AIM/FAR 2007

AERONAUTICAL INFORMATION MANUAL/FEDERAL AVIATION REGULATIONS

Edited by Charles F. Spence

McGraw-Hill

New York Chicago San Francisco Lisbon London Madrid
Mexico City Milan New Delhi San Juan Seoul
Singapore Sydney Toronto

Copyright © 2007 by The McGraw-Hill Companies, Inc. All rights reserved. Printed in the United States of America. Except as permitted under the United States Copyright Act of 1976, no part of this publication may be reproduced or distributed in any form or by any means, or stored in a data base or retrieval system, without the prior written permission of the publisher.

1 2 3 4 5 6 7 8 9 0   DOC/DOC   0 9 8 7 6

ISBN-13: 978-0-07-147924-0
ISBN-10: 0-07-147924-4

The sponsoring editor for this book was Stephen S. Chapman and the production supervisor was Pamela A. Pelton. It was set in Times Roman by TechBooks. The art director for the cover was Anthony Landi.

Printed and bound by RR Donnelley.

This book is printed on acid-free paper.

McGraw-Hill books are available at special quantity discounts to use as premiums and sales promotions, or for use in corporate training programs. For more information, please write to the Director of Special Sales, McGraw-Hill Professional, Two Penn Plaza, New York, NY 10121-2298. Or contact your local bookstore.

From the Editor

New technologies, new regulations, new subjects for tests, new, new, new seem to make changes in flight activities come more often and quicker than in the years when most of today's pilots earned their certificates. It is a trite saying, but now more than ever before it is essential that pilots keep up to date. This is true not only for our own and other's safety but also to help avoid stiffer and more restrictive regulations. Knowing and following today's regulations can help to defuse actions by regulators and proposals from others to place more clamps on flight based on possibly inadvertent violations by a few.

This is why each issue of McGraw-Hill's AIM/FAR is prepared and edited to make keeping current as convenient as possible. We do it both by what is included in the book and by its presentation. We are regularly in touch with what changes are made. Every working day of the year the Federal Register is checked to see if a government agency has issued any new rules, regulations, or proposals. There are many. These changes and additions are indicated by a screen tint over the copy so you can recognize them quickly. It is recommended that you check through these to determine how they affect your flying.

Changes are a passing parade. Merely looking at the sections with screen tints over them document that things are changing fast. Some may be just a word or two; others are long, involved changes. For instance, this 2007 edition has information about changes in lighting on taxiways and different holding position markings. Some may consider this not important but lack of knowledge of it possibly could result in a deadly runway collision. Major changes include a large section where helicopter operators will find information about night VFR operations and landing sites. The FAA issues a basic Aeronautical Information Manual every two years and provides changes three times before issuing the next basic manual. The next change will be in March 2007. Federal Aviation Regulations may be issued at any time. To keep you current between printed editions of AIM/FAR, an update is posted on a McGraw-Hill web site: *www.books.mcgraw-hill.com/engineering/update-zone.html.*

The FAA forecasts flight activity to increase three-fold in the next 20 years and that changes in operations and equipment will be necessary to handle this increase safely. For pilots, this means to always keep current. Knowledge will be freedom from error.

C. Spence

Charles F. Spence

Key to McGraw-Hill AIM/FAR features

Shaded areas over text indicate where changes or additions have been made from that published in the 2006 edition. In some cases, a segment may have numerous changes or additions in a paragraph but the entire portion has been shaded to provide ease of reading.

Contents at a glance

FEDERAL AVIATION ADMINISTRATION (FAA)

The Federal Aviation Administration is responsible for insuring the safe, efficient, and secure use of the Nation's airspace, by military as well as civil aviation, for promoting safety in air commerce, for encouraging and developing civil aeronautics, including new aviation technology, and for supporting the requirements of national defense.

The activities required to carry out these responsibilities include: safety regulations; airspace management and the establishment, operation, and maintenance of a civil-military common system of air traffic control and navigation facilities; research and development in support of the fostering of a national system of airports, promulgation of standards and specifications for civil airports, and administration of Federal grants-in-aid for developing public airports; various joint and cooperative activities with the Department of Defense; and technical assistance (under State Department auspices) to other countries.

FLIGHT INFORMATION PUBLICATION POLICY

a. The following is in essence, the statement issued by the FAA Administrator and published in the December 10, 1964, issue of the Federal Register, concerning the FAA policy as pertaining to the type of information that will be published as NOTAMs and in the Aeronautical Information Manual.

1. It is a pilot's inherent responsibility to be alert at all times for and in anticipation of all circumstances, situations, and conditions affecting the safe operation of the aircraft. For example, a pilot should expect to find air traffic at any time or place. At or near both civil and military airports and in the vicinity of known training areas, a pilot should expect concentrated air traffic and realize concentrations of air traffic are not limited to these places.

b. It is the general practice of the agency to advertise by NOTAM or other flight information publications such information it may deem appropriate; information which the agency may from time to time make available to pilots is solely for the purpose of assisting them in executing their regulatory responsibilities. Such information serves the aviation community as a whole and not pilots individually.

c. The fact that the agency under one particular situation or another may or may not furnish information does not serve as a precedent of the agency's responsibility to the aviation community; neither does it give assurance that other information of the same or similar nature will be advertised, nor does it guarantee that any and all information known to the agency will be advertised.

d. This publication, while not regulatory, provides information which reflects examples of operating techniques and procedures which may be requirements in other federal publications or regulations. It is made available solely to assist pilots in executing their responsibilities required by other publications.

Consistent with the foregoing, it shall be the policy of the Federal Aviation Administration to furnish information only when, in the opinion of the agency, a unique situation should be advertised and not to furnish routine information such as concentrations of air traffic, either civil or military. The Aeronautical Information Manual will not contain informative items concerning everyday circumstances that pilots should, either by good practices or regulation, expect to encounter or avoid.

AERONAUTICAL INFORMATION MANUAL (AIM) BASIC FLIGHT INFORMATION AND ATC PROCEDURES

This manual is designed to provide the aviation community with basic flight information and ATC procedures for use in the National Airspace System (NAS) of the United States. An international version called the Aeronautical Information Publication contains parallel information, as well as specific information on the international airports for use by the international community.

This manual contains the fundamentals required in order to fly in the United States NAS. It also contains items of interest to pilots concerning health and medical facts, factors affecting flight safety, a pilot/controller glossary of terms used in the Air Traffic Control System, and information on safety, accident, and hazard reporting.

This manual is complemented by other operational publications which are available via separate subscriptions. These publications are:

Notices to Airmen publication—A publication containing current Notices to Airmen (NOTAMs) which are considered essential to the safety of flight as well as supplemental data affecting the other operational publications listed here. It also includes current Flight Data Center NOTAMs, which are regulatory in nature, issued to establish restrictions to flight or to amend charts or published Instrument Approach Procedures. This publication is issued every four weeks and is available through subscription from the Superintendent of Documents.

The Airport/Facilities Directory, the Alaska Supplement, and the Pacific Chart Supplement—These publications contain information on airports, communications, navigation aids, instrument landing systems, VOR receiver check points, preferred routes, Flight Service Station/Weather Service telephone numbers, Air Route Traffic Control Center (ARTCC) frequencies, part-time surface areas, and various other pertinent special notices essential to air navigation. These publications are available upon subscription from the National Aeronautical Charting Office (NACO) Distribution Division, Federal Aviation Admin, Riverdale, Maryland 20737.

AERONAUTICAL INFORMATION MANUAL (AIM) CODE OF FEDERAL AVIATION REGULATIONS ADVISORY CIRCULARS

Code of Federal Aviation Regulations—The FAA publishes the Code of Federal Aviation Regulations (CFR's) to make readily available to the aviation community the regulatory requirements placed upon them. These *Regulations* are sold as individual *Parts* by the Superintendent of Documents.

The more frequently amended Parts are sold on subscription service with subscribers receiving changes automatically as issued. Less active Parts are sold on a single-sale basis. Changes to single-sale parts will be sold separately as issued. Information concerning these changes will be furnished by the FAA through its Status of Federal Aviation Regulations, AC 00-44.

Advisory Circulars—The FAA issues advisory circulars (ACs) to inform the aviation public in a systematic way of nonregulatory material. Unless incorporated into a regulation by reference, the contents of an advisory circular are not binding on the public. Advisory Circulars are issued in a numbered subject system corresponding to the subject areas of the Code of Federal Aviation Regulations (CFR's) (Title 14 Chapter 1 FAA).

The AC 00-2 checklist and Status of other FAA publications, contains advisory circulars that are for sale as well as those distributed free of charge by the Federal Aviation Administration.

Note—The above information relating to CFARs and ACs is extracted from AC 00-2 and are cross-referenced in the AIM. Many of the FARs and ACs listed in AC 00-2 are cross-referenced in the AIM. These regulatory and nonregulatory references cover a wide range of subjects and are a source of detailed information of value to the aviation community. AC 00-2 is issued annually and can be obtained free-of-charge from:

U.S. Department of Transportation
Subsequent Distribution Office, SVC-121.23
Ardmore East Business Center
3341 Q 75th Avenue
Landover, MD 20785
Telephone : 301-322-4961
AC 00-2 is also available at :
http://www.faa.gov/aba/html_policies/ac 00_2html

External References—All references to Advisory Circulars and other FAA publications in the Aeronautical Information Manual include the FAA Advisory Circular or identification numbers (when available). However due to varied publication dates, the basic publication letter is not included (Example: The Order 7110.65L, Air Traffic Control is referenced as 7110.65).

AERONAUTICAL INFORMATION MANUAL
EXPLANATION OF CHANGES

a. 1-1-19. Global Positioning System (GPS)

Updates the GPS overview to reflect the current status and adds language to explain slight differences between the course information portrayed on navigational charts and GPS navigation displays.

b. 1-1-20 Wide Area Augmentation System

Needed to reflect new criteria

c. 2-1-5 In-runway Lighting

Modifies existing taxiway centerline lead-in light patters beyond the holding position marking from solid green to alternating green and yellow.

d. 2-3-4 Taxiway Markings

New subparagraph and graphic provide information on Enhanced Taxiway Centerline Markings.

e. 2-3-5. Holding Position Markings

Clarifies subparagraph on Taxiways Located on Runway Approach Areas.

f. 4-1-9. Traffic Advisory Practices at Airports Without Operating Control Towers

Adds phraseology for improving pilot broadcast of position, intended flight activity, or ground operation on the designated Common Traffic Advisory Frequency (CTAF) at airports without a FSS on the airport.

g. 4-1-11. Designated UNICOM/MULTICOM Frequencies

Eliminates caution information that is no longer valid.

h. 4-3-3. Traffic Patterns (Figures 4-3-2 and 4-3-3).

Modifies diagrams to graphically improve the traffic pattern entry points on the downwind leg of the traffic pattern.

i. 4-6-4. Flight Planning into RVSM Airspace

Updates Domestic Reduced Vertical Separation Minimum (DRVSM) material and makes necessary editorial updates.

j. 4-6-7. Guidance on Wake Turbulence

Updates DRVSM material and makes necessary editorial updates.

k. 4-6-8. Pilot/Controller Phraseology

Updates DRVSM material and makes necessary editorial updates.

l. 4-6-10. Procedures for Accommodation of Non-RVSM Aircraft

Updates DRVSM material and makes necessary editorial updates.

m. 4-6-11. Non-RVSM Aircraft Requesting Climb to and Descent from Flight Levels Above RVSM Airspace Without Intermediate Level Off

Updates DRVSM material and makes necessary editorial updates.

n. 5-2-2. Taxi Clearance

Provides an explanation of the Pre-Departure Clearance (PDC) functions of the Tower Data Link System (TDLS)

and its limitations. Also, changes the title of the paragraph to Pre-departure Clearance Procedures.

o. 5-2-5 Departure Restrictions, Clearance Void Times, Hold For Release, and Release Times

Change allows Expect Departure Clearance Times (EDCTs) to be applied to Airspace Flow Programs (AFPs) as well as Ground Delay Programs (GDPs).

p. 5-2-6. Instrument Departure Procedures (DP) – Obstacle Departure Procedures (ODP) and Standard Instrument Departures (SID)

Adds language explaining responsibilities in the case where climb gradients exceed 200 ft/NM.

q. 5-2-7 Instrument Departure Proceedures (DPs) – Obstacle Departure Procedures (ODP and Standard Instrument Departures (SID).

Paragraph change to remove the regulatory language that was more appropriate for Operations Specifications (OpSpec) and not the AIM

r. 5-4-5. Instrument Approach Procedure Charts

Adds information concerning meaning of navigation aid and radar approach chart notes based on location on the chart.

s. 5-4-7. Instrument Approach Procedures

Advises pilots that air traffic control may clear Advanced RNAV aircraft direct to the intermediate fix. It provides pilots information about the air traffic control procedure.

t. 5-4-9 Procedure Turn

The text was revised to avoid misinterpretation and misunderstanding on when a Procedure Turn is to be flown.

u. 5-4-18. RNP SAAAR Instrument Approach Procedures

Adds new paragraph explaining RNP Special Aircraft and Aircrew Authorization Required (SAAAR) Instrument Approach Procedures.

v. 5-4-19. Approach and Landing Minimums

Adds language explaining the Precision Object Free Zone (POFZ).

w. 7-1-2. FAA Weather Services

Transcribed Weather Broadcasts (TWEB) will be provided only in Alaska.

x. 7-1-4. Preflight Briefing

TWEBs will be provided only in Alaska.

y. 7-1-5 En Route Flight Advisory Service (EFAS).

Addition of a graphic to show pilots the area in the Continental United States where they would need to be above 5000' AGL in order to contact En Route Flight Watch or Flight Watch.

z. 7-1-9. Transcribed Weather Broadcast (TWEB)

TWEBs will be provided only in Alaska.

aa. 7-1-15 ATC Inflight Weather Avoidance Assistance

Change amends the climb or descent altitude during weather deviations of more than 10 nautical miles to 300 feet.

bb. 7-5-5 Unmanned Aircraft

Adds new paragraph addressing unmanned aircraft issues.

cc. 7-5-12 Light Amplification by Stimulated Emission of Radiation (Laser) Operations

Change cancels and incorporates AC No. 70-2 Reporting of Laser Illumination of Aircraft and adds words to title.

dd. 7-5-14 Operations in Ground Icing Conditions

Adds new paragraph.

ee. 10-1-3 Helicopter Point in Space (PinS) Approach Procedures

Changes reflects changes in procedures and criteria.

ff. 10-2-2 Helicopter Night VFR Operations

New paragraph provides pilots, instructors and operators information on environmental conditions that affect seeing conditions In night VFR operations.

gg. 10-2-3 Landing Zone Safety

New paragraph provides best operating practices for selection, use, and management of temporary landing zones for helicopter emergency medical service operations.

hh. Appendix 3, Laser Beam Exposure Questionnaire

TABLE OF CONTENTS

(➤ indicates paragraphs updated for this edition)

Chapter 1. NAVIGATION AIDS

Section 1. AIR NAVIGATION AIDS

Section 2. AREA NAVIGATION (RNAV) AND REQUIRED NAVIGATION PERFORMANCE (RNP)

Chapter 2. AERONAUTICAL LIGHTING AND OTHER AIRPORT VISUAL AIDS

Section 1. AIRPORT LIGHTING AIDS

Section 2. AIR NAVIGATION AND OBSTRUCTION LIGHTING

Section 3. AIRPORT MARKING AIDS AND SIGNS

Chapter 3. AIRSPACE

Section 1. GENERAL

Section 2. CONTROLLED AIRSPACE

Section 3. CLASS G AIRSPACE

Section 4. SPECIAL USE AIRSPACE

Section 5. OTHER AIRSPACE AREAS

Chapter 4. AIR TRAFFIC CONTROL
Section 1. SERVICES AVAILABLE TO PILOTS

Section 2. RADIO COMMUNICATIONS PHRASEOLOGY AND TECHNIQUES

Section 3. AIRPORT OPERATIONS

Section 4. ATC CLEARANCES/SEPARATIONS

Section 5. SURVEILLANCE SYSTEMS

Section 6. OPERATIONAL POLICY/PROCEDURES FOR REDUCED VERTICAL SEPARATION MINIMUM (RVSM) IN THE DOMESTIC U.S., ALASKA, OFFSHORE AIRSPACE AND THE SAN JUAN FIR

Chapter 5. AIR TRAFFIC PROCEDURES
Section 1. PREFLIGHT

Section 2. DEPARTURE PROCEDURES

Section 3. EN ROUTE PROCEDURES

Section 4. ARRIVAL PROCEDURES

Section 5. PILOT/CONTROLLER ROLES AND RESPONSIBILITIES

Section 6. NATIONAL SECURITY AND INTERCEPTION PROCEDURES

Chapter 6. EMERGENCY PROCEDURES

Section 1. GENERAL

Section 2. EMERGENCY SERVICES AVAILABLE TO PILOTS

Section 3. DISTRESS AND URGENCY PROCEDURES

Section 4. TWO-WAY RADIO COMMUNICATIONS FAILURE

Section 5. AIRCRAFT RESCUE AND FIRE FIGHTING COMMUNICATIONS

Chapter 7. SAFETY OF FLIGHT

Section 1. METEOROLOGY

Section 2. ALTIMETER SETTING PROCEDURES

Section 3. WAKE TURBULENCE

Section 4. BIRD HAZARDS AND FLIGHT OVER NATIONAL REFUGES, PARKS, AND FORESTS

Section 5. POTENTIAL FLIGHT HAZARDS

Section 6. SAFETY, ACCIDENT, AND HAZARD REPORTS

Chapter 8. MEDICAL FACTS FOR PILOTS
Section 1. FITNESS FOR FLIGHT

Chapter 1. Air navigation
Section 1. NAVIGATION AIDS

1-1-1. GENERAL

a. Various types of air navigation aids are in use today, each serving a special purpose. These aids have varied owners and operators, namely: the Federal Aviation Administration (FAA), the military services, private organizations, individual states and foreign governments. The FAA has the statutory authority to establish, operate, maintain air navigation facilities and to prescribe standards for the operation of any of these aids which are used for instrument flight in federally controlled airspace. These aids are tabulated in the *Airport/Facilities Directory* (A/FD).

b. Pilots should be aware of the possibility of momentary erroneous indications on cockpit displays when the primary signal generator for a ground-based navigational transmitter (for example, a glideslope, VOR, or nondirectional beacon) is inoperative. Pilots should disregard any navigation indication, regardless of its apparent validity, if the particular transmitter was identified by NOTAM or otherwise as unusable or inoperative.

1-1-2. NONDIRECTIONAL RADIO BEACON (NDB)

a. A low or medium frequency radio beacon transmits nondirectional signals whereby the pilot of an aircraft properly equipped can determine bearings and "home" on the station. These facilities normally operate in the frequency band of 190 to 535 kilohertz (kHz), according to ICAO Annex 10 the frequency range for NDBs is between 190 and 1750 kHz, and transmit a continuous carrier with either 400 or 1020 hertz (Hz) modulation. All radio beacons except the compass locators transmit a continuous three-letter identification in code except during voice transmissions.

b. When a radio beacon is used in conjunction with the Instrument Landing System markers, it is called a Compass Locator.

c. Voice transmissions are made on radio beacons unless the letter "W" (without voice) is included in the class designator (HW).

d. Radio beacons are subject to disturbances that may result in erroneous bearing information. Such disturbances result from such factors as lightning, precipitation static, etc. At night, radio beacons are vulnerable to interference from distant stations. Nearly all disturbances which affect the ADF bearing also affect the facility's identification. Noisy identification usually occurs when the ADF needle is erratic. Voice, music or erroneous identification may be heard when a steady false bearing is being displayed. Since ADF receivers do not have a "flag" to warn the pilot when erroneous bearing information is being displayed, the pilot should continuously monitor the NDB's identification.

1-1-3. VHF OMNI-DIRECTIONAL RANGE (VOR)

a. VORs operate within the 108.0 to 117.95 MHz frequency band and have a power output necessary to provide coverage within their assigned operational service volume. They are subject to line-of-sight restrictions, and the range varies proportionally to the altitude of the receiving equipment. NOTE—normal service ranges for the various classes of VORs are given in Navigational Aid (NAVAID) Service Volumes, paragraph 1-1-8.

b. Most VORs are equipped for voice transmission on the VOR frequency. VORs without voice capability are indicated by the letter "W" (without voice) included in the class designator (VORW).

c. The only positive method of identifying a VOR is by its Morse Code identification or by the recorded automatic voice identification which is always indicated by use of the word "VOR" following the range's name. Reliance on determining the identification of an omnirange should never be placed on listening to voice transmissions by the Flight Service Station (FSS) (or approach control facility) involved. Many FSS's remotely operate several omniranges with different names. In some cases, none of the VORs have the name of the "parent" FSS. During periods of maintenance, the facility may radiate a T-E-S-T code (− • ••• −) or the code may be removed.

d. Voice identification has been added to numerous VORs. The transmission consists of a voice announcement, "AIRVILLE VOR" alternating with the usual Morse Code identification.

e. The effectiveness of the VOR depends upon proper use and adjustment of both ground and airborne equipment.

1. *Accuracy:* The accuracy of course alignment of the VOR is excellent, being generally plus or minus 1 degree.

2. *Roughness:* On some VORs, minor course roughness may be observed, evidenced by course needle or brief flag alarm activity (some receivers are more susceptible to these irregularities than others). At a few stations, usually in mountainous terrain, the pilot may occasionally observe a brief course needle oscillation, similar to the indication of "approaching station." Pilots flying over unfamiliar routes are cautioned to be on the alert for these vagaries, and in particular, to use the "to/from" indicator to determine positive station passage.

(a) Certain propeller RPM settings or helicopter rotor speeds can cause the VOR Course Deviation Indicator to fluctuate as much as plus or minus six degrees. Slight changes to the RPM setting will normally smooth out this roughness. Pilots are urged to check for this modulation phenomenon prior to reporting a VOR station or aircraft equipment for unsatisfactory operation.

1-1-4. VOR RECEIVER CHECK

a. The FAA VOR test facility (VOT) transmits a test signal which provides users a convenient means to determine the operational status and accuracy of a VOR receiver while on the ground where a VOT is located. The airborne use of VOT is permitted; however, its use is strictly limited to those areas/altitudes specifically authorized in the A/FD or appropriate supplement.

b. To use the VOT service, tune in the VOT frequency on your VOR receiver. With the Course Deviation Indicator (CDI) centered, the omni-bearing selector should read 0 degrees with the to/from indication showing "from" or the omni-bearing selector should read 180 degrees with the to/from indication showing "to." Should the VOR receiver operate an RMI (Radio Magnetic Indicator), it will indicate 180 degrees on any omni-bearing selector (OBS) setting. Two means of identification are used. One is a series of dots and the other is a continuous tone. Information concerning an individual test signal can be obtained from the local FSS.

c. Periodic VOR receiver calibration is most important. If a receiver's Automatic Gain Control or modulation circuit deteriorates, it is possible for it to display acceptable accuracy and sensitivity close into the VOR or VOT and display out-of-tolerance readings when located at greater distances where weaker signal areas exist. The likelihood of this deterioration varies between receivers, and is generally considered a function of time. The best assurance of having an accurate receiver is periodic calibration. Yearly intervals are recommended at which time an authorized repair facility should recalibrate the receiver to the manufacturer's specifications.

d. Federal Aviation Regulations (FAR Part 91.171) provides for certain VOR equipment accuracy checks prior to flight under instrument flight rules. To comply with this requirement and to ensure satisfactory operation of the airborne system, the FAA has provided pilots with the following means of checking VOR receiver accuracy:

1. VOT or a radiated test signal from an appropriately rated radio repair station.

2. Certified airborne check points.

3. Certified check points on the airport surface.

e. A radiated VOT from an appropriately rated radio repair station serves the same purpose as an FAA VOR signal and the check is made in much the same manner as a VOT with the following differences:

1. The frequency normally approved by the Federal Communications Commission is 108.0 MHz.

2. Repair stations are not permitted to radiate the VOR test signal continuously; consequently, the owner or operator must make arrangements with the repair station to have the test signal transmitted. This service is not provided by all radio repair stations. The aircraft owner or operator must determine which repair station in the local area provides this service. A representative of the repair station must make an entry into the aircraft logbook or other permanent record certifying to the radial accuracy and the date of transmission. The owner, operator or representative of the repair station may accomplish the necessary checks in the aircraft and make a logbook entry stating the results. It is necessary to verify which test radial is being transmitted and whether you should get a "to" or "from" indication.

f. Airborne and ground check points consist of certified radials that should be received at specific points on the airport surface or over specific landmarks while airborne in the immediate vicinity of the airport.

1. Should an error in excess of plus or minus 4 degrees be indicated through use of a ground check, or plus or minus 6 degrees using the airborne check, IFR flight shall not be attempted without first correcting the source of the error.

CAUTION—NO CORRECTION OTHER THAN THE CORRECTION CARD FIGURES SUPPLIED BY THE MANUFACTURER SHOULD BE APPLIED IN MAKING THESE VOR RECEIVER CHECKS.

2. Locations of airborne check points, ground check points and VOTs are published in the A/FD and are depicted on the A/G voice communications panel on the FAA IFR area chart and IFR enroute low altitude chart.

3. If a dual system VOR (units independent of each other except for the antenna) is installed in the aircraft, one system may be checked against the other. Turn both systems to the same VOR ground facility and note the indicated bearing to that station. The maximum permissible variations between the two indicated bearings is 4 degrees.

1-1-5. TACTICAL AIR NAVIGATION (TACAN)

a. For reasons peculiar to military or naval operations (unusual siting conditions, the pitching and rolling of a naval vessel, etc.) the civil VOR/Distance Measuring Equipment (DME) system of air navigation was considered unsuitable for military or naval use. A new navigational system, TACAN, was therefore developed by the military and naval forces to more readily lend itself to military and naval requirements. As a result, the FAA has integrated TACAN facilities with the civil VOR/DME program. Although the theoretical, or technical principles of operation of TACAN equipment are quite different from those of VOR/DME facilities, the end result, as far as the navigating pilot is concerned, is the same. These integrated facilities are called VORTACs.

b. TACAN ground equipment consists of either a fixed or mobile transmitting unit. The airborne unit in conjunction with the ground unit reduces the transmitted signal to a visual presentation of both azimuth and distance information. TACAN is a pulse system and operates in the UHF band of frequencies. Its use requires TACAN airborne equipment and does not operate through conventional VOR equipment.

1-1-6. VHF OMNI-DIRECTIONAL RANGE/TACTICAL AIR NAVIGATION (VORTAC)

a. A VORTAC is a facility consisting of two components, VOR and TACAN, which provides three individual services: VOR azimuth, TACAN azimuth and TACAN distance (DME) at one site. Although consisting of more than one component, incorporating more than one operating frequency, and using more than one antenna system, a VORTAC is considered to be a unified navigational aid. Both components of a VORTAC are envisioned as operating simultaneously and providing the three services at all times.

b. Transmitted signals of VOR and TACAN are each identified by three-letter code transmission and are interlocked so that pilots using VOR azimuth with TACAN distance can be assured that both signals being received are definitely from the same ground station. The frequency channels of the VOR and the TACAN at each VORTAC facility are "paired" in accordance with a national plan to simplify airborne operation.

1-1-7. DISTANCE MEASURING EQUIPMENT (DME)

a. In the operation of DME, paired pulses at a specific spacing are sent out from the aircraft (this is the interrogation) and are received at the ground station. The ground station (transponder) then transmits paired pulses back to the aircraft at the same pulse spacing but on a different frequency. The time required for the round trip of this signal exchange is measured in the airborne DME unit and is translated into distance (Nautical Miles) from the aircraft to the ground station.

b. Operating on the line-of-sight principle, DME furnishes distance information with a very high degree of accuracy. Reliable signals may be received at distances up to 199 NM at line-of-sight altitude with an accuracy of better than ½ mile or 3 percent of the distance, whichever is greater. Distance information received from DME equipment is SLANT RANGE distance and not actual horizontal distance.

c. Operating frequency range of a DME according to ICAO Annex 10 is from 960 MHz to 1215 MHz. Aircraft equipped with TACAN equipment will receive distance information from a VORTAC automatically, while aircraft equipped with VOR must have a separate DME airborne unit.

d. VOR/DME, VORTAC, ILS/DME, and LOC/DME navigation facilities established by the FAA provide course and distance information from collocated components under a frequency pairing plan. Aircraft receiving equipment which provides for automatic DME selection assures reception of azimuth and distance information from a common source when designated VOR/DME, VORTAC, ILS/DME, and LOC/DME are selected.

e. Due to the limited number of available frequencies, assignment of paired frequencies is required for certain military noncollocated VOR and TACAN facilities which serve the same area but which may be separated by distances up to a few miles. The military is presently undergoing a program to collocate VOR and TACAN facilities or to assign nonpaired frequencies to those that cannot be collocated.

f. VOR/DME, VORTAC, ILS/DME, and LOC/DME facilities are identified by synchronized identifications which are transmitted on a time share basis. The VOR or localizer portion of the facility is identified by a coded tone modulated at 1020 Hz or a combination of code and voice. The TACAN or DME is identified by a coded tone modulated at 1350 Hz. The DME or TACAN coded identification is transmitted one time for each three or four times that the VOR or localizer coded identification is transmitted. When either the VOR or the DME is inoperative, it is important to recognize which identifier is retained for the operative facility. A single coded identification with a repetition interval of approximately 30 seconds indicates that the DME is operative.

g. Aircraft equipment which provides for automatic DME selection assures reception of azimuth and distance information from a common source when designated VOR/DME, VORTAC and ILS/DME navigation facilities are selected. Pilots are cautioned to disregard any distance displays from automatically selected DME equipment when VOR or ILS facilities, which do not have the DME feature installed, are being used for position determination.

1-1-8. NAVAID SERVICE VOLUMES

a. Most air navigation radio aids which provide positive course guidance have a designated standard service volume (SSV). The SSV defines the reception limits of unrestricted NAVAIDs which are usable for random/unpublished route navigation.

b. A NAVAID will be classified as restricted if it does not conform to flight inspection signal strength and course quality standards throughout the published SSV. However, the NAVAID should not be considered usable at altitudes below that which could be flown while operating under random route IFR conditions (FAR Part 91.177), even though these altitudes may lie within the designated SSV. Service volume restrictions are first published in the Notices to Airman (NOTAM) and then with the alphabetical listing of the NAVAIDs in the A/FD.

c. Standard Service Volume limitations do not apply to published IFR routes or procedures.

d. VOR/DME/TACAN STANDARD SERVICE VOLUMES (SSV)

1. Standard service volumes (SSVs) are graphically shown in FIG 1-1-1, FIG 1-1-2, FIG 1-1-3, FIG 1-1-4, and FIG 1-1-5. The SSV of a station is indicated by using the class designator as a prefix to the station type designation.

EXAMPLE—TVOR, LDME, AND HVORTAC.

2. Within 25 NM, the bottom of the T service volume is defined by the curve in FIG 1-1-4. Within 40 NM, the bottoms of the L and H service volumes are defined by the curve in FIG 1-1-5. (See TBL 1-1-1).

e. NONDIRECTIONAL RADIO BEACON (NDB)

1. NDBs are classified according to their intended use.

2. The ranges of NDB service volumes are shown in TBL 1-1-2. The distances (radius) are the same at all altitudes.

STANDARD HIGH-ALTITUDE
SERVICE VOLUME

(See FIG 1-1-5 for altitudes
below 1000 feet.)

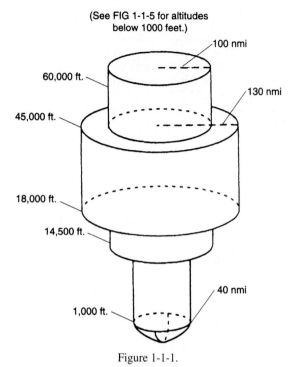

Figure 1-1-1.

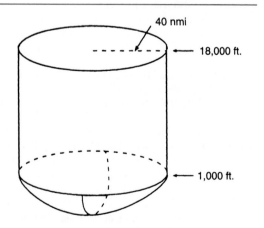

NOTE: All elevations shown are with respect
to the station's site elevation (AGL).
Coverage is not available in a cone of
airspace directly above the facility.

Figure 1-1-2.

(Refer to FIG 1-1-4 for altitudes
below 1000 feet.)

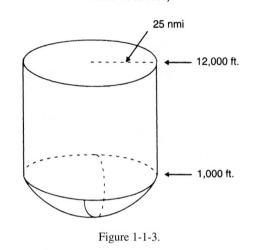

Figure 1-1-3.

Table 1-1-1
VOR/DME/TACAN Standard Service Volumes

SSV CLASS DESIGNATOR	ALTITUDE AND RANGE BOUNDARIES
T (Terminal)	From 1000 feet above ground level (AGL) up to and including 12,000 feet AGL at radial distances out to 25 NM.
L (Low Altitude)	From 1000 feet AGL up to and including 18,000 feet AGL at radial distances out to 40 NM.
H (High Altitude)	From 1000 feet AGL up to and including 14,500 feet AGL at radial distances out to 40 NM. From 14,500 feet AGL up to and including 60,000 feet at radial distances out to 100 NM. From 18,000 feet AGL up to and including 45,000 feet AGL at radial distances out to 130 NM.

Table 1-1-2
NDB Service Volumes

CLASS	DISTANCE (RADIUS)
Compass Locator	15 NM
MH	25 NM
H	50 NM*
HH	75 NM

* Service ranges of individual facilities may
be less than 50 nautical miles (NM).
Restrictions to service volumes are first
published as a Notice to Airmen and then
with the alphabetical listing of the
NAVAID in the Airport/Facility Directory.

SERVICE VOLUME LOWER EDGE TERMINAL

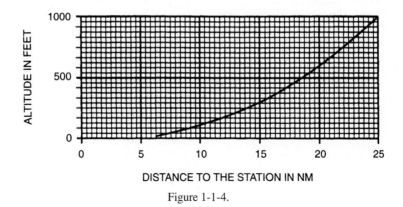

DISTANCE TO THE STATION IN NM

Figure 1-1-4.

SERVICE VOLUME LOWER EDGE
STANDARD HIGH AND LOW

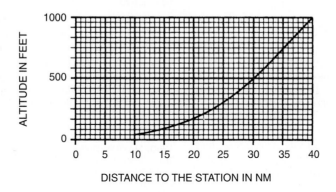

DISTANCE TO THE STATION IN NM

Figure 1-1-5.

1-1-9. INSTRUMENT LANDING SYSTEM (ILS)

a. GENERAL

1. The ILS is designed to provide an approach path for exact alignment and descent of an aircraft on final approach to a runway.

2. The ground equipment consists of two highly directional transmitting systems and, along the approach, three (or fewer) marker beacons. The directional transmitters are known as the localizer and glide slope transmitters.

3. The system may be divided functionally into three parts:

(a) *Guidance information:* localizer, glide slope

(b) *Range information:* marker beacon, DME

(c) *Visual information:* approach lights, touchdown and centerline lights, runway lights

4. Precision radar, or compass locators located at the Outer Marker (OM) or Middle Marker (MM), may be substituted for marker beacons. DME, when specified in the procedure, may be substituted for the OM.

5. Where a complete ILS system is installed on each end of a runway; (i.e., the approach end of Runway 4 and the approach end of Runway 22) the ILS systems are not in service simultaneously.

b. LOCALIZER

1. The localizer transmitter operates on one of 40 ILS channels within the frequency range of 108.10 to 111.95

MHz. Signals provide the pilot with course guidance to the runway centerline.

2. The approach course of the localizer is called the front course and is used with other functional parts, e.g., glide slope, marker beacons, etc. The localizer signal is transmitted at the far end of the runway. It is adjusted for a course width of (full scale fly-left to a full scale fly-right) of 700 feet at the runway threshold.

3. The course line along the extended centerline of a runway, in the opposite direction to the front course is called the back course.

CAUTION—UNLESS THE AIRCRAFT ILS EQUIPMENT INCLUDES REVERSE SENSING CAPABILITY, WHEN FLYING INBOUND ON THE BACK COURSE IT IS NECESSARY TO STEER THE AIRCRAFT IN THE DIRECTION OPPOSITE THE NEEDLE DEFLECTION WHEN MAKING CORRECTIONS FROM OFF-COURSE TO ON-COURSE. THIS "FLYING AWAY FROM THE NEEDLE" IS ALSO REQUIRED WHEN FLYING OUTBOUND ON THE FRONT COURSE OF THE LOCALIZER. DO NOT USE BACK COURSE SIGNALS FOR APPROACH UNLESS A BACK COURSE APPROACH PROCEDURE IS PUBLISHED FOR THAT PARTICULAR RUNWAY AND THE APPROACH IS AUTHORIZED BY ATC.

4. Identification is in International Morse Code and consists of a three-letter identifier preceded by the letter I (••) transmitted on the localizer frequency.

EXAMPLE—I-DIA

5. The localizer provides course guidance throughout the descent path to the runway threshold from a distance of 18 NM from the antenna between an altitude of 1,000 feet above the highest terrain along the course line and 4,500 feet above the elevation of the antenna site. Proper off-course indications are provided throughout the following angular areas of the operational service volume:

(a) To 10 degrees either side of the course along a radius of 18 NM from the antenna, and

(b) From 10 to 35 degrees either side of the course along a radius of 10 NM. (See FIG 1-1-6)

6. Unreliable signals may be received outside these areas.

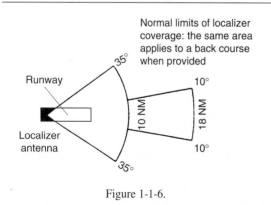

Figure 1-1-6.

c. LOCALIZER-TYPE DIRECTIONAL AID

1. The Localizer-type Directional Aid (LDA) is of comparable use and accuracy to a localizer but is not part of a complete ILS. The LDA course usually provides a more precise approach course than the similar Simplified Directional Facility (SDF) installation, which may have a course width of 6 or 12 degrees.

2. The LDA is not aligned with the runway. Straight-in minimums may be published where alignment does not exceed 30 degrees between the course and runway. Circling minimums only are published where this alignment exceeds 30 degrees.

3. A very limited number of LDA approaches also incorporate a glideslope. These are annotated in the plan view of the instrument approach chart with a note, "LDA/Glideslope." These procedures fall under a newly defined category of approaches called Approach with Vertical Guidance (APV) described in paragraph 5-4-5, Instrument Approach Procedure Charts, subparagraph a7(b), Approach with Vertical Guidance (APV). LDA minima for with and without glideslope is provided and annotated on the minima lines of the approach chart as S-LDA/GS and S-LDA. Because the final approach course is not aligned with the runway centerline, additional maneuvering will be required compared to an ILS approach.

d. GLIDE SLOPE/GLIDE PATH

1. The UHF glide slope transmitter, operating on one of the 40 ILS channels within the frequency range 329.15 MHz, to 335.00 MHz radiates its signals in the direction of the localizer front course. The term "glide path" means that portion of the glide slope that intersects the localizer.

CAUTION—FALSE GLIDE SLOPE SIGNALS MAY EXIST IN THE AREA OF THE LOCALIZER BACK COURSE APPROACH WHICH CAN CAUSE THE GLIDE SLOPE FLAG ALARM TO DISAPPEAR AND PRESENT UNRELIABLE GLIDE SLOPE INFORMATION. DISREGARD ALL GLIDE SLOPE SIGNAL INDICATIONS WHEN MAKING A LOCALIZER BACK COURSE APPROACH UNLESS A GLIDE SLOPE IS SPECIFIED ON THE APPROACH AND LANDING CHART.

2. The glide slope transmitter is located between 750 feet and 1,250 feet from the approach end of the runway (down the runway) and offset 250 to 650 feet from the runway centerline. It transmits a glide path beam 1.4 degrees (vertically) wide. The signal provides descent information for navigation down to the lowest authorized decision height (DH) specified in the approved ILS approach procedure. The glide path may not be suitable for navigation below the lowest authorized DH and any reference to glidepath indications below that height must be supplemented by visual reference to the runway environment. Glide paths with no published DH are usable to runway threshold.

3. The glidepath projection angle is normally adjusted to 3 degrees above horizontal so that it intersects the MM at about 200 feet and the OM at about 1,400 feet above the runway elevation. The glide slope is normally usable to the distance of 10 NM. However, at some locations, the glide slope has been certified for an extended service volume which exceeds 10 NM.

4. Pilots must be alert when approaching the glidepath interception. False courses and reverse sensing will occur at angles considerably greater than the published path.

5. Make every effort to remain on the indicated glide path. **CAUTION**—AVOID FLYING BELOW THE GLIDE PATH TO ASSURE OBSTACLE/TERRAIN CLEARANCE IS MAINTAINED.

6. The published glide slope threshold crossing height (TCH) DOES NOT represent the height of the actual glide path on-course indication above the runway threshold. It is used as a reference for planning purposes which represents the height above the runway threshold that an aircraft's glide slope antenna should be, if that aircraft remains on a trajectory formed by the four-mile-to-middle marker glidepath segment.

7. Pilots must be aware of the vertical height between the aircraft's glide slope antenna and the main gear in the landing configuration and, at the DH, plan to adjust the descent angle accordingly if the published TCH indicates the wheel crossing height over the runway threshold may not be satisfactory. Tests indicate a comfortable wheel crossing height is approximately 20 to 30 feet, depending on the type of aircraft.

NOTE—The TCH for a runway is established based on several factors including the largest aircraft category that normally uses the runway, how airport layout effects the glide slope antenna placement, and terrain. A higher than optimum TCH, with the same glide path angle, may cause the aircraft to touch down further from the threshold if the trajectory of the approach is maintained until the flare. Pilots should consider the effect of a high TCH on the runway available for stopping the aircraft.

e. DISTANCE MEASURING EQUIPMENT (DME)

1. When installed with the ILS and specified in the approach procedure, DME may be used:

(a) In lieu of the OM.

(b) As a back course (BC) final approach fix (FAF).

(c) To establish other fixes on the localizer course.

2. In some cases, DME from a separate facility may be used within Terminal Instrument Procedures (TERPS) limitations:

(a) To provide ARC initial approach segments.

(b) As a FAF for BC approaches.

(c) As a substitute for the OM.

f. MARKER BEACON

1. ILS marker beacons have a rated power output of 3 watts or less and an antenna array designed to produce an elliptical pattern with dimensions, at 1,000 feet above the antenna, of approximately 2,400 feet in width and 4,200 feet in length. Airborne marker beacon receivers with a selective sensitivity feature should always be operated in the "low" sensitivity position for proper reception of ILS marker beacons.

2. Ordinarily, there are two marker beacons associated with an ILS, the OM and MM. Locations with a Category II ILS also have an Inner Marker (IM). When an aircraft passes over a marker, the pilot will receive the following indications: (See TBL 1-1-3).

Table 1-1-3
Marker Passage Indications

MARKER	CODE	LIGHT
OM	– – –	BLUE
MM	• – • –	AMBER
IM	• • • •	WHITE
BC	• • • •	WHITE

(a) The OM normally indicates a position at which an aircraft at the appropriate altitude on the localizer course will intercept the ILS glide path.

(b) The MM indicates a position approximately 3,500 feet from the landing threshold. This is also the position where an aircraft on the glide path will be at an altitude of approximately 200 feet above the elevation of the touchdown zone.

(c) The inner marker (IM) will indicate a point at which an aircraft is at a designated decision height (DH) on the glide path between the MM and landing threshold.

3. A back course marker normally indicates the ILS back course final approach fix where approach descent is commenced.

g. COMPASS LOCATOR

1. Compass locator transmitters are often situated at the MM and OM sites. The transmitters have a power of less than 25 watts, a range of at least 15 miles and operate between 190 and 535 kHz. At some locations, higher powered radio beacons, up to 400 watts, are used as OM compass locators. These generally carry Transcribed Weather Broadcast (TWEB) information.

2. Compass locators transmit two letter identification groups. The outer locator transmits the first two letters of the localizer identification group, and the middle locator transmits the last two letters of the localizer identification group.

h. ILS FREQUENCY

1. See TBL 1-1-4 for frequency pairs allocated for ILS.

i. ILS MINIMUMS

1. The lowest authorized ILS minimums, with all required ground and airborne systems components operative, are

(a) *Category I*—Decision Height (DH) 200 feet and Runway Visual Range (RVR) 2,400 feet (with touchdown zone and centerline lighting, RVR 1,800 feet).

(b) *Category II*—DH 100 feet and RVR 1,200 feet.

(c) *Category IIIa*—No DH or DH below 100 feet and RVR not less than 700 feet.

(d) *Category IIIb*—No DH or DH below 50 feet and RVR less than 700 feet but not less than 150 feet.

(e) *Category IIIc*—No DH and no RVR limitation.
NOTE—Special authorization and equipment required for Category II and III.

j. INOPERATIVE ILS COMPONENTS

1. *Inoperative localizer:* When the localizer fails, an ILS approach is not authorized.

2. *Inoperative glide slope:* When the glide slope fails, the ILS reverts to a nonprecision localizer approach.

REFERENCE—SEE THE INOPERATIVE COMPONENT TABLE IN THE U.S. GOVERNMENT TERMINAL PROCEDURES PUBLICATION (TPP), FOR ADJUSTMENTS TO MINIMUMS DUE TO INOPERATIVE AIRBORNE OR GROUND SYSTEM EQUIPMENT.

k. ILS COURSE DISTORTION

1. All pilots should be aware that disturbances to ILS localizer and glide slope courses may occur when surface vehicles or aircraft are operated near the localizer or glide slope antennas. Most ILS installations are subject to signal interference by either surface vehicles, aircraft or both. ILS CRITICAL AREAS are established near each localizer and glide slope antenna.

Table 1-1-4
Frequency Pairs Allocated for ILS

Localizer MHz	Glide Slope
108.10	334.70
108.15	334.55
108.3	334.10
108.35	333.95
108.5	329.90
108.55	329.75
108.7	330.50
108.75	330.35
108.9	329.30
108.95	329.15
109.1	331.40
109.15	331.25
109.3	332.00
109.35	331.85
109.50	332.60
109.55	332.45
109.70	333.20
109.75	333.05
109.90	333.80
109.95	333.65
110.1	334.40
110.15	334.25
110.3	335.00
110.35	334.85
110.5	329.60
110.55	329.45
110.70	330.20
110.75	330.05
110.90	330.80
110.95	330.65
111.10	331.70
111.15	331.55
111.30	332.30
111.35	332.15
111.50	332.9
111.55	332.75
111.70	333.5
111.75	333.35
111.90	331.1
111.95	330.95

2. ATC issues control instructions to avoid interfering operations within ILS critical areas at controlled airports

during the hours the Airport Traffic Control Tower (ATCT) is in operation as follows:

(a) *Weather Conditions*—Less than ceiling 800 feet and/or visibility 2 miles.

(1) *LOCALIZER CRITICAL AREA*—Except for aircraft that land, exit a runway, depart or miss approach, vehicles and aircraft are not authorized in or over the critical area when an arriving aircraft is between the ILS final approach fix and the airport. Additionally, when the ceiling is less than 200 feet and/or the visibility is RVR 2,000 or less, vehicle and aircraft operations in or over the area are not authorized when an arriving aircraft is inside the ILS MM.

(2) *GLIDE SLOPE CRITICAL AREA*—Vehicles and aircraft are not authorized in the area when an arriving aircraft is between the ILS final approach fix and the airport unless the aircraft has reported the airport in sight and is circling or side stepping to land on a runway other than the ILS runway.

(b) *Weather Conditions*—At or above ceiling 800 feet and/or visibility 2 miles.

(1) No critical area protective action is provided under these conditions

(2) A flight crew, under these conditions, should advise the tower that it will conduct an AUTOLAND or COUPLED approach to ensure that the ILS critical areas are protected when the aircraft is inside the ILS MM.

EXAMPLE—GLIDE SLOPE SIGNAL NOT PROTECTED.

3. Aircraft holding below 5,000 feet between the outer marker and the airport may cause localizer signal variations for aircraft conducting the ILS approach. Accordingly, such holding is not authorized when weather or visibility conditions are less than ceiling 800 feet and/or visibility 2 miles.

4. Pilots are cautioned that vehicular traffic not subject to ATC may cause momentary deviation to ILS course or glide slope signals. Also, critical areas are not protected at uncontrolled airports or at airports with an operating control tower when weather or visibility conditions are above those requiring protective measures. Aircraft conducting coupled or autoland operations should be especially alert in monitoring automatic flight control systems. (See FIG 1-1-7).

NOTE—UNLESS OTHERWISE COORDINATED THROUGH FLIGHT STANDARDS, ILS SIGNALS TO CATEGORY I RUNWAYS ARE NOT FLIGHT INSPECTED BELOW 100 FEET AGL. GUIDANCE SIGNAL ANOMALIES MAY BE ENCOUNTERED BELOW THIS ALTITUDE.

1-1-10. SIMPLIFIED DIRECTIONAL FACILITY (SDF)

a. The SDF provides a final approach course similar to that of the ILS localizer. It does not provide glide slope information. A clear understanding of the ILS localizer and the additional factors listed below completely describe the operational characteristics and use of the SDF.

b. The SDF transmits signals within the range of 108.10 to 111.95 MHz.

c. The approach techniques and procedures used in an SDF instrument approach are essentially the same as

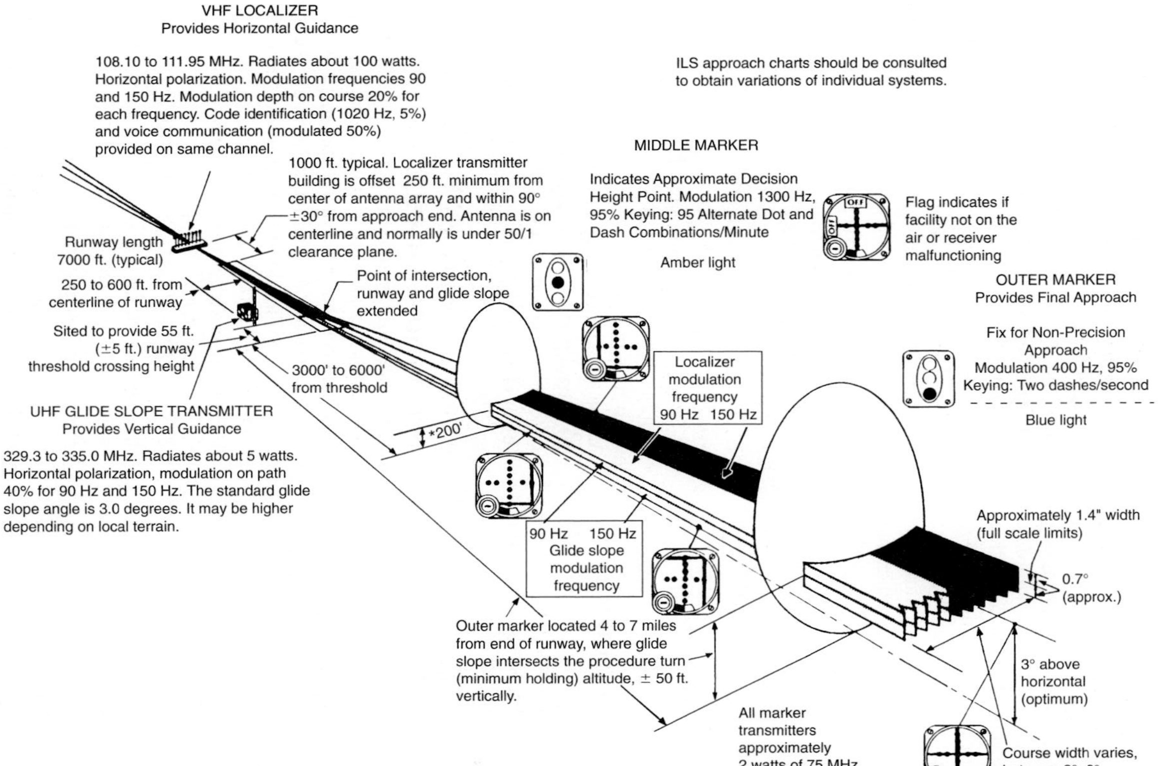

VHF LOCALIZER
Provides Horizontal Guidance

108.10 to 111.95 MHz. Radiates about 100 watts. Horizontal polarization. Modulation frequencies 90 and 150 Hz. Modulation depth on course 20% for each frequency. Code identification (1020 Hz, 5%) and voice communication (modulated 50%) provided on same channel.

ILS approach charts should be consulted to obtain variations of individual systems.

MIDDLE MARKER

Indicates Approximate Decision Height Point. Modulation 1300 Hz, 95% Keying: 95 Alternate Dot and Dash Combinations/Minute

Amber light

Flag indicates if facility not on the air or receiver malfunctioning

OUTER MARKER
Provides Final Approach

Fix for Non-Precision Approach
Modulation 400 Hz, 95%
Keying: Two dashes/second

Blue light

1000 ft. typical. Localizer transmitter building is offset 250 ft. minimum from center of antenna array and within 90°±30° from approach end. Antenna is on centerline and normally is under 50/1 clearance plane.

Point of intersection, runway and glide slope extended

Localizer modulation frequency
90 Hz   150 Hz

Approximately 1.4° width (full scale limits)

0.7° (approx.)

3° above horizontal (optimum)

Course width varies, between 3°–6° tailored to provide 700 ft. at threshold (full scale limited)

Runway length 7000 ft. (typical)

250 to 600 ft. from centerline of runway

Sited to provide 55 ft. (±5 ft.) runway threshold crossing height

3000' to 6000' from threshold

*200'

90 Hz   150 Hz
Glide slope modulation frequency

All marker transmitters approximately 2 watts of 75 MHz modulated about 95%

Outer marker located 4 to 7 miles from end of runway, where glide slope intersects the procedure turn (minimum holding) altitude, ± 50 ft. vertically.

UHF GLIDE SLOPE TRANSMITTER
Provides Vertical Guidance

329.3 to 335.0 MHz. Radiates about 5 watts. Horizontal polarization, modulation on path 40% for 90 Hz and 150 Hz. The standard glide slope angle is 3.0 degrees. It may be higher depending on local terrain.

NOTE:
Compass locators, rated at 25 watts output 190 to 535 kHz, are installed at many outer and some middle markers. A 400 Hz or a 1020 Hz tone, modulating the carrier about 95%, is keyed with the first two letters of the ILS identification on the outer locator and the last two letters on the middle locator. At some locations, simultaneous voice transmissions from the control tower are provided, with appropriate reduction in identification percentage.

* Figures marked with asterisk are typical. Actual figures vary with deviations in distances to markers, glide angles and localizer widths.

RATE OF DESCENT CHART (feet per minute)

Speed (knots)	Angle		
	2½°	2¾°	3°
90	400	440	475
110	485	535	585
130	575	630	690
150	665	730	795
160	707	778	849

Figure 1-1-7.

those employed in executing a standard localizer approach except the SDF course may not be aligned with the runway and the course may be wider, resulting in less precision.

d. Usable off-course indications are limited to 35 degrees either side of the course centerline. Instrument indications received beyond 35 degrees should be disregarded.

e. The SDF antenna may be offset from the runway centerline. Because of this, the angle of convergence between the final approach course and the runway bearing should be determined by reference to the instrument approach procedure chart. This angle is generally not more than 3 degrees. However, it should be noted that inasmuch as the approach course originates at the antenna site, an approach which is continued beyond the runway threshold will lead the aircraft to the SDF offset position rather than along the runway centerline.

f. The SDF signal is fixed at either 6 degrees or 12 degrees as necessary to provide maximum flyability and optimum course quality.

g. Identification consists of a three-letter identifier transmitted in Morse Code on the SDF frequency. The appropriate instrument approach chart will indicate the identifier used at a particular airport.

1-1-11. MICROWAVE LANDING SYSTEM (MLS)

a. GENERAL

1. The MLS provides precision navigation guidance for exact alignment and descent of aircraft on approach to a runway. It provides azimuth, elevation, and distance.

2. Both lateral and vertical guidance may be displayed on conventional course deviation indicators or incorporated into multipurpose cockpit displays. Range information can be displayed by conventional DME indicators and also incorporated into multipurpose displays.

3. The MLS supplements the ILS as the standard landing system in the United States for civil, military, and international civil aviation. At international airports, ILS service is protected to 2010.

4. The system may be divided into five functions:

(a) Approach azimuth.

(b) Back azimuth.

(c) Approach elevation.

(d) Range.

(e) Data communications.

5. The standard configuration of MLS ground equipment includes:

(a) An azimuth station to perform functions (a) and (e) above. In addition to providing azimuth navigation guidance, the station transmits basic data which consists of information associated directly with the operation of the landing system, as well as advisory data on the performance of the ground equipment.

(b) An elevation station to perform function (c).

(c) Distance Measuring Equipment (DME) to perform range guidance, both standard DME (DME/N) and precision DME (DME/P).

6. MLS Expansion Capabilities: The standard configuration can be expanded by adding one or more of the following functions or characteristics.

(a) Back azimuth: Provides lateral guidance for missed approach and departure navigation.

(b) Auxiliary data transmissions: Provides additional data, including refined airborne positioning, meteorological information, runway status, and other supplementary information.

(c) Expanded Service Volume (ESV) proportional guidance to 60 degrees.

7. MLS identification is a four-letter designation starting with the letter M. It is transmitted in International Morse Code at least six times per minute by the approach azimuth (and back azimuth) ground equipment.

b. APPROACH AZIMUTH GUIDANCE

1. The azimuth station transmits MLS angle and data on one of 200 channels within the frequency range of 5031 to 5091 MHz.

2. The equipment is normally located about 1,000 feet beyond the stop end of the runway, but there is considerable flexibility in selecting sites. For example, for heliport operations the azimuth transmitter can be collocated with the elevation transmitter.

3. The azimuth coverage extends: (See FIG 1-1-8).

(a) Laterally, at least 40 degrees on either side of the runway centerline in a standard configuration.

(b) In elevation, up to an angle of 15 degrees and to at least 20,000 feet.

(c) In range, to at least 20 NM.

c. ELEVATION GUIDANCE

1. The elevation station transmits signals on the same frequency as the azimuth station. A single frequency is time-shared between angle and data functions.

2. The elevation transmitter is normally located about 400 feet from the side of the runway between runway threshold and the touchdown zone.

3. Elevation coverage is provided in the same airspace as the azimuth guidance signals:

(a) In elevation, to at least +15 degrees.

(b) Laterally, to fill the Azimuth lateral coverage and,

(c) In range, to at least 20 NM. (See FIG 1-1-9).

d. RANGE GUIDANCE

1. The MLS Precision Distance Measuring Equipment (DME/P) functions the same as the navigation DME described in AIM paragraph 1-1-7, but there are some technical differences. The beacon transponder operates in the frequency band 962 to 1105 MHz and responds to an aircraft interrogator. The MLS DME/P accuracy is improved to be consistent with the accuracy provided by the MLS azimuth and elevation stations.

2. A DME/P channel is paired with the azimuth and elevation channel. A complete listing of the 200 paired channels

COVERAGE VOLUMES AZIMUTH

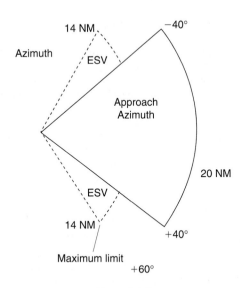

Figure 1-1-8.

COVERAGE VOLUMES ELEVATION

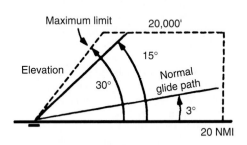

Figure 1-1-9.

of the DME/P with the angle functions is contained in FAA Standard 022 (MLS Interoperability and Performance Requirements).

3. The DME/N or DME/P is an integral part of the MLS and is installed at all MLS facilities unless a waiver is obtained. This occurs infrequently and only at outlying, low density airports where marker beacons or compass locators are already in place.

e. DATA COMMUNICATIONS

1. The data transmission can include both basic and auxiliary data words. All MLS facilities transmit basic data. Where needed, auxiliary data can be transmitted.

2. Coverage limits: MLS data are transmitted throughout the azimuth (and back azimuth when provided) coverage sectors.

3. Basic data content: Representative data include:

(a) Station identification.

(b) Exact locations of azimuth, elevation and DME/P stations (for MLS receiver processing functions).

(c) Ground equipment performance level.

(d) DME/P channel and status.

4. Auxiliary data content—Representative data include:

(a) 3-D locations of MLS equipment.

(b) Waypoint coordinates.

(c) Runway conditions.

(d) Weather (e.g., RVR, ceiling, altimeter setting, wind, wake vortex, wind shear).

f. OPERATIONAL FLEXIBILITY

1. The MLS has the capability to fulfill a variety of needs in the approach, landing, missed approach and departure phases of flight. For example:

(a) curved and segmented approaches;

(b) selectable glide path angles;

(c) accurate 3-D positioning of the aircraft in space; and

(d) the establishment of boundaries to ensure clearance from obstructions in the terminal area.

2. While many of these capabilities are available to any MLS-equipped aircraft, the more sophisticated capabilities (such as curved and segmented approaches) are dependent upon the particular capabilities of the airborne equipment.

g. SUMMARY

1. Accuracy: The MLS provides precision three-dimensional navigation guidance accurate enough for all approach and landing maneuvers.

2. Coverage: Accuracy is consistent throughout the coverage volumes. (See FIG 1-1-10).

3. Environment: The system has low susceptibility to interference from weather conditions and airport ground traffic.

4. Channels: MLS has 200 channels—enough for any foreseeable need.

5. Data: The MLS transmits ground-air data messages associated with the systems operation.

6. Range information: Continuous range information is provided with an accuracy of about 100 feet.

1-1-12. NAVAID IDENTIFIER REMOVAL DURING MAINTENANCE

During periods of routine or emergency maintenance, coded identification (or code and voice, where applicable) is removed from certain FAA NAVAIDs. Removal of identification serves as a warning to pilots that the facility is officially

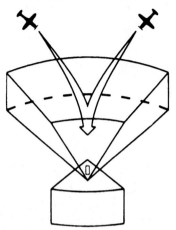

COVERAGE VOLUMES
3D REPRESENTATION

Figure 1-1-10.

off the air for tune-up or repair and may be unreliable even though intermittent or constant signals are received.

NOTE: During periods of maintenance VHF ranges may radiate a T-E-S-T code (– • ••• –).

NOTE: —DO NOT attempt to fly a procedure that *is NOTAMed out of service even if the identification is present. In certain cases, the identification may be transmitted for short periods as part of the testing.*

1-1-13. NAVAIDS WITH VOICE

a. Voice equipped en route radio navigational aids are under the operational control of either an FAA Automated Flight Service Station (AFSS) or an approach control facility. The voice communication is available on some facilities. Hazardous Inflight Weather Advisory Service (HIWAS) broadcast capability is available on selected VOR sites throughout the conterminous U.S. and does not provide two-way voice communication. The availability of two-way voice communication and HIWAS is indicated in the A/FD and aeronautical charts.

b. Unless otherwise noted on the chart, all radio navigation aids operate continuously except during shutdowns for maintenance. Hours of operation of facilities not operating continuously are annotated on charts and in the A/FD.

1-1-14. USER REPORTS ON NAVAID PERFORMANCE

a. Users of the National Airspace System (NAS) can render valuable assistance in the early correction of NAVAID malfunctions by reporting their observations of undesirable NAVAID performance. Although NAVAIDs are monitored by electronic detectors, adverse effects of electronic interference, new obstructions or changes in terrain near the

NAVAID can exist without detection by the ground monitors. Some of the characteristics of malfunction or deteriorating performance which should be reported are: erratic course or bearing indications; intermittent, or full, flag alarm; garbled, missing or obviously improper coded identification; poor quality communications reception; or, in the case of frequency interference, an audible hum or tone accompanying radio communications or NAVAID identification.

b. Reporters should identify the NAVAID, location of the aircraft, time of the observation, type of aircraft and describe the condition observed; the type of receivers in use is also useful information. Reports can be made in any of the following ways:

1. Immediate report by direct radio communication to the controlling Air Route Traffic Control Center (ARTCC), Control Tower, or FSS. This method provides the quickest result.

2. By telephone to the nearest FAA facility.

3. By FAA Form 8000-7, Safety Improvement Report, a postage-paid card designed for this purpose. These cards may be obtained at FAA FSSs, Flight Standards District Offices, and General Aviation Fixed Base Operations.

c. In aircraft that have more than one receiver, there are many combinations of possible interference between units. This can cause either erroneous navigation indications or, complete or partial blanking out of the communications. Pilots should be familiar enough with the radio installation of the particular airplanes they fly to recognize this type of interference.

1-1-15. LORAN

a. INTRODUCTION

1. The LOng RAnge Navigation-C (LORAN) system is a hyperbolic, terrestrial—based navigation system operating in the 90-110 kHz frequency band. LORAN, operated by the U.S. Coast Guard (USCG), has been in service for over 50 years and is used for navigation by the various transportation modes, as well as, for precise time and frequency applications. The system is configured to provide reliable, all weather navigation for marine users along the U.S. coasts and in the Great Lakes.

2. In the 1980's, responding to aviation user and industry requests, the USCG and FAA expanded LORAN coverage to include the entire continental U.S. This work was completed in late 1990, but the LORAN system failed to gain significant user acceptance and primarily due to transmitter and user equipment performance limitations, attempts to obtain FAA certification of nonprecision approach capable receivers were unsuccessful. More recently, concern regarding the vulnerability of Global Positioning System (GPS) and the consequences of losing GPS on the critical U.S. infrastructure (e.g., NAS) has renewed and refocused attention on LORAN.

3. LORAN is also supported in the Canadian airspace system. Currently, LORAN receivers are only certified for en route navigation.

U.S. and Canadian LORAN System Architecture

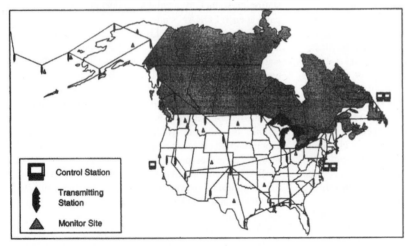

Figure 1-1-11.

LORAN Chain Based Coverage

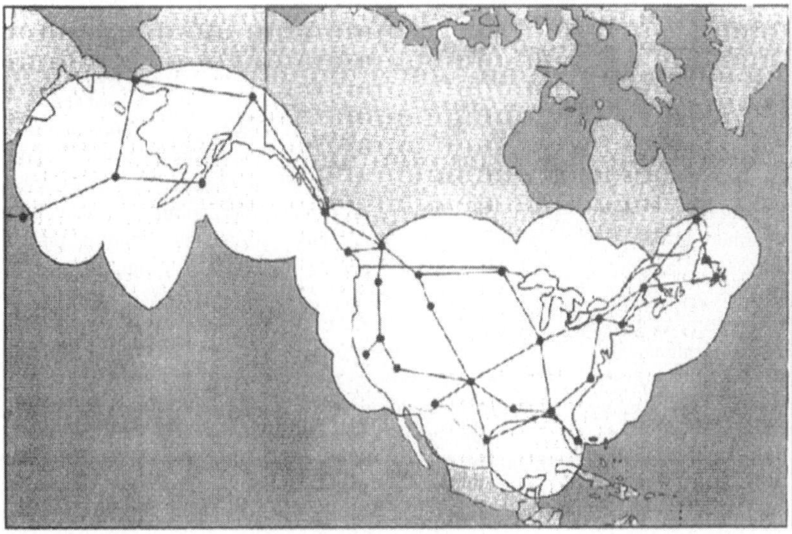

Figure 1-1-12.

4. Additional information can be found in the "LORAN-C User Handbook," COMDT PUBP16562.6, or the website **www.navcen.uscg. gov.**

b. LORAN CHAIN

1. The locations of the U.S. and Canadian LORAN transmitters and monitor sites are illustrated in FIG 1-1-11. Station operations are organized into subgroups of four to six stations called "chains." One station in the chain is designated the "Master" and the others are "secondary" stations. The resulting chain based coverage is seen in FIG 1-1-12.

2. The LORAN navigation signal is a carefully structured sequence of brief radio frequency pulses centered at

100 kHz. The sequence of signal transmissions consists of a pulse group from the Master (M) station followed at precise time intervals by groups from the secondary stations, which are designated by the U.S. Coast Guard with the letters V, W, X, Y and Z. All secondary stations radiate pulses in groups of eight, but for identification the Master signal has an additional ninth pulse. (See FIG 1-1-13.) The timing of the LORAN system is tightly controlled and synchronized to Coordinated Universal Time (UTC). Like the GPS, this is a Stratum 1 timing standard.

3. The time interval between the reoccurrence of the Master pulse group is called the Group Repetition Interval (GRI). The GRI is the same for all stations in a chain and each LORAN chain has a unique GRI. Since all stations in a particular chain operate on the same radio frequency, the GRI is the key by which a LORAN receiver can identify and isolate signal groups from a specific chain.

EXAMPLE—TRANSMITTERS IN THE NORTHEAST U.S. CHAIN (FIG 1-1-14) OPERATE WITH A GRI OF 99,600 MICROSECONDS WHICH IS SHORTENED TO 9960 FOR CONVENIENCE. THE MASTER STATION (M) AT SENECA, NEW YORK, CONTROLS SECONDARY STATIONS (W) AT CARIBOU, MAINE; (X) AT NANTUCKET, MASSACHUSETTS; (Y) AT CAROLINA BEACH, NORTH CAROLINA, AND (Z) AT DANA, INDIANA. IN ORDER TO KEEP CHAIN OPERATIONS PRECISE, MONITOR RECEIVERS ARE LOCATED AT CAPE ELIZABETH, ME; SANDY HOOK NJ; DUNBAR FOREST, MI, AND PLUMBROOLÇ OH. MONITOR RECEIVERS CONTINUOUSLY MEASURE VARIOUS ASPECTS OF THE QUALITY (E.G., PULSE SHAPE) AND ACCURACY (E.G., TIMING) OF LORAN SIGNALS AND REPORT SYSTEM STATUS TO A CONTROL STATION.

4. The line between the Master and each secondary station is the "baseline" for a pair of stations. Typical baselines are from 600 to 1,000 nautical miles in length. The continuation of the baseline in either direction is a "baseline extension."

5. At the LORAN transmitter stations there are cesium oscillators, transmitter time and control equipment, a transmitter, primary power (e.g., commercial or generator) and auxiliary power equipment (e.g., uninterruptible power supplies and generators), and a transmitting antenna (configurations may either have 1 or 4 towers) with the tower heights ranging from 700 to 1350 feet tall. Depending on the coverage area requirements a LORAN station transmits from 400 to 1,600 kilowatts of peak signal power.

6. The USCG operates the LORAN transmitter stations under a reduced staffing structure that is made possible by the remote control and monitoring of the critical station and signal parameters. The actual control of the transmitting station is accomplished remotely at Coast Guard Navigation Center (NAVCEN) located in Alexandria, Virginia. East Coast and Midwest stations are controlled by the NAVCEN. Stations on the West Coast and in Alaska are controlled by the NAVCEN Detachment (Det), located in Petaluma, California. In the event of a problem at one of these two 24 hour-a-day staffed sites, monitoring and control of the entire LORAN system can be done at either

location. If both NACEN and NAVCEN Det are down or if there is an equipment problem at a specific station, local station personnel are available to operate and perform repairs at each LORAN station.

7. The transmitted signal is also monitored in the service areas (i.e., area of published LORAN coverage) and its status provided to NAVCEN and NAVCEN Det. The System Area Monitor (SAM) is a single site used to observe the transmitted signal (signal strength, time difference, and pulse shape). If an out-of-tolerance situation that could affect navigation accuracy is detected, an alert signal called "Blink" is activated. Blink is a distinctive change in the group of eight pulses that can be recognized automatically by a receiver so the user is notified instantly that the LORAN system should not be used for navigation. Out-of-tolerance situations which only the local station can detect are also monitored. These situations when detected cause signal transmissions from a station to be halted.

8. Each individual LORAN chain provides navigation-quality signal coverage over an identified area as shown in FIG 1-1-15 for the West Coast chain, GRI 9940. The chain Master station is at Fallon, Nevada, and secondary stations are at George, Washington; Middletown, California; and Search-light, Nevada. In a signal coverage area the signal strength relative to the normal ambient radio noise must be adequate to assure successful reception. Similar coverage area charts are available for all chains.

c. THE LORAN RECEIVER

1. For a currently certified LORAN aviation receiver to provide navigation information for a pilot, it must successfully receive, or "acquire," signals from three or more stations in a chain. Acquisition involves the time synchronization of the receiver with the chain GRI, identification of the Master station signals from among those checked, identification of secondary station signals, and the proper selection of the tracking point on each signal at which measurements are made. However, a new generation of receivers has been developed that use pulses from all stations that can be received at the pilot's location. Use of "all-in-view" stations by a receiver is made possible due to the synchronization of LORAN stations signals to UTC. This new generation of receivers, along with improvements at the transmitting stations and changes in system policy and operations doctrine may allow for LORAN's use in nonprecision approaches. At this time these receivers are available for purchase, but none have been certified for aviation use.

2. The basic measurements made by certified LORAN receivers are the differences in time-of-arrival between the Master signal and the signals from each of the secondary stations of a chain. Each "time difference" (TD) value is measured to a precision of about 0.1 microseconds. As a rule of thumb, 0.1 microsecond is equal to about 100 feet.

3. An aircraft's LORAN receiver must recognize three signal conditions:

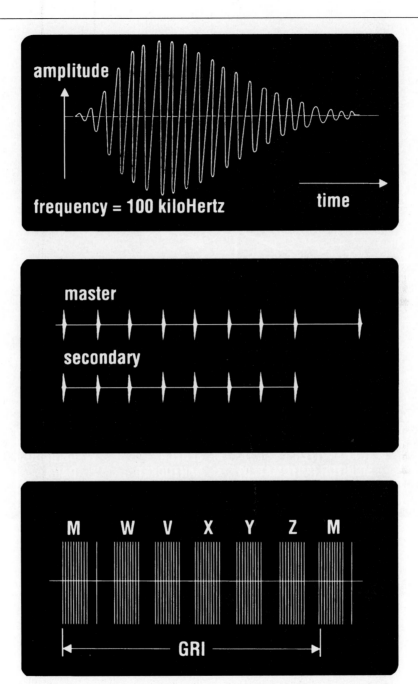

Figure 1-1-13. The LORAN Pulse and Pulse Group

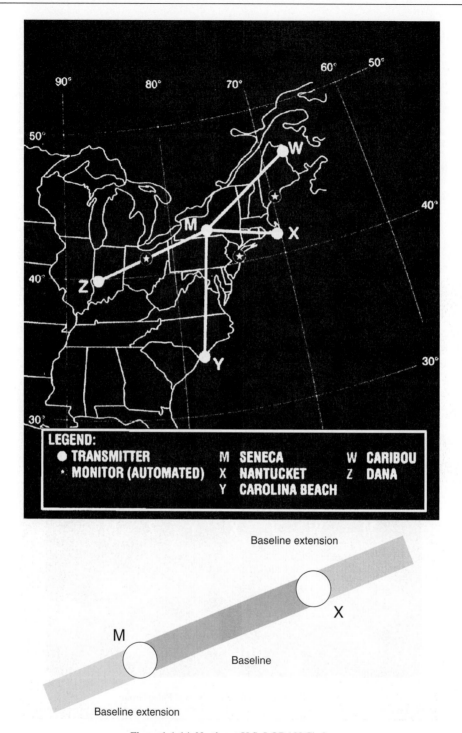

Figure 1-1-14. Northeast U.S. LORAN Chain

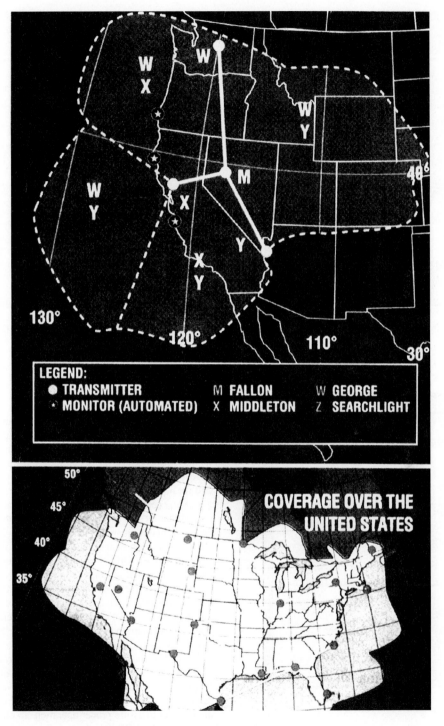

Figure 1-1-15. West Coast U.S. LORAN Chain

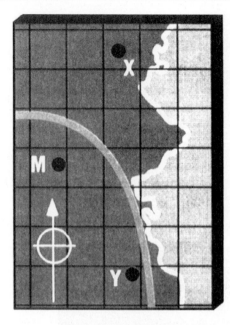

Figure 1-1-16. First line-of-position

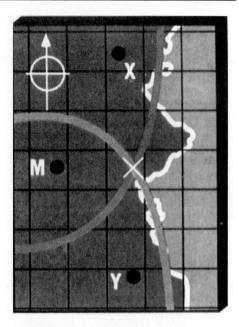

Figure 1-1-18. Intersection of lines of position

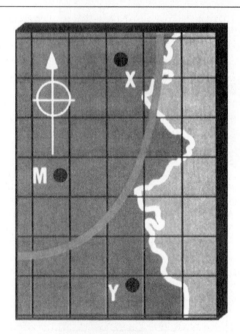

Figure 1-1-17. Second line of position

(a) Usable signals;

(b) Absence of signals, and

(c) Signal blink.

4. The most critical phase of flight is during the approach to landing at an airport. During the approach phase the receiver must detect a lost signal, or a signal Blink, within 10 seconds of the occurrence and warn the pilot of the event. At this time there are no receivers that are certified for nonprecision approaches.

5. Most certificated receivers have various internal tests for estimating the probably accuracy of the current TD values and consequent navigation solutions. Tests may include verification of the timing alignment of the receiver clock with the LORAN pulse, or a continuous measurement of the signal-to-noise ratio (SNR). SNR is the relative strength of the LORAN signals compared to the local ambient noise level. If any of the tests fail, or if the quantities measured are out of the limits set for reliable navigation, then an alarm will be activated to alert the pilot.

6. LORAN signals operate in the low frequency band (90-110 kHz) that has been reserved for marine navigation signals. Adjacent to the band, however, are numerous low frequency communications transmitters. Nearby signals can distort the LORAN signals and must be eliminated by the receiver to assure proper operation. To eliminate interfering signals, LORAN receivers have selective internal filters. These filters, commonly known as "notch filters," reduce the effect of interfering signals.

7. Careful installation of antennas, good metal-to-metal electrical bonding, and provisions for precipitation noise discharge on the aircraft are essential for the successful operation of LORAN receivers. A LORAN antenna should be installed on an aircraft in accordance with the manufacturer's instructions. Corroded bonding straps should be replaced, and static discharge devices installed at points indicated by the aircraft manufacturer.

d. LORAN NAVIGATION

1. An airborne LORAN receiver has four major parts:

(a) Signal processor;

(b) Navigation computer;

(c) Control/display, and

(d) Antenna.

2. The signal processor acquires LORAN signals and measures the difference between the time-of-arrival of each secondary station pulse group and the Master station pulse group. The measured TDs depend on the location of the receiver in relation to the three or more transmitters.

(a) The first TD will locate an aircraft somewhere on a line-of-position (LOP) on which the receiver will measure the same TD value.

(b) A second LOP is defined by a TD measurement between the Master station signal and the signal from another secondary station.

(c) The intersection of the measured LOPs is the position of the aircraft.

3. The navigation computer converts TD values to corresponding latitude and longitude. Once the time and position of the aircraft are established at two points, distance to destination, cross track error, ground speed, estimated time of arrival, etc., can be determined. Cross track error can be displayed as the vertical needle of a course deviation indicator, or digitally, as decimal parts of a mile left or right of course.

e. NOTICES TO AIRMEN (NOTAMs) are issued for LORAN chain or station outages. Domestic NOTAM (D)s are issued under the identifier "LRN." International NOTAMs are issued under the KNMH series. Pilots may obtain these NOTAMs from FSS briefers upon request.

f. LORAN STATUS INFORMATION. To find out more information on the LORAN system and its operational status you can visit **http://www.navcen.uscg.gov/loran/default.htm** or contact NAVCEN's Navigation Information Service (NIS) watchstander, phone (703) 313-5900, fax (703) 313-5920.

g. LORAN'S FUTURE. The U.S. will continue to operate the LORAN system in the short term. During this time, the FAA LORAN evaluation program, being conducted with the support of a team comprising government, academia, and industry, will identify and assess LORAN's potential contributions to required navigation services for the National Airspace System (NAS), and support decisions regarding continued operation of the system. If the government concludes LORAN should not be kept as part of the mix of federally provided radio navigation systems, it will give the users of LORAN reasonable notice so that they will have the opportunity to transition to alternative navigation aids.

1-1-16. VHF DIRECTION FINDER

a. The VHF Direction Finder (VHF/DF) is one of the common systems that helps pilots without their being aware of its operation. It is a ground based radio receiver used by the operator of the ground station. FAA facilities that provide VHF/DF service are identified in the A/FD.

b. The equipment consists of a directional antenna system and a VHF radio receiver.

c. The VHF/DF receiver display indicates the magnetic direction of the aircraft from the ground station each time the aircraft transmits.

d. DF equipment is of particular value in locating lost aircraft and in helping to identify aircraft on radar.

REFERENCE—AIM, DIRECTION FINDING INSTRUMENT APPROACH PROCEDURE, PARAGRAPH 6-2-3.

1-1-17. INERTIAL REFERENCE UNIT (IRU), INERTIAL NAVIGATION SYSTEM (INS), AND ATTITUDE HEADING REFERENCE SYSTEM (AHRS)

a. IRU's are self-contained systems comprised of gyros and accelerometers that provide aircraft attitude (pitch, roll, and heading), position, and velocity information in response to signals resulting from inertial effects on system components. Once aligned with a known position, IRU's continuously calculate position and velocity. IRU position accuracy decays with time. This degradation is known as "drift."

b. INS's combine the components of an IRU with an internal navigation computer. By programming a series of waypoints, these systems will navigate along a predetermined track.

c. AHRS's are electronic devices that attitude information to aircraft systems weather radar and autopilot, but do not compute position information.

1-1-18. DOPPLER RADAR

Doppler Radar is a semiautomatic self-contained dead reckoning navigation system (radar sensor plus computer) which is not continuously dependent on information derived from ground based or external aids. The system employs radar signals to detect and measure ground speed and drift angle, using the aircraft compass system as its directional reference. Doppler is less accurate than INS or OMEGA, however, and the use of an external reference is required for periodic updates if acceptable position accuracy is to be achieved on long range flights.

1-1-19. GLOBAL POSITIONING SYSTEM (GPS)

a. SYSTEM OVERVIEW

1. System Description. The Global Positioning System is a satellite-based radio navigation system, which broadcasts a signal that is used by receivers to determine precise position anywhere in the world. The receiver tracks multiple satellites and determines a pseudorange measurement that is then used to determine the user location. A minimum of four satellites is necessary to establish an accurate three-dimensional position. The Department of Defense (DOD) is responsible for operating the GPS satellite constellation and monitors the GPS satellites to ensure proper operation. Every satellite's orbital parameters (ephemeris data) are sent to each satellite for broadcast as part of the data message embedded in the GPS signal. The GPS coordinate system is the Cartesian earth-centered earth-fixed coordinates as specified in the World Geodetic System 1984 (WGS-84).

2. System Availability and Reliability

(a) The status of GPS satellites is broadcast as part of the data message transmitted by the GPS satellites. GPS status information is also available by means of the U.S. Coast Guard navigation information service: (703) 313-5907, Internet: http://www.navcen.uscg.gov/. Additionally, satellite status is available through the Notice to Airmen (NOTAM) system.

(b) The operational status of GNSS operations depends upon the type of equipment being used. For GPS-only equipment TSO-C 129(a), the operational status of nonprecision approach capability for flight planning purposes is provided through a prediction program that is embedded in the receiver or provided separately.

3. Receiver Autonomous Integrity Monitoring (RAIM). When GNSS equipment is not using integrity information from WAAS or LAAS, the GPS navigation receiver using RAIM provides GPS signal integrity monitoring. RAIM is necessary since delays of up to two hours can occur before an erroneous satellite transmission can be detected and corrected by the satellite control segment. The RAIM function is also referred to as fault detection. Another capability, fault exclusion, refers to the ability of the receiver to exclude a failed satellite from the position solution and is provided by some GPS receivers and by WAAS receivers.

4. The GPS receiver verifies the integrity (usability) of the signals received from the GPS constellation through receiver autonomous integrity monitoring (RAIM) to determine if a satellite is providing corrupted information. At least one satellite, in addition to those required for navigation, must be in view for the receiver to perform the RAIM function; thus, RAIM needs a minimum of 5 satellites in view, or 4 satellites and a barometric altimeter (baro-aiding) to detect an integrity anomaly. For receivers capable of doing so, RAIM needs 6 satellites in view (or 5 satellites with baro-aiding) to isolate the corrupt satellite signal and remove it from the navigation solution. Baro-aiding is a method of augmenting the GPS integrity solution by using a nonsatellite input source. GPS derived altitude should not be relied upon to determine aircraft altitude since the vertical error can be quite large and no integrity is provided. To ensure that baro-aiding is available, the current altimeter setting must be entered into the receiver as described in the operating manual.

5. RAIM messages vary somewhat between receivers; however, generally there are two types. One type indicates that there are not enough satellites available to provide RAIM integrity monitoring and another type indicates that the RAIM integrity monitor has detected a potential error that exceeds the limit for the current phase of flight. Without RAIM capability, the pilot has no assurance of the accuracy of the GPS position.

6. Selective Availability. Selective Availability (SA) is a method by which the accuracy of GPS is intentionally degraded. This feature is designed to deny hostile use of precise GPS positioning data. SA was discontinued on May 1, 2000, but many GPS receivers are designed to assume that SA is still active. New receivers may take advantage of the discontinuance of SA based on the performance values in ICAO Annex 10, and do not need to be designed to operate outside of that performance.

7. The GPS constellation of 24 satellites is designed so that a minimum of five is always observable by a user anywhere on earth. The receiver uses data from a minimum of four satellites above the mask angle (the lowest angle above the horizon at which it can use a satellite).

8. The DOD declared initial operational capability (IOC) of the U.S. GPS on December 8, 1993. The FAA has granted approval for U.S. civil operators to use properly certified GPS equipment as a primary means of navigation in oceanic airspace and certain remote areas. Properly certified GPS equipment may be used as a supplemental means of IFR navigation for domestic en route, terminal operations, and certain instrument approach procedures (IAPs). This approval permits the use of GPS in a manner that is consistent with current navigation requirements as well as approved air carrier operations specifications.

b. VFR USE OF GPS

1. GPS navigation has become a great asset to VFR pilots, providing increased navigation capability and enhanced situational awareness, while reducing operating costs due to greater ease in flying direct routes. While GPS has many benefits to the VFR pilot, care must be exercised to ensure that system capabilities are not exceeded.

2. Types of receivers used for GPS navigation under VFR are varied, from a full IFR installation being used to support a VFR flight, to a VFR only installation (in either a VFR or IFR capable aircraft) to a hand-held receiver. The limitations of each type of receiver installation or use must be understood by the pilot to avoid misusing navigation information. (See TBL 1-1-8.) In all cases, VFR pilots should never rely solely on one system of navigation. GPS navigation must be integrated with other forms of electronic navigation (when possible), as well as pilotage and dead reckoning. Only through the integration of these techniques can the VFR pilot ensure accuracy in navigation.

3. Some critical concerns in VFR use of GPS include RAIM capability, data base currency and antenna location.

(a) RAIM Capability. Many VFR GPS receivers and all hand-held units have no RAIM alerting capability. Loss of the required number of satellites in view, or the detection of a position error, cannot be displayed to the pilot by such receivers. In receivers with no RAIM capability, no alert would be provided to the pilot that the navigation solution had deteriorated, and an undetected navigation error could occur. A systematic cross-check with other navigation techniques would identify this failure, and prevent a serious deviation. See subparagraphs a8 and a9 for more information on RAIM.

(b) Database Currency

(1) In many receivers, an up-datable database is used for navigation fixes, airports, and instrument procedures. These databases must be maintained to the current update for IFR operation, but no such requirement exists for VFR use.

(2) However, in many cases, the database drives a moving map display which indicates Special Use Airspace and the various classes of airspace, in addition to other operational information. Without a current database the moving map display may be outdated and offer erroneous information to VFR pilots wishing to fly around critical airspace areas, such as a Restricted Area or a Class B airspace segment. Numerous pilots have ventured into airspace they were trying to avoid by using an outdated database. If you don't have a current database in the receiver, disregard the moving map display for critical navigation decisions.

(3) In addition, waypoints are added, removed, relocated, or re-named as required to meet operational needs. When using GPS to navigate relative to a named fix, a current database must be used to properly locate a named waypoint. Without the update, it is the pilot's responsibility to verify the waypoint location referencing to an official current source, such as the Airport/Facility Directory, Sectional Chart, or En Route Chart.

(c) Antenna Location

(1) In many VFR installations of GPS receivers, antenna location is more a matter of convenience than performance. In IFR installations, care is exercised to ensure that an adequate clear view is provided for the antenna to see satellites. If an alternate location is used, some portion of the aircraft may block the view of the antenna, causing a greater opportunity to lose navigation signal.

(2) This is especially true in the case of hand-helds. The use of hand-held receivers for VFR operations is a growing trend, especially among rental pilots. Typically, suction cups are used to place the GPS antennas on the inside of cockpit windows. While this method has great utility, the antenna location is limited to the cockpit or cabin only and is rarely optimized to provide a clear view of available satellites. Consequently, signal losses may occur in certain situations of aircraft-satellite geometry,

causing a loss of navigation signal. These losses, coupled with a lack of RAIM capability, could present erroneous position and navigation information with no warning to the pilot.

(3) While the use of a hand-held GPS for VFR operations is not limited by regulation, modification of the aircraft, such as installing a panel- or yoke-mounted holder, is governed by 14 CFR Part 43. Consult with your mechanic to ensure compliance with the regulation, and a safe installation.

4. As a result of these and other concerns, here are some tips for using GPS for VFR operations:

(a) Always check to see if your unit has RAIM capability. If no RAIM capability exists, be suspicious of your GPS position when any disagreement exists with the position derived from other radio navigation systems, pilotage, or dead reckoning.

(b) Check the currency of the database, if any. If expired, update the database using the current revision. If an update of an expired database is not possible, disregard any moving map display of airspace for critical navigation decisions. Be aware that named waypoints may no longer exist or may have been relocated since the database expired. At a minimum, the waypoints planned to be used should be checked against a current official source, such as the Airport/Facility Directory, or a Sectional Aeronautical Chart.

(c) While hand-helds can provide excellent navigation capability to VFR pilots, be prepared for intermittent loss of navigation signal, possibly with no RAIM warning to the pilot. If mounting the receiver in the aircraft, be sure to comply with 14 CFR Part 43.

(d) Plan flights carefully before taking off. If you wish to navigate to user-defined waypoints, enter them before flight, not on-the-fly. Verify your planned flight against a current source, such as a current sectional chart. There have been cases in which one pilot used waypoints created by another pilot that were not where the pilot flying was expecting. This generally resulted in a navigation error. Minimize head-down time in the aircraft and keep a sharp look out for traffic, terrain, and obstacles. Just a few minutes of preparation and planning on the ground will make a great difference in the air.

(e) Another way to minimize head-down time is to become very familiar with your receiver's operation. Most receivers are not intuitive. The pilot must take the time to learn the various keystrokes, knob functions, and displays that are used in the operation of the receiver. Some manufacturers provide computer-based tutorials or simulations of their receivers. Take the time to learn about your particular unit before you try to use in flight.

5. In summary, be careful not to rely on GPS to solve all your VFR navigational problems. Unless an IFR receiver is installed in accordance with IFR requirements, no standard of accuracy or integrity has been assured. While the practicality of GPS is compelling, the fact remains that only the pilot can navigate the aircraft, and GPS is just one of the pilot's tools to do the job.

c. VFR WAYPOINTS

1. VFR waypoints provide VFR pilots with a supplementary tool to assist with position awareness while navigating visually in aircraft equipped with area navigation receivers. VFR waypoints should be used as a tool to supplement current navigation procedures. The uses of VFR waypoints include providing navigational aids for pilots unfamiliar with an area, waypoint definition of existing reporting points, enhanced navigation in and around Class B and Class C airspace, and enhanced navigation around Special Use Airspace. VFR pilots should rely on appropriate and current aeronautical charts published specifically for visual navigation. If operating in a terminal area, pilots should take advantage of the Terminal Area Chart available for that area, if published. The use of VFR waypoints does not relieve the pilot of any responsibility to comply with the operational requirements of 14 CFR Part 91.

2. VFR waypoint names (for computer-entry and flight plans) consist of five letters beginning with the letters "VP" and are retrievable from navigation databases. The VFR waypoint names are not intended to be pronounceable, and they are not for use in ATC communications. On VFR charts, stand-alone VFR waypoints will be portrayed using the same four-point star symbol used for IFR waypoints. VFR waypoints collocated with visual check points on the chart will be identified by small magenta flag symbols. VFR waypoints collocated with visual check points will be pronounceable based on the name of the visual check point and may be used for ATC communications. Each VFR waypoint name will appear in parentheses adjacent to the geographic location on the chart. Latitude/longitude data for all established VFR waypoints may be found in the appropriate regional Airport/Facility Directory (A/FD).

3. VFR waypoints shall not be used to plan flights under IFR. VFR waypoints will not be recognized by the IFR system and will be rejected for IFR routing purposes.

4. When filing VFR flight plans, pilots may use the five letter identifier as a waypoint in the route of flight section if there is an intended course change at that point or if used to describe the planned route of flight. This VFR filing would be similar to how a VOR would be used in a route of flight. Pilots must use the VFR waypoints only when operating under VFR conditions.

5. Any VFR waypoints intended for use during a flight should be loaded into the receiver while on the ground and prior to departure. Once airborne, pilots should avoid programming routes or VFR waypoint chains into their receivers.

6. Pilots should be especially vigilant for other traffic while operating near VFR waypoints. The same effort to see and avoid other aircraft near VFR waypoints will be necessary, as was the case with VOR's and NDB's in the past. In fact, the increased accuracy of navigation through the use of GPS will demand even greater vigilance, as off-course deviations among different pilots and receivers will be less. When operating near a VFR waypoint, use whatever ATC services are available, even if outside a class of airspace where communications are required. Regardless of the class of airspace, monitor the available ATC frequency closely for information on other aircraft operating in the vicinity. It is also a good idea to turn on your landing light(s) when operating near a VFR waypoint to make your aircraft more conspicious to other pilots, especially when visibility is reduced. See paragraph 7-5-2, VFR in Congested Areas, for more information.

d. GENERAL REQUIREMENTS

1. Authorization to conduct any GPS operation under IFR requires that:

(a) GPS navigation equipment used must be approved in accordance with the requirements specified in Technical Standard Order (TSO) C-129, or equivalent, and the installation must be done in accordance with Advisory Circular AC 20-138, Airworthiness Approval of Global Positioning System (GPS) Navigation Equipment for Use as a VFR and IFR Supplemental Navigation System, or Advisory Circular AC 20-130A, Airworthiness Approval of Navigation or Flight Management Systems Integrating Multiple Navigation Sensors, or equivalent. Equipment approved in accordance with TSO C-115a does not meet the requirements of TSO C-129. Visual flight rules (VFR) and hand-held GPS systems are not authorized for IFR navigation, instrument approaches, or as a principal instrument flight reference. During IFR operations they may be considered only an aid to situational awareness.

(b) Aircraft using GPS navigation equipment under IFR must be equipped with an approved and operational alternate means of navigation appropriate to the flight. Active monitoring of alternative navigation equipment is not required if the GPS receiver uses RAIM for integrity monitoring. Active monitoring of an alternate means of navigation is required when the RAIM capability of the GPS equipment is lost.

(c) Procedures must be established for use in the event that the loss of RAIM capability is predicted to occur. In situations where this is encountered, the flight must rely on other approved equipment, delay departure, or cancel the flight.

(d) The GPS operation must be conducted in accordance with the FAA-approved aircraft flight manual (AFM) or flight manual supplement. Flight crew members must be thoroughly familiar with the particular GPS equipment installed in the aircraft, the receiver operation manual, and the AFM or flight manual supplement. Unlike ILS and VOR, the basic operation, receiver presentation to the pilot, and some capabilities of the equipment can vary

Table 1-1-5   GPS IFR Equipment Classes/Categories

Equipment Class	RAIM	INT. NAV SYS. TO PROV. RAIM EQUIV.	OCEANIC	ENROUTE	TERMINAL	NON-PREC. APPROACH CAPABLE
Class A—GPS sensor and navigation capability.						
A1	yes		yes	yes	yes	yes
A2	yes		yes	yes	yes	no
Class B—GPS sensor data to an integrated navigation system (i.e., FMS, multi-sensor navigation system, etc.).						
B1	yes		yes	yes	yes	yes
B2	yes		yes	yes	yes	no
B3		yes	yes	yes	yes	yes
B4		yes	yes	yes	yes	no
Class C—GPS sensor data to an integrated navigation system (as in Class B) which provides enhanced guidance to an autopilot, or flight director, to reduce flight tech. errors. Limited to FAR 121 or equivalent criteria.						
C1	yes		yes	yes	yes	yes
C2	yes		yes	yes	yes	no
C3		yes	yes	yes	yes	yes
C4		yes	yes	yes	yes	no

greatly. Due to these differences, operation of different brands, or even models of the same brand, of GPS receiver under IFR should not be attempted without thorough study of the operation of that particular receiver and installation. Most receivers have a built-in simulator mode which will allow the pilot to become familiar with operation prior to attempting operation in the aircraft. Using the equipment in flight under VFR conditions prior to attempting IFR operation will allow further familiarization.

(e) Aircraft navigating by IFR approved GPS are considered to be RNAV aircraft and have special equipment suffixes. File the appropriate equipment suffix in accordance with TBL 5-1-2, on the ATC flight plan. If GPS avionics become inoperative, the pilot should advise ATC and amend the equipment suffix.

(f) Prior to any GPS IFR operation, the pilot must review appropriate NOTAM's and aeronautical information. (See GPS NOTAM's/Aeronautical Information.)

(g) Air carrier and commercial operators must meet the appropriate provisions of their approved operations specifications.

e. USE OF GPS FOR IFR OCEANIC, DOMESTIC EN ROUTE, AND TERMINAL AREA OPERATIONS

1. GPS IFR operations in oceanic areas can be conducted as soon as the proper avionics systems are installed, provided all general requirements are met. A GPS installation with TSO C-129 authorization in class A1, A2, B1, B2, C1, or C2 may be used to replace one of the other approved means of long-range navigation, such as dual INS. (See TBL 1-1-5 and TBL 1-1-6.) A single GPS installation

Table 1-1-6   GPS Approval Required/Authorized Use

EQUIPMENT TYPE[1]	INSTALLATION APPROVAL REQUIRED	OPERATIONAL APPROVAL REQUIRED	IFR EN ROUTE[2]	IFR TERMINAL[2]	IFR APPROACH[3]	OCEANIC REMOTE	IN LIEU OF ADF AND/OR DME[3]
Hand held[4]	X[5]						
VFR Panel Mount[4]	X						
IFR En Route and Terminal	X	X	X	X			X
IFR Oceanic/ Remote	X	X	X	X		X	X
IFR En Route, Terminal, and Approach	X	X	X	X	X		X

NOTE-

[1]*To determine equipment approvals and limitations, refer to the AFM, AFM supplements, or pilot guides.*

[2]*Requires verification of data for correctness if database is expired.*

[3]*Requires current database.*

[4]*VFR and hand-held GPS systems are not authorized for IFR navigations, instrument approaches, or as a primary instrument flight reference. During IFR operations they may be considered only an aid to situational awareness.*

[5]*Hand-held receivers require no approval. However, any aircraft modification to support the hand-held receiver, i.e., installation of an external antenna or a permanent mounting bracket, does require approval.*

with these classes of equipment which provide RAIM for integrity monitoring may also be used on short oceanic routes which have only required one means of long-range navigation.

2. GPS domestic en route and terminal IFR operations can be conducted as soon as proper avionics systems are installed, provided all general requirements are met. The avionics necessary to receive all of the ground-based facilities appropriate for the route to the destination airport and any required alternate airport must be installed and operational. Ground-based facilities necessary for these routes must also be operational.

(a) GPS en route IFR RNAV operations may be conducted in Alaska outside the operational service volume of ground-based navigation aids when a TSO-C145a or TSO-C146a GPS/WAAS system is installed and operating. Ground-based navigation equipment is not required to be installed and operating for en route IFR RNAV operations when using GPS WAAS navigation systems. All operators should ensure that an alternate means of navigation is available in the unlikely event the GPS WAAS navigation system becomes inoperative.

3. The GPS Approach Overlay Program is an authorization for pilots to use GPS avionics under IFR for flying designated nonprecision instrument approach procedures, except LOC, LDA, and simplified directional facility (SDF) procedures. These procedures are now identified by the name of the procedure and "or GPS" (e.g., VOR/DME or GPS RWY 15). Other previous types of overlays have either been converted to this format or replaced with stand-alone procedures. Only approaches contained in the current onboard navigation database are authorized. The navigation database may contain information about nonoverlay approach procedures that is intended to be used to enhance position orientation, generally by providing a map, while flying these approaches using conventional NAVAID's. This approach information should not be confused with a GPS overlay approach (see the receiver operating manual, AFM, or AFM Supplement for details on how to identify these approaches in the navigation database).

NOTE—
OVERLAY APPROACHES ARE PREDICATED UPON THE DESIGN CRITERIA OF THE GROUND-BASED NAVAID USED AS THE BASIS OF THE APPROACH. AS SUCH, THEY DO NOT ADHERE TO THE DESIGN CRITERIA DESCRIBED IN PARAGRAPH 5-4-54I, AREA NAVIGATION (RNAV) INSTRUMENT APPROACH CHARTS FOR STAND-ALONE GPS APPROACHES.

4. GPS IFR approach operations can be conducted as soon as proper avionics systems are installed and the following requirements are met:

(a) The authorization to use GPS to fly instrument approaches is limited to U.S. airspace.

(b) The use of GPS in any other airspace must be expressly authorized by the FAA Administrator.

(c) GPS instrument approach operations outside the United States must be authorized by the appropriate sovereign authority.

F. USE OF GPS IN LIEU OF ADF AND DME

Subject to the restrictions below, operators in the U.S. NAS are authorized to use GPS equipment certified for IFR operations in place of ADF and/or DME equipment for en route and terminal operations. For some operations there is no requirement for the aircraft to be equipped with an ADF or DME receiver, see subparagraphs c. 6, (g) and (h) below. The ground based NDB or DME facility may be temporarily out of service during these operations. Charting will not change to support these operations.

(a) Operations allowed: (1) Determining the aircraft position over a DME fix. GPS satisfies the 14 CFR Section 91.205(e) requirement for DME at and above 24,000 feet MSL (FL 240).

(2) Flying a DME arc.

(3) Navigating to/from an NDB/compass locator.

(4) Determining the aircraft position over an NDB/compass locator.

(5) Determining the aircraft position over a fix defined by an NDB/compass locator bearing crossing a VOR/LOC course.

(6) Holding over an NDB/compass locator.

NOTE—
THIS APPROVAL DOES NOT ALTER THE CONDITIONS AND REQUIREMENTS FOR USE OF GPS TO FLY EXISTING NONPRECISION INSTRUMENT APPROACH PROCEDURES AS DEFINED IN THE GPS APPROACH OVERLAY PROGRAM.

(b) Restrictions

(1) GPS avionics approved for terminal IFR operations may be used in lieu of ADF and/or DME. Included in this approval are both stand-alone and multi-sensor systems actively employing GPS as a sensor. This equipment must be installed in accordance with appropriate airworthiness installation requirements and the provisions of the applicable FAA approved Aircraft Flight Manual (AFM), AFM supplement, or pilot's guide must be met. The required integrity for these operations must be provided by at least en route Receiver Autonomous Integrity Monitoring (RAIM), or an equivalent method; i.e., Wide Area Augmentation System (WAAS).

(2) For air carriers and operators for compensation or hire, POI and operations specification approval is required for any use of GPS.

(3) Waypoints, fixes, intersections and facility locations to be used for these operations must be retrieved from the GPS airborne database. The database must be current. If the required positions cannot be retrieved from the airborne database, the substitution of GPS for ADF and/or DME is not authorized.

(4) The aircraft GPS system must be operated within the guidelines contained in the AFM, AFM supplement, or pilot's guide.

(5) The Course Deviation Indicator (CDI) must be set to terminal sensitivity (normally 1 or 1¼ NM) when tracking GPS course guidance in the terminal area. This is to ensure that small deviations from course are displayed to the pilot in order to keep the aircraft within the smaller terminal protected areas.

(6) Charted requirements for ADF and/or DME can be met using the GPS system, except for use as the principal instrument approach navigation source.

(7) Procedures must be established for use in the event that GPS integrity outages are predicted or occur (RAIM annunciation). In these situations, the flight must rely on other approved equipment; this may require the aircraft to be equipped with operational NDB and/or DME receivers. Otherwise, the flight must be rerouted, delayed, canceled or conducted VFR.

(8) For TSO-C129/129A users, any required alternate airport must still have an approved instrument approach procedure other than GPS that is anticipated to be operational and available at the estimated time of arrival, and which the aircraft is equipped to fly. If the non-GPS approaches on which the pilot must rely require DME or ADF, the aircraft must be equipped with DME or ADF avionics as appropriate.

NOTE—
COINCIDENT WITH WAAS COMMISSIONING, THE FAA WILL BEGIN REMOVING THE ▲NA (ALTERNATE MINIMUMS NOT AUTHORIZED) SYMBOL FROM SELECT RNAV (GPS) AND GPS APPROACH PROCEDURES SO THEY MAY BE USED BY APPROACH APPROVED WAAS RECEIVERS AT ALTERNATE AIRPORTS. THIS DOES NOT CHANGE THE ABOVE ALTERNATE AIRPORT REQUIREMENTS FOR USERS OF GPS TSO-C129/129A, AIRBORNE SUPPLEMENTAL NAVIGATION EQUIPMENT USING THE GLOBAL POSITIONING SYSTEM (GPS), RECEIVERS.

(c) Guidance. The following provides general guidance which is not specific to any particular aircraft GPS system. For specific system guidance refer to the AFM, AFM supplement, pilot's guide, or contact the manufacturer of your system.

(1) To determine the aircraft position over a DME fix:

[a] Verify aircraft GPS system integrity monitoring is functioning properly and indicates satisfactory integrity.

[b] If the fix is identified by a five letter name which is contained in the GPS airborne database, you may select either the named fix as the active GPS waypoint (WP) or the facility establishing the DME fix as the active GPS WP.

NOTE—
WHEN USING A FACILITY AS THE ACTIVE WP, THE ONLY ACCEPTABLE FACILITY IS THE DME FACILITY WHICH IS CHARTED AS THE ONE USED TO ESTABLISH THE DME FIX. IF THIS FACILITY IS NOT IN YOUR AIRBORNE DATABASE, YOU ARE NOT AUTHORIZED TO USE A FACILITY WP FOR THIS OPERATION.

[c] If the fix is identified by a five letter name which is not contained in the GPS airborne database, or if the fix is not named, you must select the facility establishing the DME fix or another named DME fix as the active GPS WP.

NOTE—
AN ALTERNATIVE, UNTIL ALL DME SOURCES ARE IN THE DATABASE, IS USING A NAMED DME FIX AS THE ACTIVE WAYPOINT TO IDENTIFY UNNAMED DME FIXES ON THE SAME COURSE AND FROM THE SAME DME SOURCE AS THE ACTIVE WAYPOINT.

CAUTION—
PILOTS SHOULD BE EXTREMELY CAREFUL TO ENSURE THAT CORRECT DISTANCE MEASUREMENTS ARE USED WHEN UTILIZING THIS INTERIM METHOD. IT IS STRONGLY RECOMMENDED THAT PILOTS REVIEW DISTANCES FOR DME FIXING DURING PREFLIGHT PREPARATION.

[d] If you select the named fix as your active GPS WP, you are over the fix when the GPS system indicates you are at the active WP.

[e] If you select the DME providing facility as the active GPS WP, you are over the fix when the GPS distance from the active WP equals the charted DME value and you are on the appropriate bearing or course.

(2) To fly a DME arc:

[a] Verify aircraft GPS system integrity monitoring is functioning properly and indicates satisfactory integrity.

[b] You must select, from the airborne database, the facility providing the DME arc as the active GPS WP.

NOTE—
THE ONLY ACCEPTABLE FACILITY IS THE DME FACILITY ON WHICH THE ARC IS BASED. IF THIS FACILITY IS NOT IN YOUR AIRBORNE DATABASE, YOU ARE NOT AUTHORIZED TO PERFORM THIS OPERATION.

[c] Maintain position on the arc by reference to the GPS distance in lieu of a DME readout.

(3) To navigate to or from an NDB/compass locator:

NOTE—
IF THE CHART DEPICTS THE COMPASS LOCATOR COLLOCATED WITH A FIX OF THE SAME NAME, USE OF THAT FIX AS THE ACTIVE WP IN PLACE OF THE COMPASS LOCATOR FACILITY IS AUTHORIZED.

[a] Verify aircraft GPS system integrity monitoring is functioning properly and indicates satisfactory integrity.

[b] Select terminal CDI sensitivity in accordance with the AFM, AFM supplement, or pilot's guide if in the terminal area.

[c] Select the NDB/compass locator facility from the airborne database as the active WP.

[d] Select and navigate on the appropriate course to or from the active WP.

(4) To determine the aircraft position over an NDB/compass locator:

[a] Verify aircraft GPS system integrity monitoring is functioning properly and indicates satisfactory integrity.

[b] Select the NDB/compass locator facility from the airborne database as the active WP.

NOTE—
WHEN USING AN NDB/COMPASS LOCATOR, THAT FACILITY MUST BE CHARTED AND BE IN THE AIRBORNE DATABASE. IF THIS FACILITY IS NOT IN YOUR AIRBORNE DATABASE, YOU ARE NOT AUTHORIZED TO USE A FACILITY WP FOR THIS OPERATION.

[c] You are over the NDB/compass locator when the GPS system indicates you are at the active WP.

(5) To determine the aircraft position over a fix made up of an NDB/compass locator bearing crossing a VOR/LOC course:

[a] Verify aircraft GPS system integrity monitoring is functioning properly and indicates satisfactory integrity.

[b] A fix made up by crossing NDB/compass locator bearing will be identified by a five letter fix name. You may select either the named fix or the NDB/compass locator facility providing the crossing bearing to establish the fix as the active GPS WP.

NOTE—
WHEN USING AN NDB/COMPASS LOCATOR, THAT FACILITY MUST BE CHARTED AND BE IN THE AIRBORNE DATABASE. IF THIS FACILITY IS NOT IN YOUR AIRBORNE DATABASE, YOU ARE NOT AUTHORIZED TO USE A FACILITY WP FOR THIS OPERATION.

[c] If you select the named fix as your active GPS WP, you are over the fix when the GPS system indicates you are at the WP as you fly the prescribed track from the non-GPS navigation source.

[d] If you select the NDB/compass locator facility as the active GPS WP, you are over the fix when the GPS bearing to the active WP is the same as the charted NDB/compass locator bearing for the fix as you fly the prescribed track from the non-GPS navigation source.

(6) To hold over an NDB/compass locator:

[a] Verify aircraft GPS system integrity monitoring is functioning properly and indicates satisfactory integrity.

[b] Select terminal CID sensitivity in accordance with the AFM, AFM supplement, or pilot's guide if in the terminal area.

[c] Select the NDB/compass locator facility from the airborne database as the active WP.

NOTE—
WHEN USING A FACILITY AS THE ACTIVE WP, THE ONLY ACCEPTABLE FACILITY IS THE NDB/COMPASS LOCATOR FACILITY WHICH IS CHARTED. IF THIS FACILITY IS NOT IN YOUR AIRBORNE DATABASE, YOU ARE NOT AUTHORIZED TO USE A FACILITY WP FOR THIS OPERATION.

[d] Select nonsequencing (e.g., "HOLD" or "OBS") mode and the appropriate course in accordance with the AFM, AFM supplement, or pilot's guide.

[e] Hold using the GPS system in accordance with the AFM, AFM supplement, or pilot's guide.

(d) Planning. Good advance planning and intimate knowledge of your navigational systems are vital to safe and successful use of GPS in lieu of ADF and/or DME.

(1) You should plan ahead before using GPS systems as a substitute for ADF and/or DME. You will have several alternatives in selecting waypoints and system configuration. After you are cleared for the approach is not the time to begin programming your GPS. In the flight planning process you should determine whether you will use the equipment in the automatic sequencing mode or in the nonsequencing mode and select the waypoints you will use.

(2) When you are using your aircraft GPS system to supplement other navigation systems, you may need to bring your GPS control panel into your navigation scan to see the GPS information. Some GPS aircraft installations will present localizer information on the CDI whenever a localizer frequency is tuned, removing the GPS information from the CDI display. Good advance planning and intimate knowledge of your navigation systems are vital to safe and successful use of GPS.

(3) The following are some factors to consider when preparing to install a GPS receiver in an aircraft. Installation of the equipment can determine how easy or how difficult it will be to use the system.

[a] Consideration should be given to installing the receiver within the primary instrument scan to facilitate using the GPS in lieu of ADF and/or DME. This will preclude breaking the primary instrument scan while flying the aircraft and tuning, and identifying waypoints. This becomes increasingly important on approaches, and missed approaches.

[b] Many GPS receivers can drive an ADF type bearing pointer. Such an installation will provide the pilot with an enhanced level of situational awareness by providing GPS navigation information while the CDI is set to VOR or ILS.

[c] The GPS receiver may be installed so that when an ILS frequency is tuned, the navigation display defaults to the VOR/ILS mode, preempting the GPS mode. However, if the receiver installation requires a manual selection from GPS to ILS, it allows the ILS to be tuned and identified while navigating on the GPS. Additionally, this prevents the navigation display from automatically switching back to GPS when a VOR frequency is selected. If the navigation display automatically switches to GPS mode when a VOR is selected, the change may go unnoticed and could result in erroneous navigation and departing obstruction protected airspace.

[d] GPS is a supplemental navigation system in part due to signal availability. There will be times when your system will not receive enough satellites with proper geometry to provide accurate positioning or sufficient integrity. Procedures should be established by the pilot in the event that GPS outages occur. In these situations, the pilot should rely on other approved equipment, delay departure, or discontinue IFR operations.

g. EQUIPMENT AND DATABASE REQUIREMENTS.

1. Authorization to fly approaches under IFR using GPS avionics systems requires that:

(a) A pilot use GPS avionics with TSO C-129, or equivalent, authorization in class A1, B1, B3, C1, or C3; and

(b) All approach procedures to be flown must be retrievable from the current airborne navigation data base supplied by the TSO C-129 equipment manufacturer or other FAA approved source.

(c) Prior to using a procedure or waypoint retrieved from the airborne navigation database, the pilot should verify the validity of the database. This verification should include the following preflight and in-flight steps:

(1) Preflight:

[a] Determine the date of database issuance, and verify that the date/time of proposed use is before the expiration date/time.

[b] Verify that the database provider has not published a notice limiting the use of the specific waypoint or procedure.

(2) Inflight:

[a] Determine that the waypoints and transition names coincide with names found on the procedure chart. Do not use waypoints, which do not exactly match the spelling shown on published procedure charts.

[b] Determine that the waypoints are generally logical in location, in the correct order, and that their orientation to each other is as found on the procedure chart, both laterally and vertically.

NOTE—
THERE IS NO SPECIFIC REQUIREMENT TO CHECK EACH WAYPOINT LATITUDE AND LONGITUDE, TYPE OF WAYPOINT AND/OR ALTITUDE CONSTRAINT, ONLY THE GENERAL RELATIONSHIP OF WAYPOINTS IN THE PROCEDURE, OR THE LOGIC OF AN INDIVIDUAL WAYPOINTS LOCATION.

[c] If the cursory check of procedure logic or individual waypoint location, specified in [b] above, indicates a potential error, do not use the retrieved procedure or waypoint until a verification of latitude and longitude, waypoint type, and altitude constraints indicate full conformity with the published data.

h. GPS APPROACH PROCEDURES

As the production of stand-alone GPS approaches has progressed, many of the original overlay approaches have been replaced with stand-alone procedures specifically designed for use by GPS systems. The title of the remaining GPS overlay procedure has been revised on the approach chart to "or GPS" (e.g., VOR or GPS RWY 24). Therefore, all the approaches that can be used by GPS now contain "GPS" in the title (e.g., "VOR or GPS RWY 24," "GPS RWY 24," or "RNAV (GPS) RWY 24"). During these GPS approaches, underlying ground-based NAVAID's are not required to be operational and associated aircraft avionics need not be installed, operational, turned on or monitored (monitoring of the underlying approach is suggested when equipment is available and functional). Existing overlay approaches may be requested using the GPS title, such as "GPS RWY 24" for the VOR or GPS RWY 24.

NOTE—
ANY REQUIRED ALTERNATE AIRPORT MUST HAVE AN APPROVED INSTRUMENT APPROACH PROCEDURE OTHER THAN GPS THAT IS ANTICIPATED TO BE OPERATIONAL AND AVAILABLE AT THE ESTIMATED TIME OF ARRIVAL, AND WHICH THE AIRCRAFT IS EQUIPPED TO FLY.

i. GPS NOTAM'S/AERONAUTICAL INFORMATION

1. GPS satellite outages are issued as GPS NOTAM's both domestically and internationally. However, the effect of an outage on the intended operation cannot be determined unless the pilot has a RAIM availability prediction program which allows excluding a satellite which is predicted to be out of service based on the NOTAM information.

2. The term UNRELIABLE is used in conjunction with GPS NOTAMs. The term UNRELIABLE is an advisory to pilots indicating the expected level of service may not be available. GPS operation may be NOTAMed UNRELIABLE due to testing or anomalies. Air Traffic Control will advise pilots requesting a GPS or RNAV (GPS) approach of GPS UNRELIABLE for:

(a) NOTAMs not contained in the ATIS broadcast.

(b) Pilot reports of GPS anomalies received within the preceding 15 minutes.

3. Civilian pilots may obtain GPS RAIM availability information for nonprecision approach procedures by specifically requesting GPS aeronautical information from an Automated Flight Service Station during preflight briefings. GPS RAIM aeronautical information can be obtained for a period of 3 hours (ETA hour and 1 hour before to 1 hour after the ETA hour) or a 24 hour time frame at a particular airport. FAA briefers will provide RAIM information for a period of 1 hour before to 1 hour after the ETA, unless a specific time frame is requested by the pilot. If flying a published GPS departure, a RAIM prediction should also be requested for the departure airport.

4. The military provides airfield specific GPS RAIM NOTAM's for nonprecision approach procedures at military airfields. The RAIM outages are issued as M-series NOTAM's and may be obtained for up to 24 hours from the time of request.

5. Receiver manufacturers and/or database suppliers may supply "NOTAM" type information concerning database errors. Pilots should check these sources, when available, to ensure that they have the most current information concerning their electronic database.

j. RECEIVER AUTONOMOUS INTEGRITY MONITORING (RAIM)

1. RAIM outages may occur due to an insufficient number of satellites or due to unsuitable satellite geometry which causes the error in the position solution to become too large. Loss of satellite reception and RAIM warnings may occur due to aircraft dynamics (changes in pitch or bank angle). Antenna location on the aircraft, satellite position relative to the horizon, and aircraft attitude may affect reception of one or more satellites. Since the relative positions of the satellites are constantly changing, prior experience with the airport does not guarantee reception at all times, and RAIM availability should always be checked.

2. If RAIM is not available, another type of navigation and approach system must be used, another destination selected, or the trip delayed until RAIM is predicted to be available on arrival. On longer flights, pilots should consider rechecking the RAIM prediction for the destination during the flight. This may provide early indications that an unscheduled satellite outage has occurred since take-off.

3. If a RAIM failure/status annunciation occurs prior to the final approach waypoint (FAWP), the approach should not be completed since GPS may no longer provide the required accuracy. The receiver performs a RAIM prediction by 2 NM prior to the FAWP to ensure that RAIM is available at the FAWP as a condition for entering the approach mode. **The pilot should ensure that the receiver has sequenced from "Armed" to "Approach" prior to the FAWP** (normally occurs 2 NM prior). Failure to sequence may be an indication of the detection of a satellite anomaly, failure to arm the receiver (if required), or other problems which preclude completing the approach.

4. If the receiver does not sequence into the approach mode or a RAIM failure/status annunciation occurs prior to the FAWP, the pilot should not descend to MDA, but should proceed to the missed approach waypoint (MAWP) via the FAWP, perform a missed approach, and contact ATC as soon as practical. Refer to the receiver operating manual for specific indications and instructions associated with loss of RAIM prior to the FAF.

5. If a RAIM failure occurs after the FAWP, the receiver is allowed to continue operating without an annuciation for up to 5 minutes to allow completion of the approach (see receiver operating manual). **If the RAIM flag/status annunciation appears after the FAWP, the missed approach should be executed immediately.**

k. WAYPOINTS

1. GPS approaches make use of both fly-over and fly-by waypoints. Fly-by waypoints are used when an aircraft should begin a turn to the next course prior to reaching the waypoint separating the two route segments. This is known as turn anticipation and is compensated for in the airspace and terrain clearances. Approach waypoints, except for the MAWP and the missed approach holding waypoint (MAHWP), are normally fly-by waypoints. Fly-over waypoints are used when the aircraft must fly over the point prior to starting a turn. New approach charts depict fly-over waypoints as a circled waypoint symbol. Overlay approach charts and some early stand alone GPS approach charts may not reflect this convention.

2. Since GPS receivers are basically "To-To" navigators, they must always be navigating to a defined point. On overlay approaches, if no pronounceable five-character name is published for an approach waypoint or fix, it was given a data base identifier consisting of letters and numbers. These points will appear in the list of waypoints in the approach procedure data base, but may not appear on the approach chart. A point used for the purpose of defining the navigation track for an airborne computer system (i.e., GPS or FMS) is called a Computer Navigation Fix (CNF). CNF's include unnamed DME fixes, beginning and ending points of DME arcs and sensor final

approach fixes (FAF's) on some GPS overlay approaches. To aid in the approach chart/data base correlation process, the FAA has begun a program to assign five-letter names to CNF's and to chart CNF's on various National Oceanic Service aeronautical products. These CNF's are not to be used for any air traffic control (ATC) application, such as holding for which the fix has not already been assessed. CNF's will be charted to distinguish them from conventional reporting points, fixes, intersections, and waypoints. The CNF name will be enclosed in parenthesis, e.g., (MABEE), and the name will be placed next to the CNF it defines. If the CNF is not at an existing point defined by means such as crossing radials or radial/DME, the point will be indicated by an "X." The CNF name will not be used in filing a flight plan or in aircraft/ATC communications. Use current phraseology, e.g., facility name, radial, distance, to describe these fixes.

3. Unnamed waypoints in the data base will be uniquely identified for each airport but may be repeated for another airport (e.g., RW36 will be used at each airport with a runway 36 but will be at the same location for all approaches at a given airport).

4. The runway threshold waypoint, which is normally the MAWP, may have a five letter identifier (e.g., SNEEZ) or be coded as RW## (e.g., RW36, RW36L). Those thresholds which are coded as five letter identifiers are being changed to the RW## designation. This may cause the approach chart and data base to differ until all changes are complete. The runway threshold waypoint is also used as the center of the MSA on most GPS approaches. MAWP's not located at the threshold will have a five letter identifier.

l. POSITION ORIENTATION

As with most RNAV systems, pilots should pay particular attention to position orientation while using GPS. Distance and track information are provided to the next active waypoint, not to a fixed navigation aid. Receivers may sequence when the pilot is not flying along an active route, such as when being vectored or deviating for weather, due to the proximity to another waypoint in the route. This can be prevented by placing the receiver in the nonsequencing mode. When the receiver is in the nonsequencing mode, bearing and distance are provided to the selected waypoint and the receiver will not sequence to the next waypoint in the route until placed back in the auto sequence mode or the pilot selects a different waypoint. On overlay approaches, the pilot may have to compute the along track distance to stepdown fixes and other points due to the receiver showing along track distance to the next waypoint rather than DME to the VOR or ILS ground station.

m. CONVENTIONAL VERSUS GPS NAVIGATION DATA

There may be slight differences between the course information portrayed on navigational charts and a GPS navigation display when flying authorized GPS instrument procedures or along an airway. All magnetic tracks defined by any conventional navigation aids are determined by the application of the station magnetic variation. In contrast, GPS RNAV systems may use an algorithm, which applies the local magnetic variation and may produce small differences in the displayed course. However, both methods of navigation should produce the same desired ground track when using approved, IFR navigation system. Should significant differences between the approach chart and the GPS avionics' application of the navigation database arise, the published approach chart, supplemented by NOTAMs, holds precedence.

Due to the GPS avionics' computation of great circle courses, and the variations in magnetic variation, the bearing to the next waypoint and the course from the last waypoint (if available) may not be exactly 180° apart when long distances are involved. Variations in distances will occur since GPS distance-to-waypoint values are along-track distances (ATD) computed to the next waypoint and the DME values published on underlying procedures are slant-range distances measured to the station. This difference increases with aircraft altitude and proximity to the NAVAID.

n. DEPARTURE AND INSTRUMENT DEPARTURE PROCEDURES (DPS)

The GPS receiver must be set to terminal (±1 NM) course deviation indicator (CDI) sensitivity and the navigation routes contained in the data base in order to fly published IFR charted departures and DP's. Terminal RAIM should be automatically provided by the receiver. (Terminal RAIM for departure may not be available unless the waypoints are part of the active flight plan rather than proceeding direct to the first destination.) Certain segments of a DP may require some manual intervention by the pilot, especially when radar vectored to a course or required to intercept a specific course to a waypoint. The data base may not contain all of the transitions or departures from all runways and some GPS receivers do not contain DP's in the data base. It is necessary that helicopter procedures be flown at 70 knots or less since helicopter departure procedures and missed approaches use a 20:1 obstacle clearance surface (OCS), which is double the fixed-wing OCS, and turning areas are based on this speed as well.

o. FLYING GPS APPROACHES

1. Determining which area of the TAA the aircraft will enter when flying a "T" with a TAA must be accomplished using the bearing and distance to the IF(IAF). This is most critical when entering the TAA in the vicinity of the extended runway centerline and determining whether you will be entering the right or left base area. Once inside the TAA, all sectors and stepdowns are based on the bearing and distance to the IAF for that area, which the aircraft should be proceeding direct to at that time, unless on vectors. (See FIG 5-4-4 and FIG 5-4-5.)

2. Pilots should fly the full approach from an Initial Approach Waypoint (IAWP) or feeder fix unless specifically cleared otherwise. Randomly joining an approach at an intermediate fix does not assure terrain clearance.

3. When an approach has been loaded in the flight plan, GPS receivers will give an "arm" annunciation 30 NM straight line distance from the airport/heliport reference point. Pilots should arm the approach mode at this time, if it is has not already been armed (some receivers arm automatically). Without arming, the receiver will not change from en route CDI and RAIM sensitivity of ±5 NM either side of center line to ±1 NM terminal sensitivity. Where the IAWP is inside this 30 mile point, a CDI sensitivity change will occur once the approach mode is armed and the aircraft is inside 30 NM. Where the IAWP is beyond 30 NM from the airport/heliport reference point, CDI sensitivity will not change until the aircraft is within 30 miles of the airport/heliport reference point even if the approach is armed earlier. Feeder route obstacle clearance is predicated on the receiver being in terminal (±1 NM) CDI sensitivity and RAIM within 30 NM of the airport/heliport reference point, therefore, the receiver should always be armed (if required) not later than the 30 NM annunciation.

4. The pilot must be aware of what bank angle/turn rate the particular receiver uses to compute turn anticipation, and whether wind and airspeed are included in the receivers' calculations. This information should be in the receiver operating manual. Over or under banking the turn onto the final approach course may significantly delay getting on course and may result in high descent rates to achieve the next segment altitude.

5. When within 2 NM of the FAWP with the approach mode armed, the approach mode will switch to active, which results in RAIM changing to approach sensitivity and a change in CDI sensitivity. Beginning 2 NM prior to the FAWP, the full scale CDI sensitivity will smoothly change from ±1 NM, to ±3 NM at the FAWP. As sensitivity changes from ±1 NM to ±0.3 NM approaching the FAWP, with the CDI not centered, the corresponding increase in CDI displacement may give the impression that the aircraft is moving further away from the intended course even though it is on an acceptable intercept heading. Referencing the digital track displacement information (cross track error), if it is available in the approach mode, may help the pilot remain position oriented in this situation. Being established on the final approach course prior to the beginning of the sensitivity change at 2 NM will help prevent problems in interpreting the CDI display during ramp down. Therefore, requesting or accepting vectors which will cause the aircraft to intercept the final approach course within 2 NM of the FAWP is not recommended.

6. When receiving vectors to final, most receiver operating manuals suggest placing the receiver in the nonsequencing mode on the FAWP and manually setting the course. This provides an extended final approach course in cases where the aircraft is vectored onto the final approach course outside of any existing segment which is aligned with the runway. Assigned altitudes must be maintained until established on a published segment of the approach. Required altitudes at waypoints outside the FAWP or stepdown fixes must be considered. Calculating the distance to the FAWP may be required in order to descend at the proper location.

7. Overriding an automatically selected sensitivity during an approach will cancel the approach mode annunciation. If the approach mode is not armed by 2 NM prior to the FAWP, the approach mode will not become active at 2 NM prior to the FAWP, and the equipment will flag. In these conditions, the RAIM and CDI sensitivity will not ramp down, and the pilot should not descend to MDA, but fly to the MAWP and execute a missed approach. The approach active annunciator and/or the receiver should be checked to ensure the approach mode is active prior to the FAWP.

8. Do not attempt to fly an approach unless the procedure is contained in the current, on-board navigation database and identified as "GPS" on the approach chart. The navigation database may contain information about nonoverlay approach procedures that is intended to be used to enhance position orientation, generally by providing a map, while flying these approaches using conventional NAVAID's. This approach information should not be confused with a GPS overlay approach (see the receiver operating manual, AFM, or AFM Supplement for details on how to identify these procedures in the navigation database). Flying point to point on the approach does not assure compliance with the published approach procedure. The proper RAIM sensitivity will not be available and the CDI sensitivity will not automatically change to ±0.3 NM. Manually setting CDI sensitivity does not automatically change the RAIM sensitivity on some receivers. Some existing nonprecision approach procedures cannot be coded for use with GPS and will not be available as overlays.

9. Pilots should pay particular attention to the exact operation of their GPS receivers for performing holding patterns and in the case of overlay approaches, operations such as procedure turns. These procedures may require manual intervention by the pilot to stop the sequencing of waypoints by the receiver and to resume automatic GPS navigation sequencing once the maneuver is complete. The same waypoint may appear in the route of flight more than once consecutively (e.g., IAWP, FAWP, MAHWP on a procedure turn). Care must be exercised to ensure that the receiver is sequenced to the appropriate waypoint for the segment of the procedure being flown, especially if one or more fly-overs are skipped (e.g., FAWP rather than IAWP if the procedure turn is not flown). The pilot may have to sequence past one or more fly-overs of the same waypoint in order to start GPS

automatic sequencing at the proper place in the sequence of waypoints.

10. Incorrect inputs into the GPS receiver are especially critical during approaches. In some cases, an incorrect entry can cause the receiver to leave the approach mode.

11. A fix on an overlay approach identified by a DME fix will not be in the waypoint sequence on the GPS receiver unless there is a published name assigned to it. When a name is assigned, the along track to the waypoint may be zero rather than the DME stated on the approach chart. The pilot should be alert for this on any overlay procedure where the original approach used DME.

12. If a visual descent point (VDP) is published, it will not be included in the sequence of waypoints. Pilots are expected to use normal piloting techniques for beginning the visual descent. In addition, unnamed step-down fixes in the final approach segment will not be coded in the waypoint sequence and must be identified using ATD.

13. Unnamed stepdown fixes in the final approach segment will not be coded in the waypoint sequence of the aircraft's navigation database and must be identified using AFD. Stepdown fixes in the final approach segment of RNAV (GPS) approaches are being named, in addition to being identified by ATD. However, since most GPS avionics do not accommodate waypoints between the FAF and MAP, even when the waypoint is named, the waypoints for these stepdown fixes may not appear in the sequence of waypoints in the navigation database. Pilots must continue to identify these stepdown fixes using ATD.

p. MISSED APPROACH

1. A GPS missed approach requires pilot action to sequence the receiver past the MAWP to the missed approach portion of the procedure. The pilot must be thoroughly familiar with the activation procedure for the particular GPS receiver installed in the aircraft and must **initiate appropriate action after the MAWP**. Activating the missed approach prior to the MAWP will cause CDI sensitivity to immediately change to terminal (±1 NM) sensitivity and the receiver will continue to navigate to the MAWP. The receiver will not sequence past the MAWP. Turns should not begin prior to the MAWP. If the missed approach is not activated, the GPS receiver will display an extension of the inbound final approach course and the ATD will increase from the MAWP until it is manually sequenced after crossing the MAWP.

2. Missed approach routings in which the first track is via a course rather than direct to the next waypoint **require additional action by the pilot** to set the course. Being familiar with all of the inputs required is especially critical during this phase of flight.

q. GPS FAMILIARIZATION

Pilots should practice GPS approaches under visual meteorological conditions (VMC) until thoroughly proficient with all aspects of their equipment (receiver and installation) prior to attempting flight by IFR in instrument meteorological conditions (IMC). Some of the areas which the pilot should practice are:

1. Utilizing the receiver autonomous integrity monitoring (RAIM) prediction function;

2. Inserting a DP into the flight plan, including setting terminal CDI sensitivity, if required, and the conditions under which terminal RAIM is available for departure (some receivers are not DP or STAR capable);

3. Programming the destination airport;

4. Programming and flying the overlay approaches (especially procedure turns and arcs);

5. Changing to another approach after selecting an approach;

6. Programming and flying "direct" missed approaches;

7. Programming and flying "routed" missed approaches;

8. Entering, flying and exiting holding patterns, particularly on overlay approaches with a second waypoint on the holding pattern;

9. Programming and flying a "route" from a holding pattern;

10. Programming and flying an approach with radar vectors to the intermediate segment;

11. Indication of the actions required for RAIM failure both before and after the FAWP; and

12. Programming a radial and (distance from a VOR (often used in departure instructions).

1-1-20. Wide Area Augmentation System (WAAS)

a. General

1. The FAA developed the Wide Area Augmentation System (WAAS) to improve the accuracy, integrity and availability of GPS signals. WAAS will allow GPS to be used, as the aviation navigation system, from takeoff through Category I precision approach when it is complete. WAAS is a critical component of the FAA'S strategic objective for a seamless satellite navigation system for civil aviation, improving capacity and safety.

2. The International Civil Aviation Organization (ICAO) has defined Standards and Recommended Practices (SARP's) for satellite-based augmentation systems (SBAS) such as WAAS. Japan and Europe are building similar systems that are planned to be interoperable with WAAS: EGNOS, the European Geostationary Navigation Overlay System, and MSAS, the Japan Multifunctional Transport Satellite (MTSAT) Satellite-based Augmentation System. The merging of these systems will create a worldwide seamless navigation capability similar to GPS but with greater accuracy, availability and integrity.

3. Unlike traditional ground-based navigation aids, WAAS will cover a more extensive service area. Precisely surveyed wide-area ground reference stations (WRS) are linked to form the U.S. WAAS network. Signals from the GPS satellites are monitored by these WRS's to determine satellite clock and ephemeris corrections and to model the propagation effects of the ionosphere. Each station in the network relays the data to a wide-area master station (WMS) where the correction information is computed. A correction message is prepared and uplinked to a geostationary satellite (GEO) via a ground uplink station (GUS). The message is then broadcast on the same frequency as GPS (L1, 1575.42 MHz) to WAAS receivers within the broadcast coverage area of the WAAS GEO.

4. In addition to providing the correction signal, the WAAS GEO provides an additional pseudorange measurement to the aircraft receiver, improving the availability of GPS by providing, in effect, an additional GPS satellite in view. The integrity of GPS is improved through real-time monitoring, and the accuracy is improved by providing differential corrections to reduce errors. The performance improvement is sufficient to enable approach procedures with GPS/WAAS glide paths (vertical guidance).

5. The FAA has completed installation of 25 WRS's, 2 WMS's, 4 GUS's, and the required terrestrial communications to support the WAAS network. Prior to the commissioning of the WAAS for public use, the FAA has been conducting a series of test and validation activities. Enhancements to the initial phase of WAAS will include additional master and reference stations, communication satellites, and transmission frequencies as needed.

6. GNSS navigation, including GPS and WAAS, is referenced to the WGS-84 coordinate system. It should only be used where the Aeronautical Information Publications (including electronic data and aeronautical charts) conform to WGS-84 or equivalent. Other countries civil aviation authorities may impose additional limitations on the use of their SBAS systems.

b. Instrument Approach Capabilities

1. A new class of approach procedures which provide vertical guidance, but which do not meet the ICAO Annex 10 requirements for precision approaches has been developed to support satellite navigation use for aviation applications worldwide. These new procedures called Approach with Vertical Guidance (APV), are defined in ICAO Annex 6, and include approaches such as the LNAV/VNAV procedures presently being flown with barometric vertical navigation (Baro-VNAV). These approaches provide vertical guidance, but do not meet the more stringent standards of a precision approach. Properly certified WAAS receivers will be able to fly these LNAV/VNAV procedures using a WAAS electronic glide path, which eliminates the errors that can be introduced by using Barometric altimetery.

2. A new type of APV approach procedure, in addition to LNAV/VNAV, is being implemented to take advantage of the lateral precision provided by WAAS. This angular lateral precision, combined with an electronic glidepath allows the use of TERPS approach criteria very similar to that used for present precision approaches, with adjustments for the larger vertical containment limit. The resulting approach procedure minima, titled LPV, (localizer performance with vertical guidance) may have decision altitudes as low as 250 feet height above touchdown with visibility minimums as low as 1/2 mile, when the terrain and airport infrastructure support the lowest minima. LPV minima are published on the RNAV (GPS) approach charts (see paragraph, 5-4-5, Instrument Approach Procedure Charts).

3. WAAS initial operating capability provides a level of service that supports all phases of flight including LNAV, LNAV/VNAV and LPV approaches. In the long term, WAAS will provide Category I precision approach services in conjunction with modernized UPS.

c. General Requirements.

1. WAAS avionics must be certified in accordance with Technical Standard Order (TSO) TSO-C145A, Airborne Navigation Sensors Using the (GPS) Augmented by the Wide Area Augmentation System (WAAS); or TSO-146A, Stand-Alone Airborne Navigation Equipment Using the Global Positioning System (GPS) Augmented by the wide Area Augmentation System (WAAS), and installed in accordance with Advisory Circular (AC) 20-130A, Airworthiness Approval of Navigation or Flight Management Systems Integrating Multiple Navigation Sensors, or AC 20-138A, Airworthiness Approval of Global Positioning System (GPS) Navigation Equipment for Use as a VFR and IFR Navigation System.

2. GPS/WAAS operation must be conducted in accordance with the FAA-approved aircraft flight manual (AFM) and flight manual supplements. Flight manual supplements will state the level of approach procedure that the receiver supports. IFR approved WAAS receivers support all GPS only operations as long as lateral capability at the appropriate level is functional. WAAS monitors both UPS and WAAS satellites and provides integrity.

3. GPS/WAAS equipment is inherently capable of supporting oceanic and remote operations if the operator obtains a fault detection and exclusion (FDE) prediction program.

4. Air carrier and commercial operators must meet the appropriate provisions of their approved operations specifications.

5. Prior to GPS/WAAS IFR operation, the pilot must review appropriate Notices to Airmen (NOTAMs) and aeronautical information. This information is available on request from an Automated Flight Service Station. The FAA will provide NOTAMs to advise pilots of the status of the WAAS and level of service available.

(a) The term UNRELIABLE is used in conjunction with GPS and WAAS NOTAMs. The term UNRELIABLE is an advisory to pilots indicating the expected level of WAAS service (LNAV/VNAV, LPV) may not be available; e.g., **!BOS BOS WAAS LPV AND LNAV/VNAV MNM UNREL WEF 0305231700 - 0305231815.** WAAS UNRELIABLE NOTAMs are predictive in nature and published for

flight planning purposes. Upon commencing an approach at locations NOTAMed WAAS UNRELIABLE, if the WAAS avionics indicate LNAV/VNAV or LPV service is available, then vertical guidance may be used to complete the approach using the displayed level of service. Should an outage occur during the approach, reversion to LNAV minima may be required.

(1) Area-wide WAAS UNAVAILABLE NOTAMs indicate loss or malfunction of the WAAS system. In flight, Air Traffic Control will advise pilots requesting a GPS or RNAV (GPS) approach of WAAS UNAVAILABLE NOTAMs if not contained in the ATIS broadcast.

(2) Site-specific WAAS UNRELIABLE NOTAMs indicate an expected level of service, e.g., LNAV/VNAV or LPV may not be available. Pilots must request site-specific WAAS NOTAMs during flight planning. In flight, Air Traffic Control will not advise pilots of WAAS UNRELIABLE NOTAMs.

(3) When the approach chart is annotated with the ▲ symbol, site-specific WAAS UNRELIABLE NOTAMs or Air Traffic advisiories are not provided for outages in WAAS LNAV/VNAV and LPV vertical service.

NOTE—
AREA-WIDE WAAS UNAVAILABLE NOTAMS APPLY TO ALL AIRPORTS IN THE WAS UNAVAILABLE AREA DESIGNATED IN THE NOTAM, INCLUDING APPROACHES AT AIRPORTS WHERE AN APPROACH CHART IS ANNOTATED WITH THE ▲ SYMBOL.

6. GPS/WAAS was developed to be used within SBAS GEO coverage (WAAS or other interoperable system) without the need for other radio navigation equipment appropriate to the route of flight to be flown. Outside the SBAS coverage or in the event of a WAAS failure, GPS/WAAS equipment reverts to GPS-only operation and satisfies the requirements for basic GPS equipment.

7. Unlike TSO-C129 avionics, which were certified as a supplement to other means of navigation, WAAS avionics are evaluated without reliance on other navigation systems. As such, installation of WAAS avionics does not require the aircraft to have other equipment appropriate to the route to be flown.

(a) Due to initial system limitation, there are certain restrictions on WAAS operations. Pilots may plan to use any instrument approach authorized for use with WAAS avionics at a required alternate. However, when using WAAS at an alternate airport, flight planning must be based on flying the RNAV (GPS) LNAV minima line, or minima on a GPS approach procedure, or conventional approach procedure with "or UPS" in the title. Code of Federal Regulation (CFA) Part 91 nonprecision weather requirements must be used for planning. Upon arrival at an alternate, when the WAAS navigation system indicates that LNAV/VNAV or LPV service is available, then vertical guidance may be used to complete the approach using the displayed level of service. The FAA has begun removing the ▲ NA (Alternate Minimums Not Authorized) symbol from select RNAV

(GPS) and GPS approach procedures so they may be used by approach approved WAAS receivers at alternate airports. Some approach procedures will still require the ▲ NA for other reasons, such as no weather reporting, so it cannot be removed from all procedures. Since every procedure must be individually evaluated, removal of the ▲ NA from RNAV (GPS) and GPS procedures will take some time.

d. flying Procedures with WAAS

1. WAAS receivers support all basic GPS approach functions and will provide additional capabilities. One of the major improvements is the ability to generate an electronic glide path, independent of ground equipment or barometric aiding. This eliminates several problems such as cold temperature effects, incorrect altimeter setting or lack of a local altimeter source and allows approach procedures to be built without the cost of installing ground stations at each airport. Some approach certified receivers will only support a glide path with performance similar to Baro-VNAV, and are authorized to fly the LNAV/VNAV line of minima on the RNAV (GPS) approach charts. Receivers with additional capability which support the performance requirements for precision approaches (including update rates and integrity limits) will be authorized to fly the LPV line of minima. The lateral integrity changes dramatically from the 0.3 NM (556 meter) limit for UPS, LNAV and LNAV/VNAV approach mode, to 40 meters for LPV. It also adds vertical integrity monitoring, which for LNAV/VNAV and LPV approaches bounds the vertical error to *50* meters.

2. When an approach procedure is selected and active, the receiver will notify the pilot of the most accurate level of service supported by the combination of the WAAS signal, the receiver, and the selected approach, using the naming conventions on the minima lines of the selected approach procedure. For example, if an approach is published with LPV minima and the receiver is only certified for LNAV/VNAV, the equipment would indicate "LPV not available use LNAV/VNAV minima," even though the WAAS signal would support LPV. If flying an existing LNAV/VNAV procedure, the receiver will notify the pilot "LNAV/VNAV available" even if the receiver is certified for LPV and the WAAS signal supports LPV. If the WAAS signal does not support published minima lines which the receiver is certified to fly, the receiver will notify the pilot with a message such as "LPV not available use LNAV/VNAV minima" or "LPV not available use LNAV minima." Once this notification has been given, the receiver will operate in this mode for the duration of that approach procedure. The receiver cannot change back to a more accurate level of service until the next time an approach is activated.

3. Another additional feature of WAAS receivers is the ability to exclude a bad GPS signal and continue operating normally. This is normally accomplished by the WAAS correction information. Outside WAAS coverage or when WAAS is not available, it is accomplished through a receiver algorithm called FDE. In most cases this operation will be invisible to the pilot since the receiver will continue to operate with other available satellites after excluding the "bad" signal. This capability increases the reliability of navigation.

4. Both lateral and vertical scaling for the LNAV/VNAV and LPV approach procedures are different than the linear scaling of basic GPS. When the complete published procedure is flown, +/−1 NM linear scaling is provided until two (2) NM prior to the FAF, where the sensitivity increases to be similar to the angular scaling of an ILS. There are two differences in the WAAS scaling and ILS: 1) on long final approach segments, the initial scaling will be +/−0.3 NM to achieve equivalent performance to GPS (and better than ILS, which is less sensitive far from the runway); 2) close to the runway threshold, the scaling changes to linear instead of continuing to become more sensitive. The width of the final approach course is tailored so that the total width is usually 700 feet at the runway threshold. Since the origin point of the lateral splay for the angular portion of the final is not fixed due to antenna placement like localizer, the splay angle can remain fixed, making a consistent width of final for aircraft being vectored onto the final approach course on different length runways. When the complete published procedure is not flown, and instead the aircraft needs to capture the extended final approach course similar to ILS, the vector to final (VTF) mode is used. Under VTF the Area Augmentation System (WAAS), and installed in accordance with Advisory Circular (AC) 20-130A, Airworthiness Approval of Navigation or Flight Mangement Integrating Multiple Navigation Sensors, or AC 20-138A, Airworthiness Approval of Globan Positioning System (GPS) Navigation Equipment for Use as a VFR and IFR Navigation System.

5. The WAAS scaling is also different than GPS TSO-C129 in the initial portion of the missed approach. Two differences occur here. First, the scaling abruptly changes from the approach scaling to the missed approach scaling, at approximately the departure end of the runway or when the pilot requests missed approach guidance rather than ramping as GPS does. Second, when the first leg of the missed approach is a Track to Fix (TF) leg aligned within 3 degrees of the inbound course, the receiver will change to 0.3 NM linear sensitivity until the turn initiation point for the first waypoint in the missed approach procedure, at which time it will abruptly change to terminal (+/−1 NM) sensitivity. This allows the elimination of close in obstacles in the early part of the missed approach that may cause the DA to be raised.

6. A new method has been added for selecting the final approach segment of an instrument approach. Along with the current method used by most receivers using menus where the pilot selects the airport, the runway, the specific approach procedure and finally the IAF, there is also a channel number selection method. The pilot enters a unique 5-digit number provided on the approach chart, and the receiver recalls the matching final approach segment from the aircraft database. A list of information including the available IAFs is displayed and the pilot selects the appropriate IAF. The pilot should confirm that the correct final approach segment was loaded by cross checking the Approach ID, which is also provided on the approach chart.

7. The Along Track Distance (ATD) during the final approach segment of an LNAV procedure (with a minimum descent altitude) will be to the MAWP. On LNAV/VNAV and LPV approaches to a decision altitude, there is no missed approach waypoint so the along-track distance is displayed to a point normally located at the runway threshold. In most cases the MAWP for the LNAV approach is located on the runway threshold at the centerline, so these distances will be the same. This distance will always vary slightly from any ILS DME that may be present, since the ILS DME is located further down the runway. Initiation of the missed approach on the LNAV/ VNAV and LPV approaches is still based on reaching the decision altitude without any of the items listed in 14 CFR Section 91.175 being visible, and must not be delayed until the ATD reaches zero. The WAAS receiver, unlike a GPS receiver, will automatically sequence past the MAWP if the missed approach procedure has been designed for RNAV. The pilot may also select missed approach prior to the MAWP, however, navigation will continue to the MAWP prior to waypoint sequencing taking place.

1-1-21. GNSS LANDING SYSTEM (GLS)

a. General

1. The GLS provides precision navigation guidance for exact alignment and descent of aircraft on approach to a runway. It provides differential augmentation to the Global Navigation Satellite System (GNSS).

2. The United States plans to provide augmentation services to the GPS for the first phase of GNSS. This section will be revised and updated to reflect international standards and GLS services as they are provided.

1-1-22. PRECISION APPROACH SYSTEMS OTHER THAN ILS, GLS, AND MLS

a. General

Approval and use of precision approach systems other than ILS, GLS and MLS require the issuance of special instrument approach procedures.

b. Special Instrument Approach Procedure

1. Special instrument approach procedures must be issued to the aircraft operator if pilot training, aircraft equipment, and/or aircraft performance is different than published procedures. Special instrument approach procedures are not distributed for general public use. These procedures are

issued to an aircraft operator when the conditions for operations approval are satisfied.

2. General aviation operators requesting approval for special procedures should contact the local Flight Standards District Office to obtain a letter of authorization. Air carrier operators requesting approval for use of special procedures should contact their Certificate Holding District Office for authorization through their Operations Specification.

c. Transponder Landing System (TLS)

1. The TLS is designed to provide approach guidance utilizing existing airborne ILS localizer, glide slope, and transponder equipment.

2. Ground equipment consists of a transponder interrogator, sensor arrays to detect lateral and vertical position, and ILS frequency transmitters. The TLS detects the aircraft's position by interrogating its transponder. It then broadcasts ILS frequency signals to guide the aircraft along the desired approach path.

3. TLS instrument approach procedures are designated Special Instrument Approach Procedures. Special aircrew training is required. TLS ground equipment provides approach guidance for only one aircraft at a time. Even though the TLS signal is received using the ILS receiver, no fixed course or glidepath is generated. The concept of operation is very similar to an air traffic controller providing radar vectors, and just as with radar vectors, the guidance is valid only for the intended aircraft. The TLS ground equipment tracks one aircraft, based on its transponder code, and provides correction signals to course and glidepath based on the position of the tracked aircraft. Flying the TLS corrections computed for another aircraft will not provide guidance relative to the approach; therefore, aircrews must not use the TLS signal for navigation unless they have received approach clearance and completed the required coordination with the TLS ground equipment operator. Navigation fixes based on conventional NAVAID's or GPS are provided in the special instrument approach procedure to allow aircrews to verify the TLS guidance.

d. Special Category I Differential GPS (SCAT-I DGPS)

1. The SCAT-I DGPS is designed to provide approach guidance by broadcasting differential correction to GPS.

2. SCAT-I DGPS procedures require aircraft equipment and pilot training.

3. Ground equipment consists of GPS receivers and a VHF digital radio transmitter. The SCAT-I DGPS detects the position of GPS satellites relative to GPS receiver equipment and broadcasts differential corrections over the VHF digital radio.

4. Category I Ground Based Augmentation Systems (GBAS) will displace SCAT-I DGPS as the public use service.

REFERENCE—AIM, INSTRUMENT APPROACH PROCEDURES, PARAGRAPH 5-4-7f.

Section 2. AREA NAVIGATION (RNAV) AND REQUIRED NAVIGATION PERFORMANCE (RNP)

1-2-1. AREA NAVIGATION (RNAV)

a. General. RNAV is a method of navigation that permits aircraft operation on any desired flight path within the coverage of station-referenced navigation aids or within the limits of the capability of self—contained aids, or a combination of these. In the future, there will be an increased dependence on the use of RNAV in lieu of routes defined by ground-based navigation aids.

RNAV routes and terminal procedures, including departure procedures (DPs) and standard terminal arrivals (STARS), are designed with RNAV systems in mind. There are several potential advantages of RNAV routes and procedures:

1. Time and fuel savings,

2. Reduced dependence on radar vectoring, altitude, and speed assignments allowing a reduction in required ATC radio transmissions, and

3. More efficient use of airspace.

In addition to information found in this manual, guidance for domestic RNAV DPs, STARS, and routes may also be found in Advisory Circular 90-100, U.S. Terminal and En Route Area Navigation (RNAV) Operations.

b. RNAV Operations. RNAV procedures, such as DPs and STARS, demand strict pilot awareness and maintenance of the procedure centerline. Pilots should possess a working knowledge of their aircraft navigation system to ensure RNAV procedures are flown in an appropriate manner. In addition, pilots should have an understanding of the various waypoint and leg types used in RNAV procedures; these are discussed in more detail below.

1. Waypoints. A waypoint is a predetermined geographical position that is defined in terms of latitude/longitude coordinates. Waypoints may be a simple named point in space or associated with existing navaids, intersections, or fixes. A waypoint is most often used to indicate a change in direction, speed, or altitude along the desired path. RNAV procedures make use of both fly-over and fly-by waypoints.

(a) Fly-by waypoints. Fly-by waypoints are used when an aircraft should begin a turn to the next course prior to reaching the waypoint separating the two route segments. This is known as turn anticipation.

(b) Fly-over waypoints. Fly-over way-points are used when the aircraft must fly over the point prior to starting a turn.

NOTE—FIG 1-2-1 ILLUSTRATES SEVERAL DIFFERENCES BETWEEN A FLY-BY AND A FLY-OVER WAYPOINT.

2. RNAV Leg Types. A leg type describes the desired path proceeding, following, or between waypoints on an RNAV procedure. Leg types are identified by a two-letter code that describes the path (e.g., heading, course, track, etc.) and the termination point (e.g., the path terminates at an altitude, distance, fix, etc.). Leg types used for procedure design are included in the aircraft navigation database, but not normally provided on the procedure chart. The narrative depiction of the RNAV chart describes how a procedure is flown. The "path and terminator concept" defines that every leg of a procedure has a termination point and some kind of path into that termination point. Some of the available leg types are described below.

(a) Track to Fix. A Track to Fix (TF) leg is intercepted and acquired as the flight track to the following waypoint. Track to a Fix legs are sometimes called point-to-point legs for this reason.

NARRATIVE–*"VIA 087° TRACK TO CHEZZ WP"* SEE FIG 1-2-2.

(b) Direct to Fix. A Direct to Fix (DF) (DF) leg is a path described by an aircraft's track from an initial area direct to the next waypoint.

NARRATIVE–*"LEFT TURN DIRECT BARGN WP."* SEE FIG 1-2-3.

(c) Course to Fix. A Course to Fix (CF) leg is a path that terminates at a fix with a specified course at that fix.

NARRATIVE–*"VIA 078° COURSE TO PRIMY WP."* SEE FIG 1-2-4.

(d) Radius to Fix. A radius to Fix (RF) leg is defined as a constant radius circular path around a defined turn center that terminates at a fix. See FIG. 1-2-5.

(e) Heading. A Heading leg may be defined as, but not limited to, a Heading to Altitude (VA), Heading to DME range (VD), and Heading to Manual Termination, i.e., Vector (VM).

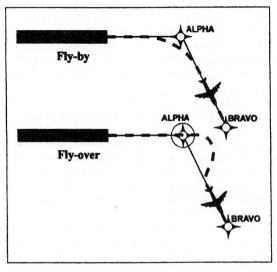

Figure 1-2-1. Fly-by and Fly-over Waypoints

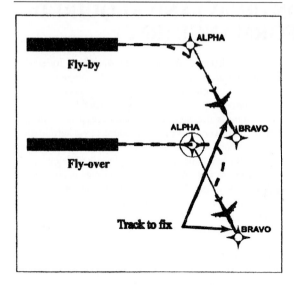

Figure 1-2-2. Track to Fix Leg Type

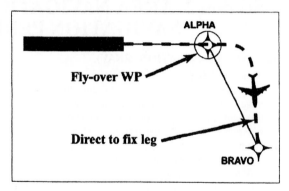

Figure 1-2-3. Direct to Fix Leg Type

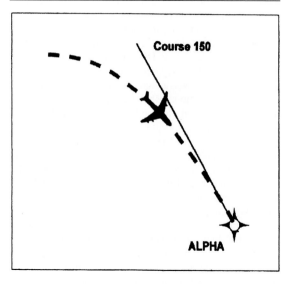

Figure 1-2-4. Course to Fix Leg Type

NARRATIVE–*"CLIMB RUNWAY HEADING TO 1500", "HEADING 265°, AT 9DME WEST OF PXR VORTAC, RIGHT TURN HEADING 360°", "FLY HEADING 090°, EXPECT RADAR VECTORS TO DRYHTINT."*

3. Navigation Issues. Pilots should be aware of their navigation system inputs, alerts, and annunciations in order to make better-informed decisions. In addition, the availability and suitability of particular sensors/systems should be considered.

(a) GPS. Operators using TSO-C129 systems should ensure departure and arrival airports are entered to ensure proper RAIM availability and CDI sensitivity.

(b) DME/DME. Operators should be aware that DME/DME position updating is dependent on FMS logic and DME facility proximity, availability, geometry, and signal masking.

(c) VOR/DME. Unique VOR characteristics may result in less accurate values from VOR/DME position updating than from GPS or DME/DME position updating.

(d) Inertial Navigation. Inertial reference units and inertial navigation systems are often coupled with other types of navigation inputs, e.g., DME/DME or GPS, to improve overall navigation system performance.

NOTE—SPECIFIC INERTIAL POSITION UPDATING REQUIREMENTS MAY APPLY.

4. Flight Management System (FMS). An FMS is an integrated suite of sensors, receivers, and computers, coupled with a navigation database. These systems generally provide performance and RNAV guidance to displays and automatic flight control systems.

Inputs can be accepted from multiple sources such as GPS, DME, VOR, LOC and IRU. These inputs may be applied to a navigation solution one at a time or in combination. Some FMSs provide for the detection and isolation of faulty navigation information.

When appropriate navigation signals are available, FMSs will normally rely on GPS and/or DME/DME (that is, the use of distance information from two or more DME stations) for position updates. Other inputs may also be incorporated based on FMS system architecture and navigation source geometry.

NOTE—*DME/DME* INPUTS COUPLED WITH ONE OR MORE *IRU(s)* ARE OFTEN ABBREVIATED AS *DME/DME/IRU* OR *D/D/I*.

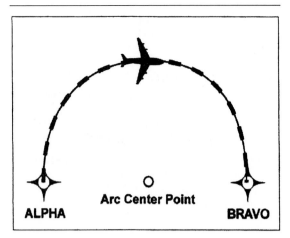

Figure 1-2-5. Radius to Fix Leg Type

1-2-2. REQUIRED NAVIGATION PERFORMANCE (RNP)

a. General. RNP is intended to provide a single performance standard for aircraft manufacturers, airspace designers, pilots, controllers, and international aviation authorities. Some RNP procedures will take advantage of improved navigation capabilities and will result in increased flight path predictability and repeatability.

Typically, various sensor inputs are processed by an RNAV system to arrive at a position estimate having a high-statistical degree of accuracy and confidence. When RNP is specified, a combination of systems may be used, provided the aircraft can achieve the required navigation performance.

While it has been a goal for RNP to be sensor-generic, this goal is unachievable as long as the aircraft capability is in any way dependent on external signals. The aircraft navigation system always consists of specific sensors or sensor combinations and the navigation infrastructure consists of specific systems.

The RNP capability of an aircraft will vary depending upon the aircraft equipment and the navigation infrastructure. For example, an aircraft may be equipped and certified for RNP 1.0, but may not be capable of RNP 1.0 operations due to limited navaid coverage.

b. RNP Operations

1. RNP Levels. An RNP "level" or "type" is applicable to a selected airspace, route, or procedure. ICAO has defined RNP values for the four typical navigation phases of flight: oceanic, en route, terminal, and approach. As defined in the Pilot/Controller Glossary, the RNP Level or Type is a value typically expressed as a distance in nautical miles from the intended centerline of a procedure, route, or path. RNP applications also account

for potential errors at some multiple of RNP level (e.g., twice the RNP level).

(a) Standard RNP Levels. U.S. standard values supporting typical RNP airspace are as specified in TBL 1-2-1. Other RNP levels as identified by ICAO, other states and the FAA may also be used.

Table 1-2-1
U.S. Standard RNP Levels

RNP Level	Typical Application	Primary Route Width (NM) Centerline to Boundary
0.3	Approach	0.3
1	Terminal	1.0
2	Terminal and en route	2.0

NOTE—

1. *THE "PERFORMANCE" OF THE NAVIGATION EQUIPMENT IN RNP REFERS NOT ONLY TO THE LEVEL OF ACCURACY OF A PARTICULAR SENSOR OR AIRCRAFT NAVIGATION SYSTEM, BUT ALSO TO THE DEGREE OF PRECISION WITH WHICH THE AIRCRAFT WILL BE FLOWN.*

2. *SPECIFIC REQUIRED FLIGHT PROCEDURES MAY VARY FOR DIFFERENT RNP LEVELS.*

(b) Application of Standard RNP Levels. U.S. standard levels of RNP typically used for various routes and procedures supporting RNAV operations may be based on use of a specific navigational system or sensor such as GPS, or on multi-sensor RNAV systems having suitable performance.

(c) Depiction of Standard RNP Levels. The applicable RNP level will be depicted on affected charts and procedures. For example, an RNAV departure procedure may contain a notation referring to eligible aircraft by equipment suffix and a phrase "or RNP-1.O." A typical RNAV approach procedure may include a notation referring to eligible aircraft by specific navigation sensor(s), equipment suffix, and a phrase "or RNP-O.3." Specific guidelines for the depiction of RNP levels will be provided through chart bulletins and accompany affected charting changes.

c. Other RNP Applications Outside the U.S. The FAA and ICAO member states have led initiatives in implementing the RNP concept to oceanic operations. For example, RNP-10 routes have been established in the northern Pacific (NOPAC) which has increased capacity and efficiency by reducing the distance between tracks to 50 NM. Additionally, the FAA has assisted those U.S. air carriers operating in Europe where the routes have been designated as RNP-5. TBL 1-2-2 shows examples of current and future RNP levels of airspace.

d. Aircraft and Airborne Equipment Eligibility for RNP Operations. Aircraft meeting RNP criteria will have an appropriate entry including special conditions and limitations in its Aircraft Flight Manual (AFM), or supplement.

RNAV installations with AFM-RNP certification based on GPS or systems integrating GPS are considered to meet U.S. standard RNP levels for all phases of flight. Aircraft with AFM-RNP certification without GPS may be limited to certain RNP levels, or phases of flight. For example, RNP based on DME/DME without other augmentation may not be appropriate for phases of flight outside the certified DME service volume. Operators of aircraft not having specific AFM-RNP certification may be issued operational approval including special conditions and limitations for specific RNP levels. Aircraft navigation systems eligible for RNP airspace will be indicated on charts or announced through other FAA media such as NOTAM's and chart bulletins.

NOTE—*SOME AIRBORNE SYSTEMS USE ESTIMATED POSITION UNCERTAINTY (EPU) AS A MEASURE OF THE CURRENT ESTIMATED NAVIGATIONAL PERFORMANCE. EPU MAY ALSO BE REFERRED TO AS ACTUAL NAVIGATION PERFORMANCE (ANP) OR ESTIMATED POSITION ERROR (EPE).*

Table 1-2-2

RNP Levels Supported for International Operations

RNP LEVEL	TYPICAL APPLICATION
1	European Precision RNAV (P-NAV)
4	Projected for oceanic/remote areas where 30 NM horizontal separation is required.
5	European Basic RNAV (B-RNAV)
10	Oceanic/remote areas where 50 NM lateral separation is applied

NOTE—*SPECIFIC OPERATIONAL AND EQUIPMENT PERFORMANCE REQUIREMENTS APPLY FOR P-RNAV AND B-RNAV.*

Chapter 2. Aeronautical lighting and other airport visual aids
Section 1. AIRPORT LIGHTING AIDS

2-1-1. APPROACH LIGHT SYSTEMS (ALS)

a. Approach light systems provide the basic means to transition from instrument flight to visual flight for landing. Operational requirements dictate the sophistication and configuration of the approach light system for a particular runway.

b. Approach light systems are a configuration of signal lights starting at the landing threshold and extending into the approach area a distance of 2400–3000 feet for precision instrument runways and 1400–1500 feet for nonprecision instrument runways. Some systems include sequenced flashing lights which appear to the pilot as a ball of light traveling towards the runway at high speed (twice a second). (See FIG 2-1-1).

2-1-2. VISUAL GLIDESLOPE INDICATORS

a. Visual Approach Slope Indicator (VASI)

1. The VASI is a system of lights so arranged to provide visual descent guidance information during the approach to a runway. These lights are visible from 3–5 miles during the day and up to 20 miles or more at night. The visual glide path of the VASI provides safe obstruction clearance within plus or minus 10 degrees of the extended runway centerline and to 4 NM from the runway threshold. Descent, using the

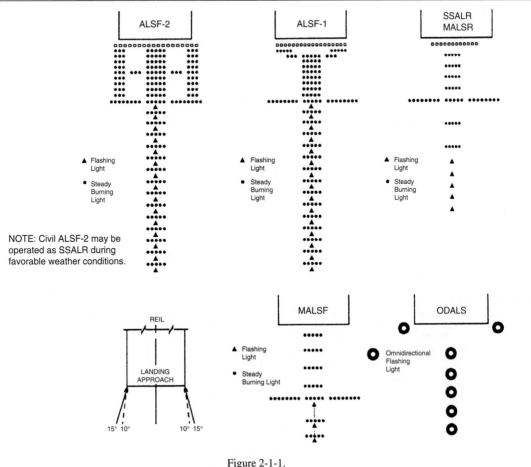

Figure 2-1-1.

VASI, should not be initiated until the aircraft is visually aligned with the runway. Lateral course guidance is provided by the runway or runway lights.

2. VASI installations may consist of either 2, 4, 6, 12, or 16 light units arranged in bars referred to as near, middle, and far bars. Most VASI installations consist of 2 bars, near and far, and may consist of 2, 4, or 12 light units. Some VASIs consist of three bars, near, middle, and far, which provide an additional visual glide path to accommodate high cockpit aircraft. This installation may consist of either 6 or 16 light units. VASI installations consisting of 2, 4, or 6 light units are located on one side of the runway, usually the left. Where the installation consists of 12 or 16 light units, the units are located on both sides of the runway.

3. Two-bar VASI installations provide one visual glide path which is normally set at 3 degrees. Three-bar VASI installations provide two visual glide paths. The lower glide path is provided by the near and middle bars and is normally set at 3 degrees while the upper glide path, provided by the middle and far bars, is normally ¼ degree higher. This higher glide path is intended for use only by high cockpit aircraft to provide a sufficient threshold crossing height. Although normal glide path angles are three degrees, angles at some locations may be as high as 4.5 degrees to give proper obstacle clearance. Pilots of

high performance aircraft are cautioned that use of VASI angles in excess of 3.5 degrees may cause an increase in runway length required for landing and rollout.

4. The basic principle of the VASI is that of color differentiation between red and white. Each light unit projects a beam of light having a white segment in the upper part of the beam and red segment in the lower part of the beam. The light units are arranged so that the pilot using the VASIs during an approach will see the combination of lights shown below.

5. For 2-bar VASI (4 light units) see FIG. 2-1-2.
6. For 3-bar VASI (6 light units) see FIG. 2-1-3.
7. For other VASI configurations see FIG. 2-1-4.

b. **Precision Approach Path Indicator (PAPI):** The precision approach path indicator (PAPI) uses light units similar to the VASI but are installed in a single row of either two or four light units. These systems have an effective visual range of about 5 miles during the day and up to 20 miles at night. The row of light units is normally installed on the left side of the runway and the glide path indications are as depicted. (See FIG 2-1-5).

c. **Tri-color Systems:** Tri-color visual approach slope indicators normally consist of a single light unit projecting a three-color visual approach path into the final approach area of the runway upon which the indicator is installed.

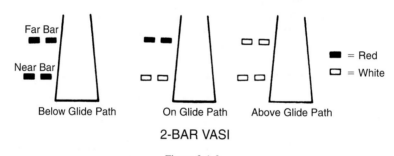

2-BAR VASI

Figure 2-1-2.

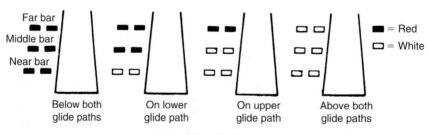

3-BAR VASI

Figure 2-1-3.

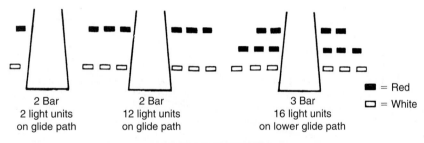

VASI VARIATIONS

Figure 2-1-4.

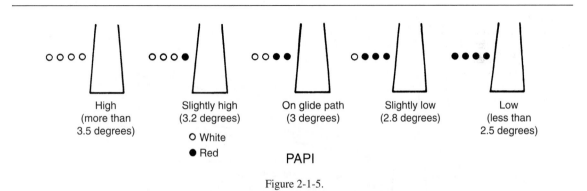

PAPI

Figure 2-1-5.

The below glide path indication is red, the above glide path indication is amber, and the on glide path indication is green. These types of indicators have a useful range of approximately one-half to one mile during the day and up to five miles at night depending upon the visibility conditions. (See FIG 2-1-6).

d. Pulsating Systems: Pulsating visual approach slope indicators normally consist of a single light unit projecting a two-color visual approach path into the final approach area of the runway upon which the indicator is installed. The on glide path indication is a steady white light. The slightly below glide path indication is a steady red light. If the aircraft descends further below the glide path, the red light starts to pulsate. The above glide path indication is a pulsating white light. The pulsating rate increases as the aircraft gets further above or below the desired glide slope. The useful range of the system is about four miles during the day and up to ten miles at night. (See FIG 2-1-7).

e. Alignment of Elements Systems: Alignment of elements systems are installed on some small general aviation airports and are a low-cost system consisting of painted plywood panels, normally black and white or fluorescent orange. Some of these systems are lighted for night use. The useful range of these systems is approximately three-quarter miles. To use the system the pilot positions the aircraft so the elements are in alignment. The glide path indications are shown in FIG 2-1-8.

2-1-3. RUNWAY END IDENTIFIER LIGHTS (REIL)

REILs are installed at many airfields to provide rapid and positive identification of the approach end of a particular runway. The system consists of a pair of synchronized flashing lights located laterally on each side of the runway threshold. REILs may be either omnidirectional or unidirectional facing the approach area. They are effective for:

a. Identification of a runway surrounded by a preponderance of other lighting.

b. Identification of a runway which lacks contrast with surrounding terrain.

c. Identification of a runway during reduced visibility.

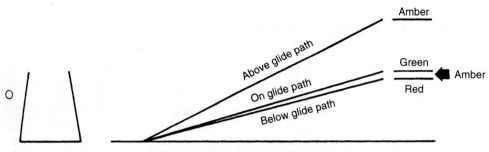

TRI-COLOR VISUAL APPROACH SLOPE INDICATOR

NOTE—
1. Since the tri-color VASI consists of a single light source which could possibly be confused with other light sources, pilots should exercise care to properly locate and identify the light signal.
2. When the aircraft descends from green to red, the pilot may see a dark amber color during the transition from green to red.

Figure 2-1-6.

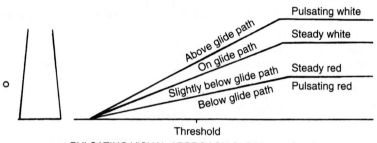

PULSATING VISUAL APPROACH SLOPE INDICATOR

NOTE—
Since the PLASI consists of a single light source which could possibly be confused with other light sources, pilots should exercise care to properly locate and identify the light signal.

Figure 2-1-7.

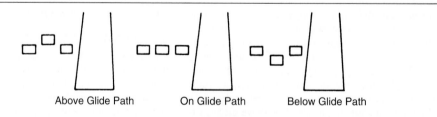

ALIGNMENT OF ELEMENTS

Figure 2-1-8.

2-1-4. RUNWAY EDGE LIGHT SYSTEMS

a. Runway edge lights are used to outline the edges of runways during periods of darkness or restricted visibility conditions. These light systems are classified according to the intensity or brightness they are capable of producing: they are the High Intensity Runway Lights (HIRL), Medium Intensity Runway Lights (MIRL), and the Low Intensity Runway Lights (LIRL). The HIRL and MIRL systems have variable intensity controls, whereas the LIRLs normally have one intensity setting.

b. The runway edge lights are white, except on instrument runways yellow replaces white on the last 2,000 feet or half the runway length, whichever is less, to form a caution zone for landings.

c. The lights marking the ends of the runway emit red light toward the runway to indicate the end of runway to a departing aircraft and emit green outward from the runway end to indicate the threshold to landing aircraft.

2-1-5. IN-RUNWAY LIGHTING

a. Runway Centerline Lighting System (RCLS): Runway centerline lights are installed on some precision approach runways to facilitate landing under adverse visibility conditions. They are located along the runway centerline and are spaced at 50-foot intervals. When viewed from the landing threshold, the runway centerline lights are white until the last 3,000 feet of the runway. The white lights begin to alternate with red for the next 2,000 feet, and for the last 1,000 feet of the runway, all centerline lights are red.

b. Touchdown Zone Lights (TDZL): Touchdown zone lights are installed on some precision approach runways to indicate the touchdown zone when landing under adverse visibility conditions. They consist of two rows of transverse light bars disposed symmetrically about the runway centerline. The system consists of steady-burning white lights which start 100 feet beyond the landing threshold and extend to 3,000 feet beyond the landing threshold or to the midpoint of the runway, whichever is less.

c. Taxiway Centerline Lead-Off Lights. Taxiway centerline lead-off lights provide visual guidance to persons exiting the runway. They are color-coded to warn pilots and vehicle drivers that they are within the runway environment or instrument landing system/microwave landing system (ILS/MLS) critical area, whichever is more restrictive. Alternate green and yellow lights are installed, beginning with green, from the runway centerline to one centerline light position beyond the runway holding position or ILS/MLS critical area holding position.

d. Taxiway Centerline Lead-On Lights. Taxiway centerline lead-on lights provide visual guidance to persons entering the runway. These "lead-on" lights are also color-coded with the same color pattern as lead-off lights to warn pilots and vehicle drivers that they are within the runway environment or instrument landing system/microwave landing system (ILS/MLS) critical area, whichever is more conservative.

Figure 2-1-9. Taxiway Lead-On Light Configuration

The fixtures used for lead-on lights are bidirectional, i.e., one side emits light for the lead-on function while the other side emits light for the lead-off function. Any fixture that emits yellow light for the lead-off function shall also emit yellow light for the lead-on function. (See FIG 2-1-9.)

REFERENCE—AIM, PILOT RESPONSIBILITIES WHEN CONDUCTING LAND AND HOLD SHORT OPERATIONS (LAHSO), PARAGRAPH 4-3-11.

2-1-6. CONTROL OF LIGHTING SYSTEMS

a. Operation of approach light systems and runway lighting is controlled by the control tower (ATCT). At some locations the FSS may control the lights where there is no control tower in operation.

b. Pilots may request that lights be turned on or off. Runway edge lights, in-pavement lights and approach lights also have intensity controls which may be varied to meet the pilots request. Sequenced flashing lights (SFL) may be turned on and off. Some sequenced flashing light systems also have intensity control.

2-1-7. PILOT CONTROL OF AIRPORT LIGHTING

Radio control of lighting is available at selected airports to provide airborne control of lights by keying the aircraft's microphone. Control of lighting systems is often available at locations without specified hours for lighting and where there is no control tower or FSS or when the tower or FSS is closed (locations with a part-time tower or FSS) or specified hours. All lighting systems which are radio controlled at an airport, whether on a single runway or multiple runways, operate on the same radio frequency. (See TBL 2-1-1 and TBL 2-1-2).

a. With FAA approved systems, various combinations of medium intensity approach lights, runway lights, taxiway lights, VASI and/or REIL may be activated by radio control.

Table 2-1-1 Runways with Approach Lights

Lighting System	No. of Int. Steps	Status During Nonuse Period	Intensity Step Selected Per No. of Mike Clicks		
			3 Clicks	5 Clicks	7 Clicks
Approach Lights (Med. Int.)	2	Off	Low	Low	High
Approach Lights (Med. Int.)	3	Off	Low	Med	High
MIRL	3	Off or Low	◆	◆	◆
HIRL	5	Off or Low	◆	◆	◆
VASI	2	Off	★	★	★

NOTES:
◆Predetermined intensity step.
★ Low intensity for night use. High intensity for day use as determined by photocell control.

Table 2-1-2 Runways without Approach Lights

Lighting System	No. of Int. Steps	Status During Nonuse Period	Intensity Step Selected Per No. of Mike Clicks		
			3 Clicks	5 Clicks	7 Clicks
MIRL	3	Off or Low	Low	Med.	High
HIRL	5	Off or Low	Step 1 or 2	Step 3	Step 5
LIRL	1	Off	On	On	On
VASI★	2	Off	◆	◆	◆
REIL★	1	Off	Off	On/Off	On
REIL★	3	Off	Low	Med.	High

NOTES:
◆ Low intensity for night use. High intensity for day use as determined by photocell control.
★ The control of VASI and/or REIL may be independent of other lighting systems.

On runways with both approach lighting and runway lighting (runway edge lights, taxiway lights, etc.) systems, the approach lighting system takes precedence for air-to-ground radio control over the runway lighting system which is set at a predetermined intensity step, based on expected visibility conditions. Runways without approach lighting may provide radio controlled intensity adjustments of runway edge lights. Other lighting systems, including VASI, REIL, and taxiway lights may be either controlled with the runway edge lights or controlled independently of the runway edge lights.

b. The control system consists of a 3-step control responsive to 7, 5, and/or 3 microphone clicks. This 3-step control will turn on lighting facilities capable of either 3-step, 2-step or 1-step operation. The 3-step and 2-step lighting facilities can be altered in intensity, while the 1-step cannot. All lighting is illuminated for a period of 15 minutes from the most recent time of activation and may not be extinguished prior to end of the 15 minute period (except for 1-step and 2-step REILs which may be turned off when desired by keying the mike 5 or 3 times respectively).

c. Suggested use is to always initially key the mike 7 times; this assures that all controlled lights are turned on to the maximum available intensity. If desired, adjustment can then be made, where the capability is provided, to a lower intensity (or the REIL turned off) by keying 5 and/or 3 times. Due to the close proximity of airports using the same frequency,

radio controlled lighting receivers may be set at a low sensitivity requiring the aircraft to be relatively close to activate the system. Consequently, even when lights are on, always key mike as directed when overflying an airport of intended landing or just prior to entering the final segment of an approach. This will assure the aircraft is close enough to activate the system and a full 15 minutes lighting duration is available. Approved lighting systems may be activated by keying the mike (within 5 seconds) as indicated in TBL 2-1-3.

Table 2-1-3 Radio Control System

Key Mike	Function
7 times within 5 seconds	Highest intensity available
5 times within 5 seconds	Medium or lower intensity (Lower REIL or REIL-off)
3 times within 5 seconds	Lowest intensity available (Lower REIL or REIL-off)

d. For all public use airports with FAA standard systems the Airport/Facility Directory contains the types of lighting, runway and the frequency that is used to activate the system. Airports with IAPs include data on the approach chart identifying the light system, the runway on which they are installed, and the frequency that is used to activate the system.

NOTE—ALTHOUGH THE CTAF IS USED TO ACTIVATE THE LIGHTS AT MANY AIRPORTS, OTHER FREQUENCIES MAY ALSO BE USED. THE APPROPRIATE FREQUENCY FOR ACTIVATING THE LIGHTS ON THE AIRPORT IS PROVIDED IN THE AIRPORT/FACILITY DIRECTORY AND THE STANDARD INSTRUMENT APPROACH PROCEDURES PUBLICATIONS. IT IS NOT IDENTIFIED ON THE SECTIONAL CHARTS.

e. Where the airport is not served by an IAP, it may have either the standard FAA approved control system or an independent type system of different specification installed by the airport sponsor. The Airport/Facility Directory contains descriptions of pilot controlled lighting systems for each airport having other than FAA approved systems, and explains the type lights, method of control, and operating frequency in clear text.

2-1-8. AIRPORT/HELIPORT BEACONS

a. Airport and heliport beacons have a vertical light distribution to make them most effective from one to ten degrees above the horizon; however, it can be seen well above and below this peak spread. The beacon may be an omnidirectional capacitor-discharge device, or it may rotate at a constant speed which produces the visual effect of flashes at regular intervals. Flashes may be one or two colors alternately. The total number of flashes are:

1. 24 to 30 per minute for beacons marking airports, landmarks, and points on Federal airways.

2. 30 to 45 per minute for beacons marking heliports.

b. The colors and color combinations of beacons are:

1. White and Green—Lighted land airport

2. *Green alone—Lighted land airport

3. White and Yellow—Lighted water airport

4. *Yellow alone—Lighted water airport

5. Green, Yellow, and White—Lighted heliport

NOTE—*GREEN ALONE OR YELLOW ALONE IS USED ONLY IN CONNECTION WITH A WHITE-AND-GREEN OR WHITE-AND-YELLOW BEACON DISPLAY, RESPECTIVELY.

c. Military airport beacons flash alternately white and green, but are differentiated from civil beacons by dual-peaked (two quick) white flashes between the green flashes.

d. In Class B, Class C, Class D and Class E surface areas, operation of the airport beacon during the hours of daylight often indicates that the ground visibility is less than 3 miles and/or the ceiling is less than 1,000 feet. ATC clearance in accordance with FAR Part 91 is required for landing, take-off and flight in the traffic pattern. Pilots should not rely solely on the operation of the airport beacon to indicate if weather conditions are IFR or VFR. At some locations with operating control towers, ATC personnel turn the beacon on or off when controls are in the tower. At many airports the airport beacon is turned on by a photoelectric cell or time clocks and ATC personnel cannot control them. There is no regulatory requirement for daylight operation and it is the pilot's responsibility to comply with proper preflight planning as required by FAR Part 91.103.

2-1-9. TAXIWAY LIGHTS

a. Taxiway Edge Lights: Taxiway edge lights are used to outline the edges of taxiways during periods of darkness or restricted visibility conditions. These fixtures emit blue light.

NOTE—AT MOST MAJOR AIRPORTS THESE LIGHTS HAVE VARIABLE INTENSITY SETTINGS AND MAY BE ADJUSTED AT PILOT REQUEST OR WHEN DEEMED NECESSARY BY THE CONTROLLER.

b. Taxiway Centerline Lights: Taxiway centerline lights are used to facilitate ground traffic under low visibility conditions. They are located along the taxiway centerline in a straight line on straight portions, on the centerline of curved portions, and along designated taxiing paths in portions of runways, ramp, and apron areas. Taxiway centerline lights are steady burning and emit green light.

c. Clearance Bar Lights: Clearance bar lights are installed at holding positions on taxiways in order to increase the conspicuity of the holding position in low visibility conditions. They may also be installed to indicate the location of an intersecting taxiway during periods of darkness. Clearance bars consist of three in-pavement steady-burning yellow lights.

d. Runway Guard Lights: Runway guard lights are installed at taxiway/runway intersections. They are primarily used to enhance the conspicuity of taxiway/runway intersections during low visibility conditions, but may be used in all weather conditions. Runway guard lights consist of either a pair of elevated flashing yellow lights installed on either side of the taxiway, or a row of in-pavement yellow lights installed across the entire taxiway, at the runway holding position marking.

NOTE—SOME AIRPORTS MAY HAVE A ROW OF THREE OR FIVE IN-PAVEMENT YELLOW LIGHTS INSTALLED AT TAXIWAY/RUNWAY INTERSECTIONS. THEY SHOULD NOT BE CONFUSED WITH CLEARANCE BAR LIGHTS DESCRIBED IN PARAGRAPH 2-1-9C ABOVE.

e. Stop Bar Lights: Stop bar lights, when installed, are used to confirm the ATC clearance to enter or cross the active runway in low visibility conditions (below 1,200 ft Runway Visual Range). A stop bar consists of a row of red, unidirectional, steady-burning in-pavement lights installed across the entire taxiway at the runway holding position, and elevated steady-burning red lights on each side. A controlled stop bar is operated in conjunction with the taxiway centerline lead-on lights which extend from the stop bar toward the runway. Following the ATC clearance to proceed, the stop bar is turned off and the lead-on lights are turned on. The stop bar and lead-on lights are automatically reset by a sensor or backup timer.

CAUTION—PILOTS SHOULD NEVER CROSS A RED ILLUMINATED STOP BAR, EVEN IF AN ATC CLEARANCE HAS BEEN GIVEN TO PROCEED ONTO OR ACROSS THE RUNWAY.

NOTE—IF AFTER CROSSING A STOP BAR, THE TAXIWAY CENTERLINE LEAD-ON LIGHTS INADVERTENTLY EXTINGUISH, PILOTS SHOULD HOLD THEIR POSITION AND CONTACT ATC FOR FURTHER INSTRUCTIONS.

Section 2. AIR NAVIGATION AND OBSTRUCTION LIGHTING

2-2-1. AERONAUTICAL LIGHT BEACONS

a. An aeronautical light beacon is a visual NAVAID displaying flashes of white and/or colored light to indicate the location of an airport, a heliport, a landmark, a certain point of a Federal airway in mountainous terrain, or an obstruction. The light used may be a rotating beacon or one or more flashing lights. The flashing lights may be supplemented by steady burning lights of lesser intensity.

b. The color or color combination displayed by a particular beacon and/or its auxiliary lights tell whether the beacon is indicating a landing place, landmark, point of the Federal airways, or an obstruction. Coded flashes of the auxiliary lights, if employed, further identify the beacon site.

2-2-2. CODE BEACONS AND COURSE LIGHTS

a. Code Beacons: The code beacon, which can be seen from all directions, is used to identify airports and landmarks. The code beacon flashes the three- or four-character airport identifier in International Morse Code six to eight times per minute. Green flashes are displayed for land airports while yellow flashes indicate water airports.

b. Course Lights: The course light, which can be seen clearly from only one direction, is used only with rotating beacons of the Federal Airway System: two course lights, back to back, direct coded flashing beams of light in either direction along the course of airway.

NOTE—AIRWAY BEACONS ARE REMNANTS OF THE "LIGHTED" AIRWAYS WHICH ANTEDATED THE PRESENT ELECTRONICALLY EQUIPPED FEDERAL AIRWAYS SYSTEM. ONLY A FEW OF THESE BEACONS EXIST TODAY TO MARK AIRWAY SEGMENTS IN REMOTE MOUNTAIN AREAS. FLASHES IN MORSE CODE IDENTIFY THE BEACON SITE.

2-2-3. OBSTRUCTION LIGHTS

a. Obstructions are marked/lighted to warn airmen of their presence during daytime and nighttime conditions. They may be marked/lighted in any of the following combinations:

1. *Aviation Red Obstruction Lights:* Flashing aviation red beacons (20 to 40 flashes per minute) and steady burning aviation red lights during nighttime operation. Aviation orange and white paint is used for daytime marking.

2. *Medium Intensity Flashing White Obstruction Lights:* Medium intensity flashing white obstruction lights may be used during daytime and twilight with automatically selected reduced intensity for nighttime operation. When this system is used on structures 500 feet (153m) AGL or less in height, other methods of marking and lighting the structure may be omitted. Aviation orange and white paint is always required for daytime marking on structures exceeding 500 feet 153m) AGL. This system is not normally installed on structures less than 200 feet (61m) AGL.

3. *High Intensity White Obstruction Lights:* Flashing high intensity white lights during daytime with reduced intensity for twilight and nighttime operation. When this type system is used, the marking of structures with red obstruction lights and aviation orange and white paint may be omitted.

4. *Dual Lighting:* A combination of flashing aviation red beacons and steady burning aviation red lights for nighttime operation and flashing high intensity white lights for daytime operation. Aviation orange and white paint may be omitted.

5. *Catenary Lighting:* Lighted markers are available for increased night conspicuity of high-voltage (69KV or higher) transmission line catenary wires. Lighted markers provide conspicuity both day and night.

b. Medium intensity omnidirectional flashing white lighting system provides conspicuity both day and night on catenary support structures. The unique sequential/simultaneous flashing light system alerts pilots of the associated catenary wires.

c. High intensity flashing white lights are being used to identify some supporting structures of overhead transmission lines located across rivers, chasms, gorges, etc. These lights flash in a middle, top, lower light sequence at approximately 60 flashes per minute. The top light is normally installed near the top of the supporting structure, while the lower light indicates the approximate lower portion of the wire span. The lights are beamed towards the companion structure and identify the area of the wire span.

d. High intensity flashing white lights are also employed to identify tall structures, such as chimneys and towers, as obstructions to air navigation. The lights provide a 360 degree coverage about the structure at 40 flashes per minute and consist of from one to seven levels of light depending upon the height of the structure. Where more than one level is used the vertical banks flash simultaneously.

Section 3. AIRPORT MARKING AIDS AND SIGNS

2-3-1. GENERAL

a. Airport pavement markings and signs provide information that is useful to a pilot during takeoff, landing, and taxiing.

b. Uniformity in airport markings and signs from one airport to another enhances safety and improves efficiency. Pilots are encouraged to work with the operators of the airports they use to achieve the marking and sign standards described in this section.

Table 2-3-1   Runway Marking Elements

Marking Element	Visual Runway	Nonprecision Instrument Runway	Precision Instrument Runway
Designation (par. 6)	X	X	X
Centerline (par. 7)	X	X	X
Threshold (par. 8)	X[1]	X	X
Aiming Point (par. 9)	X[2]	X	X
Touchdown Zone (par. 10)			X
Side Stripes (par. 11)			X

1   On runways used, or intended to be used, by international commercial transports.
2   On runways 4,000 feet (1,200 m) or longer used by jet aircraft.

c. Pilots who encounter ineffective, incorrect, or confusing markings or signs on an airport should make the operator of the airport aware of the problem. These situations may also be reported under the Aviation Safety Reporting Program as described in AIM paragraph 7-6-1. Pilots may also report these situations to the FAA regional airports division.

d. The markings and signs described in this section of the AIM reflect the current FAA recommended standards.

REFERENCE—AC 150/5340-1, STANDARDS FOR AIRPORT MARKINGS.

AC 150/5340-18, STANDARDS FOR AIRPORT SIGN SYSTEMS.

2-3-2. AIRPORT PAVEMENT MARKINGS

a. General: For the purpose of this presentation the Airport Pavement Markings have been grouped into four areas:

1. *Runway Markings.*

2. *Taxiway Markings.*

3. *Holding Position Markings.*

4. *Other Markings.*

b. Marking Colors: Markings for runways are white. Markings defining the landing area on a heliport are also white except for hospital heliports which use a red "H" on a white cross. Markings for taxiways, areas not intended for use by aircraft (closed and hazardous areas), and holding positions (even if they are on a runway) are yellow.

2-3-3. RUNWAY MARKINGS

a. General: There are three types of markings for runways: visual, nonprecision instrument and precision instrument. TBL 2-3-1 identifies the marking elements for each type of runway and TBL 2-3-2 identifies runway threshold markings.

Table 2-3-2
Number of Runway Threshold Stripes

Runway Width	Number of Stripes
60 feet (18 m)	4
75 feet (23 m)	6
100 feet (30 m)	8
150 feet (45 m)	12
200 feet (60 m)	16

b. Runway Designators: Runway numbers and letters are determined from the approach direction. The runway number is the whole number nearest one-tenth the magnetic azimuth of the centerline of the runway, measured clockwise from the magnetic north. The letters, differentiate between left (L), right (R), or center (C), parallel runways, as applicable:

1. For two parallel runways "L" "R"

2. For three parallel runways "L" "C" "R"

c. Runway Centerline Marking: The runway centerline identifies the center of the runway and provides alignment guidance during takeoff and landings. The centerline consists of a line of uniformly spaced stripes and gaps.

d. Runway Aiming Point Marking: The aiming point marking serves as a visual aiming point for a landing aircraft. These two rectangular markings consist of a broad white stripe located on each side of the runway centerline and approximately 1,000 feet from the landing threshold, as shown in FIG 2-3-1 Precision Instrument Runway Markings.

e. Runway Touchdown Zone Markers: The touchdown zone markings identify the touchdown zone for landing operations and are coded to provide distance information in 500 feet (150m) increments. These markings consist of groups of one, two, and three rectangular bars symmetrically arranged in pairs about the runway centerline, as shown in FIG 2-3-1 Precision Instrument Runway Markings. For runways having touchdown zone markings on both ends, those pairs of markings which extend to within 900 feet (270m) of the midpoint between the thresholds are eliminated.

f. Runway Side Strip Marking: Runway side stripes delineate the edges of the runway. They provide a visual contrast between runway and the abutting terrain or shoulders. Side stripes consist of continuous white stripes located, on each side of the runway as shown in FIG 2-3-5.

g. Runway Shoulder Markings: Runway shoulder stripes may be used to supplement runway side stripes to identify pavement areas contiguous to the runway sides that are not intended for use by aircraft. Runway Shoulder stripes are Yellow. (See FIG 2-3-3).

h. Runway Threshold Markings: Runway threshold markings come in two configurations. They either consist of eight longitudinal stripes of uniform dimensions disposed

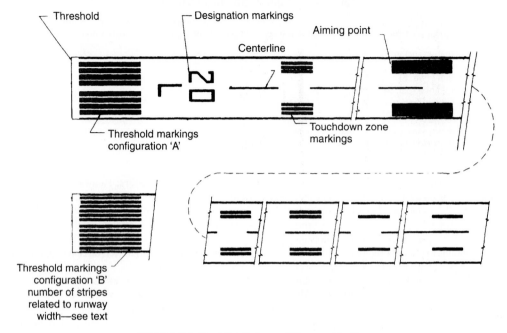

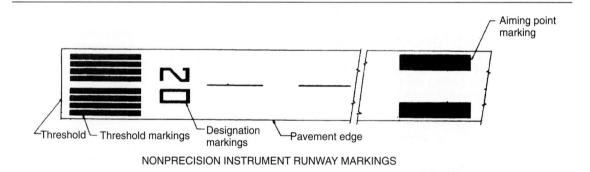

Figure 2-3-1. Precision Instrument Runway Markings

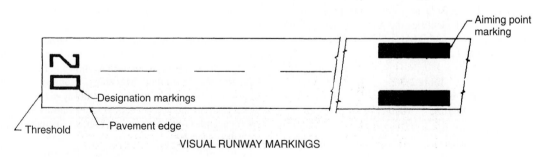

NONPRECISION INSTRUMENT RUNWAY MARKINGS

VISUAL RUNWAY MARKINGS

Figure 2-3-2. Nonprecision Instrument Runway and Visual Runway Markings.

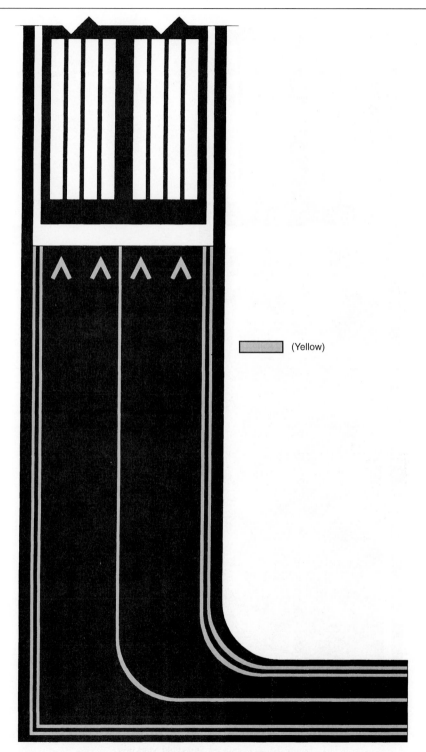

(Yellow)

Figure 2-3-3. Relocation of a Threshold with Markings for Taxiway Aligned with Runway

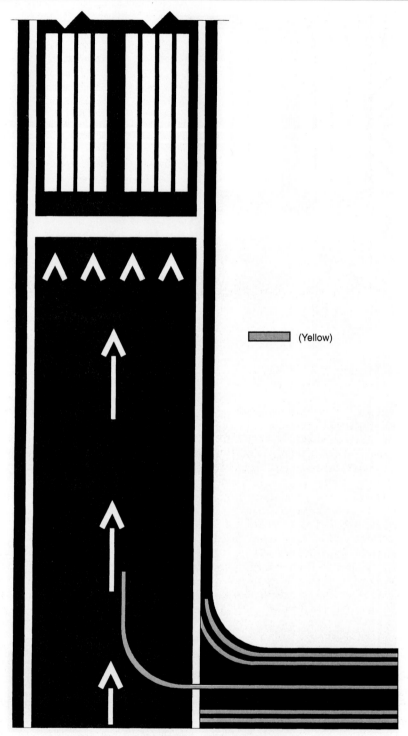

(Yellow)

Figure 2-3-4. Displaced Threshold Markings

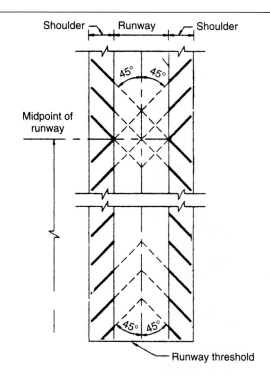

Figure 2-3-5. Runway Shoulder Markings

symmetrically about the runway centerline, as shown in FIG 2-3-1, or the number of stripes is related to the runway width as indicated in TBL 2-3-2. A threshold marking helps identify the beginning of the runway that is available for landing. In some instances the landing threshold may be relocated or displaced.

1. Relocation of a Threshold: Sometimes construction, maintenance, or other activities require the threshold to be relocated towards the rollout end of the runway. (See FIG 2-3-4). When a threshold is relocated, it closes not only a set portion of the approach end of a runway, but also shortens the length of the opposite direction runway. In these cases, a NOTAM should be issued by the airport operator identifying the portion of the runway that is closed, e.g., 10/28 W 900 CLSD. Because the duration of the relocation can vary from a few hours to several months, methods identifying the new threshold may vary. One common practice is to use a ten feet wide white threshold bar across the width of the runway. Although the runway lights in the area between the old threshold and new threshold will not be illuminated, the runway markings in this area may or may not be obliterated, removed, or covered.

2. *Displaced Threshold:* A displaced threshold is a threshold located at a point on the runway other than the designated beginning of the runway. Displacement of a threshold reduces the length of runway available for landings. The

portion of runway behind a displaced threshold is available for takeoffs in either direction and landings from the opposite direction. A ten feet wide white threshold bar is located across the width of the runway at the displaced threshold. White arrows are located along the centerline in the area between the beginning of the runway and displaced threshold. White arrow heads are located across the width of the runway just prior to the threshold bar, as shown in FIG 2-3-5.

NOTE—AIRPORT OPERATOR. WHEN REPORTING THE RELOCATION OR DISPLACEMENT OF A THRESHOLD, THE AIRPORT OPERATOR SHOULD AVOID LANGUAGE WHICH CONFUSES THE TWO.

i. Demarcation Bar: A demarcation bar delineates a runway with a displaced threshold from a blast pad, stopway or taxiway that precedes the runway. A demarcation bar is 3 feet (1m) wide and yellow, since it is not located on the runway as shown in FIG 2-3-6.

1. *Chevrons:* These markings are used to show pavement areas aligned with the runway that are unusable for landing, takeoff, and taxiing. Chevrons are yellow. (See FIG 2-3-7).

j. Runway Threshold BAR: A threshold bar delineates the beginning of the runway that is available for landing when the threshold has been relocated or displaced. A threshold bar is 10 feet (3m) in width and extends across the width of the runway, as shown in FIG 2-3-5.

2-3-4. TAXIWAY MARKINGS

a. General: All taxiways should have centerline markings and runway holding position markings whenever they intersect a runway. Taxiway edge markings are present whenever there is a need to separate the taxiway from a pavement that is not intended for aircraft use or to delineate the edge of the taxiway. Taxiways may also have shoulder markings and holding position markings for Instrument Landing System/Microwave Landing System (ILS/MLS) critical areas, and taxiway/taxiway intersection markings.

REFERENCE—AIM, PARAGRAPH 2-3-5.

b. Taxiway Centerline.

1. Normal Centerline. The taxiway centerline is a single continuous yellow line, 6 inches (15 cm) to 12 inches (30 cm) in width. This provides a visual cue to permit taxiing along a designated path. Ideally, the aircraft should be kept centered over this line during taxi. However, being centered on the taxiway centerline does not guarantee wingtip clearance with other aircraft or other objects.

2. Enhanced Centerline. At some airports, mostly the larger commercial service airports, an enhanced taxiway centerline will be used. The enhanced taxiway centerline marking consists of a parallel line of yellow dashes on either side of the normal taxiway centerline. The taxiway centerlines are enhanced for a maximum of 150 feet prior to a runway holding position marking. The purpose of this enhancement is to warn the pilot that he/she is approaching a runway holding position marking and should prepare

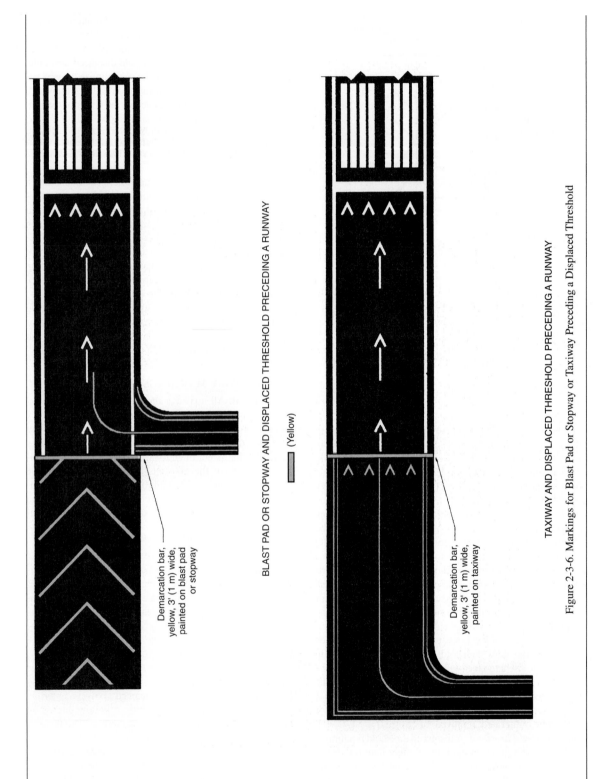

BLAST PAD OR STOPWAY AND DISPLACED THRESHOLD PRECEDING A RUNWAY

(Yellow)

Demarcation bar, yellow, 3' (1 m) wide, painted on blast pad or stopway

TAXIWAY AND DISPLACED THRESHOLD PRECEDING A RUNWAY

Demarcation bar, yellow, 3' (1 m) wide, painted on taxiway

Figure 2-3-6. Markings for Blast Pad or Stopway or Taxiway Preceding a Displaced Threshold

(Yellow)

Figure 2-3-7. Markings for Blast Pads and Stopways

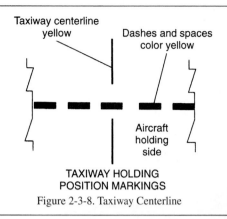

TAXIWAY HOLDING
POSITION MARKINGS

Figure 2-3-8. Taxiway Centerline

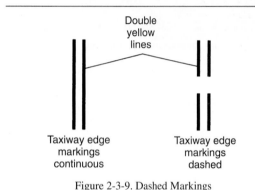

Figure 2-3-9. Dashed Markings

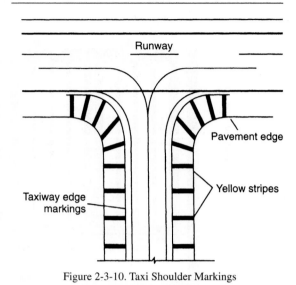

Figure 2-3-10. Taxi Shoulder Markings

to stop unless he/she has been cleared onto or across the runway by ATC. (See FIG 2-3-8.)

c. Taxiway Edge Markings: Taxiway edge markings are used to define the edge of the taxiway. They are primarily used when the taxiway edge does not correspond with the edge of the pavement. There are two types of markings depending upon whether the aircraft is suppose to cross the taxiway edge:

1. Continuous Markings: These consist of a continuous double yellow line, with each line being at least 6 inches (15 cm) in width spaced 6 inches (15 cm) apart. They are used to define the taxiway edge from the shoulder or some other abutting paved surface not intended for use by aircraft.

2. Dashed Markings: These markings are used when there is an operational need to define the edge of a taxiway or taxilane on a paved surface where the adjoining pavement to the taxiway edge is intended for use by aircraft. e.g., an apron. Dashed taxiway edge markings consist of a broken double yellow line, with each line being at least 6 inches (15 cm) in width, spaced 6 inches (15 cm) apart

(edge to edge). These lines are 15 feet (4.5 m) in length with 25 foot (7.5 m) gaps. (See FIG 2-3-9).

d. Taxi Shoulder Markings: Taxiways, holding bays, and aprons are sometimes provided with paved shoulders to prevent blast and water erosion. Although shoulders may have the appearance of full strength pavement they are not intended for use by aircraft, and may be unable to support an aircraft. Usually the taxiway edge marking will define this area. Where conditions exist such as islands or taxiway curves that may cause confusion as to which side of the edge stripe is for use by aircraft, taxiway shoulder markings may be used to indicate the pavement is unusable. Taxiway shoulder markings are yellow. (See FIG 2-3-10).

e. Surface Painted Taxiway Direction Signs: Surface painted taxiway direction signs have a yellow background with a black inscription, and are provided when it is not possible to provide taxiway direction signs at intersections, or when necessary to supplement such signs. These markings are located adjacent to the centerline with signs indicating turns to the left being on the left side of the taxiway centerline and signs indicating turns to the right being on the right side of the centerline. (See FIG 2-3-11).

f. Surface Painted Location Signs: Surface painted location signs have a black background with a yellow inscription. When necessary, these markings are used to supplement location signs located along side the taxiway and assist the pilot in confirming the designation of the taxiway on which the aircraft is located. These markings are located on the right side of the centerline. (See FIG 2-3-11).

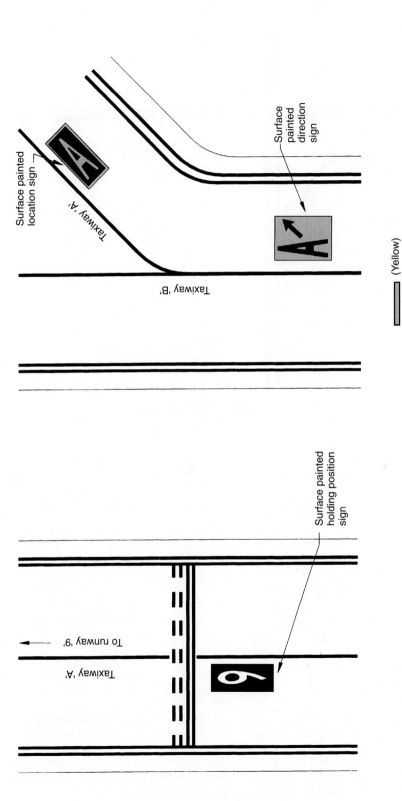

Figure 2-3-11. Surface Painted Signs

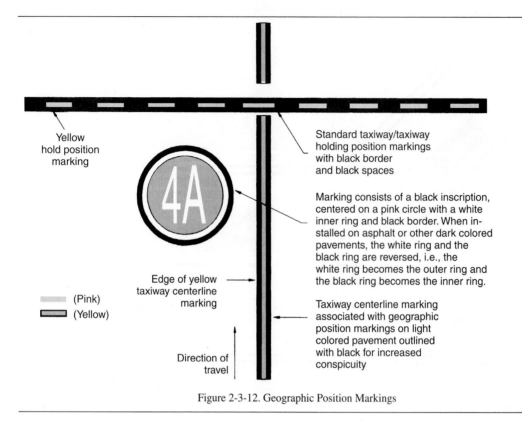

Figure 2-3-12. Geographic Position Markings

g. Geographic Position Markings: These markings are located at points along low visibility taxi routes designated in the airport's Surface Movement Guidance Control System (SMGCS) plan. They are used to identify the location of taxiing aircraft during low visibility operations. Low visibility operations are those that occur when the runway visible range (RVR) is below 1200 feet (360m). They are positioned to the left of the taxiway centerline in the direction of taxiing. (See FIG 2-3-12). The Geographic position marking is a circle comprised of an outer black ring contiguous to a white ring with a pink circle in the middle. When installed on asphalt or other dark-colored pavements, the white ring and the black ring are reversed, i.e., the white ring becomes the outer ring and the black ring becomes the inner ring. It is designated with either a number or a number and letter. The number corresponds to the consecutive position of the marking on the route.

2-3-5. HOLDING POSITION MARKINGS

a. Runway Holding Position Markings: For runways these markings indicate where an aircraft is supposed to stop. They consist of four yellow lines two solid and two dashed, spaced six inches apart and extending across the width of the taxiway or runway. The solid lines are always

on the side where the aircraft is to hold. There are three locations where runway holding position markings are encountered.

1. Runway Holding Position Markings on Taxiways: These markings identify the locations on a taxiway where an aircraft is supposed to stop when it does not have clearance to proceed onto the runway. The runway holding position markings are shown in FIG 2-3-13 and FIG 2-3-16. When instructed by ATC "Hold short of (runway "XX")" the pilot should stop so no part of the aircraft extends beyond the holding position marking. When approaching the holding position marking, a pilot should not cross the marking without ATC clearance at a controlled airport or without making sure of adequate separation from other aircraft at uncontrolled airports. An aircraft exiting a runway is not clear of the runway until all parts of the aircraft have crossed the applicable holding position marking.

2. Runway Holding Position Markings on Runways: These markings are installed on runways only if the runway is normally used by air traffic control for "land, hold short" operations or taxiing operations and have operational significance only for those two types of operations. A sign with a white inscription on a red background is

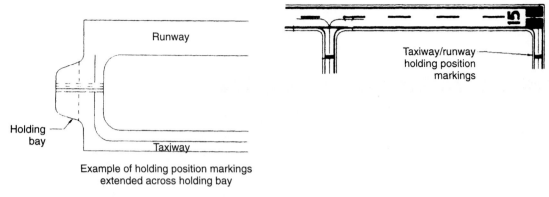

Example of holding position markings
extended across holding bay

Figure 2-3-13. Runway Holding Position Markings on Taxiways

installed adjacent to these holding position markings. (See FIG 2-3-14). The hold position markings are placed on runways prior to the intersection with another runway, or some designated point. Pilots receiving instructions "Clear to land, runway "XX"" from air traffic control are authorized to use the entire landing length of the runway and should disregard any holding position markings located on the runway. Pilots receiving and accepting instructions "clear to land runway "XX,"" hold short of runway "YY" from air traffic control must either exit runway "XX," or stop at the holding position prior to runway "YY."

3. Taxiways Located in Runway Approach Areas. These markings are used at some airports where it is necessary to hold an aircraft on a taxiway located in the approach or departure area of a runway so that the aircraft does not interfere with the operations on that runway. This marking is collocated with the runway approach area holding position sign. When specifically instructed by ATC "Hold short of (runway xx approach area)" the pilot should stop so no part of the aircraft extends beyond the holding position marking. (See subparagraph 2-3-8b2, Runway Approach Area Holding Position Sign, and FIG 2-3-15.)

b. Holding Position Markings for Instrument Landing System (ILS): Holding position markings for ILS/MLS critical areas consist of two yellow solid lines spaced two feet apart connected by pairs of solid lines spaced ten feet apart extending across the width of the taxiway as shown. (See FIG 2-3-16). A sign with an inscription in white on a red background is installed adjacent to these hold position markings. When the ILS critical area is being protected, the pilot should stop so no part of the aircraft extends beyond the holding position marking. When approaching the holding position marking, a pilot should not cross the marking without ATC clearance. ILS critical area is not clear until all parts of the aircraft have crossed the applicable holding position marking.

REFERENCE—AIM, INSTRUMENT LANDING SYSTEM (ILS) PARAGRAPH 1-1-9.

c. Holding Position Markings for Taxiway/Taxiway Intersections: Holding position markings for taxiway/taxiway intersections consist of a single dashed line extending across the width of the taxiway as shown. (See FIG 2-3-17). They are installed on taxiways where air traffic control normally holds aircraft short of a taxiway intersection. When instructed by ATC "hold short of (taxiway)" the pilot should stop so no part of the aircraft extends beyond the holding position marking. When the marking is not present the pilot should stop the aircraft at a point which provides adequate clearance from an aircraft on the intersecting taxiway.

d. Surface Painted Holding Position Signs: Surface painted holding position signs have a red background with a white inscription and supplement the signs located at the holding position. This type of marking is normally used where the width of the holding position on the taxiway is greater than 200 feet (60m). It is located to the left side of the taxiway centerline on the holding side and prior to the holding position marking. (See FIG 2-3-11).

2-3-6. OTHER MARKINGS

a. Vehicle Roadway Markings: The vehicle roadway markings are used when necessary to define a pathway for vehicle operations on or crossing areas that are also intended for aircraft. These markings consist of a white solid line to delineate each edge of the roadway and a dashed line to separate lanes within the edges of the roadway. In lieu of the solid lines, zipper markings may be used to delineate the edges of the vehicle roadway. (See FIG 2-13-18). Details of the zipper markings are shown in FIG 2-3-19.

b. VOR Receiver Checkpoint Markings: The VOR receiver checkpoint marking allows the pilot to check aircraft instruments with navigational aid signals. It consists

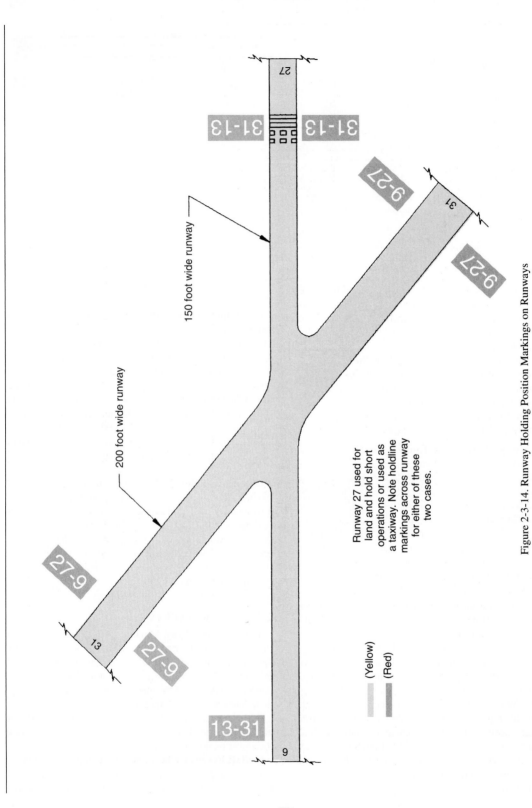

150 foot wide runway

200 foot wide runway

Runway 27 used for land and hold short operations or used as a taxiway. Note holdline markings across runway for either of these two cases.

(Yellow)

(Red)

Figure 2-3-14. Runway Holding Position Markings on Runways

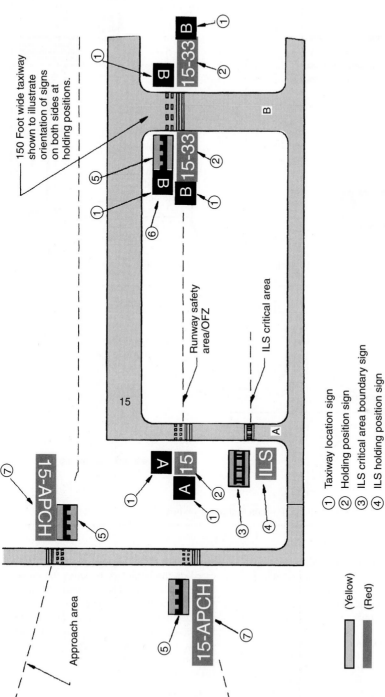

Figure 2-3-15. Taxiways Located in Runway Approach Area

① Taxiway location sign
② Holding position sign
③ ILS critical area boundary sign
④ ILS holding position sign
⑤ Runway safety area/ofz and runway approach area boundary
⑥ Taxiway location sign—optional, depending on operational need
⑦ Holding position sign for approach areas

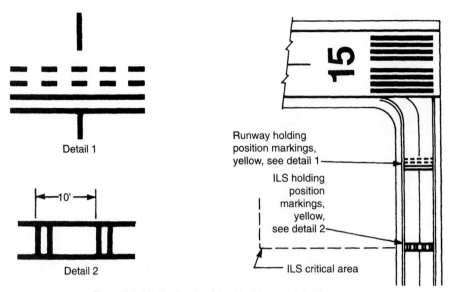

Figure 2-3-16. Holding Position Markings: ILS Critical Area

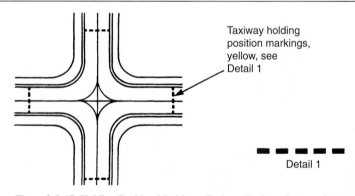

Figure 2-3-17. Holding Position Markings: Taxiway/Taxiway Intersections

of a painted circle with an arrow in the middle; the arrow is aligned in the direction of the checkpoint azimuth. This marking, and an associated sign, is located on the airport apron or taxiway at a point selected for easy access by aircraft but where other airport traffic is not to be unduly obstructed. (See FIG 2-3-20).

NOTE—THE ASSOCIATED SIGN CONTAINS THE VOR STATION IDENTIFICATION LETTER AND COURSE SELECTED (PUBLISHED) FOR THE CHECK, THE WORDS "VOR CHECK COURSE," AND DME DATA (WHEN APPLICABLE). THE COLOR OF THE LETTERS AND NUMERALS ARE BLACK ON A YELLOW BACKGROUND.

c. Nonmovement Area Boundary Markings: These markings delineate the movement area, i.e., area under air traffic control. These markings are yellow and located on the boundary between the movement and nonmovement area. The nonmovement area boundary markings consist of two yellow lines (one solid and one dashed) 6 inches (15cm) in width. The solid line is located on the nonmovement area side while the dashed yellow line is located on the movement area side. The nonmovement boundary marking area is shown in FIG 2-3-21.

EXAMPLE—
DCA 176-356
VOR CHECK COURSE
DME XXX

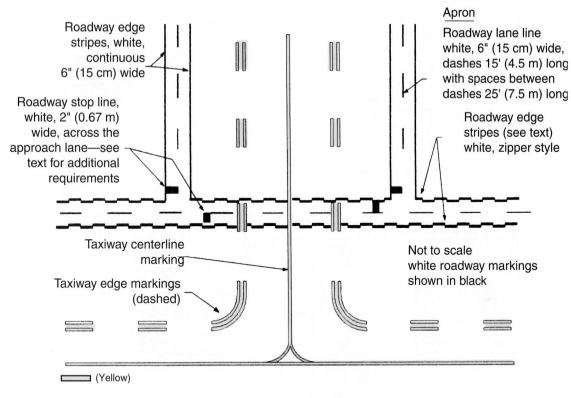

Roadway edge stripes, white, continuous 6" (15 cm) wide

Apron
Roadway lane line white, 6" (15 cm) wide, dashes 15' (4.5 m) long with spaces between dashes 25' (7.5 m) long

Roadway stop line, white, 2" (0.67 m) wide, across the approach lane—see text for additional requirements

Roadway edge stripes (see text) white, zipper style

Taxiway centerline marking

Not to scale
white roadway markings shown in black

Taxiway edge markings (dashed)

☐ (Yellow)

Figure 2-3-18. Vehicle Roadway Markings

Figure 2-3-19. Roadway Edge Stripes, White, Zipper Style

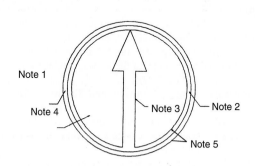

Note 1
Note 4
Note 3
Note 2
Note 5

Notes—
1 White
2 Yellow
3 Yellow. Arrow to be aligned toward the facility
4 Interior of circle black (concrete surfaces only)
5 Circle may be bordered on inside and outside with 6" black band if necessary for contrast

Figure 2-3-20. Ground Receiver Checkpoint Markings

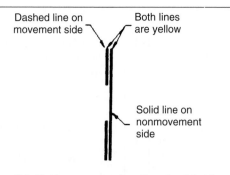

Figure 2-3-21. Nonmovement Area Boundary Markings

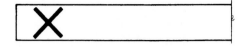

Figure 2-3-22. Closed or Temporarily Closed Runway and Taxiway Markings

HELICOPTER LANDING AREA

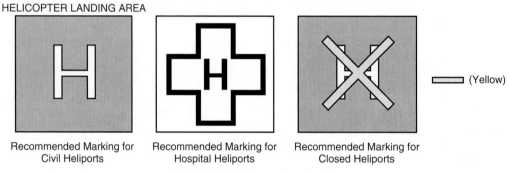

Recommended Marking for Civil Heliports   Recommended Marking for Hospital Heliports   Recommended Marking for Closed Heliports   (Yellow)

Figure 2-3-23. Helicopter Landing Areas

d. Marking and Lighting of Permanently Closed Runways and Taxiways: For runways and taxiways which are permanently closed, the lighting circuits will be disconnected. The runway threshold, runway designation, and touchdown markings are obliterated and yellow crosses are placed at each end of the runway and at 1,000 foot intervals. (See FIG 2-3-22).

e. Temporarily Closed Runways and Taxiways: To provide a visual indication to pilots that a runway is temporarily closed, crosses are placed on the runway only at each end of the runway. The crosses are yellow in color. (See FIG 2-3-22).

1. A raised lighted yellow cross may be placed on each runway end in lieu of the markings described in subparagraph "e." (above) to indicate the runway is closed.

2. A visual indication may not be present depending on the reason for the closure, duration of the closure, airfield configuration and the existence and the hours of operation of an airport traffic control tower. Pilots should check NOTAMs and the Automated Terminal Information System (ATIS) for local runway and taxiway closure information.

3. Temporarily closed taxiways are usually treated as hazardous areas, in which no part of an aircraft may enter, and

are blocked with barricades. However, as an alternative a yellow cross may be installed at each entrance to the taxiway.

f. Helicopter Landing Areas: The markings illustrated in FIG 2-3-23 are used to identify the landing and takeoff area at a public use heliport and hospital heliport. The letter "H" in the markings is oriented to align with the intended direction of approach. FIG 2-3-23 also depicts the markings for a closed airport.

Figure 2-3-24. Runway Holding Position Sign

Figure 2-3-25. Holding Position Sign at Beginning of
Takeoff Runway

2-3-7. AIRPORT SIGNS

There are six types of signs installed on airfields: mandatory instruction signs, location signs, direction signs, destination signs, information signs, and runway distance remaining signs. The characteristics and use of these signs are discussed in AIM paragraph 2-3-8 through AIM paragraph 2-3-13.

REFERENCE—AC150/5340-18, STANDARDS FOR AIRPORT SIGN SYSTEMS FOR DETAILED INFORMATION ON AIRPORT SIGNS

2-3-8. MANDATORY INSTRUCTION SIGNS

a. These signs have a red background with a white inscription and are used to denote:

1. An entrance to a runway or critical area and;

2. Areas where an aircraft is prohibited from entering.

b. Typical mandatory signs and applications are:

1. *Runway Holding Position Sign:* This sign is located at the holding position on taxiways that intersect a runway or on runways that intersect other runways. The inscription on the sign contains the designation of the intersecting runway as shown in FIG 2-3-24. The runway numbers on the sign are arranged to correspond to the respective runway threshold. For example, "15–33" indicates that the threshold for Runway 15 is to the left and the threshold for Runway 33 is to the right.

(a) On taxiways that intersect the beginning of the takeoff runway, only the designation of the takeoff runway may appear on the sign as shown in FIG 2-3-25, while all other signs will have the designation of both runway directions.

(b) If the sign is located on a taxiway that intersects the intersection of two runways, the designations for both runways will be shown on the sign along with arrows showing the approximate alignment of each runway as shown in FIG 2-3-26. In addition to showing the approximate runway alignment, the arrow indicates the direction to the threshold of the runway whose designation is immediately next to the arrow.

(c) A runway holding position sign on a taxiway will be installed adjacent to holding position markings on the taxiway pavement. On runways, holding position markings will be located only on the runway pavement adjacent to the sign, if the runway is normally used by air traffic control for "Land, Hold Short" operations or as a taxiway. The holding position markings are described in AIM paragraph 2-3-5.

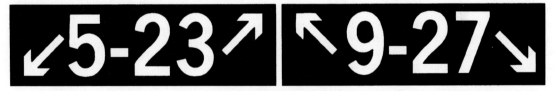

Figure 2-3-26. Holding Position Sign for a Taxiway that Intersects the Intersection of Two Runways

Figure 2-3-27. Holding Position Sign for a Runway Approach Area

Figure 2-3-28. Holding Position Sign for ILS Critical Area

Figure 2-3-29. Sign Prohibiting Aircraft Entry into an Area

2. Runway Approach Area Holding Position Sign: At some airports, it is necessary to hold an aircraft on a taxiway located in the approach or departure area for a runway so that the aircraft does not interfere with operations on that runway. In these situations, a sign with the designation of the approach end of the runway followed by a "dash" (–) and letters "APCH" will be located at the holding position on the taxiway. Holding position markings in accordance with AIM paragraph 2-3-5 will be located on the taxiway pavement. An example of this sign is shown in FIG 2-3-27. In this example, the sign may protect the approach to Runway 15 and/or the departure for Runway 33.

3. ILS Critical Area Holding Position Sign: At some airports, when the instrument landing system is being used, it is necessary to hold an aircraft on a taxiway at a location other than the holding position described in AIM paragraph 2-3-5. In these situations the holding position sign for these operations will have the inscription "ILS" and be located adjacent to the holding position marking on the taxiway described in AIM paragraph 2-3-5. An example of this sign is shown in FIG 2-3-28.

4. No Entry Sign: This sign, shown in FIG 2-3-29, prohibits an aircraft from entering an area. Typically, this sign would be located on a taxiway intended to be used in only one direction or at the intersection of vehicle roadways with runways, taxiways or aprons where the roadway may be mistaken as a taxiway or other aircraft movement surface.

NOTE—THE HOLDING POSITION SIGN PROVIDES THE PILOT WITH A VISUAL CUE AS TO THE LOCATION OF THE HOLDING POSITION MARKING. THE OPERATIONAL SIGNIFICANCE OF HOLDING POSITION MARKINGS ARE DESCRIBED IN THE NOTES FOR AIM PARAGRAPH 2-3-5.

2-3-9. LOCATION SIGNS

a. Location signs are used to identify either a taxiway or runway on which the aircraft is located. Other location signs provide a visual cue to pilots to assist them in determining when they have exited an area. The various location signs are described below.

1. Taxiway Location Sign: This sign has a black background with a yellow inscription and yellow border as shown in FIG 2-3-30. The inscription is the designation of the taxiway on which the aircraft is located. These signs are installed along taxiways either by themselves or in conjunction with direction signs (See FIG 2-3-35, or runway holding position signs, FIG 2-3-31.)

2. Runway Location Sign: This sign has a black background with a yellow inscription and yellow border as shown in FIG 2-3-32. The inscription is the designation of the runway on which the aircraft is located. These signs are intended to complement the information available to pilots through their magnetic compass and typically are installed where the proximity of two or more runways to one another could cause pilots to be confused as to which runway they are on.

3. Runway Boundary Sign: This sign has a yellow background with a black inscription with a graphic depicting the pavement holding position marking as shown in FIG 2-3-33. This sign, which faces the runway and is visible to the pilot exiting the runway, is located adjacent to the holding position marking on the pavement. The sign is intended to provide pilots with another visual cue which they can use as a guide in deciding when they are "clear of the runway."

4. ILS Critical Area Boundary Sign: This sign has a yellow background with a black inscription with a graphic depicting the ILS pavement holding position marking as shown in FIG 2-3-34. This sign is located adjacent to the ILS holding position marking on the pavement and can be seen by pilots leaving the critical area. The sign is intended

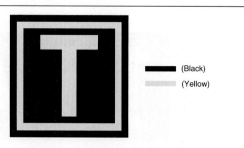

Figure 2-3-30. Taxiway Location Sign

(Red)
(Black)
(Yellow)

Figure 2-3-31. Taxiway Location Sign Collocated with Runway Holding Position Sign

(Black)
(Yellow)

Figure 2-3-32. Runway Location Sign

to provide pilots with another visual cue which they can use as a guide in deciding when they are "clear of the ILS critical area."

2-3-10. DIRECTION SIGNS

a. Direction signs have a yellow background with a black inscription. The inscription identifies the designation(s) of the intersecting taxiway(s) leading out of the intersection that a pilot would normally be expected to turn onto or hold short of. Each designation is accompanied by an arrow indicating the direction of the turn.

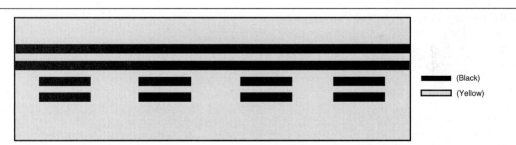

(Black)
(Yellow)

Figure 2-3-33. Runway Boundary Sign

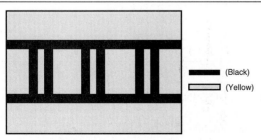

(Black)
(Yellow)

Figure 2-3-34. ILS Critical Area Boundary Sign

b. Except as noted in subparagraph e, each taxiway designation shown on the sign is accompanied by only one arrow. When more than one taxiway designation is shown on the sign each designation and its associated arrow is separated from the other taxiway designations by either a vertical message divider or a taxiway location sign as shown in FIG 2-3-35.

c. Direction signs are normally located on the left prior to the intersection. When used on a runway to indicate an exit, the sign is located on the same side of the runway as the exit. FIG 2-3-36 shows a direction sign used to indicate a runway exit.

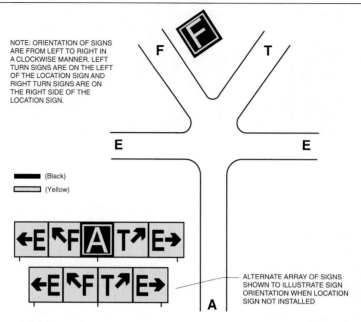

NOTE: ORIENTATION OF SIGNS
ARE FROM LEFT TO RIGHT IN
A CLOCKWISE MANNER. LEFT
TURN SIGNS ARE ON THE LEFT
OF THE LOCATION SIGN AND
RIGHT TURN SIGNS ARE ON
THE RIGHT SIDE OF THE
LOCATION SIGN.

ALTERNATE ARRAY OF SIGNS
SHOWN TO ILLUSTRATE SIGN
ORIENTATION WHEN LOCATION
SIGN NOT INSTALLED

Figure 2-3-35. Direction Sign Array with Location Sign on Far Side of Intersection

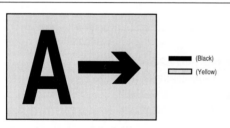

Figure 2-3-36. Direction Sign for Runway Exit

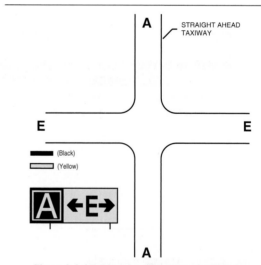

Figure 2-3-37. Direction Sign Array for Simple

d. The taxiway designations and their associated arrows on the sign are arranged clockwise starting from the first taxiway on the pilot's left. (See FIG 2-3-35).

e. If a location sign is located with the direction signs, it is placed so that the designations for all turns to the left will be to the left of the location sign; the designations for continuing straight ahead or for all turns to the right would be located to the right of the location sign. (See FIG 2-3-35).

f. When the intersection is comprised of only one crossing taxiway, it is permissible to have two arrows associated with the crossing taxiway as shown in FIG 2-3-37. In this case, the location sign is located to the left of the direction sign.

2-3-11. DESTINATION SIGNS

a. Destination signs also have a yellow background with a black inscription indicating a destination on the airport. These signs always have an arrow showing the direction of the taxiing route to that destination. FIG 2-3-38 is an exam-

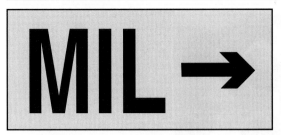

Figure 2-3-38. Destination Sign for Military Area

■ (Black)
▭ (Yellow)

Figure 2-3-39. Destination Sign for Common Taxiing Route to Two Runways

■ (Black)
▭ (Yellow)

Figure 2-3-40. Destination Sign for Different Taxiing Routes to Two Runways

ple of a typical destination sign. When the arrow on the destination sign indicates a turn, the sign is located prior to the intersection.

b. Destinations commonly shown on these types of signs include runways, aprons, terminals, military areas, civil aviation areas, cargo areas, international areas, and fixed base operators. An abbreviation may be used as the inscription on the sign for some of these destinations.

c. When the inscription for two or more destinations having a common taxiing route are placed on a sign, the destinations are separated by a "dot" (•) and one arrow would be used as shown in FIG 2-3-39. When the inscription on a sign contains two or more destinations having different taxiing routes, each destination will be accompanied by an arrow and will be separated from the other destinations on

the sign with a vertical black message divider as shown in FIG 2-3-40.

2-3-12. INFORMATION SIGNS

Information signs have a yellow background with a black inscription. They are used to provide the pilot with information on such things as areas that cannot be seen from the control tower, applicable radio frequencies, and noise abatement procedures. The airport operator determines the need, size, and location for these signs.

2-3-13. RUNWAY DISTANCE REMAINING SIGNS

Runway distance remaining signs have a black background with a white numeral inscription and may be in-

Figure 2-3-41. Runway Distance Remaining Sign Indicating 3000 feet of Runway Remaining

stalled along one or both side(s) of the runway. The number on the signs indicates the distance (in thousands of feet) of landing runway remaining. The last sign, i.e., the sign with the numeral "1," will be located at least 950 feet from the runway end. FIG 2-3-41 shows an example of a runway distance remaining sign.

2-3-14. AIRCRAFT ARRESTING DEVICES

a. Certain airports are equipped with a means of rapidly stopping military aircraft on a runway. This equipment, normally referred to as EMERGENCY ARRESTING GEAR, generally consists of pendant cables supported over the runway surface by rubber "donuts." Although most devices are located in the overrun areas, a few of these arresting systems have cables stretched over the operational areas near the ends of a runway.

b. Arresting cables which cross over a runway require special markings on the runway to identify the cable location. These markings consist of 10 feet diameter solid circles painted "identification yellow," 30 feet on center, perpendicular to the runway centerline across the entire runway width. Additional details are contained in AC 150/5220-9, Aircraft Arresting Systems for Joint Civil/ Military Airports.

NOTE—AIRCRAFT OPERATIONS ON THE RUNWAY ARE NOT RESTRICTED BY THE INSTALLATION OF AIRCRAFT ARRESTING DEVICES.

Chapter 3. Airspace
Section 1. GENERAL

3-1-1. GENERAL

a. There are two categories of airspace or airspace areas:

1. Regulatory (Class A, B, C, D and E airspace areas, restricted and prohibited areas); and

2. Nonregulatory (military operations areas (MOA's), warning areas, alert areas, and controlled firing areas).

NOTE—ADDITIONAL INFORMATION ON SPECIAL USE AIRSPACE (PROHIBITED AREAS, RESTRICTED AREAS, WARNING AREAS, MOA'S, ALERT AREAS AND CONTROLLED FIRING AREAS) MAY BE FOUND IN CHAPTER 3, AIRSPACE, SECTION 4, SPECIAL USE AIRSPACE, PARAGRAPHS 3-4-1 THROUGH 3-4-7.

b. Within these two categories, there are four types:

1. controlled,

2. uncontrolled,

3. special use, and

4. other airspace.

c. The categories and types of airspace are dictated by:

1. the complexity or density of aircraft movements;

2. the nature of the operations conducted within the airspace;

3. the level of safety required; and

4. the national and public interest.

d. It is important that pilots be familiar with the operational requirements for each of the various types or classes of airspace. Subsequent sections will cover each class in sufficient detail to facilitate understanding.

3-1-2. GENERAL DIMENSIONS OF AIRSPACE SEGMENTS

Refer to Federal Aviation Regulations (FAR) for specific dimensions, exceptions, geographical areas covered, exclusions, specific transponder or equipment requirements, and flight operations.

3-1-3. HIERARCHY OF OVERLAPPING AIRSPACE DESIGNATIONS

a. When overlapping airspace designations apply to the same airspace, the operating rules associated with the more restrictive airspace designation apply.

b. For the purpose of clarification:

1. Class A airspace is more restrictive than Class B, Class C, Class D, Class E, or Class G airspace;

2. Class B airspace is more restrictive than Class C, Class D, Class E, or Class G airspace;

3. Class C airspace is more restrictive than Class D, Class E, or Class G airspace;

4. Class D airspace is more restrictive than Class E or Class G airspace; and

5. Class E is more restrictive than Class G airspace.

3-1-4. BASIC VFR WEATHER MINIMUMS

a. No person may operate an aircraft under basic VFR when the flight visibility is less, or at a distance from clouds that is less, than that prescribed for the corresponding altitude and class of airspace. (See TBL 3-1-1).

NOTE—STUDENT PILOTS MUST COMPLY WITH FAR PART 61.89(A) (6) AND (7).

b. Except as provided in FAR Part 91.157, Special VFR Weather Minimums, no person may operate an aircraft beneath the ceiling under VFR within the lateral boundaries of controlled airspace designated to the surface for an airport when the ceiling is less than 1,000 feet. See FAR Part 91.155(c).

3-1-5. VFR CRUISING ALTITUDES AND FLIGHT LEVELS

(See TBL 3-1-2).

Section 2. CONTROLLED AIRSPACE

3-2-1. GENERAL

a. Controlled Airspace: A generic term that covers the different classification of airspace (Class A, Class B, Class C, Class D, and Class E airspace) and defined dimensions within which air traffic control service is provided to IFR flights and to VFR flights in accordance with the airspace classification. (See FIG 3-2-1).

b. IFR Requirements: IFR operations in any class of controlled airspace requires that a pilot must file an IFR flight plan and receive an appropriate ATC clearance.

c. IFR Separation: Standard IFR separation is provided to all aircraft operating under IFR in controlled airspace.

d. VFR Requirements: It is the responsibility of the pilot to insure that ATC clearance or radio communication requirements are met prior to entry into Class B, Class C, or Class D airspace. The pilot retains this responsibility when receiving ATC radar advisories. See FAR Part 91.

e. Traffic Advisories: Traffic advisories will be provided to all aircraft as the controller's work situation permits.

f. Safety Alerts: Safety Alerts are mandatory services and are provided to ALL aircraft. There are two types of Safety Alerts, Terrain/Obstruction Alert and Aircraft Conflict/Mode C Intruder Alert.

Table 3-1-1   Basic VFR Weather Minimums

Airspace	Flight Visibility	Distance from Clouds
Class A	Not Applicable	Not Applicable
Class B	3 statute miles	Clear of Clouds
Class C	3 statute miles	500 feet below 1,000 feet above 2,000 feet horizontal
Class D	3 statute miles	500 feet below 1,000 feet above 2,000 feet horizontal
Class E		
Less than 10,000 feet MSL	3 statute miles	500 feet below 1,000 feet above 2,000 feet horizontal
At or above 10,000 feet MSL	5 statute miles	1,000 feet below 1,000 feet above 1 statute mile horizontal
Class G		
1,200 feet or less above the surface (regardless of MSL altitude).		
Day, except as provided in section 91.155(b).	1 statute mile	Clear of clouds
Night, except as provided in section 91.155(b).	3 statute miles	500 feet below 1,000 feet above 2,000 feet horizontal
More than 1,200 feet above the surface but less than 10,000 feet MSL.		
Day	1 statute mile	500 feet below 1,000 feet above 2,000 feet horizontal
Night	3 statute miles	500 feet below 1,000 feet above 2,000 feet horizontal
More than 1,200 feet above the surface and at or above 10,000 feet MSL.	5 statute miles	1,000 feet below 1,000 feet above 1 statute mile horizontal

Table 3-1-2   VFR Cruising Altitudes and Flight Levels

If your magnetic course (ground track) is:	And you are more than 3,000 feet above the surface but below 18,000 feet MSL, fly:	And you are above 18,000 feet MSL to FL 290, fly:
0° to 179°	Odd thousands MSL, plus 500 feet (3,500, 5,500, 7,500, etc.)	Odd Flight Levels plus 500 feet (FL 195, FL 215, FL 235, etc.)
180° to 359°	Even thousands MSL, plus 500 feet (4,500, 6,500, 8,500, etc.)	Even Flight Levels plus 500 feet (FL 185, FL 205, FL 225, etc.)

1. *Terrain/Obstruction Alert:* A Terrain/Obstruction Alert is issued when, in the controller's judgment, an aircraft's altitude places it in unsafe proximity to terrain and/or obstructions.

2. *Aircraft Conflict/Mode C Intruder Alert:* An Aircraft Conflict/Mode C Intruder Alert is issued if the controller observes another aircraft which places it in an unsafe proximity. When feasible, the controller will offer the pilot an alternative course of action.

g. Ultralight Vehicles: No person may operate an ultralight vehicle within Class A, Class B, Class C, or Class D airspace or within the lateral boundaries of the surface area of Class E airspace designated for an airport unless that person has prior authorization from the ATC facility having jurisdiction over that airspace. See FAR Part 103.

h. Unmanned Free Balloons: Unless otherwise authorized by ATC, no person may operate an unmanned free balloon below 2,000 feet above the surface within the lateral boundaries of Class B, Class C, Class D, or Class E airspace designated for an airport. See FAR Part 101.

i. Parachute Jumps: No person may make a parachute jump, and no pilot in command may allow a parachute jump to be made from that aircraft, in or into Class A, Class B, Class C, or Class D airspace without, or in viola-

Airspace Classes

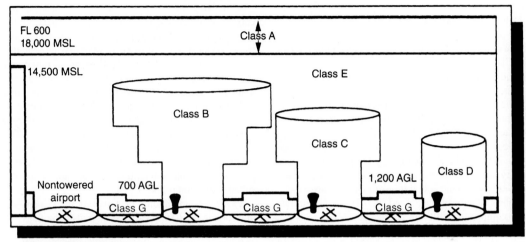

MSL—mean sea level
AGL—above ground level
FL—flight level

Figure 3-2-1.

tion of, the terms of an ATC authorization issued by the ATC facility having jurisdiction over the airspace. See FAR Part 105.

3-2-2. CLASS A AIRSPACE

a. Definition: Generally, that airspace from 18,000 feet MSL up to and including FL 600, including the airspace overlying the waters within 12 nautical miles of the coast of the 48 contiguous States and Alaska; and designated international airspace beyond 12 nautical miles of the coast of the 48 contiguous States and Alaska within areas of domestic radio navigational signal or ATC radar coverage, and within which domestic procedures are applied.

b. Operating Rules and Pilot/Equipment Requirements: Unless otherwise authorized, all persons must operate their aircraft under IFR. See FAR Part 71.33 and FAR Part 91.167 through FAR Part 91.193.

c. Charts: Class A airspace is not specifically charted.

3-2-3. CLASS B AIRSPACE

a. Definition: Generally, that airspace from the surface to 10,000 feet MSL surrounding the nation's busiest airports in terms of IFR operations or passenger enplanements. The configuration of each Class B airspace area is individually tailored and consists of a surface area and two or more layers (some Class B airspace areas resemble upside-down

wedding cakes), and is designed to contain all published instrument procedures once an aircraft enters the airspace. An ATC clearance is required for all aircraft to operate in the area, and all aircraft that are so cleared receive separation services within the airspace. The cloud clearance requirement for VFR operations is "clear of clouds."

b. Operating Rules and Pilot/Equipment Requirements for VFR Operations: Regardless of weather conditions, an ATC clearance is required prior to operating within Class B airspace. Pilots should not request a clearance to operate within Class B airspace unless the requirements of FAR Part 91.215 and FAR Part 91.131 are met. Included among these requirements are:

1. Unless otherwise authorized by ATC, aircraft must be equipped with an operable two-way radio capable of communicating with ATC on appropriate frequencies for that Class B airspace.

2. No person may take off or land a civil aircraft at the following primary airports within Class B airspace unless the pilot in command holds at least a private pilot certificate:

(a) Andrews Air Force Base, MD
(b) Atlanta Hartsfield Airport, GA
(c) Boston Logan Airport, MA
(d) Chicago O'Hare Intl Arpt, IL
(e) Los Angeles Intl Airport, CA
(f) Miami Intl Airport, FL
(g) Newark Intl Airport, NJ

(h) New York Kennedy Airport, NY

(i) New York La Guardia Arpt, NY

(j) San Francisco Intl Airport, CA

(k) Washington National Arpt, DC

(l) Dallas/Fort Worth Intl Arpt, TX

3. No person may take off or land a civil aircraft at an airport within Class B airspace or operate a civil aircraft within Class B airspace unless:

(a) The pilot in command holds at least a private pilot certificate; or,

(b) The aircraft is operated by a student pilot or recreational pilot who seeks private pilot certification and has met the requirements of FAR Part 61.95.

4. Unless otherwise authorized by ATC, each person operating a large turbine engine-powered airplane to or from a primary airport shall operate at or above the designated floors while within the lateral limits of Class B airspace.

5. Unless otherwise authorized by ATC, each aircraft must be equipped as follows:

(a) For IFR operations, an operable VOR or TACAN receiver; and

(b) For all operations, a two-way radio capable of communications with ATC on appropriate frequencies for that area; and

(c) Unless otherwise authorized by ATC, an operable radar beacon transponder with automatic altitude reporting equipment.

NOTE—ATC MAY, UPON NOTIFICATION, IMMEDIATELY AUTHORIZE A DEVIATION FROM THE ALTITUDE REPORTING EQUIPMENT REQUIREMENT; HOWEVER, A REQUEST FOR A DEVIATION FROM THE 4096 TRANSPONDER EQUIPMENT REQUIREMENT MUST BE SUBMITTED TO THE CONTROLLING ATC FACILITY AT LEAST ONE HOUR BEFORE THE PROPOSED OPERATION.

REFERENCE—AIM, TRANSPONDER OPERATION, PARAGRAPH 4-1-19.

6. *Mode C Veil:* The airspace within 30 nautical miles of an airport listed in Appendix D, Section 1 of FAR Part 91 (generally primary airports within Class B airspace areas), from the surface upward to 10,000 feet MSL. Unless otherwise authorized by air traffic control, aircraft operating within this airspace must be equipped with automatic pressure altitude reporting equipment having Mode C capability.

However, aircraft that was not originally certificated with an engine-driven electrical system or which has not subsequently been certified with a system installed, may conduct operations within a Mode C veil provided the aircraft remains outside Class A, B or C airspace; and below the altitude of the ceiling of a Class B or Class C airspace area designated for an airport or 10,000 feet MSL, whichever is lower.

c. Charts: Class B airspace is charted on Sectional Charts, IFR En Route Low Altitude, and Terminal Area Charts.

d. Flight Procedures:

1. *Flights:* Aircraft within Class B airspace are required to operate in accordance with current IFR procedures. A clearance for a visual approach to a primary airport is not authorization for turbine powered airplanes to operate below the designated floors of the Class B airspace.

2. *VFR Flights:*

(a) Arriving aircraft must obtain an ATC clearance prior to entering Class B airspace and must contact ATC on the appropriate frequency, and in relation to geographical fixes shown on local charts. Although a pilot may be operating beneath the floor of the Class B airspace on initial contact, communications with ATC should be established in rela-tion to the points indicated for spacing and sequencing purposes.

(b) Departing aircraft require a clearance to depart Class B airspace and should advise the clearance delivery position of their intended altitude and route of flight. ATC will normally advise VFR aircraft when leaving the geographical limits of the Class B airspace. Radar service is not automatically terminated with this advisory unless specifically stated by the controller.

(c) Aircraft not landing or departing the primary airport may obtain an ATC clearance to transit the Class B airspace when traffic conditions permit and provided the requirements of FAR Part 91.131 are met. Such VFR aircraft are encouraged, to the extent possible, to operate at altitudes above or below the Class B airspace or transit through established VFR corridors. Pilots operating in VFR corridors are urged to use frequency 122.750 MHz for the exchange of aircraft position information.

e. ATC Clearances and Separation: An ATC clearance is required to enter and operate within Class B airspace. VFR pilots are provided sequencing and separation from other aircraft while operating within Class B airspace.

REFERENCE—AIM, TERMINAL RADAR SERVICES FOR VFR AIRCRAFT, PARAGRAPH 4-1-17.

NOTE—

① SEPARATION AND SEQUENCING OF VFR AIRCRAFT WILL BE SUSPENDED IN THE EVENT OF A RADAR OUTAGE AS THIS SERVICE IS DEPENDENT ON RADAR. THE PILOT WILL BE ADVISED THAT THE SERVICE IS NOT AVAILABLE AND ISSUED WIND, RUNWAY INFORMATION AND THE TIME OR PLACE TO CONTACT THE TOWER.

② SEPARATION OF VFR AIRCRAFT WILL BE SUSPENDED DURING CENRAP OPERATIONS. TRAFFIC ADVISORIES AND SEQUENCING TO THE PRIMARY AIRPORT WILL BE PROVIDED ON A WORKLOAD PERMITTING BASIS. THE PILOT WILL BE ADVISED WHEN CENTER RADAR PRESENTATION (CENRAP) IS IN USE.

1. VFR aircraft are separated from all VFR/IFR aircraft which weigh 19,000 pounds or less by a minimum of:

(a) Target resolution, or

(b) 500 feet vertical separation, or

(c) Visual separation

2. VFR aircraft are separated from all VFR/IFR aircraft which weigh more than 19,000 and turbojets by no less than:

(a) 1½ miles lateral separation, or

(b) 500 feet vertical separation, or

(c) Visual separation

3. This program is not to be interpreted as relieving pilots of their responsibilities to see and avoid other traffic operating in basic VFR weather conditions, to adjust their operations and flight path as necessary to preclude serious wake encounters, to maintain appropriate terrain and obstruction clearance or to remain in weather conditions equal to or better than the minimums required by FAR Part 91.155. Approach control should be advised and a revised clearance or instruction obtained when compliance with an assigned route, heading and/or altitude is likely to compromise pilot responsibility with respect to terrain and obstruction clearance, vortex exposure, and weather minimums.

4. ATC may assign altitudes to VFR aircraft that do not conform to FAR Part 91.159. **"RESUME APPROPRIATE VFR ALTITUDES"** will be broadcast when the altitude assignment is no longer needed for separation or when leaving Class B airspace. Pilots must return to an altitude that conforms to FAR Part 91.159.

f. Proximity operations: VFR aircraft operating in proximity to Class B airspace are cautioned against operating too closely to the boundaries, especially where the floor of the Class B airspace is 3,000 feet or less or where VFR cruise altitudes are at or near the floor of higher levels. Observance of this precaution will reduce the potential for encountering an aircraft operating at the altitudes of Class B floors. Additionally, VFR aircraft are encouraged to utilize the VFR Planning Chart as a tool for planning flight in proximity to Class B airspace. Charted VFR Flyway Planning Charts are published on the back of the existing VFR Terminal Area Charts.

3-2-4. CLASS C AIRSPACE

a. Definition: Generally, that airspace from the surface to 4,000 feet above the airport elevation (charted in MSL) surrounding those airports that have an operational control tower, are serviced by a radar approach control, and that have a certain number of IFR operations or passenger enplanements. Although the configuration of each Class C airspace area is individually tailored, the airspace usually consists of a 5 NM radius core surface area that extends from the surface up to 4,000 feet above the airport elevation, and a 10 NM radius shelf area that extends no lower than 1,200 feet to 4,000 feet above the airport elevation.

b. Charts: Class C airspace is charted on Sectional Charts, IFR En Route Low Altitude, and Terminal Area Charts where appropriate.

c. Operating Rules and Pilot / Equipment Requirements:

1. *Pilot Certification:* No specific certification required.

2. *Equipment:*

(a) Two-way radio, and

(b) Unless otherwise authorized by ATC, an operable radar beacon transponder with automatic altitude reporting equipment.

NOTE—*see paragraph 4-1-19, TRANSPONDER OPERATION, Subparagraph f2(c) for Mode C transponder requirements for operating above Class C airspace.*

3. *Arrival or Through Flight Entry Requirements:* Two-way radio communication must be established with the ATC facility providing ATC services prior to entry and thereafter maintain those communications while in Class C airspace. Pilots of arriving aircraft should contact the Class C airspace ATC facility on the publicized frequency and give their position, altitude, radar beacon code, destination, and request Class C service. Radio contact should be initiated far enough from the Class C airspace boundary to preclude entering Class C airspace before two-way radio communications are established.

NOTE—

⎕ IF THE CONTROLLER RESPONDS TO A RADIO CALL WITH, "(AIRCRAFT CALLSIGN) STANDBY," RADIO COMMUNICATIONS HAVE BEEN ESTABLISHED AND THE PILOT CAN ENTER THE CLASS C AIRSPACE.

⎕ IF WORKLOAD OR TRAFFIC CONDITIONS PREVENT IMMEDIATE PROVISION OF CLASS C SERVICES, THE CONTROLLER WILL INFORM THE PILOT TO REMAIN OUTSIDE THE CLASS C AIRSPACE UNTIL CONDITIONS PERMIT THE SERVICES TO BE PROVIDED.

⎕ IT IS IMPORTANT TO UNDERSTAND THAT IF THE CONTROLLER RESPONDS TO THE INITIAL RADIO CALL WITHOUT USING THE AIRCRAFT IDENTIFICATION, RADIO COMMUNICATIONS HAVE NOT BEEN ESTABLISHED AND THE PILOT MAY NOT ENTER THE CLASS C AIRSPACE.

4. Though not requiring regulatory action, Class C airspace areas have a procedural Outer Area. Normally this area is 20 NM from the primary Class C airspace airport. Its vertical limit extends from the lower limits of radio/radar coverage up to the ceiling of the approach control's delegated airspace, excluding the Class C airspace itself, and other airspace as appropriate.

5. Pilots approaching an airport with Class C service should be aware that if they descend below the base altitude of thes 5 to 10 mile shelf during an instrument or visual approach, they may encounter nontransponder, VFR aircraft. (This outer area is not charted.)

EXAMPLE—

⎕ [aircraft callsign] "REMAIN OUTSIDE THE CLASS CHARLIE AIRSPACE AND STANDBY."

⎕ "AIRCRAFT CALLING DULLES APPROACH CONTROL, STANDBY."

4. *Departures from:*

(a) A primary or satellite airport with an operating control tower. Two-way radio communications must be established and maintained with the control tower, and thereafter as instructed by ATC while operating in Class C airspace.

(b) A satellite airport without an operating control tower: Two-way radio communications must be established as soon as practicable after departing with the ATC facility having jurisdiction over the Class C airspace.

5. *Aircraft Speed:* Unless otherwise authorized or required by ATC, no person may operate an aircraft at or below 2,500 feet above the surface within 4 nautical miles of the primary airport of a Class C airspace area at an indicated airspeed of more than 200 knots (230 mph).

e. Air Traffic Services: When two-way radio communications and radar contact are established, all participating VFR aircraft are:

1. Sequenced to the primary airport.

2. Provided Class C services within the Class C airspace and the Outer Area.

3. Provided basic radar services beyond the outer area on a workload permitting basis. This can be terminated by the controller if workload dictates.

f. Aircraft Separation: Separation is provided within the Class C airspace and the Outer Area after two-way radio communications and radar contact are established. VFR aircraft are separated from IFR aircraft within the Class C airspace by any of the following:

1. Visual separation.

2. 500 feet vertical; except when operating beneath a heavy jet.

3. Target resolution.

NOTE—

⒈ SEPARATION AND SEQUENCING OF VFR AIRCRAFT WILL BE SUSPENDED IN THE EVENT OF A RADAR OUTAGE AS THIS SERVICE IS DEPENDENT ON RADAR. THE PILOT WILL BE ADVISED THAT THE SERVICE IS NOT AVAILABLE AND ISSUED WIND, RUNWAY INFORMATION AND THE TIME OR PLACE TO CONTACT THE TOWER.

⒉ SEPARATION OF VFR AIRCRAFT WILL BE SUSPENDED DURING CENRAP OPERATIONS. TRAFFIC ADVISORIES AND SEQUENCING TO THE PRIMARY AIRPORT WILL BE PROVIDED ON A WORKLOAD PERMITTING BASIS. THE PILOT WILL BE ADVISED WHEN CENRAP IS IN USE.

⒊ PILOT PARTICIPATION IS VOLUNTARY WITHIN THE OUTER AREA AND CAN BE DISCONTINUED, WITHIN THE OUTER AREA, AT THE PILOTS REQUEST. CLASS C SERVICES WILL BE PROVIDED IN THE OUTER AREA UNLESS THE PILOT REQUESTS TERMINATION OF THE SERVICE.

⒋ SOME FACILITIES PROVIDE CLASS C SERVICES ONLY DURING PUBLISHED HOURS. AT OTHER TIMES, TERMINAL IFR RADAR SERVICE WILL BE PROVIDED. IT IS IMPORTANT TO NOTE THAT THE COMMUNICATIONS AND TRANSPONDER REQUIREMENTS ARE DEPENDENT OF THE CLASS OF AIRSPACE ESTABLISHED OUTSIDE OF THE PUBLISHED HOURS.

g. Secondary Airports:

1. In some locations Class C airspace may overlie the Class D surface area of a secondary airport. In order to allow that control tower to provide service to aircraft, portions of the overlapping Class C airspace may be procedurally excluded when the secondary airport tower is in operation. Aircraft operating in these procedurally excluded areas will only be provided airport traffic control services when in communication with the secondary airport tower.

2. Aircraft proceeding inbound to a satellite airport will be terminated at a sufficient distance to allow time to change to the appropriate tower or advisory frequency. Class C services to these aircraft will be discontinued when the aircraft is instructed to contact the tower or change to advisory frequency.

3. Aircraft departing secondary controlled airports will not receive Class C services until they have been radar identified and two-way communications have been established with the Class C airspace facility.

4. This program is not to be interpreted as relieving pilots of their responsibilities to see and avoid other traffic operating in basic VFR weather conditions, to adjust their operations and flight path as necessary to preclude serious wake encounters, to maintain appropriate terrain and obstruction clearance or to remain in weather conditions equal to or better than the minimums required by FAR Part

Table 3-2-1   Class C Airspace by State

These states currently have designated Class C airspace areas that are depicted on Sectional Charts. Pilots should consult current sectional charts and NOTAM's for the latest information on services available. Pilots should be aware that some Class C airspace underlies or is adjacent to Class B airspace.

State/City	Airport	State/City	Airport
ALABAMA		Burbank	Burbank-Glendale-Pasadena
Birmingham	International	Fresno	Air Terminal
Huntsville	Carl T Jones Field	Monterey	Peninsula
Mobile	Regional	Oakland	Metropolitan International
		Ontario	International
ALASKA		Riverside	March AFB
Anchorage	International	Sacramento	International
		San Jose	International
ARIZONA		Santa Ana	John Wayne/Orange County
Davis-Monthan	AFB	Santa Barbara	Municipal
Tucson	International		
		COLORADO	
ARKANSAS		Colorado Springs	Municipal Airport
Fayetteville (Springdale)	Northwest Arkansas Regional		
Little Rock	Adams Field		
		CONNECTICUT	
CALIFORNIA		Windsor Locks	Bradley International
Beale	AFB		

State/City	Airport	State/City	Airport
FLORIDA		**MISSOURI**	
Daytona Beach	Regional	Springfield	Springfield-Branson Regional
Fort Lauderdale	Hollywood International	**MONTANA**	
Fort Myers	SW Florida Regional	Billings	Logan International
Jacksonville	International	**NEBRASKA**	
Palm Beach	International	Lincoln	Municipal
Pensacola	NAS	Omaha	Eppley Airfield
Pensacola	Regional	Offutt	AFB
Sarasota	Bradenton	**NEVADA**	
Tallahassee	Regional	Reno	Cannon International
Whiting	NAS	**NEW HAMPSHIRE**	
GEORGIA		Manchester	Manchester
Columbus	Metropolitan	**NEW JERSEY**	
Savannah	International	Atlantic City	International
HAWAII		**NEW MEXICO**	
Kahului	Kahului	Albuquerque	International
IDAHO		**NEW YORK**	
Boise	Air Terminal	Albany	County
ILLINOIS		Buffalo	Greater Buffalo International
Champaign	U of Illinois-Willard	Islip	Long Island MacArthur
Chicago	Midway	Rochester	Greater Rochester International
Moline	Quad City	Syracuse	Hancock International
Peoria	Greater Peoria	**NORTH CAROLINA**	
Springfield	Capital	Asheville	Regional
INDIANA		Fayetteville	Regional/Grannis Field
Evansville	Regional	Greensboro	Piedmont Triad International
Fort Wayne	International	Pope	AFB
Indianapolis	International	Raleigh	Raleigh-Durham International
South Bend	Michiana Regional	**OHIO**	
IOWA		Akron	Arkon-Canton Regional
Cedar Rapids	The Eastern Iowa	Columbus	Port Columbus International
Des Moines	International	Dayton	James M. Cox International
KANSAS		Toledo	Express
Wichita	Mid-Continent	**OKLAHOMA**	
KENTUCKY		Oklahoma City	Will Rogers World
Lexington	Blue Grass	Tinker	AFB
Louisville	Standiford Field	Tulsa	International
LOUISIANA		**OREGON**	
Baton Rouge	BTR Metro, Ryan Field	Portland	International
Lafayette	Regional	**PENNSYLVANIA**	
Shreveport	Barksdale AFB	Allentown	Allentown Bethlehem-Easton
Shreveport	Regional	**PUERTO RICO**	
MAINE		San Juan	Luis Munez Marin International
Bangor	International	**RHODE ISLAND**	
Portland	International Jetport	Providence	Theodore Francis Green State
MICHIGAN		**SOUTH CAROLINA**	
Flint	Bishop International	Charleston	AFB/International
Grand Rapids	Kent County International	Columbia	Metropolitan
Lansing	Capital City	Greer	Greenville-Spartanburg
MISSISSIPPI		Myrtle Beach	Myrtle Beach International
Columbus	AFB	SHAW	AFB
Jackson	International		

Table 3-2-1 Continued

State/City	Airport	State/City	Airport
TENNESSEE		**VIRGINIA**	
Chattanooga	Lovell Field	Richmond	Richard Evelyn Byrd
Knoxville	McGhee Tyson		International
Nashville	International	Norfolk	International
		Roanoke	Regional/Woodrum Field
TEXAS			
Abilene	Regional	**WASHINGTON**	
Amarillo	International	Point Roberts	Vancouver International
Austin	Austin-Bergstrom International	Spokane	International
Corpus Christi	International	Spokane	Fairchild AFB
Dyess	AFB	Whidbey Island	NAS, Ault Field
El Paso	International		
Harlingen	Rio Grande Valley International	**WEST VIRGINIA**	
Laughlin	AFB	Charleston	Yeager
Lubbock	International		
Midland	International	**WISCONSIN**	
San Antonio	International	Green Bay	Austin Straubel International
		Madison	Dane County Regional-Traux Field
UTAH		Milwaukee	General Mitchell International
VERMONT		**WYOMING**	
Burlington	International		
VIRGIN ISLANDS			
St. Thomas	Charlotte Amalie Cyril E. King		

91.155. Approach control should be advised and a revised clearance or instruction obtained when compliance with an assigned route, heading and/or altitude is likely to compromise pilot responsibility with respect to terrain and obstruction clearance, vortex exposure, and weather minimums. (See TBL 3-2-1)

3-2-5. CLASS D AIRSPACE

a. Definition: Generally, that airspace from the surface to 2,500 feet above the airport elevation (charted in MSL) surrounding those airports that have an operational control tower. The configuration of each Class D airspace area is individually tailored and when instrument procedures are published, the airspace will normally be designed to contain the procedures.

b. Operating Rules and Pilot/Equipment Requirements:

1. *Pilot Certification:* No specific certification required.

2. *Equipment:* Unless otherwise authorized by ATC, an operable two-way radio is required.

3. *Arrival or Through Flight Entry Requirements:* Two-way radio communication must be established with the ATC facility providing ATC services prior to entry and thereafter maintain those communications while in the Class D airspace. Pilots of arriving aircraft should contact the control tower on the publicized frequency and give their position, altitude, destination, and any request(s). Radio contact should be initiated far enough from the Class D

airspace boundary to preclude entering the Class D airspace before two-way radio communications are established.

NOTE—

⚊ IF THE CONTROLLER RESPONDS TO A RADIO CALL WITH, "[AIRCRAFT CALLSIGN] STANDBY," RADIO COMMUNICATIONS HAVE BEEN ESTABLISHED AND THE PILOT CAN ENTER THE CLASS D AIR SPACE.

⚌ IF WORKLOAD OR TRAFFIC CONDITIONS PREVENT IMMEDIATE ENTRY INTO CLASS D AIRSPACE, THE CONTROLLER WILL INFORM THE PILOT TO REMAIN OUTSIDE THE CLASS D AIRSPACE UNTIL CONDITIONS PERMIT ENTRY.

EXAMPLE—

⚊ "[aircraft callsign] REMAIN OUTSIDE THE CLASS AIRSPACE AND STANDBY."

☰ IT IS IMPORTANT TO UNDERSTAND THAT IF THE CONTROLLER RESPONDS TO THE INITIAL RADIO CALL WITHOUT USING THE AIRCRAFT CALLSIGN, RADIO COMMUNICATIONS HAVE NOT BEEN ESTABLISHED AND THE PILOT MAY NOT ENTER THE CLASS D AIRSPACE.

EXAMPLE—

⚌ "AIRCRAFT CALLING MANASSAS TOWER STANDBY."

☷ AT THOSE AIRPORTS WHERE THE CONTROL TOWER DOES NOT OPERATE 24 HOURS A DAY, THE OPERATING HOURS OF THE TOWER WILL BE LISTED ON THE APPROPRIATE CHARTS AND IN THE A/FD. DURING THE HOURS THE TOWER IS NOT IN OPERATION THE CLASS E SURFACE AREA RULES OR A COMBINATION OF CLASS E RULES TO 700 FEET ABOVE GROUND LEVEL AND CLASS G RULES TO THE SURFACE WILL BECOME APPLICABLE. CHECK THE A/FD FOR SPECIFICS.

4. *Departures from:*

(a) A primary or satellite airport with an operating control tower: Two-way radio communications must be established and maintained with the control tower, and thereafter as instructed by ATC while operating in the Class D airspace.

(b) A satellite airport without an operating control tower: Two-way radio communications must be established as soon as practicable after departing with the ATC facility having jurisdiction over the Class D airspace as soon as practicable after departing.

5. *Aircraft Speed:* Unless otherwise authorized or required by ATC, no person may operate an aircraft at or below 2,500 feet above the surface within 4 nautical miles of the primary airport of a Class D airspace area at an indicated airspeed of more than 200 knots (230 mph).

c. Class D airspace areas are depicted on Sectional and Terminal charts with blue segmented lines, and on IFR En Route Lows with a boxed [D].

d. Arrival extensions for instrument approach procedures may be Class D or Class E airspace. As a general rule, if all extensions are 2 miles or less, they remain part of the Class D surface area. However, if any one extension is greater than 2 miles, then all extensions become Class E.

e. Separation for VFR Aircraft: No separation services are provided to VFR aircraft.

3-2-6. CLASS E AIRSPACE

a. Definition: Generally, if the airspace is not Class A, Class B, Class C, or Class D, and it is controlled airspace, it is Class E airspace.

b. Operating Rules and Pilot/Equipment Requirements:

1. *Pilot Certification:* No specific certification required.

2. *Equipment:* No specific equipment required by the airspace.

3. *Arrival or Through Flight Entry Requirements:* No specific requirements.

c. Charts: Class E airspace below 14,500 feet MSL is charted on Sectional, Terminal, and IFR En Route Low Altitude charts.

d. Vertical limits: Except for 18,000 feet MSL, Class E airspace has no defined vertical limit but rather it extends upward from either the surface or a designated altitude to the overlying or adjacent controlled airspace.

e. Types of Class E Airspace:

1. *Surface area designated for an airport:* When designated as a surface area for an airport, the airspace will be configured to contain all instrument procedures.

2. *Extension to a surface area:* There are Class E airspace areas that serve as extensions to Class B, Class C, and Class D surface areas designated for an airport. Such airspace provides controlled airspace to contain standard instrument approach procedures without imposing a communications requirement on pilots operating under VFR.

3. *Airspace used for transition:* There are Class E airspace areas beginning at either 700 or 1,200 feet AGL used to transition to/from the terminal or en route environment.

4. *En Route Domestic Areas:* There are Class E airspace areas that extend upward from a specified altitude and are en route domestic airspace areas that provide controlled airspace in those areas where there is a requirement to provide IFR en route ATC services but the Federal airway system is inadequate.

5. *Federal Airways:* The Federal airways are Class E airspace areas and, unless otherwise specified, extend upward from 1,200 feet to, but not including, 18,000 feet MSL. The colored airways are Green, Red, Amber, and Blue. The VOR airways are classified as Domestic, Alaskan, and Hawaiian.

6. *Offshore Airspace Areas:* There are Class E airspace areas that extend upward from a specified altitude to, but not including, 18,000 feet MSL and are designated as offshore airspace areas. These areas provide controlled airspace beyond 12 miles from the coast of the United States in those areas where there is a requirement to provide IFR en route ATC services and within which the United States is applying domestic procedures.

7. Unless designated at a lower altitude, Class E airspace begins at 14,500 feet MSL to, but not including 18,000 feet MSL overlying: the 48 contiguous States including the waters within 12 miles from the coast of the 48 contiguous States; the District of Columbia; Alaska, including the waters within 12 miles from the coast of Alaska, and that airspace above FL 600; excluding the Alaska peninsula west of long. 160°00'00" W, and the airspace below 1,500 feet above the surface of the earth unless specifically so designated.

f. Separation for VFR Aircraft: No separation services are provided to VFR aircraft.

Section 3. CLASS G AIRSPACE

3-3-1. GENERAL

Class G airspace (uncontrolled) is that portion of the airspace that has not been designated as Class A, Class B, Class C, Class D, or Class E airspace.

3-3-2. VFR REQUIREMENTS

Rules governing VFR flight have been adopted to assist the pilot in meeting the responsibility to see and avoid other aircraft. Minimum flight visibility and distance from

Table 3-3-1   IFR Altitudes
Class G Airspace

If your magnetic course (ground track) is:	And you are below 18,000 feet MSL, fly:
0° to 179°	Odd thousands MSL, (3,000; 5,000; 7,000, etc.)
180° to 359°	Even thousands MSL, (2,000; 4,000; 6,000, etc.)

clouds required for VFR flight are contained in FAR Part 91.155. (See TBL 3-1-1).

3-3-3. IFR REQUIREMENTS

a. Title 14 CFR specifies the pilot and aircraft equipment requirements for IFR flight. Pilots are reminded that in addition to altitude or flight level requirements, 14 CFR section 91.177 includes a requirement to remain at least 1,000 feet (2,000 feet in designated mountainous terrain) above the highest obstacle within a horizontal distance of 4 nautical miles from the course to be flown.

b. IFR Altitudes (See TBL 3-3-1).

Section 4. SPECIAL USE AIRSPACE

3-4-1. GENERAL

a. Special use airspace consists of that airspace wherein activities must be confined because of their nature, or wherein limitations are imposed upon aircraft operations that are not a part of those activities, or both. Except for Controlled Firing Areas, special use airspace areas are depicted on aeronautical charts.

b. Prohibited and Restricted Areas are regulatory special use airspace and are established in FAR Part 73 through the rulemaking process.

c. Warning Areas, Military Operations Areas (MOA), Alert Areas, and Controlled Firing Areas (CFA) are nonregulatory special use airspace.

d. Special use airspace descriptions (except CFAs) are contained in FAA Order 7400.8.

e. Special use airspace (except CFAs) are charted on IFR or Visual charts and include the hours of operation, altitudes, and the controlling agency.

3-4-2. PROHIBITED AREAS

Prohibited Areas contain airspace of defined dimensions identified by an area on the surface of the earth within which the flight of aircraft is prohibited. Such areas are established for security or other reasons associated with the national welfare. These areas are published in the Federal Register and are depicted on aeronautical charts.

3-4-3. RESTRICTED AREAS

a. Restricted Areas contain airspace identified by an area on the surface of the earth within which the flight of aircraft, while not wholly prohibited, is subject to restrictions. Activities within these areas must be confined because of their nature or limitations imposed upon aircraft operations that are not a part of those activities or both. Restricted Areas denote the existence of unusual, often invisible, hazards to aircraft such as artillery firing, aerial gunnery, or guided missiles. Penetration of Restricted Areas without authorization from the using or controlling agency may be extremely hazardous to the aircraft and its occupants. Restricted Areas are published in the Federal Register and constitute FAR Part 73.

b. ATC facilities apply the following procedures when aircraft are operating on an IFR clearance (including those cleared by ATC to maintain VFR-ON-TOP) via a route which lies within joint-use restricted airspace.

1. If the restricted area is not active and has been released to the controlling agency (FAA), the ATC facility will allow the aircraft to operate in the restricted airspace without issuing specific clearance for it to do so.

2. If the restricted area is active and has not been released to the controlling agency (FAA), the ATC facility will issue a clearance which will ensure the aircraft avoids the restricted airspace unless it is on an approved altitude reservation mission or has obtained its own permission to operate in the airspace and so informs the controlling facility.

NOTE—THE ABOVE APPLY ONLY TO JOINT-USE RESTRICTED AIRSPACE AND NOT TO PROHIBITED AND NONJOINT-USE AIRSPACE. FOR THE LATTER CATEGORIES, THE ATC FACILITY WILL ISSUE A CLEARANCE SO THE AIRCRAFT WILL AVOID THE RESTRICTED AIRSPACE UNLESS IT IS ON AN APPROVED ALTITUDE RESERVATION MISSION OR HAS OBTAINED ITS OWN PERMISSION TO OPERATE IN THE AIRSPACE AND SO INFORMS THE CONTROLLING FACILITY.

c. Restricted airspace is depicted on the En Route Chart appropriate for use at the altitude or flight level being flown. For joint-use restricted areas, the name of the controlling agency is shown on these charts. For all prohibited areas and nonjoint-use restricted areas, unless otherwise requested by the using agency, the phrase "NO A/G" is shown.

3-4-4. WARNING AREAS

A Warning Area is airspace of defined dimensions, extending from three nautical miles outward from the coast of the United States, that contains activity that may be hazardous to nonparticipating aircraft. The purpose of such warning areas is to warn nonparticipating pilots of the potential danger. A warning area may be located over domestic or international waters or both.

3-4-5. MILITARY OPERATIONS AREAS

a. MOAs consist of airspace of defined vertical and lateral limits established for the purpose of separating certain military training activities from IFR traffic. Whenever an MOA is being used, nonparticipating IFR traffic may be cleared through an MOA if IFR separation can be provided by ATC. Otherwise, ATC will reroute or restrict nonparticipating IFR traffic.

b. Examples of activities conducted in MOA's include, but are not limited to: air combat tactics, air intercepts, aerobatics, formation training, and low-altitude tactics. Military pilots flying in an active MOA are exempted from the provisions of 14 CFR Section 91.303(c) and (d) which prohibits acrobatic flight within Class D and Class E surface areas, and within Federal airways. Additionally, the Department of Defense has been issued an authorization to operate aircraft at indicated airspeeds in excess of 250 knots below 10,000 feet MSL within active MOA's.

c. Pilots operating under VFR should exercise extreme caution while flying within an MOA when military activity is being conducted. The activity status (active/inactive) of MOAs may change frequently. Therefore, pilots should contact any FSS within 100 miles of the area to obtain accurate real-time information concerning the MOA hours of operation. Prior to entering an active MOA, pilots should contact the controlling agency for traffic advisories.

d. MOAs are depicted on Sectional, VFR Terminal Area, and Enroute Low Altitude Charts.

3-4-6. ALERT AREAS

Alert Areas are depicted on aeronautical charts to inform nonparticipating pilots of areas that may contain a high volume of pilot training or an unusual type of aerial activity. Pilots should be particularly alert when flying in these areas. All activity within an Alert Area shall be conducted in accordance with FARs, without waiver, and pilots of participating aircraft as well as pilots transiting the area shall be equally responsible for collision avoidance.

3-4-7. CONTROLLED FIRING AREAS

CFAs contain activities which, if not conducted in a controlled environment, could be hazardous to nonparticipating aircraft. The distinguishing feature of the CFA, as compared to other special use airspace, is that its activities are suspended immediately when spotter aircraft, radar, or ground lookout positions indicate an aircraft might be approaching the area. There is no need to chart CFAs since they do not cause a nonparticipating aircraft to change its flight path.

Section 5. OTHER AIRSPACE AREAS

3-5-1. AIRPORT ADVISORY/INFORMATION SERVICES

a. There are three advisory type services available at selected airports.

1. Local Airport Advisory (LAA) service is operated within 10 statute miles of an airport where a control tower is not operating but where a FSS is located on the airport. At such locations, the FSS provides a complete local airport advisory service to arriving and departing aircraft. During periods of fast changing weather the FSS will automatically provide Final Guard as part of the service from the time the aircraft reports "on-final" or "taking-the-active-runway" until the aircraft reports "on-the-ground" or "airborne."

NOTE—
CURRENT POLICY, WHEN REQUESTING REMOTE ATC SERVICES, REQUIRES THAT A PILOT MONITOR THE AUTOMATED WEATHER BROADCAST AT THE LANDING AIRPORT PRIOR TO REQUESTING ATC SERVICES. THE FSS AUTOMATICALLY PROVIDES FINAL GUARD, WHEN APPROPRIATE, DURING LAA/REMOTE AIRPORT ADVISORY (RAA) OPERATIONS. FINAL GUARD IS A VALUE ADDED WIND/ALTIMETER MONITORING SERVICE, WHICH PROVIDES AN AUTOMATIC WIND AND ALTIMETER CHECK DURING ACTIVE WEATHER SITUATIONS WHEN THE PILOT REPORTS ON-FINAL OR TAKING THE ACTIVE RUNWAY. DURING THE LANDING OR TAKE-OFF OPERATION WHEN THE WINDS OR ALTIMETER ARE ACTIVELY CHANGING THE FSS WILL BLIND BROADCAST SIGNIFICANT CHANGES WHEN THE SPECIALIST BELIEVES THE CHANGE MIGHT AFFECT THE OPERATION. PILOTS SHOULD ACKNOWLEDGE THE FIRST WIND/ALTIMETER CHECK BUT DUE TO COCKPIT ACTIVITY NO ACKNOWLEDGEMENT IS EXPECTED FOR THE BLIND BROADCASTS. IT IS PRUDENT FOR A PILOT TO REPORT ON-THE-GROUND OR AIRBORNE TO END THE SERVICE.

2. RAA service is operated within 10 statute miles of specified high activity GA airports where a control tower is not operating. Airports offering this service are listed in the A/FD and the published service hours may be changed by NOTAM D. Final Guard is automatically provided with RAA.

3. Remote Airport Information Service (RAIS) is provided in support of short term special events like small to medium fly-in's. The service is advertised by NOTAM D only. The FSS will not have access to a continuous readout of the current winds and altimeter; therefore, RAIS does not include weather and/or Final Guard service. However, known traffic, special event instructions, and all other services are provided.

NOTE—
THE AIRPORT AUTHORITY AND/OR MANAGER SHOULD REQUEST RAIS SUPPORT ON OFFICIAL LETTERHEAD DIRECTLY WITH THE MANAGER OF THE FSS THAT WILL PROVIDE THE SERVICE AT LEAST 60 DAYS IN ADVANCE.

APPROVAL AUTHORITY RESTS WITH THE FSS MANAGER AND IS BASED ON WORKLOAD AND RESOURCE AVAILABILITY.

REFERENCE—AIM, TRAFFIC ADVISORY PRACTICES AT AIRPORTS WITHOUT OPERATING CONTROL TOWERS, PARAGRAPH 4-1-9.

b. It is not mandatory that pilots participate in the Airport Advisory programs. Participation enhances safety for everyone operating around busy GA airports; therefore, everyone is encouraged to participate and provide feedback that will help improve the program.

3-5-2. MILITARY TRAINING ROUTES

a. National security depends largely on the deterrent effect of our airborne military forces. To be proficient, the military services must train in a wide range of airborne tactics. One phase of this training involves "low level" combat tactics. The required maneuvers and high speeds are such that they may occasionally make the see-and-avoid aspect of VFR flight more difficult without increased vigilance in areas containing such operations. In an effort to ensure the greatest practical level of safety for all flight operations, the Military Training Route (MTR) program was conceived.

b. The MTR program is a joint venture by the FAA and the Department of Defense (DOD). MTR's are mutually developed for use by the military for the purpose of conducting low-altitude, high-speed training. The routes above 1,500 feet AGL are developed to be flown, to the maximum extent possible, under IFR. The routes at 1,500 feet AGL and below are generally developed to be flown under VFR.

c. Generally, MTRs are established below 10,000 feet MSL for operations at speeds in excess of 250 knots. However, route segments may be defined at higher altitudes for purposes of route continuity. For example, route segments may be defined for descent, climbout, and mountainous terrain. There are IFR and VFR routes as follows:

1. *IFR Military Training Routes-(IR):* Operations on these routes are conducted in accordance with IFR regardless of weather conditions.

2. *VFR Military Training Routes-(VR):* Operations on these routes are conducted in accordance with VFR except, flight visibility shall be 5 miles or more; and flights shall not be conducted below a ceiling of less than 3,000 feet AGL

d. Military training routes will be identified and charted as follows:

1. *Route identification.*

(a) MTRs with no segment above 1,500 feet AGL shall be identified by four number characters; e.g., IR1206, VR1207.

(b) MTRs that include one or more segments above 1,500 feet AGL shall be identified by three number characters; e.g., IR206, VR207.

(c) Alternate IR/VR routes or route segments are identified by using the basic/principal route designation followed by a letter suffix, e.g., IR008A, VR1007B, etc.

2. *Route charting.*

(a) *IFR Low Altitude En Route Chart:* This chart will depict all IR routes and all VR routes that accommodate operations above 1,500 feet AGL.

(b) *VFR Sectional Charts:* These charts will depict military training activities such as IR, VR, MOA, Restricted Area, Warning Area, and Alert Area information.

(c) *Area Planning (AP/1B) Chart (DOD Flight Information Publication-FLIP):* This chart is published by the DOD primarily for military users and contains detailed information on both IR and VR routes.

REFERENCE—AIM, AUXILIARY CHARTS, PARAGRAPH 9-1-6.

e. The FLIP contains charts and narrative descriptions of these routes. This publication is available to the general public by single copy or annual subscription from:

National Aeronautical Charting Office (NACO) Distribution Division
Federal Aviation Administration
6501 Lafayette Avenue
Riverdale, MD 20737-1199
Toll free phone: 1-800-638-8972
Commercial: 301-436-8301

This DOD FLIP is available for pilot briefings at FSS and many airports.

f. Nonparticipating aircraft are not prohibited from flying within an MTR; however, extreme vigilance should be exercised when conducting flight through or near these routes. Pilots should contact FSSs within 100 NM of a particular MTR to obtain current information or route usage in their vicinity. Information available includes times of scheduled activity, altitudes in use on each route segment, and actual route width. Route width varies for each MTR and can extend several miles on either side of the charted MTR centerline. Route width information for IR and VR MTRs is also available in the FLIP AP/1B along with additional MTR (SR/AR) information. When requesting MTR information, pilots should give the FSS their position, route of flight, and destination in order to reduce frequency congestion and permit the FSS specialist to identify the MTR which could be a factor.

3-5-3. TEMPORARY FLIGHT RESTRICTIONS

a. General: This paragraph describes the types of conditions under which the FAA may impose temporary flight restrictions. It also explains which FAA elements have been delegated authority to issue a temporary flight restrictions NOTAM and lists the types of responsible agencies/offices from which the FAA will accept requests to establish temporary flight restrictions. The 14 FAR is explicit as to what operations are prohibited, restricted, or allowed in a temporary flight restrictions area. Pilots are responsible to comply with 14 FAR Part 91.137, 91.138, 91.141 and 91.143 when conducting flight in an area where a temporary flight restrictions area is in effect, and should check appropriate NOTAMs during flight planning.

b. The purpose for establishing a temporary flight restrictions area is to:

1. Protect persons and property in the air or on the surface from an existing or imminent hazard associated with an incident on the surface when the presence of low flying aircraft would magnify, alter, spread, or compound that hazard (14 FAR Part 91.137(a)(1));

2. Provide a safe environment for the operation of disaster relief aircraft (14 FAR Part 91.137(a)(2)); or

3. Prevent an unsafe congestion of sightseeing aircraft above an incident or event which may generate a high degree of public interest (14 FAR Part 91.137(a)(3)).

4. Protect declared national disasters for humanitarian reasons in the State of Hawaii (14 CFR Section 91.138).

5. Protect the President, Vice President, or other public figures (14 CFR Section 91.141).

6. Provide a save environment for space agency operations (14 CFR Section 91.143).

c. Except for hijacking situations, when the provisions of 14 FAR Part 91.137(a)(1) or (a)(2) are necessary, a temporary flight restrictions area will only be established by or through the area manager at the Air Route Traffic Control Center (ARTCC) having jurisdiction over the area concerned. A temporary flight restrictions NOTAM involving the conditions of 14 FAR Part 91.137(a)(3) will be issued at the direction of the service area office director having oversight of the airspace concerned. When hijacking situations are involved, a temporary flight restrictions area will be implemented through the TSA Aviation Command Center. The appropriate FAA air traffic element, upon receipt of such a request, will establish a temporary flight restrictions area under 14 FAR Part 91.137(a)(1).

d. The FAA accepts recommendations for the establishment of a temporary flight restrictions area under 14 FAR Part 91.137(a)(1) from military major command headquarters, regional directors of the Office of Emergency Planning, Civil Defense State Directors, State Governors, or other similar authority. For the situations involving 14 FAR Part 91.137(a)(2), the FAA accepts recommendations from military commanders serving as regional, subregional, or Search and Rescue (SAR) coordinators; by military commanders directing or coordinating air operations associated with disaster relief; or by civil authorities directing or coordinating organized relief air operations (includes representatives of the Office of Emergency Planning, U.S. Forest Service, and State aeronautical agencies). Appropriate authorities for a temporary flight restrictions establishment under 14 FAR Part 91.137(a)(3) are any of those listed above or by State, county, or city government entities.

e. The type of restrictions issued will be kept to a minimum by the FAA consistent with achievement of the necessary objective. Situations which warrant the extreme restrictions of 14 FAR Part 91.137(a)(1) include, but are not limited to: toxic gas leaks or spills, flammable agents, or fumes which if fanned by rotor or propeller wash could endanger persons or property on the surface, or if entered by an aircraft could endanger persons or property in the air; imminent volcano eruptions which could endanger airborne aircraft and occupants; nuclear accident or incident; and hijackings. Situations which warrant the restrictions associated with 14 FAR Part 91.137(a)(2) include: forest fires which are being fought by releasing fire retardants from aircraft; and aircraft relief activities following a disaster (earthquake, tidal wave, flood, etc.). 14 FAR Part 91.137(a)(3) restrictions are established for events and incidents that would attract an unsafe congestion of sightseeing aircraft.

f. The amount of airspace needed to protect persons and property or provide a safe environment for rescue/relief aircraft operations is normally limited to within 2,000 feet above the surface and within a three-nautical-mile radius. Incidents occurring within Class B, Class C, or Class D airspace will normally be handled through existing procedures and should not require the issuance of temporary flight restrictions NOTAM. Temporary flight restrictions affecting airspace outside the United States and its territories and possessions are issued with verbiage excluding that airspace outside of the 12-mile coastal limits.

g. The FSS nearest the incident site is normally the "coordination facility." When FAA communications assistance is required, the designated FSS will function as the primary communications facility for coordination between emergency control authorities and affected aircraft. The ARTCC may act as liaison for the emergency control authorities if adequate communications cannot be established between the designated FSS and the relief organization. For example, the coordination facility may relay authorizations from the on-scene emergency response official in cases where news media aircraft operations are approved at the altitudes used by relief aircraft.

h. ATC may authorize operations in a temporary flight restrictions area under its own authority only when flight restrictions are established under 14 FAR Part 91.137(a)(2) and (a)(3). The appropriate ARTCC/air traffic control tower manager will, however, ensure that such authorized flights do not hamper activities or interfere with the event for which restrictions were implemented. However, ATC will not authorize local IFR flights into the temporary flight restrictions area.

i. To preclude misunderstanding, the implementing NOTAM will contain specific and formatted information. The facility establishing a temporary flight restrictions area will format a NOTAM beginning with the phrase "FLIGHT RESTRICTIONS" followed by: the location of the temporary flight restrictions area; the effective period; the area defined in statute miles; the altitudes affected; the FAA coordination facility and commercial telephone number; the reason for the temporary flight restrictions; the agency directing any relief activities and its commercial telephone number; and other information considered appropriate by the issuing authority.

EXAMPLE—

① 14 FAR PART 91.137(a)(1):

THE FOLLOWING NOTAM PROHIBITS ALL AIRCRAFT OPERATIONS EXCEPT THOSE SPECIFIED IN THE NOTAM.

FLIGHT RESTRICTIONS MATTHEWS, VIRGINIA, EFFECTIVE IMMEDIATELY UNTIL 9610211200. PURSUANT TO 14 FAR 91.137(A)(1) TEMPORARY FLIGHT RESTRICTIONS ARE IN EFFECT. RESCUE OPERATIONS IN PROGRESS. ONLY RELIEF AIRCRAFT OPERATIONS UNDER THE DIRECTION OF THE DEPARTMENT OF DEFENSE ARE AUTHORIZED IN THE AIRSPACE AT AND BELOW 5,000 FEET MSL WITHIN A TWO-NAUTICAL-MILE-RADIUS OF LASER AFB, MATTHEWS, VIRGINIA. COMMANDER, LASER AFB, IN CHARGE (897) 946-5543 (122.4). STEENSON FSS (792) 555-6141 (123.1) IS THE FAA COORDINATION FACILITY.

② 14 FAR PART 91.137(a)(2):

THE FOLLOWING NOTAM PERMITS FLIGHT OPERATIONS IN ACCORDANCE WITH 14CFR SECTION 91.137 (A)(2). THE ON-SITE EMERGENCY RESPONSE OFFICIAL TO AUTHORIZE MEDIA AIRCRAFT OPERATIONS BELOW THE ALTITUDES USED BY THE RELIEF AIRCRAFT.

FLIGHT RESTRICTIONS 25 MILES EAST OF BRANSOME, IDAHO, EFFECTIVE IMMEDIATELY UNTIL 9601202359 UTC. PURSUANT TO 14 FAR 91.137(A)(2) TEMPORARY FLIGHT RESTRICTIONS ARE IN EFFECT WITHIN A FOUR-MILE-RADIUS OF THE INTERSECTION OF COUNTY ROADS 564 AND 315 AT AND BELOW 3,500 FEET MSL TO PROVIDE A SAFE ENVIRONMENT FOR FIRE FIGHTING AIRCRAFT OPERATIONS. DAVIS COUNTY SHERIFF'S DEPARTMENT (792) 555-8122 (122.9) IS IN CHARGE OF ON-SCENE EMERGENCY RESPONSE ACTIVITIES. GLIVINGS FSS (792) 555-1618 (122.2) IS THE FAA COORDINATION FACILITY.

③ 14 FAR PART 91.137(a)(3):

THE FOLLOWING NOTAM PROHIBITS SIGHTSEEING AIRCRAFT OPERATIONS.

FLIGHT RESTRICTIONS BROWN, TENNESSEE, DUE TO OLYMPIC ACTIVITY, EFFECTIVE 9606181100 UTC UNTIL 9607190200 UTC. PURSUANT TO 14 CFR SECTION 91.137(A)(3) TEMPORARY FLIGHT RESTRICTIONS ARE IN EFFECT WITHIN A 3-NAUTICAL-MILE RADIUS OF N355783/W835242 AND VOLUNTEER VORTAC 019 DEGREE RADIAL 3.7 DME FIX AT AND BELOW 2,500 FEET MSL. NORTON FSS (423) 555-6742 (126.6) IS THE FAA COORDINATION FACILITY.

④ 14 CFR SECTION 91.138:

THE FOLLOWING NOTAM PROHIBITS ALL AIRCRAFT EXCEPT THOSE OPERATING UNDER THE AUTHORIZATION OF THE OFFICIAL IN CHARGE OF ASSOCIATED EMERGENCY OR DISASTER RELIEF RESPONSE ACTIVITIES, AIRCRAFT CARRYING LAW ENFORCEMENT OFFICIALS, AIRCRAFT CARRYING PERSONNEL INVOLVED IN AN EMERGENCY OR LEGITIMATE SCIENTIFIC PURPOSES, CARRYING PROPERLY ACCREDITED NEWS MEDIA, AND AIRCRAFT OPERATING IN ACCORDANCE WITH AN ATC CLEARANCE OR INSTRUCTION.

FLIGHT RESTRICTIONS KAPALUA, HAWAII, EFFECTIVE 9605101200 UTC UNTIL 9605151500 UTC. PURSUANT TO 14 CFR SECTION 91.138 TEMPORARY FLIGHT RESTRICTIONS ARE IN EFFECT WITHIN A 3-NAUTICAL-MILE RADIUS OF N205778/W1564038 AND MAUI/OGG/VORTAC 275 DEGREE RADIAL AT 14.1 NAUTICAL MILES. JOHN DOE 808-757-4469 OR 122.4 IS IN CHARGE OF THE OPERATION. HONOLULU /HNL/ 808-757-4470 (123.6) AFSS IS THE FAA COORDINATION FACILITY.

⑤ 14 CFR SECTION 91.141:

THE FOLLOWING NOTAM PROHIBITS ALL AIRCRAFT.

FLIGHT RESTRICTIONS STILLWATER, OKLAHOMA, JUNE 21, 1996. PURSUANT TO 14 CFR SECTION 91.141 AIRCRAFT FLIGHT OPERATIONS ARE PROHIBITED WITHIN A 3-NAUTICAL-MILE RADIUS, BELOW 2000 FEET AGL OF N360962/WI70515 AND THE STILLWATER/SWO/VOR/DME 176 DEGREE RADIAL 3.8-NAUTICAL-MILE FIX FROM 1400 LOCAL TIME TO 1700 LOCAL TIME JUNE 21, 1996 UNLESS OTHERWISE AUTHORIZED BY ATC.

⑥ 14 CFR SECTION 91.143:

THE FOLLOWING NOTAM PROHIBITS ANY AIRCRAFT OF U.S. REGISTRY, OR PILOT ANY AIRCRAFT UNDER THE AUTHORITY OF AN AIRMAN CERTIFICATE ISSUED BY THE FAA.

KENNEDY SPACE CENTER SPACE OPERATIONS AREA EFFECTIVE IMMEDIATELY UNTIL 9610152100 UTC. PURSUANT TO SECTION 91.143, FLIGHT OPERATIONS CONDUCTED BY FAA CERTIFICATED PILOTS OR CONDUCTED IN AIRCRAFT OF U.S. REGISTRY ARE PROHIBITED AT ANY ALTITUDE FROM SURFACE TO UNLIMITED, WITHIN THE FOLLOWING AREA 30-NAUTICAL-MILE RADIUS OF THE MELBOURNE /MLB/ VORTAC 010 DEGREE RADIAL 21-NAUTICAL-MILE FIX. ST. PETERSBURG, FLORIDA, /PIE/ AFSS 813-545-1645 (122.2) IS THE FAA COORDINATION FACILITY AND SHOULD BE CONTACTED FOR THE CURRENT STATUS OF ANY AIRSPACE ASSOCIATED WITH THE SPACE SHUTTLE OPERATIONS. THIS AIRSPACE ENCOMPASSES R2933, R2932, R2931, R3934, R2935, AREAS WILL BE ACTIVE IN CONJUNCTION WITH THE OPERATIONS. PILOTS SHALL CONSULT ALL NOTAMS REGARDING THIS OPERATION.

3-5-4. PARACHUTE JUMP AIRCRAFT OPERATIONS

a. Procedures relating to parachute jump areas are contained in FAR Part 105. Tabulations of parachute jump areas in the U.S. are contained in the A/FD.

b. Pilots of aircraft engaged in parachute jump operations are reminded that all reported altitudes must be with reference to mean sea level, or flight level, as appropriate, to enable ATC to provide meaningful traffic information.

c. Parachute operations in the vicinity of an airport without an operating control tower—There is no substitute for alertness while in the vicinity of an airport. It is essential that pilots conducting parachute operations be alert, look for other traffic, and exchange traffic information as recommended in AIM, paragraph 4-1-9, *TRAFFIC ADVISORY PRACTICES AT AIRPORTS WITHOUT OPERATING CONTROL TOWERS*. In addition, pilots should avoid releasing parachutes while in an airport traffic pattern when there are other aircraft in that pattern. Pilots should make appropriate broadcasts on the designated Common Traffic Advisory Frequency (CTAF), and monitor that CTAF until all parachute activity has terminated or the aircraft has left the area. Prior to commencing a jump operation, the pilot should broadcast the aircraft's altitude and position in relation to the airport, the approximate relative time when the jump will commence and terminate, and listen to the position reports of other aircraft in the area.

3-5-5. PUBLISHED VFR ROUTES

Published VFR routes for transitioning around, under and through complex airspace such as Class B airspace were developed through a number of FAA and industry initiatives. All of the following terms, i.e., "VFR Flyway" "VFR Corridor" and "Class B Airspace VFR Transition Route" have been used when referring to the same or different types of routes or airspace. The following paragraphs identify and clarify the functionality of each type of route, and specify where and when an ATC clearance is required.

a. VFR Flyways:

1. VFR Flyways and their associated Flyway Planning Charts were developed from the recommendations of a National Airspace Review Task Group. A VFR Flyway is defined as a general flight path not defined as a specific course, for use by pilots in planning flights into, out of, through or near complex terminal airspace to avoid Class B airspace. An ATC clearance is NOT required to fly these routes.

2. VFR Flyways are depicted on the reverse side of some of the VFR Terminal Area Charts (TAC), commonly referred to as Class B airspace charts. (See FIG 3-5-1). Eventually all TACs will include a VFR Flyway Planning

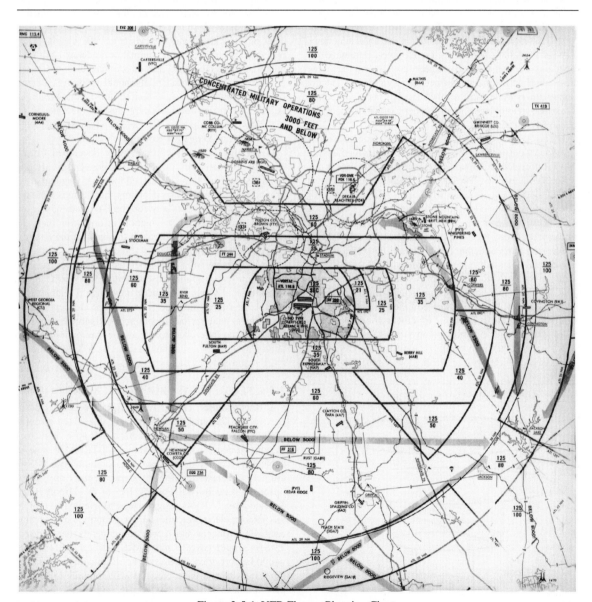

Figure 3-5-1. VFR Flyway Planning Chart.

Chart. These charts identify VFR flyways designed to help VFR pilots avoid major controlled traffic flows. They may further depict multiple VFR routings throughout the area which may be used as an alternative to flight within Class B airspace. The ground references provide a guide for improved visual navigation. These routes are not intended to discourage requests for VFR operations within Class B airspace but are designed solely to assist pilots in planning for flights under and around busy Class B airspace without actually entering Class B airspace.

3. It is very important to remember that these suggested routes are not sterile of other traffic. The entire Class B airspace, and the airspace underneath it, may be heavily congested with many different types of aircraft. Pilot adherence to VFR rules must be exercised at all times. Further, when operating beneath Class B airspace, communications must be established and maintained between your aircraft and any control tower while transiting the Class B, Class C, and Class D surface areas of those airports under Class B Airspace.

b. VFR Corridors:

1. The design of a few of the first Class B airspace areas provided a corridor for the passage of uncontrolled traffic. A VFR corridor is defined as airspace through Class B airspace, with defined vertical and lateral boundaries, in which aircraft may operate without an ATC clearance or communication with air traffic control.

2. These corridors are, in effect, a "hole" through Class B airspace. (See FIG 3-5-2). A classic example would be the corridor through the Los Angeles Class B airspace, which has been subsequently changed to Special Flight Rules airspace (SFR). A corridor is surrounded on all sides by Class B airspace and does not extend down to the surface like a VFR Flyway. Because of their finite lateral and vertical limits, and the volume of VFR traffic using a corridor, extreme caution and vigilance must be exercised.

3. Because of the heavy traffic volume and the procedures necessary to efficiently manage the flow of traffic, it has not been possible to incorporate VFR corridors in the development or modifications of Class B airspace in recent years.

c. Class B Airspace VFR Transition Routes:

1. To accommodate VFR traffic through certain Class B airspace, such as Seattle, Phoenix and Los Angeles, Class B Airspace VFR Transition Routes were developed. A Class B Airspace VFR Transition Route is defined as a specific flight course depicted on a Terminal Area Chart (TAC) for transiting a specific Class B airspace. These routes include specific ATC-assigned altitudes, and pilots must obtain an ATC clearance prior to entering Class B airspace on the route.

2. These routes, as depicted in FIG 3-5-3, are designed to show the pilot where to position the aircraft outside of, or clear of, the Class B airspace where an ATC clearance can normally be expected with minimal or no delay. Until ATC authorization is received, pilots must remain clear of Class B airspace. On initial contact, pilots should advise ATC of their position, altitude, route name desired, and direction of flight. After a clearance is received, pilot must fly the route as depicted and, most importantly, adhere to ATC instructions.

d. Terminal Area VFR Routes:

1. Terminal Area VFR Routes were developed from a concept evaluated in the Los Angeles Basin area in 1988–89, and are being developed for other terminal areas around the country. Charts depicting these routes were developed in a joint effort between the FAA and industry to provide more specific navigation information than the VFR Flyway Planning Charts on the back of the Class B airspace charts. (See FIG 3-5-4).

2. A Terminal Area VFR Route is defined as a specific flight course for optional use by pilots to avoid Class B, Class C, and Class D airspace areas while operating in complex terminal airspace. These routes are depicted on the chart(s), may include recommnended altitudes, and are described by reference to electronic navigational aidsa and/or prominent visual landmarks. An ATC clearance is NOT required to fly these routes.

3-5-6. TERMINAL RADAR SERVICE AREA (TRSA)

a. Background: TRSAs were originally established as part of the Terminal Radar Program at selected airports. TRSAs were never controlled airspace from a regulatory standpoint because the establishment of TRSAs was never subject to the rulemaking process; consequently, TRSAs are not contained in FAR Part 71 nor are there any TRSA operating rules in FAR Part 91. Part of the Airport Radar Service Area (ARSA) program was to eventually replace all TRSAs. However, the ARSA requirements became relatively stringent and it was subsequently decided that TRSAs would have to meet ARSA criteria before they would be converted. TRSAs do not fit into any of the U.S. Airspace Classes; therefore, they will continue to be non-part 71 airspace areas where participating pilots can receive additional radar services which have been redefined as TRSA Service.

CLASS B AIRSPACE

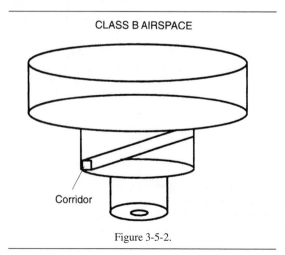

Figure 3-5-2.

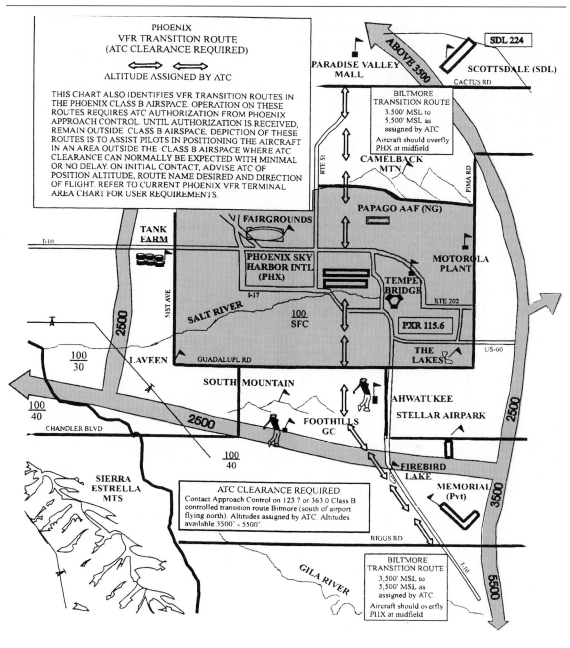

(Not to be used for navigation)

Figure 3-5-3.

b. TRSA Areas: The primary airport(s) within the TRSA become(s) Class D airspace. The remaining portion of the TRSA overlies other controlled airspace which is normally Class E airspace beginning at 700 or 1,200 feet and established to transition to/from the enroute/terminal environment.

c. Participation: Pilots operating under VFR are encouraged to contact the radar approach control and avail themselves of the TRSA Services. However, participation is voluntary on the part of the pilot. See AIM Chapter 4 for details and procedures.

d. Charts: TRSA's are depicted on VFR sectional and terminal area charts with a solid black line and altitudes for each segment. The Class D portion is charted with a blue segmented line.

3-5-7. NATIONAL SECURITY AREAS

National Security Areas consist of airspace of defined vertical and lateral dimensions established at locations where there is a requirement for increased security and safety of ground facilities. Pilots are requested to voluntarily avoid flying through the depicted NSA. When it is necessary to provide a greater level of security and safety, flight in NSAs may be temporarily prohibited by regulation under the provisions of FAR Part 99.7. Regulatory prohibitions will be issued by System Operations, System Operations Airspace and AIM Office, Airspace Rules, and disseminated via NOTAM. Inquiries about NSAs should be directed to Airspace and Rules.

Chapter 4. Air traffic control
Section 1. SERVICES AVAILABLE TO PILOTS

4-1-1. AIR ROUTE TRAFFIC CONTROL CENTERS

Centers are established primarily to provide Air Traffic Service to aircraft operating on IFR flight plans within controlled airspace, and principally during the en route phase of flight.

4-1-2. CONTROL TOWERS

Towers have been established to provide for a safe, orderly and expeditious flow of traffic on and in the vicinity of an airport. When the responsibility has been so delegated, towers also provide for the separation of IFR aircraft in the terminal areas.

REFERENCE—AIM, APPROACH CONTROL, PARAGRAPH 5-4-3.

4-1-3. FLIGHT SERVICE STATIONS

a. Flight Service Stations (FSS) are air traffic facilities which provide pilot briefings, en route communications and VFR search and rescue services, assist lost aircraft and aircraft in emergency situations, relay ATC clearances, originate Notices to Airmen, broadcast aviation weather and National Airspace System (NAS) information, receive and process IFR flight plans, and monitor NAVAIDs. In addition, at selected locations FSSs provide En Route Flight Advisory Service (Flight Watch), take weather observations, issue airport advisories, and advise Customs and Immigration of transborder flights.

b. Supplemental Weather Service Locations (SWSLs) are airport facilities staffed with contract personnel who take weather observations and provide current local weather to pilots via telephone or radio. All other services are provided by the parent FSS.

4-1-4. RECORDING AND MONITORING

a. Calls to air traffic control (ATC) facilities (ARTCCs, Towers, FSSs, Central Flow, and Operations Centers) over radio and ATC operational telephone lines (lines used for operational purposes such as controller instructions, briefings, opening and closing flight plans, issuance of IFR clearances and amendments, counter hijacking activities, etc.) may be monitored and recorded for operational uses such as accident investigations, accident prevention, search and rescue purposes, specialist training and evaluation, and technical evaluation and repair of control and communications systems.

b. Where the public access telephone is recorded, a beeper tone is not required. In place of the "beep" tone the FCC has substituted a mandatory requirement that persons to be recorded be given notice they are to be recorded and give consent. Notice is given by this entry, consent to record is assumed by the individual placing a call to the operational facility.

4-1-5. COMMUNICATIONS RELEASE OF IFR AIRCRAFT LANDING AT AN AIRPORT WITHOUT AN OPERATING CONTROL TOWER

Aircraft operating on an IFR flight plan, landing at an airport without an operating control tower will be advised to change to the airport advisory frequency when direct communications with ATC are no longer required. Towers and centers do not have nontower airport traffic and runway in use information. The instrument approach may not be aligned with the runway in use; therefore, if the information has not already been obtained, pilots should make an expeditious change to the airport advisory frequency when authorized.

REFERENCE—AIM, ADVANCE INFORMATION ON INSTRUMENT APPROACH, PARAGRAPH 5-4-4.

4-1-6. PILOT VISITS TO AIR TRAFFIC FACILITIES

Pilots are encouraged to visit air traffic facilities (Towers, Centers and FSSs) and familiarize themselves with the ATC system. On rare occasions, facilities may not be able to approve a visit because of ATC workload or other reasons. It is therefore requested that pilots contact the facility prior to the visit and advise of the number of persons in the group, the time and date of the proposed visit and the primary interest of the group. With this information available, the facility can prepare an itinerary and have someone available to guide the group through the facility.

4-1-7. OPERATION TAKE-OFF AND OPERATION RAINCHECK

Operation Take-off is a program that educates pilots in how best to utilize the FSS modernization efforts and services available in Automated Flight Service Stations (AFSS), as stated in FAA Order 7230.17. Operation Raincheck is a program designed to familiarize pilots with the ATC system, its functions, responsibilities and benefits.

4-1-8. APPROACH CONTROL SERVICE FOR VFR ARRIVING AIRCRAFT

a. Numerous approach control facilities have established programs for arriving VFR aircraft to contact approach control for landing information. This information includes: wind, runway, and altimeter setting at the airport of intended landing. This information may be omitted if contained in the ATIS broadcast and the pilot states the appropriate ATIS code.

NOTE—PILOT USE OF "HAVE NUMBERS" DOES NOT INDICATE RECEIPT OF THE ATIS BROADCAST. IN ADDITION, THE CONTROLLER WILL PROVIDE TRAFFIC ADVISORIES ON A WORKLOAD PERMITTING BASIS.

b. Such information will be furnished upon initial contact with concerned approach control facility. The pilot will be requested to change to the *tower* frequency at a predetermined time or point, to receive further landing information.

c. Where available, use of this procedure will not hinder the operation of VFR flights by requiring excessive spacing between aircraft or devious routing.

d. Compliance with this procedure is not mandatory but pilot participation is encouraged.

REFERENCE—AIM, Terminal Radar Services for VFR Aircraft, paragraph 4-1-17.

NOTE—APPROACH CONTROL SERVICES FOR VFR AIRCRAFT ARE NORMALLY DEPENDENT ON AIR TRAFFIC CONTROL RADAR. THESE SERVICES ARE NOT AVAILABLE DURING PERIODS OF A RADAR OUTAGE. APPROACH CONTROL SERVICES FOR VFR AIRCRAFT ARE LIMITED WHEN CENRAP IS IN USE.

4-1-9. TRAFFIC ADVISORY PRACTICES AT AIRPORTS WITHOUT OPERATING CONTROL TOWERS
(see TBL 4-1-1)

a. Airport Operations without Operating Control Tower:

1. There is no substitute for alertness while in the vicinity of an airport. It is essential that pilots be alert and look for other traffic and exchange traffic information when approaching or departing an airport without an operating control tower. This is of particular importance since other aircraft may not have communication capability or, in

Table 4-1-1 Summary of Recommended Communication Procedures

COMMUNICATION/BROADCAST PROCEDURES

FACILITY AT AIRPORT	FREQUENCY USE	OUTBOUND	INBOUND	PRACTICE INSTRUMENT APPROACH
1. UNICOM (No Tower or FSS)	Communicate with UNICOM station on published CTAF frequency (122.7, 122.8, 122.725, 122.975, or 123.0). If unable to contact UNICOM station, use self-announce procedures on CTAF.	Before taxiing and before taxiing on the runway for departure.	10 miles out. Entering downwind, base, and final. Leaving the runway.	
2. No Tower, FSS, or UNICOM	Self-announce on MULTICOM frequency 122.9.	Before taxiing and before taxiing on the runway for departure.	10 miles out. Entering downwind, base, and final. Leaving the runway.	Departing final approach fix (name) or on final approach segment inbound.
3. No Tower in operation, FSS open	Communicate with FSS on CTAF frequency.	Before taxiing and before taxiing on the runway for departure.	10 miles out. Entering downwind, base, and final. Leaving the runway.	Approach completed/ terminated.
4. FSS closed (No Tower)	Self-announce on CTAF.	Before taxiing and before taxiing on the runway for departure.	10 miles out. Entering downwind, base, and final. Leaving the runway.	
5. Tower or FSS not in operation	Self-announce on CTAF.	Before taxiing and before taxiing on the runway for departure.	10 miles out. Entering downwind, base, and final. Leaving the runway.	

some cases, pilots may not communicate their presence or intentions when operating into or out of such airports. To achieve the greatest degree of safety, it is essential that all radio-equipped aircraft transmit/receive on a common frequency identified for the purpose of airport advisories.

2. An airport may have a full or part-time tower or FSS located on the airport, a full or part-time UNICOM station or no aeronautical station at all. There are three ways for pilots to communicate their intention and obtain airport/traffic information when operating at an airport that does not have an operating tower: by communicating with an FSS, a UNICOM operator, or by making a self-announce broadcast.

3. Many airports are now providing completely automated weather, radio check capability and airport advisory information on an automated UNICOM system. These systems offer a variety of features, typically selectable by microphone clicks, on the UNICOM frequency. Availability of the automated UNICOM will be published in the Airport/Facility Directory and approach charts.

b. Communicating on a Common Frequency:

1. The key to communicating at an airport without an operating control tower is selection of the correct common frequency. The acronym CTAF which stands for Common Traffic Advisory Frequency, is synonymous with this program. A CTAF is a frequency designated for the purpose of carrying out airport advisory practices while operating to or from an airport without an operating control tower. The CTAF may be a UNICOM, MULTICOM, FSS, or tower frequency and is identified in appropriate aeronautical publications.

2. The CTAF frequency for a particular airport is contained in the A/FD, Alaska Supplement, Alaska Terminal Publication, Instrument Approach Procedure Charts, and Standard Instrument Departure (DP) Charts. Also, the CTAF frequency can be obtained by contacting any FSS. Use of the appropriate CTAF, combined with a visual alertness and application of the following recommended good operating practices, will enhance safety of flight into and out of all uncontrolled airports.

c. Recommended Traffic Advisory Practices:

1. Pilots of inbound traffic should monitor and communicate as appropriate on the designated CTAF from 10 miles to landing. Pilots of departing aircraft should monitor/communicate on the appropriate frequency from start-up, during taxi, and until 10 miles from the airport unless the FARs or local procedures require otherwise.

2. Pilots of aircraft conducting other than arriving or departing operations at altitudes normally used by arriving and departing aircraft should monitor/communicate on the appropriate frequency while within 10 miles of the airport unless required to do otherwise by the FARs or local procedures. Such operations include parachute jumping/dropping, en route, practicing maneuvers, etc.

REFERENCE—AIM, PARACHUTE JUMP AIRCRAFT OPERATIONS, PARAGRAPH 3-5-4.

d. Airport Advisory/Information Services Provided by a FSS

1. There are three advisory type services provided at selected airports.

(a) Local Airport Advisory (LAA) is provided at airports that have a FSS physically located on the airport, which does not have a control tower or where the tower is operated on a part-time basis. The CTAF for LAA airports is disseminated in the appropriate aeronautical publications.

(b) Remote Airport Advisory (RAA) is provided at selected very busy GA airports, which do not have an operating control tower. The CTAF for RAA airports is disseminated in the appropriate aeronautical publications.

(c) Remote Airport Information Service (RATS) is provided in support of special events at nontowered airports by request from the airport authority.

2. In communicating with a CTAF FSS, check the airport's automated weather and establish two-way communications before transmitting out bound/inbound intentions or information. An inbound aircraft should initiate contact approximately 10 miles from the airport, reporting aircraft identification and type, altitude, location relative to the airport, intentions (landing or over flight), possession of the automated weather, and request airport advisory or airport information service. A departing aircraft should initiate contact before taxiing, reporting aircraft identification and type, VFR or IFR, location on the airport, intentions, direction of takeoff, possession of the automated weather, and request airport advisory or information service. Also, report intentions before taxiing onto the active runway for departure. If you must change frequencies for other service after initial report to FSS, return to FSS frequency for traffic update.

(a) Inbound

EXAMPLE—VERO BEACH RADIO, CENTURION SIX NINER DELTA DELTA IS TEN MILES SOUTH, TWO THOUSAND, LANDING VERO BEACH. I HAVE THE AUTOMATED WEATHER, REQUEST AIRPORT ADVISORY.

(b) Outbound

EXAMPLE—VERO BEACH RADIO, CENTURION SIX NINER DELTA DELTA, READY TO TAXI TO RUNWAY 22, VFR, DEPARTING TO THE SOUTHWEST. I HAVE THE AUTOMATED WEATHER, REQUEST AIRPORT ADVISORY.

3. Airport advisory service includes wind direction and velocity, favored or designated runway, altimeter setting, known airborne and ground traffic, NOTAM's, airport taxi routes, airport traffic pattern information, and instrument approach procedures. These elements are varied so as to best serve the current traffic situation. Some airport managers have specified that under certain wind or other conditions designated runways be used. Pilots should advise the FSS of the runway they intend to use.

CAUTION—ALL AIRCRAFT IN THE VICINITY OF AN AIRPORT MAY NOT BE IN COMMUNICATION WITH THE FSS.

e. Information provided by Aeronautical Advisory Stations (UNICOM):

1. UNICOM is a nongovernment air/ground radio communication station which may provide airport information at public use airports where there is no tower or FSS.

2. On pilot request, UNICOM stations may provide pilots with weather information, wind direction, the recommended runway, or other necessary information. If the UNICOM frequency is designated as the CTAF, it will be identified in appropriate aeronautical publications.

f. Unavailability of Information from FSS or UNICOM:

Should LAA by an FSS or Aeronautical Advisory Station UNICOM be unavailable, wind and weather information may be obtainable from nearby controlled airports via Automatic Terminal Information Service (ATIS) or Automated Weather Observing System (AWOS) frequency.

g. Self Announce position or intentions

1. General. Self-announce is a procedure whereby pilots broadcast their position or intended flight activity or ground operation on the designated CTAF. The procedure is used primarily at airports which do not have an FSS on the airport. The self-announce procedure should also be used if a pilot is unable to communicate with the FSS on the designated CTAF. Pilots stating, "Traffic in the area please advise" is not a recognized Self-Announce Position and/or Intention phrase and should not be used under any condition.

2. If an airport has a tower and it is temporarily closed, or operated on a part-time basis and there is no FSS on the airport or the FSS is closed, use the CTAF to self-announce your position or intentions.

3. Where there is no tower, FSS, or UNICOM station on the airport, use MULTICOM frequency 122.9 for self-announce procedures. Such airports will be identified in appropriate aeronautical information publications.

4. *Practice Approaches.* Pilots conducting practice instrument approaches should be particularly alert for other aircraft that may be departing in the opposite direction. When conducting any practice approach, regardless of its direction relative to other airport operations, pilots should make announcements on the CTAF as follows:

(a) departing the final approach fix, inbound (nonprecision approach) or departing the Outer Marker or fix used in lieu of the outer marker, inbound (precision approach);

(b) established on the final approach segment or immediately upon being released by ATC;

(c) upon completion or termination of the approach; and

(d) upon executing the missed approach procedure.

5. Departing aircraft should always be alert for arrival aircraft coming from the opposite direction.

6. Recommended Self-Announce Phraseologies: It should be noted that aircraft operating to or from another nearby airport may be making self-announce broadcasts on the same UNICOM or MULTICOM frequency. To help

identify one airport from another, the airport name should be spoken at the beginning and end of each self-announce transmission.

(a) Inbound
EXAMPLE—STRAWN TRAFFIC, APACHE TWO TWO FIVE ZULU, (POSITION), (ALTITUDE), (DESCENDING) OR ENTERING DOWNWIND/BASE/FINAL (AS APPROPRIATE) RUNWAY ONE SEVEN FULL STOP, TOUCH-AND-GO, STRAWN.

STRAWN TRAFFIC, APACHE TWO TWO FIVE ZULU CLEAR OF RUNWAY ONE SEVEN STRAWN.

(b) Outbound
EXAMPLE—STRAWN TRAFFIC, QUEEN AIR SEVEN ONE FIVE FIVE BRAVO (LOCATION ON AIRPORT) TAXIING TO RUNWAY TWO SIX STRAWN.

STRAWN TRAFFIC, QUEEN AIR SEVEN ONE FIVE FIVE BRAVO DEPARTING RUNWAY TWO SIX DEPARTING THE PATTERN TO THE (DIRECTION), CLIMBING TO (ALTITUDE) STRAWN.

(c) Practice Instrument Approach
EXAMPLE—STRAWN TRAFFIC, CESSNA TWO ONE FOUR THREE QUEBEC (POSITION FROM AIRPORT) INBOUND DESCENDING THROUGH (ALTITUDE) PRACTICE (NAME OF APPROACH) APPROACH RUNWAY THREE FIVE STRAWN.

STRAWN TRAFFIC, CESSNA TWO ONE FOUR THREE QUEBEC PRACTICE (TYPE) APPROACH COMPLETED OR TERMINATED RUNWAY THREE FIVE STRAWN.

h. UNICOM Communications Procedures:

1. In communicating with a UNICOM station, the following practices will help reduce frequency congestion, facilitate a better understanding of pilot intentions, help identify the location of aircraft in the traffic pattern, and enhance safety of flight:

(a) Select the correct UNICOM frequency.

(b) State the identification of the UNICOM station you are calling in each transmission.

(c) Speak slowly and distinctly.

(d) Report approximately 10 miles from the airport, reporting altitude, and state your aircraft type, aircraft identification, location relative to the airport, state whether landing or overflight, and request wind information and runway in use.

(e) Report on downwind, base, and final approach.

(f) Report leaving the runway.

2. Recommended UNICOM Phraseologies:

(a) Inbound
PHRASEOLOGY—FREDERICK UNICOM CESSNA EIGHT ZERO ONE TANGO FOXTROT 10 MILES SOUTHEAST DESCENDING THROUGH (ALTITUDE) LANDING FREDERICK, REQUEST WIND AND RUNWAY INFORMATION FREDERICK.

FREDERICK TRAFFIC CESSNA EIGHT ZERO ONE TANGO FOXTROT ENTERING DOWNWIND/BASE/FINAL (AS APPROPRIATE) FOR RUNWAY ONE NINER (FULL STOP/TOUCH-AND-GO) FREDERICK.

FREDERICK TRAFFIC CESSNA EIGHT ZERO ONE TANGO FOXTROT CLEAR OF RUNWAY ONE NINER FREDERICK.

(b) Outbound

PHRASEOLOGY—FREDERICK UNICOM CESSNA EIGHT ZERO ONE TANGO FOXTROT (LOCATION ON AIRPORT) TAXIING TO RUNWAY ONE NINER, REQUEST WIND AND TRAFFIC INFORMATION FREDERICK. FREDERICK TRAFFIC CESSNA EIGHT ZERO ONE TANGO FOXTROT DEPARTING RUNWAY ONE NINER. "REMAINING IN THE PATTERN" OR "DEPARTING THE PATTERN TO THE (DIRECTION) (AS APPROPRIATE)" FREDERICK.

4-1-10. IFR APPROACHES/GROUND VEHICLE OPERATIONS

a. IFR Approaches: When operating in accordance with an IFR clearance and ATC approves a change to the advisory frequency, make an expeditious change to the CTAF and employ the recommended traffic advisory procedures.

b. Ground Vehicle Operation: Airport ground vehicles equipped with radios should monitor the CTAF frequency when operating on the airport movement area and remain clear of runways/taxiways being used by aircraft. Radio transmissions from ground vehicles should be confined to safety-related matters.

c. Radio Control of Airport Lighting Systems: Whenever possible, the CTAF will be used to control airport lighting systems at airports without operating control towers. This eliminates the need for pilots to change frequencies to turn the lights on and allows a continuous listening watch on a single frequency. The CTAF is published on the instrument approach chart and in other appropriate aeronautical information publications. For further details concerning radio controlled lights, see AC 150/5340-27.

4-1-11. DESIGNATED UNICOM/MULTICOM FREQUENCIES

a. Frequency Use.

1. The following listing depicts UNICOM and MULTICOM frequency uses as designated by the Federal Communications Commission (FCC). (See TBL 4-1-2).

Table 4-1-2 Unicom/Multicom Frequency Usage

USE	FREQUENCY
Airports without an operating control tower.	122.700
	122.725
	122.800
	122.975
	123.000
	123.050
	123.075
(MULTICOM FREQUENCY) Activities of a temporary seasonal, emergency nature or search and rescue, as well as, airports with no tower, FSS, or UNICOM.	122.900
(MULTICOM FREQUENCY) Forestry management and fire suppression, fish and game management and protection, and environmental monitoring and protection.	122.925

Airports with a control tower or FSS on airport	122.950

NOTE—

[1] IN SOME AREAS OF THE COUNTRY, FREQUENCY INTERFERENCE MAY BE ENCOUNTERED FROM NEARBY AIRPORTS USING THE SAME UNICOM FREQUENCY. WHERE THERE IS A PROBLEM, UNICOM OPERATORS ARE ENCOURAGED TO DEVELOP A "LEAST INTERFERENCE" FREQUENCY ASSIGNMENT PLAN FOR AIRPORTS CONCERNED USING THE FREQUENCIES DESIGNATED FOR AIRPORTS WITHOUT OPERATING CONTROL TOWERS. UNICOM LICENSEES ARE ENCOURAGED TO APPLY FOR UNICOM 25 KHZ SPACED CHANNEL FREQUENCIES DUE TO THE EXTREMELY LIMITED NUMBER OF FREQUENCIES WITH 50 KHZ CHANNEL SPACING, 25 KHZ CHANNEL SPACING SHOULD BE IMPLEMENTED. UNICOM LICENSEES MAY THEN REQUEST FCC TO ASSIGN FREQUENCIES IN ACCORDANCE WITH THE PLAN, WHICH FCC WILL REVIEW AND CONSIDER FOR APPROVAL.

[2] WIND DIRECTION AND RUNWAY INFORMATION MAY NOT BE AVAILABLE ON UNICOM FREQUENCY 122.950.

2. The following listing depicts other frequency uses as designated by the Federal Communications Commission (FCC). (See TBL 4-1-3).

Table 4-1-3
Other Frequency Usage Designated by FCC

USE	FREQUENCY
Air-to-air communications and private airports (not open to the public).	122.750
	122.850
Air-to-air communications (general aviation helicopters).	123.025
Aviation instruction, Glider, Hot Air Balloon (not to be used for advisory service).	123.300
	123.500

4-1-12. USE OF UNICOM FOR ATC PURPOSES

UNICOM service may be used for air traffic control purposes, only under the following circumstances:

1. Revision to proposed departure time.

2. Takeoff, arrival, or flight plan cancellation time.

3. ATC clearance, provided arrangements are made between the ATC facility and the UNICOM licensee to handle such messages.

4-1-13. AUTOMATIC TERMINAL INFORMATION SERVICE (ATIS)

a. ATIS is the continuous broadcast of recorded non-control information in selected high activity terminal areas. Its purpose is to improve controller effectiveness and to relieve frequency congestion by automating the repetitive transmission of essential but routine information. The information is continuously broadcast over a discrete VHF radio frequency or the voice portion of a local NAVAID. ATIS transmissions on a discrete VHF radio frequency are engineered to be receivable to a maximum of 60 NM from the ATIS site and a maximum altitude of 25,000 feet AGL. At most locations, ATIS signals may be received on the surface of the airport, but local conditions may limit the maximum ATIS reception distance and/or altitude. Pilots are urged to cooperate in the ATIS program as it relieves frequency congestion on approach control, ground control, and local control frequencies. The A/FD indicates airports for which ATIS is provided.

b. ATIS information includes the time of the latest weather sequence, ceiling, visibility, obstructions to visibility, temperature, dew point (if available), wind direction (magnetic), and velocity, altimeter, other pertinent remarks, instrument approach and runway in use. The ceiling/sky condition, visibility, and obstructions to vision may be omitted from the ATIS broadcast if the ceiling is above 5,000 feet and the visibility is more than 5 miles. The departure runway will only be given if different from the landing runway except at locations having a separate ATIS for departure. The broadcast may include the appropriate frequency and instructions for VFR arrivals to make initial contact with approach control. Pilots of aircraft arriving or departing the terminal area can receive the continuous ATIS broadcast at times when cockpit duties are least pressing and listen to as many repeats as desired. ATIS broadcast shall be updated upon the receipt of any official hourly and special weather. A new recording will also be made when there is a change in other pertinent data such as runway change, instrument approach in use, etc.

EXAMPLE—DULLES INTERNATIONAL INFORMATION SIERRA. 1300 ZULU WEATHER. MEASURED CEILING THREE THOUSAND OVERCAST. VISIBILITY THREE, SMOKE. TEMPERATURE SIX EIGHT. WIND THREE FIVE ZERO AT EIGHT. ALTIMETER TWO NINER NINER TWO. ILS RUNWAY ONE RIGHT APPROACH IN USE. LANDING RUNWAY ONE RIGHT AND LEFT. DEPARTURE RUNWAY THREE ZERO. ARMEL VORTAC OUT OF SERVICE. ADVISE YOU HAVE SIERRA.

c. Pilots should listen to ATIS broadcasts whenever ATIS is in operation.

d. Pilots should notify controllers on initial contact that they have received the ATIS broadcast by repeating the alphabetical code word appended to the broadcast.

EXAMPLE—"INFORMATION SIERRA RECEIVED."

e. When a pilot acknowledges receipt of the ATIS broadcast, controllers may omit those items contained in the broadcast if they are current. Rapidly changing conditions will be issued by ATC and the ATIS will contain words as follows:

EXAMPLE—"LATEST CEILING/VISIBILITY/ALTIMETER/ WIND/OTHER CONDITIONS) WILL BE ISSUED BY APPROACH CONTROL/TOWER."

NOTE—THE ABSENCE OF A SKY CONDITION OR CEILING AND/OR VISIBILITY ON ATIS INDICATES A SKY CONDITION OR CEILING OF 5,000 FEET OR ABOVE AND VISIBILITY OF 5 MILES OR MORE. A REMARK MAY BE MADE ON THE BROADCAST, "THE WEATHER IS BETTER THAN 5000 AND 5," OR THE EXISTING WEATHER MAY BE BROADCAST.

f. Controllers will issue pertinent information to pilots who do not acknowledge receipt of a broadcast or who acknowledge receipt of a broadcast which is not current.

g. To serve frequency limited aircraft, FSSs are equipped to transmit on the omnirange frequency at most en route VORs used as ATIS voice outlets. Such communication interrupts the ATIS broadcast. Pilots of aircraft equipped to receive on other FSS frequencies are encouraged to do so in order that these override transmissions may be kept to an absolute minimum.

h. While it is a good operating practice for pilots to make use of the ATIS broadcast where it is available, some pilots use the phrase "Have Numbers" in communications with the control tower. Use of this phrase means that the pilot has received wind, runway, and altimeter information ONLY and the tower does not have to repeat this information. It does not indicate receipt of the ATIS broadcast and should never be used for this purpose.

4-1-14. RADAR TRAFFIC INFORMATION SERVICE

This is a service provided by radar ATC facilities. Pilots receiving this service are advised of any radar target observed on the radar display which may be in such proximity to the position of their aircraft or its intended route of flight that it warrants their attention. This service is not intended to relieve the pilot of the responsibility for continual vigilance to see and avoid other aircraft.

a. Purpose of the Service:

1. The issuance of traffic information as observed on a radar display is based on the principle of assisting and advising a pilot that a particular radar target's position and track indicates it may intersect or pass in such proximity to that pilot's intended flight path that it warrants attention. This is to alert the pilot to the traffic, to be on the lookout for it, and thereby be in a better position to take appropriate action should the need arise.

2. Pilots are reminded that the surveillance radar used by ATC does not provide altitude information unless the aircraft is equipped with MODE C and the Radar Facility is capable of displaying altitude information.

b. Provisions of the Service:

1. Many factors, such as limitations of the radar, volume of traffic, controller workload and communications frequency congestion, could prevent the controller from providing this service. Controllers possess complete discretion for determining whether they are able to provide or continue to provide this service in a specific case. The controller's reason against providing or continuing to provide the service in a particular case is not subject to question nor need it be communicated to the pilot. In other words, the provision of this service is entirely dependent upon whether controllers believe they are in a position to provide it. Traffic information is routinely provided to all aircraft operating on IFR Flight Plans except when the pilot declines the service, or the pilot is operating within Class A airspace. Traffic information may be provided to flights not operating on IFR Flight Plans when requested by pilots of such flights.

NOTE—RADAR ATC FACILITIES NORMALLY DISPLAY AND MONITOR BOTH PRIMARY AND SECONDARY RADAR WHEN IT IS AVAILABLE, EXCEPT THAT SECONDARY RADAR MAY BE USED AS THE SOLE DISPLAY SOURCE IN CLASS A AIRSPACE, AND UNDER SOME CIRCUMSTANCES OUTSIDE OF CLASS A AIRSPACE (BEYOND PRIMARY COVERAGE AND IN EN ROUTE AREAS WHERE ONLY SECONDARY IS AVAILABLE). SECONDARY RADAR MAY ALSO BE USED OUTSIDE CLASS A AIRSPACE AS THE SOLE DISPLAY SOURCE WHEN THE PRIMARY RADAR IS TEMPORARILY UNUSABLE OR OUT OF SERVICE. PILOTS IN CONTACT WITH THE AFFECTED ATC FACILITY ARE NORMALLY ADVISED WHEN A TEMPORARY OUTAGE OCCURS; I.E., "PRIMARY RADAR OUT OF SERVICE; TRAFFIC ADVISORIES AVAILABLE ON TRANSPONDER AIRCRAFT ONLY." THIS MEANS SIMPLY THAT ONLY THE AIRCRAFT WHICH HAVE TRANSPONDERS INSTALLED AND IN USE WILL BE DEPICTED ON ATC RADAR INDICATORS WHEN THE PRIMARY RADAR IS TEMPORARILY OUT OF SERVICE.

2. When receiving VFR radar advisory service, pilots should monitor the assigned frequency at all times. This is to preclude controllers' concern for radio failure or emergency assistance to aircraft under the controller's jurisdiction. VFR radar advisory service does not include vectors away from conflicting traffic unless requested by the pilot. When advisory service is no longer desired, advise the controller before changing frequencies and then change your transponder code to 1200, if applicable. Pilots should also inform the controller when changing VFR cruising altitude. Except in programs where radar service is automatically terminated, the controller will advise the aircraft when radar is terminated.

NOTE—PARTICIPATION BY VFR PILOTS IN FORMAL PROGRAMS IMPLEMENTED AT CERTAIN TERMINAL LOCATIONS CONSTITUTES PILOT REQUEST. THIS ALSO APPLIES TO PARTICIPATING PILOTS AT THOSE LOCATIONS WHERE ARRIVING VFR FLIGHTS ARE ENCOURAGED TO MAKE THEIR FIRST CONTACT WITH THE TOWER ON THE APPROACH CONTROL FREQUENCY.

c. Issuance of Traffic Information: Traffic information will include the following concerning a target which may constitute traffic for an aircraft that is:

1. *Radar identified.*

(a) Azimuth from the aircraft in terms of the 12 hour clock, or

(b) When rapidly maneuvering civil test or military aircraft prevent accurate issuance of traffic as in (a) above,

specify the direction from an aircraft's position in terms of the eight cardinal compass points (N, NE, E, SE, S, SW, W, NW). This method shall be terminated at the pilot's request.

(c) Distance from the aircraft in nautical miles;

(d) Direction in which the target is proceeding; and

(e) Type of aircraft and altitude if known.

EXAMPLE—TRAFFIC 10 O'CLOCK, 3 MILES, WESTBOUND (TYPE AIRCRAFT AND ALTITUDE, IF KNOWN, OF THE OBSERVED TRAFFIC). THE ALTITUDE MAY BE KNOWN, BY MEANS OF MODE C, BUT NOT VERIFIED WITH THE PILOT FOR ACCURACY. (TO BE VALID FOR SEPARATION PURPOSES BY ATC, THE ACCURACY OF MODE C READOUTS MUST BE VERIFIED. THIS IS USUALLY ACCOMPLISHED UPON INITIAL ENTRY INTO THE RADAR SYSTEM BY A COMPARISON OF THE READOUT TO PILOT STATED ALTITUDE, OR THE FIELD ELEVATION IN THE CASE OF CONTINUOUS READOUT BEING RECEIVED FROM AN AIRCRAFT ON THE AIRPORT.) WHEN NECESSARY TO ISSUE TRAFFIC ADVISORIES CONTAINING UNVERIFIED ALTITUDE INFORMATION, THE CONTROLLER WILL ISSUE THE ADVISORY IN THE SAME MANNER AS IF IT WERE VERIFIED DUE TO THE ACCURACY OF THESE READOUTS. THE PILOT MAY UPON RECEIPT OF TRAFFIC INFORMATION, REQUEST A VECTOR (HEADING) TO AVOID SUCH TRAFFIC. THE VECTOR WILL BE PROVIDED TO THE EXTENT POSSIBLE AS DETERMINED BY THE CONTROLLER PROVIDED THE AIRCRAFT TO BE VECTORED IS WITHIN THE AIRSPACE UNDER THE JURISDICTION OF THE CONTROLLER.

2. *Not radar identified.*

(a) Distance and direction with respect to a fix;

(b) Direction in which the target is proceeding; and

(c) Type of aircraft and altitude if known.

EXAMPLE—TRAFFIC 8 MILES SOUTH OF THE AIRPORT NORTHEASTBOUND, (TYPE AIRCRAFT AND ALTITUDE IF KNOWN).

d. The examples depicted in the following figures point out the possible error in the position of this traffic when it is necessary for a pilot to apply drift correction to maintain this track. This error could also occur in the event a change in course is made at the time radar traffic information is issued.

EXAMPLE—IN FIG 4-1-1 TRAFFIC INFORMATION WOULD BE ISSUED TO THE PILOT OF AIRCRAFT "A" AS 12 O'CLOCK. THE ACTUAL POSITION OF THE TRAFFIC AS SEEN BY THE PILOT OF AIRCRAFT "A" WOULD BE 2 O'CLOCK. TRAFFIC INFORMATION ISSUED TO AIRCRAFT "B" WOULD ALSO BE GIVEN AS 12 O'CLOCK, BUT IN THIS CASE, THE PILOT OF "B" WOULD SEE THE TRAFFIC AT 10 O'CLOCK.

EXAMPLE—IN FIG 4-1-2 TRAFFIC INFORMATION WOULD BE ISSUED TO THE PILOT OF AIRCRAFT "C" AS 2 O'CLOCK. THE ACTUAL POSITION OF THE TRAFFIC AS SEEN BY THE PILOT OF AIRCRAFT "C" WOULD BE 3 O'CLOCK. TRAFFIC INFORMATION ISSUED TO AIRCRAFT "D" WOULD BE AT AN 11 O'CLOCK POSITION. SINCE IT IS NOT NECESSARY FOR THE PILOT OF AIRCRAFT "D" TO APPLY WIND CORRECTION (CRAB) TO REMAIN ON TRACK, THE ACTUAL POSITION OF THE TRAFFIC ISSUED WOULD BE CORRECT. SINCE THE RADAR CONTROLLER CAN ONLY OBSERVE AIRCRAFT TRACK (COURSE) ON THE

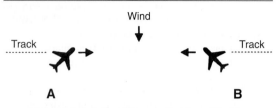

Figure 4-1-1. Induced Error in Position of Traffic

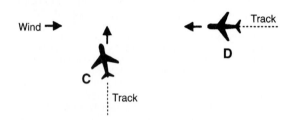

Figure 4-1-2. Induced Error in Position of Traffic

RADAR DISPLAY, TRAFFIC ADVISORIES ARE ISSUED ACCORDINGLY, AND PILOTS SHOULD GIVE DUE CONSIDERATION TO THIS FACT WHEN LOOKING FOR REPORTED TRAFFIC.

4-1-15. SAFETY ALERT

A safety alert will be issued to pilots of aircraft being controlled by ATC if the controller is aware the aircraft is at an altitude which, in the controller's judgment, places the aircraft in unsafe proximity to terrain, obstructions or other aircraft. The provision of this service is contingent upon the capability of the controller to have an awareness of a situation involving unsafe proximity to terrain, obstructions and uncontrolled aircraft. The issuance of a safety alert cannot be mandated, but it can be expected on a reasonable, though intermittent basis. Once the alert is issued, it is solely the pilot's prerogative to determine what course of action, if any, to take. This procedure is intended for use in time critical situations where aircraft safety is in question. Noncritical situations should be handled via the normal traffic alert procedures.

a. Terrain or Obstruction Alert:

1. Controllers will immediately issue an alert to the pilot of an aircraft under their control when they recognize that the aircraft is at an altitude which, in their judgment, may be in an unsafe proximity to terrain/obstructions. The primary method of detecting unsafe proximity is through MODE C automatic altitude reports.

EXAMPLE—LOW ALTITUDE ALERT, CHECK YOUR ALTITUDE IMMEDIATELY. THE, AS APPROPRIATE, MEA/MVA/MOCA IN YOUR AREA IS (ALTITUDE) OR, IF PAST THE FINAL APPROACH FIX (NONPRECISION APPROACH) OR THE OUTER MARKER OR FIX USED IN LIEU

OF THE OUTER MARKER (PRECISION APPROACH), THE, AS APPROPRIATE, MDA/DH (IF KNOWN) IS (ALTITUDE).

2. Terminal Automated Radar Terminal System (ARTS) IIIA, Common ARTS (to include ARTS IIIE and ARTS IIE) (CARTS), Micro En Route Automated Radar Tracking System (MEARTS), and Standard Terminal Automation Replacement System (STARS) facilities have an automated function which, if operating, alerts controllers when a tracked Mode C equipped aircraft under their control is below or is predicted to be below a predetermined minimum safe altitude. This function, called Minimum Safe Altitude Warning (MSAW), is designed solely as a controller aid in detecting potentially unsafe aircraft proximity to terrain/obstructions. The ARTS IIIA, CARTS, MEARTS, and STARS facility will, when MSAW is operating, provide MSAW monitoring for all aircraft with an operating Mode C altitude encoding transponder that are tracked by the system and are:

(a) Operating on an IFR flight plan, or

(b) Operating VFR and have requested MSAW monitoring.

3. Terminal AN/TPX-42A (number beacon decoder system) facilities have an automated function called Low Altitude Alert System (LAAS). Although not as sophisticated as MSAW, LAAS alerts the controller when a MODE C transponder equipped aircraft operating on an IFR flight plan is below a predetermined Minimum Safe Altitude.

NOTE—PILOTS OPERATING VFR MAY REQUEST MSAW OR LAAS MONITORING IF THEIR AIRCRAFT ARE EQUIPPED WITH MODE C TRANSPONDERS.

EXAMPLE—APACHE THREE THREE PAPA REQUEST MSAW/LAAS.

b. Aircraft Conflict Alert:

1. Controllers will immediately issue an alert to the pilot of an aircraft under their control if they are aware of another aircraft which is not under their control, at an altitude which, in the controller's judgment, places both aircraft in unsafe proximity to each other. With the alert, when feasible, the controller will offer the pilot the position of the traffic if time permits and an alternate course(s) of action. Any alternate course(s) of action the controller may recommend to the pilot will be predicated only on other traffic being worked by the controller.

EXAMPLE—AMERICAN THREE, TRAFFIC ALERT, (POSITION OF TRAFFIC, IF TIME PERMITS), ADVISE YOU TURN RIGHT/LEFT HEADING (DEGREES) AND/OR CLIMB/DESCEND TO (ALTITUDE) IMMEDIATELY

4-1-16. RADAR ASSISTANCE TO VFR AIRCRAFT

a. Radar equipped FAA ATC facilities provide radar assistance and navigation service (vectors) to VFR aircraft provided the aircraft can communicate with the facility, are within radar coverage, and can be radar identified.

b. Pilots should clearly understand that authorization to proceed in accordance with such radar navigational assistance does not constitute authorization for the pilot to violate FARs. In effect, assistance provided is on the basis that navigational guidance information issued is advisory in nature and the job of flying the aircraft safely, remains with the pilot.

c. In many cases, controllers will be unable to determine if flight into instrument conditions will result from their instructions. To avoid possible hazards resulting from being vectored into IFR conditions, pilots should keep controllers advised of the weather conditions in which they are operating and along the course ahead.

d. Radar navigation assistance (vectors) may be initiated by the controller when one of the following conditions exist:

1. The controller suggests the vector and the pilot concurs.

2. A special program has been established and vectoring service has been advertised.

3. In the controller's judgment the vector is necessary for air safety.

e. Radar navigation assistance (vectors) and other radar derived information may be provided in response to pilot requests. Many factors, such as limitations of radar, volume of traffic, communications frequency, congestion, and controller workload could prevent the controller from providing it. Controllers have complete discretion for determining if they are able to provide the service in a particular case. Their decision not to provide the service in a particular case is not subject to question.

4-1-17. TERMINAL RADAR SERVICES FOR VFR AIRCRAFT

a. Basic Radar Service:

1. In addition to the use of radar for the control of IFR aircraft, all commissioned radar facilities provide the following basic radar services for VFR aircraft:

(a) Safety alerts.

(b) Traffic advisories.

(c) Limited radar vectoring (on a workload permitting basis).

(d) Sequencing at locations where procedures have been established for this purpose and/or when covered by a letter of agreement.

NOTE—WHEN THE STAGE SERVICES WERE DEVELOPED, TWO BASIC RADAR SERVICES (TRAFFIC ADVISORIES AND LIMITED VECTORING) WERE IDENTIFIED AS "STAGE I." THIS DEFINITION BECAME UNNECESSARY AND THE TERM "STAGE I" WAS ELIMINATED FROM USE. THE TERM "STAGE II" HAS BEEN ELIMINATED IN CONJUNCTION WITH THE AIRSPACE RECLASSIFICATION, AND SEQUENCING SERVICES TO LOCATIONS WITH LOCAL PROCEDURES AND/OR LETTERS OF AGREEMENT TO PROVIDE THIS SERVICE HAVE BEEN INCLUDED IN BASIC SERVICES TO VFR AIRCRAFT. THESE BASIC SERVICES WILL STILL BE PROVIDED BY ALL TERMINAL RADAR FACILITIES WHETHER THEY INCLUDE CLASS B, CLASS C, CLASS D OR CLASS E AIRSPACE. "STAGE III" SERVICES HAVE BEEN REPLACED WITH "CLASS B" AND "TRSA" SERVICE WHERE APPLICABLE.

2. Vectoring service may be provided when requested by the pilot or with pilot concurrence when suggested by ATC.

3. Pilots of arriving aircraft should contact approach control on the publicized frequency and give their position, altitude, aircraft callsign, type aircraft, radar beacon code (if transponder equipped), destination, and request traffic information.

4. Approach control will issue wind and runway, except when the pilot states "have numbers" or this information is contained in the ATIS broadcast and the pilot states that the current ATIS information has been received. Traffic information is provided on a workload permitting basis. Approach control will specify the time or place at which the pilot is to contact the tower on local control frequency for further landing information. Radar service is automatically terminated and the aircraft need not be advised of termination when an arriving VFR aircraft receiving radar services to a tower-controlled airport where basic radar service is provided has landed, or to all other airports, is instructed to change to tower or advisory frequency. (See FAA Order 7110.65, Air Traffic Control, paragraph 5-1-13, Radar Service Termination.)

5. Sequencing for VFR aircraft is available at certain terminal locations (see locations listed in the Airport/Facility Directory). The purpose of the service is to adjust the flow of arriving VFR and IFR aircraft into the traffic pattern in a safe and orderly manner and to provide radar traffic information to departing VFR aircraft. Pilot participation is urged but is not mandatory. Traffic information is provided on a workload permitting basis. Standard radar separation between VFR or between VFR and IFR aircraft is not provided.

(a) Pilots of arriving VFR aircraft should initiate radio contact on the publicized frequency with approach control when approximately 25 miles from the airport at which sequencing services are being provided. On initial contact by VFR aircraft, approach control will assume that sequencing service is requested. After radar contact is established, the pilot may use pilot navigation to enter the traffic pattern or, depending on traffic conditions, approach control may provide the pilot with routings or vectors necessary for proper sequencing with other participating VFR and IFR traffic en route to the airport. When a flight is positioned behind a preceding aircraft and the pilot reports having that aircraft in sight, the pilot will be instructed to follow the preceding aircraft. THE ATC INSTRUCTION TO FOLLOW THE PRECEDING AIRCRAFT DOES NOT AUTHORIZE THE PILOT TO COMPLY WITH ANY ATC CLEARANCE OR INSTRUCTION ISSUED TO THE PRECEDING AIRCRAFT. If other "nonparticipating" or "local" aircraft are in the traffic pattern, the tower will issue a landing sequence. Radar service will be continued to the runway. If an arriving aircraft does not want the service, the pilot should state "NEGATIVE RADAR SERVICE" or make a similar comment, on initial contact with approach control.

(b) Pilots of departing VFR aircraft are encouraged to request radar traffic information by notifying ground control on initial contact with their request and proposed direction of flight.

EXAMPLE—XRAY GROUND CONTROL, NOVEMBER ONE EIGHT SIX, CESSNA ONE SEVENTY TWO, READY TO TAXI, VFR SOUTHBOUND AT 2,500, HAVE INFORMATION BRAVO AND REQUEST RADAR TRAFFIC INFORMATION.

NOTE—FOLLOWING TAKEOFF, THE TOWER WILL ADVISE WHEN TO CONTACT DEPARTURE CONTROL.

(c) Pilots of aircraft transiting the area and in radar contact/communication with approach control will receive traffic information on a controller workload permitting basis. Pilots of such aircraft should give their position, altitude, aircraft callsign, aircraft type, radar beacon code (if transponder equipped), destination, and/or route of flight.

b. TRSA Service (Radar Sequencing and Separation Service for VFR Aircraft in a TRSA).

1. This service has been implemented at certain terminal locations. The service is advertised in the Airport/Facility Directory. The purpose of this service is to provide separation between all participating VFR aircraft and all IFR aircraft operating within the airspace defined as the Terminal Radar Service Area (TRSA). Pilot participation is urged but is not mandatory.

2. If any aircraft does not want the service, the pilot should state "NEGATIVE TRSA SERVICE" or make a similar comment, on initial contact with approach control or ground control, as appropriate.

3. TRSAs are depicted on sectional aeronautical charts and listed in the Airport/Facility Directory.

4. While operating within a TRSA, pilots are provided TRSA service and separation as prescribed in this paragraph. In the event of a radar outage, separation and sequencing of VFR aircraft will be suspended as this service is dependent on radar. The pilot will be advised that the service is not available and issued wind, runway information, and the time or place to contact the tower. Traffic information will be provided on a workload permitting basis.

5. Visual separation is used when prevailing conditions permit and it will be applied as follows:

(a) When a VFR flight is positioned behind a preceding aircraft and the pilot reports having that aircraft in sight, the pilot will be instructed by ATC to follow the preceding aircraft. Radar service will be continued to the runway. THE ATC INSTRUCTION TO FOLLOW THE PRECEDING AIRCRAFT DOES NOT AUTHORIZE THE PILOT TO COMPLY WITH ANY ATC CLEARANCE OR INSTRUCTION ISSUED TO THE PRECEDING AIRCRAFT.

(b) If other "nonparticipating" or "local" aircraft are in the traffic pattern, the tower will issue a landing sequence.

(c) Departing VFR aircraft may be asked if they can visually follow a preceding departure out of the TRSA. The pilot will be instructed to follow the other aircraft provided that the pilot can maintain visual contact with that aircraft.

6. VFR aircraft will be separated from VFR/IFR aircraft by one of the following:

(a) 500 feet vertical separation.

(b) Visual separation.

(c) Target resolution (a process to ensure that correlated radar targets do not touch) when using broadband radar systems.

7. Participating pilots operating VFR in a TRSA:

(a) Must maintain an altitude when assigned by ATC unless the altitude assignment is to maintain at or below a specified altitude. ATC may assign altitudes for separation that do not conform to FAR Part 91.159. When the altitude assignment is no longer needed for separation or when leaving the TRSA, the instruction will be broadcast, "RESUME APPROPRIATE VFR ALTITUDES." Pilots must then return to an altitude that conforms to FAR Part 91.159 as soon as practicable.

(b) When not assigned an altitude, the pilot should coordinate with ATC prior to any altitude change.

8. Within the TRSA, traffic information on observed but unidentified targets will, to the extent possible, be provided all IFR and participating VFR aircraft. The pilot will be vectored upon request to avoid the observed traffic, provided the aircraft to be vectored is within the airspace under the jurisdiction of the controller.

9. Departing aircraft should inform ATC of their intended destination and/or route of flight and proposed cruising altitude.

10. ATC will normally advise participating VFR aircraft when leaving the geographical limits of the TRSA. Radar service is not automatically terminated with this advisory unless specifically stated by the controller.

c. Class C Service: This service provides, in addition to basic radar service, approved separation between IFR and VFR aircraft, and sequencing of VFR arrivals to the primary airport.

d. Class B Service: This service provides, in addition to basic radar service, approved separation of aircraft based on IFR, VFR, and/or weight, and sequencing of VFR arrivals to the primary airport(s).

e. PILOT RESPONSIBILITY: THESE SERVICES ARE NOT TO BE INTERPRETED AS RELIEVING PILOTS OF THEIR RESPONSIBILITIES TO SEE AND AVOID OTHER TRAFFIC OPERATING IN BASIC VFR WEATHER CONDITIONS, TO ADJUST THEIR OPERATIONS AND FLIGHT PATH AS NECESSARY TO PRECLUDE SERIOUS WAKE ENCOUNTERS, TO MAINTAIN APPROPRIATE TERRAIN AND OBSTRUCTION CLEARANCE, OR TO REMAIN IN WEATHER CONDITIONS EQUAL TO OR BETTER THAN THE MINIMUMS REQUIRED BY FAR PART 91.155. WHENEVER COMPLIANCE WITH AN ASSIGNED ROUTE, HEADING AND/OR ALTITUDE IS LIKELY TO COMPROMISE PILOT RESPONSIBILITY RESPECTING TERRAIN AND OBSTRUCTION CLEARANCE, VORTEX EXPOSURE, AND WEATHER MINIMUMS, APPROACH CONTROL SHOULD BE SO

ADVISED AND A REVISED CLEARANCE OR IN-STRUCTION OBTAINED.

f. ATC services for VFR aircraft participating in terminal radar services are dependent on air traffic control radar. Services for VFR aircraft are not available during periods of a radar outage and are limited during CENRAP operations. The pilot will be advised when VFR services are limited or not available.

NOTE—CLASS B AND CLASS C AIRSPACE ARE AREAS OF REGULATED AIRSPACE. THE ABSENCE OF ATC RADAR DOES NOT NEGATE THE REQUIREMENT OF AN ATC CLEARANCE TO ENTER CLASS B AIRSPACE OR TWO WAY RADIO CONTACT WITH ATC TO ENTER CLASS C AIRSPACE.

4-1-18. TOWER EN ROUTE CONTROL (TEC)

a. TEC is an ATC program to provide a service to aircraft proceeding to and from metropolitan areas. It links designated Approach Control Areas by a network of identified routes made up of the existing airway structure of the National Airspace System. The FAA initiated an expanded TEC program to include as many facilities as possible. The program's intent is to provide an overflow resource in the low altitude system which would enhance ATC services. A few facilities have historically allowed turbojets to proceed between certain city pairs, such as Milwaukee and Chicago, via tower en route and these locations may continue this service. However, the expanded TEC program will be applied, generally, for nonturbojet aircraft operating at and below 10,000 feet. The program is entirely within the approach control airspace of multiple terminal facilities. Essentially, it is for relatively short flights. Participating pilots are encouraged to use TEC for flights of two hours duration or less. If longer flights are planned, extensive coordination may be required within the multiple complex which could result in unanticipated delays.

b. Pilots requesting TEC are subject to the same delay factor at the destination airport as other aircraft in the ATC system. In addition, departure and en route delays may occur depending upon individual facility workload. When a major metropolitan airport is incurring significant delays, pilots in the TEC program may want to consider an alternative airport experiencing no delay.

c. There are no unique requirements upon pilots to use the TEC program. Normal flight plan filing procedures will ensure proper flight plan processing. Pilots should include the acronym "TEC" in the remarks section of the flight plan when requesting tower en route control.

d. All approach controls in the system may not operate up to the maximum TEC altitude of 10,000 feet. IFR flight may be planned to any satellite airport in proximity to the major primary airport via the same routing.

4-1-19. TRANSPONDER OPERATION

a. GENERAL

1. Pilots should be aware that proper application of transponder operating procedures will provide both VFR and IFR aircraft with a higher degree of safety in the environment where high-speed closure rates are possible. Transponders substantially increase the capability of radar to see an aircraft and the MODE C feature enables the controller to quickly determine where potential traffic conflicts may exist. Even VFR pilots who are not in contact with ATC will be afforded greater protection from IFR aircraft and VFR aircraft which are receiving traffic advisories. Nevertheless, pilots should never relax their visual scanning vigilance for other aircraft.

2. Air Traffic Control Radar Beacon System (ATCRBS) is similar to and compatible with military coded radar beacon equipment. Civil MODE A is identical to military MODE 3.

3. Civil and military transponders should be adjusted to the "on" or normal operating position as late as practicable prior to takeoff and to "off" or "standby" as soon as practicable after completing landing roll, unless the change to "standby" has been accomplished previously at the request of ATC. IN ALL CASES, WHILE IN CONTROLLED AIRSPACE EACH PILOT OPERATING AN AIRCRAFT EQUIPPED WITH AN OPERABLE ATC TRANSPONDER MAINTAINED IN ACCORDANCE WITH FAR PART 91.413 SHALL OPERATE THE TRANSPONDER, INCLUDING MODE C IF INSTALLED, ON THE APPROPRIATE CODE OR AS ASSIGNED BY ATC. IN CLASS G AIRSPACE, THE TRANSPONDER SHOULD BE OPERATING WHILE AIRBORNE UNLESS OTHERWISE REQUESTED BY ATC.

4. A pilot on an IFR flight who elects to cancel the IFR flight plan prior to reaching destination, should adjust the transponder according to VFR operations.

5. If entering a U.S. OFFSHORE AIRSPACE AREA from outside the U.S., the pilot should advise on first radio contact with a U.S. radar ATC facility that such equipment is available by adding "transponder" to the aircraft identification.

6. It should be noted by all users of ATC transponders that the coverage they can expect is limited to "line of sight." Low altitude or aircraft antenna shielding by the aircraft itself may result in reduced range. Range can be improved by climbing to a higher altitude. It may be possible to minimize antenna shielding by locating the antenna where dead spots are only noticed during abnormal flight attitudes.

7. If operating at an airport with Airport Surface Detection Equipment—Model X (ASFE-X), transponders should be transmitting "on" with altitude reporting continuously while moving on the airport surface if so equipped.

b. TRANSPONDER CODE DESIGNATION

1. For ATC to utilize one or a combination of the 4096 discrete codes FOUR DIGIT CODE DESIGNATION will be used, e.g., code 2100 will be expressed as TWO ONE ZERO ZERO. Due to the operational characteristics of the rapidly expanding automated air traffic control system, THE LAST TWO DIGITS OF THE SELECTED TRAN-SPONDER CODE SHOULD ALWAYS READ "00" UNLESS SPECIFICALLY REQUESTED BY ATC TO BE OTHERWISE.

c. AUTOMATIC ALTITUDE REPORTING (MODE C)

1. Some transponders are equipped with a MODE C automatic altitude reporting capability. This system converts aircraft altitude in 100-foot increments to coded digital information which is transmitted together with MODE C framing pulses to the interrogating radar facility. The manner in which transponder panels are designed differs, therefore, a pilot should be thoroughly familiar with the operation of the transponder so that ATC may realize its full capabilities.

2. Adjust transponder to reply on the MODE A/3 code specified by ATC and, if equipped, to reply on MODE C with altitude reporting capability activated unless deactivation is directed by ATC or unless the installed aircraft equipment has not been tested and calibrated as required by FAR Part 91.217. If deactivation is required by ATC, turn off the altitude reporting feature of your transponder. An instruction by ATC to "STOP ALTITUDE SQUAWK, ALTITUDE DIFFERS (number of feet) FEET," may be an indication that your transponder is transmitting incorrect altitude information or that you have an incorrect altimeter setting. While an incorrect altimeter setting has no effect on the MODE C altitude information transmitted by your transponder (transponders are preset at 29.92), it would cause you to fly at an actual altitude different from your assigned altitude. When a controller indicates that an altitude readout is invalid, the pilot should initiate a check to verify that the aircraft altimeter is set correctly.

3. Pilots of aircraft with operating MODE C altitude reporting transponders should report exact altitude or Flight Level to the nearest hundred foot increment when establishing initial contact with an ATC facility. Exact altitude or flight level reports on initial contact provide ATC with information that is required prior to using MODE C altitude information for separation purposes. This will significantly reduce altitude verification requests.

d. TRANSPONDER IDENT FEATURE

1. The transponder shall be operated only as specified by ATC. Activate the "IDENT" feature only upon request of the ATC controller.

e. CODE CHANGES

1. When making routine code changes, pilots should avoid inadvertent selection of codes 7500, 7600 or 7700 thereby causing momentary false alarms at automated ground facilities. For example when switching from code 2700 to code 7200, switch first to 2200 then to 7200, NOT to 7700 and then 7200. This procedure applies to nondiscrete code 7500 and all discrete codes in the 7600 and 7700 series (i.e. 7600–7677, 7700–7777) which will trigger special indicators in automated facilities. Only nondiscrete code 7500 will be decoded as the hijack code.

2. Under no circumstances should a pilot of a civil aircraft operate the transponder on Code 7777. This code is reserved for military interceptor operations.

3. Military pilots operating VFR or IFR within restricted/warning areas should adjust their transponders to code 4000 unless another code has been assigned by ATC.

f. MODE C TRANSPONDER REQUIREMENTS

1. Specific details concerning requirements to carry and operate Mode C transponders, as well as exceptions and ATC authorized deviations from the requirements are found in FAR Part 91.215 and FAR Part 99.12.

2. In general, the FAR requires aircraft to be equipped with Mode C transponders when operating:

(a) at or above 10,000 feet MSL over the 48 contiguous states or the District of Columbia, excluding that airspace below 2,500 feet AGL;

(b) within 30 miles of a Class B airspace primary airport, below 10,000 feet MSL. Balloons, gliders, and aircraft not equipped with an engine driven electrical system are excepted from the above requirements when operating below the floor of Class A airspace and/or; outside of a Class B airspace and below the ceiling of the Class B Airspace (or 10,000 feet MSL, whichever is lower);

(c) within and above all Class C airspace, up to 10,000 feet MSL;

(d) within 10 miles of certain designated airports, excluding that airspace which is both outside the Class D surface area and below 1,200 feet AGL. Balloons, gliders and aircraft not equipped with an engine driven electrical system are excepted from this requirement.

3. FAR Part 99.12 requires all aircraft flying into, within, or across the contiguous U.S. ADIZ be equipped with a Mode C or Mode S transponder. Balloons, gliders and aircraft not equipped with an engine driven electrical system are excepted from this requirement.

4. Pilots shall ensure that their aircraft transponder is operating on an appropriate ATC assigned VFR/IFR code and MODE C when operating in such airspace. If in doubt about the operational status of either feature of your transponder while airborne, contact the nearest ATC facility or FSS and they will advise you what facility you should contact for determining the status of your equipment.

5. In-flight requests for "immediate" deviation from the transponder requirement may be approved by controllers only when the flight will continue IFR or when weather conditions prevent VFR descent and continued VFR flight in airspace not affected by the FAR. All other requests for deviation should be made by contacting the nearest Flight Service or Air Traffic facility in person or by telephone. The nearest ARTCC will normally be the controlling agency and is responsible for coordinating requests involving deviations in other ARTCC areas.

g. TRANSPONDER OPERATION UNDER VISUAL FLIGHT RULES (VFR)

1. Unless otherwise instructed by an Air Traffic Control Facility, adjust Transponder to reply on MODE 3/A code 1200 regardless of altitude.

2. Adjust transponder to reply on MODE C, with altitude reporting *capability activated* if the aircraft is so equipped, unless deactivation is directed by ATC or unless the installed equipment has not been tested and calibrated as required by FAR Part 91.217. If deactivation is required and your transponder is so designed, turn off the altitude reporting switch and continue to transmit MODE C framing pulses. If this capability does not exist, turn off MODE C.

h. RADAR BEACON PHRASEOLOGY

Air traffic controllers, both civil and military, will use the following phraseology when referring to operation of the Air Traffic Control Radar Beacon System (ATCRBS). Instructions by ATC refer only to MODE A/3 or MODE C operation and do not affect the operation of the transponder on other modes.

1. *SQUAWK (number)*: Operate radar beacon transponder on designated code in MODE A/3.

2. *IDENT:* Engage the "IDENT" feature (military I/P) of the transponder.

3. *SQUAWK (number) and IDENT:* Operate transponder on specified code in MODE A/3 and engage the "IDENT" (military I/P) feature.

4. *SQUAWK STANDBY:* Switch transponder to standby position.

5. *SQUAWK LOW/NORMAL:* Operate transponder on low or normal sensitivity as specified. Transponder is operated in "NORMAL" position unless ATC specifies "LOW" ("ON" is used instead of "NORMAL" as a master control label on some types of transponders.)

6. *SQUAWK ALTITUDE:* Activate MODE C with automatic altitude reporting.

7. *STOP ALTITUDE SQUAWK:* Turn off altitude reporting switch and continue transmitting MODE C framing pulses. If your equipment does not have this capability, turn off MODE C.

8. *STOP SQUAWK (mode in use):* Switch off specified mode. (Used for military aircraft when the controller is unaware of military service requirements for the aircraft to continue operation on another MODE.)

9. *STOP SQUAWK:* Switch off transponder.

10. *SQUAWK MAYDAY:* Operate transponder in the emergency position (MODE A Code 7700 for civil transponder. MODE 3 Code 7700 and emergency feature for military transponder.)

11. *SQUAWK VFR:* Operate radar beacon transponder on code 1200 in the MODE A/3, or other appropriate VFR code.

4-1-20. HAZARDOUS AREA REPORTING SERVICE

a. Selected FSSs provide flight monitoring where regularly traveled VFR routes cross large bodies of water, swamps, and mountains. This service is provided for the purpose of expeditiously alerting Search and Rescue facilities when required. (See FIG 4-1-3.)

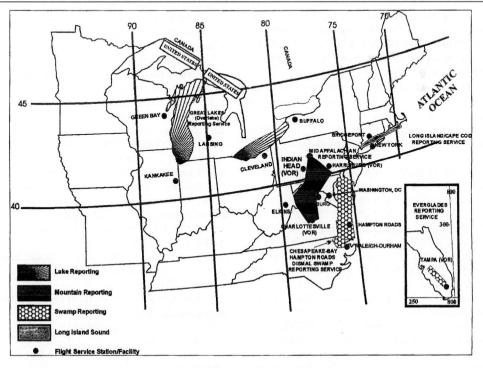

Figure 4-1-3. Hazardous Area Reporting Service

1. When requesting the service either in person, by telephone or by radio, pilots should be prepared to give the following information: type of aircraft, altitude, indicated airspeed, present position, route of flight, heading.

2. Radio contacts are desired at least every 10 minutes. If contact is lost for more than 15 minutes, Search and Rescue will be alerted. Pilots are responsible for canceling their request for service when they are outside the service area boundary. Pilots experiencing two-way radio failure are expected to land as soon as practicable and cancel their request for the service. FIG 4-1-3 depicts the areas and the FSS facilities involved in this program.

b. LONG ISLAND SOUND REPORTING SERVICE:
The New York and Bridgeport AFSSs provide Long Island Sound Reporting service on request for aircraft traversing Long Island Sound.

1. When requesting the service pilots should ask for SOUND REPORTING SERVICE and should be prepared to provide the following appropriate information:

(a) type and color of aircraft,

(b) the specific route and altitude across the sound including the shore crossing point,

(c) the overwater crossing time,

(d) number of persons on board,

(e) true airspeed.

2. Radio contacts are desired at least every 10 minutes; however, for flights of shorter duration a midsound report is requested. If contact is lost for more than 15 minutes Search and Rescue will be alerted. Pilots are responsible for canceling their request for the Long Island Sound Reporting Service when outside the service area boundary. Aircraft experiencing radio failure will be expected to land as soon as practicable and cancel their request for the service.

3. *COMMUNICATIONS:* Primary communications—pilots are to transmit on 122.1 MHz and listen on one of the following VOR frequencies:

(a) *NEW YORK AFSS CONTROLS:*

(1) Hampton RCO (FSS transmits and receives on 122.6 MHz).

(2) Calverton VORTAC VOR (FSS transmits on 117.2 and receives on standard FSS frequencies).

(3) Kennedy VORTAC (FSS transmits on 115.9 and receives on 122.1 MHz).

(b) *BRIDGEPORT AFSS CONTROLS:*

(1) Madison VORTAC (FSS transmits on 110.4 and receives on 122.1 MHz).

(2) Groton VOR (FSS transmits on 110.85 and receives on 122.1 MHz).

(3) Bridgeport VOR (FSS transmits on 108.8 and receives on 122.1 MHz).

c. BLOCK ISLAND REPORTING SERVICE:
Within the Long Island Reporting Service, the New York FSS also provides an additional service for aircraft operating between Montauk Point and Block Island. When requesting this service, pilots should ask for BLOCK

ISLAND REPORTING SERVICE and should be prepared to provide the same flight information as required for the Long Island Sound Reporting Service.

1. A minimum of three position reports are mandatory for this service; these are:

(a) Reporting leaving either Montauk Point or Block Island.

(b) Midway report.

(c) Report when over either Montauk Point or Block Island. At this time, the overwater service is canceled.

2. *COMMUNICATIONS:* Pilots are to transmit and receive on 122.6 MHz.

NOTE—PILOTS ARE ADVISED THAT 122.6 MHZ IS A REMOTE RECEIVER LOCATED AT THE HAMPTON VORTAC SITE AND DESIGNED TO PROVIDE RADIO COVERAGE BETWEEN HAMPTON AND BLOCK ISLAND. FLIGHTS PROCEEDING BEYOND BLOCK ISLAND MAY CONTACT THE BRIDGEPORT AFSS BY TRANSMITTING ON 122.1 MHZ AND LISTENING ON GROTON VOR (TMU) FREQUENCY 111.85 MHZ.

d. CAPE COD AND ISLANDS RADAR OVERWATER FLIGHT FOLLOWING:
In addition to normal VFR radar advisory services, traffic permitting, Cape Approach Control provides a radar overwater flight following service for aircraft traversing the Cape Cod and adjacent Island area. Pilots desiring this service may contact Cape RAPCON on 118.2 MHz.

1. Pilots requesting this service should be prepared to give the following information:

(a) type and color of aircraft,

(b) altitude,

(c) position and heading,

(d) route of flight, and

(e) true airspeed.

2. For best radar coverage, pilots are encouraged to fly at 1,500 feet MSL or above.

3. Pilots are responsible for canceling their request for overwater flight following when they are over the mainland and/or outside the service area boundary.

e. LAKE REPORTING SERVICE:
Cleveland and Lansing AFSSs provide Lake Reporting Service on request for aircraft traversing the western half of Lake Erie; Green Bay, Kankakee, Lansing, and Terre Haute AFSSs provide Lake Reporting Service on request for aircraft traversing Lake Michigan.

1. When requesting the service, pilots should ask for LAKE REPORTING SERVICE.

2. Pilots not on a VFR flight plan should be prepared to provide all information that is normally provided for a complete VFR flight plan.

3. Pilots already on a VFR flight plan should be prepared to provide the following information:

(a) Aircraft or flight identification.

(b) Type of aircraft.

(c) Near-shore crossing point or last fix before crossing.

(d) Proposed time over near-shore crossing point or last fix before crossing.

(e) Proposed altitude.

(f) Proposed route of flight.

(g) Estimated time over water.

(h) Next landing point.

(i) AFSS/FSS having complete VFR flight plan information.

4. Radio contacts must not exceed 10 minutes when pilots fly at an altitude that affords continuous communications. If radio contact is lost for more than 15 minutes (5 minutes after a scheduled reporting time), Search and Rescue (SAR) will be alerted.

5. The estimated time for crossing the far shore will be the scheduled reporting time for aircraft that fly at an altitude that does not afford continuous communication coverage while crossing the lake. If radio contact is not established within 5 minutes of that time, SAR will be alerted.

6. Pilots are responsible for canceling their request for Lake Reporting Service when outside the service area boundary. Aircraft experiencing radio failure will be expected to land as soon as practicable and cancel their Lake Reporting Service flight plan.

7. *COMMUNICATIONS:* Primary communications— Pilots should communicate with the following facilities on the indicated frequencies:

(a) *CLEVELAND AFSS CONTROLS:*

(1) Cleveland RCO (FSS transmits and receives on 122.35 or 122.55 MHz).

(2) Sandusky VOR (FSS transmits on 109.2 and receives on 122.1 MHz).

(b) *GREEN BAY AFSS CONTROLS:*

(1) Escanaba VORTAC (FSS transmits on 110.8 and receives on 122.1 MHz).

(2) Green Bay RCO (FSS transmits and receives on 122.55 MHz).

(3) Manistique RCO (FSS transmits and receives on 122.25 MHz).

(4) Manitowoc VOR (FSS transmits on 111.0 and receives on 122.1 MHz).

(5) Menominee VOR (FSS transmits on 109.6 and receives on 122.1 MHz).

(6) Milwaukee RCO (FSS transmits and receives on 122.65 MHz).

(7) Falls VOR (FSS transmits on 110.0 and receives on 122.1 MHz).

(c) *KANKAKEE AFSS CONTROLS:*

(1) Chicago Heights VORTAC (FSS transmits on 114.2 and receives on 122.1 MHz).

(2) Meigs RCO (FSS transmits and receives on 122.15 MHz).

(3) Waukegan RCO transmits and receives on 122.55 MHz).

(d) *LANSING AFSS CONTROLS:*

(1) *LAKE ERIE:* Detroit City RCO (FSS transmits and receives on 122.55 MHz).

(2) *LAKE MICHIGAN:*

(a) Keeler VORTAC (FSS transmits on 116.6 and receives on 122.1 MHz).

(b) Ludington RCO (FSS transmits and receives on 122.45 MHz).

(c) Manistee VORTAC (FSS transmits on 111.4 and receives on 122.1 MHz).

(d) Muskegon RCO (FSS transmits and receives on 122.5 MHz).

(e) Pellston RCO (FSS transmits and receives on 122.3 MHz).

(f) Pullman VORTAC (FSS transmits on 112.1 and receives on 122.1 MHz).

(g) Traverse City RCO (FSS transmits and receives on 122.6 MHz).

(e) *TERRE HAUTE AFSS CONTROLS:*

(1) South Bend RCOS (FSS transmits and receives on 123.65/primary and 122.6/secondary MHz).

f. Everglades Reporting Service.

This service is offered by Miami Automated International Flight Service Station (MIA AIFSS), in extreme southern Florida. The service is provided to aircraft crossing the Florida Everglades, between Lee County (Ft. Myers, FL) VORTAC (RSW) on the northwest side, and Dolphin (Miami, FL) VOR (DHP) on the southeast side.

1. The pilot must request the service from Miami AIFSS.

2. MIA AIFSS frequency information, 122.2, 122.3, and 122.65.

3. The pilot must file a VFR flight plan with the remark: ERS.

4. The pilot must maintain 2000 feet of altitude.

5. The pilot must make position reports every ten (10) minutes. SAR begins fifteen (15) minutes after position report is not made on time.

6. The pilot is expected to land as soon as is practical, in the event of two-way radio failure, and advise MIA AIFSS that the service is terminated.

7. The pilot must notify Miami AIFSS when the flight plan is cancelled or the service is suspended.

4-1-21. AIRPORT RESERVATION OPERATIONS AND SPECIAL TRAFFIC MANAGEMENT PROGRAMS

This section describes procedures for obtaining required airport reservations at high density traffic airports and for airports operating under Special Traffic Management Programs.

a. High Density Traffic Airports (HDTA).

1. The FAA, by 14 CFR Part 93, Subpart K, has designated the John F. Kennedy International (JFK), LaGuardia (LGA), Ronald Reagan Washington National (DCA), and Newark International (EWR) Airports as high density airports and has prescribed air traffic rules and requirements for operating aircraft to and from these airports. (The quota for EWR has been suspended indefinitely. Effective July 2, 2002, the slot requirements at ORD were eliminated.) Reservations for

JFK are required between 3:00 p.m. and 7:59 p.m. local time. Reservations for LGA and DCA are required between 6:00 a.m. and 11:59 p.m. local time. Helicopter operations are excluded from the requirement for a reservation.

2. The FAA has established an Airport Reservations Office (ARO) to receive and process all Instrument Flight Rules (IFR) requests for nonscheduled operations at the designated HDTA's. This office monitors operation of the high density rule and allocates reservations on a "first-come-first-served" basis determined by the time the request is received at the reservation office. Standby lists are not maintained. The ARO utilizes the Enhanced Computer Voice Reservation System (e-CVRS) to make all reservations. Users may access the computer system using a touch-tone telephone or via the Internet. Requests for IFR reservations will be accepted starting 72 hours prior to the proposed time of operation at the affected airport.

3. The toll-free telephone number for obtaining IFR reservations through e-CVRS at HDTA's is 1-800-875-9694. This number is valid for calls originating within the United States, Canada, and the Caribbean. The toll number for other areas is (703) 707-0568. The Internet address for the e-CVRS Web interface is: **http://www.fly.faa.gov/ecvrs.**

For more detailed information on operations and reservation procedures at an HDTA, please see Advisory Circular 93-1, Reservations for Unscheduled Operations at High Density Traffic Airports. A copy of the Advisory Circular may be obtained via the Internet at: **http://www.faa.gov.**

b. Special Traffic Management Programs (STMP).

1. Special procedures may be established when a location requires special traffic handling to accommodate above normal traffic demand (e.g., the Indianapolis 500, Super Bowl, etc.) or reduced airport capacity (e.g., airport runway/taxiway closures for airport construction). The special procedures may remain in effect until the problem has been resolved or until local traffic management procedures can handle the situation and a need for special handling no longer exists.

2. There will be two methods available for obtaining slot reservations at the ATCSCC: the web interface and the touch-tone interface. If these methods are used, a NOTAM will be issued relaying the web site address and toll—free telephone number. Be sure to check current NOTAM's to determine: what airports are included in the STMP; the dates and times reservations are required; the time limits for reservation requests; the point of contact for reservations; and any other instructions.

c. Users may contact the ARO at 703-904-4452 if they have a problem making a reservation or have a question concerning the HDTA/STMP regulations or procedures.

d. Making Reservations.

1. Internet Users. Detailed information and User Instruction Guides for using the Web Interface to the reservation systems are available on the websites for the HDTA (e-CVRS) and STMP's (e-STMP).

2. Telephone users. When using the telephone to make a reservation, you are prompted for input of information about what you wish to do. All input is accomplished using the keypad on the telephone. The only problem with a telephone is that most keys have a letter and number associated with them. When the system asks for a date or time, it is expecting an input of numbers. A problem arises when entering an aircraft call sign or tail number. The system does not detect if you are entering a letter (alpha character) or a number. Therefore, when entering an aircraft call sign or tail number two keys are used to represent each letter or number. When entering a number, precede the number you wish by the number 0 (zero) i.e. 01,02,03,04. If you wish to enter a letter, first press the key on which the letter appears and then press 1,2, or 3, depending upon whether the letter you desire is the first, second, or third letter on that key. For example to enter the letter "N" first press the "6" key because "N" is on that key, then press the "2" key because the letter "N" is the second letter on the "6" key. Since there are no keys for the letters "0" and "Z" e-CVRS pretends they are on the number "1" key. Therefore, to enter the letter "Q", press 11, and to enter the letter "Z" press 12.

NOTE—USERS ARE REMINDED TO ENTER THE "N" CHARACTER WITH THEIR TAIL NUMBERS. (SEE TBL 4-1-4).

Table 4-1-4

CODES FOR CALL SIGN TAIL NUMBER INPUT ONLY

A-21	J-51	S-73	1-01
B-22	K-52	T-81	2-02
C-23	L-53	U-82	3-03
D-31	M-61	V-83	4-04
E-32	N-62	W-91	5-05
F-33	O-63	X-92	6-06
G-41	P-71	Y-93	7-07
H-42	Q-11	Z-12	8-08
I-43	R-72	0-00	9-09

3. Additional helpful key entries: (See TBL 4-1-5).

Table 4-1-5 Helpful Key Entries

#	After entering a call sign tail number, depressing the "pound key" (#) twice will indicate the end of the tail number.
*2	Will take the user back to the start of the process.
*3	Will repeat the call sign tail number used in a previous reservation.
*5	Will repeat the previous question.
*8	Tutorial mode: In the tutorial mode each prompt for input includes a more detailed description of what is expected as input. *8 is a toggle on/off switch. If you are in tutorial mode and enter *8, you will return to the normal mode.
*0	Expert mode: In the expert mode each prompt for input is brief with little or no explanation. Expert mode is also an on/off toggle.

4-1-22. REQUESTS FOR WAIVERS AND AUTHORIZATIONS FROM TITLE 14, CODE OF FEDERAL REGULATIONS (14 CFR)

a. Requests for a Certificate of Waiver or Authorization (FAA Form 7711-2), or requests for renewal of a waiver or authorization, may be accepted by any FAA facility and will be forwarded, if necessary, to the appropriate office having waiver authority.

b. The grant of a Certificate of Waiver or Authorization from 14 CFR constitutes relief from specific regulations, to the degree and for the period of time specified in the certificate, and does not waive any state law or local ordinance. Should the proposed operations conflict with any state law or local ordinance, or require permission of local authorities or property owners, it is the applicant's responsibility to resolve the matter. The holder of a waiver is responsible for compliance with the terms of the waiver and its provisions.

c. A waiver may be canceled at any time by the Administrator, the person authorized to grant the waiver, or the representative designated to monitor a specific operation. In such case either written notice of cancellation, or written confirmation of a verbal cancellation will be provided to the holder.

4-1-23. Weather System Processor

The Weather System Processor (WSP) was developed for use in the National Airspace System to provide weather processor enhancements to selected Airport Surveillance Radar (ASR)-9 facilities. The WSP provides Air Traffic with warnings of hazardous wind shear and microbursts. The WSP also provides users with terminal area 6-level weather, storm cell locations and movement, as well as the location and predicted future position and intensity of wind shifts that may affect airport operations.

Section 2. RADIO COMMUNICATIONS PHRASEOLOGY AND TECHNIQUES

4-2-l. GENERAL

a. Radio communications are a critical link in the ATC system. The link can be a strong bond between pilot and controller or it can be broken with surprising speed and disastrous results. Discussion herein provides basic procedures for new pilots and also highlights safe operating concepts for all pilots.

b. The single, most important thought in pilot-controller communications is understanding. It is essential, therefore, that pilots acknowledge each radio communication with ATC by using the appropriate aircraft call sign. Brevity is important, and contacts should be kept as brief as possible, but controllers must know what you want to do before they can properly carry out their control duties. And you, the pilot, must know exactly what the controller wants you to do. Since concise phraseology may not always be adequate, use whatever words are necessary to get your message across. Pilots are to maintain vigilance in monitoring air traffic control radio communications frequencies for potential traffic conflicts with their aircraft especially when operating on an active runway and/or when conducting a final approach to landing.

c. All pilots will find the *Pilot/Controller Glossary* very helpful in learning what certain words or phrases mean. Good phraseology enhances safety and is the mark of a professional pilot. Jargon, chatter, and "CB" slang have no place in ATC communications. The *Pilot/Controller Glossary* is the same glossary used in FAA Order 7110.65, *Air Traffic Control*. We recommend that it be studied and reviewed from time to time to sharpen your communication skills.

4-2-2. RADIO TECHNIQUE

a. Listen before you transmit. Many times you can get the information you want through ATIS or by monitoring the frequency. Except for a few situations where some frequency overlap occurs, if you hear someone else talking, the keying of your transmitter will be futile and you will probably jam their receivers causing them to repeat their call. If you have just changed frequencies, pause, listen, and make sure the frequency is clear.

b. Think before keying your transmitter. Know what you want to say and if it is lengthy; e.g., a flight plan or IFR position report, jot it down.

c. The microphone should be very close to your lips and after pressing the mike button, a slight pause may be necessary to be sure the first word is transmitted. Speak in a normal, conversational tone.

d. When you release the button, wait a few seconds before calling again. The controller or FSS specialist may be jotting down your number, looking for your flight plan, transmitting on a different frequency, or selecting the transmitter for your frequency.

e. Be alert to the sounds *or the lack of sounds* in your receiver. Check your volume, recheck your frequency, and *make sure that your microphone is not stuck* in the transmit position. Frequency blockage can, and has, occurred for extended periods of time due to unintentional transmitter operation. This type of interference is commonly referred to as a "stuck mike," and controllers may refer to it in this manner when attempting to assign an alternate frequency. If the assigned frequency is completely blocked by this type of interference, use the procedures described for en route IFR radio frequency outage to establish or reestablish communications with ATC.

f. Be sure that you are within the performance range of your radio equipment and the ground station equipment. Remote radio sites do not always transmit and receive on all of a facility's available frequencies, particularly with regard to VOR sites where you can hear but not reach a ground station's receiver. Remember that higher altitudes increase the range of VHF "line of sight" communications.

4-2-3. CONTACT PROCEDURES

a. Initial Contact:

1. The terms *initial contact* or *initial callup* means the first radio call you make to a given facility or the first call to a different controller or FSS specialist within a facility. Use the following format:

(a) Name of the facility being called;

(b) Your *full* aircraft identification as filed in the flight plan or as discussed under Aircraft Call Signs below;

(c) When operating on an airport surface, state your position.

(d) The type of message to follow or your request if it is short; and

(e) The word "Over" if required.

EXAMPLE—

[1]"NEW YORK RADIO, MOONEY THREE ONE ONE ECHO."

[2]"COLUMBIA GROUND, CESSNA THREE ONE SIX ZERO FOXTROT, SOUTH RAMP, I-F-R MEMPHIS."

[3]"MIAMI CENTER, BARON FIVE SIX THREE HOTEL REQUEST V-F-R TRAFFIC ADVISORIES."

2. Many FSSs are equipped with RCOs and can transmit on the same frequency at more than one location. The frequencies available at specific locations are indicated on charts above FSS communications boxes. To enable the specialist to utilize the correct transmitter, advise the location and the frequency on which you expect a reply.

EXAMPLE—ST. LOUIS FSS CAN TRANSMIT ON FREQUENCY 122.3 AT EITHER FARMINGTON, MO, OR DECATUR, IL. IF YOU ARE IN THE VICINITY OF DECATUR, YOUR CALLUP SHOULD BE "SAINT LOUIS RADIO, PIPER SIX NINER SIX YANKEE, RECEIVING DECATUR ONE TWO TWO POINT THREE."

3. If radio reception is reasonably assured, inclusion of your request, your position or altitude, and the phrase "(ATIS) Information Charlie received" in the initial contact helps decrease radio frequency congestion. Use discretion; do not overload the controller with information unneeded or superfluous. If you do not get a response from the ground station, recheck your radios or use another transmitter, but keep the next contact short.

EXAMPLE—"ATLANTA CENTER, DUKE FOUR ONE ROMEO, REQUEST V-F-R TRAFFIC ADVISORIES, TWENTY NORTHWEST ROME, SEVEN THOUSAND FIVE HUNDRED, OVER."

b. Initial Contact When your Transmitting and Receiving Frequencies are Different:

1. If you are attempting to establish contact with a ground station and you are receiving on a different frequency than

that transmitted, indicate the VOR name or the frequency on which you expect a reply. Most FSSs and control facilities can transmit on several VOR stations in the area. Use the appropriate FSS call-sign as indicated on charts.

EXAMPLE—NEW YORK FSS TRANSMITS ON THE KENNEDY, THE HAMPTON, AND THE CALVERTON VORTACS. IF YOU ARE IN THE CALVERTON AREA, YOUR CALLUP SHOULD BE "NEW YORK RADIO, CESSNA THREE ONE SIX ZERO FOXTROT, RECEIVING CALVERTON V-O-R, OVER."

2. If the chart indicates FSS frequencies above the VORTAC or in the FSS communications boxes, transmit or receive on those frequencies nearest your location.

3. When unable to establish contact and you wish to call *any* ground station, use the phrase "ANY RADIO (tower) (station), GIVE CESSNA THREE ONE SIX ZERO FOXTROT A CALL ON (frequency) OR (V-O-R)." If an emergency exists or you need assistance, so state.

c. Subsequent Contacts and Responses to Callup from a Ground Facility:

Use the same format as used for the initial contact except you should state your message or request with the callup in one transmission. The ground station name and the word "Over" may be omitted if the message requires an obvious reply and there is no possibility for misunderstandings. *You should acknowledge all callups or clearances* unless the controller or FSS specialist advises otherwise. There are some occasions when controllers must issue time-critical instructions to other aircraft, and they may be in a position to observe your response, either visually or on radar. If the situation demands your response, take appropriate action or immediately advise the facility of any problem. Acknowledge with your aircraft identification, either at the beginning or at the end of your transmission, and one of the words "Wilco," "Roger," "Affirmative," "Negative," or other appropriate remarks; e.g., "PIPER TWO ONE FOUR LIMA, ROGER." If you have been receiving services; e.g., VFR traffic advisories and you are leaving the area or changing frequencies, advise the ATC facility and terminate contact.

d. Acknowledgement of Frequency Changes:

1. When advised by ATC to change frequencies, acknowledge the instruction. If you select the new frequency without an acknowledgement, the controller's workload is increased because there is no way of knowing whether you received the instruction or have had radio communications failure.

2. At times, a controller/specialist may be working a sector with multiple frequency assignments. In order to eliminate unnecessary verbiage and to free the controller/specialist for higher priority transmissions, the controller/specialist may request the pilot "(Identification), change to my frequency 123.4." This phrase should alert the pilot that the controller/specialist is only changing frequencies, not controller/specialist, and that initial callup phraseology may be abbreviated.

EXAMPLE—"UNITED TWO TWENTY-TWO ON ONE TWO THREE POINT FOUR" OR "ONE TWO THREE POINT FOUR, UNITED TWO TWENTY-TWO."

e. Compliance with Frequency Changes:

When instructed by ATC to change frequencies, select the new frequency as soon as possible unless instructed to make the change at a specific time, fix, or altitude. A delay in making the change could result in an untimely receipt of important information. If you are instructed to make the frequency change at a specific time, fix, or altitude, monitor the frequency you are on until reaching the specified time, fix, or altitudes unless instructed otherwise by ATC.

REFERENCE—AIM, ARTCC COMMUNICATIONS, PARAGRAPH 5-3-1.

4-2-4. AIRCRAFT CALL SIGNS

a. Precautions in the Use of Call Signs:

1. Improper use of call signs can result in pilots executing a clearance intended for another aircraft. Call signs should *never be abbreviated on an initial contact or at any time when other aircraft call signs have similar numbers/ sounds or identical letters/number;* e.g., Cessna 6132F, Cessna 1622F, Baron 123F, Cherokee 7732F, etc.

EXAMPLE—ASSUME THAT A CONTROLLER ISSUES AN APPROACH CLEARANCE TO AN AIRCRAFT AT THE BOTTOM OF A HOLDING STACK AND AN AIRCRAFT WITH A SIMILAR CALL SIGN (AT THE TOP OF THE STACK) ACKNOWLEDGES THE CLEARANCE WITH THE LAST TWO OR THREE NUMBERS OF THE AIRCRAFT'S CALL SIGN. IF THE AIRCRAFT AT THE BOTTOM OF THE STACK DID NOT HEAR THE CLEARANCE AND INTERVENE, FLIGHT SAFETY WOULD BE AFFECTED, AND THERE WOULD BE NO REASON FOR EITHER THE CONTROLLER OR PILOT TO SUSPECT THAT ANYTHING IS WRONG. THIS KIND OF "HUMAN FACTORS" ERROR CAN STRIKE SWIFTLY AND IS EXTREMELY DIFFICULT TO RECTIFY.

2. Pilots, therefore, must be certain that aircraft identification is complete and clearly identified before taking action on an ATC clearance. ATC specialists will not abbreviate call signs of air carrier or other civil aircraft having authorized call signs. ATC specialists may initiate abbreviated call signs of other aircraft by using the *prefix and the last three digits/letters* of the aircraft identification after communications are established. The pilot may use the abbreviated call sign in subsequent contacts with the ATC specialist. When aware of similar/identical call signs, ATC specialists will take action to minimize errors by emphasizing certain numbers/letters, by repeating the entire call sign, by repeating the prefix, or by asking pilots to use a different call sign temporarily. Pilots should use the phrase "VERIFY CLEARANCE FOR (your complete call sign)" if doubt exists concerning proper identity.

3. Civil aircraft pilots should state the aircraft type, model or manufacturer's name, followed by the digits/letters of the registration number. When the aircraft manufacturer's name or model is stated, the prefix "N" is dropped; e.g., Aztec Two Four Six Four Alpha.

EXAMPLE—
① BONANZA SIX FIVE FIVE GOLF

② BREEZY SIX ONE THREE ROMEO EXPERIMENTAL (OMIT "EXPERIMENTAL" AFTER INITIAL CONTACT).

4. Air Taxi or other commercial operators *not* having FAA authorized call signs should prefix their normal identification with the phonetic word "Tango."

EXAMPLE—TANGO AZTEC TWO FOUR SIX FOUR ALPHA.

5. Air carriers and commuter air carriers having FAA authorized call signs should identify themselves by stating the complete call sign (using group form for the numbers) and the word "heavy" if appropriate.

EXAMPLE—
① UNITED TWENTY-FIVE HEAVY

② MIDWEST COMMUTER SEVEN ELEVEN.

6. Military aircraft use a variety of systems including serial numbers, word call signs, and combinations of letters/numbers. Examples include Army Copter 48931; Air Force 61782; REACH 31792; Pat 157; Air Evac 17652; Navy Golf Alfa Kilo 21; Marine 4 Charlie 36, etc.

b. Air Ambulance Flights:

Because of the priority afforded air ambulance flights in the ATC system, extreme discretion is necessary when using the term "LIFEGUARD." It is only intended for those missions of an urgent medical nature and to be utilized only for that portion of the flight requiring expeditious handling. When requested by the pilot, necessary notification to expedite ground handling of patients, etc., is provided by ATC; however, when possible, this information should be passed in advance through nonATC communications systems.

1. Civilian air ambulance flights responding to medical emergencies (first call to an accident scene, carrying patients, organ donors, organs, or other urgently needed lifesaving medical material) will be expedited by ATC when necessary. When expeditious handling is necessary, add the word "LIFEGUARD" in the remarks section of the flight plan. In radio communications, use the call sign "LIFEGUARD" followed by the aircraft registration letters/numbers.

2. Similar provisions have been made for the use of "AIR EVAC" and "MED EVAC" by military air ambulance flights, except that these military flights will receive priority handling only when specifically requested.

EXAMPLE—LIFEGUARD TWO SIX FOUR SIX.

3. Air carrier and Air Taxi flights responding to medical emergencies will also be expedited by ATC when necessary. The nature of these medical emergency flights usually concerns the transportation of urgently needed lifesaving medical materials or vital organs. IT IS IMPERATIVE THAT THE COMPANY/PILOT DETERMINE, BY THE NATURE/URGENCY OF THE SPECIFIC MEDICAL CARGO, IF PRIORITY ATC ASSISTANCE IS REQUIRED. Pilots shall ensure that the word "LIFEGUARD" is included in the remarks section of the flight plan and use the call sign "LIFEGUARD" followed by the company name and flight number for all transmissions when expeditious handling is required. It is important for ATC to be aware of "LIFEGUARD" status, and it is the pilot's responsibility to ensure that this information is provided to ATC.

EXAMPLE—LIFEGUARD DELTA THIRTY-SEVEN.

c. Student Pilots Radio Identification:

1. The FAA desires to help student pilots in acquiring sufficient practical experience in the environment in which they will be required to operate. To receive additional assistance while operating in areas of concentrated air traffic, student pilots need only identify themselves as a student pilot during their initial call to an FAA radio facility.

EXAMPLE—DAYTON TOWER, FLEETWING ONE TWO THREE FOUR, STUDENT PILOT

2. This special identification will alert FAA ATC personnel and enable them to provide student pilots with such extra assistance and consideration as they may need. It is recommended that student pilots identify themselves as such, on ititial contact with each clearance delivery prior to taxiing, ground control, tower, approach and departure control frequency, or FSS contact.

4-2-5. DESCRIPTION OF INTERCHANGE OR LEASED AIRCRAFT

a. Controllers issue traffic information based on familiarity with airline equipment and color/markings. When an air carrier dispatches a flight using another company's equipment and the pilot does not advise the terminal ATC facility, the possible confusion in aircraft identification can compromise safety.

b. Pilots flying an "interchange" or "leased" aircraft not bearing the colors/markings of the company operating the aircraft should inform the terminal ATC facility on first contact the name of the operating company and trip number, followed by the company name as displayed on the aircraft, and aircraft type.

EXAMPLE—AIR CAL THREE ELEVEN, UNITED (INTERCHANGE/LEASE), BOEING SEVEN TWO SEVEN

Table 4-2-1   Calling a Ground Station

Facility	Call Sign
Airport UNICOM	"Shannon UNICOM"
FAA Flight Service Station	"Chicago Radio"
FAA Flight Service Station (En Route Flight Advisory Service (Weather))	"Seattle Flight Watch"
Airport Traffic Control Tower	"Augusta Tower"
Clearance Delivery Position (IFR)	"Dallas Clearance Delivery"
Ground Control Position in Tower	"Miami Ground"
Radar or Nonradar Approach Control Position	"Oklahoma City Approach"
Radar Departure Control Position	"St. Louis Departure"
FAA Air Route Traffic Control Center	"Washington Center"

4-2-6. GROUND STATION CALL SIGNS

Pilots, when calling a ground station, should begin with the name of the facility being called followed by the type of the facility being called as indicated in TBL 4-2-1.

4-2-7. PHONETIC ALPHABET

The International Civil Aviation Organization (ICAO) phonetic alphabet is used by FAA personnel when communications conditions are such that the information cannot be readily received without their use. ATC facilities may also request pilots to use phonetic letter equivalents when aircraft with similar sounding identifications are receiving communications on the same frequency. Pilots should use the phonetic alphabet when identifying their aircraft during initial contact with air traffic control facilities. Additionally, use the phonetic equivalents for single letters and to spell out groups of letters or difficult words during adverse communications conditions. (See TBL 4-2-2.)

4-2-8. FIGURES

a. Figures indicating hundreds and thousands in round number, as for ceiling heights, and upper wind levels up to 9,900 shall be spoken in accordance with the following.

EXAMPLE—

① 500 .... FIVE HUNDRED

② 4,500 .... FOUR THOUSAND FIVE HUNDRED

b. Numbers above 9,900 shall be spoken by separating the digits preceding the word "thousand."

EXAMPLE—

① 10,000 ...... ONE ZERO THOUSAND

② 13,500 ...... ONE THREE THOUSAND FIVE HUNDRED

c. Transmit airway or jet route numbers as follows.

EXAMPLE—

① V12 .... VICTOR TWELVE

② J533 .... J FIVE THIRTY-THREE

d. All other numbers shall be transmitted by pronouncing each digit.

EXAMPLE—10 .... ONE ZERO

e. When a radio frequency contains a decimal point, the decimal point is spoken as "POINT."

EXAMPLE—122.1............ ONE TWO TWO POINT ONE

NOTE—ICAO PROCEDURES REQUIRE THE DECIMAL POINT BE SPOKEN AS "DECIMAL." THE FAA WILL HONOR SUCH USAGE BY MILITARY AIRCRAFT AND ALL OTHER AIRCRAFT REQUIRED TO USE ICAO PROCEDURES.

4-2-9. ALTITUDES AND FLIGHT LEVELS

a. Up to but not including 18,000 feet MSL, state the separate digits of the thousands plus the hundreds if appropriate.

EXAMPLE—

① 12,000 .... ONE TWO THOUSAND

② 12,500 .... ONE TWO THOUSAND FIVE HUNDRED

b. At and above 18,000 feet MSL (FL 180), state the words "flight level" followed by the separate digits of the flight level.

Table 4-2-2   Phonetic Alphabet/Morse Code

CHARACTER	MORSE CODE	TELEPHONY	PHONIC (PRONUNCIATION)
A	• —	Alfa	(AL-FAH)
B	— • • •	Bravo	(BRAH-VOH)
C	— • — •	Charlie	(CHAR-LEE) or (SHAR-LEE)
D	— • •	Delta	(DELL-TAH)
E	•	Echo	(ECK-OH)
F	• • — •	Foxtrot	(FOKS-TROT)
G	— — •	Golf	(GOLF)
H	• • • •	Hotel	(HOH-TEL)
I	• •	India	(IN-DEE-AH)
J	• — — —	Juliett	(JEW-LEE-ETT)
K	— • —	Kilo	(KEY-LOH)
L	• — • •	Lima	(LEE-MAH)
M	— —	Mike	(MIKE)
N	— •	November	(NO-VEM-BER)
O	— — —	Oscar	(OSS-CAH)
P	• — — •	Papa	(PAH-PAH)
Q	— — • —	Quebec	(KEH-BECK)
R	• — •	Romeo	(ROW-ME-OH)
S	• • •	Sierra	(SEE-AIR-RAH)
T	—	Tango	(TANG-GO)
U	• • —	Uniform	(YOU-NEE-FORM) or (OO-NEE-FORM)
V	• • • —	Victor	(VIK-TAH)
W	• — —	Whiskey	(WISS-KEY)
X	— • • —	Xray	(ECKS-RAY)
Y	— • — —	Yankee	(YANG-KEY)
Z	— — • •	Zulu	(ZOO-LOO)
1	• — — — —	One	(WUN)
2	• • — — —	Two	(TOO)
3	• • • — —	Three	(TREE)
4	• • • • —	Four	(FOW-ER)
5	• • • • •	Five	(FIFE)
6	— • • • •	Six	(SIX)
7	— — • • •	Seven	(SEV-EN)
8	— — — • •	Eight	(AIT)
9	— — — — •	Nine	(NIN-ER)
0	— — — — —	Zero	(ZEE-RO)

EXAMPLE—
1 190 .... FLIGHT LEVEL ONE NINER ZERO
2 275 .... FLIGHT LEVEL TWO SEVEN FIVE

4-2-10. DIRECTIONS

The three digits of bearing, course, heading, or wind direction should always be magnetic. The word "true" must be added when it applies.
EXAMPLE—
1 (Magnetic course) 005 .... ZERO ZERO FIVE
2 (True course) 050 .... ZERO FIVE ZERO TRUE
3 (Magnetic bearing) 360 .... THREE SIX ZERO
4 (Magnetic heading) 100 .... HEADING ONE ZERO ZERO
5 (Wind direction) 220 .... WIND TWO TWO ZERO

4-2-11. SPEEDS

The separate digits of the speed followed by the word "KNOTS." Except, controllers may omit the word "KNOTS" when using speed adjustment procedures; e.g., "REDUCE/INCREASE SPEED TO TWO FIVE ZERO."

EXAMPLE—
(Speed) 250 .... TWO FIVE ZERO KNOTS
(Speed) 190 .... ONE NINER ZERO KNOTS

The separate digits of the MACH Number preceded by "MACH."
EXAMPLE—
(Mach number) 1.5 .... MACH ONE POINT FIVE
(Mach number) 0.64 .... MACH POINT SIX FOUR
(Mach number) 0.7 .... MACH POINT SEVEN

4-2-12. TIME

a. FAA uses Coordinated Universal Time (UTC) for all operations. The word "local" or the time zone equivalent shall be used to denote local when local time is given during radio and telephone communications. The term "Zulu" may be used to denote UTC.
EXAMPLE—
0920 UTC .... ZERO NINER TWO ZERO, ZERO ONE TWO ZERO PACIFIC OR LOCAL, OR ONE TWENTY A.

b. To Convert from Standard Time to Coordinated Universal Time:

**Table 4-2-3   Standard Time
to Coordinated Universal Time**

Eastern Standard Time	Add 5 hours
Central Standard Time	Add 6 hours
Mountain Standard Time	Add 7 hours
Pacific Standard Time	Add 8 hours
Alaska Standard Time	Add 9 hours
Hawaii Standard Time	Add 10 hours

NOTE—FOR DAYLIGHT TIME, SUBTRACT 1 HOUR.

c. A reference may be made to local daylight or standard time utilizing the 24-hour clock system. The hour is indicated by the first two figures and the minutes by the last two figures.

EXAMPLE—
0000 .... ZERO ZERO ZERO ZERO
0920 .... ZERO NINER TWO ZERO

d. Time may be stated in minutes only (two figures) in radiotelephone communications when no misunderstanding is likely to occur.

e. Current time in use at a station is stated in the nearest quarter minute in order that pilots may use this information for time checks. Fractions of a quarter minute less than 8 seconds are stated as the preceding quarter minute; fractions of a quarter minute of 8 seconds or more are stated as the succeeding quarter minute.

EXAMPLE—
0929:05 .... TIME, ZERO NINER TWO NINER
0929:10 .... TIME, ZERO NINER TWO NINER AND ONE-
                          QUARTER

4-2-13. COMMUNICATIONS WITH TOWER WHEN AIRCRAFT TRANSMITTER OR RECEIVER OR BOTH ARE INOPERATIVE

a. Arriving Aircraft:

1. *Receiver inoperative:*

(a) If you have reason to believe your receiver is inoperative, remain outside or above the Class D surface area until the direction and flow of traffic has been determined; then, advise the tower of your type aircraft, position, altitude, intention to land, and request that you be controlled with light signals.

REFERENCE—AIM, TRAFFIC CONTROL LIGHT SIGNALS, PARAGRAPH 4-3-13.

(b) When you are approximately 3 to 5 miles from the airport, advise the tower of your position and join the airport traffic pattern. From this point on, watch the tower for light signals. Thereafter, if a complete pattern is made, transmit your position downwind and/or turning base leg.

2. *Transmitter inoperative:* Remain outside or above the Class D surface area until the direction and flow of traffic has been determined; then, join the airport traffic pattern.

Monitor the primary local control frequency as depicted on Sectional Charts for landing or traffic information, and look for a light signal which may be addressed to your aircraft. During hours of daylight, acknowledge tower transmissions or light signals by rocking your wings. At night, acknowledge by blinking the landing or navigation lights. To acknowledge tower transmissions during daylight hours, hovering helicopters will turn in the direction of the controlling facility and flash the landing light. While in flight, helicopters should show their acknowledgement of receiving a transmission by making shallow banks in opposite directions. At night, helicopters will acknowledge receipt of transmissions by flashing either the landing or the search light.

3. *Transmitter and receiver inoperative:* Remain outside or above the Class D surface area until the direction and flow of traffic has been determined; then, join the airport traffic pattern and maintain visual contact with the tower to receive light signals. Acknowledge light signals as noted above.

b. Departing Aircraft: If you experience radio failure prior to leaving the parking area, make every effort to have the equipment repaired. If you are unable to have the malfunction repaired, call the tower by telephone and request authorization to depart without two-way radio communications. If tower authorization is granted, you will be given departure information and requested to monitor the tower frequency or watch for light signals as appropriate. During daylight hours, acknowledge tower transmissions or light signals by moving the ailerons or rudder. At night, acknowledge by blinking the landing or navigation lights. If radio malfunction occurs after departing the parking area, watch the tower for light signals or monitor tower frequency.

REFERENCE—FAR PART 91.125 AND FAR PART 91.129.

4-2-14. COMMUNICATIONS FOR VFR FLIGHTS

a. FSSs and Supplemental Weather Service Locations (SWSLs) are allocated frequencies for different functions; for example, 122.0 MHz is assigned as the En Route Flight Advisory Service frequency at selected FSSs. In addition, certain FSSs provide Local Airport Advisory on 123.6 MHz. Frequencies are listed in the A/FD. If you are in doubt as to what frequency to use, 122.2 MHz is assigned to the majority of FSSs as a common en route simplex frequency.

NOTE—IN ORDER TO EXPEDITE COMMUNICATIONS, STATE THE FREQUENCY BEING USED AND THE AIRCRAFT LOCATION DURING INITIAL CALLUP.

EXAMPLE—DAYTON RADIO, NOVEMBER ONE TWO THREE FOUR FIVE ON ONE TWO TWO POINT TWO, OVER SPRINGFIELD V-O-R, OVER.

b. Certain VOR voice channels are being utilized for recorded broadcasts; i.e., ATIS, HIWAS, etc. These services and appropriate frequencies are listed in the A/FD. On VFR flights, pilots are urged to monitor these frequencies. When in contact with a control facility, notify the controller if you plan to leave the frequency to monitor these broadcasts.

Section 3. AIRPORT OPERATIONS

4-3-1. GENERAL

Increased traffic congestion, aircraft in climb and descent attitudes, and pilot preoccupation with cockpit duties are some factors that increase the hazardous accident potential near the airport. The situation is further compounded when the weather is marginal—that is, just meeting VFR requirements. Pilots must be particularly alert when operating in the vicinity of an airport. This section defines some rules, practices, and procedures that pilots should be familiar with and adhere to for safe airport operations.

4-3-2. AIRPORTS WITH AN OPERATING CONTROL TOWER

a. When operating at an airport where traffic control is being exercised by a control tower, pilots are required to maintain two-way radio contact with the tower while operating within the Class B, Class C, and Class D surface area unless the tower authorizes otherwise. Initial callup should be made about 15 miles from the airport. Unless there is a good reason to leave the tower frequency before exiting the Class B, Class C, and Class D surface areas, it is a good operating practice to remain on the tower frequency for the purpose of receiving traffic information. In the interest of reducing tower frequency congestion, pilots are reminded that it is not necessary to request permission to leave the tower frequency once outside of Class B, Class C, and Class D surface areas. Not all airports with an operating control tower will have Class D airspace. These airports do not have weather reporting which is a requirement for surface based controlled airspace, previously known as a control zone. The controlled airspace over these airports will normally begin at 700 feet or 1,200 feet above ground level and can be determined from the visual aeronautical charts. Pilots are expected to use good operating practices and communicate with the control tower as described in this section.

b. When necessary, the tower controller will issue clearances or other information for aircraft to generally follow the desired flight path (traffic patterns) when flying in Class B, Class C, and Class D surface areas and the proper taxi routes when operating on the ground. If not otherwise authorized or directed by the tower, pilots of fixed-wing aircraft approaching to land must circle the airport to the left. Pilots approaching to land in a helicopter must avoid the flow of fixed-wing traffic. However, in all instances, an appropriate clearance must be received from the tower before landing.

NOTE—THIS DIAGRAM IS INTENDED ONLY TO ILLUSTRATE TERMINOLOGY USED IN IDENTIFYING VARIOUS COMPONENTS OF A TRAFFIC PATTERN. *IT SHOULD NOT BE USED AS A REFERENCE OR GUIDE ON HOW TO ENTER A TRAFFIC PATTERN.*

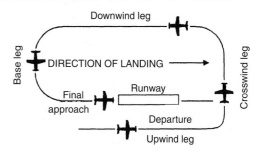

NOTE—This diagram is intended only to illustrate terminology used in identifying various components of a traffic pattern. It should not be used as a reference or guide on how to enter a traffic pattern.

Figure 4-3-1. Components of a Traffic Pattern

c. The following terminology for the various components of a traffic pattern has been adopted as standard for use by control towers and pilots (See FIG 4-3-1):

1. *Upwind leg:* A flight path parallel to the landing runway in the direction of landing.

2. *Crosswind leg:* A flight path at right angles to the landing runway off its takeoff end.

3. *Downwind leg:* A flight path parallel to the landing runway in the opposite direction of landing.

4. *Base leg:* A flight path at right angles to the landing runway off its approach end and extending from the downwind leg to the intersection of the extended runway centerline.

5. *Final approach:* A flight path in the direction of landing along the extended runway centerline from the base leg to the runway.

6. *Departure leg:* The flight plan which begins after takeoff and continues straight ahead along the extended runway centerline. The departure climb continues until reaching a point at least ½ mile beyond the departure end of the runway and within 300 feet of the traffic pattern altitude.

d. Many towers are equipped with a tower radar display. The radar uses are intended to enhance the effectiveness and efficiency of the local control, or tower, position. They are not intended to provide radar services or benefits to pilots except as they may accrue through a more efficient tower operation. The four basic uses are:

1. *To determine an aircraft's exact location:* This is accomplished by radar identifying the VFR aircraft through any of the techniques available to a radar position, such as having the aircraft *squawk ident.* Once identified, the aircraft's position and spatial relationship to other aircraft can be quickly determined, and standard instructions regarding VFR operation in Class B, Class C, and Class D surface areas will be issued. Once initial radar identification of a

VFR aircraft has been established and the appropriate instructions have been issued, radar monitoring may be discontinued; the reason being that the local controller's primary means of surveillance in VFR conditions is visually scanning the airport and local area.

2. *To provide radar traffic advisories:* Radar traffic advisories may be provided to the extent that the local controller is able to monitor the radar display. Local control has primary control responsibilities to the aircraft operating on the runways, which will normally supersede radar monitoring duties.

3. *To provide a direction or suggested heading:* The local controller may provide pilots flying VFR with generalized instructions which will facilitate operations; e.g., "PROCEED SOUTHWEST BOUND, ENTER A RIGHT DOWNWIND RUNWAY THREE ZERO," or provide a suggested heading to establish radar identification or as an advisory aid to navigation; e.g., "SUGGESTED HEADING TWO TWO ZERO, FOR RADAR IDENTIFICATION." In both cases, the instructions are advisory aids to the pilot flying VFR and are not radar vectors.

NOTE—PILOTS HAVE COMPLETE DISCRETION REGARDING ACCEPTANCE OF THE SUGGESTED HEADINGS OR DIRECTIONS AND HAVE SOLE RESPONSIBILITY FOR SEEING AND AVOIDING OTHER AIRCRAFT

4. *To provide information and instructions to aircraft operating within Class B, Class C, and Class D surface areas:* In an example of this situation, the local controller would use the radar to advise a pilot on an extended downwind when to turn base leg.

NOTE—THE ABOVE TOWER RADAR APPLICATIONS ARE INTENDED TO AUGMENT THE STANDARD FUNCTIONS OF THE LOCAL CONTROL POSITION. THERE IS NO CONTROLLER REQUIREMENT TO MAINTAIN CONSTANT RADAR IDENTIFICATION. IN FACT SUCH A REQUIREMENT COULD COMPROMISE THE LOCAL CONTROLLER'S ABILITY TO VISUALLY SCAN THE AIRPORT AND LOCAL AREA TO MEET FAA RESPONSIBILITIES TO THE AIRCRAFT OPERATING ON THE RUNWAYS AND WITHIN THE CLASS B, CLASS C, AND CLASS D SURFACE AREAS. NORMALLY, PILOTS WILL NOT BE ADVISED OF BEING IN RADAR CONTACT SINCE THAT CONTINUED STATUS CANNOT BE GUARANTEED AND SINCE THE PURPOSE OF THE RADAR IDENTIFICATION IS NOT TO ESTABLISH A LINK FOR THE PROVISION OF RADAR SERVICES.

e. A few of the radar equipped towers are authorized to use the radar to ensure separation between aircraft in specific situations, while still others may function as limited radar approach controls. The various radar uses are strictly a function of FAA operational need. The facilities may be indistinguishable to pilots since they are all referred to as tower and no publication lists the degree of radar use. THEREFORE, WHEN IN COMMUNICATION WITH A TOWER CONTROLLER WHO MAY HAVE RADAR AVAILABLE, DO NOT ASSUME THAT CONSTANT RADAR MONITORING AND COMPLETE ATC RADAR SERVICES ARE BEING PROVIDED.

4-3-3. TRAFFIC PATTERNS

At most airports and military air bases, traffic pattern altitudes for propeller-driven aircraft generally extend from 600 feet to as high as 1,500 feet above the ground. Also, traffic pattern altitudes for military turbojet aircraft sometimes extend up to 2,500 feet above the ground. Therefore, pilots of en route aircraft should be constantly on the alert for other aircraft in traffic patterns and avoid these areas whenever possible. Traffic pattern altitudes should be maintained unless otherwise required by the applicable distance from cloud criteria (FAR Part 91.155). (See FIG 4-3-2 and FIG 4-3-3).

4-3-4. VISUAL INDICATORS AT AIRPORTS WITHOUT AN OPERATING CONTROL TOWER

a. At those airports *without an operating control tower*, a segmented circle visual indicator system, if installed, is designed to provide traffic pattern information.

REFERENCE—AIM, TRAFFIC ADVISORY PRACTICES AT AIRPORTS WITHOUT OPERATING CONTROL TOWERS, PARAGRAPH 4-1-9.

b. The segmented circle system consists of the following components:

1. *The segmented circle:* Located in a position affording maximum visibility to pilots in the air and on the ground and providing a centralized location for other elements of the system.

2. *The wind direction indicator:* A wind cone, wind sock, or wind tee installed near the operational runway to indicate wind direction. The large end of the wind cone/wind sock points into the wind as does the large end (cross bar) of the wind tee. In lieu of a tetrahedron and where a wind sock or wind cone is collocated with a wind tee, the wind tee may be manually aligned with the runway in use to indicate landing direction. These signaling devices may be located in the center of the segmented circle and may be lighted for night use. Pilots are cautioned against using a tetrahedron to indicate wind direction.

3. *The landing direction indicator:* A tetrahedron is installed when conditions at the airport warrant its use. It may be used to indicate the direction of landings and takeoffs. A tetrahedron may be located at the center of a segmented circle and may be lighted for night operations. The small end of the tetrahedron points in the direction of landing. Pilots are cautioned against using a tetrahedron for any purpose other than as an indicator of landing direction. Further, pilots should use extreme caution when making runway selection by use of a tetrahedron in very light or calm wind conditions as the tetrahedron may not be aligned with the designated calm-wind runway. At airports with control towers, the tetrahedron should only be referenced when the control tower is not in operation. Tower instructions supersede tetrahedron indications.

4. *Landing strip indicators:* Installed in pairs as shown in the segmented circle diagram and used to show the alignment of landing strips.

5. *Traffic pattern indicators:* Arranged in pairs in conjunction with landing strip indicators and used to indicate the direction of turns when there is a variation from the normal left traffic pattern. (If there is no segmented circle installed at the airport, traffic pattern indicators may be installed on or near the end of the runway.)

c. Preparatory to landing at an airport without a control tower, or when the control tower is not in operation, pilots should concern themselves with the indicator for the approach end of the runway to be used. When approaching for landing, all turns must be made to the left unless a traffic pattern indicator indicates that turns should be made to the right. If the pilot will mentally enlarge the indicator for the runway to be used, the base and final approach legs of the traffic pattern to be flown immediately become apparent. Similar treatment of the indicator at the departure end of the runway will clearly indicate the direction of turn after takeoff.

d. When two or more aircraft are approaching an airport for the purpose of landing, the pilot of the aircraft at the lower altitude has the right-of-way over the pilot of the aircraft at the higher altitude. However, the pilot operating at the lower altitude should not take advantage of another aircraft, which is on final approach to land, by cutting in front of, or overtaking that aircraft.

4-3-5. UNEXPECTED MANEUVERS IN THE AIRPORT TRAFFIC PATTERN

There have been several incidents in the vicinity of controlled airports that were caused primarily by aircraft executing unexpected maneuvers. ATC service is based upon observed or known traffic and airport conditions. Controllers establish the sequence of arriving and departing aircraft by requiring them to adjust flight as necessary to achieve proper spacing. These adjustments can only be based on observed traffic, accurate pilot reports, and anticipated aircraft maneuvers. Pilots are expected to cooperate so as to preclude disrupting traffic flows or creating conflicting patterns. The pilot-in-command of an aircraft is directly responsible for and is the final authority as to the operation of the aircraft. On occasion it may be necessary for pilots to maneuver their aircraft to maintain spacing with the traffic they have been sequenced to follow. The controller can anticipate minor maneuvering such as shallow "S" turns. The controller cannot, however, anticipate a major maneuver such as a 360 degree turn. If a pilot makes a 360 degree turn after obtaining a landing sequence, the result is usually a gap in the landing interval and, more importantly, it causes a chain reaction which

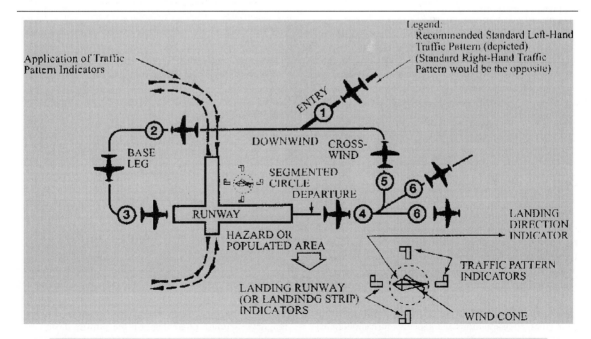

Figure 4-3-2. Traffic Pattern Operations Single Runway. (See key Traffic Pattern Operations under Fig. 4-3-3.)

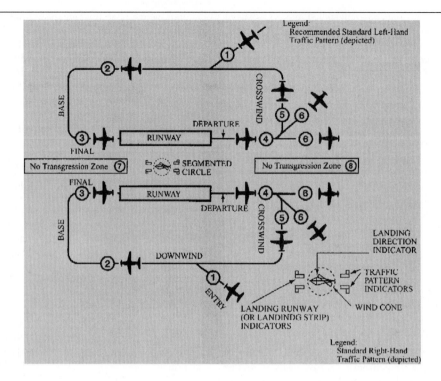

① Enter pattern in level flight, abeam the midpoint of the runway, at pattern altitude. (1000' AGL is recommended pattern altitude unless established otherwise.)

② Maintain pattern altitude until abeam approach end of the landing runway, on downwind leg.

③ Complete turn to final at least ¼ mile from the runway.

④ Continue straight ahead until beyond departure end of runway.

⑤ If remaining in the traffic pattern, commence turn to crosswind leg beyond the departure end of the runway, within 300 feet of pattern altitude.

⑥ If departing the traffic pattern, continue straight out, or exit with a 45° turn (to the left when in a left-hand traffic pattern; to the right when in a right-hand traffic pattern) beyond the departure end of the runway, after reaching pattern altitude.

⑦ Do not overshoot final or continue on a track which will penetrate the final approach of the parallel runway.

⑧ Do not continue on a track which will penetrate the departure path of the parallel runway.

Figure 4-3-3. Traffic Pattern Operations Parallel Runways

may result in a conflict with following traffic and an interruption of the sequence established by the tower or approach controller. Should a pilot decide to make maneuvering turns to maintain spacing behind a preceding aircraft, the pilot should always advise the controller if at all possible. Except when requested by the controller or in emergency situations, a 360 degree turn should never be executed in the traffic pattern or when receiving radar service without first advising the controller.

4-3-6. USE OF RUNWAYS/DECLARED DISTANCES

a. Runways are identified by numbers which indicate the nearest 10-degree increment of the azimuth of the runway centerline. For example, where the magnetic azimuth is 183 degrees, the runway designation would be 18; for a magnetic azimuth of 87 degrees, the runway designation would be 9. For a magnetic azimuth ending in the number 5, such as 185, the runway designation could be either 18 or 19. Wind direction issued by the tower is also magnetic and wind velocity is in knots.

b. Airport proprietors are responsible for taking the lead in local aviation noise control. Accordingly, they may propose specific noise abatement plans to the FAA. If approved, these plans are applied in the form of Formal or Informal Runway Use Programs for noise abatement purposes.

REFERENCE—PILOT/CONTROLLER GLOSSARY TERM-RUNWAY USE PROGRAM.

1. At airports where no runway use program is established, ATC clearances may specify:

(a) the runway most nearly aligned with the wind when it is 5 knots or more;

(b) the "calm wind" runway when wind is less than 5 knots, or;

(c) another runway if operationally advantageous.

NOTE—IT IS NOT NECESSARY FOR A CONTROLLER TO SPECIFICALLY INQUIRE IF THE PILOT WILL USE A SPECIFIC RUNWAY OR TO OFFER A CHOICE OF RUNWAYS. IF A PILOT PREFERS TO USE A DIFFERENT RUNWAY FROM THAT SPECIFIED OR THE ONE MOST NEARLY ALIGNED WITH THE WIND, THE PILOT IS EXPECTED TO INFORM ATC ACCORDINGLY.

2. At airports where a runway use program is established, ATC will assign runways deemed to have the least noise impact. If in the interest of safety a runway different from that specified is preferred, the pilot is expected to advise ATC accordingly. ATC will honor such requests and advise pilots when the requested runway is noise sensitive. When use of a runway other than the one assigned is requested, pilot cooperation is encouraged to preclude disruption of traffic flows or the creation of conflicting patterns.

c. At some airports, the airport proprietor may declare that sections of a runway at one or both ends are not available for landing or takeoff. For these airports, the declared distance of runway length available for a particular operation is published in the *Airport/Facility Directory*. Declared distances (*TORA, TODA, ASDA,* and *LDA*) are defined in the *Pilot/Controller Glossary*. These distances are calculated by adding to the full length of paved runway any applicable clearway or stopway and subtracting from that sum the sections of the runway unsuitable for satisfying the required takeoff run, takeoff, accelerate/stop, or landing distance.

4-3-7. LOW LEVEL WIND SHEAR/MICROBURST DETECTION SYSTEM

Low Level Wind Shear Alert System (LLWAS), Terminal Doppler Weather Radar (TDWR), Weather System Processor (WSP), and Integrated Terminal Weather System (ITWS) display information on hazardous wind shear and microburst activity in the vicinity of an airport to air traffic controllers who relay this information to pilots.

a. LLWAS provides wind shear alert and gust front information but does not provide microburst alerts. The LLWAS is designed to detect low level wind shear conditions around the periphery of an airport. It does not detect wind shear beyond that limitation. Controllers will provide this information to pilots by giving the pilot the airport wind followed by the boundary wind.

EXAMPLE—

WIND SHEAR ALERT; AIRPORT WIND 230 AT 5, SOUTH BOUNDARY WIND 170 AT 20.

b. LLWAS "network expansion," (LLWAS NE) and LLWAS Relocation/Sustainment (LLWAS-RS) are systems integrated with TDWR. These systems provide the capability of detecting microburst alerts and wind shear alerts. Controllers will issue the appropriate wind shear alerts or microburst alerts. In some of these systems controllers also have the ability to issue wind information oriented to the threshold or departure end of the runway.

EXAMPLE—

RUNWAY 17 ARRIVAL MICROBURST ALERT 40 KNOT LOSS 3 MILE FINAL.

REFERENCE—

AIM, MICROBURSTS, PARAGRAPH 7-1-27.

c. More advanced systems are in the field or being developed such as ITWS. ITWS provides alerts for microbursts, wind shear, and significant thunder-storm activity. ITWS displays wind information oriented to the threshold or departure end of the runway.

d. The WSP provides weather processor enhancements to selected Airport Surveillance Radar (ASR)-9 facilities. The WSP provides Air Traffic with detection and alerting of hazardous weather such as wind shear, microbursts, and significant thunderstorm activity. The WSP displays terminal area 6 level weather, storm cell locations and movement, as well as the location and predicted future position and intensity of wind shifts that may affect airport operations. Controllers will receive and issue alerts based on Areas Noted for Attention (ARENA). An ARENA extends on the runway center line from a 3 mile final to the runway to a 2 mile departure.

e. An airport equipped with the LLWAS, ITWS, or WSP is so indicated in the Airport/Facility Directory under Weather Data Sources for that particular airport.

4-3-8. BRAKING ACTION REPORTS AND ADVISORIES

a. When available, ATC furnishes pilots the quality of braking action received from pilots or airport management. The quality of braking action is described by the terms "good," "fair," "poor," and, "nil," or a combination of these terms. When pilots report the quality of braking action by using the terms noted above, they should use descriptive terms that are easily understood, such as, "braking action poor the first/last half of the runway," together with the particular type of aircraft.

b. For NOTAM purposes, braking action reports are classified according to the most critical term ("fair," "poor," or "nil") used and issued as a NOTAM(D).

c. When tower controllers have received runway braking action reports which include the terms *poor* or *nil*, or whenever weather conditions are conducive to deteriorating or rapidly changing runway braking conditions, the tower will include on the ATIS broadcast the statement, "BRAKING ACTION ADVISORIES ARE IN EFFECT."

d. During the time that braking action advisories are in effect, ATC will issue the latest braking action report for the runway in use to each arriving and departing aircraft. Pilots should be prepared for deteriorating braking conditions and should request current runway condition information if not volunteered by controllers. Pilots should also be prepared to provide a descriptive runway condition report to controllers after landing.

4-3-9. RUNWAY FRICTION REPORTS AND ADVISORIES

a. Friction is defined as the ratio of the tangential force needed to maintain uniform relative motion between two contacting surfaces (aircraft tires to the pavement surface) to the perpendicular force holding them in contact (distributed aircraft weight to the aircraft tire area). Simply stated, friction quantifies slipperiness of pavement surfaces.

b. The Greek letter MU (pronounced "myew"), is used to designate a friction value representing runway surface conditions.

c. MU (friction) values range from 0 to 100 where zero is the lowest friction value and 100 is the maximum friction value obtainable. For frozen contaminants on runway surfaces, a MU value of 40 or less is the level when the aircraft braking performance starts to deteriorate and directional control begins to be less responsive. The lower the MU value, the less effective braking performance becomes and the more difficult directional control becomes.

d. At airports with friction measuring devices, airport management should conduct friction measurements on runways covered with compacted snow and/or ice.

1. Numerical readings may be obtained by using any FAA approved friction measuring device. As these devices do not provide equal readings on contaminated surfaces, it is necessary to designate the type of friction measuring device used.

2. When the MU value for any one-third zone of an active runway is 40 or less, a report should be given to ATC by airport management for dissemination to pilots. The report will identify the runway, the time of measurement, the type of friction device used MU values for each zone, and the contaminant conditions, e.g., wet snow, dry snow, slush, deicing chemicals, etc. Measurements for each one-third zone will be given in the direction of takeoff and landing on the runway. A report should also be given when MU values rise above 40 in all zones of a runway previously reporting a MU below 40.

3. Airport management should initiate a NOTAM(D) when the friction measuring device is out of service.

e. When MU reports are provided by airport management, the ATC facility providing approach control or local airport advisory will provide the report to any pilot upon request.

f. Pilots should use MU information with other knowledge including aircraft performance characteristics, type, and weight, previous experience, wind conditions, and aircraft tire type (i.e., bias ply vs. radial constructed) to determine runway suitability.

g. No correlation has been established between MU values and the descriptive terms "good," "fair," "poor," and "nil" used in braking action reports.

4-3-10. INTERSECTION TAKEOFFS

a. In order to enhance airport capacities, reduce taxiing distances, minimize departure delays, and provide for more efficient movement of air traffic, controllers may initiate intersection takeoffs as well as approve them when the pilot requests. If for ANY reason a pilot prefers to use a different intersection or the full length of the runway or desires to obtain the distance between the intersection and the runway end, THE PILOT IS EXPECTED TO INFORM ATC ACCORDINGLY.

b. An aircraft is expected to taxi to (but not onto) the end of the assigned runway unless prior approval for an intersection departure is received from ground control.

c. Pilots should state their position on the airport when calling the tower for takeoff from a runway intersection.

EXAMPLE—CLEVELAND TOWER, APACHE THREE SEVEN TWO TWO PAPA, AT THE INTERSECTION OF TAXIWAY OSCAR AND RUNWAY TWO THREE RIGHT, READY FOR DEPARTURE.

d. Controllers are required to separate small aircraft (12,500 pounds or less, maximum certificated takeoff weight) departing (same or opposite direction) from an intersection behind a large nonheavy aircraft on the same runway, by ensuring that at least a 3-minute interval exists between the time the preceding large aircraft has taken off and the succeeding small aircraft begins takeoff roll. To inform the pilot of the required 3-minute hold, the controller will state, "Hold for wake turbulence." If after considering wake turbulence hazards, the pilot feels that a lesser time interval is appropriate, the pilot may request a waiver to the 3-minute interval. To initiate such a request, simply say "Request waiver to 3-minute interval," or a similar statement. Controllers may then issue a takeoff clearance if other traffic permits, since the pilot has accepted the responsibility for wake turbulence separation.

e. The 3-minute interval is not required when the intersection is 500 feet or less from the departure point of the preceding aircraft and both aircraft are taking off in the same direction. Controllers may permit the small aircraft to alter course after takeoff to avoid the flight path of the preceding departure.

f. The 3-minute interval is mandatory behind a heavy aircraft in all cases.

4-3-11. PILOT RESPONSIBILITIES WHEN CONDUCTING LAND AND HOLD SHORT OPERATIONS (LAHSO)

a. LAHSO is an acronym for "Land And Hold Short Operations." These operations include landing and holding short of an **intersecting runway**, an **intersecting taxiway**, or some other designated **point on a runway** other than an intersecting runway or taxiway. (See FIG 4-3-4, FIG 4-3-5, FIG 4-3-6)

b. Pilot Responsibilities and Basic Procedures:

1. LAHSO is an air traffic control procedure that requires pilot participation to balance the needs for increased airport capacity and system efficiency, consistent with safety. This procedure can be done safely **provided** pilots and controllers are knowledgeable and understand their responsibilities. The following paragraphs outline specific pilot/operator responsibilities when conducting LAHSO.

2. At controlled airports, air traffic may clear a pilot to land and hold short. Pilots may accept such a clearance provided that the pilot-in-command determines that the aircraft can safely land and stop within the Available Landing Distance (ALD). ALD data are published in the special notices section of the Airport/Facility Directory (A/FD) and in the U.S. Terminal Procedures Publications. Controllers will also provide ALD data upon request. Student pilots or pilots not familiar with LAHSO should **not** participate in the program.

3. The pilot-in-command has the final authority to accept or decline any land and hold short clearance. The safety and operation of the aircraft remain the responsibility of the pilot. Pilots are expected to decline a LAHSO clearance if they determine it will compromise safety.

4. To conduct LAHSO, pilots should become familiar with all available information concerning LAHSO at their destination airport. Pilots should have, *readily available*, the **published ALD** and runway **slope information** for **all** LAHSO runway combinations at **each** airport of intended landing. Additionally, knowledge about landing performance data permits the pilot to *readily* determine that the ALD for the assigned runway is sufficient for safe LAHSO. As part of a pilot's preflight planning process, pilots should determine if their destination airport has LAHSO. If so, their preflight planning process should include an assessment of which LAHSO combinations would work

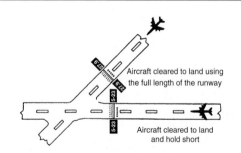

Figure 4-3-4.

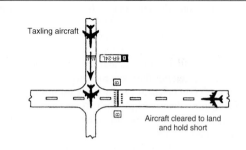

Figure 4-3-5.

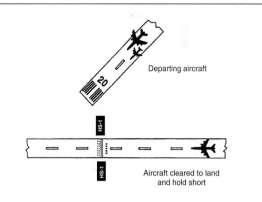

Figure 4-3-6.

for them given their aircraft's required landing distance. Good pilot decision making is knowing in advance whether one can accept a LAHSO clearance if offered.

EXAMPLE—

FIGURE 4-3-6 HOLDING SHORT AT A DESIGNATED POINT MAY BE REQUIRED TO AVOID CONFLICTS WITH THE RUNWAY SAFETY AREA/FLIGHT PATH OF A NEARBY RUNWAY.

NOTE—

EACH FIGURE SHOWS THE APPROXIMATE LOCATION OF LAHSO MARKINGS, SIGNAGE, AND IN-PAVEMENT LIGHTING WHEN INSTALLED.

REFERENCE—

AIM, CHAPTER 2, AERONAUTICAL LIGHTING AND OTHER AIRPORT VISUAL AIDS.

5. If, for any reason, such as difficulty in discerning the location of a LAHSO intersection, wind conditions, aircraft condition, etc., the pilot elects to request to land on the full length of the runway, to land on another runway, or to decline LAHSO, a pilot is expected to promptly inform air traffic, ideally even before the clearance is issued. **A LAHSO clearance, once accepted, must be adhered to, just as any other ATC clearance, unless an amended clearance is obtained or an emergency occurs. A LAHSO clearance does not preclude a rejected landing.**

6. A pilot who accepts a LAHSO clearance should land and exit the runway at the first convenient taxiway (unless directed otherwise) before reaching the hold short point. Otherwise, the pilot must stop and hold at the hold short point. **If a rejected landing becomes necessary after accepting a LAHSO clearance, the pilot should maintain safe separation from other aircraft or vehicles, and should promptly notify the controller.**

7. Controllers need a full read back of all LAHSO clearances. Pilots should read back their LAHSO clearance and include the words, "HOLD SHORT OF (RUNWAY/TAXIWAY/OR POINT)" in their acknowledgment of **all** LAHSO clearances. In order to reduce frequency congestion, pilots are encouraged to read back the LAHSO clearance without prompting. Don't make the controller have to ask for a read back!

c. LAHSO Situational Awareness

1. Situational awareness if **vital** to the success of LAHSO. Situational awareness starts with having current airport information in the cockpit, readily accessible to the pilot. (An airport diagram assists pilots in identifying their location on the airport, thus reducing requests for "progressive taxi instructions" from controllers.)

2. Situational awareness includes effective pilot-controller radio communication. ATC expects pilots to specifically acknowledge and read back all LAHSO clearances as follows:

EXAMPLE—

ATC: "(AIRCRAFT ID) CLEARED TO LAND RUNWAY SIX RIGHT, HOLD SHORT OF TAXIWAY BRAVO FOR CROSSING TRAFFIC (TYPE AIRCRAFT)."

AIRCRAFT: "(AIRCRAFT ID), WILCO, CLEARED TO LAND RUNWAY SIX RIGHT TO HOLD SHORT OF TAXIWAY BRAVO."

ATC: "(AIRCRAFT ID) CROSS RUNWAY SIX RIGHT AT TAXIWAY BRAVO, LANDING AIRCRAFT WILL HOLD SHORT."

AIRCRAFT: "(AIRCRAFT ID), WILCO, CROSS RUNWAY SIX RIGHT AT BRAVO, LANDING TRAFFIC (TYPE AIRCRAFT) TO HOLD."

3. For those airplanes flown with two crewmembers, effective **intra-cockpit** communication between cockpit crewmembers is also critical. There have been several instances where the pilot working the radios accepted a LAHSO clearance but then simply forgot to tell the pilot flying the aircraft.

4. Situational awareness also includes a thorough understanding of the airport markings, signage, and lighting associated with LAHSO. These visual aids consist of a three-part system of **yellow hold-short markings**, **red and white signage** and, in certain cases, **in-pavement lighting**. Visual aids assist the pilot in determining where to hold short. FIG 4-3-4, FIG 4-3-5, FIG 4-3-6 depict how these markings, signage, and lighting combinations will appear once installed. Pilots are cautioned that not all airports conducting LAHSO have installed any or all of the above markings, signage, or lighting.

5. Pilots should only receive a LAHSO clearance when there is a minimum ceiling of 1,000 feet and 3 statute miles visibility. The intent of having "basic" VFR weather conditions is to allow pilots to maintain visual contact with other aircraft and ground vehicles operations. Pilots should consider the effects of prevailing inflight visibility (such as landing into the sun) and how it may affect overall situational awareness. Additionally, surface vehicles and aircraft being taxied by maintenance personnel may also be participating in LAHSO, especially in those operations that involve crossing an active runway.

4-3-12. LOW APPROACH

a. A low approach (sometimes referred to as a low pass) is the go-around maneuver following an approach. Instead of landing or making a touch-and-go, a pilot may wish to go around (low approach) in order to expedite a particular operation (a series of practice instrument approaches is an example of such an operation). Unless otherwise authorized by ATC, the low approach should be made straight ahead, with no turns or climb made until the pilot has made a thorough visual check for other aircraft in the area.

b. When operating within a Class B, Class C, and Class D surface area, a pilot intending to make a low approach should contact the tower for approval. This request should be made prior to starting the final approach.

c. When operating to an airport, not within a Class B, Class C, and Class D surface area, a pilot intending to make a low approach should, prior to leaving the final approach fix inbound (non precision approach) or the outer marker or fix used in lieu of the outer marker inbound (precision approach), so advise the FSS, UNICOM, or make a broadcast as appropriate.

REFERENCE—AIM, TRAFFIC ADVISORY PRACTICES AT AIRPORTS WITHOUT OPERATING CONTROL TOWERS, PARAGRAPH 4-1-9.

4-3-13. TRAFFIC CONTROL LIGHT SIGNALS

a. The following procedures are used by ATCTs in the control of aircraft, ground vehicles, equipment, and personnel not equipped with radio. These same procedures will be used to control aircraft, ground vehicles, equipment, and personnel equipped with radio if radio contact cannot be established. ATC personnel use a directive traffic control signal which emits an intense narrow light beam of a selected color (either red, white, or green) when controlling traffic by light signals.

b. Although the traffic signal light offers the advantage that some control may be exercised over nonradio equipped aircraft, pilots should be cognizant of the disadvantages which are:

1. Pilots may not be looking at the control tower at the time a signal is directed toward their aircraft.

2. The directions transmitted by a light signal are very limited since only approval or disapproval of a pilot's anticipated actions may be transmitted. No supplement or explanatory information may be transmitted except by the use of the "General Warning Signal" which advises the pilot to be on the alert.

c. Between sunset and sunrise, a pilot wishing to attract the attention of the control tower should turn on a landing light and taxi the aircraft into a position, clear of the active runway, so that light is visible to the tower. The landing light should remain on until appropriate signals are received from the tower.

d. Air Traffic Control Tower Light Gun Signals (See TBL 4-3-1).

e. During daylight hours, acknowledge tower transmissions or light signals by moving the ailerons or rudder. At night, acknowledge by blinking the landing or navigation lights. If radio malfunction occurs after departing the parking area, watch the tower for light signals or monitor tower frequency.

4-3-14. COMMUNICATIONS

a. Pilots of departing aircraft should communicate with the control tower on the appropriate ground control/clearance delivery frequency prior to starting engines to receive engine start time, taxi and/or clearance information. Unless otherwise advised by the tower, remain on that frequency during taxiing and runup, then change to local control frequency when ready to request takeoff clearance.

NOTE—PILOTS ARE ENCOURAGED TO MONITOR THE LOCAL TOWER FREQUENCY AS SOON AS PRACTICAL CONSISTENT WITH OTHER ATC REQUIREMENTS

REFERENCE—AIM, AUTOMATIC TERMINAL INFORMATION SERVICE (ATIS) FOR CONTINUOUS BROADCAST OF TERMINAL INFORMATION, PARAGRAPH 4-1-13.

b. The tower controller will consider that pilots of turbine-powered aircraft are ready for takeoff when they reach the runway or warm-up block unless advised otherwise.

c. The majority of ground control frequencies are in the 121.6–121.9 MHz bandwidth. Ground control frequencies are provided to eliminate frequency congestion on the tower (local control) frequency and are limited to communications between the tower and aircraft on the ground and between the tower and utility vehicles on the airport, pro-viding a clear VHF channel for arriving and departing aircraft. They are used for issuance of taxi information, clearances, and other necessary contacts between the tower and aircraft or other vehicles operated on the airport. A pilot who has just landed should not change from the tower frequency to the ground control frequency until directed to do so by the controller. Normally, only one ground control frequency is assigned at an airport; however, at locations where the amount of traffic so warrants, a second ground control frequency and/or another frequency designated as a clearance delivery frequency, may be assigned.

d. A controller may omit the ground or local control frequency if the controller believes the pilot knows which frequency is in use. If the ground control frequency is in the 121 MHz bandwidth the controller may omit the numbers preceding the decimal point; e.g., 121.7, "CONTACT GROUND POINT SEVEN." However, if any doubt exists

Table 4-3-1 ATCT Light Gun Signals

MEANING

COLOR AND TYPE OF SIGNAL	MOVEMENT OF VEHICLES EQUIPMENT AND PERSONNEL	AIRCRAFT ON THE GROUND	AIRCRAFT IN FLIGHT
Steady green	Cleared to cross, proceed or go	Cleared for takeoff	Cleared to land
Flashing green	Not applicable	Cleared for taxi	Return for landing (to be followed by steady green at the proper time)
Steady red	STOP	STOP	Give way to other aircraft and continue circling
Flashing red	Clear the taxiway/runway	Taxi clear of the runway in use	Airport unsafe, do not land
Flashing white	Return to starting point on airport	Return to starting point on airport	Not applicable
Alternating red and green	Exercise extreme caution	Exercise extreme caution	Exercise extreme caution

as to what frequency is in use, the pilot should promptly request the controller to provide that information.

e. Controllers will normally avoid issuing a radio frequency change to helicopters, known to be single-piloted, which are hovering, air taxiing, or flying near the ground. At times, it may be necessary for pilots to alert ATC regarding single pilot operations to minimize delay of essential ATC communications. Whenever possible, ATC instructions will be relayed through the frequency being monitored until a frequency change can be accomplished. You must promptly advise ATC if you are unable to comply with a frequency change. Also, you should advise ATC if you must land to accomplish the frequency change unless it is clear the landing will have no impact on other air traffic; e.g., on a taxiway or in a helicopter operating area.

4-3-15. GATE HOLDING DUE TO DEPARTURE DELAYS

a. Pilots should contact ground control or clearance delivery prior to starting engines as gate hold procedures will be in effect whenever departure delays exceed or are anticipated to exceed 15 minutes. The sequence for departure will be maintained in accordance with initial call up unless modified by flow control restrictions. Pilots should monitor the ground control or clearance delivery frequency for engine startup advisories or new proposed start time if the delay changes.

b. The tower controller will consider that pilots of turbine-powered aircraft are ready for takeoff when they reach the runway or warm-up block unless advised otherwise.

4-3-16. VFR FLIGHTS IN TERMINAL AREAS

Use reasonable restraint in exercising the prerogative of VFR flight, especially in terminal areas. The weather minimums and distances from clouds are minimums. Giving yourself a greater margin in specific instances is just good judgment.

a. Approach Area: Conducting a VFR operation in a Class B, Class C, Class D and Class E surface area when the official visibility is 3 or 4 miles is not prohibited, but good judgment would dictate that you keep out of the approach area.

b. Reduced Visibility: It has always been recognized that precipitation reduces forward visibility. Consequently, although again it may be perfectly legal to cancel your IFR flight plan at any time you can proceed VFR, it is good practice, when precipitation is occurring, to continue IFR operation into a terminal area until you are reasonably close to your destination.

c. Simulated Instrument Flights: In conducting simulated instrument flights, be sure that the weather is good enough to compensate for the restricted visibility of the safety pilot and your greater concentration on your flight instruments. Give yourself a little greater margin when your flight plan lies in or near a busy airway or close to an airport.

4-3-17. VFR HELICOPTER OPERATIONS AT CONTROLLED AIRPORTS

a. General:

1. The following ATC procedures and phraseologies recognize the unique capabilities of helicopters and were developed to improve service to all users. Helicopter design characteristics and user needs often require operations from movement areas and nonmovement areas within the airport boundary. In order for ATC to properly apply these procedures, it is essential that pilots familiarize themselves with the local operations and make it known to controllers when additional instructions are necessary.

2. Insofar as possible, helicopter operations will be instructed to avoid the flow of fixed-wing aircraft to minimize overall delays; however, there will be many situations where faster/larger helicopters may be integrated with fixed-wing aircraft for the benefit of all concerned. Examples would include IFR flights, avoidance of noise sensitive areas, or use of runways/taxiways to minimize the hazardous effects of rotor downwash in congested areas.

3. Because helicopter pilots are intimately familiar with the effects of rotor downwash, they are best qualified to determine if a given operation can be conducted safely. Accordingly, the pilot has the final authority with respect to the specific airspeed/altitude combinations. ATC clearances are in no way intended to place the helicopter in a hazardous position. It is expected that pilots will advise ATC if a specific clearance will cause undue hazards to persons or property.

b. Controllers normally limit ATC ground service and instruction to *movement* areas; therefore, operations from *nonmovement* areas are conducted at pilot discretion and should be based on local policies, procedures, or letters of agreement. In order to maximize the flexibility of helicopter operations, it is necessary to rely heavily on sound pilot judgment. For example, hazards such as debris, obstructions, vehicles, or personnel must be recognized by the pilot, and action should be taken as necessary to avoid such hazards. Taxi, hover taxi, and air taxi operations are considered to be ground movements. Helicopters conducting such operations are expected to adhere to the same conditions, requirements, and practices as apply to other ground taxiing and ATC procedures in the AIM.

1. The phraseology taxi is used when it is intended or expected that the helicopter will taxi on the airport surface, either via taxiways or other prescribed routes. *Taxi* is used primarily for helicopters equipped with wheels or in response to a pilot request. Preference should be given to this procedure whenever it is necessary to minimize effects of rotor downwash.

2. Pilots may request a *hover taxi* when slow forward movement is desired or when it may be appropriate to move very short distances. Pilots should avoid this procedure if rotor downwash is likely to cause damage to parked aircraft or if blowing dust/snow could obscure visibility. If it is necessary to operate above 25 feet AGL when hover taxiing, the pilot should initiate a request to ATC.

3. *Air taxi* is the preferred method for helicopter ground movements on airports provided ground operations and conditions permit. Unless otherwise requested or instructed, pilots are expected to remain below 100 feet AGL. However, if a higher than normal airspeed or altitude is desired, the request should be made prior to lift-off. The pilot is solely responsible for selecting a safe airspeed for the altitude/operation being conducted. Use of *air taxi* enables the pilot to proceed at an optimum airspeed/altitude, minimize downwash effect, conserve fuel, and expedite movement from one point to another. Helicopters should avoid overflight of other aircraft, vehicles, and personnel during air-taxi operations. Caution must be exercised concerning active runways and pilots must be certain that air taxi instructions are understood. Special precautions may be necessary at unfamiliar airports or airports with multiple intersecting active runways. The taxi procedures given in: AIM, Taxiing, paragraph 4-3-18, Taxi During Low Visibility, paragraph 4-3-19, and Exiting the Runway After Landing, paragraph 4-3-20 also apply.

REFERENCE—
PILOT/CONTROLLER GLOSSARY TERM-TAXI.
PILOT/CONTROLLER GLOSSARY TERM-HOVER TAXI.
PILOT/CONTROLLER GLOSSARY TERM-AIR TAXI.

c. Takeoff and Landing Procedures:

1. Helicopter operations may be conducted from a runway, taxiway, portion of a landing strip, or any clear area which could be used as a landing site such as the scene of an accident, a construction site, or the roof of a building. The terms used to describe designated areas from which helicopters operate are: movement area, landing/takeoff area, apron/ramp, heliport and helipad (See *Pilot/Controller Glossary*). These areas may be improved or unimproved and may be separate from or located on an airport/heliport. ATC will issue takeoff clearances from *movement* areas other than active runways, or in diverse directions from active runways, with additional instructions as necessary. Whenever possible, takeoff clearance will be issued in lieu of extended hover/air taxi operations. Phraseology will be "CLEARED FOR TAKEOFF FROM (taxiway, helipad, runway number, etc.), MAKE RIGHT/LEFT TURN FOR (direction, heading, NAVAID radial) DEPARTURE/DEPARTURE ROUTE (number, name, etc.)." Unless requested by the pilot, downwind takeoffs will not be issued if the tailwind exceeds 5 knots.

2. Pilots should be alert to wind information as well as to wind indications in the vicinity of the helicopter. ATC should be advised of the intended method of departing. A pilot request to takeoff in a given direction indicates that the pilot is willing to accept the wind condition and controllers will honor the request if traffic permits. Departure points could be a significant distance from the control tower and it may be difficult or impossible for the controller to determine the helicopter's relative position to the wind.

3. If takeoff is requested from *nonmovement* areas, the phraseology "PROCEED AS REQUESTED" will be used. Additional instructions will be issued as necessary. The pilot is responsible for operating in a safe manner and should exercise due caution. When other known traffic is not a factor and takeoff is requested from an area not visible from the tower, an area not authorized for helicopter use, an unlighted area at night, or an area not on the airport, the phraseology "DEPARTURE FROM (location) WILL BE AT YOUR OWN RISK (with reason, and additional instructions as necessary)."

4. Similar phraseology is used for helicopter landing operations. Every effort will be made to permit helicopters to proceed direct and land as near as possible to their final destination on the airport. Traffic density, the need for detailed taxiing instructions, frequency congestion, or other factors may affect the extent to which service can be expedited. As with ground movement operations, a high degree of pilot/controller cooperation and communication is necessary to achieve safe and efficient operations.

4-3-18. TAXIING

a. General: Approval must be obtained prior to moving an aircraft or vehicle onto the movement area during the hours an Airport Traffic Control Tower is in operation.

1. Always state your position on the airport when calling the tower for taxi instructions.

2. The movement area is normally described in local bulletins issued by the airport manager or control tower. These bulletins may be found in FSSs, fixed base operators offices, air carrier offices, and operations offices.

3. The control tower also issues bulletins describing areas where they cannot provide ATC service due to nonvisibility or other reasons.

4. A clearance must be obtained prior to taxiing on a runway, taking off, or landing during the hours an Airport Traffic Control Tower is in operation.

5. When ATC clears an aircraft to "taxi to" an assigned takeoff runway, the absence of holding instructions authorizes the aircraft to "cross" all runways which the taxi route intersects except the assigned takeoff runway. It does not include authorization to "taxi onto" or "cross" the assigned takeoff runway at any point. In order to preclude misunderstandings in radio communications, ATC will not use the word "cleared" in conjunction with authorization for aircraft to taxi.

6. In the absence of holding instructions, a clearance to "taxi to" any point other than an assigned takeoff runway is a clearance to cross all runways that intersect the taxi route to that point.

7. Air traffic control will first specify the runway, issue taxi instructions, and then state any required hold short instructions, when authorizing an aircraft to taxi for departure. This does not authorize the aircraft to "enter" or "cross" the assigned departure runway at any point.

NOTE—AIR TRAFFIC CONTROLLERS ARE REQUIRED TO OBTAIN FROM THE PILOT A READBACK OF ALL RUNWAY HOLD SHORT INSTRUCTIONS.

8. If a pilot is expected to hold short of a runway approach ("APPCH") area or ILS holding position (see FIG 2-3-15, Taxiways Located in Runway Approach Area), ATC will issue instructions.

9. When taxi instructions are received from the controller, pilots should always read back:

(a) The runway assignment.

(b) Any clearance to enter a specific runway.

(c) Any instructions to hold short of a specific runway, or taxi into position and hold.

Controllers are required to request a readback of runway hold short assignment when it is not received from the pilot/vehicle.

b. ATC clearances or instructions pertaining to taxiing are predicated on known traffic and known physical airport conditions. Therefore, it is important that pilots clearly understand the clearance or instruction. Although an ATC clearance is issued for taxiing purposes, when operating in accordance with the FARs, it is the responsibility of the pilot to avoid collision with other aircraft. Since "the pilot-in-command of an aircraft is directly responsible for, and is the final authority as to, the operation of that aircraft" the pilot should obtain clarification of any clearance or instruction which is not understood.

REFERENCE—AIM, GENERAL, PARAGRAPH 7-3-1.

1. Good operating practice dictates that pilots acknowledge all runway crossing, hold short, or takeoff clearances unless there is some misunderstanding, at which time the pilot should query the controller until the clearance is understood.

NOTE—AIR TRAFFIC CONTROLLERS ARE REQUIRED TO OBTAIN FROM THE PILOT A READBACK OF ALL RUNWAY HOLD SHORT INSTRUCTIONS.

2. Pilots operating a single pilot aircraft should monitor only assigned ATC communications after being cleared onto the active runway for departure. Single pilot aircraft should not monitor other than ATC communications until flight from Class B, Class C, or Class D surface area is completed. This same procedure should be practiced from after receipt of the clearance for landing until the landing and taxi activities are complete. Proper effective scanning for other aircraft, surface vehicles, or other objects should be continuously exercised in all cases.

3. If the pilot is unfamiliar with the airport or for any reason confusion exists as to the correct taxi routing, a request may be made for progressive taxi instructions which include step-by-step routing directions. Progressive instructions may also be issued if the controller deems it necessary due to traffic or field conditions; i.e., construction or closed taxiways.

c. At those airports where the U.S. Government operates the control tower and ATC has authorized noncompliance with the requirement for two-way radio communications while operating within the Class B, Class C, or Class D surface area, or at those airports where the U.S. Government does not operate the control tower and radio communications cannot be established, pilots shall obtain a clearance by visual light signal prior to taxiing on a runway and prior to takeoff and landing.

d. The following phraseologies and procedures are used in radiotelephone communications with aeronautical ground stations.

1. *Request for taxi instructions prior to departure:* State your aircraft identification, location, type of operation planned (VFR or IFR), and the point of first intended landing.

EXAMPLE—AIRCRAFT: "WASHINGTON GROUND, BEECHCRAFT ONE THREE ONE FIVE NINER AT HANGAR EIGHT, READY TO TAXI, I-F-R TO CHICAGO."

TOWER: "BEECHCRAFT ONE THREE ONE FIVE NINER, WASHINGTON GROUND, TAXI TO RUNWAY THREE SIX, WIND ZERO THREE ZERO AT TWO FIVE. ALTIMETER THREE ZERO ZERO FOUR."

OR

TOWER: "BEECHCRAFT ONE THREE ONE FIVE NINER, WASHINGTON GROUND, RUNWAY TWO SEVEN, TAXI VIA TAXIWAYS CHARLIE AND DELTA, HOLD SHORT OF RUNWAY THREE THREE LEFT."

AIRCRAFT: "BEECHCRAFT ONE THREE ONE FIVE NINER, HOLD SHORT OF RUNWAY THREE THREE LEFT."

2. *Receipt of ATC clearance:* ARTCC clearances are relayed to pilots by airport traffic controllers in the following manner.

EXAMPLE—TOWER: "BEECHCRAFT ONE THREE ONE FIVE NINER, CLEARED TO THE CHICAGO MIDWAY AIRPORT VIA VICTOR EIGHT, MAINTAIN EIGHT THOUSAND."

AIRCRAFT: "BEECHCRAFT ONE THREE ONE FIVE NINER, CLEARED TO THE CHICAGO MIDWAY AIRPORT VIA VICTOR EIGHT, MAINTAIN EIGHT THOUSAND."

NOTE—NORMALLY, AN ATC IFR CLEARANCE IS RELAYED TO A PILOT BY THE GROUND CONTROLLER. AT BUSY LOCATIONS, HOWEVER, PILOTS MAY BE INSTRUCTED BY THE GROUND CONTROLLER TO "CONTACT CLEARANCE DELIVERY" ON A FREQUENCY DESIGNATED FOR THIS PURPOSE. NO SURVEILLANCE OR CONTROL OVER THE MOVEMENT OF TRAFFIC IS EXERCISED BY THIS POSITION OF OPERATION.

3. *Request for taxi instructions after landing:* State your aircraft identification, location, and that you request taxi instructions.

EXAMPLE—AIRCRAFT: "DULLES GROUND, BEECHCRAFT ONE FOUR TWO SIX ONE CLEARING RUNWAY ONE RIGHT ON TAXIWAY ECHO THREE, REQUEST CLEARANCE TO PAGE."

TOWER: "BEECHCRAFT ONE FOUR TWO SIX ONE, DULLES GROUND, TAXI TO PAGE VIA TAXIWAYS ECHO THREE, ECHO ONE, AND ECHO NINER."

OR

AIRCRAFT: "ORLANDO GROUND, BEECHCRAFT ONE FOUR TWO SIX ONE CLEARING RUNWAY ONE EIGHT LEFT

AT TAXIWAY BRAVO THREE, REQUEST CLEARANCE TO PAGE."

TOWER: "BEECHCRAFT ONE FOUR TWO SIX ONE, ORLANDO GROUND, HOLD SHORT OF RUNWAY ONE EIGHT RIGHT."

AIRCRAFT: "BEECHCRAFT ONE FOUR TWO SIX ONE, HOLD SHORT OF RUNWAY ONE EIGHT RIGHT."

4-3-19. TAXI DURING LOW VISIBILITY

a. Pilots and aircraft operators should be constantly aware that during certain low visibility conditions the movement of aircraft and vehicles on airports may not be visible to the tower controller. This may prevent visual confirmation of an aircraft's adherence to taxi instructions.

b. Of vital importance is the need for pilots to notify the controller when difficulties are encountered or at the first indication of becoming disoriented. Pilots should proceed with extreme caution when taxiing toward the sun. When vision difficulties are encountered pilots should immediately inform the controller.

c. Advisory Circular 120-57, Surface Movement Guidance and Control System, commonly known as SMGCS (pronounced "SMIGS") requires a low visibility taxi plan for any airport which has takeoff or landing operations in less than 1,200 feet runway visual range (RVR) visibility conditions. These plans, which affect aircrew and vehicle operators, may incorporate additional lighting, markings, and procedures to control airport surface traffic. They will be addressed at two levels; operations less than 1,200 feet RVR to 600 feet RVR and operations less than 600 feet RVR.

NOTE—SPECIFIC LIGHTING SYSTEMS AND SURFACE MARKINGS MAY BE FOUND IN AIM, PARAGRAPHS 2-1-9 AND 2-3-4.

d. When low visibility conditions exist, pilots should focus their entire attention on the safe operation of the aircraft while it is moving. Checklists and nonessential communication should be withheld until the aircraft is stopped and the brakes set.

4-3-20. EXITING THE RUNWAY AFTER LANDING

The following procedures should be followed after landing and reaching taxi speed.

a. Exit the runway without delay at the first available taxiway or on a taxiway as instructed by ATC. Pilots shall not exit the landing runway onto another runway unless authorized by ATC. At airports with an operating control tower, pilots should not stop or reverse course on the runway without first obtaining ATC approval.

b. Taxi clear of the runway unless otherwise directed by ATC. In the absence of ATC instructions the pilot is expected to taxi clear of the landing runway by clearing the hold position marking associated with the landing runway even if that requires the aircraft to protrude into or cross another taxiway, runway, or ramp area. This does not authorize an aircraft to cross a subsequent taxiway/runway/ramp after clearing the landing runway.

NOTE—THE TOWER WILL ISSUE THE PILOT WITH INSTRUCTIONS WHICH WILL NORMALLY PERMIT THE AIRCRAFT TO ENTER ANOTHER TAXIWAY, RUNWAY, OR RAMP AREA WHEN REQUIRED TO TAXI CLEAR OF THE RUNWAY BY CLEARING THE HOLD POSITION MARKING ASSOCIATED WITH THE LANDING RUNWAY.

c. Stop the aircraft after clearing the runway if instructions have not been received from ATC.

d. Immediately change to ground control frequency when advised by the tower and obtain a taxi clearance.

NOTE—

[1] THE TOWER WILL ISSUE THE PILOT INSTRUCTIONS WHICH WILL NORMALLY PERMIT THE AIRCRAFT TO ENTER ANOTHER TAXIWAY, RUNWAY, OR RAMP AREA WHEN REQUIRED TO TAXI CLEAR OF THE RUNWAY BY CLEARING THE HOLD POSITION MARKING ASSOCIATED WITH THE LANDING RUNWAY.

[2] GUIDANCE CONTAINED IN SUBPARAGRAPHS A AND B ABOVE IS CONSIDERED AN INTEGRAL PART OF THE LANDING CLEARANCE AND SATISFIES THE REQUIREMENT OF 14 CFR SECTION 91.129.

4-3-21. PRACTICE INSTRUMENT APPROACHES

a. Various air traffic incidents have indicated the necessity for adoption of measures to achieve more organized and controlled operations where practice instrument approaches are conducted. Practice instrument approaches are considered to be instrument approaches made by either a VFR aircraft not on an IFR flight plan or an aircraft on an IFR flight plan. To achieve this and thereby enhance air safety, it is Air Traffic's policy to provide for separation of such operations at locations where approach control facilities are located and, as resources permit, at certain other locations served by ARTCCs or parent approach control facilities. Pilot requests to practice instrument approaches may be approved by ATC subject to traffic and workload conditions. Pilots should anticipate that in some instances the controller may find it necessary to deny approval or withdraw previous approval when traffic conditions warrant. It must be clearly understood, however, that even though the controller may be providing separation, pilots on VFR flight plans are required to comply with basic VFR weather minimums (FAR Part 91.155). Application of ATC procedures or any action taken by the controller to avoid traffic conflictions does not relieve IFR and VFR pilots of their responsibility to see-and-avoid other traffic while operating in VFR conditions (FAR Part 91.113). In addition to the normal IFR separation minimums (which includes visual separation) during VFR conditions, 500 feet vertical separation may be applied between VFR aircraft and between a VFR aircraft and the IFR aircraft. Pilots not on IFR flight plans desiring practice instrument approaches should always state 'practice' when making requests to ATC. Controllers will instruct VFR aircraft requesting an instrument approach to maintain VFR. This is to preclude misunderstandings between the pilot and controller as to the status of the aircraft. If pilots wish to proceed in accordance with instrument flight rules, they must specifically request and obtain, an IFR clearance.

b. Before practicing an instrument approach, pilots should inform the approach control facility or the tower of the type of practice approach they desire to make and how they intend to terminate it, i.e., full-stop landing, touch-and-go, or missed or low approach maneuver. This information may be furnished progressively when conducting a series of approaches. Pilots on an IFR flight plan, who have made a series of instrument approaches to full stop landings should inform ATC when they make their final landing. The controller will control flights practicing instrument approaches so as to ensure that they do not disrupt the flow of arriving and departing itinerant IFR or VFR aircraft. The priority afforded itinerant aircraft over practice instrument approaches is not intended to be so rigidly applied that it causes grossly inefficient application of services. A minimum delay to itinerant traffic may be appropriate to allow an aircraft practicing an approach to complete that approach.

NOTE—A CLEARANCE TO LAND MEANS THAT APPROPRIATE SEPARATION ON THE LANDING RUNWAY WILL BE ENSURED. A LANDING CLEARANCE DOES NOT RELIEVE THE PILOT FROM COMPLIANCE WITH ANY PREVIOUSLY ISSUED RESTRICTION.

c. At airports without a tower, pilots wishing to make practice instrument approaches should notify the facility having control jurisdiction of the desired approach as indicated on the approach chart. All approach control facilities and ARTCCs are required to publish a Letter to Airmen depicting those airports where they provide standard separation to both VFR and IFR aircraft conducting practice instrument approaches.

d. The controller will provide approved separation between both VFR and IFR aircraft when authorization is granted to make practice approaches to airports where an approach control facility is located and to certain other airports served by approach control or an ARTCC. Controller responsibility for separation of VFR aircraft begins at the point where the approach clearance becomes effective, or when the aircraft enters Class B or Class C airspace, or a TRSA, whichever comes first.

e. VFR aircraft practicing instrument approaches are not automatically authorized to execute the missed approach procedure. This authorization must be specifically requested by the pilot and approved by the controller. Separation will not be provided unless the missed approach has been approved by ATC.

f. Except in an emergency, aircraft cleared to practice instrument approaches must not deviate from the approved procedure until cleared to do so by the controller.

g. At radar approach control locations when a full approach procedure (Procedure Turn, etc.,) cannot be approved, pilots should expect to be vectored to a final approach course for a practice instrument approach which is compatible with the general direction of traffic at that airport.

h. When granting approval for a practice instrument approach, the controller will usually ask the pilot to report to the tower prior to or over the final approach fix inbound (non-precision approaches) or over the outer marker or fix used in lieu of the outer marker inbound (precision approaches).

i. When authorization is granted to conduct practice instrument approaches to an airport with a tower, but where approved standard separation is not provided to aircraft conducting practice instrument approaches, the tower will approve the practice approach, instruct the aircraft to maintain VFR and issue traffic information, as required.

j. When an aircraft notifies a FSS providing Local Airport Advisory to the airport concerned of the intent to conduct a practice instrument approach and whether or not separation is to be provided, the pilot will be instructed to contact the appropriate facility on a specified frequency prior to initiating the approach. At airports where separation is not provided, the FSS will acknowledge the message and issue known traffic information but will neither approve or disapprove the approach.

k. Pilots conducting practice instrument approaches should be particularly alert for other aircraft operating in the local traffic pattern or in proximity to the airport.

4-3-22. OPTION APPROACH

The "Cleared for the Option" procedure will permit an instructor, flight examiner or pilot the option to make a touch-and-go, low approach, missed approach, stop-and-go, or full stop landing. This procedure can be very beneficial in a training situation in that neither the student pilot nor examinee would know what maneuver would be accomplished. The pilot should make a request for this procedure passing the final approach fix inbound on an instrument approach or entering downwind for a VFR traffic pattern. The advantages of this procedure as a training aid are that it enables an instructor or examiner to obtain the reaction of a trainee or examinee under changing conditions, the pilot would not have to discontinue an approach in the middle of the procedure due to student error or pilot proficiency requirements, and finally it allows more flexibility and economy in training programs. This procedure will only be used at those locations with an operational control tower and will be subject to ATC approval.

4-3-23. USE OF AIRCRAFT LIGHTS

a. Aircraft position lights are required to be lighted on aircraft operated on the surface and in flight from sunset to sunrise. In addition, aircraft equipped with an anti-collision light system are required to operate that light system during all types of operations (day and night). However, during any adverse meteorological conditions, the pilot-in-command may determine that the anti-collision lights should be turned off when their light output would constitute a hazard to safety (14 CFR Section 91.209). Supplementary strobe lights should be turned off on the ground when they adversely affect ground personnel or other pilots, and in flight when there are adverse reflection from clouds.

b. An aircraft anti-collision light system can use one or more rotating beacons and/or strobe lights, be colored either red or white, and have different (higher than minimum) intensities when compared to other aircraft. Many aircraft have both a rotating beacon and a strobe light system.

c. The FAA has a voluntary pilot safety program, *Operation Lights On*, to enhance the *see-and-avoid* concept. Pilots are encouraged to turn on their landing lights during takeoff; i.e., either after takeoff clearance has been received or when beginning takeoff roll. Pilots are further encouraged to turn on their landing lights when operating below 10,000 feet, day or night, especially when operating within 10 miles of any airport, or in conditions of reduced visibility and in areas where flocks of birds may be expected, i.e., coastal areas, lake areas, around refuse dumps, etc. Although turning on aircraft lights does enhance the *see-and-avoid* concept, pilots should not become complacent about keeping a sharp lookout for other aircraft. Not all aircraft are equipped with lights and some pilots may not have their lights turned on. Aircraft manufacturer's recommendations for operation of landing lights and electrical systems should be observed.

d. Prop and jet blast forces generated by large aircraft have overturned or damaged several smaller aircraft taxiing behind them. To avoid similar results, and in the interest of preventing upsets and injuries to ground personnel from such forces, the FAA recommends that air carriers and commercial operators turn on their rotating beacons anytime their aircraft engines are in operation. General Aviation pilots using rotating beacon equipped aircraft are also encouraged to participate in this program which is designed to alert others to the potential hazard. Since this is a voluntary program, exercise caution and do not rely solely on the rotating beacon as an indication that aircraft engines are in operation.

e. At the discretion of the pilot-in-command turn on all external illumination, including landing lights, when taxiing on, across, or holding in position on any runway. This increases the conspicuity of the aircraft to controllers and other pilots approaching to land, taxiing, or crossing the runway. Pilots should comply with any equipment operating limitations and consider the effects of landing and strobe lights on other aircraft in their vicinity. When cleared for takeoff pilots should turn on any remaining exterior lights.

4-3-24. FLIGHT INSPECTION/'FLIGHT CHECK' AIRCRAFT IN TERMINAL AREAS

a. *Flight check* is a call sign used to alert pilots and air traffic controllers when a FAA aircraft is engaged in flight inspection/certification of NAVAIDs and flight procedures. Flight check aircraft fly preplanned high/low altitude flight patterns such as grids, orbits, DME arcs, and tracks, including low passes along the full length of the runway to verify NAVAID performance. In most instances,

these flight checks are being automatically recorded and/or flown in an automated mode.

b. Pilots should be especially watchful and avoid the flight paths of any aircraft using the call sign "Flight Check" or "Flight Check Recorded." The latter call sign; e.g., "Flight Check 47 Recorded" indicates that automated flight inspections are in progress in terminal areas. These flights will normally receive special handling from ATC. Pilot patience and cooperation in allowing uninterrupted recordings can significantly help expedite flight inspections, minimize costly, repetitive runs, and reduce the burden on the U.S. taxpayer.

4-3-25. HAND SIGNALS

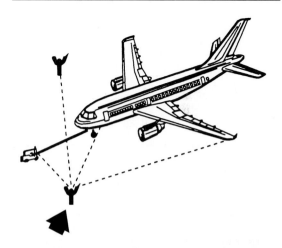

Figure 4-3-7. Signalman Directs Towing

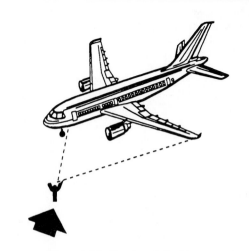

Figure 4-3-8. Signalman's Position

Figure 4-3-9. All Clear

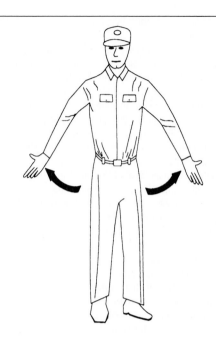

Figure 4-3-11. Pull Chocks

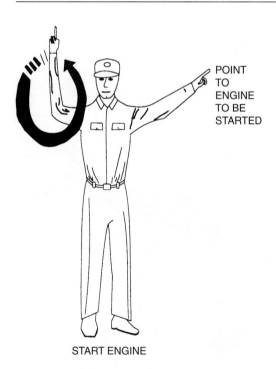

POINT
TO
ENGINE
TO BE
STARTED

START ENGINE

Figure 4-3-10 Start Engine

Figure 4-3-12. Proceed Straight Ahead

Figure 4-3-13. Left Turn

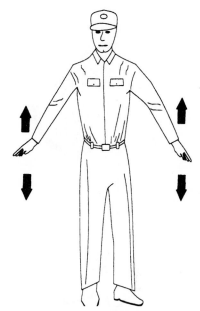

Figure 4-3-15. Slow Down

Figure 4-3-14. Right Turn

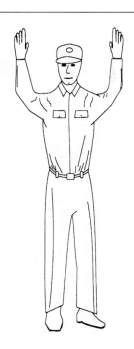

Figure 4-3-16. Flagman Directs Pilot

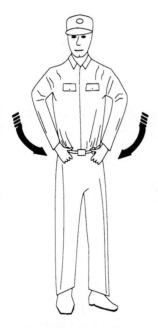

Figure 4-3-17. Insert Chocks

Use same hand
movements as day
operation

Figure 4-3-19. Night Operation

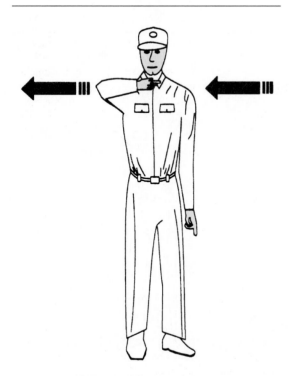

Figure 4-3-18. Cut Engines

Figure 4-3-20. Stop

4-3-26. OPERATIONS AT UNCONTROLLED AIRPORTS WITH AUTOMATED SURFACE OBSERVING SYSTEM (ASOS)/AUTOMATED WEATHER OBSERVING SYSTEM (AWOS)

a. Many airports throughout the National Airspace System are equipped with either ASOS or AWOS. At most airports with an operating control tower or human observer, the weather will be available to you in a METAR hourly or special observation format on the Automatic Terminal Information Service (ATIS) or directly transmitted from the controller/observer.

b. At uncontrolled airports that are equipped with ASOS/AWOS with ground-to-air broadcast capability, the one-minute updated airport weather should be available to you within approximately 25 NM of the airport below 10,000 feet. The frequency for the weather broadcast will be published on sectional charts and in the Airport/Facility Directory. Some part-time towered airports may also broadcast the automated weather on their ATIS frequency during the hours that the tower is closed.

c. Controllers issue SVFR or IFR clearances based on pilot request, known traffic and reported weather i.e., METAR/SPECI observations, when they are available. Pilots have access to more current weather at uncontrolled ASOS/AWOS airports than do the controllers who may be located several miles away. Controllers will rely on the pilot to determine the current airport weather from the ASOS/AWOS. All aircraft arriving or departing an ASOS/AWOS equipped uncontrolled airport should monitor the airport weather frequency to ascertain the status of the airspace. Pilots in Class E airspace must be alert for changing weather conditions which may effect the status of the airspace from IFR/VFR. If ATC service is required for IFR/SVFR approach/departure or requested for VFR service, the pilot should advise the controller that he/she has received the one-minute weather and state his/her intentions.

EXAMPLE—"I HAVE THE (AIRPORT) ONE-MINUTE WEATHER; REQUEST AN ILS RUNWAY 14 APPROACH."

REFERENCE—AIM, WEATHER OBSERVING PROGRAMS, PARAGRAPH 7-1-10.

Section 4. ATC CLEARANCES AND AIRCRAFT SEPARATIONS

4-4-1. CLEARANCE

a. A clearance issued by ATC is predicated on known traffic and known physical airport conditions. An ATC clearance means an authorization by ATC, for the purpose of preventing collision between known aircraft, for an aircraft to proceed under specified conditions within controlled airspace. IT IS NOT AUTHORIZATION FOR A PILOT TO DEVIATE FROM ANY RULE, REGULATION, OR MINIMUM ALTITUDE NOR TO CONDUCT UNSAFE OPERATION OF THE AIRCRAFT.

b. FAR Part 91.3(a) states: "The pilot-in-command of an aircraft is directly responsible for, and is the final authority as to, the operation of that aircraft." If ATC issues a clearance that would cause a pilot to deviate from a rule or regulation, or in the pilot's opinion, would place the aircraft in jeopardy, IT IS THE PILOT'S RESPONSIBILITY TO REQUEST AN AMENDED CLEARANCE. Similarly, if a pilot prefers to follow a different course of action, such as make a 360 degree turn for spacing to follow traffic when established in a landing or approach sequence, land on a different runway, takeoff from a different intersection, takeoff from the threshold instead of an intersection, or delay operation, THE PILOT IS EXPECTED TO INFORM ATC ACCORDINGLY. When the pilot requests a different course of action, however, the pilot is expected to cooperate so as to preclude disruption of traffic flow or creation of conflicting patterns. The pilot is also expected to use the appropriate aircraft call sign to acknowledge all ATC clearances, frequency changes, or advisory information.

c. Each pilot who deviates from an ATC clearance in response to a Traffic Alert and Collision Avoidance System resolution advisory shall notify ATC of that deviation as soon as possible.

REFERENCE—PILOT/CONTROLLER GLOSSARY TERM— TRAFFIC ALERT AND COLLISION AVOIDANCE SYSTEM.

d. When weather conditions permit, during the time an IFR flight is operating, it is the direct responsibility of the pilot to avoid other aircraft since VFR flights may be operating in the same area without the knowledge of ATC. Traffic clearances provide standard separation only between IFR flights.

4-4-2. CLEARANCE PREFIX

A clearance, control information, or a response to a request for information originated by an ATC facility and relayed to the pilot through an air-to-ground communication station will be prefixed by "ATC clears," "ATC advises," or "ATC requests."

4-4-3. CLEARANCE ITEMS

ATC clearances normally contain the following:

a. Clearance Limit: The traffic clearance issued prior to departure will normally authorize flight to the airport of intended landing. Under certain conditions, at some locations a short-range clearance procedure is utilized whereby a clearance is issued to a fix within or just outside of the terminal area and pilots are advised of the frequency on which they will receive the long-range clearance direct from the center controller.

b. Departure Procedure: Headings to fly and altitude restrictions may be issued to separate a departure from other air traffic in the terminal area. Where the volume of traffic warrants, DPs have been developed.

REFERENCE—AIM, ABBREVIATED IFR DEPARTURE CLEARANCE (CLEARED...AS FILED) PROCEDURES, PARAGRAPH 5-2-4.

AIM, INSTRUMENT DEPARTURE PROCEDURE (DP) PARAGRAPH 5-2-7.

c. Route of Flight:

1. Clearances are normally issued for the altitude or flight level and route filed by the pilot. However, due to traffic conditions, it is frequently necessary for ATC to specify an altitude or flight level or route different from that requested by the pilot. In addition, flow patterns have been established in certain congested areas or between congested areas whereby traffic capacity is increased by routing all traffic on preferred routes. Information on these flow patterns is available in offices where preflight briefing is furnished or where flight plans are accepted.

2. When required, air traffic clearances include data to assist pilots in identifying radio reporting points. It is the responsibility of pilots to notify ATC immediately if their radio equipment cannot receive the type of signals they must utilize to comply with their clearance.

d. Altitude Data:

1. The altitude or flight level instructions in an ATC clearance normally require that a pilot "MAINTAIN" the altitude or flight level at which the flight will operate when in controlled airspace. Altitude or flight level changes while en route should be requested prior to the time the change is desired.

2. When possible, if the altitude assigned is different from the altitude requested by the pilot, ATC will inform the pilot when to expect climb or descent clearance or to request altitude change from another facility. If this has not been received prior to crossing the boundary of the ATC facility's area and assignment at a different altitude is still desired, the pilot should reinitiate the request with the next facility.

3. The term "cruise" may be used instead of "MAINTAIN" to assign a block of airspace to a pilot from the minimum IFR altitude up to and including the altitude specified in the cruise clearance. The pilot may level off at any intermediate altitude within this block of airspace. Climb/descent within the block is to be made at the discretion of the pilot. However, once the pilot starts descent and verbally reports leaving an altitude in the block, the pilot may not return to that altitude without additional ATC clearance.

REFERENCE—PILOT/CONTROLLER GLOSSARY TERM—CRUISE.

e. Holding Instructions:

1. Whenever an aircraft has been cleared to a fix other than the destination airport and delay is expected, it is the responsibility of the ATC controller to issue complete holding instructions (unless the pattern is charted), an EFC time, and a best estimate of any additional en route/terminal delay.

2. If the holding pattern is charted and the controller doesn't issue complete holding instructions, the pilot is expected to hold as depicted on the appropriate chart. When the pattern is charted, the controller may omit all holding instructions except the charted holding direction and the statement AS PUBLISHED, e.g., "HOLD EAST AS PUBLISHED." Controllers shall always issue complete holding instructions when pilots request them.

NOTE—ONLY THOSE HOLDING PATTERNS DEPICTED ON U.S. GOVERNMENT OR COMMERCIALLY PRODUCED CHARTS WHICH MEET FAA REQUIREMENTS SHOULD BE USED.

3. If no holding pattern is charted and holding instructions have not been issued, the pilot should ask ATC for holding instructions prior to reaching the fix. This procedure will eliminate the possibility of an aircraft entering a holding pattern other than that desired by ATC. If unable to obtain holding instructions prior to reaching the fix (due to frequency congestion, stuck microphone, etc.), hold in a standard pattern on the course on which you approached the fix and request further clearance as soon as possible. In this event, the altitude/flight level of the aircraft at the clearance limit will be protected so that separation will be provided as required.

4. When an aircraft is 3 minutes or less from a clearance limit and a clearance beyond the fix has not been received, the pilot is expected to start a speed reduction so that the aircraft will cross the fix, initially, at or below the maximum holding airspeed.

5. When no delay is expected, the controller should issue a clearance beyond the fix as soon as possible and, whenever possible, at least 5 minutes before the aircraft reaches the clearance limit.

6. Pilots should report to ATC the time and altitude/flight level at which the aircraft reaches the clearance limit and report leaving the clearance limit.

NOTE—IN THE EVENT OF TWO-WAY COMMUNICATIONS FAILURE, PILOTS ARE REQUIRED TO COMPLY WITH FAR PART 91.185.

4-4-4. AMENDED CLEARANCES

a. Amendments to the initial clearance will be issued at any time an air traffic controller deems such action necessary to avoid possible confliction between aircraft. Clearances will require that a flight "hold" or change altitude prior to reaching the point where standard separation from other IFR traffic would no longer exist.

NOTE—SOME PILOTS HAVE QUESTIONED THIS ACTION AND REQUESTED "TRAFFIC INFORMATION" AND WERE AT A LOSS WHEN THE REPLY INDICATED "NO TRAFFIC REPORT." IN SUCH CASES THE CONTROLLER HAS TAKEN ACTION TO PREVENT A TRAFFIC CONFLICTION WHICH WOULD HAVE OCCURRED AT A DISTANT POINT.

b. A pilot may wish an explanation of the handling of the flight at the time of occurrence; however, controllers are not able to take time from their immediate control duties nor can they afford to overload the ATC communications channels to furnish explanations. Pilots may obtain an explanation by directing a letter or telephone call to the chief controller of the facility involved.

c. Pilots have the privilege of requesting a different clearance from that which has been issued by ATC if they feel that they have information which would make another course of action more practicable or if aircraft equipment limitations or company procedures forbid compliance with the clearance issued.

4-4-5. SPECIAL VFR CLEARANCES

a. An ATC clearance must be obtained *prior* to operating within a Class B, Class C, Class D or Class E surface area when the weather is less than that required for VFR flight. A VFR pilot may request and be given a clearance to enter, leave, or operate within most Class D and Class E surface areas and some Class B and Class C surface areas in Special VFR conditions, traffic permitting, and providing such flight will not delay IFR operations. All Special VFR flights must remain clear of clouds. The visibility requirements for Special VFR aircraft (other than helicopters) are:

1. At least 1 statute mile flight visibility for operations within Class B, Class C, Class D and Class E surface areas.

2. At least 1 statute mile ground visibility if taking off or landing. If ground visibility is not reported at that airport, the flight visibility must be at least 1 statute mile.

3. The restrictions in subparagraphs 1. and 2. do not apply to helicopters. Helicopters must remain clear of clouds and may operate in Class B, Class C, Class D and Class E surface areas with less than 1 statute mile visibility.

b. When a control tower is located within the Class B, Class C, or Class D surface area, requests for clearances should be to the tower. In a Class E surface area, a clearance may be obtained from the nearest tower, FSS, or center.

c. It is not necessary to file a complete flight plan with the request for clearance, but pilots should state their intentions in sufficient detail to permit ATC to fit their flight into the traffic flow. The clearance will not contain a specific altitude as the pilot must remain clear of clouds. The controller may require the pilot to fly at or below a certain altitude due to other traffic, but the altitude specified will permit flight at or above the Minimum Safe Altitude. In addition, at radar locations, flights may be vectored if necessary for control purposes or on pilot request.

NOTE—THE PILOT IS RESPONSIBLE FOR OBSTACLE OR TERRAIN CLEARANCE (REFERENCE FAR PART 91.119).

d. Special VFR clearances are effective within Class B, Class C, Class D and Class E surface areas only. ATC does not provide separation after an aircraft leaves the Class B, Class C, Class D or Class E surface area on a Special VFR clearance.

e. Special VFR operations by fixed-wing aircraft are prohibited in some Class B and Class C surface areas due to the volume of IFR traffic. A list of these Class B and Class C surface areas is contained in FAR Part 91, Appendix D, Section 3. They are also depicted on Sectional Aeronautical Charts.

f. ATC provides separation between Special VFR flights and between these flights and other IFR flights.

g. Special VFR operations by fixed-wing aircraft are prohibited between sunset and sunrise unless the pilot is instrument rated and the aircraft is equipped for IFR flight.

h. Pilots arrivng or departing an uncontrolled airport that has automated weather broadcast capability (ASOS/AWOS) should monitor the broadcast frequency, advise the controller that they have the "one-minute weather" and state intentions prior to operating within the Class B, Class C, Class D or Class E surface areas.

REFERENCE: Pilot/Controller Glossary TERM—ONE-MINUTE WEATHER

4-4-6. PILOT RESPONSIBILITY UPON CLEARANCE ISSUANCE

a. Record ATC clearance: When conducting an IFR operation, make a written record of your clearance. The specified conditions which are a part of your air traffic clearance may be somewhat different from those included in your flight plan. Additionally, ATC may find it necessary to ADD conditions, such as particular departure route. The very fact that ATC specifies different or additional conditions means that other aircraft are involved in the traffic situation.

b. ATC Clearance/Instruction Readback: Pilots of airborne aircraft should read back *those parts* of ATC clearances and instructions containing altitude assignments or vectors as a means of mutual verification. The readback of the "numbers" serves as a double check between pilots and controllers and reduces the kinds of communications errors that occur when a number is either "misheard" or is incorrect.

1. Include the aircraft identification in all readbacks and acknowledgements. This aids controllers in determining that the correct aircraft received the clearance or instruction. The requirement to include aircraft identification in all readbacks and acknowledgements becomes more important as frequency congestion increases and when aircraft with similar call signs are on the same frequency.

EXAMPLE—"CLIMBING TO FLIGHT LEVEL THREE THREE ZERO, UNITED TWELVE" OR "NOVEMBER FIVE CHARLIE TANGO, ROGER, CLEARED TO LAND."

2. Read back altitudes, altitude restrictions, and vectors in the same sequence as they are given in the clearance or instruction.

3. Altitudes contained in charted procedures, such as DPs, instrument approaches, etc., should not be read back unless they are specifically stated by the controller.

c. It is the responsibility of the pilot to accept or refuse the clearance issued.

4-4-7. IFR CLEARANCE VFR-ON-TOP

a. A pilot on an IFR flight plan operating in VFR weather conditions, may request VFR-ON-TOP in lieu of an assigned altitude. This permits a pilot to select an altitude or flight level of their choice (subject to any ATC restrictions.)

b. Pilots desiring to climb through a cloud, haze, smoke, or other meteorological formation and then either cancel their IFR flight plan or operate VFR-ON-TOP may request a climb to VFR-ON-TOP. The ATC authorization shall contain either a top report or a statement that no top report is available, and a request to report reaching VFR-ON-TOP. Additionally, the ATC authorization may contain a clearance limit, routing and an alternative clearance if VFR-ON-TOP is not reached by a specified altitude.

c. A pilot on an IFR flight plan, operating in VFR conditions, may request to climb/descend in VFR conditions.

d. ATC may not authorize VFR-ON-TOP/VFR CONDITIONS operations unless the pilot requests the VFR operation or a clearance to operate in VFR CONDITIONS will result in noise abatement benefits where part of the IFR departure route does not conform to an FAA approved noise abatement route or altitude.

e. When operating in VFR conditions with an ATC authorization to "MAINTAIN VFR-ON-TOP/MAINTAIN VFR CONDITIONS" pilots on IFR flight plans must:

1. Fly at the appropriate VFR altitude as prescribed in FAR Part 91.159.

2. Comply with the VFR visibility and distance from cloud criteria in FAR Part 91.155 (BASIC VFR WEATHER MINIMUMS.)

3. Comply with instrument flight rules that are applicable to this flight; i.e., minimum IFR altitudes, position reporting, radio communications, course to be flown, adherence to ATC clearance, etc.

NOTE—PILOTS SHOULD ADVISE ATC PRIOR TO ANY ALTITUDE CHANGE TO INSURE THE EXCHANGE OF ACCURATE TRAFFIC INFORMATION.

f. ATC authorization to "MAINTAIN VFR-ON-TOP" is not intended to restrict pilots so that they must operate only *above* an obscuring meteorological formation (layer). Instead, it permits operation above, below, between layers, or in areas where there is no meteorological obscuration. It is imperative, however, that pilots understand that clearance to operate "VFR-ON-TOP/VFR CONDITIONS" does not imply cancellation of the IFR flight plan.

g. Pilots operating VFR-ON-TOP/VFR CONDITIONS may receive traffic information from ATC on other pertinent IFR or VFR aircraft. However, aircraft operating in Class B airspace/TRSAs shall be separated as required by FAA Order 7110.65.

NOTE—WHEN OPERATING IN VFR WEATHER CONDITIONS, IT IS THE PILOT'S RESPONSIBILITY TO BE VIGILANT SO AS TO SEE-AND-AVOID OTHER AIRCRAFT.

h. ATC will not authorize VFR or VFR-ON-TOP operations in Class A airspace.

REFERENCE—AIM, CLASS A AIRSPACE, PARAGRAPH 3-2-2.

4-4-8. VFR/IFR FLIGHTS

A pilot departing VFR, either intending to or needing to obtain an IFR clearance en route, must be aware of the position of the aircraft and the relative terrain/obstructions. When accepting a clearance below the MEA/MIA/MVA/OROCA, pilots are responsible for their own terrain/obstruction clearance until reaching the MEA/MIA/MVA/OROCA. If pilots are unable to maintain terrain/obstruction clearance, the controller should be advised and pilots should state their intentions.

NOTE—OROCA IS AN OFF-ROUTE ALTITUDE WHICH PROVIDES OBSTRUCTION CLEARANCE WITH A 1,000 FOOT BUFFER IN NONMOUNTAINOUS TERRAIN AREAS AND A 2,000 FOOT BUFFER IN DESIGNATED MOUNTAINOUS AREAS WITHIN THE UNITED STATES. THIS ALTITUDE MAY NOT PROVIDE SIGNAL COVERAGE FROM GROUND-BASED NAVIGATIONAL AIDS, AIR TRAFFIC CONTROL RADAR, OR COMMUNICATIONS COVERAGE.

4-4-9. ADHERENCE TO CLEARANCE

a. When air traffic clearance has been obtained under either Visual or Instrument Flight Rules, the pilot-in-command of the aircraft shall not deviate from the provisions thereof unless an amended clearance is obtained. When ATC issues a clearance or instruction, pilots are expected to execute its provisions upon receipt. ATC, in certain situations, will include the word "IMMEDIATELY" in a clearance or instruction to impress urgency of an imminent situation and expeditious compliance by the pilot is expected and necessary for safety. The addition of a VFR or other restriction; i.e., climb or descent point or time, crossing altitude, etc., does not authorize a pilot to deviate from the route of flight or any other provision of the ATC clearance.

b. When a heading is assigned or a turn is requested by ATC, pilots are expected to promptly initiate the turn, to complete the turn, and maintain the new heading unless issued additional instructions.

c. The term "AT PILOT'S DISCRETION" included in the altitude information of an ATC clearance means that ATC has offered the pilot the option to start climb or descent when the pilot wishes, is authorized to conduct the climb or descent at any rate, and to temporarily level off at any intermediate altitude as desired. However, once the aircraft has vacated an altitude, it may not return to that altitude.

d. When ATC has not used the term "AT PILOT'S DISCRETION" nor imposed any climb or descent restrictions, pilots should initiate climb or descent promptly on acknowledgement of the clearance. Descend or climb at an optimum rate consistent with the operating characteristics of the aircraft to 1,000 feet above or below the assigned altitude, and then attempt to descend or climb at a rate of between 500 and 1,500 fpm until the assigned altitude is reached. If at anytime the pilot is unable to climb or descend at a rate of at least 500 feet a minute, advise ATC. If it is necessary to level off at an intermediate altitude during climb or descent, advise ATC, except when leveling off at 10,000 feet MSL on descent, or 2,500 feet above airport elevation (prior to entering a Class C or Class D surface area), when required for speed reduction.

REFERENCE—FAR PART 91.117.

NOTE—LEVELING OFF AT 10,000 FEET MSL ON DESCENT OR 2,500 FEET ABOVE AIRPORT ELEVATION (PRIOR TO ENTERING A CLASS C OR CLASS D SURFACE AREA) TO COMPLY WITH FAR PART 91.117 AIRSPEED RESTRICTIONS IS COMMONPLACE. CONTROLLERS ANTICIPATE THIS ACTION AND PLAN ACCORDINGLY. LEVELING OFF AT ANY OTHER TIME ON CLIMB OR DESCENT MAY SERIOUSLY AFFECT AIR TRAFFIC HANDLING BY ATC. CONSEQUENTLY, IT IS IMPERATIVE THAT PILOTS MAKE EVERY EFFORT TO FULFILL THE ABOVE EXPECTED ACTIONS TO AID ATC IN SAFELY HANDLING AND EXPEDITING TRAFFIC.

e. If the altitude information of an ATC DESCENT clearance includes a provision to "CROSS (fix) AT" or "AT OR ABOVE/BELOW (altitude)," the manner in which the descent is executed to comply with the crossing altitude is at the pilot's discretion. This authorization to descend at pilot's discretion is only applicable to that portion of the flight to which the crossing altitude restriction applies, and the pilot is expected to comply with the crossing altitude as a provision of the clearance. Any other clearance in which pilot execution is optional will so state "AT PILOT'S DISCRETION."

EXAMPLE—

[1] "UNITED FOUR SEVENTEEN, DESCEND AND MAINTAIN SIX THOUSAND."

NOTE—

[1] THE PILOT IS EXPECTED TO COMMENCE DESCENT UPON RECEIPT OF THE CLEARANCE AND TO DESCEND AT THE SUGGESTED RATES UNTIL REACHING THE ASSIGNED ALTITUDE OF 6,000 FEET.

EXAMPLE—

[2] "UNITED FOUR SEVENTEEN, DESCEND AT PILOT'S DISCRETION, MAINTAIN SIX THOUSAND."

NOTE—

[2] THE PILOT IS AUTHORIZED TO CONDUCT DESCENT WITHIN THE CONTEXT OF THE TERM AT PILOT'S DISCRETION AS DESCRIBED ABOVE.

EXAMPLE—

[3] "UNITED FOUR SEVENTEEN, CROSS LAKEVIEW V-O-R AT OR ABOVE FLIGHT LEVEL TWO ZERO ZERO, DESCEND AND MAINTAIN SIX THOUSAND."

NOTE—

[3] THE PILOT IS AUTHORIZED TO CONDUCT DESCENT AT PILOT'S DISCRETION UNTIL REACHING LAKEVIEW VOR AND MUST COMPLY WITH THE CLEARANCE PROVISION TO CROSS THE LAKEVIEW VOR AT OR ABOVE FL 200. AFTER PASSING LAKEVIEW VOR, THE PILOT IS EXPECTED TO DESCEND AT THE SUGGESTED RATES UNTIL REACHING THE ASSIGNED ALTITUDE OF 6,000 FEET.

EXAMPLE—

[4] "UNITED FOUR SEVENTEEN, CROSS LAKEVIEW V-O-R AT SIX THOUSAND, MAINTAIN SIX THOUSAND."

NOTE—

[4] THE PILOT IS AUTHORIZED TO CONDUCT DESCENT AT PILOT'S DISCRETION, HOWEVER, MUST COMPLY WITH THE CLEARANCE PROVISION TO CROSS THE LAKEVIEW VOR AT 6,000 FEET.

EXAMPLE—

[5] "UNITED FOUR SEVENTEEN, DESCEND NOW TO FLIGHT LEVEL TWO SEVEN ZERO, CROSS LAKEVIEW V-O-R AT OR BELOW ONE ZERO THOUSAND, DESCEND AND MAINTAIN SIX THOUSAND."

NOTE—

[5] THE PILOT IS EXPECTED TO PROMPTLY EXECUTE AND COMPLETE DESCENT TO FL 270 UPON RECEIPT OF THE CLEARANCE. AFTER REACHING FL 270 THE PILOT IS AUTHORIZED TO DESCEND "AT PILOT'S DISCRETION" UNTIL REACHING LAKEVIEW VOR. THE PILOT MUST COMPLY WITH THE CLEARANCE PROVISION TO CROSS LAKEVIEW VOR AT OR BELOW 10,000 FEET. AFTER LAKEVIEW VOR THE PILOT IS EXPECTED TO DESCEND AT THE SUGGESTED RATES UNTIL REACHING 6,000 FEET.

EXAMPLE—

[6] "UNITED THREE TEN, DESCEND NOW AND MAINTAIN FLIGHT LEVEL TWO FOUR ZERO, PILOT'S DISCRETION AFTER REACHING FLIGHT LEVEL TWO EIGHT ZERO."

NOTE—

[6] THE PILOT IS EXPECTED TO COMMENCE DESCENT UPON RECEIPT OF THE CLEARANCE AND TO DESCEND AT THE SUGGESTED RATES UNTIL REACHING FL 280. AT THAT POINT, THE PILOT IS AUTHORIZED TO CONTINUE DESCENT TO FL 240 WITHIN THE CONTEXT OF THE TERM "AT PILOT'S DISCRETION" AS DESCRIBED ABOVE.

f. In case emergency authority is used to deviate from provisions of an ATC clearance, the pilot-in-command shall notify ATC as soon as possible and obtain an amended clearance. In an emergency situation which does not result in a deviation from the rules prescribed in FAR Part 91 but which requires ATC to give priority to an aircraft, the pilot of such aircraft shall, when requested by ATC, make a report within 48 hours of such emergency situation to the manager of that ATC facility.

g. The guiding principle is that the last ATC clearance has precedence over the previous ATC clearance. When the route or altitude in a previously issued clearance is amended, the controller will restate applicable altitude restrictions. If altitude to maintain is changed or restated, whether prior to departure or while airborne, and previously issued altitude restrictions are omitted, those altitude restrictions are canceled, including DP altitude restrictions.

EXAMPLE—

[1] A DEPARTURE FLIGHT RECEIVES A CLEARANCE TO DESTINATION AIRPORT TO MAINTAIN FL 290. THE CLEARANCE INCORPORATES A DP WHICH HAS CERTAIN ALTITUDE CROSSING RESTRICTIONS. SHORTLY AFTER TAKEOFF, THE FLIGHT RECEIVES A NEW CLEARANCE CHANGING THE MAINTAINING FL FROM 290 TO 250. IF THE ALTITUDE RESTRICTIONS ARE STILL APPLICABLE, THE CONTROLLER RESTATES THEM.

☑ A DEPARTING AIRCRAFT IS CLEARED TO CROSS FLUKY INTERSECTION AT OR ABOVE 3,000 FEET, GORDONVILLE VOR AT OR ABOVE 12,000 FEET, MAINTAIN FL 200. SHORTLY AFTER DEPARTURE, THE ALTITUDE TO BE MAINTAINED IS CHANGED TO FL 240. IF THE ALTITUDE RESTRICTIONS ARE STILL APPLICABLE, THE CONTROLLER ISSUES AN AMENDED CLEARANCE AS FOLLOWS: "CROSS FLUKY INTERSECTION AT OR ABOVE THREE THOUSAND, CROSS GORDONVILLE V-O-R AT OR ABOVE ONE TWO THOUSAND, MAINTAIN FLIGHT LEVEL TWO FOUR ZERO."

☑ AN ARRIVING AIRCRAFT IS CLEARED TO THE DESTINATION AIRPORT VIA V45 DELTA VOR DIRECT; THE AIRCRAFT IS CLEARED TO CROSS DELTA VOR AT 10,000 FEET, AND THEN TO MAINTAIN 6,000 FEET. PRIOR TO DELTA VOR, THE CONTROLLER ISSUES AN AMENDED CLEARANCE AS FOLLOWS: "TURN RIGHT HEADING ONE EIGHT ZERO FOR VECTOR TO RUNWAY THREE SIX I-L-S APPROACH, MAINTAIN SIX THOUSAND."

NOTE—BECAUSE THE ALTITUDE RESTRICTION "CROSS DELTA V-O-R AT 10,000 FEET" WAS OMITTED FROM THE AMENDED CLEARANCE, IT IS NO LONGER IN EFFECT.

h. Pilots of turbojet aircraft equipped with afterburner engines should advise ATC prior to takeoff if they intend to use afterburning during their climb to the en route altitude. Often, the controller may be able to plan traffic to accommodate a high performance climb and allow the aircraft to climb to the planned altitude without restriction.

i. If an "expedite" climb or descent clearance is issued by ATC, and the altitude to maintain is subsequently changed or restated without an expedite instruction, the expedite instruction is canceled. Expedite climb/descent normally indicates to the pilot that the approximate best rate of climb/descent should be used without requiring an exceptional change in aircraft handling characteristics. Normally controllers will inform pilots of the reason for an instruction to expedite.

4-4-10. IFR SEPARATION STANDARDS

a. ATC effects separation of aircraft vertically by assigning different altitudes; longitudinally by providing an interval expressed in time or distance between aircraft on the same, converging, or crossing courses, and laterally by assigning different flight paths.

b. Separation will be provided between all aircraft operating on IFR flight plans except during that part of the flight (outside Class B airspace or a TRSA) being conducted on a VFR-ON-TOP/VFR CONDITIONS clearance. Under these conditions, ATC may issue traffic advisories, but it is the sole responsibility of the pilot to be vigilant so as to see and avoid other aircraft.

c. When radar is employed in the separation of aircraft at the same altitude, a minimum of 3 miles separation is provided between aircraft operating within 40 miles of the radar antenna site, and 5 miles between aircraft operating beyond 40 miles from the antenna site. These minima may be increased or decreased in certain specific situations.

NOTE—CERTAIN SEPARATION STANDARDS ARE INCREASED IN THE TERMINAL ENVIRONMENT WHEN CENRAP IS BEING UTILIZED.

4-4-11. SPEED ADJUSTMENTS

a. ATC will issue speed adjustments to pilots of radar-controlled aircraft to achieve or maintain required or desire spacing.

b. ATC will express all speed adjustments in terms of knots based on indicated airspeed (IAS) in 10 knot increments except that at or above FL 240 speeds may be expressed in terms of Mach numbers in 0.01 increments. The use of Mach numbers is restricted to turbojet aircraft with Mach meters.

c. Pilots complying with speed adjustments are expected to maintain a speed within plus or minus 10 knots or 0.02 Mach number of the specified speed.

d. When ATC assigns speed adjustments, it will be in accordance with the following recommended minimums:

1. To aircraft operating between FL 280 and 10,000 feet, a speed not less than 250 knots or the equivalent Mach number.

NOTE—1. ON A STANDARD DAY THE MACH NUMBERS EQUIVALENT TO 250 KNOTS CAS (SUBJECT TO MINOR VARIATIONS) ARE:

FL 240-0.6
FL 250-0.61
FL 260-0.62
FL 270-0.64
FL 280-0.65
FL 290-0.66.

2. WHEN AN OPERATIONAL/ADVANTAGE/WILL/BE/REALIZED, SPEEDS LOWER THAN THE RECOMMENDED MINIMA MAY BE APPLIED.

2. To arriving turbojet aircraft operating below 10,000 feet:

(a) A speed not less than 210 knots, except;

(b) Within 20 flying miles of the airport of intended landing, a speed not less than 170 knots.

3. To arriving reciprocating engine or turboprop aircraft within 20 flying miles of the runway threshold of the airport of intended landing, a speed not less than 150 knots.

4. To departing aircraft:

(a) Turbine-powered aircraft, a speed not less than 230 knots.

(b) Reciprocating engine aircraft, a speed not less than 150 knots.

e. When ATC combines a speed adjustment with a descent clearance, the sequence of delivery, with the word "then" between, indicates the expected order of execution;

EXAMPLE—

☐ DESCEND AND MAINTAIN (ALTITUDE); THEN, REDUCE SPEED TO (SPEED).

☑ REDUCE SPEED TO (SPEED); THEN, DESCEND AND MAINTAIN (ALTITUDE).

NOTE—THE MAXIMUM SPEEDS BELOW 10,000 FEET AS ESTABLISHED IN FAR PART 91.117 STILL APPLY. IF THERE IS ANY DOUBT CONCERNING THE MANNER IN WHICH SUCH A CLEARANCE IS TO BE EXECUTED, REQUEST CLARIFICATION FROM ATC.

f. If ATC determines (before an approach clearance is issued) that it is no longer necessary to apply speed adjustment procedures, they will inform the pilot to resume normal speed. Approach clearances supersede any prior speed adjustment assignments, and pilots are expected to make their own speed adjustments, as necessary, to complete the approach. Under certain circumstances however, it may be necessary for ATC to issue further speed adjustments after approach clearance is issued to maintain separation between successive arrivals. Under such circumstances, previously issued speed adjustments will be restated if that speed is to be maintained or additional speed adjustments are requested. ATC must obtain pilot concurrence for speed adjustments after approach clearances are issued. Speed adjustments should not be assigned inside the final approach fix on final or a point 5 miles from the runway, whichever is closer to the runway.

NOTE—AN INSTRUCTION TO "RESUME NORMAL SPEED" DOES NOT DELETE SPEED RESTRICTIONS THAT ARE CONTAINED IN A PUBLISHED PROCEDURE, UNLESS SPECIFICALLY STATED BY ATC, NOR DOES IT RELIEVE THE PILOT OF THOSE SPEED RESTRICTIONS WHICH ARE APPLICABLE TO 14 CFR SECTION 91.117.

g. The pilots retain the prerogative of rejecting the application of speed adjustment by ATC if the minimum safe airspeed for any particular operation is greater than the speed adjustment.

NOTE—IN SUCH CASES, PILOTS ARE EXPECTED TO ADVISE ATC OF THE SPEED THAT WILL BE USED.

h. Pilots are reminded that they are responsible for rejecting the application of speed adjustment by ATC if, in their opinion, it will cause them to exceed the maximum indicated airspeed prescribed by FAR Part 91.117(a), (c) and (d). IN SUCH CASES, THE PILOT IS EXPECTED TO SO INFORM ATC. Pilots operating at or above 10,000 feet MSL who are issued speed adjustments which exceed 250 knots IAS and are subsequently cleared below 10,000 feet MSL are expected to comply with FAR Part 91.117(a).

i. Speed restrictions of 250 knots do not apply to U.S. registered aircraft operating beyond 12 nautical miles from the coastline within the U.S. Flight Information Region, in Class E airspace below 10,000 feet MSL. However, in airspace underlying a Class B airspace area designated for an airport, or in a VFR corridor designated through such as a Class B airspace area, pilots are expected to comply with the 200 knot speed limit specified in 14 CFR Section 91.117(c).

j. For operations in a Class C, and Class D surface area, ATC is authorized to request or approve a speed greater than the maximum indicated airspeeds prescribed for operation within that airspace (FAR Part 91.117(b)).

NOTE—PILOTS ARE EXPECTED TO COMPLY WITH THE MAXIMUM SPEED OF 200 KNOTS WHEN OPERATING BENEATH CLASS B AIRSPACE OR IN A CLASS B VFR CORRIDOR (FAR PART 91.117(C) AND (D)).

k. When in communication with the ARTCC, or approach control facility, pilots should, as a good operating practice, state any ATC assigned speed restriction on initial radio contact associated with an ATC communications frequency change.

4-4-12. RUNWAY SEPARATION

Tower controllers establish the sequence of arriving and departing aircraft by requiring them to adjust flight or ground operation as necessary to achieve proper spacing. They may "HOLD" an aircraft short of the runway to achieve spacing between it and an arriving aircraft; the controller may instruct a pilot to "EXTEND DOWNWIND" in order to establish spacing from an arriving or departing aircraft. At times a clearance may include the word "IMMEDIATE." For example: "CLEARED FOR IMMEDIATE TAKEOFF." In such cases "IMMEDIATE" is used for purposes of *air traffic separation*. It is up to the pilot to refuse the clearance if, in the pilot's opinion, compliance would adversely affect the operation.

REFERENCE—AIM, GATE HOLDING DUE TO DEPARTURE DELAYS, PARAGRAPH 4-3-15.

4-4-13. VISUAL SEPARATION

a. Visual separation is a means employed by ATC to separate aircraft in terminal areas and en route airspace in the NAS. There are two methods employed to effect this separation:

1. The tower controller sees the aircraft involved and issues instructions, as necessary, to ensure that the aircraft avoid each other.

2. A pilot sees the other aircraft involved and upon instructions from the controller provides separation by maneuvering the aircraft to avoid it. When pilots accept responsibility to maintain visual separation, they must maintain constant visual surveillance and not pass the other aircraft until it is no longer a factor.

NOTE—TRAFFIC IS NO LONGER A FACTOR WHEN DURING APPROACH PHASE THE OTHER AIRCRAFT IS IN THE LANDING PHASE OF FLIGHT OR EXECUTES A MISSED APPROCH; AND DURING DEPARTURE OR EN ROUTE, WHEN THE OTHER AIRCRAFT TURNS AWAY OR IS ON A DIVERGING COURSE.

b. A pilot's acceptance of instructions to follow another aircraft or provide visual separation from it is an acknowledgment that the pilot will maneuver the aircraft as necessary to avoid the other aircraft or to maintain in-trail separation. In operations conducted behind heavy jet aircraft, it is also an acknowledgment that the pilot accepts the responsibility for wake turbulence separation.

NOTE—WHEN A PILOT HAS BEEN TOLD TO FOLLOW ANOTHER AIRCRAFT OR TO PROVIDE VISUAL SEPARATION FROM IT, THE PILOT SHOULD PROMPTLY NOTIFY THE CONTROLLER IF VISUAL CONTACT WITH THE OTHER AIRCRAFT IS LOST OR CANNOT BE MAINTAINED OR IF THE PILOT CANNOT ACCEPT THE RESPONSIBILITY FOR THE SEPARATION FOR ANY REASON.

c. Scanning the sky for other aircraft is a key factor in collision avoidance. Pilots and copilots (or the right seat passenger) should continuously scan to cover all areas of the sky visible from the cockpit. Pilots must develop an effective scanning technique which maximizes one's visual capabilities. Spotting a potential collision threat increases directly as more time is spent looking outside the aircraft. One must use timesharing techniques to effectively scan the surrounding airspace while monitoring instruments as well.

d. Since the eye can focus only on a narrow viewing area, effective scanning is accomplished with a series of short, regularly spaced eye movements that bring successive areas of the sky into the central visual field. Each movement should not exceed ten degrees, and each area should be observed for at least one second to enable collision detection. Although many pilots seem to prefer the method of horizontal back-and-forth scanning every pilot should develop a scanning pattern that is not only comfortable but assures optimum effectiveness. Pilots should remember, however, that they have a regulatory responsibility (FAR Part 91.113(a)) to see and avoid other aircraft when weather conditions permit.

4-4-14. USE OF VISUAL CLEARING PROCEDURES

a. Before Takeoff: Prior to taxiing onto a runway or landing area in preparation for takeoff, pilots should scan the approach areas for possible landing traffic and execute the appropriate clearing maneuvers to provide them a clear view of the approach areas.

b. Climbs and Descents: During climbs and descents in flight conditions which permit visual detection of other traffic, pilots should execute gentle banks, left and right at a frequency which permits continuous visual scanning of the airspace about them.

c. Straight and Level: Sustained periods of straight and level flight in conditions which permit visual detection of other traffic should be broken at intervals with appropriate clearing procedures to provide effective visual scanning.

d. Traffic Pattern: Entries into traffic patterns while descending create specific collision hazards and should be avoided.

e. Traffic at VOR Sites: All operators should emphasize the need for sustained vigilance in the vicinity of VORs and airway intersections due to the convergence of traffic.

f. Training Operations: Operators of pilot training programs are urged to adopt the following practices:

1. Pilots undergoing flight instruction at all levels should be requested to verbalize clearing procedures (call out "clear" left, right, above, or below) to instill and sustain the habit of vigilance during maneuvering.

2. *High-wing airplane:* momentarily raise the wing in the direction of the intended turn and look.

3. *Low-wing airplane:* momentarily lower the wing in the direction of the intended turn and look.

4. Appropriate clearing procedures should precede the execution of all turns including chandelles, lazy eights, stalls, slow flight, climbs, straight and level, spins, and other combination maneuvers.

4-4-15. TRAFFIC ALERT AND COLLISION AVOIDANCE SYSTEM (TCAS I & II)

a. TCAS I provides proximity warning only, to assist the pilot in the visual acquisition of intruder aircraft. No recommended avoidance maneuvers are provided nor authorized as a direct result of a TCAS I warning. It is intended for use by smaller commuter aircraft holding 10 to 30 passenger seats, and general aviation aircraft.

b. TCAS II provides traffic advisories (TAs) and resolution advisories (RAs). Resolution advisories provide recommended maneuvers in a vertical direction (climb or descend only) to avoid conflicting traffic. Airline aircraft, and larger commuter and business aircraft holding 31 passenger seats or more, use TCAS II equipment.

1. Each pilot who deviates from an ATC clearance in response to a TCAS II RA shall notify ATC of that deviation as soon as practicable and expeditiously return to the current ATC clearance when the traffic conflict is resolved.

2. Deviations from rules, policies, or clearances should be kept to the minimum necessary to satisfy a TCAS II RA.

3. The serving IFR air traffic facility is not responsible to provide approved standard IFR separation to an aircraft after a TCAS II RA maneuver until one of the following conditions exists:

(a) The aircraft has returned to its assigned altitude and course.

(b) Alternate ATC instructions have been issued.

c. TCAS does not alter or diminish the pilot's basic authority and responsibility to ensure safe flight. Since TCAS does not respond to aircraft which are not transponder equipped or aircraft with a transponder failure, TCAS alone does not ensure safe separation in every case.

d. At this time, no air traffic service nor handling is predicated on the availability of TCAS equipment in the aircraft.

4-4-16. TRAFFIC INFORMATION SERVICE (TIS)

a. TIS provides proximity warning only, to assist the pilot in the visual acquisition of intruder aircraft. No recommended avoidance maneuvers are provided nor authorized as a direct result of a TIS intruder display or TIS alert. It is intended for use by aircraft in which TCAS is not required.

b. TIS does not alter or diminish the pilot's basic authority and responsibility to ensure safe flight. Since TIS does not respond to aircraft which are not transponder equipped, aircraft with a transponder failure, or aircraft out of radar coverage, TIS alone does not ensure safe separation in every case.

c. At this time, no air traffic service nor handling is predicated on the availability of TIS equipment in the aircraft.

4-4-17. AUTOMATIC DEPENDENT SURVEILLANCE-BROADCAST (ADS-B)

a. ADS-B (aircraft-to-aircraft) provides proximity warning only to assist the pilot in the visual acquisition of other aircraft. No recommended avoidance maneuvers are provided nor authorized as a direct result of an ADS-B display or an ADS-B alert.

b. ADS-B does not alter or diminish the pilot's basic authority and responsibility to ensure safe flight. ADS-B only displays aircraft that are ADS-B equipped; therefore, aircraft that are not ADS-B equipped or aircraft that are experiencing an ADS-B failure will not be displayed. ADS-B alone does not ensure safe separation.

c. Presently, no air traffic services or handling is predicated on the availability of an ADS-B cockpit display. A "traffic-in-sight" reply to ATC must be based on seeing an aircraft out-the-window, NOT on the cockpit display.

4-4-18. TRAFFIC INFORMATION SERVICE-BROADCAST (TIS-B)

a. TIS-B provides traffic information to assist the pilot in the visual acquisition of other aircraft. No recommended avoidance maneuvers are provided nor authorized as the direct result of a TIS-B display or TIS-B alert.

b. TIS-B does not alter or diminish the pilot's basic authority and responsibility to ensure safe flight. TIS-B only displays aircraft with a functioning transponder; therefore, aircraft that are not transponder equipped, or aircraft that are experiencing a transponder failure, or aircraft out of radar coverage will not be displayed. TIS-B alone does not ensure safe separation.

c. Presently, no air traffic services or handling is predicated on the availability of TIS-B equipment in aircraft. A "traffic-in-sight" reply to ATC must be based on seeing an aircraft out-the-window, NOT on the cockpit display.

Section 5. SURVEILLANCE SYSTEMS

4-5-1. RADAR

a. Capabilities

1. Radar is a method whereby radio waves are transmitted into the air and are then received when they have been reflected by an object in the path of the beam. Range is determined by measuring the time it takes (at the speed of light) for the radio wave to go out to the object and then return to the receiving antenna. The direction of a detected object from a radar site is determined by the position of the rotating antenna when the reflected portion of the radio wave is received.

2. More reliable maintenance and improved equipment have reduced radar system failures to a negligible factor. Most facilities actually have some components duplicated—one operating and another which immediately takes over when a malfunction occurs to the primary component.

b. Limitations

1. It is very important for the aviation community to recognize the fact that there are limitations to radar service and that ATC controllers may not always be able to issue traffic advisories concerning aircraft which are not under ATC control and cannot be seen on radar. (See FIG 4-5-1).

(a) The characteristics of radio waves are such that they normally travel in a continuous straight line unless they are:

(1) "Bent" by abnormal atmospheric phenomena such as temperature inversions;

(2) Reflected or attenuated by dense objects such as heavy clouds, precipitation, ground obstacles, mountains, etc.; or

(3) Screened by high terrain features.

(b) The bending of radar pulses, often called anomalous propagation or ducting, may cause many extraneous blips to appear on the radar operator's display if the beam has been bent toward the ground or may decrease the detection range if the wave is bent upward. It is difficult to solve the effects of anomalous propagation, but using beacon radar and electronically eliminating stationary and slow moving targets by method called moving target indicator (MTI) usually negate the problem.

(c) Radar energy that strikes dense objects will be reflected and displayed on the operator's scope thereby blocking out aircraft at the same range and greatly weakening or completely eliminating the display of targets at a greater range. Again, radar beacon and MTI are very effectively used to combat ground clutter and weather phenomena, and a method of circularly polarizing the radar beam will eliminate some weather returns. A negative characteristic of MTI is that an aircraft flying a speed that coincides with the canceling signal of the MTI (tangential or "blind" speed) may not be displayed to the radar controller.

(d) Relatively low altitude aircraft will not be seen if they are screened by mountains or are below the radar beam due

---

PRECIPITATION ATTENUATION

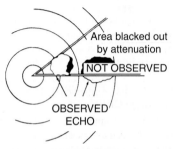

The nearby target absorbs and scatters so much of the out-going and returning energy that the radar does not detect the distant target.

Figure 4-5-1. Limitations to radar service

---

to earth curvature. The only solution to screening is the installation of strategically placed multiple radars which has been done in some areas.

(e) There are several other factors which affect radar control. The amount of reflective surface of an aircraft will determine the size of the radar return. Therefore, a small light airplane or a sleek jet fighter will be more difficult to see on radar than a large commercial jet or military bomber. Here again, the use of radar beacon is invaluable if the aircraft is equipped with an airborne transponder. All ARTCCs' radars in the conterminous U.S. and many airport surveillance radars have the capability to interrogate MODE C and display altitude information to the controller from appropriately equipped aircraft. However, there are a number of airport surveillance radars that don't have Mode C display capability and therefore altitude information must be obtained from the pilot.

(f) At some locations within the ATC en route environment, secondary-radar-only (no primary radar) gap filler radar systems are used to give lower altitude radar coverage between two larger radar systems, each of which provides both primary and secondary radar coverage. In those geographical areas served by secondary-radar only, aircraft without transponders cannot be provided with radar service. Additionally, transponder equipped aircraft cannot be provided with radar advisories concerning primary targets and weather.

REFERENCE—PILOT/CONTROLLER GLOSSARY TERM-RADAR.

(g) The controllers' ability to advise a pilot flying on instruments or in visual conditions of the aircraft's proximity to another aircraft will be limited if the unknown aircraft is not observed on radar, if no flight plan informationis available, or if the volume of traffic and workload prevent issuing traffic information. The controller's first priority is given to establishing vertical, lateral, or longitudinal separation between aircraft flying IFR under the control of ATC.

c. FAA radar units operate continuously at the locations shown in the Airport/Facility Directory, and their services are available to all pilots, both civil and military. Contact the associated FAA control tower or ARTCC on any frequency guarded for initial instructions, or in an emergency, any FAA facility for information on the nearest radar service.

4-5-2. AIR TRAFFIC CONTROL RADAR BEACON SYSTEM (ATCRBS)

a. The ATCRBS, sometimes referred to as secondary surveillance radar, consists of three main components:

1. *Interrogator.* Primary radar relies on a signal being transmitted from the radar antenna site and for this signal to be reflected or "bounced back" from an object (such as an aircraft). This reflected signal is then displayed as a "target" on the controller's radarscope. In the ATCRBS, the Interrogator, a ground based radar beacon transmitter-

receiver, scans in synchronism with the primary radar and transmits discrete radio signals which repetitively requests all transponders, on the mode being used, to reply. The replies received are then mixed with the primary returns and both are displayed on the same radarscope.

2. *Transponder.* This airborne radar beacon transmitter-receiver automatically receives the signals from the interrogator and selectively replies with a specific pulse group (code) only to those interrogations being received on the mode to which it is set. These replies are independent of, and much stronger than a primary radar return.

3. *Radarscope.* The radarscope used by the controller displays returns from both the primary radar system and the ATCRBS. These returns, called targets, are what the controller refers to in the control and separation of traffic.

b. The job of identifying and maintaining identification of primary radar targets is a long and tedious task for the controller. Some of the advantages of ATCRBS over primary radar are:

1. Reinforcement of radar targets.

2. Rapid target identification.

3. Unique display of selected codes.

c. A part of the ATCRBS ground equipment is the decoder. This equipment enables a controller to assign discrete transponder codes to each aircraft under their control. Normally only one code will be assigned for the entire flight. Assignments are made by the ARTCC computer on the basis of the National Beacon Code Allocation Plan. The equipment is also designed to receive MODE C altitude information from the aircraft.

NOTE—REFER TO FIGURES WITH EXPLANATORY LEGENDS FOR AN ILLUSTRATION OF THE TARGET SYMBOLOGY DEPICTED ON RADARSCOPES IN THE NAS STAGE A (EN ROUTE), THE ARTS III (TERMINAL) SYSTEMS, AND OTHER NONAUTOMATED (BROADBAND) RADAR SYSTEMS. (SEE FIG 4-5-2 AND FIG 4-5-3).

d. It should be emphasized that aircraft transponders greatly improve the effectiveness of radar systems.

REFERENCE—AIM, TRANSPONDER OPERATION, PARAGRAPH 4-1-19.

4-5-3. SURVEILLANCE RADAR

a. Surveillance radars are divided into two general categories: Airport Surveillance Radar (ASR) and Air Route Surveillance Radar (ARSR).

1. ASR is designed to provide relatively short-range coverage in the general vicinity of an airport and to serve as an expeditious means of handling terminal area traffic through observation of precise aircraft locations on a radarscope. The ASR can also be used as an instrument approach aid.

2. ARSR is a long-range radar system designed primarily to provide a display of aircraft locations over large areas.

3. Center Radar Automated Radar Terminal Systems (ARTS) Processing (CENRAP) was developed to provide

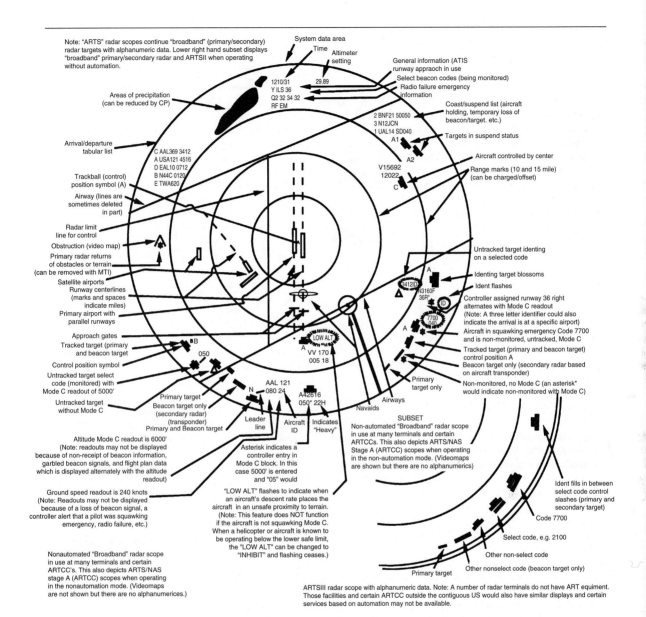

Note: "ARTS" radar scopes continue "broadband" (primary/secondary) radar targets with alphanumeric data. Lower right hand subset displays "broadband" primary/secondary radar and ARTSII when operating without automation.

System data area
Time
Altimeter setting

General information (ATIS runway approach in use)
Select beacon codes (being monitored)
Radio failure emergency information

Areas of precipitation (can be reduced by CP)

Coast/suspend list (aircraft holding, temporary loss of beacon/target. etc.)

Targets in suspend status

Aircraft controlled by center

Range marks (10 and 15 mile) (can be charged/offset)

Arrival/departure tabular list

Trackball (control) position symbol (A)

Airway (lines are sometimes deleted in part)

Radar limit line for control

Obstruction (video map)

Primary radar returns of obstacles or terrain (can be removed with MTI)

Satellite airports
Runway centerlines (marks and spaces indicate miles)
Primary airport with parallel runways

Approach gates
Tracked target (primary and beacon target)

Control position symbol
Untracked target select code (monitored) with Mode C readout of 5000'

Untracked target without Mode C

Primary target
Beacon target only (secondary radar) (transponder)
Primary and Beacon target

Leader line

Aircraft ID

Indicates "Heavy"

Untracked target identing on a selected code

Identing target blossoms

Ident flashes

Controller assigned runway 36 right alternates with Mode C readout (Note: A three letter identifier could also indicate the arrival is at a specific airport)

Aircraft in squawking emergency Code 7700 and is non-monitored, untracked, Mode C

Tracked target (primary and beacon target) control position A

Beacon target only (secondary radar based on aircraft transponder)

Non-monitored, no Mode C (an asterisk* would indicate non-monitored with Mode C)

Primary target only

Airways
Navaids

Altitude Mode C readout is 6000' (Note: readouts may not be displayed because of non-receipt of beacon information, garbled beacon signals, and flight plan data which is displayed alternately with the altitude readout)

Ground speed readout is 240 knots (Note: Readouts may not be displayed because of a loss of beacon signal, a controller alert that a pilot was squawking emergency, radio failure, etc.)

Nonautomated "Broadband" radar scope in use at many terminals and certain ARTCC's. This also depicts ARTS/NAS stage A (ARTCC) scopes when operating in the nonautomation mode. (Videomaps are not shown but there are no alphanumerices.)

Asterisk indicates a controller entry in Mode C block. In this case 5000' is entered and "05" would

"LOW ALT" flashes to indicate when an aircraft's descent rate places the aircraft in an unsafe proximity to terrain. (Note: This feature does NOT function if the aircraft is not squawking Mode C. When a helicopter or aircraft is known to be operating below the lower safe limit, the "LOW ALT" can be changed to "INHIBIT" and flashing ceases.)

SUBSET
Non-automated "Broadband" radar scope in use at many terminals and certain ARTCCs. This also depicts ARTS/NAS Stage A (ARTCC) scopes when operating in the non-automation mode. (Videomaps are shown but there are no alphanumerics.)

Ident fills in between select code control slashes (primary and secondary target)

Code 7700

Select code, e.g. 2100

Other non-select code

Other nonselect code (beacon target only)

Primary target

ARTSIII radar scope with alphanumeric data. Note: A number of radar terminals do not have ART equiment. Those facilities and certain ARTCC outside the contiguous US would also have similar displays and certain services based on automation may not be available.

NOTE: A number of radar terminals do not have arts equipment. Those facilites and certain ARTCCs outside the contiguous U.S. would have radar displays similar to the lower right hand subset. ARTS facilities and NAS stage an ARTCC's, when operating in the nonautomation mode. Would also have similar displays and certain services based on automation may not be available.

Figure 4-5-2.

an alternative to a nonradar environment at terminal facilities should an Airport Surveillance Radar (ASR) fail or malfunction. CENRAP sends aircraft radar beacon target information to the ASR terminal facility equipped with ARTS. Procedures used for the separation of aircraft may increase under certain conditions when a facility is utilizing CENRAP because radar target information updates at a slower rate than the normal ASR radar. Radar services for

VFR aircraft are also limited during CENRAP operations because of the additional workload required to provide services to IFR aircraft.

b. Surveillance radars scan through 360 degrees of azimuth and present target information on a radar display located in a tower or center. This information is used independently or in conjunction with other navigational aids in the control of air traffic.

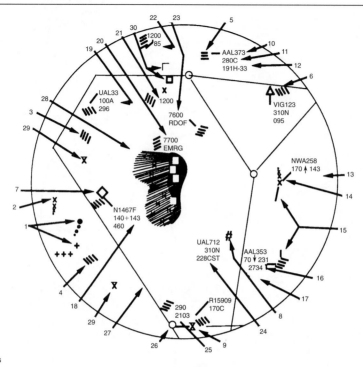

Target symbols

1 Uncorreleated primary radar target **+ ●**
2 *Correlated primary radar target **X**
3 Uncorrelated beacon target **/**
4 Correlated beacon target ****
5 Identing beacon target **≡**
   (*Correlated means the association
   of radar data with the computer projected track of an identified aricraft)

Position symbols

6 Free track(no flight plan tracking) **△**
7 First track(flight plan tracking) **◇**
8 Coast(beacon target lost) **#**
9 Present position hold **X**

Data block information

10 *Aircraft identification
11 *Assigned altitude FL280, mode C altitude same or within ±200' of asgnd
   altitude

12 *Computer ID #191, Handoff is to Sector 33
   (0-33 would mean handoff accepted)
   (*Nr's 10, 11, 12 constitute a "full data
   block")
13 Assigned altitude 17,000', aircraft is
   climbing, mode C readout was 14,300
   when last beacon interrogation was
   received
14 Leader line connecting target symbol
   and data block
15 Track velocity and direction vector line
   (projected ahead of target)
16 Assigned altitude 7000, aircraft is
   descending, last mode C readout (or
   last reported altitude was 100'
   above FL 230
17 Transponder code shows in full data block
   only when different than assigned code
18 Aircraft is 300' above assigned altitude
19 Reported altitude (no mode C readout
   same as assigned. An "N" would indicate no reported altitude)
20 Transponder set on emergency code 7700
   (EMRG flashes to attract attention)

21 Transponder code 1200 (VFR) with no
   mode C
22 Code 1200(VFR) with mode C and last
   altitude readout
23 Transponder set on Radio Failure code
   7600, (RDOF flashes)
24 Computer ID #288, CST indicates target is
   in Coast status
25 Assigned altitude FL 290, transponder

Other symbols

26 Navigational Aid
27 Airway or jet route
28 Outline of weather returns based on
   primary radar (See Chapter 4, ARTCC
   Radar Weather Display. H's represent
   areas of high density precipitation
   which might be thunderstorms. Radial
   lines indicate lower density precipitation)
29 Obstruction
30 Airports Major: **◻** Small: **Γ**

NAS Stage A Controllers View Plan Display. This figure illustrates the controller's radar scope (PVD)
when operating in the full automatic (RDP) mode, which is normally 20 hours per day. (Note: When not
in automation mode, the display is similar to the broadband mode shown in the ARTS III Radar Scope figure.
Certain ARTCC's outside the contiguous U.S. also operate in "broadband" mode.)

Figure 4-5-3.

4-5-4. PRECISION APPROACH RADAR (PAR)

a. PAR is designed to be used as a landing aid, rather than
an aid for sequencing and spacing aircraft. PAR equipment
may be used as a primary landing aid, or it may be used to
monitor other types of approaches. It is designed to display
range, azimuth and elevation information.

b. Two antennas are used in the PAR array, one scanning
a vertical plane, and the other scanning horizontally. Since

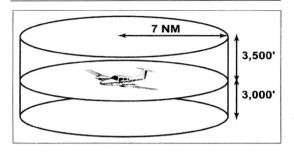

Figure 4-5-4. TIS Proximity Coverage Volume

TBL 4-5-1

STL	Lambert-St. Louis International
CLT	Charlotte Douglas International
SDF	Louisville International Standiford
DFW	Dallas/Ft. Worth International
ORD	Chicago O'Hare International
LAX	Los Angeles International
ATL	Hartsfield Atlanta International
IAD	Washington Dulles International
SEA	Seattle-Tacoma International
MKE	General Mitchell International
MCO	Orlando International
PVD	Theodore Francis Green State
PHX	Phoenix Sky Harbor International
MEM	Memphis International
RDU	Raleigh-Durham International
HOU	William P Hobby (Houston, TX)
BDL	Bradley International
SJC	San Jose International
SAT	San Antonio International
SMF	Sacramento International
FLL	Ft. Lauderdale/Hollywood
HNL	Honolulu International - Hickam AFB
OAK	Metropolitan Oakland International
IND	Indianapolis International
TPA	Tampa International
BUR	Burbank-Glendale-Pasadena
CMH	Port Columbus International
MDW	Chicago Midway
COs	Colorado Springs Municipal
SNA	John Wayne - Orange County
ONT	Ontario International
AUS	Austin-Bergstrom International
RNO	Reno/Tahoe International
ABQ	Albuquerque International Sunport
SJU	San Juan International

NOTE- The installation of ASDE-X is projected to be completed by 2009.

the range is limited to 10 miles, azimuth to 20 degrees, and elevation to 7 degrees, only the final approach area is covered. Each scope is divided into two parts. The upper half presents altitude and distance information, and the lower half presents azimuth and distance.

4-5-5. AIRPORT SURFACE DETECTION EQUIPMENT - MODEL X (ASDE-X)

a. The Airport Surface Detection Equipment - Model X (ASDE-X) is a multi-sensor surface surveillance system the FAA is acquiring for airports in the United States. This system will provide high resolution, short-range, clutter free surveillance information about aircraft and vehicles, both moving and fixed, located on or near the surface of the airport's runways and taxiways under all weather and visibility conditions. The system consists of:

1. A Primary Radar System. ASDE-X system coverage includes the airport surface and the airspace up to 200 feet above the surface. Typically located on the control tower or other strategic location on the airport, the Primary Radar antenna is able to detect and display aircraft that are not equipped with or have malfunctioning transponders.

2. Interfaces. ASDE-X contains an automation interface for flight identification via all automation platforms and interfaces with the terminal radar for position information.

3. ASDE-X Automation. A Multi-sensor Data Processor (MSDP) combines all sensor reports into a single target which is displayed to the air traffic controller.

4. Air Traffic Control Tower Display. A high resolution, color monitor in the control tower cab provides controllers with a seamless picture of airport operations on the airport surface.

b. The combination of data collected from the multiple sensors ensures that the most accurate information about aircraft location is received in the tower, thereby increasing surface safety and efficiency.

c. The following facilities have been projected to receive ASDE-X:

4-5-6. TRAFFIC INFORMATION SERVICE (TIS)

a. Introduction

The Traffic Information Service (TIS) provides information to the cockpit via data link, that is similar to VFR radar traffic advisories normally received over voice radio. Among the first FAA-provided data services, TIS is intended to improve the safety and efficiency of "see and avoid" flight through an automatic display that informs the pilot of nearby traffic and potential conflict situations. This traffic display is intended to assist the pilot in visual acquisition of these aircraft. TIS employs an enhanced capability of the terminal Mode S radar system, which contains the surveillance data, as well as the data link required to "uplink" this information to suitably-equipped aircraft (known as a TIS "client"). TIS provides estimated position, altitude, altitude trend, and ground track information for up to 8 intruder aircraft within 7 NM horizontally, +3,500 and −3,000 feet vertically of the client aircraft (see FIG 4-5-4, TIS Proximity Coverage Volume). The range of a target reported at a distance greater than 7 NM only indicates that this target will be a threat

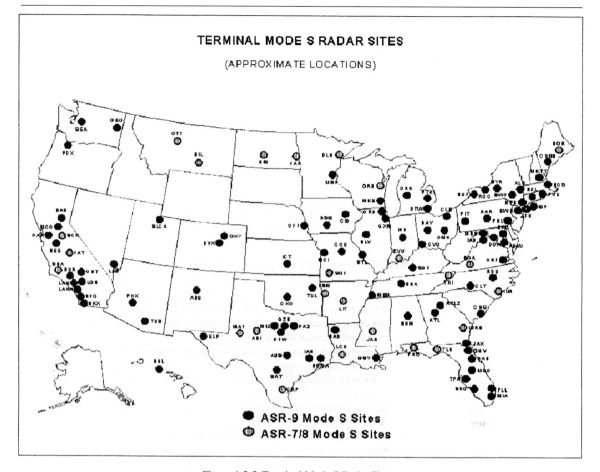

Figure 4-5-5. Terminal Mode S Radar Sites

within 34 seconds and does not display a precise distance. TIS will alert the pilot to aircraft (under surveillance of the Mode S radar) that are estimated to be within 34 seconds of potential collision, regardless of distance of altitude. TIS surveillance data is derived from the same radar used by ATC; this data is uplinked to the client aircraft on each radar scan (nominally every 5 seconds).

b. Requirements

1. In order to use TIS, the client and any intruder aircraft must be equipped with the appropriate cockpit equipment and fly within the radar coverage of a Mode S radar capable of providing TIS. Typically, this will be within 55 NM of the sites depicted in FIG 4-5-5, Terminal Mode S Radar Sites. ATC communication is not a requirement to receive TIS, although it may be required by the particular airspace or flight operations in which TIS is being used.

2. The cockpit equipment functionality required by a TIS client aircraft to receive the service consists of the following (refer to FIG 4-5-6):

(a) Mode S data link transponder with altitude encoder.

(b) Data link applications processor with TIS software installed.

(c) Control-display unit.

(d) Optional equipment includes a digital heading source to correct display errors caused by "crab angle" and turning maneuvers.

NOTE—*SOME OF THE ABOVE FUNCTIONS WILL LIKELY BE COMBINED INTO SINGLE PIECES OF AVIONICS, SUCH AS (A) AND (B).*

3. To be visible to the TIS client, the intruder aircraft must, at a minimum, have an operating transponder (Mode A, C or S). All altitude information provided by TIS from intruder aircraft is derived from Mode C reports, if appropriately equipped.

4. TIS will initially be provided by the terminal Mode S systems that are paired with ASR-9 digital primary radars. These systems are in locations with the greatest traffic densities, thus will provide the greatest initial benefit. The remaining terminal Mode S sensors, which are paired with ASR-7 or ASR-8 analog primary radars, will provide TIS pending modification or relocation of these sites. See FIG 4-5-5, Terminal Mode S Radar Sites, for site locations. There is no mechanism in place, such as NOTAM's, to provide status update on individual radar sites since TIS is a nonessential, supplemental information service.

The FAA also operates en route Mode S radars (not illustrated) that rotate once every 12 seconds. These sites will require additional development of TIS before any possible implementation. There are no plans to implement TIS in the en route Mode S radars at the present time.

c. Capabilities

1. TIS provides ground-based surveillance information over the Mode S data link to properly equipped client aircraft to aid in visual acquisition of proximate air traffic. The actual avionics capability of each installation will vary and the supplemental handbook material must be consulted prior to using TIS. A maximum of eight (8) intruder aircraft may be displayed; if more than eight aircraft match intruder parameters, the eight "most significant" intruders are uplinked. These "most significant" intruders are usually the ones in closest proximity and/or the greatest threat to the TIS client.

2. TIS, through the Mode S ground sensor, provides the following data on each intruder aircraft:

(a) Relative bearing information in 6-degree increments.

(b) Relative range information in 1/8 NM to 1 NM increments (depending on range).

(c) Relative altitude in 100-foot increments (within 1,000 feet) or 500-foot increments (from 1,000–3,500 feet) if the intruder aircraft has operating altitude reporting capability.

(d) Estimated intruder ground track in 45-degree increments.

(e) Altitude trend data (level within 500 fpm or climbing/descending >500 fpm) if the intruder aircraft has operating altitude reporting capability.

(f) Intruder priority as either an "traffic advisory" or "proximate" intruder.

3. When flying from surveillance coverage of one Mode S sensor to another, the transfer of TIS is an automatic function of the avionics system and requires no action from the pilot.

4. There are a variety of status messages that are provided by either the airborne system or ground equipment to alert the pilot of high priority intruders and data link system status. These messages include the following:

(a) Alert. Identifies a potential collision hazard within 34 seconds. This alert may be visual and/or audible, such as a flashing display symbol or a headset tone. A target is a threat if the time to the closest approach in vertical and horizontal coordinates is less than 30 seconds and the closest approach is expected to be within 500 feet vertically and *0.5* nautical miles laterally.

(b) TIS Traffic. TIS traffic data is displayed.

(c) Coasting. The TIS display is more than 6 seconds old. This indicates a missing uplink from the ground system. When the TIS display information is more than 12 seconds old, the "No Traffic" status will be indicated.

(d) No Traffic. No intruders meet proximate or alert criteria. This condition may exist when the TIS system is fully functional or may indicate "coasting" between 12 and 59 seconds old (see (c) above).

(e) TIS Unavailable. The pilot has requested TIS, but no ground system is available. This condition will also be displayed when TIS uplinks are missing for 60 seconds or more.

(f) TIS Disabled. The pilot has not requested TIS or has disconnected from TIS.

(g) Goodbye. The client aircraft has flown outside of TIS coverage.

NOTE—*DEPENDING ON THE AVIONICS MANUFACTURER IMPLEMENTATION, IT IS POSSIBLE THAT SOME OF THESE MESSAGES WILL NOT BE DIRECTLY AVAILABLE TO THE PILOT.*

5. Depending on avionics system design, TIS may be presented to the pilot in a variety of different displays, including text and/or graphics. Voice annunciation may also be used, either alone or in combination with a visual display. FIG 4-5-6, Traffic Information Service (TIS), Avionics Block Diagram, shows an example of a TIS display using symbology similar to the Traffic Alert and Collision Avoidance System (TCAS) installed on most passenger air carrier/commuter aircraft in the U.S. The small symbol in the center represents the client aircraft and the display is oriented "track up," with the 12 o'clock position at the top. The range rings indicate 2 and 5 NM. Each intruder is depicted by a symbol positioned at the approximate relative bearing and range from the client aircraft. The circular symbol near the center indicates an "alert" intruder and the diamond symbols indicate "proximate" intruders.

6. The inset in the lower right corner of FIG 4-5-6, Traffic Information Service (TIS), Avionics Block Diagram, shows a possible TIS data block display. The following information is contained in this data block:

(a) The intruder, located approximately four o'clock, three miles, is a "proximate" aircraft and currently not a collision threat to the client aircraft. This is indicated by the diamond symbol used in this example.

(b) The intruder ground track diverges to the right of the client aircraft, indicated by the small arrow.

(c) The intruder altitude is 700 feet less than or below the client aircraft, indicated by the "−07" located under the symbol.

(d) The intruder is descending >500 fpm, indicated by the downward arrow next to the "−07" relative altitude information. The absence of this arrow when an altitude tag is present indicates level flight or a climb/descent rate less than 500 fpm.

NOTE—*IF THE INTRUDER DID NOT HAVE AN OPERATING ALTITUDE ENCODER (MODE C), THE ALTITUDE AND ALTITUDE TREND "TAGS" WOULD HAVE BEEN OMITTED.*

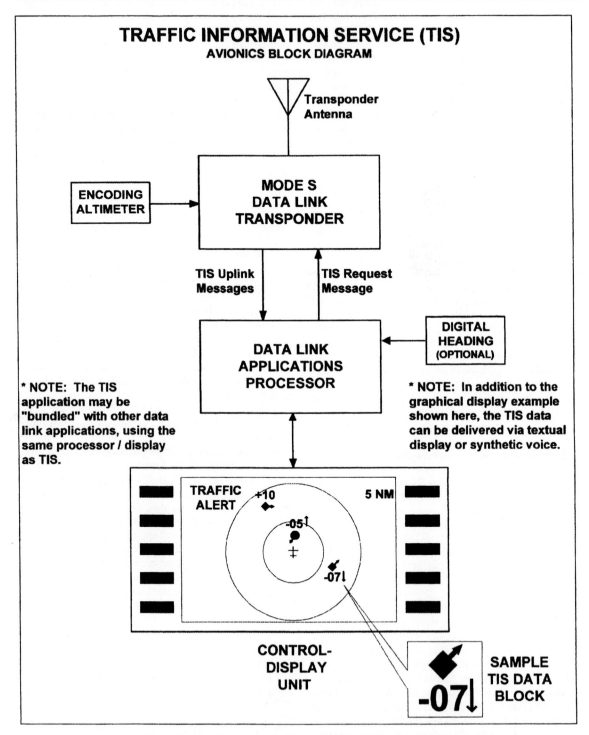

Figure 4-5-6. Traffic Information Service (TIS), Avionics Block Diagram

d. Limitations

1. TIS is <u>NOT</u> intended to be used as a collision avoidance system and does not relieve the pilot responsibility to "see and avoid" other aircraft (see paragraph 5-5-8, See and Avoid). TIS shall not be for avoidance maneuvers during IMC or other times when there is no visual contact with the intruder aircraft. TIS is intended only to assist in visual acquisition of other aircraft in VMC. <u>No recommended avoidance maneuvers are provided for. nor</u> **<u>authorized, as</u> a** <u>direct result of a TIS intruder display or TIS alert</u>.

2. While TIS is a useful aid to visual traffic avoidance, it has some system limitations that must be fully understood to ensure proper use. Many of these limitations are inherent in secondary radar surveillance. In other words, the information provided by TIS will be no better than that provided to ATC. Other limitations and anomalies are associated with the TIS predictive algorithm.

(a) Intruder Display Limitations. TIS will only display aircraft with operating transponders installed. TIS relies on surveillance of the Mode S radar, which is a "secondary surveillance" radar similar to the ATCRBS described in paragraph 4-5-2.

(b) TIS Client Altitude Reporting Requirement. Altitude reporting is required by the TIS client aircraft in order to receive TIS. If the altitude encoder is inoperative or disabled, TIS will be unavailable, as TIS requests will not be honored by the ground system. As such, TIS requires altitude reporting to determine the Proximity Coverage Volume as indicated in FIG 4-5-4. TIS users must be alert to altitude encoder malfunctions, as TIS has no mechanism to determine if client altitude reporting is correct. A failure of this nature will cause erroneous and possibly unpredictable TIS operation. If this malfunction is suspected, confirmation of altitude reporting with ATC is suggested.

(c) Intruder Altitude Reporting. Intruders without altitude reporting capability will be displayed without the accompanying altitude tag. Additionally, nonaltitude reporting intruders are assumed to be at the same altitude as the TIS client for alert computations. This helps to ensure that the pilot will be alerted to all traffic under radar coverage, but the actual altitude difference may be substantial. Therefore, visual acquisition may be difficult in this instance.

(d) Coverage Limitations. Since TIS is provided by ground-based, secondary surveillance radar, it is subject to all limitations of that radar. <u>If an aircraft is not detected by the radar, it cannot be displayed on TIS</u>. Examples of these limitations are as follows:

(1) TIS will typically be provided within **55** NM of the radars depicted in FIG 4-5-5, Terminal Mode S Radar Sites. This maximum range can vary by radar site and is always subject to "line of sight" limitations; the radar and data link signals will be blocked by obstructions, terrain, and curvature of the earth.

(2) TIS will be unavailable at low altitudes in many areas of the country, particularly in mountainous regions. Also, when flying near the "floor" of radar coverage in a particular area, intruders below the client aircraft may not be detected by TIS.

(3) TIS will be temporarily disrupted when flying directly over the radar site providing coverage if no adjacent site assumes the service. A ground-based radar, like a VOR or NDB, has a zenith cone, sometimes referred to as the cone of confusion or cone of silence. This is the area of ambiguity directly above the station where bearing information is unreliable. The zenith cone setting for TIS is 34 degrees: Any aircraft above that angle with respect to the radar horizon will lose TIS coverage from that radar until it is below this 34 degree angle. The aircraft may not actually lose service in areas of multiple radar coverage since an adjacent radar will provide TIS. If no other TIS-capable radar is available, the "Goodbye" message will be received and TIS terminated until coverage is resumed.

(e) Intermittent Operations. TIS operation may be intermittent during turns or other maneuvering, particularly if the transponder system does not include antenna diversity (antenna mounted on the top and bottom of the aircraft). As in (d) above, TIS is dependent on two-way, "line of sight" communications between the aircraft and the Mode S radar. Whenever the structure of the client aircraft comes between the transponder antenna (usually located on the underside of the aircraft) and the ground-based radar antenna, the signal may be temporarily interrupted.

(f) TIS Predictive Algorithm. TIS information is collected one radar scan prior to the scan during which the uplink occurs. Therefore, the surveillance information is approximately 5 seconds old. In order to present the intruders in a "real time" position, TIS uses a "predictive algorithm" in its tracking software. This algorithm uses track history data to extrapolate intruders to their expected positions consistent with the time of display in the cockpit. Occasionally, aircraft maneuvering will cause this algorithm to induce errors in the TIS display. These errors primarily affect relative bearing information; intruder distance and altitude will remain relatively accurate and may be used to assist in "see and avoid." Some of the more common examples of these errors are as follows:

(1) When client or intruder aircraft maneuver excessively or abruptly, the tracking algorithm will report incorrect horizontal position until the maneuvering aircraft stabilizes.

(2) When a rapidly closing intruder is on a course that crosses the client at a shallow angle (either overtaking or head on) and either aircraft abruptly changes course within 1/4 NM, TIS will display the intruder on the opposite side of the client than it actually is.

These are relatively rare occurrences and will be corrected in a few radar scans once the course has stabilized.

(g) Heading/Course Reference. Not all TIS aircraft installations will have onboard heading reference information. In these installations, aircraft course reference to the TIS display is provided by the Mode S radar. The radar only determines ground track information and has no indication of the client aircraft heading. In these installations, all intruder bearing information is referenced to ground track and does not account for wind correction. Additionally, since ground-based radar will require several scans to determine aircraft course following a course change, a lag in TIS display orientation (intruder aircraft bearing) will occur. As in (f) above, intruder distance and altitude are still usable.

(h) Closely-Spaced Intruder Errors. When operating more than 30 NM from the Mode S sensor, TIS forces any intruder within 3/8 NM of the TIS client to appear at the same horizontal position as the client aircraft. Without this feature, TIS could display intruders in a manner confusing to the pilot in critical situations (e.g. a closely-spaced intruder that is actually to the right of the client may appear on the TIS display to the left). At longer distances from the radar, TIS cannot accurately determine relative bearing/distance information on intruder aircraft that are in close proximity to the client.

Because TIS uses a ground-based, rotating radar for surveillance information, the accuracy of TIS data is dependent on the distance from the sensor (radar) providing the service. This is much the same phenomenon as experienced with ground-based navigational aids, such as VOR or NDB. As distance from the radar increases, the accuracy of surveillance decreases. Since TIS does not inform the pilot of distance from the Mode S radar, the pilot must assume that any intruder appearing at the same position as the client aircraft may actually be up to 3/8 NM away in any direction. Consistent with the operation of TIS, an alert on the display (regardless of distance from the radar) should stimulate an outside visual scan, intruder acquisition, and traffic avoidance based on outside reference.

e. Reports of TIS Malfunctions

1. Users of TIS can render valuable assistance in the early correction of malfunctions by reporting their observations of undesirable performance. Reporters should identify the time of observation, location, type and identity of aircraft, and describe the condition observed; the type of transponder processor, and software in use can also be useful information. Since TIS performance is monitored by maintenance personnel rather than ATC, it is suggested that malfunctions be reported in the following ways:

(a) By radio or telephone to the nearest Flight Service Station (FSS) facility.

(b) By FAA Form 8000-7, Safety Improvement Report, a postage-paid card designed for this purpose. These cards may be obtained at FAA FSS 's, General Aviation District Offices, Flight Standards District Offices, and General Aviation Fixed Based Operations.

4-5-7. AUTOMATIC DEPENDENT SURVEILLANCE-BROADCAST (ADS-B) SERVICES

a. Introduction

1. Automatic Dependent Surveillance-Broadcast (ADS-B) is a surveillance technology being deployed in selected areas of the NAS (see FIG 4-5-7). ADS-B broadcasts a radio transmission approximately once per second containing the aircraft's position, velocity, identification, and other information. ADS-B can also receive reports from other suitably equipped aircraft within reception range. Additionally, these broadcasts can be received by Ground Based Transceivers (GBTs) and used to provide surveillance services, along with fleet operator monitoring of aircraft. No ground infrastructure is necessary for ADS-B equipped aircraft to detect each other.

2. In the U.S., two different data links have been adopted for use with ADS-B: 1090 MHz Extended Squitter (1090 ES) and the Universal Access Transceiver (UAT). The 1090 ES link is intended for aircraft that primarily operate at FL 180 and above, whereas the UAT link is intended for use by aircraft that primarily operate at 18,000 feet and below. From a pilot's standpoint, the two links operate similarly and support ADS-B and Traffic Information Service-Broadcast (TIS-B), see paragraph 4-5-8. The UAT link additionally supports Flight Information Services-Broadcast (FIS-B), subparagraph 7-1-11d.

b. ADS-B Certification and Performance Requirements

ADS-B equipment may be certified as an air-to-air system for enhancing situational awareness and as a surveillance source for air traffic services. Refer to the aircraft's flight manual supplement for the specific aircraft installation.

c. ADS-B Capabilities

1. ADS-B enables improved surveillance services, both air-to-air and air-to-ground, especially in areas where radar is ineffective due to terrain or where it is impractical or cost prohibitive. Initial NAS applications of air-to-air ADS-B are for "advisory," use only, enhancing a pilot's visual acquisition of other nearby equipped aircraft either when airborne or on the airport surface. Additionally, ADS-B will enable ATC and fleet operators to monitor aircraft throughout the available ground station coverage area. Other applications of ADS-B may include enhanced search and rescue operations and advanced air-to-air applications such as spacing, sequencing, and merging.

2. ADS-B avionics typically allow pilots to enter the aircraft's call sign and Air Traffic Control (ATC)-assigned

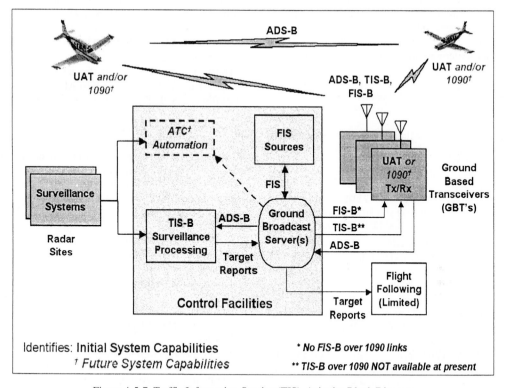

Figure 4-5-7. Traffic Information Service (TIS), Avionics Block Diagram

transponder code, which will be transmitted to other aircraft and ground receivers. Pilots are cautioned to use care when selecting and entering the aircraft's identification and transponder code. Some ADS-B avionics panels are not interconnected to the transponder. Therefore, **it is extremely important to ensure that the transponder code is identical in the ADS-B and transponder panel**. Additionally, UAT systems provide a VFR "privacy" mode switch position that may be used by pilots when not wanting to receive air traffic services. This feature will broadcast a "VFR" ID to other aircraft and ground receivers, similar to the "1200" transponder code.

3. ADS-B is intended to be used in-flight and on the airport surface. ADS-B systems should be turned "on" – and remain "on" – whenever operating in the air and on the airport surface, thus reducing the likelihood of runway incursions. Civil and military Mode A/C transponders and ADS-B systems should be adjusted to the "on" or normal operating position as soon as practical, unless the change to "standby" has been accomplished previously at the request of ATC. Mode S transponders should be left on whenever power is applied to the aircraft.

d. ATC Surveillance Services using ADS-B

Procedures and Recommended Phraseology—For Use In Alaska Only Radar procedures, with the exceptions found in this paragraph, are identical to those procedures prescribed for radar in AIM Chapter 4 and Chapter 5.

1. Preflight:

If a request for ATC services is predicated on ADS-B and such services are anticipated when either a VFR or IFR flight plan is filed, the aircraft's "N" number or callsign as filed in "Block 2" of the Flight Plan shall be entered in the ADS-B avionics as the aircraft's flight ID.

2. Inflight:

When requesting ADS-B services while airborne, pilots should ensure that their ADS-B equipment is transmitting their aircraft's "N" number or call sign prior to contacting ATC. To accomplish this, the pilot must select the ADS-B "broadcast flight ID" function.

NOTE—
THE BROADCAST "VFR" OR "STANDBY" MODE BUILT INTO SOME ADS-B SYSTEMS WILL NOT PROVIDE ATC WITH THE APPROPRIATE AIRCRAFT IDENTIFICATION INFORMATION. THIS FUNCTION SHOULD FIRST BE DISABLED BEFORE CONTACTING ATC.

3. Aircraft with an Inoperative/Malfunctioning ADS-B Transmitter or in the event of an Inoperative Ground Broadcast Transceiver (GBT).

(a) ATC will inform the flight crew when the aircraft's ADS-B transmitter appears to be inoperative or malfunctioning:

PHRASEOLOGY—

YOUR ADS-B TRANSMITTER APPEARS TO BE INOPERATIVE/MALFUNCTIONING. STOP ADS-B TRANSMISSIONS.

(b) ATC will inform the flight crew when the GBT transceiver becomes inoperative or malfunctioning, as follows:

PHRASEOLOGY—

(Name of facility) GROUND BASED TRANSCEIVER INOPER/ATIVEIMALFUNCTIONING.

(And if appropriate) RADAR CONTACT LOST.

NOTE—AN INOPERATIVE OR MALFUNCTIONING GBT MAY ALSO CAUSE A LOSS OF ATC SURVEILLANCE SERVICES.

(c) ATC will inform the flight crew if it becomes necessary to turn off the aircraft's ADS-B transmitter.

PHRASEOLOGY—

STOP ADS-B TRANSMISSIONS.

(d) Other malfunctions and considerations:

Loss of automatic altitude reporting capabilities (encoder failure) will result in loss of ATC altitude advisory services.

e. ADS-B Limitations

1. The ADS-B cockpit display of traffic is NOT intended to be used as a collision avoidance system and does not relieve the pilot's responsibility to "see and avoid" other aircraft. (See paragraph 5-5-8, See and Avoid). ADS -B shall not be used for avoidance maneuvers during IMC or other times when there is no visual contact with the intruder aircraft. ADS -B is intended only to assist in visual acquisition of other aircraft. No avoidance maneuvers are provided nor authorized, as a direct result of an ADS-B target being displayed in the cockpit.

2. Use of ADS-B radar services is limited to the service volume of the GBT.

NOTE—THE COVERAGE VOLUME OF GBTS ARE LIMITED TO LINE-OF-SIGHT.

f. Reports of ADS-B Malfunctions

Users of ADS-B can provide valuable assistance in the correction of malfunctions by reporting instances of undesirable system performance. Reporters should identify the time of observation, location, type and identity of aircraft, and describe the condition observed; the type of avionics system and its software version in use should also be included. Since ADS-B performance is monitored by maintenance personnel rather than ATC, it is suggested that malfunctions be reported in any one of the following ways:

1. By radio or telephone to the nearest Flight Service Station (FSS) facility.

2. By FAA Form 8000-7, Safety Improvement Report, a postage-paid card is designed for this purpose. These cards may be obtained from FAA FSSs, Flight Standards District Offices, and general aviation fixed-based operators.

3. By reporting the failure directly to the FAA Safe Flight 21 program at 1-877-FLYADSB or www.flyadsb.com.

4-5-8. Traffic Information Service-Broadcast (TIS-B)

a. Introduction

Traffic Information Service-Broadcast (TIS-B) is the broadcast of traffic information to ADS-B equipped aircraft from ADS-B ground stations. The source of this traffic information is derived from ground-based air traffic surveillance sensors, typically radar. TIS-B service is becoming available in selected locations where there are both adequate surveillance coverage from ground sensors and adequate broadcast coverage from Ground Based Transceivers (GBTs). The quality level of traffic information provided by TIS-B is dependent upon the number and type of ground sensors available as TIS-B sources and the timeliness of the reported data.

b. TIS-B Requirements

In order to receive TIS-B service, the following conditions must exist:

1. The host aircraft must be equipped with a UAT ADS -13 transmitter/receiver or transceiver, and a cockpit display of traffic information (CDTI). As the ground system evolves, the ADS-B data link may be either UAT or 1090 ES, or both.

2. The host aircraft must fly within the coverage volume of a compatible GBT that is configured for TIS-B uplinks. (Not all GBTs provide TIS-B due to a lack of radar coverage or because a radar feed is not available).

3. The target aircraft must be within the coverage of, and detected by, at least one of the ATC radars serving the GBT in use.

c. TIS-B Capabilities

1. TIS-B is the broadcast of traffic information to ADS -B equipped aircraft. The source of this traffic information is derived from ground-based air traffic radars. TIS-B is intended to provide ADS-B equipped aircraft with a more complete traffic picture in situations where not all nearby aircraft are equipped with ADS-B. The advisory-only application will enhance a pilot's visual acquisition of other traffic.

2. Only transponder-equipped targets (i.e., Mode A/C or Mode S transponders) are detected. Current radar citing may result in limited radar surveillance coverage at lower altitudes near some general aviation airports, with subsequently limited TIS-B service volume coverage. If there is no radar coverage in a given area, then there will be no TIS-B coverage in that area.

d. TIS-B Limitations

1. TIS-B is NOT intended to be used as collision avoidance system and does not relieve the pilot's responsibility to "see and avoid" other aircraft (See paragraph 5-5-8, See and Avoid). TIS-B shall not be used for avoidance maneuvers during time when there is no visual contact with the intruder aircraft. TIS-B is intended only to assist in the visual acquisition of other aircraft. *No avoidance maneuvers are provided for nor authorized as a direct result of a TIS-B target being displayed in the cockpit.*

2. While TIS-B is a useful aid to visual traffic avoidance, its inherent system limitations must be understood to ensure proper use.

(a) A pilot may receive an intermittent TIS-l target of themselves, typically when maneuvering (e.g., climbing turn) due to the radar not tracking the aircraft as quickly as ADS-B.

(b) The ADS-B-to-radar association process within the ground system may at times have difficulty correlating an ADS-B report with corresponding radar returns from the same aircraft. When this happens the pilot will see duplicate traffic symbols (i.e., "TIS-B shadows") on the cockpit display.

(c) Updates of TIS-B traffic reports will occur less often than ADS-B traffic updates. (TIS-B position updates will occur approximately once ever 3-13 seconds depending on the radar coverage. In comparison, the update rate for ADS-B is nominally once per second).

(d) The TIS-B system only detects and uplinks data pertaining to transponder equipped aircraft. Aircraft without a transponder will not be displayed as a TIS-B target.

(e) There is no indication provided when any aircraft is operating inside (or outside) the TIS-B service volume, therefore it is difficult to know if one is receiving uplinked TIS-B traffic information. Assume that not all aircraft are displayed as TIS-B targets.

3. Pilots and operators are reminded that the airborne equipment that displays TIS-B targets is for pilot situational awareness only and is not approved as a collision avoidance tool. Unless there is an imminent emergency requiring immediate action, any deviation from an air traffic control clearance based on TIS-B displayed cockpit information must be approved beforehand by the controlling ATC facility prior to commencing the maneuver. Uncoordinated deviations may place an aircraft in close proximity to other aircraft under ATC control not seen on the airborne equipment, and may result in a pilot deviation.

e. Reports of TIS-B Malfunctions

Users of TIS-B can provide valuable assistance in the correction of malfunctions by reporting instances of undesirable system performance. Reporters should identify the time of observation, location, type and identity of the aircraft, and describe the condition observed; the type of avionics system and its software version used. Since TIS-B performance is monitored by maintenance personnel rather than ATC, it is suggested that malfunctions be reported in anyone of the following ways:

1. By radio or telephone to the nearest Flight Service Station (FSS) facility.

2. By FAA Form 8000-7, Safety Improvement Report, a postage-paid card is designed for this purpose. These cards may be obtained from FAA FSSs, Flight Standards District Offices, and general aviation fixed-based operators.

3. By reporting the failure directly to the FAA Safe Flight 21 program at 1-877-FLYADSB or. *www.flyadsb.com*

Section 6. OPERATIONAL POLICY/PROCEDURES FOR REDUCED VERTICAL SEPARATION MINIMUM (RVSM) IN THE DOMESTIC U.S., ALASKA, OFFSHORE AIRSPACE AND THE SAN JUAN FIR

4-6-1. APPLICABILITY AND RVSM MANDATE (DATE/TIME AND AREA)

a. Applicability. The policies, guidance and direction in this section apply to RVSM operations in the airspace over the lower 48 states, Alaska, Atlantic and Gulf of Mexico High Offshore Airspace and airspace in the San Juan FIR where VHF or UHF voice direct controller-pilot communication (DCPC) is normally available. Policies, guidance and direction for RVSM operations in oceanic airspace where VHF or UHF voice DCPC is not available and the airspace of other countries are posted on the FAA "RVSM Documentation" Webpage described in paragraph 4-6-3, Aircraft and Operator Approval Policy/Procedures, RVSM Monitoring and Databases for Aircraft and Operator Approval.

b. Mandate. At 0901 UTC on January 20, 2005, the FAA implemented RVSM between flight level (FL) 290-410

(inclusive) in the following airspace: the airspace of the lower 48 states of the United States, Alaska, Atlantic and Gulf of Mexico High Offshore Airspace and the San Juan FIR. (A chart showing the location of offshore airspace is posted on the Domestic U.S. RVSM (DRVSM) Webpage. See paragraph 4-6-3.) On the same time and date, RVSM was also introduced into the adjoining airspace of Canada and Mexico to provide a seamless environment for aircraft traversing those borders. In addition, RVSM was implemented on the same date in the Caribbean and South American regions.

c. RVSM Authorization. In accordance with 14 CFR Section 91.180, with only limited exceptions, prior to operating in RVSM airspace, operators and aircraft must have received RVSM authorization from the responsible civil aviation authority. (See paragraph 4-6-10, Procedures for Accommodation of Non-RVSM Aircraft.) If the operator or aircraft or both have not been authorized for RVSM operations, the aircraft will be referred to as a "non-RVSM" aircraft. Paragraph 4-6-10 discusses ATC policies for accommodation of non-RVSM aircraft flown by the Department of Defense, Air Ambulance (Lifeguard) operators, foreign State governments and aircraft flown for certification and development. Paragraph 4-6-11, Non -RVSM Aircraft Requesting Climb to and Descent from Flight Levels Above RVSM Airspace Without Intermediate Level Off, contains policies for non-RVSM aircraft climbing and descending through RVSM airspace to/from flight levels above RVSM airspace.

d. Benefits. RVSM enhances ATC flexibility; mitigates conflict points, enhances sector throughput; reduces controller workload and enables crossing traffic. Operators gain fuel savings and operating efficiency benefits by flying at more fuel efficient flight levels and on more user preferred routings.

4-6-2. FLIGHT LEVEL ORIENTATION SCHEME

Altitude assignments for direction of flight follow a scheme of odd altitude assignment for magnetic courses 000-179 degrees and even altitudes for magnetic courses 180-359 degrees for flights up to and including FL 410, as indicated in FIG 4-6-1.

NOTE— ODD FLIGHT LEVELS: MAGNETIC COURSE 000-179 DEGREES

EVEN FLIGHT LEVELS: MAGNETIC COURSE 180-359 DEGREES

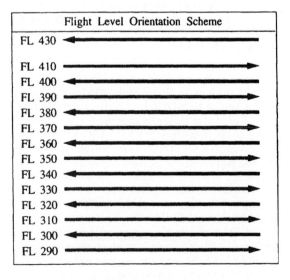

Figure 4-6-1. Flight Level Orientation Scheme

4-6-3. AIRCRAFT AND OPERATOR APPROVAL POLICY/PROCEDURES, RVSM MONITORING AND DATABASES FOR AIRCRAFT AND OPERATOR APPROVAL

a. RVSM Authority. 14 CFR Section 91.180 applies to RVSM operations within the U.S. 14 CFR Section 91.706 applies to RVSM operations outside the U.S. Both sections require that the operator obtain authorization prior to operating in RVSM airspace. 14 CFR Section 91.180 requires that, prior to conducting RVSM operations within the U.S., the operator obtain authorization from the FAA or from the responsible authority, as appropriate. In addition, it requires that the operator and the operator's aircraft comply with the standards of 14 CFR Part 91 Appendix G (Operations in RVSM Airspace).

b. Sources of Information. The FAA RVSM Website Homepage can be accessed at: www.faa.gov/ats/ato/rvsm1. htm. The "RVSM Documentation" and "Domestic RVSM" webpages are linked to the RVSM Homepage. "RVSM Documentation" contains guidance and direction for an operator to obtain aircraft and operator approval to conduct RVSM operations. It provides information for DRVSM and oceanic and international RVSM airspace. It is recommended that operators planning to operate in Domestic U.S. RVSM airspace first review the following documents to orient themselves to the approval process.

1. Under "Area of Operations Specific Information," the document, "Basic Operator Information on DRVSM Programs," provides an overview of the DRVSM program and the related aircraft and operator approval programs.

2. In the "Getting Started" section, review the "RVSM Approval Checklist - U.S. Operators" or "RVSM Approval Checklist - Non-U.S. Operators" (as applicable). These are job aids or checklists that show aircraft/operator approval process events with references to related RVSM documents published on the website.

3. Under "Documents Applicable to All RVSM Approvals," review "RVSM Area New to the Operator." This document provides a guide for operators that are conducting RVSM operations in one or more areas of operation, but are planning to conduct RVSM operations in an area where they have not previously conducted RVSM operations, such as the U.S.

c. TCAS Equipage. TCAS equipage requirements are contained in 14 CFR Sections 121.356, 125.224, 129.18 and 135.189. Part 91 Appendix G does not contain TCAS equipage requirements specific to RVSM, however, Appendix G does require that aircraft equipped with TCAS II and flown in RVSM airspace be modified to incorporate TCAS 11 Version 7.0 or a later version.

d. Aircraft Monitoring. Operators are required to participate in the RVSM aircraft monitoring program. The "Monitoring Requirements and Procedures" section of the RVSM Documentation Webpage contains policies and procedures for participation in the monitoring program. Ground-based and GPS-based monitoring systems are available for the Domestic RVSM program. Monitoring is a quality control program that enables the FAA and other civil aviation authorities to assess the in-service altitude-keeping performance of aircraft and operators.

e. Registration on RVSM Approvals Databases. The "Registration on RVSM Approvals Database" section of the RVSM Documentation Webpage provides policies/procedures for operator and aircraft registration on RVSM approvals databases.

1. Purpose of RVSM Approvals Databases. ATC does not use RVSM approvals databases to determine whether or not a clearance can be issued into RVSM airspace. RVSM program managers do regularly review the operators and aircraft that operate in RVSM airspace to identify and investigate those aircraft and operators flying in RVSM airspace, but not listed on the RVSM approvals databases.

2. Registration of U.S. Operators. When U.S. operators and aircraft are granted RVSM authority, the FAA Flight Standards office makes an input to the FAA Program Tracking and Reporting Subsystem (PTRS). The Separation Standards Group at the FAA Technical Center obtains PTRS operator and aircraft information to update the FAA maintained U.S. Operator/Aircraft RVSM Approvals Database. Basic database operator and aircraft information can be viewed on the RVSM Documentation Webpage by clicking on the appropriate database icon.

3. Registration of Non-U.S. Operators. Non-U.S. operators can find policy/procedures for registration on the North American Approvals Registry and Monitoring Organization (NAARMO) database in the "Registration on RVSM Approvals Database" section of RVSM Documentation.

4-6-4. FLIGHT PLANNING INTO RVSM AIRSPACE

a. Operators that do not file the correct aircraft equipment suffix on the FAA or ICAO Flight Plan may be denied clearance into RVSM airspace. Policies for the FAA Flight Plan are detailed in subparagraph c below. Policies for the ICAO Flight Plan are detailed in subparagraph d.

b. The operator will annotate the equipment block of the FAA or ICAO Flight Plan with an aircraft equipment suffix indicating RVSM capability only after the responsible civil aviation authority has determined that both the operator and its aircraft are RVSM-compliant and has issued RVSM authorization to the operator.

c. General Policies for FAA Flight Plan Equipment Suffix. TBL 5-1-2, Aircraft Suffixes, allows operators to indicate that the aircraft has both RVSM and Advanced Area Navigation (RNAV) capabilities or has only RVSM capability.

1. The operator will annotate the equipment block of the FAA Flight Plan with the appropriate aircraft equipment suffix from TBL 5-1-2.

2. Operators can only file one equipment suffix in block 3 of the FAA Flight Plan. Only this equipment suffix is displayed directly to the controller.

3. Aircraft with RNAV Capability. For flight in RVSM airspace, aircraft with RNAV capability, but not Advanced RNAV capability, will file "/W". Filing "/W" will not preclude such aircraft from filing and flying direct routes in en route airspace.

d. Policy for ICAO Flight Plan Equipment Suffixes.

1. Operators/aircraft that are RVSM-compliant and that file ICAO flight plans will file "/W" in block 10 (Equipment) to indicate RVSM authorization and will also file the appropriate ICAO Flight Plan suffixes to indicate navigation and communication capabilities. The equipment suffixes in "TBL 5-1-2 are for use only in an FAA Flight Plan (FAA Form 7233-1).

2. Operators/aircraft that file ICAO flight plans that include flight in Domestic U.S. RVSM airspace must file "/W" in block 10 to indicate RVSM authorization.

e. Importance of Flight Plan Equipment Suffixes. The operator must file the appropriate equipment suffix in the equipment block of the FAA Flight Plan (FAA Form 7233-1) or the ICAO Flight Plan. The equipment suffix informs ATC:

1. Whether or not the operator and aircraft are authorized to fly in RVSM airspace.

2. The navigation and/or transponder capability of the aircraft (e.g., advanced RNAV, transponder with Mode C).

f. Significant ATC uses of the flight plan equipment suffix information are:

1. To issue or deny clearance into RVSM airspace.

2. To apply a 2,000 foot vertical separation minimum in RVSM airspace to aircraft that are not authorized for RVSM, but are in one of the limited categories that the FAA has agreed to accommodate. (See paragraphs 4-6-10, Procedures for Accommodation of Non-RVSM Aircraft, and 4-6-11, Non-RVSM Aircraft Requesting Climb to and Descent from Flight Levels Above RVSM Airspace Without Intermediate Level Off, for policy on limited operation of unapproved aircraft in RVSM airspace).

3. To determine if the aircraft has "Advanced RNAV" capabilities and can be cleared to fly procedures for which that capability is required.

4-6-5. PILOT RVSM OPERATING PRACTICES AND PROCEDURES

a. RVSM Mandate. If either the operator or the aircraft or both have not received RVSM authorization (non-RVSM aircraft), the pilot will neither request nor accept a clearance into RVSM airspace unless:

1. The flight is conducted by a non-RVSM DOD, Lifeguard, certification/development or foreign State (government) aircraft in accordance with paragraph 4-6-10, Procedures for Accommodation of Non-RVSM Aircraft.

2. The pilot intends to climb to or descend from FL 430 or above in accordance with paragraph 4-6-11, Non-RVSM Aircraft Requesting Climb to and Descent from Flight Levels Above RVSM Airspace Without Intermediate Level Off.

3. An emergency situation exists.

b. Basic RVSM Operating Practices and Procedures. Appendix 4 of Guidance 91-RVSM contains pilot practices and procedures for RVSM. Operators must incorporate Appendix 4 practices and procedures, as supplemented by the applicable paragraphs of this section, into operator training or pilot knowledge programs and operator documents containing RVSM operational policies. Guidance 91-RVSM is published on the RVSM Documentation Webpage under "Documents Applicable to All RVSM Approvals."

c. Appendix 4 contains practices and procedures for flight planning, preflight procedures at the aircraft, procedures prior to RVSM airspace entry, inflight (en route) procedures, contingency procedures and post flight.

d. The following paragraphs either clarify or supplement Appendix 4 practices and procedures.

4-6-6. GUIDANCE ON SEVERE TURBULENCE AND MOUNTAIN WAVE ACTIVITY (MWA)

a. Introduction/Explanation

1. The information and practices in this paragraph are provided to emphasize to pilots and controllers the importance of taking appropriate action in RVSM airspace when aircraft experience severe turbulence and/or MWA that is of sufficient magnitude to significantly affect altitude-keeping.

2. Severe Turbulence. Severe turbulence causes large, abrupt changes in altitude and/or attitude usually accompanied by large variations in indicated airspeed. Aircraft may be momentarily out of control. Encounters with severe turbulence must be remedied immediately in any phase of flight. Severe turbulence may be associated with MWA.

3. Mountain Wave Activity (MWA)

(a) Significant MWA occurs both below and above the floor of RVSM airspace, FL 290. MWA often occurs in western states in the vicinity of mountain ranges. It may occur when strong winds blow perpendicular to mountain ranges resulting in up and down or wave motions in the atmosphere. Wave action can produce altitude excursions and airspeed fluctuations accompanied by only light turbulence. With sufficient amplitude, however, wave action can induce altitude and airspeed fluctuations accompanied by severe turbulence. MWA is difficult to forecast and can be highly localized and short lived.

(b) Wave activity is not necessarily limited to the vicinity of mountain ranges. Pilots experiencing wave activity anywhere that significantly affects altitude-keeping can follow the guidance provided below.

(c) Inflight MWA Indicators (Including Turbulence). Indicators that the aircraft is being subjected to MWA are:

(1) Altitude excursions and/or airspeed fluctuations with or without associated turbulence.

(2) Pitch and trim changes required to maintain altitude with accompanying airspeed fluctuations.

(3) Light to severe turbulence depending on the magnitude of the MWA.

4. Priority for Controller Application of Merging Target Procedures

(a) Explanation of Merging Target Procedures. As described in subparagraph c3 below, ATC will use "merging target procedures" to mitigate the effects of both severe turbulence and MWA. The procedures in subparagraph c3 have been adapted from existing procedures published in FAA Order 7110.65, Air Traffic Control, paragraph 5-1-8, Merging Target Procedures. Paragraph 5-1-8 calls for en route controllers to advise pilots of potential traffic that they perceive may fly directly above or below his/her aircraft at minimum vertical separation. In response, pilots are given the option of requesting a radar vector to ensure their radar target will not merge or overlap with the traffic's radar target.

(b) The provision of "merging target procedures" to mitigate the effects of severe turbulence and/or MWA is not optional for the controller, but rather is a priority responsibility. Pilot requests for vectors for traffic avoidance when encountering MWA or pilot reports of "Unable RVSM due turbulence or MWA" are considered first priority aircraft separation and sequencing responsibilities. (FAA Order 7110.65, paragraph 2-1-2, Duty Priority, states that the controller's first priority is to separate aircraft and issue safety alerts.)

(c) Explanation of the term "traffic permitting." The contingency actions for MWA and severe turbulence detailed in paragraph 4-6-9, Contingency Actions: Weather Encounters and Aircraft System Failures, state that the controller will "vector aircraft to avoid merging targets with traffic at adjacent flight levels, traffic permitting." The term "traffic permitting" is not intended to imply that merging target procedures are not a priority duty. The term is intended to recognize that, as stated in FAA Order 7110.65, paragraph 2-1-2, Duty Priority, there are circumstances when the controller is required to perform more than one action and must "exercise their best judgment based on the facts and circumstances known to them" to prioritize their actions. Further direction given is: "That action which is most critical from a safety standpoint is performed first."

5. TCAS Sensitivity. For both MWA and severe turbulence encounters in RVSM airspace, an additional concern is the sensitivity of collision avoidance systems when one or both aircraft operating in close proximity receive TCAS advisories in response to disruptions in altitude hold capability.

b. Pre-flight tools. Sources of observed and forecast information that can help the pilot ascertain the possibility of MWA or severe turbulence are: Forecast Winds and Temperatures Aloft (FD), Area Forecast (FA), SIGMETs and PIREPs.

c. Pilot Actions When Encountering Weather (e.g., Severe Turbulence or MWA)

1. Weather Encounters Inducing Altitude Deviations of Approximately 200 feet. When the pilot experiences weather induced altitude deviations of approximately 200 feet, the pilot will contact ATC and state "Unable RVSM Due (state reason)" (e.g., turbulence, mountain wave). See contingency actions in paragraph 4-6-9.

2. Severe Turbulence (including that associated with MWA). When pilots encounter severe turbulence, they should contact ATC and report the situation. Until the pilot reports clear of severe turbulence, the controller will apply merging target vectors to one or both passing aircraft to prevent their targets from merging:

EXAMPLE—

"YANKEE 123, FL 310, UNABLE RVSM DUE SEVERE TURBULENCE."

"YANKEE 123, FLY HEADING 290; TRAFFIC TWELVE O'CLOCK, 10 MILES, OPPOSITE DIRECTION; EASTBOUND MD-80 AT FL 320" (OR THE CONTROLLER MAY ISSUE A VECTOR TO THE MD-80 TRAFFIC TO AVOID YANKEE 123).

3. MWA. When pilots encounter MWA, they should contact ATC and report the magnitude and location of the wave activity. When a controller makes a merging targets traffic call, the pilot may request a vector to avoid flying directly over or under the traffic. In situations where the pilot is experiencing altitude deviations of 200 feet or greater, the pilot will request a vector to avoid traffic. Until the pilot reports clear of MWA, the controller will apply merging target vectors to one or both passing aircraft to prevent their targets from merging:

EXAMPLE—

"YANKEE 123, FL 310, UNABLE RVSM DUE MOUNTAIN WAVE."

"YANKEE 123, FLY HEADING 290; TRAFFIC TWELVE O'CLOCK, 10 MILES, OPPOSITE DIRECTION; EASTBOUND MD-80 AT FL 320" (OR THE CONTROLLER MAY ISSUE A VECTOR TO THE MD-80 TRAFFIC TO AVOID YANKEE 123).

4. FL Change or Re-route. To leave airspace where MWA or severe turbulence is being encountered, the pilot may request a FL change and/or re-route, if necessary.

4-6-7. Guidance on Wake Turbulence

a. Pilots should be aware of the potential for wake turbulence encounters in RVSM airspace. Experience gained since 1997 has shown that such encounters in RVSM airspace are generally moderate or less in magnitude.

b. Prior to DRVSM implementation, the FAA established provisions for pilots to report wake turbulence events in RVSM airspace using the NASA Aviation Safety Reporting System (ASRS). A "Safety Reporting" section established on the FAA RVSM Documentation webpage provides contacts, forms, and reporting procedures.

c. To date, wake turbulence has not been reported as a significant factor in DRVSM operations. European authorities also found that reports of wake turbulence encounters did not increase significantly after RVSM implementation (eight versus seven reports in a ten-month period). In addition, they found that reported wake turbulence was generally similar to moderate clear air turbulence.

d. Pilot Action to Mitigate Wake Turbulence Encounters

1. Pilots should be alert for wake turbulence when operating:

(a) In the vicinity of aircraft climbing or descending through their altitude.

(b) Approximately 10–30 miles after passing 1,000 feet below opposite-direction traffic.

(c) Approximately 10–30 miles behind and 1,000 feet below same-direction traffic.

2. Pilots encountering or anticipating wake turbulence in DRVSM airspace have the option of requesting a vector, FL change or if capable, a lateral offset.

NOTE—

1. OFFSETS OF APPROXIMATELY A WINGSPAN UPWIND GENERALLY CAN MOVE THE AIRCRAFT OUT OF THE IMMEDIATE VICINITY OF ANOTHER AIRCRAFT'S WAKE VORTEX.

2. IN DOMESTIC U.S. AIRSPACE, PILOTS MUST REQUEST CLEARANCE TO FLY A LATERAL OFFSET. STRATEGIC LATERAL OFFSETS FLOWN IN OCEANIC AIRSPACE DO NOT APPLY.

d. The FAA will track wake turbulence events as an element of its post implementation program. The FAA will advertise wake turbulence reporting procedures to the operator community and publish reporting procedures on the RVSM Documentation Webpage (See address in paragraph 4-6-3, Aircraft and Operator Approval Policy/Procedures, RVSM Monitoring and Databases for Aircraft and Operator Approval.)

4-6-8. Pilot/Controller Phraseology

TBL 4-6-1 shows standard phraseology that pilots and controllers will use to communicate in DRVSM operations.

4-6-9. Contingency Actions: Weather Encounters and Aircraft System Failures

TBL 4-2-6 provides pilot guidance on actions to take under certain conditions of aircraft system failure and weather encounters. It describes the expected ATC controller actions in these situations. It is recognized that the pilot and controller will use judgment to determine the action most appropriate to the situation.

4-6-10. Procedures for Accommodation of Non-RVSM Aircraft

a. General Policies for Accommodation of Non-RVSM Aircraft

1. The RVSM mandate calls for only RVSM authorized aircraft/operators to fly in designated RVSM airspace with limited exceptions. The policies detailed below are intended exclusively for use by aircraft that the FAA has agreed to accommodate. They are not intended to provide other operators a means to circumvent the normal RVSM approval process.

2. If either the operator or aircraft or both have not been authorized to conduct RVSM operations, the aircraft will be referred to as a "non-RVSM" aircraft. 14 CFR Section 91.180 and Part 91 Appendix G enable the FAA to authorize a deviation to operate a non-RVSM aircraft in RVSM airspace.

3. Non-RVSM aircraft flights will be handled on a workload permitting basis. The vertical separation standard applied between aircraft not approved for RVSM and all other aircraft shall be 2,000 feet.

4. Required Pilot Calls. The pilot of non-RVSM aircraft will inform the controller of the lack of RVSM approval in accordance with the direction provided in paragraph 4-6-8, Pilot/Controller Phraseology.

b. Categories of Non-RVSM Aircraft that may be Accommodated

Subject to FAA approval and clearance, the following categories of non-RVSM aircraft may operate in domestic

Table 4-6-1 Pilot/Controller Phraseology

Message	Phraseology
For a controller to ascertain the RVSM approval status of an aircraft:	(call sign) confirm RVSM approved
Pilot indication that flight is RVSM approved	Affirm RVSM

Pilot report lack of RVSM approval (non-RVSM status): Pilot will report non-RVSM status as follows:
  a. On the initial call on any frequency in the RVSM airspace and . . .
  b. In all requests for flight level changes pertaining to flight levels within the RVSM airspace and . . .
  c. In all read backs to flight level clearances pertaining to flight levels within the RVSM airspace and . . .
  d. In read back of flight level clearances involving climb and descent through RVSM airspace (FL 290 - 410)
	Negative RVSM, (supplementary information,
	
	  e.g., "Certification flight").

Pilot report of one of the following after entry into RVSM airspace: all primary altimeters, automatic altitude control systems or altitude alerters have failed. (Sec paragraph 4–6–9, Contingency Actions: Weather Encounters and Aircraft System Failures.)	Unable RVSM Due Equipment
NOTE- *This phrase is to be used to convey both the intial indication of RVSW aircraft system failure and on initial contact on all frequencies in RVSM airspace until the problem ceases to exist or the aircraft has exited RVSM airspace*	
ATC denial of clearance into RVSM airspace	Unable issue clearance into RVSM airspace, maintain FL
*Pilot reporting inability to maintain cleared flight level due to weather encounter. (See paragraph 4-6-9, Contingency Actions: Weather Encounters and Aircraft System Failures).	*Unable RVSM due (state reason) (e.g., turbulence, mountain wave)
ATC requesting pilot to confirm that an aircraft has regained RVSM-approved status or a pilot is ready to resume RVSM	Confirm able to resume RVSM
Pilot ready to resume RVSM after aircraft system or weather contingency	Ready to resume RVSM

Table 4-6-2   Contingency Actions: Weather Encounters and Aircraft System Failures

Initial Pilot Action in Contingency Situations

Initial pitot actions when unable to maintain flight level (FL) or unsure of aircraft altitude-keeping capability:
- Notify ATC and request assistance as detailed below.
- Maintain cleared flight level, to the extern possible, while evaluating the situation.
- Watch for conflicting traffic both visually and by reference to TCAS, if equipped.
- Alert nearby aircraft by illuminating exterior lights (commensurate with aircraft limitations).

Severe Turbulence and/or Mountain Wave Activity (MWA) Induced
Altitude Deviations of Approximately 208 feet

Pilot will:	Controller will:
• When experiencing severe turbulence and/or MWA induced altitude deviations of approximately 200 feet or greater, pilot will contact ATC and state "Unable RVSM Due (state reason)" (e.g., turbulence, mountain wave) • If not issued by the controller, request vector clear of traffic at adjacent FLs • If desired, request FL change or re-route • Report location and magnitude of turbulence or MWA to ATC	• Vector aircraft to avoid merging target with traffic at adjacent flight levels, traffic permitting • Advise pilot of conflicting traffic • Issue FL change or re-route, traffic permitting • Issue P1REP to other aircraft
See paragraph 4-6-6, Guidance on Severe Turbulence and Mountain Wave Activity (MWA) for detailed guidance.	Paragraph 4-6-6 explains "traffic permitting."

Mountain Wave Activity (MWA) Encounters - General

Pilot actions:	Controfler actions:
• Contact ATC and report experiencing MWA • If so desired, pilot may request a FL change or re-route • Report location and magnitude of MWA to ATC	• Advise pilot of conflicting traffic at adjacent FL • If pilot requests, vector aircraft to avoid merging target with traffic at adjacent RVSM flight levels, traffic permitting • Issue FL change or re-route, traffic permitting • Issue P1REP to other aircraft
See paragraph 4-6-6 for guidance on MWA.	Paragraph 4-6-6 explains "traffic permitting."

NOTE-
MWA encounters do not necessarily result in altitude deviations on the order of 200 feet The guidance below is intended to address itss significant MWA encounters.

Wake Turbulence Encounters

Pilot should:	Controller should:
• Contact ATC and request vector, FL change or, if capable, a lateral offset	• Issue vector, FL change or lateral offset clearance, traffic permitting
See paragraph 4-6-7, Guidance on Wake Turbulence.	Paragraph 4-6-6 explains "traffic permitting."

"Unable RVSM Due Equipment"
Failure of Automatic Altitude Control System, Attitude Alerter or AB Primary Altimeters

Pitot will:	Controller will:
• Contact ATC and state "Unable RVSM Due Equipment" • Request clearance out of RVSM airspace unless operational situation dictates otherwise	• Provide 2,000 feet vertical separation or appropriate horizontal separation • Clear aircraft out of RVSM airspace unless operational situation dictates otherwise

Table 4-6-2   (continued)
One Primary Altimeter Remains Operational

Pilot will:	Controller will:
• Cross check stand-by altimeter	• Acknowledge operation with single primary altimeter
• Notify ATC of operation with single primary altimeter	
• If unable to confirm primary altimeter accuracy, follow actions for failure of alt primary altimeters	

Transponder Failure

Pilot should:	Controller should:
• Contact ATC and request authority to continue to operate at cleared flight level	• Consider request to continue to operate at cleared flight level
• Comply with revised ATC clearance, if issued	• Issue revised clearance, if necessary

NOTE-
14 CFR Section 91.215 (ATC transponder and altitude reporting equipment and use) regulates operation with the transponder inoperative.

U.S. RVSM airspace provided they have an operational transponder:

1. Department of Defense (DOD) aircraft.

2. Flights conducted for aircraft certification and development purposes.

3. Active air ambulance flights utilizing a "Lifeguard" call sign.

4. Aircraft climbing/descending through RVSM flight levels (without intermediate level off) to/from FLs above RVSM airspace (Policies for these flights are detailed in paragraph 4-6-11, Non-RVSM Aircraft Requesting Climb to and Descent from Flight Levels Above RVSM Airspace Without Intermediate Level Off.

5. Foreign State (government) aircraft.

c. Methods for operators of non-RVSM aircraft to request access to RVSM Airspace. Operators may:

1. LOA/MOU. Enter into a Letter of Agreement (LOA)/Memorandum of Understanding (MOU) with the RVSM facility (the Air Traffic facility that provides air traffic services in RVSM airspace). Operators must comply with LOA/MOU.

2. File-and-Fly. File a flight plan to notify the FAA of their intention to request access to RVSM airspace.

NOTE—
PRIORITY FOR ACCESS TO RYSM AIRSPACE WILL BE AFFORDED TO RVSM COMPLIANT AIRCRAFT, THEN FILE-AND-FLY FLIGHTS.

3. DOD. Some DOD non-RVSM aircraft will be designated as aircraft requiring special consideration. For coordination purposes they will be referred to as STORM flights. DOD enters STORM flights on the DOD Priority Mission website and notifies the departure RVSM facility for flights that are within 60 minutes of departure.

NOTE—
SPECIAL CONSIDERATION WILL BE AFFORDED A STORM FLIGHT; HOWEVER, ACCOMMODATION OF ANY NON-RVSM FLIGHT IS WORKLOAD PERMITTING.

d. Center Phone Numbers. Center phone numbers are posted on the RVSM Documentation Webpage, North American RVSM, Domestic U.S. RVSM section. This address provides direct access to the phone number listing: http://www.faa.gov/ats/ato/150_docs/Center_Phone No_Non-RVSM Acft.doc

4-6-11. Non-RSVP Aircraft Requesting Climb and Descent from Flight Levels Above RVSM Airspace Without Intermediate Leveling Off

a. File-and-Fly. Operators of Non-RSVP aircraft climbing to and descending from RVSP flight levels should just file a flight plan.

b. Non-RVSM aircraft climbing to and descending from flight levels above RVSM airspace will be handled on a workload permitting basis. The vertical separation standard applied in RVSM airspace between non-RVSM aircraft and all other aircraft shall be 2,000 feet.

c. Non-RVSM aircraft climbing to/descending from RVSM airspace can only be considered for accommodation provided:

1. Aircraft is capable of a continuous climb/descent and does not need to level off at an intermediate altitude for any operational considerations and

2. Aircraft is capable of climb/descent at the normal rate for the aircraft.

d. Required Pilot Calls. The pilot of non-RVSM aircraft will inform the controller of the lack of RVSM approval in accordance with the direction provided in paragraph 4-6-8, Pilot/Controller Phraseology.

Chapter 5. Air traffic procedures
Section 1. PREFLIGHT

5-1-1. PREFLIGHT PREPARATION

a. Every pilot is urged to receive a preflight briefing and to file a flight plan. This briefing should consist of the latest or most current weather, airport, and en route NAVAID information. Briefing service may be obtained from an FSS either by telephone or interphone, by radio when airborne, or by a personal visit to the station. Pilots with a current medical certificate in the 48 contiguous states may access toll-free the Direct User Access Terminal System (DUATS) through a personal computer. DUATS will provide alpha-numeric preflight weather data and allow pilots to file domestic VFR or IFR flight plans.

REFERENCE—AIM, FAA WEATHER SERVICES, PARAGRAPH 7-1-2 LISTS DUATS VENDORS.

NOTE—PILOTS FILING FLIGHT PLANS VIA "FAST FILE" WHO DESIRE TO HAVE THEIR BRIEFING RECORDED, SHOULD INCLUDE A STATEMENT AT THE END OF THE RECORDING AS TO THE SOURCE OF THEIR WEATHER BRIEFING.

b. The information required by the FAA to process flight plans is contained on FAA Form 7233-1, Flight Plan. The forms are available at all flight service stations. Additional copies will be provided on request.

REFERENCE—AIM, FLIGHT PLAN-VFR FLIGHTS, PARAGRAPH 5-1-4.
AIM, FLIGHT PLAN-IFR FLIGHTS, PARAGRAPH 5-1-8.

c. Consult an FSS or a Weather Service Office (WSO) for preflight weather briefing. Supplemental Weather Service Locations (SWSLs) do not provide weather briefings.

d. FSSs are required to advise of pertinent NOTAMs if a *standard* briefing is requested, but if they are overlooked, don't hesitate to remind the specialist that you have not received NOTAM information.

NOTE—NOTAMS WHICH ARE KNOWN IN SUFFICIENT TIME FOR PUBLICATION AND ARE OF 7 DAYS DURATION OR LONGER ARE NORMALLY INCORPORATED INTO THE NOTICES TO AIRMEN PUBLICATION AND CARRIED THERE UNTIL CANCELLATION TIME. FDC NOTAMS, WHICH APPLY TO INSTRUMENT FLIGHT PROCEDURES, ARE ALSO INCLUDED IN THE NOTICES TO AIRMEN PUBLICATION UP TO AND INCLUDING THE NUMBER INDICATED IN THE FDC NOTAM LEGEND. PRINTED NOTAMS ARE NOT PROVIDED DURING A BRIEFING UNLESS SPECIFICALLY REQUESTED BY THE PILOT SINCE THE FSS SPECIALIST HAS NO WAY OF KNOWING WHETHER THE PILOT HAS ALREADY CHECKED THE NOTICES TO AIRMEN PUBLICATION PRIOR TO CALLING. REMEMBER TO ASK FOR NOTAMS IN THE NOTICES TO AIRMEN PUBLICATION. THIS INFORMATION IS NOT NORMALLY FURNISHED DURING YOUR BRIEFING.

REFERENCE—AIM, NOTICE TO AIRMEN (NOTAM) SYSTEM, PARAGRAPH 5-1-3.

e. Pilots are urged to use only the latest issue of aeronautical charts in planning and conducting flight operations.

Aeronautical charts are revised and reissued on a regular scheduled basis to ensure that depicted data are current and reliable. In the conterminous U.S., Sectional Charts are updated every 6 months, IFR En Route Charts every 56 days, and amendments to civil IFR Approach Charts are accomplished on a 56-day cycle with a change notice volume issued on the 28-day midcycle. Charts that have been superseded by those of a more recent date may contain obsolete or incomplete flight information.

REFERENCE—AIM, GENERAL DESCRIPTION OF EACH CHART SERIES, PARAGRAPH 9-1-4.

f. When requesting a preflight briefing, identify yourself as a pilot and provide the following:

1. *Type of flight planned; e.g., VFR or IFR.*
2. *Aircraft's number or pilot's name.*
3. *Aircraft type.*
4. *Departure Airport.*
5. *Route of flight.*
6. *Destination.*
7. *Flight altitude(s).*
8. *ETD and ETE.*

g. Prior to conducting a briefing, briefers are required to have the background information listed above so that they may tailor the briefing to the needs of the proposed flight. The objective is to communicate a "picture" of meteorological and aeronautical information necessary for the conduct of a safe and efficient flight. Briefers use all available weather and aeronautical information to summarize data applicable to the proposed flight. They do not read weather reports and forecasts verbatim unless specifically requested by the pilot. FSS briefers do not provide FDC NOTAM information for special instrument approach procedures unless specifically asked. Pilots authorized by the FAA to use special instrument approach procedures must specifically request FDC NOTAM information for these procedures. Pilots who receive the information electronically will receive NOTAMs for special IAPs automatically.

REFERENCE—AIM, PREFLIGHT BRIEFINGS, PARAGRAPH 7-1-3, CONTAINS THOSE ITEMS OF A WEATHER BRIEFING THAT SHOULD BE EXPECTED OR REQUESTED.

h. FAA by FAR Part 93, Subpart K, has designated High Density Traffic Airports (HDTAs) and has prescribed air traffic rules and requirements for operating aircraft (excluding helicopter operations) to and from these airports.

REFERENCE—AIRPORT/FACILITY DIRECTORY, SPECIAL NOTICES SECTION. AIM, AIRPORT RESERVATION OPERATIONS AND SPECIAL TRAFFIC MANAGEMENT PROGRAMS, PARAGRAPH 4-1-21.

i. In addition to the filing of a flight plan, if the flight will traverse or land in one or more foreign countries, it is particularly important that pilots leave a complete itinerary with someone directly concerned and keep that person advised of the flight's progress. If serious doubt arises as to the safety of the flight, that person should first contact the FSS.

REFERENCE—AIM, FLIGHTS OUTSIDE THE UNITED STATES AND U.S. TERRITORIES, PARAGRAPH 5-1-10.

j. Pilots operating under provisions of FAR Part 135 and not having an FAA assigned 3-letter designator, are urged to prefix the normal registration (N) number with the letter "T" on flight plan filing; e.g., TN1234B.

REFERENCE—AIM, AIRCRAFT CALL SIGNS, PARAGRAPH 4-2-4.

5-1-2. FOLLOW IFR PROCEDURES EVEN WHEN OPERATING VFR

a. To maintain IFR proficiency, pilots are urged to practice IFR procedures whenever possible, even when operating VFR. Some suggested practices include:

1. Obtain a complete preflight and weather briefing. Check the NOTAMs.

2. File a flight plan. This is an excellent low cost insurance policy. The cost is the time it takes to fill it out. The insurance includes the knowledge that someone will be looking for you if you become overdue at your destination.

3. Use current charts.

4. Use the navigation aids. Practice maintaining a good course-keep the needle centered.

5. Maintain a constant altitude which is appropriate for the direction of flight.

6. Estimate en route position times.

7. Make accurate and frequent position reports to the FSSs along your route of flight.

b. Simulated IFR flight is recommended (under the hood); however, pilots are cautioned to review and adhere to the requirements specified in FAR Part 91.109 before and during such flight.

c. When flying VFR at night, in addition to the altitude appropriate for the direction of flight, pilots should maintain an altitude which is at or above the minimum en route altitude as shown on charts. This is especially true in mountainous terrain, where there is usually very little ground reference. Do not depend on your eyes alone to avoid rising unlighted terrain, or even lighted obstructions such as TV towers.

5-1-3. NOTICE TO AIRMEN (NOTAM) SYSTEM

a. Time-critical aeronautical information which is of either a temporary nature or not sufficiently known in advance to permit publication on aeronautical charts or in other operational publications receives immediate dissemination via the National NOTAM System.

NOTE—

[1] NOTAM INFORMATION IS THAT AERONAUTICAL INFORMATION THAT COULD AFFECT A PILOT'S DECISION TO MAKE A FLIGHT. IT INCLUDES SUCH INFORMATION AS AIRPORT OR PRIMARY RUNWAY CLOSURES, CHANGES IN THE STATUS OF NAVIGATIONAL AIDS, ILSS, RADAR SERVICE AVAILABILITY, AND OTHER INFORMATION ESSENTIAL TO PLANNED EN ROUTE, TERMINAL, OR LANDING OPERATIONS.

[2] NOTAM INFORMATION IS TRANSMITTED USING STANDARD CONTRACTIONS TO REDUCE TRANSMISSION TIME. SEE TBL 5-1-1 FOR A LISTING OF THE MOST COMMONLY USED CONTRACTIONS.

b. NOTAM information is classified into three categories. These are NOTAM (D) or distant, NOTAM (L) or local, and Flight Data Center (FDC) NOTAMs.

1. *NOTAM (D)* information is disseminated for all navigational facilities that are part of the National Airspace System (NAS), all public use airports, seaplane bases, and heliports listed in the Airport/Facility Directory (A/FD). The complete file of all NOTAM (D) information is maintained in a computer data base at the Weather Message Switching Center (WMSC), located in Atlanta, Georgia. This category of information is distributed automatically via Service A telecommunications system. Air traffic facilities, primarily FSS's, with Service A capability have access to the entire WMSC data base of NOTAM's. These NOTAM's remain available via Service A for the duration of their validity or until published. Once published, the NOTAM data is deleted from the system.

2. *NOTAM (L)*

(a) NOTAM (L) information includes such data as taxiway closures, personnel and equipment near or crossing runways, airport lighting aids that do not affect instrument approach criteria, such as VASI.

(b) NOTAM (L) information is distributed locally only and is not attached to the hourly weather reports. A separate file of local NOTAMs is maintained at each FSS for facilities in their area only. NOTAM (L) information for other FSS areas must be specifically requested directly from the FSS that has responsibility for the airport concerned.

3. *FDC NOTAMs*

(a) On those occasions when it becomes necessary to disseminate information which is regulatory in nature, the National Flight Data Center (NFDC), in Washington, DC, will issue an FDC NOTAM. FDC NOTAMs contain such things as amendments to published IAPs and other current aeronautical charts. They are also used to advertise temporary flight restrictions caused by such things as natural disasters or large-scale public events that may generate a congestion of air traffic over a site.

(b) FDC NOTAMs are transmitted via Service A only once and are kept on file at the FSS until published or canceled. FSSs are responsible for maintaining a file of current, unpublished FDC NOTAMs concerning conditions within 400 miles of their facilities. FDC information concerning conditions that are more than 400 miles from the FSS, or that is already published, is given to a pilot only on request.

Table 5-1-1 NOTAM Contractions

Contraction:Decode

A
AADC: Approach and Departure Control
ABV: Above
A/C: Approach Control
ACCUM: Accumulate
ACFT: Aircraft
ACR: Air Carrier
ACTV/ACTVT: Active/Activate
ADF: Automatic Direction Finder
AFSS: Automated Flight Service Station
ADJ: Adjacent
ADZ/ADZD: Advise/Advised
AFD: Airport Facility Directory
AFSS: Automated Flight Service Station
ALS: Approach Light System
ALTM: Altimeter
ALTN/ALTNLY: Alternate/Alternately
ALSTG: Altimeter Setting
AMDT: Amendment
APCH: Approach
APL: Airport Lights
ARFF: Aircraft Rescue & Fire Fighting
ARPT: Airport
ARSR: Air Route Surveillance Radar
ASDE: Airport Surface Detection Equipment
ASOS: Automated Surface Observing System
ASPH: Asphalt
ASR: Airport Surveillance Radar
ATC: Air Traffic Control
ATCT: Airport Traffic Control Tower
ATIS: Automatic Terminal Information Service
AVBL: Available
AWOS: Automatic Weather Observing/Reporting System
AZM: Azimuth

B
BC: Back Course
BCN: Beacon
BERM: Snowbank/s Containing Earth/Gravel
BLO: Below
BND: Bound
BRAF: Braking Action Fair
BRAG: Braking Action Good
BRAN: Braking Action Nil
BRAP: Braking Action Poor
BYD: Beyond

C
CAAS: Class A Airspace
CAT: Category
CBAS: Class B Airspace
CBSA: Class B Surface Area
CCAS: Class C Airspace
CCLKWS: Counterclockwise
CCSA: Class C Surface Area
CD: Clearance Delivery
CDAS: Class D Airspace
CDSA: Class D Surface Area
CEAS: Class E Airspace

CESA: Class E Surface Area
CFA: Controlled Firing Area
CGAS: Class G Airspace
CHG: Change
CLKWS: Clockwise
CLNC: Clearance
CLSD: Closed
CMSN/CMSND: Commission/Commissioned
CNCL/CNCLD/CNL: Cancel/Canceled/Cancel
CNTRLN: Centerline
CONC: Concrete
CONT: Continue/Continuously
CRS: Course
CTAF: Common Traffic Advisory Frequency
CTLZ: Centered Zone

D
DALGT: Daylight
DCMSN/DCMSND: Decommission/Decommissioned
DCT: Direct
DEP: Depart/Departure
DEPT: Department
DH: Decision Height
DISABLD: Disabled
DLA/DLAD: Delay/Delayed
DLT/DLTD: Delete/Deleted
DLY:Daily
DME: Distance Measuring Equipment
DMSTN: Demonstration
DPCR: Departure Procedure
DRCT: Direct
DRFT/DRFTD: Drift/Drifted Snowbank/s Caused By
                         Wind Action
DSPLCD: Displaced
DSTC: Distance
DWPNT: Dew Point

E
E: East
EBND: Eastbound
EFAS: En Route Flight Advisory Service
EFF: Effective
ELEV: Elevate/Elevation
ENG: Engine
ENTR: Entire
EXCP: Except

F
FA: Final Approach
FAC: Facility
FAF: Final Approach Fix
FDC: Flight Data Center
FM: Fan Marker
FREQ: Frequency
FRH: Fly Runway Heading
FRZN: Frozen
FRNZ SLR: Frozen Slush on Runway/s
FSS: Flight Service Station

Table 5-1-1 NOTAM Contractions

G

GC: Ground Control
GCA: Ground Controlled Approach
GOVT: Government
GP: Glide Path
GPS: Global Positioning System
GRVL: Gravel
GS: Glide Slope

H

HAA: Height Above Airport
HAT: Height Above Touchdown
HAZ: Hazard
HEL: Helicopter
HELI: Heliport
HF: High Frequency
HIRL: High Intensity Runway Lights
HIWAS: Hazardous Inflight Weather Advisory Service
HOL: Holiday
HP: Holding Pattern

I

IAP: Instrument Approach Procedure
IBND: Inbound
ID: Identification
IDENT: Identify/Identifier/Identification
IFR: Instrument Flight Rules
ILS: Instrument Landing System
IM: Inner Marker
IN: Inch/Inches
INDEFLY: Indefinitely
INOP: Inoperative
INST: Instrument
INT: Intersection
INTST: Intensity
IR: Ice On Runway/s

L

L: Left
LAA: Local Airport Advisory
LAT: Latitude
LAWRS: Limited Aviation Weather Reporting Station
LB: Pound/Pounds
LC: Local Control
LCL: Local
LCTD: Located
LDA: Localizer Type Directional Aid
LDIN: Lead In Lighting System
LGT/LGTD/LGTS: Light/Lighted/Lights
LIRL: Low Intensity Runway Edge Lights
LLWAS: Low Level Wind Shear Alert System
LMM: Compass Locator at ILS Middle Marker
LNDG: Landing
LOC: Localizer
LOM: Compass Locator at ILS Outer Marker
LONG: Longitude
LRN: Loran
LSR: Loose Snow on Runway/s
LT: Left Turn After Take-off

M

MALS: Medium Intensity Approach Lightning System
MALSF: Medium Intensity Approach Lighting System with Sequenced Flashers
MALSR: Medium Intensity Approach Lighting System with Runway Alignment Indicator Lights
MAP: Missed Approach Point
MCA: Minimum Crossing Altitude
MDA: Minimum Descent Altitude
MEA: Minimum Enroute Altitude
MED: Medium
MIN: Minute
MIRL: Medium Intensity Runway Edge Lights
MLS: Microwave Landing System
MM: Middle Marker
MNM: Minimum
MOCA: Minimum Obstruction Clearance Altitude
MONTR: Monitor
MSA: Minimum Safe Altitude/Minimum Sector Altitude
MSAW: Minimum Safe Altitude Warning
MSL: Mean Sea Level
MU: Designate a Friction Value Representing Runway Surface Conditions
MUD: Mud
MUNI: Municipal

N

N: North
NA: Not Authorized
NBND: Northbound
NDB: Nondirectional Radio Beacon
NE: Northeast
NGT: Night
NM: Nautical Mile/s
NMR: Nautical Mile Radius
NOPT: No Procedure Turn Required
NTAP: Notice To Airmen Publication
NW: Northwest

O

OBSC: Obscured
OBSTN: Obstruction
OM: Outer Marker
OPER: Operate
OPN: Operation
ORIG: Original
OTS: Out of Service
OVR: Over

P

PAEW: Personnel and Equipment Working
PAJA: Parachute Jumping Activities
PAPI: Precision Approach Path Indicator
PAR: Precision Approach Radar
PARL: Parallel
PAT: Pattern
PCL: Pilot Controlled Lighting
PERM/PERMLY: Permanent/Permanently
PLA: Practice Low Approach
PLW: Plow/Plowed

PN: Prior Notice Required
PPR: Prior Permission Required
PREV: Previous
PRIRA: Primary Radar
PROC: Procedure
PROP: Propeller
PSGR: Passenger/s
PSR: Packed Snow on Runway/s
PT/PTN: Procedure Turn
PVT: Private

R
RAIL: Runway Alignment Indicator Lights
RCAG: Remote Communication Air/Ground Facility
RCL: Runway Centerline
RCLS: Runway Centerline Light System
RCO: Remote Communication Outlet
RCV/RCVR: Receive/Receiver
REF: Reference
REIL: Runway End Identifier Lights
RELCTD: Relocated
RMDR: Remainder
RNAV: Area Navigation
RPRT: Report
RQRD: Required
RRL: Runway Remaining Lights
RSVN: Reservation
RT: Right Turn after Take-off
RTE: Route
RTR: Remote Transmitter/Receiver
RTS: Return to Service
RUF: Rough
RVR: Runway Visual Range
RVRM: RVR Midpoint
RVRR: RVR Rollout
RVRT: RVR Touchdown
RVV: Runway Visibility Value
RY/RWY: Runway

S
S: South
SBND: Southbound
SDF: Simplified Directional Facility
SE: Southeast
SECRA: Secondary Radar
SFL: Sequenced Flashing Lights
SI: Straight-In Approach
SID: Standard Instrument Departure
SIR: Packed or Compacted Snow and Ice on Runway/s
SKED: Scheduled
SLR: Slush on Runway/s
SNBNK: Snowbank/s Caused by Plowing
SND: Sand/Sanded
SNGL: Single
SNW: Snow
SPD: Speed
SR: Sunrise
SS: Sunset
SSALF: Simplified Short Approach Lighting System with
    Sequenced Flashers
SSALR: Simplified Short Approach Lighting System with
    Runway Alignment Indicator Lights
SSALS: Simplified Short Approach Lighting System

STAR: Standard Terminal Arrival
SVC: Service
SW: Southwest
SWEPT: Swept or Broom/Broomed

T
TACAN: Tactical Air Navigational Aid
TDZ/TDZL: Touchdown Zone/Touchdown Zone Lights
TFC: Traffic
TFR: Temporary Flight Restriction
TGL: Touch and Go Landings
THN: Thin
THR: Threshold
THRU: Through
TIL: Until
TKOF: Takeoff
TMPRY: Temporary
TRML: Terminal
TRNG: Training
TRSA: Terminal Radar Service Area
TRSN: Transition
TSNT: Transient
TWEB: Transcribed Weather Broadcast
TWR: Tower
TWY: Taxiway

U
UNAVBL: Unavailable
UNLGTD: Unlighted
UNMKD: Unmarked
UNMON: Unmonitored
UNRELBL: Unreliable
UNUSBL: Unusable

V
VASI: Visual Approach Slope Indicator
VDP: Visual Descent Point
VFR: Visual Flight Rules
VIA: By Way Of
VICE: Instead/Versus
VIS/VSBY: Visibility
VMC: Visual Meteorological Conditions
VOL: Volume
VOLMET: Meterological Information for Aircraft in Flight
VOR: VHF Omni-Directional Radio Range
VORTAC: VOR and TACAN (collocated)
VOT: VOR Test Signal

W
W: West
WBND: Westbound
WEA/WX: Weather
WI: Within
WKDAYS: Monday through Friday
WKEND: Saturday and Sunday
WND: Wind
WP: Waypoint
WSR: Wet Snow on Runway/s
WTR: Water on Runway/s
WX: Weather

/: And
+: In Addition/Also

NOTE—

① DUATS VENDORS WILL PROVIDE FDC NOTAMS ONLY UPON SITE-SPECIFIC REQUESTS USING A LOCATION IDENTIFIER.

② NOTAM DATA MAY NOT ALWAYS BE CURRENT DUE TO THE CHANGEABLE NATURE OF NATIONAL AIRSPACE SYSTEM COMPONENTS, DELAYS INHERENT IN PROCESSING INFORMATION, AND OCCASIONAL TEMPORARY OUTAGES OF THE UNITED STATES NOTAM SYSTEM. WHILE EN ROUTE, PILOTS SHOULD CONTACT FSSs AND OBTAIN UPDATED INFORMATION FOR THEIR ROUTE OF FLIGHT AND DESTINATION.

c. An integral part of the NOTAM System is the bi-weekly Notices to Airmen Publication (NTAP). Data is included in this publication to reduce congestion on the telecommunications circuits and, therefore, is not available via Service A. Once published, the information is not provided during pilot weather briefings unless specifically requested by the pilot. This publication contains two sections.

1. The first section consists of notices that meet the criteria for NOTAM (D) and are expected to remain in effect for an extended period and FDC NOTAMs that are current at the time of publication. Occasionally, some NOTAM (L) and other unique information is included in this section when it will contribute to flight safety.

2. The second section contains special notices that are either too long or concern a wide or unspecified geographic area and are not suitable for inclusion in the first section. The content of these notices vary widely and there are no specific criteria for their inclusion, other than their enhancement of flight safety.

3. The number of the last FDC NOTAM included in the publication is noted on the first page to aid the user in updating the listing with any FDC NOTAMs which may have been issued between the cut-off date and the date the publication is received. All information contained will be carried until the information expires, is canceled, or in the case of permanent conditions, is published in other publications, such as the A/FD.

4. All new notices entered, excluding FDC NOTAMs, will be published only if the information is expected to remain in effect for at least 7 days after the effective date of the publication.

d. NOTAM information is not available from a Supplemental Weather Service Location (SWSL).

5-1-4. FLIGHT PLAN—VFR FLIGHTS

a. Except for operations in or penetrating a Coastal or Domestic ADIZ or DEWIZ a flight plan is not required for VFR flight.

REFERENCE—AIM, NATIONAL SECURITY, PARAGRAPH 5-6-1.

b. It is strongly recommended that a flight plan (for a VFR flight) be filed with an FAA FSS. This will ensure that you receive VFR Search and Rescue Protection.

REFERENCE—AIM, SEARCH AND RESCUE, PARAGRAPH 6-2-7 GIVES THE PROPER METHOD OF FILING A VFR FLIGHT PLAN.

c. To obtain maximum benefits from the flight plan program, flight plans should be filed directly with the nearest FSS. For your convenience, FSSs provide aeronautical and meteorological briefings while accepting flight plans. Radio may be used to file if no other means are available.

NOTE—SOME STATES OPERATE AERONAUTICAL COMMUNICATIONS FACILITIES WHICH WILL ACCEPT AND FORWARD FLIGHT PLANS TO THE FSS FOR FURTHER HANDLING.

d. When a "stopover" flight is anticipated, it is recommended that a separate flight plan be filed for each "leg" when the stop is expected to be more than 1 hour duration.

e. Pilots are encouraged to give their departure times directly to the FSS serving the departure airport or as otherwise indicated by the FSS when the flight plan is filed. This will ensure more efficient flight plan service and permit the FSS to advise you of significant changes in aeronautical facilities or meteorological conditions. When a VFR flight plan is filed, it will be held by the FSS until 1 hour after the proposed departure time unless:

1. The actual departure time is received.

2. A revised proposed departure time is received.

3. At a time of filing, the FSS is informed that the proposed departure time will be met, but actual time cannot be given because of inadequate communications (assumed departures).

f. On pilot's request, at a location having an active tower, the aircraft identification will be forwarded by the tower to the FSS for reporting the actual departure time. This procedure should be avoided at busy airports.

g. Although position reports are not required for VFR flight plans, periodic reports to FAA FSSs along the route are good practice. Such contacts permit significant information to be passed to the transiting aircraft and also serve to check the progress of the flight should it be necessary for any reason to locate the aircraft.

EXAMPLE—

① BONANZA 314K, OVER KINGFISHER AT (TIME), VFR FLIGHT PLAN, TULSA TO AMARILLO.

② CHEROKEE 5133J, OVER OKLAHOMA CITY AT (TIME), SHREVEPORT TO DENVER, NO FLIGHT PLAN.

h. Pilots not operating on an IFR flight plan and when in level cruising flight, are cautioned to conform with VFR cruising altitudes appropriate to the direction of flight.

i. When filing VFR flight plans, indicate aircraft equipment capabilities by appending the appropriate suffix to aircraft type in the same manner as that prescribed for IFR flight.

REFERENCE—AIM, FLIGHT PLAN IFR FLIGHTS, PARAGRAPH 5-1-8.

j. Under some circumstances, ATC computer tapes can be useful in constructing the radar history of a downed or crashed aircraft. In each case, knowledge of the aircraft's transponder equipment is necessary in determining whether or not such computer tapes might prove effective.

k. Flight Plan Form-(See FIG 5-1-1).

l. Explanation of VFR Flight Plan Items:

1. ***Block 1.*** Check the type flight plan. Check both the VFR and IFR blocks if composite VFR/IFR.

2. ***Block 2.*** Enter your complete aircraft identification including the prefix "N" if applicable.

3. ***Block 3.*** Enter the designator for the aircraft, or if unknown, consult an FSS briefer.

4. ***Block 4.*** Enter your true airspeed (TAS).

5. ***Block 5.*** Enter the departure airport identifier code, or if unknown, the name of the airport.

6. ***Block 6.*** Enter the proposed departure time in Coordinated Universal Time (UTC) (Z). If airborne, specify the actual or proposed departure time as appropriate.

7. ***Block 7.*** Enter the appropriate VFR altitude (to assist the briefer in providing weather and wind information).

8. ***Block 8.*** Define the route of flight by using NAVAID identifier codes and airways.

9. ***Block 9.*** Enter the destination airport identifier code, or if unknown, the airport name.

NOTE—INCLUDE THE CITY NAME (OR EVEN THE STATE NAME) IF NEEDED FOR CLARITY.

10. ***Block 10.*** Enter your estimated time en route in hours and minutes.

11. ***Block 11.*** Enter only those remarks pertinent to ATC or to the clarification of other flight plan information, such as the appropriate radiotelephony (call sign) associated with the designator filed in Block 2. Items of a personal nature are not accepted.

12. ***Block 12.*** Specify the fuel on board in hours and minutes.

13. ***Block 13.*** Specify an alternate airport if desired.

14. ***Block 14.*** Enter your complete name, address, and telephone number. Enter sufficient information to identify home base, airport, or operator.

NOTE—THIS INFORMATION IS ESSENTIAL IN THE EVENT OF SEARCH AND RESCUE OPERATIONS.

15. ***Block 15.*** Enter total number of persons on board (POB) including crew.

16. ***Block 16.*** Enter the predominant colors.

17. ***Block 17.*** Record the FSS name for closing the flight plan. If the flight plan is closed with a different FSS or facility, state the recorded FSS name that would normally have closed your flight plan.

NOTE—

◫ ***OPTIONAL*** RECORD A DESTINATION TELEPHONE NUMBER TO ASSIST SEARCH AND RESCUE CONTACT SHOULD YOU FAIL TO REPORT OR CANCEL YOUR FLIGHT PLAN WITHIN ½ HOUR AFTER YOUR ESTIMATED TIME OF ARRIVAL (ETA).

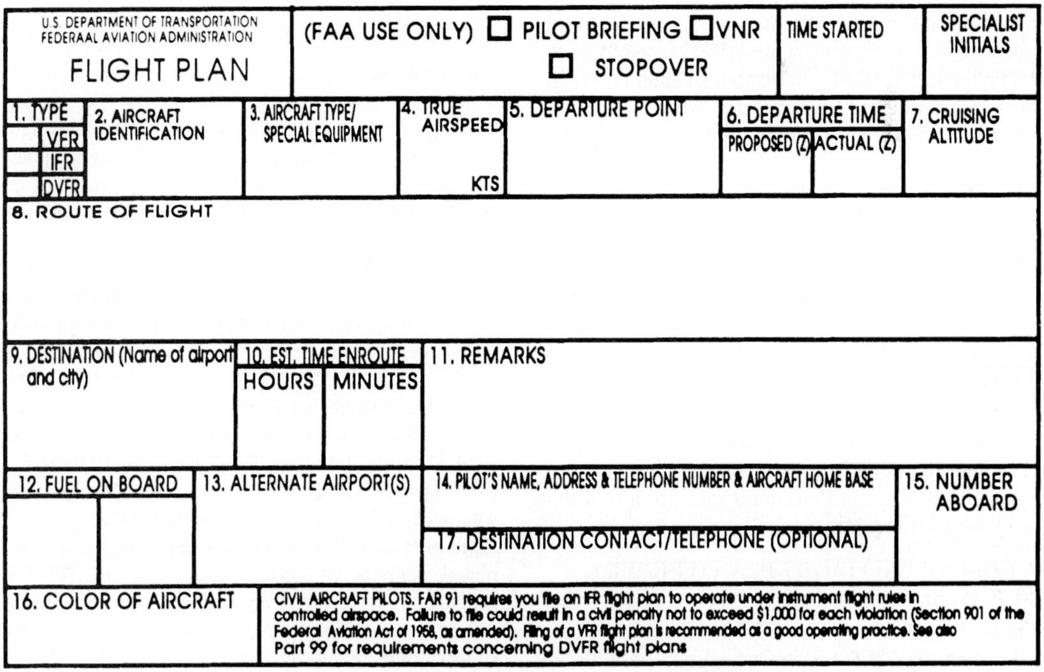
U.S. DEPARTMENT OF TRANSPORTATION
FEDERAAL AVIATION ADMINISTRATION
FLIGHT PLAN
(FAA USE ONLY) ☐ PILOT BRIEFING ☐ VNR
☐ STOPOVER
TIME STARTED
SPECIALIST INITIALS
1. TYPE
VFR
IFR
DVFR
2. AIRCRAFT IDENTIFICATION
3. AIRCRAFT TYPE/ SPECIAL EQUIPMENT
4. TRUE AIRSPEED
KTS
5. DEPARTURE POINT
6. DEPARTURE TIME
PROPOSED (Z) ACTUAL (Z)
7. CRUISING ALTITUDE
8. ROUTE OF FLIGHT
9. DESTINATION (Name of airport and city)
10. EST. TIME ENROUTE
HOURS MINUTES
11. REMARKS
12. FUEL ON BOARD
13. ALTERNATE AIRPORT(S)
14. PILOT'S NAME, ADDRESS & TELEPHONE NUMBER & AIRCRAFT HOME BASE
15. NUMBER ABOARD
17. DESTINATION CONTACT/TELEPHONE (OPTIONAL)
16. COLOR OF AIRCRAFT
CIVIL AIRCRAFT PILOTS. FAR 91 requires you file an IFR flight plan to operate under instrument flight rules in controlled airspace. Failure to file could result in a civil penalty not to exceed $1,000 for each violation (Section 901 of the Federal Aviation Act of 1958, as amended). Filing of a VFR flight plan is recommended as a good operating practice. See also Part 99 for requirements concerning DVFR flight plans

Figure 5-1-1. FAA Flight Plan Form 7233-1 (8-82)

2 THE INFORMATION TRANSMITTED TO THE DESTINATION FSS WILL CONSIST ONLY OF FLIGHT PLAN BLOCKS 2, 3, 9, AND 10. ESTIMATED TIME EN ROUTE (ETE) WILL BE CONVERTED TO THE CORRECT ETA.

5-1-5. OPERATIONAL INFORMATION SYSTEM (OIS)

a. The FAA's Air Traffic Control System Command Center (ATCSCC) maintains a web site with near real-time National Airspace System (NAS) status information. NAS operators are encouraged to access the web site at www.fly.faa.gov prior to filing their flight plan.

b. The web site consolidates information from advisories. An advisory is a message that is disseminated electronically by the ATCSCC that contains information pertinent to the NAS.

1. Advisories are normally issued for the following items:
(a) Ground Stops.
(b) Ground Delay Programs. (c) Route Information.
(d) Plan of Operations.
(e) Facility Outages and Scheduled Facility Outages.
(f) Volcanic Ash Activity Bulletins.
(g) Special Traffic Management Programs.

2. This list is not all-inclusive. Any time there is information that may be beneficial to a large number of people, an advisory may be sent. Additionally, there may be times when an advisory is not sent due to workload or the short length of time of the activity.

3. Route information is available on the web site and in specific advisories. Some route information, subject to the 56-day publishing cycle, is located on the "OIS" under "Products," Route Management Tool (RMT), and "What's New" Playbook. The RMT and Playbook contain routings for use by Air Traffic and NAS operators when they are co-ordinated "real-time" and are then published in an ATC-SCC advisory.

4. Route advisories are identified by the word "Route" in the header; the associated action is required (RQD), recommended (RMD), planned (PLN), or for your information (FYI). Operators are expected to file flight plans consistent with the Route RQD advisories.

5-1-6. FLIGHT PLAN—DEFENSE VFR (DVFR) FLIGHTS

VFR flights into a Coastal or Domestic ADIZ/DEWIZ are required to file DVFR flight plans for security purposes. Detailed ADIZ procedures are found in the National Security and Interception Procedures in Section 6 of this chapter. (See FAR Part 99.)

5-1-7. COMPOSITE FLIGHT PLAN (VFR/IFR FLIGHTS)

a. Flight plans which specify VFR operation for one portion of a flight, and IFR for another portion, will be accepted by the FSS at the point of departure. If VFR flight is conducted for the first portion of the flight, pilots should report their departure time to the FSS with whom the VFR/IFR flight plan was filed; and, subsequently, close the VFR portion and request ATC clearance from the FSS nearest the point at which change from VFR to IFR is proposed. Regardless of the type facility you are communicating with (FSS, center, or tower), it is the pilot's responsibility to request that facility to "CLOSE VFR FLIGHT PLAN." The pilot must remain in VFR weather conditions until operating in accordance with the IFR clearance.

b. When a flight plan indicates IFR for the first portion of flight and VFR for the latter portion, the pilot will normally be cleared to the point at which the change is proposed. After reporting over the clearance limit and not desiring further IFR clearance, the pilot should advise ATC to cancel the IFR portion of the flight plan. Then, the pilot should contact the nearest FSS to activate the VFR portion of the flight plan. If the pilot desires to continue the IFR flight plan beyond the clearance limit, the pilot should contact ATC at least 5 minutes prior to the clearance limit and request further IFR clearance. If the requested clearance is not received prior to reaching the clearance limit fix, the pilot will be expected to enter into a standard holding pattern on the radial or course to the fix unless a holding pattern for the clearance limit fix is depicted on a U.S. Government or commercially produced (meeting FAA requirements) Low or High Altitude En Route, Area or STAR Chart. In this case the pilot will hold according to the depicted pattern.

5-1-8. FLIGHT PLAN—IFR FLIGHTS

a. General:

1. Prior to departure from within, or prior to entering controlled airspace, a pilot must submit a complete flight plan and receive an air traffic clearance, if weather conditions are below VFR minimums. Instrument flight plans may be submitted to the nearest FSS or ATCT either in person or by telephone (or by radio if no other means are available). Pilots should file IFR flight plans at least 30 minutes prior to estimated time of departure to preclude possible delay in receiving a departure clearance from ATC. In order to provide FAA traffic management units strategic route planning capabilities, nonscheduled operators conducting IFR operations above FL 230 are requested to voluntarily file IFR flight plans at least 4 hours prior to estimated time of departure (ETD). To minimize your delay in entering Class B, Class C, Class D and Class E surface areas at destination when IFR weather conditions exist or are forecast at that airport, an IFR flight plan should be filed before departure. Otherwise, a 30 minute delay is not unusual in receiving an ATC clearance because of time spent in processing flight plan data. Traffic saturation frequently prevents control personnel from accepting flight plans by radio. In such cases, the pilot is advised to contact the nearest FSS for the purpose of filing the flight plan.

NOTE—THERE ARE SEVERAL METHODS OF OBTAINING IFR CLEARANCES AT NONTOWER, NONFSS, AND OUTLYING AIRPORTS. THE PROCEDURE MAY VARY DUE TO GEOGRAPHICAL FEATURES, WEATHER CONDITIONS, AND THE COMPLEXITY OF THE ATC SYSTEM. TO DETERMINE THE MOST EFFECTIVE MEANS OF RECEIVING AN IFR CLEARANCE, PILOTS SHOULD ASK THE NEAREST FSS THE MOST APPROPRIATE MEANS OF OBTAINING THE IFR CLEARANCE.

2. When filing an IFR flight plan, include as a prefix to the aircraft type, the number of aircraft when more than one and/or heavy aircraft indicator "H" if appropriate.

EXAMPLE—
H/DC10/A
2/F15/A

3. When filing an IFR flight plan, identify the equipment capability by adding a suffix, preceded by a slant, to the AIRCRAFT TYPE, as shown in TBL 5-1-2, Aircraft Suffixes.

NOTE—
1. ATC ISSUES CLEARANCES BASED ON FILED SUFFIXES. PILOTS SHOULD DETERMINE THE APPROPRIATE SUFFIX BASED UPON DESIRED SERVICES AND/OR ROUTING. FOR EXAMPLE, IF A DESIRED ROUTE/PROCEDURE REQUIRES GPS, A PILOT SHOULD FILE /G EVEN IF THE AIRCRAFT ALSO QUALIFIES FOR OTHER SUFFIXES.

2. FOR PROCEDURES REQUIRING GPS, IF THE NAVIGATION SYSTEM DOES NOT AUTOMATICALLY ALERT THE FLIGHT CREW OF A LOSS OF GPS, THE OPERATOR MUST DEVELOP PROCEDURES TO VERIFY CORRECT GPS OPERATION.

3. THE SUFFIX IS NOT TO BE ADDED TO THE AIRCRAFT IDENTIFICATION OR BE TRANSMITTED BY RADIO AS PART OF THE AIRCRAFT IDENTIFICATION.

4. It is recommended that pilots file the maximum transponder or navigation capability of their aircraft in the equipment suffix. This will provide ATC with the necessary information to utilize all facets of navigational equipment and transponder capabilities available.

NOTE—THE SUFFIX IS NOT TO BE ADDED TO THE AIRCRAFT IDENTIFICATION OR BE TRANSMITTED BY RADIO AS PART OF THE AIRCRAFT IDENTIFICATION.

b. Airways and Jet Routes Depiction on Flight Plan:

1. It is vitally important that the route of flight be accurately and completely described in the flight plan. To simplify definition of the proposed route, and to facilitate ATC, pilots are requested to file via airways or jet routes established for use at the altitude or Flight Level planned.

2. If flight is to be conducted via designated airways or jet routes, describe the route by indicating the type and number designators of the airway(s) or jet route(s) requested. If more than one airway or jet route is to be used, clearly indicate points of transition. If the transition is made at an unnamed intersection, show the next succeeding NAVAID or named intersection on the intended route and the complete route from that point. Reporting points may be identified by using authorized name/code

as depicted on appropriate aeronautical charts. The following two examples illustrate the need to specify the transition point when two routes share more than one transition fix.

EXAMPLE—
① ALB J37 BUMPY J14 BHM
SPELLED OUT: FROM ALBANY, NEW YORK, VIA JET ROUTE 37 TRANSITIONING TO JET ROUTE 14 AT BUMPY INTERSECTION, THENCE VIA JET ROUTE 14 TO BIRMINGHAM, ALABAMA.
② ALB J37 ENO J14 BHM
SPELLED OUT: FROM ALBANY, NEW YORK, VIA JET ROUTE 37 TRANSITIONING TO JET ROUTE 14 AT SMYRA VORTAC (ENO) THENCE VIA JET ROUTE 14 TO BIRMINGHAM, ALABAMA.

3. The route of flight may also be described by naming the reporting points or NAVAIDs over which the flight will pass, provided the points named are established for use at the altitude or flight level planned.

EXAMPLE—BWI V44 SWANN V433 DQO
SPELLED OUT: FROM BALTIMORE-WASHINGTON INTERNATIONAL, VIA VICTOR 44 TO SWANN INTERSECTION, TRANSITIONING TO VICTOR 433 AT SWANN, THENCE VIA VICTOR 433 TO DUPONT.

4. When the route of flight is defined by named reporting points, whether alone or in combination with airways or jet routes, and the navigational aids (VOR, VORTAC, TACAN, NDB) to be used for the flight are a combination of different types of aids, enough information should be included to clearly indicate the route requested.

EXAMPLE—LAX J5 LKV J3 GEG YXC FL 330 J500 VLR J515 YWG
SPELLED OUT: FROM LOS ANGELES INTERNATIONAL VIA JET ROUTE 5 LAKEVIEW, JET ROUTE 3 SPOKANE, DIRECT CRANBROOK, BRITISH COLUMBIA VOR/DME, FLIGHT LEVEL 330 JET ROUTE 500 TO LANGRUTH, MANITOBA VORTAC, JET ROUTE 515 TO WINNIPEG, MANITOBA.

5. When filing IFR, it is to the pilot's advantage to file a preferred route.

REFERENCE—PREFERRED IFR ROUTES ARE DESCRIBED AND TABULATED IN THE AIRPORT/FACILITY DIRECTORY.

6. ATC may issue a DP or a STAR, as appropriate.

REFERENCE—AIM, INSTRUMENT DEPARTURE PROCEDURES (DP) – OBSTACLE DEPARTURE PROCEDURES (ODP) AND STANDARD INSTRUMENT DEPARTURES (SID), PARAGRAPH 5-2-7
AIM STANDARD TERMINAL ARRIVAL (STAR), AREA NAVIGATION (RNAV)STAR, AND FLIGHT MANAGEMENT INSTRUMENT PROCEDURES (FMSP) FOR ARRIVALS, PARAGRAPH 5-4-1.

NOTE—PILOTS NOT DESIRING A DP OR STAR SHOULD SO INDICATE IN THE REMARKS SECTION OF THE FLIGHT PLAN AS "NO DP" OR "NO STAR."

c. Direct Flights:

1. All or any portions of the route which will not be flown on the radials or courses of established airways or routes, such as direct route flights, must be defined by indicating the radio fixes over which the flight will pass. Fixes selected to define the route shall be those over which the position of the

Table 5-1-2   Aircraft Equipment Suffixes

SUFFIX   EQUIPMENT SUFFIXES

NO DME

/X	No transponder.
/T	Transponder with no Mode C
/U	Transponder with Mode C

DME

/D	No transponder
/B	Transponder with no Mode C
/A	Transponder with Mode C

TACAN ONLY

/M	No transponder
/N	Transponder with no Mode C
/P	Transponder with Mode C

AREA NAVIGATION (RNAV)

/Y	LORAN, VOR/DME, or INS with no transponder
/C	LORAN, VOR/DME, or INS, transponder with no Mode C
/I	LORAN, VOR/DME, or INS, transponder with Mode C

ADVANCED RNAV WITH TRANSPONDER AND MODE C (If an aircraft is unable to operate with a transponder and/or Mode C, it will revert to the appropriate code listed above under Area Navigation.)

/E	Flight Management System (FMS) with DME/DME and IRU position updating
F	FMS with positioning updating <<remove all /F after this>>
/G	Global Navigation Satellite System (GNSS), including GPS or Wide Area Augmentation System (WAAS), with en route and terminal capability.
/R	Required Navigational Performance. The aircraft meets the RNP type prescribed for the route segment(s), route(s) and/or area concerned.

Reduced Vertical Separation Minimum (RVSM). Prior to conducting RVSM operations within the U.S., the operator must obtain authorization from the FAA or from the responsible authority, as appropriate.

/J	/E with RVSM
/K	/F with RVSM
/L	/G with RVSM
/Q	/R with RVSM
/W	RVSM

aircraft can be accurately determined. Such fixes automatically become compulsory reporting points for the flight, unless advised otherwise by ATC. Only those navigational aids established for use in a particular structure; i.e., in the low or high structures, may be used to define the en route phase of a direct flight within that altitude structure.

2. The azimuth feature of VOR aids and that azimuth and distance (DME) features of VORTAC and TACAN aids are assigned certain frequency protected areas of airspace which are intended for application to established airway and route use, and to provide guidance for planning flights outside of established airways or routes. These areas of airspace are expressed in terms of cylindrical service volumes of specified dimensions called "class limits" or "categories."

REFERENCE—AIM, NAVAID SERVICE VOLUMES, PARAGRAPH 1-1-8.

3. An operational service volume has been established for each class in which adequate signal coverage and frequency protection can be assured. To facilitate use of VOR, VORTAC, or TACAN aids, consistent with their operational service volume limits, pilot use of such aids for defining a direct route of flight in controlled airspace should not exceed the following:

(a) Operations above FL 450—Use aids not more than 200 NM apart. These aids are depicted on En Route High Altitude Charts.

(b) Operation off established routes from 18,000 feet MSL to FL 450—Use aids not more than 260 NM apart. These aids are depicted on En Route High Altitude Charts.

(c) Operation off established airways below 18,000 feet MSL—Use aids not more than 80 NM apart. These aids are depicted on En Route Low Altitude Charts.

(d) Operation off established airways between 14,500 feet MSL and 17,999 feet MSL in the conterminous U.S.—(H) facilities not more than 200 NM apart may be used.

4. Increasing use of self-contained airborne navigational systems which do not rely on the VOR/VORTAC/TACAN system has resulted in pilot requests for direct routes which exceed NAVAID service volume limits. These direct route requests will be approved only in a radar environment, with approval based on pilot responsibility for navigation on the authorized direct route. Radar flight following will be provided by ATC for ATC purposes.

5. At times, ATC will initiate a direct route in a radar environment which exceeds NAVAID service volume limits. In such cases ATC will provide radar monitoring and navigational assistance as necessary.

6. Airway or jet route numbers, appropriate to the stratum in which operation will be conducted, may also be included to describe portions of the route to be flown.

    EXAMPLE—MDW V262 BDF V10 BRL STJ SLN GCK SPELLED OUT: FROM CHICAGO MIDWAY AIRPORT VIA VICTOR 262 TO BRADFORD, VICTOR 10 TO BURLINGTON, IOWA, DIRECT ST. JOSEPH, MISSOURI, DIRECT SALINA, KANSAS, DIRECT GARDEN CITY, KANSAS.

    NOTE—WHEN ROUTE OF FLIGHT IS DESCRIBED BY RADIO FIXES, THE PILOT WILL BE EXPECTED TO FLY A DIRECT COURSE BETWEEN THE POINTS NAMED.

7. Pilots are reminded that they are responsible for adhering to obstruction clearance requirements on those segments of direct routes that are outside of controlled airspace. The MEAs and other altitudes shown on Low Altitude IFR Enroute Charts pertain to those route segments within controlled airspace, and those altitudes may not meet obstruction clearance criteria when operating off those routes.

d. Area Navigation (RNAV):

1. Random RNAV routes can only be approved in a radar environment. Factors that will be considered by ATC in approving random RNAV routes include the capability to provide radar monitoring and compatibility with traffic volume and flow. ATC will radar monitor each flight, however, navigation on the random RNAV route is the responsibility of the pilot.

2. Pilots of aircraft equipped with approved area navigation equipment may file for RNAV routes throughout the National Airspace System and may be filed for in accordance with the following procedures.

(a) File airport-to-airport flight plans.

(b) File the appropriate RNAV capability certification suffix in the flight plan.

(c) Plan the random route portion of the flight plan to begin and end over appropriate arrival and departure transition fixes or appropriate navigation aids for the altitude stratum within which the flight will be conducted. The use of normal preferred departure and arrival routes (DP/STAR), where established, is recommended.

(d) File route structure transitions to and from the random route portion of the flight.

(e) Define the random route by waypoints. File route description waypoints by using degree-distance fixes based on navigational aids which are appropriate for the altitude stratum.

(f) File a minimum of one route description waypoint for each ARTCC through whose area the random route will be flown. These waypoints must be located within 200 NM of the preceding center's boundary.

(g) File an additional route description waypoint for each turnpoint in the route.

(h) Plan additional route description way-points as required to ensure accurate navigation via the filed route of flight. Navigation is the pilot's responsibility unless ATC assistance is requested.

(i) Plan the route of flight so as to avoid prohibited and restricted airspace by 3 NM unless permission has been obtained to operate in that airspace and the appropriate ATC facilities are advised.

    NOTE—TO BE APPROVED FOR USE IN THE NATIONAL AIRSPACE SYSTEM, RNAV EQUIPMENT MUST MEET THE APPROPRIATE SYSTEM AVAILABILITY, ACCURACY, AND AIRWORTHINESS STANDARDS. FOR ADDITIONAL GUIDANCE ON EQUIPMENT REQUIREMENTS SEE AC 20-130, AIRWORTHINESS APPROVAL OF VERTICAL NAVIGATION (VNAV) SYSTEMS FOR USE IN THE U.S. NAS AND ALASKA, OR AC 20-138, AIRWORTHINESS APPROVAL OF GLOBAL POSITIONING SYSTEM (GPS) NAVIGATION EQUIPMENT FOR USE AS A VFR AND IFR SUPPLEMENTAL NAVIGATION SYSTEM. FOR AIRBORNE NAVIGATION DATABASE, SEE AC 90-94, GUIDELINES FOR USING GPS EQUIPMENT FOR IFR EN ROUTE AND TERMINAL OPERATIONS AND FOR NONPRECISION INSTRUMENT APPROACHES IN THE U.S. NATIONAL AIRSPACE SYSTEM, SECTION 2.

3. Pilots of aircraft equipped with latitude/longitude coordinate navigation capability, independent of VOR/ TACAN references, may file for random RNAV routes at and above FL 390 within the conterminous United States using the following procedures.

(a) File airport-to-airport flight plans prior to departure.

(b) File the appropriate RNAV capability certification suffix in the flight plan.

(c) Plan the random route portion of the flight to begin and end over published departure/arrival transition fixes or appropriate navigation aids for airports without published transition procedures. The use of preferred departure and arrival routes, such as DP and STAR where established, is recommended.

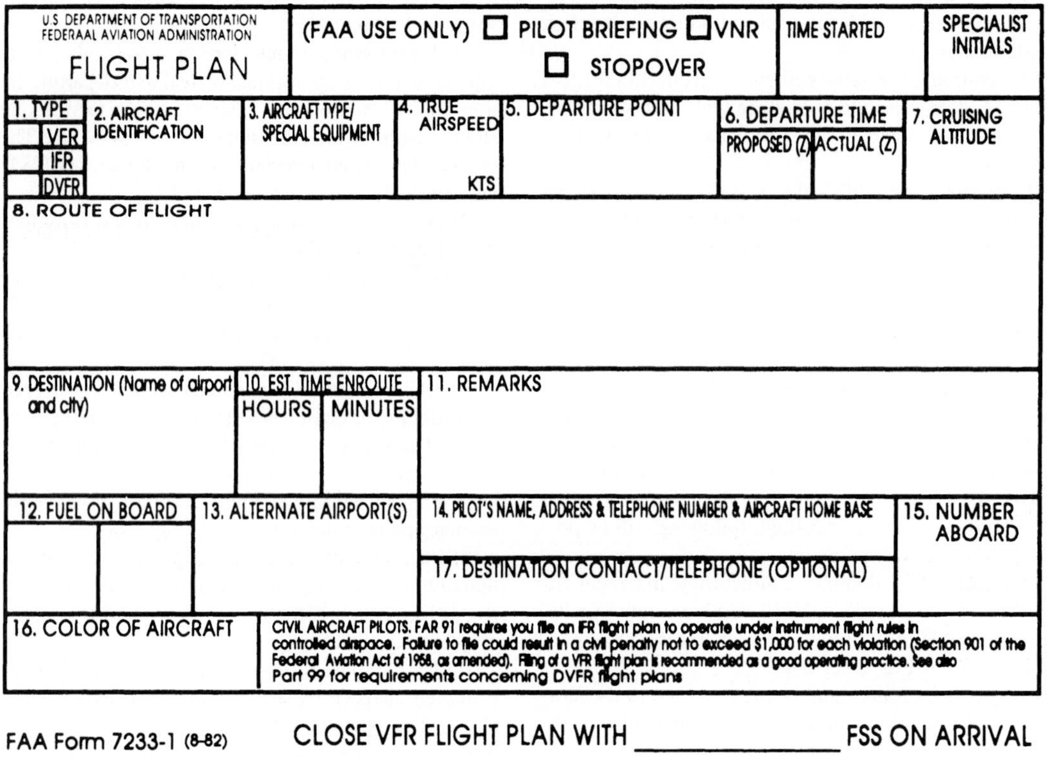

U.S. DEPARTMENT OF TRANSPORTATION FEDERAAL AVIATION ADMINISTRATION **FLIGHT PLAN**	(FAA USE ONLY) ☐ PILOT BRIEFING ☐VNR ☐ STOPOVER		TIME STARTED	SPECIALIST INITIALS

1. TYPE ☐ VFR ☐ IFR ☐ DVFR	2. AIRCRAFT IDENTIFICATION	3. AIRCRAFT TYPE/ SPECIAL EQUIPMENT	4. TRUE AIRSPEED KTS	5. DEPARTURE POINT	6. DEPARTURE TIME PROPOSED (Z) / ACTUAL (Z)	7. CRUISING ALTITUDE

8. ROUTE OF FLIGHT

9. DESTINATION (Name of airport and city)	10. EST. TIME ENROUTE HOURS / MINUTES	11. REMARKS

12. FUEL ON BOARD	13. ALTERNATE AIRPORT(S)	14. PILOT'S NAME, ADDRESS & TELEPHONE NUMBER & AIRCRAFT HOME BASE 17. DESTINATION CONTACT/TELEPHONE (OPTIONAL)	15. NUMBER ABOARD

16. COLOR OF AIRCRAFT	CIVIL AIRCRAFT PILOTS. FAR 91 requires you file an IFR flight plan to operate under instrument flight rules in controlled airspace. Failure to file could result in a civil penalty not to exceed $1,000 for each violation (Section 901 of the Federal Aviation Act of 1958, as amended). Filing of a VFR flight plan is recommended as a good operating practice. See also Part 99 for requirements concerning DVFR flight plans

FAA Form 7233-1 (8-82) CLOSE VFR FLIGHT PLAN WITH _____ FSS ON ARRIVAL

Figure 5-1-2. FAA Flight Plan Form 7233-1 (8-82)

(d) Plan the route of flight so as to avoid prohibited and restricted airspace by 3 NM unless permission has been obtained to operate in that airspace and the appropriate ATC facility is advised.

(e) Define the route of flight after the departure fix, including each intermediate fix (turnpoint) and the arrival fix for the destination airport in terms of latitude/longitude coordinates plotted to the nearest minute of in terms of Navigation Reference System (NRS) waypoints. For latitude/ longitude filing the arrival fix must be identified by both the latitude/longitude coordinates and a fix identifier.

EXAMPLE—MIA SRQ 3407/10615 407/11546 TNP LAX
1 DEPARTURE AIRPORT.
2 DEPARTURE FIX.
3 INTERMEDIATE FIX (TURNING POINT).
4.ARRIVAL FIX.
5 DESTINATION AIRPORT.
OR
ORD IOW KP49G KD34U KL160 OAL MOD2 SF0
1 DEPARTURE AIRPORT.
2 TRANSITION FIX (PITCH POINT).
3 MINNEAPOLIS ARTCC WAYPOINT.
4 DENVER ARTCC WAYPOINT.
5 LOSANGELES ARTCC WAYPOINT (CATCH POINT).
6 TRANSITION FIX.

7 ARRIVAL.
8 DESTINATION AIRPORT.

(f) Record latitude/longitude coordinates by four figures describing latitude in degrees and minutes followed by a solidus and five figures describing longitude in degrees and minutes.

(g) File at FL 390 or above for the random RNAV portion of the flight.

(h) Fly all routes/route segments on Great Circle tracks.

(i) Make any inflight requests for random RNAV clearances or route amendments to an en route ATC facility.

e. Flight Plan Form—See FIG 5-1-2.

f. Explanation of IFR Flight Plan Items:

1. *Block 1.* Check the type flight plan. Check both the VFR and IFR blocks if composite VFR/IFR.

2. *Block 2.* Enter your complete aircraft identification including the prefix "N" if applicable.

3. *Block 3.* Enter the designator for the aircraft, followed by a slant(/), and the transponder or DME equipment code letter; e.g., C-182/U. Heavy aircraft, add prefix "H" to aircraft type; example: H/DC10/U. Consult an FSS briefer for any unknown elements.

4. *Block 4.* Enter your computed true airspeed (TAS).

NOTE—IF THE AVERAGE TAS CHANGES PLUS OR MINUS 5 PERCENT OR 10 KNOTS, WHICHEVER IS GREATER, ADVISE ATC.

5. *Block 5.* Enter the departure airport identifier code (or the name if the identifier is unknown).

NOTE—USE OF IDENTIFIER CODES WILL EXPEDITE THE PROCESSING OF YOUR FLIGHT PLAN.

6. *Block 6.* Enter the proposed departure time in Coordinated Universal Time (UTC) (Z). If airborne, specify the actual or proposed departure time as appropriate.

7. *Block 7.* Enter the requested en route altitude or flight level.

NOTE—ENTER ONLY THE INITIAL REQUESTED ALTITUDE IN THIS BLOCK. WHEN MORE THAN ONE IFR ALTITUDE OR FLIGHT LEVEL IS DESIRED ALONG THE ROUTE OF FLIGHT, IT IS BEST TO MAKE A SUBSEQUENT REQUEST DIRECT TO THE CONTROLLER.

8. *Block 8.* Define the route of flight by using NAVAID identifier codes (or names if the code is unknown), airways, jet routes, and waypoints (for RNAV).

NOTE—USE NAVAIDS OR WAYPOINTS TO DEFINE DIRECT ROUTES AND RADIALS/BEARINGS TO DEFINE OTHER UNPUBLISHED ROUTES.

9. *Block 9.* Enter the destination airport identifier code (or name if the identifier is unknown).

10. *Block 10.* Enter your estimated time en route based on latest forecast winds.

11. *Block 11.* Enter only those remarks pertinent to ATC or to the clarification of other flight plan information, such as the appropriate radiotelephony (call sign) associated with the designator filed in Block 2. Items of a personal nature are not accepted. Do not assume that remarks will be automatically transmitted to every controller. Specific ATC or en route requests should be made directly to the appropriate controller.

NOTE—"DVRSN" SHOULD BE PLACED IN BLOCK 11 ONLY IF THE PILOT/COMPANY IS REQUESTING PRIORITY HANDLING TO THEIR ORIGINAL DESTINATION FROM ATC AS A RESULT OF A DIVERSION AS DEFINED IN THE PILOT/CONTROLLER GLOSSARY.

12. *Block 12.* Specify the fuel on board, computed from the departure point.

13. *Block 13.* Specify an alternate airport if desired or required, but do not include routing to the alternate airport.

14. *Block 14.* Enter the complete name, address, and telephone number of pilot-in-command, or in the case of a formation flight, the formation commander. Enter sufficient information to identify home base, airport, or operator.

NOTE—THIS INFORMATION WOULD BE ESSENTIAL IN THE EVENT OF SEARCH AND RESCUE OPERATION.

15. *Block 15.* Enter the total number of persons on board including crew.

16. *Block 16.* Enter the predominant colors.

NOTE—CLOSE IFR FLIGHT PLANS WITH TOWER, APPROACH CONTROL, OR ARTCC, OR IF UNABLE, WITH FSS. WHEN LANDING AT AN AIRPORT WITH A FUNCTIONING CONTROL TOWER, IFR FLIGHT PLANS ARE AUTOMATICALLY CANCELED.

g. The information transmitted to the ARTCC for IFR flight plans will consist of only flight plan blocks 2, 3, 4, 5, 6, 7, 8, 9, 10, and 11.

h. A description of the International Flight Plan Form is contained in the International Flight Information Manual (IFIM).

5-1-9. IFR OPERATIONS TO HIGH ALTITUDE DESTINATIONS

a. Pilots planning IFR flights to airports located in mountainous terrain are cautioned to consider the necessity for an alternate airport even when the forecast weather conditions would technically relieve them from the requirement to file one.

REFERENCE—FAR PART 91.167.
AIM, TOWER EN ROUTE CONTROL (TEC), PARAGRAPH 4-1-18.

b. The FAA has identified three possible situations where the failure to plan for an alternate airport when flying IFR to such a destination airport could result in a critical situation if the weather is less than forecast and sufficient fuel is not available to proceed to a suitable airport.

1. An IFR flight to an airport where the MDAs or landing visibility minimums for *all instrument approaches* are higher than the forecast weather minimums specified in FAR Part 91.167(b). For example, there are 3 high altitude airports in the United States with approved instrument approach procedures where all of the Minimum Descent Altitudes (MDAs) are greater than 2,000 feet and/or the landing visibility minimums are greater than 3 miles (Bishop, California; South Lake Tahoe, California. Aspen-Pitkin Co./Sardy Field, Colorado.) In the case of these 11 airports, it is possible for a pilot to elect, on the basis of forecasts, not to carry sufficient fuel to get to an alternate when the ceiling and/or visibility is actually lower than that necessary to complete the approach.

2. A small number of other airports in mountainous terrain have MDAs which are slightly (100 to 300 feet) below 2,000 feet AGL. In situations where there is an option as to whether to plan for an alternate, pilots should bear in mind that just a slight worsening of the weather conditions from those forecast could place the airport below the published IFR landing minimums.

3. An IFR flight to an airport which requires special equipment; i.e., DME, glide slope, etc., in order to make the available approaches to the lowest minimums. Pilots should be aware that all other minimums on the approach charts may require weather conditions better than those specified in FAR Part 91.167(b). An inflight equipment malfunction could result in the inability to comply with the published approach procedures or, again, in the position of having the airport below the published IFR landing minimums for all remaining instrument approach alternatives.

5-1-10. FLIGHTS OUTSIDE THE UNITED STATES AND U.S. TERRITORIES

a. When conducting flights, particularly extended flights, outside the U.S. and its territories, full account should be taken of the amount and quality of air navigation services available in the airspace to be traversed. Every effort should be made to secure information on the location and range of navigational aids, availability of communications and meteorological services, the provision of air traffic services, including alerting service, and the existence of search and rescue services.

b. Pilots should remember that there is a need to continuously guard the VHF emergency frequency 121.5 MHz when on long over-water flights, except when communications on other VHF channels, equipment limitations, or cockpit duties prevent simultaneous guarding of two channels. Guarding of 121.5 MHz is particularly critical when operating in proximity to Flight Information Region (FIR) boundaries, for example, operations on Route R220 between Anchorage and Tokyo, since it serves to facilitate communications with regard to aircraft which may experience in-flight emergencies, communications, or navigational difficulties.

REFERENCE—ICAO ANNEX 10, VOL II PARAS 5.2.2.1.1.1 AND 5.2.2.1.1.2.

c. The filing of a flight plan, always good practice, takes on added significance for extended flights outside U.S. airspace and is, in fact, usually required by the laws of the countries being visited or overflown. It is also particularly important in the case of such flights that pilots leave a complete itinerary and schedule of the flight with someone directly concerned and keep that person advised of the flight's progress. If serious doubt arises as to the safety of the flight, that person should first contact the appropriate FSS. Round Robin Flight Plans to Mexico are not accepted.

d. All pilots should review the foreign airspace and entry restrictions published in the IFIM during the flight planning process. Foreign airspace penetration without official authorization can involve both danger to the aircraft and the imposition of severe penalties and inconvenience to both passengers and crew. A flight plan on file with ATC authorities does not necessarily constitute the prior permission required by certain other authorities. The possibility of fatal consequences cannot be ignored in some areas of the world.

e. Current NOTAMs for foreign locations must also be reviewed. The publication International Notices to Airmen, published biweekly, contains considerable information pertinent to foreign flight. Current foreign NOTAMs are also available from the U.S. International NOTAM Office in Washington, D.C., through any local FSS.

f. When customs notification is required, it is the responsibility of the pilot to arrange for customs notification in a timely manner. The following guidelines are applicable:

1. When customs notification is required on flights to Canada and Mexico and a predeparture flight plan cannot be filed or an advise customs message (ADCUS) cannot be included in a predeparture flight plan, call the nearest en route domestic or International FSS as soon as radio communication can be established and file a VFR or DVFR flight plan, as required, and include as the last item the advise customs information. The station with which such a flight plan is filed will forward it to the appropriate FSS who will notify the customs office responsible for the destination airport.

2. If the pilot fails to include ADCUS in the radioed flight plan, it will be assumed that other arrangements have been made and FAA will not advise customs.

3. The FAA assumes no responsibility for any delays in advising customs if the flight plan is given too late for delivery to customs before arrival of the aircraft. It is still the pilot's responsibility to give timely notice even though a flight plan is given to FAA.

4. Air Commerce Regulations of the Treasury Department's Customs Service require all private aircraft arriving in the United States via:

(a) the United States/Mexican border or the Pacific Coast from a foreign place in the Western Hemisphere south of 33 degrees north latitude and between 97 degrees and 120 degrees west longitude; or

(b) the Gulf of Mexico and Atlantic Coasts from a foreign place in the Western Hemisphere south of 30 degrees north latitude, shall furnish a notice of arrival to the Customs service at the nearest designated airport. This notice may be furnished directly to Customs by:

(1) radio through the appropriate FAA Flight Service Station;

(2) normal FAA flight plan notification procedures (a flight plan filed in Mexico does not meet this requirement due to unreliable relay of data), or

(3) directly to the district Director of Customs or other Customs officer at place of first intended landing but must be furnished at least 1 hour prior to crossing the U.S./Mexican border or the United States coastline.

(c) This notice will be valid as long as actual arrival is within 15 minutes of the original ETA, otherwise a new notice must be given to Customs. Notices will be accepted up to 23 hours in advance. Unless an exemption has been granted by Customs, private aircraft are required to make first landing in the U.S. at one of the following designated airports nearest to the point of border of coastline crossing:

Designated Airports

ARIZONA
Bisbee Douglas Intl Airport
Douglas Municipal Airport, AZ
Nogales Intl Airport
Tucson Intl Airport
Yuma MCAS—Yuma Intl Airport

CALIFORNIA
Calexico Intl Airport
Brown Field Municipal Airport (San Diego)

FLORIDA
Fort Lauderdale Executive Airport
Fort Lauderdale/Hollywood Intl Airport)
Key West Intl Airport (Miami Intl Airport)
Opa Locka Airport (Miami)
Kendall—Tamiami Executive Airport (Miami)
St. Lucie County Intl Airport (Fort Pierce)
Tampa Intl Airport
Palm Beach Intl Airport (West Palm Beach)

LOUISIANA
New Orleans Intl Airport (Moisant Field)
New Orleans Lakefront Airport

NEW MEXICO
Las Cruces Intl Airport

NORTH CAROLINA
New Hanover Intl Airport (Wilmington)

TEXAS
Brownsville/South Padre Island Intl Airport
Corpus Christi Intl Airport
Del Rio Intl Airport
Eagle Pass Municpal Airport
El Paso Intl Airport
William P. Hobby Airport (Houston)
Laredo Intl Airport
McAllen Miller Intl Airport
Presidio Lely Intl Airport

5-1-11. CHANGE IN FLIGHT PLAN

In addition to altitude or flight level, destination and/or route changes, increasing or decreasing the speed of an aircraft constitutes a change in a flight plan. Therefore, at any time the average true airspeed at cruising altitude between reporting points varies or is expected to vary from that given in the flight plan by *plus or minus 5 percent, or 10 knots, whichever is greater,* ATC should be advised.

5-1-12. CHANGE IN PROPOSED DEPARTURE TIME

a. To prevent computer saturation in the en route environment, parameters have been established to delete proposed departure flight plans which have not been activated. Most centers have this parameter set so as to delete these flight plans a minimum of 1 hour after the proposed departure time. To ensure that a flight plan remains active, pilots whose actual departure time will be delayed 1 hour or more beyond their filed departure time, are requested to notify ATC of their departure time.

b. Due to traffic saturation, control personnel frequently will be unable to accept these revisions via radio. It is recommended that you forward these revisions to the nearest FSS.

5-1-13. CLOSING VFR/DVFR FLIGHT PLANS

Pilots are responsible for ensuring that their VFR or DVFR flight plan is canceled. You should close your flight plan with the nearest FSS, or if one is not available, you may request any ATC facility to relay your cancellation to the FSS. Control towers do not automatically close VFR or DVFR flight plans since they do not know if a particular VFR aircraft is on a flight plan. If you fail to report or cancel your flight plan within ½ hour after your ETA, search and rescue procedures are started.

REFERENCE—FAR PART 91.153.
FAR PART 91.169.

5-1-14. CANCELING IFR FLIGHT PLAN

a. FAR Part 91.153 and FAR Part 91.169 include the statement "When a flight plan has been activated, the pilot-in-command, upon canceling or completing the flight under the flight plan, shall notify an FAA Flight Service Station or ATC facility."

b. An IFR flight plan may be canceled at any time the flight is operating in VFR conditions outside Class A airspace by pilots stating "CANCEL MY IFR FLIGHT PLAN" to the controller or air/ground station with which they are communicating. Immediately after canceling an IFR flight plan, a pilot should take the necessary action to change to the appropriate air/ground frequency, VFR radar beacon code and VFR altitude or flight level.

c. ATC separation and information services will be discontinued, including radar services (where applicable). Consequently, if the canceling flight desires VFR radar advisory service, the pilot must specifically request it.

NOTE—PILOTS MUST BE AWARE THAT OTHER PROCEDURES MAY BE APPLICABLE TO A FLIGHT THAT CANCELS AN IFR FLIGHT PLAN WITHIN AN AREA WHERE A SPECIAL PROGRAM, SUCH AS A DESIGNATED TRSA, CLASS C AIRSPACE, OR CLASS B AIRSPACE, HAS BEEN ESTABLISHED.

d. If a DVFR flight plan requirement exists, the pilot is responsible for filing this flight plan to replace the canceled IFR flight plan. If a subsequent IFR operation becomes necessary, a new IFR flight plan must be filed and an ATC clearance obtained before operating in IFR conditions.

e. If operating on an IFR flight plan to an airport with a functioning control tower, the flight plan is automatically closed upon landing.

f. If operating on an IFR flight plan to an airport where there is no functioning control tower, the pilot must initiate cancellation of the IFR flight plan. This can be done after landing if there is a functioning FSS or other means of direct communications with ATC. In the event there is no FSS and/or air/ground communications with ATC is not possible below a certain altitude, the pilot should, weather conditions permitting, cancel the IFR flight plan while still airborne and able to communicate with ATC by radio. This will not only save the time and expense of canceling the flight plan by telephone but will quickly release the airspace for use by other aircraft.

5-1-15. RNAV and RNP Operations

a. During the pre-flight planning phase the availability of the navigation infrastructure required for the intended operation, including any non-RNAV contingencies, must be confirmed for the period of intended operation. Availability of the onboard navigation equipment necessary for the route to be flown must be confirmed.

b. If a pilot determines a specified RNP level cannot be achieved, revise the route or delay the operation until appropriate RNP level can be ensured.

c. The onboard navigation database must be appropriate for the region of intended operation and must include the navigation aids, waypoints, and coded terminal airspace procedures for the departure, arrival and alternate airfields.

d. During system initialization, pilots of aircraft equipped with a Flight Management System or other RNAV-certified system, must confirm that the navigation database is current, and verify that the aircraft position has been entered correctly. Flight crews should crosscheck the cleared flight plan against charts or other applicable resources, as well as the navigation system textual display and the aircraft map display. This process includes confirmation of the waypoints sequence, reasonableness of track angles and distances, any altitude or speed constraints, and identification of fly-by or fly-over waypoints. A procedure shall not be used if validity of the navigation database is in doubt.

e. Prior to commencing takeoff, the flight crew must verify that the RNAV system is operating correctly and the correct airport and runway data have been loaded.

Section 2. DEPARTURE PROCEDURES

5-2-1. PRE-TAXI CLEARANCE PROCEDURES

a. Certain airports have established Pre-taxi Clearance programs whereby pilots of departing IFR aircraft may elect to receive their IFR clearances before they start taxiing for takeoff. The following provisions are included in such procedures:

1. Pilot participation is not mandatory.

2. Participating pilots call clearance delivery or ground control not more than 10 minutes before proposed taxi time.

3. IFR clearance (or delay information, if clearance cannot be obtained) is issued at the time of this initial call-up.

4. When the IFR clearance is received on clearance delivery frequency, pilots call ground control when ready to taxi.

5. Normally, pilots need not inform ground control that they have received IFR clearance on clearance delivery frequency. Certain locations may, however, require that the pilot inform ground control of a portion of the routing or that the IFR clearance has been received.

6. If a pilot cannot establish contact on clearance delivery frequency or has not received an IFR clearance before ready to taxi, the pilot should contact ground control and inform the controller accordingly.

b. Locations where these procedures are in effect are indicated in the *Airport/Facility Directory*.

5-2-2. PRE-DEPARTURE CLEARANCE
PROCEDURES

a. Many airports in the National Airspace System are equipped with the Tower Data Link System (TDLS) that includes the Pre-departure Clearance (PDC) function. The PDC function automates the Clearance Delivery operations in the ATCT for participating users. The PDC function displays IFR clearances from the ARTCC to the ATCT. The Clearance Delivery controller in the ATCT can append local departure information and transmit the clearance via data link to participating airline/service provider computers. The airline/service provider will then deliver the clearance via the Aircraft Communications Addressing and Reporting System (ACARS) or a similar data link system or, for nondata link equipped aircraft, via a printer located at the departure gate. PDC reduces frequency congestion, controller workload and is intended to mitigate delivery/readback errors. Also, information from participating users indicates a reduction in pilot workload.

b. PDC is available only to participating aircraft that have subscribed to the service through an approved service provider.

c. Due to technical reasons, the following limitations currently exist in the PDC program:

1. Aircraft filing multiple flight plans are limited to one PDC clearance per departure airport within a 24-hour period. Additional clearances will be delivered verbally.

2. If the clearance is revised or modified prior to delivery, it will be rejected from PDC and the clearance will need to be delivered verbally.

d. No acknowledgment of receipt or readback is required for a PDC.

e. In all situations, the pilot is encouraged to contact clearance delivery if a question or concern exists regarding an automated clearance.

5-2-3. TAXI CLEARANCE

Pilots on IFR flight plans should communicate with the control tower on the appropriate ground control or clearance delivery frequency, prior to starting engines, to receive engine start time, taxi and/or clearance information.

5-2-4. ABBREVIATED IFR DEPARTURE CLEARANCE (CLEARED . . . AS FILED) PROCEDURES

a. ATC facilities will issue an abbreviated IFR departure clearance based on the ROUTE of flight filed in the IFR flight plan, provided the filed route can be approved with little or no revision. These abbreviated clearance procedures are based on the following conditions:

1. The aircraft is on the ground or it has departed VFR and the pilot is requesting IFR clearance while airborne.

2. That a pilot will not accept an abbreviated clearance if the route or destination of a flight plan filed with ATC has been changed by the pilot or the company or the operations officer before departure.

3. That it is the responsibility of the company or operations office to inform the pilot when they make a change to the filed flight plan.

4. That it is the responsibility of the pilot to inform ATC in the initial call-up (for clearance) when the filed flight plan has been either:

(a) amended, or

(b) canceled and replaced with a new filed flight plan.

NOTE—THE FACILITY ISSUING A CLEARANCE MAY NOT HAVE RECEIVED THE REVISED ROUTE OR THE REVISED FLIGHT PLAN BY THE TIME A PILOT REQUESTS CLEARANCE.

b. Controllers will issue a detailed clearance when they know that the original filed flight plan has been changed or when the pilot requests a full route clearance.

c. The clearance as issued will include the destination airport filed in the flight plan.

d. ATC procedures now require the controller to state the SID name and the current number and the DP Transition name after the phrase "Cleared to (destination) airport" and prior to the phrase, "then as filed," for ALL departure clearances when the DP or DP Transition is to be flown. The procedures apply whether or not the SID is filed in the flight plan.

e. STARs, when filed in a flight plan, are considered a part of the filed route of flight and will not normally be stated in an initial departure clearance. If the ARTCC's jurisdictional airspace includes both the departure airport and the fix where a STAR or STAR Transition begins, the STAR name, the current number and the STAR Transition name MAY be stated in the initial clearance.

f. "Cleared to (destination) airport as filed" does NOT include the en route altitude filed in a flight plan. An en route altitude will be stated in the clearance or the pilot will be advised to expect an assigned or filed altitude within a given time frame or at a certain point after departure. This may be done verbally in the departure instructions or stated in the DP.

g. In both radar and nonradar environments, the controller will state "Cleared to (destination) airport as filed" or:

1. If a DP or DP Transition is to be flown, specify the SID name, the current DP number, the DP Transition name, the assigned altitude/Flight Level, and any additional instructions (departure control frequency, beacon code assignment, etc.) necessary to clear a departing aircraft via the DP or DP Transition and the route filed.

EXAMPLE—NATIONAL SEVEN TWENTY CLEARED TO MIAMI AIRPORT INTERCONTINENTAL ONE DEPARTURE, LAKE CHARLES TRANSITION THEN AS FILED, MAINTAIN FLIGHT LEVEL TWO SEVEN ZERO.

2. When there is no DP or when the pilot cannot accept a DP, the controller will specify the assigned altitude or Flight Level, and any additional instructions necessary to clear a departing aircraft via an appropriate departure routing and the route filed.

NOTE—A DETAILED DEPARTURE ROUTE DESCRIPTION OR A RADAR VECTOR MAY BE USED TO ACHIEVE THE DESIRED DEPARTURE ROUTING.

3. If it is necessary to make a minor revision to the filed route, the controller will specify the assigned DP or DP Transition (or departure routing), the revision to the filed route, the assigned altitude or flight level and any additional instructions necessary to clear a departing aircraft.

EXAMPLE—JET STAR ONE FOUR TWO FOUR CLEARED TO ATLANTA AIRPORT, SOUTH BOSTON TWO DEPARTURE THEN AS FILED EXCEPT CHANGE ROUTE TO READ SOUTH BOSTON VICTOR 20 GREENSBORO, MAINTAIN ONE SEVEN THOUSAND.

4. Additionally, in a nonradar environment, the controller will specify one or more fixes, as necessary, to identify the initial route of flight.

EXAMPLE—CESSNA THREE ONE SIX ZERO FOXTROT CLEARED TO CHARLOTTE AIRPORT AS FILED VIA BROOKE, MAINTAIN SEVEN THOUSAND.

h. To ensure success of the program, pilots should:

1. Avoid making changes to a filed flight plan just prior to departure.

2. State the following information in the initial call-up to the facility when no change has been made to the filed flight plan: Aircraft call sign, location, type operation (IFR) and the name of the airport (or fix) to which you expect clearance.

EXAMPLE—"WASHINGTON CLEARANCE DELIVERY (OR GROUND CONTROL IF APPROPRIATE) AMERICAN SEVENTY SIX AT GATE ONE, IFR LOS ANGELES."

3. If the flight plan has been changed, state the change and request a full route clearance.

EXAMPLE—"WASHINGTON CLEARANCE DELIVERY, AMERICAN SEVENTY SIX AT GATE ONE. IFR SAN FRANCISCO. MY FLIGHT PLAN ROUTE HAS BEEN AMENDED (OR DESTINATION CHANGED). REQUEST FULL ROUTE CLEARANCE."

4. Request verification or clarification from ATC if ANY portion of the clearance is not clearly understood.

5. When requesting clearance for the IFR portion of a VFR/IFR flight, request such clearance prior to the fix where IFR operation is proposed to commence in sufficient time to avoid delay. Use the following phraseology:

EXAMPLE—"LOS ANGELES CENTER, APACHE SIX ONE PAPA, VFR ESTIMATING PASO ROBLES VOR AT THREE TWO, ONE THOUSAND FIVE HUNDRED, REQUEST IFR TO BAKERSFIELD."

5-2-5. DEPARTURE RESTRICTIONS, CLEARANCE VOID TIMES, HOLD FOR RELEASE, AND RELEASE TIMES

a. ATC may assign departure restrictions, clearance void times, hold for rase, and release times, when necessary, to separate departures from other traffic or to restrict or regulate the departure flow.

1. *CLEARANCE VOID TIMES:* A pilot may receive a clearance, when operating from an airport without a control tower, which contains a provision for the clearance to be void if not airborne by a specific time. A pilot who does not depart prior to the clearance void time must advise ATC as soon as possible of their intentions. ATC will normally advise the pilot of the time allotted to notify ATC that the aircraft did not depart prior to the clearance void time. This time cannot exceed 30 minutes. Failure of an aircraft to contact ATC within 30 minutes after the clearance void time will result in the aircraft being considered overdue and search and rescue procedures initiated.

NOTE—① OTHER IFR TRAFFIC FOR THE AIRPORT WHERE THE CLEARANCE IS ISSUED IS SUSPENDED UNTIL THE AIRCRAFT HAS CONTACTED ATC OR UNTIL 30 MINUTES AFTER THE CLEARANCE VOID TIME OR 30 MINUTES AFTER THE CLEARANCE RELEASE TIME IF NO CLEARANCE VOID TIME IS ISSUED.

② PILOTS WHO DEPART AT OR AFTER THEIR CLEARANCE VOID TIME ARE NOT AFFORDED IFR SEPARATION AND MAY BE IN VIOLATION OF FAR PART 91.173 WHICH REQUIRES THAT PILOTS RECEIVE AN APPROPRIATE ATC CLEARANCE BEFORE OPERATING IFR IN CONTROLLED AIRSPACE.

EXAMPLE—CLEARANCE VOID IF NOT OFF BY (CLEARANCE VOID TIME) AND, IF REQUIRED, IF NOT OFF BY (CLEARANCE VOID TIME) ADVISE (FACILITY) NOT LATER THAN (TIME) OF INTENTIONS.

2. *HOLD FOR RELEASE:* ATC may issue "hold for release" instructions in a clearance to delay an aircraft's departure for traffic management reasons (i.e., weather, traffic volume, etc.). When ATC states in the clearance, "hold for release," the pilot may not depart utilizing that instrument flight rules (IFR) clearance until a release time or additional instructions are issued by ATC. In addition, ATC will include departure delay information in conjunction with "hold for release" instructions. The ATC instruction, "hold for release," applies to the IFR clearance and does not prevent the pilot from departing under visual flight rules (VFR). However, prior to takeoff the pilot should cancel the IFR flight plan and operate the transponder on the appropriate VFR code. An IFR clearance may not be available after departure.

EXAMPLE—(AIRCRAFT IDENTIFICATION) CLEARED TO (DESTINATION) AIRPORT AS FILED, MAINTAIN (ALTITUDE), AND, IF REQUIRED (ADDITIONAL INSTRUCTIONS OR INFORMATION), HOLD FOR RELEASE, EXPECT (TIME IN HOURS AND/OR MINUTES) DEPARTURE DELAY.

3. *RELEASE TIMES:* A "release time" is a departure restriction issued to a pilot by ATC, specifying the earliest time an aircraft may depart. ATC will use "release times" in conjunction with traffic management procedures and/or to separate a departing aircraft from other traffic.

EXAMPLE—(AIRCRAFT IDENTIFICATION) RELEASED FOR DEPARTURE AT (TIME IN HOURS AND/OR MINUTES).

4. *EXPECT DEPARTURE CLEARANCE TIME (EDCT):* The EDCT is the runway release time assigned to an aircraft included in traffic management programs. Aircraft are expected to depart no earlier than 5 minutes before, and no later than 5 minutes after the EDCT.

b. If practical, pilots departing uncontrolled airports should obtain IFR clearances prior to becoming airborne when two-way communications with the controlling ATC facility is available.

5-2-6. DEPARTURE CONTROL

a. Departure Control is an approach control function responsible for ensuring separation between departures. So as to expedite the handling of departures, Departure Control may suggest a takeoff direction other than that which may normally have been used under VFR handling. Many times it is preferred to offer the pilot a runway that will require the fewest turns after takeoff to place the pilot on course or selected departure route as quickly as possible. At many locations particular attention is paid to the use of preferential runways for local noise abatement programs, and route departures away from congested areas.

b. Departure Control utilizing radar will normally clear aircraft out of the terminal area using DPs via radio navigation aids. When a departure is to be vectored immediately following takeoff, the pilot will be advised prior to takeoff of the initial heading to be flown but may not be advised of the purpose of the heading. Pilots operating in a radar environment are expected to associate departure headings with vectors to their planned route or flight. When given a vector taking the aircraft off a previously assigned nonradar route, the pilot will be advised briefly what the vector is to achieve. Thereafter, radar service will be provided until the aircraft has been reestablished "on-course" using an appropriate navigation aid and the pilot has been advised of the aircraft's position or a handoff is made to another radar controller with further surveillance capabilities.

c. Controllers will inform pilots of the departure control frequencies and, if appropriate, the transponder code before takeoff. Pilots should not operate their transponder until ready to start the takeoff roll, except at ASDE-X facilities where transponders should be transmitting "on" with altitude reporting continuously while operating on the airport surface if so equipped. Pilots should not change to the departure control frequency until requested. Controllers may omit the departure control frequency if a DP has or will be assigned and the departure control frequency is published on the DP.

5-2-7. INSTRUMENT DEPARTURE PROCEDURE (DP)—OBSTACLE DEPARTURE PROCEDURES (ODP) AND STANDARD INSTRUMENT DEPARTURES (SID)

Instrument departure procedures are preplanned instrument flight rule (IFR) procedures which provide obstruction clearance from the terminal area to the appropriate en route structure. There are two types of DPs, Obstacle Departure Procedures (ODPs), printed either textually or graphically, and Standard Instrument Departures (SIDS), always printed graphically. All DPs, either textual or graphic may be designed using either conventional or RNAV criteria. RNAV procedures will have RNAV printed in the title, e.g., SHEAD TWO DEPARTURE (RNAV). ODPs provide obstruction clearance via the least onerous route from the terminal area to the appropriate en route structure. ODPs are recommended for obstruction clearance and may be flown without ATC clearance unless an alternate departure procedure (SID or radar vector) has been specifically assigned by ATC. Graphic ODPs will have (OBSTACLE) printed in the procedure title, e.g., GEYSR THREE DEPARTURE (OBSTACLE), or, CROWN ONE DEPARTURE (RNAV)(OBSTACLE). Standard Instrument Departures are air traffic control (ATC) procedures printed for pilot/controller use in graphic form to provide obstruction clearance and a transition from the terminal area to the appropriate en route structure. SIDS are primarily designed for system enhancement and to reduce pilot/controller workload. ATC clearance must be received prior to flying a SID. All DPs provide the pilot with a way to depart the airport and transition to the en route structure safely. Pilots operating under 14 CFR Part 91 are strongly encouraged to file and fly a DP at night, during marginal Visual Meteorological Conditions (VMC) and Instrument Meteorological Conditions (IMC), when one is available. The following paragraphs will provide an overview of the DP program, why DPs are developed, what criteria are used, where to find them, how they are to be flown, and finally pilot and ATC responsibilities.

a. Why are DP's necessary? The primary reason is to provide obstacle clearance protection information to pilots. A secondary reason, at busier airports, is to increase efficiency and reduce communications and departure delays through the use of SID's. When an instrument approach is initially developed for an airport, the need for DP's is assessed. The procedure designer conducts an obstacle analysis to support departure operations. If an aircraft may turn in any direction from a runway, and remain clear of obstacles, that runway passes what is called a diverse departure assessment and no ODP will be published. A SID may be published if needed for air traffic control purposes. However, if an obstacle penetrates what is called the 40:1 obstacle identification surface, then the procedure designer chooses whether to:

1. Establish a steeper than normal climb gradient or

2. Establish a steeper than normal climb gradient with an alternative that increases takeoff minima to allow the pilot to visually remain clear of the obstacle(s); or

3. Design and publish a specific departure route; or

4. A combination or all of the above.

b. What criteria is used to provide obstruction clearance during departure?

1. Unless specified otherwise, required obstacle clearance for all departures, including diverse, is based on the pilot crossing the departure end of the runway at least 35 feet above the departure end of runway elevation, climbing to 400 feet above the departure end of runway elevation before making the initial turn, and maintaining a minimum climb gradient of 200 feet per nautical mile (FPNM), unless required to level off by a crossing restriction, until the minimum IFR altitude. A greater climb gradient may be specified in the DP to clear obstacles or to achieve an ATC crossing restriction. If an initial turn higher than 400 feet above the departure end of runway elevation is specified in the DP, the turn should be commenced at the higher altitude. If a turn is specified at a fix, the turn must be made at that fix. Fixes may have minimum and/or maximum crossing altitudes that must be adhered to prior to passing the fix. In rare instances, obstacles that exist on the extended runway centerline may make an "early turn" more desirable than proceeding straight ahead. In these cases, the published departure instructions will include the language "turn left(right) as soon as practicable." These departures will also include a ceiling and visibility minimum of at least 300 and 1. Pilots encountering one of these DP's should preplan the climb out to gain altitude and begin the turn as quickly as possible within the bounds of safe operating practices and operating limitations. This type of departure procedure is being phased out.

NOTE—"PRACTICAL" OR "FEASIBLE" MAY EXIST IN SOME EXISTING DEPARTURE TEXT INSTEAD OF "PRACTICABLE."

2. The 40:1 obstacle identification surface begins at the departure end of the runway and slopes upward at 152 FPNM until reaching the minimum IFR altitude or entering the en route structure.

3. Climb gradients greater than 200 FPNM are specified when required for obstacle clearance and/or ATC required crossing restrictions.

EXAMPLE—"CROSS ALPHA INTERSECTION AT OR BELOW 4000; MAINTAIN 6000." THE PILOT CLIMBS AT LEAST 200 FPNM TO 6000. IF 4000 IS REACHED BEFORE ALPHA, THE PILOT LEVELS OFF AT 4000 UNTIL PASSING ALPHA; THEN IMMEDIATELY RESUMES AT LEAST 200 FPNM CLIMB.

4. Climb gradients may be specified only to an altitude/fix, above which the normal gradient applies.

EXAMPLE—"MINIMUM CLIMB 340 FPNM TO ALPHA" THE PILOT CLIMBS AT LEAST 340 FPNM TO ALPA, THEN AT LEAST 200 FPNM TO MIA.

5. Some DPs established solely for obstacle avoidance require a climb in visual conditions to cross the airport or an on-airport NAVAID in a specified direction, at or above a specified altitude. These procedures are called Visual Climb Over the Airport (VCOA).

EXAMPLE—"CLIMB IN VISUAL CONDITIONS SO AS TO CROSS THE MCELORY AIRPORT SOUTHBOUND, AT OR ABOVE 6000, THEN CLIMB VIA KEEMMLING RADIAL ZERO THREE THREE TO KEEMMLING VORTAC."

c. Who is responsible for obstacle clearance? DPs are designed so that adherence to the procedure by the pilot will ensure obstacle protection. Additionally:

1. Obstacle clearance responsibility also rests with the pilot when he/she chooses to climb in visual conditions in lieu of flying a DP and/or depart under increased takeoff minima rather than fly the climb gradient. Standard takeoff minima are one statute mile for aircraft having two engines or less and one-half statute mile for aircraft having more than two engines. Specified ceiling and visibility minima (VCOA or increased takeoff minima) will allow visual avoidance of obstacles until the pilot enters the standard obstacle protection area. Obstacle avoidance is not guaranteed if the pilot maneuvers farther from the airport than the specified visibility minimum prior to reaching the specified altitude. DPs may also contain what are called Low Close in Obstacles. These obstacles are less than 200 feet above the departure end of runway elevation and within one NM of the runway end, and do not require increased takeoff minimums. These obstacles are identified on the SID chart or in the Take-of Minimums and (Obstacle) Departure Procedures section of the U. S. Terminal Procedure booklet These obstacles are especially critical to aircraft tha do not lift off until close to the departure end of the runway or which climb at the minimum rate. Pilot: should also consider drift following lift-off to ensure sufficient clearance from these obstacles. That segment of the procedure that requires the pilot to see and avoid obstacles ends when the aircraft crosses the specified point at the required altitude. In all cases continued obstacle clearance is based on having climbed a minimum of 200 feet per nautical mile to the specified point and then continuing to climb at least 200 foot per nautical mile during the departure until reaching the minimum enroute altitude, unless specified otherwise.

2. ATC may assume responsibility for obstacle clearance by vectoring the aircraft prior to reaching the minimum vectoring altitude by using a Diverse Vector Area (DVA). The DVA has been assessed for departures which do not follow a specific ground track. ATC may also vector an aircraft off a previously assigned DP. In all cases, the 200 FPNM climb gradient is assumed and obstacle clearance is not provided by ATC until the controller begins to provide navigational guidance in the form of radar vectors.

NOTE—WHEN USED BY THE CONTROLLER DURING DEPARTURE THE TERM "RADAR CONTACT" SHOULD NOT BE INTERPRETED AS RELIEVING PILOTS OF THEIR RESPONSIBILITY TO MAINTAIN APPROPRIATE TERRAIN AND OBSTRUCTION CLEARANCE WHICH MAY INCLUDE FLYING THE OBSTACLE DP

3. When missed approach or departure procedure climb gradients exceed 200 ft/NM, pilots must preplan that the aircraft can meet the ft/NM climb gradient requirement prescribed by the procedure.

d. Where are DP's located? DP's will be listed by airport in the IFR Takeoff Minimums and (Obstacle) Departure Procedures Section, Section C, of the Terminal Procedures Publications (TPP's). If the DP is textual, it will be described in TPP Section C. SID's and complex ODP's will be published graphically and named. The name will be listed by airport name and runway in Section C. Graphic ODP's will also have the term "(OBSTACLE)" printed in the charted procedure title, differentiating them from SID's.

1. An ODP that has been developed solely for obstacle avoidance will be indicated with the symbol 'T' on appropriate Instrument Approach Procedure (lAP) charts and DP charts for that airport. The "T" symbol will continue to refer users to TPP Section C. In the case of a graphic ODP, the TPP Section C will only contain the name of the ODP. Since there may be both a textual and a graphic DP, Section C should still be checked for additional information. The nonstandard takeoff minimums and minimum climb gradients found in TPP Section C also apply to charted DP's and radar vector departures unless different minimums are specified on the charted DP. Takeoff minimums and departure procedures apply to all runways unless otherwise specified. New graphic DP's will have all the information printed on the graphic depiction. As a general rule, ATC will only assign an ODP from a nontowered airport when compliance with the ODP is necessary for aircraft to aircraft separation. Pilots may use the ODP to help ensure separation from terrain and obstacles.

e. Responsibilities.

1. Each pilot, prior to departing an airport on an IFR flight should consider the type of terrain and other obstacles on or in the vicinity of the departure airport; and:

2. Determine whether an ODP is available; and

3. Determine if obstacle avoidance can be maintained visually or if the ODP should be flown; and

4. Consider the effect of degraded climb performance and the actions to take in the event of an engine loss during the departure.

5. After an aircraft is established on an ODP/SID and subsequently vectored or cleared off of the ODP or SID transition, pilots shall consider the ODP/SID canceled, unless the controller adds "expect to resume ODP/SID."

6. Aircraft instructed to resume a procedure which contains restrictions, such as a DP, shall be issued/reissued all applicable restrictions or shall be advised to comply with those restrictions.

7. If an altitude to "maintain" is restated, whether prior to departure or while airborne, previously issued altitude restrictions are canceled, including any DP altitude restrictions if any.

8. Pilots of civil aircraft operating from locations where SID's are established may expect ATC clearances containing a SID. Use of a DP requires pilot possession of the textual description or graphic depiction of the approved current DP, as appropriate. RNAV SID's must be retrievable by the procedure name from the aircraft database and conform to charted procedure. ATC must be immediately advised if the pilot does not possess the assigned SID, or the aircraft is not capable of flying the SID. Notification may be accomplished by filing "NO SID" in the remarks section of the filed flight plan or by the less desirable method of verbally advising ATC. Adherence to all restrictions on the DP is required unless clearance to deviate is received.

9. Controllers may omit the departure control frequency if a SID clearance is issued and the departure control frequency is published on the SID.

f. *RNAV Departure Procedures.*

1. There are two types of public RNAV SIDs and graphic ODPs:

(a) Type A. These procedures generally start with a heading or vector from the departure runway end and have an initial RNAV fix approximately 10 nautical miles (NM) from the departure airport. In addition, these procedures require system performance currently met by GPS, DME/DME, or DME/DME/IRU RNAV systems that satisfy the criteria discussed in AC 90-100, U.S. Terminal and En Route Area Navigation (RNAV) Operations. Type A terminal procedures require the aircraft's track keeping accuracy remain bounded by ±2 NM for 95% of the total flight time.

NOTE—IF NOT EQUIPPED WITH *GPS* (OR FOR MULTI-SENSOR SYSTEMS WITH *GPS* WHICH DO NOT ALERT UPON LOSS OF *GPS*), AIRCRAFT MUST BE CAPABLE OF NAVIGATION SYSTEM UPDATING USING *DME/DME* OR *DMEIDME/IRU* FOR TYPE A PROCEDURES.

(b) Type B. These procedures generally start with an initial RNAV leg near the departure runway end. In addition, these procedures require system performance currently met by GPS or DME/DME/ IRU RNAV systems that satisfy the criteria discussed in AC 90-100. Type B procedures often require the aircraft's track keeping accuracy remain bounded by ±1 NM for 95% of the total flight time.

NOTE—IF NOT EQUIPPED WITH *GPS* (OR FOR MULTI-SENSOR SYSTEMS WITH *GPS* WHICH DO NOT ALERT UPON LOSS OF *GPS*), AIRCRAFT MUST BE CAPABLE OF NAVIGATION SYSTEM UPDATING USING *DMEIDME/IRU* FOR TYPE B PROCEDURES.

2. For procedures requiring GPS, if the navigation system does not automatically alert the flight crew of a loss of GPS, the operator must develop procedures to verify correct GPS operation.

3. AC 90-100 may be used as operational and airworthiness guidance for RNAV ODPs.

4. RNAV Engagement Altitudes. For Type A procedures, the pilot must be able to engage RNAV equipment no later than 2,000 feet above airport elevation. For Type B procedures, the pilot must be able to engage RNAV equipment no later than 500 feet above airport elevation.

1. Procedure Identification.

(a) Pilots may determine the level of an RNAV DP by referring to the RNP value charted for /R aircraft. Level 2 procedures have a /R aircraft RNP requirement of 2.0. Level 1 procedures have a /R aircraft RNP requirement of 1.0.

(b) "Level 1" or "Level 2" may also be indicated on the procedure.

2. Procedure Design. The FAA provides criteria for the design of RNAV DP in FAA Order 8260.44, Civil Utilization of Area Navigation (RNAV) Departure Procedures. The two levels of procedure design provide for obstacle clearance. Level 1 criteria, with the narrower linear obstacle clearance areas, is applied in procedure design only if Level 2 criteria would not provide a usable procedure. Level 1 procedures require a higher level of overall navigation performance, as the protected obstacle clearance areas are smaller.

Section 3. EN ROUTE PROCEDURES

5-3-1. ARTCC COMMUNICATIONS

a. Direct Communications, Controllers and Pilots:

1. ARTCCs are capable of direct communications with IFR air traffic on certain frequencies. Maximum communications coverage is possible through the use of Remote Center Air/Ground (RCAG) sites comprised of both VHF and UHF transmitters and receivers. These sites are located throughout the U.S. Although they may be several hundred miles away from the ARTCC, they are remoted to the various ARTCCs by land lines or microwave links. Since IFR operations are expedited through the use of direct communications, pilots are requested to use these frequencies strictly for communications pertinent to the control of IFR aircraft. Flight plan filing, en route weather, weather forecasts, and similar data should be requested through FSSs, company radio, or appropriate military facilities capable of performing these services.

2. An ARTCC is divided into sectors. Each sector is handled by one or a team of controllers and has its own sector discrete frequency. As a flight progresses from one sector to another, the pilot is requested to change to the appropriate sector discrete frequency.

3. Controller Pilot Data Link Communications (CPDLC) is a system that supplements air/ground voice communications. As a result, it expands two-way air traffic control air/ground communications capabilities. Consequently, the air traffic system's operational capacity is increased and any associated air traffic delays become minimized. A related safety benefit is that pilot/controller read-back and hear-back errors will be significantly reduced. The CPDLC's principal operating criteria are:

(a) Voice remains the primary and controlling air/ground communications means.

(b) Participating aircraft will need to have the appropriate CPDLC avionics equipment in order to receive uplink or transmit downlink messages.

(c) CPDLC Build 1 offers four ATC data link services. These are altimeter setting (AS), transfer of communications (TC), initial contact (IC), and menu text messages (MT).

(1) Altimeter settings are usually transmitted automatically when a CPDLC session and eligibility has been established with an aircraft. A controller may also manually send an altimeter setting message.

NOTE—WHEN CONDUCTING INSTRUMENT APPROACH PROCEDURES, PILOTS ARE RESPONSIBLE TO OBTAIN AND USE THE APPROPRIATE ALTIMETER SETTING IN ACCORDANCE WITH 14 CFR SECTION 97.20. CPDLC ISSUED ALTIMETER SETTINGS ARE XCLUDED FOR THIS PURPOSE.

(2) Initial contact is a safety validation transaction that compares a pilot's initiated altitude downlink message with an aircraft's ATC host computer stored altitude. If an altitude mismatch is detected, the controller will verbally provide corrective action.

(3) Transfer of communications automatically establishes data link contact with a succeeding sector.

(4) Menu text transmissions are scripted nontrajectory altering uplink messages.

NOTE—INITIAL USE OF CPDLC WILL BE AT THE MIAMI AIR ROUTE TRAFFICCONTROL CENTER (ARTCC). AIR CARRIERS WILL BE THE FIRST USERS. SUBSEQUENTLY CPDLC WILL BE MADE AVAILABLE TO ALL NAS USERS. LATER VERSIONS WILL INCLUDE TRAJECTORY ALTERING SERVICES AND EXPANDED CLEARANCE AND ADVISORY MESSAGE CAPABILITIES.

b. ATC Frequency Change Procedures:

1. The following phraseology will be used by controllers to effect a frequency change:

EXAMPLE—(AIRCRAFT IDENTIFICATION) CONTACT (FACILITY NAME OR LOCATION NAME AND TERMINAL FUNCTION) (FREQUENCY) ATa (TIME, FIX, OR ALTITUDE).

NOTE—PILOTS ARE EXPECTED TO MAINTAIN A LISTENING WATCH ON THE TRANSFERRING CONTROLLER'S FREQUENCY UNTIL THE TIME, FIX, OR ALTITUDE SPECIFIED. ATC WILL OMIT FREQUENCY CHANGE RESTRICTIONS WHENEVER PILOT COMPLIANCE IS EXPECTED UPON RECEIPT.

2. The following phraseology should be utilized by pilots for establishing contact with the designated facility:

(a) When operating in a radar environment: On initial contact, the pilot should inform the controller of the aircraft's assigned altitude preceded by the words "level," or "climbing to," or "descending to," as appropriate; and the aircraft's present vacating altitude, if applicable.

EXAMPLE—1 (NAME) CENTER, (AIRCRAFT IDENTIFICATION), LEVEL (ALTITUDE OR FLIGHT LEVEL).

2 (NAME) CENTER, (AIRCRAFT IDENTIFICATION), LEAVING (EXACT ALTITUDE OR FLIGHT LEVEL), CLIMBING TO OR DESCENDING TO (ALTITUDE OF FLIGHT LEVEL).

NOTE—EXACT ALTITUDE OR FLIGHT LEVEL MEANS TO THE NEAREST 100 FOOT INCREMENT. EXACT ALTITUDE OR FLIGHT LEVEL REPORTS ON INITIAL CONTACT PROVIDE ATC WITH INFORMATION REQUIRED PRIOR TO USING MODE C ALTITUDE INFORMATION FOR SEPARATION PURPOSES.

(b) When operating in a nonradar environment:

(1) On initial contact, the pilot should inform the controller of the aircraft's present position, altitude and time estimate for the next reporting point.

EXAMPLE—
(NAME) CENTER, (AIRCRAFT IDENTIFICATION), (POSITION), (ALTITUDE), ESTIMATING (REPORTING POINT) AT (TIME).

(2) After initial contact, when a position report will be made, the pilot should give the controller a complete position report.

EXAMPLE—
(NAME) CENTER, (AIRCRAFT IDENTIFICATION), (POSITION), (TIME), (ALTITUDE), (TYPE OF FLIGHT PLAN), (ETA AND NAME OF NEXT REPORTING POINT), (THE NAME OF THE NEXT SUCCEEDING REPORTING POINT), AND (REMARKS).

REFERENCE—
AIM, POSITION REPORTING, PARAGRAPH 5-3-2.

3. At times controllers will ask pilots to verify that they are at a particular altitude. The phraseology used will be: "VERIFY AT (altitude)." In climbing or descending situations, controllers may ask pilots to *"VERIFY ASSIGNED ALTITUDE AS (altitude)."* Pilots should confirm that they are at the altitude stated by the controller or that the assigned altitude is correct as stated. If this is not the case, they should inform the controller of the actual altitude being maintained or the different assigned altitude.

CAUTION—PILOTS SHOULD NOT TAKE ACTION TO CHANGE THEIR ACTUAL ALTITUDE OR DIFFERENT ASSIGNED ALTITUDE TO THE ALTITUDE STATED IN THE CONTROLLERS VERIFICATION REQUEST UNLESS THE CONTROLLER SPECIFICALLY AUTHORIZES A CHANGE.

c. ARTCC Radio Frequency Outage: ARTCCs normally have at least one back-up radio receiver and transmitter system for each frequency, which can usually be placed into service quickly with little or no disruption of ATC service. Occasionally, technical problems may cause a delay but switchover seldom takes more than 60 seconds. When it appears that the outage will not be quickly remedied, the ARTCC will usually request a nearby aircraft, if there is one, to switch to the affected frequency to broadcast communications instructions. It is important, therefore, that the pilot wait at least 1 minute before deciding that the ARTCC has actually experienced a radio frequency failure. When such an outage does occur, the pilot should, if workload and equipment capability permit, maintain a listening watch on the affected frequency while attempting to comply with the following recommended communications procedures:

1. If two-way communications cannot be established with the ARTCC after changing frequencies, a pilot should attempt to recontact the transferring controller for the assignment of an alternative frequency or other instructions.

2. When an ARTCC radio frequency failure occurs after two-way communications have been established, the pilot should attempt to reestablish contact with the center on any other known ARTCC frequency, preferably that of the next responsible sector when practicable, and ask for instructions. However, when the next normal frequency change along the route is known to involve another ATC facility, the pilot should contact that facility, if feasible, for instructions. If communications cannot be reestablished by either method, the pilot is expected to request communications instructions from the FSS appropriate to the route of flight.

NOTE—THE EXCHANGE OF INFORMATION BETWEEN AN AIRCRAFT AND AN ARTCC THROUGH AN FSS IS QUICKER THAN RELAY VIA COMPANY RADIO BECAUSE THE FSS HAS DIRECT INTERPHONE LINES TO THE RESPONSIBLE ARTCC SECTOR. ACCORDINGLY, WHEN CIRCUMSTANCES DICTATE A CHOICE BETWEEN THE TWO, DURING AN ARTCC FREQUENCY OUTAGE, RELAY VIA FSS RADIO IS RECOMMENDED.

5-3-2. POSITION REPORTING

The safety and effectiveness of traffic control depends to a large extent on accurate position reporting. In order to provide the proper separation and expedite aircraft movements, ATC must be able to make accurate estimates of the progress of every aircraft operating on an IFR flight plan.

a. Position Identification:

1. When a position report is to be made passing a VOR radio facility, the time reported should be the time at which the first complete reversal of the "to/from" indicator is accomplished.

2. When a position report is made passing a facility by means of an airborne ADF, the time reported should be the time at which the indicator makes a complete reversal.

3. When an aural or a light panel indication is used to determine the time passing a reporting point, such as a fan marker, Z marker, cone of silence or intersection of range courses, the time should be noted when the signal is first received and again when it ceases. The mean of these two times should then be taken as the actual time over the fix.

4. If a position is given with respect to distance and direction from a reporting point, the distance and direction should be computed as accurately as possible.

5. Except for terminal area transition purposes, position reports or navigation with reference to aids not established for use in the structure in which flight is being conducted will not normally be required by ATC.

b. Position Reporting Points: FARs require pilots to maintain a listening watch on the appropriate frequency and, unless operating under the provisions of subparagraph c., to furnish position reports passing certain reporting points. Reporting points are indicated by symbols on en route charts. The designated compulsory reporting point symbol is a solid triangle ▲ and the "on request" reporting point symbol is the open triangle Δ. Reports passing an "on request" reporting point are only necessary when requested by ATC.

c. Position Reporting Requirements:

1. *Flights along airways or routes:* A position report is required by all flights regardless of altitude, including those operating in accordance with an ATC clearance specifying *"VFR ON TOP,"* over each designated compulsory reporting point along the route being flown.

2. *Flights Along a Direct Route:* Regardless of the altitude or flight level being flown, including flights operating in accordance with an ATC clearance specifying *"VFR ON TOP,"* pilots shall report over each reporting point used in the flight plan to define the route of flight.

3. *Flights in a Radar Environment:* When informed by ATC that their aircraft are in "Radar Contact," pilots should discontinue position reports over designated reporting points. They should resume normal position reporting when ATC advises *"RADAR CONTACT LOST"* or *"RADAR SERVICE TERMINATED."*

NOTE—ATC WILL INFORM PILOTS THAT THEY ARE IN "RADAR CONTACT"

(A) WHEN THEIR AIRCRAFT IS INITIALLY IDENTIFIED IN THE ATC SYSTEM; AND

(B) WHEN RADAR IDENTIFICATION IS REESTABLISHED AFTER RADAR SERVICE HAS BEEN TERMINATED OR RADAR CONTACT LOST.

SUBSEQUENT TO BEING ADVISED THAT THE CONTROLLER HAS ESTABLISHED RADAR CONTACT, THIS FACT WILL NOT BE REPEATED TO THE PILOT WHEN HANDED OFF TO ANOTHER CONTROLLER. AT TIMES, THE AIRCRAFT IDENTITY WILL BE CONFIRMED BY THE RECEIVING CONTROLLER; HOWEVER, THIS SHOULD NOT BE CONSTRUED TO MEAN THAT RADAR CONTACT HAS BEEN LOST. THE IDENTITY OF TRANSPONDER EQUIPPED AIRCRAFT WILL BE CONFIRMED BY ASKING THE PILOT TO "IDENT," "SQUAWK STANDBY," OR TO CHANGE CODES. AIRCRAFT WITHOUT TRANSPONDERS WILL BE ADVISED OF THEIR POSITION TO CONFIRM IDENTITY. IN THIS CASE, THE PILOT IS EXPECTED TO ADVISE THE CONTROLLER IF IN DISAGREEMENT WITH THE POSITION GIVEN. ANY PILOT WHO CANNOT CONFIRM THE ACCURACY OF THE POSITION GIVEN BECAUSE OF NOT BEING TUNED TO THE NAVAID REFERENCED BY THE CONTROLLER, SHOULD ASK FOR ANOTHER RADAR POSITION RELATIVE TO THE TUNED IN NAVAID.

d. Position Report Items:

1. *Position reports should include the following items:*

(a) Identification.

(b) Position.

(c) Time.

(d) Altitude or flight level (include actual altitude or flight level when operating on a clearance specifying VFR-ON-TOP.)

(e) Type of flight plan (not required in IFR position reports made directly to ARTCCs or approach control),

(f) ETA and name of next reporting point.

(g) The name only of the next succeeding reporting point along the route of flight, and

(h) Pertinent remarks.

5-3-3. ADDITIONAL REPORTS

a. The following reports should be made to ATC or FSS facilities without a specific ATC request:

1. *At all times:*

(a) When vacating any previously assigned altitude or flight level for a newly assigned altitude or flight level.

(b) When an altitude change will be made if operating on a clearance specifying VFR ON TOP.

(c) When *unable* to climb/descend at a rate of at least 500 feet per minute.

(d) When approach has been missed. (Request clearance for specific action; i.e., to alternative airport, another approach, etc.)

(e) Change in the average true airspeed (at cruising altitude) when it varies by 5 percent or 10 knots (whichever is greater) from that filed in the flight plan.

(f) The time and altitude or flight level upon reaching a holding fix or point to which cleared.

(g) When leaving any assigned holding fix or point.

NOTE—THE REPORTS IN SUBPARAGRAPHS (f) AND (g) MAY BE OMITTED BY PILOTS OF AIRCRAFT INVOLVED IN INSTRUMENT TRAINING AT MILITARY TERMINAL AREA FACILITIES WHEN RADAR SERVICE IS BEING PROVIDED.

(h) Any loss, in controlled airspace, of VOR, TACAN, ADF, low frequency navigation receiver capability, GPS anomalies while using installed IFR-certified GPS/GNSS receivers, complete or partial loss of ILS receiver capability or impairment of air/ground communications capability. Reports should include aircraft identification, equipment affected, degree to which the capability to operate under IFR in the ATC system is impaired, and the nature and extent of assistance desired from ATC.

NOTE—1. OTHER EQUIPMENT INSTALLED IN AN AIRCRAFT MAY EFFECTIVELY IMPAIR SAFETY AND/OR THE ABILITY TO OPERATE UNDER IFR. IF SUCH EQUIPMENT (E.G. AIRBORNE WEATHER RADAR) MALFUNCTIONS AND IN THE PILOT'S JUDGMENT EITHER SAFETY OR IFR CAPABILITIES ARE AFFECTED, REPORTS SHOULD BE MADE AS ABOVE.

2. WHEN REPORTING GPS ANOMALIES, INCLUDE THE LOCATION AND ALTITUDE OF THE ANOMALY. BE SPECIFIC WHEN DESCRIBING THE LOCATION AND INCLUDE DURATION OF THE ANOMALY IF NECESSARY.

(i) Any information relating to the safety of flight.

2. *When not in radar contact:*

(a) When leaving final approach fix inbound on final approach (nonprecision approach) or when leaving the outer marker or fix used in lieu of the outer marker inbound on final approach (precision approach).

(b) A corrected estimate at anytime it becomes apparent that an estimate as previously submitted is in error in excess of 3 minutes.

b. Pilots encountering weather conditions which have not been forecast, or hazardous conditions which have been forecast, are expected to forward a report of such weather to ATC.

REFERENCE—AIM, PILOT WEATHER REPORTS (PIREPS), PARAGRAPH 7-1-19. FAR PART 91.183(B) AND (C).

5-3-4. AIRWAYS AND ROUTE SYSTEMS

a. Two fixed route systems are established for air navigation purposes. They are the VOR and L/MF system, and the jet route system. To the extent possible, these route systems are aligned in an overlying manner to facilitate transition between each.

1. The VOR and L/MF Airway System consists of airways designated from 1,200 feet above the surface (or in some instances higher) up to but not including 18,000 feet MSL. These airways are depicted on Enroute Low Altitude Charts.

NOTE—THE ALTITUDE LIMITS OF A VICTOR AIRWAY SHOULD NOT BE EXCEEDED EXCEPT TO EFFECT TRANSITION WITHIN OR BETWEEN ROUTE STRUCTURES.

(a) Except in Alaska and coastal North Carolina, the VOR airways are predicated solely on VOR or VORTAC navigation aids; are depicted in blue on aeronautical charts; and are identified by a "V" (Victor) followed by the airway number (e.g., V12).

NOTE—SEGMENTS OF VOR AIRWAYS IN ALASKA AND NORTH CAROLINA (V56, V290) ARE BASED ON L/MF NAVIGATION AIDS AND CHARTED IN BROWN INSTEAD OF BLUE ON EN ROUTE CHARTS.

(1) A segment of an airway which is common to two or more routes carries the numbers of all the airways which coincide for that segment. When such is the case, pilots filing a flight plan need to indicate only that airway number for the route filed.

NOTE—A PILOT WHO INTENDS TO MAKE AN AIRWAY FLIGHT, USING VOR FACILITIES, WILL SIMPLY SPECIFY THE APPROPRIATE "VICTOR" AIRWAYS(S) IN THE FLIGHT PLAN. FOR EXAMPLE, IF A FLIGHT IS TO BE MADE FROM CHICAGO TO NEW ORLEANS AT 8,000 FEET, USING OMNIRANGES ONLY, THE ROUTE MAY BE INDICATED AS "DEPARTING FROM CHICAGO-MIDWAY, CRUISING 8,000 FEET VIA VICTOR 9 TO MOISANT INTERNATIONAL. " IF FLIGHT IS TO BE CONDUCTED IN PART BY MEANS OF L/MF NAVIGATION AIDS AND IN PART ON OMNIRANGES, SPECIFICATIONS OF THE APPROPRIATE AIRWAYS IN THE FLIGHT PLAN WILL INDICATE WHICH TYPES OF FACILITIES WILL BE USED ALONG THE DESCRIBED

ROUTES, AND, FOR IFR FLIGHT, PERMIT ATC TO ISSUE A TRAFFIC CLEARANCE ACCORDINGLY. A ROUTE MAY ALSO BE DESCRIBED BY SPECIFYING THE STATION OVER WHICH THE FLIGHT WILL PASS, BUT IN THIS CASE SINCE MANY VORS AND L/MF AIDS HAVE THE SAME NAME, THE PILOT MUST BE CAREFUL TO INDICATE WHICH AID WILL BE USED AT A PARTICULAR LOCATION. THIS WILL BE INDICATED IN THE ROUTE OF FLIGHT PORTION OF THE FLIGHT PLAN BY SPECIFYING THE TYPE OF FACILITY TO BE USED AFTER THE LOCATION NAME IN THE FOLLOWING MANNER: NEWARK L/MF, ALLENTOWN VOR.

(2) With respect to position reporting, reporting points are designated for VOR Airway Systems. Flights using Victor Airways will report over these points unless advised otherwise by ATC.

(b) The L/MF airways (colored airways) are predicated solely on L/MF navigation aids and are depicted in brown on aeronautical charts and are identified by color name and number (e.g., Amber One). Green and Red airways are plotted east and west. Amber and Blue airways are plotted north and south.

(c) The use of TSO-C145a or TSO-C146a GPS/WAAS navigation systems is allowed in Alaska as the only means of navigation on published air traffic routes including those Victor and colored airway segments designated with a second minimum en route altitude (MEA) depicted in blue and followed by the letter G at those lower altitudes. The altitudes so depicted are below the minimum reception altitude (MRA) of the land-based navigation facility defining the route segment, and guarantee standard en route obstacle clearance and two-way communications. Air carrier operators requiring operations specifications are authorized to conduct operations on those routes in accordance with FAA operations specifications.

NOTE—EXCEPT FOR G13 IN NORTH CAROLINA, THE COLORED AIRWAY SYSTEM EXISTS ONLY IN THE STATE OF ALASKA. ALL OTHER SUCH AIRWAYS FORMERLY SO DESIGNATED IN THE CONTERMINOUS U.S. HAVE BEEN RESCINDED.

2. The Jet Route system consists of jet routes established from 18,000 feet MSL to FL 450 inclusive.

(a) These routes are depicted on En Route High Altitude Charts. Jet routes are depicted in black on aeronautical charts and are identified by a "J" (Jet) followed by the airway number (e.g., J12). Jet routes, as VOR airways, are predicated solely on VOR or VORTAC navigation facilities (except in Alaska).

NOTE—SEGMENTS OF JET ROUTES IN ALASKA ARE BASED ON L/MF NAVIGATION AIDS AND ARE CHARTED IN BROWN COLOR INSTEAD OF BLACK ON EN ROUTE CHARTS.

(b) With respect to position reporting, reporting points are designated for Jet Route systems. Flights using Jet Routes will report over these points unless otherwise advised by ATC.

3. *Area Navigation (RNAV) Routes.*

(a) Published RNAV routes, including Q-Routes, can be flight planned for use by aircraft with RNAV capability, subject to any limitations or requirements noted on en route charts or by NOTAM.

(b) Unpublished RNAV routes are direct routes, based on area navigation capability, between waypoints defined in terms of latitude/longitude coordinates, degree-distance fixes, or offsets from established routes/airways at a specified distance and direction. Radar monitoring by ATC is required on all unpublished RNAV routes.

(c) Magnetic Reference Bearing (MRB) is the published bearing between two waypoints on an RNAV/GPS/GNSS route. The MRB is calculated by applying magnetic variation at the waypoint to the calculated true course between two waypoints. The MRB enhances situational awareness by indicating a reference bearing (no-wind heading) that a pilot should see on the compass/HSI/RMI etc., when turning prior to/over a waypoint en route to another waypoint. Pilots should use this bearing as a reference only, because their RNAV/GPS/GLASS navigation system will fly the true course between the waypoints.

b. Operation above FL 450 may be conducted on a point-to-point basis. Navigational guidance is provided on an area basis utilizing those facilities depicted on the Enroute High Altitude Charts.

c. *Radar Vectors:* Controllers may vector aircraft within controlled airspace for separation purposes, noise abatement considerations, when an operational advantage will be realized by the pilot or the controller, or when requested by the pilot. Vectors outside of controlled airspace will be provided only on pilot request. Pilots will be advised as to what the vector is to achieve when the vector is controller initiated and will take the aircraft off a previously assigned nonradar route. To the extent possible, aircraft operating on RNAV routes will be allowed to remain on their own navigation.

d. When flying in Canadian airspace, pilots are cautioned to review Canadian Air Regulations.

1. Special attention should be given to the parts which differ from U.S. FARs.

(a) The Canadian Airways Class B airspace restriction is an example. Class B airspace is all controlled low level airspace above 12,500 feet MSL or the MEA, whichever is higher, within which only IFR and controlled VFR flights are permitted. (Low level airspace means an airspace designated and defined as such in the *Designated Airspace Handbook.*)

(b) Regardless of the weather conditions or the height of the terrain, no person shall operate an aircraft under VFR conditions within Class B airspace except in accordance with a clearance for VFR flight issued by ATC.

(c) The requirement for entry into Class B airspace is a student pilot permit (under the guidance or control of a flight instructor).

(d) VFR flight requires visual contact with the ground or water at all times.

2. Segments of VOR airways and high level routes in Canada are based on L/MF navigation aids and are charted in brown color instead of blue on en route charts.

5-3-5. AIRWAY OR ROUTE COURSE CHANGES

a. Pilots of aircraft are required to adhere to airways or routes being flown. Special attention must be given to this requirement during course changes. Each course change consists of variables that make the technique applicable in each case a matter only the pilot can resolve. Some variables which must be considered are turn radius, wind effect, airspeed, degree of turn, and cockpit instrumentation. An early turn, as illustrated below, is one method of adhering to airways or routes. The use of any available cockpit instrumentation, such as Distance Measuring Equipment, may be used by the pilot to lead the turn when making course changes. This *is consistent* with the intent of FAR Part 91.181, which requires pilots to operate along the centerline of an airway and along the direct course between navigational aids or fixes.

b. Turns which begin at or after fix passage may exceed airway or route boundaries. FIG 5-3-1 contains an example flight track depicting this, together with an example of an early turn.

c. Without such actions as leading a turn, aircraft operating in excess of 290 knots true air speed (TAS) can exceed the normal airway or route boundaries depending on the amount of course change required, wind direction and velocity, the character of the turn fix (DME, overhead navigation aid, or intersection), and the pilot's technique in making a course change. For example, a flight operating at 17,000 feet MSL with a TAS of 400 knots, a 25 degree bank, and a course change of more than 40 degrees would exceed the width of the airway or route; i.e., 4 nautical miles each side of centerline. However, in the airspace below 18,000 feet MSL, operations in excess of 290 knots TAS are not prevalent and the provision of additional IFR separation in all course change situations for the occasional aircraft making a turn in excess of 290 knots TAS creates an unacceptable waste of airspace and imposes a penalty upon the preponderance of traffic which operate at low speeds. Consequently, the FAA expects pilots to lead turns and take other actions they consider necessary during course changes to adhere as closely as possible to the airways or route being flown.

d. Due to the high airspeeds used at 18,000 feet MSL and above, FAA provides additional IFR separation protection for course changes made at such altitude levels.

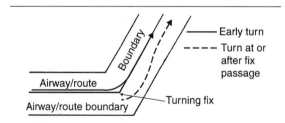

Figure 5-3-1. Adhering to Airways or Routes

5-3-6. CHANGEOVER POINTS (COPs)

a. COPs are prescribed for Federal Airways, Jet Routes, Area Navigation Routes, or other direct routes for which an MEA is designated under FAR Part 95. The COP is a point along the route or airway segment between two adjacent navigation facilities or way points where changeover in navigation guidance should occur. At this point, the pilot should change navigation receiver frequency from the station behind the aircraft to the station ahead.

b. The COP is normally located midway between the navigation facilities for straight route segments, or at the intersection of radials or courses forming a dogleg in the case of dogleg route segments. When the COP is NOT located at the midway point, aeronautical charts will depict the COP location and give the mileage to the radio aids.

c. COPs are established for the purpose of preventing loss of navigation guidance, to prevent frequency interference from other facilities, and to prevent use of different facilities by different aircraft in the same airspace. Pilots are urged to observe COPs to the fullest extent.

5-3-7. HOLDING

a. Whenever an aircraft is cleared to a fix other than the destination airport and delay is expected, it is the responsibility of the ATC controller to issue complete holding instructions (unless the pattern is charted), an EFC time and best estimate of any additional en route/terminal delay.

NOTE—ONLY THOSE HOLDING PATTERNS DEPICTED ON U.S. GOVERNMENT OR COMMERCIALLY PRODUCED (MEETING FAA REQUIREMENTS) LOW/HIGH ALTITUDE ENROUTE, AND AREA OR STAR CHARTS SHOULD BE USED.

b. If the holding pattern is charted and the controller doesn't issue complete holding instructions, the pilot is expected to hold as depicted on the appropriate chart. When the pattern is charted, the controller may omit all holding instructions except the charted holding direction and the statement *AS PUBLISHED; e.g., HOLD EAST AS PUBLISHED*. Controllers shall always issue complete holding instructions when pilots request them.

c. If no holding pattern is charted and holding instructions have not been issued, the pilot should ask ATC for holding instructions prior to reaching the fix. This procedure will eliminate the possibility of an aircraft entering a holding pattern other than that desired by ATC. If unable to obtain holding instructions prior to reaching the fix (due to frequency congestion, stuck microphone, etc.), then enter a standard pattern on the course on which the aircraft approached the fix and request further clearance as soon as possible. In this event, the altitude/flight level of the aircraft at the clearance limit will be protected so that separation will be provided as required.

d. When an aircraft is 3 minutes or less from a clearance limit and a clearance beyond the fix has not been received, the pilot is expected to start a speed reduction so that the aircraft will cross the fix, initially, at or below the maximum holding airspeed.

e. When no delay is expected, the controller should issue a clearance beyond the fix as soon as possible and, whenever possible, at least 5 minutes before the aircraft reaches the clearance limit.

f. Pilots should report to ATC the time and altitude/flight level at which the aircraft reaches the clearance limit and report leaving the clearance limit.

NOTE—IN THE EVENT OF TWO-WAY COMMUNICATIONS FAILURE, PILOTS ARE REQUIRED TO COMPLY WITH FAR PART 91.185.

g. When holding at a VOR station, pilots should begin the turn to the outbound leg at the time of the first complete reversal of the to/from indicator.

h. Patterns at the most generally used holding fixes are depicted (charted) on U.S. Government or commercially produced (meeting FAA requirements) Low or High Altitude Enroute, Area and STAR Charts. Pilots are expected to hold in the pattern depicted unless specifically advised otherwise by ATC.

NOTE—HOLDING PATTERNS THAT PROTECT FOR A MAXIMUM HOLDING AIRSPEED OTHER THAN THE STANDARD MAY BE DEPICTED BY AN ICON, UNLESS OTHERWISE DEPICTED. THE ICON IS A STANDARD HOLDING PATTERN SYMBOL (RACETRACK) WITH THE AIRSPEED RESTRICTION SHOWN IN THE CENTER. IN OTHER CASES, THE AIRSPEED RESTRICTION WILL BE DEPICTED NEXT TO THE STANDARD HOLDING PATTERN SYMBOL.

i. An ATC clearance requiring an aircraft to hold at a fix where the pattern is not charted will include the following information: (See FIG 5-3-2).

EXAMPLES OF HOLDING

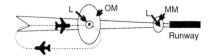

Typical procedure on an ILS outer marker

Typical procedure at intersection of VOR radials

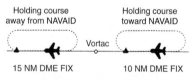

Typical procedure at DME FIX

Figure 5-3-2. Holding Patterns

1. Direction of holding from the fix in terms of the eight cardinal compass points (i.e., N, NE, E, SE, etc.).

2. Holding fix (the fix may be omitted if included at the beginning of the transmission as the clearance limit).

3. Radial, course, bearing, airway or route on which the aircraft is to hold.

4. Leg length in miles if DME or RNAV is to be used (leg length will be specified in minutes on pilot request or if the controller considers it necessary).

5. Direction of turn if left turns are to be made, the pilot requests, or the controller considers it necessary.

6. Time to expect further clearance and any pertinent additional delay information.

j. Holding pattern airspace protection is based on the following procedures.

1. *Descriptive Terms:*

(a) ***Standard Pattern:*** Right turns (See FIG 5-3-3.)

(b) ***Nonstandard Pattern:*** Left turns

2. *Airspeeds:*

(a) All aircraft may hold at the following altitudes and maximum holding airspeeds:

Table 5-3-1

Altitude (MSL)	Airspeed (KIAS)
MHA–6,000'	200
6,001'–14,000'	230
14,001' and above	265

(b) The following are exceptions to the maximum holding airspeeds:

(1) Holding patterns from 6,001' to 14,000' may be restricted to a maximum airspeed of 210 KIAS. This nonstandard pattern will be depicted by an icon.

(2) Holding patterns may be restricted to a maximum airspeed of 175 KIAS. An icon will depict this nonstandard pattern. Pilots of aircraft unable to comply with the maximum airspeed restriction should notify ATC.

(3) Holding patterns at USAF airfields only—310 KIAS maximum, unless otherwise depicted.

(4) Holding patterns at Navy fields only—230 KIAS maximum, unless otherwise depicted.

(c) The following phraseology may be used by an ATCS to advise a pilot of the maximum holding airspeed for a holding pattern airspace area.

(5) When a climb-in hold is specified by a **published procedure** (e.g., "Climb-in holding pattern to depart XYZ

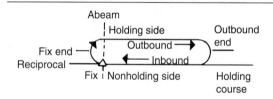

Figure 5-3-3. Holding Pattern Descriptive Terms

VORTAC at or above 10,000." or "All aircraft climb-in TRUCK holding pattern to cross TRUCK Int at or above 11,500 before proceeding on course."), additional obstacle protection area has been provided to allow for greater airspeeds in the climb for those aircraft requiring them. The holding pattern template for a maximum airspeed of 310 KIAS has been used for the holding pattern if there are no airspeed restrictions on the holding pattern as specified in subparagraph j2(b)(2) of this paragraph. Where the holding pattern is restricted to a maximum airspeed of 175 KIAS, the 200 KIAS holding pattern template has been applied for published climb-in hold procedures for altitudes 6,000 feet and below and the 230 KIAS holding pattern template has been applied for altitudes above 6,000 feet. The airspeed limitations in 14 CFR Section 91.117, Aircraft Speed, still apply.

PHRASEOLOGY—

(AIRCRAFT IDENTIFICATION)(HOLDING INSTRUCTIONS, WHEN NEEDED) MAXIMUM HOLDING AIRSPEED IS (SPEED IN KNOTS)

3. *Entry Procedures:*

(See FIG 5-3-4).

(a) *Parallel Procedure:* When approaching the holding fix from anywhere in sector (a), the parallel entry procedure would be to turn to a heading to parallel the holding course outbound on the nonholding side for one minute, turn in the direction of the holding pattern through more than 180 degrees, and return to the holding fix or intercept the holding course inbound.

(b) *Teardrop Procedure:* When approaching the holding fix from anywhere in sector (b), the teardrop entry procedure would be to fly to the fix, turn outbound to a heading for a 30 degree teardrop entry within the pattern (on the holding side) for a period of one minute, then turn in the direction of the holding pattern to intercept the inbound holding course.

(c) *Direct Entry Procedure:* When approaching the holding fix from anywhere in sector (c), the direct entry

procedure would be to fly directly to the fix and turn to follow the holding pattern.

(d) While other entry procedures may enable the aircraft to enter the holding pattern and remain within protected airspace, the parallel, teardrop and direct entries are the procedures for entry and holding recommended by the FAA.

4. *Timing:*

(a) *Inbound Leg:*

(1) At or below 14,000 feet MSL: 1 minute.

(2) Above 14,000 feet MSL: 1½ minutes.

NOTE—THE INITIAL OUTBOUND LEG SHOULD BE FLOWN FOR 1 MINUTE OR 1½ MINUTES (APPROPRIATE TO ALTITUDE). TIMING FOR SUBSEQUENT OUTBOUND LEGS SHOULD BE ADJUSTED, AS NECESSARY, TO ACHIEVE PROPER INBOUND LEG TIME. PILOTS MAY USE ANY NAVIGATIONAL MEANS AVAILABLE; IE. DME, RNAV, ETC., TO ENSURE THE APPROPRIATE INBOUND LEG TIMES.

(b) *Outbound leg* timing begins *over/abeam* the fix, whichever occurs later. If the abeam position cannot be determined, start timing when turn to outbound is completed.

5. *Distance Measuring Equipment (DME), GPS Along Track Distance (ATD).* DME/GPS holding is subject to the same entry and holding procedures except that distances (nautical miles) are used in lieu of time values. The outbound course of the DME/GPS holding pattern is called the outbound leg of the pattern. The controller or the instrument approach procedure chart will specify the length of the outbound leg. The end of the outbound leg is determined by the DME or ATD readout. The holding fix on conventional procedures, or controller defined holding based on a conventional navigation aid with DME, is a specified course or radial and distances are from the DME station for both the inbound and outbound ends of the holding pattern. When flying published GPS overlay or stand alone procedures with distance specified, the holding fix will be a waypoint in the database and the end of the outbound leg will be determined by the ATD. Some GPS overlay and early stand alone procedures may have timing specified. (See FIG 5-3-5, FIG 5-3-6 and FIG 5-3-7.) See paragraph 1-1-19, Global Positioning System (GPS), for requirements and restriction on using GPS for IFR operations.

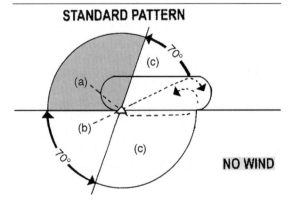

Figure 5-3-4. Holding Pattern Entry Procedures

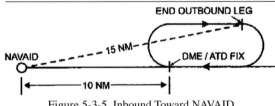

Figure 5-3-5. Inbound Toward NAVAID

NOTE—*WHEN THE INBOUND COURSE IS TOWARD THE NAVAID, THE FIX DISTANCE IS 10 NM, AND THE LEG LENGTH IS 5 NM, THEN THE END OF THE OUTBOUND LEG WILL BE REACHED WHEN THE DME/ATD READS 15 NM.*

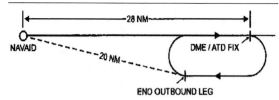

Figure 5-3-6. Inbound Leg Away from NAVAID

NOTE—*WHEN THE INBOUND COURSE IS AWAY FROM THE NAVAID AND THE FIX DISTANCE IS 28 NM, AND THE LEG LENGTH IS 8 NM, THEN THE END OF THE OUTBOUND LEG WILL BE REACHED WHEN THE DME/ATD READS 20 NM.*

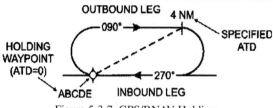

Figure 5-3-7. GPS/RNAV Holding

NOTE—*THE INBOUND COURSE IS ALWAYS TOWARD THE WAYPOINT AND THE ATD IS ZERO AT THE WAYPOINT. THE END OF THE OUTBOUND LEG OF THE HOLDING PATTERN IS REACHED WHEN THE ATD READS THE SPECIFIED DISTANCE.*

6. *Pilot Action:*

(a) Start speed reduction when 3 minutes or less from the holding fix. Cross the holding fix, initially, at or below the maximum holding airspeed.

NOTE—WHEN THE INBOUND COURSE IS TOWARD THE NAVAID AND THE FIX DISTANCE IS 10 NM, AND THE LEG LENGTH IS 5 NM, THEN THE END OF THE OUTBOUND LEG WILL BE REACHED WHEN THE DME READS 15 NM.

NOTE—WHEN THE INBOUND COURSE IS AWAY FROM THE NAVAID AND THE FIX DISTANCE IS 28 NM, AND THE LEG LENGTH IS 8 NM, THEN THE END OF THE OUTBOUND LEG WILL BE REACHED WHEN THE DME READS 20 NM.

(b) *__Make all turns during entry and while holding at:__*

(1) 3 degrees per second, or

(2) 30 degree bank angle, or

(3) 25 degree bank provided a flight director system is used.

NOTE—USE WHICHEVER REQUIRES THE LEAST BANK ANGLE.

(c) Compensate for wind effect primarily by drift correction on the inbound and outbound legs. When outbound, triple the inbound drift correction to avoid major turning adjustments; e.g., if correcting left by 8 degrees when inbound, correct right by 24 degrees when outbound.

(d) Determine entry turn from aircraft heading upon arrival at the holding fix; ±5 degrees in heading is considered to be within allowable good operating limits for determining entry.

(e) Advise ATC immediately when any increased airspeed is necessary, if any, due to turbulence, icing, etc., or if unable to accomplish any part of the holding procedures. When such higher speeds become no longer necessary, operate according to the appropriate published holding speed and notify ATC.

NOTE—AIRSPACE PROTECTION FOR TURBULENT AIR HOLDING IS BASED ON A MAXIMUM OF 280K IAS/MACH 0.8, WHICHEVER IS LOWER. CONSIDERABLE IMPACT ON TRAFFIC FLOW WILL RESULT WHEN TURBULENT AIR HOLDING PATTERNS ARE USED; THUS, PILOT DISCRETION WILL ENSURE THEIR USE IS LIMITED TO BONA FIDE CONDITIONS OR REQUIREMENTS.

7. *Nonstandard Holding Pattern:* Fix end and outbound end turns are made to the left. Entry procedures to a nonstandard pattern are oriented in relation to the 70 degree line on the holding side just as in the standard pattern.

k. When holding at a fix and instructions are received specifying the time of departure from the fix, the pilot should adjust the aircraft's flight path within the limits of the established holding pattern in order to leave the fix at the exact time specified. After departing the holding fix, normal speed is to be resumed with respect to other governing speed requirements, such as terminal area speed limits, specific ATC requests, etc. Where the fix is associated with an instrument approach and timed approaches are in effect, a procedure turn shall not be executed unless the pilot advises ATC, since aircraft holding are expected to proceed inbound on final approach directly from the holding pattern when approach clearance is received.

l. Radar surveillance of outer fix holding pattern airspace areas.

1. Whenever aircraft are holding at an outer fix, ATC will usually provide radar surveillance of the outer fix holding pattern airspace area, or any portion of it, if it is shown on the controller's radarscope.

2. The controller will attempt to detect any holding aircraft that stray outside the holding pattern airspace area and will assist any detected aircraft to return to the assigned airspace area.

NOTE—MANY FACTORS COULD PREVENT ATC FROM PROVIDING THIS ADDITIONAL SERVICE, SUCH AS WORKLOAD, NUMBER OF TARGETS, PRECIPITATION, GROUND CLUTTER, AND RADAR SYSTEM CAPABILITY. THESE CIRCUMSTANCES MAY MAKE IT UNFEASIBLE TO MAINTAIN RADAR IDENTIFICATION OF AIRCRAFT TO DETECT AIRCRAFT STRAYING FROM THE HOLDING PATTERN. THE PROVISION OF THIS SERVICE DEPENDS ENTIRELY UPON WHETHER CONTROLLERS BELIEVE THEY ARE IN A POSITION TO PROVIDE IT AND DOES NOT RELIEVE A PILOT OF THEIR RESPONSIBILITY TO ADHERE TO AN ACCEPTED ATC CLEARANCE.

3. If an aircraft is established in a published holding pattern at an assigned altitude above the published minimum holding altitude and subsequently cleared for the approach, the pilot may descend to the published minimum holding altitude. The holding pattern would only be a segment of the IAP *if* it is published on the instrument procedure chart and is used in lieu of a procedure turn.

m. For those holding patterns where there are no published minimum holding altitudes, the pilot, upon receiving an approach clearance, must maintain the last assigned altitude until leaving the holding pattern and established on the inbound course. Thereafter, the published minimum altitude of the route segment being flown will apply. It is expected that the pilot will be assigned a holding altitude that will permit a normal descent on the inbound course.

Section 4. ARRIVAL PROCEDURES

5-4-1. STANDARD TERMINAL ARRIVAL (STAR), AREA NAVIGATION (RNAV) STAR, AND FLIGHT MANAGEMENT SYSTEM PROCEDURES (FMSP) FOR ARRIVALS

a. A STAR is an ATC coded IFR arrival route established for application to arriving IFR aircraft destined for certain airports. RNAV STAR/FMSP procedures for arrivals serve the same purpose but are only used by aircraft equipped with FMS or GPS. The purpose of both is to simplify clearance delivery procedures and facilitate transition between en route and instrument approach procedures.

1. STAR/RNAV STAR/FMSP procedures may have mandatory speeds and/or crossing altitudes published. Other STARS may have planning information depicted to inform pilots what clearances or restrictions to "expect." "Expect" altitudes/speeds are not considered STAR/RNAV STAR/FMSP procedures crossing restrictions unless verbally issued by ATC.

NOTE—THE "EXPECT" ALTITUDES/SPEEDS ARE PUBLISHED SO THAT PILOTS MAY HAVE THE INFORMATION FOR PLANNING PURPOSES. THESE ALTITUDES/SPEEDS SHALL NOT BE USED IN THE EVENT OF LOST COMMUNICATIONS UNLESS ATC HAS SPECIFICALLY ADVISED THE PILOT TO EXPECT THESE ALTITUDES/SPEEDS AS PART OF A FURTHER CLEARANCE.

REFERENCE—14 CFR SECTION 91.185(c)(2)(iii).

2. Pilots navigating on STAR/RNAV STAR/ FMSP procedures shall maintain last assigned altitude until receiving authorization to descend so as to comply with all published/issued restrictions. This authorization will contain the phraseology **"DESCEND VIA."**

(a) Clearance to "descend via" authorizes pilots to:

(1) Vertically and laterally navigate on a STAR/RNAV STAR/FMSP

(2) When cleared to a waypoint depicted on a STAR/ RNAV STAR/FMSP, to descend from a previously assigned altitude at pilot's discretion to the altitude depicted for that waypoint, and once established on the depicted arrival, to navigate laterally and vertically to meet all published restrictions.

NOTE—
☐ AIR TRAFFIC IS RESPONSIBLE FOR OBSTACLE CLEARANCE WHEN ISSUING A "DESCEND VIA" INSTRUCTION TO THE PILOT. THE DESCEND VIA IS USED IN CONJUNCTION WITH STARS/RNAV STARS/FMSPS TO REDUCE PHRASEOLOGY BY NOT REQUIRING THE CONTROLLER TO RESTATE THE ALTITUDE AT THE NEXT WAYPOINT/FIX TO WHICH THE PILOT HAS BEEN CLEARED.

☐ AIR TRAFFIC WILL ASSIGN AN ALTITUDE TO CROSS THE WAYPOINT/FIX, IF NO ALTITUDE IS DEPICTED AT THE WAYPOINT/FIX FOR AIRCRAFT ON A DIRECT ROUTING TO A STAR/RNAV STAR/FMSP

☐ MINIMUM EN ROUTE ALTITUDES (MEA) ARE NOT CONSIDERED RESTRICTIONS; HOWEVER, PILOTS ARE EXPECTED TO REMAIN ABOVE MEAS.

EXAMPLE—
☐ *LATERAL/ROUTING CLEARANCE ONLY:*
"CLEARED HADLY ONE ARRIVAL."
☐ *ROUTING WITH ASSIGNED ALTITUDE:*
"CLEARED HADLY ONE ARRIVAL, DESCEND AND MAINTAIN FLIGHT LEVEL TWO FOUR ZERO."
"CLEARED HADLY ONE ARRIVAL, DESCEND AT PILOT'S DISCRETION, MAINTAIN FLIGHT LEVEL TWO FOUR ZERO."
☐ *LATERAL/ROUTING AND VERTICAL NAVIGATION CLEARANCE:*
"DESCEND VIA THE CIVIT ONE ARRIVAL."
"DESCEND VIA THE CIVIT ONE ARRIVAL, EXCEPT, CROSSARNES AT OR ABOVE ONE ONE THOUSAND."
☐ *LATERAL/ROUTING AND VERTICAL NAVIGATION CLEARANCE WHEN ASSIGNING ALTITUDE NOT PUBLISHED ON PROCEDURE:*
"DESCEND VIA THE HARIS ONE ARRIVAL, EXCEPT AFTER BRUNO, MAINTAIN ONE ZERO THOUSAND."
"DESCEND VIA THE HARIS ONE ARRIVAL, EXCEPT CROSS BRUNO AT ONE THREE THOUSAND THEN MAINTAIN ONE ZERO THOUSAND."
☐ *DIRECT ROUTING TO INTERCEPT A STAR/RNAV STAR/FMSP AND VERTICAL NAVIGATION CLEARANCE:*
"PROCEED DIRECT MAHEM, DESCEND VIA MAHEM ONE ARRIVAL."
"PROCEED DIRECT LUXOR, CROSS LUXOR AT OR ABOVE FLIGHT LEVEL TWO ZERO ZERO, THEN DESCEND VIA THE KSINO ONE ARRIVAL."

NOTE—
1. IN EXAMPLE 2, PILOTS ARE EXPECTED TO DESCEND TO FL 240 AS DIRECTED, AND MAINTAIN FL 240 UNTIL CLEARED FOR FURTHER VERTICAL NAVIGATION WITH A NEWLY ASSIGNED ALTITUDE OR A "DESCEND VIA" CLEARANCE.
2. IN EXAMPLE 4, THE AIRCRAFT SHOULD TRACK LATERALLY AND VERTICALLY ON THE HARIS ONE ARRIVAL AND SHOULD DESCEND SO AS TO COMPLY WITH ALL SPEED AND ALTITUDE RESTRICTIONS UNTIL REACHING BRUNOAND THEN MAINTAIN 10,000. UPON REACHING 10,000, AIRCRAFT SHOULD MAINTAIN 10,000 UNTIL CLEARED BY ATC TO CONTINUE TO DESCEND.

(b) Pilots cleared for vertical navigation using the phraseology **"DESCEND VIA"** shall inform ATC upon initial contact with a new frequency.

EXAMPLE—
"DELTA ONE TWENTY ONE DESCENDING VIA THE CIVIT ONE ARRIVAL."

b. Pilots of IFR aircraft destined to locations for which STARs have been published may be issued a clearance containing a STAR whenever ATC deems it appropriate.

c. Use of STAR's requires pilot possession of at least the approved chart. RNAV STAR's must be retrievable by the procedure name from the aircraft database and conform to charted procedure. As with any ATC clearance or portion thereof, it is the responsibility of each pilot to accept or refuse an issued STAR. Pilots should notify ATC if they do not wish to use a STAR by placing "NO STAR" in the remarks section of the flight plan or by the less desirable method of verbally stating the same to ATC.

d. STAR charts are published in the *Terminal Procedures Publication (TPP)* and are available on subscription from the National Aeronautical Charting Office.

e. RNAV STAR.
Procedure Design. An RNAV STAR is designed using criteria similar to Level 2 DP.

1. There are two general types of public RNAV STARS:

(a) Type A. These procedures require system performance currently met by GPS, DME/DME, or DME/DME/IRU RNAV systems that satisfy the criteria discussed in AC 90-100, U.S. Terminal and En Route Area Navigation (RNAV) Operations. Type A terminal procedures require the aircraft's track keeping accuracy remain bounded by f 2 NM for 95% of the total flight time.

NOTE—IF NOT EQUIPPED WITH *GPS* (OR FOR MULTI-SENSOR SYSTEMS WITH *GPS* WHICH DO NOT ALERT UPON LOSS OF *GPS*), AIRCRAFT MUST BE CAPABLE OF NAVIGATION SYSTEM UPDATING USING *DME/DME* OR *DME/DME/IRU* FOR TYPE *A* STARs.

(b) Type B. These procedures require system performance currently met by GPS or DME/DME/IRU RNAV systems that satisfy the criteria discussed in AC 90-100. Type B procedures may require the aircraft's track keeping accuracy remain bounded by ±1 NM for 95% of the total flight time.

NOTE—IF NOT EQUIPPED WITH *GPS* (OR FOR MULTI-SENSOR SYSTEMS WITH *GPS* WHICH DO NOT ALERT UPON LOSS OF *GPS*), AIRCRAFT MUST BE CAPABLE OF NAVIGATION SYSTEM UPDATING USING *DME/DME/IRU* FOR TYPE *B* STARS.

2. For procedures requiring GPS, if the navigation system does not automatically alert the flight crew of a loss of GPS, the operator must develop procedures to verify correct GPS operation.

5-4-2. LOCAL FLOW TRAFFIC MANAGEMENT PROGRAM

a. This program is a continuing effort by the FAA to enhance safety, minimize the impact of aircraft noise and conserve aviation fuel. The enhancement of safety and reduction of noise is achieved in this program by minimizing low altitude maneuvering of arriving turbojet and turboprop aircraft weighing more than 12,500 pounds and, by permitting departure aircraft to climb to higher altitudes sooner, as arrivals are operating at higher altitudes at the points where their flight paths cross. The application of these procedures also reduces exposure time between controlled aircraft and uncontrolled aircraft at the lower altitudes in and around the terminal environment. Fuel conservation is accomplished by absorbing any necessary arrival delays for aircraft included in this program operating at the higher and more fuel efficient altitudes.

b. A fuel efficient descent is basically an uninterrupted descent (except where level flight is required for speed adjustment) from cruising altitude to the point when level flight is necessary for the pilot to stabilize the aircraft on final approach. The procedure for a fuel efficient descent is based on an altitude loss which is most efficient for the majority of aircraft being served. This will generally result in a descent gradient window of 250–350 feet per nautical mile.

c. When crossing altitudes and speed restrictions are issued verbally or are depicted on a chart, ATC will expect the pilot to descend first to the crossing altitude and then reduce speed. Verbal clearances for descent will normally permit an uninterrupted descent in accordance with the procedure as described in paragraph b. above. Acceptance of a charted fuel efficient descent (Runway Profile Descent) clearance requires the pilot to adhere to the altitudes, speeds, and headings depicted on the charts unless otherwise instructed by ATC. PILOTS RECEIVING A CLEARANCE FOR A FUEL EFFICIENT DESCENT ARE EXPECTED TO ADVISE ATC IF THEY DO NOT HAVE RUNWAY PROFILE DESCENT CHARTS PUBLISHED FOR THAT AIRPORT OR ARE UNABLE TO COMPLY WITH THE CLEARANCE.

5-4-3. APPROACH CONTROL

a. Approach control is responsible for controlling all instrument flight operating within its area of responsibility. Approach control may serve one or more airfields, and control is exercised primarily by direct pilot and controller communications. Prior to arriving at the destination radio facility, instructions will be received from ARTCC to contact approach control on a specified frequency.

b. Radar Approach Control:
1. Where radar is approved for approach control service, it is used not only for radar approaches (ASR and PAR) but is also used to provide vectors in conjunction with published nonradar approaches based on radio NAVAIDs (ILS, MLS, VOR, NDB, TACAN). Radar vectors can provide course guidance and expedite traffic to the final approach course of any established IAP or to the traffic pattern for a visual approach. Approach control facilities that provide this radar service will operate in the following manner:

(a) Arriving aircraft are either cleared to an outer fix most appropriate to the route being flown with vertical separation and, if required, given holding information or, when radar handoffs are affected between the ARTCC and approach control, or between two approach control facilities, aircraft are cleared to the airport or to a fix so located that the handoff will be completed prior to the time the aircraft reaches the fix. When radar handoffs are utilized, successive arriving flights may be handed off to approach control with radar separation in lieu of vertical separation.

(b) After release to approach control, aircraft are vectored to the final approach course (ILS, MLS, VOR, ADF, etc.). Radar vectors and altitude or Flight Levels will be issued as required for spacing and separating aircraft. *Therefore, pilots must not deviate from the headings issued by approach control.* Aircraft will normally be informed when it is necessary to vector across the final approach course for spacing or other reasons. If approach course crossing is imminent and the pilot has not been informed that the aircraft will be vectored across the final approach course, the pilot should query the controller.

(c) The pilot is not expected to turn inbound on the final approach course unless an approach clearance has been issued. This clearance will normally be issued with the final vector for interception of the final approach course, and the vector will be such as to enable the pilot to establish the aircraft on the final approach course prior to reaching the final approach fix.

(d) In the case of aircraft already inbound on the final approach course, approach clearance will be issued prior to the aircraft reaching the final approach fix. When established inbound on the final approach course, radar separation will be maintained and the pilot will be expected to complete the approach utilizing the approach aid designated in the clearance (ILS, MLS, VOR, radio beacons, etc.) as the primary means of navigation. Therefore, once established on the final approach course, pilots must not deviate from it unless a clearance to do so is received from ATC.

(e) After passing the final approach fix on final approach, aircraft are expected to continue inbound on the final approach course and complete the approach or effect the missed approach procedure published for that airport.

2. ARTCCs are approved for and may provide approach control services to specific airports. The radar systems used by these centers do not provide the same precision as an airport surveillance radar (ASR)/precision approach radar (PAR) used by approach control facilities and towers, and the update rate is not as fast. Therefore, pilots may be requested to report established on the final approach course.

3. Whether aircraft are vectored to the appropriate final approach course or provide their own navigation on published routes to it, radar service is automatically terminated when the landing is completed or when instructed to change to advisory frequency at uncontrolled airports, whichever occurs first.

5-4-4. ADVANCE INFORMATION ON INSTRUMENT APPROACH

a. When landing at airports with approach control services and where two or more IAPs are published pilots will be provided in advance of their arrival with the type of approach to expect or that they may be vectored for a visual approach. This information will be broadcast either by a controller or on ATIS. It will not be furnished when the visibility is three miles or better and the ceiling is at or above the highest initial approach altitude established for any low altitude IAP for the airport.

b. The purpose of this information is to aid the pilot in planning arrival actions; however, it is not an ATC clearance or commitment and is subject to change. Pilots should bear in mind that fluctuating weather, shifting winds, blocked runway, etc., are conditions which may result in changes to approach information previously received. It is important that pilots advise ATC immediately they are unable to execute the approach ATC advised will be used, or if they prefer another type of approach.

c. Aircraft destined to uncontrolled airports, which have automated weather data with broadcast capability, should monitor the ASOS/AWOS frequency to ascertain the current weather for the airport. The pilot shall advise ATC when he/she has received the broadcast weather and state his/her intentions.

NOTE—

⊡ ASOS/AWOS SHOULD BE SET TO PROVIDE ONE-MINUTE BROADCAST WEATHER UPDATES AT UNCONTROLLED AIRPORTS THAT ARE WITHOUT WEATHER BROADCAST CAPABILITY BY A HUMAN OBSERVER:

② CONTROLLERS WILL CONSIDER THE LONG LINE DISSEMINATED WEATHER FROM AN AUTOMATED WEATHER SYSTEM AT AN UNCONTROLLED AIRPORT AS TREND AND PLANNING INFORMATION ONLY AND WILL RELY ON THE PILOT FOR CURRENT WEATHER INFORMATION FOR THE AIRPORT. IF THE PILOT IS UNABLE TO RECEIVE THE CURRENT BROADCAST WEATHER, THE LAST LONG LINE DISSEMINATED WEATHER WILL BE ISSUED TO THE PILOT. WHEN RECEIVING IFR SERVICES, THE PILOT/AIRCRAFT OPERATOR IS RESPONSIBLE FOR DETERMINING IF WEATHER/VISIBILITY IS ADEQUATE FOR APPROACH/LANDING.

d. When making an IFR approach to an airport not served by a tower or FSS, after ATC advises "CHANGE TO ADVISORY FREQUENCY APPROVED" you should broadcast your intentions, including the type of approach being executed, your position, and when over the final approach fix inbound (nonprecision approach) or when over the outer marker or fix used in lieu of the outer marker inbound (precision approach). Continue to monitor the appropriate frequency (UNICOM, etc.) for reports from other pilots.

5-4-5. INSTRUMENT APPROACH PROCEDURE CHARTS

(a) 14 CFR Section 91.175(a), Instrument approaches to civil airports, requires the use of SIAP's prescribed for the airport in 14 CFR Part 97 unless otherwise authorized by the Administrator (including ATC). If there are military procedures published at a civil airport, aircraft operating under 14 CFR Section 91 must use the civil procedure(s). Civil procedures are defined with "FAA" in parenthesis; e.g., (FAA), at the top, center of the procedure chart. DOD procedures are defined using the abbreviation of the applicable military service in parenthesis; e.g., (USAF), (USN), (USA). 14 CFR Section 91.175(g), Military airports, requires civil pilots flying into or out of military airports to comply with the IAP's and takeoff and landing minimums prescribed by the authority having jurisdiction at those airports. Unless an emergency exists, civil aircraft operating at military airports normally require advance authorization, commonly referred to as "Prior Permission Required" or "PPR." Information on obtaining a PPR for a particular military airport can be found in the Airport Facility Directory.

NOTE—CIVIL AIRCRAFT MAY CONDUCT PRACTICE VFR APPROACHES USING DOD INSTRUMENT APPROACH PROCEDURES WHEN APPROVED BY THE AIR TRAFFIC CONTROLLER.

1. IAP's (standard and special, civil and military) are based on joint civil and military criteria contained in the U.S. Standard for TERPS. The design of IAPs based on criteria contained in TERPS, takes into account the interrelationship between airports, facilities, and the surrounding environment, terrain, obstacles, noise sensitivity, etc. Appropriate altitudes, courses, headings, distances, and other limitations are specified and, once approved, the procedures are published and distributed by government and commercial cartographers as instrument approach charts.

2. Not all IAP's are published in chart form. Radar IAP's are established where requirements and facilities exist but they are printed in tabular form in appropriate U.S. Government Flight Information Publications.

3. The navigation equipment required to join and fly an instrument approach procedure is indicated by the title of the procedure and notes on the chart.

(a) Straight-in LAP's are identified by the navigational system providing the final approach guidance and the runway to which the approach is aligned (e.g. VOR RWY 13). Circling only approaches are identified by the navigational system providing final approach guidance and a letter (e.g., VOR A). More than one navigational system separated by a slash indicates that more than one type of equipment must be used to execute the final approach (e.g., VOR/DME RWY 31). More than one navigational system separated by the word "or" indicates either type of equipment may be used to execute the final approach (e.g., VOR or GPS RWY 15').

(b) In some cases, other types of navigation systems including radar may be required to execute other portions of the approach or to navigate to the IAF (e.g., an NDB procedure turn to an ILS, an NDB in the missed approach, or radar required to join the procedure or identify a fix). When radar or other equipment is required for procedure entry from the en route environment, a note will be charted in the planview of the approach procedure chart (e.g., RADAR REQUIRED or ADF REQUIRED). When radar or other equipment is required on portions of the procedure outside the final approach segment, including the missed approach, a note will be charted in the notes box of the pilot briefing portion of the approach chart (e.g., RADAR REQUIRED or DME REQUIRED). Notes are not charted when VOR is required outside the final approach segment. Pilots should ensure that the aircraft is equipped with the required NAVAID(s) in order to execute the approach, including the missed approach. In some cases, other types of navigation systems may be required to execute other portions of the approach (e.g., an NDB procedure turn to an ILS or an NDB in the missed approach). Pilots should ensure that the aircraft is equipped with the required NAVAID(s) in order to execute the approach, including the missed approach.

(c) The FAA has initiated a program to provide a new notation for LOC approaches when charted on an ILS approach requiring other navigational aids to fly the final approach course. The LOC minimums will be annotated with the NAVAID required e.g., "DME Required" or "RADAR Required." During the transition period, ILS approaches will still exist without the annotation.

(d) The naming of multiple approaches of the same type to the same runway is also changing. Multiple approaches with the same guidance will be annotated with an alphabetical suffix beginning at the end of the alphabet and working backwards for subsequent procedures (ILS Z RWY 28, ILS Y RWY 28, etc.). The existing annotations such as ILS 2 RWY 28 or Silver ILS RWY 28 will be phased out and replaced with the new designation. The Cat II and Cat III designations are used to differentiate between multiple ILS's to the same runway unless there are multiples of the same type.

(e) WAAS (LPV and LNAV/VNAV), and GPS (LNAV) approach procedures will be charted as RNAV (GPS) RWY (Number); e.g., RNAV (GPS) RWY 21. VOR/DME RNAV approaches will continue to be identified as VOR/DME RNAV RWY (Number); e.g., VOR/DME RNAV RWY 21. VOR/DME RNAV procedures which can be flown by GPS will be annotated with "or GPS" e.g., VOR/DME RNAV or GPS RWY 31.

4. Approach minimums are based on the local altimeter setting for that airport, unless annotated otherwise; e.g., Oklahoma City/Will Rogers World approaches are based on having a Will Rogers World altimeter setting. When a different altimeter source is required, or more than one source is authorized, it will be annotated on the approach chart; e.g., use Sidney altimeter setting, if not received, use Scottsbluff altimeter setting. Approach minimums may be raised when a nonlocal altimeter source is authorized. When more than one altimeter source is authorized, and the minima are different, they will be shown by separate lines in the approach minima box or a note; e.g., use Manhattan altimeter setting; when not available use Salina altimeter setting and increase all MDA's 40 feet. When the altimeter must be obtained from a source other than air traffic a note will indicate the source; e.g., Obtain focal altimeter setting on CTAF. When the altimeter setting(s) on which the approach is based is not available, the approach is not authorized. Baro-VNAV must be flown using the local altimeter setting only. Where no local altimeter is available, the LNAV/VNAV line will still be published for use by WAAS receivers with a note that Baro-VNAV is not authorized. When a local and at least one other altimeter setting source is authorized and the local altimeter is not available Baro-VNAV is not authorized; however, the LNAV/VNAV minima can still be used by WAAS receivers using the alternate altimeter setting source.

5. A pilot adhering to the altitudes, flight paths, and weather minimums depicted on the IAP chart or vectors and altitudes issued by the radar controller, is assured of terrain and obstruction clearance and runway or airport alignment during approach for landing.

6. IAP's are designed to provide an IFR descent from the en route environment to a point where a safe landing can be made. They are prescribed and approved by appropriate civil or military authority to ensure a safe descent during instrument flight conditions at a specific airport. It is important that pilots understand these procedures and their use prior to attempting to fly instrument approaches.

7. TERPS criteria are provided for the following types of instrument approach procedures:

(a) Precision Approach (PA). An instrument approach based on a navigation system that provides course and glidepath deviation information meeting the precision standards of ICAO Annex 10. For example, PAR, ILS, and GLS are precision approaches.

(b) Approach with Vertical Guidance (APV). An instrument approach based on a navigation system that is not required to meet the precision approach standards of ICAO Annex 10 but provides course and glidepath deviation information. For example, Baro-VNAV, LDA with glidepath, LNAV/ VNAV and LPV are APV approaches.

(c) Nonprecision Approach (NPA). An instrument approach based on a navigation system which provides course deviation information, but no glidepath deviation information. For example, VOR, NDB and LNAV. As noted in subparagraph h, Vertical Descent Angle (VDA) on Nonprecision Approaches, some approach procedures may provide a Vertical Descent Angle as an aid in flying a stabilized approach, without requiring its use in order to fly the procedure. This does not make the approach an APV procedure, since it must still be flown to an MDA and has not been evaluated with a glidepath.

b. The method used to depict prescribed altitude; on instrument approach charts differs according to techniques employed by different chart publishers Prescribed altitudes

may be depicted in four different configurations: minimum, maximum, mandatory and recommended. The U.S. Government distribute: charts produced by National Imagery and Mapping Agency (NIMA) and FAA. Altitudes are depicted of these charts in the profile view with underscore, overscore, both or none to identify them as minimum maximum, mandatory or recommended.

1. Minimum altitude will be depicted with the altitude value underscored. Aircraft are required to maintain altitude at or above the depicted value, e.g., 3000.

2. Maximum altitude will be depicted with the altitude value overscored. Aircraft are required to maintain altitude at or below the depicted value, e.g., 4000.

3. Mandatory altitude will be depicted with the altitude value both underscored and overscored. Aircraft are required to maintain altitude at the depicted value, e.g., 5000.

4. Recommended altitude will be depicted with no overscore or underscore. These altitudes are depicted for descent planning, e.g., 6000.

NOTE—PILOTS ARE CAUTIONED TO ADHERE TO ALTITUDES AS PRESCRIBED BECAUSE, IN CERTAIN INSTANCES, THEY MAY BE USED AS THE BASIS FOR VERTICAL SEPARATION OF AIRCRAFT BY ATC. WHEN A DEPICTED ALTITUDE IS SPECIFIED IN THE ATC CLEARANCE, THAT ALTITUDE BECOMES MANDATORY AS DEFINED ABOVE.

c. Minimum Safe/Sector Altitudes (MSA) are published for emergency use on IAP charts. For conventional navigation systems, the MSA is normally based on the primary omnidirectional facility on which the IAP is predicated. The MSA depiction on the approach chart contains the facility identifier of the NAVAID used to determine the MSA altitudes. For RNAV approaches, the MSA is based on the runway waypoint (RWY WP) for straight-in approaches, or the airport waypoint (APT WP) for circling approaches. For GPS approaches, the MSA center will be the missed approach waypoint (MAWP). MSAs are expressed in feet above mean sea level and normally have a 25 NM radius; however, this radius may be expanded to 30 NM if necessary to encompass the airport landing surfaces. Ideally, a single sector altitude is established and depicted on the plan view of approach charts; however, when necessary to obtain relief from obstructions, the area may be further sectored and as many as four MSAs established. When established, sectors may be no less than 90° in spread. MSAs provide 1,000 feet clearance over all obstructions but do not necessarily assure acceptable navigation signal coverage.

d. Terminal Arrival Area (TAA)

1. The objective of the TAA is to provide a seamless transition from the en route structure to the terminal environment for arriving aircraft equipped with Flight Management System (FMS) and/or Global Positioning System (GPS) navigational equipment. The underlying instrument approach procedure is an area navigation (RNAV) procedure described in this section. The TAA provides the pilot and air traffic controller with a very efficient method for routing traffic into the terminal environment with little required air traffic control interface, and with minimum altitudes depicted that provide standard obstacle clearance compatible with the instrument procedure associated with it. The TAA will not be found on all RNAV procedures, particularly in areas of heavy concentration of air traffic. When the TAA is published, it replaces the MSA for that approach procedure. See FIG 5-4-9 for a depiction of a RNAV approach chart with a TAA.

2. The RNAV procedure underlying the TAA will be the "T" design (also called the "Basic T"), or a modification of the "T." The "T" design incorporates from one to three IAF's; an intermediate fix (IF) that serves as a dual purpose IF (IAF); a final approach fix (FAF), and a missed approach point (MAP) usually located at the runway threshold. The three IAF's are normally aligned in a straight line perpendicular to the intermediate course, which is an extension of the final course leading to the runway, forming a "T." The initial segment is normally from 3–6 NM in length; the intermediate 5–7 NM, and the final segment 5 NM. Specific segment length may be varied to accomodate specific aircraft categories for which the procedure is designed. However, the published segment lengths will reflect the highest category of aircraft normally expected to use the procedure.

(a) A standard racetrack holding pattern may be provided at the center IAF, and if present may be necessary for course reversal and for altitude adjustment for entry into the procedure. In the latter case, the pattern provides an extended distance for the descent required by the procedure. Depiction of this pattern in U.S. Government publications will utilize the "hold–in–lieu–of–PT" holding pattern symbol.

(b) The published procedure will be annotated to indicate when the course reversal is not necessary when flying within a particular TAA area; e.g., "NoPT." Otherwise, the pilot is expected to execute the course reversal under the provisions of 14 CFR Part 91.175. The pilot may elect to use the course reversal pattern when it is not required by the procedure, but must inform air traffic control and receive clearance to do so. (See FIG 5-4-1 and FIG 5-4-2).

3. The "T" design may be modified by the procedure designers where required by terrain or air traffic control considerations. For instance, the "T" design may appear more like a regularly or irregularly shaped "Y", or may even have one or both outboard IAF's eliminated resulting in an upside down "L" or an "I" configuration. (See FIG 5-4-3 and FIG 5-4-10). Further, the leg lengths associated with the outboard IAF's may differ. (See FIG 5-4-5 and FIG 5-4-6).

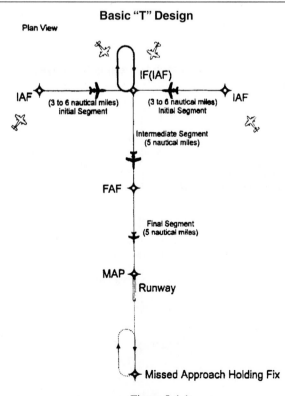

Figure 5-4-1.

Basic "T" Design

Plan View

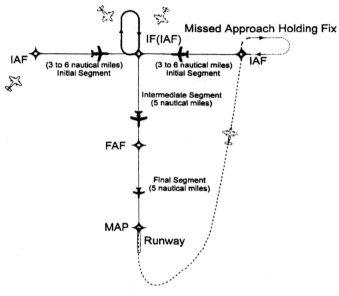

Figure 5-4-2.

Modified "T" Approach to Parallel Runways

Plan View

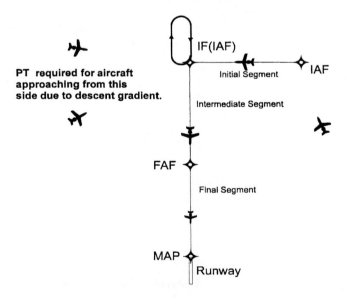

Figure 5-4-3.

Modified Basic "T" Approach to Parallel Runways

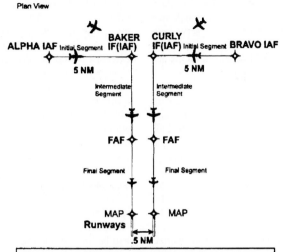

The normal "T" IAF's serve all parallel runways. Each runway will require a separate IF(IAF). Only one initial, intermediate and final segment combination will be depicted on the approach chart.

Figure 5-4-4.

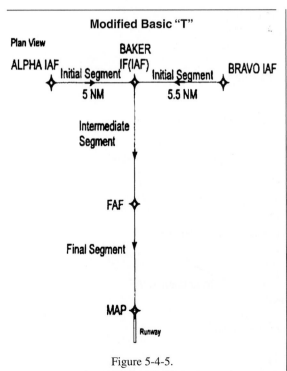

Figure 5-4-5.

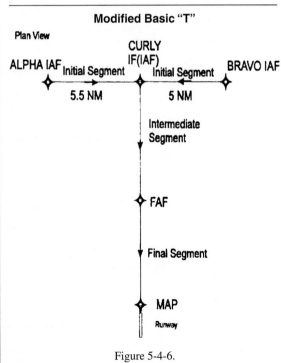

Figure 5-4-6.

4. Another modification of the "T" design may be found at airports with parallel runway configurations. Each parallel runway may be served by its own "T" IAF, IF (IAF), and FAF combination, resulting in parallel fi-nal approach courses. (See FIG 5-4-4). Common IAF's may serve both runways; however, only the intermediate and final approach segments for the landing runway will be shown on the approach chart. (See FIG 5-4-5 and FIG 5-4-6).

5. The standard TAA consists of three areas defined by the extension of the IAF legs and the intermediate segment course. These areas are called the straight-in, left-base, and right-base areas. (See FIG 5-4-7). TAA area lateral boundaries are identified by magnetic courses TO the IF (IAF). The straight-in area can be further divided into pie-shaped sectors with the boundaries identified by magnetic courses TO the IF (IAF), and may contain stepdown sections defined by arcs based on RNAV distances (DME or ATD) from the IF (IAF). The right/left–base areas can only be subdivided using arcs based on RNAV distances from the IAF's for those areas. Minimum MSL altitudes are charted within each of these defined areas/subdivisions that provide at least 1,000 feet of obstacle clearance, or more as necessary in mountainous areas.

(a) Prior to arriving at the TAA boundary, the pilot can determine which area of the TAA the aircraft will enter by selecting the IF (IAF) to determine the magnetic bearing TO the center IF (IAF). That bearing should then be compared with the published bearings that define the lateral boundaries of the TAA areas. Using the end IAF's may give a false indication of which area the aircraft will enter. This is critical when approaching the TAA near the extended boundary between the left and right-base areas, especially where these areas contain different minimum altitude requirements.

(b) Pilots entering the TAA, and cleared by air traffic control, are expected to proceed directly to the IAF associated with that area of the TAA at the altitude depicted, unless otherwise cleared by air traffic control. Pilots entering the TAA with two-way radio communications failure (14 CFR Section 91.185, IFR Operations: Two-way Radio Communications Failure), must maintain the highest altitude prescribed by Section 91.185(c)(2) until arriving at the appropriate IAF.

(c) Depiction of the TAA on U.S. Government charts will be through the use of icons located in the plan view out-side the depiction of the actual approach procedure. (See FIG 5-4-9). Use of icons is necessary to avoid obscuring any portion of the "T" procedure (altitudes, courses, minimum altitudes, etc.). The icon for each TAA area will be located and oriented on the plan view with

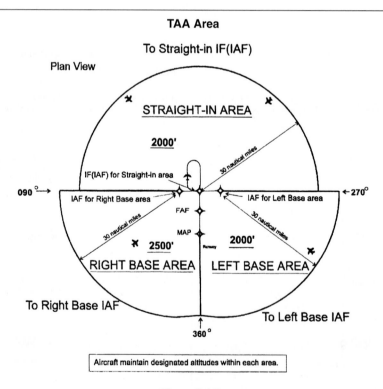

Figure 5-4-7.

Sectored TAA Areas

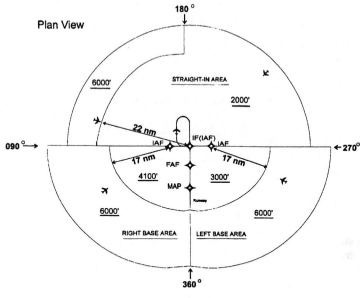

Figure 5-4-8.

respect to the direction of arrival to the approach procedure, and will show all TAA minimum altitudes and sector/radius subdivisions for that area. The IAF for each area of the TAA is included on the icon where it appears on the approach, to help the pilot orient the icon to the approach procedure. The IAF name and the distance of the TAA area boundary from the IAF are included on the outside arc of the TAA area icon. Examples here are shown with the TAA around the approach to aid pilots in visualizing how the TAA corresponds to the approach and should not be confused with the actual approach chart depiction.

(d) Each waypoint on the "T", except the missed approach waypoint, is assigned a pronounceable 5-character name used in air traffic control communications, and which is found in the RNAV databases for the procedure. The missed approach waypoint is assigned a pronounceable name when it is not located at the runway threshold.

6. Once cleared to fly the TAA, pilots are expected to obey minimum altitudes depicted within the TAA icons, unless instructed otherwise by air traffic control. In FIG 5-4-8, pilots within the left or right-base areas are expected to maintain a minimum altitude of 6,000 feet until within 17 NM of the associated IAF. After crossing the 17 NM arc, descent is authorized to the lower charted altitudes. Pilots approaching from the northwest are expected to maintain a minimum altitude of 6,000 feet, and when within

22 NM of the IF (IAF), descend to a minimum altitude of 2,000 feet MSL until reaching the IF (IAF).

7. Just as the underlying "T" approach procedure may be modified in shape, the TAA may contain modifications to the defined area shapes and sizes. Some areas may even be eliminated, with other areas expanded as needed. FIG 5-4-10 is an example of a design limitation where a course reversal is necessary when approaching the IF (IAF) from certain directions due to the amount of turn required at the IF (IAF). Design criteria require a course reversal whenever this turn exceeds 120 degrees. In this generalized example, pilots approaching on a bearing TO the IF (IAF) from 300° clockwise through 060° are expected to execute a course reversal. The term "NoPT" will be annotated on the boundary of the TAA icon for the other portion of the TAA.

8. FIG 5-4-11 depicts another TAA modification that pilots may encounter. In this generalized example, the right-base area has been eliminated. Pilots operating within the TAA between 360° clockwise to 060° bearing TO the IF (IAF) are expected to execute the course reversal in order to properly align the aircraft for entry onto the intermediate segment Aircraft operating in all other areas from 060° clockwise to 360° bearing TO the IF (IAF) need not perform the course reversal, and the term "NoPT" 'will be annotated on the TAA boundary of the icon in these areas. TAA's are no longer being produced with sections

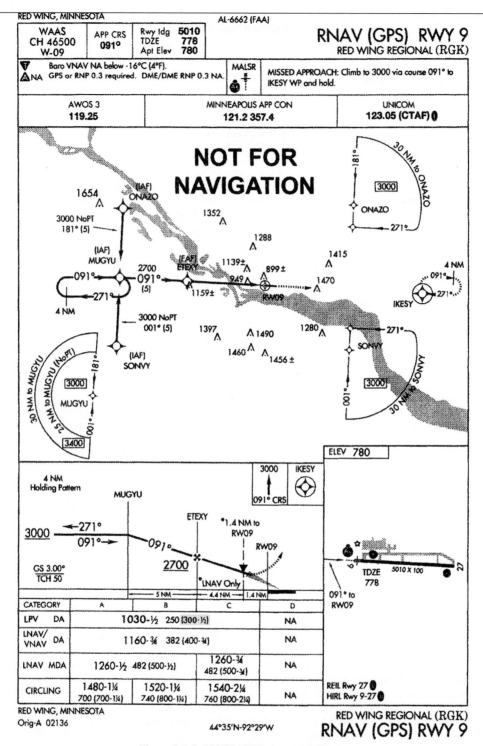

Figure 5-4-9. RNAV (GPS) Approach Chart

NOTE—This chart has been modified to depict concepts and may not reflect actual approach minima.

TAA with Left and Right Base Areas Eliminated

Plan View

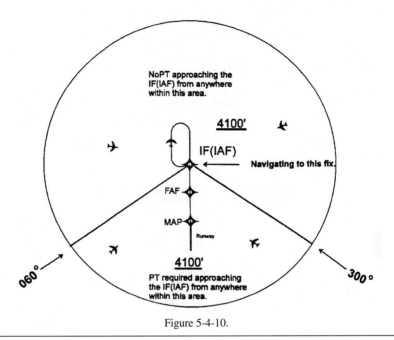

NoPT approaching the IF(IAF) from anywhere within this area.

4100'

IF(IAF)

Navigating to this fix.

FAF

MAP

Runway

4100'

PT required approaching the IF(IAF) from anywhere within this area.

060°

300°

Figure 5-4-10.

TAA with Right Base Eliminated

Plan View

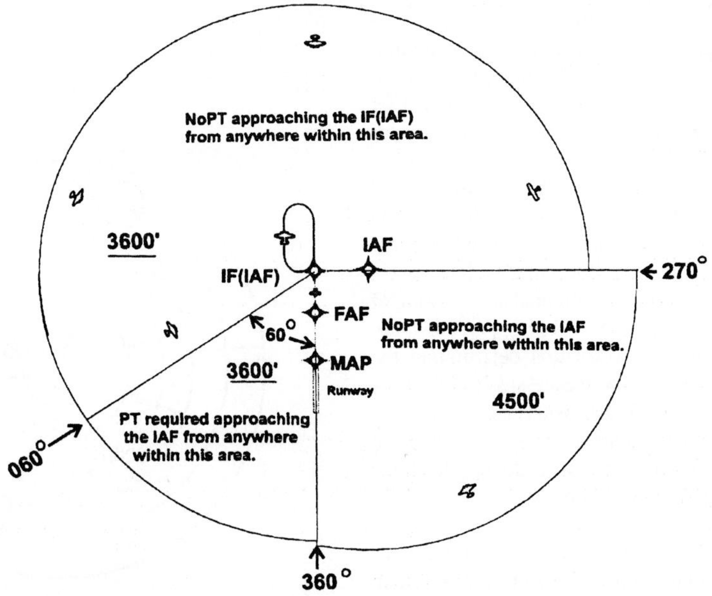

NoPT approaching the IF(IAF) from anywhere within this area.

3600'

IAF

270°

IF(IAF)

FAF

60°

NoPT approaching the IAF from anywhere within this area.

3600'

MAP

Runway

4500'

PT required approaching the IAF from anywhere within this area.

060°

360°

Figure 5-4-11.

Examples of a TAA with Feeders from Airway

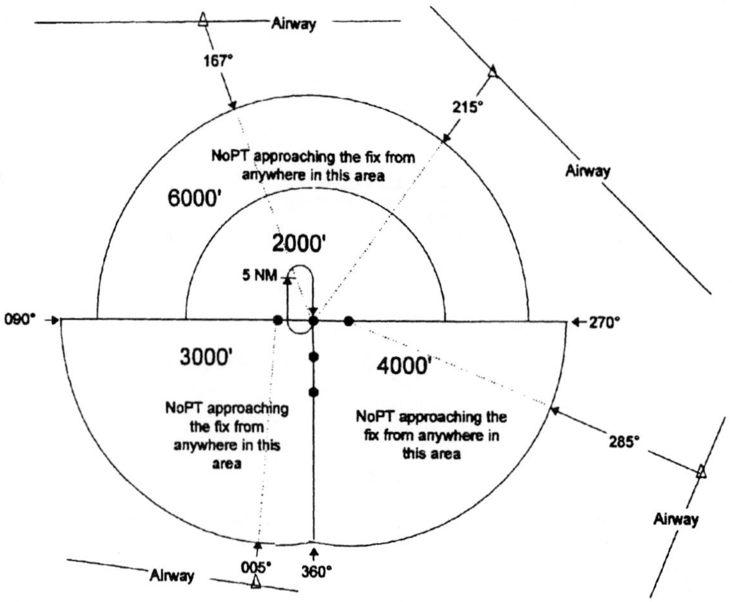

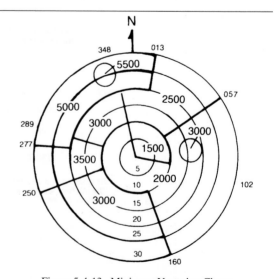

Figure 5-4-12.

removed; however, some may still exist on previously published procedures.

9. When an airway does not cross the lateral TAA boundaries, a feeder route will be established to provide a transition from the en route structure to the appropraite IAF. Each feeder route will terminate at the TAA boundary, and will be aligned along a path pointing to the associated IAF. Pilots should descend to the TAA altitude after crossing the TAA boundary and cleared by air traffic control. (See FIG 5-4-12)

e. Minimum Vectoring Altitudes (MVA) are established for use by ATC when radar ATC is exercised. MVA charts are prepared by air traffic facilities at locations where there are numerous different minimum IFR altitudes. Each MVA chart has sectors large enough to accommodate vectoring of aircraft within the sector at the MVA. Each sector boundary is at least 3 miles from the obstruction determining the MVA. To avoid a large sector with an excessively high MVA due to an isolated prominent obstruction, the obstruction may be enclosed in a buffer area whose boundaries are at least 3 miles from the obstruction. This is done to facilitate vectoring around the obstruction. (See FIG 5-4-13).

1. The minimum vectoring altitude in each sector provides 1,000 feet above the highest obstacle in nonmountainous areas and 2,000 feet above the highest obstacle in designated mountainous areas. Where lower MVAs are required in designated mountainous areas to achieve

compatibility with terminal routes or to permit vectoring to an IAP, 1,000 feet of obstacle clearance may be authorized with the use of Airport Surveillance Radar (ASR). The minimum vectoring altitude will provide at least 300 feet above the floor of controlled airspace.

Figure 5-4-13. Minimum Vectoring Charts

NOTE—OROCA IS AN OFF-ROUTE ALTITUDE WHICH PROVIDES OBSTRUCTION CLEARANCE WITH A 1,000 FOOT BUFFER IN NONMOUNTAINOUS TERRAIN AREAS AND A 2,000 FOOT BUFFER IN DESIGNATED MOUNTAINOUS AREAS WITHIN THE UNITED STATES. THIS ALTITUDE MAY NOT PROVIDE SIGNAL COVERAGE FROM GROUND-BASED NAVIGATIONAL AIDS, AIR TRAFFIC CONTROL RADAR, OR COMMUNICATIONS COVERAGE.

2. Because of differences in the areas considered for MVA, and those applied to other minimum altitudes, and the ability to isolate specific obstacles, some MVAs may be lower than the nonradar Minimum En Route Altitudes (MEA), Minimum Obstruction Clearance Altitudes (MOCA) or other minimum altitudes depicted on charts for a given location. While being radar vectored, IFR altitude assignments by ATC will be at or above MVA.

f. Visual Descent Points (VDP) are being incorporated in nonprecision approach procedures. The VDP is a defined point on the final approach course of a nonprecision straight-in approach procedure from which normal descent from the MDA to the runway touchdown point may be commenced, provided visual reference required by 14 FAR Part 91.175(c)(3) is established. The VDP will normally be identified by DME on VOR and LOC procedures and along track distance to the next waypoint for RNAV procedures. The VDP is identified on the profile view of the approach chart by the symbol: **V.**

1. VDPs are intended to provide additional guidance where they are implemented. No special technique is required to fly a procedure with a VDP. The pilot should not descend below the MDA prior to reaching the VDP and acquiring the necessary visual reference.

2. Pilots not equipped to receive the VDP should fly the approach procedure as though no VDP had been provided.

g. Visual Portion of the Final Segment. Instrument procedures designers perform a visual area obstruction evaluation off the approach end of each runway authorized for instrument landing, straight-in, or circling. Restrictions to instrument operations are imposed if penetrations of the obstruction clearance surfaces exist. These restrictions vary based on the severity of the penetrations, and may include increasing required visibility, denying VDF's and prohibiting night instrument operations to the runway.

h. Vertical Descent Angle (VDA) on Nonprecision Approaches. The FAA intends to eventually publish VDA's on all nonprecision approaches. Published along with the VDA is the threshold crossing height (TCH); i.e., the height of the descent angle above the landing threshold. The descent angle describes a computed path from the final approach fix (FAF) and altitude to the runway threshold at the published TCH. The optimum descent angle is 3.00 degrees; and whenever possible the approach will be designed to accommodate this angle.

1. The VDA provides the pilot with information not previously available on nonprecision approaches. It provides the means for the pilot to establish a stabilized approach descent from the FAF or stepdown fix to the TCH. Stabilized descent along this path is a key factor in the reduction of controlled flight into terrain (CFIT) incidents. Pilots can use the published angle and estimated/actual groundspeed to find a target rate of descent from a rate of descent table published in the back of the U.S. Terminal Procedures Publication.

2. Normally, the VDA will first appear on the nonprecision chart as the procedure is amended through the normal process. However, in some cases, pilots can expect to see this data provided via a D-NOTAM.
EXAMPLE—
GPS RWY 9L, AMDT 2...
ADD: AWZAC WP TO RW09L: 2.96 DEGREES, TCH 50.
THIS IS GPS RWY 9L, AMDT 2A

Translated, this means that the currently published (GPS RWY 9L procedure, Amendment 2, is changed by the addition of a 2.96-degree descent angle from AWZAC WP to a point 50 feet above the RWY 9L threshold. This constitutes Amendment 2A to the published procedure.

3. Pilots should be aware that the published angle is **for information only**—it is strictly *advisory in nature*. There is no implicit additional obstacle protection below the MDA. Pilots must still respect the published minimum descent altitude (MDA) unless the visual cues stated in 14 CFR Section 91.175 are present. In rare cases, the published procedure descent angle will not coincide with the Visual Glide Slope Indicator (VGSI); VASI or PAPI. In these cases, the procedure will be annotated: "VGSI and descent angle not coincident."

i. Pilot Operational Considerations When Flying Nonprecision Approaches. The missed approach point (MAP) on a nonprecision approach is not designed with any consideration to where the aircraft must begin descent to execute a safe landing. It is developed based on terrain, obstructions, NAVAID location and possibly air traffic considerations. Because the MAP may be located anywhere from well prior to the runway threshold to past the opposite end of the runway, the descent from the Minimum Descent Altitude (MDA) to the runway threshold cannot be determined based on the MAP location. Descent from MDA at the MAP when the MAP is located close to the threshold would require an excessively steep descent gradient to land in the normal touchdown zone. Any turn from the final approach course to the runway heading may also be a factor in when to begin the descent.

1. Pilots are cautioned that descent to a straight-in landing from the MDA at the MAP may be inadvisable or impossible, on a nonprecision approach, even if current weather conditions meet the published ceiling and visibility. Aircraft speed, height above the runway, descent rate, amount of turn and runway length are some of the factors which must be considered by the pilot to determine if a landing can be accomplished.

2. Visual descent points (VDP's) provide pilots with a reference for the optimal location to begin descent from the MDA, based on the designed vertical descent angle (VDA) for the approach procedure, assuming required visual references are available. Approaches without VDP's have not been assessed for terrain clearance below the MDA, and may not provide a clear vertical path to the runway at the normally expected descent angle. Therefore, pilots must be especially vigilant when descending below the MDA at locations without VDP's. This does not necessarily prevent flying the normal angle; it only means that obstacle clearance in the visual segment could be less and greater care should be exercised in looking for obstacles in the visual segment. Use of visual glide slope indicator (VGSI) systems can aid the pilot in determining if the aircraft is in a position to make the descent from the MDA. However, when the visibility is close to minimums, the VGSI may not be visible at the start descent point for a "normal" glide path, due to its location down the runway.

3. Accordingly, pilots are advised to carefully review approach procedures, prior to initiating the approach, to identify the optimum position(s), and any unacceptable positions, from which a descent to landing can be initiated (in accordance with 14 CFR Section 91.175(c)).

j. Area Navigation (RNAV) Instrument Approach Charts. Reliance on RNAV systems for instrument approach operations is becoming more commonplace as new systems such as GPS, Wide Area Augmentation System (WAAS) and Local Area Augmentation System (LAAS) are developed and deployed. In order to foster and support full integration of RNAV into the National Airspace System (NAS), the FAA has developed a new charting format for LAP's. (See FIG 5-4-9). This format avoids unnecessary duplication and proliferation of instrument approach charts. The approach minimums for unaugmented GPS (the present GPS approaches) and WAAS augmented GPS will be published on the same approach chart. The approach chart will be titled "RNAV (GPS) RWY XX." The first new RNAV approach charts appeared as stand alone "GPS" procedures, prior to WAAS becoming operational, with only a LNAV minima line. The follow-on charts contained as many as four lines of approach minimums: GLS (Global Navigation Satellite System [GNSS] Landing System); LNAV/VNAV (lateral navigation/vertical navigation); LNAV; and CIRCLING. GLS was a placeholder for WAAS and LAAS when they became available and is marked N/A. LNAV/VNAV is a new type of instrument approach called APV, with lateral and vertical navigation. The vertical portion can be flown by approach certified Baro-VNAV and by WAAS electronic VNAV as well. A new line will be added to these charts titled LPV. This will replace the GLS N/A line. The decision of whether the WAAS and LAAS precision lines of minima will remain on this chart or be moved to a precision only chart will be determined later. RNAV procedures which incorporate a final approach stepdown fix may be published without vertical navigation, on a separate chart, titled RNAV (GPS) RWY XX and then Z, Y, X, etc., as indicated in subparagraph 5-4-5a3. During a transition period until all GPS procedures are retitled both "RNAV (GPS)" and "GPS" approach charts and formats will be published. ATC clearance for the RNAV procedure authorizes a properly certified pilot to utilize any landing minimums for which the aircraft is certified. The RNAV chart includes formatted information required for quick pilot or flight crew reference located at the top of the chart. This portion of the chart, developed based on a study by the Department of Transportation, Volpe National Transportation Systems Center, is commonly referred to as the pilot briefing.

1. The minima lines are:

(a) GLS. "GLS" is the acronym for GNSS Landing System; GNSS is the acronym for Global Navigation Satellite System. The minimums line labeled GLS will accommodate aircraft equipped with precision approach certified WAAS receivers operating to their fullest capability or LAAS and may support precision (GLS) approach minimums as low as 200-foot height above touchdown (HAT) and 1/2 statute mile (SM) visibility. This line has been published as N/A as a place holder for the minima when published. The WAAS and LAAS precision minima may be moved to a different chart in the future.

(b) LPV. "LPV" is the acronym for localizer performance with vertical guidance. LPV identifies the APV minimums with electronic lateral and vertical guidance. The lateral guidance is equivalent to localizer, and the protected area is considerably smaller than the protected area for the present LNAV and LNAV/VNAV lateral protection. Aircraft can fly this minima line with a statement in the Aircraft Flight Manual that the installed equipment supports LPV approaches. This includes Class 3 and 4 TSO-C146 WAAS equipment, and future LAAS equipment.

(c) LNAV/VNAV identifies APV minimums developed to accommodate an RNAV IAP with vertical guidance, usually provided by approach certified Baro-VNAV, but with lateral and vertical integrity limits larger than a precision approach or LPV. LNAV stands for Lateral Navigation; VNAV stands for vertical Navigation. This minima line can be flown by aircraft with a statement in the Aircraft Flight Manual that the installed equipment supports GPS approaches and has an approach-approved barometric VNAV, or if the aircraft has been demonstrated to support LNAV/VNAV approaches. This includes Class 2, 3 and 4 TSO-C146 WAAS equipment Aircraft using LNAV/VNAV minimums will descend to landing via an internally generated descent path based on satellite or other approach approved VNAV systems. WAAS equipment may revert to this mode of operation when the signal does not support precision or LPV integrity. Since electronic vertical guidance is provided, the minima will be published as a DA. Other navigation systems may be specifically authorized to use this line of minima, see Section A, Terms/Landing Minima Data, of the U.S. Terminal Procedures books.

(d) LNAV. This minima is for lateral navigation only, and the approach minimum altitude will be published as a minimum descent altitude (MDA). LNAV provides the same level of service as the present GPS stand alone approaches. LNAV minimums support the following navigation systems: WAAS, when the navigation solution will not support vertical navigation; and, GPS navigation systems which are presently authorized to conduct GPS approaches. Existing GPS approaches continue to be converted to the RNAV (GPS) *format* as they am revised or reviewed.

NOTE—GPS RECEIVERS APPROVED FOR APPROACH OPERATIONS IN ACCORDANCE WITH: AC 20-138, AIRWORTHINESS APPROVAL OF GLOBAL POSITIONING SYSTEM (GPS) NAVIGATION EQUIPMENT FOR USE AS A VFR AND IFR SUPPLEMENTAL NAVIGATION SYSTEM, FOR STAND-ALONE TECHNICAL STANDARD ORDER (TSO) TSO-C129 CLASS A(1) SYSTEMS; OR AC 20-130A, AIRWORTHINESS APPROVAL OF NAVIGATION OR FLIGHT MANAGEMENT SYSTEMS INTEGRATING MULTIPLE NAVIGATION SENSORS, FOR GPS AS PART OF A MULTI-SENSOR SYSTEM, QUALIFY FOR THIS MINIMA. WAAS NAVIGATION EQUIPMENT MUST BE APPROVED IN ACCORDANCE WITH THE REQUIREMENTS SPECIFIED IN TSO-C145 OR TSO-C146 AND INSTALLED IN ACCORDANCE WITH ADVISOSY CIRCULAR AC 20-138A, AIRWORTHINESS APPROVAL OF GLOBAL NAVIGATION SATELLITE SYSTEM (GNSS) EQUIPMENT.

2. Other systems may be authorized to utilize these approaches. See the description in Section A of the U.S. Terminal Procedures books for details. These systems may include aircraft equipped with an FMS that can file /E or /F. Operational approval must also be obtained for Baro-VNAV systems to operate to the LNAV/VNAV minimums. Baro-VNAV may not be authorized on some approaches due to other factors, such as no local altimeter source being available. Baro-VNAV is not authorized on LPV procedures. Pilots are directed to their local Flight Standards District Office (FSDO) for additional information.

NOTE—RNAV AND BARO-VNAV SYSTEMS MUST HAVE A MANUFACTURER SUPPLIED ELECTRONIC DATABASE WHICH SHALL INCLUDE THE WAYPOINTS, ALTITUDES, AND VERTICAL DATA FOR THE PROCEDURE TO BE FLOWN. THE SYSTEM SHALL ALSO BE ABLE TO EXTRACT THE PROCEDURE IN ITS ENTIRETY, NOT JUST AS A MANUALLY ENTERED SERIES OF WAYPOINTS.

3. Required Navigation Performance (RNP)

(a) Pilots are advised to refer to the "TERMS/LANDING MINIMUMS DATA" (Section A) of the U.S. Government Terminal Procedures books for aircraft approach eligibility requirements by specific RNP level requirements.

(b) Some aircraft have RNP approval in their AFM without a GPS sensor. The lowest level of sensors that the FAA will support for RNP service is DME/DME. However, necessary DME signal may not be available at the airport of intended operations.

For those locations having an RNAV chart published with LNAV/VNAV minimums, a procedure note may be provided such as "DME/DME RNP-O.3 NA." This means

that RNP aircraft dependent on DME/DME to achieve RNP-O.3 are not authorized to conduct this approach. Where DME facility availability is a factor, the note may read "DME/DME RNP-O.3 Authorized; ABC and XYZ Required." This means that ABC and XYZ facilities have been determined by flight inspection to be required in the navigation solution to assure RNP-O.3. VOR/DME updating must not be used for approach procedures.

4. CHART TERMINOLOGY

(a) **Decision Altitude (DA)** replaces the familiar term Decision Height (DH). DA conforms to the international convention where altitudes relate to MSL and heights relate to AGL. DA will eventually be published for other types of instrument approach procedures with vertical guidance, as well. DA indicates to the pilot that the published descent profile is flown to the DA (MSL), where a missed approach will be initiated if visual references for landing are not established. Obstacle clearance is provided to allow a momentary descent below DA while transitioning from the final approach to the missed approach. The aircraft is expected to follow the missed instructions while continuing along the published final approach course to at least the published runway threshold waypoint or MAP (if not at the threshold) before executing any turns.

(b) **Minimum Descent Altitude (MDA)** has been in use for many years, and will continue to be used for the LNAV only and circling procedures.

(c) **Threshold Crossing Height (TCH)** has been traditionally used in "precision" approaches as the height of the glide slope above threshold. With publication of LNAV/VNAV minimums and RNAV descent angles, including graphically depicted descent profiles, TCH also applies to the height of the "descent angle," or glidepath, at the threshold. Unless otherwise required for larger type aircraft which may be using the IAP, the typical TCH is 30 to 50 feet.

5. The Minima Format will also change slightly

(a) Each line of minima on the RNAV IAP is titled to reflect the level of service available, e.g., GLS, LPV, LNAV/VNAV, and LNAV. CIRCLING minima will also be provided.

(b) The minima title box indicates the nature of the minimum altitude for the IAP. For example:

(1) DA will be published next to the minima line title for minimums supporting vertical guidance such as for GLS, LPV or LNAV/VNAV.

(2) MDA will be published where the minima line was designed to support aircraft with only lateral guidance available, such as LNAV. Descent below the MDA, including during the missed approach, is not authorized unless the visual conditions stated in 14 CFR Section 91.175 exist.

(3) Where two or more systems, such as LPV and LNAV/VNAV, share the same minima, each line of minima will be displayed separately.

6. Chart Symbology changed slightly to include:

(a) Descent Profile. The published descent profile and a graphical depiction of the vertical path to the runway will be shown. Graphical depiction of the RNAV vertical guidance will differ from the traditional depiction of an ILS glide slope (feather) through the use of a shorter vertical track beginning at the decision altitude.

(1) It is FAA policy to design IAP's with minimum altitudes established at fixes/waypoints to achieve optimum stabilized (constant rate) descents within each procedure segment. This design can enhance the safety of the operations and contribute toward reduction in the occurrence of controlled flight into terrain (CFIT) accidents. Additionally, the National Transportation Safety Board (NTSB) recently emphasized that pilots could benefit from publication of the appropriate IAP descent angle for a stabilized descent on final approach. The RNAV IAP format includes the descent angle to the hundredth of a degree; e.g., 3.00 degrees. The angle will be provided in the graphically depicted descent profile.

(2) The stabilized approach may be performed by reference to vertical navigation information provided by WAAS or LNAV/VNAV systems; or for LNAV-only systems, by the pilot determining the appropriate aircraft attitude/groundspeed combination to attain a constant rate descent which best emulates the published angle. To aid the pilot, U.S. Government Terminal Procedures Publication charts publish an expanded Rate of Descent Table on the inside of the back hard cover for use in planning and executing precision descents under known or approximate groundspeed conditions.

(b) Visual **Descent Point (VDP).** A VDP will be published on most RNAV IAP's. VDP's apply only to aircraft utilizing LNAV minima. not LPV or LNAVIVNAV minimums.

(c) Missed Approach Symbology. In order to make missed approach guidance more readily understood, a method has been developed to display missed approach guidance in the profile view through the use of quick reference icons. Due to limited space in the profile area, only four or fewer icons can be shown. However, the icons may not provide representation of the entire missed approach procedure. The entire set of textual missed approach instructions are provided at the top of the approach chart in the pilot briefing. (See FIG 5-4-9).

(d) Waypoints. All RNAV or GPS standalone IAP's are flown using data pertaining to the particular IAP obtained from an onboard database, including the sequence of all WP's used for the approach and missed approach, except that step down waypoints may not be included in some TSO C-129 receiver databases. Included in the database, in most receivers, is coding that informs the navigation system of which WP's are fly-over (FO) or fly-by (FB). The navigation system may provide guidance appropriately - including leading the turn prior to a fly-by WP; or causing overflight of a fly-over WP Where the navigation system does not provide such guidance, the pilot must accomplish the turn lead or waypoint overflight manually. Chart symbology for the FB WP provides pilot awareness of expected actions. Refer to the legend of the U.S. Terminal Procedures books.

(e) TAA's are described in paragraph 5-4-5d, Terminal Arrival Area (TAA). When published, the RNAV chart depicts the TAA areas through the use of "icons" representing each TAA area associated with the RNAV procedure (See FIG 5-4-9). These icons are depicted in the plan view of the approach chart, generally arranged on the chart in accordance with their position relative to the aircraft's arrival from the en route structure. The WP, to which navigation is appropriate and expected within each specific TAA area, will be named and depicted on the associated TAA icon. Each depicted named WP is the IAF for arrivals from within that area. TAA's may not be used on all RNAV procedures because of airspace congestion or other reasons.

(f) Cold Temperature Limitations. A minimum temperature limitation is published on procedures which authorize Baro-VNAV operation. This temperature represents the airport temperature below which use of the Baro-VNAV is not authorized to the LNAV/VNAV minimums. An example limitation will read: "Baro-VNAV NA below −20°C(−4° F)." This information will be found in the upper left hand box of the pilot briefing.

NOTE—TEMPERATURE LIMITATIONS DO NOT APPLY TO FLYING THE LNAV/VNAV LINE OF MINIMA USING APPROACH CERTIFIED WAAS RECEIVERS.

(g) WAAS~Channel Nember/Approach ID. The WAAS Channel Number is an equipment optional capability that allows the use of a 5-digit number to select a specific final approach segment. The Approach ID is an airport unique 4-letter combination for verifying selection of the correct final approach segment, e.g. W-35L, where W stands for WAAS and 35L is runway 35 left. The WAAS Channel Number and Approach ID will be displayed in the upper left corner of the approach procedure pilot briefing.

(h) At locations where outages of WAAS vertical guidance may occur daily due to initial system limitations, a negative W symbol (Ⓦ) will be placed on RNAV (GPS) approach charts. Many of these outages will be very short in duration, but may result in the disruption of the vertical portion of the approach. The Ⓦ symbol indicates that NOTAMs or Air Traffic advisories are not provided for outages which occur in the WAAS LNAV/VNAV or LPV vertical service. Use LNAV minima for flight planning at these locations, whether as a destination or alternate. For flight operations at these locations, when the WAAS avionics indicate that LNAV/VNAV or LPV service is available, then vertical guidance may be used to complete the approach using the displayed level of service. Should an outage occur during the procedure, reversion to LNAV minima may be required. As the WAAS coverage is expanded, the Ⓦ will be removed.

5-4-6. APPROACH CLEARANCE

a. An aircraft which has been cleared to a holding fix and subsequently "cleared . . . approach" has not received new routing. Even though clearance for the approach may have been issued prior to the aircraft reaching the holding fix, ATC would expect the pilot to proceed via the holding fix (his last assigned route), and the feeder route associated with that fix (if a feeder route is published on the approach chart) to the initial approach fix (IAF) to commence the approach. WHEN CLEARED FOR THE APPROACH, THE PUBLISHED OFF AIRWAY (FEEDER) ROUTES THAT LEAD FROM THE EN ROUTE STRUCTURE TO THE IAF ARE PART OF THE APPROACH CLEARANCE.

b. If a feeder route to an IAF begins at a fix located along the route of flight prior to reaching the holding fix, and clearance for an approach is issued, a pilot should commence the approach via the published feeder route; i.e., the aircraft would not be expected to overfly the feeder route and return to it. The pilot is expected to commence the approach in a similar manner at the IAF, if the IAF for the procedure is located along the route of flight to the holding fix.

c. If a route of flight directly to the initial approach fix is desired, it should be so stated by the controller with phraseology to include the words "direct . . .," "proceed direct" or a similar phrase which the pilot can interpret without question. When uncertain of the clearance, immediately query ATC as to what route of flight is desired.

d. The name of an instrument approach, as published, is used to identify the approach, even though a component of the approach aid, such as the glide slope on an Instrument Landing System, is inoperative or unreliable. The controller will use the name of the approach as published, but must advise the aircraft at the time an approach clearance is issued that the inoperative or unreliable approach aid component is unusable.

5-4-7. INSTRUMENT APPROACH PROCEDURES

a. Aircraft approach category means a grouping of aircraft based on a speed of V_{REF}, if specified, or if V_{REF} not specified, 1.3 Vso at the maximum certificated landing weight. V_{REF}, Vso, and the maximum certificated landing weight are those values as established for the aircraft by the certification authority of the country of registry. Helicopters are Category A aircraft. An aircraft must fit in only one category. Pilots are responsible for determining and briefing which category minimums will be used for each instrument approach. If a higher approach speed is used on final that places the aircraft in a higher approach category, the minimums for the higher category must be used. Approaches made with inoperative flaps, circling approaches at higher-than normal straight-in approach speeds, and approaches made in icing conditions for some types of airplanes are all examples of situations that can necessitate

the use of a higher approach category. See the following category limits:

1. Category A: Speed less than 91 knots.

2. Category B: Speed 91 knots or more but less than 121 knots.

3. Category C: Speed 121 knots or more but less than 141 knots.

4. Category D: Speed 141 knots or more but less than 166 knots.

5. Category E: Speed 166 knots or more.

NOTE—V_{REF} IS THE REFERENCE LANDING APPROACH SPEED, USUALLY ABOUT 1.3 TIMES Vso PLUS 50 PERCENT OF THE WIND GUST SPEED IN EXCESS OF THE MEAN WIND SPEED (SEE 14 CFR SECTION 23.73). Vso IS THE STALLING SPEED OR THE MINIMUM STEADY FLIGHT SPEED IN THE LANDING CONFIGURATION AT MAXIMUM WEIGHT (SEE 14 CFR SECTION 23.49).

b. When operating on an unpublished route or while being radar vectored, the pilot, when an approach clearance is received, shall, in addition to complying with the minimum altitudes for IFR operations (FAR Part 91.177), maintain the last assigned altitude unless a different altitude is assigned by ATC, or until the aircraft is established on a segment of a published route or IAP. After the aircraft is so established, published altitudes apply to descent within each succeeding route or approach segment unless a different altitude is assigned by ATC. Notwithstanding this pilot responsibility, for aircraft operating on unpublished routes or while being radar vectored, ATC will, except when conducting a radar approach, issue an IFR approach clearance only after the aircraft is established on a segment of a published route or IAP, or assign an altitude to maintain until the aircraft is established on a segment of a published route or instrument approach procedure. For this purpose, the procedure turn of a published IAP shall not be considered a segment of that IAP until the aircraft reaches the initial fix or navigation facility upon which the procedure turn is predicated.

EXAMPLE—CROSS REDDING VOR AT OR ABOVE FIVE THOUSAND, CLEARED VOR RUNWAY THREE FOUR APPROACH.

OR

FIVE MILES FROM OUTER MARKER, TURN RIGHT HEADING THREE THREE ZERO, MAINTAIN TWO THOUSAND UNTIL ESTABLISHED ON THE LOCALIZER, CLEARED ILS RUNWAY THREE SIX APPROACH.

NOTE—THE ALTITUDE ASSIGNED WILL ASSURE IFR OBSTRUCTION CLEARANCE FROM THE POINT AT WHICH THE APPROACH CLEARANCE IS ISSUED UNTIL ESTABLISHED ON A SEGMENT OF A PUBLISHED ROUTE OR IAP. IF UNCERTAIN OF THE MEANING OF THE CLEARANCE, IMMEDIATELY REQUEST CLARIFICATION FROM ATC.

c. Several IAPs, using various navigation and approach aids may be authorized for an airport. ATC may advise that a particular approach procedure is being used, primarily to expedite traffic. If issued a clearance that specifies a particular approach procedure, notify ATC immediately if a different one is desired. In this event it may be necessary for ATC to withhold clearance for the different approach until such time as traffic conditions permit. However, a pilot involved in an emergency situation will be given priority. If the pilot is not familiar with the specific approach procedure, ATC should be advised and they will provide detailed information on the execution of the procedure.

REFERENCE—AIM, ADVANCE INFORMATION ON INSTRUMENT APPROACH, PARAGRAPH 5-4-4.

d. At times ATC may not specify a particular approach procedure in the clearance, but will state "CLEARED APPROACH." Such clearance indicates that the pilot may execute any one of the authorized IAPs for that airport. This clearance does not constitute approval for the pilot to execute a contact approach or a visual approach.

e. Except when being radar vectored to the final approach course, when cleared for a specifically prescribed IAP; i.e., "cleared ILS runway one niner approach" or when "cleared approach" i.e., execution of any procedure prescribed for the airport, pilots shall execute the entire procedure at an IAF or an associated feeder route as described on the IAP Chart unless an appropriate new or revised ATC clearance is received, or the IFR flight plan is canceled.

f. Pilots planning flights to locations served by special IAPs should obtain advance approval from the owner of the procedure. Approval by the owner is necessary because special procedures are for the exclusive use of the single interest unless otherwise authorized by the owner. Additionally, some special approach procedures require certain crew qualifications training, or other special considerations in order to execute the approach. Also, some of these approach procedures are based on privately owned navigational aids. Owners of aids that are not for public use may elect to turn off the aid for whatever reason they may have; i.e., maintenance, conservation, etc. Air traffic controllers are not required to question pilots to determine if they have permission to use the procedure. Controllers presume a pilot has obtained approval and is aware of any details of the procedure if an IFR flight plan was filed to that airport.

g. Pilots should not rely on radar to identify a fix unless the fix is indicated as "RADAR" on the IAP. Pilots may request radar identification of an OM, but the controller may not be able to provide the service due either to workload or not having the fix on the video map.

h. If a missed approach is required, advise ATC and include the reason (unless initiated by ATC). Comply with the missed approach instructions for the instrument approach procedure being executed, unless otherwise directed by ATC.

REFERENCE—AIM, MISSED APPROACH, PARAGRAPH 5-4-21 , AIM, MISSED APPROACH, PARAGRAPH 5-5-5.

i. ATC may clear aircraft that have filed an Advanced RNAV equipment suffix to the intermediate fix when clearing aircraft for an instrument approach procedure. ATC will take the following actions when clearing Advanced RNAV aircraft to the intermediate fix:

1. Provide radar monitoring to the intermediate fix.

2. Advise the pilot to expect clearance direct to the intermediate fix at least 5 miles from the fix.

NOTE—THIS IS TO ALLOW THE PILOT TO PROGRAM THE RNAV EQUIPMENT TO ALLOW THE AIRCRAFT TO FLY TO THE INTERMEDIATE FIX WHEN CLEARED BY ATC.

3. Assign an altitude to maintain until the intermediate fix.

4. Insure the aircraft is on a course that will intercept the intermediate segment at an angle not greater than 90 degrees and is at an altitude that will permit normal descent from the intermediate fix to the final approach fix.

REFERENCE—AIM, MISSED APPROACH, PARAGRAPH 5-4-20. AIM, MISSED APPROACH, PARAGRAPH 5-5-5.

5-4-8. SPECIAL INSTRUMENT APPROACH PROCEDURES

Instrument Approach Procedure (IAP) charts reflect the criteria associated with the U.S. Standard for Terminal Instrument [Approach] Procedures (TERPs), which prescribes standardized methods for use in developing IAPs. Standard IAPs are published in the Federal Register (FR) in accordance with Title 14 of the Code of Federal Regulations, Part 97, and are available for use by appropriately qualified pilots operating properly equipped and airworthy aircraft in accordance with operating rules and procedures acceptable to the FAA. Special IAPs are also developed using TERPS but are not given public notice in the FR. The FAA authorizes only certain individual pilots and/or pilots in individual organizations to use special IAPs, and may require additional crew training and/or aircraft equipment or performance, and may also require the use of landing aids, communications, or weather services not available for public use. Additionally, IAPs that service private use airports or heliports are generally special IAPs.

5-4-9. PROCEDURE TURN

a. A procedure turn is the maneuver prescribed when it is necessary to reverse direction to establish the aircraft inbound on an intermediate or final approach course. The procedure turn or hold-in-lieu-of-PT is a required maneuver when it is depicted on the approach chart. However, the procedure turn or hold-in-lieu-of-PT is not permitted when the symbol "No PT" is depicted on the initial segment being used, when a RADAR VECTOR to the final approach course is provided, or when conducting a timed approach from a holding fix. The altitude prescribed for the procedure turn is a minimum altitude until the aircraft is established on the inbound course. The maneuver must be completed within the distance specified in the profile view.

NOTE—THE PILOT MAY ELECT TO USE THE PROCEDURE TURN OR HOLD-IN-LIEU-OF-PT WHEN IT IS

NOT REQUIRED BY THE PROCEDURE, BUT MUST FIRST RECEIVE AN AMENDED CLEARANCE FROM ATC. WHEN ATC IS RADAR VECTORING TO THE FINAL APPROACH COURSE OR TO THE INTERMEDIATE FIX, ATC MAY SPECIFY IN THE APPROACH CLEARANCE "CLEARED STRAIGHT-IN (TYPE) APPROACH" TO ENSURE THE PROCEDURE TURN OR HOLD-IN-LIEU-OF-PT IS NOT TO BE FLOWN. IF THE PILOT IS UNCERTAIN WHETHER THE ATC CLEARANCE INTENDS FOR A PROCEDURE TURN TO BE CONDUCTED OR TO ALLOW FOR A STRAIGHT-IN APPROACH, THE PILOT SHALL IMMEDIATELY REQUEST CLARIFICATION FROM ATC (14 CER SECTION 91.123).

1. On U.S. Government charts, a barbed arrow indicates the direction or side of the outbound course on which the procedure turn is made. Headings are provided for course reversal using the 45 degree type procedure turn. However, the point at which the turn may be commenced and the type and rate of turn is left to the discretion of the pilot. Some of the options are the 45 degree procedure turn, the racetrack pattern, the teardrop procedure turn, or the 80 degree ↔ 260 degree course reversal. Some procedure turns are specified by procedural track. These turns must be flown exactly as depicted.

2. When the approach procedure involves a procedure turn, a maximum speed of not greater than 200 knots (IAS) should be observed from first overheading the course reversal IAF through the procedure turn maneuver to ensure containment within the obstruction clearance area. Pilots should begin the outbound turn immediately after passing the procedure turn fix. The procedure turn maneuver must be executed within the distance specified in the profile view. The normal procedure turn distance is 10 miles. This may be reduced to a minimum of 5 miles where only Category A or helicopter aircraft are to be operated or increased to as much as 15 miles to accommodate high performance aircraft.

3. A teardrop procedure or penetration turn may be specified in some procedures for a required course reversal. The teardrop procedure consists of departure from an initial approach fix on an outbound course followed by a turn toward and intercepting the inbound course at or prior to the intermediate fix or point. Its purpose is to permit an aircraft to reverse direction and lose considerable altitude within reasonably limited airspace. Where no fix is available to mark the beginning of the intermediate segment, it shall be assumed to commence at a point 10 miles prior to the final approach fix. When the facility is located on the airport, an aircraft is considered to be on final approach upon completion of the penetration turn. However, the final approach segment begins on the final approach course 10 miles from the facility.

4. A procedure turn need not be established when an approach can be made from a properly aligned holding pattern. In such cases, the holding pattern is established over an intermediate fix or a final approach fix. The holding pattern maneuver is completed when the aircraft is established on the inbound course after executing the appropriate entry. If cleared for the approach prior to returning to the holding fix, and the aircraft is at the prescribed altitude, additional circuits of the holding pattern are not necessary nor expected by ATC. If pilots elect to make additional circuits to lose excessive altitude or to become better established on course, it is their responsibility to so advise ATC upon receipt of their approach clearance.

5. A procedure turn is not required when an approach can be made directly from a specified intermediate fix to the final approach fix. In such cases, the term "NoPT" is used with the appropriate course and altitude to denote that the procedure turn is not required. If a procedure turn is desired, and when cleared to do so by ATC, descent below the procedure turn altitude should not be made until the aircraft is established on the inbound course, since some NoPT altitudes may be lower than the procedure turn altitudes.

b. Limitations on Procedure Turns:

1. In the case of a radar initial approach to a final approach fix or position, or a timed approach from a holding fix, or where the procedure specifies NoPT, no pilot may make a procedure turn unless, when final approach clearance is received, the pilot so advises ATC and a clearance is received to execute a procedure turn.

2. When a teardrop procedure turn is depicted and a course reversal is required, this type turn must be executed.

3. When holding pattern replaces the procedure turn, the holding pattern must be followed, except when RADAR VECTORING is provided or when NoPT is shown on the approach course. The recommended entry procedures will ensure the aircraft remains within the holding pattern's protected airspace. As in the procedure turn, the descent from the minimum holding pattern altitude to the final approach fix altitude (when lower) may not commence until the aircraft is established on the inbound course. Where a holding pattern is established in-lieu-of a procedure turn, the maximum holding pattern airspeeds apply.

REFERENCE—
AIM, HOLDING, PARAGRAPH 5-3-7j2.

4. The absence of the procedure turn barb in the Plan View indicates that a procedure turn is not authorized for that procedure.

5-4-10. TIMED APPROACHES FROM A HOLDING FIX

a. TIMED APPROACHES may be conducted when the following conditions are met:

1. A control tower is in operation at the airport where the approaches are conducted.

2. Direct communications are maintained between the pilot and the center or approach controller until the pilot is instructed to contact the tower.

3. If more than one missed approach procedure is available, none require a course reversal.

4. If only one missed approach procedure is available, the following conditions are met:

(a) Course reversal is not required; and,

(b) Reported ceiling and visibility are equal to or greater than the highest prescribed circling minimums for the IAP.

5. When cleared for the approach, pilots shall not execute a procedure turn. (FAR Part 91.175.)

b. Although the controller will not specifically state that "timed approaches are in progress," the assigning of a time to depart the final approach fix inbound (nonprecision approach) or the outer marker or fix used in lieu of the outer marker inbound (precision approach) is indicative that timed approach procedures are being utilized, or in lieu of holding, the controller may use radar vectors to the Final Approach Course to establish a mileage interval between aircraft that will ensure the appropriate time sequence between the final approach fix/outer marker or fix used in lieu of the outer marker and the airport.

c. Each pilot in an approach sequence will be given advance notice as to the time they should leave the holding point on approach to the airport. When a time to leave the holding point has been received, the pilot should adjust the flight path to leave the fix as closely as possible to the designated time. (See FIG 5-4-14.)

5-4-11. RADAR APPROACHES

a. The only airborne radio equipment required for radar approaches is a functioning radio transmitter and receiver. The radar controller vectors the aircraft to align it with the runway centerline. The controller continues the vectors to keep the aircraft on course until the pilot can complete the approach and landing by visual reference to the surface. There are two types of radar approaches: Precision (PAR) and Surveillance (ASR).

b. A radar approach may be given to any aircraft upon request and may be offered to pilots of aircraft in distress or to expedite traffic, however, an ASR might not be approved unless there is an ATC operational requirement, or in an unusual or emergency situation. Acceptance of a PAR or ASR by a pilot does not waive the prescribed weather minimums for the airport or for the particular aircraft operator concerned. The decision to make a radar approach when the reported weather is below the established minimums rests with the pilot.

c. PAR and ASR minimums are published on separate pages in the FAA Terminal Procedures Publication (TPP).

1. A PRECISION APPROACH (PAR) is one in which a controller provides highly accurate navigational guidance in azimuth and elevation to a pilot. Pilots are given headings to fly, to direct them to, and keep their aircraft aligned with the extended centerline of the landing runway. They are told to anticipate glide path interception

---

Example—

At 12:03 local time, in the example shown, a pilot holding, receives instructions to leave the fix inbound at 12:07. These instructions are received just as the pilot has completed turn at the outbound end of the holding pattern and is proceeding inbound towards the fix. Arriving back over the fix, the pilot notes that the time is 12:04 and that he has 3 minutes to lose in order to leave the fix at the assigned time. Since the time remaining is more than 2 minutes, the pilot plans to fly a race track pattern rather than a 360 degree turn, which would use up 2 minutes. The runs at the ends of the race track pattern will consume approximately 2 minutes. Three minutes to go, minus 2 minutes required for turns, leaves 1 minute for level flight. Since two portions of level flight will be required to get back to the fix inbound, the pilot halves the 1 minute remaining and plans to fly level for 30 seconds outbound before starting his turn back toward the fix on final approach. If the winds were negligible at flight altitude, this procedure would bring the pilot inbound across the fix precisely at the specified time of 12:07. However, if the pilot expected a headwind on final approach, he should shorten his 30 seconds outbound course somewhat, knowing that the wind will carry him away from the fix faster while outbound and decrease his ground speed while returning to the fix. On the other hand, if the pilot knew he would have a tailwind on final approach, he should lengthen his calculated 30-second outbound heading somewhat, knowing that the wind would tend to hold him closer to the fix while outbound and increase his ground speed while returning to the fix.

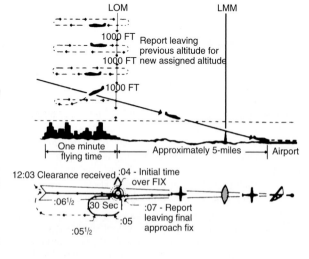

Time Approach Example — a final approach procedure from a holding pattern at a final approach fix (FAF).

Figure 5-4-14. Timed Approaches from a Holding Fix

approximately 10 to 30 seconds before it occurs and when to start descent. The published Decision Height will be given only if the pilot requests it. If the aircraft is observed to deviate above or below the glide path, the pilot is given the relative amount of deviation by use of terms "slightly" or "well" and is expected to adjust the aircraft's rate of descent/ascent to return to the glide path. Trend information is also issued with respect to the elevation of the aircraft and may be modified by the terms "rapidly" and "slowly"; e.g., "well above glide path, coming down rapidly." Range from touchdown is given at least once each mile. If an aircraft is observed by the controller to proceed outside of specified safety zone limits in azimuth and/or elevation and continue to operate outside these prescribed limits, the pilot will be directed to execute a missed approach or to fly a specified course unless the pilot has the runway environment (runway, approach lights, etc.) in sight. Navigational guidance in azimuth and elevation is provided the pilot until the aircraft reaches the published Decision Height (DH). Advisory course and glidepath information is furnished by the controller until the aircraft passes over the landing threshold, at which point the pilot is advised of any deviation from the runway centerline. Radar service is automatically terminated upon completion of the approach.

2. A SURVEILLANCE APPROACH (ASR) is one in which a controller provides navigational guidance in azimuth only. The pilot is furnished headings to fly to align the aircraft with the extended centerline of the landing runway. Since the radar information used for a surveillance approach is considerably less precise than that used for a precision approach, the accuracy of the approach will not be as great and higher minimums will apply. Guidance in elevation is not possible but the pilot will be advised when to commence descent to the Minimum Descent Altitude (MDA) or, if appropriate, to an intermediate step-down fix Minimum Crossing Altitude and subsequently to the prescribed MDA. In addition, the pilot will be advised of the location of the Missed Approach Point (MAP) prescribed for the procedure and the aircraft's position each mile on final from the runway, airport or heliport or MAP, as appropriate. If requested by the pilot, recommended altitudes will be issued at each mile, based on the descent gradient established for the procedure, down to the last mile that is at or above the MDA. Normally, navigational guidance will be provided until the aircraft reaches the MAP. Controllers will terminate guidance and instruct the pilot to execute a missed approach unless at the MAP the pilot has the runway, airport or heliport in sight or, for a helicopter point-in-space approach, the prescribed visual reference with the surface is established. Also, if, at any time during the approach the controller considers that safe guidance for the remainder of the approach cannot be provided, the controller will terminate guidance and instruct the pilot to execute a missed approach. Similarly, guidance termination and missed

approach will be affected upon pilot request and, for civil aircraft only, controllers may terminate guidance when the pilot reports the runway, airport/heliport or visual surface route (point-in-space approach) in sight or otherwise indicates that continued guidance is not required. Radar service is automatically terminated at the completion of a radar approach.

NOTE—

⑴ THE PUBLISHED MDA FOR STRAIGHT-IN APPROACHES WILL BE ISSUED TO THE PILOT BEFORE BEGINNING DESCENT. WHEN A SURVEILLANCE APPROACH WILL TERMINATE IN A CIRCLE-TO-LAND MANEUVER, THE PILOT MUST FURNISH THE AIRCRAFT APPROACH CATEGORY TO THE CONTROLLER. THE CONTROLLER WILL THEN PROVIDE THE PILOT WITH THE APPROPRIATE MDA.

⑵ ASR APPROACHES ARE NOT AVAILABLE WHEN AN ATC FACILITY IS USING CENRAP.

3. A NO-GYRO APPROACH is available to a pilot under radar control who experiences circumstances wherein the directional gyro or other stabilized compass is inoperative or inaccurate. When this occurs, the pilot should so advise ATC and request a No-Gyro vector or approach. Pilots of aircraft not equipped with a directional gyro or other stabilized compass who desire radar handling may also request a No-Gyro vector or approach. The pilot should make all turns at standard rate and should execute the turn immediately upon receipt of instructions. For example, "TURN RIGHT," "STOP TURN." When a surveillance or precision approach is made, the pilot will be advised after the aircraft has been turned onto final approach to make turns at half standard rate.

5-4-12. RADAR MONITORING OF INSTRUMENT APPROACHES

a. PAR facilities operated by the FAA and the military services at some joint-use (civil and military) and military installations monitor aircraft on instrument approaches and issue radar advisories to the pilot when weather is below VFR minimums (1,000 and 3), at night, or when requested by a pilot. This service is provided only when the PAR Final Approach Course coincides with the final approach of the navigational aid and only during the operational hours of the PAR. The radar advisories serve only as a secondary aid since the pilot has selected the navigational aid as the primary aid for the approach.

b. Prior to starting final approach, the pilot will be advised of the frequency on which the advisories will be transmitted. If, for any reason, radar advisories cannot be furnished, the pilot will be so advised.

c. Advisory information, derived from radar observations, includes information on:

1. Passing the final approach fix inbound (nonprecision approach) or passing the outer marker or fix used in lieu of the outer marker inbound (precision approach).

NOTE—AT THIS POINT, THE PILOT MAY BE REQUESTED TO REPORT SIGHTING THE APPROACH LIGHTS OR THE RUNWAY.

2. Trend advisories with respect to elevation and/or azimuth radar position and movement will be provided.

NOTE—WHENEVER THE AIRCRAFT NEARS THE PAR SAFETY LIMIT, THE PILOT WILL BE ADVISED THAT THE AIRCRAFT IS WELL ABOVE OR BELOW THE GLIDEPATH OR WELL LEFT OR RIGHT OF COURSE. GLIDEPATH INFORMATION IS GIVEN ONLY TO THOSE AIRCRAFT EXECUTING A PRECISION APPROACH, SUCH AS ILS OR MLS. ALTITUDE INFORMATION IS NOT TRANSMITTED TO AIRCRAFT EXECUTING OTHER THAN PRECISION APPROACHES BECAUSE THE DESCENT PORTIONS OF THESE APPROACHES GENERALLY DO NOT COINCIDE WITH THE DEPICTED PAR GLIDEPATH. AT LOCATIONS WHERE THE MLS GLIDEPATH AND PAR GLIDEPATH ARE NOT COINCIDENTAL, ONLY AZIMUTH MONITORING WILL BE PROVIDED.

3. If, after repeated advisories, the aircraft proceeds outside the PAR safety limit or if a radical deviation is observed, the pilot will be advised to execute a missed approach unless the prescribed visual reference with the surface is established.

d. Radar service is automatically terminated upon completion of the approach.

5-4-13. ILS/MLS APPROACHES TO PARALLEL RUNWAYS

a. ATC procedures permit ILS instrument approach operations to dual or triple parallel runway configurations. ILS/MLS approaches to parallel runways are grouped into three classes: Parallel (dependent) ILS/MLS Approaches; Simultaneous Parallel (independent) ILS/MLS Approaches; and Simultaneous Close Parallel (independent) ILS Precision Runway Monitor (PRM) Approaches (See FIG 5-4-15). The classification of a parallel runway

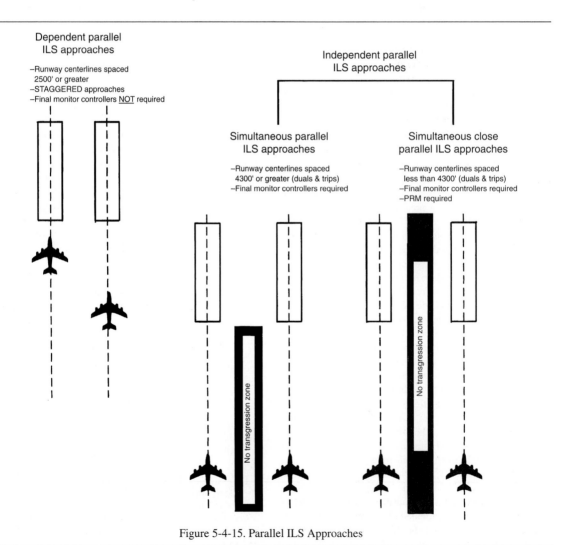

Figure 5-4-15. Parallel ILS Approaches

approach procedure is dependent on adjacent parallel runway centerline separation, ATC procedures, and airport ATC radar monitoring and communications capabilities. At some airports one or more parallel localizer courses may be offset up to 3 degrees. Offset localizer configurations result in loss of Category II capabilities and an increase in decision height (50').

b. Parallel approach operations demand heightened pilot situational awareness. A thorough Approach Procedure Chart review should be conducted with, as a minimum, emphasis on the following approach chart information: name and number of the approach, localizer frequency, inbound localizer/ azimuth course, glide slope intercept altitude, decision height, missed approach instructions, special notes/procedures, and the assigned runway location/ proximity to adjacent runways. Pilots will be advised that simultaneous ILS/MLS or simultaneous close parallel ILS PRM approaches are in use. This information may be provided through the ATIS.

c. The close proximity of adjacent aircraft conducting simultaneous parallel and simultaneous close parallel ILS PRM approaches mandates strict pilot compliance with all ATC clearances. ATC assigned airspeeds, altitudes, and headings must be complied with in a timely manner. Autopilot coupled ILS/MLS approaches require pilot knowledge of procedures necessary to comply with ATC instructions. Simultaneous parallel and simultaneous close parallel ILS/ MLS and simultaneous close parallel ILS PRM approaches necessitate precise localizer tracking to minimize final monitor controller intervention, and unwanted No Transgression Zone (NTZ) penetration. In the unlikely event of a breakout, ATC will not assign altitudes lower than the minimum vectoring altitude. Pilots should notify ATC immediately if there is a degradation of aircraft or navigation systems.

d. Strict radio discipline is mandatory during parallel ILS/MLS approach operations. This includes an alert listening watch and the avoidance of lengthy, unnecessary radio transmissions. Attention must be given to proper call sign usage to prevent the inadvertent execution of clearances intended for another aircraft. Use of abbreviated call signs must be avoided to preclude confusion of aircraft with similar sounding call signs. Pilots must be alert to unusually long periods of silence or any unusual background sounds in their radio receiver. A stuck microphone may block the issuance of ATC instructions by the final monitor controller during simultaneous parallel and simultaneous close parallel ILS/MLS and simultaneous close parallel ILS PRM approaches.

REFERENCE—AIM, CHAPTER 4, Section 2, RADIO COMMUNICATIONS PHRASEOLOGY AND TECHNIQUES, GIVES ADDITIONAL COMMUNICATIONS INFORMATION.

e. Use of Traffic Collision Avoidance Systems (TCAS) provides an additional element of safety to parallel approach operations. Pilots should follow recommended TCAS operating procedures presented in approved flight manuals, original equipment manufacturer recommendations, professional newsletters, and FAA publications.

5-4-14. PARALLEL ILS/MLS APPROACHES (DEPENDENT)

a. Parallel approaches are an ATC procedure permitting parallel ILS/MLS approaches to airports having parallel runways separated by at least 2,500 feet between centerlines. Integral parts of a total system are ILS/MLS, radar, com-munications, ATC procedures, and required airborne equipment.

b. A parallel (dependent) approach differs from a simultaneous (independent) approach in that, the minimum distance between parallel runway centerlines is reduced; there is no requirement for radar monitoring or advisories; and a staggered separation of aircraft on the adjacent localizer/azimuth course is required.

c. Aircraft are afforded a minimum of 1.5 miles radar separation diagonally between successive aircraft on the adjacent localizer/azimuth course when runway centerlines are at least 2,500 feet but no more than 4,300 feet apart. When runway centerlines are more than 4,300 feet but no more than 9,000 feet apart a minimum of 2 miles diagonal radar separation is provided. Aircraft on the same localizer/azimuth course within 10 miles of the runway end are provided a minimum of 2.5 miles radar separation. In addition, a minimum of 1,000 feet vertical or a minimum of three miles radar separation is provided between aircraft during turn on to the parallel final approach course.

d. Whenever parallel ILS/MLS approaches are in progress, pilots are informed that approaches to both runways are in use. In addition, the radar controller will have the interphone capability of communicating with the tower controller where separation responsibility has not been delegated to the tower.

5-4-15. SIMULTANEOUS PARALLEL ILS/MLS APPROACHES (INDEPENDENT)

a. System: An approach system permitting simultaneous ILS/MLS approaches to parallel runways with centerlines separated by 4,300 to 9,000 feet, and equipped with final monitor controllers. Simultaneous parallel ILS/MLS approaches require radar monitoring to ensure separation between aircraft on the adjacent parallel approach course. Aircraft position is tracked by final monitor controllers who will issue instructions to aircraft observed deviating from the assigned localizer course. Staggered radar separation procedures are not utilized. Integral parts of a total system are ILS/MLS, radar, communications, ATC procedures, and required airborne equipment. The Approach Procedure Chart permitting simultaneous parallel ILS/MLS approaches will contain the note "simultaneous approaches authorized RWYS 14L and 14R," identifying the appropriate runways as the case may be. When advised that simultaneous parallel ILS/MLS approaches are in progress, pilots shall advise approach control immediately

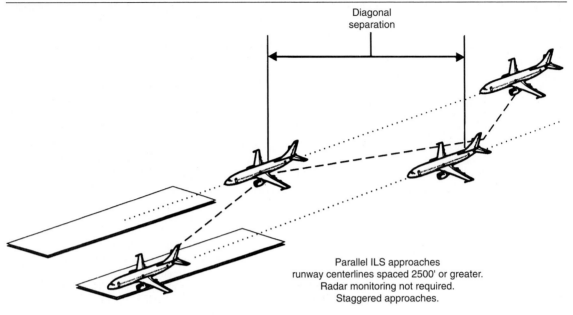

Diagonal separation

Parallel ILS approaches
runway centerlines spaced 2500' or greater.
Radar monitoring not required.
Staggered approaches.

Figure 5-4-16. Staggered ILS Approaches

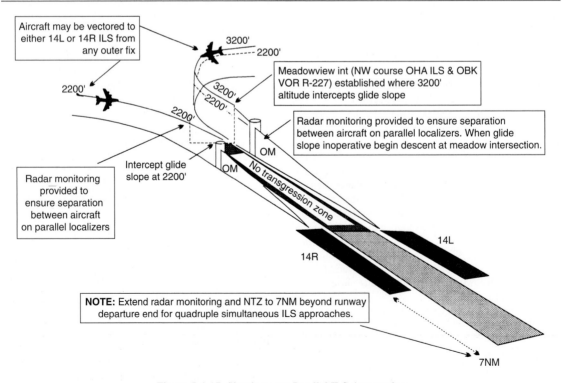

Aircraft may be vectored to either 14L or 14R ILS from any outer fix

3200'
2200'

2200'

Meadowview int (NW course OHA ILS & OBK VOR R-227) established where 3200' altitude intercepts glide slope

3200'
2200'

2200'

Radar monitoring provided to ensure separation between aircraft on parallel localizers. When glide slope inoperative begin descent at meadow intersection.

OM

Intercept glide slope at 2200'

OM

No transgression zone

Radar monitoring provided to ensure separation between aircraft on parallel localizers

14L

14R

NOTE: Extend radar monitoring and NTZ to 7NM beyond runway departure end for quadruple simultaneous ILS approaches.

7NM

Figure 5-4-17. Simultaneous Parallel ILS Approaches

of malfunctioning or inoperative receivers, or if a simultaneous parallel ILS/ MLS approach is not desired.

b. Radar Monitoring: This service is provided for each simultaneous parallel ILS/MLS approach to ensure aircraft do not deviate from the final approach course. Radar monitoring includes instructions if an aircraft nears or penetrates the prescribed NTZ (an area 2,000 feet wide located equidistant between parallel final approach courses). This service will be provided as follows:

1. During turn on to parallel final approach aircraft will be provided 3 miles radar separation or a minimum of 1,000 feet vertical separation. Aircraft will not be vectored to intercept the final approach course at an angle greater than thirty degrees.

2. The final monitor controller will have the capability of overriding the tower controller on the tower frequency.

3. Pilots will be instructed to monitor the tower frequency to receive advisories and instructions.

4. Aircraft observed to overshoot the turn-on or to continue on a track which will penetrate the NTZ will be instructed to return to the correct final approach course immediately. The final monitor controller may also issue missed approach or breakout instructions to the deviating aircraft.

PHRASEOLOGY—"(AIRCRAFT CALL SIGN) YOU HAVE CROSSED THE FINAL APPROACH COURSE. TURN (LEFT/RIGHT) IMMEDIATELY AND RETURN TO THE LOCALIZER/AZIMUTH COURSE,"

OR

"(AIRCRAFT CALL SIGN) TURN (LEFT/RIGHT) AND RETURN TO THE LOCALIZER/AZIMUTH COURSE."

5. If a deviating aircraft fails to respond to such instructions or is observed penetrating the NTZ, the aircraft on the adjacent final approach course may be instructed to alter course.

PHRASEOLOGY—"(TRAFFIC ALERT (AIRCRAFT CALL SIGN) TURN (LEFT/RIGHT) IMMEDIATELY HEADING (DEGREES), (CLIMB/DESCEND) AND MAINTAIN (ALTITUDE)."

6. Radar monitoring will automatically be terminated when visual separation is applied, the aircraft reports the approach lights or runway in sight, or the aircraft is 1 mile or less from the runway threshold (for runway centerlines spaced 4300' or greater). Final monitor controllers will not advise pilots when radar monitoring is terminated.

5-4-16. SIMULTANEOUS CLOSE PARALLEL ILS PRM APPROACHES (INDEPENDENT) AND SIMULTANEOUS OFFSET INSTRUMENT APPROACHES (SOIA) (SEE FIG 5-4-18.)

a. System

1. ILS/PRM is an acronym for Instrument Landing System/Precision Runway Monitor.

(a) An approach system that permits simultaneous ILS/PRM approaches to dual runways with centerlines separated by less than 4,300 feet but at least 3,400 feet for parallel approach courses, and at least 3,000 feet if one ILS if offset by 2.5 to 3.0 degrees. The airspace between the final approach courses contains a No Transgression Zone (NTZ) with surveillance provided by two PRM monitor controllers, one for each approach course. To qualify for reduced lateral runway separation, monitor controllers must be equipped with high update radar and high resolution ATC radar displays, collectively called a PRM system. The PRM system displays almost instantaneous radar information. Automated tracking software provides PRM monitor controllers with aircraft identification, position, speed and a ten-second projected position, as well as visual and aural controller alerts. The PRM system is a supplemental requirement for simultaneous close parallel approaches in addition to the system requirements for simultaneous parallel ILS/MLS approaches described in paragraph 5-4-15, Simultaneous Parallel ILS/MLS Approaches (Independent).

(b) Simultaneous close parallel ILS/PRM approaches are depicted on a separate Approach Procedure Chart titled ILS/PRM Rwy XXX (Simultaneous Close Parallel).

2. SOIA is an acronym for Simultaneous Offset Instrument Approach, a procedure used to conduct simultaneous approaches to runways spaced less than 3,000 feet, but at least 750 feet apart. The SOIA procedure utilizes an ILS/PRM approach to one runway and an offset Localizer Type Directional Aid (LDA)/PRM approach with glide slope to the adjacent runway.

(a) The ILS/PRM approach plates used in SOIA operations are identical to other ILS/PRM approach plates, with an additional note, which provides the separation between the two runways used for simultaneous approaches. The LDA/PRM approach plate displays the required notations for closely spaced approaches as well as depicting the visual segment of the approach, and a note that provides the separation between the two runways used for simultaneous operations.

(b) Controllers monitor the SOIA ILS/PRM and LDA/PRM approaches with a PRM system using high update radar and high-resolution ATC radar displays in exactly the same manner as is done for ILS/PRM approaches. The procedures and system requirements for SOIA ILS/PRM and LDA/PRM approaches are identical with those used for simultaneous close parallel ILS/PRM approaches until near the LDA/PRM approach missed approach point (MAP)—where visual acquisition of the ILS aircraft by the LDA aircraft must be accomplished. Since the ILS/PRM and LDA/PRM approaches are identical except for the visual segment in the SOIA concept, an understanding of the procedures for conducting ILS/PRM approaches is essential before conducting a SOIA ILS/PRM or LDA/PRM operation.

(c) In SOIA, the approach course separation (instead of the runway separation) meets established close parallel approach criteria. Refer to FIG 5-4-19 for the generic SOIA approach geometry. A visual segment of the LDA/PRM approach is established between the LDA MAP and the runway threshold. Aircraft transition in visual conditions

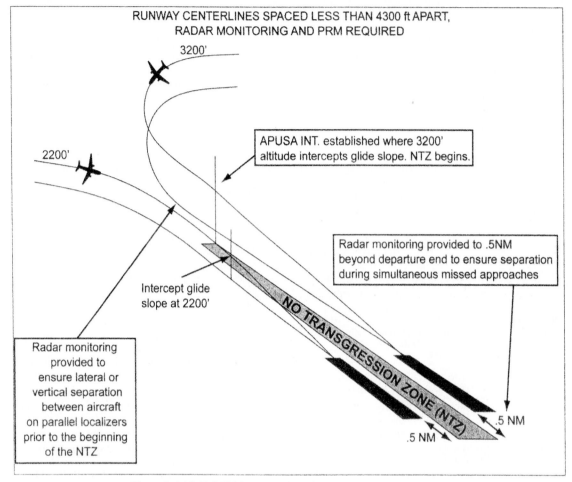

RUNWAY CENTERLINES SPACED LESS THAN 4300 ft APART,
RADAR MONITORING AND PRM REQUIRED

3200'

APUSA INT. established where 3200'
altitude intercepts glide slope. NTZ begins.

2200'

Radar monitoring provided to .5NM
beyond departure end to ensure separation
during simultaneous missed approaches

Intercept glide
slope at 2200'

NO TRANSGRESSION ZONE (NTZ)

Radar monitoring
provided to
ensure lateral or
vertical separation
between aircraft
on parallel localizers
prior to the beginning
of the NTZ

.5 NM

.5 NM

Figure 5-4-18. ILS PRM Approaches (Simultaneous Close Parallel)

from the LDA course, beginning at the LDA MAP, to align with the runway and can be stabilized by 500 feet above ground level (AGL) on the extended runway centerline. Aircraft will be "paired" in SOIA operations, with the ILS aircraft ahead of the LDA aircraft prior to the LDA aircraft reaching the LDA MAP. A cloud ceiling for the approach is established so that the LDA aircraft has nominally 30 seconds to acquire the leading ILS aircraft prior to the LDA aircraft reaching the LDA MAP. If visual acquisition is not accomplished, a missed approach must be executed.

b. Requirements

Besides system requirements as identified in subparagraph a above all pilots must have completed special training before accepting a clearance to conduct ILS/PRM or LDA/PRM Simultaneous Close Parallel Approaches.

1. Pilot Training Requirement. Pilots must complete special pilot training, as outlined below, before accepting a clearance for a simultaneous close parallel ILS/PRM or LDA/PRM approach.

(a) For operations under 14 CFR Parts 121, 129, and 135 pilots must comply with FAA approved company training as identified in their Operations Specifications. Training, at a minimum, must require pilots to view the FAA video "ILS PRM AND SOIA APPROACHES: INFORMATION FOR AIR CARRIER PILOTS." Refer to http://www.faa.gov/ for additional information and to view or download the video.

(b) For operations under Part 91:

(1) Pilots operating transport category aircraft must be familiar with PRM operations as contained in this section of the Aeronautical Information Manual (AIM). In addition, pilots operating transport category aircraft must view the FAA video "ILS PRM AND SOIA APPROACHES: INFORMATION FOR AIR CARRIER PILOTS." Refer to

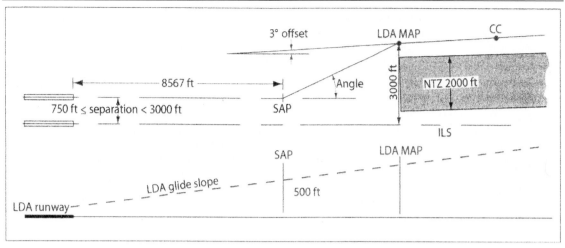

Figure 5-4-19. SOIA Approach Geometry

NOTE—

SAP	*THE SAP IS A DESIGN POINT ALONG THE EXTENDED CENTERLINE OF THE INTENDED LANDING RUNWAY ON THE4 GLIDE SLOPE AT 500 FEET ABOVE THE LANDING THRESHOLD. IT IS USED TO VERIFY A SUFFICIENT DISTANCE IS PROVIDED FOR THE VISUAL MANEUVER AFTER THE MISSED APPROACH POINT (MAP) TO PERMIT THE PILOTS TO CONFORM TO APPROVED, STABILIZED APPROACH CRITERIA.*
MAP	*THE POINT ALONG THE LDA WHERE THE COURSE SEPARATION WITH THE ADJACENT ILS REACHES 3,000 FEET. THE ALTITUDE OF THE GLIDE SLOPE AT THAT POINT DETERMINES THE APPROACH MINIMUM DESCENT ALTITUDE AND IS WHERE THE NTZ TERMINATES. MANEUVERING INSIDE THE MAP IS DONE IN VISUAL CONDITIONS.*
ANGLE	*ANGLE FORMED AT THE INTERSECTION OF THE EXTENDED LDA RUNWAY CENTERLINE AND A LINE DRAWN BETWEEN THE LDA MAP AND THE SAP. THE SIZE OF THE ANGLE IS DETERMINED BY THE FAA SOIA COMPUTER DESIGN PROGRAM, AND IS DEPENDENT ON WHETHER HEAVY AIRCRAFT USE THE LDA AND THE SPACING BETWEEN THE RUNWAYS.*
VISIBILITY	*DISTANCE FROM MAP TO RUNWAY THRESHOLD IN STATUTE MILES (LIGHT CREDIT APPLIES).*
PROCEDURE	*LDA AIRCRAFT MUST SEE THE RUNWAY LANDING ENVIRONMENT AND, IF LESS THAN STANDARD RADAR SEPARATION EXISTS BETWEEN THE AIRCRAFT ON THE ADJACENT ILS COURSE, THE LDA AIRCRAFT MUST VISUALLY ACQUIRE THE ILS AIRCRAFT AND REPORT IT IN SIGHT TO ATC PRIOR TO THE LDA MAP.*
CC	*CLEAR CLOUDS.*

http://www.faa.gov/ for additional information and to view or download the video

(2) Pilots not operating transport category aircraft must be familiar with PRM and SOIA operations as contained in this section of the AIM. The FAA strongly recommends that pilots not involved in transport category aircraft operations view the FAA video, "ILS PRM AND SOIA APPROACHES: INFORMATION FOR GENERAL AVIATION PILOTS." Refer to additional information and to view or download the video. Refer to http://www.faa.gov/ for additional information and to view or download the video.

2. ATC Directed Breakout. An ATC directed "breakout" is defined as a vector off the ILS or LDA approach course in response to another aircraft penetrating the NTZ, the 2,000 foot wide area located equidistant between the two approach courses that is monitored by the PRM monitor controllers.

3. Dual Communications. The aircraft flying the ILS/PRM or LDA/PRM approach must have the capability of enabling the pilot/s to listen to two communications frequencies simultaneously.

c. Radar Monitoring. Simultaneous close parallel ILS/PRM and LDA/PRM approaches require that final monitor controllers utilize the PRM system to ensure prescribed separation standards are met. Procedures and communications phraseology are also described in paragraph 5-4-15, Simultaneous Parallel ILS/MLS Approaches (Independent). A minimum of 3 miles radar separation or 1,000 feet vertical separation will be provided during the turn-on to close parallel final approach courses. To ensure separation is maintained, and in order to avoid an imminent situation during simultaneous close parallel ILS/PRM or SOIA ILS/PRM and LDA/PRM approaches, pilots must immediately comply with PRM monitor controller instructions. In the event of a missed approach, radar monitoring is provided to one-half mile beyond the most distant of the two runway departure ends for ILS/RPM approaches. In SOIA, PRM radar monitoring terminates at the LDA MAP. Final monitor controllers will not notify pilots when radar monitoring is terminated.

d. Attention All Users Page (AAUP). ILS/PRM and LDA/PRM approach charts have an AAUP associated with them that must be referred to in preparation for conducting the approach. This page contains the following instructions that must be followed if the pilot is unable to accept an ILS/PRM or LDA/PRM approach.

1. At airports that conduct PRM operations, (ILS/PRM or, in the case of airports where SOIAs are conducted, ILS/PRM and LDA/PRM approaches) pilots not qualified to except PRM approaches must contact the FAA Command Center prior to departure (1-800-333-4286) to obtain an arrival reservation (see FAA Advisory Circular 90-98, Simultaneous Closely Spaced Parallel Operations at Airports Using Precision Runway Monitor (PRM) Systems). Arriving flights that are unable to participate in ILS/PRM or LDA/PRM approaches and have not received an arrival reservation are subject to diversion to another airport or delays. Pilots en route to a PRM airport designated as an alternate, unable to reach their filed destination, and who are not qualified to participate in ILS/PRM or LDA/PRM approaches must advise ATC as soon as practical that they are unable to participate. Pilots who are qualified to participate but experience an en route equipment failure that would preclude participation in PRM approaches should notify ATC as soon as practical.

2. The AAUP covers the following operational topics:

(a) ATIS. When the ATIS broadcast advises ILS/PRM approaches are in progress (or ILS PRM and LDA PRM approaches in the case of SOIA), pilots should brief to fly the ILS/PRM or LDA/PRM approach. If later advised to expect the ILS or LDA approach (should one be published), the ILS/PRM or LDA/PRM chart may be used after completing the following briefing items:

(1) Minimums and missed approach procedures are unchanged.

(2) PRM Monitor frequency no longer required.

(3) ATC may assign a lower altitude for glide slope intercept.

NOTE—IN THE CASE OF THE LDAIPRM APPROACH, THIS BRIEFING PROCEDURE ONLY APPLIES IF AN LDA APPROACH IS ALSO PUBLISHED.

IN THE CASE OF THE SOIA ILS/PRM AND LDA/PRM PROCEDURE, THE AAUP DESCRIBES THE WEATHER CONDITIONS IN WHICH SIMULTANEOUS APPROACHES ARE AUTHORIZED:

SIMULTANEOUS APPROACH WEATHER MINIMUMS ARE X,XXX FEET (CEILING), X MILES (VISIBILITY).

(b) Dual VHF Communications Required. To avoid blocked transmissions, each runway will have two frequencies, a primary and a monitor frequency. The tower controller will transmit on both frequencies. The monitor controller's transmissions, if needed, will override both frequencies. Pilots will ONLY transmit on the tower controller's frequency, but will listen to both frequencies. Begin to monitor the PRM monitor controller when instructed by ATC to contact the tower. The volume levels should be set about the same on both radios so that the pilots will be able to hear transmissions on at least one frequency if the other is blocked. Site specific procedures take precedence over the general information presented in this paragraph. Refer to the AAUP for applicable procedures at specific airports.

(c) Breakouts. Breakouts differ from other types of abandoned approaches in that they can happen anywhere and unexpectedly. Pilots directed by ATC to break off an approach must assume that an aircraft is blundering toward them and a breakout must be initiated immediately.

(1) Hand-fly breakouts. All breakouts are to be hand-flown to ensure the maneuver is accomplished in the shortest amount of time.

(2) ATC Directed "Breakouts." ATC directed breakouts will consist of a turn and a climb or descent. Pilots must always initiate the breakout in response to an air traffic controller's instruction. Controllers will give a descending breakout only when there are no other reasonable options available, but in no case will the descent be below the minimum vectoring altitude (MVA) which provides at least 1,000 feet required obstruction clearance. The AAUP provides the MVA in the final approach segment as X,XXX feet at (Name) Airport.

NOTE—"TRAFFIC ALERT." IF AN AIRCRAFT ENTERS THE "NO TRANSGRESSION ZONE" (NTZ), THE CONTROLLER WILL BREAKOUT THE THREATENED AIRCRAFT ON THE ADJACENT APPROACH. THE PHRASEOLOGY FOR THE BREAKOUT WILL BE:

PHRASEOLOGY—
TRAFFIC ALERT, (AIRCRAFT CALL SIGN) TURN (LEFT/RIGHT) IMMEDIATELY, HEADING (DEGREES), CLIMB/DESCEND AND MAINTAIN (ALTITUDE).

(d) ILS/PRM Navigation. The pilot may find crossing altitudes along the final approach course. The pilot is advised that descending on the ILS glideslope ensures complying with any charted crossing restrictions.

SOIA AAUP differences from ILS PRM AAUP

(e) ILS/PRM LDA Traffic (only published on ILS/PRM AAUP when the ILS PRM approach is used in conjunctions with an LDA/PRM approach to the adjacent runway). To provide better situational awareness, and because traffic on the LDA may be visible on the ILS aircraft's TCAS, pilots are reminded of the fact that aircraft will be maneuvering behind them to align with the adjacent runway. While conducting the ILS/PRM approach to Runway XXX, other aircraft may be conducting the offset LDA/PRM approach to Runway XXX. These aircraft will approach from the (left/right)-rear and will realign with runway XXX after making visual contact with the ILS traffic. Under normal circumstances these aircraft will not pass the ILS traffic.

SOIA LDA/PRM AAUP Items. The AAUP for the SOIA LDA/PRM approach contains most information found on ILS/PRM AAUUs. It replaces certain information as seen below and provides pilots with the procedures to be used in the visual segment of the LDA/PRM approach, from the time the ILS aircraft is visually acquired until landing.

(f) SOIA LDA/PRM Navigation (replaces ILS/PRM (d) and (e) above). The pilot may find crossing altitudes along the final approach course. The pilot is advised that descending on the LDA glideslope ensures complying with any charted crossing restrictions. Remain on the LDA course until passing XXXXX (LDA MAP name) intersection prior to maneuvering to align with the centerline of runway XXX.

(g) SOIA (Name) Airport Visual Segment (replaces ILS/PRM (e) above). Pilot procedures for navigating beyond the LDA MAP are spelled out. If ATC advises that there is traffic on the adjacent ILS, pilots are authorized to continue past the LDA MAP to align with runway centerline when:

(1) the ILS traffic is in sight and is expected to remain in sight,

(2) ATC has been advised that "traffic is in sight,"

(3) the runway environment is in sight.

Otherwise, a missed approach must be executed. Between the LDA MAP and the runway threshold, pilots of the LDA aircraft are responsible for separating themselves visually from traffic on the ILS approach, which means maneuvering the aircraft as necessary to avoid the ILS traffic until landing, and providing wake turbulence avoidance, if applicable. Pilots should advise ATC, as soon as practical, if visual contact with the ILS traffic is lost and execute a missed approach unless otherwise instructed by ATC.

e. SOIA LDA Approach Wake Turbulence. Pilots are responsible for wake turbulence avoidance when maneuvering between the LDA missed approach point and the runway threshold.

f. Differences between ILS and ILS/PRM approaches of importance to the pilot.

1. Runway Spacing. Prior to ILS/PRM and LDA/PRM approaches, most ATC directed breakouts were the result of two aircraft in-trail on the same final approach course getting too close together. Two aircraft going in the same direction did not mandate quick reaction times. With PRM approaches, two aircraft could be along side each other, navigating on courses that are separated by less than 4,300 feet. In the unlikely event that an aircraft "blunders" off its course and makes a worst case turn of 30 degrees toward the adjacent final approach course, closing speeds of 135 feet per second could occur that constitute the need for quick reaction. A blunder has to be recognized by the monitor controller, and breakout instructions issued to the endangered aircraft. The pilot will not have any warning that a breakout is eminent because the blundering aircraft will be on another frequency. It is important that, when a pilot receives breakout instructions, he/she assumes that a blundering aircraft is about to or has penetrated the NTZ and is heading toward his/her approach course. The pilot must initiate a breakout as soon as safety allows. While conducting PRM approaches, pilots must maintain an increased sense of awareness in order to immediately react to an ATC instruction (breakout) and maneuver as instructed by ATC, away from a blundering aircraft.

2. Communications. To help in avoiding communication problems caused by stuck microphones and two parties talking at the same time, two frequencies for each runway will be in use during ILS/PRM and LDA/PRM approach operations, the primary tower frequency and the PRM monitor frequency. The tower controller transmits and receives in a normal fashion on the primary frequency and also transmits on the PRM monitor frequency. The monitor controller's transmissions override on both frequencies. The pilots flying the approach will listen to both frequencies but only transmit on the primary tower frequency. If the PRM monitor controller initiates a breakout and the primary frequency is blocked by another transmission, the breakout instruction will still be heard on the PRM monitor frequency.

3. Hand-flown Breakouts. The use of the autopilot is encouraged while flying an ILS/PRM or LDA/PRM approach, but the autopilot must be disengaged in the rare event that a breakout is issued. Simulation studies of breakouts have shown that a hand-flown breakout can be initiated consistently faster than a breakout performed using the autopilot.

4. TCAS. The ATC breakout instruction is the primary means of conflict resolution. TCAS, if installed, provides another form of conflict resolution in the unlikely event other separation standards would fail. TCAS is not required to conduct a closely spaced approach.

The TCAS provides only vertical resolution of aircraft conflicts, while the ATC breakout instruction provides both vertical and horizontal guidance for conflict resolutions. Pilots should always immediately follow the TCAS Resolution Advisory (RA), whenever it is received. Should a TCAS RA be received before, during, or after an ATC breakout instruction is issued, the pilot should follow the RA, even if it conflicts with the climb/descent portion of the breakout maneuver. If following an RA requires deviating from an ATC clearance, the pilot shall advise ATC as soon as practical. While following an RA, it is <u>extremely important</u> that the pilot also comply with the turn portion of the ATC breakout instruction unless the pilot determines safety to be factor. Adhering to these procedures assures the pilot that acceptable "breakout" separation margins will always be provided, even in the face of a normal procedural or system failure.

5. Breakouts. The probability is extremely low that an aircraft will "blunder" from its assigned approach course and enter the NTZ, causing ATC to "breakout" the aircraft approaching on the adjacent ILS course. However, because of the close proximity of the final approach courses, it is essential that pilots follow the ATC breakout instructions precisely and expeditiously. The controller's "breakout" instructions provide conflict resolution for the threatened aircraft, with the turn portion of the "breakout" being the single most important element in achieving maximum protection. A descending breakout will only be issued when it is the only controller option. In no case will the controller descend an aircraft below the MVA, which will provide at least 1,000 feet clearance above obstacles. The pilot is not expected to exceed 1,000 feet per minute rate of descent in the event a descending breakout is issued.

5-4-17. SIMULTANEOUS CONVERGING INSTRUMENT APPROACHES

a. ATC may conduct instrument approaches simultaneously to converging runways; i.e., runways having an included angle from 15 to 100 degrees, at airports where a program has been specifically approved to do so.

b. The basic concept requires that dedicated, separate standard instrument approach procedures be developed for each converging runway included. Missed Approach Points must be at least 3 miles apart and missed approach procedures ensure that missed approach protected airspace does not overlap.

c. Other requirements are: radar availability, nonintersecting final approach courses, precision (ILS/MLS) approach systems on each runway and, if runways intersect, controllers must be able to apply visual separation as well as intersecting runway separation criteria. Intersecting runways also require minimums of at least 700 foot ceilings and 2 miles visibility. Straight in approaches and landings must be made.

d. Whenever simultaneous converging approaches are in progress, aircraft will be informed by the controller as soon as feasible after initial contact or via ATIS. Additionally, the radar controller will have direct communications capability with the tower controller where separation responsibility has not been delegated to the tower.

5-4-18. RNP SAAAR INSTRUMENT APPROACH PROCEDURES

These procedures require authorization analogous to the special authorization required for Category II or III ILS procedures. Special aircraft and aircrew authorization required (SAAAR) procedures are to be conducted by aircrews meeting special training requirements in aircraft that meet the specified performance and functional requirements.

a. Unique characteristics of RNP SAAAR Approaches

l. RNP value. Each published line of minima has an associated RNP value. The indicated value defines the lateral and vertical performance requirements. A minimum RNP type is documented as part of the RNP SAAAR authorization for each operator and may vary depending on aircraft configuration or operational procedures (e.g., GPS inoperative, use of flight director vice autopilot).

2. Curved path procedures. Some RNP approaches have a curved path, also called a radius-to-a-fix (RF) leg. Since not all aircraft have the capability to fly these arcs, pilots are responsible for knowing if they can conduct an RNP approach with an arc or not. Aircraft speeds, winds and bank angles have been taken into consideration in the development of the procedures.

3. RNP required for extraction or not. Where required, the missed approach procedure may use RNP values less than RNP-1. The reliability of the navigation system has to be very high in order to conduct these approaches. Operation on these procedures generally requires redundant equipment, as no single point of failure can cause loss of both approach and missed approach navigation.

4. Non-standard speeds or climb gradients. RNP SAAAR approaches are developed based on standard approach speeds and a 200 ft/NM climb gradient in the missed approach. Any exceptions to these standards will be indicated on the approach procedure, and the operator should ensure they can comply with any published restrictions before conducting the operation.

5. Temperature Limits. For aircraft using barometric vertical navigation (without temperature compensation) to conduct the approach, low and high-temperature limits are identified on the procedure. Cold temperatures reduce the glidepath angle while high temperatures increase the glidepath angle. Aircraft using baro VNAV with temperature compensation or aircraft using an alternate means for vertical guidance (e.g., SBAS) may disregard the temperature restrictions. The charted temperature limits are evaluated for the final approach segment only. Regardless of charted temperature limits or temperature compensation by the FMS, the pilot may need to manually compensate for cold temperature on minimum altitudes and the decision altitude.

6. Aircraft size. The achieved minimums may be dependent on aircraft size. Large aircraft may require higher minimums due to gear height and/or wingspan. Approach procedure charts will be annotated with applicable aircraft size restrictions.

b. Types of RNP SAAAR Approach Operations

1. RNP Stand-alone Approach Operations. RNP SAAAR procedures can provide access to runways regardless of the ground-based NAVAID infrastructure, and can be designed to avoid obstacles, terrain, airspace, or resolve environmental constraints.

2. RNP Parallel Approach (RPA) Operations. RNP SAAAR procedures can be used for parallel approaches where the runway separation is adequate (See FIG 5-4-20). Parallel approach procedures can be used either simultaneously or as stand-alone operations. They may be part of either independent or dependent operations depending on the ATC ability to provide radar monitoring.

3. RNP Parallel Approach Runway Transitions (RPAT) Operations. RPAT approaches begin as a parallel IFR approach operation using simultaneous independent

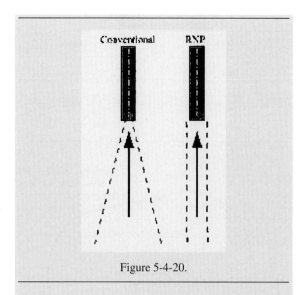

Figure 5-4-20.

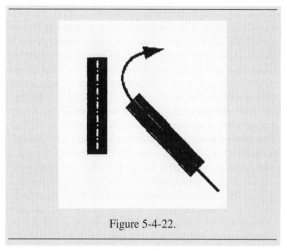

Figure 5-4-22.

or dependent procedures. (See FIG 5-4-21). Visual separation standards are used in the final segment of the approach after the final approach fix, to permit the RPAT aircraft to transition in visual conditions along a predefined lateral and vertical path to align with the runway centerline.

4. RNP Converging Runway Operations. At airports where runways converge, but may or may not intersect, an RNP SAAAR approach can provide a precise curved missed approach path that conforms to aircraft separation minimums for simultaneous operations (See FIG 5-4-22). By flying this curved missed approach path with high accuracy and containment provided by RNP, dual runway operations may continue to be used to lower ceiling and visibility values than currently available. This type of operation allows greater capacity at airports where it can be applied.

5-4-19. SIDE-STEP MANEUVER

a. ATC may authorize a standard instrument approach procedure which serves either one of parallel runways that

are separated by 1,200 feet or less followed by a straight-in landing on the adjacent runway.

b. Aircraft that will execute a side-step maneuver will be cleared for a specified approach procedure and landing on the adjacent parallel runway. Example, "cleared ILS runway 7 left approach, side-step to runway 7 right." Pilots are expected to commence the side-step maneuver as soon as possible after the runway or runway environment is in sight.

NOTE—SIDE-STEP MINIMA ARE FLOWN TO A MINIMUM DESCENT ALTITUDE (MDA) REGARDLESS OF THE APPROACH AUTHORIZED.

c. Landing minimums to the adjacent runway will be based on nonprecision criteria and therefore higher than the precision minimums to the pimary runway but will normally be lower than the published willing minimums.

5-4-20. APPROACH AND LANDING MINIMUMS

a. Landing Minimums: The rules applicable to landing minimums are contained in 14 CFR Section 91.175. TBL 5-4-1 may be used to convert RVR to ground or flight visibility. For converting RVR values that fall between listed values, use the next higher RVR value; do not interpolate. For example, when converting 1800 RVR, use 2400 RVR with the resultant visibility of ½ mile.

Table 5-4-1

RVR	Value Conversions RVR Visibility (statute miles)
1600	¼
2400	½
3200	⅝
4000	¾
4500	⅞
5000	1
6000	1¼

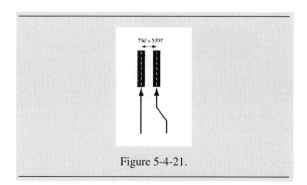

Figure 5-4-21.

b. Published Approach Minimums: Approach minimums are published for different aircraft categories and consist of a minimum altitude (DA, DH, MDA) and required visibility. These minimums are determined by applying the appropriate TERPS criteria. When a fix is incorporated in a nonprecision final segment, two sets of minimums may be published: one, for the pilot that is able to identify the fix, and a second for the pilot that cannot. Two sets of minimums may also be published when a second altimeter source is used in the procedure. When a nonprecision procedure incorporates a stepdown fix in the final segment and a second altimeter source, two sets of minimums are published to account for the stepdown fix and a note addresses minimums for the second altimeter source.

c. Obstacle Clearance: Final approach obstacle clearance is provided from the start of the final segment to the runway or Missed Approach Point, whichever occurs last. Side-step obstacle protection is provided by increasing the width of the final approach obstacle clearance area.

1. Circling approach protected areas are defined by the tangential connection of arcs drawn from each runway end. The arc radii distance differs by aircraft approach category. Because of obstacles near the airport, a portion of the circling area may be restricted by a procedural note: e.g., "Circling NA E of RWY 17-35." Obstacle clearance is provided at the published minimums for the pilot that makes a straight-in approach, side-steps, circles, or executes the missed approach. Missed approach obstacle clearance requirements may dictate the published minimums for the approach. (See FIG 5-4-23).

2. Precision Object Free Zone (POFZ). A volume of airspace above an area beginning at the runway threshold, at the threshold elevation, and centered on the extended runway centerline. The POFZ is 200 feet (60m) long and 800 feet (240m) wide. The POFZ must be clear when an aircraft on a vertically guided final approach is within 2 nautical miles of the runway threshold and the reported ceiling is below 250 feet or visibility less than 3/4 statute mile(SM) (or runway visual range below 4,000 feet). If the POFZ is not clear, the MINIMUM authorized height above touchdown (HAT) and visibility is 250 feet and 3/4 SM. The POFZ is considered clear even if the wing of the aircraft holding on a taxiway waiting for runway clearance penetrates the POFZ; however, neither the fuselage nor the tail may infringe on the POFZ. The POFZ is applicable at all runway ends including displaced thresholds.

NOTE—THE TARGET DATE FOR MANDATORY POFZ COMPLIANCE FOR EVERY AIRPORT NATIONALLY IS JANUARY 1, 2007.

d. Straight-in Minimums: Are shown on the IAP when the final approach course is within 30 degrees of the runway alignment (15 degrees for GPS IAP's) and a normal descent can be made from the IFR altitude shown on the IAP to the runway surface. When either the normal rate of descent or the runway alignment factor of 30 degrees (15 degrees for GPS IAP's) is exceeded, a straight-in minimum

Approach category	Radius (miles)
A	1.3
B	1.5
C	1.7
D	2.3
E	4.5

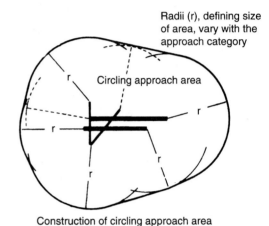

Radii (r), defining size of area, vary with the approach category

Circling approach area

Construction of circling approach area

Figure 5-4-23. Final Approach Obstacle Clearance

is not published and a circling minimum applies. The fact that a straight-in minimum is not published does not preclude pilots from landing straight-in if they have the active runway in sight and have sufficient time to make a normal approach for landing. Under such conditions and when

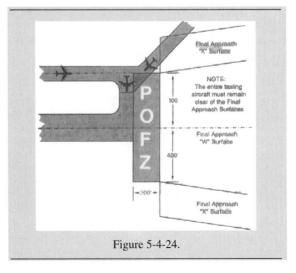

Figure 5-4-24.

ATC has cleared them for landing on that runway, pilots are not expected to circle even though only circling minimums are published. If they desire to circle, they should advise ATC.

e. Side-Step Maneuver Minimums: Landing minimums for a side-step maneuver to the adjacent runway will normally be higher than the minimums to the primary runway.

f. Circling Minimums: In some busy terminal areas, ATC may not allow circling and circling minimums will not be published. Published circling minimums provide obstacle clearance when pilots remain within the appropriate area of protection. Pilots should remain at or above the circling altitude until the aircraft is continuously in a position from which a descent to a landing on the intended runway can be made at a normal rate of descent using normal maneuvers. Circling may require maneuvers at low altitude, at low airspeed, and in marginal weather conditions. Pilots must use sound judgment, have an indepth knowledge of their capabilities, and fully understand the aircraft performance to determine the exact circling maneuver since weather, unique airport design, and the aircraft position, altitude, and airspeed must all be considered. The following basic rules apply:

1. Maneuver the shortest path to the base or downwind leg, as appropriate, considering existing weather conditions. There is no restriction from passing over the airport or other runways.

2. It should be recognized that circling maneuvers may be made while VFR or other flying is in progress at the airport. Standard left turns or specific instruction from the controller for maneuvering must be considered when circling to land.

3. At airports without a control tower, it may be desirable to fly over the airport to observe wind and turn indicators and other traffic which may be on the runway or flying in the vicinity of the airport.

g. Instrument Approach At a Military Field: When instrument approaches are conducted by civil aircraft at military airports, they shall be conducted in accordance with the procedures and minimums approved by the military agency having jurisdiction over the airport.

5-4-21. MISSED APPROACH

a. When a landing cannot be accomplished, advise ATC and, upon reaching the Missed Approach Point defined on the approach procedure chart, the pilot must comply with the missed approach instructions for the procedure being used or with an alternate missed approach procedure specified by ATC.

b. Protected obstacle clearance areas for missed approach are predicated on the assumption that the missed approach is initiated at the decision height (DH) or at the missed approach point and not lower than minimum descent altitude (MDA). A climb of at least 200 feet per nautical mile is required, (except for Copter approaches, where a climb of at least 400 feet per nautical mile is required), unless a higher climb gradient is published on the approach chart. Reasonable buffers are provided for normal maneuvers. However, no consideration is given to an abnormally early turn. Therefore, when an early missed approach is executed, pilots should, unless otherwise cleared by ATC, fly the IAP as specified on the approach plate to the missed approach point at or above the MDA or DH before executing a turning maneuver.

c. If visual reference is lost while circling-to-land from an instrument approach, the missed approach specified for that particular procedure must be followed (unless an alternate missed approach procedure is specified by ATC). To become established on the prescribed missed approach course, the pilot should make an initial climbing turn toward the landing runway and continue the turn until established on the missed approach course. Inasmuch as the circling maneuver may be accomplished in more than one direction, different patterns will be required to become established on the prescribed missed approach course, depending on the aircraft position at the time visual reference is lost. Adherence to the procedure will assure that an aircraft will remain within the circling and missed approach obstruction clearance areas. (See FIG 5-4-25).

d. At locations where ATC Radar Service is provided, the pilot should conform to radar vectors when provided by ATC in lieu of the published missed approach procedure. (See FIG 5-4-26).

e. When approach has been missed, request clearance for specific action; i.e., to alternative airport, another approach, etc.

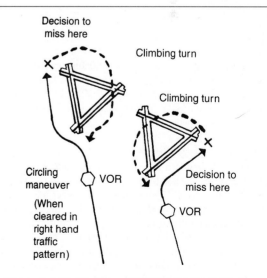

Figure 5-4-25. Circling and Missed Approach Obstruction Clearance Areas

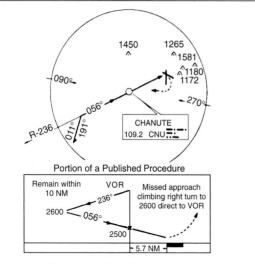

Portion of a Published Procedure

Figure 5-4-26. Missed Approach

f. Pilots must ensure that they have climbed to a safe altitude prior to proceeding off the published missed approach, especially in nonradar environments. Abandoning the missed approach prior to reaching the published altitude may not provide adequate terrain clearance. Additional climb may be required after reaching the holding pattern before proceeding back to the IAF or to an alternate.

g. Missed approach obstacle clearance is predicated on beginning the missed approach procedure at the Missed Approach Point (MAP) from MDA or DA and then climbing 200 feet/NM or greater. Initiating a go-around after passing the published MAP may result in total loss of obstacle clearance. To compensate for the possibility of reduced obstacle clearance during a go-around, a pilot should apply procedures used in takeoff planning. Pilots should refer to airport obstacle and departure data prior to initiating an instrument approach procedure. Such information may be found in the "TAKE-OFF MINIMUMS AND (OBSTACLE) DEPARTURE PROCEDURES" section of the U.S. TERMINAL PROCEDURES publication.

5-4-22. VISUAL APPROACH

a. A visual approach is conducted on an IFR flight plan and authorizes a pilot to proceed visually and clear of clouds to the airport. The pilot must have either the airport or the preceding identified aircraft in sight. This approach must be authorized and controlled by the appropriate air traffic control facility. Reported weather at the airport must have a ceiling at or above 1,000 feet and visibility 3 miles or greater. ATC may authorize this type approach when it will be operationally beneficial. Visual approaches are an IFR procedure conducted under IFR in visual meteorological conditions. Cloud clearance requirements of 14 CFR 91.155 are not applicable, unless required by operation specifications.

b. OPERATING TO AN AIRPORT WITHOUT WEATHER REPORTING SERVICE: ATC will advise the pilot when weather is not available at the destination airport. ATC may initiate a visual approach provided there is a reasonable assurance that weather at the airport is a ceiling at or above 1,000 feet and visibility 3 miles or greater (e.g. area weather reports, PIREPS, etc.).

c. OPERATING TO AN AIRPORT WITH AN OPERATING CONTROL TOWER: Aircraft may be authorized to conduct a visual approach to one runway while other aircraft are conducting IFR or VFR approaches to another parallel, intersecting, or converging runway. When operating to airports with parallel runways separated by less than 2,500 feet, the succeeding aircraft must report sighting the preceding aircraft unless standard separation is being provided by ATC. When operating to parallel runways separated by at least 2,500 feet but less than 4,300 feet, controllers will clear/vector aircraft to the final at an angle not greater than 30 degrees unless radar, vertical, or visual separation is provided during the turn-on. The purpose of the 30 degree intercept angle is to reduce the potential for overshoots of the final and to preclude side-by-side operations with one or both aircraft in a belly-up configuration during the turn-on. Once the aircraft are established within 30 degrees of final, or on the final, these operations may be conducted simultaneously. When the parallel runways are separated by 4,300 feet or more, or intersecting/converging runways are in use, ATC may authorize a visual approach after advising all aircraft involved that other aircraft are conducting operations to the other runway. This may be accomplished through use of the ATIS.

d. SEPARATION RESPONSIBILITIES: If the pilot has the airport in sight but cannot see the aircraft to be followed, ATC may clear the aircraft for a visual approach; however, ATC retains both separation and wake vortex separation responsibility. When visually following a preceding aircraft, acceptance of the visual approach clearance constitutes acceptance of pilot responsibility for maintaining a safe approach interval and adequate wake turbulence separation.

e. A visual approach is not an IAP and therefore has no missed approach segment. If a go around is necessary for any reason, aircraft operating at controlled airports will be issued an appropriate advisory/clearance/instruction by the tower. At uncontrolled airports, aircraft are expected to remain clear of clouds and complete a landing as soon as possible. If a landing cannot be accomplished, the aircraft is expected to remain clear of clouds and contact ATC as soon as possible for further clearance. Separation from other IFR aircraft will be maintained under these circumstances.

f. Visual approaches reduce pilot/controller workload and expedite traffic by shortening flight paths to the airport. It is the pilot's responsibility to advise ATC as soon as possible if a visual approach is not desired.

g. Authorization to conduct a visual approach is an IFR authorization and does not alter IFR flight plan cancellation responsibility.

REFERENCE—AIM, CANCELING IFR FLIGHT PLAN, PARAGRAPH 5-1-14.

h. Radar service is automatically terminated, without advising the pilot, when the aircraft is instructed to change to advisory frequency.

5-4-23. CHARTED VISUAL FLIGHT PROCEDURES (CVFP)

a. CVFPs are charted visual approaches established for environmental/noise considerations, and/or when necessary for the safety and efficiency of air traffic operations. The approach charts depict prominent landmarks, courses, and recommended altitudes to specific runways. CVFPs are designed to be used primarily for turbojet aircraft.

b. These procedures will be used only at airports with an operating control tower.

c. Most approach charts will depict some NAVAID information which is for supplemental navigational guidance only.

d. Unless indicating a Class B airspace floor, all depicted altitudes are for noise abatement purposes and are recommended only. Pilots are not prohibited from flying other than recommended altitudes if operational requirements dictate.

e. When landmarks used for navigation are not visible at night, the approach will be annotated *"PROCEDURE NOT AUTHORIZED AT NIGHT."*

f. CVFPs usually begin within 20 flying miles from the airport.

g. Published weather minimums for CVFPs are based on minimum vectoring altitudes rather than the recommended altitudes depicted on charts.

h. CVFPs are not instrument approaches and do not have missed approach segments.

i. ATC will not issue clearances for CVFPs when the weather is less than the published minimum.

j. ATC will clear aircraft for a CVFP after the pilot reports siting a charted landmark or a preceding aircraft. If instructed to follow a preceding aircraft, pilots are responsible for maintaining a safe approach interval and wake turbulence separation.

k. Pilots should advise ATC if at any point they are unable to continue an approach or lose sight of a preceding aircraft. Missed approaches will be handled as a go-around.

5-4-24. CONTACT APPROACH

a. Pilots operating in accordance with an IFR flight plan, provided they are clear of clouds and have at least 1 mile flight visibility and can reasonably expect to continue to the destination airport in those conditions, may request ATC authorization for a contact approach.

b. Controllers may authorize a contact approach provided:

1. The Contact Approach is specifically requested by the pilot. ATC cannot initiate this approach.

EXAMPLE—REQUEST CONTACT APPROACH.

2. The reported ground visibility at the destination airport is at least 1 statute mile.

3. The contact approach will be made to an airport having a standard or special instrument approach procedure.

4. Approved separation is applied between aircraft so cleared and between these aircraft and other IFR or special VFR aircraft.

EXAMPLE—CLEARED CONTACT APPROACH (AND, IF REQUIRED) AT OR BELOW (ALTITUDE) (ROUTING) IF NOT POSSIBLE (ALTERNATIVE PROCEDURES) AND ADVISE.

c. A contact approach is an approach procedure that may be used by a pilot (with prior authorization from ATC) in lieu of conducting a standard or special IAP to an airport. It is not intended for use by a pilot on an IFR flight clearance to operate to an airport not having a published and functioning IAP. Nor is it intended for an aircraft to conduct an instrument approach to one airport and then, when "in the clear," to discontinue that approach and proceed to another airport. In the execution of a contact approach, the pilot assumes the responsibility for obstruction clearance. If radar service is being received, it will automatically terminate when the pilot is instructed to change to advisory frequency.

5-4-25. LANDING PRIORITY

A clearance for a specific type of approach (ILS, MLS, ADF, VOR or Straight-in Approach) to an aircraft operating on an IFR flight plan does not mean that landing priority will be given over other traffic. ATCTs handle all aircraft, regardless of the type of flight plan, on a "first-come, first-served" basis. Therefore, because of local traffic or runway in use, it may be necessary for the controller in the interest of safety, to provide a different landing sequence. In any case, a landing sequence will be issued to each aircraft as soon as possible to enable the pilot to properly adjust the aircraft's flight path.

5-4-26. OVERHEAD APPROACH MANEUVER

a. Pilots operating in accordance with an IFR flight plan in visual meteorological conditions (VMC) may request ATC authorization for an overhead maneuver. An overhead maneuver is not an instrument approach procedure. Overhead maneuver patterns are developed at airports where aircraft have an operational need to conduct the maneuver. An aircraft conducting an overhead maneuver is considered to be VFR and the IFR flight plan is canceled when the aircraft reaches the initial approach portion of the maneuver. (See FIG 5-4-27). The existence of a standard overhead

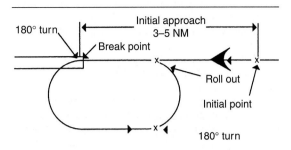

Figure 5-4-27. Overhead Maneuver

maneuver pattern does not eliminate the possible requirement for an aircraft to conform to conventional rectangular patterns if an overhead maneuver cannot be approved. Aircraft operating to an airport without a functioning control tower must initiate cancellation of an IFR flight plan prior to executing the overhead maneuver. Cancellation of the IFR flight plan must be accomplished after crossing the landing threshold on the initial portion of the maneuver or after landing. Controllers may authorize an overhead maneuver and issue the following to arriving aircraft:

1. Pattern altitude and direction of traffic. This information may be omitted if either is standard.

PHRASEOLOGY—PATTERN ALTITUDE (ALTITUDE). RIGHT TURNS.

2. Request for a report on initial approach.

PHRASEOLOGY—REPORT INITIAL.

3. "Break" information and a request for the pilot to report. The "Break Point" will be specified if nonstandard. Pilots may be requested to report "break" if required for traffic or other reasons.

PHRASEOLOGY—BREAK AT (SPECIFIED POINT). REPORT BREAK.

Section 5. PILOT/CONTROLLER ROLES AND RESPONSIBILITIES

5-5-1. GENERAL

a. The roles and responsibilities of the pilot and controller for effective participation in the ATC system are contained in several documents. Pilot responsibilities are in the FARs and the air traffic controller's are in the Air Traffic Control Order (FAA Order 7110.65) and supplemental FAA directives. Additional and supplemental information for pilots can be found in the current Aeronautical Information Manual (AIM), Notices to Airmen, Advisory Circulars and aeronautical charts. Since there are many other excellent publications produced by nongovernment organizations, as well as other government organizations, with various updating cycles, questions concerning the latest or most current material can be resolved by cross-checking with the above mentioned documents.

b. The pilot in command of an aircraft is directly responsible for, and is the final authority as to the safe operation of that aircraft. In an emergency requiring immediate action, the pilot in command may deviate from any rule in the General Subpart A and Flight Rules Subpart B in accordance with FAR Part 91.3.

c. The air traffic controller is responsible to give first priority to the separation of aircraft and to the issuance of radar safety alerts, second priority to other services that are required, but do not involve separation of aircraft and third priority to additional services to the extent possible.

d. In order to maintain a safe and efficient air traffic system, it is necessary that each party fulfill their responsibilities to the fullest.

e. The responsibilities of the pilot and the controller intentionally overlap in many areas providing a degree of redundancy. Should one or the other fail in any manner, this overlapping responsibility is expected to compensate, in many cases, for failures that may affect safety.

f. The following, while not intended to be all inclusive, is a brief listing of pilot and controller responsibilities for some commonly used procedures or phases of flight. More detailed explanations are contained in other portions of this publication, the appropriate FARs', ACs and similar publications. The information provided is an overview of the principles involved and is not meant as an interpretation of the rules nor is it intended to extend or diminish responsibilities.

5-5-2. AIR TRAFFIC CLEARANCE

a. Pilot:

1. Acknowledges receipt and understanding of an ATC clearance.

2. Reads back any hold short of runway instructions issued by ATC.

3. Requests clarification or amendment, as appropriate, any time a clearance is not fully understood or considered unacceptable from a safety standpoint.

4. Promptly complies with an air traffic clearance upon receipt except as necessary to cope with an emergency. Advises ATC as soon as possible and obtains an amended clearance, if deviation is necessary.

NOTE—A CLEARANCE TO LAND MEANS THAT APPROPRIATE SEPARATION ON THE LANDING RUNWAY WILL BE ENSURED. A LANDING CLEARANCE DOES NOT

RELIEVE THE PILOT FROM COMPLIANCE WITH ANY PREVIOUSLY ISSUED ALTITUDE CROSSING RESTRICTION.

b. Controller:

1. Issues appropriate clearances for the operation to be conducted, or being conducted, in accordance with established criteria.

2. Assigns altitudes in IFR clearances that are at or above the Minimum IFR Altitudes in controlled airspace.

3. Ensures acknowledgement by the pilot for issued information, clearances, or instructions.

4. Ensures that readbacks by the pilot of altitude, heading, or other items are correct. If incorrect, distorted, or incomplete, makes corrections as appropriate.

5-5-3. CONTACT APPROACH

a. Pilot:

1. Must request a contact approach and makes it in lieu of a standard or special instrument approach.

2. By requesting the contact approach, indicates that the flight is operating clear of clouds, has at least one mile flight visibility, and reasonably expects to continue to the destination airport in those conditions.

3. Assumes responsibility for obstruction clearance while conducting a contact approach.

4. Advises ATC immediately if unable to continue the contact approach or if encounters less than 1 mile flight visibility.

5. Is aware that if radar service is being received, it may be automatically terminated when told to contact the tower.

REFERENCE—PILOT/CONTROLLER GLOSSARY TERM-RADAR SERVICE TERMINATED.

b. Controller:

1. Issues clearance for a contact approach only when requested by the pilot. Does not solicit the use of this procedure.

2. Before issuing the clearance, ascertains that reported ground visibility at destination airport is at least 1 mile.

3. Provides approved separation between the aircraft cleared for a contact approach and other IFR or special VFR aircraft. When using vertical separation, does not assign a fixed altitude, but clears the aircraft at or below an altitude which is at least 1,000 feet below any IFR traffic but not below Minimum Safe Altitudes prescribed in FAR Part 91.119.

4. Issues alternative instructions if, in their judgment, weather conditions may make completion of the approach impracticable.

5-5-4. INSTRUMENT APPROACH

a. Pilot:

1. Be aware that the controller issues clearance for approach based only on known traffic.

2. Follows the procedure as shown on the IAP, including all restrictive notations, such as:

(a) Procedure not authorized at night;

(b) Approach not authorized when local area altimeter not available;

(c) Procedure not authorized when control tower not in operation;

(d) Procedure not authorized when glide slope not used;

(e) Straight-in minimums not authorized at night; etc.

(f) Radar required; or

(g) The circling minimums published on the instrument approach chart provide adequate obstruction clearance and pilots should not descend below the circling altitude until the aircraft is in a position to make final descent for landing. Sound judgment and knowledge of the pilot's and the aircraft's capabilities are the criteria for determining the exact maneuver in each instance since airport design and the aircraft position, altitude and airspeed must all be considered.

REFERENCE—AIM, APPROACH AND LANDING MINIMUMS, PARAGRAPH 5-4-20

3. Upon receipt of an approach clearance while on an unpublished route or being radar vectored:

(a) Complies with the minimum altitude for IFR, and

(b) Maintains the last assigned altitude until established on a segment of a published route or IAP, at which time published altitudes apply.

b. Controller:

1. Issues an approach clearance based on known traffic.

2. Issues an IFR approach clearance only after the aircraft is established on a segment of published route or IAP, or assigns an appropriate altitude for the aircraft to maintain until so established.

5-5-5. MISSED APPROACH

a. Pilot:

1. Executes a missed approach when one of the following conditions exist:

(a) Arrival at the Missed Approach Point (MAP) or the Decision Height (DH) and visual reference to the runway environment is insufficient to complete the landing.

(b) Determines that a safe approach of landing is not possible (See subparagraph 5-4-21g).

(c) Instructed to do so by ATC.

2. Advises ATC that a missed approach will be made. Include the reason for the missed approach unless the missed approach is initiated by ATC.

3. Complies with the missed approach instructions for the IAP being executed from the MAP, unless other missed approach instructions are specified by ATC.

4. If executing a missed approach prior to reaching the MAP, fly the lateral navigation path of the instrument procedure to the MAP. Climb to the altitude specified in the missed approach procedure, except when a maximum altitude is specified between the final approach fix (FAF) and the MAP In that case, comply with the maximum altitude restriction. Note, this may require a continued descent on the final approach.

5. Following a missed approach, requests clearance for specific action; i.e., another approach, hold for improved conditions, proceed to an alternate airport, etc.

6. If making a missed approach from a radar approach, executes the missed approach procedure previously given or climbs to the altitude and flies the heading specified by the controller.

7. Following a missed approach, requests clearance for specific action; i.e., another approach, hold for improved conditions, proceed to an alternate airport, etc.

b. Controller:

1. Issues an approved alternate missed approach procedure if it is desired that the pilot execute a procedure other than as depicted on the instrument approach chart.

2. May vector a radar identified aircraft executing a missed approach when operationally advantageous to the pilot or the controller.

3. In response to the pilot's stated intentions, issues a clearance to an alternate airport, to a holding fix, or for reentry into the approach sequence, as traffic conditions permit.

5-5-6. RADAR VECTORS

a. Pilot:

1. Promptly complies with headings and altitudes assigned to you by the controller.

2. Questions any assigned heading or altitude believed to be incorrect.

3. If operating VFR and compliance with any radar vector or altitude would cause a violation of any FAR, advises ATC and obtains a revised clearance or instructions.

b. Controller:

1. Vectors aircraft in Class A, Class B, Class C, Class D and Class E airspace:

(a) For separation.

(b) For noise abatement.

(c) To obtain an operational advantage for the pilot or controller.

2. Vectors aircraft in Class A, Class B, Class C, Class D, Class E and Class G airspace when requested by the pilot.

3. Vectors IFR aircraft at or above Minimum Vectoring Altitudes.

4. May vector VFR aircraft, not at an ATC assigned altitude, at any altitude. In these cases, terrain separation is the pilot's responsibility.

5-5-7. SAFETY ALERT

a. Pilot:

1. Initiates appropriate action if a safety alert is received from ATC.

2. Be aware that this service is not always available and that many factors affect the ability of the controller to be aware of a situation in which unsafe proximity to terrain, obstructions, or another aircraft may be developing.

b. Controller:

1. Issues a safety alert if aware an aircraft under their control is at an altitude which, in the controller's judgment, places the aircraft in unsafe proximity to terrain, obstructions or another aircraft. Types of safety alerts are:

(a) *Terrain or Obstruction Alert:* Immediately issued to an aircraft under their control if aware the aircraft is at an altitude believed to place the aircraft in unsafe proximity to terrain or obstructions.

(b) *Aircraft Conflict Alert:* Immediately issued to an aircraft under their control if aware of an aircraft not under their control at an altitude believed to place the aircraft in unsafe proximity to each other. With the alert, they offer the pilot an alternative, if feasible.

2. Discontinue further alerts if informed by the pilot action is being taken to correct the situation or that the other aircraft is in sight.

5-5-8. SEE AND AVOID

a. Pilot: When meteorological conditions permit, regardless of type of flight plan or whether or not under control of a radar facility, the pilot is responsible to see and avoid other traffic, terrain, or obstacles.

b. Controller:

1. Provides radar traffic information to radar identified aircraft operating outside positive control airspace on a workload permitting basis.

2. Issues safety alerts to aircraft under their control if aware the aircraft is at an altitude believed to place the aircraft in unsafe proximity to terrain, obstructions, or other aircraft.

5-5-9. SPEED ADJUSTMENTS

a. Pilot:

1. Advises ATC any time cruising airspeed varies plus or minus 5 percent or 10 knots, whichever is greater, from that given in the flight plan.

2. Complies with speed adjustments from ATC unless:

(a) The minimum or maximum safe airspeed for any particular operation is greater or less than the requested airspeed. In such cases, advises ATC.

NOTE—IT IS THE PILOT'S RESPONSIBILITY AND PREROGATIVE TO REFUSE SPEED ADJUSTMENTS CONSIDERED EXCESSIVE OR CONTRARY TO THE AIRCRAFT'S OPERATING SPECIFICATIONS.

(b) Operating at or above 10,000 feet MSL on an ATC assigned SPEED ADJUSTMENT of more than 250 knots IAS and subsequent clearance is received for descent below 10,000 feet MSL. In such cases, pilots are expected to comply with FAR Part 91.117(a).

3. When complying with speed adjustment assignments, maintains an indicated airspeed within plus or minus 10 knots or 0.02 Mach number of the specified speed.

b. Controller:

1. Assigns speed adjustments to aircraft when necessary but not as a substitute for good vectoring technique.

2. Adheres to the restrictions published in the Air Traffic Control Order (FAA Order 7110.65) as to when speed adjustment procedures may be applied.

3. Avoids speed adjustments requiring alternate decreases and increases.

4. Assigns speed adjustments to a specified IAS (KNOTS)/Mach number or to increase or decrease speed using increments of 10 knots or multiples thereof.

5. Advises pilots to resume normal speed when speed adjustments are no longer required.

6. Gives due consideration to aircraft capabilities to reduce speed while descending.

7. Does not assign speed adjustments to aircraft at or above FL 390 without pilot consent.

5-5-10. TRAFFIC ADVISORIES (Traffic Information)

a. Pilot:

1. Acknowledges receipt of traffic advisories.

2. Informs controller if traffic in sight.

3. Advises ATC if a vector to avoid traffic is desired.

4. Does not expect to receive radar traffic advisories on all traffic. Some aircraft may not appear on the radar display. Be aware that the controller may be occupied with higher priority duties and unable to issue traffic information for a variety of reasons.

5. Advises controller if service not desired.

b. Controller:

1. Issues radar traffic to the maximum extent consistent with higher priority duties except in Class A airspace.

2. Provides vectors to assist aircraft to avoid observed traffic when requested by the pilot.

3. Issues traffic information to aircraft in the Class B, Class C, and Class D surface areas for sequencing purposes.

5-5-11. VISUAL APPROACH

a. Pilot:

1. If a visual approach is not desired, advises ATC.

2. Complies with controller's instructions for vectors toward the airport of intended landing or to a visual position behind a preceding aircraft.

3. The pilot must, at all times, have either the airport or the preceding aircraft in sight. After being cleared for a visual approach, proceed to the airport in a normal manner or follow the preceding aircraft. Remain clear of clouds while conducting a visual approach.

4. If the pilot accepts a visual approach clearance to visually follow a preceding aircraft, you are required to establish a safe landing interval behind the aircraft you were instructed to follow. You are responsible for wake turbulence separation.

5. Advise ATC immediately if the pilot is unable to continue following the preceding aircraft, cannot remain clear of clouds, or lose sight of the airport.

6. Be aware that radar service is automatically terminated, without being advised by ATC, when the pilot is instructed to change to advisory frequency.

7. Be aware that there may be other traffic in the traffic pattern and the landing sequence may differ from the traffic sequence assigned by approach control or ARTCC.

b. Controller:

1. Do not clear an aircraft for a visual approach unless reported weather at the airport is ceiling at or above 1,000 feet and visibility is 3 miles or greater. When weather is not available for the destination airport, inform the pilot and do not initiate a visual approach to that airport unless there is reasonable assurance that descent and flight to the airport can be made visually.

2. Issue visual approach clearance when the pilot reports sighting either the airport or a preceding aircraft which is to be followed.

3. Provide separation except when visual separation is being applied by the pilot.

4. Continue flight following and traffic information until the aircraft has landed or has been instructed to change to advisory frequency.

5. Inform the pilot when the preceding aircraft is a heavy.

6. When weather is available for the destination airport, do not initiate a vector for a visual approach unless the reported ceiling at the airport is 500 feet or more above the MVA and visibility is 3 miles or more. If vectoring weather minima are not available but weather at the airport is ceiling at or above 1,000 feet and visibility of 3 miles or greater, visual approaches may still be conducted.

7. Informs the pilot conducting the visual approach of the aircraft class when pertinent traffic is known to be a heavy aircraft.

5-5-12. VISUAL SEPARATION

a. Pilot:

1. Acceptance of instructions to follow another aircraft or to provide visual separation from it is an acknowledgment that the pilot will maneuver the aircraft as necessary to avoid the other aircraft or to maintain in-trail separation. Pilots are responsible to maintain visual separation until flight paths (altitudes and/or courses) diverge.

2. If instructed by ATC to follow another aircraft or to provide visual separation from it, promptly notify the controller if you lose sight of that aircraft, are unable to maintain continued visual contact with it, or cannot accept the responsibility for your own separation for any reason.

3. The pilot also accepts responsibility for wake turbulence separation under these conditions.

b. Controller: Applies visual separation only:

1. Within the terminal area when a controller has both aircraft in sight or by instructing a pilot who sees the other aircraft to maintain visual separation from it.

2. Pilots are responsible to maintain visual separation until flight paths (altitudes and/or courses) diverge.

5-5-13. VFR-ON-TOP

a. Pilot:

1. This clearance must be requested by the pilot on an IFR flight plan, and if approved, allows the pilot the choice (subject to any ATC restrictions) to select an altitude or Flight Level in lieu of an assigned altitude.

NOTE—VFR-ON-TOP IS NOT PERMITTED IN CERTAIN AIRSPACE AREAS, SUCH AS CLASS A AIRSPACE, CERTAIN RESTRICTED AREAS, ETC. CONSEQUENTLY, IFR FLIGHTS OPERATING VFR-ON-TOP WILL AVOID SUCH AIRSPACE.

REFERENCE—AIM, IFR CLEARANCE VFR-ON-TOP, PARAGRAPH 4-4-7.
AIM, IFR SEPARATION STANDARDS, PARAGRAPH 4-4-10.
AIM, POSITION REPORTING, PARAGRAPH 5-3-2.
AIM, ADDITIONAL REPORTS, PARAGRAPH 5-3-3.

2. By requesting a VFR-ON-TOP clearance, the pilot assumes the sole responsibility to be vigilant so as to see and avoid other aircraft and to:

(a) Fly at the appropriate VFR altitude as prescribed in 14 CFR Part 91.159.

(b) Comply with the VFR visibility and distance from criteria in 14 CFR Part 91.155 (Basic VFR Weather Minimums).

(c) Comply with instrument flight rules that are applicable to this flight; i.e., minimum IFR altitudes, position reporting, radio communications, course to be flown, adherence to ATC clearance, etc.

3. Should advise ATC prior to any altitude change to ensure the exchange of accurate traffic information.

b. Controller:

1. May clear an aircraft to maintain VFR-ON-TOP if the pilot of an aircraft on an IFR flight plan requests the clearance.

2. Informs the pilot of an aircraft cleared to climb to VFR-ON-TOP the reported height of the tops or that no top report is available; issues an alternate clearance if necessary; and once the aircraft reports reaching VFR-ON-TOP, reclears the aircraft to maintain VFR-ON-TOP.

3. Before issuing clearance, ascertains that the aircraft is not in or will not enter Class A airspace.

5-5-14. INSTRUMENT DEPARTURES

a. Pilot:

1. Prior to departure considers the type of terrain and other obstructions on or in the vicinity of the departure airport.

2. Determines if obstruction avoidance can be maintained visually or that the departure procedure should be followed.

3. Determines whether a departure procedure and/or DP is available for obstruction avoidance.

4. At airports where IAPs have not been published, hence no published departure procedure, determines what action will be necessary and takes such action that will assure a safe departure.

b. Controller:

1. At locations with airport traffic control service, when necessary, specifies direction of takeoff, turn, or initial heading to be flown after takeoff.

2. At locations without airport traffic control service but within Class E surface area when necessary to specify direction of takeoff, turn, or initial heading to be flown, obtains pilot's concurrence that the procedure will allow the pilot to comply with local traffic patterns, terrain, and obstruction avoidance.

3. Includes established departure procedures as part of the ATC clearance when pilot compliance is necessary to ensure separation.

5-5-15. MINIMUM FUEL ADVISORY

a. Pilot:

1. Advise ATC of your minimum fuel status when your fuel supply has reached a state where, upon reaching destination, you cannot accept any undue delay.

2. Be aware this is not an emergency situation, but merely an advisory that indicates an emergency situation is possible should any undue delay occur.

3. On initial contact the term "minimum fuel" should be used after stating call sign.

EXAMPLE—SALT LAKE APPROACH, UNITED 621, "MINIMUM FUEL"

4. Be aware a minimum fuel advisory does not imply a need for traffic priority.

5. If the remaining usable fuel supply suggests the need for traffic priority to ensure a safe landing, you should declare an emergency due to low fuel and report fuel remaining in minutes.

REFERENCE—PILOT/CONTROLLER GLOSSARY ITEM— FUEL REMAINING.

b. Controller:

1. When an aircraft declares a state of minimum fuel, relay this information to the facility to whom control jurisdiction is transferred.

2. Be alert for any occurrence which might delay the aircraft.

5-5-16. RNAV AND RNP OPERATIONS

a. Pilot.

1. If unable to comply with the requirements of an RNAV or RNP procedure, pilots shall advise air traffic control as soon as possible. For example, . . .N1234, failure of GPS system, unable RNAV, request amended clearance."

2. Pilots are not authorized to fly a published RNAV or RNP procedure unless it is retrievable by the procedure name from the aircraft navigation database and conforms to the charted procedure.

3. Pilots shall not change any database waypoint type from a fly-by to fly-over, or vice versa. No other modification of database waypoints or the creation of user-defined waypoints on published RNAV or RNP procedures is permitted, except to:

(a) Change altitude and/or airspeed waypoint constraints to comply with an ATC clearance/instruction.

(b) Insert a waypoint along the published route to assist in complying with ATC instruction, example, "Descend via the WILMS arrival except cross 30 north of BRUCE at/or below FL 210." This is limited only to systems that allow along track waypoint construction.

4. Pilots of aircraft utilizing DME/DME for primary radio updating shall ensure any published *required* DME stations are in service as determined by NOTAM, ATIS, or ATC advisory. No pilot monitoring of FMS navigation source(s) is required.

5. Pilots of FMS-equipped aircraft, who are assigned an RNAV DP or STAR procedure and subsequently receive a change of runway, transition or procedure, shall verify that the appropriate changes are loaded and available for navigation.

6. While operating on RNAV segments, pilots are encouraged to use flight director, in lateral navigation mode.

7. For Type B RNAV DPs and STARS, pilots must use a CDI/flight director and/or autopilot, in lateral navigation mode. For Type A RNAV DPs and STARS, these procedures are recommended. Other methods providing an equivalent level of performance may also be acceptable.

8. For Type B RNAV DPs and STARs, pilots of aircraft without GPS, using DME/DME/IRU, must ensure the aircraft navigation system position is confirmed, within 1,000 feet, at the start point of take-off roll. The use of an automatic or manual runway update is an acceptable means of compliance with this requirement. Other methods providing an equivalent level of performance may also be acceptable.

9. RNAV terminal procedures may be amended by ATC issuing radar vectors and/or clearances direct to a waypoint. Pilots should avoid premature manual deletion of waypoints from their active "legs" page to allow for rejoining procedures.

10. While operating on RNAV segments, pilots operating /R aircraft shall adhere to any flight manual limitation or operating procedure required to maintain the RNP value specified for the procedure.

Section 6. NATIONAL SECURITY AND INTERCEPTION PROCEDURES

5-6-1. NATIONAL SECURITY

a. National security in the control of air traffic is governed by (FAR Part 99).

b. All aircraft entering domestic U.S. airspace from points outside must provide for identification prior to entry. To facilitate early aircraft identification of all aircraft in the vicinity of U.S. and international airspace boundaries, Air Defense Identification Zones (ADIZ) have been established.

REFERENCE—AIM, ADIZ BOUNDARIES AND DESIGNATED MOUNTAINOUS AREAS, PARAGRAPH 5-6-5.

c. Operational requirements for aircraft operations associated with an ADIZ are as follows:

1. *Flight Plan:* Except as specified in subparagraphs d. and e. below, an IFR or DVFR flight plan must be filed with an appropriate aeronautical facility as follows:

(a) Generally, for all operations that enter an ADIZ.

(b) For operations that will enter or exit the United States and which will operate into, within or across the Contiguous U.S. ADIZ regardless of true airspeed.

(c) The flight plan must be filed before departure except for operations associated with the Alaskan ADIZ when the airport of departure has no facility for filing a flight plan, in which case the flight plan may be filed immediately after takeoff or when within range of the aeronautical facility.

2. *Two-way Radio:* For the majority of operations associated with an ADIZ, an operating two-way radio is required. See FAR Part 99.1 for exceptions.

3. *Transponder Requirements:* Unless otherwise authorized by ATC, each aircraft conducting operations into, within, or across the Contiguous U.S. ADIZ must be equipped with an operable radar beacon transponder having altitude reporting capability (Mode C), and that transponder must be turned on and set to reply on the appropriate code or as assigned by ATC.

4. *Position Reporting.*

(a) *For IFR flight:* Normal IFR position reporting.

(b) *For DVFR flights:* The estimated time of ADIZ penetration must be filed with the aeronautical facility at least 15 minutes prior to penetration except for flight in the Alaskan ADIZ, in which case report prior to penetration.

(c) For inbound aircraft of foreign registry: The pilot must report to the aeronautical facility at least one hour prior to ADIZ penetration.

5. *Aircraft Position Tolerances:*

(a) Over land, the tolerance is within plus or minus five minutes from the estimated time over a reporting point or point of penetration and within 10 NM from the centerline of an intended track over an estimated reporting point or penetration point.

(b) Over water, the tolerance is plus or minus five minutes from the estimated time over a reporting point or point of penetration and within 20 NM from the centerline of the intended track over an estimated reporting point or point of penetration (to include the Aleutian Islands).

6. *Land-Based ADIZ:* Land-Based ADIZ are activated and deactivated over U.S. metropolitan areas as needed, with dimensions, activation dates and other relevant information disseminated via NOTAM.

(a) In addition to requirements outlined in subparagraphs c1 through c3, pilots operating within a Land-Based ADIZ must report landing or leaving the Land-Based ADIZ if flying too low for radar coverage.

(b) Pilots unable to comply with all requirements shall remain clear of Land-Based ADIZ. Pilots entering a Land-Based ADIZ without authorization or who fail to follow all requirements risk interception by military fighter aircraft.

d. Except when applicable under FAR Part 99.7, FAR Part 99 does not apply to aircraft operations:

1. Within the 48 contiguous states and the District of Columbia, or within the State of Alaska, and remains within 10 miles of the point of departure;

2. Over any island, or within three nautical miles of the coastline of any island, in the Hawaii ADIZ; or

3. Associated with any ADIZ other than the Contiguous U.S. ADIZ, when the aircraft true airspeed is less than 180 knots.

e. Authorizations to deviate from the requirements of Part 99 may also be granted by the ARTCC, on a local basis, for some operations associated with an ADIZ.

f. An Airfiled VFR Flight Plan makes an aircraft subject to interception for positive identification when entering an ADIZ. Pilots are therefore urged to file the required DVFR flight plan either in person or by telephone prior to departure.

g. Special Security Instructions:

1. During defense emergency or air defense emergency conditions, additional special security instructions may be issued in accordance with the Security Control of Air Traffic and Air Navigation Aids (SCATANA) Plan.

2. Under the provisions of the SCATANA Plan, the military will direct the action to be taken-in regard to landing, grounding, diversion, or dispersal of aircraft and the control of air navigation aids in the defense of the U.S. during emergency conditions.

3. At the time a portion or all of SCATANA is implemented, ATC facilities will broadcast appropriate instructions received from the military over available ATC frequencies. Depending on instructions received from the military, VFR flights may be directed to land at the nearest available airport, and IFR flights will be expected to proceed as directed by ATC.

4. Pilots on the ground may be required to file a flight plan and obtain an approval (through FAA) prior to conducting flight operation.

5. In view of the above, all pilots should guard an ATC or FSS frequency at all times while conducting flight operations.

5-6-2. INTERCEPTION PROCEDURES

a. General

1. Identification intercepts during peacetime operations are vastly different than those conducted under increased states of readiness. Unless otherwise directed by the control agency, intercepted aircraft will be identified by type only. When specific information is required (i.e. markings, serial numbers, etc.) the interceptor aircrew will respond only if the request can be conducted in a safe manner. During hours of darkness or Instrument Meteorological Conditions (IMC), identification of unknown aircraft will be by type only. The interception pattern described below is the typical peacetime method used by air interceptor aircrews. In all situations, the interceptor aircrew will use caution to avoid startling the intercepted aircrew and/or passengers.

2. All aircraft operating in the U.S. national airspace, if capable, will maintain a listening watch on VHF guard 121.5 or UHF 243.0. It is incumbent on all aviators to know and understand their responsibilities if intercepted. Additionally, if the U.S. military intercepts an aircraft and flares are dispensed in the area of that aircraft, aviators will pay strict attention, contact air traffic control immediately on the local frequency or on VHF guard 121.5 or UHF 243.0 and follow the intercept's visual ICAO signals. Be advised that noncompliance may result in the use of force.

b. Intercept phases (See FIG 5-6-1).

1. *Phase One-Approach Phase:*

During peacetime, intercepted aircraft will be approached from the stern. Generally two interceptor aircraft will be employed to accomplish the identification. The flight leader and wingman will coordinate their individual positions in conjunction with the ground controlling agency. Their relationship will resemble a line abreast formation. At night or in IMC, a comfortable radar trail tactic will be used. Safe vertical separation between interceptor aircraft and unknown aircraft will be maintained at all times.

2. *Phase Two-Identification Phase:*

The intercepted aircraft should expect to visually acquire the lead interceptor and possibly the wingman during this phase in visual meteorological conditions (VMC). The wingman will assume a surveillance position while the flight leader approaches the unknown aircraft. Intercepted aircraft personnel may observe the use of different drag devices to allow for speed and position stabilization during this phase. The flight leader will then initiate a gentle closure toward the intercepted aircraft, stopping at a distance no closer than absolutely necessary to obtain the information needed. The interceptor aircraft will use every possible precaution to avoid startling intercepted aircrew or passengers. Additionally, the interceptor aircrews will constantly keep in mind that maneuvers considered normal to a fighter aircraft may be considered hazardous to passengers and crews of nonfighter aircraft. When interceptor aircrews know or believe that an unsafe condition exists, the identification phase will be terminated. As previously stated, during darkness or IMC identification of unknown aircraft will be by type only. Positive vertical separation will be maintained by interceptor aircraft throughout this phase.

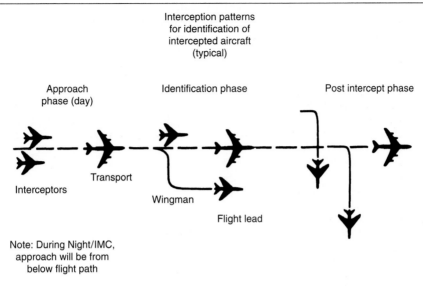

Figure 5-6-1. Interception Procedures

3. *Phase Three-Post Intercept Phase:*

Upon identification phase completion, the flight leader will turn away from the intercepted aircraft. The wingman will remain well clear and accomplish a rejoin with the leader.

c. Communication interface between interceptor aircrews and the ground controlling agency is essential to ensure successful intercept completion. Flight Safety is paramount. An aircraft which is intercepted by another aircraft shall immediately:

1. Follow the instructions given by the intercepting aircraft, interpreting and responding to the visual signals.

2. Notify, if possible, the appropriate air traffic services unit.

3. Attempt to establish radio communication with the intercepting aircraft or with the appropriate intercept control unit, by making a general call on the emergency frequency 243.0 MHz and repeating this call on the emergency frequency 121.5 MHz, if practicable, giving the identity and position of the aircraft and the nature of the flight.

4. If equipped with SSR transponder, select MODE 3/A Code 7700, unless otherwise instructed by the appropriate air traffic services unit. If any instructions received by radio from any sources conflict with those given by the intercepting aircraft by visual or radio signals, the intercepted aircraft shall request immediate clarification while continuing to comply with the instructions given by the intercepting aircraft.

5-6-3. LAW ENFORCEMENT OPERATIONS BY CIVIL AND MILITARY ORGANIZATIONS

a. Special law enforcement operations.

1. Special law enforcement operations include in-flight identification, surveillance, interdiction, and pursuit activities performed in accordance with official civil and/or military mission responsibilities.

2. To facilitate accomplishment of these special missions, exemptions from specified sections of the FAR have been granted to designated departments and agencies. However, it is each organization's responsibility to apprise ATC of their intent to operate under an authorized exemption before initiating actual operations.

3. Additionally, some departments and agencies that perform special missions have been assigned coded identifiers to permit them to apprise ATC of ongoing mission activities and solicit special air traffic assistance.

5-6-4. INTERCEPTION SIGNALS

a. TBL 5-6-1 and TBL 5-6-2.

5-6-5. ADIZ BOUNDARIES AND DESIGNATED MOUNTAINOUS AREAS

a. (See FIG 5-6-2)

Table 5-6-1 INTERCEPTION SIGNALS
Signals initiated by intercepting aircraft and responses by intercepted aircraft (as set forth in ICAO Annex 2—Appendix A, 2.1)

Series	INTERCEPTING Aircraft Signals	Meaning	INTERCEPTED Aircraft Responds	Meaning
1	DAY—Rocking wings from a position slightly above and ahead of, and normally to the left of, the intercepted aircraft and, after acknowledgment, a slow level turn, normally to the left, on to the desired heading.	You have been intercepted. Follow me.	AEROPLANES: 　DAY—Rocking wings and following:	Understood, will comply.
	NIGHT—Same and, in addition, flashing navigational lights at irregular intervals.		NIGHT—Same and, in addition, flashing navigational lights at irregular intervals.	
	Note 1.—Meterological conditions terramin may require the intercepting aircraft to take up a position slightly above and ahead of, and to the right of, the intercepted aircraft and to make the subsequent turn to the right.			
	Note 2.—If the intercepted aircraft is not able to keep pace with the intercepting aircraft, the latter is expected to fly a series of race-track patterns and to rock its wings each time it passes the intercepted aircraft.		HELICOPTERS: 　DAY or NIGHT—Rocking aircraftt, flashing navigational lights at irregular intervals and following.	
2	DAY or NIGHT—An abrupt break-away maneuver from the intercepted intercepted aircraft consisting of a climbing turn of 90 degrees or more without crossing the line of flight of the intercepted aircraft.	You may proceed.	AEROPLANES: 　DAY or NIGHT—Rocking wings.	Understood, will comply.
			HELICOPTERS: 　DAY or NIGHT —Rocking aircraftt/	
3	DAY—Circling aerodrome, lowering landing gear and overllying runway in direction of landing or, if the intercepted aircraft is a helicipter. overflying the helicopter landing area.	Land at this aerodrome.	AEROPLANES: 　DAY—Lowering landing gear, following the intercepting aircraft and, if after overflying the runway landing is considered safe, proceeding to land. 　NIGHT—Same and, in addition, showing steady landing lights (if carried).	Understood, will comply.
			HELICOPTERS: 　DAY or NIGHT—Following the intercepting aircraft and proceeding to land, showing a steady landing light (if carried).	
	NIGHT—Same in, and in addition, showing steady landing lights.			

TBL 5-6-2
Intercepting Signals

INTERCEPTING SIGNALS Signals and Responses During Aircraft Intercept Signals initiated by intercepted aircraft and responses by intercepted aircraft (as set forth in ICAO Annex 2-Appendix 1, 2.2)				
Series	INTERCEPTED Aircraft Signals	Meaning	INTERCEPTING Aircraft Responds	Meaning
4	DAY or NIGHT–Raising landing gear (if fitted) and flashing landing lights while passing over runway in use or helicopter landing area at a height exceeding 300m (1,000 ft) but not exceeding 600m (2,000 ft) (in the case of a helicopter, at a height exceeding 50m (170 ft) but not exceeding 100m (330 ft) above the aerodrome level, and continuing to circle runway in use or helicopter landing area. If unable to flash landing lights, flash any other lights available.	Aerodrome you have designated is inadequate.	DAY or NIGHT–If it is desired that the intercepted aircraft follow the intercepting aircraft to an alternate aerodrome, the intercepting aircraft raises its landing gear (if fitted) and uses the Series 1 signals prescribed for intercepting aircraft. If it is decided to release the intercepted aircraft, the intercepting aircraft uses the Series 2 signals prescribed for intercepting aircraft.	Understood, follow me. Understood, you may proceed.
5	DAY or NIGHT–Regular switching on and off of all available lights but in such a manner as to be distinct from flashing lights.	Cannot comply.	DAY or NIGHT-Use Series 2 signals prescribed for intercepting aircraft.	Understood.
6	DAY or NIGHT–Irregular flashing of all available lights.	In distress.	DAY or NIGHT-Use Series 2 signals prescribed for intercepting aircraft.	Understood.

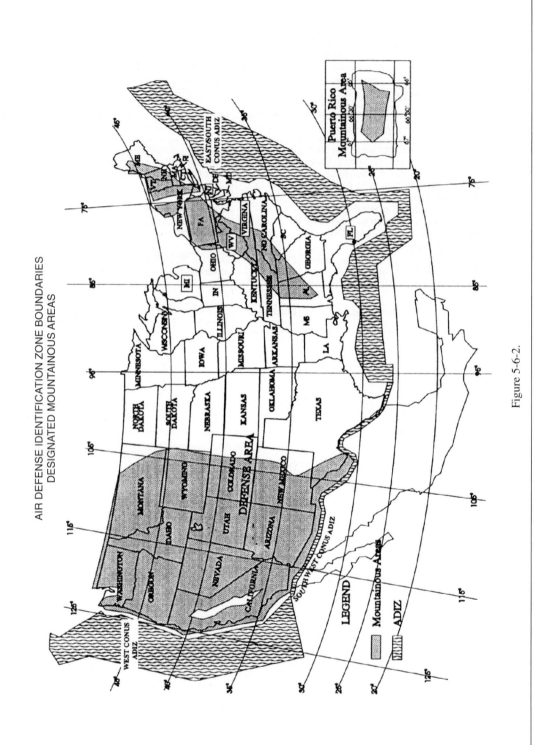

AIR DEFENSE IDENTIFICATION ZONE BOUNDARIES
DESIGNATED MOUNTAINOUS AREAS

Figure 5-6-2.

Chapter 6. Emergency procedures
Section 1. GENERAL

6-1-1. PILOT RESPONSIBILITY AND AUTHORITY

a. The pilot in command of an aircraft is directly responsible for and is the final authority as to the operation of that aircraft. In an emergency requiring immediate action, the pilot in command may deviate from any rule in the FAR, Subpart A, General, and Subpart B, Flight Rules, to the extent required to meet that emergency.

REFERENCE—FAR PART 91.3(B).

b. If the emergency authority of FAR Part 91.3.(b) is used to deviate from the provisions of an ATC clearance, the pilot in command must notify ATC as soon as possible and obtain an amended clearance.

c. Unless deviation is necessary under the emergency authority of FAR Part 91.3, pilots of IFR flights experiencing two-way radio communications failure are expected to adhere to the procedures prescribed under "IFR operations, two-way radio communications failure."

REFERENCE—FAR PART 91.185.

6-1-2. EMERGENCY CONDITION-REQUEST ASSISTANCE IMMEDIATELY

a. An emergency can be either a *distress* or *urgency* condition as defined in the *Pilot/Controller Glossary*. Pilots do not hesitate to declare an emergency when they are faced with *distress* conditions such as fire, mechanical failure, or structural damage. However, some are reluctant to report an *urgency* condition when they encounter situations which may not be immediately perilous, but are potentially catastrophic. An aircraft is in at least an *urgency* condition the moment the pilot becomes doubtful about position, fuel endurance, weather, or any other condition that could adversely affect flight safety. This is the time to ask for help, not after the situation has developed into a *distress* condition.

b. Pilots who become apprehensive for their safety for any reason should *request assistance immediately*. Ready and willing help is available in the form of radio, radar, direction finding stations and other aircraft. Delay has caused accidents and cost lives. *Safety is not a luxury! Take action!*

Section 2. EMERGENCY SERVICES AVAILABLE TO PILOTS

6-2-1. RADAR SERVICE FOR VFR AIRCRAFT IN DIFFICULTY

a. Radar equipped ATC facilities can provide radar assistance and navigation service (vectors) to VFR aircraft in difficulty when the pilot can talk with the controller, and the aircraft is within radar coverage. Pilots should clearly understand that authorization to proceed in accordance with such radar navigational assistance does not constitute authorization for the pilot to violate FARs. In effect, assistance is provided on the basis that navigational guidance information is advisory in nature, and the responsibility for flying the aircraft safely remains with the pilot.

b. Experience has shown that many pilots who are not qualified for instrument flight cannot maintain control of their aircraft when they encounter clouds or other reduced visibility conditions. In many cases, the controller will not know whether flight into instrument conditions will result from ATC instructions. To avoid possible hazards resulting from being vectored into IFR conditions, a pilot in difficulty should keep the controller advised of the current weather conditions being encountered and the weather along the course ahead and observe the following:

1. If a course of action is available which will permit flight and a safe landing in VFR weather conditions, noninstrument rated pilots should choose the VFR condition rather than requesting a vector or approach that will take them into IFR weather conditions; or

2. If continued flight in VFR conditions is not possible, the noninstrument rated pilot should so advise the controller and indicating the lack of an instrument rating, declare a *distress* condition, or

3. If the pilot is instrument rated and current, and the aircraft is instrument equipped, the pilot should so indicate by requesting an IFR flight clearance. Assistance will then be provided on the basis that the aircraft can operate safely in IFR weather conditions.

6-2-2. TRANSPONDER EMERGENCY OPERATION

a. When a *distress* or *urgency* condition is encountered, the pilot of an aircraft with a coded radar beacon transponder, who desires to alert a ground radar facility, should squawk MODE 3/A, Code 7700/Emergency and MODE C altitude reporting and then immediately establish communications with the ATC facility.

b. Radar facilities are equipped so that Code 7700 normally triggers an alarm or special indicator at all control positions. Pilots should understand that they might not be within a radar coverage area. Therefore, they should continue squawking Code 7700 and establish radio communications as soon as possible.

6-2-3. DIRECTION FINDING INSTRUMENT APPROACH PROCEDURE

a. Direction Finder (DF) equipment has long been used to locate lost aircraft and to guide aircraft to areas of good weather or to airports. Now at most DF equipped airports, DF instrument approaches may be given to aircraft in a *distress* or *urgency* condition.

b. Experience has shown that most emergencies requiring DF assistance involve pilots with little flight experience. With this in mind, DF approach procedures provide maximum flight stability in the approach by using small turns, and wings-level descents. The DF specialist will give the pilot headings to fly and tell the pilot when to begin descent.

c. DF IAPs are for emergency use only and will not be used in IFR weather conditions unless the pilot has declared a *distress* or *urgency* condition.

d. To become familiar with the procedures and other benefits of DF, pilots are urged to request practice DF guidance and approaches in VFR weather conditions. DF specialists welcome the practice and will honor such requests, workload permitting.

6-2-4. INTERCEPT AND ESCORT

a. The concept of airborne intercept and escort is based on the Search and Rescue (SAR) aircraft establishing visual and/or electronic contact with an aircraft in difficulty, providing in-flight assistance, and escorting it to a safe landing. If bailout, crash landing or ditching becomes necessary, SAR operations can be conducted without delay. For most incidents, particularly those occurring at night and/or during instrument flight conditions, the availability of intercept and escort services will depend on the proximity of SAR units with suitable aircraft on alert for immediate dispatch. In limited circumstances, other aircraft flying in the vicinity of an aircraft in difficulty can provide these services.

b. If specifically requested by a pilot in difficulty or if a *distress* condition is declared, SAR coordinators *will* take steps to intercept and escort an aircraft. Steps may be initiated for intercept and escort if an *urgency* condition is declared and unusual circumstances make such action advisable.

c. It is the pilot's prerogative to refuse intercept and escort services. Escort services will normally be provided to the nearest adequate airport. Should the pilot receiving escort services continue onto another location after reaching a safe airport, or decide not to divert to the nearest safe airport, the escort aircraft is not obligated to continue and further escort is discretionary. The decision will depend on the circumstances of the individual incident.

6-2-5. EMERGENCY LOCATOR TRANSMITTERS

a. GENERAL:

1. Emergency Locator Transmitters (ELT) are required for most General Aviation airplanes.

REFERENCE—FAR PART 91.207.

2. ELTs of various types were developed as a means of locating downed aircraft. These electronic, battery operated transmitters operate on one of three frequencies. These operating frequencies are 121.5 MHz, 243.0 MHz, and the newer 406 MHz. ELTs operating on 121.5 MHz and 243.0 MHz are analog devices. The newer 406 MHz ELT is a digital transmitter that can be encoded with the owner's contact information or aircraft data. The latest 406 MHz ELT models can also be encoded with the aircraft's position data which can help SAR forces locate the aircraft much more quickly after a crash. The 406 MHz ELTs also transmits a stronger signal when activated than the older 121.5 MHz ELTs.

(a) The Federal Communications Commission (FCC) requires 406 MHz ELTs be registered with the National Oceanic and Atmospheric Administration (NOAA) as outlined in the ELTs documentation. The FAA's 406 MHz ELT Technical Standard Order (TSO) TSO-C126 also requires that each 406 MHz ELT be registered with NOAA. The reason is NOAA maintains the owner registration database for U.S. registered 406 MHz alerting devices, which includes ELTs. NOAA also operates the United States' portion of the Cospas-Sarsat satellite distress alerting system designed to detect activated ELTs and other distress alerting devices.

(b) In the event that a properly registered 406 MHz ELT activates, the Cospas-Sarsat satellite system can decode the owner's information and provide that data to the appropriate search and rescue (SAR) center. In the United States, NOAA provides the alert data to the appropriate U.S. Air Force Rescue Coordination Center (RCC) or U.S. Coast Guard Rescue Coordination Center. That RCC can then telephone or contact the owner to verify the status of the aircraft. If the aircraft is safely secured in a hangar, a costly ground or airborne search is avoided. In the case of an inadvertent 406 MHz ELT activation, the owner can deactivate the 406 MHz ELT. If the 406 MHz ELT equipped aircraft is being flown, the RCC can quickly activate a search. 406 MHz ELTs permit the Cospas-Sarsat satellite system to narrow the search area to a more confined area compared to that of a 121.5 MHz or 243.0 MHz ELT. 406 MHz ELTs also include a low-power 121.5 MHz homing transmitter to aid searchers in finding the aircraft in the terminal search phase.

(c) Each analog ELT emits a distinctive downward swept audio tone on 121.5 MHz and 243.0 MHz.

(d) If "armed" and when subject to crash-generated forces, ELTs are designed to automatically activate and continuously emit their respective signals, analog or digital. The transmitters will operate continuously for at least 48 hours over a wide temperature range. A properly installed, maintained, and functioning ELT can expedite search and rescue operations and save lives if it survives the crash and is activated.

(e) Pilots and their passengers should know how to activate the aircraft's ELT if manual activation is required. They should also be able to verify the aircraft's ELT is functioning and transmitting an alert after a crash or manual activation.

(f) Because of the large number of 121.5 MHz ELT false alerts and the lack of a quick means of verifying the actual status of an activated 121.5 MHz or 243.0 MHz analog ELT through an owner registration database, U.S. SAR forces do not respond as quickly to initial 121.5/243.0 MHz ELT alerts as the SAR forces do to 406 MHz ELT alerts. Compared to the almost instantaneous detection of a 406 MHz ELT, SAR forces' normal practice is to wait for either a confirmation of a 121.5/243.0 MHz alert by additional satellite passes or through confirmation of an overdue aircraft or similar notification. In some cases, this confirmation process can take hours. SAR forces can initiate a response to 406 MHz alerts in minutes compared to the potential delay of hours for a 121.5/243.0 MHz ELT.

3. The Cospas-Sarsat system has announced the termination of satellite monitoring and reception of the 121.5 MHz and 243.0 MHz frequencies in 2009. The Cospas-Sarsat system will continue to monitor the 406 MHz frequency. What this means for pilots is that after the termination date, those aircraft with only 121.5 MHz or 243.0 MHz ELTs onboard will have to depend upon either a nearby Air Traffic Control facility receiving the alert signal or an overflying aircraft monitoring 121.5 MHz or 243.0 MHz detecting the alert. To ensure adequate monitoring of these frequencies and timely alerts after 2009, all airborne pilots should periodically monitor these frequencies to try and detect an activated 121.5/243.0 MHz ELT.

b. TESTING:

1. ELTs should be tested in accordance with the manufacturer's instructions, preferably in a shielded or screened room or specially designed test container to prevent the broadcast of signals which could trigger a false alert.

2. When this cannot be done, aircraft operational testing is authorized as follows:

(a) Analog 121.5/243 MHz ELTs should only be tested during the first 5 minutes after any hour. If operational tests must be made outside of this period, they should be coordinated with the nearest FAA Control Tower or FSS. Tests should be no longer than three audible sweeps. If the antenna is removable, a dummy load should be substituted during test procedures.

(b) Digital 406 MHz ELTs should only be tested in accordance with the unit's manufacturer's instructions.

(c) Airborne tests are not authorized.

c. FALSE ALARMS:

1. Caution should be exercised to prevent the inadvertent activation of ELTs in the air or while they are being handled on the ground. Accidental or unauthorized activation

will generate an emergency signal that cannot be distinguished from the real thing, leading to expensive and frustrating searches. A false ELT signal could also interfere with genuine emergency transmissions and hinder or prevent the timely location of crash sites. Frequent false alarms could also result in complacency and decrease the vigorous reaction that must be attached to all ELT signals.

2. Numerous cases of inadvertent activation have occurred as a result of aerobatics, hard landings, movement by ground crews and aircraft maintenance. These false alarms can be minimized by monitoring 121.5 MHz and/or 243.0 MHz as follows:

(a) In flight when a receiver is available.

(b) Before engine shut down at the end of each flight.

(c) When the ELT is handled during installation or maintenance.

(d) When maintenance is being performed in the vicinity of the ELT.

(e) When ground crew moves the aircraft.

(f) If an ELT signal is heard, turn off the ELT to determine if it is transmitting. If it has been activated, maintenance might be required before the unit is returned to the "ARMED" position. You should contact the nearest Air Traffic facility and notify it of the inadvertent activation.

d. IN-FLIGHT MONITORING AND REPORTING:

1. Pilots are encouraged to monitor 121.5 MHz and/or 243.0 MHz while in-flight to assist in identifying possible emergency ELT transmissions. On receiving a signal, report the following information to the nearest air traffic facility:

(a) Your position at the time the signal was first heard.

(b) Your position at the time the signal was last heard.

(c) Your position at maximum signal strength.

(d) Your flight altitudes and frequency on which the emergency signal was heard—121.5 MHz or 243.0 MHz. If possible, positions should be given relative to a navigation aid. If the aircraft has homing equipment, provide the bearing to the emergency signal with each reported position.

6-2-6. FAA K-9 EXPLOSIVES DETECTION TEAM PROGRAM

a. The FAA's Office of Civil Aviation Security Operations manages the FAA K-9 Explosive Detection Team Program which was established in 1972. Through a unique agreement with law enforcement agencies and airport authorities, the FAA has strategically placed FAA-certified K-9 teams (a team is one handler and one dog) at airports throughout the country. If a bomb threat is received while an aircraft is in flight, the aircraft can be directed to an airport with this capability.

The FAA provides initial and refresher training for all handlers, provides single purpose explosive detector dogs, and requires that each team is annually evaluated in five areas for FAA certification: aircraft (widebody and narrowbody), vehicles, terminal, freight (cargo), and luggage. **If**

you desire this service, notify your company or an FAA air traffic control facility.

b. The following list shows the locations of current FAA K-9 teams:

Table 6-2-1   FAA Sponsored Explosives Detection Dog/Handler Team Locations

Airport Symbol	Location
ATL	Atlanta, Georgia
BHM	Birmingham, Alabama
BOS	Boston, Massachusetts
BUF	Buffalo, New York
CLT	Charlotte, North Carolina
ORD	Chicago, Illinois
CVG	Cincinnati, Ohio
DFW	Dallas, Texas
DEN	Denver, Colorado
DTW	Detroit, Michigan
IAH	Houston, Texas
JAX	Jacksonville, Florida
MCI	Kansas City, Missouri
LAX	Los Angeles, California
MEM	Memphis, Tennessee
MIA	Miami, Florida
MKE	Milwaukee, Wisconsin
MSY	New Orleans, Louisiana
MCO	Orlando, Florida
PHX	Phoenix, Arizona
PIT	Pittsburgh, Pennsylvania
PDX	Portland, Oregon
SLC	Salt Lake City, Utah
SFO	San Francisco, California
SJU	San Juan, Puerto Rico
SEA	Seattle, Washington
STL	St. Louis, Missouri
TUS	Tucson, Arizona
TUL	Tulsa, Oklahoma

c. If due to weather or other considerations an aircraft with a suspected hidden explosive problem were to land or intended to land at an airport other than those listed in b. above, it is recommended that they call the FAA's Washington Operations Center (telephone 202-267-3333, if appropriate) or have an air traffic facility with which you can communicate contact the above center requesting assistance.

6-2-7. SEARCH AND RESCUE

a. GENERAL: SAR is a lifesaving service provided through the combined efforts of the federal agencies signatory to the National SAR Plan, and the agencies responsible for SAR within each state. Operational resources are provided by the U.S. Coast Guard, DOD components, the Civil Air Patrol, the Coast Guard Auxiliary, state, county and local law enforcement and other public safety agencies, and private volunteer organizations. Services include search for missing aircraft, survival aid, rescue, and emergency medical help for the occupants after an accident site is located.

b. NATIONAL SEARCH AND RESCUE PLAN: By federal interagency agreement, the National Search and Rescue Plan provides for the effective use of all available facilities in all types of SAR missions. These facilities include aircraft, vessels, pararescue and ground rescue teams, and emergency radio fixing. Under the Plan, the U.S. Coast Guard is responsible for the coordination of SAR in the Maritime Region, and the USAF is responsible in the Inland Region. To carry out these responsibilities, the Coast Guard and the Air Force have established Rescue Coordination Centers (RCCs) to direct SAR activities within their regions. For aircraft emergencies, distress, and urgency, information normally will be passed to the appropriate RCC through an ARTCC or FSS.

c. COAST GUARD RESCUE COORDINATION CENTERS. (See TBL 6-2-2).

Table 6-2-2   Coast Guard Rescue Coordination Centers

Alameda, CA 510-437-3701	Cleveland, OH 216-902-6117
Boston, MA 617-223-8555	Seattle, WA 206-220-7001
New York, NY 212-668-7055	Juneau, AK 907-463-2000
Portsmouth, VA 757-398-6390	Honolulu, HI 808-541-2500
Miami, FL 305-415-6800	San Juan, Puerto Rico 809-729-6770
New Orleans, LA 504-589-6225	

d. AIR FORCE RESCUE COORDINATION CENTERS. (See TBL 6-2-3 and TBL 6-2-4).

Table 6-2-3   Air Force Rescue Coordination Center 48 Contiguous States

Langley AFB, Virginia		Phone
	Commercial	804-764-8112
	WATS	1-800-851-3051
	DSN	574-8112

Table 6-2-4 Air Command Rescue Coordination Center—Alaska

Fort Richardson, Alaska	Phone
Commercial	907-428-7230
DSN	317-384-6726
(Outside Anchorage)	800-420-7230

e. JOINT RESCUE COORDINATION CENTER (See TBL 6-2-5).

Table 6-2-5 Joint Rescue Coordination Center Hawaii

HQ 14th CG District Honolulu	Phone
Commercial	808-541-2500
DSN	448-0301

f. EMERGENCY AND OVERDUE AIRCRAFT:

1. ARTCCs and FSSs will alert the SAR system when information is received from any source that an aircraft is in difficulty, overdue, or missing.

(a) Radar facilities providing radar flight following or advisories consider the loss of radar and radios, without service termination notice, to be a possible emergency. Pilots receiving VFR services from radar facilities should be aware that SAR may be initiated under these circumstances.

(b) A filed flight plan is the most timely and effective indicator that an aircraft is overdue. Flight plan information is invaluable to SAR forces for search planning and executing search efforts.

2. Prior to departure on every flight, local or otherwise, someone at the departure point should be advised of your destination and route of flight if other than direct. Search efforts are often wasted and rescue is often delayed because of pilots who thoughtlessly take off without telling anyone where they are going. File a flight plan for *your* safety.

3. According to the National Search and Rescue Plan, "The life expectancy of an injured survivor decreases as much as 80 percent during the first 24 hours, while the chances of survival of uninjured survivors rapidly diminishes after the first 3 days."

4. An Air Force Review of 325 SAR missions conducted during a 23-month period revealed that "Time works against people who experience a *distress* but are not on a flight plan, since 36 hours normally pass before family concern initiates an (alert)."

g. VFR SEARCH AND RESCUE PROTECTION:

1. To receive this valuable protection, *file a VFR or DVFR Flight Plan* with an FAA FSS. For maximum protection, file only to the point of first intended landing, and refile for each leg to final destination. When a lengthy flight plan is filed, with several stops en route and an ETE to final destination, a mishap could occur on any leg, and unless other information is received, it is probable that no one would start looking for you until 30 minutes after your ETA at your final destination.

2. If you land at a location other than the intended destination, report the landing to the nearest FAA FSS and advise them of your original destination.

3. If you land en route and are delayed more than 30 minutes, report this information to the nearest FSS and give them your original destination.

4. If your ETE changes by 30 minutes or more, report a new ETA to the nearest FSS and give them your original destination. Remember that if you fail to respond within one-half hour after your ETA at final destination, a search will be started to locate you.

5. It is important that you *close your flight plan IMMEDIATELY AFTER ARRIVAL AT YOUR FINAL DESTINATION WITH THE FSS DESIGNATED WHEN YOUR FLIGHT PLAN WAS FILED. The pilot is responsible for closure of a VFR or DVFR flight plan; they are not closed automatically.* This will prevent needless search efforts.

6. The rapidity of rescue on land or water will depend on how accurately your position may be determined. If a flight plan has been followed and your position is on course, rescue will be expedited.

h. SURVIVAL EQUIPMENT:

1. For flight over uninhabited land areas, it is wise to take and know how to use survival equipment for the type of climate and terrain.

2. If a forced landing occurs at sea, chances for survival are governed by the degree of crew proficiency in emergency procedures and by the availability and effectiveness of water survival equipment.

i. BODY SIGNAL ILLUSTRATIONS:

1. If you are forced down and are able to attract the attention of the pilot of a rescue airplane, the body signals illustrated on these pages can be used to transmit messages to the pilot circling over your location.

2. Stand in the open when you make the signals.

3. Be sure the background, as seen from the air, is not confusing.

4. Go through the motions slowly and repeat each signal until you are positive that the pilot understands you.

j. OBSERVANCE OF DOWNED AIRCRAFT:

1. Determine if crash is marked with a yellow cross; if so, the crash has already been reported and identified.

2. If possible, determine type and number of aircraft and whether there is evidence of survivors.

3. Fix the position of the crash as accurately as possible with reference to a navigational aid. If possible, provide geographic or physical description of the area to aid ground search parties.

4. Transmit the information to the nearest FAA or other appropriate radio facility.

5. If circumstances permit, orbit the scene to guide in other assisting units until their arrival or until you are relieved by another aircraft.

6. Immediately after landing, make a complete report to the nearest FAA facility, or Air Force or Coast Guard Rescue Coordination Center. The report can be made by a long distance collect telephone call.

NO.	MESSAGE	CODE SYMBOL
1	Require assistance	V
2	Require medical assistance	X
3	No or negative	N
4	Yes or affirmative	Y
5	Proceeding in this direction	↑

IF IN DOUBT, USE INTERNATIONAL SYMBOL **SOS**

INSTRUCTIONS

1. Lay out symbols by using strips of fabric or parachutes, pieces of wood, stones, or any available material.
2. Provide as much color contrast as possible between material used for symbols and background against which symbols are exposed.
3. Symbols should be at least 10 feet high or larger. Care should be taken to lay out symbols exactly as shown.
4. In addition to using symbols, every effort is to be made to attract attention by means of radio, flares, smoke, or other available means.
5. On snow covered ground, signals can be made by dragging, shoveling or tramping. Depressed areas forming symbols will appear black from the air.
6. Pilot should acknowledge message by rocking wings from side to side.

Figure 6-2-1.

GROUND-AIR VISUAL CODE FOR USE BY GROUND SEARCH PARTIES

NO.	MESSAGE	CODE SYMBOL
1	Operation completed.	L L L
2	We have found all personnel.	LL
3	We have found only some personnel.	 ╫
4	We are not able to continue. Returning to base.	X X
5	Have divided into two groups. Each proceeding in direction indicated.	 ⇆
6	Information received that aircraft is in this direction	→ →
7	Nothing found. Will continue search.	N N

NOTE— These visual signals have been accepted for international use and appear in Annex 12 to the Convention on International Civil Aviation.

Figure 6-2-2.

NEED MEDICAL ASSISTANCE—URGENT

Used only when life is at stake

Figure 6-2-3. Urgent Medical Assistance

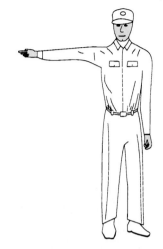

CAN PROCEED SHORTLY
WAIT IF PRACTICABLE

One arm horizontal

Figure 6-2-5. Short Delay

ALL OK—DO NOT WAIT

Wave one arm overhead

Figure 6-2-4. All OK

NEED MECHANICAL HELP
OR PARTS—LONG DELAY

Both arms horizontal

Figure 6-2-6. Long Delay

Figure 6-2-7. Drop Message

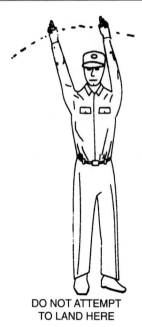

DO NOT ATTEMPT
TO LAND HERE

Both arms waved across face

Figure 6-2-9. Do Not Land Here

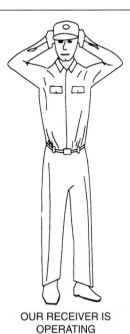

OUR RECEIVER IS
OPERATING

Cup hands over ears

Figure 6-2-8. Receiver Operates

LAND HERE

Both arms forward horizontally,
squatting and point in direction
of landing—Repeat

Figure 6-2-10. Land Here

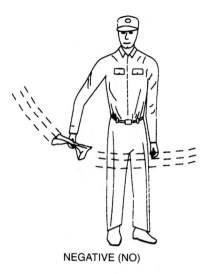

NEGATIVE (NO)

White cloth waved horizontally

Figure 6-2-11. Negative (Ground)

PICK US UP—
PLANE ABANDONED

Both arms vertical

Figure 6-2-13. Pick Us Up

AFFIRMATIVE (YES)

White cloth waved vertically

Figure 6-2-12. Affirmative (Ground)

Affirmative reply from aircraft:

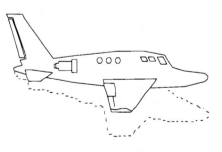

AFFIRMATIVE (YES)

Dip nose of plane several times

Figure 6-2-14. Affirmative (Aircraft)

Negative reply from aircraft:

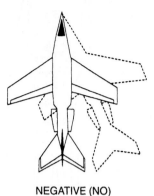

NEGATIVE (NO)

Fishtail plane

Figure 6-2-15. Negative (Aircraft)

Message received and understood by aircraft:
Day or moonlight–rocking wings
Night–green flashed from signal lamp

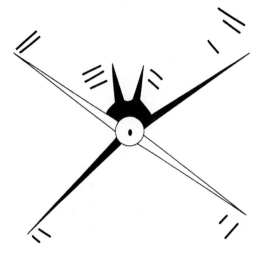

Figure 6-2-16. Message received and understood (Aircraft)

Message received and NOT understood by aircraft:
Day or moonlight–making a complete right-hand circle
Night–red flashes from signal lamp.

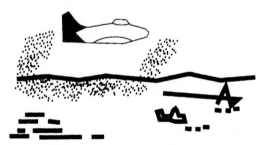

Figure 6-2-17. Message received and
NOT understood (Aircraft)

Section 3. DISTRESS AND URGENCY PROCEDURES

6-3-1. DISTRESS AND URGENCY COMMUNICATIONS

a. A pilot who encounters a *distress* or *urgency* condition can obtain assistance simply by contacting the air traffic facility or other agency in whose area of responsibility the aircraft is operating, stating the nature of the difficulty, pilot's intentions and assistance desired. *Distress* and *urgency* communications procedures are prescribed by the International Civil Aviation Organization (ICAO), however, and have decided advantages over the informal procedure described above.

b. *Distress* and *urgency* communications procedures discussed in the following paragraphs relate to the use of air ground voice communications.

c. The initial communication, and if considered necessary, any subsequent transmissions by an aircraft in *distress* should begin with the signal MAYDAY, preferably repeated three times. The signal PAN-PAN should be used in the same manner for an *urgency* condition.

d. *Distress* communications have absolute priority over all other communications, and the word MAYDAY commands radio silence on the frequency in use. *Urgency* communications have priority over all other communications except *distress*, and the word PAN-PAN warns other stations not to interfere with *urgency* transmissions.

e. Normally, the station addressed will be the air traffic facility or other agency providing air traffic services, on the frequency in use at the time. If the pilot is not communicating and receiving services, the station to be called will normally be the air traffic facility or other agency in whose area of responsibility the aircraft is operating, on the appropriate assigned frequency. If the station addressed does not respond, or if time or the situation dictates, the *distress* or *urgency* message may be broadcast, or a collect call may be used, addressing "Any Station (Tower)(Radio)(Radar)."

f. The station addressed should immediately acknowledge a *distress* or *urgency* message, provide assistance, coordinate and direct the activities of assisting facilities, and alert the appropriate search and rescue coordinator if warranted. Responsibility will be transferred to another station only if better handling will result.

g. All other stations, aircraft and ground, will continue to listen until it is evident that assistance is being provided. If any station becomes aware that the station being called either has not received a *distress* or *urgency* message, or cannot communicate with the aircraft in difficulty, it will attempt to contact the aircraft and provide assistance.

h. Although the frequency in use or other frequencies assigned by ATC are preferable, the following emergency frequencies can be used for distress or urgency communications, if necessary or desirable:

1. *121.5 MHz and 243.0 MHz:* Both have a range generally limited to line of sight. 121.5 MHz is guarded by direction finding stations and some military and civil aircraft. 243.0 MHz is guarded by military aircraft. Both 121.5 MHz and 243.0 MHz are guarded by military towers, most civil towers, FSSs, and radar facilities. Normally ARTCC emergency frequency capability does not extend to radar coverage limits. If an ARTCC does not respond when called on 121.5 MHz or 243.0 MHz, call the nearest tower or FSS.

2. *2182 kHz:* The range is generally less than 300 miles for the average aircraft installation. It can be used to request assistance from stations in the maritime service. 2182 kHz is guarded by major radio stations serving Coast Guard Rescue Coordination Centers, and Coast Guard units along the sea coasts of the U.S. and shores of the Great Lakes. The call "Coast Guard" will alert all Coast Guard Radio Stations within range. 2182 kHz is also guarded by most commercial coast stations and some ships and boats.

6-3-2. OBTAINING EMERGENCY ASSISTANCE

a. A pilot in any *distress* or *urgency* condition should *immediately* take the following action, not necessarily in the order listed, to obtain assistance:

1. Climb, if possible, for improved communications, and better radar and direction finding detection. However, it must be understood that unauthorized climb or descent under IFR conditions within controlled airspace is prohibited, except as permitted by FAR Part 91.3(b).

2. If equipped with a radar beacon transponder (civil) or IFF/SIF (military):

(a) Continue squawking assigned MODE A/3 discrete code/VFR code and MODE C altitude encoding when in radio contact with an air traffic facility or other agency providing air traffic services, unless instructed to do otherwise.

(b) If unable to immediately establish communications with an air traffic facility/agency, squawk MODE A/3, Code 7700/Emergency and MODE C.

3. Transmit a *distress* or *urgency* message consisting of *as many* as necessary of the following elements, preferably in the order listed:

(a) If *distress*, MAYDAY, MAYDAY, MAYDAY; if *urgency*, PAN-PAN, PAN-PAN, PAN-PAN.

(b) Name of station addressed.

(c) Aircraft identification and type.

(d) Nature of *distress* or *urgency*.

(e) Weather.

(f) Pilot's intentions and request.

(g) Present position, and heading; or if *lost*, last known position, time, and heading since that position.

(h) Altitude or Flight Level.

(i) Fuel remaining in minutes.

(j) Number of people on board.

(k) Any other useful information.

REFERENCE—PILOT/CONTROLLER GLOSSARY TERM—FUEL REMAINING.

b. After establishing radio contact, comply with advice and instructions received. Cooperate. Do not hesitate to ask questions or clarify instructions when you do not understand or if you cannot comply with clearance. Assist the ground station to control communications on the frequency in use. Silence interfering radio stations. Do not change frequency or change to another ground station unless absolutely necessary. If you do, advise the ground station of the new frequency and station name prior to the change, transmitting in the blind if necessary. If two-way communications cannot be established on the new frequency, return immediately to the frequency or station where two-way communications last existed.

c. When in a distress condition with bailout, crash landing or ditching imminent, take the following additional actions to assist search and rescue units:

1. Time and circumstances permitting, transmit as many as necessary of the message elements in subparagraph a. 3.

above, and any of the following that you think might be helpful:

(a) ELT status.

(b) Visible landmarks.

(c) Aircraft color.

(d) Number of persons on board.

(e) Emergency equipment on board.

2. Actuate your ELT if the installation permits.

3. For bailout, and for crash landing or ditching if risk of fire is not a consideration, set your radio for continuous transmission.

4. If it becomes necessary to ditch, make every effort to ditch near a surface vessel. If time permits, an FAA facility should be able to get the position of the nearest commercial or Coast Guard vessel from a Coast Guard Rescue Coordination Center.

5. After a crash landing, unless you have good reason to believe that you will not be located by search aircraft or ground teams, it is best to remain with your aircraft and prepare means for signaling search aircraft.

6-3-3. DITCHING PROCEDURES

a. A successful aircraft ditching is dependent on three primary factors. In order of importance they are:

1. *Sea conditions and wind.*

2. *Type of aircraft.*

3. *Skill and technique of pilot.*

b. Common oceanographic terminology:

1. *Sea.* The condition of the surface that is the result of both waves and swells.

2. *Wave (or Chop).* The condition of the surface caused by the local winds.

3. *Swell.* The condition of the surface which has been caused by a distant disturbance.

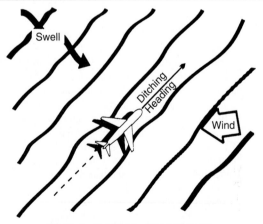

Single swell system—Wind 15 knots

Figure 6-3-1. Single Swell (15 knot wind)

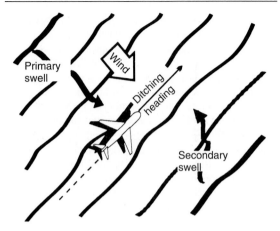

Double swell system—Wind 15 knots

Figure 6-3-2. Double Swell (15 knot wind)

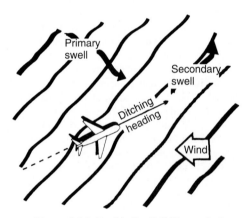

Figure 6-3-3. Double Swell (30 knot wind)

4. *Swell Face.* The side of the swell toward the observer. The backside is the side away from the observer. These definitions apply regardless of the direction of swell movement.

5. *Primary Swell.* The swell system having the greatest height from trough to crest.

6. *Secondary Swells.* Those swell systems of less height than the primary swell.

7. *Fetch.* The distance the waves have been driven by a wind blowing in a constant direction, without obstruction.

8. *Swell Period.* The time interval between the passage of two successive crests at the same spot in the water, measured in seconds.

9. *Swell Velocity.* The speed and direction of the swell with relation to a fixed reference point, measured in knots. There is little movement of water in the horizontal direction.

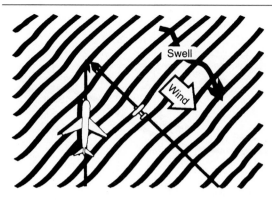

Wind—50 knots

Aircraft with low landing speeds—lands into the wind.

Aircraft with high landing speeds—choose compromise heading between wind and swell.

Both—Land on back side of swell.

Figure 6-3-4. Double swell (50 knot wind)

Swells move primarily in a vertical motion, similar to the motion observed when shaking out a carpet.

10. *Swell Direction.* The direction *from* which a swell is moving. This direction is not necessarily the result of the wind present at the scene. The swell may be moving into or across the local wind. Swells, once set in motion, tend to maintain their original direction for as long as they continue in deep water, regardless of changes in wind direction.

11. *Swell Height.* The height between crest and trough, measured in feet. The vast majority of ocean swells are lower than 12 to 15 feet, and swells over 25 feet are not common at any spot on the oceans. Successive swells may differ considerably in height.

c. In order to select a good heading when ditching an aircraft, a basic evaluation of the sea is required. Selection of a good ditching heading may well minimize damage and could save your life. It can be extremely dangerous to land into the wind without regard to sea conditions; the swell system, or systems, must be taken into consideration. Remember one axiom—*AVOID THE FACE OF A SWELL.*

1. In ditching parallel to the swell, it makes little difference whether touchdown is on the top of the crest or in the trough. It is preferable, however, to land on the top or back side of the swell, if possible. After determining which heading (and its reciprocal) will parallel the swell, select the heading with the most into the wind component.

2. If only one swell system exists, the problem is relatively simple—even with a high, fast system. Unfortunately, most cases involve two or more swell systems running in different directions. With more than one system present, the sea presents a confused appearance. One of the most difficult situations occurs when two swell systems are at right angles. For example, if one system is eight feet high, and the other three feet, plan to land parallel to the primary system, and on the down swell of the secondary system. If both systems are of equal height, a compromise may be advisable—select an intermediate heading at 45 degrees down swell to both systems. When landing down a secondary swell, attempt to touch down on the back side, not on the face of the swell.

3. If the swell system is formidable, it is considered advisable, in landplanes, to accept more crosswind in order to avoid landing directly into the swell.

4. The secondary swell system is often from the same direction as the wind. Here, the landing may be made parallel to the primary system, with the wind and secondary system at an angle. There is a choice to two directions paralleling the primary system. One direction is downwind and down the secondary swell, and the other is into the wind and into the secondary swell, the choice will depend on the velocity of the wind versus the velocity and height of the secondary swell.

d. The simplest method of estimating the wind direction and velocity is to examine the windstreaks on the water. These appear as long streaks up and down wind. Some persons may have difficulty determining wind direction after seeing the streaks on the water. Whitecaps fall forward with the wind but are overrun by the waves thus producing the illusion that the foam is sliding backward. Knowing this, and by observing the direction of the streaks, the wind direction is easily determined. Wind velocity can be estimated by noting the appearance of the whitecaps, foam and wind streaks.

1. The behavior of the aircraft on making contact with the water will vary within wide limits according to the state of the sea. If landed parallel to a single swell system, the behavior of the aircraft may approximate that to be expected on a smooth sea. If landed into a heavy swell or into a confused sea, the deceleration forces may be extremely great—resulting in breaking up of the aircraft. Within certain limits, the pilot is able to minimize these forces by proper sea evaluation and selection of ditching heading.

2. When on final approach the pilot should look ahead and observe the surface of the sea. There may be shadows and whitecaps—signs of large seas. Shadows and whitecaps close together indicate short and rough seas. Touchdown in these areas is to be avoided. Select and touch down in any area (only about 500 feet is needed) where the shadows and whitecaps are not so numerous.

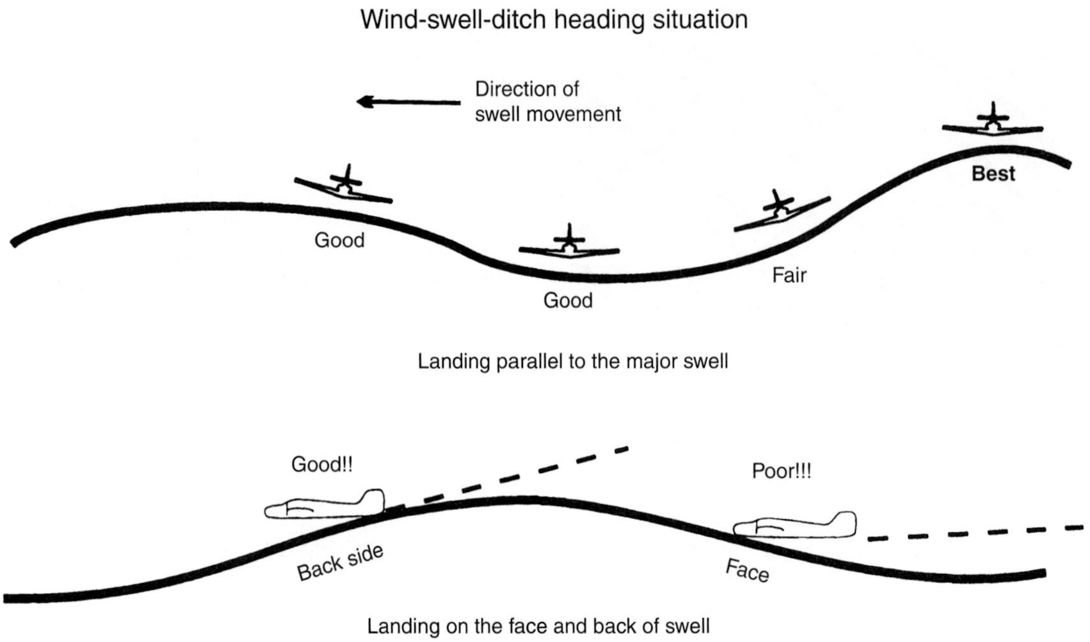

Wind-swell-ditch heading situation

Figure 6-3-5. Wind-Swell-Ditch Heading

3. Touchdown should be at the lowest speed and rate of descent which permit safe handling and optimum nose up attitude on impact. Once first impact has been made, there is often little the pilot can do to control a landplane.

e. Once preditching preparations are completed, the pilot should turn to the ditching heading and commence let-down. The aircraft should be flown low over the water, and slowed down until ten knots or so above stall. At this point, additional power should be used to overcome the increased drag caused by the nose up attitude. When a smooth stretch of water appears ahead, cut power, and touch down at the best recommended speed as fully stalled as possible. By cutting power when approaching a relatively smooth area, the pilot will prevent overshooting and will touch down with less chance of planing off into a second uncontrolled landing. Most experienced seaplane pilots prefer to make contact with the water in a semi-stalled attitude, cutting power as the tail makes contact. This technique eliminates the chance of misjudging altitude with a resultant heavy drop in a fully stalled condition. Care must be taken not to drop the aircraft from too high altitude or to balloon due to excessive speed. The altitude above water depends on the aircraft. Over glassy smooth water, or at night without sufficient light, it is very easy, for even the most experienced pilots to misjudge altitude by 50 feet or more.

Under such conditions, carry enough power to maintain nine to twelve degrees nose up attitude, and 10 to 20 percent over stalling speed until contact is made with the water. The proper use of power on the approach is of great importance. If power is available on one side only, a little power should be used to flatten the approach; however, the engine should not be used to such an extent that the aircraft cannot be turned against the good engines right down to the stall with a margin of rudder movement available. When near the stall, sudden application of excessive unbalanced power may result in loss of directional control. If power is available on one side only, a slightly higher than normal glide approach speed should be used. This will ensure good control and some margin of speed after leveling off without excessive use of power. The use of power in ditching is so important that when it is certain that the coast cannot be reached, the pilot should, if possible, ditch before fuel is exhausted. The use of power in a night or instrument ditching is far more essential than under daylight contact conditions.

1. If no power is available, a greater than normal approach speed should be used down to the flare-out. This speed margin will allow the glide to be broken early and more gradually, thereby giving the pilot time and distance to feel for the surface—decreasing the possibility of stalling high or flying into the water. When landing parallel to a swell system, little difference is noted between landing on top of a crest or in the trough. If the wings of aircraft are trimmed to the surface of the sea rather than the horizon, there is little need to worry about a wing hitting a swell crest. The actual slope of a swell is very gradual. If forced to land into a swell, touchdown should be made just after passage of the crest. If contact is made on the face of the swell, the aircraft may be swamped or thrown violently into the air, dropping heavily into the next swell. If control surfaces remain intact, the pilot should attempt to maintain the proper nose above the horizon attitude by rapid and positive use of the controls.

f. After Touchdown: In most cases drift, caused by crosswind can be ignored; the forces acting on the aircraft after touchdown are of such magnitude that drift will be only a secondary consideration. If the aircraft is under good control, the "crab" may be kicked out with rudder just prior to touchdown. This is more important with high wing aircraft, for they are laterally unstable on the water in a crosswind and may roll to the side in ditching.

REFERENCE—THIS INFORMATION HAS BEEN EXTRACTED FROM APPENDIX H OF THE "NATIONAL SEARCH AND RESCUE MANUAL."

6-3-4. SPECIAL EMERGENCY (AIR PIRACY)

a. A special emergency is a condition of air piracy, or other hostile act by a person(s) aboard an aircraft, which threatens the safety of the aircraft or its passengers.

b. The pilot of an aircraft reporting a special emergency condition should:

1. If circumstances permit, apply *distress* or *urgency* radio-telephony procedures. Include the details of the special emergency.

REFERENCE—AIM, DISTRESS AND URGENCY COMMUNICATIONS, PARAGRAPH 6-3-1.

2. If circumstances do not permit the use of prescribed *distress* or *urgency* procedures, transmit:

(a) On the air/ground frequency in use at the time.

(b) As many as possible of the following elements spoken distinctly and in the following order:

(1) Name of the station addressed (time and circumstances permitting).

(2) The identification of the aircraft and present position.

(3) The nature of the special emergency condition and pilot intentions (circumstances permitting).

(4) If unable to provide this information, use code words and/or transponder as follows: state "TRANSPONDER SEVEN FIVE ZERO ZERO." Meaning: "I am being hijacked/forced to a new destination;" and/or use Transponder Setting MODE 3/A, Code 7500.

NOTE—CODE 7500 WILL NEVER BE ASSIGNED BY ATC WITHOUT PRIOR NOTIFICATION FROM THE PILOT THAT THE AIRCRAFT IS BEING SUBJECTED TO UNLAWFUL INTERFERENCE. THE PILOT SHOULD REFUSE THE ASSIGNMENT OF CODE 7500 IN ANY OTHER SITUATION AND INFORM THE CONTROLLER ACCORDINGLY. CODE 7500 WILL TRIGGER THE SPECIAL EMERGENCY INDICATOR IN ALL RADAR ATC FACILITIES.

c. Air traffic controllers will acknowledge and confirm receipt of transponder Code 7500 by asking the pilot to verify it. If the aircraft is not being subjected to unlawful interference, the pilot should respond to the query by broadcasting in the clear that the aircraft is not being subjected to unlawful interference. Upon receipt of this information, the controller will request the pilot to verify the code selection depicted in the code selector windows in the transponder control panel and change the code to the appropriate setting. If the pilot replies in the affirmative or does not reply, the controller will not ask further questions but will flight follow, respond to pilot requests and notify appropriate authorities.

d. If it is possible to do so without jeopardizing the safety of the flight, the pilot of a hijacked passenger aircraft, after departing from the cleared routing over which the aircraft was operating, will attempt to do one or more of the following things, insofar as circumstances may permit:

1. Maintain a true airspeed of no more than 400 knots, and preferably an altitude of between 10,000 and 25,000 feet.

2. Fly a course toward the destination which the hijacker has announced.

e. If these procedures result in either radio contact or air intercept, the pilot will attempt to comply with any instructions received which may direct the aircraft to an appropriate landing field.

6-3-5. FUEL DUMPING

a. Should it become necessary to dump fuel, the pilot should immediately advise ATC. Upon receipt of information that an aircraft will dump fuel, ATC will broadcast or cause to be broadcast immediately and every 3 minutes thereafter the following on appropriate ATC and FSS radio frequencies:

EXAMPLE—ATTENTION ALL AIRCRAFT—FUEL DUMPING IN PROGRESS OVER—(LOCATION) AT (ALTITUDE) BY (TYPE AIRCRAFT) (FLIGHT DIRECTION).

b. Upon receipt of such a broadcast, pilots of aircraft affected, which are not on IFR flight plans or special VFR clearances, should clear the area specified in the advisory. Aircraft on IFR flight plans or special VFR clearances will be provided specific separation by ATC. At the termination of the fuel dumping operation, pilots should advise ATC. Upon receipt of such information, ATC will issue, on the appropriate frequencies, the following:

EXAMPLE—ATTENTION ALL AIRCRAFT—FUEL DUMPING BY—(TYPE AIRCRAFT)—TERMINATED.

Section 4. TWO-WAY
RADIO COMMUNICATIONS FAILURE

6-4-1. TWO-WAY RADIO COMMUNICATIONS FAILURE

a. It is virtually impossible to provide regulations and procedures applicable to all possible situations associated with two-way radio communications failure. During two-way radio communications failure, when confronted by a situation not covered in the regulation, pilots are expected to exercise good judgment in whatever action they elect to take. Should the situation so dictate they should not be reluctant to use the emergency action contained in FAR Part 91.3(b).

b. Whether two-way communications failure constitutes an emergency depends on the circumstances, and in any event, it is a determination made by the pilot. FAR Part 91.3(b) authorizes a pilot to deviate from any rule in Subparts A and B to the extent required to meet an emergency.

c. In the event of two-way radio communications failure, ATC service will be provided on the basis that the pilot is operating in accordance with FAR Part 91.185. A pilot experiencing two-way communications failure should (unless emergency authority is exercised) comply with FAR Part 91.185 quoted below:

NOTE—CAPITALIZATION, PRINT AND EXAMPLES CHANGED/ADDED FOR EMPHASIS.

1. *General.* Unless otherwise authorized by ATC, each pilot who has two-way radio communications failure when operating under IFR shall comply with the rules of this section.

2. *VFR conditions.* If the failure occurs in VFR conditions, or if VFR conditions are encountered after the failure, each pilot shall continue the flight under VFR and land as soon as practicable.

NOTE—This procedure also applies when two-way radio failure occurs while operating in Class A airspace. The primary objective of this provision in 14 CFR Section 91.185 is to preclude extended IFR operation by these aircraft within the ATC system. Pilots should recognize that operation under these conditions may unnecessarily as well as adversely affect other users of the airspace, since ATC may be required to reroute or delay other users in order to protect the failure aircraft. However, it is not intended that the requirement to "land as soon as practicable" be construed to mean "as soon as possible." Pilots retain the prerogative of exercising their best judgment and are not required to land at an unauthorized airport, at an airport unsuitable for the type of aircraft flown, or to land only minutes short of their intended destination.

3. *IFR conditions.* If the failure occurs in IFR conditions, or if subparagraph 2. above cannot be complied with, each pilot shall continue the flight according to the following:

(a) *Route:*

(1) By the route assigned in the last ATC clearance received;

(2) If being radar vectored, by the direct route from the point of radio failure to the fix, route, or airway specified in the vector clearance;

(3) In the absence of an assigned route, by the route that ATC has advised may be expected in a further clearance; or

(4) In the absence of an assigned route or a route that ATC has advised may be expected in a further clearance by the route filed in the flight plan.

(b) *Altitude:* At the HIGHEST of the following altitudes or Flight Levels FOR THE ROUTE SEGMENT BEING FLOWN:

(1) The altitude or flight level assigned in the last ATC clearance received;

(2) The minimum altitude (converted, if appropriate, to minimum flight level as prescribed in FAR Part 91.121.(c)) for IFR operations; or

(3) The altitude or flight level ATC has advised may be expected in a further clearance.

NOTE—THE INTENT OF THE RULE IS THAT A PILOT WHO HAS EXPERIENCED TWO-WAY RADIO FAILURE SHOULD SELECT THE APPROPRIATE ALTITUDE FOR THE PARTICULAR ROUTE SEGMENT BEING FLOWN AND MAKE THE NECESSARY ALTITUDE ADJUSTMENTS FOR SUBSEQUENT ROUTE SEGMENTS IF THE PILOT RECEIVED AN "EXPECT FURTHER CLEARANCE" CONTAINING A HIGHER ALTITUDE TO EXPECT AT A SPECIFIED TIME OR FIX, MAINTAIN THE HIGHEST OF THE FOLLOWING ALTITUDES UNTIL THAT TIME/FIX:

(1) THE LAST ASSIGNED ALTITUDE, OR

(2) THE MINIMUM ALTITUDE/FLIGHT LEVEL FOR IFR OPERATIONS.

UPON REACHING THE TIME/FIX SPECIFIED, THE PILOT SHOULD COMMENCE CLIMBING TO THE ALTITUDE ADVISED TO EXPECT IF THE RADIO FAILURE OCCURS AFTER THE TIME/FIX SPECIFIED, THE ALTITUDE TO BE EXPECTED IS NOT APPLICABLE AND THE PILOT SHOULD MAINTAIN AN ALTITUDE CONSISTENT WITH 1 OR 2 ABOVE.

IF THE PILOT RECEIVES AN "EXPECT FURTHER CLEARANCE" CONTAINING A LOWER ALTITUDE, THE PILOT SHOULD MAINTAIN THE HIGHEST OF 1 OR 2 ABOVE UNTIL THAT TIME/FIX SPECIFIED IN SUBPARAGRAPH (c) BELOW.

EXAMPLE—

☐ A PILOT EXPERIENCING TWO-WAY RADIO FAILURE AT AN ASSIGNED ALTITUDE OF 7,000 FEET IS CLEARED ALONG A DIRECT ROUTE WHICH WILL REQUIRE A CLIMB TO A MINIMUM IFR ALTITUDE OF 9,000 FEET, SHOULD CLIMB TO REACH 9,000 FEET AT THE TIME OR PLACE WHERE IT BECOMES NECESSARY (SEE FAR PART 91.177(B)). LATER WHILE PROCEEDING ALONG AN AIRWAY WITH AN MEA OF 5,000 FEET, THE PILOT WOULD DESCEND TO 7,000 FEET (THE LAST ASSIGNED ALTITUDE), BECAUSE THAT ALTITUDE IS HIGHER THAN THE MEA.

☐ A PILOT EXPERIENCING TWO-WAY RADIO FAILURE WHILE BEING PROGRESSIVELY DESCENDED TO LOWER

ALTITUDES TO BEGIN AN APPROACH IS ASSIGNED 2,700 FEET UNTIL CROSSING THE VOR AND THEN CLEARED FOR THE APPROACH. THE MOCA ALONG THE AIRWAY IS 2,700 FEET AND MEA IS 4,000 FEET. THE AIRCRAFT IS WITHIN 22 NM OF THE VOR. THE PILOT SHOULD REMAIN AT 2,700 FEET UNTIL CROSSING THE VOR BECAUSE THAT ALTITUDE IS THE MINIMUM IFR ALTITUDE FOR THE ROUTE SEGMENT BEING FLOWN.

3 THE MEA BETWEEN A AND B—5,000 FEET. THE MEA BETWEEN B AND C—5,000 FEET. THE MEA BETWEEN C AND D—11,000 FEET. THE MEA BETWEEN D AND E—7,000 FEET. A PILOT HAD BEEN CLEARED VIA A, B, C, D, TO E. WHILE FLYING BETWEEN A AND B THE ASSIGNED ALTITUDE WAS 6,000 FEET AND THE PILOT WAS TOLD TO EXPECT A CLEARANCE TO 8,000 FEET AT B. PRIOR TO RECEIVING THE HIGHER ALTITUDE ASSIGNMENT, THE PILOT EXPERIENCED TWO-WAY FAILURE. THE PILOT WOULD MAINTAIN 6,000 TO B, THEN CLIMB TO 8,000 FEET (THE ALTITUDE ADVISED TO EXPECT). THE PILOT WOULD MAINTAIN 8,000 FEET, THEN CLIMB TO 11,000 AT C, OR PRIOR TO C IF NECESSARY TO COMPLY WITH AN MCA AT C (FAR PART 91.177(B)). UPON REACHING D, THE PILOT WOULD DESCEND TO 8,000 FEET (EVEN THOUGH THE MEA WAS 7,000 FEET), AS 8,000 WAS THE HIGHEST OF THE ALTITUDE SITUATIONS STATED IN THE RULE (FAR PART 91.185).

(c) *Leave clearance limit:*

(1) When the clearance limit is a fix from which an approach begins, commence descent or descent and approach as close as possible to the expect further clearance time if one has been received, or if one has not been received, as close as possible to the Estimated Time of Arrival (ETA) as calculated from the filed or amended (with ATC) Estimated Time en Route (ETE).

(2) If the clearance limit is not a fix from which an approach begins, leave the clearance limit at the expect further clearance time if one has been received, or if none has been received, upon arrival over the clearance limit, and proceed to a fix from which an approach begins and commence descent or descent and approach as close as possible to the estimated time of arrival as calculated from the filed or amended (with ATC) estimated time en route.

6-4-2. TRANSPONDER OPERATION DURING TWO-WAY COMMUNICATIONS FAILURE

a. If an aircraft with a coded radar beacon transponder experiences a loss of two-way radio capability, the pilot should adjust the transponder to reply on MODE A/3, Code 7600.

b. The pilot should understand that the aircraft may not be in an area of radar coverage.

6-4-3. REESTABLISHING RADIO CONTACT

a. In addition to monitoring the NAVAID voice feature, the pilot should attempt to reestablish communications by attempting contact:

1. on the previously assigned frequency, or

2. with an FSS or *ARINC.

b. If communications are established with an FSS or ARINC, the pilot should advise that radio communications on the previously assigned frequency has been lost giving the aircraft's position, altitude, last assigned frequency and then request further clearance from the controlling facility. The preceding does not preclude the use of 121.5 MHz. There is no priority on which action should be attempted first. If the capability exists, do all at the same time.

NOTE—*AERONAUTICAL RADIO/INCORPORATED (ARINC) IS A COMMERCIAL COMMUNICATIONS CORPORATION WHICH DESIGNS, CONSTRUCTS, OPERATES, LEASES OR OTHERWISE ENGAGES IN RADIO ACTIVITIES SERVING THE AVIATION COMMUNITY. ARINC HAS THE CAPABILITY OF RELAYING INFORMATION TO/FROM ATC FACILITIES THROUGHOUT THE COUNTRY.

Section 5. AIRCRAFT RESCUE AND FIRE FIGHTING COMMUNICATIONS

6-5-1. DISCRETE EMERGENCY FREQUENCY

a. Direct contact between an emergency aircraft flight crew, Aircraft Rescue and Fire Fighting Incident Commander (ARFF IC), and the Airport Traffic Control Tower (ATCT), is possible on an aeronautical radio frequency (Discrete Emergency Frequency [DEF]), designated by Air Traffic Control (ATC) from the operational frequencies assigned to that facility.

b. Emergency aircraft at airports without an ATCT, (or when the ATCT is closed), may contact the ARFF IC (if ARFF service is provided), on the Common Traffic Advisory Frequency (CTAF) published for the airport or the civil emergency frequency 121.5 MHz.

6-5-2. RADIO CALL SIGNS

Preferred radio call sign for the ARFF IC is "(location/facility) Command" when communicating with the flight crew and the FAA ATCT.

EXAMPLE—
LAX COMMAND.
WASHINGTON COMMAND.

6-5-3. ARFF EMERGENCY HAND SIGNALS

In the event that electronic communications cannot be maintained between the ARF IC and the flight crew, standard emergency hand signals as depicted in FIG 6-5-1 through FIG 6-5-3 should be used. These hand signals should be known and understood by all cockpit and cabin aircrew, and all ARFF firefighters.

Figure 6-5-1. RECOMMEND EVACUATION. Evacuation recommended based on ARFF IC's assessment of external situation. Arm extended from body, and held horizontal with hand upraised at eye level. Execute beckoning arm motion angled backward. Nonbeckoning arm held against body. Night - same with wands.

Figure 6-5-2. RECOMMEND STOP. Recommend evacuation in progress be halted. Stop aircraft movement or other activity in progress. Arms in front of head - Crossed at wrists.
NIGHT - same with wands.

Figure 6-5-3. EMERGENCY CONTAINED. No outside evidence of dangerous condition or "all clear."
Arms extended outward and down at a 45 degree angle. Arms moved inward below waistline simultaneously until wrists crossed, then extended outward to starting position (umpire's "safe" signal).
NIGHT - same with wands.

Chapter 7. Safety of flight
Section 1. Meteorology

7-1-1. NATIONAL WEATHER SERVICE AVIATION PRODUCTS

a. Weather service to aviation is a joint effort of the National Weather Service (NWS), the Federal Aviation Administration (FAA), the military weather services, and other aviation oriented groups and individuals. The NWS maintains an extensive surface, upper air, and radar weather observing program; a nationwide aviation weather forecasting service; and provides limited pilot briefing service (interpretational). The majority of pilot weather briefings are provided by FAA personnel at Flight Service Stations (AFSSs/FSSs). Aviation routine weather reports (METAR) are taken manually by NWS, FAA, contractors, or supplemental observers. METAR reports are also provided by Automated Weather Observing Systems (AWOS) and Automated Surface Observing Systems (ASOS).

REFERENCE—AIM, WEATHER OBSERVING PROGRAMS, PARAGRAPH 7-1-10.

b. Aerodrome forecasts are prepared by approximately 100 Weather Forecast Offices (WFOs). These offices prepare and distribute approximately 525 aerodrome forecasts 4 times daily for specific airports in the 50 States, Puerto Rico, the Caribbean and Pacific Islands. These forecasts are valid for 24 hours and amended as required. WFOs prepare over 300 route forecasts and 39 synopses for Transcribed Weather Broadcasts (TWEB), and briefing purposes. The route forecasts are issued 4 times daily, each forecast is valid for 12 hours. A centralized aviation forecast program originating from the Aviation Weather Center (AWC) in Kansas City was implemented in October 1995. In the conterminous U.S., all Inflight Advisories Significant Meteorological Information (SIGMETs), Convective SIGMETs, and Airmen's Meteorological Information (AIRMETs) and all Area Forecasts (FA) (6 areas) are now issued by AWC. FAs are prepared 3 times a day in the conterminous U.S. and Alaska (4 times in Hawaii), and amended as required. Inflight Advisories are issued only when conditions warrant. Winds aloft forecasts are provided for 176 locations in the 48 contiguous States and 21 locations in Alaska for flight planning purposes. (Winds aloft forecasts for Hawaii are prepared locally.)

All the aviation weather forecasts are given wide distribution through the Weather Message Switching Center Replacement (WMSCR) in Atlanta, Georgia, and Salt Lake City, Utah.

REFERENCE—AIM, INFLIGHT WEATHER ADVISORIES, PARAGRAPH 7-1-6.

c. Weather element values may be expressed by using different measurement systems depending on several factors, such as whether the weather products will be used by the general public, aviation interests, international services, or a combination of these users. FIG 7-1-1 provides conversion tables for the most used weather elements that will be encountered by pilots.

REFERENCE—AIM, INFLIGHT AVIATION WEATHER ADVISORIES, PARAGRAPH 7-1-6.

7-1-2. FAA WEATHER SERVICES

a. The FAA maintains a nationwide network of Automated Flight Service Stations (AFSSs/FSSs) to serve the weather needs of pilots. In addition, NWS meteorologists are assigned to most ARTCCs as part of the Center Weather Service Unit (CWSU). They provide Center Weather Advisories (CWA) and gather weather information to support the needs of the FAA and other users of the system.

b. The primary source of preflight weather briefings is an individual briefing obtained from a briefer at the AFSS/FSS. These briefings, which are tailored to your specific flight, are available 24 hours a day through the use of the toll free number (1-800-WX BRIEF). Numbers for these services can be found in the Airport/Facility Directory (A/FD) under "FAA and NWS Telephone Numbers" section. They may also be listed in the U.S. Government section of your local telephone directory under Department of Transportation, Federal Aviation Administration, or Department of Commerce, National Weather Service. NWS pilot weather briefers do not provide aeronautical information (NOTAMs, flow control advisories, etc.) nor do they accept flight plans.

REFERENCE—AIM, PREFLIGHT BRIEFING, PARAGRAPH 7-1-3, EXPLAINS THE TYPES OF PREFLIGHT BRIEFINGS AVAILABLE AND THE INFORMATION CONTAINED IN EACH.

c. Other sources of weather information are as follows:

1. Telephone Information Briefing Service (TIBS) (AFSS); and in Alaska, Transcribed Weather Broadcast (TWEB) locations, and telephone access to the TWEB (TEL-TWEB) provide continuously updated recorded weather information for short or local flights. Separate paragraphs in this section give additional information regarding these services.

REFERENCE—AIM, TELEPHONE INFORMATION BRIEFING SERVICE (TIBS), PARAGRAPH 7-1-8.

AIM, TRANSCRIBED WEATHER BROADCAST (TWEB) (Alaska only) PARAGRAPH 7-1-9.

2. Weather and aeronautical information are also available from numerous private industry sources on an individual or contract pay basis. Information on how to obtain this service should be available from local pilot organizations.

3. The Direct User Access System (DUATS) can be accessed by pilots with a current medical certificate toll-free

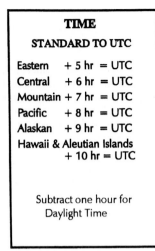

TIME

STANDARD TO UTC

Eastern	+ 5 hr = UTC
Central	+ 6 hr = UTC
Mountain	+ 7 hr = UTC
Pacific	+ 8 hr = UTC
Alaskan	+ 9 hr = UTC
Hawaii & Aleutian Islands	+ 10 hr = UTC

Subtract one hour for
Daylight Time

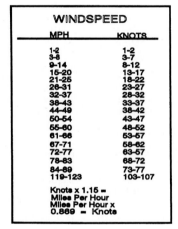

WINDSPEED

MPH	KNOTS
1-2	1-2
3-8	3-7
9-14	8-12
15-20	13-17
21-25	18-22
26-31	23-27
32-37	28-32
38-43	33-37
44-49	38-42
50-54	43-47
55-60	48-52
61-66	53-57
67-71	58-62
72-77	63-67
78-83	68-72
84-89	73-77
119-123	103-107

Knots x 1.15 =
Miles Per Hour
Miles Per Hour x
0.869 = Knots

Speed - Distance (M/H, KTS, KM/H) — Temperature (°F, °C) — Pressure - Altitude (INS. MBS./hPas., 100'S FT*) — Altimeter Setting (INS., MBS./hPas.)

STANDARD CONVERSIONS

* Standard Atmosphere

Figure 7-1-1. Weather Elements Conversion Tables

in the 48 contiguous States via personal computer. Pilots can receive alpha-numeric preflight weather data and file domestic VFR and IFR flight plans. The following are the contract DUATS vendors:

GTE Federal Systems
15000 Conference Center Drive
Chantilly, VA 22021-3808
Computer Modem Access Number: For filing flight plans and obtaining weather briefings:
(800) 767-9989
For customer service: (800) 345-3828
Data Transformation Corporation
108-D Greentree Road
Turnersville, NJ 08012

Computer Modem Access Number: For filing flight plans and obtaining weather briefings:
(800) 245-3828
For customer service: (800) 243-3828

d. Inflight weather information is available from any FSS within radio range. The common frequency for all AFSS's is 122.2. Discrete frequencies for individual stations are listed in the Airport/Facility Directory.

1. Information on In-Flight Weather broadcasts.
REFERENCE—AIM, INFLIGHT WEATHER BROADCASTS, PARAGRAPH 7-1-9.

2. En Route Flight Advisory Service (EFAS) is provided to serve the nonroutine weather needs of pilots in flight.

REFERENCE—AIM, EN ROUTE FLIGHT ADVISORY SERVICE (EFAS), PARAGRAPH 7-1-4 GIVES DETAILS ON THIS SERVICE.

7-1-3. USE OF AVIATION WEATHER PRODUCTS

a. Air carriers and operators certificated under the provisions of 14 CFR Part 119 are required to use the aeronautical weather information systems defined in the Operations Specifications issued to that certificate holder by the FAA. These systems may utilize basic FAA/National Weather Service (NWS) weather services, contractor- or operator-proprietary weather services and/or Enhanced Weather Information System (EWINS) when approved in the Operations Specifications. As an integral part of this system approval, the procedures for collecting, producing and disseminating aeronautical weather information, as well as the crew member and dispatcher training to support the use of system weather products, must be accepted or approved.

b. Operators not certificated under the provisions of 14 CFR Part 119 are encouraged to use FAA/NWS products through Flight Service Stations, Direct User Access Terminal System (DUATS), and/or Flight Information Services Data Link (FISDL).

c. The suite of available aviation weather product types is expanding, with the development of new sensor systems, algorithms and forecast models. The FAA and NWS, supported by the National Center for Atmospheric Research and the Forecast Systems Laboratory, develop and implement new aviation weather product types through a comprehensive process known as the Aviation Weather Technology Transfer process. This process ensures that user needs, and technical and operational readiness requirements are met as experimental product types mature to operational application.

d. The development of new weather products coupled with increased access to these products via the public Internet, created confusion within the aviation community regarding the relationship between regulatory requirements and new weather products. Consequently, FAA differentiates between those weather products that may be utilized to comply with regulatory requirements and those that may only be used to improve situational awareness. To clarify the proper use of aviation weather products to meet the requirements of 14 CFR, FAA defines weather products as follows:

1. Primary Weather Product. An aviation weather product that meets all the regulatory requirements and safety needs for use in making flight related, aviation weather decisions.

2. Supplementary Weather Product. An aviation weather product that may be used for enhanced situational awareness. If utilized, a supplementary weather product must only be used in conjunction with one or more primary weather product. In addition, the FAA may further restrict the use of supplementary aviation weather products through limitations described in the product label.

NOTE—AN AVIATION WEATHER PRODUCT PRODUCED BY THE FEDERAL GOVERNMENT IS A PRIMARY PRODUCT UNLESS DESIGNATED AS A SUPPLEMENTARY PRODUCT BY FAA.

e. In developing the definitions of primary and supplementary weather products, it is not the intent of FAA to change or increase the regulatory burden. Rather, the definitions are meant to eliminate confusion by differentiating between weather products that may be utilized to meet regulatory requirements and other weather products that may only be used to improve situational awareness.

f. All flight-related, aviation weather decisions must be based on primary weather products. Supplementary weather products augment the primary products by providing additional weather information but may not be used as stand-alone weather products to meet aviation weather regulatory requirements or without the relevant primary products. When discrepancies exist between primary and supplementary weather products describing the same weather phenomena, users must base flight-related decisions on the primary weather product. Furthermore, multiple primary products may be necessary to meet all aviation weather regulatory requirements.

g. The development of enhanced communications capabilities, most notably the Internet, has allowed pilots access to an ever-increasing range of weather service providers and proprietary products. The FAA has identified three distinct types of weather information available to pilots and operators.

1. Observations. Raw weather data collected by some type of sensor suite including surface and airborne observations, radar, lightning, satellite imagery, and profilers.

2. Analysis. Enhanced depiction and/or interpretation of observed weather data.

3. Forecasts. Predictions of the development and/or movement of weather phenomena based on meteorological observations and various mathematical models.

h. Not all sources of aviation weather information are able to provide all three types of weather information. The FAA has determined that operators and pilots may utilize the following approved sources of aviation weather information:

1. Federal Government. The FAA and NWS collect raw weather data, analyze the observations, and produce forecasts. The FAA and NWS disseminate meteorological observations, analyses, and forecasts through a variety of systems. In addition, the Federal Government is the only approval authority for sources of weather observations; for example, contract towers and airport operators may be approved by the Federal Government to provide weather observations.

2. Enhanced Weather Information System (EWINS). An EWINS is an FAA approved, proprietary system for tracking, evaluating, reporting, and forecasting the presence or lack of adverse weather phenomena. An EWINS is authorized to produce flight movement forecasts, adverse weather phenomena forecasts, and other meteorological advisories.

3. Commercial Weather Information Providers. In general, commercial providers produce proprietary weather products based on NWS/FAA products with formatting and layout modifications but no material changes to the weather information itself. This is also referred to as "repackaging." In addition, commercial providers may produce analyses, forecasts, and other proprietary weather products that substantially alter the information contained in government-produced products. However, those proprietary weather products that substantially alter government-produced weather products or information, may only be approved for use by Part 121 and Part 135 certificate holders if the commercial provider is EWINS qualified.

NOTE—COMMERCIAL WEATHER INFORMATION PROVIDERS CONTRACTED BY FAA TO PROVIDE WEATHER OBSERVATIONS, ANALYSES, AND FORECASTS (E.G., CONTRACT TOWERS) ARE INCLUDED IN THE FEDERAL GOVERNMENT CATEGORY OF APPROVED SOURCES BY VIRTUE OF MAINTAINING REQUIRED TECHNICAL AND QUALITY ASSURANCE STANDARDS UNDER FEDERAL GOVERNMENT OVERSIGHT.

i. Pilots and operators should be aware that weather services provided by entities other than FAA, NWS or their contractors (such as the DUATS and FISDL providers) may not meet FAA/NWS quality control standards. Hence, operators and pilots contemplating using such services should request and/or review an appropriate description of services and provider disclosure. This should include, but is not limited to, the type of weather product (e.g., current weather or forecast weather), the currency of the product (i.e., product issue and valid times), and the relevance of the product. Pilots and operators should be cautious when using unfamiliar products, or products not supported by FAA/NWS technical specifications.

NOTE—WHEN IN DOUBT, CONSULT WITH A FAA FLIGHT SERVICE STATION SPECIALIST.

j. As a point of clarification, Advisory Circular 00-62, Internet Communications of Aviation Weather and NOTAMS, describes the process for a weather information provider to become a Qualified Internet Communications Provider (QICP) and only applies to 14 CFR Part 121 and Part 135 certificate holders. Therefore, pilots conducting operations under 14 CFR Part 91 may access weather products via DUATS and the public Internet.

7-1-4. PREFLIGHT BRIEFING

a. Flight Service Stations (AFSSs/FSSs) are the primary source for obtaining preflight briefings and inflight weather information. Flight Service Specialists are qualified and certificated by the NWS as Pilot Weather Briefers. They are not authorized to make original forecasts, but are authorized to translate and interpret available forecasts and reports directly into terms describing the weather conditions which you can expect along your flight route and at your destination. Available aviation weather reports, forecasts and aviation weather charts are displayed at each AFSS/FSS, for pilot use. Pilots should feel free to use these selfbriefing displays where available, or to ask for a briefing or assistance from the specialist on duty. Three basic types of preflight briefings are available to serve your specific needs. These are: Standard Briefing, Abbreviated Briefing, and Outlook Briefing. You should specify to the briefer the type of briefing you want, along with your appropriate background information. This will enable the briefer to tailor the information to your intended flight. The following paragraphs describe the types of briefings available and the information provided in each briefing.

REFERENCE—AIM, PREFLIGHT PREPARATION, PARAGRAPH 5-1-1 FOR ITEMS THAT ARE REQUIRED.

b. Standard Briefing: You should request a Standard Briefing any time you are planning a flight and you have not received a previous briefing or have not received preliminary information through mass dissemination media; e.g., TIBS, TWEB, (Alaska only) etc. International data may be inaccurate or incomplete. If you are planning a flight outside of United States Controlled airspace, the briefer will advise you to check data as soon as practical after entering foreign airspace, unless you advise that you have the international cautionary advisory. The briefer will automatically provide the following information in the sequence listed, except as noted, when it is applicable to your proposed flight.

1. *Adverse Conditions:* Significant meteorological and aeronautical information that might influence the pilot to alter the proposed flight; e.g., hazardous weather conditions, airport closures, air traffic delays, etc.

2. *VFR Flight Not Recommended:* When VFR flight is proposed and sky conditions or visibilities are present or forecast, surface or aloft, that in the briefer's judgment would make flight under visual flight rules doubtful, the briefer will describe the conditions, affected locations, and use the phrase *"VFR flight not recommended."* This recommendation is advisory in nature. The final decision as to whether the flight can be conducted safely rests solely with the pilot.

3. *Synopsis:* A brief statement describing the type, location and movement of weather systems and/or air masses which might affect the proposed flight.

NOTE—THESE FIRST 3 ELEMENTS OF A BRIEFING MAY BE COMBINED IN ANY ORDER WHEN THE BRIEFER BELIEVES IT WILL HELP TO MORE CLEARLY DESCRIBE CONDITIONS.

4. *Current Conditions:* Reported weather conditions applicable to the flight will be summarized from all available sources; e.g., METARs/SPECIs, PIREPs, RAREPs. This element will be omitted if the proposed time of departure is beyond 2 hours, unless the information is specifically requested by the pilot.

5. *En Route Forecast:* Forecast en route conditions for the proposed route are summarized in logical order; i.e., departure/climbout, en route, and descent. (Heights are

ASL, unless the contractions "AGL" or "CIG" are denoted indicating that heights are above ground).

6. *Destination Forecast:* The destination forecast for the planned ETA. Any significant changes within 1 hour before and after the planned arrival are included.

7. *Winds Aloft:* Forecast winds aloft will be provided using degrees of the compass. The briefer will interpolate wind directions and speeds between levels and stations as necessary to provide expected conditions at planned altitudes. (Heights are MSL). Temperature information will be provided on request.

8. *Notices to Airmen (NOTAMs):*

(a) Available NOTAM (D) information pertinent to the proposed flight.

(b) NOTAM (L) information pertinent to the departure and/or local area, if available, and pertinent FDC NOTAMs within approximately 400 miles of the FSS providing the briefing. AFSS facilities will provide FDC NOTAMs for the entire route of flight.

NOTE—NOTAM INFORMATION MAY BE COMBINED WITH CURRENT CONDITIONS WHEN THE BRIEFER BELIEVES IT IS LOGICAL TO DO SO.

(c) FSS briefers do not provide FDC NOTAM information for special instrument approach procedures unless specifically asked. Pilots authorized by the FAA to use special instrument approach procedures must specifically request FDC NOTAM information for these procedures.

NOTE—NOTAM (D) INFORMATION AND FDC NOTAMS WHICH HAVE BEEN PUBLISHED IN THE NOTICES TO AIRMEN PUBLICATION ARE NOT INCLUDED IN PILOT BRIEFINGS UNLESS A REVIEW OF THIS PUBLICATION IS SPECIFICALLY REQUESTED BY THE PILOT. FOR COMPLETE FLIGHT INFORMATION YOU ARE URGED TO REVIEW THE PRINTED NOTAMS IN THE NOTICES TO AIRMEN PUBLICATION AND THE A/FD IN ADDITION TO OBTAINING A BRIEFING.

9. *ATC Delays:* Any known ATC delays and flow control advisories which might affect the proposed flight.

10. *Pilots may obtain the following from AFSS/FSS briefers upon request:*

(a) Information on Special Use Airspace (SUA), SUA related airspace and Military Training Routes (MTRs) activity within the flight plan area and a 100 NM extension around the flight plan area.

NOTE—1. SUA AND RELATED AIRSPACE INCLUDES THE FOLLOWING TYPES OF AIRSPACE: ALERT AREA, MILITARY OPERATIONS AREA (MOA), PROHIBITED AREA, RESTRICTED AREA, REFUELING ANCHOR, WARNING AREA AND AIR TRAFFIC CONTROL ASSIGNED AIRSPACE (ATCAA). MTR DATA INCLUDES THE FOLLOWING TYPES OF AIRSPACE: IFR MILITARY TRAINING ROUTE (IR), VFR MILITARY TRAINING ROUTE (VR), SLOW TRAINING ROUTE (SR) AND AERIAL REFUELING TRACK (AR).
2. PILOTS ARE ENCOURAGED TO REQUEST UPDATED INFORMATION FROM ATC FACILITIES WHILE IN FLIGHT.

(b) A review of the Notices to Airmen Publication for pertinent NOTAMs and Special Notices.

(c) Approximate density altitude data.

(d) Information regarding such items as air traffic services and rules, customs/immigration procedures, ADIZ rules, search and rescue, etc.

(e) LORAN-C NOTAMS, available military NOTAMS, and runway friction measurement value NOTAMS.

(f) GPS RAIM availability for 1 hour before to 1 hour after ETA or a time specified by the pilot.

(g) Other assistance as required.

c. Abbreviated Briefing: Request an Abbreviated Briefing when you need information to supplement mass disseminated data, update a previous briefing, or when you need only one or two specific items. Provide the briefer with appropriate background information, the time you received the previous information, and/or the specific items needed. You should indicate the source of the information already received so that the briefer can limit the briefing to the information that you have not received, and/or appreciable changes in meteorological/aeronautical conditions since your previous briefing. To the extent possible, the briefer will provide the information in the sequence shown for a Standard Briefing. If you request only one or two specific items, the briefer will advise you if adverse conditions are present or forecast. (Adverse conditions contain both meteorological and/or aeronautical information). Details on these conditions will be provided at your request. International data may be inaccurate or incomplete. If you are planning a flight outside of United States controlled airspace, the briefer will advise you to check data as soon as practical after entering foreign airspace, unless you advise that you have the international cautionary advisory.

d. Outlook Briefing: You should request an Outlook Briefing whenever your proposed time of departure is six or more hours from the time of the briefing. The briefer will provide available forecast data applicable to the proposed flight. This type of briefing is provided for planning purposes only. You should obtain a Standard or Abbreviated Briefing prior to departure in order to obtain such items as adverse conditions, current conditions, updated forecasts, winds aloft and NOTAMs, etc.

e. When filing a flight plan only, you will be asked if you require the latest information on adverse conditions pertinent to the route of flight.

f. Inflight Briefing: You are encouraged to obtain your preflight briefing by telephone or in person before departure. In those cases where you need to obtain a preflight briefing or an update to a previous briefing by radio, you should contact the nearest AFSS/FSS to obtain this information. After communications have been established, advise the specialist of the type briefing you require and provide appropriate background information. You will be provided information as specified in the above paragraphs, depending on the type of briefing requested. In addition, the specialist will recommend shifting to the Flight Watch frequency when conditions along the intended route indicate that it would be advantageous to do so.

g. Following any briefing, feel free to ask for any information that you or the briefer may have missed or are not understood. This way, the briefer is able to present the information in a logical sequence and lessens the chance of important items being overlooked.

7-1-5. EN ROUTE FLIGHT ADVISORY SERVICE (EFAS)

a. EFAS is a service specifically designed to provide en route aircraft with timely and meaningful weather advisories pertinent to the type of flight intended, route of flight, and altitude. In conjunction with this service, EFAS is also a central collection and distribution point for pilot reported weather information. EFAS is provided by specially trained specialists in selected AFSSs controlling multiple Remote Communications Outlets covering a large geographical area and is normally available throughout the conterminous U.S. and Puerto Rico from 6 a.m. to 10 p.m. EFAS provides communications capabilities for aircraft flying at 5,000 feet above ground level to 17,500 feet MSL on a common frequency of 122.0 MHz. Discrete EFAS frequencies have been established to ensure communications coverage from 18,000 through 45,000 MSL serving in each specific ARTCC area. These discrete frequencies may be used below 18,000 feet when coverage permits reliable communication.

NOTE—WHEN AN EFAS OUTLET IS LOCATED IN A TIME ZONE DIFFERENT FROM THE ZONE IN WHICH THE FLIGHT WATCH CONTROL STATION IS LOCATED, THE AVAILABILITY OF SERVICE MAY BE PLUS OR MINUS ONE HOUR FROM THE NORMAL OPERATING HOURS.

b. In some regions of the contiguous U.S. especially those that are mountainous, it is necessary to be above 5000 feet AGL in order to be at an altitude where the EFAS frequency, 122.0 MHz, is available Pilots should take this into account when fligh planning. Other AFSS communication frequencies may be available at lower altitudes. See FIG 7-1-2

c. Contact flight watch by using the name of the ARTCC facility identification serving the area of your location, followed by your aircraft identification, and the name of the nearest VOR to your position. The specialist needs to know this approximate location to select the most appropriate transmitter/receiver outlet for communications coverage.

EXAMPLE—CLEVELAND FLIGHT WATCH, CESSNA ONE TWO THREE FOUR KILO, MANSFIELD V-O-R, OVER.

d. Charts depicting the location of the flight watch control stations (parent facility) and the outlets they use are contained in the A/FD. If you do not know in which flight watch area you are flying, initiate contact by using the words "Flight Watch," your aircraft identification, and the name of the nearest VOR. The facility will respond using the name of the flight watch facility.

EXAMPLE—FLIGHT WATCH, CESSNA ONE TWO THREE FOUR KILO, MANSFIELD V-O-R, OVER.

e. AFSSs that provide En Route Flight Advisory Service are listed regionally in the A/FDs.

f. EFAS is not intended to be used for filing or closing flight plans, position reporting, getting complete preflight briefings, or obtaining random weather reports and forecasts. En route flight advisories are tailored to the phase of flight that begins after climb-out and ends with descent to land. Immediate destination weather and terminal aerodrome forecasts will be provided on request. Pilots requesting information not within the scope of flight watch will be advised of the appropriate AFSS/FSS frequency to obtain the information. Pilot participation is essential to the success of EFAS by providing a continuous exchange of information on weather, winds, turbulence, flight visibility, icing, etc., between pilots and flight watch specialists. Pilots are encouraged to report good weather as well as bad, and to confirm expected conditions as well as unexpected to EFAS facilities.

7-1-6. INFLIGHT AVIATION WEATHER ADVISORIES

a. Background

1. Inflight Aviation Weather Advisories are forecasts to advise en route aircraft of development of potentially hazardous weather. All inflight aviation weather advisories in the conterminous U.S. are issued by the Aviation Weather Center (AWC) in Kansas City, Missouri. The Weather Forecast Office (WFO) in Honolulu issues advisories for the Hawaiian Islands. In Alaska, the Alaska Aviation Weather Unit (AAWU) issues inflight aviation weather advisories. All heights are referenced MSL, except in the case of ceilings (CIG) which indicate AGL.

2. There are three types of inflight aviation weather advisories: the Significant Meteorological Information (SIGMET), the Convective SIGMET and the Airmen's

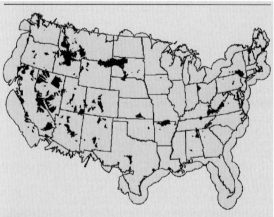

Note—
EFAS Coverage area at 5000 feet AGL. The shaded area depict the limited coverage areas in which altitude above 5000 feet AGL would be required to contact EFAS

Figure 7-1-2. EFAS Radio Coverage Areas

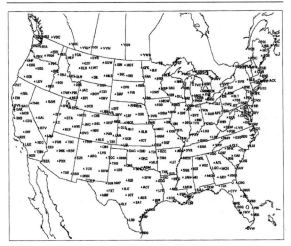

Figure 7-1-3.

Figure 7-1-4.

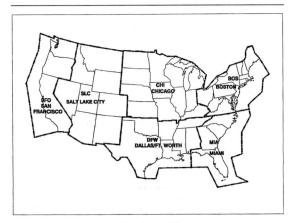

Figure 7-1-5.

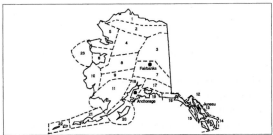

Figure 7-1-6.

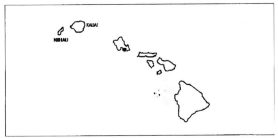

Figure 7-1-7.

Meteorological Information (AIRMET). All of these advisories use the same location identifiers (either VORs, airports, or well-known geographic areas) to describe the hazardous weather areas. See FIG 7-1-2 and FIG 7-1-3. Graphics with improved clarity can be found in Advisory Circular AC OO-45E, Aviation Weather Services, which is available on the following web site: http://www.faa.gov.

3. Two other weather products supplement these Inflight Aviation Weather Advisories:

(a) The Severe Weather Watch Bulletins (WWs), (with associated Alert Messages) (AWW), and

(b) The Center Weather Advisories (CWAs).

b. SIGMET (WS)/AIRMET (WA)

SIGMETs/AIRMET's are issued corresponding to the Area Forecast (FA) areas described in FIG 7-1-5, FIG 7-1-6, and FIG 7-1-7. The maximum forecast period is 4 hours for SIGMETs and 6 hours for AIRMETs. Both advisories are considered "widespread" because they must be either affecting or be forecasted to affect an area of at least 3,000 square miles at any one time. However, if the total area to be affected during the forecast period is very large, it could be that in actuality only a small portion of this total area would be affected at any one time.

c. SIGMET (WS)

1. A SIGMET advises of nonconvective weather that is potentially hazardous to all aircraft. SIGMETs are unscheduled products that are valid for 4 hours. However, conditions that are associated with hurricanes are valid for 6 hours.

Unscheduled updates and corrections are issued as necessary. In the conterminous U.S., SIGMETs are issued when the following phenomena occur or are expected to occur:

(a) Severe icing not associated with thunderstorms.

(b) Severe or extreme turbulence or clear air turbulence (CAT) not associated with thunderstorms.

(c) Dust storms or sandstorms lowering surface or in-flight visibilities to below 3 miles.

(d) Volcanic ash.

2. In Alaska and Hawaii, SIGMETs are also issued for:

(a) Tornadoes.

(b) Lines of thunderstorms.

(c) Embedded thunderstorms.

(d) Hail greater than or equal to 3/4 inch in diameter.

3. SIGMETs are identified by an alphabetic designator from November through Yankee excluding Sierra and Tango. (Sierra, Tango, and Zulu are reserved for AIRMETs:) The first issuance of a SIGMET will be labeled as UWS (Urgent Weather SIGMET). Subsequent issuances are at the forecaster's discretion. Issuance for the same phenomenon will be sequentially numbered, using the original designator until the phenomenon ends. For example, the first issuance in the Chicago (CHI) FA area for phenomenon moving from the Salt Lake City (SLC) FA area will be SIGMET Papa 3, if the previous two issuances, Papa 1 and Papa 2, had been in the SLC FA area. Note that no two different phenomena across the country can have the same alphabetic designator at the same time.

EXAMPLE—*Example of a SIGMET:*
BOSR WS 050600
SIGMETROMEO 2 VALID UNTIL 051000
ME NH VT
FROM CAR TO YSJ TO CON TO MPV TO CAR
MOD TO OCNL SEV TURB BLW 080 EKP DUE TO
STG NWLYFLOW CONDS CONTG BYD
1000Z

d. Convective SIGMET (WST)

1. Convective SIGMETs are issued in the conterminous U.S. for any of the following:

(a) Severe thunderstorm due to:

(1) Surface winds greater than or equal to 50 knots.

(2) Hail at the surface greater than or equal to 3/4 inches in diameter.

(3) Tornadoes.

(b) Embedded thunderstorms.

(c) A line of thunderstorms.

(d) Thunderstorms producing precipitation greater than or equal to heavy precipitation affecting 40 percent or more of an area at least 3,000 square miles.

2. Any convective SIGMET implies severe or greater turbulence, severe icing, and low-level wind shear. A convective SIGMET may be issued for any convective situation that the forecaster feels is hazardous to all categories of aircraft.

3. Convective SIGMET bulletins are issued for the western (W), central (C), and eastern (E) United States.

(Convective SIGMETs are not issued for Alaska or Hawaii.) The areas are separated at 87 and 107 degrees west longitude with sufficient overlap to cover most cases when the phenomenon crosses the boundaries. Bulletins are issued hourly at H+55. Special bulletins are issued at any time as required and updated at H+55. If no criteria meeting convective SIGMET requirements are observed or forecasted, the message "CONVECTIVE SIGMET... NONE" will be issued for each area at H+55. Individual convective SIGMETs for each area (W, C, E) are numbered sequentially from number one each day, beginning at 00Z. A convective SIGMET for a continuing phenomenon will be reissued every hour at H+55 with a new number. The text of the bulletin consists of either an observation and a forecast or just a forecast. The forecast is valid for up to 2 hours.

EXAMPLE—*Example of a Convective SIGMET:*
MKCC WST 251655
CONVECTIVE SIGMET 54C
VALID UNTIL 1855Z
WI IL
FROM 30E MSN-40ESE DBQ
DMSHG LINE TS 15 NM WIDE MOV FROM 30025KL
TOPS TO FL450. WIND GUSTS TO 50 KT POSS.

CONVECTIVE SIGMET 55C
VALID UNTIL 1855Z
WI IA
FROM 30NNW MSN-30SSE MCW
DVLPG LINE TS 10 NM WIDE MOV FROM 30015KT.
TOPS TO FL300.
CONVECTIVE SIGMET 56C
VALID UNTIL 1855Z
MT ND SD MN IA MI
LINE TS 15NM WIDE MOV FROM 27020KT. TOPS TO
FL380.
OUTLOOK VALID 151855-252255
FROM 60NW ISN-INL-TVC-SBN-BRL-FSD -BIL-60NW ISN

IR STL IMGRY SHOWS CNVTV CLD TOP TEMPS
OVER SRN WI HAVE BEEN WARMING STEADILY
INDCG A WKNG TREND. THIS ALSO REFLECTED
BY LTST RADAR AND LTNG DATA. WKNG TREND
OF PRESENT LN MAY CONT...HWVR NEWDVLPMT
IS PSBL ALG OUTFLOW BDR YAND/OR OVR NE
IA/SW W IBHD CURRENT ACT.
A SCND TS IS CONTG TO MOVE WD THRU ERN MT
WITH NEWDVLPMT OCRG OVR CNTRL ND. MT
ACT IS MOVG TWD MORE FVRBL AMS OVR THE
WRN DAKS WHERE DWPTS ARE IN THE UPR 60S
WITH LIFTED INDEX VALUES TO MS 6. TS EXPD TO
INCR IN COVERAGE AND INTSTY DURG AFTN HRS.
WST ISSUANCES EXPD TO BE RQRD THRUT AFTN
HRS WITH INCRG PTNTL FOR STGR CELLS TO
CONTAIN LRG HAIL AND PSBLY DMGG SFC WNDS.

e. International SIGMET

1. Some NWS offices have been designated by the ICAO as Meteorological Watch Offices (MWOs). These offices are responsible for issuing International SIGMETs for designated areas that include Alaska, Hawaii, portions of the Atlantic and Pacific Oceans, and the Gulf of Mexico.

2. The offices which issue International SIGMETs are:

(a) The AWC in Kansas City, Missouri.

(b) The AAWU in Anchorage, Alaska.

(c) The WFO in Honolulu, Hawaii.

(d) The WFO on Guam Island in the Pacific Ocean.

3. These SIGMETs are considered "wide spread" because they must be either affecting or be forecasted to affect an area of at least 3,000 square miles at any one time. The International SIGMET is issued for 12 hours for volcanic ash events, 6 hours for hurricanes and tropical storms, and 4 hours for all other events. Like the domestic SIGMETs, International SIGMETs are also identified by an alphabetic designator from Alpha through Mike and are numbered sequentially until that weather phenome-non ends. The criteria for an International SIGMET are:

(a) Thunderstorms occurring in lines, embedded in clouds, or in large areas producing tornadoes or large hail.

(b) Tropical cyclones.

(c) Severe icing.

(d) Severe or extreme turbulence.

(e) Dust storms and sandstorms lowering visibilities to less than 3 miles.

(f) Volcanic ash.

EXAMPLE—*Example of an International SIGMET:*
WSNT06 KKCI 022014
SIGAOF
KZMA KZNY TJZS SIGMET FOXTROT 3 VALID
022015/030015 KKCI- MIAMI OCEANIC FIR NEW
YORK OCEANIC FIR SAN JUAN FIR FRQ TS WI
AREA BOUNDED BY 2711N6807W 2156N6654W
2220N7040W 2602N7208W 2711N6807W TOPS TO
FL470. MOV NE 15KT. WKN. BASED ON SAT AND
LTG OBS.
MOSHER

f. AIRMET (WA)

1. AIRMETs (WAs) are advisories of significant weather phenomena but describe conditions at intensities lower than those which require the issuance of SIGMETs. AIRMETs are intended for dissemination to all pilots in the preflight and en route phase of flight to enhance safety. AIRMET Bulletins are issued on a scheduled basis every 6 hours beginning at 0145 UTC during Central Daylight Time and at 0245 UTC during Central Standard Time. Unscheduled updates and corrections are issued as necessary. Each AIRMET Bulletin contains any current AIRMETs in effect and an outlook for conditions expected after the AIRMET valid period. AIRMETs contain details about IFR, extensive mountain obscuration, turbulence, strong surface winds, icing, and freezing levels.

2. There are three AIRMETs: Sierra, Tango, and Zulu. After the first issuance each day, scheduled or unscheduled bulletins are numbered sequentially for easier identification.

(a) AIRMET Sierra describes IFR conditions and/or extensive mountain obscurations.

(b) AIRMET Tango describes moderate turbulence, sustained surface winds of 30 knots or greater, and/or non-convective low-level wind shear.

(c) AIRMET Zulu describes moderate icing and provides freezing level heights.

EXAMPLE—*Example of AIRMET Sierra issued for the Chicago FA area:*
CHIS WA 121345
AIRMET SIERR4 UPDT 3 FOR IFR AND MTN OBSCN
VALID UNTIL 122000.
AIR MET IFR...SD NE MN IA MO WI LM MI IL IN KY
FROM 70NW RAP TO 50W RWF TO 50W MSN TO
GRB TO MBS TO FWA TO CVG TO HNN TO TRI TO
ARG TO 40SSW BRL TO OMA TO BFF TO 70NW RAP
OCNL CIG BLW 010//VIS BLW 3SM FG/BR. CONDS
ENDG 15Z-17Z

AIRMET MTN OBSCN...KY TN
FROM HNN TO TM TO CHA TO LOZ TO HNN
MTNS OCNL OBSC CLDS/PCPN/BR. CONDS ENDG
TN PTN AREA 18Z- 20Z..CONTG KY BYD 20Z..ENDG 02Z

EXAMPLE—*Example of AIRMET Tango issued for the Salt Lake City FA area:*
SLCT WA 121345
AIRMET TANGO UPDT 2 FOR TURB VALID UNTIL 122000.
AIRMET TURB...NV UT CO AZ NM
FROM LKV TO CHE TO ELF TO 60S TUS TO YUM TO
EED TO RNO TO LKV OCNL MOD TURB BLW FL18O
DUE TO MOD SWLY/WLY WNDS. CONDS CONTG BYD
20Z THRU 021

AIRMET TURB...NV WA OR CA CSTL WTRS
FROM BLI TO REO TO BTY TO DAG TO SBA TO 120W
FOT TO 12OW TOU TO BLI
OCNL MOD TURB BTWN FLI8O AND FL400 DUE TO
WNDSHR ASSOCD WITH JTSTR. CONDS CONTG BYD
20Z THRU 02Z.

EXAMPLE—*Example of AIRMET Zu1u issued for the San Francisco FA area:*
SFOZ WA 121345
AIR MET ZULU UPDT 2 FOR ICE AND FRZ LVL VALID
UNTIL 122000.
AIR MET ICE... WA OR ID MTNV UT
FROMYQL TO SLC TO WMC TO LKV TO PDT TO YDC
TO YQL
LGT OCNL MOD RIME/MXD ICGICIP BTWN FRZLVL
AND FL 220. FRZLVL 080-120. CONDS CONTG BYD 20Z
THRU 02Z

AIRMET ICE...WA OR
FROM YDC TO PDT TO LKV TO 80 W MFR TO ONP TO
TOU TO YDC
LGT OCNL MOD RIME/MXD ICGICIP BTWN FRZLVL
 AND FL 180. FRZLVL 060-080. CONDS CONTG BYD
20z THRU 02Z.
FRZLVL...WA...060 CSTLN SLPG 100 XTRM E.
OR...060...070 CASCDS WWD. 070-095 RMNDR.
NRN CA...060-100 N OF A 30n FOT-40N RNO LN SLPG
100-110 RMNDR.

g. Severe Weather Watch Bulletins (WW's) and Alert Messages (AWW's)

1. WWs define areas of possible severe thunderstorms or tornado activity. The bulletins are issued by the Storm Prediction Center (SPC) in Norman, OK. WWs are unscheduled and are issued as required.

2. A severe thunderstorm watch describes areas of expected severe thunderstorms. (Severe thunderstorm criteria are 3/4-inch hail or larger and/or wind gusts of 50 knots [58 mph] or greater.)

3. A tornado watch describes areas where the threat of tornadoes exists.

4. In order to alert the WFOs, CWSUs, FSSs, and other users, a preliminary notification of a watch called the Alert Severe Weather Watch bulletin (AWW) is sent before the WW. (WFOs know this product as a SAW.)

EXAMPLE—*Example of an AWW*
MKC AWW 011734
WW 75 TORNADO TX OK AR 011800Z-020000Z
AXIS..80 STATUTE MILES EAST AND WEST OF A
LINE..6OESE DAL/DALLAS TX! - 3ONW ARG/ WALNUT
RIDGE AR/
.AVIATION COORDS.. 7ONM EIW /58W GGG - 25NW
ARG/
HAIL SURFACE AND ALOFT..1 3/4 INCHES. WIND
GUSTS..70 KNOTS. MAX TOPS TO 450. MEAN WIND
VECTOR 24045

5. Soon after the AWW goes out, the actual watch bulletin itself is issued. A WW is in the following format:

(a) Type of severe weather watch, watch area, valid time period, type of severe weather possible, watch axis, meaning of a watch, and a statement that persons should be on the lookout for severe weather.

(b) Other watch information; i.e., references to previous watches.

(c) Phenomena, intensities, hail size, wind speed (knots), maximum cumulonimbus (CB) tops, and estimated cell movement (mean wind vector).

(d) Cause of severe weather.

(e) Information on updating Convective Outlook (CA) products.

EXAMPLE—*Example of a WW:*
BULLETIN - IMMEDIATE BROADCASTREQUESTED
TORNADO WATCHNUMBER 381

STORM PREDICTION CENTER NORMAN OK
556 PM CDT MON JUN 2 1997
THE STORM PREDICTION CENTER HAS ISSUED A
TORNADO WATCH FOR PORTIONS OF NORTHEAST
NEW MEXICO TEXAS PANHANDLE
EFFECTIVE THIS MONDAY NIGHT AND TUESDAY
MORNING FROM 630 PM UNTIL MIDNIGHT CDT.
TORNADOES...HAIL TO 2 3/4 INCHES IN
DIAMETER...THUNDERSTORM WIND GUSTS TO 80
MPH..AND DANGEROUS LIGHTNING ARE
POSSIBLE IN THESE AREAS.
THE TORNADO WATCH AREA IS ALONG AND 60
STATUTE MILES NORTH AND SOUTH OF A LINE
FROM 50 MILES SOUTHWEST OF RATON NEW
MEXICO TO 50 MILES EAST OF AMARILLO TEXAS.
REMEMBER..A TORNADO WATCH MEANS
CONDITIONS ARE FAVORABLE FOR TORNADOES
AND SEVERE THUNDERSTORMS IN AND CLOSE TO
THE WATCH AREA. PERSONS IN THESE AREAS
SHOULD BE ON THE LOOKOUT FOR
THREATENING WEATHER CONDITIONS AND
LISTEN FOR LATER STATEMENTS AND POSSIBLE
WARNINGS.
OTHER WATCH INFORMATION...CONTINUE...
WW 378...WW 379...WW 380
DISCUSSION...THUNDERSTORMS ARE INCREASING
OVER NE NM IN MOIS TSOUTHEASTERLY UPSLOPE
FLOW OUTFLOW BOUNDARY EXTENDS
EASTWARD INTO THE TEXAS PANHANDLE AND
EXPECT STORMS TO MOVE ESE ALONG AND
NORTH OF THE BOUNDARY ON THE N EDGE OF
THE CAP. VEERING WINDS WITH HEIGHT ALONG
WITH INCREASING MID LVL FLOW INDICATE A
THREAT FOR SUPER CELLS.
AVIATION...TORNADOES AND A FEW SEVERE
THUNDERSTORMS WITH HAIL SURFACE AND
ALOFT TO 2 3/4 INCHES. EXTREME TURBULENCE
AND SURFACE WIND GUSTS TO 70 KNOTS. A FEW
CUMULONIMBI WITH MAXIMUM TOPS TO 550.
MEAN STORM MOTION VECTOR 28025

6. Status reports are issued as needed to show progress of storms and to delineate areas no longer under the threat of severe storm activity. Cancellation bulletins are issued when it becomes evident that no severe weather will develop or that storms have subsided and are no longer severe.

7. When tornadoes or severe thunderstorms have developed, the local WFO office will issue the warnings covering those areas.

h. Center Weather Advisories (CWAs)

1. CWAs are unscheduled inflight, flow control, air traffic, and air crew advisory. By nature of its short lead time, the CWA is not a flight planning product. It is generally a nowcast for conditions beginning within the next two hours. CWAs will be issued:

(a) As a supplement to an existing SIGMET, Convective SIGMET or AIRMET.

(b) When an Inflight Advisory has not been issued but observed or expected weather conditions meet SIGMET/AIRMET criteria based on current pilot reports and reinforced by other sources of information about existing meteorological conditions.

(c) When observed or developing weather conditions do not meet SIGMET, Convective SIGMET, or AIRMET criteria; e.g., in terms of intensity or area coverage, but current pilot reports or other weather information sources indicate that existing or anticipated meteorological phenomena will adversely affect the safe flow of air traffic within the ARTCC area of responsibility.

2. The following example is a CWA issued from the Kansas City, Missouri, ARTCC. The "3" after ZKC in the first line denotes this CWA has been issued for the third weather phenomena to occur for the day. The "301" in the second line denotes the phenomena number again (3) and the issuance number (01) for this phenomena. The CWA was issued at 2140Z and is valid until 2340Z.

EXAMPLE—ZKC3 CWA 032140

ZKC CWA 301 VALID UNTIL 032340

ISOLD SVR TSTM over KCOU MOVG SWWD 10 KTS ETC.

The outlook is a forecast and meteorological discussion for thunderstorm systems that are expected to require Convective SIGMET issuances during a time period 2–6 hours into the future. Furthermore, an outlook will always be made for each of the three regions, even if it is a negative statement.

e. SIGMETs (WS's) within the conterminous U.S. are issued by the Aviation Weather Center (AWC) when the following phenomena occur or are expected to occur:

1. Severe or extreme turbulence or clear air turbulence (CAT) not associated with thunderstorms.

2. Severe icing not associated with thunderstorms.

3. Dust storms, sandstorms, or volcanic ash lowering surface or inflight visibilities to below three miles.

4. Volcanic eruption.

f. Volcanic eruption SIGMETs are identified by an alphanumeric designator which consists of an alphabetic identifier and issuance number. The first time an advisory is issued for a phenomenon associated with a particular weather system, it will be given the next alphabetic designator in the series and will be numbered as the first for that designator. Subsequent advisories will retain the same alphabetic designator until the phenomenon ends. In the conterminous U.S., this means that a phenomenon that is assigned an alphabetic designator in one area will retain that designator as it moves within the area or into one or more other areas. Issuances for the same phenomenon will be sequentially numbered, using the same alphabetic designator until the phenomenon no longer exists. Alphabetic designa-

tors NOVEMBER through YANKEE, except SIERRA and TANGO are only used for SIGMETS, while designators SIERRA, TANGO and ZULU are used for AIRMETs.

g. Center Weather Advisories (CWA's):

1. CWA's are unscheduled inflight, flow control, air traffic, and air crew advisory. By nature of its short lead time, the CWA is not a flight planning product. It is generally a Nowcast for conditions beginning within the next two hours. CWA's will be issued:

(a) As a supplement to an existing SIGMET, Convective SIGMET or AIRMET.

(b) When an Inflight Advisory has not been issued but observed or expected weather conditions meet SIGMET/AIRMET criteria based on current pilot reports and reinforced by other sources of information about existing meteorological conditions.

(c) When observed or developing weather conditions do not meet SIGMET, Convective SIGMET, or AIRMET criteria; e.g., in terms of intensity or area coverage, but current pilot reports or other weather information sources indicate that existing or anticipated meteorological phenomena will adversely affect the safe flow of air traffic within the ARTCC area of responsibility.

2. The following example is a CWA issued from the Kansas City, Missouri, ARTCC. The "3" after ZKC in the first line denotes this CWA has been issued for the third weather phenomena to occur for the day. The "301" in the second line denotes the phenomena number again (3) and the issuance number (01) for this phenomena. The CWA was issued at 2140Z and is valid until 2340Z.

EXAMPLE—

ZKC3 CWA 032140

ZKC CWA 301 VALID UNTIL 032340

ISOLD SVR TSTM over KCOU MOVG SWWD 10 KTS ETC.

h. AIRMETS (WAs) may be of significance to any pilot or aircraft operator and are issued for all domestic airspace. They are of particular concern to operators and pilots of aircraft sensitive to the phenomena described and to pilots without instrument ratings and are issued by the AWC for the following weather phenomena which are potentially hazardous to aircraft:

1. Moderate icing.

2. Moderate turbulence.

3. Sustained winds of 30 knots or more at the surface.

4. Widespread area of ceilings less than 1,000 feet and/or visibility less than three miles.

5. Extensive mountain obscurement.

i. AIRMETs are issued on a scheduled basis every six hours, with unscheduled amendments issued as required. AIRMETs have fixed alphanumeric designator with ZULU for icing and freezing level data, TANGO for turbulence, strong surface winds, and windshear, and SIERRA for instrument flight rules and mountain obscuration.

7-1-7. CATEGORICAL OUTLOOKS

a. Categorical outlook terms, describing general ceiling and visibility conditions for advanced planning purposes are used only in area forecasts and are defined as follows:

1. *LIFR (Low IFR):* Ceiling less than 500 feet and/or visibility less than 1 mile.

2. *IFR:* Ceiling 500 to less than 1,000 feet and/or visibility 1 to less than 3 miles.

3. *MVFR (Marginal VFR):* Ceiling 1,000 to 3,000 feet and/or visibility 3 to 5 miles inclusive.

4. *VFR:* Ceiling greater than 3,000 feet and visibility greater than 5 miles; includes sky clear.

b. The cause of LIFR, IFR, or MVFR is indicated by either ceiling, or visibility restrictions, or both. The contraction "CIG" and/or weather and obstruction to vision symbols are used. If winds or gusts of 25 knots or greater are forecast for the outlook period, the word "WIND" is also included for all categories including VFR.

EXAMPLE—

1 LIFR CIG—LOW IFR DUE TO LOW CEILING.

2 IFR FG—IFR DUE TO VISIBILITY RESTRICTED BY FOG.

3 MVFR CIG HZ FU—MARGINAL VFR DUE TO BOTH CEILING AND VISIBILITY RESTRICTED BY HAZE AND SMOKE.

4 IFR CIG RA WIND—IFR DUE TO BOTH LOW CEILING AND VISIBILITY RESTRICTED BY RAIN; WIND EXPECTED TO BE 25 KNOTS OR GREATER.

7-1-8. TELEPHONE INFORMATION BRIEFING SERVICE (TIBS)

a. TIBS, provided by automated flight service stations (AFSSs) is a continuous recording of meteorological and aeronautical information, available by telephone. Each AFSS provides at least four route and/or area briefings. In addition, airspace procedures and special announcements (if applicable) concerning aviation interests may also be available. Depending on user demand, other items may be provided; i.e., METAR observations, terminal aerodrome forecasts, wind/temperatures aloft forecasts, etc.

b. TIBS is not intended to substitute for specialist-provided preflight briefings. It is, however, recommended for use as a preliminary briefing, and often will be valuable in helping you to make a "go or no go" decision.

c. TIBS is provided by Automated Flight Service Stations (AFSS) and provides continuous telephone recordings of meteorological and/or aeronautical information. Specifically, TIBS provides area and/or route briefings, airspace procedures, and special announcements (if applicable) concerning aviation interests.

d. Depending on user demand, other items may be provided; i.e., surface observations, terminal forecasts, winds/temperatures aloft forecasts, etc. A TOUCH-TONE™ telephone is necessary to fully utilize the TIBS program.

e. Pilots are encouraged to avail themselves of this service. TIBS locations are found at AFSS sites and can be accessed by use of 1-800-WX BRIEF toll free number.

7-1-9. TRANSCRIBED WEATHER BROADCAST (TWEB) (Alaska only)

Equipment is provided in Alaska by which meteorological and aeronautical data are recorded on tapes and broadcast continuously over selected L/MF and VOR facilities. Broadcasts are made from a series of individual tape recordings, and changes, as they occur, are transcribed onto the tapes. The information provided varies depending on the type equipment available. Generally, the broadcast contains a summary of adverse conditions, surface weather observations, pilot weather reports, and a density altitude statement (if applicable). At the discretion of the broadcast facility, recordings may also include a synopsis, winds aloft forecast, en route and terminal forecast data, and radar reports. At selected locations, telephone access to the TWEB has been provided (TEL-TWEB). Telephone numbers for this service are found in the Supplement Alaska A/FD. These broadcasts are made available primarily for preflight and inflight planning, and as such, should not be considered as a substitute for specialist-provided preflight briefings.

7-1-10. INFLIGHT WEATHER BROADCASTS

a. Weather Advisory Broadcasts: ARTCCs broadcast a Severe Weather Forecast Alert (AWW), Convective SIGMET, SIGMET, or CWA alert once on all frequencies, except emergency, when any part of the area described is within 150 miles of the airspace under their jurisdiction. These broadcasts contain SIGMET or CWA (identification) and a brief description of the weather activity and general area affected.

EXAMPLE—

1 ATTENTION ALL AIRCRAFT, SIGMET DELTA THREE, FROM MYTON TO TUBA CITY TO MILFORD, SEVERE TURBULENCE AND SEVERE CLEAR ICING BELOW ONE ZERO THOUSAND FEET. EXPECTED TO CONTINUE BEYOND ZERO THREE ZERO ZERO ZULU.

2 ATTENTION ALL AIRCRAFT, CONVECTIVE SIGMET TWO SEVEN EASTERN. FROM THE VICINITY OF ELMIRA TO PHILLIPSBURG. SCATTERED EMBEDDED THUNDERSTORMS MOVING EAST AT ONE ZERO KNOTS. A FEW INTENSE LEVEL FIVE CELLS, MAXIMUM TOPS FOUR FIVE ZERO.

3 ATTENTION ALL AIRCRAFT, KANSAS CITY CENTER WEATHER ADVISORY ONE ZERO THREE. NUMEROUS REPORTS OF MODERATE TO SEVERE ICING FROM EIGHT TO NINER THOUSAND FEET IN A THREE ZERO MILE RADIUS OF ST. LOUIS. LIGHT OR NEGATIVE ICING REPORTED FROM FOUR THOUSAND TO ONE TWO THOUSAND FEET REMAINDER OF KANSAS CITY CENTER AREA.

NOTE—TERMINAL CONTROL FACILITIES HAVE THE OPTION TO LIMIT THE AWW, CONVECTIVE SIGMET, SIGMET, OR CWA BROADCAST AS FOLLOWS: LOCAL

CONTROL AND APPROACH CONTROL POSITIONS MAY OPT TO BROADCAST SIGMET OR CWA ALERTS ONLY WHEN ANY PART OF THE AREA DESCRIBED IS WITHIN 50 MILES OF THE AIRSPACE UNDER THEIR JURISDICTION.

b. Hazardous InFlight Weather Advisory Service (HIWAS): This is a continuous broadcast of inflight weather advisories including summarized AWW, SIG-METs, Convective SIGMETs, CWAs, AIRMETs, and urgent PIREPS. HIWAS has been adopted as a national program and will be implemented throughout the conterminous U.S. as resources permit. In those areas where HIWAS is commissioned, ARTCC, Terminal ATC, and AFSS/FSS facilities have discontinued the broadcast of inflight advisories as described in the preceding paragraph. HIWAS is an additional source of hazardous weather information which makes these data available on a continuous basis. It is not, however, a placement for preflight or inflight briefings or real-time weather updates from Flight Watch (EFAS). As HIWAS is implemented in individual center areas, the commissioning will be advertised in the Notices to Airmen Publication.

1. Where HIWAS has been implemented, a HIWAS alert will be broadcast on all except emergency frequencies once upon receipt by ARTCC and terminal facilities, which will include an alert announcement, frequency instruction, number, and type of advisory updated; e.g., AWW, SIGMET, Convective SIGMET, or CWA.

EXAMPLE—
ATTENTION ALL AIRCRAFT. HAZARDOUS WEATHER INFORMATION (SIGMET, CONVECTIVE SIGMET, AIRMET, URGENT PILOT WEATHER REPORT (UUA), OR CENTER WEATHER ADVISORY (CWA) NUMBER OR NUMBERS) FOR (GEOGRAPHICAL AREA) AVAILABLE ON HIWAS, FLIGHT WATCH, OR FLIGHT SERVICE FREQUENCIES.

2. In HIWAS ARTCC areas, FSSs will broadcast a HIWAS update announcement once on all except emergency frequencies upon completion of recording an update to the HIWAS broadcast. Included in the broadcast will be the type of advisory updated; e.g. AWW, SIGMET, Convective SIGMET, CWA, etc.

EXAMPLE—
ATTENTION ALL AIRCRAFT. HAZARDOUS WEATHER INFORMATION FOR (GEOGRAPHICAL AREA) AVAILABLE FROM FLIGHT WATCH OR FLIGHT SERVICE.

3. HIWAS availability is shown on IFR Enroute Low Altitude Charts and VFR Sectional Charts. The symbol depiction is identified in the chart legend.

7-1-11. FLIGHT INFORMATION SERVICES DATA LINK (FISDL)

a. FIS. Aviation weather and other operational information may be displayed in the cockpit through the use of FIS. FIS systems are of two basic types: Broadcast only systems (called FIS-B) and two-way request/reply systems. Broadcast system components include a ground-or

space-based transmitter, an aircraft receiver, and a portable or installed cockpit display device. Two-way systems utilize transmitter/receivers at both the ground- or space-based site and the aircraft.

1. Broadcast FIS (i.e., FIS-B) allows the pilot to passively collect weather and other operational data and to display that data at the appropriate time. In addition to textual weather products such as Aviation Routine Weather Reports (METARs)/Aviation Selected Special Weather Reports (SPECIs) and Terminal Area Forecasts (TAFs), graphical weather products such as radar composite/mosaic images, temporary flight restricted airspace and other NOTAMs may be provided to the cockpit. Two-way FIS services permit the pilot to make specific weather and other operational information requests for cockpit display. A FIS service provider will then prepare a reply in response to that specific request and transmit the product to that specific aircraft.

2. FIS services are available from four types of service providers:

(a) A private sector FIS provider operating under service agreement with the FAA using broadcast data link over VHF aeronautical spectrum and whose products have been reviewed and accepted by the FAA prior to transmission. (Products and services are defined under subparagraph c.)

(b) Through an FAA operated service using a broadcast data link on the ADS-B UAT network. (Products and services are defined under subparagraph d.)

(c) Private sector FIS providers operating under customer contracts using aeronautical spectrum.

(d) Private sector FIS providers operating under customer contract using methods other than aeronautical spectrum, including Internet data-to-the-cockpit service providers.

3. FIS is a method of receiving aviation weather and other operational data in the cockpit that augments traditional pilot voice communication with FAA's Flight Service Stations (FSSs), ATC facilities or Airline Operations Control Centers (AOCCs). FIS: is not intended to replace traditional pilot and controller/flight service specialist/aircraft dispatcher pre-flight briefings or inflight voice communications. FIS, however, can provide textual and graphical background information that can help abbreviate and improve the usefulness of such communications. FIS enhances pilot situational awareness and improves safety.

4. To ensure airman compliance with Federal Aviation Regulations, manufacturer's operating manuals should remind airmen to contact ATC controllers, FSS specialists, operator dispatchers, or airline operations control centers for general and mission critical aviation weather information and/or NAS status conditions (such as NOTAMs, Special Use Airspace status, and other government flight information). If FIS products are systemically modified (for example, are displayed as abbreviated plain text and/or graphical depictions), the modification process and limitations of the resultant product should be clearly described in the vendor's user guidance.

b. Operational Use of FIS. Regardless of the type of FIS system being used, several factors must be considered when using FIS:

1. Before using FIS for inflight operations, pilots and other flight crewmembers should become familiar with the operation of the FIS system to be used, the airborne equipment to be used, including its system architecture, airborne system components, coverage service volume and other limitations of the particular system, modes of operation and indications of various system failures. Users should also be familiar with the specific content and format of the services available from the FIS provider(s). Sources of information that may provide this specific guidance include manufacturer's manuals, training programs and reference guides.

2. FIS should not serve as the sole source of aviation weather and other operational information. ATC, AFSSs and, if applicable, AOCC VHF/HF voice remain as a redundant method of communicating aviation weather, NOTAMs, and other operational information to aircraft in flight. FIS augments these traditional ATC/FSS/AOCC services and, for some products, offers the advantage of being displayed as graphical information. By using FIS for orientation, the usefulness of information received from conventional means may be enhanced. For example, FIS may alert the pilot to specific areas of concern that will more accurately focus requests made to FSS or AOCC for inflight updates or similar queries made to ATC.

3. The airspace and aeronautical environment is constantly changing. These changes occur quickly and without warning. Critical operational decisions should be based on use of the most current and appropriate data available. When differences exist between FIS and information obtained by voice communication with ATC, FSS, and/or AOCC (if applicable), pilots are <u>cautioned to use the most recent data from the most authoritative source.</u>

4. FIS aviation weather products (e.g., graphical ground-based radar precipitation depictions) are not appropriate for tactical avoidance of severe weather such as negotiating a path through a weather hazard area. FIS supports strategic weather decision making such as route selection to avoid a weather hazard area in its entirety. The misuse of information beyond its applicability may place the pilot and aircraft in jeopardy. In addition, FIS should never be used in lieu of an individual pre-flight weather and flight planning briefing.

5. FIS NOTAM products, including Temporary Flight Restriction (TFR) information, are advisory-use information and are intended for situational awareness purposes only. Cockpit displays of this information are not appropriate for tactical navigation - pilots should stay clear of any geographic area displayed as a TFR NOTAM. Pilots should contact FSSs and/or ATC while en route to obtain updated information and to verify the cockpit display of NOTAM information.

6. FIS supports better pilot decision making by increasing situational awareness. Better decision making is based on using information from a variety of sources. In addition to FIS, pilots should take advantage of other weather/NAS status sources, including, briefings from Flight Service Stations, FAA's en route "Flight Watch" service, data from other air traffic control facilities, airline operation control centers, pilot reports, as well as their own observations.

c. EAA FISDL (VHF) Service. The FAA's FISDL (VHF datalink) system is a VHF Data Link (VDL) Mode 2 implementation that provides pilots and flight crews of properly equipped aircraft with a cockpit display of certain aviation weather and flight operational information. This information may be displayed in both textual and graphical formats. The system is operated under a service agreement with the FAA, using broadcast data link on VHF aeronautical spectrum on two 25 KHz spaced frequencies (136.450 and 136.475 MHz). The FAA FISDL (VHF) service is designed to provide coverage throughout the continental U.S. from 5,000 feet AGL to 17,500 feet MSL, except in areas where this is not feasible due to mountainous terrain. Aircraft operating near transmitter sites may receive useable FISDL signals at altitudes lower than 5,000 feet AGL, including on the surface in some locations, depending on transmitter/aircraft line of sight geometry. Aircraft operating above 17,500 feet MSL may also receive useable FISDL signals under certain circumstances.

1. FAA FISDL (VHF) service provides, free of charge, the following basic text products:

(a) Aviation Routine Weather Reports (METARs).

(b) Aviation Selected Special Weather Reports (SPECIs).

(c) Terminal Area Forecasts (TAFs), and their amendments.

(d) Significant Meteorological Information (SIGMETs).

(e) Convective SIGMETs.

(f) Airman's Meteorological Information (AIRMETs)

(g) Pilot Reports (both urgent and routine) (PIREPs); and,

(h) Severe Weather Forecast Alerts and Warnings (AWWs/WW) issued by the NOAA Storm Prediction Center (SPC).

2. The format and coding of these text products are described in Advisory Circular AC-00-45, Aviation Weather Services, and paragraph 7-1-31, Key to Aerodrome Forecast (TAF) and Aviation Routine Weather Report (METAR).

3. Additional products, called "Value-Added Products," are also available from the vendor on a paid subscription basis. Details concerning the content, format, symbology and cost of these products may be obtained from the vendor.

d. FAA's Flight Information Service-Broadcast (FIS-B) Service. FIS-B is a ground broadcast service provided through the FAA's Universal Access Transceiver (UAT) "ADS-B Broadcast Services" network. The UAT network is an ADS-B data link that operates on 978 MHz. The FAA FIS-B system provides pilots and flight crews of properly equipped aircraft with a cockpit display of certain aviation weather and flight operational information. The FAA's FIS-B service is being introduced in certain regional implementations within the NAS (e.g., in Alaska and in other areas of implementation).

1. FAA's UAT FIS-B provides the initial products listed below with additional products planned for future implementation. FIS-B reception is line of sight and can be expected within 200 NM (nominal range) of each ground transmitting site. The following services are provided free of charge.

(a) Text: Aviation Routine Weather Reports (METARs).

(b) Text: Special Aviation Reports (SPECIs).

(c) Text: Terminal Area Forecasts (TAFs), and their amendments.

(d) Graphic: NEXRAD precipitation maps.

2. The format and coding of the above text weather-related products are described in Advisory Circular AC-00-45, Aviation Weather Services, and paragraph 7-1-31, Key to Aerodrome Forecast (TAF) and Aviation Routine Weather Report (METAR).

3. Details concerning the content, format, and symbology of the various data link products provided may be obtained from the specific avionics manufacturer.

e. Non-FAA FISDL Systems. Several commercial vendors also provide customers with FIS data over both the aeronautical spectrum and on other frequencies using a variety of data link protocols. In some cases, the vendors provide only the communications system that carries customer messages, such as the Aircraft Communications Addressing and Reporting System (ACARS) used by many air carrier and other operators.

1. Operators using non-FAA FIS data for inflight weather and other operational information should ensure that the products used conform to FAA/NWS standards. Specifically, aviation weather and NAS status information should meet the following criteria:

(a) The products should be either FAA/NWS "accepted" aviation weather reports or products, or based on FAA/NWS accepted aviation weather reports or products. If products are used which do not meet this criteria, they should be so identified. The operator must determine the applicability of such products to their particular flight operations.

(b) In the case of a weather product which is the result of the application of a process which alters the form, function or content of the base FAA/NWS accepted weather product(s), that process, and any limitations to the application of the resultant product, should be described in the vendor's user guidance material.

2. An example would be a NEXRAD radar composite/mosaic map, which has been modified by changing the scaling resolution. The methodology of assigning reflectivity values to the resultant image components should be described in the vendor's guidance material to ensure that the user can accurately interpret the displayed data.

7-1-12. WEATHER OBSERVING PROGRAMS

a. Manual Observations: With only a few exceptions, these reports are from airport locations staffed by FAA or NWS personnel who manually observe, perform calculations, and enter these observations into the (WMSCR) communication system. The format and coding of these observations are contained in *Key to Aviation Weather Observations and Forecasts*, AIM, paragraph 7-1-28.

b. Automated Weather Observing System (AWOS):

1. Automated weather reporting systems are increasingly being installed at airports. (See Key to AWOS: NO TAG). These systems consist of various sensors, a processor, a computer-generated voice subsystem, and a transmitter to broadcast local, minute- by-minute weather data directly to the pilot.

NOTE—WHEN THE BAROMETRIC PRESSURE EXCEEDS 31.00 INCHES HG., SEE AIM, PROCEDURES, PARAGRAPH 7-2-2, FOR THE ALTIMETER SETTING PROCEDURES.

2. The AWOS observations will include the prefix "AUTO" to indicate that the data are derived from an automated system. Some AWOS locations will be augmented by certified observers who will provide weather and obstruction to vision information in the remarks of the report when the reported visibility is less than 7 miles. These sites, along with the hours of augmentation, are to be published in the A/FD. Augmentation is identified in the observation as "OBSERVER WEATHER." The AWOS wind speed, direction and gusts, temperature, dew point, and altimeter setting are exactly the same as for manual observations. The AWOS will also report density altitude when it exceeds the field elevation by more than 1,000 feet. The reported visibility is derived from a sensor near the touchdown of the primary instrument runway. The visibility sensor output is converted to a visibility value using a 10-minute harmonic average. The reported sky condition/ceiling is derived from the ceilometer located next to the visibility sensor. The AWOS algorithm integrates the last 30 minutes of ceilometer data to derive cloud layers and heights. This output may also differ from the observer sky condition in that the AWOS is totally dependent upon the cloud advection over the sensor site.

3. These real-time systems are operationally classified into four basic levels:

(a) *AWOS-A*: only reports altimeter setting,

NOTE—ANY OTHER INFORMATION IS ADVISORY ONLY.

(b) *AWOS-1*: usually reports altimeter setting, wind data, temperature, dew point, and density altitude,

(c) *AWOS-2*: provides the information provided by AWOS-1 plus visibility, and

(d) *AWOS-3*: provides the information provided by AWOS-2 plus cloud/ceiling data.

4. The information is transmitted over a discrete VHF radio frequency or the voice portion of a local NAVAID. AWOS transmissions on a discrete VHF radio frequency are engineered to be receivable to a maximum of 25 NM from the AWOS site and a maximum altitude of 10,000 feet AGL. At many locations, AWOS signals may be received on the surface of the airport, but local conditions may limit the maximum AWOS reception distance and/or altitude. The system transmits a 20 to 30 second weather message updated each minute. Pilots should monitor the designated frequency for the automated weather broadcast. A description of the broadcast is contained in subparagraph c. There is no two-way communication capability. Most AWOS sites also have a dial-up capability so that the minute-by-minute weather messages can be accessed via telephone.

5. AWOS information (system level, frequency, phone number, etc.) concerning specific locations is published, as the systems become operational, in the A/FD, and where applicable, on published Instrument Approach Procedures. Selected individual systems may be incorporated into nationwide data collection and dissemination networks in the future.

c. AWOS Broadcasts: Computer-generated voice is used in AWOS to automate the broadcast of the minute-by-minute weather observations. In addition, some systems are configured to permit the addition of an operator-generated voice message; e.g., weather remarks following the automated parameters. The phraseology used generally follows that used for other weather broadcasts. Following are explanations and examples of the exceptions.

1. *Location and Time:* The location/name and the phrase "AUTOMATED WEATHER OBSERVATION," followed by the time are announced.

(a) If the airport's specific location is included in the airport's name, the airport's name is announced.

EXAMPLE—"BREMERTON NATIONAL AIRPORT AUTOMATED WEATHER OBSERVATION, ONE FOUR FIVE SIX ZULU;"

"RAVENSWOOD JACKSON COUNTY AIRPORT AUTOMATED WEATHER OBSERVATION, ONE FOUR FIVE SIX ZULU."

(b) If the airport's specific location is not included in the airport's name, the location is announced followed by the airport's name.

EXAMPLE—"SAULT STE MARIE, CHIPPEWA COUNTY INTERNATIONAL AIRPORT AUTOMATED WEATHER OBSERVATION;"

"SANDUSKY, COWLEY FIELD AUTOMATED WEATHER OBSERVATION."

(c) The word "TEST" is added following "OBSERVATION" when the system is not in commissioned status.

EXAMPLE—"BREMERTON NATIONAL AIRPORT AUTOMATED WEATHER OBSERVATION TEST, ONE FOUR FIVE SIX ZULU."

(d) The phrase "TEMPORARILY INOPERATIVE" is added when the system is inoperative.

EXAMPLE—"BREMERTON NATIONAL AIRPORT AUTOMATED WEATHER OBSERVING SYSTEM TEMPORARILY INOPERATIVE."

2. *Visibility:*

(a) The lowest reportable visibility value in AWOS is "less than ¼." It is announced as "VISIBILITY LESS THAN ONE QUARTER."

(b) A sensor for determining visibility is not included in some AWOS. In these systems, visibility is not announced. "VISIBILITY MISSING" is announced only if the system is configured with a visibility sensor and visibility information is not available.

3. *Weather:* In the future, some AWOSs are to be configured to determine the occurrence of precipitation. However, the type and intensity may not always be determined. In these systems, the word "PRECIPITATION" will be announced if precipitation is occurring, but the type and intensity are not determined.

4. *Ceiling and Sky Cover:*

(a) Ceiling is announced as either "CEILING" or "INDEFINITE CEILING." With the exception of indefinite ceilings, all automated ceiling heights are measured.

EXAMPLE—"BREMERTON NATIONAL AIRPORT AUTOMATED WEATHER OBSERVATION, ONE FOUR FIVE SIX ZULU. CEILING TWO THOUSAND OVERCAST;"

"BREMERTON NATIONAL AIRPORT AUTOMATED WEATHER OBSERVATION, ONE FOUR FIVE SIX ZULU. INDEFINITE CEILING TWO HUNDRED, SKY OBSCURED."

(b) The word "Clear" is not used in AWOS due to limitations in the height ranges of the sensors. No clouds detected is announced as "NO CLOUDS BELOW XXX" or, in newer systems as "CLEAR BELOW XXX" (where XXX is the range limit of the sensor).

EXAMPLE—"NO CLOUDS BELOW ONE TWO THOUSAND."

"CLEAR BELOW ONE TWO THOUSAND."

(c) A sensor for determining ceiling and sky cover is not included in some AWOS. In these systems, ceiling and sky cover are not announced. "SKY CONDITION MISSING" is announced only if the system is configured with a ceilometer and the ceiling and sky cover information is not available.

5. *Remarks:* If remarks are included in the observation, the word "REMARKS" is announced following the altimeter setting.

(a) Automated "Remarks."

(1) Density Altitude.

(2) Variable Visibility.

(3) Variable Wind Direction.

(b) Manual Input Remarks: Manual input remarks are prefaced with the phrase "OBSERVER WEATHER." As a general rule the manual remarks are limited to:

(1) Type and intensity of precipitation,

(2) Thunderstorms and direction, and

(3) Obstructions to vision when the visibility is 3 miles or less.

EXAMPLE—"REMARKS ... DENSITY ALTITUDE, TWO THOUSAND FIVE HUNDRED ... VISIBILITY VARIABLE BETWEEN ONE AND TWO ... WIND DIRECTION VARIABLE BETWEEN TWO FOUR ZERO AND THREE ONE ZERO ... OBSERVED WEATHER ... THUNDERSTORM MODERATE RAIN SHOWERS AND FOG... THUNDERSTORM OVERHEAD."

(c) If an automated parameter is "missing" and no manual input for that parameter is available, the parameter is

KEY TO DECODE AN ASOS (METAR) OBSERVATION

METAR KABC 121755Z AUTO 21016G24KT 180V240 1SM R11/P6000FT -RA BR BKN015 OVC025 06/04 A2990
RMK AO2 PK WND 20032/25 WSHFT 1715 VIS 3/4V1 1/2 VIS 3/4 RWY11 RAB07 CIG 013V017 CIG 017 RWY11 PRESFR
SLP125 P0003 60009 T00640036 10066 21012 58033 TSNO $

TYPE OF REPORT	METAR: hourly (scheduled) report; SPECI: special (unscheduled) report.	METAR
STATION IDENTIFIER	Four alphabetic characters; ICAO location identifier.	KABC
DATE/TIME	All dates and times in UTC using a 24-hour clock; two-digit date and four-digit time; always appended with Z to indicate UTC.	121755Z
REPORT MODIFIER	Fully automated report, no human intervention; removed when observer signed-on.	AUTO
WIND DIRECTION AND SPEED	Direction in tens of degrees from true north (first three digits); next two digits: speed in whole knots; as needed Gusts (character) followed by maximum observed speed; always appended with KT to indicate knots; 00000KT for calm; if direction varies by 60° or more a Variable wind direction group is reported.	21016G24KT 180V240
VISIBILITY	Prevailing visibility in statute miles and fractions (space between whole miles and fractions); always appended with SM to indicate statute miles.	1SM
RUNWAY VISUAL RANGE	10-minute RVR value in hundreds of feet; reported if prevailing visibility is ≤ one mile or RVR ≤ 6000 feet; always appended with FT to indicate feet; value prefixed with M or P to indicate value is lower or higher than the reportable RVR value .	R11/P6000FT
WEATHER PHENOMENA	RA: liquid precipitation that does not freeze; SN: frozen precipitation other than hail; UP: precipitation of unknown type; intensity prefixed to precipitation: light (-), moderate (no sign), heavy (+); FG: fog; FZFG: freezing fog (temperature below 0°C); BR: mist; HZ: haze; SQ: squall; maximum of three groups reported; augmented by observer: FC (funnel cloud/tornado/waterspout); TS (thunderstorm); GR (hail); GS (small hail; <1/4 inch); FZRA (intensity; freezing rain); VA (volcanic ash).	-RA BR
SKY CONDITION	Cloud amount and height: CLR (no clouds detected below 12000 feet); FEW (few); SCT (scattered); BKN (broken); OVC (overcast); followed by 3-digit height in hundreds of feet; or vertical visibility (VV) followed by height for indefinite ceiling.	BKN015 OVC025
TEMPERATURE/DEW POINT	Each is reported in whole degrees Celsius using two digits; values are separated by a solidus; sub-zero values are prefixed with an M (minus).	06/04
ALTIMETER	Altimeter always prefixed with an A indicating inches of mercury; reported using four digits: tens, units, tenths, and hundredths.	A2990

Figure 7-1-8.

	Example
REMARKS IDENTIFIER: RMK	RMK
TORNADIC ACTIVITY: Augmented; report should include TORNADO, FUNNEL, CLOUD, or WATERSPOUT, time begin/end, location, movement, e.g., TORNADO B25 N MOV E.	
TYPE OF AUTOMATED STATION: AO2; automated station with precipitation discriminator.	AO2
PEAK WIND: PK WND dddff(f)/(hh)mm; direction in tens of degrees, speed in whole knots, and time.	PK WND 20032/25
WIND SHIFT: WSHFT (hh)mm	WSHFT 1715
TOWER OR SURFACE VISIBILITY: TWR VIS vvvvv: visibility reported by tower personnel, e.g., TWR VIS 2; SFC VIS vvvvv: visibility reported by ASOS, e.g., SFC VIS 2.	
VARIABLE PREVAILING VISIBILITY: VIS $v_nv_nv_nv_nv_nVv_xv_xv_xv_xv_x$; reported if prevailing visibility is < 3 miles and variable.	VIS 3/4V1 1/2
VISIBILITY AT SECOND LOCATION: VIS vvvvv [LOC]; reported if different than the reported prevailing visibility in body of report.	VIS 3/4 RWY11
LIGHTNING: [FREQ] LTG [LOC]; when detected the frequency and location is reported, e.g., FRQ LTG NE.	
BEGINNING AND ENDING OF PRECIPITATION AND THUNDERSTORMS: w'w'B(hh)mmE(hh)mm; TSB(hh)mmE(hh)mm	RAB07
VIRGA: Augmented; precipitation not reaching the ground, e.g., VIRGA.	
VARIABLE CEILING HEIGHT: CIG $h_nh_nh_nVh_xh_xh_x$; reported if ceiling in body of report is < 3000 feet and variable.	CIG 013V017
CEILING HEIGHT AT SECOND LOCATION: CIG hhh [LOC]; Ceiling height reported if secondary ceilometer site is different than the ceiling height in the body of the report.	CIG 017 RWY11
PRESSURE RISING OR FALLING RAPIDLY: PRESRR or PRESFR; pressure rising or falling rapidly at time of observation.	PRESFR
SEA-LEVEL PRESSURE: SLPppp; tens, units, and tenths of SLP in hPa.	SLP125
HOURLY PRECIPITATION AMOUNT: Prrrr; in .01 inches since last METAR; a trace is P0000.	P0003
3- AND 6-HOUR PRECIPITATION AMOUNT: 6RRRR; precipitation amount in .01 inches for past 6 hours reported in 00, 06, 12, and 18 UTC observations and for past 3 hours in 03, 09, 15, and 21 UTC observations; a trace is 60000.	60009
24-HOUR PRECIPITATION AMOUNT: $7R_{24}R_{24}R_{24}R_{24}$; precipitation amount in .01 inches for past 24 hours reported in 12 UTC observation, e.g., 70015.	
HOURLY TEMPERATURE AND DEW POINT: $T s_nT_aT_aT_as_nT_a'T_a'T_a'$; tenth of degree Celsius; s_n: 1 if temperature below 0°C and 0 if temperature 0°C or higher.	T00640036
6-HOUR MAXIMUM TEMPERATURE: $1s_nT_xT_xT_x$; tenth of degree Celsius; 00, 06, 12, 18 UTC; s_n: 1 if temperature below 0°C and 0 if temperature 0°C or higher.	10066
6-HOUR MINIMUM TEMPERATURE: $2s_nT_nT_nT_n$; tenth of degree of Celsius; 00, 06, 12, 18 UTC; s_n: 1 if temperature below 0°C and 0 if temperature 0°C or higher.	21012
24-HOUR MAXIMUM AND MINIMUM TEMPERATURE: $4s_nT_xT_xT_xs_nT_nT_nT_n$; tenth of degree Celsius; reported at midnight local standard time; 1 if temperature below 0°C and 0 if temperature 0°C or higher, e.g., 400461006.	58033
PRESSURE TENDENCY: 5appp; the character (a) and change in pressure (ppp: tenths of hPa) the past 3 hours.	
SENSOR STATUS INDICATORS: RVRNO: RVR missing; PWINO: precipitation identifier information not available; PNO: precipitation amount not available; FZRANO: freezing rain information not available; TSNO: thunderstorm information not available; VISNO [LOC]: visibility a secondary location not available; e.g., VISNO RWY06; CHINO [LOC]: (cloud-height-indicator) sky condition at secondary location not available, e.g., CHINO RWY06.	TSNO
MAINTENANCE CHECK INDICATOR: Maintenance needed on the system.	$

If an element or phenomena does not occur, is missing, or cannot be observed, the corresponding group and space are omitted (body and/or remarks) from that particular report, except for Sea-Level Pressure (SLPppp). SLPNO shall be reported in a METAR when the SLP is not available.

U.S. DEPARTMENT OF TRANSPORTATION • FEDERAL AVIATION ADMINISTRATION • AVIATION WEATHER DIRECTORATE, 400 7th street, SW, Rooms 8200–8326, WASHINGTON, D.C. 20591

Figure 7-1-9.

announced as "MISSING." For example, a report with the dew point "missing" and no manual input available, would be announced as follows:

> **EXAMPLE**—"CEILING ONE THOUSAND OVERCAST ... VISIBILITY THREE ... PRECIPITATION ... TEMPERATURE THREE ZERO, DEW POINT MISSING ... WIND CALM ... ALTIMETER THREE ZERO ZERO ONE."

(d) "REMARKS" are announced in the following order of priority:

(1) Automated "REMARKS."
[a] Density Altitude
[b] Variable Visibility
[c] Variable Wind Direction
(2) Manual Input "REMARKS."
[a] Sky Condition,
[b] Visibility,
[c] Weather and Obstructions to Vision,
[d] Temperature,
[e] Dew Point,
[f] Wind, and
[g] Altimeter Setting.

> **EXAMPLE**—"REMARKS ... DENSITY ALTITUDE, TWO THOUSAND FIVE HUNDRED ... VISIBILITY VARIABLE BETWEEN ONE AND TWO ... WIND DIRECTION VARIABLE BETWEEN TWO FOUR ZERO AND THREE ONE ZERO ... OBSERVER CEILING ESTIMATED TWO THOUSAND BROKEN ... OBSERVER TEMPERATURE TWO, DEW POINT MINUS FIVE."

d. Automated Surface Observation (ASOS)/Automated Weather Sensor System (AWSS). The ASOS/AWSS is the primary surface weather observing system on the U.S. (See Key to Decode an ASOS/AWSS (METAR) Observation, FIG 7-1-8 and FIG 7-1-9.) The program to install and operate these systems throughout the U.S. is a joint effort of the NWS, the FAA and the Department of Defense. AWSS is a follow-on program that provides identical data as ASOS. ASOS/AWSS is designed to support aviation operations and weather forecast activities. The ASOS/AWSS will provide continuous minute-by-minute observations and perform the basic observing functions necessary to generate an aviation routine weather report (METAR) and other aviation weather information. The information may be transmitted over a discrete VHF radio frequency or the voice portion of a local NAVAID. ASOS/AWSS transmissions on a discrete VHF radio frequency are engineered to be receivable to a maximum of 25 NM from the ASOS/AWSS site and a maximum altitude of 10,000 feet AGL. At many locations, ASOS/AWSS signals may be received on the surface of the airport, but local conditions may limit the maximum reception distance and/or altitude. While the automated system and the human may differ in their methods of data collection and interpretation, both produce an observation quite similar in form and content. For the "objective" elements such as pressure, ambient temperature, dew point temperature, wind, and precipitation accumulation, both the automated system and the

observer use a fixed location and time-averaging technique. The quantitative differences between the observer and the automated observation of these elements are negligible. For the "subjective" elements, however, observers use a fixed time, spatial averaging technique to describe the visual elements (sky condition, visibility and present weather), while the automated systems use a fixed location, time averaging technique. Although this is a fundamental change, the manual and automated techniques yield remarkably similar results within the limits of their respective capabilities.

1. System Description.

(a) The ASOS/AWSS at each airport location consists of four main components:

(1) Individual weather sensors.
(2) Data collection and processing units.
(3) Peripherals and displays.

(b) The ASOS/AWSS sensors perform the basic function of data acquisition. They continuously sample and measure the ambient environment, derive raw sensor data and make them available to the collection and processing units.

2. *Every ASOS/AWSS will contain the following basic set of sensors:*

(a) Cloud height indicator (one or possibly three).
(b) Visibility sensor (one or possibly three).
(c) Precipitation identification sensor.
(d) Freezing rain sensor (at select sites).
(e) Pressure sensors (two sensors at small airports; three sensors at large airports).
(f) Ambient temperature/Dew point temperature sensor.
(g) Anemometer (wind direction and speed sensor).
(h) Rainfall accumulation sensor.

3. *The ASOS data outlets include:*

(a) Those necessary for on-site airport users.
(b) National communications networks.
(c) Computer-generated voice (available through FAA radio broadcast to pilots, and dial-in telephone line).

> **NOTE**—WIND DIRECTION BROADCASTS OVER FAA RADIOS IS IN REFERENCE TO MAGNETIC NORTH.

4. An ASOS/AWOS/AWSS report without human intervention will contain only that weather data capable of being reported automatically. The modifier for this METAR report is "AUTO." When an observer augments or backs-up an ASOS/AWOS/AWSS site, the "AUTO" modifier disappears.

5. There are two types of automated stations, AO1 for automated weather reporting stations without a precipitation discriminator, and AO2 for automated stations with a precipitation discriminator. As appropriate, "AO1" and "AO2" shall appear in remarks. (A precipitation discriminator can determine the difference between liquid and frozen/freezing precipitation).

> **NOTE**—TO DECODE AN ASOS REPORT REFER TO FIG 7-1-8 AND FIG 7-1-9.

> **REFERENCE**—A COMPLETE EXPLANATION OF METAR TERMINOLOGY IS LOCATED IN AIM, PARAGRAPH 7-1-27.

Table 7-1-1 Weather Observing Programs

ELEMENT REPORTED	AWOS-A	AWOS-1	AWOS-2	AWOS-3	ASOS	MANUAL
Altimeter	X	X	X	X	X	X
Wind		X	X	X	X	X
Temperature/Dew Point		X	X	X	X	X
Density Altitude		X	X	X	X	
Visibility			X	X	X	X
Clouds/Ceiling				X	X	X
Precipitation					X	X
Remarks					X	X

e. TBL 7-1-1 contains a comparison of weather observing programs and the elements reported.

f. Service Standards: During 1995, a government/industry team worked to comprehensively reassess the requirements for surface observations at the nation's airports. That work resulted in agreement on a set of service standards, and the FAA and NWS ASOS Sites to which the standards would apply. The term "Service Standards" refers to the level of detail in weather observation. The service standards consist of four different levels of service (A, B, C and D) as described below. Specific observational elements included in each service level are listed in TBL 7-1-2.

1. Service Level D defines the minimum acceptable level of service. It is a completely automated service in which the ASOS observation will constitute the entire observation, i.e., no additional weather information is added by a human observer. This service is referred to as a stand alone D site.

2. Service Level C is a service in which the human observer, usually an air traffice controller, augments or adds information to the automated observation. Service Level

Table 7-1-2

SERVICE LEVEL A	
Service Level A consists of all the elements of Service Levels B, C and D plus the elements listed to the right, if observed.	10 minute longline RVR at precedented sites or additional visibility increments of 1/8, 1/16 and 0 Sector visibility Variable sky condition Cloud layers above 12,000 feet and cloud types Widespread dust, sand and other obscurations Volcanic eruptions

SERVICE LEVEL B	
Service Level B consists of all the elements of Service Levels C and D plus the elements listed to the right, if observed.	Longline RVR at precedented sites (may be instantaneous readout) Freezing drizzle versus freezing rain Ice pellets Snow depth & snow increasing rapidly remarks Thunderstorm and lightning location remarks Observed significant weather not at the station remarks

SERVICE LEVEL C	
Service Level C consists of all the elements of Service Level D plus augmentation and backup by a human observer or an air traffic control specialist on location nearby. Backup consists of inserting the correct value if the system malfunctions or is unrepresentative. Augmentation consists of adding the elements listed to the right, if observed. During hours that the observing facility is closed, the site reverts to Service Level D.	Thunderstorms Tornadoes Hail Virga Volcanic ash Tower visibility Operationally significant remarks as deemed appropriate by the observer

SERVICE LEVEL D	
This level of service consists of an ASOS continually measuring the atmosphere at a point near the runway. The ASOS senses and measures the weather parameters listed to the right.	Wind Visibility Precipitation/Obstruction to vision Cloud height Sky cover Temperature Dew point Altimeter

C also includes backup of ASOS elements in the event of an ASOS malfunction or an unrepresentative ASOS report. In backup, the human observer inserts the correct or missing value for the automated ASOS elements. This service is provided by air traffic controllers under the Limited Aviation Weather Reporting Station (LAWRS) process, FSS and NWS observers, and, at selected sites, Non-Federal Observation Program observers.

Two categories of airports require detail beyond Service Level C in order to enhance air traffic control efficiency and increase system capacity. Services at these airports are typically provided by contract weather observers, NWS observers, and, at some locations, FSS observers.

3. Service Level B is a service in which weather observations consist of all elements provided under Service Level C, plus augmentation of additional data beyond the capability of the ASOS. This category of airports includes smaller hubs or special airports in other ways that have worse than average bad weather operations for thunderstorms and/or freezing/frozen precipitation, and/or that are remote airports.

4. Service Level A, the highest and most demanding category, includes all the data reported in Service Standard B, plus additional requirements as specified. Service Level A covers major aviation hubs and/or high volume traffic airports with average or worse weather.

7-1-13. WEATHER RADAR SERVICES

a. The National Weather Service operates a network of radar sites for detecting coverage, intensity, and movement of precipitation. The network is supplemented by FAA and DOD radar sites in the western sections of the country. Local warning radar sites augment the network by operating on an as needed basis to support warning and forecast programs.

b. Scheduled radar observations are taken hourly and transmitted in alpha-numeric format on weather telecommunications circuits for flight planning purposes. Under certain conditions, special radar reports are issued in addition to the hourly transmittals. Data contained in the reports are also collected by the National Center for Environmental Prediction and used to prepare national radar summary charts for dissemination on facsimile circuits.

c. A clear radar display (no echoes) does not mean that there is no significant weather within the coverage of the radar site. Clouds and fog are not detected by the radar. However, when echoes are present, turbulence can be implied by the intensity of the precipitation, and icing is implied by the presence of the precipitation at temperatures at or below zero degrees Celsius. Used in conjunction with other weather products, radar provides invaluable information for weather avoidance and flight planning.

d. All En Route Flight Advisory Service facilities and AFSSs have equipment to directly access the radar displays from the individual weather radar sites. Specialists at these locations are trained to interpret the display for pilot briefing and inflight advisory services. The Center Weather

Service Units located in ARTCCs also have access to weather radar displays and provide support to all air traffic facilities within their center's area.

e. Additional information on weather radar products and services can be found in AC 00-45, AVIATION WEATHER SERVICES.

REFERENCE—PILOT/CONTROLLER GLOSSARY, RADAR WEATHER ECHO INTENSITY LEVELS.

AIM, THUNDERSTORMS, PARAGRAPH 7-1-29.

A/FD, CHARTS, NWS UPPER AIR OBSERVING STATIONS AND WEATHER NETWORK FOR THE LOCATION OF SPECIFIC RADAR SITES.

7-1-14. NATIONAL CONVECTIVE WEATHER FORECAST (NCWF)

a. Description.

1. The NCWF is an automatically generated depiction *of*: (1) current convection and (2) extrapolated significant current convection. It is a supplement to, but does NOT substitute for, the report and forecast information contained in Convective SIGMETs (see paragraph 7-1-6d). Convection, particularly significant convection, is typically associated with thunderstorm activity.

2. The National Weather Service Aviation Weather Center (AWC) updates the NCWF based on input from the Next Generation Weather Radar (NEXRAD) and cloud-to-ground lightning data.

3. The NCWF is most accurate for long-lived mature multistorm systems such as organized line storms. NCWF does not forecast initiation, growth or decay of thunderstorms. Therefore, NCWF tends to under-warn on new and growing storms and over-warn on dying storms. Forecast positions of small, isolated or weaker thunderstorms are not displayed.

4. The NCWF area of coverage is limited to the 48 contiguous states.

b. Attributes.

1. The NCWF is updated frequently (every 5 minutes) using the most current available data.

2. The NCWF is able to detect the existence of convective storm locations that agree very well with concurrent radar and lightning observations.

3. The NCWF is a high-resolution forecast impacting a relatively small volume of airspace rather than covering large boxed areas. The location, speeds and directions of movement of multiple convective storms are depicted individually.

4. The NCWF extrapolation forecasts are more accurate when predicting the location and size of well organized, unchanging convective storms moving at uniform speeds. The NCWF does not work well with sporadic, explosive cells developing and dissipating in minutes.

5. In displaying forecast cell locations, the NCWF does NOT distinguish among level 3 through level 6 on the NCWF hazard scale (see TBL 7-1-3).

6. The NCWF may not detect or forecast:

(a) Some embedded convection.

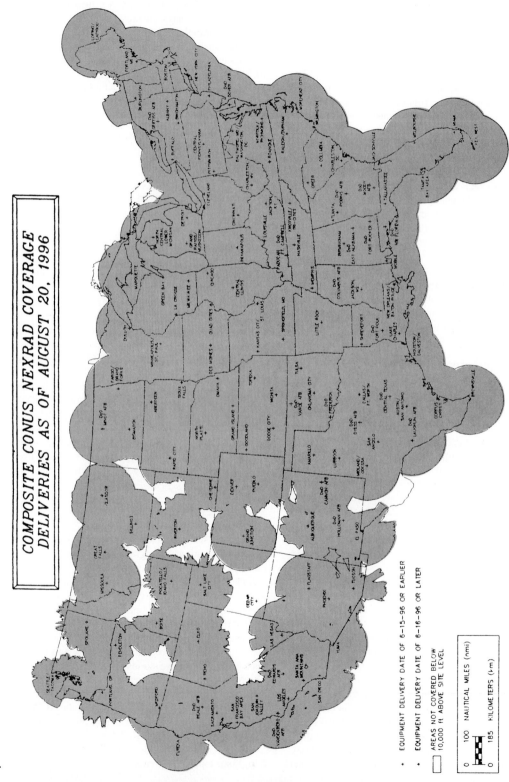

Figure 7-1-10. NOAA National Weather Service Radar Network

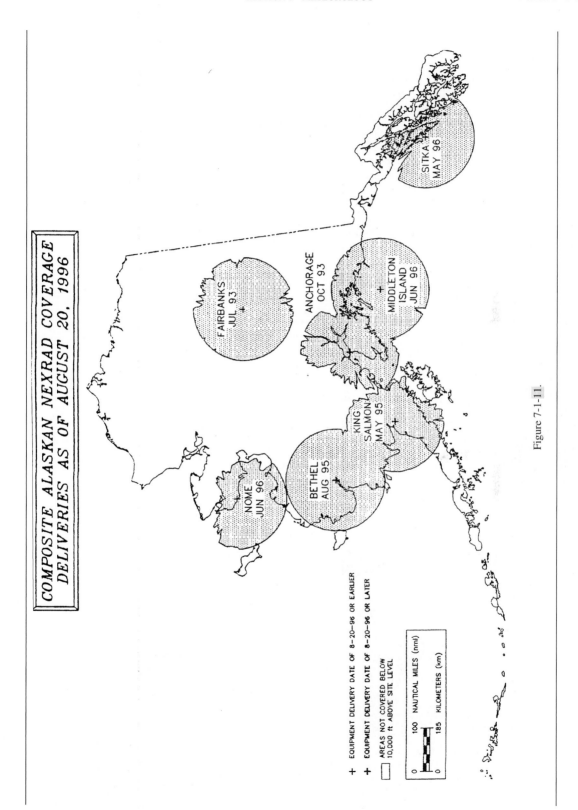

COMPOSITE ALASKAN NEXRAD COVERAGE
DELIVERIES AS OF AUGUST 20, 1996

SITKA
MAY 96

FAIRBANKS
JUL 93

ANCHORAGE
OCT 93

MIDDLETON
ISLAND
JUN 96

KING
SALMON
MAY 95

NOME
JUN 96

BETHEL
AUG 95

+ EQUIPMENT DELIVERY DATE OF 8-20-96 OR EARLIER

+ EQUIPMENT DELIVERY DATE OF 8-20-96 OR LATER

AREAS NOT COVERED BELOW
10,000 ft ABOVE SITE LEVEL

100   NAUTICAL MILES (nmi)

185   KILOMETERS (km)

0

0

Figure 7-1-11.

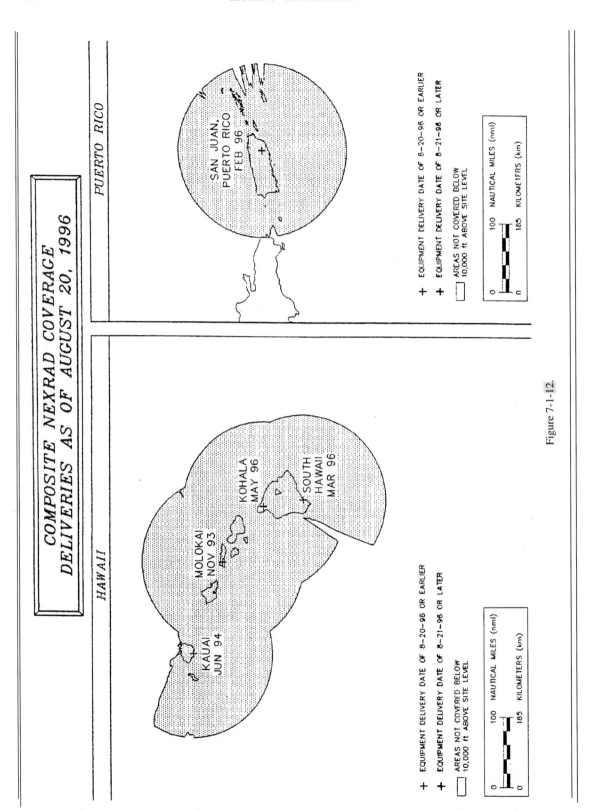

COMPOSITE NEXRAD COVERAGE
DELIVERIES AS OF AUGUST 20, 1996

Figure 7-1-12.

(b) Low-topped convection containing little or no cloud-to-ground lightning (such as may occur in cool air masses).

(c) Rapidly evolving convection.

7. The NCWF cannot provide information on specific storm hazards such as hail, high winds or tornadoes.

c. Availability and Use.

1. The NCWF is available primarily via the Internet from the AWC Aviation Digital Data Service (ADDS) at http://adds.aviationweather.gov. Used in conjunction with other weather products such as Convective SIGMETs, the NCWF provides additional information for convective weather avoidance and flight planning.

2. The NCWF access by Automated Flight Service Stations and their associated En Route Flight Advisory Service Facilities, Air Route Traffic Control Centers (ARTCC's) or Terminal Radar Approach Controls is planned but NOT currently available.

NOTE—
SEE AIM, PARAGRAPH 7-1-15, ATC INFLIGHT WEATHER AVOIDANCE ASSISTANCE, FOR FURTHER INFORMATION.

d. Display Summary.

1. Existing convective hazards (based on NEXRAD and lightning data) are depicted using the color-coded 6-level NCWF hazard scale shown in TBL 7-1-3. In displaying forecast cell locations, the NCWF does NOT distinguish among level 3 through level 6.

Table 7-1-3 NCWF Hazard Scale

Level	Color	Effect
5-6	Red	Thunderstorms may contain any or all of the following: severe turbulence, severe icing, hail, frequent lightning, tornadoes and low-level wind shear. The risk of hazardous weather generally increases with levels on the NCWF hazard scale.
3-4	Yellow Orange	
1-2	Green	

NOTE—
ALTHOUGH SIMILAR, THE NCWF HAZARD SCALE LEVELS ARE NOT IDENTICAL TO VIP LEVELS.

REFERENCE—
PILOT/CONTROLLER GLOSSARY TERM- RADAR WEATHER ECHO INTENSITY LEVELS.

2. One-hour forecast locations of signification convection (NCWF hazard scale levels of 3 or greater) are depicted with blue polygons. Their directions of movement and storm tops are also shown.

3. The Java display permits some degree of customization. Other means of viewing the NCWF may not offer these display options. Java display options include the following (see FIG 7-1-13):

(a) "Current Convective Interest Grid."

(b) "One-Hour Extrapolation Polygons."

(c) "Previous Performance Polygons."

(d) Storm speed and altitude of tops.

(e) Overlays of:

(1) Airport locations.

(2) County boundaries.

(3) ARTCC boundaries.

(f) Aviation Routine Weather Reports (METARs).

(g) Unlimited customized zooms (by holding down the left mouse button and dragging to select the rectangle of coverage desired).

4. The JavaScript display options include the following (see FIG 7-1-14):

(a) Current convective hazard "Detection" field.

(b) 1-hour extrapolation "Forecast" polygons.

(c) Previous hour "Performance" polygons.

(d) "2 hr Movie" loops of convective hazard detection fields (with forecast polygons included on the last frame).

(e) "24 hr Movie" loops of convective hazard detection fields.

(f) Zoomed views of:

(1) ARTCC boundaries.

(2) Certain major airports.

(3) Seven geographical regions: Northwest, North Central, Northeast, Southwest, South Central, Southwest, the 48 contiguous states.

5. Additional information is available via the "FYI/Help" or "i" links on the Java and JavaScript displays, respectively, field.

7-1-15. ATC INFLIGHT WEATHER AVOIDANCE ASSISTANCE

a. ATC Radar Weather Display:

1. Areas of radar weather clutter result from rain or moisture. *Radars cannot detect turbulence.* The determination of the intensity of the weather displayed is based on its precipitation density. Generally, the turbulence associated with a very heavy rate of rainfall will normally be more severe than any associated with a very light rainfall rate.

2. ARTCCs use Narrowband Radar which provides the controller with two distinct levels of weather intensity by assigning radar display symbols for specific precipitation densities measured by the narrowband system.

b. Weather Avoidance Assistance:

1. To the extent possible, controllers will issue pertinent information on weather or chaff areas and assist pilots in avoiding such areas when requested. Pilots should respond to a weather advisory by either acknowledging the advisory or by acknowledging the advisory and requesting an alternative course of action as follows:

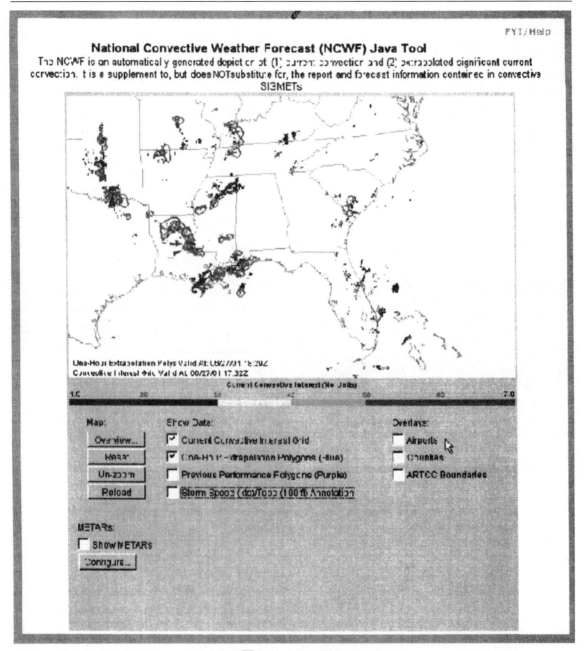

Figure 7-1-13. Example NCWF Java Display

(a) Request to deviate off course by stating the number of miles and the direction of the requested deviation. In this case, when the requested deviation is approved, navigation is at the pilot's prerogative, but must maintain the altitude assigned by ATC and to remain within the specified mileage of the original course.

(b) Request a new route to avoid the affected area.

(c) Request a change of altitude.

(d) Request radar vectors around the affected areas.

2. For obvious reasons of safety, an IFR pilot must not deviate from the course or altitude or Flight Level without a proper ATC clearance. When weather conditions

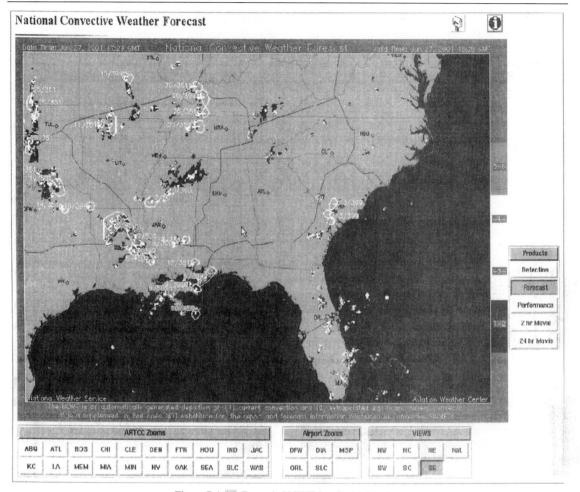

Figure 7-1-14. Example NCWF JavaScript Display

encountered are so severe that an immediate deviation is determined to be necessary and time will not permit approval by ATC, the pilot's emergency authority may be exercised.

3. When the pilot requests clearance for a route deviation or for an ATC radar vector, the controller must evaluate the air traffic picture in the affected area, and coordinate with other controllers (if ATC jurisdictional boundaries may be crossed) before replying to the request.

4. It should be remembered that the controller's primary function is to provide safe separation between aircraft. Any additional service, such as weather avoidance assistance, can only be provided to the extent that it does not derogate the primary function. It's also worth noting that the separation workload is generally greater than normal when weather disrupts the usual flow of traffic. ATC radar limitations and frequency congestion may also be a factor

in limiting the controller's capability to provide additional service.

5. It is very important, therefore, that the request for deviation or radar vector be forwarded to ATC as far in advance as possible. Delay in submitting it may delay or even preclude ATC approval or require that additional restrictions be placed on the clearance. Insofar as possible the following information should be furnished to ATC when requesting clearance to detour around weather activity:

(a) Proposed point where detour will commence.

(b) Proposed route and extent of detour (direction and distance).

(c) Point where original route will be resumed.

(d) Flight conditions (IFR or VFR).

(e) Any further deviation that may become necessary as the flight progresses.

(f) Advise if the aircraft is equipped with functioning airborne radar.

6. To a large degree, the assistance that might be rendered by ATC will depend upon the weather information available to controllers. Due to the extremely transitory nature of severe weather situations, the controller's weather information may be of only limited value if based on weather observed on radar only. Frequent updates by pilots giving specific information as to the area affected, altitudes intensity and nature of the severe weather can be of considerable value. Such reports are relayed by radio or phone to other pilots and controllers and also receive widespread teletypewriter dissemination.

7. Obtaining IFR clearance or an ATC radar vector to circumnavigate severe weather can often be accommodated more readily in the en route areas away from terminals because there is usually less congestion and, therefore, offer greater freedom of action. In terminal areas, the problem is more acute because of traffic density, ATC coordination requirements, complex departure and arrival routes, adjacent airports, etc. As a consequence, controllers are less likely to be able to accommodate all requests for weather detours in a terminal area or be in a position to volunteer such routing to the pilot. Nevertheless, pilots should not hesitate to advise controllers of any observed severe weather and should specifically advise controllers if they desire circumnavigation of observed weather.

c. Procedures for Weather Deviations and Other Contingencies in Oceanic Controlled Airspace:

1. When the pilot initiates communications with ATC, rapid response may be obtained by stating "WEATHER DEVIATION REQUIRED" to indicate priority is desired on the frequency and for ATC response.

2. The pilot still retains the option of initiating the communications using the urgency call "PAN-PAN" 3 times to alert all listening parties of a special handling condition which will receive ATC priority for issuance of a clearance or assistance.

3. ATC will:

(a) Approve the deviation,

(b) Provide vertical separation and then approve the deviation, or

(c) If ATC is unable to establish vertical separation, ATC shall advise the pilot that standard separation cannot be applied; provide essential traffic information for all affected aircraft, to the extent practicable; and if possible, suggest a course of action. ATC may suggest that the pilot climb or descend to a contingency altitude (1,000 feet above or below that assigned if operating above FL 290; 500 feet above or below that assigned if operating at or below FL 290).

PHRASEOLOGY—STANDARD SEPARATION NOT AVAILABLE, DEVIATE AT PILOT'S DISCRETION; SUGGEST CLIMB (OR DESCENT) TO (APPROPRIATE ALTITUDE); TRAFFIC (POSITION AND ALTITUDE); REPORT DEVIATION COMPLETE.

4. The pilot will follow the ATC advisory altitude when approximately 10 NM from track as well as execute the procedures detailed in paragraph 7-1-12c.5.

5. If contact cannot be established or revised ATC clearance or advisory is not available and deviation from track is required, the pilot shall take the following actions:

(a) If possible, deviate away from an organized track or route sytem.

(b) Broadcast aircraft position and intentions on the frequency in use, as well as on frequency 121.5 MHz at suitable intervals stating: flight identification (operator call sign), flight level, track code or ATS route designator, and extent of deviation expected.

(c) Watch for conflicting traffic both visually and by reference to TCAS (if equipped).

(d) Turn on aircraft exterior lights.

(e) Deviations of less than 10 NM or operations within COMPOSITE (NOPAC and CEPAC) Airspace, should REMAIN at ASSIGNED altitude. Otherwise, when the aircraft is approximately 10 NM from track, initiate an altitude change based on the following criteria:

Table 7-1-4

Route Centerline/ Track	Deviations >10 NM	Altitude Change
EAST 000–179°	LEFT RIGHT	DESCEND 300 FEET CLIMB 300 FEET
WEST 180–359°	LEFT RIGHT	CLIMB 300 FEET DESCEND 300 FEET

Pilot Memory Slogan: "East right up, West right down."

(f) When returning to track, be at assigned flight level when the aircraft is within approximately 10 NM of centerline.

(g) If contact was not established prior to deviating, continue to attempt to contact ATC to obtain a clearance. If contact was established, continue to keep ATC advised of intentions and obtain essential traffic information.

7-1-16. RUNWAY VISUAL RANGE (RVR)

There are currently two configurations of RVR in the NAS commonly identified as Taskers and New Generation RVR. The Taskers are the existing configuration which uses transmissometer technology. The New Generation RVR's were deployed in November 1994 and use forward scatter technology. The New Generation RVR's are currently being deployed in the NAS to replace the existing Taskers.

a. RVR values are measured by transmissometers mounted on 14-foot towers along the runway. A full RVR system consists of:

1. Transmissometer projector and related items.

2. Transmissometer receiver (detector) and related items.

3. Analogue recorder.

4. Signal data converter and related items.

5. Remote digital or remote display programmer.

b. The transmissometer projector and receiver are mounted on towers 250 feet apart. A known intensity of light is emitted from the projector and is measured by the receiver. Any obscuring matter such as rain, snow, dust, fog, haze or smoke reduces the light intensity arriving at the receiver. The resultant intensity measurement is then converted to an RVR value by the signal data converter. These values are displayed by readout equipment in the associated air traffic facility and updated approximately once every minute for controller issuance to pilots.

c. The signal data converter receives information on the high intensity runway edge light setting in use (step 3, 4, or 5); transmission values from the transmissometer, and the sensing of day or night conditions. From the three data sources, the system will compute appropriate RVR values. Due to variable conditions, the reported RVR values may deviate somewhat from the true observed visual range due to the slant range consideration, brief time delays between the observed RVR conditions and the time they are transmitted to the pilot, and rapidly changing visibility conditions.

d. An RVR transmissometer established on a 250 foot baseline provides digital readouts to a minimum of 600 feet, which are displayed in 200 foot increments to 3,000 feet and in 500 foot increments from 3,000 feet to a maximum value of 6,000 feet.

e. RVR values for Category IIIa operations extend down to 700 feet RVR; however, only 600 and 800 feet are reportable RVR increments. The 800 RVR reportable value covers a range of 701 feet to 900 feet and is therefore a valid minimum indication of Category IIIa operations.

f. Approach categories with the corresponding minimum RVR values: (See TBL 7-1-5.)

Table 7-1-5 Approach Category/Minimum RVR Table}}

Category	Visibility (RVR)
Nonprecision	2,400 feet
Category I	1,800 feet
Category II	1,200 feet
Category IIIa	700 feet
Category IIIb	150 feet
Category IIIc	0 feet

g. Ten minute maximum and minimum RVR values for the designated RVR runway are reported in the body of the aviation weather report when the prevailing visibility is less than one mile and/or the RVR is 6,000 feet or less. ATCTs report RVR when the prevailing visibility is 1 mile or less and/or the RVR is 6,000 feet or less.

h. Details on the requirements for the operational use of RVR are contained in FAA AC 97-1, "Runway Visual Range (RVR)." Pilots are responsible for compliance with minimums prescribed for their class of operations in the appropriate FARs and/or operations specifications.

i. RVR values are also measured by forward scatter meters mounted on 14-foot frangible fiberglass poles. A full RVR system consists of:

1. Forward scatter meter with a transmitter, receiver and associated items.

2. A runway light intensity monitor (RLIM).

3. An ambient light sensor (ALS).

4. A data processor unit (DPU).

5. Controller display (CD).

j. The forward scatter meter is mounted on a 14-foot frangible pole. Infrared light is emitted from the transmitter and received by the receiver. Any obscuring matter such as rain, snow, dust, fog, haze or smoke increases the amount of scattered light reaching the receiver. The resulting measurement along with inputs from the runway light intensity monitor and the ambient light sensor are forwarded to the DPU which calculates the proper RVR value. The RVR values are displayed locally and remotely on controller displays.

k. The runway light intensity monitors both the runway edge and centerline light step settings (steps 1 through 5). Centerline light step settings are used for CAT IIIb operations. Edge Light step settings are used for CAT I, II, and IIIa operations.

l. New Generation RVR's can measure and display RVR values down to the lowest limits of Category IIIb operations (150 feet RVR). RVR values are displayed in 100 feet increments and are reported as follows:

1. 100-feet increments for products below 800 feet.

2. 200-feet increments for products between 800 feet and 3,000 feet.

3. 500-feet increments for products between 3,000 feet and 6,500 feet.

4. 25-meter increments for products below 150 meters.

5. 50-meter increments for products between 150 meters and 800 meters.

6. 100-meter increments for products between 800 meters and 1,200 meters.

7. 200-meter increments for products between 1,200 meters and 2,000 meters.

7-1-17. REPORTING OF CLOUD HEIGHTS

a. Ceiling, by definition in the FARs and as used in Aviation Weather Reports and Forecasts, is the height above ground (or water) level of the lowest layer of clouds or obscuring phenomenon that is reported as "broken," "overcast," or "obscuration," e.g., an Aerodrome forecast (TAF) which reads "BKN030" refers to height above ground level. An Area forecast which reads "BKN030" indicates that the height is above mean sea level.

REFERENCE—AIM, KEY TO AERODROME FORECAST (TAF) AND AVIATION ROUTINE WEATHER REPORT (METAR), PARAGRAPH 7-1-31, DEFINES "BROKEN" "OVERCAST" AND "OBSCURATION."

b. Pilots usually report height values above MSL, since they determine heights by the altimeter. This is taken in account when disseminating and otherwise applying information received from pilots. ("Ceiling" heights are always above ground level.) In reports disseminated as PIREPS, height references are given the same as received from pilots, that is, above MSL.

c. In area forecasts or inflight advisories, ceilings are denoted by the contraction "CIG" when used with sky cover symbols as in "LWRG to CIG OVC005," or the contraction "AGL" after, the forecast cloud height value. When the cloud base is given in height above MSL, it is so indicated by the contraction "MSL" or "ASL" following the height value. The heights of clouds tops, freezing level, icing, and turbulence are always given in heights above ASL or MSL.

7-1-18. REPORTING PREVAILING VISIBILITY

a. Surface (horizontal) visibility is reported in METAR reports in terms of statute miles and increments thereof; e.g., 1/16, 1/8, 3/16, 1/4, 5/16, 3/8, 1/2, 5/8, 3/4, 7/8, 1, 1 1/8, etc. (Visibility reported by an unaugmented automated site is reported differently than in a manual report, i.e., ASOS: 0, 1/16, 1/8, 1/4, 1/2, 3/4, 1, 1 1/4, 1 1/2, 1 3/4, 2, 2 1/2, 3, 4, 5, etc., AWOS: M 1/4, 1/2, 3/4, 1, 1 1/4, 1 1/2, 1 3/4, 2, 2 1/2, 3, 4, 5, etc.) Visibility is determined through the ability to see and identify preselected and prominent objects at a known distance from the usual point of observation. Visibilities which are determined to be less than 7 miles, identify the obscuring atmospheric condition; e.g., fog, haze, smoke, etc., or combinations thereof.

b. Prevailing visibility is the greatest visibility equalled or exceeded throughout at least one half of the horizon circle, not necessarily continuous. Segments of the horizon circle, which may have a significantly different visibility may be reported in the remarks section of the weather report; i.e., the southeastern quadrant of the horizon circle may be determined to be 2 miles in mist while the remaining quadrants are determined to be 3 miles in mist.

c. When the prevailing visibility at the usual point of observation, or at the tower level, is less than 4 miles, certificated tower personnel will take visibility observations in addition to those taken at the usual point of observation. The lower of these two values will be used as the prevailing visibility for aircraft operations.

7-1-19. ESTIMATING INTENSITY OF RAIN AND ICE PELLETS

a. RAIN:

1. Light: From scattered drops that, regardless of duration, do not completely wet an exposed surface <u>up to</u> a condition where individual drops are easily seen.

2. Moderate: Individual drops are not clearly identifiable; spray is observable just above pavements and other hard surfaces.

3. Heavy: Rain seemingly falls in sheets; individual drops are not identifiable; heavy spray to height of several inches is observed over hard surfaces.

b. ICE PELLETS:

1. Light: Scattered pellets that do not completely cover an exposed surface regardless of duration. Visibility is not affected.

2. Moderate: Slow accumulation on ground. Visibility reduced by ice pellets to less than 7 statute miles.

3. Heavy: Rapid accumulation on ground. Visibility reduced by ice pellets to less than 3 statute miles.

7-1-20. ESTIMATING INTENSITY OF SNOW OR DRIZZLE (BASED ON VISIBILITY)

a. Light: Visibility more than 1/2 statute mile.

b. Moderate: Visibility from more than 1/4 statute mile to 1/2 statute mile.

c. Heavy: Visibility less than 1/4 statute mile or less.

7-1-21. PILOT WEATHER REPORTS (PIREPS)

a. FAA air traffic facilities are required to solicit PIREPs when the following conditions are reported or forecast: ceilings at or below 5,000 feet; visibility at or below 5 miles (surface or aloft); thunderstorms and related phenomena; icing of light degree or greater; turbulence of moderate degree or greater; windshear and reported or forecast volcanic ash clouds.

b. Pilots are urged to cooperate and promptly volunteer reports of these conditions and other atmospheric data such as: cloud bases, tops and layers; flight visibility; precipitation; visibility restrictions such as haze, smoke and dust; wind at altitude; and temperature aloft.

c. PIREPs should be given to the ground facility with which communications are established; i.e., EFAS, AFSS/FSS, ARTCC, or terminal ATC. One of the primary duties of EFAS facilities, radio call "FLIGHT WATCH," is to serve as a collection point for the exchange of PIREPs with en route aircraft.

d. If pilots are not able to make PIREPs by radio, reporting upon landing of the inflight conditions encountered to the nearest AFSS/FSS or Weather Forecast Office will be helpful. Some of the uses made of the reports are:

1. The ATCT uses the reports to expedite the flow of air traffic in the vicinity of the field and for hazardous weather avoidance procedures.

2. The AFSS/FSS uses the reports to brief other pilots, to provide inflight advisories, and weather avoidance information to en route aircraft.

3. The ARTCC uses the reports to expedite the flow of en route traffic, to determine most favorable altitudes, and to issue hazardous weather information within the center's area.

Table 7-1-6 PIREP Element Code Chart

PIREP ELEMENTS	PIREP CODE	CONTENTS
1. 3-letter station identifier	XXX	Nearest weather reporting location to the reported phenomenon.
2. Report type	UA or UUA	Routine or Urgent PIREP
3. Location	/OV	In relation to a VOR
4. Time	/TM	Coordinated Universal Time
5. Altitude	/FL	Essential for turbulence and icing reports
6. Type Aircraft	/TP	Essential for turbulence and icing reports
7. Sky cover	/SK	Cloud height and coverage (sky clear, few, scattered, broken, or overcast)
8. Weather	/WX	Flight visibility, precipitation, restrictions to visibility, etc.
9. Temperature	/TA	Degrees Celsius
10. Wind	/WV	Direction in degrees magnetic north and speed in knots
11. Turbulence	/TB	See AIM paragraph 7-1-21
12. Icing	/IC	See AIM paragraph 7-1-19
13. Remarks	/RM	For reporting elements not included or to clarify previously reported items

4. The NWS uses the reports to verify or amend conditions contained in aviation forecast and advisories. In some cases, pilot reports of hazardous conditions are the triggering mechanism for the issuance of advisories. They also use the reports for pilot weather briefings.

5. The NWS, other government organizations, the military, and private industry groups use PIREPs for research activities in the study of meteorological phenomena.

6. All air traffic facilities and the NWS forward the reports received from pilots into the weather distribution system to assure the information is made available to all pilots and other interested parties.

e. The FAA, NWS, and other organizations that enter PIREPs into the weather reporting system use the format listed in TBL 7-1-6. Items 1 through 6 are included in all transmitted PIREPs along with one or more of items 7 through 13. Although the PIREP should be as complete and concise as possible, pilots should not be overly concerned with strict format or phraseology. The important thing is that the information is relayed so other pilots may benefit from your observation. If a portion of the report needs clarification, the ground station will request the information. Completed PIREPs will be transmitted to weather circuits as in the following examples:

EXAMPLE—
① KCMH UA /OV APE 230010/TM 1516/FL085/TP BE20/SK BKN 065/WX FV03SM HZ FU/TA 20/TB LGT

NOTE—
① ONE ZERO MILES SOUTHWEST OF APPLETON VOR; TIME 1516 UTC, ALTITUDE EIGHT THOUSAND FIVE HUNDRED; AIRCRAFT TYPE BE200; BASES OF THE BROKEN CLOUD LAYER IS SIX THOUSAND FIVE HUNDRED; FLIGHT VISIBILITY 3 MILES WITH HAZE AND SMOKE; AIR TEMPERATURE 20 DEGREES CELSIUS; LIGHT TURBULENCE.

EXAMPLE—
② KCRW UV /OV KBKW 360015-KCRW/TM 1815/FL120/TP BE99/SK IMC/WX RA/TA M08 /WV 290030/TB LGT-MDT/IC LGT RIME/RM MDT MXD ICG DURC KROA NWBND FL080-100 1750Z

NOTE—
② FROM 15 MILES NORTH OF BECKLEY VOR TO CHARLESTON VOR; TIME 1815 UTC; ALTITUDE 12,000 FEET; TYPE AIRCRAFT, BE-99; IN CLOUDS; RAIN; TEMPERATURE MINUS 8 CELSIUS; WIND 290 DEGREES TRUE AT 30 KNOTS; LIGHT TO MODERATE TURBULENCE; LIGHT RIME ICING; ENCOUNTERED MODERATE MIXED ICING DURING CLIMB NORTHWESTBOUND FROM ROANOKE, VA, BETWEEN 8,000 AND 10,000 FEET AT 1750 UTC.

7-1-22. PIREPS RELATING TO AIRFRAME ICING

a. The effects of ice on aircraft are cumulative-thrust is reduced, drag increases, lift lessens, and weight increases. The results are an increase in stall speed and a deterioration of aircraft performance. In extreme cases, 2 to 3 inches of ice can form on the leading edge of the airfoil in less than 5 minutes. It takes but ½ inch of ice to reduce the lifting power of some aircraft by 50 percent and increases the frictional drag by an equal percentage.

b. A pilot can expect icing when flying in visible precipitation, such as rain or cloud droplets, and the temperature is between +02 and –10 degrees Celsius. When icing is detected, a pilot should do one of two things, particularly if the aircraft is not equipped with deicing equipment; get out of the area of precipitation; or go to an altitude where the temperature is above freezing. This "warmer" altitude may not always be a lower altitude. Proper preflight action includes obtaining information on the freezing level and the above freezing levels in precipitation areas. Report icing to ATC, and if operating IFR, request new routing or altitude if icing will be a hazard. Be sure to give the type of aircraft to ATC when reporting icing. The following describes how to report icing conditions.

1. *Trace:* Ice becomes perceptible. Rate of accumulation is slightly greater than the rate of sublimation. Deicing/anti-icing equipment is not utilized unless encountered for an extended period of time (over 1 hour).

2. *Light:* The rate of accumulation may create a problem if flight is prolonged in this environment (over 1 hour).

Occasional use of deicing/anti-icing equipment removes/ prevents accumulation. It does not present a problem if the deicing/anti-icing equipment is used.

3. *Moderate:* The rate of accumulation is such that even short encounters become potentially hazardous and use of deicing/anti-icing equipment or flight diversion is necessary.

4. *Severe:* The rate of accumulation is such that deicing/anti-icing equipment fails to reduce or control the hazard. Immediate flight diversion is necessary.

EXAMPLE—PILOT REPORT: GIVE AIRCRAFT IDENTIFICATION, LOCATION, TIME (UTC), INTENSITY OF TYPE, ALTITUDE/FL, AIRCRAFT TYPE, INDICATED AIR SPEED (IAS), AND OUTSIDE AIR TEMPERATURE (OAT).

NOTE—
① RIME ICE: ROUGH, MILKY OPAQUE ICE FORMED BY THE INSTANTANEOUS FREEZING OF SMALL SUPERCOOLED WATER DROPLETS.

② CLEAR ICE: A GLOSSY CLEAR, OR TRANSLUCENT ICE FORMED BY THE RELATIVELY SLOW FREEZING OF LARGE SUPERCOOLED WATER DROPLETS.

③ THE OAT SHOULD BE REQUESTED BY THE AFSS/FSS OR ATC IF NOT INCLUDED IN THE PIREP.

7-1-23 Definitions of Inflight Icing Terms

See TBL 7-1-7, Icing Types, and TBL 7-1-8, Icing Conditions

7-1-24. PIREPS RELATING TO TURBULENCE

a. When encountering turbulence, pilots are urgently requested to report such conditions to ATC as soon as practicable. PIREPs relating to turbulence should state:

1. *Aircraft location.*

2. *Time of occurrence in UTC.*

3. *Turbulence intensity.*

4. *Whether the turbulence occurred in or near clouds.*

5. *Aircraft altitude or Flight Level.*

6. *Type of Aircraft.*

7. *Duration of turbulence.*

EXAMPLE—
① OVER OMAHA, 1232Z, MODERATE TURBULENCE IN CLOUDS AT FLIGHT LEVEL THREE ONE ZERO, BOEING 707.

② FROM FIVE ZERO MILES SOUTH OF ALBUQUERQUE TO THREE ZERO MILES NORTH OF PHOENIX, 1250Z, OCCASIONAL MODERATE CHOP AT FLIGHT LEVEL THREE THREE ZERO, DC8.

b. Duration and classification of intensity should be made using TBL 7-1-9.

7-1-25. WIND SHEAR PIREPS

a. Because unexpected changes in wind speed and direction can be hazardous to aircraft operations at low altitudes on approach to and departing from airports, pilots are urged to promptly volunteer reports to con-

Table 7-1-7   Icing Types

Clear Ice	See Glaze Ice.
Glaze Ice	Ice, sometimes clear and smooth, but usually containing some air pockets, which results in a lumpy translucent appearance. Glaze ice results from supercooled drops/droplets striking a surface but not freezing rapidly on contact. Glaze ice is denser, harder, and sometimes more transparent than rime ice. Factors, which favor glaze formation, are those that favor slow dissipation of the heat of fusion (i.e., slight supercooling and rapid accretion). With larger accretions, the ice shape typically includes "horns" protruding from unprotected leading edge surfaces. It is the ice shape, rather than the clarity or color of the ice, which is most likely to be accurately assessed from the cockpit. The terms "clear" and "glaze" have been used for essentially the same type of ice accretion, although some reserve "clear" for thinner accretions which lack horns and conform to the airfoil.
Intercycle Ice	Ice which accumulates on a protected surface between actuation cycles of a deicing system.
Known or Observed or Detected Ice Accretion	Actual ice observed visually to be on the aircraft by the flight crew or identified by on-board sensors.
Mixed Ice	Simultaneous appearance or a combination of rime and glaze ice characteristics. Since the clarity, color, and shape of the ice will be a mixture of rime and glaze characteristics, accurate identification of mixed ice from the cockpit may be difficult.
Residual Ice	Ice which remains on a protected surface immediately after the actuation of a deicing system.
Rime Ice	A rough, milky, opaque ice formed by the rapid freezing of supercooled drops/droplets after they strike the aircraft. The rapid freezing results in air being trapped, giving the ice its paque appearance and making it porous and brittle. Time ice typically accretes along the stagnation line of an airfoil and is more regular in shape and conformal to the airfoil than glaze ice. It is the ice shape, rather than the clarity of color of the ice, which is most likely to be accurately assess from the cockpit.
Runback Ice	Ice which forms from the freezing or refreezing of water leaving protected surfaces and running back to unprotected surfaces.

Note—
Ice types are difficult for the pilot to discern and have uncertain effects on an airplane in flight. Ice type definitions will be included in the AIM for use in the "Remarks" section of the PIREP and for use in forecasting.

Table 7-1-8   Icing Conditions

Appendix C Icing Conditions	Appendix C (14 CFR, Part 25 and 29) is the certification icing condition standard for approving ice protection provisions on aircraft. The conditions are specified in terms of altitude, temperature, liquid water content (LWC), representative droplet size (mean effective drop diameter [MED]), and cloud horizontal extent.
Forecast Icing Conditions	Environmental conditions expected by a National Weather Service or an FAA-approved weather provider to be conducive to the formation of inflight icing on aircraft.
Freezing Drizzle (FZDZ)	Drizzle is precipitation at ground level or aloft in the form of liquid water drops which have diameters less than 0.5 mm and greater than 0.05 mm. Freezing drizzle is drizzle that exists at air temperatures less than 0°C (supercooled), remains in liquid form, and freezes upon contact with objects on the surface or airborne. Freezing Precipitation Freezing precipitation is freezing rain or freezing drizzle falling through or outside of visible cloud.
Freezing Rain (FZRA)	Rain is precipitation at ground level or aloft in the form of liquid water drops which have diameters greater than 0.5 mm. Freezing rain is rain that exists at air temperatures less than 0°C (supercooled), remains in liquid form, and freezes upon contact with objects on the ground or in the air.
Icing in Cloud	Icing occurring within visible cloud. Cloud droplets (diameter < 0.05 mm) will be present; freezing drizzle and/or freezing rain may or may not be present.
Icing in Precipitation	Icing occurring from an encounter with freezing precipitation, that is, supercooled drops with diameters exceeding 0.05 mm, within or outside of visible cloud. Known Icing Conditions Atmospheric conditions in which the formation of ice is observed or detected in flight. *Note—* *Because of the variability in space and time of atmospheric conditions, the existence of a report of observed icing does not assure the presence or intensity of icing conditions at a later time, nor can a report of no icing assure the absence of icing conditions at a later time.*
Potential Icing Conditions	Atmospheric icing conditions that are typically defined by airframe manufacturers relative to temperature and visible moisture that may result in aircraft ice accretion on the ground or in flight. The potential icing conditions are typically defined in the Airplane Flight Manual or in the Airplane Operation Manual.
Supercooled Drizzle Drops (SCDD)	Synonymous with freezing dr. (SCDD)
Supercooled Drops or /Droplets	Water drops/droplets which remain unfrozen at temperatures below 0°C. Supercooled drops are found in clouds, freezing drizzle, and freezing rain in the atmosphere. These drops may impinge and freeze after contact on aircraft surfaces.
Supercooled Large Drops (SLD)	Liquid droplets with diameters greater than 0.05 mm at temperatures less than 0°C, i.e., freezing rain or freezing drizzle.

trollers of wind shear conditions they encounter. An advance warning of this information will assist other pilots in avoiding or coping with a wind shear on approach or departure.

b. When describing conditions, use of the terms "negative" or "positive" wind shear should be avoided. PIREPs of "negative wind shear on final," intended to describe loss of airspeed and lift, have been interpreted to mean that no wind shear was encountered. The recommended method for wind shear reporting is to state the loss or gain of airspeed and the altitudes at which it was encountered.

EXAMPLE—

① DEWER TOWER, CESSNA 1234 ENCOUNTERED WIND SHEAR, LOSS OF 20 KNOTS AT 400.

② TULSA TOWER, AMERICAN 721 ENCOUNTERED WIND SHEAR ON FINAL, GAINED 25 KNOTS BETWEEN 600 AND 400 FEET FOLLOWED BY LOSS OF 40 KNOTS BETWEEN 400 FEET AND SURFACE.

1. Pilots who are not able to report wind shear in these specific terms are encouraged to make reports in terms of the effect upon their aircraft.

EXAMPLE—MIAMI TOWER, GULFSTREAM 403 CHARLIE ENCOUNTERED AN ABRUPT WIND SHEAR AT 800 FEET ON FINAL, MAX THRUST REQUIRED.

2. Pilots using Inertial Navigation Systems (INS) should report the wind and altitude both above and below the shear level.

7-1-26. CLEAR AIR TURBULENCE (CAT) PIREPS

CAT has become a very serious operational factor to flight operations at all levels and especially to jet traffic flying in excess of 15,000 feet. The best available information on this phenomenon must come from pilots via the PIREP reporting procedures. All pilots encountering CAT conditions are urgently requested to report time, location, and intensity (light, moderate, severe, or extreme) of the element to the FAA facility with which they are maintaining radio contact. If time and conditions permit, elements should be reported according to the standards for other PIREPs and position reports.

Table 7-1-9 Turbulence Reporting Criteria Table

Intensity	Aircraft Reaction	Reaction Inside Aircraft	Reporting Term—Definition
Light	Turbulence that momentarily causes slight, erratic changes in altitude and/or attitude (pitch, roll, yaw). Report as **Light Turbulence;**[1] *or* Turbulence that causes slight, rapid and somewhat rhythmic bumpiness without appreciable changes in altitude or attitude. Report as **Light Chop.**	Occupants may feel a slight strain against seatbelts or shoulder straps. Unsecured objects may be displaced slightly. Food service may be conducted and little or no difficulty is encountered in walking.	Occasional—Less than 1/3 of the time. Intermittent—1/3 to 2/3. Continuous—More than 2/3.
Moderate	Turbulence that is similar to Light Turbulence but of greater intensity. Changes in altitude and/or attitude occur but the aircraft remains in positive control at all times. It usually causes variations in indicated airspeed. Report as **Moderate Turbulence;**[1] *or* Turbulence that is similar to Light Chop but of greater intensity. It causes rapid bumps or jolts without appreciable changes in aircraft altitude or attitude. Report as **Moderate Chop.**[1]	Occupants feel definite strains against seatbelts or shoulder straps. Unsecured objects are dislodged. Food service and walking are difficult.	**NOTE** 1. Pilots should report location(s), time (UTC), intensity, whether in or near clouds, altitude, type of aircraft and, when applicable, duration of turbulence. 2. Duration may be based on time between two locations or over a single location. All locations should be readily identifiable.
Severe	Turbulence that causes large, abrupt changes in altitude and/or attitude. It usually causes large variations in indicated airspeed. Aircraft may be momentarily out of control. Report as **Severe Turbulence.**[1]	Occupants are forced violently against seatbelts or shoulder straps. Unsecured objects are tossed about. Food service and walking are impossible.	**EXAMPLES:** a. Over Omaha. 1232Z, Moderate Turbulence, in cloud, Flight Level 310, B707. b. From 50 miles south of Albuquerque to 30 miles north of Phoenix, 1210Z to 1250Z, occasional Moderate Chop, Flight Level 330, DC8.
Extreme	Turbulence in which the aircraft is violently tossed about and is practically impossible to control. It may cause structural damage. Report as **Extreme Turbulence.**[1]		

[1] High level turbulence (normally above 15,000 feet ASL) not associated with cumuliform cloudiness, including thunderstorms, should be reported as CAT (clear air turbulence) preceded by the appropriate intensity, or light or moderate chop.

REFERENCE—AIM, PIREPS RELATING TO TURBULENCE, PARAGRAPH 7-1-20.

7-1-27. MICROBURSTS

a. Relatively recent meteorological studies have confirmed the existence of microburst phenomenon. Microbursts are small scale intense downdrafts which, on reaching the surface, spread outward in all directions from the downdraft center. This causes the presence of both vertical and horizontal wind shears that can be extremely hazardous to all types and categories of aircraft, especially at low altitudes. Due to their small size, short life span, and the fact that they can occur over areas without surface precipitation, microbursts are not easily detectable using conventional weather radar or wind shear alert systems.

b. Parent clouds producing microburst activity can be any of the low or middle layer convective cloud types. Note, however, that microbursts commonly occur within the heavy rain portion of thunderstorms, and in much

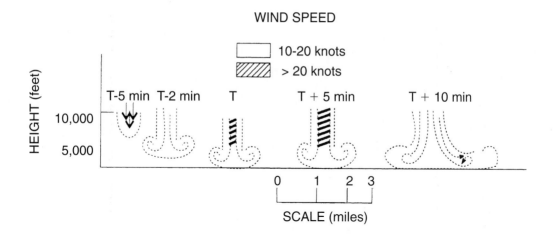

Vertical cross section of the evolution of a microburst wind field. T is the time of initial divergence at the surface. The shading refers to the vector wind speeds. Figure adapted from Wilson et al., 1984, Microburst Wind Structure and Evaluation of Doppler Radar for Wind Shear Detection, DOT/FAA Report No. DOT/FAA/PM-84/29, National Technical Information Service, Springfield, VA 37 pp.

Figure 7-1-15. Evolution of a Microburst

weaker, benign appearing convective cells that have little or no precipitation reaching the ground.

c. The life cycle of a microburst as it descends in a convective rain shaft is seen in FIG 7-1-15. An important consideration for pilots is the fact that the microburst intensifies for about 5 minutes after it strikes the ground.

d. Characteristics of microbursts include:

1. Size: The microburst downdraft is typically less than 1 mile in diameter as it descends from the cloud base to about 1,000–3,000 feet above the ground. In the transition zone near the ground, the downdraft changes to a horizontal outflow that can extend to approximately 2½ miles in diameter.

2. Intensity: The downdrafts can be as strong as 6,000 feet per minute. Horizontal winds near the surface can be as strong as 45 knots resulting in a 90 knot shear (headwind to tailwind change for a traversing aircraft) across the microburst. These strong horizontal winds occur within a few hundred feet of the ground.

3. Visual Signs: Microbursts can be found almost anywhere that there is convective activity. They may be embedded in heavy rain associated with a thunderstorm or in light rain in benign appearing virga. When there is little or no precipitation at the surface accompanying the microburst, a ring of blowing dust may be the only visual clue of its existence.

4. Duration: An individual microburst will seldom last longer than 15 minutes from the time it strikes the ground

until dissipation. The horizontal winds continue to increase during the first 5 minutes with the maximum intensity winds lasting approximately 2–4 minutes. Sometimes microbursts are concentrated into a line structure, and under these conditions, activity may continue for as long as an hour. Once microburst activity starts, multiple microbursts in the same general area are not uncommon and should be expected.

e. Microburst wind shear may create a severe hazard for aircraft within 1,000 feet of the ground, particularly during the approach to landing and landing and take-off phases. The impact of a microburst on aircraft which have the unfortunate experience of penetrating one is characterized in FIG 7-1-16. The aircraft may encounter a headwind (performance increasing) followed by a downdraft and tailwind (both performance decreasing), possibly resulting in terrain impact.

f. Detection of Microbursts, Wind Shear and Gust Fronts

1. FAA's Integrated Wind Shear Detection Plan

(a) The FAA currently employs an integrated plan for wind shear detection that will significantly improve both the safety and capacity of the majority of the airports currently served by the air carriers. This plan integrates several programs, such as the Integrated Terminal Weather System (ITWS), Terminal Doppler Weather Radar (TDWR), Weather System Processor (WSP), and Low Level Wind

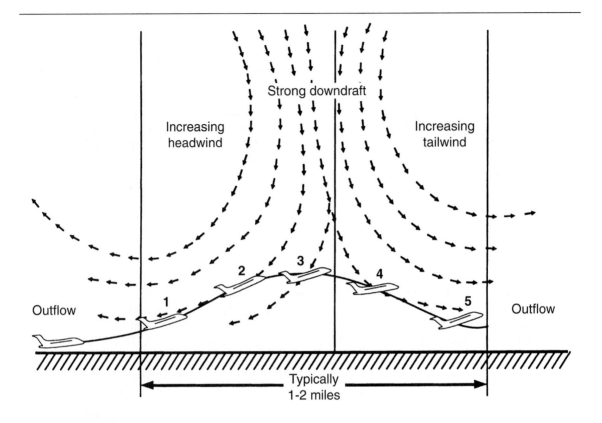

A microburst encounter during takeoff. The airplane first encounters a headwind and experiences increasing performance (1), this followed in short succession by a decreasing headwind component (2), a downdraft (3), and finally a strong tailwind (4), where 2 through 5 all result in decreasing performance of the airplane. Position (5) represents an extreme situation just prior to impact. Figure courtesy of Walter Frost, FWG Associates, Inc., Tullahoma, Tennessee.

Figure 7-1-16. Microburst Encounter During Takeoff

Shear Alert Systems (LLWAS) into a single strategic concept that significantly improves the aviation weather information in the terminal area. (See FIG 7-1-17.)

(b) The wind shear/microburst information and warnings are displayed on the ribbon display terminals (RBDT) located in the tower cabs. They are identical (and standardized) in the LLWAS, TDWR and WSP systems, and so designed that the controller does not need to interpret the data, but simply read the displayed information to the pilot. The RBDT's are constantly monitored by the controller to ensure the rapid and timely dissemination of any hazardous event(s) to the pilot.

(c) The early detection of a wind shear/microburst event, and the subsequent warning(s) issued to an aircraft on approach or departure, will alert the pilot/crew to the potential of, and to be prepared for, a situation that could become very dangerous! Without these warnings, the aircraft may NOT be able to climb out of, or safely transition, the event, resulting in a catastrophe. The air carriers, working with the FAA, have developed specialized training programs using their simulators to train and prepare their pilots on the demanding aircraft procedures required to escape these very dangerous wind shear and/or microburst encounters.

2. Low Level Wind Shear Alert System (LLWAS)

(a) The LLWAS provides wind data and software processes to detect the presence of hazardous wind shear and microbursts in the vicinity of an airport. Wind sensors,

NAS Wind Shear Product Systems

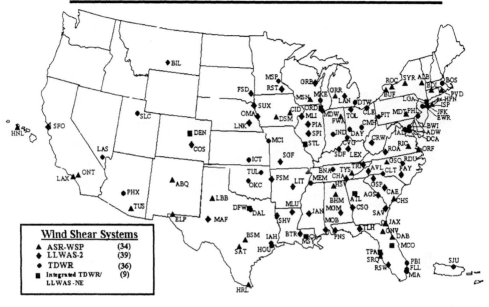

Figure 7-1-17.

mounted on poles sometimes as high as 150 feet, are (ideally) located 2,000–3,500 feet, but not more than 5,000 feet, from the centerline of the runway. (See FIG 7-1-18.)

(b) LLWAS was fielded in 1988 at 110 airports across the nation. Many of these systems have been replaced by new TDWR and WSP technology. Eventually all LLWAS systems will be phased out; however, 39 airports will be upgraded to the LLWAS-NE (Network Expansion) system, which employs the very latest software and sensor technology. The new LLWAS-NE systems will not only provide the controller with wind shear warnings and alerts, including wind shear/microburst detection at the airport wind sensor location, but will also provide the location of the hazards relative to the airport runway(s). It will also have the flexibility and capability to grow with the airport as new runways are built. As many as 32 sensors, strategically located around the airport and in relationship to its runway configuration, can be accommodated by the LLWAS-NE network.

3. Terminal Doppler Weather Radar (TDWR)

(a) TDWR's are being deployed at 45 locations across the United States. Optimum locations for TDWR's are 8 to 12 miles off of the airport proper, and designed to look at the airspace around and over the airport to detect microbursts, gust fronts, wind shifts and precipitation intensities. TDWR products advise the controller of wind shear and microburst events impacting all runways and the areas ½ mile on either side of the extended centerline of the runways out to 3 miles on final approach and 2 miles out on departure. (FIG 7-1-19 is a theoretical view of the warning boxes, including the runway, that the software uses in determining the location(s) of wind shear or microbursts). These warnings are displayed (as depicted in the examples of subparagraph 5) on the RBDT.

(b) It is very important to understand what TDWR does NOT DO:

it **DOES NOT** warn of wind shear outside of the alert boxes (on the arrival and departure ends of the runways);

it **DOES NOT** detect wind shear that is NOT a microburst or a gust front;

it **DOES NOT** detect gusty or cross wind conditions;

and, it **DOES NOT** detect turbulence.

However, research and development is continuing on these systems. Future improvements may include such areas as storm motion (movement), improved gust front detection, storm growth and decay, microburst prediction, and turbulence detection.

(c) TDWR also provides a geographical situation display (GSD) for supervisors and traffic management specialists for planning purposes. The GSD displays (in color) 6 levels of weather (precipitation), gust fronts and predicted

LLWAS SITING CRITERIA

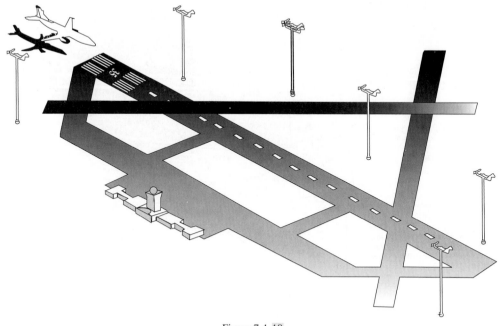

Figure 7-1-18.

storm movement(s). This data is used by the tower supervisor(s), traffic management specialists and controllers to plan for runway changes and arrival/departure route changes in order to both reduce aircraft delays and increase airport capacity.

4. Weather Systems Processor (WSP)

(a) The WSP provides the controller, supervisor, traffic management specialist, and ultimately the pilot, with the same products as the terminal doppler weather radar (TDWR) at a fraction of the cost of a TDWR. This is

WARNING BOXES

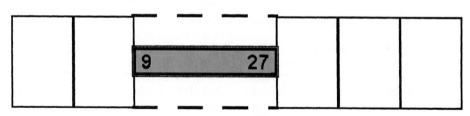

Figure 7-1-19.

MICROBURST ALERT

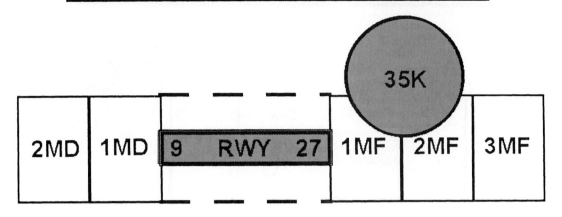

27A MBA 35K- 2MF 250 20

Figure 7-1-20.

accomplished by utilizing new technologies to access the weather channel capabilities of the existing ASR-9 radar located on or near the airport, thus eliminating the requirements for a separate radar location, land acquisition, support facilities and the associated communication landlines and expenses.

(b) The WSP utilizes the same RBDT display as the TDWR and LLWAS, and, just like the TDWR, also has a GSD for planning purposes by supervisors, traffic management specialists and controllers. The WSP GSD emulates the TDWR display, i.e., it also depicts 6 levels of precipitation, gust fronts and predicted storm movement, and like the TDWR GSD, is used to plan for runway changes and arrival/departure route changes in order to reduce aircraft delays and to increase airport capacity.

(c) This system is currently under development and is operating in a developmental test status at the Albuquerque, New Mexico, airport. When fielded, the WSP is expected to be installed at 34 airports across the nation, substantially increasing the safety of the American flying public.

5. Operational aspects of LLWAS, TDWR and WSP

To demonstrate how this data is used by both the controller and the pilot, 3 ribbon display examples and their explanations are presented:

(a) MICROBURST ALERTS

EXAMPLE

THIS IS WHAT THE CONTROLLER SEES ON HIS/HER RIBBON DISPLAY IN THE TOWER CAB.

27A MBA 35K–2MF 250 20

NOTE—

(SEE FIG 7-1-20 TO SEE HOW THE TDWR/WSP DETERMINES THE MICROBURST LOCATION).

This is what the controller will say when issuing the alert.

PHRASEOLOGY—

RUNWAY 27 ARRIVAL, MICROBURST ALERT, 35 KT LOSS 2 MILE FINAL, THRESHOLD WIND 250 AT 20.

In plain language, the controller is telling the pilot that on approach to runway 27, there is a microburst alert on the approach lane to the runway, and to anticipate or expect a 35 knot loss of airspeed at approximately 2 miles out on final approach (where it will first encounter the phenomena). With that information, the aircrew is forewarned, and should be prepared to apply wind shear/ microburst escape procedures should they decide to continue the approach. Additionally, the surface winds at the airport for landing runway 27 are reported as 250 degrees at 20 knots.

NOTE—

THRESHOLD WIND IS AT PILOT'S REQUEST OR AS DEEMED APPROPRIATE BY THE CONTROLLER.

REFERENCE—

FAA ORDER 7110.65, AIR TRAFFIC CONTROL, PARAGRAPH 3-1-8b2(a).

(b) WIND SHEAR ALERTS

EXAMPLE

THIS IS WHAT THE CONTROLLER SEES ON HIS/HER RIBBON DISPLAY IN THE TOWER CAB.

WEAK MICROBURST ALERT

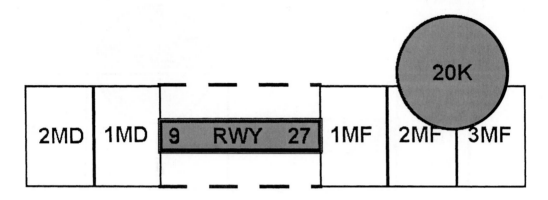

Figure 7-1-21.

27A WSA 20K–3MF 200 15

NOTE—

(SEE FIG 7-1-21 TO SEE HOW THE TDWR/WSP DETERMINES THE WIND SHEAR LOCATION).

This is what the controller will say when issuing the alert.

PHRASEOLOGY—

RUNWAY 27 ARRIVAL, WIND SHEAR ALERT, 20 KT LOSS 3 MILE FINAL, THRESHOLD WIND 200 AT 15.

In plain language, the controller is advising the aircraft arriving on runway 27 that at about 3 miles out they can expect to encounter a wind shear condition that will decrease their airspeed by 20 knots and possibly encounter turbulence. Additionally, the airport surface winds for landing runway 27 are reported as 200 degrees at 15 knots.

NOTE—

THRESHOLD WIND IS AT PILOT'S REQUEST OR AS DEEMED APPROPRIATE BY THE CONTROLLER.

REFERENCE—

FAA ORDER 7110.65, AIR TRAFFIC CONTROL, LOW LEVEL WIND SHEAR/MICROBURST ADVISORIES, PARAGRAPH 3-1-8b2(a).

(c) **MULTIPLE WIND SHEAR ALERTS**

EXAMPLE

THIS IS WHAT THE CONTROLLER SEES ON HIS/HER RIBBON DISPLAY IN THE TOWER CAB.

27A WSA 20K+ RWY 250 20

27D WSA 20K+ RWY 250 20

NOTE—

(SEE FIG 7-1-22 TO SEE HOW THE TDWR/WSP DETERMINES THE GUST FRONT/WIND SHEAR LOCATION).

This is what the controller will say when issuing the alert.

PHRASEOLOGY—

MULTIPLE WIND SHEAR ALERTS.
RUNWAY 27 ARRIVAL, WIND SHEAR ALERT, 20 KT GAIN ON RUNWAY; RUNWAY 27 DEPARTURE, WIND SHEAR ALERT, 20 KT GAIN ON RUNWAY, WINDS 250 AT 20.

EXAMPLE—

IN THIS EXAMPLE, THE CONTROLLER IS ADVISING ARRIVING AND DEPARTING AIRCRAFT THAT THEY COULD ENCOUNTER A WIND SHEAR CONDITION RIGHT ON THE RUNWAY DUE TO A GUST FRONT (SIGNIFICANT CHANGE OF WIND DIRECTION) WITH THE POSSIBILITY OF A 20 KNOT GAIN IN AIRSPEED ASSOCIATED WITH THE GUST FRONT. ADDITIONALLY, THE AIRPORT SURFACE WINDS (FOR THE RUNWAY IN USE) ARE REPORTED AS 250 DEGREES AT 20 KNOTS.

REFERENCE—

FAA ORDER 7110.65, AIR TRAFFIC CONTROL, PARAGRAPH 3-1-8b2(d).

6. The Terminal Weather Information For Pilots System (TWIP)

(a) With the increase in the quantity and quality of terminal weather information available through TDWR, the next step is to provide this information directly to pilots rather

GUST FRONT ALERT

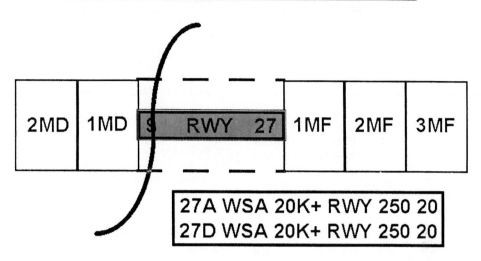

Figure 7-1-22.

than relying on voice communications from ATC. The National Airspace System has long been in need of a means of delivering terminal weather information to the cockpit more efficiently in terms of both speed and accuracy to enhance pilot awareness of weather hazards and reduce air traffic controller workload. With the TWIP capability, terminal weather information, both alphanumerically and graphically, is now available directly to the cockpit on a test basis at 9 locations.

(b) TWIP products are generated using weather data from the TDWR or the Integrated Terminal Weather System (ITWS) testbed. TWIP products are generated and stored in the form of text and character graphic messages. Software has been developed to allow TDWR or ITWS to format the data and send the TWIP products to a database resident at Aeronautical Radio, Inc. (ARINC). These products can then be accessed by pilots using the ARINC Aircraft Communications Addressing and Reporting System (ACARS) data link services. Airline dispatchers can also access this database and send messages to specific aircraft whenever wind shear activity begins or ends at an airport.

(c) TWIP products include descriptions and character graphics of microburst alerts, wind shear alerts, significant precipitation, convective activity within 30 NM surrounding the terminal area, and expected weather that will impact airport operations. During inclement weather, i.e., whenever a pre-determined level of precipitation or wind shear is detected within 15 miles of the terminal area,

TWIP products are updated once each minute for text messages and once every five minutes for character graphic messages. During good weather (below the pre-determined precipitation or wind shear parameters) each message is updated every 10 minutes. These products are intended to improve the situational awareness of the pilot/flight crew, and to aid in flight planning prior to arriving or departing the terminal area. It is important to understand that, in the context of TWIP, the pre-determined levels for inclement versus good weather has nothing to do with the criteria for VFR/MVFR/IFR/LIFR; it only deals with precipitation, wind shears and microbursts.

7-1-28. PIREPS RELATING TO VOLCANIC ASH ACTIVITY

a. Volcanic eruptions which send ash into the upper atmosphere occur somewhere around the world several times each year. Flying into a volcanic ash cloud can be extremely dangerous. At least two B747's have lost all power in all four engines after such an encounter. Regardless of the type aircraft, some damage is almost certain to ensue after an encounter with a volcanic ash cloud.

b. While some volcanoes in the United States are monitored, many in remote areas are not. These unmonitored volcanoes may erupt without prior warning to the aviation community. A pilot observing a volcanic eruption who has not had previous notification of it may be the only witness to the eruption. Pilots are strongly encouraged to transmit a PIREP regarding volcanic eruptions and any observed volcanic ash clouds.

c. Pilots should submit PIREPs regarding volcanic activity using the Volcanic Activity Reporting Form (VAR) as illustrated in Appendix 2. If a VAR form is not immediately available, relay enough information to identify the position and type of volcanic activity.

d. Pilots should verbally transmit the data required in items 1 through 8 of the VAR as soon as possible. The data required in items 9 through 16 of the VAR should be relayed after landing if possible.

7-1-29. THUNDERSTORMS

a. Turbulence, hail, rain, snow, lightning, sustained updrafts and downdrafts, icing conditions—all are present in thunderstorms. While there is some evidence that maximum turbulence exists at the middle level of a thunderstorm, recent studies show little variation of turbulence intensity with altitude.

b. There is no useful correlation between the external visual appearance of thunderstorms and the severity or amount of turbulence or hail within them. The visible thunderstorm cloud is only a portion of a turbulent system whose updrafts and downdrafts often extend far beyond the visible storm cloud. Severe turbulence can be expected up to 20 miles from severe thunderstorms. This distance decreases to about 10 miles in less severe storms.

c. Weather radar, airborne or ground based, will normally reflect the areas of moderate to heavy precipitation (radar does not detect turbulence). The frequency and severity of turbulence generally increases with the radar reflectivity which is closely associated with the areas of highest liquid water content of the storm. NO FLIGHT PATH THROUGH AN AREA OF STRONG OR VERY STRONG RADAR ECHOES SEPARATED BY 20–30 MILES OR LESS MAY BE CONSIDERED FREE OF SEVERE TURBULENCE.

d. Turbulence beneath a thunderstorm should not be minimized. This is especially true when the relative humidity is low in any layer between the surface and 15,000 feet. Then the lower altitudes may be characterized by strong out flowing winds and severe turbulence.

e. The probability of lightning strikes occurring to aircraft is greatest when operating at altitudes where temperatures are between minus 5 degrees Celsius and plus 5 degrees Celsius. Lightning can strike aircraft flying in the clear in the vicinity of a thunderstorm.

f. METAR reports do not include a descriptor for severe thunderstorms. However, by understanding severe thunderstorm criteria, i.e., 50 knot winds or ¾" hail, the information is available in the report to know that one is occurring.

g. NWS radar systems are able to objectively determine radar weather echo intensity levels by use of Video Integrator Processor (VIP) equipment. These thunderstorm intensity levels are on a scale of one to six.

REFERENCE—PILOT/CONTROLLER GLOSSARY, RADAR WEATHER ECHO INTENSITY LEVELS.
EXAMPLE—
① ALERT PROVIDED BY AN ATC FACILITY TO AN AIRCRAFT: (AIRCRAFT IDENTIFICATION) LEVEL FIVE INTENSE WEATHER ECHO BETWEEN TEN O'CLOCK AND TWO O'CLOCK, ONE ZERO MILES, MOVING EAST AT TWO ZERO KNOTS, TOPS FLIGHT LEVEL THREE NINE ZERO.
② ALERT PROVIDED BY AN AFSS/FSS: (AIRCRAFT IDENTIFICATION) LEVEL FIVE INTENSE WEATHER ECHO, TWO ZERO MILES WEST OF ATLANTA V-O-R, TWO FIVE MILES WIDE, MOVING EAST AT TWO ZERO KNOTS, TOPS FLIGHT LEVEL THREE NINE ZERO.

7-1-30. THUNDERSTORM FLYING

a. Above all, remember this: never regard any thunderstorm "lightly" even when radar observers report the echoes are of light intensity. Avoiding thunderstorms is the best policy. Following are some Do's and Don'ts of thunderstorm avoidance:

1. Don't land or takeoff in the face of an approaching thunderstorm. A sudden gust front of low level turbulence could cause loss of control.

2. Don't attempt to fly under a thunderstorm even if you can see through to the other side. Turbulence and wind shear under the storm could be disastrous.

3. Don't fly without airborne radar into a cloud mass containing scattered embedded thunderstorms. Scattered thunderstorms not embedded usually can be visually circumnavigated.

4. Don't trust the visual appearance to be a reliable indicator of the turbulence inside a thunderstorm.

5. Do avoid by at least 20 miles any thunderstorm identified as severe or giving an intense radar echo. This is especially true under the anvil of a large cumulonimbus.

6. Do clear the top of a known or suspected severe thunderstorm by at least 1,000 feet altitude for each 10 knots of wind speed at the cloud top. This should exceed the altitude capability of most aircraft.

7. Do circumnavigate the entire area if the area has ⁶⁄₁₀ thunderstorm coverage.

8. Do remember that vivid and frequent lightning indicates the probability of a strong thunderstorm.

9. Do regard as extremely hazardous any thunderstorm with tops 35,000 feet or higher whether the top is visually sighted or determined by radar.

b. If you cannot avoid penetrating a thunderstorm, following are some Do's before entering the storm:

1. Tighten your safety belt, put on your shoulder harness if you have one and secure all loose objects.

2. Plan and hold your course to take you through the storm in a minimum time.

3. To avoid the most critical icing, establish a penetration altitude below the freezing level or above the level of minus 15 degrees Celsius.

4. Verify that pitot heat is on and turn on carburetor heat or jet engine anti-ice. Icing can be rapid at any altitude and cause almost instantaneous power failure and/or loss of airspeed indication.

5. Establish power settings for turbulence penetration airspeed recommended in your aircraft manual.

6. Turn up cockpit lights to highest intensity to lessen temporary blindness from lightning.

7. If using automatic pilot, disengage altitude hold mode and speed hold mode. The automatic altitude and speed controls will increase maneuvers of the aircraft thus increasing structural stress.

8. If using airborne radar, tilt the antenna up and down occasionally. This will permit you to detect other thunderstorm activity at altitudes other than the one being flown.

c. Following are some Do's and Don'ts during the thunderstorm penetration:

1. Do keep your eyes on your instruments. Looking outside the cockpit can increase danger of temporary blindness from lightning.

2. Don't change power settings; maintain settings for the recommended turbulence penetration airspeed.

3. Don't attempt to maintain a constant altitude; let the aircraft "ride the waves."

4. Don't turn back once you are in the thunderstorm. A straight course through the storm most likely will get you out of the hazards most quickly. In addition, turning maneuvers increase stress on the aircraft.

7-1-31. KEY TO AERODROME FORECAST (TAF) AND AVIATION ROUTINE WEATHER REPORT (METAR)

U.S. Department of Transportation
Federal Aviation Administration

KEY to AERODROME FORECAST (TAF) and AVIATION ROUTINE WEATHER REPORT (METAR) (FRONT)

TAF KPIT 091730Z 091818 15005KT 5SM HZ FEW020 WS010/31022KT
    FM 1930 30015G25KT 3SM SHRA OVC015 TEMPO 2022 1/2SM +TSRA
    OVC008CB
    FM0100 27008KT 5SM SHRA BKN020 OVC040 PROB40 0407 1SM -RA BR
    FM1015 18005KT 6SM -SHRA OVC020 BECMG 1315 P6SM NSW SKC

METAR KPIT 091955Z COR 22015G25KT 3/4SM R28L/2600FT TSRA OVC010CB
18/16 A2992 RMK SLP045 T01820159

FORECAST	EXPLANATION	REPORT
TAF	Message type : TAF-routine or TAF AMD-amended forecast, METAR-hourly, SPECI-special or TESTM-non-commissioned ASOS report	METAR
KPIT	ICAO location indicator	KPIT
091730Z	Issuance time: ALL times in UTC "Z", 2-digit date, 4-digit time	091955z
091818	Valid period: 2-digit date, 2-digit beginning, 2-digit ending times	
	in U.S. METAR: CORrected of; or AUTOmated ob for automated report with no human intervention; omitted when observer logs on	COR
15005KT	Wind: 3 digit true-north direction , nearest 10 degrees (or VaRiaBle); next 2-3 digits for speed and unit, KT (KMH or MPS); as needed, Gust and maximum speed; 00000KT for calm; for METAR. if direction varies 60 degrees or more. Variability appended, e.g. 180V260	22015G25KT
5SM	Prevailing visibility; in U.S.. Statute Miles & fractions; above 6 miles in TAF Plus6SM. (Or, 4-digit minimum visibility in meters and as required, lowest value with direction)	3/4SM
	Runway Visual Range: R; 2-digit runway designator Left, Center, or Right as needed: "/", Minus or Plus in U.S., 4-digit value, FeeT in U.S., (usually meters elsewhere); 4-digit value Variability 4-digit value (and tendency Down, Up or No change)	R28L/2600FT
HZ FEW020	Significant present, forecast and recent weather: see table (on back)	TSRA OVC 010CB
	Cloud amount. height and type: Sky Clear 0/8, FEW >0/8-2/8, SCaTtered 3/8-4/8, BroKeN 5/8-7/8, OVerCast 8/8; 3-digit height in hundreds of ft; Towering Cumulus or CumulonimBus in METAR; in TAF. only CB. Vertical Visibility for obscured sky and height "VV004". More than 1 layer may be reported or forecast. In automated METAR reports only, CLeaR for "clear below 12,000 feet"	
	Temperature: degrees Celsius; first 2 digits, temperature "/" last 2 digits. dew-point temperature; Minus for below zero, e.g., M06	18/16
	Altimeter setting: indicator and 4 digits; in U.S., A-inches and hundredths: (Q-hectoPascals, e.g. Q1013)	A2992

Figure 7-1-23.

U.S. Department of Transportation
Federal Aviation Administration

KEY to AERODROME FORECAST (TAF) and AVIATION ROUTINE WEATHER REPORT (METAR) (BACK)

FORECAST	EXPLANATION	REPORT
WS010/31022KT	In U.S. TAF, non-convective low-level (≤ 2,000 ft) Wind Shear; 3-digit height (hundreds of ft); "/"; 3-digit wind direction and 2-3 digit wind speed above the indicated height, and unit, KT	
	In METAR. ReMarK indicator & remarks. For example: Sea-Level Pressure in hectoPascals & tenths, as shown: 1004.5 hPa; Temp/dew-point in tenths °C, as shown: temp. 18.2°C, dew-point 15.9°C	RMK SLP045 T01820159
FM1930	FroM and 2-digit hour and 2-digit minute beginning time: indicates significant change. Each FM starts on a new line, indented 5 spaces	
TEMPO 2022	TEMPOrary: changes expected for <1 hour and in total, < half of 2-digit hour beginning and 2-digit hour ending time period	
PROB40 0407	PROBability and 2-digit percent (30 or 40): probable condition during 2-digit hour beginning and 2-digit hour ending time period	
BECMG 1315	BECoMinG: change expected during 2-digit hour beginning and 2-digit hour ending time period	

Table of Significant Present, Forecast and Recent Weather- Grouped in categories and used in the order listed below; or as needed in TAF, No Significant Weather.

QUALIFIER

INTENSITY OR PROXIMITY

'-' Light   "no sign" Moderate   '+' Heavy

VC Vicinity: but not at aerodrome; in U.S. METAR, between 5 and 10SM of the point(s) of observation; in U.S. TAF, 5 to 10SM from center of runway complex (elsewhere within 8000m)

DESCRIPTOR

MI	Shallow	BC	Patches	PR	Partial	TS	Thunderstorm
BL	Blowing	SH	Showers	DR	Drifting	FZ	Freezing

WEATHER PHENOMENA

PRECIPITATION

DZ	Drizzle	RA	Rain	SN	Snow	SG	Snow grains
IC	Ice Crystals	PL	Ice Pellets	GR	Hail	GS	Small hail/snow pellets
UP	Unknown precipitation in automated observations						

OBSCURATION

BR	Mist (≥5/8SM)	FG	Fog (<5/8SM)	FU	Smoke	VA	Volcanic ash
SA	Sand	HZ	Haze	PY	Spray	DU	Widespread dust

OTHER

SQ	Squall	SS	Sandstorm	DU	Duststorm	PO	Well developed dust/sand whirls
FC	Funnel cloud	+FC	tornado/waterspout				

-Explanations in parentheses "( )" indicate different worldwide practices.
- - Ceiling is not specified; defined as the lowest broken or overcast layer, or the vertical visibility.
- NWS TAFs exclude turbulence, icing & temperature forecasts; NWS METARs exclude trend forecasts

January 1999   Department of Transportation
Aviation Weather Directorate   FEDERAL AVIATION ADMINISTRATION

Figure 7-1-24.

7-1-32. INTERNATIONAL CIVIL AVIATION ORGANIZATION (ICAO) WEATHER FORMATS

The United States uses the ICAO world standard for aviation weather reporting and forecasting. The utilization of terminal forecasts affirms our commitment to a single global format for aviation weather. The World Meteorological Organization's (WMO) publication No. 782 "Aerodrome Reports and Forecasts" contains the base METAR and TAF code as adopted by the WMO member countries.

a. Although the METAR code is adopted worldwide, each country is allowed to make modifications or exceptions to the code for use in their particular country, e.g., the U.S. will continue to use statute miles for visibility, feet for RVR values, knots for wind speed, and inches of mercury for altimetry. However, temperature and dew point will be reported in degrees Celsius. The U.S. will continue reporting prevailing visibility rather than lowest sector visibility. Most of the current U.S. observing procedures and policies will continue after the METAR conversion date, with the information disseminated in the METAR code and format. The elements in the body of a METAR report are separated with a space. The only exceptions are RVR, temperature and dew point, which are separated with a solidus (/). When an element does not occur, or cannot be observed, the preceding space and that element are omitted from that particular report. A METAR report contains the following sequence of elements in the following order:

1. *Type of report*
2. *ICAO Station Identifier*
3. *Date and Time of report*
4. *Modifier (as required)*
5. *Wind*
6. *Visibility*
7. *Runway Visual Range (RVR)*
8. *Weather Phenomena*
9. *Sky conditions*
10. *Temperature/Dew Point Group*
11. *Altimeter*
12. *Remarks (RMK)*

b. The following paragraphs describe the elements in a METAR report.

1. *Type of Report:* There are two types of report:

① Aviation Routine Weather Report (METAR)

② Non-Routine (special) Aviation Weather Report (SPECI).

The type of report (METAR or SPECI) will always appear as the lead element of the report.

2. *ICAO Station Identifier:* The METAR code uses ICAO 4-letter station identifiers. In the contiguous 48 states, the 3-letter domestic station identifier is prefixed with a "K;" i.e., the domestic identifier for Seattle is SEA while the ICAO identifier is KSEA. Elsewhere, the first two letters of the ICAO identifier indicate what region of the world and country (or state) the station is in. For Alaska, all station identifiers start with "PA;" for Hawaii, all station identifiers start with "PH." Canadian station identifiers start with "CU," "CW," "CY" and "CZ." Mexican station identifiers start with "MM." The identifier for the western Caribbean is "M" followed by the individual country's letter; i.e., Cuba is "MU;" Dominican Republic "MD;" the Bahamas "MY." The identifier for the eastern Caribbean is "T" followed by the individual country's letter; i.e., Puerto Rico is "TJ." For a complete worldwide listing see ICAO Document 7910, Location Indicators.

3. *Date and Time of Report.* The date and time the observation is taken are transmitted as a six-digit date/

time group appended with Z to denote Coordinated Universal Time (UTC). The first two digits are the date followed with two digits for hour and two digits for minutes.

EXAMPLE—172345Z (the 17th day of the month at 2345Z)

4. *Modifier (As Required):* "AUTO" identifies a METAR/SPECI report as an automated weather report with no human intervention. If "AUTO" is shown in the body of the report, the type of sensor equipment used at the station will be encoded in the remarks section of the report. The absence of "AUTO" indicates that a report was made manually by an observer or that an automated report had human augmentation/backup. The modifier "COR" indicates a corrected report that is sent out to replace an earlier report with an error.

NOTE—THERE ARE TWO TYPES OF AUTOMATED STATIONS, AO1 FOR AUTOMATED WEATHER REPORTING STATIONS WITHOUT A PRECIPITATION DISCRIMINATOR, AND AO2 FOR AUTOMATED STATIONS WITH A PRECIPITATION DISCRIMINATOR. (A PRECIPITATION DISCRIMINATOR CAN DETERMINE THE DIFFERENCE BETWEEN LIQUID AND FROZEN/FREEZING PRECIPITATION.) THIS INFORMATION APPEARS IN THE REMARKS SECTON OF AN AUTOMATED REPORT.

5. *Wind:* The wind is reported as a five digit group (six digits if speed is over 99 knots). The first three digits are the direction the wind is blowing from, in tens of degrees referenced to true north, or "VRB" if the direction is variable. The next two digits is the wind speed in knots, or if over 99 knots, the next three digits. If the wind is gusty, it is reported as a "G" after the speed followed by the highest gust reported. The abbreviation "KT" is appended to denote the use of knots for wind speed.

EXAMPLE—13008KT-WIND FROM 130 DEGREES AT 8 KNOTS
08032G45KT-WIND FROM 080 DEGREES AT 32 KNOTS WITH GUSTS TO 45 KNOTS
VRB04KT-WIND VARIABLE IN DIRECTION AT 4 KNOTS
00000KT-WIND CALM
210103G130KT-WIND FROM 210 DEGREES AT 103 KNOTS WITH GUSTS TO 130 KNOTS
IF THE WIND DIRECTION IS VARIABLE BY 60 DEGREES OR MORE AND THE SPEED IS GREATER THAN 6 KNOTS, A VARIABLE GROUP CONSISTING OF THE EXTREMES OF THE WIND DIRECTION SEPARATED BY A "V" WILL FOLLOW THE PREVAILING WIND GROUP.
32012G22KT 280V350

(a) Peak Wind: Whenever the peak wind exceeds 25 knots "PK WND" will be included in Remarks, e.g., PK WND 28045/1955 "Peak wind two eight zero at four five occurred at one niner five five." If the hour can be inferred from the report time, only the minutes will be appended, e.g., PK WND 34050/38 "Peak wind three four zero at five zero occurred at three eight past the hour."

(b) Wind shift: Whenever a wind shift occurs, "WSHFT" will be included in remarks followed by the time the wind shift began, e.g., WSHFT 30 FROPA "Wind shift at three zero due to frontal passage."

6. *Visibility:* Prevailing visibility is reported in statute miles with "SM" appended to it.

EXAMPLE—7SM—SEVEN STATUTE MILES
15SM—FIFTEEN STATUTE MILES
1/2 SM—ONE-HALF STATUTE MILE

(a) Tower/surface visibility: If either visibility (tower or surface) is below four statute miles, the lesser of the two will be reported in the body of the report; the greater will be reported in remarks.

(b) Automated visibility: ASOS visibility stations will show visibility ten or greater than ten miles as "10SM." AWOS visibility stations will show visibility less than ¼ statute mile as "M¼SM" and visibility ten or greater than ten miles as "10SM."

(c) Variable visibility: Variable visibility is shown in remarks (when rapid increase or decrease by ½ statute mile or more and the average prevailing visibility is less than three miles) e.g., VIS 1V2 "visibility variable between one and two."

(d) Sector visibility: Sector visibility is shown in remarks when it differs from the prevailing visibility, and either the prevailing or sector visibility is less than three miles.

EXAMPLE—VIS N2—VISIBILITY NORTH TWO

7. *Runway Visual Range (When Reported):* "R" identifies the group followed by the runway heading (and parallel runway designator, if needed) "/" and the visual range in feet (meters in other countries) followed with "FT" (feet is not spoken).

(a) Variability Values: When RVR varies (by more than one reportable value), the lowest and highest values are shown with "V" between them.

(b) Maximum/Minimum Range: P indicates an observed RVR is above the maximum value for this system (spoken as "more than"). "M" indicates an observed RVR is below the minimum value which can be determined by the system (spoken as "less than").

EXAMPLE—R32L/1200FT—RUNWAY THREE TWO LEFT R-V-R ONE THOUSAND TWO HUNDRED

R27R/M1000V4000FT—RUNWAY TWO SEVEN RIGHT R-V-R VARIABLE FROM LESS THAN ONE THOUSAND TO FOUR THOUSAND

8. *Weather Phenomena:* The weather as reported in the METAR code represents a significant change in the way weather is currently reported. In METAR, weather is reported in the format:

Intensity / Proximity / Descriptor / Precipitation / Obstruction to visibility / Other

NOTE—THE "/" ABOVE AND IN THE FOLLOWING DESCRIPTIONS (EXCEPT AS THE SEPARATOR BETWEEN THE TEMPERATURE AND DEW POINT) ARE FOR SEPARATION PURPOSES IN THIS PUBLICATION AND DO NOT APPEAR IN THE ACTUAL METARS.

(a) Intensity: applies only to the first type of precipitation reported. A "–" denotes light, no symbol denotes moderate, and a "+" denotes heavy.

(b) Proximity: applies to and reported only for weather occurring in the vicinity of the airport (between 5 and 10 miles of the point(s) of observation). It is denoted by the letters "VC." (Intensity and "VC" will not appear together in the weather group.)

(c) Descriptor: these eight descriptors apply to the precipitation or obstructions to visibility:

TS ....thunderstorm
DR ....low drifting
SH ....showers
MI ....shallow
FZ ....freezing
BC ....patches
BL ....blowing
PR ....partial

NOTE—ALTHOUGH "TS" AND "SH" ARE USED WITH PRECIPITATION AND MAY BE PRECEDED WITH AN INTENSITY SYMBOL, THE INTENSITY STILL APPLIES TO THE PRECIPITATION *NOT* THE DESCRIPTOR.

(d) Precipitation: there are nine types of precipitation in the METAR code:

RA ....rain
DZ ....drizzle
SN ....snow
GR ....hail (¼" or greater)
GS ....small hail/snow pellets
PL ....ice pellets
SG ....snow grains
IC ....ice crystals (diamond dust)
UP ....unknown precipitation
(automated stations only)

(e) Obstructions to visibility: there are eight types of obscuration phenomena in the METAR code (obscurations are any phenomena in the atmosphere, other than precipitation, that reduce horizontal visibility):

FG ....fog (vsby less than ⅝ mile)
HZ ....haze
FU ....smoke
PY ....spray
BR ....mist (vsby ⅝–6 miles)
SA ....sand
DU ....dust
VA ....volcanic ash

NOTE—FOG (FG) IS OBSERVED OR FORECAST ONLY WHEN THE VISIBILITY IS LESS THAN FIVE-EIGHTHS OF MILE, OTHERWISE MIST (BR) IS OBSERVED OR FORECAST.

(f) Other: there are five categories of other weather phenomena which are reported when they occur:

SQ .... squall
SS .... sandstorm
DS .... dust storm
PO .... dust/sand whirls
FC .... funnel cloud
+FC .... tornado / waterspout

Examples:

TSRA .... thunderstorm with moderate rain

+SN .... heavy snow

–RA FG .... light rain and fog

BRHZ .... mist and haze (visibility ⅝ mile or greater)

FZDZ .... freezing drizzle

VCSH .... rain shower in the vicinity

+SHRASNPl.............heavy rain showers, snow, ice pellets (intensity indicator refers to the predominant rain)

9. *Sky Condition:* The sky condition as reported in METAR represents a significant change from the way sky condition is currently reported. In METAR, sky condition is reported in the format:

Amount/Height/(Type) or Indefinite Ceiling/Height

(a) Amount: the amount of sky cover is reported in eighths of sky cover, using the contractions:

SKC .... clear (no clouds)

FEW .... 0 to ⅝

SCT .... scattered (⅛ to ½'s of clouds)

BKN .... broken (⅝'s to ⅞'s of clouds)

OVC .... overcast (⅞'s clouds)

CB .... Cumulonimbus when present

TCU .... Towering Cumulus when present

NOTE—

⛛ "SKC" WILL BE REPORTED AT MANUAL STATIONS. "CLR" WILL BE USED AT AUTOMATED STATIONS WHEN NO CLOUDS BELOW 12,000 FEET ARE REPORTED.

⛛ A CEILING LAYER IS NOT DESIGNATED IN THE METAR CODE. FOR AVIATION PURPOSES, THE CEILING IS THE LOWEST BROKEN OR OVERCAST LAYER, OR VERTICAL VISIBILITY INTO AN OBSCURATION. ALSO THERE IS NO PROVISION FOR REPORTING THIN LAYERS IN THE METAR CODE. WHEN CLOUDS ARE THIN, THAT LAYER SHALL BE REPORTED AS IF IT WERE OPAQUE.

(b) Height: cloud bases are reported with three digits in hundreds of feet. (Clouds above 12,000 feet cannot be reported by an automated station.)

(c) (Type): if towering cumulus clouds (TCU) or cumulonimbus clouds (CB) are present, they are reported after the height which represents their base.

EXAMPLE—(REPORTED AS) SCT025TCU BKN080 BKN250 (SPOKEN AS) "TWO THOUSAND FIVE HUNDRED SCATTERED TOWERING CUMULUS, CEILING EIGHT THOUSAND BROKEN, TWO FIVE THOUSAND BROKEN."

(REPORTED AS) SCT008 OVC012CB (SPOKEN AS) "EIGHT HUNDRED SCATTERED CEILING ONE THOUSAND TWO HUNDRED OVERCAST CUMULONIMBUS CLOUDS."

(d) Vertical Visibility (Indefinite ceiling height): The height into an indefinite ceiling is preceded by "VV" and followed by three digits indicating the vertical visibility in hundreds of feet. This layer indicates <u>total obscuration</u>.

EXAMPLE—1/8 SM FG VV006—VISIBILITY ONE EIGHTH, FOG, INDEFINITE CEILING SIX HUNDRED.

(e) Obscurations: are reported when the sky is <u>partially obscured</u> by a ground-based phenomena by indicating the amount of obscuration as FEW, SCT, BKN followed by

three zeros (000). In remarks, the obscuring phenomenon precedes the amount of obscuration and three zeros.

EXAMPLE—BKN000 (in body)....."SKY PARTIALLY OBSCURED"

FU BKN000 (IN REMARKS)....."SMOKE OBSCURING FIVE-TO SEVEN-EIGHTHS OF THE SKY"

(f) When sky conditions include a layer aloft, other than clouds, such as smoke or haze the type of phenomena, sky cover and height are shown in remarks.

EXAMPLE—BKN020 (IN BODY)....."CEILING TWO THOUSAND BROKEN"

RMK FU BKN020....."BROKEN LAYER OF SMOKE ALOFT, BASED AT TWO THOUSAND"

(g) Variable ceiling. When a ceiling is below three thousand and is variable, the remark "CIG" will be shown followed with the lowest and highest ceiling heights separated by a "V."

EXAMPLE—CIG 005V010....."CEILING VARIABLE BETWEEN FIVE HUNDRED AND ONE THOUSAND"

(h) Second site sensor. When an automated station uses meteorological discontinuity sensors, remarks will be shown to identify site specific sky conditions which differ and are lower than conditions reported in the body.

EXAMPLE—CIG 020 RY11....."CEILING TWO THOUSAND AT RUNWAY ONE ONE"

(i) Variable cloud layer. When a layer is varying in sky cover, remarks will show the variability range. If there is more than one cloud layer, the variable layer will be identified by including the layer height.

EXAMPLE—SCT V BKN....."SCATTERED LAYER VARIABLE TO BROKEN"

BKN025 V OVC....."BROKEN LAYER AT TWO THOUSAND FIVE HUNDRED VARIABLE TO OVERCAST"

(j) Significant clouds. When significant clouds are observed, they are shown in remarks, along with the specified information as shown below:

(1) Cumulonimbus (CB), or Cumulonimbus Mammatus (CBMAM), distance (if known), direction from the station, and direction of movement, if known. If the clouds are beyond 10 miles from the airport, DSNT will indicate distance.

EXAMPLE—CB W MOV E....."CUMULONIMBUS WEST MOVING EAST"

CBMAM DSNT S....."CUMULONIMBUS MAMMATUS DISTANT SOUTH"

(2) Towering Cumulus (TCU), location, (if known), or direction from the station.

EXAMPLE—TCU OHD....."TOWERING CUMULUS OVERHEAD"

TCU W....."TOWERING CUMULUS WEST"

(3) Altocumulus Castellanus (ACC), Stratocumulus Standing Lenticular (SCSL), Altocumulus Standing Lenticular (ACSL), Cirrocumulus Standing Lenticular (CCSL) or rotor clouds, describing the clouds (if needed) and the direction from the station.

EXAMPLE—ACC W....."ALTOCUMULUS CASTELLANUS WEST"

ACSL SW-S....."STANDING LENTICULAR ALTOCUMULUS SOUTHWEST THROUGH SOUTH"

APRNT ROTOR CLD S....."APPARENT ROTOR CLOUD SOUTH"

CCSL OVR MT E....."STANDING LENTICULAR CIRROCUMULUS OVER THE MOUNTAINS EAST

10. *Temperature/Dew Point:* Temperature and dew point are reported in two, two-digit groups in degrees Celsius, separated by a solidus ("/"). Temperatures below zero are prefixed with an "M." If the temperature is available but the dew point is missing, the temperature is shown followed by a solidus. If the temperature is missing, the group is omitted from the report.

EXAMPLE—15/08....."TEMPERATURE ONE FIVE, DEW POINT 8"

00/M02....."TEMPERATURE ZERO, DEW POINT MINUS 2"

M05/....."TEMPERATURE MINUS FIVE, DEW POINT MISSING"

11. *Altimeter:* Altimeter settings are reported in a four-digit format in inches of mercury prefixed with an "A" to denote the units of pressure.

EXAMPLE—A2995....."ALTIMETER TWO NINER NINER FIVE"

12. *Remarks:* Remarks will be included in all observations, when appropriate. The contraction "RMK" denotes the start of the remarks section of a METAR report.

Except for precipitation, phenomena located within 5 statute miles of the point of observation will be reported as at the station. Phenomena between 5 and 10 statute miles will be reported in the vicinity, "VC." Phenomena not occurring at the point of observation but within 10 statute miles will be shown as distant, "DSNT." Distances are in statute miles except for automated lightning remarks which are in nautical miles. Movement of clouds or weather will be indicated by the direction toward which the phenomena is moving.

a. There are two categories of remarks:
1. Automated, manual, and plain language.
2. Additive and automated maintenance data.

b. Automated, Manual, and Plain Language.

This group of remarks may be generated from either manual or automated weather reporting stations and generally elaborate on parameters reported in the body of the report. (Plain language remarks are only provided by manual stations.)

1. Volcanic eruptions
2. Tornado, Funnel Cloud, Waterspout
3. Station Type (AO1 or AO2)
4. PK WND
5. WSHFT (FROPA)
6. TWR VIS or SFC VIS
7. VRB VIS
8. Sector VIS

9. VIS @ 2nd Site
10. (freq) LTG (type) (loc)
11. Beginning/Ending of Precipitation/TSTMS
12. TSTM Location MVMT
13. Hailstone Size (GR)
14. Virga
15. VRB CIG (height)
16. Obscuration
17. VRB Sky Condition
18. Significant Cloud Types
19. Ceiling Height 2nd Location
20. PRESFR PRESRR
21. Sea-Level Pressure
22. ACFT Mishap (not transmitted)
23. NOSPECI
24. SNINCR
25. Other SIG Info

c. Additive And Automated Maintenance Data.

1. Hourly Precipitation
2. Precipitation Amount
3. 3- and 6-Hour Precipitation Amount
4. Snow Depth on Ground
5. Water Equivalent of Snow
6. Cloud Type
7. Duration of Sunshine
8. Hourly Temperature/Dew Point (Tenths)
9. 6-Hour Maximum Temperature
10. 6-Hour Minimum Temperature
11. 24-Hour Maximum/Minimum Temperature
12. Pressure Tendency
13. Sensor Status:
    PWINO
    FZRANO
    TSNO
    RVRNO
    PNO
    VISNO

Examples of METAR reports and explanation:

METAR KBNA 281250Z 33018KT 290V360 1/2SM R31/2700FT SN BLSN FG VV008 00/M03 A2991 RMK RAE42SNB42

METAR .... aviation routine weather report
KBNA .... Nashville, TN
281250Z ..... Date 28th, time 1250 UTC
(no modifier) This is a manually generated report, due to the absence of "AUTO" and " AO1 or AO2" in remarks
33018KT ..... wind three three zero at one eight
290V360 ..... wind variable between two nine zero and three six zero
1/2SM .... visibility one half
R31/2700FT... Runway three one RVR two thousand seven hundred
SN ....moderate snow
BLSN FG ..... visibility obscured by blowing snow and fog

VV008 .... indefinite ceiling eight hundred
00/M03 ...... temperature zero, dew point minus three
A2991 .... altimeter two niner niner one
RMK .... remarks
RAE42 .... rain ended at four two
SNB42 .... snow began at four two

METAR KSFO 041453Z AUTO VRB02KT 3SM IBR CLR 15/12 A3012 RMK AO2

METAR ...... aviation routine weather report
KSFO .... San Francisco, CA
041453Z .... Date 4th, time 1453 UTC
AUTO .... fully automated; no human intervention
VRB02KT .... wind variable at two
3SM .... visibility three
BR .... visibility obscured by mist
CLR .... no clouds below one two thousand
15/12 ...... temperature one five, dew point one two
A3012 ...... altimeter three zero one two
RMK .... remarks
AO2 .... this automated station has a weather discriminator (for precipitation)

SPECI KCVG 152224Z 28024G36KT 3/4SM +TSRA BKN008 OVC020CB 28/23 A3000 RMK TSRAB24 TS W MOV E

SPECI .... (non-routine) aviation special weather report
KCVG .... Cincinnati, OH
152228Z ...... Date 15th, time 2228 UTC
(no modifier) This is a manually generated report due to the absence of "AUTO" and "AO1" or AO2" in remarks
28024G36KT wind two eight zero at two four gusts three six
3/4SM .... visibility three fourths
+TSRA .... thunderstorms, heavy rain
BKN008 .... ceiling eight hundred broken
OVC020CB ..... two thousand overcast cumulonimbus clouds
28/23 .... temperature two eight, dew point two three
A3000 .... altimeter three zero zero zero
RMK .... remarks
TSRAB24 .... thunderstorm began at two four
TS W MOV E .... thunderstorm west moving east

Aerodrome Forecast (TAF): a concise statement of the expected meteorological conditions at an airport during a specified period (usually 24 hours). TAFs use the same codes as METAR weather reports. They are scheduled four times daily for 24-hour periods beginning at 0000Z, 0600Z, 1200Z, and 1800Z. TAFs are issued in the following format:

TYPE OF REPORT / ICAO STATION IDENTIFIER / DATE AND TIME OF ORIGIN/ VALID PERIOD DATE AND TIME / FORECAST METEOROLOGICAL CONDITIONS

NOTE—THE "/" ABOVE AND IN THE FOLLOWING DESCRIPTIONS ARE FOR SEPARATION PURPOSES IN THIS PUBLICATION AND DO NOT APPEAR IN THE ACTUAL TAFS.

TAF
 KOKC 051130Z 051212 14008KT 5SM BR BKN030
TEMPO 1316 1 1/2SM BR
 FM1600 16010KT P6SM NSW SKC
 FM2300 20013G20KT 4SM SHRA OVC020
PROB40 0006 2SM TSRA OVC008CB BECMG 0608
21015KT P6SM NSW SCT040

TAF format observed in the above example:
TAF = TYPE OF REPORT
KOKC = ICAO STATION IDENTIFIER
051130Z = DATE AND TIME OF ORIGIN
051212 = VALID PERIOD DATE AND TIMES
 14008KT 5SM BR BKN030 = FORECAST METEOROLOGICAL CONDITIONS

Explanation of TAF elements:
1. *Type of report:* There are two types of TAF issuances, a routine forecast issuance (TAF) and an amended forecast (TAF AMD). An amended TAF is issued when the current TAF no longer adequately describes the ongoing weather or the forecaster feels the TAF is not representative of the current or expected weather. Corrected (COR) or delayed (RTD) TAFs are identified only in the communications header which precedes the actual forecasts.

2. *ICAO Station Identifier:* The TAF code uses ICAO 4-letter location identifiers as described in the METAR section.

3. *Date and Time of Origin:* This element is the date and time the forecast is actually prepared. The format is a two-digit date and four-digit time followed, without a space, by the letter " Z."

4. *Valid Period Date and Time:* The UTC valid period of the forecast is a two-digit date followed by the two-digit beginning hour and two-digit ending hour. In the case of an amended forecast, or a forecast which is corrected or delayed, the valid period may be for less than 24 hours. Where an airport or terminal operates on a part-time basis (less than 24 hours/day), the TAFs issued for those locations will have the abbreviated statement "NIL AMD SKED AFT (closing time) Z" added to the end of the forecasts. For the TAFs issued while these locations are closed, the word "NIL" will appear in place of the forecast text. A delayed (RTD) forecast will then be issued for these locations after two complete observations are received.

5. *Forecast Meteorological Conditions:* This is the body of the TAF. The basic format is:

WIND / VISIBILITY / WEATHER / SKY CONDITION / OPTIONAL DATA (WIND SHEAR)

The wind, visibility, and sky condition elements are always included in the initial tune group of the forecast.

Weather is included only if significant to aviation. If a significant, lasting change in any of the elements is expected during the valid period, a new time period with the changes is included. It should be noted that with the exception of a "FM" group the new time period will include only those elements which are expected to change, i.e., if a lowering of the visibility is expected but the wind is expected to remain the same, the new time period reflecting the lower visibility would not include a forecast wind. The forecast wind would remain the same as in the previous time period.

Any temporary conditions expected during a specific time period are included with that time period. The following describes the elements in the above format.

(a) Wind: This five (or six) digit group includes the expected wind direction (first 3 digits) and speed (last 2 digits or 3 digits if 100 knots or greater). The contraction "KT" follows to denote the units of wind speed. Wind gusts are noted by the letter "G" appended to the wind speed followed by the highest expected gust.

A variable wind direction is noted by "VRB" where the three digit direction usually appears. A calm wind (3 knots or less) is forecast as "00000KT."

EXAMPLE—18010KT—WIND ONE EIGHT ZERO AT ONE ZERO (WIND IS BLOWING FROM 180).

35012G20KT—WIND THREE FIVE ZERO AT ONE TWO GUST TWO ZERO

(b) Visibility: The expected prevailing visibility up to and including 6 miles is forecast in statute miles, including fractions of miles, followed by "SM" to note the units of measure. Expected visibilities greater than 6 miles are forecast as P6SM (Plus six statute miles).

EXAMPLE—1/2 SM—VISIBILITY ONE-HALF

4SM—VISIBILITY FOUR

P6SM—VISIBILITY MORE THAN SIX

(c) Weather Phenomena: The expected weather phenomena is coded in TAF reports using the same format, qualifiers, and phenomena contractions as METAR reports (except UP).

Obscurations to vision will be forecast whenever the prevailing visibility is forecast to be 6 statute miles or less.

If no significant weather is expected to occur during a specific time period in the forecast, the weather phenomena group is omitted for that time period. If, after a time period in which significant weather phenomena has been forecast, a change to a forecast of no significant weather phenomena occurs, the contraction NSW (No Significant Weather) will appear as the weather group in the new time period. (NSW is included only in BECMG or TEMPO groups).

NOTE—IT IS VERY IMPORTANT THAT PILOTS UNDERSTAND THAT NSW *ONLY* REFERS TO WEATHER PHENOMENA, I.E., RAIN, SNOW, DRIZZLE, ETC. OMITTED CONDITIONS, SUCH AS SKY CONDITIONS, VISIBILITY, WINDS, ETC., ARE CARRIED OVER FROM THE PREVIOUS TIME GROUP.

(d) Sky Condition: TAF sky condition forecasts use the METAR format described in the METAR section. Cumulonimbus clouds (CB) are the only cloud type forecast in TAFs.

When clear skies are forecast, the contraction "SKC" will always be used. The contraction "CLR" is never used in the aerodrome forecast (TAF).

When the sky is obscured due to a surface-based phenomenon, vertical visibility (VV) into the obscuration is forecast. The format for vertical visibility is "VV" followed by a three-digit height in hundreds of feet.

NOTE—AS IN METAR, CEILING LAYERS ARE NOT DESIGNATED IN THE TAF CODE. FOR AVIATION PURPOSES, THE CEILING IS THE LOWEST BROKEN OR OVERCAST LAYER OR VERTICAL VISIBILITY INTO A COMPLETE OBSCURATION.

SKC.....'"SKY CLEAR"
SCT005 BKN025CB.....'"FIVE HUNDRED SCATTERED, CEILING TWO THOUSAND FIVE HUNDRED BROKEN CUMULONIMBUS CLOUDS"
VV008.....'"INDEFINITE CEILING EIGHT HUNDRED"

(e) Optional Data (Wind Shear): Wind Shear is the forecast of non-convective low level winds (up to 2,000 feet). The forecast includes the letters "WS" followed by the height of the wind shear, the wind direction and wind speed at the indicated height and the ending letters "KT" (knots). Height is given in hundreds of feet (AGL) up to and including 2,000 feet. Wind shear is encoded with the contraction "WS," followed by a three-digit height, slant character "/," and winds at the height indicated in the same format as surface winds. Then wind shear element is omitted if not expected to occur.

WS010/18040KT—"LOW LEVEL WIND SHEAR AT ONE THOUSAND, WIND ONE EIGHT ZERO AT FOUR ZERO"

d. Probability Forecast: The probability or chance of thunderstorms or other precipitation events occurring, along with associated weather conditions (wind, visibility, and sky conditions).

The PROB30 group is used when the occurrence of thunderstorms or precipitation is 30–39% and the PROB40 group is used when the occurrence of thunderstorms or precipitation is 40–49%. This is followed by a four-digit group giving the beginning hour and ending hour of the time period during which the thunderstorms or precipitation are expected.

NOTE—NEITHER PROB30 NOR PROB40 WILL BE SHOWN DURING THE FIRST SIX HOURS OF A FORECAST.

EXAMPLE—

PROB40 2102 1/2SM +TSRA—"CHANCE BETWEEN 2100Z AND 0200Z OF VISIBILITY ONE-HALF STATUTE MILE IN THUNDERSTORMS AND HEAVY RAIN"

PROB30 1014 1SM RASN—"CHANCE BETWEEN 1000Z AND 1400Z OF VISIBILITY ONE STATUTE MILE IN MIXED RAIN AND SNOW."

e. Forecast Change Indicators: The following change indicators are used when either a rapid, gradual, or temporary change is expected in some or all of the forecast meteorological conditions. Each change indicator marks a time group within the TAF report.

1. From (FM) group: The FM group is used when a rapid change, usually occurring in less than one hour, in prevailing conditions is expected. Typically, a rapid change of prevailing conditions to more or less a completely new set of prevailing conditions is associated with a synoptic feature passing through the terminal area (cold or warm frontal passage). Appended to the "FM" indicator is the four-digit hour and minute the change is expected to begin and continues until the next change group or until the end of the current forecast.

A "FM" group will mark the beginning of a new line in a TAF report (indented 5 spaces). Each "FM" group contains all the required elements—wind, visibility, weather, and sky condition. Weather will be omitted in "FM" groups when it is not significant to aviation. FM groups will not include the contraction NSW.

EXAMPLE—FM0100 14010KT P6SM SKC—"AFTER 0100Z, WIND ONE FOUR ZERO AT ONE ZERO, VISIBILITY MORE THAN SIX, SKY CLEAR"

2. Becoming (BECMG) group: The BECMG group is used when a gradual change in conditions is expected over a longer time period, usually two hours. The time period when the change is expected is a four-digit group with the beginning hour and ending hour of the change period which follows the BECMG indicator. The gradual change will occur at an unspecified time within this time period. Only the changing forecast meteorological conditions are included in BECMG groups. The omitted conditions are carried over from the previous time group.

EXAMPLE—OVC012 BECMG 1416 BKN020—"CEILING ONE THOUSAND TWO HUNDRED OVERCAST. THEN A GRADUAL CHANGE TO CEILING TWO THOUSAND BROKEN BETWEEN 1400Z AND 1600Z."

3. Temporary (TEMPO) group: The TEMPO group is used for any conditions in wind, visibility, weather, or sky condition which are expected to last for generally less than an hour at a time (occasional), and are expected to occur during less than half the time period. The TEMPO indicator is followed by a four-digit group giving the beginning hour and ending hour of the time period during which the temporary conditions are expected. Only the changing forecast meteorological conditions are included in TEMPO groups. The omitted conditions are carried over from the previous time group.

EXAMPLE—

① SCT030 TEMPO 1923 BKN030—"THREE THOUSAND SCATTERED WITH OCCASIONAL CEILINGS THREE THOUSAND BROKEN BETWEEN 1900Z AND 2300Z"

② 4SM HZ TEMPO 0006 2SM BR HZ—"VISIBILITY FOUR IN HAZE WITH OCCASIONAL VISIBILITY TWO IN MIST AND HAZE BETWEEN 0000Z AND 0600Z"

Section 2. ALTIMETER SETTING PROCEDURES

7-2-1. GENERAL

a. The accuracy of aircraft altimeters is subject to the following factors:

1. Nonstandard temperatures of the atmosphere.

2. Nonstandard atmospheric pressure.

3. Aircraft static pressure systems (position error), and

4. Instrument error.

b. EXTREME CAUTION SHOULD BE EXERCISED WHEN FLYING IN PROXIMITY TO OBSTRUCTIONS OR TERRAIN IN LOW TEMPERATURES AND PRESSURES. This is especially true in extremely cold temperatures that cause a large differential between the Standard Day temperature and actual temperature. This circumstance can cause serious errors that result in the aircraft being significantly lower than the indicated altitude.

NOTE—STANDARD TEMPERATURE AT SEA LEVEL IS 15 DEGREES CELSIUS (59 DEGREES FAHRENHEIT). THE TEMPERATURE GRADIENT FROM SEA LEVEL IS MINUS 2 DEGREES CELSIUS (3.6 DEGREES FAHRENHEIT) PER 1,000 FEET. PILOTS SHOULD APPLY CORRECTIONS FOR STATIC PRESSURE SYSTEMS AND/OR INSTRUMENTS, IF APPRECIABLE ERRORS EXIST.

c. The adoption of a standard altimeter setting at the higher altitudes eliminates station barometer errors, some altimeter instrument errors, and errors caused by altimeter settings derived from different geographical sources.

7-2-2. PROCEDURES

The cruising altitude or flight level of aircraft shall be maintained by reference to an altimeter which shall be set, when operating:

a. Below 18,000 feet MSL:

1. *When the barometric pressure is 31.00 inches Hg. or less:* to the current reported altimeter setting of a station along the route and within 100 NM of the aircraft, or if there is no station within this area, the current reported altimeter setting of an appropriate available station. When an aircraft is en route on an instrument flight plan, air traffic controllers will furnish this information to the pilot at least once while the aircraft is in the controllers area of jurisdiction. In the case of an aircraft not equipped with a radio, set to the elevation of the departure airport or use an appropriate altimeter setting available prior to departure.

2. *When the barometric pressure exceeds 31.00 inches Hg.:* the following procedures will be placed in effect by NOTAM defining the geographic area affected:

(a) *For all aircraft:* Set 31.00 inches for en route operations below 18,000 feet MSL. Maintain this setting until beyond the affected area or until reaching final approach segment. At the beginning of the final approach segment, the current altimeter setting will be set, if possible. If not possible, 31.00 inches will remain set throughout the approach. Aircraft on departure or missed approach will set 31.00 inches prior to reaching any mandatory/crossing altitude or 1,500 feet AGL, whichever is lower. (Air traffic control will issue actual altimeter settings and advise pilots to set 31.00 inches in their altimeters for en route operations below 18,000 feet MSL in affected areas.)

(b) During preflight, barometric altimeters shall be checked for normal operation to the extent possible.

(c) For aircraft with the capability of setting the current altimeter setting and operating into airports with the capability of measuring the current altimeter setting, no additional restrictions apply.

(d) For aircraft operating VFR, there are no additional restrictions, however, extra diligence in flight planning and in operating in these conditions is essential.

(e) Airports unable to accurately measure barometric pressures above 31.00 inches of Hg. will report the barometric pressure as "missing" or "in excess of 31.00 inches of Hg." Flight operations to and from those airports are restricted to VFR weather conditions.

(f) For aircraft operating IFR and unable to set the current altimeter setting, the following restrictions apply:

(1) To determine the suitability of departure alternate airports, destination airports, and destination alternate airports, increase ceiling requirements by 100 feet and visibility requirements by ¼ statute mile for each ¹⁄₁₀ of an inch of Hg., or any portion thereof, over 31.00 inches. These adjusted values are then applied in accordance with the requirements of the applicable operating regulations and operations specifications.

EXAMPLE—DESTINATION ALTIMETER IS 31.28 INCHES, ILS DH 250 FEET (200½). WHEN FLIGHT PLANNING, ADD 300¾ TO THE WEATHER REQUIREMENTS WHICH WOULD BECOME 500-1¼.

(2) On approach, 31.00 inches will remain set. Decision Height or minimum descent altitude shall be deemed to have been reached when the published altitude is displayed on the altimeter.

NOTE—ALTHOUGH VISIBILITY IS NORMALLY THE LIMITING FACTOR ON AN APPROACH, PILOTS SHOULD BE AWARE THAT WHEN REACHING DH THE AIRCRAFT WILL BE HIGHER THAN INDICATED. USING THE EXAMPLE ABOVE THE AIRCRAFT WOULD BE APPROXIMATELY 300 FEET HIGHER.

(3) These restrictions do not apply to authorized Category II and III ILS operations nor do they apply to

Table 7-2-1 Lowest Usable Flight Level

Altimeter Setting (Current Reported)	Lowest Usable Flight Level
29.92 or higher	180
29.91 to 29.42	185
29.41 to 28.92	190
28.91 to 28.42	195
28.41 to 27.92	200

Table 7-2-2 Lowest Flight Level Correction Factor

Altimeter Setting	Correction Factor
29.92 or higher	none
29.91 to 29.42	500 feet
29.41 to 28.92	1000 feet
28.91 to 28.42	1500 feet
28.41 to 27.92	2000 feet
27.91 to 27.42	2500 feet

certificate holders using approved QFE altimetry systems.

(g) The FAA Regional Flight Standards Division Manager of the Affected area is authorized to approve temporary waivers to permit emergency resupply or emergency medical service operation.

b. At or above 18,000 feet MSL: to 29.92 inches of mercury (standard setting). The lowest usable flight level is determined by the atmospheric pressure in the area of operation as shown in TBL 7-2-1.

c. Where the minimum altitude, as prescribed in FAR Part 91.159 and FAR Part 91.177, is above 18,000 feet MSL, the lowest usable flight level shall be the flight level equivalent of the minimum altitude plus the number of feet specified in TBL 7-2-2.

EXAMPLE—THE MINIMUM SAFE ALTITUDE OF A ROUTE IS 19,000 FEET MSL AND THE ALTIMETER SETTING IS REPORTED BETWEEN 29.92 AND 29.42 INCHES OF MERCURY, THE LOWEST USABLE FLIGHT LEVEL WILL BE 195, WHICH IS THE FLIGHT LEVEL EQUIVALENT OF 19,500 FEET MSL (MINIMUM ALTITUDE PLUS 500 FEET).

7-2-3. ALTIMETER ERRORS

a. Most pressure altimeters are subject to mechanical, elastic, temperature, and installation errors. (Detailed information regarding the use of pressure altimeters is found in the Instrument Flying Handbook, Chapter IV.) Although manufacturing and installation specifications, as well as the periodic test and inspections required by regulations (FAR Part 43, Appendix E), act to reduce these errors, any scale error may be observed in the following manner:

1. Set the current reported altimeter setting on the altimeter setting scale.

Table 7-2-3   ICAO Cold Temperature Error Table
Height Above Airport in Feet

Reported Temp °C	200	300	400	500	600	700	800	900	1000	1500	2000	3000	4000	5000
+10	10	10	10	10	20	20	20	20	20	30	40	60	80	90
0	20	20	30	30	40	40	50	50	60	90	120	170	230	280
−10	20	30	40	50	60	70	80	90	100	150	200	290	390	490
−20	30	50	60	70	90	100	120	130	140	210	280	420	570	710
−30	40	60	80	100	120	140	150	170	190	280	380	570	760	950
−40	50	80	100	120	150	170	190	220	240	360	480	720	970	1210
−50	60	90	120	150	180	210	240	270	300	450	590	890	1190	1500

EXAMPLE–

Temperature − 10 degrees Celsius, and the aircraft altitude is 1,000 feet above the airport elevation. The chart shows that the reported current altimeter setting may place the aircraft as much as 100 feet below the altitude indicated by the altimeter.

2. Altimeter should now read field elevation if you are located on the same reference level used to establish the altimeter setting.

3. Note the variation between the known field elevation and the altimeter indication. If this variation is in the order of plus or minus 75 feet, the accuracy of the altimeter is questionable and the problem should be referred to an appropriately rated repair station for evaluation and possible correction.

b. Once in flight, it is very important to obtain frequently current altimeter settings en route. If you do not reset your altimeter when flying *from* an area of high pressure into an area of low pressure, *your aircraft will be closer to the surface than your altimeter indicates.* An inch error in the altimeter setting equals 1,000 feet of altitude. To quote an old saying: **"GOING FROM A HIGH TO A LOW, LOOK OUT BELOW."**

c. Temperature also has an effect on the accuracy of altimeters and your altitude. The crucial values to consider are standard temperature versus the ambient (at altitude) temperature. It is this "difference" that causes the error in indicated altitude. When the air is warmer than standard, you are higher than your altimeter indicates. Subsequently, when the air is colder than standard you are lower than indicated. It is the magnitude of this "difference" that determines the magnitude of the error. When flying into a cooler air mass while maintaining a constant indicated altitude, you are losing true altitude. However, flying into a cooler air mass does not necessarily mean you will be lower than indicated if the *difference* is still on the plus side. For example, while flying at 10,000 feet (where **STANDARD** temperature is –5 degrees Celsius (C)), the outside air temperature cools from

+5 degrees C to 0 degrees C, the temperature error will nevertheless cause the aircraft to be **HIGHER** than indicated. It is the extreme "cold" difference that normally would be of concern to the pilot. Also, when flying in cold conditions over mountainous country, the pilot should exercise caution in flight planning both in regard to route and altitude to ensure adequate en route terrain clearance.

d. TBL 7-2-3, derived from ICAO formulas, indicates how much error can exist when the temperature is extremely cold. To use the table, find the reported temperature in the left column, then read across the top row to locate the height above the airport/reporting station (i.e., subtract the airport/reporting elevation from the intended flight altitude). The intersection of the column and row is how much lower the aircraft may actually be as a result of the possible cold temperature induced error.

e. The possible result of the above example should be obvious, particularly if operating at the minimum altitude or when conducting an instrument approach. When operating in extreme cold temperatures, pilots may wish to compensate for the reduction in terrain clearance by adding a cold temperature correction.

7-2-4. HIGH BAROMETRIC PRESSURE

a. Cold, dry air masses may produce barometric pressures in excess of 31.00 inches of Mercury, and many altimeters do not have an accurate means of being adjusted for settings of these levels. When the altimeter cannot be set to the higher pressure setting, the aircraft actual altitude will be higher than the altimeter indicates.

REFERENCE—AIM, ALTIMETER ERRORS, PARAGRAPH 7-2-3.

b. When the barometric pressure exceeds 31.00 inches, air traffic controllers will issue the actual altimeter setting, and:

1. *En Route/Arrivals:* Advise pilots to remain set on 31.00 inches until reaching the final approach segment.

2. *Departures:* Advise pilots to set 31.00 inches prior to reaching any mandatory/crossing altitude or 1,500 feet, whichever is lower.

c. The altimeter error caused by the high pressure will be in the opposite direction to the error caused by the cold temperature.

7-2-5. LOW BAROMETRIC PRESSURE

When abnormally low barometric pressure conditions occur (below 28.00), flight operations by aircraft unable to set the actual altimeter setting are not recommended.

NOTE—THE TRUE ALTITUDE OF THE AIRCRAFT IS LOWER THAN THE INDICATED ALTITUDE IF THE PILOT IS UNABLE TO SET THE ACTUAL ALTIMETER SETTING.

Section 3. WAKE TURBULENCE

7-3-1. GENERAL

a. Every aircraft generates a wake while in flight. Initially, when pilots encountered this wake in flight, the disturbance was attributed to "prop wash." It is known, however, that this disturbance is caused by a pair of counter rotating vortices trailing from the wing tips. The vortices from larger aircraft pose problems to encountering aircraft. For instance, the wake of these aircraft can impose rolling moments exceeding the roll-control authority of the encountering aircraft. Further, turbulence generated within the vortices can damage aircraft components and equipment if encountered at close range. The pilot must learn to envision the location of the vortex wake generated by larger (transport category) aircraft and adjust the flight path accordingly.

b. During ground operations and during takeoff, jet engine blast (thrust stream turbulence) can cause damage and upsets if encountered at close range. Exhaust velocity versus distance studies at various thrust levels have shown a need for light aircraft to maintain an adequate separation behind large turbojet aircraft. Pilots of larger aircraft should be particularly careful to consider the effects of their "jet blast" on other aircraft, vehicles, and maintenance equipment during ground operations.

7-3-2. VORTEX GENERATION

Lift is generated by the creation of a pressure differential over the wing surface. The lowest pressure occurs over the upper wing surface and the highest pressure under the wing. This pressure differential triggers the roll up of the airflow aft of the wing resulting in swirling air masses trailing downstream of the wing tips. After the roll up is completed, the wake consists of two counter rotating cylindrical vortices. (See FIG 7-3-1.) Most of the energy is within a few feet of the center of each vortex, but pilots should avoid a region within about 100 feet of the vortex core.

7-3-3. VORTEX STRENGTH

a. The strength of the vortex is governed by the weight, speed, and shape of the wing of the generating aircraft.

The vortex characteristics of any given aircraft can also be changed by extension of flaps or other wing configuring devices as well as by change in speed. However, as the basic factor is weight, the vortex strength increases proportionately. Peak vortex tangential speeds exceeding 300 feet per second have been recorded. The greatest vortex strength occurs when the generating aircraft is HEAVY, CLEAN, and SLOW.

b. INDUCED ROLL

1. In rare instances a wake encounter could cause in-flight structural damage of catastrophic proportions. However, the usual hazard is associated with induced rolling moments which can exceed the roll-control authority of the encountering aircraft. In flight experiments, aircraft have been intentionally flown directly up trailing vortex cores of larger aircraft. It was shown that the capability of an aircraft to counteract the roll imposed by the wake vortex primarily depends on the wingspan and counter-control responsiveness of the encountering aircraft.

2. Counter control is usually effective and induced roll minimal in cases where the wingspan and ailerons of the encountering aircraft extend beyond the rotational flow

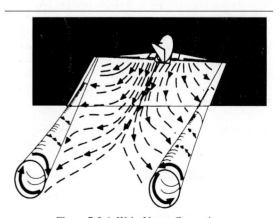

Figure 7-3-1. Wake Vortex Generation

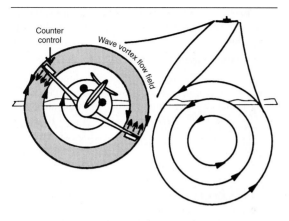

Figure 7-3-2. Wake Encounter Counter Control

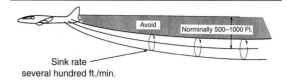

Figure 7-3-4. Vortex Flow Field

field of the vortex. It is more difficult for aircraft with short wingspan (relative to the generating aircraft) to counter the imposed roll induced by vortex flow. Pilots of short span aircraft, even of the high performance type, must be especially alert to vortex encounters. (See FIG 7-3-2.)

3. The wake of larger aircraft requires the respect of all pilots.

7-3-4. VORTEX BEHAVIOR

a. Trailing vortices have certain behavioral characteristics which can help a pilot visualize the wake location and thereby take avoidance precautions.

1. Vortices are generated from the moment aircraft leave the ground, since trailing vortices are a by-product of wing lift. Prior to takeoff or touchdown pilots should note the rotation or touchdown point of the preceding aircraft. (See FIG 7-3-3.)

2. The vortex circulation is outward, upward and around the wing tips when viewed from either ahead or behind the aircraft. Tests with large aircraft have shown that the vortices remain spaced a bit less than a wingspan apart, drifting with the wind, at altitudes greater than a wingspan from the ground. In view of this, if persistent vortex turbulence is encountered, a slight change of altitude and lateral position (preferably upwind) will provide a flight path clear of the turbulence.

3. Flight tests have shown that the vortices from larger (transport category) aircraft sink at a rate of several hundred feet per minute, slowing their descent and diminishing in strength with time and distance behind the generating aircraft. Atmospheric turbulence hastens breakup. Pilots should fly at or above the preceding aircraft's flight path, altering course as necessary to avoid the area behind and below the generating aircraft. (See FIG 7-3-4.) However vertical separation of 1,000 feet may be considered safe.

4. When the vortices of larger aircraft sink close to the ground (within 100 to 200 feet), they tend to move laterally over the ground at a speed of 2 or 3 knots. (See FIG 7-3-5.)

5. There is a small segment of the aviation community that have become convinced that wake vortices may "bounce" up to twice their nominal steady state height. With a 200-foot span aircraft, the "bounce" height could reach approximately 200 feet AGL. This conviction is based on a single unsubstantiated report of an apparent coherent vortical flow that was seen in the volume scan of a research sensor. No one can say what conditions cause vortex bouncing, how high they bounce, at what angle they bounce, or how many times a vortex may bounce. On the other hand, no one can say for certain that vortices never "bounce." Test data have shown that vortices can rise with the air mass in which they are embedded. Wind shear, particularly, can cause vortex flow field "tilting." Also, ambient thermal lifting and orographic effects (rising terrain or tree lines) can cause a vortex flow field to rise. Notwithstanding the foregoing, pilots are reminded that they should be alert at all times for possible wake vortex encounters when conducting approach and landing operations.

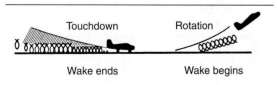

Figure 7-3-3.

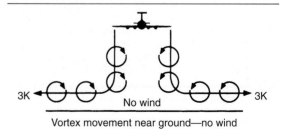

Figure 7-3-5.

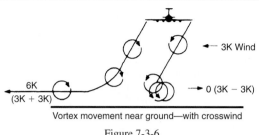

Vortex movement near ground—with crosswind

Figure 7-3-6.

The pilot has the ultimate responsibility for ensuring appropriate separations and positioning of the aircraft in the terminal area to avoid the wake turbulence created by a preceding aircraft.

b. A crosswind will decrease the lateral movement of the upwind vortex and increase the movement of the downwind vortex. Thus a light wind with a cross runway component of 1 to 5 knots could result in the upwind vortex remaining in the touchdown zone for a period of time and hasten the drift of the downwind vortex toward another runway. (See FIG 7-3-6.) Similarly, a tailwind condition can move the vortices of the preceding aircraft forward into the touchdown zone. THE LIGHT QUARTERING TAILWIND REQUIRES MAXIMUM CAUTION. Pilots should be alert to large aircraft upwind from their approach and takeoff flight paths. (See FIG 7-3-7.)

7-3-5. OPERATIONS PROBLEM AREAS

a. A wake encounter can be catastrophic. In 1972 at Fort Worth a DC-9 got too close to a DC-10 (two miles back), rolled, caught a wingtip, and cartwheeled coming to rest in an inverted position on the runway. All aboard were killed. Serious and even fatal GA accidents induced by wake vortices are not uncommon. However, a wake encounter is not necessarily hazardous. It can be one or more jolts with varying severity depending upon the direction of the encounter, weight of the generating aircraft, size of the encountering aircraft, distance from the generating aircraft, and point of vortex encounter. The probability of induced roll increases when the encountering aircraft's heading is generally aligned with the flight path of the generating aircraft.

b. AVOID THE AREA BELOW AND BEHIND THE GENERATING AIRCRAFT, ESPECIALLY AT LOW ALTITUDE WHERE EVEN A MOMENTARY WAKE ENCOUNTER COULD BE HAZARDOUS. This is not easy to do. Some accidents have occurred even though the pilot of the trailing aircraft had carefully noted that the aircraft in front was at a considerably lower altitude. Unfortunately, this does not ensure that the flight path of the lead aircraft will be below that of the trailing aircraft.

c. Pilots should be particularly alert in calm wind conditions and situations where the vortices could:

1. Remain in the touchdown area.

2. Drift from aircraft operating on a nearby runway.

3. Sink into the takeoff or landing path from a crossing runway.

4. Sink into the traffic pattern from other airport operations.

5. Sink into the flight path of VFR aircraft operating on the hemispheric altitude 500 feet below.

d. Pilots of all aircraft should visualize the location of the vortex trail behind larger aircraft and use proper vortex avoidance procedures to achieve safe operation. It is equally important that pilots of larger aircraft plan or adjust their flight paths to minimize vortex exposure to other aircraft.

7-3-6. VORTEX AVOIDANCE PROCEDURES

a. Under certain conditions, airport traffic controllers apply procedures for separating IFR aircraft. If a pilot accepts a clearance to visually follow a preceding aircraft, the pilot accepts responsibility for separation and wake turbulence avoidance. The controllers will also provide to VFR aircraft, with whom they are in communication and which in the tower's opinion may be adversely affected by wake turbulence from a larger aircraft, the position, altitude and direction of flight of larger aircraft followed by the phrase "CAUTION—WAKE TURBULENCE." After issuing the caution for wake turbulence, the airport traffic controllers generally do not provide additional information to the following aircraft unless the airport traffic controllers know the following aircraft is overtaking the preceding aircraft. WHETHER OR NOT A WARNING OR INFORMATION HAS BEEN GIVEN, HOWEVER, THE PILOT IS EXPECTED TO ADJUST AIRCRAFT OPERATIONS AND FLIGHT PATH AS NECESSARY TO PRECLUDE SERIOUS WAKE ENCOUNTERS. When any doubt exists about maintaining safe separation distances between aircraft during approaches, pilots should ask the control tower for updates on separation distance and aircraft groundspeed.

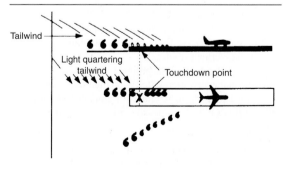

Figure 7-3-7. Vortex
Movement in Ground Effect—Tailwind

b. The following vortex avoidance procedures are recommended for the various situations:

1. *Landing behind a larger aircraft—same runway:* Stay at or above the larger aircraft's final approach flight path—note its touchdown point—land beyond it.

2. *Landing behind a larger aircraft—when parallel runway is closer than 2,500 feet:* Consider possible drift to your runway. Stay at or above the larger aircraft's final approach flight path—note its touchdown point.

3. *Landing behind a larger aircraft—crossing runway:* Cross above the larger aircraft's flight path.

4. *Landing behind a departing larger aircraft—same runway:* Note the larger aircraft's rotation point—land well prior to rotation point.

5. *Landing behind a departing larger aircraft—crossing runway:* Note the larger aircraft's rotation point—if past the intersection—continue the approach—land prior to the intersection. If larger aircraft rotates prior to the intersection, avoid flight below the larger aircraft's flight path. Abandon the approach unless a landing is ensured well before reaching the intersection.

6. *Departing behind a larger aircraft:* Note the larger aircraft's rotation point and rotate prior to the larger aircraft's rotation point. Continue climbing above the larger aircraft's climb path until turning clear of the larger aircraft's wake. Avoid subsequent headings which will cross below and behind a larger aircraft. Be alert for any critical takeoff situation which could lead to a vortex encounter.

7. *Intersection takeoffs—same runway:* Be alert to adjacent larger aircraft operations, particularly upwind of your runway. If intersection takeoff clearance is received, avoid subsequent heading which will cross below a larger aircraft's path.

8. *Departing or landing after a larger aircraft executing a low approach, missed approach or touch-and-go landing:* Because vortices settle and move laterally near the ground, the vortex hazard may exist along the runway and in your flight path after a larger aircraft has executed a low approach, missed approach or a touch-and-go landing, particular in light quartering wind conditions. You should ensure that an interval of at least 2 minutes has elapsed before your takeoff or landing.

9. *En route VFR (thousand-foot altitude plus 500 feet):* Avoid flight below and behind a large aircraft's path. If a larger aircraft is observed above on the same track (meeting or overtaking) adjust your position laterally, preferably upwind.

7-3-7. HELICOPTERS

In a slow hover taxi or stationary hover near the surface, helicopter main rotor(s) generate downwash producing high velocity outwash vortices to a distance approximately three times the diameter of the rotor. When rotor downwash hits the surface, the resulting outwash vortices have behavioral characteristics similar to wing tip vortices produced by fixed wing aircraft. However, the vortex circulation is outward, upward, around, and away from the main rotor(s) in all directions. Pilots of small aircraft should avoid operating within three rotor diameters of any helicopter in a slow hover taxi or stationary hover. In forward flight, departing or landing helicopters produce a pair of strong, high-speed trailing vortices similar to wing tip vortices of larger fixed wing aircraft. Pilots of small aircraft should use caution when operating behind or crossing behind landing and departing helicopters.

7-3-8. PILOT RESPONSIBILITY

a. Government and industry groups are making concerted efforts to minimize or eliminate the hazards of trailing vortices. However, the flight disciplines necessary to ensure vortex avoidance during VFR operations must be exercised by the pilot. Vortex visualization and avoidance procedures should be exercised by the pilot using the same degree of concern as in collision avoidance.

b. Wake turbulence may be encountered by aircraft in flight as well as when operating on the airport movement area.

REFERENCE—PILOT/CONTROLLER GLOSSARY TERM—WAKE TURBULENCE.

c. Pilots are reminded that in operations conducted behind all aircraft, acceptance of instructions from ATC in the following situations is an acknowledgment that the pilot will ensure safe takeoff and landing intervals and accepts the responsibility for providing wake turbulence separation.

1. Traffic information,

2. Instructions to follow an aircraft, and

3. The acceptance of a visual approach clearance.

d. For operations conducted behind **heavy** aircraft, ATC will specify the word **"heavy"** when this information is known. Pilots of **heavy** aircraft should always use the word **"heavy"** in radio communications.

e. Heavy and large jet aircraft operators should use the following procedures during an approach to landing. These procedures establish a dependable baseline from which pilots of in-trail, lighter aircraft may reasonably expect to make effective flight path adjustments to avoid serious wake vortex turbulence.

1. Pilots of aircraft that produce strong wake vortices should make every attempt to fly on the established glidepath, not above it; or, if glidepath guidance is not available, to fly as closely as possible to a "3-1" glidepath, not above it.

EXAMPLE—FLY 3,000 FEET AT 10 MILES FROM TOUCHDOWN, 1,500 FEET AT 5 MILES, 1,200 FEET AT 4 MILES, AND SO ON TO TOUCHDOWN.

2. Pilots of aircraft that produce strong wake vortices should fly as closely as possible to the approach course centerline or to the extended centerline of the runway of intended landing as appropriate to conditions.

f. Pilots operating lighter aircraft on visual approaches in-trail to aircraft producing strong wake vortices should use the following procedures to assist in avoiding wake turbulence. These procedures apply only to those aircraft that are on visual approaches.

1. Pilots of lighter aircraft should fly on or above the glidepath. Glidepath reference may be furnished an ILS, by a visual approach slope system, by other ground-based approach slope guidance systems, or by other means. In the absence of visible glidepath guidance, pilots may very nearly duplicate a 3-degree glideslope by adhering to the "3 to 1" glidepath principle.

EXAMPLE—FLY 3,000 FEET AT 10 MILES FROM TOUCHDOWN, 1,500 FEET AT 5 MILES, 1,200 FEET AT 4 MILES, AND SO ON TO TOUCHDOWN.

2. If the pilot of the lighter following aircraft has visual contact with the preceding heavier aircraft and also with the runway, the pilot may further adjust for possible wake vortex turbulence by the following practices:

(a) Pick a point of landing no less than 1,000 feet from the arrival end of the runway.

(b) Establish a line-of-sight to that landing point that is above and in front of the heavier preceding aircraft.

(c) When possible, note the point of landing of the heavier preceding aircraft and adjust point of intended landing as necessary.

EXAMPLE—A PUFF OF SMOKE MAY APPEAR AT THE 1,000-FOOT MARKINGS OF THE RUNWAY, SHOWING THAT TOUCHDOWN WAS THAT POINT; THEREFORE, ADJUST POINT OF INTENDED LANDING TO THE 1,500-FOOT MARKINGS.

(d) Maintain the line-of-sight to the point of intended landing above and ahead of the heavier preceding aircraft; maintain it to touchdown.

(e) Land beyond the point of landing of the preceding heavier aircraft.

3. During visual approaches pilots may ask ATC for updates on separation and groundspeed with respect to heavier preceding aircraft, especially when there is any question of safe separation from wake turbulence.

7-3-9. AIR TRAFFIC WAKE TURBULENCE SEPARATIONS

a. Because of the possible effects of wake turbulence, controllers are required to apply no less than specified minimum separation for aircraft operating behind a **heavy** jet and, in certain instances, behind large **non-heavy** aircraft (i.e., B757 aircraft).

1. Separation is applied to aircraft operating directly behind a **heavy/B757** jet at the same altitude or less than 1,000 feet below:

(a) Heavy jet behind **heavy** jet—4 miles.

(b) Large/heavy behind **B757**—4 miles.

(c) Small behind B757—5 miles.

(d) Small/large aircraft behind heavy jet—5 miles.

2. Also, separation, measured at the time the preceding aircraft is over the landing threshold, is provided to small aircraft:

(a) Small aircraft landing behind **heavy** jet—6 miles.

(b) Small aircraft landing behind **B757**—5 miles.

(c) Small aircraft landing behind **large** aircraft—4 miles.

REFERENCE—PILOT/CONTROLLER GLOSSARY TERM— AIRCRAFT CLASSES.

3. Additionally, appropriate time or distance intervals are provided to departing aircraft:

(a) Two minutes or the appropriate 4 or 5 mile radar separation when takeoff behind a heavy/B757 jet will be:

(1) from the same threshold

(2) on a crossing runway and projected flight paths will cross

(3) from the threshold of a parallel runway when staggered ahead of that of the adjacent runway by less than 500 feet and when the runways are separated by less than 2,500 feet.

NOTE— CONTROLLERS MAY NOT REDUCE OR WAIVE THESE INTERVALS.

b. A 3-minute interval will be provided when a **small** aircraft will take off:

1. From an intersection on the same runway (same or opposite direction) behind a departing **large** aircraft,

2. In the opposite direction on the same runway behind a large aircraft takeoff or low/missed approach.

NOTE—THIS 3-MINUTE INTERVAL MAY BE WAIVED UPON SPECIFIC PILOT REQUEST.

c. A 3-minute interval will be provided for all aircraft taking off when the operations are as described in subparagraph b. 1. and 2. above, the preceding aircraft is a **heavy/B757** jet, and the operations are on either the same runway or parallel runways separated by less than 2,500 feet. Controllers may not reduce or waive this interval.

d. Pilots may request additional separation, i.e., 2 minutes instead of 4 or 5 miles for wake turbulence avoidance. This request should be made as soon as practical on ground control and at least before taxiing onto the runway.

NOTE—FAR PART 91.3(A) STATES: "THE PILOT IN COMMAND OF AN AIRCRAFT IS DIRECTLY RESPONSIBLE FOR AND IS THE FINAL AUTHORITY AS TO THE OPERATION OF THAT AIRCRAFT."

e. Controllers may anticipate separation and need not withhold a takeoff clearance for an aircraft departing behind a **large/heavy** aircraft if there is reasonable assurance the required separation will exist when the departing aircraft starts takeoff roll.

Section 4. BIRD HAZARDS AND FLIGHT OVER NATIONAL REFUGES, PARKS, AND FORESTS

7-4-1. MIGRATORY BIRD ACTIVITY

a. Bird strike risk increases because of bird migration during the months of March through April, and August through November.

b. The altitudes of migrating birds vary with winds aloft, weather fronts, terrain elevations, cloud conditions, and other environmental variables. While over 90 percent of the reported bird strikes occur at or below 3,000 feet AGL, strikes at higher altitudes are common during migration. Ducks and geese are frequently observed up to 7,000 feet AGL and pilots are cautioned to minimize en route flying at lower altitudes during migration.

c. Considered the greatest potential hazard to aircraft because of their size, abundance, or habit of flying in dense flocks are gulls, waterfowl, vultures, hawks, owls, egrets, blackbirds, and starlings. Four major migratory flyways exist in the United States. The Atlantic flyway parallels the Atlantic Coast. The Mississippi Flyway stretches from Canada through the Great Lakes and follows the Mississippi River. The Central Flyway represents a broad area east of the Rockies, stretching from Canada through Central America. The Pacific Flyway follows the west coast and overflies major parts of Washington, Oregon, and California. There are also numerous smaller flyways which cross these major north-south migratory routes.

7-4-2. REDUCING BIRD STRIKE RISKS

a. The most serious strikes are those involving ingestion into an engine (turboprops and turbine jet engines) or windshield strikes. These strikes can result in emergency situations requiring prompt action by the pilot.

b. Engine ingestions may result in sudden loss of power or engine failure. Review engine out procedures, especially when operating from airports with known bird hazards or when operating near high bird concentrations.

c. Windshield strikes have resulted in pilots experiencing confusion, disorientation, loss of communications, and aircraft control problems. Pilots are encouraged to review their emergency procedures before flying in these areas.

d. When encountering birds en route, climb to avoid collision, because birds in flocks generally distribute themselves downward, with lead birds being at the highest altitude.

e. Avoid overflight of known areas of bird concentration and flying at low altitudes during bird migration. Charted wildlife refuges and other natural areas contain unusually high local concentration of birds which may create a hazard to aircraft.

7-4-3. REPORTING BIRD STRIKES

Pilots are urged to report any bird or other wildlife strike using FAA Form 5200-7, Bird/Other Wildlife Strike Report. (Form available on pages 300–301.) Additional forms are available at any FSS, or at any FAA Regional Office or at http://www.faa.gov/arp/hazard.htm. The data derived from these reports are used to develop standards to cope with this potential hazard to aircraft and for documentation of necessary habitat control on airports.

7-4-4. REPORTING BIRD AND OTHER WILDLIFE ACTIVITIES

If you observe birds or other animals on or near the runway, request airport management to disperse the wildlife before taking off. Also contact the nearest FAA ARTCC, FSS, or tower (including nonfederal towers) regarding large flocks of birds and report the:

1. Geographic location

2. Bird type (geese, ducks, gulls, etc.)

3. Approximate numbers

4. Altitude

5. Direction of bird flight path

7-4-5. PILOT ADVISORIES ON BIRD AND OTHER WILDLIFE HAZARDS

Many airports advise pilots of other wildlife hazards caused by large animals on the runway through the A/FD and the NOTAM system. Collisions of landing and departing aircraft and animals on the runway are increasing and are not limited to rural airports. These accidents have also occurred at several major airports. Pilots should exercise extreme caution when warned of the presence of wildlife on and in the vicinity of airports. If you observe deer or other large animals in close proximity to movement areas, advise the FSS, tower, or airport management.

7-4-6. FLIGHTS OVER CHARTED U.S. WILDLIFE REFUGES, PARKS, AND FOREST SERVICE AREAS

a. The landing of aircraft is prohibited on lands or waters administered by the National Park Service, U.S. Fish and Wildlife Service, or U.S. Forest Service without authorization from the respective agency. Exceptions, including

1. when forced to land due to an emergency beyond the control of the operator,

2. at officially designated landing sites, or

3. an approved official business of the Federal Government.

b. Pilots are requested to maintain a minimum altitude of 2,000 feet above the surface of the following: National Parks, Monuments, Seashores, Lakeshores, Recreation

Areas and Scenic Riverways administered by the National Park Service, National Wildlife Refuges, Big Game Refuges, Game Ranges and Wildlife Ranges administered by the U.S. Fish and Wildlife Service, and Wilderness and Primitive areas administered by the U.S. Forest Service.

NOTE—FAA ADVISORY CIRCULAR AC 91-36, VISUAL FLIGHT RULES (VFR) FLIGHT NEAR NOISE-SENSITIVE AREAS, DEFINES THE SURFACE OF A NATIONAL PARK AREA (INCLUDING PARKS, FORESTS, PRIMITIVE AREAS, WILDERNESS AREAS, RECREATIONAL AREAS, NATIONAL SEASHORES, NATIONAL MONUMENTS, NATIONAL LAKESHORES, AND NATIONAL WILDLIFE REFUGE AND RANGE AREAS) AS: THE HIGHEST TERRAIN WITHIN 2,000 FEET LATERALLY OF THE ROUTE OF FLIGHT, OR THE UPPER-MOST RIM OF A CANYON OR VALLEY.

c. Federal statutes prohibit certain types of flight activity and/or provide altitude restrictions over designated U.S. Wildlife Refuges, Parks, and Forest Service Areas. These designated areas, for example: Boundary Waters Canoe Wilderness Areas, Minnesota; Haleakala National Park, Hawaii; Yosemite National Park, California; and Grand Canyon National Park, Arizona are charted on Sectional Charts.

d. Federal regulations also prohibit airdrops by parachute or other means of persons, cargo, or objects from aircraft on lands administered by the three agencies without authorization from the respective agency. Exceptions include:

1. emergencies involving the safety of human life, or

2. threat of serious property loss.

Section 5. POTENTIAL FLIGHT HAZARDS

7-5-1. ACCIDENT CAUSE FACTORS

a. The 10 most frequent cause factors for General Aviation Accidents that involve the pilot-in-command are:

1. *Inadequate preflight preparation and/or planning.*
2. *Failure to obtain and/or maintain flying speed.*
3. *Failure to maintain direction control.*
4. *Improper level off.*
5. *Failure to see and avoid objects or obstructions.*
6. *Mismanagement of fuel.*
7. *Improper inflight decisions or planning.*
8. *Misjudgment of distance and speed.*
9. *Selection of unsuitable terrain.*
10. *Improper operation of flight controls.*

b. This list remains relatively stable and points out the need for continued refresher training to establish a higher level of flight proficiency for all pilots. A part of the FAA's continuing effort to promote increased aviation safety is the Aviation Safety Program. For information on Aviation Safety Program activities contact your nearest Flight Standards District Office.

c. ALERTNESS: Be alert at all times, especially when the weather is good. Most pilots pay attention to business when they are operating in full IFR weather conditions, but strangely, air collisions almost invariably have occurred under ideal weather conditions. Unlimited visibility appears to encourage a sense of security which is not at all justified. Considerable information of value may be obtained by listening to advisories being issued in the terminal area, even though controller workload may prevent a pilot from obtaining individual service.

d. GIVING WAY: If you think another aircraft is too close to you, give way instead of waiting for the other pilot to respect the right-of-way to which you may be entitled. It is a lot safer to pursue the right-of-way angle after you have completed your flight.

7-5-2. VFR IN CONGESTED AREAS

A high percentage of near midair collisions occur below 8,000 feet AGL and within 30 miles of an airport. When operating VFR in these highly congested areas, whether you intend to land at an airport within the area or are just flying through, it is recommended that extra vigilance be maintained and that you monitor an appropriate control frequency. Normally the appropriate frequency is an approach control frequency. By such monitoring action you can "get the picture" of the traffic in your area. When the approach controller has radar, radar traffic advisories may be given to VFR pilots upon request.
    REFERENCE—AIM, RADAR TRAFFIC INFORMATION SERVICE, PARAGRAPH 4-1-14.

7-5-3. OBSTRUCTIONS TO FLIGHT

a. General:

Many structures exist that could significantly affect the safety of your flight when operating below 500 feet AGL, and particularly below 200 feet AGL. While FAR Part 91.119 allows flight below 500 AGL when over sparsely populated areas or open water, such operations are very dangerous. At and below 200 feet AGL there are numerous power lines, antenna towers, etc., that are not marked and lighted as obstructions and therefore may not be seen in time to avoid a collision. Notices to Airmen (NOTAMs) are issued on those lighted structures experiencing temporary light outages. However, some time may pass before the FAA is notified of these outages, and the NOTAM issued, thus pilot vigilance is imperative.

b. Antenna Towers:

Extreme caution should be exercised when flying less than 2,000 feet AGL because of numerous skeletal structures, such as radio and television antenna towers, that exceed 1,000 feet AGL with some extending higher than 2,000 feet AGL. Most skeletal structures are supported by guy wires which are very difficult to see in good weather and can be invisible at dusk or during periods of reduced visibility. These wires can extend about 1,500 feet horizontally from a structure; therefore, all skeletal structures should be avoided horizontally by at least 2,000 feet. Additionally, new towers may not be on your current chart because the information was not received prior to the printing of the chart.

c. Overhead Wires:

Overhead transmission and utility lines often span approaches to runways, natural flyways such as lakes, rivers, gorges, and canyons, and cross other landmarks pilots frequently follow such as highways, railroad tracks, etc. As with antenna towers, these high voltage/power lines or the supporting structures of these lines may not always be readily visible and the wires may be virtually impossible to see under certain conditions. In some locations, the supporting structures of overhead transmission lines are equipped with unique sequence flashing white strobe light systems to indicate that there are wires between the structures. However, many power lines do not require notice to the FAA and, therefore, are not marked and/or lighted. Many of those that do require notice do not exceed 200 feet AGL or meet the Obstruction Standard of FAR Part 77 and, therefore, are not marked and/or lighted. All pilots are cautioned to remain extremely vigilant for these power lines or their supporting structures when following natural flyways or during the approach and landing phase. This is particularly important for seaplane and/or float equipped aircraft when landing on, or departing from, unfamiliar lakes or rivers.

d. Other Objects/Structures:

There are other objects or structures that could adversely affect your flight such as construction cranes near an airport, newly constructed buildings, new towers, etc. Many of these structures do not meet charting requirements or may not yet be charted because of the charting cycle. Some structures do not require obstruction marking and/or lighting and some may not be marked and lighted even though the FAA recommended it.

7-5-4. AVOID FLIGHT BENEATH UNMANNED BALLOONS

a. The majority of unmanned free balloons currently being operated have, extending below them, either a suspension device to which the payload or instrument package is attached, or a trailing wire antenna, or both. In many instances these balloon subsystems may be invisible to the pilot until the aircraft is close to the balloon, thereby creating a potentially dangerous situation. Therefore, good judgment on the part of the pilot dictates that aircraft should remain well clear of all unmanned free balloons and flight below them should be avoided at all times.

b. Pilots are urged to report any unmanned free balloons sighted to the nearest FAA ground facility with which communication is established. Such information will assist FAA ATC facilities to identify and flight follow unmanned free balloons operating in the airspace.

7-5-5. Unmanned Aircraft

a. Unmanned aircraft (UA), commonly referred to as "Unmanned Aerial Vehicles" (UAVs), are having an increasing operational presence in the national airspace system (NAS). Once the exclusive domain of the military, UAs are now being operated by various entities. Although these aircraft are "unmanned," UAs are controlled by a ground-based pilot and crew. Physical and performance characteristics of UAs vary greatly, and unlike model aircraft that typically operate lower than 400 feet above ground level, UAs may be found operating at virtually any altitude and any speed. Sizes of UAs can be as small as several pounds to as large as a commercial transport aircraft. UAs come in various categories including airplane, rotorcraft, powered-lift (tilt-rotor), and lighter-than-air. Propulsion systems of UAs include piston-powered propeller as well as turbojet.

b. To ensure segregation of UA operations from manned aircraft, the military typically conducts UA operations within restricted or other special use airspace. However, UA operations are now being approved in the NAS outside of special use airspace through the use of FAA-issued Certificates of Waiver or Authorization (COA) or through the issuance of an experimental airworthiness certificate. COA and experimental airworthiness approvals authorize UA flight operations to be contained within specific geographic boundaries, usually require coordination with an air traffic control (ATC) facility, and typically require issuance of a notice to airmen (NOTAM) describing the operation to be conducted. UA approvals also require observers to provide "see-and-avoid" capability to the UA crew and to provide necessary guidance to maneuver the UA away from any detected manned aircraft. For UA operations approved above flight level 180, UAs are operated under instrument flight rules, are in communication with ATC, and are equipped with a transponder.

c. There are several things a pilot should consider regarding UA activity in an effort to reduce potential flight hazards. Pilots are urged to exercise increased vigilance when operating in the vicinity of restricted or other special use airspace, military operations areas, and any military installation. Since the size of a UA can be very small, they may be difficult to see and track. If a UA is encountered during flight, don't assume that the pilot or crew of the UA can see you, maintain increased vigilance with the UA. Always check NOTAMs for potential UA activity along the intended route of flight and exercise increased vigilance in areas specified in the NOTAM.

7-5-6. MOUNTAIN FLYING

a. Your first experience of flying over mountainous terrain (particularly if most of your flight time has been over the flatlands of the Midwest) could be a *never-to-be-forgotten nightmare* if proper planning is not done and if you are not aware of the potential hazards awaiting. Those familiar section lines are not present in the mountains; those flat, level fields for forced landings are practically nonexistent; abrupt changes in wind direction and velocity occur; severe updrafts and downdrafts are common, particularly near or above abrupt changes of terrain such as cliffs or rugged areas; even the clouds look different and can build up with startling rapidity. Mountain flying need not be hazardous if you follow the recommendations below:

b. File a flight plan: Plan your route to avoid topography which would prevent a safe forced landing. The route should be over populated areas and well known mountain passes. Sufficient altitude should be maintained to permit gliding to a safe landing in the event of engine failure.

c. Don't fly a light aircraft when the winds aloft, at your proposed altitude, exceed 35 miles per hour. Expect the winds to be of much greater velocity over mountain passes than reported a few miles from them. Approach mountain passes with as much altitude as possible. Downdrafts of from 1,500 to 2,000 feet per minute are not uncommon on the leeward side.

d. Don't fly near or above abrupt changes in terrain. Severe turbulence can be expected, especially in high wind conditions.

e. Understand Mountain Obscuration. The term Mountain Obscuration (MTOS) is used to describe a visibility condition that is distinguished from IFR because ceilings, by definition, are described as "above ground level" (AGL). In mountainous terrain clouds can form at altitudes significantly higher than the weather reporting station and at the same time nearby mountaintops may be obscured by low visibility. In these areas the ground level can also vary greatly over a small area. Beware if operating VFR-on-top. You could be operating closer to the terrain than you think because the tops of mountains are hidden in a cloud deck below. MTOS areas are identified daily on The Aviation Weather Center located at www.awckc.noaa.gov under Official Forecast Products, AIRMETs (WA), IFR/Mountain Obscuration.

f. VFR flight operations may be conducted at night in mountainous terrain with the application of sound judgment and common sense. Proper preflight planning, giving ample consideration to winds and weather, knowledge of the terrain and pilot experience in mountain flying are prerequisites for safety of flight. Continuous visual contact with the surface and obstructions is a major concern and flight operations under an overcast or in the vicinity of clouds should be approached with extreme caution.

g. When landing at a high altitude field, the same indicated airspeed should be used as at low elevation fields. *Remember:* that due to the less dense air at altitude, this same indicated airspeed actually results in higher true airspeed, a faster landing speed, and more important, a longer landing distance. During gusty wind conditions which often prevail at high altitude fields, a power approach and power landing is recommended. Additionally, due to the faster groundspeed, your takeoff distance will increase considerably over that required at low altitudes.

h. Effects of Density Altitude: Performance figures in the aircraft owner's handbook for length of takeoff run, horsepower, rate of climb, etc., are generally based on standard atmosphere conditions (59 degrees Fahrenheit (15 degrees Celsius), pressure 29.92 inches of mercury) at sea level. However, inexperienced pilots, as well as experienced pilots, may run into trouble when they encounter an altogether different set of conditions. This is particularly true in hot weather and at higher elevations. Aircraft operations at altitudes above sea level and at higher than standard temperatures are commonplace in mountainous areas. Such operations quite often result in a drastic reduction of aircraft performance capabilities because of the changing air density. Density altitude is a measure of air density. It is not to be confused with pressure altitude, true altitude or absolute altitude. It is not to be used as a height reference, but as a determining criterion in the performance capability of an aircraft. Air density decreases with altitude. As air density decreases, density altitude increases. The further effects of high temperature and high humidity are cumulative, resulting in an increasing high density altitude condition. High density altitude reduces all aircraft performance parameters. To the pilot, this means that the normal horsepower output is reduced, propeller efficiency is reduced and a higher true airspeed is required to sustain the aircraft throughout its operating parameters. It means an increase in runway length requirements for takeoff and landings, and decreased rate of climb. An average small airplane, for example, requiring 1,000 feet for takeoff at sea level under standard atmospheric conditions will require a takeoff run of approximately 2,000 feet at an operational altitude of 5,000 feet.

NOTE—A TURBO-CHARGED AIRCRAFT ENGINE PROVIDES SOME SLIGHT ADVANTAGE IN THAT IT PROVIDES SEA LEVEL HORSEPOWER UP TO A SPECIFIED ALTITUDE ABOVE SEA LEVEL.

1. *Density Altitude Advisories:* at airports with elevations of 2,000 feet and higher, control towers and FSSs will broadcast the advisory "Check Density Altitude" when the temperature reaches a predetermined level. These advisories will be broadcast on appropriate tower frequencies or, where available, ATIS. FSSs will broadcast these advisories as a part of Local Airport Advisory, and on TWEB.

2. These advisories are provided by air traffic facilities, as a reminder to pilots that high temperatures and high field elevations will cause significant changes in aircraft characteristics. The pilot retains the responsibility to compute density altitude, when appropriate, as a part of preflight duties.

NOTE—ALL FSSs WILL COMPUTE THE CURRENT DENSITY ALTITUDE UPON REQUEST.

i. Mountain Wave: Many pilots go all their lives without understanding what a mountain wave is. Quite a few have lost their lives because of this lack of understanding. One need not be a licensed meteorologist to understand the mountain wave phenomenon.

1. Mountain waves occur when air is being blown over a mountain range or even the ridge of a sharp bluff area. As the air hits the upwind side of the range, it starts to climb, thus creating what is generally a smooth updraft which turns into a turbulent downdraft as the air passes the crest of the ridge. From this point, for many miles downwind, there will be a series of downdrafts and updrafts. Satellite photos of the Rockies have shown mountain waves extending as far as 700 miles downwind of the range. Along the east coast area, such photos of the Appalachian chain have picked up the mountain wave phenomenon over a hundred miles eastward. All it takes to form a mountain wave is wind blowing across the range at 15 knots or better at an intersection angle of not less than 30 degrees.

2. Pilots from flatland areas should understand a few things about mountain waves in order to stay out of trouble. When approaching a mountain range from the upwind side (generally the west), there will usually be a smooth updraft; therefore, it is not quite as dangerous an area as the lee of the range. From the leeward side, it is always a good idea to add an extra thousand feet or so of altitude because downdrafts can exceed the climb capability of the aircraft. Never expect an updraft when approaching a mountain chain from the leeward. Always be prepared to cope with a downdraft and turbulence.

3. When approaching a mountain ridge from the downwind side, it is recommended that the ridge be approached at approximately a 45 degree angle to the horizontal direction of the ridge. This permits a safer retreat from the ridge with less stress on the aircraft should severe turbulence and downdraft be experienced. If severe turbulence is encountered, simultaneously reduce power and adjust pitch until aircraft approaches maneuvering speed, then adjust power and trim to maintain maneuvering speed and fly away from the turbulent area.

7-5-7. USE OF RUNWAY HALF-WAY SIGNS AT UNIMPROVED AIRPORTS

When installed, runway half-way signs provide the pilot with a reference point to judge takeoff acceleration trends. Assuming that the runway length is appropriate for takeoff (considering runway condition and slope, elevation, aircraft weight, wind, and temperature), typical takeoff acceleration should allow the airplane to reach 70% of lift-off airspeed by the midpoint of the runway. The "rule of thumb" is that should airplane acceleration not allow the airspeed to reach this value by the midpoint, the takeoff should be aborted, as it may not be possible to liftoff in the remaining runway.

Several points are important when considering using this "rule of thumb";

Figure 7-5-1. Typical Runway Half-way Sign

NOTE—NO FAA STANDARD EXISTS FOR THE APPEARANCE OF THE RUNWAY HALF-WAY SIGN. FIG 7-5-1 SHOWS A GRAPHICAL DEPICTION OF A TYPICAL RUNWAY HALF-WAY SIGN.

a. Airspeed indicators in small airplanes are not required to be evaluated at speeds below stalling, and may not be usable at 70% of liftoff airspeed.

b. This "rule of thumb" is based on a uniform surface condition. Puddles, soft spots, areas of tall and/or wet grass, loose gravel, etc., may impede acceleration or even cause deceleration. Even if the airplane achieves 70% of liftoff airspeed by the midpoint, the condition of the remainder of the runway may not allow further acceleration. The entire length of the runway should be inspected prior to takeoff to ensure a usable surface.

c. This "rule of thumb" applies only to runway required for actual liftoff. In the event that obstacles affect the takeoff climb path, appropriate distance must be available after liftoff to accelerate to best angle of climb speed and to clear the obstacles. This will, in effect, require the airplane to accelerate to a higher speed by midpoint, particularly if the obstacles are close to the end of the runway. In addition, this technique does not take into account the effects of upslope or tailwinds on takeoff performance. These factors will also require greater acceleration than normal and, under some circumstances, prevent takeoff entirely.

d. Use of this "rule of thumb" does not alleviate the pilot's responsibility to comply with applicable Federal Aviation Regulations, the limitations and performance data provided in the FAA approved Airplane Flight Manual (AFM), or, in the absence of an FAA approved AFM, other data provided by the aircraft manufacturer.

In addition to their use during takeoff, runway half-way signs offer the pilot increased awareness of his or her position along the runway during landing operations.

Table 7-5-1   Jurisdictions Controlling Navigable Bodies of Water
AUTHORITY TO CONSULT FOR USE OF A BODY OF WATER

Location	Authority	Contact
Wilderness Area	U.S. Department of Agriculture, Forest Service	Local forest ranger
National Forest	USDA Forest Service	Local forest ranger
National Park	U.S. Department of the Interior, National Park Service	Local park ranger
Indian Reservation	USDI, Bureau of Indian Affairs	Local Bureau office
State Park	State government or state forestry or park service	Local state aviation office for further information
Canadian National and Provincial Parks	Supervised and restricted on an individual basis from province to province and by different departments of the Canadian government; consult Canadian Flight Information Manual and/or Water Aerodrome Supplement	Park Superintendent in an emergency

7-5-8. SEAPLANE SAFETY

a. Acquiring a seaplane class rating affords access to many areas not available to landplane pilots. Adding a seaplane class rating to your pilot certificate can be relatively uncomplicated and inexpensive. However, more effort is required to become a safe, efficient, competent "bush" pilot. The natural hazards of the backwoods have given way to modern man-made hazards. Except for the far north, the available bodies of water are no longer the exclusive domain of the airman. Seaplane pilots must be vigilant for hazards such as electric power lines, power, sail and rowboats, rafts, mooring lines, water skiers, swimmers, etc.

b. Seaplane pilots must have a thorough understanding of the right-of-way rules as they apply to aircraft versus other vessels. Seaplane pilots are expected to know and adhere to both the United States Coast Guard's (USCG) Navigation Rules, International-Inland, and 14 CFR Section and 91.115, Right-of-Way Rules; Water Operations. The navigation rules of the road are a set of collision avoidance rules as they apply to aircraft on the water. A seaplane is considered a vessel when on the water for the purposes of these collision avoidance rules. In general, a seaplane on the water

shall keep well clear of all vessels and avoid impeding their navigation. The FAR requires, in part, that aircraft operating on the water "...shall, insofar as possible, keep clear of all vessels and avoid impeding their navigation, and shall give way to any vessel or other aircraft that is given the right-of-way..." This means that a seaplane should avoid boats and commercial shipping when on the water. If on a collision course, the seaplane should slow, stop, or maneuver to the right, away from the bow of the oncoming vessel. Also, while on the surface with an engine running, an aircraft must give way to all nonpowered vessels. Since a seaplane in the water may not be as maneuverable as one in the air, the aircraft on the water has right-of-way over one in the air, and one taking off has right-of-way over one landing. A seaplane is exempt from the USCG safety equipment requirements, including the requirements for Personal Flotation Devices (PFD). Requiring seaplanes on the water to comply with USCG equipment requirements in addition to the FAA equipment requirements would be an unnecessary burden on seaplane owners and operators.

c. Unless they are under Federal jurisdiction, navigable bodies of water are under the jurisdiction of the state, or in a few cases, privately owned. Unless they are specifically restricted, aircraft have as much right to operate on these bodies of water as other vessels. To avoid problems, check with Federal or local officials in advance of operating on unfamiliar waters. In addition to the agencies listed in TBL 7-5-1, the nearest Flight Standards District Office can usually offer some practical suggestions as well as regulatory information. If you land on a restricted body of water because of an inflight emergency, or in ignorance of the restrictions you have violated, report as quickly as practical to the nearest local official having jurisdiction and explain your situation.

d. When operating a seaplane over or into remote areas, appropriate attention should be given to survival gear. Minimum kits are recommended for summer and winter, and are required by law for flight into sparsely settled areas of Canada and Alaska. Alaska State Department of Transportation and Canadian Ministry of Transport officials can provide specific information on survival gear requirements. The kit should be assembled in one container and be easily reachable and preferably floatable.

e. The FAA recommends that each seaplane owner or operator provide flotation gear for occupants any time a seaplane operates on or near water. 14 CFR Section 91.205(b) (11) requires approved flotation gear for aircraft operated for hire over water and beyond power-off gliding distance from shore. FAA-approved gear differs from that required for navigable waterways under USCG rules. FAA-approved life vests are inflatable designs as compared to the USCG's noninflatable PFDs that may consist of solid, bulky material. Such USCG PFDs are impractical for seaplanes and other aircraft because they may block passage through the relatively narrow exits available to pilots and passengers. Life vests approved under Technical Standard

Order (TSO) C13E contain fully inflatable compartments. The wearer inflates the compartments (AFTER exiting the aircraft) primarily by independent CO_2 cartridges, with an oral inflation tube as a backup. The flotation gear also contains a water-activated, self-illuminating signal light. The fact that pilots and passengers can easily don and wear inflatable life vests (when not inflated) provides maximum effectiveness and allows for unrestricted movement. It is imperative that passengers are briefed on the location and proper use of available PFDs prior to leaving the dock.

f. The FAA recommends that seaplane owners and operators obtain Advisory Circular AC 91-69, Seaplane Safety, for FAR Part 91 Operations free from the U.S. Department of Transportation, Subsequent Distribution office, SVC-121.23 Ardmore East Business Center, 3341 Q 75th Avenue, Landover, MD 20785; fax (301) 386-5394. The USCG Navigation Rules International-Inland (COMDTINSTM 16672.2B) is available for a fee from the government printing office facsimile request to (202) 512-2250, and can be ordered using Mastercard or Visa.

7-5-9. FLIGHT OPERATIONS IN VOLCANIC ASH

a. Severe volcanic eruptions which send ash into the upper atmosphere occur somewhere around the world several times each year. Flying into a volcanic ash cloud can be exceedingly dangerous. A B747-200 lost all four engines after such an encounter and a B747-400 had the same nearly catastrophic experience. Piston-powered aircraft are less likely to lose power but severe damage is almost certain to ensue after an encounter with a volcanic ash cloud which is only a few hours old.

b. Most important is to avoid any encounter with volcanic ash. The ash plume may not be visible, especially in instrument conditions or at night; and even if visible, it is difficult to distinguish visually between an ash cloud and an ordinary weather cloud. Volcanic ash clouds are not displayed on airborne or ATC radar. The pilot must rely on reports from air traffic controllers and other pilots to determine the location of the ash cloud and use that information to remain well clear of the area. Every attempt should be made to remain on the upwind side of the volcano.

c. It is recommended that pilots encountering an ash cloud should immediately reduce thrust to idle (altitude permitting), and reverse course in order to escape from the cloud. Ash clouds may extend for hundreds of miles and pilots should not attempt to fly through or climb out of the cloud. In addition, the following procedures are recommended:

1. Disengage the autothrottle if engaged. This will prevent the autothrottle from increasing engine thrust;

2. Turn on continuous ignition;

3. Turn on all accessory airbleeds including all air conditioning packs, nacelles, and wing anti-ice. This will provide an additional engine stall margin by reducing engine pressure.

d. The following has been reported by flightcrews who have experienced encounters with volcanic dust clouds:

1. Smoke or dust appearing in the cockpit;

2. An acrid odor similar to electrical smoke;

3. Multiple engine malfunctions, such as compressor stalls, increasing EGT, torching from tailpipe, and flameouts;

4. At night, St. Elmo's fire or other static discharges accompanied by a bright orange glow in the engine inlets;

5. A fire warning in the forward cargo area.

e. It may become necessary to shut down and then restart engines to prevent exceeding EGT limits. Volcanic ash may block the pitot system and result in unreliable airspeed indications.

f. If you see a volcanic eruption and have not been previously notified of it, you may have been the first person to observe it. In this case, immediately contact ATC and alert them to the existence of the eruption. If possible, use the Volcanic Activity Reporting form (VAR) depicted (on page 355) of this manual. Items 1 through 8 of the VAR should be transmitted immediately. The information requested in items 9 through 16 should be passed after landing. If a VAR form is not immediately available, relay enough information to identify the position and nature of the volcanic activity. Do not become unnecessarily alarmed if there is merely steam or very low-level eruptions of ash.

g. When landing at airports where volcanic ash has been deposited on the runway be aware that even a thin layer of dry ash can be detrimental to braking action. Wet ash on the runway may also reduce effectiveness of braking. It is recommended that reverse thrust be limited to minimum practical to reduce the possibility of reduced visibility and engine ingestion of airborne ash.

h. When departing from airports where volcanic ash has been deposited it is recommended that pilots avoid operating in visible airborne ash. Allow ash to settle before initiating takeoff roll. It is also recommended that flap extension be delayed until initiating the before takeoff checklist and that a rolling takeoff be executed to avoid blowing ash back into the air.

7-5-10. EMERGENCY AIRBORNE INSPECTION OF OTHER AIRCRAFT

a. Providing airborne assistance to another aircraft may involve flying in very close proximity to that aircraft. Most pilots receive little, if any, formal training or instruction in this type of flying activity. Close proximity flying without sufficient time to plan (i.e., in an emergency situation), coupled with the stress involved in a perceived emergency can be hazardous.

b. The pilot in the best position to assess the situation should take the responsibility of coordinating the airborne intercept and inspection, and take into account the unique flight characteristics and differences of the category(s) of aircraft involved.

c. Some of the safety considerations are:

1. Area, direction and speed of the intercept;

2. Aerodynamic effects (i.e., rotorcraft downwash);

3. Minimum safe separation distances;

4. Communications requirements, lost communications procedures, coordination with ATC;

5. Suitability of diverting the distressed aircraft to the nearest safe airport; and

6. Emergency actions to terminate the intercept.

d. Close proximity, inflight inspection of another aircraft is uniquely hazardous. The pilot in command of the aircraft experiencing the problem/emergency must not relinquish control of the situation and/or jeopardize the safety of their aircraft. The maneuver must be accomplished with minimum risk to both aircraft.

7-5-11. PRECIPITATION STATIC

a. Precipitation Static is caused by aircraft in flight coming in contact with uncharged particles. These particles can be rain, snow, fog, sleet, hail, volcanic ash, dust; any solid or liquid particles. When the aircraft strikes these neutral particles the positive element of the particle is reflected away from the aircraft and the negative particles adheres to the skin of the aircraft. In a very short period of time a substantial negative charge will develop on the skin of the aircraft. If the aircraft is not equipped with static dischargers, or has an ineffective static discharger system, when a sufficient negative voltage level is reached, the aircraft may go into "CORONA." That is, it will discharge the static electricity from the extremities of the aircraft, such as the wing tips, horizontal stabilizer, vertical stabilizer, antenna, propeller tips, etc. This discharge of static electricity is what you will hear in your headphones and is what we call P-static.

b. A review of pilot reports often shows different symptoms with each problem that is encountered. The following list of problems is a summary of many pilot reports from many different aircraft. Each problem was caused by P-static:

1. Complete loss of VHF communications

2. Erroneous magnetic compass readings (30% in error)

3. High pitched squeal on audio

4. Motor boat sound on audio

5. Loss of all avionics in clouds

6. VLF navigation system inoperative most of the time

7. Erratic instrument readouts

8. Weak transmissions and poor receptivity of radios

9. "St. Elmo's Fire" on windshield

c. Each of these symptoms is caused by one general problem on the airframe. This problem is the inability of the accumulated charge to flow easily to the wing tips and tail of the airframe, and properly discharge to the airstream.

d. Static dischargers work on the principal of creating a relatively easy path for discharging negative charges that develop on the aircraft by using a discharger with fine metal points, carbon coated rods, or carbon wicks rather than wait until a large charge is developed and discharged off the trailing edges of the aircraft that will interfere with avionics equipment. This process offers approximately 50 decibels (dB) static noise reduction which is adequate in most cases to be below the threshold of noise that would cause interference in avionics equipment.

e. It is important to remember that precipitation static problems can only be corrected with the proper number of quality static dischargers, properly installed on a properly bonded aircraft. P-static is indeed a problem in the all weather operation of the aircraft, but there are effective ways to combat it. All possible methods of reducing the effects of P-static should be considered so as to provide the best possible performance in the flight environment.

f. A wide variety of discharger designs is available on the commercial market. The inclusion of well-designed dischargers may be expected to improve airframe noise in P-static conditions by as much as 50 dB. Essentially, the discharger provides a path by which accumulated charge may leave the airframe quietly. This is generally accomplished by providing a group of tiny corona points to permit onset of corona-current flow at a low aircraft potential. Additionally, aerodynamic design of dischargers to permit corona to occur at the lowest possible atmospheric pressure also lowers the corona threshold. In addition to permitting a low-potential discharge, the discharger will minimize the radiation of radio frequency (RF) energy which accompanies the corona discharge, in order to minimize effects of RF components at communications and navigation frequencies on avionics performance. These effects are reduced through resistive attachment of the corona point(s) to the airframe, preserving direct current connection but attenuating the higher-frequency components of the discharge.

g. Each manufacturer of static dischargers offers information concerning appropriate discharger location on specific airframes. Such locations emphasize the trailing outboard surfaces of wings and horizontal tail surfaces, plus the tip of the vertical stabilizer, where charge tends to accumulate on the airframe. Sufficient dischargers must be provided to allow for current-carrying capacity which will maintain airframe potential below the corona threshold of the trailing edges.

h. In order to achieve full performance of avionic equipment, the static discharge system will require periodic maintenance. A pilot knowledgable of P-static causes and effects is an important element in assuring optimum performance by early recognition of these types of problems.

7-5-12. LIGHT AMPLIFICATION BY STIMULATED EMISSION OF RADIATION (LASER) OPERATIONS AND REPORTING ILLUMINATION OF AIRCRAFT

a. Lasers have many applications. Of concern to users of the National Airspace System are those laser events that may affect pilots, e.g., outdoor laser light shows or demonstrations for entertainment and advertisements at special events and theme parks. Generally, the beams from these events appear as bright blue-green in color; however, they may be red, yellow, or white. However, some laser systems produce light which is invisible to the human eye.

b. FAA regulations prohibit the disruption of aviation activity by any person on the ground or in the air. The FAA and the Food and Drug Administration (the Federal agency that has the responsibility to enforce compliance with Federal requirements for laser systems and laser light show products) are working together to ensure that operators of these devices do not pose a hazard to aircraft operators.

c. Pilots should be aware that illumination from theselaser operations are able to create temporary vision impairment miles from the actual location. In addition, these operations can produce permanent eye damage. Pilots should make themselves aware of where these activities are being conducted and avoid these areas if possible.

d. Recent and increasing incidents of unauthorized illumination of aircraft by lasers, as well as the proliferation and increasing sophistication of laser devices available to the general public, dictates that the FAA, in coordination with other government agencies, take action to safeguard flights from these unauthorized illuminations.

e. Pilots should report laser illumination activity to the controlling Air Traffic Control facilities, Federal Contract Towers or Flight Service Stations as soon as possible after the event. The following information should be included:

1. UTC Date and Time of Event.

2. Call Sign or Aircraft Registration Number. 3. Type Aircraft.

4. Nearest Major City. 5. Altitude.

6. Location of Event (Latitude/Longitude and/ or Fixed Radial Distance (FRD)).

7. Brief Description of the Event and any other Pertinent Information.

f. Pilots are also encouraged to complete the Laser Beam Exposure Questionnaire (See Appendix 3), and fax it to the Washington Operations Center Complex (WOCC) as soon as possible after landing.

g. When a laser event is reported to an air traffic facility, a general caution warning will be broadcast on all appropriate frequencies every five minutes for 20 minutes and broadcasted on the ATIS for one hour following the report.

PHRASEOLOGY

UNAUTHORIZED LASER ILLUMINATION EVENT, (UTC time), (location), (altitude), (color), (direction).

EXAMPLE—"UNAUTHORIZED LASER ILLUMINATION EVENT, AT 0100Z, 8 MILE FINAL RUNWAY 18R AT 3,000 FEET, GREEN LASER FROM THE SOUTHWEST."

REFERENCE—FAA O 7110.65, UNAUTHORIZED LASER ILLUMINATION OF AIRCRAFT, PARA 10-2-14.

FARO 7210.3, REPORTING LASER ILLUMINATION OF AIRCRAFTI PARA 2-1-27.

h. When these activities become known to the FAA, Notices to Airmen (NOTAMs) are issued to inform the aviation community of the events. Pilots should consult NOTAMs or the Special Notices section of the Airport/Facility Directory for information regarding these activities.

7-5-13. FLYING IN FLAT LIGHT AND WHITE OUT CONDITIONS

a. Flat Light. Flat light is an optical illusion, also known as "sector or partial white out" It is not as severe as "white out" but the condition causes pilots to lose their depth-of-field and contrast in vision. Flat light conditions are usually accompanied by overcast skies inhibiting any visual clues. Such conditions can occur anywhere in the world, primarily in snow covered areas but can occur in dust, sand, mud flats, or on glassy water. Flat light can completely obscure features of the terrain, creating an inability to distinguish distances and closure rates. As a result of this reflected light, it can give pilots the illusion that they are ascending or descending when they may actually be flying level. However, with good judgment and proper training and planning, it is possible to safely operate an aircraft in flat light conditions.

b. White Out. As defined in meteorological terms, white out occurs when a person becomes engulfed in a uniformly white glow. The glow is a result of being surrounded by blowing snow, dust, sand, mud or water. There are no shadows, no horizon or clouds and all depth-of-field and orientation are lost. A white out situation is severe in that there are no visual references. Flying is not recommended in any white out situation. Flat light conditions can lead to a white out environment quite rapidly, and both atmospheric conditions are insidious; they sneak up on you as your visual references slowly begin to disappear. White out has been the cause of several aviation accidents.

c. Self Induced White Out. This effect typically occurs when a helicopter takes off or lands on a snow-covered area. The rotor down wash picks up particles and recirculates them through the rotor down wash. The effect can vary in intensity depending upon the amount of light on the surface. This can happen on the sunniest, brightest day with good contrast everywhere. However, when it happens, there can be a complete loss of visual clues. If the pilot has not prepared for this immediate loss of visibility, the results can be disastrous. Good planning does not prevent one from encountering flat light or white out conditions.

d. Never Take Off in a White Out Situation.

1. Realize that in flat light conditions it may be possible to depart but not to return to that site. During takeoff, make sure you have a reference point. Do not lose sight of it until you have a departure reference point in view. Be prepared to return to the takeoff reference if the departure reference does not come into view.

2. Flat light is common to snow skiers. One way to compensate for the lack of visual contrast and depth-of-field loss is by wearing amber tinted lenses (also known as blue blockers). Special note of caution: Eyewear is not ideal for every pilot. Take into consideration personal factors - age, light sensitivity, and ambient lighting conditions.

3. So what should a pilot do when all visual references are lost?

(a) Trust the cockpit instruments.

(b) Execute a 180 degree turnaround and start looking for outside references.

(c) Above all - fly the aircraft.

e. Landing in Low Light Conditions. When landing in a low light condition - use extreme caution. Look for intermediate reference points, in addition to checkpoints along each leg of the route for course confirmation and timing. The lower the ambient light becomes, the more reference points a pilot should use.

f. Airport Landings.

1. Look for features around the airport or approach path that can be used in determining depth perception. Buildings, towers, vehicles or other aircraft serve well for this measurement. Use something that will provide you with a sense of height above the ground, in addition to orienting you to the runway.

2. Be cautious of snowdrifts and snow banks - anything that can distinguish the edge of the runway. Look for subtle changes in snow texture or shading to identify ridges or changes in snow depth.

g. Off-Airport Landings.

1. In the event of an off-airport landing, pilots have used a number of different visual cues to gain reference. Use whatever you must to create the contrast you need. Natural references seem to work best (trees, rocks, snow ribs, etc.)

(a) Over flight.

(b) Use of markers.

(c) Weighted flags.

(d) Smoke bombs.

(e) Any colored rags.

(f) Dye markers.

(g) Kool-aid.

(h) Trees or tree branches.

2. It is difficult to determine the depth of snow in areas that are level. Dropping items from the aircraft to use as reference points should be used as a visual aid only and not as a primary landing reference. Unless your marker is biodegradable, be sure to retrieve it after landing. Never put yourself in a position where no visual references exist.

3. Abort landing if blowing snow obscures your reference. Make your decisions early. Don't assume you can pick up a lost reference point when you get closer.

4. Exercise extreme caution when flying from sunlight into shade. Physical awareness may tell you that you are flying straight but you may actually be in a spiral dive with centrifugal force pressing against you. Having no visual references enhances this illusion. Just because you have a good visual reference does not mean that it's safe to continue. There may be snow-covered terrain not visible in the direction that you are traveling. Getting caught in a no visual reference situation can be fatal.

h. Flying Around a Lake.

1. When flying along lakeshores, use them as a reference point. Even if you can see the other side, realize that your depth perception may be poor. It is easy to fly into the surface. If you must cross the lake, check the altimeter frequently and maintain a safe altitude while you still have a good reference. Don't descend below that altitude.

2. The same rules apply to seemingly flat areas of snow. If you don't have good references, avoid going there.

i. Other Traffic. Be on the look out for other traffic in the area. Other aircraft may be using your same reference point. Chances are greater of colliding with someone traveling in the same direction as you, than someone flying in the opposite direction.

j. Ceilings. Low ceilings have caught many pilots off guard. Clouds do not always form parallel to the surface, or at the same altitude. Pilots may try to compensate for this by flying with a slight bank and thus creating a descending turn.

k. Glaciers. Be conscious of your altitude when flying over glaciers. The glaciers may be rising faster than you are climbing.

7-5-14. Operations in Ground icing Conditions

a. The presence of aircraft airframe icing during takeoff, typically caused by improper or no deicing of the aircraft being accomplished prior to flight has contributed to many recent accidents in turbine aircraft. The General Aviation Joint Steering Committee (GAJSC) is the primary vehicle for government-industry cooperation, communication, and coordination on GA accident mitigation. The Turbine Aircraft Operations Subgroup (TAOS) works to mitigate accidents in turbine accident aviation. While there is sufficient information and guidance currently available regarding the effects of icing on aircraft and methods for deicing, the TAOS has developed a list of recommended actions to further assist pilots and operators in this area.

While the efforts of the TAOS specifically focus of turbine aircraft, it is recognized that their recommendations are applicable to and can be adapted for the pilot of a small, piston powered aircraft too.

b. The following recommendations are offered:

1. Ensure that your aircraft's lift-generating surfaces are COMPLETELY free of contamination before flight through a tactile (hands on) check of the critical surfaces when feasible. Even when otherwise permitted, operators should avoid smooth or polished frost on lift-generating surfaces as an acceptable preflight condition.

2. Review and refresh your cold weather standard operating procedures.

3. Review and be familiar with the Airplane Flight Manual (AFM) limitations and procedures necessary to deal with icing conditions prior to flight, as well as in flight.

4. Protect your aircraft while on the ground, if possible, from sleet and freezing rain by taking advantage of aircraft hangars.

5. Take full advantage of the opportunities available at airports for deicing. Do not refuse deicing services simply because of cost.

6. Always consider canceling or delaying a flight if weather conditions do not support a safe operation.

c. If you haven't already developed a set of Standard Operating Procedures for cold weather operations, they should include:

1. Procedures based on information that is applicable to the aircraft operated, such as AFM limitations and procedures;

2. Concise and easy to understand guidance that outlines best operational practices;

3. A systematic procedure for recognizing, evaluating and addressing the associated icing risk, and offer clear guidance to mitigate this risk;

4. An aid (such as a checklist or reference cards) that is readily available during normal day-to-day aircraft operations.

d. There are several sources for guidance relating to airframe icing, including:

1. http://aircrafticing.gre.nasa.gov/index. html

2. http://www.ibac.org/is-bao/isbao.htm

3. http://www.natasafetylst.org/bus deice.htm

4. Advisory Circular (AC) 91-74, Pilot Guide, Flight in Icing Conditions.

5. AC 135-17, Pilot Guide Small Aircraft Ground Deicing.

6. AC 135-9, FAR Part 135 Icing Limitations.

7. AC 120-60, Ground Deicing and Anti-icing Program.

8. AC 135-16, Ground Deicing and Anti-icing Training and Checking.

The FAA Approved Deicing Program Updates is published annually as a Flight Standards Information Bulletin for Air Transportation and contains detailed information on deicing and anti-icing procedures and holdover times. It may be accessed at the following web site by selecting the current year's information bulletins:

http://www.faa.gov/library/manuals/examiners_inspectors/8400/fsat

Section 6. SAFETY, ACCIDENT, AND HAZARD REPORTS

7-6-1. AVIATION SAFETY REPORTING PROGRAM

a. The FAA has established a voluntary Aviation Safety Reporting Program designed to stimulate the free and unrestricted flow of information concerning deficiencies and discrepancies in the aviation system. This is a positive program intended to ensure the safest possible system by identifying and correcting unsafe conditions before they lead to accidents. The primary objective of the program is to obtain information to evaluate and enhance the safety and efficiency of the present system.

b. This cooperative safety reporting program invites pilots, controllers, flight attendants, maintenance personnel and other users of the airspace system, or any other person, to file written reports of actual or potential discrepancies and deficiencies involving the safety of aviation operations. The operations covered by the program include departure, en route, approach, and landing operations and procedures, air traffic control procedures and equipment, crew and air traffic control communications, aircraft cabin operations, aircraft movement on the airport, near midair collisions, aircraft maintenance and record keeping and airport conditions or services.

c. The report should give the date, time, location, persons and aircraft involved (if applicable), nature of the event, and all pertinent details.

d. To ensure receipt of this information, the program provides for the waiver of certain disciplinary actions against persons, including pilots and air traffic controllers, who file timely written reports concerning potentially unsafe incidents. To be considered timely, reports must be delivered or postmarked within 10 days of the incident unless that period is extended for good cause. Reports should be submitted on NASA ARC Forms 277, which are available free of charge, postage prepaid, at FAA Flight Standards District Offices and Flight Service Stations, and from NASA, ASRS, PO Box 189, Moffet Field, CA 94035.

e. The FAA utilizes the National Aeronautics and Space Administration (NASA) to act as an independent third party to receive and analyze reports submitted under the program. This program is described in AC 00-46, Aviation Safety Reporting Program.

7-6-2. AIRCRAFT ACCIDENT AND INCIDENT REPORTING

a. Occurrences Requiring Notification: The operator of an aircraft shall immediately, and by the most expeditious means available, notify the nearest National Transportation Safety Board (NTSB) Field Office when:

1. An aircraft accident or any of the following listed incidents occur:

(a) Flight control system malfunction or failure;

(b) Inability of any required flight crew member to perform their normal flight duties as a result of injury or illness;

(c) Failure of structural components of a turbine engine excluding compressor and turbine blades and vanes;

(d) Inflight fire;

(e) Aircraft collide in flight;

(f) Damage to property, other than the aircraft, estimated to exceed $25,000 for repair (including materials and labor) or fair market value in the event of total loss, whichever is less.

(g) For large multi-engine aircraft (more than 12,500 pounds maximum certificated takeoff weight):

(1) Inflight failure of electrical systems which requires the sustained use of an emergency bus powered by a back-up source such as a battery, auxiliary power unit, or air-driven generator to retain flight control or essential instruments;

(2) Inflight failure of hydraulic systems that results in sustained reliance on the sole remaining hydraulic or mechanical system for movement of flight control surfaces;

(3) Sustained loss of the power or thrust produced by two or more engines; and

(4) An evacuation of aircraft in which an emergency egress system is utilized.

2. An aircraft is overdue and is believed to have been involved in an accident.

b. Manner of Notification:

1. The most expeditious method of notification to the NTSB by the operator will be determined by the circumstances existing at that time. The NTSB has advised that any of the following would be considered examples of the type of notification that would be acceptable:

(a) Direct telephone notification.

(b) Telegraphic notification.

(c) Notification to the FAA who would in turn notify the NTSB by direct communication; i.e., dispatch or telephone.

c. Items To Be Notified: The notification required above shall contain the following information, if available:

1. Type, nationality, and registration marks of the aircraft;

2. Name of owner and operator of the aircraft;

3. Name of the pilot-in-command;

4. Date and time of the accident, or incident;

5. Last point of departure, and point of intended landing of the aircraft;

6. Position of the aircraft with reference to some easily defined geographical point;

7. Number of persons aboard, number killed, and number seriously injured;

8. Nature of the accident, or incident, the weather, and the extent of damage to the aircraft so far as is known; and

9. A description of any explosives, radioactive materials, or other dangerous articles carried.

d. Follow-up Reports:

1. The operator shall file a report on NTSB Form 6120.1or 6120.2, available from NTSB Field Offices or from the NTSB, Washington, DC, 20594:

(a) Within 10 days after an accident;

(b) When, after 7 days, an overdue aircraft is still missing;

(c) A report on an incident for which notification is required as described in subparagraph a (1) shall be filed only as requested by an authorized representative of the NTSB.

2. Each crewmember, if physically able at the time the report is submitted, shall attach a statement setting forth the facts, conditions and circumstances relating to the accident or incident as they appeared. If the crewmember is incapacitated, a statement shall be submitted as soon as physically possible.

e. Where To File the Reports:

1. The operator of an aircraft shall file with the NTSB Field Office nearest the accident or incident any report required by this section.

2. The NTSB Field Offices are listed under U.S. Government in the telephone directories in the following cities: Anchorage, AK; Atlanta, GA; Chicago, IL; Denver, CO; Fort Worth, TX; Los Angeles, CA; Miami, FL; Parsippany, NJ; Seattle, WA.

7-6-3. NEAR MIDAIR COLLISION REPORTING

a. Purpose and Data Uses: The primary purpose of the Near Midair Collision (NMAC) Reporting Program is to provide information for use in enhancing the safety and efficiency of the National Airspace System. Data obtained from NMAC reports are used by the FAA to improve the quality of FAA services to users and to develop programs, policies, and procedures aimed at the reduction of NMAC occurrences. All NMAC reports are thoroughly investigated by Flight Standards Facilities in coordination with Air Traffic Facilities. Data from these investigations are transmitted to FAA Headquarters in Washington, DC, where they are compiled and analyzed, and where safety programs and recommendations are developed.

b. Definition: A near midair collision is defined as an incident associated with the operation of an aircraft in which a possibility of collision occurs as a result of proximity of less than 500 feet to another aircraft, or a report is received from a pilot or a flight crew member stating that a collision hazard existed between two or more aircraft.

c. Reporting Responsibility: It is the responsibility of the pilot and/or flight crew to determine whether a near midair collision did actually occur and, if so, to initiate an NMAC report. Be specific, as ATC will not interpret a casual remark to mean that an NMAC is being reported. The pilot should state "I wish to report a near midair collision."

d. Where To File Reports: Pilots and/or flight crew members involved in NMAC occurrences are urged to report each incident immediately:

1. By radio or telephone to the nearest FAA ATC facility or FSS.

2. In writing, in lieu of the above, to the nearest Flight Standards District Office (FSDO).

e. Items To Be Reported:

1. Date and Time (UTC) of incident.

2. Location of incident and altitude.

3. Identification and type of reporting aircraft, aircrew destination, name and home base of pilot.

4. Identification and type of other aircraft, aircrew destination, name and home base of pilot.

5. Type of flight plans; station altimeter setting used.

6. Detailed weather conditions at altitude or Flight Level.

7. Approximate courses of both aircraft: indicate if one or both aircraft were climbing or descending.

8. Reported separation in distance at first sighting, proximity at closest point horizontally and vertically, and length of time in sight prior to evasive action.

9. Degree of evasive action taken, if any (from both aircraft, if possible).

10. Injuries, if any.

f. Investigation: The FSDO in whose area the incident occurred is responsible for the investigation and reporting of NMACs.

g. Existing radar, communication, and weather data will be examined in the conduct of the investigation.

When possible, all cockpit crew members will be interviewed regarding factors involving the NMAC incident. Air Traffic controllers will be interviewed in cases where one or more of the involved aircraft was provided ATC service. Both flight and ATC procedures will be evaluated. When the investigation reveals a violation of an FAA regulation, enforcement action will be pursued.

7-6-4. UNIDENTIFIED FLYING OBJECTS (UFO) REPORTS

a. Persons wanting to report UFO/Unexplained phenomena activity should contact an UFO/Unexplained Phenomena Reporting Data Collection Center, such as the National Institute for Discovery Sciences (NIDS), the National UFO Reporting Centers, etc.

b. If concern is expressed that life or property might be endangered, report the activity to the local law enforcement department.

c. NIDS will ask a series of questions (verbal and/or via questionnaire) concerning the event.

NOTE—NIDS IS THE SINGLE POINT OF CONTACT RECOGNIZED BY THE FAA IN REGARD TO UFO INFORMATION. THEY WILL MAINTAIN A NATIONAL DATABASE ON ANOMALOUS PHENOMENA AND PERIODICALLY SHARE THAT INFORMATION WITH THE FAA.

d. If concern is expressed that life or property might be endangered, refer the individual to the local police department.

318

Chapter 8. Medical facts for pilots
Section 1. FITNESS FOR FLIGHT

8-1-1. FITNESS FOR FLIGHT

a. Medical Certification:

1. All pilots except those flying gliders and free air balloons must possess valid medical certificates in order to exercise the privileges of their airman certificates. The periodic medical examinations required for medical certification are conducted by designated Aviation Medical Examiners, who are physicians with a special interest in aviation safety and training in aviation medicine.

2. The standards for medical certification are contained in FAR Part 67. Pilots who have a history of certain medical conditions described in these standards are mandatorily disqualified from flying. These medical conditions include a personality disorder manifested by overt acts, a psychosis, alcoholism, drug dependence, epilepsy, an unexplained disturbance of consciousness, myocardial infarction, angina pectoris and diabetes requiring medication for its control. Other medical conditions may be temporarily disqualifying, such as acute infections, anemia, and peptic ulcer. Pilots who do not meet medical standards may still be qualified under special issuance provisions or the exemption process. This may require that either additional medical information be provided or practical flight tests be conducted.

3. Student pilots should visit an Aviation Medical Examiner as soon as possible in their flight training in order to avoid unnecessary training expenses should they not meet the medical standards. For the same reason, the student pilot who plans to enter commercial aviation should apply for the highest class of medical certificate that might be necessary in the pilot's career.

CAUTION—THE FARs PROHIBIT A PILOT WHO POSSESSES A CURRENT MEDICAL CERTIFICATE FROM PERFORMING CREWMEMBER DUTIES WHILE THE PILOT HAS A KNOWN MEDICAL CONDITION OR INCREASE OF A KNOWN MEDICAL CONDITION THAT WOULD MAKE THE PILOT UNABLE TO MEET THE STANDARDS FOR THE MEDICAL CERTIFICATE.

b. Illness:

1. Even a minor illness suffered in day-to-day living can seriously degrade performance of many piloting tasks vital to safe flight. Illness can produce fever and distracting symptoms that can impair judgment, memory, alertness, and the ability to make calculations. Although symptoms from an illness may be under adequate control with a medication, the medication itself may decrease pilot performance.

2. The safest rule is not to fly while suffering from any illness. If this rule is considered too stringent for a particular illness, the pilot should contact an Aviation Medical Examiner for advice.

c. Medication:

1. Pilot performance can be seriously degraded by both prescribed and over-the-counter medications, as well as by the medical conditions for which they are taken. Many medications, such as tranquilizers, sedatives, strong pain relievers, and cough-suppressant preparations, have primary effects that may impair judgment, memory, alertness, coordination, vision, and the ability to make calculations. Others, such as antihistamines, blood pressure drugs, muscle relaxants, and agents to control diarrhea and motion sickness, have side effects that may impair the same critical functions. Any medication that depresses the nervous system, such as a sedative, tranquilizer or antihistamine, can make a pilot much more susceptible to hypoxia.

2. The FARs prohibit pilots from performing crewmember duties while using any medication that affects the faculties in any way contrary to safety. The safest rule is not to fly as a crewmember while taking any medication, unless approved to do so by the FAA.

d. Alcohol:

1. Extensive research has provided a number of facts about the hazards of alcohol consumption and flying. As little as one ounce of liquor, one bottle of beer or four ounces of wine can impair flying skills, with the alcohol consumed in these drinks being detectable in the breath and blood for at least 3 hours. Even after the body completely destroys a moderate amount of alcohol, a pilot can still be severely impaired for many hours by hangover. There is simply no way of increasing the destruction of alcohol or alleviating a hangover. Alcohol also renders a pilot much more susceptible to disorientation and hypoxia.

2. A consistently high alcohol related fatal aircraft accident rate serves to emphasize that alcohol and flying are a potentially lethal combination. The FARs prohibit pilots from performing crewmember duties within 8 hours after drinking any alcoholic beverage or while under the influence of alcohol. However, due to the slow destruction of alcohol, a pilot may still be under influence 8 hours after drinking a moderate amount of alcohol. Therefore, an excellent rule is to allow at least 12 to 24 hours between "bottle and throttle," depending on the amount of alcoholic beverage consumed.

e. Fatigue:

1. Fatigue continues to be one of the most treacherous hazards to flight safety, as it may not be apparent to a pilot until serious errors are made. Fatigue is best described as either acute (short-term) or chronic (long-term).

2. A normal occurrence of everyday living, acute fatigue is the tiredness felt after long periods of physical and mental strain, including strenuous muscular effort, immobility, heavy mental workload, strong emotional pressure, monotony, and lack of sleep. Consequently, coordination and alertness, so vital to safe pilot performance, can be reduced. Acute fatigue is prevented by adequate rest and sleep, as well as by regular exercise and proper nutrition.

3. Chronic fatigue occurs when there is not enough time for full recovery between episodes of acute fatigue. Performance continues to fall off, and judgment becomes impaired so that unwarranted risks may be taken. Recovery from chronic fatigue requires a prolonged period of rest.

f. Stress:

1. Stress from the pressures of everyday living can impair pilot performance, often in very subtle ways. Difficulties, particularly at work, can occupy thought processes enough to markedly decrease alertness. Distraction can so interfere with judgment that unwarranted risks are taken, such as flying into deteriorating weather conditions to keep on schedule. Stress and fatigue (see above) can be an extremely hazardous combination.

2. Most pilots do not leave stress "on the ground." Therefore, when more than usual difficulties are being experienced, a pilot should consider delaying flight until these difficulties are satisfactorily resolved.

g. Emotion:

Certain emotionally upsetting events, including a serious argument, death of a family member, separation or divorce, loss of job, and financial catastrophe, can render a pilot unable to fly an aircraft safely. The emotions of anger, depression, and anxiety from such events not only decrease alertness but also may lead to taking risks that border on self-destruction. Any pilot who experiences an emotionally upsetting event should not fly until satisfactorily recovered from it.

h. Personal Checklist: Aircraft accident statistics show that pilots should be conducting preflight checklists on themselves as well as their aircraft for pilot impairment contributes to many more accidents than failures of aircraft systems. A personal checklist, which includes all of the categories of pilot impairment as discussed in this section, that can be easily committed to memory is being distributed by the FAA in the form of a wallet-sized card.

i. PERSONAL CHECKLIST: *I'm physically and mentally safe to fly, not being impaired by:*

Illness

Medication

Stress

Alcohol

Fatigue

Emotion

8-1-2. EFFECTS OF ALTITUDE

a. Hypoxia:

1. Hypoxia is a state of oxygen deficiency in the body sufficient to impair functions of the brain and other organs.

Hypoxia from exposure to altitude is due only to the reduced barometric pressures encountered at altitude, for the concentration of oxygen in the atmosphere remains about 21 percent from the ground out to space.

2. Although a deterioration in night vision occurs at a cabin pressure altitude as low as 5,000 feet, other significant effects of altitude hypoxia usually do not occur in the normal healthy pilot below 12,000 feet. From 12,000 to 15,000 feet of altitude, judgment, memory, alertness, coordination and ability to make calculations are impaired, and headache, drowsiness, dizziness and either a sense of well-being (euphoria) or belligerence occur. The effects appear following increasingly shorter periods of exposure to increasing altitude. In fact, pilot performance can seriously deteriorate within 15 minutes at 15,000 feet.

3. At cabin pressure altitudes above 15,000 feet, the periphery of the visual field grays out to a point where only central vision remains (tunnel vision). A blue coloration (cyanosis) of the fingernails and lips develops. The ability to take corrective and protective action is lost in 20 to 30 minutes at 18,000 feet and 5 to 12 minutes at 20,000 feet, followed soon thereafter by unconsciousness.

4. The altitude at which significant effects of hypoxia occur can be lowered by a number of factors. Carbon monoxide inhaled in smoking or from exhaust fumes, lowered hemoglobin (anemia), and certain medications can reduce the oxygen-carrying capacity of the blood to the degree that the amount of oxygen provided to body tissues will already be equivalent to the oxygen provided to the tissues when exposed to a cabin pressure altitude of several thousand feet. Small amounts of alcohol and low doses of certain drugs, such as antihistamines, tranquilizers, sedatives and analgesics can, through their depressant action, render the brain much more susceptible to hypoxia. Extreme heat and cold, fever, and anxiety increase the body's demand for oxygen, and hence its susceptibility to hypoxia.

5. The effects of hypoxia are usually quite difficult to recognize, especially when they occur gradually. Since symptoms of hypoxia do not vary in an individual, the ability to recognize hypoxia can be greatly improved by experiencing and witnessing the effects of hypoxia during an altitude chamber "flight." The FAA provides this opportunity through aviation physiology training, which is conducted at the FAA Civil Aeromedical Institute and at many military facilities across the United States, to attend the Physiological Training Program at the Civil Aeromedical Institute, Mike Monroney Aeronautical Center, Oklahoma City, OK, contact by telephone (405) 954-6212, or by writing Airmen Education Program Branch, AAM-400, CAMI, Mike Monroney Aeronautical Center, P.O. Box 25082, Oklahoma City, OK 73125.

NOTE—TO ATTEND THE PHYSIOLOGICAL TRAINING PROGRAM AT ONE OF THE MILITARY INSTALLATIONS HAVING THE TRAINING CAPABILITY, AN APPLICATION FORM AND A FEE MUST BE SUBMITTED. FULL

PARTICULARS ABOUT LOCATION, FEES, SCHEDULING PROCEDURES, COURSE CONTENT, INDIVIDUAL REQUIREMENTS, ETC. ARE CONTAINED IN THE PHYSIOLOGICAL TRAINING APPLICATION, FORM NUMBER AC 3150-7, WHICH IS OBTAINED BY CONTACTING THE ACCIDENT PREVENTION SPECIALIST OR THE OFFICE FORMS MANAGER IN THE NEAREST FAA OFFICE.

6. Hypoxia is prevented by heeding factors that reduce tolerance to altitude, by enriching the inspired air with oxygen from an appropriate oxygen system, and by maintaining a comfortable, safe cabin pressure altitude. For optimum protection, pilots are encouraged to use supplemental oxygen above 10,000 feet during the day, and above 5,000 feet at night. The FARs require that at the minimum, flight crew be provided with and use supplemental oxygen after 30 minutes of exposure to cabin pressure altitudes between 12,500 and 14,000 feet and immediately on exposure to cabin pressure altitudes above 14,000 feet. Every occupant of the aircraft must be provided with supplemental oxygen at cabin pressure altitudes above 15,000 feet.

b. Ear Block:

1. As the aircraft cabin pressure decreases during ascent, the expanding air in the middle ear pushes the eustachian tube open, and by escaping down it to the nasal passages, equalizes in pressure with the cabin pressure. But during descent, the pilot must periodically open the eustachian tube to equalize pressure. This can be accomplished by swallowing, yawning, tensing muscles in the throat, or if these do not work, by a combination of closing the mouth, pinching the nose closed, and attempting to blow through the nostrils (Valsalva maneuver).

2. Either an upper respiratory infection, such as a cold or sore throat, or a nasal allergic condition can produce enough congestion around the eustachian tube to make equalization difficult. Consequently, the difference in pressure between the middle ear and aircraft cabin can build up to a level that will hold the eustachian tube closed, making equalization difficult if not impossible. The problem is commonly referred to as an "ear block."

3. An ear block produces severe ear pain and loss of hearing that can last from several hours to several days. Rupture of the ear drum can occur in flight or after landing. Fluid can accumulate in the middle ear and become infected.

4. An ear block is prevented by not flying with an upper respiratory infection or nasal allergic condition. Adequate protection is usually not provided by decongestant sprays or drops to reduce congestion around the eustachian tubes. Oral decongestants have side effects that can significantly impair pilot performance.

5. If an ear block does not clear shortly after landing, a physician should be consulted.

c. Sinus Block:

1. During ascent and descent, air pressure in the sinuses equalizes with the aircraft cabin pressure through small openings that connect the sinuses to the nasal passages. Either an upper respiratory infection, such as a cold or sinusitis, or a nasal allergic condition can produce enough congestion around an opening to slow equalization, and as the difference in pressure between the sinus and cabin mounts, eventually plug the opening. This "sinus block" occurs most frequently during descent.

2. A sinus block can occur in the frontal sinuses, located above each eyebrow, or in the maxillary sinuses, located in each upper cheek. It will usually produce excruciating pain over the sinus area. A maxillary sinus block can also make the upper teeth ache. Bloody mucus may discharge from the nasal passages.

3. A sinus block is prevented by not flying with an upper respiratory infection or nasal allergic condition. Adequate protection is usually not provided by decongestant sprays or drops to reduce congestion around the sinus openings. Oral decongestants have side effects that can impair pilot performance.

4. If a sinus block does not clear shortly after landing, a physician should be consulted.

d. Decompression Sickness After Scuba Diving:

1. A pilot or passenger who intends to fly after scuba diving should allow the body sufficient time to rid itself of excess nitrogen absorbed during diving. If not, decompression sickness due to evolved gas can occur during exposure to low altitude and create a serious inflight emergency.

2. The recommended waiting time before going to flight altitudes of up to 8,000 feet is at least 12 hours after diving which has not required controlled ascent (nondecompression stop diving), and at least 24 hours after diving which has required controlled ascent (decompression stop diving). The waiting time before going to flight altitudes above 8,000 feet should be at least 24 hours after any SCUBA dive. These recommended altitudes are actual flight altitudes above mean sea level (AMSL) and not pressurized cabin altitudes. This takes into consideration the risk of decompression of the aircraft during flight.

8-1-3. HYPERVENTILATION IN FLIGHT

a. Hyperventilation or an abnormal increase in the volume of air breathed in and out of the lungs, can occur subconsciously when a stressful situation is encountered in flight. As hyperventilation "blows off" excessive carbon dioxide from the body, a pilot can experience symptoms of lightheartedness, suffocation, drowsiness, tingling in the extremities, and coolness and react to them with even greater hyperventilation. Incapacitation can eventually result from incoordination, disorientation, and painful muscle spasms. Finally, unconsciousness can occur.

b. The symptoms of hyperventilation subside within a few minutes after the rate and depth of breathing are consciously brought back under control. The buildup of carbon dioxide in the body can be hastened by controlled breathing in and out of a paper bag held over the nose and mouth.

c. Early symptoms of hyperventilation and hypoxia are similar. Moreover, hyperventilation and hypoxia can occur at the same time. Therefore, if a pilot is using an oxygen system when symptoms are experienced, the oxygen regulator should immediately be set to deliver 100 percent oxygen, and then the system checked to assure that it has been functioning effectively before giving attention to rate and depth of breathing.

8-1-4. CARBON MONOXIDE POISONING IN FLIGHT

a. Carbon monoxide is a colorless, odorless, and tasteless gas contained in exhaust fumes. When breathed even in minute quantities over a period of time, it can significantly reduce the ability of the blood to carry oxygen. Consequently, effects of hypoxia occur.

b. Most heaters in light aircraft work by air flowing over the manifold. Use of these heaters while exhaust fumes are escaping through manifold cracks and seals is responsible every year for several nonfatal and fatal aircraft accidents from carbon monoxide poisoning.

c. A pilot who detects the odor of exhaust or experiences symptoms of headache, drowsiness, or dizziness while using the heater should suspect carbon monoxide poisoning, and immediately shut off the heater and open air vents. If symptoms are severe or continue after landing, medical treatment should be sought.

8-1-5. ILLUSIONS IN FLIGHT

a. Introduction: Many different illusions can be experienced in flight. Some can lead to spatial disorientation. Others can lead to landing errors. Illusions rank among the most common factors cited as contributing to fatal aircraft accidents.

b. Illusions Leading to Spatial Disorientation:

1. Various complex motions and forces and certain visual scenes encountered in flight can create illusions of motion and position. Spatial disorientation from these illusions can be prevented only by visual reference to reliable, fixed points on the ground or to flight instruments.

2. *The leans:* An abrupt correction of a banked attitude, which has been entered too slowly to stimulate the motion sensing system in the inner ear, can create the illusion of banking in the opposite direction. The disoriented pilot will roll the aircraft back into its original dangerous attitude, or if level flight is maintained, will feel compelled to lean in the perceived vertical plane until this illusion subsides.

(a) *Coriolis illusion:* An abrupt head movement in a prolonged constant-rate turn that has ceased stimulating the motion sensing system can create the illusion of rotation or movement in an entirely different axis. The disoriented pilot will maneuver the aircraft into a dangerous attitude in an attempt to stop rotation. This most overwhelming of all illusions in flight may be prevented by not making sudden, extreme head movements, particularly while making prolonged constant-rate turns under IFR conditions.

(b) *Graveyard spin:* A proper recovery from a spin that has ceased stimulating the motion sensing system can create the illusion of spinning in the opposite direction. The disoriented pilot will return the aircraft to its original spin.

(c) *Graveyard spiral:* An observed loss of altitude during a coordinated constant-rate turn that has ceased stimulating the motion sensing system can create the illusion of being in a descent with the wings level. The disoriented pilot will pull back on the controls, tightening the spiral and increasing the loss of altitude.

(d) *Somatogravic illusion:* A rapid acceleration during takeoff can create the illusion of being in a nose up attitude. The disoriented pilot will push the aircraft into a nose low, or dive attitude. A rapid deceleration by a quick reduction of the throttles can have the opposite effect, with the disoriented pilot pulling the aircraft into a nose up, or stall attitude.

(e) *Inversion illusion:* An abrupt change from climb to straight and level flight can create the illusion of tumbling backwards. The disoriented pilot win push the aircraft abruptly into a nose low attitude, possibly intensifying this illusion.

(f) *Elevator illusion:* An abrupt upward vertical acceleration, usually by an updraft, can create the illusion of being in a climb. The disoriented pilot will push the aircraft into a nose low attitude. An abrupt downward vertical acceleration, usually by a downdraft, has the opposite effect, with the disoriented pilot pulling the aircraft into a nose up attitude.

(g) *False horizon:* Sloping cloud formations, an obscured horizon, a dark scene spread with ground lights and stars, and certain geometric patterns of ground light can create illusions of not being aligned correctly with the actual horizon. The disoriented pilot will place the aircraft in a dangerous attitude.

(h) *Autokinesis:* In the dark, a static light will appear to move about when stared at for many seconds. The disoriented pilot will lose control of the aircraft in attempting to align it with the light.

3. *Illusions Leading to Landing Errors:*

(a) Various surface features and atmospheric conditions encountered in landing can create illusions of incorrect height above and distance from the runway threshold. Landing errors from these illusions can be prevented by anticipating them during approaches, aerial visual inspection of unfamiliar airports before landing, using electronic glide slope or VASI systems when available, and maintaining optimum proficiency in landing procedures.

(b) *Runway width illusion:* A narrower-than-usual runway can create the illusion that the aircraft is at a higher altitude than it actually is. The pilot who does not recognize this illusion will fly a lower approach, with the risk of striking objects along the approach path or landing short. A wider-than-usual runway can have the opposite effect, with the risk of leveling out high and landing hard or overshooting the runway.

(c) *Runway and terrain slopes illusion:* An upsloping runway, upsloping terrain, or both, can create the illusion that the aircraft is at a higher altitude than it actually is. The pilot who does not recognize this illusion will fly a lower approach. A downsloping runway, downsloping approach terrain, or both, can have the opposite effect.

(d) *Featureless terrain illusion:* An absence of ground features, as when landing over water, darkened areas, and terrain made featureless by snow, can create the illusion that the aircraft is at a higher altitude than it actually is. The pilot who does not recognize this illusion will fly a lower approach.

(e) *Atmospheric illusions:* Rain on the windscreen can create the illusion of greater height, and atmospheric haze the illusion of being at a greater distance from the runway. The pilot who does not recognize these illusions will fly a lower approach. Penetration of fog can create the illusion of pitching up. The pilot who does not recognize this illusion will steepen the approach, often quite abruptly.

(f) *Ground lighting illusions:* Lights along a straight path, such as a road, and even lights on moving trains can be mistaken for runway and approach lights. Bright runway and approach lighting systems, especially where few lights illuminate the surrounding terrain, may create the illusion of less distance to the runway. The pilot who does not recognize this illusion will fly a higher approach. Conversely, the pilot overflying terrain which has few lights to provide height cues may make a lower than normal approach.

8-1-6. VISION IN FLIGHT

a. Introduction: Of the body senses, vision is the most important for safe flight. Major factors that determine how effectively vision can be used are the level of illumination and the technique of scanning the sky for other aircraft.

b. Vision Under Dim and Bright Illumination:

1. Under conditions of dim illumination, small print and colors on aeronautical charts and aircraft instruments become unreadable unless adequate cockpit lighting is available. Moreover, another aircraft must be much closer to be seen unless its navigation lights are on.

2. In darkness, vision becomes more sensitive to light, a process called dark adaptation. Although exposure to total darkness for at least 30 minutes is required for complete dark adaptation, a pilot can achieve a moderate degree of dark adaptation within 20 minutes under dim red cockpit lighting. Since red light severely distorts colors, especially on aeronautical charts, and can cause serious difficulty in focusing the eyes on objects inside the aircraft, its use is advisable only where optimum outside night vision capa-

bility is necessary. Even so, white cockpit lighting must be available when needed for map and instrument reading, especially under IFR conditions. Dark adaptation is impaired by exposure to cabin pressure altitudes above 5,000 feet, carbon monoxide inhaled in smoking and from exhaust fumes, deficiency of Vitamin A in the diet, and by prolonged exposure to bright sunlight. Since any degree of dark adaptation is lost within a few seconds of viewing a bright light, a pilot should close one eye when using a light to preserve some degree of night vision.

3. Excessive illumination, especially from light reflected off the canopy, surfaces inside the aircraft, clouds, water, snow, and desert terrain, can produce glare, with uncomfortable squinting, watering of the eyes, and even temporary blindness. Sunglasses for protection from glare should absorb at least 85 percent of visible light (15 percent transmittance) and all colors equally (neutral transmittance), with negligible image distortion from refractive and prismatic errors.

c. Scanning for Other Aircraft:

1. Scanning the sky for other aircraft is a key factor in collision avoidance. It should be used continuously by the pilot and copilot (or right seat passenger) to cover all areas of the sky visible from the cockpit. Although pilots must meet specific visual acuity requirements, the ability to read an eye chart does not ensure that one will be able to efficiently spot other aircraft. Pilots must develop an effective scanning technique which maximizes one's visual capabilities. The probability of spotting a potential collision threat obviously increases with the time spent looking outside the cockpit. Thus, one must use timesharing techniques to efficiently scan the surrounding airspace while monitoring instruments as well.

2. While the eyes can observe an approximate 200 degree arc of the horizon at one glance, only a very small center area called the fovea, in the rear of the eye, has the ability to send clear, sharply focused messages to the brain. All other visual information that is not processed directly through the fovea will be of less detail. An aircraft at a distance of 7 miles which appears in sharp focus within the foveal center of vision would have to be as close as $7/10$ of a mile in order to be recognized if it were outside of foveal vision. Because the eyes can focus only on this narrow viewing area, effective scanning is accomplished with a series of short, regularly spaced eye movements that bring successive areas of the sky into the central visual field. Each movement should not exceed 10 degrees, and each area should be observed for at least 1 second to enable detection. Although horizontal back-and-forth eye movements seem preferred by most pilots, each pilot should develop a scanning pattern that is most comfortable and then adhere to it to assure optimum scanning.

3. Studies show that the time a pilot spends on visual tasks inside the cabin should represent no more that ¼ to ⅓ of the scan time outside, or no more than 4 to 5 seconds on the instrument panel for every 16 seconds outside. Since the brain is already trained to process sight information that is presented from left to right, one may find it easier to start scanning over the left shoulder and proceed across the windshield to the right.

4. Pilots should realize that their eyes may require several seconds to refocus when switching views between items in the cockpit and distant objects. The eyes will also tire more quickly when forced to adjust to distances immediately after close-up focus, as required for scanning the instrument panel. Eye fatigue can be reduced by looking from the instrument panel to the left wing past the wing tip to the center of the first scan quadrant when beginning the exterior scan. After having scanned from left to right, allow the eyes to return to the cabin along the right wing from its tip inward. Once back inside, one should automatically commence the panel scan.

5. Effective scanning also helps avoid "empty-field myopia." This condition usually occurs when flying above the clouds or in a haze layer that provides nothing specific to focus on outside the aircraft. This causes the eyes to relax and seek a comfortable focal distance which may range from 10 to 30 feet. For the pilot, this means looking without seeing, which is dangerous.

8-1-7. AEROBATIC FLIGHT

a. Pilots planning to engage in aerobatics should be aware of the physiological stresses associated with accelerative forces during aerobatic maneuvers. Many prospective aerobatic trainees enthusiastically enter aerobatic instruction but find their first experiences with G forces to be unanticipated and very uncomfortable. To minimize or avoid potential adverse effects, the aerobatic instructor and trainee must have a basic understanding of the physiology of G force adaptation.

b. Forces experienced with a rapid push-over maneuver result in the blood and body organs being displaced toward the head. Depending on forces involved and individual tolerance, a pilot may experience discomfort, headache, "red-out," and even unconsciousness.

c. Forces experienced with a rapid pull-up maneuver result in the blood and body organ displacement toward the lower part of the body away from the head. Since the brain requires continuous blood circulation for an adequate oxygen supply, there is a physiologic limit to the time the pilot can tolerate higher forces before losing consciousness. As the blood circulation to the brain decreases as a result of forces involved, a pilot will experience "narrowing" of visual fields, "gray-out," "black-out," and unconsciousness. Even a brief loss of consciousness in a maneuver can lead to improper control movement causing structural failure of the aircraft or collision with another object or terrain.

d. In steep turns, the centrifugal forces tend to push the pilot into the seat, thereby resulting in blood and body organ displacement toward the lower part of the body as in the case of rapid pull-up maneuvers and with the same physiologic effects and symptoms.

e. Physiologically, humans progressively adapt to imposed strains and stress, and with practice, any maneuver will have decreasing effect. Tolerance to G forces is dependent on human physiology and the individual pilot. These factors include the skeletal anatomy, the cardiovascular architecture, the nervous system, the quality of the blood, the general physical state, and experience and recency of exposure. The pilot should consult an Aviation Medical Examiner prior to aerobatic training and be aware that poor physical condition can reduce tolerance to accelerative forces.

f. The above information provides pilots with a brief summary of the physiologic effects of G forces. It does not address methods of "counteracting" these effects. There are numerous references on the subject of G forces during aerobatics available to pilots. Among these are "G Effects on the Pilot During Aerobatics," FAA-AM-72-28, and "G Incapacitation in Aerobatic Pilots: A Flight Hazard" FAA-AM-82-13. These are available from the National Technical Information Service, Springfield, Virginia 22161.

REFERENCE—FAA AC 91-61, A HAZARD IN AEROBATICS. EFFECTS OF G-FORCES ON PILOTS.

8-1-8. JUDGMENT ASPECTS OF COLLISION AVOIDANCE

a. Introduction: The most important aspects of vision and the techniques to scan for other aircraft are described in AIM, Vision in Flight, paragraph 8-1-6. Pilots should also be familiar with the following information to reduce the possibility of mid-air collisions.

b. Determining Relative Altitude: Use the horizon as a reference point. If the other aircraft is above the horizon, it is probably on a higher flight path. If the aircraft appears to be below the horizon, it is probably flying at a lower altitude.

c. Taking Appropriate Action: Pilots should be familiar with rules on right-of-way, so if an aircraft is on an obvious collision course, one can take immediate evasive action, preferably in compliance with applicable Federal Aviation Regulations.

d. Consider Multiple Threats: The decision to climb, descend, or turn is a matter of personal judgment, but one should anticipate that the other pilot may also be making a quick maneuver. Watch the other aircraft during the maneuver and begin your scanning again immediately since there may be other aircraft in the area.

e. Collision Course Targets: Any aircraft that appears to have no relative motion and stays in one scan quadrant is likely to be on a collision course. Also, if a target shows no lateral or vertical motion, but increases in size, *take evasive action.*

f. Recognize High Hazard Areas:

1. Airways, especially near VORs, and Class B, Class C, Class D, and Class E surface areas are places where aircraft tend to cluster.

2. Remember, most collisions occur during days when the weather is good. Being in a "radar environment" still requires vigilance to avoid collisions.

g. Cockpit Management: Studying maps, checklists, and manuals before flight, with other proper preflight planning; e.g., noting necessary radio frequencies and organizing cockpit materials, can reduce the amount of time required to look at these items during flight, permitting more scan time.

h. Windshield Conditions: Dirty or bug-smeared windshields can greatly reduce the ability of pilots to see other aircraft. Keep a clean windshield.

i. Visibility Conditions: Smoke, haze, dust, rain, and flying towards the sun can also greatly reduce the ability to detect targets.

j. Visual Obstructions in the Cockpit:

1. Pilots need to move their heads to see around blind spots caused by fixed aircraft structures, such as door posts, wings, etc. It will be necessary at times to maneuver the aircraft; e.g., lift a wing, to facilitate seeing.

2. Pilots must insure curtains and other cockpit objects; e.g., maps on glare shield, are removed and stowed during flight.

k. Lights On:

1. Day or night, use of exterior lights can greatly increase the conspicuity of any aircraft.

2. Keep interior lights low at night.

l. ATC support: ATC facilities often provide radar traffic advisories on a workload-permitting basis. Flight through Class C and Class D airspace requires communication with ATC. Use this support whenever possible or when required.

Chapter 9. Aeronautical charts and related publications
Section 1. TYPES OF CHARTS AVAILABLE

9-1-1. GENERAL

Civil aeronautical charts for the U.S. and its territories, and possessions are produced by the National Aeronautical Charting Office (NACO), www.naco.faa.gov, which is part of FAA's Office of Technical Operations Aviation Systems Standards.

9-1-2. OBTAINING AERONAUTICAL CHARTS

a. Most charts and publications described in this Chapter can be obtained by subscription or one-time sales from:

National Aeronautical Charting Office (NACO)
  Distribution Division
Federal Aviation Administration
6303 Ivy Lane, Suite 400
Greenbelt, MD 20770
Telephone: 1-800-638-8972 (Toll free within U.S.)
  301-436-8301/6990
  301-436-6829 (FAX)
e-mail: 9-AMC-Chartsales@faa.gov

b. Public sales of charts and publications are also available through a network of FAA chart agents primarily located at or near major civil airports. A listing of products and agents is printed in the free FAA catalog, aeronautical Charts and Related Products. (FAA Stock No. ACATSET). A free quarterly bulletin, Dates of Latest Editions, (FAA Stock No. 5318), is also available from NACO.

9-1-3. SELECTED CHARTS AND PRODUCTS AVAILABLE

VFR Navigation Charts
IFR Navigation Charts
Planning Charts
Supplementary Charts and Publications
Digital Products

9-1-4. GENERAL DESCRIPTION OF EACH CHART SERIES

a. VFR Navigation Charts.

1. Sectional Aeronautical Charts: Sectional Charts are designed for visual navigation of slow to medium speed aircraft. The topographic information consists of contour lines, shaded relief, drainage patterns, and an extensive selection of visual checkpoints and landmarks used for flight under VFR. Cultural features include cities and towns, roads, railroads, and other distinct landmarks. The aeronautical information includes visual and radio aids to navigation, airports, controlled airspace, special-use airspace, obstructions, and related data. Scale 1 inch = 6.86nm/1:500,000. 60 × 20 inches folded to 5 × 10 inches. Revised semiannually, except most Alaskan charts are revised annually. (See FIG 9-1-1 and FIG 9-1-11.)

2. VFR Terminal Area Charts (TAC): TAC's depict the airspace designated as Class B airspace. While similar to sectional charts, TAC's have more detail because the scale is larger. The TAC should be used by pilots intending to operate to or from airfields within or near Class B or Class C airspace. Areas with TAC coverage are indicated by a • on the Sectional Chart indexes. Scale 1 inch = 3.43nm/1:250,000. Charts are revised semiannually, except Puerto Rico-Virgin Islands revised annually. (See FIG 9-1-1 and FIG 9-1-11.)

3. World Aeronautical Charts (WAC): WAC's cover land areas for navigation by moderate speed aircraft operating at high altitudes. Included are city tints, principal roads, railroads, distinctive landmarks, drainage patterns, and relief. Aeronautical information includes visual and radio aids to navigation, airports, airways, special-use airspace, and obstructions. Because of a smaller scale, WAC's do not show as much detail as sectional or TAC's, and therefore are not recommended for exclusive use by pilots of low speed, low altitude aircraft. Scale 1 inch = 13.7nm/1:1,000,000. 60 × 20 inches folded to 5 × 10 inches. WAC's are revised annually, except for a few in Alaska and the Caribbean, which are revised biennially. (See FIG 9-1-12 and FIG 9-1-13.)

4. U.S. Gulf Coast VFR Aeronautical Chart: The Gulf Coast Chart is designed primarily for helicopter operation in the Gulf of Mexico area. Information depicted includes offshore mineral leasing areas and blocks, oil drilling platforms, and high density helicopter activity areas. Scale 1 inch = 13.7nm/1:1,000,000. 55 × 27 inches folded to 5 × 10 inches. Revised annually.

5. Grand Canyon VFR Aeronautical Chart: Covers the Grand Canyon National Park area and is designed to promote aviation safety, flight free zones, and facilitate VFR navigation in this popular area. The chart contains aeronautical information for general aviation VFR pilots on one side and commercial VFR air tour operators on the other side.

6. Helicopter Route Charts: A three-color chart series which shows current aeronautical information useful to helicopter pilots navigating in areas with high concentrations of helicopter activity. Information depicted includes helicopter routes, four classes of heliports with associated frequency and lighting capabilities, NAVAID's, and obstructions. In addition, pictorial symbols, roads, and easily identified geographical features are portrayed. Helicopter charts have a longer life span than other chart products and may be current for several years. All new editions of these charts are printed on a durable plastic material. Helicopter Route Charts are updated

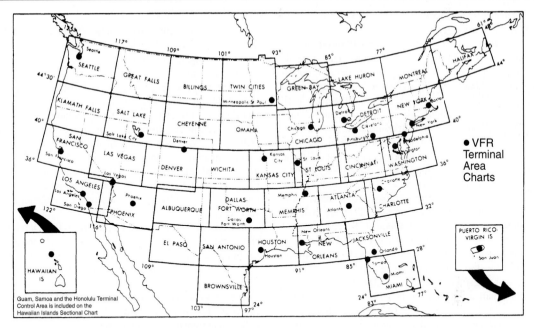

Sectional and VFR Terminal Area Charts for the Conterminous
United States, Hawaiian Islands, Puerto Rico and the Virgin Islands

Figure 9-1-1.

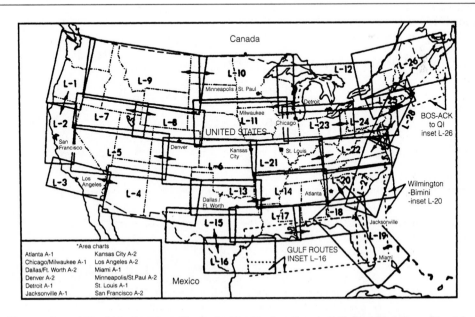

Enroute Low Altitude Instrument Charts for the Conterminous U.S. (Includes Area Charts)

Figure 9-1-2.

as requested by the FAA. Scale 1 inch = 1.71nm/1:125,000. 34 × 30 inches folded to 5 × 10 inches.

b. IFR Navigation Charts.

1. IFR Enroute Low Altitude Charts (Conterminous U.S. and Alaska): Enroute Low altitude charts provide aeronautical information for navigation under IFR conditions below 18,000 feet MSL. This four-color chart series includes airways; limits of controlled airspace; VHF NAVAID's with frequency, identification, channel, geographic coordinates; airports with terminal air/ground communications; minimum en route and obstruction clearance altitudes; airway distances; reporting points; special use airspace; and military training routes. Scales vary from 1 inch = 5nm to 1 inch = 20nm. 50 × 20 inches folded to 5 × 10 inches. Charts revised every 56 days. Area charts show congested terminal areas at a large scale. They are included with subscriptions to any conterminous U.S. Set Low (Full set, East or West sets). (See FIG 9-1-2 and FIG 9-1-4.)

2. IFR Enroute High Altitude Charts (Conterminous U.S. and Alaska): Enroute high altitude charts are designed for navigation at or above 18,000 feet MSL. This four-color chart series includes the jet route structure; VHF NAVAID's with frequency, identification, channel, geographic coordinates; selected airports; reporting points, Scales vary from 1 inch = 45nm to 1 inch = 18nm. 55 × 20

inches folded to 5 × 10 inches. Revised every 56 days. (See FIG 9-1-3 and FIG 9-1-5.)

3. U.S. Terminal Procedures Publication (TPP): TPP's are published in 24 loose-leaf or perfect bound volumes covering the conterminous U.S., Puerto Rico and the Virgin Islands. A Change Notice is published at the midpoint between revisions in bound volume format and is available on the internet for free download at the NACO website. (See FIG 9-1-9.) The TPP's include:

(a) Instrument Approach Procedure (IAP) Charts: IAP charts portray the aeronautical data that is required to execute instrument approaches to airports. Each chart depicts the IAP, all related navigation data, communications information, and an airport sketch. Each procedure is designated for use with a specific electronic navigational aid, such as ILS, VOR, NDB, RNAV, etc.

(b) Standard Instrument Departure (DP) Charts: DP charts are designed to expedite clearance delivery and to facilitate transition between takeoff and en route operations. They furnish pilots' departure routing clearance information in graphic and textual form.

(c) Standard Terminal Arrival (STAR) Charts: STAR charts are designed to expedite ATC arrival procedures and to facilitate transition between en route and instrument approach operations. They depict preplanned IFR ATC arrival

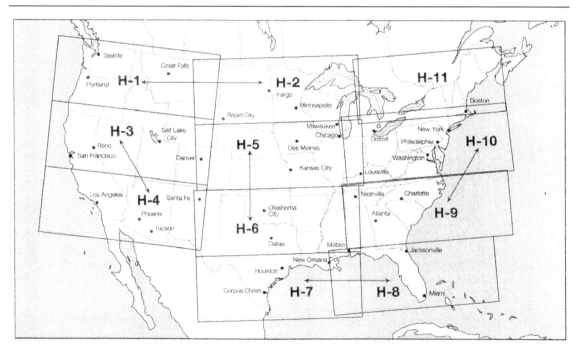

Enroute High Altitude Charts for the Conterminous U.S.

Figure 9-1-3.

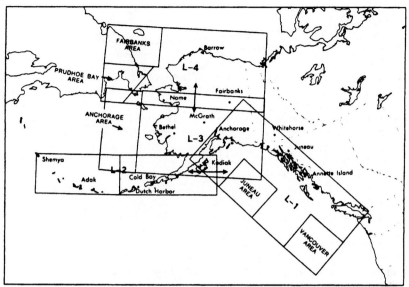

Alaska Enroute Low Altitude Chart

Figure 9-1-4.

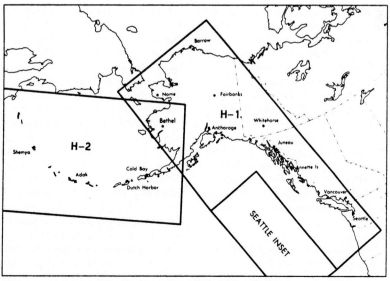

Alaska Enroute High Altitude Chart

Figure 9-1-5.

Figure 9-1-6. Planning Charts

procedures in graphic and textual form. Each STAR procedure is presented as a separate chart and may serve either a single airport or more than one airport in a given geographic area.

(d) Airport Diagrams: Full page airport diagrams are designed to assist in the movement of ground traffic at locations with complex runway/taxiway configurations and provide information for updating geodetic position navigational systems aboard aircraft. Airport diagrams are available for free download at the NACO website.

4. Alaska Terminal Procedures Publication: This publication contains all terminal flight procedures for civil and military aviation in Alaska. Included are IAP charts, DP charts, STAR charts, airport diagrams, radar minimums, and supplementary support data such as IFR alternate minimums, take-off minimums, rate of descent tables, rate of climb tables and inoperative components tables Volume is 5⅜ × 8¼ inch top bound. Publication revised every 56 days with provisions for a Terminal Change Notice, as required.

c. Planning Charts.

1. U.S. IFR/VFR Low Altitude Planning Chart: This chart is designed for preflight and en route flight planning for IFR/VFR flights. Depiction includes low altitude airways and mileage, NAVAID's, airports, special use airspace, cities, time zones, major drainage directory of airports with their airspace classification and a mileage

table showing great circle distances between major airports. Scale 1 inch = 47m 1:3,400,000. Chart revised annually, and is available either folded or unfolded for wall mounting. (See FIG 9-1-6.)

2. Gulf of Mexico and Caribbean Planning Chart: This is a VFR planning chart on the reverse side of the Puerto Rico-Virgin Islands VFR Terminal Area Chart. Information shown includes mileage between airports of entry, a selection of special use airspace and a directory of airports with their available services. Scale 1 inch = 85nm/1:6,192,178. 60 × 20 inches folded to 5 × 10 inches. Chart revised annually. (See FIG 9-1-6.)

3. Charted VFR Flyway Planning Charts: This chart is printed on the reverse side of selected charts. The coverage is the same as the associated TAC. Flyway planning charts depict flight paths and altitudes recommended for use to bypass high traffic areas. Ground references are provided as a guide for visual orientation. Flyway planning charts are designed for use in conjunction with TAC's and sectional charts and not to be used for navigation. Chart scale 1 inch = 3.43nm/1:250,000.

d. Supplementary Charts and Publications.

1. Airport/Facility Directory (A/FD). This 7-volume booklet series contains data on airports, seaplane bases, heliports, NAVAID's, communications data, weather data sources, airspace, special notices, and operational procedures. Coverage includes the conterminous United

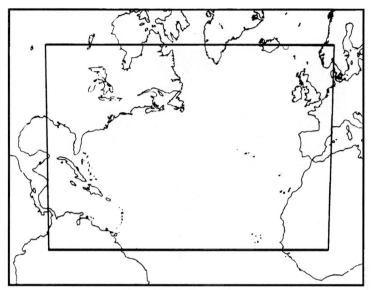

North Atlantic Route Charts
Figure 9-1-7.

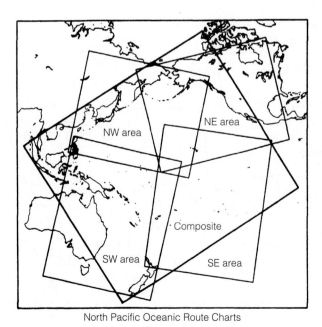

North Pacific Oceanic Route Charts

Figure 9-1-8.

U.S. Terminal Publication Volumes

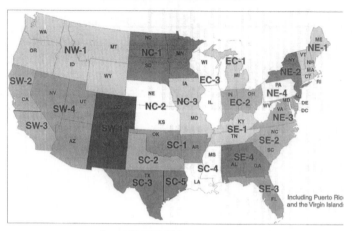

Figure 9-1-9.

Airport/Facility Geographic Areas
Figure 9-1-10.

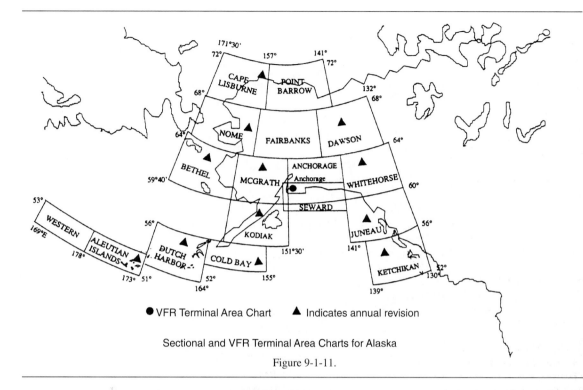

● VFR Terminal Area Chart   ▲ Indicates annual revision

Sectional and VFR Terminal Area Charts for Alaska

Figure 9-1-11.

States, Puerto Rico, and the Virgin Islands. The A/FD shows data that cannot be readily depicted in graphic form; e.g. airport hours of operations, types of fuel available, runway widths, lighting codes, etc. The A/FD also provides a means for pilots to update visual charts between edition dates (AFD is published every 56 days while sectional and Terminal Area Charts are generally revised every six months). and Terminal Area Charts are generally revised every six months). The VFR Chart Update Bulletins are available for free download from the NACO website. Volumes are side-bound 5⅜ × 8¼ inches. (See FIG 9-1-10.)

2. Supplement Alaska: This is a civil/military flight information publication issued by FAA every 56 days. It is a single volume booklet designed for use with appropriate IFR or VFR charts. The Supplement Alaska contains an airport/facility directory, airport sketches, communications data, weather data sources, airspace, listing of navigational facilities, and special notices and procedures. Volume is side-bound 5⅜ × 8¼ inches.

3. Chart Supplement Pacific: This supplement is designed for use with appropriate VFR or IFR enroute charts. Included in this one-volume booklet are the airport/facility directory, communications data, weather data sources, airspace, navigational facilities, special notices, and Pacific area procedures. IAP charts, DP charts, STAR charts, airport

diagrams, radar minimums, and supporting data for the Hawaiian and Pacific Islands are included. The manual is published every 56 days. Volume is side-bound 5⅜ × 8¼ inches.

4. North Pacific Route Charts: These charts are designed for FAA controllers to monitor transoceanic flights. They show established intercontinental air routes, including reporting points with geographic positions. Composite Chart: Scale 1 inch 164nm/1:12,000,000. 48 × 41½ inches. Area Charts: Scale 1 inch = 95.9nm/1:7,000,000. 52 × 40½ inches. All charts shipped unfolded. Charts revised every 56 days. (See FIG 9-1-8.)

5. North Atlantic Route Chart: Designed for FAA controllers to monitor transatlantic flights, this 5-color chart shows oceanic control areas, coastal navigation aids, oceanic reporting points, and NAVAID geographic coordinates. Full Size Chart: Scale 1 inch = 113.1nm/1:8,250,000. Chart is shipped flat only. Half Size Chart: Scale 1 inch = 150.8nm/1:11,000,000. Chart is 29¾ × 20½ inches, shipped folded to 5 × 10 inches only. Chart revised every 56 weeks. (See FIG 9-1-7.)

6. Airport Obstruction Charts (OC): The OC is a 1:12,000 scale graphic depicting FAR Part 77, Objects affecting Navigable Airspace surfaces, a representation of objects that penetrate these surfaces, aircraft movement and apron areas, navigational aids, prominent airport

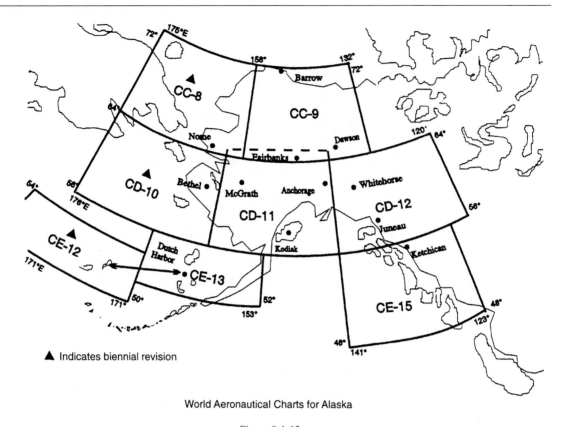

World Aeronautical Charts for Alaska

Figure 9-1-12.

buildings, and a selection of roads and other planimetric detail in the airport vicinity. Also included are tabulations of runway and other operational data.

7. FAA Aeronautical Chart User's Guide; A booklet designed to be used as a teaching aid and reference document. It describes the substantial amount of information provided on FAA's aeronautical charts and publications. It includes explanations and illustrations of chart terms and symbols organized by chart type. The users guide is available for free download at the NACO website.

e. Digital Products.

1. The Digital Aeronautical Information CD (DAICD). The DAICD is a combination of the NAVAID Digital Data File, the Digital Chart Supplement, and the Digital Obstacle File on one Compact Disk. These three digital products are no longer sold separately. The files are updated every 56 days and are available by subscription only.

(a) The NAVAID Digital Data File. This file contains a current listing of NAVAIDs that are compatible with the National Airspace System. This file contains all NAVAIDs including ILS and its components, in the U.S., Puerto Rico,

and the Virgin Islands plus bordering facilities in Canada, Mexico, and the Atlantic and Pacific areas.

(b) The Digital Obstacle File. This file describes all obstacles of interest to aviation users in the U.S., with limited coverage of the Pacific, Caribbean, Canada, and Mexico. The obstacles are assigned unique numerical identifiers, accuracy codes, and listed in order of ascending latitude within each state or area.

(c) The Digital Aeronautical Chart Supplement (DACS). The DACS is specifically designed to provide digital airspace data not otherwise readily available. The supplement includes a *Change Notice* for IAPFIX.dat at the mid-point between revisions. The *Change Notice* is available only by free download from the NACO website.

The DACS individual data tiles are:

ENHIGH.DAT: High altitude airways (conterminous U.S.)

ENLOW.DAT: Low altitude airways (conterminous U.S.)

APFIX.DAT: Selected instrument approach procedure NAVAID and fix data.

MTRFIX.DAT: Military training routes data.

ALHIGH.DAT: Alaska high altitude airways data.

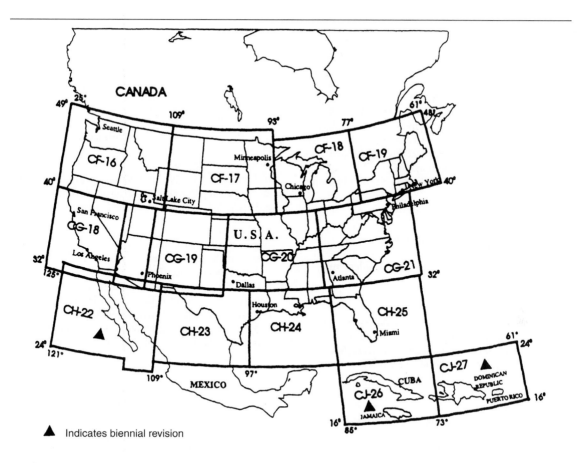

World Aeronautical Charts for the Conterminous United States, Mexico, and the Caribbean Areas

Figure 9-1-13.

ALLOWDAT: Alaska low altitude airways data.

PR.DAT: Puerto Rico airways data.

HAWAILDAT: Hawaii airways data.

BAHAMA.DAT: Bahamas routes data.

OCEANIC.DAT: Oceanic routes data.

STARS.DAT: Standard terminal arrivals data.

DP.DAT: Instrument departure procedures data.

LOPREF.DAT: Preferred low altitude IFR routes data.

HIPREF.DAT: Preferred high altitude IFR routes data.

ARF.DAT: Air route radar facilities data.

ASR.DAT: Airport surveillance radar facilities data.

2. The National Flight Database (NFD) (ARINC 424 [Ver 13 &15]). The NFD is a basic digital dataset, modeled to an international standard, which can be used as a basis to support GPS navigation. Initial data elements included are: Airport and Helicopter Records, VHF and NDB Navigation aids, en route waypoints and airways. Additional data elements will be added in subsequent releases to include: de-parture procedures, standard terminal arrivals, and GPS/RNAV instrument approach procedures. The database is updated every 28 days. The data is available by subscription only and is distributed on CD-ROM or by ftp download.

3. Sectional Raster Aeronautical Charts (SRAC). These digital VFR charts are geo-referenced scanned images of FAA sectional charts. Additional digital data may easily be overlaid on the raster image using commonly available Geographic Information System software. Data such as weather, temporary flight restrictions, obstacles, or other geospatial data can be combined with SRAC data to support a variety of needs. Most SRACs are provided in two halves, a north side and a south side. The file resolution is 200 dots per inch and the data is 8-bit color. The data is provided as a GeoTIFF and distributed on DVD-R media. The root mean square error of the transformation will not exceed two pixels. SRACs DVDs are updated every 28 days and are available by subscription only.

9-1-5. WHERE AND HOW TO GET CHARTS OF FOREIGN AREAS

a. National Imagery and Mapping Agency (NIMA) Products: An FAA catalog of NIMA Public Sale Aeronautical Charts and Publications (FAA Stock No. DMAACATSET), is available from the NOAA Distribution Division. The catalog describes available charts and publications primarily covering areas outside the U.S. A free quarterly bulletin, Dates of Latest Editions—NIMA Aeronautical Charts and Publications (FAA Stock No. DADOLE), is also available from NACO.

1. Flight Information Publication (FLIP) Planning Documents

General Planning (GP)
Area Planning
Area Planning-Special Use Airspace
Planning Charts

2. FLIP Enroute Charts and Chart Supplements

Pacific, Australasia, and Antarctica
United States—IFR and VFR Supplements
Flight Information Handbook
Caribbean and South America—Low Altitude
Caribbean and South America—High Altitude
Europe, North Africa, and Middle East—Low Altitude
Europe, North Africa, and Middle East—High Altitude
Africa
Eastern Europe and Asia
Area Arrival Charts

3. FLIP Instrument Approach Procedures (IAP's)

Africa
Canada and North Atlantic
Caribbean and South America
Eastern Europe and Asia
Europe, North Africa, and Middle East
Pacific, Australasia, and Antarctica
VFR Arrival/Departure Routes—Europe and Korea
United States

4. Miscellaneous DoD Charts and Products

Aeronautical Chart Updating Manual (CHUM)
DoD Weather Plotting Charts (WPC)
Tactical Pilotage Charts (TPC)
Operational Navigation Charts (ONC)
Global Navigation and Planning Charts (GNC)
Global LORAN-C Navigation Charts (GLCC)
LORAN-C Coastal Navigation Charts (LCNC)
Jet Navigation Charts (JNC) and Universal Jet Navigation Charts (JNU)
Jet Navigation Charts (JNCA)
Aerospace Planning Charts (ASC)
Oceanic Planning Charts (OPC)
Joint Operations Graphics—Air (JOG-A)
Standard Index Charts (SIC)
Universal Plotting Sheet (VP-OS)
Sight Reduction Tables for Air Navigation (PUB249)
Plotting Sheets (VP-30)
Dial-Up Electronic CHUM

b. Canadian Charts: Information on available Canadian charts and publications may be obtained from designated FAA chart agents or by contacting the:

NAV CANADA
Aeronautical Publications
Sales and Distribution Unit
P.O. Box 9840 Station T
Ottawa, Ontario K1G 6S8 Canada
Telephone: 613-744-6393 or 1-866-731-7827
Fax: 613-744-7120 or 1-866-740-9992

c. Mexican Charts: Information on available Mexican charts and publications may be obtained by contacting:

Dirección de Navigacion Aereo
Blvd. Puerto Aereo 485
Zona Federal Del Aeropuerto Int'l
15620 Mexico D.F.
Mexico

d. International Civil Aviation Organization: (ICAO): A free ICAO Publications and Audio=Visual Training Aids Catalogue is available from:

International Civil Aviation Organization
ATTN.: Document Sales Unit
999 University Street
Montreal, Quebec
H3C 5H7, Canada
Telephone: (514) 954-8022
Fax: (514) 954-6769
E-mail: sales_unit@icao.org
Internet: http://www.icao.org/cgi/goto.pl?icao/en/sales.htm
Sitatex: YULCAYA
Telex: 05-24513

Chapter 10. Helicopter Operations
Section 1. HELICOPTER IFR OPERATIONS

10-1-1. HELICOPTER FLIGHT CONTROL SYSTEMS

a. The certification requirements for helicopters to operate under Instrument Flight Rules (IFR) are contained in 14 CFR Part 27, Airworthiness Standards: Normal Category Rotorcraft, and 14 CFR Part 29, Airworthiness Standards: Transport Category Rotorcraft. To meet these requirements, helicopter manufacturers usually utilize a set of stabilization and/or Automatic Flight Control Systems (AFCS's).

b. Typically, these systems fall into the following categories:

1. Aerodynamic surfaces, which impart some stability or control capability not found in the basic VFR configuration.

2. Trim systems, which provide a cyclic centering effect. These systems typically involve a magnetic brake/spring device, and may also be controlled by a four-way switch on the cyclic. This is a system that supports "hands on" flying of the helicopter by the pilot.

3. Stability Augmentation Systems (SAS's), which provide short-term rate damping control inputs to increase helicopter stability. Like trim systems, SAS supports "hands on" flying.

4. Attitude Retention Systems (ATT's), which return the helicopter to a selected attitude after a disturbance. Changes in desired attitude can be accomplished usually through a four-way "beep" switch, or by actuating a "force trim" switch on the cyclic, setting the attitude manually, and releasing. Attitude retention may be a SAS function, or may be the basic "hands off" autopilot function.

5. Autopilot Systems (AP's), which provide for "hands off" flight along specified lateral and vertical paths, including heading, altitude, vertical speed, navigation tracking, and approach. These systems typically have a control panel for mode selection, and system for indication of mode status. Autopilots may or may not be installed with an associated Flight Director System (FD). Autopilots typically control the helicopter about the roll and pitch axes (cyclic control) but may also include yaw axis (pedal control) and collective control servos.

6. FD's, which provide visual guidance to the pilot to fly specific selected lateral and vertical modes of operation. The visual guidance is typically provided as either a "dual cue" (commonly known as a "cross-pointer") or "single cue" (commonly known as a "vee-bar") presentation superimposed over the attitude indicator. Some FD's also include a collective cue. The pilot manipulates the helicopter's controls to satisfy these commands, yielding the desired flight path, or may couple the flight director to the autopilot to perform automatic flight along the desired flight path. Typically, flight director mode control and indication is shared with the autopilot.

c. In order to be certificated for IFR operation, a specific helicopter may require the use of one or more of these systems, in any combination.

d. In many cases, helicopters are certificated for IFR operations with either one or two pilots. Certain equipment is required to be installed and functional for two pilot operations, and typically, additional equipment is required for single pilot operation. These requirements are usually described in the limitations section of the Rotorcraft Flight Manual.

e. In addition, the Rotorcraft Flight Manual also typically defines systems and functions that are required to be in operation or engaged for IFR flight in either the single or two pilot configuration. Often, particularly in two pilot operation, this level of augmentation is less than the full capability of the installed systems. Likewise, single pilot operation may require a higher level of augmentation.

f. The Rotorcraft Flight Manual also identifies other specific limitations associated with IFR flight. Typically, these limitations include, but are not limited to:

1. Minimum equipment required for IFR flight (in some cases, for both single pilot and two pilot operations).

2. Vmini (minimum speed-IFR).

NOTE—THE MANUFACTURER MAY ALSO RECOMMEND A MINIMUM IFR AIRSPEED DURING INSTRUMENT APPROACH.

3. Vnei (never exceed speed-IFR).

4. Maximum approach angle.

5. Weight and center of gravity limits.

6. Aircraft configuration limitations (such as aircraft door positions and external loads).

7. Aircraft system limitations (generators, inverters, etc.).

8. System testing requirements (many avionics and AFCS/AP/FD systems incorporate a self-test feature).

9. Pilot action requirements (such as the pilot must have his/her hands and feet on the controls during certain operations, such as during instrument approach below certain altitudes).

g. It is very important that pilots be familiar with the IFR requirements for their particular helicopter. Within the same make, model and series of helicopter, variations in the installed avionics may change the required equipment or the level of augmentation for a particular operation.

h. During flight operations, pilots must be aware of the mode of operation of the augmentation systems, and the control logic and functions employed. For example, during an ILS approach using a particular system in the three-cue mode (lateral, vertical and collective cues), the flight director *collective cue* responds to glideslope deviation, while the horizontal bar of the "cross-pointer" responds to airspeed deviations. The same system, while flying an ILS in the two-cue mode, provides for the *horizontal bar* to respond to glideslope deviations. This concern is particularly significant when operating using two pilots. Pilots should have an established set of procedures and responsibilities for the control of flight director/autopilot modes for the various phases of flight. Not only does a full understanding of the system modes provide for a higher degree of accuracy in control of the helicopter, it is the basis for crew identification of a faulty system.

i. Relief from the prohibition to takeoff with any inoperative instruments or equipment may be provided through a Minimum Equipment List (see 14 CFR Section 91.213 and 14 CFR Section 135.179, Inoperative Instruments and Equipment). In many cases, a helicopter configured for single pilot IFR may depart IFR with certain equipment inoperative, provided a crew of two pilots is used. Pilots are cautioned to ensure the pilot-in-command and second-in-command meet the requirements of 14 CFR Section 61.58, Pilot-in-Command Proficiency Check: Operation of Aircraft Requiring More Than One Pilot Flight Crewmember, and 14 CFR Section 61.55, Second-in-Command Qualifications, or 14 CFR Part 135, Operating Requirements: Commuter and On-Demand Operations, Subpart E, Flight Crewmember Requirements, and Subpart G, Crewmember Testing Requirements, as appropriate.

j. Experience has shown that modern AFCS/AP/FD equipment installed in IFR helicopters can, in some cases, be very complex. This complexity requires the pilot(s) to obtain and maintain a high level of knowledge of system operation, limitations, failure indications and reversionary modes. In some cases, this may only be reliably accomplished through formal training.

10-1-2. Helicopter Instrument Approaches

a. Helicopters are capable of flying any published 14 CFR Part 97, Standard Instrument Approach Procedures, for which they are properly equipped, subject to the following limitations and conditions:

1. Helicopters flying conventional (non-Copter) SIAP's may reduce the visibility minima to not less than one half the published Category A landing visibility minima, or 1/4 statue mile visibility/1200 RVR, whichever is greater. No reduction in MDA/DA is permitted. The reference for this is 14 CFR Section 97.3, Symbols and Terms used in Procedures, (d-1). The helicopter may initiate the final approach segment at speeds up to the upper limit of the highest approach category authorized by the procedure, but must be slowed to no more than 90 KIAS at the missed approach point (MAP) in order to apply the visibility reduction. Pilots are cautioned that such a decelerating approach may make early identification of wind shear on the approach path difficult or impossible. If required, use the Inoperative Components and Visual Aids Table provided in the front cover of the U.S. Terminal Procedures Volume to derive the Category A minima before applying the 14 CFR Section 97.3(d-1) rule.

2. Helicopters flying Copter SIAP's may use the published minima, with no reductions allowed. The maximum airspeed is 90 KIAS on any segment of the approach or missed approach.

3. Helicopters flying GPS Copter SIAP's must limit airspeed to 90 KIAS or less when flying any segment of the procedure, except speeds must be limited to no more than 70 KIAS on the final and missed approach segments. Military GPS Copter SIAP's are limited to no more than 90 KIAS throughout the procedure. If annotated, holding may also be limited to no more than 70 KIAS. Use the published minima, no reductions allowed.

NOTE—

OBSTRUCTION CLEARANCE SURFACES ARE BASED ON THE AIRCRAFT SPEED AND HAVE BEEN DESIGNED ON THESE APPROACHES FOR 70 KNOTS. IF THE HELICOPTER IS FLOWN AT HIGHER SPEEDS, IT MAY FLY OUTSIDE OF PROTECTED AIRSPACE. SOME HELICOPTERS HAVE A V_{MINI} GREATER THAN 70 KNOTS; THEREFORE, THEY CAN NOT MEET THE 70 KNOT LIMITATION TO CONDUCT THIS TYPE OF PROCEDURE. SOME HELICOPTER AUTOPILOTS, WHEN USED IN THE "GO-AROUND" MODE, ARE PROGRAMMED WITH A V_{YI} GREATER THAN 70 KNOTS, THEREFORE WHEN USING THE AUTOPILOT "GO-AROUND" MODE, THEY CAN NOT MEET THE 70 KNOT LIMITATION TO CONDUCT THIS TYPE OF APPROACH. IT MAY BE POSSIBLE TO USE THE AUTOPILOT FOR THE MISSED APPROACH IN THE OTHER THAN THE "GO-AROUND" MODE AND MEET THE 70 KNOT LIMITATION TO CONDUCT THIS TYPE OF APPROACH. WHEN OPERATING AT SPEEDS OTHER THAN V_{YI} OR V_Y, PERFORMANCE DATA MAY NOT BE AVAILABLE IN THE RFM TO PREDICT COMPLIANCE WITH CLIMB GRADIENT REQUIREMENTS. PILOTS MAY USE OBSERVED PERFORMANCE IN SIMILAR WEIGHT/ALTITUDE/TEMPERATURE/SPEED CONDITIONS TO EVALUATE THE SUITABILITY OF PERFORMANCE. PILOTS ARE CAUTIONED TO MONITOR CLIMB PERFORMANCE TO ENSURE COMPLIANCE WITH PROCEDURE REQUIREMENTS.

4. TBL 10-1-1 summarizes these requirements.

5. Even with weather conditions reported at or above landing minima, some combinations of reduced cockpit cutoff angle, minimal approach/runway lighting, and high MDA/DH coupled with a low visibility minima, the pilot may not be able to identify the required visual reference(s) during the approach, or those references may only be visible in a very small portion of the pilot's available field of view. Even if identified by the pilot, these visual references may not support normal maneuvering and normal rates of descent to landing. The effect of such a combination may be exacerbated by other conditions such as rain on the windshield, or incomplete windshield defogging coverage.

Table 10-1-1 Helicopter Use of Standard Instrument Approach Procedures

PROCEDURE	HELICOPTER VISIBILITY MINIMA	HELICOPTER MDA/DA	MAXIMUM SPEED LIMITATIONS
Conventional (non-Copter)	The greater of: one half the Category A visibility minima, 1/4 statute mile visibility, or 1200 RVR	As published for Category A	The helicopter may initiate the final approach segment at speeds up to the upper limit of the highest Approach Category authorized by the procedure, but must be slowed to no more than 90 KIAS at the MAP in order to apply the visibility reduction.
Copter Procedure	As published	As published	90 KIAS when on a published route/track.
GPS Copter Procedure	As published	As published	90 KIAS when on a published route or track, EXCEPT 70 KIAS when on the final approach or missed approach segment and, if annotated, in holding. Military procedures are limited to 90 KIAS for all segments.

NOTE—

SEVERAL FACTORS EFFECT THE ABILITY OF THE PILOT TO ACQUIRE AND MAINTAIN THE VISUAL REFERENCES SPECIFIED IN 14 CFR SECTION 91.175(C), EVEN IN CASES WHERE THE FLIGHT VISIBILITY MAY BE AT THE MINIMUM DERIVED BY TBL 10-1-1. THESE FACTORS INCLUDE, BUT ARE NOT LIMITED TO:

1. *Cockpit cutoff angle (the angle at which the cockpit or other airframe structure limits downward visibility below the horizon).*

2. *Combinations of high MDA/DH and low visibility minimum, such as a conventional nonprecision approach with a reduced helicopter visibility minima (per 14 CFR Section 97.3).*

3. *Type, configuration, and intensity of approach and runway lighting systems.*

4. *Type of obscuring phenomenon and/or windshield contamination.*

6. Pilots are cautioned to be prepared to execute a missed approach even though weather conditions may be reported at or above landing minima.

NOTE—SEE PARAGRAPH 5-4-21, MISSED APPROACH, FOR ADDITIONAL INFORMATION ON MISSED APPROACH PROCEDURES.

10-1-3. Helicopter Point-In-Space (PInS) Approach Procedures

a. PInS nonprecision approaches are normally developed for heliports that do not meet the design standards for an IFR heliport. A helicopter PinS approach can be developed from conventional NAVAIDs or area navigation systems (including GPS). These procedures involve a visual segment between the MAP and the landing area. There are two types of notes associated with a PinS approach:

1. To a location 10,500 feet or less from the MAP: "PROCEED VISUALLY FROM (NAMED MAP) OR CONDUCT THE SPECIFIED MISSED APPROACH."

(a) This phrase requires the pilot to acquire and maintain visual contact with the landing site at or prior to the MAP, or execute a missed approach. The visibility minimum is based on the distance from the MAP to the landing site, among other factors.

(b) The pilot is required to maintain the published minimum visibility throughout the visual segment.

(c) IFR obstruction clearance areas are not applied to the visual segment of the approach and missed approach segment protection is not provided between the MAP and the landing site.

(d) Obstacle or terrain avoidance from the MAP to the landing site is the responsibility of the pilot.

(e) Upon reaching the MAP defined on the approach procedure, or as soon as practicable after reaching the MAP, the pilot advises ATC whether proceeding visually and canceling IFR or complying with the missed approach instructions. See paragraph 5-1-13, Canceling IFR Flight Plan.

(f) In those cases where *proceed visually* cannot be approved, the procedure will be annotated *proceed VFR* . The visual requirements are contained in subpara 2(b) below.

2. In those cases where proceed visually cannot be approved, the procedure will be annotated: "PROCEED VFR

FROM (NAMED MAP) OR CONDUCT THE SPECI-FIED MISSED APPROACH."

(a) This phrase requires the pilot, at or prior to the MAP, to determine if the published minimum visibility, or the visibility required by the operating rule, or operations specifications (whichever is higher) is available to safely transition from IFR to VFR flight. If not, the pilot must execute a missed approach.

(b) Visual contact with the landing site is no required; however, the pilot must maintain the higher of the VFR weather minimums throughout the visual segment (as required by the class of airspace operating rule and/or operations specifications) provided the visibility is limited to no lower than tha published in the procedure, until canceling IFR.

(c) IFR obstruction clearance areas are not applied to the VFR segment between the MAP and the landing site. Obstacle or terrain avoidance from the MAP to the landing site is the responsibility of the pilot.

(d) Upon reaching the MAP defined on the approach procedure, or as soon as practicable after reaching the MAP, the pilot advises ATC whether proceeding VFR and canceling IFR, or complying with the missed approach instructions. See paragraph 5-1-13, Canceling IFR Flight Plan.

(e) If the visual segment penetrates Class B, C, or D airspace, pilots are responsible for obtaining a Special VFR clearance, when required.

NOTE—IN BOTH CASES, A SUBSTANTIAL VISUAL SEGMENT MAY EXIST. PILOTS ARE CAUTIONED TO REDUCE GROUND SPEED DURING THE APPROACH SO AS TO ARRIVE AT THE MAP AT A GROUND SPEED WHICH WILL PROMOTE A SAFE TRANSITION FROM IFR TO VFR FLIGHT.

10-1-4. The Gulf of Mexico Grid System

l. On October 8, 1998, the Southwest Region of the FAA, with assistance from the Helicopter Safety Advisory Conference (HSAC), implemented the world's first Instrument Flight Rules (IFR) Grid System in the Gulf of Mexico. This navigational route structure is completely independent of ground-based navigation aids, (NAVAIDs) and was designed to facilitate helicopter IFR operations to offshore destinations. The Grid System is defined by over 300 offshore waypoints located 20 minutes apart (latitude and longitude). Flight plan routes are routinely defined by just 4 segments; departure point (lat/long), first en route grid waypoint, last en route grid waypoint prior to approach procedure, and destination point (lat/long). There are over 4,000 possible offshore landing sites. Upon reaching the waypoint prior to the destination, the pilot may execute an Offshore Standard Approach Procedure (OSAP), a Helicopter En Route Descent Areas (HEDA)

approach, or an Airborne Radar Approach (ARA). For more information on these helicopter instrument procedures, refer to FAA AC 90-80B, Approval of Offshore Standard Approach Procedure (OSAP), Airborne Radar Approaches (ARA), and Helicopter En Route Areas (HEDA) Criteria, on the Flight Standards web site **http://av-info.fan.gov/terps.** The return flight plan is just the reverse with the requested stand-alone GPS approach contained in the remarks section.

2. The large number (over 300) of waypoints in the grid system makes it difficult to assign phonetically pronounceable names to the waypoints that would be meaningful to pilots and controllers. A unique naming system was adopted that enables pilots and controllers to derive the fix position from the name. The five-letter names are derived as follows:

(a) The waypoints are divided into sets of 3 columns each. A three-letter identifier, identifying a geographical area or a NAVAID to the north, represents each set.

(b) Each column in a set is named after its position, i.e., left (L), center (C), and right (R).

(c) The rows of the grid are named alphabetically from north to south, starting with A for the northern most row.

EXAMPLE—LCHRC WOULD HE PRONOUNCED "LAKE CHARLES ROMEO CHARLIE." THE WAYPOINT IS IN THE RIGHT-HAND COLUMN OF THE LAKE CHARLES VOR SET, IN ROW C (THIRD SOUTH FROM THE NORTHERN MOST ROW).

3. Since the grid system's implementation, IFR delays (frequently over 1 hour in length) for operations in this environment have been effectively eliminated. The comfort level of the pilots, knowing that they will be given a clearance quickly, plus the mileage savings in this near free-flight environment, is allowing the operators to carry less fuel. Less fuel means they can transport additional passengers, which is a substantial fiscal and operational benefit, considering the limited seating on board helicopters.

4. There are 3 requirements for operators to meet before filing IFR flight plans utilizing the grid:

(a) The helicopter must be IFR certified and equipped with IFR certified TSO C-129 CPS navigational units.

(b) The operator must obtain prior written approval from the appropriate Flight Standards District Office through a Certificate of Authorization or revision to their Operations Specifications, as appropriate.

(c) The operator must be a signatory to the Houston ARTCC Letter of Agreement.

5. FAA/NACO publishes the grid system waypoints on the IFR Gulf of Mexico Vertical Flight Reference Chart. A commercial equivalent is also available. The chart is updated annually and is available from a FAA chart agent or FAA directly, website address: **http://naco.faa.gov.**

Section 2. SPECIAL OPERATIONS

10-2-1. Offshore Helicopter Operations

a. Introduction

The offshore environment offers unique applications and challenges for helicopter pilots. The mission demands, the nature of oil and gas exploration and production facilities, and the flight environment (weather, terrain, obstacles, traffic), demand special practices, techniques and procedures not found in other flight operations. Several industry organizations have risen to the task of reducing risks in offshore operations, including the Helicopter Safety Advisory Conference (HSAC) (www.hsac.org), and the Offshore Committee of the Helicopter Association International (HAI) (www.rotor.com). The following recommended practices for offshore helicopter operations are based on guidance developed by HSAC for use in the Gulf of Mexico, and provided here with their permission. While not regulatory, these recommended practices provide aviation and oil and gas industry operators with useful information in developing procedures to avoid certain hazards of offshore helicopter operations.

NOTE—
LIKE ALL AVIATION PRACTICES, THESE RECOMMENDED PRACTICES ARE UNDER CONSTANT REVIEW. IN ADDITION TO NORMAL PROCEDURES FOR COMMENTS, SUGGESTED CHANGES, OR CORRECTIONS TO THE AIM (CONTAINED IN THE PREFACE), ANY QUESTIONS OR FEEDBACK CONCERNING THESE RECOMMENDED PROCEDURES MAY ALSO BE DIRECTED TO THE HSAC THROUGH THE FEEDBACK FEATURE OF THE HSAC WEB SITE (WWW.HSAC.ORG).

b. Passenger Management on and about Heliport Facilities

1. Background. Several incidents involving offshore helicopter passengers have highlighted the potential for incidents and accidents on and about the heliport area. The following practices will minimize risks to passengers and others involved in heliport operations.

2. Recommended Practices

(a) Heliport facilities should have a designated and posted passenger waiting area which is clear of the heliport, heliport access points, and stairways.

(b) Arriving passengers and cargo should be unloaded and cleared from the heliport and access route prior to loading departing passengers and cargo.

(c) Where a flight crew consists of more than one pilot, one crewmember should supervise the unloading/loading process from outside the aircraft.

(d) Where practical, a designated facility employee should assist with loading/unloading, etc.

c. Crane-Helicopter Operational Procedures

1. Background. Historical experience has shown that catastrophic consequences can occur when industry safe practices for crane/helicopter operations are not observed. The following recommended practices are designed to minimize risks during crane and helicopter operations.

2. Recommended Practices

(a) Personnel awareness

(1) Crane operators and pilots should develop a mutual understanding and respect of the others' operational limitations and cooperate in the spirit of safety;

(2) Pilots need to be aware that crane operators sometimes cannot release the load to cradle the crane boom, such as when attached to wire line lubricators or supporting diving bells; and

(3) Crane operators need to be aware that helicopters require warm up before takeoff, a two-minute cool down before shutdown, and cannot circle for extended lengths of time because of fuel consumption.

(b) It is recommended that when helicopters are approaching, maneuvering, taking off, or running on the heliport, cranes be shut down and the operator leave the cab. Cranes not in use shall have their booms cradled, if feasible. If in use, the crane's boom(s) are to be pointed away from the heliport and the crane shut down for helicopter operations.

(c) Pilots will not approach, land on, takeoff, or have rotor blades turning on heliports of structures not complying with the above practice.

(d) It is recommended that cranes on offshore platforms, rigs, vessels, or any other facility, which could interfere with helicopter operations (including approach/departure paths):

(1) Be equipped with a red rotating beacon or red high intensity strobe light connected to the system powering the crane, indicating the crane is under power;

(2) Be designed to allow the operator a maximum view of the helideck area and should be equipped with wide-angle mirrors to eliminate blind spots; and

(3) Have their boom tips, headache balls, and hooks painted with high visibility international orange.

d. Helicopter/Tanker Operations

1. Background. The interface of helicopters and tankers during shipboard helicopter operations is complex and may be hazardous unless appropriate procedures are coordinated among all parties. The following recommended practices are designed to minimize risks during helicopter/tanker operations:

2. Recommended Practices

(a) Management, flight operations personnel, and pilots should be familiar with and apply the operating safety standards set forth in "Guide to Helicopter/Ship Operations," International Chamber of Shipping, Third Edition, 5-89 (as amended), establishing operational guidelines/standards and safe practices sufficient to safeguard helicopter/tanker operations.

(b) Appropriate plans, approvals, and communications must be accomplished prior to reaching the vessel, allowing tanker crews sufficient time to perform required safety preparations and position crew members to receive or dispatch a helicopter safely.

(c) Appropriate approvals and direct communications with the bridge of the tanker must be maintained throughout all helicopter/tanker operations.

(d) Helicopter/tanker operations, including landings/departures, shall not be conducted until the helicopter pilot-in-command has, received and acknowledged permission from the bridge of the tanker.

(e) Helicopter/tanker operations shall not be conducted during product/cargo transfer.

(f) Generally, permission will not be granted to land on tankers during mooring operations or while maneuvering alongside another tanker.

e. Helideck/Heliport Operational Hazard Warning(s) Procedures

1. Background

(a) A number of operational hazards can develop on or near offshore helidecks or onshore heliports that can be minimized through procedures for proper notification or visual warning to pilots. Examples of hazards include but are not limited to:

(1) Perforating operations: subparagraph f.

(2) H₂S gas presence: subparagraph g.

(3) Gas venting: subparagraph h; or,

(4) Closed helidecks or heliports: subparagraph i (unspecified cause).

(b) These and other operational hazards are currently minimized through timely dissemination of a written Notice to Airmen (NOTAM) for pilots by helicopter companies and operators. A NOTAM provides a written description of the hazard, time and duration of occurrence, and other pertinent information. ANY POTENTIAL HAZARD should be communicated to helicopter operators or company aviation departments as early as possible to allow the NOTAM to be activated.

(c) To supplement the existing NOTAM procedure and further assist in reducing these hazards, a standardized visual signal(s) on the helideck/heliport will provide a positive indication to an approaching helicopter of the status of the landing area. Recommended Practice(s) have been developed to reinforce the NOTAM procedures and standardize visual signals.

f. Drilling Rig Perforating Operations: Helideck/ Heliport Operational Hazard Warning(s)/Procedure(s)

1. Background. A critical step in the oil well completion process is perforation, which involves the use of explosive charges in the drill pipe to open the pipe to oil or gas deposits. Explosive charges used in conjunction with perforation operations offshore can potentially be prematurely detonated by radio transmissions, including those from helicopters. The following practices are recommended.

2. Recommended Practices

(a) Personnel Conducting Perforating Operations. Whenever perforating operations are scheduled and operators are concerned that radio transmissions from helicopters in the vicinity may jeopardize the operation, personnel conducting perforating operations should take the following precautionary measures:

(1) Notify company aviation departments, helicopter operators or bases, and nearby manned platforms of the pending perforation operation so the Notice to Airmen (NOTAM) system can be activated for the perforation operation and the temporary helideck closure.

(2) Close the deck and make the radio warning clearly visible to passing pilots, install a temporary marking (described in subparagraph 10-2-1i1(b)) with the words "NO RADIO" stenciled in red on the legs of the diagonals. The letters should be 24 inches high and 12 inches wide. (See FIG 10-2-1.)

(3) The marker should be installed during the time that charges may be affected by radio transmissions.

Figure 10-2-1
Closed Helideck Marking - No Radio

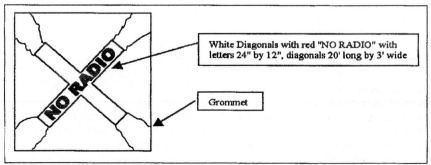

White Diagonals with red "NO RADIO" with letters 24" by 12", diagonals 20' long by 3' wide

Grommet

NO RADIO

(b) Pilots

(1) Pilots when operating within 1,000 feet of a known perforation operation or observing the white X with red "NO RADIO" warning indicating perforation operations are underway will avoid radio transmissions from or near the helideck (within 1,000 feet) and will not land on the deck if the X is present. In addition to communications radios, radio transmissions are also emitted by aircraft radar, transponders, radar altimeters, and DME equipment, and ELT's.

(2) Whenever possible, make radio calls to the platform being approached or to the Flight Following Communications Center at least one mile out on approach. Ensure all communications are complete outside the 1,000 foot hazard distance. If no response is received, or if the platform is not radio equipped, further radio transmissions should not be made until visual contact with the deck indicates it is open for operation (no white "X").

g. Hydrogen Sulfide Gas Helideck/Heliport Operational Hazard Warning(s)/Procedures

1. Background. Hydrogen sulfide (H_2S) gas:

Hydrogen sulfide gas in higher concentrations (300-500 ppm) can cause loss of consciousness within a few seconds and presents a hazard to pilots on/near offshore helidecks. When operating in offshore areas that have been identified to have concentrations of hydrogen sulfide gas, the following practices are recommended.

2. Recommended Practices

(a) Pilots

(1) Ensure approved protective air packs are available for emergency use by the crew on the helicopter.

(2) If shutdown on a helideck, request the supervisor in charge provide a briefing on location of protective equipment and safety procedures.

(3) If while flying near a helideck and the visual red beacon alarm is observed or an unusually strong odor of "rotten eggs" is detected, immediately don the protective air pack, exit to an area upwind, and notify the suspected source field of the hazard.

(b) Oil Field Supervisors

(1) If presence of hydrogen sulfide is detected, <u>a red rotating beacon</u> or red high intensity strobe light adjacent to the primary helideck stairwell or wind indicator on the structure should be turned on to provide visual warning of hazard. If the beacon is to be located near the stairwell, the State of Louisiana "Offshore Heliport Design Guide" and FAA Advisory Circular AC 150/5390-2A, "Heliport Design Guide," should be reviewed to ensure proper clearance on the helideck.

(2) Notify nearby helicopter operators and bases of the hazard and advise when hazard is cleared.

(3) Provide a safety briefing to include location of protective equipment to all arriving personnel.

(4) Wind socks or indicator should be clearly visible to provide upwind indication for the pilot.

h. Gas Venting Helideck/Heliport Operational Hazard Warning(s)/Procedures - Operations Near Gas Vent Booms

1. Background. Ignited flare booms can release a large volume of natural gas and create a hot fire and intense heat with little time for the pilot to react. Likewise, unignited gas vents can release reasonably large volumes of methane gas under certain conditions. Thus, operations conducted very near unignited gas vents require precautions to prevent inadvertent ingestion of combustible gases by the helicopter engine(s). The following practices are recommended.

2. Pilots

(a) Gas will drift upwards and downwind of the vent. Plan the approach and takeoff to observe and <u>avoid the area downwind of the vent</u>, remaining as far away as practicable from the open end of the vent boom.

(b) Do not attempt to start or land on an offshore helideck when the deck is downwind of a gas vent unless properly trained personnel verify conditions are safe.

3. Oil Field Supervisors

(a) During venting of large amounts of unignited raw gas, a red rotating beacon or red high intensity strobe light adjacent to the primary helideck stairwell or wind indicator should be turned on to provide visible warning of hazard. If the beacon is to be located near the stairwell, the State of Louisiana "Offshore Heliport Design Guide" and FAA Advisory Circular AC 150/5390-2A, Heliport Design Guide, should be reviewed to ensure proper clearance from the helideck.

(b) Notify nearby helicopter operators and bases of the hazard for planned operations.

(c) Wind socks or indicator should be clearly visible to provide upward indication for the pilot.

i. Helideck/Heliport Operational Warning(s)/ Procedure(s) - Closed Helidecks or Heliports

1. Background. A white "X" marked diagonally from corner to corner across a helideck or heliport touchdown area is the universally accepted visual indicator that the landing area is closed for safety or other reasons and that helicopter operations are not permitted. The following practices are recommended.

(a) Permanent Closing. If a helideck or heliport is to be permanently closed, X diagonals of the same size and location as indicated above should be used, but the markings should be painted on the landing area.

NOTE—
WHITE DECKS: IF A HELIDECK IS PAINTED WHITE, THEN INTERNATIONAL ORANGE OR YELLOW MARKINGS CAN BE USED FOR THE TEMPORARY OR PERMANENT DIAGONALS.

(b) Temporary Closing. A temporary marker can be used for hazards of an interim nature. This marker could be made

from vinyl or other durable material in the shape of a diagonal "X." The marker should be white with legs at least 20 feet long and 3 feet in width. This marker is designed to be quickly secured and removed from the deck using grommets and rope ties. The duration, time, location, and nature of these temporary closings should be provided to and coordinated with company aviation departments, nearby helicopter bases, and helicopter operators supporting the area. These markers MUST be removed when the hazard no longer exists.
(See FIG 10-2-2.)

Figure 10-2-2
Closed Helideck Marking

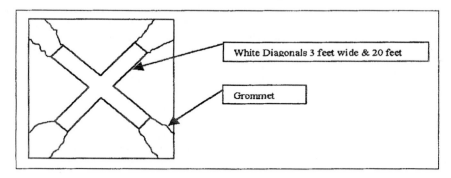

j. Offshore (VFR) Operating Altitudes for Helicopters

1. Background. Mid-air collisions constitute a significant percentage of total fatal offshore helicopter accidents. A method of reducing this risk is the use of coordinated VFR cruising altitudes. To enhance safety through standardized vertical separation of helicopters when flying in the offshore environment, it is recommended that helicopter operators flying in a particular area establish a co-operatively developed Standard Operating Procedure (SOP) for VFR operating altitudes. An example of such an SOP is contained in this example.

2. Recommended Practice Example

(a) Field Operations. Without compromising minimum safe operating altitudes, helicopters working within an offshore field "constituting a cluster" should use altitudes not to exceed 500 feet.

(b) En Route Operations

(1) Helicopters operating below *750' AGL* should avoid transitioning through offshore fields.

(2) Helicopters en route to and from offshore locations, below 3,000 feet, weather permitting, should use en route altitudes as outlined in TBL 10-2-1.

Table 10-2-1

Magnetic Heading	Altitude
00 to 179°	750'
	1750'
	2750'
180° 359°	1250'
	2250'

(c) Area Agreements. See HSAC Area Agreement Maps for operating procedures for onshore high density traffic locations.

NOTE—
PILOTS OF HELICOPTERS OPERATING VFR ABOVE 3,000 FEET ABOVE THE SURFACE SHOULD REFER TO THE CURRENT FEDERAL AVIATION REGULATIONS (14 CFR PART 91), AND PARAGRAPH 3-1-4, BASIC VFR WEATHER MINIMUMS, OF THE AIM.

(d) Landing Lights. Aircraft landing lights should be on to enhance aircraft identification:

(1) During takeoff and landings;

(2) In congested helicopter or fixed wing traffic areas;

(3) During reduced visibility; or,

(4) Any time safety could be enhanced.

k. Offshore Helidecks/Landing Communications

1. Background. To enhance safety, and provide appropriate time to prepare for helicopter operations, the following is recommended when anticipating a landing on an offshore helideck.

2. Recommended Practices

(a) Before landing on an offshore helideck, pilots are encouraged to establish communications with the company owning or operating the helideck if frequencies exist for that purpose.

(b) When impracticable, or if frequencies do not exist, pilots or operations personnel should attempt to contact the company owning or operating the helideck by telephone. Contact should be made before the pilot departs home base/point of departure to advise of intentions and obtain landing permission if necessary.

NOTE: *It is recommended that communications be established a minimum of 10 minutes prior to planned arrival time. This practice may be a requirement of some offshore owner/operators.*

NOTE—

1. SEE SUBPARAGRAPH 10-2-ID FOR TANKER OPERATIONS.

2. PRIVATE USE HELIPORT. OFFSHORE HELIPORTS ARE PRIVATELY OWNED/OPERATED FACILITIES AND THEIR USE IS LIMITED TO PERSONS HAVING PRIOR AUTHORIZATION TO UTILIZE THE FACILITY.

l. Two (2) Helicopter Operations on Offshore Helidecks

1. Background. Standardized procedures can enhance the safety of operating a second helicopter on an offshore helideck, enabling pilots to determine/maintain minimum operational parameters. Orientation of the parked helicopter on the helideck, wind and other factors may prohibit multi-helicopter operations. More conservative Rotor Diameter (RD) clearances may be required under differing conditions, i.e. temperature, wet deck, wind (velocity/direction/gusts), obstacles, approach/ departure angles, etc. Operations are at the pilot's discretion.

2. Recommended Practice. Helideck size, structural weight capability, and type of main rotor on the parked and operating helicopter will aid in determining accessibility by a second helicopter. Pilots should determine that multi-helicopter deck operations are permitted by the helideck owner/operator.

3. Recommended Criteria

(a) Minimum one-third rotor diameter clearance (1/3 RD). The landing helicopter maintains a minimum 1/3 RD clearance between the tips of its turning rotor and the closest part of a parked and secured helicopter (rotors stopped and tied down).

(b) Three foot parking distance from deck edge (3'). Helicopters operating on an offshore helideck land or park the helicopter with a skid/wheel assembly no closer than 3 feet from helideck edge.

(c) Tiedowns. Main rotors on all helicopters that are shut down be properly secured (tied down) to prevent the rotor blades from turning.

(d) Medium (transport) and larger helicopters should not land on any offshore helideck where a light helicopter is parked unless the light helicopter is property secured to the helideck and has main rotor tied down.

(e) Helideck owners/operators should ensure that the helideck has a serviceable anti-skid surface.

4. Weight and limitations markings on helideck. The helideck weight limitations should be displayed by markings visible to the pilot (see State of Louisiana "Offshore Heliport Design Guide" and FAA Advisory Circular AC 150/5390-2A, Heliport Design Guide).

NOTE—

SOME OFFSHORE HELIDECK OWNERS/OPERATORS HAVE RESTRICTIONS ON THE NUMBER OF HELICOPTERS ALLOWED ON A HELIDECK. WHEN HELIDECK SIZE PERMITS, MULTIPLE (MORE THAN TWO) HELICOPTER OPERATIONS ARE PERMITTED BY SOME OPERATORS.

m. Helicopter Rapid Refueling Procedures (HRR)

1. Background. Helicopter Rapid Refueling (HRR), engine(s)/rotors operating, can be conducted safely when utilizing trained personnel and observing safe practices. This recommended practice provides minimum guidance for HRR as outlined in National Fire Protection Association (NFPA) and industry practices. For detailed guidance, please refer to National Fire Protection Association (NFPA) Document 407, "Standard for Aircraft Fuel Servicing," 1990 edition, including 1993 HRR Amendment.

NOTE—

CERTAIN OPERATORS PROHIBIT HRR, OR "HOT REFUELING," OR MAY HAVE SPECIFIC PROCEDURES FOR CERTAIN AIRCRAFT OR REFUELING LOCATIONS. SEE THE GENERAL OPERATIONS MANUAL AND/OR OPERATIONS SPECIFICATIONS TO DETERMINE THE APPLICABLE PROCEDURES OR LIMITATIONS.

2. Recommended Practices

(a) Only turbine-engine helicopters fueled with JET A or JET A-1 with fueling ports located below any engine exhausts may be fueled while an onboard engine(s) is (are) operating.

(b) Helicopter fueling while an onboard engine(s) is (are) operating should only be conducted under the following conditions:

(1) A properly certificated and current pilot is at the controls and a trained refueler attending the fuel nozzle during the entire fuel servicing process. The pilot monitors the fuel quantity and signals the refueler when quantity is reached.

(2) No electrical storms (thunderstorms) are present within 10 nautical miles. Lightning can travel great distances beyond the actual thunderstorm.

(3) Passengers disembark the helicopter and move to a safe location prior to HRR operations. When the pilot-in-command deems it necessary for passenger safety that they remain onboard, passengers should be briefed on the evacuation route to follow to clear the area.

(4) Passengers not board or disembark during HRR operations nor should cargo be loaded or unloaded.

(5) Only designated personnel, trained in HRR operations should conduct HRR written authorization to include safe handling of the fuel and equipment. (See your Company Operations/Safety Manual for detailed instructions.)

(6) All doors, windows, and access points allowing entry to the interior of the helicopter that are adjacent to or in the immediate vicinity of the fuel inlet ports kept closed during HRR operations.

(7) Pilots insure that appropriate electrical/electronic equipment is placed in standby-off position, to preclude the possibility of electrical discharge or other fire hazard, such as radar is on standby and no radio transmissions are made (keying of the microphone/transmitter)]. Remember, in addition to communications radios, radio transmissions are also emitted by aircraft radar, transponders, radar altimeters, DME equipment, and ELT's.

(8) Smoking be prohibited in and around the helicopter during all HRR operations.

The HRR procedures are critical and present associated hazards requiring attention to detail regarding quality control, weather conditions, static electricity, bonding, and spill/fires potential.

Any activity associated with rotors turning (i.e.; refueling embarking/disembarking, loading/unloading baggage/freight; etc.) personnel should approach the aircraft when authorized to do so. Approach should be made via safe approach path/walkway or "arc"- **remain clear of all rotors.**

NOTE—

1. MARINE VESSELS, BARGES ETC.: VESSEL MOTION PRESENTS ADDITIONAL POTENTIAL HAZARDS TO HELICOPTER OPERATIONS (BLADE FLEX, AIRCRAFT MOVEMENT).

2. SEE NATIONAL FIRE PROTECTION ASSOCIATION (NFR4) DOCUMENT 407, "STANDARD FOR AIRCRAFT FUEL SERVICING" FOR SPECIFICS REGARDING NON-HRR (ROUTINE REFUELING OPERATIONS).

10-2-2. Helicopter Night VFR Operations a. Effect of Lighting on Seeing Conditions in Night VFR Helicopter Operations

NOTE—

THIS GUIDANCE WAS DEVELOPED TO SUPPORT SAFE NIGHT VFR HELICOPTER EMERGENCY MEDICAL SERVICES (HEMS) OPERATIONS. THE PRINCIPLES OF LIGHTING AND SEEING CONDITIONS ARE USEFUL IN ANY NIGHT VFR OPERATION.

While ceiling and visibility significantly affect safety in night VFR operations, lighting conditions also have a profound effect on safety. Even in conditions in which visibility and ceiling are determined to be visual meteorological conditions, the ability to discern unlighted or low contrast objects and terrain at night may be compromised. The ability to discern these objects and terrain is the seeing condition, and is related to the amount of natural and man made lighting available, and the contrast, reflectivity, and texture of surface terrain and obstruction features. In order to conduct operations safely, seeing conditions must be accounted for in the planning and execution of night VFR operations.

Night VFR seeing conditions can be described by identifying "high lighting conditions" and "low lighting conditions."

1. High lighting conditions exist when one of two sets of conditions are present:

(a) The sky cover is less than broken (less than 5/8 cloud cover), the time is between the local Moon rise and Moon set, and the lunar disk is at least 50% illuminated; or

(b) The aircraft is operated over surface lighting which, at least, provides for the lighting of prominent obstacles, the identification of terrain features (shorelines, valleys, hills, mountains, slopes) and a horizontal reference by which the pilot may control the helicopter. For example, this surface lighting may be the result of:

(1) Extensive cultural lighting (man-made, such as a built-up area of a city),

(2) Significant reflected cultural lighting (such as the illumination caused by the reflection of a major metropolitan area's lighting reflecting off a cloud ceiling), or

(3) Limited cultural lighting combined with a high level of natural reflectivity of celestial illumination, such as that provided by a surface covered by snow or a desert surface.

2. Low lighting conditions are those that do not meet the high lighting conditions requirements.

3. Some areas maybe considered a highlighting environment only in specific circumstances. For example, some surfaces, such as a forest with limited cultural lighting, normally have little reflectivity, requiring dependence on significant moonlight to achieve a high lighting condition. However, when that same forest is covered with snow, its reflectivity may support a high lighting condition based only on starlight. Similarly, a desolate area, with little cultural lighting, such as a desert, may have such inherent natural reflectivity that it may be considered a high lighting conditions area regardless of season, provided the cloud cover does not prevent starlight from being reflected from the surface. Other surfaces, such as areas of open water, may never have enough reflectivity or cultural lighting to ever be characterized as a high lighting area.

4. Through the accumulation of night flying experience in a particular area, the operator will develop the ability to determine, prior to departure, which areas can be considered supporting high or low lighting conditions. Without that operational experience, low lighting considerations should be applied by operators for both pre-flight planning and operations until high lighting conditions are observed or determined to be regularly available.

b. Astronomical Definitions and Background Information for Night Operations
1. Definitions

(a) Horizon. Wherever one is located on or near the Earth's surface, the Earth is perceived as essentially flat and, therefore, as a plane. If there are no visual obstructions, the apparent intersection of the sky with the Earth's (plane) surface is the horizon, which appears as a circle centered at the observer. For rise/set computations, the observer's eye is considered to be on the surface of the Earth, so that the horizon is geometrically exactly 90 degrees from the local vertical direction.

(b) Rise, Set. During the course of a day the Earth rotates once on its axis causing the phenomena of rising and setting. All celestial bodies, the Sun, Moon, stars and planets, seem to appear in the sky at the horizon to the East of any particular place, then to cross the sky and again disappear at the horizon to the West. Because the Sun and Moon appear as circular disks and not as points of light, a definition of rise or set must be very specific, because not all of either body is seen to rise or set at once.

(c) Sunrise and sunset refer to the times when the upper edge of the disk of the Sun is on the horizon, considered unobstructed relative to the location of interest. Atmospheric conditions are assumed to be average, and the location is in a level region on the Earth's surface.

(d) Moonrise and moonset times are computed for exactly the same circumstances as for sunrise and sunset. However, moonrise and moonset may occur at any time during a 24 hour period and, consequently, it is often possible for the Moon to be seen during daylight, and to have moonless nights. It is also possible that a moonrise or moonset does not occur relative to a specific place on a given date.

Figure 10-2-3.
Phases of the Moon

New Moon - The Moon's unilluminated side is facing the Earth. The Moon is not visible (except during a solar eclipse).

Waxing Crescent - The Moon appears to be partly but less than one-half illuminated by direct sunlight. The fraction of the Moon's disk that is illuminated is increasing.

First Quarter - One-half of the Moon appears to be illuminated by direct sunlight. The fraction of the Moon's disk that is illuminated is increasing.

Waxing Gibbous - The Moon appears to be more than one-half but not fully illuminated by direct sunlight. The fraction of the Moon's disk that is illuminated is increasing.

Full Moon - The Moon's illuminated side is facing the Earth. The Moon appears to be completely illuminated by direct sunlight.

Waning Gibbous - The Moon appears to be more than one-half but not fully illuminated by direct sunlight. The fraction of the Moon's disk that is illuminated is decreasing.

Last Quarter - One-half of the Moon appears to be illuminated by direct sunlight. The fraction of the Moon's disk that is illuminated is decreasing.

Waning Crescent - The Moon appears to be partly but less than one-half illuminated by direct sunlight. The fraction of the Moon's disk that is illuminated is decreasing.

(e) Transit. The transit time of a celestial body refers to the instant that its center crosses an imaginary line in the sky—the observer's meridian—running from north to south.

(f) Twilight. Before sunrise and again after sunset there are intervals of time, known as "twilight," during which there is natural light provided by the upper atmosphere, which does receive direct sunlight and reflects part of it toward the Earth's surface.

(g) Civil twilight is defined to begin in the morning, and to end in the evening when the center of the Sun is geometrically 6 degrees below the horizon. This is the limit at which twilight illumination is sufficient, under good weather conditions, for terrestrial objects to be clearly distinguished.

2. Title 14 of the Code of Federal Regulations applies these concepts and definitions in addressing the definition of night (Section 1.1), the requirement for aircraft lighting (Section 91.209) and pilot recency of night experience (Section 61.67).

c. Information on Moon Phases and Changes in the Percentage of the Moon Illuminated

From any location on the Earth, the Moon appears to be a circular disk which, at any specific time, is illuminated to some degree by direct sunlight. During each lunar orbit (a lunar month), we see the Moon's appearance change from not visibly illuminated through partially illuminated to fully illuminated, then back through partially illuminated to not illuminated again. There are eight distinct, traditionally recognized stages, called phases. The phases designate both the degree to which the Moon is illuminated and the geometric appearance of the illuminated part. These phases of the Moon, in the sequence of their occurrence (starting from New Moon), are listed in FIG 10-2-3.

1. The percent of the Moon's surface illuminated is a more refined, quantitative description of the Moon's appearance than is the phase. Considering the Moon as a circular disk, at New Moon the percent illuminated is 0; at First and Last Quarters it is 50% and at Full Moon it is 100%. During the crescent phases the percent illuminated is between 0 and 50% and during gibbous phases it is between 50% any 100%.

2. For practical purposes, phases of the Moon and the percent of the Moon illuminated are independent of the location on the Earth from when the Moon is observed. That is, all the phases occur at the same time regardless of the observer's position.

3. For more detailed information, refer to the United States Naval Observatory site referenced below.

d. Access to Astronomical Data for Determination of Moon Rise, Moon Set, and Percentage of Lunar Disk Illuminated

1. Astronomical data for the determination of Moon rise and set and Moon phase may be obtained from the United States Naval Observatory using an interactive query available at: http://aa.usno.navy.mil/

2. Click on "Data Services," and then on "Complete Sun and Moon Data for One Day."

3. You can obtain the times of sunrise, sunset, moonrise, moonset, transits of the Sun and Moon, and the beginning and end of civil twilight, along with information on the Moon's phase by specifying the date and location in one of the two forms on this web page and clicking on the "Get data" button at the end of the form. Form "A" is used for cities or towns in the U.S. or its territories. Form "B" for all other locations. An example of the data available from this site is shown in TBL 10-2-2.

Table 10-2-2   Sample of Astronomical Data Available from the Naval Observatory

Tuesday 29 May 2007	Central Daylight Time
SUN	
Begin civil twilight	5:34 a.m.
Sunrise	6:01 a.m.
Sun transit	12:58 p.m.
Sunset	7:55 p.m.
End civil twilight	8:22 p.m.
MOON	
Moonrise	5:10 p.m. on preceding day
Moonset	4:07 a.m.
Moonrise	6:06 p.m.
Moon transit	11:26 p.m.
Moonset	4:41 a.m. on following day

Phase of the Moon on 29 May: waxing gibbous with 95% of the Moon's visible disk illuminated.
Full Moon on 31 May 2007 at 8:04 p.m. Central Daylight Time.

4. Additionally, a yearly table may be constructed for a particular location by using the "Table of Sunrise/Sunset, Moonrise/Moonset, or Twilight Times for an Entire Year" selection.

10-2-3. Landing Zone Safety

a. This information is provided for use by helicopter emergency medical services (HEMS) pilots, program managers, medical personnel, law enforcement, fire, and rescue personnel to further their understanding of the safety issues concerning Landing Zones (LZs). It is recommended that HEMS operators establish working relationships with the ground responder organizations they may come in contact with in their flight operations and share this information in order to establish a common frame of reference for LZ selection, operations, and safety.

b. The information provided is largely based on the booklet, LZ—Preparing the Landing Zone, issued by National Emergency Medical Services Pilots Association (NEMSPA), and the guidance developed by the University of Tennessee Medical Center's LIFESTAR program,

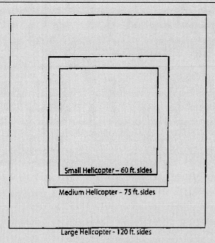

Figure 10-2-4. Recommended Minimum Landing
Zone Dimensions

Figure 10-2-5. Landing Zone Hazards

and is used with their permission. For additional information, go to http://www.nemspa.org/.

c. Information concerning the estimation of wind velocity is based on the Beaufort Scale. See http://www.spc.noaa.gov/faq/tornado/beaufort.html for more information.

d. Selecting a Scene LZ

1. If the situation requires the use of a helicopter, first check to see if there is an area large enough to land a helicopter safely.

2. For the purposes of FIG 10-2-4 the following are provided as examples of relative helicopter size:

(a) Small Helicopter: Bell 206/407, Eurocopter AS-350/355, BO-105, BK-117.

(b) Medium Helicopter: Bell UH-1 (Huey) and derivatives (Bell 212/412), Bell 222/230/430 Sikorsky S-76, Eurocopter SA-365.

(c) Large Helicopter: Boeing Chinook, Eurocopter Puma, Sikorsky H-60 series (Blackhawk), SK-92.

3. The LZ should be level, firm and free of loose debris that could possibly blow up into the rotor system.

4. The LZ should be clear of people, vehicles and obstructions such as trees, poles and wires. Remember that wires are difficult to see from the air. The LZ must also be free of stumps, brush, post and lamp rocks (see FIG 10-2-5).

5. Keep spectators back at least 200 feet. Keep emergency vehicles 100 feet away and have fire equipment (if available) standing by. Ground personnel should wear eye protection, if available, during landing and takeoff operations. To avoid loose objects being blown around in the LZ, hats should be removed; if helmets are worn, chin straps must be securely fastened.

6. Fire fighters (if available) should wet down the LZ if it is extremely dusty.

e. Helping the Flightcrew Locate the Scene

1. If the LZ coordinator has access to a GPS unit, the exact latitude and longitude of the LZ should be relayed to the HEMS pilot. If unable to contact the pilot directly, relay the information to the HEMS ground communications specialist for relaying to the pilot, so that they may locate your scene more efficiently. Recognize that the aircraft may approach from a direction different than the direct path from the takeoff point to the scene, as the pilot may have to detour around terrain, obstructions or weather en route.

2. Especially in daylight hours, mountainous and densely populated areas can make sighting a scene from the air difficult. Often, the LZ coordinator on the ground will be asked if she or he can see or hear the helicopter.

3. Flightcrews use a clock reference method for directing one another's attention to a certain direction from the aircraft. The nose of the aircraft is always 12 o'clock, the right side is 3 o'clock, etc. When the LZ coordinator sees the aircraft, he/she should use this method to assist the flightcrew by indicating the scene's clock reference position from the nose of the aircraft. For example, "Accident scene is located at your 2 o'clock position." See FIG 10-2-6.

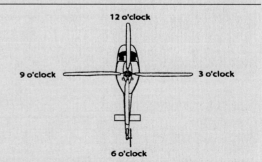

Figure 10-2-6. "Clock" System for Identifying Positions
Relative to the Nose of the Aircraft

4. When the helicopter approaches the scene, it will normally orbit at least one time as the flight crew observes the wind direction and obstacles that could interfere with the landing. This is often referred to as the "high reconnaissance" maneuver.

f. Wind Direction and Touchdown Area

1. Determine from which direction the wind is blowing. Helicopters normally land and takeoff into the wind.

2. If contact can be established with the pilot, either directly or indirectly through the HEMS ground communications specialist, describe the wind in terms of the direction the wind is *from* and the speed.

3. Common natural sources of wind direction information are smoke, dust, vegetation movement, water streaks and waves. Flags, pennants, streamers can also be used. When describing the direction, use the compass direction from which the wind is blowing (example: from the North-West).

4. Wind speed can be measured by small hand-held measurement devices, or an observer's estimate can be used to provide velocity information. The wind value should be reported in knots (nautical miles per hour). If unable to numerically measure wind speed, use TBL 10-2-3 to estimate velocity. Also, report if the wind conditions are gusty, or if the wind direction or velocity is variable or has changed recently.

5. If any obstacle(s) exist, insure their description, position and approximate height are communicated to the pilot on the initial radio call.

Table 10-2-3
Table of Common References for Estimating Wind Velocity

Wind (Knots)	Wind Classification	Appearance of Wind Effects	
		On the Water	On Land
Less than 1	Calm	Sea surface smooth and mirror-like	Calm, smoke rises vertically
1–3	Light Air	Scaly ripples, no foam crests	Smoke drift indicates wind direction, wind vanes are still
4–6	Light Breeze	Small wavelets, crests glassy, no breaking	Wind felt on face, leaves rustle, vanes begin to move
7–10	Gentle Breeze	Large wavelets, crests begin to break, scattered whitecaps	Leaves and small twigs constantly moving, light flags extended
11–16	Moderate Breeze	Small waves 1–4 ft. becoming longer, numerous whitecaps	Dust, leaves, and loose paper lifted, small tree branches move
17–21	Fresh Breeze	Moderate waves 4–8 ft. taking longer form, many whitecaps, some spray	Small trees in leaf begin to sway
22–27	Strong Breeze	Larger waves 8–13 ft., whitecaps common, more spray	Larger tree branches moving, whistling in wires
28–33	Near Gale	Sea beaps up, waves 13–20 ft., white foam streaks off breakers	Whole trees moving, resistance felt walking against wind
34–40	Gale	Moderately high (13–20 ft.) waves of greater length, edges of crests begin to break into spindrift, foam blown in streaks	Whole trees in motion, resistance felt walking against wind
41–47	Strong Gale	High waves (20 ft.) sea begins to roll, dense streaks of foam, spray may reduce visibility	Slight structural damage occurs, slate blows off roofs
48–55	Storm	Very high waves (20–30 ft.) with over-hanging crests, sea white with densely blows foam, heavy rolling, lowered visibility	Seldom experienced on land, trees broken or uprooted, "considerable structural damage."
56–63	Violent Storm	Exceptionally high (30–45 ft.) waves, foam patches cover sea, visibility more reduced	
64+	Hurricane	Air filled with foam, waves over 45 ft., sea completely white with driving spray, visibility greatly reduced	

EXAMPLE—

Wind from the South–East, estimated speed 15 knots. Wind shifted from North–East about fifteen minutes ago, and is gusty.

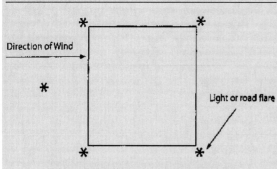

Figure 10-2-7. Recommended Lighting for
Landing Zone Operations at Night

g. Night LZs

1. There are several ways to light a night LZ:

(a) Mark the touchdown area with five lights or road flares, one in each corner and one indicating the direction of the wind. See FIG 10-2-7.

NOTE—
ROAD FLARES ARE AN INTENSE SOURCE OF IGNITION AND MAY BE UNSUITABLE OR DANGEROUS IN CERTAIN CONDITIONS. IN ANY CASE, THEY MUST BE CLOSELY MANAGED AND FIREFIGHTING EQUIPMENT SHOULD BE PRESENT WHEN USED. OTHER LIGHT SOURCES ARE PREFERRED IF AVAILABLE.

(b) If chemical light sticks may be used, care should be taken to assure they are adequately secured against being dislodged by the helicopter's rotor wash.

(c) Another method of marking a LZ uses four emergency vehicles with their low beam headlights aimed toward the intended landing area.

(d) A third method for marking a LZ uses two vehicles. Have the vehicles direct their headlight beams into the wind, crossing at the center of the LZ. (If fire/rescue personnel are available, the reflective stripes on their bunker gear will assist the pilot greatly.)

2. At night, spotlights, flood lights and hand lights used to define the LZ are not to be pointed at the helicopter. However, they are helpful when pointed toward utility poles, trees or other hazards to the landing aircraft. White lights such as spotlights, flashbulbs and hi-beam headlights ruin the pilot's night vision and temporarily blind him. Red lights, however, are very helpful in finding accident locations and do not affect the pilot's night vision as significantly.

3. As in Day LZ operations, ensure radio contact is accomplished between ground and air, if possible.

h. Ground Guide

1. When the helicopter is in sight, one person should assist the LZ Coordinator by guiding the helicopter into a safe landing area. In selecting an LZ Coordinator, recognize that medical personnel usually are very busy with the patient at this time. It is recommended that the LZ Coordinator be someone other than a medical responder, if possible. Eye protection should be worn. The ground guide should stand with his/her back to the wind and his/her arms raised over his/her head (flashlights in each hand for night operations.)

2. The pilot will confirm the LZ sighting by radio. If possible, once the pilot has identified the LZ, the ground guide should move out of the LZ.

3. As the helicopter turns into the wind and begins a descent, the LZ coordinator should provide assistance by means of radio contact, or utilize the "unsafe signal" to wave off the helicopter if the LZ is not safe (see FIG 10-2-8). The LZ Coordinator should be far enough from the touchdown area that he/she can still maintain visual contact with the pilot.

i. Assisting the Crew

1. After the helicopter has landed, do not approach the helicopter. The crew will approach you.

2. Be prepared to assist the crew by providing security for the helicopter. If asked to provide security, allow no one but the crew to approach the aircraft.

3. Once the patient is prepared and ready to load, allow the crew to open the doors to the helicopter and guide the loading of the patient.

4. When approaching or departing the helicopter, always be aware of the tail rotor and always follow the directions of the crew. Working around a running helicopter can be

Figure 10-2-8. Recommended Landing Zone
Ground Signals

potentially dangerous. The environment is very noisy and, with exhaust gases and rotor wash, often windy. In scene operations, the surface may be uneven, soft, or slippery which can lead to tripping. Be very careful of your footing in this environment.

5. The tail rotor poses a special threat to working around a running helicopter. The tail rotor turns many times faster than the main rotor, and is often invisible even at idle engine power. Avoid walking towards the tail of a helicopter beyond the end of the cabin, unless specifically directed by the crew.

NOTE—

HELICOPTERS TYPICALLY HAVE DOORS ON THE SIDES OF THE CABIN, BUT MANY USE AFT MOUNTED "CLAMSHELL" TYPE DOORS FOR LOADING AND UNLOADING PATIENTS ON LITTERS OR STRETCHERS. WHEN USING THESE DOORS, IT IS IMPORTANT TO AVOID MOVING ANY FURTHER AFT THAN NECESSARY TO OPERATE THE DOORS AND LOAD/UNLOAD THE PATIENT. AGAIN, ALWAYS COMPLY WITH THE CREW'S INSTRUCTIONS.

j. General Rules

1. When working around helicopters, always approach and depart from the front, never from the rear. Approaching from the rear can increase your risk of being struck by the tail rotor, which, when at operating engine speed, is nearly invisible.

2. To prevent injury or damage from the main rotor, never raise anything over your head.

3. If the helicopter landed on a slope, approach and depart from the down slope side only.

4. When the helicopter is loaded and ready for take off, keep the departure path free of vehicles and spectators. In an emergency, this area is needed to execute a landing.

k. Hazardous Chemicals and Gases

1. Responding to accidents involving hazardous materials requires special handling by fire/rescue units on the ground. Equally important are the preparations and considerations for helicopter operations in these areas.

2. Hazardous materials of concern are those which are toxic, poisonous, flammable, explosive, irritating, or radioactive in nature. Helicopter ambulance crews normally don't carry protective suits or breathing apparatuses to protect them from hazardous materials.

3. The helicopter ambulance crew must be told of hazardous materials on the scene in order to avoid the contamination of the crew. Patients/victims contaminated by hazardous materials may require special precautions in packaging before loading on the aircraft for the medical crew's protection, or may be transported by other means.

4. Hazardous chemicals and gases may be fatal to the unprotected person if inhaled or absorbed through the skin.

5. Upon initial radio contact, the helicopter crew must be made aware of any hazardous gases in the area. Never assume that the crew has already been informed. If the aircraft were to fly through the hazardous gases, the crew could be poisoned and/or the engines could develop mechanical problems.

6. Poisonous or irritating gases may cling to a victim's clothing and go unnoticed until the patient is loaded and the doors of the helicopter are closed. To avoid possible compromise of the crew, all of these patients must be decontaminated prior to loading.

l. Hand Signals

1. If unable to make radio contact with the HEMS pilot, use the following signals:

m. Emergency Situations

1. In the event of a helicopter accident in the vicinity of the LZ, consider the following:

(a) Emergency Exits:

(1) Doors and emergency exits are typically prominently marked. If possible, operators should familiarize ground responders with the door system on their helicopter in preparation for an emergency event.

(2) In the event of an accident during the LZ operation, be cautious of hazards such as sharp and jagged metal, plastic windows, glass, any rotating components, such as the rotors, and fire sources, such as the fuel tank(s) and the engine.

(b) Fire Suppression:

Helicopters used in HEMS operations are usually powered by turboshaft engines, which use jet fuel. Civil HEMS aircraft typically carry between 50 and 250 gallons of fuel, depending upon the size of the helicopter, and planned flight duration, and the fuel remaining after flying to the scene. Use water to control heat and use foam over fuel to keep vapors from ignition sources.

Appendix 1.

Bird/Other Wildlife Strike Report

Form Approved OMB NO. 2120-0018

BIRD/OTHER WILDLIFE STRIKE REPORT

U.S. Department of Transportation
Federal Aviation Administration

1. Name of Operator	2. Aircraft Make/Model	3. Engine Make/Model

4. Aircraft Registration	5. Date of Incident ___/___/___ Month / Day / Year	6. Local Time of Incident ☐ Dawn ☐ Dusk ___ HR ___ MIN ☐ Day ☐ Night ☐ AM ☐ PM

7. Airport Name	8. Runway Used	9. Location If En Route *(Nearest Town/Reference & State)*

10. Height *(AGL)*	11. Speed *(IAS)*	

12. Phase of Flight

☐ A. Parked
☐ B. Taxi
☐ C. Take-off Run
☐ D. Climb
☐ E. En Route
☐ F. Descent
☐ G. Approach
☐ H. Landing Roll

13. Part(s) of Aircraft Struck or Damaged

	Struck	Damaged		Struck	Damaged
A. Radome	☐	☐	H. Propeller	☐	☐
B. Windshield	☐	☐	I. Wing/Rotor	☐	☐
C. Nose	☐	☐	J. Fuselage	☐	☐
D. Engine No. 1	☐	☐	K. Landing Gear	☐	☐
E. Engine No. 2	☐	☐	L. Tail	☐	☐
F. Engine No. 3	☐	☐	M. Lights	☐	☐
G. Engine No. 4	☐	☐	N. Other:	☐	☐

(Specify, if "N. Other" is checked)

14. Effect on Flight
☐ None
☐ Aborted Take-Off
☐ Precautionary Landing
☐ Engines Shut Down
☐ Other: *(Specify)*

15. Sky Condition
☐ No Cloud
☐ Some Cloud
☐ Overcast

16. Precipitation
☐ Fog
☐ Rain
☐ Snow
☐ None

17. Bird/Other Wildlife Species

18. Number or birds seen and/or struck

Number of Birds	Seen	Struck
1	☐	☐
2-10	☐	☐
11-100	☐	☐
more than 100	☐	☐

19. Size of Bird(s)
☐ Small
☐ Medium
☐ Large

20. Pilot Warned of Birds ☐ Yes ☐ No

21. Remarks *(Describe damage, injuries and other pertinent information)*

DAMAGE / COST INFORMATION

22. Aircraft time out of service: _____ hours	23. Estimated cost of repairs or replacement *(U.S. $):* $	24. Estimated other cost *(U.S. $) (e.g. loss of revenue, fuel, hotels):* $

Reported by *(Optional)*	Title	Date

Paperwork Reduction Act Statement: The information collected on this form is necessary to allow the Federal Aviation Administration to assess the magnitude and severity of the wildlife-aircraft strike problem in the U.S. The information is used in determining the best management practices for reducing the hazard to aviation safety caused by wildlife-aircraft strikes. We estimate that it will take approximately **6 minutes** to complete the form. If you wish to make any comments concerning the accuracy of this burden estimate and any suggestions for reducing this burden, send those comments to the Federal Aviation Administration, Management Staff, ARP-10, 800 Independence Avenue, SW, Washington, DC 20591. The information collected is voluntary. Please note that an agency may not conduct or sponsor, and a person is not required to respond to, a collection of information unless it displays a currently valid OMB control number. The OMB control number associated with this collection is 2120-0045.

FAA Form 5200-7 (3-97) Supersedes Previous Edition ☆ U.S. GPO: 1997-418-084/64203 NSN:0052-00-651-9005

U.S. Department
of Transportation

**Federal Aviation
Administration**

800 Independence Ave., S.W.
Washington, D C 20591

Official Business
Penalty for Private Use, $300

NO POSTAGE
NECESSARY
IF MAILED IN
THE UNITED
STATES

BUSINESS REPLY MAIL
FIRST CLASS   PERMIT NO. 12438   WASHINGTON D.C.

POSTAGE WILL BE PAID BY THE FEDERAL AVIATION ADMINISTRATION

**Federal Aviation Administration
Office of Airport Safety and Standards, AAS-310
800 Independence Avenue, SW
WASHINGTON, DC 20591**

Appendix 2

VOLCANIC ACTIVITY REPORTING FORM (VAR)

Date _____

1. Aircraft Identification	
2. Position	
3. Time (UTC)	
4. Flight level or altitude	
5. Position/location of VOLCANIC ACTIVITY	
6. Air temperature	
7. Wind	
8. Supplementary Information (Brief description of activity including vertical and lateral extent of the ash cloud, horizontal movement, rate of growth, etc., as available.)	

Mark the appropriate box (s)

9. Density of ash cloud	☐ wispy	☐ moderately dense	☐ very dense
10. Color of ash	☐ white ☐ black	☐ light gray	☐ gray
11. Eruption	☐ continuous	☐ intermittent	☐ not visible
12. Position of activity	☐ summit ☐ multiple	☐ side ☐ not observed	☐ single
13. Other observed features of eruption	☐ lightning ☐ ash fallout	☐ glow ☐ mushroom cloud	☐ large rocks ☐ none
14. Effect on aircraft	☐ communications ☐ pitot static ☐ none	☐ nav. system ☐ windscreen	☐ engines ☐ other windows
15. Other effects	☐ turbulence ☐ ash deposits	☐ St. Elmo's fire	☐ fumes

16. Other information

Forward completed form via mail to: Global Volcanism Program NHB-119 Smithsonian Institution Washington, DC 20560 E-mail:GVN@volcano.si.edu **Or Fax to:** Global Volcanism Program (202) 357-2476

Appendix 3

LASER BEAM EXPOSURE QUESTIONNAIRE

FAX TO WASHINGTON OPERATIONS CENTER COMPLEX (WOCC) at (202) 267-5289 ATTN: DEN

PILOT NAME: _____ PHONE NUMBER: _____

COMPANY: _____ FLIGHT NUMBER: _____

1. Date and time (UTC)? _____ 2. Position of event (lat/long and/or FRD)? _____

3. Altitude? _____ 4. What was the visibility? _____

5. What were the atmospheric conditions? (Check those which apply) ❑ Clear ❑ Overcast ❑ Rainy ❑ Foggy ❑ Hazy ❑ Sunny

6. What was the color(s) of the light? _____ 7. Did the color(s) change during the exposure? ❑ Yes ❑ No

8. Did you attempt an evasive maneuver? ❑ Yes ❑ No   If yes, did the beam follow you as you tried to move away? ❑ Yes ❑ No

9. Can you estimate how far away the light source was from your location? _____

10. What was the position of the light relative to the aircraft? _____

11. Was the source moving? ❑ Yes ❑ No

12. Was the light coming directly from its source or did it appear to be reflected off other surfaces? _____

13. Were there multiple sources of light? ❑ Yes ❑ No   14. How long was the exposure? _____

15. Did the light seem to track your path or was there incidental contact? _____

16. What tasks were you performing when the exposure occurred? _____

17. Did the light prevent or hamper you from doing those tasks, or was the light more of an annoyance? _____

18. What were the visual effects you experienced (after-image, blind spot, flash-blindness, glare*)? _____

19. Did you report the incident by radio to ATC? ❑ Yes ❑ No

Any other pertinent information: _____

_____

_____

_____

_____

_____

This questionnaire may be filled out by the competent authority during interviews with aircrews exposed to unauthorized laser illumination. This information will be used to aid in subsequent investigation by ATC, law enforcement and other governmental agencies to safeguard the safety and efficiency of civil aviation operation in the NAS.

*Examples of common visual effects:

After-image. An image that remains in the visual field after an exposure to a bright light.
Blind spot. A temporary or permanent loss of vision of part of the visual field.
Flash-blindness. The inability to see (either temporarily or permanently) caused by bright light entering the eye and persisting after the illumination has ceased.
Glare. A temporary disruption in vision caused by the presence of a bright light (such as an oncoming car's headlights) within an individual's field of vision. Glare lasts only as long as the bright light is actually present within the individual's field of vision.

Appendix 4.

Abbreviations/Acronyms

As used in this manual, the following abbreviations/acronyms have the meanings indicated.

Abbreviation/ Acronym	Meaning
AAWU	Alaskan Aviation Weather Unit
AC	Advisory Circular
ACAR	Aircraft Communications Addressing and Reporting System
ADCUS	Advise Customs
ADDS	Aviation Digital Data Service
ADF	Automatic Direction Finder
ADIZ	Air Defense Identification Zone
ADS-B	(See Automatic Dependent Surveillance Broadcast)
ADS [ICAO]	(See ICAO term Automatic Dependent Surveillance)
ADS-C	(See Automatic Dependent Surveillance-Contact)
AFB	Air Force Base
AFCS	Automatic Flight Control System
A/FD	Airport/Facility Directory
AFM	Aircraft Flight Manual
AFSS	Automated Flight Service Station
AHRS	Attitude Heading Reference System
AIM	Aeronautical Information Manual
AIRMET	Airmen's Meterological Information
ALD	Available Landing Distance
ALS	Approach Light System
AMSL	Above Mean Sea Level
ANP	Actual Navigation Performance
AOCC	Airline Operations Control Center
AP	Autopilot System
APV	Approach and Vertical Guidance
ARENA	Areas Noted for Attention
ARFF IC	Aircraft Rescue and Fire Fighting Incident Commander
ARINC	Aeronautical Radio Incorporated
ARO	Airport Reservations Office

Abbreviation/ Acronym	Meaning
ARSA	Airport Radar Service Area
ARSR	Air Route Surveillance Radar
ARTCC	Air Route Traffic Control Center
ARTS	Automated Radar Terminal System
ASDE	Airport Surface Detection Equipment-Model X
ASOS	Automated Surface Observing System
ASR	Airport Surveillance Radar
ATC	Air Traffic Control
ATCRBS	Air Traffic Control Radar Beacon System
ATCSCC	Air Traffic Control System Command Center
ATCT	Airport Traffic Control Tower
ATD	Along-Track Distance
ATIS	Automatic Terminal Information Service
ATT	Attitude Retention System
AWC	Aviation Weather Center
AWOS	Automated Weather Observing System
AWW	Severe Weather Forecast Alert
BBS	Bulletin Board System
BC	Back Course
CA	Coarse Acquisition
CARTS	Common Automated Radar Terminal System (ARTS) (to include ARTS IIIE and ARTS IIE)
CAT	Clear Air Turbulenece
CD	Controller Display
CDI	Course Deviation Indicator
CERAP	Combined Center/RAPCON
CFA	Controlled Firing Area
CFR	Code of Federal Regulations
COA	Certificate of Waiver or Authorization
CPDLC	Controller Pilot Data Link Communications
CTAF	Common Traffic Advisory Frequency
CVFP	Charted Visual Flight Procedure

Abbreviation/ Acronym	Meaning
CVRS	Computerized Voice Reservation System
CWA..........	Center Weather Advisory
CSWU.........	Center Weather Service Unit
DA...........	Decision Altitude
DCA	Ronald Regan Washington National Airport
DCP..........	Data Collection Package
DF	Direction Finder
DH...........	Decision Height
DME	Distance Measuring Equipment
DME/N	Standard DME
DME/P	Precision DME
DOD	Department of Defense
DP	Instrument Departure Procedure
DRT..........	Diversion Recovery Tool
DRVSM	Domestic Reduced Vertical Separation Minimum
DUATS........	Direct User Access Terminal System
DVA	Diverse Vector Area
DVFR	Defense Visual Flight Rules
DVRSN........	Diversion
EDCT	Expect Departure Clearance Time
EFAS.........	En Route Flight Advisory Service
ELT..........	Emergency Locator Transmitter
EPE	Estimate of Position Error
ESV..........	Expanded Service Volume
ETA..........	Estimated Time of Arrival
ETD..........	Estimated Time of Departure
ETE..........	Estimated Time En Route
EWR	Newark International Airport
FA	Area Forecast
FAA..........	Federal Aviation Administration
FAF..........	Final Approach Fix
FAWP	Final Approach Waypoint
FB	Fly-by
FCC..........	Federal Communications Commission
FD	Flight Director System
FDC..........	Flight Data Center
FDE..........	Fault Detection and Exclusion

Abbreviation/ Acronym	Meaning
FIR	Flight Information Region
FISDL	Flight Information Services Data Link
FLIP	Flight Information Publication
FMS	Flight Management System
FMSP	Flight Management System Procedure
FO	Fly-over
FPNM	Feet Per Nautical Mile
FSDO	Flight Standards District Office
FSS	Flight Service Station
GBAS	Ground Based Augmentation System
GEO	Geostationary Satellite
GLS..........	GNSS Landing System
GNSS	Global Navigation Satellite System
GNSSP	Global Navigation Satellite System Panel
GPS..........	Global Positioning System
GRI	Group Repetition Interval
GSD..........	Geographical Situation Display
GUS..........	Ground Uplink Station
HAT	Height Above Touchdown
HDTA	High Density Traffic Airports
HEMS	Helicopter Emergency Medical Services
HIRL.........	High Intensity Runway Lights
HIWAS	Hazardous Inflight Weather Advisory Service
Hz	Hertz
IAF	Initial Approach Fix
IAP	Instrument Approach Procedure
IAS	Indicated Air Speed
IAWP	Initial Approach Waypoint
ICAO.........	International Civil Aviation Organization
IF............	Intermediate Fix
IFIM	International Flight Information Manual
IFR	Instrument Flight Rules
ILS...........	Instrument Landing System
ILS/PRM.......	Instrument Landing System/Precision Runway Monitor
IM	Inner Marker
IMC..........	Instrument Meterological Conditions
INS	Inertial Navigation System
IOC	Initial Operational Capability
IR............	IFR Military Training Route

Abbreviation/ Acronym	Meaning	Abbreviation/ Acronym	Meaning
IRU	Inertial Reference Unit	MSL	Mean Sea Level
ITWS	Integrated Terminal Weather System	MTI	Moving Target Indicator
JFK	John F. Kennedy International Airport	MTR	Military Training Route
kHz	Kilohertz	MVA	Minimum Vectoring Altitude
LAA	Local Airport Advisory	NACO	National Aeronautical Charting Office
LAAS	Local Augmentation System	NAS	National Airspace System
LAHSO	Land and Hold Short Operations	NASA	National Aeronautics and Space Administration
LAWRS	Limited Aviation Weather Reporting Station	NAVAID	Navigational Aid
LDA	Localizer Type Directional Aid	NAVCEN	Coast Guard Navigation Center
LGA	LaGuardia Airport	NDB	Nondirectional Radio Beacon
LIRL	Low Intensity Runway Lights	NFDC	National Flight Data Center
LLWAS	Low Level Wind Shear Alert System	NIDS	National Institute for Discovery Sciences
LLWAS NE	Low Level Wind Shear Alert System Network Expansion	NIMA	National Imagery and Mapping Agency
		NM	Nautical Mile
LLWAS RS	Low Level Wind Shear Alert System Relocation/Sustainment	NMAC	Near Midair Collision
LNAV	Lateral Navigation	NOPAC	North Pacific
LOC	Localizer	NoPT	No Procedure Turn Required
LOP	Line-of-position	NOTAM	Notices to Airmen
LORAN	Long Range Navigation System	NPA	Non Precision Approach
LPV	Localizer Performance with Vertical Guidance	NRS	Navigation Reference System
		NSA	National Security Area
LZ	Landing Zone	NSW	No Significant Weather
MAHWP	Missed Approach Holding Waypoint	NTAP	Notices to Airmen Publication
MAP	Missed Approach Point	NTSB	National Transportation Safety Board
MAWP	Missed Approach Waypoint	NWS	National Weather Service
MDA	Minimum Descent Altitude	OAT	Outside Air Temperature
MEA	Minimum En Route Altitude	OBS	Omni-bearing Selector
MEARTS	Micro En Route Automated Radar Tracking System	ODP	Obstacle Departure Procedure
		OIS	Operational Information System
METAR	Aviation Routine Weather Report	OM	Outer Marker
MHz	Megahzertz	ORD	Chicago O'Hare International Airport
MIRL	Medium Intensity Runway Lights	PA	Precision Approach
MLS	Microwave Landing System	PAPI	Precision Approach Path Indicator
MM	Middle Marker	PAR	Precision Approach Radar
MOA	Military Operations Area	PC	Personal Computer
MOCA	Minimum Obstruction Clearance Altitude	P/CG	Pilot/Controller/Glossary
MRA	Minimum Reception Altitude	PDC	Pre-departure Clearance
MRB	Magnetic Reference Bearing	PFD	Personal Flotation Device
MSA	Minimum Safe Altitude	PinS	Point-in-Space
MSAW	Minimum Safe Altitude Warning	PIREP	Pilot Weather Report

Abbreviation/ Acronym	Meaning
POB..........	Persons on Board
POFC.........	Precision Object Free Zone
POI..........	Principal Operations Inspector
PPS..........	Precise Positioning Service
PRM.........	Precision Runway Marker
PT...........	Procedure Turn
RA...........	Resolution Advisory
RAA.........	Remote Advisory Airport
RAIM........	Receiver Autonomous Integrity Monitoring
RAIS........	Remote Airport Information Service
RBDT.......	Ribbon Display Terminals
RCAG.......	Remote Center Air/Ground
RCC........	Rescue Coordination Center
RCLS.......	Runway Centerline Lighting System
RCO........	Remote Communications Outlet
REIL.......	Runway End Identifier Lights
RFM.......	Rotorcraft Flight Manual
RLIM.......	Runway Light Intensity Monitor
RMI........	Radio Magnetic Indicator
RNAV......	Area Navigation
RNP........	Required Navigation Performance
RPAT.......	RNP Parallel Approach Runway Transitions
RVR.......	Runway Visual Range
RVSM.......	Reduced Vertical Separation Minimum
SAAAR......	Special Aircraft and Aircrew Authorization Required
SAM........	System Area Monitor
SAR.........	Search and Rescue
SAS........	Stability Augmentation System
SBAS........	Satellite-based Augmentation System
SCAT-1 DGPS..	Special Category I Differential GPS
SDF.........	Simplified Directional Facility
SFL.........	Sequenced Flashing Lights
SFR.........	Special Flight Rules
SIAP........	Standard Instrument Approach Procedure
SID.........	Standard Instrument Departure
SIGMET.....	Significant Meteorological Information
SM..........	Statute Mile
SMGCS......	Surface Movement Guidance Control System

Abbreviation/ Acronym	Meaning
SNR..........	Signal-to-noise Ratio
SOIA.........	Simultaneous Offset Instrument Approaches
SOP..........	Standard Operating Procedure
SPC..........	Storm Prediction Center
SPS..........	Standard Positioning Service
STAR.......	Standard Terminal Arrival
STARS.......	Standard Terminal Automation Replacement System
STMP........	Specific Traffic Management Program
SWSL.......	Supplemental Weather Service Locations
TA...........	Traffic Advisory
TAA.........	Terminal Airrival Area
TAC..........	Terminal Area Chart
TACAN.......	Tactical Air Navigation
TAF.........	Aerodrome Forecast
TAS.........	True Air Speed
TCAS........	Traffic Alert and Collision Avoidance System
TCH..........	Threshold Crossing Height
TD...........	Time Difference
TDWR........	Terminal Doppler Weather Radar
TDZL........	Touchdown Zone Lights
TEC..........	Tower En Route Control
TIBS.........	Telephone Information Briefing Service
TIS..........	Traffic Information Service
TLS..........	Transponder Landing System
TPP..........	Terminal Procedures Publications
TRSA........	Terminal Radar Service Area
TSO..........	Technical Standard Order
TWEB........	Transcribed Weather Broadcast
TWIB........	Terminal Weather Information for Pilots System
UAV.........	Unmanned Aerial Vehicle
UFO..........	Unidentified Flying Object
UHF..........	Ultrahigh Frequency
U.S..........	United States
USCG........	United States Coast Guard
UTC..........	Coordinated Universal Time
VAR.........	Volcanic Activity Reporting
VASI.........	Visual Approach Slope Indicator

Abbreviation/ Acronym	Meaning
VCOA	Visual Climb Over Airport
VDA	Vertical Descent Angle
VDP	Visual Descent Point
VFR	Visual Flight Rules
VGSI	Visual Glide Slope Indicator
VHF	Very High Frequency
VIP	Video Integrator Processor
VMC	Visual Meterorological Conditions
Vmini	Instrument flight minimum speed, utilized in complying with minimum limit speed requirements for instrument flight.
VNAV	Vertical Navigation
VNE	Never Exceed Speed
Vnei	Instrument flight never exceed speed, utilized instead of VNE for compliance with maximum limit speed requirements for instrument flight.
VOR	Very High Frequency Omni-directional Range
VORTAC	VHF Omni-directional Range/Tactical Air Navigation
VOT	VOR Test Facility
VR	VFR Military Training Route

Abbreviation/ Acronym	Meaning
VTF	Vector to Final
VV	Vertical Visibility
Vy	Speed for best rate of climb
Vyi	Instrument climb speed, utilized instead of Vy for compliance with the climb requirements for instrument flight.
WA	AIRMET
WAAS	Wide Area Augmentation System
WAC	World Aeronautical Chart
WFO	Weather Forecast Office
WGS-84	World Geodetic System of 1984
WMO	World Meterorological Organization
WMS	Wide-Area Master Station
WMSC	Weather Message Switcing Center
WMSCR	Weather Message Switching Center Replacement
WP	Waypoint
WRS	Wide-Area Ground Reference Station
WSO	Weather Service Office
WSP	Weather System Processor
WST	Convective Significant Meterological Information

Pilot/Controller Glossary

This Glossary was compiled to promote a common understanding of the terms used in the Air Traffic Control system. It includes those terms which are intended for pilot/controller communications. Those terms most frequently used in pilot/controller communications are printed in **bold italics**. The definitions are primarily defined in an operational sense applicable to both users and operators of the National Airspace System. Use of the Glossary will preclude any misunderstandings concerning the system's design, function, and purpose.

Because of the international nature of flying, terms used in the Lexicon, published by the International Civil Aviation Organization (ICAO), are included when they differ from FAA definitions. These terms are followed by "[ICAO]." For the reader's convenience, there are also cross references to related terms in other parts of the Glossary and to other documents, such as the Federal Aviation Regulations (FARs) and the Aeronautical Information Manual (AIM).

This Glossary will be revised, as necessary, to maintain a common understanding of the system.

A

AAI (See ARRIVAL AIRCRAFT INTERVAL.)

AAR (See AIRPORT ARRIVAL RATE.)

ABBREVIATED IFR FLIGHT PLANS An authorization by ATC requiring pilots to submit only that information needed for the purpose of ATC. It includes only a small portion of the usual IFR flight plan information. In certain instances, this may be only aircraft identification, location, and pilot request. Other information may be requested if needed by ATC for separation/control purposes. It is frequently used by aircraft which are airborne and desire an instrument approach or by aircraft which are on the ground and desire a climb to VFR-on-top.
(See VFR-ON-TOP.) (Refer to AIM.)

ABEAM An aircraft is "abeam" a fix, point, or object when that fix, point, or object is approximately 90 degrees to the right or left of the aircraft track. Abeam indicates a general position rather than a precise point.

ABORT To terminate a preplanned aircraft maneuver; e.g., an aborted takeoff.

ACC [ICAO] (See AREA CONTROL CENTER.)

ACCELERATE-STOP DISTANCE AVAILABLE The runway plus stopway length declared available and suitable for the acceleration and deceleration of an airplane aborting a takeoff.

ACCELERATE-STOP DISTANCE AVAILABLE [ICAO] The length of the take-off run available plus the length of the stopway if provided.

ACDO (See AIR CARRIER DISTRICT OFFICE.)

ACKNOWLEDGE Let me know that you have received my message.
(See ICAO term ACKNOWLEDGE.)

ACKNOWLEDGE [ICAO] Let me know that you have received and understood this message.

ACL (See AIRCRAFT LISI.)

ACLS (See AUTOMATIC CARRIER LANDING SYSTEM.)

ACLT (See ACTUAL CALCULATED LANDING TIME.)

ACROBATIC FLIGHT An intentional maneuver involving an abrupt change in an aircraft's attitude, an abnormal attitude, or abnormal acceleration not necessary for normal flight.
(Refer to Part 91.) (See ICAO term ACROBATIC FLIGHT.)

ACROBATIC FLIGHT [ICAO] Manoeuvres intentionally performed by an aircraft involving an abrupt change in its attitude, an abnormal attitude, or an abnormal variation in speed.

ACTIVE RUNWAY (See RUNWAY IN USE/ACTIVE RUNWAY/DUTY RUNWAY.)

ACTUAL CALCULATED LANDING TIME ACLT is a flight's frozen calculated landing time. An actual time determined at freeze calculated landing time (FCLT) or meter list display interval (MLDI) for the adapted vertex for each arrival aircraft based upon runway configuration, airport acceptance rate, airport arrival delay period, and other metered arrival aircraft. This time is either the vertex time of arrival (VTA) of the aircraft or the tentative calculated landing time (TCLT)/ACLT of the previous aircraft plus the arrival aircraft interval (AAI), whichever is later. This time will not be updated in response to the aircraft's progress.

ACTUAL NAVIGATION PERFORMANCE (ANP) (See REQUIRED NAVIGATION PERFORMANCE)

ADDITIONAL SERVICES Advisory information provided by ATC which includes but is not limited to the following:
  a. Traffic advisories.
  b. Vectors, when requested by the pilot, to assist aircraft receiving traffic advisories to avoid observed traffic.

c. Altitude deviation information of 300 feet or more from an assigned altitude as observed on a verified (reading correctly) automatic altitude readout (Mode C).

d. Advisories that traffic is no longer a factor.

e. Weather and chaff information.

f. Weather assistance.

g. Bird activity information.

h. Holding pattern surveillance. Additional services are provided to the extent possible contingent only upon the controller's capability to fit them into the performance of higher priority duties and on the basis of limitations of the radar, volume of traffic, frequency congestion, and controller workload. The controller has complete discretion for determining if he is able to provide or continue to provide a service in a particular case. The controller's reason not to provide or continue to provide a service in a particular case is not subject to question by the pilot and need not be made known to him.

(See TRAFFIC ADVISORIES.) (Refer to AIM.)

ADF   (See AUTOMATIC DIRECTION FINDER.)

ADIZ   (See AIR DEFENSE IDENTIFICATION ZONE.)

ADLY   (See ARRIVAL DELAY.)

ADMINISTRATOR   The Federal Aviation Administrator or any person to whom he/she has delegated his/her authority in the matter concerned.

ADR   (See AIRPORT DEPARTURE RATE.)

ADS [ICAO] (See ICAO term AUTOMATIC DEPENDENT SURVEILLANCE.)

ADS-B (See AUTOMATIC DEPENDENT SURVEILLANCE-BROADCAST.)

ADS-C (See AUTOMATIC DEPENDENT SURVEILLANCE-CONTRACT.)

ADVISE INTENTIONS   Tell me what you plan to do.

ADVISORY   Advice and information provided to assist pilots in the safe conduct of flight and aircraft movement.
(See ADVISORY SERVICE.)

ADVISORY FREQUENCY   The appropriate frequency to be used for Airport Advisory Service.
(See LOCAL AIRPORT ADVISORY.) (See UNICOM.) (Refer to ADVISORY CIRCULAR NO. 90-42.) (Refer to AIM.)

ADVISORY SERVICE   Advice and information provided by a facility to assist pilots in the safe conduct of flight and aircraft movement.

(See LOCAL AIRPORT ADVISORY.) (See TRAFFIC ADVISORIES.) (See SAFETY ALERT.) (See ADDITIONAL SERVICES.) (See RADAR ADVISORY.) (See EN ROUTE FLIGHT ADVISORY SERVICE.) (Refer to AIM.)

AERIAL REFUELING   A procedure used by the military to transfer fuel from one aircraft to another during flight.
(Refer to VFR/IFR Wall Planning Charts.)

AERODROME   A defined area on land or water (including any buildings, installations and equipment) intended to be used either wholly or in part for the arrival, departure, and movement of aircraft.

AERODROME BEACON [ICAO]   Aeronautical beacon used to indicate the location of an aerodrome from the air.

AERODROME CONTROL SERVICE [ICAO]   Air traffic control service for aerodrome traffic.

AERODROME CONTROL TOWER [ICAO]   A unit established to provide air traffic control service to aerodrome traffic.

AERODROME ELEVATION [ICAO]   The elevation of the highest point of the landing area.

AERODROME TRAFFIC CIRCUIT [ICAO]   The specified path to be flown by aircraft operating in the vicinity of an aerodrome.

AERONAUTICAL BEACON   A visual NAVAID displaying flashes of white and/or colored light to indicate the location of an airport, a heliport, a landmark, a certain point of a Federal airway in mountainous terrain, or an obstruction.
(See AIRPORT ROTATING BEACON.) (Refer to AIM.)

AERONAUTICAL CHART   A map used in air navigation containing all or part of the following: Topographic features, hazards and obstructions, navigation aids, navigation routes, designated airspace, and airports. Commonly used aeronautical charts are:

a. Sectional Aeronautical Charts (1:500,000). Designed for visual navigation of slow or medium speed aircraft. Topographic information on these charts features the portrayal of relief and a judicious selection of visual check points for VFR flight. Aeronautical information includes visual and radio aids to navigation, airports, controlled airspace, restricted areas, obstructions, and related data.

b. VFR Terminal Area Charts (1:250,000). Depict Class B airspace which provides for the control or segregation of all the aircraft within Class B airspace. The chart depicts topographic information and aeronautical information which includes visual and radio aids to navigation, airports, controlled airspace, restricted areas, obstructions, and related data.

c. World Aeronautical Charts (WAC) (1:1,000,000). Provide a standard series of aeronautical charts covering land areas of the world at a size and scale convenient for navigation by moderate speed aircraft. Topographic information includes cities and towns, principal roads, railroads, distinctive landmarks, drainage, and relief. Aeronautical information includes visual and radio aids to navigation, airports, airways, restricted areas, obstructions, and other pertinent data.

d. En Route Low Altitude Charts. Provide aeronautical information for en route instrument navigation (IFR) in the low altitude stratum. Information includes the portrayal of airways, limits of controlled airspace, position identification and frequencies of radio aids, selected airports, minimum en route and minimum obstruction clearance altitudes, airway distances, reporting points, restricted areas, and related data. Area charts, which are a part of this series, furnish terminal data at a larger scale in congested areas.

e. En Route High Altitude Charts. Provide aeronautical information for en route instrument navigation (IFR) in the high altitude stratum. Information includes the portrayal of jet routes, identification and frequencies of radio aids, selected airports, distances, time zones, special use airspace, and related information.

f. Instrument Approach Procedures (IAP) Charts. Portray the aeronautical data which is required to execute an instrument approach to an airport. These charts depict the procedures, including all related data, and the airport diagram. Each procedure is designated for use with a specific type of electronic navigation system including NDB, TACAN, VOR, ILS/MLS, and RNAV. These charts are identified by the type of navigational aid(s) which provide final approach guidance.

g. Instrument Departure Procedure (DP) Charts. Designed to expedite clearance delivery and to facilitate transition between takeoff and en route operations. Each DP is presented as a separate chart and may serve a single airport or more than one airport in a given geographical location.

h. Standard Terminal Arrival (STAR) Charts. Designed to expedite air traffic control arrival procedures and to facilitate transition between en route and instrument approach operations. Each STAR procedure is presented as a separate chart and may serve a single airport or more than one airport in a given geographical location.

i. Airport Taxi Charts. Designed to expedite the efficient and safe flow of ground traffic at an airport. These charts are identified by the official airport name; e.g., Washington National Airport.

(See ICAO term AERONAUTICAL CHART.)

AERONAUTICAL CHART [ICAO]   A representation of a portion of the Earth, its culture and relief, specifically designated to meet the requirements of air navigation.

AERONAUTICAL INFORMATION MANUAL   A primary FAA publication whose purpose is to instruct airmen about operating in the National Airspace System of the U.S. It provides basic flight information, ATC Procedures and general instructional information concerning health, medical facts, factors affecting flight safety, accident and hazard reporting, and types of aeronautical charts and their use.

AERONAUTICAL INFORMATION PUBLICATION [AIP] [ICAO]   A publication issued by or with the authority of a State and containing aeronautical information of a lasting character essential to air navigation.

A/FD   (See AIRPORT/FACILITY DIRECTORY.)

AFFIRMATIVE   Yes.

AFP   (See AIRSPACE FLOW PROGRAM.)

AIM   (See AERONAUTICAL INFORMATION MANUAL.)

AIP [ICAO]   (See AERONAUTICAL INFORMATION PUBLICATION.)

AIRBORNE DELAY   Amount of delay to be encountered in airborne holding.

AIR CARRIER DISTRICT OFFICE   An FAA field office serving an assigned geographical area, staffed with Flight Standards personnel serving the aviation industry and the general public on matters related to the certification and operation of scheduled air carriers and other large aircraft operations.

AIRCRAFT   Device(s) that are used or intended to be used for flight in the air, and when used in air traffic control terminology, may include the flight crew.
(See ICAO term AIRCRAFT.)

AIRCRAFT [ICAO]   Any machine that can derive support in the atmosphere from the reactions of the air other than the reactions of the air against the Earth's surface.

AIRCRAFT APPROACH CATEGORY   A grouping of aircraft based on a speed of 1.3 times the stall speed in the landing configuration at maximum gross landing weight. An aircraft must fit in only one category. If it is necessary to maneuver at speeds in excess of the upper limit of a speed range for a category, the minimums for the category for that speed must be used. For example, an aircraft which falls in Category A, but is circling to land at a speed in excess of 91 knots, must use the approach Category B minimums when circling to land. The categories are as follows:

a. Category A. Speed less than 91 knots.

b. Category B. Speed 91 knots or more but less than 121 knots.

c. Category C. Speed 121 knots or more but less than 141 knots.

d. Category D. Speed 141 knots or more but less than 166 knots.

e. Category E. Speed 166 knots or more.
(Refer to Part 97.)

AIRCRAFT CLASSES   For the purposes of Wake Turbulence Separation Minima, ATC classifies aircraft as Heavy, Large, and Small as follows:

a. Heavy. Aircraft capable of takeoff weights of more than 255,000 pounds whether or not they are operating at this weight during a particular phase of flight.

b. Large. Aircraft of more than 41,000 pounds, maximum certificated takeoff weight, up to 255,000 pounds.

c. Small. Aircraft of 41,000 pounds or less maximum certificated takeoff weight.
(Refer to AIM.)

AIRCRAFT CONFLICT Predicted conflict, within URET, of two aircraft, or between aircraft and airspace. A Red alert is used for conflicts when the predicted minimum separation is 5 nautical miles or less. A Yellow alert is used when the predicted minimum separation is between 5 and approximately 12 nautical miles. A Blue alert is used for conflicts between an aircraft and predefined airspace.

(See USER REQUEST EVALUATION TOOL.)

AIRCRAFT LIST (ACL) A view available with URET that lists aircraft currently in or predicted to be in a particular sector's airspace. The view contains textual flight data information in line format and may be sorted into various orders based on the specify needs of the sector team.

(See USER REQUEST EVALUATION TOOL.)

AIRCRAFT SURGE LAUNCH AND RECOVERY Procedures used at USAF bases to provide increased launch and recovery rates in instrument flight rules conditions. ASLAR is based on:

a. Reduced separation between aircraft which is based on time or distance. Standard arrival separation applies between participants including multiple flights until the DRAG point. The DRAG point is a published location on an ASLAR approach where aircraft landing second in a formation slows to a predetermined airspeed. The DRAG point is the reference point at which MARSA applies as expanding elements affect separation within a flight or between subsequent participating flights.

b. ASLAR procedures shall be covered in a Letter of Agreement between the responsible USAF military ATC facility and the concerned Federal Aviation Administration facility. Initial Approach Fix spacing requirements are normally addressed as a minimum.

AIR DEFENSE EMERGENCY (ADIZ) A military emergency condition declared by a designated authority. This condition exists when an attack upon the continental U.S., Alaska, Canada, or U.S. installations in Greenland by hostile aircraft or missiles is considered probable, is imminent, or is taking place.

(Refer to AIM.)

AIR DEFENSE IDENTIFICATION ZONE The area of airspace over land or water, extending upward from the surface, within which the ready identification, the location, and the control of aircraft are required in the interest of national security.

a. Domestic Air Defense Identification Zone. An ADIZ within the United States along an international boundary of the United States.

b. Coastal Air Defense Identification Zone. An ADIZ over the coastal waters of the United States.

c. Distant Early Warning Identification Zone (DEWIZ). An ADIZ over the coastal waters of the State of Alaska.

d. Land-Based Air Defense Identification Zone. An ADIZ over U.S. metropolitan areas, which is activated and deactivated as needed, with dimensions, activation dates and other relevant information disseminated via NOTAM.

ADIZ locations and operating and flight plan requirements for civil aircraft operations are specified in FAR Part 99. (Refer to AIM.)

AIRMAN'S METEOROLOGICAL INFORMATION (See AIRMET.)

AIRMET In-flight weather advisories issued only to amend the area forecast concerning weather phenomena which are of operational interest to all aircraft and potentially hazardous to aircraft having limited capability because of lack of equipment, instrumentation, or pilot qualifications. AIRMETs concern weather of less severity than that covered by SIGMETs or Convective SIGMETs. AIRMETs cover moderate icing, moderate turbulence, sustained winds of 30 knots or more at the surface, widespread areas of ceilings less than 1,000 feet and/or visibility less than 3 miles, and extensive mountain obscurement.

(See AWW.) (See SIGMET.) (See CONVECTIVE SIGMET.) (See CWA.) (Refer to AIM.)

AIR NAVIGATION FACILITY Any facility used in, available for use in, or designed for use in, aid of air navigation, including landing areas, lights, any apparatus or equipment for disseminating weather information, for signaling, for radio-directional finding, or for radio or other electrical communication, and any other structure or mechanism having a similar purpose for guiding or controlling flight in the air or the landing and takeoff of aircraft.

(See NAVIGATIONAL AID.)

AIRPORT An area on land or water that is used or intended to be used for the landing and takeoff of aircraft and includes its buildings and facilities, if any.

AIRPORT ADVISORY AREA The area within ten miles of an airport without a control tower or where the tower is not in operation, and on which a Flight Service Station is located.

(See LOCAL AIRPORT ADVISORY.) (Refer to AIM.)

AIRPORT ARRIVAL RATE (AAR) A dynamic input parameter specifying the number of arriving aircraft which an airport or airspace can accept from the ARTCC per hour. The AAR is used to calculate the desired interval between successive arrival aircraft.

AIRPORT DEPARTURE RATE (ADR) A dynamic parameter specifying the number of aircraft which can depart an airport and the airspace can accept per hour.

AIRPORT ELEVATION The highest point of an airport's usable runways measured in feet from mean sea level.

(See TOUCHDOWN ZONE ELEVATION.) (See ICAO term AERODROME ELEVATION.)

AIRPORT/FACILITY DIRECTORY   A publication designed primarily as a pilot's operational manual containing all airports, seaplane bases, and heliports open to the public including communications data, navigational facilities, and certain special notices and procedures. This publication is issued in seven volumes according to geographical area.

AIRPORT INFORMATION AID   (See AIRPORT INFORMATION DESK.)

AIRPORT INFORMATION DESK   An airport unmanned facility designed for pilot self-service briefing, flight planning, and filing of flight plans.

(Refer to AIM.)

AIRPORT LIGHTING   Various lighting aids that may be installed on an airport. Types of airport lighting include:

a. Approach Light System (ALS). An airport lighting facility which provides visual guidance to landing aircraft by radiating light beams in a directional pattern by which the pilot aligns the aircraft with the extended centerline of the runway on his final approach for landing. Condenser-Discharge Sequential Flashing Lights/Sequenced Flashing Lights may be installed in conjunction with the ALS at some airports. Types of Approach Light Systems are:

1. ALSF-1. Approach Light System with Sequenced Flashing Lights in ILS Cat-I configuration.

2. ALSF-2. Approach Light System with Sequenced Flashing Lights in ILS Cat-II configuration. The ALSF-2 may operate as an SSALR when weather conditions permit.

3. SSALF. Simplified Short Approach Light System with Sequenced Flashing Lights.

4. SSALR. Simplified Short Approach Light System with Runway Alignment Indicator Lights.

5. MALSF. Medium Intensity Approach Light System with Sequenced Flashing Lights.

6. MALSR. Medium Intensity Approach Light System with Runway Alignment Indicator Lights.

7. LDIN. Lead-in-light system: Consists of one or more series of flashing lights installed at or near ground level that provides positive visual guidance along an approach path, either curving or straight, where special problems exist with hazardous terrain, obstructions, or noise abatement procedures.

8. RAIL. Runway Alignment Indicator Lights (Sequenced Flashing Lights which are installed only in combination with other light systems).

9. ODALS. Omnidirectional Approach Lighting System consists of seven omnidirectional flashing lights located in the approach area of a nonprecision runway. Five lights are located on the runway centerline extended with the first light located 300 feet from the threshold and extending at equal intervals up to 1,500 feet from the threshold. The other two lights are located, one on each side of the runway threshold, at a lateral distance of 40 feet from the runway edge, or 75 feet from the runway edge when installed on a runway equipped with a VASI.

(Refer to Order 6850.2.)

b. Runway Lights/Runway Edge Lights. Lights having a prescribed angle of emission used to define the lateral limits of a runway. Runway lights are uniformly spaced at intervals of approximately 200 feet, and the intensity may be controlled or preset.

c. Touchdown Zone Lighting. Two rows of transverse light bars located symmetrically about the runway centerline normally at 100 foot intervals. The basic system extends 3,000 feet along the runway.

d. Runway Centerline Lighting. Flush centerline lights spaced at 50-foot intervals beginning 75 feet from the landing threshold and extending to within 75 feet of the opposite end of the runway.

e. Threshold Lights. Fixed green lights arranged symmetrically left and right of the runway centerline, identifying the runway threshold.

f. Runway End Identifier Lights (REIL). Two synchronized flashing lights, one on each side of the runway threshold, which provide rapid and positive identification of the approach end of a particular runway.

g. Visual Approach Slope Indicator (VASI). An airport lighting facility providing vertical visual approach slope guidance to aircraft during approach to landing by radiating a directional pattern of high intensity red and white focused light beams which indicate to the pilot that he is "on path" if he sees red/white, "above path" if white/white, and "below path" if red/red. Some airports serving large aircraft have three-bar VASIs which provide two visual glide paths to the same runway.

h. Precision Approach Path Indicator (PAPI)   An airport lighting facility, similar to VASI, providing vertical approach slope guidance to aircraft during approach to landing. PAPIs consist of a single row of either two or four lights, normally installed on the left side of the runway, and have an effective visual range of about 5 miles during the day and up to 20 miles at night. PAPIs radiate a directional pattern of high intensity red and white focused light beams which indicate that the pilot is "on path" if the pilot sees an equal number of white lights and red lights, with white to the left of the red; "above path" if the pilot sees more white than red lights; and "below path" if the pilot sees more red than white lights.

i. Boundary Lights. Lights defining the perimeter of an airport or landing area.

(Refer to AIM.)

AIRPORT MARKING AIDS   Markings used on runway and taxiway surfaces to identify a specific runway, a runway threshold, a centerline, a hold line, etc. A runway should be marked in accordance with its present usage such as:

a. Visual.

b. Nonprecision instrument.

c. Precision instrument.
(Refer to AIM.)

AIRPORT MOVEMENT AREA SAFETY SYSTEM (AMASS) A software enhancement to ASDE radar which provides logic predicting the path of aircraft landing and/or departing, and aircraft and/or vehicular movements on runways. Visual and aural alarms are activated when logic projects a potential collision.

AIRPORT REFERENCE POINT (ARP) The approximate geometric center of all usable runway surfaces.

AIRPORT RESERVATION OFFICE Office responsible for monitoring the operation of the high density rule. Receives and processes requests for IFR operations at high density traffic airports.

AIRPORT ROTATING BEACON A visual NAVAID operated at many airports. At civil airports, alternating white and green flashes indicate the location of the airport. At military airports, the beacon flashes alternately white and green, but are differentiated from civil beacons by dual-peaked (two quick) white flashes between the green flashes.
(See SPECIAL VFR OPERATIONS.) (See INSTRUMENT FLIGHT RULES.) (Refer to AIM.) (See ICAO term AERODROME BEACON.)

AIRPORT SURFACE DETECTION EQUIPMENT Radar equipment specifically designed to detect all principal features on the surface of an airport, including aircraft and vehicular traffic, and to present the entire image on a radar indicator console in the control tower. Used to augment visual observation by tower personnel of aircraft and/or vehicular movements on runways and taxiways.

AIRPORT SURVEILLANCE RADAR Approach control radar used to detect and display an aircraft's position in the terminal area. ASR provides range and azimuth information but does not provide elevation data. Coverage of the ASR can extend up to 60 miles.

AIRPORT TAXI CHARTS (See AERONAUTICAL CHART.)

AIRPORT TRAFFIC CONTROL SERVICE A service provided by a control tower for aircraft operating on the movement area and in the vicinity of an airport.
(See MOVEMENT AREA.) (See TOWER.) (See ICAO term AERODROME CONTROL SERVICE.)

AIRPORT TRAFFIC CONTROL TOWER (See TOWER.)

AIR ROUTE SURVEILLANCE RADAR Air route traffic control center (ARTCC) radar used primarily to detect and display an aircraft's position while en route between terminal areas. The ARSR enables controllers to provide radar air traffic control service when aircraft are within the ARSR coverage. In some instances, ARSR may enable an ARTCC to provide terminal radar services similar to but usually more limited than those provided by a radar approach control.

AIR ROUTE TRAFFIC CONTROL CENTER A facility established to provide air traffic control service to aircraft operating on IFR flight plans within controlled airspace and principally during the en route phase of flight. When equipment capabilities and controller workload permit, certain advisory/assistance services may be provided to VFR aircraft.
(See NAS STAGE A.) (See EN ROUTE AIR TRAFFIC CONTROL SERVICES.) (Refer to AIM.)

AIRSPACE CONFLICT Predicted conflict of an aircraft and active Special Activity Airspace (SAA).

AIRSPACE FLOW PROGRAM (AFP) AFP is a Traffic Management (TM) process administered by the Air Traffic Control System Command Center (ATCSCC) where aircraft are assigned an Expect Departure Clearance Time (EDCT) in order to manage capacity and demand for a specific area of the National Airspace System (NAS). The purpose of the program is to mitigate the effects of en route constraints. It is a flexible program and may be implemented in various forms depending upon the needs of the air traffic system.

AIRSPACE HIERARCHY Within the airspace classes, there is a hierarchy and, in the event of an overlap of airspace: Class A preempts Class B, Class B preempts Class C, Class C preempts Class D, Class D preempts Class E, and Class E preempts Class G.

AIRSPEED The speed of an aircraft relative to its surrounding air mass. The unqualified term "airspeed" means one of the following:
a. Indicated Airspeed. The speed shown on the aircraft airspeed indicator. This is the speed used in pilot/controller communications under the general term "airspeed."
(Refer to FAR Part 1.)
b. True Airspeed. The airspeed of an aircraft relative to undisturbed air. Used primarily in flight planning and en route portion of flight. When used in pilot/controller communications, it is referred to as "true airspeed" and not shortened to "airspeed."

AIRSTART The starting of an aircraft engine while the aircraft is airborne, preceded by engine shutdown during training flights or by actual engine failure.

AIR TAXI Used to describe a helicopter/VTOL aircraft movement conducted above the surface but normally not above 100 feet AGL. The aircraft may proceed either via hover taxi or flight at speeds more than 20 knots. The pilot is solely responsible for selecting a safe airspeed/altitude for the operation being conducted.
(See HOVER TAXI.) (Refer to AIM.)

AIR TRAFFIC Aircraft operating in the air or on an airport surface, exclusive of loading ramps and parking areas.
(See ICAO term AIR TRAFFIC.)

AIR TRAFFIC [ICAO] All aircraft in flight or operating on the manoeuvring area of an aerodrome.

AIR TRAFFIC CLEARANCE An authorization by air traffic control for the purpose of preventing collision between known aircraft, for an aircraft to proceed under specified traffic conditions within controlled airspace. The pilot-in-command of an aircraft may not deviate from the provisions of a visual flight rules (VFR) or instrument flight rules (IFR) air traffic clearance except in an emergency or unless an amended clearance has been obtained. Additionally, the pilot may request a different clearance from that which has been issued by air traffic control (ATC) if information available to the pilot makes another course of action more practicable or if aircraft equipment limitations or company procedures forbid compliance with the clearance issued. Pilots may also request clarification or amendment, as appropriate, any time a clearance is not fully understood, or considered unacceptable because of safety of flight. Controllers should, in such instances and to the extent of operational practicality and safety, honor the pilot's request. FAR Part 91.3(a) states: "The pilot in command of an aircraft is directly responsible for, and is the final authority as to, the operation of that aircraft." THE PILOT IS RESPONSIBLE TO REQUEST AN AMENDED CLEARANCE if ATC issues a clearance that would cause a pilot to deviate from a rule or regulation, or in the pilot's opinion, would place the aircraft in jeopardy.

(See ATC INSTRUCTIONS.) (See ICAO term AIR TRAFFIC CONTROL CLEARANCE.)

AIR TRAFFIC CONTROL A service operated by appropriate authority to promote the safe, orderly and expeditious flow of air traffic.

(See ICAO term AIR TRAFFIC CONTROL SERVICE.)

AIR TRAFFIC CONTROL CLEARANCE [ICAO] Authorization for an aircraft to proceed under conditions specified by an air traffic control unit.

Note 1: For convenience, the term air traffic control clearance is frequently abbreviated to clearance when used in appropriate contexts.

Note 2: The abbreviated term clearance may be prefixed by the words taxi, takeoff, departure, en route, approach or landing to indicate the particular portion of flight to which the air traffic control clearance relates.

AIR TRAFFIC CONTROL SERVICE (See AIR TRAFFIC CONTROL.)

AIR TRAFFIC CONTROL SERVICE [ICAO] A service provided for the purpose of:
  a. Preventing collisions:
    1. Between aircraft; and
    2. On the maneuvering area between aircraft and obstructions; and
  b. Expediting and maintaining an orderly flow of air traffic.

AIR TRAFFIC CONTROL SPECIALIST A person authorized to provide air traffic control service.

(See AIR TRAFFIC CONTROL.) (See FLIGHT SERVICE STATION.) (See ICAO term CONTROLLER.)

AIR TRAFFIC CONTROL SYSTEM COMMAND CENTER An Air Traffic Tactical Operations facility responsible for monitoring and managing the flow of air traffic throughout the NAS, producing a safe, orderly, and expeditious flow of traffic while minimizing delays. The following functions are located at the ATCSCC:

  a. Central Altitude Reservation Function (CARF). Responsible for coordinating, planning, and approving special user requirements under the Altitude Reservation (ALTRV) concept.

  (See ALTITUDE RESERVATION.)

  b. Airport Reservation Office (ARO). Responsible for approving IFR flights at designated high density traffic airports (John F. Kennedy, LaGuardia, and Ronald Reagan Washington National) during specified hours.

  (Refer to 14 CFR Part 93.)
  (Refer to AIRPORT/FACILITY DIRECTORY.)

  c. U.S. Notice to Airmen (NOTAM) Office. Responsible for collecting, maintaining, and distributing NOTAMs for the U.S. civilian and military, as well as international aviation communities.

  (See NOTICE TO AIRMEN.)

  d. Weather Unit. Monitor all aspects of weather for the U.S. that might affect aviation including cloud cover, visibility, winds, precipitation, thunderstorms, icing, turbulence, and more. Provide forecasts based on observations and on discussions with meteorologists from various National Weather Service offices, FAA facilities, airlines, and private weather services.

AIR TRAFFIC SERVICE A generic term meaning:
  a. Flight Information Service:
  b. Alerting Service:
  c. Air Traffic Advisory Service:
  d. Air Traffic Control Service:
    1. Area Control Service,
    2. Approach Control Service, or
    3. Airport Control Service.

AIR TRAFFIC SERVICE (ATS) ROUTES The term "ATS route" is a generic term that includes "VOR Federal airways," "colored Federal airways," "alternate airways," "jet routes," "Military Training Routes," "named routes," and "RNAV routes." The term "ATS route" does not replace these more familiar route names, but serves only as an overall title when listing the types of routes that comprise the United States route structure.

AIRWAY A Class E airspace area established in the form of a corridor, the centerline of which is defined by radio navigational aids.

(See FEDERAL AIRWAYS.) (Refer to FAR Part 71.) (Refer to AIM.) (See ICAO term AIRWAY.)

AIRWAY [ICAO] A control area or portion thereof established in the form of a corridor equipped with radio navigational aids.

AIRWAY BEACON   Used to mark airway segments in remote mountain areas. The light flashes Morse Code to identify the beacon site.
(Refer to AIM.)

AIT   (See AUTOMATED INFORMATION TRANSFER.)

ALERFA (Alert Phase) [ICAO]   A situation wherein apprehension exists as to the safety of an aircraft and its occupants.

ALERT   A notification to a position that there is an aircraft-to-aircraft or aircraft-to-airspace conflict, as detected by Automated Problem Detection (APD).

ALERT AREA   (See SPECIAL USE AIRSPACE.)

ALERT NOTICE   A request originated by a flight service station (FSS) or an air route traffic control center (ARTCC) for an extensive communication search for overdue, unreported, or missing aircraft.

ALERTING SERVICE   A service provided to notify appropriate organizations regarding aircraft in need of search and rescue aid and assist such organizations as required.

ALNOT   (See ALERT NOTICE.)

ALONG TRACK DISTANCE (ATD)   The distance measured from a point-in-space by systems using area navigation reference capabilities that are not subject to slant range errors.

ALPHANUMERIC DISPLAY   Letters and numerals used to show identification, altitude, beacon code, and other information concerning a target on a radar display.
(See AUTOMATED RADAR TERMINAL SYSTEMS.) (See NAS STAGE A.)

ALTERNATE AERODROME [ICAO]   An aerodrome to which an aircraft may proceed when it becomes either impossible or inadvisable to proceed to or to land at the aerodrome of intended landing.
Note: The aerodrome from which a flight departs may also be an en-route or a destination alternate aerodrome for the flight.

ALTERNATE AIRPORT   An airport at which an aircraft may land if a landing at the intended airport becomes inadvisable.
(See ICAO term ALTERNATE AERODROME.)

ALTIMETER SETTING   The barometric pressure reading used to adjust a pressure altimeter for variations in existing atmospheric pressure or to the standard altimeter setting (29.92).
(Refer to FAR Part 91.) (Refer to AIM.)

ALTITUDE   The height of a level, point, or object measured in feet Above Ground Level (AGL) or from Mean Sea Level (MSL).
(See FLIGHT LEVEL.)

a. MSL Altitude. Altitude expressed in feet measured from mean sea level.

b. AGL Altitude. Altitude expressed in feet measured above ground level.

c. Indicated Altitude. The altitude as shown by an altimeter. On a pressure or barometric altimeter it is altitude as shown uncorrected for instrument error and uncompensated for variation from standard atmospheric conditions.
(See ICAO term ALTITUDE.)

ALTITUDE [ICAO]   The vertical distance of a level, a point or an object considered as a point, measured from mean sea level (MSL).

ALTITUDE READOUT   An aircraft's altitude, transmitted via the Mode C transponder feature, that is visually displayed in 100-foot increments on a radarscope having readout capability.
(See AUTOMATED RADAR TERMINAL SYSTEMS.) (See NAS STAGE A.) (See ALPHANUMERIC DISPLAY.) (Refer to AIM.)

ALTITUDE RESERVATION   Airspace utilization under prescribed conditions normally employed for the mass movement of aircraft or other special user requirements which cannot otherwise be accomplished. ALTRVs are approved by the appropriate FAA facility.
(See AIR TRAFFIC CONTROL SYSTEM COMMAND CENTER.)

ALTITUDE RESTRICTION   An altitude or altitudes, stated in the order flown, which are to be maintained until reaching a specific point or time. Altitude restrictions may be issued by ATC due to traffic, terrain, or other airspace considerations.

ALTITUDE RESTRICTIONS ARE CANCELED   Adherence to previously imposed altitude restrictions is no longer required during a climb or descent.

ALTRV   (See ALTITUDE RESERVATION.)

AMASS   (See AIRPORT MOVEMENT AREA SAFETY SYSTEM).

AMVER   (See AUTOMATED MUTUAL-ASSISTANCE VESSEL RESCUE SYSTEM.)

APB   (See AUTOMATED PROBLEM DETECTION BOUNDARY.)

APD   (See AUTOMATED PROBLEM DETECTION.)

APDIA   (See AUTOMATED PROBLEM DETECTION INHIBITED AREA.)

APPROACH CLEARANCE   Authorization by ATC for a pilot to conduct an instrument approach. The type of instrument approach for which a clearance and other pertinent information is provided in the approach clearance when required.
(See INSTRUMENT APPROACH PROCEDURE.) (See CLEARED APPROACH.) (Refer to AIM and FAR Part 91.)

APPROACH CONTROL FACILITY A terminal ATC facility that provides approach control service in a terminal area.

(See APPROACH CONTROL SERVICE.) (See RADAR APPROACH CONTROL FACILITY.)

APPROACH CONTROL SERVICE Air traffic control service provided by an approach control facility for arriving and departing VFR/IFR aircraft and, on occasion, en route aircraft. At some airports not served by an approach control facility, the ARTCC provides limited approach control service.

(Refer to AIM.) (See ICAO term APPROACH CONTROL SERVICE.)

APPROACH CONTROL SERVICE [ICAO] Air traffic control service for arriving or departing controlled flights.

APPROACH GATE An imaginary point used within ATC as a basis for vectoring aircraft to the final approach course. The gate will be established along the final approach course 1 mile from the final approach fix on the side away from the airport and will be no closer than 5 miles from the landing threshold.

APPROACH LIGHT SYSTEM (See AIRPORT LIGHTING.)

APPROACH SEQUENCE The order in which aircraft are positioned while on approach or awaiting approach clearance.

(See LANDING SEQUENCE.) (See ICAO term APPROACH SEQUENCE.)

APPROACH SEQUENCE [ICAO] The order in which two or more aircraft are cleared to approach to land at the aerodrome.

APPROACH SPEED The recommended speed contained in aircraft manuals used by pilots when making an approach to landing. This speed will vary for different segments of an approach as well as for aircraft weight and configuration.

APPROPRIATE ATS AUTHORITY [ICAO] The relevant authority designated by the State responsible for providing air traffic services in the airspace concerned. In the United States, the "appropriate ATS authority" is the Program Director for Air Traffic Planning and Procedures, ATP-1.

APPROPRIATE AUTHORITY
a. Regarding flight over the high seas: the relevant authority is the State of Registry.
b. Regarding flight over other than the high seas: the relevant authority is the State having sovereignty over the territory being overflown.

APPROPRIATE OBSTACLE CLEARANCE MINIMUM ALTITUDE Any of the following:

(See Minimum IFR Altitude MIA.) (See Minimum En Route Altitude MEA.) (See Minimum Obstruction Clearance Altitude MOCA.) (See Minimum Vectoring Altitude MVA.)

APPROPRIATE TERRAIN CLEARANCE MINIMUM ALTITUDE Any of the following:

(See Minimum IFR Altitude MIA.) (See Minimum En Route Altitude MEA.) (See Minimum Obstruction Clearance Altitude MOCA.) (See Minimum Vectoring Altitude MVA.)

APRON A defined area on an airport or heliport intended to accommodate aircraft for purposes of loading or unloading passengers or cargo, refueling, parking, or maintenance. With regard to seaplanes, a ramp is used for access to the apron from the water.

(See ICAO term APRON.)

APRON [ICAO] A defined area, on a land aerodrome, intended to accommodate aircraft for purposes of loading or unloading passengers, mail or cargo, refueling, parking or maintenance.

ARC The track over the ground of an aircraft flying at a constant distance from a navigational aid by reference to distance measuring equipment (DME).

AREA CONTROL CENTER [ICAO] An ICAO term for an air traffic control facility primarily responsible for ATC services being provided IFR aircraft during the en route phase of flight. The U.S. equivalent facility is an air route traffic control center (ARTCC).

AREA NAVIGATION Area Navigation (RNAV) provides enhanced navigational capability to the pilot. RNAV equipment can compute the airplane position, actual track and ground speed and then provide meaningful information relative to a route of flight selected by the pilot. Typical equipment will provide the pilot with distance, time, bearing and crosstrack error relative to the selected "TO" or "active" waypoint and the selected route. Several distinctly different navigational systems with different navigational performance characteristics are capable of providing area navigational functions. Present day RNAV includes INS, LORAN, VOR/DME, and GPS systems. Modern multi-sensor systems can integrate one or more of the above systems to provide a more accurate and reliable navigational system. Due to the different levels of performance, area navigational capabilities can satisfy different levels of required navigational performance (RNP). The major types of equipment are:

a. VORTAC referenced or Course Line Computer (CLC) systems, which account for the greatest number of RNAV units in use. To function, the CLC must be within the service range of a VORTAC.

b. OMEGA / VLF, although two separate systems, can be considered as one operationally. A long-range navigation system based upon Very Low Frequency radio signals transmitted from a total of 17 stations worldwide.

c. Inertial (INS) systems, which are totally self-contained and require no information from external references. They provide aircraft position and navigation information in response to signals resulting from inertial effects on components within the system.

d. MLS Area Navigation (MLS/RNAV), which provides area navigation with reference to an MLS ground facility.

e. LORAN-C is a long-range radio navigation system that uses ground waves transmitted at low frequency to provide user position information at ranges of up to 600 to 1,200 nautical miles at both en route and approach altitudes. The usable signal coverage areas are determined by the signal-to-noise ratio, the envelope-to-cycle difference, and the geometric relationship between the positions of the user and the transmitting stations.

f. GPS is a space-base radio positioning, navigation, and time-transfer system. The system provides highly accurate position and velocity information, and precise time, on a continuous global basis, to an unlimited number of properly equipped users. The system is unaffected by weather, and provides a worldwide common grid reference system.

(See ICAO term AREA NAVIGATION.)

AREA NAVIGATION [ICAO]   A method of navigation which permits aircraft operation on any desired flight path within the coverage of station-referenced navigation aids or within the limits of the capability of self-contained aids, or a combination of these.

AREA NAVIGATION (RNAV) APPROACH CONFIGURATION:

a. STANDARD T–An RNAV approach whose design allows direct flight to any one of three initial approach fixes (IAF) and eliminates the need for procedure turns. The standard design is to align the procedure on the extended centerline with the missed approach point (MAP) at the runway threshold, the final approach fix (FAF), and the initial approach/intermediate fix (IAF/IF). The other two IAF's will be established perpendicular to the IF.

b. MODIFIED T–An RNAV approach design for single or multiple runways where terrain or operational constraints do not allow for the standard T. The "T" may be modified by increasing or decreasing the angle from the corner IAF(s) to the IF or by eliminating one or both corner IAF's.

c. STANDARD I–An RNAV approach design for a single runway with both corner IAF's eliminated. Course reversal or radar vectoring may be required at busy terminals with multiple runways.

d. TERMINAL ARRIVAL AREA (TAA)–The TAA is controlled airspace established in conjunction with the Standard or Modified T and I RNAV approach configurations. In the standard TAA, there are three areas: straight-in, left base, and right base. The arc boundaries of the three areas of the TAA are published portions of the approach and allow aircraft to transition from the en route structure direct to the nearest IAF. TAA's will also eliminate or reduce feeder routes, departure extensions, and procedure turns or course reversal.

1. STRAIGHT-IN AREA. A 30NM arc centered on the IF bounded by a straight line extending through the IF perpendicular to the intermediate course.

2. LEFT BASE AREA. A 30NM arc centered on the right corner IAF. The area shares a boundary with the straight-in area except that it extends out for 30 NM from the IAF and is bounded on the other side by a line extending from the IF through the FAF to the arc.

3. RIGHT BASE AREA. A 30NM arc centered on the left corner IAF. The area shares a boundary with the straight-in area except that it extends out for 30NM from the IAF and is bounded on the other side by a line extending from the IF through the FAF to the arc.

ARINC   An acronym for Aeronautical Radio, Inc., a corporation largely owned by a group of airlines. ARINC is licensed by the FCC as an aeronautical station and contracted by the FAA to provide communications support for air traffic control and meteorological services in portions of international airspace.

ARMY AVIATION FLIGHT INFORMATION BULLETIN   A bulletin that provides air operation data covering Army, National Guard, and Army Reserve aviation activities.

ARO   (See AIRPORT RESERVATION OFFICE.)

ARRESTING SYSTEM   A safety device consisting of two major components, namely, engaging or catching devices and energy absorption devices for the purpose of arresting both tailhook and/or nontailhook-equipped aircraft. It is used to prevent aircraft from overrunning runways when the aircraft cannot be stopped after landing or during aborted takeoff. Arresting systems have various names; e.g., arresting gear, hook device, wire barrier cable.
(See ABORT.) (Refer to AIM.)

ARRIVAL AIRCRAFT INTERVAL   An internally generated program in hundredths of minutes based upon the AAR. AAI is the desired optimum interval between successive arrival aircraft over the vertex.

ARRIVAL CENTER   The ARTCC having jurisdiction for the impacted airport.

ARRIVAL DELAY   A parameter which specifies a period of time in which no aircraft will be metered for arrival at the specified airport.

ARRIVAL SECTOR   An operational control sector containing one or more meter fixes.

ARRIVAL SECTOR ADVISORY LIST   An ordered list of data on arrivals displayed at the PVD/MDM of the sector which controls the meter fix.

ARRIVAL SEQUENCING PROGRAM   The automated program designed to assist in sequencing aircraft destined for the same airport.

ARRIVAL STREAM FILTER (ASF)   An on/off filter that allows the conflict notification function to be inhibited for arrival streams into single or multiple airports to prevent nuisance alerts.

ARRIVAL TIME   The time an aircraft touches down on arrival.

ARSR   (See AIR ROUTE SURVEILLANCE RADAR.)

ARTCC   (See AIR ROUTE TRAFFIC CONTROL CENTER.)

ARTS   (See AUTOMATED RADAR TERMINAL SYSTEMS.)

ASD   (See AIRCRAFT SITUATION DISPLAY.)

ASDA   (See ACCELERATE-STOP DISTANCE AVAILABLE.)

ASDA [ICAO]   (See ICAO Term ACCELERATE-STOP DISTANCE AVAILABLE.)

ASDE   (See AIRPORT SURFACE DETECTION EQUIPMENT.)

ASF   (See ARRIVAL STREAM FILTER.)

ASLAR   (See AIRCRAFT SURGE LAUNCH AND RECOVERY.)

ASP   (See ARRIVAL SEQUENCING PROGRAM.)

ASR   (See AIRPORT SURVEILLANCE RADAR.)

ASR APPROACH   (See SURVEILLANCE APPROACH.)

ATC   (See AIR TRAFFIC CONTROL.)

ATCAA   (See ATC ASSIGNED AIRSPACE.)

ASSOCIATED   A radar target displaying a data block with flight identification and altitude information.

ATC ADVISES   Used to prefix a message of noncontrol information when it is relayed to an aircraft by other than an air traffic controller.
   (See ADVISORY.)

ATC ASSIGNED AIRSPACE   Airspace of defined vertical/lateral limits, assigned by ATC, for the purpose of providing air traffic segregation between the specified activities being conducted within the assigned airspace and other IFR air traffic.
   (See SPECIAL USE AIRSPACE.)

ATC CLEARANCE   (See AIR TRAFFIC CLEARANCE.)

ATC CLEARS   Used to prefix an ATC clearance when it is relayed to an aircraft by other than an air traffic controller.

ATC INSTRUCTIONS   Directives issued by air traffic control for the purpose of requiring a pilot to take specific actions; e.g., "Turn left heading two five zero," "Go around," "Clear the runway."
   (Refer to FAR Part 91.)

ATC PREFERRED ROUTE NOTIFICATION URET notification to the appropriate controller of the need to determine if an ATC preferred route needs to be applied, based on destination airport.
   (See ROUTE ACTION NOTIFICATION.)
   (See USER REQUEST EVALUATION TOOL.)

ATC PREFERRED ROUTES   Preferred routes that are not automatically applied by Host.

ATCRBS   (See RADAR.)

ATC REQUESTS   Used to prefix an ATC request when it is relayed to an aircraft by other than an air traffic controller.

ATCSCC   (See AIR TRAFFIC CONTROL SYSTEM COMMAND CENTER.)

ATCT   (See TOWER.)

ATD   (See ALONG-TRACK DISTANCE)

ATIS   (See AUTOMATIC TERMINAL INFORMATION SERVICE.)

ATIS [ICAO]   (See ICAO Term AUTOMATIC TERMINAL INFORMATION SERVICE.)

ATS Route [ICAO]   A specified route designed for channeling the flow of traffic as necessary for the provision of air traffic services.

Note: The term "ATS Route" is used to mean variously, airway, advisory route, controlled or uncontrolled route, arrival or departure, etc.

ATTS   (See AUTOMATED TERMINAL TRACKING SYSTEM.)

AUTOLAND APPROACH   An autoland approach is a precision instrument approach to touchdown and, in some cases, through the landing rollout. An autoland approach is performed by the aircraft autopilot which is receiving position information and/or steering commands from onboard navigation equipment. (See COUPLED APPROACH.)

Note: Autoland and coupled approaches are flown in VFR and IFR. It is common for carriers to require their crews to fly coupled approaches and autoland approaches (if certified) when the weather conditions are less than approximately 4,000 RVR.

AUTOMATED INFORMATION TRANSFER   A precoordinated process, specifically defined in facility directives, during which a transfer of altitude control and/or radar identification is accomplished without verbal coordination between controllers using information communicated in a full data block.

AUTOMATED MUTUAL-ASSISTANCE VESSEL RESCUE SYSTEM   A facility which can deliver, in a matter of minutes, a surface picture (SURPIC) of vessels in the area of a potential or actual search and rescue incident, including their predicted positions and their characteristics.
   (See FAAO Order 7110.65, paragraph 10-6-4, In-Flight Contingencies.)

AUTOMATED PROBLEM DETECTION (APD)   An Automation Processing capability that compares trajectories in order to predict conflicts.

AUTOMATED PROBLEM DETECTION BOUNDARY (APB)   The adapted distance beyond a facilities boundary defining the airspace within which URET performs conflict detection.

(See USER REQUEST EVALUATION TOOL CORE CAPABILITY LIMITED DEPLOLYMENT.)

AUTOMATED PROBLEM DETECTION INHIBITED AREA (APDIA) Airspace surrounding a terminal area within which APD is inhibited for all flights within that airspace.

AUTOMATED RADAR TERMINAL SYSTEMS (ARTS) A generic term for several tracking systems included in the Terminal Automation Systems (TAS). ARTS plus a suffix roman numeral denotes a major modification to that system.

a. ARTS IIIA. The Radar Tracking and Beacon Tracking Level (RT&BTL) of the modular, programmable automated radar terminal system, ARTS IIIA detects, tracks, and predicts primary m well as secondary radar-derived aircraft targets. This more sophisticated computer-driven system upgrades the existing ARTS III system by providing improved tracking, continuous data recording, and fail-soft capabilities.

b. Common ARTS. Includes ARTS IIE, ARTS IIIE; and ARTS IIIE with ACD (see DTAS) which combines functionalities of the previous ARTS systems.

c. Programmable Indicator Data Processor (PIDP). The PIDP is a modification to the AN/TPX-42 interrogator system currently installed in fixed RAPCONs. The PIDP detects, tracks, and predicts secondary radar aircraft targets. These are displayed by means of computer-generated symbols and alphanumeric characters depicting flight identification, aircraft altitude, ground speed, and flight plan data. Although primary radar targets are not tracked, they are displayed coincident with the secondary radar targets as well as with the other symbols and alphanumerics. The system has the capability of interfacing with ARTCCs.

AUTOMATED WEATHER SYSTEM Any of the automated weather sensor platforms that collect weather data at airports and disseminate the weather information via radio and/or landline. The systems currently consist of the Automated Surface Observing System (ASOS), Automated Weather Sensor System (AWSS) and Automated Weather Observation System (AWOS).

AUTOMATIC ALTITUDE REPORT (See ALTITUDE READOUT.)

AUTOMATIC ALTITUDE REPORTING That function of a transponder which responds to Mode C interrogations by transmitting the aircraft's altitude in 100-foot increments.

AUTOMATIC CARRIER LANDING SYSTEM U.S. Navy final approach equipment consisting of precision tracking radar coupled to a computer data link to provide continuous information to the aircraft, monitoring capability to the pilot, and a backup approach system.

AUTOMATIC DEPENDENT SURVEILLANCE (ADS) [ICAO] A surveillance technique in which aircraft automatically provide, via a data link, data derived from onboard navigation and position fixing systems, including aircraft identification, four dimensional position and additional data as appropriate.

AUTOMATIC DEPENDENT SURVEILLANCE-BROADCAST (ADS-B) A surveillance system in which an aircraft or vehicle to be detected is fitted with cooperative equipment in the form of a data link transmitter. The aircraft or vehicle periodically broadcasts its GPS-derived position and other infor-mation such as velocity over the data link, which is received by a ground-based transmitter/receiver (transceiver) for processing and display at an air traffic control facility.

(See GLOBAL POSITIONING SYSTEM.) (See GROUND-BASED TRANSCEIVER.)

AUTOMATIC DEPENDENT SURVEILLANCE CONTRACT (ADS-C) A data link position reporting system, controlled by a ground station, that establishes contracts with an aircraft's avionics that occur automatically whenever specific events occur, or specific time intervals are reached.

AUTOMATIC DIRECTION FINDER An aircraft radio navigation system which senses and indicates the direction to an L/MF nondirectional radio beacon (NDB) ground transmitter. Direction is indicated to the pilot as a magnetic bearing or as a relative bearing to the longitudinal axis of the aircraft depending on the type of indicator installed in the aircraft. In certain applications, such as military, ADF operations may be based on airborne and ground transmitters in the VHF/UHF frequency spectrum.

(See BEARING.) (See NONDIRECTIONAL BEACON.)

AUTOMATIC TERMINAL INFORMATION SERVICE The continuous broadcast of recorded noncontrol information in selected terminal areas. Its purpose is to improve controller effectiveness and to relieve frequency congestion by automating the repetitive transmission of essential but routine information; e.g., "Los Angeles information Alfa. One three zero zero Coordinated Universal Time. Weather, measured ceiling two thousand overcast, visibility three, haze, smoke, temperature seven one, dew point five seven, wind two five zero at five, altimeter two niner niner six. I-L-S Runway Two Five Left approach in use, Runway Two Five Right closed, advise you have Alfa."

(Refer to AIM.) (See ICAO term AUTOMATIC TERMINAL INFORMATION SERVICE.)

AUTOMATIC TERMINAL INFORMATION SERVICE [ICAO] The provision of current, routine information to arriving and departing aircraft by means of continuous and repetitive broadcasts throughout the day or a specified portion of the day.

AUTOROTATION A rotorcraft flight condition in which the lifting rotor is driven entirely by action of the air when the rotorcraft is in motion.

a. Autorotative Landing/Touchdown Autorotation. Used by a pilot to indicate that he will be landing without applying power to the rotor.

b. Low Level Autorotation. Commences at an altitude well below the traffic pattern, usually below 100 feet AGL and is used primarily for tactical military training.

c. 180 degrees Autorotation. Initiated from a downwind heading and is commenced well inside the normal traffic pattern. "Go around" may not be possible during the latter part of this maneuver.

AVAILABLE LANDING DISTANCE (ALD) The portion of a runway available for landing and roll-out for aircraft cleared for LAHSO. The distance is measured from the landing threshold to the hold-short point.

AVIATION WEATHER SERVICE A service provided by the National Weather Service (NWS) and FAA which collects and disseminates pertinent weather information for pilots, aircraft operators, and ATC. Available aviation weather reports and forecasts are displayed at each NWS office and FAA FSS.

(See EN ROUTE FLIGHT ADVISORY SERVICE.) (See TRANSCRIBED WEATHER BROADCAST.) (See WEATHER ADVISORY.) (Refer to AIM.)

AWW (See SEVERE WEATHER FORECAST ALERTS.)

AZIMUTH (MLS) A magnetic bearing extending from an MLS navigation facility.

Note: Azimuth bearings are described as magnetic and are referred to as "azimuth" in radio telephone communications.

———— **B** ————

BACK-TAXI A term used by air traffic controllers to taxi an aircraft on the runway opposite to the traffic flow. The aircraft may be instructed to back-taxi to the beginning of the runway or at some point before reaching the runway end for the purpose of departure or to exit the runway.

BASE LEG (See TRAFFIC PATTERN.)

BEACON (See RADAR.)
(See NONDIRECTIONAL BEACON.) (See MARKER BEACON.) (See AIRPORT ROTATING BEACON.) (See AERONAUTICAL BEACON.) (See AIRWAY BEACON.)

BEARING The horizontal direction to or from any point, usually measured clockwise from true north, magnetic north, or some other reference point through 360 degrees.
(See NONDIRECTIONAL BEACON.)

BELOW MINIMUMS Weather conditions below the minimums prescribed by regulation for the particular action involved; e.g., landing minimums, takeoff minimums.

BLAST FENCE A barrier that is used to divert or dissipate jet or propeller blast.

BLIND SPEED The rate of departure or closing of a target relative to the radar antenna at which cancellation of the primary radar target by moving target indicator (MTI) circuits in the radar equipment causes a reduction or complete loss of signal.
(See ICAO term BLIND VELOCITY.)

BLIND SPOT An area from which radio transmissions and/or radar echoes cannot be received. The term is also used to describe portions of the airport not visible from the control tower.

BLIND TRANSMISSION (See TRANSMITTING IN THE BLIND.)

BLIND VELOCITY [ICAO] The radial velocity of a moving target such that the target is not seen on primary radars fitted with certain forms of fixed echo suppression.

BLIND ZONE (See BLIND SPOT.)

BLOCKED Phraseology used to indicate that a radio transmission has been distorted or interrupted due to multiple simultaneous radio transmissions.

BOUNDARY LIGHTS (See AIRPORT LIGHTING.)

BRAKING ACTION (GOOD, FAIR, POOR, OR NIL) A report of conditions on the airport movement area providing a pilot with a degree/quality of braking that he might expect. Braking action is reported in terms of good, fair, poor, or nil.
(See RUNWAY CONDITION READING.)

BRAKING ACTION ADVISORIES When tower controllers have received runway braking action reports which include the terms "poor" or "nil," or whenever weather conditions are conducive to deteriorating or rapidly changing runway braking conditions, the tower will include on the ATIS broadcast the statement, "BRAKING ACTION ADVISORIES ARE IN EFFECT." During the time Braking Action Advisories are in effect, ATC will issue the latest braking action report for the runway in use to each arriving and departing aircraft. Pilots should be prepared for deteriorating braking conditions and should request current runway condition information if not volunteered by controllers. Pilots should also be prepared to provide a descriptive runway condition report to controllers after landing.

BREAKOUT A technique to direct aircraft out of the approach stream. In the context of close parallel operations, a breakout is used to direct threatened aircraft away from a deviating aircraft.

BROADCAST Transmission of information for which an acknowledgement is not expected.
(See ICAO term BROADCAST.)

BROADCAST [ICAO] A transmission of information relating to air navigation that is not addressed to a specific station or stations.

————— **C** —————

CALCULATED LANDING TIME A term that may be used in place of tentative or actual calculated landing time, whichever applies.

CALL UP Initial voice contact between a facility and an aircraft, using the identification of the unit being called and the unit initiating the call.
(Refer to AIM.)

CALL FOR RELEASE Wherein the overlying ARTCC requires a terminal facility to initiate verbal coordination to secure ARTCC approval for release of a departure into the en route environment.

CANADIAN MINIMUM NAVIGATION PERFORMANCE SPECIFICATION AIRSPACE That portion of Canadian domestic airspace within which MNPS separation may be applied.

CARDINAL ALTITUDES "Odd" or "Even" thousand-foot altitudes or flight levels; e.g., 5,000, 6,000, 7,000, FL 250, FL 260, FL 270.
(See ALTITUDE.) (See FLIGHT LEVEL.)

CARDINAL FLIGHT LEVELS (See CARDINAL ALTITUDES.)

CAT (See CLEAR-AIR TURBULENCE.)

CATCH POINT A fix/waypoint that serves as a transition point from the high altitude waypoint navigation structure to an arrival procedure (STAR) or the low altitude ground-based navigation structure.

CEILING The heights above the Earth's surface of the lowest layer of clouds or obscuring phenomena that is reported as "broken," "overcast," or "obscuration," and not classified as "thin" or "partial."
(See ICAO term CEILING.)

CEILING [ICAO] The height above the ground or water of the base of the lowest layer of cloud below 6,000 metres (20,000 feet) covering more than half the sky.

CENRAP (See CENTER RADAR ARTS PRESENTATION/PROCESSING.)

CENRAP-PLUS (See CENTER RADAR ARTS PRESENTATION/PROCESSING-PLUS.)

CENTER (See AIR ROUTE TRAFFIC CONTROL CENTER.)

CENTER'S AREA The specified airspace within which an air route traffic control center (ARTCC) provides air traffic control and advisory service.
(See AIR ROUTE TRAFFIC CONTROL CENTER.) (Refer to AIM.)

CENTER RADAR ARTS PRESENTATION/PROCESSING A computer program developed to provide a back-up system for airport surveillance radar in the event of a failure or malfunction. The program uses air route traffic control center radar for the processing and presentation of data on the ARTS IIA or IIIA displays.

CENTER RADAR ARTS PRESENTATION/PROCESSING-PLUS A computer program developed to provide a back-up system for airport surveillance radar in the event of a terminal secondary radar system failure. The program uses a combination of Air Route Traffic Control Center Radar and terminal airport surveillance radar primary targets displayed simultaneously for the processing and presentation of data on the ARTS IIA or IIIA displays.

CENTER TRACON AUTOMATION SYSTEM (CTAS) A computerized set of programs designed to aid Air Route Traffic Control Centers and TRACON's in the management and control of air traffic.

CENTER WEATHER ADVISORY An unscheduled weather advisory issued by Center Weather Service Unit meteorologists for ATC use to alert pilots of existing or anticipated adverse weather conditions within the next 2 hours. A CWA may modify or redefine a SIGMET.
(See AWW.) (See SIGMET.) (See CONVECTIVE SIGMET.) (See AIRMET.) (Refer to AIM.)

CENTRAL EAST PACIFIC An organized route system between the U.S. West Coast and Hawaii.

CEP (See CENTRAL EAST PACIFIC.)

CERAP (See COMBINED CENTER-RAPCON.)

CERTIFIED TOWER RADAR DISPLAY (CTRD) A FAA radar display certified for use in the NAS.

CFR (See CALL FOR RELEASE.)

CHAFF Thin, narrow metallic reflectors of various lengths and frequency responses, used to reflect radar energy. These reflectors when dropped from aircraft and allowed to drift downward result in large targets on the radar display.

CHARTED VFR FLYWAYS Charted VFR Flyways are flight paths recommended for use to bypass areas heavily traversed by large turbine-powered aircraft. Pilot compliance with recommended flyways and associated altitudes is strictly voluntary. VFR Flyway Planning charts are published on the back of existing VFR Terminal Area charts.

CHARTED VISUAL FLIGHT PROCEDURE APPROACH An approach conducted while operating on an instrument flight rules (IFR) flight plan which authorizes the pilot of an aircraft to proceed visually and clear of clouds to the airport via visual landmarks and other information depicted on a charted visual flight procedure. This approach must be authorized and under the control of the appropriate air traffic control facility. Weather minimums required are depicted on the chart.

CHASE   An aircraft flown in proximity to another aircraft normally to observe its performance during training or testing.

CHASE AIRCRAFT   (See CHASE.)

CIRCLE-TO-LAND MANEUVER   A maneuver initiated by the pilot to align the aircraft with a runway for landing when a straight-in landing from an instrument approach is not possible or is not desirable. At tower controlled airports, this maneuver is made only after ATC authorization has been obtained and the pilot has established required visual reference to the airport.

(See CIRCLE TO RUNWAY.) (See LANDING MINIMUMS.) (Refer to AIM.)

CIRCLE TO RUNWAY (RUNWAY NUMBER)   Used by ATC to inform the pilot that he must circle to land because the runway in use is other than the runway aligned with the instrument approach procedure. When the direction of the circling maneuver in relation to the airport/runway is required, the controller will state the direction (eight cardinal compass points) and specify a left or right downwind or base leg as appropriate; e.g., "Cleared VOR Runway Three Six Approach circle to Runway Two Two," or "Circle northwest of the airport for a right downwind to Runway Two Two."

(See CIRCLE-TO-LAND MANEUVER.) (See LANDING MINIMUMS.) (Refer to AIM.)

CIRCLING   APPROACH   (See   CIRCLE-TO-LAND MANEUVER.)

CIRCLING   MANEUVER   (See   CIRCLE-TO-LAND MANEUVER.)

CIRCLING MINIMA   (See LANDING MINIMUMS.)

CLASS A AIRSPACE   (See Controlled Airspace.)

CLASS B AIRSPACE   (See Controlled Airspace.)

CLASS C AIRSPACE   (See Controlled Airspace.)

CLASS D AIRSPACE   (See Controlled Airspace.)

CLASS E AIRSPACE   (See Controlled Airspace.)

CLASS G AIRSPACE   That airspace not designated as Class A, B, C, D or E.

CLEAR-AIR TURBULENCE   Turbulence encountered in air where no clouds are present. This term is commonly applied to high-level turbulence associated with wind shear. CAT is often encountered in the vicinity of the jet stream.

(See WIND SHEAR.) (See JET STREAM.)

CLEAR OF THE RUNWAY

a. A taxiing aircraft, which is approaching a runway, is clear of the runway when all parts of the aircraft are held short of the applicable holding position marking.

b. A pilot or controller may consider an aircraft, which is exiting or crossing a runway, to be clear of the runway when all parts of the aircraft are beyond the runway edge

and there is no ATC restriction to its continued movement beyond the applicable holding position marking.

c. Pilots and controllers shall exercise good judgement to ensure that adequate separation exists between all aircraft on runways and taxiways at airports with inadequate runway edge lines or holding position markings.

CLEARANCE   (See AIR TRAFFIC CLEARANCE.)

CLEARANCE LIMIT   The fix, point, or location to which an aircraft is cleared when issued an air traffic clearance.

(See ICAO term CLEARANCE LIMIT.)

CLEARANCE LIMIT [ICAO]   The point of which an aircraft is granted an air traffic control clearance.

CLEARANCE VOID IF NOT OFF BY (TIME)   Used by ATC to advise an aircraft that the departure clearance is automatically canceled if takeoff is not made prior to a specified time. The pilot must obtain a new clearance or cancel his IFR flight plan if not off by the specified time.

(See ICAO term CLEARANCE VOID TIME.)

CLEARANCE VOID TIME [ICAO]   A time specified by an air traffic control unit at which a clearance ceases to be valid unless the aircraft concerned has already taken action to comply therewith.

CLEARED AS FILED   Means the aircraft is cleared to proceed in accordance with the route of flight filed in the flight plan. This clearance does not include the altitude, DP, or DP Transition.

(See REQUEST FULL ROUTE CLEARANCE.) (Refer to AIM.)

CLEARED (Type Of) APPROACH   ATC authorization for an aircraft to execute a specific instrument approach procedure to an airport; e.g., "Cleared ILS Runway Three Six Approach."

(See INSTRUMENT APPROACH PROCEDURE.) (See APPROACH CLEARANCE.) (Refer to AIM.) (Refer to FAR Part 91.)

CLEARED APPROACH   ATC authorization for an aircraft to execute any standard or special instrument approach procedure for that airport. Normally, an aircraft will be cleared for a specific instrument approach procedure.

(See INSTRUMENT APPROACH PROCEDURE.) (See CLEARED (TYPE OF) APPROACH.) (Refer to AIM.) (Refer to FAR Part 91.)

CLEARED FOR TAKEOFF   ATC authorization for an aircraft to depart. It is predicated on known traffic and known physical airport conditions.

CLEARED FOR THE OPTION   ATC authorization for an aircraft to make a touch-and-go, low approach, missed approach, stop and go, or full stop landing at the discretion of the pilot. It is normally used in training so that an instructor can evaluate a student's performance under changing situations.

(See OPTION APPROACH.) (Refer to AIM.)

CLEARED THROUGH   ATC authorization for an aircraft to make intermediate stops at specified airports without refiling a flight plan while en route to the clearance limit.

CLEARED TO LAND   ATC authorization for an aircraft to land. It is predicated on known traffic and known physical airport conditions.

CLEARWAY   An area beyond the takeoff runway under the control of airport authorities within which terrain or fixed obstacles may not extend above specified limits. These areas may be required for certain turbine-powered operations and the size and upward slope of the clearway will differ depending on when the aircraft was certificated.
   (Refer to FAR Part 1.)

CLIMBOUT   That portion of flight operation between takeoff and the initial cruising altitude.

CLIMB TO VFR   ATC authorization for an aircraft to climb to VFR conditions within Class B, C, D, and E surface areas when the only weather limitation is restricted visibility. The aircraft must remain clear of clouds while climbing to VFR.
   (See Special VFR.) (Refer to AIM.)

CLOSE PARALLEL RUNWAYS   Two parallel runways whose extended centerlines are separated by less than 4,300 feet, having a Precision Runway Monitoring (PRM) system that permits simultaneous independent ILS approaches.

CLOSED RUNWAY   A runway that is unusable for aircraft operations. Only the airport management/military operations office can close a runway.

CLOSED TRAFFIC   Successive operations involving takeoffs and landings or low approaches where the aircraft does not exit the traffic pattern.

CLOUD   A cloud is a visible accumulation of minute water droplets and/or ice particles in the atmosphere above the Earth's surface. Cloud differs from ground fog, fog, or ice fog only in that the latter are, by definition, in contact with the Earth's surface.

CLT   (See CALCULATED LANDING TIME.)

CLUTTER   In radar operations, clutter refers to the reception and visual display of radar returns caused by precipitation, chaff, terrain, numerous aircraft targets, or other phenomena. Such returns may limit or preclude ATC from providing services based on radar.
   (See GROUND CLUTTER.) (See CHAFF.) (See PRECIPITATION.) (See TARGET.) (See ICAO term Radar Clutter.)

CMNPS   (See CANADIAN MINIMUM NAVIGATION PERFORMANCE SPECIFICATION AIRSPACE.)

COASTAL FIX   A navigation aid or intersection where an aircraft transitions between the domestic route structure and the oceanic route structure.

CODES   The number assigned to a particular multiple pulse reply signal transmitted by a transponder.
   (See DISCRETE CODE.)

COMBINED CENTER-RAPCON   An air traffic facility which combines the functions of an ARTCC and a radar approach control facility.
   (See AIR ROUTE TRAFFIC CONTROL CENTER.) (See RADAR APPROACH CONTROL FACILITY.)

COMMON POINT   A significant point over which two or more aircraft will report passing or have reported passing before proceeding on the same or diverging tracks. To establish/maintain longitudinal separation, a controller may determine a common point not originally in the aircraft's flight plan and then clear the aircraft to fly over the point. (See SIGNIFICANT POINT.)

COMMON PORTION   (See COMMON ROUTE.)

COMMON ROUTE   That segment of a North American Route between the inland navigation facility and the coastal fix.

OR

COMMON ROUTE   Typically the portion of a RNAV STAR between the en route transition end point and the runway transition start point; however, the common route may only consist of a single point that joins the en route and runway transitions.

COMMON TRAFFIC ADVISORY FREQUENCY (CTAF) A frequency designed for the purpose of carrying out airport advisory practices while operating to or from an airport without an operating control tower. The CTAF may be a UNICOM, Multicom, FSS, or tower frequency and is identified in appropriate aeronautical publications.
   (Refer to AC 90-42.)

COMPASS LOCATOR   A low power, low or medium frequency (L/MF) radio beacon installed at the site of the outer or middle marker of an instrument landing system (ILS). It can be used for navigation at distances of approximately 15 miles or as authorized in the approach procedure.
   a. Outer Compass Locator (LOM). A compass locator installed at the site of the outer marker of an instrument landing system.
   (See OUTER MARKER.)
   b. Middle Compass Locator (LMM). A compass locator installed at the site of the middle marker of an instrument landing system.
   (See MIDDLE MARKER.) (See ICAO term LOCATOR.)

COMPASS ROSE   A circle, graduated in degrees, printed on some charts or marked on the ground at an airport. It is used as a reference to either true or magnetic direction.

COMPOSITE FLIGHT PLAN   A flight plan which specifies VFR operation for one portion of flight and IFR for another portion. It is used primarily in military operations.
   (Refer to AIM.)

COMPOSITE ROUTE SYSTEM   An organized oceanic route structure, incorporating reduced lateral spacing between routes, in which composite separation is authorized.

COMPOSITE SEPARATION   A method of separating aircraft in a composite route system where, by management of route and altitude assignments, a combination of half the lateral minimum specified for the area concerned and half the vertical minimum is applied.

COMPULSORY REPORTING POINTS   Reporting points which must be reported to ATC. They are designated on aeronautical charts by solid triangles or filed in a flight plan as fixes selected to define direct routes. These points are geographical locations which are defined by navigation aids/fixes. Pilots should discontinue position reporting over compulsory reporting points when informed by ATC that their aircraft is in "radar contact."

CONFLICT ALERT   A function of certain air traffic control automated systems designed to alert radar controllers to existing or pending situations between tracked targets (known IFR or VFR aircraft) that require his immediate attention/action.
   (See MODE C INTRUDER ALERT.)

CONFLICT RESOLUTION   The resolution of potential conflictions between aircraft that are radar identified and in communication with ATC by ensuring that radar targets do not touch. Pertinent traffic advisories shall be issued when this procedure is applied.

Note: This procedure shall not be provided utilizing mosaic radar systems.

CONFORMANCE   The condition established when an aircraft's actual position is within the conformance region constructed around that aircraft at its position, according to the trajectory associated with the aircraft's Current Plan.

CONFORMANCE REGION   A volume, bounded laterally, vertically, and longitudinally, within which an aircraft must be at a given time in order to be in conformance with the Current Plan Trajectory for that aircraft. At a given time, the conformance region is determined by the simultaneous application of the lateral, vertical, and longitudinal conformance bounds for the aircraft at the position defined by time and aircraft's trajectory.

CONSOLAN   A low frequency, long-distance NAVAID used principally for transoceanic navigations.

CONTACT
   a. Establish communication with (followed by the name of the facility and, if appropriate, the frequency to be used).
   b. A flight condition wherein the pilot ascertains the attitude of his aircraft and navigates by visual reference to the surface.
   (See CONTACT APPROACH.) (See RADAR CONTACT.)

CONTACT APPROACH   An approach wherein an aircraft on an IFR flight plan, having an air traffic control authorization, operating clear of clouds with at least 1 mile flight visibility and a reasonable expectation of continuing to the destination airport in those conditions, may deviate from the instrument approach procedure and proceed to the destination airport by visual reference to the surface. This approach will only be authorized when requested by the pilot and the reported ground visibility at the destination airport is at least 1 statute mile.
   (Refer to AIM.)

CONTAMINATED RUNWAY   A runway is considered contaminated whenever standing water; ice, snow, slush, frost in any form, heavy rubber other substances are present. A runway is contaminated with respect to rubber deposits or other friction-degrading substances when the average friction value for any 500-foot segment of the runway within the ALD fails below the recommended minimum friction level and the average friction value in the adjacent 500-foot segments fails below the maintenance planning friction level.

CONTERMINOUS U.S.   The 48 adjoining States and the District of Columbia.

CONTINENTAL UNITED STATES   The 49 States located on the continent of North America and the District of Columbia.

CONTINUE   When used as a control instruction should be followed by another word or words clarifying what is expected of the pilot. Example: "continue taxi", "continue descent," "continue inbound" etc.

CONTROL AREA [ICAO]   A controlled airspace extending upwards from a specified limit above the Earth.

CONTROLLED AIRSPACE   An airspace of defined dimensions within which air traffic control service is provided to IFR flights and to VFR flights in accordance with the airspace classification.
   a. Controlled airspace is a generic term that covers Class A, Class B, Class C, Class D, and Class E airspace.
   b. Controlled airspace is also that airspace within which all aircraft operators are subject to certain pilot qualifications, operating rules, and equipment requirements in FAR Part 91 (for specific operating requirements, please refer to FAR Part 91). For IFR operations in any class of controlled airspace, a pilot must file an IFR flight plan and receive an appropriate ATC clearance. Each Class B, Class C, and

Class D airspace area designated for an airport contains at least one primary airport around which the airspace is designated (for specific designations and descriptions of the airspace classes, please refer to FAR Part 71).

c. Controlled airspace in the United States is designated as follows:

1. CLASS A: Generally, that airspace from 18,000 feet MSL up to and including FL 600, including the airspace overlying the waters within 12 nautical miles of the coast of the 48 contiguous States and Alaska. Unless otherwise authorized, all persons must operate their aircraft under IFR.

2. CLASS B: Generally, that airspace from the surface to 10,000 feet MSL surrounding the nation's busiest airports in terms of airport operations or passenger enplanements. The configuration of each Class B airspace area is individually tailored and consists of a surface area and two or more layers (some Class B airspace areas resemble upside-down wedding cakes), and is designed to contain all published instrument procedures once an aircraft enters the airspace. An ATC clearance is required for all aircraft to operate in the area, and all aircraft that are so cleared receive separation services within the airspace. The cloud clearance requirement for VFR operations is "clear of clouds."

3. CLASS C: Generally, that airspace from the surface to 4,000 feet above the airport elevation (charted in MSL) surrounding those airports that have an operational control tower, are serviced by a radar approach control, and that have a certain number of IFR operations or passenger enplanements. Although the configuration of each Class C area is individually tailored, the airspace usually consists of a surface area with a 5 nautical mile (NM) radius, a circle with a 10NM radius that extends no lower than 1,200 feet up to 4,000 feet above the airport elevation and an outer area that is not charted. Each person must establish two-way radio communications with the ATC facility providing air traffic services prior to entering the airspace and thereafter maintain those communications while within the airspace. VFR aircraft are only separated from IFR aircraft within the airspace. (See OUTER AREA)

4. CLASS D: Generally, that airspace from the surface to 2,500 feet above the airport elevation (charted in MSL) surrounding those airports that have an operational control tower. The configuration of each Class D airspace area is individually tailored and when instrument procedures are published, the airspace will normally be designed to contain the procedures. Arrival extensions for instrument approach procedures may be Class D or Class E airspace. Unless otherwise authorized, each person must establish two-way radio communications with the ATC facility providing air traffic services prior to entering the airspace and thereafter maintain those communications while in the airspace. No separation services are provided to VFR aircraft.

5. CLASS E: Generally, if the airspace is not Class A, Class B, Class C, or Class D, and it is controlled airspace, it is Class E airspace. Class E airspace extends upward from either the surface or a designated altitude to the overlying or adjacent controlled airspace. When designated as a surface area, the airspace will be configured to contain all instrument procedures. Also in this class are Federal airways, airspace beginning at either 700 or 1,200 feet AGL used to transition to/from the terminal or en route environment, en route domestic, and offshore airspace areas designated below 18,000 feet MSL. Unless designated at a lower altitude, Class E airspace begins at 14,500 MSL over the United States, including that airspace overlying the waters within 12 nautical miles of the coast of the 48 contiguous States and Alaska, up to, but not including 18,000 feet MSL, and the airspace above FL 600.

CONTROLLED AIRSPACE [ICAO]   An airspace of defined dimensions within which air traffic control service is provided to IFR flights and to VFR flights in accordance with the airspace classification.

Note: Controlled airspace is a generic term which covers ATS airspace Classes A, B, C, D, and E.

CONTROLLED TIME OF ARRIVAL   Arrival time assigned during a traffic management program. This may be modified due to adjustments or user options.

CONTROLLER   (See AIR TRAFFIC CONTROL SPECIALIST.)

CONTROLLER [ICAO]   A person authorized to provide air traffic control services.

CONTROLLER PILOT DATA LINK COMMUNICATION (CPDLC)   A two-way digital very high frequency (VHF) air/ground communications system that conveys textual air traffic control messages between controllers and pilots.

CONTROL SECTOR   An airspace area of defined horizontal and vertical dimensions for which a controller or group of controllers has air traffic control responsibility, normally within an air route traffic control center or an approach control facility. Sectors are established based on predominant traffic flows, altitude strata, and controller workload. Pilot-communications during operations within a sector are normally maintained on discrete frequencies assigned to the sector.
(See DISCRETE FREQUENCY.)

CONTROL SLASH   A radar beacon slash representing the actual position of the associated aircraft. Normally, the control slash is the one closest to the interrogating radar beacon site. When ARTCC radar is operating in narrowband (digitized) mode, the control slash is converted to a target symbol.

CONVECTIVE SIGMET   A weather advisory concerning convective weather significant to the safety of all aircraft. Convective SIGMETs are issued for tornadoes, lines of thunderstorms, embedded thunderstorms of any intensity level, areas of thunderstorms greater than or equal to VIP level 4 with an area coverage of ⁴⁄₁₀ (40%) or more, and hail ¾ inch or greater.

(See AWW.) (See SIGMET.) (See CWA.) (See AIRMET.) (Refer to AIM.)

CONVECTIVE SIGNIFICANT METEOROLOGICAL INFORMATION   (See CONVECTIVE SIGMET.)

COORDINATES   The intersection of lines of reference, usually expressed in degrees/minutes/seconds of latitude and longitude, used to determine position or location.

COORDINATION FIX   The fix in relation to which facilities will hand off, transfer control of an aircraft, or coordinate flight progress data. For terminal facilities, it may also serve as a clearance for arriving aircraft.

COPTER   (See HELICOPTER.)

CORRECTION   An error has been made in the transmission and the correct version follows.

COUPLED APPROACH   A coupled approach is an instrument approach performed by the aircraft autopilot which is receiving position information and/or steering commands from onboard navigation equipment. In general, coupled nonprecision approaches must be discontinued and flown manually at altitudes lower than 50 feet below the minimum descent altitude, and coupled precision approaches must be flown manually below 50 feet AGL.

(See AUTOLAND APPROACH.)

Note: Coupled and autoland approaches are flown in VFR and IFR. It is common for carriers to require their crews to fly coupled approaches and autoland approaches (if certified) when the weather conditions are less than approximately 4,000 RVR.

COURSE
a. The intended direction of flight in the horizontal plane measured in degrees from north.
b. The ILS localizer signal pattern usually specified as the front course or the back course.
c. The intended track along a straight, curved, or segmented MLS path.
(See BEARING.) (See RADIAL.) (See INSTRUMENT LANDING SYSTEM.) (See MICROWAVE LANDING SYSTEM.)

CPL [ICAO]   (See CURRENT FLIGHT PLAN.)

CPDLIC   (See CONTROLLER PILOT DATA LINK COMMUNICATIONS)

CRITICAL ENGINE   The engine which, upon failure, would most adversely affect the performance or handling qualities of an aircraft.

CROSS (FIX) AT (ALTITUDE)   Used by ATC when a specific altitude restriction at a specified fix is required.

CROSS (FIX) AT OR ABOVE (ALTITUDE)   Used by ATC when an altitude restriction at a specified fix is required. It does not prohibit the aircraft from crossing the fix at a higher altitude than specified; however, the higher altitude may not be one that will violate a succeeding altitude restriction or altitude assignment.

(See ALTITUDE RESTRICTION.) (Refer to AIM.)

CROSS (FIX) AT OR BELOW (ALTITUDE)   Used by ATC when a maximum crossing altitude at a specific fix is required. It does not prohibit the aircraft from crossing the fix at a lower altitude; however, it must be at or above the minimum IFR altitude.

(See MINIMUM IFR ALTITUDES.) (See ALTITUDE RESTRICTION.) (Refer to FAR Part 91.)

CROSSWIND
a. When used concerning the traffic pattern, the word means "crosswind leg."
(See TRAFFIC PATTERN.)
b. When used concerning wind conditions, the word means a wind not parallel to the runway or the path of an aircraft.
(See CROSSWIND COMPONENT.)

CROSSWIND COMPONENT   The wind component measured in knots at 90 degrees to the longitudinal axis of the runway.

CRUISE   Used in an ATC clearance to authorize a pilot to conduct flight at any altitude from the minimum IFR altitude up to and including the altitude specified in the clearance. The pilot may level off at any intermediate altitude within this block of airspace. Climb/descent within the block is to be made at the discretion of the pilot. However, once the pilot starts descent and verbally reports leaving an altitude in the block, he may not return to that altitude without additional ATC clearance. Further, it is approval for the pilot to proceed to and make an approach at destination airport and can be used in conjunction with:
a. An airport clearance limit at locations with a standard/special instrument approach procedure. The FARs require that if an instrument letdown to an airport is necessary, the pilot shall make the letdown in accordance with a standard/special instrument approach procedure for that airport, or
b. An airport clearance limit at locations that are within/below/outside controlled airspace and without a standard/special instrument approach procedure. Such a clearance is NOT AUTHORIZATION for the pilot to descend under IFR conditions below the applicable minimum IFR altitude nor does it imply that ATC is exercising control over aircraft in Class G airspace; however, it provides a means for the aircraft to proceed to destination airport, descend, and land in accordance with applicable FARs governing VFR flight operations. Also, this provides search and rescue protection until such time as the IFR flight plan is closed.
(See INSTRUMENT APPROACH PROCEDURE.)

CRUISING ALTITUDE   An altitude or flight level maintained during en route level flight. This is a constant altitude and should not be confused with a cruise clearance.

(See ALTITUDE.) (See ICAO term CRUISING LEVEL.)

CRUISING LEVEL [ICAO]   A level maintained during a significant portion of a flight.

CRUISE CLIMB   A climb technique employed by aircraft, usually at a constant power setting, resulting in an increase of altitude as the aircraft weight decreases.

CRUISING LEVEL   (See CRUISING ALTITUDE.)

CT MESSAGE   An EDCT time generated by the ATCSCC to regulate traffic at arrival airports. Normally, a CT message is automatically transferred from the Traffic Management System computer to the NAS en route computer and appears as an EDCT. In the event of a communication failure between the TMS and the NAS, the CT message can be manually entered by the TMC at the en route facility.

CTA   (See CONTROLLED TIME OF ARRIVAL.)

CTA   (See CONTROL AREA [ICAO].)

CTAF   (See COMMON TRAFFIC ADVISORY FREQUENCY.)

CTAS   (See CENTER TRACON AUTOMATION SYSTEM)

CTRD   (See CERTIFIED TOWER RADAR DISPLAY.)

CURRENT FLIGHT PLAN [ICAO]   The flight plan, including changes, if any, brought about by subsequent clearances.

CURRENT PLAN   The ATC clearance the aircraft has received and is expected to fly.

CVFP APPROACH   (See CHARTED VISUAL FLIGHT PROCEDURE APPROACH.)

CWA   (See CENTER WEATHER ADVISORY and *WEATHER ADVISORY.)*

———— **D** ————

DA [ICAO] (See ICAO Term DECISION ALTITUDE/DECISION HEIGHT.)

DAIR   (See DIRECT ALTITUDE AND IDENTITY READOUT.)

DANGER AREA [ICAO]   An airspace of defined dimensions within which activities dangerous to the flight of aircraft may exist at specified times.

Note: The term "Danger Area" is not used in reference to areas within the United States or any of its possessions or territories.

DAS   *(See Delay Assignment)*

DATA BLOCK   (See ALPHANUMERIC DISPLAY.)

D-ATIS   (See DIGITAL AUTOMATIC TERMINAL INFORMATION SERVICE.)

DEAD RECKONING   Dead reckoning, as applied to flying, is the navigation of an airplane solely by means of computations based on airspeed, course, heading, wind direction, and speed, ground speed, and elapsed time.

DECISION ALTITUDE/DECISION HEIGHT [ICAO]   A specified altitude or height (A/H) in the precision approach at which a missed approach must be initiated if the required visual reference to continue the approach has not been established.

Note 1: Decision altitude [DA] is referenced to mean sea level [MSL] and decision height [DH] is referenced to the threshold elevation.

Note 2: The required visual reference means that section of the visual aids or of the approach area which should have been in view for sufficient time for the pilot to have made an assessment of the aircraft position and rate of change of position, in relation to the desired flight path.

DECISION HEIGHT   With respect to the operation of aircraft, means the height at which a decision must be made during an ILS, MLS, or PAR instrument approach to either continue the approach or to execute a missed approach.

(See ICAO term DECISION ALTITUDE/DECISION HEIGHT.)

DECODER   The device used to decipher signals received from ATCRBS transponders to effect their display as select codes.

(See CODES.) (See RADAR.)

DEFENSE VISUAL FLIGHT RULES   Rules applicable to flights within an ADIZ conducted under the visual flight rules in FAR Part 91.

(See AIR DEFENSE IDENTIFICATION ZONE.) (Refer to FAR Part 91.) (Refer to FAR Part 99.)

DELAY ASSIGNMENT (DAS) Delays are distributed to aircraft based on the traffic management program parameters. The delay assignment is calculated in 15-minute increments and appears as a table in ETMS.

DELAY INDEFINITE (REASON IF KNOWN) EXPECT FURTHER CLEARANCE (TIME)   Used by ATC to inform a pilot when an accurate estimate of the delay time and the reason for the delay cannot immediately be determined; e.g., a disabled aircraft on the runway, terminal or center area saturation, weather below landing minimums, etc.

(See EXPECT FURTHER CLEARANCE (TIME).)

DELAY TIME   The amount of time that the arrival must lose to cross the meter fix at the assigned meter fix time. This is the difference between ACLT and VTA.

DEPARTURE CENTER The ARTCC having jurisdiction for the airspace that generates a flight to the impacted airport.

DEPARTURE CONTROL A function of an approach control facility providing air traffic control service for departing IFR and, under certain conditions, VFR aircraft.
(See APPROACH CONTROL FACILITY.) (Refer to AIM.)

DEPARTURE SEQUENCING PROGRAM A program designed to assist in achieving a specified interval over a common point for departures.

DEPARTURE TIME The time an aircraft becomes airborne.

DESCENT SPEED ADJUSTMENTS Speed deceleration calculations made to determine an accurate VTA. These calculations start at the transition point and use arrival speed segments to the vertex.

DESIRED COURSE
a. True—A predetermined desired course direction to be followed (measured in degrees from true north).
b. Magnetic—A predetermined desired course direction to be followed (measured in degrees from local magnetic north).

DESIRED TRACK The planned or intended track between two waypoints. It is measured in degrees from either magnetic or true north. The instantaneous angle may change from point to point along the great circle track between waypoints.

DETRESFA (DISTRESS PHASE) [ICAO] The code word used to designate an emergency phase wherein there is reasonable certainty that an aircraft and its occupants are threatened by grave and imminent danger or require immediate assistance.

DEVIATIONS
a. A departure from a current clearance, such as an off course maneuver to avoid weather or turbulence.
b. Where specifically authorized in the FARs and requested by the pilot, ATC may permit pilots to deviate from certain regulations.
(Refer to AIM.)

DF (See DIRECTION FINDER.)

DF APPROACH PROCEDURE Used under emergency conditions where another instrument approach procedure cannot be executed. DF guidance for an instrument approach is given by ATC facilities with DF capability.
(See DF GUIDANCE.) (See DIRECTION FINDER.) (Refer to AIM.)

DF FIX The geographical location of an aircraft obtained by one or more direction finders.
(See DIRECTION FINDER.)

DF GUIDANCE Headings provided to aircraft by facilities equipped with direction finding equipment. These headings, if followed, will lead the aircraft to a predetermined point such as the DF station or an airport. DF guidance is given to aircraft in distress or to other aircraft which request the service. Practice DF guidance is provided when workload permits.
(See DIRECTION FINDER.) (See DF FIX.) (Refer to AIM.)

DF STEER (See DF GUIDANCE.)

DH (See DECISION HEIGHT.)

DH [ICAO] (See ICAO Term DECISION ALTITUDE/ DECISION HEIGHT.)

DIGITAL-AUTOMATIC TERMINAL INFORMATION SERVICE (D-ATIS) The service provides text messages to aircraft, airlines, and other users outside the standard reception range of conventional ATIS via landline and data link communications to the cockpit. Also, the service provides a computer synthesized voice message that can be transmitted to all aircraft within range of existing transmitters. The Terminal Data Link System (TDLS) D-ATIS application uses weather inputs from local automated weather sources or manually entered meteorological data together with preprogrammed menus to provide standard information to users. Airports with D-ATIS capability are listed in the Airport/Facility Directory.

DIGITAL TARGET A computer-generated symbol representing an aircraft's position, based on a primary return or radar beacon reply, shown on a digital display.

DIGITAL TERMINAL AUTOMATION SYSTEM (DTAS) A system where digital radar and beacon data is presented on digital displays and the operational program monitors the system performance on a real-time basis.

DIGITIZED TARGET A computer-generated indication shown on an analog radar display resulting from a primary radar return or a radar beacon reply.

DIRECT Straight line flight between two navigational aids, fixes, points, or any combination thereof. When used by pilots in describing off-airway routes, points defining direct route segments become compulsory reporting points unless the aircraft is under radar contact.

DIRECT ALTITUDE AND IDENTITY READOUT The DAIR System is a modification to the AN/TPX-42 Interrogator System. The Navy has two adaptations of the DAIR System-Carrier Air Traffic Control Direct Altitude and Identification Readout System for Aircraft Carriers and Radar Air Traffic Control Facility Direct Altitude and Identity Readout System for land-based terminal operations. The DAIR detects, tracks, and predicts secondary radar aircraft targets. Targets are displayed by means of computer-generated symbols and alphanumeric characters depicting flight identification, altitude, ground speed, and flight plan data. The DAIR System is capable of interfacing with ARTCCs.

DIRECTION FINDER   A radio receiver equipped with a directional sensing antenna used to take bearings on a radio transmitter. Specialized radio direction finders are used in aircraft as air navigation aids. Others are ground-based, primarily to obtain a "fix" on a pilot requesting orientation assistance or to locate downed aircraft. A location "fix" is established by the intersection of two or more bearing lines plotted on a navigational chart using either two separately located Direction Finders to obtain a fix on an aircraft or by a pilot plotting the bearing indications of his DF on two separately located ground-based transmitters, both of which can be identified on his chart. UDFs receive signals in the ultra high frequency radio broadcast band; VDFs in the very high frequency band; and UVDFs in both bands. ATC provides DF service at those air traffic control towers and flight service stations listed in the Airport/Facility Directory and the DOD FLIP IFR En Route Supplement.
   (See DF GUIDANCE.) (See DF FIX.)

DISCRETE BEACON CODE   (See DISCRETE CODE.)

DISCRETE CODE   As used in the Air Traffic Control Radar Beacon System (ATCRBS), any one of the 4096 selectable Mode 3/A aircraft transponder codes except those ending in zero zero; e.g., discrete codes: 0010, 1201, 2317, 7777; nondiscrete codes: 0100, 1200, 7700. Nondiscrete codes are normally reserved for radar facilities that are not equipped with discrete decoding capability and for other purposes such as emergencies (7700), VFR aircraft (1200), etc.
   (See RADAR.) (Refer to AIM.)

DISCRETE FREQUENCY   A separate radio frequency for use in direct pilot-controller communications in air traffic control which reduces frequency congestion by controlling the number of aircraft operating on a particular frequency at one time. Discrete frequencies are normally designated for each control sector in en route/terminal ATC facilities. Discrete frequencies are listed in the Airport/Facility Directory and the DOD FLIP IFR En Route Supplement.
   (See CONTROL SECTOR.)

DISPLACED THRESHOLD   A threshold that is located at a point on the runway other than the designated beginning of the runway.
   (See THRESHOLD.) (Refer to AIM.)

DISTANCE (ALD)   That portion of a runway available for Available Landing Distance and roll-out for aircraft cleared for LAHSO. This distance is measured from the landing threshold to the hold-short point.

DISTANCE MEASURING EQUIPMENT   Equipment (airborne and ground) used to measure, in nautical miles, the slant range distance of an aircraft from the DME navigational aid.
   (See TACAN.) (See VORTAC.) (See MICROWAVE LANDING SYSTEM.)

DISTRESS   A condition of being threatened by serious and/or imminent danger and of requiring immediate assistance.

DIVE BRAKES   (See SPEED BRAKES.)

DIVERSE VECTOR AREA   In a radar environment, that area in which a prescribed departure route is not required as the only suitable route to avoid obstacles. The area in which random radar vectors below the MVA/MIA, established in accordance with the TERPS criteria for diverse departures, obstacles and terrain avoidance, may be issued to departing aircraft.

DIVERSION (DVRSN)   Flights that are required to land at other than their original destination for reasons beyond the control of the pilot/company, e.g., periods of significant weather.

DME   (See DISTANCE MEASURING EQUIPMENT.)

DME FIX   A geographical position determined by reference to a navigational aid which provides distance and azimuth information. It is defined by a specific distance in nautical miles and a radial, azimuth, or course (i.e., localizer) in degrees magnetic from that aid.
   (See DISTANCE MEASURING EQUIPMENT.) (See FIX.) (See MICROWAVE LANDING SYSTEM.)

DME SEPARATION   Spacing of aircraft in terms of distances (nautical miles) determined by reference to distance measuring equipment (DME).
   (See DISTANCE MEASURING EQUIPMENT.)

DOD FLIP   Department of Defense Flight Information Publications used for flight planning, en route, and terminal operations. FLIP is produced by the National Imagery and Mapping Agency (NIMA) for world-wide use. United States Government Flight Information Publications (en route charts and instrument approach procedure charts) are incorporated in DOD FLIP for use in the National Airspace System (NAS).

DOMESTIC AIRSPACE   Airspace which overlies the continental land mass of the United States plus Hawaii and U.S. possessions. Domestic airspace extends to 12 miles offshore.

DOWNBURST   A strong downdraft which induces an outburst of damaging winds on or near the ground. Damaging winds, either straight or curved, are highly divergent. The sizes of downbursts vary from ½ mile or less to more than 10 miles. An intense downburst often causes widespread damage. Damaging winds, lasting 5 to 30 minutes, could reach speeds as high as 120 knots.

DOWNWIND LEG   (See TRAFFIC PATTERN.)

DP   (See INSTRUMENT DEPARTURE PROCEDURE.)

DRAG CHUTE   A parachute device installed on certain aircraft which is deployed on landing roll to assist in deceleration of the aircraft.

DSP   (See DEPARTURE SEQUENCING PROGRAM.)

DT   (See DELAY TIME.)

DTAS (See DIGITAL TERMINAL AUTOMATION SYSTEM.)

DUE REGARD A phase of flight wherein an aircraft commander of a State-operated aircraft assumes responsibility to separate his aircraft from all other aircraft.
(See also FAA Order 7110.65, Chapter 1, Word Meanings.)

DUTY RUNWAY (See RUNWAY IN USE/ACTIVE RUNWAY/DUTY RUNWAY.)

DVA (See DIVERSE VECTOR AREA.)

DVFR (See DEFENSE VISUAL FLIGHT RULES.)

DVFR FLIGHT PLAN A flight plan filed for a VFR aircraft which intends to operate in airspace within which the ready identification, location, and control of aircraft are required in the interest of national security.

DVRSN (See DIVERSION.)

DYNAMIC Continuous review, evaluation, and change to meet demands.

DYNAMIC RESTRICTIONS Those restrictions imposed by the local facility on an "as needed" basis to manage unpredictable fluctuations in traffic demands.

──────── **E**

EARTS (See EN ROUTE AUTOMATED RADAR TRACKING SYSTEM.)

EAS (See ENROUTE AUTOMATION SYSTEM.)

EDCT (See EXPECTED DEPARTURE CLEARANCE TIME.)

EFC (See EXPECT FURTHER CLEARANCE (TIME).)

ELT (See EMERGENCY LOCATOR TRANSMITTER.)

EMERGENCY A distress or an urgency condition.

EMERGENCY LOCATOR TRANSMITTER A radio transmitter attached to the aircraft structure which operates from its own power source on 121.5 MHz and 243.0 MHz. It aids in locating downed aircraft by radiating a downward sweeping audio tone, 2-4 times per second. It is designed to function without human action after an accident.
(Refer to FAR Part 91.) (Refer to AIM.)

E-MSAW (See EN ROUTE MINIMUM SAFE ALTITUDE WARNING.)

EN ROUTE AIR TRAFFIC CONTROL SERVICES Air traffic control service provided aircraft on IFR flight plans, generally by centers, when these aircraft are operating between departure and destination terminal areas. When equipment, capabilities, and controller workload permit, certain advisory/assistance services may be provided to VFR aircraft.
(See NAS STAGE A.) (See AIR ROUTE TRAFFIC CONTROL CENTER.) (Refer to AIM.)

EN ROUTE AUTOMATED RADAR TRACKING SYSTEM An automated radar and radar beacon tracking system. Its functional capabilities and design are essentially the same as the terminal ARTS IIIA system except for the EARTS capability of employing both short-range (ASR) and long-range (ARSR) radars, use of full digital radar displays, and fail-safe design.
(See AUTOMATED RADAR TERMINAL SYSTEMS.)

EN ROUTE AUTOMATION SYSTEM (EAS) The complex integrated environment consisting of situation display systems, surveillance systems and flight data processing, remote devices, decision support tools, and the related communications equipment that form the heart of the automated IFR air traffic control system. It interfaces with automated terminal systems and is used in the control of en route IFR aircraft.
(Refer to AIM.)

EN ROUTE CHARTS (See AERONAUTICAL CHART.)

EN ROUTE DESCENT Descent from the en route cruising altitude which takes place along the route of flight.

EN ROUTE FLIGHT ADVISORY SERVICE A service specifically designed to provide, upon pilot request, timely weather information pertinent to his type of flight, intended route of flight, and altitude. The FSSs providing this service are listed in the Airport/Facility Directory.
(See FLIGHT WATCH.) (Refer to AIM.)

EN ROUTE HIGH ALTITUDE CHARTS (See AERONAUTICAL CHART.)

EN ROUTE LOW ALTITUDE CHARTS (See AERONAUTICAL CHART.)

EN ROUTE MINIMUM SAFE ALTITUDE WARNING A function of the EAS that aids the controller by providing an alert when a tracked aircraft is below or predicted by the computer to go below a predetermined minimum IFR altitude (MIA).

EN ROUTE SPACING PROGRAM A program designed to assist the exit sector in achieving the required in-trail spacing.

EN ROUTE TRANSITION
a. Conventional STARs/SIDs. The portion of a SID/STAR that connects to one or more en route airway/jet route.
b. RNAV STARs/SIDs. The portion of a STAR preceding the common route or point, or for a SID the portion following, that is coded for a specific en route fix, airway or jet route.

EPS (See ENGINEERED PERFORMANCE STANDARDS.)

ESP (See EN ROUTE SPACING PROGRAM.)

ESTABLISHED   To be stable or fixed on a route, route segment, altitude, heading, etc.

ESTIMATED ELAPSED TIME [ICAO]   The estimated time required to proceed from one significant point to another.
(See ICAO Term TOTAL ESTIMATED ELAPSED TIME.)

ESTIMATED OFF-BLOCK TIME [ICAO]   The estimated time at which the aircraft will commence movement associated with departure.

ESTIMATED POSITION ERROR (EPE)   (See REQUIRED NAVIGATION PERFORMANCE)

ESTIMATED TIME OF ARRIVAL   The time the flight is estimated to arrive at the gate (scheduled operators) or the actual runway on times for nonscheduled operators.

ESTIMATED TIME EN ROUTE   The estimated flying time from departure point to destination (lift-off to touchdown).

ETA   (See ESTIMATED TIME OF ARRIVAL.)

ETE   (See ESTIMATED TIME EN ROUTE.)

EXECUTE MISSED APPROACH   Instructions issued to a pilot making an instrument approach which means continue inbound to the missed approach point and execute the missed approach procedure as described on the Instrument Approach Procedure Chart or as previously assigned by ATC. The pilot may climb immediately to the altitude specified in the missed approach procedure upon making a missed approach. No turns should be initiated prior to reaching the missed approach point. When conducting an ASR or PAR approach, execute the assigned missed approach procedure immediately upon receiving instructions to "execute missed approach."
(Refer to AIM.)

EXPECT (ALTITUDE) AT (TIME) or (FIX)   Used under certain conditions to provide a pilot with an altitude to be used in the event of two-way communications failure. It also provides altitude information to assist the pilot in planning.
(Refer to AIM.)

EXPECT DEPARTURE CLEARANCE TIME   The runway release time assigned to an aircraft in a traffic management program and shown on the flight progress strip as an EDCT.
(See GROUND DELAY PROGRAM.)

EXPECT FURTHER CLEARANCE (TIME)   The time a pilot can expect to receive clearance beyond a clearance limit.

EXPECT FURTHER CLEARANCE VIA (AIRWAYS, ROUTES OR FIXES)   Used to inform a pilot of the routing he can expect if any part of the route beyond a short range clearance limit differs from that filed.

EXPEDITE   Used by ATC when prompt compliance is required to avoid the development of an imminent situation. Expedite climb/descent normally indicates to a pilot that the approximate best rate of climb/descent should be used without requiring an exceptional change in aircraft handling characteristics.

────────   **F**   ────────

FAF   (See FINAL APPROACH FIX.)

FAST FILE   A system whereby a pilot files a flight plan via telephone that is tape recorded and then transcribed for transmission to the appropriate air traffic facility. Locations having a fast file capability are contained in the Airport/Facility Directory.
(Refer to AIM.)

FAWP   Final Approach Waypoint

FCLT   (See FREEZE CALCULATED LANDING TIME.)

FEATHERED PROPELLER   A propeller whose blades have been rotated so that the leading and trailing edges are nearly parallel with the aircraft flight path to stop or minimize drag and engine rotation. Normally used to indicate shutdown of a reciprocating or turboprop engine due to malfunction.

FEDERAL AIRWAYS   (See LOW ALTITUDE AIRWAY STRUCTURE.)

FEEDER FIX   The fix depicted on Instrument Approach Procedure Charts which establishes the starting point of the feeder route.

FEEDER ROUTE   A route depicted on instrument approach procedure charts to designate routes for aircraft to proceed from the en route structure to the initial approach fix (IAF).
(See INSTRUMENT APPROACH PROCEDURE.)

FERRY FLIGHT   A flight for the purpose of:
　a. Returning an aircraft to base.
　b. Delivering an aircraft from one location to another.
　c. Moving an aircraft to and from a maintenance base. Ferry flights, under certain conditions, may be conducted under terms of a special flight permit.

FIELD ELEVATION   (See AIRPORT ELEVATION.)

FILED   Normally used in conjunction with flight plans, meaning a flight plan has been submitted to ATC.

FILED EN ROUTE DELAY   Any of the following preplanned delays at points/areas along the route of flight which require special flight plan filing and handling techniques.
　a. Terminal Area Delay. A delay within a terminal area for touch-and-go, low approach, or other terminal area activity.

b. Special Use Airspace Delay. A delay within a Military Operations Area, Restricted Area, Warning Area, or ATC Assigned Airspace.

c. Aerial Refueling Delay. A delay within an Aerial Refueling Track or Anchor.

FILED FLIGHT PLAN   The flight plan as filed with an ATS unit by the pilot or his designated representative without any subsequent changes or clearances.

FINAL   Commonly used to mean that an aircraft is on the final approach course or is aligned with a landing area.
   (See FINAL APPROACH COURSE.) (See FINAL APPROACH IFR.) (See SEGMENTS OF AN INSTRUMENT APPROACH PROCEDURE.)

FINAL APPROACH [ICAO]   That part of an instrument approach procedure which commences at the specified final approach fix or point, or where such a fix or point is not specified,
   a. At the end of the last procedure turn, base turn or inbound turn of a racetrack procedure, if specified; or
   b. At the point of interception of the last track specified in the approach procedure; and ends at a point in the vicinity of an aerodrome from which:
      1. A landing can be made; or
      2. A missed approach procedure is initiated.

FINAL APPROACH COURSE   A bearing/radial/track of an instrument approach leading to a runway or an extended runway centerline all without regard to distance.

FINAL APPROACH FIX   The fix from which the final approach (IFR) to an airport is executed and which identifies the beginning of the final approach segment. It is designated on Government charts by the Maltese Cross symbol for nonprecision approaches and the lightning bolt symbol for precision approaches; or when ATC directs a lower-than-published Glideslope/path Intercept Altitude, it is the resultant actual point of the glide-slope/path intercept.
   (See FINAL APPROACH POINT.) (See GLIDESLOPE INTERCEPT ALTITUDE.) (See SEGMENTS OF AN INSTRUMENT APPROACH PROCEDURE.)

FINAL APPROACH-IFR   The flight path of an aircraft which is inbound to an airport on a final instrument approach course, beginning at the final approach fix or point and extending to the airport or the point where a circle-to-land maneuver or a missed approach is executed.
   (See SEGMENTS OF AN INSTRUMENT APPROACH PROCEDURE.) (See FINAL APPROACH FIX.) (See FINAL APPROACH COURSE.) (See FINAL APPROACH POINT.) (See ICAO term FINAL APPROACH.)

FINAL APPROACH POINT   The point, applicable only to a nonprecision approach with no depicted FAF (such as an on airport VOR), where the aircraft is established inbound on the final approach course from the procedure turn and where the final approach descent may be

commenced. The FAP serves as the FAF and identifies the beginning of the final approach segment.
   (See FINAL APPROACH FIX.) (See SEGMENTS OF AN INSTRUMENT APPROACH PROCEDURE.)

FINAL APPROACH SEGMENT   (See SEGMENTS OF AN INSTRUMENT APPROACH PROCEDURE.)

FINAL APPROACH SEGMENT [ICAO]   That segment of an instrument approach procedure in which alignment and descent for landing are accomplished.

FINAL CONTROLLER   The controller providing information and final approach guidance during PAR and ASR approaches utilizing radar equipment.
   (See RADAR APPROACH.)

FINAL GUARD SERVICE   A value added service provided in conjunction with LAA/RAA only during periods of significant and fast changing weather conditions that may affect landing and takeoff operations.

FINAL MONITOR AID   A high resolution color display that is equipped with the controller alert system hardware/ software which is used in the precision runway monitor (PRM) system. The display includes alert algorithms providing the target predictors, a color change alert when a target penetrates or is predicted to penetrate the no transgression zone (NTZ), a color change alert if the aircraft transponder becomes inoperative, synthesized voice alerts, digital mapping, and like features contained in the PRM system.
   (See RADAR APPROACH.)

FINAL MONITOR CONTROLLER   Air Traffic Control Specialist assigned to radar monitor the flight path of aircraft during simultaneous parallel and simultaneous close parallel ILS approach operations. Each runway is assigned a final monitor controller during simultaneous parallel and simultaneous close parallel ILS approaches. Final monitor controllers shall utilize the Precision Runway Monitor (PRM) system during simultaneous close parallel ILS approaches.

FIR   (See FLIGHT INFORMATION REGION.)

FIRST TIER CENTER   The ARTCC immediately adjacent to the impacted center.

FIX   A geographical position determined by visual reference to the surface, by reference to one or more radio NAVAIDs, by celestial plotting, or by another navigational device.

FIX BALANCING   A process whereby aircraft are evenly distributed over several available arrival fixes reducing delays and controller workload.

FLAG   A warning device incorporated in certain airborne navigation and flight instruments indicating that:

a. Instruments are inoperative or otherwise not operating satisfactorily, or

b. Signal strength or quality of the received signal falls below acceptable values.

FLAG ALARM   (See FLAG.)

FLAMEOUT   An emergency condition caused by a loss of engine power.

FLAMEOUT PATTERN   An approach normally conducted by a single-engine military aircraft experiencing loss or anticipating loss of engine power or control. The standard overhead approach starts at a relatively high altitude over a runway ("high key") followed by a continuous 180 degree turn to a high, wide position ("low key") followed by a continuous 180 degree turn final. The standard straight-in pattern starts at a point that results in a straight-in approach with a high rate of descent to the runway. Flameout approaches terminate in the type approach requested by the pilot (normally fullstop).

FLIGHT CHECK   A call-sign prefix used by FAA aircraft engaged in flight inspection/certification of navigational aids and flight procedures. The word "recorded" may be added as a suffix; e.g., "Flight Check 320 recorded" to indicate that an automated flight inspection is in progress in terminal areas.
(See FLIGHT INSPECTION.) (Refer to AIM.)

FLIGHT FOLLOWING   (See TRAFFIC ADVISORIES.)

FLIGHT INFORMATION REGION   An airspace of defined dimensions within which Flight Information Service and Alerting Service are provided.

a. Flight Information Service. A service provided for the purpose of giving advice and information useful for the safe and efficient conduct of flights.

b. Alerting Service. A service provided to notify appropriate organizations regarding aircraft in need of search and rescue aid and to assist such organizations as required.

FLIGHT INFORMATION SERVICE   A service provided for the purpose of giving advice and information useful for the safe and efficient conduct of flights.

FLIGHT INSPECTION   Inflight investigation and evaluation of a navigational aid to determine whether it meets established tolerances.
(See NAVIGATIONAL AID.) (See FLIGHT CHECK.)

FLIGHT LEVEL   A level of constant atmospheric pressure related to a reference datum of 29.92 inches of mercury. Each is stated in three digits that represent hundreds of feet. For example, flight level (FL) 250 represents a barometric altimeter indication of 25,000 feet; FL 255, an indication of 25,500 feet.
(See ICAO term FLIGHT LEVEL.)

FLIGHT LEVEL [ICAO]   A surface of constant atmospheric pressure which is related to a specific pressure datum, 1013.2 hPa (1013.2 mb), and is separated from other such surfaces by specific pressure intervals.

Note 1: A pressure type altimeter calibrated in accordance with the standard atmosphere:

a. When set to a QNH altimeter setting, will indicate altitude;

b. When set to a QFE altimeter setting, will indicate height above the QFE reference datum; and

c. When set to a pressure of 1013.2 hPa (1013.2 mb), may be used to indicate flight levels.

Note 2: The terms 'height' and 'altitude,' used in Note 1 above, indicate altimetric rather than geometric heights and altitudes.

FLIGHT LINE   A term used to describe the precise movement of a civil photogrammetric aircraft along a predetermined course(s) at a predetermined altitude during the actual photographic run.

FLIGHT MANAGEMENT SYSTEMS   A computer system that uses a large data base to allow routes to be preprogrammed and fed into the system by means of a data loader. The system is constantly updated with respect to position accuracy by reference to conventional navigation aids. The sophisticated program and its associated data base ensures that the most appropriate aids are automatically selected during the information update cycle.

FLIGHT MANAGEMENT SYSTEM PROCEDURE   An arrival, departure, or approach procedure developed for use by aircraft with a slant (/) E or slant (/) F equipment suffix.

FLIGHT PATH   A line, course, or track along which an aircraft is flying or intended to be flown.
(See TRACK.) (See COURSE.)

FLIGHT PLAN   Specified information relating to the intended of an aircraft that is filed orally or in writing with an FSS or an ATC facility.
(See FAST FILE.) (See FILED.) (Refer to AIM.)

FLIGHT PLAN AREA   The geographical area assigned by regional air traffic divisions to a flight service station for the purpose of search and rescue for VFR aircraft, issuance of NOTAMs, pilot briefing, in-flight services, broadcast, emergency services, flight data processing, international operations, and aviation weather services. Three letter identifiers are assigned to every flight service station and are annotated in AFDs and FAAO 7350.7, LOCATION IDENTIFIERS as tie-in-facilities.
(See FAST FILE.) (See FILED.) (Refer to AIM.)

FLIGHT RECORDER   A general term applied to any instrument or device that records information about the performance of an aircraft in flight or about conditions encountered in flight. Flight recorders may make records of airspeed, outside air temperature, vertical acceleration, engine RPM, manifold pressure, and other pertinent variables for a given flight.
(See ICAO term FLIGHT RECORDER.)

FLIGHT RECORDER [ICAO]   Any type of recorder installed in the aircraft for the purpose of complementing accident/incident investigation.

Note: See Annex 6 Part I, for specifications relating to flight recorders.

FLIGHT SERVICE STATION   Air traffic facilities which provide pilot briefing, en route communications and VFR search and rescue services, assist lost aircraft and aircraft in emergency situations, relay ATC clearances, originate Notices to Airmen, broadcast aviation weather and NAS information, and receive and process IFR flight plans. In addition, at selected locations, FSSs provide En Route Flight Advisory Service (Flight Watch), issue airport advisories, and advise Customs and Immigration of transborder flights. Selected flight service stations in Alaska also provide TWEB recordings and take weather observations.

(Refer to AIM.)

FLIGHT STANDARDS DISTRICT OFFICE   An FAA field office serving an assigned geographical area and staffed with Flight Standards personnel who serve the aviation industry and the general public on matters relating to the certification and operation of air carrier and general aviation aircraft. Activities include general surveillance of operational safety, certification of airmen and aircraft, accident prevention, investigation, enforcement, etc.

FLIGHT TEST   A flight for the purpose of:

a. Investigating the operation/flight characteristics of an aircraft or aircraft component.

b. Evaluating an applicant for a pilot certificate or rating.

FLIGHT VISIBILITY   (See VISIBILITY.)

FLIGHT WATCH   A shortened term for use in air-ground contacts to identify the flight service station providing En Route Flight Advisory Service; e.g., "Oakland Flight Watch."

(See EN ROUTE FLIGHT ADVISORY SERVICE.)

FLIP   (See DOD FLIP.)

FLY-BY WAYPOINT   A fly-by waypoint requires the use of turn anticipation to avoid overshoot of the next flight segment.

FLY HEADING (DEGREES)   Informs the pilot of the heading he should fly. The pilot may have to turn to, or continue on, a specific compass direction in order to comply with the instructions. The pilot is expected to turn in the shorter direction to the heading unless otherwise instructed by ATC.

FLY-OVER WAYPOINT   A fly-over waypoint precludes any turn until the waypoint is overflown and is followed by an intercept maneuver of the next flight segment.

FMA   (See FINAL MONITOR AID.)

FMS   (See FLIGHT MANAGEMENT SYSTEM.)

FMSP   (See FLIGHT MANAGEMENT SYSTEM PROCEDURE.)

FORMATION FLIGHT   More than one aircraft which, by prior arrangement between the pilots, operate as a single aircraft with regard to navigation and position reporting. Separation between aircraft within the formation is the responsibility of the flight leader and the pilots of the other aircraft in the flight. This includes transition periods when aircraft within the formation are maneuvering to attain separation from each other to effect individual control and during join-up and breakaway.

a. A standard formation is one in which a proximity of no more than 1 mile laterally or longitudinally and within 100 feet vertically from the flight leader is maintained by each wingman.

b. Nonstandard formations are those operating under any of the following conditions:

1. When the flight leader has requested and ATC has approved other than standard formation dimensions.

2. When operating within an authorized altitude reservation (ALTRV) or under the provisions of a letter of agreement.

3. When the operations are conducted in airspace specifically designed for a special activity.

(See ALTITUDE RESERVATION.) (Refer to FAR Part 91.)

FRC   (See REQUEST FULL ROUTE CLEARANCE.)

FREEZE/FROZEN   Terms used in referring to arrivals which have been assigned ACLTs and to the lists in which they are displayed.

FREEZE CALCULATED LANDING TIME   A dynamic parameter number of minutes prior to the meter fix calculated time of arrival for each aircraft when the TCLT is frozen and becomes an ACLT (i.e., the VTA is updated and consequently the TCLT is modified as appropriate until FCLT minutes prior to meter fix calculated time of arrival, at which time updating is suspended and an ACLT and a frozen meter fix crossing time (MFT) is assigned).

FREEZE HORIZON   The time or point at which an aircraft's STA becomes fixed and no longer fluctuates with each radar update. This setting insures a constant time for each aircraft, necessary for the metering controller to plan his/her delay technique. This setting can be either in distance from the meter fix or a prescribed flying time to the meter fix.

FREEZE SPEED PARAMETER   A speed adapted for each aircraft to determine fast and slow aircraft. Fast aircraft freeze on parameter FCLT and slow aircraft freeze on parameter MLDI.

FRICTION MEASUREMENT   A measurement of the friction characteristics of the runway pavement surface using continuous self-watering friction measurement equipment in accordance with the specifications, procedures and schedules contained in AC 150/5320-12, Measurement, Construction, and Maintenance of Skid-Resistant Airport Pavement Surfaces.

FSDO (See FLIGHT STANDARDS DISTRICT OFFICE.)

FSPD (See FREEZE SPEED PARAMETER.)

FSS (See FLIGHT SERVICE STATION.)

FUEL DUMPING Airborne release of usable fuel. This does not include the dropping of fuel tanks.
(See JETTISONING OF EXTERNAL STORES.)

FUEL REMAINING A phrase used by either pilots or controllers when relating to the fuel remaining on board until actual fuel exhaustion. When transmitting such information in response to either a controller question or pilot initiated cautionary advisory to air traffic control, pilots will state the APPROXIMATE NUMBER OF MINUTES the flight can continue with the fuel remaining. All reserve fuel SHOULD BE INCLUDED in the time stated, as should an allowance for established fuel gauge system error.

FUEL SIPHONING Unintentional release of fuel caused by overflow, puncture, loose cap, etc.

FUEL VENTING (See FUEL SIPHONING.)

——————— **G** ———————

GATE HOLD PROCEDURES Procedures at selected airports to hold aircraft at the gate or other ground location whenever departure delays exceed or are anticipated to exceed 15 minutes. The sequence for departure will be maintained in accordance with initial call-up unless modified by flow control restrictions. Pilots should monitor the ground control/clearance delivery frequency for engine start/taxi advisories or new proposed start/taxi time if the delay changes.
(See FLOW CONTROL.)

GBT (See GROUND-BASED TRANSCEIVER.)

GCA (See GROUND CONTROLLED APPROACH.)

GDP (See GROUND DELAY PROGRAM.)

GENERAL AVIATION That portion of civil aviation which encompasses all facets of aviation except air carriers holding a certificate of public convenience and necessity from the Civil Aeronautics Board and large aircraft commercial operators.
(See ICAO term GENERAL AVIATION.)

GENERAL AVIATION [ICAO] All civil aviation operations other than scheduled air services and nonscheduled air transport operations for remuneration or hire.

GENERAL AVIATION DISTRICT OFFICE An FAA field office serving a designated geographical area and staffed with Flight Standards personnel who have the responsibility for serving the aviation industry and the general public on all matters relating to the certification and operation of general aviation aircraft.

GEO MAP The digitized map markings associated with the ASR-9 Radar System.

GLIDEPATH (See GLIDESLOPE.)

GLIDEPATH INTERCEPT ALTITUDE (See GLIDESLOPE INTERCEPT ALTITUDE.)

GLIDESLOPE Provides vertical guidance for aircraft during approach and landing. The glideslope/glidepath is based on the following:

a. Electronic components emitting signals which provide vertical guidance by reference to airborne instruments during instrument approaches such as ILS/MLS, or

b. Visual ground aids, such as VASI, which provide vertical guidance for a VFR approach or for the visual portion of an instrument approach and landing.

c. PAR. Used by ATC to inform an aircraft making a PAR approach of its vertical position (elevation) relative to the descent profile.
(See ICAO term GLIDEPATH.)

GLIDEPATH [ICAO] A descent profile determined for vertical guidance during a final approach.

GLIDESLOPE INTERCEPT ALTITUDE The minimum altitude to intercept the glideslope/path on a precision approach. The intersection of the published intercept altitude with the glideslope/path, designated on Government charts by the lightning bolt symbol, is the precision FAF; however, when the approach chart shows an alternative lower glide-slope intercept altitude, and ATC directs a lower altitude, the resultant lower intercept position is then the FAF.
(See FINAL APPROACH FIX.) (See SEGMENTS OF AN INSTRUMENT APPROACH PROCEDURE.)

GLOBAL POSITIONING SYSTEM (GPS) A space-base radio positioning, navigation, and time-transfer system. The system provides highly accurate position and velocity information, and precise time, on a continuous global basis, to an unlimited number of properly equipped users. The system is unaffected by weather, and provides a worldwide common grid reference system. The GPS concept is predicated upon accurate and continuous knowledge of the spatial position of each satellite in the system with respect to time and distance from a transmitting satellite to the user. The GPS receiver automatically selects appropriate signals from the satellites in view and translates these into three-dimensional position, velocity, and time. System accuracy for civil users is normally 100 meters horizontally.

GO AHEAD Proceed with your message. Not to be used for any other purpose.

GO AROUND Instructions for a pilot to abandon his approach to landing. Additional instructions may follow. Unless otherwise advised by ATC, a VFR aircraft or an aircraft conducting visual approach should overfly the runway while climbing to traffic pattern altitude and enter the traffic pattern via the crosswind leg. A pilot on an IFR flight plan making an instrument approach should execute the published missed approach procedure or proceed as instructed by ATC; e.g., "Go around" (additional instructions if required).
(See LOW APPROACH.) (See MISSED APPROACH.)

GPD (See GRAPHIC PLAN DISPLAY.)

GPS (See Global Positioning System.)

GRAPHIC PLAN DISPLAY (GPD) A view available with URET CCLD that provides a graphic display of aircraft, traffic, and notification of predicted conflicts. Graphic routes for Current Plans and Trial Plans are displayed upon controller request.
(See *USER REQUEST EVALUATION TOOL.*)

GROUND-BASED TRANSCEIVER (GBT) The ground-based transmitter/receiver (transceiver) receives automatic dependent surveillance-broadcast messages, which are forwarded to an air traffic control facility for processing and display with other radar targets on the plan position indicator (radar display).
(See *AUTOMATIC DEPENDENT SURVEILLANCE-BROADCAST.*)

GROUND CLUTTER A pattern produced on the radar scope by ground returns which may degrade other radar returns in the affected area. The effect of ground clutter is minimized by the use of moving target indicator (MTI) circuits in the radar equipment resulting in a radar presentation which displays only targets which are in motion.
(See CLUTTER.)

GROUND COMMUNICATION OUTLET (GCO) An unstaffed, remotely controlled, ground/ground communications facility. Pilots at uncontrolled airports may contact ATC and FSS via VHF to a telephone connection to obtain an instrument clearance or close a VFR or IFR flight plan. They may also get an updated weather briefing prior to takeoff. Pilots will use four "key clicks" on the VHF radio to contact the appropriate ATC facility or six "key clicks" to contact the FSS. The GCO system is intended to be used only on the ground.

GROUND CONTROLLED APPROACH A radar approach system operated from the ground by air traffic control personnel transmitting instructions to the pilot by radio. The approach may be conducted with surveillance radar (ASR) only or with both surveillance and precision approach radar (PAR). Usage of the term "GCA" by pilots is discouraged except when referring to a GCA facility. Pilots should specifically request a "PAR" approach when a precision radar approach is desired or request an "ASR" or "surveillance" approach when a nonprecision radar approach is desired.
(See RADAR APPROACH.)

GROUND DELAY PROGRAM (GDP) A traffic management process administered by the ATCSCC; when aircraft are held on the ground. The purpose of the program is to support the TM mission and limit airborne holding. It is a flexible program and may be implemented in various forms depending upon the needs of the AT system. Ground delay programs provide for equitable assignment of delays to all system users.

GROUND SPEED The speed of an aircraft relative to the surface of the Earth.

GROUND STOP (GS) The GS is a process that requires aircraft that meet a specific criteria to remain on the ground. The criteria may be airport specific, airspace specific, or equipment specific; for example, all departures to San Francisco, or all departures entering Yorktown sector, or all Category I and II aircraft going to Charlotte. GSs normally occur with little or no warning.

GROUND VISIBILITY (See VISIBILITY.)

GS (See GROUND STOP)

H

HAA (See HEIGHT ABOVE AIRPORT.)

HAL (See HEIGHT ABOVE LANDING.)

HANDOFF An action taken to transfer the radar identification of an aircraft from one controller to another if the aircraft will enter the receiving controller's airspace and radio communications with the aircraft will be transferred.

HAR (See HIGH ALTITUDE REDESIGN.)

HAT (See HEIGHT ABOVE TOUCHDOWN.)

HAVE NUMBERS Used by pilots to inform ATC that they have received runway, wind, and altimeter information only.

HAZARDOUS INFLIGHT WEATHER ADVISORY SERVICE Continuous recorded hazardous inflight weather forecasts broadcasted to airborne pilots over selected VOR outlets defined as an HIWAS BROADCAST AREA.

HAZARDOUS WEATHER INFORMATION Summary of significant meteorological information (SIGMET/WS), convective significant meteorological information (convective SIGMET/WST), urgent pilot weather reports (urgent PIREP/UUA), center weather advisories (CWA), airmen's meteorological information (AIRMET/WA) and any other weather such as isolated thunderstorms that are rapidly developing and increasing in intensity, or low ceilings and visibilities that are becoming widespread which is considered significant and are not included in a current hazardous weather advisory.

HEAVY (AIRCRAFT) (See AIRCRAFT CLASSES.)

HEIGHT ABOVE AIRPORT   The height of the Minimum Descent Altitude above the published airport elevation. This is published in conjunction with circling minimums.
(See MINIMUM DESCENT ALTITUDE.)

HEIGHT ABOVE LANDING   The height above a designated helicopter landing area used for helicopter instrument approach procedures.
(Refer to FAR Part 97.)

HEIGHT ABOVE TOUCHDOWN   The height of the Decision Height or Minimum Descent Altitude above the highest runway elevation in the touchdown zone (first 3,000 feet of the runway). HAT is published on instrument approach charts in conjunction with all straight-in minimums.
(See DECISION HEIGHT.) (See MINIMUM DESCENT ALTITUDE.)

HELICOPTER   Rotorcraft that, for its horizontal motion, depends principally on its engine-driven rotors.
(See ICAO term HELICOPTER.)

HELICOPTER [ICAO]   A heavier-than-air aircraft supported in flight chiefly by the reactions of the air on one or more power-driven rotors on substantially vertical axes.

HELIPAD   A small, designated area, usually with a prepared surface, on a heliport, airport, landing/takeoff area, apron/ramp, or movement area used for takeoff, landing, or parking of helicopters.

HELIPORT   An area of land, water, or structure used or intended to be used for the landing and takeoff of helicopters and includes its buildings and facilities if any.

HELIPORT REFERENCE POINT (HRP)   The geographic center of a heliport.

HERTZ   The standard radio equivalent of frequency in cycles per second of an electromagnetic wave. Kilohertz (kHz) is a frequency of one thousand cycles per second. Megahertz (MHz) is a frequency of one million cycles per second.

HF   (See HIGH FREQUENCY.)

HF COMMUNICATIONS   (See HIGH FREQUENCY COMMUNICATIONS.)

HIGH ALTITUDE REDESIGN (HAR)   A level of non-restrictive routing (NRR) service for aircraft that have all waypoints associated with the HAR program in their flight management systems of RNAV equipage.

HIGH FREQUENCY   The frequency band between 3 and 30 MHz.
(See HIGH FREQUENCY COMMUNICATIONS.)

HIGH FREQUENCY COMMUNICATIONS   High radio frequencies (HF) between 3 and 30 MHz used for air-to-ground voice communication in overseas operations.

HIGH SPEED EXIT   (See HIGH SPEED TAXIWAY.)

HIGH SPEED TAXIWAY   A long radius taxiway designed and provided with lighting or marking to define the path of aircraft, traveling at high speed (up to 60 knots), from the runway center to a point on the center of a taxiway. Also referred to as long radius exit or turn-off taxiway. The high speed taxiway is designed to expedite aircraft turning off the runway after landing, thus reducing runway occupancy time.

HIGH SPEED TURNOFF   (See HIGH SPEED TAXIWAY.)

HIWAS   (See HAZARDOUS INFLIGHT WEATHER ADVISORY SERVICE.)

HIWAS AREA   (See HAZARDOUS INFLIGHT WEATHER ADVISORY SERVICE.)

HIWAS BROADCAST AREA   A geographical area of responsibility including one or more HIWAS outlet areas assigned to an AFSS/FSS for hazardous weather advisory broadcasting.

HIWAS OUTLET AREA   An area defined as a 150 NM radius of a HIWAS outlet, expanded as necessary to provide coverage.

HOLDING PROCEDURE   (See HOLD PROCEDURE.)

HOLD PROCEDURE   A predetermined maneuver which keeps aircraft within a specified airspace while awaiting further clearance from air traffic control. Also used during ground operations to keep aircraft within a specified area or at a specified point while awaiting further clearance from air traffic control.
(See HOLDING FIX.) (Refer to AIM.)

HOLDING FIX   A specified fix identifiable to a pilot by NAVAIDs or visual reference to the ground used as a reference point in establishing and maintaining the position of an aircraft while holding.
(See FIX.) (See VISUAL HOLDING.) (Refer to AIM.)

HOLDING POINT [ICAO]   A specified location, identified by visual or other means, in the vicinity of which the position of an aircraft in flight is maintained in accordance with air traffic control clearances.

HOLD FOR RELEASE   Used by ATC to delay an aircraft for traffic management reasons; i.e., weather, traffic volume, etc. Hold for release instructions (including departure delay information) are used to inform a pilot or a controller (either directly or through an authorized relay) that an IFR departure clearance is not valid until a release time or additional instructions have been received.
(See ICAO term HOLDING POINT.)

HOLD IN LIEU OF PROCEDURE TURN   A hold in lieu of procedure turn shall be established over a final or intermediate fix when an approach can be made from a properly aligned holding pattern. The hold in lieu of procedure turn permits the pilot to align with the final or intermediate segment of the approach and/or descend in the holding pattern to an altitude that will permit a normal descent to the final approach fix altitude. The hold in lieu of procedure turn is a required maneuver (the same as a procedure turn) unless the aircraft is being radar vectored to the final approach course, when "NoPT" is shown on the approach chart, or when the pilot requests or the controller advises the pilot to make a "straight-in" approach.

HOLD PROCEDURE A predetermined maneuver which keeps aircraft within a specified airspace while awaiting further clearance from air traffic control. Also used during ground operations to keep aircraft within a specified area or at a specified point while awaiting further clearance from air traffic control.

(See HOLDING FIX.)
(Refer to AIM.)

HOLDING FIX A specified fix identifiable to a pilot by NAVAIDs or visual reference to the ground used as a reference point in establishing and maintaining the position of an aircraft while holding.

(See FIX.)
(See VISUAL HOLDING.)
(Refer to AIM.)

HOLDING POINT [ICAO] A specified location, identified by visual or other means, in the vicinity of which the position of an aircraft in flight is maintained in accordance with air traffic control clearances.

HOLD-SHORT POINT A point on the runway beyond which a landing aircraft with a LAHSO clearance is not authorized to proceed. This point may be located prior to an intersecting runway, taxiway, predetermined point, or approach/departure flight path.

HOLD-SHORT POSITION MARKING The painted runway marking located at the hold-short point on all LAHSO runways.

HOLD-SHORT POSITION LIGHTS Flashing in-pavement white lights located at specified hold-short points.

HOLD-SHORT POSITION SIGNS Red and White holding position signs located alongside the hold-short point.

HOMING Flight toward a NAVAID, without correcting for wind, by adjusting the aircraft heading to maintain a relative bearing of zero degrees.

(See BEARING.) (See ICAO term HOMING.)

HOMING [ICAO] The procedure of using the direction-finding equipment of one radio station with the emission of another radio station, where at least one of the stations is mobile, and whereby the mobile station proceeds continuously towards the other station.

HOVER CHECK Used to describe when a helicopter/VTOL aircraft requires a stabilized hover to conduct a performance/power check prior to hover taxi, air taxi, or takeoff. Altitude of the hover will vary based on the purpose of the check.

HOVER TAXI Used to describe a helicopter/VTOL aircraft movement conducted above the surface and in ground effect at airspeeds less than approximately 20 knots. The actual height may vary, and some helicopters may require hover taxi above 25 feet AGL to reduce ground effect turbulence or provide clearance for cargo slingloads.

(See AIR TAXI.) (See HOVER CHECK.) (Refer to AIM.)

HOW DO YOU HEAR ME? A question relating to the quality of the transmission or to determine how well the transmission is being received.

HZ (See HERTZ.)

---------- **I** ----------

IAF (See INITIAL APPROACH FIX.)

IAP (See INSTRUMENT APPROACH PROCEDURE.)

IAWP Initial Approach Waypoint

ICAO (See INTERNATIONAL CIVIL AVIATION ORGANIZATION.)

ICAO [ICAO] (See ICAO Term INTERNATIONAL CIVIL AVIATION ORGANIZATION.)

ICING The accumulation of airframe ice.

Types of icing are:

 a. Rime Ice—Rough, milky, opaque ice formed by the instantaneous freezing of small supercooled water droplets.

 b. Clear Ice—A glossy, clear, or translucent ice formed by the relatively slow freezing or large supercooled water droplets.

 c. Mixed—A mixture of clear ice and rime ice.

Intensity of icing:

 a. Trace—Ice becomes perceptible. Rate of accumulation is slightly greater than the rate of sublimation. Deicing/anti-icing equipment is not utilized unless encountered for an extended period of time (over 1 hour).

 b. Light—The rate of accumulation may create a problem if flight is prolonged in this environment (over 1 hour). Occasional use of deicing/anti-icing equipment removes/prevents accumulation. It does not present a problem if the deicing/anti-icing equipment is used.

 c. Moderate—The rate of accumulation is such that even short encounters become potentially hazardous and use of deicing/anti-icing equipment or flight diversion is necessary.

 d. Severe—The rate of accumulation is such that deicing/anti-icing equipment fails to reduce or control the hazard. Immediate flight diversion is necessary.

IDENT A request for a pilot to activate the aircraft transponder identification feature. This will help the controller to confirm an aircraft identity or to identify an aircraft.
  (Refer to AIM.)

IDENT FEATURE The special feature in the Air Traffic Control Radar Beacon System (ATCRBS) equipment. It is used to immediately distinguish one displayed beacon target from other beacon targets.
  (See IDENT.)

IF (See INTERMEDIATE FIX.)

IFIM (See INTERNATIONAL FLIGHT INFORMATION MANUAL.)

IF NO TRANSMISSION RECEIVED FOR (TIME) Used by ATC in radar approaches to prefix procedures which should be followed by the pilot in event of lost communications.
  (See LOST COMMUNICATIONS.)

IFR (See INSTRUMENT FLIGHT RULES.)

IFR AIRCRAFT An aircraft conducting flight in accordance with instrument flight rules.

IFR CONDITIONS Weather conditions below the minimum for flight under visual flight rules.
  (See INSTRUMENT METEOROLOGICAL CONDITIONS.)

IFR DEPARTURE PROCEDURE (See IFR TAKEOFF MINIMUMS AND DEPARTURE PROCEDURES.)
  (Refer to AIM.)

IFR FLIGHT (See IFR AIRCRAFT.)

IFR LANDING MINIMUMS (See LANDING MINIMUMS.)

IFR MILITARY TRAINING ROUTES (IR) Routes used by the Department of Defense and associated Reserve and Air Guard units for the purpose of conducting low-altitude navigation and tactical training in both IFR and VFR weather conditions below 10,000 feet MSL at airspeeds in excess of 250 knots IAS.

IFR TAKEOFF MINIMUMS AND DEPARTURE PROCEDURES Federal Aviation Regulations, Part 91, prescribes standard takeoff rules for certain civil users. At some airports, obstructions or other factors require the establishment of nonstandard takeoff minimums, departure procedures, or both to assist pilots in avoiding obstacles during climb to the minimum en route altitude. Those airports are listed in NOS/DOD Instrument Approach Charts (IAPs) under a section entitled "IFR Takeoff Minimums and Departure Procedures." The NOS/DOD IAP chart legend illustrates the symbol used to alert the pilot to nonstandard takeoff minimums and departure procedures. When departing IFR from such airports or from any airports where there are no departure procedures, DP's, or ATC facilities available, pilots should advise ATC of any departure limitations. Controllers may query a pilot to determine acceptable departure directions, turns, or headings after takeoff. Pilots should be familiar with the departure procedures and must assure that their aircraft can meet or exceed any specified climb gradients.

IF/IAWP Intermediate Fix/Initial Approach Waypoint. The waypoint where the final approach course of a T approach meets the crossbar of the T. When designated (in conjunction with a TAA) this waypoint will be used as an IAWP when approaching the airport from certain directions, and as an IFWP when beginning the approach from another IAWP.

IFWP Intermediate Fix Waypoint.

ILS (See INSTRUMENT LANDING SYSTEM.)

ILS CATEGORIES 1. ILS Category I. An ILS approach procedure which provides for approach to a height above touchdown of not less than 200 feet and with runway visual range of not less than 1,800 feet. 2. ILS Category II. An ILS approach procedure which provides for approach to a height above touchdown of not less than 100 feet and with runway visual range of not less than 1,200 feet. 3. ILS Category III:
  a. IIIA. An ILS approach procedure which provides for approach without a decision height minimum and with runway visual range of not less than 700 feet.
  b. IIIB. An ILS approach procedure which provides for approach without a decision height minimum and with runway visual range of not less than 150 feet.
  c. IIIC. An ILS approach procedure which provides for approach without a decision height minimum and without runway visual range minimum.

ILS PRM APPROACH An instrument landing system (ILS) approach conducted to parallel runways whose extended centerlines are separated by less than 4,300 feet and the parallel runways have a Precision Runway Monitoring (PRM) system that permits simultaneous independent approaches.

IM (See INNER MARKER.)

IMC (See INSTRUMENT METEOROLOGICAL CONDITIONS.)

IMMEDIATELY Used by ATC or pilots when such action compliance is required to avoid an imminent situation.

INCERFA (Uncertainty Phase) [ICAO] A situation wherein uncertainty exists as to the safety of an aircraft and its occupants.

INCREASE SPEED TO (SPEED) (See SPEED ADJUSTMENT.)

INERTIAL NAVIGATION SYSTEM An RNAV system which is a form of self-contained navigation.
(See Area Navigation/RNAV.)

INFLIGHT REFUELING (See AERIAL REFUELING.)

INFLIGHT WEATHER ADVISORY (See WEATHER ADVISORY.)

INFORMATION REQUEST A request originated by an FSS for information concerning an overdue VFR aircraft.

INITIAL APPROACH FIX The fixes depicted on instrument approach procedure charts that identify the beginning of the initial approach segment(s).
(See FIX.) (See SEGMENTS OF AN INSTRUMENT APPROACH PROCEDURE.)

INITIAL APPROACH SEGMENT (See SEGMENTS OF AN INSTRUMENT APPROACH PROCEDURE.)

INITIAL APPROACH SEGMENT [ICAO] That segment of an instrument approach procedure between the initial approach fix and the intermediate approach fix or, where applicable, the final approach fix or point.

INLAND NAVIGATION FACILITY A navigation aid on a North American Route at which the common route and/or the noncommon route begins or ends.

INNER MARKER A marker beacon used with an ILS (CAT II) precision approach located between the middle marker and the end of the ILS runway, transmitting a radiation pattern keyed at six dots per second and indicating to the pilot, both aurally and visually, that he is at the designated decision height (DH), normally 100 feet above the touchdown zone elevation, on the ILS CAT II approach. It also marks progress during a CAT III approach.
(See INSTRUMENT LANDING SYSTEM.) (Refer to AIM.)

INNER MARKER BEACON (See INNER MARKER.)

INREQ (See INFORMATION REQUEST.)

INS (See INERTIAL NAVIGATION SYSTEM.)

INSTRUMENT APPROACH (See INSTRUMENT APPROACH PROCEDURE.)

INSTRUMENT APPROACH PROCEDURE A series of predetermined maneuvers for the orderly transfer of an aircraft under instrument flight conditions from the beginning of the initial approach to a landing or to a point from which a landing may be made visually. It is prescribed and approved for a specific airport by competent authority.
(See SEGMENTS OF AN INSTRUMENT APPROACH PROCEDURE.) (Refer to FAR Part 91.) (See AIM.)
    a. U.S. civil standard instrument approach procedures are approved by the FAA as prescribed under Part 97 and are available for public use.

    b. U.S. military standard instrument approach procedures are approved and published by the Department of Defense.
    c. Special instrument approach procedures are approved by the FAA for individual operators but are not published in Part 97 for public use.
(See ICAO term INSTRUMENT APPROACH PROCEDURE.)

INSTRUMENT APPROACH PROCEDURE [ICAO] A series of predetermined manoeuvres by reference to flight instruments with specified protection from obstacles from the initial approach fix, or where applicable, from the beginning of a defined arrival route to a point from which a landing can be completed and thereafter, if a landing is not completed, to a position at which holding or en route obstacle clearance criteria apply.

INSTRUMENT DEPARTURE PROCEDURE (DP) A preplanned instrument flight rule (IFR) departure procedure published for pilot use, in graphic or textual format, that provides obstruction clearance from the terminal area to the appropriate en route structure. There are two types of DP, Obstacle Departure Procedure (ODP), printed either textually or graphically, and, Standard Instrument Departure (SID), which is always printed graphically.
(See IFR TAKEOFF MINIMUMS AND DEPARTURE PROCEDURES.)
(See OBSTACLE DEPARTURE PROCEDURES.) (See STANDARD INSTRUMENT DEPARTURES.)
(Refer to AIM.)

INSTRUMENT APPROACH PROCEDURE (DP) CHARTS
(See AERONAUTICAL CHART.)

INSTRUMENT FLIGHT RULES Rules governing the procedures for conducting instrument flight. Also a term used by pilots and controllers to indicate type of flight plan.
(See VISUAL FLIGHT RULES.) (See INSTRUMENT METEOROLOGICAL CONDITIONS.) (See VISUAL METEOROLOGICAL CONDITIONS.) (Refer to AIM.) (See ICAO term INSTRUMENT FLIGHT RULES.)

INSTRUMENT FLIGHT RULES [ICAO] A set of rules governing the conduct of flight under instrument meteorological conditions.

INSTRUMENT LANDING SYSTEM A precision instrument approach system which normally consists of the following electronic components and visual aids:
    a. Localizer. *(See LOCALIZER.)*
    b. Glideslope. *(See GLIDESLOPE.)*
    c. Outer Marker. *(See OUTER MARKER.)*
    d. Middle Marker. *(See MIDDLE MARKER.)*
    e. Approach Lights. *(See AIRPORT LIGHTING.) (Refer to FAR Part 91.) (See AIM.)*

INSTRUMENT METEOROLOGICAL CONDITIONS Meteorological conditions expressed in terms of visibility, distance from cloud, and ceiling less than the minima specified for visual meteorological conditions.

(See VISUAL METEOROLOGICAL CONDITIONS.) (See INSTRUMENT FLIGHT RULES.) (See VISUAL FLIGHT RULES.)

INSTRUMENT RUNWAY A runway equipped with electronic and visual navigation aids for which a precision or nonprecision approach procedure having straight-in landing minimums has been approved.

(See ICAO term INSTRUMENT RUNWAY.)

INSTRUMENT RUNWAY [ICAO] One of the following types of runways intended for the operation of aircraft using instrument approach procedures:

a. Nonprecision Approach Runway An instrument runway served by visual aids and a nonvisual aid providing at least directional guidance adequate for a straight-in approach.

b. Precision Approach Runway, Category I An instrument runway served by ILS and visual aids intended for operations down to 60 m (200 feet) decision height and down to an RVR of the order of 800 m.

c. Precision Approach Runway, Category II An instrument runway served by ILS and visual aids intended for operations down to 30 m (100 feet) decision height and down to an RVR of the order of 400 m.

d. Precision Approach Runway, Category III An instrument runway served by ILS to and along the surface of the runway and:

1. Intended for operations down to an RVR of the order of 200 m (no decision height being applicable) using visual aids during the final phase of landing;

2. Intended for operations down to an RVR of the order of 50 m (no decision height being applicable) using visual aids for taxiing;

3. Intended for operations without reliance on visual reference for landing or taxiing.

Note 1: See Annex 10 Volume 1, Part I Chapter 3, for related ILS specifications.

Note 2: Visual aids need not necessarily be matched to the scale of nonvisual aids provided. The criterion for the selection of visual aids is the conditions in which operations are intended to be conducted.

INTEGRITY The ability of a system to provide timely warnings to users when the system should not be used for navigation.

INTERMEDIATE APPROACH SEGMENT (See SEGMENTS OF AN INSTRUMENT APPROACH PROCEDURE.)

INTERMEDIATE APPROACH SEGMENT [ICAO] That segment of an instrument approach procedure between either the intermediate approach fix and the final approach fix or point, or between the end of a reversal, race

track or dead reckoning track procedure and the final approach fix or point, as appropriate.

INTERMEDIATE FIX The fix that identifies the beginning of the intermediate approach segment of an instrument approach procedure. The fix is not normally identified on the instrument approach chart as an intermediate fix (IF).

(See SEGMENTS OF AN INSTRUMENT APPROACH PROCEDURE.)

INTERMEDIATE LANDING On the rare occasion that this option is requested, it should be approved. The departure center, however, must advise the ATCSCC so that the appropriate delay is carried over and assigned at the intermediate airport. An intermediate landing airport within the arrival center will not be accepted without coordination with and the approval of the ATCSCC.

INTERNATIONAL AIRPORT Relating to international flight, it means:

a. An airport of entry which has been designated by the Secretary of Treasury or Commissioner of Customs as an international airport for customs service.

b. A landing rights airport at which specific permission to land must be obtained from customs authorities in advance of contemplated use.

c. Airports designated under the Convention on International Civil Aviation as an airport for use by international commercial air transport and/or international general aviation.

(Refer to AIRPORT/FACILITY DIRECTORY.) (Refer to AIM.) (See ICAO term INTERNATIONAL AIRPORT.)

INTERNATIONAL AIRPORT [ICAO] Any airport designated by the Contracting State in whose territory it is situated as an airport of entry and departure for international air traffic, where the formalities incident to customs, immigration, public health, animal and plant quarantine and similar procedures are carried out.

INTERNATIONAL CIVIL AVIATION ORGANIZATION [ICAO] A specialized agency of the United Nations whose objective is to develop the principles and techniques of international air navigation and to foster planning and development of international civil air transport.

a. Regions include:

1. African-Indian Ocean Region
2. Caribbean Region
3. European Region
4. Middle East/Asia Region
5. North American Region
6. North Atlantic Region
7. Pacific Region
8. South American Region

INTERNATIONAL FLIGHT INFORMATION MANUAL A publication designed primarily as a pilot's preflight planning guide for flights into foreign airspace and for flights returning to the U.S. from foreign locations.

INTERROGATOR The ground-based surveillance radar beacon transmitter-receiver, which normally scans in synchronism with a primary radar, transmitting discrete radio signals which repetitively request all transponders on the mode being used to reply. The replies received are mixed with the primary radar returns and displayed on the same plan position indicator (radarscope). Also, applied to the airborne element of the TACAN/DME system.
(See TRANSPONDER.) (Refer to AIM.)

INTERSECTING RUNWAYS Two or more runways which cross or meet within their lengths.
(See INTERSECTION.)

INTERSECTION
a. A point defined by any combination of courses, radials, or bearings of two or more navigational aids.
b. Used to describe the point where two runways, a runway and a taxiway, or two taxiways cross or meet.

INTERSECTION DEPARTURE A departure from any runway intersection except the end of the runway.
(See INTERSECTION.)

INTERSECTION TAKEOFF (See INTERSECTION DEPARTURE.)

IR (See IFR MILITARY TRAINING ROUTES.)

I SAY AGAIN The message will be repeated.

----------- **J** -----------

JAMMING Electronic or mechanical interference which may disrupt the display of aircraft on radar or the transmission / reception of radio communications / navigation.

JET BLAST Jet engine exhaust (thrust stream turbulence).
(See WAKE TURBULENCE.)

JET ROUTE A route designed to serve aircraft operations from 18,000 feet MSL up to and including flight level 450. The routes are referred to as "J" routes with numbering to identify the designated route; e.g., J105.
(See Class A airspace.) (Refer to FAR Part 71.)

JET STREAM A migrating stream of high-speed winds present at high altitudes.

JETTISONING OF EXTERNAL STORES Airborne release of external stores; e.g., tiptanks, ordnance.
(See FUEL DUMPING.) (Refer to FAR Part 91.)

JOINT USE RESTRICTED AREA (See RESTRICTED AREA.)

----------- **K** -----------

KNOWN TRAFFIC With respect to ATC clearances, means aircraft whose altitude, position, and intentions are known to ATC.

----------- **L** -----------

LAA (See LOCAL AIRPORT ADVISORY.)

LAAS (See LOW ALTITUDE ALERT SYSTEM.)

LAHSO An acronym for "Land and Hold Short Operation." These operations include landing and holding short of an intersecting runway, a taxiway, a predetermined point, or an approach/departure flightpath.

LAHSO-DRY. Land and hold short operations on runways that are dry.

LAHSO-WET. Land and hold short operations on runways that are wet (but not contaminated).

LAND AND HOLD SHORT OPERATIONS Operations which include simultaneous takeoffs and landings and/or simultaneous landings when a landing aircraft is able and is instructed by the controller to hold-short of the intersecting runway/taxiway or designated hold-short point. Pilots are expected to promptly inform the controller if the hold short clearance cannot be accepted.
(See PARALLEL RUNWAYS.) (Refer to AIM.)

LANDING AREA Any locality either on land, water, or structures, including airports/heliports and intermediate landing fields, which is used, or intended to be used, for the landing and takeoff of aircraft whether or not facilities are provided for the shelter, servicing, or for receiving or discharging passengers or cargo.
(See ICAO term LANDING AREA.)

LANDING AREA [ICAO] That part of a movement area intended for the landing or takeoff of aircraft.

LANDING DIRECTION INDICATOR A device which visually indicates the direction in which landings and takeoffs should be made.
(See TETRAHEDRON.) (Refer to AIM.)

LANDING DISTANCE AVAILABLE [ICAO] The length of runway which is declared available and suitable for the ground run of an aeroplane landing.

LANDING MINIMUMS The minimum visibility prescribed for landing a civil aircraft while using an instrument approach procedure. The minimum applies with other limitations set forth in FAR Part 91 with respect to the Minimum Descent Altitude (MDA) or Decision Height (DH) prescribed in the instrument approach procedures as follows:
a. Straight-in landing minimums. A statement of MDA and visibility, or DH and visibility, required for a straight-in landing on a specified runway, or

b. Circling minimums. A statement of MDA and visibility required for the circle-to-land maneuver.

Note: Descent below the established MDA or DH is not authorized during an approach unless the aircraft is in a position from which a normal approach to the runway of intended landing can be made and adequate visual reference to required visual cues is maintained.

(See STRAIGHT-IN LANDING.) (See CIRCLE-TO-LAND MANEUVER.) (See DECISION HEIGHT.) (See MINIMUM DESCENT ALTITUDE.) (See VISIBILITY.) (See INSTRUMENT APPROACH PROCEDURE.) (Refer to FAR Part 91.)

LANDING ROLL   The distance from the point of touchdown to the point where the aircraft can be brought to a stop or exit the runway.

LANDING SEQUENCE   The order in which aircraft are positioned for landing.

(See APPROACH SEQUENCE.)

LAST ASSIGNED ALTITUDE   The last altitude/flight level assigned by ATC and acknowledged by the pilot.

(See MAINTAIN.) (Refer to FAR Part 91.)

LATERAL NAVIGATION (LNAV)   A function of area navigation (RNAV) equipment which calculates, displays, and provides lateral guidance to a profile or path.

LATERAL SEPARATION   The lateral spacing of aircraft at the same altitude by requiring operation on different routes or in different geographical locations.

(See SEPARATION.)

LDA   (See LOCALIZER TYPE DIRECTIONAL AID.)

LDA [ICAO]   (See ICAO Term LANDING DISTANCE AVAILABLE.)

LF   (See LOW FREQUENCY.)

LIGHTED AIRPORT   An airport where runway and obstruction lighting is available.

(See AIRPORT LIGHTING.) (Refer to AIM.)

LIGHT GUN   A handheld directional light signaling device which emits a brilliant narrow beam of white, green, or red light as selected by the tower controller. The color and type of light transmitted can be used to approve or disapprove anticipated pilot actions where radio communication is not available. The light gun is used for controlling traffic operating in the vicinity of the airport and on the airport movement area.

(Refer to AIM.)

LOCALIZER   The component of an ILS which provides course guidance to the runway.

(See INSTRUMENT LANDING SYSTEM.) (Refer to AIM.) (See ICAO term LOCALIZER COURSE.)

LOCALIZER COURSE [ICAO]   The locus of points, in any given horizontal plane, at which the DDM (difference in depth of modulation) is zero.

LOCALIZER OFFSET   An angular offset of the localizer from the runway extended centerline in a direction away from the no transgression zone (NTZ) that increases the normal operating zone (NOZ) width. An offset requires a 50 foot increase in DH and is not authorized for CAT II and CAT III approaches.

LOCALIZER TYPE DIRECTIONAL AID   A NAVAID used for nonprecision instrument approaches with utility and accuracy comparable to a localizer but which is not a part of a complete ILS and is not aligned with the runway.

(Refer to AIM.)

LOCALIZER USABLE DISTANCE   The maximum distance from the localizer transmitter at a specified altitude, as verified by flight inspection, at which reliable course information is continuously received.

(Refer to AIM.)

LOCAL AIRPORT ADVISORY [LAA]   A service provided by facilities, which are located on the landing airport, have a discrete ground-to-air communication frequency or the tower frequency when the tower is closed, automated weather reporting with voice broadcasting, and a continuous ASOS/AWOS data display, other continuous direct reading instruments, or manual observations available to the specialist.

(See AIRPORT ADVISORY AREA.)

LOCAL TRAFFIC   Aircraft operating in the traffic pattern or within sight of the tower, or aircraft known to be departing or arriving from flight in local practice areas, or aircraft executing practice instrument approaches at the airport.

(See TRAFFIC PATTERN.)

LOCATOR [ICAO]   An LM/MF NDB used as an aid to final approach.

Note: A locator usually has an average radius of rated coverage of between 18.5 and 46.3 km (10 and 25 NM).

LONGITUDINAL SEPARATION   The longitudinal spacing of aircraft at the same altitude by a minimum distance expressed in units of time or miles.

(See SEPARATION.) (Refer to AIM.)

LONG RANGE NAVIGATION   (See LORAN.)

LORAN   An electronic navigational system by which hyperbolic lines of position are determined by measuring the difference in the time of reception of synchronized pulse signals from two fixed transmitters. Loran A operates in the 1750–1950 kHz frequency band. Loran C and D operate in the 100–110 kHz frequency band.

(Refer to AIM.)

LOST COMMUNICATIONS   Loss of the ability to communicate by radio. Aircraft are sometimes referred to as NORDO (No Radio). Standard pilot procedures are specified in Part 91. Radar controllers issue procedures for pilots to follow in the event of lost communications during a radar approach when weather reports indicate that an aircraft will likely encounter IFR weather conditions during the approach.

(Refer to FAR Part 91.) (See AIM.)

LOW ALTITUDE AIRWAY STRUCTURE The network of airways serving aircraft operations up to but not including 18,000 feet MSL.

(See AIRWAY.) (Refer to AIM.)

LOW ALTITUDE ALERT, CHECK YOUR ALTITUDE IMMEDIATELY (See SAFETY ALERT.)

LOW ALTITUDE ALERT SYSTEM An automated function of the TPX-42 that alerts the controller when a Mode C transponder-equipped aircraft on an IFR flight plan is below a predetermined minimum safe altitude. If requested by the pilot, LAAS monitoring is also available to VFR Mode C transponder-equipped aircraft.

LOW APPROACH An approach over an airport or runway following an instrument approach or a VFR approach including the go-around maneuver where the pilot intentionally does not make contact with the runway.

(Refer to AIM.)

LOW FREQUENCY The frequency band between 30 and 300 kHz.

(Refer to AIM.)

LPV A type of approach with vertical guidance (APV) based on WAAS, published on RNAV (GPS) approach charts. This procedure takes advantage of the precise lateral guidance available from WAAS. The minima is published as a decision altitude (DA).

———————— **M** ————————

M-EARTS (See MICRO-EN ROUTE AUTOMATED RADAR TRACKING SYSTEM.)

MAA (See MAXIMUM AUTHORIZED ALTITUDE.)

MACH NUMBER The ratio of true airspeed to the speed of sound; e.g., MACH .82, MACH 1.6.

(See AIRSPEED.)

MACH TECHNIQUE [ICAO] Describes a control technique used by air traffic control whereby turbojet aircraft operating successively along suitable routes are cleared to maintain appropriate MACH numbers for a relevant portion of the en route phase of flight. The principle objective is to achieve improved utilization of the airspace and to ensure that separation between successive aircraft does not decrease below the established minima.

MAHWP Missed Approach Holding Waypoint.

MAINTAIN

a. Concerning altitude/flight level, the term means to remain at the altitude/flight level specified. The phrase "climb and" or "descend and" normally precedes "maintain" and the altitude assignment; e.g., "descend and maintain 5,000."

b. Concerning other ATC instructions, the term is used in its literal sense; e.g., maintain VFR.

MAINTENANCE PLANNING FRICTION LEVEL The friction level specified in AC 150/5320-12, Measurement, Construction, and Maintenance of Skid Resistant Airport Pavement Surfaces, which represents the friction value below which the runway pavement surface remains acceptable for any category or class of aircraft operations but which is beginning to show signs of deterioration. This value will vary depending on the particular friction measurement equipment used.

MAKE SHORT APPROACH Used by ATC to inform a pilot to alter his traffic pattern so as to make a short final approach.

(See TRAFFIC PATTERN.)

MANDATORY ALTITUDE An altitude depicted on an instrument Approach Procedure Chart requiring the aircraft to maintain altitude at the depicted value.

MAN PORTABLE AIR DEFENSE SYSTEMS (MANPADS) MANPADS are lightweight, shoulder-launched, missile systems used to bring down aircraft and create mass casualties. The potential for MANPADS use against airborne aircraft is real and requires familiarity with the subject. Terrorists choose MANPADS because the weapons are low cost, highly mobile, require minimal set-up time, and are easy to use and maintain. Although the weapons have limited range, and their accuracy is affected by poor visibility and adverse weather, they can be fired from anywhere on land or from boats where there is unrestricted visibility to the target.

MANPADS (See MAN PORTABLE AIR DEFENSE SYSTEMS.)

MAP (See MISSED APPROACH POINT.)

MARKER BEACON An electronic navigation facility transmitting a 75 MHz vertical fan or boneshaped radiation pattern. Marker beacons are identified by their modulation frequency and keying code, and when received by compatible airborne equipment, indicate to the pilot, both aurally and visually, that he is passing over the facility.

(See OUTER MARKER.) See MIDDLE MARKER.) (See INNER MARKER.) (Refer to AIM.)

MARSA (See MILITARY AUTHORITY ASSUMES RESPONSIBILITY FOR SEPARATION OF AIRCRAFT.)

MAXIMUM AUTHORIZED ALTITUDE A published altitude representing the maximum usable altitude or flight level for an airspace structure or route segment. It is the highest altitude on a Federal airway, jet route, area navigation low or high route, or other direct route for which an MEA is designated in Part 95 at which adequate reception of navigation aid signals is assured.

MAWP Missed Approach Waypoint.

MAYDAY The international radiotelephony distress signal. When repeated three times, it indicates imminent and grave danger and that immediate assistance is requested.
(See PAN-PAN-PAN.) (Refer to AIM.)

MCA (See MINIMUM CROSSING ALTITUDE.)

MDA (See MINIMUM DESCENT ALTITUDE.)

MEA (See MINIMUM EN ROUTE IFR ALTITUDE.)

MEARTS (See MICRO-EN ROUTE AUTOMATED RADAR TRACKING SYSTEM.)

METEOROLOGICAL IMPACT STATEMENT An unscheduled planning forecast describing conditions expected to begin within 4 to 12 hours which may impact the flow of air traffic in a specific center's (ARTCC) area.

METER FIX ARC A semicircle, equidistant from a meter *fix*, usually in low altitude relatively close to the meter fix, used to help CTAS/HOST calculate a meter time, and determine appropriate sector meter list assignments for aircraft not on an established arrival route or assigned a meter fix.

METER FIX TIME/SLOT TIME A calculated time to depart the meter fix in order to cross the vertex at the ACLT. This time reflects descent speed adjustment and any applicable time that must be absorbed prior to crossing the meter fix.

METER LIST (See ARRIVAL SECTOR ADVISORY LIST.)

METER LIST DISPLAY INTERVAL A dynamic parameter which controls the number of minutes prior to the flight plan calculated time of arrival at the meter fix for each aircraft, at which time the TCLT is frozen and becomes an ACLT, i.e., the VTA is updated and consequently the TCLT modified as appropriate until frozen at which time updating is suspended and an ACLT is assigned. When frozen, the flight entry is inserted into the arrival sector's meter list for display on the sector PVD/MDM. MLDI is used if filed true airspeed is less than or equal to freeze speed parameters (FSPD).

METERING A method of time-regulating arrival traffic flow into a terminal area so as not to exceed a predetermined terminal acceptance rate.

METERING AIRPORTS Airports adapted for metering and for which optimum flight paths are defined. A maximum of 15 airports may be adapted.

METERING FIX A fix along an established route from over which aircraft will be metered prior to entering terminal airspace. Normally, this fix should be established at a distance from the airport which will facilitate a profile descent 10,000 feet above airport elevation [AAE] or above.

METERING POSITION(S) Adapted PVDs/MDM's and associated "D" positions eligible for display of a metering position list. A maximum of four PVDs/MDM's may be adapted.

METERING POSITION LIST An ordered list of data on arrivals for a selected metering airport displayed on a metering position PVD/MDM.

MFT (See METER FIX TIME/SLOT TIME.)

MHA (See MINIMUM HOLDING ALTITUDE.)

MIA (See MINIMUM IFR ALTITUDES.)

MICROBURST A small downburst with outbursts of damaging winds extending 2.5 miles or less. In spite of its small horizontal scale, an intense microburst could induce wind speeds as high as 150 knots.
(Refer to AIM.)

MICRO-EN ROUTE AUTOMATED RADAR TRACKING SYSTEM (MEARTS) An automated radar and radar beacon tracking system capable of employing both short-range (ASR) and long-range (ARSR) radars. This microcomputer driven system provides improved tracking, continuous data recording, and use of full digital radar displays.

MICROWAVE LANDING SYSTEM A precision instrument approach system operating in the microwave spectrum which normally consists of the following components:
a. Azimuth Station.
b. Elevation Station.
c. Precision Distance Measuring Equipment.
(See MLS CATEGORIES.)

MIDDLE COMPASS LOCATOR (See COMPASS LOCATOR.)

MIDDLE MARKER A marker beacon that defines a point along the glideslope of an ILS normally located at or near the point of decision height (ILS Category I). It is keyed to transmit alternate dots and dashes, with the alternate dots and dashes keyed at the rate of 95 dot/dash combinations per minute on a 1300 Hz tone, which is received aurally and visually by compatible airborne equipment.
(See MARKER BEACON.) (See INSTRUMENT LANDING SYSTEM.) (Refer to AIM.)

MID RVR (See VISIBILITY.)

MILES-IN-TRAIL A specified distance between aircraft, normally, in the same stratum associated with the same destination or route of flight.

MILITARY AUTHORITY ASSUMES RESPONSIBILITY FOR SEPARATION OF AIRCRAFT A condition whereby the military services involved assume responsibility for separation between participating military aircraft in the ATC system. It is used only for required IFR operations which are specified in letters of agreement or other appropriate FAA or military documents.

MILITARY LANDING ZONE A landing strip used exclusively for military training. A military landing zone does not carry a runway designation.

MILITARY OPERATIONS AREA (See SPECIAL USE AIRSPACE.)

MILITARY TRAINING ROUTES Airspace of defined vertical and lateral dimensions established for the conduct of military flight training at airspeeds in excess of 250 knots IAS.
(See IFR MILITARY TRAINING ROUTES.) (See VFR MILITARY TRAINING ROUTES.)

MINIMA (See MINIMUMS.)

MINIMUM CROSSING ALTITUDE The lowest altitude at certain fixes at which an aircraft must cross when proceeding in the direction of a higher minimum en route IFR altitude (MEA).
(See MINIMUM EN ROUTE IFR ALTITUDE.)

MINIMUM DESCENT ALTITUDE The lowest altitude, expressed in feet above mean sea level, to which descent is authorized on final approach or during circle-to-land maneuvering in execution of a standard instrument approach procedure where no electronic glideslope is provided.
(See NONPRECISION APPROACH PROCEDURE.)

MINIMUM EN ROUTE IFR ALTITUDE The lowest published altitude between radio fixes which assures acceptable navigational signal coverage and meets obstacle clearance requirements between those fixes. The MEA prescribed for a Federal airway or segment thereof, area navigation low or high route, or other direct route applies to the entire width of the airway, segment, or route between the radio fixes defining the airway, segment, or route.
(Refer to Part 91.) (Refer to Part 95.) (Refer to AIM.)

MINIMUM FRICTION LEVEL The friction level specified in AC 150/5320-12, Measurement, Construction, and Maintenance of Skid Resistant Airport Pavement Surfaces, that represents the minimum recommended wet pavement surface friction value for any turbojet aircraft engaged in LAHSO. This value will vary with the particular friction measurement equipment used.

MINIMUM FUEL Indicates that an aircraft's fuel supply has reached a state where, upon reaching the destination, it can accept little or no delay. This is not an emergency situation but merely indicates an emergency situation is possible should any undue delay occur.
(Refer to AIM.)

MINIMUM HOLDING ALTITUDE The lowest altitude prescribed for a holding pattern which assures navigational signal coverage, communications, and meets obstacle clearance requirements.

MINIMUM IFR ALTITUDES Minimum altitudes for IFR operations as prescribed in Part 91. These altitudes are published on aeronautical charts and prescribed in Part 95 for airways and routes, and in Part 97 for standard instrument approach procedures. If no applicable minimum altitude is prescribed in FAR 95 or FAR 97, the following minimum IFR altitude applies:

a. In designated mountainous areas, 2,000 feet above the highest obstacle within a horizontal distance of 4 nautical miles from the course to be flown; or

b. Other than mountainous areas, 1,000 feet above the highest obstacle within a horizontal distance of 4 nautical miles from the course to be flown; or

c. As otherwise authorized by the Administrator or assigned by ATC.
(See MINIMUM EN ROUTE IFR ALTITUDE.) (See MINIMUM OBSTRUCTION CLEARANCE ALTITUDE.) (See MINIMUM CROSSING ALTITUDE.) (See MINIMUM SAFE ALTITUDE.) (See MINIMUM VECTORING ALTITUDE.) (Refer to Part 91.)

MINIMUM NAVIGATION PERFORMANCE SPECIFICATION A set of standards which require aircraft to have a minimum navigation performance capability in order to operate in MNPS designated airspace. In addition, aircraft must be certified by their State of Registry for MNPS operation.

MINIMUM NAVIGATION PERFORMANCE SPECIFICATION AIRSPACE Designated airspace in which MNPS procedures are applied between MNPS certified and equipped aircraft. Under certain conditions, non-MNPS aircraft can operate in MNPSA. However, standard oceanic separation minima is provided between the non-MNPS aircraft and other traffic. Currently, the only designated MNPSA is described as follows:

a. Between FL 285 and FL 420;

b. Between latitudes 27°N and the North Pole;

c. In the east, the eastern boundaries of the CTA's Santa Maria Oceanic, Shanwick Oceanic, and Reykjavik;

d. In the west, the western boundaries of CTA's Reykjavik and Gander Oceanic, Shannon and New York Oceanic excluding the area west of 60°W and south of 38°–30' N.

MINIMUM OBSTRUCTION CLEARANCE ALTITUDE The lowest published altitude in effect between radio fixes on VOR airways, off-airway routes, or route segments which meets obstacle clearance requirements for the entire route segment and which assures acceptable navigational signal coverage only within 25 statute (22 nautical) miles of a VOR.

(Refer to Part 91.) (Refer to Part 95.)

MINIMUM RECEPTION ALTITUDE   The lowest altitude at which an intersection can be determined.

(Refer to Part 95.)

MINIMUM SAFE ALTITUDE

a. The minimum altitude specified in Part 91 for various aircraft operations.

b. Altitudes depicted on approach charts which provide at least 1,000 feet of obstacle clearance for emergency use within a specified distance from the navigation facility upon which a procedure is predicated. These altitudes will be identified as Minimum Sector Altitudes or Emergency Safe Altitudes and are established as follows:

1. Minimum Sector Altitudes. Altitudes depicted on approach charts which provide at least 1,000 feet of obstacle clearance within a 25-mile radius of the navigation facility upon which the procedure is predicated. Sectors depicted on approach charts must be at least 90 degrees in scope. These altitudes are for emergency use only and do not necessarily assure acceptable navigational signal coverage.

(See ICAO term Minimum Sector Altitude.)

2. Emergency Safe Altitudes. Altitudes depicted on approach charts which provide at least 1,000 feet of obstacle clearance in nonmountainous areas and 2,000 feet of obstacle clearance in designated mountainous areas within a 100-mile radius of the navigation facility upon which the procedure is predicated and normally used only in military procedures. These altitudes are identified on published procedures as "Emergency Safe Altitudes."

MINIMUM SAFE ALTITUDE WARNING   A function of the ARTS III computer that aids the controller by alerting him when a tracked Mode C-equipped aircraft is below or is predicted by the computer to go below a predetermined minimum safe altitude.

(Refer to AIM.)

MINIMUM SECTOR ALTITUDE [ICAO]   The lowest altitude which may be used under emergency conditions which will provide a minimum clearance of 300 m (1,000 feet) above all obstacles located in an area contained within a sector of a circle of 46 km (25 NM) radius centered on a radio aid to navigation.

MINIMUMS   Weather condition requirements established for a particular operation or type of operation; e.g., IFR takeoff or landing, alternate airport for IFR flight plans, VFR flight, etc.

(See LANDING MINIMUMS.) (See IFR TAKEOFF MINIMUMS AND DEPARTURE PROCEDURES.) (See VFR CONDITIONS.) (See IFR CONDITIONS.) (Refer to Part 91.) (Refer to AIM.)

MINIMUM VECTORING ALTITUDE   The lowest MSL altitude at which an IFR aircraft will be vectored by a radar controller, except as otherwise authorized for radar approaches, departures, and missed approaches. The altitude meets IFR obstacle clearance criteria. It may be lower than the published MEA along an airway or J-route segment. It may be utilized for radar vectoring only upon the controller's determination that an adequate radar return is being received from the aircraft being controlled. Charts depicting minimum vectoring altitudes are normally available only to the controllers and not to pilots.

(Refer to AIM.)

MINUTES-IN-TRAIL   A specified interval between aircraft expressed in time. This method would more likely be utilized regardless of altitude.

MIS   (See METEOROLOGICAL IMPACT STATEMENT.)

MISSED APPROACH

a. A maneuver conducted by a pilot when an instrument approach cannot be completed to a landing. The route of flight and altitude are shown on instrument approach procedure charts. A pilot executing a missed approach prior to the Missed Approach Point (MAP) must continue along the final approach to the MAP.

b. A term used by the pilot to inform ATC that he is executing the missed approach.

c. At locations where ATC radar service is provided, the pilot should conform to radar vectors when provided by ATC in lieu of the published missed approach procedure.

(See Missed Approach Point) (Refer to AIM.)

MISSED APPROACH POINT   A point prescribed in each instrument approach procedure at which a missed approach procedure shall be executed if the required visual reference does not exist.

(See MISSED APPROACH.) (See SEGMENTS OF AN INSTRUMENT APPROACH PROCEDURE.)

MISSED APPROACH PROCEDURE [ICAO]   The procedure to be followed if the approach cannot be continued.

MISSED APPROACH SEGMENT   (See SEGMENTS OF AN INSTRUMENT APPROACH PROCEDURE.)

MLDI   (See METER LIST DISPLAY INTERVAL.)

MLS   (See MICROWAVE LANDING SYSTEM.)

MLS CATEGORIES

a. MLS Category I. An MLS approach procedure which provides for an approach to a height above touchdown of not less than 200 feet and a runway visual range of not less than 1,800 feet.

b. MLS Category II. Undefined until data gathering/analysis completion.

c. MLS Category III. Undefined until data gathering/analysis completion.

MM   (See MIDDLE MARKER.)

MNPS (See MINIMUM PERFORMANCE SPECIFICATION.)

MNPSA (See MINIMUM PERFORMANCE SPECIFICATIONS AIRSPACE.)

MOA (See MILITARY OPERATIONS AREA.)

MOCA (See MINIMUM OBSTRUCTION CLEARANCE ALTITUDE.)

MODE The letter or number assigned to a specific pulse spacing of radio signals transmitted or received by ground interrogator or airborne transponder components of the Air Traffic Control Radar Beacon System (ATCRBS). Mode A (military Mode 3) and Mode C (altitude reporting) are used in air traffic control.
(See TRANSPONDER.) (See INTERROGATOR.) (See RADAR.) (Refer to AIM.) (See ICAO term MODE.)

MODE (SSR MODE) [ICAO] The letter or number assigned to a specific pulse spacing of the interrogation signals transmitted by an interrogator. There are 4 modes, A, B, C and D specified in Annex 10, corresponding to four different interrogation pulse spacings.

MODE C INTRUDER ALERT A function of certain air traffic control automated systems designed to alert radar controllers to existing or pending situations between a tracked target (known IFR or VFR aircraft) and an untracked target (unknown IFR or VFR aircraft) that requires immediate attention/action.
(See CONFLICT ALERT.)

MONITOR (When used with communication transfer) listen on a specific frequency and stand by for instructions. Under normal circumstances do not establish communications.

MONITOR ALERT (MA) A function of the ETMS that provides traffic management personnel with a tool for predicting potential capacity problems in individual operational sectors. The MA is an indication that traffic management personnel need to analyze a particular sector for actual activity and to determine the required action(s), if any, needed to control the demand.

MONITOR ALERT PARAMETER (MAP) The number designated for use in monitor alert processing by the ETMS. The MAP is designated for each operational sector for increments of 15 minutes.

MOVEMENT AREA The runways, taxiways, and other areas of an airport/heliport which are utilized for taxiing/hover taxiing, air taxiing, takeoff, and landing of aircraft, exclusive of loading ramps and parking areas. At those airports/heliports with a tower, specific, approval for entry onto the movement area must be obtained from ATC.
(See ICAO term MOVEMENT AREA.)

MOVEMENT AREA [ICAO] That part of an aerodrome to be used for the takeoff, landing and taxiing of aircraft, consisting of the manoeuvring area and the apron(s).

MOVING TARGET INDICATOR An electronic device which will permit radar scope presentation only from targets which are in motion. A partial remedy for ground clutter.

MRA (See MINIMUM RECEPTION ALTITUDE.)

MSA (See MINIMUM SAFE ALTITUDE.)

MSAW (See MINIMUM SAFE ALTITUDE WARNING.)

MTI (See MOVING TARGET INDICATOR.)

MTR (See MILITARY TRAINING ROUTES.)

MULTICOM A mobile service not open to public correspondence used to provide communications essential to conduct the activities being performed by or directed from private aircraft.

MULTIPLE RUNWAYS The utilization of a dedicated arrival runway(s) for departures and a dedicated departure runway(s) for arrivals when feasible to reduce delays and enhance capacity.

MVA (See MINIMUM VECTORING ALTITUDE.)

--------- N ---------

NAS (See NATIONAL AIRSPACE SYSTEM.)

NAS STAGE A The en route ATC system's radar, computers and computer programs, controller plan view displays (PVDs/Radarscopes), input/output devices, and the related communications equipment which are integrated to form the heart of the automated IFR air traffic control system. This equipment performs Flight Data Processing (FDP) and Radar Data Processing (RDP). It interfaces with automated terminal systems and is used in the control of en route IFR aircraft.
(Refer to AIM).

NATIONAL AIRSPACE SYSTEM The common network of U.S. airspace; air navigation facilities, equipment and services, airports or landing areas; aeronautical charts, information and services; rules, regulations and procedures, technical information, and manpower and material. Included are system components shared jointly with the military.

NATIONAL BEACON CODE ALLOCATION PLAN AIRSPACE Airspace over United States territory located within the North American continent between Canada and Mexico, including adjacent territorial waters outward to about boundaries of oceanic control areas (CTA)/Flight Information Regions (FIR).
(See FLIGHT INFORMATION REGION.)

NATIONAL FLIGHT DATA CENTER A facility in Washington D.C., established by FAA to operate a central aeronautical information service for the collection, validation, and dissemination of aeronautical data in support of the activities of government, industry, and the aviation community. The information is published in the National Flight Data Digest.
(See NATIONAL FLIGHT DATA DIGEST.)

NATIONAL FLIGHT DATA DIGEST A daily (except weekends and Federal holidays) publication of flight information appropriate to aeronautical charts, aeronautical publications, Notices to Airmen, or other media serving the purpose of providing operational flight data essential to safe and efficient aircraft operations.

NATIONAL ROUTE PROGRAM (NRP) The NRP is a set of rules and procedures which are designed to increase the flexibility of user flight planning within published guidelines.

NATIONAL SEARCH AND RESCUE PLAN An interagency agreement which provides for the effective utilization of all available facilities in all types of search and rescue missions.

NAVAID (See NAVIGATIONAL AID.)

NAVAID CLASSES VOR, VORTAC, and TACAN aids are classed according to their operational use. The three classes of NAVAIDs are:
   a. T—Terminal.
   b. L—Low altitude.
   c. H—High altitude.

Note: The normal service range for T, L, and H class aids is found in the AIM. Certain operational requirements make it necessary to use some of these aids at greater service ranges than specified. Extended range is made possible through flight inspection determinations. Some aids also have lesser service range due to location, terrain, frequency protection, etc. Restrictions to service range are listed in Airport/Facility Directory.

NAVIGABLE AIRSPACE Airspace at and above the minimum flight altitudes prescribed in the FARs including airspace needed for safe takeoff and landing.
(Refer to FAR Part 91.)

NAVIGATIONAL AID Any visual or electronic device airborne or on the surface which provides point-to-point guidance information or position data to aircraft in flight.
(See AIR NAVIGATION FACILITY.)

NAVIGATION REFERENCE SYSTEM (NRS) The NRS is a system of waypoints developed for use within the United States for flight planning and navigation without reference to ground based navigational aids. The NRS waypoints are located in a grid pattern along defined latitude and longitude lines. The initial use of the NRS will be in the high altitude environment in conjunction with the High Altitude Redesign initiative. The NRS waypoints are intended for use by aircraft capable of point-to-point navigation.

NBCAP AIRSPACE (See NATIONAL BEACON CODE ALLOCATION PLAN AIRSPACE.)

NDB (See NONDIRECTIONAL BEACON.)

NEGATIVE "No," or "permission not granted," or "that is not correct."

NEGATIVE CONTACT Used by pilots to inform ATC that:
   a. Previously issued traffic is not in sight. It may be followed by the pilot's request for the controller to provide assistance in avoiding the traffic.
   b. They were unable to contact ATC on a particular frequency.

NFDC (See NATIONAL FLIGHT DATA CENTER.)

NFDD (See NATIONAL FLIGHT DATA DIGEST.)

NIGHT The time between the end of evening civil twilight and the beginning of morning civil twilight, as published in the American Air Almanac, converted to local time.
(See ICAO term NIGHT.)

NIGHT [ICAO] The hours between the end of evening civil twilight and the beginning of morning civil twilight or such other period between sunset and sunrise as may be specified by the appropriate authority.

Note: Civil twilight ends in the evening when the centre of the sun's disk is 6 degrees below the horizon and begins in the morning when the centre of the sun's disk is 6 degrees below the horizon.

NO GYRO APPROACH A radar approach/vector provided in case of a malfunctioning gyro-compass or directional gyro. Instead of providing the pilot with headings to be flown, the controller observes the radar track and issues control instructions "turn right/left" or "stop turn" as appropriate.
(Refer to AIM.)

NO GYRO VECTOR (See NO GYRO APPROACH.)

NO TRANSGRESSION ZONE (NTZ) The NTZ is a 2,000 foot wide zone, located equidistant between parallel runway final approach courses in which flight is not allowed.

NONAPPROACH CONTROL TOWER Authorizes aircraft to land or takeoff at the airport controlled by the tower or to transit the Class D airspace. The primary function of a nonapproach control tower is the sequencing of aircraft in the traffic pattern and on the landing area. Nonapproach control towers also separate aircraft operating under instrument flight rules clearances from approach controls and centers. They provide ground control services to aircraft, vehicles, personnel, and equipment on the airport movement area.

NONCOMMON ROUTE/PORTION   That segment of a North American Route between the inland navigation facility and a designated North American terminal.

NONCOMPOSITE SEPARATION   Separation in accordance with minima other than the composite separation minimum specified for the area concerned.

NONDIRECTIONAL BEACON   An L/MF or UHF radio beacon transmitting nondirectional signals whereby the pilot of an aircraft equipped with direction finding equipment can determine his bearing to or from the radio beacon and "home" on or track to or from the station. When the radio beacon is installed in conjunction with the Instrument Landing System marker, it is normally called a Compass Locator.

(See COMPASS LOCATOR.) (See AUTOMATIC DIRECTION FINDER.)

NONMOVEMENT AREAS   Taxiways and apron (ramp) areas not under the control of air traffic.

NONPRECISION APPROACH   (See NONPRECISION APPROACH PROCEDURE.)

NONPRECISION APPROACH PROCEDURE   A standard instrument approach procedure in which no electronic glideslope is provided; e.g., VOR, TACAN, NDB, LOC, ASR, LDA, or SDF approaches.

NONRADAR   Precedes other terms and generally means without the use of radar, such as:

a. Nonradar Approach. Used to describe instrument approaches for which course guidance on final approach is not provided by ground-based precision or surveillance radar. Radar vectors to the final approach course may or may not be provided by ATC. Examples of nonradar approaches are VOR, NDB, TACAN, and ILS/MLS approaches.

(See FINAL APPROACH—IFR=FINAL APPROACH—IFR) (See FINAL APPROACH COURSE.) (See RADAR APPROACH.) (See INSTRUMENT APPROACH PROCEDURE.)

b. Nonradar Approach Control. An ATC facility providing approach control service without the use of radar.

(See APPROACH CONTROL FACILITY.) (See APPROACH CONTROL SERVICE.)

c. Nonradar Arrival. An aircraft arriving at an airport without radar service or at an airport served by a radar facility and radar contact has not been established or has been terminated due to a lack of radar service to the airport.

(See RADAR ARRIVAL.) (See RADAR SERVICE.)

d. Nonradar Route. A flight path or route over which the pilot is performing his own navigation. The pilot may be receiving radar separation, radar monitoring, or other ATC services while on a nonradar route.

(See RADAR ROUTE.)

e. Nonradar Separation. The spacing of aircraft in accordance with established minima without the use of radar; e.g., vertical, lateral, or longitudinal separation.

(See RADAR SEPARATION.) (See ICAO term NONRADAR SEPARATION.)

NONRADAR SEPARATION [ICAO]   The separation used when aircraft position information is derived from sources other than radar.

NON-RESTRICTIVE ROUTING (NRR)   Portions of a proposed route of flight where a user can flight plan the most advantageous flight path with no requirement to make reference to ground-based NAVAIDs.

NOPAC   (See NORTH PACIFIC.)

NORDO   (See LOST COMMUNICATIONS.)

NORMAL OPERATING ZONE (NOZ)   The NOZ is the operating zone within which aircraft flight remains during normal independent simultaneous parallel ILS approaches.

NORTH AMERICAN ROUTE   A numerically coded route preplanned over existing airway and route systems to and from specific coastal fixes serving the North Atlantic. North American Routes consist of the following:

a. Common Route/Portion. That segment of a North American Route between the inland navigation facility and the coastal fix.

b. NonCommon Route/Portion. That segment of a North American Route between the inland navigation facility and a designated North American terminal.

c. Inland Navigation Facility. A navigation aid on a North American Route at which the common route and/or the noncommon route begins or ends.

d. Coastal Fix. A navigation aid or intersection where an aircraft transition between the domestic route structure and the oceanic route structure.

NORTH MARK   A beacon data block sent by the host computer to be displayed by the ARTS on a 360 degree bearing at a locally selected radar azimuth and distance. The North Mark is used to ensure correct range/azimuth orientation during periods of CENRAP.

NORTH PACIFIC   An organized route system between the Alaskan west coast and Japan.

NOTAM   (See NOTICE TO AIRMEN.)

NOTICE TO AIRMEN   A notice containing information (not known sufficiently in advance to publicize by other means) concerning the establishment, condition, or change in any component (facility, service, or procedure of, or hazard in the National Airspace System) the timely knowledge of which is essential to personnel concerned with flight operations.

a. NOTAM(D). A NOTAM given (in addition to local dissemination) distant dissemination beyond the area of responsibility of the Flight Service Station. These NOTAMs will be stored and available until canceled.

b. NOTAM(L). A NOTAM given local dissemination by voice and other means, such as telautograph and telephone, to satisfy local user requirements.

c. FDC NOTAM. A NOTAM regulatory in nature, transmitted by USNOF and given system wide dissemination.

(See ICAO term NOTAM.)

NOTAM [ICAO]   A notice containing information concerning the establishment, condition or change in any aeronautical facility, service, procedure or hazard, the timely knowledge of which is essential to personnel concerned with flight operations.

a. I Distribution—Distribution by means of telecommunication.

b. II Distribution—Distribution by means other than telecommunications.

NOTICES TO AIRMEN PUBLICATION   A publication issued every 28 days, designed primarily for the pilot, which contains current NOTAM information considered essential to the safety of flight as well as supplemental data to other aeronautical publications. The contraction NTAP is used in NOTAM text.

(See NOTICE TO AIRMEN.)

NRR   (See NON-RESTRICTIVE ROUTING.)

NRS   (See Navigation Reference System.)

NTAP   (See NOTICES TO AIRMEN PUBLICATION.)

NUMEROUS TARGETS VICINITY (LOCATION)   A traffic advisory issued by ATC to advise pilots that targets on the radarscope are too numerous to issue individually.

(See TRAFFIC ADVISORIES.)

O

OBSTACLE   An existing object, object of natural growth, or terrain at a fixed geographical location or which may be expected at a fixed location within a prescribed area with reference to which vertical clearance is or must be provided during flight operation.

OBSTACLE DEPARTURE PROCEDURE (ODP)   A preplanned instrument flight rule (IFR) departure procedure printed for pilot use in textual or graphic form to provide obstruction clearance via the least onerous route from the terminal area to the appropriate en route structure. ODPs are recommended for obstruction clearance and may be flown without ATC clearance unless an alternate departure procedure (SID or radar vector) has been specifically assigned by ATC.

(See IFR TAKEOFF MINIMUMS AND DEPARTURE PROCEDURES.)
(See STANDARD INSTRUMENT DEPARTURES.)
(Refer to AIM.)

OBSTACLE FREE ZONE   The OFZ is a three dimensional volume of airspace which protects for the transition of aircraft to and from the runway. The OFZ clearing standard precludes taxiing and parked airplanes and object penetrations, except for frangible NAVAID locations that are fixed by function. Additionally, vehicles, equipment, and personnel may be authorized by air traffic control to enter the area using the provisions of FAA Order 7110.65, Air Traffic Control, paragraph 3-1-5. The runway OFZ and when applicable, the inner-approach OFZ, and the inner-transitional OFZ, comprise the OFZ.

a. Runway OFZ. The runway OFZ is a defined volume of airspace centered above the runway. The runway OFZ is the airspace above a surface whose elevation at any point is the same as the elevation of the nearest point on the runway centerline. The runway OFZ extends 200 feet beyond each end of the runway. The width is as follows:

1. For runways serving large airplanes, the greater of:
(a) 400 feet, or
(b) 180 feet, plus the wingspan of the most demanding airplane, plus 20 feet per 1,000 feet of airport elevation.

2. For runways serving only small airplanes:
(a) 300 feet for precision instrument runways.
(b) 250 feet for other runways serving small airplanes with approach speeds of 50 knots, or more.
(c) 120 feet for other runways serving small airplanes with approach speeds of less than 50 knots.

b. Inner-approach OFZ. The inner-approach OFZ is a defined volume of airspace centered on the approach area. The inner-approach OFZ applies only to runways with an approach lighting system. The inner-approach OFZ begins 200 feet from the runway threshold at the same elevation as the runway threshold and extends 200 feet beyond the last light unit in the approach lighting system. The width of the inner-approach OFZ is the same as the runway OFZ and rises at a slope of 50 (horizontal) to 1 (vertical) from the beginning.

c. Inner-transitional OFZ. The inner transitional surface OFZ is a defined volume of airspace along the sides of the runway and inner-approach OFZ and applies only to precision instrument runways. The inner-transitional surface OFZ slopes 3 (horizontal) to 1 (vertical) out from the edges of the runway OFZ and inner-approach OFZ to a height of 150 feet above the established airport elevation.

(Refer to AC 150/5300-13, Chapter 3 and FAA Order 7110.65 paragraph 3-1-5.)

OBSTRUCTION Any object/obstacle exceeding the obstruction standards specified by FAR Part 77, Subpart C.

OBSTRUCTION LIGHT A light or one of a group of lights, usually red or white, frequently mounted on a surface structure or natural terrain to warn pilots of the presence of an obstruction.

OCEANIC AIRSPACE Airspace over the oceans of the world, considered international airspace, where oceanic separation and procedures per the International Civil Aviation Organization are applied. Responsibility for the provisions of air traffic control service in this airspace is delegated to various countries, based generally upon geographic proximity and the availability of the required resources.

OCEANIC DISPLAY AND PLANNING SYSTEM An automated digital display system which provides flight data processing, conflict probe, and situation display for oceanic air traffic control.

OCEANIC NAVIGATIONAL ERROR REPORT A report filed when an aircraft exiting oceanic airspace has been observed by radar to be off course. ONER reporting parameters and procedures are contained in FAA Order 7110.82, Monitoring of Navigational Performance In Oceanic Areas.

OCEANIC PUBLISHED ROUTE A route established in international airspace and charted or described in flight information publications, such as Route Charts, DOD En route Charts, Chart Supplements, NOTAMs, and Track Messages.

OCEANIC TRANSITION ROUTE An ATS route established for the purpose of transitioning aircraft to/from an organized track system.

ODAPS (See OCEANIC DISPLAY AND PLANNING SYSTEM.)

ODP (See OBSTACLE DEPARTURE PROCEDURE.)

OFF COURSE A term used to describe a situation where an aircraft has reported a position fix or is observed on radar at a point not on the ATC-approved route of flight.

OFFSHORE/CONTROL AIRSPACE AREA That portion of airspace between the U.S. 12 NM limit and the oceanic CTA/FIR boundary within which air traffic control is exercised. These areas are established to provide air traffic control services. Offshore/Control Airspace Areas may be classified as either Class A airspace or Class E airspace.

OFF-ROUTE VECTOR A vector by ATC which takes an aircraft off a previously assigned route. Altitudes assigned by ATC during such vectors provide required obstacle clearance.

OFFSET PARALLEL RUNWAYS Staggered runways having centerlines which are parallel.

OFFSHORE CONTROL AREA That portion of airspace between the U.S 12-mile limit and the oceanic CTA/FIR boundary within which air traffic control is exercised. These areas are established to permit the application of domestic procedures in the provision of air traffic control services. Offshore control area is generally synonymous with Federal Aviation Regulations, FAR Part 71, Subpart E, "Control Areas and Control Area Extensions."

OFT (See OUTER FIX TIME.)

OM (See OUTER MARKER.)

OMEGA An RNAV system designed for long-range navigation based upon ground-based electronic navigational aid signals.

ONE-MINUTE WEATHER The most recent one minute update weather broadcast received by a pilot from an uncrossed airport ASOS/AWOS.

ONER (See OCEANIC NAVIGATIONAL ERROR REPORT.)

OPERATIONAL (See DUE REGARD.)

ON COURSE

a. Used to indicate that an aircraft is established on the route centerline.

b. Used by ATC to advise a pilot making a radar approach that his aircraft is lined up on the final approach course.

(See ON-COURSE INDICATION.)

ON-COURSE INDICATION An indication on an instrument, which provides the pilot a visual means of determining that the aircraft is located on the centerline of a given navigational track, or an indication on a radar scope that an aircraft is on a given track.

OPERATIONAL ACCEPTABLE LEVEL OF TRAFFIC An air traffic activity level associated with the designed capacity for a sector or airport. The OALT considers dynamic changes in staffing, personnel experience levels, equipment outages, operational configurations, weather, traffic complexity, aircraft performance mixtures, transitioning flights, adjacent airspace, handoff/point-out responsibilities, and other factors that may affect an air traffic operational position or system element. The OALT is normally considered to be the total number of aircraft that any air traffic functional position can accommodate for a defined period of time under a given set of circumstances.

OPPOSITE DIRECTION AIRCRAFT Aircraft are operating in opposite directions when:

a. They are following the same track in reciprocal directions; or

b. Their tracks are parallel and the aircraft are flying in reciprocal directions; or,

c. Their tracks intersect at an angle of more than 135°.

OPTION APPROACH   An approach requested and conducted by a pilot which will result in either a touch-and-go, missed approach, low approach, stop-and-go, or full stop landing.
   (See CLEARED FOR THE OPTION.) (Refer to AIM.)

ORGANIZED TRACK SYSTEM   A movable system of oceanic tracks that traverses the North Atlantic between Europe and North America the physical position of which is determined twice daily taking the best advantage of the winds aloft.

ORGANIZED TRACK SYSTEM   A series of ATS routes which are fixed and charted; i.e., CEP, NOPAC, or flexible and described by NOTAM; i.e., NAT TRACK MESSAGE.

OROCA   An off-route altitude which provides obstruction clearance with a 1,000 foot buffer in nonmountainous terrain areas and a 2,000 foot buffer in designated mountainous areas within the United States. This altitude may not provide signal coverage from ground-based navigational aids, air traffic control radar, or communications coverage.

OTR   (See OCEANIC TRANSITION ROUTE.)

OTS   (See ORGANIZED TRACK SYSTEM.)

OUT   The conversation is ended and no response is expected.

OUTER AREA (associated with Class C airspace)   Nonregulatory airspace surrounding designated Class C airspace airports wherein ATC provides radar vectoring and sequencing on a full-time basis for all IFR and participating VFR aircraft. The service provided in the outer area is called Class C service which includes: IFR/IFR-standard IFR separation; IFR/VFR-traffic advisories and conflict resolution; and VFR/VFR-traffic advisories and, as appropriate, safety alerts. The normal radius will be 20 nautical miles with some variations based on site-specific requirements. The outer area extends outward from the primary Class C airspace airport and extends from the lower limits of radar/radio coverage up to the ceiling of the approach control's delegated airspace excluding the Class C charted area and other airspace as appropriate.
   (See CONTROLLED AIRSPACE.) (See CONFLICT RESOLUTION.)

OUTER COMPASS LOCATOR   (See COMPASS LOCATOR.)

OUTER FIX   A general term used within ATC to describe fixes in the terminal area, other than the final approach fix. Aircraft are normally cleared to these fixes by an Air Route Traffic Control Center or an Approach Control Facility. Aircraft are normally cleared from these fixes to the final approach fix or final approach course.

OR

OUTER FIX   An adapted fix along the converted route of flight, prior to the meter fix, for which crossing times are calculated and displayed in the metering position list.

OUTER FIX ARC   A semicircle, usually about a 50-70 mile radius from a meter fix, usually in high altitude, which is used by CTAS/HOST to calculate outer fix times and determine appropriate sector meter list assignments for aircraft on an established arrival route that will traverse the arc.

OUTER FIX TIME   A calculated time to depart the outer fix in order to cross the vertex at the ACLT. The time reflects descent speed adjustments and any applicable delay time that must be absorbed prior to crossing the meter fix.

OUTER MARKER   A marker beacon at or near the glide-slope intercept altitude of an ILS approach. It is keyed to transmit two dashes per second on a 400 Hz tone, which is received aurally and visually by compatible airborne equipment. The OM is normally located four to seven miles from the runway threshold on the extended centerline of the runway.
   (See MARKER BEACON.) (See INSTRUMENT LANDING SYSTEM.) (Refer to AIM.)

OVER   My transmission is ended; I expect a response.

OVERHEAD MANEUVER   A series of predetermined maneuvers prescribed for aircraft (often in formation) for entry into the visual flight rules (VFR) traffic pattern and to proceed to a landing. An overhead maneuver is not an instrument flight rules (IFR) approach procedure. An aircraft executing an overhead maneuver is considered VFR and the IFR flight plan is canceled when the aircraft reaches the "initial point" on the initial approach portion of the maneuver. The pattern usually specifies the following:
   a. The radio contact required of the pilot.
   b. The speed to be maintained.
   c. An initial approach 3 to 5 miles in length.
   d. An elliptical pattern consisting of two 180 degree turns.
   e. A break point at which the first 180 degree turn is started.
   f. The direction of turns.
   g. Altitude (at least 500 feet above the conventional pattern).
   h. A "Roll-out" on final approach not less than ¼ mile from the landing threshold and not less than 300 feet above the ground.

OVERLYING CENTER   The ARTCC facility that is responsible for arrival/departure operations at a specific terminal.

———— **P** ————

P TIME   (See PROPOSED DEPARTURE TIME.)

P-ACP   (See PREARRANGED COORDINATION PROCEDURES.)

PAN-PAN The international radio-telephony urgency signal. When repeated three times, indicates uncertainty or alert followed by the nature of the urgency.
(See MAYDAY.) (Refer to AIM.)

PAR (See PRECISION APPROACH RADAR.)

PAR [ICAO] (See ICAO Term PRECISION APPROACH RADAR.)

PARALLEL ILS APPROACHES Approaches to parallel runways by IFR aircraft which, when established inbound toward the airport on the adjacent final approach courses, are radar-separated by at least 2 miles.
(See FINAL APPROACH COURSE.) (See SIMULTANE-OUS ILS APPROACHES.)

PARALLEL MLS APPROACHES (See PARALLEL ILS APPROACHES.)

PARALLEL OFFSET ROUTE A parallel track to the left or right of the designated or established airway/ route. Normally associated with Area Navigation (RNAV) operations.
(See AREA NAVIGATION.)

PARALLEL RUNWAYS Two or more runways at the same airport whose centerlines are parallel. In addition to runway number, parallel runways are designated as L (left) and R (right) or, if three parallel runways exist, L (left), C (center), and R (right).

PBCT (See PROPOSED BOUNDARY CROSSING TIME.)

PDC (See PRE-DEPARTURE CLEARANCE.)

PERMANENT ECHO Radar signals reflected from fixed objects on the Earth's surface; e.g., buildings, towers, terrain. Permanent echoes are distinguished from "ground clutter" by being definable locations rather than large areas. Under certain conditions they may be used to check radar alignment.

PHOTO RECONNAISSANCE Military activity that requires locating individual photo targets and navigating to the targets at a preplanned angle and altitude. The activity normally requires a lateral route width of 16 NM and altitude range of 1,500 feet to 10,000 feet AGL.

PIDP (See PROGRAMMABLE INDICATOR DATA PROCESSOR.)

PILOT BRIEFING A service provided by the FSS to assist pilots in flight planning. Briefing items may include weather information, NOTAMS, military activities, flow control information, and other items as requested.
(Refer to AIM.)

PILOT IN COMMAND The pilot responsible for the operation and safety of an aircraft during flight time.
(Refer to FAR Part 91.)

PILOT'S DISCRETION When used in conjunction with altitude assignments, means that ATC has offered the pilot the option of starting climb or descent whenever he wishes and conducting the climb or descent at any rate he wishes. He may temporarily level off at any intermediate altitude. However, once he has vacated an altitude, he may not return to that altitude.

PILOT WEATHER REPORT A report of meteorological phenomena encountered by aircraft in flight.
(Refer to AIM.)

PIREP (See PILOT WEATHER REPORT.)

PITCH POINT A fix/waypoint that serves as a transition point from a departure procedure or the low altitude ground-based navigation structure into the high altitude waypoint system.

PLANS DISPLAY A display available in URET that provides detailed flight plan and predicted conflict information in textual format for requested Current Plans and all Trial Plans.
(See USER REQUEST EVALUATION TOOL.)

POINT OUT (See RADAR POINT OUT.)

POINT TO POINT (PTP) A level of NRR service for aircraft that is based on traditional waypoints in their FMSs or RNAV equipage.

POLAR TRACK STRUCTURE A system of organized routes between Iceland and Alaska which overlie Canadian MNPS Airspace.

POSITION AND HOLD Used by ATC to inform a pilot to taxi onto the departure runway in takeoff position and hold. It is not authorization for takeoff. It is used when takeoff clearance cannot immediately be issued because of traffic or other reasons.
(See CLEARED FOR TAKEOFF.)

POSITION REPORT A report over a known location as transmitted by an aircraft to ATC.
(Refer to AIM.)

POSITION SYMBOL A computer-generated indica-tion shown on a radar display to indicate the mode of tracking.

POSITIVE CONTROL The separation of all air traffic within designated airspace by air traffic control.

PRACTICE INSTRUMENT APPROACH An instrument approach procedure conducted by a VFR or an IFR aircraft for the purpose of pilot training or proficiency demonstrations.

PREARRANGED COORDINATION A standardized procedure which permits an air traffic controller to enter the airspace assigned to another air traffic controller without verbal coordination. The procedures are defined in a facility directive which ensures standard separation between aircraft.

PRECIPITATION Any or all forms of water particles (rain, sleet, hail, or snow) that fall from the atmosphere and reach the surface.

PRECISION APPROACH (See PRECISION AP-PROACH PROCEDURE.)

PRECISION APPROACH PROCEDURE A standard instrument approach procedure in which an electronic glideslope/glidepath is provided; e.g., ILS/MLS and PAR.

(See INSTRUMENT LANDING SYSTEM.) (See MI-CROWAVE LANDING SYSTEM.) (See PRECISION AP-PROACH RADAR.)

PRECISION APPROACH RADAR Radar equipment in some ATC facilities operated by the FAA and/or the military services at joint-use civil/military locations and separate military installations to detect and display azimuth, elevation, and range of aircraft on the final approach course to a runway. This equipment may be used to monitor certain nonradar approaches, but is primarily used to conduct a precision instrument approach (PAR) wherein the controller issues guidance instructions to the pilot based on the aircraft's position in relation to the final approach course (azimuth), the glidepath (elevation), and the distance (range) from the touchdown point on the runway as displayed on the radar scope.

(Note: The abbreviation "PAR" is also used to denote preferential arrival routes in ARTCC computers.)
(See GLIDEPATH.) (See PAR.) (See PREFERENTIAL ROUTES.) (See ICAO term PRECISION APPROACH RADAR.) (Refer to AIM.)

PRECISION APPROACH RADAR [ICAO] Primary radar equipment used to determine the position of an aircraft during final approach, in terms of lateral and vertical deviations relative to a nominal approach path, and in range relative to touchdown.

Note: Precision approach radars are designed to enable pilots of aircraft to be given guidance by radio communication during the final stages of the approach to land.

PRECISION RUNWAY MONITOR Provides air traffic controllers with high precision secondary surveillance data for aircraft on final approach to parallel runways that have extended centerlines separated by less than 4,300 feet. High resolution color monitoring displays (FMA) are required to present surveillance track data to controllers along with detailed maps depicting approaches and no transgression zone.

PRE-DEPARTURE CLEARANCE An application with the Terminal Data Link System (TDLS) that provides clearance information to subscribers, through a service provider, in text to the cockpit or gate printer.

PREFERENTIAL ROUTES Preferential routes (PDRs, PARs, and PDARs) are adapted in ARTCC computers to accomplish inter/intrafacility controller coordination and to assure that flight data is posted at the proper control positions. Locations having a need for these specific inbound and outbound routes normally publish such routes in local facility bulletins, and their use by pilots minimizes flight plan route amendments. When the workload or traffic situation permits, controllers normally provide radar vectors or assign requested routes to minimize circuitous routing. Preferential routes are usually confined to one ARTCC's area and are referred to by the following names or acronyms:

a. Preferential Departure Route (PDR). A specific departure route from an airport or terminal area to an en route point where there is no further need for flow control. It may be included in a Instrument Departure Procedure (DP) or a Preferred IFR Route.

b. Preferential Arrival Route (PAR). A specific arrival route from an appropriate en route point to an airport or terminal area. It may be included in a Standard Terminal Arrival (STAR) or a Preferred IFR Route. The abbreviation "PAR" is used primarily within the ARTCC and should not be confused with the abbreviation for Precision Approach Radar.

c. Preferential Departure and Arrival Route (PDAR). A route between two terminals which are within or immediately adjacent to one ARTCC's area. PDARs are not synonymous with Preferred IFR Routes but may be listed as such as they do accomplish essentially the same purpose.

(See PREFERRED IFR ROUTES.) (See NAS STAGE A.)

PREFERRED IFR ROUTES Routes established between busier airports to increase system efficiency and capacity. They normally extend through one or more ARTCC areas and are designed to achieve balanced traffic flows among high density terminals. IFR clearances are issued on the basis of these routes except when severe weather avoidance procedures or other factors dictate otherwise. Preferred IFR Routes are listed in the Airport/Facility Directory. If a flight is planned to or from an area having such routes but the departure or arrival point is not listed in the Airport/Facility Directory, pilots may use that part of a Preferred IFR Route which is appropriate for the departure or arrival point that is listed. Preferred IFR Routes are correlated with DP's and STARs and may be defined by airways, jet routes, direct routes between NAVAIDs, Waypoints, NAVAID radials/DME, or any combinations thereof.

(See INSTRUMENT DEPARTURE PROCEDURE.) (See STANDARD TERMINAL ARRIVAL.) (See PREFERENTIAL ROUTES.) (See CENTER'S AREA.) (Refer to AIRPORT/FACILITY DIRECTORY.) (Refer to NOTICES TO AIRMEN PUBLICATION.)

PRE-FLIGHT PILOT BRIEFING (See PILOT BRIEFING.)

PREVAILING VISIBILITY (See VISIBILITY.)

PRM (See ILS PRM APPROACH AND PRECISION RUNWAY MONITOR.)

PROCEDURE TURN The maneuver prescribed when it is necessary to reverse direction to establish an aircraft on the intermediate approach segment or final approach course. The outbound course, direction of turn, distance within which the turn must be completed, and minimum altitude are specified in the procedure. However, unless otherwise restricted, the point at which the turn may be commenced and the type and rate of turn are left to the discretion of the pilot.
(See ICAO term PROCEDURE TURN.)

PROCEDURE TURN [ICAO] A manoeuvre in which a turn is made away from a designated track followed by a turn in the opposite direction to permit the aircraft to intercept and proceed along the reciprocal of the designated track.

Note 1: Procedure turns are designated "left" or "right" according to the direction of the initial turn.

Note 2: Procedure turns may be designated as being made either in level flight or while descending, according to the circumstances of each individual approach procedure.

PROCEDURE TURN INBOUND That point of a procedure turn maneuver where course reversal has been completed and an aircraft is established inbound on the intermediate approach segment or final approach course. A report of "procedure turn inbound" is normally used by ATC as a position report for separation purposes.
(See FINAL APPROACH COURSE.) (See PROCEDURE TURN.) (See SEGMENTS OF AN INSTRUMENT APPROACH PROCEDURE.)

PROFILE DESCENT An uninterrupted descent (except where level flight is required for speed adjustment; e.g., 250 knots at 10,000 feet MSL) from cruising altitude/level to interception of a glideslope or to a minimum altitude specified for the initial or intermediate approach segment of a nonprecision instrument approach. The profile descent normally terminates at the approach gate or where the glideslope or other appropriate minimum altitude is intercepted.

PROGRAMMABLE INDICATOR DATA PROCESSOR The PIDP is a modification to the AN/TPX-42 interrogator system currently installed in fixed RAPCONs. The PIDP detects, tracks, and predicts secondary radar aircraft targets. These are displayed by means of computer-generated symbols and alphanumeric characters depicting flight identification, aircraft altitude, ground speed, and flight plan data. Although primary radar targets are not tracked, they are displayed coincident with the secondary radar targets as well as with the other symbols and alphanumerics. The system has the capability of interfacing with ARTCCs.

PROGRESS REPORT (See POSITION REPORT.)

PROGRESSIVE TAXI Precise taxi instructions given to a pilot unfamiliar with the airport or issued in stages as the aircraft proceeds along the taxi route.

PROHIBITED AREA (See SPECIAL USE AIRSPACE.)
(See ICAO term PROHIBITED AREA.)

PROHIBITED AREA [ICAO] An airspace of defined dimensions, above the land areas or territorial waters of a State, within which the flight of aircraft is prohibited.

PROPOSED BOUNDARY CROSSING TIME Each center has a PBCT parameter for each internal airport. Proposed internal flight plans are transmitted to the adjacent center if the flight time along the proposed route from the departure airport to the center boundary is less than or equal to the value of PBCT or if airport adaptation specifies transmission regardless of PBCT.

PROPOSED DEPARTURE TIME The time that the aircraft expects to become airborne.

PROTECTED AIRSPACE The airspace on either side of an oceanic route/track that is equal to one-half the lateral separation minimum except where reduction of protected airspace has been authorized.

PT (See PROCEDURE TURN.)

PTP (See POINT TO POINT.)

PTS (See POLAR TRACK STRUCTURE.)

PUBLISHED ROUTE A route for which an IFR altitude has been established and published; e.g., Federal Airways, Jet Routes, Area Navigation Routes, Specified Direct Routes.

————— **Q** —————

Q ROUTE 'Q' is the designator assigned to published RNAV routes used by the United States.

QNE The barometric pressure used for the standard altimeter setting (29.92 inches Hg.).

QNH The barometric pressure as reported by a particular station.

QUADRANT A quarter part of a circle, centered on a NAVAID, oriented clockwise from magnetic north as follows: NE quadrant 000-089, SE quadrant 090-179, SW quadrant 180-269, NW quadrant 270-359.

QUEUING (See STAGING/QUEUING.)

QUICK LOOK A feature of EAS and ARTS which provides the controller the capability to display full data blocks of tracked aircraft from other control positions.

R

RAA   (See REMOTE AIRPORT ADVISORY.)

RADAR   A device which, by measuring the time interval between transmission and reception of radio pulses and correlating the angular orientation of the radiated antenna beam or beams in azimuth and/or elevation, provides information on range, azimuth, and/or elevation of objects in the path of the transmitted pulses.

 a. Primary Radar. A radar system in which a minute portion of a radio pulse transmitted from a site is reflected by an object and then received back at that site for processing and display at an air traffic control facility.

 b. Secondary Radar/Radar Beacon (ATCRBS). A radar system in which the object to be detected is fitted with cooperative equipment in the form of a radio receiver/transmitter (transponder). Radar pulses transmitted from the searching transmitter/receiver (interrogator) site are received in the cooperative equipment and used to trigger a distinctive transmission from the transponder. This reply transmission, rather than a reflected signal, is then received back at the transmitter/receiver site for processing and display at an air traffic control facility.

 (See INTERROGATOR.) (See TRANSPONDER.) (See ICAO term PRIMARY RADAR.) (See ICAO term RADAR.) (See ICAO term SECONDARY RADAR.) (Refer to AIM.)

RADAR [ICAO]   A radio detection device which provides information on range, azimuth and/or elevation of objects.

 a. <u>Primary Radar</u>—Radar system which uses reflected radio signals.

 b. <u>Secondary Radar</u>—Radar system wherein a radio signal transmitted from a radar station initiates the transmission of a radio signal from another station.

RADAR ADVISORY   The provision of advice and information based on radar observations.

 (See ADVISORY SERVICE.)

RADAR ALTIMETER   (See RADIO ALTIMETER.)

RADAR APPROACH   An instrument approach procedure which utilizes Precision Approach Radar (PAR) or Airport Surveillance Radar (ASR).

 (See AIRPORT SURVEILLANCE RADAR.) (See INSTRUMENT APPROACH PROCEDURE.) (See PRECISION APPROACH RADAR.) (See SURVEILLANCE APPROACH.) (See ICAO term RADAR APPROACH.) (Refer to AIM.)

RADAR APPROACH [ICAO]   An approach, executed by an aircraft, under the direction of a radar controller.

RADAR APPROACH CONTROL FACILITY   A terminal ATC facility that uses radar and nonradar capabilities to provide approach control services to aircraft arriving, departing, or transiting airspace controlled by the facility.

 (See APPROACH CONTROL SERVICE.)

 a. Provides radar ATC services to aircraft operating in the vicinity of one or more civil and/or military airports in a terminal area. The facility may provide services of a ground controlled approach (GCA); i.e., ASR and PAR approaches. A radar approach control facility may be operated by FAA, USAF, US Army, USN, USMC, or jointly by FAA and a military service. Specific facility nomenclatures are used for administrative purposes only and are related to the physical location of the facility and the operating service generally as follows:

 1. Army Radar Approach Control (ARAC) (Army).

 2. Radar Air Traffic Control Facility (RATCF) (Navy/FAA).

 3. Radar Approach Control (RAPCON) (Air Force/FAA).

 4. Terminal Radar Approach Control (TRACON) (FAA).

 5. Air Traffic Control Tower (ATCT) (FAA). (Only those towers delegated approach control authority.).

RADAR ARRIVAL   An aircraft arriving at an airport served by a radar facility and in radar contact with the facility.

 (See NONRADAR.)

RADAR BEACON   (See RADAR.)

RADAR CLUTTER [ICAO]   The visual indication on a radar display of unwanted signals.

RADAR CONTACT

 a. Used by ATC to inform an aircraft that it is identified in the radar display and radar flight following will be provided until radar identification is terminated. Radar service may also be provided within the limits of necessity and capability. When a pilot is informed of "radar contact," he automatically discontinues reporting over compulsory reporting points.

 (See RADAR CONTACT LOST.) (See RADAR FLIGHT FOLLOWING.) (See RADAR SERVICE.) (See RADAR SERVICE TERMINATED.) (Refer to AIM.)

 b. The term used to inform the controller that the aircraft is identified and approval is granted for the aircraft to enter the receiving controllers airspace.

 (See ICAO term RADAR CONTACT.)

RADAR CONTACT [ICAO]   The situation which exists when the radar blip or radar position symbol of a particular aircraft is seen and identified on a radar display.

RADAR CONTACT LOST   Used by ATC to inform a pilot that radar data used to determine the aircraft's position is no longer being received, or is no longer reliable and radar service is no longer being provided. The loss may be attributed to several factors including the aircraft merging with weather or ground clutter, the aircraft operating below radar line of sight coverage, the aircraft entering an area of poor radar return, failure of the aircraft transponder, or failure of the ground radar equipment.

 (See CLUTTER.) (See RADAR CONTACT.)

RADAR ENVIRONMENT An area in which radar service may be provided.

(See ADDITIONAL SERVICES.) (See RADAR CONTACT.) (See RADAR SERVICE.) (See TRAFFIC ADVISORIES.)

RADAR FLIGHT FOLLOWING The observation of the progress of radar identified aircraft, whose primary navigation is being provided by the pilot, wherein the controller retains and correlates the aircraft identity with the appropriate target or target symbol displayed on the radar scope.

(See RADAR CONTACT.) (See RADAR SERVICE.) (Refer to AIM.)

RADAR IDENTIFICATION The process of ascertaining that an observed radar target is the radar return from a particular aircraft.

(See RADAR CONTACT.) (See RADAR SERVICE.) (See ICAO term RADAR IDENTIFICATION.)

RADAR IDENTIFICATION [ICAO] The process of correlating a particular radar blip or radar position symbol with a specific aircraft.

RADAR IDENTIFIED AIRCRAFT An aircraft, the position of which has been correlated with an observed target or symbol on the radar display.

(See RADAR CONTACT.) (See RADAR CONTACT LOST.)

RADAR MONITORING (See RADAR SERVICE.)

RADAR NAVIGATIONAL GUIDANCE (See RADAR SERVICE.)

RADAR POINT OUT An action taken by a controller to transfer the radar identification of an aircraft to another controller if the aircraft will or may enter the airspace or protected airspace of another controller and radio communications will not be transferred.

RADAR REQUIRED A term displayed on charts and approach plates and included in FDC Notams to alert pilots that segments of either an instrument approach procedure or a route are not navigable because of either the absence or unusability of a NAVAID. The pilot can expect to be provided radar navigational guidance while transiting segments labeled with this term.

(See RADAR ROUTE.) (See RADAR SERVICE.)

RADAR ROUTE A flight path or route over which an aircraft is vectored. Navigational guidance and altitude assignments are provided by ATC.

(See FLIGHT PATH.) (See ROUTE.)

RADAR SEPARATION (See RADAR SERVICE.)

RADAR SERVICE A term which encompasses one or more of the following services based on the use of radar which can be provided by a controller to a pilot of a radar identified aircraft.

a. Radar Monitoring.—The radar flight-following of aircraft, whose primary navigation is being performed by the pilot, to observe and note deviations from its authorized flight path, airway, or route. When being applied specifically to radar monitoring of instrument approaches; i.e., with precision approach radar (PAR) or radar monitoring of simultaneous ILS/MLS approaches, it includes advice and instructions whenever an aircraft nears or exceeds the prescribed PAR safety limit or simultaneous ILS/MLS no transgression zone.

(See ADDITIONAL SERVICES.) (See TRAFFIC ADVISORIES.)

b. Radar Navigational Guidance. Vectoring aircraft to provide course guidance.

c. Radar Separation. Radar spacing of aircraft in accordance with established minima.

(See ICAO term RADAR SERVICE.)

RADAR SERVICE [ICAO] Term used to indicate a service provided directly by means of radar.

a. Monitoring.—The use of radar for the purpose of providing aircraft with information and advice relative to significant deviations from nominal flight path.

b. Separation.—The separation used when aircraft position information is derived from radar sources.

RADAR SERVICE TERMINATED Used by ATC to inform a pilot that he will no longer be provided any of the services that could be received while in radar contact. Radar service is automatically terminated, and the pilot is not advised in the following cases:

a. An aircraft cancels its IFR flight plan, except within Class B airspace, Class C airspace, a TRSA, or where Basic Radar service is provided.

b. An aircraft conducting an instrument, visual, or contact approach has landed or has been instructed to change to advisory frequency.

c. An arriving VFR aircraft, receiving radar service to a tower-controlled airport within Class B airspace, Class C airspace, a TRSA, or where sequencing service is provided, has landed; or to all other airports, is instructed to change to tower or advisory frequency.

d. An aircraft completes a radar approach.

RADAR SURVEILLANCE The radar observation of a given geographical area for the purpose of performing some radar function.

RADAR TRAFFIC ADVISORIES Advisories issued to alert pilots to known or observed radar traffic which may affect the intended route of flight of their aircraft.

(See TRAFFIC ADVISORIES.)

RADAR TRAFFIC INFORMATION SERVICE (See TRAFFIC ADVISORIES.)

RADAR VECTORING [ICAO] Provision of navigational guidance to aircraft in the form of specific headings, based on the use of radar.

RADAR WEATHER ECHO INTENSITY LEVELS
Existing radar systems cannot detect turbulence. However, there is a direct correlation between the degree of turbulence and other weather features associated with thunderstorms and the radar weather echo intensity. The National Weather Service has categorized radar weather echo intensity for precipitation into six levels. These levels are sometimes expressed during communications as "VIP LEVEL" 1 through 6 (derived from the component of the radar that produces the information-Video Integrator and Processor). The following list gives the "VIP LEVELS" in relation to the precipitation intensity within a thunderstorm:
- a. Level 1. WEAK
- b. Level 2. MODERATE
- c. Level 3. STRONG
- d. Level 4. VERY STRONG
- e. Level 5. INTENSE
- f. Level 6. EXTREME

(See AC 00-45.)

RADIAL A magnetic bearing extending from a VOR/VORTAC/TACAN navigation facility.

RADIO
- a. A device used for communication.
- b. Used to refer to a flight service station; e.g., "Seattle Radio" is used to call Seattle FSS.

RADIO ALTIMETER Aircraft equipment which makes use of the reflection of radio waves from the ground to determine the height of the aircraft above the surface.

RADIO BEACON (See NONDIRECTIONAL BEACON.)

RADIO DETECTION AND RANGING (See RADAR.)

RADIO MAGNETIC INDICATOR An aircraft navigational instrument coupled with a gyro compass or similar compass that indicates the direction of a selected NAVAID and indicates bearing with respect to the heading of the aircraft.

RAIS (See REMOTE AIRPORT INFORMATION SERVICE.)

RAMP (See APRON.)

RANDOM ALTITUDE An altitude inappropriate for direction of flight and/or not in accordance with FAA Order 7110.65, paragraph 4-5-1.

RANDOM ROUTE Any route not established or charted/published or not otherwise available to all users.

RC (See ROAD RECONNAISSANCE.)

RCAG (See REMOTE COMMUNICATIONS AIR/GROUND FACILITY.)

RCC (See RESCUE COORDINATION CENTER.)

RCO (See REMOTE COMMUNICATIONS OUTLET.)

RCR (See RUNWAY CONDITION READING.)

READ BACK Repeat my message back to me.

RECEIVER AUTONOMOUS INTEGRITY MONITORING (RAIM) A technique whereby a civil GNSS receiver/processor determines the integrity of the GNSS navigation signals without reference to sensors or non-DoD integrity systems other than the receiver itself. This determination is achieved by a consistency check among redundant pseudorange measurements.

RECEIVING CONTROLLER A controller/facility receiving control of an aircraft from another controller/facility.

RECEIVING FACILITY (See RECEIVING CONTROLLER.)

RECONFORMANCE The automated process of bringing an aircraft's Current Plan Trajectory into conformance with its track.

REDUCE SPEED TO (SPEED) (See SPEED ADJUSTMENT.)

REIL (See RUNWAY END IDENTIFIER LIGHTS.)

RELEASE TIME A departure time restriction issued to a pilot by ATC (either directly or through an authorized relay) when necessary to separate a departing aircraft from other traffic.

(See ICAO term RELEASE TIME.)

RELEASE TIME [ICAO] Time prior to which an aircraft should be given further clearance or prior to which it should not proceed in case of radio failure.

REMOTE AIRPORT ADVISORY (RAA) A remote service which may be provided by facilities, which are not located on the landing airport, but have ground-to-air communication on frequency 123.6 or 123.65 or tower frequency when the tower is closed, automated weather reporting with voice available to the pilot at the landing airport, and a continuous ASOS/AWOS data display, other direct reading instruments, or manual observation is available to the AFSS specialist.

REMOTE AIRPORT INFORMATION SERVICE (RAIS) A temporary service provided by facilities, which are not located on the landing airport, but have communication capability and automated weather reporting available to the pilot at the landing airport.

REMOTE COMMUNICATIONS AIR/GROUND FACILITY An unmanned VHF/UHF transmitter/receiver facility which is used to expand ARTCC air/ground communications coverage and to facilitate direct contact between pilots and controllers. RCAG facilities are sometimes not equipped with emergency frequencies 121.5 MHz and 243.0 MHz.

(Refer to AIM.)

REMOTE COMMUNICATIONS OUTLET An unmanned communications facility remotely controlled by air traffic personnel. RCOs serve FSSs. RTRs serve terminal ATC facilities. An RCO or RTR may be UHF or VHF and will extend the communication range of the air traffic facility. There are several classes of RCOs and RTRs. The class is determined by the number of transmitters or receivers. Classes A through G are used primarily for air/ground purposes. RCO and RTR class O facilities are nonprotected outlets subject to undetected and prolonged outages. RCO (Os) and RTR (Os) were established for the express purpose of providing ground-to-ground communications between air traffic control specialists and pilots located at a satellite airport for delivering en route clearances, issuing departure authorizations, and acknowledging instrument flight rules cancellations or departure/landing times. As a secondary function, they may be used for advisory purposes whenever the aircraft is below the coverage of the primary air/ground frequency.

REMOTE TRANSMITTER/RECEIVER (See REMOTE COMMUNICATIONS OUTLET.)

REPORT Used to instruct pilots to advise ATC of specified information; e.g., "Report passing Hamilton VOR."

REPORTING POINT A geographical location in relation to which the position of an aircraft is reported.
(See COMPULSORY REPORTING POINTS.) (See ICAO term REPORTING POINT.) (Refer to AIM.)

REPORTING POINT [ICAO] A specified geographical location in relation to which the position of an aircraft can be reported.

REQUEST FULL ROUTE CLEARANCE Used by pilots to request that the entire route of flight be read verbatim in an ATC clearance. Such request should be made to preclude receiving an ATC clearance based on the original filed flight plan when a filed IFR flight plan has been revised by the pilot, company, or operations prior to departure.

REQUIRED NAVIGATION PERFORMANCE (RNP) A statement of the navigational performance necessary for operation within a defined airspace. The following terms are commonly associated with RNP:
a. Required Navigation Performance Level or Type (RNP-X). A value, in nautical miles (NM), from the intended horizontal position within which an aircraft would be at least 95-percent of the total flying time.
b. Required Navigation Performance (RNP) Airspace. A generic term designating airspace, route(s), leg(s), operation(s), or procedure(s) where minimum required navigational performance (RNP) have been established.
c. Actual Navigation Performance (ANP). A measure of the current estimated navigational performance. Also referred to as Estimated Position Error (EPE).

d. Estimated Position Error (EPE). A measure of the current estimated navigational performance. Also referred to as Actual Navigation Performance (ANP).
e. Lateral Navigation (LNAV). A function of area navigation (RNAV) equipment which calculates, displays, and provides lateral guidance to a profile or path.
f. Vertical Navigation (VNAV). A function of area navigation (RNAV) equipment which calculates, displays, and provides vertical guidance to a profile or path.

RESCUE CO-ORDINATION CENTER A search and rescue (SAR) facility equipped and manned to coordinate and control SAR operations in an area designated by the SAR plan. The U.S. Coast Guard and the U.S. Air Force have responsibility for the operation of RCCs.
(See ICAO term RESCUE CO-ORDINATION CENTRE.)

RESCUE COORDINATION CENTRE [ICAO] A unit responsible for promoting efficient organization of search and rescue service and for co-ordinating the conduct of search and rescue operations within a search and rescue region.

RESOLUTION ADVISORY A display indication given to the pilot by the traffic alert and collision avoidance systems(TCAS II) recommending a maneuver to increase vertical separation relative to an intruding aircraft. Positive, negative, and vertical speed limit (VSL) advisories constitute the resolution advisories. A resolution advisory is also classified as corrective or preventive.

RESTRICTED AREA (See SPECIAL USE AIRSPACE.)
(See ICAO term RESTRICTED AREA.)

RESTRICTED AREA [ICAO] An airspace of defined dimensions, above the land areas or territorial waters of a State, within which the flight of aircraft is restricted in accordance with certain specified conditions.

RESUME NORMAL SPEED Used by ATC to advise a pilot that previously issued speed control restrictions are deleted. An instruction to "resume normal speed" does not delete speed restrictions that are applicable to published procedures of upcoming segments of flight, unless specifically stated by ATC. This does not relieve the pilot of those speed restrictions which are applicable to FAR 91.117.

RESUME OWN NAVIGATION Used by ATC to advise a pilot to resume his own navigational responsibility. It is issued after completion of a radar vector or when radar contact is lost while the aircraft is being radar vectored.
(See RADAR CONTACT LOST.) (See RADAR SERVICE TERMINATED.)

RMI (See RADIO MAGNETIC INDICATOR.)

RNAV (See AREA NAVIGATION.)

RNAV [ICAO] (See ICAO Term AREA NAVIGATION.)

RNAV APPROACH An instrument approach procedure which relies on aircraft area navigation equipment for navigational guidance.

(See AREA NAVIGATION.) (See INSTRUMENT APPROACH PROCEDURE.)

ROAD RECONNAISSANCE Military activity requiring navigation along roads, railroads, and rivers. Reconnaissance route/route segments are seldom along a straight line and normally require a lateral route width of 10 NM to 30 NM and an altitude range of 500 feet to 10,000 feet AGL.

ROGER I have received all of your last transmission. It should not be used to answer a question requiring a yes or a no answer.

(See AFFIRMATIVE.) (See NEGATIVE.)

ROLLOUT RVR (See VISIBILITY.)

ROUTE A defined path, consisting of one or more courses in a horizontal plane, which aircraft traverse over the surface of the Earth.

(See AIRWAY.) (See JET ROUTE.) (See PUBLISHED ROUTE.) (See UNPUBLISHED ROUTE.)

ROUTE ACTION NOTIFICATION URET notification that a PAR/PDR/PDAR has been applied to the flight plan.

(See ATC PREFERRED ROUTE NOTIFICATION.)

(See USER REQUEST EVALUATION TOOL.)

ROUTE SEGMENT As used in Air Traffic Control, a part of a route that can be defined by two navigational fixes, two NAVAIDs, or a fix and a NAVAID.

(See FIX.) (See ROUTE.) (See ICAO term ROUTE SEGMENT.)

ROUTE SEGMENT [ICAO] A portion of a route to be flown, as defined by two consecutive significant points specified in a flight plan.

RSA (See RUNWAY SAFETY AREA.)

RTR (See REMOTE TRANSMITTER/RECEIVER.)

RUNWAY A defined rectangular area on a land airport prepared for the landing and takeoff run of aircraft along its length. Runways are normally numbered in relation to their magnetic direction rounded off to the nearest 10 degrees; e.g., Runway 1, Runway 25.

(See PARALLEL RUNWAYS.) (See ICAO term RUNWAY.)

RUNWAY [ICAO] A defined rectangular area on a land aerodrome prepared for the landing and takeoff of aircraft.

RUNWAY CENTERLINE LIGHTING (See AIRPORT LIGHTING.)

RUNWAY CONDITION READING Numerical decelerometer readings relayed by air traffic controllers at USAF and certain civil bases for use by the pilot in determining runway braking action. These readings are routinely relayed only to USAF and Air National Guard Aircraft.

(See BRAKING ACTION.)

RUNWAY END IDENTIFIER LIGHTS (See AIRPORT LIGHTING.)

RUNWAY GRADIENT The average slope, measured in percent, between two ends or points on a runway. Runway gradient is depicted on Government aerodrome sketches when total runway gradient exceeds 0.3%.

RUNWAY HEADING The magnetic direction that corresponds with the runway centerline extended, not the painted runway number. When cleared to "fly or maintain runway heading," pilots are expected to fly or maintain the heading that corresponds with the extended centerline of the departure runway. Drift correction shall not be applied, e.g., Runway 4, actual magnetic heading of the runway centerline 044, fly 044.

RUNWAY IN USE/ACTIVE RUNWAY/DUTY RUNWAY Any runway or runways currently being used for takeoff or landing. When multiple runways are used, they are all considered active runways. In the metering sense, a selectable adapted item which specifies the landing runway configuration or direction of traffic flow. The adapted optimum flight plan from each transition fix to the vertex is determined by the runway configuration for arrival metering processing purposes.

RUNWAY LIGHTS (See AIRPORT LIGHTING.)

RUNWAY MARKINGS (See AIRPORT MARKING AIDS.)

RUNWAY OVERRUN In military aviation exclusively, a stabilized or paved area beyond the end of a runway, of the same width as the runway plus shoulders, centered on the extended runway centerline.

RUNWAY PROFILE DESCENT An instrument flight rules (IFR) air traffic control arrival procedure to a runway published for pilot use in graphic and/or textual form and may be associated with a STAR. Runway Profile Descents provide routing and may depict crossing altitudes, speed restrictions, and headings to be flown from the en route structure to the point where the pilot will receive clearance for and execute an instrument approach procedure. A Runway Profile Descent may apply to more than one runway if so stated on the chart.

(Refer to AIM.)

RUNWAY SAFETY AREA A defined surface surrounding the runway prepared, or suitable, for reducing the risk of damage to airplanes in the event of an undershoot, overshoot, or excursion from the runway. The dimensions of the RSA vary and can be determined by using the criteria contained within AC 150/5300-13, Chapter 3. Figure 3-1 in AC

150/5300-13 depicts the RSA. The design standards dictate that the RSA shall be:

a. Cleared, graded, and have no potentially hazardous ruts, humps, depressions, or other surface variations;

b. Drained by grading or storm sewers to prevent water accumulation;

c. Capable, under dry conditions, of supporting snow removal equipment, aircraft rescue and firefighting equipment, and the occasional passage of aircraft without causing structural damage to the aircraft; and,

d. Free of objects, except for objects that need to be located in the runway safety area because of their function. These objects shall be constructed on low impact resistant supports (frangible mounted structures) to the lowest practical height with the frangible point no higher than 3 inches above grade.

(Refer to AC 150/5300-13, Chap. 3.)

RUNWAY TRANSITION

a. Conventional STARS/SIDS. The portion of a STAR/SID that serves a particular runway or runways at an airport.

b. RNAV STARS/SIDs. Defines a path(s) from the common route to the final point(s) on a STAR. For a SID, the common route that serves a particular runway or runways at an airport.

RUNWAY USE PROGRAM   A noise abatement runway selection plan designed to enhance noise abatement efforts with regard to airport communities for arriving and departing aircraft. These plans are developed into runway use programs and apply to all turbojet aircraft 12,500 pounds or heavier; turbojet aircraft less than 12,500 pounds are included only if the airport proprietor determines that the aircraft creates a noise problem. Runway use programs are coordinated with FAA offices, and safety criteria used in these programs are developed by the Office of Flight Operations. Runway use programs are administered by the Air Traffic Service as "Formal" or "Informal" programs.

a. Formal Runway Use Program. An approved noise abatement program which is defined and acknowledged in a Letter of Understanding between Flight Operations, Air Traffic Service, the airport proprietor, and the users. Once established, participation in the program is mandatory for aircraft operators and pilots as provided for in FAR Part 91.129.

b. Informal Run Use Program. An approved noise abatement program which does not require a Letter of Understanding, and participation in the program is voluntary for aircraft operators/pilots.

RUNWAY VISIBILITY VALUE   (See VISIBILITY.)

RUNWAY VISUAL RANGE   (See VISIBILITY.)

———— **S** ————

SAA   (See SPECIAL ACTIVITY AIRSPACE.)

SAFETY ALERT   A safety alert issued by ATC to aircraft under their control if ATC is aware the aircraft is at an altitude which, in the controller's judgment, places the aircraft in unsafe proximity to terrain, obstructions, or other aircraft. The controller may discontinue the issuance of further alerts if the pilot advises he is taking action to correct the situation or has the other aircraft in sight.

a. Terrain/Obstruction Alert. A safety alert issued by ATC to aircraft under their control if ATC is aware the aircraft is at an altitude which, in the controller's judgment, places the aircraft in unsafe proximity to terrain/obstructions; e.g., "Low Altitude Alert, check your altitude immediately."

b. Aircraft Conflict Alert. A safety alert issued by ATC to aircraft under their control if ATC is aware of an aircraft that is not under their control at an altitude which, in the controller's judgment, places both aircraft in unsafe proximity to each other. With the alert, ATC will offer the pilot an alternate course of action when feasible; e.g., "Traffic Alert, advise you turn right heading zero niner zero or climb to eight thousand immediately."

The issuance of a safety alert is contingent upon the capability of the controller to have an awareness of an unsafe condition. The course of action provided will be predicated on other traffic under ATC control. Once the alert is issued, it is solely the pilot's prerogative to determine what course of action, if any, he will take.

SAFETY LOGIC SYSTEM   A software enhancement to ASDE-3, ASDE-X, and ASDE-3X, that predicts the path of aircraft landing and/or departing, and/or vehicular movements on runways. Visual and aural alarms are activated when the safety logic projects a potential collision. The Airport Movement Area Safety System (AMASS) is a safety logic system enhancement to the ASDE-3. The Safety Logic System for ASDE-X and ASDE-3X is an integral part of the software program.

SAFETY LOGIC SYSTEM ALERTS

a. ALERT   An actual situation involving two real safety logic tracks (aircraft/aircraft, aircraft/vehicle, or aircraft/other tangible object) that safety logic has predicted will result in an imminent collision, based upon the current set of Safety Logic parameters.

b. FALSE ALERT

1. Alerts generated by one or more false surface-radar targets that the system has interpreted as real tracks and placed into safety logic.

2. Alerts in which the safety logic software did not perform correctly, based upon the design specifications and the current set of Safety Logic parameters.

c. NUISANCE ALERT   An alert in which one or more of the following is true:

1. The alert is generated by a known situation that is not considered an unsafe operation, such as LAHSO or other approved operations.

2. The alert is generated by inaccurate secondary radar data received by the Safety Logic System.

3. The alert is generated by surface radar targets caused by moderate or greater precipitation.

4. One or more of the aircraft involved in the alert is not intending to use a runway (i.e., helicopter, pipeline patrol, non-Mode C overflight, etc.).

d. VALID NON-ALERT   A situation in which the safety logic software correctly determines that an alert is not required, based upon the design specifications and the current set of Safety Logic parameters.

e. INVALID NON-ALERT   A situation in which the safety logic software did not issue an alert when an alert was required, based upon the design specifications.

SAIL BACK   A maneuver during high wind conditions (usually with power off) where float plane movement is controlled by water rudders/opening and closing cabin doors.

SAME DIRECTION AIRCRAFT   Aircraft are operating in the same direction when:

a. They are following the same track in the same direction; or

b. Their tracks are parallel and the aircraft are flying in the same direction; or

c. Their tracks intersect at an angle of less than 45 degrees.

SAR   (See SEARCH AND RESCUE.)

SAY AGAIN   Used to request a repeat of the last transmission. Usually specifies transmission or portion thereof not understood or received; e.g., "Say again all after ABRAM VOR."

SAY ALTITUDE   Used by ATC to ascertain an aircraft's specific altitude/flight level. When the aircraft is climbing or descending, the pilot should state the indicated altitude rounded to the nearest 100 feet.

SAY HEADING   Used by ATC to request an aircraft heading. The pilot should state the actual heading of the aircraft.

SCHEDULED TIME OF ARRIVAL (STA)   A STA is the desired time that an aircraft should cross a certain point (landing or metering fix). It takes other traffic and airspace configuration into account. A STA time shows the results of the TMA scheduler that has calculated an arrival time according to parameters such as optimized spacing, aircraft performance, and weather.

SDF   (See SIMPLIFIED DIRECTIONAL FACILITY.)

SEA LANE   A designated portion of water outlined by visual surface markers for and intended to be used by aircraft designed to operate on water.

SEARCH AND RESCUE   A service which seeks missing aircraft and assists those found to be in need of assistance.

It is a cooperative effort using the facilities and services of available Federal, state and local agencies. The U.S. Coast Guard is responsible for coordination of search and rescue for the Maritime Region, and the U.S. Air Force is responsible for search and rescue for the Inland Region. Information pertinent to search and rescue should be passed through any air traffic facility or be transmitted directly to the Rescue Coordination Center by telephone.

(See FLIGHT SERVICE STATION.) (See RESCUE CO-ORDINATION CENTER.) (Refer to AIM.)

SEARCH AND RESCUE FACILITY   A facility responsible for maintaining and operating a search and rescue (SAR) service to render aid to persons and property in distress. It is any SAR unit, station, NET, or other operational activity which can be usefully employed during an SAR Mission; e.g., a Civil Air Patrol Wing, or a Coast Guard Station.

(See SEARCH AND RESCUE.)

SECTIONAL AERONAUTICAL CHARTS   (See AERONAUTICAL CHART.)

SECTOR LIST DROP INTERVAL   A parameter number of minutes after the meter fix time when arrival aircraft will be deleted from the arrival sector list.

SEE AND AVOID   When weather conditions permit, pilots operating IFR or VFR are required to observe and maneuver to avoid other aircraft. Right-of-way rules are contained in FAR Part 91.

SEGMENTED CIRCLE   A system of visual indicators designed to provide traffic pattern information at airports without operating control towers.

(Refer to AIM.)

SEGMENTS OF AN INSTRUMENT APPROACH PROCEDURE   An instrument approach procedure may have as many as four separate segments depending on how the approach procedure is structured.

a. Initial Approach. The segment between the initial approach fix and the intermediate fix or the point where the aircraft is established on the intermediate course or final approach course.

(See ICAO term INITIAL APPROACH SEGMENT.)

b. Intermediate Approach. The segment between the intermediate fix or point and the final approach fix.

(See ICAO term INTERMEDIATE APPROACH SEGMENT.)

c. Final Approach. The segment between the final approach fix or point and the runway, airport, or missed approach point.

(See ICAO term FINAL APPROACH SEGMENT.)

d. Missed Approach. The segment between the missed approach point or the point of arrival at decision height and the missed approach fix at the prescribed altitude.

(Refer to FAR Part 97.) (See ICAO term MISSED APPROACH PROCEDURE.)

SEPARATION   In air traffic control, the spacing of aircraft to achieve their safe and orderly movement in flight and while landing and taking off.
(See SEPARATION MINIMA.) (See ICAO term SEPARATION.)

SEPARATION [ICAO]   Spacing between aircraft, levels or tracks.

SEPARATION MINIMA   The minimum longitudinal, lateral, or vertical distances by which aircraft are spaced through the application of air traffic control procedures.
(See SEPARATION.)

SERVICE   A generic term that designates functions or assistance available from or rendered by air traffic control. For example, Class C service would denote the ATC services provided within a Class C airspace area.

SEVERE WEATHER AVOIDANCE PLAN   An approved plan to minimize the affect of severe weather on traffic flows in impacted terminal and/or ARTCC areas. SWAP is normally implemented to provide the least disruption to the ATC system when flight through portions of airspace is difficult or impossible due to severe weather.

SEVERE WEATHER FORECAST ALERTS   Preliminary messages issued in order to alert users that a Severe Weather Watch Bulletin (WW) is being issued. These messages define areas of possible severe thunderstorms or tornado activity. The messages are unscheduled and issued as required by the National Severe Storm Forecast Center at Kansas City, Missouri.
(See AIRMET.) (See SIGMET.) (See CONVECTIVE SIGMET.) (See CWA.)

SFA   (See SINGLE FREQUENCY APPROACH.)

SFO   (See SIMULATED FLAMEOUT.)

SHF   (See SUPER HIGH FREQUENCY.)

SHORT RANGE CLEARANCE   A clearance issued to a departing IFR flight which authorizes IFR flight to a specific fix short of the destination while air traffic control facilities are coordinating and obtaining the complete clearance.

SHORT TAKEOFF AND LANDING AIRCRAFT   An aircraft which, at some weight within its approved operating weight, is capable of operating from a STOL runway in compliance with the applicable STOL characteristics, airworthiness, operations, noise, and pollution standards.
(See VERTICAL TAKEOFF AND LANDING AIRCRAFT.)

SIAP   (See STANDARD INSTRUMENT APPROACH PROCEDURE.)

SID   (See STANDARD INSTRUMENT DEPARTURE.)

SIDESTEP MANEUVER   A visual maneuver accomplished by a pilot at the completion of an instrument approach to permit a straight-in landing on a parallel runway not more than 1,200 feet to either side of the runway to which the instrument approach was conducted.
(Refer to AIM.)

SIGMET   A weather advisory issued concerning weather significant to the safety of all aircraft. SIGMET advisories cover severe and extreme turbulence, severe icing, and widespread dust or sandstorms that reduce visibility to less than 3 miles.
(See AIRMET.) (See AWW.) (See CONVECTIVE SIGMET.) (See CWA.) (See ICAO term SIGMET INFORMATION.) (Refer to AIM.)

SIGMET INFORMATION [ICAO]   Information issued by a meteorological watch office concerning the occurrence or expected occurrence of specified en-route weather phenomena which may affect the safety of aircraft operations.

SIGNIFICANT METEOROLOGICAL INFORMATION (See SIGMET.)

SIGNIFICANT POINT   A point, whether a named intersection, a NAVAID, a fix derived from a NAVAID(s), or geographical coordinate expressed in degrees of latitude and longitude, which is established for the purpose of providing separation, as a reporting point, or to delineate a route of flight.

SIMPLIFIED DIRECTIONAL FACILITY   A NAVAID used for nonprecision instrument approaches. The final approach course is similar to that of an ILS localizer except that the SDF course may be offset from the runway, generally not more than 3 degrees, and the course may be wider than the localizer, resulting in a lower degree of accuracy.
(Refer to AIM.)

SIMULATED FLAMEOUT   A practice approach by a jet aircraft (normally military) at idle thrust to a runway. The approach may start at a relatively high altitude over a runway (high key) and may continue on a relatively high and wide downwind leg with a high rate of descent and a continuous turn to final. It terminates in a landing or low approach. The purpose of this approach is to simulate a flameout.
(See FLAME OUT.)

SIMULTANEOUS ILS APPROACHES   An approach system permitting simultaneous ILS/MLS approaches to airports having parallel runways separated by at least 4,300 feet between centerlines. Integral parts of a total system are ILS/MLS, radar, communications, ATC procedures, and appropriate airborne equipment.
(See PARALLEL RUNWAYS.) (Refer to AIM.)

SIMULTANEOUS MLS APPROACHES   (See SIMULTANEOUS ILS APPROACHES.)

SINGLE DIRECTION ROUTES Preferred IFR Routes which are sometimes depicted on high altitude en route charts and which are normally flown in one direction only.
(See PREFERRED IFR ROUTES.) (Refer to AIRPORT/ FACILITY DIRECTORY.)

SINGLE FREQUENCY APPROACH A service provided under a letter of agreement to military single-piloted turbojet aircraft which permits use of a single UHF frequency during approach for landing. Pilots will not normally be required to change frequency from the beginning of the approach to touchdown except that pilots conducting an en route descent are required to change frequency when control is transferred from the air route traffic control center to the terminal facility. The abbreviation "SFA" in the DOD FLIP IFR Supplement under "Communications" indicates this service is available at an aerodrome.

SINGLE-PILOTED AIRCRAFT A military turbojet aircraft possessing one set of flight controls, tandem cockpits, or two sets of flight controls but operated by one pilot is considered single-piloted by ATC when determining the appropriate air traffic service to be applied.
(See SINGLE FREQUENCY APPROACH.)

SKYSPOTTER A pilot who has received specialized training in observing and reporting inflight weather phenomena.

SLASH A radar beacon reply displayed as an elongated target.

SLDI (See SECTOR LIST DROP INTERVAL.)

SLOT TIME (See METER FIX TIME/SLOT TIME.)

SLOW TAXI To taxi a float plane at low power or low RPM.

SN (See SYSTEM STRATEGIC NAVIGATION.)

SPEAK SLOWER Used in verbal communications as a request to reduce speech rate.

SPECIAL INSTRUMENT APPROACH PROCEDURE (See INSTRUMENT APPROACH PROCEDURE.)

SPECIAL ACTIVITY AIRSPACE (SAA) Any airspace with defined dimensions within the National Airspace System wherein limitations may be imposed upon aircraft operations. This airspace may be restricted areas, prohibited areas, military operations areas, air ATC assigned airspace, and any other designated airspace areas. The dimensions of this airspace are programmed into URET and can be designated as either active or inactive by screen entry. Aircraft trajectories are constantly tested against the dimensions of active areas and alerts issued to the applicable sectors when violations are predicted.
(See USER REQUEST EVALUATION TOOL.)

SPECIAL EMERGENCY A condition of air piracy or other hostile act by a person(s) aboard an aircraft which threatens the safety of the aircraft or its passengers.

SPECIAL INSTRUMENT APPROACH PROCEDURE (See INSTRUMENT APPROACH PROCEDURE.)

SPECIAL USE AIRSPACE Airspace of defined dimensions identified by an area on the surface of the Earth wherein activities must be confined because of their nature and/or wherein limitations may be imposed upon aircraft operations that are not a part of those activities. Types of special use airspace are:

a. Alert Area. Airspace which may contain a high volume of pilot training activities or an unusual type of aerial activity, neither of which is hazardous to aircraft. Alert Areas are depicted on aeronautical charts for the information of nonparticipating pilots. All activities within an Alert Area are conducted in accordance with Federal Aviation Regulations, and pilots of participating aircraft as well as pilots transiting the area are equally responsible for collision avoidance.

b. Controlled Firing Area. Airspace wherein activities are conducted under conditions so controlled as to eliminate hazards to nonparticipating aircraft and to ensure the safety of persons and property on the ground.

c. Military Operations Area (MOA). A MOA is airspace established outside of Class A airspace area to separate or segregate certain nonhazardous military activities from IFR traffic and to identify for VFR traffic where these activities are conducted.
(Refer to AIM.)

d. Prohibited Area. Airspace designated under part 73 within which no person may operate an aircraft without the permission of the using agency.
(Refer to En Route Charts, AIM.)

e. Restricted Area. Airspace designated under FAR Part 73, within which the flight of aircraft, while not wholly prohibited, is subject to restriction. Most restricted areas are designated joint use and IFR/VFR operations in the area may be authorized by the controlling ATC facility when it is not being utilized by the using agency. Restricted areas are depicted on en route charts. Where joint use is authorized, the name of the ATC controlling facility is also shown.
(Refer to FAR Part 73.) (Refer to AIM.)

f. Warning Area. A warning area is airspace of defined dimensions extending from 3 nautical miles outward from the coast of the United States, that contains activity that may be hazardous to nonparticipating aircraft. The purpose of such warning area is to warn nonparticipating pilots of the potential danger. A warning area may be located over domestic or international waters or both.

SPECIAL VFR CONDITIONS Meteorological conditions that are less than those required for basic VFR flight in Class B, C, D, or E surface areas and in which some aircraft are permitted flight under visual flight rules.

(See SPECIAL VFR OPERATIONS.) (Refer to FAR Part 91.)

SPECIAL VFR FLIGHT [ICAO]   A VFR flight cleared by air traffic control to operate within Class B, C, D, and E surface areas in meteorological conditions below VMC.

SPECIAL VFR OPERATIONS   Aircraft operating in accordance with clearances within Class B, C, D, and E surface areas in weather conditions less than the basic VFR weather minima. Such operations must be requested by the pilot and approved by ATC.

(See SPECIAL VFR CONDITIONS.) (See ICAO term SPECIAL VFR FLIGHT.)

SPEED   (See AIRSPEED.)
(See GROUND SPEED.)

SPEED ADJUSTMENT   An ATC procedure used to request pilots to adjust aircraft speed to a specific value for the purpose of providing desired spacing. Pilots are expected to maintain a speed of plus or minus 10 knots or 0.02 Mach number of the specified speed.

Examples of speed adjustments are:
  1. "Increase/reduce speed to Mach point (number)."
  2. "Increase/reduce speed to (speed in knots)" or "Increase/reduce speed (number of knots) knots."

SPEED BRAKES   Movable aerodynamic devices on aircraft that reduce airspeed during descent and landing.

SPEED SEGMENTS   Portions of the arrival route between the transition point and the vertex along the optimum flight path for which speeds and altitudes are specified. There is one set of arrival speed segments adapted from each transition point to each vertex. Each set may contain up to six segments.

SQUAWK (Mode, Code, Function)   Activate specific modes/codes/functions on the aircraft transponder; e.g., "Squawk three/alpha, two one zero five, low."

(See TRANSPONDER.)

STA   (See SCHEDULED TIME OF ARRIVAL.)

STAGING/QUEUING   The placement, integration, and segregation of departure aircraft in designated movement areas of an airport by departure fix, EDCT, and/or restriction.

STANDARD INSTRUMENT APPROACH PROCEDURE   (See INSTRUMENT APPROACH PROCEDURE.)

STANDARD INSTRUMENT APPROACH PROCEDURE (SIAP)   (See INSTRUMENT APPROACH PROCEDURE.)

STANDARD INSTRUMENT DEPARTURE   A preplanned instrument flight rule (IFR) air traffic control departure procedure printed for pilot use in graphic and/or textual form. SIDs provide transition from the terminal to the appropriate en route structure.

(See IFR TAKEOFF MINIMUMS AND DEPARTURE PROCEDURES.) (Refer to AIM.)

STANDARD INSTRUMENT DEPARTURE CHARTS   (See AERONAUTICAL CHART.)

STANDARD INSTRUMENT DEPARTURE (SID)   A preplanned instrument flight rule (IFR) air traffic control (ATC) departure procedure printed for pilot controller use in graphic form to provide obstacle clearance and a transition from the terminal area to the appropriate en route structure. SIDs are primarily designed for system enhancement to expedite traffic flow and to reduce pilot/controller workload. ATC clearance must always be received prior to flying a SID.

(See IFR TAKEOFF MINIMUMS AND DEPARTURE PROCEDURES.)
(See OBSTACLE DEPARTURE PROCEDURE.)
(Refer to AIM.)

STANDARD RATE TURN   A turn of three degrees per second.

STANDARD TERMINAL ARRIVAL   A preplanned instrument flight rule (IFR) air traffic control arrival procedure published for pilot use in graphic and/or textual form. STARs provide transition from the en route structure to an outer fix or an instrument approach fix/arrival waypoint in the terminal area.

STANDARD TERMINAL ARRIVAL CHARTS   (See AERONAUTICAL CHART.)

STAND BY   Means the controller or pilot must pause for a few seconds, usually to attend to other duties of a higher priority. Also means to wait as in "stand by for clearance." The caller should reestablish contact if a delay is lengthy. "Stand by" is not an approval or denial.

STAR   (See STANDARD TERMINAL ARRIVAL.)

STATE AIRCRAFT   Aircraft used in military, customs and police service, in the exclusive service of any government, or of any political subdivision, thereof including the government of any state, territory, or possession of the United States or the District of Columbia, but not including any government-owned aircraft engaged in carrying persons or property for commercial purposes.

STATIC RESTRICTIONS   Those restrictions that are usually not subject to change, fixed, in place, and/or published.

STATIONARY RESERVATIONS   Altitude reservations which encompass activities in a fixed area. Stationary reservations may include activities, such as special tests of weapons systems or equipment, certain U.S. Navy carrier, fleet, and anti-submarine operations, rocket, missile and drone operations, and certain aerial refueling or similar operations.

STEPDOWN FIX A fix permitting additional descent within a segment of an instrument approach procedure by identifying a point at which a controlling obstacle has been safely overflown.

STEP TAXI To taxi a float plane at full power or high RPM.

STEP TURN A maneuver used to put a float plane in a planing configuration prior to entering an active sea lane for takeoff. The STEP TURN maneuver should only be used upon pilot request.

STEREO ROUTE A routinely used route of flight established by users and ARTCCs identified by a coded name; e.g., ALPHA 2. These routes minimize flight plan handling and communications.

STOL AIRCRAFT (See SHORT TAKEOFF AND LANDING AIRCRAFT.)

STOP ALTITUDE SQUAWK Used by ATC to inform an aircraft to turn-off the automatic altitude reporting feature of its transponder. It is issued when the verbally reported altitude varies 300 feet or more from the automatic altitude report.
(See ALTITUDE READOUT.) (See TRANSPONDER.)

STOP AND GO A procedure wherein an aircraft will land, make a complete stop on the runway, and then commence a takeoff from that point.
(See LOW APPROACH.) (See OPTION APPROACH.)

STOP BURST (See STOP STREAM.)

STOP BUZZER (See STOP STREAM.)

STOPOVER FLIGHT PLAN A flight plan format which permits in a single submission the filing of a sequence of flight plans through interim full-stop destinations to a final destination.

STOP SQUAWK (Mode or Code) Used by ATC to tell the pilot to turn specified functions of the aircraft transponder off.
(See STOP ALTITUDE SQUAWK.) (See TRANSPONDER.)

STOP STREAM Used by ATC to request a pilot to suspend electronic attack activity.
(See JAMMING.)

STOPWAY An area beyond the takeoff runway no less wide than the runway and centered upon the extended centerline of the runway, able to support the airplane during an aborted takeoff, without causing structural damage to the airplane, and designated by the airport authorities for use in decelerating the airplane during an aborted takeoff.

STRAIGHT-IN APPROACH IFR An instrument approach wherein final approach is begun without first having executed a procedure turn, not necessarily completed with a straight-in landing or made to straight-in landing minimums.
(See STRAIGHT-IN LANDING.) (See LANDING MINIMUMS.) (See STRAIGHT-IN APPROACH VFR.)

STRAIGHT-IN APPROACH VFR Entry into the traffic pattern by interception of the extended runway centerline (final approach course) without executing any other portion of the traffic pattern.
(See TRAFFIC PATTERN.)

STRAIGHT-IN LANDING A landing made on a runway aligned within 30° of the final approach course following completion of an instrument approach.
(See STRAIGHT-IN APPROACH-IFR.)

STRAIGHT-IN LANDING MINIMUMS (See LANDING MINIMUMS.)

STRAIGHT-IN MINIMUMS (See STRAIGHT-IN LANDING MINIMUMS.)

STRATEGIC PLANNING Planning whereby solutions are sought to resolve potential conflicts.

SUBSTITUTE ROUTE A route assigned to pilots when any part of an airway or route is unusable because of NAVAID status. These routes consist of:
a. Substitute routes which are shown on U.S. Government charts.
b. Routes defined by ATC as specific NAVAID radials or courses.
c. Routes defined by ATC as direct to or between NAVAIDs.

SUNSET AND SUNRISE The mean solar times of sunset and sunrise as published in the Nautical Almanac, converted to local standard time for the locality concerned. Within Alaska, the end of evening civil twilight and the beginning of morning civil twilight, as defined for each locality.

SUPER HIGH FREQUENCY The frequency band between 3 and 30 gigahertz (gHz). The elevation and azimuth stations of the microwave landing system operate from 5031 MHz to 5091 MHz in this spectrum.

SUPPLEMENTAL WEATHER SERVICE LOCATION Airport facilities staffed with contract personnel who take weather observations and provide current local weather to pilots via telephone or radio. (All other services are provided by the parent FSS.)

SUPPS Refers to ICAO Document 7030 Regional Supplementary Procedures. SUPPS contain procedures for each ICAO Region which are unique to that Region and are not covered in the worldwide provisions identified in the ICAO Air Navigation Plan. Procedures contained in chapter 8 are based in part on those published in SUPPS.

SURFACE AREA   The airspace contained by the lateral boundary of the Class B, C, D, or E airspace designated for an airport that begins at the surface and extends upward.

SURPIC   A description of surface vessels in the area of a Search and Rescue incident including their predicted positions and their characteristics.

(See FAA Order 7110.65, Paragraph 10-7-4, In-Flight Contingencies.)

SURVEILLANCE APPROACH   An instrument approach wherein the air traffic controller issues instructions, for pilot compliance, based on aircraft position in relation to the final approach course (azimuth), and the distance (range) from the end of the runway as displayed on the controller's radar scope. The controller will provide recommended altitudes on final approach if requested by the pilot.

(Refer to AIM.)

SWAP   (See SEVERE WEATHER AVOIDANCE PLAN.)

SWSL   (See SUPPLEMENTAL WEATHER SERVICE LOCATION.)

SYSTEM STRATEGIC NAVIGATION   Military activity accomplished by navigating along a preplanned route using internal aircraft systems to maintain a desired track. This activity normally requires a lateral route width of 10 NM and altitude range of 1,000 feet to 6,000 feet AGL with some route segments that permit terrain following.

———— **T** ————

TACAN   (See TACTICAL AIR NAVIGATION.)

TACAN-ONLY AIRCRAFT   An aircraft, normally military, possessing TACAN with DME but no VOR navigational system capability. Clearances must specify TACAN or VORTAC fixes and approaches.

TACTICAL AIR NAVIGATION   An ultra-high frequency electronic rho-theta air navigation aid which provides suitably equipped aircraft a continuous indication of bearing and distance to the TACAN station.

(See VORTAC.) (Refer to AIM.)

TAILWIND   Any wind more than 90 degrees to the longitudinal axis of the runway. The magnetic direction of the runway shall be used as the basis for determining the longitudinal axis.

TAKEOFF AREA   (See LANDING AREA.)

TAKE-OFF DISTANCE AVAILABLE [ICAO]   The length of the take-off run available plus the length of the clearway, if provided.

TAKE-OFF RUN AVAILABLE [ICAO]   The length of runway declared available and suitable for the ground run of an aeroplane take-off.

TARGET   The indication shown on a radar display resulting from a primary radar return or a radar beacon reply.

(See RADAR.) (See TARGET SYMBOL.) (See ICAO term TARGET.)

TARGET [ICAO]   In radar:

a. Generally, any discrete object which reflects or retransmits energy back to the radar equipment.

b. Specifically, an object of radar search or surveillance.

TARGET RESOLUTION   A process to ensure that correlated radar targets do not touch. Target resolution shall be applied as follows:

a. Between the edges of two primary targets or the edges of the ASR-9 primary target symbol.

b. Between the end of the beacon control slash and the edge of a primary target.

c. Between the ends of two beacon control slashes.

MANDATORY TRAFFIC ADVISORIES AND SAFETY ALERTS SHALL BE ISSUED WHEN THIS PROCEDURE IS USED.

Note: This procedure shall not be provided utilizing mosaic radar systems.

TARGET SYMBOL   A computer-generated indication shown on a radar display resulting from a primary radar return or a radar beacon reply.

TAXI   The movement of an airplane under its own power on the surface of an airport (FAR Part 135.100 [Note]). Also, it describes the surface movement of helicopters equipped with wheels.

(See AIR TAXI.) (See HOVER TAXI.) (Refer to AIM.) (Refer to FAR Part 135.100.)

TAXI INTO POSITION AND HOLD   Used by ATC to inform a pilot to taxi onto the departure runway in takeoff position and hold. It is not authorization for takeoff. It is used when takeoff clearance cannot immediately be issued because of traffic or other reasons.

(See CLEARED FOR TAKEOFF.)

TAXI PATTERNS   Patterns established to illustrate the desired flow of ground traffic for the different runways or airport areas available for use.

TCAS   (See TRAFFIC ALERT AND COLLISION AVOIDANCE SYSTEM.)

TCH   (See THRESHOLD CROSSING HEIGHT.)

TCLT   (See TENTATIVE CALCULATED LANDING TIME.)

TDLS   (See TERMINAL DATA LINK SYSTEM.)

TDZE   (See TOUCHDOWN ZONE ELEVATION.)

TELEPHONE INFORMATION BRIEFING SERVICE
A continuous telephone recording of meteorological and/or aeronautical information.
(Refer to AIM.)

TENTATIVE CALCULATED LANDING TIME   A projected time calculated for adapted vertex for each arrival aircraft based upon runway configuration, airport acceptance rate, airport arrival delay period, and other metered arrival aircraft. This time is either the VTA of the aircraft or the TCLT/ACLT of the previous aircraft plus the AAI, whichever is later. This time will be updated in response to an aircraft's progress and its current relationship to other arrivals.

TERMINAL AREA   A general term used to describe airspace in which approach control service or airport traffic control service is provided.

TERMINAL AREA FACILITY   A facility providing air traffic control service for arriving and departing IFR, VFR, Special VFR, and on occasion en route aircraft.
(See APPROACH CONTROL FACILITY.) (See TOWER.)

TERMINAL DATA LINK SYSTEM (TDLS)   A system that provides Digital Automatic Terminal Information Service (D-ATIS) both on a specified radio frequency and also, for subscribers, in a text message via data link to the cockpit or to a gate printer. TDLS also provides Pre-departure Clearances (PDC), at selected airports, to subscribers, through a service provider, in text to the cockpit or to a gate printer. In addition, TDLS will emulate the Flight Data Input/Output (FDIO) information within the control tower.

TERMINAL VFR RADAR SERVICE   A national program instituted to extend the terminal radar services provided instrument flight rules (IFR) aircraft to visual flight rules (VFR) aircraft. The program is divided into four types service referred to as basic radar service, terminal radar service area (TRSA) service, Class B service and Class C service. The type of service provided at a particular location is contained in the Airport/Facility Directory.
    a. Basic Radar Service: These services are provided for VFR aircraft by all commissioned terminal radar facilities. Basic radar service includes safety alerts, traffic advisories, limited radar vectoring when requested by the pilot, and sequencing at locations where procedures have been established for this purpose and/or when covered by a letter of agreement. The purpose of this service is to adjust the flow of arriving IFR and VFR aircraft into the traffic pattern in a safe and orderly manner and to provide traffic advisories to departing VFR aircraft.
    b. TRSA Service: This service provides, in addition to basic radar service, sequencing of all IFR and participating VFR aircraft to the primary airport and separation between all participating VFR aircraft. The purpose of this service is to provide separation between all participating VFR aircraft and all IFR aircraft operating within the area defined as a TRSA.
    c. Class C Service: This service provides, in addition to basic radar service, approved separation between IFR and VFR aircraft, and sequencing of VFR aircraft, and sequencing of VFR arrivals to the primary airport.
    d. Class B Service: This service provides, in addition to basic radar service, approved separation of aircraft based on IFR, VFR, and/or weight, and sequencing of VFR arrivals to the primary airport(s).
(See CONTROLLED AIRSPACE.) (See TERMINAL RADAR SERVICE AREA.) (Refer to AIM.) (Refer to AIRPORT/FACILITY DIRECTORY.)

TERMINAL RADAR SERVICE AREA   Airspace surrounding designated airports wherein ATC provides radar vectoring, sequencing, and separation on a full-time basis for all IFR and participating VFR aircraft. The AIM contains an explanation of TRSA. TRSAs are depicted on VFR aeronautical charts. Pilot participation is urged but is not mandatory.
(See TERMINAL RADAR PROGRAM.) (Refer to AIM.) (Refer to AIRPORT/FACILITY DIRECTORY.)

TERMINAL—VERY HIGH FREQUENCY OMNIDIRECTIONAL RANGE STATION   A very high frequency terminal omnirange station located on or near an airport and used as an approach aid.
(See NAVIGATIONAL AID.) (See VOR.)

TERRAIN FOLLOWING   The flight of a military aircraft maintaining a constant AGL altitude above the terrain or the highest obstruction. The altitude of the aircraft will constantly change with the varying terrain and/or obstruction.

TETRAHEDRON   A device normally located on uncontrolled airports and used as a landing direction indicator. The small end of a tetrahedron points in the direction of landing. At controlled airports, the tetrahedron, if installed, should be disregarded because tower instructions supersede the indicator.
(See SEGMENTED CIRCLE.) (Refer to AIM.)

TF   (See TERRAIN FOLLOWING.)

THAT IS CORRECT   The understanding you have is right.

360 OVERHEAD   (See OVERHEAD APPROACH.)

THRESHOLD   The beginning of that portion of the runway usable for landing.
(See AIRPORT LIGHTING.) (See DISPLACED THRESHOLD.)

THRESHOLD CROSSING HEIGHT   The theoretical height above the runway threshold at which the aircraft's glideslope antenna would be if the aircraft maintains the trajectory established by the mean ILS glideslope or MLS glidepath.
(See GLIDESLOPE.) (See THRESHOLD.)

THRESHOLD LIGHTS   (See AIRPORT LIGHTING.)

TIBS   (See TELEPHONE INFORMATION BRIEFING SERVICE.)

TIME GROUP   Four digits representing the hour and minutes from the Coordinated Universal Time (UTC) clock. FAA uses UTC for all operations. The term "ZULU" may be used to denote UTC. The word "local" or the time zone equivalent shall be used to denote local when local time is given during radio and telephone communications. When written, a time zone designator is used to indicate local time; e.g., "0205M" (Mountain). The local time may be based on the 24-hour clock system. The day begins at 0000 and ends at 2359.

TMA   (See TRAFFIC MANAGEMENT ADVISOR.)

TMPA   (See TRAFFIC MANAGEMENT PROGRAM ALERT.)

TMU   (See TRAFFIC MANAGEMENT UNIT.)

TODA [ICAO]   (See ICAO Term TAKE-OFF DISTANCE AVAILABLE.)

TORA [ICAO]   (See ICAO Term TAKE-OFF RUN AVAILABLE.)

TORCHING   The burning of fuel at the end of an exhaust pipe or stack of a reciprocating aircraft engine, the result of an excessive richness in the fuel air mixture.

TOTAL ESTIMATED ELAPSED TIME [ICAO]   For IFR flights, the estimated time required from take-off to arrive over that designated point, defined by reference to navigation aids, from which it is intended that an instrument approach procedure will be commenced, or, if no navigation aid is associated with the destination aerodrome, to arrive over the destination aerodrome. For VFR flights, the estimated time required from takeoff to arrive over the destination aerodrome.
(See ESTIMATED ELAPSED TIME.)

TOUCH-AND-GO   An operation by an aircraft that lands and departs on a runway without stopping or exiting the runway.

TOUCH-AND-GO LANDING   (See TOUCH-AND-GO.)

TOUCHDOWN
   a. The point at which an aircraft first makes contact with the landing surface.
   b. Concerning a precision radar approach (PAR), it is the point where the glide path intercepts the landing surface.
(See ICAO term TOUCHDOWN.)

TOUCHDOWN [ICAO]   The point where the nominal glide path intercepts the runway.

Note: Touchdown as defined above is only a datum and is not necessarily the actual point at which the aircraft will touch the runway.

TOUCHDOWN RVR   (See VISIBILITY.)

TOUCHDOWN ZONE   The first 3,000 feet of the runway beginning at the threshold. The area is used for determination of Touchdown Zone Elevation in the development of straight-in landing minimums for instrument approaches.
(See ICAO term TOUCHDOWN ZONE.)

TOUCHDOWN ZONE [ICAO]   The portion of a runway, beyond the threshold, where it is intended landing aircraft first contact the runway.

TOUCHDOWN ZONE ELEVATION   The highest elevation in the first 3,000 feet of the landing surface. TDZE is indicated on the instrument approach procedure chart when straight-in landing minimums are authorized.
(See TOUCHDOWN ZONE.)

TOUCHDOWN ZONE LIGHTING   (See AIRPORT LIGHTING.)

TOWER   A terminal facility that uses air/ground communications, visual signaling, and other devices to provide ATC services to aircraft operating in the vicinity of an airport or on the movement area. Authorizes aircraft to land or takeoff at the airport controlled by the tower or to transit the Class D airspace area regardless of flight plan or weather conditions (IFR or VFR). A tower may also provide approach control services (radar or nonradar).
(See AIRPORT TRAFFIC CONTROL SERVICE.) (See APPROACH CONTROL FACILITY.) (See APPROACH CONTROL SERVICE.) (See MOVEMENT AREA.) (See TOWER EN ROUTE CONTROL SERVICE.) (Refer to AIM.) (See ICAO term AERODROME CONTROL TOWER.)

TOWER EN ROUTE CONTROL SERVICE   The control of IFR en route traffic within delegated airspace between two or more adjacent approach control facilities. This service is designed to expedite traffic and reduce control and pilot communication requirements.

TOWER TO TOWER   (See TOWER EN ROUTE CONTROL SERVICE.)

TPX-42   A numeric beacon decoder equipment/system. It is designed to be added to terminal radar systems for beacon decoding. It provides rapid target identification, reinforcement of the primary radar target, and altitude information from Mode C.
(See AUTOMATED RADAR TERMINAL SYSTEMS.) (See TRANSPONDER.)

TRACEABLE PRESSURE STANDARD   The facility station pressure instrument, with certification/calibration traceable to the National Institute of Standards and Technology. Traceable pressure standards may be mercurial barometers, commissioned ASOS or dual transducer AWOS, or portable pressure standards or DASI.

TRACK   The actual flight path of an aircraft over the surface of the Earth.

(See COURSE.) (See ROUTE.) (See FLIGHT PATH.) (See ICAO term TRACK.)

TRACK [ICAO]   The projection on the Earth's surface of the path of an aircraft, the direction of which path at any point is usually expressed in degrees from North (True, Magnetic, or Grid).

TRAFFIC

a. A term used by a controller to transfer radar identification of an aircraft to another controller for the purpose of coordinating separation action. Traffic is normally issued:

1. in response to a handoff or point out,

2. in anticipation of a handoff or point out, or

3. in conjunction with a request for control of an aircraft.

b. A term used by ATC to refer to one or more aircraft.

TRAFFIC ADVISORIES   Advisories issued to alert pilots to other known or observed air traffic which may be in such proximity to the position or intended route of flight of their aircraft to warrant their attention. Such advisories may be based on:

a. Visual observation.

b. Observation of radar identified and nonidentified aircraft targets on an ATC radar display, or

c. Verbal reports from pilots or other facilities.

Note 1: The word "traffic" followed by additional information, if known, is used to provide such advisories; e.g., "Traffic, 2 o'clock, one zero miles, southbound, eight thousand."

Note 2: Traffic advisory service will be provided to the extent possible depending on higher priority duties of the controller or other limitations; e.g., radar limitations, volume of traffic, frequency congestion, or controller workload. Radar/nonradar traffic advisories do not relieve the pilot of his responsibility to see and avoid other aircraft. Pilots are cautioned that there are many times when the controller is not able to give traffic advisories concerning all traffic in the aircraft's proximity; in other words, when a pilot requests or is receiving traffic advisories, he should not assume that all traffic will be issued.

(Refer to AIM.)

TRAFFIC ALERT   (aircraft call sign) TURN (left/right) IMMEDIATELY, (climb/descend) AND MAINTAIN (altitude).

TRAFFIC ALERT AND COLLISION AVOIDANCE SYSTEM   An airborne collision avoidance system based on radar beacon signals which operates independent of ground-based equipment. TCAS-I generates traffic advisories only. TCAS-II generates traffic advisories, and resolution (collision avoidance) advisories in the vertical plane.

TRAFFIC INFORMATION   (See TRAFFIC ADVISORIES.)

TRAFFIC IN SIGHT   Used by pilots to inform a controller that previously issued traffic is in sight.

(See NEGATIVE CONTACT.) (See TRAFFIC ADVISORIES.)

TRAFFIC MANAGEMENT ADVISOR (TMA)   A computerized tool which assists Traffic Management Coordinators to efficiently schedule arrival traffic to a metered airport, by calculating meter fix times and delays then sending that information to the sector controllers.

TRAFFIC MANAGEMENT PROGRAM ALERT   A term used in a Notice to Airmen (NOTAM) issued in conjunction with a special traffic management program to alert pilots to the existence of the program and to refer them to either the Notices to Airmen publication or a special traffic management program advisory message for program details. The contraction TMPA is used in NOTAM text.

TRAFFIC MANAGEMENT UNIT   The entity in ARTCCs and designated terminals directly involved in the active management of facility traffic. Usually under the direct supervision of an assistant manager for traffic management.

TRAFFIC NO FACTOR   Indicates that the traffic described in a previously issued traffic advisory is no factor.

TRAFFIC NO LONGER OBSERVED   Indicates that the traffic described in a previously issued traffic advisory is no longer depicted on radar, but may still be a factor.

TRAFFIC PATTERN   The traffic flow that is prescribed for aircraft landing at, taxiing on, or taking off from an airport. The components of a typical traffic pattern are upwind leg, crosswind leg, downwind leg, base leg, and final approach.

a. Upwind Leg. A flight path parallel to the landing runway in the direction of landing.

b. Crosswind Leg. A flight path at right angles to the landing runway off its upwind end.

c. Downwind Leg. A flight path parallel to the landing runway in the direction opposite to landing. The downwind leg normally extends between the crosswind leg and the base leg.

d. Base Leg. A flight path at right angles to the landing runway off its approach end. The base leg normally extends from the downwind leg to the intersection of the extended runway centerline.

e. Final Approach. A flight path in the direction of landing along the extended runway centerline. The final approach normally extends from the base leg to the runway. An aircraft making a straight-in approach VFR is also considered to be on final approach.

(See STRAIGHT-IN APPROACH VFR.) (See TAXI PATTERNS.) (Refer to AIM.) (Refer to FAR Part 91.) (See ICAO term AERODROME TRAFFIC CIRCUIT.)

TRAFFIC SITUATION DISPLAY (TSD)   TSD is a computer system that receives radar track data from all 20 CONUS ARTCC's, organizes this data into a mosaic display, and presents it on a computer screen. The display allows the traffic management coordinator multiple methods of selection and highlighting of individual aircraft or groups of aircraft. The user has the option of superimposing these aircraft positions over any number of background displays. These background options include ARTCC boundaries, any stratum of en route sector boundaries, fixes, airways, military and other special use airspace, airports, and geopolitical boundaries. By using the TSD, a coordinator can monitor any number of traffic situations or the entire systemwide traffic flows.

TRAJECTORY   A URET representation of the path an aircraft is predicted to fly based upon a Current Plan or Trial Plan.
   (See USER REQUEST EVALUATION TOOL.)

TRAJECTORY MODELING   The automated process of calculating a trajectory.

TRANSCRIBED WEATHER BROADCAST   A continuous recording of meteorological and aeronautical information that is broadcast on L/MF and VOR facilities for pilots.
   (Refer to AIM.)

TRANSFER OF CONTROL   That action whereby the responsibility for the separation of an aircraft is transferred from one controller to another.
   (See ICAO term TRANSFER OF CONTROL.)

TRANSFER OF CONTROL [ICAO]   Transfer of responsibility for providing air traffic control service.

TRANSFERRING CONTROLLER   A controller/facility transferring control of an aircraft to another controller/facility.
   (See ICAO term TRANSFERRING UNIT/CONTROLLER.)

TRANSFERRING FACILITY   (See TRANSFERRING CONTROLLER.)

TRANSFERRING UNIT/CONTROLLER [ICAO]   Air traffic control unit/air traffic controller in the process of transferring the responsibility for providing air traffic control service to an aircraft to the next air traffic control unit/air traffic controller along the route of flight.

Note: See definition of accepting unit/controller.

TRANSITION
   a. The general term that describes the change from one phase of flight or flight condition to another; e.g., transition from en route flight to the approach or transition from instrument flight to visual flight.
   b. A published procedure (DP Transition) used to connect the basic DP to one of several en route airways/jet routes, or a published procedure (STAR Transition) used to connect one of several en route airways/jet routes to the basic STAR.
   (Refer to DP/STAR Charts.)

TRANSITIONAL AIRSPACE   That portion of controlled airspace wherein aircraft change from one phase of flight or flight condition to another.

TRANSITION POINT   A point at an adapted number of miles from the vertex at which an arrival aircraft would normally commence descent from its en route altitude. This is the first fix adapted on the arrival speed segments.

TRANSITION WAYPOINT   The waypoint that defines the beginning of a runway or en route transition on an RNAV SID or STAR.

TRANSMISSOMETER   An apparatus used to determine visibility by measuring the transmission of light through the atmosphere. It is the measurement source for determining runway visual range (RVR) and runway visibility value (RVV).
   (See VISIBILITY.)

TRANSMITTING IN THE BLIND   A transmission from one station to other stations in circumstances where two-way communication cannot be established, but where it is believed that the called stations may be able to receive the transmission.

TRANSPONDER   The airborne radar beacon receiver/transmitter portion of the Air Traffic Control Radar Beacon System (ATCRBS) which automatically receives radio signals from interrogators on the ground, and selectively replies with a specific reply pulse or pulse group only to those interrogations being received on the mode to which it is set to respond.
   (See INTERROGATOR.) (Refer to AIM.) (See ICAO term TRANSPONDER.)

TRANSPONDER [ICAO]   A receiver/transmitter which will generate a reply signal upon proper interrogation; the interrogation and reply being on different frequencies.

TRANSPONDER CODES   (See CODES.)

TRIAL PLAN   A proposed amendment which utilizes automation to analyze and display potential conflicts along the predicted trajectory of the selected aircraft.

TRSA   (See TERMINAL RADAR SERVICE AREA.)

TSD   (See TRAFFIC SITUATION DISPLAY.)

TURBOJET AIRCRAFT   An aircraft having a jet engine in which the energy of the jet operates a turbine which in turn operates the air compressor.

TURBOPROP AIRCRAFT   An aircraft having a jet engine in which the energy of the jet operates a turbine which drives the propeller.

TURN ANTICIPATION   (Maneuver anticipation.)

TVOR   (See TERMINAL—VERY HIGH FREQUENCY OMNIDIRECTIONAL RANGE STATION.)

TWEB   (See TRANSCRIBED WEATHER BROADCAST.)

TWO-WAY RADIO COMMUNICATIONS FAILURE (See LOST COMMUNICATIONS.)

——————— U ———————

UDF   (See DIRECTION FINDER.)

UHF   (See ULTRAHIGH FREQUENCY.)

ULTRAHIGH FREQUENCY   The frequency band between 300 and 3,000 MHz. The bank of radio frequencies used for military air/ground voice communications. In some instances this may go as low as 225 MHz and still be referred to as UHF.

ULTRALIGHT VEHICLE   An aeronautical vehicle operated for sport or recreational purposes which does not require FAA registration, an airworthiness certificate, nor pilot certification. They are primarily single occupant vehicles, although some two-place vehicles are authorized for training purposes. Operation of an ultralight vehicle in certain airspace requires authorization from ATC.
   (See FAR Part 103.)

UNABLE   Indicates inability to comply with a specific instruction, request, or clearance.

UNDER THE HOOD   Indicates that the pilot is using a hood to restrict visibility outside the cockpit while simulating instrument flight. An appropriately rated pilot is required in the other control seat while this operation is being conducted.
   (Refer to FAR Part 91.)

UNFROZEN   The Scheduled Time of Arrival (STA) tags, which are still being rescheduled by traffic management advisor (TMA) calculations. The aircraft will remain unfrozen until the time the corresponding estimated time of arrival (ETA) tag passes the preset freeze horizon for that aircraft's stream class. At this point the automatic rescheduling will stop, and the STA becomes "frozen."

UNICOM   A nongovernment communication facility which may provide airport information at certain airports. Locations and frequencies of UNICOMs are shown on aeronautical charts and publications.
   (See AIRPORT/FACILITY DIRECTORY.) (Refer to AIM.)

UNPUBLISHED ROUTE   A route for which no minimum altitude is published or charted for pilot use. It may include a direct route between NAVAIDS, a radial, a radar vector, or a final approach course beyond the segments of an instrument approach procedure.
   (See PUBLISHED ROUTE.) (See ROUTE.)

UNRELIABLE   An advisory to pilots indicating the expected level of service of the GPS and/or WAAS may not be available. Pilots must then determine the adequacy of the signal for desired use.

UNRELIABLE (GPS/WAAS)   An advisory to pilots indicating the expected level of service of the GPS and/or WAAS may not be available. Pilots must then determine the adequancy of the signal for desired use.

UPWIND LEG   (See TRAFFIC PATTERN.)

URET
   (See USER REQUEST EVALUATION TOOL CORE CAPABILITY LIMITED DEPLOYMENT.)

URET CCLD   (See USER REQUEST EVALUATION TOOL.)

URGENCY   A condition of being concerned about safety and of requiring timely but not immediate assistance; a potential distress condition.
   (See ICAO term URGENCY.)

URGENCY [ICAO]   A condition concerning the safety of an aircraft or other vehicle, or of person on board or in sight, but which does not require immediate assistance.

USAFIB   (See ARMY AVIATION FLIGHT INFORMATION BULLETIN.)

USER REQUEST EVALUATION TOOL CORE CAPABILITY LIMITED DEPLOYMENT (URET)   User Request Evaluation Tool Core Capability Limited Deployment is an automated tool provided at each Radar Associate position in selected En Route facilities. This tool utilizes flight and radar data to determine present and future trajectories for all active and proposal aircraft and provides enhanced, automated flight data management.

UVDF   (See DIRECTION FINDER.)

——————— V ———————

VASI   (See VISUAL APPROACH SLOPE INDICATOR.)

VDF   (See DIRECTION FINDER.)

VDP   (See VISUAL DESCENT POINT.)

VECTOR   A heading issued to an aircraft to provide navigational guidance by radar.

   (See ICAO term RADAR VECTORING.)

VERIFY   Request confirmation of information; e.g., "verify assigned altitude."

VERIFY SPECIFIC DIRECTION OF TAKEOFF (OR TURNS AFTER TAKEOFF)   Used by ATC to ascertain an aircraft's direction of takeoff and/or direction of turn after takeoff. It is normally used for IFR departures from an airport not having a control tower. When direct communication with the pilot is not possible, the request and information may be relayed through an FSS, dispatcher, or by other means.

(See IFR TAKEOFF MINIMUMS AND DEPARTURE PROCEDURES.)

VERTEX The last fix adapted on the arrival speed segments. Normally, it will be the outer marker of the runway in use. However, it may be the actual threshold or other suitable common point on the approach path for the particular runway configuration.

VERTEX TIME OF ARRIVAL A calculated time of aircraft arrival over the adapted vertex for the runway configuration in use. The time is calculated via the optimum flight path using adapted speed segments.

VERTICAL NAVIGATION (VNAV) A function of area navigation (RNAV) equipment which calculates, displays, and provides vertical guidance to a profile or path.

VERTICAL SEPARATION Separation established by assignment of different altitudes or flight levels.

(See SEPARATION.) (See ICAO term VERTICAL SEPARATION.)

VERTICAL SEPARATION [ICAO] Separation between aircraft expressed in units of vertical distance.

VERTICAL TAKEOFF AND LANDING AIRCRAFT Aircraft capable of vertical climbs and/or descents and of using very short runways or small areas for takeoff and landings. These aircraft include, but are not limited to, helicopters.

(See SHORT TAKEOFF AND LANDING AIRCRAFT.)

VERY HIGH FREQUENCY The frequency band between 30 and 300 MHz. Portions of this band, 108 to 118 MHz, are used for certain NAVAIDS; 118 to 136 MHz are used for civil air/ground voice communications. Other frequencies in this band are used for purposes not related to air traffic control.

VERY HIGH FREQUENCY OMNIDIRECTIONAL RANGE STATION (See VOR.)

VERY LOW FREQUENCY The frequency band between 3 and 30 kHz.

VFR (See VISUAL FLIGHT RULES.)

VFR AIRCRAFT An aircraft conducting flight in accordance with visual flight rules.

(See VISUAL FLIGHT RULES.)

VFR CONDITIONS Weather conditions equal to or better than the minimum for flight under visual flight rules. The term may be used as an ATC clearance/instruction only when:

a. An IFR aircraft requests a climb/descent in VFR conditions.

b. The clearance will result in noise abatement benefits where part of the IFR departure route does not conform to an FAA approved noise abatement route or altitude.

c. A pilot has requested a practice instrument approach and is not on an IFR flight plan.

Note: All pilots receiving this authorization must comply with the VFR visibility and distance from cloud criteria in FAR Part 91. Use of the term does not relieve controllers of their responsibility to separate aircraft in Class B and Class C airspace or TRSAs as required by FAA Order 7110.65. When used as an ATC clearance/instruction, the term may be abbreviated "VFR;" e.g., "MAINTAIN VFR," "CLIMB/DESCEND VFR," etc.

VFR FLIGHT (See VFR AIRCRAFT.)

VFR MILITARY TRAINING ROUTES Routes used by the Department of Defense and associated Reserve and Air Guard units for the purpose of conducting low-altitude navigation and tactical training under VFR below 10,000 feet MSL at airspeeds in excess of 250 knots IAS.

VFR NOT RECOMMENDED An advisory provided by a flight service station to a pilot during a preflight or in-flight weather briefing that flight under visual flight rules is not recommended. To be given when the current and/or forecast weather conditions are at or below VFR minimums. It does not abrogate the pilot's authority to make his own decision.

VFR-ON-TOP ATC authorization for an IFR aircraft to operate in VFR conditions at any appropriate VFR altitude (as specified in FAR and as restricted by ATC). A pilot receiving this authorization must comply with the VFR visibility, distance from cloud criteria, and the minimum IFR altitudes specified in FAR Part 91. The use of this term does not relieve controllers of their responsibility to separate aircraft in Class B and Class C airspace or TRSAs as required by FAA Order 7110.65.

VFR TERMINAL AREA CHARTS (See AERONAUTICAL CHART.)

VFR WAYPOINT (See WAYPOINT.)

VHF (See VERY HIGH FREQUENCY.)

VHF OMNIDIRECTIONAL RANGE/TACTICAL AIR NAVIGATION (See VORTAC.)

VIDEO MAP An electronically displayed map on the radar display that may depict data such as airports, heliports, runway centerline extensions, hospital emergency landing areas, NAVAIDs and fixes, reporting points, airway/route centerlines, boundaries, handoff points, special use tracks, obstructions, prominent geographic features, map alignment indicators, range accuracy marks, minimum vectoring altitudes.

VISIBILITY The ability, as determined by atmospheric conditions and expressed in units of distance, to see and identify prominent unlighted objects by day and prominent lighted objects by night. Visibility is reported as statute miles, hundreds of feet or meters.

(Refer to FAR Part 91.) (See AIM.)

a. Flight Visibility. The average forward horizontal distance, from the cockpit of an aircraft in flight, at which prominent unlighted objects may be seen and identified by day and prominent lighted objects may be seen and identified by night.

b. Ground Visibility. Prevailing horizontal visibility near the Earth's surface as reported by the United States National Weather Service or an accredited observer.

c. Prevailing Visibility. The greatest horizontal visibility equaled or exceeded throughout at least half the horizon circle which need not necessarily be continuous.

d. Runway Visibility Value (RVV). The visibility determined for a particular runway by a transmissometer. A meter provides a continuous indication of the visibility (reported in miles or fractions of miles) for the runway. RVV is used in lieu of prevailing visibility in determining minimums for a particular runway.

e. Runway Visual Range (RVR). An instrumentally derived value, based on standard calibrations, that represents the horizontal distance a pilot will see down the runway from the approach end. It is based on the sighting of either high intensity runway lights or on the visual contrast of other targets whichever yields the greater visual range. RVR, in contrast to prevailing or runway visibility, is based on what a pilot in a moving aircraft should see looking down the runway. RVR is horizontal visual range, not slant visual range. It is based on the measurement of a transmissometer made near the touchdown point of the instrument runway and is reported in hundreds of feet. RVR is used in lieu of RVV and/or prevailing visibility in determining minimums for a particular runway.

1. Touchdown RVR. The RVR visibility readout values obtained from RVR equipment serving the runway touchdown zone.

2. Mid-RVR. The RVR readout values obtained from RVR equipment located midfield of the runway.

3. Rollout RVR. The RVR readout values obtained from RVR equipment located nearest the rollout end of the runway.

(See ICAO term VISIBILITY.) (See ICAO term FLIGHT VISIBILITY.) (See ICAO term GROUND VISIBILITY.) (See ICAO term RUNWAY VISUAL RANGE.)

VISIBILITY [ICAO]   The ability, as determined by atmospheric conditions and expressed in units of distance, to see and identify prominent unlighted objects by day and prominent lighted objects by night.

a. Flight Visibility. The visibility forward from the cockpit of an aircraft in flight.

b. Ground Visibility. The visibility at an aerodrome as reported by an accredited observer.

c. Runway Visual range [RVR]. The range over which the pilot of an aircraft on the centre line of a runway can see the runway surface markings or the lights delineating the runway or identifying its centre line.

VISUAL APPROACH   An approach conducted on an instrument flight rules (IFR) flight plan which authorizes the pilot to proceed visually and clear of clouds to the airport. The pilot must, at all times, have either the airport or the preceding aircraft in sight. This approach must be authorized and under the control of the appropriate air traffic control facility. Reported weather at the airport must be ceiling at or above 1,000 feet and visibility of 3 miles or greater.

(See ICAO term VISUAL APPROACH.)

VISUAL APPROACH [ICAO]   An approach by an IFR flight when either part or all of an instrument approach procedure is not completed and the approach is executed in visual reference to terrain.

VISUAL APPROACH SLOPE INDICATOR   (See AIRPORT LIGHTING.)

VISUAL DESCENT POINT   A defined point on the final approach course of a nonprecision straight-in approach procedure from which normal descent from the MDA to the runway touchdown point may be commenced, provided the approach threshold of that runway, or approach lights, or other markings identifiable with the approach end of that runway are clearly visible to the pilot.

VISUAL FLIGHT RULES   Rules that govern the procedures for conducting flight under visual conditions. The term "VFR" is also used in the United States to indicate weather conditions that are equal to or greater than minimum VFR requirements. In addition, it is used by pilots and controllers to indicate type of flight plan.

(See INSTRUMENT FLIGHT RULES.) (See INSTRUMENT METEOROLOGICAL CONDITIONS.) (See VISUAL METEOROLOGICAL CONDITIONS.) (Refer to FAR Part 91.) (Refer to AIM.)

VISUAL HOLDING   The holding of aircraft at selected, prominent geographical fixes which can be easily recognized from the air.

(See HOLDING FIX.)

VISUAL METEOROLOGICAL CONDITIONS   Meteorological conditions expressed in terms of visibility, distance from cloud, and ceiling equal to or better than specified minima.

(See INSTRUMENT FLIGHT RULES.) (See INSTRUMENT METEOROLOGICAL CONDITIONS.) (See VISUAL FLIGHT RULES.)

VISUAL SEPARATION A means employed by ATC to separate aircraft in terminal areas and en route airspace in the NAS. There are two ways to effect this separation:

a. The tower controller sees the aircraft involved and issues instructions, as necessary, to ensure that the aircraft avoid each other.

b. A pilot sees the other aircraft involved and upon instructions from the controller provides his own separation by maneuvering his aircraft as necessary to avoid it. This may involve following another aircraft or keeping it in sight until it is no longer a factor.

(See and Avoid.) (Refer to FAR Part 91.)

VLF (See VERY LOW FREQUENCY.)

VMC (See VISUAL METEOROLOGICAL CONDITIONS.)

VOICE SWITCHING AND CONTROL SYSTEM The VSCS is a computer controlled switching system that provides air traffic controllers with all voice circuits (air to ground and ground to ground) necessary for air traffic control.
(See VOICE SWITCHING AND CONTROL SYSTEM.) (Refer to AIM.)

VOR A ground-based electronic navigation aid transmitting very high frequency navigation signals, 360 degrees in azimuth, oriented from magnetic north. Used as the basis for navigation in the National Airspace System. The VOR periodically identifies itself by Morse Code and may have an additional voice identification feature. Voice features may be used by ATC or FSS for transmitting instructions/information to pilots.
(See NAVIGATIONAL AID.) (Refer to AIM.)

VORTAC A navigation aid providing VOR azimuth, TACAN azimuth, and TACAN distance measuring equipment (DME) at one site.
(See DISTANCE MEASURING EQUIPMENT.) (See NAVIGATIONAL AID.) (See TACAN.) (See VOR.) (Refer to AIM.)

VORTICES Circular patterns of air created by the movement of an airfoil through the air when generating lift. As an airfoil moves through the atmosphere in sustained flight, an area of low pressure is created above it. The air flowing from the high pressure area to the low pressure area around and about the tips of the airfoil tends to roll up into two rapidly rotating vortices, cylindrical in shape. These vortices are the most predominant parts of aircraft wake turbulence and their rotational force is dependent upon the wing loading, gross weight, and speed of the generating aircraft. The vortices from medium to heavy aircraft can be of extremely high velocity and hazardous to smaller aircraft.
(See AIRCRAFT CLASSES.) (See WAKE TURBULENCE.) (Refer to AIM.)

VOR TEST SIGNAL (See VOT.)

VOT A ground facility which emits a test signal to check VOR receiver accuracy. Some VOTs are available to the user while airborne, and others are limited to ground use only.
(Refer to FAR Part 91.) (See AIM.) (See AIRPORT/FACILITY DIRECTORY.)

VR (See VFR MILITARY TRAINING ROUTES.)

VSCS (See VOICE SWITCHING AND CONTROL SYSTEM.)

VTA (See VERTEX TIME OF ARRIVAL.)

VTOL AIRCRAFT (See VERTICAL TAKEOFF AND LANDING AIRCRAFT.)

——————— **W** ———————

WA (See AIRMET.)
(See WEATHER ADVISORY.)

WAKE TURBULENCE Phenomena resulting from the passage of an aircraft through the atmosphere. The term includes vortices, thrust stream turbulence, jet blast, jet wash, propeller wash, and rotor wash both on the ground and in the air.
(See AIRCRAFT CLASSES.) (See JET BLAST.) (See VORTICES.) (Refer to AIM.)

WARNING AREA (See SPECIAL USE AIRSPACE.)

WAAS (See WIDE-AREA AUGMENTATION SYSTEM.)

WAYPOINT A predetermined geographical position used for route/instrument approach definition, progress reports published VFR routes, visual reporting points or points for transitioning and/or circumnavigating controlled and/or special use airspace, that is defined relative to a VORTAC station or in terms of latitude/longitude coordinates.

WEATHER ADVISORY In aviation weather forecast practice, an expression of hazardous weather conditions not predicted in the area forecast, as they affect the operation of air traffic and as prepared by the NWS.
(See SIGMET.) (See AIRMET.)

WHEN ABLE When used in conjunction with ATC instructions, gives the pilot the latitude to delay compliance until a condition or event has been reconciled. Unlike "pilot discretion," when instructions are prefaced "when able," the pilot is expected to seek the first opportunity to comply. Once a maneuver has been initiated, the pilot is expected to continue until the specifications of the instructions have been met. "When able," should not be used when expeditious compliance is required.

WIDE-AREA AUGMENTATION SYSTEM (WAAS) The WAAS is a satellite navigation system consisting of the equipment and software which augments the GSP Standard Positioning Service (SPS). The WAAS provides enhanced integrity, accuracy, availability, and continuity over and above GPS SPS. The differential correction function provides improved accuracy required for precision approach.

WILCO I have received your message, understand it, and will comply with it.

WIND GRID DISPLAY A display that presents the latest forecasted wind data overlaid on a map of the ARTCC area. Wind data is automatically entered and updated periodically by transmissions from the National Weather Service. Winds at specific altitudes, along with temperatures and air pressure can be viewed.

WIND SHEAR A change in wind speed and/or wind direction in a short distance resulting in a tearing or shearing effect. It can exist in a horizontal or vertical direction and occasionally in both.

WING TIP VORTICES (See VORTICES.)

WORDS TWICE
a. As a request: "Communication is difficult. Please say every phrase twice."
b. As information: "Since communications are difficult, every phrase in this message will be spoken twice."

WORLD AERONAUTICAL CHARTS (See AERONAUTICAL CHART.)

WS (See SIGMET.)
(See WEATHER ADVISORY.)

WST (See CONVECTIVE SIGMET.)
(See WEATHER ADVISORY.)

Federal Aviation Regulations / Part

Part 1—Definitions and abbreviations

Authority: 49 U.S.C. 106(g), 40113, 44701

§ 1.1 General definitions.

As used in subchapters A through K of this chapter unless the context requires otherwise:

"Administrator" means the Federal Aviation Administrator or any person to whom he has delegated his authority in the matter concerned.

"Aerodynamic coefficients" means nondimensional coefficients for aerodynamic forces and moments.

"Air carrier" means a person who undertakes directly by lease, or other arrangement, to engage in air transportation.

Air commerce" means interstate, overseas, or foreign air commerce or the transportation of mail by aircraft or any operation or navigation of aircraft within the limits of any Federal airway or any operation or navigation of aircraft which directly affects, or which may endanger safety in, interstate, overseas, or foreign air commerce.

"Aircraft" means a device that is used or intended to be used for flight in the air.

"Aircraft engine" means an engine that is used or intended to be used for propelling aircraft. It includes turbosuperchargers, appurtenances, and accessories necessary for its functioning, but does not include propellers.

"Airframe" means the fuselage, booms, nacelles, cowlings, fairings, airfoil surfaces (including rotors but excluding propellers and rotating airfoils of engines), and landing gear of an aircraft and their accessories and controls.

"Airplane" means an engine-driven fixed-wing aircraft heavier than air, that is supported in flight by the dynamic reaction of the air against its wings.

"Airport" means an area of land or water that is used or intended to be used for the landing and takeoff of aircraft, and includes its buildings and facilities, if any.

"Airship" means an engine-driven lighter-than-air aircraft that can be steered.

"Air traffic" means aircraft operating in the air or on an airport surface, exclusive of loading ramps and parking areas.

"Air traffic clearance" means an authorization by air traffic control, for the purpose of preventing collision between known aircraft, for an aircraft to proceed under specified traffic conditions within controlled airspace.

"Air traffic control" means a service operated by appropriate authority to promote the safe, orderly, and expeditious flow of air traffic.

Air Traffic Service (ATS) route is a specified route designated for channeling the flow of traffic as necessary for the provision of air traffic services. The term "ATS route" refers to a variety of airways, including jet routes, area navigation (RNAV) routes, and arrival and departure routes. An ATS route is defined by route specifications, which may include:

(1) An ATS route designator;

(2) The path to or from significant points;

(3) Distance between significant points;

(4) Reporting requirements; and

(5) The lowest safe altitude determined by the appropriate authority.

"Air transportation" means interstate, overseas, or foreign air transportation or the transportation of mail by aircraft.

"Alert Area." An alert area is established to inform pilots of a specific area wherein a high volume of pilot training or an unusual type of aeronautical activity is conducted.

"Alternate airport" means an airport at which an aircraft may land if a landing at the intended airport becomes inadvisable.

"Altitude engine" means a reciprocating aircraft engine having a rated takeoff power that is producible from sea level to an established higher altitude.

"Appliance" means any instrument, mechanism, equipment, part, apparatus, appurtenance, or accessory, including communications equipment, that is used or intended to be used in operating or controlling an aircraft in flight, is installed in or attached to the aircraft, and is not part of an airframe, engine, or propeller.

"Approved," unless used with reference to another person, means approved by the Administrator.

"Area navigation (RNAV)" means a method of navigation that permits aircraft operations on any desired course within the coverage of station-referenced navigation signals or within the limits of self-contained system capability.

Area navigation (RNAV) is a method of navigation that permits aircraft operations on any desired flight path.

Area navigation route (KNAV) is an ATS route based on RNAV that can be used by suitably equipped aircraft.

"Armed Forces" means the Army, Navy, Air Force, Marine Corps, and Coast Guard, including their regular and reserve components and members serving without component status.

"Autorotation" means a rotorcraft flight condition in which the lifting rotor is driven entirely by action of the air when the rotorcraft is in motion.

"Auxiliary rotor" means a rotor that serves either to counteract the effect of the main rotor torque on a rotorcraft or to maneuver the rotorcraft about one or more of its three principal axes.

"Balloon" means a lighter-than-air aircraft that is not engine driven, and sustains flight through the use of either gas buoyancy or an airborne heater.

"Brake horsepower" means the power delivered at the propeller shaft (main drive or main output) of an aircraft engine.

"Calibrated airspeed" means indicated airspeed of an aircraft, corrected for position and instrument error. Calibrated airspeed is equal to true airspeed in standard atmosphere at sea level.

"Canard" means the forward wing of a canard configuration and may be a fixed, movable, or variable geometry surface, with or without control surfaces.

"Canard configuration" means a configuration in which the span of the forward wing is substantially less than that of the main wing.

"Category":

(1) As used with respect to the certification, ratings, privileges, and limitations of airmen, means a broad classification of aircraft. Examples include: airplane; rotorcraft; glider; and lighter-than-air; and

(2) As used with respect to the certification of aircraft, means a grouping of aircraft based upon intended use of operating limitations. Examples include: transport, normal, utility, acrobatic, limited, restricted, and provisional.

"Category A," with respect to transport category rotorcraft, means multiengine rotorcraft designed with engine and system isolation features specified in Part 29 and utilizing scheduled takeoff and landing operations under a critical engine failure concept which assures adequate designated surface area and adequate performance capability for continued safe flight in the event of engine failure.

"Category B," with respect to transport category rotorcraft, means single-engine or multiengine rotorcraft which do not fully meet all Category A standards. Category B rotorcraft have no guaranteed stay-up ability in the event of engine failure and unscheduled landing is assumed.

"Category II operations," with respect to the operation of aircraft, means an ILS approach to the runway of an airport under a Category II ILS instrument approach procedure issued by the Administrator or other appropriate authority.

"Category III operations," with respect to the operation of aircraft, means an ILS approach to, and landing on, the runway of an airport using Category III ILS instrument approach procedure issued by the Administrator or other appropriate authority.

Category IIIa operations, an ILS approach and landing with no decision height (DH), or a DH below 100 feet (30 meters), and controlling runway visual range not less than 700 feet (200 meters).

Category IIIb operations, an ILS approach and landing with no DH, or with a DH below 50 feet (15 meters), and controlling runway visual range less than 700 feet (200 meters), but not less than 150 feet (50 meters).

Category IIIc operations, an ILS approach and landing with no DH and no runway visual range limitation.

"Ceiling" means the height above the earth's surface of the lowest layer of clouds or obscuring phenomena that is reported as "broken," "overcast," or "obscuration," and not classified as "thin" or "partial."

"Civil aircraft" means aircraft other than public aircraft.

"Class":

(1) As used with respect to the certification, ratings, privileges, and limitations of airmen, means a classification of aircraft within a category having similar operating characteristics. Examples include: single engine; multiengine; land; water; gyroplane, helicopter; airship; and free balloon; and

(2) As used with respect to the certification of aircraft, means a broad grouping of aircraft having similar characteristics of propulsion, flight, or landing. Examples include: airplane; rotorcraft; glider; balloon; landplane; and seaplane.

"Clearway" means:

(1) For turbine engine powered airplanes certificated after August 29, 1959, an area beyond the runway, not less than 500 feet wide, centrally located about the extended centerline of the runway, and under the control of the airport authorities. The clearway is expressed in terms of a clearway plane, extending from the end of the runway with an upward slope not exceeding 1.25 percent, above which no object nor any terrain protrudes. However, threshold lights may protrude above the plane if their height above the end of the runway is 26 inches or less and if they are located to each side of the runway.

(2) For turbine engine powered airplanes certificated after September 30, 1958, but before August 30, 1959, an area beyond the takeoff runway extending no less than 300 feet on either side of the extended centerline of the runway, at an elevation no higher than the elevation of the end of the runway, clear of all fixed obstacles, and under the control of the airport authorities.

"Climbout speed," with respect to rotorcraft, means a referenced airspeed which results in a flight path clear of the height-velocity envelope during initial climbout.

"Commercial operator" means a person who, for compensation or hire, engages in the carriage by aircraft in air commerce of persons or property, other than as an air carrier or foreign air carrier or under the authority of Part 375 of this Title. Where it is doubtful that an operation is for "compensation or hire," the test applied is whether the carriage by air is merely incidental to the person's other business or is, in itself, a major enterprise for profit.

"Consensus standard" means, for the purpose of certificating light-sport aircraft, an industry-developed consensus standard that applies to aircraft design, production, and airworthiness. It includes, but is not limited to, standards for aircraft design and performance, required equipment, manufacturer quality assurance systems, production acceptance test procedures, operating instructions, maintenance and inspection procedures, identification and recording of major repairs and major alterations, and continued airworthiness.

"Controlled airspace" means an airspace of defined dimensions within which air traffic control service is provided to IFR flights and to VFR flights in accordance with the airspace classification.

[NOTE—CONTROLLED AIRSPACE IS A GENERIC TERM THAT COVERS CLASS A CLASS B, CLASS C, CLASS D, AND CLASS E AIRSPACE.]

"Controlled Firing Area." A controlled firing area is established to contain activities, which if not conducted in a controlled environment, would be hazardous to nonparticipating aircraft.

"Crewmember" means a person assigned to perform duty in an aircraft during flight time.

"Critical altitude" means the maximum altitude at which, in standard atmosphere, it is possible to maintain, at a specified rotational speed, a specified power or a specified manifold pressure. Unless otherwise stated, the critical altitude is the maximum altitude at which it is possible to maintain, at the maximum continuous rotational speed, one of the following:

(1) The maximum continuous power, in the case of engines for which this power rating is the same at sea level and at the rated altitude.

(2) The maximum continuous rated manifold pressure, in the case of engines the maximum continuous power of which, is governed by a constant manifold pressure.

"Critical engine" means the engine whose failure would most adversely affect the performance or handling qualities of an aircraft.

"Decision height," with respect to the operation of aircraft, means the height at which a decision must be made, during an ILS or PAR instrument approach, to either continue the approach or to execute a missed approach.

Enhanced flight visibility (EFV) means the average forward horizontal distance, from the cockpit of an aircraft in flight, at which prominent topographical objects may be clearly distinguished and identified by day or night by a pilot using an enhanced flight vision system. *Enhanced flight vision system (EFVS)* means an electronic means to provide a display of the forward external scene topography (the natural or manmade features of a place or region especially in a way to show their relative positions and elevation) through the use of imaging sensors, such as a forward looking infrared, millimeter wave radiometry, millimeter wave radar, low light level image intensifying.

"Equivalent airspeed" means the calibrated airspeed of an aircraft corrected for adiabatic compressible flow for the particular altitude. Equivalent airspeed is equal to calibrated airspeed in standard atmosphere at sea level.

"Extended over-water operation" means—

(1) With respect to aircraft other than helicopters, and operation over water at a horizontal distance of more than 50 nautical miles from the nearest shoreline; and

(2) With respect to helicopters, an operation over water at a horizontal distance of more than 50 nautical miles from the nearest shoreline and more than 50 nautical miles from an off-shore heliport structure.

"External load" means a load that is carried, or extends, outside of the aircraft fuselage.

"External-load attaching means" means the structural components used to attach an external load to an aircraft, including external-load containers, the backup structure at the attachment points, and any quick-release device used to jettison the external load.

"Final takeoff speed" means the speed of the airplane that exists at the end of the takeoff path in the en route configuration with one engine inoperative.

"Fireproof"—

(1) With respect to materials and parts used to confine fire in a designated fire zone, means the capacity to withstand at least as well as steel in dimensions appropriate for the purpose for which they are used, the heat produced when there is a severe fire of extended duration in that zone; and

(2) With respect to other materials and parts, means the capacity to withstand the heat associated with fire at least as well as steel in dimensions appropriate for the purpose for which they are used.

"Fire resistant"—

(1) With respect to sheet or structural members means the capacity to withstand the heat associated with fire at least as well as aluminum alloy in dimensions appropriate for the purpose for which they are used; and

(2) With respect to fluid-carrying lines, fluid system parts, wiring, air ducts, fittings, and powerplant controls, means the capacity to perform the intended functions under the heat and other conditions likely to occur when there is a fire at the place concerned.

"Flame resistant" means not susceptible to combustion to the point of propagating a flame, beyond safe limits, after the ignition source is removed.

"Flammable," with respect to a fluid or gas, means susceptible to igniting readily or to exploding.

"Flap extended speed" means the highest speed permissible with wing flaps in a prescribed extended position.

"Flash resistant" means not susceptible to burning violently when ignited.

"Flight crewmember" means a pilot, flight engineer, or flight navigator assigned to duty in an aircraft during flight time.

"Flight level" means a level of constant atmospheric pressure related to a reference datum of 29.92 inches of mercury. Each is stated in three digits that represent hundreds of feet. For example, flight level 250 represents a barometric altimeter indication of 25,000 feet; flight level 255, an indication of 25,500 feet.

"Flight plan" means specified information, relating to the intended flight of an aircraft, that is filed orally or in writing with air traffic control.

Flight time" means:

(1) Pilot time that commences when an aircraft moves under its own power for purpose of flight and ends when the aircraft comes to rest after landing, or

(2) For a glider without self-launch capability, pilot time that commences when the glider is towed for the purpose

of flight and ends when the glider comes to rest after landing.

"Flight visibility" means the average forward horizontal distance, from the cockpit of an aircraft in flight, at which prominent unlighted objects may be seen and identified by day and prominent lighted objects may be seen and identified by night.

"Foreign air carrier" means any person other than a citizen of the United States, who undertakes directly, by lease or other arrangement, to engage in air transportation.

"Foreign air commerce" means the carriage by aircraft of persons or property for compensation or hire, or the carriage of mail by aircraft, or the operation or navigation of aircraft in the conduct or furtherance of a business or vocation, in commerce between a place in the United States and any place outside thereof; whether such commerce moves wholly by aircraft or partly by aircraft and partly by other forms of transportation.

"Foreign air transportation" means the carriage by aircraft of persons or property as a common carrier for compensation or hire, or the carriage of mail by aircraft, in commerce between a place in the United States and any place outside of the United States, whether that commerce moves wholly by aircraft or partly by aircraft and partly by, other forms of transportation.

"Forward wing" means a forward lifting surface of a canard configuration or tandem-wing configuration airplane. The surface may be a fixed, movable, or variable geometry surface, with or without control surfaces.

"Glider" means a heavier-than-air aircraft, that is supported in flight by the dynamic reaction of the air against its lifting surfaces and whose free flight does not depend principally on an engine.

"Go-around power or thrust setting" means the maximum allowable in-flight power or thrust setting identified in the performance data.

"Ground visibility" means prevailing horizontal visibility near the earth's surface as reported by the United States National Weather Service or an accredited observer.

"Gyrodyne" means a rotorcraft whose rotors are normally engine-driven for takeoff, hovering, and landing, and for forward flight through part of its speed range, and whose means of propulsion, consisting usually of conventional propellers, is independent of the rotor system.

"Gyroplane" means a rotorcraft whose rotors are not engine-driven except for initial starting, but are made to rotate by action of the air when the rotorcraft is moving; and whose means of propulsion, consisting usually of conventional propellers, is independent of the rotor system.

"Helicopter" means a rotorcraft that, for its horizontal motion, depends principally on its engine-driven rotors.

"Heliport" means an area of land, water, or structure used or intended to be used for the landing and takeoff of helicopters.

"Idle thrust" means the jet thrust obtained with the engine power control lever set at the stop for the least thrust position at which it can be placed.

"IFR conditions" means weather conditions below the minimum for flight under visual flight rules.

"IFR over-the-top," with respect to the operation of aircraft, means the operation of an aircraft over-the-top on an IFR flight plan when cleared by air traffic control to maintain "VFR conditions" or "VFR conditions on top".

"Indicated airspeed" means the speed of an aircraft as shown on its pitot static airspeed indicator calibrated to reflect standard atmosphere adiabatic compressible flow, at sea level uncorrected for airspeed system errors.

"Instrument" means a device using an internal mechanism to show visually or aurally the attitude, altitude, or operation of an aircraft or aircraft part. It includes electronic devices for automatically controlling an aircraft in flight.

"Interstate air commerce" means the carriage by aircraft of persons or property for compensation or hire, or the carriage of mail by aircraft, or the operation or navigation of aircraft in the conduct or furtherance of a business or vocation, in commerce between a place in any State of the United States, or the District of Columbia, and a place in any other State of the United States, or the District of Columbia; or between places in the same State of the United States through the airspace over any place outside thereof, or between places in the same territory or possession of the United States, or the District of Columbia.

"Interstate air transportation" means the carriage by aircraft of persons or property as a common carrier for compensation or hire, or the carriage of mail by aircraft in commerce:

(1) Between a place in a State or the District of Columbia and another place in another State or the District of Columbia;

(2) Between places in the same State through the airspace over any place outside that State; or

(3) Between places in the same possession of the United States; whether that commerce moves wholly by aircraft or partly by aircraft and partly by other forms of transportation.

"Intrastate air transportation" means the carriage of persons or property as a common carrier for compensation or hire, by turbojet-powered aircraft capable of carrying thirty or more persons, wholly within the same State of the United States.

"Kite" means a framework, covered with paper, cloth, metal, or other material, intended to be flown at the end of a rope or cable, and having as its only support the force of the wind moving past its surfaces.

"Landing gear extended speed" means the maximum speed at which an aircraft can be safely flown with the landing gear extended.

"Landing gear operating speed" means the maximum speed at which the landing gear can be safely extended or retracted.

"Large aircraft" means aircraft of more than 12,500 pounds, maximum certificated takeoff weight.

"Lighter-than-air aircraft" means aircraft that can rise and remain suspended by using contained gas weighing less than the air that is displaced by the gas.

"Light-sport aircraft" means an aircraft, other than a helicopter or powered-lift that, since its original certification, has continued to meet the following:

(1) A maximum takeoff weight of not more than

(i) 660 pounds (300 kilograms) for lighter-than-air aircraft;

(ii) 1,320 pounds (600 kilograms) for aircraft not intended for operation on water;

or

(iii) 1,430 pounds (650 kilograms) for an aircraft intended for operation on water.

(2) A maximum airspeed in level flight with maximum continuous power (VH) of not more than 120 knots CAS under standard atmospheric conditions at sea level.

(3) A maximum never-exceed speed (VNE) of not more than 120 knots CAS for a glider.

(4) A maximum stalling speed or minimum steady flight speed without the use of lift-enhancing devices (VsI) of not more than 45 knots CAS at the aircraft's maximum certificated takeoff weight and most critical center of gravity.

(5) A maximum seating capacity of no more than two persons, including the pilot.

(6) A single, reciprocating engine, if powered.

(7) A fixed or ground-adjustable propeller if a powered aircraft other than a powered glider.

(8) A fixed or autofeathering propeller system if a powered glider.

(9) A fixed-pitch, semi-rigid, teetering, two-blade rotor system, if a gyroplane.

(10) A nonpressurized cabin, if equipped with a cabin.

(11) Fixed landing gear, except for an aircraft intended for operation on water or a glider.

(12) Fixed or repositionable landing gear, or a hull, for an aircraft intended for operation on water.

(13) Fixed or retractable landing gear for a glider.

"Load factor" means the ratio of a specified load to the total weight of the aircraft. The specified load is expressed in terms of any of the following: aerodynamic forces, inertia forces, or ground or water reactions.

"Long-range communication system (LRCS)." A system that uses satellite relay, data link, high frequency, or another approved communication system which extends beyond line of sight.

"Long-range navigation system (LRNS)." An electronic navigation unit that is approved for use under instrument flight rules as a primary means of navigation, and has at least one source of navigational imput, such as inertial navigation system, global positioning system, Omega/very low frequency, or Loran C.

"Mach number" means the ratio of true airspeed to the speed of sound.

"Main rotor" means the rotor that supplies the principal lift to a rotorcraft.

"Maintenance" means inspection, overhaul, repair, preservation, and the replacement of parts, but excludes preventive maintenance.

"Major alteration" means an alteration not listed in the aircraft, aircraft engine, or propeller specifications—

(1) That might appreciably affect weight, balance, structural strength, performance, powerplant operation, flight characteristics, or other qualities affecting airworthiness; or

(2) That is not done according to accepted practices or cannot be done by elementary operations.

"Major repair" means a repair—

(1) That, if improperly done, might appreciably affect weight, balance, structural strength, performance, powerplant operation, flight characteristics, or other qualities affecting airworthiness; or

(2) That is not done according to accepted practices or cannot be done by elementary operations.

"Manifold pressure" means absolute pressure as measured at the appropriate point in the induction system and usually expressed in inches of mercury.

Maximum speed for stability characteristics, "V_{FC}/M_{FC} means a speed that may not be less than a speed midway between maximum operating limit speed (V_{MO}/M_{MO}) and demonstrated flight diving speed (V_{DF}/M_{DF}), except that, for altitudes where the Mach number is the limiting factor, M_{FC} need not exceed the Mach number at which effective speed warning occurs.

"Medical certificate" means acceptable evidence of physical fitness on a form prescribed by the Administrator.

"Military operations area." A military operations area (MOA) is airspace established outside Class A airspace to separate or segregate certain nonhazardous military activities from IFR Traffic and to identify for VFR traffic where these activities are conducted.

"Minimum descent altitude" means the lowest altitude, expressed in feet above mean sea level, to which descent is authorized on final approach or during circle-to-land maneuvering in execution of a standard instrument approach procedure, where no electronic glide slope is provided.

"Minor alteration" means an alteration other than a major alteration.

"Minor repair" means a repair other than a major repair.

"Navigable airspace" means airspace at and above the minimum flight altitudes prescribed by or under this chapter, including airspace needed for safe takeoff and landing.

"Night" means the time between the end of evening civil twilight and the beginning of morning civil twilight, as published in the American Air Almanac, converted to local time.

"Nonprecision approach procedure" means a standard instrument approach procedure in which no electronic glide slope is provided.

"Operate," with respect to aircraft, means use, cause to use or authorize to use aircraft, for the purpose (except as provided in § 91.13 of this chapter) of air navigation including the piloting of aircraft, with or without the right of legal control (as owner, lessee, or otherwise).

"Operational control," with respect to a flight, means the exercise of authority over initiating, conducting, or terminating a flight.

"Overseas air commerce" means the carriage by aircraft of persons or property for compensation or hire, or the carriage of mail by aircraft, or the operation or navigation of aircraft in the conduct or furtherance of a business or vocation, in commerce between a place in any State of the United States, or the District of Columbia, and any place in a territory or possession of the United States; or between a place in a territory or possession of the United States, and a place in any other territory or possession of the United States.

"Overseas air transportation" means the carriage by aircraft of persons or property as a common carrier or compensation or hire, or the carriage of mail by aircraft, in commerce—

(1) Between a place in a State or the District of Columbia and a place in a possession of the United States; or

(2) Between a place in a possession of the United States and a place in another possession of the United States; whether that commerce moves wholly by aircraft or partly by aircraft and partly by other forms of transportation.

"Over-the-top" means above the layer of clouds or other obscuring phenomena forming the ceiling.

"Parachute" means a device used or intended to be used to retard the fall of a body or object through the air.

"Person" means an individual, firm, partnership, corporation, company, association, joint-stock association, or governmental entity. It includes a trustee, receiver, assignee, or similar representative of any of them.

"Pilotage" means navigation by visual reference to landmarks.

"Pilot in command" means the person who:

(1) Has final authority and responsibility for the operation and safety of the flight;

(2) Has been designated as pilot in command before or during the flight; and

(3) Holds the appropriate category, class, and type rating, if appropriate, for the conduct of the flight.

"Pitch setting" means the propeller blade setting as determined by the blade angle measured in a manner, and at a radius, specified by the instruction manual for the propeller.

"Positive control" means control of all air traffic, within designated airspace, by air traffic control.

"Powered-lift" means a heavier-than-air aircraft capable of vertical takeoff, vertical landing, and low speed flight that depends principally on engine-driven lift devices or engine thrust for lift during these flight regimes and on nonrotating airfoil(s) for lift during horizontal flight.

"Powered parachute" means a powered aircraft comprised of a flexible or semi-rigid wing connected to a fuselage so that the wing is not in position for flight until the aircraft is in motion. The fuselage of a powered parachute contains the aircraft engine, a seat for each occupant and is attached to the aircraft's landing gear.

"Precision approach procedure" means a standard instrument approach procedure in which an electronic glide slope is provided, such as ILS and PAR.

"Preventive maintenance" means simple or minor preservation operations and the replacement of small standard parts not involving complex assembly operations.

"Prohibited area." A prohibited area is airspace designated under Part 73 within which no person may operate an aircraft without the permission of the using agency.

"Propeller" means a device for propelling an aircraft that has blades on an engine-driven shaft and that, when rotated, produces by its action on the air, a thrust approximately perpendicular to its plane of rotation. It includes control components normally supplied by its manufacturer, but does not include main and auxiliary rotors or rotating airfoils of engines.

"Public aircraft" means any of the following aircraft when not being used for a commercial purpose or to carry an individual other than a crewmember or qualified noncrewmember:

(1) An aircraft used only for the United States Government; an aircraft owned by the Government and operated by any person for purposes related to crew training, equipment development, or demonstration; an aircraft owned and operated by the government of a State, the District of Columbia, or a territory or possession of the United States or a political subdivision of one of these governments; or an aircraft exclusively leased for at least 90 continuous days by the government of a State, the District of Columbia, or a territory or possession of the United States or a political subdivision of one of these governments.

(i) For the sole purpose of determining public aircraft status, commercial purposes means the transportation of persons or property for compensation or hire, but does not include the operation of an aircraft by the armed forces for reimbursement when that reimbursement is required by any Federal statute, regulation, or directive, in effect on November 1, 1999, or by one government on behalf of another government under a cost reimbursement agreement if the government on whose behalf the operation is conducted certifies to the Administrator of the Federal Aviation Administration that the operation is necessary to respond to a significant and imminent threat to life or property (including natural resources) and that no service by a private operator is reasonably available to meet the threat.

(ii) For the sole purpose of determining public aircraft status, govern mental function means an activity undertaken by a government, such as national defense, intelligence missions, firefighting, search and rescue, law enforcement (including transport of prisoners, detainees, and illegal aliens), aeronautical research, or biological or geological resource management.

(iii) For the sole purpose of determining public aircraft status, qualified non-crewmember means an individual, other than a member of the crew, aboard an aircraft operated by the armed forces or an intelligence agency of the United States Government, or whose presence is required to perform, or is associated with the performance of, a governmental function.

(2) An aircraft owned or operated by the armed forces or chartered to provide transportation to the armed forces if—

(i) The aircraft is operated in accordance with title 10 of the United States Code;

(ii) The aircraft is operated in the performance of a governmental function under title 14, 31, 32, or 50 of the United States Code and the aircraft is not used for commercial purposes; or

(iii) The aircraft is chartered to provide transportation to the armed forces and the Secretary of Defense (or the Secretary of the department in which the Coast Guard is operating) designates the operation of the aircraft as being required in the national interest.

(3) An aircraft owned or operated by the National Guard of a State, the District of Columbia, or any territory or possession of the United States, and that meets the criteria of paragraph (2) of this definition, qualifies as a public aircraft only to the extent that it is operated under the direct control of the Department of Defense.

"Rated continuous OEI power," with respect to rotorcraft turbine engines, means the approved brake horsepower developed under static conditions at specified altitudes and temperatures within the operating limitations established for the engine under Part 33 of this chapter, and limited in use to the time required to complete the flight after the failure of one engine of a multiengine rotorcraft.

"Rated maximum continuous augmented thrust," with respect to turbojet engine type certification, means the approved jet thrust that is developed statically or in flight, in standard atmosphere at a specified altitude, with fluid injection or with the burning of fuel in a separate combustion chamber, within the engine operating limitations established under Part 33 of this chapter, and approved for unrestricted periods of use.

"Rated maximum continuous power," with respect to reciprocating, turbopropeller, and turboshaft engines, means the approved brake horsepower that is developed statically or in flight, in standard atmosphere at a specified altitude, within the engine operating limitations established under Part 33, and approved for unrestricted periods of use.

"Rated maximum continuous thrust," with respect to turbojet engine type certification, means the approved jet thrust that is developed statically or in flight, in standard

atmosphere at a specified altitude, without fluid injection and without the burning of fuel in a separate combustion chamber, within the engine operating limitations established under Part 33 of this chapter, and approved for unrestricted periods of use.

"Rated takeoff augmented thrust," with respect to turbojet engine type certification, means the approved jet thrust that is developed statically under standard sea level conditions, with fluid injection or with the burning of fuel in a separate combustion chamber, within the engine operating limitations established under Part 33 of this chapter, and limited in use to periods of not over 5 minutes for takeoff operation.

"Rated takeoff power," with respect to reciprocating, turbopropeller, and turboshaft engine type certification, means the approved brake horsepower that is developed statically under standard sea level conditions, within the engine operating limitations established under Part 33, and limited in use to periods of not over 5 minutes for takeoff operation.

"Rated takeoff thrust," with respect to turbojet engine type certification, means the approved jet thrust that is developed statically under standard sea level conditions, without fluid injection and without the burning of fuel in a separate combustion chamber, within the engine operating limitations established under Part 33 of this chapter, and limited in use to periods of not over 5 minutes for takeoff operation.

"Rated $2\frac{1}{2}$-minute OEI power," with respect to rotorcraft turbine engines, means the approved brake horsepower developed under static conditions at specified altitudes and temperatures within the operating limitations established for the engine under Part 33 of this chapter, and limited to three periods of use no longer than $2\frac{1}{2}$ minutes each in any one flight, and followed by mandatory inspection and prescribed maintenance action.

"Rated 30-second OEI power," with respect to rotorcraft turbine engines, means the approved brake horsepower-developed under static conditions at specified altitudes and temperatures within the operating limitations established for the engine under Part 33 of this chapter, for continued one-flight operation after the failure of one engine in multiengine rotorcraft, limited to three periods of use no longer than 30 seconds each in any one flight, and followed by mandatory inspection and prescribed maintenance action.

"Rating" means a statement that, as a part of a certificate, sets forth special conditions, privileges, or limitations.

"Reference landing speed" means the speed of the airplane, in a specified landing configuration, at the point where it descends through the 50 foot height in the determination of the landing distance.

"Reporting point" means a geographical location in relation to which the position of an aircraft is reported.

"Restricted area." A restricted area is airspace designated under Part 73 within which the flight of aircraft, while not wholly prohibited, is subject to restriction.

"RNAV way point (W/P)" means a predetermined geographical position used for route or instrument approach definition or progress reporting purposes that is defined relative to a VORTAC station position.

"Rocket" means an aircraft propelled by ejected expanding gases generated in the engine from self-contained propellants and not dependent on the intake of outside substances. It includes any part which becomes separated during the operation.

"Rotorcraft" means a heavier-than-air aircraft that depends principally for its support in flight on the lift generated by one or more rotors.

"Rotorcraft-load combination" means the combination of a rotorcraft and an external-load, including the external-load attaching means. Rotorcraft-load combinations are designated as Class A, Class B, Class C, and Class D, as follows:

(1) "Class A rotorcraft-load combination" means one in which the external load cannot move freely, cannot be jettisoned, and does not extend below the landing gear.

(2) "Class B rotorcraft-load combination" means one in which the external load is jettisonable and is lifted free of land or water during the rotorcraft operation.

(3) "Class C rotorcraft-load combination" means one in which the external load is jettisonable and remains in contact with land or water during the rotorcraft operation.

(4) "Class D rotorcraft-load combination" means one in which the external-load is other than a Class A, B, or C and has been specifically approved by the Administrator for that operation.

Route Segment is a portion of a route bounded on each end by a fix or navigation aid (NAVAID).

"Sea level engine" means a reciprocating aircraft engine having a rated takeoff power that is producible only at sea level.

"Second in command" means a pilot who is designated to be second in command of an aircraft during flight time.

"Show," unless the context otherwise requires, means to show to the satisfaction of the Administrator.

"Small aircraft" means aircraft of 12,500 pounds or less, maximum certificated takeoff weight.

Special VFR conditions mean meteorological conditions that are less than those required for basic VFR flight in controlled airspace and in which some aircraft are permitted flight under visual flight rules.

Special VFR operations means aircraft operating in accordance with clearances within controlled airspace in meteorological conditions less than the basic VFR weather minima. Such operations must be requested by the pilot and approved by ATC.

"Standard atmosphere" means the atmosphere defined in U.S. Standard Atmosphere, 1962 (Geo-potential altitude tables).

"Stopway" means an area beyond the takeoff runway, no less wide than the runway and centered upon the extended centerline of the runway, able to support the airplane during an aborted takeoff, without causing structural damage to the airplane, and designated by the airport authorities for use in decelerating the airplane during an aborted takeoff.

Synthetic vision means a computer-generated image of the external scene topography from the perspective of the flight deck that is derived from aircraft attitude, high-precision navigation solution, and database of terrain, obstacles and relevant cultural features. *Synthetic vision* system means an electronic means to display a synthetic vision image of the external scene topography to the flight crew.

"Takeoff power"—

(1) With respect to reciprocating engines, means the brake horsepower that is developed under standard sea level conditions, and under the maximum conditions of crankshaft rotational speed and engine manifold pressure approved for the normal takeoff, and limited in continuous use to the period of time shown in the approved engine specification; and

(2) With respect to turbine engines, means the brake horsepower that is developed under static conditions at a specified altitude and atmospheric temperature, and under the maximum conditions of rotorshaft rotational speed and gas temperature approved for the normal take off, and limited in continuous use to the period of time shown in the approved engine specification.

"Takeoff safety speed" means a referenced airspeed obtained after lift-off at which the required one-engine-inoperative climb performance can be achieved.

"Takeoff thrust," with respect to turbine engines, means the jet thrust that is developed under static conditions at a specific altitude and atmospheric temperature under the maximum conditions of rotorshaft rotational speed and gas temperature approved for the normal takeoff, and limited in continuous use to the period of time shown in the approved engine specification.

"Tandem wing configuration" means a configuration having two wings of similar span, mounted in tandem.

"Time in service," with respect to maintenance time records, means the time from the moment an aircraft leaves the surface of the earth until it touches it at the next point of landing.

"True airspeed" means the airspeed of an aircraft relative to undisturbed air. True airspeed is equal to equivalent airspeed multiplied by $(0)^{1/2}$.

"Traffic pattern" means the traffic flow that is prescribed for aircraft landing at, taxiing on, or taking off from, an airport.

"Type":

(1) As used with respect to the certification, ratings, privileges, and limitations of airmen, means a specific make and basic model of aircraft, including modifications thereto that do not change its handling or flight characteristics. Examples include: DC-7, 1049, and F-27; and

(2) As used with respect to the certification of aircraft, means those aircraft which are similar in design. Examples

include: DC-7 and DC-7C; 1049G and 1049H; and F-27 and F-27F.

(3) As used with respect to the certification of aircraft engines means those engines which are similar in design. For example, JT8D and JT8D-7 are engines of the same type, and JT9D-3A and JT9D-7 are engines of the same type.

"United States," in a geographical sense, means (1) the States, the District of Columbia, Puerto Rico, and the possessions, including the territorial waters, and (2) the airspace of those areas.

"United States air carrier" means a citizen of the United States who undertakes directly by lease, or other arrangement, to engage in air transportation.

"VFR over-the-top," with respect to the operation of aircraft, means the operation of an aircraft over-the-top under VFR when it is not being operated on an IFR flight plan.

"Warning area." A warning area is airspace of defined dimensions, extending from 3 nautical miles outward from the coast of the aunited States, that contains activity that may be hazardous to nonparticipating pilots of the potential danger. A warning area may be located over domestic or international waters or both.

"Weight-shift-control aircraft" means a powered aircraft with a framed pivoting wing and a fuselage controllable only in pitch and roll by the pilot's ability to change the aircraft's center of gravity with respect to the wing. Flight control of the aircraft depends on the wing's ability to flexibly deform rather than the use of control surfaces.

"Winglet or tip fin" means an out-of-plane surface extending from a lifting surface. The surface may or may not have control surfaces.

[Doc. No. 1150, 27 FR 4588, May 15, 1962, as amended by Amdt. 1–36, 54 FR 34389, Aug. 18, 1989; Amdt. 1–37, 56 FR 351, Jan. 3, 1991; Amdt. 1-38, 56 FR 65653, Dec. 17, 1991; Amdt. 1-39, 60 FR 5067, Jan. 25, 1995; Amdt. 1–40, 60 FR 30749, June 9, 1995; Amdt. 1–42, 61 FR 2081, Jan. 24, 1996; Amdt. 1–43, 61 FR 5183, Feb. 9, 1996; Amdt. 1–44, 61 FR 7190, Feb. 26, 1996; Amdt. 1–46, 61 FR 31328, June 19, 1996; Amdt. 1–45, 61 FR 34547, July 2, 1996; Amdt. 1–47, 62 FR 16298, Apr. 4, 1997]

§ 1.2 Abbreviations and symbols.

In Subchapters A through K of this chapter:

"AGL" means above ground level.

"ALS" means approach light system.

"ASR" means airport surveillance radar.

"ATC" means air traffic control.

"CAS" means calibrated airspeed.

"CAT II" means Category II.

"CONSOL or CONSOLAN" means a kind of low or medium frequency long range navigational aid.

"DH" means decision height.

"DME" means distance measuring equipment compatible with TACAN.

"EAS" means equivalent airspeed.

"EFVS" means enhanced flight vision system.

"FAA" means Federal Aviation Administration.

"FM" means fan marker.

"GS" means glide slope.

"HIRL" means high-intensity runway light system.

"IAS" means indicated airspeed.

"ICAO" means International Civil Aviation Organization.

"IFR" means instrument flight rules.

"ILS" means instrument landing system.

"IM" means ILS inner marker.

"INT" means intersection.

"LDA" means localizer-type directional aid.

"LFR" means low-frequency radio range.

"LMM" means compass locator at middle marker.

"LOC" means ILS localizer.

"LOM" means compass locator at outer marker.

"M" means mach number.

"MAA" means maximum authorized IFR altitude.

"MALS" means medium intensity approach light system.

"MALSR" means medium intensity approach light system with runway alignment indicator lights.

"MCA" means minimum crossing altitude.

"MDA" means minimum descent altitude.

"MEA" means minimum en route IFR altitude.

"MM" means ILS middle marker.

"MOCA" means minimum obstruction clearance altitude.

"MRA" means minimum reception altitude.

"MSL" means mean sea level.

"NDB(ADF)" means nondirectional beacon (automatic direction finder).

"NOPT" means no procedure turn required.

"OEI" means one engine inoperative.

"OM" means ILS outer marker.

"PAR" means precision approach radar.

"RAIL" means runway alignment indicator light system.

"RBN" means radio beacon.

"RCLM" means runway centerline marking.

"RCLS" means runway centerline light system.

"REIL" means runway end identification lights.

"RR" means low or medium frequency radio range station.

"RVR" means runway visual range as measured in the touchdown zone area.

"SALS" means short approach light system.

"SSALS" means simplified short approach light system.

"SSALSR" means simplified short approach light system with runway alignment indicator lights.

"TACAN" means ultra-high frequency tactical air navigational aid.

"TAS" means true airspeed.

"TCAS" means a traffic alert and collision avoidance system.

"TDZL" means touchdown zone lights.

"TVOR" means very high frequency terminal omnirange station.

V_A means design maneuvering speed.

V_B means design speed for maximum gust intensity.

V_C means design cruising speed.

V_D means design diving speed.

V_{DF}/M_{DF} means demonstrated flight diving speed.

V_{EF} means the speed at which the critical engine is assumed to fail during a takeoff.

V_F means design flap speed.

V_{FC}/M_{FC} means maximum speed for stability characteristics.

V_{FE} means maximum flap extended speed.

V_{FTO} means final takeoff speed.

V_H means maximum speed in level flight with maximum continuous power.

V_{LE} means maximum landing gear extended speed.

V_{LO} means maximum landing gear operating speed.

V_{LOF} means lift-off speed.

V_{MC} means minimum control speed with the critical engine inoperative.

V_{MO}/M_{MO} means maximum operating limit speed.

V_{MU} means minimum unstick speed.

V_{NE} means never-exceed speed.

V_{NO} means maximum structural cruising speed.

V_R means rotation speed.

V_{REF} means reference landing speed.

V_S means the stalling speed or the minimum steady flight speed at which the airplane is controllable.

V_{SO} means the stalling speed or the minimum steady flight speed in the landing configuration.

V_{SR} means reference stall speed.

V_{SRO} means reference stall speed in the landing configuration.

V_{S1} means the stalling speed or the minimum steady flight speed obtained in a specific configuration.

V_{SR1} means reference stall speed in a specific configuration.

V_{SW} means speed at which onset of natural or artificial stall warning occurs.

V_{TOSS} means takeoff safety speed for Category A rotorcraft.

V_X means speed for best angle of climb.

V_Y means speed for best rate of climb.

V_1 means the maximum speed in the takeoff at which the pilot must take the first action (e.g., apply brakes, reduce thrust, deploy speed brakes) to stop the airplane within the accelerate-stop distance. V_1 also means the minimum speed in the takeoff, following a failure of the critical engine at V_{EF}, at which the pilot can continue the takeoff and achieve the required height above the takeoff surface within the takeoff distance.

V_2 means takeoff safety speed.

$V_{2\ MIN}$ means minimum takeoff safety speed.

"VFR" means visual flight rules.

"VHF" means very high frequency.

"VOR" means very high frequency omnirange station.

"VORTAC" means collocated VOR and TACAN.

(Doc. No 1150, 27 FR 4590, May 15, 1962, as amended by Amdt. 1–35, 54 FR 950, Jan. 10, 1989; Amdt. 1–48, 63 FR 8318, Feb. 18, 1998)

§ 1.3 Rules of construction.

(a) In Subchapters A through K of this chapter, unless the context requires otherwise:

(1) Words importing the singular include the plural;

(2) Words importing the plural include the singular; and

(3) Words importing the masculine gender include the feminine.

(b) In Subchapters A through K of this chapter, the word:

(1) "Shall" is used in an imperative sense;

(2) "May" is used in a permissive sense to state authority or permission to do the act prescribed, and the words "no person may . . ." or "a person may not . . ." mean that no person is required, authorized, or permitted to do the act prescribed; and

(3) "Includes" means "includes but is not limited to."

(Doc. No 1150, 27 FR 4590, May 15, 1962, as amended by Amdt. 1–10, 31 FR 5055, Mar. 29, 1966

Part 43—Maintenance, preventive maintenance, rebuilding, and alteration

Source: Docket No. 1993 (29 FR 5451, 4/23/64) unless otherwise noted.

§ 43.1 Applicability.

(a) Except as provided in paragraphs (b) and (d) of this section, this part prescribes rules governing the maintenance, preventive maintenance, rebuilding, and alteration of any

(1) Aircraft having a U.S. airworthiness certificate;

(2) Foreign-registered civil aircraft used in common carriage or carriage of mail under the provisions of Part 121, 127, or 135 of this chapter; and

(3) Airframe, aircraft engines, propellers, appliances, and component parts of such aircraft.

(b) This part does not apply to any aircraft for which the FAA has issued an experimental certificate, unless the FAA has previously issued a different kind of airworthiness certificate for that aircraft.

(c) This part applies to all life-limited parts that are removed from a type certificated product, segregated, or controlled as provided in § 43.10.

(d) This part applies to any aircraft issued a special airworthiness certificate in the light-sport category except:

(1) The repair or alteration form specified in §§43.5 (b) and 43.9 (d) is not required to be completed for products not produced under an FAA approval;

(2) Major repairs and major alterations for products not produced under an FAA approval are not required to be recorded in accordance with appendix B of this part; and

(3) The listing of major alterations and major repairs specified in paragraphs (a) and (b) of appendix A of this part is not applicable to products not produced under an FAA approval.

(Amdt. 43-23, 47 FR 41084 Eff. 10/15/82)

§ 43.2 Records of overhaul and rebuilding.

(a) No person may describe in any required maintenance entry or form an aircraft, airframe, aircraft engine, propeller, appliance, or component part as being overhauled unless—

(1) Using methods, techniques, and practices acceptable to the Administrator, it has been disassembled, cleaned, inspected, repaired as necessary, and reassembled; and

(2) It has been tested in accordance with approved standards and technical data, or in accordance with current standards and technical data acceptable to the Administrator, which have been developed and documented by the holder of the type certificate, supplemental type certificate, or a material, part, process, or appliance approval under § 21.305 of this chapter.

(b) No person may describe in any required maintenance entry or form an aircraft, airframe, aircraft engine, propeller, appliance, or component part as being rebuilt unless it has been disassembled, cleaned, inspected, repaired as necessary, reassembled, and tested to the same tolerances and limits as a new item, using either new parts or used parts that either conform to new part tolerances and limits or to approved oversized or undersized dimensions.

Docket No. 21071 (47 FR 41076, 9/16/82)

(Amdt. 43-23, Eff. 10/15/82)

§ 43.3 Persons authorized to perform maintenance, preventive maintenance, rebuilding, and alterations.

(a) Except as provided in this section and § 43.17, no person may maintain, rebuild, alter, or perform preventive maintenance on an aircraft, airframe, aircraft engine, propeller, appliance, or component part to which this part applies. Those items, the performance of which is a major alteration, a major repair, or preventive maintenance, are listed in Appendix A.

(b) The holder of a mechanic certificate may perform maintenance, preventive maintenance, and alterations as provided in Part 65 of this chapter.

(c) The holder of a repairman certificate may perform maintenance, preventive maintenance, and alterations as provided in part 65 of this chapter.

(d) A person working under the supervision of a holder of a mechanic or repairman certificate may perform the maintenance, preventive maintenance, and alterations that his supervisor is authorized to perform, if the supervisor personally observes the work being done to the extent necessary to ensure that it is being done properly and if the supervisor is readily available, in person, for consultation. However, this paragraph does not authorize the performance of any inspection required by Part 91 or Part 125 of this chapter or any inspection performed after a major repair or alteration.

(e) The holder of a repair station certificate may perform maintenance, preventive maintenance, and alterations as provided in Part 145 of this chapter.

(f) The holder of an air carrier operating certificate or an operating certificate issued under Part 121, 127, or 135, may perform maintenance, preventive maintenance, and alterations as provided in Part 121, 127, or 135.

(g) Except for holders of a sport pilot certificate, the holder of a pilot certificate issued under part 61 may perform preventive maintenance on any aircraft owned or operated by that pilot which is not used under part 121, 129, or 135 of this chapter. The holder of a sport pilot certificate may perform preventive maintenance on an aircraft owned or operated by that pilot and issued a special airworthiness certificate in the light-sport category.

(h) Notwithstanding the provisions of paragraph (g) of this section, the Administrator may approve a certificate holder under Part 135 of this chapter, operating rotorcraft in a remote area, to allow a pilot to perform specific preventive maintenance items provided—

(1) The items of preventive maintenance are a result of a known or suspected mechanical difficulty or malfunction that occurred en route to or in a remote area;

(2) The pilot has satisfactorily completed an approved training program and is authorized in writing by the certificate holder for each item of preventive maintenance that the pilot is authorized to perform;

(3) There is no certificated mechanic available to perform preventive maintenance;

(4) The certificate holder has procedures to evaluate the accomplishment of a preventive maintenance item that requires a decision concerning the airworthiness of the rotorcraft; and

(5) The items of preventive maintenance authorized by this section are those listed in paragraph (c) of Appendix A of this part.

(i) Notwithstanding the provisions of paragraph (g) of this section, in accordance with an approval issued to the holder of a certificate issued under part 135 of this chapter, a pilot of an aircraft type-certificated for 9 or fewer passenger seats, excluding any pilot seat, may perform the removal and reinstallation of approved aircraft cabin seats, approved cabin-mounted stretchers, and when no tools are required, approved cabin-mounted medical oxygen bottles, provided—

(1) The pilot has satisfactorily completed an approved training program and is authorized in writing by the certificate holder to perform each task; and

(2) The certificate holder has written procedures available to the pilot to evaluate the accomplishment of the task.

(j) A manufacturer may—

(1) Rebuild or alter any aircraft, aircraft engine, propeller, or appliance manufactured by him under a type or production certificate;

(2) Rebuild or alter any appliance or part of aircraft, aircraft engines, propellers, or appliances manufactured by him under a Technical Standard Order Authorization, an FAA-Parts Manufacturer Approval, or Product and Process Specification issued by the Administrator; and

(3) Perform any inspection required by Part 91 or Part 125 of this chapter on aircraft it manufactures, while currently operating under a production certificate or under a currently approved production inspection system for such aircraft.

Docket No 1993, 29 FR 5451, Apr. 23, 1964, as amended by Amdt. 43-4, 31 FR 5249, Apr. 1, 1966; Amdt. 43-23, 47 FR 41084, Sept. 16, 1982; Amdt. 43–25, 51 FR 40702, Nov. 7, 1986; Amdt. 43–36, 61 FR 19501, May 1, 1996)

§ 43.5 Approval for return to service after maintenance, preventive maintenance, rebuilding, or alteration.

No person may approve for return to service any aircraft, airframe, aircraft engine, propeller, or appliance, that has undergone maintenance, preventive maintenance, rebuilding, or alteration unless—

(a) The maintenance record entry required by § 43.9 or § 43.11, as appropriate, has been made;

(b) The repair or alteration form authorized by or furnished by the Administrator has been executed in a manner prescribed by the Administrator; and

(c) If a repair or an alteration results in any change in the aircraft operating limitations or flight data contained in the approved aircraft flight manual, those operating limitations or flight data are appropriately revised and set forth as prescribed in § 91.9 of this chapter.

(Amdt. 43-23, Eff. 10/15/82); (Amdt. 43-31, Eff. 8/18/90)

§ 43.7 Persons authorized to approve aircraft, airframes, aircraft engines, propellers, appliances, or component parts for return to service after maintenance, preventive maintenance, rebuilding, or alteration.

(a) Except as provided in this section and § 43.17, no person, other than the Administrator, may approve an aircraft, airframe, aircraft engine, propeller, appliance, or component part for return to service after it has undergone maintenance, preventive maintenance, rebuilding, or alteration.

(b) The holder of a mechanic certificate or an inspection authorization may approve an aircraft, airframe, aircraft engine, propeller, appliance, or component part for return to service as provided in Part 65 of this chapter.

(c) The holder of a repair station certificate may approve an aircraft, airframe, aircraft engine, propeller, appliance, or component part for return to service as provided in Part 145 of this chapter.

(d) A manufacturer may approve for return to service any aircraft, airframe, aircraft engine, propeller, appliance, or component part which that manufacturer has worked on under § 43.3(j). However, except for minor alterations, the work must have been done in accordance with technical data approved by the Administrator.

(e) The holder of an air carrier operating certificate or an operating certificate issued under Part 121, 127, or 135, may approve an aircraft, airframe, aircraft engine, propeller, appliance, or component part for return to service as provided in Part 121, 127, or 135 of this chapter, as applicable.

(f) A person holding at least a private pilot certificate may approve an aircraft for return to service after performing preventive maintenance under the provisions of § 43.3(g).

(g) The holder of a repairman certificate (light-sport aircraft) with a maintenance rating may approve an aircraft issued a special airworthiness certificate in light-sport category for return to service, as provided in part 65 of this chapter.

(h) The holder of at least a sport pilot certificate may approve an aircraft owned or operated by that pilot and issued a special airworthiness certificate in the light-sport category for return to service after performing preventive maintenance under the provisions of §43.3 (g).

(Amdt. 43-6, Eff. 7/6/66); (Amdt. 43-12, Eff. 11/15/69); (Amdt. 43-23, Eff. 10/15/82); (Amdt. 43-36, Eff. 5/31/96).

§ 43.9 Content, form, and disposition of maintenance, preventive maintenance, rebuilding, and alteration records (except inspections performed in accordance with Part 91, Part 125, § 135.411(a)(1), and § 135.419 of this chapter).

(a) *Maintenance record entries.* Except as provided in paragraphs (b) and (c) of this section, each person who maintains, performs preventive maintenance, rebuilds, or alters an aircraft, airframe, aircraft engine, propeller, appliance, or component part shall make an entry in the maintenance record of that equipment containing the following information:

(1) A description (or reference to data acceptable to the Administrator) of work performed.

(2) The date of completion of the work performed.

(3) The name of the person performing the work if other than the person specified in paragraph (a)(4) of this section.

(4) If the work performed on the aircraft, airframe, aircraft engine, propeller, appliance, or component part has been performed satisfactorily, the signature, certificate number, and kind of certificate held by the person approving the work. The signature constitutes the approval for return to service only for the work performed.

In addition to the entry required by this paragraph, major repairs and major alterations shall be entered on a form, and the form disposed of, in the manner prescribed in Appendix B, by the person performing the work.

(b) Each holder of an air carrier operating certificate or an operating certificate issued under Part 121, 127, or 135, that is required by its approved operations specifications to provide for a continuous airworthiness maintenance program, shall make a record of the maintenance, preventive maintenance, rebuilding, and alteration, on aircraft, airframes, aircraft engines, propellers, appliances, or component parts which it operates in accordance with the applicable provisions of Part 121, 127, or 135 of this chapter, as appropriate.

(c) This section does not apply to persons performing inspections in accordance with Part 91, 125, §135.411 (a)(1), or § 135.419 of this chapter.

(d) In addition to the entry required by paragraph (a) of this section, major repairs and major alterations shall be entered on a form, and the form disposed of, in the manner prescribed in appendix B, by the person performing the work.

(Amdt. 43-1, Eff. 4/1/65); (Amdt. 43-3, Eff. 4/2/66); (Amdt. 43-11, Eff. 10/16/69); (Amdt. 43-15, Eff. 10/23/72); (Amdt. 43-16, Eff. 9/8/72); (Amdt. 43-23, Eff. 10/15/82)

§ 43.10 Disposition of life-limited aircraft parts.

(a) *Definitions used in this section.* For the purposes of this section the following definitions apply.

Life-limited part means any part for which a mandatory replacement limit is specified in the type design, the Instructions for Continued Airworthiness, or the maintenance manual.

Life status means the accumulated cycles, hours, or any other mandatory replacement limit of a life-limited part.

(b) *Temporary removal of parts from type-certificated products.* When a life-limited part is temporarily removed and reinstalled for the purpose of performing maintenance, no disposition under paragraph (c) of this section is required if—

(1) The life status of the part has not changed;

(2) The removal and reinstallation is performed on the same serial numbered product; and

(3) That product does not accumulate time in service while the part is removed.

(c) *Disposition of parts removed from type-certificated products.* Except as provided in paragraph (b) of this section, after April 15, 2002 each person who removes a life-limited part from a type-certificated product must ensure that the part is controlled using one of the methods in this paragraph. The method must deter the installation of the part after it has reached its life limit. Acceptable methods include:

(1) *Record keeping system.* The part may be controlled using a record keeping system that substantiates the part number, serial number, and current life status of the part. Each time the part is removed from a type certificated product, the record must be updated with the current life status. This system may include electronic, paper, or other means of record keeping.

(2) *Tag or record attached to part.* A tag or other record may be attached to the part. The tag or record must include the part number, serial number, and current life status of the part. Each time the part is removed from a type certificated product, either a new tag or record must be created, or the existing tag or record must be updated with the current life status.

(3) *Non-permanent marking.* The part may be legibly marked using a nonpermanent method showing its current life status. The life status must be updated each time the part is removed from a type certificated product, or if the mark is removed, another method in this section may be used. The mark must be accomplished in accordance with the instructions under § 45.16 of this chapter in order to maintain the integrity of the part.

(4) *Permanent marking.* The part may be legibly marked using a permanent method showing its current life status. The life status must be updated each time the part is removed from a type certificated product. Unless the part is permanently removed from use on type certificated products, this permanent mark must be accomplished in accordance with the instructions under § 45.16 of this chapter in order to maintain the integrity of the part.

(5) The part may be segregated using methods that deter its installation on a type-certificated product. These methods must include, at least—

(i) Maintaining a record of the part number, serial number, and current life status, and

(ii) Ensuring the part is physically stored separately from parts that are currently eligible for installation.

(6) *Mutilation.* The part may be mutilated to deter its installation in a type certificated produce. The mutilation must render the part beyond repair and incapable of being reworked to appear to be airworthy.

(7) *Other methods.* Any other method approved or accepted by the FAA.

(d) *Transfer of life-limited parts.* Each person who removes a life-limited part from a type certificated product and later sells or otherwise transfers that part must transfer with the part the mark, tag, or other record used to comply with this section, unless the part is mutilated before it is sold or transferred.

§ 43.11 Content, form, and disposition of records for inspections conducted under Parts 91 and 125 and §§ 135.411(a)(1) and 135.419 of this chapter.

(a) *Maintenance record entries.* The person approving or disapproving for return to service an aircraft, airframe, aircraft engine, propeller, appliance, or component part after any inspection performed in accordance with Part 91, 125, § 135.411(a)(1), or § 135.419 shall make an entry in the maintenance record of that equipment containing the following information:

(1) The type of inspection and a brief description of the extent of the inspection.

(2) The date of the inspection and aircraft total time in service.

(3) The signature, the certificate number, and kind of certificate held by the person approving or disapproving for return to service the aircraft, airframe, aircraft engine, propeller, appliance, component part, or portions thereof.

(4) Except for progressive inspections, if the aircraft is found to be airworthy and approved for return to service, the following or a similarly worded statement—"I certify

that this aircraft has been inspected in accordance with (insert type) inspection and was determined to be in airworthy condition."

(5) Except for progressive inspections, if the aircraft is not approved for return to service because of needed maintenance, noncompliance with applicable specifications, airworthiness directives, or other approved data, the following or a similarly worded statement—"I certify that this aircraft has been inspected in accordance with (insert type) inspection and a list of discrepancies and unairworthy items dated (date) has been provided for the aircraft owner or operator."

(6) For progressive inspections, the following or a similarly worded statement—"I certify that in accordance with a progressive inspection program, a routine inspection of (identify whether aircraft or components) and a detailed inspection of (identify components) were performed and the (aircraft or components) are (approved or disapproved) for return to service." If disapproved, the entry will further state "and a list of discrepancies and unairworthy items dated (date) has been provided to the aircraft owner or operator."

(7) If an inspection is conducted under an inspection program provided for in Part 91, 125, or § 135.411(a)(1), the entry must identify the inspection program, that part of the inspection program accomplished, and contain a statement that the inspection was performed in accordance with the inspections and procedures for that particular program.

(b) *Listing of discrepancies and placards.* If the person performing any inspection required by Part 91 or 125 or § 135.411(a)(1) of this chapter finds that the aircraft is unairworthy or does not meet the applicable type certificate data, airworthiness directives, or other approved data upon which its airworthiness depends, that person must give the owner or lessee a signed and dated list of those discrepancies. For those items permitted to be inoperative under § 91.213(d)(2) of this chapter, that person shall place a placard, that meets the aircraft's airworthiness certification regulations, on each inoperative instrument and the cockpit control of each item of inoperative equipment, marking it "Inoperative," and shall add the items to the signed and dated list of discrepancies given to the owner or lessee.

[Amdt. 43-23, 47 FR 41085, Sept. 16, 1982, as amended by Amdt. 43-30, 53 FR 50195, Dec. 13, 1988; Amdt 43-36, 61 FR 19501, May 1, 1996]

§ 43.12 Maintenance records: Falsification, reproduction, or alteration.

(a) No person may make or cause to be made:

(1) Any fraudulent or intentionally false entry in any record or report that is required to be made, kept, or used to show compliance with any requirement under this part;

(2) Any reproduction, for fraudulent purpose, of any record or report under this part; or

(3) Any alteration, for fraudulent purpose, of any record or report under this part.

(b) The commission by any person of an act prohibited under paragraph (a) of this section is a basis for suspending or revoking the applicable airman, operator, or production certificate, Technical Standard Order Authorization, FAA-Parts Manufacturer Approval, or Product and Process Specification issued by the Administrator and held by that person.

[Amdt. 43–19, 43 FR 22639, May 25, 1978, as amended by Amdt. 43–23, 47 FR 41085, Sept. 16, 1982]

§ 43.13 Performance rules (general).

(a) Each person performing maintenance, alteration, or preventive maintenance on an aircraft, engine, propeller, or appliance shall use the methods, techniques, and practices prescribed in the current manufacturer's maintenance manual or Instructions for Continued Airworthiness prepared by its manufacturer, or other methods, techniques, and practices acceptable to the Administrator, except as noted in § 43.16. He shall use the tools, equipment, and test apparatus necessary to assure completion of the work in accordance with accepted industry practices. If special equipment or test apparatus is recommended by the manufacturer involved, he must use that equipment or apparatus or its equivalent acceptable to the Administrator.

(b) Each person maintaining or altering, or performing preventive maintenance, shall do that work in such a manner and use materials of such a quality, that the condition of the aircraft, airframe, aircraft engine, propeller, or appliance worked on will be at least equal to its original or properly altered condition (with regard to aerodynamic function, structural strength, resistance to vibration and deterioration, and other qualities affecting airworthiness).

(c) *Special provisions for holders of air carrier operating certificates and operating certificates issued under the provisions of Part 121, 127, or 135 and Part 129 operators holding operations specifications.* Unless otherwise notified by the administrator, the methods, techniques, and practices contained in the maintenance manual or the maintenance part of the manual of the holder of an air carrier operating certificate or an operating certificate under Part 121, 127, or 135 and Part 129 operators holding operations specifications (that is required by its operating specifications to provide a continuous airworthiness maintenance and inspection program) constitute acceptable means of compliance with this section.

[Doc. No. 1993, 29 FR 5451, Apr. 23, 1964, as amended by Amdt 43-20, 45 FR 60182, Sept. 11, 1980; Amdt. 43-23, 47 FR 41085, Sept. 16, 1982; Amdt. 43-28 52 FR 20028, June 16, 1987]

§ 43.15 Additional performance rules for inspections.

(a) *General.* Each person performing an inspection required by Part 91, 123, 125, or 135 of this chapter, shall—

(1) Perform the inspection so as to determine whether the aircraft, or portion(s) thereof under inspection, meets all applicable airworthiness requirements; and

(2) If the inspection is one provided for in Part 123, 125, 135, or § 91.409(e) of this chapter, perform the inspection in accordance with the instructions and procedures set forth in the inspection program for the aircraft being inspected.

(b) *Rotorcraft.* Each person performing an inspection required by Part 91 on a rotorcraft shall inspect the following systems in accordance with the maintenance manual or Instructions for Continued Airworthiness of the manufacturer concerned:

(1) The drive shafts or similar systems.

(2) The main rotor transmission gear box for obvious defects.

(3) The main rotor and center section (or the equivalent area).

(4) The auxiliary rotor on helicopters.

(c) *Annual and 100-hour inspections.* (1) Each person performing an annual or 100-hour inspection shall use a checklist while performing the inspection. The checklist may be of the person's own design, one provided by the manufacturer of the equipment being inspected or one obtained from another source. This checklist must include the scope and detail of the items contained in Appendix D to this part and paragraph (b) of this section.

(2) Each person approving a reciprocating-engine-powered aircraft for return to service after an annual or 100-hour inspection shall, before that approval, run the aircraft engine or engines to determine satisfactory performance in accordance with the manufacturer's recommendations of—

(i) Power output (static and idle r.p.m.);

(ii) Magnetos;

(iii) Fuel and oil pressure; and

(iv) Cylinder and oil temperature.

(3) Each person approving a turbine-engine-powered aircraft for return to service after an annual, 100-hour, or progressive inspection shall, before that approval, run the aircraft engine or engines to determine satisfactory performance in accordance with the manufacturer's recommendations.

(d) *Progressive inspection.* (1) Each person performing a progressive inspection shall, at the start of a progressive inspection system, inspect the aircraft completely. After this initial inspection, routine and detailed inspections must be conducted as prescribed in the progressive inspection schedule. Routine inspections consist of visual examination or check of the appliances, the aircraft, and its components and systems, insofar as practicable without disassembly. Detailed inspections consist of a thorough examination of the appliances, the aircraft, and its components and systems, with such disassembly as is necessary. For the purposes of this subparagraph, the overhaul of a component or system is considered to be a detailed inspection.

(2) If the aircraft is away from the station where inspections are normally conducted, an appropriately rated mechanic, a certificated repair station, or the manufacturer of the aircraft may perform inspections in accordance with the procedures and using the forms of the person who would otherwise perform the inspection.

[Doc. No. 1993, 29 FR 5451, Apr. 23, 1964, as amended by Amdt. 43-23, 47 FR 41086, Sept. 16, 1982; Amdt. 43-25, 51 FR 40702, Nov. 7, 1986; Amdt. 43-31, 54 FR 34336, August 18, 1989]

§ 43.16 Airworthiness limitations.

Each person performing an inspection or other maintenance specified in an Airworthiness Limitations section of a manufacturer's maintenance manual or Instructions for Continued Airworthiness shall perform the inspection or other maintenance in accordance with that section, or in accordance with operations specifications approved by the Administrator under Parts 121, 123, 127, or 135, or an inspection program approved under § 91.409(e).

Docket No. 8444 (33 FR 14104, 9/18/68)

(Amdt. 43-9, Eff. 10/17/68); (Amdt. 43-20, Eff. 10/14/80); (Amdt. 43-23, Eff. 10/15/82); (Amdt. 43-31, Eff. 8/18/90)

§ 43.17 Maintenance, preventive maintenance, and alterations performed on U.S. aeronautical products by certain Canadian persons.

(a) *Definitions.* For purposes of this section:

Aeronautical product means any civil aircraft or airframe, aircraft engine, propeller, appliance, component, or part to be installed thereon.

Canadian aeronautical product means any aeronautical product under airworthiness regulation by Transport Canada Civil Aviation.

U.S. aeronautical product means any aeronautical product under airworthiness regulation by the FAA.

[(b) *Applicability.* This section does not apply to any U.S. aeronautical products maintained or altered under any bilateral agreement made between Canada and any other than the United States.

(c) *Authorized persons.* (1) A person holding a valid Transport Canada Civil Aviation Maintenance Engineer license and appropriate ratings may, with respect to a U.S.-registered aircraft located in Canada, perform maintenance, preventive maintenance, and alterations in accordance with the requirements of paragraph (d) of this section and approve the affected aircraft for return to service in accordance with the requirements of paragraph (e) of this section.

(2) A Transport Canada Civil Aviation Approved Maintenance Organization (AMO) holding appropriate ratings may, with respect to a U.S.-registered aircraft or other U.S. aeronautical products located in Canada, perform maintenance, preventive maintenance, and alterations in accordance with the requirements of paragraph (d) of this section and approve the affected products for return to service in accordance with the requirements of paragraph (e) of this section.

(d) *Performance requirements.* A person authorized in paragraph (c) of this section may perform maintenance (including any inspection required by Sec. 91.409 of this chapter, except an annual inspection), preventive maintenance, and alterations, provided

(1) The person performing the work is authorized by Transport Canada Civil Aviation to perform the same type of work with respect to Canadian aeronautical products;

(2) The maintenance, preventive maintenance, or alteration is performed in accordance with a Bilateral Aviation Safety Agreement between the United States and Canada and associated Maintenance Implementation Procedures that provide a level of safety equivalent to that provided by the provisions of this chapter;

(3) The maintenance, preventive maintenance, or alteration is performed such that the affected product complies with the applicable requirements of part 36 of this chapter; and

(4) The maintenance, preventive maintenance, or alteration is recorded in accordance with a Bilateral Aviation Safety Agreement between the United States and Canada and associated Maintenance Implementation Procedures that provide a level of safety equivalent to that provided by the provisions of this chapter.

[(e) *Approval requirements.*

[(1) To return an affected product to service, a person authorized in paragraph (c) of this section must approve (certify) maintenance, preventive maintenance, and alterations performed under this section, except that an Aircraft Maintenance Engineer may not approve a major repair or major alteration.

(2) An AMO whose system of quality control for the maintenance, preventive maintenance, alteration, and inspection of aeronautical products has been approved by Transport Canada Civil Aviation, or an authorized employee performing work for such an AMO, may approve (certify) a major repair or major alteration performed under this section if the work was performed in accordance with technical data approved by the FAA.

[(f) No person may operate in air commerce an aircraft, airframe, aircraft engine, propeller, or appliance on which maintenance, preventive maintenance, or alteration has been performed under this section unless it has been approved for return to service by a person authorized in this section.]

[Amdt. 43-33, 56 FR 57571, Nov. 12, 1991]

Appendix A—Major alterations, major repairs, and preventive maintenance

(a) *Major alterations*—(1) *Airframe major alterations.* Alterations of the following parts and alterations of the following types, when not listed in the aircraft specifications issued by the FAA, are airframe major alterations:

(i) Wings.

(ii) Tail surfaces.

(iii) Fuselage.

(iv) Engine mounts.

(v) Control system.

(vi) Landing gear.

(vii) Hull or floats.

(viii) Elements of an airframe including spars, ribs, fittings, shock absorbers, bracing, cowling, fairings, and balance weights.

(ix) Hydraulic and electrical actuating system of components.

(x) Rotor blades.

(xi) Changes to the empty weight or empty balance which result in an increase in the maximum certificated weight or center of gravity limits of the aircraft.

(xii) Changes to the basic design of the fuel, oil, cooling, heating, cabin pressurization, electrical, hydraulic, de-icing, or exhaust systems.

(xiii) Changes to the wing or to fixed or movable control surfaces which affect flutter and vibration characteristics.

(2) *Powerplant major alterations.* The following alterations of a powerplant when not listed in the engine specifications issued by the FAA, are powerplant major alterations.

(i) Conversion of an aircraft engine from one approved model to another, involving any changes in compression ratio, propeller reduction gear, impeller gear ratios or the substitution of major engine parts which requires extensive rework and testing of the engine.

(ii) Changes to the engine by replacing aircraft engine structural parts with parts not supplied by the original manufacturer or parts not specifically approved by the Administrator.

(iii) Installation of an accessory which is not approved for the engine.

(iv) Removal of accessories that are listed as required equipment on the aircraft or engine specification.

(v) Installation of structural parts other than the type of parts approved for the installation.

(vi) Conversions of any sort for the purpose of using fuel of a rating or grade other than that listed in the engine specifications.

(3) *Propeller major alterations.* The following alterations of a propeller when not authorized in the propeller specifications issued by the FAA are propeller major alterations:

(i) Changes in blade design.

(ii) Changes in hub design.

(iii) Changes in the governor or control design.

(iv) Installation of a propeller governor or feathering system.

(v) Installation of propeller de-icing system.

(vi) Installation of parts not approved for the propeller.

(4) *Appliance major alterations.* Alterations of the basic design not made in accordance with recommendations of the appliance manufacturer or in accordance with an FAA Airworthiness Directive are appliance major alterations. In addition, changes in the basic design of radio communication and navigation equipment approved under type certification or a Technical Standard Order that have an effect on frequency stability, noise level, sensitivity, selectivity, distortion, spurious radiation, AVC characteristics, or ability to meet environmental test conditions and other changes that have an effect on the performance of the equipment are also major alterations.

(b) *Major repairs—(1) Airframe major repairs.* Repairs to the following parts of an airframe and repairs of the following types, involving the strengthening, reinforcing, splicing, and manufacturing of primary structural members or their replacement, when replacement is by fabrication such as riveting or welding, are airframe major repairs.

(i) Box beams.

(ii) Monocoque or semimonocoque wings or control surfaces.

(iii) Wing stringers or chord members.

(iv) Spars.

(v) Spar flanges.

(vi) Members of truss-type beams.

(vii) Thin sheet webs of beams.

(viii) Keel and chine members of boat hulls or floats.

(ix) Corrugated sheet compression members which act as flange material of wings or tail surfaces.

(x) Wing main ribs and compression members.

(xi) Wing or tail surface brace struts.

(xii) Engine mounts.

(xiii) Fuselage longerons.

(xiv) Members of the side truss, horizontal truss, or bulkheads.

(xv) Main seat support braces and brackets.

(xvi) Landing gear brace struts.

(xvii) Axles.

(xviii) Wheels.

(xix) Skis, and ski pedestals.

(xx) Parts of the control system such as control columns, pedals, shafts, brackets, or horns.

(xxi) Repairs involving the substitution of material.

(xxii) The repair of damaged areas in metal or plywood stressed covering exceeding six inches in any direction.

(xxiii) The repair of portions of skin sheets by making additional seams.

(xxiv) The splicing of skin sheets.

(xxv) The repair of three or more adjacent wing or control surface ribs or the leading edge of wings and control surfaces, between such adjacent ribs.

(xxvi) Repair of fabric covering involving an area greater than that required to repair two adjacent ribs.

(xxvii) Replacement of fabric on fabric covered parts such as wings, fuselages, stabilizers, and control surfaces.

(xxviii) Repairing, including rebottoming, of removable or integral fuel tanks and oil tanks.

(2) *Powerplant major repairs.* Repairs of the following parts of an engine and repairs of the following types, are powerplant major repairs:

(i) Separation or disassembly of a crankcase or crankshaft of a reciprocating engine equipped with an integral supercharger.

(ii) Separation or disassembly of a crankcase or crankshaft of a reciprocating engine equipped with other than spur-type propeller reduction gearing.

(iii) Special repairs to structural engine parts by welding, plating, metalizing, or other methods.

(3) *Propeller major repairs.* Repairs of the following types to a propeller are propeller major repairs:

(i) Any repairs to, or straightening of steel blades.

(ii) Repairing or machining of steel hubs.

(iii) Shortening of blades.

(iv) Retipping of wood propellers.

(v) Replacement of outer laminations on fixed pitch wood propellers.

(vi) Repairing elongated bolt holes in the hub of fixed pitch wood propellers.

(vii) Inlay work on wood blades.

(viii) Repairs to composition blades.

(ix) Replacement of tip fabric.

(x) Replacement of plastic covering.

(xi) Repair of propeller governors.

(xii) Overhaul of controllable pitch propellers.

(xiii) Repairs to deep dents, cuts, scars, nicks, etc., and straightening of aluminum blades.

(xiv) The repair or replacement of internal elements of blades.

(4) *Appliance major repairs.* Repairs of the following types to appliances are appliance major repairs:

(i) Calibration and repair of instruments.

(ii) Calibration of radio equipment.

(iii) Rewinding the field coil of an electrical accessory.

(iv) Complete disassembly of complex hydraulic power valves.

(v) Overhaul of pressure type carburetors, and pressure type fuel, oil and hydraulic pumps.

(c) *Preventive maintenance.* Preventive maintenance is limited to the following work, provided it does not involve complex assembly operations: (1) Removal, installation, and repair of landing gear tires.

(2) Replacing elastic shock absorber cords on landing gear.

(3) Servicing landing gear shock struts by adding oil, air, or both.

(4) Servicing landing gear wheel bearings, such as cleaning and greasing.

(5) Replacing defective safety wiring or cotter keys.

(6) Lubrication not requiring disassembly other than removal of nonstructural items such as cover plates, cowlings, and fairings.

(7) Making simple fabric patches not requiring rib stitching or the removal of structural parts or control surfaces. In the case of balloons, the making of small fabric repairs to envelopes (as defined in, and in accordance with, the balloon manufacturers' instructions) not requiring load tape repair or replacement.

(8) Replenishing hydraulic fluid in the hydraulic reservoir.

(9) Refinishing decorative coating of fuselage, balloon baskets, wings tail group surfaces (excluding balanced control surfaces), fairings, cowlings, landing gear, cabin, or cockpit interior when removal or disassembly of any primary structure or operating system is not required.

(10) Applying preservative or protective material to components where no disassembly of any primary structure or operating system is involved and where such coating is not prohibited or is not contrary to good practices.

(11) Repairing upholstery and decorative furnishings of the cabin, cockpit, or balloon basket interior when the repairing does not require disassembly of any primary structure or operating system or interfere with an operating system or affect the primary structure of the aircraft.

(12) Making small simple repairs to fairings, nonstructural cover plates, cowlings, and small patches and reinforcements not changing the contour so as to interfere with proper air flow.

(13) Replacing side windows where that work does not interfere with the structure or any operating system such as controls, electrical equipment, etc.

(14) Replacing safety belts.

(15) Replacing seats or seat parts with replacement parts approved for the aircraft, not involving disassembly of any primary structure or operating system.

(16) Troubleshooting and repairing broken circuits in landing light wiring circuits.

(17) Replacing bulbs, reflectors, and lenses of position and landing lights.

(18) Replacing wheels and skis where no weight and balance computation is involved.

(19) Replacing any cowling not requiring removal of the propeller or disconnection of flight controls.

(20) Replacing or cleaning spark plugs and setting of spark plug gap clearance.

(21) Replacing any hose connection except hydraulic connections.

(22) Replacing prefabricated fuel lines.

(23) Cleaning or replacing fuel and oil strainers or filter elements.

(24) Replacing and servicing batteries.

(25) Cleaning of balloon burner pilot and main nozzles in accordance with the balloon manufacturer's instructions.

(26) Replacement or adjustment of nonstructural standard fasteners incidental to operations.

(27) The interchange of balloon baskets and burners on envelopes when the basket or burner is designated as interchangeable in the balloon type certificate data and the baskets and burners are specifically designed for quick removal and installation.

(28) The installations of anti-misfueling devices to reduce the diameter of fuel tank filler openings provided the specific device has been made a part of the aircraft type certificate data by the aircraft manufacturer, the aircraft manufacturer has provided FAA-approved instructions for installation of the specific device, and installation does not involve the disassembly of the existing tank filler opening.

(29) Removing, checking, and replacing magnetic chip detectors.

(30) The inspection and maintenance tasks prescribed and specifically identified as preventive maintenance in a primary category aircraft type certificate or supplemental type certificate holder's approved special inspection and preventive maintenance program when accomplished on a primary category aircraft provided:

(i) They are performed by the holder of at least a private pilot certificate issued under part 61 who is the registered owner (including co-owners) of the affected aircraft and who holds a certificate of competency for the affected aircraft (1) issued by a school approved under § 147.21(e) of this chapter; (2) issued by the holder of the production certificate for that primary category aircraft that has a special training program approved under § 21.24 of this subchapter; or (3) issued by another entity that has a course approved by the Administrator; and

(ii) The inspections and maintenance tasks are performed in accordance with instructions contained by the special inspection and preventive maintenance program approved as part of the aircraft's type design or supplemental type design.

(31) Removing and replacing self-contained, front instrument panel-mounted navigation and communication devices that employ tray-mounted connectors that connect the unit when the unit is installed into the instrument panel, (excluding automatic flight control systems, transponders, and microwave frequency distance measuring equipment (DME)). The approved unit must be designed to be readily and repeatedly removed and replaced, and pertinent instructions must be provided. Prior to the unit's intended use, and operational check must be performed in accordance with the applicable sections of part 91 of this chapter.

(32) Updating self-contained, front instrument panel-mounted Air Traffic Control (ATC) navigational software data bases (excluding those of automatic flight control systems, transponders, and microwave frequency distance measuring equipment (DME) provided no disassembly of

the unit is required and pertinent instructions are provided. Prior to the unit's intended use, and operational check must be performed in accordance with applicable sections of part 91 of this chapter.

(Amdt. 43-14, Eff. 8/18/72); (Amdt. 43-23, Eff. 10/15/82); (Amdt. 43-24, Eff. 11/7/84); (Amdt. 43-25, Eff. 1/6/87); (Amdt. 43-27, Eff. 6/5/87); [(Admt. 43-34, Eff. 12/31/92)]; (Amdt. 43-36, Eff. 5/31/96).

Appendix B—Recording of major repairs and major alterations

(a) Except as provided in paragraphs (b), (c), and (d) of this appendix, each person performing a major repair or major alteration shall—

(1) Execute FAA Form 337 at least in duplicate;

(2) Give a signed copy of that form to the aircraft owner; and

(3) Forward a copy of that form to the local Flight Standards District Office within 48 hours after the aircraft, airframe, aircraft engine, propeller, or appliance is approved for return to service.

(b) For major repairs made in accordance with a manual or specifications acceptable to the Administrator, a certificated repair station may, in place of the requirements of paragraph (a)—

(1) Use the customer's work order upon which the repair is recorded;

(2) Give the aircraft owner a signed copy of the work order and retain a duplicate copy for at least two years from the date of approval for return to service of the aircraft, airframe, aircraft engine, propeller, or appliance;

(3) Give the aircraft owner a maintenance release signed by an authorized representative of the repair station and incorporating the following information:

(i) Identity of the aircraft, airframe, aircraft engine, propeller or appliance.

(ii) If an aircraft, the make, model, serial number, nationality and registration marks, and location of the repaired area.

(iii) If an airframe, aircraft engine, propeller, or appliance, give the manufacturer's name, name of the part, model, and serial numbers (if any); and

(4) Include the following or a similarly worded statement—

"The aircraft, airframe, aircraft engine, propeller, or appliance identified above was repaired and inspected in accordance with current Regulations of the Federal Aviation Agency and is approved for return to service.

Pertinent details of the repair are on file at this repair station under Order No._____,

No._____   Date_____

Signed _____ for
        (signature of authorized representative)

_____   _____
      (repair station name)   (certificate number)

_____."
                    (address)

(c) For a major repair or major alteration made by a person authorized in § 43.17, the person who performs the major repair or major alteration and the person authorized by § 43.17 to approve that work shall execute a FAA Form 337 at least in duplicate. A completed copy of that form shall be—

(1) Given to the aircraft owner; and

(2) Forwarded to the Federal Aviation Administration, Aircraft Registration Branch, Post Office Box 25082, Oklahoma City, Okla. 73125, within 48 hours after the work is inspected.

(d) For extended-range fuel tanks installed within the passenger compartment or a baggage compartment, the person who performs the work and the person authorized to approve the work by § 43.7 of this part shall execute an FAA Form 337 in at least triplicate. One (1) copy of the FAA Form 337 shall be placed on board the aircraft as specified in § 91.417 of this chapter. The remaining forms shall be distributed as required by paragraph (a) (2) and (3) or (c) (1) and (2) of this paragraph as appropriate.

(Amdt. 43-10, Eff. 11/29/68); (Amdt. 43-29, Eff. 12/8/87); (Amdt. 43-31, Eff. 8/18/90)

Appendix C—[Reserved]

Appendix D—Scope and detail of items (as applicable to the particular aircraft) to be included in annual and 100-hour inspections

(a) Each person performing an annual or 100-hour inspection shall, before that inspection, remove or open all necessary inspection plates, access doors, fairing, and cowling. He shall thoroughly clean the aircraft and aircraft engine.

(b) Each person performing an annual or 100-hour inspection shall inspect (where applicable) the following components of the fuselage and hull group:

(1) Fabric and skin—for deterioration, distortion, other evidence of failure, and defective or insecure attachment of fittings.

(2) Systems and component—for improper installation, apparent defects, and unsatisfactory operation.

(3) Envelope, gas bags, ballast tanks, and related parts—for poor condition.

(c) Each person performing an annual or 100-hour inspection shall inspect (where applicable) the following components of the cabin and cockpit group:

(1) Generally—for uncleanliness and loose equipment that might foul the controls.

(2) Seats and safety belts—for poor condition and apparent defects.

(3) Windows and windshields—for deterioration and breakage.

(4) Instruments—for poor condition, mounting, marking, and (where practicable) improper operation.

(5) Flight and engine controls—for improper installation and improper operation.

(6) Batteries—for improper installation and improper charge.

(7) All systems—for improper installation, poor general condition, apparent and obvious defects, and insecurity of attachment.

(d) Each person performing an annual or 100-hour inspection shall inspect (where applicable) components of the engine and nacelle group as follows:

(1) Engine section—for visual evidence of excessive oil, fuel, or hydraulic leaks, and sources of such leaks.

(2) Studs and nuts—for improper torquing and obvious defects.

(3) Internal engine—for cylinder compression and for metal particles or foreign matter on screens and sump drain plugs. If there is weak cylinder compression, for improper internal condition and improper internal tolerances.

(4) Engine mount—for cracks, looseness of mounting, and looseness of engine to mount.

(5) Flexible vibration dampeners—for poor condition and deterioration.

(6) Engine controls—for defects, improper travel, and improper safetying.

(7) Lines, hoses, and clamps—for leaks, improper condition and looseness.

(8) Exhaust stacks—for cracks, defects, and improper attachment.

(9) Accessories—for apparent defects in security of mounting.

(10) All systems—for improper installation, poor general condition, defects, and insecure attachment.

(11) Cowling—for cracks, and defects.

(e) Each person performing an annual or 100-hour inspection shall inspect (where applicable) the following components of the landing gear group:

(1) All units—for poor condition and insecurity of attachment.

(2) Shock absorbing devices—for improper oleo fluid level.

(3) Linkages, trusses, and members—for undue or excessive wear fatigue, and distortion.

(4) Retracting and locking mechanism—for improper operation.

(5) Hydraulic lines—for leakage.

(6) Electrical system—for chafing and improper operation of switches.

(7) Wheels—for cracks, defects, and condition of bearings.

(8) Tires—for wear and cuts.

(9) Brakes—for improper adjustment.

(10) Floats and skis—for insecure attachment and obvious or apparent defects.

(f) Each person performing an annual or 100-hour inspection shall inspect (where applicable) all components of the wing and center section assembly for poor general condition, fabric or skin deterioration, distortion, evidence of failure, and insecurity of attachment.

(g) Each person performing an annual or 100-hour inspection shall inspect (where applicable) all components and systems that make up the complete empennage assembly for poor general condition, fabric or skin deterioration, distortion, evidence of failure, insecure attachment, improper component installation, and improper component operation.

(h) Each person performing an annual or 100-hour inspection shall inspect (where applicable) the following components of the propeller group:

(1) Propeller assembly—for cracks, nicks, binds, and oil leakage.

(2) Bolts—for improper torquing and lack of safetying.

(3) Anti-icing devices—for improper operations and obvious defects.

(4) Control mechanisms—for improper operation, insecure mounting, and restricted travel.

(i) Each person performing an annual or 100-hour inspection shall inspect (where applicable) the following components of the radio group:

(1) Radio and electronic equipment—for improper installation and insecure mounting.

(2) Wiring and conduits—for improper routing, insecure mounting, and obvious defects.

(3) Bonding and shielding—for improper installation and poor condition.

(4) Antenna including trailing antenna—for poor condition, insecure mounting, and improper operation.

(j) Each person performing an annual or 100-hour inspection shall inspect (where applicable) each installed miscellaneous item that is not otherwise covered by this listing for improper installation and improper operation.

(Amdt. 43-3, Eff. 4/2/66)

Appendix E—Altimeter system test and inspection

Each person performing the altimeter system tests and inspections required by § 91.411 shall comply with the following:

(a) Static pressure system:

(1) Ensure freedom from entrapped moisture and restrictions.

(2) Determine that leakage is within the tolerances established in § 23.1325 or § 25.1325, whichever is applicable.

(3) Determine that the static port heater, if installed, is operative.

(4) Ensure that no alterations or deformations of the airframe surface have been made that would affect the relationship between air pressure in the static pressure system and true ambient static air pressure for any flight condition.

(b) Altimeter:

(1) Test by an appropriately rated repair facility in accordance with the following subparagraphs. Unless otherwise specified, each test for performance may be conducted with the instrument subjected to vibration. When tests are conducted with the temperature substantially different from ambient temperature of approximately 25 degrees C., allowance shall be made for the variation from the specified condition.

(i) *Scale error.* With the barometric pressure scale at 29.92 inches of mercury, the altimeter shall be subjected successively to pressures corresponding to the altitude specified in Table I up to the maximum normally expected operating altitude of the airplane in which the altimeter is to be installed. The reduction in pressure shall be made at a rate not in excess of 20,000 feet per minute to within approximately 2,000 feet of the test point. The test point shall be approached at a rate compatible with the test equipment. The altimeter shall be kept at the pressure corresponding to each test point for at least 1 minute, but not more than 10 minutes, before a reading is taken. The error at all test points must not exceed the tolerances specified in Table I.

(ii) *Hysteresis.* The hysteresis test shall begin not more than 15 minutes after the altimeter's initial exposure to the pressure corresponding to the upper limit of the scale error test prescribed in subparagraph (i); and while the altimeter is at this pressure, the hysteresis test shall commence. Pressure shall be increased at a rate simulating a descent in altitude at the rate of 5,000 to 20,000 feet per minute until within 3,000 feet of the first test point (50

TABLE I

Altitude	Equivalent pressure (inches of mercury)	Tolerance ± (feet)
—1,000	31.018	20
0	29.921	20
500	29.385	20
1,000	28.856	20
1,500	28.335	25
2,000	27.821	30
3,000	26.817	30
4,000	25.842	35
6,000	23.978	40
8,000	22.225	60
10,000	20.577	80
12,000	19.029	90
14,000	17.577	100
16,000	16.216	110
18,000	14.942	120
20,000	13.750	130
22,000	12.636	140
25,000	11.104	155
30,000	8.885	180
35,000	7.041	205
40,000	5.538	230
45,000	4.355	255
50,000	3.425	280

percent of maximum altitude). The test point shall then be approached at a rate of approximately 3,000 feet per minute. The altimeter shall be kept at this pressure for at least 5 minutes, but not more than 15 minutes, before the test reading is taken. After the reading has been taken, the pressure shall be increased further, in the same manner as before, until the pressure corresponding to the second test point (40 percent of maximum altitude) is reached. The altimeter shall be kept at this pressure for at least 1 minute, but not more than 10 minutes, before the test reading is taken. After the reading has been taken, the pressure shall be increased further, in the same manner as before, until atmospheric pressure is reached. The reading of the altimeter at either of the two test points shall not differ by more than the tolerance specified in Table II from the reading of the altimeter for the corresponding altitude recorded during the scale error test prescribed in paragraph (b)(i).

(iii) *After effect.* Not more than 5 minutes after the completion of the hysteresis test prescribed in paragraph (b)(ii), the reading of the altimeter (corrected for any change in atmospheric pressure) shall not differ from the original atmospheric pressure reading by more than the tolerance specified in Table II.

TABLE II—TEST TOLERANCES

Test	Tolerance (feet)
Case Leak Test	±100
Hysteresis Test:	
First Test Point (50 percent of maximum altitude)	75
Second Test Point (40 percent of maximum altitude)	75
After Effect Test	30

TABLE III—FRICTION

Altitude (feet)	Tolerance (feet)
1,000	±70
2,000	70
3,000	70
5,000	70
10,000	80
15,000	90
20,000	100
25,000	120
30,000	140
35,000	160
40,000	180
50,000	250

(iv) *Friction.* The altimeter shall be subjected to a steady rate of decrease of pressure approximating 750 feet per minute. At each altitude listed in Table III, the change in reading of the pointers after vibration shall not exceed the corresponding tolerance listed in Table III.

(v) *Case leak.* The leakage of the altimeter case, when the pressure within it corresponds to an altitude of 18,000 feet, shall not change the altimeter reading by more than the tolerance shown in Table II during an interval of 1 minute.

TABLE IV—PRESSURE-ALTITUDE DIFFERENCE

Pressure (inches of Hg)	Altitude difference (feet)
28.10	−1,727
28.50	−1,340
29.00	−863
29.50	−392
29.92	0
30.50	+531
30.90	+893
30.99	+974

(vi) *Barometric scale error.* At constant atmospheric pressure, the barometric pressure scale shall be set at each of the pressures (falling within its range of adjustment) that are listed in Table IV, and shall cause the pointer to indicate the equivalent altitude difference shown in Table IV with a tolerance of 25 feet.

(2) Altimeters which are the air data computer type with associated computing systems, or which incorporate air data correction internally, may be tested in a manner and to specifications developed by the manufacturer which are acceptable to the Administrator.

(c) Automatic Pressure Altitude Reporting Equipment and ATC Transponder System Integration Test. The test must be conducted by an appropriately rated person under the conditions specified in paragraph (a). Measure the automatic pressure altitude at the output of the installed ATC transponder when interrogated on Mode C at a sufficient number of test points to ensure that the altitude reporting equipment, altimeters, and ATC transponders perform their intended functions as installed in the aircraft. The difference between the automatic reporting output and the altitude displayed at the altimeter shall not exceed 125 feet.

(d) Records: Comply with the provisions of § 43.9 of this chapter as to content, form, and disposition of the records. The person performing the altimeter tests shall record on the altimeter the date and maximum altitude to which the altimeter has been tested and the persons approving the airplane for return to service shall enter that data in the airplane log or other permanent record.

(Amdt. 43-2, Eff. 7/29/65); (Amdt. 43-7, Eff. 8/1/67); (Amdt. 43-19, Eff. 6/26/78); (Amdt. 43-23, Eff. 10/15/82); (Amdt. 43-31, Eff. 8/18/90)

Appendix F—ATC transponder tests and inspections

The ATC transponder tests required by § 91.413 of this chapter may be conducted using a bench check or portable test equipment and must meet the requirements prescribed in paragraphs (a) through (j) of this appendix. If portable test equipment with appropriate coupling to the aircraft antenna system is used, operate the test equipment for the

ATCRBS transponders at a nominal rate of 235 interrogations per second to avoid possible ATCRBS interference. Operate the test equipment at a nominal rate of 50 Mode S interrogations per second for Mode S. An additional 3 dB loss is allowed to compensate for antenna coupling errors during receiver sensitivity measurements conducted in accordance with paragraph (c)(1) when using portable test equipment.

(a) Radio Reply Frequency:

(1) For all classes of ATCRBS transponders, interrogate the transponder and verify that the reply frequency is 1090±3 Megahertz (MHz).

(2) For classes 1B, 2B, and 3B Mode S transponders, interrogate the transponder and verify that the reply frequency is 1090±3 MHz.

(3) For classes 1B, 2B, and 3B Mode S transponders that incorporate the optional 1090±1 MHz reply frequency, interrogate the transponder and verify that the reply frequency is correct.

(4) For classes 1A, 2A, 3A, and 4 Mode S transponders, interrogate the transponder and verify that the reply frequency is 1090±1 MHz.

(b) Suppression: When Classes 1B and 2B ATCRBS Transponders, or Classes 1B, 2B, and 3B Mode S transponders are interrogated Mode 3/A at an interrogation rate between 230 and 1,000 interrogations per second; or when Classes 1A and 2A ATCRBS Transponders, or Classes 1B, 2A, 3A, and 4 Mode S transponders are interrogated at a rate between 230 and 1,200 Mode 3/A interrogations per second:

(1) Verify that the transponder does not respond to more than 1 percent of ATCRBS interrogations when the amplitude of P_2 pulse is equal to the P_1 pulse.

(2) Verify that the transponder replies to at least 90 percent of ATCRBS interrogations when the amplitude of the P_2 pulse is 9 dB less than the P_1 pulse. If the test is conducted with a radiated test signal, the interrogation rate shall be 235±5 interrogations per second unless a higher rate has been approved for the test equipment used at that location.

(c) Receiver Sensitivity:

(1) Verify that for any class of ATCRBS Transponder, the receiver minimum triggering level (MTL) of the system is –73±4 dbm, or that for any class of Mode S transponder the receiver MTL for Mode S format (P6 type) interrogations is –74±3 dbm by use of a test set either:

(i) Connected to the antenna end of the transmission line;

(ii) Connected to the antenna terminal of the transponder with a correction for transmission line loss; or

(iii) Utilized radiated signal.

(2) Verify that the difference in Mode 3/A and Mode C receiver sensitivity does not exceed 1 db for either any class of ATCRBS transponder or any class of Mode S transponder.

(d) Radio Frequency (RF) Peak Output Power:

(1) Verify that the transponder RF output power is within specifications for the class of transponder. Use the same conditions as described in (c)(1) (i), (ii), and (iii) above.

(i) For Class 1A and 2A ATCRBS transponders, verify that the minimum RF peak output power is at least 21.0 dbw (125 watts).

(ii) For Class 1B and 2B ATCRBS Transponders, verify that the minimum RF peak output power is at least 18.5 dbw (70 watts).

(iii) For Class 1A, 2A, 3A, and 4 and those Class 1B, 2B, and 3B Mode S transponders that include the optional high RF peak output power, verify that the minimum RF peak output power is at least 21.0 dbw (125 watts).

(iv) For Classes 1B, 2B, and 3B Mode S transponders, verify that the minimum RF peak output power is at least 18.5 dbw (70 watts).

(v) For any class of ATCRBS or any class of Mode S transponders, verify that the maximum RF peak output power does not exceed 27.0 dbw (500 watts).

Note: The tests in (e) through (j) apply only to Mode S transponders.

(e) Mode S Diversity Transmission Channel Isolation: For any class of Mode S transponder that incorporates diversity operation, verify that the RF peak output power transmitted from the selected antenna exceeds the power transmitted from the nonselected antenna by at least 20 db.

(f) Mode S Address: Interrogate the Mode S transponder and verify that it replies only to its assigned address. Use the correct address and at least two incorrect addresses. The interrogations should be made at a nominal rate of 50 interrogations per second.

(g) Mode S Formats: Interrogate the Mode S transponder with uplink formats (UF) for which it is equipped and verify that the replies are made in the correct format. Use the surveillance formats UF = 4 and 5. Verify that the altitude reported in the replies to UF = 4 are the same as that reported in a valid ATCRBS Mode C reply. Verify that the identity reported in the replies to UF = 5 are the same as that reported in a valid ATCRBS Mode 3/A reply. If the transponder is so equipped, use the communication formats UF = 20, 21, and 24.

(h) Mode S All-Call Interrogations: Interrogate the Mode S transponder with the Mode S only all-call format UF = 11, and the ATCRBS/Mode S all-call formats (1.6 microsecond P_4 pulse) and verify that the correct address and capability are reported in the replies (downlink format DF = 11).

(i) ATCRBS-Only All-Call Interrogation: Interrogate the Mode S transponder with the ATCRBS-only all-call interrogation (0.8 microsecond P_4 pulse) and verify that no reply is generated.

(j) Squitter: Verify that the Mode S transponder generates a correct squitter approximately once per second.

(k) Records: Comply with the provisions of § 43.9 of this chapter as to content, form, and disposition of the records.

(Amdt. 43-17, Eff. 1/26/73); (Amdt. 43-18, Eff. 12/31/73); (Amdt. 43-19, Eff. 6/26/78); (Amdt. 43-26, Eff. 4/6/87); (Amdt. 43-31, Eff. 8/18/90).

Part 61—Certification: Pilots, flight instructors, and ground instructors

Subpart K-Flight Instructors with a Sport Pilot Rating

Authority: 49 U.S.C. 106(g), 40113, 44701-44703, 44707, 44709-44711, 45102-45103, 45301-45302.

Source: Amdt. 61-102, 62 FR 16298, Apr. 4, 1997, unless otherwise noted.

Subpart A—General

§ 61.1 Applicability and definitions.

(a) This part prescribes:

(1) The requirements for issuing pilot, flight instructor, and ground instructor certificates and ratings; the conditions under which those certificates and ratings are necessary; and the privileges and limitations of those certificates and ratings.

(2) The requirements for issuing pilot, flight instructor, and ground instructor authorizations; the conditions under which those authorizations are necessary; and the privileges and limitations of those authorizations.

(3) The requirements for issuing pilot, flight instructor, and ground instructor certificates and ratings for persons who have taken courses approved by the Administrator under other parts of this chapter.

(b) For the purpose of this part:

(1) Aeronautical experience means pilot time obtained in an aircraft, flight simulator, or flight training device for meeting the appropriate training and flight time requirements for an airman certificate, rating, flight review, or recency of flight experience requirements of this part.

(2) Authorized instructor means—

(i) A person who holds a valid ground instructor certificate issued under part 61 or part 143 of this chapter when conducting ground training in accordance with the privileges and limitations of his or her ground instructor certificate;

(ii) A person who holds a current flight instructor certificate issued under part 61 of this chapter when conducting ground training or flight training in accordance with the privileges and limitations of his or her flight instructor certificate; or

(iii) A person authorized by the Administrator to provide ground training or flight training under SFAR No. 58, or part 61, 121, 135, or 142 of this chapter when conducting ground training or flight training in accordance with that authority.

(3) Cross-country time means—

(i) Except as provided in paragraphs (b)(3)(ii) through (b)(3)(vi) of this section, time acquired during flight—

(A) Conducted by a person who holds a pilot certificate;

(B) Conducted in an aircraft;

(C) That includes a landing at a point other than the point of departure; and

(D) That involves the use of dead reckoning, pilotage, electronic navigation aids, radio aids, or other navigation systems to navigate to the landing point.

(ii) For the purpose of meeting the aeronautical experience requirements (except for a rotorcraft category rating), for a private pilot certificate (except for a powered parachute category rating), a commercial pilot certificate, or an instrument rating, or for the purpose of exercising recreational pilot privileges (except in a rotorcraft) under §61.101 (c), time acquired during a flight—

(A) Conducted in an appropriate aircraft;

(B) That includes a point of landing that was at least a straight-line distance of more than 50 nautical miles from the original point of departure; and

(C) That involves the use of dead reckoning, pilotage, electronic navigation aids, radio aids, or other navigation systems to navigate to the landing point.

(iii) For the purpose of meeting the aeronautical experience requirements for a sport pilot certificate (except for powered parachute privileges), time acquired during a flight conducted in an appropriate aircraft that—

(A) Includes a point of landing at least a straight line distance of more than 25 nautical miles from the original point of departure; and

(B) Involves, as applicable, the use of dead reckoning; pilotage; electronic navigation aids; radio aids; or other navigation systems to navigate to the landing point.

(iv) For the purpose of meeting the aeronautical experience requirements for a sport pilot certificate with powered parachute privileges or a private pilot certificate with a powered parachute category rating, time acquired during a flight conducted in an appropriate aircraft that—

(A) Includes a point of landing at least a straight line distance of more than 15 nautical miles from the original point of departure; and

(B) Involves, as applicable, the use of dead reckoning; pilotage; electronic navigation aids; radio aids; or other navigation systems to navigate to the landing point.

(v) For the purpose of meeting the aeronautical experience requirements for any pilot certificate with a rotorcraft category rating or an instrument-helicopter rating, or for the purpose of exercising recreational pilot privileges, in a rotorcraft, under § 61.101(c), time acquired during a flight—

(A) Conducted in an appropriate aircraft;

(B) That includes a point of landing that was at least a straight-line distance of more than 25 nautical miles from the original point of departure; and

(C) That involves the use of dead reckoning, pilotage, electronic navigation aids, radio aids, or other navigation systems to navigate to the landing point.

(vi) For the purpose of meeting the aeronautical experience requirements for an airline transport pilot certificate (except with a rotorcraft category rating), time acquired during a flight—

(A) Conducted in an appropriate aircraft;

(B) That is at least a straight-line distance of more than 50 nautical miles from the original point of departure; and

(C) That involves the use of dead reckoning, pilotage, electronic navigation aids, radio aids, or other navigation systems.

(vii) For a military pilot who qualifies for a commercial pilot certificate (except with a rotorcraft category rating) under § 61.73 of this part, time acquired during a flight—

(A) Conducted in an appropriate aircraft;

(B) That is at least a straight-line distance of more than 50 nautical miles from the original point of departure; and

(C) That involves the use of dead reckoning, pilotage, electronic navigation aids, radio aids, or other navigation systems.

(4) *Examiner* means any person who is authorized by the Administrator to conduct a pilot proficiency test or a practical test for an airman certificate or rating issued under this part, or a person who is authorized to conduct a knowledge test under this part.

(5) *Flight simulator* means a device that—

(i) Is a full-size aircraft cockpit replica of a specific type of aircraft, or make, model, and series of aircraft;

(ii) Includes the hardware and software necessary to represent the aircraft in ground operations and flight operations;

(iii) Uses a force cueing system that provides cues at least equivalent to those cues provided by a 3 degree freedom of motion system;

(iv) Uses a visual system that provides at least a 45 degree horizontal field of view and a 30 degree vertical field of view simultaneously for each pilot; and

(v) Has been evaluated, qualified, and approved by the Administrator.

(6) *Flight training* means that training, other than ground training, received from an authorized instructor in flight in an aircraft.

(7) *Flight training device* means a device that—

(i) Is a full-size replica of the instruments, equipment, panels, and controls of an aircraft, or set of aircraft, in an open flight deck area or in an enclosed cockpit, including the hardware and software for the systems installed, that is necessary to simulate the aircraft in ground and flight operations;

(ii) Need not have a force (motion) cueing or visual system; and

(iii) Has been evaluated, qualified, and approved by the Administrator.

(8) *Ground training* means that training, other than flight training, received from an authorized instructor.

(9) *Instrument approach* means an approach procedure defined in part 97 of this chapter.

(10) *Instrument training* means that time in which instrument training is received from an authorized instructor under actual or simulated instrument conditions.

(11) *Knowledge test* means a test on the aeronautical knowledge areas required for an airman certificate or rating that can be administered in written form or by a computer.

(12) *Pilot time* means that time in which a person—

(i) Serves as a required pilot flight crewmember;

(ii) Receives training from an authorized instructor in an aircraft, flight simulator, or flight training device; or

(iii) Gives training as an authorized instructor in an aircraft, flight simulator, or flight training device.

(13) *Practical test* means a test on the areas of operations for an airman certificate, rating, or authorization that is conducted by having the applicant respond to questions and demonstrate maneuvers in flight, in a flight simulator, or in a flight training device.

(14) *Set of aircraft* means aircraft that share similar performance characteristics, such as similar airspeed and altitude operating envelopes, similar handling characteristics, and the same number and type of propulsion systems.

(15) *Student pilot seeking a sport pilot certificate* means a person who has received an endorsement—

(i) To exercise student pilot privileges from a certificated flight instructor with a sport pilot rating; or

(ii) That includes a limitation for the operation of a light-sport aircraft specified in §61.89 (c) issued by a certificated flight instructor with other than a sport pilot rating.

(16) *Training time* means training received—

(i) In flight from an authorized instructor;

(ii) On the ground from an authorized instructor; or

(iii) In a flight simulator or flight training device from an authorized instructor.

[Amdt. 61-102, 62 FR 16298, Apr. 4, 1997; Amdt. 61-103, 62 FR 40893, July 30, 1997]

§ 61.2 Removed.

§ 61.3 Requirement for certificates, ratings, and authorizations.

(a) *Pilot certificate.* A person may not act as pilot in command or in any other capacity as a required pilot flight crewmember of a civil aircraft of U.S. registry, unless that person—

(1) Has a valid pilot certificate or special purpose pilot authorization issued under this part in that person's physical possession or readily accessible in the aircraft when exercising the privileges of that pilot certificate or authorization. However, when the aircraft is operated within a foreign country, a current pilot license issued by the country in which the aircraft is operated may be used; and

(2) Has a photo identification that is in that person's physical possession or readily accessible in the aircraft when exercising the privileges of that pilot certificate or authorization. The photo identification must be a:

(i) Valid driver's license issued by a State, the District of Columbia, or territory or possession of the United States;

(ii) Government identification card issued by the Federal government, a State, the District of Columbia, or a territory or possession of the United States;

(iii) U.S. Armed Forces' identification card;

(iv) Official passport;

(v) Credential that authorizes unescorted access to a security identification display area at an airport regulated under 49 CFR part 1542; or

(vi) Other form of identification that the Administrator finds acceptable.

(b) *Required pilot flight crewmember certificate for operating a foreign-registered aircraft.* A person may not act as pilot in command or in any other capacity as a required pilot of a civil aircraft of foreign registry within the United States, unless that person's pilot certificate:

(1) Is valid and in that person's physical possession, or readily accessible in the aircraft when exercising the privileges of that pilot certificate; and

(2) Has been issued under this part, or has been issued or validated by the country in which the aircraft is registered.

(c) *Medical certificate.* (1) Except as provided for in paragraph (c)(2) of this section, a person may not act as pilot in command or in any other capacity as a required pilot flight crewmember of an aircraft, under a certificate issued to that person under this part, unless that person has a current and appropriate medical certificate that has been issued under part 67 of this chapter, or other documentation acceptable to the Administrator, which is in that person's physical possession or readily accessible in the aircraft.

(2) A person is not required to meet the requirements of paragraph (c)(1) of this section if that person—

(i) Is exercising the privileges of a student pilot certificate while seeking a pilot certificate with a glider category rating, balloon class rating, or glider or balloon privileges;

(ii) Is exercising the privileges of a student pilot certificate while seeking a sport pilot certificate with other than glider or balloon privileges and holds a current and valid U.S. driver's license;

(iii) Is exercising the privileges of a student pilot certificate while seeking a pilot certificate with a weight-shift-control aircraft category rating or a powered parachute category rating and holds a current and valid U.S. driver's license;

(iv) Is exercising the privileges of a sport pilot certificate with glider or balloon privileges;

(v) Is exercising the privileges of a sport pilot certificate with other than glider or balloon privileges and holds a current and valid U.S. driver's license. A person who has applied for or held a medical certificate may exercise the privileges of a sport pilot certificate using a current and valid U.S. driver's license only if that person—

(A) Has been found eligible for the issuance of at least a third-class airman medical certificate at the time of his or her most recent application; and

(B) Has not had his or her most recently issued medical certificate suspended or revoked or most recent Authorization for a Special Issuance of a Medical Certificate withdrawn.

(vi) Is holding a pilot certificate with a balloon class rating and is piloting or providing training in a balloon as appropriate;

(vii) Is holding a pilot certificate or a flight instructor certificate with a glider category rating, and is piloting or providing training in a glider, as appropriate;

(viii) Except as provided in paragraph (c)(2)(iii) of this section, is exercising the privileges of a flight instructor certificate, provided the person is not acting as pilot in command or as a required pilot flight crewmember;

(ix) Is exercising the privileges of a ground instructor certificate;

(x) Is operating an aircraft within a foreign country using a pilot license issued by that country and possesses evidence of current medical qualification for that license; or

(xi) Is operating an aircraft with a U.S. pilot certificate, issued on the basis of a foreign pilot license, issued under § 61.75 of this part, and holds a current medical certificate issued by the foreign country that issued the foreign pilot license, which is in that person's physical possession or readily accessible in the aircraft when exercising the privileges of that airman certificate.

(d) *Flight instructor certificate.* (1) A person who holds a flight instructor certificate issued under this part must have that certificate, or other documentation acceptable to the Administrator, in that person's physical possession or readily accessible in the aircraft when exercising the privileges of that flight instructor certificate.

(2) Except as provided in paragraph (d)(3) of this section, no person other than the holder of a flight instructor certificate issued under this part with the appropriate rating on that certificate may—

(i) Give training required to qualify a person for solo flight and solo cross-country flight;

(ii) Endorse an applicant for a—

(A) Pilot certificate or rating issued under this part;

(B) Flight instructor certificate or rating issued under this part; or

(C) Ground instructor certificate or rating issued under this part;

(iii) Endorse a pilot logbook to show training given; or

(iv) Endorse a student pilot certificate and logbook for solo operating privileges.

(3) A flight instructor certificate issued under this part is not necessary—

(i) Under paragraph (d)(2) of this section, if the training is given by the holder of a commercial pilot certificate with a lighter-than-air rating, provided the training is given in accordance with the privileges of the certificate in a lighter-than-air aircraft;

(ii) Under paragraph (d)(2) of this section, if the training is given by the holder of an airline transport pilot certificate with a rating appropriate to the aircraft in which the training is given, provided the training is given in accordance with the privileges of the certificate and conducted in accordance with an approved air carrier training program approved under part 121 or part 135 of this chapter;

(iii) Under paragraph (d)(2) of this section, if the training is given by a person who is qualified in accordance with subpart C of part 142 of this chapter, provided the training is conducted in accordance with an approved part 142 training program;

(iv) Under paragraphs (d)(2)(i), (d)(2)(ii)(C), and (d)(2)(iii) of this section, if the training is given by the holder of a ground instructor certificate in accordance with the privileges of the certificate; or

(v) Under paragraph (d)(2)(iii) of this section, if the training is given by an authorized flight instructor under § 61.41 of this part.

(e) Instrument rating. No person may act as pilot in command of a civil aircraft under IFR or in weather conditions less than the minimums prescribed for VFR flight unless that person holds:

(1) The appropriate aircraft category, class, type (if required), and instrument rating on that person's pilot certificate for any airplane, helicopter, or powered-lift being flown;

(2) An airline transport pilot certificate with the appropriate aircraft category, class, and type rating (if required) for the aircraft being flown;

(3) For a glider, a pilot certificate with a glider category rating and an airplane instrument rating; or

(4) For an airship, a commercial pilot certificate with a lighter-than-air category rating and airship class rating.

(f) Category II pilot authorization. Except for a pilot conducting Category II operations under part 121 or part 135, a person may not:

(1) Act as pilot in command of a civil aircraft during Category II operations unless that person—

(i) Holds a current Category II pilot authorization for that category or class of aircraft, and the type of aircraft, if applicable; or

(ii) In the case of a civil aircraft of foreign registry, is authorized by the country of registry to act as pilot in command of that aircraft in Category II operations.

(2) Act as second in command of a civil aircraft during Category II operations unless that person—

(i) Holds a valid pilot certificate with category and class ratings for that aircraft and a current instrument rating for that category aircraft;

(ii) Holds an airline transport pilot certificate with category and class ratings for that aircraft; or

(iii) In the case of a civil aircraft of foreign registry, is authorized by the country of registry to act as second in command of that aircraft during Category II operations.

(g) Category III pilot authorization. Except for a pilot conducting Category III operations under part 121 or part 135, a person may not:

(1) Act as pilot in command of a civil aircraft during Category III operations unless that person—

(i) Holds a current Category III pilot authorization for that category or class of aircraft, and the type of aircraft, if applicable; or

(ii) In the case of a civil aircraft of foreign registry, is authorized by the country of registry to act as pilot in command of that aircraft in Category III operations.

(2) Act as second in command of a civil aircraft during Category III operations unless that person—

(i) Holds a valid pilot certificate with category and class ratings for that aircraft and a current instrument rating for that category aircraft;

(ii) Holds an airline transport pilot certificate with category and class ratings for that aircraft; or

(iii) In the case of a civil aircraft of foreign registry, is authorized by the country of registry to act as second in command of that aircraft during Category III operations.

(h) Category A aircraft pilot authorization. The administrator may issue a certificate of authorization for a Category II or Category III operation to the pilot of a small aircraft that is a Category A aircraft, as identified in § 97.3(b)(1) of this chapter if:

(1) The Administrator determines that the Category II or Category III operation can be performed safely by that pilot under the terms of the certificate of authorization; and

(2) The Category II or Category III operation does not involve the carriage of persons or property for compensation or hire.

(i) Ground instructor certificate.

(1) Each person who holds a ground instructor certificate issued under this part or part 143 must have that certificate in that person's physical possession or immediately accessible when exercising the privileges of that certificate.

(2) Except as provided in paragraph (i)(3) of this section, no person other than the holder of a ground instructor certificate, issued under this part or part 143, with the appropriate rating on that certificate may—

(i) Give ground training required to qualify a person for solo flight and solo cross-country flight;

(ii) Endorse an applicant for a knowledge test required for a pilot, flight instructor, or ground instructor certificate or rating issued under this part; or

(iii) Endorse a pilot logbook to show ground training given.

(3) A ground instructor certificate issued under this part is not necessary—

(i) Under paragraph (i)(2) of this section, if the training is given by the holder of a flight instructor certificate issued under this part in accordance with the privileges of that certificate;

(ii) Under paragraph (i)(2) of this section, if the training is given by the holder of a commercial pilot certificate with a lighter-than-air rating, provided the training is given in accordance with the privileges of the certificate in a lighter-than-air aircraft;

(iii) Under paragraph (i)(2) of this section, if the training is given by the holder of an airline transport pilot certificate with a rating appropriate to the aircraft in which the training is given, provided the training is given in accordance with the privileges of the certificate and conducted in accordance with an approved air carrier training program approved under part 121 or part 135 of this chapter;

(iv) Under paragraph (i)(2) of this section, if the training is given by a person who is qualified in accordance with subpart C of part 142 of this chapter, provided the training is conducted in accordance with an approved part 142 training program; or

(v) Under paragraph (i)(2)(iii) of this section, if the training is given by an authorized flight instructor under § 61.41 of this part.

(j) Age limitation for certain operations.

(1) Age limitation. Except as provided in paragraph (j)(3) of this section, no person who holds a pilot certificate issued under this part shall serve as a pilot on a civil airplane of U.S. registry in the following operations if the person has reached his or her 60th birthday—

(i) Scheduled international air services carrying passengers in turbojet-powered airplanes;

(ii) Scheduled international air services carrying passengers in airplanes having a passenger-seat configuration of more than nine passenger seats, excluding each crewmember seat;

(iii) Nonscheduled international air transportation for compensation or hire in airplanes having a passenger-seat configuration of more than 30 passenger seats, excluding each crewmember seat; or

(iv) Scheduled international air services, or nonscheduled international air transportation for compensation or hire, in airplanes having a payload capacity of more than 7,500 pounds.

(2) Definitions. (i) "International air service," as used in paragraph (j) of this section, means scheduled air service performed in airplanes for the public transport of passengers, mail, or cargo, in which the service passes through the airspace over the territory of more than one country.

(ii) "International air transportation," as used in paragraph (j) of this section, means air transportation performed in airplanes for the public transport of passengers, mail, or cargo, in which the service passes through the airspace over the territory of more than one country.

(3) Delayed pilot age limitation. Until December 20, 1999, a person may serve as a pilot in operations covered by this paragraph after that person has reached his or her 60th birthday if, on March 20, 1997, that person was employed as a pilot in operations covered by this paragraph.

(k) Special purpose pilot authorization. Any person that is required to hold a special purpose pilot authorization, issued in accordance with § 61.77 of this part, must have that authorization and the person's foreign pilot license in that person's physical possession or have it readily accessible

in the aircraft when exercising the privileges of that authorization.

(l) Inspection of certificate. Each person who holds an airman certificate, medical certificate, authorization, or license required by this part must present it and their photo identification as described in paragraph (a)(2) of this section for inspection upon a request from:

(1) The Administrator;

(2) An authorized representative of the National Transportation Safety Board; or

(3) Any Federal, State, or local law enforcement officer.

(4) An authorized representative of the Transportation Security Administration.

[Amdt. 61-102, 62 FR 16298, Apr. 4, 1997; Amdt. 61-103, 62 FR 40894, July 30, 1997; Amdt. 61-107, 67 FR 65861, Oct. 28, 2002]

§ 61.4 Qualification and approval of flight simulators and flight training devices.

(a) Except as specified in paragraph (b) or (c) of this section, each flight simulator and flight training device used for training, and for which an airman is to receive credit to satisfy any training, testing, or checking requirement under this chapter, must be qualified and approved by the Administrator for—

(1) The training, testing, and checking for which it is used;

(2) Each particular maneuver, procedure, or crewmember function performed; and

(3) The representation of the specific category and class of aircraft, type of aircraft, particular variation within the type of aircraft, or set of aircraft for certain flight training devices.

(b) Any device used for flight training, testing, or checking that has been determined to be acceptable to or approved by the Administrator prior to August 1, 1996, which can be shown to function as originally designed, is considered to be a flight training device, provided it is used for the same purposes for which it was originally accepted or approved and only to the extent of such acceptance or approval.

(c) The Administrator may approve a device other than a flight simulator or flight training device for specific purposes.

[Amdt. 60-102, 62 FR 16298, Apr. 4, 1997; Amdt. 61-103, 62 FR 40895, July 30, 1997]

§ 61.5 Certificates and ratings issued under this part.

(a) The following certificates are issued under this part to an applicant who satisfactorily accomplishes the training and certification requirements for the certificate sought:

(1) Pilot certificates—

(i) Student pilot.

(ii) Sport Pilot.

(iii) Recreational pilot.

(iv) Private pilot.

(v) Commercial pilot.

(vi) Airline transport pilot.

(vii) Powered parachute,

(viii) Weight-shift control aircraft.

(2) Flight instructor certificates.

(3) Ground instructor certificates.

(b) The following ratings are placed on a pilot certificate (other than student pilot) when an applicant satisfactorily accomplishes the training and certification requirements for the rating sought:

(1) Aircraft category ratings—

(i) Airplane.

(ii) Rotorcraft.

(iii) Glider.

(iv) Lighter-than-air.

(v) Powered-lift.

(2) Airplane class ratings—

(i) Single-engine land.

(ii) Multiengine land.

(iii) Single-engine sea.

(iv) Multiengine sea.

(3) Rotorcraft class ratings—

(i) Helicopter.

(ii) Gyroplane.

(4) Lighter-than-air class ratings—

(i) Airship.

(ii) Balloon.

(5) Weight-shift control aircraft class rating—

(i) Weight-shift control aircraft land,

(ii) Weight-shift control aircraft sea.

(6) Powered parachute class ratings—

(i) Powered parachute land,

(ii) Powered parachute sea.

(7) Aircraft type ratings—

(i) Large aircraft other than lighter-than-air.

(ii) Turbojet-powered airplanes.

(iii) Other aircraft type ratings specified by the Administrator through the aircraft type certification procedures.

(iv) Second-in-command pilot type rating for aircraft that is certificated for operations with a minimum crew of at least two pilots.

(8) Instrument ratings (on private and commercial pilot certificates only)—

(i) Instrument—Airplane.

(ii) Instrument—Helicopter.

(iii) Instrument—Powered-lift.

(c) The following ratings are placed on a flight instructor certificate when an applicant satisfactorily accomplishes the training and certification requirements for the rating sought:

(1) Aircraft category ratings—

(i) Airplane.

(ii) Rotorcraft.

(iii) Glider.

(iv) Powered-lift.

(2) Airplane class ratings—

(i) Single-engine.

(ii) Multiengine.

(3) Rotorcraft class ratings—

(i) Helicopter.

(ii) Gyroplane.

(4) Instrument ratings—

(i) Instrument—Airplane.

(ii) Instrument—Helicopter.

(iii) Instrument—Powered-lift.

(5) Sport pilot rating.

(d) The following ratings are placed on a ground instructor certificate when an applicant satisfactorily accomplishes the training and certification requirements for the rating sought:

(1) Basic.

(2) Advanced.

(3) Instrument.

§ 61.7 Obsolete certificates and ratings.

(a) The holder of a free-balloon pilot certificate issued before November 1, 1973, may not exercise the privileges of that certificate.

(b) The holder of a pilot certificate that bears any of the following category ratings without an associated class rating may not exercise the privileges of that category rating:

(1) Rotorcraft.

(2) Lighter-than-air.

(3) Helicopter.

(4) Autogyro.

§ 61.9 [Reserved.]

§ 61.11 Expired pilot certificates and reissuance.

(a) No person who holds an expired pilot certificate or rating may:

(1) Exercise the privileges of that pilot certificate or rating; or

(2) Act as pilot in command or as a required pilot flight crewmember of an aircraft of the same category and class specified on the expired pilot certificate or rating.

(b) The following pilot certificates and ratings have expired and may will not be reissued:

(1) An airline transport pilot certificate issued before May 1, 1949, or an airline transport pilot certificate that contains a horsepower limitation;

(2) A private or commercial pilot certificate issued before July 1, 1945; and

(3) A pilot certificate with a lighter-than-air or free-balloon rating issued before July 1, 1945.

(c) A pilot certificate issued on the basis of a foreign pilot license will expire on the date the foreign license expires unless otherwise specified on the U.S. pilot certificate. A certificate only if that person meets the requirements of § 61.75 for the issuance of a pilot license.

(d) An airline transport pilot certificate issued after April 30, 1949, that bears an expiration date but does not contain a horsepower limitation may be reissued without an expiration date.

(e) A private or commercial pilot certificate issued after June 30, 1945, that bears an expiration date may be reissued without an expiration date.

(f) A pilot certificate with a lighter-than-air or free-balloon rating issued after June 30, 1945, that bears an expiration date may be reissued without an expiration date.

[Amdt. 61-102, 62 FR 16298, Apr. 4, 1997; Amdt. 61-103, 62 FR 40895, July 30, 1997]

§ 61.13 Issuance of airman certificates, ratings, and authorizations.

(a) Application.

(1) An applicant for an airman certificate, rating, or authorization under this part must make that application on a form and in a manner acceptable to the Administrator.

(2) An applicant who is neither a citizen of the United States nor a resident alien of the United States—

(i) Must show evidence that the appropriate fee prescribed in appendix A to part 187 of this chapter has been paid when that person applies for a—

(A) Student pilot certificate that is issued outside the United States; or

(B) Knowledge test or practical test for an airman certificate or rating issued under this part, if the test is administered outside the United States.

(ii) May be refused issuance of any U.S. airman certificate, rating, or authorization by the Administrator.

(3) Except as provided in paragraph (a)(2)(ii) of this section, an applicant who satisfactorily accomplishes the training and certification requirements for the certificate, rating, or authorization sought is entitled to receive that airman certificate, rating, or authorization.

(b) Limitations.

(1) An applicant who cannot comply with certain areas of operation required on the practical test because of physical limitations may be issued an airman certificate, rating, or authorization with the appropriate limitation placed on the applicant's airman certificate provided the—

(i) Applicant is able to meet all other certification requirements for the airman certificate, rating, or authorization sought;

(ii) Physical limitation has been recorded with the FAA on the applicant's medical records; and

(iii) Administrator determines that the applicant's inability to perform the particular area of operation will not adversely affect safety.

(2) A limitation placed on a person's airman certificate may be removed, provided that person demonstrates for an examiner satisfactory proficiency in the area of operation appropriate to the airman certificate, rating, or authorization sought.

(c) Additional requirements for Category II and Category III pilot authorizations. (1) A Category II or Category

III pilot authorization is issued by a letter of authorization as part of an applicant's instrument rating or airline transport pilot certificate.

(2) Upon original issue, the authorization contains the following limitations:

(i) For Category II operations, the limitation is 1,600 feet RVR and a 150-foot decision height; and

(ii) For Category III operations, each initial limitation is specified in the authorization document.

(3) The limitations on a Category II or Category III pilot authorization may be removed as follows:

(i) In the case of Category II limitations, a limitation is removed when the holder shows that, since the beginning of the sixth preceding month, the holder has made three Category II ILS approaches with a 150-foot decision height to a landing under actual or simulated instrument conditions.

(ii) In the case of Category III limitations, a limitation is removed as specified in the authorization.

(4) To meet the experience requirements of paragraph (c)(3) of this section, and for the practical test required by this part for a Category II or a Category III pilot authorization, a flight simulator or flight training device may be used if it is approved by the Administrator for such use.

(d) Application during suspension or revocation.

(1) Unless otherwise authorized by the Administrator, a person whose pilot, flight instructor, or ground instructor certificate has been suspended may not apply for any certificate, rating, or authorization during the period of suspension.

(2) Unless otherwise authorized by the Administrator, a person whose pilot, flight instructor, or ground instructor certificate has been revoked may not apply for any certificate, rating, or authorization for 1 year after the date of revocation.

[Amdt. 61-102, 62 FR 16298, Apr. 4, 1997; Amdt. 61-103, 62 FR 40895, July 30, 1997]

§ 61.14 Refusal to submit to a drug or alcohol test.

(a) This section applies to an individual who holds a certificate under this part and is subject to the types of testing required under appendix I to part 121 or appendix J of part 21 of this chapter.

(b) Refusal by the holder of a certificate issued under this part to take a drug test required under the provisions of appendix I to part 121 or an alcohol test required under the provisions of appendix J to part 121 is grounds for:

(1) Denial of an application for any certificate, rating, or authorization issued under this part for a period of up to 1 year after the date of such refusal; and

(2) Suspension or revocation of any certificate, rating, or authorization issued under this part.

§ 61.15 Offenses involving alcohol or drugs.

(a) A conviction for the violation of any Federal or State statute relating to the growing, processing, manufacture, sale, disposition, possession, transportation, or importation of narcotic drugs, marijuana, or depressant or stimulant drugs or substances is grounds for:

(1) Denial of an application for any certificate, rating, or authorization issued under this part for a period of up to 1 year after the date of final conviction; or

(2) Suspension or revocation of any certificate, rating, or authorization issued under this part.

(b) Committing an act prohibited by § 91.17(a) or § 91.19(a) of this chapter is grounds for:

(1) Denial of an application for a certificate, rating, or authorization issued under this part for a period of up to 1 year after the date of that act; or

(2) Suspension or revocation of any certificate, rating, or authorization issued under this part.

(c) For the purposes of paragraphs (d), (e), and (f) of this section, a motor vehicle action means:

(1) A conviction after November 29, 1990, for the violation of any Federal or State statute relating to the operation of a motor vehicle while intoxicated by alcohol or a drug, while impaired by alcohol or a drug, or while under the influence of alcohol or a drug;

(2) The cancellation, suspension, or revocation of a license to operate a motor vehicle after November 29, 1990, for a cause related to the operation of a motor vehicle while intoxicated by alcohol or a drug, while impaired by alcohol or a drug, or while under the influence of alcohol or a drug; or

(3) The denial after November 29, 1990, of an application for a license to operate a motor vehicle for a cause related to the operation of a motor vehicle while intoxicated by alcohol or a drug, while impaired by alcohol or a drug, or while under the influence of alcohol or a drug.

(d) Except for a motor vehicle action that results from the same incident or arises out of the same factual circumstances, a motor vehicle action occurring within 3 years of a previous motor vehicle action is grounds for:

(1) Denial of an application for any certificate, rating, or authorization issued under this part for a period of up to 1 year after the date of the last motor vehicle action; or

(2) Suspension or revocation of any certificate, rating, or authorization issued under this part.

(e) Each person holding a certificate issued under this part shall provide a written report of each motor vehicle action to the FAA, Civil Aviation Security Division (AMC-700), P.O. Box 25810, Oklahoma City, OK 73125, not later than 60 days after the motor vehicle action. The report must include:

(1) The person's name, address, date of birth, and airman certificate number;

(2) The type of violation that resulted in the conviction or the administrative action;

(3) The date of the conviction or administrative action;

(4) The State that holds the record of conviction or administrative action; and

(5) A statement of whether the motor vehicle action resulted from the same incident or arose out of the same factual circumstances related to a previously reported motor vehicle action.

(f) Failure to comply with paragraph (e) of this section is grounds for:

(1) Denial of an application for any certificate, rating, or authorization issued under this part for a period of up to 1 year after the date of the motor vehicle action; or

(2) Suspension or revocation of any certificate, rating, or authorization issued under this part.

§ 61.16 Refusal to submit to an alcohol test or to furnish test results.

A refusal to submit to a test to indicate the percentage by weight of alcohol in the blood, when requested by a law enforcement officer in accordance with § 91.17(c) of this chapter, or a refusal to furnish or authorize the release of the test results requested by the Administrator in accordance with § 91.17(c) or (d) of this chapter, is grounds for:

(a) Denial of an application for any certificate, rating, or authorization issued under this part for a period of up to 1 year after the date of that refusal; or

(b) Suspension or revocation of any certificate, rating, or authorization issued under this part.

§ 61.17 Temporary certificate.

(a) A temporary pilot, flight instructor, or ground instructor certificate or rating is issued for up to 120 days, at which time a permanent certificate will be issued to a person whom the Administrator finds qualified under this part.

(b) A temporary pilot, flight instructor, or ground instructor certificate or rating expires:

(1) On the expiration date shown on the certificate;

(2) Upon receipt of the permanent certificate; or

(3) Upon receipt of a notice that the certificate or rating sought is denied or revoked.

§ 61.18 Security disqualification.

(a) Eligibility standard. No person is eligible to hold a certificate, rating, or authorization issued under this part when the Transportation Security Administration (TSA) has notified the FAA in writing that the person poses a security threat.

(b) Effect of the issuance by the TSA of an Initial Notification of Threat Assessment. (1) The FAA will hold in abeyance pending the outcome of the TSA's final threat assessment review an application for any certificate, rating, or authorization under this part by any person who has been issued an Initial Notification of Threat Assessment by the TSA.

(1) The FAA will suspend any certificate, rating, or authorization issued under this part after the TSA issues to the holder an Initial Notification of Threat Assessment.

(c) *Effect of the issuance by the TSA of Final Notification of Threat Assessment.* (1) The FAA will deny an application for any certificate, rating, or authorization under this part to any person who has been issued a Final Notification of Threat Assessment.

(1) The FAA will revoke any certificate, rating, or authorization issued under this part after the TSA has issued to the holder a Final Notification of Threat Assessment.

§ 61.19   Duration of pilot and instructor certificates.

(a) *General.* The holder of a certificate with an expiration date may not, after that date, exercise the privileges of that certificate.

(b) *Student pilot certificate.* A student pilot certificate expires 24 calendar months from the month in which it is issued.

(c) *Other pilot certificates.* A pilot certificate (other than a student pilot certificate) issued under this part is issued without a specific expiration date. The holder of a pilot certificate issued on the basis of a foreign pilot license may exercise the privileges of that certificate only while that person's foreign pilot license is effective.

(d) *Flight instructor certificate.* A flight instructor certificate:

(1) Is effective only while the holder has a current pilot certificate; and

(2) Except as specified in § 61.197(b) of this part, expires 24 calendar months from the month in which it was issued or renewed.

(e) *Ground instructor certificate.* A ground instructor certificate issued under this part is issued without a specific expiration date.

(f) *Surrender, suspension, or revocation.* Any certificate issued under this part ceases to be effective if it is surrendered, suspended, or revoked.

(g) *Return of certificates.* The holder of any certificate issued under this part that has been suspended or revoked must return that certificate to the FAA when requested to do so by the Administrator.

§ 61.21   Duration of a Category II and a Category III pilot authorization (for other than part 121 and part 135 use).

(a) A Category II pilot authorization or a Category III pilot authorization expires at the end of the sixth calendar month after the month in which it was issued or renewed.

(b) Upon passing a practical test for a Category II or Category III pilot authorization, the authorization may be renewed for each type of aircraft for which the authorization is held.

(c) A Category II or Category III pilot authorization for a specific type aircraft for which an authorization is held will not be renewed beyond 12 calendar months from the month the practical test was accomplished in that type aircraft.

(d) If the holder of a Category II or Category III pilot authorization passes the practical test for a renewal in the month before the authorization expires, the holder is considered to have passed it during the month the authorization expired.

§ 61.23   Medical certificates: Requirement and duration.

(a) *Operations requiring a medical certificate.* Except as provided in paragraph (b) and (c) of this section, a person:

(1) Must hold a first-class medical certificate when exercising the privileges of an airline transport pilot certificate;

(2) Must hold at least a second-class medical certificate when exercising the privileges of a commercial pilot certificate; or

(3) Must hold at least a third-class medical certificate—

(i) When exercising the privileges of a private pilot certificate;

(ii) When exercising the privileges of a recreational pilot certificate;

(iii) When exercising the privileges of a student pilot certificate;

(iv) When exercising the privileges of a flight instructor certificate, except for a flight instructor certificate with a glider category rating or sport pilot rating, if the person is acting as pilot in command or is serving as a required flight crewmember; or

(v) Except for a glider category rating or a balloon class rating, prior to taking a practical test that is performed in an aircraft for a certificate or rating at the recreational, private, commercial, or airline transport pilot certificate level.

(b) *Operations not requiring a medical certificate.* A person is not required to hold a valid medical certificate—

(1) When exercising the privileges of a student pilot certificate while seeking

(i) A sport pilot certificate with glider or balloon privileges; or

(ii) A pilot certificate with a glider category rating or balloon class rating;

(2) When exercising the privileges of a sport pilot certificate with privileges in a glider or balloon;

(3) When exercising the privileges of a pilot certificate with a glider category or balloon class rating;

(4) When exercising the privileges of a flight instructor certificate with—

(i) A sport pilot rating in a glider or balloon; or

(ii) A glider category rating;

(5) When exercising the privileges of a flight instructor certificate if the person is not acting as pilot in command or serving as a required pilot flight crewmember;

(6) When exercising the privileges of a ground instructor certificate;

(7) When serving as an examiner or check airman during the administration of a test or check for a certificate, rating, or authorization conducted in a flight simulator or flight training device; or

(8) When taking a test or check for a certificate, rating, or authorization conducted in a flight simulator or flight training device.

(c) *Operations requiring either a medical certificate or U.S. driver's license.*

(1) A person must hold and possess either a valid medical certificate issued under part 67 of this chapter or a current and valid U.S. driver's license when exercising the privileges of—

(i) A student pilot certificate while seeking sport pilot privileges in a light-sport aircraft other than a glider or balloon;

(ii) A sport pilot certificate in a light-sport aircraft other than a glider or balloon; or

(iii) A flight instructor certificate with a sport pilot rating while acting as pilot in command or serving as a required flight crewmember of a light-sport aircraft other than a glider or balloon.

(2) A person using a current and valid U.S. driver's license to meet the requirements of this paragraph must—

(i) Comply with each restriction and limitation imposed by that person's U.S. driver's license and any judicial or administrative order applying to the operation of a motor vehicle;

(ii) Have been found eligible for the issuance of at least a third-class airman medical certificate at the time of his or her most recent application (if the person has applied for a medical certificate);

(iii) Not have had his or her most recently issued medical certificate (if the person has held a medical certificate) suspended or revoked or most recent Authorization for a Special Issuance of a Medical Certificate withdrawn; and

(iv) Not know or have reason to know of any medical condition that would make that person unable to operate a light-sport aircraft in a safe manner.

(d) *Duration of a medical certificate.*

(1) A first-class medical certificate expires at the end of the last day of—

(i) The sixth month after the month of the date of examination shown on the certificate for operations requiring an airline transport pilot certificate;

(ii) The 12th month after the month of the date of examination shown on the certificate for operations requiring a commercial pilot certificate or an air traffic control tower operator certificate; and

(iii) The period specified in paragraph (c)(3) of this section for operations requiring a recreational pilot certificate, a private pilot certificate, a flight instructor certificate (when acting as pilot in command or a required pilot flight

crewmember in operations other than glider or balloon), or a student pilot certificate.

(2) A second-class medical certificate expires at the end of the last day of—

(i) The 12th month after the month of the date of examination shown on the certificate for operations requiring a commercial pilot certificate or an air traffic control tower operator certificate; and

(ii) The period specified in paragraph (c)(3) of this section for operations requiring a recreational pilot certificate, a private pilot certificate, a flight instructor certificate (when acting as pilot in command or a required pilot flight crewmember in operations other than glider or balloon), or a student pilot certificate.

(3) A third-class medical certificate for operations requiring a recreational pilot certificate, a private pilot certificate, a flight instructor certificate (when acting as pilot in command or a required pilot flight crewmember in operations other than glider or balloon), or a student pilot certificate issued—

(i) Before September 16, 1996, expires at the end of the 24th month after the month of the date of examination shown on the certificate; or

(ii) On or after September 16, 1996, expires at the end of:

(A) The 36th month after the month of the date of the examination shown on the certificate if the person has not reached his or her 40th birthday on or before the date of examination; or

(B) The 24th month after the month of the date of the examination shown on the certificate if the person has reached his or her 40th birthday on or before the date of the examination.

[Amdt. 61-102, 62 FR 16298, Apr. 4, 1997; Amdt. 61-103, 62 FR 40895, July 30, 1997]

§ 61.25 Change of name.

(a) An application to change the name on a certificate issued under this part must be accompanied by the applicant's:

(1) Current airman certificate; and

(2) A copy of the marriage license, court order, or other document verifying the name change.

(b) The documents in paragraph (a) of this section will be returned to the applicant after inspection.

§ 61.27 Voluntary surrender or exchange of certificate.

(a) The holder of a certificate issued under this part may voluntarily surrender it for:

(1) Cancellation;

(2) Issuance of a lower grade certificate; or

(3) Another certificate with specific ratings deleted.

(b) Any request made under paragraph (a) of this section must include the following signed statement or its equivalent: "This request is made for my own reasons, with full knowledge that my (insert name of certificate or rating, as appropriate) may not be reissued to me unless I again pass the tests prescribed for its issuance."

§ 61.29 Replacement of a lost or destroyed airman or medical certificate or knowledge test report.

(a) A request for the replacement of a lost or destroyed airman certificate issued under this part must be made by letter to the Department of Transportation, FAA, Airman Certification Branch, P.O. Box 25082, Oklahoma City, OK 73125, and must be accompanied by a check or money order for the appropriate fee payable to the FAA.

(b) A request for the replacement of a lost or destroyed medical certificate must be made by letter to the Department of Transportation, FAA, Aeromedical Certification Branch, P.O. Box 25082, Oklahoma City, OK 73125, and must be accompanied by a check or money order for the appropriate fee payable to the FAA.

(c) A request for the replacement of a lost or destroyed knowledge test report must be made by letter to the Department of Transportation, FAA, Airman Certification Branch, P.O. Box 25082, Oklahoma City, OK 73125, and must be accompanied by a check or money order for the appropriate fee payable to the FAA.

(d) The letter requesting replacement of a lost or destroyed airman certificate, medical certificate, or knowledge test report must state:

(1) The name of the person;

(2) The permanent mailing address (including ZIP code), or if the permanent mailing address includes a post office box number, then the person's current residential address;

(3) The social security number;

(4) The date and place of birth of the certificate holder; and

(5) Any available information regarding the—

(i) Grade, number, and date of issuance of the certificate, and the ratings, if applicable;

(ii) Date of the medical examination, if applicable; and

(iii) Date the knowledge test was taken, if applicable.

(e) A person who has lost an airman certificate, medical certificate, or knowledge test report may obtain a facsimile from the FAA Aeromedical Certification Branch or the Airman Certification Brancsh, as appropriate confirming that it was issued and the:

(1) Facsimile may be carried as an airman certificate, medical certificate, or knowledge test report, as appropriate, for up to 60 days pending the person's receipt of a duplicate under paragraph (a), (b), or (c) of this section, unless the person has been notified that the certificate has been suspended or revoked.

(2) Request for such a facsimile must include the date on which a duplicate certificate or knowledge test report was previously requested.

[Amdt. 61-102, 62 FR 16298, Apr. 4, 1997; Amdt. 61-103, 62 FR 40896, July 30, 1997]

§ 61.31 Type rating requirements, additional training, and authorization requirements.

(a) Type ratings required. A person who acts as a pilot in command of any of the following aircraft must hold a type rating for that aircraft:

(1) Large aircraft (except lighter-than-air).

(2) Turbojet-powered airplanes.

(3) Other aircraft specified by the Administrator through aircraft type certificate procedures.

(b) Authorization in lieu of a type rating. A person may be authorized to operate without a type rating for up to 60 days an aircraft requiring a type rating, provided—

(1) The Administrator has authorized the flight or series of flights;

(2) The Administrator has determined that an equivalent level of safety can be achieved through the operating limitations on the authorization;

(3) The person shows that compliance with paragraph (a) of this section is impracticable for the flight or series of flights; and

(4) The flight—

(i) Involves only a ferry flight, training flight, test flight, or practical test for a pilot certificate or rating;

(ii) Is within the United States;

(iii) Does not involve operations for compensation or hire unless the compensation or hire involves payment for the use of the aircraft for training or taking a practical test; and

(iv) Involves only the carriage of flight crewmembers considered essential for the flight.

(5) If the flight or series of flights cannot be accomplished within the time limit of the authorization, the Administrator may authorize an additional period of up to 60 days to accomplish the flight or series of flights.

(c) Aircraft category, class, and type ratings: Limitations on the carriage of persons, or operating for compensation or hire. Unless a person holds a category, class, and type rating (if a class and type rating is required) that applies to the aircraft, that person may not act as pilot in command of an aircraft that is carrying another person, or is operated for compensation or hire. That person also may not act as pilot in command of that aircraft for compensation or hire.

(d) Aircraft category, class, and type ratings: Limitations on operating an aircraft as the pilot in command. To serve as the pilot in command of an aircraft, a person must—

(1) Hold the appropriate category, class, and type rating (if a class rating and type rating are required) for the aircraft to be flown;

(2) Be receiving training for the purpose of obtaining an additional pilot certificate and rating that are appropriate

to that aircraft, and be under the supervision of an authorized instructor; or

(3) Have received training required by this part that is appropriate to the aircraft category, class, and type rating (if a class or type rating is required) for the aircraft to be flown, and have received the required endorsements from an instructor who is authorized to provide the required endorsements for solo flight in that aircraft.

(e) Additional training required for operating complex airplanes.

(1) Except as provided in paragraph (e)(2) of this section, no person may act as pilot in command of a complex airplane (an airplane that has a retractable landing gear, flaps, and a controllable pitch propeller; or, in the case of a seaplane, flaps and a controllable pitch propeller), unless the person has—

(i) Received and logged ground and flight training from an authorized instructor in a complex airplane, or in a flight simulator or flight training device that is representative of a complex airplane, and has been found proficient in the operation and systems of the airplane; and

(ii) Received a one-time endorsement in the pilot's logbook from an authorized instructor who certifies the person is proficient to operate a complex airplane.

(2) The training and endorsement required by paragraph (e)(1) of this section is not required if the person has logged flight time as pilot in command of a complex airplane, or in a flight simulator or flight training device that is representative of a complex airplane prior to August 4, 1997.

(f) Additional training required for operating high-performance airplanes.

(1) Except as provided in paragraph (f)(2) of this section, no person may act as pilot in command of a high-performance airplane (an airplane with an engine of more than 200 horsepower), unless the person has—

(i) Received and logged ground and flight training from an authorized instructor in a high-performance airplane, or in a flight simulator or flight training device that is representative of a high-performance airplane, and has been found proficient in the operation and systems of the airplane; and

(ii) Received a one-time endorsement in the pilot's logbook from an authorized instructor who certifies the person is proficient to operate a high-performance airplane.

(2) The training and endorsement required by paragraph (f)(1) of this section is not required if the person has logged flight time as pilot in command of a high-performance airplane, or in a flight simulator or flight training device that is representative of a high-performance airplane prior to August 4, 1997.

(g) Additional training required for operating pressurized aircraft capable of operating at high altitudes.

(1) Except as provided in paragraph (g)(3) of this section, no person may act as pilot in command of a pressurized aircraft (an aircraft that has a service ceiling or maximum operating altitude, whichever is lower, above 25,000 feet MSL), unless that person has received and logged ground training from an authorized instructor and obtained an endorsement in the person's logbook or training record from an authorized instructor who certifies the person has satisfactorily accomplished the ground training. The ground training must include at least the following subjects:

(i) High-altitude aerodynamics and meteorology;

(ii) Respiration;

(iii) Effects, symptoms, and causes of hypoxia and any other high-altitude sickness;

(iv) Duration of consciousness without supplemental oxygen;

(v) Effects of prolonged usage of supplemental oxygen;

(vi) Causes and effects of gas expansion and gas bubble formation;

(vii) Preventive measures for eliminating gas expansion, gas bubble formation, and high-altitude sickness;

(viii) Physical phenomena and incidents of decompression; and

(ix) Any other physiological aspects of high-altitude flight.

(2) Except as provided in paragraph (g)(3) of this section, no person may act as pilot in command of a pressurized aircraft unless that person has received and logged training from an authorized instructor in a pressurized aircraft, or in a flight simulator or flight training device that is representative of a pressurized aircraft, and obtained an endorsement in the person's logbook or training record from an authorized instructor who found the person proficient in the operation of a pressurized aircraft. The flight training must include at least the following subjects:

(i) Normal cruise flight operations while operating above 25,000 feet MSL;

(ii) Proper emergency procedures for simulated rapid decompression without actually depressurizing the aircraft; and

(iii) Emergency descent procedures.

(3) The training and endorsement required by paragraphs (g)(1) and (g)(2) of this section are not required if that person can document satisfactory accomplishment of any of the following in a pressurized aircraft, or in a flight simulator or flight training device that is representative of a pressurized aircraft:

(i) Serving as pilot in command before April 15, 1991;

(ii) Completing a pilot proficiency check for a pilot certificate or rating before April 15, 1991;

(iii) Completing an official pilot-in-command check conducted by the military services of the United States; or

(iv) Completing a pilot-in-command proficiency check under part 121, 125, or 135 of this chapter conducted by the Administrator or by an approved pilot check airman.

(h) Additional training required by the aircraft's type certificate. No person may serve as pilot in command of an aircraft that the Administrator has determined requires aircraft type-specific training unless that person has—

(1) Received and logged type-specific training in the aircraft, or in a flight simulator or flight training device that is representative of that type of aircraft; and

(2) Received a logbook endorsement from an authorized instructor who has found the person proficient in the operation of the aircraft and its systems.

(i) Additional training required for operating tailwheel airplanes.

(1) Except as provided in paragraph (i)(2) of this section, no person may act as pilot in command of a tailwheel airplane unless that person has received and logged flight training from an authorized instructor in a tailwheel airplane and received an endorsement in the person's logbook from an authorized instructor who found the person proficient in the operation of a tailwheel airplane. The flight training must include at least the following maneuvers and procedures:

(i) Normal and crosswind takeoffs and landings;

(ii) Wheel landings (unless the manufacturer has recommended against such landings); and

(iii) Go-around procedures.

(2) The training and endorsement required by paragraph (i)(1) of this section is not required if the person logged pilot-in-command time in a tailwheel airplane before April 15, 1991.

(j) Additional training required for operating a glider.

(1) No person may act as pilot in command of a glider—

(i) Using ground-tow procedures, unless that person has satisfactorily accomplished ground and flight training on ground-tow procedures and operations, and has received an endorsement from an authorized instructor who certifies in that pilot's logbook that the pilot has been found proficient in ground-tow procedures and operations;

(ii) Using aerotow procedures, unless that person has satisfactorily accomplished ground and flight training on aerotow procedures and operations, and has received an endorsement from an authorized instructor who certifies in that pilot's logbook that the pilot has been found proficient in aerotow procedures and operations; or

(iii) Using self-launch procedures, unless that person has satisfactorily accomplished ground and flight training on self-launch procedures and operations, and has received an endorsement from an authorized instructor who certifies in that pilot's logbook that the pilot has been found proficient in self-launch procedures and operations.

(2) The holder of a glider rating issued prior to August 4, 1997, is considered to be in compliance with the training and logbook endorsement requirements of this paragraph for the specific operating privilege for which the holder is already qualified.

(k) Exceptions.

(1) This section does not require a category and class rating for aircraft not type certificated as airplanes, rotorcraft,

gliders, lighter-than-air aircraft, powered-lifts, powered parachutes, or weight-shift-control aircraft.

(2) The rating limitations of this section do not apply to—

(i) An applicant when taking a practical test given by an examiner;

(ii) The holder of a student pilot certificate; —

(iii) The holder of a pilot certificate when operating an aircraft under the authority of—

(A) A provisional type certificate; or

(B) An experimental certificate, unless the operation involves carrying a passenger;

(iv) The holder of a pilot certificate with a lighter-than-air category rating when operating a balloon; or

(v) The holder of a recreational pilot certificate operating under the provisions of § 61.101(h).

(vi) The holder of a sport pilot certificate when operating a light-sport aircraft.

[Amdt. 61-102, 62 FR 16298, Apr. 4, 1997; Amdt. 61-103, 62 FR 40896, July 30, 1997]

§ 61.33 Tests: General procedure.

Tests prescribed by or under this part are given at times and places, and by persons designated by the Administrator.

§ 61.35 Knowledge test: Prerequisites and passing grades.

(a) An applicant for a knowledge test must have:

(1) Received an endorsement, if required by this part, from an authorized instructor certifying that the applicant accomplished a ground-training or a home-study course required by this part for the certificate or rating sought and is prepared for the knowledge test; and

(2) Proper identification at the time of application that contains the applicant's—

(i) Photograph;

(ii) Signature;

(iii) Date of birth, which shows the applicant meets or will meet the age requirements of this part for the certificate sought before the expiration date of the airman knowledge test report; and

(iv) Actual residential address, if different from the applicant's mailing address.

(b) The Administrator shall specify the minimum passing grade for the knowledge test.

§ 61.37 Knowledge tests: Cheating or other unauthorized conduct.

(a) An applicant for a knowledge test may not:

(1) Copy or intentionally remove any knowledge test;

(2) Give to another applicant or receive from another applicant any part or copy of a knowledge test;

(3) Give assistance on, or receive assistance on, a knowledge test during the period that test is being given;

(4) Take any part of a knowledge test on behalf of another person;

(5) Be represented by, or represent, another person for a knowledge test;

(6) Use any material or aid during the period that the test is being given, unless specifically authorized to do so by the Administrator; and

(7) Intentionally cause, assist, or participate in any act prohibited by this paragraph.

(b) An applicant who the Administrator finds has committed an act prohibited by paragraph (a) of this section is prohibited, for 1 year after the date of committing that act, from:

(1) Applying for any certificate, rating, or authorization issued under this chapter; and

(2) Applying for and taking any test under this chapter.

(c) Any certificate or rating held by an applicant may be suspended or revoked if the Administrator finds that person has committed an act prohibited by paragraph (a) of this section.

§ 61.39 Prerequisites for practical tests.

(a) Except as provided in paragraphs (b) and (c) of this section, to be eligible for a practical test for a certificate or rating issued under this part, an applicant must:

(1) Pass the required knowledge test within the 24-calendar-month period preceding the month the applicant completes the practical test, if a knowledge test is required;

(2) Present the knowledge test report at the time of application for the practical test, if a knowledge test is required;

(3) Have satisfactorily accomplished the required training and obtained the aeronautical experience prescribed by this part for the certificate or rating sought;

(4) Hold at least a current third-class medical certificate, if a medical certificate is required;

(5) Meet the prescribed age requirement of this part for the issuance of the certificate or rating sought;

(6) Have an endorsement, if required by this part, in the applicant's logbook or training record that has been signed by an authorized instructor who certifies that the applicant—

(i) Has received and logged training time within 60 days preceding the date of application in preparation for the practical test;

(ii) Is prepared for the required practical test; and

(iii) Has demonstrated satisfactory knowledge of the subject areas in which the applicant was deficient on the airman knowledge test; and

(7) Have a completed and signed application form.

(b) Notwithstanding the provisions of paragraphs (a)(1) and (2) of this section, an applicant for an airline transport pilot certificate or an additional rating to an airline transport certificate may take the practical test for that certificate or rating with an expired knowledge test report, provided that the applicant:

(1) Is employed as a flight crewmember by a certificate holder under part 121, 125, or 135 of this chapter at the time of the practical test and has satisfactorily accomplished that operator's approved—

(i) Pilot in command aircraft qualification training program that is appropriate to the certificate and rating sought; and

(ii) Qualification training requirements appropriate to the certificate and rating sought; or

(2) Is employed as a flight crewmember in scheduled U.S. military air transport operations at the time of the practical test, and has accomplished the pilot in command aircraft qualification training program that is appropriate to the certificate and rating sought.

(c) A person is not required to comply with the provisions of paragraph (a)(6) of this section if that person:

(1) Holds a foreign-pilot license issued by a contracting State to the Convention on International Civil Aviation that authorizes at least the pilot privileges of the airman certificate sought;

(2) Is applying for a type rating only, or a class rating with an associated type rating; or

(3) Is applying for an airline transport pilot certificate or an additional rating to an airline transport pilot certificate in an aircraft that does not require an aircraft type rating practical test.

(d) If all increments of the practical test for a certificate or rating are not completed on one date, all remaining increments of the test must be satisfactorily completed not more than 60 calendar days after the date on which the applicant began the test.

(e) If all increments of the practical test for a certificate or a rating are not satisfactorily completed within 60 calendar days after the date on which the applicant began the test, the applicant must retake the entire practical test, including those increments satisfactorily completed.

[Amdt. 61-102, 62 FR 16298, Apr. 4, 1997; Amdt. 61-103, 62 FR 40897, July 30, 1997]

§ 61.41 Flight training received from flight instructors not certificated by the FAA.

(a) A person may credit flight training toward the requirements of a pilot certificate or rating issued under this part, if that person received the training from:

(1) A flight instructor of an Armed Force in a program for training military pilots of either—

(i) The United States; or

(ii) A foreign contracting State to the Convention on International Civil Aviation.

(2) A flight instructor who is authorized to give such training by the licensing authority of a foreign contracting State to the Convention on International Civil Aviation, and the flight training is given outside the United States.

(b) A flight instructor described in paragraph (a) of this section is only authorized to give endorsements to show training given.

§ 61.43 Practical tests: General procedures.

(a) Except as provided in paragraph (b) of this section, the ability of an applicant for a certificate or rating issued under this part to perform the required tasks on the practical test is based on that applicant's ability to safely:

(1) Perform the tasks specified in the areas of operation for the certificate or rating sought within the approved standards;

(2) Demonstrate mastery of the aircraft with the successful outcome of each task performed never seriously in doubt;

(3) Demonstrate satisfactory proficiency and competency within the approved standards;

(4) Demonstrate sound judgment; and

(5) Demonstrate single-pilot competence if the aircraft is type certificated for single-pilot operations.

(b) If an applicant does not demonstrate single pilot proficiency, as required in paragraph (a)(5) of this section, a limitation of "Second in Command Required" will be placed on the applicant's airman certificate. The limitation may be removed if the applicant passes the appropriate practical test by demonstrating single-pilot competency in the aircraft in which single-pilot privileges are sought.

(c) If an applicant fails any area of operation, that applicant fails the practical test.

(d) An applicant is not eligible for a certificate or rating sought until all the areas of operation are passed.

(e) The examiner or the applicant may discontinue a practical test at any time:

(1) When the applicant fails one or more of the areas of operation; or

(2) Due to inclement weather conditions, aircraft airworthiness, or any other safety-of-flight concern.

(f) If a practical test is discontinued, the applicant is entitled credit for those areas of operation that were passed, but only if the applicant:

(1) Passes the remainder of the practical test within the 60-day period after the date the practical test was discontinued;

(2) Presents to the examiner for the retest the original notice of disapproval form or the letter of discontinuance form, as appropriate;

(3) Satisfactorily accomplishes any additional training needed and obtains the appropriate instructor endorsements, if additional training is required; and

(4) Presents to the examiner for the retest a properly completed and signed application.

§ 61.45 Practical tests: Required aircraft and equipment.

(a) General. Except as provided in paragraph (a)(2) of this section or when permitted to accomplish the entire flight increment of the practical test in a flight simulator or a flight training device, an applicant for a certificate or rating issued under this part must furnish:

(1) An aircraft of U.S. registry for each required test that—

(i) Is of the category, class, and type, if applicable, for which the applicant is applying for a certificate or rating; and

(ii) Has a current standard airworthiness certificate or special airworthiness certificate in the limited, primary, or light-sport category.

(2) At the discretion of the examiner who administers the practical test, the applicant may furnish—

(i) An aircraft that has a current airworthiness certificate other than a standard airworthiness certificate or special airworthiness certificate in the limited, primary, or light-sport category, but that otherwise meets the requirements of paragraph (a)(1) of this section;

(ii) An aircraft of the same category, class, and type, if applicable, of foreign registry that is properly certificated by the country of registry; or

(iii) A military aircraft of the same category, class, and type, if applicable, for which the applicant is applying for a certificate or rating.

(b) *Required equipment (other than controls).* (1) Except as provided in paragraph (b)(2) of this section, an aircraft used for a practical test must have—

(i) The equipment for each area of operation required for the practical test;

(ii) No prescribed operating limitations that prohibit its use in any of the areas of operation required for the practical test;

(iii) Except as provided in paragraphs (e) and (f) of this section, at least two pilot stations with adequate visibility for each person to operate the aircraft safely; and

(iv) Cockpit and outside visibility adequate to evaluate the performance of the applicant when an additional jump seat is provided for the examiner.

(2) An applicant for a certificate or rating may use an aircraft with operating characteristics that preclude the applicant from performing all of the tasks required for the practical test. However, the applicant's certificate or rating, as appropriate, will be issued with an appropriate limitation.

(3) Except as provided in paragraph (e) of this section, at least two pilot stations with adequate visibility for each person to operate the aircraft safely; and

(4) Cockpit and outside visibility adequate to evaluate the performance of the applicant when an additional jump seat is provided for the examiner.

(c) Required controls. An aircraft (other than a lighter-than-air aircraft) used for a practical test must have engine power controls and flight controls that are easily reached and operable in a conventional manner by both pilots, unless the examiner determines that the practical test can be conducted safely in the aircraft without the controls being easily reached.

(d) Simulated instrument flight equipment. An applicant for a practical test that involves maneuvering an aircraft solely by reference to instruments must furnish:

(1) Equipment on board the aircraft that permits the applicant to pass the areas of operation that apply to the rating sought; and

(2) A device that prevents the applicant from having visual reference outside the aircraft, but does not prevent the examiner from having visual reference outside the aircraft, and is otherwise acceptable to the Administrator.

(e) *Aircraft with single controls.* A practical test may be conducted in an aircraft having a single set of controls, provided the:

(1) Examiner agrees to conduct the test;

(2) Test does not involve a demonstration of instrument skills; and

(3) Proficiency of the applicant can be observed by an examiner who is in a position to observe the applicant.

(f) *Light-sport aircraft with a single seat.* A practical test for a sport pilot certificate may be conducted in a light-sport aircraft having a single seat provided that the—

(1) Examiner agrees to conduct the test;

(2) Examiner is in a position to observe the operation of the aircraft and evaluate the proficiency of the applicant; and

(3) Pilot certificate of an applicant successfully passing the test is issued a pilot certificate with a limitation "No passenger carriage and flight in a single-seat light-sport aircraft only."

[Amdt. 61-102, 62 FR 16298, Apr. 4, 1997; Amdt. 61-103, 62 FR 40897, July 30, 1997]

§ 61.47 Status of an examiner who is authorized by the Administrator to conduct practical tests.

(a) An examiner represents the Administrator for the purpose of conducting practical tests for certificates and ratings issued under this part and to observe an applicant's ability to perform the areas of operation on the practical test.

(b) The examiner is not the pilot in command of the aircraft during the practical test unless the examiner agrees to act in that capacity for the flight or for a portion of the flight by prior arrangement with:

(1) The applicant; or

(2) A person who would otherwise act as pilot in command of the flight or for a portion of the flight.

(c) Notwithstanding the type of aircraft used during the practical test, the applicant and the examiner (and any other occupants authorized to be on board by the examiner) are not subject to the requirements or limitations for the carriage of passengers that are specified in this chapter.

[Amdt. 61-102, 62 FR 16298, Apr. 4, 1997; Amdt. 61-103, 62 FR 40897, July 30, 1997]

§ 61.49 Retesting after failure.

(a) An applicant for a knowledge or practical test who fails that test may reapply for the test only after the applicant has received:

(1) The necessary training from an authorized instructor who has determined that the applicant is proficient to pass the test; and

(2) An endorsement from an authorized instructor who gave the applicant the additional training.

(b) An applicant for a flight instructor certificate with an airplane category rating or, for a flight instructor certificate with a glider category rating, who has failed the practical test due to deficiencies in instructional proficiency on stall awareness, spin entry, spins, or spin recovery must:

(1) Comply with the requirements of paragraph (a) of this section before being retested;

(2) Bring an aircraft to the retest that is of the appropriate aircraft category for the rating sought and is certificated for spins; and

(3) Demonstrate satisfactory instructional proficiency on stall awareness, spin entry, spins, and spin recovery to an examiner during the retest.

§ 61.51 Pilot logbooks.

(a) *Training time and aeronautical experience.* Each person must document and record the following time in a manner acceptable to the Administrator:

(1) Training and aeronautical experience used to meet the requirements for a certificate, rating, or flight review of this part.

(2) The aeronautical experience required for meeting the recent flight experience requirements of this part.

(b) *Logbook entries.* For the purposes of meeting the requirements of paragraph (a) of this section, each person must enter the following information for each flight or lesson logged:

(1) General—

(i) Date.

(ii) Total flight time or lesson time.

(iii) Location where the aircraft departed and arrived, or for lessons in a flight simulator or flight training device, the location where the lesson occurred.

(iv) Type and identification of aircraft, flight simulator, or flight training device, as appropriate.

(v) The name of a safety pilot, if required by § 91.109(b) of this chapter.

(2) Type of pilot experience or training—

(i) Solo.

(ii) Pilot in command.

(iii) Second in command.

(iv) Flight and ground training received from an authorized instructor.

(v) Training received in a flight simulator or flight training device from an authorized instructor.

(3) Conditions of flight—

(i) Day or night.

(ii) Actual instrument.

(iii) Simulated instrument conditions in flight, a flight simulator, or a flight training device.

(c) *Logging of pilot time.* The pilot time described in this section may be used to:

(1) Apply for a certificate or rating issued under this part; or a privilege authorized under this part; or

(2) Satisfy the recent flight experience requirements of this part.

(d) *Logging of solo flight time.* Except for a student pilot performing the duties of pilot in command of an airship requiring more than one flight crewmember, a pilot may log as solo flight time only that flight time when the pilot is the sole occupant of the aircraft.

(e) *Logging pilot-in-command flight time.*

(1) A sport, recreational, private, or commercial pilot may log pilot-in-command time only for that flight time during which that person—

(i) Is the sole manipulator of the controls of an aircraft for which the pilot is rated or has privileges;

(ii) Is the sole occupant of the aircraft; or

(iii) Except for a recreational pilot, is acting as pilot in command of an aircraft on which more than one pilot is required under the type certification of the aircraft or the regulations under which the flight is conducted.

(2) An airline transport pilot may log as pilot-in-command time all of the flight time while acting as pilot-in-command of an operation requiring an airline transport pilot certificate.

(3) An authorized instructor may log as pilot-in-command time all flight time while acting as an authorized instructor.

(4) A student pilot may log pilot-in-command time only when the student pilot—

(i) Is the sole occupant of the aircraft or is performing the duties of pilot in command of an airship requiring more than one pilot flight crewmember

(ii) Has a current solo flight endorsement as required under § 61.87 of this part; and

(iii) Is undergoing training for a pilot certificate or rating

(f) *Logging second-in-command time.* A person may log second-in-command flight time only for that flight time during which that person:

(1) Is qualified in accordance with the second-in-command requirements of § 61.55 of this part, and occupies a crewmember station in an aircraft that requires more than one pilot by the aircraft's type certificate; or

(2) Holds the appropriate category, class, and instrument rating (if an instrument rating is required for the flight) for the aircraft being flown, and more than one pilot is required under the type certification of the aircraft or the regulations under which the flight is being conducted.

(g) *Logging instrument flight time.*

(1) A person may log instrument time only for that flight time when the person operates the aircraft solely by reference to instruments under actual or simulated instrument flight conditions.

(2) An authorized instructor may log instrument time when conducting instrument flight instruction in actual instrument flight conditions.

(3) For the purposes of logging instrument time to meet the recent instrument experience requirements of § 61.57(c) of this part, the following information must be recorded in the person's logbook—

(i) The location and type of each instrument approach accomplished; and

(ii) The name of the safety pilot, if required.

(4) A flight simulator or flight training device may be used by a person to log instrument flight time, provided an authorized instructor is present during the simulated flight.

(h) *Logging training time.*

(1) A person may log training time when that person receives training from an authorized instructor in an aircraft, flight simulator, or flight training device.

(2) The training time must be logged in a logbook and must:

(i) Be endorsed in a legible manner by the authorized instructor; and

(ii) Include a description of the training given, the length of the training lesson, and the instructor's authorized signature, certificate number, and certificate expiration date.

(i) *Presentation of required documents.*

(1) Persons must present their pilot certificate, medical certificate, logbook, or any other record required by this part for inspection upon a reasonable request by—

(i) The Administrator;

(ii) An authorized representative from the National Transportation Safety Board; or

(iii) Any Federal, State, or local law enforcement officer.

(2) A student pilot must carry the following items in the aircraft on all solo cross-country flights as evidence of the required authorized instructor clearances and endorsements—

(i) Pilot logbook;

(ii) Student pilot certificate; and

(iii) Any other record required by this section.

(3) A sport pilot must carry his or her logbook or other evidence of required authorized instructor endorsements on all flights

(4) A recreational pilot must carry his or her logbook with the required authorized instructor endorsements on all solo flights—

(i) That exceed 50 nautical miles from the airport at which training was received;

(ii) Within airspace that requires communication with air traffic control;

(iii) Conducted between sunset and sunrise; or

(iv) In an aircraft for which the pilot does not hold an appropriate category or class rating.

(5) A flight instructor with a sport pilot rating must carry his or her logbook or other evidence of required authorized instructor endorsements on all flights when providing flight training.

[Amdt. 61-102, 62 FR 16298, Apr. 4, 1997; Amdt. 61-103, 62 FR 40897, July 30, 1997, and amended at Amdt. 61-103, 62 FR 40897, July 30, 1997]

§ 61.52 Use of aeronautical experience obtained in ultralight vehicles.

(a) A person may use aeronautical experience obtained in an ultralight vehicle to meet the requirements for the following certificates and ratings issued under this part:

(1) A sport pilot certificate.

(2) A flight instructor certificate with a sport pilot rating;

(3) A private pilot certificate with a weight-shift-control or powered parachute category rating.

(b) A person may use aeronautical experience obtained in an ultralight vehicle to meet the provisions of §§61.69 and 61.415 (e).

(c) A person using aeronautical experience obtained in an ultralight vehicle to meet the requirements for a certificate or rating specified in paragraph (a) of this section or the requirements of paragraph (b) of this section must

(1) Have been a registered ultralight pilot with an FAA-recognized ultralight organization when that aeronautical experience was obtained;

(2) Document and log that aeronautical experience in accordance with the provisions for logging aeronautical experience specified by an FAA-recognized ultralight organization and in accordance with provisions for logging pilot time in aircraft as specified in §61.51; and—

(3) Obtain the experience in a category and class of vehicle corresponding to the rating or privileges sought.

§ 61.53 Prohibition on operations during medical deficiency.

(a) Operations that require a medical certificate. Except as provided for in paragraph (b) of this section, a person who holds a current medical certificate issued under part 67 of this chapter shall not act as pilot in command, or in any other capacity as a required pilot flight crewmember, while that person:

(1) Knows or has reason to know of any medical condition that would make the person unable to meet the requirements for the medical certificate necessary for the pilot operation; or

(2) Is taking medication or receiving other treatment for a medical condition that results in the person being unable to meet the requirements for the medical certificate necessary for the pilot operation.

(b) Operations that do not require a medical certificate. For operations provided for in § 61.23(b) of this part, a person shall not act as pilot in command, or in any other capacity as a required pilot flight crewmember, while that person knows or has reason to know of any medical condition that would make the person unable to operate the aircraft in a safe manner.

(c) Operations requiring a medical certificate or a U.S. driver's license. For operations provided for in §61.23 (c), a person must meet the provisions of—

(1) Paragraph (a) of this section if that person holds a valid medical certificate issued under part 67 of this chapter and does not hold a current and valid U.S. driver's license.

(2) Paragraph (b) of this section if that person holds a current and valid U.S. driver's license.

§ 61.55 Second-in-command qualifications.

(a) A person may serve as a second-in-command of an aircraft type certificated for more than one required pilot flight crewmember or in operations requiring a second-in-command pilot flight crewmember only if that person holds:

(1) At least a current private pilot certificate with the appropriate category and class rating; and

(2) An instrument rating or privilege that applies to the aircraft being flown if the flight is under IFR; and

(3) The appropriate pilot type rating for the aircraft unless the flight will be conducted as domestic flight operations within United States airspace.

(b) Except as provided in paragraph (e) of this section, no person may serve as a second in command of an aircraft type certificated for more than one required pilot flight crewmember or in operations requiring a second in command unless that person has within the previous 12 calendar months:

(1) Become familiar with the following information for the specific type aircraft for which second-in-command privileges are requested—

(i) Operational procedures applicable to the powerplant, equipment, and systems.

(ii) Performance specifications and limitations.

(iii) Normal, abnormal, and emergency operating procedures.

(iv) Flight manual.

(v) Placards and markings.

(2) Except as provided in paragraph (g) of this section, performed and logged pilot time in the type of aircraft or in a flight simulator that represents the type of aircraft for which second-in-command privileges are requested, which includes—

(i) Three takeoffs and three landings to full stop as the sole manipulator of the flight controls;

(ii) Engine-out procedures and maneuvering with an engine out while executing the duties of pilot in command; and

(iii) Crew resource management training.

(c) If a person complies with the requirements in paragraph (b) of this section in the calendar month before or the calendar month after the month in which compliance with this section is required, then that person is considered to have accomplished the training and practice in the month it is due.

(d) A person may receive a second-in command pilot type rating for an aircraft after satisfactorily completing the second-in-command familiarization training requirements under paragraph (b) of this section in that type of aircraft. Provided the training was completed within 12 calendar months before the month of application for the SIC pilot type rating. The person must comply with the following application and pilot certification procedures:

(1) The person who provided the training must sign the applicant's logbook or training record after each lesson in accordance with § 61.51(h)(2) of this part. In lieu of the trainer, it is permissible for a qualified management official within the organization to sign the applicant's training records or logbook and make the required endorsement. The qualified management official must hold the position of Chief Pilot, Director of Training, Director of Operations, or another comparable management position within the organization that provided the training and must be in a position to verify the applicant's training records and that the training was given.

(2) The trainer or qualified management official must make an endorsement in the applicant's logbook that states "[Applicant's Name and Pilot Certificate Number] has demonstrated the skill and knowledge required for the safe operation of the [Type of Aircraft], relevant to the duties and responsibilities of a second in command."

(3) If the applicant's flight experience and/or training records are in an electronic form, the applicant must present a paper copy of those records containing the signature of the trainer or qualified management official to an FAA Flight Standards District Office or Examiner.

(4) The applicant must complete and sign an Airman Certificate and/or Rating Application, FAA Form 8710-1, and present the application to an FAA Flight Standards District Office or to an Examiner.

(5) The person who provided the ground and flight training to the applicant must sign the "Instructor's Recommendation" section of the Airman Certificate and/or Rating Application, FAA Form 8710-1. In lieu of the trainer, it is permissible for a qualified management official within the organization to sign the applicant's FAA Form 8710-1.

(6) The applicant must appear in person at a FAA Flight Standards District Office or to an Examiner with his or her logbook/training records and with the completed and signed FAA Form 8710-1.

(7) There is no practical test required for the issuance of the "SIC Privileges Only" pilot type rating.

(e) A person may receive a second-in-command pilot type rating for the type of aircraft after satisfactorily completing an approved second-in-command training program, proficiency check, or competency check under subpart K of part 91, part 121, part 125, or part 135, as appropriate, in that type of aircraft provided the training was completed within the 12 calendar months before the month of application for the SIC pilot type rating. The person must comply with the following application and pilot certification procedures:

(1) The person who provided the training must sign the applicant's logbook or training record after each lesson in

accordance with § 61.51(h)(2) of this part. In lieu of the trainer, it is permissible for a qualified management official within the organization to sign the applicant's training records or logbook and make the required endorsement. The qualified management official must hold the position of Chief Pilot, Director of Training, Director of Operations, or another comparable management position within the organization that provided the training and must be in a position to verify the applicant's training records and that the training was given.

(2) The trainer or qualified management official must make an endorsement in the applicant's logbook that states "[Applicant's Name and Pilot Certificate Number] has demonstrated the skill and knowledge required for the safe operation of the [Type of Aircraft], relevant to the duties and responsibilities of a second in command."

(3) If the applicant's flight experience and/or training records are in an electronic form, the applicant must provide a paper copy of those records containing the signature of the trainer or qualified management official to an FAA Flight Standards District Office, an Examiner, or an Aircrew Program Designee.

(4) The applicant must complete and sign an Airman Certificate and/or Rating Application, FAA Form 8710-1, and present the application to an FAA Flight Standards District Office or to an Examiner or to an authorized Aircrew Program Designee.

(5) The person who provided the ground and flight training to the applicant must sign the "Instructor's Recommendation" section of the Airman Certificate and/or Rating Application, FAA Form 8710-1. In lieu of the trainer, it is permissible for a qualified management official within the organization to sign the applicant's FAA Form 8710-1.

(6) The applicant must appear in person at an FAA Flight Standards District Office or to an Examiner or to an authorized Aircrew Program Designee with his or her logbook/ training records and with the completed and signed FAA Form 8710-1.

(7) There is no practical test required for the issuance of the "SIC Privileges Only" pilot type rating.

(f) The familiarization training requirements of paragraph (b) of this section does not apply to a person who is:

(1) Designated and qualified as pilot in command under subpart K of part 91, part 121, 125, or 135 of this chapter in that specific type of aircraft;

(2) Designated as the second in command under subpart K of part 91, part 121, 125, or 135 of this chapter in that specific type of aircraft;

(3) Designated as the second in command in that specific type of aircraft for the purpose of receiving flight training required by this section, and no passengers or cargo are carried on the aircraft; or

(4) Designated as a safety pilot for purposes required by § 91.109(b) of this chapter.

(g) The holder of a commercial or airline transport pilot certificate with the appropriate category and class rating is not required to meet the requirements of paragraph (b)(2) of this section, provided the pilot:

(1) Is conducting a ferry flight, aircraft flight test, or evaluation flight of an aircraft's equipment; and

(2) Is not carrying any person or property on board the aircraft, other than necessary for conduct of the flight.

(h) For the purpose of meeting the requirements of paragraph (b) of this section, a person may serve as second in command in that specific type aircraft, provided:

(1) The flight is conducted under day VFR or day IFR; and

(2) No person or property is carried on board the aircraft, other than necessary for conduct of the flight.

(i) The training under paragraphs (b) and (d) of this section and the training, proficiency check, and competency check under paragraph (e) of this section may be accomplished in a flight simulator that is used in accordance with an approved training course conducted by a training center certificated under part 142 of this chapter or under subpart K of part 91, part 121 or part 135 of this chapter.

(j) When an applicant for an initial second-in-command qualification for a particular type of aircraft receives all the training in a flight simulator, that applicant must satisfactorily complete one takeoff and one landing in an aircraft of the same type for which the qualification is sought. This requirement does not apply to an applicant who completes a proficiency check under part 121 or competency check under subpart K, part 91, part 125, or part 135 for the particular type of aircraft.

[Doc. No. 25910, 62 FR 16298, Apr. 4,1997; Amdt. 61-103, 62 FR 40898, July 30, 1997; Amdt. 61-109, 68 FR 54559, Sept. 17, 2003; Amdt. 61-113, 70 FR 45271, Aug. 4, 2005; Amdt. 61-109, 70 FR 61890, Oct. 27, 2005]

§ 61.56 Flight review.

(a) Except as provided in paragraphs (b) and (f) of this section, a flight review consists of a minimum of 1 hour of flight training and 1 hour of ground training. The review must include:

(1) A review of the current general operating and flight rules of part 91 of this chapter; and

(2) A review of those maneuvers and procedures that, at the discretion of the person giving the review, are necessary for the pilot to demonstrate the safe exercise of the privileges of the pilot certificate.

(b) Glider pilots may substitute a minimum of three instructional flights in a glider, each of which includes a flight to traffic pattern altitude, in lieu of the 1 hour of flight training required in paragraph (a) of this section.

(c) Except as provided in paragraphs (d), (e), and (g) of this section, no person may act as pilot in command of an

aircraft unless, since the beginning of the 24th calendar month before the month in which that pilot acts as pilot in command, that person has:

(1) Accomplished a flight review given in an aircraft for which that pilot is rated by an authorized instructor; and

(2) A logbook endorsed from an authorized instructor who gave the review certifying that the person has satisfactorily completed the review.

(d) A person who has, within the period specified in paragraph (c) of this section, passed a pilot proficiency check conducted by an examiner, an approved pilot check airman, or a U.S. Armed Force, for a pilot certificate, rating, or operating privilege need not accomplish the flight review required by this section.

(e) A person who has, within the period specified in paragraph (c) of this section, satisfactorily accomplished one or more phases of an FAA-sponsored pilot proficiency award program need not accomplish the flight review required by this section.

(f) A person who holds a current flight instructor certificate who has, within the period specified in paragraph (c) of this section, satisfactorily completed a renewal of a flight instructor certificate under the provisions in § 61.197 need not accomplish the 1 hour of ground training specified in paragraph (a) of this section.

(g) A student pilot need not accomplish the flight review required by this section provided the student pilot is undergoing training for a certificate and has a current solo flight endorsement as required under § 61.87 of this part.

(h) The requirements of this section may be accomplished in combination with the requirements of § 61.57 and other applicable recent experience requirements at the discretion of the authorized instructor conducting the flight review.

(i) A flight simulator or flight training device may be used to meet the flight review requirements of this section subject to the following conditions:

(1) The flight simulator or flight training device must be used in accordance with an approved course conducted by a training center certificated under part 142 of this chapter.

(2) Unless the flight review is undertaken in a flight simulator that is approved for landings, the applicant must meet the takeoff and landing requirements of § 61.57(a) or § 61.57(b) of this part.

(3) The flight simulator or flight training device used must represent an aircraft, or set of aircraft, for which the pilot is rated.

[Amdt. 61-102, 62 FR 16298, Apr. 4, 1997; Amdt. 61-103, 62 FR 40898, July 30, 1997]

§ 61.57 Recent flight experience: Pilot in command.

(a) General experience.

(1) Except as provided in paragraph (e) of this section, no person may act as a pilot in command of an aircraft

carrying passengers or of an aircraft certificated for more than one pilot flight crewmember unless that person has made at least three takeoffs and three landings within the preceding 90 days, and—

(i) The person acted as the sole manipulator of the flight controls; and

(ii) The required takeoffs and landings were performed in an aircraft of the same category, class, and type (if a type rating is required), and, if the aircraft to be flown is an airplane with a tailwheel, the takeoffs and landings must have been made to a full stop in an airplane with a tailwheel.

(2) For the purpose of meeting the requirements of paragraph (a)(1) of this section, a person may act as a pilot in command of an aircraft under day VFR or day IFR, provided no persons or property are carried on board the aircraft, other than those necessary for the conduct of the flight.

(3) The takeoffs and landings required by paragraph (a)(1) of this section may be accomplished in a flight simulator or flight training device that is—

(i) Approved by the Administrator for landings; and

(ii) Used in accordance with an approved course conducted by a training center certificated under part 142 of this chapter.

(b) *Night takeoff and landing experience.*

(1) Except as provided in paragraph (e) of this section, no person may act as pilot in command of an aircraft carrying passengers during the period beginning 1 hour after sunset and ending 1 hour before sunrise, unless within the preceding 90 days that person has made at least three takeoffs and three landings to a full stop during the period beginning 1 hour after sunset and ending 1 hour before sunrise, and—

(i) That person acted as sole manipulator of the flight controls; and

(ii) The required takeoffs and landings were performed in an aircraft of the same category, class, and type (if a type rating is required).

(2) The takeoffs and landings required by paragraph (b)(1) of this section may be accomplished in a flight simulator that is—

(i) Approved by the Administrator for takeoffs and landings, if the visual system is adjusted to represent the period described in paragraph (b)(1) of this section; and

(ii) Used in accordance with an approved course conducted by a training center certificated under part 142 of this chapter.

(c) *Instrument experience.* Except as provided in paragraph (e) of this section, no person may act as pilot in command under IFR or in weather conditions less than the minimums prescribed for VFR, unless within the preceding 6 calendar months, that person has:

(1) For the purpose of obtaining instrument experience in an aircraft (other than a glider), performed and logged under actual or simulated instrument conditions, either in flight in the appropriate category of aircraft for the instrument privileges sought or in a flight simulator or flight training device that is representative of the aircraft category for the instrument privileges sought—

(i) At least six instrument approaches;

(ii) Holding procedures; and

(iii) Intercepting and tracking courses through the use of navigation systems.

(2) For the purpose of obtaining instrument experience in a glider, performed and logged under actual or simulated instrument conditions—

(i) At least 3 hours of instrument time in flight, of which 1½ hours may be acquired in an airplane or a glider if no passengers are to be carried; or

(ii) 3 hours of instrument time in flight in a glider if a passenger is to be carried.

(d) *Instrument proficiency check.* Except as provided in paragraph (e) of this section, a person who does not meet the instrument experience requirements of paragraph (c) of this section within the prescribed time, or within 6 calendar months after the prescribed time, may not serve as pilot in command under IFR or in weather conditions less than the minimums prescribed for VFR until that person passes an instrument proficiency check consisting of a representative number of tasks required by the instrument rating practical test.

(1) The instrument proficiency check must be—

(i) In an aircraft that is appropriate to the aircraft category;

(ii) For other than a glider, in a flight simulator or flight training device that is representative of the aircraft category; or

(iii) For a glider, in a single-engine airplane or a glider.

(2) The instrument proficiency check must be given by—

(i) An examiner;

(ii) A person authorized by the U.S. Armed Forces to conduct instrument flight tests, provided the person being tested is a member of the U.S. Armed Forces;

(iii) A company check pilot who is authorized to conduct instrument flight tests under part 121, 125, or 135 of this chapter or subpart K of part 91 of this chapter, and provided that both the check pilot and the pilot being tested are employees of that operator or fractional ownership program manager, as applicable;

(iv) An authorized flight instructor; or

(v) A person approved by the Administrator to conduct instrument practical tests.

(e) *Exceptions.*

(1) Paragraphs (a) and (b) of this section do not apply to a pilot in command who is employed by a certificate holder under part 125 and engaged in a flight operation for that certificate holder if the pilot is in compliance with §§ 125.281 and 125.285 of this chapter.

(2) This section does not apply to a pilot in command who is employed by an air carrier certificated under part 121 or 135 and is engaged in a flight operation under part 91, 121, or 135 for that air carrier if the pilot is in compliance

with §§ 121.437 and 121.439, or §§ 135.243 and 135.247 of this chapter, as appropriate.

(3) Paragraph (b) of this section does not apply to a pilot in command of a turbine-powered airplane that is type certificated for more than one pilot crewmember, provided that pilot has complied with the requirements of paragraph (e)(3)(i) or (ii) of this section:

(i) The pilot in command must hold at least a commercial pilot certificate with the appropriate category, class, and type rating for each airplane that is type certificated for more than one pilot crewmember that the pilot seeks to operate under this alternative, and:

(A) That pilot must have logged at least 1,500 hours of aeronautical experience as a pilot;

(B) In each airplane that is type certificated for more than one pilot crewmember that the pilot seeks to operate under this alternative, that pilot must have accomplished and logged the daytime takeoff and landing recent flight experience of paragraph (a) of this section, as the sole manipulator of the flight controls;

(C) Within the preceding 90 days prior to the operation of that airplane that is type certificated for more than one pilot crewmember, the pilot must have accomplished and logged at least 15 hours of flight time in the type of airplane that the pilot seeks to operate under this alternative; and

(D) That pilot has accomplished and logged at least 3 takeoffs and 3 landings to a full stop, as the sole manipulator of the flight controls, in a turbine-powered airplane that requires more than one pilot crewmember. The pilot must have performed the takeoffs and landings during the period beginning 1 hour after sunset and ending 1 hour before sunrise within the preceding 6 months prior to the month of the flight.

(ii) The pilot in command must hold at least a commercial pilot certificate with the appropriate category, class, and type rating for each airplane that is type certificated for more than one pilot crewmember that the pilot seeks to operate under this alternative, and:

(A) That pilot must have logged at least 1,500 hours of aeronautical experience as a pilot;

(B) In each airplane that is type certificated for more than one pilot crewmember that the pilot seeks to operate under this alternative, that pilot must have accomplished and logged the daytime takeoff and landing recent flight experience of paragraph (a) of this section, as the sole manipulator of the flight controls;

(C) Within the preceding 90 days prior to the operation of that airplane that is type certificated for more than one pilot crewmember, the pilot must have accomplished and logged at least 15 hours of flight time in the type of airplane that the pilot seeks to operate under this alternative; and

(D) Within the preceding 12 months prior to the month of the flight, the pilot must have completed a training program that is approved under part 142 of this chapter. The approved training program must have required and the

pilot must have performed, at least 6 takeoffs and 6 landings to a full stop as the sole manipulator of the controls in a flight simulator that is representative of a turbine-powered airplane that requires more than one pilot crewmember. The flight simulator's visual system must have been adjusted to represent the period beginning 1 hour after sunset and ending 1 hour before sunrise.

§ 61.58 Pilot-in-command proficiency check: Operation of aircraft requiring more than one pilot flight crewmember.

(a) Except as otherwise provided in this section, to serve as pilot in command of an aircraft that is type certificated for more than one required pilot flight crewmember, a person must—

(1) Within the preceding 12 calendar months, complete a pilot-in-command proficiency check in an aircraft that is type certificated for more than one required pilot flight crewmember; and

(2) Within the preceding 24 calendar months, complete a pilot-in-command proficiency check in the particular type of aircraft in which that person will serve as pilot in command.

(b) This section does not apply to persons conducting operations under subpart K of part 91, part 121, 125, 133, 135, or 137 of this chapter, or persons maintaining continuing qualification under an Advanced Qualification program approved under SFAR 58.

(c) The pilot-in-command proficiency check given in accordance with the provisions of subpart K of part 91, part 121, 125, or 135 of this chapter may be used to satisfy the requirements of this section.

(d) The pilot-in-command proficiency check required by paragraph (a) of this section may be accomplished by satisfactory completion of one of the following:

(1) A pilot-in-command proficiency check conducted by a person authorized by the Administrator, consisting of the maneuvers and procedures required for a type rating, in an aircraft type certificated for more than one required pilot flight crewmember;

(2) The practical test required for a type rating, in an aircraft type certificated for more than one required pilot flight crewmember;

(3) The initial or periodic practical test required for the issuance of a pilot examiner or check airman designation, in an aircraft type certificated for more than one required pilot flight crewmember; or

(4) A military flight check required for a pilot in command with instrument privileges, in an aircraft that the military requires to be operated by more than one pilot flight crewmember.

(e) A check or test described in paragraphs (d)(1) through (d)(4) of this section may be accomplished in a flight simulator under part 142 of this chapter, subject to the following:

(1) Except as provided for in paragraphs (e)(2) and (e)(3) of this section, if an otherwise qualified and approved flight simulator used for a pilot-in-command proficiency check is not qualified and approved for a specific required maneuver—

(i) The training center must annotate, in the applicant's training record, the maneuver or maneuvers omitted; and

(ii) Prior to acting as pilot in command, the pilot must demonstrate proficiency in each omitted maneuver in an aircraft or flight simulator qualified and approved for each omitted maneuver.

(2) If the flight simulator used pursuant to paragraph (e) of this section is not qualified and approved for circling approaches—

(i) The applicant's record must include the statement, "Proficiency in circling approaches not demonstrated"; and

(ii) The applicant may not perform circling approaches as pilot in command when weather conditions are less than the basic VFR conditions described in § 91.155 of this chapter, until proficiency in circling approaches has been successfully demonstrated in a flight simulator qualified and approved for circling approaches or in an aircraft to a person authorized by the Administrator to conduct the check required by this section.

(3) If the flight simulator used pursuant to paragraph (e) of this section is not qualified and approved for landings, the applicant must—

(i) Hold a type rating in the airplane represented by the simulator; and

(ii) Have completed within the preceding 90 days at least three takeoffs and three landings (one to a full stop) as the sole manipulator of the flight controls in the type airplane for which the pilot-in-command proficiency check is sought.

(f) For the purpose of meeting the pilot-in-command proficiency check requirements of paragraph (a) of this section, a person may act as pilot in command of a flight under day VFR conditions or day IFR conditions if no person or property is carried, other than as necessary to demonstrate compliance with this part.

(g) If a pilot takes the pilot-in-command proficiency check required by this section in the calendar month before or the calendar month after the month in which it is due, the pilot is considered to have taken it in the month in which it was due for the purpose of computing when the next pilot-in-command proficiency check is due.

[Amdt. 61-102, 62 FR 16298, Apr. 4, 1997; Amdt. 61-103, 62 FR 40899, July 30, 1997]

§ 61.59   Falsification, reproduction, or alteration of applications, certificates, logbooks, reports, or records.

(a) No person may make or cause to be made:

(1) Any fraudulent or intentionally false statement on any application for a certificate, rating, authorization, or duplicate thereof, issued under this part;

(2) Any fraudulent or intentionally false entry in any logbook, record, or report that is required to be kept, made, or used to show compliance with any requirement for the issuance or exercise of the privileges of any certificate, rating, or authorization under this part;

(3) Any reproduction for fraudulent purpose of any certificate, rating, or authorization, under this part; or

(4) Any alteration of any certificate, rating, or authorization under this part.

(b) The commission of an act prohibited under paragraph (a) of this section is a basis for suspending or revoking any airman certificate, rating, or authorization held by that person.

§ 61.60   Change of address.

The holder of a pilot, flight instructor, or ground instructor certificate who has made a change in permanent mailing address may not, after 30 days from that date, exercise the privileges of the certificate unless the holder has notified in writing the FAA, Airman Certification Branch, P.O. Box 25082, Oklahoma City, OK 73125, of the new permanent mailing address, or if the permanent mailing address includes a post office box number, then the holder's current residential address.

Subpart B—Aircraft Ratings and Pilot Authorizations

§ 61.61   Applicability.

This subpart prescribes the requirements for the issuance of additional aircraft ratings after a pilot certificate is issued, and the requirements for and limitations of pilot authorizations issued by the Administrator.

§ 61.63   Additional aircraft ratings (other than airline transport pilot certificate).

(a) General. To be eligible for an additional aircraft rating to a pilot certificate, for other than an airline transport pilot certificate, an applicant must meet the appropriate requirements of this section for the additional aircraft rating sought.

(b) Additional category rating. An applicant who holds a pilot certificate and applies to add a category rating to that pilot certificate:

(1) Must have received the required training and possess the aeronautical experience prescribed by this part that applies to the pilot certificate for the aircraft category and, if applicable, class rating sought;

(2) Must have an endorsement in his or her logbook or training record from an authorized instructor, and that

endorsement must attest that the applicant has been found competent in the aeronautical knowledge areas appropriate to the pilot certificate for the aircraft category and, if applicable, class rating sought;

(3) Must have an endorsement in his or her logbook or training record from an authorized instructor, and that endorsement must attest that the applicant has been found proficient on the areas of operation that are appropriate to the pilot certificate for the aircraft category and, if applicable, class rating sought;

(4) Must pass the required practical test that is appropriate to the pilot certificate for the aircraft category and, if applicable, class rating sought; and

(5) Need not take an additional knowledge test, provided the applicant holds an airplane, rotorcraft, powered-lift, or airship rating at that pilot certificate level.

(c) *Additional class rating.* Any person who applies for an additional class rating to be added on a pilot certificate:

(1) Must have an endorsement in his or her logbook or training record from an authorized instructor and that endorsement must attest that the applicant has been found competent in the aeronautical knowledge areas appropriate to the pilot certificate for the aircraft class rating sought;

(2) Must have an endorsement in his or her logbook or training record from an authorized instructor, and that endorsement must attest that the applicant has been found proficient in the areas of operation appropriate to the pilot certificate for the aircraft class rating sought;

(3) Must pass the required practical test that is appropriate to the pilot certificate for the aircraft class rating sought;

(4) Need not meet the specified training time requirements prescribed by this part that apply to the pilot certificate for the aircraft class rating sought unless the person holds a lighter-than-air category rating with a balloon class rating and is seeking an airship class rating and

(5) Need not take an additional knowledge test, provided the applicant holds an airplane, rotorcraft, powered-lift, or airship rating at that pilot certificate level.

(d) *Additional type rating.* Except as specified in paragraph (d)(7) of this section, a person who applies for an additional aircraft type rating to be added on a pilot certificate, or the addition of an aircraft type rating that is accomplished concurrently with an additional aircraft category or class rating:

(1) Must hold or concurrently obtain an instrument rating that is appropriate to the aircraft category, class, or type rating sought;

(2) Must have an endorsement in his or her logbook or training record from an authorized instructor, and that endorsement must attest that the applicant has been found competent in the aeronautical knowledge areas appropriate to the pilot certificate for the aircraft category, class, or type rating sought;

(3) Must have an endorsement in his or her logbook, or training record from an authorized instructor, and that endorsement must attest that the applicant has been found proficient in the areas of operation required for the issuance of an airline transport pilot certificate for the aircraft category, class, and type rating sought;

(4) Must pass the required practical test appropriate to the airline transport pilot certificate for the aircraft category, class, and type rating sought;

(5) Must perform the practical test in actual or simulated instrument conditions, unless the aircraft's type certificate makes the aircraft incapable of operating under instrument flight rules. If the practical test cannot be accomplished for this reason, the person may obtain a type rating limited to "VFR only." The "VFR only" limitation may be removed for that aircraft type when the person passes the practical test in actual or simulated instrument conditions. When an instrument rating is issued to a person who holds one or more type ratings, the type ratings on the amended pilot certificate shall bear the "VFR only" limitation for each aircraft type rating for which the person has not demonstrated instrument competency;

(6) Need not take an additional knowledge test, provided the applicant holds an airplane, rotorcraft, powered-lift, or airship rating on their pilot certificate; and

(7) In the case of a pilot employee of a certificate holder operating under part 121 or 135 of this chapter or of a fractional ownership program manager under subpart K of part 91 of this chapter, must have—

(i) Met the appropriate requirements of paragraphs (d)(1), (d)(4), and (d)(5) of this section for the aircraft type rating sought; and

(ii) Received an endorsement in his or her flight training record from the certificate holder or program manager attesting that the applicant has completed the certificate holder's or program manager's approved ground and flight training program appropriate to the aircraft type rating sought.

(e) *Use of a flight simulator or, flight training device for an additional rating in an airplane.* The areas of operation required to be performed by paragraphs (b), (c), and (d) of this section shall be performed as follows:

(1) Except as provided in paragraph (e)(2) of this section, the areas of operation must be performed in an airplane of the same category, class, and type, if applicable, as the airplane for which the additional rating is sought.

(2) Subject to the limitations of paragraph (e)(3) through (e)(12) of this section, the areas of operation may be performed in a flight simulator or flight training device that represents the airplane for which the additional rating is sought.

(3) The use of a flight simulator or flight training device permitted by paragraph (e)(2) of this section shall be conducted in accordance with an approved course at a training center certificated under part 142 of this chapter.

(4) To complete all training and testing (except preflight inspection) for an additional airplane rating without limitations when using a flight simulator—

(i) The flight simulator must be qualified and approved as Level C or Level D; and

(ii) The applicant must meet at least one of the following:

(A) Hold a type rating for a turbojet airplane of the same class of airplane for which the type rating is sought, or have been appointed by a military service as a pilot in command of an airplane of the same class of airplane for which the type rating is sought, if a type rating in a turbojet airplane is sought.

(B) Hold a type rating for a turbopropeller airplane of the same class of airplane for which the type rating is sought, or have been designated by a military service as a pilot in command of an airplane of the same class of airplane for which the type rating is sought, if a type rating in a turbopropeller airplane is sought.

(C) Have at least 2,000 hours of flight time, of which 500 hours is in turbine-powered airplanes of the same class of airplane for which the type rating is sought.

(D) Have at least 500 hours of flight time in the same type airplane as the airplane for which the rating is sought.

(E) Have at least 1,000 hours of flight time in at least two different airplanes requiring a type rating.

(5) Subject to the limitation of paragraph (e)(6) of this section, an applicant who does not meet the requirements of paragraph (e)(4) of this section may complete all training and testing (except for preflight inspection) for an additional rating when using a flight simulator if—

(i) The flight simulator is qualified and approved as Level C or Level D; and

(ii) The applicant meets at least one of the following:

(A) Holds a type rating in a propeller-driven airplane if a type rating in a turbojet airplane is sought, or holds a type rating in a turbojet airplane if a type rating in a propeller-driven airplane is sought; or

(B) Since the beginning of the 12th calendar month before the month in which the applicant completes the practical test for an additional airplane rating, has logged:

(1) At least 100 hours of flight time in airplanes of the same class for which the type rating is sought and which requires a type rating; and

(2) At least 25 hours of flight time in airplanes of the same type for which the rating is sought.

(6) An applicant meeting only the requirements of paragraph (e)(5) of this section will be issued an additional rating with a limitation.

(7) The limitation on a certificate issued under the provisions of paragraph (e)(6) of this section shall state, "This certificate is subject to pilot-in-command limitations for the additional rating."

(8) An applicant who has been issued a pilot certificate with the limitation specified in paragraph (e)(7) of this section—

(i) May not act as pilot in command of that airplane for which the additional rating was obtained under the provisions of this section until the limitation is removed from the pilot certificate; and

(ii) May have the limitation removed by accomplishing 15 hours of supervised operating experience as pilot in command under the supervision of a qualified and current pilot in command, in the seat normally occupied by the pilot in command, in the same type of airplane to which the limitation applies.

(9) An applicant who does not meet the requirements of paragraph (e)(4) or paragraph (e)(5) of this section may be issued an additional rating after successful completion of one of the following requirements:

(i) Compliance with paragraphs (e)(2) and (e)(3) of this section and the following tasks, which must be successfully completed on a static airplane or in flight, as appropriate:

(A) Preflight inspection;

(B) Normal takeoff;

(C) Normal ILS approach;

(D) Missed approach; and

(E) Normal landing.

(ii) Compliance with paragraphs (e)(2), (e)(3), and (e)(10) through (e)(12) of this section.

(10) An applicant meeting only the requirements of paragraph (e)(9)(ii) of this section will be issued an additional rating with a limitation.

(11) The limitation on a certificate issued under the provisions of paragraph (e)(10) of this section shall state, "This certificate is subject to pilot-in-command limitations for the additional rating."

(12) An applicant who has been issued a pilot certificate with the limitation specified in paragraph (e)(11) of this section—

(i) May not act as pilot in command of that airplane for which the additional rating was obtained under the provisions of this section until the limitation is removed from the pilot certificate; and

(ii) May have the limitation removed by accomplishing 25 hours of supervised operating experience as pilot in command under the supervision of a qualified and current pilot in command, in the seat normally occupied by the pilot in command, in that airplane of the same type to which the limitation applies.

(f) Use of a flight simulator or flight training device for an additional rating in a helicopter. The areas of operation required to be performed by paragraphs (b), (c), and (d) of this section shall be performed as follows:

(1) Except as provided in paragraph (f)(2) of this section, the areas of operation must be performed in a helicopter of the same type for the additional rating sought.

(2) Subject to the limitations of paragraph (f)(3) through (f)(12) of this section, the areas of operation may be performed in a flight simulator or flight training device that represents that helicopter for the additional rating sought.

(3) The use of a flight simulator or flight training device permitted by paragraph (f)(2) of this section shall be conducted in accordance with an approved course at a training center certificated under part 142 of this chapter.

(4) To complete all training and testing (except preflight inspection) for an additional helicopter rating without limitations when using a flight simulator—

(i) The flight simulator must be qualified and approved as Level C or Level D; and

(ii) The applicant must meet at least one of the following if a type rating is sought in a turbine-powered helicopter:

(A) Hold a type rating in a turbine-powered helicopter or have been appointed by a military service as a pilot in command of a turbine-powered helicopter.

(B) Have at least 2,000 hours of flight time that includes at least 500 hours in turbine-powered helicopters.

(C) Have at least 500 hours of flight time in turbine-powered helicopters.

(D) Have at least 1,000 hours of flight time in at least two different turbine-powered helicopters.

(5) Subject to the limitation of paragraph (f)(6) of this section, an applicant who does not meet the requirements of paragraph (f)(4) of this section may complete all training and testing (except for preflight inspection) for an additional rating when using a flight simulator if—

(i) The flight simulator is qualified and approved as Level C or Level D; and

(ii) The applicant meets at least one of the following:

(A) Holds a type rating in a turbine-powered helicopter if a type rating in a turbine-powered helicopter is sought; or

(B) Since the beginning of the 12th calendar month before the month in which the applicant completes the practical test for an additional helicopter rating, has logged at least 25 hours of flight time in helicopters of the same type for which the rating is sought.

(6) An applicant meeting only the requirements of paragraph (f)(5) of this section will be issued an additional rating with a limitation.

(7) The limitation on a certificate issued under the provisions of paragraph (f)(6) of this section shall state, "This certificate is subject to pilot-in-command limitations for the additional rating."

(8) An applicant who is issued a pilot certificate with the limitation specified in paragraph (f)(7) of this section—

(i) May not act as pilot in command of that helicopter for which the additional rating was obtained under the provisions of this section until the limitation is removed from the pilot certificate; and

(ii) May have the limitation removed by accomplishing 15 hours of supervised operating experience as pilot in command under the supervision of a qualified and current pilot in command, in the seat normally occupied by the pilot in command, in the same type of helicopter to which the limitation applies.

(9) An applicant who does not meet the requirements of paragraph (f)(4) or paragraph (f)(5) of this section may be issued an additional rating after successful completion of one of the following requirements:

(i) Compliance with paragraphs (f)(2) and (f)(3) of this section and the following tasks, which must be successfully completed on a static helicopter or in flight, as appropriate:

(A) Preflight inspection;

(B) Normal takeoff;

(C) Normal ILS approach;

(D) Missed approach; and

(E) Normal landing.

(ii) Compliance with paragraphs (f)(2), (f)(3), and (f)(10) through (f)(12) of this section.

(10) An applicant meeting only the requirements of paragraph (g)(9)(ii) of this section will be issued an additional rating with a limitation.

(11) The limitation on a certificate issued under the provisions of paragraph (f)(10) of this section shall state, "This certificate is subject to pilot-in-command limitations for the additional rating."

(12) An applicant who has been issued a pilot certificate with the limitation specified in paragraph (f)(11) of this section—

(i) May not act as pilot in command of that helicopter for which the additional rating was obtained under the provisions of this section until the limitation is removed from the pilot certificate; and

(ii) May have the limitation removed by accomplishing 25 hours of supervised operating experience as pilot in command under the supervision of a qualified and current pilot in command, in the seat normally occupied by the pilot in command, in that helicopter of the same type as to which the limitation applies.

(g) Use of a flight simulator or flight training device for an additional rating in a powered-lift. The areas of operation required to be performed by paragraphs (b), (c), and (d) of this section shall be performed as follows:

(1) Except as provided in paragraph (g)(2) of this section, the areas of operation must be performed in a powered-lift of the same type for the additional rating sought.

(2) Subject to the limitations of paragraphs (g)(3) through (g)(12) of this section, the areas of operation may be performed in a flight simulator or flight training device that represents that powered-lift for the additional rating sought.

(3) The use of a flight simulator or flight training device permitted by paragraph (g)(2) of this section shall be conducted in accordance with an approved course at a training center certificated under part 142 of this chapter.

(4) To complete all training and testing (except preflight inspection) for an additional powered-lift rating without limitations when using a flight simulator—

(i) The flight simulator must be qualified and approved as Level C or Level D; and

(ii) The applicant must meet at least one of the following if a type rating is sought in a turbine powered-lift:

(A) Hold a type rating in a turbine powered-lift or have been appointed by a military service as a pilot in command of a turbine powered-lift.

(B) Have at least 2,000 hours of flight time that includes at least 500 hours in turbine powered-lifts.

(C) Have at least 500 hours of flight time in turbine powered-lifts.

(D) Have at least 1,000 hours of flight time in at least two different turbine powered-lifts.

(5) Subject to the limitation of paragraph (g)(6) of this section, an applicant who does not meet the requirements of paragraph (g)(4) of this section may complete all training and testing (except for preflight inspection) for an additional rating when using a flight simulator if—

(i) The flight simulator is qualified and approved as Level C or Level D; and

(ii) The applicant meets at least one of the following:

(A) Holds a type rating in a turbine powered-lift if a type rating in a turbine powered-lift is sought; or

(B) Since the beginning of the 12th calendar month before the month in which the applicant completes the practical test for an additional powered-lift rating, has logged at least 25 hours of flight time in powered-lifts of the same type for which the rating is sought.

(6) An applicant meeting only the requirements of paragraph (g)(5) of this section will be issued an additional rating with a limitation.

(7) The limitation on a certificate issued under the provisions of paragraph (g)(6) of this section shall state, "This certificate is subject to pilot-in-command limitations for the additional rating."

(8) An applicant who is issued a pilot certificate with the limitation specified in paragraph (g)(7) of this section—

(i) May not act as pilot in command of that powered-lift for which the additional rating was obtained under the provisions of this section until the limitation is removed from the pilot certificate; and

(ii) May have the limitation removed by accomplishing 15 hours of supervised operating experience as pilot in command under the supervision of a qualified and current pilot in command, in the seat normally occupied by the pilot in command, in the same type of powered-lift to which the limitation applies.

(9) An applicant who does not meet the requirements of paragraph (g)(4) or paragraph (g)(5) of this section may be issued an additional rating after successful completion of one of the following requirements:

(i) Compliance with paragraphs (g)(2) and (g)(3) of this section and the following tasks, which must be successfully completed on a static powered-lift or in flight, as appropriate:

(A) Preflight inspection;

(B) Normal takeoff;

(C) Normal ILS approach;

(D) Missed approach; and

(E) Normal landing.

(ii) Compliance with paragraphs (g)(2), (g)(3), and (g)(10) through (g)(12) of this section.

(10) An applicant meeting only the requirements of paragraph (g)(9) of this section will be issued an additional rating with a limitation.

(11) The limitation on a certificate issued under the provisions of paragraph (g)(10) of this section shall state, "This certificate is subject to pilot-in-command limitations for the additional rating."

(12) An applicant who has been issued a pilot certificate with the limitation specified in paragraph (g)(11) of this section—

(i) May not act as pilot in command of that powered-lift for which the additional rating was obtained under the provisions of this section until the limitation is removed from the pilot certificate; and

(ii) May have the limitation removed by accomplishing 25 hours of supervised operating experience as pilot in command under the supervision of a qualified and current pilot in command, in the seat normally occupied by the pilot in command, in that powered-lift of the same type as to which the limitation applies.

(h) *Aircraft not capable of instrument maneuvers and procedures.* An applicant for a type rating who provides an aircraft not capable of the instrument maneuvers and procedures required by the appropriate requirements contained in § 61.157 of this part for the practical test may—

(1) Obtain a type rating limited to "VFR only"; and

(2) Remove the "VFR only" limitation for each aircraft type in which the applicant demonstrates compliance with the appropriate instrument requirements contained in § 61.157 or § 61.73 of this part.

(i) *Multiengine, single-pilot station airplane.* An applicant for a type rating in a multiengine, single-pilot station airplane may meet the requirements of this part in a multiseat version of that multiengine airplane.

(j) *Single-engine, single-pilot station airplane.* An applicant for a type rating in a single-engine, single-pilot station airplane may meet the requirements of this part in a multiseat version of that single-engine airplane.

(k) *Category class ratings for the operation of aircraft with experimental certificates:* Notwithstanding the provisions of paragraphs (b) and (c) of this section, a person holding at least a recreational pilot certificate may apply for a category and class rating limited to a specific make and model of experimental aircraft, provided

(1) The person has logged at least 5 hours flight time while acting as pilot in command in the same category, class, make, and model of aircraft that has been issued an experimental certificate;

(2) The person has received a logbook endorsement from an authorized instructor who has determined that he

or she is proficient to act as pilot in command of the same category, class, make, and model of aircraft for which application is made; and

(3) The flight time specified in paragraph (k)(1) of this section must be logged between September 1, 2004 and August 31, 2005.

(l) Waivers. Unless the Administrator requires certain or all tasks to be performed, the examiner who conducts the practical test may waive any of the tasks for which the Administrator approves waiver authority.

[Amdt. 61-102, 62 FR 19298, Apr. 4, 1997; Amdt. 61-103, 62 FR 40899, July 30, 1997]

§ 61.64 [Reserved.]

§ 61.65 Instrument rating requirements.

(a) General. A person who applies for an instrument rating must:

(1) Hold at least a current private pilot certificate with an airplane, helicopter, or powered-lift rating appropriate to the instrument rating sought;

(2) Be able to read, speak, write, and understand the English language. If the applicant is unable to meet any of these requirements due to a medical condition, the Administrator may place such operating limitations on the applicant's pilot certificate as are necessary for the safe operation of the aircraft;

(3) Receive and log ground training from an authorized instructor or accomplish a home-study course of training on the aeronautical knowledge areas of paragraph (b) of this section that apply to the instrument rating sought;

(4) Receive a logbook or training record endorsement from an authorized instructor certifying that the person is prepared to take the required knowledge test;

(5) Receive and log training on the areas of operation of paragraph (c) of this section from an authorized instructor in an aircraft, flight simulator, or flight training device that represents an airplane, helicopter, or powered-lift appropriate to the instrument rating sought;

(6) Receive a logbook or training record endorsement from an authorized instructor certifying that the person is prepared to take the required practical test;

(7) Pass the required knowledge test on the aeronautical knowledge areas of paragraph (b) of this section; however, an applicant is not required to take another knowledge test when that person already holds an instrument rating; and

(8) Pass the required practical test on the areas of operation in paragraph (c) of this section in—

(i) An airplane, helicopter, or powered-lift, appropriate to the rating sought; or

(ii) A flight simulator or a flight training device appropriate to the rating sought and approved for the specific maneuver or procedure performed. If a flight training device is used for the practical test, the instrument approach procedures conducted in that flight training device are limited to one precision and one nonprecision approach, provided the flight training device is approved for the procedure performed.

(b) Aeronautical knowledge. A person who applies for an instrument rating must have received and logged ground training from an authorized instructor or accomplished a home-study course on the following aeronautical knowledge areas that apply to the instrument rating sought:

(1) Federal Aviation Regulations of this chapter that apply to flight operations under IFR;

(2) Appropriate information that applies to flight operations under IFR in the "Aeronautical Information Manual;"

(3) Air traffic control system and procedures for instrument flight operations;

(4) IFR navigation and approaches by use of navigation systems;

(5) Use of IFR en route and instrument approach procedure charts;

(6) Procurement and use of aviation weather reports and forecasts and the elements of forecasting weather trends based on that information and personal observation of weather conditions;

(7) Safe and efficient operation of aircraft under instrument flight rules and conditions;

(8) Recognition of critical weather situations and windshear avoidance;

(9) Aeronautical decision making and judgment; and

(10) Crew resource management, including crew communication and coordination.

(c) Flight proficiency. A person who applies for an instrument rating must receive and log training from an authorized instructor in an aircraft, or in a flight simulator or flight training device, in accordance with paragraph (e) of this section, that includes the following areas of operation:

(1) Preflight preparation;

(2) Preflight procedures;

(3) Air traffic control clearances and procedures;

(4) Flight by reference to instruments;

(5) Navigation systems;

(6) Instrument approach procedures;

(7) Emergency operations; and

(8) Postflight procedures.

(d) Aeronautical experience. A person who applies for an instrument rating must have logged the following:

(1) At least 50 hours of cross-country flight time as pilot in command, of which at least 10 hours must be in airplanes for an instrument—airplane rating; and

(2) A total of 40 hours of actual or simulated instrument time on the areas of operation of this section, to include—

(i) At least 15 hours of instrument flight training from an authorized instructor in the aircraft category for which the instrument rating is sought;

(ii) At least 3 hours of instrument training that is appropriate to the instrument rating sought from an authorized instructor in preparation for the practical test within the 60 days preceding the date of the test;

(iii) For an instrument—airplane rating, instrument training on cross-country flight procedures specific to airplanes that includes at least one cross-country flight in an airplane that is performed under IFR, and consists of—

(A) A distance of at least 250 nautical miles along airways or ATC-directed routing;

(B) An instrument approach at each airport; and

(C) Three different kinds of approaches with the use of navigation systems;

(iv) For an instrument—helicopter rating, instrument training specific to helicopters on cross-country flight procedures that includes at least one cross-country flight in a helicopter that is performed under IFR, and consists of—

(A) A distance of at least 100 nautical miles along airways or ATC-directed routing;

(B) An instrument approach at each airport; and

(C) Three different kinds of approaches with the use of navigation systems; and

(v) For an instrument—powered-lift rating, instrument training specific to a powered-lift on cross-country flight procedures that includes at least one cross-country flight in a powered-lift that is performed under IFR and consists of—

(A) A distance of at least 250 nautical miles along airways or ATC-directed routing;

(B) An instrument approach at each airport; and

(C) Three different kinds of approaches with the use of navigation systems.

(e) Use of flight simulators or flight training devices. If the instrument training was provided by an authorized instructor in a flight simulator or flight training device—

(1) A maximum of 30 hours may be performed in that flight simulator or flight training device if the training was accomplished in accordance with part 142 of this chapter; or

(2) A maximum of 20 hours may be performed in that flight simulator or flight training device if the training was not accomplished in accordance with part 142 of this chapter.

[Amdt. 61-102, 62 FR 16298, Apr. 4, 1997; Amdt. 61-103, 62 FR 40900, July 30, 1997]

§ 61.67 Category II pilot authorization requirements.

(a) General. A person who applies for a Category II pilot authorization must hold:

(1) At least a private or commercial pilot certificate with an instrument rating or an airline transport pilot certificate;

(2) A type rating for the aircraft for which the authorization is sought if that aircraft requires a type rating; and

(3) A category and class rating for the aircraft for which the authorization is sought.

(b) Experience requirements. An applicant for a Category II pilot authorization must have at least—

(1) 50 hours of night flight time as pilot in command.

(2) 75 hours of instrument time under actual or simulated instrument conditions that may include not more than—

(i) A combination of 25 hours of simulated instrument flight time in a flight simulator or flight training device; or

(ii) 40 hours of simulated instrument flight time if accomplished in an approved course conducted by an appropriately rated training center certificated under part 142 of this chapter.

(3) 250 hours of cross-country flight time as pilot in command.

(c) Practical test requirements.

(1) A practical test must be passed by a person who applies for—

(i) Issuance or renewal of a Category II pilot authorization; and

(ii) The addition of another type aircraft to the applicant's Category II pilot authorization.

(2) To be eligible for the practical test for an authorization under this section, an applicant must—

(i) Meet the requirements of paragraphs (a) and (b) of this section; and

(ii) If the applicant has not passed a practical test for this authorization during the 12 calendar months preceding the month of the test, then that person must—

(A) Meet the requirements of § 61.57(c); and

(B) Have performed at least six ILS approaches during the 6 calendar months preceding the month of the test, of which at least three of the approaches must have been conducted without the use of an approach coupler.

(3) The approaches specified in paragraph (c)(2)(ii)(B) of this section—

(i) Must be conducted under actual or simulated instrument flight conditions;

(ii) Must be conducted to the decision height for the ILS approach in the type aircraft in which the practical test is to be conducted;

(iii) Need not be conducted to the decision height authorized for Category II operations;

(iv) Must be conducted to the decision height authorized for Category II operations only if conducted in a flight simulator or flight training device; and

(v) Must be accomplished in an aircraft of the same category and class, and type, as applicable, as the aircraft in which the practical test is to be conducted or in a flight simulator that—

(A) Represents an aircraft of the same category and class, and type, as applicable, as the aircraft in which the authorization is sought; and

(B) Is used in accordance with an approved course conducted by a training center certificated under part 142 of this chapter.

(4) The flight time acquired in meeting the requirements of paragraph (c)(2)(ii)(B) of this section may be used to meet the requirements of paragraph (c)(2)(ii)(A) of this section.

(d) Practical test procedures. The practical test consists of an oral increment and a flight increment.

(1) Oral increment. In the oral increment of the practical test an applicant must demonstrate knowledge of the following:

(i) Required landing distance;

(ii) Recognition of the decision height;

(iii) Missed approach procedures and techniques using computed or fixed attitude guidance displays;

(iv) Use and limitations of RVR;

(v) Use of visual clues, their availability or limitations, and altitude at which they are normally discernible at reduced RVR readings;

(vi) Procedures and techniques related to transition from nonvisual to visual flight during a final approach under reduced RVR;

(vii) Effects of vertical and horizontal windshear;

(viii) Characteristics and limitations of the ILS and runway lighting system;

(ix) Characteristics and limitations of the flight director system, auto approach coupler (including split axis type if equipped), auto throttle system (if equipped), and other required Category II equipment;

(x) Assigned duties of the second in command during Category II approaches, unless the aircraft for which authorization is sought does not require a second in command; and

(xi) Instrument and equipment failure warning systems.

(2) Flight increment. The following requirements apply to the flight increment of the practical test:

(i) The flight increment must be conducted in an aircraft of the same category, class, and type, as applicable, as the aircraft in which the authorization is sought or in a flight simulator that—

(A) Represents an aircraft of the same category and class, and type, as applicable, as the aircraft in which the authorization is sought; and

(B) Is used in accordance with an approved course conducted by a training center certificated under part 142 of this chapter.

(ii) The flight increment must consist of at least two ILS approaches to 100 feet AGL including at least one landing and one missed approach.

(iii) All approaches performed during the flight increment must be made with the use of an approved flight control guidance system, except if an approved auto approach coupler is installed, at least one approach must be hand flown using flight director commands.

(iv) If a multiengine airplane with the performance capability to execute a missed approach with one engine inoperative is used for the practical test, the flight increment must include the performance of one missed approach with an engine, which shall be the most critical engine, if applicable, set at idle or zero thrust before reaching the middle marker.

(v) If a multiengine flight simulator or multiengine flight training device is used for the practical test, the applicant must execute a missed approach with the most critical engine, if applicable, failed.

(vi) For an authorization for an aircraft that requires a type rating, the practical test must be performed in coordination with a second in command who holds a type rating in the aircraft in which the authorization is sought.

(vii) Oral questioning may be conducted at any time during a practical test.

[Amdt. 61-102, 62 FR 16298, Apr. 4, 1997; Amdt. 61-103, 62 FR 40900, July 30, 1997]

§ 61.68 Category III pilot authorization requirements.

(a) General. A person who applies for a Category III pilot authorization must hold:

(1) At least a private pilot certificate or commercial pilot certificate with an instrument rating or an airline transport pilot certificate;

(2) A type rating for the aircraft for which the authorization is sought if that aircraft requires a type rating; and

(3) A category and class rating for the aircraft for which the authorization is sought.

(b) Experience requirements. An applicant for a Category III pilot authorization must have at least—

(1) 50 hours of night flight time as pilot in command.

(2) 75 hours of instrument flight time during actual or simulated instrument conditions that may include not more than—

(i) A combination of 25 hours of simulated instrument flight time in a flight simulator or flight training device; or

(ii) 40 hours of simulated instrument flight time if accomplished in an approved course conducted by an appropriately rated training center certificated under part 142 of this chapter.

(3) 250 hours of cross-country flight time as pilot in command.

(c) Practical test requirements. (1) A practical test must be passed by a person who applies for—

(i) Issuance or renewal of a Category III pilot authorization; and

(ii) The addition of another type of aircraft to the applicant's Category III pilot authorization.

(2) To be eligible for the practical test for an authorization under this section, an applicant must—

(i) Meet the requirements of paragraphs (a) and (b) of this section; and

(ii) If the applicant has not passed a practical test for this authorization during the 12 calendar months preceding the month of the test, then that person must—

(A) Meet the requirements of § 61.57(c); and

(B) Have performed at least six ILS approaches during the 6 calendar months preceding the month of the test, of which at least three of the approaches must have been conducted without the use of an approach coupler.

(3) The approaches specified in paragraph (c)(2)(ii)(B) of this section—

(i) Must be conducted under actual or simulated instrument flight conditions;

(ii) Must be conducted to the alert height or decision height for the ILS approach in the type aircraft in which the practical test is to be conducted;

(iii) Need not be conducted to the decision height authorized for Category III operations;

(iv) Must be conducted to the alert height or decision height, as applicable, authorized for Category III operations only if conducted in an approved flight simulator or approved flight training device; and

(v) Must be accomplished in an aircraft of the same category and class, and type, as applicable, as the aircraft in which the practical test is to be conducted or in a flight simulator that—

(A) Represents an aircraft of the same category and class, and type, as applicable, as the aircraft for which the authorization is sought; and

(B) Is used in accordance with an approved course conducted by a training center certificated under part 142 of this chapter.

(4) The flight time acquired in meeting the requirements of paragraph (c)(2)(ii)(B) of this section may be used to meet the requirements of paragraph (c)(2)(ii)(A) of this section.

(d) Practical test procedures. The practical test consists of an oral increment and a flight increment.

(1) Oral increment. In the oral increment of the practical test an applicant must demonstrate knowledge of the following:

(i) Required landing distance;

(ii) Determination and recognition of the alert height or decision height, as applicable, including use of a radar altimeter;

(iii) Recognition of and proper reaction to significant failures encountered prior to and after reaching the alert height or decision height, as applicable;

(iv) Missed approach procedures and techniques using computed or fixed attitude guidance displays and expected height loss as they relate to manual go-around or automatic go-around, and initiation altitude, as applicable;

(v) Use and limitations of RVR, including determination of controlling RVR and required transmissometers;

(vi) Use, availability, or limitations of visual cues and the altitude at which they are normally discernible at reduced RVR readings including—

(A) Unexpected deterioration of conditions to less than minimum RVR during approach, flare, and rollout;

(B) Demonstration of expected visual references with weather at minimum conditions;

(C) The expected sequence of visual cues during an approach in which visibility is at or above landing minima; and

(D) Procedures and techniques for making a transition from instrument reference flight to visual flight during a final approach under reduced RVR.

(vii) Effects of vertical and horizontal windshear;

(viii) Characteristics and limitations of the ILS and runway lighting system;

(ix) Characteristics and limitations of the flight director system auto approach coupler (including split axis type if equipped), auto throttle system (if equipped), and other Category III equipment;

(x) Assigned duties of the second in command during Category III operations, unless the aircraft for which authorization is sought does not require a second in command;

(xi) Recognition of the limits of acceptable aircraft position and flight path tracking during approach, flare, and, if applicable, rollout; and

(xii) Recognition of, and reaction to, airborne or ground system faults or abnormalities, particularly after passing alert height or decision height, as applicable.

(2) Flight increment. The following requirements apply to the flight increment of the practical test—

(i) The flight increment may be conducted in an aircraft of the same category and class, and type, as applicable, as the aircraft for which the authorization is sought, or in a flight simulator that—

(A) Represents an aircraft of the same category and class, and type, as applicable, as the aircraft in which the authorization is sought; and

(B) Is used in accordance with an approved course conducted by a training center certificated under part 142 of this chapter.

(ii) The flight increment must consist of at least two ILS approaches to 100 feet AGL, including one landing and one missed approach initiated from a very low altitude that may result in a touchdown during the go-around maneuver;

(iii) All approaches performed during the flight increment must be made with the approved automatic landing system or an equivalent landing system approved by the Administrator;

(iv) If a multiengine aircraft with the performance capability to execute a missed approach with one engine inoperative is used for the practical test, the flight increment must include the performance of one missed approach with the most critical engine, if applicable, set at idle or zero thrust before reaching the middle or outer marker;

(v) If a multiengine flight simulator or multiengine flight training device is used, a missed approach must be executed with an engine, which shall be the most critical engine, if applicable, failed;

(vi) For an authorization for an aircraft that requires a type rating, the practical test must be performed in coordination with a second in command who holds a type rating in the aircraft in which the authorization is sought;

(vii) Oral questioning may be conducted at any time during the practical test;

(viii) Subject to the limitations of this paragraph, for Category IIIb operations predicated on the use of a fail-passive rollout control system, at least one manual rollout using visual reference or a combination of visual and instrument references must be executed. The maneuver required by this paragraph shall be initiated by a fail-passive disconnect of the rollout control system—

(A) After main gear touchdown;

(B) Prior to nose gear touchdown;

(C) In conditions representative of the most adverse lateral touchdown displacement allowing a safe landing on the runway; and

(D) In weather conditions anticipated in Category IIIb operations.

[Amdt. 61-102, 62 FR 16298, Apr. 4, 1997; Amdt. 61-103, 62 FR 40900, July 30, 1997]

§ 61.69 Glider and unpowered ultralight vehicle towing: Experience and training requirements.

(a) No person may act as pilot in command for towing a glider or unpowered ultralight vehicle unless that person

(1) Holds at least a private pilot certificate with a category rating for powered aircraft;

(2) Has logged at least 100 hours of pilot-in-command time in the aircraft category, class and type, if required, that the pilot is using to tow a glider or unpowered ultralight vehicle;

(3) Has a logbook endorsement from an authorized instructor who certifies that the person has received ground and flight training in gliders or unpowered ultralight vehicles and is proficient in—

(i) The techniques and procedures essential to the safe towing of gliders or unpowered ultralight vehicles, including airspeed limitations;

(ii) Emergency procedures;.

(iii) Signals used; and

(iv) Maximum angles of bank.

(4) Except as provided in paragraph (b) of this section, has logged at least three flights as the sole manipulator of the controls of an aircraft towing a glider or unpowered ultralight vehicle simulating towing flight procedures while accompanied by a pilot who meets the requirements of paragraphs (c) and (d) of this section;

(5) Except as provided in paragraph (b) of this section, has received a logbook endorsement from the pilot, described in paragraph (a)(4) of this section, certifying that the person has accomplished at least 3 flights in an aircraft while towing a glider or unpowered ultralight vehicle, or while simulating towing flight procedures; and

(6) Within the preceding 12 months has—

(i) Made at least three actual or simulated tows of a glider or unpowered ultralight vehicle while accompanied by a qualified pilot who meets the requirements of this section; or

(ii) Made at least three flights as pilot in command of a glider or unpowered ultralight vehicle towed by an aircraft.

(b) Any person who, before May 17, 1967, has made and logged 10 or more flights as pilot in command of an aircraft towing a glider or unpowered ultralight vehicle in accordance with a certificate of waiver need not comply with paragraphs (a)(4) and (a)(5) of this section.

(c) The pilot, described in paragraph (a)(4) of this section, who endorses the logbook of a person seeking towing privileges must have—

(1) Met the requirements of this section prior to endorsing the logbook of the person seeking towing privileges; and

(2) Logged at least 10 flights as pilot in command of an aircraft while towing a glider or unpowered ultralight vehicle.

(d) If the pilot described in paragraph (a)(4) of this section holds only a private pilot certificate, then that pilot must have—

(1) Logged at least 100 hours of pilot-in-command time in airplanes, or 200 hours of pilot-in-command time in a combination of powered and other-than-powered aircraft; and

(2) Performed and logged at least three flights within the 12 calendar months preceding the month that pilot accompanies or endorses the logbook of a person seeking towing privileges—

(i) In an aircraft while towing a glider or unpowered ultralight vehicle accompanied by another pilot who meets the requirements of this section; or

(ii) As pilot in command of a glider or unpowered ultralight vehicle being towed by another aircraft.

§ 61.71 Graduates of an approved training program other than under this part: Special rules.

(a) A person who graduates from an approved training program under part 141 or part 142 of this chapter is considered to have met the applicable aeronautical experience, aeronautical knowledge, and areas of operation requirements of this part if that person presents the graduation certificate and passes the required practical test within the 60-day period after the date of graduation.

(b) A person may apply for an airline transport pilot certificate, type rating, or both under this part, and will be considered to have met the applicable requirements under § 61.157 of this part for that certificate and rating, if that person has:

(1) Satisfactorily accomplished an approved training program and the pilot-in-command proficiency check for that airplane type, in accordance with the pilot-in-command requirements under subparts N and O of part 121 of this chapter; and

(2) Applied for the airline transport pilot certificate, type rating, or both within the 60-day period from the date the person satisfactorily accomplished the approved training program and pilot-in-command proficiency check for that airplane type.

[Amdt. 61-102, 62 FR 16298, Apr. 4, 1997; Amdt. 61-103, 62 FR 40901, July 30, 1997]

§ 61.73 Military pilots or former military pilots: Special rules.

(a) General. Except for a rated military pilot or former rated military pilot who has been removed from flying status for lack of proficiency, or because of disciplinary action involving aircraft operations, a rated military pilot or former rated military pilot who meets the applicable requirements of this section may apply, on the basis of his or her military training, for:

(1) A commercial pilot certificate;

(2) An aircraft rating in the category and class of aircraft for which that military pilot is qualified;

(3) An instrument rating with the appropriate aircraft rating for which that military pilot is qualified; or

(4) A type rating, if appropriate.

(b) Military pilots on active flying status within the past 12 months. A rated military pilot or former rated military pilot who has been on active flying status within the 12 months before applying must:

(1) Pass a knowledge test on the appropriate parts of this chapter that apply to pilot privileges and limitations, air traffic and general operating rules, and accident reporting rules;

(2) Present documentation showing compliance with the requirements of paragraph (d) of this section for at least one aircraft category rating; and

(3) Present documentation showing that the applicant is or was, at any time during the 12 calendar months before the month of application—

(i) A rated military pilot on active flying status in an armed force of the United States; or

(ii) A rated military pilot of an armed force of a foreign contracting State to the Convention on International Civil Aviation, assigned to pilot duties (other than flight training) with an armed force of the United States and holds, at the time of application, a current civil pilot license issued by that contracting State authorizing at least the privileges of the pilot certificate sought.

(c) Military pilots not on active flying status during the 12 calendar months before the month of application. A rated military pilot or former rated military pilot who has not been on active flying status within the 12 calendar months before the month of application must:

(1) Pass the appropriate knowledge and practical tests prescribed in this part for the certificate or rating sought; and

(2) Present documentation showing that the applicant was, before the beginning of the 12th calendar month

before the month of application, a rated military pilot as prescribed by paragraph (b)(3)(i) or paragraph (b)(3)(ii) of this section.

(d) Aircraft category, class, and type ratings. A rated military pilot or former rated military pilot who applies for an aircraft category, class, or type rating, if applicable, is issued that rating at the commercial pilot certificate level if the pilot presents documentary evidence that shows satisfactory accomplishment of:

(1) An official U.S. military pilot check and instrument proficiency check in that aircraft category, class, or type, if applicable, as pilot in command during the 12 calendar months before the month of application;

(2) At least 10 hours of pilot-in-command time in that aircraft category, class, or type, if applicable, during the 12 calendar months before the month of application; or

(3) An FAA practical test in that aircraft after—

(i) Meeting the requirements of paragraphs (b)(1) and (b)(2) of this section; and

(ii) Having received an endorsement from an authorized instructor who certifies that the pilot is proficient to take the required practical test, and that endorsement is made within the 60-day period preceding the date of the practical test.

(e) Instrument rating. A rated military pilot or former rated military pilot who applies for an airplane instrument rating, a helicopter instrument rating, or a powered-lift instrument rating to be added to his or her commercial pilot certificate may apply for an instrument rating if the pilot has, within the 12 calendar months preceding the month of application:

(1) Passed an instrument proficiency check by a U.S. Armed Force in the aircraft category for the instrument rating sought; and

(2) Received authorization from a U.S. Armed Force to conduct IFR flights on Federal airways in that aircraft category and class for the instrument rating sought.

(f) Aircraft type rating. An aircraft type rating is issued only for aircraft types that the Administrator has certificated for civil operations.

(g) Aircraft type rating placed on an airline transport pilot certificate. A rated military pilot or former rated military pilot who holds an airline transport pilot certificate and who requests an aircraft type rating to be placed on that person's airline transport pilot certificate may be issued that aircraft type rating at the airline transport pilot certificate level, provided that person:

(1) Holds a category and class rating for that type of aircraft at the airline transport pilot certificate level; and

(2) Passed an official U.S. military pilot check and instrument proficiency check in that type of aircraft as pilot in command during the 12 calendar months before the month of application.

(h) Evidentiary documents. The following documents are satisfactory evidence for the purposes indicated:

(1) An official identification card issued to the pilot by an armed force may be used to demonstrate membership in the armed forces.

(2) An original or a copy of a certificate of discharge or release may be used to demonstrate discharge or release from an armed force or former membership in an armed force.

(3) Current or previous status as a rated military pilot with a U.S. Armed Force may be demonstrated by—

(i) An official U.S. Armed Force order to flight status as a military pilot;

(ii) An official U.S. Armed Force form or logbook showing military pilot status; or

(iii) An official order showing that the rated military pilot graduated from a U.S. military pilot school and received a rating as a military pilot.

(4) A certified U.S. Armed Force logbook or an appropriate official U.S. Armed Force form or summary may be used to demonstrate flight time in military aircraft as a member of a U.S. Armed Force.

(5) An official U.S. Armed Force record of a military checkout as pilot in command may be used to demonstrate pilot in command status.

(6) A current instrument grade slip that is issued by a U.S. Armed Force, or an official record of satisfactory accomplishment of an instrument proficiency check during the 12 calendar months preceding the month of the application may be used to demonstrate instrument pilot qualification.

[Amdt. 61-102, 62 FR 16298, Apr. 4, 1997; Amdt. 61-103, 62 FR 40901, July 30, 1997]

§ 61.75 Private pilot certificate issued on the basis of a foreign pilot license.

(a) *General.* A person who holds a current foreign pilot license issued by a contracting State to the Convention on International Civil Aviation may apply for and be issued a private pilot certificate with the appropriate ratings when the application is based on the foreign pilot license that meets the requirements of this section.

(b) *Certificate issued.* A U.S. private pilot certificate that is issued under this section shall specify the person's foreign license number and country of issuance. A person who holds a current foreign pilot license issued by a contracting State to the Convention on International Civil Aviation may be issued a private pilot certificate based on the foreign pilot license without any further showing of proficiency, provided the applicant:

(1) Meets the requirements of this section;

(2) Holds a foreign pilot license that—

(i) Is not under an order of revocation or suspension by the foreign country that issued the foreign pilot license; and

(ii) Does not contain an endorsement stating that the applicant has not met all of the standards of ICAO for that license;

(3) Does not currently hold a U.S. pilot certificate;

(4) Holds a current medical certificate issued under part 67 of this chapter or a current medical certificate issued by the country that issued the person's foreign pilot license; and

(5) Is able to read, speak, write, and understand the English language. If the applicant is unable to meet one of these requirements due to medical reasons, then the Administrator may place such operating limitations on that applicant's pilot certificate as are necessary for the safe operation of the aircraft.

(c) *Aircraft ratings issued.* Aircraft ratings listed on a person's foreign pilot license, in addition to any issued after testing under the provisions of this part, may be placed on that person's U.S. pilot certificate.

(d) *Instrument ratings issued.* A person who holds an instrument rating on the foreign pilot license issued by a contracting State to the Convention on International Civil Aviation may be issued an instrument rating on a U.S. private pilot certificate provided:

(1) The person's foreign pilot license authorizes instrument privileges;

(2) Within 24 months preceding the month in which the person applies for the instrument rating, the person passes the appropriate knowledge test; and

(3) The person is able to read, speak, write, and understand the English language. If the applicant is unable to meet one of these requirements due to medical reasons, then the Administrator may place such operating limitations on that applicant's pilot certificate as are necessary for the safe operation of the aircraft.

(e) *Operating privileges and limitations.* A person who receives a U.S. private pilot certificate that has been issued under the provisions of this section:

(1) May act as a pilot of a civil aircraft of U.S. registry in accordance with the private pilot privileges authorized by this part;

(2) Is limited to the privileges placed on the certificate by the Administrator;

(3) Is subject to the limitations and restrictions on the person's U.S. certificate and foreign pilot license when exercising the privileges of that U.S. pilot certificate in an aircraft of U.S. registry operating within or outside the United States; and

(4) Shall not exercise the privileges of that U.S. private pilot certificate when the person's foreign pilot license has been revoked or suspended.

(f) *Limitation on licenses used as the basis for a U.S. certificate.* Only one foreign pilot license may be used as a basis for issuing a U.S. private pilot certificate. The foreign pilot license and medical certification used as a basis for issuing a U.S. private pilot certificate under this section must be in the English language or accompanied by an English language transcription that has been signed by an official or representative of the foreign aviation authority that issued the foreign pilot license.

(g) Limitation placed on a U.S. private pilot certificate. A U.S. private pilot certificate issued under this section is valid only when the holder has the foreign pilot license upon which the issuance of the U.S. private pilot certificate was based in the holder's personal possession or readily accessible in the aircraft.

§ 61.77   Special purpose pilot authorization: Operation of U.S.-registered civil aircraft leased by a person who is not a U.S. citizen.

(a) General. The holder of a foreign pilot license issued by a contracting State to the Convention on International Civil Aviation who meets the requirements of this section may be issued a special purpose pilot authorization by the Administrator for the purpose of performing pilot duties.

(1) On a civil aircraft of U.S. registry that is leased to a person who is not a citizen of the United States, and

(2) For carrying persons or property for compensation or hire on that aircraft.

(b) Eligibility. To be eligible for the issuance or renewal of a special purpose pilot authorization, an applicant must present the following to an FAA Flight Standards District Office:

(1) A current foreign pilot license that has been issued by the aeronautical authority of a contracting State to the Convention on International Civil Aviation from which the person holds citizenship or resident status and that contains the appropriate aircraft category, class, instrument rating, and type rating, if appropriate, for the aircraft to be flown;

(2) A current certification by the lessee of the aircraft—

(i) Stating that the applicant is employed by the lessee;

(ii) Specifying the aircraft type on which the applicant will perform pilot duties; and

(iii) Stating that the applicant has received ground and flight instruction that qualifies the applicant to perform the duties to be assigned on the aircraft.

(3) Documentation showing when the applicant will reach the age of 60 years (an official copy of the applicant's birth certificate or other official documentation);

(4) Documentation that the applicant meets the medical standards for the issuance of the foreign pilot license from the aeronautical authority of the contracting State to the Convention on International Civil Aviation where the applicant holds citizenship or resident status;

(5) Documentation that the applicant meets the recent flight experience requirements of this part (a logbook or flight record); and

(6) A statement that the applicant does not already hold a special purpose pilot authorization; however, if the applicant already holds a special purpose pilot authorization, then that special purpose pilot authorization must be surrendered to either the FAA Flight Standards District Office that issued it, or the FAA Flight Standards District Office processing the application for the authorization, prior to being issued another special purpose pilot authorization.

(c) Privileges. A person issued a special purpose pilot authorization under this section—

(1) May exercise the privileges prescribed on the special purpose pilot authorization; and

(2) Must comply with the limitations specified in this section and any additional limitations specified on the special purpose pilot authorization.

(d) General limitations. A special purpose pilot authorization is valid only—

(1) For flights between foreign countries or for flights in foreign air commerce within the time period allotted on the authorization;

(2) If the foreign pilot license required by paragraph (b)(1) of this section, the medical documentation required by paragraph (b)(4) of this section, and the special purpose pilot authorization issued under this section are in the holder's physical possession or immediately accessible in the aircraft;

(3) While the holder is employed by the person to whom the aircraft described in the certification required by paragraph (b)(2) of this section is leased;

(4) While the holder is performing pilot duties on the U.S.-registered aircraft described in the certification required by paragraph (b)(2) of this section; and

(5) If the holder has only one special purpose pilot authorization as provided in paragraph (b)(6) of this section.

(e) Age limitation. Except as provided in paragraph (g) of this section, no person who holds a special purpose pilot authorization issued under this part, and no person who holds a special purpose pilot certificate issued under this part before August 4, 1997, shall serve as a pilot on a civil airplane of U.S. registry if the person has reached his or her 60th birthday, in the following operations:

(1) Scheduled international air services carrying passengers in turbojet-powered airplanes;

(2) Scheduled international air services carrying passengers in airplanes having a passenger-seat configuration of more than nine passenger seats, excluding each crewmember seat;

(3) Nonscheduled international air transportation for compensation or hire in airplanes having a passenger-seat configuration of more than 30 passenger seats, excluding each crewmember seat; or

(4) Scheduled international air services, or nonscheduled international air transportation for compensation or hire, in airplanes having a payload capacity of more than 7,500 pounds.

(f) Definitions.

(1) International air service, as used in paragraph (e) of this section, means scheduled air service performed in airplanes for the public transport of passengers, mail, or cargo, in which the service passes through the air space over the territory of more than one country.

(2) International air transportation, as used in paragraph (e) of this section, means air transportation performed in

airplanes for the public transport of passengers, mail, or cargo, in which service passes through the air space over the territory of more than one country.

(g) Delayed pilot age limitations for certain operations. Until December 20, 1999, a person may serve as a pilot in the operations specified in paragraph (e) of this section after that person has reached his or her 60th birthday, if, on March 20, 1997, that person was employed as a pilot in any of the following operations:

(1) Scheduled international air services carrying passengers in nontransport category turbopropeller-powered airplanes type certificated after December 31, 1964, that have a passenger-seat configuration of 10 to 19 seats;

(2) Scheduled international air services carrying passengers in transport category turbopropeller-powered airplanes that have a passenger-seat configuration of 20 to 30 seats; or

(3) Scheduled international air services carrying passengers in turbojet-powered airplanes having a passenger-seat configuration of 1 to 30 seats.

(h) Expiration date. Each special purpose pilot authorization issued under this section expires—

(1) 60 calendar months from the month it was issued, unless sooner suspended or revoked;

(2) When the lease agreement for the aircraft expires or the lessee terminates the employment of the person who holds the special purpose pilot authorization;

(3) Whenever the person's foreign pilot license has been suspended, revoked, or is no longer valid; or

(4) When the person no longer meets the medical standards for the issuance of the foreign pilot license.

(i) Renewal. A person exercising the privileges of a special purpose pilot authorization may apply for a 60-calendar-month extension of that authorization, provided the person—

(1) Continues to meet the requirements of this section; and

(2) Surrenders the expired special purpose pilot authorization upon receipt of the new authorization.

(j) Surrender. The holder of a special purpose pilot authorization must surrender the authorization to the Administrator within 7 days after the date the authorization terminates.

[Amdt. 61–102, 62 FR 16298, Apr. 4, 1997; Amdt. 61–103, 62 FR 40902, July 30, 1997]

Subpart C—Student Pilots

§ 61.81 Applicability.

This subpart prescribes the requirements for the issuance of student pilot certificates, the conditions under which those certificates are necessary, and the general operating rules and limitations for the holders of those certificates.

§ 61.83 Eligibility requirements for student pilots.

To be eligible for a student pilot certificate, an applicant must:

(a) Be at least 16 years of age for other than the operation of a glider or balloon.

(b) Be at least 14 years of age for the operation of a glider or balloon.

(c) Be able to read, speak, write, and understand the English language. If the applicant is unable to meet one of these requirements due to medical reasons, then the Administrator may place such operating limitations on that applicant's pilot certificate as are necessary for the safe operation of the aircraft.

§ 61.85 Application.

An application for a student pilot certificate is made on a form and in a manner provided by the Administrator and is submitted to:

(a) A designated aviation medical examiner if applying for an FAA medical certificate under part 67 of this chapter;

(b) An examiner; or

(c) A Flight Standards District Office.

§ 61.87 Solo requirements for student pilots.

(a) General. A student pilot may not operate an aircraft in solo flight unless that student has met the requirements of this section. The term "solo flight," as used in this subpart, means that flight time during which a student pilot is the sole occupant of the aircraft, or that flight time during which the student performs the duties of a pilot in command of a gas balloon or an airship requiring more than one pilot flight crewmember.

(b) Aeronautical knowledge. A student pilot must demonstrate satisfactory aeronautical knowledge on a knowledge test that meets the requirements of this paragraph:

(1) The test must address the student pilot's knowledge of—

(i) Applicable sections of parts 61 and 91 of this chapter;

(ii) Airspace rules and procedures for the airport where the solo flight will be performed; and

(iii) Flight characteristics and operational limitations for the make and model of aircraft to be flown.

(2) The student's authorized instructor must—

(i) Administer the test; and

(ii) At the conclusion of the test, review all incorrect answers with the student before authorizing that student to conduct a solo flight.

(c) Pre-solo flight training. Prior to conducting a solo flight, a student pilot must have:

(1) Received and logged flight training for the maneuvers and procedures of this section that are appropriate to the make and model of aircraft to be flown; and

(2) Demonstrated satisfactory proficiency and safety, as judged by an authorized instructor, on the maneuvers and procedures required by this section in the make and model of aircraft or similar make and model of aircraft to be flown.

(d) Maneuvers and procedures for pre-solo flight training in a single-engine airplane. A student pilot who is receiving training for a single-engine airplane rating or privileges must receive and log flight training for the following maneuvers and procedures:

(1) Proper flight preparation procedures, including pre-flight planning and preparation, powerplant operation, and aircraft systems;

(2) Taxiing or surface operations, including runups;

(3) Takeoffs and landings, including normal and cross-wind;

(4) Straight and level flight, and turns in both direc-tions;

(5) Climbs and climbing turns;

(6) Airport traffic patterns, including entry and depar-ture procedures;

(7) Collision avoidance, windshear avoidance, and wake turbulence avoidance;

(8) Descents, with and without turns, using high and low drag configurations;

(9) Flight at various airspeeds from cruise to slow flight;

(10) Stall entries from various flight attitudes and power combinations with recovery initiated at the first indication of a stall, and recovery from a full stall;

(11) Emergency procedures and equipment malfunctions;

(12) Ground reference maneuvers;

(13) Approaches to a landing area with simulated en-gine malfunctions;

(14) Slips to a landing; and

(15) Go-arounds.

(e) Maneuvers and procedures for pre-solo flight training in a multiengine airplane. A student pilot who is receiving training for a multiengine airplane rating must receive and log flight training for the following maneuvers and procedures:

(1) Proper flight preparation procedures, including pre-flight planning and preparation, powerplant operation, and aircraft systems;

(2) Taxiing or surface operations, including runups;

(3) Takeoffs and landings, including normal and cross-wind;

(4) Straight and level flight, and turns in both direc-tions;

(5) Climbs and climbing turns;

(6) Airport traffic patterns, including entry and depar-ture procedures;

(7) Collision avoidance, windshear avoidance, and wake turbulence avoidance;

(8) Descents, with and without turns, using high and low drag configurations;

(9) Flight at various airspeeds from cruise to slow flight;

(10) Stall entries from various flight attitudes and power combinations with recovery initiated at the first indication of a stall, and recovery from a full stall;

(11) Emergency procedures and equipment malfunctions;

(12) Ground reference maneuvers;

(13) Approaches to a landing area with simulated en-gine malfunctions; and

(14) Go-arounds.

(f) Maneuvers and procedures for pre-solo flight training in a helicopter. A student pilot who is receiving training for a helicopter rating must receive and log flight training for the following maneuvers and procedures:

(1) Proper flight preparation procedures, including pre-flight planning and preparation, powerplant operation, and aircraft systems;

(2) Taxiing or surface operations, including runups;

(3) Takeoffs and landings, including normal and cross-wind;

(4) Straight and level flight, and turns in both directions;

(5) Climbs and climbing turns;

(6) Airport traffic patterns, including entry and departure procedures;

(7) Collision avoidance, windshear avoidance, and wake turbulence avoidance;

(8) Descents with and without turns;

(9) Flight at various airspeeds;

(10) Emergency procedures and equipment malfunc-tions;

(11) Ground reference maneuvers;

(12) Approaches to the landing area;

(13) Hovering and hovering turns;

(14) Go-arounds;

(15) Simulated emergency procedures, including au-torotational descents with a power recovery and power re-covery to a hover;

(16) Rapid decelerations; and

(17) Simulated one-engine-inoperative approaches and landings for multiengine helicopters.

(g) Maneuvers and procedures for pre-solo flight training in a gyroplane. A student pilot who is receiving training for a gyroplane rating or privileges must receive and log flight training for the following maneuvers and procedures:

(1) Proper flight preparation procedures, including pre-flight planning and preparation, powerplant operation, and aircraft systems;

(2) Taxiing or surface operations, including runups;

(3) Takeoffs and landings, including normal and cross-wind;

(4) Straight and level flight, and turns in both direc-tions;

(5) Climbs and climbing turns;

(6) Airport traffic patterns, including entry and depar-ture procedures;

(7) Collision avoidance, windshear avoidance, and wake turbulence avoidance;

(8) Descents with and without turns;

(9) Flight at various airspeeds;

(10) Emergency procedures and equipment malfunctions;

(11) Ground reference maneuvers;

(12) Approaches to the landing area;

(13) High rates of descent with power on and with simulated power off, and recovery from those flight configurations;

(14) Go-arounds; and

(15) Simulated emergency procedures, including simulated power-off landings and simulated power failure during departures.

(h) Maneuvers and procedures for pre-solo flight training in a powered-lift. A student pilot who is receiving training for a powered-lift rating must receive and log flight training in the following maneuvers and procedures:

(1) Proper flight preparation procedures, including preflight planning and preparation, powerplant operation, and aircraft systems;

(2) Taxiing or surface operations, including runups;

(3) Takeoffs and landings, including normal and crosswind;

(4) Straight and level flight, and turns in both directions;

(5) Climbs and climbing turns;

(6) Airport traffic patterns, including entry and departure procedures;

(7) Collision avoidance, windshear avoidance, and wake turbulence avoidance;

(8) Descents with and without turns;

(9) Flight at various airspeeds from cruise to slow flight;

(10) Stall entries from various flight attitudes and power combinations with recovery initiated at the first indication of a stall, and recovery from a full stall;

(11) Emergency procedures and equipment malfunctions;

(12) Ground reference maneuvers;

(13) Approaches to a landing with simulated engine malfunctions;

(14) Go-arounds;

(15) Approaches to the landing area;

(16) Hovering and hovering turns; and

(17) For multiengine powered-lifts, simulated one-engine-inoperative approaches and landings.

(i) Maneuvers and procedures for pre-solo flight training in a glider. A student pilot who is receiving training for a glider rating or privileges must receive and log flight training for the following maneuvers and procedures:

(1) Proper flight preparation procedures, including preflight planning, preparation, aircraft systems, and, if appropriate, powerplant operations;

(2) Taxiing or surface operations, including runups, if applicable;

(3) Launches, including normal and crosswind;

(4) Straight and level flight, and turns in both directions, if applicable

(5) Airport traffic patterns, including entry procedures;

(6) Collision avoidance, windshear avoidance, and wake turbulence avoidance;

(7) Descents with and without turns using high and low drag configurations;

(8) Flight at various airspeeds;

(9) Emergency procedures and equipment malfunctions;

(10) Ground reference maneuvers, if applicable

(11) Inspection of towline rigging and review of signals and release procedures, if applicable

(12) Aerotow, ground tow, or self-launch procedures;

(13) Procedures for disassembly and assembly of the glider;

(14) Stall entry, stall, and stall recovery;

(15) Straight glides, turns, and spirals;

(16) Landings, including normal and crosswind;

(17) Slips to a landing;

(18) Procedures and techniques for thermalling; and

(19) Emergency operations, including towline break procedures.

(j) Maneuvers and procedures for pre-solo flight training in an airship. A student pilot who is receiving training for an airship rating or privileges must receive and log flight training for the following maneuvers and procedures:

(1) Proper flight preparation procedures, including preflight planning and preparation, powerplant operation, and aircraft systems;

(2) Taxiing or surface operations, including runups;

(3) Takeoffs and landings, including normal and crosswind;

(4) Straight and level flight, and turns in both directions;

(5) Climbs and climbing turns;

(6) Airport traffic patterns, including entry and departure procedures;

(7) Collision avoidance, windshear avoidance, and wake turbulence avoidance;

(8) Descents with and without turns;

(9) Flight at various airspeeds from cruise to slow flight;

(10) Emergency procedures and equipment malfunctions;

(11) Ground reference maneuvers;

(12) Rigging, ballasting, and controlling pressure in the ballonets, and superheating; and

(13) Landings with positive and with negative static trim.

(k) Maneuvers and procedures for pre-solo flight training in a balloon. A student pilot who is receiving training in a balloon must receive and log flight training for the following maneuvers and procedures:

(1) Layout and assembly procedures;

(2) Proper flight preparation procedures, including preflight planning and preparation, and aircraft systems;

(3) Ascents and descents;

(4) Landing and recovery procedures;

(5) Emergency procedures and equipment malfunctions;

(6) Operation of hot air or gas source, ballast, valves, vents, and rip panels, as appropriate;

(7) Use of deflation valves or rip panels for simulating an emergency;

(8) The effects of wind on climb and approach angles; and

(9) Obstruction detection and avoidance techniques.

(l) Maneuvers and procedures for pre-solo flight training in a powered parachute. A student pilot who is receiving training for a powered parachute rating or privileges must receive and log flight training for the following maneuvers and procedures:

(1) Proper flight preparation procedures, including pre-flight planning and preparation, preflight assembly and rigging, aircraft systems, and powerplant operations.

(2) Taxiing or surface operations, including run-ups.

(3) Takeoffs and landings, including normal and cross-wind.

(4) Straight and level flight, and turns in both directions.

(5) Climbs, and climbing turns in both directions.

(6) Airport traffic patterns, including entry and departure procedures.

(7) Collision avoidance, windshear avoidance, and wake turbulence avoidance.

(8) Descents, and descending turns in both directions.

(9) Emergency procedures and equipment malfunctions.

(10) Ground reference maneuvers.

(11) Straight glides, and gliding turns in both directions.

(12) Go-arounds.

(13) Approaches to landing areas with a simulated engine malfunction.

(14) Procedures for canopy packing and aircraft disassembly.

(m) Maneuvers and procedures for pre-solo flight training in a weight-shift-control aircraft. A student pilot who is receiving training for a weight-shift-control aircraft rating or privileges must receive and log flight training for the following maneuvers and procedures:

(1) Proper flight preparation procedures, including pre-flight planning and preparation, preflight assembly and rigging, aircraft systems, and powerplant operations.

(2) Taxiing or surface operations, including run-ups.

(3) Takeoffs and landings, including normal and cross-wind.

(4) Straight and level flight, and turns in both directions.

(5) Climbs, and climbing turns in both directions.

(6) Airport traffic patterns, including entry and departure procedures.

(7) Collision avoidance, windshear avoidance, and wake turbulence avoidance.

(8) Descents, and descending turns in both directions.

(9) Flight at various airspeeds from maximum cruise to slow flight.

(10) Emergency procedures and equipment malfunctions.

(11) Ground reference maneuvers.

(12) Stall entry, stall, and stall recovery.

(13) Straight glides, and gliding turns in both directions.

(14) Go-arounds.

(15) Approaches to landing areas with a simulated engine malfunction. (16) Procedures for disassembly.

(n) Limitations on student pilots operating an aircraft in solo flight. A student pilot may not operate an aircraft in solo flight unless that student pilot has received:

(1) An endorsement from an authorized instructor on his or her student pilot certificate for the specific make and model aircraft to be flown; and

(2) An endorsement in the student's logbook for the specific make and model aircraft to be flown by an authorized instructor, who gave the training within the 90 days preceding the date of the flight.

(o) Limitations on student pilots operating an aircraft in solo flight at night. A student pilot may not operate an aircraft in solo flight at night unless that student pilot has received:

(1) Flight training at night on night flying procedures that includes takeoffs, approaches, landings, and go-arounds at night at the airport where the solo flight will be conducted;

(2) Navigation training at night in the vicinity of the airport where the solo flight will be conducted; and

(3) An endorsement in the student's logbook for the specific make and model aircraft to be flown for night solo flight by an authorized instructor who gave the training within the 90-day period preceding the date of the flight.

(p) Limitations on flight instructors authorizing solo flight.

(1) No instructor may authorize a student pilot to perform a solo flight unless that instructor has—

(i) Given that student pilot training in the make and model of aircraft or a similar make and model of aircraft in which the solo flight is to be flown;

(ii) Determined the student pilot is proficient in the maneuvers and procedures prescribed in this section;

(iii) Determined the student pilot is proficient in the make and model of aircraft to be flown;

(iv) Ensured that the student pilot's certificate has been endorsed by an instructor authorized to provide flight training for the specific make and model aircraft to be flown; and

(v) Endorsed the student pilot's logbook for the specific make and model aircraft to be flown, and that endorsement remains current for solo flight privileges, provided an authorized instructor updates the student's logbook every 90 days thereafter.

(2) The flight training required by this section must be given by an instructor authorized to provide flight training who is appropriately rated and current.

[Amdt. 61-102, 62 FR 16298, Apr. 4, 1997; Amdt. 61-103, 62 FR 40902, July 30, 1997]

§ 61.89 General limitations.

(a) A student pilot may not act as pilot in command of an aircraft:

(1) That is carrying a passenger;

(2) That is carrying property for compensation or hire;

(3) For compensation or hire;

(4) In furtherance of a business;

(5) On an international flight, except that a student pilot may make solo training flights from Haines, Gustavus, or Juneau, Alaska, to White Horse, Yukon, Canada, and return over the province of British Columbia;

(6) With a flight or surface visibility of less than 3 statute miles during daylight hours or 5 statute miles at night;

(7) When the flight cannot be made with visual reference to the surface; or

(8) In a manner contrary to any limitations placed in the pilot's logbook by an authorized instructor.

(b) A student pilot may not act as a required pilot flight crewmember on any aircraft for which more than one pilot is required by the type certificate of the aircraft or regulations under which the flight is conducted, except when receiving flight training from an authorized instructor on board an airship, and no person other than a required flight crewmember is carried on the aircraft.

(c) A student pilot seeking a sport pilot certificate must comply with the provisions of paragraphs (a) and (b) of this section and may not act as pilot in command—

(1) Of an aircraft other than a light-sport aircraft;

(2) At night;

(3) At an altitude of more than 10,000 feet MSL; and

(4) In Class B, C, and D airspace, at an airport located in Class B, C, or D airspace, and to, from, through, or on an airport having an operational control tower without having received the ground and flight training specified in §61.94 and an endorsement from an authorized instructor.

§ 61.91 [Reserved]

§ 61.93 Solo cross-country flight requirements.

(a) General.

(1) Except as provided in paragraph (b) of this section, a student pilot must meet the requirements of this section before—

(i) Conducting a solo cross-country flight, or any flight greater than 25 nautical miles from the airport from where the flight originated.

(ii) Making a solo flight and landing at any location other than the airport of origination.

(2) Except as provided in paragraph (b) of this section, a student pilot who seeks solo cross-country flight privileges must:

(i) Have received flight training from an instructor authorized to provide flight training on the maneuvers and procedures of this section that are appropriate to the make and model of aircraft for which solo cross-country privileges are sought;

(ii) Have demonstrated cross-country proficiency on the appropriate maneuvers and procedures of this section to an authorized instructor;

(iii) Have satisfactorily accomplished the pre-solo flight maneuvers and procedures required by § 61.87 of this part in the make and model of aircraft or similar make and model of aircraft for which solo cross-country privileges are sought; and

(iv) Comply with any limitations included in the authorized instructor's endorsement that are required by paragraph (c) of this section.

(3) A student pilot who seeks solo cross-country flight privileges must have received ground and flight training from an authorized instructor on the cross-country maneuvers and procedures listed in this section that are appropriate to the aircraft to be flown.

(b) Authorization to perform certain solo flights and cross-country flights. A student pilot must obtain an endorsement from an authorized instructor to make solo flights from the airport where the student pilot normally receives training to another location. A student pilot who receives this endorsement must comply with the requirements of this paragraph.

(1) Solo flights may be made to another airport that is within 25 nautical miles from the airport where the student pilot normally receives training, provided—

(i) An authorized instructor has given the student pilot flight training at the other airport, and that training includes flight in both directions over the route, entering and exiting the traffic pattern, and takeoffs and landings at the other airport;

(ii) The authorized instructor who gave the training endorses the student pilot's logbook authorizing the flight;

(iii) The student pilot has current solo flight endorsements in accordance with § 61.87 of this part;

(iv) The authorized instructor has determined that the student pilot is proficient to make the flight; and

(v) The purpose of the flight is to practice takeoffs and landings at that other airport.

(2) Repeated specific solo cross-country flights may be made to another airport that is within 50 nautical miles of the airport from which the flight originated, provided—

(i) The authorized instructor has given the student flight training in both directions over the route, including entering and exiting the traffic patterns, takeoffs, and landings at the airports to be used;

(ii) The authorized instructor who gave the training has endorsed the student's logbook certifying that the student is proficient to make such flights;

(iii) The student has current solo flight endorsements in accordance with § 61.87 of this part; and

(iv) The student has current solo cross-country flight endorsements in accordance with paragraph (c) of this section; however, for repeated solo cross-country flights to another airport within 50 nautical miles from which the flight originated, separate endorsements are not required to be made for each flight.

(c) Endorsements for solo cross-country flights. Except as specified in paragraph (b)(2) of this section, a student pilot must have the endorsements prescribed in this paragraph for each cross-country flight:

(1) Student pilot certificate endorsement. A student pilot must have a solo cross-country endorsement from the authorized instructor who conducted the training, and that endorsement must be placed on that person's student pilot certificate for the specific category of aircraft to be flown.

(2) Logbook endorsement. (i) A student pilot must have a solo cross-country endorsement from an authorized instructor that is placed in the student pilot's logbook for the specific make and model of aircraft to be flown.

(ii) For each cross-country flight, the authorized instructor who reviews the cross-country planning must make an endorsement in the person's logbook after reviewing that person's cross-country planning, as specified in paragraph (d) of this section. The endorsement must—

(A) Specify the make and model of aircraft to be flown;

(B) State that the student's preflight planning and preparation is correct and that the student is prepared to make the flight safely under the known conditions; and

(C) State that any limitations required by the student's authorized instructor are met.

(d) Limitations on authorized instructors to permit solo cross-country flights. An authorized instructor may not permit a student pilot to conduct a solo cross-country flight unless that instructor has:

(1) Determined that the student's cross-country planning is correct for the flight;

(2) Reviewed the current and forecast weather conditions and has determined that the flight can be completed under VFR;

(3) Determined that the student is proficient to conduct the flight safely;

(4) Determined that the student has the appropriate solo cross-country endorsement for the make and model of aircraft to be flown; and

(5) Determined that the student's solo flight endorsement is current for the make and model aircraft to be flown.

(e) Maneuvers and procedures for cross-country flight training in a single-engine airplane. A student pilot who is receiving training for cross-country flight in a single-engine airplane must receive and log flight training in the following maneuvers and procedures:

(1) Use of aeronautical charts for VFR navigation using pilotage and dead reckoning with the aid of a magnetic compass;

(2) Use of aircraft performance charts pertaining to cross-country flight;

(3) Procurement and analysis of aeronautical weather reports and forecasts, including recognition of critical weather situations and estimating visibility while in flight;

(4) Emergency procedures;

(5) Traffic pattern procedures that include area departure, area arrival, entry into the traffic pattern, and approach;

(6) Procedures and operating practices for collision avoidance, wake turbulence precautions, and windshear avoidance;

(7) Recognition, avoidance, and operational restrictions of hazardous terrain features in the geographical area where the cross-country flight will be flown;

(8) Procedures for operating the instruments and equipment installed in the aircraft to be flown, including recognition and use of the proper operational procedures and indications;

(9) Use of radios for VFR navigation and two-way communications;

(10) Takeoff, approach, and landing procedures, including short-field, soft-field, and crosswind takeoffs, approaches, and landings;

(11) Climbs at best angle and best rate; and

(12) Control and maneuvering solely by reference to flight instruments, including straight and level flight, turns, descents, climbs, use of radio aids, and ATC directives.

(f) Maneuvers and procedures for cross-country flight training in a multiengine airplane. A student pilot who is receiving training for cross-country flight in a multiengine airplane must receive and log flight training in the following maneuvers and procedures:

(1) Use of aeronautical charts for VFR navigation using pilotage and dead reckoning with the aid of a magnetic compass;

(2) Use of aircraft performance charts pertaining to cross-country flight;

(3) Procurement and analysis of aeronautical weather reports and forecasts, including recognition of critical weather situations and estimating visibility while in flight;

(4) Emergency procedures;

(5) Traffic pattern procedures that include area departure, area arrival, entry into the traffic pattern, and approach;

(6) Procedures and operating practices for collision avoidance, wake turbulence precautions, and windshear avoidance;

(7) Recognition, avoidance, and operational restrictions of hazardous terrain features in the geographical area where the cross-country flight will be flown;

(8) Procedures for operating the instruments and equipment installed in the aircraft to be flown, including recognition and use of the proper operational procedures and indications;

(9) Use of radios for VFR navigation and two-way communications;

(10) Takeoff, approach, and landing procedures, including short-field, soft-field, and crosswind takeoffs, approaches, and landings;

(11) Climbs at best angle and best rate; and

(12) Control and maneuvering solely by reference to flight instruments, including straight and level flight, turns, descents, climbs, use of radio aids, and ATC directives.

(g) *Maneuvers and procedures for cross-country flight training in a helicopter.* A student pilot who is receiving training for cross-country flight in a helicopter must receive and log flight training for the following maneuvers and procedures:

(1) Use of aeronautical charts for VFR navigation using pilotage and dead reckoning with the aid of a magnetic compass;

(2) Use of aircraft performance charts pertaining to cross-country flight;

(3) Procurement and analysis of aeronautical weather reports and forecasts, including recognition of critical weather situations and estimating visibility while in flight;

(4) Emergency procedures;

(5) Traffic pattern procedures that include area departure, area arrival, entry into the traffic pattern, and approach;

(6) Procedures and operating practices for collision avoidance, wake turbulence precautions, and windshear avoidance;

(7) Recognition, avoidance, and operational restrictions of hazardous terrain features in the geographical area where the cross-country flight will be flown;

(8) Procedures for operating the instruments and equipment installed in the aircraft to be flown, including recognition and use of the proper operational procedures and indications;

(9) Use of radios for VFR navigation and two-way communications; and

(10) Takeoff, approach, and landing procedures.

(h) *Maneuvers and procedures for cross-country flight training in a gyroplane.* A student pilot who is receiving training for cross-country flight in a gyroplane must receive and log flight training in the following maneuvers and procedures:

(1) Use of aeronautical charts for VFR navigation using pilotage and dead reckoning with the aid of a magnetic compass;

(2) Use of aircraft performance charts pertaining to cross-country flight;

(3) Procurement and analysis of aeronautical weather reports and forecasts, including recognition of critical weather situations and estimating visibility while in flight;

(4) Emergency procedures;

(5) Traffic pattern procedures that include area departure, area arrival, entry into the traffic pattern, and approach;

(6) Procedures and operating practices for collision avoidance, wake turbulence precautions, and windshear avoidance;

(7) Recognition, avoidance, and operational restrictions of hazardous terrain features in the geographical area where the cross-country flight will be flown;

(8) Procedures for operating the instruments and equipment installed in the aircraft to be flown, including recog-

nition and use of the proper operational procedures and indications;

(9) Use of radios for VFR navigation and two-way communications; and

(10) Takeoff, approach, and landing procedures, including short-field and soft-field takeoffs, approaches, and landings.

(i) *Maneuvers and procedures for cross-country flight training in a powered-lift.* A student pilot who is receiving training for cross-country flight training in a powered-lift must receive and log flight training in the following maneuvers and procedures:

(1) Use of aeronautical charts for VFR navigation using pilotage and dead reckoning with the aid of a magnetic compass;

(2) Use of aircraft performance charts pertaining to cross-country flight;

(3) Procurement and analysis of aeronautical weather reports and forecasts, including recognition of critical weather situations and estimating visibility while in flight;

(4) Emergency procedures;

(5) Traffic pattern procedures that include area departure, area arrival, entry into the traffic pattern, and approach;

(6) Procedures and operating practices for collision avoidance, wake turbulence precautions, and windshear avoidance;

(7) Recognition, avoidance, and operational restrictions of hazardous terrain features in the geographical area where the cross-country flight will be flown;

(8) Procedures for operating the instruments and equipment installed in the aircraft to be flown, including recognition and use of the proper operational procedures and indications;

(9) Use of radios for VFR navigation and two-way communications;

(10) Takeoff, approach, and landing procedures that include high-altitude, steep, and shallow takeoffs, approaches, and landings; and

(11) Control and maneuvering solely by reference to flight instruments, including straight and level flight, turns, descents, climbs, use of radio aids, and ATC directives.

(j) *Maneuvers and procedures for cross-country flight training in a glider.* A student pilot who is receiving training for cross-country flight in a glider must receive and log flight training in the following maneuvers and procedures:

(1) Use of aeronautical charts for VFR navigation using pilotage and dead reckoning with the aid of a magnetic compass;

(2) Use of aircraft performance charts pertaining to cross-country flight;

(3) Procurement and analysis of aeronautical weather reports and forecasts, including recognition of critical weather situations and estimating visibility while in flight;

(4) Emergency procedures;

(5) Traffic pattern procedures that include area departure, area arrival, entry into the traffic pattern, and approach;

(6) Procedures and operating practices for collision avoidance, wake turbulence precautions, and windshear avoidance;

(7) Recognition, avoidance, and operational restrictions of hazardous terrain features in the geographical area where the cross-country flight will be flown;

(8) Procedures for operating the instruments and equipment installed in the aircraft to be flown, including recognition and use of the proper operational procedures and indications;

(9) Landings accomplished without the use of the altimeter from at least 2,000 feet above the surface; and

(10) Recognition of weather and upper air conditions favorable for cross-country soaring, ascending and descending flight, and altitude control.

(k) Maneuvers and procedures for cross-country flight training in an airship. A student pilot who is receiving training for cross-country flight in an airship must receive and log flight training for the following maneuvers and procedures:

(1) Use of aeronautical charts for VFR navigation using pilotage and dead reckoning with the aid of a magnetic compass;

(2) Use of aircraft performance charts pertaining to cross-country flight;

(3) Procurement and analysis of aeronautical weather reports and forecasts, including recognition of critical weather situations and estimating visibility while in flight;

(4) Emergency procedures;

(5) Traffic pattern procedures that include area departure, area arrival, entry into the traffic pattern, and approach;

(6) Procedures and operating practices for collision avoidance, wake turbulence precautions, and windshear avoidance;

(7) Recognition, avoidance, and operational restrictions of hazardous terrain features in the geographical area where the cross-country flight will be flown;

(8) Procedures for operating the instruments and equipment installed in the aircraft to be flown, including recognition and use of the proper operational procedures and indications;

(9) Use of radios for VFR navigation and two-way communications;

(10) Control of air pressure with regard to ascending and descending flight and altitude control;

(11) Control of the airship solely by reference to flight instruments; and

(12) Recognition of weather and upper air conditions conducive for the direction of cross-country flight.

(l) Maneuvers and procedures for cross-country flight training in a powered parachute. A student pilot who is receiving training for cross-country flight in a powered parachute must receive and log flight training in the following maneuvers and procedures:

(1) Use of aeronautical charts for VFR navigation using pilotage and dead reckoning with the aid of a magnetic compass, as appropriate.

(2) Use of aircraft performance charts pertaining to cross-country flight.

(3) Procurement and analysis of aeronautical weather reports and forecasts, including recognizing critical weather situations and estimating visibility while in flight.

(4) Emergency procedures.

(5) Traffic pattern procedures that include area departure, area arrival, entry into the traffic pattern, and approach.

(6) Procedures and operating practices for collision avoidance, wake turbulence precautions. and windshear avoidance.

(7) Recognition, avoidance, and operational restrictions of hazardous terrain features in the geographical area where the cross-country flight will be flown.

(8) Procedures for operating the instruments and equipment installed in the aircraft to be flown, including recognition and use of the proper operational procedures and indications.

(9) If equipped for flight with navigation radios, the use of radios for VFR navigation.

(10) Recognition of weather and upper air conditions favorable for the cross-country flight.

(11) Takeoff, approach and landing procedures.

(m) Maneuvers and procedures for cross-country flight training in a weight-shift-control aircraft. A student pilot who is receiving training for cross-country flight in a weight-shift-control aircraft must receive and log flight training for the following maneuvers and procedures:

(1) Use of aeronautical charts for VFR navigation using pilotage and dead reckoning with the aid of a magnetic compass, as appropriate.

(2) Use of aircraft performance charts pertaining to cross-country flight.

(3) Procurement and analysis of aeronautical weather reports and forecasts, including recognizing critical weather situations and estimating visibility while in flight.

(4) Emergency procedures.

(5) Traffic pattern procedures that include area departure, area arrival, entry into the traffic pattern, and approach.

(6) Procedures and operating practices for collision avoidance, wake turbulence precautions, and windshear avoidance.

(7) Recognition, avoidance, and operational restrictions of hazardous terrain features in the geographical area where the cross-country flight will be flown.

(8) Procedures for operating the instruments and equipment installed in the aircraft to be flown, including recognition and use of the proper operational procedures and indications.

(9) If equipped for flight using navigation radios, the use of radios for VFR navigation.

(10) Recognition of weather and upper air conditions favorable for the cross-country flight.

(11) Takeoff, approach and landing procedures, including crosswind approaches and landings.

[Amdt. 61-102, 62 FR 16298, Apr. 4, 1997; Amdt. 61-103, 62 FR 40902, July 30, 1997]

§ 61.94 Student pilot seeking a sport pilot certificate or a recreational pilot certificate: Operations at airports within, and in airspace located within, Class B, C, and D airspace, or at airports with an operational control tower in other airspace.

(a) A student pilot seeking a sport pilot certificate or a recreational pilot certificate who wants to obtain privileges to operate in Class B, C, and D airspace, at an airport located in Class B, C, or D airspace, and to, from, through, or at an airport having an operational control tower, must receive and log ground and flight training from an authorized instructor in the following aeronautical knowledge areas and areas of operation:

(1) The use of radios, communications, navigation systems and facilities, and radar services.

(2) Operations at airports with an operating control tower, to include three takeoffs and landings to a full stop, with each landing involving a flight in the traffic pattern, at an airport with an operating control tower.

(3) Applicable flight rules of part 91 of this chapter for operations in Class B, C, and D airspace and air traffic control clearances.

(4) Ground and flight training for the specific Class B, C, or D airspace for which the solo flight is authorized, if applicable, within the 90-day period preceding the date of the flight in that airspace. The flight training must be received in the specific airspace area for which solo flight is authorized.

(5) Ground and flight training for the specific airport located in Class B, C, or D airspace for which the solo flight is authorized, if applicable, within the 90-day period preceding the date of the flight at that airport. The flight and ground training must be received at the specific airport for which solo flight is authorized.

(b) The authorized instructor who provides the training specified in paragraph (a) of this section must provide a logbook endorsement that certifies the student has received that training and is proficient to conduct solo flight in that specific airspace or at that specific airport and in those aeronautical knowledge areas and areas of operation specified in this section.

§ 61.95 Operations in Class B airspace and at airports located within Class B airspace.

(a) A student pilot may not operate an aircraft on a solo flight in Class B airspace unless:

(1) The student pilot has received both ground and flight training from an authorized instructor on that Class B airspace area, and the flight training was received in the specific Class B airspace area for which solo flight is authorized;

(2) The logbook of that student pilot has been endorsed by the authorized instructor who gave the student pilot flight training, and the endorsement is dated within the 90-day period preceding the date of the flight in that Class B airspace area; and

(3) The logbook endorsement specifies that the student pilot has received the required ground and flight training, and has been found proficient to conduct solo flight in that specific Class B airspace area.

(b) A student pilot may not operate an aircraft on a solo flight to, from, or at an airport located within Class B airspace pursuant to § 91.131(b) of this chapter unless:

(1) The student pilot has received both ground and flight training from an instructor authorized to provide training to operate at that airport, and the flight and ground training has been received at the specific airport for which the solo flight is authorized;

(2) The logbook of that student pilot has been endorsed by an authorized instructor who gave the student pilot flight training, and the endorsement is dated within the 90-day period preceding the date of the flight at that airport; and

(3) The logbook endorsement specifies that the student pilot has received the required ground and flight training, and has been found proficient to conduct solo flight operations at that specific airport.

(c) This section does not apply to a student pilot seeking a sport pilot certificate or a recreational pilot certificate.

[Amdt. 61-102, 62 FR 16298, Apr. 4, 1997; Amdt. 61-103, 62 FR 40902, July 30, 1997]

Subpart D—Recreational Pilots

§ 61.96 Applicability and eligibility requirements: General.

(a) This subpart prescribes the requirement for the issuance of recreational pilot certificates and ratings, the conditions under which those certificates and ratings are necessary, and the general operating rules for persons who hold those certificates and ratings.

(b) To be eligible for a recreational pilot certificate, a person who applies for that certificate must:

(1) Be at least 17 years of age;

(2) Be able to read, speak, write, and understand the English language. If the applicant is unable to meet one of these requirements due to medical reasons, then the Administrator may place such operating limitations on that applicant's pilot certificate as are necessary for the safe operation of the aircraft;

(3) Receive a logbook endorsement from an authorized instructor who—

(i) Conducted the training or reviewed the applicant's home study on the aeronautical knowledge areas listed in

§ 61.97(b) of this part that apply to the aircraft category and class rating sought; and

(ii) Certified that the applicant is prepared for the required knowledge test.

(4) Pass the required knowledge test on the aeronautical knowledge areas listed in § 61.97(b) of this part;

(5) Receive flight training and a logbook endorsement from an authorized instructor who—

(i) Conducted the training on the areas of operation listed in § 61.98(b) of this part that apply to the aircraft category and class rating sought; and

(ii) Certified that the applicant is prepared for the required practical test.

(6) Meet the aeronautical experience requirements of § 61.99 of this part that apply to the aircraft category and class rating sought before applying for the practical test;

(7) Pass the required practical test on the areas of operation listed in § 61.98(b) of this part that apply to the aircraft category and class rating sought; and

(8) Comply with the sections of this part that apply to the aircraft category and class rating sought.

[Amdt. 61-102, 62 FR 16298, Apr. 4, 1997; Amdt. 61-103, 62 FR 40902, July 30, 1997]

§ 61.97 Aeronautical knowledge.

(a) General. A person who applies for a recreational pilot certificate must receive and log ground training from an authorized instructor or complete a home-study course on the aeronautical knowledge areas of paragraph (b) of this section that apply to the aircraft category and class rating sought.

(b) Aeronautical knowledge areas.

(1) Applicable Federal Aviation Regulations of this chapter that relate to recreational pilot privileges, limitations, and flight operations;

(2) Accident reporting requirements of the National Transportation Safety Board;

(3) Use of the applicable portions of the "Aeronautical Information Manual" and FAA advisory circulars;

(4) Use of aeronautical charts for VFR navigation using pilotage with the aid of a magnetic compass;

(5) Recognition of critical weather situations from the ground and in flight, windshear avoidance, and the procurement and use of aeronautical weather reports and forecasts;

(6) Safe and efficient operation of aircraft, including collision avoidance, and recognition and avoidance of wake turbulence;

(7) Effects of density altitude on takeoff and climb performance;

(8) Weight and balance computations;

(9) Principles of aerodynamics, powerplants, and aircraft systems;

(10) Stall awareness, spin entry, spins, and spin recovery techniques, if applying for an airplane single-engine rating;

(11) Aeronautical decision making and judgment; and

(12) Preflight action that includes—

(i) How to obtain information on runway lengths at airports of intended use, data on takeoff and landing distances, weather reports and forecasts, and fuel requirements; and

(ii) How to plan for alternatives if the planned flight cannot be completed or delays are encountered.

[Amdt. 61-102, 62 FR 16298, Apr. 4, 1997; Amdt. 61-103, 62 FR 40902, July 30, 1997]

§ 61.98 Flight proficiency.

(a) General. A person who applies for a recreational pilot certificate must receive and log ground and flight training from an authorized instructor on the areas of operation of this section that apply to the aircraft category and class rating sought.

(b) Areas of operation.

(1) For a single-engine airplane rating:

(i) Preflight preparation;

(ii) Preflight procedures;

(iii) Airport operations;

(iv) Takeoffs, landings, and go-arounds;

(v) Performance maneuvers;

(vi) Ground reference maneuvers;

(vii) Navigation;

(viii) Slow flight and stalls;

(ix) Emergency operations; and

(x) Postflight procedures.

(2) For a helicopter rating: (i) Preflight preparation;

(ii) Preflight procedures;

(iii) Airport and heliport operations;

(iv) Hovering maneuvers;

(v) Takeoffs, landings, and go-arounds;

(vi) Performance maneuvers;

(vii) Ground reference maneuvers;

(viii) Navigation;

(ix) Emergency operations; and

(x) Postflight procedures.

(3) For a gyroplane rating:

(i) Preflight preparation;

(ii) Preflight procedures;

(iii) Airport operations;

(iv) Takeoffs, landings, and go-arounds;

(v) Performance maneuvers;

(vi) Ground reference maneuvers;

(vii) Navigation;

(viii) Flight at slow airspeeds;

(ix) Emergency operations; and

(x) Postflight procedures.

[Amdt. 61-102, 62 FR 16298, Apr. 4, 1997; Amdt. 61-103, 62 FR 40902, July 30, 1997]

§ 61.99 Aeronautical experience.

A person who applies for a recreational pilot certificate must receive and log at least 30 hours of flight time that includes at least:

(a) 15 hours of flight training from an authorized instructor on the areas of operation listed in § 61.98 of this part that consists of at least:

(1) Except as provided in § 61.100 of this part, 2 hours of flight training en route to an airport that is located more than 25 nautical miles from the airport where the applicant normally trains, which includes at least three takeoffs and three landings at the airport located more than 25 nautical miles from the airport where the applicant normally trains; and

(2) 3 hours of flight training in the aircraft for the rating sought in preparation for the practical test within the 60 days preceding the date of the practical test.

(b) 3 hours of solo flying in the aircraft for the rating sought, on the areas of operation listed in § 61.98 of this part that apply to the aircraft category and class rating sought.

§ 61.100 Pilots based on small islands.

(a) An applicant located on an island from which the flight training required in § 61.99(a)(1) of this part cannot be accomplished without flying over water for more than 10 nautical miles from the nearest shoreline need not comply with the requirements of that section. However, if other airports that permit civil operations are available to which a flight may be made without flying over water for more than 10 nautical miles from the nearest shoreline, the applicant must show completion of a dual flight between two airports, which must include three landings at the other airport.

(b) An applicant who complies with paragraph (a) of this section and meets all requirements for the issuance of a recreational pilot certificate, except the requirements of § 61.99(a)(1) of this part, will be issued a pilot certificate with an endorsement containing the following limitation, "Passenger carrying prohibited on flights more than 10 nautical miles from (the appropriate island)." The limitation may be subsequently amended to include another island if the applicant complies with the requirements of paragraph (a) of this section for another island.

(c) Upon meeting the requirements of § 61.99(a)(1) of this part, the applicant may have the limitation(s) in paragraph (b) of this section removed.

§ 61.101 Recreational pilot privileges and limitations.

(a) A person who holds a recreational pilot certificate may:

(1) Carry no more than one passenger; and

(2) Not pay less than the pro rata share of the operating expenses of a flight with a passenger, provided the expenses involve only fuel, oil, airport expenses, or aircraft rental fees.

(b) A person who holds a current and valid recreational pilot certificate may act as pilot in command of an aircraft on a flight that is within 50 nautical miles from the departure airport, provided that person has:

(1) Received ground and flight training for takeoff, departure, arrival, and landing procedures at the departure airport;

(2) Received ground and flight training for the area, terrain, and aids to navigation that are in the vicinity of the departure airport;

(3) Been found proficient to operate the aircraft at the departure airport and the area within 50 nautical miles from that airport; and

(4) Received from an authorized instructor a logbook endorsement, which is carried in the person's possession in the aircraft, that permits flight within 50 nautical miles from the departure airport.

(c) A person who holds a current and valid recreational pilot certificate may act as pilot in command of an aircraft on a flight that exceeds 50 nautical miles from the departure airport, provided that person has:

(1) Received ground and flight training from an authorized instructor on the cross-country training requirements of subpart E of this part that apply to the aircraft rating held;

(2) Been found proficient in cross-country flying; and

(3) Received from an authorized instructor a logbook endorsement, which is carried on the person's possession in the aircraft, that certifies the person has received and been found proficient in the cross-country training requirements of subpart E of this part that apply to the aircraft rating held.

(d) A person who holds a current and valid recreational pilot certificate may act as pilot in command of an aircraft in Class B, C, and D airspace, at an airport located in Class B, C, or D airspace, and to, from, through, or at an airport having an operational control tower, provided that person has—

(1) Received and logged ground and flight training from an authorized instructor on the following aeronautical knowledge areas and areas of operation, as appropriate to the aircraft rating held:

(i) The use of radios, communications, navigation system and facilities, and radar services.

(ii) Operations at airports with an operating control tower to include three takeoffs and landings to a full stop, with each landing involving a flight in the traffic pattern at an airport with an operating control tower.

(iii) Applicable flight rules of part 91 of this chapter for operations in Class B, C, and D airspace and air traffic control clearances;

(2) Been found proficient in those aeronautical knowledge areas and areas of operation specified in paragraph (d)(1) of this section; and

(3) Received from an authorized instructor a logbook endorsement, which is carried on the person's possession or readily accessible in the aircraft, that certifies the person has received and been found proficient in those aeronautical knowledge areas and areas of operation specified in paragraph (d)(1) of this section.

(e) Except as provided in paragraphs (d) and (i) of this section, a recreational pilot may not act as pilot in command of an aircraft

(1) That is certificated—

(i) For more than four occupants;

(ii) With more than one powerplant;

(iii) With a powerplant of more than 180 horsepower; or

(iv) With retractable landing gear;

(2) That is classified as a multiengine airplane, powered-lift, glider, airship, balloon, powered parachute, or weight-shift-control aircraft;

(3) That is carrying a passenger or property for compensation or hire;

(4) For compensation or hire;

(5) In furtherance of a business;

(6) Between sunset and sunrise;

(7) In Class A, B, C, and D airspace, at an airport located in Class B, C, or D airspace, or to, from, through, or at an airport having an operational control tower;

(8) At an altitude of more than 10,000 feet MSL or 2,000 feet AGL, whichever is higher;

(9) When the flight or surface visibility is less than 3 statute miles;

(10) Without visual reference to the surface;

(11) On a flight outside the United States, unless authorized by the country in which the flight is conducted;

(12) To demonstrate that aircraft in flight as an aircraft salesperson to a prospective buyer;

(13) That is used in a passenger-carrying airlift and sponsored by a charitable organization; and

(14) That is towing any object.

(e) A recreational pilot may not act as a pilot flight crewmember on any aircraft for which more than one pilot is required by the type certificate of the aircraft or the regulations under which the flight is conducted, except when:

(1) Receiving flight training from a person authorized to provide flight training on board an airship; and

(2) No person other than a required flight crewmember is carried on the aircraft.

(f) A person who holds a recreational pilot certificate, has logged fewer than 400 flight hours, and has not logged pilot-in-command time in an aircraft within the 180 days preceding the flight shall not act as pilot in command of an aircraft until the pilot receives flight training and a logbook endorsement from an authorized instructor, and the instructor certifies that the person is proficient to act as pilot in command of the aircraft. This requirement can be met in combination with the requirements of §§ 61.56 and 61.57 of this part, at the discretion of the authorized instructor.

(g) A recreational pilot certificate issued under this subpart carries the notation, "Holder does not meet ICAO requirements."

(h) For the purpose of obtaining additional certificates or ratings while under the supervision of an authorized instructor, a recreational pilot may fly as the sole occupant of an aircraft:

(1) For which the pilot does not hold an appropriate category or class rating;

(2) Within airspace that requires communication with air traffic control; or

(3) Between sunset and sunrise, provided the flight or surface visibility is at least 5 statute miles.

(i) In order to fly solo as provided in paragraph (h) of this section, the recreational pilot must meet the appropriate aeronautical knowledge and flight training requirements of § 61.87 for that aircraft. When operating an aircraft under the conditions specified in paragraph (h) of this section, the recreational pilot shall carry the logbook that has been endorsed for each flight by an authorized instructor who:

(1) Has given the recreational pilot training in the make and model of aircraft in which the solo flight is to be made;

(2) Has found that the recreational pilot has met the applicable requirements of § 61.87; and

(3) Has found that the recreational pilot is competent to make solo flights in accordance with the logbook endorsement.

Subpart E—Private Pilots

§ 61.102 Applicability.

This subpart prescribes the requirements for the issuance of private pilot certificates and ratings, the conditions under which those certificates and ratings are necessary, and the general operating rules for persons who hold those certificates and ratings.

§ 61.103 Eligibility requirements: General.

To be eligible for a private pilot certificate, a person must:

(a) Be at least 17 years of age for a rating in other than a glider or balloon.

(b) Be at least 16 years of age for a rating in a glider or balloon.

(c) Be able to read, speak, write, and understand the English language. If the applicant is unable to meet one of these requirements due to medical reasons, then the Administrator may place such operating limitations on that applicant's pilot certificate as are necessary for the safe operation of the aircraft.

(d) Receive a logbook endorsement from an authorized instructor who:

(1) Conducted the training or reviewed the person's home study on the aeronautical knowledge areas listed in § 61.105 (b) of this part that apply to the aircraft rating sought; and

(2) Certified that the person is prepared for the required knowledge test.

(e) Pass the required knowledge test on the aeronautical knowledge areas listed in § 61.105(b) of this part.

(f) Receive flight training and a logbook endorsement from an authorized instructor who:

(1) Conducted the training in the areas of operation listed in § 61.107(b) of this part that apply to the aircraft rating sought; and

(2) Certified that the person is prepared for the required practical test.

(g) Meet the aeronautical experience requirements of this part that apply to the aircraft rating sought before applying for the practical test.

(h) Pass a practical test on the areas of operation listed in § 61.107(b) of this part that apply to the aircraft rating sought.

(i) Comply with the appropriate sections of this part that apply to the aircraft category and class rating sought.

§ 61.105 Aeronautical knowledge.

(a) General. A person who is applying for a private pilot certificate must receive and log ground training from an authorized instructor or complete a home-study course on the aeronautical knowledge areas of paragraph (b) of this section that apply to the aircraft category and class rating sought.

(b) Aeronautical knowledge areas.

(1) Applicable Federal Aviation Regulations of this chapter that relate to private pilot privileges, limitations, and flight operations;

(2) Accident reporting requirements of the National Transportation Safety Board;

(3) Use of the applicable portions of the "Aeronautical Information Manual" and FAA advisory circulars;

(4) Use of aeronautical charts for VFR navigation using pilotage, dead reckoning, and navigation systems;

(5) Radio communication procedures;

(6) Recognition of critical weather situations from the ground and in flight, windshear avoidance, and the procurement and use of aeronautical weather reports and forecasts;

(7) Safe and efficient operation of aircraft, including collision avoidance, and recognition and avoidance of wake turbulence;

(8) Effects of density altitude on takeoff and climb performance;

(9) Weight and balance computations;

(10) Principles of aerodynamics, powerplants, and aircraft systems;

(11) Stall awareness, spin entry, spins, and spin recovery techniques for the airplane and glider category ratings;

(12) Aeronautical decision making and judgment; and

(13) Preflight action that includes—

(i) How to obtain information on runway lengths at airports of intended use, data on takeoff and landing distances, weather reports and forecasts, and fuel requirements; and

(ii) How to plan for alternatives if the planned flight cannot be completed or delays are encountered.

[Amdt. 61-102, 62 FR 16298, Apr. 4, 1997; Amdt. 61-103, 62 FR 40902, July 30, 1997]

§ 61.107 Flight proficiency.

(a) General. A person who applies for a private pilot certificate must receive and log ground and flight training from an authorized instructor on the areas of operation of this section that apply to the aircraft category and class rating sought.

(b) Areas of operation.

(1) For an airplane category rating with a single-engine class rating:

(i) Preflight preparation;

(ii) Preflight procedures;

(iii) Airport and seaplane base operations;

(iv) Takeoffs, landings, and go-arounds;

(v) Performance maneuvers;

(vi) Ground reference maneuvers;

(vii) Navigation;

(viii) Slow flight and stalls;

(ix) Basic instrument maneuvers;

(x) Emergency operations;

(xi) Night operations, except as provided in § 61.110 of this part; and

(xii) Postflight procedures.

(2) For an airplane category rating with a multiengine class rating:

(i) Preflight preparation;

(ii) Preflight procedures;

(iii) Airport and seaplane base operations;

(iv) Takeoffs, landings, and go-arounds;

(v) Performance maneuvers;

(vi) Ground reference maneuvers;

(vii) Navigation;

(viii) Slow flight and stalls;

(ix) Basic instrument maneuvers;

(x) Emergency operations;

(xi) Multiengine operations;

(xii) Night operations, except as provided in § 61.110 of this part; and

(xiii) Postflight procedures.

(3) For a rotorcraft category rating with a helicopter class rating:

(i) Preflight preparation;

(ii) Preflight procedures;

(iii) Airport and heliport operations;

(iv) Hovering maneuvers;

(v) Takeoffs, landings, and go-arounds;

(vi) Performance maneuvers;

(vii) Navigation;

(viii) Emergency operations;

(ix) Night operations, except as provided in § 61.110 of this part; and

(x) Postflight procedures.

(4) For a rotorcraft category rating with a gyroplane class rating:

(i) Preflight preparation;

(ii) Preflight procedures;

(iii) Airport operations;

(iv) Takeoffs, landings, and go-arounds;

(v) Performance maneuvers;

(vi) Ground reference maneuvers;

(vii) Navigation;

(viii) Flight at slow airspeeds;

(ix) Emergency operations;

(x) Night operations, except as provided in § 61.110 of this part; and

(xi) Postflight procedures.

(5) For a powered-lift category rating:

(i) Preflight preparation;

(ii) Preflight procedures;

(iii) Airport and heliport operations;

(iv) Hovering maneuvers;

(v) Takeoffs, landings, and go-arounds;

(vi) Performance maneuvers;

(vii) Ground reference maneuvers;

(viii) Navigation;

(ix) Slow flight and stalls;

(x) Basic instrument maneuvers;

(xi) Emergency operations;

(xii) Night operations, except as provided in § 61.110 of this part; and

(xiii) Postflight procedures.

(6) For a glider category rating:

(i) Preflight preparation;

(ii) Preflight procedures;

(iii) Airport and gliderport operations;

(iv) Launches and landings;

(v) Performance speeds;

(vi) Soaring techniques;

(vii) Performance maneuvers;

(viii) Navigation;

(ix) Slow flight and stalls;

(x) Emergency operations; and

(xi) Postflight procedures.

(7) For a lighter-than-air category rating with an airship class rating:

(i) Preflight preparation;

(ii) Preflight procedures;

(iii) Airport operations;

(iv) Takeoffs, landings, and go-arounds;

(v) Performance maneuvers;

(vi) Ground reference maneuvers;

(vii) Navigation;

(viii) Emergency operations; and

(ix) Postflight procedures.

(8) For a lighter-than-air category rating with a balloon class rating:

(i) Preflight preparation;

(ii) Preflight procedures;

(iii) Airport operations;

(iv) Launches and landings;

(v) Performance maneuvers;

(vi) Navigation;

(vii) Emergency operations; and

(viii) Postflight procedures.

(9) For a powered parachute category rating

(i) Preflight preparation;

(ii) Preflight procedures;

(iii) Airport and seaplane base operations, as applicable;

(iv) Takeoffs, landings, and go-arounds;

(v) Performance maneuvers;

(vi) Ground reference maneuvers;

(vii) Navigation;

(viii) Night operations, except as provided in §61.110;

(ix) Emergency operations; and

(x) Post-flight procedures.

(10) For a weight-shift-control aircraft category rating

(i) Preflight preparation;

(ii) Preflight procedures;

(iii) Airport and seaplane base operations, as applicable;

(iv) Takeoffs, landings, and go-arounds;

(v) Performance maneuvers;

(vi) Ground reference maneuvers;

(vii) Navigation;

(viii) Slow flight and stalls;

(ix) Night operations, except as provided in §61.110;

(x) Emergency operations; and

(xi) Post-flight procedures.

§ 61.109 Aeronautical experience.

(a) *For an airplane single-engine rating.* Except as provided in paragraph (k) of this section, a person who applies for a private pilot certificate with an airplane category and single-engine class rating must log at least 40 hours of flight time that includes at least 20 hours of flight training from an authorized instructor and 10 hours of solo flight training in the areas of operation listed in § 61.107(b)(1) of this part, and the training must include at least—

(1) 3 hours of cross-country flight training in a single-engine airplane;

(2) Except as provided in § 61.110 of this part, 3 hours of night flight training in a single-engine airplane that includes—

(i) One cross-country flight of over 100 nautical miles total distance; and

(ii) 10 takeoffs and 10 landings to a full stop (with each landing involving a flight in the traffic pattern) at an airport.

(3) 3 hours of flight training in a single-engine airplane on the control and maneuvering of an airplane solely by reference to instruments, including straight and level flight, constant airspeed climbs and descents, turns to a heading, recovery from unusual flight attitudes, radio communications, and the use of navigation systems/facilities and radar services appropriate to instrument flight;

(4) 3 hours of flight training in preparation for the practical test in a single-engine airplane, which must have been performed within 60 days preceding the date of the test; and

(5) 10 hours of solo flight time in a single-engine airplane, consisting of at least—

(i) 5 hours of solo cross-country time;

(ii) One solo cross-country flight of at least 150 nautical miles total distance, with full-stop landings at a minimum of three points, and one segment of the flight consisting of a straight-line distance of at least 50 nautical miles between the takeoff and landing locations; and

(iii) Three takeoffs and three landings to a full stop (with each landing involving a flight in the traffic pattern) at an airport with an operating control tower.

(b) For an airplane multiengine rating. Except as provided in paragraph (k) of this section, a person who applies for a private pilot certificate with an airplane category and multiengine class rating must log at least 40 hours of flight time that includes at least 20 hours of flight training from an authorized instructor and 10 hours of solo flight training in the areas of operation listed in § 61.107(b)(2) of this part, and the training must include at least—

(1) 3 hours of cross-country flight training in a multiengine airplane;

(2) Except as provided in § 61.110 of this part, 3 hours of night flight training in a multiengine airplane that includes—

(i) One cross-country flight of over 100 nautical miles total distance; and

(ii) 10 takeoffs and 10 landings to a full stop (with each landing involving a flight in the traffic pattern) at an airport.

(3) 3 hours of flight training in a multiengine airplane on the control and maneuvering of an airplane solely by reference to instruments, including straight and level flight, constant airspeed climbs and descents, turns to a heading, recovery from unusual flight attitudes, radio communications, and the use of navigation systems/facilities and radar services appropriate to instrument flight;

(4) 3 hours of flight training in preparation for the practical test in a multiengine airplane, which must have been performed within the 60-day period preceding the date of the test; and

(5) 10 hours of solo flight time in an airplane consisting of at least—

(i) 5 hours of solo cross-country time;

(ii) One solo cross-country flight of at least 150 nautical miles total distance, with full-stop landings at a minimum of three points, and one segment of the flight consisting of a straight-line distance of at least 50 nautical miles between the takeoff and landing locations; and

(iii) Three takeoffs and three landings to a full stop (with each landing involving a flight in the traffic pattern) at an airport with an operating control tower.

(c) For a helicopter rating. Except as provided in paragraph (k) of this section, a person who applies for a private pilot certificate with rotorcraft category and helicopter class rating must log at least 40 hours of flight time that includes at least 20 hours of flight training from an autho-

rized instructor and 10 hours of solo flight training in the areas of operation listed in § 61.107(b)(3) of this part, and the training must include at least—

(1) 3 hours of cross-country flight training in a helicopter;

(2) Except as provided in § 61.110 of this part, 3 hours of night flight training in a helicopter that includes—

(i) One cross-country flight of over 50 nautical miles total distance; and

(ii) 10 takeoffs and 10 landings to a full stop (with each landing involving a flight in the traffic pattern) at an airport.

(3) 3 hours of flight training in preparation for the practical test in a helicopter, which must have been performed within 60 days preceding the date of the test; and

(4) 10 hours of solo flight time in a helicopter, consisting of at least—

(i) 3 hours cross-country time;

(ii) One solo cross-country flight of at least 75 nautical miles total distance, with landings at a minimum of three points, and one segment of the flight being a straight-line distance of at least 25 nautical miles between the takeoff and landing locations; and

(iii) Three takeoffs and three landings to a full stop (with each landing involving a flight in the traffic pattern) at an airport with an operating control tower.

(d) For a gyroplane rating. Except as provided in paragraph (k) of this section, a person who applies for a private pilot certificate with rotorcraft category and gyroplane class rating must log at least 40 hours of flight time that includes at least 20 hours of flight training from an authorized instructor and 10 hours of solo flight training in the areas of operation listed in § 61.107(b)(4) of this part, and the training must include at least—

(1) 3 hours of cross-country flight training in a gyroplane;

(2) Except as provided in § 61.110 of this part, 3 hours of night flight training in a gyroplane that includes—

(i) One cross-country flight of over 50 nautical miles total distance; and

(ii) 10 takeoffs and 10 landings to a full stop (with each landing involving a flight in the traffic pattern) at an airport.

(3) 3 hours of flight training in preparation for the practical test in a gyroplane, which must have been performed within the 60-day period preceding the date of the test; and

(4) 10 hours of solo flight time in a gyroplane, consisting of at least—

(i) 3 hours of cross-country time;

(ii) One solo cross-country flight of over 75 nautical miles total distance, with landings at a minimum of three points, and one segment of the flight being a straight-line distance of at least 25 nautical miles between the takeoff and landing locations; and

(iii) Three takeoffs and three landings to a full stop (with each landing involving a flight in the traffic pattern) at an airport with an operating control tower.

(e) For a powered-lift rating. Except as provided in paragraph (k) of this section, a person who applies for a private pilot certificate with a powered-lift category rating must log at least 40 hours of flight time that includes at least 20 hours of flight training from an authorized instructor and 10 hours of solo flight training in the areas of operation listed in § 61.107(b)(5) of this part, and the training must include at least—

(1) 3 hours of cross-country flight training in a powered-lift;

(2) Except as provided in § 61.110 of this part, 3 hours of night flight training in a powered-lift that includes—

(i) One cross-country flight of over 100 nautical miles total distance; and

(ii) 10 takeoffs and 10 landings to a full stop (with each landing involving a flight in the traffic pattern) at an airport.

(3) 3 hours of flight training in a powered-lift on the control and maneuvering of a powered-lift solely by reference to instruments, including straight and level flight, constant airspeed climbs and descents, turns to a heading, recovery from unusual flight attitudes, radio communications, and the use of navigation systems/facilities and radar services appropriate to instrument flight;

(4) 3 hours of flight training in preparation for the practical test in a powered-lift, which must have been performed within the 60-day period preceding the date of the test; and

(5) 10 hours of solo flight time in an airplane or powered-lift consisting of at least—

(i) 5 hours cross-country time;

(ii) One cross-country flight of at least 150 nautical miles total distance, with landings at a minimum of three points, and one segment of the flight being a straight-line distance of at least 50 nautical miles between the takeoff and landing locations; and

(iii) Three takeoffs and three landings to a full stop (with each landing involving a flight in the traffic pattern) at an airport with an operating control tower.

(f) *For a glider category rating.* (1) If the applicant for a private pilot certificate with a glider category rating has not logged at least 40 hours of flight time as a pilot in a heavier-than-air aircraft, the applicant must log at least 10 hours of flight time in a glider in the areas of operation listed in § 61.107(b)(6) of this part, and that flight time must include at least—

(i) 20 flights in a glider in the areas of operations listed in § 61.107(b)(6) of this part, including at least 3 training flights in a glider with an authorized instructor in preparation for the practical test that must have been performed within the 60-day period preceding the date of the test; and

(ii) 2 hours of solo flight time in a glider in the areas of operation listed in § 61.107(bj)(6) of this part, with not less than 10 launches and landing being performed.

(2) If the applicant has logged at least 40 hours of flight time in a heavier-than-air aircraft, the applicant must log at least 3 hours of flight time in a glider in the areas of operation listed in § 61.107(b)(6) of this part, and that flight time must include at least—

(i) 10 solo flights in a glider in the areas of operation listed in § 61.107(b)(6) of this part; and

(ii) 3 training flights in a glider with an authorized instructor in preparation for the practical test that must have been performed within the 60-day period preceding the date of the test.

(g) For an airship rating. A person who applies for a private pilot certificate with a lighter-than-air category and airship class rating must log at least:

(1) 25 hours of flight training in airships on the areas of operation listed in § 61.107(b)(7) of this part, which consists of at least:

(i) 3 hours of cross-country flight training in an airship;

(ii) Except as provided in § 61.110 of this part, 3 hours of night flight training in an airship that includes:

(A) A cross-country flight of over 25 nautical miles total distance; and

(B) Five takeoffs and five landings to a full stop (with each landing involving a flight in the traffic pattern) at an airport.

(2) 3 hours of flight training in an airship on the control and maneuvering of an airship solely by reference to instruments, including straight and level flight, constant airspeed climbs and descents, turns to a heading, recovery from unusual flight attitudes, radio communications, and the use of navigation systems/ facilities and radar services appropriate to instrument flight;

(3) 3 hours of flight training in an airship in preparation for the practical test within the 60 days preceding the date of the test; and

(4) 5 hours of performing the duties of pilot in command in an airship and with an authorized instructor.

(h) For a balloon rating. A person who applies for a private pilot certificate with a lighter-than-air category and balloon class rating must log at least 10 hours of flight training that includes at least six training flights with an authorized instructor in the areas of operation listed in § 61.107(b)(8) of this part, that includes—

(1) Gas balloon. If the training is being performed in a gas balloon, at least two flights of 2 hours each that consists of—

(i) At least one training flight with an authorized instructor within 60 days prior to application for the rating on the areas of operation for a gas balloon;

(ii) At least one flight performing the duties of pilot in command in a gas balloon; and

(iii) At least one flight involving a controlled ascent to 3,000 feet above the launch site.

(2) Balloon with an airborne heater. If the training is being performed in a balloon with an airborne heater, at least—

(i) Two flights of 1 hour each within 60 days prior to application for the rating on the areas of operation appropriate to a balloon with an airborne heater;

(ii) One solo flight in a balloon with an airborne heater; and

(iii) At least one flight involving a controlled ascent to 2,000 feet above the launch site.

(i) For a powered parachute rating. A person who applies for a private pilot certificate with a powered parachute category rating must log at least 25 hours of flight time in a powered parachute that includes at least 10 hours of flight training with an authorized instructor, including 30 takeoffs and landings, and 10 hours of solo flight training in the areas of operation listed in §61.107 (b)(9) and the training must include at least—

(1) One hour of cross-country flight training in a powered parachute that includes a 1-hour cross-country flight with a landing at an airport at least 25 nautical miles from the airport of departure;

(2) Except as provided in §61.110, 3 hours of night flight training in a powered parachute that includes 10 takeoffs and landings (with each landing involving a flight in the traffic pattern) at an airport;

(3) Three hours of flight training in preparation for the practical test in a powered parachute, which must have been performed within the 60-day period preceding the date of the test; and

(4) Three hours of solo flight time in a powered parachute, consisting of at least—

(i) One solo cross-country flight with a landing at an airport at least 25 nautical miles from the departure airport; and

(ii) Twenty solo takeoffs and landings to a full stop (with each landing involving a flight in a traffic pattern) at an airport, with at least 3 takeoffs and landings at an airport with an operating control tower.

(j) For a weight-shift-control aircraft rating. A person who applies for a private pilot certificate with a weight-shift-control rating must log at least 40 hours of flight time that includes at least 20 hours of flight training with an authorized instructor and 10 hours of solo flight training in the areas listed in §61.107 (b)(10) and the training must include at least—

(1) Three hours of cross-country flight training in a weight-shift-control aircraft;

(2) Except as provided in §61.110, 3 hours of night flight training in a weight-shift-control aircraft that includes—

(i) One cross-country flight over 75 nautical miles total distance; and

(ii) Ten takeoffs and landings (with each landing involving a flight in the traffic pattern) at an airport;

(3) Three hours of flight training in preparation for the practical test in a weight-shift-control aircraft, which must have been performed within the 60-day period preceding the date of the test; and

(4) Ten hours of solo flight time in a weight-shift-control aircraft, consisting of at least—

(i) Five hours of solo cross-country time;

(ii) One solo cross-country flight over 100 nautical miles total distance, with landings at a minimum of three points, and one segment of the flight being a straight line distance of at least 50 nautical miles between takeoff and landing locations; and

(iii) Three takeoffs and landings (with each landing involving a flight in the traffic pattern) at an airport with an operating control tower.

(k) Permitted credit for use of a flight simulator or flight training device. (1) Except as provided in paragraphs (k)(2) of this section, a maximum of 2.5 hours of training in a flight simulator or flight training device representing the category, class, and type, if applicable, of aircraft appropriate to the rating sought, may be credited toward the flight training time required by this section, if received from an authorized instructor.

(2) A maximum of 5 hours of training in a flight simulator or flight training device representing the category, class, and type, if applicable, of aircraft appropriate to the rating sought, may be credited toward the flight training time required by this section if the training is accomplished in a course conducted by a training center certificated under part 142 of this chapter.

(3) Except when fewer hours are approved by the Administrator, an applicant for a private pilot certificate with an airplane, rotorcraft, or powered-lift rating, who has satisfactorily completed an approved private pilot course conducted by a training center certificated under part 142 of this chapter, need only have a total of 35 hours of aeronautical experience to meet the requirements of this section.

[Amdt. 61-102, 62 FR 16298, Apr. 4, 1997; Amdt. 61-103, 62 FR 40902, July 30, 1997]

§ 61.110   Night flying exceptions.

(a) Subject to the limitations of paragraph (b) of this section, a person is not required to comply with the night flight training requirements of this subpart if the person receives flight training in and resides in the State of Alaska.

(b) A person who receives flight training in and resides in the State of Alaska but does not meet the night flight training requirements of this section:

(1) May be issued a pilot certificate with a limitation "Night flying prohibited;" and

(2) Must comply with the appropriate night flight training requirements of this subpart within the 12-calendar-month period after the issuance of the pilot certificate. At the end of that period, the certificate will become invalid use until the person complies with the appropriate night training requirements of this subpart. The person may have the "Night flying prohibited" limitation removed if the person—

(i) Accomplishes the appropriate night flight training requirements of this subpart; and

(ii) Presents to an examiner a logbook or training record endorsement from an authorized instructor that verifies accomplishment of the appropriate night flight training requirements of this subpart.

(c) A person who does not meet the night flying requirements in §61.109 (d)(2), (i)(2), or (J)(2) may be issued a private pilot certificate with the limitation "Night flying prohibited." This limitation may be removed by an examiner if the holder complies with the requirements of §61.109 (d)(2), (i)(2), or (j)(2), as appropriate.

[Amdt. 61-102, 62 FR 16298, Apr. 4, 1997; Amdt. 61-103, 62 FR 40904, July 30, 1997]

§ 61.111 Cross-country flights: Pilots based on small islands.

(a) Except as provided in paragraph (b) of this section, an applicant located on an island from which the cross-country flight training required in § 61.109 of this part cannot be accomplished without flying over water for more than 10 nautical miles from the nearest shoreline need not comply with the requirements of that section.

(b) If other airports that permit civil operations are available to which a flight may be made without flying over water for more than 10 nautical miles from the nearest shoreline, the applicant must show completion of two round-trip solo flights between those two airports that are farthest apart, including a landing at each airport on both flights.

(c) An applicant who complies with paragraph (a) or paragraph (b) of this section, and meets all requirements for the issuance of a private pilot certificate, except the cross-country training requirements of § 61.109 of this part, will be issued a pilot certificate with an endorsement containing the following limitation, "Passenger carrying prohibited on flights more than 10 nautical miles from (the appropriate island)." The limitation may be subsequently amended to include another island if the applicant complies with the requirements of paragraph (b) of this section for another island.

(d) Upon meeting the cross-country training requirements of § 61.109 of this part, the applicant may have the limitation in paragraph (c) of this section removed.

[Amdt. 61-102, 62 FR 16298, Apr. 4, 1997; Amdt. 61-103, 62 FR 40904, July 30, 1997]

§ 61.113 Private pilot privileges and limitations: Pilot in command.

(a) Except as provided in paragraphs (b) through (g) of this section, no person who holds a private pilot certificate may act as pilot in command of an aircraft that is carrying passengers or property for compensation or hire; nor may that person, for compensation or hire, act as pilot in command of an aircraft.

(b) A private pilot may, for compensation or hire, act as pilot in command of an aircraft in connection with any business or employment if:

(1) The flight is only incidental to that business or employment; and

(2) The aircraft does not carry passengers or property for compensation or hire.

(c) A private pilot may not pay less than the pro rata share of the operating expenses of a flight with passengers, provided the expenses involve only fuel, oil, airport expenditures, or rental fees.

(d) A private pilot may act as pilot in command of an aircraft used in a passenger-carrying airlift sponsored by a charitable organization described in paragraph (d)(7) of this section, and for which the passengers make a donation to the organization, when the following requirements are met:

(1) The sponsor of the airlift notifies the FAA Flight Standards District Office with jurisdiction over the area concerned at least 7 days before the event and furnishes—

(i) A signed letter from the sponsor that shows the name of the sponsor, the purpose of the charitable event, the date and time of the event, and the location of the event; and

(ii) A photocopy of each pilot in command's pilot certificate, medical certificate, and logbook entries that show the pilot is current in accordance with §§ 61.56 and 61.57 of this part and has logged at least 200 hours of flight time.

(2) The flight is conducted from a public airport that is adequate for the aircraft to be used, or from another airport that has been approved by the FAA for the operation.

(3) No aerobatic or formation flights are conducted.

(4) Each aircraft used for the charitable event holds a standard airworthiness certificate.

(5) Each aircraft used for the charitable event is airworthy and complies with the applicable requirements of subpart E of part 91 of this chapter.

(6) Each flight for the charitable event is made during day VFR conditions.

(7) The charitable organization is an organization identified as such by the U.S. Department of Treasury.

(e) A private pilot may be reimbursed for aircraft operating expenses that are directly related to search and location operations, provided the expenses involve only fuel, oil, airport expenditures, or rental fees, and the operation is sanctioned and under the direction and control of:

(1) A local, State, or Federal agency; or

(2) An organization that conducts search and location operations.

(f) A private pilot who is an aircraft salesman and who has at least 200 hours of logged flight time may demonstrate an aircraft in flight to a prospective buyer.

(g) A private pilot who meets the requirements of § 61.69 of this part may act as pilot in command of an aircraft towing a glider or unpowered ultralight vehicle.

§ 61.115 Balloon rating: Limitations.

(a) If a person who applies for a private pilot certificate with a balloon rating takes a practical test in a balloon with an airborne heater:

(1) The pilot certificate will contain a limitation restricting the exercise of the privileges of that certificate to a balloon with an airborne heater; and

(2) The limitation may be removed when the person obtains the required aeronautical experience in a gas balloon and receives a logbook endorsement from an authorized instructor who attests to the person's accomplishment of the required aeronautical experience and ability to satisfactorily operate a gas balloon.

(b) If a person who applies for a private pilot certificate with a balloon rating takes a practical test in a gas balloon:

(1) The pilot certificate will contain a limitation restricting the exercise of the privilege of that certificate to a gas balloon; and

(2) The limitation may be removed when the person obtains the required aeronautical experience in a balloon with an airborne heater and receives a logbook endorsement from an authorized instructor who attests to the person's accomplishment of the required aeronautical experience and ability to satisfactorily operate a balloon with an airborne heater.

§ 61.117 Private pilot privileges and limitations: Second in command of aircraft requiring more than one pilot.

Except as provided in § 61.113 of this part, no private pilot may, for compensation or hire, act as second in command of an aircraft that is type certificated for more than one pilot, nor may that pilot act as second in command of such an aircraft that is carrying passengers or property for compensation or hire.

[Amdt. 61-102, 62 FR 16298, Apr. 4, 1997; Amdt. 61-103, 62 FR 40904, July 30, 1997]

§§ 61.118-61.120 [Reserved.]

Subpart F—Commercial Pilots

§ 61.121 Applicability.

This subpart prescribes the requirements for the issuance of commercial pilot certificates and ratings, the conditions under which those certificates and ratings are necessary, and the general operating rules for persons who hold those certificates and ratings.

§ 61.123 Eligibility requirements: General.

To be eligible for a commercial pilot certificate, a person must:

(a) Be at least 18 years of age;

(b) Be able to read, speak, write, and understand the English language. If the applicant is unable to meet one of these requirements due to medical reasons, then the Administrator may place such operating limitations on that applicant's pilot certificate as are necessary for the safe operation of the aircraft.

(c) Receive a logbook endorsement from an authorized instructor who:

(1) Conducted the required ground training or reviewed the person's home study on the aeronautical knowledge areas listed in § 61.125 of this part that apply to the aircraft category and class rating sought; and

(2) Certified that the person is prepared for the required knowledge test that applies to the aircraft category and class rating sought.

(d) Pass the required knowledge test on the aeronautical knowledge areas listed in § 61.125 of this part;

(e) Receive the required training and a logbook endorsement from an authorized instructor who:

(1) Conducted the training on the areas of operation listed in § 61.127(b) of this part that apply to the aircraft category and class rating sought; and

(2) Certified that the person is prepared for the required practical test.

(f) Meet the aeronautical experience requirements of this subpart that apply to the aircraft category and class rating sought before applying for the practical test;

(g) Pass the required practical test on the areas of operation listed in § 61.127(b) of this part that apply to the aircraft category and class rating sought;

(h) Hold at least a private pilot certificate issued under this part or meet the requirements of § 61.73; and

(i) Comply with the sections of this part that apply to the aircraft category and class rating sought.

§ 61.125 Aeronautical knowledge.

(a) General. A person who applies for a commercial pilot certificate must receive and log ground training from an authorized instructor, or complete a home-study course, on the aeronautical knowledge areas of paragraph (b) of this section that apply to the aircraft category and class rating sought.

(b) Aeronautical knowledge areas.

(1) Applicable Federal Aviation Regulations of this chapter that relate to commercial pilot privileges, limitations, and flight operations;

(2) Accident reporting requirements of the National Transportation Safety Board;

(3) Basic aerodynamics and the principles of flight;

(4) Meteorology to include recognition of critical weather situations, windshear recognition and avoidance, and the use of aeronautical weather reports and forecasts;

(5) Safe and efficient operation of aircraft;

(6) Weight and balance computations;

(7) Use of performance charts;

(8) Significance and effects of exceeding aircraft performance limitations;

(9) Use of aeronautical charts and a magnetic compass for pilotage and dead reckoning;

(10) Use of air navigation facilities;

(11) Aeronautical decision making and judgment;

(12) Principles and functions of aircraft systems;

(13) Maneuvers, procedures, and emergency operations appropriate to the aircraft;

(14) Night and high-altitude operations;

(15) Procedures for operating within the National Airspace System; and

(16) Procedures for flight and ground training for lighter-than-air ratings.

§ 61.127   Flight proficiency.

(a) *General.* A person who applies for a commercial pilot certificate must receive and log ground and flight training from an authorized instructor on the areas of operation of this section that apply to the aircraft category and class rating sought.

(b) *Areas of operation.*

(1) For an airplane category rating with a single-engine class rating:

(i) Preflight preparation;

(ii) Preflight procedures;

(iii) Airport and seaplane base operations;

(iv) Takeoffs, landings, and go-arounds;

(v) Performance maneuvers;

(vi) Ground reference maneuvers;

(vii) Navigation;

(viii) Slow flight and stalls;

(ix) Emergency operations;

(x) High-altitude operations; and

(xi) Postflight procedures.

(2) For an airplane category rating with a multiengine class rating:

(i) Preflight preparation;

(ii) Preflight procedures;

(iii) Airport and seaplane base operations;

(iv) Takeoffs, landings, and go-arounds;

(v) Performance maneuvers;

(vi) Navigation;

(vii) Slow flight and stalls;

(viii) Emergency operations;

(ix) Multiengine operations;

(x) High-altitude operations; and

(xi) Postflight procedures.

(3) For a rotorcraft category rating with a helicopter class rating:

(i) Preflight preparation;

(ii) Preflight procedures;

(iii) Airport and heliport operations;

(iv) Hovering maneuvers;

(v) Takeoffs, landings, and go-arounds;

(vi) Performance maneuvers;

(vii) Navigation;

(viii) Emergency operations;

(ix) Special operations; and

(x) Postflight procedures.

(4) For a rotorcraft category rating with a gyroplane class rating:

(i) Preflight preparation;

(ii) Preflight procedures;

(iii) Airport operations;

(iv) Takeoffs, landings, and go-arounds;

(v) Performance maneuvers;

(vi) Navigation;

(vii) Flight at slow airspeeds;

(viii) Emergency operations; and

(ix) Postflight procedures.

(5) For a powered-lift category rating:

(i) Preflight preparation;

(ii) Preflight procedures;

(iii) Airport and heliport operations;

(iv) Hovering maneuvers;

(v) Takeoffs, landings, and go-arounds;

(vi) Performance maneuvers;

(vii) Ground reference maneuvers;

(viii) Navigation;

(ix) Slow flight and stalls;

(x) Emergency operations;

(xi) High-altitude operations;

(xii) Special operations; and

(xiii) Postflight procedures.

(6) For a glider category rating:

(i) Preflight preparation;

(ii) Preflight procedures;

(iii) Airport and gliderport operations;

(iv) Launches and landings;

(v) Performance speeds;

(vi) Soaring techniques;

(vii) Performance maneuvers;

(viii) Navigation;

(ix) Slow flight and stalls;

(x) Emergency operations; and

(xi) Postflight procedures.

(7) For a lighter-than-air category rating with an airship class rating:

(i) Fundamentals of instructing;

(ii) Technical subjects;

(iii) Preflight preparation;

(iv) Preflight lesson on a maneuver to be performed in flight;

(v) Preflight procedures;

(vi) Airport operations;

(vii) Takeoffs, landings, and go-arounds;

(viii) Performance maneuvers;

(ix) Navigation;

(x) Emergency operations; and

(xi) Postflight procedures.

(8) For a lighter-than-air category rating with a balloon class rating:

(i) Fundamentals of instructing;

(ii) Technical subjects;

(iii) Preflight preparation;

(iv) Preflight lesson on a maneuver to be performed in flight;

(v) Preflight procedures;

(vi) Airport operations;

(vii) Launches and landings;

(viii) Performance maneuvers;

(ix) Navigation;

(x) Emergency operations; and

(xi) Postflight procedures.

§ 61.129 Aeronautical experience.

(a) For an airplane single-engine rating. Except as provided in paragraph (i) of this section, a person who applies for a commercial pilot certificate with an airplane category and single-engine class rating must log at least 250 hours of flight time as a pilot that consists of at least:

(1) 100 hours in powered aircraft, of which 50 hours must be in airplanes.

(2) 100 hours of pilot-in-command flight time, which includes at least—

(i) 50 hours in airplanes; and

(ii) 50 hours in cross-country flight of which at least 10 hours must be in airplanes.

(3) 20 hours of training on the areas of operation listed in § 61.127(b)(1) of this part that includes at least—

(i) 10 hours of instrument training of which at least 5 hours must be in a single-engine airplane;

(ii) 10 hours of training in an airplane that has a retractable landing gear, flaps, and a controllable pitch propeller, or is turbine-powered, or for an applicant seeking a single-engine seaplane rating, 10 hours of training in a seaplane that has flaps and a controllable pitch propeller;

(iii) One cross-country flight of at least 2 hours in a single-engine airplane in day VFR conditions, consisting of a total straight-line distance of more than 100 nautical miles from the original point of departure;

(iv) One cross-country flight of at least 2 hours in a single-engine airplane in night VFR conditions, consisting of a total straight-line distance of more than 100 nautical miles from the original point of departure; and

(v) 3 hours in a single-engine airplane in preparation for the practical test within the 60-day period preceding the date of the test.

(4) 10 hours of solo flight in a single-engine airplane on the areas of operation listed in § 61.127(b)(1) of this part, which includes at least—

(i) One cross-country flight of not less than 300 nautical miles total distance, with landings at a minimum of three points, one of which is a straight-line distance of at least 250 nautical miles from the original departure point. However, if this requirement is being met in Hawaii, the longest segment need only have a straight-line distance of at least 150 nautical miles; and

(ii) 5 hours in night VFR conditions with 10 takeoffs and 10 landings (with each landing involving a flight in the traffic pattern) at an airport with an operating control tower.

(b) For an airplane multiengine rating. Except as provided in paragraph (i) of this section, a person who applies for a commercial pilot certificate with an airplane category and multiengine class rating must log at least 250 hours of flight time as a pilot that consists of at least:

(1) 100 hours in powered aircraft, of which 50 hours must be in airplanes.

(2) 100 hours of pilot-in-command flight time, which includes at least—

(i) 50 hours in airplanes; and

(ii) 50 hours in cross-country flight of which at least 10 hours must be in airplanes.

(3) 20 hours of training on the areas of operation listed in § 61.127(b)(2) of this part that includes at least—

(i) 10 hours of instrument training of which at least 5 hours must be in a multiengine airplane;

(ii) 10 hours of training in a multiengine airplane that has a retractable landing gear, flaps, and controllable pitch propellers, or is turbine-powered, or for an applicant seeking a multiengine seaplane rating, 10 hours of training in a multiengine seaplane that has flaps and a controllable pitch propeller;

(iii) One cross-country flight of at least 2 hours in a multiengine airplane in day VFR conditions, consisting of a total straight-line distance of more than 100 nautical miles from the original point of departure;

(iv) One cross-country flight of at least 2 hours in a multiengine airplane in night VFR conditions, consisting of a total straight-line distance of more than 100 nautical miles from the original point of departure; and

(v) 3 hours in a multiengine airplane in preparation for the practical test within the 60-day period preceding the date of the test.

(4) 10 hours of solo flight time in a multiengine airplane or 10 hours of flight time performing the duties of pilot in command in a multiengine airplane with an authorized instructor (either of which may be credited towards the flight time requirement in paragraph (b)(2) of this section), on the areas of operation listed in § 61.127(b)(2) of this part that includes at least—

(i) One cross-country flight of not less than 300 nautical miles total distance with landings at a minimum of three points, one of which is a straight-line distance of at least 250 nautical miles from the original departure point. However, if this requirement is being met in Hawaii, the longest segment need only have a straight-line distance of at least 150 nautical miles; and

(ii) 5 hours in night VFR conditions with 10 takeoffs and 10 landings (with each landing involving a flight with a traffic pattern) at an airport with an operating control tower.

(c) For a helicopter rating. Except as provided in paragraph (i) of this section, a person who applies for a commercial pilot certificate with a rotorcraft category and helicopter class rating must log at least 150 hours of flight time as a pilot that consists of at least:

(1) 100 hours in powered aircraft, of which 50 hours must be in helicopters.

(2) 100 hours of pilot-in-command flight time, which includes at least—

(i) 35 hours in helicopters; and

(ii) 10 hours in cross-country flight in helicopters.

(3) 20 hours of training on the areas of operation listed in § 61.127(b)(3) of this part that includes at least—

(i) 10 hours of instrument training in an aircraft;

(ii) One cross-country flight of at least 2 hours in a helicopter in day VFR conditions, consisting of a total straight-line distance of more than 50 nautical miles from the original point of departure;

(iii) One cross-country flight of at least 2 hours in a helicopter in night VFR conditions, consisting of a total straight-line distance of more than 50 nautical miles from the original point of departure; and

(iv) 3 hours in a helicopter in preparation for the practical test within the 60-day period preceding the date of the test.

(4) 10 hours of solo flight in a helicopter on the areas of operation listed in § 61.127(b)(3) of this part, which includes at least—

(i) One cross-country flight with landings at a minimum of three points, with one segment consisting of a straight-line distance of at least 50 nautical miles from the original point of departure; and

(ii) 5 hours in night VFR conditions with 10 takeoffs and 10 landings (with each landing involving a flight in the traffic pattern).

(d) For a gyroplane rating. A person who applies for a commercial pilot certificate with a rotorcraft category and gyroplane class rating must log at least 150 hours of flight time as a pilot (of which 5 hours may have been accomplished in a flight simulator or flight training device that is representative of a gyroplane) that consists of at least:

(1) 100 hours in powered aircraft, of which 25 hours must be in gyroplanes.

(2) 100 hours of pilot-in-command flight time, which includes at least—

(i) 10 hours in gyroplanes; and

(ii) 3 hours in cross-country flight in gyroplanes.

(3) 20 hours of training on the areas of operation listed in § 61.127(b)(4) of this part that includes at least—

(i) 5 hours of instrument training in an aircraft;

(ii) One cross-country flight of at least 2 hours in a gyroplane in day VFR conditions, consisting of a total

straight-line distance of more than 50 nautical miles from the original point of departure;

(iii) One cross-country flight of at least 2 hours in a gyroplane in night VFR conditions, consisting of a total straight-line distance of more than 50 nautical miles from the original point of departure; and

(iv) 3 hours in a gyroplane in preparation for the practical test within the 60-day period preceding the date of the test.

(4) 10 hours of solo flight in a gyroplane on the areas of operation listed in § 61.127(b)(4) of this part, which includes at least—

(i) One cross-country flight with landings at a minimum of three points, with one segment consisting of a straight-line distance of at least 50 nautical miles from the original point of departure; and

(ii) 5 hours in night VFR conditions with 10 takeoffs and 10 landings (with each landing involving a flight in the traffic pattern).

(e) For a powered-lift rating. Except as provided in paragraph (i) of this section, a, person who applies for a commercial pilot certificate with a powered-lift category rating must log at least 250 hours of flight time as a pilot that consists of at least:

(1) 100 hours in powered aircraft, of which 50 hours must be in a powered-lift.

(2) 100 hours of pilot-in-command flight time, which includes at least—

(i) 50 hours in a powered-lift; and

(ii) 50 hours in cross-country flight of which 10 hours must be in a powered-lift.

(3) 20 hours of training on the areas of operation listed in § 61.127(b)(5) of this part that includes at least—

(i) 10 hours of instrument training, of which at least 5 hours must be in a powered-lift;

(ii) One cross-country flight of at least 2 hours in a powered-lift in day VFR conditions, consisting of a total straight-line distance of more than 100 nautical miles from the original point of departure;

(iii) One cross-country flight of at least 2 hours in a powered-lift in night VFR conditions, consisting of a total straight-line distance of more than 100 nautical miles from the original point of departure; and

(iv) 3 hours in a powered-lift in preparation for the practical test within the 60-day period preceding the date of the test.

(4) 10 hours of solo flight in a powered-lift on the areas of operation listed in § 61.127(b)(5) of this part, which includes at least—

(i) One cross-country flight of not less than 300 nautical miles total distance with landings at a minimum of three points, one of which is a straight-line distance of at least 250 nautical miles from the original departure point. However, if this requirement is being met in Hawaii the longest segment need only have a straight-line distance of at least 150 nautical miles; and

(ii) 5 hours in night VFR conditions with 10 takeoffs and 10 landings (with each landing involving a flight in the traffic pattern) at an airport with an operating control tower.

(f) *For a glider rating.* A person who applies for a commercial pilot certificate with a glider category rating must log at least—

(1) 25 hours of flight time as a pilot in a glider and that flight time must include at least 100 flights in a glider as pilot in command, including at least—

(i) 3 hours of flight training in a glider or 10 training flights in a glider with an authorized instructor on the areas of operation listed in § 61.127(b)(6) of this part, including at least 3 training flights in a glider with an authorized instructor in preparation for the practical test within the 60-day period preceding the date of the test; and

(ii) 2 hours of solo flight that include not less than 10 solo flights in a glider on the areas of operation listed in § 61.27(b)(6) of this part; or

(2) 200 hours of flight time as a pilot in heavier-than-air aircraft and at least 20 flights in a glider as pilot in command, including at least—

(i) 3 hours of flight training in a glider or 10 training flights in a glider with an authorized instructor on the areas of operation listed in § 61.127(b)(6) of this part including at least 3 training flights in a glider with an authorized instructor in preparation for the practical test within the 60-day period preceding the date of the test; and

(ii) 5 solo flights in a glider on the areas of operation listed in § 61.127(b)(6) of this part.

(iii) Three training flights in preparation for the practical test within the 60-day period preceding the date of the test.

(g) For an airship rating. A person who applies for a commercial pilot certificate with a lighter-than-air category and airship class rating must log at least 200 hours of flight time as a pilot, which includes at least the following hours:

(1) 50 hours in airships.

(2) 30 hours of pilot-in-command time in airships, which consists of at least—

(i) 10 hours of cross-country flight time in airships; and

(ii) 10 hours of night flight time in airships.

(3) 40 hours of instrument time, which consists of at least 20 hours in flight, of which 10 hours must be in flight in airships.

(4) 20 hours of flight training in airships on the areas of operation listed in § 61.127(b)(7) of this part, which includes at least—

(i) 3 hours in an airship in preparation for the practical test within the 60-day period preceding the date of the test;

(ii) One cross-country flight of at least 1 hour in duration in an airship in day VFR conditions, consisting of a total straight-line distance of more than 25 nautical miles from the original point of departure; and

(iii) One cross-country flight of at least 1 hour in duration in an airship in night VFR conditions, consisting of a total straight-line distance of more than 25 nautical miles from the original point of departure.

(5) 10 hours of flight training performing the duties of pilot in command with an authorized instructor on the areas of operation listed in § 61.127(b)(7) of this part, which includes at least—

(i) One cross-country flight with landings at a minimum of three points, with one segment consisting of a straight-line distance of at least 25 nautical miles from the original point of departure; and

(ii) 5 hours in night VFR conditions with 10 takeoffs and 10 landings (with each landing involving a flight in the traffic pattern).

(h) For a balloon rating. A person who applies for a commercial pilot certificate with a lighter-than-air category and a balloon class rating must log at least 35 hours of flight time as a pilot, which includes at least the following requirements:

(1) 20 hours in balloons;

(2) 10 flights in balloons;

(3) Two flights in balloons as the pilot in command; and

(4) 10 hours of flight training that includes at least 10 training flights with an authorized instructor in balloons on the areas of operation listed in § 61.127(b)(8) of this part, which consists of at least—

(i) For a gas balloon—

(A) Two training flights of 2 hours each with an authorized instructor in a gas balloon on the areas of operation appropriate to a gas balloon within 60 days prior to application for the rating;

(B) Two flights performing the duties of pilot in command in a gas balloon with an authorized instructor on the appropriate areas of operation; and

(C) One flight involving a controlled ascent to 5,000 feet above the launch site.

(ii) For a balloon with an airborne heater—

(A) Two training flights of 1 hour each with an authorized instructor in a balloon with an airborne heater on the areas of operation appropriate to a balloon with an airborne heater within 60 days prior to application for the rating;

(B) Two solo flights in a balloon with an airborne heater on the appropriate areas of operation; and

(C) One flight involving a controlled ascent to 3,000 feet above the launch site.

(i) Permitted credit for use of a flight simulator or flight training device.

(1) Except as provided in paragraph (i)(2) of this section, an applicant who has not accomplished the training required by this section in a course conducted by a training center certificated under part 142 of this chapter may:

(i) Credit a maximum of 50 hours toward the total aeronautical experience requirements for an airplane or powered-lift rating, provided the aeronautical experience was obtained from an authorized instructor in a flight simulator or flight training device that represents that class of airplane or powered-lift category and type, if applicable, appropriate to the rating sought; and

(ii) Credit a maximum of 25 hours toward the total aeronautical experience requirements of this section for a helicopter rating, provided the aeronautical experience was obtained from an authorized instructor in a flight simulator or flight training device that represents a helicopter and type, if applicable, appropriate to the rating sought.

(2) An applicant who has accomplished the training required by this section in a course conducted by a training center certificated under part 142 of this chapter may:

(i) Credit a maximum of 100 hours toward the total aeronautical experience requirements of this section for an airplane and powered-lift rating, provided the aeronautical experience was obtained from an authorized instructor in a flight simulator or flight training device that represents that class of airplane or powered-lift category and type, if applicable, appropriate to the rating sought; and

(ii) Credit a maximum of 50 hours toward the total aeronautical experience requirements of this section for a helicopter rating, provided the aeronautical experience was obtained from an authorized instructor in a flight simulator or flight training device that represents a helicopter and type, if applicable, appropriate to the rating sought.

(3) Except when fewer hours are approved by the Administrator an applicant for a commercial pilot certificate with an airplane or a powered-lift rating who has satisfactorily completed an approved commercial pilot course conducted by a training center certificated under part 142 of this chapter need only have 190 hours of total aeronautical experience to meet the requirements of this section.

(i) 190 hours for an airplane or powered-lift rating; and

(ii) 150 hours for a helicopter rating.

[Amdt. 61-102, 62 FR 16298, Apr. 4, 1997; Amdt. 61-103, 62 FR 40904, July 30, 1997]

§ 61.131 Exceptions to the night flying requirements.

(a) Subject to the limitations of paragraph (b) of this section, a person is not required to comply with the night flight training requirements of this subpart if the person receives flight training in and resides in the State of Alaska.

(b) A person who receives flight training in and resides in the State of Alaska but does not meet the night flight training requirements of this section:

(1) May be issued a pilot certificate with the limitation "night flying prohibited."

(2) Must comply with the appropriate night flight training requirements of this subpart within the 12-calendar-month period after the issuance of the pilot certificate. At the end of that period, the certificate will become invalid for use until the person complies with the appropriate night flight training requirements of this subpart. The person may have the "night flying prohibited" limitation removed if the person—

(i) Accomplishes the appropriate night flight training requirements of this subpart; and

(ii) Presents to an examiner a logbook or training record endorsement from an authorized instructor that verifies accomplishment of the appropriate night flight training requirements of this subpart.

[Amdt. 61-102, 62 FR 16298, Apr. 4, 1997; Amdt. 61-103, 62 FR 40905, July 30, 1997]

§ 61.133 Commercial pilot privileges and limitations.

(a) Privileges.

(1) General. A person who holds a commercial pilot certificate may act as pilot in command of an aircraft—

(i) Carrying persons or property for compensation or hire, provided the person is qualified in accordance with this part and with the applicable parts of this chapter that apply to the operation; and

(ii) For compensation or hire, provided the person is qualified in accordance with this part and with the applicable parts of this chapter that apply to the operation.

(2) Commercial pilots with lighter-than-air category ratings. A person with a commercial pilot certificate with a lighter-than-air category rating may—

(i) For an airship—

(A) Give flight and ground training in an airship for the issuance of a certificate or rating;

(B) Give an endorsement for a pilot certificate with an airship rating;

(C) Endorse a student pilot certificate or logbook for solo operating privileges in an airship;

(D) Act as pilot in command of an airship under IFR or in weather conditions less than the minimum prescribed for VFR flight; and

(E) Give flight and ground training and endorsements that are required for a flight review, an operating privilege, or recency-of-experience requirements of this part.

(ii) For a balloon—

(A) Give flight and ground training in a balloon for the issuance of a certificate or rating;

(B) Give an endorsement for a pilot certificate with a balloon rating;

(C) Endorse a student pilot certificate or logbook for solo operating privileges in a balloon; and

(D) Give ground and flight training and endorsements that are required for a flight review, an operating privilege, or recency-of-experience requirements of this part.

(b) Limitations.

(1) A person who applies for a commercial pilot certificate with an airplane category or powered-lift category rating and does not hold an instrument rating in the same category and class will be issued a commercial pilot certificate that contains the limitation, "The carriage of passengers for hire in (airplanes) (powered-lifts) on cross-country flights in excess of 50 nautical miles or at night is prohibited."

The limitation may be removed when the person satisfactorily accomplishes the requirements listed in § 61.65 of this part for an instrument rating in the same category and class of aircraft listed on the person's commercial pilot certificate.

(2) If a person who applies for a commercial pilot certificate with a balloon rating takes a practical test in a balloon with an airborne heater—

(i) The pilot certificate will contain a limitation restricting the exercise of the privileges of that certificate to a balloon with an airborne heater.

(ii) The limitation specified in paragraph (b)(2)(i) of this section may be removed when the person obtains the required aeronautical experience in a gas balloon and receives a logbook endorsement from an authorized instructor who attests to the person's accomplishment of the required aeronautical experience and ability to satisfactorily operate a gas balloon.

(3) If a person who applies for a commercial pilot certificate with a balloon rating takes a practical test in a gas balloon—

(i) The pilot certificate will contain a limitation restricting the exercise of the privileges of that certificate to a gas balloon.

(ii) The limitation specified in paragraph (b)(3)(i) of this section may be removed when the person obtains the required aeronautical experience in a balloon with an airborne heater and receives a logbook endorsement from an authorized instructor who attests to the person's accomplishment of the required aeronautical experience and ability to satisfactorily operate a balloon with an airborne heater.

[Amdt. 61-102, 62 FR 16298, Apr. 4, 1997; Amdt. 61-103, 62 FR 40905, July 30, 1997]

§§ 61.135—61.141 [Reserved.]

Subpart G—Airline Transport Pilots

§ 61.151 Applicability.

This subpart prescribes the requirements for the issuance of airline transport pilot certificates and ratings, the conditions under which those certificates and ratings are necessary, and the general operating rules for persons who hold those certificates and ratings.

§ 61.153 Eligibility requirements: General.

To be eligible for an airline transport pilot certificate, a person must:

(a) Be at least 23 years of age;

(b) Be able to read, speak, write, and understand the English language. If the applicant is unable to meet one of these requirements due to medical reasons, then the Administrator may place such operating limitations on that applicant's pilot certificate as are necessary for the safe operation of the aircraft;

(c) Be of good moral character;

(d) Meet at least one of the following requirements:

(1) Hold at least a commercial pilot certificate and an instrument rating;

(2) Meet the military experience requirements under § 61.73 of this part to qualify for a commercial pilot certificate, and an instrument rating if the person is a rated military pilot or former rated military pilot of an Armed Force of the United States; or

(3) Hold either a foreign airline transport pilot or foreign commercial pilot license and an instrument rating, without limitations issued by a contracting State to the Convention on International Civil Aviation.

(e) Meet the aeronautical experience requirements of this subpart that apply to the aircraft category and class rating sought before applying for the practical test;

(f) Pass a knowledge test on the aeronautical knowledge areas of § 61.155(c) of this part that apply to the aircraft category and class rating sought;

(g) Pass the practical test on the areas of operation listed in § 61.157(e) of this part that apply to the aircraft category and class rating sought; and

(h) Comply with the sections of this part that apply to the aircraft category and class rating sought.

[Amdt. 61-102, 62 FR 16298, Apr. 4, 1997; Amdt. 61-103, 62 FR 40905, July 30, 1997]

§ 61.155 Aeronautical knowledge.

(a) General. The knowledge test for an airline transport pilot certificate is based on the aeronautical knowledge areas listed in paragraph (c) of this section that are appropriate to the aircraft category and class rating sought.

(b) Aircraft type rating. A person who is applying for an additional aircraft type rating to be added to an airline transport pilot certificate is not required to pass a knowledge test if that person's airline transport pilot certificate lists the aircraft category and class rating that is appropriate to the type rating sought.

(c) Aeronautical knowledge areas.

(1) Applicable Federal Aviation Regulations of this chapter that relate to airline transport pilot privileges, limitations, and flight operations;

(2) Meteorology, including knowledge of and effects of fronts, frontal characteristics, cloud formations, icing, and upper-air data;

(3) General system of weather and NOTAM collection, dissemination, interpretation, and use;

(4) Interpretation and use of weather charts, maps, forecasts, sequence reports, abbreviations, and symbols;

(5) National Weather Service functions as they pertain to operations in the National Airspace System;

(6) Windshear and microburst awareness, identification, and avoidance;

(7) Principles of air navigation under instrument meteorological conditions in the National Airspace System;

(8) Air traffic control procedures and pilot responsibilities as they relate to en route operations, terminal area and radar operations, and instrument departure and approach procedures;

(9) Aircraft loading, weight and balance, use of charts, graphs, tables, formulas, and computations, and their effect on aircraft performance;

(10) Aerodynamics relating to an aircraft's flight characteristics and performance in normal and abnormal flight regimes;

(11) Human factors;

(12) Aeronautical decision making and judgment; and

(13) Crew resource management to include crew communication and coordination.

§ 61.157   Flight proficiency.

(a) *General.* (1) The practical test for an airline transport pilot certificate is given for—

(i) An airplane category and single-engine class rating;

(ii) An airplane category and multiengine class rating;

(iii) A rotorcraft category and helicopter class rating;

(iv) A powered-lift category rating; and

(v) An aircraft type rating for the category and class ratings listed in paragraphs (a)(1)(i) through (a)(1)(iv) of this section.

(2) A person who is applying for an airline transport pilot practical test must meet—

(i) The eligibility requirements of § 61.153 of this part; and

(ii) The aeronautical knowledge and aeronautical experience requirements of this subpart that apply to the aircraft category and class rating sought.

(b) *Aircraft type rating.* Except as provided in paragraph (c) of this section, a person who is applying for an aircraft type rating to be added to an airline transport pilot certificate:

(1) Must receive and log ground and flight training from an authorized instructor on the areas of operation in this section that apply to the aircraft type rating sought;

(2) Must receive a logbook endorsement from an authorized instructor certifying that the applicant completed the training on the areas of operation listed in paragraph (e) of this section that apply to the aircraft type rating sought; and

(3) Must perform the practical test in actual or simulated instrument conditions, unless the aircraft's type certificate makes the aircraft incapable of operating under instrument flight rules. If the practical test cannot be accomplished for this reason, the person may obtain a type rating limited to "VFR only." The "VFR only" limitation

may be removed for that aircraft type when the person passes the practical test under instrument flight rules.

(c) *Exceptions.* A person who is applying for an aircraft type rating to be added to an airline transport pilot certificate or an aircraft type rating concurrently with an airline transport pilot certificate, and who is an employee of a certificate holder operating under part 121 or 135 of this chapter or of a fractional ownership program manager operating under subpart K of part 91 of this chapter, need not comply with the requirements of paragraph (b) of this section if the applicant presents a training record that shows satisfactory completion of that certificate holder's or program manager's approved pilot-in-command training program for the aircraft type rating sought.

(d) *Upgrading type ratings.* Any type rating(s) on the pilot certificate of an applicant who successfully completes an airline transport pilot practical test shall be included on the airline transport pilot certificate with the privileges and limitations of the airline transport pilot certificate, provided the applicant passes the practical test in the same category and class of aircraft for which the applicant holds the type rating(s). However, if a type rating for that category and class of aircraft on the superseded pilot certificate is limited to VFR, that limitation shall be carried forward to the person's airline transport pilot certificate level.

(e) Areas of operation.

(1) For an airplane category—single-engine class rating:

(i) Preflight preparation;

(ii) Preflight procedures;

(iii) Takeoff and departure phase;

(iv) In-flight maneuvers;

(v) Instrument procedures;

(vi) Landings and approaches to landings;

(vii) Normal and abnormal procedures;

(viii) Emergency procedures; and

(ix) Postflight procedures.

(2) For an airplane category—multiengine class rating

(i) Preflight preparation;

(ii) Preflight procedures;

(iii) Takeoff and departure phase;

(iv) In-flight maneuvers;

(v) Instrument procedures;

(vi) Landings and approaches to landings;

(vii) Normal and abnormal procedures;

(viii) Emergency procedures; and

(ix) Postflight procedures.

(3) For a powered-lift category rating:

(i) Preflight preparation;

(ii) Preflight procedures;

(iii) Takeoff and departure phase;

(iv) In-flight maneuvers;

(v) Instrument procedures;

(vi) Landings and approaches to landings;

(vii) Normal and abnormal procedures;

(viii) Emergency procedures; and

(ix) Postflight procedures.

(4) For a rotorcraft category—helicopter class rating: (i) Preflight preparation;

(ii) Preflight procedures;

(iii) Takeoff and departure phase;

(iv) In-flight maneuvers;

(v) Instrument procedures;

(vi) Landings and approaches to landings;

(vii) Normal and abnormal procedures;

(viii) Emergency procedures; and

(ix) Postflight procedures.

(f) Proficiency and competency checks conducted under part 121, part 135, or subpart K of part 91.

(1) Successful completion of any of the following checks satisfy the requirements of this section for the appropriate aircraft rating:

(i) A proficiency check under § 121.441 of this chapter.

(ii) Both a competency check under § 135.293 of this chapter and a pilot-in-command instrument proficiency check under § 135.297 of this chapter.

(iii) Both a competency check under § 91.1065 of this chapter and a pilot-in-command instrument proficiency check under § 91.1069 of this chapter.

(2) The checks specified in paragraph (f)(1) of this section must be conducted by an authorized designated pilot examiner or FAA aviation safety inspector.

(g) Use of a flight simulator or flight training device for an airplane rating. If a flight simulator or flight training device is used for accomplishing all of the training and the required practical test for an airline transport pilot certificate with an airplane category, class, and type rating, if applicable, the applicant, flight simulator, and flight training device are subject to the following requirements:

(1) The flight simulator and flight training device must represent that airplane type if the rating involves a type rating in an airplane, or is representative of an airplane if the applicant is only seeking an airplane class rating and does not require a type rating.

(2) The flight simulator and flight training device must be used in accordance with an approved course at a training center certificated under part 142 of this chapter.

(3) All training and testing (except preflight inspection) must be accomplished by the applicant to receive an airplane class rating and type rating, if applicable, without limitations and—

(i) The flight simulator must be qualified and approved as Level C or Level D; and

(ii) The applicant must meet the aeronautical experience requirements of § 61.159 of this part and at least one of the following—

(A) Hold a type rating for a turbojet airplane of the same class of airplane for which the type rating is sought, or have been designated by a military service as a pilot in command of an airplane of the same class of airplane for which the type rating is sought, if a turbojet type rating is sought;

(B) Hold a type rating for a turbopropeller airplane of the same class as the airplane for which the type rating is sought, or have been appointed by a military service as a pilot in command of an airplane of the same class of airplane for which the type rating is sought, if a turbopropeller airplane type rating is sought;

(C) Have at least 2,000 hours of flight time, of which 500 hours must be in turbine-powered airplanes of the same class as the airplane for which the type rating is sought;

(D) Have at least 500 hours of flight time in the same type of airplane as the airplane for which the type rating is sought; or

(E) Have at least 1,000 hours of flight time in at least two different airplanes requiring a type rating.

(4) Subject to the limitation of paragraph (g)(5) of this section, an applicant who does not meet the requirements of paragraph (g)(3) of this section may complete all training and testing (except for preflight inspection) for an additional rating if—

(i) The flight simulator is qualified and approved as Level C or Level D; and

(ii) The applicant meets the aeronautical experience requirements of § 61.159 of this part and at least one of the following—

(A) Holds a type rating in a propeller-driven airplane if a type rating in a turbojet airplane is sought, or holds a type rating in a turbojet airplane if a type rating in a propeller-driven airplane is sought;

(B) Since the beginning of the 12th calendar month before the month in which the applicant completes the practical test for the additional rating, has logged—

(1) At least 100 hours of flight time in airplanes in the same class as the airplane for which the type rating is sought and which requires a type rating; and

(2) At least 25 hours of flight time in airplanes of the same type for which the type rating is sought.

(5) An applicant meeting only the requirements of paragraph (g)(4)(ii)(A) and (B) of this section will be issued an additional rating, or an airline transport pilot certificate with an added rating, as applicable, with a limitation. The limitation shall state: "This certificate is subject to pilot-in-command limitations for the additional rating."

(6) An applicant who has been issued a certificate with the limitation specified in paragraph (g)(5) of this section—

(i) May not act as pilot in command of the aircraft for which an additional rating was obtained under the provisions of this section until the limitation is removed from the certificate; and

(ii) May have the limitation removed by accomplishing 15 hours of supervised operating experience as pilot in command under the supervision of a qualified and current pilot in command, in the seat normally occupied by the pilot in command, in an airplane of the same type for which the limitation applies.

(7) An applicant who does not meet the requirements of paragraph (g)(3)(ii)(A) through (E) or (g)(4)(ii)(A) and (B) of this section may be issued an airline transport pilot certificate or an additional rating to that pilot certificate after successful completion of one of the following requirements—

(i) An approved course at a part 142 training center that includes all training and testing for that certificate or rating, followed by training and testing on the following tasks, which must be successfully completed on a static airplane or in flight, as appropriate—

(A) Preflight inspection;

(B) Normal takeoff;

(C) Normal ILS approach;

(D) Missed approach; and

(E) Normal landing.

(ii) An approved course at a part 142 training center that complies with paragraphs (g)(8) and (g)(9) of this section and includes all training and testing for a certificate or rating.

(8) An applicant meeting only the requirements of paragraph (g)(7)(ii) of this section will be issued an additional rating or an airline transport pilot certificate with an additional rating, as applicable, with a limitation. The limitation shall state: "This certificate is subject to pilot-in-command limitations for the additional rating."

(9) An applicant issued a pilot certificate with the limitation specified in paragraph (g)(8) of this section—

(i) May not act as pilot in command of the aircraft for which an additional rating was obtained under the provisions of this section until the limitation is removed from the certificate; and

(ii) May have the limitation removed by accomplishing 25 hours of supervised operating experience as pilot in command under the supervision of a qualified and current pilot in command, in the seat normally occupied by the pilot in command, in an airplane of the same type for which the limitation applies.

(h) *Use of a flight simulator or a flight training device for a helicopter rating.* If a flight simulator or flight training device is used for accomplishing all of the training and the required practical test for an airline transport pilot certificate with a helicopter class rating and type rating, if applicable, the applicant, flight simulator, and flight training device are subject to the following requirements:

(1) The flight simulator and flight training device must represent that helicopter type if the rating involves a type rating in a helicopter, or is representative of a helicopter if the applicant is only seeking a helicopter class rating and does not require a type rating.

(2) The flight simulator and flight training device must be used in accordance with an approved course at a training center certificated under part 142 of this chapter.

(3) All training and testing requirements (except preflight inspection) must be accomplished by the applicant to receive a helicopter class rating and type rating, if applicable, without limitations and—

(i) The flight simulator must be qualified and approved as a Level C or Level D; and

(ii) The applicant must meet the aeronautical experience requirements of § 61.161 of this part and at least one of the following—

(A) Hold a type rating for a turbine-powered helicopter, or have been designated by a military service as a pilot in command of a turbine-powered helicopter, if a turbine-powered helicopter type rating is sought;

(B) Have at least 1,200 hours of flight time, of which 500 hours must be in turbine-powered helicopters;

(C) Have at least 500 hours of flight time in the same type helicopter as the helicopter for which the type rating is sought; or

(D) Have at least 1,000 hours of flight time in at least two different helicopters requiring a type rating.

(4) Subject to the limitation of paragraph (h)(5) of this section, an applicant who does not meet the requirements of paragraph (h)(3) of this section may complete all training and testing (except for preflight inspection) for an additional rating if—

(i) The flight simulator is qualified and approved as Level C or Level D; and

(ii) The applicant meets the aeronautical experience requirements of § 61.161 of this part and, since the beginning of the 12th calendar month before the month in which the applicant completes the practical test for the additional rating, has logged—

(A) At least 100 hours of flight time in helicopters; and

(B) At least 15 hours of flight time in helicopters of the same type of helicopter for which the type rating is sought.

(5) An applicant meeting only the requirements of paragraph (h)(4)(ii) (A) and (B) of this section will be issued an additional rating or an airline transport pilot certificate with a limitation. The limitation shall state: "This certificate is subject to pilot-in-command limitations for the additional rating."

(6) An applicant who has been issued a certificate with the limitation specified in paragraph (h)(5) of this section—

(i) May not act as pilot in command of the helicopter for which an additional rating was obtained under the provisions of this section until the limitation is removed from the certificate; and

(ii) May have the limitation removed by accomplishing 15 hours of supervised operating experience as pilot in command under the supervision of a qualified and current pilot in command, in the seat normally occupied by the pilot in command, in a helicopter of the same type for which the limitation applies.

(7) An applicant who does not meet the requirements of paragraph (h)(3)(ii) (A) through (D), or (h)(4)(ii) (A)

and (B) of this section may be issued an airline transport pilot certificate or an additional rating to that pilot certificate after successful completion of one of the following requirements—

(i) An approved course at a part 142 training center that includes all training and testing for that certificate or rating, followed by training and testing on the following tasks, which must be successfully completed on a static aircraft or in flight, as appropriate—

(A) Preflight inspection;

(B) Normal takeoff from a hover;

(C) Manually flown precision approach; and

(D) Steep approach and landing to an off-airport heliport; or

(ii) An approved course at a training center that includes all training and testing for that certificate or rating and compliance with paragraphs (h)(8) and (h)(9) of this section.

(8) An applicant meeting only the requirements of paragraph (h)(7)(ii) of this section will be issued an additional rating or an airline transport pilot certificate with an additional rating, as applicable, with a limitation. The limitation shall state: "This certificate is subject to pilot-in-command limitations for the additional rating."

(9) An applicant issued a certificate with the limitation specified in paragraph (h)(8) of this section—

(i) May not act as pilot in command of the aircraft for which an additional rating was obtained under the provisions of this section until the limitation is removed from the certificate; and

(ii) May have the limitation removed by accomplishing 25 hours of supervised operating experience as pilot in command under the supervision of a qualified and current pilot in command, in the seat normally occupied by the pilot in command, in an aircraft of the same type for which the limitation applies.

(i) Use of a flight simulator or flight training device for a powered-lift rating. If a flight simulator or flight training device is used for accomplishing all of the training and the required practical test for an airline transport pilot certificate with a powered-lift category rating and type rating, if applicable, the applicant, flight simulator, and flight training device are subject to the following requirements:

(1) The flight simulator and flight training device must represent that powered-lift type, if the rating involves a type rating in a powered-lift, or is representative of a powered-lift if the applicant is only seeking a powered-lift category rating and does not require a type rating.

(2) The flight simulator and flight training device must be used in accordance with an approved course at a training center certificated under part 142 of this chapter.

(3) All training and testing requirements (except preflight inspection) must be accomplished by the applicant to receive a powered-lift category rating and type rating, if applicable, without limitations; and—

(i) The flight simulator must be qualified and approved as Level C or Level D; and

(ii) The applicant must meet the aeronautical experience requirements of § 61.163 of this part and at least one of the following—

(A) Hold a type rating for a turbine-powered powered-lift, or have been designated by a military service as a pilot in command of a turbine-powered powered-lift, if a turbine-powered powered-lift type rating is sought;

(B) Have at least 1,200 hours of flight time, of which 500 hours must be in turbine-powered powered-lifts;

(C) Have at least 500 hours of flight time in the same type of powered-lift for which the type rating is sought; or

(D) Have at least 1,000 hours of flight time in at least two different powered-lifts requiring a type rating.

(4) Subject to the limitation of paragraph (i)(5) of this section, an applicant who does not meet the requirements of paragraph (i)(3) of this section may complete all training and testing (except for preflight inspection) for an additional rating if—

(i) The flight simulator is qualified and approved as Level C or Level D; and

(ii) The applicant meets the aeronautical experience requirements of § 61.163 of this part and, since the beginning of the 12th calendar month before the month in which the applicant completes the practical test for the additional rating, has logged—

(A) At least 100 hours of flight time in powered-lifts; and

(B) At least 15 hours of flight time in powered-lifts of the same type of powered-lift for the type rating sought.

(5) An applicant meeting only the requirements of paragraph (i)(4)(ii) (A) and (B) of this section will be issued an additional rating or an airline transport pilot certificate with a limitation. The limitation shall state: "This certificate is subject to pilot-in-command limitations for the additional rating."

(6) An applicant who has been issued a certificate with the limitation specified in paragraph (i)(5) of this section—

(i) May not act as pilot in command of the powered-lift for which an additional rating was obtained under the provisions of this section until the limitation is removed from the certificate; and

(ii) May have the limitation removed by accomplishing 15 hours of supervised operating experience as pilot in command under the supervision of a qualified and current pilot in command, in the seat normally occupied by the pilot in command, in a powered-lift of the same type for which the limitation applies.

(7) An applicant who does not meet the requirements of paragraph (i)(3)(ii) (A) through (D) or (i)(4)(ii) (A) and (B) of this section may be issued an airline transport pilot certificate or an additional rating to that pilot certificate after successful completion of one of the following requirements—

(i) An approved course at a part 142 training center that includes all training and testing for that certificate or rating, followed by training and testing on the following tasks, which must be successfully completed on a static aircraft or in flight, as appropriate—

(A) Preflight inspection;

(B) Normal takeoff from a hover;

(C) Manually flown precision approach; and

(D) Steep approach and landing to an off-airport site; or

(ii) An approved course at a training center that includes all training and testing for that certificate or rating and is in compliance with paragraphs (i)(8) and (i)(9) of this section.

(8) An applicant meeting only the requirements of paragraph (i)(7)(ii) of this section will be issued an additional rating or an airline transport pilot certificate with an additional rating, as applicable, with a limitation. The limitation shall state: "This certificate is subject to pilot-in-command limitations for the additional rating."

(9) An applicant issued a pilot certificate with the limitation specified in paragraph (i)(8) of this section—

(i) May not act as pilot in command of the aircraft for which an additional rating was obtained under the provisions of this section until the limitation is removed from the certificate; and

(ii) May have the limitation removed by accomplishing 25 hours of supervised operating experience as pilot in command under the supervision of a qualified and current pilot in command, in the seat normally occupied by the pilot in command, in a powered-lift of the same type for which the limitation applies.

(j) *Waiver authority.* Unless the Administrator requires certain or all tasks to be performed, the examiner who conducts the practical test for an airline transport pilot certificate may waive any of the tasks for which the Administrator approves waiver authority.

[Amdt. 61-102, 62 FR 16298, Apr. 4, 1997; Amdt. 61-103, 62 FR 40905, July 30, 1997]

§ 61.158 [Reserved.]

§ 61.159 Aeronautical experience: Airplane category rating.

(a) Except as provided in paragraphs (b), (c), and (d) of this section, a person who is applying for an airline transport pilot certificate with an airplane category and class rating must have at least 1,500 hours of total time as a pilot that includes at least:

(1) 500 hours of cross-country flight time.

(2) 100 hours of night flight time.

(3) 75 hours of instrument flight time, in actual or simulated instrument conditions, subject to the following:

(i) Except as provided in paragraph (a)(3)(ii) of this section, an applicant may not receive credit for more than a total of 25 hours of simulated instrument time in a flight simulator or flight training device.

(ii) A maximum of 50 hours of training in a flight simulator or flight training device may be credited toward the instrument flight time requirements of paragraph (a)(3) of this section if the training was accomplished in a course conducted by a training center certificated under part 142 of this chapter.

(iii) Training in a flight simulator or flight training device must be accomplished in a flight simulator or flight training device, representing an airplane.

(4) 250 hours of flight time in an airplane as a pilot in command, or as second in command performing the duties of a pilot in command while under the supervision of a pilot in command, or any combination thereof, which includes at least—

(i) 100 hours of cross-country flight time; and

(ii) 25 hours of night flight time.

(5) Not more than 100 hours of the total aeronautical experience requirements of paragraph (a) of this section may be obtained in a flight simulator or flight training device that represents an airplane, provided the aeronautical experience was obtained in an approved course conducted by a training center certificated under part 142 of this chapter.

(b) A person who has performed at least 20 night takeoffs and landings to a full stop may substitute each additional night takeoff and landing to a full stop for 1 hour of night flight time to satisfy the requirements of paragraph (a)(2) of this section; however, not more than 25 hours of night flight time may be credited in this manner.

(c) A commercial pilot may credit the following second-in-command flight time or flight-engineer flight time toward the 1,500 hours of total time as a pilot required by paragraph (a) of this section:

(1) Second-in-command time, provided the time is acquired in an airplane—

(i) Required to have more than one pilot flight crewmember by the airplane's flight manual, type certificate, or the regulations under which the flight is being conducted;

(ii) Engaged in operations under subpart K of part 91, part 121, or part 135 of this chapter for which a second in command is required; or

(iii) That is required by the operating rules of this chapter to have more than one pilot flight crewmember.

(2) Flight-engineer time, provided the time—

(i) Is acquired in an airplane required to have a flight engineer by the airplane's flight manual or type certificate;

(ii) Is acquired while engaged in operations under part 121 of this chapter for which a flight engineer is required;

(iii) Is acquired while the person is participating in a pilot training program approved under part 121 of this chapter; and

(iv) Does not exceed more than 1 hour for each 3 hours of flight engineer flight time for a total credited time of no more than 500 hours.

(d) An applicant may be issued an airline transport pilot certificate with the endorsement, "Holder does not meet the pilot in command aeronautical experience requirements of ICAO," as prescribed by Article 39 of the Convention on International Civil Aviation, if the applicant:

(1) Credits second-in-command or flight-engineer time under paragraph (c) of this section toward the 1,500 hours total flight time requirement of paragraph (a) of this section;

(2) Does not have at least 1,200 hours of flight time as a pilot, including no more than 50 percent of his or her second-in-command time and none of his or her flight-engineer time; and

(3) Otherwise meets the requirements of paragraph (a) of this section.

(e) When the applicant specified in paragraph (d) of this section presents satisfactory evidence of the accumulation of 1,200 hours of flight time as a pilot including no more than 50 percent of his or her second-in-command flight time and none of his or her flight-engineer time, the applicant is entitled to an airline transport pilot certificate without the endorsement prescribed in that paragraph.

[Amdt. 61-102, 62 FR 16298, Apr. 4, 1997; Amdt. 61-103, 62 FR 40906, July 30, 1997]

§ 61.161 Aeronautical experience: Rotorcraft category and helicopter class rating.

(a) A person who is applying for an airline transport pilot certificate with a rotorcraft category and helicopter class rating, must have at least 1,200 hours of total time as a pilot that includes at least:

(1) 500 hours of cross-country flight time;

(2) 100 hours of night flight time, of which 15 hours are in helicopters;

(3) 200 hours of flight time in helicopters, which includes at least 75 hours as a pilot in command, or as second in command performing the duties of a pilot in command under the supervision of a pilot in command, or any combination thereof; and

(4) 75 hours of instrument flight time in actual or simulated instrument meteorological conditions, of which at least 50 hours are obtained in flight with at least 25 hours in helicopters as a pilot in command, or as second in command performing the duties of a pilot in command under the supervision of a pilot in command, or any combination thereof.

(b) Training in a flight simulator or flight training device may be credited toward the instrument flight time requirements of paragraph (a)(4) of this section, subject to the following:

(1) Training in a flight simulator or a flight training device must be accomplished in a flight simulator or flight training device that represents a rotorcraft.

(2) Except as provided in paragraph (b)(3) of this section, an applicant may receive credit for not more than a total of 25 hours of simulated instrument time in a flight simulator and flight training device.

(3) A maximum of 50 hours of training in a flight simulator or flight training device may be credited toward the instrument flight time requirements of paragraph (a)(4) of this section if the aeronautical experience is accomplished

in an approved course conducted by a training center certificated under part 142 of this chapter.

[Amdt. 61-102, 62 FR 16298, Apr. 4, 1997; Amdt. 61-103, 62 FR 40906, July 30, 1997]

§ 61.163 Aeronautical experience: Powered-lift category rating.

(a) A person who is applying for an airline transport pilot certificate with a powered-lift category rating must have at least 1,500 hours of total time as a pilot that includes at least:

(1) 500 hours of cross-country flight time;

(2) 100 hours of night flight time;

(3) 250 hours in a powered-lift as a pilot in command, or as a second in command performing the duties of a pilot in command under the supervision of a pilot in command, or any combination thereof, which includes at least:

(i) 100 hours of cross-country flight time; and

(ii) 25 hours of night flight time.

(4) 75 hours of instrument flight time in actual or simulated instrument conditions, subject to the following:

(i) Except as provided in paragraph (a)(4)(ii) of this section, an applicant may not receive credit for more than a total of 25 hours of simulated instrument time in a flight simulator or flight training device.

(ii) A maximum of 50 hours of training in a flight simulator or flight training device may be credited toward the instrument flight time requirements of paragraph (a)(4) of this section if the training was accomplished in a course conducted by a training center certificated under part 142 of this chapter.

(iii) Training in a flight simulator or flight training device must be accomplished in a flight simulator or flight training device that represents a powered-lift.

(b) Not more than 100 hours of the total aeronautical experience requirements of paragraph (a) of this section may be obtained in a flight simulator or flight training device that represents a powered-lift, provided the aeronautical experience was obtained in an approved course conducted by a training center certificated under part 142 of this chapter.

[Amdt. 61-102, 62 FR 16298, Apr. 4, 1997; Amdt. 61-103, 62 FR 40906, July 30, 1997]

§ 61.165 Additional aircraft category and class ratings.

(a) Rotorcraft category and helicopter class rating. A person applying for an airline transport certificate with a rotorcraft category and helicopter class rating who holds an airline transport pilot certificate with another aircraft category rating must:

(1) Meet the eligibility requirements of § 61.153 of this part;

(2) Pass a knowledge test on the aeronautical knowledge areas of § 61.155(c) of this part;

(3) Comply with the requirements in § 61.157(b) of this part, if appropriate;

(4) Meet the applicable aeronautical experience requirements of § 61.161 of this part; and

(5) Pass the practical test on the areas of operation of § 61.157(e)(4) of this part.

(b) *Airplane category rating with a single-engine class rating.* A person applying for an airline transport certificate with an airplane category and single-engine class rating who holds an airline transport pilot certificate with another aircraft category rating must:

(1) Meet the eligibility requirements of § 61.153 of this part;

(2) Pass a knowledge test on the aeronautical knowledge areas of § 61.155(c) of this part;

(3) Comply with the requirements in § 61.157(b) of this part, if appropriate;

(4) Meet the applicable aeronautical experience requirements of § 61.159 of this part; and

(5) Pass the practical test on the areas of operation of § 61.157(e)(1) of this part.

(c) *Airplane category rating with a multiengine class rating.* A person applying for an airline transport certificate with an airplane category and multiengine class rating who holds an airline transport certificate with another aircraft category rating must:

(1) Meet the eligibility requirements of § 61.153 of this part;

(2) Pass a knowledge test on the aeronautical knowledge areas of § 61.155(c) of this part;

(3) Comply with the requirements in § 61.157(b) of this part, if appropriate;

(4) Meet the applicable aeronautical experience requirements of § 61.159 of this part; and

(5) Pass the practical test on the areas of operation of § 61.157(e)(2) of this part.

(d) *Powered-lift category.* A person applying for an airline transport pilot certificate with a powered-lift category rating who holds an airline transport certificate with another aircraft category rating must:

(1) Meet the eligibility requirements of § 61.153 of this part;

(2) Pass a required knowledge test on the aeronautical knowledge areas of § 61.155(c) of this part;

(3) Comply with the requirements in § 61.157(b) of this part, if appropriate;

(4) Meet the applicable aeronautical experience requirements of § 61.163 of this part; and

(5) Pass the required practical test on the areas of operation of § 61.157(e)(3) of this part.

(e) *Additional class rating within the same aircraft category.* A person applying for an airline transport certificate with an additional class rating who holds an airline transport certificate in the same aircraft category must—

(1) Meet the eligibility requirements of § 61.153, except paragraph (f) of that section;

(2) Comply with the requirements in § 61.157(b) of this part, if applicable;

(3) Meet the applicable aeronautical experience requirements of subpart G of this part; and

(4) Pass a practical test on the areas of operation of § 61.157(e) appropriate to the aircraft rating sought.

(f) *Category class ratings for the operation of aircraft with experimental certificates.* Notwithstanding the provisions of paragraphs (a) through (e) of this section, a person holding an airline transport certificate may apply for a category and class rating limited to a specific make and model of experimental aircraft, provided—

(1) The person has logged at least 5 hours flight time while acting as pilot in command in the same category, class, make, and model of aircraft that has been issued an experimental certificate;

(2) The person has received a logbook endorsement from an authorized instructor who has determined that he or she is proficient to act as pilot in command of the same category, class, make, and model of aircraft for which application is made; and

(3) The flight time specified in paragraph (f)(1) of this section must be logged between September 1, 2004 and August 31, 2005.

[Amdt. 61-102, 62 FR 16298, Apr. 4, 1997; Amdt. 61-103, 62 FR 40906, July 30, 1997, and amended at Amdt. 61-103, 62 FR 40906, July 30, 1997]

§ 61.167 Privileges.

(a) A person who holds an airline transport pilot certificate is entitled to the same privileges as those afforded a person who holds a commercial pilot certificate with an instrument rating.

(b) An airline transport pilot may instruct—

(1) Other pilots in air transportation service in aircraft of the category, class, and type, as applicable, for which the airline transport pilot is rated and endorse the logbook or other training record of the person to whom training has been given;

(2) In flight simulators, and flight training devices representing the aircraft referenced in paragraph (b)(1) of this section, when instructing under the provisions of this section and endorse the logbook or other training record of the person to whom training has been given;

(3) Only as provided in this section, unless the airline transport pilot also holds a flight instructor certificate, in which case the holder may exercise the instructor privileges of subpart H of part 61 for which he or she is rated; and

(4) In an aircraft, only if the aircraft has functioning dual controls, when instructing under the provisions of this section.

(c) Excluding briefings and debriefings, an airline transport pilot may not instruct in aircraft, flight simulators, and flight training devices under this section—

(1) For more than 8 hours in any 24-consecutive-hour period; or

(2) For more than 36 hours in any 7-consecutive-day period.

(d) An airline transport pilot may not instruct in Category II or Category III operations unless he or she has been trained and successfully tested under Category II or Category III operations, as applicable.

[Amdt. 61-102, 62 FR 16298, Apr. 4, 1997; Amdt. 61-103, 62 FR 40907, July 30, 1997]

§§ 61.169—61.171 [Reserved.]

[Amdt. 61-102, 62 FR 16298, Apr. 4, 1997; Amdt. 61-103, 62 FR 40907, July 30, 1997]

Subpart H—Flight Instructors Other Than Flight Instructors with a Sport Pilot Rating

§ 61.181 Applicability.

This subpart prescribes the requirements for the issuance of flight instructor certificates and ratings (except for flight instructor certificates with a sport pilot rating), the conditions under which those certificates and ratings are necessary, and the limitations on those certificates and ratings.

§ 61.183 Eligibility requirements.

To be eligible for a flight instructor certificate or rating a person must:

(a) Be at least 18 years of age;

(b) Be able to read, speak, write, and understand the English language. If the applicant is unable to meet one of these requirements due to medical reasons, then the Administrator may place such operating limitations on that applicant's flight instructor certificate as are necessary;

(c) Hold either a commercial pilot certificate or airline transport pilot certificate with:

(1) An aircraft category and class rating that is appropriate to the flight instructor rating sought; and

(2) An instrument rating or privileges on that person's pilot certificate that is appropriate to the flight instructor rating sought, if applying for—

(i) A flight instructor certificate with an airplane category and single-engine class rating;

(ii) A flight instructor certificate with an airplane category and multiengine class rating;

(iii) A flight instructor certificate with a powered-lift rating; or

(iv) A flight instructor certificate with an instrument rating.

(d) Receive a logbook endorsement from an authorized instructor on the fundamentals of instructing listed in § 61.185 of this part appropriate to the required knowledge test;

(e) Pass a knowledge test on the areas listed in § 61.185(a)(1) of this part, unless the applicant:

(1) Holds a flight instructor certificate or ground instructor certificate issued under this part;

(2) Holds a current teacher's certificate issued by a State, county, city, or municipality that authorizes the person to teach at an educational level of the 7th grade or higher; or

(3) Is employed as a teacher at an accredited college or university.

(f) Pass a knowledge test on the aeronautical knowledge areas listed in § 61.185(a)(2) and (a)(3) of this part that are appropriate to the flight instructor rating sought;

(g) Receive a logbook endorsement from an authorized instructor on the areas of operation listed in § 61.187(b) of this part, appropriate to the flight instructor rating sought;

(h) Pass the required practical test that is appropriate to the flight instructor rating sought in an:

(1) Aircraft that is representative of the category and class of aircraft for the aircraft rating sought; or

(2) Flight simulator or flight training device that is representative of the category and class of aircraft for the rating sought, and used in accordance with an approved course at a training center certificated under part 142 of this chapter.

(i) Accomplish the following for a flight instructor certificate with an airplane or a glider rating:

(1) Receive a logbook endorsement from an authorized instructor indicating that the applicant is competent and possesses instructional proficiency in stall awareness, spin entry, spins, and spin recovery procedures after providing the applicant with flight training in those training areas in an airplane or glider, as appropriate, that is certificated for spins; and

(2) Demonstrate instructional proficiency in stall awareness, spin entry, spins, and spin recovery procedures. However, upon presentation of the endorsement specified in paragraph (i)(1) of this section an examiner may accept that endorsement as satisfactory evidence of instructional proficiency in stall awareness, spin entry, spins, and spin recovery procedures for the practical test, provided that the practical test is not a retest as a result of the applicant failing the previous test for deficiencies in the knowledge or skill of stall awareness, spin entry, spins, or spin recovery instructional procedures. If the retest is a result of deficiencies in the ability of an applicant to demonstrate knowledge or skill of stall awareness, spin entry, spins, or spin recovery instructional procedures, the examiner must test the person on stall awareness, spin entry, spins, and spin recovery instructional procedures in an airplane or glider, as appropriate, that is certificated for spins;

(j) Log at least 15 hours as pilot in command in the category and class of aircraft that is appropriate to the flight instructor rating sought; and

(k) Comply with the appropriate sections of this part that apply to the flight instructor rating sought.

[Amdt. 61-102, 62 FR 16298, Apr. 4, 1997; Amdt. 61-103, 62 FR 40907, July 30, 1997]

§ 61.185 Aeronautical knowledge.

(a) A person who is applying for a flight instructor certificate must receive and log ground training from an authorized instructor on:

(1) Except as provided in paragraph (b) of this section, the fundamentals of instructing, including:

(i) The learning process;

(ii) Elements of effective teaching;

(iii) Student evaluation and testing;

(iv) Course development;

(v) Lesson planning; and

(vi) Classroom training techniques.

(2) The aeronautical knowledge areas for a recreational, private, and commercial pilot certificate applicable to the aircraft category for which flight instructor privileges are sought; and

(3) The aeronautical knowledge areas for the instrument rating applicable to the category for which instrument flight instructor privileges are sought.

(b) The following applicants do not need to comply with paragraph (a)(1) of this section:

(1) The holder of a flight instructor certificate or ground instructor certificate issued under this part;

(2) The holder of a current teacher's certificate issued by a State, county, city, or municipality that authorizes the person to teach at an educational level of the 7th grade or higher; or

(3) A person employed as a teacher at an accredited college or university.

[Amdt. 61-102, 62 FR 16298, Apr. 4, 1997; Amdt. 61-103, 62 FR 40907, July 30, 1997]

§ 61.187 Flight proficiency.

(a) General. A person who is applying for a flight instructor certificate must receive and log flight and ground training from an authorized instructor on the areas of operation listed in this section that apply to the flight instructor rating sought. The applicant's logbook must contain an endorsement from an authorized instructor certifying that the person is proficient to pass a practical test on those areas of operation.

(b) Areas of operation. (1) For an airplane category rating with a single- engine class rating: (i) Fundamentals of instructing;

(ii) Technical subject areas;

(iii) Preflight preparation;

(iv) Preflight lesson on a maneuver to be performed in flight;

(v) Preflight procedures;

(vi) Airport and seaplane base operations;

(vii) Takeoffs, landings, and go-arounds;

(viii) Fundamentals of flight;

(ix) Performance maneuvers;

(x) Ground reference maneuvers;

(xi) Slow flight, stalls, and spins;

(xii) Basic instrument maneuvers;

(xiii) Emergency operations; and

(xiv) Postflight procedures.

(2) For an airplane category rating with a multiengine class rating:

(i) Fundamentals of instructing;

(ii) Technical subject areas;

(iii) Preflight preparation;

(iv) Preflight lesson on a maneuver to be performed in flight;

(v) Preflight procedures;

(vi) Airport and seaplane base operations;

(vii) Takeoffs, landings, and go-arounds;

(viii) Fundamentals of flight;

(ix) Performance maneuvers;

(x) Ground reference maneuvers;

(xi) Slow flight and stalls;

(xii) Basic instrument maneuvers;

(xiii) Emergency operations;

(xiv) Multiengine operations; and

(xv) Postflight procedures.

(3) For a rotorcraft category rating with a helicopter class rating:

(i) Fundamentals of instructing;

(ii) Technical subject areas;

(iii) Preflight preparation;

(iv) Preflight lesson on a maneuver to be performed in flight;

(v) Preflight procedures;

(vi) Airport and heliport operations;

(vii) Hovering maneuvers;

(viii) Takeoffs, landings, and go-arounds;

(ix) Fundamentals of flight;

(x) Performance maneuvers;

(xi) Emergency operations;

(xii) Special operations; and

(xiii) Postflight procedures.

(4) For a rotorcraft category rating with a gyroplane class rating:

(i) Fundamentals of instructing;

(ii) Technical subject areas;

(iii) Preflight preparation;

(iv) Preflight lesson on a maneuver to be performed in flight;

(v) Preflight procedures;

(vi) Airport operations;

(vii) Takeoffs, landings, and go-arounds;

(viii) Fundamentals of flight;

(ix) Performance maneuvers;

(x) Flight at slow airspeeds;

(xi) Ground reference maneuvers;

(xii) Emergency operations; and

(xiii) Postflight procedures.

(5) For a powered-lift category rating:

(i) Fundamentals of instructing;

(ii) Technical subject areas;

(iii) Preflight preparation;

(iv) Preflight lesson on a maneuver to be performed in flight;

(v) Preflight procedures;

(vi) Airport and heliport operations;

(vii) Hovering maneuvers;

(viii) Takeoffs, landings, and go-arounds;

(ix) Fundamentals of flight;

(x) Performance maneuvers;

(xi) Ground reference maneuvers;

(xii) Slow flight and stalls;

(xiii) Basic instrument maneuvers;

(xiv) Emergency operations;

(xv) Special operations; and

(xvi) Postflight procedures.

(6) For a glider category rating: (i) Fundamentals of instructing;

(ii) Technical subject areas;

(iii) Preflight preparation;

(iv) Preflight lesson on a maneuver to be performed in flight;

(v) Preflight procedures;

(vi) Airport and gliderport operations;

(vii) Launches, landings, and go-arounds;

(viii) Fundamentals of flight;

(ix) Performance speeds;

(x) Soaring techniques;

(xi) Performance maneuvers;

(xii) Slow flight, stalls, and spins;

(xiii) Emergency operations; and

(xiv) Postflight procedures.

(7) For an instrument rating with the appropriate aircraft category and class rating:

(i) Fundamentals of instructing;

(ii) Technical subject areas;

(iii) Preflight preparation;

(iv) Preflight lesson on a maneuver to be performed in flight;

(v) Air traffic control clearances and procedures;

(vi) Flight by reference to instruments;

(vii) Navigation aids;

(viii) Instrument approach procedures;

(ix) Emergency operations; and

(x) Postflight procedures.

(c) The flight training required by this section may be accomplished:

(1) In an aircraft that is representative of the category and class of aircraft for the rating sought; or

(2) In a flight simulator or flight training device representative of the category and class of aircraft for the rating sought, and used in accordance with an approved course at a training center certificated under part 142 of this chapter. [Amdt. 61-102, 62 FR 16298, Apr. 4, 1997; Amdt. 61-103, 62 FR 40907, July 30, 1997]

§ 61.189 Flight instructor records.

(a) A flight instructor must sign the logbook of each person to whom that instructor has given flight training or ground training.

(b) A flight instructor must maintain a record in a logbook or a separate document that contains the following:

(1) The name of each person whose logbook or student pilot certificate that instructor has endorsed for solo flight privileges, and the date of the endorsement; and

(2) The name of each person that instructor has endorsed for a knowledge test or practical test, and the record shall also indicate the kind of test, the date, and the results.

(c) Each flight instructor must retain the records required by this section for at least 3 years.

§ 61.191 Additional flight instructor ratings.

(a) A person who applies for an additional flight instructor rating on a flight instructor certificate must meet the eligibility requirements listed in § 61.183 of this part that apply to the flight instructor rating sought.

(b) A person who applies for an additional rating on a flight instructor certificate is not required to pass the knowledge test on the areas listed in § 61.185(a)(1) of this part. [Amdt. 61-102, 62 FR 16298, Apr. 4, 1997; Amdt. 61-103, 62 FR 40907, July 30, 1997]

§ 61.193 Flight instructor privileges.

A person who holds a flight instructor certificate is authorized within the limitations of that person's flight instructor certificate and ratings to give training and endorsements that are required for, and relate to:

(a) A student pilot certificate;

(b) A pilot certificate;

(c) A flight instructor certificate;

(d) A ground instructor certificate;

(e) An aircraft rating;

(f) An instrument rating;

(g) A flight review, operating privilege, or recency of experience requirement of this part;

(h) A practical test; and

(i) A knowledge test. [Amdt. 61-102, 62 FR 16298, Apr. 4, 1997; Amdt. 61-103, 62 FR 40907, July 30, 1997]

§ 61.195 Flight instructor limitations and qualifications.

A person who holds a flight instructor certificate is subject to the following limitations:

(a) Hours of training. In any 24-consecutive-hour period, a flight instructor may not conduct more than 8 hours of flight training.

(b) Aircraft ratings. A flight instructor may not conduct flight training in any aircraft for which the flight instructor does not hold:

(1) A pilot certificate and flight instructor certificate with the applicable category and class rating; and

(2) If appropriate, a type rating.

(c) Instrument Rating. A flight instructor who provides instrument flight training for the issuance of an instrument rating or a type rating not limited to VFR must hold an instrument rating on his or her flight instructor certificate and pilot certificate that is appropriate to the category and class of aircraft in which instrument training is being provided.

(d) Limitations on endorsements. A flight instructor may not endorse a:

(1) Student pilot's certificate or logbook for solo flight privileges, unless that flight instructor has—

(i) Given that student the flight training required for solo flight privileges required by this part; and

(ii) Determined that the student is prepared to conduct the flight safely under known circumstances, subject to any limitations listed in the student's logbook that the instructor considers necessary for the safety of the flight.

(2) Student pilot's certificate and logbook for a solo cross-country flight, unless that flight instructor has determined the student's flight preparation, planning, equipment, and proposed procedures are adequate for the proposed flight under the existing conditions and within any limitations listed in the logbook that the instructor considers necessary for the safety of the flight;

(3) Student pilot's certificate and logbook for solo flight in a Class B airspace area or at an airport within Class B airspace unless that flight instructor has—

(i) Given that student ground and flight training in that Class B airspace or at that airport; and

(ii) Determined that the student is proficient to operate the aircraft safely.

(4) Logbook of a recreational pilot, unless that flight instructor has—

(i) Given that pilot the ground and flight training required by this part; and

(ii) Determined that the recreational pilot is proficient to operate the aircraft safely.

(5) Logbook of a pilot for a flight review, unless that instructor has conducted a review of that pilot in accordance with the requirements of § 61.56(a) of this part; or

(6) Logbook of a pilot for an instrument proficiency check, unless that instructor has tested that pilot in accordance with the requirements of § 61.57(d) of this part.

(e) Training in an aircraft that requires a type rating. A flight instructor may not give flight training in an aircraft that requires the pilot in command to hold a type rating unless the flight instructor holds a type rating for that aircraft on his or her pilot certificate.

(f) Training received in a multiengine airplane, a helicopter, or a powered-lift. A flight instructor may not give training required for the issuance of a certificate or rating in a multiengine airplane, a helicopter, or a powered-lift unless that flight instructor has at least 5 flight hours of pilot-in-command time in the specific make and model of multiengine airplane, helicopter, or powered-lift, as appropriate.

(g) Position in aircraft and required pilot stations for providing flight training.

(1) A flight instructor must perform all training from in an aircraft that complies with the requirements of § 91.109 of this chapter.

(2) A flight instructor who provides flight training for a pilot certificate or rating issued under this part must provide that flight training in an aircraft that meets the following requirements—

(i) The aircraft must have at least two pilot stations and be of the same category, class, and type, if appropriate, that applies to the pilot certificate or rating sought.

(ii) For single-place aircraft, the pre-solo flight training must have been provided in an aircraft that has two pilot stations and is of the same category, class, and type, if appropriate.

(h) Qualifications of the flight instructor for training first-time flight instructor applicants. (1) The ground training provided to an initial applicant for a flight instructor certificate must be given by an authorized instructor who—

(i) Holds a current ground or flight instructor certificate with the appropriate rating, has held that certificate for at least 24 months, and has given at least 40 hours of ground training; or

(ii) Holds a current ground or flight instructor certificate with the appropriate rating, and has given at least 100 hours of ground training in an FAA-approved course.

(2) Except for an instructor who meets the requirements of paragraph (h)(3)(ii) of this section, a flight instructor who provides training to an initial applicant for a flight instructor certificate must—

(i) Meet the eligibility requirements prescribed in § 61.183 of this part;

(ii) Hold the appropriate flight instructor certificate and rating;

(iii) Have held a flight instructor certificate for at least 24 months;

(iv) For training in preparation for an airplane, rotorcraft, or powered-lift rating, have given at least 200 hours of flight training as a flight instructor; and

(v) For training in preparation for a glider rating, have given at least 80 hours of flight training as a flight instructor.

(3) A flight instructor who serves as a flight instructor in an FAA-approved course for the issuance of a flight instructor rating must hold a current flight instructor certificate with the appropriate rating and pass the required initial and recurrent flight instructor proficiency tests, in accordance with the requirements of the part under which the FAA-approved course is conducted, and must—

(i) Meet the requirements of paragraph (h)(2) of this section; or

(ii) Have trained and endorsed at least five applicants for a practical test for a pilot certificate, flight instructor certificate, ground instructor certificate, or an additional rating, and at least 80 percent of those applicants passed that test on their first attempt; and

(A) Given at least 400 hours of flight training as a flight instructor for training in an airplane, a rotorcraft, or for a powered-lift rating; or

(B) Given at least 100 hours of flight training as a flight instructor, for training in a glider rating.

(i) Prohibition against self-endorsements. A flight instructor shall not make any self-endorsement for a certificate, rating, flight review, authorization, operating privilege, practical test, or knowledge test that is required by this part.

(j) A flight instructor may not give training in Category II or Category III operations unless the flight instructor has been trained and tested in Category II or Category III operations, pursuant to § 61.67 or § 61.68 of this part, as applicable.

§ 61.197 Renewal of flight instructor certificates.

(a) A person who holds a flight instructor certificate that has not expired may renew that certificate by—

(1) Passing a practical test for—

(i) One of the ratings listed on the current flight instructor certificate; or

(ii) An additional flight instructor rating; or

(2) Presenting to an authorized FAA Flight Standards Inspector—

(i) A record of training students showing that, during the preceding 24 calendar months, the flight instructor has endorsed at least five students for a practical test for a certificate or rating and at least 80 percent of those students passed that test on the first attempt;

(ii) A record showing that, within the preceding 24 calendar months, the flight instructor has served as a company check pilot, chief flight instructor, company check airman, or flight instructor in a part 121 or part 135 operation, or in a position involving the regular evaluation of pilots; or

(iii) A graduation certificate showing that, within the preceding 3 calendar months, the person has successfully completed an approved flight instructor refresher course consisting of ground training or flight training, or a combination of both.

(b) The expiration month of a renewed flight instructor certificate shall be 24 calendar months from—

(1) The month the renewal requirements of paragraph (a) of this section are accomplished; or

(2) The month of expiration of the current flight instructor certificate provided—

(i) The renewal requirements of paragraph (a) of this section are accomplished within the 3 calendar months preceding the expiration month of the current flight instructor certificate, and

(ii) If the renewal is accomplished under paragraph (a)(2)(iii) of this section, the approved flight instructor refresher course must be completed within the 3 calendar months preceding the expiration month of the current flight instructor certificate.

(c) The practical test required by paragraph (a)(1) of this section may be accomplished in a flight simulator or flight training device if the test is accomplished pursuant to an approved course conducted by a training center certificated under part 142 of this chapter.

[Amdt. 61-102, 62 FR 16298, Apr. 4, 1997; Amdt. 61-103, 62 FR 40907, July 30, 1997]

§ 61.199 Expired flight instructor certificates and ratings.

(a) Flight instructor certificates. The holder of an expired flight instructor certificate may exchange that certificate for a new certificate with the same ratings by passing a practical test prescribed in § 61.183(h) of this part for one of the ratings listed on the expired flight instructor certificate.

(b) Flight instructor ratings.

(1) A flight instructor rating or a limited flight instructor rating on a pilot certificate is no longer valid and may not be exchanged for a similar rating or a flight instructor certificate.

(2) The holder of a flight instructor rating or a limited flight instructor rating on a pilot certificate may be issued a flight instructor certificate with the current ratings, but only if the person passes the required knowledge and practical test prescribed in this subpart for the issuance of the current flight instructor certificate and rating.

§ 61.201 [Reserved.]

Subpart I—Ground Instructors

§ 61.211 Applicability.

This subpart prescribes the requirements for the issuance of ground instructor certificates and ratings, the conditions under which those certificates and ratings are necessary, and the limitations upon those certificates and ratings.

§ 61.213 Eligibility requirements.

(a) To be eligible for a ground instructor certificate or rating a person must:

(1) Be at least 18 years of age;

(2) Be able to read, write, speak, and understand the English language. If the applicant is unable to meet one

of these requirements due to medical reasons, then the Administrator may place such operating limitations on that applicant's ground instructor certificate as are necessary;

(3) Except as provided in paragraph (b) of this section, pass a knowledge test on the fundamentals of instructing to include—

(i) The learning process;

(ii) Elements of effective teaching;

(iii) Student evaluation and testing;

(iv) Course development;

(v) Lesson planning; and

(vi) Classroom training techniques.

(4) Pass a knowledge test on the aeronautical knowledge areas in—

(i) For a basic ground instructor rating §§61.97, 61.105, and 61.309;

(ii) For an advanced ground instructor rating §§61.97, 61.105, 61.125, 61.155, and 61.309; and

(iii) For an instrument ground instructor rating, § 61.65.

(b) The knowledge test specified in paragraph (a)(3) of this section is not required if the applicant:

(1) Holds a ground instructor certificate or flight instructor certificate issued under this part;

(2) Holds a current teacher's certificate issued by a State, county, city, or municipality that authorizes the person to teach at an educational level of the 7th grade or higher; or

(3) Is employed as a teacher at an accredited college or university.

§ 61.215 Ground instructor privileges.

(a) A person who holds a basic ground instructor rating is authorized to provide:

(1) Ground training in the aeronautical knowledge areas required for the issuance of a sport pilot certificate, recreational pilot certificate, private pilot certificate, or associated ratings under this part;

(2) Ground training required for a sport pilot, recreational pilot, and private pilot flight review; and

(3) A recommendation for a knowledge test required for the issuance of a sport pilot certificate, recreational

pilot certificate, or private pilot certificate under this part.

(b) A person who holds an advanced ground instructor rating is authorized to provide:

(1) Ground training in the aeronautical knowledge areas required for the issuance of any certificate or rating under this part;

(2) Ground training required for any flight review; and

(3) A recommendation for a knowledge test required for the issuance of any certificate under this part.

(c) A person who holds an instrument ground instructor rating is authorized to provide:

(1) Ground training in the aeronautical knowledge areas required for the issuance of an instrument rating under this part;

(2) Ground training required for an instrument proficiency check; and

(3) A recommendation for a knowledge test required for the issuance of an instrument rating under this part.

(d) A person who holds a ground instructor certificate is authorized, within the limitations of the ratings on the ground instructor certificate, to endorse the logbook or other training record of a person to whom the holder has provided the training or recommendation specified in paragraphs (a) through (c) of this section.

§ 61.217 Recent experience requirements.

The holder of a ground instructor certificate may not perform the duties of a ground instructor unless, within the preceding 12 months:

(a) The person has served for at least 3 months as a ground instructor; or

(b) The person has received an endorsement from an authorized ground or flight instructor certifying that the person has demonstrated satisfactory proficiency in the subject areas prescribed in § 61.213 (a)(3) and (a)(4), as applicable.

[Amdt. 61-102, 62 FR 16298, Apr. 4, 1997; Amdt. 61-103, 62 FR 40907, July 30, 1997]

Subpart J—Sport Pilots

§ 61.301 What is the purpose of this subpart and to whom does it apply?

(a) This subpart prescribes the following requirements that apply to a sport pilot certificate:

(1) Eligibility.

(2) Aeronautical knowledge.

(3) Flight proficiency.

(4) Aeronautical experience.

(5) Endorsements.

(6) Privileges and limits.

(7) Transition provisions for registered ultralight pilots.

(b) Other provisions of this part apply to the logging of flight time and testing.

(c) This subpart applies to applicants for, and holders of, sport pilot certificates. It also applies to holders of recreational pilot certificates and higher, as provided in § 61.303.

§ 61.303 If I want to operate a light-sport aircraft, what operating limits and endorsement requirements in this subpart must I comply with?

(a) Use the following table to determine what operating limits and endorsement requirements in this subpart, if any, apply to you when you operate a light-sport aircraft. The medical certificate specified in this table must be valid. If you hold a recreational pilot certificate, but not a medical certificate, you must comply with cross-country requirements in §61.101 (c), even if your flight does not exceed 50 nautical miles from your departure airport. You must also comply with requirements in other subparts of this part that apply to your certificate and the operation you conduct.

If you hold	And you hold	Then you may operate	And
(1) A medical certificate,	(i) A sport pilot certificate,	(A) Any light sport aircraft for which you hold the endorsements required for its category, class, make and model,	(1) You must hold any other endorsements required by this subpart, and comply with the limitations in §61.315.
	(ii) At least a recreational pilot certificate with a category and class rating,	(A) Any light sport aircraft in that category and class,	(1) You do not have to hold any of the endorsements required by this subpart, nor do you have to comply with the limitations in §61.315.
	(iii) At least a recreational pilot certificate but not a rating for the category and class of light sport aircraft you operate,	(A) That light sport aircraft, only if you hold the endorsements required in §61.321 for its category and class,	(1) You must comply with the limitations in §61.315, except §61.315 (c)(14) and, if a private pilot or higher, §61.315 (c)(7).
(2) Only a U.S. driver's license,	(i) A sport pilot certificate,	(A) Any light sport aircraft for which you hold the endorsements required for its category, class, make and model,	(1) You must hold any other endorsements required by this subpart, and comply with the limitations in §61.315.
	(ii) At least a recreational pilot certificate with a category and class rating,	(A) Any light sport aircraft in that category and class,	(1) You do not have to hold any of the endorsements required by this subpart, but you must comply with the limitations in §61.315.
	(iii) At least a recreational pilot certificate but not a rating for the category and class of light-sport aircraft you operate,	(A) That light sport aircraft, only if you hold the endorsements required in §61.321 for its category and class,	(1) You must comply with the limitations in §61.315, except §61.315 (c)(14) and, if a private pilot or higher, §61.315 (c)(7).
(3) Neither a medical certificate nor a U.S. driver's license,	(i) A sport pilot certificate,	(A) Only a light sport glider or balloon for which you hold the endorsements required for its category, class, make and model,	(1) You must hold any other endorsements required by this subpart, and comply with the limitations in §61.315.
	(ii) At least a private pilot certificate with a category and class rating for glider or balloon,	(A) Only a light sport glider or balloon in that category and class,	(1) You do not have to hold any of the endorsements required by this subpart, but you must comply with the limitations in §61.315.
	(iii) At least a private pilot certificate but not a rating for glider or balloon,	(A) Only a light sport glider or balloon, if you hold the endorsements required in §61.321 for its category and class,	(1) You must comply with the limitations in §61.315, except §61.315 (c)(14) and, if a private pilot or higher, §61.315 (c)(7).

(b) A person using a current and valid U.S. driver's license to meet the requirements of this paragraph must—

(1) Comply with each restriction and limitation imposed by that person's U.S. driver's license and any judicial or administrative order applying to the operation of a motor vehicle;

(2) Have been found eligible for the issuance of at least a third-class airman medical certificate at the time of his or her most recent application (if the person has applied for a medical certificate);

(3) Not have had his or her most recently issued medical certificate (if the person has held a medical certificate) sus-

pended or revoked or most recent Authorization for a Special Issuance of a Medical Certificate withdrawn; and

(4) Not know or have reason to know of any medical condition that would make that person unable to operate a light-sport aircraft in a safe manner.

§ 61.305   What are the age and language requirements for a sport pilot certificate?

(a) To be eligible for a sport pilot certificate you must:

(1) Be at least 17 years old (or 16 years old if you are applying to operate a glider or balloon).

(2) Be able to read, speak, write, and understand English. If you cannot read, speak, write, and understand English because of medical reasons, the FAA may place limits on your certificate as are necessary for the safe operation of light-sport aircraft.

§ 61.307   What tests do I have to take to obtain a sport pilot certificate?

To obtain a sport pilot certificate, you must pass the following tests:

(a) Knowledge test. You must pass a knowledge test on the applicable aeronautical knowledge areas listed in §61.309. Before you may take the knowledge test for a sport pilot certificate, you must receive a logbook endorsement from the authorized instructor who trained you or reviewed and evaluated your home-study course on the aeronautical knowledge areas listed in §61.309 certifying you are prepared for the test.

(b) Practical test. You must pass a practical test on the applicable areas of operation listed in §§61.309 and 61.311. Before you may take the practical test for a sport pilot certificate, you must receive a logbook endorsement from the authorized instructor who provided you with flight training on the areas of operation specified in §§61.309 and 61.311 in preparation for the practical test. This endorsement certifies that you meet the applicable aeronautical knowledge and experience requirements and are prepared for the practical test.

§ 61.309   What aeronautical knowledge must I have to apply for a sport pilot certificate?

Except as specified in §61.329, to apply for a sport pilot certificate you must receive and log ground training from an authorized instructor or complete a home-study course on the following aeronautical knowledge areas:

(a) Applicable regulations of this chapter that relate to sport pilot privileges, limits, and flight operations.

(b) Accident reporting requirements of the National Transportation Safety Board.

If you are applying for a sport pilot certificate with...	Then you must log at least...	Which must include at least...
(a) Airplane category and single-engine land or sea class privileges,	(1) 20 hours of flight time, including at least 15 hours of flight training from an authorized instructor in a single-engine airplane and at least 5 hours of solo flight training in the areas of operation listed in §61.311,	(i) 2 hours of cross-country flight training, (ii) 10 takeoffs and landings to a full stop (with each landing involving a flight in the traffic pattern) at an airport; (iii) One solo cross-country flight of at least 75 nautical miles total distance, with a full-stop landing at a minimum of two points and one segment of the flight consisting of a straight-line distance of at least 25 nautical miles between the takeoff and landing locations, and (iv) 3 hours of flight training on those areas of operation specified in 61.311 preparing for the practical test within 60 days defore the date of the test.
(b) Glider category privileges, and you have not logged at least 20 hours of flight time in a heavier-than-air aircraft,	(1) 10 hours of flight time in a glider, including 10 flights in a glider receiving flight training from an authorized instructor and at least 2 hours of solo flight training in the areas of operation listed in §61.311,	(i) Five solo launches and landings, and (ii) 3 hours of flight training on those areas of operation specified in §61.311 preparing for the practical test within 60 days before the date of the test.
(c) Glider category privileges, and you have logged 20 hours flight time in a heavier-than-air aircraft,	(1) 3 hours of flight time in a glider, including five flights in a glider while receiving flight training from an authorized instructor and at least 1 hour of solo flight training in the areas of operation listed in §61.311,	(i) Three solo launches and landing, and (ii) 3 hours of flight training on those areas of operation specified in §61.311 preparing for the practical test within 60 days before the date of the test.

If you are applying for a sport pilot certificate with...	Then you must log at least...	Which must include at least...
(d) Rotorcraft category and gyroplane class privileges,	(1) 20 hours of flight time, including 15 hours of flight training from an authorized instructor in a gyroplane and at least 5 hours of solo flight training in the areas of operation listed in §61.311,	(i) 2 hours of cross-country flight training, (ii) 10 takeoffs and landings to a full stop (with each landing involving a flight in the traffic pattern) at an airport, (iii) One solo cross-country flight of at least 50 nautical miles total distance, with a full-stop landing at a minimum of two points, and one segment of the flight consisting of a straight-line distance of at least 25 nautical miles between the takeoff and landing locations, and (iv) 3 hours of flight training on those areas of operation specified in §61.311 preparing for the practical test within 60 days before the date of the test.
(e) Lighter-than-air category and airship class privileges,	(1) 20 hours of flight time, including 15 hours of flight training from an authorized instructor in an airship and at least 3 hours performing the duties of pilot in command in an airship with an authorized instructor in the areas of operation listed in §61.311,	(i) 2 hours of cross-country flight training, (ii) Three takeoffs and landings to a full stop (with each landing involving a flight in the traffic pattern) at an airport, (iii) One cross-country flight of at least 25 nautical miles between the takeoff and landing locations, and (iv) 3 hours of flight training on those areas of operation specified in §61.311 preparing for the practical test within 60 days before the date of the test.
(f) Lighter-than-air category and balloon class privileges,	(1) 7 hours of flight time in a balloon, including three flights with an authorized instructor and one flight performing the duties of pilot in command in a balloon with an authorized instructor in the areas of operation listed in §61.311,	(i) 2 hours of cross-country flight training, and (ii) 3 hours of flight training on those areas of operation specified in §61.311 preparing for the practical test within 60 days before the date of the test.
(g) Powered parachute category land or sea class privileges,	(1) 12 hours of flight time in a powered parachute, including 10 hours flight training and, and at least 2 hours solo flight training in the areas of operation listed in §61.311,	(i) 1 hour of cross-country flight training, (ii) 20 takeoffs and landings to a full stop in a powered parachute with each landing involving flight in the traffic pattern at an airport; (iii) 10 solo takeoffs and landings to a full stop (with each landing involving a flight in the traffic pattern) at an airport, (iv) One solo flight with a landing at a different airport and one segment of the flight consisting of a straight-line distance of at least 10 nautical miles between takeoff and landing locations, and (v) 3 hours of flight training on those areas of operation specified in §61.311 preparing for the practical test within 60 days before the date of the test.
(h) Weight-shift-control aircraft category land or sea class privileges,	(1) 20 hours of flight time, including 15 hours of flight training from an authorized instructor in a weight-shift-control aircraft and at least 5 hours of solo flight training in the areas of operation listed in §61.311,	(i) 2 hours of cross-country flight training; (ii) 10 takeoffs and landings to a full stop (with each landing involving a flight in the traffic pattern) at an airport, (iii) One solo cross-country flight of at least 50 nautical miles total distance, with a full-stop landing at a minimum of two points, and one segment of the flight consisting of a straight-line distance of at least 25 nautical miles between takeoff and landing locations, and (iv) 3 hours of flight training on those areas of operation specified in §61.311 preparing for the practical test within 60 days before the date of the test.

(c) Use of the applicable portions of the aeronautical information manual and FAA advisory circulars.

(d) Use of aeronautical charts for VFR navigation using pilotage, dead reckoning, and navigation systems, as appropriate.

(e) Recognition of critical weather situations from the ground and in flight, windshear avoidance, and the procurement and use of aeronautical weather reports and forecasts.

(f) Safe and efficient operation of aircraft, including collision avoidance, and recognition and avoidance of wake turbulence.

(g) Effects of density altitude on takeoff and climb performance. (h) Weight and balance computations.

(h) Principles of aerodynamics, powerplants, and aircraft systems.

(i) Stall awareness, spin entry, spins, and spin recovery techniques, as applicable.

(j) Aeronautical decision making and risk management.

(k) Preflight actions that include—

(1) How to get information on runway lengths at airports of intended use, data on takeoff and landing distances, weather reports and forecasts, and fuel requirements; and

(2) How to plan for alternatives if the planned flight cannot be completed or if you encounter delays.

§ 61.311   What flight proficiency requirements must I meet to apply for a sport pilot certificate?

Except as specified in §61.329, to apply for a sport pilot certificate you must receive and log ground and flight training from an authorized instructor on the following areas of operation, as appropriate, for airplane single-engine land or sea, glider, gyroplane, airship, balloon, powered parachute land or sea, and weight-shift-control aircraft land or sea privileges:

(a) Preflight preparation.

(b) Preflight procedures.

(c) Airport, seaplane base, and gliderport operations, as applicable.

(d) Takeoffs (or launches), landings, and go-arounds.

(e) Performance maneuvers, and for gliders, performance speeds.

(f) Ground reference maneuvers (not applicable to gliders and balloons).

(g) Soaring techniques (applicable only to gliders).

(h) Navigation.

(i) Slow flight (not applicable to lighter-than-air aircraft and powered parachutes).

(j) Stalls (not applicable to lighter-than-air aircraft, gyroplanes, and powered parachutes).

(k) Emergency operations.

(l) Post-flight procedures.

§ 61.313   What aeronautical experience must I have to apply for a sport pilot certificate?

Except as specified in §61.329, use the following table to determine the aeronautical experience you must have to apply for a sport pilot certificate:

§ 61.315   What are the privileges and limits of my sport pilot certificate?

(a) If you hold a sport pilot certificate you may act as pilot in command of a light-sport aircraft, except as specified in paragraph (c) of this section.

(b) You may share the operating expenses of a flight with a passenger, provided the expenses involve only fuel, oil, airport expenses, or aircraft rental fees. You must pay at least half the operating expenses of the flight.

(c) You may not act as pilot in command of a light-sport aircraft:

(1) That is carrying a passenger or property for compensation or hire.

(2) For compensation or hire.

(3) In furtherance of a business.

(4) While carrying more than one passenger.

(5) At night.

(6) In Class A airspace.

(7) In Class B, C, and D airspace, at an airport located in Class B, C, or D airspace, and to, from, through, or at an airport having an operational control tower unless you have met the requirements specified in §61.325.

(8) Outside the United States, unless you have prior authorization from the country in which you seek to operate. Your sport pilot certificate carries the limit "Holder does not meet ICAO requirements."

(9) To demonstrate the aircraft in flight to a prospective buyer if you are an aircraft salesperson.

(10) In a passenger-carrying airlift sponsored by a charitable organization.

(11) At an altitude of more than 10,000 feet MSL.

(12) When the flight or surface visibility is less than 3 statute miles.

(13) Without visual reference to the surface.

(14) If the aircraft has a V_H that exceeds 87 knots CAS, unless you have met the requirements of §61.327.

(15) Contrary to any operating limitation placed on the airworthiness certificate of the aircraft being flown.

(16) Contrary to any limit or endorsement on your pilot certificate, airman medical certificate, or any other limit or endorsement from an authorized instructor.

(17) Contrary to any restriction or limitation on your U.S. driver's license or any restriction or limitation imposed by judicial or administrative order when using your driver's license to satisfy a requirement of this part.

(18) While towing any object.

(19) As a pilot flight crewmember on any aircraft for which more than one pilot is required by the type certificate of the aircraft or the regulations under which the flight is conducted.

§ 61.317   Is my sport pilot certificate issued with aircraft category and class ratings?

Your sport pilot certificate does not list aircraft category and class ratings. When you successfully pass the practical test for a sport pilot certificate, regardless of the light-sport aircraft privileges you seek, the FAA will issue you a sport pilot certificate without any category and class ratings. The FAA will provide you with a logbook endorsement for the category, class, and make and model aircraft of aircraft in which you are authorized to act as pilot in command.

§ 61.319 Can I operate a make and model of aircraft other than the make and model aircraft for which I have received an endorsement?

If you hold a sport pilot certificate you may operate any make and model of light-sport aircraft in the same category and class and within the same set of aircraft as the make and model of aircraft for which you have received an endorsement.

§ 61.321 How do I obtain privileges to operate an additional category or class of light-sport aircraft?

If you hold a sport pilot certificate and seek to operate an additional category or class of light-sport aircraft, you must

(a) Receive a logbook endorsement from the authorized instructor who trained you on the applicable aeronautical knowledge areas specified in §61.309 and areas of operation specified in §61.311. The endorsement certifies you have met the aeronautical knowledge and flight proficiency requirements for the additional light-sport aircraft privilege you seek;

(b) Successfully complete a proficiency check from an authorized instructor other than the instructor who trained you on the aeronautical knowledge areas and areas of operation specified in §§61.309 and 61.311 for the additional light-sport aircraft privilege you seek;

(c) Complete an application for those privileges on a form and in a manner acceptable to the FAA and present this application to the authorized instructor who conducted the proficiency check specified in paragraph (b) of this section; and

(d) Receive a logbook endorsement from the instructor who conducted the proficiency check specified in paragraph (b) of this section certifying you are proficient in the applicable areas of operation and aeronautical knowledge areas, and that you are authorized for the additional category and class light-sport aircraft privilege.

§ 61.323 How do I obtain privileges to operate a make and model of light-sport aircraft in the same category and class within a different set of aircraft?

If you hold a sport pilot certificate and seek to operate a make and model of light-sport aircraft in the same category and class but within a different set of aircraft as the make and model of aircraft for which you have received an endorsement, you must

(a) Receive and log ground and flight training from an authorized instructor in a make and model of light-sport aircraft that is within the same set of aircraft as the make and model of aircraft you intend to operate;

(b) Receive a logbook endorsement from the authorized instructor who provided you with the aircraft specific training specified in paragraph (a) of this section certifying you are proficient to operate the specific make and model of light-sport aircraft.

§ 61.325 How do I obtain privileges to operate a light-sport aircraft at an airport within, or in airspace within, Class B, C, and D airspace, or in other airspace with an airport having an operational control tower?

If you hold a sport pilot certificate and seek privileges to operate a light-sport aircraft in Class B, C, or D airspace, at an airport located in Class B, C, or D airspace, or to, from, through, or at an airport having an operational control tower, you must receive and log ground and flight training. The authorized instructor who provides this training must provide a logbook endorsement that certifies you are proficient in the following aeronautical knowledge areas and areas of operation:

(a) The use of radios, communications, navigation system/facilities, and radar services.

(b) Operations at airports with an operating control tower to include three takeoffs and landings to a full stop, with each landing involving a flight in the traffic pattern, at an airport with an operating control tower.

(c) Applicable flight rules of part 91 of this chapter for operations in Class B, C, and D airspace and air traffic control clearances.

§ 61.327 How do I obtain privileges to operate a light-sport aircraft that has a V_H greater than 87 knots CAS?

If you hold a sport pilot certificate and you seek to operate a light-sport aircraft that has a V_H greater than 87 knots CAS you must

(a) Receive and log ground and flight training from an authorized instructor in an aircraft that has a V_H greater than 87 knots CAS; and

(b) Receive a logbook endorsement from the authorized instructor who provided the training specified in paragraph (a) of this section certifying that you are proficient in the operation of light-sport aircraft with a V_H greater than 87 knots CAS.

§ 61.329 Are there special provisions for obtaining a sport pilot certificate for persons who are registered ultralight pilots with an FAA-recognized ultralight organization?

(a) If you are a registered ultralight pilot with an FAA-recognized ultralight organization use the following table to determine how to obtain a sport pilot certificate.

If you are...	Then you must...
(1) A registered ultralight pilot with an FAA-recognized ultralight organization on or before September 1, 2004, and you want to apply for a sport pilot certificate,	(i) Not later than January 31, 2007— (A) Meet the eligibility requirements in §61.305 and 61.23, but not the aeronautical knowledge requirements specified in §61.309, the flight proficiency requirements specified in §61.311, and the aeronautical experience requirements specified in §61.313, (B) Pass the knowledge test for a sport pilot certificate specified in §61.307 or the knowledge test for a flight instructor certificate with a sport pilot rating specified in §61.405, (C) Pass the practical test for a sport pilot certificate specified in §61.307, (D) Provide the FAA with a certified copy of your ultralight pilot records from an FAA-recognized ultralight organization, and those records must— (1) Document that you are a registered ultralight pilot with that FAA-recognized ultralight organization, and (2) Indicate that you are recognized to operate each category and class of aircraft for which you seek sport pilot privileges.
(2) A registered ultralight pilot with an FAA-recognized ultralight organization after September 1, 2004, and you want to apply for a sport pilot certificate,	(i) Meet the eligibility requirements in §§61.305 and 61.23, (ii) Meet the aeronautical knowledge requirements specified in §61.309, the flight proficiency requirements specified in §61.311, and aeronautical experience requirements specified in §61.313; however, you may credit your ultralight aeronautical experience in accordance with §61.52 toward the requirements in §§61.309, 61.311, and 61.313, (iii) Pass the knowledge and practical tests for a sport pilot certificate specified in §61.307, and (iv) Provide the FAA with a certified copy of your ultralight pilot records from an FAA-recognized ultralight organization, and those records must— (A) Document that you are a registered ultralight pilot with that FAA-recognized ultralight organization, and (B) Indicate that you are recognized to operate the category and class of aircraft for which you seek sport pilot privileges.

(b) When you successfully pass the practical test for a sport pilot certificate, the FAA will issue you a sport pilot certificate without any category and class ratings. The FAA will provide you with a logbook endorsement for the category, class, and make and model of aircraft in which you have successfully passed the practical test and for which you are authorized to act as pilot in command. If you meet the provisions of paragraph (a)(1) of this section, the FAA will provide you with a logbook endorsement for each category, class, and make and model of aircraft listed on the ultralight pilot records you provide to the FAA.

Subpart K—Flight Instructors with a Sport Pilot Rating

§ 61.401   What is the purpose of this subpart?

(a) This part prescribes the following requirements that apply to a flight instructor certificate with a sport pilot rating:

(1) Eligibility.

(2) Aeronautical knowledge.

(3) Flight proficiency.

(4) Endorsements.

(5) Privileges and limits.

(6) Transition provisions for registered ultralight flight instructors.

(b) Other provisions of this part apply to the logging of flight time and testing.

§ 61.403   What are the age, language, and pilot certificate requirements for a flight instructor certificate with a sport pilot rating?

To be eligible for a flight instructor certificate with a sport pilot rating you must:

(a) Be at least 18 years old.

(b) Be able to read, speak, write, and understand English. If you cannot read, speak, write, and understand English because of medical reasons, the FAA may place limits on your certificate as are necessary for the safe operation of light-sport aircraft.

(c) Hold at least a current and valid sport pilot certificate with category and class ratings or privileges, as applicable, that are appropriate to the flight instructor privileges sought.

§ 61.405   What tests do I have to take to obtain a flight instructor certificate with a sport pilot rating?

To obtain a flight instructor certificate with a sport pilot rating you must pass the following tests:

(a) Knowledge test. Before you take a knowledge test, you must receive a logbook endorsement certifying you are prepared for the test from an authorized instructor who trained you or evaluated your home-study course on the aeronautical knowledge areas listed in §61.407. You must pass knowledge tests on

(1) The fundamentals of instructing listed in §61.407 (a), unless you meet the requirements of §61.407 (c); and

(2) The aeronautical knowledge areas for a sport pilot certificate applicable to the aircraft category and class for which flight instructor privileges are sought.

(b) Practical test.

(1) Before you take the practical test, you must-

(i) Receive a logbook endorsement from the authorized instructor who provided you with flight training on the areas of operation specified in §61.409 that apply to the category and class of aircraft privileges you seek. This endorsement certifies you meet the applicable aeronautical knowledge and experience requirements and are prepared for the practical test;

(ii) If you are seeking privileges to provide instruction in an airplane or glider, receive a logbook endorsement from an authorized instructor indicating that you are competent and possess instructional proficiency in stall awareness, spin entry, spins, and spin recovery procedures after you have received flight training in those training areas in an airplane or glider, as appropriate, that is certificated for spins;

(2) You must pass a practical test—

(i) On the areas of operation listed in §61.409 that are appropriate to the category and class of aircraft privileges you seek;

(ii) In an aircraft representative of the category and class of aircraft for the privileges you seek;

(iii) In which you demonstrate that you are able to teach stall awareness, spin entry, spins, and spin recovery procedures if you are seeking privileges to provide instruction in an airplane or glider. If you have not failed a practical test based on deficiencies in your ability to demonstrate knowledge or skill in these areas and you provide the endorsement required by paragraph (b)(1)(ii) of this section, an examiner may accept the endorsement instead of the demonstration required by this paragraph. If you are taking a test because you previously failed a test based on not meeting the requirements of this paragraph, you must pass a practical test on stall awareness, spin entry, spins, and spin recovery instructional competency and proficiency in the applicable category and class of aircraft that is certificated for spins.

§ 61.407 What aeronautical knowledge must I have to apply for a flight instructor certificate with a sport pilot rating?

(a) Except as specified in paragraph (c) of this section you must receive and log ground training from an authorized instructor on the fundamentals of instruction that includes:

(1) The learning process.
(2) Elements of effective teaching.
(3) Student evaluation and testing.
(4) Course development.
(5) Lesson planning.
(6) Classroom training techniques.

(b) You must receive and log ground training from an authorized instructor on the aeronautical knowledge areas applicable to a sport pilot certificate for the aircraft category and class in which you seek flight instructor privileges.

(c) You do not have to meet the requirements of paragraph (a) of this section if you—

(1) Hold a flight instructor certificate or ground instructor certificate issued under this part;

(2) Hold a current teacher's certificate issued by a State, county, city, or municipality; or

(3) Are employed as a teacher at an accredited college or university.

§ 61.409 What flight proficiency requirements must I meet to apply for a flight instructor certificate with a sport pilot rating?

You must receive and log ground and flight training from an authorized instructor on the following areas of operation for the aircraft category and class in which you seek flight instructor privileges:

(a) Technical subject areas.
(b) Preflight preparation.
(c) Preflight lesson on a maneuver to be performed in flight.
(d) Preflight procedures.
(e) Airport, seaplane base, and gliderport operations, as applicable.
(f) Takeoffs (or launches), landings, and go-arounds.
(g) Fundamentals of flight.
(h) Performance maneuvers and for gliders, performance speeds.
(i) Ground reference maneuvers (except for gliders and lighter-than-air).
(j) Soaring techniques.
(k) Slow flight (not applicable to lighter-than-air and powered parachutes).
(l) Stalls (not applicable to lighter-than-air, powered parachutes, and gyroplanes).
(m) Spins (applicable to airplanes and gliders).
(n) Emergency operations.
(o) Tumble entry and avoidance techniques (applicable to weight-shift-control aircraft).
(p) Post-flight procedures.

§ 61.411 What aeronautical experience must I have to apply for a flight instructor certificate with a sport pilot rating?

Use the following table to determine the experience you must have for each aircraft category and class:

If you are applying for a flight instructor certificate with a sport pilot rating for...	Then you must log at least...	Which must include at least...
(a) Airplane category and single-engine class privileges,	(1) 150 hours of flight time as a pilot,	(i) 100 hours of flight time as pilot in command in powered aircraft, (ii) 50 hours of flight time in a single-engine airplane, (iii) 25 hours of cross-country flight time, (iv) 10 hours of cross-country flight time in a single-engine airplane, and (v) 15 hours of flight time as pilot in command in a single-engine airplane that is a light-sport aircraft.
(b) Glider category privileges,	(1) 25 hours of flight time as pilot in command in a glider, 100 flights in a glider, and 15 flights as pilot in command in a glider that is a light-sport aircraft, or (2) 100 hours in heavier-than-air aircraft, 20 flights in a glider, and 15 flights as pilot in command in a glider that is a light-sport aircraft.	
(c) Rotorcraft category and gyroplane class privileges,	(1) 125 hours of flight time as a pilot,	(i) 100 hours of flight time as pilot in command in powered aircraft, (ii) 50 hours of flight time in a gyroplane, (iii) 10 hours of cross-country flight time, (iv) 3 hours of cross-country flight time in a gyroplane, and (v) 15 hours of flight time as pilot in command in a gyroplane that is a light-sport aircraft.
(d) Lighter-than-air category and airship class privileges,	(1) 100 hours of flight time as a pilot,	(i) 40 hours of flight time in an airship, (ii) 20 hours of pilot in command time in an airship, (iii) 10 hours of cross-country flight time, (iv) 5 hours of cross-country flight time in an airship, and (v) 15 hours of flight time as pilot in command in an airship that is a light-sport aircraft.
(e) Lighter-than-air category and balloon class privileges,	(1) 35 hours of flight time as pilot-in-command,	(i) 20 hours of flight time in a balloon, (ii) 10 flights in a balloon, and (iii) 5 flights as pilot in command in a balloon that is a light-sport aircraft.
(f) Weight-shift-control aircraft category privileges,	(1) 150 hours of flight time as a pilot,	(i) 100 hours of flight time as pilot in command in powered aircraft, (ii) 50 hours of flight time in a weight-shift-control aircraft, (iii) 25 hours of cross-country flight time, (iv) 10 hours of cross-country flight time in a weight-shift-control aircraft, and (v) 15 hours of flight time as pilot in command in a weight-shift-control aircraft that is a light-sport aircraft.
(g) Powered-parachute category privileges,	(1) 100 hours of flight time as a pilot,	(i) 75 hours of flight time as pilot in command in powerd aircraft, (ii) 50 hours of flight time in a powered parachute, (iii) 15 hours of cross-country flight time, (iv) 5 hours of cross-country flight time in a powered parachute, and (v) 15 hours of flight time as pilot in command in a powered parachute that is a light-sport aircraft.

§ 61.413   What are the privileges of my flight instructor certificate with a sport pilot rating?

If you hold a fight flight instructor certificate with a sport pilot rating, you are authorized, within the limits of your certificate and rating, to provide training and logbook endorsements for:

(a) A student pilot seeking a sport pilot certificate;

(b) A sport pilot certificate;

(c) A flight instructor certificate with a sport pilot rating;

(d) A powered parachute or weight-shift-control aircraft rating;

(e) Sport pilot privileges;

(f) A flight review or operating privilege for a sport pilot;

(g) A practical test for a sport pilot certificate, a private pilot certificate with a powered parachute or weight-shift-control aircraft rating or a flight instructor certificate with a sport pilot rating;

(h) A knowledge test for a sport pilot certificate, a private pilot certificate with a powered parachute or weight-shift-control aircraft rating or a flight instructor certificate with a sport pilot rating; and

(i) A proficiency check for an additional category, class, or make and model privilege for a sport pilot certificate or a flight instructor certificate with a sport pilot rating.

§ 61.415   What are the limits of a flight instructor certificate with a sport pilot rating?

If you hold a flight instructor certificate with a sport pilot rating, you are subject to the following limits:

(a) You may not provide ground or flight training in any aircraft for which you do not hold:

(1) A sport pilot certificate with applicable category and class privileges and make and model privileges or a pilot certificate with the applicable category and class rating; and

(2) Applicable category and class privileges for your flight instructor certificate with a sport pilot rating.

(b) You may not provide ground or flight training for a private pilot certificate with a powered parachute or weight-shift-control aircraft rating unless you hold:

(1) At least a private pilot certificate with the applicable category and class rating; and

(2) Applicable category and class privileges for your flight instructor certificate with a sport pilot rating.

(c) You may not conduct more than 8 hours of flight training in any 24-consecutive-hour period.

(d) You may not endorse a:

(1) Student pilot's certificate or logbook for solo flight privileges, unless you have—

(i) Given that student the flight training required for solo flight privileges required by this part; and

(ii) Determined that the student is prepared to conduct the flight safely under known circumstances, subject to any limitations listed in the student's logbook that you consider necessary for the safety of the flight.

(2) Student pilot's certificate and logbook for a solo cross-country flight, unless you have determined the student's flight preparation, planning, equipment, and proposed procedures are adequate for the proposed flight under the existing conditions and within any limitations listed in the logbook that you consider necessary for the safety of the flight.

(3) Student pilot's certificate and logbook for solo flight in Class B, C and D airspace areas, at an airport within Class B, C, or D airspace and to from, through or on an airport having an operational control tower, unless that you have

(i) Given that student ground and flight training in that airspace or at that airport; and

(ii) Determined that the student is proficient to operate the aircraft safely.

(4) Logbook of a pilot for a flight review, unless you have conducted a review of that pilot in accordance with the requirements of §61.56.

(e) You may not provide flight training in an aircraft unless you have at least 5 hours of flight time in a make and model of light-sport aircraft within the same set of aircraft as the aircraft in which you are providing training.

(f) You may not provide training to operate a light-sport aircraft in Class B, C, and D airspace, at an airport located in Class B, C, or D airspace, and to, from, through, or at an airport having an operational control tower, unless you have the endorsement specified in §61.325, or are otherwise authorized to conduct operations in this airspace and at these airports.

(g) You may not provide training in a light-sport aircraft with a VH greater than 87 knots CAS unless you have the endorsement specified in §61.327, or are otherwise authorized to operate that light-sport aircraft.

(h) You must perform all training in an aircraft that complies with the requirements of §91.109 of this chapter.

(i) If you provide flight training for a certificate, rating or privilege, you must provide that flight training in an aircraft that meets the following:

(1) The aircraft must have at least two pilot stations and be of the same category and class appropriate to the certificate, rating or privilege sought.

(2) For single place aircraft, pre-solo flight training must be provided in an aircraft that has two pilot stations and is of the same category and class appropriate to the certificate, rating, or privilege sought.

§ 61.417   Will my flight instructor certificate with a sport pilot rating list aircraft category and class ratings?

Your flight instructor certificate does not list aircraft category and class ratings. When you successfully pass the practical test for a flight instructor certificate with a sport pilot rating, regardless of the light-sport aircraft privileges you seek, the FAA will issue you a flight instructor certificate with a sport pilot rating without any category and class ratings. The FAA will provide you with a logbook endorsement for the category and class of light-sport aircraft you are authorized to provide training in.

§ 61.419 How do I obtain privileges to provide training in an additional category or class of light-sport aircraft?

If you hold a flight instructor certificate with a sport pilot rating and seek to provide training in an additional category or class of light-sport aircraft you must—

(a) Receive a logbook endorsement from the authorized instructor who trained you on the applicable areas of operation specified in §61.409 certifying you have met the aeronautical knowledge and flight proficiency requirements for the additional category and class flight instructor privilege you seek;

(b) Successfully complete a proficiency check from an authorized instructor other than the instructor who trained you on the areas specified in §61.409 for the additional category and class flight instructor privilege you seek;

(c) Complete an application for those privileges on a form and in a manner acceptable to the FAA and present this application to the authorized instructor who conducted the proficiency check specified in paragraph (b) of this section; and

(d) Receive a logbook endorsement from the instructor who conducted the proficiency check specified in paragraph (b) of this section certifying you are proficient in the areas of operation and authorized for the additional category and class flight instructor privilege.

§ 61.421 May I give myself an endorsement?

No. If you hold a flight instructor certificate with a sport pilot rating, you may not give yourself an endorsement for any certificate, privilege, rating, flight review, authorization, practical test, knowledge test, or proficiency check required by this part.

§ 61.423 What are the recordkeeping requirements for a flight instructor with a sport pilot rating?

(a) As a flight instructor with a sport pilot rating you must:

(1) Sign the logbook of each person to whom you have given flight training or ground training.

(2) Keep a record of the name, date, and type of endorsement for:

(i) Each person whose logbook or student pilot certificate you have endorsed for solo flight privileges.

(ii) Each person for whom you have provided an endorsement for a knowledge test, practical test, or proficiency check, and the record must indicate the kind of test or check, and the results.

(iii) Each person whose logbook you have endorsed as proficient to operate(A) An additional category or class of light-sport aircraft;

(B) An additional make and model of light-sport aircraft;

(C) In Class B, C, and D airspace; at an airport located in Class B, C, or D airspace; and to, from, through, or at an airport having an operational control tower; and (D) A light-sport aircraft with a V_H greater than 87 knots CAS.

(iv) Each person whose logbook you have endorsed as proficient to provide flight training in an additional—

(A) Category or class of light-sport aircraft; and

(B) Make and model of light-sport aircraft.

(b) Within 10 days after providing an endorsement for a person to operate or provide training in an additional category and class of light-sport aircraft you must—

(1) Complete, sign, and submit to the FAA the application presented to you to obtain those privileges; and

(2) Retain a copy of the form.

(c) You must keep the records listed in this section for 3 years. You may keep these records in a logbook or a separate document.

§ 61.425 How do I renew my flight instructor certificate?

If you hold a flight instructor certificate with a sport pilot rating you may renew your certificate in accordance with the provisions of §61.197.

§ 61.427 What must I do if my flight instructor certificate with a sport pilot rating expires?

You may exchange your expired flight instructor certificate with a sport pilot rating for a new certificate with a sport pilot rating and any other rating on that certificate by passing a practical test as prescribed in §61.405 (b) or §61.183 (h) for one of the ratings listed on the expired flight instructor certificate. The FAA will reinstate any privilege authorized by the expired certificate.

§ 61.429 May I exercise the privileges of a flight instructor certificate with a sport pilot rating if I hold a flight instructor certificate with another rating?

If you hold a current and valid flight instructor certificate, a commercial pilot certificate with an airship rating, or a commercial pilot certificate with a balloon rating issued under this part, and you seek to exercise the privileges of a flight instructor certificate with a sport pilot rating, you may do so without any further showing of proficiency, subject to the following limits:

(a) You are limited to the aircraft category and class ratings listed on your flight instructor certificate, commercial pilot certificate with an airship rating, or commercial pilot certificate with a balloon rating, as appropriate, when exercising your flight instructor privileges and the privileges specified in §61.413.

(b) You must comply with the limits specified in §61.415 and the recordkeeping requirements of §61.423.

(c) If you want to exercise the privileges of your flight instructor certificate, commercial pilot certificate with an airship rating, or commercial pilot certificate with a balloon rating, as appropriate, in a category, class, or make and model of light-sport aircraft for which you are not currently rated, you must meet all applicable requirements to provide training in an additional category or class of light-sport aircraft specified in §61.419.

§ 61.431 Are there special provisions for obtaining a flight instructor certificate with a sport pilot rating for persons who are registered ultralight instructors with an FAA-recognized ultralight organization?

If you are a registered ultralight instructor with an FAA-recognized ultralight organization on or before September 1, 2004, and you want to apply for a flight instructor certificate with a sport pilot rating, not later than January 31, 2008—

(a) You must hold either a current and valid sport pilot certificate, a current recreational pilot certificate and meet the requirements §61.101 (c), or at least a current and valid private pilot certificate issued under this part.

(b) You must meet the eligibility requirements in §§61.403 and 61.23. You do not have to meet the aeronautical knowledge requirements specified in §61.407, the flight proficiency requirements specified in §61.409 and the aeronautical experience requirements specified in §61.411, except you must meet the minimum total flight time requirements in the category and class of light-sport aircraft specified in §61.411.

(c) You do not have to meet the aeronautical knowledge requirement specified in §61.407 (a) if you have passed an FAA-recognized ultralight organization's fundamentals of instruction knowledge test.

(d) You must submit a certified copy of your ultralight pilot records from the FAA-recognized ultralight organization. Those records must—

(1) Document that you are a registered ultralight flight instructor with that FAA-recognized ultralight organization; and

(2) Indicate that you are recognized to operate and provide training in the category and class of aircraft for which you seek privileges.

(e) You must pass the knowledge test and practical test for a flight instructor certificate with a sport pilot rating applicable to the aircraft category and class for which you seek flight instructor privileges.

Special Federal Aviation Regulations

SFAR NO. 73—ROBINSON R-22/R-44 SPECIAL TRAINING AND EXPERIENCE REQUIREMENTS

Sections

1. Applicability.

2. Required training, aeronautical experience, endorsements, and flight review.

3. Expiration date.

1. Applicability. Under the procedures prescribed herein, this SFAR applies to all persons who seek to manipulate the controls or act as pilot in command of a Robinson model R-22 or R-44 helicopter. The requirements stated in this SFAR are in addition to the current requirements of part 61.

2. Required training, aeronautical experience, endorsements, and flight review.

(a) Awareness Training:

(1) Except as provided in paragraph (a)(2) of this section, no person may manipulate the controls of a Robinson model R-22 or R-44 helicopter after March 27, 1995, for the purpose of flight unless the awareness training specified in paragraph (a)(3) of this section is completed and the person's logbook has been endorsed by a certified flight instructor authorized under paragraph (b)(5) of this section.

(2) A person who holds a rotorcraft category and helicopter class rating on that person's pilot certificate and meets the experience requirements of paragraph (b)(1) or paragraph (b)(2) of this section may not manipulate the controls of a Robinson model R-22 or R-44 helicopter for the purpose of flight after April 26, 1995, unless the awareness training specified in paragraph (a)(3) of this section is completed and the person's logbook has been endorsed by a certified flight instructor authorized under paragraph (b)(5) of this section.

(3) Awareness training must be conducted by a certified flight instructor who has been endorsed under paragraph (b)(5) of this section and consists of instruction in the following general subject areas:

(i) Energy management;

(ii) Mast bumping;

(iii) Low rotor RPM (blade stall);

(iv) Low G hazards; and

(v) Rotor RPM decay.

(4) A person who can show satisfactory completion of the manufacturer's safety course after January 1, 1994, may obtain an endorsement from an FAA aviation safety inspector in lieu of completing the awareness training required in paragraphs (a)(1) and (a)(2) of this section.

(b) Aeronautical Experience:

(1) No person may act as pilot in command of a Robinson model R-22 unless that person:

(i) Has had at least 200 flight hours in helicopters, at least 50 flight hours of which were in the Robinson -22; or

(ii) Has had at least 10 hours dual instruction in the Robinson R-22 and has received an endorsement from a certified flight instructor authorized under paragraph (b)(5) of this section that the individual has been given the training required by this paragraph and is proficient to act as pilot in command of an R-22. Beginning 12 calendar months after the date of the endorsement, the individual may not act as pilot in command unless the individual has completed a flight review in an R-22 within the preceding 12 calendar months and obtained an endorsement for that flight review. The dual instruction must include at least the following abnormal and emergency procedures flight training:

(A) Enhanced training in autorotation procedures,

(B) Engine rotor RPM control without the use of the governor,

(C) Low rotor RPM recognition and recovery, and

(D) Effects of low G maneuvers and proper recovery procedures.

(2) No person may act as pilot in command of a Robinson model R-44 unless that person:

(i) Has had at least 200 flight hours in helicopters, at least 50 flight hours of which were in the Robinson R-44. The pilot in command may credit up to 25 flight hours in the Robinson 22 toward the 50 hour requirement in the Robinson R-44 or

(ii) Has had at least 10 hours dual instruction in the Robinson helicopter, at least 5 hours of which must have been accomplished in the Robinson R-44 helicopter and has received an endorsement from a certified flight instructor authorized under paragraph (b) (5) of this section that the individual has been given the training required by this paragraph and is proficient to act as pilot in command of an R44. Beginning 12 calendar months after the date of the endorsement, the individual may not act as pilot in command unless the individual has completed a flight review in a Robinson R-44 within the preceding 12 calendar months and obtained an endorsement for that flight review. The dual instruction must include at least the following abnormal and emergency procedures flight training:

(A) Enhanced training in autorotation procedures,

(B) Engine rotor RPM control without the use of the governor,

(C) Low rotor RPM recognition and recovery, and

(D) Effects of low G maneuvers and proper recovery procedures.

(3) A person who does not hold a rotorcraft category and helicopter class rating must have had at least 20 hours of dual instruction in a Robinson R-22 helicopter prior to operating it in solo flight. In addition, the person must obtain an endorsement from a certified flight instructor authorized under paragraph (b)(5) of this section that instruction has been given in those maneuvers and procedures, and the instructor has found the applicant proficient to solo a Robinson R-22. This endorsement is valid for a period of 90 days. The dual instruction must include at least the following abnormal and emergency procedures flight training:

(i) Enhanced training in autorotation procedures,

(ii) Engine rotor RPM control without the use of the governor,

(iii) Low rotor RPM recognition and recovery, and

(iv) Effects of low G maneuvers and proper recovery procedures.

(4) A person who does not hold a rotorcraft category and helicopter class rating must have had at least 20 hours of dual instruction in a Robinson R-44 helicopter prior to operating it in solo flight. In addition, the person must obtain an endorsement from a certified flight instructor authorized under paragraph (b)(5) of this section that instruction has been given in those maneuvers and procedures, and the instructor has found the applicant proficient to solo a Robinson R-44. This endorsement is valid for a period of 90 days. The dual instruction must include at least the following abnormal and emergency procedures flight training:

(i) Enhanced training in autorotation procedures,

(ii) Engine rotor RPM control without the use of the governor,

(iii) Low rotor RPM recognition and recovery, and

(iv) Effects of low G maneuvers and proper recovery procedures.

(5) No certified flight instructor may provide instruction or conduct a flight review in a Robinson R-22 or R-44 unless that instructor:

(i) Completes the awareness training in paragraph (2)(a) of this SFAR;

(ii) and for the R-22, has had at least 200 flight hours in helicopters, at least 50 flight hours of which were in the Robinson R-22, or for the Robinson R-44, has had at least 200 flight hours in helicopters, 50 flight hours of which were in Robinson helicopters. Up to 25 flight hours of Robinson R-22 flight time may be credited toward the 50 hour requirement.

(iii) Has completed flight training in a Robinson R-22, R-44, or both, on the following abnormal and emergency procedures:

(A) Enhanced training in autorotation procedures,

(B) Engine rotor RPM control without the use of the governor,

(C) Low rotor RPM recognition and recovery, and

(D) Effects of low G maneuvers and proper recovery procedures.

(iv) Has been authorized by endorsement from an FAA aviation safety inspector or authorized designated examiner that the instructor has completed the appropriate training, meets the experience requirements, and has satisfactorily demonstrated an ability to provide instruction on the general subject areas of paragraph 2(a)(3) of this SFAR, and the flight training identified in paragraph 2(b)(5)(iii) of this SFAR.

(c) Flight Review:

(1) No flight review completed to satisfy § 61.56 by an individual after becoming eligible to function as pilot in command in a Robinson R-22 helicopter shall be valid for the operation of R-22 helicopter unless that flight review was taken in an R-22.

(2) No flight review completed to satisfy § 61.56 by individual after becoming eligible to function as pilot in command in a Robinson R-44 helicopter shall be valid for the operation of R-44 helicopter unless that flight review was taken in the R-44.

(d) Currency Requirements: No person may act as pilot in command of a Robinson model R-22 or R-44 helicopter carrying passengers unless the pilot in command has met the recency of flight experience requirements of § 61.57 in an R-22 or R-44, as appropriate.

3. Expiration date. This SFAR expires on March 1, 2008, unless sooner superseded or rescinded.

SFAR No. 100–1 Relief for U.S. Military and Civilian Personnel Who Are Assigned Outside the United States in Support of U.S. Armed Forces Operations

1. *Applicability*. Flight Standards District Offices are authorized to accept from an eligible person, as described in paragraph 2 of this SFAR, the following:

(a) An expired flight instructor certificate to show eligibility for renewal of a flight instructor certificate under § 61.197, or an expired written test report to show eligibility under part 61 to take a practical test;

(b) An expired written test report to show eligibility under §§ 63.33 and 63.57 to take a practical test; and

(c) An expired written test report to show eligibility to take a practical test required under part 65 or an expired inspection authorization to show eligibility for renewal under § 65.93.

2. *Eligibility*. A person is eligible for the relief described in paragraph 1 of this SFAR if:

(a) The person served in a U.S. military or civilian capacity outside the United States in support of the U.S. Armed Forces' operation during some period of time from September 11,2001, to June 20, 2010;

(b) The person's flight instructor certificate, airman written test report, or inspection authorization expired some time between September 11, 2001, and 6 calendar months after returning to the United States, or June 20, 2010, whichever is earlier; and

(c) The person complies with § 61.197 or § 65.93 of this chapter, as appropriate, or completes the appropriate practical test within 6 calendar months after returning to the United States, or June 20, 2010, whichever is earlier.

3. *Required documents*. The person must send the Airman Certificate and/or Rating Application (FAA Form 8710-1) to the appropriate Flight Standards District Office. The person must include with the application one of the following documents, which must show the date of assignment outside the United States and the date of return to the United States:

(a) An official U.S. Government notification of personnel action, or equivalent document, showing the person was a civilian on official duty for the U.S. Government outside the United States and was assigned to a U.S. Armed Forces' operation some time between September 11, 2001, and June 20, 2010;

(b) Military orders showing the person was assigned to duty outside the United States and was assigned to a U.S. Armed Forces' operation some time between September 11, 2001, and June 20, 2010; or

(c) A letter from the person's military commander or civilian supervisor providing the dates during which the person served outside the United States and was assigned to a U.S. Armed Forces' operation some time between September 11, 2001, and June 20, 2010.

4. *Expiration date*. This Special Federal Aviation Regulation No. 100 expires June 20, 2010, unless sooner superseded or rescinded.

(Also applies to Parts 63 and 65)

Part 65—Certification: Airmen other than flight crewmembers

Authority: 49 U.S.C. 106(g), 40113, 44701–44703, 44707,
    44709–44711, 45102–45103, 45301–45302.
Source: Docket No. 1179, 27 FR 7973, Aug. 10, 1962,
    unless otherwise noted.

Subpart A—General

§65.1 Applicability.

This part prescribes the requirements for issuing the following certificates and associated ratings and the general operating rules for the holders of those certificates and ratings:

(a) Air-traffic control-tower operators.

(b) Aircraft dispatchers.

(c) Mechanics.

(d) Repairmen.

(e) Parachute riggers.

§65.3 Certification of foreign airmen other than flight crewmembers.

A person who is neither a U.S. citizen nor a resident alien is issued a certificate under subpart D of this part, outside the United States, only when the Administrator finds that the certificate is needed for the operation or continued airworthiness of a U.S.-registered civil aircraft.

[Doc. 65-28, 47 FR 35693, Aug. 16, 1982]

§65.11 Application and issue.

(a) Application for a certificate and appropriate class rating, or for an additional rating, under this part must be made on a form and in a manner prescribed by the Administrator. Each person who is neither a U.S. citizen nor a resident alien and who applies for a written or practical test to be administered outside the United States or for any certificate or rating issued under this part must show evidence that the fee prescribed in appendix A of part 187 of this chapter has been paid.

(b) An applicant who meets the requirements of this part is entitled to an appropriate certificate and rating.

(c) Unless authorized by the Administrator, a person whose air traffic control tower operator, mechanic, or parachute rigger certificate is suspended may not apply for any rating to be added to that certificate during the period of suspension.

(d) Unless the order of revocation provides otherwise—

(1) A person whose air traffic control tower operator, aircraft dispatcher, or parachute rigger certificate is revoked may not apply for the same kind of certificate for 1 year after the date of revocation; and

(2) A person whose mechanic or repairman certificate is revoked may not apply for either of those kinds of certificates for 1 year after the date of revocation.

[Doc. No. 1179, 27 FR 7973, Aug. 10, 1962, as amended by Amdt. 65-9, 31 FR 13524, Oct. 20, 1966; Amdt. 65-28, 47 FR 35693, Aug. 16, 1982]

§65.12 Offenses involving alcohol or drugs.

(a) A conviction for the violation of any Federal or state statute relating to the growing, processing, manufacture, sale, disposition, possession, transportation, or importation of narcotic drugs, marihuana, or depressant or stimulant drugs or substances is grounds for—

(1) Denial of an application for any certificate or rating issued under this part for a period of up to 1 year after the date of final conviction; or

(2) Suspension or revocation of any certificate or rating issued under this part.

(b) The commission of an act prohibited by §91.19(a) of this chapter is grounds for—

(1) Denial of an application for a certificate or rating issued under this part for a period of up to 1 year after the date of that act; or

(2) Suspension or revocation of any certificate or rating issued under this part.

[Doc. No. 21956, 50 FR 15379, Apr. 17, 1985, as amended by Amdt. 65-34, 54 FR 34330, Aug. 18, 1989]

§65.13 Temporary certificate.

A certificate and ratings effective for a period of not more than 120 days may be issued to a qualified applicant, pending review of his application and supplementary documents and the issue of the certificate and ratings for which he applied.

[Doc. No. 1179, 27 FR 7973, Aug. 10, 1962, as amended by Amdt. 65-23, 43 FR 22640, May 25, 1978]

§ 65.14 Security disqualification.

(a) Eligibility standard. No person is eligible to hold a certificate, rating, or authorization issued under this part when the Transportation Security Administration (TSA) has notified the FAA in writing that the person poses a security threat.

(b) Effect of the issuance by the TSA of an Initial Notification of Threat Assessment. (1) The FAA will hold in abeyance pending the outcome of the TSA's final threat assessment review an application for any certificate, rating, or authorization under this part by any person who has been issued an Initial Notification of Threat Assessment by the TSA.

(1) The FAA will suspend any certificate, rating, or authorization issued under this part after the TSA issues to the holder an Initial Notification of Threat Assessment.

(c) Effect of the issuance by the TSA of a Final Notification of Threat Assessment. (1) The FAA will deny an application for any certificate, rating, or authorization under this part to any person who has been issued a Final Notification of Threat Assessment.

(1) The FAA will revoke any certificate, rating, or authorization issued under this part after the TSA has issued to the holder a Final Notification of Threat Assessment.

§65.15 Duration of certificates.

(a) Except for repairman certificates, a certificate or rating issued under this part is effective until it is surrendered, suspended, or revoked.

(b) Unless it is sooner surrendered, suspended, or revoked, a repairman certificate is effective until the holder is relieved from the duties for which the holder was employed and certificated.

(c) The holder of a certificate issued under this part that is suspended, revoked, or no longer effective shall return it to the Administrator.

[Doc. No. 22052, 47 FR 35693, Aug. 16, 1982]

§65.16 Change of name: Replacement of lost or destroyed certificate.

(a) An application for a change of name on a certificate issued under this part must be accompanied by the applicant's current certificate and the marriage license, court order, or other document verifying the change. The documents are returned to the applicant after inspection.

(b) An application for a replacement of a lost or destroyed certificate is made by letter to the Department of Transportation, Federal Aviation Administration, Airman Certification Branch, Post Office Box 25082, Oklahoma City, OK 73125. The letter must—

(1) Contain the name in which the certificate was issued, the permanent mailing address (including zip code), social security number (if any), and date and place of birth of the certificate holder, and any available information regarding the grade, number, and date of issue of the certificate, and the ratings on it; and

(2) Be accompanied by a check or money order for $2, payable to the Federal Aviation Administration.

(c) An application for a replacement of a lost or destroyed medical certificate is made by letter to the Department of Transportation, Federal Aviation Administration, Civil Aeromedical Institute, Aeromedical Certification Branch, Post Office Box 25082, Oklahoma City, OK 73125, accompanied by a check or money order for $2.00.

(d) A person whose certificate issued under this part or medical certificate, or both, has been lost may obtain a telegram from the FAA confirming that it was issued. The telegram may be carried as a certificate for a period not to exceed 60 days pending his receiving a duplicate certificate under paragraph (b) or (c) of this section, unless he has been notified that the certificate has been suspended or revoked. The request for such a telegram may be made by prepaid telegram, stating the date upon which a duplicate certificate was requested, or including the request for a duplicate and a money order for the necessary amount. The request for a telegraphic certificate should be sent to the office prescribed in paragraph (b) or (c) of this section, as appropriate. However, a request for both at the same time should be sent to the office prescribed in paragraph (b) of this section.

[Doc. No. 7258, 31 FR 13524, Oct. 20, 1966, as amended by Doc. No. 8084, 32 FR 5769, Apr. 11, 1967; Amdt. 65-16, 35 FR 14075, Sept. 4, 1970; Amdt. 65-17, 36 FR 2865, Feb. 11, 1971]

§65.17 Tests: General procedure.

(a) Tests prescribed by or under this part are given at times and places, and by persons, designated by the Administrator.

(b) The minimum passing grade for each test is 70 percent.

§65.18 Written tests: Cheating or other unauthorized conduct.

(a) Except as authorized by the Administrator, no person may—

(1) Copy, or intentionally remove, a written test under this part;

(2) Give to another, or receive from another, any part or copy of that test;

(3) Give help on that test to, or receive help on that test from, any person during the period that test is being given;

(4) Take any part of that test in behalf of another person;

(5) Use any material or aid during the period that test is being given; or

(6) Intentionally cause, assist, or participate in any act prohibited by this paragraph.

(b) No person who commits an act prohibited by paragraph (a) of this section is eligible for any airman or ground instructor certificate or rating under this chapter for a period of 1 year after the date of that act. In addition, the commission of that act is a basis for suspending or revoking any airman or ground instructor certificate or rating held by that person.

[Doc. No. 4086, 30 FR 2196, Feb. 18, 1965]

§65.19 Retesting after failure.

An applicant for a written, oral, or practical test for a certificate and rating, or for an additional rating under this part, may apply for retesting—

(a) After 30 days after the date the applicant failed the test; or

(b) Before the 30 days have expired if the applicant presents a signed statement from an airman holding the certificate and rating sought by the applicant, certifying that the airman has given the applicant additional instruction in each of the subjects failed and that the airman considers the applicant ready for retesting.

[Doc. No. 16383, 43 FR 22640, May 25, 1978]

§65.20 Applications, certificates, logbooks, reports, and records: Falsification, reproduction, or alteration.

(a) No person may make or cause to be made—

(1) Any fraudulent or intentionally false statement on any application for a certificate or rating under this part;

(2) Any fraudulent or intentionally false entry in any logbook, record, or report that is required to be kept, made, or used, to show compliance with any requirement for any certificate or rating under this part;

(3) Any reproduction, for fraudulent purpose, of any certificate or rating under this part; or

(4) Any alteration of any certificate or rating under this part.

(b) The commission by any person of an act prohibited under paragraph (a) of this section is a basis for suspending or revoking any airman or ground instructor certificate or rating held by that person.

[Doc. No. 4086, 30 FR 2196, Feb. 18, 1965]

§65.21 Change of address.

Within 30 days after any change in his permanent mailing address, the holder of a certificate issued under this part shall notify the Department of Transportation, Federal Aviation Administration, Airman Certification Branch, Post Office Box 25082, Oklahoma City, OK 73125, in writing, of his new address.

[Doc. No. 10536, 35 FR 14075, Sept. 4, 1970]

§65.23 Refusal to submit to a drug or alcohol test.

(a) *General.* This section applies to an individual who holds a certificate under this part and is subject to the types of testing required under appendix I to part 121 or appendix J of part 21 of this chapter.

(b) Refusal by the holder of a certificate issued under this part to take a drug test required under the provisions of appendix I to part 121 or an alcohol test required under the provisions of appendix J to part 121 is grounds for—

(1) Denial of an application for any certificate or rating issued under this part for a period of up to 1 year after the date of such refusal; and

(2) Suspension or revocation of any certificate or rating issued under this part.

[Amdt. 65-37, 59 FR 7389, Feb. 15, 1994]

Subpart B—Air Traffic Control Tower Operators

Source: Docket No. 10193, 35 FR 12326, Aug. 1, 1970, unless otherwise noted.

§65.31 Required certificates, and rating or qualification.

No person may act as an air traffic control tower operator at an air traffic control tower in connection with civil aircraft unless he—

(a) Holds an air traffic control tower operator certificate issued to him under this subpart;

(b) Holds a facility rating for that control tower issued to him under this subpart, or has qualified for the operating position at which he acts and is under the supervision of the holder of a facility rating for that control tower; and For the purpose of this subpart, operating position means an air traffic control function performed within or directly associated with the control tower;

(c) Except for a person employed by the FAA or employed by, or on active duty with, the Department of the Air Force, Army, or Navy or the Coast Guard, holds at least a second-class medical certificate issued under part 67 of this chapter.

[Doc. No. 10193, 35 FR 12326, Aug. 1, 1970, as amended by Amdt. 65-25, 45 FR 18911, Mar. 24, 1980; Amdt. 65-31, 52 FR 17518, May 8, 1987]

§65.33 Eligibility requirements: General.

To be eligible for an air traffic control tower operator certificate a person must --

(a) Be at least 18 years of age;

(b) Be of good moral character;

(c) Be able to read, write, and understand the English language and speak it without accent or impediment of speech that would interfere with two-way radio conversation;

(d) Except for a person employed by the FAA or employed by, or on active duty with, the Department of the Air Force, Army, or Navy or the Coast Guard, hold at least a second-class medical certificate issued under part 67 of this chapter within the 12 months before the date application is made; and

(e) Comply with §65.35.

[Doc. No. 10193, 35 FR 12326, Aug. 1, 1970, as amended by Amdt. 65-25, 45 FR 18911, Mar. 24, 1980; Amdt. 65-31, 52 FR 17518, May 8, 1987]

§65.35 Knowledge requirements.

Each applicant for an air traffic control tower operator certificate must pass a written test on --

(a) The flight rules in part 91 of this chapter:

(b) Airport traffic control procedures, and this subpart:

(c) En route traffic control procedures;

(d) Communications operating procedures;

(e) Flight assistance service;

(f) Air navigation, and aids to air navigation; and

(g) Aviation weather.

§65.37 Skill requirements: Operating positions.

No person may act as an air traffic control tower operator at any operating position unless he has passed a practical test on—

(a) Control tower equipment and its use;

(b) Weather reporting procedures and use of reports;

(c) Notices to Airmen, and use of the Airman's Information Manual;

(d) Use of operational forms;

(e) Performance of noncontrol operational duties; and

(f) Each of the following procedures that is applicable to that operating position and is required by the person performing the examination:

(1) The airport, including rules, equipment, runways, taxiways, and obstructions.

(2) The terrain features, visual checkpoints, and obstructions within the lateral boundaries of the surface areas of Class B, Class C, Class D, or Class E airspace designated for the airport.

(3) Traffic patterns and associated procedures for use of preferential runways and noise abatement.

(4) Operational agreements.

(5) The center, alternate airports, and those airways, routes, reporting points, and air navigation aids used for terminal air traffic control.

(6) Search and rescue procedures.

(7) Terminal air traffic control procedures and phraseology.

(8) Holding procedures, prescribed instrument approach, and departure procedures.

(9) Radar alignment and technical operation.

(10) The application of the prescribed radar and non-radar separation standard, as appropriate.

[Doc. No. 10193, 35 FR 12326, Aug. 1, 1991, as amended by Amdt. 65-36, 56 FR 65653, Dec. 17, 1991]

§65.39 Practical experience requirements: Facility rating.

Each applicant for a facility rating at any air traffic control tower must have satisfactorily served—

(a) As an air traffic control tower operator at that control tower without a facility rating for at least 6 months; or

(b) As an air traffic control tower operator with a facility rating at a different control tower for at least 6 months before the date he applies for the rating.

However, an applicant who is a member of an Armed Force of the United States meets the requirements of this section if he has satisfactorily served as an air traffic control tower operator for at least 6 months.

[Doc. No. 1179, 27 FR 7973, Aug. 10, 1962, as amended by Amdt. 65-19, 36 FR 21280, Nov. 5, 1971]

§65.41 Skill requirements: Facility ratings.

Each applicant for a facility rating at an air traffic control tower must have passed a practical test on each item listed in §65.37 of this part that is applicable to each operating position at the control tower at which the rating is sought.

§65.43 Rating privileges and exchange.

(a) The holder of a senior rating on August 31, 1970, may at any time after that date exchange his rating for a facility rating at the same air traffic control tower. However, if he does not do so before August 31, 1971, he may not thereafter exercise the privileges of his senior rating at the control tower concerned until he makes the exchange.

(b) The holder of a junior rating on August 31, 1970, may not control air traffic, at any operating position at the control tower concerned, until he has met the applicable requirements of §65.37 of this part. However, before meeting those requirements he may control air traffic under the supervision, where required, of an operator with a senior rating (or facility rating) in accordance with §65.41 of this part in effect before August 31, 1970.

§65.45 Performance of duties.

(a) An air traffic control tower operator shall perform his duties in accordance with the limitations on his certificate and the procedures and practices prescribed in air traffic control manuals of the FAA, to provide for the safe, orderly, and expeditious flow of air traffic.

(b) An operator with a facility rating may control traffic at any operating position at the control tower at which he holds a facility rating. However, he may not issue an air traffic clearance for IFR flight without authorization from the appropriate facility exercising IFR control at that location.

(c) An operator who does not hold a facility rating for a particular control tower may act at each operating position for which he has qualified, under the supervision of an operator holding a facility rating for that control tower.

[Doc. No. 10193, 35 FR 12326, Aug. 1, 1970, as amended by Amdt. 65-16, 35 FR 14075, Sept. 4, 1970]

§65.46 Use of prohibited drugs.

(a) The following definitions apply for the purposes of this section:

(1) An employee is a person who performs an air traffic control function for an employer. For the purpose of this section, a person who performs such a function pursuant to a contract with an employer is considered to be performing that function for the employer.

(2) An "employer" means an air traffic control facility not operated by the FAA or by or under contract to the U.S. military that employs a person to perform an air traffic control function.

(b) Each employer shall provide each employee performing a function listed in appendix I to part 121 of this chapter and his or her supervisor with the training specified in that appendix. No employer may use any contractor to perform an air traffic control function unless that contractor provides each of its employees performing that function for the employer and his or her supervisor with the training specified in that appendix.

(c) No employer may knowingly use any person to perform, nor may any person perform for an employer, either directly or by contract, any air traffic control function while that person has a prohibited drug, as defined in appendix I to part 121 of this chapter, in his or her system.

(d) No employer shall knowingly use any person to perform, nor may any person perform for an employer, either directly or by contract, any air traffic control function if the person has a verified positive drug test result on or has refused to submit to a drug test required by appendix I to part 121 of this chapter and the person has not met the requirements of appendix I to part 121 of this chapter for returning to the performance of safety-sensitive duties.

(e) Each employer shall test each of its employees who performs any air traffic control function in accordance with appendix I to part 121 of this chapter. No employer may use any contractor to perform any air traffic control function unless that contractor tests each employee performing such a function for the employer in accordance with that appendix.

[Doc. No. 25148, 53 FR 47056, Nov. 21, 1988, as amended by Amdt. 65-38, 59 FR 42927, Aug. 19, 1994]

§65.46a Misuse of alcohol.

(a) This section applies to employees who perform air traffic control duties directly or by contract for an employer that is an air traffic control facility not operated by the FAA or the U.S. military (covered employees).

(b) Alcohol concentration. No covered employee shall report for duty or remain on duty requiring the performance of safety-sensitive functions while having an alcohol concentration of 0.04 or greater. No employer having actual knowledge that an employee has an alcohol concentration of 0.04 or greater shall permit the employee to perform or continue to perform safety-sensitive functions.

(c) On-duty use. No covered employee shall use alcohol while performing safety-sensitive functions. No employer having actual knowledge that a covered employee is using alcohol while performing safety-sensitive functions shall permit the employee to perform or continue to perform safety-sensitive functions.

(d) Pre-duty use. No covered employee shall perform air traffic control duties within 8 hours after using alcohol. No

employer having actual knowledge that such an employee has used alcohol within 8 hours shall permit the employee to perform or continue to perform air traffic control duties.

(e) Use following an accident. No covered employee who has actual knowledge of an accident involving an aircraft for which he or she performed a safety-sensitive function at or near the time of the accident shall use alcohol for 8 hours following the accident, unless he or she has been given a post-accident test under appendix J to part 121 of this chapter, or the employer has determined that the employee's performance could not have contributed to the accident.

(f) Refusal to submit to a required alcohol test. A covered employee may not refuse to submit to any alcohol test required under appendix J to part 121 of this chapter. An employer may not permit an employee who refuses to submit to such a test to perform or continue to perform safety-sensitive functions.

[Amdt. 65-37, 59 FR 7389, Feb. 15, 1994]

§65.46b Testing for alcohol.

(a) Each air traffic control facility not operated by the FAA or the U.S. military (hereinafter employer) must establish an alcohol misuse prevention program in accordance with the provisions of appendix J to part 121 of this chapter.

(b) No employer shall use any person who meets the definition of covered employee in appendix J to part 121 to perform a safety-sensitive function listed in that appendix unless such person is subject to testing for alcohol misuse in accordance with the provisions of appendix J.

[Amdt. 65-37, 59 FR 7389, Feb. 15, 1994]

§65.47 Maximum hours.

Except in an emergency, a certificated air traffic control tower operator must be relieved of all duties for at least 24 consecutive hours at least once during each 7 consecutive days. Such an operator may not serve or be required to serve—

(a) For more than 10 consecutive hours; or

(b) For more than 10 hours during a period of 24 consecutive hours, unless he has had a rest period of at least 8 hours at or before the end of the 10 hours of duty.

§65.49 General operating rules.

(a) Except for a person employed by the FAA or employed by, or on active duty with, the Department of the Air Force, Army, or Navy, or the Coast Guard, no person may act as an air traffic control tower operator under a certificate issued to him or her under this part unless he or she has in his or her personal possession an appropriate current medical certificate issued under part 67 of this chapter.

(b) Each person holding an air traffic control tower operator certificate shall keep it readily available when performing duties in an air traffic control tower, and shall present that certificate or his medical certificate or both for inspection upon the request of the Administrator or an authorized representative of the National Transportation Safety Board, or of any Federal, State, or local law enforcement officer.

(c) A certificated air traffic control tower operator who does not hold a facility rating for a particular control tower may not act at any operating position at the control tower concerned unless there is maintained at that control tower, readily available to persons named in paragraph (b) of this section, a current record of the operating positions at which he has qualified.

(d) An air traffic control tower operator may not perform duties under his certificate during any period of known physical deficiency that would make him unable to meet the physical requirements for his current medical certificate.

However, if the deficiency is temporary, he may perform duties that are not affected by it whenever another certificated and qualified operator is present and on duty.

(e) A certificated air traffic control tower operator may not control air traffic with equipment that the Administrator has found to be inadequate.

(f) The holder of an air traffic control tower operator certificate, or an applicant for one, shall, upon the reasonable request of the Administrator, cooperate fully in any test that is made of him.

[Doc. No. 1179, 27 FR 7973, Aug. 10, 1962, as amended by Amdt. 65-31, 52 FR 17519, May 8, 1987]

§65.50 Currency requirements.

The holder of an air traffic control tower operator certificate may not perform any duties under that certificate unless—

(a) He has served for at least three of the preceding 6 months as an air traffic control tower operator at the control tower to which his facility rating applies, or at the operating positions for which he has qualified; or

(b) He has shown that he meets the requirements for his certificate and facility rating at the control tower concerned, or for operating at positions for which he has previously qualified.

Subpart C—Aircraft Dispatchers (Eff. 4-6-00)

Source: Docket No. FAA-1998-4553, 64 FR 68923, Dec. 8, 1999, unless otherwise noted.

§65.51 Certificate required.

(a) No person may act as an aircraft dispatcher (exercising responsibility with the pilot in command in the operational control of a flight) in connection with any civil aircraft in air commerce unless that person has in his or her personal possession an aircraft dispatcher certificate issued under this subpart.

(b) Each person who holds an aircraft dispatcher certificate must present it for inspection upon the request of the Administrator or an authorized representative of the National Transportation Safety Board, or of any Federal, State, or local law enforcement officer.

§65.53 Eligibility requirements: General.

(a) To be eligible to take the aircraft dispatcher knowledge test, a person must be at least 21 years of age.

(b) To be eligible for an aircraft dispatcher certificate, a person must—

(1) Be at least 23 years of age;

(2) Be able to read, speak, write, and understand the English language;

(3) Pass the required knowledge test prescribed by §65.55 of this part;

(4) Pass the required practical test prescribed by §65.59 of this part; and

(5) Comply with the requirements of §65.57 of this part.

§65.55 Knowledge requirements.

(a) A person who applies for an aircraft dispatcher certificate must pass a knowledge test on the following aeronautical knowledge areas:

(1) Applicable Federal Aviation Regulations of this chapter that relate to airline transport pilot privileges, limitations, and flight operations;

(2) Meteorology, including knowledge of and effects of fronts, frontal characteristics, cloud formations, icing, and upper-air data;

(3) General system of weather and NOTAM collection, dissemination, interpretation, and use;

(4) Interpretation and use of weather charts, maps, forecasts, sequence reports, abbreviations, and symbols;

(5) National Weather Service functions as they pertain to operations in the National Airspace System;

(6) Windshear and microburst awareness, identification, and avoidance;

(7) Principles of air navigation under instrument meteorological conditions in the National Airspace System;

(8) Air traffic control procedures and pilot responsibilities as they relate to enroute operations, terminal area and radar operations, and instrument departure and approach procedures;

(9) Aircraft loading, weight and balance, use of charts, graphs, tables, formulas, and computations, and their effect on aircraft performance;

(10) Aerodynamics relating to an aircraft's flight characteristics and performance in normal and abnormal flight regimes;

(11) Human factors;

(12) Aeronautical decision making and judgment; and

(13) Crew resource management, including crew communication and coordination.

(b) The applicant must present documentary evidence satisfactory to the administrator of having passed an aircraft dispatcher knowledge test within the preceding 24 calendar months.

§65.57 Experience or training requirements.

An applicant for an aircraft dispatcher certificate must present documentary evidence satisfactory to the Administrator that he or she has the experience prescribed in paragraph (a) of this section or has accomplished the training described in paragraph (b) of this section as follows:

(a) A total of at least 2 years experience in the 3 years before the date of application, in any one or in any combination of the following areas:

(1) In military aircraft operations as a—

(i) Pilot;

(ii) Flight navigator; or

(iii) Meteorologist.

(2) In aircraft operations conducted under part 121 of this chapter as—

(i) An assistant in dispatching air carrier aircraft, under the direct supervision of a dispatcher certificated under this subpart;

(ii) A pilot;

(iii) A flight engineer; or

(iv) A meteorologist.

(3) In aircraft operations as—

(i) An Air Traffic Controller; or

(ii) A Flight Service Specialist.

(4) In aircraft operations, performing other duties that the Administrator finds provide equivalent experience.

(b) A statement of graduation issued or revalidated in accordance with §65.70(b) of this part, showing that the person has successfully completed an approved aircraft dispatcher course.

§65.59 Skill requirements.

An applicant for an aircraft dispatcher certificate must pass a practical test given by the Administrator, with respect to any one type of large aircraft used in air carrier operations. The practical test must be based on the aircraft dispatcher practical test standards, as published by the FAA, on the items outlined in appendix A of this part.

§65.61 Aircraft dispatcher certification courses: Content and minimum hours.

(a) An approved aircraft dispatcher certification course must:

(1) Provide instruction in the areas of knowledge and topics listed in appendix A of this part;

(2) Include a minimum of 200 hours of instruction.

(b) An applicant for approval of an aircraft dispatcher course must submit an outline that describes the major topics and subtopics to be covered and the number of hours proposed for each.

(c) Additional subject headings for an aircraft dispatcher certification course may also be included, however the hours proposed for any subjects not listed in appendix A of this part must be in addition to the minimum 200 course hours required in paragraph (a) of this section.

(d) For the purpose of completing an approved course, a student may substitute previous experience or training for a portion of the minimum 200 hours of training. The course operator determines the number of hours of credit based on an evaluation of the experience or training to determine if it is comparable to portions of the approved course curriculum. The credit allowed, including the total hours and the basis for it, must be placed in the student's record required by §65.70(a) of this part.

§65.63 Aircraft dispatcher certification courses: Application, duration, and other general requirements.

(a) Application. Application for original approval of an aircraft dispatcher certification course or the renewal of approval of an aircraft dispatcher certification course under this part must be:

(1) Made in writing to the Administrator;

(2) Accompanied by two copies of the course outline required under §65.61(b) of this part, for which approval is sought;

(3) Accompanied by a description of the equipment and facilities to be used; and

(4) Accompanied by a list of the instructors and their qualifications.

(b) Duration. Unless withdrawn or canceled, an approval of an aircraft dispatcher certification course of study expires:

(1) On the last day of the 24th month from the month the approval was issued; or

(2) Except as provided in paragraph (f) of this section, on the date that any change in ownership of the school occurs.

(c) Renewal. Application for renewal of an approved aircraft dispatcher certification course must be made within 30 days preceding the month the approval expires, provided the course operator meets the following requirements:

(1) At least 80 percent of the graduates from that aircraft dispatcher certification course, who applied for the practical test required by §65.59 of this part, passed the practical test on their first attempt; and

(2) The aircraft dispatcher certification course continues to meet the requirements of this subpart for course approval.

(d) Course revisions. Requests for approval of a revision of the course outline, facilities, or equipment must be in accordance with paragraph (a) of this section. Proposed revisions of the course outline or the description of facilities and equipment must be submitted in a format that will allow an entire page or pages of the approved outline or description to be removed and replaced by any approved revision. The list of instructors may be revised at any time without request for approval, provided the minimum requirements of §65.67 of this part are maintained and the Administrator is notified in writing.

(e) Withdrawal or cancellation of approval. Failure to continue to meet the requirements of this subpart for the approval or operation of an approved aircraft dispatcher certification course is grounds for withdrawal of approval of the course. A course operator may request cancellation of course approval by a letter to the Administrator. The operator must forward any records to the FAA as requested by the Administrator.

(f) Change in ownership. A change in ownership of a part 65, appendix A-approved course does not terminate that aircraft dispatcher certification course approval if, within 10 days after the date that any change in ownership of the school occurs:

(1) Application is made for an appropriate amendment to the approval; and

(2) No change in the facilities, personnel, or approved aircraft dispatcher certification course is involved.

(g) Change in name or location. A change in name or location of an approved aircraft dispatcher certification course does not invalidate the approval if, within 10 days after the date that any change in name or location occurs, the course operator of the part 65, appendix A-approved course notifies the Administrator, in writing, of the change.

§65.65 Aircraft dispatcher certification courses: Training facilities.

An applicant for approval of authority to operate an aircraft dispatcher course of study must have facilities, equipment, and materials adequate to provide each student the theoretical and practical aspects of aircraft dispatching. Each room, training booth, or other space used for instructional purposes must be temperature controlled, lighted, and ventilated to conform to local building, sanitation, and health codes. In addition, the training facility must be so located that the students in that facility are not distracted by the instruction conducted in other rooms.

§65.67 Aircraft dispatcher certification courses: Personnel.

(a) Each applicant for an aircraft dispatcher certification course must meet the following personnel requirements:

(1) Each applicant must have adequate personnel, including one instructor who holds an aircraft dispatcher certificate and is available to coordinate all training course instruction.

(2) Each applicant must not exceed a ratio of 25 students for one instructor.

(b) The instructor who teaches the practical dispatch applications area of the appendix A course must hold an aircraft dispatchers certificate

§65.70 Aircraft dispatcher certification courses: Records.

(a) The operator of an aircraft dispatcher course must maintain a record for each student, including a chronological log of all instructors, subjects covered, and course examinations and results. The record must be retained for at least 3 years after graduation. The course operator also must prepare, for its records, and transmit to the Administrator not later than January 31 of each year, a report containing the following information for the previous year:

(1) The names of all students who graduated, together with the results of their aircraft dispatcher certification courses.

(2) The names of all the students who failed or withdrew, together with the results of their aircraft dispatcher certification courses or the reasons for their withdrawal.

(b) Each student who successfully completes the approved aircraft dispatcher certification course must be given a written statement of graduation, which is valid for 90 days. After 90 days, the course operator may revalidate the graduation certificate for an additional 90 days if the course operator determines that the student remains proficient in the subject areas listed in appendix A of this part.

Subpart D—Mechanics

§65.71 Eligibility requirements: General.

(a) To be eligible for a mechanic certificate and associated ratings, a person must—

(1) Be at least 18 years of age;

(2) Be able to read, write, speak, and understand the English language, or in the case of an applicant who does not meet this requirement and who is employed outside of the United States by a U.S. air carrier, have his certificate endorsed "Valid only outside the United States";

(3) Have passed all of the prescribed tests within a period of 24 months; and

(4) Comply with the sections of this subpart that apply to the rating he seeks.

(b) A certificated mechanic who applies for an additional rating must meet the requirements of §65.77 and, within a period of 24 months, pass the tests prescribed by §§65.75 and 65.79 for the additional rating sought.

[Doc. No. 1179, 27 FR 7973, Aug. 10, 1962, as amended by Amdt. 65-6, 31 FR 5950, Apr. 19, 1966]

§65.73 Ratings.

(a) The following ratings are issued under this subpart:

(1) Airframe.

(2) Powerplant.

(b) A mechanic certificate with an aircraft or aircraft engine rating, or both, that was issued before, and was valid on, June 15, 1952, is equal to a mechanic certificate with an airframe or powerplant rating, or both, as the case may be, and may be exchanged for such a corresponding certificate and rating or ratings.

§65.75 Knowledge requirements.

(a) Each applicant for a mechanic certificate or rating must, after meeting the applicable experience requirements of §65.77, pass a written test covering the construction and maintenance of aircraft appropriate to the rating he seeks, the regulations in this subpart, and the applicable provisions of parts 43 and 91 of this chapter. The basic principles covering the installation and maintenance of propellers are included in the powerplant test.

(b) The applicant must pass each section of the test before applying for the oral and practical tests prescribed by §65.79. A report of the written test is sent to the applicant.

[Doc. No. 1179, 27 FR 7973, Aug. 10, 1962, as amended by Amdt. 65-1, 27 FR 10410, Oct. 25, 1962; Amdt. 65-6, 31 FR 5950, Apr. 19, 1966]

§65.77 Experience requirements.

Each applicant for a mechanic certificate or rating must present either an appropriate graduation certificate or certificate of completion from a certificated cated aviation maintenance technician school or documentary evidence, satisfactory to the Administrator, of—

(a) At least 18 months of practical experience with the procedures, practices, materials, tools, machine tools, and equipment generally used in constructing, maintaining, or altering airframes, or powerplants appropriate to the rating sought; or

(b) At least 30 months of practical experience concurrently performing the duties appropriate to both the airframe and powerplant ratings.

[Doc. No. 1179, 27 FR, 7973, Aug. 10, 1962, as amended by Amdt. 65-14, 35 FR, 5533, Apr. 3, 1970]

§65.79 Skill requirements.

Each applicant for a mechanic certificate or rating must pass an oral and a practical test on the rating he seeks. The tests cover the applicant's basic skill in performing practical projects on the subjects covered by the written test for that rating. An applicant for a powerplant rating must show his ability to make satisfactory minor repairs to, and minor alterations of, propellers.

§65.80 Certificated aviation maintenance technician school students.

Whenever an aviation maintenance technician school certificated under part 147 of this chapter shows to an FAA inspector that any of its students has made satisfactory progress at the school and is prepared to take the oral and practical tests prescribed by §65.79, that student may take those tests during the final subjects of his training in the approved curriculum, before he meets the applicable experience requirements of §65.77 and before he passes each section of the written test prescribed by §65.75.

[Doc. No. 9444, 35 FR 5533, Apr. 3, 1970]

§65.81 General privileges and limitations.

(a) A certificated mechanic may perform or supervise the maintenance, preventive maintenance or alteration of an aircraft or appliance, or a part thereof, for which he is rated (but excluding major repairs to, and major alterations of, propellers, and any repair to, or alteration of, instruments), and may perform additional duties in accordance with §§65.85, 65.87, and 65.95. However, he may not supervise the maintenance, preventive maintenance, or alteration of, or approve and return to service, any aircraft or appliance, or part thereof, for which he is rated unless he has satisfactorily performed the work concerned at an earlier date. If he has not so performed that work at an earlier date, he may show his ability to do it by performing it to the satisfaction of the Administrator or under the direct supervision of a certificated and appropriately rated mechanic, or a certificated repairman, who has had previous experience in the specific operation concerned.

(b) A certificated mechanic may not exercise the privileges of his certificate and rating unless he understands the current instructions of the manufacturer, and the maintenance manuals, for the specific operation concerned.

[Doc. No. 1179, 27 FR 7973, Aug. 10, 1962, as amended by Amdt. 65-2, 29 FR 5451, Apr. 23, 1964; Amdt. 65-26, 45 FR 46737, July 10, 1980]

§65.83 Recent experience requirements.

A certificated mechanic may not exercise the privileges of his certificate and rating unless, within the preceding 24 months—

(a) The Administrator has found that he is able to do that work; or

(b) He has, for at least 6 months—

(1) Served as a mechanic under his certificate and rating;

(2) Technically supervised other mechanics;

(3) Supervised, in an executive capacity, the maintenance or alteration of aircraft; or

(4) Been engaged in any combination of paragraph (b) (1), (2), or (3) of this section.

§65.85 Airframe rating; additional privileges.

(a) A certificated mechanic with an airframe rating may approve and return to service an airframe, or any related part or appliance, after he has performed, supervised, or inspected its maintenance or alteration (excluding major repairs and major alterations). In addition, he may perform the 100-hour inspection required by part 91 of this chapter on an airframe, or any related part or appliance, and approve and return it to service.

(b) A certificated mechanic with an airframe rating can approve and return to service an airframe, or any related part or appliance, of an aircraft with a special airworthiness certificate in the light-sport category after performing and inspecting a major repair or major alteration for products that are not produced under an FAA approval provided the work was performed in accordance with instructions developed by the manufacturer or a person acceptable to the FAA.

[Doc. No. 1179, 27 FR 7973, Aug. 10, 1962, as amended by Amdt. 65-10, 32 FR 5770, Apr. 11, 1967]

§65.87 Powerplant rating; additional privileges.

(a) A certificated mechanic with a powerplant rating may approve and return to service a powerplant or propeller or any related part or appliance, after he has performed, supervised, or inspected its maintenance or alteration (excluding major repairs and major alterations). In addition, he may perform the 100-hour inspection required by part 91 of this chapter on a powerplant or propeller, or any part thereof, and approve and return it to service.

(b) A certificated mechanic with a powerplant rating can approve and return to service a powerplant or propeller, or any related part or appliance, of an aircraft with a special airworthiness certificate in the light-sport category after performing and inspecting a major repair or major alteration for products that are not produced under an FAA approval, provided the work was performed in accordance with instructions developed by the manufacturer or a person acceptable to the FAA.

[Doc. No. 1179, 27 FR 7973, Aug. 10, 1962, as amended by Amdt. 65-10, 32 FR 5770, Apr. 11, 1967]

§65.89 Display of certificate.

Each person who holds a mechanic certificate shall keep it within the immediate area where he normally exercises the privileges of the certificate and shall present it for inspection upon the request of the Administrator or an authorized representative of the National Transportation Safety Board, or of any Federal, State, or local law enforcement officer.

[Doc. No. 7258, 31 FR 13524, Oct. 20, 1966, as amended by Doc. No. 8084, 32 FR 5769, Apr. 11, 1967]

§65.91 Inspection authorization.

(a) An application for an inspection authorization is made on a form and in a manner prescribed by the Administrator.

(b) An applicant who meets the requirements of this section is entitled to an inspection authorization.

(c) To be eligible for an inspection authorization, an applicant must—

(1) Hold a currently effective mechanic certificate with both an airframe rating and a powerplant rating, each of which is currently effective and has been in effect for a total of at least 3 years;

(2) Have been actively engaged, for at least the 2-year period before the date he applies, in maintaining aircraft certificated and maintained in accordance with this chapter;

(3) Have a fixed base of operations at which he may be located in person or by telephone during a normal working week but it need not be the place where he will exercise his inspection authority;

(4) Have available to him the equipment, facilities, and inspection data necessary to properly inspect airframes, powerplants, propellers, or any related part or appliance; and

(5) Pass a written test on his ability to inspect according to safety standards for returning aircraft to service after major repairs and major alterations and annual and progressive inspections performed under part 43 of this chapter. An applicant who fails the test prescribed in paragraph (c)(5) of this section may not apply for retesting until at least 90 days after the date he failed the test.

[Doc. No. 1179, 27 FR 7973, Aug. 10, 1962, as amended by Amdt. 65-5, 31 FR 3337, Mar. 3, 1966; Amdt. 65-22, 42 FR 46279, Sept. 15, 1977; Amdt. 65-30, 50 FR 15700, Apr. 19, 1985]

§65.92 Inspection authorization: Duration.

(a) Each inspection authorization expires on March 31 of each year. However, the holder may exercise the privileges of that authorization only while he holds a currently effective mechanic certificate with both a currently effective airframe rating and a currently effective powerplant rating.

(b) An inspection authorization ceases to be effective whenever any of the following occurs:

(1) The authorization is surrendered, suspended, or revoked.

(2) The holder no longer has a fixed base of operation.

(3) The holder no longer has the equipment, facilities, and inspection data required by §65.91(c) (3) and (4) for issuance of his authorization.

(c) The holder of an inspection authorization that is suspended or revoked shall, upon the Administrator's request, return it to the Administrator.

[Doc. No. 12537, 42 FR 46279, Sept. 15, 1977]

§65.93 Inspection authorization: Renewal.

(a) To be eligible for renewal of an inspection authorization for a 1-year period an applicant must present evidence annually, during the month of March, at an FAA Flight Standards District Office or an International Field Office that the applicant still meets the requirements of §65.91(c)(1) through (4) and must show that, during the current period that the applicant held the inspection authorization, the applicant—

(1) Has performed at least one annual inspection for each 90 days that the applicant held the current authority; or

(2) Has performed inspections of at least two major repairs or major alterations for each 90 days that the applicant held the current authority; or

(3) Has performed or supervised and approved at least one progressive inspection in accordance with standards prescribed by the Administrator; or

(4) Has attended and successfully completed a refresher course, acceptable to the Administrator, of not less than 8 hours of instruction during the 12-month period preceding the application for renewal; or

(5) Has passed on oral test by an FAA inspector to determine that the applicant's knowledge of applicable regulations and standards is current.

(b) The holder of an inspection authorization that has been in effect for less than 90 days before the expiration date need not comply with paragraphs (a) (1) through (5) of this section.

[Doc. No. 18241, 45 FR 46738, July 10, 1980, as amended by Amdt. 65-35, 54 FR 39292, Sept. 25, 1989]

§65.95 Inspection authorization: Privileges and limitations.

(a) The holder of an inspection authorization may—

(1) Inspect and approve for return to service any aircraft or related part or appliance (except any aircraft maintained in accordance with a continuous airworthiness program under part 121 of this chapter) after a major repair or major alteration to it in accordance with part 43 [New] of this chapter, if the work was done in accordance with technical data approved by the Administrator; and

(2) Perform an annual, or perform or supervise a progressive inspection according to §§43.13 and 43.15 of this chapter.

(b) When he exercises the privileges of an inspection authorization the holder shall keep it available for inspection by the aircraft owner, the mechanic submitting the aircraft, repair, or alteration for approval (if any), and shall present it upon the request of the Administrator or an authorized representative of the National Transportation Safety Board, or of any Federal, State, or local law enforcement officer.

(c) If the holder of an inspection authorization changes his fixed base of operation, he may not exercise the privileges of the authorization until he has notified the FAA Flight Standards District Office or International Field Office for the area in which the new base is located, in writing, of the change.

[Doc. No. 1179, 27 FR 7973, Aug. 10, 1962, as amended by Amdt. 65-2, 29 FR 5451, Apr. 23, 1964; Amdt. 65-4, 30 FR 3638, Mar. 14, 1965; Amdt. 65-5, 31 FR 3337, Mar. 3, 1966; Amdt. 65-9, 31 FR 13524, Oct. 20, 1966; 32 FR 5769, Apr. 11, 1967; Amdt. 65-35, 54 FR 39292, Sept. 25, 1989; Amdt. 65-41, 66 FR 21066, Apr. 27, 2001]

Subpart E—Repairmen

§65.101 Eligibility requirements: General.

(a) To be eligible for a repairman certificate a person must—

(1) Be at least 18 years of age;

(2) Be specially qualified to perform maintenance on aircraft or components thereof, appropriate to the job for which he is employed;

(3) Be employed for a specific job requiring those special qualifications by a certificated repair station, or by a certificated commercial operator or certificated air carrier, that is required by its operating certificate or approved operations specifications to provide a continuous airworthiness maintenance program according to its maintenance manuals;

(4) Be recommended for certification by his employer, to the satisfaction of the Administrator, as able to satisfactorily maintain aircraft or components, appropriate to the job for which he is employed;

(5) Have either—

(i) At least 18 months of practical experience in the procedures, practices, inspection methods, materials, tools, machine tools, and equipment generally used in the maintenance duties of the specific job for which the person is to be employed and certificated; or

(ii) Completed formal training that is acceptable to the Administrator and is specifically designed to qualify the applicant for the job on which the applicant is to be employed; and

(6) Be able to read, write, speak, and understand the English language, or, in the case of an applicant who does not meet this requirement and who is employed outside the United States by a certificated repair station, a certificated U.S. commercial operator, or a certificated U.S. air carrier, described in paragraph

(c) of this section, have his certificate endorsed "Valid only outside the United States."

(b) This section does not apply to the issuance of a repairman certificate (experimental aircraft builder) under §65.104 or to a repairman certificate (light-sport aircraft) under §65.107.

[Doc. No. 1179, 27 FR 7973, Aug. 10, 1962, as amended by Amdt. 65-11, 32 FR 13506, Sept. 27, 1967; Amdt. 65-24, 44 FR 46781, Aug. 9, 1979; Amdt. 65-27, 47 FR 13316, Mar. 29, 1982]

§65.103 Repairman certificate: Privileges and limitations.

(a) A certificated repairman may perform or supervise the maintenance, preventive maintenance, or alteration of aircraft or aircraft components appropriate to the job for which the repairman was employed and certificated, but only in connection with duties for the certificate holder by whom the repairman was employed and recommended.

(b) A certificated repairman may not perform or supervise duties under the repairman certificate unless the repairman understands the current instructions of the certificate holder by whom the repairman is employed and the manufacturer's instructions for continued airworthiness relating to the specific operations concerned.

(c) This section does not apply to the holder of a repairman certificate (light-sport aircraft) while that repairman is performing work under that certificate.

[Doc. No. 18241, 45 FR 46738, July 10, 1980]

§65.104 Repairman certificate -- experimental aircraft builder -- Eligibility, privileges and limitations.

(a) To be eligible for a repairman certificate (experimental aircraft builder), an individual must --

(1) Be at least 18 years of age;

(2) Be the primary builder of the aircraft to which the privileges of the certificate are applicable;

(3) Show to the satisfaction of the Administrator that the individual has the requisite skill to determine whether the aircraft is in a condition for safe operations; and

(4) Be a citizen of the United States or an individual citizen of a foreign country who has lawfully been admitted for permanent residence in the United States.

(b) The holder of a repairman certificate (experimental aircraft builder) may perform condition inspections on the aircraft constructed by the holder in accordace with the operating limitations of that aircraft.

(c) Section 65.103 does not apply to the holder of a repairman certificate (experimental aircraft builder) while performing under that certificate.

[Doc. No. 18739, 44 FR 46781, Aug. 9, 1979]

§65.105 Display of certificate.

Each person who holds a repairman certificate shall keep it within the immediate area where he normally exercises the privileges of the certificate and shall present it for inspection upon the request of the Administrator or an authorized representative of the National Transportation Safety Board, or of any Federal, State, or local law enforcement officer.

[Doc. No. 7258, 31 FR 13524, Oct. 20, 1966, as amended by Doc. No. 8084, 32 FR 5769, Apr. 11, 1967]

§65.107 Repairman certificate (light-sport aircraft): Eligibility, privileges, and limits.

(a) Use the following table to determine your eligibility for a repairman certificate (light-sport aircraft) and appropriate rating:

To be eligible for . . .	You must . . .
(1) A repairman certificate (light-sport aircraft),	(i) Be at least 18 years old, (ii) Be able to read, speak, write, and understand English. If for medical reasons you cannot meet one of these requirements, the FAA may place limits on your repairman certificate necessary to safely perform the actions authorized by the certificate and rating, (iii) Demonstrate the requisite skill to determine whether a light-sport aircraft is in a condition for safe operation, and (iv) Be a citizen of the United States, or a citizen of a foreign country who has been lawfully admitted for permanent residence in the United States.
(2) A repairman certificate (light-sport aircraft) with an inspection rating,	(i) Meet the requirements of paragraph (a)(I) of this section, and (ii) Complete a 16-hour training course acceptable to the FAA on inspecting the particular class of experimental light-sport aircraft for which you intend to exercise the privileges of this rating.
(3) A repairman certificate (light-sport aircraft) with a maintenance rating,	(i) Meet the requirements of paragraph (a)(I) of this section, and

To be eligible for . . .	You must . . .
	(ii) Complete a training course acceptable to the FAA on maintaining the particular class of light-sport aircraft for which you intend to exercise the privileges of this rating. The training course must, at a minimum, provide the following number of hours of instruction: (A) For airplane class privileges—120 hours, (B) For weight-shift control aircraft class privileges—104 hours, (C) For powered parachute class privileges—104 hours, (D) For lighter than air class privileges—80 hours, (E) For glider class privileges—80 hours.

(b) The holder of a repairman certificate (light-sport aircraft) with an inspection rating may perform the annual condition inspection on a light-sport aircraft:

(1) That is owned by the holder;

(2) That has been issued an experimental certificate for operating a light-sport aircraft under §21.191 (i) of this chapter; and

(3) That is in the same class of light-sport-aircraft for which the holder has completed the training specified in paragraph (a)(2)(ii) of this section.

(c) The holder of a repairman certificate (light-sport aircraft) with a maintenance rating may—

(1) Approve and return to service an aircraft that has been issued a special airworthiness certificate in the light-sport category under §21.190 of this chapter, or any part thereof, after performing or inspecting maintenance (to include the annual condition inspection and the 100-hour inspection required by §91.327 of this chapter), preventive maintenance, or an alteration (excluding a major repair or a major alteration on a product produced under an FAA approval);

(2) Perform the annual condition inspection on a light-sport aircraft that has been issued an experimental certificate for operating a light-sport aircraft under §21.191 (i) of this chapter; and

(3) Only perform maintenance, preventive maintenance, and an alteration on a light-sport aircraft that is in the same class of light-sport aircraft for which the holder has completed the training specified in paragraph (a)(3)(ii) of this section. Before performing a major repair, the holder must complete additional training acceptable to the FAA and appropriate to the repair performed.

(d) The holder of a repairman certificate (light-sport aircraft) with a maintenance rating may not approve for return to service any aircraft or part thereof unless that person has previously performed the work concerned satisfactorily. If that person has not previously performed that work, the person may show the ability to do the work by performing it to the satisfaction of the FAA, or by performing it under the direct supervision of a certificated and appropriately rated mechanic, or a certificated repairman, who has had previous experience in the specific operation concerned. The repairman may not exercise the privileges of the certificate unless the repairman understands the current instructions of the manufacturer and the maintenance manuals for the specific operation concerned.

Subpart F—Parachute Riggers

§65.111 Certificate required.

(a) No person may pack, maintain, or alter any personnel-carrying parachute intended for emergency use in connection with civil aircraft of the United States (including the reserve parachute of a dual parachute system to be used for intentional parachute jumping) unless that person holds an appropriate current certificate and type rating issued under this subpart and complies with §§65.127 through 65.133.

(b) No person may pack, maintain, or alter any main parachute of a dual-parachute system to be used for intentional parachute jumping in connection with civil aircraft of the United States unless that person—

(1) Has an appropriate current certificate issued under this subpart;

(2) Is under the supervision of a current certificated parachute rigger;

(3) Is the person making the next parachute jump with that parachute in accordance with §105.43(a) of this chapter; or

(4) Is the parachutist in command making the next parachute jump with that parachute in a tandem parachute operation conducted under §105.45(b)(1) of this chapter.

(c) Each person who holds a parachute rigger certificate shall present it for inspection upon the request of the Administrator or an authorized representative of the National

Transportation Safety Board, or of any Federal, State, or local law enforcement officer.

(d) The following parachute rigger certificates are issued under this part:

(1) Senior parachute rigger.

(2) Master parachute rigger.

(e) Sections 65.127 through 65.133 do not apply to parachutes packed, maintained, or altered for the use of the armed forces.

[Doc. No. 1179, 27 FR 7973, Aug. 10, 1962, as amended by Amdt. 65-9, 31 FR 13524, Oct. 20, 1966; 32 FR 5769, Apr. 11, 1967; Amdt. 65-42, 66 FR 23553, May 9, 2001]

§65.113 Eligibility requirements: General.

(a) To be eligible for a parachute rigger certificate, a person must—

(1) Be at least 18 years of age;

(2) Be able to read, write, speak, and understand the English language, or, in the case of a citizen of Puerto Rico, or a person who is employed outside of the United States by a U.S. air carrier, and who does not meet this requirement, be issued a certificate that is valid only in Puerto Rico or while he is employed outside of the United States by that air carrier, as the case may be; and

(3) Comply with the sections of this subpart that apply to the certificate and type rating he seeks.

(b) Except for a master parachute rigger certificate, a parachute rigger certificate that was issued before, and was valid on, October 31, 1962, is equal to a senior parachute rigger certificate, and may be exchanged for such a corresponding certificate.

§65.115 Senior parachute rigger certificate: Experience, knowledge, and skill requirements.

Except as provided in §65.117, an applicant for a senior parachute rigger certificate must—

(a) Present evidence satisfactory to the Administrator that he has packed at least 20 parachutes of each type for which he seeks a rating, in accordance with the manufacturer's instructions and under the supervision of a certificated parachute rigger holding a rating for that type or a person holding an appropriate military rating;

(b) Pass a written test, with respect to parachutes in common use, on—

(1) Their construction, packing, and maintenance;

(2) The manufacturer's instructions;

(3) The regulations of this subpart; and

(c) Pass an oral and practical test showing his ability to pack and maintain at least one type of parachute in common use, appropriate to the type rating he seeks.

[Doc. No. 10468, 37 FR 13251, July 6, 1972]

§65.117 Military riggers or former military riggers: Special certification rule.

In place of the procedure in §65.115, an applicant for a senior parachute rigger certificate is entitled to it if he passes a written test on the regulations of this subpart and presents satisfactory documentary evidence that he—

(a) Is a member or civilian employee of an Armed Force of the United States, is a civilian employee of a regular armed force of a foreign country, or has, within the 12 months before he applies, been honorably discharged or released from any status covered by this paragraph;

(b) Is serving, or has served within the 12 months before he applies, as a parachute rigger for such an Armed Force; and

(c) Has the experience required by §65.115(a).

§65.119 Master parachute rigger certificate: Experience, knowledge, and skill requirements.

An applicant for a master parachute rigger certificate must meet the following requirements:

(a) Present evidence satisfactory to the Administrator that he has had at least 3 years of experience as a parachute rigger and has satisfactorily packed at least 100 parachutes of each of two types in common use, in accordance with the manufacturer's instructions—

(1) While a certificated and appropriately rated senior parachute rigger; or

(2) While under the supervision of a certificated and appropriately rated parachute rigger or a person holding appropriate military ratings. An applicant may combine experience specified in paragraphs (a) (1) and (2) of this section to meet the requirements of this paragraph.

(b) If the applicant is not the holder of a senior parachute rigger certificate, pass a written test, with respect to parachutes in common use, on—

(1) Their construction, packing, and maintenance;

(2) The manufacturer's instructions; and

(3) The regulations of this subpart.

(c) Pass an oral and practical test showing his ability to pack and maintain two types of parachutes in common use, appropriate to the type ratings he seeks.

[Doc. No. 10468, 37 FR 13252, July 6, 1972]

§65.121 Type ratings.

(a) The following type ratings are issued under this subpart:

(1) Seat.

(2) Back.

(3) Chest.

(4) Lap.

(b) The holder of a senior parachute rigger certificate who qualifies for a master parachute rigger certificate is entitled to have placed on his master parachute rigger certificate the ratings that were on his senior parachute rigger certificate.

§65.123 Additional type ratings: Requirements.

A certificated parachute rigger who applies for an additional type rating must—

(a) Present evidence satisfactory to the Administrator that he has packed at least 20 parachutes of the type for which he seeks a rating, in accordance with the manufacturer's instructions and under the supervision of a certificated parachute rigger holding a rating for that type or a person holding an appropriate military rating; and

(b) Pass a practical test, to the satisfaction of the Administrator, showing his ability to pack and maintain the type of parachute for which he seeks a rating.

[Doc. No. 1179, 27 FR 7973, Aug. 10, 1962, as amended by Amdt. 65-20, 37 FR 13251, July 6, 1972]

§65.125 Certificates: Privileges.

(a) A certificated senior parachute rigger may --

(1) Pack or maintain (except for major repair) any type of parachute for which he is rated; and

(2) Supervise other persons in packing any type of parachute for which that person is rated in accordance with §105.43(a) or §105.45(b)(1) of this chapter.

(b) A certificated master parachute rigger may—

(1) Pack, maintain, or alter any type of parachute for which he is rated; and

(2) Supervise other persons in packing, maintaining, or altering any type of parachute for which the certificated parachute rigger is rated in accordance with §105.43(a) or §105.45(b)(1) of this chapter.

(c) A certificated parachute rigger need not comply with §§65.127 through 65.133 (relating to facilities, equipment, performance standards, records, recent experience, and seal) in packing, maintaining, or altering (if authorized) the main parachute of a dual parachute pack to be used for intentional jumping.

[Doc. No. 1179, 27 FR 7973, Aug. 10, 1962, as amended by Amdt. 65-20, 37 FR 13252, July 6, 1972; Amdt. 65-42, 66 FR 23553, May 9, 2001]

§65.127 Facilities and equipment.

No certificated parachute rigger may exercise the privileges of his certificate unless he has at least the following facilities and equipment available to him:

(a) A smooth top table at least three feet wide by 40 feet long.

(b) Suitable housing that is adequately heated, lighted, and ventilated for drying and airing parachutes.

(c) Enough packing tools and other equipment to pack and maintain the types of parachutes that he services.

(d) Adequate housing facilities to perform his duties and to protect his tools and equipment.

[Doc. No. 1179, 27 FR 7973, Aug. 10, 1962, as amended by Amdt. 65-27, 47 FR 13316, Mar. 29, 1982]

§65.129 Performance standards.

No certificated parachute rigger may—

(a) Pack, maintain, or alter any parachute unless he is rated for that type;

(b) Pack a parachute that is not safe for emergency use;

(c) Pack a parachute that has not been thoroughly dried and aired;

(d) Alter a parachute in a manner that is not specifically authorized by the Administrator or the manufacturer;

(e) Pack, maintain, or alter a parachute in any manner that deviates from procedures approved by the Administrator or the manufacturer of the parachute; or

(f) Exercise the privileges of his certificate and type rating unless he understands the current manufacturer's instructions for the operation involved and has—

(1) Performed duties under his certificate for at least 90 days within the preceding 12 months; or

(2) Shown the Administrator that he is able to perform those duties.

§65.131 Records.

(a) Each certificated parachute rigger shall keep a record of the packing, maintenance, and alteration of parachutes performed or supervised by him. He shall keep in that record, with respect to each parachute worked on, a statement of—

(1) Its type and make;

(2) Its serial number;

(3) The name and address of its owner;

(4) The kind and extent of the work performed;

(5) The date when and place where the work was performed; and

(6) The results of any drop tests made with it.

(b) Each person who makes a record under paragraph (a) of this section shall keep it for at least 2 years after the date it is made.

(c) Each certificated parachute rigger who packs a parachute shall write, on the parachute packing record attached to the parachute, the date and place of the packing and a notation of any defects he finds on inspection. He shall sign that record with his name and the number of his certificate.

§65.133 Seal.

Each certificated parachute rigger must have a seal with an identifying mark prescribed by the Administrator, and a seal press. After packing a parachute he shall seal the pack with his seal in accordance with the manufacturer's recommendation for that type of parachute.

Appendix A to Part 65—Aircraft Dispatcher Courses

Overview

This appendix sets forth the areas of knowledge necessary to perform dispatcher functions. The items listed below indicate the minimum set of topics that must be covered in a training course for aircraft dispatcher certification. The order of coverage is at the discretion of the approved school. For the latest technological advancements refer to the Practical Test Standards as published by the FAA.

I. Regulations

A. Subpart C of this part;

B. Parts 1, 25, 61, 71, 91, 121, 139, and 175, of this chapter;

C. 49 CFR part 830;

D. General Operating Manual.

II. Meteorology

A. Basic Weather Studies

(1) The earth's motion and its effects on weather.

(2) Analysis of the following regional weather types, characteristics, and structures, or combinations thereof:

(a) Maritime.

(b) Continental.

(c) Polar.

(d) Tropical.

(3) Analysis of the following local weather types, characteristics, and structures or combinations thereof:

(a) Coastal.

(b) Mountainous.

(c) Island.

(d) Plains.

(4) The following characteristics of the atmosphere:

(a) Layers.

(b) Composition.

(c) Global Wind Patterns.

(d) Ozone.

(5) Pressure:

(a) Units of Measure.

(b) Weather Systems Characteristics.

(c) Temperature Effects on Pressure.

(d) Altimeters.

(e) Pressure Gradient Force.

(f) Pressure Pattern Flying Weather.

(6) Wind:

(a) Major Wind Systems and Coriolis Force.

(b) Jetstreams and their Characteristics.

(c) Local Wind and Related Terms.

(7) States of Matter:

(a) Solids, Liquid, and Gases.

(b) Causes of change of state.

(8) Clouds:

(a) Composition, Formation, and Dissipation.

(b) Types and Associated Precipitation.

(c) Use of Cloud Knowledge in Forecasting.

(9) Fog

(a) Causes, Formation, and Dissipation.

(b) Types.

(10) Ice:

(a) Causes, Formation, and Dissipation.

(b) Types.

(11) Stability/Instability:

(a) Temperature Lapse Rate, Convection.

(b) Adiabatic Processes.

(c) Lifting Processes.

(d) Divergence.

(e) Convergence.

(12) Turbulence:

(a) Jetstream Associated.

(b) Pressure Pattern Recognition.

(c) Low Level Windshear.

(d) Mountain Waves.

(e) Thunderstorms.

(f) Clear Air Turbulence.

(13) Airmasses:

(a) Classification and Characteristics.

(b) Source Regions.

(c) Use of Airmass Knowledge in Forecasting.

(14) Fronts:

(a) Structure and Characteristics, Both Vertical and Horizontal.

(b) Frontal Types.

(c) Frontal Weather Flying.

(15) Theory of Storm Systems:

(a) Thunderstorms.

(b) Tornadoes.

(c) Hurricanes and Typhoons.

(d) Microbursts.

(e) Causes, Formation, and Dissipation.

B. Weather, Analysis, and Forecasts

(1) Observations:

(a) Surface Observations.

(i) Observations made by certified weather observer.

(ii) Automated Weather Observations.

(b) Terminal Forecasts.

(c) Significant En route Reports and Forecasts.

(i) Pilot Reports.

(ii) Area Forecasts.

(iii) Sigmets, Airmets.

(iv) Center Weather Advisories.

(d) Weather Imagery.

(i) Surface Analysis.

(ii) Weather Depiction.

(iii) Significant Weather Prognosis.

(iv) Winds and Temperature Aloft.

(v) Tropopause Chart.

(vi) Composite Moisture Stability Chart.

(vii) Surface Weather Prognostic Chart.

(viii) Radar Meteorology.

(ix) Satellite Meteorology.

(x) Other charts as applicable.

(e) Meteorological Information Data Collection Systems.

(2) Data Collection, Analysis, and Forecast Facilities.

(3) Service Outlets Providing Aviation Weather Products.

C. Weather Related Aircraft Hazards

(1) Crosswinds and Gusts.

(2) Contaminated Runways.

(3) Restrictions to Surface Visibility.

(4) Turbulence and Windshear.

(5) Icing.

(6) Thunderstorms and Microburst.

(7) Volcanic Ash.

III. Navigation

A. Study of the Earth

(1) Time reference and location (0 Longitude, UTC).

(2) Definitions.

(3) Projections.

(4) Charts.

B. Chart Reading, Application, and Use.

C. National Airspace Plan.

D. Navigation Systems.

E. Airborne Navigation Instruments.

F. Instrument Approach Procedures.

(1) Transition Procedures.

(2) Precision Approach Procedures.

(3) Non-precision Approach Procedures.

(4) Minimums and the relationship to weather.

G. Special Navigation and Operations.

(1) North Atlantic.

(2) Pacific.

(3) Global Differences.

IV. AIRCRAFT

A. Aircraft Flight Manual.

B. Systems Overview.

(1) Flight controls.

(2) Hydraulics.

(3) Electrical.

(4) Air Conditioning and Pressurization.

(5) Ice and Rain protection.

(6) Avionics, Communication, and Navigation.

(7) Powerplants and Auxiliary Power Units.

(8) Emergency and Abnormal Procedures.

(9) Fuel Systems and Sources.

C. Minimum Equipment List/Configuration Deviation List (MEL/CDL) and Applications.

D. Performance.

(1) Aircraft in general.

(2) Principles of flight:

(a) Group one aircraft.

(b) Group two aircraft.

(3) Aircraft Limitations.

(4) Weight and Balance.

(5) Flight instrument errors.

(6) Aircraft performance:

(a) Take-off performance.

(b) En route performance.

(c) Landing performance.

V. Communications

A. Regulatory requirements.

B. Communication Protocol.

C. Voice and Data Communications.

D. Notice to Airmen (NOTAMS).

E. Aeronautical Publications.

F. Abnormal Procedures.

VI. Air Traffic Control

A. Responsibilities.

B. Facilities and Equipment.

C. Airspace classification and route structure.

D. Flight Plans.

(1) Domestic.

(2) International.

E. Separation Minimums.

F. Priority Handling.

G. Holding Procedures.

H. Traffic Management.

VII. Emergency and Abnormal Procedures

A. Security measures on the ground.

B. Security measures in the air.

C. FAA responsibility and services.

D. Collection and dissemination of information on overdue or missing aircraft.

E. Means of declaring an emergency.

F. Responsibility for declaring an emergency.

G. Required reporting of an emergency.

H. NTSB reporting requirements.

VIII. Practical Dispatch Applications

A. Human Factors.

(1) Decisionmaking:

(a) Situation Assessment.

(b) Generation and Evaluation of Alternatives.

(i) Tradeoffs and Prioritization.

(ii) Contingency Planning.

(c) Support Tools and Technologies.

(2) Human Error:

(a) Causes.

(i) Individual and Organizational Factors.

(ii) Technology-Induced Error.

(b) Prevention.

(c) Detection and Recovery.

(3) Teamwork:

(a) Communication and Information Exchange.

(b) Cooperative and Distributed Problem-Solving.

(c) Resource Management.

(i) Air Traffic Control (ATC) activities and workload.

(ii) Flightcrew activities and workload.

(iii) Maintenance activities and workload.

(iv) Operations Control Staff activities and workload.

B. Applied Dispatching.

(1) Briefing techniques, Dispatcher, Pilot.

(2) Preflight:

(a) Safety.

(b) Weather Analysis.

(i) Satellite imagery.

(ii) Upper and lower altitude charts.

(iii) Significant en route reports and forecasts.

(iv) Surface charts.

(v) Surface observations.

(vi) Terminal forecasts and orientation to Enhanced Weather Information System (EWINS).

(c) NOTAMS and airport conditions.

(d) Crew.

(i) Qualifications.

(ii) Limitations.

(e) Aircraft.

(i) Systems.

(ii) Navigation instruments and avionics systems.

(iii) Flight instruments.

(iv) Operations manuals and MEL/CDL.

(v) Performance and limitations.

(f) Flight Planning.

(i) Route of flight.

1. Standard Instrument Departures and Standard Terminal Arrival Routes.

2. En route charts.

3. Operational altitude.

4. Departure and arrival charts.

(ii) Minimum departure fuel.

1. Climb.

2. Cruise.

3. Descent.

(g) Weight and balance.

(h) Economics of flight overview (Performance, Fuel Tankering).

(i) Decision to operate the flight.

(j) ATC flight plan filing.

(k) Flight documentation.

(i) Flight plan.

(ii) Dispatch release.

(3) Authorize flight departure with concurrence of pilot in command.

(4) In-flight operational control:

(a) Current situational awareness.

(b) Information exchange.

(c) Amend original flight release as required.

(5) Post-Flight:

(a) Arrival verification.

(b) Weather debrief.

(c) Flight irregularity reports as required.

[Doc. No. FAA-1998–4553, 64 FR 68925, Dec. 8, 1999]

Part 67—Medical standards and certification

Authority: 49 U.S.C. 106(g), 40113, 44701-44703, 44707, 44709–44711, 45102-45103, 45301-45303.

Subpart A—General

§ 67.1 Applicability.

This part prescribes the medical standards and certification procedures for issuing medical certificates for airmen and for remaining eligible for a medical certificate.

§ 67.3 Issue.

Except as provided in § 67.5, a person who meets the medical standards prescribed in this part, based on medical examination and evaluation of the person's history and condition, is entitled to an appropriate medical certificate.

§ 67.7 Access to the National Driver Register.

At the time of application for a certificate issued under this part, each person who applies for a medical certificate shall execute an express consent form authorizing the Administrator to request the chief driver licensing official of any state designated by the Administrator to transmit information contained in the National Driver Register about the person to the Administrator. The Administrator shall make information received from the National Driver Register, if any, available on request to the person for review and written comment.

Subpart B—First-Class Airman Medical Certificate

§ 67.101 Eligibility.

To be eligible for a first-class airman medical certificate, and to remain eligible for a first-class airman medial certificate, a person must meet the requirements of this subpart.

§ 67.103 Eye.

Eye standards for a first-class airman medical certificate are:

(a) Distant visual acuity of 20/20 or better in each eye separately, with or without corrective lenses. If corrective lenses (spectacles or contact lenses) are necessary for 20/20 vision, the person may be eligible only on the condition that corrective lenses are worn while exercising the privileges of an airman certificate.

(b) Near vision of 20/40 or better, Snellen equivalent, at 16 inches in each eye separately, with or without corrective leases. If age 50 or older, near vision of 20/40 or better, Snellen equivalent, at both 16 inches and 32 inches in each eye separately, with or without corrective lenses.

(c) Ability to perceive those colors necessary for the safe performance of airman duties.

(d) Normal fields of vision.

(e) No acute or chronic pathological condition of either eye or adnexa that interferes with the proper function of an eye, that may reasonably be expected to progress to that degree, or that may reasonably be expected to be aggravated by flying.

(f) Bifoveal fixation and vergence-phoria relationship sufficient to prevent a break in fusion under conditions that may reasonably be expected to occur in performing airman duties. Tests for the factors named in this paragraph are not, required except for persons found to have more than 1 prism diopter of hyperphoria, 6 prison diopters of esophoria, or 6 prism diopters of exophoria. If any of these values are exceeded, the Federal Air Surgeon may require the person to be examined by a qualified eye specialist to determine if there is bifoveal fixation and an adequate vergence-phoria relationship. However, if otherwise eligible, the person is issued a medical certificate pending the results of the examination.

§ 67.105 Ear, nose, throat and equilibrium.

Ear, nose, throat, and equilibrium standards for a first-class airman medical certificate are:

(a) The person shall demonstrate acceptable hearing by at least one of the following tests:

(1) Demonstrate an ability to hear an average conversational voice in a quiet room. using both ears, at a distance of 6 feet from the examiner, with the back turned to the examiner.

(2) Demonstrate an acceptable understanding of speech as determined by audiometric speech discrimination testing to a score of at least 70 percent obtained in one ear or in a sound field environment.

(3) Provide acceptable results of pure tone audiometric testing of unaided hearing acuity according to the following table of worst acceptable thresholds, using the calibration standards of the American National Standards Institute, 1969 (11 West 42d Street, New York, NY 10036):

Frequency (Hz)	500 Hz	1000 Hz	2000 Hz	3000 Hz
Better ear (Db)	35	30	30	40
Poorer ear (Db)	35	50	50	60

(b) No disease or condition of the middle or internal ear, nose, oral cavity, pharynx, or larynx that—

(1) Interferes with, or is aggravated by, flying or may reasonably be expected to do so; or

(2) Interferes with, or may reasonably be expected to interfere with, clear and effective speech communication.

(c) No disease or condition manifested by, or that may reasonably be expected to be manifested by, vertigo or a disturbance of equilibrium.

§ 67.107 Mental.

Mental standards for a first-class airman medical certificate are:

(a) No established medical history or clinical diagnosis of any of the following:

(1) A personality disorder that is severe enough to have repeatedly manifested itself by overt acts.

(2) A psychosis. As used in this section, "psychosis" refers to a mental disorder in which:

(i) The individual has manifested delusions, hallucinations, grossly bizarre or disorganized behavior, or other commonly accepted symptoms of this condition; or

(ii) The individual may reasonably be expected to manifest delusions, hallucinations, grossly bizarre or disorganized behavior, or other commonly accepted symptoms of this condition.

(3) A bipolar disorder.

(4) Substance dependence, except where there is established clinical evidence, satisfactory to the Federal Air Surgeon, of recovery, including sustained total abstinence from the substance(s) for not less than the preceding 2 years. As used in this section—

(i) "Substance" includes: Alcohol; other sedatives and hypnotics; anxiolytics; opioids; central nervous system stimulants such as cocaine, amphetamines, and similarly acting sympathomimetics; hallucinogens; phencyclidine or similarly acting arylcyclohexylamines; cannabis; inhalants; and other psychoactive drugs and chemicals; and

(ii) "Substance dependence" means a condition in which a person is dependent on a substance, other than tobacco or ordinary xanthine-containing (e.g., caffeine) beverages, as evidenced by—

(A) Increased tolerance;

(B) Manifestation of withdrawal symptoms;

(C) Impaired control of use; or

(D) Continued use despite damage to physical health or impairment of social, personal, or occupational functioning.

(b) No substance abuse within the preceding 2 years defined as:

(1) Use of a substance in a situation in which that use was physically hazardous, if there has been at any other time an instance of the use of a substance also in a situation in which that use was physically hazardous;

(2) A verified positive drug test result, an alcohol test result of 0.04 or greater alcohol concentration, or a refusal to submit to a drug or alcohol test required by the U.S. Department of Transportation or an agency of the U.S. Department of transportation; or

(3) Misuse of a substance that the Federal Air Surgeon, based on case history and appropriate, qualified medical judgment relating to the substance involved, finds—

(i) Makes the person unable to safely perform the duties or exercise the privileges of the airman certificate applied for or held; or

(ii) May reasonably be expected, for the maximum duration of the airman medical certificate applied for or held, to make the person unable to perform those duties or exercise those privileges.

(c) No other personality disorder, neurosis, or other mental condition that the Federal Air Surgeon, based on the case history and appropriate, qualified medical judgment relating to the condition involved, finds—

(1) Makes the person unable to safely perform the duties or exercise the privileges of the airman certificate applied for or held; or

(2) May reasonably be expected, for the maximum duration of the airman medical certificate applied for or held, to make the person unable to perform those duties or exercise those privileges.

§ 67.109 Neurologic.

Neurologic standards for a first-class airman medical certificate are:

(a) No established medical history or clinical diagnosis of any of the following:

(1) Epilepsy:

(2) A disturbance of consciousness without satisfactory medical explanation of the cause; or

(3) A transient loss of control of nervous system function(s) without satisfactory medical explanation of the cause.

(b) No other seizure disorder, disturbance of consciousness, or neurologic condition that the Federal Air Surgeon, based on the case history and appropriate, qualified medical judgment relating to the condition involved, finds—

(1) Makes the person unable to safely perform the duties or exercise the privileges of the airman certificate applied for or held; or

(2) May reasonably be expected, for the maximum duration of the airman medical certificate applied for or hold, to make the person unable to perform those duties or exercise those privileges.

§ 67.111 Cardiovascular

Cardiovascular standards for a first-class airman medical certificate are:

(a) No established medical history or clinical diagnosis of any of the following:

(1) Myocardial infarction;

(2) Angina pectoris;

(3) Coronary heart disease that has required treatment or, if untreated, that has been symptomatic or clinically significant;

(4) Cardiac valve replacement;

(5) Permanent cardiac pacemaker implantation; or

(6) Heart replacement;

(b) A person applying for first-class medical certification must demonstrate an absence of myocardial infarction and other clinically significant abnormality on electrocardiographic examination:

(1) At the first application after reaching the 35th birthday; and

(2) On an annual basis after reaching the 40th birthday.

(c) An electrocardiogram will satisfy a requirement of paragraph (b) of this section if it is dated no earlier than 60 days before the date of the application it is to accompany and was performed and transmitted according to acceptable standards and techniques.

§ 67.113 General medical condition.

The general medical standards for a first-class airman medical certificate are:

(a) No established medical history or clinical diagnosis of diabetes mellitus that requires insulin or any other hypoglycemic drug for control.

(b) No other organic, functional, or structural disease, defect, or limitation that the Federal Air Surgeon, based on the case history and appropriate, qualified medical judgment relating to the condition involved, finds—

(1) Makes the person unable to safely perform the duties or exercise the privileges of the airman certificate applied for or held; or

(2) May reasonably be expected, for the maximum duration of the airman medical certificate applied for or held, to make the person unable to perform those duties or exercise those privileges.

(c) No medication or other treatment that the Federal Air Surgeon, based on the case history and appropriate, qualified medical judgment relating to the medication or other treatment involved, finds—

(1) Makes the person unable to safely perform the duties or exercise the privileges of the airman certificate applied for or held; or

(2) May reasonably be expected, for the maximum duration of the airman medical certificate applied for or held, to make the person unable to perform those duties or exercise those privileges.

§ 67.115 Discretionary issuance.

A person who does not meet the provisions of §§ 67.103 through 67.113 may apply for the discretionary issuance of a certificate under § 67.401.

Subpart C—Second-Class Airman Medical Certificate

§ 67.201 Eligibility.

To be eligible for a second-class airman medical certificate, and to remain eligible for a second-class airman medical certificate, a person must meet the requirements of this subpart.

§ 67.203 Eye.

Eye standards for a second-class airman medical certificate are:

(a) Distant visual acuity of 20/20 or better in each eye separately, with or without corrective lenses. If corrective lenses (spectacles or contact lenses) are necessary for 20/20 vision, the person may be eligible only on the condition that corrective leases are worn while exercising the privileges of an airman certificate.

(b) Near vision of 20/40 or better, Snellen equivalent, at 16 inches in each eye separately, with or without corrective lenses. If age 50 or older, near vision of 20/40 or better, Snellen equivalent, at both 16 inches and 32 inches in each eye separately, with or without corrective lenses.

(c) Ability to perceive those colors necessary for the safe performance of airman duties.

(d) Normal fields of vision.

(e) No acute or chronic pathological condition of either eye or adnexa that interferes with the proper function of an eye, that may reasonably be expected to progress to that degree, or that may reasonably be expected to be aggravated by flying.

(f) Bifoveal fixation and vergence-phoria relationship sufficient to prevent a break in fusion under conditions that may reasonably be expected to occur in performing airman duties. Tests for the factors named in this paragraph are not required except for persons found to have more than 1 prism diopter of hyperphoria, 6 prism diopters of esophoria, or 6 prism diopters of exophoria. If any of these values are exceeded, the Federal Air Surgeon may require the person to be examined by a qualified eye specialist to determine if there is bifoveal fixation and an adequate vergence-phoria relationship. However, if otherwise eligible, the person is issued a medical certificate pending the results of the examination.

§ 67.205 Ear, nose, throat, and equilibrium

Ear, nose, throat, and equilibrium standards for a second-class airman medical certificate are:

(a) The person shall demonstrate acceptable hearing by at least one of the following tests:

(1) Demonstrate an ability to hear an average conversational voice in a quiet room, using both ears, at a distance of 6 feet from the examiner, with the back turned to the examiner.

(2) Demonstrate an acceptable understanding of speech as determined by audiometric speech discrimination testing to a score of at least 70 percent obtained in one ear or in a sound field environment.

(3) Provide acceptable results of pure tone audiometric testing of unaided hearing acuity according to the following table of worst acceptable thresholds, using the calibration standards of the American National Standards Institute, 1969:

Frequency (Hz)	500 Hz	1000 Hz	2000 Hz	3000 Hz
Better ear (Db)	35	30	30	40
Poorer ear (Db)	35	50	50	60

(b) No disease or condition of the middle or internal ear, nose, oral cavity, pharynx, or larynx that—

(1) Interferes with, or is aggravated by, flying or may reasonably be expected to do so; or

(2) Interferes with, or may reasonably be expected to interfere with clear and effective speech communication.

(c) No disease or condition manifested by, or that may reasonably be expected to be manifested by, vertigo or a disturbance of equilibrium.

§ 67.207 Mental.

Mental standards for a second-class airman medical certificate are:

(a) No established medical history or clinical diagnosis of any of the following:

(1) A personality disorder that is severe enough to have repeatedly manifested itself by overt acts.

(2) A psychosis. As used in this section, "psychosis" refers to a mental disorder in which:

(i) The individual has manifested delusions, hallucinations, grossly bizarre or disorganized behavior, or other commonly accepted symptoms of this condition; or

(ii) The individual may reasonably be expected to manifest delusions, hallucinations, grossly bizarre or disorganized behavior, or other commonly accepted symptoms of this condition.

(3) A bipolar disorder.

(4) Substance dependence, except where there is established clinical evidence, satisfactory to the Federal Air Surgeon, of recovery, including sustained total abstinence from the substance(s) for not less than the preceding 2 years. As used in this section—

(i) "Substance" includes: Alcohol; other sedatives and hypnotics; anxiolytics; opioids; central nervous system stimulants such as cocaine, amphetamines, and similarly acting sympathomimetics; hallucinogens; phencyclidine or similarly acting arylcyclohexylamines; cannabis; inhalants; and other psychoactive drugs and chemicals; and

(ii) "Substance dependence" means a condition in which a person is dependent on a substance, other than tobacco or ordinary xanthine-containing (e.g., caffeine) beverages, as evidenced by—

(A) Increased tolerance;

(B) Manifestation of withdrawal symptoms;

(C) Impaired control of use; or

(D) Continued use despite damage to physical health or impairment of social, personal, or occupational functioning.

(b) No substance abuse within the preceding 2 years defined as.

(1) Use of a substance in a situation in which that use was physically hazardous, if there has been at any other time an instance of the use of a substance also in a situation in which that use was physically hazardous;

(2) A verified positive drug test result, an alcohol test result of 0.04 or greater alcohol concentration, or a refusal to submit to a drug or alcohol test required by the U.S. Department of Transportation or an agency of the U.S. Department of transportation; or

(3) Misuse of a substance that the Federal Air Surgeon, based on case history and appropriate, qualified medical judgment relating to the substance involved, finds—

(i) Makes the person unable to safely perform the duties or exercise the privileges of the airmen certificate applied for or held; or

(ii) May reasonably be expected, for the maximum duration of the airman medical certificate applied for or held, to make the person unable to perform those duties or exercise those privileges.

(c) No other personality disorder, neurosis, or other mental condition that the Federal Air Surgeon, based on the case history and appropriate, qualified medical judgment relating to the condition involved, finds—

(1) Makes the person unable to safely perform the duties or exercise the privileges of the airman certificate applied for or held; or

(2) May reasonably be expected, for the maximum duration of the airman certificate applied for or held, to make the person unable to perform those duties or exercise those privileges.

§ 67.209 Neurologic.

Neurologic standards for a second-class airman medical certificate are:

(a) No established medical history or clinical diagnosis of any of the following:

(1) Epilepsy;

(2) A disturbance of consciousness without satisfactory medical explanation of the cause; or

(3) A transient loss of control of nervous system function(s) without satisfactory medical explanation of the cause;

(b) No other seizure disorder, disturbance of consciousness, or neurologic condition that the Federal Air Surgeon, based on the case history and appropriate, qualified medical judgment relating to the condition involved, finds—

(1) Makes the person unable to safely perform the duties or exercise the privileges of the airman certificate applied for or held; or

(2) May reasonably be expected, for the maximum duration of the airman medical certificate applied for or held, to make the person unable to perform those duties or exercise those privileges.

§ 67.211 Cardiovascular.

Cardiovascular standards for a second-class medical certificate are no established medical history or clinical diagnosis of any of the following:

(a) Myocardial infarction;

(b) Angina pectoris;

(c) Coronary heart disease that has required treatment or, if untreated, that has been symptomatic or clinically significant;

(d) Cardiac valve replacement;

(e) Permanent cardiac pacemaker implantation; or

(f) Heart replacement.

§ 67.213 General medical condition.

The general medical standards for a second-class airman medical certificate are:

(a) No established medical history or clinical diagnosis of diabetes mellitus that requires insulin or any other hypoglycemic drug for control.

(b) No other organic, functional, or structural disease, defect, or limitation that the Federal Air Surgeon, based on the case history and appropriate, qualified medical judgment relating to the condition involved, finds—

(1) Makes the person unable to safely perform duties or exercise the privileges of the airman certificate applied for or held; or

(2) May reasonably be expected, for the maximum duration of the airman medical certificate applied for or held, to make the person unable to perform those duties or exercise those privileges.

(c) No medication or other treatment that the Federal Air Surgeon, based on the case history and appropriate, qualified medical judgment relating to the medication or other treatment involved, finds—

(1) Makes the person unable to safely perform the duties or exercise the privileges of the airman certificate applied for or held; or

(2) May reasonably be expected, for the maximum duration of the airman medical certificate applied for or held, to make the person unable to perform those duties or exercise those privileges.

§ 67.215 Discretionary issuance.

A person who does not meet the provisions of §§ 67.203 through 67.213 may apply for the discretionary issuance of a certificate under § 67.401.

Subpart D—Third-Class Airman Medical Certificate

§ 67.301 Eligibility.

To be eligible for a third-class airman medial certificate, or to remain eligible for a third-class airman medical certificate, a person must meet the requirements of this subpart.

§ 67.303 Eye.

Eye standards for a third-class airman medical certificate are:

(a) Distant visual acuity of 20/40 or better in each eye separately, with or without corrective lenses. If corrective lenses (spectacles or contact lenses) are necessary for 20/40 vision, the person may be eligible only on the condition that corrective lenses are worn while exercising the privileges of an airman certificate.

(b) Near vision of 20/40 or better, Snellen equivalent, at 16 inches in each eye separately, with or without corrective lenses.

(c) Ability to perceive those colors necessary for the safe performance of airman duties.

(d) No acute or chronic pathological condition of either eye or adnexa that interferes with the proper function of an eye, that may reasonably be expected to progress to that degree, or that may reasonably be expected to be aggravated by flying.

§ 67.305 Ear, nose, throat, and equilibrium.

Ear, nose, throat, and equilibrium standards for a third-class airman medical certificate are:

(a) The person shall demonstrate acceptable hearing by at least one of the following tests:

(1) Demonstrate an ability to hear an average conversation voice in a quiet room, using both ears, at a distance of 6 feet from the examiner, with the back turned to the examiner.

(2) Demonstrate an acceptable understanding of speech as determined by audiometric speech discrimination testing to a score of at least 70 percent obtained in one ear or in a sound field environment

(3) Provide acceptable results of pure tone audiometric testing of unaided hearing acuity according to the following table of worst acceptable thresholds, using the calibration standards of the American National Standards Institute, 1969:

Frequency (Hz)	500 Hz	1000 Hz	2000 Hz	3000 Hz
Better ear (Db)	35	30	30	40
Poorer ear (Db)	35	50	50	60

(b) No disease or condition of the middle or internal ear, nose, oral cavity, pharynx, or larynx that—

(1) Interferes with, or is aggravated by, flying or may reasonably be expected to do so; or

(2) Interferes with clear and effective speech communication.

(c) No disease or condition manifested by, or that may reasonably be expected to be manifested by, vertigo or a disturbance of equilibrium.

§ 67.307 Mental.

Mental standards for a third-class airman medical certificate are:

(a) No established medical history or clinical diagnosis of any of the following:

(1) A personality disorder that is severe enough to have repeatedly manifested itself by overt acts.

(2) A psychosis. As used in this section, "psychosis" refers to a mental disorder in which—

(i) The individual has manifested delusions, hallucinations, grossly bizarre or disorganized behavior, or other commonly accepted symptoms of this condition; or

(ii) The individual may reasonably be expected to manifest delusions, hallucinations, grossly bizarre or disorganized behavior, or other commonly accepted symptoms of this condition.

(3) A bipolar disorder.

(4) Substance dependence, except where there is established clinical evidence, satisfactory to the Federal Air Surgeon, of recovery, including sustained total abstinence from the substance(s) for not less than the preceding 2 years. As used in this section—

(i) "Substance" includes: alcohol; other sedatives and hypnotics; anxiolytics; opioids; central nervous system stimulants such as cocaine, amphetamines, and similarly acting sympathomimetics; hallucinogens; phencyclidine or similarly acting arylcyclohexylamines; cannabis; inhalants; and other psychoactive drugs and chemicals; and

(ii) "Substance dependence" means a condition in which a person is dependent on a substance, other than tobacco or ordinary xanthine-containing (e.g., caffeine) beverages, as evidenced by—

(A) Increased tolerance;

(B) Manifestation of withdrawal symptoms;

(C) Impaired control of use; or

(D) Continued use despite damage to physical health or impairment of social, personal, or occupational functioning.

(b) No substance abuse within the preceding 2 years defined as:

(1) Use of a substance in a situation in which that use was physically hazardous, if there has been at any other time an instance of the use of a substance also in a situation in which that use was physically hazardous;

(2) A verified positive drug test result, an alcohol test result of 0.04 or greater alcohol concentration, or a refusal to submit to a drug or alcohol test required by the U.S. Department of Transportation or an agency of the U.S. Department of transportation; or

(3) Misuse of a substance that the Federal Air Surgeon, based on case history and appropriate, qualified medical judgment relating to the substance involved, finds—

(i) Makes the person unable to safely perform the duties or exercise the privileges of the airman certificate applied for or held; or

(ii) May reasonably be expected, for the maximum duration of the airman medical certificate applied for or held, to make the person unable to perform those duties or exercise those privileges.

(c) No other personality disorder, neurosis, or other mental condition that, the Federal Air Surgeon, based on the case history and appropriate, qualified medical judgment relating to the condition involved, finds—

(1) Makes the person unable to safely perform the duties or exercise the privileges of the airman certificate applied for or held; or

(2) May reasonably be expected, for the maximum duration of the airman medical certificate applied for or held, to make the person unable to perform those duties or exercise those privileges.

§ 67.309 Neurologic.

Neurologic standards for a third-class airman medical certificate are:

(a) No established medical history or clinical diagnosis of any of the following:

(1) Epilepsy;

(2) A disturbance of consciousness without satisfactory medical explanation of the cause; or

(3) A transient loss of control of nervous system function(s) without satisfactory medical explanation of the cause.

(b) No other seizure disorder, disturbance of consciousness, or neurologic condition that the Federal Air Surgeon, band on the case history and appropriate, qualified medical judgment relating to the condition involved finds—

(1) Makes the person unable to safely perform the duties or exercise the privileges of the airman certificate applied for or held; or

(2) May reasonably be expected, for the maximum duration of the airman medical certificate applied for or held, to make the person unable to perform those duties or exercise those privileges.

§ 67.311 Cardiovascular.

Cardiovascular standards for a third-class airman medical certificate are no established medical history or clinical diagnosis of any of the following:

(a) Myocardial infarction;

(b) Angina pectoris;

(c) Coronary heart disease that has required treatment or, if untreated, that has been symptomatic or clinically significant:

(d) Cardiac valve replacement;

(e) Permanent cardiac pacemaker implantation; or

(f) Heart replacement.

§ 67.313 General medical condition.

The general medical standards for a third-class airman medical certificate are:

(a) No established medical history or clinical diagnosis of diabetes mellitus that requires insulin or any other hypoglycemic drug for control.

(b) No other organic, functional, or structural disease, defect, or limitation that the Federal Air Surgeon, based on the case history and appropriate, qualified medical judgment relating to the condition involved, finds—

(1) Makes the person unable to safely perform the duties or exercise the privileges of the airman certificate applied for or held; or

(2) May reasonably be expected, for the maximum duration of the airman medical certificate applied for or held, to

make the person unable to perform those duties or exercise those privileges.

(c) No medication or other treatment that the Federal Air Surgeon, based on the case history and appropriate, qualified medical judgment relating to the medication or other treatment involved, finds—

(1) Makes the person unable to safely perform the duties or exercise the privileges of the airman certificate applied for or held; or

(2) May reasonably be expected, for the maximum duration of the airman medical certificate applied for or held, to make the person unable to perform those duties or exercise those privileges.

§ 67.315 Discretionary issuance.

A person who does not meet the provisions of §§ 67.303 through 67.313 may apply for the discretionary issuance of a certificate under § 67.401.

Subpart E—Certification Procedures

§ 67.401 Special issuance of medical certificates.

(a) At the discretion of the Federal Air Surgeon, an Authorization for Special Issuance of a Medical Certificate (Authorization), valid for a specified period, may be granted to a person who does not meet the provisions of subparts B, C, or D of this part if the person shows to the satisfaction of the Federal Air Surgeon that the duties authorized by the class of medical certificate applied for can be performed without endangering public safety during the period in which the Authorization would be in force. The Federal Air Surgeon may authorize a special medical flight test, practical test, or medical evaluation for this purpose. A medical certificate of the appropriate class may be issued to a person who does not meet the provisions of subparts B, C, or D of this part if that person possesses a valid Authorization and is otherwise eligible. An airman medical certificate issued in accordance with this section shall expire no later than the end of the validity period or upon the withdrawal of the Authorization upon which it is based. At the end of its specified validity period, for grant of a new Authorization, the person must again show to the satisfaction of the Federal Air Surgeon that the duties authorized by the class of medical certificate applied for can be performed without endangering public safety during the period in which the Authorization would be in force.

(b) At the discretion of the Federal Air Surgeon, a Statement of Demonstrated Ability (SODA) may be granted, instead of an Authorization, to a person whose disqualifying condition is static or nonprogressive and who has been found capable of performing airman duties without endangering public safety. A SODA does not expire and authorizes a designated aviation medical examiner to issue a medical certificate of a specified class if the examiner finds that the condition described on its face has not adversely changed.

(c) In granting an Authorization or SODA, the Federal Air Surgeon may consider the person's operational experience and any medical facts that may affect the ability of the person to perform airman duties including—

(1) The combined effect on the person of failure to meet more than one requirement of this part; and -

(2) The prognosis derived from professional consideration of an available information regarding the person.

(d) In granting an Authorization or SODA under this section, the Federal Air Surgeon specifies the class of medical certificate authorized to be issued and may do any or all of the following:

(1) Limit the duration of an Authorization;

(2) Condition the granting of a new Authorization on the results of subsequent medical tests, examinations, or evaluations;

(3) State on the Authorization or SODA, and any medical certificate based upon it, any operational limitation needed for safety; or

(4) Condition the continued effect of an Authorization or SODA, and any second- or third-class medical certificate based upon it, on compliance with a statement of functional limitations issued to the person in coordination with the Director of Flight Standards or the Director's designee.

(e) In determining whether an Authorization or SODA should be granted to an applicant for a third-class medical certificate, the Federal Air Surgeon considers the freedom of an airman, exercising the privileges of a private pilot certificate, to accept reasonable risks to his or her person and property that are not acceptable in the exercise of commercial or airline transport pilot privileges, and, at the same time, considers the need to protect the safety of persons and property in other aircraft and on the ground.

(f) An Authorization or SODA granted under the provisions of this section to a person who does not meet the applicable provisions of subparts B, C, or D of this part may be withdrawn, at the discretion of the Federal Air Surgeon, at any time if—

(1) There is adverse change in the holder's medical condition;

(2) The holder fails to comply with a statement of functional limitations or operational limitations issued as a condition of certification under this section;

(3) Public safety would be endangered by the holder's exercise of airman privileges;

(4) The holder fails to provide medical information reasonably needed by the Federal Air Surgeon for certification under this section; or

(5) The holder makes or causes to be made a statement or entry that is the basis for withdrawal of an Authorization or SODA under § 67.403.

(g) A person who has been granted an Authorization or SODA under this section based on a special medical flight or practical test need not take the test again during later physical examinations unless the Federal Air Surgeon determines or has reason to believe that the physical deficiency has or may have degraded to a degree to require another special medical flight test or practical test.

(h) The authority of the Federal Air Surgeon under this section is also exercised by the Manager, Aeromedical Certification Division, and each Regional Flight Surgeon.

(i) If an Authorization or SODA is withdrawn under paragraph (f) of this section the following procedures apply:

(1) The holder of the Authorization or SODA will be served a letter of withdrawal, stating the reason for the action;

(2) By not later than 60 days after the service of the letter of withdrawal, the holder of the Authorization or SODA may request, in writing, that the Federal Air Surgeon provide for review of the decision to withdraw. The request for review may be accompanied by supporting medical evidence;

(3) Within 60 days of receipt of a request for review, a written final decision either affirming or reversing the decision to withdraw will be issued; and

(4) A medical certificate rendered invalid pursuant to a withdrawal, in accordance with paragraph (a) of this section, shall be surrendered to the Administrator upon request.

(j) No grant of a special issuance made prior to September 16, 1996, may be used to obtain a medical certificate after the earlier of the following dates:

(1) September 16, 1997; or

(2) The date on which the holder of such special issuance is required to provide additional information to the FAA as a condition for continued medical certification.

§ 67.403 Applications, certificates, logbooks, reports, and records: Falsification, reproduction, or alteration; Incorrect statements.

(a) No person may make or cause to be made—

(1) A fraudulent or intentionally false statement on any application for a medical certificate or on a request for any Authorization for Special Issuance of a Medical Certificate (Authorization) or Statement of Demonstrated Ability (SODA) under this part;

(2) A fraudulent or intentionally false entry in any logbook, record, or report that is kept, made, or used, to show compliance with any requirement for any medical certificate or for any Authorization or SODA under this part;

(3) A reproduction, for fraudulent purposes, of any medical certificate under this part; or

(4) An alteration of any medical certificate under this part.

(b) The commission by any person of an act prohibited under paragraph (a) of this section is a basis for—

(1) Suspending or revoking all airman, ground instructor, and medical certificates and ratings held by that person;

(2) Withdrawing all Authorizations or SODA's held by that person; and

(3) Denying all applications for medical certification and requests for Authorizations or SODA'S.

(c) The following may serve as a basis for suspending or revoking a medical certificate; withdrawing an Authorization or SODA; or denying an application for a medical certificate or request for an authorization or SODA:

(1) An incorrect statement, upon which the FAA relied, made in support of an application for a medical certificate or request for an Authorization or SODA.

(2) An incorrect entry, upon which the FAA relied, made in any logbook, record, or report that is kept, made, or used to show compliance with any requirement for a medical certificate or an Authorization or SODA.

§ 67.405 Medical examinations: Who may give.

(a) *First-class.* Any aviation medical examiner who is specifically designated for the purpose may give the examination for the first-class medical certificate. Any interested person may obtain a list of these aviation medical examiners, in any area, from the FAA Regional Flight Surgeon of the region in which the area is located.

(b) *Second- and third-class.* Any aviation medical examiner may give the examination for the second- or third-class medical certificate. Any interested person may obtain a list of aviation medical examiners, in any area, from the FAA Regional Flight Surgeon of the region in which the area is located.

§ 67.407 Delegation of authority.

(a) The authority of the Administrator under 49 U.S.C. 44703 to issue or deny medical certificates is delegated to the Federal Air Surgeon to the extent necessary to—

(1) Examine applicants for and holders of medical certificates to determine whether they meet applicable medical standards; and

(2) Issue, renew, and deny medical certificates, and issue, renew, deny, and withdraw Authorizations for Special Issuance of a Medical Certificate and Statements of Demonstrated Ability to a person based upon meeting or failing to meet applicable medical standards.

(b) Subject to limitations in this chapter, the delegated functions of the Federal Air Surgeon to examine applicants for and holders of medical certificates for compliance with applicable medical standards and to issue, renew, and deny medical certificates are also delegated to aviation medical examiners and to authorized representatives of the Federal Air Surgeon within the FAA.

(c) The authority of the Administrator under 49 U.S.C. 44702, to reconsider the action of an aviation medical examiner is delegated to the Federal Air Surgeon; the Manager, Aeromedical Certification Division; and each Regional Flight Surgeon. Where the person does not meet the standards of §§ 67.107(b)(3) and (c), 67.109(b), 67.113(b) and (c), 67.207(b)(3) and (c), 67.209(b), 67.213(b) and (c), 67.307(b)(3) and (c), 67.309(b), or 67.313(b) and (c), any action taken under this paragraph other than by the Federal Air Surgeon is subject to reconsideration by the Federal Air Surgeon. A certificate issued by an aviation medical examiner is considered to be affirmed as issued unless an FAA official named in this paragraph (authorized official) reverses that issuance within 60 days after the date of issuance. However, if within 60 days after the date of issuance an authorized official requests the certificate holder to submit additional medical information, an authorized official may reverse the issuance within 60 days after receipt of the requested information.

(d) The authority of the Administrator under 49 U.S.C. 44709 to re-examine any civil airman to the extent necessary to determine an airman's qualification to continue to hold an airman medical certificate, is delegated to the Federal Air Surgeon and his or her authorized representatives within the FAA.

§ 67.409 Denial of medical certificate.

(a) Any person who is denied a medical certificate by an aviation medical examiner may, within 30 days after the date of the denial, apply in writing and in duplicate to the Federal Air Surgeon, Attention: Manager, Aeromedical Certification Division, AAM-300, Federal Aviation Administration, P.O. Box 26080, Oklahoma City, Oklahoma 73126, for reconsideration of that denial. If the person does not ask for reconsideration during the 30-day period

after the date of the denial, he or she is considered to have withdrawn the application for a medical certificate.

(b) The denial of a medical certificate

(1) By an aviation medical examiner is not a denial by the Administrator under 49 U.S.C. 44703.

(2) By the Federal Air Surgeon is considered to be a denial by the Administrator under 49 U.S.C. 44703.

(3) By the Manager, Aeromedical Certification Division, or a Regional Flight Surgeon is considered to be a denial by the Administrator under 49 U.S.C. 44703 except where the person does not meet the standards of §§ 67.107(b)(3) and (c), 67.109(b), or 67.113(b) and (c); 67.207(b)(3) and (c), 67.209(b), or 67.213(b) and (c); or 67.307(b)(3) and (c), 67.309(b), or 67.313(b) and (c).

(c) Any action taken under § 67.407 (c) that wholly or partly reverses the issue of a medical certificate by an aviation medical examiner is the denial of a medical certificate under paragraph (b) of this section.

(d) If the issue of a medical certificate is wholly or partly reversed by the Federal Air Surgeon; the Manager, Aeromedical Certification Division; or a Regional Flight Surgeon, the person holding that certificate shall surrender it, upon request of the FAA.

§ 67.411 Medical certificates by flight surgeons of Armed Forces.

(a) The FAA has designated flight surgeons of the Armed Forces on specified military posts, stations, and facilities, as aviation medical examiners.

(b) An aviation medical examiner described in paragraph (a) of this section may give physical examinations for the FAA medical certificates to persons who are on active duty or who are, under Department of Defense medical programs, eligible for FAA medical certification as civil airmen. In addition, such an examiner may issue or deny an appropriate FAA medical certificate in accordance with the regulations of this chapter and the policies of the FAA.

(c) Any interested person may obtain a list of the military posts, stations, and facilities at which a flight surgeon has been designated as an aviation medical examiner from the Surgeon General of the Armed Force concerned or from the Manager, Aeromedical Education Division, AAM-400, Federal Aviation Administration, P.O. Box 26082, Oklahoma City, Oklahoma 73125.

§ 67.413 Medical records.

(a) Whenever the Administrator finds that additional medical information or history is necessary to determine whether an applicant for or the holder of a medical certificate meets the medical standards for it, the Administrator requests that person to furnish that information or to authorize any clinic, hospital, physician, or other person to release to the Administrator all available information or records concerning that history. If the applicant or holder fails to provide the requested medical information or history or to authorize the release so requested, the Administrator may suspend, modify, or revoke all medical certificates the airman holds or may, in the case of an applicant, deny the application for an airman medical certificate.

(b) If an airman medical certificate is suspended or modified under paragraph (a) of this section, that suspension or modification remains in effect until the requested information, history, or authorization is provided to the FAA and until the Federal Air Surgeon determines whether the person meets the medical standards under this part.

§ 67.415 Return of medical certificate after suspension or revocation.

The holder of any medical certificate issued under this part that is suspended or revoked shall, upon the Administrator's request, return it to the Administrator.

Part 71—Designation of Class A, B, C, D, and E airspace areas; air traffic service routes; and reporting points

---

Source: Docket No. 24456, (56 FR 65638), December 17, 1991.

§ 71.1 [Applicability]

The complete listing for all Class A, Class B, Class C, Class D, and Class E airspace areas and for all reporting points can be found in FAA Order 7400.9O, Airspace Designations and Reporting Points, dated August 30, 2002. This incorporation by reference was approved by the Director of the Federal Register in accordance with 5 U.S.C. 552(a) and 1 CFR part 51. The approval to incorporate by reference FAA Order 7400.9O is effective September 1, 2005, through September 15, 2006. During the incorporation by reference period, proposed changes to the listings of Class A, Class B, Class C, Class D, and Class E airspace areas and to reporting points will be published in full text as proposed rule documents in the *Federal Register*. Amendments to the listings of Class A, Class B, Class C, Class D, and Class E airspace areas and to reporting points will be published in full text as final rules in the *Federal Register*.

Periodically, the final rule amendments will be integrated into a revised edition of the order and submitted to the Director of the Federal Register for approval for incorporation by reference in this section. Copies of FAA Order 7400.9O may be obtained from the Library of Chief Counsel, Federal Aviation Administration, 800 Independence Avenue, SW., Washington, DC 20591, (202) 267-3174. Copies of FAA Order 7400.9O may be inspected in Docket No. 29334 at the Federal Aviation Administration, at the same address above, weekdays between 8:30 a.m. and 5:00 p.m., or at the Office of the Federal Register, 800 North Capitol Street, NW., Suite 700, Washington, DC. This section is applicable September 16, 2005, through September 15, 2006.

[Amdt. 71-14, 56 FR 65657, Dec. 17, 1991, as amended by Amdt. 71-20, 58 FR 36299, July 6, 1993; Amdt. 71-23, 59 FR 43035, Aug. 22, 1994; Amdt. 71–26, 60 FR 47266, Sept. 12, 1995; Amdt. 71-28, 61 Fr 48403, Sept. 13, 1996; Amdt. 71-29, 62 FR 52492, Oct. 8, 1997; Amdt. 71-30, 63 FR 50139, Sept. 21, 1998, September 17, 1999, September 16, 2000]

§ 71.3 [Reserved]
§ 71.5 Reporting points.

The reporting points listed in subpart H of FAA Order 7400.9N (incorporated by reference, see § 71.1) consist of geographic locations at which the position of an aircraft must be reported in accordance with part 91 of this chapter. [Amdt. 71-14, 56 FR 65655, Dec. 17, 1991, as amended by Amdt. 71-20, 58 FR 36299, July 6, 1993; Amdt. 71-23, 59 FR 43035, Aug. 22, 1994; Amdt. 71-26, 60 FR 47266, Sept. 12, 1995; Amdt. 71-28, 61 FR 48403, Sept. 13, 1996; Amdt. 71-29, 62 FR 52492, Oct. 8, 1997; Amdt. 71-30, 63 FR 50139, Sept. 21, 1998, September 17, 1999, September 19, 2000]

§ 71.7 Bearings, radials, and mileages.

All bearings and radials in this part are true and are applied from point of origin and all mileages in this part are stated as nautical miles.

§ 71.9 Overlapping airspace designations.

(a) When overlapping airspace designations apply to the same airspace, the operating rules associated with the more restrictive airspace designation apply.

(b) For the purpose of this section—

(1) Class A airspace is more restrictive than Class B, Class C, Class D, Class E, or Class G airspace;

(2) Class B airspace is more restrictive than Class C, Class D, Class E, or Class G airspace;

(3) Class C airspace is more restrictive than Class D, Class E, or Class G airspace;

(4) Class D airspace is more restrictive than Class E or Class G airspace; and

(5) Class E is more restrictive than Class G airspace.

§ 71.11 Air Traffic Service (ATS) routes.

Unless otherwise specified, the following apply:

(a) An Air Traffic Service (ATS) route is based on a centerline that extends from one navigation aid, fix, or intersection, to another navigation aid, fix, or intersection (or through several navigation aids, fixes, or intersections) specified for that route.

(b) ATS routes include the primary protected airspace dimensions defined in FAA Order 8260.3, "United States Standard For Terminal Instrument Procedures (TERPS)," Order 8260.3 *is* incorporated by reference in § 97.20 of this chapter.

(c) An ATS route does not include the airspace of a prohibited area.

§ 71.13 Classification of Air Traffic Services (ATS) routes

Unless otherwise specified, ATS routes are classified as follows:

(a) In subpart A of this part:

(1) Jet routes.

(2) Area navigation (RNAV) routes.

(b) In subpart E of this part:

(1) VOR Federal airways.

(2) Colored Federal airways.

(i) Green Federal airways.

(ii) Amber Federal airways.

(iii) Red Federal airways.

(iv) Blue Federal airways.

(3) Area navigation (RNAV) routes.

§ 71.15 Designation of jet routes and VOR Federal airways.

Unless otherwise specified, the place names appearing in the descriptions of airspace areas designated as jet routes in subpart A of FAA Order 7400.9, and as VOR Federal airways in subpart E of FAA Order 7400.9N, are the names of VOR or VORTAC navigation aids. FAA Order 7400.9 Is incorporated by reference in § 71.1.

Subpart A—Class A airspace

§ 71.31 Class A airspace.

The airspace descriptions contained in § 71.33 of this part and the routes contained in subpart A of FAA Order 7400.9N (incorporated by reference, see § 71.1) are designated as Class A airspace within which all pilots and aircraft are subject to the rating requirements, operating rules, and equipment requirements of part 91 of this chapter.

§ 71.33 Class A airspace areas.

(a) That airspace of the United States, including that airspace overlying the waters within 12 nautical miles of the coast of the 48 contiguous states, from 18,000 feet MSL to and including FL600 excluding the states of Alaska and Hawaii, Santa Barbara Island, Farallon Island, and the airspace south of latitude 25°04'00" North.

(b) That airspace of the State of Alaska, including that airspace overlying the waters within 12 nautical miles of the coast, from 18,000 feet MSL to and including FL600 but not including the airspace less than 1,500 feet above the surface of the earth and the Alaska Peninsula west of longitude 160°00'00" West.

(c) The airspace areas listed as offshore airspace areas in subpart A of FAA Order 7400.9N (incorporated by reference, see § 71.1) that are designated in international airspace within areas of domestic radio navigational signal or ATC radar coverage, and within which domestic ATC procedures are applied.

[Amdt. 71-14, 56 FR 65655, Dec. 17, 1991, as amended by Amdt. 71-19, 58 FR 12137, Mar. 2, 1993; Amdt. 71-23, 59 FR 43035, Aug. 22, 1994; Amdt. 71-26, 60 FR 47266, Sept. 12, 1995; Amdt. 71-28, 61 FR 48403, Sept. 13, 1996; Amdt. 71-29, 62 FR 52492, Oct 8, 1997; Amdt. 71-30, 63 FR 50139, Sept. 21, 1998]

Subpart B—Class B airspace

§ 71.41 Class B airspace.

The Class B airspace areas listed in subpart B of FAA Order 7400.9N (incorporated by reference, see § 71.1) consist of specified airspace within which all aircraft operators are subject to the minimum pilot qualification requirements, operating rules, and aircraft equipment requirements of part 91 of this chapter. Each Class B airspace area designated for an airport in subpart B of FAA Order 7400.9H (incorporated by reference, see § 71.1) contains at least one primary airport around which the airspace is designated. [Amdt. 71-14, 56 FR 65655, Dec. 17, 1991, as amended by Amdt. 71-20, 58 FR 36299, July 6, 1993; Amdt. 71-23, 59 FR 43035, Aug. 22, 1994; Amdt. 71-26, 60 FR 47266, Sept. 12, 1995; Amdt. 71-28, 61 FR 48403, Sept. 13, 1996; Amdt. 71-29, 62 FR 52492, Oct 8, 1997; Amdt. 71-30, 63 FR 50139, Sept. 21, 1998, September 17, 1999, September 16, 2000]

Subpart C—Class C airspace

§ 71.51 Class C airspace.

The Class C airspace areas listed in subpart C of FAA Order 7400.9K (incorporated by reference, see § 71.1) consist of specified airspace within which all aircraft operators are subject to operating rules and equipment requirements specified in part 91 of this chapter. Each Class C airspace area designated for an airport in subpart C of FAA Order 7400.9N (incorporated by reference, see § 71.1) contains at least one primary airport around which the airspace is designated. [Amdt. 71-14, 56 FR 65655, Dec. 17, 1991, as amended by Amdt. 71-20, 58 FR 36299, July 6, 1993; Amdt. 71-23, 59 FR 43035, Aug. 22, 1994; Amdt. 71-26, 60 FR 47266, Sept. 12, 1995; Amdt. 71-28, 61 FR 48403, Sept. 13, 1996; Amdt. 71-29, 62 FR 52492, Oct. 8, 1997; Amdt. 71-30, 63 FR 50139, Sept. 21, 1998, September 17, 1999, September 16, 2000]

Subpart D—Class D airspace

§ 71.61 Class D airspace.

The Class D airspace areas listed in subpart D of FAA Order 7400.9N (incorporated by reference, see § 71.1) consist of specified airspace within which all aircraft operators are subject to operating rules and equipment requirements specified in part 91 of this chapter. Each Class D airspace area designated for an airport in subpart D of FAA Order 7400.9H (incorporated by reference, see § 71.1) contains at least one primary airport around which the airspace is designated. [Amdt. 71-14, 56 FR 65655, Dec. 17, 1991, as amended by Amdt. 71-20, 58 FR 36299, July 6, 1993; Amdt. 71-23, 59 FR 43035, Aug. 22, 1994; Amdt. 71-26, 60 FR 47266, Sept. 12, 1995; Amdt. 71-28, 61 FR 48403, Sept. 13, 1996; Amdt. 71-29, 62 FR 52492, Oct. 8, 1997; Amdt. 71-30, 63 FR 50139, Sept. 21, 1998, September 17, 1999, September 16, 2000]

Subpart E—Class E airspace

§ 71.71 Class E airspace.

Class E Airspace consists of:

(a) [The airspace of the United States, including that airspace overlying the waters within 12 nautical miles of the coast of the 48 contiguous states and Alaska, extending upward from 14,500 feet MSL up to, but not including 18,000 feet MSL, and the airspace above FL600, excluding—*]

(1) The Alaska peninsula west of longitude 160°00'00"W.; and

(2) The airspace below 1,500 feet above the surface of the earth.

(b) The airspace areas designated for an airport in subpart E of FAA Order 7400.9N (incorporated by reference, see § 71.1) within which all aircraft operators are subject to the operating rules specified in part 91 of this chapter.

(c) The airspace areas listed as domestic airspace areas in subpart E of FAA Order 7400.9N (incorporated by reference, see § 71.1) which extend upward from 700 feet or more above the surface of the earth when designated in conjunction with an airport for which an approved instrument approach procedure has been prescribed, or from 1,200 feet or more above the surface of the earth for the purpose of transitioning to or from the terminal or en route environment. When such areas are designated in conjunction with airways or routes, the extent of such designation has the lateral extent identical to that of a Federal airway and extends upward from 1,200 feet or higher. Unless otherwise specified, the airspace areas in the paragraph extend upward from 1,200 feet or higher above the surface to, but not including, 14,500 feet MSL.

(d) The Federal airways described in subpart E of FAA Order 7400.9H (incorporated by reference, see § 71.1).

(e) The airspace areas listed as en route domestic airspace areas in subpart E of FAA Order 7400.9H (incorporated by reference, see § 71.1). Unless otherwise specified, each airspace area has a lateral extent identical to that of a Federal airway and extends upward from 1,200 feet above the surface of the earth to the overlying or adjacent controlled airspace.

(f) The airspace areas listed as offshore airspace areas in subpart E of FAA Order 7400.9K (incorporated by reference, see § 71.1) that are designated in international airspace within areas of domestic radio navigational signal or ATC radar coverage, and within which domestic ATC procedures are applied. Unless otherwise specified, each airspace area extends upward from a specified altitude up to, but not including, 18,000 feet MSL. (Amdt. 71-16, Eff. 9/16/93); (Amdt. 71-19, Eff. 9/1/93; [Amdt. 71-14, 56 FR 65655, Dec. 17, 1991, as amended by Amdt. 71-19, 58 FR 12137, Mar. 2, 1993; Amdt. 71-16, 58 FR 15259 Mar. 19, 1993; Amdt. 71-20, 58 FR 36299, July 6, 1993; Amdt. 71-21. 58 FR 44121, Aug.19, 1993; Amdt. 71-23, 59 FR 43055, Aug. 22, 1994; Amdt. 71-26, 60 FR 47266, Sept. 12, 1995; Amdt. 71-28, 61 FR 48403, Sept. 13, 1996; Amdt. 71-29, 62 FR 52492, Oct. 8, 1997; Amdt, 71-30, 63 FR 50139, Sept. 21, 1998, September 17, 1999, September 16, 2000]

Subpart F—[Reserved]
Subpart G—[Reserved]
Subpart H—Reporting points

§ 71.901 Applicability.

Unless otherwise designated:

(a) Each reporting point listed in subpart H of FAA Order 7400.9N (incorporated by reference, see § 71.1) applies to all directions of flight. In any case where a geographic location is designated as a reporting point for less than all airways passing through that point, or for a particular direction of flight along an airway only, it is so indicated by including the airways or direction of flight in the designation of geographical location.

(b) Place names appearing in the reporting point descriptions indicate VOR or VORTAC facilities identified by those names.

[Amdt. 71-14, 56 FR 65657, Dec. 17, 1991, as amended by Amdt. 71-20, 58 FR 36299, July 6, 1993; Amdt. 71-23, 59 FR 43035, Aug. 22, 1994; Amdt. 71-26, 60 FR 47266, Sept. 12, 1995; Amdt. 71-28, 61 FR 48403, Sept. 13, 1996; Amdt. 71-29, 62 FR 52492, Oct. 8, 1997; Amdt. 71-30, 63 FR 50139, Sept. 21, 1998, September 17, 1999, September 16, 2000]

Special Federal Aviation Regulation No. 45-1

Contrary provisions of the Federal Aviation Regulations notwitstanding,

1. The Robert Mueller Municipal Airport, Austin, Tx, and the Port Columbus International Airport, Columbus, OH, are designated Airport Radar Service Aeas (ARSA) airports at which, unless otherwise provided for in a Notice to Airmen issued by the ARSA facility, radar services will be provided continuously.

2. For purposes of this Special Federal Aviation Regulation:

(a) The Austin, TX, ARSA is that airspace extending upward from the surface to and including 4,600 feet MSL within a 5 nautical mile radius of the Robert Mueller Municipal Airport and that airspace extending upward from 2,000 feet MSL to 4,600 feet MSL, within a 10 nautical mile radius of Robert Mueller Municipal Airport from the 027 deg. true bearing from the airport clockwise to the 207 deg. true bearing from the airport and that airspace extending upward from 2,300 feet MSL to 4,600 feet MSL within 10 nautical mile radius of the airport from the 207 deg. true bearing from the airport clockwise to the 027 deg. true bearing from the airport.

(b) The Columbus, OH, ARSA is that airspace extending upward from the surface to and including 4,800 feet MSL within a 5 nautical mile radius of the Port Columbus International Airport and that airspace extending upward from 2,500 feet MSL to 4,800 feet MSL within a 10 natical mile radius of Port Columbus International Airport from the 008 deg. true bearing from the airport clockwise to the 127 deg. true bearing from the airport and that airspace extending upward from 2,200 feet MSL to 4,800 feet MSL within a 10 nautical mile radius of the airport from the 127 deg. true bearing from the airport clockwise to the 088 deg. true bearing from the airport.

(c) The primary airport is the airport for which the ARSA is designated; a satellite airport is any other airport, heliport, helipad, etc., within that ARSA.

3. Unless otherwise authorized or required by ATC—

(a) Arrivals and Overflights—no person may operate an aircraft in an ARSA unless two-way radio communication is established with ATC prior to entering that ARSA and is thereafter maintained with ATC while witin that ARSA;

(b) Depatures—no person may operate an aircraft within an ARSA unless two-way radio communication is maintained with ATC while within that ARSA, except that for aircraft departing a satellite airport, two-way radio communication is established as soon as practicable and thereafter maintained with ATC while within that ARSA;

(c) ATC Instructions and Clearances—except in an emergency, no person may, while within an ARSA, operate an aircraft contrary to an ATC clearnace or instruction;

(d) Traffic patterns—no person may takeoff or land an aircraft within an ARSA except in compliance with FAA arrival and departure traffic patterns; and

(e) Ultralight Vehicle and Parachute Jump Operations— no person may operate an ultralight vehicle within, or make a parachute jump within or into, an ARSA except under the terms of an ATC authorization issued by the ATC facility having jurisdiction over that ARSA.

4. In addition to the required preflight actions of FAR Section 91.103, each pilot in command shall review NOTAM's pertaining to flight within an ARSA prior to conducting flight in an ARSA.

This SFAR terminates at 12:01 a.m. local time on June 21, 1985, unless sooner superseded or recinded.

(Secs. 307 and 313(a), Federal Aviation Act of 1985, as amended (49 U.S.C. 1348, 1354(a)); 49 U.S.C. 106(g) (Revised, Pub. L. 97-449, January 12, 1983); 14 CRF 11.45; and 14 CFR 11.65)

[Doc. No. 23708, 49 FR 47177, Nov. 30, 1984; 49 FR 49089, Dec. 18, 1984, as amended by Amdt. 71-13, 54 FR 34331, Aug. 18, 1989]

Part 73—Special use airspace

Subpart A—General

§ 73.1 Applicability.

The airspace that is described in Subpart B and Subpart C of this part is designated as special use airspace. This part prescribes the requirements for the use of that airspace.

§ 73.3 Special use airspace.

(a) Special use airspace consists of airspace of defined dimensions identified by an area on the surface of the earth wherein activities must be confined because of their nature, or wherein limitations are imposed upon aircraft operations that are not a part of those activities, or both.

(b) The vertical limits of special use airspace are measured by designated altitude floors and ceilings expressed as flight levels or as feet above mean sea level. Unless otherwise specified, the word "to" (an altitude or flight level) means "to and including" (that altitude or flight level).

(c) The horizontal limits of special use airspace are measured by boundaries described by geographic coordinates or other appropriate references that clearly define their perimeter.

(d) The period of time during which a designation of special use airspace is in effect is stated in the designation.

§ 73.5 Bearings; radials; miles.

(a) All bearings and radials in this part are true from point of origin.

(b) Unless otherwise specified, all mileages in this part are stated as statute miles.

Subpart B—Restricted areas

§ 73.11 Applicability.

This subpart designates restricted areas and prescribes limitations on the operation of aircraft within them.

§ 73.13 Restrictions.

No person may operate an aircraft within a restricted area between the designated altitudes and during the time of designation, unless he has the advance permission of—

(a) The using agency described in § 73.15; or
(b) The controlling agency described in § 73.17.

§ 73.15 Using agency.

(a) For the purposes of this subpart, the following are using agencies:

(1) The agency, organization, or military command whose activity within a restricted area necessitated the area being so designated.

[(2) [Reserved]]

(b) Upon the request of the FAA, the using agency shall execute a letter establishing procedures for joint use of a restricted area by the using agency and the controlling agency, under which the using agency would notify the controlling agency whenever the controlling agency may grant permission for transit through the restricted area in accordance with the terms of the letter.

(c) The using agency shall—

(1) Schedule activities within the restricted area;

(2) Authorize transit through, or flight within, the restricted area as feasible; and

(3) Contain within the restricted area all activities conducted therein in accordance with the purpose for which it was designated.

§ 73.17 Controlling agency.

For the purposes of this part, the controlling agency is the FAA facility that may authorize transit through or flight within a restricted area in accordance with a joint-use letter issued under § 73.15.

§ 73.19 Reports by using agency.

(a) Each using agency shall prepare a report on the use of each restricted area assigned thereto during any part of the preceding 12-month period ended September 30, and transmit it by the following January 31 of each year to the Manager, Air Traffic Division in the regional office of the Federal Aviation Administration having jurisdiction over the area in which the restricted area is located, with a copy to the Program Director for Air Traffic Airspace Management, Federal Aviation Administration, Washington, D.C. 20591.

(b) In the report under this section the using agency shall:

(1) State the name and number of the restricted area as published in this part, and the period covered by the report.

(2) State the activities (including average daily number of operations if appropriate) conducted in the area, and any other pertinent information concerning current and future electronic monitoring devices.

(3) State the number of hours daily, the days of the week, and the number of weeks during the year that the area was used.

(4) For restricted areas having a joint-use designation, also state the number of hours daily, the days of the week, and the number of weeks during the year that the restricted area was released to the controlling agency for public use.

(5) State the mean sea level altitudes or flight levels (whichever is appropriate) used in aircraft operations and the maximum and average ordinate of surface firing (expressed in feet, mean sea level altitude) used on a daily, weekly, and yearly basis.

(6) Include a chart of the area (of optional scale and design) depicting, if used, aircraft operating areas, flight patterns, ordnance delivery areas, surface firing points, and target, fan, and impact areas. After once submitting an appropriate chart, subsequent annual charts are not required unless there is a change in the area, activity or altitude (or flight levels) used, which might alter the depiction of the activities originally reported. If no change is to be submitted, a statement indicating "no change" shall be included in the report.

(7) Include any other information not otherwise required under this part which is considered pertinent to activities carried on in the restricted area.

(c) If it is determined that the information submitted under paragraph (b) of this section is not sufficient to evaluate the nature and extent of the use of a restricted area, the FAA may request the using agency to submit supplementary reports. Within 60 days after receiving a request for additional information, the using agency shall submit such information as the Program Director for Air Traffic Airspace Management considers appropriate. Supplementary reports must be sent to the FAA officials designated in paragraph (a) of this section.

§§ 73.21 through 73.72

[Redesignations.] [§§ 608.21 through 608.72 of the Regulations of the Administrator are hereby redesignated as §§ 73.21 through 73.72, respectively.]*

Subpart C—Prohibited areas

§ 73.81 Applicability.

This subpart designates prohibited areas and prescribes limitations on the operation of aircraft therein.

§ 73.83 Restrictions.

No person may operate an aircraft within a prohibited area unless authorization has been granted by the using agency.

§ 73.85 Using agency.

For the purpose of this subpart, the using agency is the agency, organization or military command that established the requirements for the prohibited area.

§§ 73.87 through 73.99*

These sections are reserved for descriptions of designated prohibited areas.

Ch. 4 (Amdt. 73-5, Eff. 10/25/89)

---

*The airspace descriptions in this part and their subsequent changes are published in the Federal Register. Due to their complexity and length, they will not be included in this publication of Part 73.

Special Federal Aviation Regulation No. 53.
Establishment of warning areas in the airspace
overlying the waters between 3 and 12 nautical
miles from the United States coast

[**1. Applicability.** This rule establishes warning areas in the same location as nonregulatory warning areas previously designated over international waters. This special regulation does not affect the validity of any nonregulatory warning area which is designated over international waters beyond 12 nautical miles from the coast of the United States. This special regulation expires on January 15, 1996.]

2. [Definition-Warning area. A warning area established under this special rule is airspace of defined dimensions, extending from 3 to 12 nautical miles from the coast of the United States, that contains activity which may be hazardous to nonparticipating aircraft. The purpose of such warning areas is to warn nonparticipating pilots of the potential danger. Part 91 is applicable within the airspace designated under this special rule.]

3. Participating aircraft. Each person conducting an aircraft operation within a warning area designated under this special rule and operating with the approval of the using agency may deviate from the rules of Part 91, Subpart B, to the extent that the rules are not compatible with approved operations.

4. Nonparticipating aircraft. Nonparticipating pilots, while not excluded from the warning areas established by this SFAR, are on notice that military activity, which may be hazardous to nonparticipating aircraft, is conducted in these areas.

Ch. 6 (SFAR 53-2, Eff. 12/27/90)

Part 91—General operating and flight rules

Appendix A
Category II operations: Manual, instruments, equipment, and maintenance

Appendix B
Authorizations to exceed Mach 1 (section 91.817)

Appendix C
Operations in the North Atlantic (NAT) Minimum Navigation Performance Specifications (MNPS) airspace

Appendix D
Airports/locations: special operating restrictions

Appendix E
Airplane flight recorder specifications

Appendix F
Helicopter flight recorder specifications

Appendix G
Operation in Reduced Vertical Separation Minimum (RVSM) Airspace

Special Federal Aviation Regulations

Special Federal Aviation Regulation No. 51-1. Special flight rules in the vicinity of Los Angeles International Airport *(Rule)*

Special Federal Aviation Regulation No. 60. Air Traffic Control System emergency operation *(Rule)*

Special Federal Aviation Regulation No. 61-2. Prohibition Against Certain Flights Between the United States and Iraq

Special Federal Aviation Regulation No. 71. Special operating Rules for Air Tour Operators in the State of Hawaii

Special Federal Aviation Regulation No. 76. Prohibition Against Certain Flights within the Territory and Airspace of Iran.

Special Federal Aviation Regulation No. 79. Prohibition Against Certain Flights within the Flight Information Region (FIR) of the Democratic People's Republic of Korea (DPRK).

Special Federal Aviation Regulation No. 87. Prohibition Against Certain Flights Within the Territory and Airspace of Ethiopia

Special Federal Aviation Regulation No. 94. Enhanced Security Procedures for Operating at Certain Airports in the Washington DC Metropolitan Area Special Flight Rules Area

Special Federal Aviation Regulation No. 97. Special Operating Rules for the Conduct of Instrument Flight Rules (IFR) Area Navigation (RNAV) Operations using Global Positioning Systems (GPS) in Alaska

Subpart A—General

§ 91.1 Applicability.

(a) Except as provided in paragraph (b) and (c) of this section and §91.701 and § 91.703, this part prescribes rules governing the operation of aircraft (other than moored balloons, kites, unmanned rockets, and unmanned free balloons, which are governed by part 101 of this chapter, and ultralight vehicles operated in accordance with part 103 of this chapter) within the United States, including the waters within 3 nautical miles of the U.S. coast.

(b) Each person operating an aircraft in the airspace overlying the waters between 3 and 12 nautical miles from the coast of the United States must comply with §§91.1 through 91.21; §§91.101 through 91.143; §§91.151 through 91.159; §§91.167 through 91.193; §91.203; §91.205; §§ 91.209 through 91.217; §91.221; §§91.303 through 91.319; §§91.323 through 91.327; §91.605; §91.609; §§91.703 through 91.715; and §91.903.

(c) This part applies to each person on board an aircraft being operated under this part unless otherwise specified.

§ 91.3 Responsibility and authority of the pilot in command.

(a) The pilot in command of an aircraft is directly responsible for, and is the final authority as to, the operation of that aircraft.

(b) In an in-flight emergency requiring immediate action, the pilot in command may deviate from any rule of this part to the extent required to meet that emergency.

(c) Each pilot in command who deviates from a rule under paragraph (b) of this section shall, upon the request of the Administrator, send a written report of that deviation to the Administrator.

(Approved by the Office of Management and Budget under OMB control number 2120-0005).

§ 91.5 Pilot in command of aircraft requiring more than one required pilot.

No person may operate an aircraft that is type certificated for more than one required pilot flight crewmember unless the pilot in command meets the requirements of § 61.58 of this chapter.

§ 91.7 Civil aircraft airworthiness.

(a) No person may operate a civil aircraft unless it is in an airworthy condition.

(b) The pilot in command of a civil aircraft is responsible for determining whether that aircraft is in condition for safe flight. The pilot in command shall discontinue the flight when unairworthy mechanical, electrical, or structural conditions occur.

§ 91.9 Civil aircraft flight manual, marking, and placard requirements.

(a) Except as provided in paragraph (d) of this section, no person may operate a civil aircraft without complying with the operating limitations specified in the approved Airplane or Rotorcraft Flight Manual, markings, and placards, or as otherwise prescribed by the certificating authority of the country of registry.

(b) No person may operate a U.S.-registered civil aircraft—

(1) For which an Airplane or Rotorcraft Flight Manual is required by § 21.5 of this chapter unless there is available in the aircraft a current, approved Airplane or Rotorcraft Flight Manual or the manual provided for in § 121.141(b); and

(2) For which an Airplane or Rotorcraft Flight Manual is not required by § 21.5 of this chapter, unless there is available in the aircraft a current approved Airplane or Rotorcraft Flight Manual, approved manual material, markings, and placards, or any combination thereof.

(c) No person may operate a U.S.-registered civil aircraft unless that aircraft is identified in accordance with part 45 of this chapter.

(d) Any person taking off or landing a helicopter certificated under part 29 of this chapter at a heliport constructed over water may make such momentary flight as is necessary for takeoff or landing through the prohibited range of the limiting height-speed envelope established for the helicopter if that flight through the prohibited range takes place over water on which a safe ditching can be accomplished and if the helicopter is amphibious or is equipped with floats or other emergency flotation gear adequate to accomplish a safe emergency ditching on open water.

§ 91.11 Prohibition against interference with crewmembers.

No person may assault, threaten, intimidate, or interfere with a crewmember in the performance of the crewmember's duties aboard an aircraft being operated.

§ 91.13 Careless or reckless operation.

(a) *Aircraft operations for the purpose of air navigation.* No person may operate an aircraft in a careless or reckless manner so as to endanger the life or property of another.

(b) *Aircraft operations other than for the purpose of air navigation.* No person may operate an aircraft, other than for the purpose of air navigation, on any part of the surface of an airport used by aircraft for air commerce (including areas used by those aircraft for receiving or discharging persons or cargo), in a careless or reckless manner so as to endanger the life or property of another.

§ 91.15 Dropping objects.

No pilot in command of a civil aircraft may allow any object to be dropped from that aircraft in flight that creates a hazard to persons or property. However, this section does not prohibit the dropping of any object if reasonable precautions are taken to avoid injury or damage to persons or property.

§ 91.17 Alcohol or drugs.

(a) No person may act or attempt to act as a crewmember of a civil aircraft—

(1) Within 8 hours after the consumption of any alcoholic beverage;

(2) While under the influence of alcohol;

(3) While using any drug that affects the person's faculties in any way contrary to safety; or

(4) While having an alcohol concentration of 0.04 or greater in a blood or breath specimen. Alcohol concentration means grams of alcohol per deciliter of blood or grams of alcohol per 210 liters of breath.

(b) Except in an emergency, no pilot of a civil aircraft may allow a person who appears to be intoxicated or who demonstrates by manner or physical indications that the individual is under the influence of drugs (except a medical patient under proper care) to be carried in that aircraft.

(c) A crewmember shall do the following:

(1) On request of a law enforcement officer, submit to a test to indicate the alcohol concentration in the blood or breath, when—

(i) The law enforcement officer is authorized under State or local law to conduct the test or to have the test conducted; and

(ii) The law enforcement officer is requesting submission to the test to investigate a suspected violation of State or local law governing the same or substantially similar conduct prohibited by paragraph (a)(1), (a)(2), or (a)(4) of this section.

(2) Whenever the FAA has a reasonable basis to believe that a person may have violated paragraph (a)(1), (a)(2), or (a)(4) of this section, on request of the FAA, that person must furnish to the FAA the results, or authorize any clinic, hospital, or doctor, or other person to release to the FAA, the results of each test taken within 4 hours after acting or attempting to act as a crewmember that indicates an alcohol concentration in the blood or breath specimen.

(d) Whenever the Administrator has a reasonable basis to believe that a person may have violated paragraph (a)(3) of this section, that person shall, upon request by the Administrator, furnish the Administrator, or authorize any clinic, hospital, doctor, or other person to release to the Administrator, the results of each test taken within 4 hours after acting or attempting to act as a crewmember that indicates the presence of any drugs in the body.

(e) Any test information obtained by the Administrator under paragraph (c) or (d) of this section may be evaluated in determining a person's qualifications for any airman certificate or possible violations of this chapter and may be used as evidence in any legal proceeding under section 602, 609, or 901 of the Federal Aviation Act of 1958.

§ 91.19 Carriage of narcotic drugs, marijuana, and depressant or stimulant drugs or substances.

(a) Except as provided in paragraph (b) of this section, no person may operate a civil aircraft within the United States with knowledge that narcotic drugs, marijuana, and depressant or stimulant drugs or substances as defined in Federal or State statutes are carried in the aircraft.

(b) Paragraph (a) of this section does not apply to any carriage of narcotic drugs, marijuana, and depressant or stimulant drugs or substances authorized by or under any Federal or State statute or by any Federal or State agency.

§ 91.21 Portable electronic devices.

(a) Except as provided in paragraph (b) of this section, no person may operate, nor may any operator or pilot in command of an aircraft allow the operation of, any portable electronic device on any of the following U.S.-registered civil aircraft:

(1) Aircraft operated by a holder of an air carrier operating certificate or an operating certificate; or

(2) Any other aircraft while it is operated under IFR.

(b) Paragraph (a) of this section does not apply to—

(1) Portable voice recorders;

(2) Hearing aids;

(3) Heart pacemakers;

(4) Electric shavers; or

(5) Any other portable electronic device that the operator of the aircraft has determined will not cause interference with the navigation or communication system of the aircraft on which it is to be used.

(c) In the case of an aircraft operated by a holder of an air carrier operating certificate or an operating certificate, the determination required by paragraph (b)(5) of this section shall be made by that operator of the aircraft on which the particular device is to be used. In the case of other aircraft, the determination may be made by the pilot in command or other operator of the aircraft.

§ 91.23 Truth-in-leasing clause requirement in leases and conditional sales contracts.

(a) Except as provided in paragraph (b) of this section, the parties to a lease or contract of conditional sale involving a U.S.-registered large civil aircraft and entered into after January 2, 1973, shall execute a written lease or contract and include therein a written truth-in-leasing clause as a concluding paragraph in large print, immediately preceding the space for the signature of the parties, which contains the following with respect to each such aircraft:

(1) Identification of the Federal Aviation Regulations under which the aircraft has been maintained and inspected

during the 12 months preceding the execution of the lease or contract of conditional sale, and certification by the parties thereto regarding the aircraft's status of compliance with applicable maintenance and inspection requirements in this part for the operation to be conducted under the lease or contract of conditional sale.

(2) The name and address (printed or typed) and the signature of the person responsible for operational control of the aircraft under the lease or contract of conditional sale, and certification that each person understands that person's responsibilities for compliance with applicable Federal Aviation Regulations.

(3) A statement that an explanation of factors bearing on operational control and pertinent Federal Aviation Regulations can be obtained from the nearest FAA Flight Standards district office.

(b) The requirements of paragraph (a) of this section do not apply—

(1) To a lease or contract of conditional sale when—

(i) The party to whom the aircraft is furnished is a foreign air carrier or certificate holder under part 121, 125, 127, 135, or 141 of this chapter, or

(ii) The party furnishing the aircraft is a foreign air carrier or a person operating under part 121, 125, and 141 of this chapter, or a person operating under part 135 of this chapter having authority to engage in on-demand operations with large aircraft.

(2) To a contract of conditional sale, when the aircraft involved has not been registered anywhere prior to the execution of the contract, except as a new aircraft under a dealer's aircraft registration certificate issued in accordance with § 47.61 of this chapter.

(c) No person may operate a large civil aircraft of U.S. registry that is subject to a lease or contract of conditional sale to which paragraph (a) of this section applies, unless—

(1) The lessee or conditional buyer, or the registered owner if the lessee is not a citizen of the United States, has mailed a copy of the lease or contract that complies with the requirements of paragraph (a) of this section, within 24 hours of its execution, to the Aircraft Registration Branch, Attn: Technical Section, P.O. Box 25724, Oklahoma City, Oklahoma 73125;

(2) A copy of the lease or contract that complies with the requirements of paragraph (a) of this section is carried in the aircraft. The copy of the lease or contract shall be made available for review upon request by the Administrator, and

(3) The lessee or conditional buyer, or the registered owner if the lessee is not a citizen of the United States, has notified by telephone or in person the FAA Flight Standards district office nearest the airport where the flight will originate. Unless otherwise authorized by that office, the notification shall be given at least 48 hours before takeoff in the case of the first flight of that aircraft under that lease or contract and inform the FAA of—

(i) The location of the airport of departure;

(ii) The departure time; and

(iii) The registration number of the aircraft involved.

(d) The copy of the lease or contract furnished to the FAA under paragraph (c) of this section is commercial or financial information obtained from a person. It is, therefore, privileged and confidential and will not be made available by the FAA for public inspection or copying under 5 U.S.C. 552(b)(4) unless recorded with the FAA under part 49 of this chapter.

(e) For the purpose of this section, a lease means any agreement by a person to furnish an aircraft to another person for compensation or hire, whether with or without flight crewmembers, other than an agreement for the sale of an aircraft and a contract of conditional sale under section 101 of the Federal Aviation Act of 1958. The person furnishing the aircraft is referred to as the lessor, and the person to whom it is furnished the lessee.

(Approved by the Office of Management and Budget under OMB control number 2120-0005)

(Amdt. 91-212, Eff. 8/18/90)

§ 91.25 Aviation Safety Reporting Program: Prohibition against use of reports for enforcement purposes.

The Administrator of the FAA will not use reports submitted to the National Aeronautics and Space Administration under the Aviation Safety Reporting Program (or information derived therefrom) in any enforcement action except information concerning accidents or criminal offenses which are wholly excluded from the Program.

§§ 91.27–91.99 [Reserved]

[Amdt. 91–240, Corrected, Eff. 5/12/94]

Subpart B—Flight rules general

§ 91.101 Applicability.

This subpart prescribes flight rules governing the operation of aircraft within the United States and within 12 nautical miles from the coast of the United States.

§ 91.103 Preflight action.

Each pilot in command shall, before beginning a flight, become familiar with all available information concerning that flight. This information must include—

(a) For a flight under IFR or a flight not in the vicinity of an airport, weather reports and forecasts, fuel requirements, alternatives available if the planned flight cannot be completed, and any known traffic delays of which the pilot in command has been advised by ATC;

(b) For any flight, runway lengths at airports of intended use, and the following takeoff and landing distance information:

(1) For civil aircraft for which an approved Airplane or Rotorcraft Flight Manual containing takeoff and landing distance data is required, the takeoff and landing distance data contained therein; and

(2) For civil aircraft other than those specified in paragraph (b)(1) of this section, other reliable information appropriate to the aircraft, relating to aircraft performance under expected values of airport elevation and runway slope, aircraft gross weight, and wind and temperature.

§ 91.105 Flight crewmembers at stations.

(a) During takeoff and landing, and while en route, each required flight crewmember shall—

(1) Be at the crewmember station unless the absence is necessary to perform duties in connection with the operation of the aircraft or in connection with physiological needs; and

(2) Keep the safety belt fastened while at the crewmember station.

(b) [Each required flight crewmember of a U.S.-registered civil aircraft shall, during takeoff and landing, keep his or her shoulder harness fastened while at his or her assigned duty station. This paragraph does not apply if—]

(1) The seat at the crewmember's station is not equipped with a shoulder harness; or

(2) The crewmember would be unable to perform required duties with the shoulder harness fastened.

[(Admt. 91-231, Eff. 10/15/92)]

§ 91.107 [Use of safety belts, shoulder harnesses, and child restraint systems.]

(a) Unless otherwise authorized by the Administrator—

(1) No pilot may take off a U.S.-registered civil aircraft (except a free balloon that incorporates a basket or gondola, or an airship type certificated before November 2, 1987) unless the pilot in command of that aircraft ensures that each person on board is briefed on how to fasten and unfasten that person's safety belt and, if installed, shoulder harness.

(2) No pilot may cause to be moved on the surface, take off, or land a U.S.-registered civil aircraft (except a free balloon that incorporates a basket or gondola, or an airship type certificated before November 2, 1987) unless the pilot in command of that aircraft ensures that each person on board has been notified to fasten his or her safety belt and, if installed, his or her shoulder harness.

(3) Except as provided in this paragraph, each person on board a U.S.-registered civil aircraft (except a free balloon

that incorporates a basket or gondola or an airship type certificated before November 2, 1987) must occupy an approved seat or berth with a safety belt and, if installed, shoulder harness, properly secured about him or her during movement on the surface, takeoff, and landing. For seaplane and float equipped rotorcraft operations during movement on the surface, the person pushing off the seaplane or rotorcraft from the dock and the person mooring the seaplane or rotorcraft at the dock are excepted from preceding seating and safety belt requirements. Notwithstanding the preceding requirements of this paragraph, a person may:

(i) Be held by an adult who is occupying an approved seat or berth, provided that the person being held has not reached his or her second birthday and does not occupy or use any restraining device;

(ii) Use the floor of the aircraft as a seat, provided that the person is on board for the purpose of engaging in sport parachuting; or

(iii) Notwithstanding any other requirement of this chapter, occupy an approved child restraint system furnished by the operator or one of the persons described in paragraph (a)(3)(iii)(A) of this section provided that:

(A) The child is accompanied by a parent, guardian, or attendant designated by the child's parent or guardian to attend to the safety of the child during the flight;

(B) Except as provided in paragraph (a)(3)(iii)(B)(4) of this section, the approved child restraint system bears one or more labels as follows:

(1) Seats manufactured to U.S. standards between January 1, 1981, and February 25, 1985, must bear the label: "This child restraint system conforms to all applicable Federal motor vehicle safety standards."

(2) Seats manufactured to U.S. standards on or after February 26, 1985, must bear two labels:

(i) "This child restraint system conforms to all applicable Federal motor vehicle safety standards"; and

(ii) THIS RESTRAINT IS CERTIFIED FOR USE IN MOTOR VEHICLES AND AIRCRAFT" in red lettering;

(3) Seats that do not qualify under paragraphs (a)(3)(iii)(B)(1) and (a)(3)(iii)(B)(2) of this section must bear a label or markings showing:

(i) That the seat was approved by a foreign government;

(ii) That the seat was manufactured under the standards of the United Nations; or

(iii) That the seat or child restraint device furnished by the operator was approved by the FAA through Type Certificate, Supplemental Type Certificate, or applicable Technical Standard Order.

(iv) That the seat or child restraint device furnished by the operator, or one of the persons described in paragraph (a) (3) (iii) (A) of this section, was approved by the FAA in accordance with § 21.305(d) or Technical Standard Order C-100b, or a later version,

(4) Except as provided in § 91.107(a)(3)(iii)(B)(3)(iii), notwithstanding any other provision of this section,

booster-type child restraint systems (as defined in Federal Motor Vehicle Safety Standard No. 213 (49 CFR 571.213)), vest- and harness-type child restraint systems, and lap held child restraints are not approved for use in aircraft; and

(C) The operator complies with the following requirements:

(1) The restraint system must be properly secured to an approved forward-facing seat or berth;

(2) The child must be properly secured in the restraint system and must not exceed the specified weight limit for the restraint system; and

(3) The restraint system must bear the appropriate label(s).

(b) Unless otherwise stated, this section does not apply to operations conducted under Part 121, 125, or 135 of this chapter. Paragraph (a)(3) of this section does not apply to persons subject to § 91.105.]

(Admt. 91-231, Eff. 10/15/92)

§ 91.109 Flight instruction; Simulated instrument flight and certain flight tests.

(a) No person may operate a civil aircraft (except a manned free balloon) that is being used for flight instruction unless that aircraft has fully functioning dual controls. However, instrument flight instruction may be given in a single-engine airplane equipped with a single, functioning throwover control wheel in place of fixed, dual controls of the elevator and ailerons when—

(1) The instructor has determined that the flight can be conducted safely; and

(2) The person manipulating the controls has at least a private pilot certificate with appropriate category and class ratings.

(b) No person may operate a civil aircraft in simulated instrument flight unless—

(1) The other control seat is occupied by a safety pilot who possesses at least a private pilot certificate with category and class ratings appropriate to the aircraft being flown.

(2) The safety pilot has adequate vision forward and to each side of the aircraft, or a competent observer in the aircraft adequately supplements the vision of the safety pilot; and

(3) Except in the case of lighter-than-air aircraft, that aircraft is equipped with fully functioning dual controls. However, simulated instrument flight may be conducted in a single-engine airplane, equipped with a single, functioning, throwover control wheel, in place of fixed, dual controls of the elevator and ailerons, when—

(i) The safety pilot has determined that the flight can be conducted safely; and

(ii) The person manipulating the controls has at least a private pilot certificate with appropriate category and class ratings.

(c) No person may operate a civil aircraft that is being used for a flight test for an airline transport pilot certificate or a class or type rating on that certificate, or for a part 121 proficiency flight test, unless the pilot seated at the controls, other than the pilot being checked, is fully qualified to act as pilot in command of the aircraft.

§ 91.111 Operating near other aircraft.

(a) No person may operate an aircraft so close to another aircraft as to create a collision hazard.

(b) No person may operate an aircraft in formation flight except by arrangement with the pilot in command of each aircraft in the formation.

(c) No person may operate an aircraft, carrying passengers for hire, in formation flight.

§ 91.113 Right-of-way rules: Except water operations.

(a) *Inapplicability.* This section does not apply to the operation of an aircraft on water.

(b) *General.* When weather conditions permit, regardless of whether an operation is conducted under instrument flight rules or visual flight rules, vigilance shall be maintained by each person operating an aircraft so as to see and avoid other aircraft. When a rule of this section gives another aircraft the right-of-way, the pilot shall give way to that aircraft and may not pass over, under, or ahead of it unless well clear.

(c) *In distress.* An aircraft in distress has the right-of-way over all other air traffic.

(d) *Converging.* When aircraft of the same category are converging at approximately the same altitude (except head-on, or nearly so), the aircraft to the other's right has the right-of-way. If the aircraft are of different categories—

(1) A balloon has the right-of-way over any other category of aircraft;

(2) A glider has the right-of-way over an airship, powered parachute, weight-shift-control aircraft, airplane, or rotorcraft.

(3) An airship has the right-of-way over a powered parachute, weight-shift-control aircraft, airplane, or rotorcraft.

However, an aircraft towing or refueling other aircraft has the right-of-way over all other engine-driven aircraft.

(e) *Approaching head-on.* When aircraft are approaching each other head-on, or nearly so, each pilot of each aircraft shall alter course to the right.

(f) *Overtaking.* Each aircraft that is being overtaken has the right-of-way and each pilot of an overtaking aircraft shall alter course to the right to pass well clear.

(g) *Landing.* Aircraft, while on final approach to land or while landing, have the right-of-way over other aircraft in flight or operating on the surface, except that they shall not take advantage of this rule to force an aircraft off the runway surface which has already landed and is attempting to make way for an aircraft on final approach. When two or more aircraft are approaching an airport for the purpose of landing, the aircraft at the lower altitude has the right-of-way, but it shall not take advantage of this rule to cut in front of another which is on final approach to land or to overtake that aircraft.

§ 91.115 Right-of-way rules: Water operations.

(a) *General.* Each person operating an aircraft on the water shall, insofar as possible, keep clear of all vessels and avoid impeding their navigation, and shall give way to any vessel or other aircraft that is given the right-of-way by any rule of this section.

(b) *Crossing.* When aircraft, or an aircraft and a vessel, are on crossing courses, the aircraft or vessel to the other's right has the right-of-way.

(c) *Approaching head-on.* When aircraft, or an aircraft and a vessel, are approaching head-on, or nearly so, each shall alter its course to the right to keep well clear.

(d) *Overtaking.* Each aircraft or vessel that is being overtaken has the right-of-way, and the one overtaking shall alter course to keep well clear.

(e) *Special circumstances.* When aircraft, or an aircraft and a vessel, approach so as to involve risk of collision, each aircraft or vessel shall proceed with careful regard to existing circumstances, including the limitations of the respective craft.

§ 91.117 Aircraft speed.

(a) Unless otherwise authorized by the Administrator, no person may operate an aircraft below 10,000 feet MSL at an indicated airspeed of more than 250 knots (288 m.p.h.).]

(b) Unless otherwise authorized or required by ATC, no person may operate an aircraft at or below 2,500 feet above the surface within 4 nautical miles of the primary airport of a Class C or Class D airspace area at an indicated airspeed of more than 200 knots (230 mph.). This paragraph (b) does not apply to any operations within a Class B airspace area. Such operations shall comply with paragraph (a) of this section.]

(c) No person may operate an aircraft in the airspace underlying a Class B airspace area designated for an airport or in a VFR corridor designated through such a Class B airspace area, at an indicated airspeed of more than 200 knots (230 mph).

(d) If the minimum safe airspeed for any particular operation is greater than the maximum speed prescribed in this section, the aircraft may be operated at that minimum speed.

(Amdt. 91-219, Eff. 8/24/90); (Amdt. 91-227, Eff. 9/16/93); (Amdt. 91-227, Corrected, Eff. 9/16/93); [*(Amdt. 91-233, Eff. 9/16/93*]

§ 91.119 Minimum safe altitudes: General.

Except when necessary for takeoff or landing, no person may operate an aircraft below the following altitudes:

(a) *Anywhere.* An altitude allowing, if a power unit fails, an emergency landing without undue hazard to persons or property on the surface.

(b) *Over congested areas.* Over any congested area of a city, town, or settlement, or over any open air assembly of persons, an altitude of 1,000 feet above the highest obstacle within a horizontal radius of 2,000 feet of the aircraft.

(c) *Over other than congested areas.* An altitude of 500 feet above the surface, except over open water or sparsely populated areas. In those cases, the aircraft may not be operated closer than 500 feet to any person, vessel, vehicle, or structure.

(d) *Helicopters.* Helicopters may be operated at less than the minimums prescribed in paragraph (b) or (c) of this section if the operation is conducted without hazard to persons or property on the surface. In addition, each person operating a helicopter shall comply with any routes or altitudes specifically prescribed for helicopters by the Administrator.

§ 91.121 Altimeter settings.

(a) Each person operating an aircraft shall maintain the cruising altitude or flight level of that aircraft, as the case may be, by reference to an altimeter that is set, when operating—

(1) Below 18,000 feet MSL, to—

(i) The current reported altimeter setting of a station along the route and within 100 nautical miles of the aircraft;

(ii) If there is no station within the area prescribed in paragraph (a)(1)(i) of this section, the current reported altimeter setting of an appropriate available station; or

(iii) In the case of an aircraft not equipped with a radio, the elevation of the departure airport or an appropriate altimeter setting available before departure; or

(2) At or above 18,000 feet MSL, to 29.92" Hg.

(b) The lowest usable flight level is determined by the atmospheric pressure in the area of operation as shown in the following table:

Current altimeter setting	Lowest usable flight level
29.92" (or higher)	180
29.91" through 29.42"	185
29.41" through 28.92"	190
28.91" through 28.42"	195
28.41" through 27.92"	200
27.91" through 27.42"	205
27.41" through 26.92"	210

(c) To convert minimum altitude prescribed under § 91.119 and § 91.177 to the minimum flight level, the pilot shall take the flight level equivalent of the minimum altitude in feet and add the appropriate number of feet specified below, according to the current reported altimeter setting:

Current altimeter setting	Adjustment factor
29.92" (or higher)	None
29.91" through 29.42"	500
29.41" through 28.92"	1,000
28.91" through 28.42"	1,500
28.41" through 27.92"	2,000
27.91" through 27.42"	2,500
27.41" through 26.92"	3,000

§ 91.123 Compliance with ATC clearances and instructions.

(a) When an ATC clearance has been obtained, no pilot in command may deviate from that clearance unless an amended clearance is obtained, an emergency exists, or the deviation is in response to a traffic alert and collision avoidance system resolution advisory. However, except in Class A airspace, a pilot may cancel an IFR flight plan if the operation is being conducted in VFR weather conditions. When a pilot is uncertain of an ATC clearance, that pilot shall immediately request clarification from ATC.

(b) Except in an emergency, no person may operate an aircraft contrary to an ATC instruction in an area in which air traffic control is exercised.

(c) Each pilot in command who, in an emergency, or in response to a traffic alert and collision avoidance system resolution advisory, deviates from an ATC clearance or instruction shall notify ATC of that deviation as soon as possible.

(d) Each pilot in command who (though not deviating from a rule of this subpart) is given priority by ATC in an emergency, shall submit a detailed report of that emergency within 48 hours to the manager of that ATC facility, if requested by ATC.

(e) Unless otherwise authorized by ATC, no person operating an aircraft may operate that aircraft according to any clearance or instruction that has been issued to the pilot of another aircraft for radar air traffic control purposes.

(Approved by the Office of Management and Budget under OMB control number 2120-0005).

[(Amdt. 91-227, Eff. 9/16/93]

§ 91.125 ATC light signals.

ATC light signals have the meaning shown in the following table:

Color and type of signal	Meaning with respect to aircraft on the surface	Meaning with respect to aircraft in flight
Steady green	Cleared for takeoff.	Cleared to land.
Flashing green	Cleared to taxi.	Return for landing (to be followed by steady green at proper time).
Steady red	Stop.	Give way to other aircraft and continue circling.
Flashing red	Taxi clear of runway in use.	Airport unsafe—do not land.
Flashing white	Return to starting point on airport.	Not applicable.
Alternating red and green	Exercise extreme caution.	Exercise extreme caution.

[§ 91.126 Operating on or in the vicinity of an airport in Class G airspace.

[(a) *General.* Unless otherwise authorized or required, each person operating an aircraft on or in the vicinity of an airport in a Class G airspace area must comply with the requirements of this section.

(b) *Direction of turns.* [(When approaching to land at an airport without an operating control tower in Class G airspace—]

(1) Each pilot of an airplane must make all turns of that airplane to the left unless the airport displays approved light signals or visual markings indicating that turns should be made to the right, in which case the pilot must make all turns to the right; and

(2) Each pilot of a helicopter or a powered parachute must avoid the flow of fixed-wing aircraft.

(c) *Flap settings.* Except when necessary for training or certification, the pilot in command of a civil turbojet-powered aircraft must use, as a final flap setting, the minimum certificated landing flap setting set forth in the approved performance information in the Airplane Flight Manual for the applicable conditions. However, each pilot in command has the final authority and responsibility for the safe operation of the pilot's airplane, and may use a different flap setting for that airplane if the pilot determines that it is necessary in the interest of safety.]

[(d) *Communications with control towers.* Unless otherwise authorized or required by ATC, no person may operate an aircraft to, from, through, or on an airport having an operational control tower unless two-way radio communications are maintained between that aircraft and the control tower. Communications must be established prior to 4 nautical miles from the airport, up to and including 2,500 feet AGL. However, if the aircraft radio fails in flight, the pilot in command may operate that aircraft and land if weather conditions are at or above basic VFR weather minimums, visual contact with the tower is maintained, and a clearance to land is received. If the aircraft radio fails while in flight under IFR, the pilot must comply with § 91.185.]

(Amdt. 91-227, Eff. 9/16/93); [(Admt. 91–239, Eff. 3/11/94)]

§ 91.127 Operating on or in the vicinity of an airport in Class E airspace.

(a) Unless otherwise required by Part 93 of this chapter or unless otherwise authorized or required by the ATC facility having jurisdiction over the Class E airspace area, each person operating an aircraft on or in the vicinity of an airport in a Class E airspace area must comply with the requirements of § 91.126.

(b) *Departures.* Each pilot of an aircraft must comply with any traffic patterns established for that airport in Part 93 of this chapter.

[(c) *Communications with control towers.* Unless otherwise authorized or required by ATC, no person may operate an aircraft to, from, through, or on an airport having an operational control tower unless two-way radio communications are maintained between that aircraft and the control tower. Communications must be established prior to 4 nautical miles from the airport, up to and including 2,500 feet AGL. However, if the aircraft radio fails in flight, the pilot in command may operate that aircraft and land if weather conditions are at or above basic VFR weather minimums, visual contact with the tower is maintained, and a clearance to land is received. If the aircraft radio fails while in flight under IFR, the pilot must comply with §91.185.]

(Amdt. 91-227, Eff. 9/16/93); (Amdt. 91-239, Eff. 3/11/94)]

§ 91.129 Operations in Class D airspace.

(a) *General.* Unless otherwise authorized or required by the ATC facility having jurisdiction over the Class D airspace area, each person operating an aircraft in Class D airspace must comply with the applicable provisions of this section. In addition, each person must comply with §§ 91.126 and 91.127. For the purpose of this section, the primary airport is the airport for which the Class D airspace area is designated. A satellite airport is any other airport within the Class D airspace area.

(b) *Deviations.* An operator may deviate from any provision of this section under the provisions of an ATC authorization issued by the ATC facility having jurisdiction over the airspace concerned. ATC may authorize a deviation on a continuing basis or for an individual flight, as appropriate.

(c) *Communications.* Each person operating an aircraft in Class D airspace must meet the following two-way radio communications requirements:

(1) Arrival or through flight. Each person must establish two-way radio communications with the ATC facility (including foreign ATC in the case of foreign airspace designated in the United States) providing air traffic services prior to entering that airspace and thereafter maintain those communications while within that airspace.

(2) Departing flight. Each person—

(i) From the primary airport or satellite airport with an operating control tower must establish and maintain two-way radio communications with the control tower, and thereafter as instructed by ATC while operating in the Class D airspace area; or

(ii) From a satellite airport without an operating control tower, must establish and maintain two-way radio communications with the ATC facility having jurisdiction over the Class D airspace area as soon as practicable after departing.

(d) *Communications failure.* Each person who operates an aircraft in a Class D airspace area must maintain two-way radio communications with the ATC facility having jurisdiction over that area.

(1) If the aircraft radio fails in flight under IFR, the pilot must comply with § 91.185 of the part.

(2) If the aircraft radio fails in flight under VFR, the pilot in command may operate that aircraft and land if—

(i) Weather conditions are at or above basic VFR weather minimums;

(ii) Visual contact with the tower is maintained; and

(iii) A clearance to land is received.

[(e) *Minimum Altitudes.* When operating to an airport in Class D airspace, each pilot of—

(1) [A large or turbine-powered airplane shall, unless otherwise required by the applicable distance from cloud criteria, enter the traffic pattern at an altitude of at least 1,500 feet above the elevation of the airport and maintain at least 1,500 feet until further descent is required for a safe landing;

(2) [A large or turbine-powered airplane approaching to land on a runway served by an instrument landing system (ILS), if the airplane is ILS equipped, shall fly that airplane at an altitude at or above the glide slope between the outer marker (or point of interception of glide slope, if compliance with the applicable distance from cloud criteria requires interception closer in) and the middle marker; and

(3) [An airplane approaching to land on a runway served by a visual approach slope indicator shall maintain an altitude at or above the glide slope until a lower altitude is necessary for a safe landing.]

Paragraphs (e)(2) and (e)(3) of this section do not prohibit normal bracketing maneuvers above or below the glide slope that are conducted for the purpose of remaining on the glide slope.

(f) *Approaches.* Except when conducting a circling approach under Part 97 of this chapter or unless otherwise required by ATC, each pilot must—

(1) Circle the airport to the left, if operating an airplane; or

(2) Avoid the flow of fixed-wing aircraft, if operating a helicopter.

(g) *Departures.* No person may operate an aircraft departing from an airport except in compliance with the following:

(1) Each pilot must comply with any departure procedures established for that airport by the FAA.

(2) Unless otherwise required by the prescribed departure procedure for that airport or the applicable distance from clouds criteria, each pilot of a turbine-powered airplane and each pilot of a large airplane must climb to an altitude of 1,500 feet above the surface as rapidly as practicable.

(h) *Noise abatement.* Where a formal runway use program has been established by the FAA, each pilot of a large or turbine-powered airplane assigned a noise abatement runway by ATC must use that runway. However, consistent with the final authority of the pilot in command concerning the safe operation of the aircraft as prescribed in § 91.3(a), ATC may assign a different runway if requested by the pilot in the interest of safety.

(i) *Takeoff, landing, taxi clearance.* No person may, at any airport with an operating control tower, operate an aircraft on a runway or taxiway, or take off or land an aircraft, unless an appropriate clearance is received from ATC. A clearance to "taxi to" the takeoff runway assigned to the aircraft is not a clearance to cross that assigned takeoff runway, or to taxi on that runway at any point, but is a clearance to cross other runways that intersect the taxi route to that assigned takeoff runway. A clearance to "taxi to" any point other than an assigned takeoff runway is clearance to cross all runways that intersect the taxi route to that point.

(Amdt. 91-227, Eff. 9/16/93); (Amdt. 91-234, Eff. 9/16/93)]

§ 91.130 Operations in Class C airspace.

(a) *General*. Unless otherwise authorized by ATC, each aircraft operation in Class C airspace must be conducted in compliance with this section and § 91.129. For the purpose of this section, the primary airport is the airport for which the Class C airspace area is designated. A satellite airport is any other airport within the Class C airspace area.

(b) *Traffic patterns*. No person may take off or land an aircraft at a satellite airport within a Class C airspace area except in compliance with FAA arrival and departure traffic patterns.

(c) *Communications*. Each person operating an aircraft in Class C airspace must meet the following two-way radio communications requirements:

(1) Arrival or through flight. Each person must establish two-way radio communications with the ATC facility (including foreign ATC in the case of foreign airspace designated in the United States) providing air traffic services prior to entering that airspace and thereafter maintain those communications while within that airspace.

(2) Departing flight. Each person—

(i) From the primary airport or satellite airport with an operating control tower must establish and maintain two-way radio communications with the control tower, and thereafter as instructed by ATC while operating in the Class C airspace area; or

(ii) From a satellite airport without an operating control tower, must establish and maintain two-way radio communications with the ATC facility having jurisdiction over the Class C airspace area as soon as practicable after departing.

(d) *Equipment requirements*. Unless otherwise authorized by the ATC having jurisdiction over the Class C airspace area, no person may operate an aircraft within a Class C airspace area designated for an airport unless that aircraft is equipped with the applicable equipment specified in § 91.215.

[(e) *Deviations*. An operator may deviate from any provision of this section under the provisions of an ATC authorization issued by the ATC facility having jurisdiction over the airspace concerned. ATC may authorize a deviation on a continuing basis or for an individual flight, as appropriate.]

(Amdt. 91-215, Eff. 8/18/90); (Amdt. 91-227, Eff. 9/16/93); (Amdt. 91-232, Eff. 9/16/93; [(Amdt. 91-239, Eff. 3/11/94)]

[§ 91.131 Operations in Class B airspace.

[(a) *Operating rules*. No person may operate an aircraft within a Class B airspace area except in compliance with § 91.129 and the following rules:

[(1) The operator must receive an ATC clearance from the ATC facility having jurisdiction for that area before operating an aircraft in that area.

[(2) Unless otherwise authorized by ATC, each person operating a large turbine engine-powered airplane to or from a primary airport for which a Class B airspace area is designated must operate at or above the designated floors of the Class B airspace area while within the lateral limits of that area.

[(3) Any person conducting pilot training operations at an airport within a Class B airspace area must comply with any procedures established by ATC for such operations in that area.

[(b) *Pilot requirements*.

[(1) No person may take off or land a civil aircraft at an airport within a Class B airspace area or operate a civil aircraft within a Class B airspace area unless—

(i) The pilot in command holds at least a private pilot certificate;

(ii) The pilot in command holds a recreational pilot certificate and has met—

(A) The requirements of §61.101 (d) of this chapter; or

(B) The requirements for a student pilot seeking a recreational pilot certificate in §61.94 of this chapter;

(iii) The pilot in command holds a sport pilot certificate and has met—

(A) The requirements of §61.325 of this chapter; or

(B) The requirements for a student pilot seeking a recreational pilot certificate in §61.94 of this chapter; or

(iv) The aircraft is operated by a student pilot who has met the requirements of §61.94 or §61.95 of this chapter, as applicable.

(2) Notwithstanding the provisions of paragraphs (b)(1)(ii), (b)(1)(iii) and (b)(1)(iv) of this section, no person may take off or land a civil aircraft at those airports listed in section 4 of appendix D to this part unless the pilot in command holds at least a private pilot certificate.

(c) *Communications and navigation equipment requirements*. Unless otherwise authorized by ATC, no person may operate an aircraft within a Class B airspace area unless that aircraft is equipped with—

(1) For IFR operation. An operable VOR or TACAN receiver; and

(2) For all operations. An operable two-way radio capable of communications with ATC on appropriate frequencies for that Class B airspace area.

(d) *Transponder requirements.* No person may operate an aircraft in a Class B airspace area unless the aircraft is equipped with the applicable operating transponder and automatic altitude reporting equipment specified in paragraph (a) of § 91.215, except as provided in paragraph (d) of that section.]

(Amdt. 91-214, Eff. 8/18/90); (Amdt. 91-216, Eff. 8/18/90); [(Amdt. 91-227, Eff. 9/16/93)]

§ 91.133 Restricted and prohibited areas.

(a) No person may operate an aircraft within a restricted area (designated in Part 73) contrary to the restrictions imposed, or within a prohibited area, unless that person has the permission of the using or controlling agency, as appropriate.

(b) Each person conducting, within a restricted area, an aircraft operation (approved by the using agency) that creates the same hazards as the operations for which the restricted area was designated may deviate from the rules of this subpart that are not compatible with the operation of the aircraft.

[§ 91.135 Operations in Class A airspace.

[Except as provided in paragraph (d) of this section, each person operating an aircraft in Class A airspace must conduct that operation under instrument flight rules (IFR) and in compliance with the following:

[(a) *Clearance.* Operations may be conducted only under an ATC clearance received prior to entering the airspace.

[(b) *Communications.* Unless otherwise authorized by ATC, each aircraft operating in Class A airspace must be equipped with a two-way radio capable of communicating with ATC on a frequency assigned by ATC. Each pilot must maintain two-way radio communications with ATC while operating in Class A airspace.

[(c) *Transponder requirement.* Unless otherwise authorized by ATC, no person may operate an aircraft within Class A airspace unless that aircraft is equipped with the applicable equipment specified in § 91.215.

[(d) *ATC authorizations.* An operator may deviate from any provision of this section under the provisions of an ATC authorization issued by the ATC facility having jurisdiction of the airspace concerned. In the case of an inoperative transponder, ATC may immediately approve an operation within a Class A airspace area allowing flight to continue, if desired, to the airport of ultimate destination, including any intermediate stops, or to proceed to a place where suitable repairs can be made, or both. Requests for deviation from any provision of this section must be submitted in writing, at least 4 days before the proposed operation. ATC may authorize a deviation on a continuing basis or for an individual flight.]

[(Amdt. 91-227, Eff. 9/16/93)]

§ 91.137 Temporary flight restrictions.

(a) The Administrator will issue a Notice to Airmen (NOTAM) designating an area within which temporary flight restrictions apply and specifying the hazard or condition requiring their imposition, whenever he determines it is necessary in order to—

(1) Protect persons and property on the surface or in the air from a hazard associated with an incident on the surface;

(2) Provide a safe environment for the operation of disaster relief aircraft; or

(3) Prevent an unsafe congestion of sightseeing and other aircraft above an incident or event which may generate a high degree of public interest.

The Notice to Airmen will specify the hazard or condition that requires the imposition of temporary flight restrictions.

(b) When a NOTAM has been issued in accordance with this section, no person may operate an aircraft within the designated area unless at least one of the following conditions is met:

(1) That person has obtained authorization from the official in charge of associated emergency or disaster relief response activities, and is operating the aircraft under the conditions of that authorization.

(2) The aircraft is carrying law enforcement officials.

(3) The aircraft is carrying persons involved in an emergency or a legitimate scientific purpose.

(4) The aircraft is carrying properly accredited news persons, and that prior to entering the area, a flight plan is filed with the appropriate FAA or ATC facility specified in the NOTAM and the operation is conducted in compliance with the conditions and restrictions established by the official in charge of on-scene emergency response activities.

(5) The aircraft is operating in accordance with an ATC clearance or instruction.

(c) When a NOTAM has been issued under paragraph (a)(2) of this section, no person may operate an aircraft within the designated area unless at least one of the following conditions are met:

(1) The aircraft is participating in hazard relief activities and is being operated under the direction of the official in charge of on scene emergency response activities.

(2) The aircraft is carrying law enforcement officials.

(3) The aircraft is operating under the ATC approved IFR flight plan.

(4) The operation is conducted directly to or from an airport within the area, or is necessitated by the impracticability of VFR flight above or around the area due to weather, or terrain; notification is given to the Flight Service Station (FSS) or ATC facility specified in the NOTAM to receive advisories concerning disaster relief aircraft operations; and the operation does not hamper or endanger relief activities and is not conducted for the purpose of observing the disaster.

(5) The aircraft is carrying properly accredited news representatives, and, prior to entering the area, a flight plan is filed with the appropriate FAA or ATC facility specified

in the Notice to Airmen and the operation is conducted above the altitude used by the disaster relief aircraft, unless otherwise authorized by the official in charge of on scene emergency response activities.

(d) When a NOTAM has been issued under paragraph (a)(3) of this section, no person may operate an aircraft within the designated area unless at least one of the following conditions is met:

(1) The operation is conducted directly to or from an airport within the area, or is necessitated by the impracticability of VFR flight above or around the area due to weather or terrain, and the operation is not conducted for the purpose of observing the incident or event.

(2) The aircraft is operating under an ATC approved IFR flight plan.

(3) The aircraft is carrying incident or event personnel, or law enforcement officials.

(4) The aircraft is carrying properly accredited news representatives and, prior to entering that area, a flight plan is filed with the appropriate FSS or ATC facility specified in the NOTAM.

(e) Flight plans filed and notifications made with an FSS or ATC facility under this section shall include the following information:

(1) Aircraft identification, type and color.

(2) Radio communications frequencies to be used.

(3) Proposed times of entry of, and exit from, the designated area.

(4) Name of news media or organization and purpose of flight.

(5) Any other information requested by ATC.

[§ 91.138 Temporary flight restrictions in national disaster areas in the State of Hawaii.

[(a) When the Administrator has determined, pursuant to a request and justification provided by the Governor of the State of Hawaii, or the Governor's designee, that an inhabited area within a declared national disaster area in the State of Hawaii is in need of protection for humanitarian reasons, the Administrator will issue a Notice to Airmen (NOTAM) designating an area within which temporary flight restrictions apply. The Administrator will designate the extent and duration of the temporary flight restrictions necessary to provide for the protection of persons and property on the surface.

(b) When a NOTAM has been issued in accordance with this section, no person may operate an aircraft within the designated area unless at least one of the following conditions is met:

(1) That person has obtained authorization from the official in charge of associated emergency or disaster relief response activities, and is operating the aircraft under the conditions of that authorization.

(2) The aircraft is carrying law enforcement officials.

(3) The aircraft is carrying persons involved in an emergency or a legitimate scientific purpose.

(4) The aircraft is carrying properly accredited newspersons, and that prior to entering the area, a flight plan is filed with the appropriate FAA or ATC facility specified in the NOTAM and the operation is conducted in compliance with the conditions and restrictions established by the official in charge of on-scene emergency response activities.

(5) The aircraft is operating in accordance with an ATC clearance or instruction.

[(c) A NOTAM issued under this section is effective for 90 days or until the national disaster designation is terminated, whichever comes first, unless terminated by notice or extended by the Administrator at the request of the Governor of the State of Hawaii or the Governor's designee.]

[(Amdt. 91-222, Eff. 5/20/91)]

§ 91.139 Emergency air traffic rules.

(a) This section prescribes a process for utilizing Notices to Airmen (NOTAMs) to advise of the issuance and operations under emergency air traffic rules and regulations and designates the official who is authorized to issue NOTAMs on behalf of the Administrator in certain matters under this section.

(b) Whenever the Administrator determines that an emergency condition exists, or will exist, relating to the FAA's ability to operate the air traffic control system and during which normal flight operations under this chapter cannot be conducted consistent with the required levels of safety and efficiency—

(1) The Administrator issues an immediately effective air traffic rule or regulation in response to that emergency condition; and

(2) The Administrator or the Associate Administrator for Air Traffic may utilize the NOTAM system to provide notification of the issuance of the rule or regulation.

Those NOTAMs communicate information concerning the rules and regulations that govern flight operations, the use of navigation facilities, and designation of that airspace in which the rules and regulations apply.

(c) When a NOTAM has been issued under this section, no person may operate an aircraft, or other device governed by the regulation concerned, within the designated airspace except in accordance with the authorizations, terms, and conditions prescribed in the regulation covered by the NOTAM.

§ 91.141 Flight restrictions in the proximity of the Presidential and other parties.

No person may operate an aircraft over or in the vicinity of any area to be visited or traveled by the President, the Vice President, or other public figures contrary to the restrictions established by the Administrator and published in a Notice to Airmen (NOTAM).

§ 91.143 Flight limitation in the proximity of space flight operations.

When a Notice to Airmen (NOTAM) is issued in accordance with this section, no person may operate any aircraft of U.S. registry, or pilot any aircraft under the authority of an airman certificate issued by the Federal Aviation Administration, within areas designated in a NOTAM for space flight operation except when authorized by ATC.

§ 91.144 Temporary restriction on flight operations during abnormally high barometric pressure conditions.

(a) *Special flight restrictions.* When any information indicates that barometric pressure on the route of flight currently exceeds or will exceed 31 inches of mercury, no person may operate an aircraft or initiate a flight contrary to the requirements established by the Administrator and published in a Notice to Airmen issued under this section.

(b) *Waivers.* The Administrator is authorized to waive any restriction issued under paragraph (a) of this section to permit emergency supply, transport, or medical services to be delivered to isolated communities, where the operation can be conducted with an acceptable level of safety.]

[(Amdt. 91-240, Eff. 5/12/94)]

§ 91.145 Management of aircraft operations in the vicinity of aerial demonstrations and major sporting events.

(a) The FAA will issue a Notice to Airmen (NOTAM) designating an area of airspace in which a temporary flight restriction applies when it determines that a temporary flight restriction is necessary to protect persons or property on the surface or in the air, to maintain air safety and efficiency, or to prevent the unsafe congestion of aircraft in the vicinity of an aerial demonstration or major sporting event. These demonstrations and events may include:

(1) United States Naval Flight Demonstration Team (Blue Angels);

(2) United States Air Force Air Demonstration Squadron (Thunderbirds);

(3) United States Army Parachute Team (Golden Knights);

(4) Summer/Winter Olympic Games;

(5) Annual Tournament of Roses Football Game;

(6) World Cup Soccer;

(7) Major League Baseball All-Star Game;

(8) World Series;

(9) Kodak Albuquerque International Balloon Fiesta;

(10) Sandia Classic Hang Gliding Competition;

(11) Indianapolis 500 Mile Race;

(12) Any other aerial demonstration or sporting event the FAA determines to need a temporary flight restriction in accordance with paragraph (b) of this section.

(b) In deciding whether a temporary flight restriction is necessary for an aerial demonstration or major sporting event not listed in paragraph (a) of this section, the FAA considers the following factors:

(1) Area where the event will be held.

(2) Effect flight restrictions will have on known aircraft operations.

(3) Any existing ATC airspace traffic management restrictions.

(4) Estimated duration of the event.

(5) Degree of public interest.

(6) Number of spectators.

(7) Provisions for spectator safety.

(8) Number and types of participating aircraft.

(9) Use of mixed high and low performance aircraft.

(10) Impact on non-participating aircraft.

(11) Weather minimums.

(12) Emergency procedures that will be in effect.

(c) A NOTAM issued under this section will state the name of the aerial demonstration or sporting event and specify the effective dates and times, the geographic features or coordinates, and any other restrictions or procedures governing flight operations in the designated airspace.

(d) When a NOTAM has been issued in accordance with this section, no person may operate an aircraft or device, or engage in any activity within the designated airspace area, except in accordance with the authorizations, terms, and conditions of the temporary flight restriction published in the NOTAM, unless otherwise authorized by:

(1) Air traffic control; or

(2) A Flight Standards Certificate of Waiver or Authorization issued for the demonstration or event.

(e) For the purpose of this section:

(1) *Flight restricted airspace area for an aerial demonstration*—The amount of airspace needed to protect persons and property on the surface or in the air, to maintain air safety and efficiency, or to prevent the unsafe congestion of aircraft will vary depending on the aerial demonstration and the factors listed in paragraph (b) of this section. The restricted airspace area will normally be limited to a 5 nautical mile radius from the center of the demonstration and an altitude 17,000 mean sea level (for high performance aircraft) or 13,000 feet above the surface (for certain parachute operations), but will be no greater than the minimum airspace necessary for the management of aircraft operations in the vicinity of the specified area.

(2) *Flight restricted area for a major sporting event*—The amount of airspace needed to protect persons and property on the surface or in the air, to maintain air safety and efficiency, or to prevent the unsafe congestion of aircraft will vary depending on the size of the event and the factors listed in paragraph (b) of this section. The restricted airspace will normally be limited to a 3 nautical mile radius from the center of the event and 2,500 feet above the surface but will not be greater than the minimum airspace necessary for the management of aircraft operations in the vicinity of the specified area.

(f) A NOTAM issued under this section will be issued at least 30 days in advance of an aerial demonstration or a major sporting event, unless the FAA finds good cause for a shorter period and explains this in the NOTAM.

(g) When warranted, the FAA Administrator may exclude the following flights from the provisions of this section:

(1) Essential military.

(2) Medical and rescue.

(3) Presidential and Vice Presidential.

(4) Visiting heads of state.

(5) Law enforcement and security.

(6) Public health and welfare.

§§ 91.146 – 91.149 [Reserved]

Visual flight rules (Subpart B—Flight rules)

§ 91.151 Fuel requirements for flight in VFR conditions.

(a) No person may begin a flight in an airplane under VFR conditions unless (considering wind and forecast weather conditions) there is enough fuel to fly to the first point of intended landing and, assuming normal cruising speed—

(1) During the day, to fly after that for at least 30 minutes; or

(2) At night, to fly after that for at least 45 minutes.

(b) No person may begin a flight in a rotorcraft under VFR conditions unless (considering wind and forecast weather conditions) there is enough fuel to fly to the first point of intended landing and, assuming normal cruising speed, to fly after that for at least 20 minutes.

§ 91.153 VFR flight plan: Information required.

(a) *Information required.* Unless otherwise authorized by ATC, each person filing a VFR flight plan shall include in it the following information:

(1) The aircraft identification number and, if necessary, its radio call sign.

(2) The type of the aircraft or, in the case of a formation flight, the type of each aircraft and the number of aircraft in the formation.

(3) The full name and address of the pilot in command or, in the case of a formation flight, the formation commander.

(4) The point and proposed time of departure.

(5) The proposed route, cruising altitude (or flight level), and true airspeed at that altitude.

(6) The point of first intended landing and the estimated elapsed time until over that point.

(7) The amount of fuel on board (in hours).

(8) The number of persons in the aircraft, except where that information is otherwise readily available to the FAA.

(9) Any other information the pilot in command or ATC believes is necessary for ATC purposes.

(b) *Cancellation.* When a flight plan has been activated, the pilot in command, upon canceling or completing the flight under the flight plan, shall notify an FAA Flight Service Station or ATC facility.

§ 91.155 Basic VFR weather minimums.

(a) Except as provided in paragraph (b) of this section and § 91.157, no person may operate an aircraft under VFR when the flight visibility is less, or at a distance from clouds that is less, than that prescribed for the corresponding altitude and class of airspace in the table on page 503:

(b) *Class G Airspace.* Notwithstanding the provisions of paragraph (a) of this section, the following operations may be conducted in Class G airspace below 1,200 feet above the surface:

(1) Helicopter. A helicopter may be operated clear of clouds if operated at a speed that allows the pilot adequate opportunity to see any air traffic or obstruction in time to avoid a collision.

(2) *Airplane, powered parachute, or weight-shift-control aircraft.* If the visibility is less than 3 statute miles but not less than 1 statute mile during night hours and you are operating in an airport traffic pattern within 1/2 mile of the runway, you may operate an airplane, powered parachute, or weight-shift-control aircraft clear of clouds.

(c) Except as provided in § 91.157, no person may operate an aircraft beneath the ceiling under VFR within the lateral boundaries of controlled airspace designated to the surface for an airport when the ceiling is less than 1,000 feet.

(d) Except as provided in § 91.157 of this part, no person may take off or land an aircraft, or enter the traffic pattern of an airport, under VFR, within the lateral boundaries of the surface areas of Class B, Class C, Class D, or Class E airspace designated for an airport—

(1) Unless ground visibility at that airport is at least 3 statute miles; or

(2) If ground visibility is not reported at that airport, unless flight visibility during landing or takeoff, or while operating in the traffic pattern is at least 3 statute miles.

(e) For the purpose of this section, an aircraft operating at the base altitude of a Class E airspace area is considered to be within the airspace directly below that area.

(Amdt. 91-213, Eff. 8/18/90); (Amdt. 91-224, Eff. 9/23/91); (Amdt. 91-227, Eff. 9/16/93); [(Amdt. 91-235, Eff. 10/5/93)]

[§ **91.157 Special VFR weather minimums.**

[(a) Except as provided in Appendix D, Section 3, of this part, special VFR operations may be conducted under the weather minimums and requirements of this section, instead of those contained in § 91.155, below 10,000 feet MSL within the airspace contained by the upward extension of the lateral boundaries of the controlled airspace designated to the surface for an airport.

[(b) Special VFR operations may only be conducted—

[(1) With an ATC clearance;

[(2) Clear of clouds;

[(3) Except for helicopters, when flight visibility is at least 1 statute mile; and

[(4) Except for helicopters, between sunrise and sunset (or in Alaska, when the sun is 6° or more below the horizon) unless—

[(i) The person being granted the ATC clearance meets the applicable requirements for instrument flight under part 61 of this chapter; and

[(ii) The aircraft is equipped as required in § 91.205(d).

[(c) No person may take off or land an aircraft (other than a helicopter) under special VFR—

Airspace	Flight visibility	Distance from clouds
Class A	Not Applicable	Not Applicable.
Class B	3 statute miles	Clear of Clouds.
Class C	3 statute miles	500 feet below. 1,000 feet above. 2,000 feet horizontal.
Class D	3 statute miles	500 feet below. 1,000 feet above. 2,000 feet horizontal.
Class E		
Less than 10,000 feet MSL	3 statute miles	500 feet below. 1,000 feet above. 2,000 feet horizontal.
At or above 10,000 feet MSL	5 statute miles	1,000 feet below. 1,000 feet above. 1 statute mile horizontal.
Class G		
1,200 feet or less above the surface (regardless of MSL altitude)		
Day, except as provided in §91.155(b)	1 statute mile	Clear of clouds.
Night, except as provided in §91.155(b)	3 statute miles	500 feet below. 1,000 feet above. 2,000 feet horizontal.
More than 1,200 feet above the surface but less than 10,000 feet MSL		
Day	1 statute mile	500 feet below. 1,000 feet above. 2,000 feet horizontal.
Night	3 statute miles	500 feet below. 1,000 feet above. 2,000 feet horizontal.
More than 1,200 feet above the surface and at or above 10,000 feet MSL	5 statute miles	1,000 feet below. 1,000 feet above. 1 statute mile horizontal.

(1) Unless ground visibility is at least 1 statute mile; or

(2) If ground visibility is not reported, unless flight visibility is at least 1 statute mile. For the purposes of this paragraph, the term flight visibility includes the visibility from the cockpit of an aircraft in takeoff position if:

(i) The flight is conducted under this part 91; and

(ii) The airport at which the aircraft is located is a satellite airport that does not have weather reporting capabilities.

(d) The determination of visibility by a pilot in accordance with paragraph (c)(2) of this section is not an official weather report or an official ground visibility report.

§ 91.159   VFR cruising altitude or flight level.

Except while holding in a holding pattern of 2 minutes or less, or while turning, each person operating an aircraft under VFR in level cruising flight more than 3,000 feet above the surface shall maintain the appropriate altitude or flight level prescribed below, unless otherwise authorized by ATC:

(a) When operating below 18,000 feet MSL and—

(1) On a magnetic course of zero degrees through 179 degrees, any odd thousand foot MSL altitude +500 feet (such as 3,500, 5,500, or 7,500); or

(2) On a magnetic course of 180 degrees through 359 degrees, any even thousand foot MSL altitude +500 feet (such as 4,500, 6,500, or 8,500).

(b) When operating above 18,000 feet MSL maintain the altitude or flight level assigned by ATC and—

(1) On a magnetic course of zero degrees through 179 degrees, any odd flight level +500 feet (such as 195, 215, or 235); or

(2) On a magnetic course of 180 degrees through 359 degrees, any even flight level +500 feet (such as 185, 205, or 225).

(c) When operating above flight level 290 and—

(1) On a magnetic course of zero degrees through 179 degrees, any flight level, at 4,000-foot intervals, beginning at and including flight level 300 (such as flight level 300, 340, or 380); or

(2) On a magnetic course of 180 degrees through 359 degrees, any flight level, at 4,000-foot intervals, beginning at and including flight level 320 (such as flight level 320, 360, or 400).

§§ 91.161–91.165   [Reserved]

Instrument flight rules (Subpart B—Flight rules)

§ 91.167 Fuel requirements for flight in IFR conditions.

(a) No person may operate a civil aircraft in IFR conditions unless it carries enough fuel (considering weather reports and forecasts and weather conditions) to—

(1) Complete the flight to the first airport of intended landing;

(2) Except as provided in paragraph (b) of this section, fly from that airport to the alternate airport; and

(3) Fly after that for 45 minutes at normal cruising speed or, for helicopters, fly after that for 30 minutes at normal cruising speed.

(b) Paragraph (a)(2) of this section does not apply if:

(1) Part 97 of this chapter prescribes a standard instrument approach procedure to, or a special instrument approach procedure has been issued by the Administrator to the operator for, the first airport of intended landing; and

(2) Appropriate weather reports or weather forecasts, or a combination of them, indicate the following:

(i) *For aircraft other than helicopters.* For at least 1 hour before and for 1 hour after the estimated time of arrival, the ceiling will be at least 2,000 feet above the airport elevation and the visibility will be at least 3 statute miles.

(ii) *For helicopters.* At the estimated time of arrival and for 1 hour after the estimated time of arrival, the ceiling will be at least 1,000 feet above the airport elevation, or at least 400 feet above the lowest applicable approach minima, whichever is higher, and the visibility will be at least 2 statute miles.

§ 91.169 IFR flight plan: Information required.

(a) *Information required.* Unless otherwise authorized by ATC, each person filing an IFR flight plan must include in it the following information:

(1) Information required under § 91.153 (a) of this part;

(2) Except as provided in paragraph (b) of this section, an alternate airport.

(b) Paragraph (a)(2) of this section does not apply if:

(1) Part 97 of this chapter prescribes a standard instrument approach procedure to, or a special instrument approach procedure has been issued by the Administrator to the operator for, the first airport of intended landing; and

(2) Appropriate weather reports or weather forecasts, or a combination of them, indicate the following:

(i) *For aircraft other than helicopters.* For at least 1 hour before and for 1 hour after the estimated time of arrival, the ceiling will be at least 2,000 feet above the airport elevation and the visibility will be at least 3 statute miles.

(ii) *For helicopters.* At the estimated time of arrival and for 1 hour after the estimated time of arrival, the ceiling will be at least 1,000 feet above the airport elevation, or at least 400 feet above the lowest applicable approach minima, whichever is higher, and the visibility will be at least 2 statute miles.

(c) *IFR alternate airport weather minima.* Unless otherwise authorized by the Administrator, no person may include an alternate airport in an IFR flight plan unless appropriate weather reports or weather forecasts, or a combination of them, indicate that, at the estimated time of arrival at the alternate airport, the ceiling and visibility at that airport will be at or above the following weather minima:

(1) If an instrument approach procedure has been published in part 97 of this chapter, or a special instrument approach procedure has been issued by the Administrator to the operator, for that airport, the following minima:

(i) *For aircraft other than helicopters:* The alternate airport minima specified in that procedure, or if none are specified the following standard approach minima:

(A) *For a precision approach procedure.* Ceiling 600 feet and visibility 2 statute miles.

(B) *For a nonprecision approach procedure.* Ceiling 800 feet and visibility 2 statute miles.

(ii) *For helicopters:* Ceiling 200 feet above the minimum for the approach to be flown, and visibility at least 1 statute mile but never less than the minimum visibility for the approach to be flown, and

(2) If no instrument approach procedure has been published in part 97 of this chapter and no special instrument approach procedure has been issued by the Administrator to the operator, for the alternate airport, the ceiling and visibility minima are those allowing descent from the MEA, approach, and landing under basic VFR.

§ 91.171 VOR equipment check for IFR operations.

(a) No person may operate a civil aircraft under IFR using the VOR system of radio navigation unless the VOR equipment of that aircraft—

(1) Is maintained, checked, and inspected under an approved procedure; or

(2) Has been operationally checked within the preceding 30 days, and was found to be within the limits of the permissible indicated bearing error set forth in paragraph (b) or (c) of this section.

(b) Except as provided in paragraph (c) of this section, each person conducting a VOR check under paragraph (a)(2) of this section shall—

(1) Use, at the airport of intended departure, an FAA-operated or approved test signal or a test signal radiated by a certificated and appropriately rated radio repair station or, outside the United States, a test signal operated or approved by an appropriate authority to check the VOR equipment (the maximum permissible indicated bearing error is plus or minus 4 degrees); or

(2) Use, at the airport of intended departure, a point on the airport surface designated as a VOR system checkpoint

by the Administrator, or, outside the United States, by an appropriate authority (the maximum permissible bearing error is plus or minus 4 degrees);

(3) If neither a test signal nor a designated checkpoint on the surface is available, use an airborne checkpoint designated by the Administrator or, outside the United States, by an appropriate authority (the maximum permissible bearing error is plus or minus 6 degrees); or

(4) If no check signal or point is available, while in flight—

(i) Select a VOR radial that lies along the centerline of an established VOR airway;

(ii) Select a prominent ground point along the selected radial preferably more than 20 nautical miles from the VOR ground facility and maneuver the aircraft directly over the point at a reasonably low altitude; and

(iii) Note the VOR bearing indicated by the receiver when over the ground point (the maximum permissible variation between the published radial and the indicated bearing is 6 degrees).

(c) If dual system VOR (units independent of each other except for the antenna) is installed in the aircraft, the person checking the equipment may check one system against the other in place of the check procedures specified in paragraph (b) of this section. Both systems shall be tuned to the same VOR ground facility and note the indicated bearings to that station. The maximum permissible variation between the two indicated bearings is 4 degrees.

(d) Each person making the VOR operational check, as specified in paragraph (b) or (c) of this section, shall enter the date, place, bearing error, and sign the aircraft log or other record. In addition, if a test signal radiated by a repair station, as specified in paragraph (b)(1) of this section, is used, an entry must be made in the aircraft log or other record by the repair station certificate holder or the certificate holder's representative certifying to the bearing transmitted by the repair station for the check and the date of transmission.

(Approved by the Office of Management and Budget under OMB control number 2120-0005).

§ 91.173 ATC clearance and flight plan required.

No person may operate an aircraft in controlled airspace under IFR unless that person has—

(a) Filed an IFR flight plan; and

(b) Received an appropriate ATC clearance.

§ 91.175 Takeoff and landing under IFR.

(a) *Instrument approaches to civil airports.* Unless otherwise authorized by the Administrator, when an instrument letdown to a civil airport is necessary, each person operating an aircraft, except a military aircraft of the United States, shall use a standard instrument approach procedure prescribed for the airport in part 97 of this chapter.

(b) *Authorized DH or MDA.* For the purpose of this section, when the approach procedure being used provides for and requires the use of a DH or MDA, the authorized DH or MDA is the highest of the following:

(1) The DH or MDA prescribed by the approach procedure.

(2) The DH or MDA prescribed for the pilot in command.

(3) The DH or MDA for which the aircraft is equipped.

(c) *Operation below DH or MDA.* Except as provided in paragraph (1) of this section, where a DH or MDA is applicable, no pilot may operate an aircraft, except a military aircraft of the United States, at any airport below the authorized MDA or continue an approach below the authorized DH unless—

(1) The aircraft is continuously in a position from which a descent to a landing on the intended runway can be made at a normal rate of descent using normal maneuvers, and for operations conducted under Part 121 or Part 135 unless that descent rate will allow touchdown to occur within the touchdown zone of the runway of intended landing;

(2) The flight visibility is not less than the visibility prescribed in the standard instrument approach being used; and

(3) Except for a Category II or Category III approach where any necessary visual reference requirements are specified by the Administrator, at least one of the following visual references for the intended runway is distinctly visible and identifiable to the pilot:

(i) The approach light system, except that the pilot may not descend below 100 feet above the touchdown zone elevation using the approach lights as a reference unless the red terminating bars or the red side row bars are also distinctly visible and identifiable.

(ii) The threshold.

(iii) The threshold markings.

(iv) The threshold lights.

(v) The runway end identifier lights.

(vi) The visual approach slope indicator.

(vii) The touchdown zone or touchdown zone markings.

(viii) The touchdown zone lights.

(ix) The runway or runway markings.

(x) The runway lights.

(d) *Landing.* No pilot operating an aircraft, except a military aircraft of the United States, may land that aircraft when—

(1) For operations conducted under paragraph (1) of this section, the requirements of (l)(4) of this section are not met; or

(2) For all other part 91 operations and parts 121, 125, 129, and 135 operations, the flight visibility is less than the visibility prescribed in the standard instrument approach procedure being used.

(e) *Missed approach procedures.* Each pilot operating an aircraft, except a military aircraft of the United States,

shall immediately execute an appropriate missed approach procedure when either of the following conditions exist:

(1) Whenever operating an aircraft pursuant to paragraph (c) or (1) of this section and the requirements of that paragraph are not met at either of the following times:

(i) When the aircraft is being operated below MDA; or

(ii) Upon arrival at the missed approach point, including a DH where a DH is specified and its use is required, and at any time after that until touchdown.

(2) Whenever an identifiable part of the airport is not distinctly visible to the pilot during a circling maneuver at or above MDA, unless the inability to see an identifiable part of the airport results only from a normal bank of the aircraft during the circling approach.

(f) *Civil airport takeoff minimums.* Unless otherwise authorized by the Administrator, no pilot operating an aircraft under Parts 121, 125, 127, 129, or 135 of this chapter may take off from a civil airport under IFR unless weather conditions are at or above the weather minimum for IFR takeoff prescribed for that airport under Part 97 of this chapter. If takeoff minimums are not prescribed under Part 97 of this chapter for a particular airport, the following minimums apply to takeoffs under IFR for aircraft operating under those parts:

(1) For aircraft, other than helicopters, having two engines or less—1 statute mile visibility.

(2) For aircraft having more than two engines—½ statute mile visibility.

(3) For helicopters—½ statute mile visibility.

(g) *Military airports.* Unless otherwise prescribed by the Administrator, each person operating a civil aircraft under IFR into or out of a military airport shall comply with the instrument approach procedures and the takeoff and landing minimum prescribed by the military authority having jurisdiction of that airport.

(h) *Comparable values of RVR and ground visibility.*

(1) Except for Category II or Category III minimums, if RVR minimums for takeoff or landing are prescribed in an instrument approach procedure, but RVR is not reported for the runway of intended operation, the RVR minimum shall be converted to ground visibility in accordance with the table in paragraph (h)(2) of this section and shall be the visibility minimum for takeoff or landing on that runway.

(2) RVR Table:

RVR (feet)	Visibility (statute miles)
1,600	¼
2,400	½
3,200	⅝
4,000	¾
4,500	⅞
5,000	1
6,000	1¼

(i) *Operations on unpublished routes and use of radar in instrument approach procedures.* When radar is approved at certain locations for ATC purposes, it may be used not only for surveillance and precision radar approaches, as applicable, but also may be used in conjunction with instrument approach procedures predicated on other types of radio navigational aids. Radar vectors may be authorized to provide course guidance through the segments of an approach to the final course or fix. When operating on an unpublished route or while being radar vectored, the pilot, when an approach clearance is received, shall, in addition to complying with § 91.177, maintain the last altitude assigned to that pilot until the aircraft is established on a segment of a published route or instrument approach procedure unless a different altitude is assigned by ATC. After the aircraft is so established, published altitudes apply to descent within each succeeding route or approach segment unless a different altitude is assigned by ATC. Upon reaching the final approach course or fix, the pilot may either complete the instrument approach in accordance with a procedure approved for the facility or continue a surveillance or precision radar approach to a landing.

(j) *Limitation on procedure turns.* In the case of a radar vector to a final approach course or fix, a timed approach from a holding fix, or an approach for which the procedure specifies "No PT," no pilot may make a procedure turn unless cleared to do so by ATC.

(k) *ILS components.* The basic ground components of an ILS are the localizer, glide slope, outer marker, middle marker, and, when installed for use with Category II or Category III instrument approach procedures, an inner marker. A compass locator or precision radar may be substituted for the outer or middle marker. DME, VOR, or nondirectional beacon fixes authorized in the standard instrument approach procedure or surveillance radar may be substituted for the outer marker. Applicability of, and substitution for, the inner marker for Category II or III approaches is determined by the appropriate Part 97 approach procedure, letter of authorization, or operations specification pertinent to the operations.

(l) *Approach to straight-in landing operations below DH, or MDA using an enhanced flight vision system (EFVS).* For straight-in instrument approach procedures other than Category II or Category III, no pilot operating under this section or §§ 121.651, 125.381, and 135.225 of this chapter may operate an aircraft at any airport below the authorized MDA or continue an approach below the authorized DH and land unless—

(1) The aircraft is continuously in a position from which a descent to a landing on the intended runway can be made at a normal rate of descent using normal maneuvers, and, for operations conducted under part 121 or part 135 of this chapter, the descent rate will allow touchdown to occur within the touchdown zone of the runway of intended landing;

(2) The pilot determines that the enhanced flight visibility observed by use of a certified enhanced flight vision system is not less than the visibility prescribed in the standard instrument approach procedure being used;

(3) The following visual references for the intended runway are distinctly visible and identifiable to the pilot using the enhanced flight vision system:

(i) The approach light system (if installed); or

(ii) The following visual references in both paragraphs (l)(3)(ii)(A) and (B) of this section:

(A) The runway threshold, identified by at least one of the following:

(1) The beginning of the runway landing surface;

(2) The threshold lights; or

(3) The runway end identifier lights.

(B) The touchdown zone, identified by at least one of the following:

(1) The runway touchdown zone landing surface;

(2) The touchdown zone lights;

(3) The touchdown zone markings; or

(4) The runway lights.

(4) At 100 feet above the touchdown zone elevation of the runway of intended landing and below that altitude, the flight visibility must be sufficient for the following to be distinctly visible and identifiable to the pilot without reliance on the enhanced flight vision system to continue to a landing:

(i) The lights or markings of the threshold; or

(ii) The lights or markings of the touchdown zone;

(5) The pilot(s) is qualified to use an EFVS as follows—

(i) For parts 119 and 125 certificate holders, the applicable training, testing and qualification provisions of parts 121, 125, and 135 of this chapter;

(ii) For foreign persons, in accordance with the requirements of the civil aviation authority of the State of the operator; or

(iii) For persons conducting any other operation, in accordance with the applicable currency and proficiency requirements of part 61 of this chapter;

(6) For parts 119 and 125 certificate holders, and part 129 operations specifications holders, their operations specifications authorize use of EFVS; and

(7) The aircraft is equipped with, and the pilot uses, an enhanced flight vision system, the display of which is suitable for maneuvering the aircraft and has either an FAA type design approval or, for a foreign-registered aircraft, the EFVS complies with all of the EFVS requirements of this chapter.

(m) For purposes of this section, "enhanced flight vision system" (EFVS) is an installed airborne system comprised of the following features and characteristics:

(1) An electronic means to provide a display of the forward external scene topography (the natural or manmade features of a place or region especially in a way to show their relative positions and elevation) through the use of imaging sensors, such as a forward-looking infrared, millimeter wave radiometry, millimeter wave radar, and low-light level image intensifying;

(2) The EFVS sensor imagery and aircraft flight symbology (*i.e.*, at least airspeed, vertical speed, aircraft attitude, heading, altitude, command guidance as appropriate for the approach to be flown, path deviation indications, and flight path vector, and flight path angle reference cue) are presented on a head-up display, or an equivalent display, so that they are clearly visible to the pilot flying in his or her normal position and line of vision and looking forward along the flight path, to include:

(i) The displayed EFVS imagery, attitude symbology, flight path vector, and flight path angle reference cue, and other cues, which are referenced to this imagery and external scene topography, must be presented so that they are aligned with and scaled to the external view; and

(ii) The flight path angle reference cue must be displayed with the pitch scale, selectable by the pilot to the desired descent angle for the approach, and suitable for monitoring the vertical flight path of the aircraft on approaches without vertical guidance; and

(iii) The displayed imagery and aircraft flight symbology do not adversely obscure the pilot's outside view or field of view through the cockpit window;

(3) The EFVS includes the display element, sensors, computers and power supplies, indications, and controls. It may receive inputs from an airborne navigation system or flight guidance system; and

(4) The display characteristics and dynamics are suitable for manual control of the aircraft.

§ 91.177 Minimum altitudes for IFR operations.

(a) *Operation of aircraft at minimum attitudes.* Except when necessary for takeoff or landing, no person may operate an aircraft under IFR below—

(1) The applicable minimum altitudes prescribed in Parts 95 and 97 of this chapter; or

(2) If no applicable minimum altitude is prescribed in those parts—

(i) In the case of operations over an area designated as a mountainous area in Part 95, an altitude of 2,000 feet above the highest obstacle within a horizontal distance of 4 nautical miles from the course to be flown; or

(ii) In any other case, an altitude of 1,000 feet above the highest obstacle within a horizontal distance of 4 nautical miles from the course to be flown.

However, if both a MEA and a MOCA are prescribed for a particular route or route segment, a person may operate an aircraft below the MEA down to, but not below, the MOCA, when within 22 nautical miles of the VOR concerned (based on the pilot's reasonable estimate of that distance).

(b) *Climb.* Climb to a higher minimum IFR altitude shall begin immediately after passing the point beyond which that minimum altitude applies, except that when ground obstructions intervene, the point beyond which that higher minimum altitude applies shall be crossed at or above the applicable MCA.

§ 91.179 IFR cruising altitude or flight level.

(a) *In controlled airspace.* Each person operating an aircraft under IFR in level cruising flight in controlled airspace shall maintain the altitude or flight level assigned that aircraft by ATC. However, if the ATC clearance assigns "VFR conditions on-top," that person shall maintain an altitude or flight level as prescribed by § 91.159.

(b) *In uncontrolled airspace.* Except while in a holding pattern of 2 minutes or less or while turning, each person operating an aircraft under IFR in level cruising flight in uncontrolled airspace shall maintain an appropriate altitude as follows:

(1) When operating below 18,000 feet MSL and—

(i) On a magnetic course of zero degrees through 179 degrees, any odd thousand foot MSL altitude (such as 3,000, 5,000, or 7,000); or

(ii) On a magnetic course of 180 degrees through 359 degrees, any even thousand foot MSL altitude (such as 2,000, 4,000, or 6,000).

(2) When operating at or above 18,000 feet MSL but below flight level 290, and—

(i) On a magnetic course of zero degrees through 179 degrees, any odd flight level (such as 190, 210, or 230); or

(ii) On a magnetic course of 180 degrees through 359 degrees, any even flight level (such as 180, 200, or 220).

(3) When operating at flight level 290 and above in non-RVSM airspace, and—

(i) On a magnetic course of zero degrees through 179 degrees, any flight level, at 4,000-foot intervals, beginning at and including flight level 290 (such as flight level 290, 330, or 370); or

(ii) On a magnetic course of 180 degrees through 359 degrees, any flight level, at 4,000-foot intervals, beginning at and including flight level 310 (such as flight level 310, 350, or 390).

(4) When operating at flight level 290 and above in airspace designated as Reduced Vertical Separation Minimum (RVSM) airspace and—

(i) On a magnetic course of zero degrees through 179 degrees, any odd flight level, at 2,000-foot intervals beginning at and including flight level 290 (such as flight level 290, 310, 330, 350, 370, 390, 410); or

(ii) On a magnetic course of 180 degrees through 359 degrees, any even flight level, at 2000-foot intervals beginning at and including flight level 300 (such as 300, 320, 340, 360, 380, 400).

§ 91.180 Operations within airspace designated as Reduced Vertical Separation Minimum airspace.

(a) Except as provided in paragraph (b) of this section, no person may operate a civil aircraft in airspace designated as Reduced Vertical Separation Minimum (RVSM) airspace unless:

(1) The operator and the operator's aircraft comply with the minimum standards of appendix C of this part; and

(2) The operator is authorized by the Administrator or the country of registry to conduct such operations.

(b) The Administrator may authorize a deviation from the requirements of this section.

§ 91.181 Course to be flown.

Unless otherwise authorized by ATC, no person may operate an aircraft within controlled airspace under IFR except as follows:

(a) On a Federal airway, along the centerline of that airway.

(b) On any other route, along the direct course between the navigational aids or fixes defining that route. However, this section does not prohibit maneuvering the aircraft to pass well clear of other air traffic or the maneuvering of the aircraft in VFR conditions to clear the intended flight path both before and during climb or descent.

§ 91.183 IFR radio communications.

The pilot in command of each aircraft operated under IFR in controlled airspace shall have a continuous watch maintained on the appropriate frequency and shall report by radio as soon as possible—

(a) The time and altitude of passing each designated reporting point, or the reporting points specified by ATC, except that while the aircraft is under radar control, only the passing of those reporting points specifically requested by ATC need be reported;

(b) Any unforecast weather conditions encountered; and

(c) Any other information relating to the safety of flight.

§ 91.185 IFR operations: Two-way radio communications failure.

(a) *General.* Unless otherwise authorized by ATC, each pilot who has two-way radio communications failure when operating under IFR shall comply with the rules of this section.

(b) *VFR conditions.* If the failure occurs in VFR conditions, or if VFR conditions are encountered after the failure, each pilot shall continue the flight under VFR and land as soon as practicable.

(c) *IFR conditions.* If the failure occurs in IFR conditions, or if paragraph (b) of this section cannot be complied with, each pilot shall continue the flight according to the following:

(1) *Route.*

(i) By the route assigned in the last ATC clearance received;

(ii) If being radar vectored, by the direct route from the point of radio failure to the fix, route, or airway specified in the vector clearance;

(iii) In the absence of an assigned route, by the route that ATC has advised may be expected in a further clearance; or

(iv) In the absence of an assigned route or a route that ATC has advised may be expected in a further clearance, by the route filed in the flight plan.

(2) *Altitude.* At the highest of the following altitudes or flight levels for the route segment being flown:

(i) The altitude or flight level assigned in the last ATC clearance received;

(ii) The minimum altitude (converted, if appropriate, to minimum flight level as prescribed in § 91.121(c)) for IFR operations; or

(iii) The altitude or flight level ATC has advised may be expected in a further clearance.

(3) *Leave clearance limit.*

(i) When the clearance limit is a fix from which an approach begins, commence descent or descent and approach as close as possible to the expect-further-clearance time if one has been received, or if one has not been received, as close as possible to the estimated time of arrival as calculated from the filed or amended (with ATC) estimated time en route.

(ii) If the clearance limit is not a fix from which an approach begins, leave the clearance limit at the expect-further-clearance time if one has been received, or if none has been received, upon arrival over the clearance limit, and proceed to a fix from which an approach begins and commence descent or descent and approach as close as possible to the estimated time of arrival as calculated from the filed or amended (with ATC) estimated time en route.

§ 91.187 Operation under IFR in controlled airspace: Malfunction reports.

(a) The pilot in command of each aircraft operated in controlled airspace under IFR shall report as soon as practical to ATC any malfunctions of navigational, approach, or communication equipment occurring in flight.

(b) In each report required by paragraph (a) of this section, the pilot in command shall include the—

(1) Aircraft identification;

(2) Equipment affected;

(3) Degree to which the capability of the pilot to operate under IFR in the ATC system is impaired; and

(4) Nature and extent of assistance desired from ATC.

§ 91.189 Category II and III operations: General operating rules.

(a) No person may operate a civil aircraft in a Category II or III operation unless—

(1) The flight crew of the aircraft consists of a pilot in command and a second in command who hold the appropriate authorizations and ratings prescribed in § 61.3 of this chapter;

(2) Each flight crewmember has adequate knowledge of, and familiarity with, the aircraft and the procedures to be used; and

(3) The instrument panel in front of the pilot who is controlling the aircraft has appropriate instrumentation for the type of flight control guidance system that is being used.

(b) Unless otherwise authorized by the Administrator, no person may operate a civil aircraft in a Category II or

Category III operation unless each ground component required for that operation and the related airborne equipment is installed and operating.

(c) *Authorized DH.* For the purpose of this section, when the approach procedure being used provides for and requires the use of a DH, the authorized DH is the highest of the following:

(1) The DH prescribed by the approach procedure.

(2) The DH prescribed for the pilot in command.

(3) The DH for which the aircraft is equipped.

(d) Unless otherwise authorized by the Administrator, no pilot operating an aircraft in a Category II or Category III approach that provides and requires use of a DH may continue the approach below the authorized decision height unless the following conditions are met:

(1) The aircraft is in a position from which a descent to a landing on the intended runway can be made at a normal rate of descent using normal maneuvers, and where that descent rate will allow touchdown to occur within the touchdown zone of the runway of intended landing.

(2) At least one of the following visual references for the intended runway is distinctly visible and identifiable to the pilot:

(i) The approach light system, except that the pilot may not descend below 100 feet above the touchdown zone elevation using the approach lights as a reference unless the red terminating bars or the red side row bars are also distinctly visible and identifiable.

(ii) The threshold.

(iii) The threshold markings.

(iv) The threshold lights.

(v) The touchdown zone or touchdown zone markings.

(vi) The touchdown zone lights.

(e) Unless otherwise authorized by the Administrator, each pilot operating an aircraft shall immediately execute an appropriate missed approach whenever, prior to touchdown, the requirements of paragraph (d) of this section are not met.

(f) No person operating an aircraft using a Category III approach without decision height may land that aircraft except in accordance with the provisions of the letter of authorization issued by the Administrator.

(g) Paragraphs (a) through (f) of this section do not apply to operations conducted by certificate holders operating under part 121, 125, 129, or 135 of this chapter, or holders of management specifications issued in accordance

with subpart K of this part. Holders of operations specifications or management specifications may operate a civil aircraft in a Category II or Category III operation only in accordance with their operations specifications or management specifications, as applicable.

§ 91.191 Category II and Category III manual.

(a) Except as provided in paragraph (c) of this section, after August 4, 1997, no person may operate a U.S.-registered civil aircraft in a Category II or a Category III operation unless—

(1) There is available in the aircraft a current and approved Category II or Category III manual, as appropriate, for that aircraft;

(2) The operation is conducted in accordance with the procedures, instructions, and limitations in the appropriate manual; and

(3) The instruments and equipment listed in the manual that are required for a particular Category II or Category III operation have been inspected and maintained in accordance with the maintenance program contained in the manual.

(b) Each operator must keep a current copy of each approved manual at its principal base of operations and must make each manual available for inspection upon request by the Administrator.

(c) This section does not apply to operations conducted by a certificate holder operating under part 121 or part 135 of this chapter or a holder of management specifications issued in accordance with subpart K of this part.

(Approved by the Office of Management and Budget under OMB control number 2120-0005).

§ 91.193 Certificate of authorization for certain Category II operations.

The Administrator may issue a certificate of authorization authorizing deviations from the requirements of § 91.189, § 91.191, and § 91.205(f) for the operation of small aircraft identified as Category A aircraft in § 97.3 of this chapter in Category II operations if the Administrator finds that the proposed operation can be safely conducted under the terms of the certificate. Such authorization does not permit operation of the aircraft carrying persons or property for compensation or hire.

Subpart C—Equipment, instrument, and certificate requirements

§ 91.201 [Reserved]

§ 91.203 Civil aircraft: Certifications required.

(a) Except as provided in § 91.715, no person may operate a civil aircraft unless it has within it the following:

(1) An appropriate and current airworthiness certificate. Each U.S. airworthiness certificate used to comply with this subparagraph (except a special flight permit, a copy of the applicable operations specifications issued under § 21.197(c) of this chapter, appropriate sections of the air

carrier manual required by Parts 121 and 135 of this chapter containing that portion of the operations specifications issued under § 21.197(c), or an authorization under § 91.611) must have on it the registration number assigned to the aircraft under Part 47 of this chapter. However, the airworthiness certificate need not have on it an assigned special identification number before 10 days after that number is first affixed to the aircraft. A revised airworthiness certificate having on it an assigned special identification number, that has been affixed to an aircraft, may only be obtained upon application to an FAA Flight Standards district office.

(2) An effective U.S. registration certificate issued to its owner or, for operation within the United States, the second duplicate copy (pink) of the Aircraft Registration Application as provided for in § 47.31(b), or a registration certificate issued under the laws of a foreign country.

(b) No person may operate a civil aircraft unless the airworthiness certificate required by paragraph (a) of this section or a special flight authorization issued under § 91.715 is displayed at the cabin or cockpit entrance so that it is legible to passengers or crew.

(c) No person may operate an aircraft with a fuel tank installed within the passenger compartment or a baggage compartment unless the installation was accomplished pursuant to Part 43 of this chapter, and a copy of FAA Form 337 authorizing that installation is on board the aircraft.

[(d) No person may operate a civil airplane (domestic or foreign) into or out of an airport in the United States unless it complies with the fuel venting and exhaust emissions requirements of Part 34 of this chapter.]

[(Amdt. 91-218, Eff. 9/10/90)]

§ 91.205 Powered civil aircraft with standard category U.S. airworthiness certificates: Instrument and equipment requirements.

(a) *General.* Except as provided in paragraphs (c)(3) and (e) of this section, no person may operate a powered civil aircraft with a standard category U.S. airworthiness certificate in any operation described in paragraphs (b) through (f) of this section unless that aircraft contains the instruments and equipment specified in those paragraphs (or FAA-approved equivalents) for that type of operation, and those instruments and items of equipment are in operable condition.

(b) *Visual-flight rules (day).* For VFR flight during the day, the following instruments and equipment are required:

(1) Airspeed indicator.

(2) Altimeter.

(3) Magnetic direction indicator.

(4) Tachometer for each engine.

(5) Oil pressure gauge for each engine using pressure system.

(6) Temperature gauge for each liquid-cooled engine.

(7) Oil temperature gauge for each air-cooled engine.

(8) Manifold pressure gauge for each altitude engine.

(9) Fuel gauge indicating the quantity of fuel in each tank.

(10) Landing gear position indicator, if the aircraft has a retractable landing gear.

(11) For small civil airplanes certificated after March 11, 1996, in accordance with part 23 of this chapter, an approved aviation red or aviation white anticollision light system. In the event of failure of any light of the anticollision light systems, operation of the aircraft may continue to a location were repairs or replacement can be made.

(12) If the aircraft is operated for hire over water and beyond power-off gliding distance from shore, approved flotation gear readily available to each occupant and, unless the aircraft is operating under part 121 of this chapter, at least one pyrotechnic signaling device. As used in this section, "shore" means that area of the land adjacent to the water which is above the high water mark and excludes land areas which are intermittently under water.

(13) [An approved safety belt with an approved metal-to-metal latching device for each occupant 2 years of age or older.]

(14) For small civil airplanes manufactured after July 18, 1978, an approved shoulder harness for each front seat. The shoulder harness must be designed to protect the occupant from serious head injury when the occupant experiences the ultimate inertia forces specified in § 23.561(b)(2) of this chapter. Each shoulder harness installed at a flight crewmember station must permit the crewmember, when seated and with the safety belt and shoulder harness fastened, to perform all functions necessary for flight operations. For purposes of this paragraph—

(i) The date of manufacture of an airplane is the date the inspection acceptance records reflect that the airplane is complete and meets the FAA-approved type design data; and

(ii) A front seat is a seat located at a flight crewmember station or any seat located alongside such a seat.

(15) An emergency locator transmitter, if required by § 91.207.

(16) For normal, utility, and acrobatic category airplanes with a seating configuration, excluding pilot seats, of 9 or less, manufactured after December 12, 1986, a shoulder harness for—

(i) Each front seat that meets the requirements of § 23.785(g) and (h) of this chapter in effect on December 12, 1985;

(ii) Each additional seat that meets the requirements of § 23.785(g) of this chapter in effect on December 12, 1985.

(17) For rotorcraft manufactured after September 16, 1992, a shoulder harness for each seat that meets the requirements of § 27.2 or § 29.2 of this chapter in effect on September 16, 1991.

(c) *Visual flight rules (night).* For VFR flight at night, the following instruments and equipment are required:

(1) Instruments and equipment specified in paragraph (b) of this section.

(2) Approved position lights.

(3) An approved aviation red or aviation white anticollision light system on all U.S.-registered civil aircraft. Anticollision light systems initially installed after August 11, 1971, on aircraft for which a type certificate was issued or applied for before August 11, 1971, must at least meet the anticollision light standards of Part 23, 25, 27, or 29 of this chapter, as applicable, that were in effect on August 10, 1971, except that the color may be either aviation red or aviation white. In the event of failure of any light of the anticollision light system, operations with the aircraft may be continued to a stop where repairs or replacement can be made.

(4) If the aircraft is operated for hire, one electric landing light.

(5) An adequate source of electrical energy for all installed electrical and radio equipment.

(6) One spare set of fuses, or three spare fuses of each kind required, that are accessible to the pilot in flight.

(d) *Instrument flight rules.* For IFR flight, the following instruments and equipment are required:

(1) Instruments and equipment specified in paragraph (b) of this section, and, for night flight, instruments and equipment specified in paragraph (c) of this section.

(2) Two-way radio communications system and navigational equipment appropriate to the ground facilities to be used.

(3) Gyroscopic rate-of-turn indicator, except on the following aircraft:

(i) Airplanes with a third attitude instrument system usable through flight attitudes of 360 degrees of pitch and roll and installed in accordance with the instrument requirements prescribed in § 121.305(j) of this chapter; and:

(ii) Rotorcraft with a third attitude instrument system usable through flight attitudes of ±80 degrees of pitch and ±120 degrees of roll and installed in accordance with § 29.1303(g) of this chapter.

(4) Slip-skid indicator.

(5) Sensitive altimeter adjustable for barometric pressure.

(6) A clock displaying hours, minutes, and seconds with a sweep-second pointer or digital presentation.

(7) Generator or alternator of adequate capacity.

(8) Gyroscopic pitch and bank indicator (artificial horizon).

(9) Gyroscopic direction indicator (directional gyro or equivalent).

(e) *Flight at and above 24,000 ft. MSL (FL 240).* If VOR navigational equipment is required under paragraph (d)(2) of this section, no person may operate a U.S.-registered civil aircraft within the 50 states and the District of Columbia at or above FL 240 unless that aircraft is equipped with approved distance measuring equipment (DME). When DME required by this paragraph fails at and above FL 240, the pilot in command of the aircraft shall notify ATC immediately, and then may continue operations at and above FL 240 to the next airport of intended landing at which repairs or replacement of the equipment can be made.

(f) *Category II operations.* The requirements for Category II operations are the instruments and equipment specified in—

(1) Paragraph (d) of this section; and

(2) Appendix A to this part.

(g) *Category III operations.* The instruments and equipment required for Category III operations are specified in paragraph (d) of this section.

(h) *Exclusions.* Paragraphs (f) and (g) of this section do not apply to operations conducted by a holder of a certificate issued under part 121 or part 135 of this chapter.

(Amdt. 91-220, Eff. 11/26/90); (Amdt. 91-223, Eff. 9/16/91); [(Amdt. 91-231, Eff. 10/15/92)]

§ 91.207 Emergency locator transmitters.

(a) Except as provided in paragraphs (e) and (f) of this section, no person may operate a U.S.-registered civil airplane unless—

(1) There is attached to the airplane an approved automatic type emergency locator transmitter that is in operable condition for the following operations, except that after June 21, 1995, an emergency locator transmitter that meets the requirements of TSO-C91 may not be used for new installations:

(i) Those operations governed by the supplemental air carrier and commercial operator rules of Parts 121 and 125;

(ii) Charter flights governed by the domestic and flag air carrier rules of Part 121 of this chapter; and

(iii) Operations governed by Part 135 of this chapter; or

(2) For operations other than those specified in paragraph (a)(1) of this section, there must be attached to the airplane an approved personal type or an approved automatic type emergency locator transmitter that is in operable condition, except that after June 21, 1995, an emergency locator transmitter that meets the requirements of TSO-C91 may not be used for new installations.

(b) Each emergency locator transmitter required by paragraph (a) of this section must be attached to the airplane in such a manner that the probability of damage to the transmitter in the event of crash impact is minimized. Fixed and deployable automatic type transmitters must be attached to the airplane as far aft as practicable.

(c) Batteries used in the emergency locator transmitters required by paragraphs (a) and (b) of this section must be replaced (or recharged, if the batteries are rechargeable)—

(1) When the transmitter has been in use for more than 1 cumulative hour; or

(2) When 50 percent of their useful life (or, for rechargeable batteries, 50 percent of their useful life of charge) has expired, as established by the transmitter manufacturer under its approval.

The new expiration date for replacing (or recharging) the battery must be legibly marked on the outside of the transmitter and entered in the aircraft maintenance record. Paragraph (c)(2) of this section does not apply to batteries (such as water-activated batteries) that are essentially unaffected during probable storage intervals.

(d) Each emergency locator transmitter required by paragraph (a) of this section must be inspected within 12 calendar months after the last inspection for—

(1) Proper installation;

(2) Battery corrosion;

(3) Operation of the controls and crash sensor; and

(4) The presence of a sufficient signal radiated from its antenna.

(e) Notwithstanding paragraph (a) of this section, a person may—

(1) Ferry a newly acquired airplane from the place where possession of it was taken to a place where the emergency locator transmitter is to be installed; and

(2) Ferry an airplane with an inoperative emergency locator transmitter from a place where repairs or replacements cannot be made to a place where they can be made.

No person other than required crewmembers may be carried aboard an airplane being ferried under paragraph (e) of this section.

(f) Paragraph (a) of this section does not apply to—

(1) Before January 1, 2004, turbojet-powered aircraft;

(2) Aircraft while engaged in scheduled flights by scheduled air carriers;

(3) Aircraft while engaged in training operations conducted entirely within a 50-nautical mile radius of the airport from which such local flight operations began;

(4) Aircraft while engaged in flight operations incident to design and testing;

(5) New aircraft while engaged in flight operations incident to their manufacture, preparation, and delivery;

(6) Aircraft while engaged in flight operations incident to the aerial application of chemicals and other substances for agricultural purposes;

(7) Aircraft certificated by the Administrator for research and development purposes;

(8) Aircraft while used for showing compliance with regulations, crew training, exhibition, air racing, or market surveys;

(9) Aircraft equipped to carry not more than one person.

(10) An aircraft during any period for which the transmitter has been temporarily removed for inspection, repair, modification, or replacement, subject to the following:

(i) No person may operate the aircraft unless the aircraft records contain an entry which includes the date of initial removal, the make, model, serial number, and reason for removing the transmitter, and a placard located in view of the pilot to show "ELT not installed."

(ii) No person may operate the aircraft more than 90 days after the ELT is initially removed from the aircraft

and, (11) On and after January 1, 2004, aircraft with a maximum payload of more than 18,000 pounds when used in air transportation..

§ 91.209 Aircraft lights.

No person may:

(a) During the period from sunset to sunrise (or, in Alaska, during the period a prominent unlighted object cannot be seen from a distance of 3 statute miles or the the sun is more than 6 degrees below the horizon)—

(1) Operate an aircraft unless it has lighted position lights;

(2) Park or move an aircraft in, or in dangerous proximity to, a night flight operations area of an airport unless the aircraft—

(i) Is clearly illuminated;

(ii) Has lighted position lights: or

(iii) Is in an area that is marked by obstruction lights;

(3) Anchor an aircraft unless the aircraft—

(i) Has lighted anchor lights; or

(ii) Is in an area where anchor lights are not required on vessels; or

(b) Operate an aircraft that is equipped with an anticollisions light system, unlesss it has lighted anticollision lights. However, the anticollision lights need not be lighted when the pilot-in-command determines that, because of operating conditions, it would be in the interest of safety to turn the lights off.

§ 91.211 Supplemental oxygen.

(a) *General.* No person may operate a civil aircraft of U.S. registry—

(1) At cabin pressure altitudes above 12,500 feet (MSL) up to and including 14,000 feet (MSL) unless the required minimum flight crew is provided with and uses supplemental oxygen for that part of the flight at those altitudes that is of more than 30 minutes duration;

(2) At cabin pressure altitudes above 14,000 feet (MSL) unless the required minimum flight crew is provided with and uses supplemental oxygen during the entire flight time at those altitudes; and

(3) At cabin pressure altitudes above 15,000 feet (MSL) unless each occupant of the aircraft is provided with supplemental oxygen.

(b) *Pressurized cabin aircraft.*

(1) No person may operate a civil aircraft of U.S. registry with a pressurized cabin—

(i) At flight altitudes above flight level 250 unless at least a 10-minute supply of supplemental oxygen, in addition to any oxygen required to satisfy paragraph (a) of this section, is available for each occupant of the aircraft for use in the event that a descent is necessitated by loss of cabin pressurization; and

(ii) At flight altitudes above flight level 350 unless one pilot at the controls of the airplane is wearing and using an oxygen mask that is secured and sealed and that either supplies oxygen at all times or automatically supplies oxygen whenever the cabin pressure altitude of the airplane exceeds 14,000 feet (MSL), except that the one pilot need not wear and use an oxygen mask while at or below flight level 410 if there are two pilots at the controls and each pilot has a quick-donning type of oxygen mask that can be placed on the face with one hand from the ready position within 5 seconds, supplying oxygen and properly secured and sealed.

(2) Notwithstanding paragraph (b)(1)(ii) of this section, if for any reason at any time it is necessary for one pilot to leave the controls of the aircraft when operating at flight altitudes above flight level 350, the remaining pilot at the controls shall put on and use an oxygen mask until the other pilot has returned to that crewmember's station.

§ 91.213 Inoperative instruments and equipment.

(a) Except as provided in paragraph (d) of this section, no person may take off an aircraft with inoperative instruments or equipment installed unless the following conditions are met:

(1) An approved Minimum Equipment List exists for that aircraft.

(2) The aircraft has within it a letter of authorization, issued by the FAA Flight Standards district office having jurisdiction over the area in which the operator is located, authorizing operation of the aircraft under the Minimum Equipment List. The letter of authorization may be obtained by written request of the airworthiness certificate holder. The Minimum Equipment List and the letter of authorization constitute a supplemental type certificate for the aircraft.

(3) The approved Minimum Equipment List must—

(i) Be prepared in accordance with the limitations specified in paragraph (b) of this section; and

(ii) Provide for the operation of the aircraft with the instruments and equipment in an inoperable condition,

(4) The aircraft records available to the pilot must include an entry describing the inoperable instruments and equipment.

(5) The aircraft is operated under all applicable conditions and limitations contained in the Minimum Equipment List and the letter authorizing the use of the list.

(b) The following instruments and equipment may not be included in a Minimum Equipment List:

(1) Instruments and equipment that are either specifically or otherwise required by the airworthiness requirements under which the aircraft is type certificated and which are essential for safe operations under all operating conditions.

(2) Instruments and equipment required by an airworthiness directive to be in operable condition unless the airworthiness directive provides otherwise.

(3) Instruments and equipment required for specific operations by this part.

(c) A person authorized to use an approved Minimum Equipment List issued for a specific aircraft under subpart K of this part, part 121, 125, or 135 of this chapter must use that Minimum Equipment List to comply with the requirements in this section.

(d) Except for operations conducted in accordance with paragraph (a) or (c) of this section, a person may take off an aircraft in operations conducted under this part with inoperative instruments and equipment without an approved Minimum Equipment List provided—

(1) The flight operation is conducted in a—

(i) Rotorcraft, non-turbine-powered airplane, glider, lighter-than-air aircraft, powered parachute, or weight-shift-control aircraft, for which a master minimum equipment list has not been developed; or

(ii) Small rotorcraft, nonturbine-powered small airplane, glider, or lighter-than-air aircraft for which a Master Minimum Equipment List has been developed; and

(2) The inoperative instruments and equipment are not—

(i) Part of the VFR-day type certification instruments and equipment prescribed in the applicable airworthiness regulations under which the aircraft was type certificated;

(ii) Indicated as required on the aircraft's equipment list, or on the Kinds of Operations Equipment List for the kind of flight operation being conducted;

(iii) Required by § 91.205 or any other rule of this part for the specific kind of flight operation being conducted; or

(iv) Required to be operational by an airworthiness directive; and

(3) The inoperative instruments and equipment are—

(i) Removed from the aircraft, the cockpit control placarded, and the maintenance recorded in accordance with § 43.9 of this chapter; or

(ii) Deactivated and placarded "Inoperative." If deactivation of the inoperative instrument or equipment involves maintenance, it must be accomplished and recorded in accordance with Part 43 of this chapter; and

(4) A determination is made by a pilot, who is certificated and appropriately rated under Part 61 of this chapter, or by a person, who is certificated and appropriately rated to perform maintenance on the aircraft, that the inoperative instrument or equipment does not constitute a hazard to the aircraft. An aircraft with inoperative instruments or equipment as provided in paragraph (d) of this section is considered to be in a properly altered condition acceptable to the Administrator.

(e) Notwithstanding any other provision of this section, an aircraft with inoperable instruments or equipment may be operated under a special flight permit issued in accordance with § 21.197 and § 21.199 of this chapter.

§ 91.215 ATC transponder and altitude reporting equipment and use.

(a) *All airspace: U.S.-registered civil aircraft.* [For operations not conducted under part 121, 127 or 135 of this chapter, ATC transponder equipment installed must meet the performance and environmental requirements of any class of TSO-C74b (Mode A) or any class of TSO-C74c (Mode A with altitude reporting capability) as appropriate, or the appropriate class of TSO-C112 (Mode S).]

(b) *All airspace.* Unless otherwise authorized or directed by ATC, no person may operate an aircraft in the airspace described in paragraphs (b)(1) through (b)(5) of this section, unless that aircraft is equipped with an operable coded radar beacon transponder having either Mode 3/A 4096 code capability, replying to Mode 3/A interrogations with the code specified by ATC, or a Mode S capability, replying to Mode 3/A interrogations with the code specified by ATC and intermode and Mode S interrogations in accordance with the applicable provisions specified in TSO C-112, and that aircraft is equipped with automatic pressure altitude reporting equipment having a Mode C capability that automatically replies to Mode C interrogations by transmitting pressure altitude information in 100-foot increments. This requirement applies—

(1) All aircraft. In Class A, Class B, and Class C airspace areas;

(2) All aircraft. In all airspace within 30 nautical miles of an airport listed in Appendix D, Section 1 of this part from the surface upward to 10,000 feet MSL;

(3) Notwithstanding paragraph (b)(2) of this section, any aircraft which was not originally certificated with an engine-driven electrical system or which has not subsequently been certified with such a system installed, balloon or glider may conduct operations in the airspace within 30 nautical miles of an airport listed in Appendix D, Section 1 of this part provided such operations are conducted—

(i) Outside any Class A, Class B, or Class C airspace area; and

(ii) Below the altitude of the ceiling of a Class B or Class C airspace area designated for an airport or 10,000 feet MSL, whichever is lower; and

(4) All aircraft in all airspace above the ceiling and within the lateral boundaries of a Class B or Class C airspace area designated for an airport upward to 10,000 feet MSL; and

(5) All aircraft except any aircraft which was not originally certificated with an engine-driven electrical system or which has not subsequently been certified with such a system installed, balloon, or glider—

(i) In all airspace of the 48 contiguous states and the District of Columbia at and above 10,000 feet MSL, excluding the airspace at and below 2,500 feet above the surface; and

(ii) In the airspace from the surface to 10,000 feet MSL within a 10-nautical-mile radius of any airport listed in Appendix D, Section 2 of this part, excluding the airspace below 1,200 feet outside of the lateral boundaries of the surface area of the airspace designated for that airport.

(c) *Transponder-on operation.* While in the airspace as specified in paragraph (b) of this section or in all controlled airspace, each person operating an aircraft equipped with an operable ATC transponder maintained in accordance with § 91.413 of this part shall operate the transponder, including Mode C equipment if installed, and shall reply on the appropriate code or as assigned by ATC.

(d) *ATC authorized deviations.* Requests for ATC authorized deviations must be made to the ATC facility having jurisdiction over the concerned airspace within the time periods specified as follows:

(1) For operation of an aircraft with an operating transponder but without operating automatic pressure altitude reporting equipment having a Mode C capability, the request may be made at any time.

(2) For operation of an aircraft with an inoperative transponder to the airport of ultimate destination, including any intermediate stops, or to proceed to a place where suitable repairs can be made or both, the request may be made at any time.

(3) For operation of an aircraft that is not equipped with a transponder, the request must be made at least one hour before the proposed operation.

(Amdt. 91-221, Eff. 1/4/91); (Amdt. 91-227, Eff. 12/12/91); [(*Amdt. 91-229, Eff. 7/30/92)*]

§ 91.217 Data correspondence between automatically reported pressure altitude data and the pilot's altitude reference.

No person may operate any automatic pressure altitude reporting equipment associated with a radar beacon transponder—

(a) When deactivation of that equipment is directed by ATC;

(b) Unless, as installed, that equipment was tested and calibrated to transmit altitude data corresponding within 125 feet (on a 95 percent probability basis) of the indicated or calibrated datum of the altimeter normally used to maintain flight altitude, with that altimeter referenced to 29.92" of mercury for altitudes from sea level to the maximum operating altitude of the aircraft; or

(c) Unless the altimeters and digitizers in that equipment meet the standards of TSO C10b and TSO C88, respectively.

§ 91.219 Altitude alerting system or device: Turbojet-powered civil airplanes.

(a) Except as provided in paragraph (d) of this section, no person may operate a turbojet-powered U.S.-registered civil airplane unless that airplane is equipped with an approved altitude alerting system or device that is in operable condition and meets the requirements of paragraph (b) of this section.

(b) Each altitude alerting system or device required by paragraph (a) of this section must be able to—

(1) Alert the pilot—

(i) Upon approaching a preselected altitude in either ascent or descent, by a sequence of both aural and visual signals in sufficient time to establish level flight at that preselected altitude; or

(ii) Upon approaching a preselected altitude in either ascent or descent, by a sequence of visual signals in sufficient time to establish level flight at that preselected altitude, and when deviating above and below that preselected altitude, by an aural signal;

(2) Provide the required signals from sea level to the highest operating altitude approved for the airplane in which it is installed;

(3) Preselect altitudes in increments that are commensurate with the altitudes at which the aircraft is operated;

(4) Be tested without special equipment to determine proper operation of the alerting signals; and

(5) Accept necessary barometric pressure settings if the system or device operates on barometric pressure.

However, for operation below 3,000 feet AGL, the system or device need only provide one signal, either visual or aural, to comply with this paragraph. A radio altimeter may be included to provide the signal if the operator has an approved procedure for its use to determine DH or MDA, as appropriate.

(c) Each operator to which this section applies must establish and assign procedures for the use of the altitude alerting system or device and each flight crewmember must comply with those procedures assigned to him.

(d) Paragraph (a) of this section does not apply to any operation of an airplane that has an experimental certificate or to the operation of any airplane for the following purposes:

(1) Ferrying a newly acquired airplane from the place where possession of it was taken to a place where the altitude alerting system or device is to be installed.

(2) Continuing a flight as originally planned, if the altitude alerting system or device becomes inoperative after the airplane has taken off; however, the flight may not depart from a place where repair or replacement can be made.

(3) Ferrying an airplane with any inoperative altitude alerting system or device from a place where repairs or replacements cannot be made to a place where it can be made.

(4) Conducting an airworthiness flight test of the airplane.

(5) Ferrying an airplane to a place outside the United States for the purpose of registering it in a foreign country.

(6) Conducting a sales demonstration of the operation of the airplane.

(7) Training foreign flight crews in the operation of the airplane before ferrying it to a place outside the United States for the purpose of registering it in a foreign country.

§ 91.221 Traffic alert and collision avoidance system equipment and use.

(a) *All airspace: U.S.-registered civil aircraft.* Any traffic alert and collision avoidance system installed in a U.S.-registered civil aircraft must be approved by the Administrator.

(b) *Traffic alert and collision avoidance system, operation required.* Each person operating an aircraft equipped with an operable traffic alert and collision avoidance system shall have that system on and operating.

§ 91.223 Terrain awareness and warning system.

(a) *Airplanes manufactured after March 29, 2002.* Except as provided in paragraph (d) of this section, no person may operate a turbine-powered U.S.-registered airplane configured with six or more passenger seats, excluding any pilot seat, unless that airplane is equipped with an approved terrain awareness and warning system that as a minimum meets the requirements for Class B equipment in Technical Standard Order (TSO)–C151.

(b) *Airplanes manufactured on or before March 29, 2002.* Except as provided in paragraph (d) of this section, no person may operate a turbine-powered U.S.-registered airplane configured with six or more passenger seats, excluding any pilot seat, after March 29, 2005, unless that airplane is equipped with an approved terrain awareness and warning system that as a minimum meets the requirements for Class B equipment in Technical Standard Order (TSO)–C151.

(c) *Airplane Flight Manual.* The Airplane Flight Manual shall contain appropriate procedures for—

(1) The use of the terrain awareness and warning system; and

(2) Proper flight crew reaction in response to the terrain awareness and warning system audio and visual warnings.

(d) *Exceptions.* Paragraphs (a) and (b) of this section do not apply to—

(1) Parachuting operations when conducted entirely within a 50 nautical mile radius of the airport from which such local flight operations began.

(2) Firefighting operations.

(3) Flight operations when incident to the aerial application of chemicals and other substances.

§§ 91.225–9.229 [Reserved]

Subpart D—Special flight operations

§ 91.301 [Reserved]

§ 91.303 Aerobatic flight.

No person may operate an aircraft in aerobatic flight—

(a) Over any congested area of a city, town, or settlement;

(b) Over an open air assembly of persons;

(c) [*Within the lateral boundaries of the surface areas of Class B, Class C, Class D, or Class E airspace designated for an airport;*]

(d) [*Within 4 nautical miles of the center line of any Federal airway;*]

(e) [*Below an altitude of 1,500 feet above the surface; or*]

[(*f*) *When flight visibility is less than 3 statute miles. For the purposes of this section, aerobatic flight means an intentional maneuver involving an abrupt change in an aircraft's attitude, an abnormal attitude, or abnormal acceleration, not necessary for normal flight.*]

For the purposes of this section, aerobatic flight means an intentional maneuver involving an abrupt change in an aircraft's attitude, an abnormal attitude, or abnormal acceleration, not necessary for normal flight.

[*(Amdt. 91-227, Eff. 9/16/93)*]

§ 91.305 Flight test areas.

No person may flight test an aircraft except over open water, or sparsely populated areas, having light air traffic.

§ 91.307 Parachutes and parachuting.

(a) No pilot of a civil aircraft may allow a parachute that is available for emergency use to be carried in that aircraft unless it is an approved type and—

(1) If a chair type (canopy in back), it has been packed by a certificated and appropriately rated parachute rigger within the preceding 120 days; or

(2) If any other type, it has been packed by a certificated and appropriately rated parachute rigger—

(i) Within the preceding 120 days, if its canopy, shrouds, and harness are composed exclusively of nylon, rayon, or other similar synthetic fiber or materials that are substantially resistant to damage from mold, mildew, or other fungi and other rotting agents propagated in a moist environment; or

(ii) Within the preceding 60 days, if any part of the parachute is composed of silk, pongee, or other natural fiber, or materials not specified in paragraph (a)(2)(i) of this section.

(b) Except in an emergency, no pilot in command may allow, and no person may conduct a parachute jump from an aircraft within the United States except in accordance with Part 105.

(c) Unless each occupant of the aircraft is wearing an approved parachute, no pilot of a civil aircraft carrying any person (other than a crewmember) may execute any intentional maneuver that exceeds—

(1) A bank of 60 degrees relative to the horizon; or

(2) A nose-up or nose-down attitude of 30 degrees relative to the horizon.

(d) Paragraph (c) of this section does not apply to—

(1) Flight tests for pilot certification or rating; or

(2) Spins and other flight maneuvers required by the regulations for any certificate or rating when given by—

(i) A certificated flight instructor; or

(ii) An airline transport pilot instructing in accordance with §61.67 of this chapter.

(e) For the purposes of this section, "approved parachute" means—

(1) A parachute manufactured under a type certificate or a technical standard order (C-23 series); or

(2) A personnel-carrying military parachute identified by an NAF, AAF, or AN drawing number, an AAF order number, or any other military designation or specification number.

§ 91.309 Towing: Gliders.

(a) No person may operate a civil aircraft towing glider or unpowered ultralight vehicle unless—

(1) The pilot in command of the towing aircraft is qualified under § 61.69 of this chapter;

(2) The towing aircraft is equipped with a tow-hitch of a kind, and installed in a manner, that is approved by the Administrator;

(3) The towline used has breaking strength not less than 80 percent of the maximum certificated operating weight of the glider or unpowered ultralight vehicle and not more than twice this operating weight. However, the towline used may have a breaking strength more than twice the maximum certificated operating weight of the glider or unpowered ultralight vehicle if—

(i) A safety link is installed at the point of attachment of the towline to the glider or unpowered ultralight vehicle with a breaking strength not less than 80 percent of the maximum certificated operating weight of the glider or unpowered ultralight vehicle and not greater than twice this operating weight;

(ii) A safety link is installed at the point of attachment of the towline to the towing aircraft with a breaking strength greater, but not more than 25 percent greater, than that of the safety link at the towed glider or unpowered ultralight vehicle end of the towline and not greater than twice the maximum certificated operating weight of the glider or unpowered ultralight vehicle;

(4) [*Before conducting any towing operation within the lateral boundaries of the surface areas of Class B, Class C, Class D, or Class E airspace designated for an airport, or before making each towing flight within such controlled airspace if required by ATC, the pilot in command notifies the control tower. If a control tower does*

not exist or is not in operation, the pilot in command must notify the FAA flight service station serving that controlled airspace before conducting any towing operations in that airspace; and]

(5) The pilots of the towing aircraft and the glider or unpowered ultralight vehicle have agreed upon a general course of action, including takeoff and release signals, airspeeds, and emergency procedures for each pilot.

(b) No pilot of a civil aircraft may intentionally release a towline, after release of a glider or unpowered ultralight vehicle, in a manner that endangers the life or property of another.

[*(Amdt. 91-227, Eff. 9/16/93)*]

§ 91.311 Towing: Other than under § 91.309

No pilot of a civil aircraft may tow anything with that aircraft (other than under § 91.309) except in accordance with the terms of a certificate of waiver issued by the Administrator.

§ 91.313 Restricted category civil aircraft: Operating limitations.

(a) No person may operate a restricted category civil aircraft—

(1) For other than the special purpose for which it is certificated; or

(2) In an operation other than one necessary to accomplish the work activity directly associated with that special purpose.

(b) For the purpose of paragraph (a) of this section, operating a restricted category civil aircraft to provide flight crewmember training in a special purpose operation for which the aircraft is certificated is considered to be an operation for that special purpose.

(c) No person may operate a restricted category civil aircraft carrying persons or property for compensation or hire. For the purposes of this paragraph, a special purpose operation involving the carriage of persons or material necessary to accomplish that operation, such as crop dusting, seeding, spraying, and banner towing (including the carrying of required persons or material to the location of that operation), and operation for the purpose of providing flight crewmember training in a special purpose operation, are not considered to be the carriage of persons or property for compensation or hire.

(d) No person may be carried on a restricted category civil aircraft unless that person—

(1) Is a flight crewmember;

(2) Is a flight crewmember trainee;

(3) Performs an essential function in connection with a special purpose operation for which the aircraft is certificated; or

(4) Is necessary to accomplish the work activity directly associated with that special purpose.

(e) Except when operating in accordance with the terms and conditions of a certificate of waiver or special operating limitations issued by the Administrator, no person may operate a restricted category civil aircraft within the United States—

(1) Over a densely populated area;

(2) In a congested airway; or

(3) Near a busy airport where passenger transport operations are conducted.

(f) This section does not apply to nonpassenger-carrying civil rotorcraft external-load operations conducted under Part 133 of this chapter.

(g) No person may operate a small restricted-category civil airplane manufactured after July 18, 1978, unless an approved shoulder harness is installed for each front seat. The shoulder harness must be designed to protect each occupant from serious head injury when the occupant experiences the ultimate inertia forces specified in § 23.561 (b)(2) of this chapter. The shoulder harness installation at each flight crewmember station must permit the crewmember, when seated and with the safety belt and shoulder harness fastened, to perform all functions necessary for flight operation. For purposes of this paragraph—

(1) The date of manufacture of an airplane is the date the inspection acceptance records reflect that the airplane is complete and meets the FAA-approved type design data; and

(2) A front seat is a seat located at a flight crewmember station or any seat located alongside such a seat.

§ 91.315 Limited category civil aircraft: Operating limitations.

No person may operate a limited category civil aircraft carrying persons or property for compensation or hire.

§ 91.317 Provisionally certificated civil aircraft: Operating limitations.

(a) No person may operate a provisionally certificated civil aircraft unless that person is eligible for a provisional airworthiness certificate under § 21.213 of this chapter.

(b) No person may operate a provisionally certificated civil aircraft outside the United States unless that person has specific authority to do so from the Administrator and each foreign country involved.

(c) Unless otherwise authorized by the Director, Flight Standards Service, no person may operate a provisionally certificated civil aircraft in air transportation.

(d) Unless otherwise authorized by the Administrator, no person may operate a provisionally certificated civil aircraft except—

(1) In direct conjunction with the type or supplemental type certification of that aircraft;

(2) For training flight crews, including simulated air carrier operations;

(3) Demonstration flight by the manufacturer for prospective purchasers;

(4) Market surveys by the manufacturer;

(5) Flight checking of instruments, accessories, and equipment that do not affect the basic airworthiness of the aircraft; or

(6) Service testing of the aircraft.

(e) Each person operating a provisionally certificated civil aircraft shall operate within the prescribed limitations displayed in the aircraft or set forth in the provisional aircraft flight manual or other appropriate document. However, when operating in direct conjunction with the type or supplemental type certification of the aircraft, that person shall operate under the experimental aircraft limitations of § 21.191 of this chapter and when flight testing, shall operate under the requirement of § 91.305 of this part.

(f) Each person operating a provisionally certificated civil aircraft shall establish approved procedures for—

(1) The use and guidance of flight and ground personnel in operating under this section; and

(2) Operating in and out of airports where takeoffs or approaches over populated areas are necessary. No person may operate that aircraft except in compliance with the approved procedures.

(g) Each person operating a provisionally certificated civil aircraft shall ensure that each flight crewmember is properly certificated and has adequate knowledge of, and familiarity with, the aircraft and procedures to be used by that crewmember.

(h) Each person operating a provisionally certificated civil aircraft shall maintain it as required by applicable regulations and as may be specially prescribed by the Administrator.

(i) Whenever the manufacturer, or the Administrator, determines that a change in design, construction, or operation is necessary to ensure safe operation, no person may operate a provisionally certificated civil aircraft until that change has been made and approved. Section § 21.99 of this chapter applies to operations under this section.

(j) Each person operating a provisionally certificated civil aircraft—

(1) May carry in that aircraft only persons who have a proper interest in the operations allowed by this section or who are specifically authorized by both the manufacturer and the Administrator; and

(2) Shall advise each person carried that the aircraft is provisionally certificated.

(k) The Administrator may prescribe additional limitations or procedures that the Administrator considers necessary, including limitations on the number of persons who may be carried in the aircraft.

(Approved by the Office of Management and Budget under OMB control number 2120-0005)

(Amdt. 91-212, Eff. 8/18/90)

§ 91.319 Aircraft having experimental certificates: Operating limitations.

(a) No person may operate an aircraft that has an experimental certificate—

(1) For other than the purpose for which the certificate was issued; or

(2) Carrying persons or property for compensation or hire.

(b) No person may operate an aircraft that has an experimental certificate outside of an area assigned by the Administrator until it is shown that—

(1) The aircraft is controllable throughout its normal range of speeds and throughout all the maneuvers to be executed; and

(2) The aircraft has no hazardous operating characteristics or design features.

(c) Unless otherwise authorized by the Administrator in special operating limitations, no person may operate an aircraft that has an experimental certificate over a densely populated area or in a congested airway. The Administrator may issue special operating limitations for particular aircraft to permit takeoffs and landings to be conducted over a densely populated area or in a congested airway, in accordance with terms and conditions specified in the authorization in the interest of safety in air commerce.

(d) Each person operating an aircraft that has an experimental certificate shall—

(1) Advise each person carried of the experimental nature of the aircraft;

(2) Operate under VFR, day only, unless otherwise specifically authorized by the Administrator; and

(3) Notify the control tower of the experimental nature of the aircraft when operating the aircraft into or out of airports with operating control towers.

(e) No person may operate an aircraft that is issued an experimental certificate under §21.191 (i) of this chapter for compensation or hire, except a person may operate an aircraft issued an experimental certificate under §21.191 (i)(1) for compensation or hire to—

(1) Tow a glider that is a light-sport aircraft or unpowered ultralight vehicle in accordance with §91.309; or

(2) Conduct flight training in an aircraft which that person provides prior to January 31, 2010.

(f) No person may lease an aircraft that is issued an experimental certificate under §21.191 (i) of this chapter, except in accordance with paragraph (e)(1) of this section.

(g) No person may operate an aircraft issued an experimental certificate under §21.191 (i)(1) of this chapter to tow a glider that is a light-sport aircraft or unpowered ultralight vehicle for compensation or hire or to conduct flight training for compensation or hire in an aircraft which that persons provides unless within the preceding 100 hours of time in service the aircraft has—

(1) Been inspected by a certificated repairman (light-sport aircraft) with a maintenance rating, an appropriately

rated mechanic, or an appropriately rated repair station in accordance with inspection procedures developed by the aircraft manufacturer or a person acceptable to the FAA; or

(2) Received an inspection for the issuance of an airworthiness certificate in accordance with part 21 of this chapter.

(h) The FAA may issue deviation authority providing relief from the provisions of paragraph (a) of this section for the purpose of conducting flight training. The FAA will issue this deviation authority as a letter of deviation authority.

(1) The FAA may cancel or amend a letter of deviation authority at any time.

(2) An applicant must submit a request for deviation authority to the FAA at least 60 days before the date of intended operations. A request for deviation authority must contain a complete description of the proposed operation and justification that establishes a level of safety equivalent to that provided under the regulations for the deviation requested.

(i) The Administrator may prescribe additional limitations that the Administrator considers necessary, including limitations on the persons that may be carried in the aircraft.

(Approved by the Office of Management and Budget under OMB control number 2120-0005).

§ 91.321 Carriage of candidates in elections.

(a) As an aircraft operator, you may receive payment for carrying a candidate, agent of a candidate, or person traveling on behalf of a candidate, running for Federal, State, or local election, without having to comply with the rules in parts 121, 125 or 135 of this chapter, under the following conditions:

(1) Your primary business is not as an air carrier or commercial operator;

(2) You carry the candidate, agent, or person traveling on behalf of a candidate, under the rules of a part 91; and

(3) By Federal, state or local law, you are required to receive payment for carrying the candidate, agent, or person traveling on behalf of a candidate. For federal elections, the payment may not exceed the amount required by the Federal Election Commission. For a state or local election, the payment may not exceed the amount required under the applicable state or local law.

(b) For the purposes of this section, for Federal elections, the terms *candidate* and *election* have the same meaning as set forth in the regulations of the Federal Election Commission. For State or local elections, the terms *candidate* and *election* have the same meaning as provided by the applicable State or local law and those terms relate to candidates for election to public office in State and local government elections.

§ 91.323 Increased maximum certificated weights for certain airplanes operated in Alaska.

(a) Notwithstanding any other provision of the Federal Aviation Regulations, the Administrator will approve, as provided in this section, an increase in the maximum certificated weight of an airplane type certificated under Aeronautics Bulletin No. 7 A of the U.S. Department of Commerce dated January 1, 1931, as amended, or under the normal category of part 4a of the former Civil Air Regulations (14 CFR Part 4a, 1964 ed.) if that airplane is operated in the State of Alaska by—

(1) A certificate holder conducting operations under part 121 or part 135 of this chapter; or

(2) The U.S. Department of Interior in conducting its game and fish law enforcement activities or its management, fire detection, and fire suppression activities concerning public lands.

(b) The maximum certificated weight approved under this section may not exceed—

(1) 12,500 pounds;

(2) 115 percent of the maximum weight listed in the FAA aircraft specifications;

(3) The weight at which the airplane meets the positive maneuvering load factor requirement for the normal category specified in § 23.337 of this chapter; or

(4) The weight at which the airplane meets the climb performance requirements under which it was type certificated,

(c) In determining the maximum certificated weight, the Administrator considers the structural soundness of the airplane and the terrain to be traversed.

(d) The maximum certificated weight determined under this section is added to the airplane's operation limitations and is identified as the maximum weight authorized for operations within the State of Alaska.

[§ 91.325 Primary category aircraft: Operating limitations.

[(a) No person may operate a primary category aircraft carrying persons or property for compensation or hire.

[(b) No person may operate a primary category aircraft that is maintained by the pilot-owner under an approved special inspection and maintenance program except—

[(1) The pilot-owner; or

[(2) A designee of the pilot-owner, provided that the pilot-owner does not receive compensation for the use of the aircraft.]

[(Amdt. 91-230, Eff. 12/31/92)]

§91.327 Aircraft having a special airworthiness certificate in the light-sport category: Operating limitations.

(a) No person may operate an aircraft that has a special airworthiness certificate in the light-sport category for compensation or hire except—

(1) To tow a glider or an unpowered ultralight vehicle in accordance with §91.309 of this chapter; or

(2) To conduct flight training.

(b) No person may operate an aircraft that has a special airworthiness certificate in the light-sport category unless—

(1) The aircraft is maintained by a certificated repairman with a light-sport aircraft maintenance rating, an appropriately rated mechanic, or an appropriately rated repair station in accordance with the applicable provisions of part 43 of this chapter and maintenance and inspection procedures developed by the aircraft manufacturer or a person acceptable to the FAA;

(2) A condition inspection is performed once every 12 calendar months by a certificated repairman (light-sport aircraft) with a maintenance rating, an appropriately rated mechanic, or an appropriately rated repair station in accordance with inspection procedures developed by the aircraft manufacturer or a person acceptable to the FAA;

(3) The owner or operator complies with all applicable airworthiness directives;

(4) The owner or operator complies with each safety directive applicable to the aircraft that corrects an existing unsafe condition. In lieu of complying with a safety directive an owner or operator may—

(i) Correct the unsafe condition in a manner different from that specified in the safety directive provided the person issuing the directive concurs with the action; or

(ii) Obtain an FAA waiver from the provisions of the safety directive based on a conclusion that the safety directive was issued without adhering to the applicable consensus standard;

(5) Each alteration accomplished after the aircraft's date of manufacture meets the applicable and current consensus standard and has been authorized by either the manufacturer or a person acceptable to the FAA;

(6) Each major alteration to an aircraft product produced under a consensus standard is authorized, performed and inspected in accordance with maintenance and inspection procedures developed by the manufacturer or a person acceptable to the FAA; and

(7) The owner or operator complies with the requirements for the recording of major repairs and major alterations performed on type-certificated products in accordance with §43.9 (d) of this chapter, and with the retention requirements in §91.417.

(c) No person may operate an aircraft issued a special airworthiness certificate in the light-sport category to tow a glider or unpowered ultralight vehicle for compensation or hire or conduct flight training for compensation or hire in an aircraft which that persons provides unless within the preceding 100 hours of time in service the aircraft has—

(1) Been inspected by a certificated repairman with a light-sport aircraft maintenance rating, an appropriately rated mechanic, or an appropriately rated repair station in accordance with inspection procedures developed by the aircraft manufacturer or a person acceptable to the FAA and been approved for return to service in accordance with part 43 of this chapter; or

(2) Received an inspection for the issuance of an airworthiness certificate in accordance with part 21 of this chapter.

(d) Each person operating an aircraft issued a special airworthiness certificate in the light-sport category must operate the aircraft in accordance with the aircraft's operating instructions, including any provisions for necessary operating equipment specified in the aircraft's equipment list.

(e) Each person operating an aircraft issued a special airworthiness certificate in the light-sport category must advise each person carried of the special nature of the aircraft and that the aircraft does not meet the airworthiness requirements for an aircraft issued a standard airworthiness certificate.

(f) The FAA may prescribe additional limitations that it considers necessary.

Subpart E—Maintenance, preventive maintenance, and alterations

§ 91.401 Applicability.

(a) This subpart prescribes rules governing the maintenance, preventive maintenance, and alterations of U.S.-registered civil aircraft operating within or outside of the United States.

(b) Sections 91.409, 91.411, 91.417, and 91.419 of this subpart do not apply to an aircraft maintained in accordance with a continuous airworthiness maintenance program as provided in Part 121, 129, or §91.1411 or 135.411(a) (2) of this chapter.

(c) Sections 91.405 and 91.409 of this part do not apply to an airplane inspected in accordance with Part 125 of this chapter.

§ 91.403 General.

(a) The owner or operator of an aircraft is primarily responsible for maintaining that aircraft in an airworthy condition, including compliance with Part 39 of this chapter.

(b) No person may perform maintenance, preventive maintenance, or alterations on an aircraft other than as prescribed in this subpart and other applicable regulations, including Part 43 of this chapter.

(c) No person may operate an aircraft for which a manufacturer's maintenance manual or instructions for continued airworthiness has been issued that contains an airworthiness limitations section unless the mandatory replacement times, inspection intervals, and related procedures specified in that section or alternative inspection intervals and related procedures set forth in an operations specification approved by the Administrator under Part 121, 127, or 135 of this chapter or in accordance with an inspection program approved under § 91.409(e) have been compiled with.

(d) A person must not alter an aircraft based on a supplemental type certificate unless the owner or operator of the aircraft is the holder of the supplemental type certificate, or has written permission from the holder.

§ 91.405 Maintenance required.

Each owner or operator of an aircraft—

(a) Shall have that aircraft inspected as prescribed in Subpart E of this part and shall between required inspections, except as provided in paragraph (c) of this section, have discrepancies repaired as prescribed in Part 43 of this chapter.

(b) Shall ensure that maintenance personnel make appropriate entries in the aircraft maintenance records indicating the aircraft has been approved for return to service;

(c) Shall have any inoperative instrument or item of equipment, permitted to be inoperative by § 91.213(d) (2) of this part, repaired, replaced, removed, or inspected at the next required inspection; and

(d) When listed discrepancies include inoperative instruments or equipment, shall ensure that a placard has been installed as required by § 43.11 of this chapter.

§ 91.407 Operation after maintenance, preventive maintenance, rebuilding, or alteration.

(a) No person may operate any aircraft that has undergone maintenance, preventive maintenance, rebuilding, or alteration unless—

(1) It has been approved for return to service by a person authorized under § 43.7 of this chapter; and

(2) The maintenance record entry required by § 43.9 or § 43.11, as applicable, of this chapter has been made.

(b) No person may carry any person (other than crewmembers) in an aircraft that has been maintained, rebuilt, or altered in a manner that may have appreciably changed its flight characteristics or substantially affected its operation in flight until an appropriately rated pilot with at least a private pilot certificate flies the aircraft, makes an operational check of the maintenance performed or alteration made, and logs the flight in the aircraft records.

(c) The aircraft does not have to be flown as required by paragraph (b) of this section, if, prior to flight, ground tests, inspections, or both show conclusively that the maintenance, preventive maintenance, rebuilding, or alteration has not appreciably changed the flight characteristics or substantially affected the flight operation of the aircraft.

(Approved by the Office of Management and Budget under OMB control number 2120-0005)

§ 91.409 Inspections.

(a) Except as provided in paragraph (c) of this section, no person may operate an aircraft unless within the preceding 12 calendar months, it has had—

(1) An annual inspection in accordance with Part 43 of this chapter and has been approved for return to service by a person authorized by § 43.7 of this chapter, or

(2) An inspection for the issue of an airworthiness certificate in accordance with Part 21 of this chapter.

No inspection performed under paragraph (b) of the section may be substituted for any inspection required by this paragraph unless it is performed by a person authorized to perform annual inspections and is entered as an "annual" inspection in the required maintenance records.

(b) Except as provided in paragraph (c) of this section, no person may operate an aircraft carrying any person (other than a crewmember) for hire, and no person may give flight instruction for hire in an aircraft which that person provides, unless within the preceding 100 hours of time in service the aircraft has received an annual or 100-hour inspection and been approved for return to service in accordance with Part 43 of this chapter or has received an inspection for the issuance of an airworthiness certificate in accordance with Part 21 of this chapter. The 100-hour limitation may be exceeded by not more than 10 hours while en route to reach a place where the inspection can be done. The excess time used to reach a place where the inspection can be done must be included in computing the next 100 hours of time in service.

(c) Paragraphs (a) and (b) of this section do not apply to—

(1) An aircraft that carries a special flight permit, a current experimental certificate, or a light-sport or provisional airworthiness certificate:

(2) An aircraft inspected in accordance with an approved aircraft inspection program under Part 125, 127, or 135 of this chapter and so identified by the registration number in the operations specifications of the certificate holder having the approved inspection program;

(3) An aircraft subject to the requirements of paragraph (d) or (e) of this section; or

(4) Turbine-powered rotorcraft when the operator elects to inspect that rotorcraft in accordance with paragraph (e) of this section.

(d) *Progressive inspection.* Each registered owner or operator of an aircraft desiring to use a progressive inspection program must submit a written request to the FAA Flight Standards district office having jurisdiction over the area in which the applicant is located, and shall provide—

(1) A certificated mechanic holding an inspection authorization, a certificated airframe repair station, or the manufacturer of the aircraft to supervise or conduct the progressive inspection;

(2) A current inspection procedures manual available and readily understandable to pilot and maintenance personnel containing, in detail—

(i) An explanation of the progressive inspection, including the continuity of inspection responsibility, the making of reports, and the keeping of records and technical reference material;

(ii) An inspection schedule, specifying the intervals in hours or days when routine and detailed inspections will be performed and including instructions for exceeding an inspection interval by not more than 10 hours while en route and for changing an inspection interval because of service experience;

(iii) Sample routine and detailed inspection forms and instructions for their use; and

(iv) Sample reports and records and instructions for their use;

(3) Enough housing and equipment for necessary disassembly and proper inspection of the aircraft; and

(4) Appropriate current technical information for the aircraft.

The frequency and detail of the progressive inspection shall provide for the complete inspection of the aircraft within each 12 calendar months and be consistent with the manufacturer's recommendations, field service experience, and the kind of operation in which the aircraft is engaged. The progressive inspection schedule must ensure that the aircraft, at all times, will be airworthy and will conform to all applicable FAA aircraft specifications, type certificate data sheets, airworthiness directives, and other approved data. If the progressive inspection is discontinued, the owner or operator shall immediately notify the local FAA Flight Standards district office, in writing, of the discontinuance. After the discontinuance, the first annual inspection under § 91.409(a) (1) is due within 12 calendar months after the last complete inspection of the aircraft under the progressive inspection. The 100-hour inspection under § 91.409(b) is due within 100 hours after that complete inspection. A complete inspection of the aircraft, for the purpose of determining when the annual and 100-hour inspections are due, requires a detailed inspection of the aircraft and all its components in accordance with the progressive inspection. A routine inspection of the aircraft and a detailed inspection of several components is not considered to be a complete inspection.

(e) *Large airplanes (to which Part 125 is not applicable), turbojet multiengine airplanes, turbopropeller powered multiengine airplanes, and turbine-powered rotor-craft.* No person may operate a large airplane, turbojet multiengine airplane, or turbopropeller-powered multiengine airplane, or turbine-powered rotorcraft unless the replacement times for life-limited parts specified in the aircraft specifications, type data sheets, or other documents approved by the Administrator are complied with and the airplane or turbine-powered rotorcraft, including the airframe, engines, propellers, rotors, appliances, survival equipment, and emergency equipment, is inspected in accordance with an inspection program selected under

the provisions of paragraph (f) of this section, except that, the owner or operator of a turbine-powered rotorcraft may elect to use the inspection provisions of § 91.409(a), (b), (c), or (d) in lieu of an inspection option of § 91.409(f).

(f) *Selection of inspection program under paragraph (e) of this section.* The registered owner or operator of each airplane or turbine-powered rotorcraft described in paragraph (e) of this section must select, identify in the aircraft maintenance records, and use one of the following programs for the inspection of the aircraft:

(1) A continuous airworthiness inspection program that is part of a continuous airworthiness maintenance program currently in use by a person holding an air carrier operating certificate or an operating certificate issued under Part 121, 127, or 135 of this chapter and operating that make and model aircraft under Part 121 of this chapter or operating that make and model under Part 135 of this chapter and maintaining it under § 135.411(a) (2) of this chapter.

(2) An approved aircraft inspection program approved under § 135.419 of this chapter and currently in use by a person holding an operating certificate issued under Part 135 of this chapter.

(3) A current inspection program recommended by the manufacturer.

(4) Any other inspection program established by the registered owner or operator of that airplane or turbine-powered rotorcraft and approved by the Administrator under paragraph (g) of this section. However, the Administrator may require revision of this inspection program in accordance with the provisions of § 91.415.

Each operator shall include in the selected program the name and address of the person responsible for scheduling the inspections required by the program and make a copy of that program available to the person performing inspections on the aircraft and, upon request, to the Administrator.

(g) *Inspection program approved under paragraph (e), of this section.* Each operator of an airplane or turbine-powered rotorcraft desiring to establish or change an approved inspection program under paragraph (f) (4) of this section must submit the program for approval to the local FAA Flight Standards district office having jurisdiction over the area in which the aircraft is based. The program must be in writing and include at least the following information:

(1) Instructions and procedures for the conduct of inspections for the particular make and model airplane or turbine-powered rotorcraft, including necessary tests and checks. The instructions and procedures must set forth in detail the parts and areas of the airframe, engines, propellers, rotors, and appliances, including survival and emergency equipment required to be inspected.

(2) A schedule for performing the inspections that must be performed under the program expressed in terms of the time in service, calendar time, number of system operations, or any combination of these.

(h) *Changes from one inspection program to another.* When an operator changes from one inspection program under paragraph (f) of this section to another, the time in service, calendar times, or cycles of operation accumulated under the previous program must be applied in determining inspection due times under the new program.

(Approved by the Office of Management and Budget under OMB control number 2120-0005)

§91.410 Special maintenance program requirements

(a) No person may operate an Airbus Model A300 (excluding the -600 series), British Aerospace Model BAC 1-11, Boeing Model, 707, 720, 727, 737 or 747, McDonnell Douglas Model DC-8, DC-9/MD-80 or DC-10, Fokker Model F28, or Lockheed Model L-1011 airplane beyond applicable flight cycle implementation time specified below, or May 25, 2001, whichever occurs later, unless repair assessment guidelines applicable to the fuselage pressure boundary (fuselage skin, door skin, and bulkhead webs) that have been approved by the FAA Aircraft Certification Office (ACO), or office of the Transport Airplane Directorate, having cognizance over the type certificate for the affected airplane are incorporated within its inspection program:

(a)(1) For the Airbus Model A300 (excluding the -600 series), the flight cycle implementation time is:

(a)(1)(i) Model B2: 36,000 flights.

(a)(1)(ii) Model B4-100 (including Model B4-2C): 30,000 flights above the window line, and 36,000 flights below the window line.

(a)(1)(iii) Model B4-200: 25,000 flights above the window line, and 34,000 flights below the window line.

(a)(2) For all models of the British Aerospace BAC 1-11, the flight cycle implementation time is 60,000 flights.

(a)(3) For all models of the Boeing 707, the flight cycle implementation time is 15,000 flights.

(a)(4) For all models of the Boeing 720, the flight cycle implementation time is 23,000 flights.

(a)(5) For all models of the Boeing 727, the flight cycle implementation time is 45,000 flights.

(a)(6) For all models of the Boeing 737, the flight cycle implementation time is 60,000 flights.

(a)(7) For all models of the Boeing 747, the flight cycle implementation time is 15,000 flights.

(a)(8) For all models of the McDonnell Douglas DC-8, the flight cycle implementation time is 30,000 flights.

(a)(9) For all models of the McDonnell Douglas DC-9/MD-80, the flight cycle implementation time is 60,000 flights.

(a)(10) For all models of the McDonnell Douglas DC-10, the flight cycle implementation time is 30,000 flights.

(a)(11) For all models of the Lockheed L-1011, the flight cycle implementation time is 27,000 flights.

(a)(12) For the Fokker F-28 Mark 1000, 2000, 3000, and 4000, the flight cycle implementation time is 27,000 flights.

(b) After December 16, 2008, no person may operate a turbine-powered transport category airplane with a type certificate issued after January 1, 1958, *and* either a maximum type certificated passenger capacity of 30 or more, or a maximum type certificated payload capacity of 7,500 pounds or more, unless instructions for maintenance and inspection of the fuel tank system are incorporated into its inspection program. Those instructions must address the actual configuration of the fuel tank systems of each affected airplane, and must be approved by the FAA Aircraft Certification Office (ACO), or office of the Transport Airplane Directorate, having cognizance over the type certificate for the affected airplane. Operators must submit their request through the cognizant Flight Standards District Office, who may add comments and then send it to the manager of the appropriate office. Thereafter, the approved instructions can be revised only with the approval of the FAA Aircraft Certification Office (ACO), or office of the Transport Airplane Directorate, having cognizance over the type certificate for the affected airplane. Operators must submit their request for revisions through the cognizant Flight Standards District Office, who may add comments and then send it to the manager of the appropriate office.

§ 91.411 Altimeter system and altitude reporting equipment tests and inspections.

(a) No person may operate an airplane, or helicopter, in controlled airspace under IFR unless—

(1) Within the preceding 24 calendar months, each static pressure system, each altimeter instrument, and each automatic pressure altitude reporting system has been tested and inspected and found to comply with Appendix E of Part 43 of this chapter;

(2) Except for the use of system drain and alternate static pressure valves, following any opening and closing of the static pressure system, that system has been tested and inspected and found to comply with paragraph (a), Appendices E and F, of Part 43 of this chapter; and

(3) Following installation or maintenance on the automatic pressure altitude reporting system of the ATC transponder where data correspondence error could be introduced, the integrated system has been tested, inspected, and found to comply with paragraph (c), Appendix E, of Part 43 of this chapter.

(b) The tests required by paragraph (a) of this section must be conducted by—

(1) The manufacturer of the airplane, or helicopter, on which the tests and inspections are to be performed;

(2) A certificated repair station properly equipped to perform those functions and holding—

(i) An instrument rating, Class I;

(ii) A limited instrument rating appropriate to the make and model of appliance to be tested;

(iii) A limited rating appropriate to the test to be performed;

(iv) An airframe rating appropriate to the airplane, or helicopter, to be tested; or

(3) A certificated mechanic with an airframe rating (static pressure system tests and inspections only).

(c) Altimeter and altitude reporting equipment approved under Technical Standard Orders are considered to be tested and inspected as of the date of their manufacture.

(d) No person may operate an airplane, or helicopter, in controlled airspace under IFR at an altitude above the maximum altitude at which all altimeters and the automatic altitude reporting system of that airplane, or helicopter, have been tested.

§ 91.413 ATC transponder tests and inspections.

(a) No person may use an ATC transponder that is specified in § 91.215(a), § 121.345(c), § 127.123(b) or § 135.143(c) of this chapter unless, within the preceding 24 calendar months, the ATC transponder has been tested and inspected and found to comply with Appendix F of Part 43 of this chapter; and

(b) Following any installation or maintenance on an ATC transponder where data correspondence error could be introduced, the integrated system has been tested, inspected, and found to comply with paragraph (c), Appendix E, of Part 43 of this chapter.

(c) The tests and inspections specified in this section must be conducted by—

(1) A certificated repair station properly equipped to perform those functions and holding—

(i) A radio rating, Class III;

(ii) A limited radio rating appropriate to the make and model transponder to be tested;

(iii) A limited rating appropriate to the test to be performed;

(2) A holder of a continuous airworthiness maintenance program as provided in Part 121, 127, or § 135.411(a) (2) of this chapter; or

(3) The manufacturer of the aircraft on which the transponder to be tested is installed, if the transponder was installed by that manufacturer.

§ 91.415 Changes to aircraft inspection programs.

(a) Whenever the Administrator finds that revisions to an approved aircraft inspection program under § 91.409(f) (4) are necessary for the continued adequacy of the program, the owner or operator must, after notification by the Administrator, make any changes in the program found to be necessary by the Administrator.

(b) The owner or operator may petition the Administrator to reconsider the notice to make any changes in a program in accordance with paragraph (a) of this section.

(c) The petition must be filed with the Director, Flight Standards Service within 30 days after the certificate holder or fractional ownership program manager receives the notice.

(d) Except in the case of an emergency requiring immediate action in the interest of safety, the filing of the petition stays the notice pending a decision by the Administrator.

§ 91.417 Maintenance records.

(a) Except for work performed in accordance with § 91.411 and 91.413, each registered owner or operator shall keep the following records for the periods specified in paragraph (b) of this section:

(1) Records of the maintenance, preventive maintenance, and alteration and records of the 100-hour, annual, progressive, and other required or approved inspections, as appropriate, for each aircraft (including the airframe) and each engine, propeller, rotor, and appliance of an aircraft. The records must include—

(i) A description (or reference to data acceptable to the Administrator) of the work performed; and

(ii) The date of completion of the work performed; and

(iii) The signature, and certificate number of the person approving the aircraft for return to service.

(2) Records containing the following information:

(i) The total time in service of the airframe, each engine, each propeller, and each rotor.

(ii) The current status of life-limited parts of each airframe, engine, propeller, rotor, and appliance.

(iii) The time since last overhaul of all items installed on the aircraft which are required to be overhauled on a specified time basis.

(iv) The current inspection status of the aircraft, including the time since the last inspection required by the inspection program under which the aircraft and its appliances are maintained.

(v) The current status of applicable airworthiness directives (AD) including, for each, the method of compliance, the AD number, and revision date. If the AD involves recurring action, the time and date when the next action is required.

(vi) Copies of the forms prescribed by § 43.9(a) of this chapter for each major alteration to the airframe and currently installed engines, rotors, propellers, and appliances.

(b) The owner or operator shall retain the following records for the periods prescribed:

(1) The records specified in paragraph (a) (1) of this section shall be retained until the work is repeated or superseded by other work or for 1 year after the work is performed.

(2) The records specified in paragraph (a) (2) of this section shall be retained and transferred with the aircraft at the time the aircraft is sold.

(3) A list of defects furnished to a registered owner or operator under § 43.11 of this chapter shall be retained until the defects are repaired and the aircraft is approved for return to service.

(c) The owner or operator shall make all maintenance records required to be kept by this section available for inspection by the Administrator or any authorized representative of the National Transportation Safety Board (NTSB). In addition, the owner or operator shall present the Form 337 described in paragraph (d) of this section for inspection upon request of any law enforcement officer.

(d) When a fuel tank is installed within the passenger compartment or a baggage compartment pursuant to Part 43 of this chapter, a copy of the FAA Form 337 shall be kept on board the modified aircraft by the owner or operator.

(Approved by the Office of Management and Budget under OMB control number 2120-0005)

§ 91.419 Transfer of maintenance records.

Any owner or operator who sells a U.S.-registered aircraft shall transfer to the purchaser, at the time of sale, the following records of that aircraft, in plain language form or in coded form at the election of the purchaser, if the coded form provides for the preservation and retrieval of information in a manner acceptable to the Administrator:

(a) The records specified in § 91.417(a) (2).

(b) The records specified in § 91.417(a) (1) which are not included in the records covered by paragraph (a) of this section, except that the purchaser may permit the seller to keep physical custody of such records. However, custody of records by the seller does not relieve the purchaser of the responsibility under § 91.417(c) to make the records available for inspection by the Administrator or any authorized representative of the National Transportation Safety Board (NTSB).

§ 91.421 Rebuilt engine maintenance records.

(a) The owner or operator may use a new maintenance record, without previous operating history, for an aircraft engine rebuilt by the manufacturer or by an agency approved by the manufacturer.

(b) Each manufacturer or agency that grants zero time to an engine rebuilt by it shall enter in the new record—

(1) A signed statement of the date the engine was rebuilt;

(2) Each change made as required by airworthiness directives; and

(3) Each change made in compliance with manufacturer's service bulletins, if the entry is specifically requested in that bulletin.

(c) For the purposes of this section, a rebuilt engine is a used engine that has been completely disassembled, inspected, repaired as necessary, reassembled, tested, and approved in the same manner and to the same tolerances and limits as a new engine with either new or used parts. However, all parts used in it must conform to the production drawing tolerances and limits for new parts or be of approved oversized or undersized dimensions for a new engine.

§§ 91.423–91.499 [Reserved]

Subpart F—Large and turbine-powered multiengine airplanes

§ 91.501 Applicability.

(a) This subpart prescribes operating rules, in addition to those prescribed in other subparts of this part, governing the operation of large airplanes of U.S. registry, turbojet-powered multiengine civil airplanes of U.S. registry, and fractional ownership program aircraft of U.S. registry that are operating under subpart K of this part in operations not involving common carriage. The operating rules in this subpart do not apply to those aircraft when they are required to be operated under parts 121, 125, 129, 135, and 137 of this chapter. (Section 91.409 prescribes an inspection program for large and for turbine-powered (turbojet and turboprop) multiengine airplanes and turbine-powered rotorcraft of U.S. registry when they are operated under this part or part 129 or 137.)

(b) Operations that may be conducted under the rules in this subpart instead of those in parts 121, 129, 135, and 137 of this chapter when common carriage is not involved, include—

(1) Ferry or training flights;

(2) Aerial work operations such as aerial photography or survey, or pipeline patrol, but not including fire fighting operations;

(3) Flights for the demonstration of an airplane to prospective customers when no charge is made except for those specified in paragraph (d) of this section;

(4) Flights conducted by the operator of an airplane for his personal transportation, or the transportation of his guests when no charge, assessment, or fee is made for the transportation;

(5) Carriage of officials, employees, guests, and property of a company on an airplane operated by that company, or the parent or a subsidiary of the company or a subsidiary of the parent, when the carriage is within the scope of, and incidental to, the business of the company (other than transportation by air) and no charge, assessment or fee is made for the carriage in excess of the cost of owning, operating, and maintaining the airplane,

except that no charge of any kind may be made for the carriage of a guest of a company, when the carriage is not within the scope of, and incidental to, the business of that company;

(6) The carriage of company officials, employees, and guests of the company on an airplane operated under a time sharing, interchange, or joint ownership agreement as defined in paragraph (c) of this section;

(7) The carriage of property (other than mail) on an airplane operated by a person in the furtherance of a business or employment (other than transportation by air) when the carriage is within the scope of, and incidental to, that business or employment and no charge, assessment, or fee is made for the carriage other than those specified in paragraph (d) of this section;

(8) The carriage on an airplane of an athletic team, sports group, choral group, or similar group having a common purpose or objective when there is no charge, assessment, or fee of any kind made by any person for that carriage; and

(9) The carriage of persons on an airplane operated by a person in the furtherance of a business other than transportation by air for the purpose of selling them land, goods, or property, including franchises or distributorships, when the carriage is within the scope of, and incidental to, that business and no charge, assessment, or fee is made for that carriage.

(c) As used in this section—

(1) A "time sharing agreement" means an arrangement whereby a person leases his airplane with flight crew to another person, and no charge is made for the flights conducted under that arrangement other than those specified in paragraph (d) of this section;

(2) An "interchange agreement" means an arrangement whereby a person leases his airplane to another person in exchange for equal time, when needed, on the other person's airplane, and no charge, assessment, or fee is made, except that a charge may be made not to exceed the difference between the cost of owning, operating, and maintaining the two airplanes;

(3) A "joint ownership agreement" means an arrangement whereby one of the registered joint owners of an airplane employs and furnishes the flight crew for that airplane and each of the registered joint owners pays a share of the charge specified in the agreement.

(d) The following may be charged, as expenses of a specific flight, for transportation as authorized by paragraphs (b)(3) and (7) and (c)(1) of this section:

(1) Fuel, oil, lubricants, and other additives.

(2) Travel expenses of the crew, including food, lodging, and ground transportation.

(3) Hangar and tie-down costs away from the aircraft's base of operation.

(4) Insurance obtained for the specific flight.

(5) Landing fees, airport taxes, and similar assessments.

(6) Customs, foreign permit, and similar fees directly related to the flight.

(7) In flight food and beverages.

(8) Passenger ground transportation.

(9) Flight planning and weather contract services.

(10) Any operation identified in paragraphs (b)(1) through (b)(9) of this section when conducted—

(i) By a fractional ownership program manager, or

(ii) By a fractional owner in a fractional ownership program aircraft operated under subpart K of this part, except that a flight under a joint ownership arrangement under paragraph (b)(6) of this section may not be conducted. For a flight under an interchange agreement under paragraph (b)(6) of this section, the exchange of equal time for the operation must be properly accounted for as part of the total hours associated with the fractional owner's share of ownership.

§ 91.503 Flying equipment and operating information.

(a) The pilot in command of an airplane shall ensure that the following flying equipment and aeronautical charts and data, in current and appropriate form, are accessible for each flight at the pilot station of the airplane:

(1) A flashlight having at least two size "D" cells, or the equivalent, that is in good working order.

(2) A cockpit checklist containing the procedures required by paragraph (b) of this section.

(3) Pertinent aeronautical charts.

(4) For IFR, VFR over-the-top, or night operations, each pertinent navigational en route, terminal area, and approach and letdown chart.

(5) In the case of multiengine airplanes, one-engine inoperative climb performance data.

(b) Each cockpit checklist must contain the following procedures and shall be used by the flight crewmembers when operating the airplane:

(1) Before starting engines.

(2) Before takeoff.

(3) Cruise.

(4) Before landing.

(5) After landing.

(6) Stopping engines.

(7) Emergencies.

(c) Each emergency cockpit checklist procedure required by paragraph (b)(7) of this section must contain the following procedures, as appropriate:

(1) Emergency operation of fuel, hydraulic, electrical, and mechanical systems.

(2) Emergency operation of instruments and controls.

(3) Engine inoperative procedures.

(4) Any other procedures necessary for safety.

(d) The equipment, charts, and data prescribed in this section shall be used by the pilot in command and other members of the flight crew, when pertinent.

§ 91.505 Familiarity with operating limitations and emergency equipment.

(a) Each pilot in command of an airplane shall, before beginning a flight, become familiar with the Airplane Flight Manual for that airplane, if one is required, and with any placards, listings, instrument markings, or any combination thereof, containing each operating limitation prescribed for that airplane by the Administrator, including those specified in § 91.9(b).

(b) Each required member of the crew shall, before beginning a flight, become familiar with the emergency equipment installed on the airplane to which that crewmember is assigned and with the procedures to be followed for the use of that equipment in an emergency situation.

§ 91.507 Equipment requirements: Over-the-top or night VFR operations.

No person may operate an airplane over-the-top or at night under VFR unless that airplane is equipped with the instruments and equipment required for IFR operations under § 91.205(d) and one electric landing light for night operations. Each required instrument and item of equipment must be in operable condition.

§ 91.509 Survival equipment for overwater operations.

(a) No person may take off an airplane for a flight over water more than 50 nautical miles from the nearest shore unless that airplane is equipped with a life preserver or an approved flotation means for each occupant of the airplane.

(b) Except as provided in paragraph (c) of this section, no person may take off an airplane for flight over water more than 30 minutes flying time or 100 nautical miles from the nearest shore, whichever is less, unless it has on board the following survival equipment:

(1) A life preserver, equipped with an approved survivor locator light, for each occupant of the airplane.

(2) Enough liferafts (each equipped with an approved survival locator light) of a rated capacity and buoyancy to accommodate the occupants of the airplane.

(3) At least one pyrotechnic signaling device for each liferaft.

(4) One self-buoyant, water-resistant, portable emergency radio signaling device that is capable of transmission on the appropriate emergency frequency or frequencies and not dependent upon the airplane power supply.

(5) A lifeline stored in accordance with § 25.1411(g) of this chapter.

(c) A fractional ownership program manager under subpart K of this part may apply for a deviation from paragraphs (b)(2) through (5) of this section for a particular over water operation or the Administrator may amend the management specifications to require the carriage of all or any specific items of the equipment listed in paragraphs (b)(2) through (5) of this section.

(d) The required life rafts, life preservers, and signaling devices must be installed in conspicuously marked locations and easily accessible in the event of a ditching without appreciable time for preparatory procedures.

(e) A survival kit, appropriately equipped for the route to be flown, must be attached to each required life raft.

(f) As used in this section, the term shore means that area of the land adjacent to the water that is above the high water mark and excludes land areas that are intermittently under water.

§ 91.511 Radio equipment for overwater operations.

(a) Except as provided in paragraphs (c), (d), and (f) of this section, no person may take off an airplane for a flight over water more than 30 minutes flying time or 100 nautical miles from the nearest shore unless it has at least the following operable equipment:

(1) Radio communication equipment appropriate to the facilities to be used and able to transmit to, and receive from, any place on the route, at least one surface facility:

(i) Two transmitters.

(ii) Two microphones.

(iii) Two headsets or one headset and one speaker.

(iv) Two independent receivers.

(2) Appropriate electronic navigational equipment consisting of at least two independent electronic navigation units capable of providing the pilot with the information necessary to navigate the airplane within the airspace assigned by air traffic control. However, a receiver that can receive both communications and required navigational signals may be used in place of a separate communications receiver and a separate navigational signal receiver or unit.

(b) For the purposes of paragraphs (a)(1)(iv) and (a)(2) of this section, a receiver or electronic navigation unit is independent if the function of any part of it does not depend on the functioning of any part of another receiver or electronic navigation unit.

(c) Notwithstanding the provisions of paragraph (a) of this section, a person may operate an airplane on which no passengers are carried from a place where repairs or replacement cannot be made to a place where they can be made, if not more than one of each of the dual items of radio communication and navigational equipment specified in paragraphs (a)(1)(i) through (iv) and (a)(2) of this section malfunctions or becomes inoperative,

(d) Notwithstanding the provisions of paragraph (a) of this section, when both VHF and HF communications equipment are required for the route and the airplane has two VHF transmitters and two VHF receivers for communications, only one HF transmitter and one HF receiver is required for communications.

(e) As used in this section, the term "shore" means that area of the land adjacent to the water which is above the high-water mark and excludes land areas which are intermittently under water.

(f) Notwithstanding the requirements in paragraph (a)(2) of this section, a person may operate in the Gulf of Mexico, the Caribbean Sea, and the Atlantic Ocean west of a line which extends from 44°47'00" N/67°00'00" W to 39°00'00" N /67°00'00" W to 38°30'00" N / 60°00'00Æ W south along the 60°00'00" W longitude line to the point were the line intersects with the northern coast of South America, when:

A single long-range navigation system is installed, operational and appropriate for the route, and

Flight conditions and the aircraft's capabilities are such that no more than a 30-minute gap in two-way radio very high frequency communications is expected to exist.

§ 91.513 Emergency equipment.

(a) No person may operate an airplane unless it is equipped with the emergency equipment listed in this section.

(b) Each item of equipment—

(1) Must be inspected in accordance with § 91.409 to ensure its continued serviceability and immediate readiness for its intended purposes;

(2) Must be readily accessible to the crew;

(3) Must clearly indicate its method of operation; and

(4) When carried in a compartment or container, must have that compartment or container marked as to contents and date of last inspection.

(c) Hand fire extinguishers must be provided for use in crew, passenger, and cargo compartments in accordance with the following:

(1) The type and quantity of extinguishing agent must be suitable for the kinds of fires likely to occur in the compartment where the extinguisher is intended to be used.

(2) At least one hand fire extinguisher must be provided and located on or near the flight deck in a place that is readily accessible to the flight crew.

(3) At least one hand fire extinguisher must be conveniently located in the passenger compartment of each airplane accommodating more than six but less than 31 passengers, and at least two hand fire extinguishers must be conveniently located in the passenger compartment of each airplane accommodating more than 30 passengers.

(4) Hand fire extinguishers must be installed and secured in such a manner that they will not interfere with the safe operation of the airplane or adversely affect the safety of the crew and passengers. They must be readily accessible and, unless the locations of the fire extinguishers are obvious, their stowage provisions must be properly identified.

(d) First aid kits for treatment of injuries likely to occur in flight or in minor accidents must be provided.

(e) Each airplane accommodating more than 19 passengers must be equipped with a crash axe.

(f) Each passenger-carrying airplane must have a portable battery-powered megaphone or megaphones readily accessible to the crewmembers assigned to direct emergency evacuation, installed as follows:

(1) One megaphone on each airplane with a seating capacity of more than 60 but less than 100 passengers, at the most rearward location in the passenger cabin where it would be readily accessible to a normal flight attendant seat. However, the Administrator may grant a deviation from the requirements of this subparagraph if the Administrator finds that a different location would be more useful for evacuation of persons during an emergency.

(2) On each airplane with a seating capacity of 100 or more passengers, one megaphone installed at the forward end and one installed at the most rearward location where it would be readily accessible to a normal flight attendant seat.

§ 91.515 Flight altitude rules.

(a) Notwithstanding § 91.119, and except as provided in paragraph (b) of this section, no person may operate an airplane under VFR at less than—

(1) One thousand feet above the surface, or 1,000 feet from any mountain, hill, or other obstruction to flight, for day operations; and

(2) The altitudes prescribed in § 91.177, for night operations.

(b) This section does not apply—

(1) During takeoff or landing;

(2) When a different altitude is authorized by a waiver to this section under subpart J of this part; or

(3) When a flight is conducted under the special VFR weather minimums of § 91.157 with an appropriate clearance from ATC.

§ 91.517 [Passenger information.

[(a) Except as provided in paragraph (b) of this section, no person may operate an airplane carrying passengers unless it is equipped with signs that are visible to passengers and flight attendants to notify them when smoking is prohibited and when safety belts must be fastened. The signs must be so constructed that the crew can turn them on and off. They must be turned on during airplane movement on the surface, for each takeoff, for each landing, and when otherwise considered to be necessary by the pilot in command.

[(b) The pilot in command of an airplane that is not required, in accordance with applicable aircraft and equipment requirements of this chapter, to be equipped as provided in paragraph (a) of this section shall ensure that the passengers are notified orally each time that it is necessary to fasten their safety belts and when smoking is prohibited.

[(c) If passenger information signs are installed, no passenger or crewmember may smoke while any "no smoking" sign is lighted nor may any passenger or crewmember smoke in any lavatory.

[(d) Each passenger required by § 91.107(a)(3) to occupy a seat or berth shall fasten his or her safety belt about him or her and keep it fastened while any "fasten seat belt" sign is lighted.

[(e) Each passenger shall comply with instructions given him or her by crewmembers regarding compliance with paragraphs (b), (c), and (d) of this section.]

[(Amdt. 91-231, Eff. 10/15/92)]

§ 91.519 Passenger briefing.

(a) Before each takeoff the pilot in command of an airplane carrying passengers shall ensure that all passengers have been orally briefed on—

(1) [Smoking: Each passenger shall be briefed on when, where, and under what condition smoking is prohibited. This briefing shall include a statement, as appropriate, that the Federal Aviation Regulations require passenger compliance with lighted passenger information signs and no smoking placards, prohibit smoking in lavatories, and require compliance with crewmember instructions with regard to these items;

(2) [Use of safety belts and shoulder harnesses: Each passenger shall be briefed on when, where, and under what conditions it is necessary to have his or her safety belt and, if installed, his or her shoulder harness fastened about him or her. This briefing shall include a statement, as appropriate, that Federal Aviation Regulations require passenger compliance with the lighted passenger sign and/or crewmember instructions with regard to these items;]

(3) Location and means for opening the passenger entry door and emergency exits;

(4) Location of survival equipment;

(5) Ditching procedures and the use of flotation equipment required under § 91.509 for a flight over water; and

(6) The normal and emergency use of oxygen equipment installed on the airplane.

(b) The oral briefing required by paragraph (a) of this section shall be given by the pilot in command or a member of the crew, but need not be given when the pilot in command determines that the passengers are familiar with the contents of the briefing. It may be supplemented by printed cards for the use of each passenger containing—

(1) A diagram of, and methods of operating, the emergency exits; and

(2) Other instructions necessary for use of emergency equipment.

(c) Each card used under paragraph (b) must be carried in convenient locations on the airplane for the use of each passenger and must contain information that is pertinent only to the type and model airplane on which it is used.

(d) For operations under subpart K of this part, the passenger briefing requirements of § 91.1035 apply, instead of the requirements of paragraphs (a) through (c) of this section.

[(Amdt. 91-231, Eff. 10/15/92)]

§ 91.521 Shoulder harness.

(a) No person may operate a transport category airplane that was type certificated after January 1, 1958, unless it is equipped at each seat at a flight deck station with a combined safety belt and shoulder harness that meets the applicable requirements specified in § 25.785 of this chapter, except that—

(1) Shoulder harnesses and combined safety belt and shoulder harnesses that were approved and installed before March 6, 1980, may continue to be used; and

(2) Safety belt and shoulder harness restraint systems may be designed to the inertia load factors established under the certification basis of the airplane.

(b) No person may operate a transport category airplane unless it is equipped at each required flight attendant seat in the passenger compartment with a combined safety belt and shoulder harness that meets the applicable requirements specified in § 25.785 of this chapter, except that—

(1) Shoulder harnesses and combined safety belt and shoulder harnesses that were approved and installed before March 6, 1980, may continue to be used; and

(2) Safety belt and shoulder harness restraint systems may be designed to the inertia load factors established under the certification basis of the airplane.

§ 91.523 Carry-on baggage.

No pilot in command of an airplane having a seating capacity of more than 19 passengers may permit a passenger to stow baggage aboard that airplane except—

(a) In a suitable baggage or cargo storage compartment, or as provided in § 91.525; or

(b) Under a passenger seat in such a way that it will not slide forward under crash impacts severe enough to induce the ultimate inertia forces specified in § 25.561(b)(3) of this chapter, or the requirements of the regulations under which the airplane was type certificated. Restraining devices must also limit sideward motion of under-seat baggage and be designed to withstand crash impacts severe enough to induce sideward forces specified in § 25.561(b)(3) of this chapter.

§ 91.525 Carriage of cargo.

(a) No pilot in command may permit cargo to be carried in any airplane unless—

(1) It is carried in an approved cargo rack, bin, or compartment installed in the airplane;

(2) It is secured by means approved by the Administrator; or

(3) It is carried in accordance with each of the following:

(i) It is properly secured by a safety belt or other tiedown having enough strength to eliminate the possibility of shifting under all normally anticipated flight and ground conditions.

(ii) It is packaged or covered to avoid possible injury to passengers.

(iii) It does not impose any load on seats or on the floor structure that exceeds the load limitation for those components.

(iv) It is not located in a position that restricts the access to or use of any required emergency or regular exit, or the use of the aisle between the crew and the passenger compartment.

(v) It is not carried directly above seated passengers.

(b) When cargo is carried in cargo compartments that are designed to require the physical entry of a crewmember to extinguish any fire that may occur during flight, the cargo must be loaded so as to allow a crewmember to effectively reach all parts of the compartment with the contents of a hand fire extinguisher.

§ 91.527 Operating in icing conditions.

(a) No pilot may take off an airplane that has—

(1) Frost, snow, or ice adhering to any propeller, windshield, or powerplant installation or to an airspeed, altimeter, rate of climb, or flight attitude instrument system;

(2) Snow or ice adhering to the wings or stabilizing or control surfaces; or

(3) Any frost adhering to the wings or stabilizing or control surfaces, unless that frost has been polished to make it smooth.

(b) Except for an airplane that has ice protection provisions that meet the requirements in section 34 of Special Federal Aviation Regulation No. 23, or those for transport category airplane type certification, no pilot may fly—

(1) Under IFR into known or forecast moderate icing conditions; or

(2) Under VFR into known light or moderate icing conditions unless the aircraft has functioning de-icing or anti-icing equipment protecting each propeller, windshield, wing, stabilizing or control surface, and each airspeed, altimeter, rate of climb, or flight attitude instrument system.

(c) Except for an airplane that has ice protection provisions that meet the requirements in section 34 of Special Federal Aviation Regulation No. 23, or those for transport category airplane type certification, no pilot may fly an airplane into known or forecast severe icing conditions.

(d) If current weather reports and briefing information relied upon by the pilot in command indicate that the forecast icing conditions that would otherwise prohibit the flight will not be encountered during the flight because of changed weather conditions since the forecast, the restrictions in paragraphs (b) and (c) of this section based on forecast conditions do not apply.

§ 91.529 Flight engineer requirements.

(a) No person may operate the following airplanes without a flight crewmember holding a current flight engineer certificate:

(1) An airplane for which a type certificate was issued before January 2, 1964, having a maximum certificated takeoff weight of more than 80,000 pounds.

(2) An airplane type certificated after January 1, 1964, for which a flight engineer is required by the type certification requirements.

(b) No person may serve as a required flight engineer on an airplane unless, within the preceding 6 calendar months, that person has had at least 50 hours of flight time as a flight engineer on that type airplane or has been checked by the Administrator on that type airplane and is found to be familiar and competent with all essential current information and operating procedures.

§ 91.531 Second in command requirements.

(a) Except as provided in paragraph (b) and (d) of this section, no person may operate the following airplanes without a pilot who is designated as second in command of that airplane:

(1) A large airplane, except that a person may operate an airplane certificated under SFAR 41 without a pilot who is designated as second in command if that airplane is certificated for operation with one pilot.

(2) A turbojet-powered multiengine airplane for which two pilots are required under the type certification requirements for that airplane.

(3) A commuter category airplane, except that a person may operate a commuter category airplane notwithstanding paragraph (a)(1) of this section, that has a passenger seating configuration, excluding pilot seats, of nine or less without a pilot who is designated as second in command if that airplane is type certificated for operations with one pilot.

(b) The Administrator may issue a letter of authorization for the operation of an airplane without compliance with the requirements of paragraph (a) of this section if that airplane is designed for and type certificated with only one pilot station. The authorization contains any conditions that the Administrator finds necessary for safe operation.

(c) No person may designate a pilot to serve as second in command, nor may any pilot serve as second in command, of an airplane required under this section to have two pilots unless that pilot meets the qualifications for second in command prescribed in § 61.55 of this chapter.

(d) No person may operate an aircraft under subpart K of this part without a pilot who is designated as second in command of that aircraft in accordance with § 9 1.1049(d). The second in command must meet the experience requirements of § 91.1053.

§ 91.533 Flight attendant requirements.

(a) No person may operate an airplane unless at least the following number of flight attendants are on board the airplane:

(1) For airplanes having more than 19 but less than 51 passengers on board, one flight attendant.

(2) For airplanes having more than 50 but less than 101 passengers on board, two flight attendants.

(3) For airplanes having more than 100 passengers on board, two flight attendants plus one additional flight attendant for each unit (or part of a unit) of 50 passengers above 100.

(b) No person may serve as a flight attendant on an airplane when required by paragraph (a) of this section unless that person has demonstrated to the pilot in command familiarity with the necessary functions to be performed in an emergency or a situation requiring emergency evacuation and is capable of using the emergency equipment installed on that airplane.

[§ 91.535 Stowage of food, beverage, and passenger service equipment during aircraft movement on the surface, takeoff, and landing.

[(a) No operator may move an aircraft on the surface, take off, or land when any food, beverage, or tableware furnished by the operator is located at any passenger seat.

[(b) No operator may move an aircraft on the surface, take off, or land unless each food and beverage tray and seat back tray table is secured in its stowed position.

[(c) No operator may permit an aircraft to move on the surface, take off, or land unless each passenger serving cart is secured in its stowed position.

[(d) No operator may permit an aircraft to move on the surface, take off, or land unless each movie screen that extends into the aisle is stowed.

[(e) Each passenger shall comply with instructions given by a crewmember with regard to compliance with this section.]

[(Amdt. 91-231, Eff. 10/15/92)]

§§ 91.537–91.599 [Reserved]

Subpart G—Additional equipment and operating requirements for large and transport category aircraft

§ 91.601 Applicability.

This subpart applies to operation of large and transport category U.S.-registered civil aircraft.

§ 91.603 Aural speed warning device.

No person may operate a transport category airplane in air commerce unless that airplane is equipped with an aural speed warning device that complies with § 25.1303(c)(1).

§ 91.605 Transport category civil airplane weight limitations.

(a) No person may take off any transport category airplane (other than a turbine-engine-powered airplane certificated after September 30, 1958) unless—

(1) The takeoff weight does not exceed the authorized maximum takeoff weight for the elevation of the airport of takeoff;

(2) The elevation of the airport of takeoff is within the altitude range for which maximum takeoff weights have been determined;

(3) Normal consumption of fuel and oil in flight to the airport of intended landing will leave a weight on arrival not in excess of the authorized maximum landing weight for the elevation of that airport; and

(4) The elevations of the airport of intended landing and of all specified alternate airports are within the altitude range for which the maximum landing weights have been determined.

(b) No person may operate a turbine-engine-powered transport category airplane certificated after September 30, 1958, contrary to the Airplane Flight Manual, or take off that airplane unless—

(1) The takeoff weight does not exceed the takeoff weight specified in the Airplane Flight Manual for the elevation of the airport and for the ambient temperature existing at the time of takeoff;

(2) Normal consumption of fuel and oil in flight to the airport of intended landing and to the alternate airports will leave a weight on arrival not in excess of the landing weight specified in the Airplane Flight Manual for the elevation of each of the airports involved and for the ambient temperatures expected at the time of landing;

(3) The takeoff weight does not exceed the weight shown in the Airplane Flight Manual to correspond with the minimum distances required for takeoff considering the elevation of the airport, the runway to be used, the effective runway gradient, and the ambient temperature and wind component existing at the time of takeoff; and

(4) Where the takeoff distance includes a clearway, the clearway distance is not greater than one-half of—

(i) The takeoff run, in the case of airplanes certificated after September 30, 1958, and before August 30, 1959; or

(ii) The runway length, in the case of airplanes certificated after August 29, 1959.

(c) No person may take off a turbine-engine-powered transport category airplane certificated after August 29, 1959, unless, in addition to the requirements of paragraph (b) of this section—

(1) The accelerate-stop distance is no greater than the length of the runway plus the length of the stopway (if present); and

(2) The takeoff distance is no greater than the length of the runway plus the length of the clearway (if present); and

(3) The takeoff run is no greater than the length of the runway.

§ 91.607 Emergency exits for airplanes carrying passengers for hire.

(a) Notwithstanding any other provision of this chapter, no person may operate a large airplane (type certificated under the Civil Air Regulations effective before April 9, 1957) in passenger-carrying operations for hire, with more than the number of occupants—

(1) Allowed under Civil Air Regulations 4b.362 (a), (b), and (c) as in effect on December 20, 1951; or

(2) Approved under Special Civil Air Regulations SR 387, SR 389, SR 389A, or SR 389B, or under this section as in effect.

However, an airplane type listed in the following table may be operated with up to the listed number of occupants (including crewmembers) and the corresponding number of exits (including emergency exits and doors) approved for the emergency exit of passengers or with an occupant-exit configuration approved under paragraph (b) or (c) of this section.

Airplane type	Max. number of occupants including all crewmembers	Corresponding number of exits authorized for passenger use
B-307	61	4
B-377	96	9
C-46	67	4
CV-240	53	6
CV-340 and CV-440.	53	6
DC-3	35	4
DC-3 (Super)	39	5
DC-4	86	5
DC-6	87	7
DC-6B	112	11
L-18	17	3
L-049, L-649, L-749.	87	7
L-1049	96	9
M-202	53	6
M-404	53	7
Viscount 700 Series.	53	7

(b) Occupants in addition to those authorized under paragraph (a) of this section may be carried as follows:

(1) For each additional floor-level exit at least 24 inches wide by 48 inches high, with an unobstructed 20-inch-wide access aisleway between the exit and the main passenger aisle, 12 additional occupants.

(2) For each additional window exit located over a wing that meets the requirements of the airworthiness standards under which the airplane was type certificated or that is large enough to inscribe an ellipse 19 × 26 inches, eight additional occupants.

(3) For each additional window exit that is not located over a wing but that otherwise complies with paragraph (b)(2) of this section, five additional occupants.

(4) For each airplane having a ratio (as computed from the table in paragraph (a) of this section) of maximum number of occupants to number of exits greater than 14:1, and for each airplane that does not have at least one full-size, door-type exit in the side of the fuselage in the rear part of the cabin, the first additional exit must be a floor-level exit that complies with paragraph (b)(1) of this section and must be located in the rear part of the cabin on the opposite side of the fuselage from the main entrance door. However, no person may operate an airplane under this section carrying more than 115 occupants unless there is such an exit on each side of the fuselage in the rear part of the cabin.

(c) No person may eliminate any approved exit except in accordance with the following:

(1) The previously authorized maximum number of occupants must be reduced by the same number of additional occupants authorized for that exit under this section.

(2) Exits must be eliminated in accordance with the following priority schedule: First, non-over-wing window exits; second, over-wing window exits; third, floor-level exits located in the forward part of the cabin; and fourth, floor-level exits located in the rear of the cabin.

(3) At least one exit must be retained on each side of the fuselage regardless of the number of occupants.

(4) No person may remove any exit that would result in a ratio of maximum number of occupants to approved exits greater than 14:1.

(d) This section does not relieve any person operating under Part 121 of this chapter from complying with § 121.291.

§ 91.609 Flight recorders and cockpit voice recorders.

(a) No holder of an air carrier operating certificate or an operating certificate may conduct any operation under this part with an aircraft listed in the holder's operations specifications or current list of aircraft used in air transportation unless that aircraft complies with any applicable flight recorder and cockpit voice recorder requirements of the part under which its certificate is issued except that the operator may—

(1) Ferry an aircraft with an inoperative flight recorder or cockpit voice recorder from a place where repair or replacement cannot be made to a place where they can be made;

(2) Continue a flight as originally planned, if the flight recorder or cockpit voice recorder becomes inoperative after the aircraft has taken off;

(3) Conduct an airworthiness flight test during which the flight recorder or cockpit voice recorder is turned off to test it or to test any communications or electrical equipment installed in the aircraft; or

(4) Ferry a newly acquired aircraft from the place where possession of it is taken to a place where the flight recorder or cockpit voice recorder is to be installed.

(b) [Notwithstanding paragraphs (c) and (e) of this section, an operator other than the holder of an air carrier or a commercial operator certificate may—

[(1) Ferry an aircraft with an inoperative flight recorder or cockpit voice recorder from a place where repair or replacement cannot be made to a place where they can be made;

[(2) Continue a flight as originally planned if the flight recorder or cockpit voice recorder becomes inoperative after the aircraft has taken off;

[(3) Conduct an airworthiness flight test during which the flight recorder or cockpit voice recorder is turned off to test it or to test any communications or electrical equipment installed in the aircraft;

[(4) Ferry a newly acquired aircraft from a place where possession of it was taken to a place where the flight recorder or cockpit voice recorder is to be installed; or

[(5) Operate an aircraft:

[(i) For not more than 15 days while the flight recorder and/or cockpit voice recorder is inoperative and/or removed for repair provided that the aircraft maintenance records contain an entry that indicates the date of failure, and a placard is located in view of the pilot to show that the flight recorder or cockpit voice recorder is inoperative.

[(ii) For not more than an additional 15 days, provided that the requirements in paragraph (b)(5)(i) are met and that a certificated pilot, or a certificated person authorized to return an aircraft to service under § 43.7 of this chapter, certifies in the aircraft maintenance records that additional time is required to complete repairs or obtain a replacement unit.]

(c) No person may operate a U.S. civil registered, multiengine, turbine-powered airplane or rotorcraft having a passenger seating configuration, excluding any pilot seats of 10 or more that has been manufactured after October 11, 1991, unless it is equipped with one or more approved flight recorders that utilize a digital method of recording and storing data and a method of readily retrieving that data from the storage medium, that are capable of recording the data specified in Appendix E to this part, for an airplane, or Appendix F to this part, for a rotorcraft, of this part within the range, accuracy, and recording interval specified, and that are capable of retaining no less than 8 hours of aircraft operation.

(d) Whenever a flight recorder, required by this section, is installed, it must be operated continuously from the

instant the airplane begins the takeoff roll or the rotorcraft begins lift-off until the airplane has completed the landing roll or the rotorcraft has landed at its destination.

(e) Unless otherwise authorized by the Administrator, after October 11, 1991, no person may operate a U.S. civil registered multiengine, turbine-powered airplane or rotorcraft having a passenger seating configuration of six passengers or more and for which two pilots are required by type certification or operating rule unless it is equipped with an approved cockpit voice recorder that:

(1) Is installed in compliance with § 23.1457(a) (1) and (2), (b), (c), (d), (e), (f), and (g); § 25.1457(a) (1) and (2), (b), (c), (d), (e), (f), and (g); 27.1457(a) (1) and (2), (b), (c), (d), (e), (f), and (g); or § 29.1457(a) (1) and (2), (b), (c), (d), (e), (f), and (g) of this chapter, as applicable; and

(2) Is operated continuously from the use of the checklist before the flight to completion of the final checklist at the end of the flight.

(f) In complying with this section, an approved cockpit voice recorder having an erasure feature may be used, so that at any time during the operation of the recorder, information recorded more than 15 minutes earlier may be erased or otherwise obliterated.

(g) In the event of an accident or occurrence requiring immediate notification to the National Transportation Safety Board under Part 830 of its regulations that results in the termination of the flight, any operator who has installed approved flight recorders and approved cockpit voice recorders shall keep the recorded information for at least 60 days or, if requested by the Administrator or the Board, for a longer period. Information obtained from the record is used to assist in determining the cause of accidents or occurrences in connection with the investigation under Part 830. The Administrator does not use the cockpit voice recorder record in any civil penalty or certificate action.

(Amdt. 91-226, Eff. 10/11/91); [(Amdt. 91-228, Eff. 5/5/92)]

§ 91.611 Authorization for ferry flight with one engine inoperative.

(a) *General.* The holder of an air carrier operating certificate or an operating certificate issued under Part 125 may conduct a ferry flight of a four-engine airplane or a turbine-engine-powered airplane equipped with three engines, with one engine inoperative, to a base for the purpose of repairing that engine subject to the following:

(1) The airplane model has been test flown and found satisfactory for safe flight in accordance with paragraph (b) or (c) of this section, as appropriate. However, each operator who before November 19, 1966, has shown that a model of airplane with an engine inoperative is satisfactory for safe flight by a test flight conducted in accordance with performance data contained in the applicable Airplane Flight Manual under paragraph (a)(2)

of this section need not repeat the test flight for that model.

(2) The approved Airplane Flight Manual contains the following performance data and the flight is conducted in accordance with that data:

(i) Maximum weight.

(ii) Center of gravity limits.

(iii) Configuration of the inoperative propeller (if applicable).

(iv) Runway length for takeoff (including temperature accountability).

(v) Altitude range.

(vi) Certificate limitations.

(vii) Ranges of operational limits.

(viii) Performance information.

(ix) Operating procedures.

(3) The operator has FAA approved procedures for the safe operation of the airplane, including specific requirements for—

(i) Limiting the operating weight on any ferry flight to the minimum necessary for the flight plus the necessary reserve fuel load;

(ii) A limitation that takeoffs must be made from dry runways unless, based on a showing of actual operating takeoff techniques on wet runways with one engine inoperative, takeoffs with full controllability from wet runways have been approved for the specific model aircraft and included in the Airplane Flight Manual;

(iii) Operations from airports where the runways may require a takeoff or approach over populated areas; and

(iv) Inspection procedures for determining the operating condition of the operative engines.

(4) No person may take off an airplane under this section if—

(i) The initial climb is over thickly populated areas; or

(ii) Weather conditions at the takeoff or destination airport are less than those required for VFR flight.

(5) Persons other than required flight crewmembers shall not be carried during the flight.

(6) No person may use a flight crewmember for flight under this section unless that crewmember is thoroughly familiar with the operating procedures for one-engine inoperative ferry flight contained in the certificate holder's manual and the limitations and performance information in the Airplane Flight Manual.

(b) *Flight tests: reciprocating-engine-powered airplanes.* The airplane performance of a reciprocating-engine-powered airplane with one engine inoperative must be determined by flight test as follows:

(1) A speed not less than $1.3 \ V_{S1}$ must be chosen at which the airplane may be controlled satisfactorily in a climb with the critical engine inoperative (with its propeller removed or in a configuration desired by the operator and with all other engines operating at the maximum power determined in paragraph (b)(3) of this section.

(2) The distance required to accelerate to the speed listed in paragraph (b)(1) of this section and to climb to 50 feet must be determined with—

(i) The landing gear extended;

(ii) The critical engine inoperative and its propeller removed or in a configuration desired by the operator; and

(iii) The other engines operating at not more than maximum power established under paragraph (b)(3) of this section.

(3) The takeoff, flight and landing procedures, such as the approximate trim settings, method of power application, maximum power, and speed must be established.

(4) The performance must be determined at a maximum weight not greater than the weight that allows a rate of climb of at least 400 feet per minute in the en route configuration set forth in § 25.67(d) of this chapter in effect on January 31, 1977, at an altitude of 5,000 feet.

(5) The performance must be determined using temperature accountability for the takeoff field length, computed in accordance with § 25.61 of this chapter in effect on January 31, 1977.

(c) *Flight tests: Turbine-engine-powered airplanes.* The airplane performance of a turbine-engine-powered airplane with one engine inoperative must be determined by flight tests, including at least three takeoff tests, in accordance with the following:

(1) Takeoff speeds V_R and V_2, not less than the corresponding speeds under which the airplane was type certificated under § 25.107 of this chapter, must be chosen at which the airplane may be controlled satisfactorily with the critical engine inoperative (with its propeller removed or in a configuration desired by the operator, if applicable) and with all other engines operating at not more than the power selected for type certification as set forth in § 25.101 of this chapter.

(2) The minimum takeoff field length must be the horizontal distance required to accelerate and climb to the 35-foot height at V_2 speed (including any additional speed increment obtained in the tests) multiplied by 115 percent and determined with—

(i) The landing gear extended;

(ii) The critical engine inoperative and its propeller removed or in a configuration desired by the operator (if applicable); and

(iii) The other engine operating at not more than the power selected for type certification as set forth in § 25.101 of this chapter.

(3) The takeoff, flight, and landing procedures such as the approximate trim setting, method of power application, maximum power, and speed must be established. The airplane must be satisfactorily controllable during the entire takeoff run when operated according to these procedures.

(4) The performance must be determined at a maximum weight not greater than the weight determined under § 25.121(c) of this chapter but with—

(i) The actual steady gradient of the final takeoff climb requirement not less than 1.2 percent at the end of the takeoff path with two critical engines inoperative; and

(ii) The climb speed not less than the two-engine inoperative trim speed for the actual steady gradient of the final takeoff climb prescribed by paragraph (c)(4)(i) of this section.

(5) The airplane must be satisfactorily controllable in a climb with two critical engines inoperative. Climb performance may be shown by calculations based on, and equal in accuracy to, the results of testing.

(6) The performance must be determined using temperature accountability for takeoff distance and final takeoff climb computed in accordance with § 25.101 of this chapter.

For the purpose of paragraphs (c)(4) and (5) of this section, "two critical engines" means two adjacent engines on one side of an airplane with four engines, and the center engine and one outboard engine on an airplane with three engines.

§ 91.613 Materials for compartment interiors.

(a) No person may operate an airplane that conforms to an amended or supplemental type certificate issued in accordance with SFAR No. 41 for a maximum certificated takeoff weight in excess of 12,500 pounds unless within 1 year after issuance of the initial airworthiness certificate under that SFAR the airplane meets the compartment interior requirements set forth in § 25.853(a), (b), (b-1), (b-2), and (b-3) of this chapter in effect on September 26, 1978.

(b) Thermal/acoustic insulation materials. For transport category airplanes type certificated after January 1, 1958:

(1) For airplanes manufactured before September 2, 2005, when thermal/acoustic insulation is installed in the fuselage as replacements after September 2, 2005, the insulation must meet the flame propagation requirements of § 25.856 of this chapter, effective September 2, 2003, if it is:

(i) Of a blanket construction or

(ii) Installed around air ducting.

(2) For airplanes manufactured after September 2, 2005, thermal/acoustic insulation materials installed in the fuselage must meet the flame propagation requirements of § 25.856 of this chapter, effective September 2, 2003.

§§ 91.615–91.699 [Reserved]

Subpart H—Foreign aircraft operations and operations of U.S.-registered civil aircraft outside of the United States

§ 91.701 Applicability.

(a) This subpart applies to the operations of civil aircraft of U.S. registry outside of the United States and the operations of foreign civil aircraft within the United States.

(b) Section 91.702 of this subpart also applies to each person on board an aircraft operated as follows:

(1) A U.S. registered civil aircraft operated outside the United States;

(2) Any aircraft operated outside the United States--

(i) That has its next scheduled destination or last place of departure in the United States if the aircraft next lands in the United States; or

(ii) If the aircraft lands in the United States with the individual still on the aircraft regardless of whether it was a scheduled or otherwise planned landing site.

§91.702 Persons on board.

Section 91.11 of this part (Prohibitions on interference with crewmembers) applies to each person on board an aircraft.

§ 91.703 Operations of civil aircraft of U.S. registry outside of the United States.

(a) Each person operating a civil aircraft of U.S. registry outside of the United States shall—

(1) When over the high seas, comply with annex 2 (Rules of the Air) to the Convention on International Civil Aviation and with §§ 91.117(c), 91.130, and 91.131;

(2) When within a foreign country, comply with the regulations relating to the flight and maneuver of aircraft there in force;

(3) Except for §§ 91.307(b), 91.309, 91.323, and 91.711, comply with this part so far as it is not inconsistent with applicable regulations of the foreign country where the aircraft is operated or annex 2 of the Convention on International Civil Aviation; and

(4) When over the North Atlantic within airspace designated as Minimum Navigation Performance Specifications airspace, comply with § 91.705.

(b) Annex 2 to the Convention on International Civil Aviation, Eighth Edition—July 1986, with amendments through Amendment 28 effective November 1987, to which reference is made in this part, is incorporated into this part and made a part hereof as provided in 5 U.S.C. 552 and pursuant to 1 CFR Part 51. Annex 2 (including a complete historic file of changes thereto) is available for public inspection at the Rules Docket, AGC-10, Federal Aviation Administration, 800 Independence Avenue SW., Washington, DC 20591. In addition, Annex 2 may be purchased from the International Civil Aviation Organization (Attention: Distribution Officer), P.O. Box 400, Succursale, Place de L'Aviation Internationale, 1000 Sherbrooke Street West, Montreal, Quebec, Canada H3A 2R2.

[*(Amdt. 91-227, Eff. 9/16/93)*]

§ 91.705 Operations within the North Atlantic Minimum Navigation Performance Specifications airspace.

No person may operate a civil aircraft of U.S. registry in North Atlantic (NAT) airspace designated as Minimum Navigation Performance Specifications (MNPS) airspace unless—

(a) The aircraft has approved navigation performance capability which complies with the requirements of Appendix C of this part; and

(b) The operator is authorized by the Administrator to perform such operations.

(c) The Administrator authorizes deviations from the requirements of this section in accordance with section 3 of Appendix C to this part.

§ 91.707 Flights between Mexico or Canada and the United States.

Unless otherwise authorized by ATC, no person may operate a civil aircraft between Mexico or Canada and the United States without filing an IFR or VFR flight plan, as appropriate.

§ 91.709 Operations to Cuba.

No person may operate a civil aircraft from the United States to Cuba unless—

(a) Departure is from an international airport of entry designated in § 6.13 of the Air Commerce Regulations of the Bureau of Customs (19 CFR 6.13); and

(b) In the case of departure from any of the 48 contiguous States or the District of Columbia, the pilot in command of the aircraft has filed—

(1) A DVFR or IFR flight plan as prescribed in §§ 99.11 or 99.13 of this chapter; and

(2) A written statement, within 1 hour before departure, with the Office of Immigration and Naturalization Service at the airport of departure, containing—

(i) All information in the flight plan;

(ii) The name of each occupant of the aircraft;

(iii) The number of occupants of the aircraft; and

(iv) A description of the cargo, if any.

This section does not apply to the operation of aircraft by a scheduled air carrier over routes authorized in operations specifications issued by the Administrator.

(Approved by the Office of Management and Budget under OMB control number 2120-0005).

§ 91.711 Special rules for foreign civil aircraft.

(a) *General.* In addition to the other applicable regulations of this part, each person operating a foreign civil aircraft within the United States shall comply with this section.

(b) *VFR.* No person may conduct VFR operations which require two-way radio communications under this part unless at least one crewmember of that aircraft is able to conduct two-way radio communications in the English language and is on duty during that operation.

(c) *IFR.* No person may operate a foreign civil aircraft under IFR unless— (1) That aircraft is equipped with—

(i) **[Radio equipment allowing two-way radio communication with ATC when it is operated in controlled airspace; and]**

(ii) Radio navigational equipment appropriate to the navigational facilities to be used;

(2) Each person piloting the aircraft—

(i) Holds a current United States instrument rating or is authorized by his foreign airman certificate to pilot under IFR; and

(ii) Is thoroughly familiar with the United States en route, holding, and letdown procedures; and

(3) At least one crewmember of that aircraft is able to conduct two-way radiotelephone communications in the English language and that crewmember is on duty while the aircraft is approaching, operating within, or leaving the United States.

(d) *Over water.* Each person operating a foreign civil aircraft over water off the shores of the United States shall give flight notification or file a flight plan in accordance with the Supplementary Procedures for the ICAO region concerned.

(e) *Flight at and above FL 240.* If VOR navigational equipment is required under paragraph (c)(1)(ii) of this section, no person may operate a foreign civil aircraft within the 50 States and the District of Columbia at or above FL 240, unless the aircraft is equipped with distance measuring equipment (DME) capable of receiving and indicating distance information from the VORTAC facilities to be used. When DME required by this paragraph fails at and above FL 240, the pilot in command of the aircraft shall notify ATC immediately and may then continue operations at and above FL 240 to the next airport of intended landing at which repairs or replacement of the equipment can be made. However, paragraph (e) of this section does not apply to foreign civil aircraft that are not equipped with DME when operated for the following purposes and if ATC is notified prior to each takeoff:

(1) Ferry flights to and from a place in the United States where repairs or alterations are to be made.

(2) Ferry flights to a new country of registry.

(3) Flight of a new aircraft of U.S. manufacture for the purpose of—

(i) Flight testing the aircraft;

(ii) Training foreign flight crews in the operation of the aircraft; or

(iii) Ferrying the aircraft for export delivery outside the United States.

(4) Ferry, demonstration, and test flight of an aircraft brought to the United States for the purpose of demonstration or testing the whole or any part thereof.

[(Amdt. 91-227, Eff. 9/16/93)]

§ 91.713　Operation of civil aircraft of Cuban registry.

No person may operate a civil aircraft of Cuban registry except in controlled airspace and in accordance with air traffic clearance or air traffic control instructions that may require use of specific airways or routes and landings at specific airports.

§ 91.715　Special flight authorizations for foreign civil aircraft.

(a) Foreign civil aircraft may be operated without airworthiness certificates required under § 91.203 if a special flight authorization for that operation is issued under this section. Application for a special flight authorization must be made to the [Flight Standards Division Manager or Aircraft Certification Directorate Manager] of the FAA region in which the applicant is located or to the region within which the U.S. point of entry is located. However, in the case of an aircraft to be operated in the U.S. for the purpose of demonstration at an airshow, the application may be made to the [Flight Standards Division Manager or Aircraft Certification Directorate Manager] of the FAA region in which the airshow is located.

(b) The Administrator may issue a special flight authorization for a foreign civil aircraft subject to any conditions and limitations that the Administrator considers necessary for safe operation in the U.S. airspace.

(c) No person may operate a foreign civil aircraft under a special flight authorization unless that operation also complies with Part 375 of the Special Regulations of the Department of Transportation (14 CFR Part 375).

(Approved by the Office of Management and Budget under OMB control number 2120-0005).

(Amdt. 91-212, Eff. 8/18/90)

§§ 91.717–91.799　[Reserved]

Subpart I—Operating noise limits

§ 91.801　Applicability: Relation to Part 36.

(a) This subpart prescribes operating noise limits and related requirements that apply, as follows, to the operation of civil aircraft in the United States.

(1) Sections 91.803, 91.805, 91.807, 91.809, and 91.811 apply to civil subsonic turbojet airplanes with maximum weights of more than 75,000 pounds and—

(i) If U.S. registered, that have standard airworthiness certificates; or

(ii) If foreign registered, that would be required by this chapter to have a U.S. standard airworthiness certificate in order to conduct the operations intended for the airplane were it registered in the United States. Those sections apply to operations to or from airports in the United States under this part and Parts 121, 125, 129, and 135 of this chapter.

(2) Section 91.813 applies to U.S. operators of civil subsonic turbojet airplanes covered by this subpart. This section applies to operators operating to or from airports in the United States under this part and Parts 121, 125, and 135, but not to those operating under Part 129 of this chapter.

(3) Sections 91.803, 91.819, and 91.821 apply to U.S.-registered civil supersonic airplanes having standard airworthiness certificates and to foreign-registered civil supersonic airplanes that, if registered in the United States, would be required by this chapter to have U.S. standard airworthiness certificates in order to conduct the operations intended for the airplane. Those sections apply to operations under this part and under Parts 121, 125, 129, and 135 of this chapter.

(b) Unless otherwise specified, as used in this subpart "Part 36" refers to 14 CFR Part 36, including the noise levels under Appendix C of that part, notwithstanding the provisions of that part excepting certain airplanes from the specified noise requirements. For purposes of this subpart, the various stages of noise levels, the terms used to describe airplanes with respect to those levels, and the terms "subsonic airplane" and "supersonic airplane" have the meanings specified under Part 36 of this chapter. For purposes of this subpart, for subsonic airplanes operated in foreign air commerce in the United States, the Administrator may accept compliance with the noise requirements under annex 16 of the International Civil Aviation Organization when those requirements have been shown to be substantially compatible with, and achieve results equivalent to those achievable under, Part 36 for that airplane. Determinations made under these provisions are subject to the limitations of § 36.5 of this chapter as if those noise levels were Part 36 noise levels.

(c) Sections 91.851 through 91.877 of this subpart prescribe operating noise limits and related requirements that apply to any civil subsonic turbojet airplane (for which an airworthiness certificated other than an experimental certificate has been issued by the Administrator) with a maximum certificated weight of more than 75,000 pounds operating to or from an airport in the 48 contiguous United States and the District of Columbia under this part, Part 121, 125, 129, or 135 of this chapter on and after September 25, 1991).

(d) Section 91.877 prescribes reporting requirements that apply to any civil subsonic turbojet airplane with a maximum weight of more than 75,000 pounds operated by an air carrier or foreign air carrier between the contiguous

United States and the State of Hawaii, between the State of Hawaii and any point outside of the 48 contiguous United States, or between the islands of Hawaii in turnaround service, under part 121 or 129 of this chapter on or after November 5, 1990.

(Amdt. 91-225, Eff. 9/25/91)

§91.803 Part 125 operators: Designation of applicable regulations.

For airplanes covered by this subpart and operated under Part 125 of this chapter, the following regulations apply as specified:

(a) For each airplane operation to which requirements prescribed under this subpart applied before November 29, 1980, those requirements of this subpart continue to apply.

(b) For each subsonic airplane operation to which requirements prescribed under this subpart did not apply before November 29, 1980, because the airplane was not operated in the United States under this part or Part 121, 129, or 135 of this chapter, the requirements prescribed under §§ 91.805, of this subpart apply.

(c) For each supersonic airplane operation to which requirements prescribed under this subpart did not apply before November 29, 1980, because the airplane was not operated in the United States under this part or Part 121, 129, or 135 of this chapter, the requirements of §§ 91.819 and 91.821 of this subpart apply.

(d) For each airplane required to operate under Part 125 for which a deviation under that part is approved to operate, in whole or in part, under this part or Part 121, 129, or 135 of this chapter, notwithstanding the approval, the requirements prescribed under paragraphs (a), (b), and (c) of this section continue to apply.

§91.805 Final compliance: Subsonic airplanes.

Except as provided in §§ 91.809 and 91.811, on and after January 1, 1985, no person may operate to or from an airport in the United States any subsonic airplane covered by this subpart unless that airplane has been shown to comply with Stage 2 or Stage 3 noise levels under Part 36 of this chapter.

§91.807 [Removed and Reserved]

§91.809 [Removed and Reserved]

§91.811 [Removed and Reserved]

§91.815 Agricultural and fire fighting airplanes: Noise operating limitations.

(a) This section applies to propeller-driven, small airplanes having standard airworthiness certificates that are designed for "agricultural aircraft operations" (as defined in 137.3 of this chapter, as effective on January 1, 1966) or for dispensing fire fighting materials.

(b) If the Airplane Flight Manual, or other approved manual material information, markings, or placards for the airplane indicate that the airplane has not been shown to comply with the noise limits under Part 36 of this chapter, no person may operate that airplane, except—

(1) To the extent necessary to accomplish the work activity directly associated with the purpose for which it is designed;

(2) To provide flight crewmember training in the special purpose operation for which the airplane is designed; and

(3) To conduct "nondispensing aerial work operations" in accordance with the requirements under § 137.29(c) of this chapter.

§91.817 Civil aircraft sonic boom.

(a) No person may operate a civil aircraft in the United States at a true flight Mach number greater than 1 except in compliance with conditions and limitations in an authorization to exceed Mach 1 issued to the operator under Appendix B of this part.

(b) In addition, no person may operate a civil aircraft for which the maximum operating limit speed M_{M0} exceeds a Mach number of 1, to or from an airport in the United States, unless—

(1) Information available to the flight crew includes flight limitations that ensure that flights entering or leaving the United States will not cause a sonic boom to reach the surface within the United States; and

(2) The operator complies with the flight limitations prescribed in paragraph (b)(1) of this section or complies with conditions and limitations in an authorization to exceed Mach 1 issued under Appendix B of this part.

(Approved by the Office of Management and Budget under OMB control number 2120-0005).

§91.819 Civil supersonic airplanes that do not comply with Part 36.

(a) *Applicability.* This section applies to civil supersonic airplanes that have not been shown to comply with the Stage 2 noise limits of Part 36 in effect on October 13, 1977, using applicable trade-off provisions, and that are operated in the United States, after July 31, 1978.

(b) *Airport use.* Except in an emergency, the following apply to each person who operates a civil supersonic airplane to or from an airport in the United States:

(1) Regardless of whether a type design change approval is applied for under Part 21 of this chapter, no person may land or take off an airplane covered by this section for which the type design is changed, after July 31, 1978, in a manner constituting an "acoustical change" under § 21.93 unless the acoustical change requirements of Part 36 are complied with.

(2) No flight may be scheduled, or otherwise planned, for takeoff or landing after 10 p.m. and before 7 a.m. local time.

§ 91.821 Civil supersonic airplanes: Noise limits.

Except for Concorde airplanes having flight time before January 1, 1980, no person may operate in the United States, a civil supersonic airplane that does not comply with Stage 2 noise limits of Part 36 in effect on October 13, 1977, using applicable trade-off provisions.

§§ 91.823–91.849 [Reserved]

[§ 91.851 Definitions.

[For the purposes of §§ 91.851 through 91.875 of this subpart:

[*Contiguous United States* means the area encompassed by the 48 contiguous United States and the District of Columbia.

[*Fleet* means those civil subsonic turbojet airplanes with a maximum certificated weight of more than 75,000 pounds that are listed on an operator's operations specifications as eligible for operation in the contiguous United States.

[*Import* means a change in ownership of an airplane from a non-U.S. person to a U.S. person when the airplane is brought into the United States for operation.

[*Operations specifications* means an enumeration of airplanes by type, model, series, and serial number operated by the operator or foreign air carrier on a given day, regardless of how or whether such airplanes are formally listed or designated by the operator.

[*Owner* means any person that has indicia of ownership sufficient to register the airplane in the United States pursuant to Part 47 of this chapter.

[*New entrant* means an air carrier or foreign air carrier that, on or before November 5, 1990, did not conduct operations under Part 121 or 129 of this chapter using an airplane covered by this subpart to or from any airport in the contiguous United States, but that initiates such operation after that date.

[*Stage 2 noise levels* means the requirements for Stage 2 noise levels as defined in Part 36 of this chapter in effect on November 5, 1990.

[*Stage 3 noise levels* means the requirements for Stage 3 noise levels as defined in Part 36 of this chapter in effect on November 5, 1990.

[*Stage 2 airplane* means a civil subsonic turbojet airplane with a maximum certificated weight of 75,000 pounds or more that complies with Stage 2 noise levels as defined in Part 36 of this chapter.

[*Stage 3 airplane* means a civil subsonic turbojet airplane with a maximum certificated weight of 75,000 pounds or more that complies with Stage 3 noise levels as defined in Part 36 of this chapter.]

[(Amdt. 91-225, Eff. 9/25/91)]

[Chapter 4 noise level means a noise level at or below the maximum noise level prescribed in Chapter 4, Paragraph 4.4, Maximum Noise Levels, of the International Civil Aviation Organization (ICAO) Annex 16, Volume I,

Amendment 7, effective March 21, 2002. The Director of the Federal Register in accordance with 5 U.S.C. 552 (a) and 1 CFR part 51 approved the incorporation by reference of this document, which can be obtained from the International Civil Aviation Organization (ICAO), Document Sales Unit, 999 University Street, Montreal, Quebec H3C 5H7, Canada. Also, you may obtain documents on the Internet at *http://www.ICAO.int/eshop/index.cfm.* Copies may be reviewed at the U.S. Department of Transportation, Docket Management System, 400 7th Street, SW., Room PL 401, Washington, DC or at the National Archives and Records Administration (NARA). For information on the availability of this material at NARA, call 202-741-6030, or go to: *http://www.archives.gov/federalregister/code*_of_ federal_regulations/ibr/_locations.html.]

[Stage 4 noise level means a noise level at or below the Stage 4 noise limit prescribed in part 36 of this chapter.]

[Stage 4 airplane means an airplane that has been shown not to exceed the Stage 4 noise limit prescribed in part 36 of this chapter. A Stage 4 airplane complies with all of the noise operating rules of this part.]

§ 91.853 Final compliance: Civil subsonic airplanes.

[Except as provided in § 91.873, after December 31, 1999, no person shall operate to or from any airport in the contiguous United States any airplane subject to § 91.801(c) of this subpart, unless that airplane has been shown to comply with Stage 3 or Stage 4 noise levels.]

[(Amdt. 91-225, Eff. 9/25/91)]

[§ 91.855 Entry and nonaddition rule.

[No person may operate any airplane subject to § 91.801(c) of this subpart to or from an airport in the contiguous United States unless one or more of the following apply:

[(a) The airplane complies with Stage 3 or Stage 4 noise levels.

[(b) The airplane complies with Stage 2 noise levels and was owned by a U.S. person on and since November 5, 1990. Stage 2 airplanes that meet these criteria and are leased to foreign airlines are also subject to the return provisions of paragraph (e) of this section.

[(c) The airplane complies with Stage 2 noise levels, is owned by a non-U.S. person, and is the subject of a binding lease to a U.S. person effective before and on September 19, 1991. Any such airplane may be operated for the term of the lease in effect on that date, and any extensions thereof provided for in that lease.

[(d) The airplane complies with Stage 2 noise levels and is operated by a foreign air carrier.

[(e) The airplane complies with Stage 2 noise levels and is operated by a foreign operator other than for the purpose of foreign air commerce.

[(f) The airplane complies with Stage 2 noise levels and—

[(1) On November 5, 1990, was owned by:

[(i) A corporation, trust, or partnership organized under the laws of the United States or any State (including individual States, territories, possessions, and the District of Columbia);

[(ii) An individual who is a citizen of the United States; or

[(iii) An entity owned or controlled by a corporation, trust, partnership, or individual described in paragraph (g)(1)(i) or (ii) of this section; and

[(2) Enters into the United States not later than 6 months after the expiration of a lease agreement (including any extensions thereof) between an owner described in paragraph (f)(1) of this section and a foreign airline.

[(g) The airplane complies with Stage 2 noise levels and was purchased by the importer under a written contract executed before November 5, 1990.

[(h) Any Stage 2 airplane described in this section is eligible for operation in the contiguous United States only as provided under § 91.865 or § 91.867.]

[(Amdt. 91-225, Eff. 9/25/91)]

§ 91.857 Stage 2 operations outside of the 48 contiguous United States, and authorization for maintenance.

An operator of a Stage 2 airplane that is operating only between points outside the contiguous United States on or after November 5, 1990, must include a statement that such airplane may not be used to provide air transportation to or from any airport in the contiguous United States.

§ 91.858 Special Flight authorizations for non-revenue Stage 2 operations.

(a) After December 31, 1999, any operator of a Stage 2 airplane over 75,000 pounds may operate that airplane in nonrevenue service in the contiguous United States only for the following purposes:

(1) Sell, lease, or scrap the airplane;

(2) Obtain modifications to meet Stage 3 noise levels;

(3) Obtain scheduled heavy maintenance or significant modifications;

(4) Deliver the airplane to a lessee or return it to a lessor;

(5) Park or store the airplane; and

(6) Prepare the airplane for any of the purposes listed in paragraph (a)(1) thru (a)(5) of this section.

(b) An operator of a Stage 2 airplane that needs to operate in the contiguous United States for any of the purposes listed above may apply to FAA's Office of Environment and Energy for a special flight authorization. The applicant must file in advance. Applications are due 30 days in advance of the planned flight and must provide the information necessary for the FAA to determine that the planned flight is within the limits prescribed in the law.

[§ 91.859 Modification to meet Stage 3 or Stage 4 noise levels.

For an airplane subject to § 91.801(c) of this subpart and otherwise prohibited from operation to or from an airport in the contiguous United States by § 91.855, any person may apply for a special flight authorization for that airplane to operate in the contiguous United States for the purpose of obtaining modifications to meet Stage 3 or Stage 4 noise levels.]

[§ 91.861 Base level.

[(a) U.S. Operators. The base level of a U.S. operator is equal to the number of owned or leased Stage 2 airplanes subject to § 91.801(c) of this subpart that were listed on that operator's operations specifications for operations to or from airports in the contiguous United States on any one day selected by the operator during the period January 1, 1990 through July 1, 1991, plus or minus adjustments made pursuant to paragraphs (a)(1) and (2).

[(1) The base level of a U.S. operator shall be increased by a number equal to the total of the following—

[(i) The number of Stage 2 airplanes returned to service in the United States pursuant to § 91.855(f);

[(ii) The number of Stage 2 airplanes purchased pursuant to § 91.855(g); and

[(iii) Any U.S. operator base level acquired with a Stage 2 airplane transferred from another person under § 91.863.

[(2) The base level of a U.S. operator shall be decreased by the amount of U.S. operator base level transferred with the corresponding number of Stage 2 airplanes to another person under § 91.863.

[(b) Foreign air carriers. The base level of a foreign air carrier is equal to the number of owned or leased Stage 2 airplanes that were listed on that carrier's U.S. operations specifications on any one day during the period January 1, 1990, through July 1, 1991, plus or minus any adjustments to the base levels made pursuant to paragraphs (b)(1) and (2).

[(1) The base level of a foreign air carrier shall be increased by the amount of foreign air carrier base level acquired with a Stage 2 airplane from another person under § 91.863.

[(2) The base level of a foreign air carrier shall be decreased by the amount of foreign air carrier base level transferred with a Stage 2 airplane to another person under § 91.863.

[(c) New entrants do not have a base level.]

[(Amdt. 91-225, Eff. 9/25/91)]

[§ 91.863 Transfers of Stage 2 airplanes with base level.

[(a) Stage 2 airplanes may be transferred with or without the corresponding amount of base level. Base level may not be transferred without the corresponding number of Stage 2 airplanes.

[(b) No portion of a U.S. operator's base level established under § 91.861(a) may be used for operations by a foreign air carrier. No portion of a foreign air carrier's base level established under § 91.861(b) may be used for operations by a U.S. operator.

[(c) Whenever a transfer of Stage 2 airplanes with base level occurs, the transferring and acquiring parties shall, within 10 days, jointly submit written notification of the transfer to the FAA, Office of Environment and Energy. Such notification shall state:

[(1) The names of the transferring and acquiring parties;

[(2) The name, address, and telephone number of the individual responsible for submitting the notification on behalf of the transferring and acquiring parties;

[(3) The total number of Stage 2 airplanes transferred, listed by airplane type, model, series, and serial number;

[(4) The corresponding amount of base level transferred and whether it is U.S. operator or foreign air carrier base level; and

[(5) The effective date of the transaction.

[(d) If, taken as a whole, a transaction or series of transactions made pursuant to this section does not produce an increase or decrease in the number of Stage 2 airplanes for either the acquiring or transferring operator, such transaction or series of transactions may not be used to establish compliance with the requirements of § 91.865.]

[(Amdt. 91-225, Eff. 9/25/91)]

[§ 91.865 Phased compliance for operators with base level.

[Except as provided in paragraph (a) of this section, each operator that operates an airplane under Part 91, 121, 125, 129, or 135 of this chapter, regardless of the national registry of the airplane, shall comply with paragraph (b) or (d) of this section at each interim compliance date with regard to its subsonic airplane fleet covered by § 91.801(c) of this subpart.

[(a) This section does not apply to new entrants covered by § 91.867 or to foreign operators not engaged in foreign air commerce.

[(b) Each operator that chooses to comply with this paragraph pursuant to any interim compliance requirement shall reduce the number of Stage 2 airplanes it operates that are eligible for operation in the contiguous United States to a maximum of:

[(1) After December 31, 1994, 75 percent of the base level held by the operator;

[(2) After December 31, 1996, 50 percent of the base level held by the operator;

[(3) After December 31, 1998, 25 percent of the base level held by the operator.

[(c) Except as provided under § 91.871, the number of Stage 2 airplanes that must be reduced at each compliance date contained in paragraph (b) of this section shall be determined by reference to the amount of base level held by the operator on that compliance date, as calculated under § 91.861.

[(d) Each operator that chooses to comply with this paragraph pursuant to any interim compliance requirement shall operate a fleet that consists of:

[(1) After December 31, 1994, not less than 55 percent Stage 3 airplanes;

[(2) After December 31, 1996, not less than 65 percent Stage 3 airplanes;

[(3) After December 31, 1998, not less than 75 percent Stage 3 airplanes.

[(e) Calculations resulting in fractions may be rounded to permit the continued operation of the next whole number of Stage 2 airplanes.]

[(Amdt. 91-225, Eff. 9/25/91)]

[§ 91.867 Phased compliance for new entrants.

[(a) New entrant U.S. air carriers.

[(1) A new entrant initiating operations under Part 121 of this chapter on or before December 31, 1994, may initiate service without regard to the percentage of its fleet composed of Stage 3 airplanes.

[(2) After December 31, 1994, at least 25 percent of the fleet of a new entrant must comply with Stage 3 noise levels.

[(3) After December 31, 1996, at least 50 percent of the fleet of a new entrant must comply with Stage 3 noise levels.

[(4) After December 31, 1998, at least 75 percent of the fleet of a new entrant must comply with Stage 3 noise levels.

[(b) New entrant foreign air carriers.

[(1) A new entrant foreign air carrier initiating Part 129 operations on or before December 31, 1994, may initiate service without regard to the percentage of its fleet composed of Stage 3 airplanes.

[(2) After December 31, 1994, at least 25 percent of the fleet on U.S. operations specifications of a new entrant foreign air carrier must comply with Stage 3 noise levels.

[(3) After December 31, 1996, at least 50 percent of the fleet on U.S. operations specifications of a new entrant foreign air carrier must comply with Stage 3 noise levels.

[(4) After December 31, 1998, at least 75 percent of the fleet on U.S. operations specifications of a new entrant foreign air carrier must comply with Stage 3 noise levels.

[(c) Calculations resulting in fractions may be rounded to permit the continued operation of the next whole number of Stage 2 airplanes.]

[(Amdt. 91-225, Eff. 9/25/91)]

[§ 91.869 Carry-forward compliance.

[(a) Any operator that exceeds the requirements of paragraph (b) of § 91.865 of this part on or before December 31, 1994, or on or before December 31, 1996, may claim a credit that may be applied at a subsequent interim compliance date.

[(b) Any operator that eliminates or modifies more Stage 2 airplanes pursuant to § 91.865(b) than required as of December 31, 1994, or December 31, 1996, may count the number of additional Stage 2 airplanes reduced as a credit toward—

[(1) The number of Stage 2 airplanes that it would otherwise be required to reduce following a subsequent interim compliance date specified in § 91.865(b); or

[(2) The number of Stage 3 airplanes it would otherwise be required to operate in its fleet following a subsequent interim compliance date to meet the percentage requirements specified in § 91.865(d).]

[Amdt. 91-225, Eff. 9/25/91)]

[§ 91.871 Waivers from interim compliance requirements.

[(a) Any U.S. operator or foreign air carrier subject to the requirements of § 91.865 or § 91.867 of this subpart may request a waiver from any individual compliance requirement.

[(b) Applications must be filed with the Secretary of Transportation at least 120 days prior to the compliance date from which the waiver is requested.

[(c) Applicants must show that a grant of waiver would be in the public interest, and must include in its application its plans and activities for modifying its fleet, including evidence of good faith efforts to comply with the requirements of § 91.865 or § 91.867. The application should contain all information the applicant considers relevant, including, as appropriate, the following:

[(1) The applicant's balance sheet and cash flow positions;

[(2) The composition of the applicant's current fleet; and

[(3) The applicant's delivery position with respect to new airplanes or noise-abatement equipment.

[(d) Waivers will be granted only upon a showing by the applicant that compliance with the requirements of § 91.865 or § 91.867 at a particular interim compliance date is financially onerous, physically impossible, or technologically infeasible, or that it would have an adverse effect on competition or on service to small communities.

[(e) The conditions of any waiver granted under this section shall be determined by the circumstances presented in the application, but in no case may the term extend beyond the next interim compliance date.

[(f) A summary of any request for a waiver under this section will be published in the *Federal Register*, and public comment will be invited. Unless the Secretary finds that circumstances require otherwise, the public comment period will be at least 14 days.]

[(Amdt. 91-225, Eff. 9/25/91)]

[§ 91.873 Waivers from final compliance.

[(a) A U.S. air carrier or a foreign air carrier may apply for a waiver from the prohibition contained in § 91.853 of this part for its remaining Stage 2 airplanes, provided that, by July 1, 1999, at least 85 percent of the airplanes used by the carrier to provide service to or from an airport in the contiguous United States will comply with the Stage 3 noise levels.

[(b) An application for the waiver described in paragraph (a) of this section must be filed with the Secretary of Transportation no later than January 1, 1999, or, in the case of a foreign air carrier no later than April 20, 2000. Such application must include a plan with firm orders for replacing or modifying all airplanes to comply with Stage 3 levels at the earliest practical time.

[(c) To be eligible to apply for the waiver under this section, a new entrant U.S. air carrier must initiate service no later than January 1, 1999, and must comply fully with all provisions of this section.

[(d) The Secretary may grant a waiver under this section if the Secretary finds that granting such waiver is in the public interest. In making such a finding, the Secretary shall include consideration of the effect of granting such a waiver on competition in the air carrier industry and the effect on small community air service, and any other information submitted by the applicant that the Secretary considers relevant.

[(e) The term of any waiver granted under this section shall be determined by the circumstances presented in the application, but in no case will the waiver permit the operation of any Stage 2 airplane covered by this subchapter in the contiguous United States after December 31, 2003.

[(f) A summary of any request for a waiver under this section will be published in the *Federal Register*, and public comment will be invited. Unless the Secretary finds that circumstances require otherwise, the public comment period will be at least 14 days.]

(Amdt. 91-225, Eff. 9/25/91)]

[§ 91.875 Annual progress reports.

[(a) Each operator subject to § 91.865 or § 91.867 of this chapter shall submit an annual report to the FAA, Office of Environment and Energy, on the progress it has made toward complying with the requirements of that section. Such reports shall be submitted no later than 45 days after the end of a calendar year. All progress reports must provide the information through the end of the calendar year, be certified by the operator as true and complete (under penalty of 18 U.S.C. 1001), and include the following information:

[(1) The name and address of the operator;

[(2) The name, title, and telephone number of the person designated by the operator to be responsible for ensuring the accuracy of the information in the report;

[(3) The operator's progress during the reporting period toward compliance with the requirements of § 91.853, § 91.865 or § 91.867. For airplanes on U.S. operations specifications, each operator shall identify the airplanes by type, model, series, and serial number.

[(i) Each Stage 2 airplane added or removed from operation or U.S. operations specifications (grouped separately by those airplanes acquired with and without base level);

[(ii) Each Stage 2 airplane modified to Stage 3 noise levels (identifying the manufacturer and model of noise abatement retrofit equipment);

[(iii) Each Stage 3 airplane on U.S. operations specifications as of the last day of the reporting period; and

[(iv) For each Stage 2 airplane transferred or acquired, the name and address of the recipient or transferor; and if base level was transferred, the person to or from whom base level was transferred or acquired pursuant to Section 91.863 along with the effective date of each base level transaction, and the type of base level transferred or acquired.

[(b) Each operator subject to § 91.865 or § 91.867 of this chapter shall submit an initial progress report covering the period from January 1, 1990, through December 31, 1991, and provide:

[(1) For each operator subject to § 91.865:

[(i) The date used to establish its base level pursuant to § 91.861 (a); and

[(ii) a list of those Stage 2 airplanes (by type, model, series, and serial number) in its base level, including adjustments made pursuant to § 91.861 after the date its base level was established.

[(2) For each U.S. operator:

[(i) A plan to meet the compliance schedules in § 91.865 or § 91.867 and the final compliance date of § 91.853, including the schedule for delivery of replacement Stage 3 airplanes or the installation of noise abatement retrofit equipment; and

[(ii) A separate list (by type, model, series, and serial number) of those airplanes included in the operator's base level, pursuant to § 91.861 (a)(1)(i) and (ii), under the categories "returned" or "purchased," along with the date each was added to its operations specifications.

[(c) Each operator subject to § 91.865 or § 91.867 of this chapter shall submit subsequent annual progress reports covering the calendar year preceding the report and including any changes in the information provided in paragraphs (a) and (b) of this section; including the use of any carry-forward credits pursuant to § 91.869.

[(d) An operator may request, in any report, that specific planning data be considered proprietary.

[(e) If an operator's actions during any reporting period cause it to achieve compliance with § 91.853, the report should include a statement to that effect. Further progress reports are not required unless there is any change in the information reported pursuant to (a) of this section.

[(f) For each U.S. operator subject to § 91.865, progress reports submitted for calendar years 1994, 1996, and 1998, shall also state how the operator achieved compliance with the requirements of that section, i.e.—

[(1) By reducing the number of Stage 2 airplanes in its fleet to no more than the maximum permitted percentage of its base level under § 91.865(b), or

[(2) By operating a fleet that consists of at least the minimum required percentage of Stage 3 airplanes required under § 91.865(d).]

§91.877 Annual reporting of Hawaiian operations.

(a) Each air carrier or foreign air carrier subject to § 91.865 or § 91.867 of this part that conducts operations between the contiguous United States and the State of Hawaii, between the State of Hawaii and any point outside of the contiguous United States, or between the islands of Hawaii in turnaround service, on or since November 5, 1990, shall include in its annual report the information described in paragraph (c) of this section.

(b) Each air carrier or foreign air carrier not subject to § 91.865 or § 91.867 of this part that conducts operations between the contiguous U.S. and the State of Hawaii, between the State of Hawaii, between the State of Hawaii and any point outside of the contiguous United States, or between the islands of Hawaii in turnaround service, on or since November 5, 1990, shall submit an annual report to the FAA, Office of Environment and Energy, on its compliance with the Hawaiian operations provisions of 49 U.S.C. 47528. Such reports shall be submitted no later than 45 days after the end of a calendar year. All progress reports must provide the information through the end of the calendar year, be certified by the operator as true and complete (under penalty of 18 U.S.C. 1001), and include the following information—

(1) The name and address of the air carrier or foreign air carrier;

(2) The name, title, and telephone number of the person designated by the air carrier or foreign air carrier to be responsible for ensuring the accuracy of the information in the report; and

(3) The information specified in paragraph (c) of this section.

(c) The following information must be included in reports filed pursuant to this section—

(1) For operations conducted between the contiguous United States and the State of Hawaii—

(i) The number of Stage 2 airplanes used to conduct such operations as of November 5, 1990;

(ii) Any change to that number during the calendar year being reported, including the date of such change;

(2) For air carriers that conduct inter island turnaround service in the State of Hawaii—

(i) The number of Stage 2 airplanes used to conduct such operations as of November 5, 1990;

(ii) Any change to that number during the calendar year being reported, including the date of such change;

(iii) For an air carrier that provided inter-island turnaround service within the state of Hawaii on November 5,

1990, the number reported under paragraph (c)(2)(i) of this section may include all Stage 2 airplanes with a maximum certificated takeoff weight of more than 75,000 pounds that were owned or leased by the air carrier on November 5, 1990, regardless of whether such airplanes were operated by that air carrier or foreign air carrier on that date.

(3) For operations conducted between the State of Hawaii and a point outside the contiguous United States—

(i) The number of Stage 2 airplanes used to conduct such operations as of November 5, 1990; and

(ii) Any change to that number during the calendar year being reported, including the date of such change.

(d) Reports or amended reports for years predating this regulation are required to be filed concurrently with the next annual report.

§ 91.899 [Reserved]

Subpart J—Waivers

§ 91.901 [Reserved]

§ 91.903 Policy and procedures.

(a) The Administrator may issue a certificate of waiver authorizing the operation of aircraft in deviation from any rule listed in this subpart if the Administrator finds that the proposed operation can be safely conducted under the terms of that certificate of waiver.

(b) An application for a certificate of waiver under this part is made on a form and in a manner prescribed by the Administrator and may be submitted to any FAA office.

(c) A certificate of waiver is effective as specified in that certificate of waiver.

§ 91.905 List of rules subject to waivers.

Sec

91.107	Use of safety belts.
91.111	Operating near other aircraft.
91.113	Right-of-way rules: Except water operations.
91.115	Right-of-way rules: Water operations.
91.117	Aircraft speed.
91.119	Minimum safe altitudes: General.
91.121	Altimeter settings.
91.123	Compliance with ATC clearances and instructions.
91.125	ATC light signals.
[91.126	*Operating on or in the vicinity of an airport in Class G airspace.]*
[91.127	*Operating on or in the vicinity of an airport in Class E airspace.]*
[91.129	*[Operations in Class D airspace.]*
[91.130	*Operations in Class C airspace.]*
[91.131	*Operations in Class B airspace.]*
91.133	Restricted and prohibited areas.
[91.135	*Operations in Class A airspace.]*
91.137	Temporary flight restrictions.
91.141	Flight restrictions in the proximity of the Presidential and other parties.
91.143	Flight limitation in the proximity of space flight operations.
91.153	VFR flight plan: Information required.
91.155	Basic VFR weather minimums.
91.157	Special VFR weather minimums.
91.159	VFR cruising altitude or flight level.
91.169	IFR flight plan: Information required.
91.173	ATC clearance and flight plan required.
91.175	Takeoff and landing under IFR.
91.177	Minimum altitudes for IFR operations.
91.179	IFR cruising altitude or flight level.
91.181	Course to be flown.
91.183	IFR radio communications.
91.185	IFR operations: Two-way radio communications failure.
91.187	Operations under IFR in controlled airspace: Malfunction reports.
91.209	Aircraft lights.
91.303	Aerobatic flights.
91.305	Flight test areas.
91.311	Towing: Other than under §91.309
91.313(e)	Restricted category civil aircraft: Operating limitations.
91.515	Flight altitude rules.
91.705	Operations within the North Atlantic Minimum Navigation Performance Specifications Airspace.
91.707	Flights between Mexico or Canada and the United States.
91.713	Operation of civil aircraft of Cuban registry.

[(Amdt. 91-227, Eff. 9/16/93)]

§§91.907-91.999 [Reserved}

Subpart K—Fractional Ownership Operations

§ 91.1001 Applicability.

(a) This subpart prescribes rules, in addition to those prescribed in other subparts of this part, that apply to fractional owners and fractional ownership program managers governing—

(1) The provision of program management services in a fractional ownership program;

(2) The operation of a fractional ownership program aircraft in a fractional ownership program; and

(3) The operation of a program aircraft included in a fractional ownership program managed by an affiliate of the manager of the program to which the owner belongs.

(b) As used in this part—

(1) *Affiliate of a program manager* means a manager that, directly, or indirectly, through one or more intermediaries, controls, is controlled by, or is under common control with, another program manager. The holding of at least forty percent (40 percent) of the equity and forty percent (40 percent) of the voting power of an entity will be presumed to constitute control for purposes of determining an affiliation under this subpart.

(2) A *dry-lease aircraft exchange* means an arrangement, documented by the written program agreements, under which the program aircraft are available, on an as needed basis without crew, to each fractional owner.

(3) A *fractional owner or owner* means an individual or entity that possesses a minimum fractional ownership interest in a program aircraft and that has entered into the applicable program agreements; provided, however, that in the case of the flight operations described in paragraph (b)(6)(ii) of this section, and solely for purposes of requirements pertaining to those flight operations, the fractional owner operating the aircraft will be deemed to be a fractional owner in the program managed by the affiliate.

(4) A *fractional ownership interest* means the ownership of an interest or holding of a multi-year leasehold interest and/or a multi-year leasehold interest that is convertible into an ownership interest in a program aircraft.

(5) A *fractional ownership program or program* means any system of aircraft ownership and exchange that consists of all of the following elements.

(i) The provision for fractional ownership program management services by a single fractional ownership program manager on behalf of the fractional owners.

(ii) Two or more airworthy aircraft.

(iii) One or more fractional owners per program aircraft, with at least one program aircraft having more than one owner.

(iv) Possession of at least a minimum fractional ownership interest in one or more program aircraft by each fractional owner.

(v) A dry-lease aircraft exchange arrangement among all of the fractional owners.

(vi) Multi-year program agreements covering the fractional ownership, fractional ownership program management services, and dry-lease aircraft exchange aspects of the program.

(6) A *fractional ownership program* aircraft *or program aircraft* means:

(i) An aircraft in which a fractional owner has a minimal fractional ownership interest and that has been included in the dry-lease aircraft exchange pursuant to the program agreements, or

(ii) In the case of a fractional owner from one program operating an aircraft in a different fractional ownership program managed by an affiliate of the operating owner's program manager, the aircraft being operated by the fractional owner, so long as the aircraft is:

(A) Included in the fractional ownership program managed by the affiliate of the operating owner's program manager, and

(B) Included in the operating owner's program's dry-lease aircraft exchange pursuant to the program agreements of the operating owner's program.

(iii) An aircraft owned in whole or in part by the program manager that has been included in the dry-lease aircraft exchange and is used to supplement program operations.

(7) A *Fractional Ownership Program Flight or Program Flight* means a flight under this subpart when one or more passengers or property designated by a fractional owner are on board the aircraft.

(8) *Fractional ownership program management services or program management services* mean administrative and aviation support services furnished in accordance with the applicable requirements of this subpart or provided by the program manager on behalf of the fractional owners, including, but not limited to, the—

(i) Establishment and implementation of program safety guidelines;

(ii) Employment, furnishing, or contracting of pilots and other crewmembers;

(iii) Training and qualification of pilots and other crewmembers and personnel;

(iv) Scheduling and coordination of the program aircraft and crews;

(v) Maintenance of program aircraft;

(vi) Satisfaction of recordkeeping requirements;

(vii) Development and use of a program operations manual and procedures; and

(viii) Application for and maintenance of management specifications and other authorizations and approvals.

(9) A *fractional ownership program manager or program manager* means the entity that offers fractional ownership program management services to fractional owners,

and is designated in the multi-year program agreements referenced in paragraph (b)(1)(v) of this section to fulfill the requirements of this chapter applicable to the manager of the program containing the aircraft being flown. When a fractional owner is operating an aircraft in a fractional ownership program managed by an affiliate of the owner's program manager, the references in this subpart to the flight-related responsibilities of the program manager apply, with respect to that particular flight, to the affiliate of the owner's program manager rather than to the owner's program manager.

(10) A *minimum fractional ownership interest* means—

(i) A fractional ownership interest equal to, or greater than, one-sixteenth (1/16) of at least one subsonic, fixed-wing or powered-lift program aircraft; or

(ii) A fractional ownership interest equal to, or greater than, one-thirty-second (1/32) of at least one rotorcraft program aircraft.

(c) The rules in this subpart that refer to a fractional owner or a fractional ownership program manager also apply to any person who engages in an operation governed by this subpart without the management specifications required by this subpart.

§ 91.1002 Compliance date.

No person that conducted flights before November 17, 2003 under a program that meets the definition of fractional ownership program in § 91.1001 may conduct such flights after February 17, 2005 unless it has obtained management specifications under this subpart.

§ 91.1003 Management contract between owner and program manager.

Each owner must have a contract with the program manager that—

(a) Requires the program manager to ensure that the program conforms to all applicable requirements of this chapter.

(b) Provides the owner the right to inspect and to audit, or have a designee of the owner inspect and audit, the records of the program manager pertaining to the operational safety of the program and those records required to show compliance with the management specifications and all applicable regulations. These records include, but are not limited to, the management specifications, authorizations, approvals, manuals, log books, and maintenance records maintained by the program manager.

(c) Designates the program manager as the owner's agent to receive service of notices pertaining to the program that the FAA seeks to provide to owners and authorizes the FAA to send such notices to the program manager in its capacity as the agent of the owner for such service.

(d) Acknowledges the FAA's right to contact the owner directly if the Administrator determines that direct contact is necessary.

§ 91.1005 Prohibitions and limitations.

(a) Except as provided in § 91.321 or § 91.501, no owner may carry persons or property for compensation or hire on a program flight.

(b) During the term of the multi-year program agreements under which a fractional owner has obtained a minimum fractional ownership interest in a program aircraft, the flight hours used during that term by the owner on program aircraft must not exceed the total hours associated with the fractional owner's share of ownership.

(c) No person may sell or lease an aircraft interest in a fractional ownership program that is smaller than that prescribed in the definition of "minimum fractional ownership interest" in §91.1001(b)(10) unless flights associated with that interest are operated under part 121 or 135 of this chapter and are conducted by an air carrier or commercial operator certificated under part **119** of this chapter.

§ 91.1007 Flights conducted under part 121 or part 135 of this chapter.

(a) Except as provided in § 91.501(b), when a nonprogram aircraft is used to substitute for a program flight, the flight must be operated in compliance with part 121 or part 135 of this chapter, as applicable.

(b) A program manager who holds a certificate under part 119 of this chapter may conduct a flight for the use of a fractional owner under part 121 or part 135 of this chapter if the aircraft is listed on that certificate holder's operations specifications for part 121 or part 135, as applicable.

(c) The fractional owner must be informed when a flight is being conducted as a program flight or is being conducted under part 121 or part 135 of this chapter.

Operational Control

§ 91.1009 Clarification of operational control.

(a) An owner is in operational control of a program flight when the owner—

(1) Has the rights and is subject to the limitations set forth in §§ 91.1003 through 91.1013;

(2) Has directed that a program aircraft carry passengers or property designated by that owner; and

(3) The aircraft is carrying those passengers or property.

(b) An owner is not in operational control of a flight in the following circumstances:

(1) A program aircraft is used for a flight for administrative purposes such as demonstration, positioning, ferrying, maintenance, or crew training, and no passengers or property designated by such owner are being carried; or

(2) The aircraft being used for the flight is being operated under part 121 or 135 of this chapter.

§ 91.1011 Operational control responsibilities and delegation.

(a) Each owner in operational control of a program flight is ultimately responsible for safe operations and for complying with all applicable requirements of this chapter, including those related to airworthiness and operations in connection with the flight. Each owner may delegate some or all of the performance of the tasks associated with carrying out this responsibility to the program manager, and may rely on the program manager for aviation expertise and program management services. When the owner delegates performance of tasks to the program manager or relies on the program manager's expertise, the owner and the program manager are jointly and individually responsible for compliance.

(b) The management specifications, authorizations, and approvals required by this subpart are issued to, and in the sole name of, the program manager on behalf of the fractional owners collectively. The management specifications, authorizations, and approvals will not be affected by any change in ownership of a program aircraft, as long as the aircraft remains a program aircraft in the identified program.

§ 91.1013 Operational control briefing and acknowledgment.

(a) Upon the signing of an initial program management services contract, or a renewal or extension of a program management services contract, the program manager must brief the fractional owner on the owner's operational control responsibilities, and the owner must review and sign an acknowledgment of these operational control responsibilities. The acknowledgment must be included with the program management services contract. The acknowledgment must define when a fractional owner is in operational control and the owner's responsibilities and liabilities under the program. These include:

(1) Responsibility for compliance with the management specifications and all applicable regulations.

(2) Enforcement actions for any noncompliance.

(3) Liability risk in the event of a flight-related occurrence that causes personal injury or property damage.

(b) The fractional owner's signature on the acknowledgment will serve as the owner's affirmation that the owner has read, understands, and accepts the operational control responsibilities described in the acknowledgment.

(c) Each program manager must ensure that the fractional owner or owner's representatives have access to the acknowledgments for such owner's program aircraft. Each program manager must ensure that the FAA has access to the acknowledgments for all program aircraft.

Program Management

§ 91.1014 Issuing or denying management specifications.

(a) A person applying to the Administrator for management specifications under this subpart must submit an application—

(1) In a form and manner prescribed by the Administrator; and

(2) Containing any information the Administrator requires the applicant to submit.

(b) Management specifications will be issued to the program manager on behalf of the fractional owners if, after investigation, the Administrator finds that the applicant:

(1) Meets the applicable requirements of this subpart; and

(2) Is properly and adequately equipped in accordance with the requirements of this chapter and is able to conduct safe operations under appropriate provisions of part 91 of this chapter and management specifications issued under this subpart.

(c) An application for management specifications will be denied if the Administrator finds that the applicant is not properly or adequately equipped or is not able to conduct safe operations under this part.

§ 91.1015 Management specifications.

(a) Each person conducting operations under this subpart or furnishing fractional ownership program management services to fractional owners must do so in accordance with management specifications issued by the Administrator to the fractional ownership program manager under this subpart. Management specifications must include:

(1) The current list of all fractional owners and types of aircraft, registration markings and serial numbers;

(2) The authorizations, limitations, and certain procedures under which these operations are to be conducted,

(3) Certain other procedures under which each class and size of aircraft is to be operated;

(4) Authorization for an inspection program approved under § 91.1109, including the type of aircraft, the registration markings and serial numbers of each aircraft to be operated under the program. No person may conduct any program flight using any aircraft not listed.

(5) Time limitations, or standards for determining time limitations, for overhauls, inspections, and checks for airframes, engines, propellers, rotors, appliances, and emergency equipment of aircraft.

(6) The specific location of the program manager's principal base of operations and, if different, the address that will serve as the primary point of contact for correspondence between the FAA and the program manager and the name and mailing address of the program manager's agent for service;

(7) Other business names the program manager may use;

(8) Authorization for the method of controlling weight and balance of aircraft;

(9) Any authorized deviation and exemption granted from any requirement of this chapter; and

(10) Any other information the Administrator determines is necessary.

(b) The program manager may keep the current list of all fractional owners required by paragraph (a)(1) of this section at its principal base of operation or other location approved by the Administrator and referenced in its management specifications. Each program manager shall make this list of owners available for inspection by the Administrator.

(c) Management specifications issued under this subpart are effective unless—

(1) The management specifications are amended as provided in § 91.1017; or

(2) The Administrator suspends or revokes the management specifications.

(d) At least 30 days before it proposes to establish or change the location of its principal base of operations, its main operations base, or its main maintenance base, a program manager must provide written notification to the Flight Standards District Office that issued the program manager's management specifications.

(e) Each program manager must maintain a complete and separate set of its management specifications at its principal base of operations, or at a place approved by the Administrator, and must make its management specifications available for inspection by the Administrator and the fractional owner(s) to whom the program manager furnishes its services for review and audit.

(f) Each program manager must insert pertinent excerpts of its management specifications, or references thereto, in its program manual and must—

(1) Clearly identify each such excerpt as a part of its management specifications; and

(2) State that compliance with each management specifications requirement is mandatory.

(g) Each program manager must keep each of its employees and other persons who perform duties material to its operations informed of the provisions of its management specifications that apply to that employee's or person's duties and responsibilities.

§ 91.1017 Amending program manager's management specifications.

(a) The Administrator may amend any management specifications issued under this subpart if—

(1) The Administrator determines that safety and the public interest require the amendment of any management specifications; or

(2) The program manager applies for the amendment of any management specifications, and the Administrator determines that safety and the public interest allows the amendment.

(b) Except as provided in paragraph (e) of this section, when the Administrator initiates an amendment of a program manager's management specifications, the following procedure applies:

(1) The Flight Standards District Office that issued the program manager's management specifications will notify the program manager in writing of the proposed amendment.

(2) The Flight Standards District Office that issued the program manager's management specifications will set a reasonable period (but not less than 7 days) within which the program manager may submit written information, views, and arguments on the amendment.

(3) After considering all material presented, the Flight Standards District Office that issued the program manager's management specifications will notify the program manager of—

(i) The adoption of the proposed amendment,

(ii) The partial adoption of the proposed amendment, or

(iii) The withdrawal of the proposed amendment.

(4) If the Flight Standards District Office that issued the program manager's management specifications issues an amendment of the management specifications, it becomes effective not less than 30 days after the program manager receives notice of it unless—

(i) The Flight Standards District Office that issued the program manager's management specifications finds under paragraph (e) of this section that there is an emergency requiring immediate action with respect to safety; or

(ii) The program manager petitions for reconsideration of the amendment under paragraph (d) of this section.

(c) When the program manager applies for an amendment to its management specifications, the following procedure applies:

(1) The program manager must file an application to amend its management specifications—

(i) At least 90 days before the date proposed by the applicant for the amendment to become effective, unless a shorter time is approved, in cases such as mergers, acquisitions of operational assets that require an additional showing of safety (for example, proving tests or validation tests), and resumption of operations following a suspension of operations as a result of bankruptcy actions.

(ii) At least 15 days before the date proposed by the applicant for the amendment to become effective in all other cases.

(2) The application must be submitted to the Flight Standards District Office that issued the program manager's management specifications in a form and manner prescribed by the Administrator.

(3) After considering all material presented, the Flight Standards District Office that issued the program manager's management specifications will notify the program manager of—

(i) The adoption of the applied for amendment;

(ii) The partial adoption of the applied for amendment; or

(iii) The denial of the applied for amendment. The program manager may petition for reconsideration of a denial under paragraph (d) of this section.

(4) If the Flight Standards District Office that issued the program manager's management specifications approves

the amendment, following coordination with the program manager regarding its implementation, the amendment is effective on the date the Administrator approves it.

(d) When a program manager seeks reconsideration of a decision of the Flight Standards District Office that issued the program manager's management specifications concerning the amendment of management specifications, the following procedure applies:

(1) The program manager must petition for reconsideration of that decision within 30 days of the date that the program manager receives a notice of denial of the amendment of its management specifications, or of the date it receives notice of an FAA-initiated amendment of its management specifications, whichever circumstance applies.

(2) The program manager must address its petition to the Director, Flight Standards Service.

(3) A petition for reconsideration, if filed within the 30-day period, suspends the effectiveness of any amendment issued by the Flight Standards District Office that issued the program manager's management specifications unless that District Office has found, under paragraph (e) of this section, that an emergency exists requiring immediate action with respect to safety.

(4) If a petition for reconsideration is not filed within 30 days, the procedures of paragraph (c) of this section apply.

(e) If the Flight Standards District Office that issued the program manager's management specifications finds that an emergency exists requiring immediate action with respect to safety that makes the procedures set out in this section impracticable or contrary to the public interest—

(1) The Flight Standards District Office amends the management specifications and makes the amendment effective on the day the program manager receives notice of it; and

(2) In the notice to the program manager, the Flight Standards District Office will articulate the reasons for its finding that an emergency exists requiring immediate action with respect to safety or that makes it impracticable or contrary to the public interest to stay the effectiveness of the amendment.

§ 91.1019 Conducting tests and inspections.

(a) At any time or place, the Administrator may conduct an inspection or test, other than an en route inspection, to determine whether a program manager under this subpart is complying with title 49 of the United States Code, applicable regulations, and the program manager's management specifications.

(b) The program manager must—

(1) Make available to the Administrator at the program manager's principal base of operations, or at a place approved by the Administrator, the program manager's management specifications; and

(2) Allow the Administrator to make any test or inspection, other than an en route inspection, to determine compliance respecting any matter stated in paragraph (a) of this section.

(c) Each employee of, or person used by, the program manager who is responsible for maintaining the program manager's records required by or necessary to demonstrate compliance with this subpart must make those records available to the Administrator.

(d) The Administrator may determine a program manager's continued eligibility to hold its management specifications on any grounds listed in paragraph (a) of this section, or any other appropriate grounds.

(e) Failure by any program manager to make available to the Administrator upon request, the management specifications, or any required record, document, or report is grounds for suspension of all or any part of the program manager's management specifications.

§ 91.1021 Internal safety reporting and incident/accident response.

(a) Each program manager must establish an internal anonymous safety reporting procedure that fosters an environment of safety without any potential for retribution for filing the report.

(b) Each program manager must establish procedures to respond to an aviation incident/accident.

§ 91.1023 Program operating manual requirements.

(a) Each program manager must prepare and keep current a program operating manual setting forth procedures and policies acceptable to the Administrator. The program manager's management, flight, ground, and maintenance personnel must use this manual to conduct operations under this subpart. However, the Administrator may authorize a deviation from this paragraph if the Administrator finds that, because of the limited size of the operation, part of the manual is not necessary for guidance of management, flight ground, or maintenance personnel.

(b) Each program manager must maintain at least one copy of the manual at its principal base of operations.

(c) No manual may be contrary to any applicable U.S. regulations, foreign regulations applicable to the program flights in foreign countries, or the program manager's management specifications.

(d) The program manager must make a copy of the manual, or appropriate portions of the manual (and changes and additions), available to its maintenance and ground operations personnel and must furnish the manual to—

(1) Its crewmembers; and

(2) Representatives of the Administrator assigned to the program manager.

(e) Each employee of the program manager to whom a manual or appropriate portions of it are furnished under paragraph (d)(i) of this section must keep it up-to-date with the changes and additions furnished to them.

(f) Except as provided in paragraph (h) of this section, the appropriate parts of the manual must be carried on each aircraft when away from the principal operations base. The appropriate parts must be available for use by ground or flight personnel.

(g) For the purpose of complying with paragraph (d) of this section, a program manager may furnish the persons listed therein with all or part of its manual in printed form or other form, acceptable to the Administrator, that is retrievable in the English language. If the program manager furnishes all or part of the manual in other than printed form, it must ensure there is a compatible reading device available to those persons that provides a legible image of the maintenance information and instructions, or a system that is able to retrieve the maintenance information and instructions in the English language.

(h) If a program manager conducts aircraft inspections or maintenance at specified facilities where the approved aircraft inspection program is available, the program manager is not required to ensure that the approved aircraft inspection program is carried aboard the aircraft en route to those facilities.

(i) Program managers that are also certificated to operate under part 121 or 135 of this chapter may be authorized to use the operating manual required by those parts to meet the manual requirements of subpart K, provided:

(1) The policies and procedures are consistent for both operations, or

(2) When policies and procedures are different, the applicable policies and procedures are identified and used.

§ 91.1025 Program operating manual contents.

Each program operating manual must have the date of the last revision on each revised page. Unless otherwise authorized by the Administrator, the manual must include the following:

(a) Procedures for ensuring compliance with aircraft weight and balance limitations;

(b) Copies of the program manager's management specifications or appropriate extracted information, including area of operations authorized, category and class of aircraft authorized, crew complements, and types of operations authorized;

(c) Procedures for complying with accident notification requirements;

(d) Procedures for ensuring that the pilot in command knows that required airworthiness inspections have been made and that the aircraft has been approved for return to service in compliance with applicable maintenance requirements;

(e) Procedures for reporting and recording mechanical irregularities that come to the attention of the pilot in command before, during, and after completion of a flight;

(f) Procedures to be followed by the pilot in command for determining that mechanical irregularities or defects reported for previous flights have been corrected or that correction of certain mechanical irregularities or defects have been deferred;

(g) Procedures to be followed by the pilot in command to obtain maintenance, preventive maintenance, and servicing of the aircraft at a place where previous arrangements have not been made by the program manager or owner, when the pilot is authorized to so act for the operator;

(h) Procedures under § 91.213 for the release of, and continuation of flight if any item of equipment required for the particular type of operation becomes inoperative or unserviceable en route;

(i) Procedures for refueling aircraft, eliminating fuel contamination, protecting from fire (including electrostatic protection), and supervising and protecting passengers during refueling;

(j) Procedures to be followed by the pilot in command in the briefing under § 91.1035.

(k) Procedures for ensuring compliance with emergency procedures, including a list of the functions assigned each category of required crewmembers in connection with an emergency and emergency evacuation duties;

(l) The approved aircraft inspection program, when applicable;

(m) Procedures for the evacuation of persons who may need the assistance of another person to move expeditiously to an exit if an emergency occurs;

(n) Procedures for performance planning that take into account takeoff, landing and en route conditions;

(o) An approved Destination Airport Analysis, when required by § 91.1037(c), that includes the following elements, supported by aircraft performance data supplied by the aircraft manufacturer for the appropriate runway conditions—

(1) Pilot qualifications and experience;

(2) Aircraft performance data to include normal, abnormal and emergency procedures as supplied by the aircraft manufacturer;

(3) Airport facilities and topography;

(4) Runway conditions (including contamination);

(5) Airport or area weather reporting;

(6) Appropriate additional runway safety margins, if required;

(7) Airplane inoperative equipment;

(8) Environmental conditions; and

(9) Other criteria that affect aircraft performance.

(p) A suitable system (which may include a coded or electronic system) that provides for preservation and retrieval of maintenance recordkeeping information required by § 91.1113 in a manner acceptable to the Administrator that provides—

(1) A description (or reference to date acceptable to the Administrator) of the work performed:

(2) The name of the person performing the work if the work is performed by a person outside the organization of the program manager; and

(3) The name or other positive identification of the individual approving the work.

(q) Flight locating and scheduling procedures; and

(r) Other procedures and policy instructions regarding program operations that are issued by the program manager or required by the Administrator.

§ 91.1027 Recordkeeping.

(a) Each program manager must keep at its principal base of operations or at other places approved by the Administrator, and must make available for inspection by the Administrator all of the following:

(1) The program manager's management specifications.

(2) A current list of the aircraft used or available for use in operations under this subpart, the operations for which each is equipped (for example, MNPS, RNP5/10, RVSM.).

(3) An individual record of each pilot used in operations under this subpart, including the following information:

(i) The full name of the pilot.

(ii) The pilot certificate (by type and number) and ratings that the pilot holds.

(iii) The pilot's aeronautical experience in sufficient detail to determine the pilot's qualifications to pilot aircraft in operations under this subpart.

(iv) The pilot's current duties and the date of the pilot's assignment to those duties.

(v) The effective date and class of the medical certificate that the pilot holds.

(vi) The date and result of each of the initial and recurrent competency tests and proficiency checks required by this subpart and the type of aircraft flown during that test or check.

(vii) The pilot's flight time in sufficient detail to determine compliance with the flight time limitations of this subpart.

(viii) The pilot's check pilot authorization, if any.

(ix) Any action taken concerning the pilot's release from employment for physical or professional disqualification; and

(x) The date of the satisfactory completion of initial, transition, upgrade, and differences training and each recurrent training phase required by this subpart.

(4) An individual record for each flight attendant used in operations under this subpart, including the following information:

(i) The full name of the flight attendant, and

(ii) The date and result of training required by § 91.1063, as applicable.

(5) A current list of all fractional owners and associated aircraft. This list or a reference to its location must be included in the management specifications and should be of sufficient detail to determine the minimum fractional ownership interest of each aircraft.

(b) Each program manager must keep each record required by paragraph (a)(2) of this section for at least 6 months, and must keep each record required by paragraphs (a)(3) and (a)(4) of this section for at least 12 months. When an employee is no longer employed or affiliated with the program manager or fractional owner, each record required by paragraphs (a)(3) and (a)(4) of this section must be retained for at least 12 months.

(c) Each program manager is responsible for the preparation and accuracy of a load manifest in duplicate containing information concerning the loading of the aircraft. The manifest must be prepared before each takeoff and must include—

(1) The number of passengers;

(2) The total weight of the loaded aircraft;

(3) The maximum allowable takeoff weight for that flight;

(4) The center of gravity limits;

(5) The center of gravity of the loaded aircraft, except that the actual center of gravity need not be computed if the aircraft is loaded according to a loading schedule or other approved method that ensures that the center of gravity of the loaded aircraft is within approved limits. In those cases, an entry must be made on the manifest indicating that the center of gravity is within limits according to a loading schedule or other approved method;

(6) The registration number of the aircraft or flight number;

(7) The origin and destination; and

(8) Identification of crewmembers and their crew position assignments.

(d) The pilot in command of the aircraft for which a load manifest must be prepared must carry a copy of the completed load manifest in the aircraft to its destination. The program manager must keep copies of completed load manifest for at least 30 days at its principal operations base, or at another location used by it and approved by the Administrator.

(e) Each program manager is responsible for providing a written document that states the name of the entity having operational control on that flight and the part of this chapter under which the flight is operated. The pilot in command of the aircraft must carry a copy of the document in the aircraft to its destination. The program manager must keep a copy of the document for at least 30 days at its principal operations base, or at another location used by it and approved by the Administrator.

(f) Records may be kept either in paper or other form acceptable to the Administrator.

(g) Program managers that are also certificated to operate under part 121 or 135 of this chapter may satisfy the recordkeeping requirements of this section and of § 91.1113 with records maintained to fulfill equivalent obligations under part 121 or 135 of this chapter.

§ 91.1029 Flight scheduling and locating requirements.

(a) Each program manager must establish and use an adequate system to schedule and release program aircraft.

(b) Except as provided in paragraph (d) of this section, each program manager must have adequate procedures established for locating each flight, for which a flight plan is not filed, that—

(1) Provide the program manager with at least the information required to be included in a VFR flight plan;

(2) Provide for timely notification of an FAA facility or search and rescue facility, if an aircraft is overdue or missing; and

(3) Provide the program manager with the location, date, and estimated time for reestablishing radio or telephone communications, if the flight will operate in an area where communications cannot be maintained.

(c) Flight locating information must be retained at the program manager's principal base of operations, or at other places designated by the program manager in the flight locating procedures, until the completion of the flight.

(d) The flight locating requirements of paragraph (b) of this section do not apply to a flight for which an FAA flight plan has been filed and the flight plan is canceled within 25 nautical miles of the destination airport.

§ 91.1031 Pilot in command or second in command: Designation required.

(a) Each program manager must designate a—

(1) Pilot in command for each program flight; and

(2) Second in command for each program flight requiring two pilots.

(b) The pilot in command, as designated by the program manager, must remain the pilot in command at all times during that flight.

§ 91.1033 Operating information required.

(a) Each program manager must, for all program operations, provide the following materials, in current and appropriate form, accessible to the pilot at the pilot station, and the pilot must use them—

(1) A cockpit checklist;

(2) For multiengine aircraft or for aircraft with retractable landing gear, an emergency cockpit checklist containing the procedures required by paragraph (c) of this section, as appropriate;

(3) At least one set of pertinent aeronautical charts; and

(4) For IFR operations, at least one set of pertinent navigational en route, terminal area, and instrument approach procedure charts.

(b) Each cockpit checklist required by paragraph (a)(1) of this section must contain the following procedures:

(1) Before starting engines;

(2) Before takeoff;

(3) Cruise;

(4) Before landing;

(5) After landing; and

(6) Stopping engines.

(c) Each emergency cockpit checklist required by paragraph (a)(2) of this section must contain the following procedures, as appropriate:

(1) Emergency operation of fuel, hydraulic, electrical, and mechanical systems.

(2) Emergency operation of instruments and controls.

(3) Engine inoperative procedures.

(4) Any other emergency procedures necessary for safety.

§ 91.1035 Passenger awareness.

(a) Prior to each takeoff, the pilot in command of an aircraft carrying passengers on a program flight must ensure that all passengers have been orally briefed on—

(1) *Smoking:* Each passenger must be briefed on when, where, and under what conditions smoking is prohibited. This briefing must include a statement, as appropriate, that the regulations require passenger compliance with lighted passenger information signs and no smoking placards, prohibit smoking in lavatories, and require compliance with crewmember instructions with regard to these items;

(2) *Use of safety belts, shoulder harnesses, and child restraint systems:*

Each passenger must be briefed on when, where and under what conditions it is necessary to have his or her safety belt and, if installed, his or her shoulder harness fastened about him or her, and if a child is being transported, the appropriate use of child restraint systems, if available. This briefing must include a statement, as appropriate, that the regulations require passenger compliance with the lighted passenger information sign and/or crewmember instructions with regard to these items;

(3) The placement of seat backs in an upright position before takeoff and landing;

(4) Location and means for opening the passenger entry door and emergency exits;

(5) Location of survival equipment;

(6) Ditching procedures and the use of flotation equipment required under § 91.509 for a flight over water;

(7) The normal and emergency use of oxygen installed in the aircraft; and

(8) Location and operation of fire extinguishers.

(b) Prior to each takeoff, the pilot in command of an aircraft carrying passengers on a program flight must ensure that each person who may need the assistance of another person to move expeditiously to an exit if an emergency occurs and that person's attendant, if any, has received a briefing as to the procedures to be followed if an evacuation occurs. This paragraph does not apply to a person who has been given a briefing before a previous leg of that flight in the same aircraft.

(c) Prior to each takeoff, the pilot in command must advise the passengers of the name of the entity in operational control of the flight.

(d) The oral briefings required by paragraphs (a), (b), and (c) of this section must be given by the pilot in command or another crewmember.

(e) The oral briefing required by paragraph (a) of this section may be delivered by means of an approved recording playback device that is audible to each passenger under normal noise levels.

(f) The oral briefing required by paragraph (a) of this section must be supplemented by printed cards that must be carried in the aircraft in locations convenient for the use of each passenger. The cards must—

(1) Be appropriate for the aircraft on which they are to be used;

(2) Contain a diagram of, and method of operating, the emergency exits; and

(3) Contain other instructions necessary for the use of emergency equipment on board the aircraft.

§ 91.1037 Large transport category airplanes: Turbine engine powered; Limitations; Destination and alternate airports.

(a) No program manager or any other person may permit a turbine engine powered large transport category airplane on a program flight to take off that airplane at a weight that (allowing for normal consumption of fuel and oil in flight to the destination or alternate airport) the weight of the airplane on arrival would exceed the landing weight in the Airplane Flight Manual for the elevation of the destination or alternate airport and the ambient temperature expected at the time of landing.

(b) Except as provided in paragraph (c) of this section, no program manager or any other person may permit a turbine engine powered large transport category airplane on a program flight to take off that airplane unless its weight on arrival, allowing for normal consumption of fuel and oil in flight (in accordance with the landing distance in the Airplane Flight Manual for the elevation of the destination airport and the wind conditions expected there at the time of landing), would allow a full stop landing at the intended destination airport within 60 percent of the effective length of each runway described below from a point 50 feet above the intersection of the obstruction clearance plane and the runway. For the purpose of determining the allowable landing weight at the destination airport, the following is assumed:

(1) The airplane is landed on the most favorable runway and in the most favorable direction, in still air.

(2) The airplane is landed on the most suitable runway considering the probable wind velocity and direction and the ground handling characteristics of that airplane, and considering other conditions such as landing aids and terrain.

(c) A program manager or other person flying a turbine engine powered large transport category airplane on a program flight may permit that airplane to take off at a weight in excess of that allowed by paragraph (b) of this section if all of the following conditions exist:

(1) The operation is conducted in accordance with an approved Destination Airport Analysis in that person's program operating manual that contains the elements listed in § 91.1025(o).

(2) The airplane's weight on arrival, allowing for normal consumption of fuel and oil in flight (in accordance with the landing distance in the Airplane Flight Manual for the elevation of the destination airport and the wind conditions expected there at the time of landing), would allow a full stop landing at the intended destination airport within 80 percent of the effective length of each runway described below from a point 50 feet above the intersection of the obstruction clearance plane and the runway. For the purpose of determining the allowable landing weight at the destination airport, the following is assumed:

(i) The airplane is landed on the most favorable runway and in the most favorable direction, in still air.

(ii) The airplane is landed on the most suitable runway considering the probable wind velocity and direction and the ground handling characteristics of that airplane, and considering other conditions such as landing aids and terrain.

(3) The operation is authorized by management specifications.

(d) No program manager or other person may select an airport as an alternate airport for a turbine engine powered large transport category airplane unless (based on the assumptions in paragraph (b) of this section) that airplane, at the weight expected at the time of arrival, can be brought to a full stop landing within 80 percent of the effective length of the runway from a point 50 feet above the intersection of the obstruction clearance plane and the runway.

(e) Unless, based on a showing of actual operating landing techniques on wet runways, a shorter landing distance (but never less than that required by paragraph (b) or (c) of this section) has been approved for a specific type and model airplane and included in the Airplane Flight Manual, no person may take off a turbojet airplane when the appropriate weather reports or forecasts, or any combination of them, indicate that the runways at the destination or alternate airport may be wet or slippery at the estimated time of arrival unless the effective runway length at the destination airport is at least 115 percent of the runway length required under paragraph (b) or (c) of this section.

§ 91.1039 IFR takeoff, approach and landing minimums.

(a) No pilot on a program aircraft operating a program flight may begin an instrument approach procedure to an airport unless—

(1) Either that airport or the alternate airport has a weather reporting facility operated by the U.S. National

Weather Service, a source approved by the U.S. National Weather Service, or a source approved by the Administrator; and

(2) The latest weather report issued by the weather reporting facility includes a current local altimeter setting for the destination airport. If no local altimeter setting is available at the destination airport, the pilot must obtain the current local altimeter setting from a source provided by the facility designated on the approach chart for the destination airport.

(b) For flight planning purposes, if the destination airport does not have a weather reporting facility described in paragraph (a)(1) of this section, the pilot must designate as an alternate an airport that has a weather reporting facility meeting that criteria.

(c) The MDA or Decision Altitude and visibility landing minimums prescribed in part 97 of this chapter or in the program manager's management specifications are increased by 100 feet and 1/2 mile respectively, but not to exceed the ceiling and visibility minimums for that airport when used as an alternate airport, for each pilot in command of a turbine-powered aircraft who has not served at least 100 hours as pilot in command in that type of aircraft.

(d) No person may take off an aircraft under IFR from an airport where weather conditions are at or above takeoff minimums but are below authorized IFR landing minimums unless there is an alternate airport within one hour's flying time (at normal cruising speed, in still air) of the airport of departure.

(e) Each pilot making an IFR takeoff or approach and landing at an airport must comply with applicable instrument approach procedures and take off and landing weather minimums prescribed by the authority having jurisdiction over the airport. In addition, no pilot may, at that airport take off when the visibility is less than 600 feet.

§91.1041 Aircraft proving and validation tests.

(a) No program manager may permit the operation of an aircraft, other than a turbojet aircraft, for which two pilots are required by the type certification requirements of this chapter for operations under VFR, if it has not previously proved such an aircraft in operations under this part in at least 25 hours of proving tests acceptable to the Administrator including—

(1) Five hours of night time, if night flights are to be authorized;

(2) Five instrument approach procedures under simulated or actual conditions, if IFR flights are to be authorized; and

(3) Entry into a representative number of en route airports as determined by the Administrator.

(b) No program manager may permit the operation of a turbojet airplane if it has not previously proved a turbojet airplane in operations under this part in at least

25 hours of proving tests acceptable to the Administrator including—

(1) Five hours of night time, if night flights are to be authorized;

(2) Five instrument approach procedures under simulated or actual conditions, if IFR flights are to be authorized; and

(3) Entry into a representative number of en route airports as determined by the Administrator.

(c) No program manager may carry passengers in an aircraft during proving tests, except those needed to make the tests and those designated by the Administrator to observe the tests. However, pilot flight training may be conducted during the proving tests.

(d) Validation testing is required to determine that a program manager is capable of conducting operations safely and in compliance with applicable regulatory standards. Validation tests are required for the following authorizations:

(1) The addition of an aircraft for which two pilots are required for operations under VFR or a turbojet airplane, if that aircraft or an aircraft of the same make or similar design has not been previously proved or validated in operations under this part.

(2) Operations outside U.S. airspace.

(3) Class II navigation authorizations.

(4) Special performance or operational authorizations.

(e) Validation tests must be accomplished by test methods acceptable to the Administrator. Actual flights may not be required when an applicant can demonstrate competence and compliance with appropriate regulations without conducting a flight.

(f) Proving tests and validation tests may be conducted simultaneously when appropriate.

(g) The Administrator may authorize deviations from this section if the Administrator finds that special circumstances make full compliance with this section unnecessary.

§ 91.1043 (Reserved)

§ 91.1045 Additional Equipment Requirements

No person may operate a program aircraft on a program flight unless the aircraft is equipped with the following—

(a) Airplanes having a passenger-seat configuration of more than 30 seats or a payload capacity of more than 7,500 pounds:

(1) A cockpit voice recorder as required by § 121.359 of this chapter as applicable to the aircraft specified in this section.

(2) A flight recorder as required by § 121.343 or § 121.344 of this chapter as applicable to aircreaft specified in that section.

(3) A terrain awareness and warning system as required by § 121.354 of this chapter as applicable to the aircraft specified in that section.

(4) A traffic alert and collision avoidance system as required by § 121.356 of this chapter as applicable to the aircraft specified in that section.

(5) Airborne weather radar as required by § 121.357 of this chapter, as applicable to aircraft specified in that section.

(b) Airplanes having a passenger-seat configuaration of 30 seats or fewer, excluding each crewmember, and a payload capacity of 7,500 pounds or less, and any rotorcraft as applicable:

(1) A cockpit voice recorder as required by § 135.151 of this chapter as applicable to the aircraft specified in that section.

(2) A flight recorder as required by § 135.152 of this chapter as applicable to the aircraft specified in that section.

(3) A terrain awareness and warning system as required by § 135.154 of this chapter as applicable to the aircraft specified in that section.

(4) A traffic alert and collision avoidance system as required by § 135.180 of this chapter as applicable to the aircraft specified in that section.

(5) As applicable to the aircraft specified in that section, either:

(i) Airborne thunderstorm detection equipment as required by § 135.173 of this chapter; or

(ii) Airborne weather radar as required by § 135.175 of this chapter.

§ 91.1047   Drug and alcohol misuse education program.

(a) Each program manager must provide each direct employee performing flight crewmember, flight attendant, flight instructor, or aircraft maintenance duties with drug and alcohol misuse education.

(b) No program manager may use any contract employee to perform flight crewmember, flight attendant, flight instructor, or aircraft maintenance duties for the program manager unless that contract employee has been provided with drug and alcohol misuse education.

(c) Program managers must disclose to their owners and prospective owners the existence of a company drug and alcohol misuse testing program. If the program manager has implemented a company testing program, the program manager's disclosure must include the following:

(1) Information on the substances that they test for, for example, alcohol and a list of the drugs;

(2) The categories of employees tested, the types of tests, for example, pre-employment, random, reasonable cause/suspicion, post accident, return to duty and followup; and

(3) The degree to which the program manager's company testing program is comparable to the federally mandated drug and alcohol misuse prevention program required under part 121, appendices I and J, of this chapter, regarding the information in paragraphs (c)(1) and (c)(2) of this section.

(d) If a program aircraft is operated on a program flight into an airport at which no maintenance personnel are available that are subject to the requirements of paragraphs (a) or (b) of this section and emergency maintenance is required, the program manager may use persons not meeting the requirements of paragraphs (a) or (b) of this section to provide such emergency maintenance under both of the following conditions:

(1) The program manager must notify the Drug Abatement Program Division, AAM—800, 800 Independence Avenue, SW., Washington, DC 20591 in writing within 10 days after being provided emergency maintenance in accordance with this paragraph. The program manager must retain copies of all such written notifications for two years.

(2) The aircraft must be reinspected by maintenance personnel who meet the requirements of paragraph (a) or (b) of this section when the aircraft is next at an airport where such maintenance personnel are available.

(e) For purposes of this section, emergency maintenance means maintenance that—

(1) Is not scheduled, and

(2) Is made necessary by an aircraft condition not discovered prior to the departure for that location.

(f) Notwithstanding paragraphs (a) and (b) of this section, drug and alcohol misuse education conducted under an FAA-approved drug and alcohol misuse prevention program may be used to satisfy these requirements.

§ 91.1049   Personnel.

(a) Each program manager and each fractional owner must use in program operations on program aircraft flight crews meeting § 91.1053 criteria and qualified under the appropriate regulations. The program manager must provide oversight of those crews.

(b) Each program manager must employ (either directly or by contract) an adequate number of pilots per program aircraft. Flight crew staffing must be determined based on the following factors, at a minimum:

(1) Number of program aircraft.

(2) Program manager flight, duty, and rest time considerations, and in all cases within the limits set forth in §§ 91.1057 through 91.1061.

(3) Vacations.

(4) Operational efficiencies.

(5) Training.

(6) Single pilot operations, if authorized by deviation under paragraph (d) of this section.

(c) Each program manager must publish pilot and flight attendant duty schedules sufficiently in advance to follow the flight, duty, and rest time limits in §§ 91.1057 through 91.1061 in program operations.

(d) Unless otherwise authorized by the Administrator, when any program aircraft is flown in program operations with passengers onboard, the crew must consist of at least two qualified pilots employed or contracted by the program manager or the fractional owner.

(e) The program manager must ensure that trained and qualified scheduling or flight release personnel are on duty to schedule and release program aircraft during all hours that such aircraft are available for program operations.

§ 91.1051 Pilot safety background check.

Within 90 days of an individual beginning service as a pilot, the program manager must request the following information:

(a) FAA records pertaining to—

(1) Current pilot certificates and associated type ratings.

(2) Current medical certificates.

(3) Summaries of legal enforcement actions resulting in a finding by the Administrator of a violation.

(b) Records from all previous employers during the five years preceding the date of the employment application where the applicant worked as a pilot. If any of these firms are in bankruptcy, the records must be requested from the trustees in bankruptcy for those employees. If the previous employer is no longer in business, a documented good faith effort must be made to obtain the records. Records from previous employers must include, as applicable—

(1) Crew member records.

(2) Drug testing—collection, testing, and rehabilitation records pertaining to the individual.

(3) Alcohol misuse prevention program records pertaining to the individual.

(4) The applicant's individual record that includes certifications, ratings, aeronautical experience, effective date and class of the medical certificate.

§ 91.1053 Crewmember experience.

(a) No program manager or owner may use any person, nor may any person serve, as a pilot in command or second in command of a program aircraft, or as a flight attendant on a program aircraft, in program operations under this subpart unless that person has met the applicable requirements of part 61 of this chapter and has the following experience and ratings:

(1) Total flight time for all pilots:

(i) Pilot in command—A minimum of 1,500 hours.

(ii) Second in command—A minimum of 500 hours.

(2) For multi-engine turbine-powered fixed-wing and powered-lift aircraft, the following FAA certification and ratings requirements:

(i) Pilot in command—Airline transport pilot and applicable type ratings.

(ii) Second in command—Commercial pilot and instrument ratings.

(iii) Flight attendant (if required or used)—Appropriately trained personnel.

(3) For all other aircraft, the following FAA certification and rating requirements:

(i) Pilot in command—Commercial pilot and instrument ratings.

(ii) Second in command—Commercial pilot and instrument ratings.

(iii) Flight attendant (if required or used)—Appropriately trained personnel.

(b) The Administrator may authorize deviations from paragraph (a)(1) of this section if the Flight Standards District Office that issued the program manager's management specifications finds that the crewmember has comparable experience, and can effectively perform the functions associated with the position in accordance with the requirements of this chapter. Grants of deviation under this paragraph may be granted after consideration of the size and scope of the operation, the qualifications of the intended personnel and the circumstances set forth in § 91.1055(b)(1) through (3). The Administrator may, at any time, terminate any grant of deviation authority issued under this paragraph.

§ 91.1055 Pilot operating limitations and pairing requirement

(a) If the second in command of a fixed-wing program aircraft has fewer than 100 hours of flight time as second in command flying in the aircraft make and model and, if a type rating is required, in the type aircraft being flown, and the pilot in command is not an appropriately qualified check pilot, the pilot in command shall make all takeoffs and landings in any of the following situations:

(1) Landings at the destination airport when a Destination Airport Analysis is required by § 91.1037(c); and

(2) In any of the following conditions:

(i) The prevailing visibility for the airport is at or below 3/4 mile.

(ii) The runway visual range for the runway to be used is at or below 4,000 feet.

(iii) The runway to be used has water, snow, slush, ice or similar contamination that may adversely affect aircraft performance.

(iv) The braking action on the runway to be used is reported to be less than "good."

(v) The crosswind component for the runway to be used is in excess of 15 knots.

(vi) Windshear is reported in the vicinity of the airport.

(vii) Any other condition in which the pilot in command determines it to be prudent to exercise the pilot in command's authority.

(b) No program manager may release a program flight under this subpart unless, for that aircraft make or model and, if a type rating is required, for that type aircraft, either the pilot in command or the second in command has at least 75 hours of flight time, either as pilot in command or second in command. The Administrator may, upon application by the program manager, authorize deviations from the requirements of this paragraph by an appropriate amendment to the management specifications in any of the following circumstances:

(1) A newly authorized program manager does not employ any pilots who meet the minimum requirements of this paragraph.

(2) An existing program manager adds to its fleet a new category and class aircraft not used before in its operation.

(3) An existing program manager establishes a new base to which it assigns pilots who will be required to become qualified on the aircraft operated from that base.

(c) No person may be assigned in the capacity of pilot in command in a program operation to more than two aircraft types that require a separate type rating.

§ 91.1057 Flight, duty and rest time requirements: All crewmembers.

(a) For purposes of this subpart—

Augmented flight crew means at least three pilots.

Calendar day means the period of elapsed time, using Coordinated Universal Time or local time that begins at midnight and ends 24 hours later at the next midnight.

Duty period means the period of elapsed time between reporting for an assignment involving flight time and release from that assignment by the program manager. All time between these two points is part of the duty period, even if flight time is interrupted by nonflight-related duties. The time is calculated using either Coordinated Universal Time or local time to reflect the total elapsed time.

Extension of flight time means an increase in the flight time because of circumstances beyond the control of the program manager or flight crewmember (such as adverse weather) that are not known at the time of departure and that prevent the flightcrew from reaching the destination within the planned flight time.

Flight attendant means an individual, other than a flight crewmember, who is assigned by the program manager, in accordance with the required minimum crew complement under the program manager's management specifications or in addition to that minimum complement, to duty in an aircraft during flight time and whose duties include but are not necessarily limited to cabin-safety-related responsibilities.

Multi-time zone flight means an easterly or westerly flight or multiple flights in one direction in the same duty period that results in a time zone difference of 5 or more hours and is conducted in a geographic area that is south of 60 degrees north latitude and north of 60 degrees south latitude.

Reserve status means that status in which a flight crewmember, by arrangement with the program manager:

Holds himself or herself fit to fly to the extent that this is within the control of the flight crewmember; remains within a reasonable response time of the aircraft as agreed between the flight crewmember and the program manager; and maintains a ready means whereby the flight crewmember may be contacted by the program manager. Reserve status is not part of any duty period or rest period.

Rest period means a period of time required pursuant to this subpart that is free of all responsibility for work or duty prior to the commencement of, or following completion of, a duty period, and during which the flight crewmember or flight attendant cannot be required to receive contact from the program manager. A rest period does not include any time during which the program manager imposes on a flight crewmember or flight attendant any duty or restraint, including any actual work or present responsibility for work should the occasion arise.

Standby means that portion of a duty period during which a flight crewmember is subject to the control of the program manager and holds himself or herself in a condition of readiness to undertake a flight. Standby is not part of any rest period.

(b) A program manager may assign a crewmember and a crewmember may accept an assignment for flight time only when the applicable requirements of this section and §§ 91.1059—91.1062 are met.

(c) No program manager may assign any crewmember to any duty during any required rest period.

(d) Time spent in transportation, not local in character, that a program manager requires of a crewmember and provides to transport the crewmember to an airport at which he or she is to serve on a flight as a crewmember, or from an airport at which he or she was relieved from duty to return to his or her home station, is not considered part of a rest period.

(e) A flight crewmember may continue a flight assignment if the flight to which he or she is assigned would normally terminate within the flight time limitations, but because of circumstances beyond the control of the program manager or flight crewmember (such as adverse weather conditions), is not at the time of departure expected to reach its destination within the planned flight time. The extension of flight time under this paragraph may not exceed the maximum time limits set forth in § 91.1059.

(f) Each flight assignment must provide for at least 10 consecutive hours of rest during the 24-hour period that precedes the completion time of the assignment.

(g) The program manager must provide each crewmember at least 13 rest periods of at least 24 consecutive hours each in each calendar quarter.

(h) A flight crewmember may decline a flight assignment if, in the flight crewmember's determination, to do so would not be consistent with the standard of safe operation required under this subpart, this part, and applicable provisions of this title.

(i) Any rest period required by this subpart may occur concurrently with any other rest period.

(j) If authorized by the Administrator, a program manager may use the applicable unscheduled flight time limitations, duty period limitations, and rest requirements of part 121 or part 135 of this chapter instead of the flight time limitations, duty period limitations, and rest requirements of this subpart.

§ 91.1059 Flight time limitations and rest requirements: One or two pilot crews.

(a) No program manager may assign any flight crewmember, and no flight crewmember may accept an assignment, for flight time as a member of a one- or two-pilot crew if that crewmember's total flight time in all commercial flying will exceed—

(1) 500 hours in any calendar quarter;

(2) 800 hours in any two consecutive calendar quarters;

(3) 1,400 hours in any calendar year.

(b) Except as provided in paragraph (c) of this section, during any 24 consecutive hours the total flight time of the assigned flight, when added to any commercial flying by that flight crewmember, may not exceed—

(1) 8 hours for a flight crew consisting of one pilot; or

(2) 10 hours for a flight crew consisting of two pilots qualified under this subpart for the operation being conducted.

(c) No program manager may assign any flight crewmember, and no flight crewmember may accept an assignment, if that crewmember's flight time or duty period will exceed, or rest time will be less than—

	Normal duty	Extension of flight time
(1) Minimum Rest Immediately Before Duty	10 Hours	10 Hours.
(2) Duty Period	Up to 14 Hours	Up to 14 Hours.
(3) Flight Time For 1 Pilot	Up to 8 Hours	Exceeding 8 Hours up to 9 Hours.
(4) Flight Time For 2 Pilots	Up to 10 Hours	Exceeding 10 Hours up to 12 Hours.
(5) Minimum After Duty Rest	10 Hours	12 Hours.
(6) Minimum After Duty Rest Period for Multi-Time Zone Flights	14 Hours	18 Hours.

§ 91.1061 Augmented flight crews

(a) No program manager may assign any flight crewmember, and no flight crewmember may accept an assignment, for flight time as a member of an augmented crew if that crewmember's total flight time in all commercial flying will exceed—

(1) 500 hours in any calendar quarter;

(2) 800 hours in any two consecutive quarters

(3) 1,400 hours in any calendar year.

(1) Adequate sleeping facilities are installed on the aircraft for pilots.

(2) No more than 8 hours of flight deck duty is accrued in any 24 consecutive hours.

(3) For a three-pilot crew, the crew must consist of at least the following:

(i) A pilot in command (PIC) who meets the applicable flight crewmember requirements of this subpart and § 61.57 of this chapter,

(ii) A PIC qualified pilot who meets the applicable flight crewmember requirements of this subpart and § 61.57 (c) and (d) of this chapter.

(iii) A second in command (SIC) who meets the SIC qualifications of this subpart. For flight under IFR, that person must also meet the recent instrument experience requirements of part 61 of this chapter.

(4) For a four-pilot crew, at least three pilots who meet the conditions of paragraph (b)(3) of this section, plus a fourth pilot who meets the SIC qualifications of this subpart.For flight under IFR, that person must also meet the recent instrument experience requirements of part 61 of this chapter.

(c) No program manager may assign any flight crewmember, and no flight crewmember may accept an assignment, if that crewmember's flight time or duty period will be less than—

	3-pilot crew	4-Pilot crew
(1) Minimum rest immediately before Duty	10 hours	10 Hours
(2) Duty Period	Up to 16 Hours	Up to 18 Hours
(3) Flight Time	Up to 12 Hours	Up to 16 Hours
(4) Minimum after duty rest	12 Hours	18 Hours
(5) Minimum after duty rest for Multi-Time Zone Flights	18 Hours	24 hours

§ 91.1062 Duty periods and rest requirements: Flight attendants.

(a) Except as provided in paragraph (b) of this section, a program manager may assign a duty period to a flight attendant only when the assignment meets the applicable duty period limitations and rest requirements of this paragraph.

(1) Except as provided in paragraphs (a)(4), (a)(5), and (a)(6) of this section, no program manager may assign a flight attendant to a scheduled duty period of more than 14 hours.

(2) Except as provided in paragraph (a)(3) of this section, a flight attendant scheduled to a duty period of 14 hours or less as provided under paragraph (a)(1) of this section must be given a scheduled rest period of at least 9

consecutive hours. This rest period must occur between the completion of the scheduled duty period and the commencement of the subsequent duty period.

(3) The rest period required under paragraph (a)(2) of this section may be scheduled or reduced to 8 consecutive hours if the flight attendant is provided a subsequent rest period of at least 10 consecutive hours; this subsequent rest period must be scheduled to begin no later than 24 hours after the beginning of the reduced rest period and must occur between the completion of the scheduled duty period and the commencement of the subsequent duty period.

(4) A program manager may assign a flight attendant to a scheduled duty period of more than 14 hours, but no more than 16 hours, if the program manager has assigned to the flight or flights in that duty period at least one flight attendant in addition to the minimum flight attendant complement required for the flight or flights in that duty period under the program manager's management specifications.

(5) A program manager may assign a flight attendant to a scheduled duty period of more than 16 hours, but no more than 18 hours, if the program manager has assigned to the flight or flights in that duty period at least two flight attendants in addition to the minimum flight attendant complement required for the flight or flights in that duty period under the program manager's management specifications.

(6) A program manager may assign a flight attendant to a scheduled duty period of more than 18 hours, but no more than 20 hours, if the scheduled duty period includes one or more flights that land or take off outside the 48 contiguous states and the District of Columbia, and if the program manager has assigned to the flight or flights in that duty period at least three flight attendants in addition to the minimum flight attendant complement required for the flight or flights in that duty period under the program manager's management specifications.

(7) Except as provided in paragraph (a)(8) of this section, a flight attendant scheduled to a duty period of more than 14 hours but no more than 20 hours, as provided in paragraphs (a)(4), (a)(5), and (a)(6) of this section, must be given a scheduled rest period of at least 12 consecutive hours. This rest period must occur between the completion of the scheduled duty period and the commencement of the subsequent duty period.

(8) The rest period required under paragraph (a)(7) of this section may be scheduled or reduced to 10 consecutive hours if the flight attendant is provided a subsequent rest period of at least 14 consecutive hours; this subsequent rest period must be scheduled to begin no later than 24 hours after the beginning of the reduced rest period and must occur between the completion of the scheduled duty period and the commencement of the subsequent duty period.

(9) Notwithstanding paragraphs (a)(4), (a)(5), and (a)(6) of this section, if a program manager elects to reduce the rest period to 10 hours as authorized by paragraph (a)(8) of this section, the program manager may not schedule a flight attendant for a duty period of more than 14 hours during the 24-hour period commencing after the beginning of the reduced rest period.

(b) Notwithstanding paragraph (a) of this section, a program manager may apply the flight crewmember flight time and duty limitations and rest requirements of this part to flight attendants for all operations conducted under this part provided that the program manager establishes written procedures that—

(1) Apply to all flight attendants used in the program manager's operation;

(2) Include the flight crewmember rest and duty requirements of §§ 91.1057, 91.1059, and 91.1061, as appropriate to the operation being conducted, except that rest facilities on board the aircraft are not required;

(3) Include provisions to add one flight attendant to the minimum flight attendant complement for each flight crewmember who is in excess of the minimum number required in the aircraft type certificate data sheet and who is assigned to the aircraft under the provisions of § 91.1061; and

(4) Are approved by the Administrator and described or referenced in the program manager's management specifications.

§ 91.1063 Testing and training: Applicability and terms used.

(a) Sections 91.1065 through 91.1107:

(1) Prescribe the tests and checks required for pilots and flight attendant crewmembers and for the approval of check pilots in operations under this subpart;

(2) Prescribe the requirements for establishing and maintaining an approved training program for crewmembers, check pilots and instructors, and other operations personnel employed or used by the program manager in program operations;

(3) Prescribe the requirements for the qualification, approval and use of aircraft simulators and flight training devices in the conduct of an approved training program; and

(4) Permits training center personnel authorized under part 142 of this chapter who meet the requirements of § 91.1075 to conduct training, testing and checking under contract or other arrangements to those persons subject to the requirements of this subpart.

(b) If authorized by the Administrator, a program manager may comply with the applicable training and testing sections of subparts N and O of part 121 of this chapter instead of §§ 91.1065 through 91.1107, except for the operating experience requirements of § 121.434 of this chapter.

(c) If authorized by the Administrator, a program manager may comply with the applicable training and testing sections of subparts G and H of part 135 of this chapter instead of §§ 91.1065 through 91.1107, except for the operating experience requirements of § 135.244 of this chapter.

(d) For the purposes of this subpart, the following terms and definitions apply:

(1) *Initial training.* The training required for crewmembers who have not qualified and served in the same capacity on an aircraft.

(2) *Transition training.* The training required for crewmembers who have qualified and served in the same capacity on another aircraft.

(3) *Upgrade training.* The training required for crewmembers who have qualified and served as second in command on a particular aircraft type, before they serve as pilot in command on that aircraft.

(4) *Differences training.* The training required for crewmembers who have qualified and served on a particular type aircraft, when the Administrator finds differences training is necessary before a crewmember serves in the same capacity on a particular variation of that aircraft.

(5) *Recurrent training.* The training required for crewmembers to remain adequately trained and currently proficient for each aircraft crewmember position, and type of operation in which the crewmember serves.

(6) *In flight.* The maneuvers, procedures, or functions that will be conducted in the aircraft.

(7) *Training center.* An organization governed by the applicable requirements of part 142 of this chapter that conducts training, testing, and checking under contract or other arrangement to program managers subject to the requirements of this subpart.

(8) *Requalification training.* The training required for crewmembers previously trained and qualified, but who have become unqualified because of not having met within the required period any of the following:

(i) Recurrent crewmember training requirements of § 91.1107.

(ii) Instrument proficiency check requirements of § 91.1069.

(iii) Testing requirements of § 91.1065.

(iv) Recurrent flight attendant testing requirements of § 91.1067.

§ 91.1065 Initial and recurrent pilot testing requirements.

(a) No program manager or owner may use a pilot, nor may any person serve as a pilot, unless, since the beginning of the 12th month before that service, that pilot has passed either a written or oral test (or a combination), given by the Administrator or an authorized check pilot, on that pilot's knowledge in the following areas—

(1) The appropriate provisions of parts 61 and 91 of this chapter and the management specifications and the operating manual of the program manager;

(2) For each type of aircraft to be flown by the pilot, the aircraft powerplant, major components and systems, major appliances, performance and operating limitations, standard and emergency operating procedures, and the contents of the accepted operating manual or equivalent, as applicable;

(3) For each type of aircraft to be flown by the pilot, the method of determining compliance with weight and balance limitations for takeoff, landing and en route operations;

(4) Navigation and use of air navigation aids appropriate to the operation or pilot authorization, including, when applicable, instrument approach facilities and procedures;

(5) Air traffic control procedures, including WR procedures when applicable;

(6) Meteorology in general, including the principles of frontal systems, icing, fog, thunderstorms, and windshear, and, if appropriate for the operation of the program manager, high altitude weather;

(7) Procedures for—

(i) Recognizing and avoiding severe weather situations;

(ii) Escaping from severe weather situations, in case of inadvertent encounters, including low-altitude windshear (except that rotorcraft aircraft pilots are not required to be tested on escaping from low-altitude windshear); and

(iii) Operating in or near thunderstorms (including best penetration altitudes), turbulent air (including clear air turbulence), icing, hail, and other potentially hazardous meteorological conditions; and

(8) New equipment, procedures, or techniques, as appropriate.

(b) No program manager or owner may use a pilot, nor may any person serve as a pilot, in any aircraft unless, since the beginning of the 12th month before that service, that pilot has passed a competency check given by the Administrator or an authorized check pilot in that class of aircraft, if single-engine aircraft other than turbojet, or that type of aircraft, if rotorcraft, multiengine aircraft, or turbojet airplane, to determine the pilot's competence in practical skills and techniques in that aircraft or class of aircraft. The extent of the competency check will be determined by the Administrator or authorized check pilot conducting the competency check. The competency check may include any of the maneuvers and procedures currently required for the original issuance of the particular pilot certificate required for the operations authorized and appropriate to the category, class and type of aircraft involved. For the purposes of this paragraph, type, as to an airplane, means any one of a group of airplanes determined by the Administrator to have a similar means of propulsion, the same manufacturer, and no significantly different handling or flight characteristics. For the purposes of this paragraph, type, as to a rotorcraft, means a basic make and model.

(c) The instrument proficiency check required by § 91.1069 may be substituted for the competency check required by this section for the type of aircraft used in the check.

(d) For the purpose of this subpart, competent performance of a procedure or maneuver by a person to be used as a pilot requires that the pilot be the obvious master of the aircraft, with the successful outcome of the maneuver never in doubt.

(e) The Administrator or authorized check pilot certifies the competency of each pilot who passes the knowledge or flight check in the program manager's pilot records.

(f) All or portions of a required competency check may be given in an aircraft simulator or other appropriate training device, if approved by the Administrator.

§ 91.1067 Initial and recurrent flight attendant crewmember testing requirements.

No program manager or owner may use a flight attendant crewmember, nor may any person serve as a flight attendant crewmember unless, since the beginning of the 12th month before that service, the program manager has determined by appropriate initial and recurrent testing that the person is knowledgeable and competent in the following areas as appropriate to assigned duties and responsibilities—

(a) Authority of the pilot in command;

(b) Passenger handling, including procedures to be followed in handling deranged persons or other persons whose conduct might jeopardize safety;

(c) Crewmember assignments, functions, and responsibilities during ditching and evacuation of persons who may need the assistance of another person to move expeditiously to an exit in an emergency;

(d) Briefing of passengers;

(e) Location and operation of portable fire extinguishers and other items of emergency equipment;

(f) Proper use of cabin equipment and controls;

(g) Location and operation of passenger oxygen equipment;

(h) Location and operation of all normal and emergency exits, including evacuation slides and escape ropes; and

(i) Seating of persons who may need assistance of another person to move rapidly to an exit in an emergency as prescribed by the program manager's operations manual.

§ 91.1069 Flight crew: Instrument proficiency check requirements.

(a) No program manager or owner may use a pilot, nor may any person serve, as a pilot in command of an aircraft under IFR unless, since the beginning of the 6th month before that service, that pilot has passed an instrument proficiency check under this section administered by the Administrator or an authorized check pilot.

(b) No program manager or owner may use a pilot, nor may any person serve, as a second command pilot of an aircraft under IFR unless, since the beginning of the 12th

month before that service, that pilot has passed an instrument proficiency check under this section administered by the Administrator or an authorized check pilot.

(c) No pilot may use any type of precision instrument approach procedure under IFR unless, since the beginning of the 6th month before that use, the pilot satisfactorily demonstrated that type of approach procedure. No pilot may use any type of nonprecision approach procedure under IFR unless, since the beginning of the 6th month before that use, the pilot has satisfactorily demonstrated either that type of approach procedure or any other two different types of nonprecision approach procedures. The instrument approach procedure or procedures must include at least one straight-in approach, one circling approach, and one missed approach. Each type of approach procedure demonstrated must be conducted to published minimums for that procedure.

(d) The instrument proficiency checks required by paragraphs (a) and (b) of this section consists of either an oral or written equipment test (or a combination) and a flight check under simulated or actual IFR conditions. The equipment test includes questions on emergency procedures, engine operation, fuel and lubrication systems, power settings, stall speeds, best engine-out speed, propeller and supercharger operations, and hydraulic, mechanical, and electrical systems, as appropriate. The flight check includes navigation by instruments, recovery from simulated emergencies, and standard instrument approaches involving navigational facilities which that pilot is to be authorized to use.

(e) Each pilot taking the instrument proficiency check must show that standard of competence required by § 91.1065(d).

(1) The instrument proficiency check must—

(i) For a pilot in command of an aircraft requiring that the PIC hold an airline transport pilot certificate, include the procedures and maneuvers for an airline transport pilot certificate in the particular type of aircraft, if appropriate; and

(ii) For a pilot in command of a rotorcraft or a second in command of any aircraft requiring that the SIC hold a commercial pilot certificate include the procedures and maneuvers for a commercial pilot certificate with an instrument rating and, if required, for the appropriate type rating.

(2) The instrument proficiency check must be given by an authorized check pilot or by the Administrator.

(f) If the pilot is assigned to pilot only one type of aircraft, that pilot must take the instrument proficiency check required by paragraph (a) of this section in that type of aircraft.

(g) If the pilot in command is assigned to pilot more than one type of aircraft, that pilot must take the instrument proficiency check required by paragraph (a) of this section in each type of aircraft to which that pilot is assigned, in rotation, but not more than one flight check during each period described in paragraph (a) of this section.

(h) If the pilot in command is assigned to pilot both single-engine and multiengine aircraft, that pilot must initially take the instrument proficiency check required by paragraph (a) of this section in a multiengine aircraft, and each succeeding check alternately in single-engine and multiengine aircraft, but not more than one flight check during each period described in paragraph (a) of this section.

(i) All or portions of a required flight check may be given in an aircraft simulator or other appropriate training device, if approved by the Administrator.

§ 91.1071 Crewmember: Tests and checks, grace provisions, training to accepted standards.

(a) If a crewmember who is required to take a test or a flight check under this subpart, completes the test or flight check in the month before or after the month in which it is required, that crewmember is considered to have completed the test or check in the month in which it is required.

(b) If a pilot being checked under this subpart fails any of the required maneuvers, the person giving the check may give additional training to the pilot during the course of the check. In addition to repeating the maneuvers failed, the person giving the check may require the pilot being checked to repeat any other maneuvers that are necessary to determine the pilot's proficiency. If the pilot being checked is unable to demonstrate satisfactory performance to the person conducting the check, the program manager may not use the pilot, nor may the pilot serve, as a flight crewmember in operations under this subpart until the pilot has satisfactorily completed the check If a pilot who demonstrates unsatisfactory performance is employed as a pilot for a certificate holder operating under part 121, 125, or 135 of this chapter, he or she must notify that certificate holder of the unsatisfactory performance.

§ 91.1073 Training program: General.

(a) Each program manager must have a training program and must:

(1) Establish, obtain the appropriate initial and final approval of, and provide a training program that meets this subpart and that ensures that each crewmember, including each flight attendant if the program manager uses a flight attendant crewmember, flight instructor, check pilot, and each person assigned duties for the carriage and handling of hazardous materials (as defined in 49 CFR 171.8) is adequately trained to perform these assigned duties.

(2) Provide adequate ground and flight training facilities and properly qualified ground instructors for the training required by this subpart.

(3) Provide and keep current for each aircraft type used and, if applicable, the particular variations within the aircraft type, appropriate training material, examinations,

forms, instructions, and procedures for use in conducting the training and checks required by this subpart.

(4) Provide enough flight instructors, check pilots, and simulator instructors to conduct required flight training and flight checks, and simulator training courses allowed under this subpart.

(b) Whenever a crewmember who is required to take recurrent training under this subpart completes the training in the month before, or the month after, the month in which that training is required, the crewmember is considered to have completed it in the month in which it was required.

(c) Each instructor, supervisor, or check pilot who is responsible for a particular ground training subject, segment of flight training, course of training, flight check, or competence check under this subpart must certify as to the proficiency and knowledge of the crewmember, flight instructor, or check pilot concerned upon completion of that training or check. That certification must be made a part of the crewmember's record. When the certification required by this paragraph is made by an entry in a computerized recordkeeping system, the certifying instructor, supervisor, or check pilot, must be identified with that entry. However, the signature of the certifying instructor, supervisor, or check pilot is not required for computerized entries.

(d) Training subjects that apply to more than one aircraft or crewmember position and that have been satisfactorily completed during previous training while employed by the program manager for another aircraft or another crewmember position, need not be repeated during subsequent training other than recurrent training.

(e) Aircraft simulators and other training devices may be used in the program manager's training program if approved by the Administrator.

(f) Each program manager is responsible for establishing safe and efficient crew management practices for all phases of flight in program operations including crew resource management training for all crewmembers used in program operations.

(g) If an aircraft simulator has been approved by the Administrator for use in the program manager's training program, the program manager must ensure that each pilot annually completes at least one flight training session in an approved simulator for at least one program aircraft. The training session may be the flight training portion of any of the pilot training or check requirements of this subpart, including the initial, transition, upgrade, requalification, differences, or recurrent training, or the accomplishment of a competency check or instrument proficiency check. If there is no approved simulator for that aircraft type in operation, then all flight training and checking must be accomplished in the aircraft.

§ 91.1075 Training program: Special rules.

Other than the program manager, only the following are eligible under this subpart to conduct training, testing, and checking under contract or other arrangement to those persons subject to the requirements of this subpart.

(a) Another program manager operating under this subpart:

(b) A training center certificated under part 142 of this chapter to conduct training, testing, and checking required by this subpart if the training center—

(1) Holds applicable training specifications issued under part 142 of this chapter;

(2) Has facilities, training equipment, and courseware meeting the applicable requirements of part 142 of this chapter;

(3) Has approved curriculums, curriculum segments, and portions of curriculum segments applicable for use in training courses required by this subpart; and

(4) Has sufficient instructors and check pilots qualified under the applicable requirements of §§ 91.1089 through 91.1095 to conduct training, testing, and checking to persons subject to the requirements of this subpart.

(c) A part 119 certificate holder operating under part 121 or part 135 of this chapter.

(d) As authorized by the Administrator, a training center that is not certificated under part 142 of this chapter.

§ 91.1077 Training program and revision: Initial and final approval.

(a) To obtain initial and final approval of a training program, or a revision to an approved training program, each program manager must submit to the Administrator—

(1) An outline of the proposed or revised curriculum, that provides enough information for a preliminary evaluation of the proposed training program or revision; and

(2) Additional relevant information that may be requested by the Administrator.

(b) If the proposed training program or revision complies with this subpart, the Administrator grants initial approval in writing after which the program manager may conduct the training under that program. The Administrator then evaluates the effectiveness of the training program and advises the program manager of deficiencies, if any, that must be corrected.

(c) The Administrator grants final approval of the proposed training program or revision if the program manager shows that the training conducted under the initial approval in paragraph (b) of this section ensures that each person who successfully completes the training is adequately trained to perform that person's assigned duties.

(d) Whenever the Administrator finds that revisions are necessary for the continued adequacy of a training program that has been granted final approval, the program manager must, after notification by the Administrator, make any changes in the program that are found necessary by the Administrator. Within 30 days after the program manager receives the notice, it may file a petition to reconsider the notice with the Administrator. The filing of a petition to reconsider stays the notice pending a decision by the Administrator. However, if the Administrator finds that there is an emergency that requires immediate action in the interest of safety, the Administrator may, upon a statement of the reasons, require a change effective without stay.

§ 91.1079 Training program: Curriculum.

(a) Each program manager must prepare and keep current a written training program curriculum for each type of aircraft for each crewmember required for that type aircraft. The curriculum must include ground and flight training required by this subpart.

(b) Each training program curriculum must include the following:

(1) A list of principal ground training subjects, including emergency training subjects, that are provided.

(2) A list of all the training devices, mock-ups, systems trainers, procedures trainers, or other training aids that the program manager will use.

(3) Detailed descriptions or pictorial displays of the approved normal, abnormal, and emergency maneuvers, procedures and functions that will be performed during each flight training phase or flight check, indicating those maneuvers, procedures and functions that are to be performed during the inflight portions of flight training and flight checks.

§ 91.1081 Crewmember training requirements.

(a) Each program manager must include in its training program the following initial and transition ground training as appropriate to the particular assignment of the crewrnember:

(1) Basic indoctrination ground training for newly hired crewmembers including instruction in at least the—

(i) Duties and responsibilities of crewmembers as applicable;

(ii) Appropriate provisions of this chapter;

(iii) Contents of the program manager's management specifications (not required for flight attendants); and

(iv) Appropriate portions of the program manager's operating manual.

(2) The initial and transition ground training in §§ 91.1101 and 91.1105, as applicable.

(3) Emergency training in § 91.1083.

(b) Each training program must provide the initial and transition flight training in § 91.1103, as applicable.

(c) Each training program must provide recurrent ground and flight training as provided in § 91.1107.

(d) Upgrade training in § 91.1101 and 91.1103 for a particular type aircraft may be included in the training program for crewmembers who have qualified and served as second in command on that aircraft.

(e) In addition to initial, transition, upgrade and recurrent training, each training program must provide ground and flight training, instruction, and practice necessary to ensure that each crewmember—

(1) Remains adequately trained and currently proficient for each aircraft, crewmember position, and type of operation in which the crewmember serves; and

(2) Qualifies in new equipment, facilities, procedures, and techniques, including modifications to aircraft.

§ 91.1083 Crewmember emergency training.

(a) Each training program must provide emergency training under this section for each aircraft type, model, and configuration, each crewmember, and each kind of operation conducted, as appropriate for each crewmember and the program manager.

(b) Emergency training must provide the following:

(1) Instruction in emergency assignments and procedures, including coordination among crewmembers.

(2) Individual instruction in the location, function, and operation of emergency equipment including—

(i) Equipment used in ditching and evacuation;

(ii) First aid equipment and its proper use; and

(iii) Portable fire extinguishers, with emphasis on the type of extinguisher to be used on different classes of fires.

(3) Instruction in the handling of emergency situations including—

(i) Rapid decompression;

(ii) Fire in flight or on the surface and smoke control procedures with emphasis on electrical equipment and related circuit breakers found in cabin areas;

(iii) Ditching and evacuation;

(iv) Illness, injury, or other abnormal situations involving passengers or crewmembers; and

(v) Hijacking and other unusual situations.

(4) Review and discussion of previous aircraft accidents and incidents involving actual emergency situations.

(c) Each crewmember must perform at least the following emergency drills, using the proper emergency equipment and procedures, unless the Administrator finds that, for a particular drill, the crewmember can be adequately trained by demonstration:

(1) Ditching, If applicable.

(2) Emergency evacuation.

(3) Fire extinguishing and smoke control.

(4) Operation and use of emergency exits, including deployment and use of evacuation slides, if applicable.

(5) Use of crew and passenger oxygen.

(6) Removal of life rafts from the aircraft, inflation of the life rafts, use of lifelines, and boarding of passengers and crew, if applicable.

(7) Donning and inflation of life vests and the use of other individual flotation devices, if applicable.

(d) Crewmembers who serve in operations above 25,000 feet must receive instruction in the following:

(1) Respiration.

(2) Hypoxia.

(3) Duration of consciousness without supplemental oxygen at altitude.

(4) Gas expansion.

(5) Gas bubble formation.

(6) Physical phenomena and incidents of decompression.

§ 91.1085 Hazardous materials recognition training.

No program manager may use any person to perform, and no person may perform, any assigned duties and responsibilities for the handling or carriage of hazardous materials (as defined in 49 CFR 171.8), unless that person has received training in the recognition of hazardous materials.

§ 91.1087 Approval of aircraft simulators and other training devices.

(a) Training courses using aircraft simulators and other training devices may be included in the program manager's training program if approved by the Administrator.

(b) Each aircraft simulator and other training device that is used in a training course or in checks required under this subpart must meet the following requirements:

(1) It must be specifically approved for—

(i) The program manager; and

(ii) The particular maneuver, procedure, or crewmember function involved.

(2) It must maintain the performance, functional, and other characteristics that are required for approval.

(3) Additionally, for aircraft simulators, it must be.—

(i) Approved for the type aircraft and, if applicable, the particular variation within type for which the training or check is being conducted; and

(ii) Modified to conform with any modification to the aircraft being simulated that changes the performance, functional, or other characteristics required for approval.

(c) A particular aircraft simulator or other training device may be used by more than one program manager.

(d) In granting initial and final approval of training programs or revisions to them, the Administrator considers the training devices, methods, and procedures listed in the program manager's curriculum under § 91.1079.

§ 91.1089 Qualifications: Check pilots (aircraft) and check pilots (simulator).

(a) For the purposes of this section and §91.1093:

(1) A check pilot (aircraft) is a person who is qualified to conduct flight checks in an aircraft, in a flight simulator, or in a flight training device for a particular type aircraft.

(2) A check pilot (simulator) is a person who is qualified to conduct flight checks, but only in a flight simulator, in a flight training device, or both, for a particular type aircraft.

(3) Check pilots (aircraft) and check pilots (simulator) are those check pilots who perform the functions described in § 91.1073(a)(4) and (c).

(b) No program manager may use a person, nor may any person serve as a check pilot (aircraft) in a training program established under this subpart unless, with respect to the aircraft type involved, that person—

(1) Holds the pilot certificates and ratings required to serve as a pilot in command in operations under this subpart;

(2) Has satisfactorily completed the training phases for the aircraft, including recurrent training, that are required to serve as a pilot in command in operations under this subpart;

(3) Has satisfactorily completed the proficiency or competency checks that are required to serve as a pilot in command in operations under this subpart;

(4) Has satisfactorily completed the applicable training requirements of § 91.1093;

(5) Holds at least a Class III medical certificate unless serving as a required crewmember, in which case holds a Class I or Class II medical certificate as appropriate; and

(6) Has been approved by the Administrator for the check pilot duties involved.

(c) No program manager may use a person, nor may any person serve as a check pilot (simulator) in a training program established under this subpart unless, with respect to the aircraft type involved, that person meets the provisions of paragraph (b) of this section, or—

(1) Holds the applicable pilot certificates and ratings, except medical certificate, required to serve as a pilot in command in operations under this subpart;

(2) Has satisfactorily completed the appropriate training phases for the aircraft, including recurrent training, that are required to serve as a pilot in command in operations under this subpart;

(3) Has satisfactorily completed the appropriate proficiency or competency checks that are required to serve as a pilot in command in operations under this subpart;

(4) Has satisfactorily completed the applicable training requirements of § 91.1093; and

(5) Has been approved by the Administrator for the check pilot (simulator) duties involved.

(d) Completion of the requirements in paragraphs (b)(2), (3), and (4) or (c)(2), (3), and (4) of this section, as applicable, must be entered in the individual's training record maintained by the program manager.

(e) A check pilot who does not hold an appropriate medical certificate may function as a check pilot (simulator), but may not serve as a fiightcrew member in operations under this subpart.

(f) A check pilot (simulator) must accomplish the following—

(1) Fly at least two flight segments as a required crewmember for the type, class, or category aircraft involved within the 12-month period preceding the performance of any check pilot duty in a flight simulator; or

(2) Before performing any check pilot duty in a flight simulator, satisfactorily complete an approved line-observation program within the period prescribed by that program.

(g) The flight segments or line-observation program required in paragraph (f) of this section are considered to be completed in the month required if completed in the month before or the month after the month in which they are due.

§ 91.1091 Qualifications: Flight instructors (aircraft) and flight instructors (simulator).

(a) For the purposes of this section and § 91.1095:

(1) A flight instructor (aircraft) is a person who is qualified to instruct in an aircraft, in a flight simulator, or in a flight training device for a particular type, class, or category aircraft.

(2) A flight instructor (simulator) is a person who is qualified to instruct in a flight simulator, in a flight training device, or in both, for a particular type, class, or category aircraft.

(3) Flight instructors (aircraft) and flight instructors (simulator) are those instructors who perform the functions described in § 91.1073(a)(4) and (c).

(b) No program manager may use a person, nor may any person serve as a flight instructor (aircraft) in a training program established under this subpart unless, with respect to the type, class, or category aircraft involved, that person—

(1) Holds the pilot certificates and ratings required to serve as a pilot in command in operations under this subpart or part 121 or 135 of this chapter;

(2) Has satisfactorily completed the training phases for the aircraft, including recurrent training, that are required to serve as a pilot in command in operations under this subpart;

(3) Has satisfactorily completed the proficiency or competency checks that are required to serve as a pilot in command in operations under this subpart;

(4) Has satisfactorily completed the applicable training requirements of § 91.1095; and

(5) Holds at least a Class Ill medical certificate.

(c) No program manager may use a person, nor may any person serve as a flight instructor (simulator) in a training program established under this subpart, unless, with respect to the type, class, or category aircraft involved, that person meets the provisions of paragraph (b) of this section, or—

(1) Holds the pilot certificates and ratings, except medical certificate, required to serve as a pilot in command in operations under this subpart or part 121 or 135 of this chapter;

(2) Has satisfactorily completed the appropriate training phases for the aircraft, including recurrent training, that are required to serve as a pilot in command in operations under this subpart;

(3) Has satisfactorily completed the appropriate proficiency or competency checks that are required to serve as a pilot in command in operations under this subpart; and

(4) Has satisfactorily completed the applicable training requirements of § 91.1095.

(d) Completion of the requirements in paragraphs (b)(2), (3), and (4) or (c)(2), (3), and (4) of this section, as applicable, must be entered in the individual's training record maintained by the program manager.

(e) A pilot who does not hold a medical certificate may function as a flight instructor in an aircraft if functioning as a non-required crewmember, but may not serve as a flightcrew member in operations under this subpart.

(f) A flight instructor (simulator) must accomplish the following—

(1) Fly at least two flight segments as a required crewmember for the type, class, or category aircraft involved within the 12-month period preceding the performance of any flight instructor duty in a flight simulator; or

(2) Satisfactorily complete an approved line-observation program within the period prescribed by that program and that must precede the performance of any check pilot duty in a flight simulator.

(g) The flight segments or line-observation program required in paragraph (f) of this section are considered completed in the month required if completed in the month before, or in the month after, the month in which they are due.

§ 91.1093 Initial and transition training and checking check pilots (aircraft), check pilots (simulator).

(a) No program manager may use a person nor may any person serve as a check pilot unless—

(1) That person has satisfactorily completed initial or transition check pilot training; and

(2) Within the preceding 24 months, that person satisfactorily conducts a proficiency or competency check under the observation of an FAA inspector or an aircrew designated examiner employed by the program manager. The observation check may be accomplished in part or in full in an aircraft, in a flight simulator, or in a flight training device.

(b) The observation check required by paragraph (a)(2) of this section is considered to have been completed in the month required if completed in the month before or the month after the month in which it is due.

(c) The initial ground training for check pilots must include the following:

(1) Check pilot duties, functions, and responsibilities.

(2) The applicable provisions of the Code of Federal Regulations and the program manager's policies and procedures.

(3) The applicable methods, procedures, and techniques for conducting the required checks.

(4) Proper evaluation of student performance including the detection of—

(i) Improper and insufficient training; and

(ii) Personal characteristics of an applicant that could adversely affect safety.

(5) The corrective action in the case of unsatisfactory checks.

(6) The approved methods, procedures, and limitations for performing the required normal, abnormal, and emergency procedures in the aircraft.

(d) The transition ground training for a check pilot must include the approved methods, procedures, and limitations for performing the required normal, abnormal, and emergency procedures applicable to the aircraft to which the check pilot is in transition.

(e) The initial and transition flight training for a check pilot (aircraft) must include the following—

(1) The safety measures for emergency situations that are likely to develop during a check;

(2) The potential results of improper, untimely, or nonexecution of safety measures during a check;

(3) Training and practice in conducting flight checks from the left and right pilot seats in the required normal, abnormal, and emergency procedures to ensure competence to conduct the pilot flight checks required by this subpart; and

(4) The safety measures to be taken from either pilot seat for emergency situations that are likely to develop during checking.

(f) The requirements of paragraph (e) of this section may be accomplished in full or in part in flight, in a flight simulator, or in a flight training device, as appropriate.

(g) The initial and transition flight training for a check pilot (simulator) must include the following:

(1) Training and practice in conducting flight checks in the required normal, abnormal, and emergency procedures to ensure competence to conduct the flight checks required by this subpart. This training and practice must be accomplished in a flight simulator or in a flight training device.

(2) Training in the operation of flight simulators, flight training devices, or both, to ensure competence to conduct the flight checks required by this subpart.

§ 91.1095 Initial and transition training and checking: Right instructors (aircraft), flight instructors (simulator).

(a) No program manager may use a person nor may any person serve as a flight instructor unless—

(1) That person has satisfactorily completed initial or transition flight instructor training; and

(2) Within the preceding 24 months, that person satisfactorily conducts instruction under the observation of an FAA inspector, a program manager check pilot, or an aircrew designated examiner employed by the program manager. The observation check may be accomplished in part or in full in an aircraft, in a flight simulator, or in a flight training device.

(b) The observation chock required by paragraph (a)(2) of this section is considered to have been completed in the month required if completed in the month before, or the month after, the month in which it is due.

(c) The initial ground training for flight instructors must include the following:

(1) Flight instructor duties, functions, and responsibilities.

(2) The applicable Code of Federal Regulations and the program manager's policies and procedures.

(3) The applicable methods, procedures, and techniques for conducting flight instruction.

(4) Proper evaluation of student performance including the detection of—

(i) Improper and insufficient training; and

(ii) Personal characteristics of an applicant that could adversely affect safety.

(5) The corrective action in the case of unsatisfactory training progress.

(6) The approved methods, procedures, and limitations for performing the required normal, abnormal, and emergency procedures in the aircraft.

(7) Except for holders of a flight instructor certificate—

(i) The fundamental principles of the teaching-learning process;

(ii) Teaching methods and procedures; and

(iii) The instructor-student relationship.

(d) The transition ground training for flight instructors must include the approved methods, procedures, and limitations for performing the required normal, abnormal, and emergency procedures applicable to the type, class, or category aircraft to which the flight instructor is in transition.

(e) The initial and transition flight training for flight instructors (aircraft) must include the following—

(1) The safety measures for emergency situations that are likely to develop during instruction;

(2) The potential results of improper or untimely safety measures during instruction;

(3) Training and practice from the left and right pilot seats in the required normal, abnormal, and emergency maneuvers to ensure competence to conduct the flight instruction required by this subpart; and

(4) The safety measures to be taken from either the left or right pilot seat for emergency situations that are likely to develop during instruction.

(f) The requirements of paragraph (e) of this section may be accomplished in full or in part in flight, in a flight simulator, or in a flight training device, as appropriate.

(g) The initial and transition flight training for a flight instructor (simulator) must include the following:

(1) Training and practice in the required normal, abnormal, and emergency procedures to ensure competence to conduct the flight instruction required by this subpart. These maneuvers and procedures must be accomplished in full or in part in a flight simulator or in a flight training device.

(2) Training in the operation of flight simulators, flight training devices, or both, to ensure competence to conduct the flight instruction required by this subpart.

§ 91.1097 Pilot and flight attendant crewmember training programs.

(a) Each program manager must establish and maintain an approved pilot training program, and each program manager who uses a flight attendant crewmember must establish and maintain an approved flight attendant training program, that is appropriate to the operations to which each pilot and flight attendant is to be assigned, and will ensure that they are adequately trained to meet the applicable knowledge and practical testing requirements of §§ 91.1065 through 91.1071.

(b) Each program manager required to have a training program by paragraph (a) of this section must include in that program ground and flight training curriculums for—

(1) Initial training

(2) Transition training;

(3) Upgrade training;

(4) Differences training;

(5) Recurrent training; and

(6) Requalification training.

(c) Each program manager must provide current and appropriate study materials for use by each required pilot and flight attendant.

(d) The program manager must furnish copies of the pilot and flight attendant crewmember training program, and all changes and additions, to the assigned representative of the Administrator. If the program manager uses training facilities of other persons, a copy of those training programs or appropriate portions used for those facilities must also be furnished. Curricula that follow FAA published curricula may be cited by reference in the copy of the training program furnished to the representative of the Administrator and need not be furnished with the program.

§ 91.1099 Crewmember initial and recurrent training requirements.

No program manager may use a person, nor may any person serve, as a crewmember in operations under this subpart unless that crewmember has completed the appropriate initial or recurrent training phase of the training program appropriate to the type of operation in which the crewmember is to serve since the beginning of the 12th month before that service.

§ 91.1101 Pilots: Initial, transition, and upgrade ground training.

Initial, transition, and upgrade ground training for pilots must include instruction in at least the following, as applicable to their duties:

(a) General subjects—

(1) The program manager's flight locating procedures;

(2) Principles and methods for determining weight and balance, and runway limitations for takeoff and landing;

(3) Enough meteorology to ensure a practical knowledge of weather phenomena, including the principles of

frontal systems, icing, fog, thunderstorms, windshear and, if appropriate, high altitude weather situations;

(4) Air traffic control systems, procedures, and phraseology;

(5) Navigation and the use of navigational aids, including instrument approach procedures;

(6) Normal and emergency communication procedures;

(7) Visual cues before and during descent below Decision Altitude or MDA; and

(8) Other instructions necessary to ensure the pilot's competence.

(b) For each aircraft type—

(1) A general description;

(2) Performance characteristics;

(3) Engines and propellers;

(4) Major components;

(5) Major aircraft systems (that is, flight controls, electrical, and hydraulic), other systems, as appropriate, principles of normal, abnormal, and emergency operations, appropriate procedures and limitations;

(6) Knowledge and procedures for—

(i) Recognizing and avoiding severe weather situations;

(ii) Escaping from severe weather situations, in case of inadvertent encounters, including low-altitude windshear (except that rotorcraft pilots are not required to be trained in escaping from low-altitude windshear);

(iii) Operating in or near thunderstorms (including best penetration altitudes), turbulent air (including clear air turbulence), inflight icing, hail, and other potentially hazardous meteorological conditions; and

(iv) Operating airplanes during ground icing conditions, (that is, any time conditions are such that frost, ice, or snow may reasonably be expected to adhere to the aircraft), if the program manager expects to authorize takeoffs in ground icing conditions, including:

(A) The use of holdover times when using deicing/anti-icing fluids;

(B) Airplane deicing/anti-icing procedures, including inspection and check procedures and responsibilities;

(C) Communications;

(D) Airplane surface contamination (that is, adherence of frost, ice, or snow) and critical area identification, and knowledge of how contamination adversely affects airplane performance and flight characteristics;

(E) Types and characteristics of deicing/anti-icing fluids, if used by the program manager;

(F) Cold weather preflight inspection procedures;

(G) Techniques for recognizing contamination on the airplane;

(7) Operating limitations;

(8) Fuel consumption and cruise control;

(9) Flight planning;

(10) Each normal and emergency procedure; and

(11) The approved Aircraft Flight Manual or equivalent.

§ 91.1103 Pilots: Initial, transition, upgrade, requalification, and differences flight training.

(a) Initial, transition, upgrade, requalification, and differences training for pilots must include flight and practice in each of the maneuvers and procedures contained in each of the curriculums that are a part of the approved training program.

(b) The maneuvers and procedures required by paragraph (a) of this section must be performed in flight, except to the extent that certain maneuvers and procedures may be performed in an aircraft simulator, or an appropriate training device, as allowed by this subpart.

(c) If the program manager's approved training program includes a course of training using an aircraft simulator or other training device, each pilot must successfully complete—

(1) Training and practice in the simulator or training device in at least the maneuvers and procedures in this subpart that are capable of being performed in the aircraft simulator or training device; and

(2) A flight check in the aircraft or a check in the simulator or training device to the level of proficiency of a pilot in command or second in command, as applicable, in at least the maneuvers and procedures that are capable of being performed in an aircraft simulator or training device.

§ 91.1105 Flight attendants: Initial and transition ground training.

Initial and transition ground training for flight attendants must include instruction in at least the following—

(a) General subjects—

(1) The authority of the pilot in command; and

(2) Passenger handling, including procedures to be followed in handling deranged persons or other persons whose conduct might jeopardize safety.

(b) For each aircraft type—

(1) A general description of the aircraft emphasizing physical characteristics that may have a bearing on ditching, evacuation, and inflight emergency procedures and on other related duties;

(2) The use of both the public address system and the means of communicating with other flight crewmembers, including emergency means in the case of attempted hijacking or other unusual situations; and

(3) Proper use of electrical galley equipment and the controls for cabin heat and ventilation.

§ 91.1107 Recurrent training.

(a) Each program manager must ensure that each crewmember receives recurrent training and is adequately trained and currently proficient for the type aircraft and crewmember position involved.

(b) Recurrent ground training for crewmembers must include at least the following:

(1) A quiz or other review to determine the crewmember's knowledge of the aircraft and crewmember position involved.

(2) Instruction as necessary in the subjects required for initial ground training by this subpart, as appropriate, including low-altitude windshear training and training on operating during ground icing conditions, as prescribed in § 91.1097 and described in § 91.1101, and emergency training.

(c) Recurrent flight training for pilots must include, at least, flight training in the maneuvers or procedures in this subpart, except that satisfactory completion of the check required by § 91.1065 within the preceding 12 months may be substituted for recurrent flight training.

§ 91.1109 Aircraft maintenance: Inspection program.

Each program manager must establish an aircraft inspection program for each make and model program aircraft and ensure each aircraft is inspected in accordance with that inspection program.

(a) The inspection program must be in writing and include at least the following information:

(1) Instructions and procedures for the conduct of inspections for the particular make and model aircraft, including necessary tests and checks. The instructions and procedures must set forth in detail the parts and areas of the airframe, engines, propellers, rotors, and appliances, including survival and emergency equipment required to be inspected.

(2) A schedule for performing the inspections that must be accomplished under the inspection program expressed in terms of the time in service, calendar time, number of system operations, or any combination thereof.

(3) The name and address of the person responsible for scheduling the inspections required by the inspection program. A copy of the inspection program must be made available to the person performing inspections on the aircraft and, upon request, to the Administrator.

(b) Each person desiring to establish or change an approved inspection program under this section must submit the inspection program for approval to the Flight Standards District Office that issued the program manager's management specifications. The inspection program must be derived from one of the following programs:

(1) An inspection program currently recommended by the manufacturer of the aircraft, aircraft engines, propellers, appliances, and survival and emergency equipment;

(2) An inspection program that is part of a continuous airworthiness maintenance program currently in use by a person holding an air carrier or operating certificate issued under part 119 of this chapter and operating that make and model aircraft under part 121 or 135 of this chapter;

(3) An aircraft inspection program approved under § 135.419 of this chapter and currently in use under part 135

of this chapter by a person holding a certificate issued under part 119 of this chapter; or

(4) An airplane inspection program approved under § 125.247 of this chapter and currently in use under part 125 of this chapter.

(5) An inspection program that is part of the program manager's continuous airworthiness maintenance program under §§ 91.1411 through 91.1443.

(c) The Administrator may require revision of the inspection program approved under this section in accordance with the provisions of § 91.415.

§ 91.1111 Maintenance training.

The program manager must ensure that all employees who are responsible for maintenance related to program aircraft undergo appropriate initial and annual recurrent training and are competent to perform those duties.

§ 91.1113 Maintenance recordkeeping.

Each fractional ownership program manager must keep (using the system specified in the manual required in § 91.1025) the records specified in § 91.417(a) for the periods specified in § 91.417(b).

§ 91.1115 Inoperable instruments and equipment

(a) No person may take off an aircraft with inoperable instruments or equipment installed unless the following conditions are met:

(1) An approved Minimum Equipment List exists for that aircraft.

(2) The program manager has been issued management specifications authorizing operations in accordance with an approved Minimum Equipment List. The flight crew must have direct access at all times prior to flight to all of the information contained in the approved Minimum Equipment List through printed or other means approved by the Administrator in the program manager's management specifications. An approved Minimum Equipment List, as authorized by the management specifications, constitutes an approved change to the type design without requiring recertification.

(3) The approved Minimum Equipment List must:

(i) Be prepared in accordance with the limitations specified in paragraph (b) of this section.

(ii) Provide for the operation of the aircraft with certain instruments and equipment in an inoperable condition.

(4) Records identifying the inoperable instruments and equipment and the information required by (a)(3)(ii) of this section must be available to the pilot.

(5) The aircraft is operated under all applicable conditions and limitations contained in the Minimum Equipment List and the management specifications authorizing use of the Minimum Equipment List.

(b) The following instruments and equipment may not be included in the Minimum Equipment List:

(1) Instruments and equipment that are either specifically or otherwise required by the airworthiness requirements under which the airplane is type certificated and that are essential for safe operations under all operating conditions.

(2) Instruments and equipment required by an airworthiness directive to be in operable condition unless the airworthiness directive provides otherwise.

(3) Instruments and equipment required for specific operations by this part.

(c) Notwithstanding paragraphs (b)(1) and (b)(3) of this section, an aircraft with inoperable instruments or equipment may be operated under a special flight permit under §§ 21.197 and 21.199 of this chapter.

(d) A person authorized to use an approved Minimum Equipment List issued for a specific aircraft under part 121, 125, or 135 of this chapter must use that Minimum Equipment List to comply with this section.

§ 91.1411 Continuous airworthiness maintenance program use by fractional ownership program manager.

Fractional ownership program aircraft may be maintained under a continuous airworthiness maintenance program (CAMP) under §§91.1413 through 91.1443. Any program manager who elects to maintain the program aircraft using a continuous airworthiness maintenance program must comply with §§ 91.1413 through 91.1443.

§ 91.1413 CAMP: Responsibility for airworthiness.

(a) For aircraft maintained in accordance with a Continuous Airworthiness Maintenance Program, each program manager is primarily responsible for the following:

(1) Maintaining the airworthiness of the program aircraft, including airframes, aircraft engines, propellers, rotors, appliances, and parts.

(2) Maintaining its aircraft in accordance with the requirements of this chapter.

(3) Repairing defects that occur between regularly scheduled maintenance required under part 43 of this chapter.

(b) Each program manager who maintains program aircraft under a CAMP must—

(1) Employ a Director of Maintenance or equivalent position. The Director of Maintenance must be a certificated mechanic with airframe and powerplant ratings who has responsibility for the maintenance program on all program aircraft maintained under a continuous airworthiness maintenance program. This person cannot also act as Chief Inspector.

(2) Employ a Chief Inspector or equivalent position. The Chief Inspector must be a certificated mechanic with airframe and powerplant ratings who has overall responsibility for inspection aspects of the CAMP. This person cannot also act as Director of Maintenance.

(3) Have the personnel to perform the maintenance of program aircraft, including airframes, aircraft engines, propellers, rotors, appliances, emergency equipment and parts, under its manual and this chapter; or make arrangements with another person for the performance of maintenance. However, the program manager must ensure that any maintenance, preventive maintenance, or alteration that is performed by another person is performed under the program manager's operating manual and this chapter.

§ 91.1415 CAMP: Mechanical reliability reports.

(a) Each program manager who maintains program aircraft under a CAMP must report the occurrence or detection of each failure, malfunction, or defect in an aircraft concerning—

(1) Fires during flight and whether the related fire-warning system functioned properly;

(2) Fires during flight not protected by related fire-warning system;

(3) False fire-warning during flight;

(4) An exhaust system that causes damage during flight to the engine, adjacent structure, equipment, or components;

(5) An aircraft component that causes accumulation or circulation of smoke, vapor, or toxic or noxious fumes in the crew compartment or passenger cabin during flight;

(6) Engine shutdown during flight because of flameout;

(7) Engine shutdown during flight when external damage to the engine or aircraft structure occurs;

(8) Engine shutdown during flight because of foreign object ingestion or icing;

(9) Shutdown of more than one engine during flight;

(10) A propeller feathering system or ability of the system to control overspeed during flight;

(11) A fuel or fuel-dumping system that affects fuel flow or causes hazardous leakage during flight;

(12) An unwanted landing gear extension or retraction or opening or closing of landing gear doors during flight;

(13) Brake system components that result in loss of brake actuating force when the aircraft is in motion on the ground;

(14) Aircraft structure that requires major repair;

(15) Cracks, permanent deformation, or corrosion of aircraft structures, if more than the maximum acceptable to the manufacturer or the FAA; and

(16) Aircraft components or systems that result in taking emergency actions during flight (except action to shut down an engine).

(b) For the purpose of this section, *during flight* means the period from the moment the aircraft leaves the surface of the earth on takeoff until it touches down on landing.

(c) In addition to the reports required by paragraph (a) of this section, each program manager must report any other failure, malfunction, or defect in an aircraft that occurs or is detected at any time if, in the manager's opinion, the failure, malfunction, or defect has endangered or may endanger the safe operation of the aircraft.

(d) Each program manager must send each report required by this section, in writing, covering each 24-hour period beginning at 0900 hours local time of each day and ending at 0900 hours local time on the next day to the Flight Standards District Office that issued the program manager's management specifications. Each report of occurrences during a 24-hour period must be mailed or transmitted to that office within the next 72 hours. However, a report that is due on Saturday or Sunday may be mailed or transmitted on the following Monday and one that is due on a holiday may be mailed or transmitted on the next workday. For aircraft operated in areas where mail is not collected, reports may be mailed or transmitted within 72 hours after the aircraft returns to a point where the mail is collected.

(e) The program manager must transmit the reports required by this section on a form and in a manner prescribed by the Administrator, and must include as much of the following as is available:

(1) The type and identification number of the aircraft.

(2) The name of the program manager.

(3) The date.

(4) The nature of the failure, malfunction, or defect.

(5) Identification of the part and system involved, including available information pertaining to type designation of the major component and time since last overhaul, if known.

(6) Apparent cause of the failure, malfunction or defect (for example, wear, crack, design deficiency, or personnel error).

(7) Other pertinent information necessary for more complete identification, determination of seriousness, or corrective action.

(f) A program manager that is also the holder of a type certificate (including a supplemental type certificate), a Parts Manufacturer Approval, or a Technical Standard Order Authorization, or that is the licensee of a type certificate need not report a failure, malfunction, or defect under this section if the failure, malfunction, or defect has been reported by it under § 21.3 of this chapter or under the accident reporting provisions of part 830 of the regulations of the National Transportation Safety Board.

(g) No person may withhold a report required by this section even when not all information required by this section is available.

(h) When the program manager receives additional information, including information from the manufacturer or other agency, concerning a report required by this section, the program manager must expeditiously submit it as a supplement to the first report and reference the date and place of submission of the first report.

§ 91.1417 CAMP: Mechanical Interruption summary report.

Each program manager who maintains program aircraft under a CAMP must mail or deliver, before the end of the 10th day of the following month, a summary report of the following occurrences in multiengine aircraft for the preceding month to the Flight Standards District Office that issued the management specifications:

(a) Each interruption to a flight, unscheduled change of aircraft en route, or unscheduled stop or diversion from a route, caused by known or suspected mechanical difficulties or malfunctions that are not required to be reported under § 91.1415.

(b) The number of propeller featherings in flight, listed by type of propeller and engine and aircraft on which it was installed. Propeller featherings for training, demonstration, or flight check purposes need not be reported.

§ 91.1423 CAMP: Maintenance organization.

(a) Each program manager who maintains program aircraft under a CAMP that has its personnel perform any of its maintenance (other than required inspections), preventive maintenance, or alterations, and each person with whom it arranges for the performance of that work, must have an organization adequate to perform the work.

(b) Each program manager who has personnel perform any inspections required by the program manager's manual under § 91.1427(b) (2) or (3), (in this subpart referred to as required inspections), and each person with whom the program manager arranges for the performance of that work, must have an organization adequate to perform that work.

(c) Each person performing required inspections in addition to other maintenance, preventive maintenance, or alterations, must organize the performance of those functions so as to separate the required inspection functions from the other maintenance, preventive maintenance, or alteration functions. The separation must be below the level of administrative control at which overall responsibility for the required inspection functions and other maintenance, preventive maintenance, or alterations is exercised.

§ 91.1425 CAMP: Maintenance, preventive maintenance, and alteration programs.

Each program manager who maintains program aircraft under a CAMP must have an inspection program and a program covering other maintenance, preventive maintenance, or alterations that ensures that—

(a) Maintenance, preventive maintenance, or alterations performed by its personnel, or by other persons, are performed under the program manager's manual;

(b) Competent personnel and adequate facilities and equipment are provided for the proper performance of maintenance, preventive maintenance, or alterations; and

(c) Each aircraft released to service is airworthy and has been properly maintained for operation under this part.

§ 91.1427 CAMP: Manual requirements

(a) Each program manager who maintains program aircraft under a CAMP must put in the operating manual the chart or description of the program manager's organization required by § 91.1423 and a list of persons with whom it has arranged for the performance of any of its required inspections, and other maintenance, preventive maintenance, or alterations, including a general description of that work.

(b) Each program manager must put in the operating manual the programs required by § 91.1425 that must be followed in performing maintenance, preventive maintenance, or alterations of that program manager's aircraft, including airframes, aircraft engines, propellers, rotors, appliances, emergency equipment, and parts, and must include at least the following:

(1) The method of performing routine and nonroutine maintenance (other than required inspections), preventive maintenance, or alterations.

(2) A designation of the items of maintenance and alteration that must be inspected (required inspections) including at least those that could result in a failure, malfunction, or defect endangering the safe operation of the aircraft, if not performed properly or if improper parts or materials are used.

(3) The method of performing required inspections and a designation by occupational title of personnel authorized to perform each required inspection.

(4) Procedures for the reinspection of work performed under previous required inspection findings (buy-back procedures).

(5) Procedures, standards, and limits necessary for required inspections and acceptance or rejection of the items required to be inspected and for periodic inspection and calibration of precision tools, measuring devices, and test equipment.

(6) Procedures to ensure that all required inspections are performed.

(7) Instructions to prevent any person who performs any item of work from performing any required inspection of that work.

(8) Instructions and procedures to prevent any decision of an inspector regarding any required inspection from being countermanded by persons other than supervisory personnel of the inspection unit, or a person at the level of administrative control that has overall responsibility for the management of both the required inspection functions and the other maintenance, preventive maintenance, or alterations functions.

(9) Procedures to ensure that maintenance (including required inspections), preventive maintenance, or alterations that are not completed because of work interruptions are properly completed before the aircraft is released to service.

(c) Each program manager must put in the manual a suitable system (which may include an electronic or coded system) that provides for the retention of the following information —

(1) A description (or reference to data acceptable to the Administrator) of the work performed;

(2) The name of the person performing the work if the work is performed by a person outside the organization of the program manager; and

(3) The name or other positive identification of the individual approving the work.

(d) For the purposes of this part, the program manager must prepare that part of its manual containing maintenance information and instructions, in whole or in part, in a format acceptable to the Administrator, that is retrievable in the English language.

§ 91.1429 CAMP: Required inspection personnel.

(a) No person who maintains an aircraft under a CAMP may use any person to perform required inspections unless the person performing the inspection is appropriately certificated, properly trained, qualified, and authorized to do so.

(b) No person may allow any person to perform a required inspection unless, at the time the work was performed, the person performing that inspection is under the supervision and control of the chief inspector.

(c) No person may perform a required inspection if that person performed the item of work required to be inspected.

(d) Each program manager must maintain, or must ensure that each person with whom it arranges to perform required inspections maintains, a current listing of persons who have been trained, qualified, and authorized to conduct required inspections. The persons must be identified by name, occupational title, and the inspections that they are authorized to perform. The program manager (or person with whom it arranges to perform its required inspections) must give written information to each person so authorized, describing the extent of that person's responsibilities, authorities, and inspectional limitations. The list must be made available for inspection by the Administrator upon request.

§ 91.1431 CAMP: Continuing analysis and surveillance.

(a) Each program manager who maintains program aircraft under a CAMP must establish and maintain a system for the continuing analysis and surveillance of the performance and effectiveness of its inspection program and the program covering other maintenance, preventive maintenance, and alterations and for the correction of any deficiency in those programs, regardless of whether those programs are carried out by employees of the program manager or by another person.

(b) Whenever the Administrator finds that the programs described in paragraph (a) of this section does not contain adequate procedures and standards to meet this part, the program manager must, after notification by the Administrator, make changes in those programs requested by the Administrator.

(c) A program manager may petition the Administrator to reconsider the notice to make a change in a program. The petition must be filed with the Director, Flight Standards Service, within 30 days after the program manager receives the notice. Except in the case of an emergency requiring immediate action in the interest of safety, the filing of the petition stays the notice pending a decision by the Administrator.

§ 91.1433 CAMP: Maintenance and preventive maintenance training program.

Each program manager who maintains program aircraft under a CAMP or a person performing maintenance or preventive maintenance functions for it must have a training program to ensure that each person (including inspection personnel) who determines the adequacy of work done is fully informed about procedures and techniques and new equipment in use and is competent to perform that person's duties.

§ 91.1435 CAMP: Certificate requirements.

(a) Except for maintenance, preventive maintenance, alterations, and required inspections performed by repair stations located outside the United States certificated under the provisions of part 145 of this chapter, each person who is directly in charge of maintenance, preventive maintenance, or alterations for a CAMP, and each person performing required inspections for a CAMP must hold an appropriate airman certificate.

(b) For the purpose of this section, a person "directly in charge" is each person assigned to a position in which that person is responsible for the work of a shop or station that performs maintenance, preventive maintenance, alterations, or other functions affecting airworthiness. A person who is directly in charge need not physically observe and direct each worker constantly but must be available for consultation and decision on matters requiring instruction or decision from higher authority than that of the person performing the work.

§ 91.1437 CAMP: Authority to perform and approve maintenance.

A program manager who maintains program aircraft under a CAMP may employ maintenance personnel, or make arrangements with other persons to perform maintenance and preventive maintenance as provided in its maintenance manual. Unless properly certificated, the program manager may not perform or approve maintenance for return to service.

§ 91.1439 CAMP: Maintenance recording requirements.

(a) Each program manager who maintains program aircraft under a CAMP must keep (using the system specified in the manual required in § 91.1427) the following records for the periods specified in paragraph (b) of this section:

(1) All the records necessary to show that all requirements for the issuance of an airworthiness release under § 91.1443 have been met.

(2) Records containing the following information:

(i) The total time in service of the airframe, engine, propeller, and rotor.

(ii) The current status of life-limited parts of each airframe, engine, propeller, rotor, and appliance.

(iii) The time since last overhaul of each item installed on the aircraft that are required to be overhauled on a specified time basis.

(iv) The identification of the current inspection status of the aircraft, including the time since the last inspections required by the inspection program under which the aircraft and its appliances are maintained.

(v) The current status of applicable airworthiness directives, including the date and methods of compliance, and, if the airworthiness directive involves recurring action, the time and date when the next action is required.

(vi) A list of current major alterations and repairs to each airframe, engine, propeller, rotor, and appliance.

(b) Each program manager must retain the records required to be kept by this section for the following periods:

(1) Except for the records of the last complete overhaul of each airframe, engine, propeller, rotor, and appliance the records specified in paragraph (a)(1) of this section must be retained until the work is repeated or superseded by other work or for one year after the work is performed.

(2) The records of the last complete overhaul of each airframe, engine, propeller, rotor, and appliance must be retained until the work is superseded by work of equivalent scope and detail.

(3) The records specified in paragraph (a)(2) of this section must be retained as specified unless transferred with the aircraft at the time the aircraft is sold.

(c) The program manager must make all maintenance records required to be kept by this section available for inspection by the Administrator or any representative of the National Transportation Safety Board.

§91.1441 CAMP: Transfer of maintenance records.

When a U.S.-registered fractional ownership program aircraft maintained under a CAMP is removed from the list of program aircraft in the management specifications, the program manager must transfer to the purchaser, at the time of the sale, the following records of that aircraft, in plain language form or in coded form that provides for the preservation and retrieval of information in a manner acceptable to the Administrator:

(a) The records specified in § 91.1439(a)(2).

(b) The records specified in § 91.1439(a)(1) that are not included in the records covered by paragraph (a) of this section, except that the purchaser may allow the program manager to keep physical custody of such records. However, custody of records by the program manager does not relieve the purchaser of its responsibility under § 91.1439(c) to make the records available for inspection by the Administrator or any representative of the National Transportation Safety Board.

§ 91.1443 CAMP: Airworthiness release or aircraft maintenance log entry.

(a) No program aircraft maintained under a CAMP may be operated after maintenance, preventive maintenance, or alterations are performed unless qualified, certificated personnel employed by the program manager prepare, or cause the person with whom the program manager arranges for the performance of the maintenance, preventive maintenance, or alterations, to prepare—

(1) An airworthiness release; or

(2) An appropriate entry in the aircraft maintenance log.

(b) The airworthiness release or log entry required by paragraph (a) of this section must—

(1) Be prepared in accordance with the procedure in the program manager's manual;

(2) Include a certification that—

(i) The work was performed in accordance with the requirements of the program manager's manual;

(ii) All items required to be inspected were inspected by an authorized person who determined that the work was satisfactorily completed;

(iii) No known condition exists that would make the aircraft unairworthy;

(iv) So far as the work performed is concerned, the aircraft is in condition for safe operation; and

(3) Be signed by an authorized certificated mechanic.

(c) Notwithstanding paragraph (b)(3) of this section, after maintenance, preventive maintenance, or alterations performed by a repair station certificated under the provisions of part 145 of this chapter, the approval for return to service or log entry required by paragraph (a) of this section may be signed by a person authorized by that repair station.

(d) Instead of restating each of the conditions of the certification required by paragraph (b) of this section, the program manager may state in its manual that the signature of an authorized certificated mechanic or repairman constitutes that certification.

Appendix A
Category II operations: Manual, instruments, equipment, and maintenance

1. Category II manual

(a) *Application for approval.* An applicant for approval of a Category II manual or an amendment to an approved Category II manual must submit the proposed manual or amendment to the Flight Standards District Office having jurisdiction of the area in which the applicant is located. If the application requests an evaluation program, it must include the following:

(1) The location of the aircraft and the place where the demonstrations are to be conducted; and

(2) The date the demonstrations are to commence (at least 10 days after filing the application).

(b) *Contents.* Each Category II manual must contain:

(1) The registration number, make, and model of the aircraft to which it applies;

(2) A maintenance program as specified in Section 4 of this appendix; and

(3) The procedures and instructions related to recognition of decision height, use of runway visual range information, approach monitoring, the decision region (the region between the middle marker and the decision height), the maximum permissible deviations of the basic ILS indicator within the decision region, a missed approach,

use of airborne low approach equipment, minimum altitude for the use of the autopilot, instrument and equipment failure warning systems, instrument failure, and other procedures, instructions, and limitations that may be found necessary by the Administrator.

2. Required instruments and equipment

The instruments and equipment listed in this section must be installed in each aircraft operated in a Category II operation. This section does not require duplication of instruments and equipment required by § 91.205 or any other provisions of this chapter.

(a) *Group I.*

(1) Two localizer and glide slope receiving systems. Each system must provide a basic ILS display and each side of the instrument panel must have a basic ILS display. However, a single localizer antenna and a single glide slope antenna may be used.

(2) A communications system that does not affect the operation of at least one of the ILS systems.

(3) A marker beacon receiver that provides distinctive aural and visual indications of the outer and the middle markers.

(4) Two gyroscopic pitch and bank indicating systems.

(5) Two gyroscopic direction indicating systems.

(6) Two airspeed indicators.

(7) Two sensitive altimeters adjustable for barometric pressure, each having a placarded correction for altimeter scale error and for the wheel height of the aircraft. After June 26, 1979, two sensitive altimeters adjustable for barometric pressure, having markings at 20-foot intervals and each having a placarded correction for altimeter scale error and for the wheel height of the aircraft.

(8) Two vertical speed indicators.

(9) A flight control guidance system that consists of either an automatic approach coupler or a flight director system. A flight director system must display computed information as steering command in relation to an ILS localizer and, on the same instrument, either computed information as pitch command in relation to an ILS glide slope or basic ILS glide slope information. An automatic approach coupler must provide at least automatic steering in relation to an ILS localizer. The flight control guidance system may be operated from one of the receiving systems required by subparagraph (1) of this paragraph.

(10) For Category II operations with decision heights below 150 feet either a marker beacon receiver providing aural and visual indications of the inner marker or a radio altimeter.

(b) *Group II.*

(1) Warning systems for immediate detection by the pilot of system faults in items (1), (4), (5), and (9) of Group I and, if installed for use in Category III operations, the radio altimeter and autothrottle system.

(2) Dual controls.

(3) An externally vented static pressure system with an alternate static pressure source.

(4) A windshield wiper or equivalent means of providing adequate cockpit visibility for a safe visual transition by either pilot to touchdown and rollout.

(5) A heat source for each airspeed system pitot tube installed or an equivalent means of preventing malfunctioning due to icing of the pitot system.

3. Instruments and equipment approval

(a) *General.* The instruments and equipment required by Section 2 of this appendix must be approved as provided in this section before being used in Category II operations. Before presenting an aircraft for approval of the instruments and equipment, it must be shown that since the beginning of the 12th calendar month before the date of submission—

(1) The ILS localizer and glide slope equipment were bench checked according to the manufacturer's instructions and found to meet those standards specified in RTCA Paper 23 63/DO 117 dated March 14, 1963, "Standard Adjustment Criteria for Airborne Localizer and Glide Slope Receivers," which may be obtained from the RTCA Secretariat, 1425 K St., NW., Washington, DC 20005.

(2) The altimeters and the static pressure systems were tested and inspected in accordance with Appendix E to Part 43 of this chapter; and

(3) All other instruments and items of equipment specified in Section 2(a) of this appendix that are listed in the proposed maintenance program were bench checked and found to meet the manufacturer's specifications.

(b) *Flight control guidance system.* All components of the flight control guidance system must be approved as installed by the evaluation program specified in paragraph (e) of this section if they have not been approved for Category III operations under applicable type or supplemental type certification procedures. In addition, subsequent changes to make, model, or design of the components must be approved under this paragraph. Related systems or devices, such as the autothrottle and computed missed approach guidance system, must be approved in the same manner if they are to be used for Category II operations.

(c) *Radio altimeter.* A radio altimeter must meet the performance criteria of this paragraph for original approval and after each subsequent alteration.

(1) It must display to the flight crew clearly and positively the wheel height of the main landing gear above the terrain.

(2) It must display wheel height above the terrain to an accuracy of plus or minus 5 feet or 5 percent, whichever is greater, under the following conditions:

(i) Pitch angles of zero to plus or minus 5 degrees about the mean approach attitude.

(ii) Roll angles of zero to 20 degrees in either direction.

(iii) Forward velocities from minimum approach speed up to 200 knots.

(iv) Sink rates from zero to 15 feet per second at altitudes from 100 to 200 feet.

(3) Over level ground, it must track the actual altitude of the aircraft without significant lag or oscillation.

(4) With the aircraft at an altitude of 200 feet or less, any abrupt change in terrain representing no more than 10 percent of the aircraft's altitude must not cause the altimeter to unlock, and indicator response to such changes must not exceed 0.1 second and, in addition, if the system unlocks for greater changes, it must reacquire the signal in less than 1 second.

(5) Systems that contain a push-to-test feature must test the entire system (with or without an antenna) at a simulated altitude of less than 500 feet.

(6) The system must provide to the flight crew a positive failure warning display any time there is a loss of power or an absence of ground return signals within the designed range of operating altitudes.

(d) *Other instruments and equipment.* All other instruments and items of equipment required by § 2 of this appendix must be capable of performing as necessary for Category II operations. Approval is also required after each subsequent alteration to these instruments and items of equipment.

(e) *Evaluation program*—

(1) *Application.* Approval by evaluation is requested as a part of the application for approval of the Category II manual.

(2) *Demonstrations.* Unless otherwise authorized by the Administrator, the evaluation program for each aircraft requires the demonstrations specified in this paragraph. At least 50 ILS approaches must be flown with at least five approaches on each of three different ILS facilities and no more than one half of the total approaches on any one ILS facility. All approaches shall be flown under simulated instrument conditions to a 100-foot decision height and 90 percent of the total approaches made must be successful. A successful approach is one in which—

(i) At the 100-foot decision height, the indicated airspeed and heading are satisfactory for a normal flare and landing (speed must be plus or minus 5 knots of programmed airspeed, but may not be less than computed threshold speed if autothrottles are used);

(ii) The aircraft at the 100-foot decision height, is positioned so that the cockpit is within, and tracking so as to remain within, the lateral confines of the runway extended;

(iii) Deviation from glide slope after leaving the outer marker does not exceed 50 percent of full-scale deflection as displayed on the ILS indicator;

(iv) No unusual roughness or excessive attitude changes occur after leaving the middle marker; and

(v) In the case of an aircraft equipped with an approach coupler, the aircraft is sufficiently in trim when the approach coupler is disconnected at the decision height to allow for the continuation of a normal approach and landing.

(3) *Records.* During the evaluation program the following information must be maintained by the applicant for the aircraft with respect to each approach and made available to the Administrator upon request:

(i) Each deficiency in airborne instruments and equipment that prevented the initiation of an approach.

(ii) The reasons for discontinuing an approach, including the altitude above the runway at which it was discontinued.

(iii) Speed control at the 100-foot decision height if auto throttles are used.

(iv) Trim condition of the aircraft upon disconnecting the auto coupler with respect to continuation to flare and landing.

(v) Position of the aircraft at the middle marker and at the decision height indicated both on a diagram of the basic ILS display and a diagram of the runway extended to the middle marker. Estimated touchdown point must be indicated on the runway diagram.

(vi) Compatibility of flight director with the auto coupler, if applicable.

(vii) Quality of overall system performance.

(4) *Evaluation.* A final evaluation of the flight control guidance system is made upon successful completion of the demonstrations. If no hazardous tendencies have been displayed or are otherwise known to exist, the system is approved as installed.

4. Maintenance program

(a) Each maintenance program must contain the following:

(1) A list of each instrument and item of equipment specified in § 2 of this appendix that is installed in the aircraft and approved for Category II operations, including the make and model of those specified in § 2(a).

(2) A schedule that provides for the performance of inspections under subparagraph (5) of this paragraph within 3 calendar months after the date of the previous inspection. The inspection must be performed by a person authorized by part 43 of this chapter, except that each alternate inspection may be replaced by a functional flight check. This functional flight check must be performed by a pilot holding a Category II pilot authorization for the type aircraft checked.

(3) A schedule that provides for the performance of bench checks for each listed instrument and item of equipment that is specified in section 2(a) within 12 calendar months after the date of the previous bench check.

(4) A schedule that provides for the performance of a test and inspection of each static pressure system in accordance with Appendix E to Part 43 of this chapter within 12 calendar months after the date of the previous test and inspection.

(5) The procedures for the performance of the periodic inspections and functional flight checks to determine the ability of each listed instrument and item of equipment specified in section 2(a) of this appendix to perform as approved for Category II operations including a procedure for recording functional flight checks.

(6) A procedure for assuring that the pilot is informed of all defects in listed instruments and items of equipment.

(7) A procedure for assuring that the condition of each listed instrument and item of equipment upon which maintenance is performed is at least equal to its Category II approval condition before it is returned to service for Category II operations.

(8) A procedure for an entry in the maintenance records required by § 43.9 of this chapter that shows the date, airport, and reasons for each discontinued Category II operation because of a malfunction of a listed instrument or item of equipment.

(b) *Bench check.* A bench check required by this section must comply with this paragraph.

(1) It must be performed by a certificated repair station holding one of the following ratings as appropriate to the equipment checked:

(i) An instrument rating.

(ii) A radio rating.

(2) It must consist of removal of an instrument or item of equipment and performance of the following:

(i) A visual inspection for cleanliness, impending failure, and the need for lubrication, repair, or replacement of parts;

(ii) Correction of items found by that visual inspection; and

(iii) Calibration to at least the manufacturer's specifications unless otherwise specified in the approved Category II manual for the aircraft in which the instrument or item of equipment is installed.

(c) *Extensions.* After the completion of one maintenance cycle of 12 calendar months, a request to extend the period for checks, tests, and inspections is approved if it is shown that the performance of particular equipment justifies the requested extension.

Appendix B
Authorizations to exceed Mach 1 (section 91.817)

Section 1. Application

(a) An applicant for an authorization to exceed Mach 1 must apply in a form and manner prescribed by the Administrator and must comply with this appendix.

(b) In addition, each application for an authorization to exceed Mach 1 covered by section 2(a) of this appendix must contain all information requested by the Administrator necessary to assist him in determining whether the designation of a particular test area or issuance of a particular authorization is a "major Federal action significantly affecting the quality of the human environment" within the meaning of the National Environmental Policy Act of 1969 (42 U.S.C. 4321 et seq.), and to assist him in complying with that act and with related Executive Orders, guidelines, and orders prior to such action.

(c) In addition, each application for an authorization to exceed Mach 1 covered by section 2(a) of this appendix must contain—

(1) Information showing that operation at a speed greater than Mach 1 is necessary to accomplish one or more of the purposes specified in section 2(a) of this appendix, including a showing that the purpose of the test cannot be safely or properly accomplished by overocean testing;

(2) A description of the test area proposed by the applicant, including an environmental analysis of that area meeting the requirements of paragraph (b) of this section; and

(3) Conditions and limitations that will ensure that no measurable sonic boom overpressure will reach the surface outside of the designated test area.

(d) An application is denied if the Administrator finds that such action is necessary to protect or enhance the environment.

Section 2. Issuance

(a) For a flight in a designated test area, an authorization to exceed Mach 1 may be issued when the Administrator has taken the environmental protective actions specified in section 1(b) of this appendix and the applicant shows one or more of the following:

(1) The flight is necessary to show compliance with airworthiness requirements.

(2) The flight is necessary to determine the sonic boom characteristics of the airplane or to establish means of reducing or eliminating the effects of sonic boom.

(3) The flight is necessary to demonstrate the conditions and limitations under which speeds greater than a true flight Mach number of 1 will not cause a measurable sonic boom overpressure to reach the surface.

(b) For a flight outside of a designated test area, an authorization to exceed Mach 1 may be issued if the applicant shows conservatively under paragraph (a)(3) of this section that—

(1) The flight will not cause a measurable sonic boom overpressure to reach the surface when the aircraft is operated under conditions and limitations demonstrated under paragraph (a)(3) of this section; and

(2) Those conditions and limitations represent all foreseeable operating conditions.

Section 3. Duration

(a) An authorization to exceed Mach 1 is effective until it expires or is surrendered, or until it is suspended or terminated by the Administrator. Such an authorization may be amended or suspended by the Administrator at any time if the Administrator finds that such action is necessary to protect the environment. Within 30 days of notification of amendment, the holder of the authorization must request reconsideration or the amendment becomes final. Within 30 days of notification of suspension, the holder of the authorization must request reconsideration or the authorization is automatically terminated. If reconsideration is requested within the 30-day period, the amendment or suspension continues until the holder shows why the authorization should not be amended or terminated. Upon such showing, the Administrator may terminate or amend the authorization if the Administrator finds that such action is necessary to protect the environment, or he may reinstate the authorization without amendment if he finds that termination or amendment is not necessary to protect the environment.

(b) Findings and actions by the Administrator under this section do not affect any certificate issued under Title VI of the Federal Aviation Act of 1958.

Appendix C
Operations in the North Atlantic (NAT) Minimum Navigation Performance Specifications (MNPS) airspace

Section 1. NAT MNPS airspace is that volume of airspace between FL 275 and FL 400 extending between latitude 27 degrees north and the North Pole, bounded in the east by the eastern boundaries of control areas Santa Maria Oceanic, Shanwick Oceanic, and Reykjavik Oceanic and in the west by the western boundary of Reykjavik Oceanic Control Area, the western boundary of Gander Oceanic Control Area, and the western boundary of New York Oceanic Control Area, excluding the area west of 60 degrees west and south of 38 degrees 30 minutes north.

Section 2. The navigation performance capability required for aircraft to be operated in the airspace defined in section 1 of this appendix is as follows:

(a) The standard deviation of lateral track errors shall be less than 6.3 NM (11.7 Km). Standard deviation is a statistical measure of data about a mean value. The mean is zero nautical miles. The overall form of data is such that the plus and minus 1 standard deviation about the mean encompasses approximately 68 percent of the data and plus or minus 2 deviations encompasses approximately 95 percent.

(b) The proportion of the total flight time spent by aircraft 30 NM (55.6 Km) or more off the cleared track shall be less than 5.3×10^{-4} (less than 1 hour in 1,887 flight hours).

(c) The proportion of the total flight time spent by aircraft between 50 NM and 70 NM (92.6 Km and 129.6 Km) off the cleared track shall be less than 13×10^{-5} (less than 1 hour in 7,693 flight hours.)

Section 3. Air traffic control (ATC) may authorize an aircraft operator to deviate from the requirements of § 91.705 for a specific flight if, at the time of flight plan filing for that flight, ATC determines that the aircraft may be provided appropriate separation and that the flight will not interfere with, or impose a burden upon, the operations of other aircraft which meet the requirements of § 91.705.

Appendix D
Airports/locations: Special operating restrictions

Section 1. Locations at which the requirements of § 91.215(b)(2) apply.

The requirements of § 91.215(b)(2) apply below 10,000 feet above the surface within 30-nautical-mile radius of each location in the following list:

Atlanta, GA (The William B. Hartsfield Atlanta International Airport)

Baltimore, MD (Baltimore Washington International Airport)

Boston, MA (General Edward Lawrence Logan International Airport)

Chantilly, VA (Washington Dulles International Airport)
Charlotte, NC (Charlotte/Douglas International Airport)
Chicago, IL (Chicago-O'Hare International Airport)
Cleveland, OH (Cleveland-Hopkins International Airport)
Covington, KY (Cincinnati Northern Kentucky International Airport)

Dallas, TX (Dallas/Fort Worth Regional Airport)
Denver, CO (International Airport)
Detroit, MI (Metropolitan Wayne County Airport)
Honolulu, HI (Honolulu International Airport)
Houston, TX (George Bush Intercontinental Airport/Houston)

Kansas City, KS (Mid-Continent International Airport)
Las Vegas, NV (McCarran International Airport)
Los Angeles, CA (Los Angeles International Airport)
Memphis, TN (Memphis International Airport)
Miami, FL (Miami International Airport)
Minneapolis, MN (Minneapolis-St. Paul International Airport)

Newark, NJ (Newark International Airport)
New Orleans, LA (New Orleans International Airport-Moisant Field)
New York, NY (John F. Kennedy International Airport)

New York, NY (LaGuardia Airport)

Orlando, FL (Orlando International Airport)

Philadelphia, PA (Philadelphia International Airport)

Phoenix, AZ (Phoenix Sky Harbor International Airport)

Pittsburgh, PA (Greater Pittsburgh International Airport)

St. Louis, MO (Lambert-St. Louis International Airport)

Salt Lake City, UT (Salt Lake City International Airport)

San Diego, CA (San Diego International Airport)

San Francisco, CA (San Francisco International Airport)

Seattle, WA (Seattle-Tacoma International Airport)

Tampa, FL (Tampa International Airport)

Washington, DC(Ronald Reagan Washington National Airport and Andrews Air Force Base, MD)

Section 2. Airports at which the requirements of § 91.215(b)(5)(ii) apply.

The requirements of § 91.215(b)(5)(ii) apply to operations in the vicinity of each of the following airports:

Billings, MT (Logan International Airport)

[Reserved]

Section 3. Locations at which fixed-wing Special VFR operations are prohibited.

The Special VFR weather minimums of § 91.157 do not apply to the following airports:

Atlanta, GA (The William B. Hartsfield Atlanta International Airport)

Baltimore, MD (Baltimore/Washington International Airport)

Boston, MA (General Edward Lawrence Logan International Airport)

Buffalo, NY (Greater Buffalo International Airport)

Chicago, IL (Chicago-O'Hare International Airport)

Cleveland, OH (Cleveland-Hopkins International Airport)

Columbus, OH (Port Columbus International Airport)

Covington, KY (Greater Cincinnati International Airport)

Dallas, TX (Dallas/Fort Worth Regional Airport)

Dallas, TX (Love Field)

Denver, CO (International Airport)

Detroit, MI (Metropolitan Wayne County Airport)

Honolulu, HI (Honolulu International Airport)

Houston, TX (Houston Intercontinental Airport)

Indianapolis, IN (Indianapolis International Airport)

Los Angeles, CA (Los Angeles International Airport)

Louisville, KY (Standiford Field)

Memphis, TN (Memphis International Airport)

Miami, FL (Miami International Airport)

Minneapolis, MN (Minneapolis-St. Paul International Airport)

Newark, NJ (Newark International Airport)

New York, NY (John F. Kennedy International Airport)

New York, NY (LaGuardia Airport)

New Orleans, LA (New Orleans International Airport-Moisant Field)

Philadelphia, PA (Philadelphia International Airport)

Pittsburgh, PA (Greater Pittsburgh International Airport)

Portland, OR (Portland International Airport)

San Francisco, CA (San Francisco International Airport)

Seattle, WA (Seattle-Tacoma International Airport)

St. Louis, MO (Lambert-St. Louis International Airport)

Tampa, FL (Tampa International Airport)

Washington, DC

Section 4. Locations at which solo student, sport, and recreational pilot activity is not permitted.

Pursuant to §91.131 (b)(2), solo student, sport, and recreational pilot operations are not permitted at any of the following airports.

Atlanta, GA (The William B. Hartsfield Atlanta International Airport)

Boston, MA (General Edward Lawrence Logan International Airport)

Chicago, IL (Chicago-O'Hare International Airport)

Dallas, TX (Dallas/Fort Worth Regional Airport)

Los Angeles, CA (Los Angeles International Airport)

Miami, FL (Miami International Airport)

Newark, NJ (Newark International Airport)

New York, NY (John F. Kennedy International Airport)

New York, NY (LaGuardia Airport)

San Francisco, CA (San Francisco International Airport)

Washington, DC (Washington National Airport)

Andrews Air Force Base, MD)]

[Reserved]

(Amdt. 91-217, Eff. 7/23/90); (Amdt. 91-227, Eff. 9/16/93); (Amdt. 91-235, Eff. 10/5/93); [Amdt. 91-236 & 91-237, Eff. 3/9/94 and Amdt. 91-238, Eff. 5/15/94)]; [*(Amdt. 91-241 delays effective date of name change indefinitely.)] [(Amdt. 91-243, Eff. 1/28/95)]

Appendix E to Part 91—Airplane flight recorder specifications

Parameters	Range	Installed system[1] minimum accuracy (to recovered data)	Sampling interval (per second)	Resolution[4] read out
Relative Time (from recorded on prior to takeoff).	8 hr minimum	±0.125% per hour	1	1 sec.
Indicated Airspeed.	Vso to VD (KIAS)	±5% or ±10 kts., whichever is greater. Resolution 2 kts. below 175 KIAS	1	1%[3]
Altitude.	−1,000 ft. to max cert. alt. of A/C	±100 to ±700 ft. (see Table 1, TSO C51-a)	1	25 to 150 ft.
Magnetic Heading.	360°	±5°	1	1½
Vertical Acceleration.	−3g to +6g	±0.2g in addition to ±0.3g maximum datum	4 (or 1 per second where peaks, ref. to 1g are recorded)	0.03g
Longitudinal Acceleration.	±1.0g	±1.5% max. range excluding datum error of ±5%	2	0.01g
Pitch Attitude.	100% of usable	±2°	1	0.8°
Roll Attitude.	±60° or 100% of usable range, whichever is greater	±2°	1	0.8°
Stabilizer Trim Position, OR.	Full Range	±3% unless higher uniquely required	1	1%[3]
Pitch Control Position.				
Engine Power, Each Engine:	Full Range	±3% unless higher uniquely required	1	1%[3]
Fan or N1 Speed or EPR or Cockpit Indications. Used for Aircraft Certification OR.	Maximum Range	±5%	1	1%[3]
Prop. Speed and Torque (Sample once/sec as close together as practicable).			1 (prop Speed) 1 (torque)	1%[3] 1%[3]
Altitude Rate[2] (need depends on altitude resolution).	±8,000 fpm	±10%. Resolution 250 fpm below 12,000 ft. indicated	1	250 fpm below 12,000
Angle of Attack[2] (need depends on altitude resolution).	−20° to 40° or 100% of usable range	±2°	1	0.8%[3]
Radio Transmitter Keying (Discrete).			On/Off	1

Parameters	Range	Installed system[1] minimum accuracy (to recovered data)	Sampling interval (per second)	Resolution[4] read out
TE Flaps (Discrete or Analog).	Each discrete position (U, D, T/O, AAP) OR		1	
LE Flaps (Discrete or Analog).	Analog 0–100% range	±3°	1	1%[3]
	Each discrete position (U, D, T/O, AAP) OR		1	
Thrust Reverser, Each Engine (Discrete).	Analog 0–100% range Stowed or full reverse	±3°	1	1%[3]
Spoiler/Speedbrake (Discrete).	Stowed or out		1	
Autopilot Engaged (Discrete).	Engaged or Disengaged		1	

[1] When data sources are aircraft instruments (except altimeters) of acceptable quality to fly the aircraft the recording system excluding these sensors (but including all other characteristics of the recording system) shall contribute no more than half of the values in this column.

[2] If data from the altitude encoding altimeter (100 ft. resolution) is used, then either one of these parameters should also be recorded. If however, altitude is recorded at a minimum resolution of 25 feet, then these two parameters can be omitted.

[3] Percent of full range.

[4] This column applies to aircraft manufactured after October 11, 1991.

Appendix F to Part 91—Helicopter flight recorder specifications

Parameters	Range	Installed system[1] minimum accuracy (to recovered data)	Sampling interval (per second)	Resolution[3] read out
Relative Time (from recorded on prior to takeoff).	4 hr minimum	±0.125% per hour	1	1 sec.
Indicated Airspeed.	VM in to VD (KIAS) (minimum airspeed signal attainable with installed pilot-static system)	±5% or ±10 kts., whichever is greater	1	1 kt.
Altitude.	−1,000 ft. to 20,000 ft. pressure altitude	±100 to ±700 ft. (see Table 1, TSO C51-a)	1	25 to 150 ft.
Magnetic Heading.	360°	±5°	1	1°
Vertical Acceleration.	−3g to +6g	±0.2g in addition to ±0.3g maximum datum	4 (or 1 per second where peaks, ref. to 1g are recorded)	0.05g
Longitudinal Acceleration.	±1.0g	±1.5% max. range excluding datum error of ±5%	2	0.03g
Pitch Attitude.	100% of usable range	±2°	1	0.8°
Roll Attitude.	±60 or 100% of usable range, whichever is greater	±2°	1	0.8°

Parameters	Range	Installed system[1] minimum accuracy (to recovered data)	Sampling interval (per second)	Resolution[3] read out
Altitude Rate.	±8,000 fpm	±10%. Resolution 250 fpm below 12,000 ft. indicated	1	250 fpm below 12,000
Engine Power, Each Engine				
Main Rotor Speed.	Maximum Range	±5%	1	1%[2]
Free or Power Turbine.	Maximum Range	±5%	1	1%[2]
Engine Torque.	Maximum Range	±5%	1	1%[2]
Flight Control Hydraulic Pressure				
Primary (Discrete).	High/Low		1	
Secondary—if applicable (Discrete).	High/Low		1	
Radio Transmitter Keying (Discrete)	On/Off		1	
Autopilot Engaged (Discrete).	Engaged or Disengaged		1	
SAS Status-Engaged (Discrete).	Engaged or Disengaged		1	
SAS Fault Status (Discrete).	Fault/OK		1	
Flight Controls				
Collective.	Full range	±3%	2	1%[2]
Pedal Position.	Full range	±3%	2	1%[2]
Lat. Cyclic.	Full range	±3%	2	1%[2]
Long. Cyclic.	Full range	±3%	2	1%[2]
Controllable Stabilator Position.	Full range	±3%	2	1%[2]

[1] When data sources are aircraft instruments (except altimeters) of acceptable quality to fly the aircraft the recording system excluding these sensors (but including all other characteristics of the recording system) shall contribute no more than half of the values in this column.

[2] Percent of full range.

[3] This column applies to aircraft manufactured after October 11, 1991.

Appendix G to Part 91—Operations in Reduced Vertical Separation Minimum (RVSM) Airspace

Section 1. Special Definitions

Reduced Vertical Separation Minimum (RVSM) Airspace. Within RVSM airspace, air traffic control (ATC) separates aircraft by a minimum of 1,000 feet vertically between flight level (FL) 290 and FL 410 inclusive. RVSM airspace is special qualification airspace; the operator and the aircraft used by the operator must be approved by the Administrator. Air-traffic control notifies operators of RVSM by providing route planning information. Section 8 of this appendix identifies airspace where RVSM may be applied.

RVSM Group Aircraft. Aircraft within a group of aircraft, approved as a group by the Administrator, in which each of the aircraft satisfy each of the following:

(a) The aircraft have been manufactured to the same design, and have been approved under the same type certificate, amended type certificate, or supplemental type certificate.

(b) The static system of each aircraft is installed in a manner and position that is the same as those of the other aircraft in the group. The same static source error correction is incorporated in each aircraft of the group.

(c) The avionics units installed in each aircraft to meet the minimum RVSM equipment requirements of this appendix are:

(1) Manufactured to the same manufacturer specification and have the same part number; or

(2) Of a different manufacturer or part number, if the applicant demonstrates that the equipment provides equivalent system performance.

RVSM Nongroup Aircraft. An aircraft that is approved for RVSM operations as an individual aircraft.

RVSM Flight envelope. An RVSM flight envelope includes the range of Mach number, weight divided by atmospheric pressure ratio, and altitudes over which an aircraft is approved to be operated in cruising flight within RVSM airspace. RVSM flight envelopes are defined as follows:

(a) The *full RVSM flight envelope* is bounded as follows:

(1) The altitude flight envelope extends from FL 290 upward to the lowest altitude of the following:

(i) FL 410 (the RVSM altitude limit);

(ii) The maximum certificated altitude for the aircraft; or

(iii) The altitude limited by cruise thrust, buffet, or other flight limitations.

(2) The airspeed flight envelope extends:

(i) From the airspeed of the slats/flaps-up maximum endurance (holding) airspeed, or the maneuvering airspeed, whichever is lower;

(ii) To the maximum operating airspeed (Vmo/Mmo), or airspeed limited by cruise thrust buffet, or other flight limitations, whichever is lower.

(3) All permissible gross weights within the flight envelopes defined in paragraphs (1) and (2) of this definition.

(b) The *basic RVSM flight envelope* is the same as the full RVSM flight envelope except that the airspeed flight envelope extends:

(1) From the airspeed of the slats/flaps-up maximum endurance (holding) airspeed, or the maneuver airspeed, whichever is lower;

(2) To the upper Mach/airspeed boundary defined for the full RVSM flight envelope, or a specified lower value not less than the long-range cruise Mach number plus .04 Mach, unless further limited by available cruise thrust, buffet, or other flight limitations.

Section 2. Aircraft Approval

(a) An operator may be authorized to conduct RVSM operations if the Administrator finds that its aircraft comply with this section.

(b) The applicant for authorization shall submit the appropriate data package for aircraft approval. The package must consist of at least the following:

(1) An identification of the RVSM aircraft group or the nongroup aircraft;

(2) A definition of the RVSM flight envelopes applicable to the subject aircraft;

(3) Documentation that establishes compliance with the applicable RVSM aircraft requirements of this section; and

(4) The conformity tests used to ensure that aircraft approved with the data package meet the RVSM aircraft requirements.

(c) *Altitude-keeping equipment: All aircraft.* To approve an aircraft group or a nongroup aircraft, the Administrator must find that the aircraft meets the following requirements:

(1) The aircraft must be equipped with two operational independent altitude measurement systems.

(2) The aircraft must be equipped with at least one automatic altitude control system that controls the aircraft altitude —

(i) Within a tolerance band of ±65 feet about an acquired altitude when the aircraft is operated in straight and level flight under nonturbulent, nongust conditions; or

(ii) Within a tolerance band of ±130 feet under nonturbulent, nongust conditions for aircraft for which application for type certification occurred on or before April 9, 1997 that are equipped with an automatic altitude control system with flight management/performance system inputs.

(3) The aircraft must be equipped with an altitude alert system that signals an alert when the altitude displayed to the flight crew deviates from the selected altitude by more than:

(i) ±300 feet for aircraft for which application for type certification was made on or before April 9, 1997; or

(ii) ±200 feet for aircraft for which application for type certification is made after April 9, 1997.

(d) *Altimetry system error containment: Group aircraft for which application for type certification was made on or before April 9, 1997.* To approve group aircraft for which application for type certification was made on or before April 9, 1997, the Administrator must find that the altimetry system error (ASE) is contained as follows:

(1) At the point in the basic RVSM flight envelope where mean ASE reaches its largest absolute value, the absolute value may not exceed 80 feet.

(2) At the point in the basic RVSM flight envelope where mean ASE plus three standard deviations reaches its largest absolute value, the absolute value may not exceed 200 feet.

(3) At the point in the full RVSM flight envelope where mean ASE reaches its largest absolute value, the absolute value may not exceed 120 feet.

(4) At the point in the full RVSM flight envelope where mean ASE plus three standard deviations reaches its largest absolute value, the absolute value may not exceed 245 feet.

(5) *Necessary operating restrictions.* If the applicant demonstrates that its aircraft otherwise comply with the ASE containment requirements, the Administrator may establish an operating restriction on that applicant's aircraft to restrict the aircraft from operating in areas of the basic RVSM flight envelope where the absolute value of mean ASE exceeds 80 feet, and/or the absolute value of mean ASE plus three standard deviations exceeds 200 feet; or from operating in areas of the full RVSM flight envelope where the absolute value of the mean ASE exceeds 120 feet and/or the absolute value of the mean ASE plus three standard deviations exceeds 245 feet.

(e) *Altimetry system error containment: Group aircraft for which application for type certification is made after April 9, 1997.* To approve group aircraft for which application for type certification is made after April 9, 1997, the Administrator must find that the altimetry system error (ASE) is contained as follows:

(1) At the point in the full RVSM flight envelope where mean ASE reaches its largest absolute value, the absolute value may not exceed 80 feet.

(2) At the point in the full RVSM flight envelope where mean ASE plus three standard deviations reaches its largest absolute value, the absolute value may not exceed 200 feet.

(f) *Altimetry system error containment: Nongroup aircraft.* To approve a nongroup aircraft, the Administrator must find that the altimetry system error (ASE) is contained as follows:

(1) For each condition in the basic RVSM flight envelope, the largest combined absolute value for residual sta-

tic source error plus the avionics error may not exceed 160 feet.

(2) For each condition in the full RVSM flight envelope, the largest combined absolute value for residual static source error plus the avionics error may not exceed 200 feet.

(g) Traffic Alert and Collision Avoidance System (TCAS) Compatibility With RVSM Operations: All aircraft. After March 31, 2002, unless otherwise authorized by the Administrator, if you operate an aircraft that is equipped with TCAS II in RVSM airspace, it must be a TCAS II that meets TSO C-119b (Version 7.0), or a later version.

(h) If the Administrator finds that the applicant's aircraft comply with this section, the Administrator notifies the applicant in writing.

Section 3. Operator Authorization

(a) Authority for an operator to conduct flight in airspace where RVSM is applied is issued in operations specifications or a Letter of Authorization, or management specifications issued under subpart K of this part, as appropriate. To issue an RVSM authorization, the Administrator must find that the operator's aircraft have been approved in accordance with Section 2 of this appendix and that the operator complies with this section.

(b) An applicant for authorization to operate within RVSM airspace shall apply in a form and manner prescribed by the Administrator. The application must include the following:

(1) An approved RVSM maintenance program outlining procedures to maintain RVSM aircraft in accordance with the requirements of this appendix. Each program must contain the following:

(i) Periodic inspections, functional flight tests, and maintenance and inspection procedures, with acceptable maintenance practices, for ensuring continued compliance with the RVSM aircraft requirements.

(ii) A quality assurance program for ensuring continuing accuracy and reliability of test equipment used for testing aircraft to determine compliance with the RVSM aircraft requirements.

(iii) Procedures for returning noncompliant aircraft to service.

(2) For an applicant who operates under part 121 or 135 of this chapter or under subpart K of this part, initial and recurring pilot training requirements.

(3) Policies and Procedures. An applicant who operates under part 121 or 135 of this chapter or under subpart K of this part must submit RVSM policies and procedures that will enable it to conduct RVSM operations safely.

(c) Validation and Demonstration. In a manner prescribed by the Administrator, the operator must provide evidence that:

(1) It is capable to operate and maintain each aircraft or aircraft group for which it applies for approval to operate in RVSM airspace; and

(2) Each pilot has an adequate knowledge of RVSM requirements, policies, and procedures.

Section 4. RVSM Operations

(a) Each person requesting a clearance to operate within RVSM airspace shall correctly annotate the flight plan filed with air traffic control with the status of the operator and aircraft with regard to RVSM approval. Each operator shall verify RVSM applicability for the flight planned route through the appropriate flight planning information sources.

(b) No person may show, on the flight plan filed with air traffic control, an operator or aircraft as approved for RVSM operations, or operate on a route or in an area where RVSM approval is required, unless:

(1) The operator is authorized by the Administrator to perform such operations; and

(2) The aircraft has been approved and complies with the requirements of Section 2 of this appendix.

Section 5. Deviation Authority Approval

The Administrator may authorize an aircraft operator to deviate from the requirements of ~91.180 or §91.706 for a specific flight in RVSM airspace if that operator has not been approved in accordance with Section 3 of this appendix, if:

(a) The operator submits a request in a time and manner acceptable to the Administrator, and

(b) At the time of filing the flight plan for that flight, ATC determines that the aircraft may be provided appropriate separation and that the flight will not interfere with, or impose a burden on, the operations of operators who have been approved for RVSM operations in accordance with Section 3 of this appendix.

Section 6. Reporting Altitude-Keeping Errors

Each operator shall report to the Administrator each event in which the operator's aircraft has exhibited the following altitude-keeping performance:

(a) Total vertical error of 300 feet or more;

(b) Altimetry system error of 245 feet or more; or

(c) Assigned altitude deviation of 300 feet or more.

Section 7. Removal or Amendment of Authority

The Administrator may amend operations specifications or management specifications issued under subpart K of this part to revoke or restrict an RVSM authorization, or may revoke or restrict an RVSM letter of authorization, if the Administrator determines that the operator is not complying, or is unable to comply, with this appendix or subpart H of this part. Examples of reasons for amendment, revocation, or restriction include, but are not limited to, an operator's:

(a) Committing one or more altitude-keeping errors in RVSM airspace;

(b) Failing to make an effective and timely response to identify and correct an altitude-keeping error; or

(c) Failing to report an altitude-keeping error.

Section 8. Airspace Designation

(a) *RVSM in the North Atlantic.* (1) RVSM may be applied in the NAT in the following ICAO Flight Information Regions (FIRs): New York Oceanic, Gander Oceanic, Sondrestrom FIR, Reykjavik Oceanic, Shanwick Oceanic, and Santa Maria Oceanic.

(2) RVSM may be effective in the Minimum Navigation Performance Specification (MNPS) airspace within the NAT. The MNPS airspace within the NAT is defined by the volume of airspace between FL 285 and FL 420 (inclusive) extending between latitude 27 degrees north and the North Pole, bounded in the east by the eastern boundaries of control areas Santa Maria Oceanic, Shanwick Oceanic, and Reykjavik Oceanic and in the west by the western boundaries of control areas Reykjavik Oceanic, Gander Oceanic, and New York Oceanic, excluding the areas west of 60 degrees west and south of 38 degrees 30 minutes north.

(b) *RVSM in the Pacific.* (1) RVSM may be applied in the Pacific in the following ICAO Flight Information Regions (FIRs): Anchorage Arctic, Anchorage Continental, Anchorage Oceanic, Auckland Oceanic, Brisbane, Edmonton, Honiara, Los Angeles, Melbourne, Nadi, Naha, Nauru, New Zealand, Oakland, Oakland Oceanic, Port Moresby, Seattle, Tahiti, Tokyo, Ujung Pandang and Vancouver.

(c) *RVSM in the West Atlantic Route System (WATRS).* RVSM may be applied in the New York FIR portion of the West Atlantic Route System (WATRS). The area is defined as beginning at a point 38°30'; N/60°00'; W direct to 38°30';N/69°15'; W direct to 38°20'; N/69°57'; W direct to 37°31'; N/71°41'; W direct to 37°13'; N/72°40'; W direct to 35°05'; N/72°40'; W direct to 34°54'; N/72°57'; W direct to 34°29'; N/73°34'; W direct to 34°33'; N/73°41'; W direct to 34°19'; N/74°02'; W direct to 34°14'; N/73°57'; W direct to 32°12'; N/76°49'; W direct to 32°20'; N/77°00'; W direct to 28°08'; N/77°00'; W direct to 27°50'; N/76°32'; W direct to 27°50'; N/74°50'; W direct to 25°00'; N/73°21'; W direct to 25°00'; 05'; N/69°13';06'; W direct to 25°00'; N/69°07'; W direct to 23°30'; N/68°40'; W direct to 23°30'; N/60°00'; W to the point of beginning.

(d) *RVSM in the United States.* RVSM may be applied in the airspace of the 48 contiguous states, District of Columbia, and Alaska, including that airspace overlying the waters within 12 nautical miles of the coast.

(e) *RVSM in the Gulf of Mexico.* RVSM may be applied in the Gulf of Mexico in the following areas: Gulf of Mexico High Offshore Airspace, Houston Oceanic ICAO FIR and Miami Oceanic ICAO FIR.

(f) *RVSM in Atlantic High Offshore Airspace and the San Juan FIR.* RVSM may be applied in Atlantic High Offshore Airspace and in the San Juan ICAO FIR.

Special Federal Aviation Regulation No. 51-1. Special flight rules in the vicinity of Los Angeles International Airport.

Section 1. Applicability. [This rule establishes a special operating area for persons operating aircraft under visual flight rules (VFR) in the following airspace of the Los

Angeles Class B airspace area designated as the Los Angeles Special Flight Rules Area:]

That part of Area A of the Los Angeles TCA between 3,500 feet above mean sea level (MSL) and 4,500 feet MSL, inclusive, bounded on the north by Ballona Creek, on the east by the San Diego Freeway; on the south by Imperial Highway and on the west by the Pacific Ocean shoreline.

Section 2. *Aircraft operations, general.* Unless otherwise authorized by the Administrator, no person may operate an aircraft in the airspace described in Section 1 unless the operation is conducted under the following rules:

(a) [The flight must be conducted under VFR and only when operation may be conducted in compliance with § 91.155(a)].

(b) [The aircraft must be equipped as specified in § 91.215(b) replying on Code 1201 prior to entering and while operating in this area.]

(c) The pilot shall have a current Los Angeles Terminal Area Chart in the aircraft.

(d) The pilot shall operate on the Santa Monica very high frequency omni-directional radio range (VOR) 132 degree radial.

(e) Operations in a southeasterly direction shall be in level flight at 3,500 feet MSL.

(f) Operations in a northwesterly direction shall be in level flight at 4,500 feet MSL.

(g) Indicated airspeed shall not exceed 140 knots.

(h) Anticollision lights and aircraft position/navigation lights shall be on. Use of landing lights is recommended.

(i) Turbojet aircraft are prohibited from VFR operations in this area.

Section 3. [Notwithstanding the provisions of § 91.131 (a), an air traffic control authorization is not required in the Los Angeles Special Flight Rules Area for operations in compliance with Section 2 of this SFAR. All other provisions of § 91.131 apply to operate in the Special Flight Rules Area.]

Authority: 49 U.S.C. app. 1303, 1348, 1354(a), 1421, and 1422; 49 U.S.C. 106(g).

Special Federal Aviation Regulation No. 60. Air Traffic Control System emergency operation.

1. Each person shall, before conducting any operation under the Federal Aviation Regulations (14 CFR Chapter 1), be familiar with all available information concerning that operation, including Notices to Airmen issued under § 91.139 and, when activated, the provisions of the National Air Traffic Reduced Complement Operations Plan available for inspection at operating air traffic facilities and Regional air traffic division offices, and the General Aviation Reservation Program. No operator may change the designated airport of intended operation for any flight contained in the October 1, 1990, OAG.

2. Notwithstanding any provision of the Federal Aviation Regulations to the contrary, no person may operate an aircraft in the Air Traffic Control System:

a. Contrary to any restriction, prohibition, procedure or other action taken by the Director of the Office of Air Traffic Systems Management (Director) pursuant to Paragraph 3 of this regulation and announced in a Notice to Airmen pursuant to § 91.139 of the Federal Aviation Regulations.

b. When the National Air Traffic Reduced Complement Operations Plan is activated pursuant to Paragraph 4 of this regulation, except in accordance with the pertinent provisions of the National Air Traffic Reduced Complement Operations Plan.

3. Prior to or in connection with the implementation of the RCOP, and as conditions warrant, the Director is authorized to:

a. Restrict, prohibit, or permit VFR and/or IFR operations at any airport, Class B airspace area, Class C airspace area, or other class of controlled airspace.

b. Give priority at any airport to flights that are of military necessity, or are medical emergency flights, Presidential flights, and flights transporting critical Government employees.

c. Implement, at any airport, traffic management procedures, that may include reduction of flight operations. Reduction of flight operations will be accomplished, to the extent practical, on a pro rata basis among and between air carrier, commercial operator, and general aviation operations. Flights cancelled under this SFAR at a high density traffic airport will be considered to have been operated for purposes of Part 93 of the Federal Aviation Regulations.

4. The Director may activate the National Air Traffic Reduced Complement Operations Plan at any time he finds that it is necessary for the safety and efficiency of the National Airspace System. Upon activation of the RCOP and notwithstanding any provision of the FAR to the contrary, the Director is authorized to suspend or modify any airspace designation.

5. Notice of restrictions, prohibitions, procedures and other actions taken by the Director under this regulation with respect to the operation of the Air Traffic Control system will be announced in Notices to Airmen issued pursuant to § 91.139 of the Federal Aviation Regulations.

6. The Director may delegate his authority under this regulation to the extent he considers necessary for the safe and efficient operation of the National Air Traffic Control System.

APPENDIX 1
Key Airports
ATL Atlanta Hartsfield International
BNA Nashville International
BOS Boston Logan International
CLT Charlotte-Douglas International
CVG Greater Cincinnati International
DAL Dallas-Love
DAY Cox-Dayton International
DCA Washington National
DEN Denver International

DFW Dallas-Fort Worth International
DTW Detroit Metropolitan Wayne County
EWR Newark International
FLL Ft. Lauderdale-Hollywood International
HOU William B. Hobby
IAD Washington-Dulles International
IAH Houston Intercontinental
IND Indianapolis International
JFK John F. Kennedy International
LAX Los Angeles International
LGA La Guardia
MCO Orlando International
MDW Chicago Midway
MEM Memphis International
MIA Miami International
ORD Chicago-O'Hare International
PPI Palm Beach International
PHL Philadelphia International
PHX Phoenix Sky Harbor International
PIT Greater Pittsburgh International
RDU Raleigh-Durham International
SAN San Diego-Lindbergh International
SEA Seattle-Tacoma International
SFO San Francisco International
SJC San Jose International
SLC Salt Lake City International
STL Lambert-St. Louis International

SFAR No. 71. Special Operating Rules for Air Tour Operators in the State of Hawaii.

Section 1. Applicability. This Special Federal Aviation Regulation prescribes operating rules for airplane and helicopter visual flight rules air tour flights conducted in the State of Hawaii under 14 CFR parts 91, 121, and 135. This rules does not apply to:

(a) Operations conducted under 14 CFR part 121 in airplanes with a passenger seating configuration of more than 30 seats or a payload capacity of more than 7,500 pounds.

(b) Flights conducted in gliders or hot air balloons.

Section 2. Definitions. For the purposes of this SFAR:

"Air tour" means any sightseeing flight conducted under visual flight rules in an airplane or helicopter for compensation or hire.

"Air tour operator" means any person who conducts an air tour.

Section 3. Helicopter flotation equipment. No person may conduct an air tour in Hawaii in a single-engine helicopter beyond the shore of any island, regardless of whether the helicopter is within gliding distance of the shore, unless:

(a) The helicopter is amphibious or is equipped with floats adequate to accomplish a safe emergency ditching and approved flotation gear is easily accessible for each occupant; or

(b) Each person on board the helicopter is wearing approved flotation gear.

Section 4. Helicopter performance plan. Each operator must complete a performance plan before each helicopter air tour flight. The performance plan must be based on the information in the Rotorcraft Flight Manual (RFM), considering the maximum density altitude for which the operation is planned for the flight to determine the following:

(a) Maximum gross weight and center of gravity (CG) limitations for hovering in ground effect;

(b) Maximum gross weight and CG limitations for hovering out of ground effect; and,

(c) Maximum combination of weight, altitude, and temperature for which height-velocity information in the RFM is valid.

The pilot in command (PIC) must comply with the performance plan.

Section 5. Helicopter operating limitations. Except for approach to and transition from a hover, the PIC shall operate the helicopter at a combination of height and forward speed (including hover) that would permit a safe landing in event of engine power loss, in accordance with the height-speed envelope for that helicopter under current weight and aircraft altitude.

Section 6. Minimum flight altitudes. Except when necessary for takeoff and landing, or operating in compliance with an air traffic control clearance, or as otherwise authorized by the Administrator, no person may conduct an air tour in Hawaii:

(a) Below an altitude of 1,500 feet above the surface over all areas of the State of Hawaii, and,

(b) Closer than 1,500 feet to any person or property; or,

(c) Below any altitude prescribed by federal statute or regulation.

Section 7. Passenger briefing. Before takeoff, each PIC of an air tour flight of Hawaii with a flight segment beyond the ocean shore of any island shall ensure that each passenger has been briefed on the following, in addition to requirements set forth in 14 CFR 91.107, 121.571, or 135.117:

(a) Water ditching procedures;

(b) Use of required flotation equipment; and

(c) Emergency egress from the aircraft in event of a water landing.

Section 8. Termination date. This SFAR 71 shall remain in effect until further notice.

Special Federal Aviation Regulation No. 76. Prohibition Against Certain Flights Within the Territory and Airspace of Iran.

1. *Applicability.* This rule applies to the following persons:

(a) All U.S. air carriers and commercial operators;

(b) All persons exercising the privileges of an airman certificate issued by the FAA except such persons operating U.S.-registered aircraft for a foreign air carrier; or

(c) All operators of aircraft registered in the United States except where the operator of such aircraft is a foreign air carrier.

2. Flight Prohibition. Except as provided in paragraphs 3 and 4 of this SFAR, no person described in paragraph 1 may conduct flight operations over or within the territory and airspace of Iran.

3. Permitted Operations. This SFAR does not prohibit persons described in paragraph 1 from conducting flight operations over or within the territory and airspace of Iran where such operations are authorized by an exemption issued by the Administrator.

4. Emergency Situations. In an emergency that requires immediate decision and action for the safety of the flight, the pilot in command of an aircraft may deviate from this SFAR to the extent required by that emergency. Except for U.S. air carriers and commercial operators that are subject to the requirements of 14 CFR part 119, 121, or 135, each person who deviates from this rule shall, within ten (10) days of the deviation, excluding Saturdays, Sundays, and Federal holidays, submit to the nearest FAA Flight Standards District office a complete report of the operations of the aircraft involved in the deviation, including a description of the deviation and the reasons therefore.

5. Expiration. This Special Federal Aviation Regulation will remain in effect until further notice.

Special Federal Aviation Regulation No. 77— Prohibition Against Certain Flights Within the Territory and Airspace of Iraq

1. *Applicability.* This rule applies to the following persons:
(a) All U.S. air carriers or commercial operators;
(b) All persons exercising the privileges of an airman certificate issued by the FAA except such persons operating U.S.-registered aircraft for a foreign air carrier; or
(c) All operators of aircraft registered in the United States except where the operator of such aircraft is a foreign air carrier.

2. *Flight prohibition.* No person may conduct flight operations over or within the territory of Iraq except as provided in paragraphs 3 and 4 of this SFAR or except as follows:
(a) Overflights of Iraq may be conducted above flight level (FL) 200 subject to the approval of, and in accordance with the conditions established by, the appropriate authorities of Iraq.
(b) Flights departing from countries adjacent to Iraq whose climb performance will not permit operation above FL 200 prior to entering Iraqi airspace may operate at altitudes below FL 200 within Iraq to the extent necessary to permit a climb above FL 200, subject to the approval of, and in accordance with the conditions established by, the appropriate authorities of Iraq.
(c) [Reserved]

3. *Permitted operations.* This SFAR does not prohibit persons described in paragraph 1 from conducting flight operations within the territory and airspace of Iraq when such operations are authorized either by another agency of

the United States Government with the approval of the FAA or by an exemption issued by the Administrator.

4. *Emergency situations.* In an emergency that requires immediate decision and action for the safety of the flight, the pilot in command of an aircraft may deviate from this SFAR to the extent required by that emergency. Except for U.S. air carriers or commercial operators that are subject to the requirements of 14 CFR parts 119, 121, or 135, each person who deviates from this rule shall, within ten (10) days of the deviation, excluding Saturdays, Sundays, and Federal holidays, submit to the nearest FAA Flight Standards District Office a complete report of the operations of the aircraft involved in the deviation including a description of the deviation and the reasons therefore.

5. *Expiration.* This Special Federal Aviation Regulation will remain in effect until further notice.
[Doc. No. 28691, 61 FR 54021, Oct. 16, 1996. as amended by Doc. No. FAA–2003–14766, 68 FR 17870, Apr. 11, 2003; 68 FR 65382, Nov. 19, 2003]

SFAR No. 79. Prohibition Against Certain Flights Within the Flight Information Region (FIR) of the Democratic People's Republic of Korea (DPRK).

1. *Applicability.* This rule applies to the following persons:
(a) All U.S. air carriers or commercial operators.
(b) All persons exercising the privileges of an airman certificate issued by the FAA, except such persons operating U.S.-registered aircraft for a foreign air carrier.
(c) All operators of aircraft registered in the United States except where the operator of such aircraft is a foreign air carrier.

2. *Flight prohibitions.*
(a) Except as provided in paragraphs 2(b), 3, and 4 of this SFAR, no person described in paragraph 1 may conduct flight operations through the Pyongyang FIR.
(b) Flight operations within the Pyongyang FIR east of 132 degrees east longitude are prohibited until the FAA determines, based on information from the DPRK civil aviation authority, that the proper level of operational overflight safety can be assured. The FAA will amend this SFAR and publish a notice to airmen (NOTAM) to permit flights east of 132 degrees east longitude once this determination is made.

3. *Permitted operations.* This SFAR does not prohibit persons described in paragraph 1 from conducting flight operations within the Pyongyang FIR where such operations are authorized either by exemption issued by the Administrator or by another agency of the United States Government with FAA approval.

4. *Emergency situations.* In an emergency that requires immediate decision and action for the safety of the flight, the pilot in command on an aircraft may deviate from this SFAR to the extent required by that emergency. Except for U.S. air carriers and commercial operators that are subject to the requirements of 14 CFR parts 121, 125, or 135, each person who deviates from this rule shall, within ten

(10) days of the deviation, excluding Saturdays, Sundays, and Federal holidays, submit to the nearest FAA Flight Standards District Office a complete report of the operations of the aircraft involved in the deviation, including a description of the deviation and the reasons therefore.

5. *Expiration.* This Special Federal Aviation Regulation No. 79 will remain in effect until further notice.

Special Federal Aviation Regulation No. 87— Prohibition Against Certain Flights Within the Territory and Airspace of Ethiopia

1. *Applicability.* This Special Federal Aviation Regulation (SFAR) No. 87 applies to all U.S. air carriers or commercial operators, all persons exercising the privileges of an airman certificate issued by the FAA unless that person is engaged in the operation of a U.S.-registered aircraft for a foreign air carrier, and all operators using aircraft registered in the United States except where the operator of such aircraft is a foreign air carrier.

2. *Flight prohibition.* Except as provided in paragraphs 3 and 4 of this SFAR, no person described in paragraph 1 may conduct flight operations within the territory and airspace of Ethiopia north of 12 degrees north latitude.

3. *Permitted operations.* This SFAR does not prohibit persons described in paragraph 1 from conducting flight operations within the territory and airspace of Ethiopia where such operations are authorized either by exemption issued by the Administrator or by an authorization issued by another agency of the United States Government with the approval of the FAA.

4. *Emergency situations.* In an emergency that requires immediate decision and action for the safety of the flight, the pilot in command of an aircraft may deviate from this SFAR to the extent required by that emergency. Except for U.S. air carriers and commercial operators that are subject to the requirements of 14 CFR 121.557, 121.559, or 135.19, each person who deviates from this rule shall, within ten (10) days of the deviation, excluding Saturdays, Sundays, and Federal holidays, submit to the nearest FAA Flight Standards District Office a complete report of the operations of the aircraft involved in the deviation, including a description of the deviation and the reasons therefor.

5. *Expiration.* This Special Federal Aviation Regulation shall remain in effect until further notice.

SFAR No. 94. Enhanced Security Procedures for Operations at Certain Airports in the Washington, DC Metropolitan Area Special Flight Rules Area

1. *Applicability.* This Special Federal Aviation Regulation (SFAR) establishes rules for all persons operating an aircraft to or from the following airports located within the airspace designated as the Washington, DC Metropolitan Area

Special Flight Rules Area: (a) College Park Airport (CGS). (b) Potomac Airfield (VKX). (c) Washington Executive/Hyde Field (W32).

2. *Definitions.* For the purposes of this SFAR the following definitions apply:

Administrator means the Federal Aviation Administrator, the Under Secretary of Transportation for Security, or any person delegated the authority of the Federal Aviation Administrator or Under Secretary of Transportation for Security. Washington, DC Metropolitan Area Special Flight Rules Area means that airspace within an area from the surface up to but not including Flight Level 180, bounded by a line beginning at the Washington (DCA) VOR/DME 300 degree radial at 15 nautical miles (Lat. 38°56'55"N., Long. 77°20'08"W.); thence clockwise along the DCA 15 nautical mile arc to the DCA 022 degree radial at 15 nautical miles (Lat. 39°06'11"N., Long 76°57'51"W.); thence southeast via a line drawn to the DCA 049 degree radial at 14 nautical miles (Lat. 39°02'18"N., Long. 76°50'38"W.); thence south via a line drawn to the DCA 064 degree radial at 13 nautical miles (Lat. 38°59'01"N., Long. 76°48'32"W.); thence clockwise along the DCA 13 nautical mile arc to the DCA 282 degree radial at 13 nautical miles (Lat. 38°52'14"N., Long 77°18'48"W.); thence north via a line drawn to the point of the beginning; excluding the airspace within a one nautical mile radius of Freeway Airport (W00), Mitchellville, Md.

3. *Operating requirements.*

(a) Except as specified in paragraph 3(c) of this SFAR, no person may operate an aircraft to or from an airport to which this SFAR applies unless security procedures that meet the provisions of paragraph 4 of this SFAR have been approved by the Administrator for operations at that airport.

(b) Except as specified in paragraph 3(c) of this SFAR, each person serving as a required flightcrew member of an aircraft operating to or from an airport to which this SFAR applies must: (1) Prior to obtaining authorization to operate to or from the airport, present to the Administrator the following: (i) A current and valid airman certificate; (ii) A current medical certificate; (iii) One form of Government issued picture identification; and (iv) A list containing the make, model, and registration number of each aircraft that the pilot intends to operate to or from the airport; (2) Successfully complete a background check by a law enforcement agency, which may include submission of fingerprints and the conduct of a criminal history, records check. (3) Attend a briefing acceptable to the Administrator that describes procedures for operating to or from the airport; (4) Not have been convicted or found not guilty by reason of insanity, in any jurisdiction, during the 10 years prior to being authorized to operate to or from the airport, or while authorized to operate to or from the airport, of those crimes specified in § 108.229 (d) of this chapter; (5) Not have a record on file with the FAA of:

(i) A violation of a prohibited area designated under part 73 of this chapter, a flight restriction established under § 91.141 of this chapter, or special security instructions issued under §99.7 of this chapter; or (ii) More than one violation of a restricted area designated under part 73 of this chapter, emergency air traffic rules issued under § 91.139 of this chapter, a temporary flight restriction designated under § 91.137, § 91.138, or § 91.145 of this chapter, an area designated under § 91.143 of this chapter, or any combination thereof; (6) Be authorized by the Administrator to conduct operations to or from the airport; (7) Protect from unauthorized disclosure any identification information issued by the Administrator for the conduct of operations to or from the airport; (8) Operate an aircraft that is authorized by the Administrator for operations to or from the airport; (9) File an IFR or VFR flight plan telephonically with Leesburg AFSS prior to departure and obtain an ATC clearance prior to entering the Washington, DC Metropolitan Area Special Flight Rules Area; (10) Operate the aircraft in accordance with an open IFR or VFR flight plan while in the Washington, DC Metropolitan Area Special Flight Rules Area, unless otherwise authorized by ATC; (11) Maintain two-way communications with an appropriate ATC facility while in the Washington, DC Metropolitan Area Special Flight Rules Area; (12) Ensure that the aircraft is equipped with an operable transponder with altitude reporting capability and use an assigned discrete beacon code while operating in the Washington, DC Metropolitan Area Special Flight Rules Area; (13) Comply with any instructions issued by ATC for the flight; (14) Secure the aircraft after returning to the airport from any flight; (15) Comply with all additional safety and security requirements specified in applicable NOTAMs; and (16) Comply with any Transportation Security Administration, or law enforcement requirements to operate to or from the airport. © A person may operate a U.S. Armed Forces, law enforcement, or aeromedical services aircraft to or from an affected airport provided the operator complies with paragraphs 3(b)(10) through 3(b)(16) of this SFAR and any additional procedures specified by the Administrator necessary to provide for the security of aircraft operations to or from the airport.

4. *Airport Security Procedures.*

(a) Airport security procedures submitted to the Administrator for approval must: (1) Identify and provide contact information for the airport manager who is responsible for ensuring that the security procedures at the airport are implemented and maintained; (2) Contain procedures to identify those aircraft eligible to be authorized for operations to or from the airport; (3) Contain procedures to ensure that a current record of those persons authorized to conduct operations to or from the airport and the aircraft in which the person is authorized to conduct those operations is maintained at the airport; (4) Contain airport arrival and departure route descriptions, air traffic control clearance procedures, flight plan requirements, communications procedures, and procedures for transponder use; (5) Contain procedures to monitor the security of aircraft at the airport during operational and non-operational hours and to alert aircraft owners and operators, airport operators, and the Administrator of unsecured aircraft; (6) Contain

procedures to ensure that security awareness procedures are implemented and maintained at the airport; (7) Contain procedures to ensure that a copy of the approved security procedures is maintained at the airport and can be made available for inspection upon request of the Administrator; (8) Contain procedures to provide the Administrator with the means necessary to make any inspection to determine compliance with the approved security procedures; and (9) Contain any additional procedures necessary to provide for the security of aircraft operations to or from the airport.

(b) Airport security procedures are approved without an expiration date and remain in effect unless the Administrator makes a determination that operations at the airport have not been conducted in accordance with those procedures or that those procedures must be amended in accordance with paragraph 4.(a)(9) of this SFAR.

5. *Waivers.* The Administrator may permit an operation to or from an airport to which this SFAR applies, in deviation from the provisions of this SFAR if the Administrator finds that such action is in the public interest, provides the level of security required by this SFAR, and the operation can be conducted safely under the terms of the waiver.

6. *Delegation.* The authority of the Administrator under this SFAR is also exercised by the Associate Administrator for Civil Aviation Security and the Deputy Associate Administrator for Civil Aviation Security. This authority may be further delegated.

7. *Expiration.* This Special Federal Aviation Regulation shall remain in effect until February 13, 2005. [Doc. No. FAA-2002-11580, 67 FR 7544, Feb. 19, 2002]

Special Federal Aviation Regulation No. 97—Special Operating Rules for the Conduct of Instrument Flight Rules (IFR) Area Navigation (RNAV) Operations using Global Positioning Systems (GPS) In Alaska

Those persons identified in Section 1 may conduct IFR en route RNAV operations in the State of Alaska and its airspace on published air traffic routes using TSO C145a/C146a navigation systems as the only means of IFR navigation. Despite contrary provisions of parts 71, 91, 95, 121, 125, and 135 of this chapter, a person may operate aircraft in accordance with this SFAR if the following requirements are met.

Section 1. Purpose, use, and limitations

a. This SFAR permits TSO C145a/C146a GPS (RNAV) systems to be used for IFR en route operations in the United States airspace over and near Alaska (as set forth in paragraph c of this section) at Special Minimum En Route Altitudes (MEA) that are outside the operational service volume of ground-based navigation aids, if the aircraft operation also meets the requirements of sections 3 and 4 of this SFAR.

b. Certificate holders and part 91 operators may operate aircraft under this SFAR provided that they comply with the requirements of this SFAR.

c. Operations conducted under this SFAR are limited to United States Airspace within and near the State of Alaska as defined in the following area description: From 62°00'00.000"N, Long. 141°00'00.00"W.; to Lat. 59°47' 54.11"N., Long. 135°28'38.34"W.; to Lat. 56°00' 04.11"N., Long. 130°00'07.80"W.; to Lat. 54°43'00.00"N., Long. 130°37'00.00"W.; to Lat. 51 24'00.00"N., Long. 167°49'00.00"W.; to Lat. 50°08'00.00"N., Long. 176°34'00.00"W.; to Lat. 45°42'00.00"N., Long.—162° 55'00.00"E.; to Lat. 50°05'00.00"N., Long.—159°00' 00.00" to Lat. 54°00'00"N., Long.—169°00'00.00"E; to Lat. 60°00'00.00"N., Long. 180°00'00.00"E; to Lat. 65°00'00.00"N., Long. 168°58'23.00"W.; to Lat. 90°00' 00.00"N., Long. 00°00'0.00"W.; to Lat. 62°00'00" 000"N, Long. 141°00'00.00"W.

(d) No person may operate an aircraft under IFR during the en route portion of flight below the standard MEA or at the special MEA unless the operation is conducted in accordance with sections 3 and 4 of this SFAR.

Section 2. Definitions and abbreviations

For the purposes of this SFAR, the following definitions and abbreviations apply.

Area navigation (RNAV). RNAV is a method of navigation that permits aircraft operations on any desired flight path.

Area navigation (RNAV) route. RNAV route is a published route based on RNAV that can be used by suitably equipped aircraft.

Certificate holder. A certificate holder means a person holding a certificate issued under part 119 or part 125 of this chapter or holding operations specifications issued under part 129 of this chapter.

Global Navigation Satellite System (GNSS). GNSS is a world-wide position and time determination system that uses satellite ranging signals to determine user location. It encompasses all satellite ranging technologies, Including GPS and additional satellites. Components of the GNSS include GPS, the Global Orbiting Navigation Satellite System, and WAAS satellites.

Global Positioning System (GPS). GPS is a satellite-based radio navigational, positioning, and time transfer system. The system provides highly accurate position and velocity information and precise time on a continuous global basis to properly equipped users.

Minimum crossing altitude (MCA). The minimum crossing altitude (MCA) applies to the operation of an aircraft proceeding to a higher minimum en route altitude when crossing specified fixes.

Required navigation system. Required navigation system means navigation equipment that meets the performance requirements of TSO C145a/C146a navigation systems certified for IFR en route operations.

Route segment. Route segment is a portion of a route bounded on each end by a fix or NAVAID,

Special MEA. Special MEA refers to the minimum en route altitudes, using required navigation systems, on published routes outside the operational service volume of ground-based navigation aids and are depicted on the published Low Altitude and High Altitude En Route Charts using the color blue and with the suffix "G." For example, a GPS MEA of 4000 feet MSL would be depicted using the color blue, as 4000G.

Standard MEA. Standard MEA refers to the minimum en route IFR altitude on published routes that uses ground-based navigation aids and are depicted on the published Low Altitude and High Altitude En Route Charts using the color black.

Station referenced. Station referenced refers to radio navigational aids or fixes that are referenced by ground based navigation facilities such as VOR facilities.

Wide Area Augmentation System (WAAS). WAAS is an augmentation to GPS that calculates GPS integrity and correction data on the ground and uses geo-stationary satellites to broadcast GPS integrity and correction data to GPS/WAAS users and to provide ranging signals. It is a safety critical system consisting of a ground network of reference and integrity monitor data processing sites to assess current GPS performance, as well as a space segment that broadcasts that assessment to GNSS users to support en route through precision approach navigation. Users of the system include all aircraft applying the WAAS data and ranging signal.

Section 3. Operational Requirements

To operate an aircraft under this SFAR, the following requirements must be met:

a. Training and qualification for operations and maintenance personnel on required navigation equipment used under this SFAR.

b. Use authorized procedures for normal, abnormal, and emergency situations unique to these operations, Including degraded navigation capabilities, and satellite system outages.

c. For certificate holders, training of flight crewmembers and other personnel authorized to exercise operational control on the use of those procedures specified in paragraph b of this section.

d. Part 129 operators must have approval from the State of the operator to conduct operations in accordance with this SFAR.

e. In order to operate under this SFAR, a certificate holder must be authorized in operations specifications.

Section 4. Equipment Requirements

a. The certificate holder must have properly installed, certificated, and functional dual required navigation systems as defined in section 2 of this SFAR for the en route operations covered under this SFAR.

b. When the aircraft is being operated under part 91, the aircraft must be equipped with at least one properly installed, certificated, and functional required navigation system as defined in section 2 of this SFAR for the en route operations covered under this SFAR.

Part 93—Special air traffic rules and airport traffic patterns

§ 93.301 Applicability.

This subpart prescribes special operating rules for all persons operating aircraft in the following airspace, designated as the Grand Canyon National Park Special Flight Rules Area: That airspace extending from the surface up to but not including 18,000 feet MSL within an area bounded by a line beginning at Lat. 35°55'12" N., Long. 112°04'05" W.; east to Lat. 35°55'30" N., Long. 111°45'00" W.; to Lat. 35°59'02" N., Long. 111°36'03" W., north to Lat. 36°15'30" N., Long. 111°36'06" W.; to Lat. 36°24'49" N., Long. 111°47'45" W.; to Lat. 36°52'23" N., Long. 111°33'10" W.; west-northwest to Lat. 36°53'37" N., Long. 111°38'29" W.; southwest to Lat. 36°35'02" N., Long. 111°53'28" W.; to Lat. 36°21'30" N., Long. 112°00'03" W.; west-northwest to Lat. 36°30'30" N., Long. 112°35'59" W.; southwest to Lat. 36°24'46" N., Long. 112°51'10" W., thence west along the boundary of Grand Canyon National Park (GCNP) to Lat. 36°14'08" N., Long. 113°10'07" W.; west-southwest to Lat. 36°09'30" N., Long. 114°03'03° W.; southeast to Lat. 36°05'11" N., Long. 113°58'46" W.; thence south along the boundary of GCNP to Lat. 35°58'23" N., Long. 113°54'14" W.; north to Lat. 36°00'10" N., Long. 113°53'48" W.; northeast to Lat. 36°02'14" N., Long. 113°50'16" W.; to Lat. 36°02'17" N., Long. 113°53'48" W.; northeast to Lat. 36°02'14" N., Long. 113°50'16" W.; to Lat. 36°02'17" N., Long. 113°49'11" W.; southeast to Lat. 36°01'22" N., Long. 113°48'21" W.; to Lat. 35°59'15" N., Long. 113°47'13" W.; to Lat. 35°57'51" N., Long. 113°46'01" W.; to Lat. 35°57'45" N., Long. 113°45'23" W.; southwest to Lat. 35°54'48" N., Long. 113°50'24" W.; southeast to Lat. 35°41'01" N., Long. 113°35'27" W.; thence clockwise via the 4.2-nau-tical mile radius of the Peach Springs VORTAC to Lat. 36°38'53" N., Long. 113°27'49" W.; northeast to Lat. 35°42'58" N., Long. 113°10'57" W.; north to Lat. 35°57'51" N., Long. 113°11'06" W.; east to Lat. 35°57'44" N., Long. 112°14'04" W.; thence clockwise via the 4.3-nautical mile radius of the Grand Canyon National Park Airport reference point (Lat. 35°57'08" N., Long. 112°08'49" W.) to the point of origin.

[Amdt. 93-8065 FR 17736, April 4, 2000]

§ 93.303 Definitions.

For the purposes of this subpart:

Allocation means authorization to conduct a commercial air tour in the Grand Canyon National Park (GCNP) Special Flight Rules Area (SFRA).

Commercial air tour means any flight conducted for compensation or hire in a powered aircraft where a purpose of the flight is sightseeing. If the operator of a flight asserts that the flight is not a commercial air tour, factors that can be considered by the Administrator in making a determination of whether the flight is a commercial air tour include, but are not limited to—

(1) Whether there was a holding out to the public of willingness to conduct a sightseeing flight for compensation or hire;

(2) Whether a narrative was provided that referred to areas or points of interest on the surface;

(3) The area of operation;

(4) The frequency of flights;

(5) The route of flight;

(6) The inclusion of sightseeing flights as part of any travel arrangement package; or

(7) Whether the flight in question would or would not have been canceled based on poor visibility of the surface.

Commercial Special Flight Rules Area Operation means any portion of any flight within the Grand Canyon National Park Special Flight Rules Area that is conducted by a certificate holder that has operations specifications authorizing flights within the Grand Canyon National Park Special Flight Rules Area. This term does not include operations conducted under an FAA Form 7711-1, Certificate of Waiver or Authorization. The types of flights covered by this definition are set forth in the "Las Vegas Flight Standards District Office Grand Canyon National Park Special Flight Rules Area Procedures Manual" which is available from the Las Vegas Flight Standards District Office.

Flight Standards District Office means the FAA Flight Standards District Office with jurisdiction for the geographical area containing the Grand Canyon.

Park means Grand Canyon National Park.

Special Flight Rules Area means the Grand Canyon National Park Special Flight Rules area.

[Amdt. 93-81 65 FR 17736, April 4, 2000]

§ 93.305   Flight-free zones and flight corridors.

Except in an emergency or if otherwise necessary for safety of flight, or unless otherwise authorized by the Flight Standards District Office for a purpose listed in 93.309, no person may operate an aircraft in the Special Flight Rules Area within the following flight-free zones:

(a) *Desert View Flight-free Zone.* That airspace extending from the surface up to but not including 14,500 feet MSL within an area bounded by a line beginning at Lat. 35°59'58" N., Long. 111°52'47" W.; thence east to Lat. 36°00'00" N., Long. 111°51'04" W.; thence north to 36°00'24" N., Long. 111°51'04" W.; thence east to 36°00'24" N., Long. 111°45'44" W.; thence north along the GCNP boundary to Lat. 36°14'05" N., Long. 111°48'34" W.; thence southwest to Lat. 36°12'06" N., Long. 111°51'14" W.; to the point of origin; but not including the airspace at and above 10,500 feet MSL within 1 nautical mile of the western boundary of the zone. The corridor to the west between the Desert View and Bright Angel Flight-free Zones, is designated the "Zuni Point Corridor." This corridor is 2 nautical miles wide for commercial air tour flights and 4 nautical miles wide for transient and general aviation operations.

(b) *Bright Angel Flight-free zone.* This corridor is 2 nautical miles wide for commercial air tour flights and 4 nautical miles wide for transient and general aviation operations. The Bright Angel Flight-free Zone does not include the following airspace designated as the Bright Angel Corridor: That airspace one-half nautical mile on either side of a line extending from Lat. 36°14'57" N., Long. 112°08'45" W. and Lat. 36°15'01" N., Long. 111°55'39" W.

(c) *Toroweap/Shinumo Flight-free Zone.* That airspace extending from the surface up to but not including 14,500 feet MSL within an area bounded by a line beginning at Lat. 36°05'44" N., Long. 112°19'27" W.; north-northeast to Lat. 36°10'49" N., Long. 112°13'19" W.; to Lat. 36°21'02" N., Long. 112°08'47" W.; thence west and south along the GCNP boundary to Lat. 36°10'58" N., Long. 113°08'35" W.; south to Lat. 36°10'12" N., Long. 113°08'34" W.; thence in an easterly direction along the park boundary to the point of origin; but not including the following airspace designated as the "Tuckup Corridor": at or above 10,500 feet MSL within 2 nautical miles either side of a line extending between Lat. 36°24'42" N., Long. 112°48'47" W. and Lat. 36°14'17" N., Long. 112°48'31" W. The airspace designated as the "Fossil Canyon Corridor" is also excluded from the

Toroweap/Shinumo Flight-free Zone at or above 10,500 feet MSL within 2 nautical miles either side of a line extending between Lat. 36°16'26" N., Long. 112°34'35" W. and Lat. 36°22'51" N., Long. 112°18'18" W. The Fossil Canyon Corridor is to be used for transient and general aviation operations only.

(d) *Sanup Flight-free Zone.* That airspace extending from the surface up to but not including 8,000 feet MSL within an area bounded by a line beginning at Lat. 35°59'32" N., Long. 113°20'28" W.; west to Lat. 36°00'55"N., Long. 113°42'09" W.; southeast to Lat. 35°59'57" N., Long. 113°41'09" W.; to Lat. 35°59'09" N., Long. 113°40'53" W.; to Lat. 35°58'45" N., Long 113°40'15" W.; to Lat. 35°57'52" N., Long. 113°39'34" W.; to Lat. 35°56'44" N., Long. 113°39'07" W.; to Lat. 35°56'04" N., Long. 113°39'20" W.; to Lat. 35°55'02" N., Long. 113°40'43" W.; to Lat. 35°54'47" N., Long. 113°40'51" W.; southeast to Lat. 35°50'16" N., Long. 113°37'13" W.; thence along the park boundary to the point of origin.

§ 93.307   Minimum flight altitudes.

Except in an emergency, or if otherwise necessary for safety of flight, or unless otherwise authorized by the Flight Standards District Office for a purpose listed in 93.309, no person may operate an aircraft in the Special Flight Rules Area at an altitude lower than the following:

(a) Minimum sector altitudes.

(1) Commercial air tours.

(i) Marble Canyon Sector. Lees Ferry to Boundary Ridge: 6,000 feet MSL.

(ii) Supai Sector. Boundary Ridge to Supai Point: 7,500 feet MSL.

(iii) Diamond Creek Sector. Supai Point to Diamond Creek: 6,500 feet MSL.

(iv) Pearce Ferry Sector. Diamond Creek to the Grand Wash Cliffs: 5,000 feet MSL.

(2) Transient and general aviation operations.

(i) Marble Canyon Sector. Lees Ferry to Boundary Ridge: 8,000 feet MSL.

(ii) Supai Sector. Boundary Ridge to Supai Point: 10,000 feet MSL.

(iii) Diamond Creek Sector. Supai Point to Diamond Creek: 9,000 feet MSL.

(iv) Fossil Canyon Corridor, 10,500 feet MSL

(b) Minimum corridor altitudes.

(1) Commercial sightseeing flights.

(i) Zuni Point Corridors. 7,500 feet MSL.

(ii) Dragon Corridor. 7,500 feet MSL.

(2) Transient and general aviation operations.

(i) Zuni Point Corridor. 10,500 feet MSL.

(ii) Dragon Corridor. 10,500 feet MSL.

(iii) Tuckup Corridor. 10,500 feet MSL.

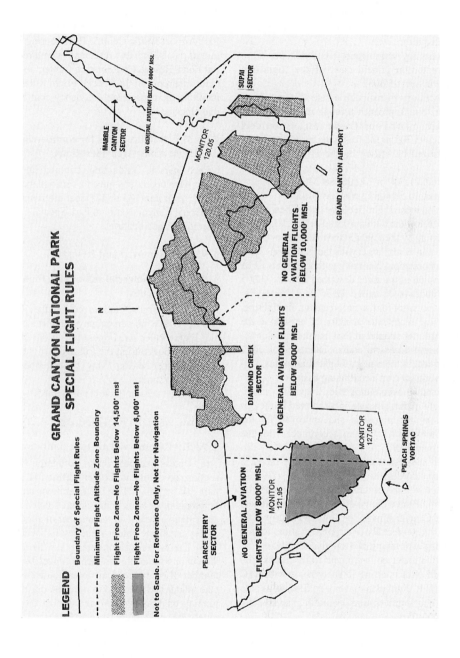

GRAND CANYON NATIONAL PARK
SPECIAL FLIGHT RULES

LEGEND

Boundary of Special Flight Rules

Minimum Flight Altitude Zone Boundary

Flight Free Zone—No Flights Below 14,500' msl

Flight Free Zones—No Flights Below 8,000' msl

Not to Scale. For Reference Only, Not for Navigation

§ 93.309   General operating procedures.

Except in an emergency, no person may operate an aircraft in the Special Flight Rules Area unless the operation is conducted in accordance with the following procedures. (Note: The following procedures do not relieve the pilot from see-and-avoid responsibility or compliance with the minimum safe altitude requirements specified in Sec. 91.119 of this chapter.):

(a) Unless necessary to maintain a safe distance from other aircraft or terrain remain clear of the flight-free zones described in Sec. 93.305;

(b) Unless necessary to maintain a safe distance from other aircraft or terrain, proceed through the Zuni Point, Dragon, and Tuckup and Fossil Canyon Flight Corridors described in Sec. 93.305 at the following altitudes unless otherwise authorized in writing by the Flight Standards District Office:

(1) Northbound. 11,500 or 13,500 feet MSL.

(2) Southbound. 10,500 or 12,500 feet MSL.

(c) For operation in the flight-free zones described in Sec. 93.305, or flight below the altitudes listed in Sec. 93.307, is authorized in writing by the Flight Standards District Office and is conducted in compliance with the conditions contained in that authorization. Normally authorization will be granted for operation in the areas described in Sec. 93.305 or below the altitudes listed in Sec. 93.307 only for operations of aircraft necessary for law enforcement, firefighting, emergency medical treatment/evacuation of persons in the vicinity of the Park; for support of Park maintenance or activities; or for aerial access to and maintenance of other property located within the Special Flight Rules Area. Authorization may be issued on a continuing basis;

(d) Is conducted in accordance with a specific authorization to operate in that airspace incorporated in the operator's operations specifications and approved by the Flight Standards District Office in accordance with the provisions of this subpart;

(e) Is a search and rescue mission directed by the U.S. Air Force Rescue Coordination Center;

(f) Is conducted within 3 nautical miles of Grand Canyon Bar Ten Airstrip, Pearce Ferry Airstrip, Cliff Dwellers Airstrip, Marble Canyon Airstrip, or Tuweep Airstrip at an altitude less than 3,000 feet above airport elevation, for the purpose of landing at or taking off from that facility; or

(g) Is conducted under an instrument flight rules (IFR) clearance and the pilot is acting in accordance with ATC instructions. An IFR flight plan may not be filed on a route or at an altitude that would require operation in an area described in Sec. 93.305.

[Amdt. 93-80, 65 FR 17736, April 4, 2000]

§ 93.311   Minimum terrain clearance.

Except in an emergency, when necessary for takeoff or landing, or unless otherwise authorized by the Flight Standards District Office for a purpose listed in Sec. 93.309(c), no person may operate an aircraft within 500 feet of any terrain or structure located between the north and south rims of the Grand Canyon.

§ 93.313   Communications.

Except when in contact with the Grand Canyon National Park Airport Traffic Control Tower during arrival or departure or on a search and rescue mission directed by the U.S. Air Force Rescue Coordination Center, no person may operate an aircraft in the Special Flight Rules Area unless he monitors the appropriate frequency continuously while in that airspace.

§ 93.315   Requirements for commercial Special Flight Rules Area operations.

Each person conducting commercial Special Flight Rules Area operations must be certified in accordance with Part 119 for Part 135 or 121 operations and hold appropriate Grand Canyon National Park Special Flight Rules Area operations specifications.

§ 93.316   Removed and Reserved.

§ 93.317   Commercial sightseeing flight reporting requirements.

Unless otherwise authorized by the Flight Standards District Office, no person may conduct a commercial Special Flight Rules Area operation in the Dragon and Zuni Point corridors during the following flight-free periods:

(a) Summer season (May 1–September 30)–6 p.m. to 8 a.m. daily; and

(b) Winter season (October 1–April 30)–5 p.m. to 9 a.m. daily.

[Amdt. 93-81, 65 FR 17736, April 4, 2000]

§ 93.319 Commercial air tour limitations.

(a) Unless excepted under paragraph (f) or (g) of this section, no certificate holder certified in accordance with part 119 for part 121 or 135 operations may conduct more commercial air tours in the Grand Canyon National Park in any calendar year than the number of allocations specified on the certificate holder's operations specifications.

(b) The Administrator determines the number of initial allocations for each certificate holder based on the total number of commercial air tours conducted by the certificate holder and reported to the FAA during the period beginning on May 1, 1997 and ending on April 30, 1998, unless excepted under paragraph (g).

(c) Certificate holders who conducted commercial air tours during the base year and reported them to the FAA receive an initial allocation.

(d) A certificate holder must use one allocation for each flight that is a commercial air tour, unless excepted under paragraph (f) or (g) of this section.

(e) Each certificate holder's operation specifications will identify the following information, as applicable:

(1) Total SFRA allocations; and

(2) Dragon corridor and Zuni Point corridor allocations.

(f) Certificate holders satisfying the requirements of §93.315 of this subpart are not required to use a commercial air tour allocation for each commercial air tour flight in the GCNP SFRA provided the following conditions are satisfied:

(1) The certificate holder conducts its operations in conformance with the routes and airspace authorizations as specified in its Grand Canyon National Park Special Flight Rules Area operations specifications;

(2) The certificate holder must have executed a written contract with the Hualapai Indian Nation which grants the certificate holder a trespass permit and specifies the maximum number of flights to be permitted to land at Grand Canyon West Airport and at other sites located in the vicinity of that airport and operates in compliance with that contract; and

(3) The certificate holder must have a valid operations specification that authorizes the certificate holder to conduct the operations specified in the contract with the Hualapai Indian Nation and specifically approves the number of operations that may transit the Grand Canyon National Park Special Flight Rules Area under this exception.

(g) Certificate holders conducting commercial air tours at or above 14,500 feel MSL but below 18,000 feet MSL who did not receive initial allocations in 1999 because they were not required to report during the base year may operate without an allocation when conducting air tours at those altitudes. Certificate holders conducting commercial air tours in the area affected by the eastward shift of the SFRA who did not receive initial allocations in 1999 because they were not required to report during the base year may continue to operate on the specified routes without an allocation in the area bounded by longitude line 111 degrees 42 minutes east and longitude line 111 degrees 36 minutes east. This exception does not include operation in the Zuni Point corridor.

7. Section 93.321 is added to read as follows:

[Amdt. 93-81, 65 FR 17736, April 4, 2000]

§ 93.321 Transfer and termination of allocations.

(a) Allocations are not a property interest; they are an operating privilege subject to absolute FAA control.

(b) Allocations are subject to the following conditions:

(1) The Administrator will re-authorize and re-distribute allocations no earlier than two years from the effective date of this rule.

(2) Allocations that are held by the FAA at the time of re-allocation may be distributed among remaining certificate holders, proportionate to the size of each certificate holder's allocation.

(3) The aggregate SFRA allocations will not exceed the number of operations reported to the FAA for the base year beginning on May 1, 1997 and ending on April 30, 1998, except as adjusted to incorporate operations occurring for the base year of April 1, 2000 and ending on March 31, 2001, that operate at or above 14,500 feet MSL and below 18,000 feet MSL and operations in the area affected by the eastward shift of the SFRA bounded by longitude line 111 degrees 42 minutes east to longitude 111 degrees 36 minutes east.

(4) Allocations may be transferred among Part 135 or Part 121 certificate holders, subject to all of the following:

(i) Such transactions are subject to all other applicable requirements of this chapter.

(ii) Allocations authorizing commercial air tours outside the Dragon and Zuni Point Corridors may not be transferred into the Dragon and Zuni Point corridors. Allocations authorizing commercial air tours within the Dragon and Zuni Point corridors may be transferred outside of the Dragon and Zuni Point corridors.

(iii) A certificate holder must notify in writing the Las Vegas Flight Standards District Office within 10 calendar days of a transfer of allocations. This notification must identify the parties involved, the type of transfer (permanent or temporary) and the number of allocations transferred. Permanent transfers are not effective until the Flight Standards District Office reissues the operations specifications reflecting the transfer. Temporary transfers are effective upon notification.

(5) An allocation will revert to the FAA upon voluntary cessation of commercial air tours within the SFRA for any consecutive 180-day period unless the certificate holder notifies the FSDO in writing, prior to the expiration of the 180-day time period, of the following: i) the reason why the certificate holder has not conducted any commercial air tours during the consecutive 180-day period; and ii) the date the certificate holder intends on resuming commercial air tours operations. The FSDO will notify the certificate holder of any extension to the consecutive 180-day time period, not to exceed an additional consecutive 180-days. A certificate holder may be granted one extension.

(6) The FAA retains the right to redistribute, reduce, or revoke allocations based on:

(i) Efficiency of airspace;

(ii) Voluntary surrender of allocations;

(iii) Involuntary cessation of operations; and

(iv) Aviation safety.

[Amdt. 93-81, 65 FR 17736, April 4, 2000]

§ 93.323 Flight plans.

Each certificate holder conducting commercial SFRA operation must file a visual flight rules (VFR) flight plan in accordance with §91.153. This section does not apply to operations conducted in accordance with § 93.309 (g). The flight plan must be on file with a FAA Flight Service Station prior to each flight. Each VFR flight plan must identify the purpose of the flight in the "remarks" section according to only the types set forth in the "Las Vegas Flight Standards District Office Grand Canyon National Park Special Flight Rules Area Procedures Manual" which is available from the Las Vegas Flight Standards District Office.

[Amdt. 93-81, 65 FR 17736, April 4, 2000]

§ 93.325 Quarterly reporting.

(a) Each certificate holder must submit in writing, within 30 days of the end of each calendar quarter, the number of commercial SFRA operations conducted for that quarter. Quarterly reports must be filed with the Las Vegas Flight Standards District Office.

(b) Each quarterly report must contain the following information.

(1) Make and model of aircraft;

(2) Identification number (registration number) for each aircraft;

(3) Departure airport for each segment flown;

(4) Departure date and actual Universal Coordinated Time, as applicable for each segment flown;

(5) Type of operation; and

(6) Route(s) flown.

Part 97—Standard instrument approach procedures

Subpart A—General

§ 97.1 Applicability.

This part prescribes standard instrument approach procedures for instrument letdown to airports in the United States and the weather minimums that apply to takeoffs and landings under IFR at those airports.

§ 97.3 Symbols and terms used in procedures.

As used in the standard terminal instrument procedures prescribed in this part—

(a) "A" means alternate airport weather minimum.

(b) Aircraft approach category means a grouping of aircraft based on a speed of V_{REF}, if specified, or if V_{REF} is not specified, 1.3 V_{SO} at the maximum certificated landing weight. V_{REF}, V_{SO}, and the maximum certificated landing weight are those values as established for the aircraft by the certification authority of the country of registry.

(1) Category A: Speed less than 91 knots.

(2) Category B: Speed 91 knots or more but less than 121 knots.

(3) Category C: Speed 121 knots or more but less than 141 knots.

(4) Category D: Speed 141 knots or more but less than 166 knots.

(5) Category E: Speed 166 knots or more.

(c) Approach procedure segments for which altitudes (all altitudes prescribed are minimum altitudes unless otherwise specified) or courses, or both, are prescribed in procedures, are as follows:

(1) "Initial approach" is the segment between the initial approach fix and the intermediate fix or the point where the aircraft is established on the intermediate course or final approach course.

(2) "Initial approach altitude" means the altitude (or altitudes, in High Altitude Procedures) prescribed for the initial approach segment of an instrument approach.

(3) "Intermediate approach" is the segment between the intermediate fix or point and the final approach fix.

(4) "Final approach" is the segment between the final approach fix or point and the runway, airport, or missed-approach point.

(5) "Missed approach" is the segment between the missed-approach point, or point of arrival at decision height, and the missed-approach fix at the prescribed altitude.

(d) "C" means circling landing minimum, a statement of ceiling and visibility values, or minimum descent altitude and visibility, required for the circle-to-land maneuver.

(d-1) "Copter procedures" means helicopter procedures, with applicable minimums as prescribed in § 97.35 of this part. Helicopters may also use other procedures prescribed in Subpart C of this part and may use the Category A minimum descent altitude (MDA) or decision height (DH). The required visibility minimum may be reduced to one-half the published visibility minimum for Category A aircraft, but in no case may it be reduced to less than one-quarter mile or 1,200 feet RVR.

(e) "Ceiling minimum" means the minimum ceiling, expressed in feet above the surface of the airport, required for takeoff or required for designating an airport as an alternate airport.

(f) "d" means day.

(g) "FAF" means final approach fix.

(h) "HAA" means height above airport.

(h-1) "HAL" means height above a designated helicopter landing area used for helicopter instrument approach procedures.

(i) "HAT" means height above touchdown.

(j) "MAP" means missed approach point.

(k) "More than 65 knots" means an aircraft that has a stalling speed of more than 65 knots (as established in an approved flight manual) at maximum certificated landing weight with full flaps, landing gear extended, and power off.

(l) "MSA" means minimum safe altitude, an emergency altitude expressed in feet above mean sea level, which provides 1,000 feet clearance over all obstructions in that sector within 25 miles of the facility on which the procedure is based (LOM in ILS procedures).

(m) "n" means night.

(n) "NA" means not authorized.

(o) "NOPT" means no procedure turn required (altitude prescribed applies only if procedure turn is not executed).

(o-1) "Point in space approach" means a helicopter instrument approach procedure to a missed approach point that is more than 2,600 feet from an associated helicopter landing area.

(p) "Procedure turn" means the maneuver prescribed when it is necessary to reverse direction to establish the aircraft on an intermediate or final approach course. The outbound course, direction of turn, distance within which the turn must be completed, and minimum altitude are specified in the procedure. However, the point at which the turn may be commenced, and the type and rate of turn, is left to the discretion of the pilot.

(q) "RA" means radio altimeter setting height.

(r) "RVV" means runway visibility value.

(s) "S" means straight-in landing minimum, a statement of ceiling and visibility, minimum descent altitude and visibility, or decision height and visibility, required for a straight-in landing on a specified runway. The number appearing with the "S" indicates the runway to which the minimum applies. If a straight-in minimum is not prescribed in the procedure, the circling minimum specified applies to a straight-in landing.

(t) "Shuttle" means a shuttle, or race-track-type, pattern with 2-minute legs prescribed in lieu of a procedure turn.

(u) "65 knots or less" means an aircraft that has a stalling speed of 65 knots or less (as established in an approved flight manual) at maximum certificated landing weight with full flaps, landing gear extended, and power off.

(v) "T" means takeoff minimum.

(w) "TDZ" means touchdown zone.

(x) "Visibility minimum" means the minimum visibility specified for approach, or landing, or takeoff, expressed in statute miles, or in feet where RVR is reported.

§ 97.5 Bearings; courses; headings; radials; miles.

(a) All bearings, courses, headings, and radials in this part are magnetic.

(b) RVR values are stated in feet. Other visibility values are stated in statute miles. All other mileages are stated in nautical miles.

Subpart B—Procedures

§ 97.10 General.

This subpart prescribes standard instrument approach procedures other than those based on the criteria contained in the U.S. Standard for Terminal Instrument Approach Procedures (TERPS). Standard instrument approach procedures adopted by the FAA and described on FAA Form 3139 are incorporated into this part and made a part hereof as provided in 5 U.S.C. 522 (a) (1) and pursuant to 1 CFR Part 20. The incorporated standard instrument approach procedures are available for examination at the Rules Docket and at the National Flight Data Center, Federal Aviation Administration, 800 Independence Avenue, S.W., Washington, D.C. 20590. Copies of SIAPs adopted in a particular FAA Region are also available for examination at the Headquarters of that Region. Moreover, copies of SIAPs originating in a particular Flight Inspection District Office are available for examination at that Office. Based on the information contained on FAA Form 3139, standard instrument approach procedures are portrayed on charts prepared for the use of pilots by the United States Coast and Geodetic Survey and other publishers of aeronautical charts.

§ 97.11 Low or medium frequency range, automatic direction finding, and very high frequency omnirange procedures.

Section 609.100 of the Regulations of the Administrator is hereby designated as § 97.11.

§ 97.13 Terminal very high frequency omnirange procedures.

Section 609.200 of the Regulations of the Administrator is hereby designated as § 97.13.

§ 97.15 Very high frequency omnirange-distance measuring equipment procedures.

Section 609.300 of the Regulations of the Administrator is hereby designated as § 97.15.

§ 97.17 Instrument landing system procedures.

Section 609.400 of the Regulations of the Administrator is hereby designated as § 97.17.

§ 97.19 Radar procedures.

Section 609.500 of the Regulations of the Administrator is hereby designated as § 97.19.

Subpart C—TERPS procedures

§ 97.20 General.

(a) This subpart prescribes standard instrument procedures based on the criteria contained in FAA Order 8260.3B, "U.S. Standard for Terminal instrument Procedures (TERPS) (July 7, 1976) and FAA Order 8260.19C, "Flight Procedures and Airspace" (September 16, 1993). These standard instrument procedures and FAA Orders were approved for incorporation by reference by the Director of the Federal Register pursuant to 5 U.S.C. 552(a) and 1 CFR part 51. They may be examined at the following locations:

(1) FAA Orders 8260.3 and 8260.19 may be examined at the Federal Aviation Administration, Flight Standards Service, Flight Technologies and Procedures Division (AFS-420), 6500 S. MacArthur Blvd., Oklahoma City, OK, and at the Office of the Federal Register, 800 North Capitol Street, NW., suite 700, Washington, DC. These Orders are available for purchase from the U.S. Government Printing Office, 710 N. Capitol Street, NW. Washington, DC 20401.

(2) Standard instrument procedures may be examined at the Federal Aviation Administration National Flight Data Center (ATA-1lO), 800 Independence Avenue, SW., Washington, DC, and at the Office of the Federal Register, 800 North Capitol Street, NW. suite 700, Washington, DC.

(b) Standard instrument procedures and associated supporting data are documented on specific forms under FAA Order 8260.19C (September 16, 1993) and are promulgated by the FAA through the National flight Data Center (NFDC) as the source for aeronautical charts and avionics databases. These procedures are then portrayed on aeronautical charts and Included in avionics databases prepared by the National Aeronautical Charting Office (AVN—500) and other publishers of aeronautical data for use by pilots using the NFDC source data. The terminal aeronautical charts published by the U.S. Government were approved for incorporation by reference by the Director of the Federal Register pursuant to 5 U.S.C. 552(a) and 1 CFR part 51. They may be examined at the Federal Aviation Administration, National Flight Data Center (ATA-1lO), 800 Independence Avenue, SW., Washington, DC, and at the Office of the Federal Register, 800 North Capitol Street, NW., suite 700, Washington. DC. These charts are available for purchase from the FAA National Aeronautical Charting Office, Distribution Division AVN—530, 6303 Ivy Lane, Suite 400, Greenbelt, MD 20770.

§ 97.21 Low or medium frequency range (L/MF) procedures.

§ 97.23 Very high frequency omnirange (VOR) and very high frequency distance measuring equipment (VOR/DME) procedures.

§ 97.25 Localizer (LOC) and localizer-type directional aid (LDA) procedures.

§ 97.27 Nondirectional beacon (automatic direction finder) (NDB(ADF)) procedures.

§ 97.29 Instrument landing system (ILS) procedures.

§ 97.31 Precision approach radar (PAR) and airport surveillance radar (ASR) procedures.

§ 97.33 Area navigation (RNAV) procedures.

§ 97.35 Helicopter procedures.

NOTE: The procedures set forth in § 97.35 are not carried in the Code of Federal Regulations. For Federal Register citations affecting these procedures see List of CFR Sections Affected.

Part 99—Security control of air traffic

Subpart A—General

§ 99.1 Applicability.

(a) This subpart prescribes rules for operating civil aircraft (except for Department of Defense and law enforcement aircraft) in a defense area, or into, within, or out of the United States through an Air Defense Identification Zone (ADIZ), designated in Subpart B.

(b) Except for §§ 99.7, 99.13, and 99.15 this subpart does not apply to the operation of any aircraft—

(1) Within the 48 contiguous States and the District of Columbia, or within the State of Alaska, on a flight which remains within 10 nautical miles of the point of departure;

(2) Operating at true airspeed of less than 180 knots in the Hawaii ADIZ or over any island, or within 12 nautical miles or the coastline of any island, in the Hawaii ADIZ;

(3) Operating at true airspeed of less than 180 knots in the Alaska ADIZ while the pilot maintains a continuous listening watch on the appropriate frequency; or

(4) Operating at true airspeed of less than 180 knots in the Guam ADIZ.]

(c) An FAA ATC center may exempt the following operations from this subpart (except § 99.7) on a local basis only, with the concurrence of the U.S. military commanders concerned, or pursuant to an agreement with a U.S. Federal security or intelligence agency:

(1) Aircraft operations that are conducted wholly within the boundaries of an ADIZ and are not currently significant to the air defense system.

(2) Aircraft operations conducted in accordance with special procedures prescribed by the military authorities concerned.

§ 99.3 Definitions.

Aeronautical facility means, for the purposes of this subpart, a communications facility where flight plans or position reports are normally filed during flight operations.

Air defense identification zone (ADIZ) means an area of airspace over land or water in which the ready identification, location, and control of all aircraft (except for Department of Defense and law enforcement aircraft) is required in the interest of national security.

Defense area means any airspace of the contiguous United States that is not an ADIZ in which the control of aircraft is required for reasons of national security.

Defense visual flight rules (DVFR) flight means, for the purposes of this subpart, a flight within an *ADIZ* conducted by any aircraft (except for Department of Defense and law enforcement aircraft) under the visual flight rules in part 91 of this title.

§ 99.5 Emergency situations.

In an emergency that requires immediate decision and action for the safety of the flight, the pilot in command of an aircraft may deviate from the rules in this part to the extent required by that emergency. He shall report the reasons for the deviation to the communications facility where flight plans or position reports are normally filed (referred to in this part as "an appropriate aeronautical facility") as soon as possible.

§ 99.7 Special security instructions.

Each person operating an aircraft in an ADIZ or Defense Area must, in addition to the applicable rules of this part, comply with special security instructions issued by the Administrator in the interest of national security, pursuant to agreement between the FAA and the Department of Defense, or between the FAA and a U.S. Federal security or intelligence agency.

§ 99.9 Radio requirements.

(a) A person who operates a civil aircraft into an ADIZ must have a functioning two-way radio, and the pilot must maintain a continuous listening watch on the appropriate aeronautical facility's frequency.

(b) No person may operate an aircraft into, within, or whose departure point is within an ADIZ unless—.

(1) The person files a DVFR flight plan containing the time and point of ADIZ penetration, and

(2) The aircraft departs within five minutes of the estimated departure time contained in the flight plan.

(c) If the pilot operating an aircraft under DVFR in an ADIZ cannot maintain two-way radio communications, the pilot may proceed, in accordance with original DVFR flight plan, or land as soon as practicable. The pilot must report the radio failure to an appropriate aeronautical facility as soon as possible.

(d) If a pilot operating an aircraft under IFR in an ADIZ cannot maintain two-way radio communications, the pilot must proceed in accordance with §91.185 of this chapter.

§ 99.11 ADIZ flight plan requirements.

(a) No person may operate an aircraft into, within, or from a departure point within an ADIZ, unless the person files, activates, and closes a flight plan with the appropriate aeronautical facility, or is otherwise authorized by air traffic control.

(b) Unless ATC authorizes an abbreviated flight plan—

(1) A flight plan for IFR flight must contain the information specified in [§ 91.169]; and

(2) A flight plan for VFR flight must contain the information specified in [§ 91.153 (a) (1) through (6)].

(3) If airport of departure is within the Alaskan ADIZ and there is no facility for filing a flight plan then:

(i) Immediately after takeoff or when within range of an appropriate aeronautical facility, comply with provisions of paragraph (b)(1) or (b)(2) as appropriate.

(ii) Proceed according to the instructions issued by the appropriate aeronautical facility.

(c) The pilot shall designate a flight plan for VFR flight as a DVFR flight plan.

(d) The pilot in command of an aircraft for which a flight plan has been filed must file an arrival or completion notice with an appropriate aeronautical facility.

§ 99.12 Reserved.

§ 99.13 Transponder-on requirements.

[(a) *Aircraft transponder-on operation.* Each person operating an aircraft into or out of the United States into, within, or across an ADIZ designated in Subpart B of this part, if that aircraft is equipped with an operable radar beacon transponder, shall operate the transponder, including altitude encoding equipment if installed, and shall reply on the appropriate code or as assigned by ATC.

[(b) *ATC transponder equipment and use.* Effective September 7, 1990, unless otherwise authorized by ATC, no person may operate a civil aircraft into or out of the United States into, within, or across the contiguous U.S. ADIZ designated in Subpart B of this part unless that aircraft is equipped with a coded radar beacon transponder.

[(c) *ATC transponder and altitude reporting equipment and use.* Effective December 30, 1990, unless otherwise authorized by ATC, no person may operate a civil aircraft into or out of the United States into, within, or across the contiguous U.S. ADIZ unless that aircraft is equipped with a coded radar beacon transponder and automatic pressure altitude reporting equipment having altitude reporting capability that automatically replies to interrogations by transmitting pressure altitude information in 100-foot increments.

[(d) Paragraphs (b) and (c) of this section do not apply to the operation of an aircraft which was not originally certificated with an engine-driven electrical system and which has not subsequently been certified with such a system installed, a balloon, or a glider.]

§ 99.15 Position reports.

(a) The pilot of an aircraft operating in or penetrating an ADIZ under IFR—

(1) In controlled airspace, must make the position reports required in § 91.183; and

(2) In uncontrolled airspace, must make the position reports required in this section.

(b) No pilot may operate an aircraft penetrating an ADIZ under DVFR unless——

(1) The pilot reports to an appropriate aeronautical facility before penetration: the time, position, and altitude at which the aircraft passed the last reporting point before penetration and the estimated time of arrival over the next appropriate reporting point along the flight route;

(2) If there is no appropriate reporting point along the flight route, the pilot reports at least 15 minutes before penetration: The estimated time, position. and altitude at which the pilot will penetrate; or

(3) If the departure airport is within an ADIZ or so close to the ADIZ boundary that it prevents the pilot from complying with paragraphs (b)(1) or (2) of this section, the pilot must report immediately after departure: the time of departure, the altitude, and the estimated time of arrival over the first reporting point along the flight route.

(c) In addition to any other reports as ATC may require, no pilot in command of a foreign civil aircraft may enter the United States through an ADIZ unless that pilot makes the reports required in this section or reports the position of the aircraft when it is not less that one hour and not more that 2 hours average direct cruising distance from the United States.

§ 99.17 Deviation from flight plans and ATC clearances and instructions.

(a) No pilot may deviate from the provisions of an ATC clearance or ATC, instruction except in accordance with § 91.123 of this chapter.

(b) No pilot may deviate from the filed IFR flight plan when operating an aircraft in uncontrolled airspace unless that pilot notifies an appropriate aeronautical facility before deviating.

(c) No pilot may deviate from the filed DVFR flight plan unless that pilot notifies an appropriate aeronautical facility before deviating,

§ 99.19 Reserved

§ 99.21 Reserved

§ 99.23 Reserved

§ 99.27 Reserved

§ 99.29 Reserved

§ 99.31 Reserved

Subpart B—Designated air defense identification zones

§ 99.41 General

The airspace above the areas described in this subpart is established as an ADIZ Defense Area. The lines between points described in this subpart are great circles except that the lines joining adjacent points on the same parallel of latitude are rhumb lines.

§ 99.43 Contiguous U.S. ADIZ.

The area bounded by a line from
43°15'N; 65°55'W; 44°21'N; 67°16'W;
43°10'N; 69°40W; 41°05'N; 69°40'W;
40°32'N; 72°15'W; 39°55'N; 73°00'W;
39°38'N; 73°00'W; 39°36'N; 73°40'W;
37°00'N; 75°30~W; 36°10'N; 75°10'W;
35°10'N; 75°10'W; 32°00'N; 80°30'W;
30°30'N; 81°00'W; 26°40'N; 79°40'W;
25°00'N; 80°05'W; 24°25'N; 81°15'W;
24°20'N; 81°45'W; 24°30'N; 82°06W';
24°41'N; 82°06'W; 24°43'N; 82°00'W;
25°00'N; 81°30'W; 25°10'N; 81°23'W;
25°35'N; 81°30'W; 26°15'N; 82°20'W;
27°50'N; 83°05'W; 28°55'N; 83°30'W;
29°42'N; 84°00'W; 29°20'N; 85°00'W;
30°00'N; 87°10'W; 30°00'N; 88°30'W;
28°45'N; 86°55'W; 26°45'N; 90°00'W;
29°25'N; 94°00'W; 28°20'N; 96°00'W;
27°30'N; 97°00'W; 26°00'N; 97°00'W;
25°58'N; 97°07'W; westward along the
US/Mexico border to 32°32'03'N;
117°07'25'W; 32°30'N; 117°25'W
32°35'N; 118°30'W; 33°05'N; 119°45'W;
33°55'N; 120°40'W; 34°50'N; 121°10'W;
38°50'N; 124°00'W; 40°00'N; 124°35'W;
40°25'N; 124°40'W; 42°50'N; 124°50'W;
46°15'N; 124°30'W; 48°30'N; 125°00'W;
48°20'N; 128°00'W; 48°20'N; 132°00'W;
37°42'N; 130°40'W; 29°00'N; 124°00'W;
30°45'N; 120°50'W; 32°00'N; 118°24'W;
32°30'N; 117°20'W; 32°32'03'N;
117°07'25'W; eastward along the U.S./
Mexico border to 25°58'N; 97°07'W;
26°00'N; 97°00'W; 26°00'N; 95°00'W;
26°30'N; 95°00'W; then via 26°30'N;
parallel to 26°30'N; 84°00'W; 24°00'N;
83°00'W; then via 24°00'N; parallel to
24°00'N; 79°25'W; 25°40'N; 79°25'W;
27°30'N; 78°50'W; 30°45'N; 74°00'W;
39°30'N; 63°45'W; 43°00'N; 65°48'W; to
point of beginning.

§ 99.45 Alaska ADIZ.

The area is bounded by a line from
54°00'N; 136°00'W; 55°57'N; 144°00'W;
57°00'N; 145°00'W; 53°00'N; 158°00'W;
50°00'N; 169°00'W; 50°00'N; 180°00';
50°00'N; 170°00'E; 53°00'N; 170°00'E;

60°00'N; 180°00'; 65°00'N; 169°00'W;
then along 169°00'W; to 75°00'N; 169°00'W;
then along the 75°00'N parallel to 75°00'N;
141°00W; 69°50'N; 141°00'W; 71°18'N;
156°44'W; 68°40'N; 167°10'W; 67°00'N;
165°00'W; 65°40'N; 168°15'\V; 63°45'N;
165°30'W; 61°20'N; 166°40'W; 59°00'N;
163°00'W; then south along 163°00'W to 54°00'N;
163°00'W; 56°30'N; 154°00"W; 59°20'N;
146°00'W; 59°30'N; 140°00'W; 57°00'N;
136°00'W; 54°35'N, 133°00'W; to point
of beginning.

§ 99.47 Guam ADIZ.

(a) *Inner boundary.* From a point 13°52'07"N, 143°59'16"E, counterclockwise along the 50-nautical-mile radius arc of the NIMITZ VORTAC located at 13°27'11"N, 144°43'51"E); to a point 13°02'08"N, 145°28'17"E; then to a point 14°49'07"N, 146°13'58"E; counterclockwise

along the 35-nautical-mile radius arc of the SAIPAN NDB (located at 15°06'46"N, 145°42'42"E); to a point 15°24'21"N, 145°11'21"E; then to point of origin.

(b) *Outer boundary.* The area bounded by a circle with a radius of 250 NM centered at latitude 13°32'41"N, longitude 144°50'30"E.

§ 99.49 Hawaii ADIZ.

(a) *Outer boundary.* The area included in the irregular octagonal figure formed by a line connecting 26°30'N, 156°00'W; 26°30'N, 161°00'W; 24°00'N, 164°00'W; 20°00'N, 164°00'W; 17°00'N, 160°00'W; 17°00'N, 156°00'W; 20°00'N, 153°00'W; 22°00'N, 153°00'W; to point of orgin.

(b) *Inner boundary.* The inner boundary to follow a line connecting 22°30'N, 157°00'W; 22°30'N, 160°00'W; 22°00'N, 161°00'W; 21°00'N, 161°00'W; 20°00'N, 160°00'W; 20°00'N, 156°30'W; 21°00'N, 155°30'W; to point of origin.

Part 103—Ultralight vehicles

Subpart A—General

Source: Docket No. 21631 (47 FR 38776, 9/2/82) effective 10/4/82, for each subpart, unless otherwise noted.

§ 103.1 Applicability.

This part prescribes rules governing the operation of ultralight vehicles in the United States. For the purposes of this part, an ultralight vehicle is a vehicle that:

(a) Is used or intended to be used for manned operation in the air by a single occupant;

(b) Is used or intended to be used for recreation or sport purposes only;

(c) Does not have any U.S. or foreign airworthiness certificate; and

(d) If unpowered, weighs less than 155 pounds; or

(e) If powered:

(1) Weighs less than 254 pounds empty weight, excluding floats and safety devices which are intended for deployment in a potentially catastrophic situation;

(2) Has a fuel capacity not exceeding 5 U.S. gallons;

(3) Is not capable of more than 55 knots calibrated airspeed at full power in level flight; and

(4) Has a power-off stall speed which does not exceed 24 knots calibrated airspeed.

§ 103.3 Inspection requirements.

(a) Any person operating an ultralight vehicle under this part shall, upon request, allow the Administrator, or his designee, to inspect the vehicle to determine the applicability of this part.

(b) The pilot or operator of an ultralight vehicle must, upon request of the Administrator, furnish satisfactory evidence that the vehicle is subject only to the provisions of this part.

§ 103.5 Waivers.

No person may conduct operations that require a deviation from this part except under a written waiver issued by the Administrator.

§ 103.7 Certification and registration.

(a) Notwithstanding any other section pertaining to certification of aircraft or their parts or equipment, ultralight vehicles and their component parts and equipment are not required to meet the airworthiness certification standards specified for aircraft or to have certificates of airworthiness.

(b) Notwithstanding any other section pertaining to airman certification, operators of ultralight vehicles are not required to meet any aeronautical knowledge, age, or experience requirements to operate those vehicles or to have airman or medical certificates.

(c) Notwithstanding any other section pertaining to registration and marking of aircraft, ultralight vehicles are not required to be registered or to bear markings of any type.

Subpart B—Operating rules

§ 103.9 Hazardous operations.

(a) No person may operate any ultralight vehicle in a manner that creates a hazard to other persons or property.

(b) No person may allow an object to be dropped from an ultralight vehicle if such action creates a hazard to other persons or property.

§ 103.11 Daylight operations.

(a) No person may operate an ultralight vehicle except between the hours of sunrise and sunset.

(b) Notwithstanding paragraph (a) of this section, ultralight vehicles may be operated during the twilight periods 30 minutes before official sunrise and 30 minutes after official sunset or, in Alaska, during the period of civil twilight as defined in the Air Almanac, if:

(1) The vehicle is equipped with an operating anticollision light visible for at least 3 statute miles; and

(2) All operations are conducted in uncontrolled airspace.

§ 103.13 Operation near aircraft; right-of-way rules.

(a) Each person operating an ultralight vehicle shall maintain vigilance so as to see and avoid aircraft and shall yield the right-of-way to all aircraft.

(b) No person may operate an ultralight vehicle in a manner that creates a collision hazard with respect to any aircraft.

(c) Powered ultralights shall yield the right-of-way to unpowered ultralights.

§ 103.15 Operations over congested areas.

No person may operate an ultralight vehicle over any congested area of a city, town, or settlement, or over any open air assembly of persons.

§ 103.17 Operations in certain airspace.

No person may operate an ultralight vehicle within Class A, Class B, Class C, or Class D airspace or within the lateral boundaries of the surface area of Class E airspace designated for an airport unless that person has prior authorization from the ATC facility having jurisdiction over that airspace.]

[(Amdt. 103-4, Eff. 9/16/93)]

§ 103.19 Operations in prohibited or restricted areas.

No person may operate an ultralight vehicle in prohibited or restricted areas unless that person has permission from the using or controlling agency, as appropriate.

§ 103.20 Flight restrictions in the proximity of certain areas designated by notice to airmen.

No person may operate an ultralight vehicle in areas designated in a Notice to Airmen under § 91.137, § 91.138, § 91.141, § 91.143 or § 91.145 of this chapter, unless authorized by:

(a) Air Traffic Control (ATC); or

(b) A Flight Standards Certificate of Waiver or Authorization issued for the demonstration or event.

§ 103.21 Visual reference with the surface.

No person may operate an ultralight vehicle except by visual reference with the surface.

§ 103.23 Flight visibility and cloud clearance requirements.

[No person may operate an ultralight vehicle when the flight visibility or distance from clouds is less than that in the table found on page 508. All operations in Class A, Class B, Class C, and Class D airspace or Class E airspace designated for an airport must receive prior ATC authorization as required in § 103.17 of this part.

See basic VFR weather minimums on page 592.

BASIC VFR WEATHER MINIMUMS

Airspace	Flight visibility	Distance from clouds
Class A	Not Applicable	Not Applicable.
Class B	3 statute miles	Clear of Clouds.
Class C	3 statute miles	500 feet below. 1,000 feet above. 2,000 feet horizontal.
Class D	3 statute miles	500 feet below. 1,000 feet above. 2,000 feet horizontal.
Class E		
Less than 10,000 feet MSL	3 statute miles	500 feet below. 1,000 feet above. 2,000 feet horizontal.
At or above 10,000 feet MSL	5 statute miles	1,000 feet below. 1,000 feet above. 1 statute mile horizontal.
Class G		
1,200 feet or less above the surface (regardless of MSL altitude)		
Day, except as provided in § 91.155(b)	1 statute mile	Clear of clouds.
Night, except as provided in § 91.155(b)	3 statute miles	500 feet below. 1,000 feet above. 2,000 feet horizontal.
More than 1,200 feet above the surface but less than 10,000 feet MSL		
Day	1 statute mile	500 feet below. 1,000 feet above. 2,000 feet horizontal.
Night	3 statute miles	500 feet below. 1,000 feet above. 2,000 feet horizontal.
More than 1,200 feet above the surface and at or above 10,000 feet MSL	5 statute miles	1,000 feet below. 1,000 feet above. 1 statute mile horizontal.

Part 105—Parachute operations

Authority: 49 U.S.C. 106(g), 40113-40114, 44701-44702, 44721.

Source: Doc. No. FAA-1999-5483, 66 FR 23553, May 9, 2001, unless otherwise noted.

Subpart A—General

§ 105.1 Applicability.

(a) Except as provided in paragraphs (b) and (c) of this section, this part prescribes rules governing parachute operations conducted in the United States.

(b) This part does not apply to a parachute operation conducted—

(1) In response to an in-flight emergency, or

(2) To meet an emergency on the surface when it is conducted at the direction or with the approval of an agency of the United States, or of a State, Puerto Rico, the District of Columbia, or a possession of the United States, or an agency or political subdivision thereof.

(c) Sections 105.5, 105.9, 105.13, 105.15, 105.17, 105.19 through 105.23, 105.25(a)(1) and 105.27 of this part do not apply to a parachute operation conducted by a member of an Armed Force—

(1) Over or within a restricted area when that area is under the control of an Armed Force.

(2) During military operations in uncontrolled airspace.

§ 105.3 Definitions.

For the purposes of this part—

Approved parachute means a parachute manufactured under a type certificate or a Technical Standard Order (C-23 series), or a personnel-carrying U.S. military parachute (other than a high altitude, high speed, or ejection type) identified by a Navy Air Facility, an Army Air Field, and Air Force-Navy drawing number, an Army Air Field order number, or any other military designation or specification number.

Automatic Activation Device means a self-contained mechanical or electro-mechanical device that is attached to the interior of the reserve parachute container, which automatically initiates parachute deployment of the reserve parachute at a pre-set altitude, time, percentage of terminal velocity, or combination thereof.

Direct Supervision means that a certificated rigger personally observes a non-certificated person packing a main parachute to the extent necessary to ensure that it is being done properly, and takes responsibility for that packing.

Drop Zone means any pre-determined area upon which parachutists or objects land after making an intentional parachute jump or drop. The center-point target of a drop zone is expressed in nautical miles from the nearest VOR facility when 30 nautical miles or less; or from the nearest airport, town, or city depicted on the appropriate Coast and Geodetic Survey World Aeronautical Chart or Sectional Aeronautical Chart, when the nearest VOR facility is more than 30 nautical miles from the drop zone.

Foreign parachutist means a parachutist who is neither a U.S. citizen or a resident alien and is participating in parachute operations within the United States using parachute equipment not manufactured in the United States. Freefall means the portion of a parachute jump or drop between aircraft exit and parachute deployment in which the parachute is activated manually by the parachutist at the parachutist's discretion or automatically, or, in the case of an object, is activated automatically.

Main parachute means a parachute worn as the primary parachute used or intended to be used in conjunction with a reserve parachute.

Object means any item other than a person that descends to the surface from an aircraft in flight when a parachute is used or is intended to be used during all or part of the descent.

Parachute drop means the descent of an object to the surface from an aircraft in flight when a parachute is used or intended to be used during all or part of that descent.

Parachute jump means a parachute operation that involves the descent of one or more persons to the surface from an aircraft in flight when a aircraft is used or intended to be used during all or part of that descent.

Parachute operation means the performance of all activity for the purpose of, or in support of, a parachute jump or a parachute drop. This parachute operation can involve, but is not limited to, the following persons: parachutist, parachutist in command and passenger in tandem parachute operations, drop zone or owner or operator, jump master, certificated parachute rigger, or pilot. Parachutist means a person who intends to exit an aircraft while in flight using a single-harness, dual parachute system to descend to the surface.

Parachutist in command means the person responsible fro the operation and safety of a tandem parachute operation.

Passenger parachutist means a person who boards an aircraft, acting as other than the parachutist in command of a tandem parachute operation, with the intent of existing the aircraft while in-flight using the forward harness of a dual harness tandem parachute system to descend to the surface.

Pilot chute means a small parachute used to initiate and/or accelerate deployment of a main or reserve parachute.

Ram-air parachute means a parachute with a canopy consisting of an upper and lower surface that is inflated by ram air entering through specially designed openings in the front of the canopy to form a gliding airfoil.

Reserve parachute means an approved parachute worn for emergency use to be activated only upon failure of the main parachute or in any other emergency where use of the main parachute is impractical or use of the main parachute would increase risk.

Single-harness, dual parachute system: means the combination of a main parachute, approved reserve parachute, and approved single person harness and dual-parachute container. This parachute system may have an operational automatic activation device installed.

Tandem parachute operation: means a parachute operation in which more than one person simultaneously uses the same tandem parachute system while descending to the surface from an aircraft in flight.

Tandem parachute system: means the combination of a main parachute, approved reserve parachute, and approved harness and dual parachute container, and a separate approved forward harness for a passenger parachutist. This parachute system must have an operational automatic activation device installed.

§ 105.5 General.

No person may conduct a parachute operation, and no pilot in command of an aircraft may allow a parachute operation to be conducted from an aircraft, if that operation creates a hazard to air traffic or to persons or property on the surface.

§ 105.7 Use of alcohol and drugs.

No person may conduct a parachute operation, and no pilot in command of an aircraft may allow a person to conduct a parachute operation from that aircraft, if that person is or appears to be under the influence of—

(a) Alcohol, or

(b) Any drug that affects that person's faculties in any way contrary to safety.

§ 105.9 Inspections.

The Administrator may inspect any parachute operation to which this part applies (including inspections at the site where the parachute operation is being conducted) to determine compliance with the regulations of this part.

Subpart B—Operating rules

§ 105.13 Radio equipment and use requirements.

(a) Except when otherwise authorized by air traffic control —

(1) No person may conduct a parachute operation, and no pilot in command of an aircraft may allow a parachute operation to be conducted from that aircraft, in or into controlled airspace unless, during that flight—

(i) The aircraft is equipped with a functioning two-way radio communication system appropriate to the air traffic control facilities being used; and

(ii) Radio communications have been established between the aircraft and the air traffic control facility having jurisdiction over the affected airspace of the first intended exit altitude at least 5 minutes before the parachute operation begins. The pilot in command must establish radio communications to receive information regarding air traffic activity in the vicinity of the parachute operation.

(2) The pilot in command of an aircraft used for any parachute operation in or into controlled airspace must, during each flight—

(i) Continuously monitor the appropriate frequency of the aircraft's radio communications system from the time radio communications are first established between the aircraft and air traffic control, until the pilot advises air traffic control that the parachute operation has ended for that flight.

(ii) Advise air traffic control when the last parachutist or object leaves the aircraft.

(b) Parachute operations must be aborted if, prior to receipt of a required air traffic control authorization, or during any parachute operation in or into controlled airspace, the required radio communications system is or becomes inoperative.

§ 105.15 Information required and notice of cancellation or postponement of a parachute operation.

(a) Each person requesting an authorization under §§ 105.21(b) and 105.25(a)(2) of this part and each person submitting a notification under § 105.25(a)(3) of this part must provide the following information (on an individual or group basis):

(1) The date and time the parachute operation will begin.

(2) The radius of the drop zone around the target expressed in nautical miles.

(3) The location of the center of the drop zone in relation to —

(i) The nearest VOR facility in terms of the VOR radial on which it is located and its distance in nautical miles from the VOR facility when that facility is 30 nautical miles or less from the drop zone target; or

(ii) the nearest airport, town, or city depicted on the appropriate Coast and Geodetic Survey World Aeronautical Chart or Sectional Aeronautical Chart, when the nearest VOR facility is more than 30 nautical miles from the drop zone target.

(4) Each altitude above mean sea level at which the aircraft will be operated when parachutists or objects exist the aircraft.

(5) The duration of the intended parachute operation.

(6) The name, address, and telephone number of the person who requests the authorization or gives notice of the parachute operation.

(7) The registration number of the aircraft to be used.

(8) The name of the air traffic control facility with jurisdiction of the airspace at the first intended exit altitude to be used for the parachute operation.

(b) Each holder of a certificate of authorization issued under §§ 105.21(b) and 105.25(b) of this part must present that certificate for inspection upon the request of the Administrator or any Federal, State, or local official.

(c) Each person requesting an authorization under §§ 105.21(b) and 105.25(a)(2) of this part and each person submitting a notice under § 105.25(a)(3) of this part must promptly notify the air traffic control facility having jurisdiction over the affected airspace if the proposed or scheduled parachute operation is canceled or postponed.

§ 105.17 Flight visibility and clearance from cloud requirements.

No person may conduct a parachute operation, and no pilot in command of an aircraft may allow a parachute operation to be conducted from that aircraft—

(a) Into or through a cloud, or

(b) When the flight visibility or the distance from any cloud is less than that prescribed in the following table:

Altitude	Flight visibility (statute miles)	Distance from clouds
1,200 feet or less above the surface regardless of the MSL altitude.	3	500 feet below, 1,000 feet above, 2,000 feet horizontal.
More than 1,200 feet above the surface but less than 10,000 feet MSL	3	500 feet below, 1,000 feet above, 2,000 feet horizontal.
More than 1,200 feet above the surface and at or above 10,000 feet MSL	5	1,000 feet below, 1,000 feet above, 1 mile horizontal.

§ 105.19 Parachute operations between sunset and sunrise.

(a) No person may conduct a parachute operation, and no pilot in command of an aircraft may allow a person to conduct a parachute operation from an aircraft between sunset and sunrise, unless the person or object descending from the aircraft displays a light that is visible for at least 3 statute miles.

(b) The light required by paragraph (a) of this section must be displayed from the time that the person or object is under a properly functioning open parachute until that person or object reaches the surface.

§ 105.21 Parachute operations over or into a congested area or an open-air assembly of persons.

(a) No person may conduct a parachute operation, and no pilot in command of an aircraft may allow a parachute operation to be conducted from that aircraft, over or into a congested area of a city, town, or settlement, or an open-air assembly of persons unless a certificate of authorization for that parachute operation has been issued under this section. However, a parachutist may drift over a congested area or an open-air assembly of persons with a fully deployed and properly functioning parachute if that parachutist is at a sufficient altitude to avoid creating a hazard to persons or property on the surface.

(b) An application for a certificate of authorization issued under this section must—

(1) Be made in the form and manner prescribed by the Administrator, and

(2) Contain the information required in § 105.15(a) of this part.

(c) Each holder of, and each person named as a participant in a certificate of authorization issued under this section must comply with all requirements contained in the certificate of authorization.

(d) Each holder of a certificate of authorization issued under this section must present that certificate for inspection upon the request of the Administrator, or any Federal, State, or local official.

§ 105.23   Parachute operations over or onto airports.

No person may conduct a parachute operation, and no pilot in command of an aircraft may allow a parachute operation to be conducted from that aircraft, over or onto any airport unless—

(a) For airports with an operating control tower:

(1) Prior approval has been obtained from the management of the airport to conduct parachute operations over or on that airport.

(2) Approval has been obtained from the control tower to conduct parachute operations over or onto that airport.

(3) Two-way radio communications are maintained between the pilot of the aircraft involved in the parachute operation and the control tower of the airport over or onto which the parachute operation is being conducted.

(b) For airports without an operating control tower, prior approval has been obtained from the management of the airport to conduct parachute operations over or on that airport.

(c) A parachutist may drift over that airport with a fully deployed and properly functioning parachute if the parachutist is at least 2,000 feet above that airport's traffic pattern, and avoids creating a hazard to air traffic or to persons and property on the ground.

§ 105.25   Parachute operations in designated airspace.

(a) No person may conduct a parachute operation, and no pilot in command of an aircraft may allow a parachute operation to be conducted from that aircraft —

(1) Over or within a restricted area or prohibited area unless the controlling agency of the area concerned has authorized that parachute operation;

(2) Within or into a Class A, B, C, D airspace area without, or in violation of the requirements of, an air traffic control authorization issued under this section;

(3) Except as provided in paragraph (c) and (d) of this section, within or into Class E or G airspace area unless the air traffic control facility having jurisdiction over the airspace at the first intended exit altitude is notified of the parachute operation no earlier than 24 hours before or no later than 1 hour before the parachute operation begins.

(b) Each request for a parachute operation authorization or notification required under this section must be submitted to the air traffic control facility having jurisdiction over the airspace at the first intended exit altitude and must include the information prescribed by § 105.15(a) of this part.

(c) For the purposes of paragraph (a)(3) of this section, air traffic control facilities may accept a written notification from an organization that conducts parachute operations and lists the scheduled series of parachute operations to be conducted over a stated period of time not longer than 12 calendar months. The notification must contain the information prescribed by § 105.15(a) of this part, identify the responsible persons associated with that parachute operation, and be submitted at least 15 days, but not more than 30 days, before the parachute operation begins. The FAA may revoke the acceptance of the notification for any failure of the organization conducting the parachute operations to comply with its requirements.

(d) Paragraph (a)(3) of this section does not apply to a parachute operation conducted by a member of an Armed Force within a restricted area that extends upward from the surface when that area is under the control of an Armed Force.

Subpart C—Parachute equipment and packing

§ 105.41   Applicability.

This subpart prescribed rules governing parachute equipment used in civil parachute operations.

§ 105.43   Use of single-harness, dual-parachute systems.

No person may conduct a parachute operation using a single-harness, dual-parachute system, and no pilot in command of an aircraft may allow any person to conduct a parachute operation from that aircraft using a single-harness, dual-parachute system, unless that system has at least one main parachute, one approved reserve parachute, and one approved single person harness and container that are packed as follows:

(a) The main parachute must have been packed within 120 days before the date of its use of a certificated parachute rigger, the person making the next jump with that parachute, or a non-certificated person under the direct supervision of a certification parachute rigger.

(b) The reserve parachute must have been packed by a certificated parachute rigger —

(1) Within 120 days before the date of its use, if its canopy, shroud, and harness are composed exclusively of

nylon, rayon, or similar synthetic fiber or material that is substantially resistant to damage from mold, mildew, and other fungi, and other rotting agents propagated in a moist environment; or

(2) Within 60 days before the date of its use, if it is composed of any amount of silk, pongee, or other natural fiber, or material not specified in paragraph

(b)(1) of this section.

(c) If installed, the automatic activation device must be maintained in accordance with manufacturer instructions for that automatic activation device.

§105.45 Use of tandem parachute systems.

(a) No person may conduct a parachute operation using a tandem parachute system, and no pilot in command of an aircraft may allow any person to conduct a parachute operation from that aircraft using a tandem parachute system, unless

(1) One of the parachutists using the tandem parachute system is the parachutist in command, and meets the following requirements:

(i) Has a minimum of 3 years of experience in parachuting, and must provide documentation that the parachutist —

(ii) Has completed a minimum of 500 freefall parachute jumps using a ram-air parachute, and

(iii) Holds a master parachute license issued by an organization recognized by the FAA, and

(iv) Has successfully completed a tandem instructor course given by the manufacturer of the tandem parachute system used in the parachute operation or a course acceptable to the Administrator.

(v) Has been certified by the appropriate parachute manufacturer or tandem course provider as being properly trained on the use of the specific tandem parachute system to be used.

(2) The person acting as parachutist in command:

(i) Has briefed the passenger parachutist before boarding the aircraft. The briefing must include the procedures to be used in case of an emergency with the aircraft or after exiting the aircraft, while preparing to exit and exiting the aircraft, freefall, operating the parachute after freefall, landing approach, and landing.

(ii) Uses the harness position prescribed by the manufacturer of the tandem parachute equipment.

(b) No person may make a parachute jump with a tandem parachute system unless —

(1) The main parachute has been packed by a certificated parachute rigger, the parachutist in command making the next jump with that parachute, or a person under the direct supervision of a certificated parachute rigger.

(2) The reserve parachute has been packed by a certificated parachute rigger in accordance with §105.43(b) of this part.

(3) The tandem parachute system contains an operational automatic activation device for the reserve parachute,

approved by the manufacturer of that tandem parachute system. The device must—

(i) Have been maintained in accordance with manufacturer instructions, and

(ii) Be armed during each tandem parachute operation.

(4) The passenger parachutist is provided with a manual main parachute activation device and instructed on the use of that device, if required by the owner/operator.

(5) The main parachute is equipped with a single-point release system.

(6) The reserve parachute meets Technical Standard Order C23 specifications.

§105.47 Use of static lines.

(a) Except as provided in paragraph (c) of this section, no person may conduct a parachute operation using a static line attached to the aircraft and the main parachute unless an assist device, described and attached as follows, is used to aid the pilot chute in performing its function, or, if no pilot chute is used, to aid in the direct deployment of the main parachute canopy. The assist device must—

(1) Be long enough to allow the main parachute container to open before a load is placed on the device.

(2) Have a static load strength of—

(i) At least 28 pounds but not more than 160 pounds if it is used to aid the pilot chute in performing its function; or

(ii) At least 56 pounds but not more than 320 pounds if it is used to aid in the direct deployment of the main parachute canopy; and

(3) Be attached as follows:

(i) At one end, to the static line above the static-line pins or, if static-line pins are not used, above the static-line ties to the parachute cone.

(ii) At the other end, to the pilot chute apex, bridle cord, or bridle loop, or, if no pilot chute is used, to the main parachute canopy.

(b) No person may attach an assist device required by paragraph (a) of this section to any main parachute unless that person is a certificated parachute rigger or that person makes the next parachute jump with that parachute.

(c) An assist device is not required for parachute operations using direct-deployed, ram-air parachutes.

§105.49 Foreign parachutists and equipment.

(a) No person may conduct a parachute operation, and no pilot in command of an aircraft may allow a parachute operation to be conducted from that aircraft with an unapproved foreign parachute system unless—

(1) The parachute system is worn by a foreign parachutist who is the owner of that system.

(2) The parachute system is of a single-harness dual parachute type.

(3) The parachute system meets the civil aviation authority requirements of the foreign parachutist's country.

(4) All foreign non-approved parachutes deployed by a foreign parachutist during a parachute operation conducted under this section shall be packed as follows—

(i) The main parachute must be packed by the foreign parachutist making the next parachute jump with that parachute, a certificated parachute rigger, or any other person acceptable to the Administrator.

(ii) The reserve parachute must be packed in accordance with the foreign parachutist's civil aviation authority requirements, by a certificated parachute rigger, or any other person acceptable to the Administrator.

Part 119—Certification:
Air carriers and commercial operators

Subpart A—General

§ 119.1 Applicability.

(a) This part applies to each person operating or intending to operate civil aircraft—

(1) As an air carrier or commercial operator, or both, in air commerce; or

(2) When common carriage is not involved, in operations of U.S.-registered civil airplanes with a seat configuration of 20 or more passengers, or a maximum payload capacity of 6,000 pounds or more.

(b) This part prescribes—

(1) The types of air operator certificates issued by the Federal Aviation Administration, including air carrier certificates and operating certificates;

(2) The certification requirements an operator must meet

in order to obtain and hold a certificate authorizing operations under part 121, 125, or 135 of this chapter and operations specifications for each kind of operation to be conducted and each class and size of aircraft to be operated under part 121 or 135 of this chapter:

(3) The requirements an operator must meet to conduct operations under part 121, 125, or 135 of this chapter and in operating each class and size of aircraft authorized in its operations specifications;

(4) Requirements affecting wet leasing of aircraft and other arrangements for transportation by air;

(5) Requirements for obtaining deviation authority to perform operations under a military contract and obtaining deviation authority to perform emergency operation; and

(6) Requirements for management personnel for operations conducted under part 121 or part 135 of this chapter.

(c) Persons subject to this part must comply with the other requirements of this chapter, except where those requirements are modified by or where additional requirements are imposed by part 119, 121, 125, or 135 of this chapter.

(d) This part does not govern operations conducted under part 91, subpart K (when common carriage is not involved) nor does it govern operations conducted under part 129, 133, 137, or 139 of this chapter.

(e) Except for operations when common carriage is not involved conducted with airplanes having a passenger-seat configuration of 20 seats or more, excluding any required crewmember seat, or a payload capacity of 6,000 pounds or more, this part does not apply to—

(1) Student instruction;

(2) Nonstop sightseeing flights conducted with aircraft having a passenger seat configuration of 30 or fewer, excluding each crewmember seat, and a payload capacity of 7,500 pounds or less, that begin and end at the same airport, and are conducted within a 25 statute mile radius of that airport; however, for nonstop sightseeing flights for compensation or hire conducted in the vicinity of the Grand Canyon National Park, Arizona, the requirements of as applicable, apply;

(3) Ferry or training flights;

(4) Aerial work operations, including—

(i) Crop dusting, seeding, spraying, and bird chasing;

(ii) Banner towing;

(iii) Aerial photography or survey;

(iv) Fire fighting;

(v) Helicopter operations in construction or repair work (but it does apply to transportation to and from the site of operations); and

(vi) Powerline or pipeline patrol;

(5) Sightseeing flights conducted in hot air balloons;

(6) Nonstop flights conducted within a 25 statute mile radius of the airport of takeoff carrying persons for the purpose of intentional parachute jumps;

(7) Helicopter flights conducted within a 25 statute mile radius of the airport of takeoff if—

(i) Not more than two passengers are carried in the helicopter in addition to the required flightcrew;

(ii) Each flight is made under day VFR conditions;

(iii) The helicopter used is certificated in the standard category and complies with the 100-hour inspection requirements of part 91 of this chapter;

(iv) The operator notifies the FAA Flight Standards District Office responsible for the geographic area concerned at least 72 hours before each flight and furnishes any essential information that the office requests;

(v) The number of flights does not exceed a total of six in any calendar year;

(vi) Each flight has been approved by the Administrator; and

(vii) Cargo is not carried in or on the helicopter;

(8) Operations conducted under part 133 of this chapter or 375 of this title;

(9) Emergency mail service conducted under 49 U.S.C. 41906; or

(10) Operations conducted under the provisions of § 91.321 of this chapter.

§ 119.3 Definitions.

For the purpose of subchapter G of this chapter, the term—

"All cargo operation" means any operation for compensation or hire that is other than a passenger-carrying operation or, if passengers are carried, they are only those, specified in § 121.583(a) or 135.85 of this chapter.

"Certificate-holding district office" means the Flight Standards District Office that has responsibility for administering the certificate and is charged with the overall inspection of the certificate holder's operations.

"Commuter operation" means any scheduled operation conducted by any person operating one of the following types of aircraft with a frequency of operations of at least five round trips per week or at least one route between two or more points according to the published flight schedules:

(1) Airplanes, other than turbojet powered airplanes, having a maximum passenger-seat configuration of 9 seats or less, excluding each crewmember seat, and a maximum payload capacity of 7,500 pounds or less; or

(2) Rotorcraft.

"Direct air carrier" means a person who provides or offers to provide air transportation and who has control over the operational functions performed in providing that transportation.

"Domestic operation" means any scheduled operation conducted by any person operating any airplane described in paragraph (1) of this definition at locations described in paragraph (2) of this definition:

(1) Airplanes:

(i) Turbojet-powered airplanes;

(ii) Airplanes having a passenger-seat configuration of more than 9 passenger seats, excluding each crewmember seat; or

(iii) Airplanes having a payload capacity of more than 7,500 pounds.

(2) Locations:

(i) Between any points within the 48 contiguous States of the United States or the District of Columbia; or

(ii) Operations solely within the 48 contiguous States of the United States or the District of Columbia; or

(iii) Operations entirely within any State, territory, or possession of the United States; or

(iv) When specifically authorized by the Administrator, operations between any point within the 48 contiguous States of the United States or the District of Columbia and any specifically authorized point located outside the 48 contiguous States of the United States or the District of Columbia.

"Empty weight" means the weight of the airframe, engines, propellers, rotors, and fixed equipment. Empty weight excludes the weight of the crew and payload, but includes the weight of all fixed ballast, unusable fuel supply, undrainable oil, total quantity of engine coolant, and total quantity of hydraulic fluid.

"Flag operation" means any, scheduled operation conducted by any person operating any airplane described in paragraph (1) of this definition at the locations described in paragraph (2) of this definition:

(1) Airplanes:

(i) Turbojet-powered airplanes;

(ii) Airplanes having a passenger-seat configuration of more than 9 passenger seats, excluding each crewmember seat; or

(iii) Airplanes having a payload capacity of more than 7,500 pounds.

(2) Locations:

(i) Between any point within the State of Alaska or the State of Hawaii or any territory or possession of the United States and any point outside the State of Alaska or the State of Hawaii or any territory or possession of the United States, respectively; or

(ii) Between any point within the 48 contiguous States of the United States or the District of Columbia and any point outside the 48 contiguous States of the United States and the District of Columbia.

(iii) Between any point outside the U.S. and another point outside the U.S.

"Justifiable aircraft" equipment means any equipment necessary for the operation of the aircraft. It does not include equipment or ballast specifically installed, permanently or otherwise, for the purpose of altering the empty weight of an aircraft to meet the maximum payload capacity.

"Kind of operation" means one of the various operations a certificate holder is authorized to conduct, as specified in its operations specifications, i.e., domestic, flag, supplemental, commuter, or on-demand operations.

"Maximum payload capacity" means:

(1) For an aircraft for which a maximum zero fuel weight is prescribed in FAA technical specifications, the maximum zero fuel weight, less empty weight, less all justifiable aircraft equipment, and less the operating load (consisting of minimum flightcrew, foods and beverages, and supplies and equipment related to foods and beverages, but not including disposable fuel or oil).

(2) For all other aircraft, the maximum certificated takeoff weight of an aircraft, less the empty weight, less all justifiable aircraft equipment, and less the operating load (consisting of minimum fuel load, oil, and flightcrew). The allowance for the weight of the crew, oil, and fuel is as follows:

(i) Crew—for each crewmember required by the Federal Aviation Regulations—

(A) For male flight crewmembers—180 pounds.

(B) For female flight crewmembers—140 pounds.

(C) For male flight attendants—180 pounds.

(D) For female flight attendants—130 pounds.

(E) For flight attendants not identified by gender—140 pounds.

(ii) Oil—350 pounds or the oil capacity as specified on the Type Certificate Data Sheet.

(iii) Fuel—the minimum weight of fuel required by the applicable Federal Aviation Regulations for a flight between domestic points 174 nautical miles apart under VFR weather conditions that does not involve extended overwater operations.

"Maximum zero fuel weight" means the maximum permissible weight of an aircraft with no disposable fuel or oil. The zero fuel weight figure may be found in either the aircraft type certificate data sheet, the approved Aircraft Flight Manual, or both.

"Noncommon carriage" means an aircraft operation for compensation or hire that does not involve a holding out to others.

"On-demand operation" means any operation for compensation or hire that is one of the following:

(1) Passenger-carrying operations conducted as a public charter under part 380 of this title or any operations in which the departure time, departure location, and arrival location are specifically negotiated with the customer or the customer's representative that are any of the following types of operations:

(i) Common carriage operations conducted with airplanes, including turbojet-powered airplanes, having a passenger-seat configuration of 30 seats or fewer, excluding each crewmember seat, and a payload capacity of 7,500 pounds or less, except that operations using a specific airplane that is also used in domestic or flag operations and that is so listed in the operations specifications as required by § 119.49(a)(4) for those operations are considered supplemental operations;

(ii) Noncommon or private carriage operations conducted with airplanes having a passenger-seat configuration of less than 20 seats, excluding each crewmember seat, or a payload capacity of less than 6,000 pounds; or

(iii) Any rotorcraft operation.

(2) Scheduled passenger-carrying operations conducted with one of the following types of aircraft with a frequency of operations of less than five round trips per week on at least one route between two or more points according to the published flight schedules:

(i) Airplanes, other than turbojet powered airplanes, having a maximum passenger-seat configuration of 9 seats or less, excluding each crewmember seat, and a maximum payload capacity of 7,500 pounds or less; or

(ii) Rotorcraft.

(3) All-cargo operations conducted with airplanes having a payload capacity of 7,500 pounds or less, or with rotorcraft.

"Passenger-carrying operation" means any aircraft operation carrying any person, unless the only persons on the aircraft are those identified in §§ 121.583(a) or 135.85 of this chapter, as applicable. An aircraft used in a passenger-carrying operation may also carry cargo or mail in addition to passengers.

"Principal base of operations" means the primary operating location of a certificate holder as established by the certificate holder.

"Provisional airport" means an airport approved by the Administrator for use by a certificate holder for the purpose of providing service to a community when the regular airport used by the certificate holder is not available.

"Regular airport" means an airport used by a certificate holder in scheduled operations and listed in its operations specifications.

Scheduled operation means any common carriage passenger-carrying operation for compensation or hire conducted by an air carrier or commercial operator for which the certificate holder or its representative offers in advance the departure location, departure time, and arrival location. It does not include any passenger-carrying operation that is conducted as a public charter operation under part 380 of this title.

"Supplemental operation" means any common carriage operation for compensation or hire conducted with any airplane described in paragraph (1) of this definition that is a type of operation described in paragraph (2) of this definition:

(1) Airplanes:

(i) Airplanes having a passenger-seat configuration of more than 30 seats, excluding each crewmember seat;

(ii) Airplanes having a payload capacity of more than 7,500 pounds; or

(iii) Each airplane having a passenger-seat configuration of more than 9 seats and less than 31 seats, excluding each crewmember seat and any turbojet powered airplane, that is also used in domestic or flag operations and that is so listed in the operations specifications as required by § 119.49(a)(4) for those operations.

(2) Types of operation:

(i) Operations for which the departure time, departure location, and arrival location are specifically negotiated with the customer or the customer's representative;

(ii) All-cargo operations; or

(iii) Passenger-carrying public charter operations conducted under part 380 of this title.

"Wet lease" means any leasing arrangement whereby a person agrees to provide an entire aircraft and at least one crewmember. A wet lease does not include a code-sharing arrangement.

When common carriage is not involved or operations not involving common carriage means any of the following:

(1) Noncommon carriage.

(2) Operations in which persons or cargo are transported without compensation or hire.

(3) Operations not involving the transportation of persons or cargo.

(4) Private carriage.

§ 119.5 Certifications, authorizations, and prohibitions.

(a) A person authorized by the Administrator to conduct operations as a direct air carrier will be issued an Air Carrier Certificate.

(b) A person who is not authorized to conduct direct air carrier operations, but who is authorized by the Administrator to conduct operations as a U.S. commercial operator, will be issued an Operating Certificate.

(c) A person who is not authorized to conduct direct air carrier operations, but who is authorized by the Administrator to conduct operations when common carriage is not involved as an operator of U.S.-registered civil airplanes with a seat configuration of 20 or more passengers, or a maximum payload capacity of 6,000 pounds or more, will be issued an Operating Certificate.

(d) A person authorized to engage in common carriage under part 121 or part 135 of this chapter, or both, shall be issued only one certificate authorizing such common carriage, regardless of the kind of operation or the class or size of aircraft to be operated.

(e) A person authorized to engage in noncommon or private carriage under part 125 or part 135 of this chapter, or both, shall be issued only one certificate authorizing such carriage, regardless of the kind of operation or the class or size of aircraft to be operated.

(f) A person conducting operations under more than one paragraph of §§ 119.21, 119.23, or 119.25 shall conduct those operations in compliance with—

(1) The requirements specified in each paragraph of those sections for the kind of operation conducted under that paragraph; and

(2) The appropriate authorizations, limitations, and procedures specified in the operations specifications for each kind of operation.

(g) No person may operate as a direct air carrier or as a commercial operator without, or in violation of, an appropriate certificate and appropriate operations specifications. No person may operate as a direct air carrier or as a commercial operator in violation of any deviation or exemption authority, if issued to that person or that person's representative.

(h) A person holding an Operating Certificate authorizing noncommon or private carriage operations shall not conduct any operations in common carriage. A person holding an Air Carrier Certificate or Operating Certificate authorizing common carriage operations shall not conduct any operations in noncommon carriage.

(i) No person may operate as a direct air carrier without holding appropriate economic authority from the Department of Transportation.

(j) A certificate holder under this part may not operate aircraft under part 121 or part 135 of this chapter in a geographical area unless its operations specifications specifically authorize the certificate holder to operate in that area.

(k) No person may advertise or otherwise offer to perform an operation subject to this part unless that person is authorized by the Federal Aviation Administration to conduct that operation.

(l) No person may operate an aircraft under this part, part 121 of this chapter, or part 135 or this chapter in violation of an air carrier operating certificate, operating certificate, or appropriate operations specifications issued under this part.

§ 119.7 Operations specifications.

(a) Each certificate holder's operations specifications must contain—

(1) The authorizations, limitations, and certain procedures under which each kind of operation, if applicable, is to be conducted; and

(2) Certain other procedures under which each class and size of aircraft is to be operated.

(b) No person may operate an aircraft under part 121 or part 135 of this chapter unless the name of the certificate holder who is operating the aircraft, or the air carrier or operating certificate number of the certificate holder who is operating the aircraft, is legibly displayed on the aircraft and is clearly visible and readable from the outside of the aircraft to a person standing on the ground at any time except during flight time. The means of displaying the name on the aircraft and its readability must be acceptable to the Administrator.

§ 119.9 Use of business names.

(a) A certificate holder under this part may not operate an aircraft under part 121 or part 135 of this chapter using a business name other than a business name appearing in the certificate holder's operations specifications.

(b) Unless otherwise authorized by the Assistant Administrator for Civil Aviation Security, no person may operate an aircraft under part 121 or part 135 of this chapter unless the name of the certificate holder who is operating the aircraft is legibly displayed on the aircraft and is clearly visible and readable from the outside of the aircraft to a person standing on the ground at any time except during flight time. The means of displaying the name on the aircraft and its readability must be acceptable to the Administrator.

Subpart B—Applicability of Operating Requirements to Different Kinds of Operations Under Part 121, 125, and 135 of This Chapter

§ 119.21 Commercial operators engaged in common carriage and direct air carriers.

(a) Each person who conducts operations as a direct air carrier or as a commercial operator engaged in intrastate common carriage of persons or property for compensation or hire in air commerce, shall comply, with the certification and operations specifications requirements in subpart C of this part, and shall conduct its:

(1) Domestic operations in accordance with the applicable requirements of part 121 of this chapter, and shall be issued operations specifications for those operations in accordance with those requirements. However, based on a showing of safety in air commerce, the Administrator may permit persons who conduct domestic operations between any point located within any of the following Alaskan islands and any point in the State of Alaska to comply with the requirements applicable to flag operations contained in subpart U of part 121 of this chapter:

(i) The Aleutian Islands.

(ii) The Pribilof Islands.

(iii) The Shumagin Islands.

(2) Flag operations in accordance with the applicable requirements of part 121 of this chapter, and shall be issued operations specifications for those operations in accordance with those requirements.

(3) Supplemental operations in accordance with the applicable requirements of part 121 of this chapter, and shall be issued operations specifications for those operations in

accordance with those requirements. However, based on a determination of safety in air commerce, the Administrator may authorize or require the following operations to be conducted under paragraph (a) (1) or (2) of this section:

(i) Passenger-carrying operations which are conducted between points that are also served by the certificate holder's domestic or flag operations.

(ii) All-cargo operations which are conducted regularly and frequently between the same two points.

(4) Commuter operations in accordance with the applicable requirements of part 135 of this chapter, and shall be issued operations specifications for those operations in accordance with those requirements.

(5) On-demand operations in accordance with the applicable requirements of part 135 of this chapter, and shall be issued operations specifications for those operations in accordance with those requirements.

(b) Persons who are subject to the requirements of paragraph (a)(4) of this section may conduct those operations in accordance with the requirements of paragraph (a)(1) or (a)(2) of this section, provided they obtain authorization from the Administrator.

(c) Persons who are subject to the requirements of paragraph (a)(5) of this section may conduct those operations in accordance with the requirements of paragraph (a)(3) of this section, provided they obtain authorization from the Administrator.

§ 119.23 Operators engaged in passenger-carrying operations, cargo operations, or both with airplanes when common carriage is not involved.

(a) Each person who conducts operations when common carriage is not involved with airplanes having a passenger-seat configuration of 20 seats or more, excluding each crewmember seat, or a payload capacity of 6,000 pounds or more, shall, unless deviation authority is issued—

(1) Comply with the certification and operations specifications requirements of part 125 of this chapter;

(2) Conduct its operations with those airplanes in accordance with the requirements of part 125 of this chapter; and

(3) Be issued operations specifications in accordance with those requirements.

(b) Each person who conducts noncommon or private carriage operations for compensation or hire with airplanes having a passenger-seat configuration of less than 20 seats, excluding each crewmember seat, and a payload capacity of less than 6,000 pounds shall—

(1) Comply with the certification and operations specifications requirements in subpart C of this part;

(2) Conduct those operations in accordance with the requirements of part 135 of this chapter, except for those requirements applicable only to commuter operations; and

(3) Be issued operations specifications in accordance with those requirements.

§ 119.25 Rotorcraft operations: Direct air carriers and commercial operators.

Each person who conducts rotorcraft operations for compensation or hire must comply with the certification and operations specifications requirements of Subpart C of this part, and shall conduct its:

(a) Commuter operations in accordance with the applicable requirements of part 135 of this chapter, and shall be issued operations specifications for those operations in accordance with those requirements.

(b) On-demand operations in accordance with the applicable requirements of part 135 of this chapter, and shall be issued operations specifications for those operations in accordance with those requirements.

Subpart C—Certification, Operations Specifications, and Certain Other Requirements for Operations Conducted Under Part 121 or Part 135 of This Chapter

§ 119.31 Applicability.

This subpart sets out certification requirements and prescribes the content of operations specifications and certain other requirements for operations conducted under part 121 or part 135 of this chapter.

§ 119.33 General requirements.

(a) A person may not operate as a direct air carrier unless that person—

(1) Is a citizen of the United States;

(2) Obtains an Air Carrier Certificate; and

(3) Obtains operations specifications that prescribe the authorizations, limitations, and procedures under which each kind of operation must be conducted.

(b) A person other than a direct air carrier may not conduct any commercial passenger or cargo aircraft operation for compensation or hire under part 121 or part 135 of this chapter unless that person—

(1) Is a citizen of the United States;

(2) Obtains an Operating Certificate; and

(3) Obtains operations specifications that prescribe the authorizations, limitations, and procedures under which each kind of operation must be conducted.

(c) Each applicant for a certificate under this part shall conduct proving tests as authorized by the Administrator during the application process for authority to conduct operations under part 121 or part 135 of this chapter. All proving tests must be conducted in a manner acceptable to the Administrator. All proving tests must be conducted under the appropriate operating and maintenance requirements of part 121 or 135 of this chapter that would apply if the applicant were fully certificated. The Administrator will issue a letter of authorization to each person stating the various authorities under which the proving tests shall be conducted.

§ 119.35 Certificate application for all operators.

(a) A person applying to the Administrator for an Air Carrier Certificate or Operating Certificate under this part (applicant) must submit an application—

(1) In a form and manner prescribed by the Administrator; and

(2) Containing any information the Administrator requires the applicant to submit.

(b) Each applicant must submit the application to the Administrator at least 90 days before the date of intended operation.

§ 119.36 Additional certificate application requirements for commercial operators.

(a) Each applicant for the original issue of an operating certificate for the purpose of conducting intrastate common carriage operations under part 121 or part 135 of this chapter must submit an application in a form and manner prescribed by the Administrator to the Flight Standards District Office in whose area the applicant proposes to establish or has established his or her principal base of operations.

(b) Each application submitted under paragraph (a) of this section must contain a signed statement showing the following:

(1) For corporate applicants:

(i) The name and address of each stockholder who owns 5 percent or more of the total voting stock of the corporation, and if that stockholder is not the sole beneficial owner of the stock, the name and address of each beneficial owner. An individual is considered to own the stock owned, directly or indirectly, by or for his or her spouse, children, grandchildren, or parents.

(ii) The name and address of each director and each officer and each person employed in a management position described in §§119.65 and 119.69, as applicable.

(iii) The name and address of each person directly or indirectly controlling or controlled by the applicant and each person under direct or indirect control with the applicant.

(2) For non-corporate applicants:

(i) The name and address of each person having a financial interest therein and the nature and extent of that interest.

(ii) The name and address of each person employed or who will be employed in a management position described in §§119.65 and 119.69, as applicable.

(c) In addition, each applicant for the original issue of an operating certificate under paragraph (a) of this section must submit with the application a signed statement showing—

(1) The nature and scope of its intended operation, including the name and address of each person, if any, with whom the applicant has a contract to provide series as a commercial operator and the scope, nature, date, and duration of each of those contracts; and

(2) For applicants intending to conduct operations under part 121 of this chapter, the financial information listed in paragraph (e) of this section.

(d) Each applicant for, or holder of, a certificate issued under paragraph (a) of this section, shall notify the Administrator within 10 days after—

(1) A change in any of the persons, or the names and addresses of any of the persons, submitted to the Administrator under paragraph (b)(1) or (b)(2) of this section; or

(2) For applicants intending to conduct operators under part 121 of this chapter, a change in the financial information submitted to the Administrator under paragraph (e) of this section that occurs while the application for the issue is pending before the FAA and that would make the applicant's financial situation substantially less favorable than originally reported.

(e) Each applicant for the original issue of an operating certificate under paragraph (a) of this section who intends to conduct operations under part 121 of this chapter must submit the following financial information:

(1) A balance sheet that shows assets, liabilities, and net worth, as of a date not more than 60 days before the date of application.

(2) An itemization of liabilities more than 60 days past due on the balance sheet date, if any, showing each creditor's name and address, a description of the liability, and the amount and due date of the liability.

(3) An itemization of claims in litigation, if any, against the applicant as of the date of application showing each claimant's name and address and a description and the amount of the claim.

(4) A detailed projection of the proposed operation covering 6 complete months after the month in which the certificate is expected to be issued including—

(i) Estimated amount and source of both operating and nonoperating revenue, including identification of its existing and anticipated income producing contracts and estimated revenue per mile or hour of operation by aircraft type:

(ii) Estimated amount of operating and nonoperating expenses by expense objective classification; and

(iii) Estimated net profit or loss for the period.

(5) An estimate of the cash that will be needed for the proposed operations during the first 6 months after the month in which the certificate is expected to be issued, including—

(i) Acquisition of property and equipment (explain);

(ii) Retirement of debt (explain);

(iii) Additional working capital (explain);

(iv) Operating losses other than depreciation and amortization (explain); and

(v) Other (explain).

(6) An estimate of the cash that will be available during the first 6 months after the month in which the certificate is expected to be issued, from—

(i) Sale of property or flight equipment (explain);

(ii) New debt (explain);

(iii) New equity (explain);

(iv) Working capital reduction (explain);

(v) Operations (profits) (explain);

(vi) Depreciation and amortization (explain); and

(vii) Other (explain).

(7) A schedule of insurance coverage in effect on the balance sheet date showing insurance companies; policy numbers; types, amounts, and period of coverage; and special conditions, exclusions, and limitations.

(8) Any other financial information that the Administrator requires to enable him or her to determine that the applicant has sufficient financial resources to conduct his or her operations with the degree of safety required in the public interest.

(f) Each financial statement containing financial information required by paragraph (e) of this section must be based on accounts prepared and maintained on an accrual basis in accordance with generally accepted accounting principles applied on a consistent basis, and must contain the name and address of the applicant's public accounting firm, if any. Information submitted must be signed by an officer, owner, or partner of the applicant or certificate holder.

§ 119.37 Contents of an Air Carrier Certificate or Operating Certificate.

The Air Carrier Certificate or Operating Certificate includes—

(a) The certificate holder's name;

(b) The location of the certificate holder's principal base of operations;

(c) The certificate number;

(d) The certificate's effective date; and

(e) The name or the designator of the certificate-holding district office.

§ 119.39 Issuing or denying a certificate.

(a) An applicant may be issued an Air Carrier Certificate or Operating Certificate if, after investigation, the Administrator finds that the applicant—

(1) Meets the applicable requirements of this part;

(2) Holds the economic authority applicable to the kinds of operations to be conducted, issued by the Department of Transportation, if required; and

(3) Is properly and adequately equipped in accordance with the requirements of this chapter and is able to conduct a safe operation under appropriate provisions of part 121 or part 135 of this chapter and operations specifications issued under this part.

(b) An application for a certificate may be denied if the Administrator finds that—

(1) The applicant is not properly or adequately equipped or is not able to conduct safe operations under this subchapter;

(2) The applicant previously held an Air Carrier Certificate or Operating Certificate which was revoked;

(3) The applicant intends to or fills a key management position listed in § 119.65(a) or § 119.69(a), as applicable, with an individual who exercised control over or who held the same or a similar position with a certificate holder whose certificate was revoked, or is in the process of being revoked, and that individual materially contributed to the circumstances causing revocation or causing the revocation process;

(4) An individual who will have control over or have a substantial ownership interest in the applicant had the same or similar control or interest in a certificate holder whose certificate was revoked, or is in the process of being revoked, and that individual materially contributed to the circumstances causing revocation or causing the revocation process; or

(5) In the case of an applicant for an Operating Certificate for intrastate common carriage, that for financial reasons the applicant is not able to conduct a safe operation.

§ 119.41 Amending a certificate.

(a) The Administrator may amend any certificate issued under this part if—

(1) The Administrator determines, under 49 U.S.C. 44709 and part 13 of this chapter, that safety in air commerce and the public interest requires the amendment; or

(2) The certificate holder applies for the amendment and the certificate-holding district office determines that safety in air commerce and the public interest allows the amendment.

(b) When the Administrator proposes to issue an order amending, suspending, or revoking all or part of any certificate, the procedure in § 13.19 of this chapter applies.

(c) When the certificate holder applies for an amendment of its certificate, the following procedure applies:

(1) The certificate holder must file an application to amend its certificate with the certificate-holding district office at least 15 days before the date proposed by the applicant for the amendment to become effective, unless the administrator approves filing within a shorter period: and

(2) The application must be submitted to the certificate-holding district office in the form and manner prescribed by the Administrator.

(d) When a certificate holder seeks reconsideration of a decision from the certificate-holding district office concerning amendments of a certificate, the following procedure applies:

(1) The petition for reconsideration must be made within 30 days after the certificate holder receives the notice of denial; and

(2) The certificate holder must petition for reconsideration to the Director, Flight Standards Service.

§ 119.43 Certificate holder's duty to maintain operations specifications.

(a) Each certificate holder shall maintain a complete and separate set of its operations specifications at its principal base of operations.

(b) Each certificate holder shall insert pertinent excerpts of its operations specifications, or references thereto, in its manual and shall—

(1) Clearly identify each such excerpt as a part of its operations specifications; and

(2) State that compliance with each operations specifications requirement is mandatory.

(c) Each certificate holder shall keep each of its employees and other persons used in its operations informed of the provisions of its operations specifications that apply to that employee's or person's duties and responsibilities.

§ 119.45 [Reserved]

§ 119.47 Maintaining a principal base of operations, main operations base, and main maintenance base; change of address.

(a) Each certificate holder must maintain a principal base of operations. Each certificate holder may also establish a main operations base and a main maintenance base which may be located at either the same location as the principal base of operations or at separate locations.

(b) At least 30 days before it proposes to establish or change the location of its principal base of operations, its main operations base, or its main maintenance base, a certificate holder must provide written notification to its certificate-holding district office.

§ 119.49 Contents of operations specifications.

(a) Each certificate holder conducting domestic, flag, or commuter operations must obtain operations specifications containing all of the following:

(1) The specific location of the certificate holder's principal base of operations and, if different, the address that shall serve as the primary point of contact for correspondence between the FAA and the certificate holder and the name and mailing address of the certificate holder's agent for service.

(2) Other business names under which the certificate holder may operate.

(3) Reference to the economic authority issued by the Department of Transportation, if required.

(4) Type of aircraft, registration markings, and serial numbers of each aircraft authorized for use, each regular and alternate airport to be used in scheduled operations, and, except for commuter operations, each provisional and refueling airport.

(i) Subject to the approval of the Administrator with regard to form and content, the certificate holder may incorporate by reference the items listed in paragraph (a)(4) of this section into the certificate holder's operations specifications by maintaining a current listing of those items and by referring to the specific list in the applicable paragraph of the operations specifications.

(ii) The certificate holder may not conduct any operation using any aircraft or airport not listed.

(5) Kinds of operations authorized.

(6) Authorization and limitations for routes and areas of operations.

(7) Airport limitations.

(8) Time limitations, or standards for determining time limitations, for overhauling, inspecting, and checking airframes, engines, propellers, rotors, appliances, and emergency equipment.

(9) Authorization for the method of controlling weight and balance of aircraft.

(10) Interline equipment interchange requirements, if relevant.

(11) Aircraft wet lease information required by § 119.53(c).

(12) Any authorized deviation and exemption granted from any requirement of this chapter.

(13) Any other item the Administrator determines is necessary.

(b) Each certificate holder conducting supplemental operations must obtain operations specifications containing all of the following:

(1) The specific location of the certificate holder's principal base of operations, and, if different, the address that shall serve as the primary point of contact for correspondence between the FAA and the certificate holder and the name and mailing address of the certificate holder's agent for service.

(2) Other business names under which the certificate holder may operate.

(3) Reference to the economic authority issued by the Department of Transportation, if required.

(4) Type of aircraft, registration markings, and serial number of each aircraft authorized for use.

(i) Subject to the approval of the Administrator with regard to form and content, the certificate holder may incorporate by reference the items listed in paragraph (b)(4) of this section into the certificate holder's operations specifications by maintaining a current listing of those items and by referring to the specific list in the applicable paragraph of the operations specifications.

(ii) The certificate holder may not conduct any operation using any aircraft not listed.

(5) Kinds of operations authorized.

(6) Authorization and limitations for routes and areas of operations.

(7) Special airport authorizations and limitations.

(8) Time limitations, or standards for determining time limitations, for overhauling, inspecting, and checking airframes, engines, propellers, appliances, and emergency equipment.

(9) Authorization for the method of controlling weight and balance of aircraft.

(10) Aircraft wet lease information required by § 119.53(c).

(11) Any authorization or requirement to conduct supplemental operations as provided by § 119.21 (a)(3) (i) or (ii).

(12) Any authorized deviation or exemption from any requirement of this chapter.

(13) Any other item the Administrator determines is necessary.

(c) Each certificate holder conducting on-demand operations must obtain operations specifications containing all of the following:

(1) The specific location of the certificate holder's principal base of operations, and if different, the address that shall serve as the primary point of contact for correspondence between the FAA and the name and mailing address of the certificate holder's agent for service.

(2) Other business names under which the certificate holder may operate.

(3) Reference to the economic authority issued by the Department of Transportation, if required.

(4) Kind and area of operations authorized.

(5) Category and class of aircraft that may be used in those operations.

(6) Type of aircraft, registration markings, and serial number of each aircraft that is subject to an airworthiness maintenance program required by § 135.411(a)(2) of this chapter.

(i) Subject to the approval of the Administrator with regard to form and content, the certificate holder may incorporate by reference the items listed in paragraph (c)(6) of this section into the certificate holder's operations specifications by maintaining a current listing of those items and by referring to the specific list in the applicable paragraph of the operations specifications.

(ii) The certificate holder may not conduct any operation using any aircraft not listed.

(7) Registration markings of each aircraft that is to be inspected under an approved aircraft inspection program under § 135.419 of this chapter.

(8) Time limitations or standards for determining time limitations, for overhauls, inspections, and checks for airframes, engines, propellers, rotors, appliances, and emergency equipment of aircraft that are subject to an airworthiness maintenance program required by § 135.411 (a)(2) of this chapter.

(9) Additional maintenance items required by the Administrator under 135.421 of this chapter.

(10) Aircraft wet lease information required by § 119.53(c).

(11) Any authorized deviation or exemption from any requirement of this chapter.

(12) Any other item the Administrator determines is necessary.

§ 119.51 Amending operations specifications.

(a) The Administrator may amend any operations specifications issued under this part if—

(1) The Administrator determines that safety in air commerce and the public interest require the amendment; or

(2) The certificate holder applies for the amendment, and the Administrator determines that safety in air commerce and the public interest allows the amendment.

(b) Except as provided in paragraph (e) of this section, when the Administrator initiates an amendment to a certificate holder's operations specifications, the following procedure applies:

(1) The certificate-holding district office notifies the certificate holder in writing of the proposed amendment.

(2) The certificate-holding district office sets a reasonable period (but not less than 7 days) within which the certificate holder may submit written information, views, and arguments on the amendment.

(3) After considering all material presented, the certificate-holding district office notifies the certificate holder of—

(i) The adoption of the proposed amendment;

(ii) The partial adoption of the proposed amendment; or

(iii) The withdrawal of the proposed amendment.

(4) If the certificate-holding district office issues an amendment to the operations specifications, it becomes effective not less than 30 days after the certificate holder receives notice of it unless—

(i) The certificate-holding district office finds under paragraph (e) of this section that there is an emergency requiring immediate action with respect to safety in air commerce; or

(ii) The certificate holder petitions for reconsideration of the amendment under paragraph (d) of this section.

(c) When the certificate holder applies for an amendment to its operations specifications, the following procedure applies:

(1) The certificate holder must file an application to amend its operations specifications—

(i) At least 90 days before the date proposed by the applicant for the amendment to become effective, unless a shorter time is approved, in cases of mergers: acquisitions of airline operational assets that require an additional showing of safety (e.g., proving tests); changes in the kind of operation as defined in § 119.3; resumption of operations following a suspension of operations as a result of bankruptcy actions; or the initial introduction of aircraft not before proven for use in air carrier or commercial operator operations.

(ii) At least 15 days before the date proposed by the applicant for the amendment to become effective in all other cases.

(2) The application must be submitted to the certificate-holding district office in a form and manner prescribed by the Administrator.

(3) After considering all material presented, the certificate-holding district office notifies the certificate holder of—

(i) The adoption of the applied for amendment;

(ii) The partial adoption of the applied for amendment; or

(iii) The denial of the applied for amendment. The certificate holder may petition for reconsideration of a denial under paragraph (d) of this section.

(4) If the certificate-holding district office approves the amendment, following coordination with the certificate holder regarding its implementation, the amendment is effective on the date the Administrator approves it.

(d) When a certificate holder seeks reconsideration of a decision from the certificate-holding district office concerning the amendment of operations specifications, the following procedure applies:

(1) The certificate holder must petition for reconsideration of that decision within 30 days of the date that the certificate holder receives a notice of denial of the amendment to its operations specifications, or of the date it receives notice of an FAA-initiated amendment to its operations specifications, whichever circumstance applies.

(2) The certificate holder must address its petition to the Director, Flight Standards Service.

(3) A petition for reconsideration, if filed within the 30-day period, suspends the effectiveness of any amendment issued by the certificate-holding district office unless the certificate-holding district office has found, under paragraph (e) of this section, that an emergency exists requiring immediate action with respect to safety in air transportation or air commerce.

(4) If a petition for reconsideration is not filed within 30 days, the procedures of paragraph (c) of this section apply.

(e) If the certificate-holding district office finds that an emergency exists requiring immediate action with respect to safety in air commerce or air transportation that makes the procedures set out in this section impracticable or contrary to the public interest:

(1) The certificate-holding district office amends the operations specifications and makes the amendment effective on the day the certificate holder receives notice of it.

(2) In the notice to the certificate holder, the certificate-holding district office articulates the reasons for its finding that an emergency exists requiring immediate action with respect to safety in air transportation or air commerce or that makes it impracticable or contrary to the public interest to stay the effectiveness of the amendment.

§ 119.53 Wet leasing of aircraft and other arrangements for transportation by air.

(a) Unless otherwise authorized by the Administrator, prior to conducting operations involving a wet lease, each certificate holder under this part authorized to conduct common carriage operations under this subchapter shall provide the Administrator with a copy of the wet lease to be executed which would lease the aircraft to any other person engaged in common carriage operations under this subchapter, including foreign air carriers, or to any other foreign person engaged in common carriage wholly outside the United States.

(b) No certificate holder under this part may wet lease from a foreign air carrier or any other foreign person or any person not authorized to engage in common carriage.

(c) Upon receiving a copy of a wet lease, the Administrator determines which party to the agreement has operational control of the aircraft and issues amendments to the operations specifications of each party to the agreement, as needed. The lessor must provide the following information to be incorporated into the operations specifications of both parties, as needed.

(1) The names of the parties to the agreement and the duration thereof.

(2) The nationality and registration markings of each aircraft involved in the agreement.

(3) The kind of operation (e.g., domestic, flag, supplemental, commuter, or on-demand).

(4) The airports or areas of operation.

(5) A statement specifying the party deemed to have operational control and the times, airports, or areas under which such operational control is exercised.

(d) In making the determination of paragraph (c) of this section, the Administrator will consider the following:

(1) Crewmembers and training.

(2) Airworthiness and performance of maintenance.

(3) Dispatch.

(4) Servicing the aircraft.

(5) Scheduling.

(6) Any other factor the Administrator considers relevant.

(e) Other arrangements for transportation by air: Except as provided in paragraph (f) of this section, a certificate holder under this part operating under part 121 or 135 of this chapter may not conduct any operation for another certificate holder under this part or a foreign air carrier under part 129 of this chapter or a foreign person engaged in common carriage wholly outside the United States unless it holds applicable Department of Transportation economic authority, if required, and is authorized under its operations specifications to conduct the same kinds of operations (as defined in § 119.3). The certificate holder conducting the substitute operation must conduct that operation in accordance with the same operations authority held by the certificate holder arranging for the substitute operation. These substitute operations must be conducted between airports for which the substitute certificate holder holds authority for scheduled operations or within areas of operations for which the substitute certificate holder has authority or supplemental or on-demand operations.

(f) A certificate holder under this part may, if authorized by the Department of Transportation under § 380.3 of this title and the Administrator in the case of interstate commuter, interstate domestic, and flag operations, or the Administrator in the case of scheduled intrastate common carriage operations, conduct one or more flights for passengers who are stranded because of the cancellation of their scheduled flights. These flights must be conducted under the rules of part 121 or part 135 of this chapter applicable to supplemental or on-demand operations.

§ 119.55 Obtaining deviation authority to perform operations under a U.S. military contract.

(a) The Administrator may authorize a certificate holder that is authorized to conduct supplemental or on-demand operations to deviate from the applicable requirements of this part, part 121, or part 135 of this chapter in order to perform operations under a U.S. military contract.

(b) A certificate holder that has a contract with the U.S. Department of Defense's Air Mobility Command (AMC) must submit a request for deviation authority to AMC. AMC will review the requests, then forward the carriers' consolidated requests, along with AMC's recommendations, to the FAA for review and action.

(c) The Administrator may authorize a deviation to perform operations under a U.S. military contract under the following conditions—

(1) The Department of Defense certifies to the Administrator that the operation is essential to the national defense;

(2) The Department of Defense further certifies that the certificate holder cannot perform the operation without deviation authority;

(3) The certificate holder will perform the operation under a contract or subcontract for the benefit of a U.S. armed service; and

(4) The Administrator finds that the deviation is based on grounds other than economic advantage either to the certificate holder or to the United States.

(d) In the case where the Administrator authorizes a deviation under this section, the Administrator will issue an appropriate amendment to the certificate holder's operations specifications.

(e) The Administrator may, at any time, terminate any grant of deviation authority issued under this section.

§ 119.57 Obtaining deviation authority to perform an emergency operation.

(a) In emergency conditions, the Administrator may authorize deviations if—

(1) Those conditions necessitate the transportation of persons or supplies for the protection of life or property; and

(2) The Administrator finds that a deviation is necessary for the expeditious conduct of the operations.

(b) When the Administrator authorizes deviations for operations under emergency conditions—

(1) The Administrator will issue an appropriate amendment to the certificate holder's operations specifications; or

(2) If the nature of the emergency does not permit timely amendment of the operations specifications—

(i) The Administrator may authorize the deviation orally; and

(ii) The certificate holder shall provide documentation describing the nature of the emergency to the certificate-holding district office within 24 hours after completing the operation.

§ 119.58 Emergencies requiring immediate decision and action.

(a) In an emergency situation that requires immediate decision and action, the pilot in command may take any action that he considers necessary under the circumstances. In such a case, he may deviate from prescribed operations procedures and methods, weather minimums, and this chapter to the extent required in the interest of safety.

(b) In an emergency situation arising during flight, that requires immediate decision and action by an aircraft dispatcher or appropriate management personnel, and that is known to him, he shall advise the pilot in command of the emergency, shall ascertain the decision of the pilot in command, and shall have the decision recorded. If he cannot communicate with the pilot, he shall declare an emergency and take any reasonable action necessary, under the circumstances.

(c) Whenever a pilot in command or a dispatcher or an appropriate management person exercises emergency authority, he shall keep the appropriate ATC facility, ground radio station, and, if applicable, dispatch centers, fully informed of the progress of the flight. The person declaring the emergency shall send a written report of any deviation through the certificate holder's management to the Administrator within 10 days of the emergency action.

§ 119.59 Conducting tests and inspections.

(a) At any time or place, the Administrator may conduct an inspection or test to determine whether a certificate holder under this part is complying with title 49 of the United States Code, applicable regulations, the certificate, or the certificate holder's operations specifications.

(b) The certificate holder must—

(1) Make available to the Administrator at the certificate holder's principal base of operations—

(i) The certificate holder's Air Carrier Certificate or the certificate holder's Operating Certificate and the certificate holder's operations specifications; and

(ii) A current listing that will include the location and persons responsible for each record, document, and report required to be kept by the certificate holder under title 49 of the United States Code applicable to the operation of the certificate holder.

(2) Allow the Administrator to make any test or inspection to determine compliance respecting any matter stated in paragraph (a) of this section.

(c) Each employee of, or person used by, the certificate holder who is responsible for maintaining the certificate holder's records must make those records available to the Administrator.

(d) The Administrator may determine a certificate-holder's continued eligibility to hold its certificate and/or operations specifications on any grounds listed in paragraph (a) of this section, or any other appropriate grounds.

(e) Failure by any certificate holder to make available to the Administrator upon request, the certificate, operations specifications, or any required record, document, or report is grounds for suspension of all or any part of the certificate holder's certificate and operations specifications.

(f) In the case of operators conducting intrastate common carriage operations, these inspections and tests include inspections and tests of financial books and records.

§ 119.61 Duration and surrender of certificate and operations specifications.

(a) An Air Carrier Certificate or Operating Certificate issued, under this part is effective until—

(1) The certificate holder surrenders it to the Administrator or

(2) The Administrator suspends, revokes, or otherwise terminates the certificate.

(b) Operations specifications issued under this part, part 121, or part 135 of this chapter are effective unless—

(1) The Administrator suspends, revokes, or otherwise terminates the certificate;

(2) The operations specifications are amended as provided in § 119.51;

(3) The certificate holder does not conduct a kind of operation for more than the time specified in § 119.63 and fails to follow the procedures of § 119.63 upon resuming that kind of operation; or

(4) The Administrator suspends or revokes the operations specifications for a kind of operation.

(c) Within 30 days after a certificate holder terminates operations under part 135 of this chapter, the operating certificate and operations specifications must be surren-

dered by the certificate holder to the certificate-holding district office.

§ 119.63 Recency of operation.

(a) Except as provided in paragraph (b) of this section, no certificate holder may conduct a kind of operation for which it holds authority in its operations specifications unless the certificate holder has conducted that kind of operation within the preceding number of consecutive calendar days specified in this paragraph:

(i) For domestic, flag, or commuter operations—30 days.

(2) For supplemental or on-demand operations—90 days, except that if the certificate holder has authority to conduct domestic, flag, or commuter operations, and has conducted domestic, flag or commuter operations within the previous 30 days, this paragraph does not apply.

(b) If a certificate holder does not conduct a kind of operation for which it is authorized in its operations specifications within the number of preceding 30 consecutive calendar days specified in paragraph (a) of this section, it shall not conduct such kind of operation unless—

(1) It advises the Administrator at least 5 consecutive calendar days before resumption of that kind of operation; and

(2) It makes itself available and accessible during the 5 consecutive calendar day period in the event that the FAA decides to conduct a full inspection reexamination to determine whether the holder remains properly and adequately equipped and able to conduct a safe operation.

§ 119.65 Management personnel required for operations conducted under part 121 of this chapter.

(a) Each certificate holder must have sufficient qualified management and technical personnel to ensure the highest degree of safety in its operations. The certificate holder must have qualified personnel serving full-time in the following or equivalent positions:

(1) Director of Safety.

(2) Director of Operations.

(3) Chief Pilot.

(4) Director of Maintenance.

(5) Chief Inspector.

(b) The Administrator may approve positions or numbers of positions other than those listed in paragraph (a) of this section for a particular operation if the certificate holder shows that it can perform the operation with the highest degree of safety under the direction of fewer or different categories of management personnel due to—

(1) The kind of operation involved;

(2) The number and type of airplanes used; and

(3) The area of operations.

(c) The title of the positions required under paragraph (a) of this section or the title and number of equivalent positions approved under paragraph (b) of this section shall be set forth in the certificate holder's operations specifications.

(d) The individuals who serve in the positions required or approved under paragraph (a) or (b) of this section and anyone in a position to exercise control over operations conducted under the operating certificate must—

(1) Be qualified through training, experience, and expertise;

(2) To the extent of their responsibilities, have a full understanding of the following materials with respect to the certificate holder's operation—

(i) Aviation safety standards and safe operating practices;

(ii) 14 CFR Chapter I (Federal Aviation Regulations);

(iii) The certificate holder's operations specifications;

(iv) All appropriate maintenance and airworthiness requirements of this chapter (e.g., parts 1, 21, 23, 25, 43, 45, 47, 65, 91, and 121 of this chapter); and

(v) The manual required by § 121.133 of this chapter; and

(3) Discharge their duties to meet applicable legal requirements and to maintain safe operations.

(e) Each certificate holder must:

(1) State in the general policy provisions of the manual required by § 121.133 of this chapter, the duties. responsibilities, and authority of personnel required under paragraph (a) of this section:

(2) List in the manual the names and business addresses of the individuals assigned to those positions; and

(3) Notify the certificate-holding district office within 10 days of any change in personnel or any vacancy in any position listed.

§ 119.67 Management personnel: Qualifications for operations conducted under part 121 of this chapter.

(a) To serve as Director of Operations under § 119.65(a) a person must—

(1) Hold an airline transport pilot certificate;

(2) Have at least 3 years supervisory or managerial experience within the last 6 years in a position that exercised operational control over any operations conducted with large airplanes under part 121 or part 135 of this chapter, or if the certificate holder uses only small airplanes in its operations, the experience may be obtained in large or small airplanes; and

(3) In the case of a person becoming a Director of Operations—

(i) For the first time ever, have at least 3 years experience, within the past 6 years, as pilot in command of a large airplane operated under part 121 or part 135 of this chapter, if the certificate holder operates large airplanes. If the certificate holder uses only small airplanes in its operation, the experience may be obtained in either large or small airplanes.

(ii) In the case of a person with previous experience as a Director of Operations, have at least 3 years experience as

pilot in command of a large airplane operated under part 121 or part 135 of this chapter, if the certificate holder operates large airplanes. If the certificate holder uses only small airplanes in its operation, the experience may be obtained in either large or small airplanes.

(b) To serve as Chief Pilot under § 119.65(a) a person must hold an airline transport pilot certificate with appropriate ratings for at least one of the airplanes used in the certificate holder's operation and:

(1) In the case of a person becoming a Chief Pilot for the first time ever, have at least 3 years experience, within the past 6 years, as pilot in command of a large airplane operated under part 121 or part 135 of this chapter, if the certificate holder operates large airplanes. If the certificate holder uses only small airplanes in its operation, the experience may be obtained in either large or small airplanes.

(2) In the case of a person with previous experience as a Chief Pilot, have at least 3 years experience, as pilot in command of a large airplane operated under part 121 or part 135 of this chapter, if the certificate holder operates large airplanes. If the certificate holder uses only small airplanes in its operation, the experience may be obtained in either large or small airplanes.

(c) To serve as Director of Maintenance under § 119.65(a) a person must—

(1) Hold a mechanic certificate with airframe and powerplant ratings;

(2) Have 1 year of experience in a position responsible for returning airplanes to service;

(3) Have at least 1 year of experience in a supervisory capacity under either paragraph (c)(4)(i) or (c)(4)(ii) of this section maintaining the same category and class of airplane as the certificate holder uses; and

(4) Have 3 years experience within the past 6 years in one or a combination of the following—

(i) Maintaining large airplanes with 10 or more passenger seats, including at the time of appointment as Director of Maintenance, experience in maintaining the same category and class of airplane as the certificate holder uses; or

(ii) Repairing airplanes in a certificated airframe repair station that is rated to maintain airplanes in the same category and class of airplane as the certificate holder uses.

(d) To serve as Chief Inspector under 119.65(a) a person must—

(1) Hold a mechanic certificate with both airframe and powerplant ratings, and have held these ratings for at least 3 years;

(2) Have at least 3 years of maintenance experience on different types of large airplanes with 10 or more passenger seats with an air carrier or certificated repair station, 1 year of which must have been as maintenance inspector; and

(3) Have at least 1 year in a supervisory capacity maintaining large aircraft with 10 or more passenger seats.

(e) A certificate holder may request a deviation to employ a person who does not meet the appropriate airman experience, managerial experieince, or supervisory experience requirements of this section if the Manager of the Air Transportation Division, AFS-200, or the Manager of the Aircraft Maintenance Division, AFS-300, as appropriate, finds that the person has comparable experience, and can effectively perform the functions associated with the position in accordance with the requirements of this chapter and the procedures outlined in the certificate holder's manual. Grants of deviation under this paragraph may be granted after consideration of the size and scope of the operation and the qualifications of the intended personnel. The administrator may, at any time, terminate any grant of deviation authority issued under this paragraph.

§ 119.69 Management personnel required for operations conducted under part 135 of this chapter.

(a) Each certificate holder must have sufficient qualified management and technical personnel to ensure the safety of its operations. Except for a certificate holder using only one pilot in its operations, the certificate holder must have qualified personnel serving in the following or equivalent positions:

(1) Director of Operations.

(2) Chief Pilot.

(3) Director of Maintenance.

(b) The Administrator may approve positions or numbers of positions other than those listed in paragraph (a) of this section for a particular operation if the certificate holder shows that it can perform the operation with the highest degree of safety under the direction of fewer or different categories of management personnel due to—

(1) The kind of operation involved;

(2) The number and type of aircraft used; and

(3) The area of operations.

(c) The title of the positions required under paragraph (a) of this section or the title and number of equivalent positions approved under paragraph (b) of this section shall be set forth in the certificate holder's operations specifications.

(d) The individuals who serve in the positions required or approved under paragraph (a) or (b) of this section and anyone in a position to exercise control over operations conducted under the operating certificate must—

(1) Be qualified through training, experience, and expertise;

(2) To the extent of their responsibilities, have a full understanding of the following material with respect to the certificate holder's operation—

(i) Aviation safety standards and safe operating practices;

(ii) 14 CFR Chapter I (Federal Aviation Regulations);

(iii) The certificate holder's operations specifications;

(iv) All appropriate maintenance and airworthiness requirements of this chapter (e.g., parts 1, 21, 23, 25, 43, 45. 47, 65, 91, and 135 of this chapter); and

(v) The manual required by § 135.21 of this chapter; and

(3) Discharge their duties to meet applicable legal requirements and, to maintain safe operations.

(e) Each certificate holder must—

(1) State in the general policy provisions of the manual required by § 135.21 of this chapter, the duties, responsibilities, and authority of personnel required or approved under paragraph (a) or (b), respectively , of this section;

(2) List in the manual the names and business addresses of the individuals assigned to those positions; and

(3) Notify the certificate-holding district office within 10 days of any change in personnel or any vacancy in any position listed.

§ 119.71 Management personnel: Qualifications for operations conducted under part 135 of this chapter.

(a) To serve as Director of Operations under § 119.69(a) for a certificate holder conducting any operations for which the pilot in command is required to hold an airline transport pilot certificate a person must hold an airline transport pilot certificate and either:

(1) Have at least 3 years supervisory or managerial experience within the last 6 years in a position that exercised operational control over any operations conducted under part 121 or part 135 of this chapter; or

(2) In the case of a person becoming Director of Operations—

(i) For the first time ever, have at least 3 years experience, within the past 6 years, as pilot in command of an aircraft operated under part 121 or part 135 of this chapter.

(ii) In the case of a person with previous experience as a Director of Operations, have at least 3 years experience, as pilot in command of an aircraft operated under part 121 or part 135 of this chapter.

(b) To serve as Director of Operations under §119.69(a) for a certificate holder that only conducts operations for which the pilot in command is required to hold a commercial pilot certificate, a person must hold at least a commercial pilot certificate. If an instrument rating is required for any pilot in command for that certificate holder, the Director of Operations must also hold an instrument rating. In addition, the Director of Operations must either—

(1) Have at least 3 years supervisory or managerial experience within the last 6 years in a position that exercised operational control over any operations conducted under part 121 or part 135 of this chapter; or

(2) In the case of a person becoming Director of Operations—

(i) For the first time ever, have at least 3 years experience, within the past 6 years, as pilot in command of an aircraft operated under part 121 or part 135 of this chapter.

(ii) In the case of a person with previous experience as a Director of Operations, have at least 3 years experience as pilot in command of an aircraft operated under part 121 or part 135 of this chapter.

(c) To serve as Chief Pilot under § 119.69(a) for a certificate holder conducting any operation for which the pilot in command is required to hold an airline transport pilot certificate a person must hold an airline transport pilot certificate with appropriate ratings and be qualified to serve as pilot in command in at least one aircraft used in the certificate holder's operation and:

(1) In the case of a person becoming a Chief Pilot for the first time ever, have at least 3 years experience, within the past 6 years, as pilot in command of an aircraft operated under part 121 or part 135 of this chapter.

(2) In the case of a person with previous experience as a Chief Pilot, have at least 3 years experience as pilot in command of an aircraft operated under part 121 or part 135 of this chapter.

(d) To serve as Chief Pilot under §119.69(a) for a certificate holder that only conducts operations for which the pilot in command is required to hold a commercial pilot certificate, a person must hold at least a commercial pilot certificate. If an instrument rating is required for any pilot in command for that certificate holder, the Chief Pilot must also hold an instrument rating. The Chief Pilot must be qualified to serve as pilot in command in at least one aircraft used in the certificate holder's operation. In addition, the Chief Pilot must:

(1) In the case of a person becoming a Chief Pilot for the first time ever, have at least 3 years experience, within the past 6 years, as pilot in command of an aircraft operated under part 121 or part 135 of this chapter.

(2) In the case of a person with previous experience as a Chief Pilot, have at least 3 years experience as pilot in command of an aircraft operated under part 121 or part 135 of this chapter.

(e) To serve as Director of Maintenance under § 119.69(a) a person must hold a mechanic certificate with airframe and powerplant ratings and either:

(1) Have 3 years of experience within the past 3 years maintaining aircraft as a certificated mechanic, including, at the time of appointment as Director of Maintenance, experience in maintaining the same category and class of aircraft as the certificate holder uses; or

(2) Have 3 years of experience within the past 3 years repairing aircraft in a certificated airframe repair station, including 1 year in the capacity of approving aircraft for return to service.

(f) A certificate holder may request a deviation to employ a person who does not meet the appropriate airman experience requirements, managerial experience requirements, or supervisory experience requirements of this section if the Manager of the Air Transportation Division, AFS-200, or the Manager of the Aircraft Maintenance Division, AFS-300, as appropriate, find that the person has comparable experience, and can effectively perform the functions associated with the position in accordance with the requirements of this chapter and the procedures outlined in the certificate holder's manual. Grants of deviation under this paragraph may be granted after consideration of the size and scope of the operation and the qualifications of the intended personnel. The Administrator may, at any time, terminate any grant of deviation authority issued under this paragraph.

Part 121—Appendix I—Drug testing program

This appendix contains the standards and components that must be included in an antidrug program required by this chapter.

I. General

A. *Purpose.* The purpose of this appendix is to establish a program designed to help prevent accidents and injuries resulting from the use of prohibited drugs by employees who perform safety-sensitive functions.

B. *DOT Procedures.* Each employer shall ensure that drug testing programs conducted pursuant to 14 CFR parts 65, 121, and 135 comply with the requirements of this appendix and the "Procedures for Transportation Workplace Drug Testing Programs" published by the Department of Transportation (DOT) (49 CFR part 40). An employer may not use or contract with any drug testing laboratory that is not certified by the Department of Health and Human Services (HHS) under the National Laboratory Certification Program.

C. *Employer Responsibility.* As an employer, you are responsible for all actions of your officials, representatives, and service agents in carrying out the requirements of this appendix and 49 CFR part 40.

D. *Applicable Federal Regulations.* The following applicable regulations appear in 49 CFR or 14 CFR:

1. 49 CFR

Part 40—Procedures for Transportation Workplace Drug Testing Programs

2. 14CFR

61.14—Refusal to submit to a drug or alcohol test.

65.46—Use of prohibited drugs.

67.107—First-Class Airman Medical Certificate, Mental.

67.207—Second-Class Airman Medical Certificate, Mental.

67.307—Third-Class Airman Medical Certificate, Mental.

135.1—Applicability.

135.249—Use of prohibited drugs.

135.251—Testing for prohibited drugs.

135.353—Prohibited drugs.

E. *Falsification.* No person may make, or cause to be made, any of the following:

1. Any fraudulent or intentionally false statement in any application of an antidrug program.

2. Any fraudulent or intentionally false entry in any record or report that is made, kept, or used to show compliance with this appendix.

3. Any reproduction or alteration, for fraudulent purposes, of any report or record required to be kept by this appendix.

II. Definitions

For the purpose of this appendix, the following definitions apply:

Accident means an occurrence associated with the operation of an aircraft which takes place between the time any person boards the aircraft with the intention of flight and all such persons have disembarked, and in which any person suffers death or serious injury, or in which the aircraft receives substantial damage.

Annualized rate for the purposes of unannounced testing of employees based on random selection means the percentage of specimen collection and testing of employees performing a safety-sensitive function during a calendar year. The employer shall determine the annualized rate by referring to the total number of employees performing a safety-sensitive function for the employer at the beginning of the calendar year.

Contractor is an individual or company that performs a safety-sensitive function by contract for an employer or another contractor.

DOT agency means an agency (or "operating administration") of the United States Department of Transportation administering regulations requiring drug testing (14 CFR part 61 et al.; 46 CFR part 16; 49 CFR parts 199, 219, and 382) in accordance with 49 CFR part 40.

Employee is a person who is hired, either directly or by contract, to perform a safety-sensitive function for an employer, as defined below. An employee is also a person who transfers into a position to perform a safety-sensitive function for an employer.

Employer is a part 121 certificate holder, a part 135 certificate holder, an operator as defined in § 135.1(c) of this chapter, or an air traffic control facility not operated by the FAA or by or under contract to the U.S. military. An employer may use a contract employee who is not included under that employer's FAA-mandated antidrug program to perform a safety-sensitive function only if that contract employee is included under the contractor's FAA-mandated antidrug program and is performing a safety-sensitive function on behalf of that contractor (*i.e.,* within the scope of employment with the contractor.)

Hire means retaining an individual for a safety-sensitive function as a paid employee, as a volunteer, or through barter or other form of compensation.

Performing (a safety-sensitive function): an employee is considered to be performing a safety-sensitive function during any period in which he or she is actually performing, ready to perform, or immediately available to perform such function.

Prohibited drug means marijuana, cocaine, opiates, phencyclidine (PCP), and amphetamines, as specified in 49 CFR 40.85.

Refusal to submit means that a covered employee engages in conduct specified in 49 CFR 40.191.

Safety-sensitive function means a function listed in section III of this appendix.

Substance abuse professional means a licensed physician (Medical Doctor or Doctor of Osteopathy), or a licensed or certified psychologist, social worker, employee assistance professional, or addiction counselor (certified by the National Association of Alcoholism and Drug Abuse Counselors Certification Commission), with knowledge of and clinical experience in the diagnosis and treatment of disorders related to drug use and abuse.

Verified negative drug test result means a drug test result from an HHS-certified laboratory that has undergone review by an MRO and has been determined by the MRO to be a negative result.

Verified positive drug test result means a drug test result from an HHS-certified laboratory that has undergone review by an MRO and has been determined by the MRO to be a positive result.

III. Employees Who Must Be Tested

Each employee, including any assistant, helper, or individual in a training status, who performs a safety-sensitive function listed in this section directly or by contract for an employer as defined in this appendix must be subject to drug testing under an antidrug program implemented in accordance with this appendix. This includes full-time, part-time, temporary, and intermittent employees regardless of the degree of supervision. The safety-sensitive functions are:

A. Flight crewmember duties.

B. Flight attendant duties.

C. Flight instruction duties.

D. Aircraft dispatcher duties.

E. Aircraft maintenance and preventive maintenance duties.

F. Ground security coordinator duties.

G. Aviation screening duties.

H. Air traffic control duties.

IV. Substances for Which Testing Must Be Conducted

Each employer shall test each employee who performs a safety-sensitive function for evidence of marijuana, cocaine, opiates, phencyclidine (PCP), and amphetamines during each test required by section V of this appendix. As part of a reasonable cause drug testing program established pursuant to this part, employers may test for drugs in addition to those specified in this pad only with approval granted by the FAA under 49 CFR part 40 and for substances for which the Department of Health and Human Services has established an approved testing protocol and positive threshold.

Substances for Which Testing Must Be Conducted. Each employer shall test each employee who performs a safety-sensitive function for evidence of marijuana, cocaine, opiates, phencyclidine (PCP), and amphetamines during each test required by section V of this appendix.

V. Types of Drug Testing Required

Each employer shall conduct the following types of testing in accordance with the procedures set forth in this appendix and the DOT "Procedures for Transportation Workplace Drug Testing Programs" (49 CFR part 40):

A. *Pre-Employment Testing.*

1. No employer may hire any individual for a safety-sensitive function listed in section III of this appendix unless the employer first conducts a pre-employment test and receives a verified negative drug test result for that individual.

2. No employer may allow an individual to transfer from a nonsafety-sensitive to a safety-sensitive function unless the employer first conducts a pre-employment test and receives a verified negative drug test result for the individual.

3. Employers must conduct another pro-employment test and receive a verified negative drug test result before hiring or transferring an individual into a safety-sensitive function if more than 180 days elapse between conducting the pre-employment test required by section V.A.1. or V.A.2. of this appendix and hiring or transferring the individual into a safety-sensitive function, resulting in that individual being brought under an FAA drug testing program.

4. If the following criteria are met, an employer is permitted to conduct a pre-employment test, and if such a test is conducted, the employer must receive a negative test result before putting the individual into a safety-sensitive function:

(a) The individual previously performed a safety-sensitive function for the employer and the employer is not required to pre-employment test the individual under section V.A.1. or V.A.2 of this appendix before putting the individual to work in a safety-sensitive function;

(b) The employer removed the individual from the employer's random testing program conducted under this appendix for reasons other than a verified positive test result on an FAA-mandated drug test or a refusal to submit to such testing; and

(c) The individual will be returning to the performance of a safety-sensitive function.

5. Before hiring or transferring an individual to a safety-sensitive function, the employer must advise each individual that the individual will be required to undergo pre-employment testing in accordance with this appendix, to determine the presence of marijuana, cocaine, opiates, phencyclidine (PCP), and amphetamines, or a metabolite of those drugs in the individual's system. The employer shall provide this same notification to undergo pre-employment testing under section V.A.4. of this appendix.

C. *Random Testing.*

1. Except as provided in paragraphs 2-4 of this section, the minimum annual percentage rate for random drug testing shall be 50 percent of covered employees.

2. The Administrator's decision to increase or decrease the minimum annual percentage rate for random drug testing is based on the reported positive rate for the entire

industry. All information used for this determination is drawn from the statistical reports required by section X of this appendix. In order to ensure reliability of the data, the Administrator considers the quality and completeness of the reported data, may obtain additional information or reports from employers, and may make appropriate modifications in calculating the industry positive rate. Each year, the Administrator will publish in the Federal Register the minimum annual percentage rate for random drug testing of covered employees. The new minimum annual percentage rate for random drug testing will be applicable starting January 1 of the calendar year following publication.

3. When the minimum annual percentage rate for random drug testing is 50 percent, the Administrator may lower this rate to 25 percent of all covered employees if the Administrator determines that the data received under the reporting requirements of this appendix for two consecutive calendar years indicate that the reported positive rate is less than 1.0 percent.

4. When the minimum annual percentage rate for random drug testing is 25 percent, and the data received under the reporting requirements of this appendix for any calendar year indicate that the reported positive rate is equal to or greater than 1.0 percent, the Administrator will increase the minimum annual percentage rate for random drug testing to 50 percent of all covered employees.

5. The selection of employees for random drug testing shall be made by a scientifically valid method, such as a random-number table or a computer-based random number generator that is matched with employees' Social Security numbers, payroll identification numbers, or other comparable identifying numbers. Under the selection process used, each covered employee shall have an equal chance of being tested each time selections are made.

6. The employer shall randomly select a sufficient number of covered employees for testing during each calendar year to equal an annual rate not less than the minimum annual percentage rate for random drug testing determined by the Administrator. If the employer conducts random drug testing through a Consortium/Third-party administrator (C/TPA), the number of employees to be tested may be calculated for each individual employer or may be based on the total number of covered employees covered by the C/TPA who are subject to random drug testing at the same minimum annual percentage rate under this part or any DOT drug testing rule.

7. Each employer shall ensure that random drug tests conducted under this appendix are unannounced and that the dates for administering random tests are spread reasonably throughout the calendar year.

8. Each employer shall require that each safety-sensitive employee who is notified of selection for random drug testing proceeds to the collection site immediately; provided, however, that if the employee is performing a safety-sensitive function at the time of the notification, the employer shall instead ensure that the employee ceases to perform the safety-sensitive function and proceeds to the collection site as soon as possible.

9. If an employer is required to conduct random drug testing under the drug testing rules of more than one DOT agency, the employer may—

(a) Establish separate pools for random selection, with each pool containing the covered employees who are subject to testing at the same required rate; or

(b) Randomly select covered employees for testing at the highest percentage rate established for the calendar year by any DOT agency to which the employer is subject.

10. An employer required to conduct random drug testing under the anti drug rules of more than one DOT agency shall provide each such agency access to the employer's records of random drug testing, as determined to be necessary by the agency to ensure the employer's compliance with the rule.

D. *Testing Based on Reasonable Cause.* Each employer must test each employee who performs a safety-sensitive function and who is reasonably suspected of having used a prohibited drug. The decision to test must be based on a reasonable and articulable belief that the employee is using a prohibited drug on the basis of specific contemporaneous physical, behavioral, or performance indicators of probable drug use. At least two of the employee's supervisors, one of whom is trained in detection of the symptoms of possible drug use, must substantiate and concur in the decision to test an employee who is reasonably suspected of drug use; except that in the case of an employer, other than a part 121 certificate holder, who employs 50 or fewer employees who perform safety-sensitive functions, one supervisor who is trained in detection of symptoms of possible drug use must substantiate the decision to test an employee who is reasonably suspected of drug use.

E. *Testing Based on Reasonable Cause.* Each employer shall test each employee who performs a safety-sensitive function and who is reasonably suspected of using a prohibited drug. Each employer shall test an employee's specimen for the presence of marijuana, cocaine, opiates, phencyclidine (PCP), and amphetamines, or a metabolite of those drugs. An employer may test an employee's specimen for the presence of other prohibited drugs or drug metabolites only in accordance with this appendix and the DOT "Procedures for Transportation Workplace Drug Testing Programs" (49 CFR part 40). At least two of the employee's supervisors, one of whom is trained in detection of the symptoms of possible drug use, shall substantiate and concur in the decision to test an employee who is reasonably suspected of drug use; provided, however, that in the case of an employer other than a part 121 certificate holder who employs 50 or fewer employees who perform safety-sensitive functions, one supervisor who is trained in detection of symptoms of possible drug use shall substantiate the decision to test an employee who is reasonably suspected of drug use. The decision to test must be based on a reasonable and articulable belief that the employee is using a prohibited drug on the basis of specific contemporaneous physical, behavioral, or performance indicators of probable drug use.

F. *Return to Duty Testing.* Each employer shall ensure that before an individual is returned to duty to perform a safety-sensitive function after refusing to submit to a drug test required by this appendix or receiving a verified positive drug test result on a test conducted under this appendix the individual shall undergo a return to duty drug test. No employer shall allow an individual required to undergo return to duty testing to perform a safety-sensitive function unless the employer has received a verified negative drug test result for the individual.

The test cannot occur until after the SAP has determined that the employee has successfully complied with the prescribed education and/or treatment.

G. *Follow-up Testing.*

1. Each employer shall implement a reasonable program of unannounced testing of each individual who has been hired to perform or who has been returned to the performance of a safety-sensitive function after refusing to submit to a drug test required by this appendix or receiving a verified positive drug test result on a test conducted under this appendix.

2. The number and frequency of such testing shall be determined by the employer's Substance Abuse Professional conducted in accordance with the provisions of 49 CFR part 40, but shall consist of at least six tests in the first 12 months following the employee's return to duty.

3. The employer must direct the employee to undergo testing for alcohol in accordance with appendix J of this part, in addition to drugs, if the Substance Abuse Professional determines that alcohol testing is necessary for the particular employee. Any such alcohol testing shall be conducted in accordance with the provisions of 49 CFR part 40.

4. Follow-up testing shall not exceed 60 months after the date the individual begins to perform or returns to the performance of a safety-sensitive function. The Substance Abuse Professional may terminate the requirement for follow-up testing at any time after the first six tests have been conducted, if the Substance Abuse Professional determines that such testing is no longer necessary.

VI. Administrative and Other Matters

A. *MRO Record Retention Requirements.*

1. Records concerning drug tests confirmed positive by the laboratory shall be maintained by the MRO for 5 years. Such records include the MRO copies of the custody and control form, medical interviews, documentation of the basis for verifying as negative test results confirmed as positive by the laboratory, any other documentation concerning the MRO's verification process.

2. Should the employer change MROs for any reason, the employer shall ensure that the former MRO forwards all records maintained pursuant to this rule to the new MRO within ten working days of receiving notice from the employer of the new MRO's name and address.

3. Any employer obtaining MRO services by contract, including a contract through a C/TPA, shall ensure that the contract includes a recordkeeping provision that is consistent with this paragraph, including requirements for transferring records to a new MRO.

B. *Access to Records.* The employer and the MRO shall permit the Administrator or the Administrator's representative to examine records required to be kept under this appendix and 49 CFR part 40. The Administrator or the Administrator's representative may require that all records maintained by the service agent for the employer must be produced at the employer's place of business.

C. *Release of Drug Testing Information.* An employer shall release information regarding an employee's drug testing results, evaluation, or rehabilitation to a third party in accordance with 49 CFR part 40. Except as required by law, this appendix, or 49 CFR part 40, no employer shall release employee information.

D. *Refusal to Submit to Testing.*

1. Each employer must notify the FAA within 5 working days of any employee who holds a certificate issued under part 61, part 63, or part 65 of this chapter who has refused to submit to a drug test required under this appendix. Send these notifications to: Federal Aviation Administration, Office of Aerospace Medicine, Drug Abatement Division (AAM—800), 800 Independence Avenue, SW, Washington, DC 20591.

2. Employers are not required to notify the above office of refusals to submit to pre-employment or return to duty testing.

E. *Permanent Disqualification from Service.* An employee who has verified positive drug test results on two drug tests required by Appendix I to part 121 of this chapter and conducted after September 19, 1994 is permanently precluded from performing for an employer the safety-sensitive duties the employee performed prior to the second drug test.

2. An employee who has engaged in prohibited drug use during the performance of a safety-sensitive function after September 19, 1994 is permanently precluded from performing that safety-sensitive function for an employer.

F. *DOT Management Information System Annual Reports.* Copies of any annual reports submitted to the FAA under this appendix must be maintained by the employer for a minimum of 5 years.

VII. Medical Review Officer/Substance Abuse Professional

The employer shall designate or appoint a Medical Review Officer (MRO) who shall be qualified in accordance with 49 CFR part 40 and shall perform the functions set forth in 49 CFR part 40 and this appendix. If the employer does not have a qualified individual on staff to serve as MRO, the employer may contract for the provision of MRO services as part of its drug testing program.

A. *Medical Review Officer (MRO).* The MRO must perform the functions set forth in 49 CFR part 40, Subpart G, and this appendix. The MRO shall not delay verification of the primary test result following a request for a split specimen test unless such delay is based on reasons other than the fact that the split specimen test result is pending. If the primary test result is verified as positive, actions required under this role (e.g., notification to the Federal Air Surgeon, removal from safety-sensitive position) are not stayed during the 72-hour request period or pending receipt of the split specimen test result.

B. *Substance Abuse Professional (SAP).* The SAP must perform the functions set forth in 49 CFR part 40, Subpart O.

C. *Additional Medical Review Officer, Substance Abuse Professional, and Employer Responsibilities Regarding 14 CFR part 67 Airman Medical Certificate Holders.*

1. As part of verifying a confirmed positive test result, the MRO shall inquire, and the individual shall disclose, whether the individual is or would be required to hold a medical certificate issued under 14 CFR part 67 to perform a safety-sensitive function for the employer, if the individual answers in the negative, the MRO shall then inquire, and the individual shall disclose whether the individual currently holds a medical certificate issued under 14 CFR part 67. If the individual answers in the affirmative to either question, in addition to notifying the employer in accordance with 49 CFR part 40, the MRO must forward to the Federal Air Surgeon, at the address listed in paragraph 5, the name of the individual, along with identifying information and supporting documentation, within 12 working days after verifying a positive drug test result.

2. The SAP shall inquire, and the individual shall disclose, whether the individual is or would be required to hold a medical certificate issued under 14 CFR part 67 of this chapter to perform a safety sensitive function for the employer, if the individual answers in the affirmative, the SAP cannot recommend that the individual be returned to a safety-sensitive function that requires the individual to hold a 14 CFR part 67 medical certificate unless and until such individual has received a medical certificate or a special issuance medical certificate from the Federal Air Surgeon. The receipt of a medical certificate or a special issuance medical certificate does not alter any obligations otherwise required by 49 CFR part 40 or this appendix.

3. The employer must forward to the Federal Air Surgeon a copy of any report provided by the SAP, if available, regarding an individual for whom the MRO has provided a report to the Federal Air Surgeon under section VII.C.1 of this appendix, within 12 working days of the employer's receipt of the report.

4. The employer cannot permit an employee who is required to hold a medical certificate under part 67 of this chapter to perform a safety-sensitive duty to resume that duty until the employee has received a medical certificate or a special issuance medical certificate from the Federal Air Surgeon unless and until the employer has ensured that the employee meets the return-to-duty requirements in accordance with 49 CFR part 40.

5. Reports required under this section shall be forwarded to the Federal Air Surgeon, Federal Aviation Administration, Office of Aerospace Medicine, Attn: Drug Abatement Division (AAM—800), 800 Independence Avenue, SW., Washington, DC 20591.

VIII. Employee Assistance Program (EAP)

The employer shall provide an EAP for employees. The employer may establish the EAP as a part of its internal personnel services or the employer may contract with an entity that will provide EAP services to an employee. Each EAP must include education and training on drug use for employees and training for supervisors making determinations for testing of employees based on reasonable cause.

A. *EAP Education Program.* Each EAP education program must include at least the following elements: display and distribution of informational material; display and distribution of a community service hot-line telephone number for employee assistance; and display and distribution of the employer's policy regarding drug use in the workplace. The employer's policy shall include information regarding the consequences under the rule of using drugs while performing safety-sensitive functions, receiving a verified positive drug test result, or refusing to submit to a drug test required under the rule.

B. *EAP Training Program.* Each employer shall implement a reasonable program of initial training for employees. The employee training program must include at least the following elements: The effects and consequences of drug use on personal health, safety, and work environment; the manifestations and behavioral cues that may indicate drug use and abuse; and documentation of training given to employees and employer's supervisory personnel. The employer's supervisory personnel who will determine when an employee is subject to testing based on reasonable cause shall receive specific training on specific, contemporaneous physical, behavioral, and performance indicators of probable drug use in addition to the training specified above. The employer shall ensure that supervisors who will make reasonable cause determinations receive at least 60 minutes of initial training. The employer shall implement a reasonable recurrent training program for supervisory personnel making reasonable cause determinations during subsequent years. The employer shall identify the employee and supervisor EAP training in the employer's drug testing plan submitted to the FAA for approval.

IX. Implementing an Antidrug Program.

A. Each company must meet the requirements of this appendix. Use the following chart to determine whether your company must obtain an Antidrug and Alcohol Misuse Prevention Program Operations Specification or whether you must register with the FAA:

If you are...	You must...
1. A part 121 or 135 certificate holder	Obtain an Antidrug and Alcohol Misuse Prevention Program Operations Specification by contacting your FAA Principal Operations Inspector.
2. A sightseeing operator as defined in § 135.1(c) of this chapter	Register with the FAA, Office of Aerospace Medicine, Drug Abatement Division (AAM—810),, 800 Independence Avenue, SW, Washington, DC 20591 by March 12, 2004.
3. An air traffic control facility not operated by the FAA or by or under contract to the U.S. Military	Register with the FAA, Office of Aerospace Medicine, Drug Abatement Division (AAM—810), 800 Independence Avenue, SW, Washington, DC 20591 by March 12, 2004
4. A part 145 certificate holder who has your own antidrug program	Obtain an Antidrug and Alcohol Misuse Prevention Program Operations Specification by contacting your Principal Maintenance Inspector.
5. A contractor who has your own antidrug program	Register with the FAA, Office of Aerospace Medicine, Drug Abatement Division (AAM—810), 800 Independence Avenue, SW, Washington, DC 20591 by March 12, 2004.

B. Use the following chart for implementing an antidrug program is you are applying for a part 121 or 135 certificate, if you intend to begin sightseeing operations as defined in § 135.1(c) of this chapter, or if you intend to begin air traffic control operations (not operated by the FAA or by or under contract to the U.S. military.) Use it to determine whether you need to have an Antidrug and Alcohol Misuse Prevention Program Operations Specifications, or whether you need to register with the FAA. Your employees who perform safety-sensitive duties must be tested in accordance with this appendix. The chart follows:

If you are...	You must...
1 Apply for a part 121 certificate or apply for a part 135 certificate	a. Have an Antidrug and Alcohol Misuse Prevention Program operations Specification. b. Implement an FAA antidrug program no later than the date you start operations, and c. Meet the requirements of this appendix.
2. Intend to begin sightseeing operations as defined in § 135.1(c) of this chapter	a. Register with the FAA Office of Aerospace Medicine, Drug Abatement Division (AAM—810), 800 Independence Avenue, SW, Washington, DC 20591 prior to starting operations. b. Implement an FAA antidrug program no later than the date you start operations, and c. Meet the requirements of this appendix.
3. Intend to begin air traffic control operations (at an air traffic control facility not operated by the FAA or by or under contract to the U.S.military.	a. Register with the FAA, Office of Aerospace Medicine, Drug Abatement Division (AAM—800), 800 Independence Avenue, SW Washington, DC 20591, b. Implement an FAA antidrug program no later than the date you start operations, and c. Meet the requirement of this appendix.

C. 1. If you are an individual or a company that intends to provide safety-sensitive services by contract to a part 121 or 135 certificate holder, a sightseeing operation as defined in § 135.1(c) of this chapter, or an air traffic control facility not operated by the FAA or by or under contract to the U.S. military, use the chart in paragraph C.2 of this section to determine what you must do if you if you opt to have your own antidrug program.

2. The following chart explains what you must do if you have your own antidrug program:

If you...	You must...
a. Are a part 145 certificate holder	i. Have an Antidrug and Alcohol Misuse Prevention Program Operations Specification, ii Implement an FAA Antidrug Program no later than the date you start performing safety-sensitive functions for a part 121 or 135 certificate holder or sightseeing operator as defined in § 135.1(c) of this chapter, and iii. Meet the requirements of this appendix as if you were an employer.
b. Are a contractor (e,g,, a security com[amy, a non-certificated repair station, a temporary employment service company or any other individual or company that provides safety-sensitive services,)	i. Register with the FAA Office of Aerospace Medicine, Drug Abatement Division (AAM—810), 800 Independence Avenue SW, Washington, DC 20591 ii. Implement an FAA Antidrug Program lo later than the date you start performing safety-sensitive functions for a part 121 0r 135 certificate holder, a sightseeing operator as defines in § 135.1(c) of this chapter, or an air traffic control facility not operated by the FAA or by or under contract to the U.S. military, and iii. Meet the requirements of this appendix as if you were an employer.

D. 1. To obtain an Antidrug and Alcohol Misuse Prevention Program Operations Specification, you must contact your FAA Principal Operations Inspector or Principal Maintenance Inspector. Provide him/her with the following information:

a. Company name.

b. Certificate number.

c. Telephone number.

d. Address where your Antidrug and Alcohol Misuse Prevention Program records are kept.

e. Whether you have 50 or more safety-sensitive employees, or 49 or fewer safety-sensitive employees. (Part 121 certificate holders are not required to provide this information.)

2. You must certify on your Antidrug and Alcohol Misuse Prevention Program Operations Specification issued by your FAA Principal Operations Inspector or Principal Maintenance Inspector that you will comply with this appendix, appendix J of this part, and 49 CFR part 40.

3. You are required to obtain only one Antidrug and Alcohol Misuse Prevention Program Operations Specification to satisfy this requirement under this appendix and appendix J of this part.

4. You must update the Antidrug and Alcohol Misuse Prevention Program Operations Specification when any changes to the information contained in the Operation Specification occur.

E. 1. To register with the FAA, submit the following information:

a. Company name.

b. Telephone number

c. Address where your Antidrug and Alcohol Misuse Prevention Program reords are kept.

d. Type of safety-sensitive functions you perform for an employer (such as flight instruction duties, aircraft dispatcher duties, maintenance or preventive maintenance duties, ground security coordinator duties, aviation screening duties, air traffic control duties).

e. Whether you have 50 or more safety-sensitive employees, or 49 or fewer covered employees.

f. A signed statement indicating that: your company will comply with this appendix, appendix J of this part, and 49 CFR part 40; and, if you are a contractor, you intend to provide safety-sensitive functions by contract to a part **121** or part **135** certificate holder, a sightseeing operator as defined in § 135.1(c) of this chapter, or an air traffic control facility not operated by the FAA or by or under contract to the U.S. military.

2. Send this information in the form and manner prescribed by the Administrator, in duplicate to: The Federal Aviation Administration, Office of Aerospace Medicine, Drug Abatement Division (AAM—810), 800 Independence Avenue, SW., Washington, DC 20591

3. Update the registration information as changes occur. Send the updates in duplicate to the address specified in paragraph 2.

4. This registration will satisfy the registration requirements for both your Antidrug Program under this appendix and your Alcohol Misuse Prevention Program under appendix J of this part.

X. Annual Reports.

A. Annual reports of testing results must be submitted to the FAA by March 15 of the succeeding calendar year for the prior calendar year (January 1 through December 31) in accordance with the provisions below.

1. Each part 121 certificate holder shall submit an annual report each year.

2. Each entity conducting an antidrug program under this part, other than a part 121 certificate holder, that has 50 or more employees performing a safety-sensitive function on January 1 of any calendar year shall submit an annual report to the FAA for that calendar year.

3. The Administrator reserves the right to require that aviation employers not otherwise required to submit annual reports prepare and submit such reports to the FAA. Employers that will be required to submit annual reports under this provision will be notified in writing by the FAA.

B. As an employer, you must use the Management Information System (MIS) form and instructions as required by 49 CFR part 40 (at 49 CFR 40.26 and appendix H to 49 CFR part 40). You may also use the electronic version of the MIS form provided by DOT. The Administrator may designate means (e.g., electronic program transmitted via the Internet) other than hard-copy, for MIS form submission. For information on where to submit M1S forms and for the electronic version of the form, see: *http://www. faa.gov/avr/aam/adap.*

C. A service agent may prepare the MIS report on behalf of an employer. However, a company official (e.g, Designated Employer Representative as defined in 49 CFR part 40) must certify the accuracy and completeness of the MIS report, no matter who prepares it.

XI. Preemption

A. The issuance of 14 CFR parts 65, 121, and 135 by the FAA preempts any state or local law, rule, regulation, order, or standard covering the subject matter of 14 CFR parts 65, 121, and 135, including but not limited to, drug testing of aviation personnel performing safety-sensitive functions.

B. The issuance of 14 CFR parts 65, 121, and 135 does not preempt provisions of state criminal law that impose sanctions for reckless conduct of an individual that leads to actual loss of life, injury, or damage to property whether such provisions apply specifically to aviation employees or generally to the public.

XII. Testing Outside the Territory of the United States.

A. No part of the testing process (including specimen collection, laboratory processing, and MRO actions) shall be conducted outside the territory of the United States.

1. Each employee who is assigned to perform safety-sensitive functions solely outside the territory of the United States shall be removed from the random testing pool upon the inception of such assignment.

2. Each covered employee who is removed from the random testing pool under this paragraph A shall be returned to the random testing pool when the employee resumes the performance of safety-sensitive functions wholly or partially within the territory of the United States.

B. The provisions of this appendix shall not apply to any person who performs a function listed in section III of this appendix by contract for an employer outside the territory of the United States.

XIII. Waivers from 49 CFR 40.21. An employer subject to this part may petition the Drug Abatement Division, Office of Aviation Medicine, for a waiver allowing the employer to stand down an employee following a report of a laboratory confirmed positive drug test or refusal, pending the outcome of the verification process.

A. Each petition for a waiver must be in writing and include substantial facts and justification to support the waiver. Each petition must satisfy the substantive requirements for obtaining a waiver, as provided in 49 CFR 40.21.

B. Each petition for a waiver must be submitted to the Federal Aviation Administration, Office of Aviation Medicine, Drug Abatement Division (AAM—800), 800 Independence Avenue, SW, Washington, DC 20591.

C. The Administrator may grant a waiver subject to 49 CFR 40.2 1(d).

Appendix J—Alcohol misuse prevention program

This appendix contains the standards and components that must be included in an alcohol misuse prevention program required by this chapter.

I. General

A. *Purpose.* The purpose of this appendix is to establish programs designed to help prevent accidents and injuries resulting from the misuse of alcohol by employees who perform safety-sensitive functions in aviation.

B. *Alcohol testing procedures.* Each employer shall ensure that all alcohol testing conducted pursuant to this appendix complies with the procedures set forth in 49 CFR part 40. The provisions of 49 CFR part 40 that address alcohol testing are made applicable to employers by this appendix.

C. *Employer Responsibility.* As an employer, you are responsible for all actions of your officials, representatives, and service agents in carrying out the requirements of the DOT agency regulations.

D. *Definitions.*

As used In this appendix—

Accident means an occurrence associated with the operation of an aircraft which takes place between the time any person boards the aircraft with the intention of flight and the time all such persons have disembarked, and in which any person suffers death or serious injury or in which the aircraft receives substantial damage.

Alcohol means the intoxicating agent in beverage alcohol, ethyl alcohol, or other low molecular weight alcohols, including methyl or isopropyl alcohol.

Alcohol concentration (or content) means the alcohol in a volume of breath expressed in terms of grams of alcohol per 210 liters of breath as indicated by an evidential breath test under this appendix.

Alcohol use means the consumption of any beverage, mixture, or preparation, including any medication, containing alcohol.

Confirmation test means a second test, following a screening test with a result 0.02 or greater, that provides quantitative data of alcohol concentration.

Consortium means an entity, including a group or association of employers or contractors, that provides alcohol testing as required by this appendix and that acts on behalf of such employers or contractors, provided that it has submitted an alcohol misuse prevention program certification statement to the FAA in accordance with this appendix.

Contractor means an individual or company that performs a safety-sensitive function by contract for an employer or another contractor.

Covered employee means a person who performs, either directly or by contract, a safety-sensitive function listed in section II of this appendix for an employer as (defined below). For purposes of pre-employment only, the term "covered employee" includes a person applying to perform a safety-sensitive function.

DOT agency means an agency (or "operating administration") of the United States Department of Transportation administering regulations requiring alcohol testing (14 CFR parts 65, 121, and 135; 49 CFR parts 199, 219, and 382) in accordance with 49 CFR part 40.

Employer means a part 121 certificate holder; a part 135 certificate holder; an air traffic control facility not operated by the FAA or by or under contract to the U.S. military; and an operator as defined in 14 CFR 135.1(c).

Performing (a safety-sensitive function): an employee is considered to be performing a safety-sensitive function during any period in which he or she is actually performing, ready to perform, or immediately available to perform such functions.

Refusal to submit means that a covered employee engages in conduct specified in 49 CFR 40.261.

Safety-sensitive function means a function listed in section II of this appendix.

Screening test means an analytical procedure to determine whether a covered employee may have a prohibited concentration of alcohol in his or her system.

Substance abuse professional means a licensed physician (Medical Doctor or Doctor of Osteopathy), or a licensed or certified psychologist, social worker, employee assistance professional, or an addiction counselor (certified by the National Association of Alcoholism and Drug Abuse Counselors Certification Commission) with knowledge of and clinical experience in the diagnosis and treatment of alcohol-related disorders.

Violation rate means the number of covered employees (as reported under section IV of this appendix) found during random tests given under this appendix to have an alcohol concentration of 0.04 or greater plus the number of employees who refused a random test required by this appendix, divided by the total reported number of employees in the industry given random alcohol tests under this appendix plus the total reported number of employees in the industry who refuse a random test required by this appendix.

E. *Preemption of State and local laws.*

1. Except as provided in subparagraph 2 of this paragraph, these regulations preempt any State or local law, rule, regulation, or order to the extent that:

(a) Compliance with both the State or local requirement and this appendix is not possible; or

(b) Compliance with the State or local requirement is an obstacle to the accomplishment and execution of any requirement in this appendix.

2. The alcohol misuse requirements of this title shall not be construed to preempt provisions of State criminal law that impose sanctions for reckless conduct leading to actual loss of life, injury, or damage to property, whether the provisions apply specifically to transportation employees or employers or to the general public.

F. *Other requirements imposed by employers.*

Except as expressly provided in these alcohol misuse requirements, nothing in these requirements shall be construed to affect the authority of employers, or the rights of employees, with respect to the use or possession of alcohol, including any authority and rights with respect to alcohol testing and rehabilitation.

G. *Requirement for notice.*

Before performing an alcohol test under this appendix, each employer shall notify a covered employee that the alcohol test is required by this appendix. No employer shall falsely represent that a test is administered under this appendix.

H. *Applicable Federal Regulations.*

The following applicable regulations appear in 49 CFR and 14 CFR:

1. 49 CFR

Part 40—Procedures for Transportation Workplace Drug Testing Programs

2. 14 CFR

61.14—Refusal to submit to a drug or alcohol test.

63.12b—Refusal to submit to a drug or alcohol test.

65.23—Refusal to submit to a drug or alcohol test.

65.46a—Misuse of Alcohol.

65.46b—Testing for Alcohol.

67.107—First-Class Airman Medical Certificate, Mental.

67.207—Second-Class Airman Medical Certificate, Mental.

67.307—Third-Class Airman Medical Certificate, Mental.

135.1—Applicability.

135.253—Misuse of alcohol.

135.255—Testing for alcohol.

I. *Falsification.*

No person may make, or cause to be made, any of the following:

1. Any fraudulent or intentionally false statement in any application of an alcohol misuse prevention program.

2. Any fraudulent or intentionally false entry in any record or report that is made, kept, or used to show compliance with this appendix.

3. Any reproduction or alteration, for fraudulent purposes, of any report or record required to be kept by this appendix.

II. Covered employees

A. Each employee, including any assistant, helper, or individual in a training status, who performs a safety-sensitive function listed in this section directly or by contract for an employer as defined in this appendix must be subject to alcohol testing under an alcohol misuse prevention program implemented in accordance with this appendix. This not only includes full-time and part-time employees, but temporary and intermittent employees regardless of the degree of supervision. The safety-sensitive functions are:

1. Flight crewmember duties.

2. Flight attendant duties.

3. Flight instruction duties.

4. Aircraft dispatcher duties.

5. Aircraft maintenance or preventive maintenance duties.

6. Ground security coordinator duties.

7. Aviation screening duties.

8. Air traffic control duties.

III. Tests required

A. *Pre-employment testing*

As an employer, you may, but are not required to, conduct pre-employment alcohol testing under this part. If you choose to conduct pre-employment alcohol testing, you must comply with the following requirements:

1. You must conduct a pre-employment alcohol test before the first performance of safety-sensitive functions by every covered employee (whether a new employee or someone who has transferred to a position involving the performance of safety-sensitive functions).

2. You must treat all safety-sensitive employees performing safety-sensitive functions the same for the purpose of pre-employment alcohol testing (i.e., you must not test some covered employees and not others).

3. You must conduct the pre-employment tests after making a contingent offer of employment or transfer, subject to the employee passing the pre-employment alcohol test.

4. You must conduct all pre-employment alcohol tests using the alcohol testing procedures of 49 CFR Part 40.

5. You must not allow a covered employee to begin performing safety-sensitive functions unless the result of the employee's test indicates an alcohol concentration of less than 0.04. If a pre-employment test result under this paragraph indicates an alcohol concentration of 0.02 or greater but less than 0.04, the provisions of paragraph F of section V of this appendix apply.

B. *Post-accident testing*

1. As soon as practicable following an accident, each employer shall test each surviving covered employee for alcohol if that employee's performance of a safety-sensitive function either contributed to the accident or cannot be completely discounted as a contributing factor to the accident. The decision not to administer a test under this section shall be based on the employer's determination, using the best available information at the time of the determination, that the covered employee's performance could not have contributed to the accident.

2. If a test required by this section is not administered within 2 hours following the accident, the employer shall prepare and maintain on file a record stating the reasons the test was not promptly administered. If a test required by this section is not administered within 8 hours following the accident, the employer shall cease attempts to administer an alcohol test and shall prepare and maintain the same record. Records shall be submitted to the FAA upon request of the Administrator or his or her designee.

3. A covered employee who is subject to post-accident testing shall remain readily available for such testing or may be deemed by the employer to have refused to submit to testing. Nothing in this section shall be construed to require the delay of necessary medical attention for injured people following an accident or to prohibit a covered employee from leaving the scene of an accident for the period necessary to obtain assistance in responding to the accident or to obtain necessary emergency medical care.

C. *Random testing*

1. Except as provided in paragraphs 2–4 of this section, the minimum annual percentage rate for random alcohol testing will be 25 percent of the covered employees.

2. The Administrator's decision to increase or decrease the minimum annual percentage rate for random alcohol testing is based on the violation rate for the entire industry. All information used for this determination is drawn from alcohol MIS reports required by this appendix. In order to ensure reliability of the data, the Administrator considers the quality and completeness of the reported data, may obtain additional information or reports from employers, and may make appropriate modifications in calculating the industry violation rate. Each year, the Administrator will publish in the **Federal Register** the minimum annual percentage rate for random alcohol testing of covered employees. The new minimum annual percentage rate for random alcohol testing will be applicable starting January 1 of the calendar year following publication.

3. (a) When the minimum annual percentage rate for random alcohol testing is 25 percent or more, the Adminis-

trator may lower this rate to 10 percent of all covered employees if the Administrator determines that the data received under the reporting requirements of this appendix for two consecutive calendar years indicate that the violation rate is less than 0.5 percent.

(b) When the minimum annual percentage rate for random alcohol testing is 50 percent, the Administrator may lower this rate to 25 percent of all covered employees if the Administrator determines that the data received under the reporting requirements of this appendix for two consecutive calendar years indicate that the violation rate is less than 1.0 percent but equal to or greater than 0.5 percent.

4. (a) When the minimum annual percentage rate for random alcohol testing is 10 percent, and the data received under the reporting requirements of this appendix for that calendar year indicate that the violation rate is equal to or greater than 0.5 percent but less than 1.0 percent, the Administrator will increase the minimum annual percentage rate for random alcohol testing to 25 percent of all covered employees.

(b) When the minimum annual percentage rate for random alcohol testing is 25 percent or less, and the data received under the reporting requirements of this appendix for that calendar year indicate that the violation rate is equal to or greater than 1.0 percent, the Administrator will increase the minimum annual percentage rate for random alcohol testing to 50 percent of all covered employees.

5. The selection of employees for random alcohol testing shall be made by a scientifically valid method, such as a random-number table or a computer-based random number generator that is matched with employees' Social Security numbers, payroll identification numbers, or other comparable identifying numbers. Under the selection process used, each covered employee shall have an equal chance of being tested each time selections are made.

6. The employer shall randomly select a sufficient number of covered employees for testing during each calendar year to equal an annual rate not less than the minimum annual percentage rate for random alcohol testing determined by the Administrator. If the employer conducts random testing through a Consortium/Third-party administrator (C/TPA), the number of employees to be tested may be calculated for each individual employer or may be based on the total number of covered employees who are subject to random alcohol testing at the same minimum annual percentage rate under this appendix or any DOT alcohol testing rule.

7. Each employer shall ensure that random alcohol tests conducted under this appendix are unannounced and that the dates for administering random tests are spread reasonably throughout the calendar year.

8. Each employer shall require that each covered employee who is notified of selection for random testing proceeds to the testing site immediately; provided, however,

that if the employee is performing a safety-sensitive function at the time of the notification, the employer shall instead ensure that the employee ceases to perform the safety-sensitive function and proceeds to the testing site as soon as possible.

9. A covered employee shall only be randomly tested while the employee is performing safety-sensitive functions; just before the employee is to perform safety-sensitive functions; or just after the employee has ceased performing such functions.

10. If a given covered employee is subject to random alcohol testing under the alcohol testing rules of more than one DOT agency, the employee shall be subject to random alcohol testing at the percentage rate established for the calendar year by the DOT agency regulating more than 50 percent of the employee's functions.

11. If an employer is required to conduct random alcohol testing under the alcohol testing rules of more than one DOT agency, the employer may—

(a) Establish separate pools for random selection, with each pool containing the covered employees who are subject to testing at the same required rate; or

(b) Randomly select such employees for testing at the highest percentage rate established for the calendar year by any DOT agency to which the employer is subject.

D. *Reasonable Suspicion Testing*

1. An employer shall require a covered employee to submit to an alcohol test when the employer has reasonable suspicion to believe that the employee has violated the alcohol misuse prohibitions in § 65.46a, 121.458, or 135.253 of this chapter.

2. The employer's determination that reasonable suspicion exists to require the covered employee to undergo an alcohol test shall be based on specific, contemporaneous, articulable observations concerning the appearance, behavior, speech or body odors of the employee. The required observations shall be made by a supervisor who is trained in detecting the symptoms of alcohol misuse. The supervisor who makes the determination that reasonable suspicion exists shall not conduct the breath alcohol test on that employee.

3. Alcohol testing is authorized by this section only if the observations required by paragraph 2 are made during, just preceding, or just after the period of the workday that the covered employee is required to be in compliance with this rule. An employee may be directed by the employer to undergo reasonable suspicion testing for alcohol only while the employee is performing safety-sensitive functions; just before the employee is to perform safety-sensitive functions; or just after the employee has ceased performing such functions.

4. (a) For the years stated in this paragraph, employers who submit MIS reports shall submit to the FAA each record of a test required by this section that is not completed within 8 hours. The employer's records of tests that are not

completed within 8 hours shall be submitted to the FAA by March 15, 1996; March 15, 1997; and March 15, 1998; for calendar years 1995, 1996, and 1997, respectively. Employers shall append these records to their MIS submissions. Each record shall include the following information:

(i) Type of test (reasonable suspicion/post-accident);

(ii) Triggering event (including date, time, and location);

(iii) Employee category (do *not* include employee name or other identifying information);

(iv) Reason(s) test could not be completed within 8 hours; and

(v) If blood alcohol testing could have been completed within eight hours, the name, address, and telephone number of the testing site where blood testing could have occurred.

(b) Notwithstanding the absence of a reasonable suspicion alcohol test under this section, no covered employee shall report for duty or remain on duty requiring the performance of safety-sensitive functions while the employee is under the influence of or impaired by alcohol, as shown by the behavioral, speech, or performance indicators of alcohol misuse, nor shall an employer permit the covered employee to perform or continue to perform safety-sensitive functions until:

(1) An alcohol test is administered and the employee's alcohol concentration measures less than 0.02; or

(2) The start of the employee's next regularly scheduled duty period, but not less than 8 hours following the determination made under paragraph 2 of this section that there is reasonable suspicion that the employee has violated the alcohol misuse provisions in § 65.46a, 121.458, or 135.253 of this chapter.

(c) No employer shall take any action under this appendix against a covered employee based solely on the employee's behavior and appearance in the absence of an alcohol test. This does not prohibit an employer with authority independent of this appendix from taking any action otherwise consistent with law.

E. *Return to Duty Testing*

Each employer shall ensure that before a covered employee returns to duty requiring the performance of a safety-sensitive function after engaging in conduct prohibited in § 65.46a, § 121.458, or § 135.253 of this chapter, the employee shall undergo a return to duty alcohol test with a result indicating an alcohol concentration of less than 0.02. The test cannot occur until after the SAP has determined that the employee has successfully complied with the prescribed education and/or treatment.

F. *Follow-up Testing*

1. Each employer shall ensure that the employee who engages in conduct prohibited by §65.46a, §121.458, or § 135.253 of this chapter is subject to unannounced follow-up alcohol testing as directed by a SAP.

2. The number and frequency of such testing shall be determined by the employer's SAP, but must consist of at least six tests in the first 12 months following the employee's return to duty.

3. The employer must direct the employee to undergo testing for drugs in accordance with appendix I of this part, in addition to alcohol, if the SAP determines that drug testing is necessary for the particular employee. Any such drug testing shall be conducted in accordance with the provisions of 49 CFR part 40.

4. Follow-up testing shall not exceed 60 months after the date the individual begins to perform or returns to the performance of a safety-sensitive function. The SAP may terminate the requirement for follow-up testing at any time after the first six tests have been conducted, if the SAP determines that such testing is no longer necessary.

5. A covered employee shall be tested for alcohol under this paragraph only while the employee is performing safety-sensitive functions, just before the employee is to perform safety-sensitive functions, or just after the employee has ceased performing such functions.

G. *Retesting of Covered Employees with an Alcohol Concentration of 0.02 or Greater but Less Than 0.04*

Each employer shall retest a covered employee to ensure compliance with the provisions of section V, paragraph F of this appendix. If the employer chooses to permit the employee to perform a safety-sensitive function within 8 hours following the administration of an alcohol test indicating an alcohol concentration of 0.02 or greater but less than 0.04.

IV. Handling of test results, record retention, and confidentiality

A. *Retention of Records*

1. *General Requirement.* In addition to the records required to be maintained under 49 CFR part 40, employers must maintain records required by this appendix in a secure location with controlled access.

2. *Period of retention.*

(a) *Five years.*

(1) Copies of any annual reports submitted to the FAA under this appendix for a minimum of 5 years.

(2) Records of notifications to the Federal Air Surgeon of violations of the alcohol misuse prohibitions in this chapter by covered employees who hold medical certificates issued under part 67 of this chapter.

(3) Documents presented by a covered employee to dispute the result of an alcohol test administered under this appendix.

(4) Records related to other violations of §65.46a, § 121.458, or § 135.253 of this chapter.

(b) *Two years.* Records related to the testing process and training required under this appendix.

(1) Documents related to the random selection process.

(2) Documents generated in connection with decisions to administer reasonable suspicion alcohol tests.

(3) Documents generated in connection with decisions on post-accident tests.

(4) Documents verifying existence of a medical explanation of the inability of a covered employee to provide adequate breath for testing.

(5) Materials on alcohol misuse awareness, including a copy of the employer's policy on alcohol misuse.

(6) Documentation of compliance with the requirements of section VI, paragraph A of this appendix.

(7) Documentation of training provided to supervisors for the purpose of qualifying the supervisors to make a determination concerning the need for alcohol testing based on reasonable suspicion.

(8) Certification that any training conducted under this appendix complies with the requirements for such training.

(1) The employer's copy of the alcohol test form, including the results of the test;

(2) Documents related to the refusal of any covered employee to submit to an alcohol test required by this appendix.

(3) Documents presented by a covered employee to dispute the result of an alcohol test administered under this appendix.

(c) Records related to other violations of §§65.46a, 121.458, or 135.253 of this chapter.

(d) Records related to evaluations:

(1) Records pertaining to a determination by a substance abuse professional concerning a covered employee's need for assistance.

(2) Records concerning a covered employee's compliance with the recommendations of the substance abuse professional.

(3) Records of notifications to the Federal Air Surgeon of violations of the alcohol misuse prohibitions in this chapter by covered employees who hold medical certificates issued under part 67 of this chapter.

(e) Records related to education and training:

(1) Materials on alcohol misuse awareness, including a copy of the employer's policy on alcohol misuse.

(2) Documentation of compliance with the requirements of section VI, paragraph A of this appendix.

(3) Documentation of training provided to supervisors for the purpose of qualifying the supervisors to make a determination concerning the need for alcohol testing based on reasonable suspicion.

(4) Certification that any training conducted under this appendix complies with the requirements for such training.

B. *Reporting of Results in a Management Information System*

1. Annual reports summarizing the results of alcohol misuse prevention programs shall be submitted to the FAA in the form and manner prescribed by the Administrator by March 15 of each year covering the previous calendar year (January 1 through December 31) in accordance with the provisions below.

(a) Each part 121 certificate holder shall submit an annual report each year.

(b) Each entity conducting an alcohol misuse prevention program under the provisions of this appendix, other than a part 121 certificate holder, that has 50 or more covered employees on January 1 of any calendar year shall submit an annual report to the FAA for that calendar year.

(c) The Administrator reserves the right to require employers not otherwise required to submit annual reports to prepare and submit such reports to the FAA. Employers that will be required to submit annual reports under this provision will be notified in writing by the FAA.

2. Each employer that is subject to more than one DOT agency alcohol rule shall identify each employee covered by the regulations of more than one DOT agency. The identification will be by the total number and category of covered function. Prior to conducting any alcohol test on a covered employee subject to the rules of more than one DOT agency, the employer shall determine which DOT agency rule or rules authorizes or requires the test. The test result information shall be directed to the appropriate DOT agency or agencies.

3. Each employer shall ensure the accuracy and timeliness of each report submitted.

4. Each report shall be submitted in the form and manner prescribed by the Administrator.

5. Each report shall be signed by the employer's alcohol misuse prevention program manager or other designated representative.

6. Each report that contains information on an alcohol screening test result of 0.02 or greater or a violation of the alcohol misuse provisions of § 65.46a, 121.458, or 135.253 of this chapter shall include the following informational elements:

(a) Number of covered employees by employee category.

(b) Number of covered employees in each category subject to alcohol testing under the alcohol misuse rule of another DOT agency, identified by each agency.

(c)(1) Number of screening tests by type of test and employee category.

(2) Number of confirmation tests, by type of test and employee category.

(d) Number of confirmation alcohol tests indicating an alcohol concentration of 0.02 or greater but less than 0.04 by type of test and employee category.

(e) Number of confirmation alcohol tests indicating an alcohol concentration of 0.04 or greater, by type of test and employee category.

(f) Number of persons denied a position as a covered employee following a pre-employment alcohol test indicating an alcohol concentration of 0.04 or greater.

(g) Number of covered employees with a confirmation alcohol test indicating an alcohol concentration of 0.04 or greater who were returned to duty in covered positions (having complied with the recommendations of a substance abuse professional as described in 49 CFR part 40).

(h) Number of covered employees who were administered alcohol and drug tests in the same time, with both a positive drug test result and an alcohol test result indicating an alcohol concentration of 0.04 or greater.

(i) Number of covered employees who were found to have violated other alcohol misuse provisions of §§ 65.46a, 121.458, or 135.253 of this chapter, and the action taken in response to the violation.

(j) Number of covered employees who refused to submit to an alcohol test required under this appendix, the number of such refusals that were not random tests, and the action taken in response to each refusal.

(k) Number of supervisors who have received required training during the reporting period in determining the existence of reasonable suspicion of alcohol misuse.

7. Each report with no screening test results of 0.02 or greater or violations of the alcohol misuse provisions of §§ 65.46a, 121.458, or 135.253 of this chapter shall include the following informational elements. (This report may only be submitted if the program results meet these criteria.)

(a) Number of covered employees by employee category.

(b) Number of covered employees in each category subject to alcohol testing under the alcohol misuse rule of another DOT agency, identified by each agency.

(c) Number of screening tests by type of test and employee category.

(d) Number of covered employees who engaged in alcohol misuse who were returned to duty in covered positions (having complied with the recommendations of a substance abuse professional as described in 49 CFR part 40).

(e) Number of covered employees who refused to submit to an alcohol test required under this appendix, and the action taken in response to each refusal.

(f) Number of supervisors who have received required training during the reporting period in determining the existence of reasonable suspicion of alcohol misuse.

8. A C/TPA may prepare reports on behalf of individual aviation employers for purposes of compliance with this reporting requirement. However, the aviation employer shall sign and submit such a report and shall remain responsible for ensuring the accuracy and timeliness of each report prepared on its behalf by a C/TPA. A C/TPA must not sign the form.

C. *Access to Records and Facilities*

1. Except as required by law or expressly authorized or required in this appendix, no employer shall release covered employee information that is contained in records required to be maintained under this appendix.

2. A covered employee is entitled, upon written request, to obtain copies of any records pertaining to the employee's use of alcohol, including any records pertaining to his or her alcohol tests in accordance with 49 CFR part 40. The employer shall promptly provide the records requested by the employee. Access to an employee's records shall not be contingent upon payment for records other than those specifically requested.

3. Each employer shall permit access to all facilities utilized in complying with the requirements of this appendix to the Secretary of Transportation or any DOT agency with regulatory authority over the employer or any of its covered employees.

4. Each employer shall make available copies of all results of alcohol testing conducted under this appendix and any other information pertaining to the employer's alcohol misuse prevention program, when requested by the Secretary of Transportation or any DOT agency with regulatory authority over the employer or covered employee.

V. Consequences for employees engaging in alcohol-related conduct

A. *Removal from Safety-sensitive Function*

1. Except as provided in 49 CFR part 40, no covered employee shall perform safety-sensitive functions if the employee has engaged in conduct prohibited by § 65.46a, 121.458, or 135.253 of this chapter or an alcohol misuse rule of another DOT agency.

2. No employer shall permit any covered employee to perform safety-sensitive functions if the employer has determined that the employee has violated this paragraph.

B. *Permanent Disqualification from Service*

An employee who violates §§ 65.46a(c), 121.458(c), or 135.253(c) of this chapter, or who engages in alcohol use that violates another alcohol misuse provision of §§65.46a, 121.458, or 135.253 of this chapter and had previously engaged in alcohol use that violated the provision of §§65.46a, 121.458, or 135.253 of this chapter after becoming subject to such prohibitions is permanently precluded from performing for an employer the safety-sensitive duties the employee performed before such violation.

C. *Notice to the Federal Air Surgeon*

1. An employer who determines that a covered employee who holds an airman medical certificate issued under part 67 of this chapter has engaged in alcohol use that violated the alcohol misuse provisions of §§65.46a, 121.458, or 135.253 of this chapter shall notify the Federal Air Surgeon within 2 working days.

2. Each such employer shall forward to the Federal Air Surgeon a copy of the report of any evaluation performed under the provisions of section VI.C. of this appendix within 2 working days of the employer's receipt of the report.

3. All documents must be sent to the Federal Air Surgeon, Office of Aviation Medicine, Federal Aviation Administration, Attn: Drug Abatement Division (AAM-800), 800 Independence Avenue, SW., Washington, DC 20591.

4. No covered employee who is required to bold a medical certificate under part 67 of this chapter to perform a safety-sensitive duty shall perform that duty following a violation of this appendix until and unless the Federal Air Surgeon has recommended that the employee be permitted to perform such duties.

5. Once the Federal Air Surgeon has recommended under paragraph C.4. of this section that the employee be permitted

to perform safety-sensitive duties, the employer cannot permit the employee to perform those safety-sensitive duties until the employer has ensured that the employee meets the return to duty requirements in accordance with 49 CFR part 40.

D. *Notice of Refusals*

1. Except as provided in subparagraph 2 of this paragraph D, each employer shall notify the FAA within 5 working days of any covered employee who holds a certificate issued under 14 CFR part 61, part 63, or part 65 who has refused to submit to an alcohol test required under this appendix. Notifications must be sent to: Federal Aviation Administration, Office of Aviation Medicine, Drug Abatement Division (AAM-800), 800 Independence Avenue, SW., Washington, DC 20591.

2. An employer is not required to notify the above office of refusals to submit to preemployment alcohol tests or refusals to duty tests.

E. *Required Evaluation and Testing*

No covered employee who has engaged in conduct prohibited by § 65.46a, 121.458, or 135.253 of this chapter shall perform safety-sensitive functions unless the employee has met the requirements of 49 CFR part 40. No employer shall permit a covered employee who has engaged in such conduct to perform safety-sensitive functions unless the employee has met the requirements of 49 CFR part 40.

F. *Other Alcohol-Related Conduct*

1. No covered employee tested under the provisions of section III of this appendix who is found to have an alcohol concentration of 0.02 or greater but less than 0.04 shall perform or continue to perform safety-sensitive functions for an employer, nor shall an employer permit the employee to perform or continue to perform safety-sensitive functions, until:

(a) The employee's alcohol concentration measures less than 0.02; or

(b) The start of the employee's next regularly scheduled duty period, but not less than 8 hours following administration of the test.

2. Except as provided in subparagraph 1 of this paragraph, no employer shall take any action under this rule against an employee based solely on test results showing an alcohol concentration less than 0.04. This does not prohibit an employer with authority independent of this rule from taking any action otherwise consistent with law.

VI. Alcohol misuse information, training, and substance abuse professional

A. *Employer Obligation to Promulgate a Policy on the Misuse of Alcohol*

1. *General requirements.* Each employer shall provide educational materials that explain these alcohol misuse requirements and the employer's policies and procedures with respect to meeting those requirements.

(a) The employer shall ensure that a copy of these materials is distributed to each covered employee prior to the start of alcohol testing under the employer's FAA-mandated alcohol misuse prevention program and to each person subsequently hired for or transferred to a covered position.

(b) Each employer shall provide written notice to representatives of employee organizations of the availability of this information.

2. *Required content.* The materials to be made available to employees shall include detailed discussion of at least the following:

(a) The identity of the person designated by the employer to answer employee questions about the materials.

(b) The categories of employees who are subject to the provisions of these alcohol misuse requirements.

(c) Sufficient information about the safety-sensitive functions performed by those employees to make clear what period of the workday the covered employee is required to be in compliance with these alcohol misuse requirements.

(d) Specific information concerning employee conduct that is prohibited by this chapter.

(e) The circumstances under which a covered employee will be tested for alcohol under this appendix.

(f) The procedures that will be used to test for the presence of alcohol, protect the employee and the integrity of the breath testing process, safeguard the validity of the test results, and ensure that those results are attributed to the correct employee.

(g) The requirement that a covered employee submit to alcohol tests administered in accordance with this appendix.

(h) An explanation of what constitutes a refusal to submit to an alcohol test and the attendant consequences.

(i) The consequences for covered employees found to have violated the prohibitions in this chapter, including the requirement that the employee be removed immediately from performing safety-sensitive functions, and the process in 49 CFR part 40, subpart O.

(j) The consequences for covered employees found to have an alcohol concentration of 0.02 or greater but less than 0.04.

(k) Information concerning the effects of alcohol misuse on an individual's health, work, and personal life: signs and symptoms of an alcohol problem; and available methods of evaluating and resolving problems associated with the misuse of alcohol; and intervening when an alcohol problem is suspected, including confrontation, referral to any available employee assistance program, and/or referral to management.

(l) *Optional provisions.* The materials supplied to covered employees may also include information on additional employer policies with respect to the use or possession of alcohol, including any consequences for an employee found to have a specified alcohol level, that are based on the employer's authority independent of this appendix. Any such additional policies or consequences must be clearly and obviously described as being based on independent authority.

B. *Training for Supervisors*

Each employer shall ensure that persons designated to determine whether reasonable suspicion exists to require a covered employee to undergo alcohol testing under section II of this appendix receive at least 60 minutes of training on the physical, behavioral, speech, and performance indicators of probable alcohol misuse.

C. *Substance Abuse Professional (SAP) Duties*

The SAP must perform the functions set forth in 49 CFR part 40, Subpart 0, and this appendix.

VII. How lo Implement an Alcohol Misuse Prevention Program

A. Each company must meet the requirements of this appendix. Use the following chart to determine whether your company must obtain an Antidrug and Alcohol Misuse Prevention Program Operations Specification or whether you must register with the FAA:

If you are...	You must...
1. A part 121 or 135 certificate holder	Obtain an Antidrug and Alcohol Misuse Prevention Program Operations Specification by contacting your FAA Principal Operations Inspector.
2. A sightseeing operator as defined in § 135.1(c)	Register with the FAA, Office of Aerospace Medicine, Drug Abatement Division (AAM—810), 800 Independence Avenue, SW, Washington, DC 20591 by March 12, 2004.
3. An air traffic control facility not operated by the FAA or by or under contract to the U. S. military	Register with the FAA, Office of Aerospace Medicine, Drug Abatement Division (AAM—810), 800 Independence Avenue, SW, Washington DC 20591 by March 12, 2004
4. A part 145 certificate holder who has your own alcohol misuse prevention program	Obtain an Antidrug and Alcohol Misuse Prevention Program Operations Specification by contacting your FAA Principal Maintenance Inspector.
5. A contractor who has your own alcohol misuse prevention program	Register with the FAA, Office of Aerospace Medicine, Drug Abatement Division (AAM—810), 800 Independence Avenue, SW, Washington, DC 20591 by March 12, 2004.

B. Use the following chart for implementing an Alcohol Misuse Prevention Program if you are applying for a part 121 or 135 certificate, if you intend to begin sightseeing operations as defined in § 135.1(c) of this chapter, or if you intend to begin air traffic control operations (not operated by the FAA or by or under contract to the U.S. military.) Use it to determine whether you need to have an Antidrug and Alcohol Misuse Prevention Program Operations Specification, or whether you need to register with the FAA. Your employees who perform safety-sensitive duties must be tested in accordance with this appendix. The chart follows:

If you...	You must...
1 Apply for a part 121 certificate or apply for a part 135 certificate	a. Have an Antidrug and Alcohol Misuse Prevention operations b. Implement an FAA Alcohol Misuse Prevention Program no later than the date you start operations, and c. Meet the requirements of this appendix.
2. Intend to begin sightseeing operations as defined in § 135.1(c) of this chapter...	a. Register with the FAA Office of Aerospace Medicine, Drug Abatement Division (AAM—810), 800 Independence Avenue, SW, Washington, DC 20591 prior to starting operations b. Implement an FAA Alcohol Misuse Prevention Program no later than the date you start operations, and c. Meet the requirements of this appendix.
3. Intend to begin air traffic control operations (at an air traffic control facility not operated by the FAA or by or under contract to the U.S. military...	a. Register with the FAA Office of Aerospace Medicine, Drug Abatement Division (AAM—800), 800 Independence Avenue, SW, 800 b. Implement the Alcohol Misuse Prevention Program no later than the date you start operations, and c. Meet the requirements of this appendix.

C. 1. If you are an individual or a company that intends to provide safety-sensitive services by contract to a part 121 or 135 certificate holder or a sightseeing operator as defined in § 135.1(c) of this chapter, use the chart in paragraph C.2 of this section to determine what you must do if you opt to have your own Alcohol Misuse Prevention Program.

2. The following chart explains what you must do if you have your own Alcohol Misuse Prevention Program:

If you...	You must...
a. Are a part 145 certificate holder........	i. Have an Antidrug and Alcohol Misuse Prevention Program Operations Specification, ii. Implement an FAA Alcohol Misuse Prevention Program no later than the date you start performing safety-sensitive functions for a part 121 or 135 certificate holder or sightseeing operator as defined in § 135.1(c) of this chapter, and iii. Meet the requirements of this appendix as if you were an employer.
b. Are a contractor (e,g., a security company, a non-certificated repair station, a temporary employment service company or any other individual or company that provides safety-sensitive services,)	i. Register with the FAA, Office of Aerospace Medicine, Drug Abatement Division (AAM—810), 800 Independence Avenue, SW Washington DC, 20591 ii. Implement an FAA Alcohol Misuse Prevention Program no later than the date you start performing safety-sensitive functions for a part 121 or 135 certificate holder or sightseeing operator as defined in § 135.1(c) of this chapter, and iii. Meet the requirements of this appendix as if you were an employer.

D. 1. To obtain an Antidrug and Alcohol Misuse Prevention Program Operations Specification, you must contact your FAA Principal Operations Inspector or Principal Maintenance Inspector. Provide him/her with the following information:

a. Company name.

b. Certificate number.

c. Telephone number.

d. Address where your Antidrug and Alcohol Misuse Prevention Program records are kept:

e. Whether you have 50 or more covered employees, or 49 or fewer covered employees. (Part 121 certificate holders are not required to provide this information.)

2. You must certify on your Antidrug and Alcohol Misuse Prevention Program Operations Specification, issued by your FAA Principal Operations Inspector or Principal Maintenance Inspector, that you will comply with appendix I of this part, this appendix, and 49 CFR part 40.

3. You are required to obtain only one Antidrug and Alcohol Misuse Prevention Program Operations Specification to satisfy this requirement under appendix I of this part and this appendix.

4. You must update the Antidrug and Alcohol Misuse Prevention Program Operations Specification when any changes to the information contained in the Operation Specification occur.

E. 1. To register with the FAA, submit the following information:

a. Company name.

b. Telephone number.

c. Address where your Antidrug and Alcohol Misuse Prevention Program records are kept.

d. Type of safety-sensitive functions you perform for an employer (such as flight instruction duties, aircraft dispatcher duties, maintenance or preventive maintenance duties, ground security coordinator duties, aviation screening duties, air traffic control duties).

e. Whether you have 50 or more covered employees, or 49 or fewer covered employees.

f. A signed statement indicating that: your company will comply with this appendix, appendix I of this part, and 49 CFR part 40; and, if you are a contractor, you intend to provide safety-sensitive functions by contract to a part 121 or part 135 certificate holder, a sightseeing operator as defined by § 135.1(c) of this chapter, or an air traffic control facility not operated by the FAA or by or under contract to the U.S. military.

2. Send this information in the form and manner prescribed by the Administrator, in duplicate to: The Federal Aviation Administration, Office of Aerospace Medicine, Drug Abatement Division (AAM—810), 800 Independence Avenue, SW., Washington, DC 20591.

3. Update the registration information as changes occur. Send the updates in duplicate to the address specified in paragraph 2.

4. This registration will satisfy the registration requirements for both your Antidrug Program under appendix I of

this part and your Alcohol Misuse Prevention Program under this appendix.

VIII. Employees located outside the U.S.

A. No covered employee shall be tested for alcohol misuse while located outside the territory of the United States.

1. Each covered employee who is assigned to perform safety-sensitive functions solely outside the territory of the United States shall be removed from the random testing pool upon the inception of such assignment.

2. Each covered employee who is removed from the random testing pool under this paragraph shall be returned to the random testing pool when the employee resumes the performance of safety-sensitive functions wholly or partially within the territory of the United States.

B. The provisions of this appendix shall not apply to any person who performs a safety-sensitive function by contract for an employer outside the territory of the United States.

Part 135—Operating requirements: commuter and on-demand operations

Subpart A—General

§ 135.1 Applicability.

(a) This part prescribes rules governing—

(1) The commuter or on-demand operations of each person who holds or is required to hold an Air Carrier Certificate or Operating Certificate under part 119 of this chapter.

(2) Each person employed or used by a certificate holder conducting operations under this part including the maintenance, preventative maintenance and alteration of an aircraft.

(3) The transportation of mail by aircraft conducted under a postal service contract awarded under 39 U.S.C. 5402c.

(4) Each person who applies for provisional approval of an Advanced Qualification program curriculum, curriculum segment, or portion of a curriculum segment under SFAR No. 58 of 14 CFR part 121 and each person employed or used by an air carrier or commercial operator under this part to perform training, qualification, or evaluation functions under an Advanced Qualification Program under SFAR No. 58 of 14 CFR part 121.

(5) Nonstop sightseeing flights for compensation or hire that begin and end at the same airport, and are conducted within a 25 statute mile radius of that airport; however, except for operations subject to SFAR 50-2; these operations, when conducted for compensation or hire, must comply only with §§135.249, 135.251, 135.253, 135.255, and 135.353.

(6) Each person who is on board an aircraft being operated under this part.

(7) Each person who is an applicant for an Air Carrier Certificate or an Operating Certificate under 119 of this chapter, when conducting proving tests.

(b) Reserved.

(c) For the purpose of §§ 135.249, 135.251, 135.253, 135.255, and 135.353, *operator* means any person or entity conducting non-stop sight-seeing flights for compensation or hire in an airplane or rotorcraft that begin and end at the same airport and are conducted within a 25 statute mile radius of that airport.

(d) [Notwithstanding the provisions of this part and Appendices I and J to Part 121 of this chapter, an operator who does not hold a Part 121 or Part 135 certificate is permitted to use a person who is otherwise authorized to perform aircraft maintenance or preventive maintenance duties and who is not subject to FAA-approved anti-drug and alcohol misuse prevention programs to perform—

(1) [Aircraft maintenance or preventive maintenance on the operator's aircraft if the operator would otherwise be required to transport the aircraft more than 50 nautical miles further than the repair point closest to operator's principal place of operation to obtain these services; or

(2) [Emergency repairs on the operator's aircraft if the aircraft cannot be safely operated to a location where an employee subject to FAA-approved programs can perform the repairs.]

(Amdt. 135-5, Eff. 7/1/80); (Amdt. 135-7, Eff. 2/1/81); (Amdt. 135-20, Eff. 1/6/87); (Amdt. 135-28, Eff. 12/21/88); (Amdt. 135-32, Eff. 8/18/90); (Amdt. 135-37, Eff. 10/1/90); (Amdt. 135-40, Eff. 10/5/91); [(Amdt. 135-48, Eff. 3/17/94)]

§ 135.2 Compliance schedule for operators that transition to part 121 of this chapter; certain new entrant operators.

(a) *Applicability.* This section applies to the following:

(1) Each certificate holder that was issued an air carrier or operating certificate and operations specifications under the requirements of part 135 of this chapter or under SFAR No. 38-2 of 14 CFR part 121 before January 19, 1996, and that conducts scheduled passenger-carrying operations with:

(i) Nontransport category turbopropeller powered airplanes type certificated after December 31, 1964, that have a passenger seat configuration of 10–19 seats:

(ii) Transport category turbopropeller powered airplanes that have a passenger seat configuration of 20–40 seats; or

(iii) Turbojet engine powered airplanes having a passenger seat configuration of 1–30 seats.

(2) Each person who, after January 19, 1996, applies for or obtains an initial air carrier or operating certificate and operations specifications to conduct scheduled passenger-carrying operations in the kinds of airplanes described in paragraphs (a)(1)(i), (a)(1)(ii), or paragraph (a)(1)(iii) of this section.

(b) *Obtaining operations specifications.* A certificate holder described in paragraph (a)(1) of this section may not, after March 20, 1997, operate an airplane described in paragraphs (a)(1)(i), (a)(1)(ii), or (a)(1)(iii) of this section in scheduled passenger-carrying operations, unless it obtains operations specifications to conduct its scheduled operations under part 121 of this chapter on or before March 20, 1997.

(c) *Regular or accelerated compliance.* Except as provided in paragraphs (d), (e), and (i) of this section, each certificate holder described in paragraphs (a)(1) of this section shall comply with each applicable requirement of part 121 of this chapter on and after March 20, 1997 or on and after the date on which the certificate holder is issued operations specifications under this part, whichever occurs first. Except as provided in paragraphs (d) and (e) of this section, each person described in paragraph (a)(2) of this section shall comply with each applicable requirement of part 121 of this chapter on and after the date on which that person is issued a certificate and operations specifications under part 121 of this chapter.

(d) *Delayed compliance dates.* Unless paragraph (e) of this section specifies an earlier compliance date, no certificate holder that is covered by paragraph (a) of this section may operate an airplane in 14 CFR part 121 operations on or after a date listed in this paragraph unless that airplane meets the applicable requirement of this paragraph:

(1) *Nontransport category turbopropeller powered airplanes type certificated after December 31, 1964, that have a passenger seat configuration of 10–19 seats.* No certificate holder may operate under this part an airplane that is described in paragraph (a)(1)(i) of this section on or after a date listed in paragraph (d)(1)(i), (ii), and (iii) of this section unless that airplane meets the applicable requirement listed in paragraph (d)(1)(i), (ii), and (iii) of this section:

(i) December 20, 1997:

(A) Section 121.289, Landing gear aural warning.

(B) Section 121.308, Lavatory fire protection.

(C) Section 121.310(e), Emergency exit handle illumination.

(D) Section 121.337(b)(8), Protective breathing equipment.

(E) Section 121.340, Emergency flotation means.

(ii) December 20, 1999: Section 121.342, Pitot heat indication system.

(iii) December 20, 2010:

(iv) March 12, 1999: Section 121.310(b)(1), Interior emergency exit locating sign.

(A) For airplanes described in § 121.157(f), the Airplane Performance Operating Limitations in §§ 121.189 through 121.197.

(B) Section 121.161(b), Ditching approval.

(C) Section 121.305(j), Third attitude indicator.

(D) Section 121.312(c), Passenger seat cushion flammability.

(2) *Transport category turbopropeller powered airplanes that have a passenger seat configuration of 20–30 seats.* No certificate holder may operate under this part an airplane that is described in paragraph (a)(1)(ii) of this section on or after a date listed in this paragraph (d) unless that airplane meets the applicable requirement listed in this paragraph (d) (2) of this section:

(i) December 20, 1997:

(A) Section 121.308, Lavatory fire protection.

(B) Section 121.337(b) (8) and (9), Protective breathing equipment.

(C) Section 121.340, Emergency flotation means.

(ii) December 20, 2010: Section 121.305(j), Third attitude indicator.

(e) *Newly manufactured airplanes.* No certificate holder that is described in paragraph (a) of this section may operate under part 121 of this chapter an airplane manufactured on or after a date listed in this paragraph (e) unless that airplane meets the applicable requirement listed in this paragraph (e).

(1) For nontransport category turbopropeller powered airplanes type certificated after December 31, 1964, that have a passenger seat configuration of 10–19 seats:

(i) Manufactured on or after March 20, 1997:

(A) Section 121.305(j), Third attitude indicator.

(B) Section 121.311(f), Safety belts and shoulder harnesses.

(ii) Manufactured on or after December 20, 1997: Section 121.317(a), Fasten seat belt light.

(iii) Manufactured on or after December 20, 1999: Section 121.293, Takeoff warning system.

(iv) Manufactured on or after March 12, 1999: Section 121.21(b)(1), Interior emergency exit locating sign.

(2) For transport category turbopropeller powered airplanes that have a passenger seat configuration of 20–30 seats manufactured on or after March 20, 1997: Section 121.305(j), Third attitude indicator.

(f) *New type certification requirements.* No person may operate an airplane for which the application for a type certificate was filed after March 29, 1995, in 14 CFR part 121 operations unless that airplane is type certificated under part 25 of this chapter.

(g) *Transition plan.* Before March 19, 1996 each certificate holder described in paragraph (a)(1) of this section must submit to the FAA a transition plan (containing a calendar of events) for moving from conducting its scheduled operations under the commuter requirements of part 135 of this chapter to the requirements for domestic or flag operations under part 121 of this chapter. Each transition plan must contain details on the following:

(1) Plans for obtaining new operations specifications authorizing domestic or flag operations;

(2) Plans for being in compliance with the applicable requirements of part 121 of this chapter on or before March 20, 1997; and

(3) Plans for complying with the compliance date schedules contained in paragraphs (d) and (e) of this section.

§ 135.3 Rules applicable to operations subject to this part.

(a) Each person operating an aircraft in operations under this part shall—

(1) While operating inside the United States, comply with the applicable rules of this chapter; and

(2) While operating outside the United States, comply with Annex 2, Rules of the Air, to the Convention on International Civil Aviation or the regulations of any foreign country, whichever applies, and with any rules of parts 61 and 91 of this chapter and this part that are more restrictive than that Annex or those regulations and that can be complied with without violating that Annex or those regulations. Annex 2 is incorporated by reference in § 91.703(b) of this chapter.

(b) After March 19, 1997, each certificate holder that conducts commuter operations under this part with airplanes in which two pilots are required by the type certification rules of this chapter, or with airplanes having a passenger seating configuration, excluding any pilot seat, of 10 seats or more, shall comply with subparts N and O of part 121 instead of the requiremens of subparts E, G, and H of this part. Each affected certificate holder must submit to the Administrator and obtain approval of a transition plan (containing a calendar of events) for moving from its present part 135 training, checking, testing, and qualification requirements to the requirements of part 121 of this chapter. Each transition plan must be submitted by March 19, 1996, and must contain details on how the certificate holder plans to be in compliance with subparts N and O of part 121 on or before March 19, 1997.

(c) If authorized by the Administrator upon application, each certificate holder that conducts operations under this part to which paragraph (b) of this section does not apply, may comply with the applicable sections of subparts N and O of part 121 instead of the requirements of subparts E, G, and H of this part, except that those authorized certificate holders may choose to comply with the operating experience requirements of § 135.244, instead of the requirements of § 121.434 of this chapter.

§ 135.4 Applicability of rules for eligible on-demand operations.

(a) An "eligible on-demand operation" is an on-demand operation conducted under this part that meets the following requirements:

(1) *Two-pilot crew.* The flightcrew must consist of at least two qualified pilots employed or contracted by the certificate holder.

(2) *Flight crew experience.* The crewmembers must have met the applicable requirements of part 61 of this chapter and have the following experience and ratings:

(i) Total flight time for all pilots:

(A) Pilot in command—A minimum of 1,500 hours.

(B) Second in command—A minimum of 500 hours.

(ii) For multi-engine turbine-powered fixed-wing and powered-lift aircraft, the following FAA certification and ratings requirements:

(A) Pilot in command—Airline transport pilot and applicable type ratings.

(B) Second in command—Commercial pilot and instrument ratings.

(iii) For all other aircraft, the following FAA certification and rating requirements:

(A) Pilot in command—Commercial pilot and instrument ratings.

(B) Second in command—Commercial pilot and instrument ratings.

(3) *Pilot operating limitations.* If the second in command of a fixed-wing aircraft has fewer than 100 hours of flight time as second in command flying in the aircraft make and model and, if a type rating is required, in the type aircraft being flown, and the pilot in command is not an appropriately qualified check pilot, the pilot in command shall make all takeoffs and landings in any of the following situations:

(i) Landings at the destination airport when a Destination Airport Analysis is required by § 135.385(f); and

(ii) In any of the following conditions:

(A) The prevailing visibility for the airport is at or below \3/4\ mile.

(B) The runway visual range for the runway to be used is at or below 4,000 feet.

(C) The runway to be used has water, snow, slush, ice, or similar contamination that may adversely affect aircraft performance.

(D) The braking action on the runway to be used is reported to be less than "good."

(E) The crosswind component for the runway to be used is in excess of 15 knots.

(F) Windshear is reported in the vicinity of the airport.

(G) Any other condition in which the pilot in command determines it to be prudent to exercise the pilot in command's authority.

(4) *Crew pairing.* Either the pilot in command or the second in command must have at least 75 hours of flight time in that aircraft make or model and, if a type rating is required, for that type aircraft, either as pilot in command or second in command.

(b) The Administrator may authorize deviations from paragraphs (a)(2)(i) or (a)(4) of this section if the Flight Standards District Office that issued the certificate holder's operations specifications finds that the crewmember has comparable experience, and can effectively perform the functions associated with the position in accordance with the requirements of this chapter. The Administrator may, at any time, terminate any grant of deviation authority issued under this paragraph. Grants of deviation under this paragraph may be granted after consideration of the size and scope of the operation, the qualifications of the intended personnel and the following circumstances:

(1) A newly authorized certificate holder does not employ any pilots who meet the minimum requirements of paragraphs (a)(2)(i) or (a)(4) of this section.

(2) An existing certificate holder adds to its fleet a new category and class aircraft not used before in its operation.

(3) An existing certificate holder establishes a new base to which it assigns pilots who will be required to become qualified on the aircraft operated from that base.

(c) An eligible on-demand operation may comply with alternative requirements specified in §§135.225(b), 135.385(f), and 135.387(b) instead of the requirements that apply to other on-demand operations.

§ 135.5 Removed

§ 135.7 Applicability of rules to unauthorized operators.

The rules in this part which apply to a person certificated under part 119 of this chapter also apply to a person who engages in any operation governed by this part without an appropriate certificate and operations specifications required by part 119 of this chapter.

§ 135.9 Removed

§ 135.10 Removed

§ 135.11 Removed

§ 135.12 Previously trained crewmembers.

A certificate holder may use a crewmember who received the certificate holder's training in accordance with subparts E, G, and H of this part before March 19, 1997 without complying with initial training and qualification requirements of subparts N and O of part 121 of this chapter. The crewmember must comply with the applicable recurrent training requirements of part 121 of this chapter.

§ 135.13 Removed

§ 135.15 Removed

§ 135.17 Removed

§ 135.19 Emergency operations.

(a) In an emergency involving the safety of persons or property, the certificate holder may deviate from the rules of this part relating to aircraft and equipment and weather minimums to the extent required to meet that emergency.

(b) In an emergency involving the safety of persons or property, the pilot in command may deviate from the rules of this part to the extent required to meet that emergency.

(c) Each person who, under the authority of this section, deviates from a rule of this part shall, within 10 days, excluding Saturdays, Sundays, and Federal holidays, after the deviation, send to the FAA Flight Standards District Office charged with the overall inspection of the certificate holder a complete report of the aircraft operation involved, including a description of the deviation and reasons for it.

§ 135.21 Manual requirements.

(a) Each certificate holder, other than one who uses only one pilot in the certificate holder's operations, shall prepare and keep current a manual setting forth the certificate holder's procedures and policies acceptable to the Administrator. This manual must be used by the certificate holder's flight, ground, and maintenance personnel in conducting its operations. However, the Administrator may authorize a deviation from this paragraph if the Administrator finds that, because of the limited size of the operation, all or part of the manual is not necessary for guidance of flight, ground, or maintenance personnel.

(b) Each certificate holder shall maintain at least one copy of the manual at its principal base of operations.

(c) The manual must not be contrary to any applicable Federal regulations, foreign regulation applicable to the certificate holder's operations in foreign countries, or the certificate holder's operating certificate or operations specifications.

(d) A copy of the manual, or appropriate portions of the manual (and changes and additions) shall be made available to maintenance and ground operations personnel by the certificate holder and furnished to—

(1) Its flight crewmembers; and

(2) Representatives of the Administrator assigned to the certificate holder.

(e) Each employee of the certificate holder to whom a manual or appropriate portions of it are furnished under paragraph (d)(1) of this section shall keep it up to date with the changes and additions furnished to them.

(f) Except as provided in paragraph (h) of this section, each certificate holder must carry appropriate parts of the manual on each aircraft when away from the principal operations base. The appropriate parts must be available for use by ground or flight personnel.

(g) For the purpose of complying with paragraph (d) of this section, a certificate holder may furnish the persons listed therein with all or part of its manual in printed form or other form, acceptable to the Administrator, that is retrievable in the English language. If the certificate holder furnishes all or part of the manual in other than printed form, it must ensure there is a compatible reading device available to those persons that provides a legible image of the information and instructions, or a system that is able to retrieve the information and instructions in the English language.

(h) If a certificate holder conducts aircraft inspections or maintenance at specified stations where it keeps the approved inspection program manual, it is not required to carry the manual aboard the aircraft en route to those stations.

(Amdt. 135-18, Eff. 8/2/82)

§ 135.23 Manual contents.

Each manual shall have the date of the last revision on each revised page. The manual must include—

(a) The name of each management person required under § 119.69(a) of this chapter who is authorized to act for the certificate holder, the person's assigned area of responsibility, the person's duties, responsibilities, and authority, and the name and title of each person authorized to exercise operational control under § 135.77;

(b) Procedures for ensuring compliance with aircraft weight and balance limitations and, for multiengine aircraft, for determining compliance with § 135.185;

(c) Copies of the certificate holder's operations specifications or appropriate extracted information, including area of operations authorized, category and class of aircraft authorized, crew complements, and types of operations authorized;

(d) Procedures for complying with accident notification requirements;

(e) Procedures for ensuring that the pilot in command knows that required airworthiness inspections have been made and that the aircraft has been approved for return to service in compliance with applicable maintenance requirements;

(g) Procedures to be followed by the pilot in command for determining that mechanical irregularities or defects reported for previous flights have been corrected or that correction has been deferred;

(h) Procedures to be followed by the pilot in command to obtain maintenance, preventive maintenance, and servicing of the aircraft at a place where previous arrangements have not been made by the operator, when the pilot is authorized to so act for the operator;

(i) Procedures under § 135.179 for the release for, or continuation of, flight if any item of equipment required for the particular type of operation becomes inoperative or unserviceable en route;

(j) Procedures for refueling aircraft, eliminating fuel contamination, protecting from fire (including electrostatic protection), and supervising and protecting passengers during refueling;

(k) Procedures to be followed by the pilot in command in the briefing under § 135.117;

(l) Flight locating procedures, when applicable;

(m) Procedures for ensuring compliance with emergency procedures, including a list of the functions assigned each category of required crewmembers in connection with an emergency and emergency evacuation duties under § 135.123;

(n) En route qualification procedures for pilots, when applicable;

(o) The approved aircraft inspection program, when applicable;

(p) (1) Procedures and information, as described in paragraph (p)(2) of this section, to assist each crewmember and person performing or directly supervising the following job functions involving items for transport on an aircraft:

(i) Acceptance;

(ii) Rejection;

(iii) Handling;

(iv) Storage incidental to transport;

(v) Packaging of company material; or

(vi) Loading.

(2) Ensure that the procedures and information described in this paragraph are sufficient to assist a person in identifying packages that are marked or labeled as containing hazardous materials or that show signs of containing undeclared hazardous materials. The procedures and information must include:

(i) Procedures for rejecting packages that do not conform to the Hazardous Materials Regulations in 49 CFR parts 171 through 180 or that appear to contain undeclared hazardous materials;

(ii) Procedures for complying with the hazardous materials incident reporting requirements of 49 CFR 171.15 and 171.16 and discrepancy reporting requirements of 49 CFR 175.31.

(iii) The certificate holder's hazmat policies and whether the certificate holder is authorized to carry, or is prohibited from carrying, hazardous materials; and

(iv) If the certificate holder's operations specifications permit the transport of hazardous materials, procedures and information to ensure the following:

(A) That packages containing hazardous materials are properly offered and accepted in compliance with 49 CFR parts 171 through 180;

(B) That packages containing hazardous materials are properly handled, stored, packaged, loaded and carried on board an aircraft in compliance with 49 CFR parts 171 through 180;

(C) That the requirements for Notice to the Pilot in Command (49 CFR 175.33) are complied with; and

(D) That aircraft replacement parts, consumable materials or other items regulated by 49 CFR parts 171 through 180 are properly handled, packaged, and transported.

(q) Procedures for the evacuation of persons who may need the assistance of another person to move expeditiously to an exit if an emergency occurs; and

(r) If required by § 135.385, an approved Destination Airport Analysis establishing runway safety margins at destination airports, taking into account the following factors as supported by published aircraft performance data supplied by the aircraft manufacturer for the appropriate runway conditions—

(1) Pilot qualifications and experience;

(2) Aircraft performance data to include normal, abnormal and emergency procedures as supplied by the aircraft manufacturer;

(3) Airport facilities and topography; (4) Runway conditions (including contamination);

(4) Airport or area weather reporting; (6) Appropriate additional runway safety margins, if required;

(5) Airplane inoperative equipment;

(6) Environmental conditions; and

(7) Other criteria affecting aircraft performance.

(s) Other procedures and policy instructions regarding the certificate holder's operations Issued by the certificate holder.

(Amdt. 135-20, Eff. 1/6/87)

§ 135.25 Aircraft requirements.

(a) Except as provided in paragraph (d) of this section, no certificate holder may operate an aircraft under this part unless that aircraft—

(1) Is registered as a civil aircraft of the United States and carries an appropriate and current airworthiness certificate issued under this chapter; and

(2) Is in an airworthy condition and meets the applicable airworthiness requirements of this chapter, including those relating to identification and equipment.

(b) Each certificate holder must have the exclusive use of at least one aircraft that meets the requirements for at least one kind of operation authorized in the certificate holder's operations specifications. In addition, for each kind of operation for which the certificate holder does not have the exclusive use of an aircraft, the certificate holder must have available for use under a written agreement (including arrangements for performing required maintenance) at least one aircraft that meets the requirements for that kind of operation. However, this paragraph does not prohibit the operator from using or authorizing the use of the aircraft for other than operations under this part and does not require the certificate holder to have exclusive use of all aircraft that the certificate holder uses.

(c) For the purposes of paragraph (b) of this section, a person has exclusive use of an aircraft if that person has the sole possession, control, and use of it for flight, as owner, or has a written agreement (including arrangements for performing required maintenance), in effect when the aircraft is operated, giving the person that possession, control, and use for at least 6 consecutive months.

(d) A certificate holder may operate in common carriage, and for the carriage of mail, a civil aircraft which is leased or chartered to it without crew and is registered in a country which is a party to the Convention on International Civil Aviation if—

(1) The aircraft carries an appropriate airworthiness certificate issued by the country of registration and meets the registration and identification requirements of that country;

(2) The aircraft is of a type design which is approved under a U.S. type certificate and complies with all of the requirements of this chapter (14 CFR Chapter 1) that would be applicable to that aircraft were it registered in the United States, including the requirements which must be met for issuance of a U.S. standard airworthiness certificate (including type design conformity, condition for safe operation, and the noise, fuel venting, and engine emission requirements of this chapter), except that a U.S. registration certificate and a U.S. standard airworthiness certificate will not be issued for the aircraft;

(3) The aircraft is operated by U.S.-certificated airmen employed by the certificate holder; and

(4) The certificate holder files a copy of the aircraft lease or charter agreement with the FAA Aircraft Registry, Department of Transportation, 6400 South MacArthur Boulevard, Oklahoma City, Oklahoma (Mailing address: P.O. Box 25504, Oklahoma City, Oklahoma 73125).

(Amdt. 135-8, Eff. 10/16/80)

§ 135.41 Carriage of narcotic drugs, marihuana, and depressant or stimulant drugs or substances.

If the holder of a certificate operating under this part allows any aircraft owned or leased by that holder to be engaged in any operation that the certificate holder knows to be in violation of § 91.19(a) of this chapter, that operation is a basis for suspending or revoking the certificate.

(Amdt 135-32, Eff. 8/18/90)

§ 135.43 Crewmember certificates: International operations.

(a) This section describes the certificates that were issued to United States citizens who were employed by air carriers at the time of issuance as flight crewmembers on United States registered aircraft engaged in international air commerce. The purpose of the certificate is to facilitate the entry and clearnace of those crewmembers into ICAO contracting states. They were issued under Annex 9, as amended, to the Convention on International Civil Aviation.

(b) The holder of a certificate issued under this section, or the air carrier by whom the holder is employed, shall surrender the certificate for cancellation at the nearest FAA Flight Standards District Office at the termination of the holder's employment with that air carrier.

[Amdt. 135-65, 61 FR 30435, June 14, 1996]

Subpart B—Flight operations

§ 135.61 General.

This subpart prescribes rules, in addition to those in Part 91 of this chapter, that apply to operations under this part.

§ 135.63 Recordkeeping requirements.

(a) Each certificate holder shall keep at its principal business office or at other places approved by the Administrator, and shall make available for inspection by the Administrator the following—

(1) The certificate holder's operating certificate;

(2) The certificate holder's operations specifications;

(3) [A current list of the aircraft used or available for use in operations under this part and the operations for which each is equipped;]

(4) An individual record of each pilot used in operations under this part, including the following information:

(i) The full name of the pilot.

(ii) The pilot certificate (by type and number) and ratings that the pilot holds.

(iii) The pilot's aeronautical experience in sufficient detail to determine the pilot's qualifications to pilot aircraft in operation under this part.

(iv) The pilot's current duties and the date of the pilot's assignment to those duties.

(v) The effective date and class of the medical certificate that the pilot holds.

(vi) The date and result of each of the initial and recurrent competency tests and proficiency and route checks required by this part and the type of aircraft flown during that test or check.

(vii) The pilot's flight time in sufficient detail to determine compliance with the flight time limitations of this part.

(viii) The pilot's check pilot authorization, if any.

(ix) Any reaction taken concerning the pilot's release from employment for physical or professional disqualification.

(x) [The date of the completion of the initial phase and each recurrent phase of the training required by this part; and

[(5) An individual record for each flight attendant who is required under this part, maintained in sufficient detail to determine compliance with the applicable portions of § 135.273 of this part.]

(b) [Each certificate holder must keep each record required by paragraph (a)(3) of this section for at least 6 months, and must keep each record required by paragraphs (a)(4) and (a)(5) of this section for at least 12 months.]

(c) For multiengine aircraft, each certificate holder is responsible for the preparation and accuracy of a load manifest in duplicate containing information concerning the loading of the aircraft. The manifest must be prepared before each takeoff and must include—

(1) The number of passengers;

(2) The total weight of the loaded aircraft;

(3) The maximum allowable takeoff weight for that flight;

(4) The center of gravity limits;

(5) The center of gravity of the loaded aircraft, except that the actual center of gravity need not be computed if the aircraft is loaded according to a loading schedule or other approved method that ensures that the center of gravity of the loaded aircraft is within approved limits. In those cases, an entry shall be made on the manifest indicating that the center of gravity is within limits according to a loading schedule or other approved method;

(6) The registration number of the aircraft or flight number;

(7) The origin and destination; and

(8) Identification of crewmembers and their crew position assignments.

(d) The pilot in command of the aircraft for which a load manifest must be prepared shall carry a copy of the completed load manifest in the aircraft to its destination. The certificate holder shall keep copies of completed load manifest for at least 30 days at its principal operations base, or at another location used by it and approved by the Administrator.

§ 135.64   Retention of contracts and amendments: Commercial operators who conduct intrastate operations for compensation or hire.

Each commercial operator who conducts intrastate operations for compensation or hire shall keep a copy of each written contract under which it provides services as a commercial operator for a period of at least one year after the date of execution of the contract. In the case of an oral contract, it shall keep a memorandum stating its elements, and of any amendments to it, for a period of at least one year after the execution of that contract or change.

§ 135.65   Reporting mechanical irregularities.

(a) Each certificate holder shall provide an aircraft maintenance log to be carried on board each aircraft for recording or deferring mechanical irregularities and their correction.

(b) The pilot in command shall enter or have entered in the aircraft maintenance log each mechanical irregularity that comes to the pilot's attention during flight time. Before each flight, the pilot in command shall, if the pilot does not already know, determine the status of each irregularity entered in the maintenance log at the end of the preceding flight.

(c) Each person who takes corrective action or defers action concerning a reported or observed failure or malfunction of an airframe, powerplant, propeller, rotor, or appliance, shall record the action taken in the aircraft maintenance log under the applicable maintenance requirements of this chapter.

(d) Each certificate holder shall establish a procedure for keeping copies of the aircraft maintenance log required by this section in the aircraft for access by appropriate personnel and shall include that procedure in the manual required by § 135.21.

§ 135.67   Reporting potentially hazardous meteorological conditions and irregularities of communications or navigational facilities.

Whenever a pilot encounters a potentially hazardous meteorological condition or an irregularity in a ground communications or navigational facility in flight, the knowledge of which the pilot considers essential to the safety of other flights, the pilot shall notify an appropriate ground radio station as soon as practicable.

(Amdt. 135-1, Eff. 5/7/79)

§ 135.69   Restriction or suspension of operations: Continuation of flight in an emergency.

(a) During operations under this part, if a certificate holder or pilot in command knows of conditions, including airport and runway conditions, that are a hazard to safe operations, the certificate holder or pilot in command, as the case may be, shall restrict or suspend operations as necessary until those conditions are corrected.

(b) No pilot in command may allow a flight to continue toward any airport of intended landing under the conditions set forth in paragraph (a) of this section, unless in the opinion of the pilot in command, the conditions that are a hazard to safe operations may reasonably be expected to be corrected by the estimated time of arrival or, unless there is no safer procedure. In the latter event, the continuation toward that airport is an emergency situation under § 135.19.

§ 135.71 Airworthiness check.

The pilot in command may not begin a flight unless the pilot determines that the airworthiness inspections required by § 91.409 of this chapter, or § 135.419, whichever is applicable, have been made.

(Amdt. 135-32, Eff. 8/18/90)

§ 135.73 Inspections and tests.

Each certificate holder and each person employed by the certificate holder shall allow the Administrator, at any time or place, to make inspections or tests (including en route inspections) to determine the holder's compliance with the Federal Aviation Act of 1958, applicable regulations, and the certificate holder's operating certificate, and operations specifications.

§ 135.75 Inspectors credentials: Admission to pilots' compartment: Forward observer's seat.

(a) Whenever, in performing the duties of conducting an inspection, an FAA inspector presents an Aviation Safety Inspector credential, FAA Form 110A, to the pilot in command of an aircraft operated by the certificate holder, the inspector must be given free and uninterrupted access to the pilot compartment of that aircraft. However, this paragraph does not limit the emergency authority of the pilot in command to exclude any person from the pilot compartment in the interest of safety.

(b) A forward observer's seat on the flight deck, or forward passenger seat with headset or speaker must be provided for use by the Administrator while conducting en route inspections. The suitability of the location of the seat and the headset or speaker for use in conducting en route inspections is determined by the Administrator.

§ 135.77 Responsibility for operational control.

Each certificate holder is responsible for operational control and shall list, in the manual required by § 135.21, the name and title of each person authorized by it to exercise operational control.

§ 135.79 Flight locating requirements.

(a) Each certificate holder must have procedures established for locating each flight, for which an FAA flight plan is not filed, that—

(1) Provide the certificate holder with at least the information required to be included in a VFR flight plan;

(2) Provide for timely notification of an FAA facility or search and rescue facility, if an aircraft is overdue or missing; and

(3) Provide the certificate holder with the location, date, and estimated time for reestablishing radio or telephone communications, if the flight will operate in an area where communications cannot be maintained.

(b) Flight locating information shall be retained at the certificate holder's principal place of business, or at other places designated by the certificate holder in the flight locating procedures, until the completion of the flight.

(c) Each certificate holder shall furnish the representative of the Administrator assigned to it with a copy of its flight locating procedures and any changes or additions, unless those procedures are included in a manual required under this part.

§ 135.81 Informing personnel of operational information and appropriate changes.

Each certificate holder shall inform each person in its employment of the operations specifications that apply to that person's duties and responsibilities and shall make available to each pilot in the certificate holder's employ the following materials in current form:

(a) Airman's Information Manual (Alaska Supplement in Alaska and Pacific Chart Supplement in Pacific-Asia Regions) or a commercial publication that contains the same information.

(b) This part and Part 91 of this chapter.

(c) Aircraft Equipment Manuals, and Aircraft Flight Manual or equivalent.

(d) For foreign operations, the International Flight Information Manual or a commercial publication that contains the same information concerning the pertinent operational and entry requirements of the foreign country or countries involved.

§ 135.83 Operating information required.

(a) The operator of an aircraft must provide the following materials, in current and appropriate form, accessible to the pilot at the pilot station, and the pilot shall use them:

(1) A cockpit checklist.

(2) For multiengine aircraft or for aircraft with retractable landing gear, an emergency cockpit checklist containing the procedures required by paragraph (c) of this section, as appropriate.

(3) Pertinent aeronautical charts.

(4) For IFR operations, each pertinent navigational en route, terminal area, and approach and letdown chart.

(5) For multiengine aircraft, one-engine-inoperative climb performance data and if the aircraft is approved for use in IFR or over-the-top operations, that data must be sufficient to enable the pilot to determine compliance with § 135.181(a)(2).

(b) Each cockpit checklist required by paragraph (a)(1) of this section must contain the following procedures: (1) Before starting engines; (2) Before takeoff; (3) Cruise; (4) Before landing; (5) After landing; (6) Stopping engines.

(c) Each emergency cockpit checklist required by paragraph (a)(2) of this section must contain the following procedures as appropriate:

(1) Emergency operation of fuel, hydraulic, electrical, and mechanical systems.

(2) Emergency operation of instruments and controls.

(3) Engine inoperative procedures.

(4) Any other emergency procedures necessary for safety.

§ 135.85 Carriage of persons without compliance with the passenger-carrying provisions of this part.

The following persons may be carried aboard an aircraft without complying with the passenger-carrying requirements of this part:

(a) A crewmember or other employee of the certificate holder.

(b) A person necessary for the safe handling of animals on the aircraft.

(c) A person necessary for the safe handling of hazardous materials (as defined in Subchapter C of Title 49 CFR).

(d) A person performing duty as a security or honor guard accompanying a shipment made by or under the authority of the U.S. Government.

(e) A military courier or a military route supervisor carried by a military cargo contract air carrier or commercial operator in operations under a military cargo contract, if that carriage is specifically authorized by the appropriate military service.

(f) An authorized representative of the Administrator conducting an en route inspection.

(g) A person, authorized by the Administrator, who is performing a duty connected with a cargo operation of the certificate holder.

§ 135.87 Carriage of cargo including carry-on baggage.

No person may carry cargo, including carry-on baggage, in or on any aircraft unless—

(a) It is carried in an approved cargo rack, bin, or compartment installed in or on the aircraft;

(b) It is secured by an approved means; or

(c) It is carried in accordance with each of the following:

(1) For cargo, it is properly secured by a safety belt or other tie-down having enough strength to eliminate the possibility of shifting under all normally anticipated flight and ground conditions, or for carry-on baggage, it is restrained so as to prevent its movement during air turbulence.

(2) It is packaged or covered to avoid possible injury to occupants.

(3) It does not impose any load on seats or on the floor structure that exceeds the load limitation for those components.

(4) It is not located in a position that obstructs the access to, or use of, any required emergency or regular exit, or the use of the aisle between the crew and the passenger compartment, or located in a position that obscures any passenger's view of the "seat belt" sign, "no smoking" sign, or any required exit sign, unless an auxiliary sign or other approved means for proper notification of the passengers is provided.

(5) It is not carried directly above seated occupants.

(6) It is stowed in compliance with this section for takeoff and landing.

(7) For cargo only operations, paragraph (c)(4) of this section does not apply if the cargo is loaded so that at least one emergency or regular exit is available to provide all occupants of the aircraft a means of unobstructed exit from the aircraft if an emergency occurs.

(d) Each passenger seat under which baggage is stowed shall be fitted with a means to prevent articles of baggage stowed under it from sliding under crash impacts severe enough to induce the ultimate inertia forces specified in the emergency landing condition regulations under which the aircraft was type certificated.

(e) When cargo is carried in cargo compartments that are designed to require the physical entry of a crewmember to extinguish any fire that may occur during flight, the cargo must be loaded so as to allow a crewmember to effectively reach all parts of the compartment with the contents of a hand fire extinguisher.

§ 135.89 Pilot requirements: Use of oxygen.

(a) *Unpressurized aircraft.* Each pilot of an unpressurized aircraft shall use oxygen continuously when flying

(1) At altitudes above 10,000 feet through 12,000 feet MSL for that part of the flight at those altitudes that is of more than 30 minutes duration; and

(2) Above 12,000 feet MSL.

(b) *Pressurized aircraft.*

(1) Whenever a pressurized aircraft is operated with the cabin pressure altitude more than 10,000 feet MSL, each pilot shall comply with paragraph (a) of this section.

(2) Whenever a pressurized aircraft is operated at altitudes above 25,000 feet through 35,000 feet MSL unless each pilot has an approved quick-donning type oxygen mask—

(i) At least one pilot at the controls shall wear, secured and sealed, an oxygen mask that either supplies oxygen at all times or automatically supplies oxygen whenever the cabin pressure altitude exceeds 12,000 feet MSL; and

(ii) During that flight, each other pilot on flight deck duty shall have an oxygen mask, connected to an oxygen supply, located so as to allow immediate placing of the mask on the pilot's face sealed and secured for use.

(3) Whenever a pressurized aircraft is operated at altitudes above 35,000 feet MSL, at least one pilot at the controls shall wear, secured and sealed, an oxygen mask required by paragraph (2)(i) of this paragraph.

(4) If one pilot leaves a pilot duty station of an aircraft when operating at altitudes above 25,000 feet MSL, the remaining pilot at the controls shall put on and use an approved oxygen mask until the other pilot returns to the pilot duty station of the aircraft.

§ 135.91 Oxygen for medical use by passengers.

(a) Except as provided in paragraphs (d) and (e) of this section, no certificate holder may allow the carriage or operation of equipment for the storage, generation or dispensing of medical oxygen unless the unit to be carried is constructed so that all valves, fittings, and gauges are protected from damage during that carriage or operation and unless the following conditions are met—

(1) The equipment must be—

(i) Of an approved type or in conformity with the manufacturing, packaging, marking, labeling and maintenance requirements of Title 49 CFR Parts 171, 172, and 173, except § 173.24(a)(1);

(ii) When owned by the certificate holder, maintained under the certificate holder's approved maintenance program;

(iii) Free of flammable contaminants on all exterior surfaces; and

(iv) Appropriately secured.

(2) When the oxygen is stored in the form of a liquid, the equipment must have been under the certificate holder's approved maintenance program since its purchase new or since the storage container was last purged.

(3) When the oxygen is stored in the form of a compressed gas as defined in Title 49 CFR § 173.300(a)—

(i) When owned by the certificate holder, it must be maintained under its approved maintenance program; and

(ii) The pressure in any oxygen cylinder must not exceed the rated cylinder pressure.

(4) The pilot in command must be advised when the equipment is on board, and when it is intended to be used.

(5) The equipment must be stowed, and each person using the equipment must be seated, so as not to restrict access to or use of any required emergency or regular exit, or of the aisle in the passenger compartment.

(b) No person may smoke and no certificate holder may allow any person to smoke within 10 feet of oxygen storage and dispensing equipment carried under paragraph (a) of this section.

(c) No certificate holder may allow any person other than a person trained in the use of medical oxygen equipment to connect or disconnect oxygen bottles or any other ancillary component while any passenger is aboard the aircraft.

(d) Paragraph (a)(1)(i) of this section does not apply when that equipment is furnished by a professional or medical emergency service for use on board an aircraft in a medical emergency when no other practical means of transportation (including any other properly equipped certificate holder) is reasonably available and the person carried under the medical emergency is accompanied by a person trained in the use of medical oxygen.

(e) Each certificate holder who, under the authority of paragraph (d) of this section, deviates from paragraph (a)(1)(i) of this section under a medical emergency shall, within 10 days, excluding Saturdays, Sundays, and Federal holidays, after the deviation, send to the certificate-holding district office a complete report of the operation involved, including a description of the deviation and the reasons for it.

§ 135.93 Autopilot: Minimum altitudes for use.

(a) Except as provided in paragraphs (b), (c), (d), and (e) of this section, no person may use an autopilot at an altitude above the terrain which is less than 500 feet or less than twice the maximum altitude loss specified in the approved Aircraft Flight Manual or equivalent for a malfunction of the autopilot, whichever is higher.

(b) When using an instrument approach facility other than ILS, no person may use an autopilot at an altitude above the terrain that is less than 50 feet below the approved minimum descent altitude for that procedure, or less than twice the maximum loss specified in the approved Airplane Flight Manual or equivalent for a malfunction of the autopilot under approach conditions, whichever is higher.

(c) For ILS approaches, when reported weather conditions are less than the basic weather conditions in § 91.155 of this chapter, no person may use an autopilot with an approach coupler at an altitude above the terrain that is less than 50 feet above the terrain, or the maximum altitude loss specified in the approved Airplane Flight Manual or equivalent for the malfunction of the autopilot with approach coupler, whichever is higher.

(d) Without regard to paragraph (a), (b), or (c) of this section, the Administrator may issue operations specifications to allow the use, to touchdown, of an approved flight control guidance system with automatic capability, if—

(1) The system does not contain any altitude loss (above zero) specified in the approved Aircraft Flight Manual or equivalent for malfunction of the autopilot with approach coupler; and

(2) The Administrator finds that the use of the system to touchdown will not otherwise adversely affect the safety standards of this section.

(e) Notwithstanding paragraph (a) of this section, the Administrator issues operations specifications to allow the use of an approved autopilot system with automatic capability during the takeoff and initial climb phase of flight provided:

(1) The Airplane Flight Manual specifies a minimum altitude engagement certification restriction;

(2) The system is not engaged prior to the minimum engagement certification restriction specified in the Airplane Flight Manual, or an altitude specified by the Administrator, whichever is higher; and

(3) The Administrator finds that the use of the system will not otherwise affect the safety standards required by this section.

(f) This section does not apply to the operations conducted in rotorcraft.

(Amdt. 135-32, Eff. 8/18/90)

§ 135.95 Airmen: Limitations on use of services.

No certificate holder may use the services of any person as an airman unless the person performing those services—

(a) Holds an appropriate and current airman certificate; and

(b) Is qualified, under this chapter, for the operation for which the person is to be used.

§ 135.97 Aircraft and facilities for recent flight experience.

Each certificate holder shall provide aircraft and facilities to enable each of its pilots to maintain and demonstrate the pilot's ability to conduct all operations for which the pilot is authorized.

§ 135.99 Composition of flight crew.

(a) No certificate holder may operate an aircraft with less than the minimum flight crew specified in the aircraft operating limitations or the Aircraft Flight Manual for that aircraft and required by this part for the kind of operation being conducted.

(b) No certificate holder may operate an aircraft without a second in command if that aircraft has a passenger seating configuration, excluding any pilot seat, of ten seats or more.

§ 135.100 Flight crewmember duties.

(a) No certificate holder shall require, nor may any flight crewmember perform, any duties during a critical phase of flight except those duties required for the safe operation of the aircraft. Duties such as company required calls made for such nonsafety related purposes as ordering galley supplies and confirming passenger connections, announcements made to passengers promoting the air carrier or pointing out sights of interest, and filling out company payroll and related records are not required for the safe operation of the aircraft.

(b) No flight crewmember may engage in, nor may any pilot in command permit, any activity during a critical phase of flight which could distract any flight crewmember from the performance of his or her duties or which could interfere in any way with the proper conduct of those duties. Activities such as eating meals, engaging in nonessential conversations within the cockpit and nonessential communications be-

tween the cabin and cockpit crews, and reading publications not related to the proper conduct of the flight are not required for the safe operation of the aircraft.

(c) For the purposes of this section, critical phases of flight includes all ground operations involving taxi, takeoff and landing, and all other flight operations conducted below 10,000 feet, except cruise flight.

NOTE: Taxi is defined as "movement of an airplane under its own power on the surface of an airport."

(Amdt. 135-11, Eff. 5/18/81); (Amdt. 135-14, Eff. 6/18/81); (Amdt. 135-15, Eff. 6/11/81)

§ 135.101 Second in command required in IFR conditions.

Except as provided in §135.105, no person may operate an aircraft carrying passengers under IFR unless there is a second in command in the aircraft.

§ 135.103 Removed and reserved

§ 135.105 Exception to second in command requirement: Approval for use of autopilot system.

(a) Except as provided in §§ 135.99 and 135.111, unless two pilots are required by this chapter for operations under VFR, a person may operate an aircraft without a second in command, if it is equipped with an operative approved autopilot system and the use of that system is authorized by appropriate operations specifications. No certificate holder may use any person, nor may any person serve, as a pilot in command under this section of an aircraft operated as a commuter operation in part 119 of this chapter unless that person has at least 100 hours pilot in command flight time in the make and model of aircraft to be flown and has met all other applicable requirements of this part.

(b) The certificate holder may apply for an amendment of its operations specifications to authorize the use of an autopilot system in place of a second in command.

(c) The Administrator issues an amendment to the operations specifications authorizing the use of an autopilot system, in place of a second in command, if—

(1) The autopilot is capable of operating the aircraft controls to maintain flight and maneuver it about the three axes; and

(2) The certificate holder shows, to the satisfaction of the Administrator, that operations using the autopilot system can be conducted safely and in compliance with this part.

The amendment contains any conditions or limitations on the use of the autopilot system that the Administrator determines are needed in the interest of safety.

(Amdt. 135-3, Eff. 3/1/80)

§ 135.107 Flight attendant crewmember requirement.

No certificate holder may operate an aircraft that has a passenger seating configuration, excluding any pilot seat, of more than 19 unless there is a flight attendant crewmember on board the aircraft.

§ 135.109 Pilot in command or second in command: Designation required.

(a) Each certificate holder shall designate a—

(1) Pilot in command for each flight; and

(2) Second in command for each flight requiring two pilots.

(b) The pilot in command, as designated by the certificate holder, shall remain the pilot in command at all times during the flight.

§ 135.111 Second in command required in Category II operations.

No person may operate an aircraft in a Category II operation unless there is a second in command of the aircraft.

§ 135.113 Passenger occupancy of pilot seat.

No certificate holder may operate an aircraft type certificate after October 15, 1971, that has a passenger seating configuration, excluding any pilot seat, of more than eight seats if any person other than the pilot in command, a second in command, a company check airman, or an authorized representative of the Administrator, the National Transportation Safety Board, or the United States Postal Service occupies a pilot seat.

§ 135.115 Manipulation of controls.

No pilot in command may allow any person to manipulate the flight controls of an aircraft during flight conducted under this part, nor may any person manipulate the controls during such flight unless that person is—

(a) A pilot employed by the certificate holder and qualified in the aircraft; or,

(b) An authorized safety representative of the Administrator who has the permission of the pilot in command, is qualified in the aircraft, and is checking flight operations.

§ 135.117 Briefing of passengers before flight.

(a) Before each takeoff each pilot in command of an aircraft carrying passengers shall ensure that all passengers have been orally briefed on—

(1) Smoking. [Each passenger shall be briefed on when, where, and under what conditions smoking is prohibited (including, but not limited to, any applicable requirements of Part 252 of this title). This briefing shall include a statement that the Federal Aviation Regulations require passenger compliance with the lighted passenger information signs (if such signs are required), posted placards, areas designated for safety purposes as no smoking areas, and crewmember instructions with regard to these items. The briefing shall also include a statement (if the aircraft is equipped with a lavatory) that Federal

law prohibits: tampering with, disabling, or destroying any smoke detector installed in an aircraft lavatory; smoking in lavatories; and, when applicable, smoking in passenger compartments.

(2) [The use of safety belts, including instructions on how to fasten and unfasten the safety belts. Each passenger shall be briefed on when, where, and under what conditions the safety belt must be fastened about that passenger. This briefing shall include a statement that the Federal Aviation Regulations require passenger compliance with lighted passenger information signs and crewmember instructions concerning the use of safety belts.]

(3) The placement of seat backs in an upright position before takeoff and landing;

(4) Location and means for opening the passenger entry door and emergency exits;

(5) Location of survival equipment;

(6) If the flight involves extended overwater operation, ditching procedures and the use of required flotation equipment;

(7) If the flight involves operations above 12,000 feet MSL, the normal and emergency use of oxygen; and

(8) Location and operation of fire extinguishers.

(b) Before each takeoff the pilot in command shall ensure that each person who may need the assistance of another person to move expeditiously to an exit if an emergency occurs and that person's attendant, if any, has received a briefing as to the procedures to be followed if an evacuation occurs. This paragraph does not apply to a person who has been given a briefing before a previous leg of a flight in the same aircraft.

(c) The oral briefing required by paragraph (a) of this section shall be given by the pilot in command or a crewmember.

(d) Notwithstanding the provisions of paragraph (c) of this section, for aircraft certificated to carry 19 passengers or less, the oral briefing required by paragraph (a) of this section shall be given by the pilot in command, a crewmember, or other qualified person designated by the certificate holder and approved by the Administrator.

(e) The oral briefing required by paragraph (a) of this section must be supplemented by printed cards which must be carried in the aircraft in locations convenient for the use of each passenger. The cards must—

(1) Be appropriate for the aircraft on which they are to be used;

(2) Contain a diagram of, and method of operating, the emergency exits;

(3) Contain other instructions necessary for the use of emergency equipment on board the aircraft; and

(4) No later than June 12, 2005, for scheduled Commuter passenger-carrying flights, include the sentence, "Final assembly of this aircraft was completed in [INSERT NAME OF COUNTRY]."

(f) The briefing required by paragraph (a) may be delivered by means of an approved recording playback device that is audible to each passenger under normal noise levels.

(Amdt. 135-20, Eff. 1/6/87); (Amdt. 135-25, Eff. 4/23/88); [(Amdt. 135-44, Eff. 10/15/92)]

§ 135.119 Prohibition against carriage of weapons.

No person may, while on board an aircraft being operated by a certificate holder, carry on or about that person a deadly or dangerous weapon, either concealed or unconcealed. This section does not apply to—

(a) Officials or employees of a municipality or a State, or of the United States, who are authorized to carry arms; or

(b) Crewmembers and other persons authorized by the certificate holder to carry arms.

§ 135.120 Prohibition on interference with crewmembers

No person may assault, threaten, intimidate, or interfere with a crewmember in the performace of the crewmember's duties aboard an aircraft being operated under this part.

§ 135.121 Alcoholic beverages.

(a) No person may drink any alcoholic beverage aboard an aircraft unless the certificate holder operating the aircraft has served that beverage.

(b) No certificate holder may serve any alcoholic beverage to any person aboard its aircraft if that person appears to be intoxicated.

(c) No certificate holder may allow any person to board any of its aircraft if that person appears to be intoxicated.

[§ 135.122 Stowage of food, beverage, and passenger service equipment during aircraft movement on the surface, takeoff, and landing.

[(a) No certificate holder may move an aircraft on the surface, take off, or land when any food, beverage, or tableware furnished by the certificate holder is located at any passenger seat.

[(b) No certificate holder may move an aircraft on the surface, take off, or land unless each food and beverage tray and seat back tray table is secured in its stowed position.

[(c) No certificate holder may permit an aircraft to move on the surface, take off, or land unless each passenger serving cart is secured in its stowed position.

[(d) Each passenger shall comply with instructions given by a crewmember with regard to compliance with this system.]

[(Amdt. 135-44, Eff. 10/15/92)]

§ 135.123 Emergency and emergency evacuation duties.

(a) Each certificate holder shall assign to each required crewmember for each type of aircraft as appropriate, the necessary functions to be performed in an emergency or in a situation requiring emergency evacuation. The certificate holder shall ensure that those functions can be practicably accomplished, and will meet any reasonably anticipated emergency including incapacitation of individual crewmembers or their inability to reach the passenger cabin because of shifting cargo in combination cargo passenger aircraft.

(b) The certificate holder shall describe in the manual required under § 135.21 the functions of each category of required crewmembers assigned under paragraph (a) of this section.

§ 135.125 Airplane security.

Certificate holders conducting operations under this part must comply with the applicable security requirements in 49 CFR chapter XII.

§ 135.127 Passenger information.

(a) No person may conduct a scheduled flight segment on which smoking is prohibited unless the "No Smoking" passenger information signs are lighted during the entire flight segment, or one or more "No Smoking" placards meeting the requirements of § 25.1541 are posted during the entire flight segment. If both the lighted signs and the placards are used, the signs must remain lighted during the entire flight segment.

Smoking is prohibited on scheduled flight segments—

(1) Between any two points within Puerto Rico, the United States Virgin Islands, the District of Columbia, or any State of the United States (other than Alaska or Hawaii) or between any two points in any one of the above-mentioned jurisdictions (other than Alaska or Hawaii);

(2) Within the State of Alaska or within the State of Hawaii; or

(3) Scheduled in the current Worldwide or North American Edition of the *Official Airline Guide* or 6 hours or less in duration and between any point listed in paragraph (a)(1) of this section and any point in Alaska or Hawaii, or between any point in Alaska and any point in Hawaii.

(b) [No person may smoke while a "No Smoking" sign is lighted or while "No Smoking" placards are posted, except that the pilot in command may authorize smoking on the flight deck (if it is physically separated from the passenger compartment) except during any movement of an aircraft on the surface, takeoff, and landing.]

(c) No person may smoke in any aircraft lavatory.

(d) No person may operate an aircraft with a lavatory equipped with a smoke detector unless there is in that lavatory a sign or placard which reads: "Federal law provides for a penalty of up to $2,000 for tampering with the smoke detector installed in this lavatory."

(e) No person may tamper with, disable, or destroy any smoke detector installed in any aircraft lavatory.

(f) [On flight segments other than those described in paragraph (a) of this section, the "No Smoking" sign required by § 135.177(a)(3) of this part must be turned on during any movement of the aircraft on the surface, for each take-off or landing, and at any other time considered necessary by the pilot in command.

[(g) The passenger information requirements prescribed in § 91.517(b) and (d) of this chapter are in addition to the requirements prescribed in this section.

[(h) Each passenger shall comply with instructions given him or her by crewmembers regarding compliance with paragraphs (b), (c), and (e) of this section.]

(Amdt. 135-25, Eff. 4/23/88); (Amdt. 135-35, Eff. 2/25/90); [(Amdt. 135-44, Eff. 10/15/92)]

§ 135.128 Use of safety belts and child restraint systems.

[(a) Except as provided in this paragraph, each person on board an aircraft operated under this part shall occupy an approved seat or berth with a separate safety belt properly secured about him or her during movement on the surface, takeoff, and landing. For seaplane and float equipped rotorcraft operations during movement on the surface, the person pushing off the seaplane or rotorcraft from the dock and the person mooring the seaplane or rotorcraft at the dock are excepted from the preceding seating and safety belt requirements. A safety belt provided for the occupant of a seat may not be used by more than one person who has reached his or her second birthday. Notwithstanding the preceding requirements, a child may:

[(1) Be held by an adult who is occupying an approved seat or berth, provided the child has not reached his or her second birthday and the child does not occupy or use any restraining device; or

[(2) Notwithstanding any other requirement of this chapter, occupy an approved child restraint system furnished by the certificate holder or one of the persons described in paragraph (a)(2)(i) of this section, provided:

[(i) The child is accompanied by a parent, guardian, or attendant designated by the child's parent or guardian to attend to the safety of the child during the flight;

[(ii) Except as provided in paragraph (a)(2)(ii)(D) of this section, the approved child restraint system bears one or more labels as follows:

[(A) Seats manufactured to U.S. standards between January 1, 1981, and February 25, 1985, must bear the label: "This child restraint system conforms to all applicable Federal motor vehicle safety standards." Vest- and harness-type child restraint systems manufactured before February 26, 1985, bearing such a label are not approved for the purposes of this section;

[(B) Seats manufactured to U.S. standards on or after February 26, 1985, must bear two labels:

[(*1*) "This child restraint system conforms to all applicable Federal motor vehicle safety standards"; and

[(*2*) "THIS RESTRAINT IS CERTIFIED FOR USE IN MOTOR VEHICLES AND AIRCRAFT" in red lettering;

(C) Seats that do not qualify under paragraph (a)(2)(ii)(A) and (a)(2)(ii)(B) of this section must bear a label or markings showing:

(1) That the seat was approved by a foreign government;

(2) That the seat was manufactured under the standards of the United Nations;

(3) That the seat or child restraint device furnished by the certificate holder was approved by the FAA through Type Certificate or Supplemental Type Certificate.

(4) That the seat or child restraint device furnished by the certificate holder, or one of the persons described in paragraph (b)(2)(i) of this section, was approved by the FAA in accordance with § 21.305(d) or Technical Standard Order C-100b, or a later version.

(D) Except as provided in § 135.128(a)(2)(C)(3) and § 135.128(a)(2)(C)(4), booster-type child restraint systems (as defined in Federal Motor Vehicle Safety Standard No. 213 (49 CFR 571.213)), vest- and harness type child restraint systems, and lap held child restraints are not approved for use in aircraft; and

[(iii) The certificate holder complies with the following requirements:

[(A) The restraint system must be properly secured to an approved forward-facing seat or berth;

[(B) The child must be properly secured in the restraint system and must not exceed the specified weight limit for the restraint system; and

[(C) The restraint system must bear the appropriate label(s).

(b) Except as provided in paragraph (b)(3) of this section, the following prohibitions apply to certificate holders:

(1) Except as provided in § 135.128 (a)(2)(ii)(C)(3) and § 135.128 (a)(2)(ii)(C)(4), no certificate holder may permit a child, in an aircraft, to occupy a booster-type child restraint system, a vest-type child restraint system, a harness-type child restraint system, or a lap held child restraint system during take off, landing, and movement on the surface.

(2) Except as required in paragraph (b)(1) of this section, no certificate holder may prohibit a child, if requested by the child's parent, guardian, or designated attendant, from occupying a child restraint system furnished by the child's parent, guardian, or designated attendant provided:

(i) The child holds a ticket for an approved seat or berth or such seat or berth is otherwise made available by the certificate holder for the child's use:

(ii) The requirements of paragraph (a)(2)(i) of this section are met;

(iii) The requirements of paragraph (a)(2)(iii) of this section are met; and

(iv) The child restraint has one or more of the labels described in paragraphs (a)(2)(ii)(A) through (a)(2)(ii)(C) of this section.

(3) This section does not prohibit the certificate holder from providing child restraint systems authorized by this

or, consistent with safe operating practices, determining the most appropriate passenger seat location for the child restraint system.}}

[(Amdt. 135-44, Eff. 10/15/92)]

§ 135.129 [Exit seating.]

(a)(1) [*Applicability.* This section applies to all certificate holders operating under this part, except for on-demand operations with aircraft having 19 or fewer passenger seats and commuter operations with aircraft having 9 or fewer passenger seats.

[(2) *Duty to make determination of suitability.* Each certificate holder shall determine, to the extent necessary to perform the applicable functions of paragraph (d) of this section, the suitability of each person it permits to occupy an exit seat. For the purpose of this section—

(i) *Exit seat* means—

(A) Each seat having direct access to an exit; and,

(B) Each seat in a row of seats through which passengers would have to pass to gain access to an exit, from the first seat inboard of the exit to the first aisle inboard of the exit.

(ii) A passenger seat having *direct access* means a seat from which a passenger can proceed directly to the exit without entering an aisle or passing around an obstruction.]

([3]) [*Persons designated to make determination.*] Each certificate holder shall make the passenger exit seating determinations required by this paragraph in a non-discriminatory manner consistent with the requirements of this section, by persons designated in the certificate holder's required operations manual.

([4]) [*Submission of designation for approval.*] Each certificate holder shall designate the exit seats for each passenger seating configuration in its fleet in accordance with the definitions in this paragraph and submit those designations for approval as part of the procedures required to be submitted for approval under paragraphs (n) and (p) of this section.

(b) No certificate holder may seat a person in a seat affected by this section if the certificate holder determines that it is likely that the person would be unable to perform one or more of the applicable functions listed in paragraph (d) of this section because—

(1) The person lacks sufficient mobility, strength, or dexterity in both arms and hands, and both legs:

(i) To reach upward, sideways, and downward to the location of emergency exit and exit-slide operating mechanisms;

(ii) To grasp and push, pull, turn, or otherwise manipulate those mechanisms;

(iii) To push, shove, pull, or otherwise open emergency exits;

(iv) To lift out, hold, deposit on nearby seats, or maneuver over the seatbacks to the next row objects the size and weight of over-wing window exit doors;

(v) To remove obstructions of size and weight similar over-wing exit doors;

(vi) To reach the emergency exit expeditiously;

(vii) To maintain balance while removing obstructions;

(viii) To exit expeditiously;

(ix) To stabilize an escape slide after deployment; or

(x) To assist others in getting off an escape slide;

(2) The person is less than 15 years of age or lacks the capacity to perform one or more of the applicable functions listed in paragraph (d) of this section without the assistance of an adult companion, parent, or other relative;

(3) The person lacks the ability to read and understand instructions required by this section and related to emergency evacuation provided by the certificate holder in printed or graphic form or the ability to understand oral crew commands.

(4) The person lacks sufficient visual capacity to perform one or more of the applicable functions in paragraph (d) of this section without the assistance of visual aids beyond contact lenses or eyeglasses;

(5) The person lacks sufficient aural capacity to hear and understand instructions shouted by flight attendants. without assistance beyond a hearing aid;

(6) The person lacks the ability adequately to impart information orally to other passengers; or,

(7) The person has:

(i) A condition or responsibilities, such as caring for small children, that might prevent the person from performing one or more of the applicable functions listed in paragraph (d) of this section; or

(ii) A condition that might cause the person harm if he or she performs one or more of the applicable functions listed in paragraph (d) of this section.

(c) Each passenger shall comply with instructions given by a crewmember or other authorized employee of the certificate holder implementing exit seating restrictions established in accordance with this section.

(d) Each certificate holder shall include on passenger information cards, presented in the language in which briefings and oral commands are given by the crew, at each exit seat affected by this section, information that, in the event of an emergency in which a crewmember is not available to assist, a passenger occupying an exit seat may use if called upon to perform the following functions:

(1) Locate the emergency exit;

(2) Recognize the emergency exit opening mechanism;

(3) Comprehend the instructions for operating the emergency exit;

(4) Operate the emergency exit;

(5) Assess whether opening the emergency exit will increase the hazards to which passengers may be exposed;

(6) Follow oral directions and hand signals given by a crewmember;

(7) Stow or secure the emergency exit door so that it will not impede use of the exit;

(8) Assess the condition of an escape slide, activate the slide, and stabilize the slide after deployment to assist others in getting off the slide;

(9) Pass expeditiously through the emergency exit; and,

(10) Assess, select, and follow a safe path away from the emergency exit.

(e) Each certificate holder shall include on passenger information cards, at each exit seat—

(1) In the primary language in which emergency commands are given by the crew, the selection criteria set forth in paragraph (b) of this section, and a request that a passenger identify himself or herself to allow reseating if he or she—

(i) Cannot meet the selection criteria set forth in paragraph (b) of this section;

(ii) Has a nondiscernible condition that will prevent him or her from performing the applicable functions listed in paragraph (d) of this section;

(iii) May suffer bodily harm as the result of performing one or more of those functions; or

(iv) Does not wish to perform those functions; and,

(2) In each language used by the certificate holder for passenger information cards, a request that a passenger identify himself or herself to allow reseating if he or she lacks the ability to read, speak, or understand the language or the graphic form in which instructions required by this section and related to emergency evacuation are provided by the certificate holder, or the ability to understand the specified language in which crew commands will be given in an emergency;

(3) May suffer bodily harm as the result of performing one or more of those functions; or,

(4) Does not wish to perform those functions.

A certificate holder shall not require the passenger to disclose his or her reason for needing reseating.

(f) Each certificate holder shall make available for inspection by the public at all passenger loading gates and ticket counters at each airport where it conducts passenger operations, written procedures established for making determinations in regard to exit row seating.

(g) No certificate holder may allow taxi or pushback unless at least one required crewmember has verified that no exit seat is occupied by a person the crewmember determines is likely to be unable to perform the applicable functions listed in paragraph (d) of this section.

(h) Each certificate holder shall include in its passenger briefings a reference to the passenger information cards, required by paragraphs (d) and (e), the selection criteria set forth in paragraph (b), and the functions to be performed, set forth in paragraph (d) of this section.

(i) Each certificate holder shall include in its passenger briefings a request that a passenger identify himself or herself to allow reseating if he or she—

(1) Cannot meet the selection criteria set forth in paragraph (b) of this section.

(2) Has a nondiscernible condition that will prevent him or her from performing the applicable functions listed in paragraph (d) of this section;

(3) May suffer bodily harm as the result of performing one or more of those functions; or,

(4) Does not wish to perform those functions.

A certificate holder shall not require the passenger to disclose his or her reason for needing reseating.

(j) Removed and [Reserved]

(k) In the event a certificate holder determines in accordance with this section that it is likely that a passenger assigned to an exit seat would be unable to perform the functions listed in paragraph (d) of this section or a passenger requests a non-exit seat, the certificate holder shall expeditiously relocate the passenger to a non-exit seat.

(l) In the event of full booking in the non-exit seats and if necessary to accommodate a passenger being relocated from an exit seat, the certificate holder shall move a passenger who is willing and able to assume the evacuation functions that may be required, to an exit seat.

(m) A certificate holder may deny transportation to any passenger under this section only because—

(1) The passenger refuses to comply with instructions given by a crewmember or other authorized employee of the certificate holder implementing exit seating restrictions established in accordance with this section, or

(2) The only seat that will physically accommodate the person's handicap is an exit seat.

(n) In order to comply with this section certificate holders shall—

(1) Establish procedures that address:

(i) The criteria listed in paragraph (b) of this section;

(ii) The functions listed in paragraph (d) of this section;

(iii) The requirements for airport information, passenger information cards, crewmember verification of appropriate seating in exit seats, passenger briefings, seat assignments, and denial of transportation as set forth in this section;

(iv) How to resolve disputes arising from implementation of this section, including identification of the certificate holder employee on the airport to whom complaints should be addressed for resolution; and,

(2) Submit their procedures for preliminary review and approval to the principal operations inspectors assigned to them at the certificate-holding district office.

(o) Certificate holders shall assign seats prior to boarding consistent with the criteria listed in paragraph (b) and the functions listed in paragraph (d) of this section, to the maximum extent feasible.

(p) The procedures required by paragraph (n) of this section will not become effective until final approval is granted by the Director, Flight Standards Service, Washington, DC. Approval will be based solely upon the safety aspects of the certificate holder's procedures.

(Amdt. 135-36, Eff. 4/5/90); (Amdt. 135-45, Eff. 10/27/92); [(Amdt. 135-50, Eff. 7/29/94)]

Subpart C—Aircraft and equipment

§ 135.141 Applicability.

This subpart prescribes aircraft and equipment requirements for operations under this part. The requirements of this subpart are in addition to the aircraft and equipment requirements of Part 91 of this chapter. However, this part does not require the duplication of any equipment required by this chapter.

§ 135.143 General requirements.

(a) No person may operate an aircraft under this part unless that aircraft and its equipment meet the applicable regulations of this chapter.

(b) Except as provided in § 135.179, no person may operate an aircraft under this part unless the required instruments and equipment in it have been approved and are in an operable condition.

(c) ATC transponder equipment installed within the time periods indicated below must meet the performance and environmental requirements of the following TSO's.

(1) *Through January 1, 1992:*

(i) Any class of TSO-C74b or any class of TSO-C74c as appropriate, provided that the equipment was manufactured before January 1, 1990; or

(ii) The appropriate class of TSO-C112 (Mode S).

(2) *After January 1, 1992:* The appropriate class of TSO-C112 (Mode S). For purposes of paragraph (c)(2) of this section, "installation" does not include—

(i) Temporary installation of TSO-C74b or TSO-C74c substitute equipment, as appropriate, during maintenance of the permanent equipment;

(ii) Reinstallation of equipment after temporary removal for maintenance; or

(iii) For fleet operations, installation of equipment in a fleet aircraft after removal of the equipment for maintenance from another aircraft in the same operator's fleet.

(Amdt. 135-22, Eff. 5/26/87)

§ 135.144 Portable electronic devices.

(a) Except as provided in paragraph (b) of this section, no person may operate, nor may any operator or pilot in command of an aircraft allow the operation of, any portable electronic device on any of the following U.S.-registered civil aircraft operating under this part.

(b) Paragraph (a) of this section does not apply to—

(1) Portable voice recorders;

(2) Hearing aids;

(3) Heart pace makers;

(4) Electric shavers; or

(5) Any other portable electronic device that the part 119 certificate holder has determined will not cause interference with the navigation or communication system of the aircraft on which it is to be used.

(c). The determination required by paragraph (b)(5) of this section shall be made by that part 119 certificate holder operating the aircraft on which the particular device is to be used.

§ 135.145 Aircraft proving and validation tests.

(a) No certificate holder may operate an aircraft, other than a turbojet aircraft, for which two pilots are required by this chapter for operations under VFR, if it has not previously proved such an aircraft in operations under this part in at least 25 hours of proving tests acceptable to the Administrator including—

(1) Five hours of night time, if night flights are to be authorized;

(2) Five instrument approach procedures under simulated or actual conditions, if IFR flights are to be authorized; and

(3) Entry into a representative number of en mute airports as determined by the Administrator.

(b) No certificate holder may operate a turbojet airplane if it has not previously proved a turbojet airplane in operations under this part in at least 25 hours of proving tests acceptable to the Administrator including—

(1) Five hours of night time, if night flights are to be authorized;

(2) Five instrument approach procedures under simulated or actual conditions, if WR flights are to be authorized; and

(3) Entry into a representative number to of en route airports as determined by the Administrator.

(c) No certificate holder may carry passengers in an aircraft during proving tests, except those needed to make the tests and those designated by the Administrator to observe the tests. However, pilot flight training may be conducted during the proving tests.

(d) Validation testing is required to determine that a certificate holder is capable of conducting operations safely and in compliance with applicable regulatory standards. Validation tests are required for the following authorizations:

(1) The addition of an aircraft for which two pilots are required for operations under VFR or a turbojet airplane, If that aircraft or an aircraft of the same make or similar design has not been previously proved or validated in operations under this part.

(2) Operations outside U.S. airspace.

(3) Class II navigation authorizations.

(4) Special performance or operational authorizations.

(e) Validation tests must be accomplished by test methods acceptable to the Administrator. Actual flights may not be required when an applicant can demonstrate competence and compliance with appropriate regulations without conducting a flight.

(f) Proving tests and validation tests may be conducted simultaneously when appropriate.

(g) The Administrator may authorize deviations from this section if the Administrator finds that special circumstances make full compliance with this section unnecessary.

§ 135.147 Dual controls required.

No person may operate an aircraft in operations requiring two pilots unless it is equipped with functioning dual controls. However, if the aircraft type certification operating limitations do not require two pilots, a throwover control wheel may be used in place of two control wheels.

§ 135.149 Equipment requirements: General.

No person may operate an aircraft unless it is equipped with—

(a) A sensitive altimeter that is adjustable for barometric pressure;

(b) Heating or deicing equipment for each carburetor or, for a pressure carburetor, an alternate air source;

(c) For turbojet airplanes, in addition to two gyroscopic bank-and-pitch indicators (artificial horizons) for use at the pilot stations, a third indicator that is installed in accordance with the instrument requirements prescribed in § 121.305(j) of this chapter.

(d) [Reserved]

(e) For turbine powered aircraft, any other equipment as the Administrator may require.

(Amdt. 135-1, Eff. 5/7/79); (Amdt. 135-34, Eff. 11/27/89); (Amdt. 135-38, Eff. 11/26/90)

§ 135.150 Public address and crewmember interphone systems.

No person may operate an aircraft having a passenger seating configuration, excluding any pilot seat, of more than 19 unless it is equipped with—

(a) A public address system which—

(1) Is capable of operation independent of the crewmember interphone system required by paragraph (b) of this section, except for handsets, headsets, microphones, selector switches, and signaling devices;

(2) Is approved in accordance with § 21.305 of this chapter;

(3) Is accessible for immediate use from each of two flight crewmember stations in the pilot compartment;

(4) For each required floor-level passenger emergency exit which has an adjacent flight attendant seat, has a microphone which is readily accessible to the seated flight attendant, except that one microphone may serve more than one exit, provided the proximity of the exits allows unassisted verbal communication between seated flight attendants;

(5) Is capable of operation within 10 seconds by a flight attendant at each of those stations in the passenger compartment from which its use is accessible;

(6) Is audible at all passenger seats, lavatories, and flight attendant seats and work stations; and

(7) For transport category airplanes manufactured on or after November 27, 1990, meets the requirements of § 25.1423 of this chapter.

(b) A crewmember interphone system which—

(1) Is capable of operation independent of the public address system required by paragraph (a) of this section, except for handsets, headsets, microphones, selector switches, and signaling devices;

(2) Is approved in accordance with § 21.305 of this chapter;

(3) Provides a means of two-way communication between the pilot compartment and—

(i) Each passenger compartment; and

(ii) Each galley located on other than the main passenger deck level;

(4) Is accessible for immediate use from each of two flight crewmember stations in the pilot compartment;

(5) Is accessible for use from at least one normal flight attendant station in each passenger compartment;

(6) Is capable of operation within 10 seconds by a flight attendant at each of those stations in each passenger compartment from which its use is accessible; and

(7) For large turbojet-powered airplanes—

(i) Is accessible for use at enough flight attendant stations so that all floor-level emergency exits (or entryways to those exits in the case of exits located within galleys) in each passenger compartment are observable from one or more of those stations so equipped;

(ii) Has an alerting system incorporating aural or visual signals for use by flight crewmembers to alert flight attendants and for use by flight attendants to alert flight crewmembers;

(iii) For the alerting system required by paragraph (b)(7)(ii) of this section, has a means for the recipient of a call to determine whether it is a normal call or an emergency call; and

(iv) When the airplane is on the ground, provides a means of two-way communication between ground personnel and either of at least two flight crewmembers in the pilot compartment. The interphone system station for use by ground personnel must be so located that personnel using the system may avoid visible detection from within the airplane.

Docket No. 24995 (54 FR 43926) Eff. 10/27/89 (Amdt. 135-34, Eff. 11/27/89)

§ 135.151 Cockpit voice recorders.

(a) No person may operate a multiengine, turbine-powered airplane or rotorcraft having a passenger seating configuration of six or more and for which two pilots are required by certification or operating rules unless it is equipped with an approved cockpit voice recorder that:

(1) Is installed in compliance with Part 23.1457 (a)(1) and (2), (b), (c), (d), (e), (f), and (g); § 25.1457(a)(1) and (2), (b), (c), (d), (e), (f), and (g); § 27.1457 (a)(1) and (2), (b), (c), (d), (e), (f), and (g); § 29.1457(a)(1) and (2), (b), (c), (d), (e), (f), and (g); of this chapter, as applicable; and

(2) Is operated continuously from the use of the check list before the flight to completion of the final check list at the end of the flight.

(b) No person may operate a multiengine, turbine-powered airplane or rotorcraft having a passenger seating configuration of 20 or more seats unless it is equipped with an approved cockpit voice recorder that—

(1) Is installed in compliance with § 23.1457, § 25.1457, § 27.1457 or § 29.1457 of this chapter, as applicable; and

(2) Is operated continuously from the use of the check list before the flight to completion of the final check list at the end of the flight.

(c) In the event of an accident, or occurrence requiring immediate notification of the National Transportation Safety Board which results in termination of the flight, the certificate holder shall keep the recorded information for at least 60 days or, if requested by the Administrator or the Board, for a longer period. Information obtained from the record may be used to assist in determining the cause of accidents or occurrences in connection with investigations. The Administrator does not use the record in any civil penalty or certificate action.

(d) For those aircraft equipped to record the uninterrupted audio signals received by a boom or a mask microphone the flight crewmembers are required to use the boom microphone below 18,000 feet mean sea level. No person may operate a large turbine engine powered airplane manufactured after October 11, 1991, or on which a cockpit voice recorder has been installed after October 11, 1991, unless it is equipped to record the uninterrupted audio signal received by a boom or mask microphone in accordance with § 25.1457(c)(5) of this chapter.

(e) In complying with this section, an approved cockpit voice recorder having an erasure feature may be used, so that during the operation of the recorder, information:

(1) Recorded in accordance with paragraph (a) of this section and recorded more than 15 minutes earlier; or

(2) Recorded in accordance with paragraph (b) of this section and recorded more than 30 minutes earlier; may be erased or otherwise obliterated.

(Amdt. 135-23, Eff. 5/26/87); (Amdt. 135-26, Eff. 10/11/88)

§ 135.152 Flight recorders.

(a) Except as provided in paragraph (k) of this section, no person may operate under this part a multi-engine, turbine-engine powered airplane or rotorcraft having a passenger seating configuration, excluding any required crewmember seat, of 10 to 19 seats, that was either brought onto the U.S. register after, or was registered outside the United States and added to the operator's U.S. operations specifications after, October 11, 1991, unless it is equipped with one or more approved flight recorders that use a digital method of recording and storing data and a method of readily retrieving that data from the storage medium. The parameters specified in either Appendix B or C of this part, as applicable must be recorded within the range, accuracy, resolution, and recording intervals as specified. The recorder shall retain no less than 25 hours of aircraft operation.

(b) After October 11, 1991, no person may operate a multiengine, turbine-powered airplane having a passenger seating configuration of 20 to 30 seats or a multiengine, turbine-powered rotorcraft having a passenger seating configuration of 20 or more seats unless it is equipped with one or more approved flight recorders that utilize a digital method of recording and storing data, and a method of readily retrieving that data from the storage medium. The parameters in Appendix D or E of this part, as applicable, that are set forth below, must be recorded within the ranges, accuracies, resolutions, and sampling intervals as specified:

(1) Except as provided in paragraph (b)(3) of this section for aircraft type certificated before October 1, 1969, the following parameters must be recorded:

(i) Time;

(ii) Altitude;

(iii) Airspeed;

(iv) Vertical acceleration;

(v) Heading;

(vi) Time of each radio transmission to or from air traffic control;

(vii) Pitch attitude;

(viii) Roll attitude;

(ix) Longitudinal acceleration;

(x) Control column or pitch control surface position; and

(xi) Thrust of each engine.

(2) Except as provided in paragraph (b)(3) of this section for aircraft type certificated after September 30, 1969, the following parameters must be recorded:

(i) Time;

(ii) Altitude;

(iii) Airspeed;

(iv) Vertical acceleration;

(v) Heading;

(vi) Time of each radio transmission either to or from air traffic control;

(vii) Pitch attitude;

(viii) Roll attitude;

(ix) Longitudinal acceleration;

(x) Pitch trim position;

(xi) Control column or pitch control surface position;

(xii) Control wheel or lateral control surface position;

(xiii) Rudder pedal or yaw control surface position;

(xiv) Thrust of each engine;

(xv) Position of each thrust reverser;

(xvi) Trailing edge flap or cockpit flap control position; and

(xvii) Leading edge flap or cockpit flap control position.

(3) For aircraft manufactured after October 11, 1991, all of the parameters listed in Appendix D or E of this part, as applicable, must be recorded.

(c) Whenever a flight recorder required by this section is installed, it must be operated continuously from the instant the airplane begins the takeoff roll or the rotorcraft begins the lift-off until the airplane has completed the landing roll or the rotorcraft has landed at its destination.

(d) Except as provided in paragraph (c) of this section, and except for recorded data erased as authorized in this paragraph, each certificate holder shall keep the recorded data prescribed in paragraph (a) of this section until the aircraft has been operating for at least 25 hours of the operating time specified in paragraph (c) of this section. In addition, each certificate holder shall keep the recorded data prescribed in paragraph (b) of this section for an airplane until the airplane has been operating for at least 25 hours, and for a rotorcraft until the rotorcraft has been operating for at least 10 hours, of the operating time specified in paragraph (c) of this section. A total of 1 hour of recorded data may be erased for the purpose of testing the flight recorder or the flight recorder system. Any erasure made in accordance with this paragraph must be of the oldest recorded data accumulated at the time of testing. Except as provided in paragraph (c) of this section, no record need be kept more than 60 days.

(e) In the event of an accident or occurrence that requires that immediate notification of the National Transportation Safety Board under 49 CFR Part 830 of its regulations and that results in termination of the flight, the certificate holder shall remove the recording media from the aircraft and keep the recorded data required by paragraphs (a) and (b) of this section for at least 60 days or for a longer period upon request of the Board or the Administrator.

(f)(1) For airplanes manufactured on or before August 18, 2000, and all other aircraft, each flight recorder required by this section must be installed in accordance with the requirements of §23.1459, 25.1459, 27.1459, or 29.1459, as appropriate, of this chapter. The correlation required by paragraph (c) of §23.1459, 25.1459, 27.1459, or 29.1459, as appropriate, of this chapter need be established only on one aircraft of a group of aircraft:

(i) That are of the same type;

(ii) On which the flight recorder models and their installations are the same; and

(iii) On which there are no differences in the type designs with respect to the installation of the first pilot's instruments associated with the flight recorder. The most recent instrument calibration, including the recording medium from which this calibration is derived, and the recorder correlation must be retained by the certificate holder.

(f)(2) For airplanes manufactured after August 18, 2000, each flight data recorder system required by this section must be installed in accordance with the requirements of §23.1459(a), (b), (d) and (e) of this chapter, or §25.1459 (a), (b), (d), and (e) of this chapter. A correlation must be established between the values recorded by the flight data recorder and the corresponding values being measured. The correlation must contain a sufficient number of correlation points to accurately establish the conversion from the recorded values to engineering units or discrete state over the full operating range of the parameter. Except for airplanes having separate altitude and airspeed sensors that are an integral part of the flight data recorder system, a single correlation may be established for any group of airplanes—

(i) That are of the same type;

(ii) On which the flight recorder system and its installation are the same; and

(iii) On which there is no difference in the type design with respect to the installation of those sensors associated with the flight data recorder system. Documentation sufficient to convert recorded data into the engineering units and discrete values specified in the applicable appendix must be maintained by the certificate holder.

(g) Each flight recorder required by this section that records the data specified in paragraphs (a) and (b) of this section must have an approved device to assist in locating that recorder under water.

(h) The operational parameters required to be recorded by digital flight data recorders required by paragraphs (i) and (j) of this section are as follows, the phrase "when an information source is installed" following a parameter indicates that recording of that parameter is not intended to require a change in installed equipment.

(1) Time;

(2) Pressure altitude;

(3) Indicated airspeed;

(4) Heading—primary flight crew reference (if selectable, record discrete, true or magnetic);

(5) Normal acceleration (Vertical);

(6) Pitch attitude;

(7) Roll attitude;

(8) Manual radio transmitter keying, or CVR/DFDR synchronization reference;

(9) Thrust/power of each engine—primary flight crew reference;

(10) Autopilot engagement status;

(11) Longitudinal acceleration;

(12) Pitch control input;

(13) Lateral control input;

(14) Rudder pedal input;

(15) Primary pitch control surface position;

(16) Primary lateral control surface position;

(17) Primary yaw control surface position;

(18) Lateral acceleration;

(19) Pitch trim surface position or parameters of paragraph (h)(82) of this section if currently recorded;

(20) Trailing edge flap or cockpit flap control selection (except when parameters of paragraph (h)(85) of this section apply);

(21) Leading edge flap or cockpit flap control selection (except when parameters of paragraph (h)(86) of this section apply);

(22) Each Thrust reverser position (or equivalent for propeller airplane);

(23) Ground spoiler position or speed brake selection (except when parameters of paragraph (h)(87) of this section apply);

(24) Outside or total air temperature;

(25) Automatic Flight Control System (AFCS) modes and engagement status, including autothrottle;

(26) Radio altitude (when an information source is installed);

(27) Localizer deviation, MLS Azimuth;

(28) Glideslope deviation, MLS Elevation;

(29) Marker beacon passage;

(30) Master warning;

(31) Air ground sensor (primary airplane system reference nose or main gear);

(32) Angle of attack (when information source is installed);

(33) Hydraulic pressure low (each system);

(34) Ground speed (when an information source is installed);

(35) Ground proximity warning system;

(36) Landing gear position or landing gear cockpit control selection;

(37) Drift angle (when an information source is installed);

(38) Wind speed and direction (when an information source is installed);

(39) Latitude and longitude (when an information source is installed);

(40) Stick shaker/pusher (when an information source is installed);

(41) Windshear (when an information source is installed);

(42) Throttle/power lever position;

(43) Additional engine parameters (as designated in appendix F of this part);

(44) Traffic alert and collision avoidance system;

(45) DME 1 and 2 distances;

(46) Nav 1 and 2 selected frequency;

(47) Selected barometric setting (when an information source is installed);

(48) Selected altitude (when an information source is installed);

(49) Selected speed (when an information source is installed);

(50) Selected mach (when an information source is installed);

(51) Selected vertical speed (when an information source is installed);

(52) Selected heading (when an information source is installed);

(53) Selected flight path (when an information source is installed);

(54) Selected decision height (when an information source is installed);

(55) EFIS display format;

(56) Multi-function/engine/alerts display format;

(57) Thrust command (when an information source is installed);

(58) Thrust target (when an information source is installed);

(59) Fuel quantity in CG trim tank (when an information source is installed);

(60) Primary Navigation System Reference;

(61) Icing (when an information source is installed);

(62) Engine warning each engine vibration (when an information source is installed);

(63) Engine warning each engine over temp. (when an information source is installed);

(64) Engine warning each engine oil pressure low (when an information source is installed);

(65) Engine warning each engine over speed (when an information source is installed);

(66) Yaw trim surface position;

(67) Roll trim surface position;

(68) Brake pressure (selected system);

(69) Brake pedal application (left and right);

(70) Yaw or sideslip angle (when an information source is installed);

(71) Engine bleed valve position (when an information source is installed);

(72) De-icing or anti-icing system selection (when an information source is installed);

(73) Computed center of gravity (when an information source is installed);

(74) AC electrical bus status:

(75) DC electrical bus status;

(76) APU bleed valve position (when an information source is installed);

(77) Hydraulic pressure (each system);

(78) Loss of cabin pressure;

(79) Computer failure;

(80) Heads-up display (when an information source is installed);

(81) Para-visual display (when an information source is installed);

(82) Cockpit trim control input position—pitch;

(83) Cockpit trim control input position—roll;

(84) Cockpit trim control input position—yaw;

(85) Trailing edge flap and cockpit flap control position;

(86) Leading edge flap and cockpit flap control position;

(87) Ground spoiler position and speed brake selection; and

(88) All cockpit flight control input forces (control wheel, control column, rudder pedal).

(i) For all turbine-engine powered airplanes with a seating configuration, excluding any required crewmember seat, of 10 to 30 passenger seats, manufactured after August 18, 2000—

(1) The parameters listed in paragraphs (h)(1) through (h)(57) of this section must be recorded within the ranges, accuracies, resolutions, and recording intervals specified in Appendix F of this part.

(2) Commensurate with the capacity of the recording system, all additional parameters for which information sources are installed and which are connected to the recording system must be recorded within the ranges, accuracies, resolutions, and sampling intervals specified in Appendix F of this part.

(j) For all turbine-engine-powered airplanes with a seating configuration, excluding any required crewmember seat, of 10 to 30 passenger seats, that are manufactured after August 19, 2002 the parameters listed in paragraph (a)(1) through (a)(88) of this section must be recorded within the ranges, accuracies, resolutions, and recording intervals specified in Appendix F of this part.

(k) For airplanes manufactured before August 18, 1997 the following airplane type need not comply with this section: deHavilland DHC–6.

§135.153 Ground proximity warning system.

(a) No person may operate a turbine-powered airplane having a passenger seat configuration of 10 seats or more, excluding any pilot seat, unless it is equipped with an approved ground proximity warning system.

(b) [Reserved]

(c) For a system required by this section, the Airplane Flight Manual shall contain—

(1) Appropriate procedures for—

(i) The use of the equipment;

(ii) Proper flight crew action with respect to the equipment; and

(iii) Deactivation for planned abnormal and emergency conditions; and

(2) An outline of all input sources that must be operating.

(d) No person may deactivate a system required by this section except under procedures in the Airplane Flight Manual.

(e) Whenever a system required by this section is deactivated, an entry shall be made in the airplane maintenance record that includes the date and time of deactivation.

(f) This section expires on March 29, 2005.

§135.154 Terrain awareness and warning system.

(a) *Airplanes manufactured after March 29, 2002:*

(1) No person may operate a turbine-powered airplane configured with 10 or more passenger seats, excluding any pilot seat, unless that airplane is equipped with an approved terrain awareness and warning system that meets the requirements for Class A equipment in Technical Standard Order (TSO)-C151. The airplane must also include an approved terrain situational awareness display.

(2) No person may operate a turbine-powered airplane configured with 6 to 9 passenger seats, excluding any pilot seat, unless that airplane is equipped with an approved terrain awareness and warning system that meets as a minimum the requirements for Class B equipment in Technical Standard Order (TSO)-C151.

(b) *Airplanes manufactured on or before March 29, 2002:*

(1) No person may operate a turbine-powered airplane configured with 10 or more passenger seats, excluding any pilot seat, after March 29, 2005, unless that airplane is equipped with an approved terrain awareness and warning system that meets the requirements for Class A equipment in Technical Standard Order (TSO)-C151. The airplane must also include an approved terrain situational awareness display.

(2) No person may operate a turbine-powered airplane configured with 6 to 9 passenger seats, excluding any pilot seat, after March 29, 2005, unless that airplane is equipped with an approved terrain awareness and warning system that meets as a minimum the requirements for Class B equipment in Technical Standard Order (TSO)-C151.

(c) *Airplane Flight Manual.* The Airplane Flight Manual shall contain appropriate procedures for—

(1) The use of the terrain awareness and warning system; and

(2) Proper flight crew reaction in response to the terrain awareness and warning system audio and visual warnings.

§135.155 Fire extinguishers: Passenger-carrying aircraft.

No person may operate an aircraft carrying passengers unless it is equipped with hand fire extinguishers of an approved type for use in crew and passenger compartments as follows—

(a) The type and quantity of extinguishing agent must be suitable for all the kinds of fires likely to occur;

(b) At least one hand fire extinguisher must be provided and conveniently located on the flight deck for use by the flight crew; and

(c) At least one hand fire extinguisher must be conveniently located in the passenger compartment of each aircraft having a passenger seating configuration, excluding any pilot seat, of at least 10 seats but less than 31 seats.

§ 135.157 Oxygen equipment requirements.

(a) *Unpressurized aircraft.* No person may operate an unpressurized aircraft at altitudes prescribed in this section unless it is equipped with enough oxygen dispensers and oxygen to supply the pilots under § 135.89(a) and to supply, when flying—

(1) At altitudes above 10,000 feet through 15,000 feet MSL, oxygen to at least 10 percent of the occupants of the aircraft, other than the pilots, for that part of the flight at those altitudes that is of more than 30 minutes duration; and

(2) Above 15,000 feet MSL oxygen to each occupant of the aircraft other than the pilots.

(b) *Pressurized aircraft.* No person may operate pressurized aircraft

(1) At altitudes above 25,000 feet MSL, unless at least a 10-minute supply of supplemental oxygen is available for each occupant of the aircraft, other than the pilots, for use when a descent is necessary due to loss of cabin pressurization; and

(2) Unless it is equipped with enough oxygen dispensers and oxygen to comply with paragraph (a) of this section whenever the cabin pressure altitude exceeds 10,000 feet MSL and, if the cabin pressurization fails, to comply with § 135.89(a) or to provide a 2-hour supply for each pilot, whichever is greater, and to supply when flying—

(i) At altitudes above 10,000 feet through 15,000 feet MSL, oxygen to at least 10 percent of the occupants of the aircraft, other than the pilots, for that part of the flight at those altitudes that is of more than 30 minutes duration; and

(ii) Above 15,000 feet MSL, oxygen to each occupant of the aircraft, other than the pilots, for one hour unless, at all times during flight above that altitude, the aircraft can safely descend to 15,000 feet MSL within four minutes, in which case only a 30-minute supply is required.

(c) The equipment required by this section must have a means—

(1) To enable the pilots to readily determine, in flight, the amount of oxygen available in each source of supply and whether the oxygen is being delivered to the dispensing units; or

(2) In the case of individual dispensing units, to enable each user to make those determinations with respect to that person's oxygen supply and delivery; and

(3) To allow the pilots to use undiluted oxygen at their discretion at altitudes above 25,000 feet MSL.

§ 135.158 Pitot heat indication systems.

(a) Except as provided in paragraph (b) of this section, after April 12, 1981, no person may operate a transport category airplane equipped with a flight instrument pitot heating system unless the airplane is also equipped with an operable pitot heat indication system that complies with § 25.1326 of this chapter in effect on April 12, 1978.

(b) A certificate holder may obtain an extension of the April 12, 1981, compliance date specified in paragraph (a) of this section, but not beyond April 12, 1983, from the Director, Flight Standards Service if the certificate holder—

(1) Shows that due to circumstances beyond its control it cannot comply by the specified compliance date; and

(2) Submits by the specified compliance date a schedule for compliance, acceptable to the Director, indicating that compliance will be achieved at the earliest practicable date.

(Amdt. 135-17, Eff. 9/30/81); (Amdt. 135-33, Eff. 10/25/89)

§ 135.159 Equipment requirements: Carrying passengers under VFR at night or under VFR over-the-top conditions.

No person may operate an aircraft carrying passengers under VFR at night or under VFR over-the-top unless it is equipped with—

(a) A gyroscopic rate-of-turn indicator except on the following aircraft:

(1) Airplanes with a third attitude instrument system usable through flight attitudes of 360 degrees of pitch-and-roll and installed in accordance with the instrument requirements prescribed in § 121.3056) of this chapter.

(2) Helicopters with a third attitude instrument system usable through flight attitudes of ±80 degrees of pitch and ±120 degrees of roll and installed in accordance with § 29.1303(g) of this chapter.

(3) Helicopters with a maximum certificated takeoff weight of 6,000 pounds or less.

(b) A slip skid indicator.

(c) A gyroscopic bank-and-pitch indicator.

(d) A gyroscopic direction indicator.

(e) A generator or generators able to supply all probable combinations of continuous in-flight electrical loads for required equipment and for recharging the battery.

(f) For night flights—

(1) An anticollision light system;

(2) Instrument lights to make all instruments, switches, and gauges easily readable, the direct rays of which are shielded from the pilot's eyes; and

(3) A flashlight having at least two size "D" cells or equivalent.

(g) For the purpose of paragraph (e) of this section, a continuous in-flight electrical load includes one that draws current continuously during flight, such as radio equipment, electrically driven instruments and lights, but does not include occasional intermittent loads.

(h) Notwithstanding provisions of paragraphs (b), (c), and (d), helicopters having a maximum certificated takeoff weight of 6,000 pounds or less may be operated until January 6, 1988, under visual flight rules at night without a slip skid indicator, a gyroscopic bank-and-pitch indicator, or a gyroscopic direction indicator.

Docket No. 24550 (51 FR 40709) Eff. 11/7/86; (Amdt. 135-20, Eff. 1/6/87); (Amdt. 135-38, Eff. 11/26/90)

§ 135.161 Radio and navigational equipment: Carrying passengers under VFR at night or under VFR over-the-top.

(a) No person may operate an aircraft carrying passengers under VFR at night, or under VFR over-the-top, unless it has two-way communications equipment able, at least in flight, to transmit to, and receive from, ground facilities 25 miles away.

(b) No person may operate an aircraft carrying passengers under VFR over-the-top unless it has radio navigational equipment able to receive radio signals from the ground facilities to be used.

(c) No person may operate an airplane carrying passengers under VFR at night unless it has radio navigational equipment able to receive radio signals from the ground facilities to be used.

§ 135.163 Equipment requirements: Aircraft carrying passengers under IFR.

No person may operate an aircraft under IFR, carrying passengers, unless it has—

(a) A vertical speed indicator;

(b) A free-air temperature indicator;

(c) A heated pitot tube for each airspeed indicator;

(d) A power failure warning device or vacuum indicator to show the power available for gyroscopic instruments from each power source;

(e) An alternate source of static pressure for the altimeter and the airspeed and vertical speed indicators;

(f) For a single-engine aircraft:

(1) Two independent electrical power generating sources each of which is able to supply all probable combinations of continuous inflight electrical loads for required instruments and equipment; or

(2) In addition to the primary electrical power generating source, a standby battery or an alternate source of electric power that is capable of supplying 150% of the electrical loads of all required instruments and equipment necessary for safe emergency operation of the aircraft for at least one hour;

(g) For multi-engine aircraft, at least two generators or alternators each of which is on a separate engine, of which any combination of one-half of the total number are rated sufficiently to supply the electrical loads of all required instruments and equipment necessary for safe emergency operation of the aircraft except that for multi-engine helicopters, the two required generators may be mounted on the main rotor drive train; and

(h) Two independent sources of energy (with means of selecting either), of which at least one is an engine-driven pump or generator, each of which is able to drive all gyroscopic instruments powered by, or to be powered by, that particular source, and installed so that failure of one instrument or source, does not interfere with the energy supply to the remaining instruments or the other energy source unless, for single-engine aircraft in all-cargo operations only, the rate-of-turn indicator has a source of energy separate from the bank and pitch and direction indicators. For the purpose of this paragraph, for multi-engine aircraft, each engine-driven source of energy must be on a different engine.

(i) For the purpose of paragraph (f) of this section, a continuous inflight electrical load includes one that draws current continuously during flight, such as radio equipment, electrically driven instruments, and lights, but does not include occasional intermittent loads.

§ 135.165 Radio and navigational equipment: Extended overwater or IFR operations.

(a) No person may operate a turbojet airplane having a passenger seating configuration, excluding any pilot seat, of 10 seats or more, or a multiengine airplane in a commuter operation, as defined in part 119 of this chapter, under IFR or in extended overwater operations unless it has at least the following radio communication and navigational equipment appropriate to the facilities to be used which are capable of transmitting to and receiving from, at any place on the route to be flown, at least one ground facility:

(1) Two transmitters, (2) two microphones, (3) two headsets or one headset and one speaker, (4) a marker beacon receiver, (5) two independent receivers for navigation, and (6) two independent receivers for communications.

(b) No person may operate an aircraft other than that specified in paragraph (a) of this section, under IFR or in extended overwater operations unless it has at least the following radio communication and navigational equipment appropriate to the facilities to be used and which are capable of transmitting to, and receiving from, at any place on the route, at least one ground facility:

(1) A transmitter, (2) two microphones, (3) two headsets or one headset and one speaker, (4) a marker beacon receiver, (5) two independent receivers for navigation, (6) two independent receivers for communications, and (7) for extended overwater operations only, an additional transmitter.

(c) For the purpose of paragraphs (a)(5), (a)(6), (b)(5), and (b)(6) of this section, a receiver is independent if the function of any part of it does not depend on the functioning of any part of another receiver. However, a receiver that can receive both communications and navigational signals may be used in place of a separate communications receiver and a separate navigational signal receiver.

Notwithstanding the requirements of paragraphs (a) and (b) of this section, installation and use of a single long-range navigation system and a single long-range communication systems, for extended overwater operations, may be authorized by the Administrator and approved in the certificate holder's operations specifications. The following are among the operational factors the Administrator may consider in granting an authorization:

(1) The ability of the flightcrew to reliably fix the position of the airplane within the degree of accuracy required by ATC,

(2) The length of the route being flown, and

(3) The duration of the very high frequency communications gap.

§ 135.167   Emergency equipment: Extended overwater operations.

(a) Except where the Administrator, by amending the operations specifications of the certificate holder, requires the carriage of all or any specific items of the equipment listed below for any overwater operation, or, upon application of the certificate holder, the Administrator allows deviation for a particular extended overwater operation, no person may operate an aircraft in extended overwater operations unless it carries, installed in conspicuously marked locations easily accessible to the occupants if a ditching occurs, the following equipment:

(1) An approved life preserver equipped with an approved survivor locator light for each occupant of the aircraft. The life preserver must be easily accessible to each seated occupant.

(2) Enough approved life rafts of a rated capacity and buoyancy to accommodate the occupants of the aircraft.

(b) Each life raft required by paragraph (a) of this section must be equipped with or contain at least the following:

(1) One approved survivor locator light.

(2) One approved pyrotechnic signaling device.

(3) Either—

(i) One survival kit, appropriately equipped for the route to be flown; or

(ii) One canopy (for sail, sunshade, or rain catcher);

(iii) One radar reflector;

(iv) One life raft repair kit;

(v) One bailing bucket;

(vi) One signaling mirror;

(vii) One police whistle;

(viii) One raft knife;

(ix) One CO± bottle for emergency inflation;

(x) One inflation pump;

(xi) Two oars;

(xii) One 75-foot retaining line;

(xiii) One magnetic compass;

(xiv) One dye marker;

(xv) One flashlight having at least two size "D" cells or equivalent;

(xvi) A two-day supply of emergency food rations supplying at least 1,000 calories a day for each person;

(xvii) For each two persons the raft is rated to carry, two pints of water or one sea water desalting kit;

(xviii) One fishing kit; and

(xix) One book on survival appropriate for the area in which the aircraft is operated.

(c) No person may operate an airplane in extended overwater operations unless there is attached to one of the life rafts required by paragraph (a) of this section, an approved survival type emergency locator transmitter. Batteries used in this transmitter must be replaced (or recharged, if the batteries are rechargeable) when the transmitter has been in use for more than 1 cumulative hour, or, when 50 percent of their useful life (or for rechargeable batteries, 50 percent of their useful life of charge) has expired, as established by the transmitter manufacturer under its approval. The new expiration date for replacing (or recharging) the battery must be legibly marked on the outside of the transmitter. The battery useful life (or useful life of charge) requirements of this paragraph do not apply to batteries (such as water-activated batteries) that are essentially unaffected during probable storage intervals.

(Amdt. 135-4, Eff. 9/9/80); (Amdt. 135-20, Eff. 1/6/87); [(Amdt. 135-49, Eff. 6/21/94)]

§ 135.169   Additional airworthiness requirements.

(a) Except for commuter category airplanes, no person may operate a large airplane unless it meets the additional airworthiness requirements of §§ 121.213 through 121.283 and 121.307 of this chapter.

(b) No person may operate a reciprocating-engine or turbopropeller-powered small airplane that has a passenger seating configuration, excluding pilot seats, of 10 seats or more unless it is type certificated—

(1) In the transport category;

(2) Before July 1, 1970, in the normal category and meets special conditions issued by the Administrator for airplanes intended for use in operations under this part;

(3) Before July 19, 1970, in the normal category and meets the additional airworthiness standards in Special Federal Aviation Regulation No. 23;

(4) In the normal category and meets the additional airworthiness standards in Appendix A;

(5) In the normal category and complies with section 1.(a) of Special Federal Aviation Regulation No. 41;

(6) In the normal category and complies with section 1.(b) of Special Federal Aviation Regulation No. 41; or

(7) In the commuter category.

(c) No person may operate a small airplane with a passenger seating configuration, excluding any pilot seat, of 10 seats or more, with a seating configuration greater than the maximum seating configuration used in that type airplane in operations under this part before August 19, 1977. This paragraph does not apply to—

(1) An airplane that is type certificated in the transport category; or

(2) An airplane that complies with—

(i) Appendix A of this part provided that its passenger seating configuration, excluding pilot seats, does not exceed 19 seats; or

(ii) Special Federal Aviation Regulation No. 41.

(d) Cargo or baggage compartments:

(1) After March 20, 1991, each Class C or D compartment, as defined in § 25.857 of Part 25 of this chapter, greater than 200 cubic feet in volume in a transport

category airplane type certificated after January 1, 1958, must have ceiling and sidewall panels which are constructed of:

(i) Glass fiber reinforced resin;

(ii) Materials which meet the test requirements of Part 25, Appendix F, Part III of this chapter; or

(iii) In the case of liner installations approved prior to March 20, 1989, aluminum.

(2) For compliance with this paragraph, the term "liner" includes any design feature, such as a joint or fastener, which would affect the capability of the liner to safely contain a fire.

(Amdt. 135-2, Eff. 10/17/79); (Amdt. 135-21, Eff. 2/17/87); (Amdt. 135-31, Eff. 3/20/89)

§ 135.170 Materials for compartment interiors.

(a) No person may operate an airplane that conforms to an amended or supplemental type certificate issued in accordance with SFAR No. 41 for a maximum certificated takeoff weight in excess of 12,500 pounds unless within one year after issuance of the initial airworthiness certificate under that SFAR, the airplane meets the compartment interior requirements set forth in § 25.853(a) in effect March 6, 1995 (formerly § 25.853 (a), (b), (b-1), (b-2), and (b-3) of this chapter in effect on September 26, 1978).

(b) Except for commuter category airplanes and airplanes certificated under Special Federal Aviation Regulation No. 41, no person may operate a large airplane unless it meets the following additional airworthiness requirements:

(1) Except for those materials covered by paragraph (b)(2) of this section, all materials in each compartment used by the crewmembers or passengers must meet the requirements of § 25.853 of this chapter in effect as follows or later amendment thereto:

(i) Except as provided in paragraph (b)(1)(iv) of this section, each airplane with a passenger capacity of 20 or more and manufactured after August 19, 1988, but prior to August 20, 1990, must comply with the heat release rate testing provisions of § 25.853(d) in effect March 6, 1995 (formerly § 25.853(a-1) in effect on August 20, 1986), except that the total heat release over the first 2 minutes of sample exposure rate must not exceed 100 kilowatt minutes per square meter and the peak heat release rate must not exceed 100 kilowatts per square meter.

(ii) Each airplane with a passenger capacity of 20 or more and manufactured after August 19, 1990, must comply with the heat release rate and smoke testing provisions of § 25.853(d) in effect March 6, 1995 (formerly § 25.853(a-1) in effect on September 26, 1988).

(iii) Except as provided in paragraph (b)(1)(v) or (vi) of this section, each airplane for which the application for type certificate was filed prior to May 1, 1972, must comply with the provisions of § 25.853 in effect on April 30, 1972, regardless of the passenger capacity, if there is a substantially complete replacement of the cabin interior after April 30, 1972.

(iv) Except as provided in paragraph (b)(1)(v) or (vi) of this section, each airplane for which the application for type certificate was filed after May 1, 1972, must comply with the material requirements under which the airplane was type certificated regardless of the passenger capacity if there is substantially complete replacement of the cabin interior after that date.

(v) Except as provided in paragraph (b)(1)(vi) of this section, each airplane that was type certificated after January 1, 1958, must comply with the heat release testing provisions of § 25.853(d) in effect March 6, 1995 (formerly § 25.853(a-1) in effect on August 20, 1986), if there is a substantially complete replacement of the cabin interior components identified in that paragraph on or after that date, except that the total heat release over the first 2 minutes of sample exposure shall not exceed 100 kilowatt-minutes per square meter and the peak heat release rate shall not exceed 100 kilowatts per square meter.

(vi) Each airplane that was type certificated after January 1, 1958, must comply with the heat release rate and smoke testing provisions of § 25.853(d) in effect March 6, 1995 (formerly § 25.853(a-1) in effect on August 20, 1986), if there is a substantially complete replacement of the cabin interior components identified in that paragraph after August 19, 1990.

(vii) Contrary provisions of this section notwithstanding, the Manager of the Transport Airplane Directorate, Aircraft Certification Service, Federal Aviation Administration, may authorize deviation from the requirements of paragraph (b)(1)(i), (b)(1)(ii), (b)(1)(v), or (b)(1)(vi) of this section for specific components of the cabin interior that do not meet applicable flammability and smoke emission requirements, if the determination is made that special circumstances exist that make compliance impractical. Such grants of deviation will be limited to those airplanes manufactured within 1 year after the applicable date specified in this section and those airplanes in which the interior is replaced within 1 year of that date. A request for such grant of deviation must include a thorough and accurate analysis of each component subject to § 25.853(d) in effect March 6, 1995 (formerly § 25.853(a-1) in effect on August 20, 1986), the steps being taken to achieve compliance, and for the few components for which timely compliance will not be achieved, credible reasons for such noncompliance.

(viii) Contrary provisions of this section notwithstanding, galley carts and standard galley containers that do not meet the flammability and smoke emission requirements of § 25.853(d) in effect March 6, 1995 (formerly § 25.853(a-1) in effect on August 20, 1986), may be used in airplanes that must meet the requirements of paragraph (b)(1)(i), (b)(1)(ii), (b)(1)(iv) or (b)(1)(vi) of this section provided the galley carts or standard containers were manufactured prior to March 6, 1995.

(c) Thermal/acoustic insulation materials. For transport category airplanes type certificated after January 1, 1958:

(1) For airplanes manufactured before September 2, 2005, when thermal/ acoustic insulation is installed in the fuselage as replacements after September 2, 2005, the insulation must meet the flame propagation requirements of § 25.856 of this chapter, effective September 2, 2003, if it is:

(i) Of a blanket construction, or

(ii) Installed around air ducting.

(2) For airplanes manufactured after September 2, 2005, thermal/acoustic insulation materials installed in the fuselage must meet the flame propagation requirements of § 25.856 of this chapter, effective September 2, 2003.

(3) For airplanes type certificated after January 1, 1958, seat cushions, except those on flight crewmember seats, in any compartment occupied by crew or passengers must comply with the requirements pertaining to fire protection of seat cushions in § 25.853(c) effective November 26, 1984.

§ 135.171 Shoulder harness installation at flight crewmember stations.

(a) No person may operate a turbojet aircraft or an aircraft having a passenger seating configuration, excluding any pilot seat, of 10 seats or more unless it is equipped with an approved shoulder harness installed for each flight crewmember station.

(b) Each flight crewmember occupying a station equip-ped with a shoulder harness must fasten the shoulder harness during takeoff and landing, except that the shoulder harness may be unfastened if the crewmember cannot perform the required duties with the shoulder harness fastened.

§ 135.173 Airborne thunderstorm detection equipment requirements.

(a) No person may operate an aircraft that has a passenger seating configuration, excluding any pilot seat, of 10 seats or more in passenger-carrying operations, except a helicopter operating under day VFR conditions, unless the aircraft is equipped with either approved thunderstorm detection equipment or approved airborne weather radar equipment.

(b) No person may operate a helicopter that has a passenger seating configuration, excluding any pilot seat, of 10 seats or more in passenger-carrying operations, under night VFR when current weather reports indicate that thunderstorms or other potentially hazardous weather conditions that can be detected with airborne thunderstorm detection equipment may reasonably be expected along the route to be flown, unless the helicopter is equipped with either approved thunderstorm detection equipment or approved airborne weather radar equipment.

(c) No person may begin a flight under IFR or night VFR conditions when current weather reports indicate that thunderstorms or other potentially hazardous weather conditions that can be detected with airborne thunderstorm detection equipment, required by paragraph (a) or (b) of this section, may reasonably be expected along the route to be flown, unless the airborne thunderstorm detection equipment is in satisfactory operating condition.

(d) If the airborne thunderstorm detection equipment becomes inoperative en route, the aircraft must be operated under the instructions and procedures specified for that event in the manual required by § 135.21.

(e) This section does not apply to aircraft used solely within the State of Hawaii, within the State of Alaska, within that part of Canada west of longitude 130 degrees W, between latitude 70 degrees N, and latitude 53 degrees N, or during any training, test, or ferry flight.

(f) Without regard to any other provision of this part, an alternate electrical power supply is not required for airborne thunderstorm detection equipment.

(Amdt. 135-20, Eff. 1/6/87)

§ 135.175 Airborne weather radar equipment requirements.

(a) No person may operate a large, transport category aircraft in passenger-carrying operations unless approved airborne weather radar equipment is installed in the aircraft.

(b) No person may begin a flight under IFR or night VFR conditions when current weather reports indicate that thunderstorms, or other potentially hazardous weather conditions that can be detected with airborne weather radar equipment, may reasonably be expected along the route to be flown, unless the airborne weather radar equipment required by paragraph (a) of this section is in satisfactory operating condition.

(c) If the airborne weather radar equipment becomes inoperative en route, the aircraft must be operated under the instructions and procedures specified for that event in the manual required by § 135.21.

(d) This section does not apply to aircraft used solely within the State of Hawaii, within the State of Alaska, within that part of Canada west of longitude 130 degrees W, between latitude 70 degrees N, and latitude 53 degrees N, or during any training, test, or ferry flight.

(e) Without regard to any other provision of this part, an alternate electrical power supply is not required for airborne weather radar equipment.

§ 135.177 Emergency equipment requirements for aircraft having a passenger seating configuration of more than 19 passengers.

(a) No person may operate an aircraft having a passenger seating configuration, excluding any pilot seat, of more than 19 seats unless it is equipped with the following emergency equipment:

(1) One approved first-aid kit for treatment of injuries likely to occur in flight or in a minor accident, which meets the following specifications and requirements:

(i) Each first-aid kit must be dust and moisture proof, and contain only materials that either meet Federal

Specifications GGK-319a, as revised, or as approved by the Administrator.

(ii) Required first-aid kits must be readily accessible to the cabin flight attendants.

(iii) Except as provided in paragraph (a)(1)(iv) of this section, at time of takeoff, each first aid kit must contain at least the following or other contents approved by the Administrator:

Contents (**Quantity**)

Adhesive bandage compressors, 1 in (**16**)

Antiseptic swabs (**20**)

Ammonia inhalents (**10**)

Bandage compressors, 4 in (**8**)

Triangular bandage compressors, 40 in (**5**)

Arm splint, noninflatable (**1**)

Leg splint, noninflatable (**1**)

Roller bandage, 4 in (**4**)

Adhesive tape, 1-in standard roll (**2**)

Bandage scissors (**1**)

Protective latex gloves or equivalent nonpermeable gloves (**1 pair**)

(iv) Protective latex gloves or equivalent nonpermeable gloves may be placed in the first aid kit or in a location that is readily accessible to crewmembers.

(2) A crash axe carried so as to be accessible to the crew but inaccessible to passengers during normal operations.

(3) Signs that are visible to all occupants to notify them when smoking is prohibited and when safety belts must be fastened. The signs must be constructed so that they can be turned on during any movement of the aircraft on the surface, for each takeoff or landing, and at other times considered necessary by the pilot in command. "No smoking" signs shall be turned on when required by § 135.127.

(4) (Reserved)

(b) Each item of equipment must be inspected regularly under inspection periods established in the operations specifications to ensure its condition for continued serviceability and immediate readiness to perform its intended emergency purposes.

(Amdt. 135-25, Eff. 4/23/88); *[(Amdt. 135-43, Eff. 6/30/92)]; [(Amdt. 135-44, Eff. 10/15/92)]; [(Amdt. 135-47, Eff. 1/12/94)]

[§ 135.178 Additional emergency equipment.

[No person may operate an airplane having a passenger seating configuration of more than 19 seats, unless it has the additional emergency equipment specified in paragraphs (a) through (l) of this section.

[(a) *Means for emergency evacuation.* Each passenger-carrying landplane emergency exit (other than over-the-wing) that is more than 6 feet from the ground, with the airplane on the ground and the landing gear extended, must have an approved means to assist the occupants in descending to the ground. The assisting means for a floor-level emergency exit must meet the requirements of § 25.809(f)(1) of this chapter in effect on April 30, 1972, except that, for any airplane for which the application for the type certificate was filed after that date, it must meet the requirements under which the airplane was type certificated. An assisting means that deploys automatically must be armed during taxiing, takeoffs, and landings; however, the Administrator may grant a deviation from the requirement of automatic deployment if he finds that the design of the exit makes compliance impractical, if the assisting means automatically erects upon deployment and, with respect to required emergency exits, if an emergency evacuation demonstration is conducted in accordance with § 121.291(a) of this chapter. This paragraph does not apply to the rear window emergency exit of Douglas DC-3 airplanes operated with fewer than 36 occupants, including crewmembers, and fewer than five exits authorized for passenger use.

[(b) *Interior emergency exit marking.* The following must be complied with for each passenger-carrying airplane:

[(1) Each passenger emergency exit, its means of access, and its means of opening must be conspicuously marked. The identity and location of each passenger emergency exit must be recognizable from a distance equal to the width of the cabin. The location of each passenger emergency exit must be indicated by a sign visible to occupants approaching along the main passenger aisle. There must be a locating sign—

[(i) Above the aisle near each over-the-wing passenger emergency exit, or at another ceiling location if it is more practical because of low headroom;

[(ii) Next to each floor level passenger emergency exit, except that one sign may serve two such exits if they both can be seen readily from that sign; and

[(iii) On each bulkhead or divider that prevents fore and aft vision along the passenger cabin, to indicate emergency exits beyond and obscured by it, except that if this is not possible, the sign may be placed at another appropriate location.

[(2) Each passenger emergency exit marking and each locating sign must meet the following:

[(i) For an airplane for which the application for the type certificate was filed prior to May 1, 1972, each passenger emergency exit marking and each locating sign must be manufactured to meet the requirements of § 25.812(b) of this chapter in effect on April 30, 1972. On these airplanes, no sign may continue to be used if its luminescence (brightness) decreases to below 100 microlamberts. The colors may be reversed if it increases the emergency illumination of the passenger compartment. However, the Administrator may authorize deviation from the 2-inch background requirements if he finds that special circumstances exist that make compliance impractical and that the proposed deviation provides an equivalent level of safety.

[(ii) For an airplane for which the application for the type certificate was filed on or after May 1, 1972, each passenger emergency exit marking and each locating sign must be manufactured to meet the interior emergency exit marking requirements under which the airplane was type certificated. On these airplanes, no sign may continue to be used if its luminescence (brightness) decreases to below 250 microlamberts.

[(c) *Lighting for interior emergency exit markings.* Each passenger-carrying airplane must have an emergency lighting system, independent of the main lighting system; however, sources of general cabin illumination may be common to both the emergency and the main lighting systems if the power supply to the emergency lighting system is independent of the power supply to the main lighting system. The emergency lighting system must—

[(1) Illuminate each passenger exit marking and locating sign;

[(2) Provide enough general lighting in the passenger cabin so that the average illumination when measured at 40-inch intervals at seat armrest height, on the centerline of the main passenger aisle, is at least 0.05 foot-candle; and

[(3) For airplanes type certificated after January 1, 1958, include floor proximity emergency escape path marking which meets the requirements of § 25.812(e) of this chapter in effect on November 26, 1984.

[(d) *Emergency light operation.* Except for lights forming part of emergency lighting subsystems provided in compliance with § 25.812(h) of this chapter (as prescribed in paragraph (h) of this section) that serve no more than one assist means, are independent of the airplane's main emergency lighting systems, and are automatically activated when the assist means is deployed, each light required by paragraphs (c) and (h) of this section must:

[(1) Be operable manually both from the flightcrew station and from a point in the passenger compartment that is readily accessible to a normal flight attendant seat;

[(2) Have a means to prevent inadvertent operation of the manual controls;

[(3) When armed or turned on at either station, remain lighted or become lighted upon interruption of the airplane's normal electric power;

[(4) Be armed or turned on during taxiing, takeoff, and landing. In showing compliance with this paragraph, a transverse vertical separation of the fuselage need not be considered;

[(5) Provide the required level of illumination for at least 10 minutes at the critical ambient conditions after emergency landing; and

[(6) Have a cockpit control device that has an "on," "off," and "armed" position.

[(e) *Emergency exit operating handles.*

[(1) For a passenger-carrying airplane for which the application for the type certificate was filed prior to May 1, 1972, the location of each passenger emergency exit operating handle, and instructions for opening the exit, must be shown by a marking on or near the exit that is readable from a distance of 30 inches. In addition, for each Type I and Type II emergency exit with a locking mechanism released by rotary motion of the handle, the instructions for opening must be shown by—

[(i) A red arrow with a shaft at least three-fourths inch wide and a head twice the width of the shaft, extending along at least 70° of arc at a radius approximately equal to three-fourths of the handle length; and

[(ii) The word "open" in red letters 1 inch high placed horizontally near the head of the arrow.

[(2) For a passenger-carrying airplane for which the application for the type certificate was filed on or after May 1, 1972, the location of each passenger emergency exit operating handle and instructions for opening the exit must be shown in accordance with the requirements under which the airplane was type certificated. On these airplanes, no operating handle or operating handle cover may continue to be used if its luminescence (brightness) decreases to below 100 microlamberts.

[(f) *Emergency exit access.* Access to emergency exits must be provided as follows for each passenger-carrying airplane:

[(1) Each passageway between individual passenger areas, or leading to a Type I or Type II emergency exit, must be unobstructed and at least 20 inches wide.

[(2) There must be enough space next to each Type I or Type II emergency exit to allow a crewmember to assist in the evacuation of passengers without reducing the unobstructed width of the passageway below that required in paragraph (f)(1) of this section; however, the Administrator may authorize deviation from this requirement for an airplane certificated under the provisions of Part 4b of the Civil Air Regulations in effect before December 20, 1951, if he finds that special circumstances exist that provide an equivalent level of safety.

[(3) There must be access from the main aisle to each Type III and Type IV exit. The access from the aisle to these exits must not be obstructed by seats, berths, or other protrusions in a manner that would reduce the effectiveness of the exit. In addition, for a transport category airplane type certificated after January 1, 1958, there must be placards installed in accordance with 25.813(c)(3) of this chapter for each Type III exit after December 3, 1992.

[(4) If it is necessary to pass through a passageway between passenger compartments to reach any required emergency exit from any seat in the passenger cabin, the passageway must not be obstructed. Curtains may, however, be used if they allow free entry through the passageway.

[(5) No door may be installed in any partition between passenger compartments.

[(6) If it is necessary to pass through a doorway separating the passenger cabin from other areas to reach a required emergency exit from any passenger seat, the door must have a means to latch it in the open position, and the door must be latched open during each takeoff and landing. The latching means must be able to withstand the loads imposed upon it when the door is subjected to the ultimate inertia forces, relative to the surrounding structure, listed in § 25.561(b) of this chapter.

[(g) *Exterior exit markings.* Each passenger emergency exit and the means of opening that exit from the outside must be marked on the outside of the airplane. There must be a 2-inch colored band outlining each passenger emergency exit on the side of the fuselage. Each outside marking, including the band, must be readily distinguishable from the surrounding fuselage area by contrast in color. The markings must comply with the following:

[(1) If the reflectance of the darker color is 15 percent or less, the reflectance of the lighter color must be at least 45 percent.

[(2) If the reflectance of the darker color is greater than 15 percent, at least a 30 percent difference between its reflectance and the reflectance of the lighter color must be provided.

[(3) Exits that are not in the side of the fuselage must have the external means of opening and applicable instructions marked conspicuously in red or, if red is inconspicuous against the background color, in bright chrome yellow and, when the opening means for such an exit is located on only one side of the fuselage, a conspicuous marking to that effect must be provided on the other side. "Reflectance" is the ratio of the luminous flux reflected by a body to the luminous flux it receives.

[(h) *Exterior emergency lighting and escape route.*

[(1) Each passenger-carrying airplane must be equipped with exterior lighting that meets the following requirements:

[(i) For an airplane for which the application for the type certificate was filed prior to May 1, 1972, the requirements of § 25.812 (f) and (g) of this chapter in effect on April 30, 1972.

[(ii) For an airplane for which the application for the type certificate was filed on or after May 1, 1972, the exterior emergency lighting requirements under which the airplane was type certificated.

[(2) Each passenger-carrying airplane must be equipped with a slip-resistant escape route that meets the following requirements:

[(i) For an airplane for which the application for the type certificate was filed prior to May 1, 1972, the requirements of § 25.803(e) of this chapter in effect on April 30, 1972.

[(ii) For an airplane for which the application for the type certificate was filed on or after May 1, 1972, the slip-resistant escape route requirements under which the airplane was type certificated.

[(i) *Floor level exits.* Each floor level door or exit in the side of the fuselage (other than those leading into a cargo or baggage compartment that is not accessible from the passenger cabin) that is 44 or more inches high and 20 or

more inches wide, but not wider than 46 inches, each passenger ventral exit (except the ventral exits on Martin 404 and Convair 240 airplanes), and each tail cone exit, must meet the requirements of this section for floor level emergency exits. However, the Administrator may grant a deviation from this paragraph if he finds that circumstances make full compliance impractical and that an acceptable level of safety has been achieved.

[(j) *Additional emergency exits.* Approved emergency exits in the passenger compartments that are in excess of the minimum number of required emergency exits must meet all of the applicable provisions of this section, except paragraphs (f)(1), (2), and (3) of this section, and must be readily accessible.

[(k) On each large passenger-carrying turbojet-powered airplane, each ventral exit and tailcone exit must be—

[(1) Designed and constructed so that it cannot be opened during flight; and

[(2) Marked with a placard readable from a distance of 30 inches and installed at a conspicuous location near the means of opening the exit, stating that the exit has been designed and constructed so that it cannot be opened during flight.

[(l) *Portable lights.* No person may operate a passenger-carrying airplane unless it is equipped with flashlight stowage provisions accessible from each flight attendant seat.]

[(Amdt. 135-43, Eff. 6/3/92)]

§ 135.179 Inoperable instruments and equipment.

(a) No person may take off an aircraft with inoperable instruments or equipment installed unless the following conditions are met:

(1) An approved Minimum Equipment List exists for that aircraft.

(2) The certificate-holding district office has issued the certificate holder operations specifications authorizing operations in accordance with an approved Minimum Equipment List. The flight crew shall have direct access at all times prior to flight to all of the information contained in the approved Minimum Equipment List through printed or other means approved by the Administrator in the certificate holders operations specifications. An approved Minimum Equipment List, as authorized by the operations specifications, constitutes an approved change to the type design without requiring recertification.

(3) The approved Minimum Equipment List must:

(i) Be prepared in accordance with the limitations specified in paragraph (b) of this section.

(ii) Provide for the operation of the aircraft with certain instruments and equipment in an inoperable condition.

(4) Records identifying the inoperable instruments and equipment and the information required by (a)(3)(ii) of this section must be available to the pilot.

(5) The aircraft is operated under all applicable conditions and limitations contained in the Minimum Equipment

List and the operations specifications authorizing use of the Minimum Equipment List.

(b) The following instruments and equipment may not be included in the Minimum Equipment List:

(1) Instruments and equipment that are either specifically or otherwise required by the airworthiness requirements under which the airplane is type certificated and which are essential for safe operations under all operating conditions.

(2) Instruments and equipment required by an airworthiness directive to be in operable condition unless the airworthiness directive provides otherwise.

(3) Instruments and equipment required for specific operations by this part.

(c) Notwithstanding paragraphs (b)(i) and (b)(3) of this section, an aircraft with inoperable instruments or equipment may be operated under a special flight permit under % 21.197 and 21.199 of this chapter.

(Amdt. 135-39, Eff. 6/20/91)

§ 135.180   Traffic alert and collision avoidance system.

(a) Unless otherwise authorized by the Administrator, after December 31, 1995, no person may operate a turbine powered airplane that has a passenger seat configuration, excluding any pilot seat, of 10 to 30 seats unless it is equipped with an approved traffic alert and collision avoidance system. If a TCAS II system is installed, it must be capable of coordinating with TCAS units that meet TSO C-119.

(b) The airplane flight manual required by § 135.21 of this part shall contain the following information on the TCAS I system required by this section:

(1) Appropriate procedures for—

(i) The use of the equipment; and

(ii) Proper flightcrew action with respect to the equipment operation.

(2) An outline of all input sources that must be operating for the TCAS to function properly.

Docket No. 25355 (54 FR 951) Eff. 1/10/89;

(Amdt. 135-30, Eff. 2/9/89)

§ 135.181   Performance requirements: Aircraft operated over-the-top or in IFR conditions.

(a) Except as provided in paragraphs (b) and (c) of this section, no person may—

(1) Operate a single-engine aircraft carrying passengers over-the-top; or

(2) Operate a multiengine aircraft carrying passengers over-the-top or in IFR conditions at a weight that will not allow it to climb, with the critical engine inoperative, at least 50 feet a minute when operating at the MEAs of the route to be flown or 5,000 feet MSL, whichever is higher.

(b) Notwithstanding the restrictions in paragraph (a)(2) of this section, multiengine helicopters carrying passengers offshore may conduct such operations in over-the-top or in IFR conditions at a weight that will allow the helicopter to climb at least 50 feet per minute with the critical engine inoperative when operating at the MEA of the route to be flown or 1,500 feet MSL, whichever is higher.

(c) Without regard to paragraph (a) of this section, if the latest weather reports or forecasts, or any combination of them, indicate that the weather along the planned route (including takeoff and landing) allows flight under VFR under the ceiling (if a ceiling exists) and that the weather is forecast to remain so until at least 1 hour after the estimated time of arrival at the destination, a person may operate an aircraft over-the-top.

(d) Without regard to paragraph (a) of this section, a person may operate an aircraft over-the-top under conditions allowing—

(1) For multiengine aircraft, descent or continuance of the flight under VFR if its critical engine fails; or

(2) For single-engine aircraft, descent under VFR if its engine fails.

(Amdt. 135-20, Eff. 1/6/87)

§ 135.183   Performance requirements: Land aircraft operated over water.

No person may operate a land aircraft carrying passengers over water unless—

(a) It is operated at an altitude that allows it to reach land in the case of engine failure;

(b) It is necessary for takeoff or landing;

(c) It is a multiengine aircraft operated at a weight that will allow it to climb, with the critical engine inoperative, at least 50 feet a minute, at an altitude of 1,000 feet above the surface; or

(d) It is a helicopter equipped with helicopter flotation devices.

§ 135.185   Empty weight and center of gravity: Currency requirement.

(a) No person may operate a multiengine aircraft unless the current empty weight and center of gravity are calculated from values established by actual weighing of the aircraft within the preceding 36 calendar months.

(b) Paragraph (a) of this section does not apply to—

(1) Aircraft issued an original airworthiness certificate within the preceding 36 calendar months; and

(2) Aircraft operated under a weight and balance system approved in the operations specifications of the certificate holder.

Subpart D—VFR/IFR operating limitations and weather requirements

§ 135.201 Applicability.

This subpart prescribes the operating limitations for VFR/IFR flight operations and associated weather requirements for operations under this part.

§ 135.203 VFR: Minimum altitudes.

Except when necessary for takeoff and landing, no person may operate under VFR—

(a) An airplane—

(1) During the day, below 500 feet above the surface or less than 500 horizontally from any obstacle; or

(2) At night, at an altitude less than 1,000 feet above the highest obstacle within a horizontal distance of 5 miles from the course intended to be flown or, in designated mountainous terrain, less than 2,000 feet above the highest obstacle within a horizontal distance of 5 miles from the course intended to be flown; or

(b) A helicopter over a congested area at an altitude less than 300 feet above the surface.

§ 135.205 VFR: Visibility requirements.

(a) No person may operate an airplane under VFR in uncontrolled airspace when the ceiling is less than 1,000 feet unless flight visibility is at least 2 miles.

(b) [No person may operate a helicopter under VFR in Class G airspace at an altitude of 1,200 feet or less above the surface or within the lateral boundaries of the surface areas of Class B, Class C, Class D, or Class E airspace designated for an airport unless the visibility is at least—]

(1) During the day—½ mile; or

(2) At night—1 mile.

[(Amdt. 135-41, Eff. 9/16/93)]

§ 135.207 VFR: Helicopter surface reference requirements.

No person may operate a helicopter under VFR unless that person has visual surface reference or, at night, visual surface light reference, sufficient to safely control the helicopter.

§ 135.209 VFR: Fuel supply.

(a) No person may begin a flight operation in an airplane under VFR unless, considering wind and forecast weather conditions, it has enough fuel to fly to the first point of intended landing and, assuming normal cruising fuel consumption—

(1) During the day, to fly after that for at least 30 minutes; or

(2) At night, to fly after that for at least 45 minutes.

(b) No person may begin a flight operation in a helicopter under VFR unless, considering wind and forecast weather conditions, it has enough fuel to fly to the first point of intended landing and, assuming normal cruising fuel consumption, to fly after that for at least 20 minutes.

§ 135.211 VFR: Over-the-top carrying passengers: Operating limitations.

Subject to any additional limitations in § 135.181, no person may operate an aircraft under VFR over-the-top carrying passengers, unless—

(a) Weather reports or forecasts, or any combination of them, indicate that the weather at the intended point of termination of over-the-top flight—

(1) Allows descent to beneath the ceiling under VFR and is forecast to remain so until at least 1 hour after the estimated time of arrival at that point; or

(2) Allows an IFR approach and landing with flight clear of the clouds until reaching the prescribed initial approach altitude over the final approach facility, unless the approach is made with the use of radar under § 91.175(f) of this chapter; or

(b) It is operated under conditions allowing—

(1) For multiengine aircraft, descent or continuation of the flight under VFR if its critical engine fails; or

(2) For single-engine aircraft, descent under VFR if its engine fails.

(Amdt. 135-32, Eff. 8/18/90)

§ 135.213 Weather \reports and forecasts.

(a) Whenever a person operating an aircraft under this part is required to use a weather report or forecast, that person shall use that of the U.S. National Weather Service, a source approved by the U.S. National Weather Service, or a source approved by the Administrator. However, for operations under VFR, the pilot in command may, if such a report is not available, use weather information based on that pilot's own observations or on those of other persons competent to supply appropriate observations.

(b) For the purposes of paragraph (a) of this section, weather observations made and furnished to pilots to conduct IFR operations at an airport must be taken at the airport where those IFR operations are conducted, unless the Administrator issues operations specifications allowing the use of weather observations taken at a location not at the airport where the IFR operations are conducted. The Administrator issues such operations specifications when, after investigation by the U.S. National Weather Service and the certificate-holding district office, it is found that the standards of safety for that operation would allow the deviation from this paragraph for a particular operation for which an air carrier operations certificate or operating certificate has been issued.

§ 135.215 IFR: Operating limitations.

(a) Except as provided in paragraphs (b), (c) and (d) of this section, no person may operate an aircraft under IFR outside of controlled airspace or at any airport that does not have an approved standard instrument approach procedure.

(b) The Administrator may issue operations specifications to the certificate holder to allow it to operate under IFR over routes outside controlled airspace if—

(1) The certificate holder shows the Administrator that the flight crew is able to navigate, without visual reference to the ground, over an intended track without deviating more than 5 degrees or 5 miles, whichever is less, from that track; and

(2) The Administrator determines that the proposed operations can be conducted safely.

(c) A person may operate an aircraft under IFR outside of controlled airspace if the certificate holder has been approved for the operations and that operation is necessary to—

(1) Conduct an instrument approach to an airport for which there is in use a current approved standard or special instrument approach procedure; or

(2) Climb into controlled airspace during an approved missed approach procedure; or

(3) Make an IFR departure from an airport having an approved instrument approach procedure.

(d) The Administrator may issue operations specifications to the certificate holder to allow it to depart at an airport that does not have an approved standard instrument approach procedure when the Administrator determines that it is necessary to make an IFR departure from that airport and that the proposed operations can be conducted safely. The approval to operate at that airport does not include an approval to make an IFR approach to that airport.

§ 135.217 IFR: Takeoff limitations.

No person may take off an aircraft under IFR from an airport where weather conditions are at or above takeoff minimums but are below authorized IFR landing minimums unless there is an alternate airport within 1 hour's flying time (at normal cruising speed, in still air) of the airport of departure.

§ 135.219 IFR: Destination airport weather minimums.

No person may take off an aircraft under IFR or being an IFR or over-the-top operation unless the latest weather reports or forecasts, or any combination of them, indicate that weather conditions at the estimated time of arrival at the next airport of intended landing will be at or above authorized IFR landing minimums.

§ 135.221 IFR: Alternate airport weather minimums.

No person may designate an alternate airport unless the weather reports or forecasts, or any combination of them, indicate that the weather conditions will be at or above authorized alternate airport landing minimums for that airport at the estimated time of arrival.

§ 135.223 IFR: Alternate airport requirements.

(a) Except as provided in paragraph (b) of this section, no person may operate an aircraft in IFR conditions unless it carries enough fuel (considering weather reports or forecasts or any combination of them) to—

(1) Complete the flight to the first airport of intended landing;

(2) Fly from that airport to the alternate airport; and

(3) Fly after that for 45 minutes at normal cruising speed, or helicopters, fly after that for 30 minutes at normal cruising speed.

(b) Paragraph (a)(2) of this section does not apply if Part 97 of this chapter prescribes a standard instrument approach procedure for the first airport of intended landing and, for at least one hour before and after the estimated time of arrival, the appropriate weather reports or forecasts, or any combination of them, indicate that—

(1) The ceiling will be at least 1,500 feet above the lowest circling approach MDA; or

(2) If a circling instrument approach is not authorized for the airport, the ceiling will be at least 1,500 feet above the lowest published minimum or 2,000 feet above the airport elevation, whichever is higher; and

(3) Visibility for that airport is forecast to be at least three miles, or two miles more than the lowest applicable visibility minimums, whichever is the greater, for the instrument approach procedure to be used at the destination airport.

(Amdt. 135-20, Eff. 1/6/87)

§ 135.225 IFR: Takeoff, approach and landing minimums.

(a) Except to the extent permitted by paragraph (b) of this section, no pilot may begin an instrument approach procedure to an airport unless—

(1) That airport has a weather reporting facility operated by the U.S. National Weather Service, a source approved by U.S. National Weather Service, or a source approved by the Administrator; and

(2) The latest weather report issued by that weather reporting facility indicates that weather conditions are at or above the authorized IFR landing minimums for that airport.

(b) A pilot conducting an eligible on- demand operation may begin an instrument approach procedure to an airport that does not have a weather reporting facility operated by the U.S. National Weather Service, a source approved by the U.S. National Weather Service, or a source approved by the Administrator if—

(1) The alternate airport has a weather reporting facility operated by the U.S. National Weather Service, a source approved by the U.S. National Weather Service, or a source approved by the Administrator; and

(2) The latest weather report issued by the weather reporting facility includes a current local altimeter setting for the destination airport. If no local altimeter setting for the destination airport is available, the pilot may use the current altimeter setting provided by the facility designated on the approach chart for the destination airport.

(c) If a pilot has begun the final approach segment of an instrument approach to an airport under paragraph (b) of this section, and the pilot receives a later weather report indicating that conditions have worsened to below the minimum requirements, then the pilot may continue the approach only if the requirements of §91.175(l) of this chapter, or both of the following conditions, are met—

(1) The later weather report is received when the aircraft is in one of the following approach phases:

(i) The aircraft is on an ILS final approach and has passed the final approach fix;

(ii) The aircraft is on an ASR or PAR final approach and has been turned over to the final approach controller; or

(iii) The aircraft is on a nonprecision final approach and the aircraft.

(A) Has passed the appropriate facility or final approach fix; or

(B) Where a final approach fix is not specified, has completed the procedure turn and is established inbound toward the airport on the final approach course within the distance prescribed in the procedure; and

(2) The pilot in command finds, on reaching the authorized MDA or DH, that the actual weather conditions are at or above the minimums prescribed for the procedure being used.

(d) If a pilot has begun the final approach segment of an instrument approach to an airport under paragraph (c) of this section and a later weather report indicating below minimum conditions is received after the aircraft is—

(1) On an ILS final approach and has passed the final approach fix; or

(2) On an ASR or PAR final approach and has been turned over to the final approach controller; or

(3) On a final approach using a VOR, NDB, or comparable approach procedure; and the aircraft-

(i) Has passed the appropriate facility or final approach fix; or

(ii) Where a final approach fix is not specified, has completed the procedure turn and is established inbound toward the airport on the final approach course within the distance prescribed in the procedure; the approach may be continued and a landing made if the pilot finds, upon reaching the authorized MDA or DH, that actual weather conditions are at least equal to the minimums prescribed for the procedure.

(e) The MDA or DH and visibility landing minimums prescribed in part 97 of this chapter or in the operator's operations specifications are increased by 100 feet and 1/2 mile respectively, but not to exceed the ceiling and visibility minimums for that airport when used as an alternate airport, for each pilot in command of a turbine-powered airplane who has not served at least 100 hours as pilot in command in that type of airplane.

(f) Each pilot making an IFR take-off or approach and landing at a military or foreign airport shall comply with applicable instrument approach procedures and weather minimums prescribed by the authority having jurisdiction over that airport. In addition, no pilot may, at that airport—

(1) Take off under IFR when the visibility is less than 1 mile; or

(2) Make an instrument approach when the visibility is less than 1/2 mile.

(g) If takeoff minimums are specified in part 97 of this chapter for the take- off airport, no pilot may take off an aircraft under IFR when the weather conditions reported by the facility described in paragraph (a)(1) of this section are less than the takeoff minimums specified for the take-off airport in part 97 or in the certificate holder's operations specifications.

(h) Except as provided in paragraph (i) of this section, if takeoff minimums are not prescribed in part 97 of this chapter for the takeoff airport, no pilot may takeoff an aircraft under IFR when the weather conditions reported by the facility described in paragraph (a) (1) of this section are less than that prescribed in part 91 of this chapter or in the certificate holder's operations specifications.

(i) At airports where straight-in instrument approach procedures are authorized, a pilot may takeoff an aircraft under IFR when the weather conditions reported by the facility described in paragraph (a)(1) of this section are equal to or better than the lowest straight-in landing minimums, unless otherwise restricted, if

(1) The wind direction and velocity at the time of take-off are such that a straight-in instrument approach can be made to the runway served by the instrument approach;

(2) The associated ground facilities upon which the landing minimums are predicated and the related airborne equipment are in normal operation; and

(3) The certificate holder has been approved for such operations.

[Doc. No. 16097, 43 FR 46783, Oct. 10, 1978, as amended by Amdt. 135-91, 68 FR 54586, Sept. 17, 2003; Amdt. 135-93, 69 FR 1641, Jan. 9, 2004]

§ 135.227 Icing conditions: Operating limitations.

(a) [No pilot may take off an aircraft that has frost, ice, or snow adhering to any rotor blade, propeller, windshield, wing, stabilizing or control surface, to a powerplant installation, or to an airspeed, altimeter, rate of climb, or flight attitude instrument system, except under the following conditions:

[(l) Takeoffs may be made with frost adhering to the wings, or stabilizing or control surfaces, if the frost has been polished to make it smooth.

[(2) Takeoffs may be made with frost under the wing in the area of the fuel tanks if authorized by the Administrator.]

[(b) No certificate holder may authorize an airplane to take off and no pilot may take off an airplane any time conditions are such that frost, ice, or snow may reasonably be expected to adhere to the airplane unless the pilot has completed all applicable training as required by § 135.341 and unless one of the following requirements is met:

[(1) A pretakeoff contamination check, that has been established by the certificate holder and approved by the Administrator for the specific airplane type, has been completed within 5 minutes prior to beginning takeoff. A pretakeoff contamination check is a check to make sure the wings and control surfaces are free of frost, ice, or snow.

[(2) The certificate holder has an approved alternative procedure and under that procedure the airplane is determined to be free of frost, ice, or snow.

[(3) The certificate holder has an approved deicing/anti-icing program that complies with § 121.629(c) of this chapter and the takeoff complies with that program.]

([c]) Except for an airplane that has ice protection provisions that meet § 34 of Appendix A, or those for transport category airplane type certificate, no pilot may fly—

(1) Under IFR into known or forecast light or moderate icing conditions; or

(2) Under VFR into known light or moderateicing conditions; unless the aircraft has functioning deicing or anti-icing equipment protecting each rotor blade, propeller, windshield, wing, stabilizing or control surface, and each airspeed, altimeter, rate of climb, or flight attitude instrument system.

([d]) No pilot may fly a helicopter under IFR into known or forecast icing conditions or under VFR into known icing conditions unless it has been type certificated and appropriately equipped for operations in icing conditions.

([e]) Except for an airplane that has ice protection provisions that meet § 34 of Appendix A, or those for transport category airplane type certification, no pilot may fly an aircraft into known or forecast severe icing conditions.

([f]) If current weather reports and briefing information relied upon by the pilot in command indicate that the forecast icing condition that would otherwise prohibit the flight will not be encountered during the flight because of changed weather conditions since the forecast, the restrictions in paragraphs (c), (d), and (e) of this section based on forecast conditions do not apply.

(Amdt. 135-20, Eff. 1/6/87); [(Amdt. 135-46, Eff. 1/31/94)]

§ 135.229 Airport requirements.

(a) No certificate holder may use any airport unless it is adequate for the proposed operation, considering such items as size, surface, obstructions, and lighting.

(b) No pilot of an aircraft carrying passengers at night may take off from, or land on, an airport unless—

(1) That pilot has determined the wind direction from an illuminated wind direction indicator or local ground communications or, in the case of takeoff, that pilot's personal observations; and

(2) The limits of the area to be used for landing or takeoff are clearly shown—

(i) For airplanes, by boundary or runway marker lights;

(ii) For helicopters, by boundary or runway marker lights or reflective material.

(c) For the purpose of paragraph (b) of this section, if the area to be used for takeoff or landing is marked by flare pots or lanterns, their use must be approved by the Administrator.

Subpart E—Flight crewmember requirements

§ 135.241 Applicability.

Except as provided in § 135.3, this subpart prescribes the flight crewmember requirements for operations under this part.

§ 135.243 Pilot in command qualifications.

(a) No certificate holder may use a person, nor may any person serve, as pilot in command in passenger-carrying operations—

(1) Of a turbojet airplane, of an airplane having a passenger-seat configuration, excluding each crewmember seat, of 10 seats or more, or of a multiengine airplane in a commuter operation as defined in part 119 of this chapter, unless that person holds an airline transport pilot certificate with appropriate category and class ratings and, if required, an appropriate type rating for that airplane.

(2) Of a helicopter in a scheduled interstate air transportation operation by an air carrier within the 48 contiguous states unless that person holds an airline transport pilot certificate, appropriate type ratings, and an instrument rating.

(b) Except as provided in paragraph (a) of this section, no certificate holder may use a person, nor may any person serve, as pilot in command of an aircraft under VFR unless that person—

(1) Holds at least a commercial pilot certificate with appropriate category and class ratings and, if required, an appropriate type rating for that aircraft; and

(2) Has had at least 500 hours of flight time as a pilot, including at least 100 hours of cross-country flight time, at least 25 hours of which were at night; and

(3) For an airplane, holds an instrument rating or an airline transport pilot certificate with an airplane category rating; or

(4) For helicopter operations conducted VFR over-the-top, holds a helicopter instrument rating, or an airline transport pilot certificate with a category and class rating for that aircraft, not limited to VFR.

(c) Except as provided in paragraph (a) of this section, no certificate holder may use a person, nor may any person serve, as pilot in command of an aircraft under IFR unless that person—

(1) Holds at least a commercial pilot certificate with appropriate category and class ratings and, if required, an appropriate type rating for that aircraft; and

(2) Has had at least 1,200 hours of flight time as a pilot, including 500 hours of cross-country flight time, 100 hours of night flight time, and 75 hours of actual or simulated instrument time at least 50 hours of which were in actual flight; and

(3) For an airplane, holds an instrument rating or an airline transport pilot certificate with an airplane category rating; or

(4) For a helicopter, holds a helicopter instrument rating, or an airline transport pilot certificate with a category and class rating for the aircraft, not limited to VFR.

(d) Paragraph (b)(3) of this section does not apply when—

(1) The aircraft used is a single reciprocating-engine-powered airplane;

(2) The certificate holder does not conduct any operation pursuant to a published flight schedule which specifies five or more round trips a week between two or more points and places between which the round trips are performed, and does not transport mail by air under a contract or contracts with the United States Postal Service having total amount estimated at the beginning of any semiannual reporting period (January 1–June 30; July 1–December 31) to be in excess of $20,000 over the 12 months commencing with the beginning of the reporting period;

(3) The area, as specified in the certificate holder's operations specifications, is an isolated area, as determined by the Flight Standards district office, if it is shown that—

(i) The primary means of navigation in the area is by pilotage, since radio navigational aids are largely ineffective; and

(ii) The primary means of transportation in the area is by air;

(4) Each flight is conducted under day VFR with a ceiling of not less than 1,000 feet and visibility not less than 3 statute miles;

(5) Weather reports or forecasts, or any combination of them, indicate that for the period commencing with the planned departure and ending 30 minutes after the planned arrival at the destination the flight may be conducted under VFR with a ceiling of not less than 1,000 feet and visibility of not less than 3 statute miles, except that if weather reports and forecasts are not available, the pilot in command may use that pilot's observations or those of other persons competent to supply weather observations if those observations indicate the flight may be conducted under VFR with the ceiling and visibility required in this paragraph;

(6) The distance of each flight from the certificate holder's base of operation to destination does not exceed 250 nautical miles for a pilot who holds a commercial pilot certificate with an airplane rating without an instrument rating, provided the pilot's certificate does not contain any limitation to the contrary; and

(7) The areas to be flown are approved by the certificate-holding FAA Flight Standards district office and are listed in the certificate holder's operations specifications.

(Amdt. 135-15, Eff. 6/11/81)

§ 135.244 Operating experience.

(a) No certificate holder may use any person, nor may any person serve, as a pilot in command of an aircraft operated in a commuter operation, as defined in part 119 of this chapter, unless that person has completed, prior to designation as pilot in command, on that make and basic model aircraft and in that crewmember position, the following operating experience in each make and basic model of aircraft to be flown:

(1) Aircraft, single engine—10 hours.

(2) Aircraft multiengine, reciprocating engine-powered—15 hours.

(3) Aircraft multiengine, turbine engine-powered—20 hours.

(4) Airplane, turbojet-powered—25 hours.

(b) In acquiring the operating experience, each person must comply with the following:

(1) The operating experience must be acquired after satisfactory completion of the appropriate ground and flight training for the aircraft and crewmember position. Approved provisions for the operating experience must be included in the certificate holder's training program.

(2) The experience must be acquired in flight during commuter passenger-carrying operations under this part. However, in the case of an aircraft not previously used by the certificate holder in operations under this part, operating experience acquired in the aircraft during proving flights or ferry flights may be used to meet this requirement.

(3) Each person must acquire the operating experience while performing the duties of a pilot in command under the supervision of a qualified check pilot.

(4) The hours of operating experience may be reduced to not less than 50 percent of the hours required by this section by the substitution of one additional takeoff and landing for each hour of flight.

Docket No. 20011 (45 FR 7541) Eff. 2/4/80;

(Amdt. 135-3, Eff. 3/1/80); (Amdt. 135-9, Eff. 12/1/80)

§ 135.245 Second in command qualifications.

(a) Except as provided in paragraph (b), no certificate holder may use any person, nor may any person serve, as second in command of an aircraft unless that person holds at least a commercial pilot certificate with appropriate category and class ratings and an instrument rating. For flight under IFR, that person must meet the recent instrument experience requirements of Part 61 of this chapter.

(b) A second in command of a helicopter operated under VFR, other than over-the-top, must have at least a commercial pilot certificate with an appropriate aircraft category and class rating.

(Amdt. 135-1, Eff. 5/7/79)

§ 135.247 Pilot qualifications: Recent experience.

(a) No certificate holder may use any person, nor may any person serve, as pilot in command of an aircraft carrying passengers unless, within the preceding 90 days, that person has—

(1) Made three takeoffs and three landings as the sole manipulator of the flight controls in an aircraft of the same category and class and, if a type rating is required, of the same type in which that person is to serve; or

(2) For operation during the period beginning 1 hour after sunset and ending 1 hour before sunrise (as published in the Air Almanac), made three takeoffs and three landings during that period as the sole manipulator of the flight controls in an aircraft of the same category and class and, if a type rating is required, of the same type in which that person is to serve.

A person who complies with paragraph (a)(2) of this section need not comply with paragraph (a)(1) of this section.

(b) For the purpose of paragraph (a) of this section, if the aircraft is a tailwheel airplane, each takeoff must be made in a tailwheel airplane and each landing must be made to a full stop in a tailwheel airplane.

(3) Paragraph (a)(2) of this section does not apply to a pilot in command of a turbine-powered airplane that is type certificated for more than one pilot crewmember, provided that pilot has complied with the requirements of paragraph (a)(3)(i) or (ii) of this section:

(i) The pilot in command must hold at least a commercial pilot certificate with the appropriate category, class, and type rating for each airplane that is type certificated for more than one pilot crewmember that the pilot seeks to operate under this alternative, and:

(A) That pilot must have logged at least 1,500 hours of aeronautical experience as a pilot;

(B) In each airplane that is type certificated for more than one pilot crewmember that the pilot seeks to operate under this alternative, that pilot must have accomplished and logged the daytime takeoff and landing recent flight experience of paragraph (a) of this section, as the sole manipulator of the flight controls;

(C) Within the preceding 90 days prior to the operation of that airplane that is type certificated for more than one pilot crewmember, the pilot must have accomplished and logged at least 15 hours of flight lime in the type of airplane that the pilot seeks to operate under this alternative; and

(D) That pilot has accomplished and logged at least 3 takeoffs and 3 landings to a full stop, as the sole manipulator of the flight controls, in a turbine-powered airplane that requires more than one pilot crewmember. The pilot must have performed the takeoffs and landings during the period beginning 1 hour after sunset and ending 1 hour before sunrise within the preceding 6 months prior to the month of the flight.

(ii) The pilot in command must hold at least a commercial pilot certificate with the appropriate category, class, and type rating for each airplane that is type certificated for more than one pilot crewmember that the pilot seeks to operate under this alternative, and:

(A) That pilot must have logged at least 1,500 hours of aeronautical experience as a pilot;

(B) In each airplane that is type certificated for more than one pilot crewmember that the pilot seeks to operate under this alternative, that pilot must have accomplished and logged the daytime takeoff and landing recent flight experience of paragraph (a) of this section, as the sole manipulator of the flight controls;

(C) Within the preceding 90 days prior to the operation of that airplane that is type certificated for more than one pilot crewmember, the pilot must have accomplished and logged at least 15 hours of flight time in the type of airplane that the pilot seeks to operate under this alternative; and

(D) Within the preceding 12 months prior to the month of the flight, the pilot must have completed a training program that is approved under part 142 of this chapter. The approved training program must have required and the pilot must have performed, at least 6 takeoffs and 6 landings to a full stop as the sole manipulator of the controls in a flight simulator that is representative of a turbine-powered airplane that requires more than one pilot crewmember. The flight simulator's visual system must have been adjusted to represent the period beginning 1 hour after sunset and ending 1 hour before sunrise.

§ 135.249 Use of prohibited drugs.

(a) This section applies to persons who perform a function listed in Appendix I to Part 121 of this chapter for a certificate holder or an operator. For the purpose of this section, a person who performs such a function pursuant to a contract with the certificate holder or the operator is considered to be performing that function for the certificate holder or the operator.

(b) No certificate holder or operator may knowingly use any person to perform, nor may any person perform for a certificate holder or an operator, either directly or by contract, any function listed in Appendix I to Part 121 of this chapter while that person has a prohibited drug, as defined in that appendix, in his or her system.

(c) No certificate holder or operator shall knowingly use any person to perform, nor shall any person perform for a certificate holder or operator, either directly or by contract, any safety-sensitive function if the person has a verified positive drug test result on or has refused to submit to a drug test required by Appendix I of Part 121 of this chapter and the person has not met the requirements of Appendix I to Part 121 of this chapter for returning to the performance of safety-sensitive duties.

Docket No. 25148 (54 FR 47061) Eff. 11/21/88; (Amdt. 135-28, Eff. 12/21/88)

§ 135.251 Testing for prohibited drugs.

(a) Each certificate holder or operator shall test each of its employees who performs a function listed in Appendix I to Part 121 of this chapter in accordance with that appendix.

(b) Except as provided in paragraph (c) of this section, no certificate holder or operator may use any contractor to perform a function listed in appendix I part 121 of this chapter unless that contractor tests each employee performing such a function for the certificate holder or operator in accordance with that appendix.

(c) If a certificate holder conducts an on-demand operation into an airport at which no maintenance providers are available that are subject to the requirements of appendix I to part 121 and emergency maintenance is required, the certificate holder may use persons not meeting the requirements of paragraph (b) of this section to provide such emergency maintenance under both of the following conditions:

(1) The certificate holder must give written notification of the emergency maintenance to the Drug Abatement Program Division, AAM—800, 800 Independence Avenue, Washington, DC, 20591, within 10 days after being provided same in accordance with this paragraph. A certificate holder must retain copies of all such written notifications for two years.

(2) The aircraft must be reinspected by maintenance personnel who meet the requirements of paragraph (b) of this section when the aircraft is next at an airport where such maintenance personnel are available.

(d) For purposes of this section, emergency maintenance means maintenance that—

(1) Is not scheduled and

(2) Is made necessary by an aircraft condition not discovered prior to the departure for that location.

§ 135.253 Misuse of alcohol.

(a) This section applies to employees who perform a function listed in Appendix J to Part 121 of this chapter for a certificate holder or operator (*covered employees*). For the purpose of this section, a person who meets the

definition of covered employee in Appendix J is considered to be performing the function for the certificate holder or operator.

(b) *Alcohol concentration.* No covered employee shall report for duty or remain on duty requiring the performance of safety-sensitive functions while having an alcohol concentration of 0.04 or greater. No certificate holder or operator having actual knowledge that an employee has an alcohol concentration of 0.04 or greater shall permit the employee to perform or continue to perform safety-sensitive functions.

(c) *On-duty use.* No covered employee shall use alcohol while performing safety-sensitive functions. No certificate holder or operator having actual knowledge that a covered employee is using alcohol while performing safety-sensitive functions shall permit the employee to perform or continue to perform safety-sensitive functions.

(d) *Pre-duty use.*

(1) No covered employee shall perform flight crewmember or flight attendant duties within 8 hours after using alcohol. No certificate holder or operator having actual knowledge that such an employee has used alcohol within 8 hours shall permit the employee to perform or continue to perform the specified duties.

(2) No covered employee shall perform safety-sensitive duties other than those specified in paragraph (d)(1) of this section within 4 hours after using alcohol. No certificate holder or operator having actual knowledge that such an employee has used alcohol within 4 hours shall permit the employee to perform or continue to perform safety-sensitive functions.

(e) *Use following an accident.* No covered employee who has actual knowledge of an accident involving an aircraft for which he or she performed a safety-sensitive function at or near the time of the accident shall use alcohol for 8 hours following the accident, unless he or she has been given a post-accident test under Appendix J of Part 121 of this chapter, or the employer has determined that the employee's performance could not have contributed to the accident.

(f) *Refusal to submit to a required alcohol test.* A covered employee may not refuse to submit to any alcohol test required under appendix J to part 121 of this chapter.

An operator or certificate holder may not permit an employee who refuses to submit to such a test to perform or continue to perform safety-sensitive functions.

§ 135.255 Testing for alcohol.

(a) Each certificate holder and operator must establish an alcohol misuse prevention program in accordance with the provisions of Appendix J to Part 121 of this chapter.

(b) Except as provided in paragraph (c) of this section, no certificate holder or operator may use any person who meets the definition of "covered employee" in appendix J to part 121 of this chapter to perform a safety-sensitive function listed in that appendix unless such person is subject to testing for alcohol misuse in accordance with the provisions of appendix J.

(c) If a certificate holder conducts an on-demand operation into an airport at which no maintenance providers are available that are subject to the requirements of appendix J to part 121 of this chapter and emergency maintenance is required, the certificate holder may use persons not meeting the requirements of paragraph (b) of this section to provide such emergency maintenance under both of the following conditions:

(1) The certificate holder must give written notification of the emergency maintenance to the Drug Abatement Program Division, AAM—800, 800 Independence Avenue, Washington, DC, 20591, within 10 days after being provided same in accordance with this paragraph. A certificate holder must retain copies of all such written notifications for two years.

(2) The aircraft must be reinspected by maintenance personnel who meet the requirements of paragraph (b) of this section when the aircraft is next at an airport where such maintenance personnel are available.

(d) For purposes of this section, emergency maintenance means maintenance that—

(1) Is not scheduled, and

(2) Is made necessary by an aircraft condition not discovered prior to the departure for that location.

Subpart F—Crewmember flight time and duty period limitations and rest requirements

§ 135.261 Applicability.

[Sections 135.263 through 135.273 of this part prescribe flight time limitations, duty period limitations, and rest requirements for operations conducted under this part as follows:]

(a) Section 135.263 applies to all operations under this subpart.

(b) Section 135.265 applies to:

(1) Scheduled passenger-carrying operations except those conducted solely within the state of Alaska. "Scheduled passenger-carrying operations" means passenger-carrying operations that are conducted in accordance with a published schedule which covers at least five round trips per week on at least one route between two or more points, includes dates or times (or both), and is openly advertised or otherwise made readily available to the general public, and

(2) Any other operation under this part, if the operator elects to comply with § 135.265 and obtains an appropriate operations specification amendment.

(c) Sections 135.267 and 135.269 apply to any operation that is not a scheduled passenger-carrying operation and to any operation conducted solely within the State of Alaska, unless the operator elects to comply with § 135.265 as authorized under paragraph (b)(2) of this section.

(d) Section 135.271 contains special daily flight time limits for operations conducted under the helicopter emergency medical evacuation service (HEMES).

[(e) Section 135.273 prescribes duty period limitations and rest requirements for flight attendants in all operations conducted under this part.]

§ 135.263 Flight time limitations and rest requirements: All certificate holders.

(a) A certificate holder may assign a flight crewmember and a flight crewmember may accept an assignment for flight time only when the applicable requirements of §§ 135.263 through 135.271 are met.

(b) No certificate holder may assign any flight crewmember to any duty with the certificate holder during any required rest period.

(c) Time spent in transportation, not local in character, that a certificate holder requires of a flight crewmember and provides to transport the crewmember to an airport at which he is to serve on a flight as a crewmember, or from an airport at which he was relieved from duty to return to his home station, is not considered part of a rest period.

(d) A flight crewmember is not considered to be assigned flight time in excess of flight time limitations if the flights to which he is assigned normally terminate within the limitations, but due to circumstances beyond the control of the certificate holder or flight crewmember (such as adverse weather conditions), are not at the time of departure expected to reach their destination within the planned flight time.

§ 135.265 Flight time limitations and rest requirements: Scheduled operations.

(a) No certificate holder may schedule any flight crewmember, and no flight crewmember may accept an assignment, for flight time in scheduled operations or in other commercial flying if that crewmember's total flight time in all commercial flying will exceed—

(1) 1,200 hours in any calendar year.

(2) 120 hours in any calendar month.

(3) 34 hours in any 7 consecutive days.

(4) 8 hours during any 24 consecutive hours for a flight crew consisting of one pilot.

(5) 8 hours between required rest periods for a flight crew consisting of two pilots qualified under this part for the operation being conducted.

(b) Except as provided in paragraph (c) of this section, no certificate holder may schedule a flight crewmember, and no flight crewmember may accept an assignment, for flight time during the 24 consecutive hours preceding the scheduled completion of any flight segment without a scheduled rest period during that 24 hours of at least the following:

(1) 9 consecutive hours of rest for less than 8 hours of scheduled flight time.

(2) 10 consecutive hours of rest for 8 or more but less than 9 hours of scheduled flight time.

(3) 11 consecutive hours of rest for 9 or more hours of scheduled flight time.

(c) A certificate holder may schedule a flight crewmember for less than the rest required in paragraph (b) of this section or may reduce a scheduled rest under the following conditions:

(1) A rest required under paragraph (b)(1) of this section may be scheduled for or reduced to a minimum of 8 hours if the flight crewmember is given a rest period of at least 10 hours that must begin no later than 24 hours after the commencement of the reduced rest period.

(2) A rest required under paragraph (b)(2) of this section may be scheduled for or reduced to a minimum of 8 hours if the flight crewmember is given a rest period of at least 11 hours that must begin no later than 24 hours after the commencement of the reduced rest period.

(3) A rest required under paragraph (b)(3) of this section may be scheduled for or reduced to a minimum of 9 hours if the flight crewmember is given a rest period of at least 12 hours that must begin no later than 24 hours after the commencement of the reduced rest period.

(d) Each certificate holder shall relieve each flight crewmember engaged in scheduled air transportation from all further duty for at least 24 consecutive hours during any 7 consecutive days.

§ 135.267 Flight time limitations and rest requirements: Unscheduled one- and two-pilot crews.

(a) No certificate holder may assign any flight crewmember, and no flight crewmember may accept an assignment, for flight time as a member of a one- or two-pilot crew if that crewmember's total flight time in all commercial flying will exceed—

(1) 500 hours in any calendar quarter.

(2) 800 hours in any two consecutive calendar quarters.

(3) 1,400 hours in any calendar year.

(b) Except as provided in paragraph (c) of this section, during any 24 consecutive hours the total flight time of the assigned flight when added to any other commercial flying by that flight crewmember may not exceed—

(1) 8 hours for a flight crew consisting of one pilot; or

(2) 10 hours for a flight crew consisting of two pilots qualified under this part for the operation being conducted.

(c) A flight crewmember's flight time may exceed the flight time limits of paragraph (b) of this section if the assigned flight time occurs during a regularly assigned duty period of no more than 14 hours and—

(1) If this duty period is immediately preceded by and followed by a required rest period of at least 10 consecutive hours of rest;

(2) If flight time is assigned during this period, that total flight time when added to any other commercial flying by the flight crewmember may not exceed—

(i) 8 hours for a flight crew consisting of one pilot; or

(ii) 10 hours for a flight crew consisting of two pilots; and

(3) If the combined duty and rest periods equal 24 hours.

(d) Each assignment under paragraph (b) of this section must provide for at least 10 consecutive hours of rest during the 24-hour period that precedes the planned completion time of the assignment.

(e) When a flight crewmember has exceeded the daily flight time limitations in this section, because of circumstances beyond the control of the certificate holder or flight crewmember (such as adverse weather conditions), that flight crewmember must have a rest period before being assigned or accepting an assignment for flight time of at least—

(1) 11 consecutive hours of rest if the flight time limitation is exceeded by not more than 30 minutes;

(2) 12 consecutive hours of rest if the flight time limitation is exceeded by more than 30 minutes, but not more than 60 minutes; and

(3) 16 consecutive hours of rest if the flight time limitation is exceeded by more than 60 minutes.

(f) The certificate holder must provide each flight crewmember at least 13 rest periods of at least 24 consecutive hours each in each calendar quarter.

(Amdt. 135-33, Eff. 10/25/89)

§ 135.269 Flight time limitations and rest requirements: Unscheduled three- and four-pilot crews.

(a) No certificate holder may assign any flight crewmember, and no flight crewmember may accept an assignment, for flight time as a member of a three- or four-pilot crew if that crewmember's total flight time in all commercial flying will exceed—

(1) 500 hours in any calendar quarter.

(2) 800 hours in any two consecutive calendar quarters.

(3) 1,400 hours in any calendar year.

(b) No certificate holder may assign any pilot to a crew of three or four pilots, unless that assignment provides—

(1) At least 10 consecutive hours of rest immediately preceding the assignment;

(2) No more than 8 hours of flight deck duty in any 24 consecutive hours;

(3) No more than 18 duty hours for a three-pilot crew or 20 duty hours for a four-pilot crew in any 24 consecutive hours;

(4) No more than 12 hours aloft for a three-pilot crew or 16 hours aloft for a four-pilot crew during the maximum duty hours specified in paragraph (b)(3) of this section;

(5) Adequate sleeping facilities on the aircraft for the relief pilot;

(6) Upon completion of the assignment, a rest period of at least 12 hours;

(7) For a three-pilot crew, a crew which consists of at least the following:

(i) A pilot in command (PC) who meets the applicable flight crewmember requirements of Subpart E of Part 135;

(ii) A PC who meets the applicable flight crewmember requirements of Subpart E of Part 135, except those prescribed in §§ 135.244 and 135.247; and

(iii) A second in command (SIC) who meets the SIC qualifications of § 135.245.

(8) For a four-pilot crew, at least three pilots who meet the conditions of paragraph (b)(7) of this section, plus a fourth pilot who meets the SIC qualifications of § 135.245.

(c) When a flight crewmember has exceeded the daily flight deck duty limitation in this section by more than 60 minutes, because of circumstances beyond the control of the certificate holder or flight crewmember, that flight crewmember must have a rest period before the next duty period of at least 16 consecutive hours.

(d) A certificate holder must provide each flight crewmember at least 13 rest periods of at least 24 consecutive hours each in each calendar quarter.

§ 135.271 Helicopter hospital emergency medical evacuation service (HEMES).

(a) No certificate holder may assign any flight crewmember, and no flight crewmember may accept an assignment for flight time if that crewmember's total flight time in all commercial flying will exceed—

(1) 500 hours in any calendar quarter.

(2) 800 hours in any two consecutive calendar quarters.

(3) 1,400 hours in any calendar year.

(b) No certificate holder may assign a helicopter flight crewmember, and no flight crewmember may accept an assignment, for hospital emergency medical evacuation service helicopter operations unless that assignment provides for at least 10 consecutive hours of rest immediately preceding reporting to the hospital for availability for flight time.

(c) No flight crewmember may accrue more than 8 hours of flight time during any 24 consecutive hour period of a HEMES assignment, unless an emergency medical evacuation operation is prolonged. Each flight crewmember who exceeds the daily 8 hour flight time limitation in this paragraph must be relieved of the HEMES assignment immediately upon the completion of that emergency medical operation and must be given a rest period in compliance with paragraph (h) of this section.

(d) Each flight crewmember must receive at least 8 consecutive hours of rest during any 24 consecutive hour period of a HEMES assignment. A flight crewmember must be relieved of the HEMES assignment if he or she has not

or cannot receive at least 8 consecutive hours of rest during any 24 consecutive hour period of a HEMES assignment.

(e) A HEMES assignment may not exceed 72 consecutive hours at the hospital.

(f) An adequate place of rest must be provided at, or in close proximity to, the hospital at which the HEMES assignment is being performed.

(g) No certificate holder may assign any other duties to a flight crewmember during a HEMES assignment.

(h) Each pilot must be given a rest period upon completion of the HEMES assignment and prior to being assigned any further duty with the certificate holder of—

(1) At least 12 consecutive hours for an assignment of less than 48 hours.

(2) At least 16 consecutive hours for an assignment of more than 48 hours.

(i) The certificate holder must provide each flight crewmember at least 13 rest periods of at least 24 consecutive hours each in each calendar quarter.

§ 135.273 Duty period limitations and rest time requirements.

[(a) For purposes of this section—

"Calendar day" means the period of elapsed time, using Coordinated Universal Time or local time, that begins at midnight and ends 24 hours later at the next midnight.

"Duty period" means the period of elapsed time between reporting for an assignment involving flight time and release from that assignment by the certificate holder. The time is calculated using either Coordinated Universal Time or local time to reflect the total elapsed time.

"Flight attendant" means an individual, other than a flight crewmember, who is assigned by the certificate holder, in accordance with the required minimum crew complement under the certificate holder's operations specifications or in addition to that minimum complement, to duty in an aircraft during flight time and whose duties include but are not necessarily limited to cabin-safety-related responsibilities.

"Rest period" means the period free of all responsibility for work or duty should the occasion arise.

(b) Except as provided in paragraph (c) of this section, a certificate holder may assign a duty period to a flight attendant only when the applicable duty period limitations and rest requirements of this paragraph are met.

(1) Except as provided in paragraphs (b)(4), (b)(5), and (b)(6) of this section, no certificate holder may assign a flight attendant to a scheduled duty period of more than 14 hours.

(2) Except as provided in paragraph (b)(3) of this section, a flight attendant scheduled to a duty period of 14 hours or less as provided under paragraph (b)(1) of this section must be given a scheduled rest period of at least 9 consecutive hours. This rest period must occur between the completion of the scheduled duty period and the commencement of the subsequent duty period.

(3) The rest period required under paragraph (b)(2) of this section may be scheduled or reduced to 8 consecutive hours if the flight attendant is provided a subsequent rest period of at least 10 consecutive hours; this subsequent rest period must be scheduled to begin no later than 24 hours after the beginning of the reduced rest period and must occur between the completion of the scheduled duty period and the commencement of the subsequent duty period.

(4) A certificate holder may assign a flight attendant to a scheduled duty period of more than 14 hours, but no more than 16 hours, if the certificate holder has assigned to the flight or flights in that duty period at least one flight attendant in addition to the minimum flight attendant complement required for the flight or flights in that duty period under the certificate holder's operations specifications.

(5) A certificate holder may assign a flight attendant to a scheduled duty period of more than 16 hours, but no more than 18 hours, if the certificate holder has assigned to the flight or flights in that duty period at least two flight attendants in addition to the minimum flight attendant complement required for the flight or flights in that duty period under the certificate holder's operations specifications.

(6) A certificate holder may assign a flight attendant to a scheduled duty period of more than 18 hours, but no more than 20 hours, if the scheduled duty period includes one or more flights that land or takeoff outside the 48 contiguous states and the District of Columbia, and if the certificate holder was assigned to the flight or flights in that duty period at least three flight attendants in addition to the minimum flight attendant complement required for the flight or flights in that duty period under the certificate holder's operations specifications.

(7) Except as provided in paragraph (b)(8) of this section, a flight attendant scheduled to a duty period of more than 14 hours but no more than 20 hours, as provided in paragraphs (b)(4), (b)(5), and (b)(6) of this section, must be given a scheduled rest period of at least 12 consecutive hours. This rest period must occur between the completion of the scheduled duty period and the commencement of the subsequent duty period.

(8) The rest period required under paragraph (b)(7) of this section may be scheduled or reduced to 10 consecutive hours if the flight attendant is provided a subsequent rest period of at least 14 consecutive hours; this subsequent rest period must be scheduled to begin no later than 24 hours after the beginning of the reduced rest period and must occur between the completion of the scheduled duty period and the commencement of the subsequent duty period.

(9) Notwithstanding paragraphs (b)(4), (b)(5), and (b)(6) of this section, if a certificate holder elects to reduce the rest period to 10 hours as authorized by paragraph (b)(8) of this section, the certificate holder may not schedule a flight attendant for a duty period of more than 14 hours during the 24-hour period commencing after the beginning of the reduced rest period.

(10) No certificate holder may assign a flight attendant any duty period with the certificate holder unless the flight attendant has had at least the minimum rest required under this section.

(11) No certificate holder may assign a flight attendant to perform any duty with the certificate holder during any required rest period.

(12) Time spent in transportation, not local in character, that a certificate holder requires of a flight attendant and provides to transport the flight attendant to an airport at which that flight attendant is to serve on a flight as a crewmember, or from an airport at which the flight attendant was relieved from duty to return to the flight attendant's home station, is not considered part of a rest period.

(13) Each certificate holder must relieve each flight attendant engaged in air transportation from all further duty for at least 24 consecutive hours during any 7 consecutive calendar days.

(14) A flight attendant is not considered to be scheduled for duty in excess of duty period limitations if the flights to which the flight attendant is assigned are scheduled and normally terminate within the limitations but due to circumstances beyond the control of the certificate holder (such as adverse weather conditions) are not at the time of departure expected to reach their destination within the scheduled time.

(c) Notwithstanding paragraph (b) of this section, a certificate holder may apply the flight crewmember flight time and duty limitations and rest requirements of this part to flight attendants for all operations conducted under this part provided that—

(1) The certificate holder establishes written procedures that—

(i) Apply to all flight attendants used in the certificate holder's operation;

(ii) Include the flight crewmember requirements contained in subpart F of this part, as appropriate to the operation being conducted, except that rest facilities on board the aircraft are not required; and

(iii) Include provisions to add one flight attendant to the minimum flight attendant complement for each flight crewmember who is in excess of the minimum number required in the aircraft type certificate data sheet and who is assigned to the aircraft under the provisions of subpart F of this part, as applicable.

(iv) Are approved by the Administrator and described or referenced in the certificate holder's operations specifications; and

(2) Whenever the Administrator finds that revisions are necessary for the continued adequacy of duty period limitation and rest requirement procedures that are required by paragraph (c)(1) of this section and that had been granted final approval, the certificate holder must, after notification by the Administrator, make any changes in the procedures that are found necessary by the Administrator.

Within 30 days after the certificate holder receives such notice, it may file a petition to reconsider the notice with the certificate-holding district office. The filing of a petition to reconsider stays the notice, pending decision by the Administrator. However, if the Administrator finds that there is an emergency that requires immediate action in the interest of safety, the Administrator may, upon a statement of the reasons, require a change effective without stay.]

Subpart G—Crewmember testing requirements

§ 135.291 Applicability.

Except as provided in § 135.3, this subpart—

(a) Prescribes the tests and checks required for pilot and flight attendant crewmembers and for the approval of check pilots in operations under this part; and

(b) Permits training center personnel authorized under part 142 of this chapter who meet the requirements of § 135.337 and 135.339 to conduct training, testing, and checking under contract or other arrangement to those persons subject to the requirements of this subpart.

§ 135.293 Initial and recurrent pilot testing requirements.

(a) No certificate holder may use a pilot, nor may any person serve as a pilot, unless, since the beginning of the 12th calendar month before that service, that pilot has passed a written or oral test, given by the Administrator or an authorized check pilot, on that pilot's knowledge in the following areas—

(1) The appropriate provisions of Parts 61, 91, and 135 of this chapter and the operations specifications and the manual of the certificate holder;

(2) For each type of aircraft to be flown by the pilot, the aircraft powerplant, major components and systems, major appliances, performance and operating limitations, standard and emergency operating procedures, and the contents of the approved Aircraft Flight Manual or equivalent, as applicable;

(3) For each type of aircraft to be flown by the pilot, the method of determining compliance with weight and balance limitations for takeoff, landing and en route operations;

(4) Navigation and use of air navigation aids appropriate to the operation or pilot authorization, including, when applicable, instrument approach facilities and procedures;

(5) Air traffic control procedures, including IFR procedures when applicable;

(6) Meteorology in general, including the principles of frontal systems, icing, fog, thunderstorms, and windshear, and, if appropriate for the operation of the certificate holder, high altitude weather;

(7) Procedures for—

(i) Recognizing and avoiding severe weather situations;

(ii) Escaping from severe weather situations, in case of inadvertent encounters, including low-altitude windshear (except that rotorcraft pilots are not required to be tested on escaping from low-altitude windshear); and

(iii) Operating in or near thunderstorms (including best penetrating altitudes), turbulent air (including clear air turbulence), icing, hail, and other potentially hazardous meteorological conditions; and

(8) New equipment, procedures, or techniques, as appropriate.

(b) No certificate holder may use a pilot, nor may any person serve as a pilot, in any aircraft unless, since the beginning of the 12th calendar month before that service, that pilot has passed a competency check given by the Administrator or an authorized check pilot in that class of aircraft, if single-engine airplane other than turbojet, or that type of aircraft, if helicopter, multiengine airplane, or turbojet airplane, to determine the pilot's competence in practical skills and techniques in that aircraft or class of aircraft. The extent of the competency check shall be determined by the Administrator or authorized check point conducting the competency check. The competency check may include any of the maneuvers and procedures currently required for the original issuance of the particular pilot certificate required for the operations authorized and appropriate to the category, class and type of aircraft involved. For the purposes of this paragraph, type, as to an airplane, means any one of a group of airplanes determined by the Administrator to have a similar means of propulsion, the same manufacturer, and no significantly different handling or flight characteristics. For the purposes of this paragraph, type, as to a helicopter, means a basic make and model.

(c) The instrument proficiency check required by § 135.297 may be substituted for the competency check required by this section for the type of aircraft used in the check.

(d) For the purpose of this part, competent performance of a procedure or maneuver by a person to be used as a pilot requires that the pilot be the obvious master of the aircraft, with the successful outcome of the maneuver never in doubt.

(e) The Administrator or authorized check pilot certifies the competency of each pilot who passes the knowledge or flight check in the certificate holder's pilot records.

(f) Portions of a required competency check may be given in an aircraft simulator for other appropriate training device, if approved by the Administrator.

(Amdt. 135-27, Eff. 1/2/89)

§ 135.295 Initial and recurrent flight attendant crewmember testing requirements.

No certificate holder may use a flight attendant crewmember, nor may any person serve as a flight attendant crewmember unless, since the beginning of the 12th calendar month before that service, the certificate holder has determined by appropriate initial and recurrent testing that the person is knowledgeable and competent in the following areas as appropriate to assigned duties and responsibilities—

(a) Authority of the pilot in command;

(b) Passenger handling, including procedures to be followed in handling deranged persons or other persons whose conduct might jeopardize safety;

(c) Crewmember assignments, functions, and responsibilities during ditching and evacuation of persons who may need the assistance of another person to move expeditiously to an exit in an emergency;

(d) Briefing of passengers;

(e) Location and operation of portable fire extinguishers and other items of emergency equipment;

(f) Proper use of cabin equipment and controls;

(g) Location and operation of passenger oxygen equipment;

(h) Location and operation of all normal and emergency exits, including evacuation chutes and escape ropes; and

(i) Seating of persons who may need assistance of another person to move rapidly to an exit in an emergency as prescribed by the certificate holder's operations manual.

§ 135.297 Pilot in command: Instrument proficiency check requirements.

(a) No certificate holder may use a pilot, nor may any person serve, as a pilot in command of an aircraft under IFR unless, since the beginning of the 6th calendar month before that service, that pilot has passed an instrument proficiency check under this section administered by the Administrator or an authorized check pilot.

(b) No pilot may use any type of precision instrument approach procedure under IFR unless, since the beginning of the 6th calendar month before that use, the pilot satisfactorily demonstrated that type of approach procedure. No pilot may use any type of nonprecision approach procedure under IFR unless, since the beginning of the 6th calendar month before that use, the pilot has satisfactorily demonstrated either that type of approach procedure or any other two different types of nonprecision approach procedures. The instrument approach procedure or procedures must include at least one straight-in approach, one circling approach, and one missed approach. Each type of approach procedure demonstrated must be conducted to published minimums for that procedure.

(c) The instrument proficiency check required by paragraph (a) of this section consists of an oral or written equipment test and a flight check under simulated or actual IFR conditions. The equipment test includes questions on emergency procedures, engine operation, fuel and lubrication systems, power settings, stall speeds, best engine-out speed, propeller and supercharger operations, and hydraulic, mechanical, and electrical systems, as appropriate. The flight check includes navigation by instruments, recovery from simulated emergencies, and standard instrument approaches involving navigational facilities which that pilot is to be authorized to use. Each pilot taking the instrument proficiency check must show that standard of competence required by § 135.293(d).

(1) The instrument proficiency check must—

(i) For a pilot in command of an airplane under § 135.243(a), include the procedures and maneuvers for an airline transport pilot certificate in the particular type of airplane, if appropriate; and

(ii) For a pilot in command of an airplane or helicopter under § 135.243(c), include the procedures and maneuvers for a commercial pilot certificate with an instrument rating and, if required, for the appropriate type rating.

(2) The instrument proficiency check must be given by an authorized check airman or by the Administrator.

(d) If the pilot in command is assigned to pilot only one type of aircraft, that pilot must take the instrument proficiency check required by paragraph (a) of this section in that type of aircraft.

(e) If the pilot in command is assigned to pilot more than one type of aircraft, that pilot must take the instrument proficiency check required by paragraph (a) of this section in each type of aircraft to which that pilot is assigned, in rotation, but not more than one flight check during each period described in paragraph (a) of this section.

(f) If the pilot in command is assigned to pilot both single-engine and multiengine aircraft, that pilot must initially take the instrument proficiency check required by paragraph (a) of this section in a multiengine aircraft, and each succeeding check alternately in single-engine and multiengine aircraft, but not more than one flight check during each period described in paragraph (a) of this section. Portions of a required flight check may be given in an aircraft simulator or other appropriate training device, if approved by the Administrator.

(g) If the pilot in command is authorized to use an autopilot system in place of a second in command, that pilot must show, during the required instrument proficiency check, that the pilot is able (without a second in command) both with and without using the autopilot to—

(1) Conduct instrument operations competently; and

(2) Properly conduct air-ground communications and comply with complex air traffic control instructions.

(3) Each pilot taking the autopilot check must show that, while using the autopilot, the airplane can be operated as proficiently as it would be if a second in command were present to handle air-ground communications and air

traffic control instructions. The autopilot check need only be demonstrated once every 12 calendar months during the instrument proficiency check required under paragraph (a) of this section.

(h) [Deleted.]

(Amdt. 135-15, Eff. 6/11/81)

§ 135.299 Pilot in command: Line checks: Routes and airports.

(a) No certificate holder may use a pilot, nor may any person serve as a pilot in command of a flight unless, since the beginning of the 12th calendar month before that service, that pilot has passed a flight check in one of the types of aircraft which that pilot is to fly. The flight check shall—

(1) Be given by an approved check pilot or by the Administrator;

(2) Consist of at least one flight over one route segment; and

(3) Include takeoffs and landings at one or more representative airports. In addition to the requirements of this paragraph, for a pilot authorized to conduct IFR operations, at least one flight shall be flown over a civil airway, an approved off-airway route, or a portion of either of them.

(b) The pilot who conducts the check shall determine whether the pilot being checked satisfactorily performs the duties and responsibilities of a pilot in command in operations under this part, and shall so certify in the pilot training record.

(c) Each certificate holder shall establish in the manual required by § 135.21 a procedure which will ensure that each pilot who has not flown over a route and into an airport within the preceding 90 days will, before beginning the flight, become familiar with all available information required for the safe operation of that flight.

§ 135.301 Crewmember: Tests and checks, grace provisions, training to accepted standards.

(a) If a crewmember who is required to take a test or a flight check under this part, completes the test or flight check in the calendar month before or after the calendar month in which it is required, that crewmember is considered to have completed the test or check in the calendar month in which it is required.

(b) If a pilot being checked under this subpart fails any of the required maneuvers, the person giving the check may give additional training to the pilot during the course of the check. In addition to repeating the maneuvers failed, the person giving the check may require the pilot being checked to repeat any other maneuvers that are necessary to determine the pilot's proficiency. If the pilot being checked is unable to demonstrate satisfactory performance to the person conducting the check, the certificate holder may not use the pilot, nor may the pilot serve, as a flight crewmember in operations under this part until the pilot has satisfactorily completed the check.

§ 135.303 [(Removed)]

[(Amdt. 135-44, Eff. 10/15/92)]

Subpart H—Training

§ 135.321 Applicability and terms used.

(a) Except as provided in § 135.3, this subpart prescribes the requirements applicable to—

(1) A certificate holder under this part which contracts with, or otherwise arranges to use the services of a training center certificated under part 142 to perform training, testing, and checking functions;

(2) Each certificate holder for establishing and maintaining an approved training program for crewmembers, check airmen and instructors, and other operations personnel employed or used by that certificate holder; and

(3) Each certificate holder for the qualifications, approval, and use of aircraft simulators and flight training devices in the conduct of the program.

(b) For the purposes of this subpart, the following terms and definitions apply:

(1) *Initial training.* The training required for crewmembers who have not qualified and served in the same capacity on an aircraft.

(2) *Transition training.* The training required for crewmembers who have qualified and served in the same capacity on another aircraft.

(3) *Upgrade training.* The training required for crewmembers who have qualified and served as second in command on a particular aircraft type, before they serve as pilot in command on that aircraft.

(4) *Differences training.* The training required for crewmembers who have qualified and served on a particular type aircraft, when the Administrator finds differences training is necessary before a crewmember serves in the same capacity on a particular variation of that aircraft.

(5) *Recurrent training.* The training required for crewmembers to remain adequately trained and currently proficient for each aircraft, crewmember position, and type of operation in which the crewmember serves.

(6) *In flight.* The maneuvers, procedures, or functions that must be conducted in the aircraft.

(7) *Training center.* An organization governed by the applicable requirements of part 142 of this chapter that conducts training, testing, and checking under contract or other arrangement to certificate holders subject to the requirements of this part.

(8) *Requalification training.* The training required for crewmembers previously trained and qualified, but who

have become unqualified due to not having met within the required period the—

(i) Recurrent pilot testing requirements of § 135.293;

(ii) Instrument proficiency check requirements of § 135.297; or

(iii) Line checks required by § 135.299.

§ 135.323 Training program: General.

(a) Each certificate holder required to have a training program under § 135.341 shall:

(1) Establish and implement a training program that satisfies the requirements of this subpart and that ensures that each crewmember, aircraft dispatcher, flight instructor and check airman is adequately trained to perform his or her assigned duties. Prior to implementation, the certificate holder must obtain initial and final FAA approval of the training program.

(2) Provide adequate ground and flight training facilities and properly qualified ground instructors for the training required by this subpart.

(3) Provide and keep current for each aircraft type used and, if applicable, the particular variations within the aircraft type, appropriate training material, examinations, forms, instructions, and procedures for use in conducting the training and checks required by this subpart.

(4) Provide enough flight instructors, check airmen, and simulator instructors to conduct required flight training and flight checks, and simulator training courses allowed under this subpart.

(b) Whenever a crewmember who is required to take recurrent training under this subpart completes the training in the calendar month before, or the calendar month after, the month in which that training is required, the crewmember is considered to have completed it in the calendar month in which it was required.

(c) Each instructor, supervisor, or check airman who is responsible for a particular ground training subject, segment of flight training, course of training, flight check, or competence check under this part shall certify as to the proficiency and knowledge of the crewmember, flight instructor, or check airman concerned upon completion of that training or check. That certification shall be made a part of the crewmember's record. When the certification required by this paragraph is made by an entry in a computerized recordkeeping system, the certifying instructor, supervisor, or check airman, must be identified with that entry. However, the signature of the certifying instructor, supervisor, or check airman, is not required for computerized entries.

(d) Training subjects that apply to more than one aircraft or crewmember position and that have been satisfactorily completed during previous training while employed by the certificate holder for another aircraft or another crewmember position, need not be repeated during subsequent training other than recurrent training.

(e) Aircraft simulators and other training devices may be used in the certificate holder's training program if approved by the Administrator.

§ 135.324 Training program: Special Rules.

(a) Other than the certificate holder, only another certificate holder certificated under this part or a training center certificated under part 142 of this chapter is eligible under this subpart to conduct training, testing, and checking under contract or other arrangement to those persons subject to the requirements of this subpart.

(b) A certificate holder may contract with, or otherwise arrange to use the services of, a training center certificated under part 142 of this chapter to conduct training, testing, and checking required by this part only if the training center—

(1) Holds applicable training specifications issued under part 142 of this chapter;

(2) Has facilities, training equipment, and courseware meeting the applicable requirements of part 142 of this chapter;

(3) Has approved curriculums, curriculum segments, and portions of curriculum segments applicable for use in training courses required by this subpart; and

(4) Has sufficient instructor and check airmen qualified under the applicable requirements of §§135.337–135.340 to provide training, testing, and checking to persons subject to the requirements of this subpart.

§ 135.325 Training program and revision: Initial and final approval.

(a) To obtain initial and final approval of a training program, or a revision to an approved training program, each certificate holder must submit to the Administrator—

(1) An outline of the proposed or revised curriculum, that provides enough information for a preliminary evaluation of the proposed training program or revision; and

(2) Additional relevant information that may be requested by the Administrator.

(b) If the proposed training program or revision complies with this subpart, the Administrator grants initial approval in writing after which the certificate holder may conduct the training under that program. The Administrator then evaluates the effectiveness of the training program and advises the certificate holder of deficiencies, if any, that must be corrected.

(c) The Administrator grants final approval of the proposed training program or revision if the certificate holder shows that the training conducted under the initial approval in paragraph (b) of this section ensures that each person who successfully completes the training is adequately trained to perform that person's assigned duties.

(d) Whenever the Administrator finds that revisions are necessary for the continued adequacy of a training program that has been granted final approval, the certificate holder shall, after notification by the Administrator, make any changes in the program that are found necessary by the Administrator. Within 30 days after the certificate holder receives the notice, it may file a petition to reconsider the notice with the Administrator. The filing of a petition to reconsider stays the notice pending a decision by the Administrator. However, if the Administrator finds that there is an emergency that requires immediate action in the interest of safety, the Administrator may, upon a statement of the reasons, require a change effective without stay.

§ 135.327 Training program: Curriculum.

(a) Each certificate holder must prepare and keep current a written training program curriculum for each type of aircraft for each crewmember required for that type aircraft. The curriculum must include ground and flight training required by this subpart.

(b) Each training program curriculum must include the following:

(1) A list of principal ground training subjects, including emergency training subjects, that are provided.

(2) A list of all the training devices, mockups, systems trainers, procedures trainers, or other training aids that the certificate holder will use.

(3) Detailed descriptions or pictorial displays of the approved normal, abnormal, and emergency maneuvers, procedures and functions that will be performed during each flight training phase or flight check, indicating those maneuvers, procedures and functions that are to be performed during the inflight portions of flight training and flight checks.

§ 135.329 Crewmember training requirements.

(a) Each certificate holder must include in its training program the following initial and transition ground training as appropriate to the particular assignment of the crewmember:

(1) Basic indoctrination ground training for newly hired crewmembers including instruction in at least the—

(i) Duties and responsibilities of crewmembers as applicable;

(ii) Appropriate provisions of this chapter;

(iii) Contents of the certificate holder's operating certificate and operations specifications (not required for flight attendants); and

(iv) Appropriate portions of the certificate holder's operating manual.

(2) The initial and transition ground training in §§ 135.345 and 135.349, as applicable.

(3) Emergency training in § 135.331.

(b) Each training program must provide the initial and transition flight training in § 135.347, as applicable.

(c) Each training program must provide recurrent ground and flight training in § 135.351.

(d) Upgrade training in §§ 135.345 and 135.347 for a particular type aircraft may be included in the training

program for crewmembers who have qualified and served as second in command on that aircraft.

(e) In addition to initial, transition, upgrade and recurrent training, each training program must provide ground and flight training, instruction, and practice necessary to ensure that each crewmember—

(1) Remains adequately trained and currently proficient for each aircraft, crewmember position, and type of operation in which the crewmember serves; and

(2) Qualifies in new equipment, facilities, procedures, and techniques, including modifications to aircraft.

§ 135.331 Crewmember emergency training.

(a) Each training program must provide emergency training under this section for each aircraft type, model, and configuration, each crewmember, and each kind of operation conducted, as appropriate for each crewmember and the certificate holder.

(b) Emergency training must provide the following:

(1) Instruction in emergency assignments and procedures, including coordination among crewmembers.

(2) Individual instruction in the location, function, and operation of emergency equipment including—

(i) Equipment used in ditching and evacuation;

(ii) First aid equipment and its proper use; and

(iii) Portable fire extinguishers, with emphasis on the type of extinguisher to be used on different classes of fires.

(3) Instruction in the handling of emergency situations including—

(i) Rapid decompression;

(ii) Fire in flight or on the surface and smoke control procedures with emphasis on electrical equipment and related circuit breakers found in cabin areas;

(iii) Ditching and evacuation;

(iv) Illness, injury, or other abnormal situations involving passengers or crewmembers; and

(v) Hijacking and other unusual situations.

(4) Review of the certificate holder's previous aircraft accidents and incidents involving actual emergency situations.

(c) Each crewmember must perform at least the following emergency drills, using the proper emergency equipment and procedures, unless the Administrator finds that, for a particular drill, the crewmember can be adequately trained by demonstration:

(1) Ditching, if applicable.

(2) Emergency evacuation.

(3) Fire extinguishing and smoke control.

(4) Operation and use of emergency exits, including deployment and use of evacuation chutes, if applicable.

(5) Use of crew and passenger oxygen.

(6) Removal of life rafts from the aircraft, inflation of the life rafts, use of life lines, and boarding of passengers and crew, if applicable.

(7) Donning and inflation of life vests and the use of other individual flotation devices, if applicable.

(d) Crewmembers who serve in operations above 25,000 feet must receive instruction in the following:

(1) Respiration.

(2) Hypoxia.

(3) Duration of consciousness without supplemental oxygen at altitude.

(4) Gas expansion.

(5) Gas bubble formation.

(6) Physical phenomena and incidents of decompression.

§ 135.133 [Removed.]

§ 135.335 Approval of aircraft simulators and other training devices.

(a) Training courses using aircraft simulators and other training devices may be included in the certificate holder's training program if approved by the Administrator.

(b) Each aircraft simulator and other training device that is used in a training course or in checks required under this subpart must meet the following requirements:

(1) It must be specifically approved for—

(i) The certificate holder; and

(ii) The particular maneuver, procedure, or crewmember function involved.

(2) It must maintain the performance, functional, and other characteristics that are required for approval.

(3) Additionally, for aircraft simulators, it must be—

(i) Approved for the type aircraft and, if applicable, the particular variation within type for which the training or check is being conducted; and

(ii) Modified to conform with any modification to the aircraft being simulated that changes the performance, functional, or other characteristics required for approval.

(c) A particular aircraft simulator or other training device may be used by more than one certificate holder.

[(d) In granting initial and final approval of training programs or revisions to them, the Administrator considers the training devices, methods, and procedures listed in the certificate holder's curriculum under § 135.327.]

(Amdt. 135-1, Eff. 5/7/79)

§ 135.337 Qualifications: Check airmen (aircraft) and check airmen (simulator)

(a) No certificate holder may use a person, nor may any person serve, as a flight instructor or check airman in a training program established under this subpart unless, for the particular aircraft type involved, that person—

(1) Holds the airman certificate and ratings that must be held to serve as a pilot in command in operations under this part;

(2) Has satisfactorily completed the appropriate training phases for the aircraft, including recurrent training, required to serve as a pilot in command in operations under this part;

(3) Has satisfactorily completed the appropriate proficiency or competency checks required to serve as a pilot in command in operations under this part;

(4) Has satisfactorily completed the applicable training requirements of § 135.339;

(5) Holds a Class I or Class II medical certificate required to serve as a pilot in command in operations under this part;

(6) In the case of a check airman, has been approved by the Administrator for the airman duties involved; and

(7) In the case of a check airman used in an aircraft simulator only, holds a Class III medical certificate.

(b) No certificate holder may use a person, nor may any person serve, as a simulator instructor for a course of training given in an aircraft simulator under this subpart unless that person—

(1) Holds at least a commercial pilot certificate; and

(2) Has satisfactorily completed the following as evidenced by the approval of a check airman—

(i) Appropriate initial pilot and flight instructor ground training under this subpart; and

(ii) A simulator flight training course in the type simulator in which that person instructs under this subpart.

§ 135.338 Qualifications: Flight instructors (aircraft) and flight instructors (simulator).

(a) For the purposes of this section and Sec. 135.340:

(1) A flight instructor (aircraft) is a person who is qualified to instruct in an aircraft, in a flight simulator, or in a flight training device for a particular type, class, or category aircraft.

(2) A flight instructor (simulator) is a person who is qualified to instruct in a flight simulator, in a flight training device, or in both, for a particular type, class, or category aircraft.

(3) Flight insructors (aircraft) and flight instructors (simulator) are those instructors who perform the functions described in Sec. 135.321 (a) and 135.323 (a) (4) and (c).

(b) No certificate holder may use a person, nor may any person serve as a flight instructor (aircraft) in a training program established under this subpart unless, with respect to the type, class, or category aircraft involved, that person—

(1) Holds the airman cerificates and ratings required to serve as a pilot in command in oeprations under this part;

(2) Has satisfactorily completed the training phases for the aircraft, including recurrent training, that are required to serve as a pilot in command in operations under this part;

(3) Has satifactorily completed the proficiency or competency checks that are required to serve as a pilot in command in opertions under this part;

(4) Has satisfactorily completed the applicable training requirements of Sec. 135.340;

(5) Holds at least a Class III medical certificate; and

(6) Has satisfied the recency of experience requirements of Sec. 135.247.

(c) No certificate holder may use a person, nor may any person serve as a flight instructor (simulator) in a training program established under this subpart, unless; with respect to the type, class, or category aircraft involved, that person meets the provisions of paragraph (b) of this section, or—

(1) Holds the airman certificates and ratings, except medical certificate, required to serve as a pilot in command in operations under this part except before March 19, 1997 that person need not hold a type rating for the type, class, or category of aircraft involved.

(2) Has satisfactorily completed the appropriate training phases for aircraft, including recurrent training, that are required to serve as a pilot in command in operations under this part;

(3) Has satisfactorily completed the appropriate proficiency or competency checks that are required to serve as a pilot in command in operations under this part; and

(4) Has satisfactorily completed the applicable training requirements of Sec. 135.340.

(d) Completion of the requirements in paragraphs (b) (2), (3), and (4) or (c) (2), (3), and (4) of this section, as applicable, shall be entered in the individual's training record maintained by the certificate holder.

(e) An airman who does not hold a medical certificate may function as a flight instructor in an aircraft if functioning as a non-required crewmember, but may not serve as a flightcrew member in operations under this part.

(f) A flight instructor (simulator) must accomplish the following—

(1) Fly at least two flight segments as a required crewmember for the type, class, or category aircraft involved within the 12-month period preceding the performance of any flight instructor duty in a flight simulator; or

(2) Satisfactorily complete an approved line-observation program within the period prescribed by that program and that must precede the performance of any check airman duty in a flight simulator.

(g) The flight segments or line-observation program required in paragragh (f) of this section are considered completed in the month required if completed in the calendar month before, or in the calendar month after, the month in which they are due.

[Amdt. 135-64, 61 FR 30744, June 17, 1996; Amdt. 39-9881, 62 FR 3739, Jan. 24, 1997]

§ 135.339 Check airmen and flight instructors: Initial and transition training.

(a) The initial and transition ground training for pilot check airmen must include the following:

(1) Pilot check airman duties, functions, and responsibilities.

(2) The applicable provisions of this chapter and certificate holder's policies and procedures.

(3) The appropriate methods, procedures, and techniques for conducting the required checks.

(4) Proper evaluation of pilot performance including the detection of—

(i) Improper and insufficient training; and

(ii) Personal characteristics that could adversely affect safety.

(5) The appropriate corrective action for unsatisfactory checks.

(6) The approved methods, procedures, and limitations for performing the required normal, abnormal, and emergency procedures in the aircraft.

(b) The initial and transition ground training for pilot flight instructors, except for the holder of a valid flight instructor certificate, must include the following:

(1) The fundamental principles of the teaching-learning process.

(2) Teaching methods and procedures.

(3) The instructor-student relationship.

(c) The initial and transition flight training for pilot check airmen and pilot flight instructors must include the following:

(1) Enough inflight training and practice in conducting flight checks from the left and right pilot seats in the required normal, abnormal, and emergency maneuvers to ensure that person's competence to conduct the pilot flight checks and flight training under this subpart.

(2) The appropriate safety measures to be taken from either pilot seat for emergency situations that are likely to develop in training.

(3) The potential results of improper or untimely safety measures during training.

The requirements of paragraphs (2) and (3) of this paragraph may be accomplished in flight or in an approved simulator.

§ 135.340 Initial and transition training and checking: Flight instructors (aircraft), flight instructors (simulator).

(a) No certificate holder may use a person nor may any person serve as a flight instructor unless—

(1) That person has satisfactorily completed initial or transition flight instructor training; and

(2) Within the preceding 24 calendar months, that person satisfactorily conducts instruction under the observation of an FFA inspector, an operator check airman, or an aircrew designated examiner employed by the operator. The observation check may be accomplished in part or in full in an aircraft, in a flight simulator, or in a flight training device. This paragraph applies after March 19, 1997.

(b) The observation check required by paragraph (a) (2) of this section is considered to have been completed in the month required if completed in the calendar month before, or the calendar month after, the month in which it is due.

(c) The initial ground training for flight instructors must include the following:

(1) Flight instructor duties, functions, and responsibilities.

(2) The applicable Code of Federal Regulations and the certificate holder's policies and procedures.

(3) The applicable methods, procedures, and techniques for conducting flight instruction.

(4) Proper evaluation of student performance including the detection of —

(i) Improper and insufficient training; and

(ii) Personal characteristics of an applicant that could adversely affect safety.

(5) The corrective action in the case of unsatisfactory training progress.

(6) The approved methods, procedures, and limitations for performing the required normal, abnormal, and emergency procedures in the aircraft.

(7) Except for holders of a flight instructor cerificate—

(i) The fundamental principles of the teaching-learning process;

(ii) Teaching methods and procedures; and

(iii) The instructor-student relationship.

(d) The transition ground training for flight instructors must include the approved, procedures, and limitations for performing the required normal, abnormal, and emergency procedures applicalbe to the type, class, or category aircraft to which the flight instructor is in transition.

(e) The initial and transition flight training for flight instructors (aircraft) must include the following—

(1) The safety measures for emergency situations that are likely to develop during instruction;

(2) The potential results of improper or untimely safety measures during instruction;

(3) Training and practice from the left and right pilot seats in the required normal, abnormal, and emergency maneuvers to ensure competence to conduct the flight instruction required by this part; and

(4) The safety measures to be taken from either the left or right pilot seat for emergency situations that are likely to develop during instruction.

(f) The requirements of paragraph (e) of this section may be accomplished in full or in part in flight, in a flight simulator, or in a flight training device, as appropriate.

(g) The initial and transition flight training for a flight instructor (simulator) must include the following:

(1) Training and practice in the required normal, abnormal, and emergency procedures to ensure competence to conduct the flight instruction required by this part. These maneuvers and procedures must be accomplished in full or in part in a flight simulator or in a flight training device.

(2) Training in the operation of flight simulators, flight training devices, or both, to ensure competence to conduct the flight instruction required by this part.

[Amdt. 135-64, 61 FR 30745, June 17, 1996; 61 FR 34927, July 3, 1996, Amdt. 39-9881, 62 FR 3739, Jan. 24, 1997]

§ 135.341 Pilot and flight attendant crewmember training programs.

(a) Each certificate holder, other than one who uses only one pilot in the certificate holder's operations, shall establish and maintain an approved pilot training program, and each certificate holder who uses a flight attendant crew-member shall establish and maintain an approved flight attendant training program, that is appropriate to the operations to which each pilot and flight attendant is to be assigned, and will ensure that they are adequately trained to meet the applicable knowledge and practical testing requirements of §§ 135.293 through 135.301. However, the Administrator may authorize a deviation from this section if the Administrator finds that, because of the limited size and scope of the operation, safety will allow a deviation from these requirements.

(b) Each certificate holder required to have a training program by paragraph (a) of this section shall include in that program ground and flight training curriculums for—

(1) Initial training;

(2) Transition training;

(3) Upgrade training;

(4) Differences training; and

(5) Recurrent training.

(c) Each certificate holder required to have a training program by paragraph (a) of this section shall provide current and appropriate study materials for use by each required pilot and flight attendant.

(d) The certificate holder shall furnish copies of the pilot and flight attendant crewmember training program, and all changes and additions, to the assigned representative of the Administrator. If the certificate holder uses training facilities of other persons, a copy of those training programs or appropriate portions used for those facilities shall also be furnished. Curricula that follow FAA published curricula may be cited by reference in the copy of the training program furnished to the representative of the Administrator and need not be furnished with the program.

(Amdt. 135-18, Eff. 8/2/82)

§ 135.343 Crewmember initial and recurrent training requirements.

No certificate holder may use a person, nor may any person serve, as a crewmember in operations under this part unless that crewmember has completed the appropriate initial or recurrent training phase of the training program appropriate to the type of operation in which the crew-member is to serve since the beginning of the 12th calendar month before that service. This section does not apply to a certificate holder that uses only one pilot in the certificate holder's operations.

(Amdt. 135-18, Eff. 8/2/82)

§ 135.345 Pilots: Initial, transition, and upgrade ground training.

Initial, transition, and upgrade ground training for pilots must include instruction in at least the following, as applicable to their duties:

(a) General subjects—

(1) The certificate holder's flight locating procedures;

(2) Principles and methods for determining weight and balance, and runway limitations for takeoff and landing;

(3) Enough meteorology to ensure a practical knowledge of weather phenomena, including the principles of frontal systems, icing, fog, thunderstorms, windshear and, if appropriate, high altitude weather situations;

(4) Air traffic control systems, procedures, and phraseology;

(5) Navigation and the use of navigational aids, including instrument approach procedures;

(6) Normal and emergency communication procedures;

(7) Visual cues before and during descent below DH or MDA; and

(8) Other instructions necessary to ensure the pilot's competence.

(b) For each aircraft type—

(1) A general description;

(2) Performance characteristics;

(3) Engines and propellers;

(4) Major components;

(5) Major aircraft systems (i.e., flight controls, electrical, and hydraulic), other systems, as appropriate, principles of normal, abnormal, and emergency operations, appropriate procedures and limitations;

(6) [Knowledge and] procedures for—

(i) Recognizing and avoiding severe weather situations;

(ii) Escaping from severe weather situations, in case of inadvertent encounters, including low-altitude windshear (except that rotorcraft pilots are not required to be trained in escaping from low-altitude windshear);

(iii) Operating in or near thunderstorms (including best penetrating altitudes), turbulent air (including clear air turbulence), icing, hail, and other potentially hazardous meteorological conditions; and

[(iv) Operating airplanes during ground icing conditions (i.e., any time conditions are such that frost, ice, or snow may reasonably be expected to adhere to the airplane), if the certificate holder expects to authorize takeoffs in ground icing conditions, including:

[(A) The use of holdover times when using deicing/anti-icing fluids;

[(B) Airplane deicing/anti-icing procedures, including inspection and check procedures and responsibilities;

[(C) Communications;

[(D) Airplane surface contamination (i.e., adherence of frost, ice, or snow) and critical area identification, and knowledge of how contamination adversely affects airplane performance and flight characteristics;

[(E) Types and characteristics of deicing/anti-icing fluids, if used by the certificate holder;

[(F) Cold weather preflight inspection procedures;

[(G) Techniques for recognizing contamination on the airplane;]

(7) Operating limitations;

(8) Fuel consumption and cruise control;

(9) Flight planning;

(10) Each normal and emergency procedure; and

(11) The approved Aircraft Flight Manual, or equivalent.

(Amdt. 135-27, Eff. 1/2/89); [(Amdt. 135-46, Eff. 1/31/94)]

§ 135.347 Pilots: Initial, transition, upgrade, and differences flight training.

(a) Initial, transition, upgrade, and differences training for pilots must include flight and practice in each of the maneuvers and procedures in the approved training program curriculum.

(b) The maneuvers and procedures required by paragraph (a) of this section must be performed in flight, except to the extent that certain maneuvers and procedures may be performed in an aircraft simulator, or an appropriate training device, as allowed by this subpart.

(c) If the certificate holder's approved training program includes a course of training using an aircraft simulator or other training device, each pilot must successfully complete—

(1) Training and practice in the simulator or training device in at least the maneuvers and procedures in this subpart that are capable of being performed in the aircraft simulator or training device; and

(2) A flight check in the aircraft or a check in the simulator or training device to the level of proficiency of a pilot in command or second in command, as applicable, in at least the maneuvers and procedures that are capable of being performed in an aircraft simulator or training device.

§ 135.349 Flight attendants: Initial and transition ground training.

Initial and transition ground training for flight attendants must include instruction in at least the following—

(a) General subjects—

(1) The authority of the pilot in command; and

(2) Passenger handling, including procedures to be followed in handling deranged persons or other persons whose conduct might jeopardize safety.

(b) For each aircraft type—

(1) A general description of the aircraft emphasizing physical characteristics that may have a bearing on ditching, evacuation, and inflight emergency procedures and on other related duties;

(2) The use of both the public address system and the means of communicating with other flight crewmembers, including emergency means in the case of attempted hijacking or other unusual situations; and

(3) Proper use of electrical galley equipment and the controls for cabin heat and ventilation.

§ 135.351 Recurrent training.

(a) Each certificate holder must ensure that each crewmember receives recurrent training and is adequately trained and currently proficient for the type aircraft and crewmember position involved.

(b) Recurrent ground training for crewmembers must include at least the following:

(1) A quiz or other review to determine the crewmember's knowledge of the aircraft and crewmember position involved.

(2) [Instruction as necessary in the subjects required for initial ground training by this subpart, as appropriate, including low-altitude windshear training and training on operating during ground icing conditions, as prescribed in §135.341 and described in §135.345, and emergency training.]

(c) Recurrent flight training for pilots must include, at least, flight training in the maneuvers or procedures in this subpart, except that satisfactory completion of the check required by § 135.293 within the preceding 12 calendar months may be substituted for recurrent flight training.

(Amdt. 135-27, Eff. 1/2/89); [(Amdt. 135-46, Eff. 1/31/94)]

§ 135.353 Prohibited drugs.

(a) Each certificate holder or operator shall provide each employee performing a function listed in Appendix I to Part 121 of this chapter and his or her supervisor with the training specified in that appendix.

(b) No certificate holder or operator may use any contractor to perform a function specified in Appendix I to Part 121 of this chapter unless that contractor provides each of its employees performing that function for the certificate holder or the operator and his or her supervisor with the training specified in that appendix.

Docket No. 25148 (53 FR 47061) Eff. 11/21/88; (Amdt. 135-28, Eff. 12/21/88)

Subpart I—Airplane performance operating limitations

§ 135.361 Applicability.

(a) This subpart prescribes airplane performance operating limitations applicable to the operation of the categories of airplanes listed in § 135.363 when operated under this part.

(b) For the purpose of this subpart, "effective length of the runway," for landing means the distance from the point at which the obstruction clearance plane associated with the approach end of the runway intersects the centerline of the runway to the far end of the runway.

(c) For the purpose of this subpart, "obstruction clearance plane" means a plane sloping upward from the runway at a slope of 1:20 to the horizontal, and tangent to or clearing all obstructions within a specified area surrounding the runway as shown in a profile view of that area. In the plan view, the centerline of the specified area coincides with the centerline of the runway, beginning at the point where the obstruction clearance plane intersects the centerline of the runway and proceeding to a point at least 1,500 feet from the beginning point. After that the centerline coincides with the takeoff path over the ground for the runway (in the case of takeoffs) or with the instrument approach counterpart (for landings), or, where the applicable one of these paths has not been established, it proceeds consistent with turns of at least 4,000-foot radius until a point is reached beyond which the obstruction clearance plane clears all obstructions. This area extends laterally 200 feet on each side of the centerline at the point where the obstruction clearance plane intersects the runway and continues at this width to the end of the runway; then it increases uniformly to 500 feet on each side of the centerline at a point 1,500 feet from the intersection of the obstruction clearance plane with the runway; after that it extends laterally 500 feet on each side of the centerline.

§ 135.363 General.

(a) Each certificate holder operating a reciprocating engine powered large transport category airplane shall comply with §§ 135.365 through 135.377.

(b) Each certificate holder operating a turbine engine powered large transport category airplane shall comply with §§ 135.379 through 135.387, except that when it operates a turbo-propeller-powered large transport category airplane certificated after August 29, 1959, but previously type certificated with the same number of reciprocating engines, it may comply with §§ 135.365 through 135.377.

(c) Each certificate holder operating a large nontransport category airplane shall comply with §§ 135.389 through 135.395 and any determination of compliance must be based only on approved performance data. For the purpose of this subpart, a large nontransport category airplane is an airplane that was type certificated before July 1, 1942.

(d) Each certificate holder operating a small transport category airplane shall comply with § 135.397.

(e) Each certificate holder operating a small nontransport category airplane shall comply with § 135.399.

(f) The performance data in the Airplane Flight Manual applies in determining compliance with §§ 135.365 through 135.387. Where conditions are different from those on which the performance data is based, compliance is determined by interpolation or by computing the effects of change in the specific variables, if the results of the interpolation or computations are substantially as accurate as the results of direct tests.

(g) No person may take off a reciprocating engine powered large transport category airplane at a weight that is more than the allowable weight for the runway being used (determined under the runway takeoff limitations of the transport category operating rules of this subpart) after taking into account the temperature operating correction factors in § 4a.749a-T or § 4b.117 of the Civil Air Regulations in effect on January 31, 1965, and in the applicable Airplane Flight Manual.

(h) The Administrator may authorize in the operations specifications deviations from this subpart if special circumstances make a literal observance of a requirement unnecessary for safety.

(i) The 10-mile width specified in §§ 135.369 through 135.373 may be reduced to 5 miles, for not more than 20 miles, when operating under VFR or where navigation facilities furnish reliable and accurate identification of high ground and obstructions located outside of 5 miles, but within 10 miles, on each side of the intended track.

(j) Each certificate holder operating a commuter category airplane shall comply with § 135.398.

(Amdt. 135-21, Eff. 2/17/87)

§ 135.365 Large transport category airplanes: Reciprocating engine powered: Weight limitations.

(a) No person may take off a reciprocating engine powered large transport category airplane from an airport located at an elevation outside of the range for which maximum takeoff weights have been determined for that airplane.

(b) No person may take off a reciprocating engine powered large transport category airplane for an airport of intended destination that is located at an elevation outside of the range for which maximum landing weights have been determined for that airplane.

(c) No person may specify, or have specified, an alternate airport that is located at an elevation outside of the range for which maximum landing weights have been determined for the reciprocating engine powered large transport category airplane concerned.

(d) No person may take off a reciprocating engine powered large transport category airplane at a weight more than the maximum authorized takeoff weight for the elevation of the airport.

(e) No person may take off a reciprocating engine powered large transport category airplane if its weight on arrival at the airport of destination will be more than the maximum authorized landing weight for the elevation of that airport, allowing for normal consumption of fuel and oil en route.

§ 135.367 Large transport category airplanes: Reciprocating engine powered: Takeoff limitations.

(a) No person operating a reciprocating engine powered large transport category airplane may take off that airplane unless it is possible—

(1) To stop the airplane safely on the runway, as shown by the accelerate-stop distance data, at any time during takeoff until reaching critical engine failure speed;

(2) If the critical engine fails at any time after the airplane reaches critical-engine failure speed V_1, to continue the takeoff and reach a height of 50 feet, as indicated by the takeoff path data, before passing over the end of the runway; and

(3) To clear all obstacles either by at least 50 feet vertically (as shown by the takeoff path data) or 200 feet horizontally within the airport boundaries and 300 feet horizontally beyond the boundaries, without banking before reaching a height of 50 feet (as shown by the takeoff path data) and after that without banking more than 15 degrees.

(b) In applying this section, corrections must be made for any runway gradient. To allow for wind effect, takeoff data based on still air may be corrected by taking into account not more than 50 percent of any reported headwind component and not less than 150 percent of any reported tailwind component.

§ 135.369 Large transport category airplanes: Reciprocating engine powered: En route limitations: All engines operating.

(a) No person operating a reciprocating engine powered large transport category airplane may take off that airplane at a weight, allowing for normal consumption of fuel and oil, that does not allow a rate of climb (in feet per minute), with all engines operating, of at least 6.90 Vs_0 (that is, the number of feet per minute obtained by multiplying the number of knots by 6.90) at an altitude of at least 1,000 feet above the highest ground or obstruction within ten miles of each side of the intended track.

(b) This section does not apply to large transport category airplanes certificated under Part 4a of the Civil Air Regulations.

§ 135.371   Large transport category airplanes: Reciprocating engine powered: En route limitations: One engine inoperative.

(a) Except as provided in paragraph (b) of this section, no person operating a reciprocating engine powered large transport category airplane may take off that airplane at a weight, allowing for normal consumption of fuel and oil, that does not allow a rate of climb (in feet per minute), with one engine inoperative, of at least $(0.079-0.106/N)$ $Vs_0{}^2$ (where N is the number of engines installed and Vs_0 is expressed in knots) at an altitude of least 1,000 feet above the highest ground or obstruction within 10 miles of each side of the intended track. However, for the purposes of this paragraph the rate of climb for transport category airplanes certificated under Part 4a of the Civil Air Regulations is $0.026\ Vs_0{}^2$.

(b) In place of the requirements of paragraph (a) of this section, a person may, under an approved procedure, operate a reciprocating engine powered large transport category airplane at an all-engine-operating altitude that allows the airplane to continue, after an engine failure, to an alternate airport where a landing can be made under § 135.377, allowing for normal consumption of fuel and oil. After the assumed failure, the flight path must clear the ground and any obstruction within five miles on each side of the intended track by at least 2,000 feet.

(c) If an approved procedure under paragraph (b) of this section is used, the certificate holder shall comply with the following:

(1) The rate of climb (as prescribed in the Airplane Flight Manual for the appropriate weight and altitude) used in calculating the airplane's flight path shall be diminished by an amount in feet per minute equal to $(0.079-0.106/N)\ Vs_0{}^2$ (when N is the number of engines installed and Vs_0 is expressed in knots) for airplanes certificated under Part 25 of this chapter and by $0.026\ Vs_0{}^2$ for airplanes certificated under Part 4a of the Civil Air Regulations.

(2) The all-engines-operating altitude shall be sufficient so that in the event the critical engine becomes inoperative at any point along the route, the flight will be able to proceed to a predetermined alternate airport by use of this procedure. In determining the takeoff weight, the airplane is assumed to pass over the critical obstruction following engine failure at a point no closer to the critical obstruction than the nearest approved radio navigational fix, unless the Administrator approves a procedure established on a different basis upon finding that adequate operational safeguards exist.

(3) The airplane must meet the provisions of paragraph (a) of this section at 1,000 feet above the airport used as an alternate in this procedure.

(4) The procedure must include an approved method of accounting for winds and temperatures that would otherwise adversely affect the flight path.

(5) In complying with this procedure, fuel jettisoning is allowed if the certificate holder shows that it has an adequate training program, that proper instructions are given to the flight crew, and all other precautions are taken to ensure a safe procedure.

(6) The certificate holder and the pilot in command shall jointly elect an alternate airport for which the appropriate weather reports or forecasts, or any combination of them, indicate that weather conditions will be at or above the alternate weather minimum specified in the certificate holder's operations specifications for that airport when the flight arrives.

§ 135.373   Part 25 transport category airplanes with four or more engines: Reciprocating engine powered: En route limitations: Two engines inoperative.

(a) No person may operate an airplane certificated under Part 25 and having four or more engines unless—

(1) There is no place along the intended track that is more than 90 minutes (with all engines operating at cruising power) from an airport that meets § 135.377; or

(2) It is operated at a weight allowing the airplane, with the two critical engines inoperative, to climb at $0.013\ Vs_0{}^2$ feet per minute (that is, the number of feet per minute obtained by multiplying the number of knots squared by 0.013) at an altitude of 1,000 feet above the highest ground or obstruction within 10 miles on each side of the intended track, or at an altitude of 5,000 feet, whichever is higher.

(b) For the purposes of paragraph (a)(2) of this section, it is assumed that—

(1) The two engines fail at the point that is most critical with respect to the takeoff weight;

(2) Consumption of fuel and oil is normal with all engines operating up to the point where the two engines fail with two engines operating beyond that point;

(3) Where the engines are assumed to fail at an altitude above the prescribed minimum altitude, compliance with the prescribed rate of climb at the prescribed minimum altitude need not be shown during the descent from the cruising altitude to the prescribed minimum altitude, if those requirements can be met once the prescribed minimum altitude is reached, and assuming descent to be along a net flight path and the rate of descent to be $0.013\ Vs_0{}^2$ greater than the rate in the approved performance data; and

(4) If fuel jettisoning is provided, the airplane's weight at the point where the two engines fail is considered to be

not less than that which would include enough fuel to proceed to an airport meeting § 135.377 and to arrive at an altitude of at least 1,000 feet directly over that airport.

§ 135.375   Large transport category airplanes: Reciprocating engine powered: Landing limitations: Destination airports.

(a) Except as provided in paragraph (b) of this section, no person operating a reciprocating engine powered large transport category airplane may take off that airplane, unless its weight on arrival, allowing for normal consumption of fuel and oil in flight, would allow a full stop landing at the intended destination within 60 percent of the effective length of each runway described below from a point 50 feet directly above the intersection of the obstruction clearance plane and the runway. For the purposes of determining the allowable landing weight at the destination airport the following is assumed:

(1) The airplane is landed on the most favorable runway and in the most favorable direction in still air.

(2) The airplane is landed on the most suitable runway considering the probable wind velocity and direction (forecast for the expected time of arrival), the ground handling characteristics of the type of airplane, and other conditions such as landing aids and terrain, and allowing for the effect of the landing path and roll of not more than 50 percent of the headwind component or not less than 150 percent of the tailwind component.

(b) An airplane that would be prohibited from being taken off because it could not meet paragraph (a)(2) of this section may be taken off if an alternate airport is selected that meets all of this section except that the airplane can accomplish a full stop landing within 70 percent of the effective length of the runway.

§ 135.377   Large transport category airplanes: Reciprocating engine powered: Landing limitations: Alternate airports.

No person may list an airport as an alternate airport in a flight plan unless the airplane (at the weight anticipated at the time of arrival at the airport), based on the assumptions in § 135.375(a)(1) and (2), can be brought to a full stop landing within 70 percent of the effective length of the runway.

§ 135.379   Large transport category airplanes: Turbine engine powered: Takeoff limitations.

(a) No person operating a turbine engine powered large transport category airplane may take off that airplane at a weight greater than that listed in the Airplane Flight Manual for the elevation of the airport and for the ambient temperature existing at takeoff.

(b) No person operating a turbine engine powered large transport category airplane certificated after August 26, 1957, but before August 30, 1959 (SR422, 422A), may take off that airplane at a weight greater than that listed in the Airplane Flight Manual for the minimum distance required for takeoff. In the case of an airplane certificated after September 30, 1958 (SR422A, 422B), the takeoff distance may include a clearway distance but the clearway distance included may not be greater than one-half of the takeoff run.

(c) No person operating a turbine engine powered large transport category airplane certificated after August 29, 1959 (SR422B), may take off that airplane at a weight greater than that listed in the Airplane Flight Manual at which compliance with the following may be shown:

(1) The accelerate-stop distance, as defined in § 25.109 of this chapter, must not exceed the length of the runway plus the length of any stopway.

(2) The takeoff distance must not exceed the length of the runway plus the length of any clearway except that the length of any clearway included must not be greater than one-half the length of the runway.

(3) The takeoff run must not be greater than the length of the runway.

(d) No person operating a turbine engine powered large transport category airplane may take off that airplane at a weight greater than that listed in the Airplane Flight Manual—

(1) For an airplane certificated after August 26, 1957, but before October 1, 1958 (SR422), that allows a takeoff path that clears all obstacles either by at least (35 + 0.01 D) feet vertically (D is the distance along the intended flight path from the end of the runway in feet), or by at least 200 feet horizontally within the airport boundaries and by at least 300 feet horizontally after passing the boundaries; or

(2) For an airplane certificated after September 30, 1958 (SR422A, 422B), that allows a net takeoff flight path that clears all obstacles either by a height of at least 35 feet vertically, or by at least 200 feet horizontally within the airport boundaries and by at least 300 feet horizontally after passing the boundaries.

(e) In determining maximum weights, minimum distances and flight paths under paragraphs (a) through (d) of this section, correction must be made for the runway to be used, the elevation of the airport, the effective runway gradient, and the ambient temperature and wind component at the time of takeoff.

(f) For the purposes of this section, it is assumed that the airplane is not banked before reaching a height of 50 feet, as shown by the takeoff path or net takeoff flight path data (as appropriate) in the Airplane Flight Manual, and after that the maximum is not more than 15 degrees.

(g) For the purposes of this section, the terms, "takeoff distance," "takeoff run," "net takeoff flight path," have the same meanings as set forth in the rules under which the airplane was certificated.

§ 135.381 Large transport category airplanes: Turbine engine powered: En route limitations: One engine inoperative.

(a) No person operating a turbine engine powered large transport category airplane may take off that airplane at a weight, allowing for normal consumption of fuel and oil, that is greater than that which (under the approved, one engine inoperative, en route net flight path data in the Airplane Flight Manual for that airplane) will allow compliance with subparagraph (1) or (2) of this paragraph, based on the ambient temperatures expected en route.

(1) There is a positive slope at an altitude of at least 1,000 feet above all terrain and obstructions within five statute miles on each side of the intended track, and, in addition, if that airplane was certificated after August 29, 1958 (SR422B), there is a positive slope at 1,500 feet above the airport where the airplane is assumed to land after an engine fails.

(2) The net flight path allows the airplane to continue flight from the cruising altitude to an airport where a landing can be made under § 135.387 clearing all terrain and obstructions within five statute miles of the intended track by at least 2,000 feet vertically and with a positive slope at 1,000 feet above the airport where the airplane lands after an engine fails, or, if that airplane was certificated after September 30, 1958 (SR422A, 422B), with a positive slope at 1,500 feet above the airport where the airplane lands after an engine fails.

(b) For the purpose of paragraph (a)(2) of this section, it is assumed that—

(1) The engine fails at the most critical point en route;

(2) The airplane passes over the critical obstruction, after engine failure at a point that is no closer to the obstruction than the approved radio navigation fix, unless the Administrator authorizes a different procedure based on adequate operational safeguards;

(3) An approved method is used to allow for adverse winds;

(4) Fuel jettisoning will be allowed if the certificate holder shows that the crew is properly instructed, that the training program is adequate, and that all other precautions are taken to ensure a safe procedure;

(5) The alternate airport is selected and meets the prescribed weather minimums; and

(6) The consumption of fuel and oil after engine failure is the same as the consumption that is allowed for in the approved net flight path data in the Airplane Flight Manual.

§ 135.383 Large transport category airplanes: Turbine engine powered: En route limitations: Two engines inoperative.

(a) Airplanes certificated after August 26, 1957, but before October 1, 1958 (SR422). No person may operate a turbine engine powered large transport category airplane along an intended route unless that person complies with either of the following:

(1) There is no place along the intended track that is more than 90 minutes (with all engines operating at cruising power) from an airport that meets § 135.387.

(2) Its weight, according to the two-engine-inoperative, en route, net flight path data in the Airplane Flight Manual, allows the airplane to fly from the point where the two engines are assumed to fail simultaneously to an airport that meets § 135.387, with a net flight path (considering the ambient temperature anticipated along the track) having a positive slope at an altitude of at least 1,000 feet above all terrain and obstructions within five statute miles on each side of the intended track, or at an altitude of 5,000 feet, whichever is higher.

For the purposes of paragraph (2) of this paragraph, it is assumed that the two engines fail at the most critical point en route, that if fuel jettisoning is provided, the airplane's weight at the point where the engines fail includes enough fuel to continue to the airport and to arrive at an altitude of at least 1,000 feet directly over the airport, and that the fuel and oil consumption after engine failure is the same as the consumption allowed for in the net flight path data in the Airplane Flight Manual.

(b) Airplanes certificated after September 30, 1958, but before August 30, 1959 (SR422A). No person may operate a turbine engine powered large transport category airplane along an intended route unless that person complies with either of the following:

(1) There is no place along the intended track that is more than 90 minutes (with all engines operating at cruising power) from an airport that meets § 135.387.

(2) Its weight, according to the two-engine-inoperative, en route, net flight path data in the Airplane Flight Manual allows the airplane to fly from the point where the two engines are assumed to fail simultaneously to an airport that meets § 135.387 with a net flight path (considering the ambient temperatures anticipated along the track) having a positive slope at an altitude of at least 1,000 feet above all terrain and obstructions within five statute miles on each side of the intended track, or at an altitude of 2,000 feet, whichever is higher.

For the purpose of paragraph (2) of this paragraph, it is assumed that the two engines fail at the most critical point en route, that the airplane's weight at the point where the engines fail includes enough fuel to continue to the airport, to arrive at an altitude of at least 1,500 feet directly over the airport, and after that to fly for 15 minutes at cruise power or thrust, or both, and that the consumption of fuel and oil after engine failure is the same as the consumption allowed for in the net flight path data in the Airplane Flight Manual.

(c) Aircraft certificated after August 29, 1959 (SR422B). No person may operate a turbine engine powered large transport category airplane along an intended route unless that person complies with either of the following:

(1) There is no place along the intended track that is more than 90 minutes (with all engines operating at cruising power) from an airport that meets § 135.387.

(2) Its weight, according to the two-engine-inoperative, en route, net flight path data in the Airplane Flight Manual, allows the airplane to fly from the point where the two engines are assumed to fail simultaneously to an airport that meets § 135.387, with the net flight path (considering the ambient temperatures anticipated along the track) clearing vertically by at least 2,000 feet all terrain and obstructions within five statute miles on each side of the intended track. For the purposes of this paragraph, it is assumed that—

(i) The two engines fail at the most critical point en route;

(ii) The net flight path has a positive slope at 1,500 feet above the airport where the landing is assumed to be made after the engines fail;

(iii) Fuel jettisoning will be approved if the certificate holder shows that the crew is properly instructed, that the training program is adequate, and that all other precautions are taken to ensure a safe procedure;

(iv) The airplane's weight at the point where the two engines are assumed to fail provides enough fuel to continue to the airport, to arrive at an altitude of at least 1,500 feet directly over the airport, and after that to fly for 15 minutes at cruise power or thrust, or both; and

(v) The consumption of fuel and oil after the engines fail is the same as the consumption that is allowed for in the net flight path data in the Airplane Flight Manual.

§ 135.385 Large transport category airplanes: Turbine engine powered: Landing limitations: Destination airports.

(a) No person operating a turbine engine powered large transport category airplane may take off that airplane at a weight that (allowing for normal consumption of fuel and oil in flight to the destination or alternate airport) the weight of the airplane on arrival would exceed the landing weight in the Airplane Flight Manual for the elevation of the destination or alternate airport and the ambient temperature anticipated at the time of landing.

(b) Except as provided in paragraph (c), (d), (e), or (f) of this section, no person operating a turbine engine powered large transport category airplane may take off that airplane unless its weight on arrival, allowing for

normal consumption of fuel and oil in flight (in accordance with the landing distance in the Airplane Flight Manual for the elevation of the destination airport and the wind conditions expected there at the time of landing), would allow a full stop landing at the intended destination airport within 60 percent of the effective length of each runway described below from a point 50 feet above the intersection of the obstruction clearance plane and the runway. For the purpose of determining the allowable landing weight at the destination airport the following is assumed:

(1) The airplane is landed on the most favorable runway and in the most favorable direction, in still air.

(2) The airplane is landed on the most suitable runway considering the probable wind velocity and direction and the ground handling characteristics of the airplane, and considering other conditions such as landing aids and terrain.

(c) A turbopropeller powered airplane that would be prohibited from being taken off because it could not meet paragraph (b)(2) of this section, may be taken off if an alternate airport is selected that meets all of this section except that the airplane can accomplish a full stop landing within 70 percent of the effective length of the runway.

(d) Unless, based on a showing of actual operating landing techniques on wet runways, a shorter landing distance (but never less than that required by paragraph (b) of this section) has been approved for a specific type and model airplane and included in the Airplane Flight Manual, no person may take off a turbojet airplane when the appropriate weather reports or forecasts, or any combination of them indicate that the runways at the destination airport may be wet or slippery at the estimated time of arrival unless the effective runway length at the destination airport is at least 115 percent of the runway length required under paragraph (b) of this section.

(e) A turbojet airplane that would be prohibited from being taken off because it could not meet paragraph (b)(2) of this section may be taken off if an alternate airport is selected that meets all of paragraph (b) of this section.

(f) An eligible on-demand operator may take off a turbine engine powered large transport category airplane on an on-demand flight if all of the following conditions exist:

(1) The operation is permitted by an approved Destination Airport Analysis in that person's operations manual.

(2) The airplane's weight on arrival, allowing for normal consumption of fuel and oil in flight (in accordance with the landing distance in the Airplane Flight Manual for the elevation of the destination airport and the wind conditions expected there at the lime of landing), would allow a full stop landing at the intended destination airport within 80 percent of the effective length of each runway described below from a point 50 feet above the intersection of the obstruction clearance plane and the

runway. For the purpose of determining the allowable landing weight at the destination airport, the following is assumed:

(i) The airplane is landed on the most favorable runway and in the most favorable direction, in still air.

(ii) The airplane is landed on the most suitable runway considering the probable wind velocity and direction and the ground handling characteristics of the airplane, and considering other conditions such as landing aids and terrain.

(3) The operation is authorized by operations specifications.

§ 135.387 Large transport category airplanes: Turbine engine powered: Landing limitations: Alternate airports.

(a) Except as provided in paragraph (b) of this section, no person may select an airport as an alternate airport for a turbine engine powered large transport category airplane unless (based on the assumptions in § 135.385(b)) that airplane, at the weight expected at the time of arrival, can be brought to a full stop landing within 70 percent of the effective length of the runway for turbopropeller-powered airplanes and 60 percent of the effective length of the runway for turbojet airplanes, from a point 50 feet above the intersection of the obstruction clearance plane and the runway.

(b) Eligible on-demand operators may select an airport as an alternate airport for a turbine engine powered large transport category airplane if (based on the assumptions in § 135.385(f)) that airplane, at the weight expected at the time of arrival, can be brought to a full stop landing within 80 percent of the effective length of the runway from a point 50 feet above the intersection of the obstruction clearance plane and the runway.

§ 135.389 Large nontransport category airplanes: Takeoff limitations.

(a) No person operating a large nontransport category airplane may take off that airplane at a weight greater than the weight that would allow the airplane to be brought to a safe stop within the effective length of the runway, from any point during the takeoff before reaching 105 percent of minimum control speed (the minimum speed at which an airplane can be safely controlled in flight after an engine becomes inoperative) or 115 percent of the power off stalling speed in the takeoff configuration, whichever is greater.

(b) For the purposes of this section—

(1) It may be assumed that takeoff power used on all engines during the acceleration;

(2) Not more than 50 percent of the reported headwind component, or not less than 150 percent of the reported tailwind component, may be taken into account;

(3) The average runway gradient (the difference between the elevations of the endpoints of the runway divided by the total length) must be considered if it is more than one-half of one percent;

(4) It is assumed that the airplane is operating in standard atmosphere; and

(5) For takeoff, "effective length of the runway" means the distance from the end of the runway at which the takeoff is started to a point at which the obstruction clearance plane associated with the other end of the runway intersects the runway centerline.

§ 135.391 Large nontransport category airplanes: En route limitations: One engine inoperative.

(a) Except as provided in paragraph (b) of this section, no person operating a large nontransport category airplane may take off that airplane at a weight that does not allow a rate of climb of at least 50 feet a minute, with the critical engine inoperative, at an altitude of at least 1,000 feet above the highest obstruction within five miles on each side of the intended track, or 5,000 feet, whichever is higher.

(b) Without regard to paragraph (a) of this section, if the Administrator finds that safe operations are not impaired, a person may operate the airplane at an altitude that allows the airplane, in case of engine failure, to clear all obstructions within five miles on each side of the intended track by 1,000 feet. If this procedure is used, the rate of descent for the appropriate weight and altitude is assumed to be 50 feet a minute greater than the rate in the approved performance data. Before approving such a procedure, the Administrator considers the following for the route, route segment, or area concerned:

(1) The reliability of wind and weather forecasting.

(2) The location and kinds of navigation aids.

(3) The prevailing weather conditions, particularly the frequency and amount of turbulence normally encountered.

(4) Terrain features.

(5) Air traffic problems.

(6) Any other operational factors that affect the operations.

(c) For the purposes of this section, it is assumed that—

(1) The critical engine is inoperative;

(2) The propeller of the inoperative engine is in the minimum drag position;

(3) The wing flaps and landing gear are in the most favorable position;

(4) The operating engines are operating at the maximum continuous power available;

(5) The airplane is operating in standard atmosphere; and

(6) The weight of the airplane is progressively reduced by the anticipated consumption of fuel and oil.

§ 135.393 Large nontransport category airplanes: Landing limitations: Destination airports.

(a) No person operating a large nontransport category airplane may take off that airplane at a weight that—

(1) Allowing for anticipated consumption of fuel and oil, is greater than the weight that would allow a full stop landing within 60 percent of the effective length of the most suitable runway at the destination airport; and

(2) Is greater than the weight allowable if the landing is to be made on the runway—

(i) With the greatest effective length in still air; and

(ii) Required by the probable wind, taking into account not more than 50 percent of the headwind component or not less than 150 percent of the tailwind component.

(b) For the purpose of this section, it is assumed that—

(1) The airplane passes directly over the intersection of the obstruction clearance plane and the runway at a height of 50 feet in a steady gliding approach at a true indicated airspeed of at least 1.3 Vs_0;

(2) The landing does not require exceptional pilot skill; and

(3) The airplane is operating in standard atmosphere.

§ 135.395 Large nontransport category airplanes: Landing limitations: Alternate airports.

No person may select an airport as an alternate airport for a large nontransport category airplane unless that airplane (at the weight anticipated at the time of arrival), based on the assumptions in § 135.393(b), can be brought to a full stop landing within 70 percent of the effective length of the runway.

§ 135.397 Small transport category airplane performance operating limitations.

(a) No person may operate a reciprocating engine powered small transport category airplane unless that person complies with the weight limitations in § 135.365, the takeoff limitations in § 135.367 (except paragraph (a)(3)), and the landing limitations in §§ 135.375 and 135.377.

(b) No person may operate a turbine engine powered small transport category airplane unless that person complies with the takeoff limitations in § 135.379 (except paragraphs (d) and (f)) and the landing limitations in §§ 135.385 and 135.387.

§ 135.398 Commuter category airplane performance operating limitations.

(a) No person may operate a commuter category airplane unless that person complies with the takeoff weight limitations in the approved Airplane Flight Manual.

(b) No person may take off an airplane type certificated in the commuter category at a weight greater than that listed in the Airplane Flight Manual that allows a net take-off flight path that clears all obstacles either by a height of at least 35 feet vertically, or at least 200 feet horizontally within the airport boundaries and by at least 300 feet horizontally after passing the boundaries.

(c) No person may operate a commuter category airplane unless that person complies with the landing limitations prescribed in §§ 135.385 and 135.387 of this part. For purposes of this paragraph, §§ 135.385 and 135.387 are applicable to all commuter category airplanes notwithstanding their stated applicability to turbine-engine-powered large transport category airplanes.

(d) In determining maximum weights, minimum distances and flight paths under paragraphs (a) through (c) of this section, correction must be made for the runway gradient, and ambient temperature, and wind component at the time of takeoff.

(e) For the purpose of this section, the assumption is that the airplane is not banked before reaching a height of 50 feet as shown by the new takeoff flight path data in the Airplane Flight Manual and thereafter the maximum bank is not more than 15 degrees.

Docket No. 23516 (52 FR 1836) Eff. 1/15/87; (Amdt. 135-21, Eff. 2/17/87)

§ 135.399 Small nontransport category airplane performance operating limitations.

(a) No person may operate a reciprocating engine or turbopropeller-powered small airplane that is certificated under § 135.169(b)(2); (3), (4), (5), or (6) unless that person complies with the takeoff weight limitations in the approved Airplane Flight Manual or equivalent for operations under this part, and, if the airplane is certificated under § 135.169(b)(4) or (5) with the landing weight limitations in the Approved Airplane Flight Manual or equivalent for operations under this part.

(b) No person may operate an airplane that is certificated under § 135.169(b)(6) unless that person complies with the landing limitations prescribed in §§ 135.385 and 135.387 of this part. For purposes of this paragraph, §§ 135.385 and 135.387 are applicable to reciprocating and turbopropeller-powered small airplanes notwithstanding their stated applicability to turbine engine powered large transport category airplanes.

(Amdt. 135-2, Eff. 10/17/79)

Subpart J—Maintenance, preventive maintenance, and alterations

§ 135.411 Applicability.

(a) This subpart prescribes rules in addition to those in other parts of this chapter for the maintenance, preventive maintenance, and alterations for each certificate holder as follows:

(1) Aircraft that are type certificated for a passenger seating configuration, excluding any pilot seat, of nine seats or less, shall be maintained under parts 91 and 43 of this chapter and §§ 135.415, 135.416, 135.417, 135.421 and 135.422. An approved aircraft inspection program may be used under § 135.419.

(2) Aircraft that are type certificated for a passenger seating configuration, excluding any pilot seat, of ten seats or more, shall be maintained under a maintenance program in §§ 135.415, 135.416, 135.417, 135.423 through 135.443.

(b) A certificate holder who is not otherwise required, may elect to maintain its aircraft under paragraph (a)(2) of this section.

(c) Single engine aircraft used in passenger-carrying IFR operations shall also be maintained in accordance with §135.421 (c), (d), and (e).

§ 135.413 Responsibility for airworthiness.

(a) Each certificate holder is primarily responsible for the airworthiness of its aircraft, including airframes, aircraft engines, propellers, rotors, appliances, and parts, and shall have its aircraft maintained under this chapter, and shall have defects repaired between required maintenance under Part 43 of this chapter.

(b) Each certificate holder who maintains its aircraft under § 135.411(a)(2) shall—

(1) Perform the maintenance, preventive maintenance, and alteration of its aircraft, including airframe, aircraft engines, propellers, rotors, appliances, emergency equipment and parts, under its manual and this chapter; or

(2) Make arrangements with another person for the performance of maintenance, preventive maintenance or alteration. However, the certificate holder shall ensure that any maintenance, preventive maintenance, or alteration that is performed by another person is performed under the certificate holder's manual and this chapter.

§ 135.415 Service Difficulty Reports (Operational)

(a) Each certificate holder shall report the occurrence or detection of each failure, malfunction, or defect concerning—

(1) Any fire and, when monitored by a related fire-warning system, whether the fire-warning system functioned properly;

(2) Any false warning of fire or smoke;

(3) An engine exhaust system that causes damage to the engine, adjacent structure, equipment, or components;

(4) An aircraft component that causes the accumulation or circulation of smoke, vapor, or toxic or noxious fumes

(5) Any engine flameout or shutdown during flight or ground operations;

(6) A propeller feathering system or ability of the system to control overspeed;

(7) A fuel or fuel-dumping system that affects fuel flow or causes hazardous leakage;

(8) A landing gear extension or retraction, or the opening or closing of landing gear doors during flight;

(9) Any brake system component that results in any detectable loss of brake actuating force when the aircraft is in motion on the ground.

(10) Any aircraft component or system that results in a rejected takeoff after initiation of the takeoff roll or the taking of emergency action, as defined by the Aircraft Flight Manual or Pilot's Operating Handbook;

(11) Any emergency evacuation system or component including any exit door, passenger emergency evacuation lighting system, or evacuation equipment found to be defective, or that fails to perform the intended function during an actual emergency or during training, testing, maintenance, demonstrations, or inadvertent deployments; and

(12) Autothrottle, autoflight, or flight control systems or components of these systems.

(13) Brake system components that result in loss of brake actuating force when the aircraft is in motion on the ground;

(14) Aircraft structure that requires major repair;

(15) Cracks, permanent deformation, or corrosion of aircraft structures, if more than the maximum acceptable to the manufacturer or the FAA; and

(16) Aircraft components or systems that result in taking emergency actions during flight (except action to shutdown an engine).

(b) For the purpose of this section, "during flight" means the period from the moment the aircraft leaves the surface of the earth on takeoff until it touches down on landing.

(c) In addition to the reports required by paragraph (a) of this section, each certificate holder shall report any other failure, malfunction, or defect in an aircraft, system, component or powerplant that occurs or is detected at any time if the failure, malfunction, or defect has endangered or may endanger the safe operation of the aircraft.

(d) Each certificate holder shall submit each report required by this section, covering each 24-hour period beginning at 0900 local time of each day and ending at 0900 local time on the next day to a centralized collection point as specified by the Administrator. Each report of occurrences during a 24-hour period shall be submitted to the FAA within the next 96 hours. However, a report due on Saturday or Sunday may be submitted on the following Monday, and a report due on a holiday may be submitted on the next workday. For aircraft operating in areas where mail is not collected, reports may be submitted within 24 hours after the aircraft returns to a point where the mail is collected. Each certificate holder also shall make the report data available for 30 days for examination by the certificate-holding district office in a form and manner acceptable to the Administrator.

(e) The certificate holder shall submit the reports required by this section on a form or in another format acceptable to the Administrator. The reports shall include the following information:

(1) The manufacturer, model, and serial number of the aircraft, engine, or propeller:

(2) The registration number of the aircraft.

(3) The operator designator;

(4) The date on which the failure, malfunction, or defect was discovered;

(5) The stage of flight or ground operation during which the failure, malfunction, or defect was discovered;

(6) The nature of the failure, malfunction, or defect;

(7) The applicable Joint Aircraft System/Component Code;

(8) The total cycles, if applicable, and total time of the aircraft, aircraft engine, propeller, or component;

(9) The manufacturer, manufacturer part number, part name, serial number, and location of the component that failed, malfunctioned, or was defective, if applicable.

(10) The manufacturer, manufacturer part number, part name, serial number, and location of the part that failed, malfunctioned, or was defective, if applicable;

(11) The precautionary or emergency action taken;

(12) Other information necessary for more complete analysis of the cause of the failure, malfunction, or defect, including available information pertaining to type designation of the major component and the time since the last maintenance overhaul, repair, or inspection; and

(13) A unique control number for the occurrence, in a form acceptable to the Administrator.

(f) A certificate holder that is also the holder of a Type Certificate (including a Supplemental Type Certificate), a Parts Manufacturer Approval, or a Technical Standard Order Authorization, or that is the licensee of Type Certificate holder, need not report a failure, malfunction, or defect under this section if the failure, malfunction, or defect has been reported by that certificate holder § 21.3 of this chapter or

under the accident reporting provisions of 49 CFR part 830.

(g) A report required by this section may be submitted by a certificated repair station when the reporting task has been assigned to that repair station by a part 135 certificate holder. However, the part 135 certificate holder remains primarily responsible for ensuring compliance with the provisions of this section. The part 135 certificate holder shall receive a copy of each report submitted by the repair station.

(h) No person may withhold a report required by this section although all information required by this section is not available.

(i) When a certificate holder gets supplemental information to complete the report required by this section, the certificate holder shall expeditiously submit that information as a supplement to the original report and use the unique control number from the original report.

§ 135.416 Service difficulty reports (structural).

(a) Each certificate holder shall report the occurrence or detection of each failure or defect related to—

(1) Corrosion, cracks, or disbonding that requires replacement of the affected part;

(2) Corrosion, cracks, or disbonding that requires rework or blendout because the corrosion, cracks, or disbonding exceeds the manufacturer's established allowable damage limits;

(3) Cracks, fractures, or disbonding in a composite structure that the equipment manufacturer has designated as a primary structure or a principal structural element; or

(4) Repairs made in accordance with approved data not contained in the manufacturer's maintenance manual.

(b) In addition to the reports required by paragraph (a) of this section, each certificate holder shall report any other failre or defect in aircraft structure that occurs or is detected at any time if that failure or defect has endangered or may endanger the safe operation of an aircraft.

(c) Each certificate holder shall submit each report required by this section, covering each 24-hour period beginning at 0900 local time of each day and ending at 0900 local time on the next day, to a centralized collection point as specified by the Administrator. Each report of occurrences during a 24-hour period shall be submitted to the FAA within the next 96 hours. However, a report due on Saturday or Sunday may be submitted on the following Monday, and a report due on a holiday may be submitted on the next workday. For aircaft operating in areas where mail is not collected, reports may be submitted within 24 hours after the aircraft returns to a point where the mail is collected. Each certificate holder also shall make the report data available for 30 days for examination by the certificate-holding district office in a form and manner acceptable to the Administrator.

(d) The certficate holder shall submit the reports required by this section on a form or in another format acceptable to the Administrator. The reports shall include the following information:

(1) The manufacturer, model, serial number, and registration number of the aircraft;

(2) The operator designator;

(3) The date on which the failure or defect was discovered;

(4) The stage of ground operation during which the failure or defect was discovered;

(5) The part name, part condition, and location of the failure or defect;

(6) The applicable Joint Aircraft System/Component Code;

(7) The total cycles, if applicable, and total time of the aircraft;

(8) Other information necessary for a more complete analysis of the cause of the failure or defect, including corrosion classification, if applicable, or crack length and available information pertaining to type designation of the major component and the time since the last maintenance overhaul, repair, or inspection; and

(9) A unique control number for the occurrence, in a form acceptable to the Administrator.

(e) A certificate holder that also is the holder of a Type Certificate (including a Supplemental Type Certificate), a Parts Manufacturer Approval, or a Technical Standard Order authorization, or that is a licensee of a Type Certificate holder, need not report a failure or defect under this section if the failure or defect has been reported by that certificate holder under § 21.3 of this chapter or under the accident reporting provisions of 49 CFR part 830.

(f) A report required by this section may be submitted by a certificated repair station when the reporting task has been assigned to that repair station by the part 135 certificate holder. However, the part 135 certificate holder remains primarily responsible for ensuring compliance with the provisions of this section. The part 135 certificate holder shall receive a copy of each report submitted by the repair station.

(g) No person may withhold a report required by this section although all information required by this section is not available.

(h) When a certificate holder gets supplemental information to complete the report required by this section, the certificate holder shall expeditiously submit that information as a supplement to the original report and use the unique control number for the the original report.

§ 135.417 Mechanical interruption summary report.

Each certificate holder shall submit to the Administrator, before the end of the 10th day of the following month, a summary report for the previous month of each interruption to a flight, unscheduled change of aircraft en route, unscheduled stop or diverison from a route, or unscheduled engine removal, caused by known or suspected mechanical difficulties or malfunctions that are not required to be reported under § 135.415 or § 135.416 of this part.

§ 135.419 Approved aircraft inspection program.

(a) Whenever the Administrator finds that the aircraft inspections required or allowed under Part 91 of this chapter are not adequate to meet this part, or upon application by a certificate holder, the Administrator may amend the certificate holder's operations specifications under §119.51, to require or allow an approved aircraft inspection program for any make and model aircraft of which the certificate holder has the exclusive use of at least one aircraft (as defined in § 135.25(b)).

(b) A certificate holder who applies for an amendment of its operations specifications to allow an approved aircraft inspection program must submit that program with its application for approval by the Administrator.

(c) Each certificate holder who is required by its operations specifications to have an approved aircraft inspection program shall submit a program for approval by the Administrator within 30 days of the amendment of its operations specifications or within any other period that the Administrator may prescribe in the operations specifications.

(d) The aircraft inspection program submitted for approval by the Administrator must contain the following:

(1) Instructions and procedures for the conduct of aircraft inspections (which must include necessary tests and checks), setting forth in detail the parts and areas of the airframe, engines, propellers, rotors and appliances, including emergency equipment, that must be inspected.

(2) A schedule for the performance of the aircraft inspections under paragraph (1) of this paragraph expressed in terms of the time in service, calendar time, number of system operations, or any combination of these.

(3) Instructions and procedures for recording discrepancies found during inspections and correction or deferral of discrepancies including form and disposition of records.

(e) After approval, the certificate holder shall include the approved aircraft inspection program in the manual required by § 135.21.

(f) Whenever the Administrator finds that revisions to an approved aircraft inspection program are necessary for the continued adequacy of the program, the certificate holder shall, after notification by the Administrator, make any changes in the program found by the Administrator to be necessary. The certificate holder may petition the Administrator to reconsider the notice to make any changes in a program. The petition must be filed with the representatives of the Administrator assigned to it within 30 days after the certificate holder receives the notice. Except in the case of an emergency requiring immediate action in the interest of safety, the filing of the petition stays the notice pending a decision by the Administrator.

(g) Each certificate holder who has an approved aircraft inspection program shall have each aircraft that is subject to the program inspected in accordance with the program.

(h) The registration number of each aircraft that is subject to an approved aircraft inspection program must be included in the operations specifications of the certificate holder.

§ 135.421 Additional maintenance requirements.

(a) Each certificate holder who operates an aircraft type certificated for a passenger seating configuration, excluding any pilot seat, of nine seats or less, must comply with the manufacturer's recommended maintenance programs, or a program approved by the Administrator, for each aircraft engine, propeller, rotor, and each item of emergency equipment required by this chapter.

(b) For the purpose of this section, a manufacturer's maintenance program is one which is contained in the maintenance manual or maintenance instructions set forth by the manufacturer as required by this chapter for the aircraft, aircraft engine, propeller, rotor or item of emergency equipment.

(c) For each single engine aircraft to be used in passenger-carrying IFR operations, each certificate holder must incorporate into its maintenance program either;

(1) the manufacturer's recommended engine trend monitoring program, which includes an oil analysis, if appropriate, or

(2) an FAA approved engine trend monitoring program that includes an oil analysis at each 100 hour interval or at the manufacturer's suggested interval, whichever is more frequent.

(d) For single engine aircraft to be used in passenger-carrying IFR operations, written maintenance instructions containing the methods, techniques, and practices necessary to maintain the equipment specified in §§135.105, and 135.163 (f) and (h) are required.

(e) No certificate holder may operate a single engine aircraft under IFR, carrying passengers, unless the certificate holder records and maintains in the engine maintenance records the results of each test, observation, and inspection required by the applicable engine trend monitoring program specified in (c)(1) and (c)(2) of this section.

§135.422 Aging airplane inspections and records reviews for multiengine airplanes certificated with nine or fewer passenger seats.

(a) Applicability. This section applies to multiengine airplanes certificated with nine or fewer passenger seats, operated by a certificate holder in a scheduled operation under this part, except for those airplanes operated by a certificate holder in a scheduled operation between any point within the State of Alaska and any other point within the State of Alaska.

(b) Operation after inspections and records review. After the dates specified in this paragraph, a certificate holder may not operate a multiengine airplane in a scheduled operation under this part unless the Administrator has notified the certificate holder that the Administrator has completed the aging airplane inspection and records review required by this section. During the inspection and records review, the certificate holder must demonstrate to the Administrator that the maintenance of age-sensitive parts and components of the airplane has been adequate and timely enough to ensure the highest degree of safety.

(1) Airplanes exceeding 24 years in service on December 8, 2003; initial one repetitive inspections and records reviews. For an airplane that has exceeded 24 years in service on December 8, 2003, no later than December 5, 2007, and thereafter at intervals not to exceed 7 years.

(2) Airplanes exceeding 14 years in service but not 24 years in service on December 8, 2003; initial and repetitive inspections and records reviews. For an airplane that has exceeded 14 years in service, but not 24 years in service, on December 8, 2003, no later than December 4, 2008, and thereafter at intervals not to exceed 7 years.

(3) Airplanes not exceeding 14 years in service on December 8, 2003; initial and repetitive inspections and records reviews. For an airplane that has not exceeded 14 years in service on December 8, 2003, no later than 5 years after the start of the airplane's 15th year in service and thereafter at intervals not to exceed 7 years.

(c) Unforeseen schedule conflict. In the event of an unforeseen scheduling conflict for a specific airplane, the Administrator may approve an extension of up to 90 days beyond an interval specified in paragraph (b) of this section.

(d) Airplane and records availability. The certificate holder must make available to the Administrator each airplane for which an inspection and records review is required under this section, in a condition for inspection specified by the Administrator, together with the records containing the following information:

(1) Total years in service of the airplane;

(2) Total time in service of the airframe;

(3) Date of the last inspection and records review required by this section;

(4) Current status of life-limited parts of the airframe;

(5) Time since the last overhaul of all structural components required to be overhauled on a specific time basis;

(6) Current inspection status of the airplane, including the time since the last inspection required by the inspection program under which the airplane is maintained;

(7) Current status of applicable airworthiness directives, including the date and methods of compliance, and, if the airworthiness directive involves recurring action, the time and date when the next action is required;

(8) A list of major structural alterations; and

(9) A report of major structural repairs and the current inspection status for these repairs.

(e) Notification to the Administrator. Each certificate holder must notify the Administrator at least 60 days before the date on which the airplane and airplane records will be made available for the inspection and records review.

§ 135.423 Maintenance, preventive maintenance, and alteration organization.

(a) Each certificate holder that performs any of its maintenance (other than required inspections), preventive maintenance, or alterations, and each person with whom it arranges for the performance of that work, must have an organization adequate to perform the work.

(b) Each certificate holder that performs any inspections required by its manual under § 135.427(b)(2) or (3) (in this subpart referred to as "required inspections"), and each person with whom it arranges for the performance of that work, must have an organization adequate to perform that work.

(c) Each person performing required inspections in addition to other maintenance, preventive maintenance, or alterations, shall organize the performance of those functions so as to separate the required inspection functions from the other maintenance, preventive maintenance, and alteration functions. The separation shall be below the level of administrative control at which overall responsibility for the required inspection functions and other maintenance, preventive maintenance, and alteration functions is exercised.

§ 135.425 Maintenance, preventive maintenance, and alteration programs.

Each certificate holder shall have an inspection program and a program covering other maintenance, preventive maintenance, and alterations, that ensures that—

(a) Maintenance, preventive maintenance, and alterations performed by it, or by other persons, are performed under the certificate holder's manual;

(b) Competent personnel and adequate facilities and equipment are provided for the proper performance of maintenance, preventive maintenance and alterations; and

(c) Each aircraft released to service is airworthy and has been properly maintained for operation under this part.

§ 135.427 Manual requirements.

(a) Each certificate holder shall put in its manual the chart or description of the certificate holder's organization required by § 135.424 and a list of persons with whom it has arranged for the performance of any of its required inspections, other maintenance, preventive maintenance, or alterations, including a general description of that work.

(b) Each certificate holder shall put in its manual the programs required by § 135.425 that must be followed in performing maintenance, preventive maintenance, and alterations of that certificate holder's aircraft, including airframes, aircraft engines, propellers, rotors, appliances, emergency equipment, and parts, and must include at least the following:

(1) The method of performing routine and nonroutine maintenance (other than required inspections), preventive maintenance, and alterations.

(2) A designation of the items of maintenance and alteration that must be inspected (required inspections) including at least those that could result in a failure, malfunction, or defect endangering the safe operation of the aircraft, if not performed properly or if improper parts or materials are used.

(3) The method of performing required inspections and a designation by occupational title of personnel authorized to perform each required inspection.

(4) Procedures for the reinspection of work performed under previous required inspection findings ("buy-back procedures").

(5) Procedures, standards, and limits necessary for required inspections and acceptance or rejection of the items required to be inspected and for periodic inspection and calibration of precision tools, measuring devices, and test equipment.

(6) Procedures to ensure that all required inspections are performed.

(7) Instructions to prevent any person who performs any item of work from performing any required inspection of that work.

(8) Instructions and procedures to prevent any decision of an inspector regarding any required inspection from being countermanded by persons other than supervisory personnel of the inspection unit, or a person at the level of administrative control that has overall responsibility for the management of both the required inspection functions and the other maintenance, preventive maintenance, and alterations functions.

(9) Procedures to ensure that required inspections, other maintenance, preventive maintenance, and alterations that are not completed as a result of work interruptions are properly completed before the aircraft is released to service.

(c) Each certificate holder shall put in its manual a suitable system (which may include a coded system) that provides for the retention of the following information—

(1) A description (or reference to data acceptable to the Administrator) of the work performed;

(2) The name of the person performing the work if the work is performed by a person outside the organization of the certificate holder; and

(3) The name or other positive identification of the individual approving the work.

(d) For the purposes of this part, the certificate holder must prepare that part of its manual containing maintenance information and instructions, in whole or in part, in printed form or other form, acceptable to the Administrator, that is retrievable in the English language.

§ 135.429 Required inspection personnel.

(a) No person may use any person to perform required inspections unless the person performing the inspection is appropriately certificated, properly trained, qualified, and authorized to do so.

(b) No person may allow any person to perform a required inspection unless, at the time, the person performing that inspection is under the supervision and control of an inspection unit.

(c) No person may perform a required inspection if that person performed the item of work to be inspected.

(d) In the case of rotorcraft that operate in remote areas or sites, the Administrator may approve procedures for the performance of required inspection items by a pilot when no other qualified person is available, provided—

(1) The pilot is employed by the certificate holder;

(2) It can be shown to the satisfaction of the Administrator that each pilot authorized to perform required inspections is properly trained and qualified;

(3) The required inspection is a result of a mechanical interruption and is not a part of a certificate holder's continuous airworthiness maintenance program;

(4) Each item is inspected after each flight until the item has been inspected by an appropriately certificated mechanic other than the one who originally performed the item of work; and

(5) Each item of work that is a required inspection item that is part of the flight control system shall be flight tested and reinspected before the aircraft is approved for return to service.

(e) Each certificate holder shall maintain, or shall determine that each person with whom it arranges to perform its required inspections maintains, a current listing of persons who have been trained, qualified, and authorized to conduct required inspections. The persons must be identified by name, occupational title and the inspections that they are authorized to perform. The certificate holder (or person with whom it arranges to perform its required inspections) shall give written information to each person so authorized, describing the extent of that person's responsibilities, authorities, and inspectional limitations. The list shall be made available for inspection by the Administrator upon request.

(Amdt. 135-20, Eff. 1/6/87)

§ 135.431 Continuing analysis and surveillance.

(a) Each certificate holder shall establish and maintain a system for the continuing analysis and surveillance of the performance and effectiveness of its inspection program and the program covering other maintenance, preventive maintenance, and alterations and for the correction of any deficiency in those programs, regardless of whether those programs are carried out by the certificate holder or by another person.

(b) Whenever the Administrator finds that either or both of the programs described in paragraph (a) of this section does not contain adequate procedures and standards to meet this part, the certificate holder shall, after notification by the Administrator, make changes in those programs requested by the Administrator.

(c) A certificate holder may petition the Administrator to reconsider the notice to make a change in a program. {{The petition must be filed with the certificate-holding district office within 30 days}} after the certificate holder receives the notice. Except in the case of an emergency requiring immediate action in the interest of safety, the filing of the petition stays the notice pending a decision by the Administrator.

§ 135.433 Maintenance and preventive maintenance training program.

Each certificate holder or a person performing maintenance or preventive maintenance functions for it shall have a training program to ensure that each person (including inspection personnel) who determines the adequacy of work done is fully informed about procedures and techniques and new equipment in use and is competent to perform that person's duties.

§ 135.435 Certificate requirements.

(a) Except for maintenance, preventive maintenance, alterations, and required inspections performed by a certificated repair station that is located outside the United States, each person who is directly in charge of maintenance, preventive maintenance, or alterations, and each person performing required inspections must hold an appropriate airman certificate.

(b) For the purpose of this section, a person "directly in charge" is each person assigned to a position in which that person is responsible for the work of a shop or station that performs maintenance, preventive maintenance, alterations, or other functions affecting airworthiness. A person who is "directly in charge" need not physically observe and direct each worker constantly but must be available for consultation and decision on matters requiring instruction or decision from higher authority than that of the person performing the work.

§ 135.437 Authority to perform and approve maintenance, preventive maintenance, and alterations.

(a) A certificate holder may perform, or make arrangements with other persons to perform, maintenance, preventive maintenance, and alterations as provided in its maintenance manual. In addition, a certificate holder may perform these functions for another certificate holder as provided in the maintenance manual of the other certificate holder.

(b) A certificate holder may approve any airframe, aircraft engine, propeller, rotor, or appliance for return to service after maintenance, preventive maintenance, or alterations that are performed under paragraph (a) of this section. However, in the case of a major repair or alteration, the work must have been done in accordance with technical data approved by the Administrator.

§ 135.439 Maintenance recording requirements.

(a) Each certificate holder shall keep (using the system specified in the manual required in § 135.427) the following records for the periods specified in paragraph (b) of this section:

(1) All the records necessary to show that all requirements for the issuance of an airworthiness release under § 135.443 have been met.

(2) Records contain the following information:

(i) The total time in service of the airframe, engine, propeller, and rotor.

(ii) The current status of life-limited parts of each airframe, engine, propeller, rotor, and appliance.

(iii) The time since last overhaul of each item installed on the aircraft which are required to be overhauled on a specified time basis.

(iv) The identification of the current inspection status of the aircraft, including the time since the last inspections required by the inspection program under which the aircraft and its appliances are maintained.

(v) The current status of applicable airworthiness directives, including the date and methods of compliance, and, if the airworthiness directive involves recurring action, the time and date when the next action is required.

(vi) A list of current major alterations and repairs to each airframe, engine, propeller, rotor, and appliance.

(b) Each certificate holder shall retain the records required to be kept by this section for the following periods:

(1) Except for the records of the last complete overhaul of each airframe, engine, propeller, rotor, and appliance the records specified in paragraph (a)(1) of this section shall be retained until the work is repeated or superseded by other work or for one year after the work is performed.

(2) The records of the last complete overhaul of each airframe, engine, propeller, rotor, and appliance shall be retained until the work is superseded by work of equivalent scope and detail.

(3) The records specified in paragraph (a)(2) of this section shall be retained and transferred with the aircraft at the time the aircraft is sold.

(c) The certificate holder shall make all maintenance records required to be kept by this section available for inspection by the Administrator or any representative of the National Transportation Safety Board.

§ 135.441 Transfer of maintenance records.

Each certificate holder who sells a United States registered aircraft shall transfer to the purchaser, at the time of the sale, the following records of that aircraft, in plain language form or in coded form which provides for the preservation and retrieval of information in a manner acceptable to the Administrator.

(a) The records specified in § 135.439(a)(2).

(b) The records specified in § 135.439(a)(1) which are not included in the records covered by paragraph (a) of this section, except that the purchaser may allow the seller to keep physical custody of such records. However, custody of records by the seller does not relieve the purchaser of its responsibility under § 135.439(c) to make the records available for inspection by the Administrator or any representative of the National Transportation Safety Board.

§ 135.443 Airworthiness release or aircraft maintenance log entry.

(a) No certificate holder may operate an aircraft after maintenance, preventive maintenance, or alterations are performed on the aircraft unless the certificate holder prepares, or causes the person with whom the certificate holder arranges for the performance of the maintenance, preventive maintenance, or alterations, to prepare—

(1) An airworthiness release; or

(2) An appropriate entry in the aircraft maintenance log.

(b) The airworthiness release or log entry required by paragraph (a) of this section must—

(1) Be prepared in accordance with the procedure in the certificate holder's manual;

(2) Include a certification that—

(i) The work was performed in accordance with the requirements of the certificate holder's manual;

(ii) All items required to be inspected were inspected by an authorized person who determined that the work was satisfactorily completed;

(iii) No known condition exists that would make the aircraft unairworthy; and

(iv) So far as the work performed is concerned, the aircraft is in condition for safe operation; and

(3) Be signed by an authorized certificated mechanic or repairman, except that a certificated repairman may sign the release or entry only for the work for which that person is employed and for which that person is certificated.

(c) Notwithstanding paragraph (b)(3) of this section, after maintenance, preventive maintenance, or alterations performed by a repair station located outside the United States, the airworthiness release or log entry required by paragraph (a) of this section may be signed by a person authorized by that repair station.

(d) Instead of restating each of the conditions of the certification required by paragraph (b) of this section, the certificate holder may state in its manual that the signature of an authorized certificated mechanic or repairman constitutes that certification.

Subpart K—Hazardous Materials Training Program

§ 135.501   Applicability and definitions.

(a) This subpart prescribes the requirements applicable to each certificate holder for training each crewmember and person performing or directly supervising any of the following job functions involving any item for transport on board an aircraft:

(1) Acceptance;

(2) Rejection;

(3) Handling;

(4) Storage incidental to transport;

(5) Packaging of company material; or

(6) Loading.

(b) *Definitions.* For purposes of this sub art, the following definitions apply: (1) Company material (COMAT)-Material owned or used by a certificate holder.

(2) *Initial hazardous materials training*—The basic training required for each newly hired person, or each person changing job functions, who performs or directly supervises any of the job functions specified in paragraph (a) of this section.

(3) *Recurrent hazardous materials training*—The training required every 24 months for each person who has satisfactorily completed the certificate holder's approved initial hazardous materials training program and performs or directly supervises any of the job functions specified in paragraph (a) of this section.

§ 135.503   Hazardous materials training: General.

(a) Each certificate holder must establish and implement a hazardous materials training program that:

(1) Satisfies the requirements of Appendix O of part 121 of this part;

(2) Ensures that each person performing or directly supervising any of the job functions specified in § 135.501 (a) is trained to comply with all applicable parts of 49 CFR parts 171 through 180 and the requirements of this subpart; and

(3) Enables the trained person to recognize items that contain, or may contain, hazardous materials regulated by 49 CFR parts 171 through 180.

(b) Each certificate holder must provide initial hazardous materials training and recurrent hazardous materials training to each crewmember and person performing or directly supervising any of the job functions specified in § 135.501(a).

(c) Each certificate holder's hazardous materials training program must be approved by the FAA prior to implementation.

§ 135.505   Hazardous materials training required.

(a) *Training requirement.* Except as provided in paragraphs (b), (c) and (f) of this section, no certificate holder may use any crewmember or person to perform any of the job functions or direct supervisory responsibilities, and no person may perform any of the job functions or direct supervisory responsibilities, specified in § 135.501 (a) unless that person has satisfactorily completed the certificate holder's FAA-approved initial or recurrent hazardous materials training program within the past 24 months.

(b) *New hire or new job function.* A person who is a new hire and has not et satisfactorily completed the required initial hazardous materials training, or a person who is changing job functions and has not received initial or recurrent training for a job function involving storage incidental to transport, or loading of items for transport on an aircraft, may perform those job functions for not more than 30 days from the date of hire or a change in job function, if the person is under the direct visual supervision of a person who is authorized by the certificate holder to supervise that person and who has successfully completed the certificate holder's FAA-approved initial or recurrent training program within the past 24 months.

(c) *Persons who work for more than one certificate holder.* A certificate holder that uses or assigns a person to perform or directly supervise a job function specified in § 135.501(a), when that person also performs or directly supervises the same job function for another certificate holder, need only train that person in its own policies and procedures regarding those job functions, if all of the following are met:

(1) The certificate holder using this exception receives written verification from the person designated to hold the training records representing the other certificate holder that the person has satisfactorily completed hazardous materials training for the specific job function under the other certificate holder's FAA approved hazardous material training program under appendix O of part 121 of this chapter; and

(2) The certificate holder who trained the person has the same operations specifications regarding the acceptance, handling, and transport of hazardous materials as the certificate holder using this exception.

(d) *Recurrent hazardous materials training-Completion date.* A person who satisfactorily completes recurrent hazardous materials training in the calendar month before, or the calendar month after, the month in which the recurrent training is due, is considered to have taken that training during the month in which it is due. If the person completes this training earlier than the month before it is due, the month of the completion date becomes his or her new anniversary month.

(e) *Repair stations.* A certificate holder must ensure that each repair station performing work for, or on the certificate holder's behalf is notified in writing of the certificate holder's policies and operations specification authorization permitting or prohibition against the acceptance, rejection, handling, storage incidental to transport, and transportation of hazardous materials, including company material. This notification requirement applies only to repair stations that are regulated by 49 CFR parts 171 through 180.

(f) *Certificate holders operating at foreign locations.* This exception applies if a certificate holder operating at a foreign location where the country requires the certificate holder to use persons working in that country to load aircraft. In such a case, the certificate holder may use those persons even if they have not been trained in accordance with the certificate holder's FAA approved hazardous materials training program. Those persons, however, must be under the direct visual supervision of someone who has successfully completed the certificate holder's approved initial or recurrent hazardous materials training program in accordance with this part. This exception applies only to those persons who load aircraft.

§ 135.507 Hazardous materials training records.

(a) *General requirement.* Each certificate holder must maintain a record of all training required by this part received within the preceding three years for each person who performs or directly supervises a job function specified in § 135.501(a). The record must be maintained during the time that the person performs or directly supervises any of those job functions, and for 90 days thereafter. These training records must be kept for direct employees of the certificate holder, as well as independent contractors, subcontractors, and any other person who performs or directly supervises these job functions for the certificate holder.

(b) *Location of records.* The certificate holder must retain the training records required by paragraph (a) of this section for all initial and recurrent training received within the preceding 3 years for all persons performing or directly supervising the job functions listed in Appendix O of part 121 of this chapter at a designated location. The records must be available upon request at the location where the trained person performs or directly supervises the job function specified in § 135.501(a). Records may be maintained electronically and provided on location electronically. When the person ceases to perform or directly supervise a hazardous materials job function, the certificate holder must retain the hazardous materials training records for an additional 90 days and make them available upon request at the last location where the person worked.

(c) *Content of records.* Each record must contain the following:

(1) The individual's name;

(2) The most recent training completion date;

(3) A description, copy or reference to training materials used to meet the training requirement;

(4) The name and address of the organization providing the training; and

(5) A copy of the certification issued when the individual was trained, which shows that a test has been completed satisfactorily.

(d) *New hire or new job function.* Each certificate holder using a person under the exception in § 135.505(b) must maintain a record for that person. The records must be available upon request at the location where the trained person performs or directly supervises the job function specified in § 135.501(a). Records may be maintained electronically and provided on location electronically. The record must include the following:

(1) A signed statement from an authorized representative of the certificate holder authorizing the use of the person in accordance with the exception;

(2) The date of hire or change in job function;

(3) The person's name and assigned job function;

(4) The name of the supervisor of the job function; and

(5) The date the person is to complete hazardous materials training in accordance with Appendix O of part 121 of this chapter.

Appendix A—
Additional airworthiness standards
for 10 or more passenger airplanes

APPLICABILITY

1. *Applicability.* This appendix prescribes the additional airworthiness standard required by § 135.169.

2. *References.* Unless otherwise provided references in this appendix to specific sections of Part 23 of the Federal Aviation Regulations (FAR Part 23) are to those sections of Part 23 in effect on March 30, 1967.

FLIGHT REQUIREMENTS

3. *General.* Compliance must be shown with the applicable requirements of Subpart B of FAR Part 23, as supplemented or modified in §§ 4 through 10.

PERFORMANCE

4. *General.* (a) Unless otherwise prescribed in this appendix, compliance with each applicable performance requirement in §§ 4 through 7 must be shown for ambient atmospheric conditions and still air.

(b) The performance must correspond to the propulsive thrust available under the particular ambient atmospheric conditions and the particular flight condition. The available propulsive thrust must correspond to engine power or thrust, not exceeding the approved power or thrust less—

(1) Installation losses; and

(2) The power or equivalent thrust absorbed by the accessories and services appropriate to the particular ambient atmospheric conditions and the particular flight condition.

(c) Unless otherwise prescribed in this appendix, the applicant must select the takeoff, en route, and landing configurations for the airplane.

(3) The airplane configuration may vary with weight, altitude, and temperature, to the extent they are compatible with the operating procedures required by paragraph (e) of this section.

(e) Unless otherwise prescribed in this appendix, in determining the critical engine inoperative takeoff performance, the accelerate-stop distance, takeoff distance, changes in the airplane's configuration, speed, power, and thrust must be made under procedures established by the applicant for operation in service.

(f) Procedures for the execution of balked landings must be established by the applicant and included in the Airplane Flight Manual.

(g) The procedures established under paragraphs (e) and (f) of this section must—

(1) Be able to be consistently executed in service by a crew of average skill;

(2) Use methods or devices that are safe and reliable; and

(3) Include allowance for any time delays, in the execution of the procedures, that may reasonably be expected in service.

5. *Takeoff*—(a) *General.* Takeoff speeds, the accelerate-stop distance, the takeoff distance, and the one-engine-inoperative takeoff flight path data (described in paragraphs (b), (c), (d), and (f) of this section), must be determined for—

(1) Each weight, altitude, and ambient temperature within the operational limits selected by the applicant;

(2) The selected configuration for takeoff;

(3) The center of gravity in the most unfavorable position;

(4) The operating engine within approved operating limitations; and

(5) Takeoff data based on smooth, dry, hard-surface runway.

(b) *Takeoff speeds.* (1) The decision speed V_1 is the calibrated airspeed on the ground at which, as a result of engine failure or other reasons, the pilot is assumed to have made a decision to continue or discontinue the takeoff. The speed V_1 must be selected by the applicant but may not be less than—

(i) 1.10 V_{S1};

(ii) 1.10 V_{MC};

(iii) A speed that allows acceleration to V_1 and stop under paragraph (c) of this section; or

(iv) A speed at which the airplane can be rotated for takeoff and shown to be adequate to safely continue the takeoff, using normal piloting skill, when the critical engine is suddenly made inoperative.

(2) The initial climb out speed V_2, in terms of calibrated airspeed, must be selected by the applicant so as to allow the gradient of climb required in § 6(b)(2), but it must not be less than V_1 or less than 1.2 V_{S1}.

(3) Other essential takeoff speeds necessary for safe operation of the airplane.

(c) *Accelerate-stop distance.* (1) The accelerate-stop distance is the sum of the distances necessary to—

(i) Accelerate the airplane from a standing start to V_1; and

(ii) Come to a full stop from the point at which V_1 is reached assuming that in the case of engine failure, failure of the critical engine is recognized by the pilot at the speed V_1.

(2) Means other than wheel brakes may be used to determine the accelerate-stop distance if that means is available with the critical engine inoperative and—

(i) Is safe and reliable;

(ii) Is used so that consistent results can be expected under normal operating conditions; and

(iii) Is such that exceptional skill is not required to control the airplane.

(d) *All engines operating takeoff distance.* The all engine operating takeoff distance is the horizontal distance required to take off and climb to a height of 50 feet above the takeoff surface under the procedures in FAR 23.51(a).

(e) *One-engine-inoperative takeoff.* Determine the weight for each altitude and temperature within the operational limits established for the airplane, at which the airplane has the capability, after failure of the critical engine at V_1 determined under paragraph (b) of this section, to take off and climb at not less than V_2, to a height 1,000 feet above the takeoff surface and attain the speed and configuration at which compliance is shown with the en route one-engine-inoperative gradient of climb specified in § 6(c).

(f) *One-engine-inoperative takeoff flight path data.* The one-engine-inoperative takeoff flight path data consist of takeoff flight paths extending from a standing start to a point in the takeoff at which the airplane reaches a height 1,000 feet above the takeoff surface under paragraph (e) of this section.

6. *Climb*—(a) *Landing climb: All-engines-operating.* The maximum weight must be determined with the airplane in the landing configuration, for each altitude, and ambient temperature within the operational limits established for the airplane, with the most unfavorable center of gravity, and out-of-ground effect in free air, at which the steady gradient of climb will not be less than 3.3 percent, with:

(1) The engines at the power that is available 8 seconds after initiation of movement of the power or thrust controls from the minimum flight idle to the takeoff position.

(2) A climb speed not greater than the approach speed established under § 7 and not less than the greater of 1.05 V_{MC} or 1.10 V_{S1}.

(b) *Takeoff climb: one-engine-inoperative.* The maximum weight at which the airplane meets the minimum climb performance specified in subparagraphs (1) and (2) of this paragraph must be determined for each altitude and ambient temperature within the operational limits established for the airplane, out of ground effect in free air, with the airplane in the takeoff configuration, with the most unfavorable center of gravity, the critical engine inoperative, the remaining engines at the maximum takeoff power or thrust, and the propeller of the inoperative engine windmilling with the propeller controls in the normal position except that, if an approved automatic feathering system is installed, the propellers may be in the feathered position:

(1) *Takeoff landing gear extended.* The minimum steady gradient of climb must be measurably positive at the speed V_1.

(2) *Takeoff landing gear retracted.* The minimum steady gradient of climb may not be less than 2 percent at speed V_2. For airplanes with fixed landing gear this requirement must be met with the landing gear extended.

(c) *En route climb: one-engine-inoperative.* The maximum weight must be determined for each altitude and ambient temperature within the operational limits established for the airplane, at which the steady gradient of climb is not less than 1.2 percent at an altitude 1,000 feet above the takeoff surface, with the airplane in the en route configuration, the critical engine inoperative, the remaining engine at the maximum continuous power or thrust, and the most unfavorable center of gravity.

7. *Landing.* (a) The landing field length described in paragraph (b) of this section must be determined for standard atmosphere at each weight and altitude within the operational limits established by the applicant.

(b) The landing field length is equal to the landing distance determined under FAR 23.75 (a) divided by a factor of 0.6 for the destination airport and 0.7 for the alternate airport. Instead of the gliding approach specified in FAR 23.75(a)(1), the landing may be preceded by a steady approach down to the 50-foot height at a gradient of descent not greater than 5.2 percent (3°) at a calibrated airspeed not less than 1.3 V_{S1}.

TRIM

8. *Trim*—(a) *Lateral and directional trim.* The airplane must maintain lateral and directional trim in level flight at a speed of V_H or V_{MO}/M_{MO}, whichever is lower, with landing gear and wing flaps retracted.

(b) *Longitudinal trim.* The airplane must maintain longitudinal trim during the following conditions, except that it need not maintain trim at a speed greater than V_{MO}/M_{MO}:

(1) In the approach conditions specified in FAR 23.161(c)(3) through (5), except that instead of the speeds specified in those paragraphs, trim must be maintained with a stick force of not more than 10 pounds down to a speed used in showing compliance with § 7 or 1.4 V_{S1} whichever is lower.

(2) In level flight at any speed from V_H or V_{MO}/M_{MO}, whichever is lower, to either V_X or 1.4 V_{S1}, with the landing gear and wing flaps retracted.

STABILITY

9. *Static longitudinal stability.* (a) In showing compliance with FAR 23.175(b) and with paragraph (b) of this section, the airspeed must return to within 7½ percent of the trim speed.

(b) *Cruise stability.* The stick force curve must have a stable slope for a speed range of ±50 knots from the trim speed except that the speeds need not exceed V_{FC}/M_{FC} or be less than 1.4 V_{S1}. This speed range will be considered to begin at the outer extremes of the friction band and the stick force may not exceed 50 pounds with—

(1) Landing gear retracted;

(2) Wing flaps retracted;

(3) The maximum cruising power as selected by the applicant as an operating limitation for turbine engines or

75 percent of maximum continuous power for reciprocating engines except that the power need not exceed that required at V_{MO}/M_{MO};

(4) Maximum takeoff weight; and

(5) The airplane trimmed for level flight with the power specified in subparagraph (3) of this paragraph.

V_{FC}/M_{FC} may not be less than a speed midway between V_{MO}/M_{MO} and V_{DF}/M_{DF}, except that, for altitudes where Mach number is the limiting factor, M_{FC} need not exceed the Mach number at which effective speed warning occurs.

(c) *Climb stability (turbopropeller powered airplanes only).* In showing compliance with FAR 23.175(a), an applicant must, instead of the power specified in FAR 23.175(a)(4), use the maximum power or thrust selected by the applicant as an operating limitation for use during climb at the best rate of climb speed, except that the speed need not be less than 1.4 V_{S1}.

STALLS

10. *Stall warning.* If artificial stall warning is required to comply with FAR 23.207, the warning device must give clearly distinguishable indications under expected conditions of flight. The use of a visual warning device that requires the attention of the crew within the cockpit is not acceptable by itself.

CONTROL SYSTEMS

11. *Electric trim tabs.* The airplane must meet FAR 23.677 and in addition it must be shown that the airplane is safely controllable and that a pilot can perform all the maneuvers and operations necessary to effect a safe landing following any probable electric trim tab runaway which might be reasonably expected in service allowing for appropriate time delay after pilot recognition of the runaway. This demonstration must be conducted at the critical airplane weights and center of gravity positions.

INSTRUMENTS: INSTALLATION

12. *Arrangement and visibility.* Each instrument must meet FAR 23.1321 and in addition:

(a) Each flight, navigation, and powerplant instrument for use by any pilot must be plainly visible to the pilot from the pilot's station with the minimum practicable deviation from the pilot's normal position and line of vision when the pilot is looking forward along the flight path.

(b) The flight instruments required by FAR 23.1303 and by the applicable operating rules must be grouped on the instrument panel and centered as nearly as practicable about the vertical plane of each pilot's forward vision. In addition—

(1) The instrument that most effectively indicates the attitude must be in the panel in the top center position;

(2) The instrument that most effectively indicates the airspeed must be on the panel directly to the left of the instrument in the top center position;

(3) The instrument that most effectively indicates altitude must be adjacent to and directly to the right of the instrument in the top center position; and

(4) The instrument that most effectively indicates direction of flight must be adjacent to and directly below the instrument in the top center position.

13. *Airspeed indicating system.* Each airspeed indicating system must meet FAR 23.1323 and in addition:

(a) Airspeed indicating instruments must be of an approved type and must be calibrated to indicate true airspeed at sea level in the standard atmosphere with a minimum practicable instrument calibration error when the corresponding pitot and static pressures are supplied to the instruments.

(b) The airspeed indicating system must be calibrated to determine the system error, i.e., the relation between IAS and CAS, in flight and during the accelerate-takeoff ground run. The ground run calibration must be obtained between 0.8 of the minimum value of V_1 and 1.2 times the maximum value of V_1, considering the approved ranges of altitude and weight. The ground run calibration is determined assuming an engine failure at the minimum value of V_1.

(c) The airspeed error of the installation excluding the instrument calibration error, must not exceed 3 percent or 5 knots whichever is greater, throughout the speed range from V_{MO} to 1.3 V_{S1}, with flaps retracted and from 1.3 V_{SO} to V_{FE} with flaps in the landing position.

(d) Information showing the relationship between IAS and CAS must be shown in the Airplane Flight Manual.

14. *Static air vent system.* The static air vent system must meet FAR 23.1325. The altimeter system calibration must be determined and shown in the Airplane Flight Manual.

OPERATING LIMITATIONS AND INFORMATION

15. *Maximum operating limit speed V_{MO}/M_{MO}.* Instead of establishing operating limitations based on V_{NE}/V_{NO}, the applicant must establish a maximum operating limit speed V_{MO}/M_{MO} as follows:

(a) The maximum operating limit speed must not exceed the design cruising speed V_C and must be sufficiently below V_D/M_D or V_{DF}/M_{DF} to make it highly improbable that the latter speeds will be inadvertently exceeded in flight.

(b) The speed V_{MO} must not exceed 0.8 V_D/M_D or 0.8 V_{DF}/M_{DF} unless flight demonstrations involving upsets as specified by the Administrator indicates a lower speed margin will not result in speeds exceeding V_D/M_D or V_{DF}. Atmospheric variations, horizontal gusts, system and equipment errors, and airframe production variations are taken into account.

16. *Minimum flight crew.* In addition to meeting FAR 23.1523, the applicant must establish the minimum number and type of qualified flight crew personnel sufficient for safe operation of the airplane considering—

(a) Each kind of operation for which the applicant desires approval;

(b) The workload on each crewmember considering the following:

(1) Flight path control.

(2) Collision avoidance.

(3) Navigation.

(4) Communications.

(5) Operation and monitoring of all essential aircraft systems.

(6) Command decisions; and

(c) The accessibility and ease of operation of necessary controls by the appropriate crewmember during all normal and emergency operations when at the crewmember flight station.

17. *Airspeed indicator.* The airspeed indicator must meet FAR 23.1545 except that, the airspeed notations and markings in terms of V_{NO} and V_{NH} must be replaced by the V_{MO}/M_{MO} notations. The airspeed indicator markings must be easily read and understood by the pilot. A placard adjacent to the airspeed indicator is an acceptable means of showing compliance with FAR 23.1545(c).

AIRPLANE FLIGHT MANUAL

18. *General.* The Airplane Flight Manual must be prepared under FARs 23.1583 and 23.1587, and in addition the operating limitations and performance information in §§ 19 and 20 must be included.

19. *Operating limitations.* The Airplane Flight Manual must include the following limitations—

(a) *Airspeed limitations.* (1) The maximum operating limit speed V_{MO}/M_{MO} and a statement that this speed limit may not be deliberately exceeded in any regime of flight (climb, cruise, or descent) unless a higher speed is authorized for flight test or pilot training;

(2) If an airspeed limitation is based upon compressibility effects, a statement to this effect and information as to any symptoms, the probable behavior of the airplane, and the recommended recovery procedures; and

(3) The airspeed limits, shown in terms of V_{MO}/M_{MO} instead of V_{NO} and V_{NE}.

(b) *Takeoff weight limitations.* The maximum takeoff weight for each airport elevation, ambient temperature, and available takeoff runway length within the range selected by the applicant may not exceed the weight at which—

(1) The all-engine-operating takeoff distance determined under § 5(b) or the accelerate-stop distance determined under § 5(c), whichever is greater, is equal to the available runway length;

(2) The airplane complies with the engine-inoperative takeoff requirements specified in § 5(e); and

(3) The airplane complies with the one-engine-inoperative takeoff and en route climb requirements specified in §§ 6(b) and (c).

(c) *Landing weight limitations.* The maximum landing weight for each airport elevation (standard temperature) and available landing runway length, within the range selected by the applicant. This weight may not exceed the weight at which the landing field length determined under § 7(b) is equal to the available runway length. In showing compliance with this operating limitation, it is acceptable to assume that the landing weight at the destination will be equal to the takeoff weight reduced by the normal consumption of fuel and oil en route.

20. *Performance information.* The Airplane Flight Manual must contain the performance information determined under the performance requirements of this appendix. The information must include the following:

(a) Sufficient information so that the takeoff weight limits specified in § 19(b) can be determined for all temperatures and altitudes within the operation limitations selected by the applicant.

(b) The conditions under which the performance information was obtained, including the airspeed at the 50-foot height used to determine landing distances.

(c) The performance information (determined by extrapolation and computed for the range of weights between the maximum landing and takeoff weights) for—

(1) Climb in the landing configuration; and

(2) Landing distance.

(d) Procedure established under § 4 related to the limitations and information required by this section in the form of guidance material including any relevant limitations or information.

(e) An explanation of significant or unusual flight or ground handling characteristics of the airplane.

(f) Airspeeds, as indicated airspeeds, corresponding to those determined for takeoff under § 5(b).

21. *Maximum operating altitudes.* The maximum operating altitude to which operation is allowed, as limited by flight, structural, powerplant, functional, or equipment characteristics, must be specified in the Airplane Flight Manual.

22. *Stowage provision for airplane flight manual.* Provision must be made for stowing the Airplane Flight Manual in a suitable fixed container which is readily accessible to the pilot.

23. *Operating procedures.* Procedures for restarting turbine engines in flight (including the effects of altitude) must be set forth in the Airplane Flight Manual.

AIRFRAME REQUIREMENTS FLIGHT LOADS

24. *Engine Torque.* (a) Each turbopropeller engine mount and its supporting structure must be designed for the torque effects of:

(1) The conditions in FAR 23.361(a).

(2) The limit engine torque corresponding to takeoff power and propeller speed multiplied by a factor accounting for propeller control system malfunction, including

quick feathering action, simultaneously with 1g level flight loads. In the absence of a rational analysis, a factor of 1.6 must be used.

(b) The limit torque is obtained by multiplying the mean torque by a factor of 1.25.

25. *Turbine engine gyroscopic loads.* Each turbopropeller engine mount and its supporting structure must be designed for the gyroscopic loads that result, with the engines at maximum continuous r.p.m. under either—

(a) The conditions in FARs 23.351 and 23.423; or

(b) All possible combinations of the following:

(1) A yaw velocity of 2.5 radians per second.

(2) A pitch velocity of 1.0 radians per second.

(3) A normal load factor of 2.5.

(4) Maximum continuous thrust.

26. *Unsymmetrical loads due to engine failure.*

(a) Turbopropeller powered airplanes must be designed for the unsymmetrical loads resulting from the failure of the critical engine including the following conditions in combination with a single malfunction of the propeller drag limiting system, considering the probable pilot corrective action on the flight controls:

(1) At speeds between V_{MO} and V_D, the loads resulting from power failure because of fuel flow interruption are considered to be limit loads.

(2) At speeds between V_{MO} and V_C, the loads resulting from the disconnection of the engine compressor from the turbine or from loss of the turbine blades are considered to be ultimate loads.

(3) The time history of the thrust decay and drag buildup occurring as a result of the prescribed engine failures must be substantiated by test or other data applicable to the particular engine-propeller combination.

(4) The timing and magnitude of the probable pilot corrective action must be conservatively estimated, considering the characteristics of the particular engine-propeller-airplane combination.

(b) Pilot corrective action may be assumed to be initiated at the time maximum yawing velocity is reached, but not earlier than 2 seconds after the engine failure. The magnitude of the corrective action may be based on the control forces in FAR 23.397 except that lower forces may be assumed where it is shown by analysis or test that these forces can control the yaw and roll resulting from the prescribed engine failure conditions.

GROUND LOADS

27. *Dual wheel landing gear units.* Each dual wheel landing gear unit and its supporting structure must be shown to comply with the following:

(a) *Pivoting.* The airplane must be assumed to pivot about one side of the main gear with the brakes on that side locked. The limit vertical load factor must be 1.0 and the coefficient of friction 0.8. This condition need apply only to the main gear and its supporting structure.

(b) *Unequal tire inflation.* A 60-40 percent distribution of the loads established under FAR 23.471 through FAR 23.483 must be applied to the dual wheels.

(c) *Flat tire.* (1) Sixty percent of the loads in FAR 23.471 through FAR 23.483 must be applied to either wheel in a unit.

(2) Sixty percent of the limit drag and side loads and 100 percent of the limit vertical load established under FARs 23.493 and 23.485 must be applied to either wheel in a unit except that the vertical load need not exceed the maximum vertical load in paragraph (c)(1) of this section.

FATIGUE EVALUATION

28. *Fatigue evaluation of wing and associated structure.* Unless it is shown that the structure, operating stress levels, materials and expected use are comparable from a fatigue standpoint to a similar design which has had substantial satisfactory service experience, the strength, detail design, and the fabrication of those parts of the wing, wing carrythrough, and attaching structure whose failure would be catastrophic must be evaluated under either—

(a) A fatigue strength investigation in which the structure is shown by analysis, tests, or both to be able to withstand the repeated loads of variable magnitude expected in service; or

(b) A fail-safe strength investigation in which it is shown by analysis, tests, or both that catastrophic failure of the structure is not probable after fatigue, or obvious partial failure, of a principal structural element, and that the remaining structure is able to withstand a static ultimate load factor of 75 percent of the critical limit load factor at V_C. These loads must be multiplied by a factor of 1.15 unless the dynamic effects of failure under static load are otherwise considered.

DESIGN AND CONSTRUCTION

29. *Flutter.* For multiengine turbopropeller powered airplanes, a dynamic evaluation must be made and must include—

(a) The significant elastic, inertia, and aerodynamic forces associated with the rotations and displacements of the plane of the propeller; and

(b) Engine-propeller-nacelle stiffness and damping variations appropriate to the particular configuration.

LANDING GEAR

30. *Flap operated landing gear warning device.* Airplanes having retractable landing gear and wing flaps must be equipped with a warning device that functions continuously when the wing flaps are extended to a flap position that activates the warning device to give adequate warning before landing, using normal landing procedures, if the landing gear is not fully extended and locked. There may not be a manual shut off for this warning device. The flap position sensing unit may be installed at any suitable

location. The system for this device may use any part of the system (including the aural warning devices).

PERSONNEL AND CARGO ACCOMMODATIONS

31. *Cargo and baggage compartments.* Cargo and baggage compartments must be designed to meet FAR 23.787 (a) and (b), and in addition means must be provided to protect passengers from injury by the contents of any cargo or baggage compartment when the ultimate forward inertia force is 9 *g*.

32. *Doors and exits.* The airplane must meet FAR 23.783 and FAR 23.807 (a)(3), (b), and (c), and in addition:

(a) There must be a means to lock and safeguard each external door and exit against opening in flight either inadvertently by persons, or as a result of mechanical failure. Each external door must be operable from both the inside and the outside.

(b) There must be means for direct visual inspection of the locking mechanism by crewmembers to determine whether external doors and exits, for which the initial opening movement is outward, are fully locked. In addition, there must be a visual means to signal to crewmembers when normally used external doors are closed and fully locked.

(c) The passenger entrance door must qualify as a floor level emergency exit. Each additional required emergency exit except floor level exits must be located over the wing or must be provided with acceptable means to assist the occupants in descending to the ground. In addition to the passenger entrance door:

(1) For a total seating capacity of 15 or less, an emergency exit as defined in FAR 23.807(b) is required on each side of the cabin.

(2) For a total seating capacity of 16 through 23, three emergency exits as defined in FAR 23.807(b) are required with one on the same side as the door and two on the side opposite the door.

(d) An evacuation demonstration must be conducted utilizing the maximum number of occupants for which certification is desired. It must be conducted under simulated night conditions utilizing only the emergency exits on the most critical side of the aircraft. The participants must be representative of average airline passengers with no previous practice or rehearsal for the demonstration. Evacuation must be completed within 90 seconds.

(e) Each emergency exit must be marked with the word "Exit" by a sign which has white letters 1 inch high on a red background 2 inches high, be self-illuminated or independently internally electrically illuminated, and have a minimum luminescence (brightness) of at least 160 microlamberts. The colors may be reserved if the passenger compartment illumination is essentially the same.

(f) Access to window type emergency exits must not be obstructed by seats or seat backs.

(g) The width of the main passenger aisle at any point between seats must equal or exceed the values in the following table:

Minimum main passenger aisle width	Total seating capacity	
	Less than 25 inches from floor	25 inches and more from floor
10 through 23	9 inches	15 inches

MISCELLANEOUS

33. *Lightning strike protection.* Parts that are electrically insulated from the basic airframe must be connected to it through lightning arrestors unless a lightning strike on the insulated part—

(a) Is improbable because of shielding by other parts; or

(b) Is not hazardous.

34. *Ice protection.* If certification with ice protection provisions is desired, compliance with the following must be shown:

(a) The recommended procedures for the use of the ice protection equipment must be set forth in the Airplane Flight Manual.

(b) An analysis must be performed to establish, on the basis of the airplane's operational needs, the adequacy of the ice protection system for the various components of the airplane. In addition, tests of the ice protection system must be conducted to demonstrate that the airplane is capable of operating safely in continuous maximum and intermittent maximum icing conditions as described in Appendix C of Part 25 of this chapter.

(c) Compliance with all or portions of this section may be accomplished by reference, where applicable because of similarity of the designs, to analysis and tests performed by the applicant for a type certificated model.

35. *Maintenance information.* The applicant must make available to the owner at the time of delivery of the airplane the information the applicant considers essential for the proper maintenance of the airplane. That information must include the following:

(a) Description of systems, including electrical, hydraulic, and fuel controls.

(b) Lubrication instructions setting forth the frequency and the lubricants and fluids which are to be used in the various systems.

(c) Pressures and electrical loads applicable to the various systems.

(d) Tolerances and adjustments necessary for proper functioning.

(e) Methods of leveling, raising, and towing.

(f) Methods of balancing control surfaces.

(g) Identification of primary and secondary structures.

(h) Frequency and extent of inspections necessary to the proper operation of the airplane.

(i) Special repair methods applicable to the airplane.

(j) Special inspection techniques, such as X-ray, ultrasonic, and magnetic particle inspection.

(k) List of special tools.

PROPULSION

GENERAL

36. *Vibration characteristics.* For turbopropeller powered airplanes, the engine installation must not result in vibration characteristics of the engine exceeding those established during the type certification of the engine.

37. *In-flight restarting of engine.* If the engine on turbopropeller powered airplanes cannot be restarted at the maximum cruise altitude, a determination must be made of the altitude below which restarts can be consistently accomplished. Restart information must be provided in the Airplane Flight Manual.

38. *Engines.* (a) *For turbopropeller powered airplanes.* The engine installation must comply with the following:

(1) *Engine isolation.* The powerplants must be arranged and isolated from each other to all operation, in at least one configuration, so that the failure or malfunction of any engine, or of any system that can affect the engine, will not—

(i) Prevent the continued safe operation of the remaining engines; or

(ii) Require immediate action by any crewmember for continued safe operation.

(2) *Control of engine rotation.* There must be a means to individually stop and restart the rotation of any engine in flight except that engine rotation need not be stopped if continued rotation could not jeopardize the safety of the airplane. Each component of the stopping and restarting system on the engine side of the firewall, and that might be exposed to fire, must be at least fire resistant. If hydraulic propeller feathering systems are used for this purpose, the feathering lines must be at least fire resistant under the operating conditions that may be expected to exist during feathering.

(3) *Engine speed and gas temperature control devices.* The powerplant systems associated with engine control devices, systems, and instrumentation must provide reasonable assurance that those engine operating limitations that adversely affect turbine rotor structural integrity will not be exceeded in service.

(b) *For reciprocating engine powered airplanes.* To provide engine isolation, the powerplants must be arranged and isolated from each other to allow operation, in at least one configuration, so that the failure or malfunction of any engine, or of any system that can affect that engine, will not—

(1) Prevent the continued safe operation of the remaining engines; or

(2) Require immediate action by any crewmember for continued safe operation.

39. *Turbopropeller reversing systems.* (a) Turbopropeller reversing systems intended for ground operation must be designed so that no single failure or malfunction of the system will result in unwanted reverse thrust under any expected operating condition. Failure of structural elements need not be considered if the probability of this kind of failure is extremely remote.

(b) Turbopropeller reversing systems intended for in flight use must be designed so that no unsafe condition will result during normal operation of the system, or from any failure (or reasonably likely combination of failures) of the reversing system, under any anticipated condition of operation of the airplane. Failure of structural elements need not be considered if the probability of this kind of failure is extremely remote.

(c) Compliance with this section may be shown by failure analysis, testing, or both for propeller systems that allow propeller blades to move from the flight low-pitch position to a position that is substantially less than that at the normal flight low-pitch stop position. The analysis may include or be supported by the analysis made to show compliance with the type certification of the propeller and associated installation components. Credit will be given for pertinent analysis and testing completed by the engine and propeller manufacturers.

40. *Turbopropeller drag-limiting systems.* Turbopropeller drag-limiting systems must be designed so that no single failure or malfunction of any of the systems during normal or emergency operation results in propeller drag in excess of that for which the airplane was designed. Failure of structure elements of the drag-limiting systems need not be considered if the probability of this kind of failure is extremely remote.

41. *Turbine engine powerplant operating characteristics.* For turbopropeller powered airplanes, the turbine engine powerplant operating characteristics must be investigated in flight to determine that no adverse characteristics (such as stall, surge, or flameout) are present to a hazardous degree, during normal and emergency operation within the range of operating limitations of the airplane and of the engine.

42. *Fuel flow.* (a) For turbopropeller powered airplanes—

(1) The fuel system must provide for continuous supply of fuel to the engines for normal operation without interruption due to depletion of fuel in any tank other than the main tank; and

(2) The fuel flow rate for turbopropeller engine fuel pump systems must not be less than 125 percent of the fuel flow required to develop the standard sea level atmospheric conditions takeoff power selected and included as an operating limitation in the Airplane Flight Manual.

(b) For reciprocating engine powered airplanes, it is acceptable for the fuel flow rate for each pump system (main and reverse supply) to be 125 percent of the takeoff fuel consumption of the engine.

FUEL SYSTEM COMPONENTS

43. *Fuel pumps.* For turbopropeller powered airplanes, a reliable and independent power source must be provided for

each pump used with turbine engines which do not have provisions for mechanically driving the main pumps. It must be demonstrated that the pump installations provide a reliability and durability equivalent to that in FAR 23.991(a).

44. *Fuel strainer or filter.* For turbopropeller powered airplanes, the following apply:

(a) There must be a fuel strainer or filter between the tank outlet and the fuel metering device of the engine. In addition, the fuel strainer or filter must be—

(1) Between the tank outlet and the engine driven positive displacement pump inlet, if there is an engine-driven positive displacement pump;

(2) Accessible for drainage and cleaning and, for the strainer screen, easily removable; and

(3) Mounted so that its weight is not supported by the connecting lines or by the inlet or outlet connections of the strainer or filter itself.

(b) Unless there are means in the fuel system to prevent the accumulation of ice on the filter, there must be means to automatically maintain the fuel-flow if ice-clogging of the filter occurs; and

(c) The fuel strainer or filter must be of adequate capacity (for operating limitations established to ensure proper service) and of appropriate mesh to ensure proper engine operation, with the fuel contaminated to a degree (for particle size and density) that can be reasonably expected in service. The degree of fuel filtering may not be less than that established for the engine type certification.

45. *Lightning strike protection.* Protection must be provided against the ignition of flammable vapors in the fuel vent system due to lightning strikes.

COOLING

46. *Cooling test procedures for turbopropeller powered airplanes.* (a) Turbopropeller powered airplanes must be shown to comply with FAR 23.1041 during takeoff, climb, en route, and landing stages of flight that correspond to the applicable performance requirements. The cooling tests must be conducted with the airplane in the configuration, and operating under the conditions that are critical relative to cooling during each stage of flight. For the cooling tests a temperature is "stabilized" when its rate of change is less than 2°F per minute.

(b) Temperatures must be stabilized under the conditions from which entry is made into each stage of flights being investigated unless the entry condition is not one during which component and engine fluid temperatures would stabilize, in which case, operation through the full entry condition must be conducted before entry into the stage of flight being investigated to allow temperatures to reach their natural levels at the time of entry. The takeoff cooling test must be preceded by a period during which the powerplant component and engine fluid temperatures are stabilized with the engines at ground idle.

(c) Cooling tests for each stage of flight must be continued until—

(1) The component and engine fluid temperatures stabilize;

(2) The stage of flight is completed; or

(3) An operating limitation is reached.

INDUCTION SYSTEM

47. *Air induction.* For turbopropeller powered airplanes—

(a) There must be means to prevent hazardous quantities of fuel leakage or overflow from drains, vents, or other components of flammable fluid systems from entering the engine intake systems; and

(b) The air inlet ducts must be located or protected so as to minimize the ingestion of foreign matter during takeoff, landing, and taxiing.

48. *Induction system icing protection.* For turbopropeller powered airplanes, each turbine engine must be able to operate throughout its flight power range without adverse effect on engine operation or serious loss of power or thrust, under the icing conditions specified in Appendix C of Part 25 of this chapter. In addition, there must be means to indicate to appropriate flight crewmembers the functioning of the powerplant ice protection system.

49. *Turbine engine bleed air systems.* Turbine engine bleed air systems of turbopropeller powered airplanes must be investigated to determine—

(a) That no hazard to the airplane will result if a duct rupture occurs. This condition must consider that a failure of the duct can occur anywhere between the engine port and the airplane bleed service; and

(b) That, if the bleed air system is used for direct cabin pressurization, it is not possible for hazardous contamination of the cabin air system to occur in event of lubrication system failure.

EXHAUST SYSTEM

50. *Exhaust system drains.* Turbopropeller engine exhaust systems having low spots or pockets must incorporate drains at those locations. These drains must discharge clear of the airplane in normal and ground attitudes to prevent the accumulation of fuel after the failure of an attempted engine start.

POWERPLANT CONTROLS AND ACCESSORIES

51. *Engine controls.* If throttles or power levers for turbopropeller powered airplanes are such that any position of these controls will reduce the fuel flow to the engine(s) below that necessary for satisfactory and safe idle operation of the engine while the airplane is in flight, a means must be provided to prevent inadvertent movement of the control into this position. The means provided must incorporate a positive lock or stop at this idle position and must require a separate and distinct operation by the crew to displace the control from the normal engine operating range.

52. *Reverse thrust controls.* For turbopropeller powered airplanes, the propeller reverse thrust controls must have a means to prevent their inadvertent operation. The means must have a positive lock or stop at the idle position and must require a separate and distinct operation by the crew to displace the control from the flight regime.

53. *Engine ignition systems.* Each turbopropeller airplane ignition system must be considered an essential electrical load.

54. *Powerplant accessories.* The powerplant accessories must meet FAR 23.1163, and if the continued rotation of any accessory remotely driven by the engine is hazardous when malfunctioning occurs, there must be means to prevent rotation without interfering with the continued operation of the engine.

POWERPLANT FIRE PROTECTION

55. *Fire detector system.* For turbopropeller powered airplanes, the following apply:

(a) There must be a means that ensures prompt detection of fire in the engine compartment. An overtemperature switch in each engine cooling air exit is an acceptable method of meeting this requirement.

(b) Each fire detector must be constructed and installed to withstand the vibration, inertia, and other loads to which it may be subjected in operation.

(c) No fire detector may be affected by any oil, water, other fluids, or fumes that might be present.

(d) There must be means to allow the flight crew to check, in flight, the functioning of each fire detector electric circuit.

(e) Wiring and other components of each fire detector system in a fire zone must be at least fire resistant.

56. *Fire protection, cowling and nacelle skin.* For reciprocating engine powered airplanes, the engine cowling must be designed and constructed so that no fire originating in the engine compartment can enter either through openings or by burn through, any other region where it would create additional hazards.

57. *Flammable fluid fire protection.* If flammable fluids or vapors might be liberated by the leakage of fluid systems in areas other than engine compartments, there must be means to—

(a) Prevent the ignition of those fluids or vapors by any other equipment; or

(b) Control any fire resulting from that ignition.

EQUIPMENT

58. *Powerplant instruments.* (a) The following are required for turbopropeller airplanes:

(1) The instruments required by FAR 23.1305(a) (1) through (4), (b)(2) and (4).

(2) A gas temperature indicator for each engine.

(3) Free air temperature indicator.

(4) A fuel flowmeter indicator for each engine.

(5) Oil pressure warning means for each engine.

(6) A torque indicator or adequate means for indicating power output for each engine.

(7) Fire warning indicator for each engine.

(8) A means to indicate when the propeller blade angle is below the low-pitch position corresponding to idle operation in flight.

(9) A means to indicate the functioning of the ice protection system for each engine.

(b) For turbopropeller powered airplanes, the turbopropeller blade position indicator must begin indicating when the blade has moved below the flight low-pitch position.

(c) The following instruments are required for reciprocating engine powered airplanes:

(1) The instruments required by FAR 23.1305.

(2) A cylinder head temperature indicator for each engine.

(3) A manifold pressure indicator for each engine.

SYSTEMS AND EQUIPMENTS
GENERAL

59. *Function and installation.* The systems and equipment of the airplane must meet FAR 23.1301, and the following:

(a) Each item of additional installed equipment must—

(1) Be of a kind and design appropriate to its intended function;

(2) Be labeled as to its identification, function, or operating limitations, or any applicable combination of these factors, unless misuse or inadvertent actuation cannot create a hazard;

(3) Be installed according to limitations specified for that equipment; and

(4) Function properly when installed.

(b) Systems and installations must be designed to safeguard against hazards to the aircraft in the event of their malfunction or failure.

(c) Where an installation, the functioning of which is necessary in showing compliance with the applicable requirements, requires a power supply, that installation must be considered an essential load on the power supply, and the power sources and the distribution system must be capable of supplying the following power loads in probable operation combinations and for probable durations:

(1) All essential loads after failure of any prime mover, power converter, or energy storage device.

(2) All essential loads after failure of any one engine on two-engine airplanes.

(3) In determining the probable operating combinations and durations of essential loads for the power failure conditions described in subparagraphs (1) and (2) of this paragraph, it is permissible to assume that the power loads are reduced in accordance with a monitoring procedure which is consistent with safety in the types of operations authorized.

60. *Ventilation.* The ventilation system of the airplane must meet FAR 23.831, and in addition, for pressurized

aircraft, the ventilating air in flight crew and passenger compartments must be free of harmful or hazardous concentrations of gases and vapors in normal operation and in the event of reasonably probable failures or malfunctioning of the ventilating, heating, pressurization, or other systems, and equipment. If accumulation of hazardous quantities of smoke in the cockpit area is reasonably probable, smoke evacuation must be readily accomplished.

ELECTRICAL SYSTEMS AND EQUIPMENT

61. *General.* The electrical systems and equipment of the airplane must meet FAR 23.1351, and the following:

(a) *Electrical system capacity.* The required generating capacity, and number and kinds of power sources must—

(1) Be determined by an electrical load analysis; and

(2) Meet FAR 23.1301.

(b) *Generating system.* The generating system includes electrical power sources, main power busses, transmission cables, and associated control, regulation and protective devices. It must be designed so that—

(1) The system voltage and frequency (as applicable) at the terminals of all essential load equipment can be maintained within the limits for which the equipment is designed, during any probable operating conditions;

(2) System transients due to switching, fault clearing, or other causes do not make essential loads inoperative, and do not cause a smoke or fire hazard;

(3) There are means, accessible in flight to appropriate crewmembers, for the individual and collective disconnection of the electrical power sources from the system; and

(4) There are means to indicate to appropriate crewmembers the generating system quantities essential for the safe operation of the system, including the voltage and current supplied by each generator.

62. *Electrical equipment and installation.* Electrical equipment, controls, and wiring must be installed so that operation of any one unit or system of units will not adversely affect the simultaneous operation of any other electrical unit or system essential to the safe operation.

63. *Distribution system.* (a) For the purpose of complying with this section, the distribution system includes the distribution busses, their associated feeders, and each control and protective device.

(b) Each system must be designed so that essential load circuits can be supplied in the event of reasonably probable faults or open circuits, including faults in heavy current carrying cables.

(c) If two independent sources of electrical power for particular equipment or systems are required under this Appendix, their electrical energy supply must be ensured by means such as duplicate electrical equipment, throwover switching, or multi-channel or loop circuits separately routed.

64. *Circuit protective devices.* The circuit protective devices for the electrical circuits of the airplane must meet FAR 23.1357, and in addition circuits for loads which are essential to safe operation must have individual and exclusive circuit protection.

Docket No. 25530 (53 FR 26152) Eff. 7/11/88; (Amdt. 135-26, Eff. 10/11/88)

Appendix B—Airplane flight recorder specifications

Parameters	Range	Installed system[1] minimum accuracy (to recovered data)	Sampling interval (per second)	Resolution[4] read out
Relative time (from recorded on prior to takeoff).	25 hr minimum	±0.125% per hour	1	1 sec.
Indicated airspeed.	V_{SO} to V_D (KIAS)	±5% or ±10 kts., whichever is greater. Resolution 2 kts. below 175 KIAS.	1	1%[3]
Altitude.	−1,000 ft. to max cert. alt. of A/C.	±100 to ±700 ft. (see Table 1, TSO C51-a).	1	25 to 150
Magnetic heading.	360°	±5°	1	1°
Vertical acceleration.	−3g to +6g	±0.2g in addition to ±0.3g maximum datum.	4 (or 1 per second where peaks, ref. to 1g are recorded).	0.03g

Parameters	Range	Installed system[1] minimum accuracy (to recovered data)	Sampling interval (per second)	Resolution[4] read out
Longitudinal acceleration.	±1.0g	±1.5% max. range excluding datum error of ±5%.	2	0.01g
Pitch attitude.	100% if usable	±2°	1	0.8°
Roll attitude.	±60° or 100% of usable range, whichever is greater.	±2°	1	0.8°
Stabilizer trim position,	Full range	±3% unless higher uniquely required.	1	1%[3]
Or				
Pitch control position.	Full range	±3% unless higher uniquely required.	1	1%[3]
Engine Power, Each Engine				
Fan or N_1 speed or EPR or cockpit indications used for aircraft certification.	Maximum range	±5%	1	1%[3]
Or				
Prop. speed and torque (sample once/sec as close together as practicable).			1 (prop speed), 1 (torque).	
Altitude rate[2] (need depends on altitude resolution).	±8,000 fpm	±10%. Resolution 250 fpm below 12,000 ft. indicated.	1	250 fpm below 12,000
Angle of attack[2] (need depends on altitude resolution).	–20° to 40° or of usable range.	±2°	1	0.8%[3]
Radio transmitter keying (discrete).	On/Off		1	
TE flaps (discrete or analog).	Each discrete position (U, D, T/O, AAP).		1	
Or				
	Analog 0–100% range	±3°	1	1%[3]
LE flaps (discrete or analog).	Each discrete position (U, D, T/O, AAP).		1	
Or				
	Analog 0–100% range	±3°	1	1%[3]
Thrust reverser, each engine (discrete).	Stowed or full reverse		1	
Spoiler/speedbrake (discrete).	Stowed or out		1	

Parameters	Range	Installed system[1] minimum accuracy (to recovered data)	Sampling interval (per second)	Resolution[4] read out
Autopilot engaged (discrete).	Engaged or disengaged		1	

[1] When data sources are aircraft instruments (except altimeters) of acceptable quality to fly the aircraft the recording system excluding these sensors (but including all other characteristics of the recording system) shall contribute no more than half of the values in this column.
[2] If data from the altitude encoding altimeter (100 ft. resolution) is used, then either one of these parameters should also be recorded. If however, altitude is recorded at a minimum resolution of 25 feet, then these two parameters can be omitted.
[3] Percent of full range.
[4] This column applies to aircraft manufactured after October 11, 1991.

Docket No. 25530 (53 FR 26152) Eff. 7/11/88;
(Amdt. 135–26, Eff. 10/11/88)

Appendix C—Helicopter flight recorder specifications

Parameters	Range	Installed system[1] minimum accuracy (to recovered data)	Sampling interval (per second)	Resolution[3] read out
Relative time (from recorded on prior to takeoff).	25 hr minimum	±0.125% per hour	1	1 sec.
Indicated airspeed.	V_m in to V_D (KIAS) (minimum airspeed signal attainable with installed pilot-static system).	±5% or ±10 kts., whichever is greater.	1	1 kt.
Altitude.	–1,000 ft. to 20,000 ft. pressure altitude.	±100 to ±700 ft. (see Table 1, TSO C51-a).	1	25 to 150 ft.
Magnetic heading.	360°	±5°	1	1°
Vertical acceleration.	–3g to +6g	±0.2g in addition to ±0.3g maximum datum.	4 (or 1 per second where peaks, ref to 1g are recorded).	0.05g
Longitudinal acceleration.	±1.0g	±1.5% max. range excluding datum error of ±5%.	2	0.03g
Pitch attitude.	100% of usable range	±2°	1	0.8°
Roll attitude.	±60% or 100% of usable range, whichever is greater.	±2°	1	0.8°
Altitude rate.	±8,000 fpm	±10% resolution 250 fpm below 12,000 ft. indicated.	1	250 fpm below 12,000
Engine Power, Each Engine				
Main rotor speed.	Maximum range	±5%	1	1%[2]

Parameters	Range	Installed system[1] minimum accuracy (to recovered data)	Sampling interval (per second)	Resolution[3] read out
Free or power turbine.	Maximum range	+5%	1	1%[2]
Engine torque.	Maximum range	±5%	1	1%[2]
Flight Control— Hydraulic Pressure				
Primary (discrete).	High/low		1	
Secondary—if applicable (discrete).	High/low		1	
Radio transmitter keying (discrete).	On/off		1	
Autopilot engaged (discrete).	Engaged or disengaged		1	
SAS status—engaged (discrete).	Engaged/disengaged		1	
SAS fault status (discrete).	Fault/OK		1	
Flight Controls				
Collective.	Full range	±3%	2	1%[2]
Pedal position.	Full range	±3%	2	1%[2]
Lat. cyclic.	Full range	±3%	2	1%[2]
Long. cyclic.	Full range	±3%	2	1%[2]
Controllable stabilator position.	Full range	±3%	2	1%[2]

[1] When data sources are aircraft instruments (except altimeters) of acceptable quality to fly the aircraft the recording system excluding these sensors (but including all other characteristics of the recording system) shall contribute no more than half of the values in this column.
[2] Percent of full range.
[3] This column applies to aircraft manufactured after October 11, 1991.

Docket No. 25530 (53 FR 26152) Eff. 7/11/88;
(Amdt. 135–26, Eff. 10/11/88)

Appendix D—Airplane flight recorder specifications

Parameters	Range	Accuracy sensor input to DFDR readout	Sampling interval (per second)	Resolution[4] read out
Time (GMT or Frame Counter) (range 0 to 4095, sampled 1 per frame).	24 hrs	±0.125% per hour	0.25 (1 per 4 seconds).	1 sec.
Altitude.	−1,000 ft to max certificated altitude of aircraft.	±100 to ±700 ft (See Table 1, TSO)–C51a).	1	5' to 35'[1]
Airspeed.	50 KIAS to V_{SO1} to V_{SO} to 1.2 V_D.	±5%, ±3%	1	1 kt.
Heading.	360°	±2°	1	0.5°

Parameters	Range	Accuracy sensor input to DFDR readout	Sampling interval (per second)	Resolution[4] read out
Normal Acceleration (Vertical).	−3g to +6g	±1% of max range excluding datum error or ±5%.	8	0.01g
Pitch Attitude.	±75°	±2°	1	0.5°
Roll Attitude.	±180°	±2°	1	0.5°
Radio Transmitter Keying.	On-Off (Discrete)		1	
Thrust/Power on Each Engine.	Full range forward	±2%	1 (per engine)	0.2%[2]
Trailing Edge Flap or Cockpit Control Selection.	Full range or each discrete position.	±3° or as pilot's indicator	0.5	0.5%[2]
Leading Edge Flap on or Cockpit Control Selection.	Full range or each discrete position.	±3° or as pilot's indicator	0.5	0.5%[2]

Parameters	Range	Accuracy sensor input to DFDR readout	Sampling interval (per second)	Resolution[4] read out
Thrust Reverser Position.	Stowed, in transit, and reverse (discrete).		1 (per 4 seconds per engine).	
Ground Spoiler Position/ Speed Brake Selection.	Full range or each discrete position.	±2% unless higher accuracy uniquely required.	1	0.22[2]
Marker Beacon Passage.	Discrete		1	
Autopilot Engagement.	Discrete		1	
Longitudinal Acceleration.	±1g	±1.5% max range excluding datum error of ±5%.	4	0.01g
Pilot Input and/or Surface Position-Primary Controls (Pitch, Roll, Yaw)[3].	Full range	±2° unless higher accuracy uniquely required.	1	0.2%[2]
Lateral Acceleration.	±1g	±1.5% max range excluding datum error of ±5%.	4	0.01g
Pitch Trim Position.	Full range	±3% unless higher accuracy uniquely required.	1	0.3%[2]
Glideslope Deviation.	±400 Microamps	±3%	1	0.3%[2]
Localizer Deviation.	±400 Microamps	±3%	1	0.3%[2]
AFCS Mode and Engagement Status.	Discrete		1	
Radio Altitude.	−20 ft to 2,500 ft	±2 ft or ±3% whichever is greater below 500 ft and ±5% above 500 ft.	1	1 ft + 5%[2] above 500'
Master Warning.	Discrete		1	
Main Gear Squat Switch Status.	Discrete		1	
Angle of Attack (if recorded directly).	As installed	As installed	2	0.3%[2]

Parameters	Range	Accuracy sensor input to DFDR readout	Sampling interval (per second)	Resolution[4] read out
Outside Air Temperature or Total Air Temperature.	−50°C to +90°C	±2°C	0.5	0.3°C
Hydraulics, Each System Low Pressure.	Discrete		0.5	Or 0.5%[2]
Groundspeed.	As installed	Most accurate systems installed (IMS equipped aircraft only).	1	0.2%[2]
Drift angle.	When available. As installed.	As installed	4	
Wind Speed and Direction.	When available. As installed.	As installed	4	
Latitude and longitude.	When available. As installed.	As installed	4	
Brake pressure/Brake pedal position.	As installed	As installed	1	
Additional engine parameters:				
EPR	As installed	As installed	1 (per engine)	
N1	As installed	As installed	1 (per engine)	
N2	As installed	As installed	1 (per engine)	
EGT	As installed	As installed	1 (per engine)	
Throttle Lever Position	As installed	As installed	1 (per engine)	
Fuel Flow	As installed	As installed	1 (per engine)	
TCAS:				
TA	As installed	As installed	1	
RA	As installed	As installed	1	
Sensitivity level (as selected by crew).	As installed	As installed	2	
GPWS (ground proximity warning system).	Discrete		1	
Landing gear or gear selector position.	Discrete		0.25 (1 per 4 seconds).	
DME 1 and 2 Distance.	0–200 NM;	As installed	0.25	1 mi.
Nav 1 and 2 Frequency Selection.	Full range	As installed	0.25	

If additional recording capacity is available, recording of the following parameters is recommended. The parameters are listed in order of significance:

[1] When altitude rate is recorded. Altitude rate must have sufficient resolution and sampling to permit the derivation of altitude to 5 feet.

[2] Percent of full range.

[3] For airplanes that can demonstrate the capability of deriving either the control input on control movement (one from the other) for all modes of operation and flight regimes, the "or" applies. For airplanes with non-mechanical control systems (fly-by-wire) the "and" applies. In airplanes with split surfaces, suitable combination of inputs is acceptable in lieu of recording each surface separately.

[4] This column applies to aircraft manufactured after October 11, 1991.

Docket No. 25530 (53 FR 26152) Eff. 7/11/88;
(Amdt. 135–26, Eff. 10/11/88)

Appendix E—Helicopter flight recorder specifications

Parameters	Range	Accuracy sensor input to DFDR readout	Sampling interval (per second)	Resolution[2] read out
Time (GMT).	24 hrs	±0.125% per hour	0.25 (1 per 4 seconds).	1 sec.
Altitude.	−1,000 ft to max certificated altitude of aircraft.	±100 to ±700 ft (See Table 1, TSO–C51a).	1	5' to 30'
Airspeed.	As the installed measuring system.	±3%	1	1 kt.
Heading.	360°	±2°	1	0.5°
Normal Acceleration (Vertical).	−3g to +6g	±1% of max range excluding datum error of ±5%.	8	0.01g
Pitch Attitude.	±75°	±2°	2	0.5°
Roll Attitude.	±180°	±2°	2	0.5°
Radio Transmitter Keying.	On-Off (Discrete)		1	0.25 sec.
Power in Each Engine: Free Power Turbine Speed *and* Engine Torque.	0–130% (power Turbine Speed) Full range (Torque).	±2%	1 speed 1 torque (per engine).	0.2%[1] to 0.4%[1]
Main Rotor Speed.	0–130%	±2%	2	0.3%[1]
Altitude Rate.	±6,000 ft/min	As installed	2	0.2%[1]
Pilot Input—Primary Controls (Collective, Longitudinal Cyclic, Lateral Cyclic, Pedal).	Full range	±3%	2	0.5%[1]
Flight Control Hydraulic Pressure Low.	Discrete, each circuit		1	
Flight Control Hydraulic Pressure Selector Switch Position, 1st and 2nd stage.	Discrete		1	
AFCS Mode and Engagement Status.	Discrete (5 bits necessary).		1	
Stability Augmentation System Engage.	Discrete		1	
SAS Fault Status.	Discrete		0.25	
Main Gearbox Temperature Low.	As installed	As installed	0.25	0.5%[1]
Main Gearbox Temperature High.	As installed	As installed	0.5	0.5%[1]
Controllable Stabilator Position.	Full Range	±3%	2	0.4%[1]
Longitudinal Acceleration.	±1g	±1.5% max range excluding datum error of ±5%.	4	0.01g
Lateral Acceleration.	±1g	±1.5% max range excluding datum of ±5%.	4	0.01g
Master Warning.	Discrete		1	
Nav 1 and 2 Frequency Selection.	Full Range	As installed	0.25	
Outside Air Temperature.	−50° C +90° C	±2° C	0.5	0.3 °C

[1] Percent of full range.
[2] This column applies to aircraft manufactured after October 11, 1991.

Docket No. 25530 (53 FR 26152) Eff. 7/11/88;
(Amdt. 135–26, Eff. 10/11/88)

Appendix F—Airplane Flight Recorder Specification

Parameters	Range	Accuracy (sensor input)	Seconds per sampling interval	Resolution	Remarks
1. Time or Relative Time Counts	24 Hrs, 0 to 4095	±0.125% per hour	4	1 sec	UTC time preferred when available. Counter increments each 4 seconds of system operation.
2. Pressure Altitude	−1000 ft to max certificated altitude of aircraft. +5000 ft.	±100 to ±700 ft (see table, TSO C124a or TSO C51a).	1	5' to 35"	Data should be obtained from the air data computer when practicable.
3. Indicated airspeed or Calibrated	50 KIAS or minimum value to Max V_{SO} and V_{SO} to 1.2 V_D.	±5% and ±3%	1	1 kt	Data should be obtained from the air data computer when practicable.
4. Heading (Primary crew reference.)	0–360° and Discrete "true" or "mag".	±2°	1	0.5°	When true or magnetic heading can be selected as the primary heading reference, a discrete indicating selection must be recorded.
5. Normal Acceleration (Vertical).	−3g to +6g	±1% of max range excluding datum error of ±5%.	0.125	0.004g.	
6. Pitch attitude	±75%	±2%	1 or 0.25 for airplanes operated under §135.152(j).	0.5°	A sampling rate of 0.25 is recommended.
7. Roll attitude	±180°	±2°	1 or 0.5 for airplanes operated under §135.152(j).	0.5°	A sampling rate of 0.5 is recommended.
8. Manual Radio Transmitter Keying or CVR/DFDR synchronization reference.	On-Off (Discrete) None		1		Preferably each crew member but one discrete acceptable for all transmission provided the CVR/FDR system complies with TSO C124a CVR synchronization requirements (paragraph 4.2.1 ED–55).
9. Thrust/Power on Each Engine— primary flight crew reference.	Full Range Forward	±2%	1 (per engine)	0.2% of full range.	Sufficient parameters (e.g., EPR, N1 or Torque, NP) as appropriate to the particular engine be recorded to determine power in forward and reverse thrust, including potential overspeed conditions.

Parameters	Range	Accuracy (sensor input)	Seconds per sampling interval	Resolution	Remarks
10. Autopilot Engagement	Discrete "on" or "off"		1		
11. Longitudinal Acceleration.	±1g	±1.5% max. range excluding datum error of ±5%.	0.25	0.004g.	
12a. Pitch Control(s) position (non-fly-by-wire systems.)	Full Range	±2° Unless Higher Accuracy Uniquely Required.	0.5 or 0.25 for air-planes operated under §135.152(j).	0.2% of full range.	For airplanes that have a flight control break away capability that allows either pilot to operate the controls independently, record both control inputs. The control inputs may be sampled alternately once per second to produce the sampling interval of 0.5 or 0.25, as applicable.
12b. Pitch Control(s) position (fly-by-wire systems).	Full Range	±2° Unless Higher Accuracy. Uniquely Required	0.5 or 0.25 for air-planes operated under §135.152(j).	0.2% of full range.	
13a. Lateral Control posi-tion(s) (non-fly-by-wire).	Full Range	±2° Unless Higher Accuracy Uniquely Required.	0.5 or 0.25 for air-planes under §135.152(j).	0.2% of full range.	For airplanes that have a flight control break away capability that allows either pilot to operate the controls independently, record both control inputs. The control inputs may be sampled alternately once per second to produce the sampling interval of 0.5 or 0.25, as applicable.
13b. Lateral Con-trol posi-tion(s) (fly-by-wire).	Full Range	±2° Unless Higher Accuracy Uniquely Required.	0.5 or 0.25 for air-planes operated under §135.152(j).	0.2% of full range.	
14a. Yaw Control position(s) (non-fly-by-wire).	Full Range	±2° Unless Higher Accuracy Uniquely Required.	0.5 or 0.25 for air-planes operated under §135.152(j)	0.2% of full range.	For airplanes that have a flight control break away capability that allows either pilot to operate the controls independently, record both control inputs. The control inputs may be sampled alternately once per second to produce the sampling interval of 0.5.

Parameters	Range	Accuracy (sensor input)	Seconds per sampling interval	Resolution	Remarks
14b. Yaw Control position(s) (fly-by-wire).	Full Range	±2° Unless Higher Accuracy Uniquely Required.	0.5	0.2% of full range.	
15. Pitch Control Surface(s) Position.	Full Range	±2° Unless Higher Accuracy Uniquely Required.	0.5 or 0.25 for airplanes operated under §135.152(j)	0.2% of full range.	For airplanes fitted with multiple or split surfaces, a suitable combination of inputs is acceptable in lieu of recording each surface separately. The control surfaces may be sampled alternately to produce the sampling interval of 0.5 or 0.25.
16. Lateral Control Surface(s) Position.	Full Range	±2% Unless Higher Accuracy Uniquely Required.	0.5 or 0.25 for airplanes operated under §135.152(j).	0.2% of full range.	A suitable combination of surface position sensors is acceptable in lieu of recording each surface separately. The control surfaces may be sampled alternately to produce the sampling interval of 0.5 or 0.25.
17. Yaw Control Surface(s) Position.	Full Range	±2° Unless Higher Accuracy Uniquely Required.	0.5	0.2% of full range.	For Airplanes with multiple or split surfaces, a suitable combination of surface position sensors is acceptable in lieu of recording each surface separately. The control surfaces may be sampled alternately to produce the sampling interval of 0.5.
18. Lateral Acceleration	±1g	±1.5% max. range excluding datum error of ±5%.	0.25	0.004g.	
19. Pitch Trim Surface Position.	Full Range	±3% Unless Higher Accuracy Uniquely Required.	1	0.3% of full range.	
20. Trailing Edge Flap or Cockpit Control Selection.	Full Range or Each Position (discrete).	±3° or as Pilot's indicator.	2	0.5% of full range.	Flap position and cockpit control may each be sampled alternately at 4 second intervals, to give a data point every 2 seconds.
21. Leading Edge Flap or Cockpit	Full Range or Each Discrete	±3° or as Pilot's indicator and suf-	2	0.5% of full range.	Left and right sides, or flap position and

Parameters	Range	Accuracy (sensor input)	Seconds per sampling interval	Resolution	Remarks
Control Selection.	Position.	ficient to determine each discrete position.			cockpit control may each be sampled at 4 second intervals, so as to give a data point every 2 seconds.
22. Each Thrust reverser Position (or equivalent for propeller airplane).	Stowed, in Transit, and reverse (Discrete).		1 (per engine)		Turbo-jet—2 discretes enable the 3 states to be determined. Turbo-prop—1 discrete.
23. Ground Spoiler Position or Speed Brake Selection.	Full Range or Each Position (discrete).	±2 Unless Higher Accuracy	1 0.5 for airplanes Uniquely Required.	0.2% of full range. operated under §135.152(j).	
24. Outside Air Temperature or Total Air Temperature.	–50°C to +90°C	±2° C	2		0.3°C.
25. Autopilot/ Autothrottle/ AFCS Mode and Engagement Status.	A suitable combination of discretes.		1		Discretes should show show which systems are engaged and which primary modes are controlling the flight path and speed of the aircraft.
26. Radio Altitude	–20 ft to 2,5000 ft	±2 ft or ±3% Whichever is Greater Below 500 ft and ±5% Above 500 ft.	1	1 ft +5% above 500 ft	For autoland/category 3 operations. Each radio altimeter should be recorded, but arranged so that at least one is recorded each second.
27. Localizer Deviation, MLS Azimuth, or GPS Lateral Deviation.	±400 Microamps or available sensor range as installed ±62°.	As installed ±3% recommended.	1	0.3% of full range.	For autoland/category 3 operations. Each system should be recorded but arranged so that at least one is recorded each second. It is not necessary to record ILS and MLS at the same time, only the approach aid in use need be recorded.
28. Glideslope Deviation, MLS Elevation, or GPS Vertical Deviation.	±400 Microamps or available sensor range as installed. 0.9 to +30°	As installed ±3% recommended.	1	0.3% of full range.	For autoland/category 3 operations. Each system should be recorded but arranged so that at least one is recorded each second. It is not necessary to record ILS and MLS at the same time, only the approach aid in use need be recorded.

Parameters	Range	Accuracy (sensor input)	Seconds per sampling interval	Resolution	Remarks
29. Marker Beacon Passage	Discrete "on" or "off"		1		A single discrete is acceptable for all markers.
30. Master Warning	Discrete		1		Record the master warning and record each "red" warning that cannot be determined from other parameters or from the cockpit voice recorder.
31. Air/ground sensor primary airplane system reference nose or main gear).	Discrete "air" or "ground."		1 (0.25 recommended.)		
32. Angle of Attack (If measured directly).	As installed	As installed	2 or 0.5 for airplanes operated under §135.152(j)	0.3% of full range.	If left and right sensors are available, each may be recorded at 4 or 1 second intervals, as appropriate, so as to give a data point at 2 seconds or 0.5 second, as required.
33. Hydraulic Pressure Low, Each System	Discrete or available sensor range, "low" or "normal."	±5%	2	0.5% of full range.	
34. Groundspeed	As installed	Most Accurate Systems Installed.	1	0.2% of full range.	
35. GPWS (ground proximity warning system).	Discrete "warning" or "off."		1		A suitable combination of discretes unless recorder capability is limited in which case a single discrete for all modes is acceptable.
36. Landing Gear Position or Landing gear cockpit control selection.	Discrete		4		A suitable combination of discretes should be recorded.
37. Drift Angle	As installed	As installed	4	0.1°	
38. Wind Speed and Direction.	As installed	As installed	4	1 knot, and 1.0°.	
39. Latitude and Longitude.	As installed	As installed	4	0.002°, or as installed.	Provided by the Primary Navigation System Reference. Where capacity permits latitude/longitude resolution should be 0.0002°.

Parameters	Range	Accuracy (sensor input)	Seconds per sampling interval	Resolution	Remarks
40. Stick shaker and pusher activation.	Discrete(s) "on" or "off."		1		A suitable combination of discretes to determine activation.
41. Windshear Detection.	Discrete "warning" or "off."		1		
42. Throttle/ power lever position.	Full range	±2%	1 for each lever	2% of full range.	For airplanes with non-mechanically linked cockpit engine controls.
43. Additional Engine Parameters.	As installed	As installed	Each engine each second.	2% of full range.	Where capacity permits, the preferred priority is indicated vibration level, N2, EGT, Fuel Flow, Fuel Cut-off lever position and N3, unless engine manufacturer recommends otherwise.
44. Traffic Alert and Collision System (TCAS).	Discretes	As installed	1		A suitable combination of discretes should be recorded to determine the status of—Combined Control, Vertical Control, Up, Advisory, and Attachment 6E, TCAS, VERTICAL RA DATA OUTPUT WORD.)
45. DME 1 and 2 Distance	0–200 NM;	As installed	4	1NM	1 mile.
46. Nav 1 and 2 Selected Frequency.	Full range	As installed	4		Sufficient to determine selected frequency.
47. Selected barometric setting.	Full Range	±5%	(1 per 64 sec.)	0.2% of full range.	
48. Selected altitude.	Full Range	±5%	1	100 ft.	
49. Selected speed	Full Range	±5%	1	1 knot.	
50. Selected Mach	Full Range	±5%	1	.01	
51. Selected vertical speed	Full Range	±5%	1	100 ft./min.	
52. Selected heading	Full Range	±5%	1	100 ft./min.	
53. Selected flight path	Full Range	±5%	1	1°.	
54. Selected decision height.	Full Range	±5%	64	1 ft.	
55. EFIS display format	Discrete(s)		4		Discretes should show the display system

Parameters	Range	Accuracy (sensor input)	Seconds per sampling interval	Resolution	Remarks
					status (e.g., off, normal, fail, composite, sector, plan, nav aids, weather radar, range, copy).
56. Multi-function/ Engine Alerts Display format.	Discrete(s)		4		Discretes should show the display system status (e.g., off, normal, fail, and the identity of display pages for emergency procedures, need not be recorded.
57. Thrust command	Full Range	±2%	2	2% of full range.	
58. Thrust target	Full Range	±2%	4	2% of full range.	
59. Fuel quantity in CG trim tank.	Full Range	±5%	(1 per 64 sec.)	1% of full range.	
60. Primary Navigation System Reference.	Discrete GPS, INS, VOR/DME, MLS, Loran C, Omega, Localizer Glidescope.		4		A suitable combination of discretes to determine the Primary Navigation System reference.
61. Ice Detection	Discrete "ice" or "no ice".		4		
62. Engine warning each engine vibration.	Discrete		1		
63. Engine warning each engine over temp.	Discrete		1		
64. Engine warning each engine oil pressure low.	Discrete		1		
65. Engine warning each engine over speed.	Discrete		1		
66. Yaw Trim Surface Position.	Full Range	±3% Unless Higher Accuracy Uniquely Required.	2	0.3% of full range.	
67. Roll Trim Surface Position.	Full Range	±3% Unless Higher Accuracy Uniquely Required.	2	0.3% of full range.	
68. Brake Pressure (left and right).	As installed	±5%	1		To determine braking effort applied by pilots or by autobrakes.
69. Brake Pedal Application (left and right).	Discrete or Analog "applied" or "off."	±5 (Analog)	1		To determine braking effort applied by pilots.

Parameters	Range	Accuracy (sensor input)	Seconds per sampling interval	Resolution	Remarks
70. Yaw or side-slip angle.	Full Range	±5%	1	0.5°.	
71. Engine bleed valve position.	Discrete "open" or "closed."	±5%	1	0.5°.	
72. De-icing or anti-icing system selection.	Discrete "on" or "off."		4		
73. Computed center of gravity.	Full Range	±5%	(1 per 64 sec.)	1% of full range.	
74. AC electrical bus status.	Discrete "power" or "off"		4		Each bus.
75. DC electrical bus status.	Discrete "power" or "off"		4		Each bus.
76. APU bleed valve position.	Discrete "open" or "closed"		4		
77. Hydraulic Pressure (each system).	Full range	±5%	2	100 psi.	
78. Loss of cabin pressure.	Discrete "loss" or "normal."		1		
79. Computer failure (critical flight and engine control systems.	Discrete "fail" or "normal."		4		
80. Heads-up display (when an information source is installed.	Discrete(s) "on" or "off"		4		
81. Para-visual display (when an information source is installed).	Discrete(s) "on" or "off"		1		
82. Cockpit trim control input position—pitch.	Full Range	±5%	1	0.2% of full range	Where mechanical means for control inputs are not available, cockpit display trim positions should be recorded.
83. Cockpit trim control input position—roll.	Full Range	±5%	1	0.7% of full range	Where mechanical means for control inputs are not available, cockpit display trim positions should be recorded.
84. Cockpit trim control input position—yaw.	Full Range	±5%	1	0.3% of full range	Where mechanical means for control inputs are not available, cockpit display trim positions should be recorded.

Parameters	Range	Accuracy (sensor input)	Seconds per sampling interval	Resolution	Remarks
85. Trailing edge flap and cockpit flap control position.	Full Range	±5%	2	0.5% of full range.	Trailing edge flaps and cockpit flap control position may each be sampled alternately at 4 second intervals to provide a sample each 0.5 second.
86. Leading edge flap and cockpit flap control position.	Full Range or Discrete	±5%	1	0.5% of full range.	
87. Ground spoiler position and speed brake selection.	Full Range or discrete	±5%	0.5	0.2% of full range.	
88. All cockpit flight control input (control wheel, control column, rudder pedal).	Full Range Control wheel ±70 lbs. Control Column ±85 lb Rudder pedal ±165 lbs	±5%	1	0.2% of full range.	For fly-by-wire flight control systems, where flight control surface position is a function of the displacement of the control input device only, it is not necessary to record this parameter. For airplanes that have a flight control break away capability that allows either pilot to operate the control independently, record both control force inputs. The control force inputs may be sampled alternately once per 2 seconds to produce the sampling interval of 1.

Docket No. 25530 (53 FR 26152) Eff. 7/11/88;
(Amdt. 135–26, Eff. 10/11/88)

Special Federal Aviation Regulation 36. Operations review program amendment no. 2A; development of major repair data

1. *Definitions.* For purposes of this Special Federal Aviation Regulation—

(a) A product is an aircraft, airframe, aircraft engine, propeller, or appliance;

(b) An article is an airframe, powerplant, propeller, instrument, radio, or accessory; and

(c) A component is a part of a product or article.

2. *General.*

(a) Contrary to provisions of § 121.379(b) and § 135.437(b) of this chapter notwithstanding, the holder of an air carrier certificate or operating certificate, that operates large aircraft, and that has been issued operations specifications for operations required to be conducted in accordance with 14 CFR part 121 or 135, may perform a major repair on a product as described in § 121.379(b) or § 135.437(a), using technical data that have not been approved by the Administrator, and approve that product for return to service, if authorized in accordance with this Special Federal Aviation Regulation.

(b) Reserved

(c) Contrary to provisions of § 145.51 of the Federal Aviation Regulations notwithstanding, the holder of a domestic repair station certificate under 14 CFR Part 145 may perform a major repair on an article for which it is rated, using technical data not approved by the Administrator, and approve that article for return to service, if authorized in accordance with this Special Federal Aviation Regulation. If the certificate holder holds a rating limited to a component of a product or article, the holder may not, by virtue of this Special Federal Aviation Regulation, approve that product or article for return to service.

3. *Major repair data and return to service.*

(a) As referenced in section 2 of this Special Federal Aviation Regulation, a certificate holder may perform a major repair on a product or article using technical data that have not been approved by the Administrator, and approve that product or article for return to service, if the certificate holder—

(1) Has been issued an authorization under, and a procedures manual that complies with, Special Federal Aviation Regulation No. 36—8, effective on January 23, 2004;

(2) Has developed the technical data in accordance with the procedures manual;

(3) Has developed the technical data specifically for the product or article being repaired; and

(4) Has accomplished the repair in accordance with the procedures manual and the procedures approved by the Administrator for the certificate.

(b) For purposes of this section, an authorization holder may develop technical data to perform a major repair on a product or article and use that data to repair a subsequent product or article of the same type as long as the holder—

(1) Evaluates each subsequent repair and the technical data to determine that performing the subsequent repair with the same data will return the product or article to its original or properly altered condition, and that the repaired product or article conforms with applicable airworthiness requirements; and

(2) Records each evaluation in the records referenced in paragraph (a) of section 13 of this Special Federal Aviation Regulation.]

4. *Application.* The applicant for an authorization under this Special Federal Aviation Regulation must submit an application, in writing and signed by an officer of the applicant, to the FAA Flight Standards District Office charged with the overall inspection of the applicant's operations under its certificate. The application must contain—

(a) If the applicant is

(1) The holder of an air carrier operating or commercial operating certificate, or the holder of an air taxi operating certificate that operates large aircraft, the—

(i) The applicant's certificate number; and

(ii) The specific product(s) the applicant is authorized to maintain under its certificate, operations specifications, and maintenance manual; or

(2) The holder of a domestic repair station certificate—

(i) The applicant's certificate number;

(ii) A copy of the applicant's operations specifications; and

(iii) The specific article(s) for which the applicant is rated;

(b) The name, signature, and title of each person for whom authorization to approve, on behalf of the authorization holder, the use of technical data for major repairs is requested; and

(c) The qualifications of the applicant's staff that show compliance with section 5 of this Special Federal Aviation Regulation.

5. *Eligibility.*

(a) To be eligible for an authorization under this Special Federal Aviation Regulation, the applicant, in addition to having the authority to repair products or articles must—

(1) Hold an air carrier certificate or operating certificate, operate large aircraft, and have been issued operations specifications for operations required to be conducted in accordance with 14 CFR Part 121 or 127, or § 135.2, or hold a domestic repair station certificate under 14 CFR Part 145;

(2) Have an adequate number of sufficiently trained personnel in the United States to develop data and repair the products that the applicant is authorized to maintain under its operating certificate or the articles for which it is rated under its domestic repair station certificate;

(3) Employ, or have available, a staff of engineering personnel that can determine compliance with the applicable airworthiness requirements of the Federal Aviation Regulations.

(b) At least one member of the staff required by paragraph (a)(3) of this section must—

(1) Have a thorough working knowledge of the applicable requirements of the Federal Aviation Regulations;

(2) Occupy a position on the applicant's staff that has the authority to establish a repair program that ensures that each repaired product or article meets the applicable requirements of the Federal Aviation Regulations;

(3) Have at least one year of satisfactory experience in processing engineering work, in direct contact with the FAA, for type certification or major repair projects; and

(4) Have at least eight years of aeronautical engineering experience (which may include the one year of experience in processing engineering work for type certification or major repair projects).

(c) The holder of an authorization issued under this Special Federal Aviation Regulation shall notify the Administrator within 48 hours of any change (including a change of personnel) that could affect the ability of the holder to meet the requirements of this Special Federal Aviation Regulation.]

[6. *Procedures Manual.*

(a) A certificate holder may not approve a product or article for return to service under section 2 of this Special Federal Aviation Regulation unless the holder—

(1) Has a procedures manual that has been approved by the Administrator as complying with paragraph (b) of this section; and

(2) Complies with the procedures contained in the procedures manual.

(b) The approved procedures manual must contain—

(1) The procedures for developing and determining the adequacy of technical data for major repairs;

(2) The identification (names, signatures, and responsibilities) of officials and of each staff member described in section 5 of this Special Federal Aviation Regulation who—

(i) Has the authority to make changes in procedures that require a revision to the procedures manual; and

(ii) Prepares or determines the adequacy of technical data, plans or conducts tests, and approves, on behalf of the authorization holder, test results; and (3) A "log of revisions" page that identifies each revised item, page, and date of revision, and contains the signature of the person approving the change for the Administrator.

(c) The holder of an authorization issued under this Special Federal Aviation Regulation may not approve a product or article for return to service after a change in staff necessary to meet the requirements of section 5 of this regulation or a change in procedures from those approved under paragraph (a) of this section, unless that change has been approved by the FAA and entered in the procedures manual.

7. *Duration of Authorization.* Each authorization issued under this Special Federal Aviation Regulation is effective from the date of issuance until January 23, 2009, unless it is earlier surrendered, suspended, revoked, or otherwise terminated. Upon termination of such authorization, the terminated authorization holder must:

(a) Surrender to the FAA all data developed pursuant to Special Federal Aviation Regulation No. 36; or

(b) Maintain indefinitely all data developed pursuant to Special Federal Aviation Regulation No. 36, and make that data available to the FAA for inspection upon request.

8. *Transferability.* An authorization issued under this special Federal Aviation Regulation is not transferable.

9. *Inspections.* Each holder of an authorization issued under this Special Federal Aviation Regulation and each applicant for an authorization must allow the Administrator to inspect its personnel, facilities, products and articles, and records upon request.

10. *Limits of Applicability.* An authorization issued under this Special Federal Aviation Regulation applies only to—

(a) A product that the air carrier, commercial, or air taxi operating certificate holder is authorized to maintain pursuant to its continuous airworthiness maintenance program or maintenance manual; or

(b) An article for which the domestic repair station certificate holder is rated. If the certificate holder is rated for a component of an article, the holder may not, in accordance with this Special Federal Aviation Regulation, approve that article for return to service.

11. *Additional Authorization Limitations.* Each holder of an authorization issued under this Special Federal Aviation Regulation must comply with any additional limitations prescribed by the Administrator and made a part of the authorization.]

[12. *Data Review and Service Experience.* If the Administrator finds that a product or article has been approved for return to service after a major repair has been performed under this Special Federal Aviation Regulation, that the product or article may not conform to the applicable airworthiness requirements or that an unsafe feature or characteristic of the product or article may exist, and that the nonconformance or unsafe feature or characteristic may be attributed to the repair performed, the holder of the authorization, upon notification by the Administrator, shall—

(a) Investigate the matter;

(b) Report to the Administrator the results of the investigation and any action proposed or taken; and

(c) If notified that an unsafe condition exists, provide within the time period stated by the Administrator, the information necessary for the FAA to issue an airworthiness directive under Part 39 of the Federal Aviation Regulations.

13. *Current Records.* Each holder of an authorization issued under this Special Federal Aviation Regulation shall maintain, at its facility, current records containing—

(a) For each product or article for which it has developed and used major repair data, a technical data file that includes all data and amendments thereto (including drawings, photographs, specifications, instructions, and reports) necessary to accomplish the major repair;

(b) A list of products or articles by make, model, manufacturer's serial number (including specific part numbers and serial numbers of components) and, if applicable, FAA Technical Standard Order (TSO) or Parts Manufacturer Approval (PMA) identification, that have been repaired under the authorization; and

(c) A file of information from all available sources on difficulties experienced with products and articles repaired under the authorization.

This Special Federal Aviation Regulation terminates January 23, 2009.

[(SFAR 36-6, Eff. 1/23/94)]

Special Federal Aviation Regulation 58. Advanced qualification program

1. *Purpose and Eligibility.*

(a) This Special Federal Aviation Regulation provides for approval of an alternate method (known as "Advanced Qualification Program" or "AQP") for qualifying, training, certifying, and otherwise ensuring competency of crewmembers, aircraft dispatchers, other operations personnel, instructors, and evaluators who are required to be trained or qualified under Parts 121 and 135 of the FAR or under this SFAR.

(b) A certificate holder is eligible under this Special Federal Aviation Regulation if the certificate holder is required to have an approved training program under § 121.401 or § 135.341 of the FAR, or elects to have an approved training program under § 135.341.

(c) A certificate holder obtains approval of each proposed curriculum under this AQP as specified in section 10 of this SFAR.

(d) A curriculum approved under the AQP may include elements of present Part 121 and Part 135 training programs. Each curriculum must specify the make, model, and series aircraft (or variant) and each crewmember position or other positions to be covered by that curriculum. Positions to be covered by the AQP must include all flight crewmember positions, instructors, and evaluators and may include other positions, such as flight attendants, aircraft dispatchers, and other operations personnel.

(e) Each certificate holder that obtains approval of an AQP under this SFAR shall comply with all of the requirements of that program.

2. *Definitions.*

As used in this SFAR:

"Curriculum" means a portion of an Advanced Qualification Program that covers one of three program areas: (1) indoctrination, (2) qualification, or (3) continuing qualification. A qualification or continuing qualification curriculum addresses the required training and qualification activities for a specific make, model, and series aircraft (or variant) and for a specific duty position.

"Evaluator" means a person who has satisfactorily completed training and evaluation that qualifies that person to evaluate the performance of crewmembers, instructors, other evaluators, aircraft dispatchers, and other operations personnel.

"Facility" means the physical environment required for training and qualification (e.g., buildings, classrooms).

"Training center" means an independent organization that provides training under contract or other arrangement to certificate holders. A training center may be a certificate holder that provides training to another certificate holder, an aircraft manufacturer that provides training to certificate holders, or any non-certificate holder that provides training to a certificate holder.

"Variant" means a specifically configured aircraft for which the FAA has identified training and qualification requirements that are significantly different from those applicable to other aircraft of the same make, model, and series.

3. *Required Curriculums.*

Each AQP must have separate curriculums for indoctrination, qualification, and continuing qualification as specified in §§ 4, 5, and 6 of this SFAR.

4. *Indoctrination Curriculums.*

Each indoctrination curriculum must include the following:

(a) For newly hired persons being trained under an AQP: Company policies and operating practices and general operational knowledge.

(b) For newly hired flight crewmembers and aircraft dispatchers: General aeronautical knowledge.

(c) For instructors: The fundamental principles of the teaching and learning process; methods and theories of instruction; and the knowledge necessary to use aircraft, flight training devices, flight simulators, and other training equipment in advanced qualification curriculums.

(d) For evaluators: Evaluation requirements specified in each approved curriculum; methods of evaluating crewmembers and aircraft dispatchers and other operations personnel; and policies and practices used to conduct the kinds of evaluations particular to an advanced qualification curriculum (e.g., proficiency and on-line).

5. *Qualification Curriculums.*

Each qualification curriculum must include the following:

(a) The certificate holder's planned hours of training, evaluation, and supervised operating experience.

(b) A list of and text describing the training, qualification, and certification activities, as applicable for specific positions subject to the AQP, as follows:

(1) *Crewmembers, aircraft dispatchers, and other operations personnel.* Training, evaluation, and certification activities which are aircraft- and equipment-specific to qualify a person for a particular duty position on, or duties

related to the operation of a specific make, model, and series aircraft (or variant); a list of and text describing the knowledge requirements, subject materials, job skills, and each maneuver and procedure to be trained and evaluated; the practical test requirements in addition to or in place of the requirements of Parts 61, 63, and 65; and a list of and text describing supervised operating experience.

(2) *Instructors.* Training and evaluation to qualify a person to impart instruction on how to operate, or on how to ensure the safe operation of a particular make, model, and series aircraft (or variant).

(3) *Evaluators.* Training, evaluation, and certification activities that are aircraft and equipment specific to qualify a person to evaluate the performance of persons who operate or who ensure the safe operation of, a particular make, model, and series aircraft (or variant).

6. *Continuing Qualification Curriculums.*

Continuing qualification curriculums must comply with the following requirements:

(a) *General.* A continuing qualification curriculum must be based on—

(1) A continuing qualification cycle that ensures that during each cycle each person qualified under an AQP, including instructors and evaluators, will receive a balanced mix of training and evaluation on all events and subjects necessary to ensure that each person maintains the minimum proficiency level of knowledge, skills, and attitudes required for original qualification; and

(2) If applicable, flight crewmember or aircraft dispatcher recency of experience requirements.

(b) *Continuing Qualification Cycle Content.* Each continuing qualification cycle must include at least the following:

(1) *Evaluation period.* An evaluation period during which each person qualified under an AQP must receive at least one training session and a proficiency evaluation at a training facility. The number and frequency of training sessions must be approved by the Administrator. A training session, including any proficiency evaluation completed at that session, that occurs any time during the two calendar months before the last date for completion of an evaluation period can be considered by the certificate holder to be completed in the last calendar month.

(2) *Training.* Continuing qualification must include training in all events and major subjects required for original qualification, as follows:

(i) For pilots in command, seconds in command, flight engineers, and instructors and evaluators: Ground training including a general review of knowledge and skills covered in qualification training, updated information on newly developed procedures, and safety information.

(ii) For crewmembers, aircraft dispatchers, instructors, evaluators, and other operation personnel who conduct their duties in flight: Proficiency training in an aircraft, flight training device, or flight simulator on nor-

mal, abnormal, and emergency flight procedures and maneuvers.

(iii) For instructors and evaluators who are limited to conducting their duties in flight simulators and flight training devices: Proficiency training in a flight training device and/or flight simulator regarding operation of this training equipment and in operational flight procedures and maneuvers (normal, abnormal, and emergency).

(3) *Evaluations.* Continuing qualification must include evaluation in all events and major subjects required for original qualification, and online evaluations for pilots in command and other eligible flight crewmembers. Each person qualified under an AQP must successfully complete a proficiency evaluation and, if applicable, an online evaluation during each evaluation period. An individual's proficiency evaluation may be accomplished over several training sessions if a certificate holder provides more than one training session in an evaluation period. The following evaluation requirements apply:

(i) Proficiency evaluations as follows:

(A) For pilots in command, seconds in command, and flight engineers: A proficiency evaluation, portions of which may be conducted in an aircraft, flight simulator, or flight training device as approved in the certificate holder's curriculum which must be completed during each evaluation period.

(B) For any other persons covered by an AQP a means to evaluate their proficiency in the performance of their duties in their assigned tasks in an operational setting.

(ii) On-line evaluations as follows:

(A) For pilots in command: An on line evaluation conducted in an aircraft during actual flight operations under Part 121 or Part 135 or during operationally (line) oriented flights, such as ferry flights or proving flights. An on-line evaluation in an aircraft must be completed in the calendar month that includes the midpoint of the evaluation period. An on-line evaluation that is satisfactorily completed in the calendar month before or the calendar month after the calendar month in which it becomes due is considered to have been completed during the calendar month it became due. However, in no case is an on-line evaluation under this paragraph required more often than once during an evaluation period.

(B) During the on-line evaluations required under paragraph (5)(3)(ii)(A) of this section, each person performing duties as a pilot in command, second in command, or flight engineer for that flight, must be individually evaluated to determine whether he or she—(1) Remains adequately trained and currently proficient with respect to the particular aircraft, crew position, and type of operation in which he or she serves; and (2) Has sufficient knowledge and skills to operate effectively as part of a crew.

(4) *Recency of experience.* For pilots in command and seconds in command, and, if the certificate holder elects,

flight engineers and aircraft dispatchers, approved recency of experience requirements.

(c) *Duration periods.* Initially the continuing qualification cycle approved for an AQP may not exceed 26 calendar months and the evaluation period may not exceed 13 calendar months. Thereafter, upon demonstration by a certificate holder that an extension is warranted, the Administrator may approve extensions of the continuing qualification cycle and the evaluation period in increments not exceeding 3 calendar months. However, a continuing qualification cycle may not exceed 39 calendar months and an evaluation period may not exceed 26 calendar months.

(d) *Requalification.* Each continuing qualification curriculum must include a curriculum segment that covers the requirements for requalifying a crewmember, aircraft dispatcher, or other operations personnel who has not maintained continuing qualification.

7. *Other Requirements.* In addition to the requirements of sections 4, 5, and 6, each AQP qualification and continuing qualification curriculum must include the following requirements:

(a) Approved Cockpit Resource Management (CRM) Training applicable to each position for which training is provided under an AQP.

(b) Approved training on and evaluation of skills and proficiency of each person being trained under an AQP to use their cockpit resource management skills and their technical (piloting or other) skills in an actual or simulated operations scenario. For flight crewmembers this training and evaluation must be conducted in an approved flight training device or flight simulator.

(c) Data collection procedures that will ensure that the certificate holder provides information from its crewmembers, instructors, and evaluators that will enable the FAA to determine whether the training and evaluations are working to accomplish the overall objectives of the curriculum.

8. *Certification.* A person enrolled in an AQP is eligible to receive a commercial or airline transport pilot, flight engineer, or aircraft dispatcher certificate or appropriate rating based on the successful completion of training and evaluation events accomplished under that program if the following requirements are met:

(a) Training and evaluation of required knowledge and skills under the AQP must meet minimum certification and rating criteria established by the Administrator in Parts 61, 63, or 65. The Administrator may accept substitutes for the practical test requirements of Parts 61, 63, or 65, as applicable.

(b) The applicant satisfactorily completes the appropriate qualification curriculum.

(c) The applicant shows competence in required technical knowledge and skills (e.g., piloting) and cockpit resource management knowledge and skills in scenarios that test both types of knowledge and skills together.

(d) The applicant is otherwise eligible under the applicable requirements of Part 61, 63, or 65.

9. *Training Devices and Simulators.*

(a) *Qualification and approval of flight training devices and flight simulators.*

(1) Any training device or simulator that will be used in an AQP for one of the following purposes must be evaluated by the Administrator for assignment of a flight training device or flight simulator qualification level:

(i) Required evaluation of individual or crew proficiency.

(ii) Training activities that determine if an individual or crew is ready for a proficiency evaluation.

(iii) Activities used to meet recency of experience requirements.

(iv) Line Operational Simulations (LOS).

(2) To be eligible to request evaluation for a qualification level of a flight training device or flight simulator an applicant must—

(i) Hold an operating certificate; or

(ii) Be a training center that has applied for authorization to the Administrator or has been authorized by the Administrator to conduct training or qualification under an AQP.

(3) Each flight training device or flight simulator to be used by a certificate holder or training center for any of the purposes set forth in paragraph (a)(1) of this section must—

(i) Be, or have been, evaluated against a set of criteria established by the Administrator for a particular qualification level of simulation;

(ii) Be approved for its intended use in a specified AQP; and

(iii) Be part of a flight simulator or flight training device continuing qualification program approved by the Administrator.

(b) *Approval of other Training Equipment.*

(1) Any training device that is intended to be used in an AQP for purposes other than those set forth in paragraph (a)(1) of this section must be approved by the Administrator for its intended use.

(2) An applicant for approval of a training device under this paragraph must identify the device by its nomenclature and describe its intended use.

(3) Each training device approved for use in an AQP must be part of a continuing program to provide for its serviceability and fitness to perform its intended function as approved by the Administrator.

10. *Approval Of Advanced Qualification Program.*

(a) *Approval Process.* Each applicant for approval of an AQP curriculum under this SFAR shall apply for approval of that curriculum. Application for approval is made to the certificate holder's FAA Flight Standards District Office.

(b) *Approval Criteria.* An application for approval of an AQP curriculum will be approved if the program meets the following requirements:

(1) It must be submitted in a form and manner acceptable to the Administrator.

(2) It must meet all of the requirements of this SFAR.

(3) It must indicate specifically the requirements of Parts 61, 63, 65, 121 or 135, as applicable, that would be replaced by an AQP curriculum. If a requirement of Parts 61, 63, 65, 121, or 135 is replaced by an AQP curriculum, the certificate holder must show how the AQP curriculum provides an equivalent level of safety for each requirement that is replaced. Each applicable requirement of Parts 61, 63, 65, 121 or 135 that is not specifically addressed in an AQP curriculum continues to apply to the certificate holder.

(c) *Application and Transition.* Each certificate holder that applies for one or more advanced qualification curriculums or for a revision to a previously approved curriculum must comply with § 121.405 or § 135.325, as applicable, and must include as part of its application a proposed transition plan (containing a calendar of events) for moving from its present approved training to the advanced qualification training.

(d) *Advanced Qualification Program Revisions or Rescissions of Approval.* If after a certificate holder begins operations under an AQP, the Administrator finds that the certificate holder is not meeting the provisions of its approved AQP, the Administrator may require the certificate holder to make revisions in accordance with § 121.405 or § 135.325, as applicable, or to submit and obtain approval for a Plan (containing a schedule of events) that the certificate holder must comply with and use to transition to an approved Part 121 or Part 135 training program, as appropriate.

11. *Approval of Training, Qualification, or Evaluation by a Person who Provides Training by Arrangement.*

(a) A certificate holder under Part 121 or Part 135 may arrange to have AQP required training, qualification, or evaluation functions performed by another person (a "training center") if the following requirements are met:

(1) The training center's training and qualification curriculums, curriculum segments, or portions of curriculum segments must be provisionally approved by the Administrator. A training center may apply for provisional approval independently or in conjunction with, a certificate application for AQP approval. Application for provisional approval must be made to the FAA's Flight Standards District Office that has responsibility for the training center.

(2) The specific use of provisionally approved curriculums, curriculum segments, or portions of curriculum segments in a certificate holder's AQP must be approved by the Administrator as set forth in Section 10 of this SFAR.

(b) An applicant for provisional approval of a curriculum, curriculum segment, or portion of a curriculum segment under this paragraph must show that the following requirements are met:

(1) The applicant must have a curriculum for the qualification and continuing qualification of each instructor or evaluator employed by the applicant.

(2) The applicant's facilities must be found by the Administrator to be adequate for any planned training, qualification, or evaluation for a Part 121 or Part 135 certificate holder.

(3) Except for indoctrination curriculums, the curriculum, curriculum segment, or portion of a curriculum segment must identify the specific make, model, and series aircraft (or variant) and crewmember or other positions for which it is designed.

(c) A certificate holder who wants approval to use a training center's provisionally approved curriculum, curriculum segment, or portion of a curriculum segment in its AQP, must show that the following requirements are met:

(1) Each instructor or evaluator used by the training center must meet all of the qualification and continuing qualification requirements that apply to employees of the certificate holder that has arranged for the training, including knowledge of the certificate holder's operations.

(2) Each provisionally approved curriculum, curriculum segment, or portion of a curriculum segment must be approved by the Administrator for use in the certificate holder's AQP. The Administrator will either provide approval or require modifications to ensure that each curriculum, curriculum segment, or portion of a curriculum segment is applicable to the certificate holder's AQP.

12. *Recordkeeping Requirements.*

Each certificate holder and each training center holding AQP provisional approval shall show that it will establish and maintain records in sufficient detail to establish the training, qualification, and certification of each person qualified under an AQP in accordance with the training, qualification, and certification requirements of this SFAR.

13. *Expiration.* This Special Federal Aviation Regulation terminates on October 2, 1995, unless sooner terminated.

SFAR No. 71—Special Operating Rules for Air Tour Operators in the State of Hawaii

Section 1. Applicability. This Special Federal Aviation Regulation prescribes operating rules for airplane and helicopter visual flight rules air tour flights conducted in the State of Hawaii under parts 91 and 135 of the Federal Aviation Regulations. This rule does not apply to flights conducted in gliders or hot air balloons.

Section 2. Definitions. For the purposes of this SFAR:

"Air tour" means any sightseeing flight conducted under visual flight rules in an airplane or helicopter for compensation or hire.

"Air tour operator" means any person who conducts an air tour.

Section 3. Helicopter flotation equipment. No person may conduct an air tour in Hawaii in a single-engine helicopter beyond the shore of any island, regardless of whether the helicopter is within gliding distance of the shore, unless:

(a) The helicopter is amphibious or is equipped with floats adequate to accomplish a safe emergency ditching and approved flotation gear is easily accessible for each occupant; or

(b) Each person on board the helicopter is wearing approved flotation gear.

Section 4: Helicopter performance plan. Each operator must complete a performance plan before each helicopter air tour flight. The performance plan must be based on the information in the Rotorcraft Flight Manual (RFM), considering the maximum density altitude for which the operation is planned for the flight to determine the following:

(a) Maximum gross weight and center of gravity (CG) limitations for hovering in ground effect;

(b) Maximum gross weight and CG limitations for hovering out of ground effect; and,

(c) Maximum combination of weight, altitude, and temperature for which height-velocity information in the RFM is valid.

The pilot in command (PIC) must comply with the performance plan.

Section 5. Helicopter operating limitations. Except for approach to and transition from a hover, the PIC shall operate the helicopter at a combination of height and forward speed (including hover) that would permit a safe landing in event of engine power loss, in accordance with the height-speed envelope for that helicopter under current weight and aircraft altitude.

Section 6. Minimum flight altitudes. Except when necessary for takeoff and landing, or operating in compliance with an air traffic control clearance, or as otherwise authorized by the Administrator, no person may conduct an air tour in Hawaii:

(a) Below an altitude of 1,500 feet above the surface over all areas of the State of Hawaii, and,

(b) Closer than 1,500 feet to any person or property; or,

(c) Below any altitude prescribed by federal statute or regulation.

Section 7. Passenger briefing. Before takeoff each PIC of an air tour flight of Hawaii with a flight segment beyond the ocean shore of any island shall ensure that each passenger has been briefed on the following, in addition to requirements set forth in §91.107 or 135.117:

(a) Water ditching procedures;

(b) Use of required flotation equipment; and

(c) Emergency egress from the aircraft in event of a water landing.

Section 8. Termination date. This Special Federal Aviation Regulation expires on October 26, 1997.

SFAR No. 81—Passenger-Carrying Single-Engine IFR Operations.

1. Purpose and Eligibility.

(a) This Special Federal Aviation Regulation provides for the approval of single-engine passenger-carrying operations under instrument flight rules (IFR) during the month prior to the effective date of the Commercial Passenger-Carrying Operations in Single-Engine Aircraft Under Instrument Flight Rules rule.

(b) This SFAR terminates on May 3, 1998.

(c) Only those single-engine, passenger-carrying operations meeting all the applicable requirements of part 135 and those requirements set forth in paragraph 2 of this SFAR may operate under IFR.

2. Contrary provisions of §§135.103 and 135.181 notwithstanding, a person may conduct passenger-carrying operations under IFR in single-engine aircraft if the following conditions are met:

(a) The aircraft has two independent electrical power generating sources each of which is able to supply all probable combinations of continuous inflight electrical loads for required instruments and equipment; or in addition to the primary electrical power generating source, a standby battery or an alternate source of electric power that is capable of supplying 150% of the electrical loads of all required instruments and equipment necessary for safe emergency operation of the aircraft for at least one hour;

(b) The aircraft has two independent sources of energy (with means of selecting either), of which at least one is an engine-driven pump or generator, each of which is able to drive all gyroscopic instruments and installed so that failure of one instrument or source does not interfere with the energy supply to the remaining instruments or the other energy source;

(c) The aircraft meets the autopilot requirements of §135.105 or has a second in command;

(d) The certificate holder's maintenance inspection program incorporates either the manufacturer's recommended engine trend monitoring program, which includes an oil analysis, if appropriate, or an FAA approved engine trend monitoring program that includes an oil analysis at each 100 hour interval or at the manufacturer's suggested interval, whichever is more frequent.

(e) The results of each test, observation, and inspection required by the applicable engine trend monitoring program are recorded and maintained in the engine maintenance records; and

(f) Written maintenance instructions containing the methods, techniques, and practices necessary to maintain the equipment specified in paragraph 2(a), (b), and (c) are prepared.

Part 137—Agricultural aircraft operations

Authority: 49 U.S.C. 106(g), 40103, 40113, 44701-44702.

Source: Docket No. 1464, 30 FR 8106, June 24, 1965, unless otherwise noted.

Subpart A—General

§ 137.1 Applicability.

(a) This part prescribes rules governing—

(1) Agricultural aircraft operations within the United States; and

(2) The issue of commercial and private agricultural aircraft operator certificates for those operations.

(b) In a public emergency, a person conducting agricultural aircraft operations under this part may, to the extent necessary, deviate from the operating rules of this part for relief and welfare activities approved by an agency of the United States or of a State or local government.

(c) Each person who, under the authority of this section, deviates from a rule of this part shall, within 10 days after the deviation send to the nearest FAA Flight Standards District Office a complete report of the aircraft operation involved, including a description of the operation and the reasons for it.

[Doc. No. 1464, 30 FR 8106, June 24, 1965, as amended by Amdt. 137-13, 54 FR 39294, Sept. 25, 1989]

§ 137.3 Definition of terms.

For the purposes of this part—

"Agricultural aircraft operation" means the operation of an aircraft for the purpose of (1) dispensing any economic poison, (2) dispensing any other substance intended for plant nourishment, soil treatment, propagation of plant life, or pest control, or (3) engaging in dispensing activities directly affecting agriculture, horticulture, or forest preservation, but not including the dispensing of live insects.

"Economic poison" means (1) any substance or mixture of substances intended for preventing, destroying, repelling, or mitigating any insects, rodents, nematodes, fungi, weeds, and other forms of plant or animal life or viruses, except viruses on or in living man or other animals, which the Secretary of Agriculture shall declare to be a pest, and (2) any substance or mixture of substances intended for use as a plant regulator, defoliant or desiccant.

[Doc. No. 1464, 30 FR 8106, June 24, 1965, as amended by Amdt. 137-3, 33 FR 9601, July 2, 1968]

Subpart B—Certification Rules

§ 137.11 Certificate required.

(a) Except as provided in paragraphs (c) and (d) of this section, no person may conduct agricultural aircraft operations without, or in violation of, an agricultural aircraft operator certificate issued under this part.

(b) Notwithstanding Part 133 of this chapter, an operator may, if he complies with this part, conduct agricultural aircraft operations with a rotorcraft with external dispensing equipment in place without a rotorcraft external-load operator certificate.

(c) A Federal, State, or local government conducting agricultural aircraft operations with public aircraft need not comply with this subpart.

(d) The holder of a rotorcraft external-load operator certificate under Part 133 of this chapter conducting an agricultural aircraft operation, involving only the dispensing of water on forest fires by rotorcraft external-load means, need not comply with this subpart.

[Doc. No. 1464, 30 FR 8106, June 24, 1965, as amended by Amdt. 137-3, 33 FR 9601, July 2, 1968; Amdt. 137-6, 41 FR 35060, Aug. 19, 1976]

§ 137.15 Application for certificate.

An application for an agricultural aircraft operator certificate is made on a form and in a manner prescribed by the Administrator, and filed with the FAA Flight Standards District Office that has jurisdiction over the area in which the applicant's home base of operations is located.

[Doc. No. 1464, 30 FR 8106, June 24, 1965, as amended by Amdt. 137-13, 54 FR 39294, Sept. 25, 1989]

§ 137.17 Amendment of certificate.

(a) An agricultural aircraft operator certificate may be amended—

(1) On the Administrator's own initiative, under § 609 of the Federal Aviation Act of 1958 (49 U.S.C. 1429) and Part 13 of this chapter; or

(2) Upon application by the holder of that certificate.

(b) An application to amend an agricultural aircraft operator certificate is submitted on a form and in a manner prescribed by the Administrator. The applicant must file the application with the FAA Flight Standards District Office having jurisdiction over the area in which the applicant's home base of operations is located at least 15 days before the date that it proposes the amendment become effective, unless a shorter filing period is approved by that office.

(c) The Flight Standards District Office grants a request to amend a certificate if it determines that safety in air commerce and the public interest so allow.

(d) Within 30 days after receiving a refusal to amend, the holder may petition the Director, Flight Standards Service, to reconsider the refusal.

[Doc. No. 1464, 30 FR 8106, June 24, 1965, as amended by Amdt. 137-9, 43 FR 52206, Nov. 9, 1978; Amdt. 137-11, 45 FR 47838, July 17, 1980; Amdt. 137-13, 54 FR 39294, Sept. 25, 1989]

§ 137.19 Certification requirements.

(a) General. An applicant for a private agricultural aircraft operator certificate is entitled to that certificate if he shows that he meets the requirements of paragraphs (b), (d), and (e) of this section. An applicant for a commercial agricultural aircraft operator certificate is entitled to that certificate if he shows that he meets the requirements of paragraphs (c), (d), and (e) of this section. However, if an applicant applies for an agricultural aircraft operator certificate containing a prohibition against the dispensing of economic poisons, that applicant is not required to demonstrate the knowledge required in paragraphs (e)(1) (ii) through (iv) of this section.

(b) Private operator—pilot. The applicant must hold a current U.S. private, commercial, or airline transport pilot certificate and be properly rated for the aircraft to be used.

(c) Commercial operator—pilots. The applicant must have available the services of at least one person who holds a current U.S. commercial or airline transport pilot certificate and who is properly rated for the aircraft to be used. The applicant himself may be the person available.

(d) Aircraft. The applicant must have at least one certificated and airworthy aircraft, equipped for agricultural operation.

(e) Knowledge and skill tests. The applicant must show, or have the person who is designated as the chief supervisor of agricultural aircraft operations for him show, that he has satisfactory knowledge and skill regarding agricultural aircraft operations, as described in paragraphs (e) (1) and (2) of this section.

(1) The test of knowledge consists of the following:

(i) Steps to be taken before starting operations, including survey of the area to be worked.

(ii) Safe handling of economic poisons and the proper disposal of used containers for those poisons.

(iii) The general effects of economic poisons and agricultural chemicals on plants, animals, and persons, with emphasis on those normally used in the areas of intended operations; and the precautions to be observed in using poisons and chemicals.

(iv) Primary symptoms of poisoning of persons from economic poisons, the appropriate emergency measures to be taken, and the location of poison control centers.

(v) Performance capabilities and operating limitations of the aircraft to be used.

(vi) Safe flight and application procedures.

(2) The test of skill consists of the following maneuvers that must be shown in any of the aircraft specified in paragraph (d) of this section, and at that aircraft's maximum certificated take-off weight, or the maximum weight established for the special purpose load, whichever is greater:

(i) Short-field and soft-field takeoffs (airplanes and gyroplanes only).

(ii) Approaches to the working area.

(iii) Flare-outs.

(iv) Swath runs.

(v) Pullups and turnarounds.

(vi) Rapid deceleration (quick stops) in helicopters only.

[Doc. No. 1464, 30 FR 8106, June 24, 1965, as amended by Amdt. 137-1, 30 FR 15143, Dec. 8, 1965; Amdt. 137-7, 43 FR 22643, May 25, 1978]

§ 137.21 Duration of certificate.

An agricultural aircraft operator certificate is effective until it is surrendered, suspended, or revoked. The holder of an agricultural aircraft operator certificate that is suspended or revoked shall return it to the Administrator.

§ 137.23 Carriage of narcotic drugs, marihuana, and depressant or stimulant drugs or substances.

If the holder of a certificate issued under this part permits any aircraft owned or leased by that holder to be engaged in any operation that the certificate holder knows to be in violation of § 91.19(a) of this chapter, that operation is a basis for suspending or revoking the certificate.

[Doc. No. 12035, Amdt. 137-4, 38 FR 17493, July 2, 1973, as amended by Amdt. 137-12, 54 FR 34332, Aug. 18, 1989]

Subpart C—Operating Rules

§ 137.29 General.

(a) Except as provided in paragraphs (d) and (e) of this section, this subpart prescribes rules that apply to persons and aircraft used in agricultural aircraft operations conducted under this part.

(b) [Reserved]

(c) The holder of an agricultural aircraft operator certificate may deviate from the provisions of Part 91 of this chapter without a certificate of waiver, as authorized in this subpart for dispensing operations, when conducting nondispensing aerial work operations related to agriculture, horticulture, or forest preservation in accordance with the operating rules of this subpart.

(d) §§ 137.31 through 137.35, §§ 137.41, and 137.53 through 137.59 do not apply to persons and aircraft used in agricultural aircraft operations conducted with public aircraft.

(e) §§ 137.31 through 137.35, §§ 137.39, 137.41, 137.51 through 137.59, and Subpart D do not apply to persons and rotorcraft used in agricultural aircraft operations conducted by a person holding a certificate under Part 133 of this chapter and involving only the dispensing of water on forest fires by rotorcraft external-load means. However, the operation shall be conducted in accordance with—

(1) The rules of Part 133 of this chapter governing rotorcraft external-load operations; and

(2) The operating rules of this subpart contained in §§ 137.29, 137.37, and §§ 137.43 through 137.49.

[Doc. No. 1464, 30 FR 8106, June 24, 1965, as amended by Amdt. 137-3, 33 FR 9601, July 2, 1968; Amdt. 137-6, 41 FR 35060, Aug. 19, 1976]

§ 137.31 Aircraft requirements.

No person may operate an aircraft unless that aircraft—

(a) Meets the requirements of § 137.19(d); and

(b) Is equipped with a suitable and properly installed shoulder harness for use by each pilot.

§ 137.33 Carrying of certificate.

(a) No person may operate an aircraft unless a facsimile of the agricultural aircraft operator certificate, under which the operation is conducted, is carried on that aircraft. The facsimile shall be presented for inspection upon the request of the Administrator or any Federal, State, or local law enforcement officer.

(b) Notwithstanding Part 91 of this chapter, the registration and airworthiness certificates issued for the aircraft need not be carried in the aircraft. However, when those certificates are not carried in the aircraft they shall be kept available for inspection at the base from which the dispensing operation is conducted.

[Doc. No. 1464, 30 FR 8106, June 24, 1965, as amended by Amdt. 137-3, 33 FR 9601, July 2, 1968]

§ 137.35 Limitations on private agricultural aircraft operator.

No person may conduct an agricultural aircraft operation under the authority of a private agricultural aircraft operator certificate—

(a) For compensation or hire;

(b) Over a congested area; or

(c) Over any property unless he is the owner or lessee of the property, or has ownership or other property interest in the crop located on that property.

§ 137.37 Manner of dispensing.

No persons may dispense, or cause to be dispensed, from an aircraft, any material or substance in a manner that creates a hazard to persons or property on the surface.

[Doc. No. 1464, 30 FR 8106, June 24, 1965, as amended by Amdt. 137-3, 33 FR 9601, July 2, 1968]

§ 137.39 Economic poison dispensing.

(a) Except as provided in paragraph (b) of this section, no person may dispense or cause to be dispensed from an aircraft, any economic poison that is registered with the U.S. Department of Agriculture under the Federal Insecticide, Fungicide, and Rodenticide Act (7 U.S.C. 135-135k)—

(1) For a use other than that for which it is registered;

(2) Contrary to any safety instructions or use limitations on its label; or

(3) In violation of any law or regulation of the United States.

(b) This section does not apply to any person dispensing economic poisons for experimental purposes under—

(1) The supervision of a Federal or State agency authorized by law to conduct research in the field of economic poisons; or

(2) A permit from the U.S. Department of Agriculture issued pursuant to the Federal Insecticide, Fungicide, and Rodenticide Act (7 U.S.C. 135-135k).

[Amdt. 137-2, 31 FR 6686, May 5, 1966]

§ 137.41 Personnel.

(a) Information. The holder of an agricultural aircraft operator certificate shall insure that each person used in the holder's agricultural aircraft operation is informed of that person's duties and responsibilities for the operation.

(b) Supervisors. No person may supervise an agricultural aircraft operation unless he has met the knowledge and skill requirements of § 137.19(e).

(c) Pilot in command. No person may act as pilot in command of an aircraft unless he holds a pilot certificate and rating prescribed by § 137.19 (b) or (c), as appropriate to the type of operation conducted. In addition, he must demonstrate to the holder of the Agricultural Aircraft Operator Certificate conducting the operation that he has met the knowledge and skill requirements of § 137.19(e). If the holder of that certificate has designated a person under § 137.19(e) to supervise his agricultural aircraft operations the demonstration must be made to the person so designated. However, a demonstration of the knowledge and skill requirement is not necessary for any pilot in command who—

(1) Is, at the time of the filing of an application by an agricultural aircraft operator, working as a pilot in command for that operator; and

(2) Has a record of operation under that applicant that does not disclose any question regarding the safety of his flight operations or his competence in dispensing agricultural materials or chemicals.

§ 137.42 Fastening of safety belts and shoulder harnesses.

No person may operate an aircraft in operations required to be conducted under Part 137 without a safety belt and shoulder harness properly secured about that person except that the shoulder harness need not be fastened if that person would be unable to perform required duties with the shoulder harness fastened.

[Amdt. 137-10, 44 FR 61325, Oct. 25, 1979]

§ 137.43 Operations in controlled airspace designated for an airport.

(a) Except for flights to and from a dispensing area, no person may operate an aircraft within the lateral boundaries of the surface area of Class D airspace designated for an airport unless authorization for that operation has been obtained from the ATC facility having jurisdiction over that area.

(b) No person may operate an aircraft in weather conditions below VFR minimums within the lateral boundaries of a Class E airspace area that extends upward from the surface unless authorization for that operation has been obtained from the ATC facility having jurisdiction over that area.

(c) Notwithstanding § 91.157(a)(2) of this chapter, an aircraft may be operated under the special VFR weather minimums without meeting the requirements prescribed therein.

[Amdt. 137-14, 56 FR 65663, Dec. 17, 1991; 58 FR 32840, June 14, 1993]

§ 137.45 Nonobservance of airport traffic pattern.

Notwithstanding Part 91 of this chapter, the pilot in command of an aircraft may deviate from an airport traffic pattern when authorized by the control tower concerned. At an airport without a functioning control tower, the pilot in command may deviate from the traffic pattern if—

(a) Prior coordination is made with the airport management concerned;

(b) Deviations are limited to the agricultural aircraft operation;

(c) Except in an emergency, landing and takeoffs are not made on ramps, taxiways, or other areas of the airport not intended for such use; and

(d) The aircraft at all times remains clear of, and gives way to, aircraft conforming to the traffic pattern for the airport.

§ 137.47 Operation without position lights.

Notwithstanding Part 91 of this chapter, an aircraft may be operated without position lights if prominent unlighted objects are visible for at least 1 mile and takeoffs and landings at—

(a) Airports with a functioning control tower are made only as authorized by the control tower operator; and

(b) Other airports are made only with the permission of the airport management and no other aircraft operations requiring position lights are in progress at that airport.

§ 137.49 Operations over other than congested areas.

Notwithstanding Part 91 of this chapter, during the actual dispensing operation, including approaches, departures, and turnarounds reasonably necessary for the operation, an aircraft may be operated over other than congested areas below 500 feet above the surface and closer than 500 feet to persons, vessels, vehicles, and structures, if the operations are conducted without creating a hazard to persons or propty on the surface.

[Amdt. 137-3, 33 FR 9601, July 2, 1968]

§ 137.51 Operation over congested areas: General.

(a) Notwithstanding Part 91 of this chapter, an aircraft may be operated over a congested area at altitudes required for the proper accomplishment of the agricultural aircraft operation if the operation is conducted—

(1) With the maximum safety to persons and property on the surface, consistent with the operation; and

(2) In accordance with the requirements of paragraph (b) of this section.

(b) No person may operate an aircraft over a congested area except in accordance with the requirements of this paragraph.

(1) Prior written approval must be obtained from the appropriate official or governing body of the political subdivision over which the operations are conducted.

(2) Notice of the intended operation must be given to the public by some effective means, such as daily newspapers, radio, television, or door-to-door notice.

(3) A plan for each complete operation must be submitted to, and approved by appropriate personnel of the FAA Flight Standards District Office having jurisdiction over the area where the operation is to be conducted. The plan must include consideration of obstructions to flight; the emergency landing capabilities of the aircraft to be used; and any necessary coordination with air traffic control.

(4) Single engine aircraft must be operated as follows:

(i) Except for helicopters, no person may take off a loaded aircraft, or make a turnaround over a congested area.

(ii) No person may operate an aircraft over a congested area below the altitudes prescribed in Part 91 of this chapter except during the actual dispensing operation, including the approaches and departures necessary for that operation.

(iii) No person may operate an aircraft over a congested area during the actual dispensing operation, including the approaches and departures for that operation, unless it is operated in a pattern and at such an altitude that the aircraft can land, in an emergency, without endangering persons or property on the surface.

(5) Multiengine aircraft must be operated as follows:

(i) No person may take off a multiengine airplane over a congested area except under conditions that will allow the airplane to be brought to a safe stop within the effective length of the runway from any point on takeoff up to the time of attaining, with all engines operating at normal takeoff power, 105 percent of the minimum control speed with the critical engine inoperative in the takeoff configuration or 115 percent of the power-off stall speed in the takeoff configuration, whichever is greater, as shown by the accelerate stop distance data. In applying this requirement, takeoff data is based upon still-air conditions, and no correction is made for any uphill gradient of 1 percent or less when the percentage is measured as the difference between elevation at the end points of the runway divided by the total length. For uphill gradients greater than 1 percent, the effective takeoff length of the runway is reduced 20 percent for each 1-percent grade.

(ii) No person may operate a multiengine airplane at a weight greater than the weight that, with the critical engine inoperative, would permit a rate of climb of at least 50 feet per minute at an altitude of at least 1,000 feet above the elevation of the highest ground or obstruction within the area to be worked or at an altitude of 5,000 feet, whichever is higher. For the purposes of this subdivision, it is assumed that the propeller of the inoperative engine is in the minimum drag position; that the wing flaps and landing gear are in the most favorable positions; and that the remaining engine or engines are operating at the maximum continuous power available.

(iii) No person may operate any multiengine aircraft over a congested area below the altitudes prescribed in Part 91 of this chapter except during the actual dispensing operation, including the approaches, departures, and turnarounds necessary for that operation.

[Doc. No. 1464, 30 FR 8106, June 24, 1965, as amended by Doc. No. 8084, 32 FR 5769, Apr. 11, 1967; Amdt. 137-13, 54 FR 39294, Sept. 25, 1989]

§ 137.53 Operation over congested areas: Pilots and aircraft.

(a) *General.* No person may operate an aircraft over a congested area except in accordance with the pilot and aircraft rules of this section.

(b) *Pilots.* Each pilot in command must have at least—

(1) 25 hours of pilot-in-command flight time in the make and basic model of the aircraft, at least 10 hours of which must have been acquired within the preceding 12 calendar months; and

(2) 100 hours of flight experience as pilot in command in dispensing agricultural materials or chemicals.

(c) *Aircraft.* (1) Each aircraft must—(i) If it is an aircraft not specified in paragraph (c)(1)(ii) of this section, have had within the preceding 100 hours of time in service a 100-hour or annual inspection by a person authorized by Part 65 or 145 of this chapter, or have been inspected under a progressive inspection system; and

(ii) If it is a large or turbine-powered multiengine civil airplane of U.S. registry, have been inspected in accordance with the applicable inspection program requirements of § 91.409 of this chapter.

(2) If other than a helicopter, it must be equipped with a device capable of jettisoning at least one-half of the aircraft's maximum authorized load of agricultural material

within 45 seconds. If the aircraft is equipped with a device for releasing the tank or hopper as a unit, there must be a means to prevent inadvertent release by the pilot or other crewmember.

[Doc. No. 1464, 30 FR 8106, June 24, 1965, as amended by Amdt. 137-5, 41 FR 16796, Apr. 22, 1976; Amdt. 137-12, 54 FR 34332, Aug. 18, 1989]

§ 137.55 Business name: Commercial agricultural aircraft operator.

No person may operate under a business name that is not shown on his commercial agricultural aircraft operator certificate.

§ 137.57 Availability of certificate.

Each holder of an agricultural aircraft operator certificate shall keep that certificate at his home base of operations and shall present it for inspection on the request of the Administrator or any Federal, State, or local law enforcement officer.

§ 137.59 Inspection authority.

Each holder of an agricultural aircraft operator certificate shall allow the Administrator at any time and place to make inspections, including on-the-job inspections, to determine compliance with applicable regulations and his agricultural aircraft operator certificate.

Subpart D—Records and Reports

§ 137.71 Records: Commercial agricultural aircraft operator.

(a) Each holder of a commercial agricultural aircraft operator certificate shall maintain and keep current, at the home base of operations designated in his application, the following records:

(1) The name and address of each person for whom agricultural aircraft services were provided;

(2) The date of the service;

(3) The name and quantity of the material dispensed for each operation conducted; and

(4) The name, address, and certificate number of each pilot used in agricultural aircraft operations and the date that pilot met the knowledge and skill requirements of § 137.19(e).

(b) The records required by this section must be kept at least 12 months and made available for inspection by the Administrator upon request.

§ 137.75 Change of address.

Each holder of an agricultural aircraft operator certificate shall notify the FAA in writing in advance of any change in the address of his home base of operations.

§ 137.77 Termination of operations.

Whenever a person holding an agricultural aircraft operator certificate ceases operations under this part, he shall surrender that certificate to the FAA Flight Standards District Office last having jurisdiction over his operation.

[Doc. No. 1464, 30 FR 8106, June 24, 1965, as amended by Amdt. 137-13, 54 FR 39294, Sept. 25, 1989; 54 FR 52872, Dec. 22, 1989]

14 CFR 137 * Amendment 137-15 * Dec. 28, 1995

Part 141—Pilot schools

Authority: 49 U.S.C. 106(g), 40113, 44701-44703, 44707, 44709, 44711, 45102-45103, 45301-45302.

Source: Amdt. 141-8, 62 FR 16347, Apr. 4, 1997, unless otherwise noted.

Subpart A—General

§ 141.1 Applicability.

This part prescribes the requirements for issuing pilot school certificates, provisional pilot school certificates, and associated ratings, and the general operating rules applicable to a holder of a certificate or rating issued under this part.

§ 141.3 Certificate required.

No person may operate as a certificated pilot school without, or in violation of, a pilot school certificate or provisional pilot school certificate issued under this part.

§ 141.5 Requirements for a pilot school certificate.

An applicant may be issued a pilot school certificate with associated ratings if the applicant:

(a) Completes the application for a pilot school certificate on a form and in a manner prescribed by the Administrator;

(b) Holds a provisional pilot school certificate, issued under this part, for at least 24 calendar months preceding the month in which the application for a pilot school certificate is made;

(c) Meets the applicable requirements of subparts A through C of this part for the school ratings sought; and

(d) Has trained and recommended for pilot certification and rating tests, within 24 calendar months preceding the month the application is made for the pilot school certificate, at least 10 students for a knowledge or practical test for a pilot certificate, flight instructor certificate, ground instructor certificate, an additional rating, an end-of-course test for a training course specified in appendix K to this part, or any combination of those tests, and at least 80 percent of all tests administered were passed on the first attempt.

[Amdt. 141-8, 62 FR 16347, Apr. 4, 1997; Amdt. 141-9, 62 FR 40907, July 30, 1997]

§ 141.7 Provisional pilot school certificate.

An applicant that meets the applicable requirements of subparts A, B, and C of this part, but does not meet the recent training activity requirements of § 141.5(d) of this part, may be issued a provisional pilot school certificate with ratings.

§ 141.9 Examining authority.

An applicant is issued examining authority for its pilot school certificate if the applicant meets the requirements of subpart D of this part.

§ 141.11 Pilot school ratings.

(a) The ratings listed in paragraph (b) of this section may be issued to an applicant for:

(1) A pilot school certificate, provided the applicant meets the requirements of § 141.5 of this part; or

(2) A provisional pilot school certificate, provided the applicant meets the requirements of § 141.7 of this part.

(b) An applicant may be authorized to conduct the following courses:

(1) Certification and rating courses. (Appendixes A through J.)

(i) Recreational pilot course.

(ii) Private pilot course.

(iii) Commercial pilot course.

(iv) Instrument rating course.

(v) Airline transport pilot course.

(vi) Flight instructor course.

(vii) Flight instructor instrument course.

(viii) Ground instructor course.

(ix) Additional aircraft category or class rating course.

(x) Aircraft type rating course.

(2) Special preparation courses. (Appendix K).

(i) Pilot refresher course.

(ii) Flight instructor refresher course.

(iii) Ground instructor refresher course.

(iv) Agricultural aircraft operations course.

(v) Rotorcraft external-load operations course.

(vi) Special operations course.

(vii) Test pilot course.

(3) Pilot ground school course. (Appendix L).

§ 141.13 Application for issuance, amendment, or renewal.

(a) Application for an original certificate and rating, an additional rating, or the renewal of a certificate under this part must be made on a form and in a manner prescribed by the Administrator.

(b) Application for the issuance or amendment of a certificate or rating must be accompanied by two copies of each proposed training course curriculum for which approval is sought.

§ 141.17 Duration of certificate and examining authority.

(a) Unless surrendered, suspended, or revoked, a pilot school's certificate or a provisional pilot school's certificate expires:

(1) On the last day of the 24th calendar month from the month the certificate was issued;

(2) Except as provided in paragraph (b) of this section, on the date that any change in ownership of the school occurs;

(3) On the date of any change in the facilities upon which the school's certificate is based occurs; or

(4) Upon notice by the Administrator that the school has failed for more than 60 days to maintain the facilities,

aircraft, or personnel required for any one of the school's approved training courses.

(b) A change in the ownership of a pilot school or provisional pilot school does not terminate that school's certificate if, within 30 days after the date that any change in ownership of the school occurs:

(1) Application is made for an appropriate amendment to the certificate; and

(2) No change in the facilities, personnel, or approved training courses is involved.

(c) An examining authority issued to the holder of a pilot school certificate expires on the date that the pilot school certificate expires, or is surrendered, suspended, or revoked.

§ 141.18 Carriage of narcotic drugs, marijuana, and depressant or stimulant drugs or substances.

If the holder of a certificate issued under this part permits any aircraft owned or leased by that holder to be engaged in any operation that the certificate holder knows to be in violation of § 91.19(a) of this chapter, that operation is a basis for suspending or revoking the certificate.

§ 141.19 Display of certificate.

(a) Each holder of a pilot school certificate or a provisional pilot school certificate must display that certificate in a place in the school that is normally accessible to the public and is not obscured.

(b) A certificate must be made available for inspection upon request by:

(1) The Administrator;

(2) An authorized representative of the National Transportation Safety Board; or

(3) A Federal, State, or local law enforcement officer.

§ 141.21 Inspections.

Each holder of a certificate issued under this part must allow the Administrator to inspect its personnel, facilities, equipment, and records to determine the certificate holder's:

(a) Eligibility to hold its certificate;

(b) Compliance with 49 U.S.C. 40101 et seq., formerly the Federal Aviation Act of 1958, as amended; and

(c) Compliance with the Federal Aviation Regulations.

§ 141.23 Advertising limitations.

(a) The holder of a pilot school certificate or a provisional pilot school certificate may not make any statement relating to its certification and ratings that is false or designed to mislead any person contemplating enrollment in that school.

(b) The holder of a pilot school certificate or a provisional pilot school certificate may not advertise that the school is certificated unless it clearly differentiates between courses that have been approved under part 141 of this chapter and those that have not been approved under part 141 of this chapter.

(c) The holder of a pilot school certificate or a provisional pilot school certificate must promptly remove:

(1) From vacated premises, all signs indicating that the school was certificated by the Administrator; or

(2) All indications (including signs), wherever located, that the school is certificated by the Administrator when its certificate has expired or has been surrendered, suspended, or revoked.

§ 141.25 Business office and operations base.

(a) Each holder of a pilot school or a provisional pilot school certificate must maintain a principal business office with a mailing address in the name shown on its certificate.

(b) The facilities and equipment at the principal business office must be adequate to maintain the files and records required to operate the business of the school.

(c) The principal business office may not be shared with, or used by, another pilot school.

(d) Before changing the location of the principal business office or the operations base, each certificate holder must notify the FAA Flight Standards District Office having jurisdiction over the area of the new location, and the notice must be:

(1) Submitted in writing at least 30 days before the change of location; and

(2) Accompanied by any amendments needed for the certificate holder's approved training course outline.

(e) A certificate holder may conduct training at an operations base other than the one specified in its certificate, if:

(1) The Administrator has inspected and approved the base for use by the certificate holder; and

(2) The course of training and any needed amendments have been approved for use at that base.

§ 141.26 Training agreements.

A training center certificated under part 142 of this chapter may provide the training, testing, and checking for pilot schools certificated under part 141 of this chapter, and is considered to meet the requirements of part 141, provided—

(a) There is a training agreement between the certificated training center and the pilot school;

(b) The training, testing, and checking provided by the certificated training center is approved and conducted under part 142;

(c) The pilot school certificated under part 141 obtains the Administrator's approval for a training course outline that includes the training, testing, and checking to be conducted under part 141 and the training, testing, and checking to be conducted under part 142; and

(d) Upon completion of the training, testing, and checking conducted under part 142, a copy of each student's training record is forwarded to the part 141 school and becomes part of the student's permanent training record.

§ 141.27 Renewal of certificates and ratings.

(a) Pilot school.

(1) A pilot school may apply for renewal of its school certificate and ratings within 30 days preceding the month the pilot school's certificate expires, provided the school meets the requirements prescribed in paragraph (a)(2) of this section for renewal of its certificate and ratings.

(2) A pilot school may have its school certificate and ratings renewed for an additional 24 calendar months if the Administrator determines the school's personnel, aircraft, facility and airport, approved training courses, training records, and recent training ability and quality meet the requirements of this part.

(3) A pilot school that does not meet the renewal requirements in paragraph (a)(2) of this section, may apply for a provisional pilot school certificate if the school meets the requirements of § 141.7 of this part.

(b) Provisional pilot school.

(1) Except as provided in paragraph (b)(3) of this section, a provisional pilot school may not have its provisional pilot school certificate or the ratings on that certificate renewed.

(2) A provisional pilot school may apply for a pilot school certificate and associated ratings provided that school meets the requirements of § 141.5 of this part.

(3) A former provisional pilot school may apply for another provisional pilot school certificate, provided 180 days have elapsed since its last provisional pilot school certificate expired.

§ 141.29 [Reserved.]

Subpart B—Personnel, Aircraft, and Facilities Requirements

§ 141.31 Applicability.

(a) This subpart prescribes:

(1) The personnel and aircraft requirements for a pilot school certificate or a provisional pilot school certificate; and

(2) The facilities that a pilot school or provisional pilot school must have available on a continuous basis.

(b) As used in this subpart, to have continuous use of a facility, including an airport, the school must have:

(1) Ownership of the facility or airport for at least 6 calendar months after the date the application for initial certification and on the date of renewal of the school's certificate; or

(2) A written lease agreement for the facility or airport for at least 6 calendar months at the time of application for initial certification and on the date of renewal of the school's certificate is made.

[Amdt. 141-8, 62 FR 16347, Apr. 4, 1997; Amdt. 141-9, 62 FR 40907, July 30, 1997]

§ 141.33 Personnel.

(a) An applicant for a pilot school certificate or for a provisional pilot school certificate must meet the following personnel requirements:

(1) Each applicant must have adequate personnel, including certificated flight instructors, certificated ground instructors, or holders of a commercial pilot certificate with a lighter-than-air rating, and a chief instructor for each approved course of training who is qualified and competent to perform the duties to which that instructor is assigned.

(2) If the school employs dispatchers, aircraft handlers, and line and service personnel, then it must instruct those persons in the procedures and responsibilities of their employment.

(3) Each instructor to be used for ground or flight training must hold a flight instructor certificate, ground instructor certificate, or commercial pilot certificate with a lighter-than-air rating, as appropriate, with ratings for the approved course of training and any aircraft used in that course.

(b) An applicant for a pilot school certificate or for a provisional pilot school certificate must designate a chief instructor for each of the school's approved training courses, who must meet the requirements of § 141.35 of this part.

(c) When necessary, an applicant for a pilot school certificate or for a provisional pilot school certificate may designate a person to be an assistant chief instructor for an approved training course, provided that person meets the requirements of § 141.36 of this part.

(d) A pilot school and a provisional pilot school may designate a person to be a check instructor for conducting student stage checks, end-of-course tests, and instructor proficiency checks, provided:

(1) That person meets the requirements of § 141.37 of this part; and

(2) That school has a student enrollment of at least 50 students at the time designation is sought.

(e) A person, as listed in this section, may serve in more than one position for a school, provided that person is qualified for each position.

[Amdt. 141-8, 62 FR 16347, Apr. 4, 1997; Amdt. 141-9, 62 FR 40907, July 30, 1997]

§ 141.35 Chief instructor qualifications.

(a) To be eligible for designation as a chief instructor for a course of training, a person must meet the following requirements:

(1) Hold a commercial pilot certificate or an airline transport pilot certificate, and, except for a chief instructor for a course of training solely for a lighter-than-air rating, a current flight instructor certificate. The certificates must contain the appropriate aircraft

category and class ratings for the category and class of aircraft used in the course and an instrument rating, if an instrument rating is required for enrollment in the course of training;

(2) Meet the pilot-in-command recent flight experience requirements of § 61.57 of this chapter;

(3) Pass a knowledge test on—

(i) Teaching methods;

(ii) Applicable provisions of the "Aeronautical Information Manual";

(iii) Applicable provisions of parts 61, 91, and 141 of this chapter; and

(iv) The objectives and approved course completion standards of the course for which the person seeks to obtain designation.

(4) Pass a proficiency test on instructional skills and ability to train students on the flight procedures and maneuvers appropriate to the course;

(5) Except for a course of training for gliders, balloons, or airships, the chief instructor must meet the applicable requirements in paragraphs (b),

(c), and (d) of this section; and

(6) A chief instructor for a course of training for gliders, balloons or airships is only required to have 40 percent of the hours required in paragraphs (b) and (d) of this section.

(b) For a course of training leading to the issuance of a recreational or private pilot certificate or rating, a chief instructor must have:

(1) At least 1,000 hours as pilot in command; and

(2) Primary flight training experience, acquired as either a certificated flight instructor or an instructor in a military pilot flight training program, or a combination thereof, consisting of at least—

(i) 2 years and a total of 500 flight hours; or

(ii) 1,000 flight hours.

(c) For a course of training leading to the issuance of an instrument rating or a rating with instrument privileges, a chief instructor must have:

(1) At least 100 hours of flight time under actual or simulated instrument conditions;

(2) At least 1,000 hours as pilot in command; and

(3) Instrument flight instructor experience, acquired as either a certificated flight instructor-instrument or an instructor in a military pilot flight training program, or a combination thereof, consisting of at least—

(i) 2 years and a total of 250 flight hours; or

(ii) 400 flight hours.

(d) For a course of training other than those leading to the issuance of a recreational or private pilot certificate or rating, or an instrument rating or a rating with instrument privileges, a chief instructor must have:

(1) At least 2,000 hours as pilot in command; and

(2) Flight training experience, acquired as either a certificated flight instructor or an instructor in a military pilot

flight training program, or a combination thereof, consisting of at least—

(i) 3 years and a total of 1,000 flight hours; or

(ii) 1,500 flight hours.

(e) To be eligible for designation as chief instructor for a ground school course, a person must have 1 year of experience as a ground school instructor at a certificated pilot school.

[Amdt. 141-8, 62 FR 16347, Apr. 4, 1997; Amdt. 141-9, 62 FR 40907, July 30, 1997]

§ 141.36 Assistant chief instructor qualifications.

(a) To be eligible for designation as an assistant chief instructor for a course of training, a person must meet the following requirements:

(1) Hold a commercial pilot or an airline transport pilot certificate and, except for the assistant chief instructor for a course of training solely for a lighter-than-air rating, a current flight instructor certificate. The certificates must contain the appropriate aircraft category, class, and instrument ratings if an instrument rating is required by the couse of training for the category and class of aircraft used in the course;

(2) Meet the pilot-in-command recent flight experience requirements of § 61.57 of this chapter;

(3) Pass a knowledge test on—

(i) Teaching methods;

(ii) Applicable provisions of the "Aeronautical Information Manual";

(iii) Applicable provisions of parts 61, 91, and 141 of this chapter; and

(iv) The objectives and approved course completion standards of the course for which the person seeks to obtain designation.

(4) Pass a proficiency test on the flight procedures and maneuvers appropriate to that course; and

(5) Meet the applicable requirements in paragraphs (b), (c), and (d) of this section. However, an assistant chief instructor for a course of training for gliders, balloons, or airships is only required to have 40 percent of the hours required in paragraphs (b) and (d) of this section.

(b) For a course of training leading to the issuance of a recreational or private pilot certificate or rating, an assistant chief instructor must have:

(1) At least 500 hours as pilot in command; and

(2) Flight training experience, acquired as either a certificated flight instructor or an instructor in a military pilot flight training program, or a combination thereof, consisting of at least—

(i) 1 year and a total of 250 flight hours; or

(ii) 500 flight hours.

(c) For a course of training leading to the issuance of an instrument rating or a rating with instrument privileges, an assistant chief flight instructor must have:

(1) At least 50 hours of flight time under actual or simulated instrument conditions;

(2) At least 500 hours as pilot in command; and

(3) Instrument flight instructor experience, acquired as either a certificated flight instructor-instrument or an instructor in a military pilot flight training program, or a combination thereof, consisting of at least—

(i) 1 year and a total of 125 flight hours; or

(ii) 200 flight hours.

(d) For a course of training other than one leading to the issuance of a recreational or private pilot certificate or rating, or an instrument rating or a rating with instrument privileges, an assistant chief instructor must have:

(1) At least 1,000 hours as pilot in command; and

(2) Flight training experience, acquired as either a certificated flight instructor or an instructor in a military pilot flight training program, or a combination thereof, consisting of at least—

(i) 1 ½ years and a total of 500 flight hours; or

(ii) 750 flight hours.

(e) To be eligible for designation as an assistant chief instructor for a ground school course, a person must have 6 months of experience as a ground school instructor at a certificated pilot school.

[Amdt. 141-8, 62 FR 16347, Apr. 4, 1997; Amdt. 141-9, 62 FR 40907, July 30, 1997]

§ 141.37 Check instructor qualifications.

(a) To be designated as a check instructor for conducting student stage checks, end-of-course tests, and instructor proficiency checks under this part, a person must meet the eligibility requirements of this section:

(1) For checks and tests that relate to either flight or ground training, the person must pass a test, given by the chief instructor, on—

(i) Teaching methods;

(ii) Applicable provisions of the "Aeronautical Information Manual";

(iii) Applicable provisions of parts 61, 91, and 141 of this chapter; and

(iv) The objectives and course completion standards of the approved training course for the designation sought.

(2) For checks and tests that relate to a flight training course, the person must—

(i) Meet the requirements in paragraph (a)(1) of this section;

(ii) Hold a commercial pilot certificate or an airline transport pilot certificate and, except for a check instructor for a course of training for a lighter-than-air rating, a current flight instructor certificate. The certificates must contain the appropriate aircraft category, class, and instrument ratings for the category and class of aircraft used in the course;

(iii) Meet the pilot-in-command recent flight experience requirements of § 61.57 of this chapter; and

(iv) Pass a proficiency test, given by the chief instructor or assistant chief instructor, on the flight procedures and maneuvers of the approved training course for the designation sought.

(3) For checks and tests that relate to ground training, the person must—

(i) Meet the requirements in paragraph (a)(1) of this section;

(ii) Except for a course of training for a lighter-than-air rating, hold a current flight instructor certificate or ground instructor certificate with ratings appropriate to the category and class of aircraft used in the course; and

(iii) For a course of training for a lighter-than-air rating, hold a commercial pilot certificate with a lighter-than-air category rating and the appropriate class rating.

(b) A person who meets the eligibility requirements in paragraph (a) of this section must:

(1) Be designated, in writing, by the chief instructor to conduct student stage checks, end-of-course tests, and instructor proficiency checks; and

(2) Be approved by the FAA Flight Standards District Office having jurisdiction over the school.

(c) A check instructor may not conduct a stage check or an end-of-course test of any student for whom the check instructor has:

(1) Served as the principal instructor; or

(2) Recommended for a stage check or end-of-course test.

[Amdt. 141-8, 62 FR 16347, Apr. 4, 1997; Amdt. 141-9, 62 FR 40907, July 30, 1997]

§ 141.38 Airports.

(a) An applicant for a pilot school certificate or a provisional pilot school certificate must show that he or she has continuous use of each airport at which training flights originate.

(b) Each airport used for airplanes and gliders must have at least one runway or takeoff area that allows training aircraft to make a normal takeoff or landing under the following conditions at the aircraft's maximum certificated takeoff gross weight:

(1) Under wind conditions of not more than 5 miles per hour;

(2) At temperatures in the operating area equal to the mean high temperature for the hottest month of the year;

(3) If applicable, with the powerplant operation, and landing gear and flap operation recommended by the manufacturer; and

(4) In the case of a takeoff—

(i) With smooth transition from liftoff to the best rate of climb speed without exceptional piloting skills or techniques; and

(ii) Clearing all obstacles in the takeoff flight path by at least 50 feet.

(c) Each airport must have a wind direction indicator that is visible from the end of each runway at ground level;

(d) Each airport must have a traffic direction indicator when:

(1) The airport does not have an operating control tower; and

(2) UNICOM advisories are not available.

(e) Except as provided in paragraph (f) of this section, each airport used for night training flights must have permanent runway lights;

(f) An airport or seaplane base used for night training flights in seaplanes is permitted to use adequate nonpermanent lighting or shoreline lighting, if approved by the Administrator.

[Amdt. 141-8, 62 FR 16347, Apr. 4, 1997; Amdt. 141-9, 62 FR 40907, July 30, 1997]

§ 141.39 Aircraft.

An applicant for a pilot school certificate or provisional pilot school certificate must show that each aircraft used by that school for flight training and solo flights meets the following requirements:

(a) Each aircraft must be registered as a civil aircraft in the United States;

(b) Each aircraft must be certificated with a standard airworthiness certificate or a primary airworthiness certificate, unless the Administrator determines that due to the nature of the approved course, an aircraft not having a standard airworthiness certificate or primary airworthiness certificate may be used;

(c) Each aircraft must be maintained and inspected in accordance with the requirements under subpart E of part 91 of this chapter that apply to aircraft operated for hire;

(d) Each aircraft used in flight training must have at least two pilot stations with engine-power controls that can be easily reached and operated in a normal manner from both pilot stations; and

(e) Each aircraft used in a course involving IFR en route operations and instrument approaches must be equipped and maintained for IFR operations. For training in the control and precision maneuvering of an aircraft by reference to instruments, the aircraft may be equipped as provided in the approved course of training.

[Amdt. 141-8, 62 FR 16347, Apr. 4, 1997; Amdt. 141-9, 62 FR 40908, July 30, 1997]

§ 141.41 Flight simulators, flight training devices, and training aids.

An applicant for a pilot school certificate or a provisional pilot school certificate must show that its flight simulators, flight training devices, training aids, and equipment meet the following requirements:

(a) Flight simulators. Each flight simulator used to obtain flight training credit allowed for flight simulators in an approved pilot training course curriculum must—

(1) Be a full-size aircraft cockpit replica of a specific type of aircraft, or make, model, and series of aircraft;

(2) Include the hardware and software necessary to represent the aircraft in ground operations and flight operations;

(3) Use a force cueing system that provides cues at least equivalent to those cues provided by a 3 degree freedom of motion system;

(4) Use a visual system that provides at least a 45-degree horizontal field of view and a 30-degree vertical field of view simultaneously for each pilot; and

(5) Have been evaluated, qualified, and approved by the Administrator.

(b) Flight training devices. Each flight training device used to obtain flight training credit allowed for flight training devices in an approved pilot training course curriculum must—

(1) Be a full-size replica of instruments, equipment panels, and controls of an aircraft, or set of aircraft, in an open flight deck area or in an enclosed cockpit, including the hardware and software for the systems installed that is necessary to simulate the aircraft in ground and flight operations;

(2) Need not have a force (motion) cueing or visual system; and

(3) Have been evaluated, qualified, and approved by the Administrator.

(c) Training aids and equipment. Each training aid, including any audiovisual aid, projector, tape recorder, mockup, chart, or aircraft component listed in the approved training course outline, must be accurate and appropriate to the course for which it is used.

[Amdt. 141-8, 62 FR 16347, Apr. 4, 1997; Amdt. 141-9, 62 FR 40908, July 30, 1997]

§ 141.43 Pilot briefing areas.

(a) An applicant for a pilot school certificate or provisional pilot school certificate must show that the applicant has continuous use of a briefing area located at each airport at which training flights originate that is:

(1) Adequate to shelter students waiting to engage in their training flights;

(2) Arranged and equipped for the conduct of pilot briefings; and

(3) Except as provided in paragraph (c) of this section, for a school with an instrument rating or commercial pilot course, equipped with private landline or telephone communication to the nearest FAA Flight Service Station.

(b) A briefing area required by paragraph (a) of this section may not be used by the applicant if it is available for use by any other pilot school during the period it is required for use by the applicant.

(c) The communication equipment required by paragraph (a)(3) of this section is not required if the briefing area and the flight service station are located on the same airport, and are readily accessible to each other.

§ 141.45 Ground training facilities.

An applicant for a pilot school or provisional pilot school certificate must show that:

(a) Each room, training booth, or other space used for instructional purposes is heated, lighted, and ventilated to conform to local building, sanitation, and health codes; and

(b) The training facility is so located that the students in that facility are not distracted by the training conducted in other rooms, or by flight and maintenance operations on the airport.

Subpart C—Training Course Outline and Curriculum

§ 141.51 Applicability.

This subpart prescribes the curriculum and course outline requirements for the issuance of a pilot school certificate or provisional pilot school certificate and ratings.

§ 141.53 Approval procedures for a training course: General.

(a) *General.* An applicant for a pilot school certificate or provisional pilot school certificate must obtain the Administrator's approval of the outline of each training course for which certification and rating is sought.

(b) *Application.*

(1) An application for the approval of an initial or amended training course must be submitted in duplicate to the FAA Flight Standards District Office having jurisdiction over the area where the school is based.

(2) An application for the approval of an initial or amended training course must be submitted at least 30 days before any training under that course, or any amendment thereto, is scheduled to begin.

(3) An application for amending a training course must be accompanied by two copies of the amendment.

(c) *Training courses.* (1) A training course submitted for approval prior to August 4, 1997 may, if approved, retain that approval until 1 year after August 4, 1997.

(2) An applicant for a pilot school certificate or provisional pilot school certificate may request approval of the training courses specified in § 141.11(b) of this part.

[Amdt. 141-8, 62 FR 16347, Apr. 4, 1997; Amdt. 141-9, 62 FR 40908, July 30, 1997]

§ 141.55 Training course: Contents.

(a) Each training course for which approval is requested must meet the minimum curriculum requirements in accordance with the appropriate appendix of this part.

(b) Except as provided in paragraphs (d) and (e) of this section, each training course for which approval is requested must meet the minimum ground and flight training time requirements in accordance with the appropriate appendix of this part.

(c) Each training course for which approval is requested must contain:

(1) A description of each room used for ground training, including the room's size and the maximum number of students that may be trained in the room at one time;

(2) A description of each type of audiovisual aid, projector, tape recorder, mockup, chart, aircraft component, and other special training aids used for ground training;

(3) A description of each flight simulator or flight training device used for training;

(4) A listing of the airports at which training flights originate and a description of the facilities, including pilot briefing areas that are available for use by the school's students and personnel at each of those airports;

(5) A description of the type of aircraft including any special equipment used for each phase of training;

(6) The minimum qualifications and ratings for each instructor assigned to ground or flight training; and

(7) A training syllabus that includes the following information—

(i) The prerequisites for enrolling in the ground and flight portion of the course that include the pilot certificate and rating (if required by this part), training, pilot experience, and pilot knowledge;

(ii) A detailed description of each lesson, including the lesson's objectives, standards, and planned time for completion;

(iii) A description of what the course is expected to accomplish with regard to student learning;

(iv) The expected accomplishments and the standards for each stage of training; and

(v) A description of the checks and tests to be used to measure a student's accomplishments for each stage of training.

(d) A pilot school may request and receive initial approval for a period of not more than 24 calendar months for any of the training courses of this part without specifying the minimum ground and flight training time requirements of this part, provided the following provisions are met:

(1) The school holds a pilot school certificate issued under this part and has held that certificate for a period of at

least 24 consecutive calendar months preceding the month of the request;

(2) In addition to the information required by paragraph (c) of this section, the training course specifies planned ground and flight training time requirements for the course;

(3) The school does not request the training course to be approved for examining authority, nor may that school hold examining authority for that course; and

(4) The practical test or knowledge test for the course is to be given by—

(i) An FAA inspector; or

(ii) An examiner who is not an employee of the school.

(e) A certificated pilot school may request and receive final approval for any of the training courses of this part without specifying the minimum ground and flight training time requirements of this part, provided the following conditions are met:

(1) The school has held initial approval for that training course for at least 24 calendar months.

(2) The school has—

(i) Trained at least 10 students in that training course within the preceding 24 calendar months and recommended those students for a pilot, flight instructor, or ground instructor certificate or rating; and

(ii) At least 80 percent of those students passed the practical or knowledge test, or any combination thereof, on the first attempt, and that test was given by—

(A) An FAA inspector; or

(B) An examiner who is not an employee of the school.

(3) In addition to the information required by paragraph (c) of this section, the training course specifies planned ground and flight training time requirements for the course.

(4) The school does not request that the training course be approved for examining authority nor may that school hold examining authority for that course.

§ 141.57 Special curricula.

An applicant for a pilot school certificate or provisional pilot school certificate may apply for approval to conduct a special course of airman training for which a curriculum is not prescribed in the appendixes of this part, if the applicant shows that the training course contains features that could achieve a level of pilot proficiency equivalent to that achieved by a training course prescribed in the appendixes of this part or the requirements of part 61 of this chapter.

Subpart D—Examining Authority

§ 141.61 Applicability.

This subpart prescribes the requirements for the issuance of examining authority to the holder of a pilot school certificate, and the privileges and limitations of that examining authority.

§ 141.63 Examining authority qualification requirements.

(a) A pilot school must meet the following prerequisites to receive initial approval for examining authority:

(1) The school must complete the application for examining authority on a form and in a manner prescribed by the Administrator;

(2) The school must hold a pilot school certificate and rating issued under this part;

(3) The school must have held the rating in which examining authority is sought for at least 24 consecutive calendar months preceding the month of application for examining authority;

(4) The training course for which examining authority is requested may not be a course that is approved without meeting the minimum ground and flight training time requirements of this part; and

(5) Within 24 calendar months before the date of application for examining authority, that school must meet the following requirements—

(i) The school must have trained at least 10 students in the training course for which examining authority is sought and recommended those students for a pilot, flight instructor, or ground instructor certificate or rating; and

(ii) At least 90 percent of those students passed the required practical or knowledge test, or any combination thereof, for the pilot, flight instructor, or ground instructor certificate or rating on the first attempt, and that test was given by—

(A) An FAA inspector; or

(B) An examiner who is not an employee of the school.

(b) A pilot school must meet the following requirements to retain approval of its examining authority:

(1) The school must complete the application for renewal of its examining authority on a form and in a manner prescribed by the Administrator;

(2) The school must hold a pilot school certificate and rating issued under this part;

(3) The school must have held the rating for which continued examining authority is sought for at least 24 calendar months preceding the month of application for renewal of its examining authority; and

(4) The training course for which examining authority is requested may not be a course that is approved without meeting the minimum ground and flight training time requirements of this part.

[Amdt. 141-8, 62 FR 16347, Apr. 4, 1997; Amdt. 141-9, 62 FR 40908, July 30, 1997]

§ 141.65 Privileges.

A pilot school that holds examining authority may recommend a person who graduated from its course for the appropriate pilot, flight instructor, or ground instructor certificate or rating without taking the FAA knowledge test or practical test in accordance with the provisions of this subpart.

§ 141.67 Limitations and reports.

A pilot school that holds examining authority may only recommend the issuance of a pilot, flight instructor, or ground instructor certificate and rating to a person who does not take an FAA knowledge test or practical test, if the recommendation for the issuance of that certificate or rating is in accordance with the following requirements:

(a) The person graduated from a training course for which the pilot school holds examining authority.

(b) Except as provided in this paragraph, the person satisfactorily completed all the curriculum requirements of that pilot school's approved training course. A person who transfers from one part 141 approved pilot school to another part 141 approved pilot school may receive credit for that previous training, provided the following requirements are met:

(1) The maximum credited training time does not exceed one-half of the receiving school's curriculum requirements;

(2) The person completes a knowledge and proficiency test conducted by the receiving school for the purpose of determining the amount of pilot experience and knowledge to be credited;

(3) The receiving school determines (based on the person's performance on the knowledge and proficiency test required by paragraph (b)(2) of this section) the amount of credit to be awarded, and records that credit in the person's training record;

(4) The person who requests credit for previous pilot experience and knowledge obtained the experience and knowledge from another part 141 approved pilot school and training course; and

(5) The receiving school retains a copy of the person's training record from the previous school.

(c) Tests given by a pilot school that holds examining authority must be approved by the Administrator and be at least equal in scope, depth, and difficulty to the comparable knowledge and practical tests prescribed by the Administrator under part 61 of this chapter.

(d) A pilot school that holds examining authority may not use its knowledge or practical tests if the school:

(1) Knows, or has reason to believe, the test has been compromised; or

(2) Is notified by an FAA Flight Standards District Office that there is reason to believe or it is known that the test has been compromised.

(e) A pilot school that holds examining authority must maintain a record of all temporary airman certificates it issues, which consist of the following information:

(1) A chronological listing that includes—

(i) The date the temporary airman certificate was issued;

(ii) The student to whom the temporary airman certificate was issued, and that student's permanent mailing address and telephone number;

(iii) The training course from which the student graduated;

(iv) The name of person who conducted the knowledge or practical test;

(v) The type of temporary airman certificate or rating issued to the student; and

(vi) The date the student's airman application file was sent to the FAA for processing for a permanent airman certificate.

(2) A copy of the record containing each student's graduation certificate, airman application, temporary airman certificate, superseded airman certificate (if applicable), and knowledge test or practical test results; and

(3) The records required by paragraph (e) of this section must be retained for 1 year and made available to the Administrator upon request. These records must be surrendered to the Administrator when the pilot school ceases to have examining authority.

(f) Except for pilot schools that have an airman certification representative, when a student passes the knowledge test or practical test, the pilot school that holds examining authority must submit that student's airman application file and training record to the FAA for processing for the issuance of a permanent airman certificate.

[Amdt. 141-8, 62 FR 16347, Apr. 4, 1997; Amdt. 141-9, 62 FR 40908, July 30, 1997]

Subpart E—Operating Rules

§ 141.71 Applicability.

This subpart prescribes the operating rules applicable to a pilot school or provisional pilot school certificated under the provisions of this part.

§ 141.73 Privileges.

(a) The holder of a pilot school certificate or a provisional pilot school certificate may advertise and conduct approved pilot training courses in accordance with the certificate and any ratings that it holds.

(b) A pilot school that holds examining authority for an approved training course may recommend a graduate of that course for the issuance of an appropriate pilot, flight instructor, or ground instructor certificate and rating, without taking an FAA knowledge test or practical test, provided the training course has been approved and meets the minimum ground and flight training time requirements of this part.

§ 141.75 Aircraft requirements.

The following items must be carried on each aircraft used for flight training and solo flights:

(a) A pretakeoff and prelanding checklist; and

(b) The operator's handbook for the aircraft, if one is furnished by the manufacturer, or copies of the handbook if furnished to each student using the aircraft.

[Amdt. 141-8, 62 FR 16347, Apr. 4, 1997; Amdt. 141-9, 62 FR 40908, July 30, 1997]

§ 141.77 Limitations.

(a) The holder of a pilot school certificate or a provisional pilot school certificate may not issue a graduation certificate to a student, or recommend a student for a pilot certificate or rating, unless the student has:

(1) Completed the training specified in the pilot school's course of training; and

(2) Passed the required final tests.

(b) Except as provided in paragraph (c) of this section, the holder of a pilot school certificate or a provisional pilot school certificate may not graduate a student from a course of training unless the student has completed all of the curriculum requirements of that course;

(c) A student may be given credit towards the curriculum requirements of a course for previous pilot experience and knowledge, provided the following conditions are met:

(1) If the credit is based upon a part 141-approved training course, the credit given that student for the previous pilot experience and knowledge may be 50 percent of the curriculum requirements and must be based upon a proficiency test or knowledge test, or both, conducted by the receiving pilot school;

(2) If the credit is not based upon a part 141-approved training course, the credit given that student for the previous pilot experience and knowledge shall not exceed more than 25 percent of the curriculum requirements and must be based upon a proficiency test or knowledge test, or both, conducted by the receiving pilot school;

(3) The receiving school determines the amount of course credit to be transferred under paragraph (c)(1) or paragraph (c)(2) of this section, based on a proficiency test or knowledge test, or both, of the student; and

(4) Credit for training specified in paragraph (c)(1) or paragraph (c)(2) of this section may be given only if the previous provider of the training has certified in writing, or other form acceptable to the Administrator as to the kind and amount of training provided, and the result of each stage check and end-of-course test, if applicable, given to the student.

[Amdt. 141-8, 62 FR 16347, Apr. 4, 1997; Amdt. 141-9, 62 FR 40908, July 30, 1997]

§ 141.79 Flight training.

(a) No person other than a certificated flight instructor or commercial pilot with a lighter-than-air rating who has the ratings and the minimum qualifications specified in the approved training course outline may give a student flight training under an approved course of training.

(b) No student pilot may be authorized to start a solo practice flight from an airport until the flight has been approved by a certificated flight instructor or commercial pilot with a lighter-than-air rating who is present at that airport.

(c) Each chief instructor and assistant chief instructor assigned to a training course must complete, at least once every 12 calendar months, an approved syllabus of training consisting of ground or flight training, or both, or an approved flight instructor refresher course.

(d) Each certificated flight instructor or commercial pilot with a lighter-than-air rating who is assigned to a flight training course must satisfactorily complete the following tasks, which must be administered by the school's chief instructor, assistant chief instructor, or check instructor:

(1) Prior to receiving authorization to train students in a flight training course, must—

(i) Accomplish a review of and receive a briefing on the objectives and standards of that training course; and

(ii) Accomplish an initial proficiency check in each make and model of aircraft used in that training course in which that person provides training; and

(2) Every 12 calendar months after the month in which the person last complied with the requirements of paragraph (d)(1)(ii) of this section, accomplish a recurrent proficiency check in one of the aircraft in which the person trains students.

[Amdt. 141-8, 62 FR 16347, Apr. 4, 1997; Amdt. 141-9, 62 FR 40908, July 30, 1997]

§ 141.81   Ground training.

(a) Except as provided in paragraph (b) of this section, each instructor who is assigned to a ground training course must hold a flight or ground instructor certificate, or a commercial pilot certificate with a lighter-than-air rating, with the appropriate rating for that course of training.

(b) A person who does not meet the requirements of paragraph (a) of this section may be assigned ground training duties in a ground training course, if:

(1) The chief instructor who is assigned to that ground training course finds the person qualified to give that training; and

(2) The training is given while under the supervision of the chief instructor or the assistant chief instructor who is present at the facility when the training is given.

(c) An instructor may not be used in a ground training course until that instructor has been briefed on to the objectives and standards of that course by the chief instructor, assistant chief instructor, or check instructor.

[Amdt. 141-8, 62 FR 16347, Apr. 4, 1997; Amdt. 141-9, 62 FR 40908, July 30, 1997]

§ 141.83   Quality of training.

(a) Each pilot school or provisional pilot school must meet the following requirements:

(1) Comply with its approved training course; and

(2) Provide training of such quality that meets the requirements of § 141.5(d) of this part.

(b) The failure of a pilot school or provisional pilot school to maintain the quality of training specified in paragraph (a) of this section may be the basis for suspending or revoking that school's certificate.

(c) When requested by the Administrator, a pilot school or provisional pilot school must allow the FAA to administer any knowledge test, practical test, stage check, or end-of-course test to its students.

(d) When a stage check or end-of-course test is administered by the FAA under the provisions of paragraph (c) of this section, and the student has not completed the training course, then that test will be based on the standards prescribed in the school's approved training course.

(e) When a practical test or knowledge test is administered by the FAA under the provisions of paragraph (c) of this section, to a student who has completed the school's training course, that test will be based upon the areas of operation approved by the Administrator.

[Amdt. 141-8, 62 FR 16347, Apr. 4, 1997; Amdt. 141-9, 62 FR 40908, July 30, 1997]

§ 141.85   Chief instructor responsibilities.

(a) Each person designated as a chief instructor for a pilot school or provisional pilot school shall be responsible for:

(1) Certifying each student's training record, graduation certificate, stage check and end-of-course test report, recommendation for course completion, and application;

(2) Ensuring that each certificated flight instructor, certificated ground instructor, or commercial pilot with a lighter-than-air rating passes an initial proficiency check prior to that instructor being assigned instructing duties in the school's approved training course and thereafter that the instructor passes a recurrent proficiency check every 12 calendar months after the month in which the initial test was accomplished;

(3) Ensuring that each student accomplishes the required stage checks and end-of-course tests in accordance with the school's approved training course; and

(4) Maintaining training techniques, procedures, and standards for the school that are acceptable to the Administrator.

(b) The chief instructor or an assistant chief instructor must be available at the pilot school or, if away from the pilot school, be available by telephone, radio, or other electronic means during the time that training is given for an approved training course.

(c) The chief instructor may delegate authority for conducting stage checks, end-of-course tests, and flight instructor proficiency checks to the assistant chief instructor or a check instructor.

[Amdt. 141-8, 62 FR 16347, Apr. 4, 1997; Amdt. 141-9, 62 FR 40908, July 30, 1997]

§ 141.87   Change of chief instructor.

Whenever a pilot school or provisional pilot school makes a change of designation of its chief instructor, that school:

(a) Must immediately provide the FAA Flight Standards District Office that has jurisdiction over the area in which the school is located with written notification of the change;

(b) May conduct training without a chief instructor for that training course for a period not to exceed 60 days while awaiting the designation and approval of another chief instructor;

(c) May, for a period not to exceed 60 days, have the stage checks and end-of-course tests administered by:

(1) The training course's assistant chief instructor, if one has been designated;

(2) The training course's check instructor, if one has been designated;

(3) An FAA inspector; or

(4) An examiner.

(d) Must, after 60 days without a chief instructor, cease operations and surrender its certificate to the Administrator; and

(e) May have its certificate reinstated, upon:

(1) Designating and approving another chief instructor;

(2) Showing it meets the requirements of § 141.27(a)(2) of this part; and

(3) Applying for reinstatement on a form and in a manner prescribed by the Administrator.

§ 141.89 Maintenance of personnel, facilities, and equipment.

The holder of a pilot school certificate or provisional pilot school certificate may not provide training to a student who is enrolled in an approved course of training unless:

(a) Each airport, aircraft, and facility necessary for that training meets the standards specified in the holder's approved training course outline and the appropriate requirements of this part; and

(b) Except as provided in § 141.87 of this part, each chief instructor, assistant chief instructor, check instructor, or instructor meets the qualifications specified in the holder's approved course of training and the appropriate requirements of this part.

§ 141.91 Satellite bases.

The holder of a pilot school certificate or provisional pilot school certificate may conduct ground training or flight training in an approved course of training at a base other than its main operations base if:

(a) An assistant chief instructor is designated for each satellite base, and that assistant chief instructor is available at that base or, if away from the premises, by telephone, radio, or other electronic means during the time that training is provided for an approved training course;

(b) The airport, facilities, and personnel used at the satellite base meet the appropriate requirements of subpart B of this part and its approved training course outline;

(c) The instructors are under the direct supervision of the chief instructor or assistant chief instructor for the appropriate training course, who is readily available for consultation in accordance with § 141.85(b) of this part; and

(d) The FAA Flight Standards District Office having jurisdiction over the area in which the school is located is notified in writing if training is conducted at a base other than the school's main operations base for more than 7 consecutive days.

[Amdt. 141-8, 62 FR 16347, Apr. 4, 1997; Amdt. 141-9, 62 FR 40908, July 30, 1997]

§ 141.93 Enrollment.

(a) The holder of a pilot school certificate or a provisional pilot school certificate must, at the time a student is enrolled in an approved training course, furnish that student with a copy of the following:

(1) A certificate of enrollment containing—

(i) The name of the course in which the student is enrolled; and

(ii) The date of that enrollment.

(2) A copy of the student's training syllabus.

(3) A copy of the safety procedures and practices developed by the school that describe the use of the school's facilities and the operation of its aircraft. Those procedures and practices shall include training on at least the following information—

(i) The weather minimums required by the school for dual and solo flights;

(ii) The procedures for starting and taxiing aircraft on the ramp;

(iii) Fire precautions and procedures;

(iv) Redispatch procedures after unprogrammed landings, on and off airports;

(v) Aircraft discrepancies and approval for return-to-service determinations;

(vi) Securing of aircraft when not in use;

(vii) Fuel reserves necessary for local and cross-country flights;

(viii) Avoidance of other aircraft in flight and on the ground;

(ix) Minimum altitude limitations and simulated emergency landing instructions; and

(x) A description of and instructions regarding the use of assigned practice areas.

(b) The holder of a pilot school certificate or provisional pilot school certificate must maintain a monthly listing of persons enrolled in each training course offered by the school.

[Amdt. 141-8, 62 FR 16347, Apr. 4, 1997; Amdt. 141-9, 62 FR 40908, July 30, 1997]

§ 141.95 Graduation certificate.

(a) The holder of a pilot school certificate or provisional pilot school certificate must issue a graduation certificate to each student who completes its approved course of training.

(b) The graduation certificate must be issued to the student upon completion of the course of training and contain at least the following information:

(1) The name of the school and the certificate number of the school;

(2) The name of the graduate to whom it was issued;

(3) The course of training for which it was issued;

(4) The date of graduation;

(5) A statement that the student has satisfactorily completed each required stage of the approved course of training including the tests for those stages;

(6) A certification of the information contained on the graduation certificate by the chief instructor for that course of training; and

(7) A statement showing the cross-country training that the student received in the course of training.

[Amdt. 141-8, 62 FR 16347, Apr. 4, 1997; Amdt. 141-9, 62 FR 40908, July 30, 1997]

Subpart F—Records

§ 141.101 Training records.

(a) Each holder of a pilot school certificate or provisional pilot school certificate must establish and maintain a current and accurate record of the participation of each student enrolled in an approved course of training conducted by the school that includes the following information:

(1) The date the student was enrolled in the approved course;

(2) A chronological log of the student's course attendance, subjects, and flight operations covered in the student's training, and the names and grades of any tests taken by the student; and

(3) The date the student graduated, terminated training, or transferred to another school.

(b) The records required to be maintained in a student's logbook will not suffice for the record required by paragraph (a) of this section.

(c) Whenever a student graduates, terminates training, or transfers to another school, the student's record must be certified to that effect by the chief instructor.

(d) The holder of a pilot school certificate or a provisional pilot school certificate must retain each student record required by this section for at least 1 year from the date that the student:

(1) Graduates from the course to which the record pertains;

(2) Terminates enrollment in the course to which the record pertains; or

(3) Transfers to another school.

(e) The holder of a pilot school certificate or a provisional pilot school certificate must make a copy of the student's training record available upon request by the student.

[Amdt. 141-8, 62 FR 16347, Apr. 4, 1997; Amdt. 141-9, 62 FR 40908, July 30, 1997]

Appendix A To Part 141—Recreational Pilot Certification Course

1. Applicability. This appendix prescribes the minimum curriculum required for a recreational pilot certification course under this part, for the following ratings:

(a) Airplane single-engine.

(b) Rotorcraft helicopter.

(c) Rotorcraft gyroplane.

2. Eligibility for enrollment. A person must hold a student pilot certificate prior to enrolling in the flight portion of the recreational pilot certification course.

3. Aeronautical knowledge training. Each approved course must include at least 20 hours of ground training on the following aeronautical knowledge areas, appropriate to the aircraft category and class for which the course applies:

(a) Applicable Federal Aviation Regulations for recreational pilot privileges, limitations, and flight operations;

(b) Accident reporting requirements of the National Transportation Safety Board;

(c) Applicable subjects in the "Aeronautical Information Manual" and the appropriate FAA advisory circulars;

(d) Use of aeronautical charts for VFR navigation using pilotage with the aid of a magnetic compass;

(e) Recognition of critical weather situations from the ground and in flight, windshear avoidance, and the procurement and use of aeronautical weather reports and forecasts;

(f) Safe and efficient operation of aircraft, including collision avoidance, and recognition and avoidance of wake turbulence;

(g) Effects of density altitude on takeoff and climb performance;

(h) Weight and balance computations;

(i) Principles of aerodynamics, powerplants, and aircraft systems;

(j) Stall awareness, spin entry, spins, and spin recovery techniques, if applying for an airplane single-engine rating;

(k) Aeronautical decision making and judgment; and

(l) Preflight action that includes—

(1) How to obtain information on runway lengths at airports of intended use, data on takeoff and landing distances, weather reports and forecasts, and fuel requirements; and

(2) How to plan for alternatives if the planned flight cannot be completed or delays are encountered.

4. Flight training.

(a) Each approved course must include at least 30 hours of flight training (of which 15 hours must be with a certificated flight instructor and 3 hours must be solo flight training) as provided in section No. 5 of this appendix on the approved areas of operation listed in paragraph (c) of this section that are appropriate to the aircraft category and class rating for which the course applies, including:

(1) Except as provided in § 61.100 of this chapter, 2 hours of dual flight training to and at an airport that is located more than 25 nautical miles from the airport where the applicant normally trains, with at least three takeoffs and three landings; and

(2) 3 hours of dual flight training in an aircraft that is appropriate to the aircraft category and class for which the course applies, in preparation for the practical test within 60 days preceding the date of the test.

(b) Each training flight must include a preflight briefing and a postflight critique of the student by the flight instructor assigned to that flight.

(c) Flight training must include the following approved areas of operation appropriate to the aircraft category and class rating—

(1) For an airplane single-engine course:

(i) Preflight preparation;

(ii) Preflight procedures;

(iii) Airport operations;

(iv) Takeoffs, landings, and go-arounds;

(v) Performance maneuvers;

(vi) Ground reference maneuvers;

(vii) Navigation;

(viii) Slow flight and stalls;

(ix) Emergency operations; and

(x) Postflight procedures.

(2) For a rotorcraft helicopter course:

(i) Preflight preparation;

(ii) Preflight procedures;

(iii) Airport and heliport operations;

(iv) Hovering maneuvers;

(v) Takeoffs, landings, and go-arounds;

(vi) Performance maneuvers;

(vii) Navigation;

(viii) Emergency operations; and

(ix) Postflight procedures.

(3) For a rotorcraft gyroplane course:

(i) Preflight preparation;

(ii) Preflight procedures;

(iii) Airport operations;

(iv) Takeoffs, landings, and go-arounds;

(v) Performance maneuvers;

(vi) Ground reference maneuvers;

(vii) Navigation;

(viii) Flight at slow airspeeds;

(ix) Emergency operations; and

(x) Postflight procedures.

5. Solo flight training. Each approved course must include at least 3 hours of solo flight training on the approved areas of operation listed in paragraph (c) of section No. 4 of this appendix that are appropriate to the aircraft category and class rating for which the course applies.

6. Stage checks and end-of-course tests.

(a) Each student enrolled in a recreational pilot course must satisfactorily accomplish the stage checks and end-of-course tests, in accordance with the school's approved training course, consisting of the approved areas of operation listed in paragraph (c) of section No. 4 of this appendix that are appropriate to the aircraft category and class rating for which the course applies.

(b) Each student must demonstrate satisfactory proficiency prior to receiving an endorsement to operate an aircraft in solo flight.

[Amdt. 141-8, 62 FR 16347, Apr. 4, 1997; Amdt. 141-9, 62 FR 40908, July 30, 1997]

Appendix B To Part 141—Private Pilot Certification Course

1. Applicability. This appendix prescribes the minimum curriculum for a private pilot certification course required under this part, for the following ratings:

(a) Airplane single-engine.

(b) Airplane multiengine.

(c) Rotorcraft helicopter.

(d) Rotorcraft gyroplane.

(e) Powered-lift.

(f) Glider.

(g) Lighter-than-air airship.

(h) Lighter-than-air balloon.

2. Eligibility for enrollment. A person must hold a recreational or student pilot certificate prior to enrolling in the flight portion of the private pilot certification course.

3. Aeronautical knowledge training.

(a) Each approved course must include at least the following ground training on the aeronautical knowledge areas listed in paragraph (b) of this section, appropriate to the aircraft category and class rating:

(1) 35 hours of training if the course is for an airplane, rotorcraft, or powered-lift category rating.

(2) 15 hours of training if the course is for a glider category rating.

(3) 10 hours of training if the course is for a lighter-than-air category with a balloon class rating.

(4) 35 hours of training if the course is for a lighter-than-air category with an airship class rating.

(b) Ground training must include the following aeronautical knowledge areas:

(1) Applicable Federal Aviation Regulations for private pilot privileges, limitations, and flight operations;

(2) Accident reporting requirements of the National Transportation Safety Board;

(3) Applicable subjects of the "Aeronautical Information Manual" and the appropriate FAA advisory circulars;

(4) Aeronautical charts for VFR navigation using pilotage, dead reckoning, and navigation systems;

(5) Radio communication procedures;

(6) Recognition of critical weather situations from the ground and in flight, windshear avoidance, and the procurement and use of aeronautical weather reports and forecasts;

(7) Safe and efficient operation of aircraft, including collision avoidance, and recognition and avoidance of wake turbulence;

(8) Effects of density altitude on takeoff and climb performance;

(9) Weight and balance computations;

(10) Principles of aerodynamics, powerplants, and aircraft systems;

(11) If the course of training is for an airplane category or glider category rating, stall awareness, spin entry, spins, and spin recovery techniques;

(12) Aeronautical decision making and judgment; and

(13) Preflight action that includes—

(i) How to obtain information on runway lengths at airports of intended use, data on takeoff and landing distances, weather reports and forecasts, and fuel requirements; and

(ii) How to plan for alternatives if the planned flight cannot be completed or delays are encountered.

4. Flight training.

(a) Each approved course must include at least the following flight training, as provided in this section and section No. 5 of this appendix, on the approved areas of operation listed in paragraph (d) of this section, appropriate to the aircraft category and class rating:

(1) 35 hours of training if the course is for an airplane, rotorcraft, powered-lift, or airship rating.

(2) 6 hours of training if the course is for a glider rating.

(3) 8 hours of training if the course is for a balloon rating.

(b) Each approved course must include at least the following flight training:

(1) For an airplane single-engine course: 20 hours of flight training from a certificated flight instructor on the approved areas of operation in paragraph (d)(1) of this section that includes at least—

(i) Except as provided in § 61.111 of this chapter, 3 hours of cross-country flight training in a single-engine airplane;

(ii) 3 hours of night flight training in a single-engine airplane that includes—

(A) One cross-country flight of more than 100-nautical-miles total distance; and

(B) 10 takeoffs and 10 landings to a full stop (with each landing involving a flight in the traffic pattern) at an airport.

(iii) 3 hours of instrument training in a single-engine airplane; and

(iv) 3 hours of flight training in a single-engine airplane in preparation for the practical test within 60 days preceding the date of the test.

(2) For an airplane multiengine course: 20 hours of flight training from a certificated flight instructor on the approved areas of operation in paragraph (d)(2) of this section that includes at least—

(i) Except as provided in § 61.111 of this chapter, 3 hours of cross-country flight training in a multiengine airplane;

(ii) 3 hours of night flight training in a multiengine airplane that includes—

(A) One cross-country flight of more than 100-nautical-miles total distance; and

(B) 10 takeoffs and 10 landings to a full stop (with each landing involving a flight in the traffic pattern) at an airport.

(iii) 3 hours of instrument training in a multiengine airplane; and

(iv) 3 hours of flight training in a multiengine airplane in preparation for the practical test within 60 days preceding the date of the test.

(3) For a rotorcraft helicopter course: 20 hours of flight training from a certificated flight instructor on the approved areas of operation in paragraph (d)(3) of this section that includes at least—

(i) Except as provided in § 61.111 of this chapter, 3 hours of cross-country flight training in a helicopter.

(ii) 3 hours of night flight training in a helicopter that includes—

(A) One cross-country flight of more than 50-nautical-miles total distance; and

(B) 10 takeoffs and 10 landings to a full stop (with each landing involving a flight in the traffic pattern) at an airport.

(iii) 3 hours of flight training in a helicopter in preparation for the practical test within 60 days preceding the date of the test.

(4) For a rotorcraft gyroplane course: 20 hours of flight training from a certificated flight instructor on the approved areas of operation in paragraph (d)(4) of this section that includes at least—

(i) Except as provided in § 61.111 of this chapter, 3 hours of cross-country flight training in a gyroplane.

(ii) 3 hours of night flight training in a gyroplane that includes—

(A) One cross-country flight over 50-nautical-miles total distance; and

(B) 10 takeoffs and 10 landings to a full stop (with each landing involving a flight in the traffic pattern) at an airport.

(iii) 3 hours of flight training in a gyroplane in preparation for the practical test within 60 days preceding the date of the test.

(5) For a powered-lift course: 20 hours of flight training from a certificated flight instructor on the approved areas of operation in paragraph (d)(5) of this section that includes at least—

(i) Except as provided in § 61.111 of this chapter, 3 hours of cross-country flight training in a powered-lift;

(ii) 3 hours of night flight training in a powered-lift that includes—

(A) One cross-country flight of more than 100-nautical-miles total distance; and

(B) 10 takeoffs and 10 landings to a full stop (with each landing involving a flight in the traffic pattern) at an airport.

(iii) 3 hours of instrument training in a powered-lift; and

(iv) 3 hours of flight training in a powered-lift in preparation for the practical test, within 60 days preceding the date of the test.

(6) For a glider course: 4 hours of flight training from a certificated flight instructor on the approved areas of operation in paragraph (d)(6) of this section that includes at least—

(i) Five training flights in a glider with a certified instructor on launch/tow procedures approved for the course and in the appropriate approved areas of operation listed in paragraph (d)(6) of this section; and

(ii) Three training flights in a glider with a certified instructor in preparation for the practical test within 60 days preceding the date of the test.

(7) For a lighter-than-air airship course: 20 hours of flight training from a commercial pilot with an airship rating on the approved areas of operation in paragraph (d)(7) of this section that includes at least—

(i) Except as provided in § 61.111 of this chapter, 3 hours of cross-country flight training in an airship;

(ii) 3 hours of night flight training in an airship that includes—

(A) One cross-country flight over 25-nautical-miles total distance; and

(B) Five takeoffs and five landings to a full stop (with each landing involving a flight in the traffic pattern) at an airport.

(iii) 3 hours of instrument training in an airship; and

(iv) 3 hours of flight training in an airship in preparation for the practical test within 60 days preceding the date of the test.

(8) For a lighter-than-air balloon course: 8 hours of flight training, including at least five training flights, from a commercial pilot with a balloon rating on the approved areas of operation in paragraph (d)(8) of this section, that includes—

(i) If the training is being performed in a gas balloon—

(A) Two flights of 1 hour each;

(B) One flight involving a controlled ascent to 3,000 feet above the launch site; and

(C) Two flights in preparation for the practical test within 60 days preceding the date of the test.

(ii) If the training is being performed in a balloon with an airborne heater—

(A) Two flights of 30 minutes each;

(B) One flight involving a controlled ascent to 2,000 feet above the launch site; and

(C) Two flights in preparation for the practical test within 60 days preceding the date of the test.

(c) For use of flight simulators or flight training devices:

(1) The course may include training in a flight simulator or flight training device, provided it is representative of the aircraft for which the course is approved, meets the requirements of this paragraph, and the training is given by an authorized instructor.

(2) Training in a flight simulator that meets the requirements of § 141.41(a) of this part may be credited for a maximum of 20 percent of the total flight training hour requirements of the approved course, or of this section, whichever is less.

(3) Training in a flight training device that meets the requirements of § 141.41(b) of this part may be credited for a maximum of 15 percent of the total flight training hour requirements of the approved course, or of this section, whichever is less.

(4) Training in flight simulators or flight training devices described in paragraphs (c)(2) and (c)(3) of this section, if used in combination, may be credited for a maximum of 20 percent of the total flight training hour re-

quirements of the approved course, or of this section, whichever is less. However, credit for training in a flight training device that meets the requirements of § 141.41(b) cannot exceed the limitation provided for in paragraph (c)(3) of this section.

(d) Each approved course must include the flight training on the approved areas of operation listed in this paragraph that are appropriate to the aircraft category and class rating—

(1) For a single-engine airplane course:
(i) Preflight preparation;
(ii) Preflight procedures;
(iii) Airport and seaplane base operations;
(iv) Takeoffs, landings, and go-arounds;
(v) Performance maneuvers;
(vi) Ground reference maneuvers;
(vii) Navigation;
(viii) Slow flight and stalls;
(ix) Basic instrument maneuvers;
(x) Emergency operations;
(xi) Night operations, and
(xii) Postflight procedures.

(2) For a multiengine airplane course:
(i) Preflight preparation;
(ii) Preflight procedures;
(iii) Airport and seaplane base operations;
(iv) Takeoffs, landings, and go-arounds;
(v) Performance maneuvers;
(vi) Ground reference maneuvers;
(vii) Navigation;
(viii) Slow flight and stalls;
(ix) Basic instrument maneuvers;
(x) Emergency operations;
(xi) Multiengine operations;
(xii) Night operations; and
(xiii) Postflight procedures.

(3) For a rotorcraft helicopter course:
(i) Preflight preparation;
(ii) Preflight procedures;
(iii) Airport and heliport operations;
(iv) Hovering maneuvers;
(v) Takeoffs, landings, and go-arounds;
(vi) Performance maneuvers;
(vii) Navigation;
(viii) Emergency operations;
(ix) Night operations; and
(x) Postflight procedures.

(4) For a rotorcraft gyroplane course:
(i) Preflight preparation;
(ii) Preflight procedures;
(iii) Airport operations;
(iv) Takeoffs, landings, and go-arounds;
(v) Performance maneuvers;
(vi) Ground reference maneuvers;
(vii) Navigation;

(viii) Flight at slow airspeeds;

(ix) Emergency operations;

(x) Night operations; and

(xi) Postflight procedures.

(5) For a powered-lift course:

(i) Preflight preparation;

(ii) Preflight procedures;

(iii) Airport and heliport operations;

(iv) Hovering maneuvers;

(v) Takeoffs, landings, and go-arounds;

(vi) Performance maneuvers;

(vii) Ground reference maneuvers;

(viii) Navigation;

(ix) Slow flight and stalls;

(x) Basic instrument maneuvers;

(xi) Emergency operations;

(xii) Night operations; and

(xiii) Postflight procedures.

(6) For a glider course:

(i) Preflight preparation;

(ii) Preflight procedures;

(iii) Airport and gliderport operations;

(iv) Launches/tows, as appropriate, and landings;

(v) Performance speeds;

(vi) Soaring techniques;

(vii) Performance maneuvers;

(viii) Navigation;

(ix) Slow flight and stalls;

(x) Emergency operations; and

(xi) Postflight procedures.

(7) For a lighter-than-air airship course:

(i) Preflight preparation;

(ii) Preflight procedures;

(iii) Airport operations;

(iv) Takeoffs, landings, and go-arounds;

(v) Performance maneuvers;

(vi) Ground reference maneuvers;

(vii) Navigation;

(viii) Emergency operations; and

(ix) Postflight procedures.

(8) For a lighter-than-air balloon course:

(i) Preflight preparation;

(ii) Preflight procedures;

(iii) Airport operations;

(iv) Launches and landings;

(v) Performance maneuvers;

(vi) Navigation;

(vii) Emergency operations; and

(viii) Postflight procedures.

5. Solo flight training. Each approved course must include at least the following solo flight training:

(a) For an airplane single-engine course: 5 hours of solo flight training in a single-engine airplane on the approved areas of operation in paragraph (d)(1) of section No. 4 of this appendix that includes at least—

(1) One solo cross-country flight of at least 100 nautical miles with landings at a minimum of three points, and one segment of the flight consisting of a straight-line distance of at least 50 nautical miles between the takeoff and landing locations; and

(2) Three takeoffs and three landings to a full stop (with each landing involving a flight in the traffic pattern) at an airport with an operating control tower.

(b) For an airplane multiengine course: 5 hours of flight training in a multiengine airplane performing the duties of a pilot in command while under the supervision of a certificated flight instructor. The training must consist of the approved areas of operation in paragraph (d)(2) of section No. 4 of this appendix, and include at least—

(1) One cross-country flight of at least 100 nautical miles with landings at a minimum of three points, and one segment of the flight consisting of a straight-line distance of at least 50 nautical miles between the takeoff and landing locations; and

(2) Three takeoffs and three landings to a full stop (with each landing involving a flight in the traffic pattern) at an airport with an operating control tower.

(c) For a rotorcraft helicopter course: 5 hours of solo flight training in a helicopter on the approved areas of operation in paragraph (d)(3) of section No. 4 of this appendix that includes at least—

(1) One solo cross-country flight of more than 50 nautical miles with landings at a minimum of three points, and one segment of the flight consisting of a straight-line distance of at least 25 nautical miles between the takeoff and landing locations; and

(2) Three takeoffs and three landings to a full stop (with each landing involving a flight in the traffic pattern) at an airport with an operating control tower.

(d) For a rotorcraft gyroplane course: 5 hours of solo flight training in gyroplanes on the approved areas of operation in paragraph (d)(4) of section No. 4 of this appendix that includes at least—

(1) One solo cross-country flight of more than 50 nautical miles with landings at a minimum of three points, and one segment of the flight consisting of a straight-line distance of at least 25 nautical miles between the takeoff and landing locations; and

(2) Three takeoffs and three landings to a full stop (with each landing involving a flight in the traffic pattern) at an airport with an operating control tower.

(e) For a powered-lift course: 5 hours of solo flight training in a powered-lift on the approved areas of operation in paragraph (d)(5) of section No. 4 of this appendix that includes at least—

(1) One solo cross-country flight of at least 100 nautical miles with landings at a minimum of three points, and one segment of the flight consisting of a straight-line distance of at least 50 nautical miles between the takeoff and landing locations; and

(2) Three takeoffs and three landings to a full stop (with each landing involving a flight in the traffic pattern) at an airport with an operating control tower.

(f) For a glider course: Two solo flights in a glider on the approved areas of operation in paragraph (d)(6) of section No. 4 of this appendix, and the launch and tow procedures appropriate for the approved course.

(g) For a lighter-than-air airship course: 5 hours of flight training in an airship performing the duties of pilot in command while under the supervision of a commercial pilot with an airship rating. The training shall consist of the approved areas of operation in paragraph (d)(7) of section No. 4 of this appendix.

(h) For a lighter-than-air balloon course: Two solo flights in a balloon with an airborne heater if the course involves a balloon with an airborne heater, or, if the course involves a gas balloon, at least two flights in a gas balloon performing the duties of pilot in command while under the supervision of a commercial pilot with a balloon rating. The training shall consist of the approved areas of operation in paragraph (d)(8) of section No. 4 of this appendix, in the kind of balloon for which the course applies.

6. Stage checks and end-of-course tests.

(a) Each student enrolled in a private pilot course must satisfactorily accomplish the stage checks and end-of-course tests in accordance with the school's approved training course, consisting of the approved areas of operation listed in paragraph (d) of section No. 4 of this appendix that are appropriate to the aircraft category and class rating for which the course applies.

(b) Each student must demonstrate satisfactory proficiency prior to receiving an endorsement to operate an aircraft in solo flight.

[Amdt. 141-8, 62 FR 16347, Apr. 4, 1997; Amdt. 141-9, 62 FR 40908, July 30, 1997]

Appendix C to Part 141—Instrument Rating Course

1. Applicability. This appendix prescribes the minimum curriculum for an instrument rating course and an additional instrument rating course, required under this part, for the following ratings:

(a) Instrument—airplane.

(b) Instrument—helicopter.

(c) Instrument—powered-lift.

2. Eligibility for enrollment. A person must hold at least a private pilot certificate with an aircraft category and class rating appropriate to the instrument rating for which the course applies prior to enrolling in the flight portion of the instrument rating course.

3. Aeronautical knowledge training.

(a) Each approved course must include at least the following ground training on the aeronautical knowledge areas listed in paragraph (b) of this section appropriate to the instrument rating for which the course applies:

(1) 30 hours of training if the course is for an initial instrument rating.

(2) 20 hours of training if the course is for an additional instrument rating.

(b) Ground training must include the following aeronautical knowledge areas:

(1) Applicable Federal Aviation Regulations for IFR flight operations;

(2) Appropriate information in the "Aeronautical Information Manual";

(3) Air traffic control system and procedures for instrument flight operations;

(4) IFR navigation and approaches by use of navigation systems;

(5) Use of IFR en route and instrument approach procedure charts;

(6) Procurement and use of aviation weather reports and forecasts, and the elements of forecasting weather trends on the basis of that information and personal observation of weather conditions;

(7) Safe and efficient operation of aircraft under instrument flight rules and conditions;

(8) Recognition of critical weather situations and windshear avoidance;

(9) Aeronautical decision making and judgment; and

(10) Crew resource management, to include crew communication and coordination.

4. Flight training.

(a) Each approved course must include at least the following flight training on the approved areas of operation listed in paragraph (d) of this section, appropriate to the instrument-aircraft category and class rating for which the course applies:

(1) 35 hours of instrument training if the course is for an initial instrument rating.

(2) 15 hours of instrument training if the course is for an additional instrument rating.

(b) For the use of flight simulators or flight training devices—

(1) The course may include training in a flight simulator or flight training device, provided it is representative of the aircraft for which the course is approved, meets the requirements of this paragraph, and the training is given by an authorized instructor.

(2) Training in a flight simulator that meets the requirements of § 141.41(a) of this part may be credited for a maximum of 50 percent of the total flight training hour requirements of the approved course, or of this section, whichever is less.

(3) Training in a flight training device that meets the requirements of § 141.41(b) of this part may be credited for a maximum of 40 percent of the total flight training hour requirements of the approved course, or of this section, whichever is less.

(4) Training in flight simulators or flight training devices described in paragraphs (b)(2) and (b)(3) of this section, if used in combination, may be credited for a maximum of 50 percent of the total flight training hour requirements of the approved course, or of this section, whichever is less. However, credit for training in a flight training device that meets the requirements of § 141.41(b) cannot exceed the limitation provided for in paragraph (b)(3) of this section.

(c) Each approved course must include the following flight training—

(1) For an instrument airplane course: Instrument training time from a certificated flight instructor with an instrument rating on the approved areas of operation in paragraph (d) of this section including at least one cross-country flight that—

(i) Is in the category and class of airplane that the course is approved for, and is performed under IFR;

(ii) Is a distance of at least 250 nautical miles along airways or ATC-directed routing with one segment of the flight consisting of at least a straight-line distance of 100 nautical miles between airports;

(iii) Involves an instrument approach at each airport; and

(iv) Involves three different kinds of approaches with the use of navigation systems.

(2) For an instrument helicopter course: Instrument training time from a certificated flight instructor with an instrument rating on the approved areas of operation in paragraph (d) of this section including at least one cross-country flight that—

(i) Is in a helicopter and is performed under IFR;

(ii) Is a distance of at least 100 nautical miles along airways or ATC-directed routing with one segment of the flight consisting of at least a straight-line distance of 50 nautical miles between airports;

(iii) Involves an instrument approach at each airport; and

(iv) Involves three different kinds of approaches with the use of navigation systems.

(3) For an instrument powered-lift course: Instrument training time from a certificated flight instructor with an instrument rating on the approved areas of operation in paragraph (d) of this section including at least one cross-country flight that—

(i) Is in a powered-lift and is performed under IFR;

(ii) Is a distance of at least 250 nautical miles along airways or ATC-directed routing with one segment of the flight consisting of at least a straight-line distance of 100 nautical miles between airports;

(iii) Involves an instrument approach at each airport; and

(iv) Involves three different kinds of approaches with the use of navigation systems.

(d) Each approved course must include the flight training on the approved areas of operation listed in this paragraph appropriate to the instrument aircraft category and class rating for which the course applies:

(1) Preflight preparation;

(2) Preflight procedures;

(3) Air traffic control clearances and procedures;

(4) Flight by reference to instruments;

(5) Navigation systems;

(6) Instrument approach procedures;

(7) Emergency operations; and

(8) Postflight procedures.

5. Stage checks and end-of-course tests. Each student enrolled in an instrument rating course must satisfactorily accomplish the stage checks and end-of-course tests, in accordance with the school's approved training course, consisting of the approved areas of operation listed in paragraph (d) of section No. 4 of this appendix that are appropriate to the aircraft category and class rating for which the course applies.

[Amdt. 141-8, 62 FR 16347, Apr. 4, 1997; Amdt. 141-9, 62 FR 40909, July 30, 1997]

Appendix D to Part 141—Commercial Pilot Ccrtification Course

1. Applicability. This appendix prescribes the minimum curriculum for a commercial pilot certification course required under this part, for the following ratings:

(a) Airplane single-engine.

(b) Airplane multiengine.

(c) Rotorcraft helicopter.

(d) Rotorcraft gyroplane.

(e) Powered-lift.

(f) Glider.

(g) Lighter-than-air airship.

(h) Lighter-than-air balloon.

2. Eligibility for enrollment. A person must hold the following prior to enrolling in the flight portion of the commercial pilot certification course:

(a) At least a private pilot certificate; and

(b) If the course is for a rating in an airplane or a powered-lift category, then the person must:

(1) Hold an instrument rating in the aircraft that is appropriate to the aircraft category rating for which the course applies; or

(2) Be concurrently enrolled in an instrument rating course that is appropriate to the aircraft category rating for which the course applies, and pass the required instrument rating practical test prior to completing the commercial pilot certification course.

3. Aeronautical knowledge training.

(a) Each approved course must include at least the following ground training on the aeronautical knowledge areas listed in paragraph (b) of this section, appropriate to the aircraft category and class rating for which the course applies:

(1) 35 hours of training if the course is for an airplane category rating or a powered-lift category rating.

(2) 65 hours of training if the course is for a lighter-than-air category with an airship class rating.

(3) 30 hours of training if the course is for a rotorcraft category rating.

(4) 20 hours of training if the course is for a glider category rating.

(5) 20 hours of training if the course is for lighter-than-air category with a balloon class rating.

(b) Ground training must include the following aeronautical knowledge areas:

(1) Federal Aviation Regulations that apply to commercial pilot privileges, limitations, and flight operations;

(2) Accident reporting requirements of the National Transportation Safety Board;

(3) Basic aerodynamics and the principles of flight;

(4) Meteorology, to include recognition of critical weather situations, windshear recognition and avoidance, and the use of aeronautical weather reports and forecasts;

(5) Safe and efficient operation of aircraft;

(6) Weight and balance computations;

(7) Use of performance charts;

(8) Significance and effects of exceeding aircraft performance limitations;

(9) Use of aeronautical charts and a magnetic compass for pilotage and dead reckoning;

(10) Use of air navigation facilities;

(11) Aeronautical decision making and judgment;

(12) Principles and functions of aircraft systems;

(13) Maneuvers, procedures, and emergency operations appropriate to the aircraft;

(14) Night and high-altitude operations;

(15) Descriptions of and procedures for operating within the National Airspace System; and

(16) Procedures for flight and ground training for lighter-than-air ratings.

4. Flight training.

(a) Each approved course must include at least the following flight training, as provided in this section and section No. 5 of this appendix, on the approved areas of operation listed in paragraph (d) of this section that are appropriate to the aircraft category and class rating for which the course applies:

(1) 120 hours of training if the course is for an airplane or powered-lift rating.

(2) 155 hours of training if the course is for an airship rating.

(3) 115 hours of training if the course is for a rotorcraft rating.

(4) 6 hours of training if the course is for a glider rating.

(5) 10 hours of training and 8 training flights if the course is for a balloon rating.

(b) Each approved course must include at least the following flight training:

(1) For an airplane single-engine course: 55 hours of flight training from a certificated flight instructor on the approved areas of operation listed in paragraph (d)(1) of this section that includes at least—

(i) 5 hours of instrument training in a single-engine airplane;

(ii) 10 hours of training in a single-engine airplane that has retractable landing gear, flaps, and a controllable pitch propeller, or is turbine-powered;

(iii) One cross-country flight in a single-engine airplane of at least a 2-hour duration, a total straight-line distance of more than 100 nautical miles from the original point of departure, and occurring in day VFR conditions;

(iv) One cross-country flight in a single-engine airplane of at least a 2-hour duration, a total straight-line distance of more than 100 nautical miles from the original point of departure, and occurring in night VFR conditions; and

(v) 3 hours in a single-engine airplane in preparation for the practical test within 60 days preceding the date of the test.

(2) For an airplane multiengine course: 55 hours of flight training from a certificated flight instructor on the approved areas of operation listed in paragraph (d)(2) of this section that includes at least—

(i) 5 hours of instrument training in a multiengine airplane;

(ii) 10 hours of training in a multiengine airplane that has retractable landing gear, flaps, and a controllable pitch propeller, or is turbine-powered;

(iii) One cross-country flight in a multiengine airplane of at least a 2-hour duration, a total straight-line distance of more than 100 nautical miles from the original point of departure, and occurring in day VFR conditions;

(iv) One cross-country flight in a multiengine airplane of at least a 2-hour duration, a total straight-line distance of more than 100 nautical miles from the original point of departure, and occurring in night VFR conditions; and

(v) 3 hours in a multiengine airplane in preparation for the practical test within 60 days preceding the date of the test.

(3) For a rotorcraft helicopter course: 30 hours of flight training from a certificated flight instructor on the approved areas of operation listed in paragraph (d)(3) of this section that includes at least—

(i) 5 hours of instrument training;

(ii) One cross-country flight in a helicopter of at least a 2-hour duration, a total straight-line distance of more than 50 nautical miles from the original point of departure and occurring in day VFR conditions;

(iii) One cross-country flight in a helicopter of at least a 2-hour duration, a total straight-line distance of more than 50 nautical miles from the original point of departure, and occurring in night VFR conditions; and

(iv) 3 hours in a helicopter in preparation for the practical test within 60 days preceding the date of the test.

(4) For a rotorcraft gyroplane course: 30 hours of flight training from a certificated flight instructor on the approved areas of operation listed in paragraph (d)(4) of this section that includes at least—

(i) 5 hours of instrument training;

(ii) One cross-country flight in a gyroplane of at least a 2-hour duration, a total straight-line distance of more than 50 nautical miles from the original point of departure, and occurring in day VFR conditions;

(iii) One cross-country flight in a gyroplane of at least a 2-hour duration, a total straight-line distance of more than 50 nautical miles from the original point of departure, and occurring in night VFR conditions; and

(iv) 3 hours in a gyroplane in preparation for the practical test within 60 days preceding the date of the test.

(5) For a powered-lift course: 55 hours of flight training from a certificated flight instructor on the approved areas of operation listed in paragraph (d)(5) of this section that includes at least—

(i) 5 hours of instrument training in a powered-lift;

(ii) One cross-country flight in a powered-lift of at least a 2-hour duration, a total straight-line distance of more than 100 nautical miles from the original point of departure, and occurring in day VFR conditions;

(iii) One cross-country flight in a powered-lift of at least a 2-hour duration, a total straight-line distance of more than 100 nautical miles from the original point of departure, and occurring in night VFR conditions; and

(iv) 3 hours in a powered-lift in preparation for the practical test within 60 days preceding the date of the test.

(6) For a glider course: 4 hours of flight training from a certificated flight instructor on the approved areas of operation in paragraph (d)(6) of this section, that includes at least—

(i) Five training flights in a glider with a certified flight instructor on launch/tow procedures approved for the course and on the appropriate approved areas of operation listed in paragraph (d)(6) of this section; and

(ii) Three training flights in a glider with a certified flight instructor in preparation for the practical test within the 60 days preceding the date of the test.

(7) For a lighter-than-air airship course: 55 hours of flight training in airships from a commercial pilot with an airship rating on the approved areas of operation in paragraph (d)(7) of this section that includes at least—

(i) 3 hours of instrument training in an airship;

(ii) One cross-country flight in an airship of at least a 1-hour duration, a total straight-line distance of more than 25 nautical miles from the original point of departure, and occurring in day VFR conditions; and

(iii) One cross-country flight in an airship of at least a 1-hour duration, a total straight-line distance of more than 25 nautical miles from the original point of departure, and occurring in night VFR conditions; and

(iv) 3 hours in an airship, in preparation for the practical test within 60 days preceding the date of the test.

(8) For a lighter-than-air balloon course: Flight training from a commercial pilot with a balloon rating on the approved areas of operation in paragraph (d)(8) of this section that includes at least—

(i) If the course involves training in a gas balloon:

(A) Two flights of 1 hour each;

(B) One flight involving a controlled ascent to at least 5,000 feet above the launch site; and

(C) Two flights in preparation for the practical test within 60 days preceding the date of the test.

(ii) If the course involves training in a balloon with an airborne heater:

(A) Two flights of 30 minutes each;

(B) One flight involving a controlled ascent to at least 3,000 feet above the launch site; and

(C) Two flights in preparation for the practical test within 60 days preceding the date of the test.

(c) For the use of flight simulators or flight training devices:

(1) The course may include training in a flight simulator or flight training device, provided it is representative of the aircraft for which the course is approved, meets the requirements of this paragraph, and is given by an authorized instructor.

(2) Training in a flight simulator that meets the requirements of § 141.41(a) of this part may be credited for a maximum of 30 percent of the total flight training hour requirements of the approved course, or of this section, whichever is less.

(3) Training in a flight training device that meets the requirements of § 141.41(b) of this part may be credited for a maximum of 20 percent of the total flight training hour requirements of the approved course, or of this section, whichever is less.

(4) Training in the flight training devices described in paragraphs (c)(2) and (c)(3) of this section, if used in combination, may be credited for a maximum of 30 percent of the total flight training hour requirements of the approved course, or of this section, whichever is less. However, credit for training in a flight training device that meets the requirements of § 141.41(b) cannot exceed the limitation provided for in paragraph (c)(3) of this section.

(d) Each approved course must include the flight training on the approved areas of operation listed in this paragraph that are appropriate to the aircraft category and class rating—

(1) For an airplane single-engine course:

(i) Preflight preparation;

(ii) Preflight procedures;

(iii) Airport and seaplane base operations;

(iv) Takeoffs, landings, and go-arounds;

(v) Performance maneuvers;

(vi) Navigation;

(vii) Slow flight and stalls;

(viii) Emergency operations;

(ix) High-altitude operations; and

(x) Postflight procedures.

(2) For an airplane multiengine course:

(i) Preflight preparation;

(ii) Preflight procedures;

(iii) Airport and seaplane base operations;

(iv) Takeoffs, landings, and go-arounds;

(v) Performance maneuvers;

(vi) Navigation;

(vii) Slow flight and stalls;

(viii) Emergency operations;

(ix) Multiengine operations;

(x) High-altitude operations; and

(xi) Postflight procedures.

(3) For a rotorcraft helicopter course:

(i) Preflight preparation;

(ii) Preflight procedures;

(iii) Airport and heliport operations;

(iv) Hovering maneuvers;

(v) Takeoffs, landings, and go-arounds;

(vi) Performance maneuvers;

(vii) Navigation;

(viii) Emergency operations;

(ix) Special operations; and

(x) Postflight procedures.

(4) For a rotorcraft gyroplane course:

(i) Preflight preparation;

(ii) Preflight procedures;

(iii) Airport operations;

(iv) Takeoffs, landings, and go-arounds;

(v) Performance maneuvers;

(vi) Navigation;

(vii) Flight at slow airspeeds;

(viii) Emergency operations; and

(ix) Postflight procedures.

(5) For a powered-lift course:

(i) Preflight preparation;

(ii) Preflight procedures;

(iii) Airport and heliport operations;

(iv) Hovering maneuvers;

(v) Takeoffs, landings, and go-arounds;

(vi) Performance maneuvers;

(vii) Navigation;

(viii) Slow flight and stalls;

(ix) Emergency operations;

(x) High altitude operations;

(xi) Special operations; and

(xii) Postflight procedures.

(6) For a glider course:

(i) Preflight preparation;

(ii) Preflight procedures;

(iii) Airport and gliderport operations;

(iv) Launches/tows, as appropriate, and landings;

(v) Performance speeds;

(vi) Soaring techniques;

(vii) Performance maneuvers;

(viii) Navigation;

(ix) Slow flight and stalls;

(x) Emergency operations; and

(xi) Postflight procedures.

(7) For a lighter-than-air airship course:

(i) Fundamentals of instructing;

(ii) Technical subjects;

(iii) Preflight preparation;

(iv) Preflight lessons on a maneuver to be performed in flight;

(v) Preflight procedures;

(vi) Airport operations;

(vii) Takeoffs, landings, and go-arounds;

(viii) Performance maneuvers;

(ix) Navigation;

(x) Emergency operations; and

(xi) Postflight procedures.

(8) For a lighter-than-air balloon course:

(i) Fundamentals of instructing;

(ii) Technical subjects;

(iii) Preflight preparation;

(iv) Preflight lesson on a maneuver to be performed in flight;

(v) Preflight procedures;

(vi) Airport operations;

(vii) Launches and landings;

(viii) Performance maneuvers;

(ix) Navigation;

(x) Emergency operations; and

(xi) Postflight procedures.

5. Solo training. Each approved course must include at least the following solo flight training:

(a) For an airplane single-engine course: 10 hours of solo flight training in a single-engine airplane on the approved areas of operation in paragraph (d)(1) of section No. 4 of this appendix that includes at least—

(1) One cross-country flight, if the training is being performed in the State of Hawaii, with landings at a minimum of three points, and one of the segments consisting of a straight-line distance of at least 150 nautical miles;

(2) One cross-country flight, if the training is being performed in a State other than Hawaii, with landings at a minimum of three points, and one segment of the flight consisting of a straight-line distance of at least 250 nautical miles; and

(3) 5 hours in night VFR conditions with 10 takeoffs and 10 landings (with each landing involving a flight with a traffic pattern) at an airport with an operating control tower.

(b) For an airplane multiengine course: 10 hours of flight training in a multiengine airplane performing the duties of pilot in command while under the supervision of a certificated flight instructor. The training must consist of the approved areas of operation in paragraph (d)(2) of section No. 4 of this appendix, and include at least—

(1) One cross-country flight, if the training is being performed in the State of Hawaii, with landings at a minimum of three points, and one of the segments consisting of a straight-line distance of at least 150 nautical miles;

(2) One cross-country flight, if the training is being performed in a State other than Hawaii, with landings at a minimum of three points and one segment of the flight consisting of straight-line distance of at least 250 nautical miles; and

(3) 5 hours in night VFR conditions with 10 takeoffs and 10 landings (with each landing involving a flight with a traffic pattern) at an airport with an operating control tower.

(c) For a rotorcraft helicopter course: 10 hours of solo flight training in a helicopter on the approved areas of operation in paragraph (d)(3) of section No. 4 of this appendix that includes at least—

(1) One cross-country flight with landings at a minimum of three points and one segment of the flight consisting of a straight-line distance of at least 50 nautical miles from the original point of departure; and

(2) 5 hours in night VFR conditions with 10 takeoffs and 10 landings (with each landing involving a flight with a traffic pattern) at an airport with an operating control tower.

(d) For a rotorcraft-gyroplane course: 10 hours of solo flight training in a gyroplane on the approved areas of operation in paragraph (d)(4) of section No. 4 of this appendix that includes at least—

(1) One cross-country flight with landings at a minimum of three points, and one segment of the flight consisting of a straight-line distance of at least 50 nautical miles from the original point of departure; and

(2) 5 hours in night VFR conditions with 10 takeoffs and 10 landings (with each landing involving a flight with a traffic pattern) at an airport with an operating control tower.

(e) For a powered-lift course: 10 hours of solo flight training in a powered-lift on the approved areas of operation in paragraph (d)(5) of section No. 4 of this appendix that includes at least—

(1) One cross-country flight, if the training is being performed in the State of Hawaii, with landings at a minimum of three points, and one segment of the flight consisting of a straight-line distance of at least 150 nautical miles;

(2) One cross-country flight, if the training is being performed in a State other than Hawaii, with landings at a minimum of three points, and one segment of the flight consisting of a straight-line distance of at least 250 nautical miles; and

(3) 5 hours in night VFR conditions with 10 takeoffs and 10 landings (with each landing involving a flight with a traffic pattern) at an airport with an operating control tower.

(f) For a glider course: 5 solo flights in a glider on the approved areas of operation in paragraph (d)(6) of section No. 4 of this appendix.

(g) For a lighter-than-air airship course: 10 hours of flight training in an airship, while performing the duties of pilot in command under the supervision of a commercial pilot with an airship rating. The training shall consist of the approved areas of operation in paragraph (d)(7) of section No. 4 of this appendix and include at least—

(1) One cross-country flight with landings at a minimum of three points, and one segment of the flight consisting of a straight-line distance of at least 25 nautical miles from the original point of departure; and

(2) 5 hours in night VFR conditions with 10 takeoffs and 10 landings (with each landing involving a flight with a traffic pattern).

(h) For a lighter-than-air balloon course: Two solo flights if the course is for a hot air balloon rating, or, if the course is for a gas balloon rating, at least two flights in a gas balloon, while performing the duties of pilot in command under the supervision of a commercial pilot with a balloon rating. The training shall consist of the approved areas of operation in paragraph (d)(8) of section No. 4 of this appendix, in the kind of balloon for which the course applies.

6. Stage checks and end-of-course tests.

(a) Each student enrolled in a commercial pilot course must satisfactorily accomplish the stage checks and end-of-course tests, in accordance with the school's approved training course, consisting of the approved areas of operation listed in paragraph (d) of section No. 4 of this appendix that are appropriate to aircraft category and class rating for which the course applies.

(b) Each student must demonstrate satisfactory proficiency prior to receiving an endorsement to operate an aircraft in solo flight.

[Amdt. 141-8, 62 FR 16347, Apr. 4, 1997; Amdt. 141-9, 62 FR 40909, July 30, 1997]

Appendix E to Part 141—Airline Transport Pilot Certification Course

1. Applicability. This appendix prescribes the minimum curriculum for an airline transport pilot certification course under this part, for the following ratings:

(a) Airplane single-engine.

(b) Airplane multiengine.

(c) Rotorcraft helicopter.

(d) Powered-lift.

2. Eligibility for enrollment. Prior to enrolling in the flight portion of the airline transport pilot certification course, a person must:

(a) Meet the aeronautical experience requirements prescribed in subpart G of part 61 of this chapter for an airline transport pilot certificate that is appropriate to the aircraft category and class rating for which the course applies;

(b) Hold at least a commercial pilot certificate and an instrument rating;

(c) Meet the military experience requirements under § 61.73 of this chapter to qualify for a commercial pilot certificate and an instrument rating, if the person is a rated military pilot or former rated military pilot of an Armed Force of the United States; or

(d) Hold either a foreign airline transport pilot license or foreign commercial pilot license and an instrument rating, if the person holds a pilot license issued by a contracting State to the Convention on International Civil Aviation.

3. Aeronautical knowledge areas.

(a) Each approved course must include at least 40 hours of ground training on the aeronautical knowledge areas listed in paragraph (b) of this section, appropriate to the aircraft category and class rating for which the course applies.

(b) Ground training must include the following aeronautical knowledge areas:

(1) Applicable Federal Aviation Regulations of this chapter that relate to airline transport pilot privileges, limitations, and flight operations;

(2) Meteorology, including knowledge of and effects of fronts, frontal characteristics, cloud formations, icing, and upper-air data;

(3) General system of weather and NOTAM collection, dissemination, interpretation, and use;

(4) Interpretation and use of weather charts, maps, forecasts, sequence reports, abbreviations, and symbols;

(5) National Weather Service functions as they pertain to operations in the National Airspace System;

(6) Windshear and microburst awareness, identification, and avoidance;

(7) Principles of air navigation under instrument meteorological conditions in the National Airspace System;

(8) Air traffic control procedures and pilot responsibilities as they relate to en route operations, terminal area and radar operations, and instrument departure and approach procedures;

(9) Aircraft loading; weight and balance; use of charts, graphs, tables, formulas, and computations; and the effects on aircraft performance;

(10) Aerodynamics relating to an aircraft's flight characteristics and performance in normal and abnormal flight regimes;

(11) Human factors;

(12) Aeronautical decision making and judgment; and

(13) Crew resource management to include crew communication and coordination.

4. Flight training.

(a) Each approved course must include at least 25 hours of flight training on the approved areas of operation listed in paragraph (c) of this section appropriate to the aircraft category and class rating for which the course applies. At least 15 hours of this flight training must be instrument flight training.

(b) For the use of flight simulators or flight training devices—

(1) The course may include training in a flight simulator or flight training device, provided it is representative of the aircraft for which the course is approved, meets the requirements of this paragraph, and the training is given by an authorized instructor.

(2) Training in a flight simulator that meets the requirements of § 141.41(a) of this part may be credited for a maximum of 50 percent of the total flight training hour requirements of the approved course, or of this section, whichever is less.

(3) Training in a flight training device that meets the requirements of § 141.41(b) of this part may be credited for a maximum of 25 percent of the total flight training hour requirements of the approved course, or of this section, whichever is less.

(4) Training in flight simulators or flight training devices described in paragraphs (b)(2) and (b)(3) of this section, if used in combination, may be credited for a maximum of 50 percent of the total flight training hour requirements of the approved course, or of this section, whichever is less. However, credit for training in a flight training device that meets the requirements of § 141.41(b) cannot exceed the limitation provided for in paragraph (b)(3) of this section.

(c) Each approved course must include flight training on the approved areas of operation listed in this paragraph appropriate to the aircraft category and class rating for which the course applies:

(1) Preflight preparation;

(2) Preflight procedures;

(3) Takeoff and departure phase;

(4) In-flight maneuvers;

(5) Instrument procedures;

(6) Landings and approaches to landings;

(7) Normal and abnormal procedures;

(8) Emergency procedures; and

(9) Postflight procedures.

5. Stage checks and end-of-course tests.

(a) Each student enrolled in an airline transport pilot course must satisfactorily accomplish the stage checks and end-of-course tests, in accordance with the school's approved training course, consisting of the approved areas of operation listed in paragraph (c) of section No. 4 of this appendix that are appropriate to the aircraft category and class rating for which the course applies.

(b) Each student must demonstrate satisfactory proficiency prior to receiving an endorsement to operate an aircraft in solo flight.

[Amdt. 141-8, 62 FR 16347, Apr. 4, 1997; Amdt. 141-9, 62 FR 40909, July 30, 1997]

Appendix F to Part 141—Flight Instructor Certification Course

1. *Applicability.* This appendix prescribes the minimum curriculum for a flight instructor certification course and an additional flight instructor rating course required under this part, for the following ratings:

(a) Airplane single-engine.

(b) Airplane multiengine.

(c) Rotorcraft helicopter.

(d) Rotorcraft gyroplane.

(e) Powered-lift.

(f) Glider category.

2. *Eligibility for enrollment.* A person must hold the following prior to enrolling in the flight portion of the flight instructor or additional flight instructor rating course:

(a) A commercial pilot certificate or an airline transport pilot certificate, with an aircraft category and class rating appropriate to the flight instructor rating for which the course applies; and

(b) An instrument rating or privilege in an aircraft that is appropriate to the aircraft category and class rating for which the course applies, if the course is for a flight instructor airplane or powered-lift instrument rating.

3. *Aeronautical knowledge training.*

(a) Each approved course must include at least the following ground training in the aeronautical knowledge areas listed in paragraph (b) of this section:

(1) 40 hours of training if the course is for an initial issuance of a flight instructor certificate; or

(2) 20 hours of training if the course is for an additional flight instructor rating.

(b) Ground training must include the following aeronautical knowledge areas:

(1) The fundamentals of instructing including—

(i) The learning process;

(ii) Elements of effective teaching;

(iii) Student evaluation and testing;

(iv) Course development;

(v) Lesson planning; and

(vi) Classroom training techniques.

(2) The aeronautical knowledge areas in which training is required for—

(i) A recreational, private, and commercial pilot certificate that is appropriate to the aircraft category and class rating for which the course applies; and

(ii) An instrument rating that is appropriate to the aircraft category and class rating for which the course applies, if the course is for an airplane or powered-lift aircraft rating.

(c) A student who satisfactorily completes 2 years of study on the principles of education at a college or university may be credited with no more than 20 hours of the training required in paragraph (a)(1) of this section.

4. *Flight training.*

(a) Each approved course must include at least the following flight training on the approved areas of operation of paragraph (c) of this section appropriate to the flight instructor rating for which the course applies:

(1) 25 hours, if the course is for an airplane, rotorcraft, or powered-lift rating; and

(2) 10 hours which must include 10 flights, if the course is for a glider category rating.

(b) For the use of flight simulators or flight training devices:

(1) The course may include training in a flight simulator or flight training device, provided it is representative of the aircraft for which the course is approved, meets the requirements of this paragraph, and the training is given by an authorized instructor.

(2) Training in a flight simulator that meets the requirements of § 141.41(a) of this part, may be credited for a maximum of 10 percent of the total flight training hour requirements of the approved course, or of this section, whichever is less.

(3) Training in a flight training device that meets the requirements of § 141.41(b) of this part, may be credited for a maximum of 5 percent of the total flight training hour requirements of the approved course, or of this section, whichever is less.

(4) Training in flight simulators or flight training devices described in paragraphs (b)(2) and (b)(3) of this section, if used in combination, may be credited for a maximum of 10 percent of the total flight training hour requirements of the approved course, or of this section, whichever is less. However, credit for training in a flight training device that meets the requirements of § 141.41(b) cannot exceed the limitation provided for in paragraph (b)(3) of this section.

(c) Each approved course must include flight training on the approved areas of operation listed in this paragraph that are appropriate to the aircraft category and class rating for which the course applies—

(1) For an airplane—single-engine course:

(i) Fundamentals of instructing;

(ii) Technical subject areas;

(iii) Preflight preparation;

(iv) Preflight lesson on a maneuver to be performed in flight;

(v) Preflight procedures;

(vi) Airport and seaplane base operations;

(vii) Takeoffs, landings, and go-arounds;

(viii) Fundamentals of flight;

(ix) Performance maneuvers;

(x) Ground reference maneuvers;

(xi) Slow flight, stalls, and spins;

(xii) Basic instrument maneuvers;

(xiii) Emergency operations; and

(xiv) Postflight procedures.

(2) For an airplane—multiengine course:

(i) Fundamentals of instructing;

(ii) Technical subject areas;

(iii) Preflight preparation;

(iv) Preflight lesson on a maneuver to be performed in flight;

(v) Preflight procedures;

(vi) Airport and seaplane base operations;

(vii) Takeoffs, landings, and go-arounds;

(viii) Fundamentals of flight;

(ix) Performance maneuvers;

(x) Ground reference maneuvers;

(xi) Slow flight and stalls;

(xii) Basic instrument maneuvers;

(xiii) Emergency operations;

(xiv) Multiengine operations; and

(xv) Postflight procedures.

(3) For a rotorcraft—helicopter course:

(i) Fundamentals of instructing;

(ii) Technical subject areas;

(iii) Preflight preparation;

(iv) Preflight lesson on a maneuver to be performed in flight;

(v) Preflight procedures;

(vi) Airport and heliport operations;

(vii) Hovering maneuvers;

(viii) Takeoffs, landings, and go-arounds;

(ix) Fundamentals of flight;

(x) Performance maneuvers;

(xi) Emergency operations;

(xii) Special operations; and

(xiii) Postflight procedures.

(4) For a rotorcraft—gyroplane course:

(i) Fundamentals of instructing;

(ii) Technical subject areas;

(iii) Preflight preparation;

(iv) Preflight lesson on a maneuver to be performed in flight;

(v) Preflight procedures;

(vi) Airport operations;

(vii) Takeoffs, landings, and go-arounds;

(viii) Fundamentals of flight;

(ix) Performance maneuvers;

(x) Flight at slow airspeeds;

(xi) Ground reference maneuvers;

(xii) Emergency operations; and

(xiii) Postflight procedures.

(5) For a powered-lift course:

(i) Fundamentals of instructing;

(ii) Technical subject areas;

(iii) Preflight preparation;

(iv) Preflight lesson on a maneuver to be performed in flight;

(v) Preflight procedures;

(vi) Airport and heliport operations;

(vii) Hovering maneuvers;

(viii) Takeoffs, landings, and go-arounds;

(ix) Fundamentals of flight;

(x) Performance maneuvers;

(xi) Ground reference maneuvers;

(xii) Slow flight and stalls;

(xiii) Basic instrument maneuvers;

(xiv) Emergency operations;

(xv) Special operations; and

(xvi) Postflight procedures.

(6) For a glider course:

(i) Fundamentals of instructing;

(ii) Technical subject areas;

(iii) Preflight preparation;

(iv) Preflight lesson on a maneuver to be performed in flight;

(v) Preflight procedures;

(vi) Airport and gliderport operations;

(vii) Tows or launches, landings, and go-arounds, if applicable;

(viii) Fundamentals of flight;

(ix) Performance speeds;

(x) Soaring techniques;

(xi) Performance maneuvers;

(xii) Slow flight, stalls, and spins;

(xiii) Emergency operations; and

(xiv) Postflight procedures.

5. Stage checks and end-of-course tests.

(a) Each student enrolled in a flight instructor course must satisfactorily accomplish the stage checks and end-of-course tests, in accordance with the school's approved training course, consisting of the appropriate approved areas of operation listed in paragraph (c) of section No. 4 of this appendix appropriate to the flight instructor rating for which the course applies.

(b) In the case of a student who is enrolled in a flight instructor-airplane rating or flight instructor-glider rating course, that student must have:

(1) Received a logbook endorsement from a certificated flight instructor certifying the student received ground and flight training on stall awareness, spin entry, spins, and spin recovery procedures in an aircraft that is certificated for spins and is appropriate to the rating sought; and

(2) Demonstrated instructional proficiency in stall awareness, spin entry, spins, and spin recovery procedures.

[Amdt. 141-8, 62 FR 16347, Apr. 4, 1997; Amdt. 141-9, 62 FR 40909, July 30, 1997]

Appendix G to Part 141—Flight Instructor Instrument (For an Airplane, Helicopter, or Powered-Lift Instrument Instructor Rating, as Appropriate) Certification Course

1. Applicability. This appendix prescribes the minimum curriculum for a flight instructor instrument certification course required under this part, for the following ratings:

(a) Flight Instructor Instrument—Airplane.

(b) Flight Instructor Instrument—Helicopter.

(c) Flight Instructor Instrument—Powered-lift aircraft.

2. Eligibility for enrollment. A person must hold the following prior to enrolling in the flight portion of the flight instructor instrument course:

(a) A commercial pilot certificate or airline transport pilot certificate with an aircraft category and class rating appropriate to the flight instructor category and class rating for which the course applies; and

(b) An instrument rating or privilege on that flight instructor applicant's pilot certificate that is appropriate to the flight instructor instrument rating (for an airplane-, helicopter-, or powered-lift-instrument rating, as appropriate) for which the course applies.

3. Aeronautical knowledge training.

(a) Each approved course must include at least 15 hours of ground training on the aeronautical knowledge areas listed in paragraph (b) of this section, appropriate to the flight instructor instrument rating (for an airplane-, helicopter-, or powered-lift-instrument rating, as appropriate) for which the course applies:

(b) Ground training must include the following aeronautical knowledge areas:

(1) The fundamentals of instructing including:

(i) The learning process;

(ii) Elements of effective teaching;

(iii) Student evaluation and testing;

(iv) Course development;

(v) Lesson planning; and

(vi) Classroom training techniques.

(2) The aeronautical knowledge areas in which training is required for an instrument rating that is appropriate to the aircraft category and class rating for the course which applies.

4. Flight training.

(a) Each approved course must include at least 15 hours of flight training in the approved areas of operation of paragraph (c) of this section appropriate to the flight instructor rating for which the course applies.

(b) For the use of flight simulators or flight training devices:

(1) The course may include training in a flight simulator or flight training device, provided it is representative of the aircraft for which the course is approved for, meets requirements of this paragraph, and the training is given by an instructor.

(2) Training in a flight simulator that meets the requirements of § 141.41(a) of this part, may be credited for a maximum of 10 percent of the total flight training hour requirements of the approved course, or of this section, whichever is less.

(3) Training in a flight training device that meets the requirements of § 141.41(b) of this part, may be credited for a maximum of 5 percent of the total flight training hour re-

quirements of the approved course, or of this section, whichever is less.

(4) Training in flight simulators or flight training devices described in paragraphs (b)(2) and (b)(3) of this section, if used in combination, may be credited for a maximum of 10 percent of the total flight training hour requirements of the approved course, or of this section, whichever is less. However, credit for training in a flight training device that meets the requirements of § 141.41(b) cannot exceed the limitation provided for in paragraph (b)(3) of this section.

(c) An approved course for the flight instructor-instrument rating must include flight training on the following approved areas of operation that are appropriate to the instrument-aircraft category and class rating for which the course applies:

(1) Fundamentals of instructing;

(2) Technical subject areas;

(3) Preflight preparation;

(4) Preflight lesson on a maneuver to be performed in flight;

(5) Air traffic control clearances and procedures;

(6) Flight by reference to instruments;

(7) Navigation systems;

(8) Instrument approach procedures;

(9) Emergency operations; and

(10) Postflight procedures.

5. Stage checks and end-of-course tests. Each student enrolled in a flight instructor instrument course must satisfactorily accomplish the stage checks and end-of-course tests, in accordance with the school's approved training course, consisting of the approved areas of operation listed in paragraph (c) of section No. 4 of this appendix that are appropriate to the flight instructor instrument rating (for an airplane-, helicopter-, or powered-lift- instrument rating, as appropriate) for which the course applies.

[Amdt. 141-8, 62 FR 16347, Apr. 4, 1997; Amdt. 141-9, 62 FR 40909, July 30, 1997]

Appendix H to Part 141—Ground Instructor Certification Course

1. Applicability. This appendix prescribes the minimum curriculum for a ground instructor certification course and an additional ground instructor rating course, required under this part, for the following ratings:

(a) Ground Instructor—Basic.

(b) Ground Instructor—Advanced.

(c) Ground Instructor—Instrument.

2. Aeronautical knowledge training.

(a) Each approved course must include at least the following ground training on the knowledge areas listed in paragraphs (b), (c), (d), and (e) of this section, appropriate to the ground instructor rating for which the course applies:

(1) 20 hours of training if the course is for an initial issuance of a ground instructor certificate; or

(2) 10 hours of training if the course is for an additional ground instructor rating.

(b) Ground training must include the following aeronautical knowledge areas:

(1) Learning process;

(2) Elements of effective teaching;

(3) Student evaluation and testing;

(4) Course development;

(5) Lesson planning; and

(6) Classroom training techniques.

(c) Ground training for a basic ground instructor certificate must include the aeronautical knowledge areas applicable to a recreational and private pilot.

(d) Ground training for an advanced ground instructor rating must include the aeronautical knowledge areas applicable to a recreational, private, commercial, and airline transport pilot.

(e) Ground training for an instrument ground instructor rating must include the aeronautical knowledge areas applicable to an instrument rating.

(f) A student who satisfactorily completed 2 years of study on the principles of education at a college or university may be credited with 10 hours of the training required in paragraph (a)(1) of this section.

3. Stage checks and end-of-course tests. Each student enrolled in a ground instructor course must satisfactorily accomplish the stage checks and end-of-course tests, in accordance with the school's approved training course, consisting of the approved knowledge areas in paragraph (b), (c), (d), and (e) of section No. 2 of this appendix appropriate to the ground instructor rating for which the course applies.

Appendix I to Part 141—Additional Aircraft Category or Class Rating Course

1. Applicability. This appendix prescribes the minimum curriculum for an additional aircraft category rating course or an additional aircraft class rating course required under this part, for the following ratings:

(a) Airplane single-engine.

(b) Airplane multiengine.

(c) Rotorcraft helicopter.

(d) Rotorcraft gyroplane.

(e) Powered-lift.

(f) Glider.

(g) Lighter-than-air airship.

(h) Lighter-than-air balloon.

2. Eligibility for enrollment. A person must hold the level of pilot certificate for the additional aircraft category and class rating for which the course applies prior to enrolling in the flight portion of an additional aircraft category or additional aircraft class rating course.

3. Aeronautical knowledge training. Each approved course for an additional aircraft category rating or additional aircraft class rating must include the ground training time requirements and ground training on the aeronautical knowledge areas that are specific to that aircraft category and class rating and pilot certificate level for which the course applies as required in appendix A, B, D, or E of this part, as appropriate.

4. Flight training.

(a) Each approved course for an additional aircraft category rating or additional aircraft class rating must include the flight training time requirements and flight training on the areas of operation that are specific to that aircraft category and class rating and pilot certificate level for which the course applies as required in appendix A, B, D, or E of this part, as appropriate.

(b) For the use of flight simulators or flight training devices:

(1) The course may include training in a flight simulator or flight training device, provided it is representative of the aircraft for which the course is approved, meets the requirements of this paragraph, and the training is given by an authorized instructor.

(2) Training in a flight simulator that meets the requirements of § 141.41(a) of this part may be credited for a maximum of 30 percent of the total flight training hour requirements of the approved course, or of this section, whichever is less.

(3) Training in a flight training device that meets the requirements of § 141.41(b) of this part may be credited for a maximum of 20 percent of the total flight training hour requirements of the approved course, or of this section, whichever is less.

(4) Training in the flight simulators or flight training devices described in paragraphs (b)(2) and (b)(3) of this section, if used in combination, may be credited for a maximum of 30 percent of the total flight training hour requirements of the approved course, or of this section, whichever is less. However, credit for training in a flight training device that meets the requirements of § 141.41(b) cannot exceed the limitation provided for in paragraph (c)(3) of this section.

5. Stage checks and end-of-course tests.

(a) Each student enrolled in an additional aircraft category rating course or an additional aircraft class rating course must satisfactorily accomplish the stage checks and end-of-course tests, in accordance with the school's approved training course, consisting of the approved areas of operation in section No. 4 of this appendix that are appropriate to the aircraft category and class rating for which the course applies at the appropriate pilot certificate level.

(b) Each student must demonstrate satisfactory proficiency prior to receiving an endorsement to operate an aircraft in solo flight.

[Amdt. 141-8, 62 FR 16347, Apr. 4, 1997; Amdt. 141-9, 62 FR 40909, July 30, 1997]

Appendix J to Part 141—Aircraft Type Rating Course, For Other Than an Airline Transport Pilot Certificate

1. Applicability. This appendix prescribes the minimum curriculum for an aircraft type rating course other than an airline transport pilot certificate, for:

(a) A type rating in an airplane category—single-engine class.

(b) A type rating in an airplane category—multiengine class.

(c) A type rating in a rotorcraft category—helicopter class.

(d) A type rating in a powered-lift category.

(e) Other aircraft type ratings specified by the Administrator through the aircraft type certificate procedures.

2. Eligibility for enrollment. Prior to enrolling in the flight portion of an aircraft type rating course, a person must hold at least a private pilot certificate and:

(a) An instrument rating in the category and class of aircraft that is appropriate to the aircraft type rating for which the course applies, provided the aircraft's type certificate does not have a VFR limitation; or

(b) Be concurrently enrolled in an instrument rating course in the category and class of aircraft that is appropriate to the aircraft type rating for which the course applies, and pass the required instrument rating practical test concurrently with the aircraft type rating practical test.

3. Aeronautical knowledge training.

(a) Each approved course must include at least 10 hours of ground training on the aeronautical knowledge areas listed in paragraph (b) of this section, appropriate to the aircraft type rating for which the course applies.

(b) Ground training must include the following aeronautical areas:

(1) Proper control of airspeed, configuration, direction, altitude, and attitude in accordance with procedures and limitations contained in the aircraft's flight manual, checklists, or other approved material appropriate to the aircraft type;

(2) Compliance with approved en route, instrument approach, missed approach, ATC, or other applicable procedures that apply to the aircraft type;

(3) Subjects requiring a practical knowledge of the aircraft type and its powerplant, systems, components, operational, and performance factors;

(4) The aircraft's normal, abnormal, and emergency procedures, and the operations and limitations relating thereto;

(5) Appropriate provisions of the approved aircraft's flight manual;

(6) Location of and purpose for inspecting each item on the aircraft's checklist that relates to the exterior and interior preflight; and

(7) Use of the aircraft's prestart checklist, appropriate control system checks, starting procedures, radio and electronic equipment checks, and the selection of proper navigation and communication radio facilities and frequencies.

4. Flight training.

(a) Each approved course must include at least:

(1) Flight training on the approved areas of operation of paragraph (c) of this section in the aircraft type for which the course applies; and

(2) 10 hours of training of which at least 5 hours must be instrument training in the aircraft for which the course applies.

(b) For the use of flight simulators or flight training devices:

(1) The course may include training in a flight simulator or flight training device, provided it is representative of the aircraft for which the course is approved, meets requirements of this paragraph, and the training is given by an authorized instructor.

(2) Training in a flight simulator that meets the requirements of § 141.41(a) of this part, may be credited for a maximum of 50 percent of the total flight training hour requirements of the approved course, or of this section, whichever is less.

(3) Training in a flight training device that meets the requirements of § 141.41(b) of this part, may be credited for a maximum of 25 percent of the total flight training hour requirements of the approved course, or of this section, whichever is less.

(4) Training in the flight simulators or flight training devices described in paragraphs (b)(2) and (b)(3) of this section, if used in combination, may be credited for a maximum of 50 percent of the total flight training hour requirements of the approved course, or of this section, whichever is less. However, credit training in a flight training device that meets the requirements of § 141.41(b) cannot exceed the limitation provided for in paragraph (b)(3) of this section.

(c) Each approved course must include the flight training on the areas of operation listed in this paragraph, that are appropriate to the aircraft category and class rating for which the course applies:

(1) A type rating for an airplane—single-engine course:
(i) Preflight preparation;
(ii) Preflight procedures;
(iii) Takeoff and departure phase;
(iv) In-flight maneuvers;
(v) Instrument procedures;
(vi) Landings and approaches to landings;
(vii) Normal and abnormal procedures;
(viii) Emergency procedures; and
(ix) Postflight procedures.

(2) A type rating for an airplane—multiengine course:
(i) Preflight preparation;
(ii) Preflight procedures;
(iii) Takeoff and departure phase;
(iv) In-flight maneuvers;

(v) Instrument procedures;

(vi) Landings and approaches to landings;

(vii) Normal and abnormal procedures;

(viii) Emergency procedures; and

(ix) Postflight procedures.

(3) A type rating for a powered-lift course:

(i) Preflight preparation;

(ii) Preflight procedures;

(iii) Takeoff and departure phase;

(iv) In-flight maneuvers;

(v) Instrument procedures;

(vi) Landings and approaches to landings;

(vii) Normal and abnormal procedures;

(viii) Emergency procedures; and

(ix) Postflight procedures.

(4) A type rating for a rotorcraft—helicopter course:

(i) Preflight preparation;

(ii) Preflight procedures;

(iii) Takeoff and departure phase;

(iv) In-flight maneuvers;

(v) Instrument procedures;

(vi) Landings and approaches to landings;

(vii) Normal and abnormal procedures;

(viii) Emergency procedures; and

(ix) Postflight procedures.

(5) Other aircraft type ratings specified by the Administrator through aircraft type certificate procedures:

(i) Preflight preparation;

(ii) Preflight procedures;

(iii) Takeoff and departure phase;

(iv) In-flight maneuvers;

(v) Instrument procedures;

(vi) Landings and approaches to landings;

(vii) Normal and abnormal procedures;

(viii) Emergency procedures; and

(ix) Postflight procedures.

5. Stage checks and end-of-course tests.

(a) Each student enrolled in an aircraft type rating course must satisfactorily accomplish the stage checks and end-of-course tests, in accordance with the school's approved training course, consisting of the approved areas of operation that are appropriate to the aircraft type rating for which the course applies at the airline transport pilot certificate level; and

(b) Each student must demonstrate satisfactory proficiency prior to receiving an endorsement to operate an aircraft in solo flight.

[Amdt. 141-8, 62 FR 16347, Apr. 4, 1997; Amdt. 141-9, 62 FR 40910, July 30, 1997]

Appendix K to Part 141—Special Preparation Courses

1. Applicability. This appendix prescribes the minimum curriculum for the special preparation courses that are listed in § 141.11 of this part.

2. Eligibility for enrollment. Prior to enrolling in the flight portion of a special preparation course, a person must hold a pilot certificate, flight instructor certificate, or ground instructor certificate that is appropriate for the exercise of the operating privileges or authorizations sought.

3. General requirements.

(a) To be approved, a special preparation course must:

(1) Meet the appropriate requirements of this appendix; and

(2) Prepare the graduate with the necessary skills, competency, and proficiency to exercise safely the privileges of the certificate, rating, or authorization for which the course is established.

(b) An approved special preparation course must include ground and flight training on the operating privileges or authorization sought, for developing competency, proficiency, resourcefulness, self-confidence, and self-reliance in the student.

4. Use of flight simulators or flight training devices.

(a) The approved special preparation course may include training in a flight simulator or flight training device, provided it is representative of the aircraft for which the course is approved, meets requirements of this paragraph, and the training is given by an authorized instructor.

(b) Training in a flight simulator that meets the requirements of § 141.41(a) of this part, may be credited for a maximum of 10 percent of the total flight training hour requirements of the approved course, or of this section, whichever is less.

(c) Training in a flight training device that meets the requirements of § 141.41(b) of this part, may be credited for a maximum of 5 percent of the total flight training hour requirements of the approved course, or of this section, whichever is less.

(d) Training in the flight simulators or flight training devices described in paragraphs (b) and (c) of this section, if used in combination, may be credited for a maximum of 10 percent of the total flight training hour requirements of the approved course, or of this section, whichever is less. However, credit for training in a flight training device that meets the requirements of § 141.41(b) cannot exceed the limitation provided for in paragraph (c) of this section.

5. Stage check and end-of-course tests. Each person enrolled in a special preparation course must satisfactorily accomplish the stage checks and end-of-course tests, in accordance with the school's approved training course, consisting of the approved areas of operation that are appropriate to the operating privileges or authorization sought, and for which the course applies.

6. Agricultural aircraft operations course. An approved special preparation course for pilots in agricultural aircraft operations must include at least the following—

(a) 25 hours of training on:

(1) Agricultural aircraft operations;

(2) Safe piloting and operating practices and procedures for handling, dispensing, and disposing agricultural and industrial chemicals, including operating in and around congested areas; and

(3) Applicable provisions of part 137 of this chapter.

(b) 15 hours of flight training on agricultural aircraft operations.

7. Rotorcraft external-load operations course. An approved special preparation course for pilots of external-load operations must include at least the following—

(a) 10 hours of training on:

(1) Rotorcraft external-load operations;

(2) Safe piloting and operating practices and procedures for external-load operations, including operating in and around congested areas; and

(3) Applicable provisions of part 133 of this chapter.

(b) 15 hours of flight training on external-load operations.

8. Test pilot course. An approved special preparation course for pilots in test pilot duties must include at least the following—

(a) Aeronautical knowledge training on:

(1) Performing aircraft maintenance, quality assurance, and certification test flight operations;

(2) Safe piloting and operating practices and procedures for performing aircraft maintenance, quality assurance, and certification test flight operations;

(3) Applicable parts of this chapter that pertain to aircraft maintenance, quality assurance, and certification tests; and

(4) Test pilot duties and responsibilities.

(b) 15 hours of flight training on test pilot duties and responsibilities.

9. Special operations course. An approved special preparation course for pilots in special operations that are mission-specific for certain aircraft must include at least the following—

(a) Aeronautical knowledge training on:

(1) Performing that special flight operation;

(2) Safe piloting operating practices and procedures for performing that special flight operation;

(3) Applicable parts of this chapter that pertain to that special flight operation; and

(4) Pilot in command duties and responsibilities for performing that special flight operation.

(b) Flight training:

(1) On that special flight operation; and

(2) To develop skills, competency, proficiency, resourcefulness, self-confidence, and self-reliance in the student for performing that special flight operation in a safe manner.

10. Pilot refresher course. An approved special preparation pilot refresher course for a pilot certificate, aircraft category and class rating, or an instrument rating must include at least the following—

(a) 4 hours of aeronautical knowledge training on:

(1) The aeronautical knowledge areas that are applicable to the level of pilot certificate, aircraft category and class rating, or instrument rating, as appropriate, that pertain to that course;

(2) Safe piloting operating practices and procedures; and

(3) Applicable provisions of parts 61 and 91 of this chapter for pilots.

(b) 6 hours of flight training on the approved areas of operation that are applicable to the level of pilot certificate, aircraft category and class rating, or instrument rating, as appropriate, for performing pilot-in-command duties and responsibilities.

11. Flight instructor refresher course. An approved special preparation flight instructor refresher course must include at least a combined total of 16 hours of aeronautical knowledge training, flight training, or any combination of ground and flight training on the following—

(a) Aeronautical knowledge training on:

(1) The aeronautical knowledge areas of part 61 of this chapter that apply to student, recreational, private, and commercial pilot certificates and instrument ratings;

(2) The aeronautical knowledge areas of part 61 of this chapter that apply to flight instructor certificates;

(3) Safe piloting operating practices and procedures, including airport operations and operating in the National Airspace System; and

(4) Applicable provisions of parts 61 and 91 of this chapter that apply to pilots and flight instructors.

(b) Flight training to review:

(1) The approved areas of operations applicable to student, recreational, private, and commercial pilot certificates and instrument ratings; and

(2) The skills, competency, and proficiency for performing flight instructor duties and responsibilities.

12. Ground instructor refresher course. An approved special preparation ground instructor refresher course must include at least 16 hours of aeronautical knowledge training on:

(a) The aeronautical knowledge areas of part 61 of this chapter that apply to student, recreational, private, and commercial pilots and instrument rated pilots;

(b) The aeronautical knowledge areas of part 61 of this chapter that apply to ground instructors;

(c) Safe piloting operating practices and procedures, including airport operations and operating in the National Airspace System; and

(d) Applicable provisions of parts 61 and 91 of this chapter that apply to pilots and ground instructors.

[Amdt. 141-8, 62 FR 16347, Apr. 4, 1997; Amdt. 141-9, 62 FR 40910, July 30, 1997]

Appendix L to Part 141—Pilot Ground School Course

1. Applicability. This appendix prescribes the minimum curriculum for a pilot ground school course required under this part.

2. General requirements. An approved course of training for a pilot ground school must include training on the aeronautical knowledge areas that are:

(a) Needed to safely exercise the privileges of the certificate, rating, or authority for which the course is established; and

(b) Conducted to develop competency, proficiency, resourcefulness, self-confidence, and self-reliance in each student.

3. Aeronautical knowledge training requirements. Each approved pilot ground school course must include:

(a) The aeronautical knowledge training that is appropriate to the aircraft rating and pilot certificate level for which the course applies; and

(b) An adequate number of total aeronautical knowledge training hours appropriate to the aircraft rating and pilot certificate level for which the course applies.

4. Stage checks and end-of-course tests. Each person enrolled in a pilot ground school course must satisfactorily accomplish the stage checks and end-of-course tests, in accordance with the school's approved training course, consisting of the approved areas of operation that are appropriate to the operating privileges or authorization that graduation from the course will permit and for which the course applies.

Part 142—Training Centers

Authority: 49 U.S.C. 106(g), 40113, 40119, 44101, 44701–44703, 44705, 44707, 44709–44711, 45102–45103, 45301–45302.

Subpart A—General

§ 142.1 Applicability.

(a) This subpart prescribes the requirements governing the certification and operation of aviation training centers. Except as provided in paragraph (b) of this section, this part provides an alternative means to accomplish training required by parts 61, 63, 91, 121, 125, 127, 135, or 137 of this chapter.

(b) Certification under this part is not required for training that is—

(1) Approved under the provisions of parts 63, 91, 121, 127, 135, or 137 of this chapter;

(2) Approved under SFAR 58, Advanced Qualification Programs, for the authorization holder's own employees;

(3) Conducted under part 61 unless that part requires certification under this part;

(4) Conducted by a part 121 certificate holder for another part 121 certificate holder;

(5) Conducted by a part 135 certificate holder for another part 135 certificate holder; or

(6) Conducted by a part 91 fractional ownership program manager for another part 91 fractional ownership program manager.

(c) Except as provided in paragraph (b) of this section, after August 3, 1998, no person may conduct training, testing, or checking in advanced flight training devices or flight simulators without, or in violation of, the certificate and training specifications required by this part.

§ 142.3 Definitions.

As used in this part:

Advanced Flight Training Device as used in this part, means a flight training device as defined in part 61 of this chapter that has a cockpit that accurately replicates a specific make, model, and type aircraft cockpit, and handling characteristics that accurately model the aircraft handling characteristics.

Core Curriculum means a set of courses approved by the Administrator, for use by a training center and its satellite training centers. The core curriculum consists of training which is required for certification. It does not include training for tasks and circumstances unique to a particular user.

Course means—

(1) A program of instruction to obtain pilot certification, qualification, authorization, or currency;

(2) A program of instruction to meet a specified number of requirements of a program for pilot training, certification, qualification, authorization, or currency; or

(3) A curriculum, or curriculum segment, as defined in SFAR 58 of part 121 of this chapter.

Courseware means instructional material developed for each course or curriculum, including lesson plans, flight event descriptions, computer software programs, audiovisual programs, workbooks, and handouts.

Evaluator means a person employed by a training center certificate holder who performs tests for certification, added ratings, authorizations, and proficiency checks that are authorized by the certificate holder's training specification, and who is authorized by the Administrator to administer such checks and tests.

Flight training equipment means flight simulators, as defined in § 61.1(b)(5) of this chapter, flight training devices, as defined in § 61.1(b) of this chapter, and aircraft.

Instructor means a person employed by a training center and designated to provide instruction in accordance with subpart C of this part.

Line-Operational Simulation means simulation conducted using operational-oriented flight scenarios that accurately replicate interaction among flightcrew members and between flightcrew members and dispatch facilities, other crewmembers, air traffic control, and ground operations. Line operational simulation simulations are conducted for training and evaluation purposes and include random, abnormal, and emergency occurrences. Line operational simulation specifically includes line-oriented flight training, special purpose operational training, and line operational evaluation.

Specialty Curriculum means a set of courses that is designed to satisfy a requirement of the Federal Aviation Regulations and that is approved by the Administrator for use by a particular training center or satellite training center. The specialty curriculum includes training requirements unique to one or more training center clients.

Training center means an organization governed by the applicable requirements of this part that provides training, testing, and checking under contract or other arrangement to airmen subject to the requirements of this chapter.

Training program consists of courses, courseware, facilities, flight training equipment, and personnel necessary to accomplish a specific training objective. It may include a core curriculum and a specialty curriculum.

Training specifications means a document issued to a training center certificate holder by the Administrator that prescribes that center's training, checking, and testing authorizations and limitations, and specifies training program requirements.

§ 142.5 Certificate and training specifications required.

(a) No person may operate a certificated training center without, or in violation of, a training center certificate and training specifications issued under this part.

(b) An applicant will be issued a training center certificate and training specifications with appropriate limitations if the applicant shows that it has adequate facilities, equipment, personnel, and courseware required by § 142.11 to conduct training approved under § 142.37.

§ 142.7 Duration of a certificate.

(a) Except as provided in paragraph (b) of this section, a training center certificate issued under this part is effective until the certificate is surrendered or until the Administrator suspends, revokes, or terminates it.

(b) Unless sooner surrendered, suspended, or revoked, a certificate issued under this part for a training center located outside the United States expires at the end of the twelfth month after the month in which it is issued or renewed.

(c) If the Administrator suspends, revokes, or terminates a training center certificate, the holder of that certificate shall return the certificate to the Administrator within 5 working days after being notified that the certificate is suspended, revoked, or terminated.

§ 142.9 Deviation or waivers.

(a) The Administrator may issue deviations or waivers from any of the requirements of this part.

(b) A training center applicant requesting a deviation or waiver under this section must provide the Administrator with information acceptable to the Administrator that shows—

(1) Justification for the deviation or waiver; and

(2) That the deviation or waiver will not adversely affect the quality of instruction or evaluation.

§ 142.11 Application for issuance or amendment.

(a) An application for a training center certificate and training specifications shall—

(1) Be made on a form and in a manner prescribed by the Administrator;

(2) Be filed with the FAA Flight Standards District Office that has jurisdiction over the area in which the applicant's principal business office is located; and

(3) Be made at least 120 calendar days before the beginning of any proposed training or 60 calendar days before effecting an amendment to any approved training, unless a shorter filing period is approved by the Administrator.

(b) Each application for a training center certificate and training specification shall provide—

(1) A statement showing that the minimum qualification requirements for each management position are met or exceeded;

(2) A statement acknowledging that the applicant shall notify the Administrator within 10 working days of any change made in the assignment of persons in the required management positions;

(3) The proposed training authorizations and training specifications requested by the applicant;

(4) The proposed evaluation authorization;

(5) A description of the flight training equipment that the applicant proposes to use;

(6) A description of the applicant's training facilities, equipment, qualifications of personnel to be used, and proposed evaluation plans;

(7) A training program curriculum, including syllabi, outlines, courseware, procedures, and documentation to support the items required in subpart B of this part, upon request by the Administrator.

(8) A description of a recordkeeping system that will identify and document the details of training, qualification, and certification of students, instructors, and evaluators;

(9) A description of quality control measures proposed; and

(10) A method of demonstrating the applicant's qualifications and ability to provide training for a certificate or rating in fewer than the minimum hours prescribed in part 61 of this chapter if the applicant proposes to do so.

(c) The facilities and equipment described in paragraph (b)(6) of this section shall—

(1) Be available for inspection and evaluation prior to approval; and

(2) Be in place and operational at the location of the proposed training center prior to issuance of a certificate under this part.

(d) An applicant who meets the requirements of this part and is approved by the Administrator is entitled to—

(1) A training center certificate containing all business names included on the application under which the certificate holder may conduct operations and the address of each business office used by the certificate holder; and

(2) Training specifications, issued by the Administrator to the certificate holder, containing—

(i) The type of training authorized, including approved courses;

(ii) The category, class, and type of aircraft that may be used for training, testing, and checking.

(iii) For each flight simulator or flight training device, the make, model, and series of airplane or the set of airplanes being simulated and the qualification level assigned, or the make, model, and series of rotorcraft, or set of rotorcraft being simulated and the qualification level assigned;

(iv) For each flight simulator and flight training device subject to qualification evaluation by the Administrator, the identification number assigned by the FAA;

(v) The name and address of all satellite training centers, and the approved courses offered at each satellite training center;

(vi) Authorized deviations or waivers from this part; and

(vii) Any other items the Administrator may require or allow.

(e) The Administrator may deny, suspend, revoke, or terminate a certificate under this part if the Administrator finds that the applicant or the certificate holder—

(1) Held a training center certificate that was revoked, suspended, or terminated within the previous 5 years; or

(2) Employs or proposes to employ a person who—

(i) Was previously employed in a management or supervisory position by the holder of a training center certificate that was revoked, suspended, or terminated within the previous 5 years;

(ii) Exercised control over any certificate holder whose certificate has been revoked, suspended, or terminated within the last 5 years; and

(iii) Contributed materially to the revocation, suspension, or termination of that certificate and who will be employed in a management or supervisory position, or who will be in control of or have a substantial ownership interest in the training center.

(3) Has provided incomplete, inaccurate, fraudulent, or false information for a training center certificate;

(4) Should not be granted a certificate if the grant would not foster aviation safety.

(f) At any time, the Administrator may amend a training center certificate—

(1) On the Administrator's own initiative, under section 609 of the Federal Aviation Act of 1958 (49 U.S.C. 1429), as amended, and part 13 of this chapter; or

(2) Upon timely application by the certificate holder.

(g) The certificate holder must file an application to amend a training center certificate at least 60 calendar days prior to the applicant's proposed effective amendment date unless a different filing period is approved by the Administrator.

§ 142.13   Management and personnel requirements.

An applicant for a training center certificate must show that—

(a) For each proposed curriculum, the training center has, and shall maintain, a sufficient number of instructors who are qualified in accordance with subpart C of this part to perform the duties to which they are assigned;

(b) The training center has designated, and shall maintain, a sufficient number of approved evaluators to provide required checks and tests to graduation candidates within 7 calendar days of training completion for any curriculum leading to airman certificates or ratings, or both;

(c) The training center has, and shall maintain, a sufficient number of management personnel who are qualified and competent to perform required duties; and

(d) A management representative, and all personnel who are designated by the training center to conduct direct student training, are able to understand, read, write, and fluently speak the English language.

§ 142.15   Facilities.

(a) An applicant for, or holder of, a training center certificate shall ensure that—

(1) Each room, training booth, or other space used for instructional purposes is heated, lighted, and ventilated to conform to local building, sanitation, and health codes; and

(2) The facilities used for instruction are not routinely subject to significant distractions caused by flight operations and maintenance operations at the airport.

(b) An applicant for, or holder of, a training center certificate shall establish and maintain a principal business office that is physically located at the address shown on its training center certificate.

(c) The records required to be maintained by this part must be located in facilities adequate for that purpose.

(d) An applicant for, or holder of, a training center certificate must have available exclusively, for adequate periods of time and at a location approved by the Administrator, adequate flight training equipment and courseware, including at least one flight simulator or advanced flight training device.

§ 142.17   Satellite training centers.

(a) The holder of a training center certificate may conduct training in accordance with an approved training program at a satellite training center located if—

(1) The facilities, equipment, personnel, and course content of the satellite training center meet the applicable requirements of this part;

(2) The instructors and evaluators at the satellite training center are under the direct supervision of management personnel of the principal training center;

(3) The Administrator is notified in writing that a particular satellite is to begin operations at least 60 days prior to proposed commencement of operations at the satellite training center; and

(4) The certificate holder's training specifications reflect the name and address of the satellite training center and the approved courses offered at the satellite training center.

(b) The certificate holder's training specifications shall prescribe the operations required and authorized at each satellite training center.

§ 142.21–142.25   [Reserved]

§ 142.27   Display of certificate.

(a) Each holder of a training center certificate must prominently display that certificate in a place accessible to the public in the principal business office of the training center.

(b) A training center certificate and training specifications must be made available for inspection upon request by—

(1) The Administrator.

(2) An authorized representative of the National Transportation Safety Board; or

(3) Any Federal, State, or local law enforcement agency.

§ 142.29   Inspections.

Each certificate holder must allow the Administrator to inspect training center facilities, equipment, and records at any reasonable time and in any reasonable place in order to determine compliance with or to determine initial or continuing eligibility under 49 U.S.C. 44701, 44707, formerly the Federal Aviation Act of 1958, as amended, and the training center's certificate and training specifications.

§ 142.31 Advertising limitations.

(a) A certificate holder may not conduct, and may not advertise to conduct, any training, testing, and checking that is not approved by the Administrator if that training is designed to satisfy any requirement of this chapter.

(b) A certificate holder whose certificate has been surrendered, suspended, revoked, or terminated must—

(1) Promptly remove all indications, including signs, wherever located, that the training center was certificated by the Administrator; and

(2) Promptly notify all advertising agents, or advertising media, or both, employed by the certificate holder to cease all advertising indicating that the training center is certificated by the Administrator.

§ 142.33 Training agreements.

A pilot school certificated under part 141 of this chapter may provide training, testing, and checking for a training center certificated under this part if—

(a) There is a training, testing, and checking agreement between the certificated training center and the pilot school;

(b) The training, testing, and checking provided by the certificated pilot school is approved and conducted in accordance with this part;

(c) The pilot school certificated under part 141 obtains the Administrator's approval for a training course outline that includes the portion of the training, testing, and checking to be conducted under part 141; and

(d) Upon completion of training, testing, and checking conducted under part 141, a copy of each student's training record is forwarded to the part 142 training center and becomes part of the student's permanent training record.

Subpart B—Aircrew Curriculum and Syllabus Requirements

§ 142.35 Applicability.

This subpart prescribes the curriculum and syllabus requirements for the issuance of a training center certificate and training specifications for training, testing, and checking conducted to meet the requirements of part 61 of this chapter.

§ 142.37 Approval of flight aircrew training program.

(a) Except as provided in paragraph (b) of this section, each applicant for, or holder of, a training center certificate must apply to the Administrator for training program approval.

(b) A curriculum approved under SFAR 58 of part 121 of this chapter is approved under this part without modifications.

(c) Application for training program approval shall be made in a form and in a manner acceptable to the Administrator.

(d) Each application for training program approval must indicate—

(1) Which courses are part of the core curriculum and which courses are part of the specialty curriculum;

(2) Which requirements of part 61 of this chapter would be satisfied by the curriculum or curriculums; and

(3) Which requirements of part 61 of this chapter would not be satisfied by the curriculum or curriculums.

(e) If, after a certificate holder begins operations under an approved training program, the Administrator finds that the certificate holder is not meeting the provisions of its approved training program, the Administrator may require the certificate holder to make revisions to that training program.

(f) If the Administrator requires a certificate holder to make revisions to an approved training program and the certificate holder does not make those required revisions, within 30 calendar days, the Administrator may suspend, revoke, or terminate the training center certificate under the provisions of § 142.11(e).

§ 142.39 Training program curriculum requirements.

Each training program curriculum submitted to the Administrator for approval must meet the applicable requirements of this part and must contain—

(a) A syllabus for each proposed curriculum;

(b) Minimum aircraft and flight training equipment requirements for each proposed curriculum;

(c) Minimum instructor and evaluator qualifications for each proposed curriculum;

(d) A curriculum for initial training and continuing training of each instructor or evaluator employed to instruct in a proposed curriculum; and

(e) For each curriculum that provides for the issuance of a certificate or rating in fewer than the minimum hours prescribed by part 61 of this chapter—

(1) A means of demonstrating the ability to accomplish such training in the reduced number of hours; and

(2) A means of tracking student performance.

Subpart C—Personnel and Flight Training Equipment Requirements

§ 142.45 Applicability.

This subpart prescribes the personnel and flight training equipment requirements for a certificate holder that is training to meet the requirements of part 61 of this chapter.

§ 142.47 Training center instructor eligibility requirements.

(a) A certificate holder may not employ a person as an instructor in a flight training course that is subject to approval by the Administrator unless that person—

(1) Is at least 18 years of age;

(2) Is able to read, write, and speak and understand in the English language;

(3) If instructing in an aircraft in flight, is qualified in accordance with subpart H of part 61 of this chapter.

(4) Satisfies the requirements of paragraph (c) of this section; and

(5) Meets at least one of the following requirements—

(i) Except as allowed by paragraph (a)(5)(ii) of this section, meets the aeronautical experience requirements of § 61.129(a), (b), (c), or (e) of this chapter, as applicable, excluding the required hours of instruction in preparation for the commercial pilot practical test;

(ii) If instructing in a flight simulator or flight training device that represents an airplane requiring a type rating or if instructing in a curriculum leading to the issuance of an airline transport pilot certificate or an added rating to an airline transport pilot certificate, meets the aeronautical experience requirements of § 61.159 or § 61.161, or § 61.163 of this chapter, as applicable; or

(iii) Is employed as a flight simulator instructor or a flight training device instructor for a training center providing instruction and testing to meet the requirements of part 61 of this chapter as applicable; or

(b) A training center must designate each instructor in writing to instruct in each approved course, prior to that person functioning as an instructor in that course.

(c) Prior to initial designation, each instructor shall:

(1) Complete at least 8 hours of ground training on the following subject matter;

(i) Instruction methods and techniques.

(ii) Training policies and procedures.

(iii) The fundamental principles of the learning process.

(iv) Instructor duties, privileges, responsibilities, and limitations.

(v) Proper operation of simulation controls and systems.

(vi) Proper operation of environmental control and warning or caution panels.

(vii) Limitations of simulation.

(viii) Minimum equipment requirements for each curriculum.

(ix) Revisions to the training courses.

(x) Cockpit resource management and crew coordination.

(2) Satisfactorily complete a written test—

(i) On the subjects specified in paragraph (c)(i) of this section; and

(ii) That is accepted by the Administrator as being of equivalent difficulty, complexity, and scope as the tests provided by the Administrator for the flight instructor airplane and instrument flight instructor knowledge tests.

§ 142.49 Training center instructor and evaluator privileges and limitations.

(a) A certificate holder may allow an instructor to provide:

(1) Instruction for each curriculum for which that instructor is qualified.

(2) Testing and checking for which that instructor is qualified.

(3) Instruction, testing, and checking intended to satisfy the requirements of any part of this chapter.

(b) A training center whose instructor or evaluator is designated in accordance with the requirements of this subpart to conduct training, testing, or checking in qualified and approved flight training equipment, may allow its instructor or evaluator to give endorsements required by part 61 of this chapter if that instructor or evaluator is authorized by the Administrator to instruct or evaluate in a part 142 curriculum that requires such endorsements.

(c) A training center may not allow an instructor to—

(1) Excluding briefings and debriefings, conduct more than 8 hours of instruction in any 24-consecutive-hour period;

(2) Provide flight training equipment instruction unless that instructor meets the requirements of § 142.53 (a)(1) through (a)(4), and § 142.53(b), as applicable; or

(3) Provide flight instruction in an aircraft unless that instructor—

(i) Meets the requirements of § 142.53 (a)(1), (a)(2), and (a)(5);

(ii) Is qualified and authorized in accordance with subpart H of part 61 of this chapter;

(iii) Holds certificates and ratings specified by part 61 of this chapter appropriate to the category, class, and type aircraft in which instructing;

(iv) If instructing or evaluating in an aircraft in flight while occupying a required crewmember seat, holds at least a valid second class medical certificate; and

(v) Meets the recency of experience requirements of part 61 of this chapter.

§ 142.51 [Reserved]

§ 142.53 Training center instructor training and testing requirements.

(a) Except as provided in paragraph (c) of this section, prior to designation and every 12 calendar months beginning the first day of the month following an instructor's initial designation, a certificate holder must ensure that each of its instructors meets the following requirements:

(1) Each instructor must satisfactorily demonstrate to an authorized evaluator knowledge of, and proficiency in, instructing in a representative segment of each curriculum for which that instructor is designated to instruct under this part.

(2) Each instructor must satisfactorily complete an approved course of ground instruction in at least—

(i) The fundamental principles of the learning process;

(ii) Elements of effective teaching, instruction methods, and techniques;

(iii) Instructor duties, privileges, responsibilities, and limitations;

(iv) Training policies and procedures;

(v) Cockpit resource management and crew coordination; and

(vi) Evaluation.

(3) Each instructor who instructs in a qualified and approved flight simulator or flight training device must satisfactorily complete an approved course of training in the operation of the flight simulator, and an approved course of ground instruction, applicable to the training courses the instructor is designated to instruct.

(4) The flight simulator training course required by paragraph (a)(3) of this section which must include—

(i) Proper operation of flight simulator and flight training device controls and systems;

(ii) Proper operation of environmental and fault panels;

(iii) Limitations of simulation; and

(iv) Minimum equipment requirements for each curriculum.

(5) Each flight instructor who provides training in an aircraft must satisfactorily complete an approved course of ground instruction and flight training in an aircraft, flight simulator, or flight training device.

(6) The approved course of ground instruction and flight training required by paragraph (a)(5) of this section which must include instruction in—

(i) Performance and analysis of flight training procedures and maneuvers applicable to the training courses that the instructor is designated to instruct;

(ii) Technical subjects covering aircraft subsystems and operating rules applicable to the training courses that the instructor is designated to instruct;

(iii) Emergency operations;

(iv) Emergency situations likely to develop during training; and

(v) Appropriate safety measures.

(7) Each instructor who instructs in qualified and approved flight training equipment must pass a written test and annual proficiency check—

(i) In the flight training equipment in which the instructor will be instructing; and

(ii) On the subject matter and maneuvers of a representative segment of each curriculum for which the instructor will be instructing.

(b) In addition to the requirements of paragraphs (a)(1) through (a)(7) of this section, each certificate holder must ensure that each instructor who instructs in a flight simulator that the Administrator has approved for all training and all testing for the airline transport pilot certification test, aircraft type rating test, or both, has met at least one of the following three requirements:

(1) Each instructor must have performed 2 hours in flight, including three takeoffs and three landings as the sole manipulator of the controls of an aircraft of the same category and class, and, if a type rating is required, of the same type replicated by the approved flight simulator in which that instructor is designated to instruct;

(2) Each instructor must have participated in an approved line-observation program under part 121 or part 135 of this chapter, and that—

(i) Was accomplished in the same airplane type as the airplane represented by the flight simulator in which that instructor is designated to instruct; and

(ii) Included line-oriented flight training of at least 1 hour of flight during which the instructor was the sole manipulator of the controls in a flight simulator that replicated the same type aircraft for which that instructor is designated to instruct; or

(3) Each instructor must have participated in an approved in-flight observation training course that—

(i) Consisted of at least 2 hours of flight time in an airplane of the same type as the airplane replicated by the flight simulator in which the instructor is designated to instruct; and

(ii) Included line-oriented flight training of at least 1 hour of flight during which the instructor was the sole manipulator of the controls in a flight simulator that replicated the same type aircraft for which that instructor is designated to instruct.

(c) An instructor who satisfactorily completes a curriculum required by paragraph (a) or (b) of this section in the calendar month before or after the month in which it is due is considered to have taken it in the month in which it was due for the purpose of computing when the next training is due.

(d) The Administrator may give credit for the requirements of paragraph (a) or (b) of this section to an instructor who has satisfactorily completed an instructor training course for a part 121 or part 135 certificate holder if the

Administrator finds such a course equivalent to the requirements of paragraph (a) or (b) of this section.

§ 142.55 Training center evaluator requirements.

(a) Except as provided by paragraph (d) of this section, a training center must ensure that each person authorized as an evaluator—

(1) Is approved by the Administrator;

(2) Is in compliance with §§ 142.47, 142.49, and 142.53 and applicable sections of part 187 of this chapter; and

(3) Prior to designation, and except as provided in paragraph (b) of this section, every 12-calendar-month period following initial designation, the certificate holder must ensure that the evaluator satisfactorily completes a curriculum that includes the following:

(i) Evaluator duties, functions, and responsibilities;

(ii) Methods, procedures, and techniques for conducting required tests and checks;

(iii) Evaluation of pilot performance; and

(iv) Management of unsatisfactory tests and subsequent corrective action; and

(4) If evaluating in qualified and approved flight training equipment must satisfactorily pass a written test and annual proficiency check in a flight simulator or aircraft in which the evaluator will be evaluating.

(b) An evaluator who satisfactorily completes a curriculum required by paragraph (a) of this section in the calendar month before or the calendar month after the month in which it is due is considered to have taken it in the month is which it was due for the purpose of computing when the next training is due.

(c) The Administrator may give credit for the requirements of paragraph (a)(3) of this section to an evaluator who has satisfactorily completed an evaluator training course for a part 121 or part 135 certificate holder if the Administrator finds such a course equivalent to the requirements of paragraph (a)(3) of this section.

(d) An evaluator who is qualified under SFAR 58 shall be authorized to conduct evaluations under the Advanced Qualification Program without complying with the requirements of this section.

§ 142.57 Aircraft requirements.

(a) An applicant for, or holder of, a training center certificate must ensure that each aircraft used for flight instruction and solo flights meets the following requirements:

(1) Except for flight instruction and solo flights in a curriculum for agricultural aircraft operations, external load operations, and similar aerial work operations, the aircraft must have an FAA standard airworthiness certificate or a foreign equivalent of an FAA standard airworthiness certificate, acceptable to the Administrator.

(2) The aircraft must be maintained and inspected in accordance with—

(i) The requirements of part 91, subpart E, of this chapter; and

(ii) An approved program for maintenance and inspection.

(3) The aircraft must be equipped as provided in the training specifications for the approved course for which it is used.

(b) Except as provided in paragraph (c) of this section, an applicant for, or holder of, a training center certificate must ensure that each aircraft used for flight instruction is at least a two-place aircraft with engine power controls and flight controls that are easily reached and that operate in a conventional manner from both pilot stations.

(c) Airplanes with controls such as nose-wheel steering, switches, fuel selectors, and engine air flow controls that are not easily reached and operated in a conventional manner by both pilots may be used for flight instruction if the certificate holder determines that the flight instruction can be conducted in a safe manner considering the location of controls and their nonconventional operation, or both.

§ 142.59 Flight simulators and flight training devices.

(a) An applicant for, or holder of, a training center certificate must show that each flight simulator and flight training device used for training, testing, and checking (except AQP) will be or is specifically qualified and approved by the Administrator for—

(1) Each maneuver and procedure for the make, model, and series of aircraft, set of aircraft, or aircraft type simulated, as applicable; and

(2) Each curriculum or training course in which the flight simulator or flight training device is used, if that curriculum or course is used to satisfy any requirement of 14 CFR chapter 1.

(b) The approval required by paragraph (a)(2) of this section must include—

(1) The set of aircraft, or type aircraft;

(2) If applicable, the particular variation within type, for which the training, testing, or checking is being conducted; and

(3) The particular maneuver, procedure, or crewmember function to be performed.

(c) Each qualified and approved flight simulator or flight training device used by a training center must—

(1) Be maintained to ensure the reliability of the performances, functions, and all other characteristics that were required for qualification;

(2) Be modified to conform with any modification to the aircraft being simulated if the modification results in changes to performance, function, or other characteristics required for qualification;

(3) Be given a functional preflight check each day before being used; and

(4) Have a discrepancy log in which the instructor or evaluator, at the end of each training session, enters each discrepancy.

(d) Unless otherwise authorized by the Administrator, each component on a qualified and approved flight simulator or flight training device used by a training center must be operative if the component is essential to, or involved in, the training, testing, or checking of airmen.

(e) Training centers shall not be restricted to specific—

(1) Route segments during line-oriented flight training scenarios; and

(2) Visual data bases replicating a specific customer's bases of operation.

(f) Training centers may request evaluation, qualification, and continuing evaluation for qualification of flight simulators and flight training devices without—

(1) Holding an air carrier certificate; or

(2) Having a specific relationship to an air carrier certificate holder.

Subpart D—Operating Rules

§ 142.61 Applicability.

This subpart prescribes the operating rules applicable to a training center certificated under this part and operating a course or training program curriculum approved in accordance with subpart B of this part.

§ 142.63 Privileges.

A certificate holder may allow flight simulator instructors and evaluators to meet recency of experience requirements through the use of a qualified and approved flight simulator or qualified and approved flight training device if that flight simulator or flight training device is—

(a) Used in a course approved in accordance with subpart B of this part; or

(b) Approved under the Advanced Qualification Program for meeting recency of experience requirements.

§ 142.65 Limitations.

(a) A certificate holder shall—

(1) Ensure that a flight simulator or flight training device freeze, slow motion, or repositioning feature is not used during testing or checking; and

(2) Ensure that a repositioning feature is used during line operational simulation for evaluation and line-ori-

ented flight training only to advance along a flight route to the point where the descent and approach phase of the flight begins.

(b) When flight testing, flight checking, or line operational simulation is being conducted, the certificate holder must ensure that one of the following occupies each crewmember position:

(1) A crewmember qualified in the aircraft category, class, and type, if a type rating is required, provided that no flight instructor who is giving instruction may occupy a crewmember position.

(2) A student, provided that no student may be used in a crewmember position with any other student not in the same specific course.

(c) The holder of a training center certificate may not recommend a trainee for a certificate or rating, unless the trainee—

(1) Has satisfactorily completed the training specified in the course approved under § 142.37; and

(2) Has passed the final tests required by § 142.37.

(d) The holder of a training center certificate may not graduate a student from a course unless the student has satisfactorily completed the curriculum requirements of that course.

Subpart E—Recordkeeping

§ 142.71 Applicability.

This subject prescribes the training center recordkeeping requirements for trainees enrolled in a course, and instructors and evaluators designated to instruct a course, approved in accordance with subpart B of this part.

§ 142.73 Recordkeeping requirements.

(a) A certificate holder must maintain a record for each trainee that contains—

(1) The name of the trainee;

(2) A copy of the trainee's pilot certificate, if any, and medical certificate;

(3) The name of the course and the make and model of flight training equipment used;

(4) The trainee's prerequisite experience and course time completed;

(5) The trainee's performance on each lesson and the name of the instructor providing instruction;

(6) The date and result of each end-of-course practical test and the name of the evaluator conducting the test; and

(7) The number of hours of additional training that was accomplished after any unsatisfactorily practical test.

(b) A certificate holder shall maintain a record for each instructor or evaluator designated to instruct a course approved in accordance with subpart B of this part that indicates that the instructor or evaluator has complied with the requirements of §§ 142.13, 142.45, 142.47, 142.49, and 142.53, as applicable.

(c) The certificate holder shall—

(1) Maintain the records required by paragraphs (a) of this section for at least 1 year following the completion of training, testing or checking;

(2) Maintain the qualification records required by paragraph (b) of this section while the instructor or evaluator is in the employ of the certificate holder and for 1 year thereafter; and

(3) Maintain the recurrent demonstration of proficiency records required by paragraph (b) of this section for at least 1 year.

(d) The certificate holder must provide the records required by this section to the Administrator, upon request and at a reasonable time, and shall keep the records required by—

(1) Paragraph (a) of this section at the training center, or satellite training center where the training, testing, or checking, if appropriate, occurred; and

(2) Paragraph (b) of this section at the training center or satellite training center where the instructor or evaluator is primarily employed.

(e) The certificate holder shall provide to a trainee, upon request and at a reasonable time, a copy of his or her training records.

Subpart F—Other Approved Courses

§ 142.81 Conduct of other approved courses.

(a) An applicant for, or holder of, a training center certificate may apply for approval to conduct a course for which a curriculum is not prescribed by this part.

(b) The course for which application is made under paragraph (a) of this section may be for flight crewmembers other than pilots, airmen other than flight crewmembers, material handlers, ground servicing personnel, and security personnel, and others approved by the Administrator.

(c) An applicant for course approval under this subpart must comply with the applicable requirements of subpart A through subpart F of this part.

(d) The Administrator approves the course for which the application is made if the training center or training center applicant shows that the course contains a curriculum that will achieve a level of competency equal to, or greater than, that required by the appropriate part of this chapter.

Issued in Washington, DC, on May 23, 1996.
David R. Hinson,
Administrator.
[FR Doc. 96-16432 Filed 7-1-96; 8:45 am]
BILLING CODE 4910-13-P

Part 143—Removed and reserved

NTSB Part 830—Notification and reporting of aircraft accidents or incidents and overdue aircraft, and preservation of aircraft wreckage, mail, cargo, and records

Authority: 49 U.S.C. 1441 and 1901 et seq.
Amended: June 21, 1989

Subpart A—General

§ 830.1 Applicability.

This part contains rules pertaining to:

(a) Notification and reporting aircraft accidents and incidents and certain other occurrences in the operation of aircraft when they involve civil aircraft of the United States wherever they occur, or foreign civil aircraft when such events occur in the United States, its territories or possessions.

(b) Reporting aircraft accidents and listed incidents in the operation of aircraft when they involve certain public aircraft.

(c) Preservation of aircraft wreckage, mail, cargo, and records involving all civil aircraft in the United States, its territories or possessions.

§ 830.2 Definitions.

As used in this part the following words or phrases are defined as follows:

"Aircraft accident" means an occurrence associated with the operation of an aircraft which takes place between the time any person boards the aircraft with the intention of flight and all such persons have disembarked, and in which any person suffers death or serious injury, or in which the aircraft receives substantial damage.

"Civil aircraft" means any aircraft other than a public aircraft.

"Fatal injury" means any injury which results in death within 30 days of the accident.

"Incident" means an occurrence other than an accident associated with the operation of an aircraft, which affects or could affect the safety of operations.

"Operator" means any person who causes or authorizes the operation of an aircraft, such as the owner, lessee, or bailee of an aircraft.

"Public aircraft" means an aircraft used exclusively in the service of any government or of any political subdivision thereof, including the government of any State, Territory, or possession of the United States, or the District of Columbia, but not including any government-owned aircraft engaged in carrying persons or property for commercial purposes. For purposes of this section "used exclusively in the service of" means, for other than the Federal Government, an aircraft which is owned and operated by a governmental entity for other than commercial purposes or which is exclusively leased by such governmental entity for not less than 90 continuous days.

"Serious injury" means any injury which: (1) Requires hospitalization for more than 48 hours, commencing within 7 days from the date of the injury was received; (2) results in a fracture of any bone (except simple fractures of fingers, toes, or nose); (3) causes severe hemorrhages, nerve, muscle, or tendon damage; (4) involves any internal organ; or (5) involves second- or third-degree burns, or any burns affecting more than 5 percent of the body surface.

"Substantial damage" means damage or failure which adversely affects the structural strength, performance, or flight characteristics of the aircraft, and which normally require major repair or replacement of the affected component. Engine failure or damage limited to an engine if only one engine fails or is damaged, bent fairings or cowling, dented skin, small punctured holes in the skin or fabric, ground damage to rotor or propeller blades, and damage to landing gear, wheels, tires, flaps, engine accessories, brakes, or wingtips are not considered "substantial damage" for the purpose of this part.

Subpart B—Initial notification of aircraft accidents, incidents and overdue aircraft

§ 830.5 Immediate notification.

The operator of an aircraft shall immediately, and by the most expeditious means available, notify the nearest National Transportation Safety Board (Board), field office[1] when:

(a) An aircraft accident or any of the following listed incidents occur:

(1) Flight control system malfunction or failure;

(2) Inability of any required flight crewmember to perform normal flight duties as a result of injury or illness;

(3) Failure of structural components of a turbine engine excluding compressor and turbine blades and vanes;

(4) In-flight fire; or

(5) Aircraft collide in flight.

(6) Damage to property, other than the aircraft, estimated to exceed $25,000 for repair (including materials and labor) or fair market value in the event of total loss, whichever is less.

(7) For large multiengine aircraft (more than 12,500 pounds maximum certificated takeoff weight):

(i) In-flight failure of electrical systems which requires the sustained use of an emergency bus powered by a back-up source such as a battery, auxiliary power unit, or air-driven generator to retain flight control or essential instruments;

(ii) In-flight failure of hydraulic systems that results in sustained reliance on the sole remaining hydraulic or mechanical system for movement of flight control surfaces;

(iii) Sustained loss of the power or thrust produced by two or more engines; and

(iv) An evacuation of aircraft in which an emergency egress system is utilized.

(b) An aircraft is overdue and is believed to have been involved in an accident.

§ 830.6 Information to be given in notification.

The notification required in § 830.5 shall contain the following information, if available;

(a) Type, nationality, and registration marks of the aircraft;

(b) Name of owner, and operator of the aircraft;

(c) Name of the pilot-in-command;

(d) Date and time of the accident;

(e) Last point of departure and point of intended landing of the aircraft;

(f) Position of the aircraft with reference to some easily defined geographical point;

(g) Number of persons aboard, number killed, and number seriously injured;

(h) Nature of the accident, the weather and the extent of damage to the aircraft, so far as is known; and

(i) A description of any explosives, radioactive materials, or other dangerous articles carried.

Subpart C—Preservation of aircraft wreckage, mail, cargo, and records

§ 830.10 Preservation of aircraft wreckage, mail, cargo, and records.

(a) The operator of an aircraft involved in an accident or incident for which notification must be given is responsible for preserving to the extent possible any aircraft wreckage, cargo, and mail aboard the aircraft, and all records, including all recording mediums of flight, maintenance, and voice recorders, pertaining to the operation and maintenance of the aircraft and to the airmen until the Board takes custody thereof or a release is granted pursuant to § 831.12(b) of this chapter.

(b) Prior to the time the Board or its authorized representative takes custody of aircraft wreckage, mail, or cargo, such wreckage, mail, or cargo may not be disturbed or moved except to the extent necessary:

(1) To remove persons injured or trapped;

(2) To protect the wreckage from further damage; or

(3) To protect the public from injury.

(c) Where it is necessary to move aircraft wreckage, mail or cargo, sketches, descriptive notes, and photographs shall be made, if possible, of the original positions and condition of the wreckage and any significant impact marks.

(d) The operator of an aircraft involved in an accident or incident shall retain all records, reports, internal documents, and memoranda dealing with the accident or incident, until authorized by the Board to the contrary.

Subpart D—Reporting of aircraft accidents, incidents, and overdue aircraft

§ 830.15 Reports and statements to be filed.

(a) *Reports.* The operator of an aircraft shall file a report on Board Form 6120.1 (OMB No. 3147-005) or Board Form 7120.2 (OMB No. 3147-0001)[2] within 10 days after an accident, or after 7 days if an overdue aircraft is still missing. A report on an incident for which notification is required by § 830.5(a) shall be filed only as requested by an authorized representative of the Board.

(b) *Crewmember statement.* Each crewmember, if physically able at the time the report is submitted, shall attach a statement setting forth the facts, conditions, and circumstances relating to the accident or incident as they appear to him. If the crewmember is incapacitated, he shall submit the statement as soon as he is physically able.

(c) *Where to file the reports.* The operator of an aircraft shall file any report with the field office of the Board nearest the accident or incident.

Subpart E—Reporting of public aircraft accidents and incidents

§ 830.20 Reports to be filed.

The operator of a public aircraft other than an aircraft of the Armed Forces or Intelligence Agencies shall file a report on NTSB Form 6120.1 (OMB No. 3147-001)[3] within 10 days after an accident or incident listed in § 830.5(a). The operator shall file the report with the filed office of the Board nearest the accident or incident.[4]

Signed at Washington, DC, on this 16th day of September 1988.

James L. Kolstad,
Acting Chairman
[FR Doc. 88-21705 Filed 9-22-88; 8:45 am]

---

[1] The National Transportation Safety Board field offices are listed under U.S. Government in the telephone directories in the following cities: Anchorage, Alaska; Atlanta Ga.; Chicago, Ill.; Denver, Colo.; Fort Worth, Tex.; Kansas City, Mo.; Los Angeles, Calif.; Miami, Fla.; New York, N.Y.; Seattle, Wash.

[2] Forms are available from the Board field offices, (see footnote 1), the National Transportation Safety Board, Washington, DC 20594, and the Federal Aviation Administration, Flight Standards District Office

[3] To obtain this form, see footnote 2.

[4] The locations or the Board's field offices are set forth in footnote 1.

Part 49—Code of Federal Regulations

Editor's note—Security matters relating to all transportation were assigned to a new Department of Homeland Security. This is the Transportation Security Administration. It assumed authority over security from the Federal Aviation Administration. The FAA's regulations come under CFR 14, the TSA under CFR 49.

49 CFR
CHAPTER XII
TRANSPORTATION SECURITY ADMINISTRATION, DEPARTMENT OF TRANSPORTATION
SUBCHAPTER A — ADMINISTRATIVE AND PROCEDURAL RULES

PART 1500—APPLICABILITY, TERMS, AND ABBREVIATIONS

Sec.
1500.1 Applicability.
1500.3 Terms and abbreviations used in this chapter.
1500.5 Rules of construction.
Authority: 49 U.S.C. 114, 5103, 40119, 44901–44907, 44913–44914, 44916–44918, 44935–44936, 44942, 46105.

Source: 67 FR 8351, Feb. 22, 2002, unless otherwise noted.

§1500.1 Applicability.

This chapter, this subchapter, and this part apply to all matters regulated by the Transportation Security Administration.

§1500.3 Terms and abbreviations used in this chapter.

As used in this chapter:

Person means an individual, corporation, company, association, firm, partnership, society, joint-stock company, or governmental authority. It includes a trustee, receiver, assignee, successor, or similar representative of any of them.

Transportation Security Regulations (TSR) means the regulations issued by the Transportation Security Administration, in title 49 of the Code of Federal Regulations, chapter XII, which includes parts 1500 through 1699.

TSA means the Transportation Security Administration.

Under Secretary means the Under Secretary of Transportation for Security.

United States, in a geographical sense, means the States of the United States, the District of Columbia, and territories and possessions of the United States, including the territorial sea and the overlying airspace.

§1500.5 Rules of construction.

(a) In this chapter, unless the context requires otherwise:
(1) Words importing the singular include the plural.
(2) Words importing the plural include the singular.
(3) Words importing the masculine gender include the feminine.
(b) In this chapter, the word:
(1) "Must" is used in an imperative sense;
(2) "May" is used in a permissive sense to state authority or permission to do the act prescribed, and the words "no person may * * *" or "a person may not * * *" mean that no person is required, authorized, or permitted to do the act prescribed; and
(3) "Includes" means "includes but is not limited to".

SUBCHAPTER C—CIVIL AVIATION SECURITY
PART 1540—CIVIL AVIATION SECURITY: GENERAL RULES

Subpart A—General

Sec.
1540.1   Applicability of this subchapter and this part.
1540.3   Delegation of authority.
1540.5   Terms used in this subchapter.

Subpart B—Responsibilities of Passengers and Other Individuals and Persons

1540.101   Applicability of this subpart.
1540.103   Fraud and intentional falsification of records.
1540.105   Security responsibilities of employees and other persons.
1540.107   Submission to screening and inspection.
1540.109   Prohibition against interference with screening personnel.
1540.111   Carriage of weapons, explosives, and incendiaries by individuals.
1540.113   Inspection of airman certificate.
1540.115   Threat assessments regarding citizens of the United States holding, or applying for certificates, ratings, or authorizations.
1540.117   Threat assessments regarding aliens holding, or applying for FAA certificates, ratings, or authorizations.
Authority: 49 U.S.C. 114, 5103, 40119, 44901-44907, 44913-44914, 44916-44918,
44935-44936, 44942, 46105.
Source: 67 FR 8353, Feb. 22, 2002, unless otherwise noted.

§1540.1 Applicability of this subchapter and this part.

This subchapter and this part apply to persons engaged in aviation-related activities.

§1540.3 Delegation of authority.

(a) Where the Under Secretary is named in this subchapter as exercising authority over a function, the authority is exercised by the Under Secretary or the Deputy Under Secretary, or any individual formally designated to act as the Under Secretary or the Deputy Under Secretary.

(b) Where TSA or the designated official is named in this subchapter as exercising authority over a function, the authority is exercised by the official designated by the Under Secretary to perform that function.

§1540.5 Terms used in this subchapter.

In addition to the terms in part 1500 of this chapter, the following terms are used in this subchapter:

Air operations area (AOA) means a portion of an airport, specified in the airport security program, in which security measures specified in this part are carried out. This area includes aircraft movement areas, aircraft parking areas, loading ramps, and safety areas, for use by aircraft regulated under 49 CFR part 1544 or 1546, and any adjacent areas (such as general aviation areas) that are not separated by adequate security systems, measures, or procedures. This area does not include the secured area.

Aircraft operator means a person who uses, causes to be used, or authorizes to be used an aircraft, with or without the right of legal control (as owner, lessee, or otherwise), for the purpose of air navigation including the piloting of aircraft, or on any part of the surface of an airport. In specific parts or sections of this subchapter, "aircraft operator" is used to refer to specific types of operators as described in those parts or sections.

Airport operator means a person that operates an airport serving a aircraft operator or a foreign air carrier required to have a security program under part 1544 or 1546 of this chapter.

Airport security program means a security program approved by TSA under §1542.101 of this chapter.

Airport tenant means any person, other than an aircraft operator or foreign air carrier that has a security program under part 1544 or 1546 of this chapter, that has an agreement with the airport operator to conduct business on airport property.

Airport tenant security program means the agreement between the airport operator and an airport tenant that specifies the measures by which the tenant will perform security functions, and approved by TSA, under §1542.113 of this chapter.

Approved, unless used with reference to another person, means approved by TSA. Cargo means property tendered for air transportation accounted for on an air waybill. All accompanied commercial courier consignments, whether or not accounted for on an air waybill, are also classified as cargo. Aircraft operator security programs further define the term "cargo."

Checked baggage means property tendered by or on behalf of a passenger and accepted by an aircraft operator for transport, which is inaccessible to passengers during flight. Accompanied commercial courier consignments are not classified as checked baggage.

Escort means to accompany or monitor the activities of an individual who does not have unescorted access authority into or within a secured area or SIDA. Exclusive area means any portion of a secured area, AOA, or SIDA, including individual access points, for which an aircraft operator or foreign air carrier that has a security program under part 1544 or 1546 of this chapter has assumed responsibility under §1542.111 of this chapter.

Exclusive area agreement means an agreement between the airport operator and an aircraft operator or a foreign air carrier that has a security program under parts 1544 or 1546 of this chapter that permits such an aircraft operator or foreign air carrier to assume responsibility for specified security measures in accordance with §1542.111 of this chapter.

FAA means the Federal Aviation Administration.

Flightcrew member means a pilot, flight engineer, or flight navigator assigned to duty in an aircraft during flight time.

Indirect air carrier means any person or entity within the United States not in possession of an FAA air carrier operating certificate, that undertakes to engage indirectly in air transportation of property, and uses for all or any part of such transportation the services of a passenger air carrier. This does not include the United States Postal Service (USPS) or its representative while acting on the behalf of the USPS.

Loaded firearm means a firearm that has a live round of ammunition, or any component thereof, in the chamber or cylinder or in a magazine inserted in the firearm.

Passenger seating configuration means the total maximum number of seats for which the aircraft is type certificated that can be made available for passenger use aboard a flight, regardless of the number of seats actually installed, and includes that seat in certain aircraft that may be used by a representative of the FAA to conduct flight checks but is available for revenue purposes on other occasions.

Private charter means any aircraft operator flight—

(1) For which the charterer engages the total passenger capacity of the aircraft for the carriage of passengers; the passengers are invited by the charterer; the cost of the flight is borne entirely by the charterer and not directly or indirectly by any individual passenger; and the flight is not advertised to the public, in any way, to solicit passengers.

(2) For which the total passenger capacity of the aircraft is used for the purpose of civilian or military air movement conducted under contract with the Government of the United States or the government of a foreign country.

Public charter means any charter flight that is not a private charter. Scheduled passenger operation means an air transportation operation (a flight) from identified air terminals at a set time, which is held out to the public and announced by timetable or schedule, published in a newspaper, magazine, or other advertising medium.

Screening function means the inspection of individuals and property for weapons, explosives, and incendiaries.

Screening location means each site at which individuals or property are inspected for the presence of weapons, explosives, or incendiaries.

Secured area means a portion of an airport, specified in the airport security program, in which certain security measures specified in part 1542 of this chapter are carried out. This area is where aircraft operators and foreign air carriers that have a security program under part 1544 or 1546 of this chapter enplane and deplane passengers and sort and load baggage and any adjacent areas that are not separated by adequate security measures.

Security Identification Display Area (SIDA) means a portion of an airport, specified in the airport security program, in which security measures specified in this part are carried out. This area includes the secured area and may include other areas of the airport.

Sterile area means a portion of an airport defined in the airport security program that provides passengers access to boarding aircraft and to which the access generally is controlled by TSA, or by an aircraft operator under part 1544 of this chapter or a foreign air carrier under part 1546 of this chapter, through the screening of persons and property.

Unescorted access authority means the authority granted by an airport operator, an aircraft operator, foreign air carrier, or airport tenant under part 1542, 1544, or 1546 of this chapter, to individuals to gain entry to, and be present without an escort in, secured areas and SIDA's of airports.

[67 FR 8353, Feb. 22, 2002, as amended at 67 FR 8209, Feb. 22, 2002]

Subpart B—Responsibilities of Passengers and Other Individuals and Persons

§1540.101 Applicability of this subpart.

This subpart applies to individuals and other persons.

§1540.103 Fraud and intentional falsification of records.

No person may make, or cause to be made, any of the following:

(a) Any fraudulent or intentionally false statement in any application for any security program, access medium, or identification medium, or any amendment thereto, under this subchapter.

(b) Any fraudulent or intentionally false entry in any record or report that is kept, made, or used to show compliance with this subchapter, or exercise any privileges under this subchapter.

(c) Any reproduction or alteration, for fraudulent purpose, of any report, record, security program, access medium, or identification medium issued under this subchapter.

§1540.105 Security responsibilities of employees and other persons.

(a) No person may:

(1) Tamper or interfere with, compromise, modify, attempt to circumvent, or cause a person to tamper or interfere with, compromise, modify, or attempt to circumvent any security system, measure, or procedure implemented under this subchapter.

(2) Enter, or be present within, a secured area, AOA, SIDA or sterile area without complying with the systems, measures, or procedures being applied to control access to, or presence or movement in, such areas.

(3) Use, allow to be used, or cause to be used, any airport-issued or airport-approved access medium or identification medium that authorizes the access, presence, or movement of persons or vehicles in secured areas, AOA's, or SIDA's in any other manner than that for which it was issued by the appropriate authority under this subchapter.

(b) The provisions of paragraph (a) of this section do not apply to conducting inspections or tests to determine compliance with this part or 49 U.S.C. Subtitle VII authorized by:

(1) TSA, or

(2) The airport operator, aircraft operator, or foreign air carrier, when acting in accordance with the procedures described in a security program approved by TSA.

§1540.107 Submission to screening and inspection.

Link to an amendment published at 67 FR 41639, June 19, 2002.

No individual may enter a sterile area without submitting to the screening and inspection of his or her person and accessible property in accordance with the procedures being applied to control access to that area under this subchapter.

§1540.109 Prohibition against interference with screening personnel.

No person may interfere with, assault, threaten, or intimidate screening personnel in the performance of their screening duties under this subchapter.

§1540.111 Carriage of weapons, explosives, and incendiaries by individuals.

Link to an amendment published at 67 FR 41639, June 19, 2002.

(a) On an individual's person or accessible property—prohibitions. Except as provided in paragraph (b) of this section, an individual may not have a weapon, explosive, or incendiary, on or about the individual's person or accessible property—

(1) When performance has begun of the inspection of the individual's person or accessible property before entering a sterile area;

(2) When the individual is entering or in a sterile area; or

(3) When the individual is attempting to board or onboard an aircraft for which screening is conducted under §1544.201 or §1546.201 of this chapter.

(b) On an individual's person or accessible property—permitted carriage of a weapon. Paragraph (a) of this section does not apply as to carriage of firearms and other weapons if the individual is one of the following:

(1) Law enforcement personnel required to carry a firearm or other weapons while in the performance of law enforcement duty at the airport.

(2) An individual authorized to carry a weapon in accordance with §§1544.219, 1544.221, 1544.223, or 1546.211 of this chapter.

(3) An individual authorized to carry a weapon in a sterile area under a security program.

(c) In checked baggage. A passenger may not transport or offer for transport in checked baggage:

(1) Any loaded firearm(s).

(2) Any unloaded firearm(s) unless—

(i) The passenger declares to the aircraft operator, either orally or in writing, before checking the baggage, that the passenger has a firearm in his or her bag and that it is unloaded;

(ii) The firearm is unloaded;

(iii) The firearm is carried in a hard-sided container; and

(iv) The container in which it is carried is locked, and only the passenger retains the key or combination.

(3) Any unauthorized explosive or incendiary.

(d) Ammunition. This section does not prohibit the carriage of ammunition in checked baggage or in the same container as a firearm. Title 49 CFR part 175 provides additional requirements governing carriage of ammunition on aircraft.

§1540.113 Inspection of airman certificate.

Each individual who holds an airman certificate, medical certificate, authorization, or license issued by the FAA must present it for inspection upon a request from TSA.

§ 1540.115 Threat assessments regarding citizens of the United States holding, or applying for certificates, ratings, or authorizations.

Applicability. This section applies when TSA that an individual who is a United States citizen and who holds, or is applying for, an airman certificate, rating, or authorization issued by the Administrator, poses a security threat.

Definitions. The following terms apply to this section:

Assistant Administrator means the Assistant Administrator for Intelligence for TSA.

Date of service means—

(1) The date of personal delivery in the case of personal service;

(2) The mailing date shown on the certificate of service;

(3) The date shown on the postmark if there is no certificate of service; or

(4) Another mailing date shown by other evidence if there is no certificate of service or postmark.

Deputy Administrator means the officer next in rank below the Under Secretary of Transportation for Security.

FAA Administrator means the Administrator of the Federal Aviation Administration.

Individual means an individual whom TSA determines poses a security threat. Under Secretary means the Under Secretary of Transportation for Security.

(c) Security threat. An individual poses a security threat when the individual is suspected of posing, or is known to pose—

(1) A threat to transportation or national security;

(2) A threat of air piracy or terrorism;

(3) A threat to airline or passenger security; or

(4) A threat to civil aviation security.

(d) Representation by counsel. The individual may, if he or she so chooses, be represented by counsel at his or her own expense.

(e) Initial Notification of Threat Assessment. (1) Issuance. If the Assistant Administrator determines that an individual poses a security threat, the Assistant Administrator serves upon the individual an Initial Notification of Threat Assessment and serves the determination upon the FAA Administrator. The Initial Notification includes—

(i) A statement that the Assistant Administrator personally has reviewed the materials upon which the Initial Notification was based; and

(ii) A statement that the Assistant Administrator has determined that the individual poses a security threat.

(2) Request for Materials. Not later than 15 calendar days after the date of service of the Initial Notification, the individual may serve a written request for copies of the releasable materials upon which the Initial Notification was based.

(3) TSA response. Not later than 30 calendar days, or such longer period as TSA may determine for good cause, after receiving the individual's request for copies of the releasable materials upon which the Initial Notification was based, TSA serves a response. TSA will not include in its response any classified information or other information described in paragraph (g) of this section.

(4) Reply. The individual may serve upon TSA a written reply to the Initial Notification of Threat Assessment not later than 15 calendar days after the date of service of the Initial Notification, or the date of service of TSA's response to the individual's request under paragraph (e)(2) if such a request was served. The reply may include any information that the individual believes TSA should consider in reviewing the basis for the Initial Notification,

(5) TSA final determination. Not later than 30 calendar days, or such longer period as TSA may determine for good cause, after TSA receives the individual's reply, TSA serves a final determination in accordance with paragraph (f) of this section.

(f) Final Notification of Threat Assessment. (1) In general. The Deputy Administrator reviews the Initial Notification, the materials upon which the Initial Notification was based, the individual's reply, if any, and any other materials or information available to him.

(2) Review and Issuance of Final Notification. If the Deputy Administrator determines that the individual poses a security threat, the Under Secretary reviews the Initial Notification, the materials upon which the Initial Notification was based, the individual's reply, if any, and any other materials or information available to him. If the Under Secretary determines that the individual poses a security threat, the Under Secretary serves upon the individual a Final Notification of Threat Assessment and serves the determination upon the FAA Administrator. The Final Notification includes a statement that the Under Secretary personally has reviewed the Initial Notification, the individual's reply, if any, and any other materials or information available to him, and has determined that the individual poses a security threat.

(3) Withdrawal of Initial Notification. If the Deputy Administrator does not determine that the individual poses a security threat, or upon review, the Under Secretary does not determine that the individual poses a security threat, TSA serves upon the individual a Withdrawal of the Initial Notification and provides a copy of the Withdrawal to the FAA Administrator.

(g) Nondisclosure of certain information, in connection with the procedures under this section, TSA does not disclose to the individual classified information, as defined in Executive Order 12988 section 1.1(d), and reserves the right not to disclose any other information or material not warranting disclosure or protected from disclosure under law.

Issued in Washington, DC on January 21, 2003.

§ 1540.117 Threat Assessments regarding aliens holding or applying for FAA certificates, ratings, or authorizations

(a) Applicability. This section applies when TSA has determined that an individual who is not a citizen of the United States and who holds, or is applying for, an airman certificate, rating, or authorization issued by the FAA Administrator, poses a security threat applying for an airman certificate, rating, or authorization issued by the FAA, poses a security threat.

(b) Definitions. The following terms apply in this section:

Assistant Administrator means the Assistant Administrator for Intelligence for TSA.

Date of service means—

(1) The date of personal delivery in the case of personal service;

(2) The mailing date shown on the certificate of service;

(3) The date shown on the postmark if there is no certificate of service; or

(4) Another mailing date shown by other evidence if there is no certificate of service or postmark.

Deputy Administrator means the officer next in rank below the Under Secretary of Transportation for Security.

FAA Administrator means the Administrator of the Federal Aviation Administration.

Individual means an individual whom TSA determines poses a security threat.

(c) Security threat. An individual poses a security threat when the individual is suspected of posing, or is known to pose—

(1) A threat to transportation or national security;

(2) A threat of air piracy or terrorism;

(3) A threat to airline or passenger security; or

(4) A threat to civil aviation security.

(d) Representation by counsel. The individual may, if he or she so chooses, be represented by counsel at his or her own expense.

(e) Initial Notification of Threat Assessment. (1) Issuance. If the Assistant Administrator determines that an individual poses a security threat, the Assistant Administrator serves upon the individual an Initial Notification of Threat Assessment and serves the determination upon the FAA Administrator. The Initial Notification includes—

(i) A statement that the Assistant Administrator personally has reviewed the materials upon which the Initial Notification was based; and

(ii) A statement that the Assistant Administrator has determined that the individual poses a security threat.

(2) Request for materials. Not later than 15 calendar days after the date of service of the Initial Notification, the individual may serve a written request for copies of the releasable materials upon which the Initial Notification was based.

(3) TSA response. Not later than 30 calendar days, or such longer period as TSA may determine for good cause, after receiving the individual's request for copies of the releasable materials upon which the Initial Notification was based, TSA serves a response. TSA will not include in its response any classified information or other information described in paragraph (g) of this section.

(4) Reply. The individual may serve upon TSA a written reply to the Initial Notification of Threat Assessment not later than 15 calendar days after the date of service of the Initial Notification, or the date of service of TSA's response to the individual's request under paragraph (e)(2) if such a request was served. The reply may include any information that the individual believes TSA should consider in reviewing the basis for the Initial Notification.

(5) TSA final determination. Not later than 30 calendar days, or such longer period as TSA may determine for good cause, after TSA receives the individual's reply, TSA serves a final determination in accordance with paragraph (f) of this section.

(f) Final Notification of Threat Assessment. (1) In general. The Deputy Administrator reviews the Initial Notification, the materials upon which the Initial Notification was based, the individual's reply, if any, and any other materials or information available to him.

(2) Issuance of Final Notification. If the Deputy Administrator determines that the individual poses a security threat, the Deputy Administrator serves upon the individual a Final Notification of Threat Assessment and serves the determination upon the FAA Administrator. The Final Notification includes a statement that the Deputy Administrator personally has reviewed the Initial Notification, the individual's reply, if any, and any other materials or information available to him, and has determined that the individual poses a security threat.

(3) Withdrawal of Initial Notification. If the Deputy Administrator does not determine that the individual poses a security threat, TSA serves upon the individual a Withdrawal of the Initial Notification and provides a copy of the Withdrawal to the FAA Administrator.

(g) Nondisclosure of certain information. In connection with the procedures under this section, TSA does not disclose to the individual classified information, as defined in Executive Order 12968 section 1.1(d), and TSA reserves the right not to disclose any other information or material not warranting disclosure or protected from disclosure under law.

Issued in Washington, DC, on January 21, 2003.

SUBCHAPTER C—CIVIL AVIATION SECURITY
PART 1544—AIRCRAFT OPERATOR SECURITY: AIR CARRIERS AND COMMERCIAL OPERATORS

Authority: 49 U.S.C. 114, 5103, 40119, 44901-44905, 44907, 44913-44914, 44916-44918, 44932, 44935-44936, 44942, 46105.

Source: 76 FR 8364, Feb. 22, 2002, unless otherwise noted.

Subpart A—General

§1544.1 Applicability of this part.

(a) This part prescribes aviation security rules governing the following:

(1) The operations of aircraft operators holding operating certificates under 14 CFR part 119 for scheduled passenger operations, public charter passenger operations, private charter passenger operations; the operations of aircraft operators holding operating certificates under 14 CFR part 119 operating aircraft with a maximum certificated takeoff weight of 12,500 pounds or more; and other aircraft operators adopting and obtaining approval of an aircraft operator security program.

(2) Each law enforcement officer flying armed aboard an aircraft operated by an aircraft operator described in paragraph (a)(1) of this section.

(3) Each aircraft operator that receives a Security Directive or Information Circular and each person who receives information from a Security Directive or Information Circular issued by TSA.

(b) As used in this part, "aircraft operator" means an aircraft operator subject to this part as described in §1544.101.

[67 FR 8364, Feb. 22, 2002, as amended at 67 FR 8209, Feb. 22, 2002]

§1544.3 TSA inspection authority.

(a) Each aircraft operator must allow TSA, at any time or place, to make any inspections or tests, including copying records, to determine compliance of an airport operator, aircraft operator, foreign air carrier, indirect air carrier, or other airport tenants with—

(1) This subchapter and any security program under this subchapter, and part 1520 of this chapter; and

(2) 49 U.S.C. Subtitle VII, as amended.

(b) At the request of TSA, each aircraft operator must provide evidence of compliance with this part and its security program, including copies of records.

(c) TSA may enter and be present within secured areas, AOA's, and SIDA's without access media or identification media issued or approved by an airport operator or aircraft operator, in order to inspect or test compliance, or perform other such duties as TSA may direct.

(d) At the request of TSA and the completion of SIDA training as required in a security program, each aircraft operator must promptly issue to TSA personnel access and identification media to provide TSA personnel with unescorted access to, and movement within, areas controlled by the aircraft operator under an exclusive area agreement.

Subpart B—Security Program

§1544.101 Adoption and implementation.

Link to an amendment published at 67 FR 41639, June 19, 2002.

(a) Full program. Each aircraft operator must carry out subparts C, D, and E of this part and must adopt and carry out a security program that meets the requirements of §1544.103 for each of the following operations:

(1) A scheduled passenger or public charter passenger operation with an aircraft having a passenger seating configuration of 61 or more seats.

(2) A scheduled passenger or public charter passenger operation with an aircraft having a passenger seating configuration of 60 or fewer seats when passengers are enplaned from or deplaned into a sterile area.

(b) Partial program—adoption. Each aircraft operator must carry out the requirements specified in paragraph (c) of this section for each of the following operations:

(1) A scheduled passenger or public charter passenger operation with an aircraft having a passenger-seating configuration of 31 or more but 60 or fewer seats that does not enplane from or deplane into a sterile area.

(2) A scheduled passenger or public charter passenger operation with an aircraft having a passenger-seating configuration of 60 or fewer seats engaged in operations to, from, or outside the United States that does not enplane from or deplane into a sterile area.

(c) Partial program-content: For operations described in paragraph (b) of this section, the aircraft operator must carry out the following, and must adopt and carry out a security program that meets the applicable requirements in §1544.103 (c):

(1) The requirements of §§1544.215, 1544.217, 1544.219, 1544.223, 1544.230, 1544.235, 1544.237, 1544.301, 1544.303, and 1544.305.

(2) Other provisions of subparts C, D, and E of this part that TSA has approved upon request.

(3) The remaining requirements of subparts C, D, and E when TSA notifies the aircraft operator in writing that a security threat exists concerning that operation.

(d) Twelve-five program-adoption: Each aircraft operator must carry out the requirements of paragraph (e) of this section for each operation that meets all of the following—

(1) Is in an aircraft with a maximum certificated takeoff weight of 12,500 pounds or more;

(2) Is in scheduled or charter service;

(3) Is carrying passengers or cargo or both; and

(4) Is not under a full program or partial program under paragraph (a) or (b) of this section.

(e) Twelve-five program-contents: For each operation described in paragraph (d) of this section, the aircraft operator must carry out the following, and must adopt and carry out a security program that meets the applicable requirements of §1544.103 (c):

(1) The requirements of §§1544.215, 1544.217, 1544.219, 1544.223, 1544.230, 1544.235, 1544.237, 1544.301(a) and (b), 1544.303, and 1544.305.

(2) Other provisions of subparts C, D, and E that TSA has approved upon request.

(3) The remaining requirements of subparts C, D, and E when TSA notifies the aircraft operator in writing that a security threat exists concerning that operation.

(f) Private charter program: In addition to paragraph (d) of this section, if applicable, each aircraft operator must carry out §§1544.201, 1544.207, 1544.209, 1544.213, 1544.215, 1544.217, 1544.219, 1544.229, 1544.230, 1544.233, 1544.235, 1544.303, 1544.305, and subpart E, and must adopt and carry out a security program that meets the applicable requirements of §1544.103 for each private charter operation in which passengers are enplaned from or deplaned into a sterile area.

(g) Limited program: In addition to paragraph (d) of this section, if applicable, TSA may approve a security program after receiving a request by an aircraft operator holding a certificate under 14 CFR part 119, other than one identified in paragraph (a), (b), (d), or (f) of this section. The aircraft operator must—

(1) Carry out selected provisions of subparts C, D, and E;

(2) Carry out the provisions of §1544.305, as specified in its security program; and

(3) Adopt and carry out a security program that meets the applicable requirements of §1544.103 (c).

[67 FR 8364, Feb. 22, 2002, as amended at 67 FR 8209, Feb. 22, 2002]

§1544.103 Form, content, and availability.

(a) General requirements. Each security program must:

(1) Provide for the safety of persons and property traveling on flights provided by the aircraft operator against acts of criminal violence and air piracy, and the introduction of explosives, incendiaries, or weapons aboard an aircraft.

(2) Be in writing and signed by the aircraft operator or any person delegated authority in this matter.

(3) Be approved by TSA.

(b) Availability. Each aircraft operator having a security program must:

(1) Maintain an original copy of the security program at its corporate office.

(2) Have accessible a complete copy, or the pertinent portions of its security program, or appropriate implementing instructions, at each airport served. An electronic version of the program is adequate.

(3) Make a copy of the security program available for inspection upon request of TSA.

(4) Restrict the distribution, disclosure, and availability of information contained in the security program to persons with a need-to-know as described in part 1520 of this chapter.

(5) Refer requests for such information by other persons to TSA.

(c) Content. The security program must include, as specified for that aircraft operator in §1544.101, the following:

(1) The procedures and description of the facilities and equipment used to comply with the requirements of §1544.201 regarding the acceptance and screening of individuals and their accessible property, including, if applicable, the carriage weapons as part of State-required emergency equipment.

(2) The procedures and description of the facilities and equipment used to comply with the requirements of §1544.203 regarding the acceptance and screening of checked baggage.

(3) The procedures and description of the facilities and equipment used to comply with the requirements of §1544.205 regarding the acceptance and screening of cargo.

(4) The procedures and description of the facilities and equipment used to comply with the requirements of §1544.207 regarding the screening of individuals and property.

(5) The procedures and description of the facilities and equipment used to comply with the requirements of §1544.209 regarding the use of metal detection devices.

(6) The procedures and description of the facilities and equipment used to comply with the requirements of §1544.211 regarding the use of x-ray systems.

(7) The procedures and description of the facilities and equipment used to comply with the requirements of §1544.213 regarding the use of explosives detection systems.

(8) The procedures used to comply with the requirements of §1544.215 regarding the responsibilities of security coordinators. The names of the Aircraft Operator Security Coordinator (AOSC) and any alternate, and the means for contacting the AOSC(s) on a 24-hour basis, as provided in §1544.215.

(9) The procedures used to comply with the requirements of §1544.217 regarding the requirements for law enforcement personnel.

(10) The procedures used to comply with the requirements of §1544.219 regarding carriage of accessible weapons.

(11) The procedures used to comply with the requirements of §1544.221 regarding carriage of prisoners under the control of armed law enforcement officers.

(12) The procedures used to comply with the requirements of §1544.223 regarding transportation of Federal Air Marshals.

(13) The procedures and description of the facilities and equipment used to perform the aircraft and facilities control function specified in §1544.225.

(14) The specific locations where the air carrier has entered into an exclusive area agreement under §1544.227.

(15) The procedures used to comply with the applicable requirements of §§1544.229 and 1544.230 regarding fingerprint-based criminal history records checks.

(16) The procedures used to comply with the requirements of §1544.231 regarding personnel identification systems.

(17) The procedures and syllabi used to accomplish the training required under §1544.233.

(18) The procedures and syllabi used to accomplish the training required under §1544.235.

(19) An aviation security contingency plan as specified under §1544.301.

(20) The procedures used to comply with the requirements of §1544.303 regarding bomb and air piracy threats.

(21) The procedures used to comply with §1544.237 regarding flight deck privileges.

[67 FR 8364, Feb. 22, 2002, as amended at 67 FR 8209, Feb. 22, 2002]

§1544.105 Approval and amendments.

(a) Initial approval of security program. Unless otherwise authorized by TSA, each aircraft operator required to have a security program under this part must submit its proposed security program to the designated official for approval at least 90 days before the intended date of passenger operations. The proposed security program must meet the requirements applicable to its operation as described in §1544.101. Such requests will be processed as follows:

(1) The designated official, within 30 days after receiving the proposed aircraft operator security program, will either approve the program or give the aircraft operator written notice to modify the program to comply with the applicable requirements of this part.

(2) The aircraft operator may either submit a modified security program to the designated official for approval, or petition the Under Secretary to reconsider the notice to modify within 30 days of receiving a notice to modify. A petition for reconsideration must be filed with the designated official.

(3) The designated official, upon receipt of a petition for reconsideration, either amends or withdraws the notice, or transmits the petition, together with any pertinent information, to the Under Secretary for reconsideration. The Under Secretary disposes of the petition within 30 days of receipt by either directing the designated official to withdraw or amend the notice to modify, or by affirming the notice to modify.

(b) Amendment requested by an aircraft operator. An aircraft operator may submit a request to TSA to amend its security program as follows:

(1) The request for an amendment must be filed with the designated official at least 45 days before the date it proposes for the amendment to become effective, unless a shorter period is allowed by the designated official.

(2) Within 30 days after receiving a proposed amendment, the designated official, in writing, either approves or denies the request to amend.

(3) An amendment to an aircraft operator security program may be approved if the designated official determines that safety and the public interest will allow it, and the proposed amendment provides the level of security required under this part.

(4) Within 30 days after receiving a denial, the aircraft operator may petition the Under Secretary to reconsider the denial. A petition for reconsideration must be filed with the designated official.

(5) Upon receipt of a petition for reconsideration, the designated official either approves the request to amend or transmits the petition, together with any pertinent information, to the Under Secretary for reconsideration. The Under Secretary disposes of the petition within 30 days of receipt by either directing the designated official to approve the amendment, or affirming the denial.

(6) Any aircraft operator may submit a group proposal for an amendment that is on behalf of it and other aircraft operators that co-sign the proposal.

(c) Amendment by TSA. If safety and the public interest require an amendment, TSA may amend a security program as follows:

(1) The designated official notifies the aircraft operator, in writing, of the proposed amendment, fixing a period of not less than 30 days within which the aircraft operator may submit written information, views, and arguments on the amendment.

(2) After considering all relevant material, the designated official notifies the aircraft operator of any amendment adopted or rescinds the notice. If the amendment is adopted, it becomes effective not less than 30 days after the aircraft operator receives the notice of amendment, unless the aircraft operator petitions the Under Secretary to reconsider no later than 15 days before the effective date of the amendment. The aircraft operator must send the petition for reconsideration to the designated official. A timely petition for reconsideration stays the effective date of the amendment.

(3) Upon receipt of a petition for reconsideration, the designated official either amends or withdraws the notice or transmits the petition, together with any pertinent information, to the Under Secretary for reconsideration. The Under Secretary disposes of the petition within 30 days of receipt by either directing the designated official to withdraw or amend the amendment, or by affirming the amendment.

(d) Emergency amendments. If the designated official finds that there is an emergency requiring immediate action with respect to safety in air transportation or in air commerce that makes procedures in this section contrary to the public interest, the designated official may issue an amendment, without the prior notice and comment procedures in paragraph (c) of this section, effective without stay on the date the aircraft operator receives notice of it. In such a case, the designated official will incorporate in the notice a brief statement of the reasons and findings for the amendment to be adopted. The aircraft operator may file a petition for reconsideration under paragraph (c) of this section; however, this does not stay the effective date of the emergency amendment.

Subpart C—Operations

§1544.201 Acceptance and screening of individuals and accessible property.

(a) Preventing or deterring the carriage of any explosive, incendiary, or deadly or dangerous weapon. Each aircraft operator must use the measures in its security program to prevent or deter the carriage of any weapon, explosive, or incendiary on or about each individual's person or accessible property before boarding an aircraft or entering a sterile area.

(b) Screening of individuals and accessible property. Except as provided in its security program, each aircraft operator must ensure that each individual entering a sterile area at each preboard screening checkpoint for which it is responsible, and all accessible property under that individual's control, are inspected for weapons, explosives, and incendiaries as provided in §1544.207.

(c) Refusal to transport. Each aircraft operator must deny entry into a sterile area and must refuse to transport—

(1) Any individual who does not consent to a search or inspection of his or her person in accordance with the system prescribed in this part; and

(2) Any property of any individual or other person who does not consent to a search or inspection of that property in accordance with the system prescribed by this part.

(d) Prohibitions on carrying a weapon, explosive, or incendiary. Except as provided in §§1544.219, 1544.221, and 1544.223, no aircraft operator may permit any individual to have a weapon, explosive, or incendiary, on or about the individual's person or accessible property when onboard an aircraft.

(e) Staffing. Each aircraft operator must staff its security screening checkpoints with supervisory and non-supervisory personnel in accordance with the standards specified in its security program.

§1544.203 Acceptance and screening of checked baggage.

(a) Preventing or deterring the carriage of any explosive or incendiary. Each aircraft operator must use the procedures, facilities, and equipment described in its security program to prevent or deter the carriage of any unauthorized explosive or incendiary onboard aircraft in checked baggage.

(b) Acceptance. Each aircraft operator must ensure that checked baggage carried in the aircraft is received by its authorized aircraft operator representative.

(c) Screening of checked baggage. Except as provided in its security program, each aircraft operator must ensure that all checked baggage is inspected for explosives and incendiaries before loading it on its aircraft, in accordance with §1544.207.

(d) Control. Each aircraft operator must use the procedures in its security program to control checked baggage that it accepts for transport on an aircraft, in a manner that:

(1) Prevents the unauthorized carriage of any explosive or incendiary aboard the aircraft.

(2) Prevents access by persons other than an aircraft operator employee or its agent.

(e) Refusal to transport. Each aircraft operator must refuse to transport any individual's checked baggage or property if the individual does not consent to a search or inspection of that checked baggage or property in accordance with the system prescribed by this part.

(f) Firearms in checked baggage. No aircraft operator may knowingly permit any person to transport in checked baggage:

(1) Any loaded firearm(s).

(2) Any unloaded firearm(s) unless—

(i) The passenger declares to the aircraft operator, either orally or in writing before checking the baggage that any firearm carried in the baggage is unloaded;

(ii) The firearm is carried in a hard-sided container;

(iii) The container in which it is carried is locked, and only the individual checking the baggage retains the key or combination; and

(iv) The checked baggage containing the firearm is carried in an area that is inaccessible to passengers, and is not carried in the flightcrew compartment,.

(3) Any unauthorized explosive or incendiary.

(g) Ammunition. This section does not prohibit the carriage of ammunition in checked baggage or in the same container as a firearm. Title 49 CFR part 175 provides additional requirements governing carriage of ammunition on aircraft.

§1544.205 Acceptance and screening of cargo.

(a) General requirements. Each aircraft operator must use the procedures, facilities, and equipment described in its security program to prevent or deter the carriage of unauthorized explosives or incendiaries in cargo onboard a passenger aircraft.

(b) Screening of cargo baggage. Each aircraft operator must ensure that, as required in its security program, cargo is inspected for explosives and incendiaries before loading it on its aircraft in accordance with §1544.207.

(c) Control. Each aircraft operator must use the procedures in its security program to control cargo that it accepts for transport on an aircraft in a manner that:

(1) Prevents the carriage of any unauthorized explosive or incendiary aboard the aircraft.

(2) Prevents access by persons other than an aircraft operator employee or its agent.

(d) Refusal to transport. Each aircraft operator must refuse to transport any cargo if the shipper does not consent to a search or inspection of that cargo in accordance with the system prescribed by this part.

§1544.207 Screening of individuals and property.

(a) Applicability of this section. This section applies to the inspection of individuals, accessible property, checked baggage, and cargo as required under this part.

(b) Locations within the United States at which TSA conducts screening. Each aircraft operator must ensure that the individuals or property have been inspected by TSA before boarding or loading on its aircraft. This paragraph applies when TSA is conducting screening using TSA employees or when using companies under contract with TSA.

(c) Aircraft operator conducting screening. Each aircraft operator must use the measures in its security program and in subpart E of this part to inspect the individual or property. This paragraph does not apply at locations identified in paragraphs (b) and (d) of this section.

(d) Locations outside the United States at which the foreign government conducts screening. Each aircraft operator must ensure that all individuals and property have been inspected by the foreign government. This paragraph applies when the host government is conducting screening using government employees or when using companies under contract with the government.

§1544.209 Use of metal detection devices.

(a) No aircraft operator may use a metal detection device within the United States or under the aircraft operator's operational control outside the United States to inspect persons, unless specifically authorized under a security program under this part. No aircraft operator may use such a device contrary to its security program.

(b) Metal detection devices must meet the calibration standards established by TSA.

§1544.211 Use of X-ray systems.

(a) TSA authorization required. No aircraft operator may use any X-ray system within the United States or under the aircraft operator's operational control outside the United States to inspect accessible property or checked baggage, unless specifically authorized under its security program. No aircraft operator may use such a system in a manner contrary to its security program. TSA authorizes aircraft operators to use X-ray systems for inspecting accessible property or checked baggage under a security program if the aircraft operator shows that—

(1) The system meets the standards for cabinet X-ray systems primarily for the inspection of baggage issued by the Food and Drug Administration (FDA) and published in 21 CFR 1020.40;

(2) A program for initial and recurrent training of operators of the system is established, which includes training in radiation safety, the efficient use of X-ray systems, and the identification of weapons, explosives, and incendiaries; and

(3) The system meets the imaging requirements set forth in its security program using the step wedge specified in American Society for Testing Materials (ASTM) Standard F792-88 (Reapproved 1993). This standard is incorporated by reference in paragraph (g) of this section.

(b) Annual radiation survey. No aircraft operator may use any X-ray system unless, within the preceding 12 calendar months, a radiation survey is conducted that shows that the system meets the applicable performance standards in 21 CFR 1020.40.

(c) Radiation survey after installation or moving. No aircraft operator may use any X-ray system after the system has been installed at a screening point or after the system has been moved unless a radiation survey is conducted which shows that the system meets the applicable performance standards in 21 CFR 1020.40. A radiation survey is not required for an X-ray system that is designed and constructed as a mobile unit and the aircraft operator shows that it can be moved without altering its performance.

(d) Defect notice or modification order. No aircraft operator may use any X-ray system that is not in full compliance with any defect notice or modification order issued for that system by the FDA, unless the FDA has advised TSA that the defect or failure to comply does not create a significant risk of injury, including genetic injury, to any person.

(e) Signs and inspection of photographic equipment and film. (1) At locations at which an aircraft operator uses an X-ray system to inspect accessible property the aircraft operator must ensure that a sign is posted in a conspicuous place at the screening checkpoint. At locations outside the United States at which a foreign government uses an X-ray system to inspect accessible property the aircraft operator must ensure that a sign is posted in a conspicuous place at the screening checkpoint.

(2) At locations at which an aircraft operator or TSA uses an X-ray system to inspect checked baggage the aircraft operator must ensure that a sign is posted in a conspicuous place where the aircraft operator accepts checked baggage.

(3) The signs required under this paragraph (e) must notify individuals that such items are being inspected by an X-ray and advise them to remove all X-ray, scientific, and high-speed film from accessible property and checked baggage before inspection. This sign must also advise individuals that they may request that an inspection be made of their photographic equipment and film packages without exposure to an X-ray system. If the X-ray system exposes any accessible property or checked baggage to more

than one milliroentgen during the inspection, the sign must advise individuals to remove film of all kinds from their articles before inspection.

(4) If requested by individuals, their photographic equipment and film packages must be inspected without exposure to an X-ray system.

(f) Radiation survey verification after installation or moving. Each aircraft operator must maintain at least one copy of the results of the most recent radiation survey conducted under paragraph (b) or (c) of this section and must make it available for inspection upon request by TSA at each of the following locations—

(1) The aircraft operator's principal business office; and

(2) The place where the X-ray system is in operation.

(g) Incorporation by reference. The American Society for Testing and Materials (ASTM) Standard F792-88 (Reapproved 1993), "Standard Practice for Design and Use of Ionizing Radiation Equipment for the Detection of Items Prohibited in Controlled Access Areas," is approved for incorporation by reference by the Director of the Federal Register pursuant to 5 U.S.C. 552(a) and 1 CFR part 51. ASTM Standard F792-88 may be examined at the Department of Transportation (DOT) Docket, 400 Seventh Street SW, Room Plaza 401, Washington, DC 20590, or on DOT's Docket Management System (DMS) web page at http://dms.dot.gov/search (under docket number FAA-2001-8725). Copies of the standard may be examined also at the Office of the Federal Register, 800 North Capitol St., NW, Suite 700, Washington, DC. In addition, ASTM Standard F792-88 (Reapproved 1993) may be obtained from the American Society for Testing and Materials, 100 Barr Harbor Drive, West Conshohocken, PA 19428-2959.

(h) Duty time limitations. Each aircraft operator must comply with the X-ray operator duty time limitations specified in its security program.

§1544.213 Use of explosives detection systems.

(a) Use of explosive detection equipment. If TSA so requires by an amendment to an aircraft operator's security program, each aircraft operator required to conduct screening under a security program must use an explosives detection system approved by TSA to screen checked baggage on international flights.

(b) Signs and inspection of photographic equipment and film. (1) At locations at which an aircraft operator or TSA uses an explosives detection system that uses X-ray technology to inspect checked baggage the aircraft operator must ensure that a sign is posted in a conspicuous place where the aircraft operator accepts checked baggage. The sign must notify individuals that such items are being inspected by an explosives detection system and advise them to remove all X-ray, scientific, and high-speed film from checked baggage before inspection. This sign must also advise individuals that they may request that an in-

spection be made of their photographic equipment and film packages without exposure to an explosives detection system.

(2) If the explosives detection system exposes any checked baggage to more than one milliroentgen during the inspection the aircraft operator must post a sign which advises individuals to remove film of all kinds from their articles before inspection. If requested by individuals, their photographic equipment and film packages must be inspected without exposure to an explosives detection system.

§1544.215 Security coordinators.

(a) Aircraft Operator Security Coordinator. Each aircraft operator must designate and use an Aircraft Operator Security Coordinator (AOSC). The AOSC and any alternates must be appointed at the corporate level and must serve as the aircraft operator's primary contact for security-related activities and communications with TSA, as set forth in the security program. Either the AOSC, or an alternate AOSC, must be available on a 24-hour basis.

(b) Ground Security Coordinator. Each aircraft operator must designate and use a Ground Security Coordinator for each domestic and international flight departure to carry out the Ground Security Coordinator duties specified in the aircraft operator's security program. The Ground Security Coordinator at each airport must conduct the following daily:

(1) A review of all security-related functions for which the aircraft operator is responsible, for effectiveness and compliance with this part, the aircraft operator's security program, and applicable Security Directives.

(2) Immediate initiation of corrective action for each instance of noncompliance with this part, the aircraft operator's security program, and applicable Security Directives. At foreign airports where such security measures are provided by an agency or contractor of a host government, the aircraft operator must notify TSA for assistance in resolving noncompliance issues.

(c) In-flight Security Coordinator. Each aircraft operator must designate and use the pilot in command as the In-flight Security Coordinator for each domestic and international flight to perform duties specified in the aircraft operator's security program.

§1544.217 Law enforcement personnel.

(a) The following applies to operations at airports within the United States that are not required to hold a security program under part 1542 of this chapter.

(1) For operations described in §1544.101(a) each aircraft operator must provide for law enforcement personnel meeting the qualifications and standards specified in §§1542.215 and 1542.217 of this chapter.

(2) For operations described in §1544.101(b) or (c) each aircraft operator must—

(i) Arrange for law enforcement personnel meeting the qualifications and standards specified in §1542.217 of this chapter to be available to respond to an incident; and

(ii) Provide its employees, including crewmembers, current information regarding procedures for obtaining law enforcement assistance at that airport.

(b) The following applies to operations at airports required to hold security programs under part 1542 of this chapter. For operations described in §1544.101(c), each aircraft operator must—

(1) Arrange with TSA and the airport operator, as appropriate, for law enforcement personnel meeting the qualifications and standards specified in §1542.217 of this chapter to be available to respond to incidents, and

(2) Provide its employees, including crewmembers, current information regarding procedures for obtaining law enforcement assistance at that airport.

§1544.219 Carriage of accessible weapons.

(a) Flights for which screening is conducted. The provisions of §1544.201(d), with respect to accessible weapons, do not apply to a law enforcement officer (LEO) aboard a flight for which screening is required if the requirements of this section are met. Paragraph (a) of this section does not apply to a Federal Air Marshal on duty status under §1544.223.

(1) Unless otherwise authorized by TSA, the armed LEO must meet the following requirements:

(i) Be a Federal law enforcement officer or a full-time municipal, county, or state law enforcement officer who is a direct employee of a government agency.

(ii) Be sworn and commissioned to enforce criminal statutes or immigration statutes.

(iii) Be authorized by the employing agency to have the weapon in connection with assigned duties.

(iv) Has completed the training program "Law Enforcement Officers Flying Armed."

(2) In addition to the requirements of paragraph (a)(1) of this section, the armed LEO must have a need to have the weapon accessible from the time he or she would otherwise check the weapon until the time it would be claimed after deplaning. The need to have the weapon accessible must be determined by the employing agency, department, or service and be based on one of the following:

(i) The provision of protective duty, for instance, assigned to a principal or advance team, or on travel required to be prepared to engage in a protective function.

(ii) The conduct of a hazardous surveillance operation.

(iii) On official travel required to report to another location, armed and prepared for duty.

(iv) Employed as a Federal LEO, whether or not on official travel, and armed in accordance with an agency-wide policy governing that type of travel established by the employing agency by directive or policy statement.

(v) Control of a prisoner, in accordance with §1544.221, or an armed LEO on a round trip ticket returning from escorting, or traveling to pick up, a prisoner.

(vi) TSA Federal Air Marshal on duty status.

(3) The armed LEO must comply with the following notification requirements:

(i) All armed LEOs must notify the aircraft operator of the flight(s) on which he or she needs to have the weapon accessible at least 1 hour, or in an emergency as soon as practicable, before departure.

(ii) Identify himself or herself to the aircraft operator by presenting credentials that include a clear full-face picture, the signature of the armed LEO, and the signature of the authorizing official of the agency, service, or department or the official seal of the agency, service, or department. A badge, shield, or similar device may not be used, or accepted, as the sole means of identification.

(iii) If the armed LEO is a State, county, or municipal law enforcement officer, he or she must present an original letter of authority, signed by an authorizing official from his or her employing agency, service or department, confirming the need to travel armed and detailing the itinerary of the travel while armed.

(iv) If the armed LEO is an escort for a foreign official then this paragraph (a)(3) may be satisfied by a State Department notification.

(4) The aircraft operator must do the following:

(i) Obtain information or documentation required in paragraphs (a)(3)(ii), (iii), and (iv) of this section.

(ii) Advise the armed LEO, before boarding, of the aircraft operator's procedures for carrying out this section.

(iii) Have the LEO confirm he/she has completed the training program "Law Enforcement Officers Flying Armed" as required by TSA, unless otherwise authorized by TSA.

(iv) Ensure that the identity of the armed LEO is known to the appropriate personnel who are responsible for security during the boarding of the aircraft.

(v) Notify the pilot in command and other appropriate crewmembers, of the location of each armed LEO aboard the aircraft. Notify any other armed LEO of the location of each armed LEO, including FAM's. Under circumstances described in the security program, the aircraft operator must not close the doors until the notification is complete.

(vi) Ensure that the information required in paragraphs (a)(3)(i) and (ii) of this section is furnished to the flight crew of each additional connecting flight by the Ground Security Coordinator or other designated agent at each location.

(b) Flights for which screening is not conducted. The provisions of §1544.201(d), with respect to accessible weapons, do not apply to a LEO aboard a flight for which screening is not required if the requirements of paragraphs (a)(1), (3), and (4) of this section are met.

(c) *Alcohol.* (1) No aircraft operator may serve any alcoholic beverage to an armed LEO.

(2) No armed LEO may:

(i) Consume any alcoholic beverage while aboard an aircraft operated by an aircraft operator.

(ii) Board an aircraft armed if they have consumed an alcoholic beverage within the previous 8 hours.

(d) *Location of weapon.* (1) Any individual traveling aboard an aircraft while armed must at all times keep their weapon:

(i) Concealed and out of view, either on their person or in immediate reach, if the armed LEO is not in uniform.

(ii) On their person, if the armed LEO is in uniform.

(2) No individual may place a weapon in an overhead storage bin.

§1544.221 Carriage of prisoners under the control of armed law enforcement officers.

(a) This section applies as follows:

(1) This section applies to the transport of prisoners under the escort of an armed law enforcement officer.

(2) This section does not apply to the carriage of passengers under voluntary protective escort.

(3) This section does not apply to the escort of non-violent detainees of the Immigration and Naturalization Service. This section does not apply to individuals who may be traveling with a prisoner and armed escort, such as the family of a deportee who is under armed escort.

(b) For the purpose of this section:

(1) "High risk prisoner" means a prisoner who is an exceptional escape risk, as determined by the law enforcement agency, and charged with, or convicted of, a violent crime.

(2) "Low risk prisoner" means any prisoner who has not been designated as "high risk."

(c) No aircraft operator may carry a prisoner in the custody of an armed law enforcement officer aboard an aircraft for which screening is required unless, in addition to the requirements in §1544.219, the following requirements are met:

(1) The agency responsible for control of the prisoner has determined whether the prisoner is considered a high risk or a low risk.

(2) Unless otherwise authorized by TSA, no more than one high risk prisoner may be carried on the aircraft.

(d) No aircraft operator may carry a prisoner in the custody of an armed law enforcement officer aboard an aircraft for which screening is required unless the following staffing requirements are met:

(1) A minimum of one armed law enforcement officer must control a low risk prisoner on a flight that is scheduled for 4 hours or less. One armed law enforcement officer may control no more than two low risk prisoners.

(2) A minimum of two armed law enforcement officers must control a low risk prisoner on a flight that is

scheduled for more than 4 hours. Two armed law enforcement officers may control no more than two low risk prisoners.

(3) For high-risk prisoners:

(i) For one high-risk prisoner on a flight: A minimum of two armed law enforcement officers must control a high risk prisoner. No other prisoners may be under the control of those two armed law enforcement officers.

(ii) If TSA has authorized more than one high-risk prisoner to be on the flight under paragraph (c)(2) of this section, a minimum of one armed law enforcement officer for each prisoner and one additional armed law enforcement officer must control the prisoners. No other prisoners may be under the control of those armed law enforcement officers.

(e) An armed law enforcement officer who is escorting a prisoner—

(1) Must notify the aircraft operator at least 24 hours before the scheduled departure, or, if that is not possible as far in advance as possible of the following—

(i) The identity of the prisoner to be carried and the flight on which it is proposed to carry the prisoner; and

(ii) Whether or not the prisoner is considered to be a high risk or a low risk.

(2) Must arrive at the check-in counter at least 1 hour before to the scheduled departure.

(3) Must assure the aircraft operator, before departure, that each prisoner under the control of the officer(s) has been searched and does not have on or about his or her person or property anything that can be used as a weapon.

(4) Must be seated between the prisoner and any aisle.

(5) Must accompany the prisoner at all times, and keep the prisoner under control while aboard the aircraft.

(f) No aircraft operator may carry a prisoner in the custody of an armed law enforcement officer aboard an aircraft unless the following are met:

(1) When practicable, the prisoner must be boarded before any other boarding passengers and deplaned after all other deplaning passengers.

(2) The prisoner must be seated in a seat that is neither located in any passenger lounge area nor located next to or directly across from any exit and, when practicable, the aircraft operator should seat the prisoner in the rearmost seat of the passenger cabin.

(g) Each armed law enforcement officer escorting a prisoner and each aircraft operator must ensure that the prisoner is restrained from full use of his or her hands by an appropriate device that provides for minimum movement of the prisoner's hands, and must ensure that leg irons are not used.

(h) No aircraft operator may provide a prisoner under the control of a law enforcement officer—

(1) With food or beverage or metal eating utensils unless authorized to do so by the armed law enforcement officer.

(2) With any alcoholic beverage.

§1544.223 Transportation of Federal Air Marshals.

(a) A Federal Air Marshal on duty status may have a weapon accessible while aboard an aircraft for which screening is required.

(b) Each aircraft operator must carry Federal Air Marshals, in the number and manner specified by TSA, on each scheduled passenger operation, and public charter passenger operation designated by TSA.

(c) Each Federal Air Marshal must be carried on a first priority basis and without charge while on duty, including positioning and repositioning flights. When a Federal Air Marshal is assigned to a scheduled flight that is canceled for any reason, the aircraft operator must carry that Federal Air Marshal without charge on another flight as designated by TSA.

(d) Each aircraft operator must assign the specific seat requested by a Federal Air Marshal who is on duty status. If another LEO is assigned to that seat or requests that seat, the aircraft operator must inform the Federal Air Marshal. The Federal Air Marshal will coordinate seat assignments with the other LEO.

(e) The Federal Air Marshal identifies himself or herself to the aircraft operator by presenting credentials that include a clear, full-face picture, the signature of the Federal Air Marshal, and the signature of the FAA Administrator. A badge, shield, or similar device may not be used or accepted as the sole means of identification.

(f) The requirements of §1544.219(a) do not apply for a Federal Air Marshal on duty status.

(g) Each aircraft operator must restrict any information concerning the presence, seating, names, and purpose of Federal Air Marshals at any station or on any flight to those persons with an operational need to know.

(h) Law enforcement officers authorized to carry a weapon during a flight will be contacted directly by a Federal Air Marshal who is on that same flight.

§1544.225 Security of aircraft and facilities.

Each aircraft operator must use the procedures included, and the facilities and equipment described, in its security program to perform the following control functions with respect to each aircraft operation:

(a) Prevent unauthorized access to areas controlled by the aircraft operator under an exclusive area agreement in accordance with §1542.111 of this chapter.

(b) Prevent unauthorized access to each aircraft.

(c) Conduct a security inspection of each aircraft before placing it into passenger operations if access has not been controlled in accordance with the aircraft operator security program and as otherwise required in the security program.

§1544.227 Exclusive area agreement.

(a) An aircraft operator that has entered into an exclusive area agreement with an airport operator, under §1542.111 of this chapter must carry out that exclusive area agreement.

(b) The aircraft operator must list in its security program the locations at which it has entered into exclusive area agreements with an airport operator.

(c) The aircraft operator must provide the exclusive area agreement to TSA upon request.

(d) Any exclusive area agreements in effect on November 14, 2001, must meet the requirements of this section and §1542.111 of this chapter no later than November 14, 2002.

§1544.229 Fingerprint-based criminal history records checks (CHRC):

Unescorted access authority, authority to perform screening functions, and authority to perform checked baggage or cargo functions.

(a) Scope. The following individuals are within the scope of this section. Unescorted access authority, authority to perform screening functions, and authority to perform checked baggage or cargo functions, are collectively referred to as "covered functions."

(1) New unescorted access authority or authority to perform screening functions.

(i) Each employee or contract employee covered under a certification made to an airport operator on or after December 6, 2001, pursuant to 14 CFR 107.209(n) in effect prior to November 14, 2001 (see 14 CFR Parts 60 to 139 revised as of January 1, 2001) or §1542.209(n) of this chapter.

(ii) Each individual issued on or after December 6, 2001, an aircraft operator identification media that one or more airports accepts as airport-approved media for unescorted access authority within a security identification display area (SIDA), as described in §1542.205 of this chapter (referred to as "unescorted access authority").

(iii) Each individual, on or after December 6, 2001, granted authority to perform the following screening functions at locations within the United States (referred to as "authority to perform screening functions")—

(A) Screening passengers or property that will be carried in a cabin of an aircraft of an aircraft operator required to screen passengers under this part.

(B) Serving as an immediate supervisor (checkpoint security supervisor (CSS)), and the next supervisory level (shift or site supervisor), to those individuals described in paragraph (a)(1)(iii)(A) of this section.

(2) Current unescorted access authority or authority to perform screening functions. (i) Each employee or contract employee covered under a certification made to an airport operator pursuant to 14 CFR 107.31(n) in effect prior to November 14, 2001 (see 14 CFR Parts 60 to 139 revised as of January 1, 2001), or pursuant to 14 CFR 107.209(n) in effect prior to December 6, 2001 (see 14 CFR Parts 60 to 139 revised as of January 1, 2001).

(ii) Each individual who holds on December 6, 2001, an aircraft operator identification media that one or more airports accepts as airport-approved media for unescorted access authority within a security identification display area (SIDA), as described in §1542.205 of this chapter.

(iii) Each individual who is performing on December 6, 2001, a screening function identified in paragraph (a)(1)(iii) of this section.

(3) *New authority to perform checked baggage or cargo functions.* Each individual who, on and after February 17, 2002, is granted the authority to perform the following checked baggage and cargo functions (referred to as "authority to perform checked baggage or cargo functions"), except for individuals described in paragraph (a)(1) of this section:

(i) Screening of checked baggage or cargo of an aircraft operator required to screen passengers under this part, or serving as an immediate supervisor of such an individual.

(ii) Accepting checked baggage for transport on behalf of an aircraft operator required to screen passengers under this part.

(4) *Current authority to perform checked baggage or cargo functions.* Each individual who holds on February 17, 2002, authority to perform checked baggage or cargo functions, except for individuals described in paragraph (a)(1) or (2) of this section.

(b) *Individuals seeking unescorted access authority, authority to perform screening functions, or authority to perform checked baggage or cargo functions.* Each aircraft operator must ensure that each individual identified in paragraph (a)(1) or (3) of this section has undergone a fingerprint-based CHRC that does not disclose that he or she has a disqualifying criminal offense, as described in paragraph (d) of this section, before—

(1) Making a certification to an airport operator regarding that individual;

(2) Issuing an aircraft operator identification medium to that individual;

(3) Authorizing that individual to perform screening functions; or

(4) Authorizing that individual to perform checked baggage or cargo functions.

(c) *Individuals who have not had a CHRC.* (1) *Deadline for conducting a CHRC.* Each aircraft operator must ensure that, on and after December 6, 2002:

(i) No individual retains unescorted access authority, whether obtained as a result of a certification to an airport operator under 14 CFR 107.31(n) in effect prior to November 14, 2001 (see 14 CFR parts 60 to 139 revised as of January 1, 2001), or under 14 CFR 107.209(n) in effect prior to December 6, 2001 (see 14 CFR Parts 60 to 139 revised as of January 1, 2001), or obtained as a result of the issuance of an aircraft operator's identification media, unless the individual has been subject to a fingerprint-based CHRC for unescorted access authority under this part.

(ii) No individual continues to have authority to perform screening functions described in paragraph (a)(1)(iii) of this section, unless the individual has been subject to a fingerprint-based CHRC under this part.

(iii) No individual continues to have authority to perform checked baggage or cargo functions described in paragraph (a)(3) of this section, unless the individual has been subject to a fingerprint-based CHRC under this part.

(2) *Lookback for individuals with unescorted access authority or authority to perform screening functions.* When a CHRC discloses a disqualifying criminal offense for which the conviction or finding was on or after December 6, 1991, the aircraft operator must immediately suspend that individual's unescorted access authority or authority to perform screening functions.

(3) *Lookback for individuals with authority to perform checked baggage or cargo functions.* When a CHRC discloses a disqualifying criminal offense for which the conviction or finding was on or after February 17, 1992, the aircraft operator must immediately suspend that individual's authority to perform checked baggage or cargo functions.

(d) *Disqualifying criminal offenses.* An individual has a disqualifying criminal offense if the individual has been convicted, or found not guilty by reason of insanity, of any of the disqualifying crimes listed in this paragraph in any jurisdiction during the 10 years before the date of the individual's application for authority to perform covered functions, or while the individual has authority to perform covered functions. The disqualifying criminal offenses are as follows:

(1) Forgery of certificates, false marking of aircraft, and other aircraft registration violation; 49 U.S.C. 46306.

(2) Interference with air navigation; 49 U.S.C. 46308.

(3) Improper transportation of a hazardous material; 49 U.S.C. 46312.

(4) Aircraft piracy; 49 U.S.C. 46502.

(5) Interference with flight crew members or flight attendants; 49 U.S.C. 46504.

(6) Commission of certain crimes aboard aircraft in flight; 49 U.S.C. 46506.

(7) Carrying a weapon or explosive aboard aircraft; 49 U.S.C. 46505.

(8) Conveying false information and threats; 49 U.S.C. 46507.

(9) Aircraft piracy outside the special aircraft jurisdiction of the United States; 49 U.S.C. 46502(b).

(10) Lighting violations involving transporting controlled substances; 49 U.S.C. 46315.

(11) Unlawful entry into an aircraft or airport area that serves air carriers or foreign air carriers contrary to established security requirements; 49 U.S.C. 46314.

(12) Destruction of an aircraft or aircraft facility; 18 U.S.C. 32.

(13) Murder.

(14) Assault with intent to murder.

(15) Espionage.

(16) Sedition.

(17) Kidnapping or hostage taking.

(18) Treason.

(19) Rape or aggravated sexual abuse.

(20) Unlawful possession, use, sale, distribution, or manufacture of an explosive or weapon.

(21) Extortion.

(22) Armed or felony unarmed robbery.

(23) Distribution of, or intent to distribute, a controlled substance.

(24) Felony arson.

(25) Felony involving a threat.

(26) Felony involving—

(i) Willful destruction of property;

(ii) Importation or manufacture of a controlled substance;

(iii) Burglary;

(iv) Theft;

(v) Dishonesty, fraud, or misrepresentation;

(vi) Possession or distribution of stolen property;

(vii) Aggravated assault;

(viii) Bribery; or

(ix) Illegal possession of a controlled substance punishable by a maximum term of imprisonment of more than 1 year.

(27) Violence at international airports; 18 U.S.C. 37.

(28) Conspiracy or attempt to commit any of the criminal acts listed in this paragraph (d).

(e) Fingerprint application and processing. (1) At the time of fingerprinting, the aircraft operator must provide the individual to be fingerprinted a fingerprint application that includes only the following—

(i) The disqualifying criminal offenses described in paragraph (d) of this section.

(ii) A statement that the individual signing the application does not have a disqualifying criminal offense.

(iii) A statement informing the individual that Federal regulations under 49 CFR 1544.229 impose a continuing obligation to disclose to the aircraft operator within 24 hours if he or she is convicted of any disqualifying criminal offense that occurs while he or she has authority to perform a covered function.

(iv) A statement reading, "The information I have provided on this application is true, complete, and correct to the best of my knowledge and belief and is provided in good faith. I understand that a knowing and willful false statement on this application can be punished by fine or imprisonment or both. (See section 1001 of Title 18 United States Code.)"

(v) A line for the printed name of the individual.

(vi) A line for the individual's signature and date of signature.

(2) Each individual must complete and sign the application prior to submitting his or her fingerprints.

(3) The aircraft operator must verify the identity of the individual through two forms of identification prior to fingerprinting, and ensure that the printed name on the fingerprint application is legible. At least one of the two forms of identification must have been issued by a government authority, and at least one must include a photo.

(4) The aircraft operator must:

(i) Advise the individual that a copy of the criminal record received from the FBI will be provided to the individual, if requested by the individual in writing; and

(ii) Identify a point of contact if the individual has questions about the results of the CHRC.

(5) The aircraft operator must collect, control, and process one set of legible and classifiable fingerprints under direct observation by the aircraft operator or a law enforcement officer.

(6) Fingerprints may be obtained and processed electronically, or recorded on fingerprint cards approved by the FBI and distributed by TSA for that purpose.

(7) The fingerprint submission must be forwarded to TSA in the manner specified by TSA.

(f) Fingerprinting fees. Aircraft operators must pay for all fingerprints in a form and manner approved by TSA. The payment must be made at the designated rate (available from the local TSA security office) for each set of fingerprints submitted. Information about payment options is available though the designated TSA headquarters point of contact. Individual personal checks are not acceptable.

(g) Determination of arrest status. (1) When a CHRC on an individual described in paragraph (a)(1) or (3) of this section discloses an arrest for any disqualifying criminal offense listed in paragraph (d) of this section without indicating a disposition, the aircraft operator must determine, after investigation, that the arrest did not result in a disqualifying offense before granting authority to perform a covered function. If there is no disposition, or if the disposition did not result in a conviction or in a finding of not guilty by reason of insanity of one of the offenses listed in paragraph (d) of this section, the individual is not disqualified under this section.

(2) When a CHRC on an individual described in paragraph (a)(2) or (4) of this section discloses an arrest for any disqualifying criminal offense without indicating a disposition, the aircraft operator must suspend the individual's authority to perform a covered function not later than 45 days after obtaining the CHRC unless the aircraft operator determines, after investigation, that the arrest did not result in a disqualifying criminal offense. If there is no disposition, or if the disposition did not result in a conviction or in a finding of not guilty by reason of insanity of one of the offenses listed in paragraph (d) of this section, the individual is not disqualified under this section.

(3) The aircraft operator may only make the determinations required in paragraphs (g)(1) and (g)(2) of this

section for individuals for whom it is issuing, or has issued, authority to perform a covered function; and individuals who are covered by a certification from an aircraft operator under §1542.209(n) of this chapter. The aircraft operator may not make determinations for individuals described in §1542.209(a) of this chapter.

(h) *Correction of FBI records and notification of disqualification.* (1) Before making a final decision to deny authority to an individual described in paragraph (a)(1) or (3) of this section, the aircraft operator must advise him or her that the FBI criminal record discloses information that would disqualify him or her from receiving or retaining authority to perform a covered function and provide the individual with a copy of the FBI record if he or she requests it.

(2) The aircraft operator must notify an individual that a final decision has been made to grant or deny authority to perform a covered function.

(3) Immediately following the suspension of authority to perform a covered function, the aircraft operator must advise the individual that the FBI criminal record discloses information that disqualifies him or her from retaining his or her authority, and provide the individual with a copy of the FBI record if he or she requests it.

(i) *Corrective action by the individual.* The individual may contact the local jurisdiction responsible for the information and the FBI to complete or correct the information contained in his or her record, subject to the following conditions—

(1) For an individual seeking unescorted access authority or authority to perform screening functions on or after December 6, 2001; or an individual seeking authority to perform checked baggage or cargo functions on or after February 17, 2002; the following applies:

(i) Within 30 days after being advised that the criminal record received from the FBI discloses a disqualifying criminal offense, the individual must notify the aircraft operator in writing of his or her intent to correct any information he or she believes to be inaccurate. The aircraft operator must obtain a copy, or accept a copy from the individual, of the revised FBI record or a certified true copy of the information from the appropriate court, prior to authority to perform a covered function.

(ii) If no notification, as described in paragraph (h)(1) of this section, is received within 30 days, the aircraft operator may make a final determination to deny authority to perform a covered function.

(2) For an individual with unescorted access authority or authority to perform screening functions before December 6, 2001; or an individual with authority to perform checked baggage or cargo functions before February 17, 2002; the following applies: Within 30 days after being advised of suspension because the criminal record received from the FBI discloses a disqualifying criminal offense, the individual must notify the aircraft operator in writing of his or her intent to correct any information he or she

believes to be inaccurate. The aircraft operator must obtain a copy, or accept a copy from the individual, of the revised FBI record, or a certified true copy of the information from the appropriate court, prior to reinstating authority to perform a covered function.

(j) *Limits on dissemination of results.* Criminal record information provided by the FBI may be used only to carry out this section and §1542.209 of this chapter. No person may disseminate the results of a CHRC to anyone other than:

(1) The individual to whom the record pertains, or that individual's authorized representative.

(2) Officials of airport operators who are determining whether to grant unescorted access to the individual under part 1542 of this chapter when the determination is not based on the aircraft operator's certification under §1542.209(n) of this chapter.

(3) Other aircraft operators who are determining whether to grant authority to perform a covered function under this part.

(4) Others designated by TSA.

(k) *Recordkeeping.* The aircraft operator must maintain the following information.

(1) *Investigation conducted before December 6, 2001.* The aircraft operator must maintain and control the access or employment history investigation files, including the criminal history records results portion, for investigations conducted before December 6, 2001.

(2) *Fingerprint application process on or after December 6, 2001.* The aircraft operator must physically maintain, control, and, as appropriate, destroy the fingerprint application and the criminal record. Only direct aircraft operator employees may carry out the responsibility for maintaining, controlling, and destroying criminal records.

(3) *Protection of records—all investigations.* The records required by this section must be maintained in a manner that is acceptable to TSA and in a manner that protects the confidentiality of the individual.

(4) *Duration—all investigations.* The records identified in this section with regard to an individual must be maintained until 180 days after the termination of the individual's authority to perform a covered function. When files are no longer maintained, the criminal record must be destroyed.

(l) *Continuing responsibilities.* (1) Each individual with unescorted access authority or the authority to perform screening functions on December 6, 2001, who had a disqualifying criminal offense in paragraph (d) of this section on or after December 6, 1991, must, by January 7, 2002, report the conviction to the aircraft operator and surrender the SIDA access medium to the issuer and cease performing screening functions, as applicable.

(2) Each individual with authority to perform a covered function who has a disqualifying criminal offense must report the offense to the aircraft operator and sur-

render the SIDA access medium to the issuer within 24 hours of the conviction or the finding of not guilty by reason of insanity.

(3) If information becomes available to the aircraft operator indicating that an individual with authority to perform a covered function has a possible conviction for any disqualifying criminal offense in paragraph (d) of this section, the aircraft operator must determine the status of the conviction. If a disqualifying criminal offense is confirmed the aircraft operator must immediately revoke any authority to perform a covered function.

(4) Each individual with authority to perform checked baggage or cargo functions on February 17, 2002, who had a disqualifying criminal offense in paragraph (d) of this section on or after February 17, 1992, must, by March 25 2002, report the conviction to the aircraft operator and cease performing check baggage or cargo functions.

(m) Aircraft operator responsibility. The aircraft operator must—

(1) Designate an individual(s) to be responsible for maintaining and controlling the employment history investigations for those whom the aircraft operator has made a certification to an airport operator under 14 CFR 107.209(n) in effect prior to November 14, 2001 (see 14 CFR Parts 60 to 139 revised as of January 1, 2001), and for those whom the aircraft operator has issued identification media that are airport-accepted. The aircraft operator must designate a direct employee to maintain, control, and, as appropriate, destroy criminal records.

(2) Designate an individual(s) to maintain the employment history investigations of individuals with authority to perform screening functions whose files must be maintained at the location or station where the screener is performing his or her duties.

(3) Designate an individual(s) at appropriate locations to serve as the contact to receive notification from individuals seeking authority to perform covered functions of their intent to seek correction of their FBI criminal record.

(4) Audit the employment history investigations performed in accordance with this section and 14 CFR 108.33 in effect prior to November 14, 2001 (see 14 CFR Parts 60 to 139 revised as of January 1, 2001). The aircraft operator must set forth the audit procedures in its security program.

§1544.230 Fingerprint-based criminal history records checks (CHRC):

Flightcrew members.

(a) Scope. This section applies to each flightcrew member for each aircraft operator, except that this section does not apply to flightcrew members who are subject to §1544.229.

(b) CHRC required. Each aircraft operator must ensure that each flightcrew member has undergone a fingerprint-based CHRC that does not disclose that he or she has a dis-

qualifying criminal offense, as described in §1544.229(d), before allowing that individual to serve as a flightcrew member.

(c) Application and fees. Each aircraft operator must ensure that each flightcrew member's fingerprints are obtained and submitted as described in §1544.229 (e) and (f).

(d) Determination of arrest status. (1) When a CHRC on an individual described in paragraph (a) of this section discloses an arrest for any disqualifying criminal offense listed in §1544.229(d) without indicating a disposition, the aircraft operator must determine, after investigation, that the arrest did not result in a disqualifying offense before the individual may serve as a flightcrew member. If there is no disposition, or if the disposition did not result in a conviction or in a finding of not guilty by reason of insanity of one of the offenses listed in §1544.229(d), the flight crewmember is not disqualified under this section.

(2) When a CHRC on an individual described in paragraph (a) of this section discloses an arrest for any disqualifying criminal offense listed in §1544.229(d) without indicating a disposition, the aircraft operator must suspend the individual's flightcrew member privileges not later than 45 days after obtaining a CHRC, unless the aircraft operator determines, after investigation, that the arrest did not result in a disqualifying criminal offense. If there is no disposition, or if the disposition did not result in a conviction or in a finding of not guilty by reason of insanity of one of the offenses listed in §1544.229(d), the flight crewmember is not disqualified under this section.

(3) The aircraft operator may only make the determinations required in paragraphs (d)(1) and (d)(2) of this section for individuals whom it is using, or will use, as a flightcrew member. The aircraft operator may not make determinations for individuals described in §1542.209(a) of this chapter.

(e) Correction of FBI records and notification of disqualification. (1) Before making a final decision to deny the individual the ability to serve as a flightcrew member, the aircraft operator must advise the individual that the FBI criminal record discloses information that would disqualify the individual from serving as a flightcrew member and provide the individual with a copy of the FBI record if the individual requests it.

(2) The aircraft operator must notify the individual that a final decision has been made to allow or deny the individual flightcrew member status.

(3) Immediately following the denial of flightcrew member status, the aircraft operator must advise the individual that the FBI criminal record discloses information that disqualifies him or her from retaining his or her flightcrew member status, and provide the individual with a copy of the FBI record if he or she requests it.

(f) Corrective action by the individual. The individual may contact the local jurisdiction responsible for the information and the FBI to complete or correct the information contained in his or her record, subject to the following conditions—

(1) Within 30 days after being advised that the criminal record received from the FBI discloses a disqualifying criminal offense, the individual must notify the aircraft operator in writing of his or her intent to correct any information he or she believes to be inaccurate. The aircraft operator must obtain a copy, or accept a copy from the individual, of the revised FBI record or a certified true copy of the information from the appropriate court, prior to allowing the individual to serve as a flightcrew member.

(2) If no notification, as described in paragraph (f)(1) of this section, is received within 30 days, the aircraft operator may make a final determination to deny the individual flightcrew member status.

(g) Limits on the dissemination of results. Criminal record information provided by the FBI may be used only to carry out this section. No person may disseminate the results of a CHRC to anyone other than—

(1) The individual to whom the record pertains, or that individual's authorized representative.

(2) Others designated by TSA.

(h) Recordkeeping. (1) Fingerprint application process. The aircraft operator must physically maintain, control, and, as appropriate, destroy the fingerprint application and the criminal record. Only direct aircraft operator employees may carry out the responsibility for maintaining, controlling, and destroying criminal records.

(2) Protection of records. The records required by this section must be maintained by the aircraft operator in a manner that is acceptable to TSA that protects the confidentiality of the individual.

(3) Duration. The records identified in this section with regard to an individual must be made available upon request by TSA, and maintained by the aircraft operator until 180 days after the termination of the individual's privileges to perform flightcrew member duties with the aircraft operator. When files are no longer maintained, the aircraft operator must destroy the CHRC results.

(i) Continuing responsibilities. (1) Each flightcrew member identified in paragraph (a) of this section who has a disqualifying criminal offense must report the offense to the aircraft operator within 24 hours of the conviction or the finding of not guilty by reason of insanity.

(2) If information becomes available to the aircraft operator indicating that a flightcrew member identified in paragraph (a) of this section has a possible conviction for any disqualifying criminal offense in §1544.229 (d), the aircraft operator must determine the status of the conviction. If a disqualifying criminal offense is confirmed, the aircraft operator may not assign that individual to flightcrew duties in operations identified in paragraph (a).

(j) Aircraft operator responsibility. The aircraft operator must—(1) Designate a direct employee to maintain, control, and, as appropriate, destroy criminal records.

(2) Designate an individual(s) to maintain the CHRC results.

(3) Designate an individual(s) at appropriate locations to receive notification from individuals of their intent to seek correction of their FBI criminal record.

(k) Compliance date. Each aircraft operator must comply with this section for each flightcrew member described in paragraph (a) of this section not later than December 6, 2002. [67 FR 8209, Feb. 22, 2002]

§1544.231 Airport-approved and exclusive area personnel identification systems.

(a) Each aircraft operator must establish and carry out a personnel identification system for identification media that are airport-approved, or identification media that are issued for use in an exclusive area. The system must include the following:

(1) Personnel identification media that—

(i) Convey a full face image, full name, employer, and identification number of the individual to whom the identification medium is issued;

(ii) Indicate clearly the scope of the individual's access and movement privileges;

(iii) Indicate clearly an expiration date; and

(iv) Are of sufficient size and appearance as to be readily observable for challenge purposes.

(2) Procedures to ensure that each individual in the secured area or SIDA continuously displays the identification medium issued to that individual on the outermost garment above waist level, or is under escort.

(3) Procedures to ensure accountability through the following:

(i) Retrieving expired identification media.

(ii) Reporting lost or stolen identification media.

(iii) Securing unissued identification media stock and supplies.

(iv) Auditing the system at a minimum of once a year, or sooner, as necessary to ensure the integrity and accountability of all identification media.

(v) As specified in the aircraft operator security program, revalidate the identification system or reissue identification media if a portion of all issued, unexpired identification media are lost, stolen, or unretrieved, including identification media that are combined with access media.

(vi) Ensure that only one identification medium is issued to an individual at a time. A replacement identification medium may only be issued if an individual declares in writing that the medium has been lost or stolen.

(b) The aircraft operator may request approval of a temporary identification media system that meets the standards in §1542.211(b) of this chapter, or may arrange with the airport to use temporary airport identification media in accordance with that section.

(c) Each aircraft operator must submit a plan to carry out this section to TSA no later than May 13, 2002. Each aircraft operator must fully implement its plan no later than November 14, 2003.

§1544.233 Security coordinators and crewmembers, training.

(a) No aircraft operator may use any individual as a Ground Security Coordinator unless, within the preceding 12-calendar months, that individual has satisfactorily completed the security training as specified in the aircraft operator's security program.

(b) No aircraft operator may use any individual as an in-flight security coordinator or crewmember on any domestic or international flight unless, within the preceding 12-calendar months or within the time period specified in an Advanced Qualifications Program approved under SFAR 58 in 14 CFR part 121, that individual has satisfactorily completed the security training required by 14 CFR 121.417(b)(3)(v) or 135.331(b)(3)(v), and as specified in the aircraft operator's security program.

(c) With respect to training conducted under this section, whenever an individual completes recurrent training within one calendar month earlier, or one calendar month after the date it was required, that individual is considered to have completed the training in the calendar month in which it was required.

§1544.235 Training and knowledge for individuals with security-related duties.

(a) No aircraft operator may use any direct or contractor employee to perform any security-related duties to meet the requirements of its security program unless that individual has received training as specified in its security program including their individual responsibilities in §1540.105 of this chapter.

(b) Each aircraft operator must ensure that individuals performing security-related duties for the aircraft operator have knowledge of the provisions of this part, applicable Security Directives and Information Circulars, the approved airport security program applicable to their location, and the aircraft operator's security program to the extent that such individuals need to know in order to perform their duties.

§1544.237 Flight deck privileges.

(a) For each aircraft that has a door to the flight deck, each aircraft operator must restrict access to the flight deck as provided in its security program.

(b) This section does not restrict access for an FAA air carrier inspector, an authorized representative of the National Transportation Safety Board, or for an Agent of the United States Secret Service, under 14 CFR parts 121, 125, or 135. This section does not restrict access for a Federal Air Marshal under this part. [67 FR 8210, Feb. 22, 2002]

Subpart D—Threat and Threat Response

§1544.301 Contingency plan.

Each aircraft operator must adopt a contingency plan and must:

(a) Implement its contingency plan when directed by TSA.

(b) Ensure that all information contained in the plan is updated annually and that appropriate persons are notified of any changes.

(c) Participate in an airport-sponsored exercise of the airport contingency plan or its equivalent, as provided in its security program.

§1544.303 Bomb or air piracy threats.

(a) Flight: Notification. Upon receipt of a specific and credible threat to the security of a flight, the aircraft operator must—

(1) Immediately notify the ground and in-flight security coordinators of the threat, any evaluation thereof, and any measures to be applied; and

(2) Ensure that the in-flight security coordinator notifies all crewmembers of the threat, any evaluation thereof, and any measures to be applied; and

(3) Immediately notify the appropriate airport operator.

(b) Flight: Inspection. Upon receipt of a specific and credible threat to the security of a flight, each aircraft operator must attempt to determine whether or not any explosive or incendiary is present by doing the following:

(1) Conduct a security inspection on the ground before the next flight or, if the aircraft is in flight, immediately after its next landing.

(2) If the aircraft is on the ground, immediately deplane all passengers and submit that aircraft to a security search.

(3) If the aircraft is in flight, immediately advise the pilot in command of all pertinent information available so that necessary emergency action can be taken.

(c) Ground facility. Upon receipt of a specific and credible threat to a specific ground facility at the airport, the aircraft operator must:

(1) Immediately notify the appropriate airport operator.

(2) Inform all other aircraft operators and foreign air carriers at the threatened facility.

(3) Conduct a security inspection.

(d) Notification. Upon receipt of any bomb threat against the security of a flight or facility, or upon receiving information that an act or suspected act of air piracy has been committed, the aircraft operator also must notify TSA. If the aircraft is in airspace under other than U.S. jurisdiction, the aircraft operator must also notify the appropriate authorities of the State in whose territory the aircraft is located and, if the aircraft is in flight, the appropriate authorities of the State in whose territory the aircraft is to land. Notification of the appropriate air traffic controlling authority is sufficient action to meet this requirement.

§1544.305 Security Directives and Information Circulars.

(a) TSA may issue an Information Circular to notify aircraft operators of security concerns. When TSA determines that additional security measures are necessary to respond to a threat assessment or to a specific threat against civil aviation, TSA issues a Security Directive setting forth mandatory measures.

(b) Each aircraft operator required to have an approved aircraft operator security program must comply with each Security Directive issued to the aircraft operator by TSA, within the time prescribed in the Security Directive for compliance.

(c) Each aircraft operator that receives a Security Directive must—

(1) Within the time prescribed in the Security Directive, verbally acknowledge receipt of the Security Directive to TSA.

(2) Within the time prescribed in the Security Directive, specify the method by which the measures in the Security Directive have been implemented (or will be implemented, if the Security Directive is not yet effective).

(d) In the event that the aircraft operator is unable to implement the measures in the Security Directive, the aircraft operator must submit proposed alternative measures and the basis for submitting the alternative measures to TSA for approval. The aircraft operator must submit the proposed alternative measures within the time prescribed in the Security Directive. The aircraft operator must implement any alternative measures approved by TSA.

(e) Each aircraft operator that receives a Security Directive may comment on the Security Directive by submitting data, views, or arguments in writing to TSA. TSA may amend the Security Directive based on comments received. Submission of a comment does not delay the effective date of the Security Directive.

(f) Each aircraft operator that receives a Security Directive or Information Circular and each person who receives information from a Security Directive or Information Circular must:

(1) Restrict the availability of the Security Directive or Information Circular, and information contained in either document, to those persons with an operational need-to-know.

(2) Refuse to release the Security Directive or Information Circular, and information contained in either document, to persons other than those with an operational need-to-know without the prior written consent of TSA.

Subpart E—Screener Qualifications When the Aircraft Operator Performs Screening

§1544.401 Applicability of this subpart.

(a) Aircraft operator screening. This subpart applies when the aircraft operator is conducting inspections as provided in §1544.207(c).

(b) Current screeners. As used in this subpart, "current screener" means each individual who first performed screening functions before the date the aircraft operator must begin use of the new screener training program provided by TSA. Until November 19, 2002, each current screener must comply with §1544.403. Until November 19, 2002, each aircraft operator must apply §1544.403 for each current screener. On and after November 19, 2002, each such current screener must comply with §§1544.405 through 1544.411, and each aircraft operator must comply with §§1544.405 through 1544.411 for such individuals.

(c) New screeners. As used in this subpart, "new screener" means each individual who first performs screening functions on and after the date the aircraft operator must begin use of the new screener training program provided by TSA. Each aircraft operator must apply §§1544.405 through 1544.411 for individuals who first perform screening functions for new screeners.

§1544.403 Current screeners.

This section applies to current screeners. This section no longer applies on and after November 19, 2002.

(a) No aircraft operator may use any person to perform any screening function, unless that person has:

(1) A high school diploma, a General Equivalency Diploma, or a combination of education and expe-

rience that the aircraft operator has determined to have equipped the person to perform the duties of the position.

(2) Basic aptitudes and physical abilities including color perception, visual and aural acuity, physical coordination, and motor skills to the following standards:

(i) Screeners operating X-ray equipment must be able to distinguish on the X-ray monitor the appropriate imaging standard specified in the aircraft operator's security program. Wherever the X-ray system displays colors, the operator must be able to perceive each color;

(ii) Screeners operating any screening equipment must be able to distinguish each color displayed on every type of screening equipment and explain what each color signifies;

(iii) Screeners must be able to hear and respond to the spoken voice and to audible alarms generated by screening equipment in an active checkpoint environment;

(iv) Screeners performing physical searches or other related operations must be able to efficiently and thoroughly manipulate and handle such baggage, containers, and other objects subject to security processing; and

(v) Screeners who perform pat-downs or hand-held metal detector searches of persons must have sufficient dexterity and capability to thoroughly conduct those procedures over a person's entire body.

(3) The ability to read, speak, and write English well enough to—

(i) Carry out written and oral instructions regarding the proper performance of screening duties;

(ii) Read English language identification media, credentials, airline tickets, and labels on items normally encountered in the screening process;

(iii) Provide direction to and understand and answer questions from English-speaking persons undergoing screening; and

(iv) Write incident reports and statements and log entries into security records in the English language.

(4) Satisfactorily completed all initial, recurrent, and appropriate specialized training required by the aircraft operator's security program, except as provided in paragraph (b) of this section.

(b) The aircraft operator may use a person who has not completed the training required by paragraph (a)(4) of this section during the on-the-job portion of training to perform security functions provided that the person:

(1) Is closely supervised, and

(2) Does not make independent judgments as to whether persons or property may enter a sterile area or aircraft without further inspection.

(c) No aircraft operator must use a person to perform a screening function after that person has failed an operational test related to that function until that person has successfully completed the remedial training specified in the aircraft operator's security program.

(d) Each aircraft operator must ensure that a Ground Security Coordinator conducts and documents an annual evaluation of each individual assigned screening duties and may continue that individual's employment in a screening capacity only upon the determination by the Ground Security Coordinator that the individual:

(1) Has not suffered a significant diminution of any physical ability required to perform a screening function since the last evaluation of those abilities;

(2) Has a satisfactory record of performance and attention to duty based on the standards and requirements in its security program; and

(3) Demonstrates the current knowledge and skills necessary to courteously, vigilantly, and effectively perform screening functions.

(e) Paragraphs (a) through (d) of this section do not apply to those screening functions conducted outside the United States over which the aircraft operator does not have operational control. In the event the aircraft operator is unable to implement paragraphs (a) through (d) of this section for screening functions outside the United States, the aircraft operator must notify TSA of those aircraft operator stations so affected.

(f) At locations outside the United States where the aircraft operator has operational control over a screening function, the aircraft operator may use screeners who do not meet the requirements of paragraph (a)(3) of this section, provided that at least one representative of the aircraft operator who has the ability to functionally read and speak English is present while the aircraft operator's passengers are undergoing security screening.

§1544.405 New screeners: Qualifications of screening personnel.

(a) No individual subject to this subpart may perform a screening function unless that individual has the qualifications described in §§1544.405 through 1544.411. No aircraft operator may use such an individual to perform a screening function unless that person complies with the requirements of §§1544.405 through 1544.411.

(b) A screener must have a satisfactory or better score on a screener selection test administered by TSA.

(c) A screener must be a citizen of the United States.

(d) A screener must have a high school diploma, a General Equivalency Diploma, or a combination of education and experience that the TSA has determined to be sufficient for the individual to perform the duties of the position.

(e) A screener must have basic aptitudes and physical abilities including color perception, visual and aural acuity, physical coordination, and motor skills to the following standards:

(1) Screeners operating screening equipment must be able to distinguish on the screening equipment monitor the appropriate imaging standard specified in the aircraft operator's security program.

(2) Screeners operating any screening equipment must be able to distinguish each color displayed on every

type of screening equipment and explain what each color signifies.

(3) Screeners must be able to hear and respond to the spoken voice and to audible alarms generated by screening equipment at an active screening location.

(4) Screeners who perform physical searches or other related operations must be able to efficiently and thoroughly manipulate and handle such baggage, containers, cargo, and other objects subject to screening.

(5) Screeners who perform pat-downs or hand-held metal detector searches of individuals must have sufficient dexterity and capability to thoroughly conduct those procedures over an individual's entire body.

(f) A screener must have the ability to read, speak, and write English well enough to—

(1) Carry out written and oral instructions regarding the proper performance of screening duties;

(2) Read English language identification media, credentials, airline tickets, documents, air waybills, invoices, and labels on items normally encountered in the screening process;

(3) Provide direction to and understand and answer questions from English-speaking individuals undergoing screening; and

(4) Write incident reports and statements and log entries into security records in the English language.

(g) At locations outside the United States where the aircraft operator has operational control over a screening function, the aircraft operator may use screeners who do not meet the requirements of paragraph (f) of this section, provided that at least one representative of the aircraft operator who has the ability to functionally read and speak English is present while the aircraft operator's passengers are undergoing security screening. At such locations the aircraft operator may use screeners who are not United States citizens.

§1544.407 New screeners: Training, testing, and knowledge of individuals who perform screening functions.

(a) Training required. Before performing screening functions, an individual must have completed initial, recurrent, and appropriate specialized training as specified in this section and the aircraft operator's security program. No aircraft operator may use any screener, screener in charge, or checkpoint security supervisor unless that individual has satisfactorily completed the required training. This paragraph does not prohibit the performance of screening functions during on-the-job training as provided in §1544.409 (b).

(b) Use of training programs. Training for screeners must be conducted under programs provided by TSA. Training programs for screeners-in-charge and checkpoint security supervisors must be conducted in accordance with the aircraft operator's security program.

(c) Classroom instruction. Each screener must complete at least 40 hours of classroom instruction or successfully complete a program that TSA determines will train individuals to a level of proficiency equivalent to the level that would be achieved by such classroom instruction.

(d) Screener readiness test. Before beginning on-the-job training, a screener trainee must pass the screener readiness test prescribed by TSA.

(e) On-the-job training and testing. Each screener must complete at least 60 hours of on-the-job training and must pass an on-the-job training test prescribed by TSA. No aircraft operator may permit a screener trainee to exercise independent judgment as a screener, until the individual passes an on-the-job training test prescribed by TSA.

(f) Knowledge requirements. Each aircraft operator must ensure that individuals performing as screeners, screeners-in-charge, and checkpoint security supervisors for the aircraft operator have knowledge of the provisions of this part, the aircraft operator's security program, and applicable Security Directives and Information Circulars to the extent necessary to perform their duties.

(g) Disclosure of sensitive security information during training. The aircraft operator may not permit a trainee to have access to sensitive security information during screener training unless a criminal history records check has successfully been completed for that individual in accordance with §1544.229, and the individual has no disqualifying criminal offense.

§1544.409 New screeners: Integrity of screener tests.

(a) Cheating or other unauthorized conduct. (1) Except as authorized by the TSA, no person may—

(i) Copy or intentionally remove a test under this part;

(ii) Give to another or receive from another any part or copy of that test;

(iii) Give help on that test to or receive help on that test from any person during the period that the test is being given; or

(iv) Use any material or aid during the period that the test is being given.

(2) No person may take any part of that test on behalf of another person.

(3) No person may cause, assist, or participate intentionally in any act prohibited by this paragraph (a).

(b) Administering and monitoring screener tests. (1) Each aircraft operator must notify TSA of the time and location at which it will administer each screener readiness test required under §1544.405(d).

(2) Either TSA or the aircraft operator must administer and monitor the screener readiness test. Where more than one aircraft operator or foreign air carrier uses a screening location, TSA may authorize an employee of one or more of the aircraft operators or foreign air carriers to monitor the test for a trainee who will screen at that location.

(3) If TSA or a representative of TSA is not available to administer and monitor a screener readiness test, the aircraft operator must provide a direct employee to administer and monitor the screener readiness test.

(4) An aircraft operator employee who administers and monitors a screener readiness test must not be an instructor, screener, screener-in-charge, checkpoint security supervisor, or other screening supervisor. The employee must be familiar with the procedures for administering and monitoring the test and must be capable of observing whether the trainee or others are engaging in cheating or other unauthorized conduct.

§1544.411 New screeners: Continuing qualifications for screening personnel.

(a) Impairment. No individual may perform a screening function if he or she shows evidence of impairment, such as impairment due to illegal drugs, sleep deprivation, medication, or alcohol.

(b) Training not complete. An individual who has not completed the training required by §1544.405 may be deployed during the on-the-job portion of training to perform security functions provided that the individual—

(1) Is closely supervised; and

(2) Does not make independent judgments as to whether individuals or property may enter a sterile area or aircraft without further inspection.

(c) Failure of operational test. No aircraft operator may use an individual to perform a screening function after that individual has failed an operational test related to that function, until that individual has successfully completed the remedial training specified in the aircraft operator's security program.

(d) Annual proficiency review. Each individual assigned screening duties shall receive an annual evaluation. The aircraft operator must ensure that a Ground Security Coordinator conducts and documents an annual evaluation of each individual who performs screening functions. An individual who performs screening functions may not continue to perform such functions unless the evaluation demonstrates that the individual—

(1) Continues to meet all qualifications and standards required to perform a screening function;

(2) Has a satisfactory record of performance and attention to duty based on the standards and requirements in the aircraft operator's security program; and

(3) Demonstrates the current knowledge and skills necessary to courteously, vigilantly, and effectively perform screening functions.

SUBCHAPTER C—CIVIL AVIATION SECURITY

PART 1550—AIRCRAFT SECURITY UNDER GENERAL OPERATING AND FLIGHT RULES

Sec.
Authority: 49 U.S.C. 114, 5103, 40119, 44901–44907, 44913–44914, 44916–44918, 44935–44936, 44942, 46105.

Source: 67 FR 8383, Feb. 22, 2002, unless otherwise noted.

§1550.1 Applicability of this part.

This part applies to the operation of aircraft for which there are no security requirements in other parts of this subchapter.

§1550.3 TSA inspection authority.

(a) Each aircraft operator subject to this part must allow TSA, at any time or place, to make any inspections or tests, including copying records, to determine compliance with—

(1) This subchapter and any security program or security procedures under this subchapter, and part 1520 of this chapter; and

(2) 49 U.S.C. Subtitle VII, as amended.

(b) At the request of TSA, each aircraft operator must provide evidence of compliance with this part and its security program or security procedures, including copies of records.

§1550.5 Operations using a sterile area.

(a) Applicability of this section. This section applies to all aircraft operations in which passengers, crewmembers, or other individuals are enplaned from or deplaned into a sterile area, except for scheduled passenger operations, public charter passenger operations, and private charter passenger operations, that are in accordance with a security program issued under part 1544 or 1546 of this chapter.

(b) Procedures. Any person conducting an operation identified in paragraph (a) of this section must conduct a search of the aircraft before departure and must screen passengers, crewmembers, and other individuals and their accessible property (carry-on items) before boarding in accordance with security procedures approved by TSA.

(c) Sensitive security information. The security program procedures approved by TSA for operations specified in paragraph (a) of this section are sensitive security information. The operator must restrict the distribution, disclosure, and availability of information contained in the security procedures to persons with a need to know as described in part 1520 of this chapter.

(d) Compliance date. Persons conducting operations identified in paragraph (a) of this section must implement security procedures on October 6, 2001.

(e) Waivers. TSA may permit a person conducting an operation under this section to deviate from the provisions of this section if TSA finds that the operation can be conducted safely under the terms of the waiver.

§1550.7 Operations in aircraft of 12,500 pounds or more.

(a) Applicability of this section. This section applies to each aircraft operation conducted in an aircraft with a maximum certificated takeoff weight of 12,500 pounds or more except for those operations specified in §1550.5 and

those operations conducted under a security program under part 1544 or 1546 of this chapter.

(b) Procedures. Any person conducting an operation identified in paragraph (a) of this section must conduct a search of the aircraft before departure and screen passengers, crewmembers, and other persons and their accessible property (carry-on items) before boarding in accordance with security procedures approved by TSA.

(c) Compliance date. Persons identified in paragraph (a) of this section must implement security procedures when notified by TSA. TSA will notify operators by NOTAM, letter, or other communication when they must implement security procedures.

(d) Waivers. TSA may permit a person conducting an operation identified in this section to deviate from the provisions of this section if TSA finds that the operation can be conducted safely under the terms of the waiver.

PART 1552—FLIGHT SCHOOLS

Subpart A-Flight Training for Aliens and Other Designated Individuals

Sec.
1552.1 Scope and definitions.
1552.3 Flight training.
1552.5 Fees.

Subpart B -Flight School Security Awareness Training

1552.21 Scope and definitions.
1552.23 Security awareness training programs.
1552.25 Documentation, recordkeeping, and inspection.
Authority: 49 U.S.C. 114, 44939.

§1552.1 Scope and definitions.

(a) Scope. This subpart applies to flight schools that provide instruction under 49 U.S.C. Subtitle VII, Part A, in the operation of aircraft or aircraft simulators, and individuals who apply to obtain such instruction or who receive such instruction.

(b) Definitions. As used in this part:

Aircraft simulator means a flight simulator or flight training device, as those terms are defined at 14 CFR 61.1.

Alien means any person not a citizen or national of the United States.

Candidate means an alien or other individual designated by TSA who applies for flight training or recurrent training. It does not include an individual endorsed by the Department of Defense for flight training.

Day means a day from Monday through Friday, including State and local holidays but not Federal holidays: for any time period less than 11 days specified in this part. For any time period greater than 11 days, day means calendar day.

Demonstration flight for marketing purposes means a flight for the purpose of demonstrating an aircraft's or aircraft simulator's capabilities or characteristics to a potential purchaser, or to an agent of a potential purchaser, of the aircraft or simulator, including any acceptance flight after an aircraft manufacturer delivers an aircraft to a purchaser.

Flight school means any pilot school, flight training center, air carrier flight training facility, or flight instructor certificated under 14 CFR part 61, 121, 135, 141, or 142; or any other person or entity that provides instruction under 49 U.S.C. Subtitle VII, Part A, in the operation of any aircraft or aircraft simulator.

Flight training means instruction received from a flight school in an aircraft or aircraft simulator. Flight training does not include recurrent training, ground training, a demonstration flight for marketing purposes, or any military training provided by the Department of Defense, the U.S. Coast Guard, or an entity under contract with the Department of Defense or U.S. Coast Guard.

Ground training means classroom or computer-based instruction in the operation of aircraft, aircraft systems, or cockpit procedures. Ground training does not include instruction in an aircraft simulator.

National of the United States means a person who, though not a citizen of the United States, owes permanent allegiance to the United States, and includes a citizen of American Samoa or Swains Island.

Recurrent training means periodic training required under 14 CFR part 61, 121,125, 135, or Subpart K of part 91. Recurrent training does not include training that would enable a candidate who has a certificate or type rating for a particular aircraft to receive a certificate or type rating for another aircraft.

§1552.3 Flight training.

This section describes the procedures a flight school must follow before providing flight training.

(a) *Category 1*–Regular processing for flight training on aircraft more than 12,500 pounds. A flight school may not provide flight training in the operation of any aircraft having a maximum certificated takeoff weight of more than 12,500 pounds to a candidate, except for a candidate who receives expedited processing under paragraph (b) of this section, unless

(1) The flight school has first notified TSA that the candidate has requested such flight training.

(2) The candidate has submitted to TSA, in a form and manner acceptable to TSA, the following:

(i) The candidate's full name, including any aliases used by the candidate or variations in the spelling of the candidate's name;

(ii) A unique candidate identification number created by TSA;

(iii) A copy of the candidate's current, unexpired passport and visa;

(iv) The candidate's passport and visa information, including all current and previous passports and visas held by the candidate and all the information necessary to obtain a passport and visa;

(v) The candidate's country of birth, current country or countries of citizenship, and each previous country of citizenship, if any;

(vi) The candidate's actual date of birth or, if the candidate does not know his or her date of birth, the approximate date of birth used consistently by the candidate for his or her passport or visa;

(vii) The candidate's requested dates of training and the location of the training;

(viii) The type of training for which the candidate is applying, including the aircraft type rating the candidate would be eligible to obtain upon completion of the training;

(ix) The candidate's current U.S. pilot certificate, certificate number, and type rating, if any;

(x) Except as provided in paragraph (k) of this section, the candidate's fingerprints, in accordance with paragraph (f) of this section;

(xi) The candidate's current address and phone number and each address for the 5 years prior to the date of the candidate's application;

(xii) The candidate's gender; and

(xiii) Any fee required under this part.

(3) The flight school has submitted to TSA, in a form and manner acceptable to TSA, a photograph of the candidate taken when the candidate arrives at the flight school for flight training.

(4) TSA has informed the flight school that the candidate does not pose a threat to aviation or national security, or more than 30 days have elapsed since TSA received all of the information specified in paragraph (a)(2) of this section.

(5) The flight school begins the candidate's flight training within 180 days of either event specified in paragraph (a)(4) of this section.

(b) *Category 2*–*Expedited processing for flight training on aircraft more than 12,500 pounds.* (1) A flight school may not provide flight training in the operation of any aircraft having a maximum certificated takeoff weight of more than 12,500 pounds to a candidate who meets any of the criteria of paragraph (b)(2) of this section unless

(i) The flight school has first notified TSA that the candidate has requested such flight training.

(ii) The candidate has submitted to TSA, in a form and manner acceptable to TSA:

(A) The information and fee required under paragraph (a)(2) of this section; and

(B) The reason the candidate is eligible for expedited processing under paragraph (b)(2) of this section and information that establishes that the candidate is eligible for expedited processing.

(iii) The flight school has submitted to TSA, in a form and manner acceptable to TSA, a photograph of the candidate taken when the candidate arrives at the flight school for flight training.

(iv) TSA has informed the flight school that the candidate does not pose a threat to aviation or national security or more than 5 days have elapsed since TSA received all of the information specified in paragraph (a)(2) of this section.

(v) The flight school begins the candidate's flight training within 180 days of either event specified in paragraph (b)(1)(iv) of this section.

(2) A candidate is eligible for expedited processing if he or she –

(i) Holds an airman's certificate from a foreign country that is recognized by the Federal Aviation Administration or a military agency of the United States, and that permits the candidate to operate a multi-engine aircraft that has a certificated takeoff weight of more than 12,500 pounds;

(ii) Is employed by a foreign air carrier that operates under 14 CFR part 129 and has a security program approved under 49 CFR part 1546;

(iii) Has unescorted access authority to a secured area of an airport under 49 U.S.C. 44936(a)(1)(A)(ii), 49 CFR 1542.209, or 49 CFR 1544.229;

(iv) Is a flightcrew member who has successfully completed a criminal history records check in accordance with 49 CFR 1544.230; or

(v) Is part of a class of individuals that TSA has determined poses a minimal threat to aviation or national security because of the flight training already possessed by that class of individuals.

(c) *Category 3*–*Flight training on aircraft 12,500 pounds or less.* A flight school may not provide flight training in the operation of any aircraft having a maximum certificated takeoff weight of 12,500 pounds or less to a candidate unless–

(1) The flight school has first notified TSA that the candidate has requested such flight training.

(2) The candidate has submitted to TSA, in a form and manner acceptable to TSA:

(i) The information required under paragraph (a)(2) of this section; and

(ii) Any other information required by TSA.

(3) The flight school has submitted to TSA, in a form and manner acceptable to TSA, a photograph of the candidate taken when the candidate arrives at the flight school for flight training.

(4) The flight school begins the candidate's flight training within 180 days of the date the candidate submitted the information required under paragraph (a)(2) of this section to TSA.

(d) *Category 4–Recurrent training for all aircraft.* Prior to beginning recurrent training for a candidate, a flight school must–

(1) Notify TSA that the candidate has requested such recurrent training; and

(2) Submit to TSA, in a form and manner acceptable to TSA:

(i) The candidate's full name, including any aliases used by the candidate or variations in the spelling of the candidate's name;

(ii) Any unique student identification number issued to the candidate by the Department of justice or TSA;

(iii) A copy of the candidate's current, unexpired passport and visa;

(iv) The candidate's current U.S. pilot certificate, certificate number, and type rating(s);

(v) The type of training for which the candidate is applying;

(vi) The date of the candidate's prior recurrent training, if any, and a copy of the training form documenting that recurrent training;

(vii) The candidate's requested dates of training; and

(viii) A photograph of the candidate taken when the candidate arrives at the flight school for flight training.

(e) *Interruption of flight training.* A flight school must immediately terminate or cancel a candidate's flight training if TSA notifies the flight school at any time that the candidate poses a threat to aviation or national security.

(f) *Fingerprints.* (1) Fingerprints submitted in accordance with this subpart must be collected—

(i) By United States Government personnel at a United States embassy or consulate; or

(ii) By another entity approved by n TSA.

(2) A candidate must confirm his or her identity to the individual or agency collecting his or her fingerprints under paragraph (f)(1) of this section by providing the individual or agency his or her:

(i) Passport;

(ii) Resident alien card; or

(iii) U.S. driver's license.

(3) A candidate must pay any fee imposed by the agency taking his or her fingerprints.

(g) General requirements. (1) False statements. If a candidate makes a knowing and willful false statement, or omits a material fact, when submitting the information required under this part, the candidate may be—-

(i) Subject to fine or imprisonment or both under 18 U.S.C. 1001;

(ii) Denied approval for flight training under this section; and

(iii) Subject to other enforcement action, as appropriate.

(2) *Preliminary approval.* For purposes of facilitating a candidate's visa process with the U.S. Department of State, TSA may inform a flight school and a candidate that the candidate has received preliminary approval for flight training based on information submitted by the flight school or the candidate under this section. A flight school may then issue an I-20 form to the candidate to present with the candidate's visa application. Preliminary approval does not initiate the waiting period under paragraph (a)(3) or (b)(1)(iii) of this section or the period in which a flight school must initiate a candidate's training after receiving TSA approval under paragraph (a)(4) or (b)(1)(iv) of this section.

(h) *U.S. citizens and nationals and Department of Defense endorsees.* A flight school must determine whether an individual is a citizen or national of the United States, or a Department of Defense endorsee, prior to providing flight training to the individual.

(1) *U.S. citizens and nationals.* To establish U.S. citizenship or nationality an individual must present to the flight school his or her:

(i) Valid, unexpired United States passport;

(ii) Original or government-issued certified birth certificate of the United States, American Samoa, or Swains Island, together with a government issued picture identification of the individual;

(iii) Original United States naturalization certificate with raised seal, or a Certificate of Naturalization issued by the U.S. Citizenship and Immigration Services (USCIS) or the U.S. Immigration and Naturalization Service (INS) (Form N-550 or Form N-570), together with a government-issued picture identification of the individual;

(iv) Original certification of birth abroad with raised seal, U.S. Department of State Form FS-545, or U.S. Department of State Form DS-1350, together with a government-issued picture identification of the individual;

(v) Original certificate of United States citizenship with raised seal, a Certificate of United States Citizenship issued by the USCIS or INS (Form N–560 or Form N-561), or a Certificate of Repatriation issued by the USCIS or INS (Form N-581), together with a government-issued picture identification of the individual; or

(vi) In the case of flight training provided to a Federal employee (including military personnel) pursuant to a contract between a Federal agency and a flight school, the agency's written certification as to its employee's United States citizenship or nationality, together with the employee's government-issued credentials or other Federally-issued picture identification.

(2) *Department of Defense endorsees.* To establish that an individual has been endorsed by the U.S. Department of Defense for flight training, the individual must present to the flight school a written statement acceptable to TSA from the U.S. Department of Defense attaché in the individual's country of residence together with a government-issued picture identification of the individual.

(i) *Recordkeeping requirements.* A flight school must

(1) Maintain the following information for a minimum of 5 years:

(i) For each candidate:

(A) A copy of the photograph required under paragraph (a)(3), (b)(1)(iii), (c)(3), or (d)(2)(viii) of this section; and

(B) A copy of the approval sent by TSA confirming the candidate's eligibility for flight training.

(ii) For a Category 1, Category 2, or Category 3 candidate, a copy of the information required under paragraph (a)(2) of this section, except the information in paragraph (a)(2)(x).

(iii) For a Category 4 candidate, a copy of the information required under paragraph (d)(2) of this section.

(iv) For an individual who is a United States citizen or national, a copy of the information required under paragraph (h)(1) of this section.

(v) For an individual who has been endorsed by the U.S. Department of Defense for flight training, a copy of the information required under paragraph (h)(2) of this section.

(vi) A record of all fees paid to TSA in accordance with this part.

(2) Permit TSA and the Federal Aviation Administration to inspect the records required by paragraph (i)(1) of this section during reasonable business hours.

(j) *Candidates subject to the Department of Justice rule.* A candidate who submits a completed Flight Training

Candidate Checks Program form and fingerprints to the Department of justice in accordance with 28 CFR part 105 before September 28, 2004, or a later date specified by TSA, is processed in accordance with the requirements of that part. If TSA specifies a date later than the compliance dates identified in this part, individuals and flight schools who comply with 28 CFR part 105 up to that date will be considered to be in compliance with the requirements of this part.

(k) Additional or missed flight training. (1) A Category 1, 2, or 3 candidate who has been approved for flight training by TSA may take additional flight training without submitting fingerprints as specified in paragraph (a)(2)(x) of this section if the candidate:

(i) Submits all other information required in paragraph (a)(2) of this section, including the fee; and

(ii) Waits for TSA approval or until the applicable waiting period expires before initiating the additional flight training.

(2) A Category 1, 2, or 3 candidate who is approved for flight training by TSA, but does not initiate that flight training within 180 days, may reapply for flight training without submitting fingerprints as specified in paragraph (a)(2)(x) of this section if the candidate submits all other information required in paragraph (a)(2) of this section, including the fee.

§ 1552.5 Fees.

(a) *Imposition of fees.* The following fee is required for TSA to conduct a security threat assessment for a candidate for flight training subject to the requirements of § 1552.3: $130.

(b) *Remittance of fees.* (1) A candidate must remit the fee required under this subpart to TSA, in a form and manner acceptable to TSA, each time the candidate or the flight school is required to submit the information required under § 1552.3 to TSA.

(2) TSA will not issue any fee refunds, unless a fee was paid in error.

Subpart B-Flight School Security Awareness Training

§ 1552.21 Scope and definitions.

(a) Scope. This subpart applies to flight schools that provide instruction under 49 U.S.C. Subtitle VII, Part A, in the operation of aircraft or aircraft simulators, and to employees of such flight schools.

(b) *Definitions:* As used in this sub art:

Flight school employee means a flight instructor or ground instructor certificated under 14 CFR part 61, 141, or 142; a chief instructor certificated under 14 CFR part 141; a director of training certificated under 14 CFR part

142; or any other person employed by a flight school, including an independent contractor, who has direct contact with a flight school student. This includes an independent or solo flight instructor certificated under 14 CFR part 61.

§ 1552.23 Security awareness training programs.

(a) *General.* A flight school must ensure that–

(1) Each of its flight school employees receives initial and recurrent security awareness training in accordance with this subpart; and

(2) If an instructor is conducting the initial security awareness training program, the instructor has first successfully completed the initial flight school security awareness training program offered by TSA or an alternative initial flight school security awareness training program that meets the criteria of paragraph (c) of this section,

(b) *Initial security awareness training program.* (1) A flight school must ensure that—

(i) Each flight school employee employed on January 18, 2005 receives initial security awareness training in accordance with this subpart by January 18, 2005; and

(ii) Each flight school employee hired after January 18, 2005 receives initial security awareness training within 60 days of being hired.

(2) In complying with paragraph (b)(2) of this section, a flight school may use either:

(i) The initial flight school security awareness training program offered by TSA; or

(ii) An alternative initial flight school security awareness training program that meets the criteria of paragraph (c) of this section.

(c)Alternative initial security awareness training program. At a minimum, an alternative initial security awareness training program must—

(1) Require active participation by the flight school employee receiving the training.

(2) Provide situational scenarios requiring the flight school employee receiving the training to assess specific situations and determine appropriate courses of action.

(3) Contain information that enables a flight school employee to identify—

(i) Uniforms and other identification, if any are required at the flight school, for flight school employees or other persons authorized to be on the flight school grounds.

(ii) Behavior by clients and customers that may be considered suspicious, including, but not limited to:

(A) Excessive or unusual interest in restricted airspace or restricted ground structures;

(B) Unusual questions or interest regarding aircraft capabilities;

(C) Aeronautical knowledge inconsistent with the client or customer's existing airman credentialing; and

(D) Sudden termination of the client or customer's instruction.

(iii) Behavior by other on-site persons that may be considered suspicious, including, but not limited to:

(A) Loitering on the flight school grounds for extended periods of time; and

(B) Entering "authorized access only" areas without permission.

(iv) Circumstances regarding aircraft that may be considered suspicious, including, but not limited to:

(A) Unusual modifications to aircraft, such as the strengthening of landing gear, changes to the tail number, or stripping of the aircraft of seating or equipment;

(B) Damage to propeller locks or other parts of an aircraft that is inconsistent with the pilot training or aircraft flight tog; and

(C) Dangerous or hazardous cargo loaded into an aircraft.

(v) Appropriate responses for the employee to specific situations, including:

(A) Taking no action, if a situation does not warrant action;

(B) Questioning an individual, if his or her behavior may be considered suspicious;

(C) Informing a supervisor, if a situation or an individual's behavior warrants further investigation;

(D) Calling the TSA General Aviation Hotline; or

(E) Calling local law enforcement, if a situation or an individual's behavior could pose an immediate threat.

(vi) Any other information relevant to security measures or procedures at the flight school, including applicable information in the TSA Information Publication "Security Guidelines for General Aviation Airports".

(d) *Recurrent security awareness training program.* (1) A flight school must ensure that each flight school employee receives recurrent security awareness training each year in the same month as the month the flight school employee received initial security awareness training in accordance with this subpart.

(2) At a minimum, a recurrent security awareness training program must contain information regarding

(i) Any new security measures or procedures implemented by the flight school;

(ii) Any security incidents at the flight school, and any lessons learned as a result of such incidents;

(iii) Any new threats posed by or incidents involving general aviation aircraft contained on the TSA Web site; and

(iv) Any new TSA guidelines or recommendations concerning the security of general aviation aircraft, airports, or flight schools.

§1552.25 Documentation, recordkeeping, and inspection.

(a) *Documentation.* A flight school must issue a document to each flight school employee each time the flight school employee receives initial or recurrent security awareness training in accordance with this subpart. The document must—

(1) Contain the flight school employee's name and a distinct identification number.

(2) Indicate the date on which the flight school employee received the security awareness training.

(3) Contain the name of the instructor who conducted the training, if any.

(4) Contain a statement certifying that the flight school employee received the security awareness training.

(5) Indicate the type of training received, initial or recurrent.

(6) Contain a statement certifying that the alternative training program used by the flight school meets the criteria in 49 CFR 1552.23(c), if the flight school uses an alternative training program to comply with this subpart.

(7) Be signed by the flight school employee and an authorized official of the flight school.

(b) *Recordkeeping requirements.* A flight school must establish and maintain the following records for one year after an individual no longer is a flight school employee:

(1) A copy of the document required by paragraph (a) of this section for the initial and each recurrent security awareness training conducted for each flight school employee in accordance with this subpart; and

(2) The alternative flight school security awareness training program used by the flight school, if the flight school uses such a program.

(c) *Inspection.* A flight school must permit TSA and the Federal Aviation Administration to inspect the records required under paragraph (b) of this section during reasonable business hours.

Issued in Arlington, Virginia, on September 16, 2004.

David M. Stone, Assistant Secretary.

Selected Aviation Web Sites

There are literally hundreds of aviation and weather web sites. The following listing touches on a few of the popular sites used by persons involved in general aviation. Many of them link to other aviation pages. The editor welcomes additional sites that you have found to be helpful. Home pages change and come and go. These addresses were all active at press time for this edition. Try some for interesting web-flying instead of web-surfing. (This listing does not include DUATS or commercial venders of software or aviation information.)

Selected Aviation Weather Sites
Aviation Digital Data Service—funded and directed by FAA offering experimentl weather products and displays: http://adds.aviationweather.gov

NATIONAL WEATHER SERVICE—Current terminal forecasts, METAR reports, weather discussions, and more. www.nws.noaa.gov

THE AVIATION HOMEPAGE—Meteorology, satellite and radar images, plus links to aviation clubs, organizations, etc: www.avhome.com

AOPA—The members only section provides weather information—including radar animations. This is the same information as is found at many FBOs: www.aopa.org

SELECTED WEB SITES OF INTEREST TO GENERAL AVIATION
AirNav—Airport data, facilities directory, etc.: www.airnav.com

Air Force—Thunderbirds, events, etc. http://events.airforce.com

Air & Space Magazine:—Information about Air and Space Museum and developing comprehensive air and space related worldwide web: www.airspacemag.com

Air Racing—News and pictures of events, results, etc. www.warbird.com

Aircraft Owners and Pilots Association—membership organization of pilots and aircraft owners. News for pilots, membership information, learning to fly information, weather, President's position on current aviation issues, AOPA Air Safety Foundation connection, and more: www.aopa.org

Aircraft Performance—Performance data on almost any general aviation aircraft in production in the past 40 years: www.risingup.com

Aircraft Shoppers Online—Aircraft listings, brokers and dealers, search for aircraft: www.aso.solid.com

American Asso. of Airport Executives—association of individuals involved in airport management: www.airportnet.org

Air Force Association—aerospace education, Air Force Magazine, links, etc :—www.afa.org

Air National Guard—Air national guard page including mission, safety, addresses/phone numbers: www.ang.af.mil

Antique Aircraft—Links to museums, antique aircraft, organizations: www.wingsofhistory.org

Aviation International News—Publication focusing on business flying: www.ainonline.com

Aviation Jobs—Airline and corporate pilot jobs, resumé listings, links to recruiters: www.flightdeckrecruitment.com

Aviation News—NOTAMs, TFRs, airports by identifiers, aviation directory, links to databases, search for Designated Examiners, etc.: www.landings.com

Aviation Safety Reporting System—NASA's site for reporting flight or system errors. Publications, reporting forms, database information. http://asrs.arc.nasa.gov

Avweb—Aviation news wire, safety information, airmanship, products, weather, databases: www.avweb.com

Be a Pilot—General aviation community's web site for students. Information about learning to fly, flight schools, etc.: www.beapilot.com

Civil Air Patrol—Membership news, search, education, etc.: www.cap.gov

Department of Transportation—News releases, speeches, etc. relating to DOT actions in aviation: www.dot.gov

Experimental Aircraft Association—Home page for the EAA with links to chapters, information about EAA, chapter areas, membership, Oshkosh fly-in, etc.: www.eaa.org

Embry-Riddle Aeronautical University—Information about the university and links to its vast stores of information: www.erau.edu

Federal Aviation Administration—FAA home page that connects to various offices and databases, airports, air traffic, certification, public affairs: www.faa.gov

Flight Planning—IFR flight planning for corporate, charter, and business pilots: www.fltplan.com

Flight Safety Foundation—non-profit organization devoted to international aviation safety: www.flightsafety.org

Federation International Aeronautic—news, world records, membership, etc.: www.fai.org

General Aviation—links to many sites—travel, shopping, aircraft merchandise, etc.: www.generalaviation.com

General Aviation Manufacturers Association—Association of manufacturers of aircraft, engines, avionics, and components: www.gama.aero/home.php

General Aviation News—Bi-weekly news and feature newspaper for general aviation: www.generalaviationnews.com

General Aviation World Wide—General aviation commission of Fedération Aéronautique Internationale. FAI's general aviation activities, world records, and competitions: www.fai.org/general_aviation

Government Printing Office—Route to Federal Register, and other printed matter from the federal government: www.access.gpo.gov

International Aircraft Owners and Pilots Association—International pilot information, flight rules, connects to web sites of AOPA organizations in other nations including Australia, Germany, Italy, Luxembourg, Netherlands, Sweden, and United Kingdom: www.iaopa.org/

Helicopter Association International—Professional trade organization for the civil helicopter industry: www.rotor.com

I FLY America—Places to fly, accident reports, other info for pilots: www.iflyamerica.org

International Civil Aviation Organization—About ICAO. regional offices, publications, etc.: www.icao.org

Lawyer Pilots Bar Association—Organization of lawyers who are also pilots: www.lpba.org

Landings—Data bases, aircraft sales, weather, pilot supplies: www.landings.com

Light Aircraft Manufacturers Association—Sport aviation news. www.lama.b3/

Links to sites—Listing of aviation-related links selected from top 500 aviation sites on the web: www.ultimateaviationlinks.com

McGraw-Hill Aviation Publications—Aviation Week, Business and Commercial Aviation, and A/C Flyer magazines: www.aviationnow.com

McGraw-Hill Companies—Company information link to aviation books: www.mcgraw-hill.com

National Aeronautic Association—U.S. aviation record organization, membership news, activities: www.naa-usa.org

National Air Transportation Association—Association for FBO's and charter operators: www.nata.aero

NASA—News of NASA activities, publications, links: www.nasa.gov

National Air Traffic Controllers Association—Traffic controller information, membership, etc.: www.natca.org

National Association of State Aviation Officials—Information about NASAO and links to home pages of state aeronautics departments and divisions: www.nasao.org

National Business Aviation Association—Association of businesses and corporations that use aircraft for business: www.nbaa.org

National Transportation Safety Board—Home page for the NTSB, to obtain sources of aviation safety information, press releases, Board meetings, etc.: www.ntsb.gov

Resource for Air and Space professional and non-professionals interested in flying, space flight, pilot training, etc.: www.generalaviation.com

Safety data—Good source of current safety data for general aviation, amateur built, and ultralight aircraft, includes daily accident reports, ADs, and other goodies. Sponsored by EAA: www.safetydata.com

Silver Wings—Organization of pilots who soloed 25 or more years ago: silverwings.org

Smilinjack—Airports, travel, aviation shopping, weather, servers, home pages, airlines, and more: www.smilinjack.com

Soaring Society of America—Information about soaring, places to soar, competitions, etc.: www.ssa.org

The CFI—Regulations, flight planning, accident prevention, test sites, and links: www.TheCFI.com

Virtual Skies—Education site for schools: http://virtualskies.arc.nasa.gov

Women in Aviation International—Education, careers, chapters on line: www.women-in-aviation.com

Flight Forum

Flight Forum is a page from FAA Aviation News, a publication of the Federal Aviation Administration. It provides an opportunity for pilots to express their views and receive answers from the Administration. Some of the subjects discussed contain interesting matter that many pilots would find useful. To get wider distribution of the views, certain subjects are reprinted here.

Proposed change to AC 90-42F

As an active flight instructor I would like to see the phase "Departing Runway____" changed to "Taking off Runway" in Section 10. (2) in Advisory Circular 90-42, Traffic Advisory Practices at Airports without Operating Control Towers. The reason for this change is that departing a runway can easily be confused. Are we:

- Taking off
- Departing the runway on the taxi way
- On the departure leg
- Departing the pattern.

Confusing, especially if you are stepped on or the radio traffic is too heavy and/or you are at a new and strange airport. I believe that at a Non-towered field the phraseology should mimic the FAA standards as set forth in Aeronautical Information Manual or the federal aviation regulations.

I have personally seen and heard pilots so confused where they were that safety issue were prevalent and compromised. Pilots maneuvering at low altitudes, trying to catch sight of aircraft that the pilot has given confusing position report continue to claim lives.

AOPA reminds us again in the CD "Maneuvering" how dangerous low and slow can be for pilots trying to set up for the landing. Location of aircraft and understanding a reported position is so critical in aviation safety, especially at a non towered field. The skies are becoming busier now that Sport Piloting is being introduced. We all must be on the same page.

Tom Evans, Master CFI

Thank you for your comments. We will forward a copy of your comments to the appropriate Air Traffic organization for consideration.

Banning Skydiving

I have been a reader of your magazine for many years and normally agree with your articles, but in your September/October 2004 issue, you have one article that is of great issue.

Your article on "Drop Zone Flying for the GA Pilot" is wrongly named, it should be "Get out of my way-Here I come." I realize this article was written by someone who is a member of USPA [Editor: United States Parachute Association], and who will do whatever is necessary to promote this very hazardous sport. Please check the USPA web site and you will find that almost three people per month are killed in USPA accidents, not to mention skydive groups that do not report or innocent people in other aircraft that are not counted on their tally sheet of death.

My concern is not for the person who elects to jump out of an airplane, but the innocent flying public who are affected by skydiving over or around active airports. Please refer to the FAA Incident Data System Reports of aircraft/parachute incidents and accidents.

When you mix student pilots and student skydivers, you are promoting accidents. When we HAD a skydiving club at our local airport, jumpers would land all over the airport, sometimes on the runways, sometimes across the fence on the highway, and occasionally in a local lake. Sometimes they actually hit the jump zone.

The FAA promotes SAFETY SAFETY SAFETY. . .I do not understand how the FAA could possibly allow; much less promote skydiving on or around active airports. Skydiving over or onto to active airports must be banned .... period.

Gary F Jones
Paris, TN

I want to thank you for being a loyal reader of FAA Aviation News for many years. I respect your opinion, but I must take exception to your comment about banning skydiving at an active airport. You are right. FAA does promote safety. But FAA also recognizes that there are many different types of aviation activities wanting access to the national airspace and public access airports. Unless there is a documented safety issue involved, FAA believes that no flight activity should be banned at public airports. However, FAA recognizes that the safe operation of a public airport is a local airport management issue.

To ensure safety for all, it is important for the various aviation groups at an airport to work together to maximize safety while minimizing risks so that everyone can enjoy a public access airport. This is especially important for those airports accepting Federal funding.

The problem with banning a type of operation today is which operation will be banned tomorrow. Will it be business jets or training aircraft or light sport aircraft or all non commercial flight activities?

GPS Update Cards

Does the FAA require any type of record entry when a front panel-mounted GPS unit is updated?

Name withheld by request.

Clinton, Maryland

The short answer is yes. A record is required with an appropriate signature.

The requirement is based upon Title 14 Code of Federal Regulations part 43, Maintenance, Preventive Maintenance, Rebuilding, and Alteration. Appendix A to part 43-Major Alterations, Major Repairs, and Preventive Maintenance, subparagraph (c) Preventive maintenance, item (32) permits the updating. Item (32) states, "Updating self-contained, front instrument panel-mounted Air Traffic Control (ATC) navigational software data bases (excluding those of automatic flight control systems, transponders, and microwave frequency distance measuring equipment (DME)) provided no disassembly of the unit is required and pertinent instructions are provided. Prior to the unit's intended use, an operational check must be performed in accordance with applicable sections of part 91 of this chapter."

Since updating is authorized as a preventive maintenance item under part 43, the part has a record keeping requirement for preventive work. That requirement is specified in section 43.9 titled, Content, form, and disposition of maintenance, preventive maintenance, rebuilding, and alteration records (except inspections performed in accordance with part 91, part 125, Sec. 735.411(a)(1), and Sec. 135.419 of this chapter). "(a) Maintenance record entries. Except as provided in paragraphs (b) and (c) of this section, each person who maintains, performs preventive maintenance, rebuilds, or alters an aircraft, airframe, aircraft engine, propeller, appliance, or component part shall make an entry in the maintenance record of that equipment containing the following information: (1) A description (or reference to data acceptable to the Administrator) of work performed. (2) The date of completion of the work performed. (3) The name of the person performing the work if other than the person specified in paragraph (a)(4) of this section. (4) If the work performed on the aircraft, airframe, aircraft engine, propeller, appliance, or component part has been performed satisfactorily, the signature, certificate number, and kind of certificate held by the person approving the work. The signature constitutes the approval for return to service only for the work performed."

Finally, section 43.7 lists the persons authorized to approve and sign off work on aircraft, airframes, aircraft engines, propellers, appliances, or component parts for return to service after maintenance, preventive maintenance, rebuilding, or alteration. Subparagraph (f) of the section says, "A person holding at least a private pilot certificate may approve an aircraft for return to service after performing preventive maintenance under the provisions of Sec. 43.3(g)."

Stick and Rudder Versus the Jetsons

I love the *FAA Aviation News* and read it faithfully... best in the business! I also detect from Dean's articles that he is a true pilot who loves flying.

As for the article on FITS and TAA by Mike Gaffney, I love the idea of scenario-based training and am currently working to incorporate this into our 141 school. This is a very powerful technique and the new FAA focus is very welcome. But what is the attraction of a "hands folded in lap aircraft," especially in primary training? I can understand this for an Airbus that was designed for automated flight, not a Cessna 172. I really like flying too much! The idea of programming my plane and watching it fly somewhere seems like a nightmare out of a Jetson's cartoon. This also seems very out of sync with the new emphasis on "stick and rudder skills" to address takeoff and landing, as well as maneuvering accidents. How proficient will a pilot trained in these planes become at landing. Will that be programmed in also?

In my perfect world all students would learn the first 20 hours or solo in sport plane that requires positive control and presents few distractions. (And why fly around $50K in superfluous avionics?) Then as they learn to navigate, they transition to the "starship," if desired.

Thanks for a great magazine.

David St. George, MCFI, DPE

Thanks for the complements on the magazine. As for stick and rudder versus scenariobased training, there is no right or wrong way to teach flying. Whatever works best for you and your students is all that matters, as long as it is done safely.

VOR Test Facilities

An aircraft's VOR must have been operationally checked within 30 days prior to flight to use that system for navigation under instrument flight rules. (See FAR 91.171.) Following is a list of ground-based locations and frequencies of VOR test facilities. The FAA's Airport Facilities Directory, published in seven regional volumes, provides additional ground and airborne checkpoints.

Akron, OH—Akron-Canton Regional	110.6
Albany, NY—Albany County	108.2
Albuquerque, NM—Int'l	111.0
Atlanta, GA—Hartsfield	111.0
Bakersfield, CA—Meadows	111.2
Bangor, ME—Int'l	111.0
Bedford, MA—Hanscom	110.0
Birmingham, AL—Birmingham	110.0
Boise, ID—Gowen	116.7
Boston, MA—Logan	111.0
Bradley, CT—Int'l	111.4
Bridgeport, CT—Sikorsky	109.25
Brunswick, GA—Glynco	111.0
Buffalo, NY—Int'l	109.0
Burlington, VT—Int'l	109.0
Charleston, SC AFB/Int'l	111.0
Charleston, WV—Yeager	108.8
Charlotte, NC—Douglas Int'l	112.0
Chicago, IL—Midway	111.0
Chicago, IL—O'Hare	112.0
Cincinnati, OH—Lunken	108.4
Cleveland, OH—Cleveland-Hopkins Int'l	110.4
Colorado Springs, CO—Municipal	110.4
Columbus, OH—Int'l	111.0
Compton, CA—Compton	113.9
Dallas, TX—Love	113.8
Davenport, IA—Municipal	111.8
Dayton, OH—Cox/Int'l	111.0
Daytona Beach, FL—Municipal	111.0
Denver, CO—Centenial	108.2
Denver, CO—Int'l	110.0
Des Moines, IA—Int'l	109.2
Detroit, MI—City	111.6
Detroit, MI—Wayne	109.8
Detroit, MI—Willow Run	112.0
El Paso, TX—Int'l	111.0
Fort Wayne, IN—Int'l	111.0
Fort Worth, TX—Meacham	108.2
Groton, CT—Groton-New London	110.25
Harrisburg, PA—Int'l	110.0
Hawthorne, CA—Municipal	113.9

Hickory, NC—Regional	110.0
Honolulu, HI—Int'l	111.0
Houston, TX—Hobby	111.6
Huntsville, AL—Jones	111.0
Indianapolis, IN—Int'l	109.6
Islip, NY—MacArthur	109.4
Jackson, MS—Int'l	111.0
Jacksonville, FL—Int'l	111.0
Jefferson City, MO—Memorial	112.0
Kansas City, MO—Downtown	108.6
Knoxville, TN—McGee Tyson	112.0
Las Vegas, NV—North Las Vegas	108.2
Long Beach, CA—Daugherty	113.9
Los Angeles, CA—Int'l	113.9
Louisville, KY—Standiford	111.0
Medford, OR—Rogue Valley Int'l	117.2
Memphis, TN—Int'l	111.0
Miami, FL—Int'l	112.0
Midland, TX	108.2
Milwaukee, WI—Mitchell	109.0
Minneapolis MN—Wold Chamberlain Int'l	111.0
Nashville, TN—Int'l	108.6
New Orleans, LA—Lakefront	111.0
New York, NY—JFK Int'l	115.1
Newark, NJ—Neward Int'l	110.0
Oklahoma City, OK—Will Rogers	108.8
Omaha, NB—Eppley	109.0
Philadelphia, PA—Int'l	109.8
Phoenix, AZ—Sky Harbor	109.0
Portland, ME—Int'l	111.0
Portland, OR—Int'l	111.0
Portland, OR—Hillsboro	115.2
Prescott, AZ—	110.0
Providence, RI—Green State	108.2
Sacramento, CA—Executive	111.4
Sacramento, CA—Int'l	111.4
St. Louis, MO—Lambert Int'l	111.0
St. Louis, MO—Spirit	112.2
St. Paul, MN—Downtown	114.4
Salt Lake City, UT—Int'l	111.0
San Antonio, TX—Int'l	110.4
San Diego, CA—Lindberg	109.0
San Diego, CA—Brown	109.0
San Diego, CA—Montgomery	109.0
San Diego, CA—Gillespie	110.0
San Francisco, CA—Int'l	111.0
Santa Ana, CA—Orange Co	110.0
Santa Monica, CA—Municiple	113.9
Savannah, GA—Int'l	111.0

Seattle, WA—Seattle/Tacoma	117.5	Topeka, KS—Forbes	111.0
Seattle, WA—Boeing/King Int'l	108.6	Torrance, CA—Zamperini	113.9
Shreveport, LA—Regional	108.2	Tulsa, OK—Int'l	109.0
Smyrna, TN	110.2	Vero Beach, FL—Municiple	111.0
Spokane, WA—Int'l	109.6	Washington, DC—Reagan National	109.4
Spokane, WA—Felts	114.0	West Palm Beach, FL—Int'l	109.0
Tallahassee, FL—Regional	111.0	Wichita-Mid-Continent	114.0
Tampa, FL—Int'l	111.0	Worchester, MA—Regional	108.2

Navigation and Communication Frequencies

Navigation frequencies

75 MHz—Transmitting frequencies of fan markers, Z markers, and ILS markers.

108.0 to 108.05 MHz—Commonly used for VOR ramp testers.

108.1 to 108.15 MHz—Commonly used for ILS localizer ramp testers.

108.2 to 111.85 MHz—Transmitting frequencies of VORs. Operated on even tenths.

108.3 to 111.95 MHz—ILS localizers with or without voice. Operated on odd tenths.

112.0 to 117.95—Transmitting frequencies of VORs.

Voice communication frequencies

118.0 to 121.4 MHz—ATC

121.5 MHz—emergency frequency (search and rescue (SAR), emergency locator transmitter (ELT) signals (five-second operational check)

121.6 to 121.925 MHz—airport ground control, ELT test

121.95 MHz—aviation support

121.975 MHz—private aircraft advisory (FSS)

122.0 to 122.05 MHz—FSS EFAS (Flight Watch)

122.075 to 122.675 MHz—private aircraft advisory (FSS)

122.7 to 122.725 MHz—unicom, nontower-controlled airports

122.75 MHz—air-to-air communications (fixed-wing aircraft)

122.775 MHz—aviation support

122.8 MHz—unicom, nontower-controlled airports

122.825 MHz—aeronautical en route (ARINC)

122.85 MHz—multicom

122.875 MHz—ARINC

122.9 MHz—multicom, SAR training, airports with no tower, FSS, or unicom

122.925 MHz—multicom, special use (forestry management/fire suppression, fish and game management/protection, etc.)

122.95 MHz—unicom, tower-controlled airports, airports with fulltime FSSs

122.975 to 123.0 MHz—unicom, nontower-controlled airports

123.025 MHz—air-to-air communications (helicopter)

123.05 to 123.075 MHz—unicom, nontower-controlled airports

123.1—SAR, temporary control towers

123.125 to 123.275 MHz—flight test stations of aircraft manufacturers

123.3 MHz—flight schools

123.325 to 123.475 MHz—flight test stations of aircraft manufacturers

123.5 MHz—flight schools

123.525 to 123.575 MHz—flight test stations of aircraft manufacturers

123.6 to 123.65 MHz—air carrier advisory (FSS)

123.675 to 126.175 MHz—ATC

126.2 MHz—military common advisory

126.225 to 128.8 MHz—ATC

128.825 to 132.0 MHz—ARINC

132.025 to 134.075 MHz—ATC

134.1 MHz—military common advisory

134.125 to 135.825 MHz—ATC

135.85 MHz—FAA flight inspection

135.875 to 135.925 MHz—ATC

135.95 MHz—FAA flight inspection

135.975 to 136.075 MHz—ATC

136.1 MHz—future unicom or AWOS

136.125 to 136.175 MHz—ATC

136.2 MHz—future unicom or AWOS

136.225 to 136.25 MHz—ATC

136.275 MHz—future unicom or AWOS

136.3 to 136.35 MHz—ATC

136.375 MHz—future unicom or AWOS

136.4 to 136.45 MHz—ATC

136.475 MHz—future unicom or AWOS

136.5 to 136.975 MHz—ARINC

Selected Addresses and Telephone Numbers

United States Congress

Members of Congress may be contacted by telephone, mail, or e-mail. To reach the U.S. Capitol switchboard for either the Senate or the House call: 202-224-3121.

To contact your Senator by mail, write to:
The Honorable (Full name)
United States Senate
Washington, DC 20510
Dear Senator (Last name)

For e-mail: www.senate.gov
From this Senate home page you will find links to specific Senators.

To contact your Representative by mail, write to:
The Honorable (Full Name)
United States House of Representatives
Washington, DC 20515
Dear Representative (Last Name)

For e-mail: www.House.gov
From this House home page you will find links to specific Representatives.

Department of Transportation
400 7th Street SW
Washington, DC 20590
Phone: 202/366-4000

Federal Aviation Administration
800 Independence Ave. SW
Washington DC 20591
Phone: 202/267-3484

Mike Monroney Aeronautical Center
6500 South MacArthur
Oklahoma City, OK 73169
(Mail Address: P.O. Box 25082
Oklahoma City, OK 73125)
405/954-3011

William J. Hughes Technical Center
Atlantic City International Airport
New Jersey 08504
609/485-4000

FAA Regional Offices

Alaska Region
707-271-5859
http://www.alaska.faa.gov

Central Region
816-329-3881

Eastern Region
718-553-3052

Great Lakes Region
847-294-7166

New England Region
781-238-7660

Northwest Mountain Region
425-227-1179

Southern Region
4040305-5784

Western Region
310-725-7563
http://www.awp.faa.gov

FAA Flight Standards District Offices
Contact a Flight Standards District Office (FSDO) for:
Aircraft noise reporting
Aircraft modifications and permits
Certification and surveillance of air operators, air agencies, and airmen
Enforcement and Investigation
Aviation safety and training.

Alabama & Northwest Florida
1500 Urban Center Drive
Suite 250
Vestavia Hills, AL 35242
Phone: (205) 731-1557 Fax: (205) 731-0939

Alaska
  Anchorage
  Phone: (907) 271-2000
  4419 Airport Way
  Fairbanks Fairbanks, AK 99709
  Phone: (907) 474-0276 Fax: (907) 479-9650
  Juneau
  Phone: (907) 586-7532

Arizona
  17777 N. Perimeter Dr.
  Suite 101
  Scottsdale, AZ 85255-5453
  Phone: (480) 419-0111 Fax: (480) 419-0800

Arkansas
1701 Bond St.
Little Rock, AR 72202
Phone: (501) 918-4400 Fax: (501) 918-4403

California
Riverside 6961 Flight Road
Riverside, CA 92504
Phone: (951) 276-6701 Fax: (951) 689-4309

Oakland
1420 Harbor Bay Parkway
Suite 280
Alameda, CA 94502-7083
Phone: (510) 748-0122 Fax: (510) 748-9559
Los Angeles
2250 E. Imperial Hwy
Suite 140
El Segundo, CA 90245
Phone: (310) 215-2150 Fax: (310) 645-3768
Long Beach
5001 Airport Plaza Drive
Long Beach, CA 90815
Phone: (562) 420-1755 Fax: (562) 420-6765
Fresno
4955 E. Anderson
Suite 110
Fresno, CA 93727
Phone: (559) 487-5306 Fax: (559) 454-8808

Colorado
26805 East 68th Avenue Suite 200
Denver, CO 80249-6339
Phone: (800) 847-3808

Connecticut
1st Floor
Building 85-214
Bradley International Airport
Windsor Locks, CT 06096-1009
Phone: (860) 654-1000 Fax: (860) 654-1009

Florida
Ft. Lauderdale
1050 Lee Wagener Blvd.
Suite 201
Ft. Lauderdale, FL 33315
Phone: (954) 635-1300 Fax: (954) 635-1260
Orlando
5950 Hazeltine National Drive, Suite #500,
Citadel International, Orlando, FL 32822-5023
Phone: (407) 812-7700 Fax: (407) 812-7710
Tampa
Phone: (813) 287-4900
South FL (Miami)
Phone: (305) 716-3400

Georgia
Atlanta
Campus Building Suite 2-110
1701 Columbia Ave.
College Park, GA 30337
Phone: (404) 305-7200 Fax: (404) 305-7215

Hawaii
Honolulu
Phone: (808) 837-8300

IDAHO
3295 Elder Street
Airport Plaza, Suite 350
Boise, ID 83705
Phone: (208) 387-4000 or (800) 453-0001
Fax: (208) 387-4020

Illinois
Springfield
1250 North Airport Drive
Suite 1
Springfield, IL 62707-8417
Phone: (217) 744-1910 Fax: (217) 744-1947
Chicago
Chicago O'Hare
9950 W. Lawrence Avenue
Suite 400
Schiller Park, IL 60176
Phone: (847) 928-8000 Fax: (847) 928-8002

Indiana
Indianapolis
8303 West Southern Avenue
Indianapolis, IN 46241
Phone: (317) 487-2400 Fax: (317) 487-2429
South Bend
5800 Nimtz Parkway
South Bend, IN 46628
Phone: (574) 245-4600 Fax: (574) 233-9387

IOWA
3753 SE Convenience Blvd.
Ankeny, IA 50021
Phone: (515) 289-3840 Fax: (515) 289-3855

Kansas
1801 Airport Road
Suite 300
Wichita, KS 67209
Phone: (316) 941-1200 Fax: (316) 941-1276

Kentucky
Watterson Towers 11th Floor
1930 Bishop Lane
Louisville, KY 40218
Switchboard: (502) 753-4200

Louisiana
6100 Corporate Blvd.
Suite 200
Baton Rouge, LA 70808
Phone: (225) 932-5900 or (800) 821-1960
Fax: (225) 932-5975

Maine
Portland International Jetport
412 Yellowbird Road
Portland, ME 04102
Phone: (207) 780-3263
Fax: (207) 780-3296

Maryland
890 Airport Park Road
Suite 101
Glen Burnie, MD 21061-2559
Phone: (410) 787-0040, Fax: (410) 787-8708

Massachusetts
Boston
One Cranberry Hill
Fourth Floor, Suite 402
Lexington, MA 02421
Phone: (781) 274-7130 Fax: (781) 274-6725
Boston
70 Everett Avenue
Chelsea, MA02150
Phone: (617) 887-9237 Fax: (617) 887-9261

Michigan
Detroit
8800 Beck Road-Eastside
Belleville, MI 48111
Phone: (734) 487-7222 Fax: (734) 487-7221
Grand Rapids
3196 Kraft Avenue SE
Grand Rapids, MI 49512
Phone: (616) 954-6657 Fax: (616) 940-3140

Minnesota
6020 28th Ave.
Minneapolis, MN 55450
Phone: (612) 713-4211 Fax: (612) 713-4290

Mississippi
100 West Cross Street
Suite C
Jackson, MS 39208
Phone: (601) 664-9800 Fax: (601) 664-9910

Missouri
Kansas City
10015 N.W. Ambassador Drive
Kansas City, MO 64153
Phone: (816) 891-2100, (800) 519-3269
Fax: (816) 891-2155

St. Louis
10801 Pear Tree Ln
Suite 200
St. Ann, MO 63074
Phone: (314) 429-1006, (800) 322-8876
Fax: (314) 429-6367

Montana
2725 Skyway Drive
Helena, MT 59602-1213
Phone: (406) 449-5270
Fax: (406) 449-5275

Nebraska
3431 Aviation Road
Suite 120
Lincoln, NE 68524
Phone: (402) 475-1738 Fax: (402) 458-7841

Nevada
Las Vegas
7181 Amigo St.
Suite 180
Las Vegas, NV 89119
Phone: (702) 269-1445 Fax: (702) 269-8013
Reno Flight
4900 Energy Way
Reno, Nevada 89502
Phone: (775) 858-7700 Fax: (775) 858-7737

New Jersey
Park 80 West
Plaza One
Saddle Brook, NJ 07663
Phone: (201) 556-6600 Fax: (201) 556-6623

New Mexico
1601 Randolph Rd. SE
Suite 200N
Albuquerque, NM 87106
Phone: (505) 764-1200 Fax: (505) 764-1233

New York
Albany
7 Airport Park Blvd.
Latham, NY 12110
Phone: (518) 785-5660, Fax: (518) 785-7165
Farmingdale
7150 Republic Airport
Suite 235
Farmingdale, NY 11735
Phone: (631) 755-1300 Fax: (631) 694-5516
Rochester
One Airport Way
Suite 110
Rochester, New York 14624
Phone: (585) 436-3880, Fax: (585) 436-2322

New York City
990 Stewart Avenue
Suite 630
Garden City, NY 11530-4858
Phone: (516) 228-8029, Fax: (516) 228-8827

North Carolina
Charlotte
3800 Arco Corporate Drive
Suite 233
Charlotte, NC 28273
Phone: (704) 319-7020 Fax: (704) 319-7021
Greensboro
6433 Bryan Blvd
Greensboro, North Carolina 27409
Phone: (336) 662-1000 Fax: (336) 662-1080

North Dakota
4620 Amber Valley Parkway
Fargo, ND 58104
Phone: (701) 277-1245
Fax: (701) 277-3682

Ohio
Cincinnati
4240 Airport Road
Cincinnati, OH 45226
Phone: (513) 979-6400 Fax: (513) 979-6420
Cleveland
Great Northern Technology Park II
25249 Country Club Blvd.
North Olmsted, OH 44070
Phone: (440) 686-2001 Fax: (440) 686-2080
Columbus
2780 Airport Drive
Suite 300
Columbus, OH 43219
Phone: (614) 255-3120 Fax: (614) 255-3159

Oklahoma
1300 S. Meridian
Ste. 601
Oklahoma City, OK 73108
Phone: (405) 951-4200 Fax: (405) 951-4282

Oregon
3180 NW 229th Avenue
Hillsboro, OR 97124

Pennsylvania
Allegheny
101 Towne Square Way
Suite 201
Pittsburgh, PA 15227
Phone: (412) 886-2580 Fax: (412) 886-2591
Allentown
961 Marcon Blvd.
Suite 111
Allentown, PA 18109
Phone: (610) 264-2888, Fax: (610) 264-3179

Harrisburg
400 Airport Road
Room 101
New Cumberland, PA 17070
Phone: (717) 774-8271,
Fax: (717) 774-8327
Philadelphia
2 International Plaza
Suite 110
Philadelphia, PA 19113-1504
Phone: (610) 595-1500 Fax: (610) 595-1519

Puerto Rico
525 F.D. Roosevelt Avenue
La Torre de Plaza, Suite 901
San Juan, Puerto Rico 00918
Phone: (787) 764-2538
Fax: (787) 764-2641

South Carolina
125-B Summer Lake Drive
West Columbia, SC 29170
Phone: (803) 765-5931 Fax: (803) 253-3999

South Dakota
909 St. Joseph St.
Suite 700
Rapid City, SD 57701-2699
Phone: (605) 737-3050 Fax: (605) 737-3069

Tennessee
Memphis
2842 Business Park Drive
Bldg G
Memphis, TN 38118
Phone: (901) 322-8600
Fax: (901) 322-8601 or (901) 322-8602
Nashville
2 International Plaza
Suite 700
Nashville, TN 37217
Phone: (615) 324-1300 Fax: (615) 324-1360

Texas
Dallas
Phone: (214) 902-1800
Alliance/Fort Worth
2221 Alliance Blvd.
Suite 400
Fort Worth, TX 76177-4300
Phone: (817) 491-5000 Fax: (817) 491-5014
Lubbock
6202 N. Interstate 27, Suite 2
Lubbock, TX 79403-9712
Phone: (806) 740-3800 or 1-800-858-4115
Fax: (806) 740-3809

Houston
12000 Aerospace Ave.
Suite 400
Houston, TX 77034-5576
Phone: (281) 929-7000 or (888) 285-2127
Fax: (281) 929-7059

Utah
1020 North Flyer Way
Salt Lake City, UT 84116-2959
Phone: (801) 257-5020
Fax: (801) 257-5066

Virginia
Richmond
5707 Huntsman Road
Suite 100
Richmond International Airport
Richmond, VA 23250-2415
Phone: (804) 222-7494
Fax: (804) 222-4843
Washington (Dulles)
44965 Aviation Drive
Suite 112
Washington Dulles Airport
Dulles, VA 20166-7524
Phone: (703) 661-8160
Fax: (703) 661-8744

Washington
Seattle
1601 Lind Avenue SW
Renton, WA 98055
Phone: (425) 227-2813 or (800) 354-1940
Fax: (425) 227-1810
Spokane
6133 E. Rutter Avenue
Spokane, WA 99212
Phone: (509) 532-2340 or (800) 341-2623
Fax: (509) 532-2380

West Virginia
Yeager Airport
301 Eagle Mountain Road, Room 144
Charleston, WV 25311-1093
Phone: (304) 347-5199
Fax: (304) 343-2011

Wisconsin
4915 South Howell Avenue
Milwaukee, WI 53207
Phone: (414) 486-2920
Fax: (414) 486-2921/22

Wyoming
951 Werner Court #320
Casper, WY 82601-1312
Phone: (800) 325-5785

National Transportation Safety Board
490 L'Enfant Plaza East SW
Washington DC 20594
202/314-6000
e-mail: ntsb.gov

Southeast Regional Office
8405 N.W. 53rd Street Suite B-103
Miami, FL 33166
305/597-4610

Southeast Field Office
Atlanta Federal Center
60 Forsyth Street, SW Suite 3M25
Atlanta, GA 30303-3104
404/562-1666

North Central Regional Office
31 West 775 North Avenue
West Chicago, IL 60185
630/377-8187

South Central Regional Office
624 Six Flags Drive Suite 150
Arlington, TX 76011
817/652-7800

South Central Field Office
4760 Oakland Street, Suite 500
Denver, CO 80239
303/361-0600

Northwest Regional Office
19518 Pacific Higfiway South Room 201
Seattle, WA 98188
206/870-2200

Northwest Field Office
222 West 7th Avenue Room 216, Box 11
Anchorage, AK 99513
907/271-5001

Southwest Regional Office
1515 W. 190th Street, Suite 555
Gardena, CA 90248
310/380-5660

Northeast Regional Office
2001 Route 46, Suite 504
Parsippany, NJ 07054
201/334-6420

Northeast Field Office
490 L'Enfant Plaza, S.W.
Washington, D.C. 20594
202/314-6320

State Aeronautic Offices

Many state aeronautic offices have web sites that offer a diversity of information about airports and other subjects of interest to pilots and aircraft owners. Their Association—the National Association of State Aviation Officials—maintains a web site and has links to those states. The NASAO web site address is: www.NASAO.org

Mr. John Eagerton, IV, D.P.A.
Bureau Chief
Alabama DOT/Aeronautics Bureau
1409 Coliseum Blvd.
Montgomery, AL 36130
Phone: 334-242-6823
FAX: 334-353-6540
E-mail: eagertonj@dot.state.al.us

John Torgerson
Deputy Commissioner for Aviation
Alaska Statewide Aviation
Alaska Dept. of Trans. & Public Facilities
P.O. Box 196900
Anchorage, AK 99519-6900
Phone: 907-269-0724
Fax: 907-269-0489
e-mail: john_torgerson@dot.state.ak.us

Barclay Dick, Director
Arizona Division of Aeronautics
Arizona Department of Transportation
P.O. Box 13588, MD 426M
Phoenix, AZ 85002-3588
phone: 602-294-9144
fax: 602-294-9141

John Knight, Director
Arkansas Department of Aeronautics
Regional Airport Terminal, 3rd floor
#1 Airport Drive
Little Rock, AR 72202
phone: 501-376-6781
fax: 501-378-0820
e-mail: john.knight@arkansas.gov

Mary Frederick, Acting Division Chief
California Aeronautics Program M.S. # 40
California Dept. of Transportation
P.O. Box 942874
Sacramento, CA 94274-0001
phone: 916-654-5470
fax: 916-653-9531
e-mail: mary_frederick@dot.ca.gov

Travis L. Vallin, Director
Colorado Department of Transportation
Division of Aeronautics
5126 Front Range Parkway
Watkins, CO 80137
phone: 303-261-7705
fax: 303-261-9608
e-mail: travis.vallin@dot.state.co.us

Steve Korta, Commissioner of Transportation
Connecticut Bureau of Aviation and Ports
Connecticut Department of Transportation
P.O. Box 317546
2800 Berlin Turnpike
Newington, CT 06131-7546
phone: 860-594-2529
fax: 860-594-2574
e-mail: skorta@bradleyairport.com

Michael Kirkpatrick, Administrator
Delaware Office of Aeronautics
Delaware Department of Transportation
P.O. Box 778
Dover, DE 19903
phone: 302-760-2111
fax: 302-739-2251
e-mail: michael.kirkpatrick@state.de.us

William J. Ashbaker, P.E., State Aviation Manager
Florida Aviation Office
Florida Department of Transportation
605 Suwannee Street, M.S. 46
Tallahassee, FL 32399-0460
phone: 850-414-4505
fax: 850-412-8050
e-mail: bj.ashbaker@dot.state.fl.us

Edward Ratigan, Manager
Georgia Off. of Intermodal Prog.-Aviation
Georgia Department of Transportation
276 Memorial Drive, SW
Atlanta, GA 30303-3743
phone: 404-651-9221
fax: 404-657-4221
e-mail: ed.ratigan@dot.state.ga.us

Brian Sekiguchi, Deputy Director, Airports
Hawaii Airports Division
Hawaii Department of Transportation
Honolulu International Airport
400 Rodgers Blvd., Suite 700
Honolulu. Hl 96819-1880

phone: 808-838-8600
fax: 808-838-8734
e-mail: airadministrator@exec.state.hi.us

Robert Martin, Administrator
Idaho Division of Aeronautics
Idaho Department of Transportation
P.O. Box 7129
Boise, ID 83707-1129
phone: 208-334-8775
fax: 208-334-8789
e-mail: bob.martin@itd.state.id.us

Susan Shea, Director
Illinois Division of Aeronautics
Illinois Department of Transportation
Capital Airport—One Langhorne Bond Dr.
Springfield, IL 62707-8415
phone: 217-785-8515
fax: 217-524-1022
email: sheasr@dot.il.gov

Jim Keefer, P.E. Manager
Indiana Aeronautics Section
Indiana Department of Transportation
100 North Senate Avenue Room N901
Indianapolis, IN 46204-2217
phone: 317-232-1477
fax: 317-232-1499
e-mail: jgkeefer@indot.state.in.us

Michelle McEnary, Director
Office of Aviation
Iowa Department of Transportation
800 Lincoln Way
Ames, IA 50010
phone: 515-239-1569
fax: 515-233-7983
e-mail: michelle.mcnary@dot.iowa.gov

Ed Young, Director
Kansas Division of Aviation
Kansas Department of Transportation
700 SW Harrison
Topeka, KS 66603-3754
phone: 785-296-7449
fax: 785-296-3833
e-mail: ceyoung@ksdot.org

Paul Steely, Commissioner
Kentucky Division of Aeronautics
Kentucky Transportation Cabinet
125 Holmes Street
Third Floor
Frankfort, KY 40622

phone: 502-564-4480
fax: 502-564-7953
e-mail: Paul.Steely@ky.gov

Phil Jones, Aviation Director
Louisiana Aviation Division
Louisiana Dept. of Trans. & Development
P.O. Box 94245
Baton Rouge, LA 70804-9245
phone: 225-274-4112
fax: 225-274-4181
e-mail: pjones@dotd.louisiana.gov

Ronald L. Roy, Director
Maine Office of Passenger Transportation
Maine Department of Transportation
16 State House Station
Augusta, ME 04333-0016
phone: 207-624-3250
fax: 207-624-3251
e-mail: ron.roy@maine.gov

Tim Campbell, Executive Director
Maryland Aviation Administration
Maryland Department of Transportation
P.O. Box 8766
BWI Airport MD 21240
phone: 410-859-7060
fax: 410-850-4729
e-mail: tcampbell@bwiairport.com

Robert Welch, Executive Director
Massachusetts Aeronautics Commission
10 Park Plaza Room 6620
Boston, MA 02116-3933
phone: 617-973-8881
fax: 617-973-8889
e-mail: bob.welch@state.ma.ur

Rob Abent, Director
Multi-Model Transportation Services Bureau
Michigan Bureau of Aeronautics
Michigan Department of Transportation
2700 East Airport Service Drive
Lansing, MI 48906-2171
phone: 517-335-9943
fax: 517-321-6522
e-mail: abentr@michighan.gov

Raymond J. Rought, Director
Minnesota Aeronautics Office
Minnesota Department of Transportation
222 East Plato Blvd.
St. Paul, MN 55107-1618
phone: 651-296-8046

fax: 651-297-5643
e-mail: ray.rought@dot.state.mn.us

Elton E. Jay, Director
Mississippi Aeronautics Division
Mississippi Dept. of Transportation
P.O. Box 1850
Jackson, MS 39215-1850
phone: 601-359-7850
fax: 601-359-7855
e-mail: ejay@mdot.state.ms.us

Joe Pestka, A.I.C.P.
Administrator of Aviation
Missouri Aviation Section
Missouri Department of Transportation
P.O. Box 270
Jefferson City, MO 65102
phone: 573-526-7912
fax: 573-526-4709
e-mail: joseph.pestka@modot.mo.us

Debbie Alke, Administrator
Montana Aeronautics Division
Montana Department of Transportation
P.O. Box 200507
Helena, MT 59620-0507
phone: 406-444-2506
fax: 406-444-2519
e-mail: dalke@state.mt.us

Stuart MacTaggart, Director
Nebraska Department of Aeronautics
P.O. Box 82088
Lincoln, NE 68501
phone: 402-471-2371
fax: 402-471-2906
e-mail: stuartm@mail.state.ne.us

Kent Cooper, Asst. Director, Planning
Nevada Department of Transportation
1263 South Stewart Street
Carson City, NV 89712
phone: 775-888-7002
fax: 775-888-7203
e-mail: kcooper@dot.state.nv.us

Jack Ferns, Director
New Hampshire Division of Aeronautics
New Hampshire Dept. of Transportation
John O. Morton Building
P.O. Box 483, Hazen Dr
Concord, NH 03302-0483
phone: 603-271-1676
fax: 603-271-3914
e-mail: jferns@dot.state.nh.us

Thomas Thatcher, Director
New Jersey Division of Aeronautics
1035 Parkway Avenue, CN 610
Trenton, NJ 08625-0610
phone: 609-530-2900
fax: 609-530-4549
e-mail: tom.thatcher@dot.state.nj.us

Tom Baca, Director
New Mexico Aviation Division
New Mexico State Highway & Trans. Dept.
P.O. Box 1149
Sante Fe, NM 87504-1149
phone: 505-476-0936
fax: 505-476-0942
e-mail: tom.baca@nmshtd.state.nm.us

Seth Edelman
New York Aviation Services Bureau
New York State Dept. of Transportation
50 Wolf Road
Albany, NY 12232
phone: 518-457-8343
fax: 518-457-9779
e-mail: sedelman@gw.dot.state.ny.us

William H. Williams, Jr. Director
North Carolina Division of Aviation
North Carolina Dept. of Transportation
1560 Mail Service Center
Raleigh, NC 27699-1560
phone: 919-840-0112
fax: 919-840-9267
e-mail: wwilliams@dot.state.nc.us

Gary R. Ness, Director
North Dakota Aeronautics Commission
Box 5020
Bismarck, ND 58502
phone: 701-328-9655
fax: 701-328-9656
e-mail: gness@state.nd.us

James Bryant, Aviation Administrator
Ohio Office of Aviation
Ohio Department of Transportation
2829 W. Dublin-Granville Road
Columbus, OH 43235-2786
Phone: 614-793-5041
fax: 614-793-8972
e-mail: james.bryant@odot.dot.oh.gov

Vic Bird, Director
Oklahoma Aeronautics Commission
3700 N. Classen Blud. Suite 240

Oklahoma City, OK 73118
phone: 405-604-6906
fax: 405-604-6919
e-mail: victor.bird@oac.state.ok.us

Richard Hidley, Director
Oregon Department of Aviation
Oregon Department of Transportation
3040 25th Street SE
Salem, OR 97302-1125
phone: 503-378-4880 × 226
fax: 503-373-1688
e-mail: robert.hidley@state.or.us

Rick Harner, Director
Pennsylvania Bureau of Aviation
Pennsylvania Dept. of Transportation
P.O. Box 3457
Harrisburg, PA 17105-3457
phone: 717-705-1200
fax: 717-705-1255
e-mail: riharner@state.pa.us

Mark Brewer, Acting Executive Director & CEO
Rhode Island Airport Corporation
Theodore Francis Green Airport
2000 Post Road
Warwick, RI 02886-1533
phone: 401-737-4000 × 221
fax: 401-732-3034
e-mail: mbrewer@pvd-ri.com

Michael O'Donnell, Acting Director
South Carolina Division of Aeronautics
South Carolina Dept. of Commerce
P.O. Box 280068
Columbia, SC 29228-0068
phone: 803-896-6261
fax: 803-896-6277
e-mail: modonnell@commerce.state.sc.us

Bruce Lindholm, Program Manager
SD Aviation Off/Dept. of Trans.
Office Air, Rail and Transit
Becker-Hansen Building
700 East Broadway Ave.
Pierre, SD 57501-2586
phone: 605-773-3574
fax: 605-773-3921
e-mail: bruce.lindholm@state.sd.us

Robert Woods, Director
Tennessee Aeronautics Division
Tennessee Dept. of Transportation

P.O. Box 17326
Nashville, TN 37217
phone: 615-741-3208
fax: 515-741-4959
e-mail: Bob.Woods@state.tn.us

David Fulton, Director
Texas Division of Aviation
Texas Department of Transportation
125 E. 11th Street Austin, TX 78701-2483
phone: 512-416-4501
fax: 512-416-4510
e-mail: dfulton@mailgw.dot.state.tx.us

M. Patrick Morley, Director
Utah Aeronautical Operations Division
Utah Department of Transportation
135 North 2400 West
Salt Lake City, UT 84116
phone: 801-715-2260
fax: 801-715-2276
e-mail: pmorley@utah.gov

Richard Turner, Aviation Program Manager
Vermont Agency of Transportation
Maintenance Division/Aviation Section
Nat'l Life Building/Drawer 33
Montpelier, VT 05633
phone: 802-828-2833
fax: 802-828-2829
e-mail: rich.turner@state.vt.us

Randall Burdette, Director
Virginia Department of Aviation
5702 Gulfstream Road
Richmond, VA 23250-2422
phone: 804-236-3625 × 108
fax: 804-236-3635
e-mail: randall.burdette@doav.virginia.gov

John Sibold, Acting Director
Washington Aviation Division
Washington Dept. of Transportation
3704 172nd Street NE Suite K2
P.O. Box 3367
Seattle, WA 98223-3367
phone: 360-651-6300
fax: 360-651-6319
e-mail: sibold@wsdot.wa.gov

Ms. Susan V. Chernonko, Director
West Virginia Aeronautics Commission
West Virginia Dept. of Transportation

1900 Kanawha Blvd. East
Building 5, Room A-503
Charleston, WV 25305
phone: 304-558-3436
fax: 304-558-0333
e-mail: schernenko@mail.dot.slate.wv.us

David Greene, Director
Wisconsin Bureau of Aeronautics
Wisconsin Department of Transportation
P.O. Box 7914
Madison, WI 53707-7914
phone: 608-266-2480
fax: 608-267-6748
e-mail: David.Greene@dot.state.wi.us

Shelly Reams, Administrator
Wyoming Aeronautics Division
Wyoming Department of Transportation
5300 Bishop Boulevard

Cheyenne, WY 82003-3340
phone: 307-777-3953
fax: 307-637-7352
e-mail: sreams@state.wy.us

Miguel Calimano, Executive Director
Puerto Rico Ports Authority
P.O. Box 362829
San Juan, PR 00936-2829
phone: 787-729-8804
fax: 707-722-7867
e-mail: mcalimano@prpa.gobierno.pr

Jesus Torres, Executive Manager
Guam Airport Authority
P.O. Box 8770
Tamuning GU 96931
phone: 671-646-0300
fax: 671-646-8823
e-mail: jest@gvamairport.net

Test Preparations

This edition of AIM/FAR contains information needed for passing the knowledge portion of testing for a FAA pilot certificate or to complete the knowledge portion of a flight review. It also contains sections that are important to training schools, repair stations, parachute jumping, and other aeronautical activities but of lesser importance to pilots.

Obviously, some parts of the AIM and FAR deal more specifically with pilot examinations than do others. The purpose of this text is to help you zero in on particular sections that probably will be covered in examinations for a recreational or private pilot certificate and for a flight review. Suggested here is only a portion of areas that might be included in any test. This should, however, help you focus on some of the known areas that tests will cover.

Private Pilot

In the AIM, check Chapter 2 for airport lighting and visual aids, such as runway/taxiway markings, hold short indicators, etc. With the FAA's current interest in reducing runway incursions, it probably will be a good idea to bone up on airport signage and communications.

Airspace in Chapter 3 is an important area to study, particularly those sections relating to classes of airspace and the requirements and restrictions relating to them. Another section of Chapter 3 to study relates to weather minimums and flight altitudes.

Also for weather, study section 1 of Part 7 relating to meteorology. This includes various types of weather forecasts, where and how to obtain them, decoding the information, severe weather, and other meteorological issues. Section 2 of Part 7 provides information about altimeter settings that are important.

In Chapter 4 pay close attention to operations at airports, those with control towers and those without, and know the differences. Also, read section 2 relating to radio phraseology.

Although no pilot wants to experience an emergency, the FAA wants each one to be prepared if one occurs and Chapter 6 provides the information.

Remember that information in the AIM is just that—information. Although the FAA will test on certain aspects of it and knowledge of it will make your flying safer, more enjoyable, and more efficient, the information does not have the binding effect of a federal aviation regulation. FARs are binding. Not know them and you can fail a test; violate them and you can face a certificate suspension, revocation, fine—or physical and aircraft injury.

These rules come under Part 14 of the Code of Federal Regulations (CFRs). This issue of AIM/FAR contains 18 of the 68 parts of the FARs that are of concern particularly to general aviation. (The other parts refer to airline operations, airworthiness standards, recording aircraft titles, security, and other subjects rarely of interest in general aviation.). Parts 61 and 91 contain the essential rules for pilots. (It's wise to get some familiarization also with Part 45 as there might be questions about who can perform what kinds of aircraft maintenance.)

The most likely sections to be covered in testing include:

61.3—Certificates required

61.14, 15, 16—Drug and alcohol testing

61.23—Medical certificates: how long they are valid for different ratings.

61.31—High performance aircraft, tailwheel aircraft: what they are and who can pilot them

61.51—Logging flight time: what must be logged, who may log what.

61.57—Recent flight experience: when you may fly and who you can take with you.

91.3—Responsibilities and authority of pilot-in-command.

91.103—Preflight action: what you must do before starting a flight.

91.111—Operating near other aircraft

91.113—Right of way rules

91.117—Aircraft speeds

91.119—Minimum safe altitudes

91.126-91.135—Operating in the various classes of airspace.

91.411—Altimeter system and transponder checks.

You might find all of these subjects as test questions. You might find only some of them. There might be others not mentioned here. Whether or not these subjects are included in the test, they are all important to know.

The Flight Review

The flight review is required by section 61.56 of the FARs. That section spells out who must take the review. It doesn't cover what questions to ask. Much of this is left to the discretion of the instructor. The review requires a minimum of one-hour flight training and one-hour ground training. The ground portion might include specifics about the particular aircraft but most questioning will be to information found in Part 91 of the FARs—general operating and flight rules. What an instructor will discuss depends upon several factors, including where most of your flying is done (flying in congested airspace vs. non-congested areas, for instance). The AOPA Air Safety Foundation recommends these sections of Part 91 as a general subject area:

Subpart A — General

91.3—Responsibility and authority of the pilot in command

91.7—Civil aircraft airworthiness

91.9—Civil aircraft flight manual, marking, and placards

91.17—Alcohol or drugs

91.21—Portable electronic devices

Subpart B — Flight Rules

91.103—Preflight action

91.107—Use of safety belts, shoulder harnesses and child restraint

91.113—Right of way rules

91.117—Aircraft speed

91.119—Minimum safe altitudes

91.123—Compliance with ATC clearances and instructions

91.126–130—Operations in Class A, B, C, D, E and G airspace

91.139—Emergency air traffic rules

91.151—Fuel requirements for flight in VFR conditions

91.155—Basic VFR weather minimums

91.157—Special VFR weather minimums

91.167—Fuel requirements for flight in IFR conditions

91.175—Takeoff and landing under IFR

91.177—Minimum altitudes for IFR operations

91.185—IFR operations: Two–way radio communications failure

Subpart C — Equipment, Instrument, and Certificate Requirements

91.207—Emergency locator transmitters

91.213—Inoperative instruments and equipment

91.215—ATC transponder and altitude reporting equipment and use

Subpart D — Special Flight Operations

91.303—Aerobatic flight

91.325—Primary Category Aircraft: Operating limitations

Subpart E — Maintenance, Preventive Maintenance and Alterations

91.409—Inspections

91.411—Altimeter system and altitude reporting equipment tests

91.413—ATC transponder tests and inspections

91.417—Maintenance records

Remember for passing tests or safe flying it is better to know more than you will be asked than to be asked more than you know.

NASA Aviation Safety Reporting System

DO NOT REPORT AIRCRAFT ACCIDENTS AND CRIMINAL ACTIVITIES ON THIS FORM.
ACCIDENTS AND CRIMINAL ACTIVITIES ARE NOT INCLUDED IN THE ASRS PROGRAM AND SHOULD NOT BE SUBMITTED TO NASA.
ALL IDENTITIES CONTAINED IN THIS REPORT WILL BE REMOVED TO ASSURE COMPLETE REPORTER ANONYMITY.

(SPACE BELOW RESERVED FOR ASRS DATE/TIME STAMP)

IDENTIFICATION STRIP: *Please fill in all blanks to ensure return of strip.*
NO RECORD WILL BE KEPT OF YOUR IDENTITY. This section will be returned to you.

TELEPHONE NUMBERS where we may reach you for further details of this occurrence:

HOME Area _____ No. _____ - _____ Hours _____

WORK Area _____ No. _____ - _____ Hours _____

NAME _____

ADDRESS/PO BOX _____

_____

CITY _____ STATE _____ ZIP _____

TYPE OF EVENT/SITUATION _____

_____

DATE OF OCCURRENCE _____

LOCAL TIME (24 hr. clock) _____

PLEASE FILL IN APPROPRIATE SPACES AND CHECK ALL ITEMS WHICH APPLY TO THIS EVENT OR SITUATION.

REPORTER	FLYING TIME	CERTIFICATES/RATINGS		ATC EXPERIENCE	
o Captain	total _____ hrs.	o student	o private	o FPL	o Developmental
o First Officer		o commercial	o ATP	radar _____ yrs.	
o pilot flying	last 90 days ____ hrs.	o instrument	o CFI	non-radar _____ yrs.	
o pilot not flying		o multiengine	o F/E	supervisory _____ yrs.	
o Other Crewmember	time in type ____ hrs.	o _____		military _____ yrs.	
o _____					

AIRSPACE		WEATHER		LIGHT/VISIBILITY		ATC/ADVISORY SERV.	
o Class A (PCA)	o Special Use Airspace	o VMC	o ice	o daylight	o night	o local	o center
o Class B (TCA)	o airway/route ____	o IMC	o snow	o dawn	o dusk	o ground	o FSS
o Class C (ARSA)	o unknown/other ____	o mixed	o turbulence	ceiling ____ feet		o apch	o UNICOM
o Class D (Control Zone/ATA)	_____	o marginal	o tstorm	visibility ____ miles		o dep	o CTAF
o Class E (General Controlled)	_____	o rain	o windshear	RVR ____ feet		Name of ATC Facility:	
o Class G (Uncontrolled)	_____	o fog	o ____			_____	

	AIRCRAFT 1			AIRCRAFT 2		
Type of Aircraft (Make/Model)	(Your Aircraft) _____	o EFIS o FMS/FMC		(Other Aircraft) _____	o EFIS o FMS/FMC	
Operator	o air carrier o military o corporate			o air carrier o military o corporate		
	o commuter o private o other ____			o commuter o private o other ____		
Mission	o passenger o training o business			o passenger o training o business		
	o cargo o pleasure o unk/other ____			o cargo o pleasure o unk/other ____		
Flight plan	o VFR o SVFR o none			o VFR o SVFR o none		
	o IFR o DVFR o unknown			o IFR o DVFR o unknown		
Flight phases at time of occurrence	o taxi o cruise o landing			o taxi o cruise o landing		
	o takeoff o descent o missed apch/GAR			o takeoff o descent o missed apch/GAR		
	o climb o approach o other ____			o climb o approach o other ____		
Control status	o visual apch o on vector o on SID/STAR			o visual apch o on vector o on SID/STAR		
	o controlled o none o unknown			o controlled o none o unknown		
	o no radio o radar advisories			o no radio o radar advisories		

If more than two aircraft were involved, please describe the additional aircraft in the "Describe Event/Situation" section.

LOCATION	CONFLICTS			
Altitude _____ o MSL o AGL	Estimated miss distance in feet: horiz ____ vert ____			
Distance and radial from airport, NAVAID, or other fix ____	Was evasive action taken?		o Yes	o No
_____	Was TCAS a factor?	o TA	o RA	o No
Nearest City/State _____	Did GPWS activate?		o Yes	o No

NATIONAL AERONAUTICS AND SPACE ADMINISTRATION

NASA has established an Aviation Safety Reporting System (ASRS) to identify issues in the aviation system which need to be addressed. The program of which this system is a part is described in detail in FAA Advisory Circular 00-46C. Your assistance in informing us about such issues is essential to the success of the program. Please fill out this form as completely as possible, enclose in an sealed envelope, affix proper postage, and and send it directly to us.

The information you provide on the identity strip will be used only if NASA determines that it is necessary to contact you for further information. THIS IDENTITY STRIP WILL BE RETURNED DIRECTLY TO YOU. The return of the identity strip assures your anonymity.

AVIATION SAFETY REPORTING SYSTEM

Section 91.25 of the Federal Aviation Regulations (14 CFR 91.25) prohibits reports filed with NASA from being used for FAA enforcement purposes. This report will not be made available to the FAA for civil penalty or certificate actions for violations of the Federal Air Regulations. Your identity strip, stamped by NASA, is proof that you have submitted a report to the Aviation Safety Reporting System. We can only return the strip to you, however, if you have provided a mailing address. Equally important, we can often obtain additional useful information if our safety analysts can talk with you directly by telephone. For this reason, we have requested telephone numbers where we may reach you.

Thank you for your contribution to aviation safety.

NOTE: *AIRCRAFT ACCIDENTS SHOULD NOT BE REPORTED ON THIS FORM. SUCH EVENTS SHOULD BE FILED WITH THE NATIONAL TRANSPORTATION SAFETY BOARD AS REQUIRED BY NTSB Regulation 830.5 (49CFR830.5).*

Please fold both pages (and additional pages if required), enclose in a sealed, stamped envelope, and mail to:

NASA AVIATION SAFETY REPORTING SYSTEM
POST OFFICE BOX 189
MOFFETT FIELD, CALIFORNIA 94035-0189

DESCRIBE EVENT/SITUATION

Keeping in mind the topics shown below, discuss those which you feel are relevant and anything else you think is important. Include what you believe really caused the problem, and what can be done to prevent a recurrence, or correct the situation. (USE ADDITIONAL PAPER IF NEEDED)

CHAIN OF EVENTS		Page 2 of 2	HUMAN PERFORMANCE CONSIDERATIONS	
- How the problem arose	- How it was discovered		- Perceptions, judgments, decisions	- Actions or inactions
- Contributing factors	- Corrective actions		- Factors affecting the quality of human performance	

Index

NOTE: words in **boldface** indicate aviation-related phrases.

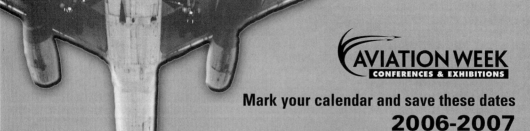

MRO April
2007 17-19
Cobb Galleria Centre
Atlanta, GA

MRO April
Military 17-19
Cobb Galleria Centre
Atlanta, GA

Get more information or register at
www.aviationweek.com/conferences
or call 1.800.240.7645 ext. 5 or
+1.212.904.3225 today!

Aviation Week Books is an imprint of McGraw-Hill Professional in conjunction with the Aviation Week division of The McGraw-Hill Companies. With nearly 50 products and services and a core audience of some one million professionals and enthusiasts, Aviation Week is the world's largest multimedia information and service provider to the global aviation and aerospace market.

For more information, use www.aviationnow.com

Editorial Director: Stanley Kandebo, Assistant Managing Editor, Aviation Week & Space Technology